Lecture Notes in Computer Science 3827

Xiaotie Deng Dingzhu Du (Eds.)

Algorithms and Computation

16th International Symposium, ISAAC 2005
Sanya, Hainan, China, December 19-21, 2005
Proceedings

Springer

Volume Editors

Xiaotie Deng
City University of Hong Kong, Department of Computer Science
83 Tat Chee Avenue, Kowloon Tong, Hong Kong SAR, China
E-mail: csdeng@cityu.edu.hk

Dingzhu Du
University of Minnesota, Department of Computer Science and Engineering
4-192 EE/CS Building, 200 Union Street S.E., Minneapolis, MN 55455, USA
E-mail: dzd@cs.umn.edu

Library of Congress Control Number: 2005937163

CR Subject Classification (1998): F.2, C.2, G.2-3, I.3.5, C.2.4, E.5

ISSN 0302-9743
ISBN-10 3-540-30935-7 Springer Berlin Heidelberg New York
ISBN-13 978-3-540-30935-2 Springer Berlin Heidelberg New York

Springer is a part of Springer Science+Business Media

springer.com

Printed in Germany

Typesetting: Camera-ready by author, data conversion by Scientific Publishing Services, Chennai, India
Printed on acid-free paper SPIN: 11602613 06/3142 5 4 3 2 1 0

Preface

ISAAC 2005, the 16th International Symposium on Algorithms and Computation, took place in Hainan, China, December 19-21, 2005. The symposium provided a forum for researchers working in algorithms and the theory of computation from all over the world. The final count of electronic submissions was 549, of which 112 were accepted. Among them, the submitting authors' emails are: 18 from edu (USA) accounts, 14 from de (Germany), 10 from jp (Japan), 8 from fr (France), 8 from hk (Hong Kong), 7 from ca (Canada), 6 from cn (China), 5 from gr (Greece), 4 from gmail, 3 from tw (Taiwan), 3 from it (Italy), 3 from se (Sweden), 3 from sg (Singapore), 2 from cz (Czech Republic), 2 from ch (Switzerland), 2 from 163.com, and 1 each from no (Norway), uk (United Kingdom), be (Belgium), au (Australia), es (Spain), nl (The Netherlands), kr (Korea), in (India), il (Israel), cy (Cyprus), cl (Chile), pl (Poland), ie (Ireland), and net. This represents a total of 30 countries or regions, including 3 Internet dot coms.

We would like to thank the two invited speaker, Mihalis Yannakakis and Frances Y. Yao, for their insightful speeches and new directions in algorithms and computation.

We received 6 nominations for the best paper and 6 nominations for the best student paper. Only papers with all co-authors students could qualify for the latter. Unlike previous years, this year we made a decision after the authors made their presentation. We hope this change allowed the Program Committee to make the most informed decision.

We would like to thank the conference Co-chairs, Francis Chin and Frances Yao, for their leadership, advice and help on crucial matters concerning the conference. We would like to thank the International Program Committee for spending their valuable time and effort in the review process. We would also like to thank the Organizing Committee, led by Xiaodong Hu, for their contribution to making this conference a success.

In addition, we would like to thank those who spoke out on matters regarding the program and the organization, as well as other important ISAAC matters, and have provided invaluable feedback to ISAAC 2005.

Finally, we would like to thank our sponsors, the Academy of Mathematics and System Sciences of the Chinese Academy of Sciences, and to thank the Department of Computer Science, City University of Hong Kong, for the clerical support in handling the enormous amount of submissions and registrations.

December 2005

Xiaotie Deng
Dingzhu Du

Preface

ISAAC 2005, the 16th International Symposium on Algorithms and Computation, took place in Hainan, China, December 19–21, 2005. The symposium provided a forum for researchers working in algorithms and the theory of computation from all over the world. The final count of electronic submissions was 549, of which 112 were accepted. [illegible]

We would like to thank the two invited speakers, Mihalis Yannakakis and Frances Yao, for their insightful speeches and new directions in algorithms and computation. [illegible]

[illegible] for the referees [illegible] We would like to thank the Program Committee members for spending their valuable time and effort in the review process. We would also like to thank the Organizing Committee [illegible]

Organization

ISAAC 2005 was jointly organized by the Academy of Mathematics and System Sciences of the Chinese Academy of Sciences, City University of Hong Kong, and the University of Hong Kong.

Conference Co-chairs

Francis Chin	The University of Hong Kong
Frances Yao	City University of Hong Kong

Organizing Committee

Xiaodong Hu (Chair)	Chinese Academy of Sciences
Anthony Yingjie Fu	City University of Hong Kong

Treasurer

Xiaohua Jia	City University of Hong Kong

Program Co-chairs

Xiaotie Deng	City University of Hong Kong
Dingzhu Du	University of Minnesota

Invited Speakers

Mihalis Yannakakis	Columbia University
Frances Yao	City University of Hong Kong

Program Committee

Sanjeev Arora	Princeton University
Tetsuo Asano	JAIST
G. Ausiello	Università di Roma
Carlo Blundo	Università di Salerno
Franz J. Brandenburg	Universität Passau
Leizhen Cai	Chinese University of Hong Kong
Mao-Cheng Cai	Chinese Academy of Sciences
Jianer Chen	University of Texas A&M
ZhiZhong Chen	Tokyo Denki University
Frank Dehne	Griffith University
Josep Diaz	Universitat Politècnica de Catalunya
Qizhi Fang	Shandong University
Haodi Feng	Shandong University
Rudolf Fleischer	Fudan University

Philippe Golle	Palo Alto Research Center
Monika Rauch Henzinger	Google Corp.
Toshihide Ibaraki	Kwansei Gakuin University
Markus Jakobsson	Indiana University
Elias Koutsoupias	University of Athens
Hao Li	Laboratoire de Recherche en Informatique
Jianping Li	Yunnan University
Hsueh-I Lu	National Taiwan University
Bin Ma	University of Western Ontario
Subhas C. Nandy	Indian Statistical Institute
Koji Nakano	Hiroshima University
Rolf Niedermeier	Universität Jena
Christos Papadimitriou	University of California, Berkeley
Kunihiko Sadakane	Kyushu University
Yaoyun Shi	University of Michigan
Xiaoming Sun	Tsinghua University
Caoan Wang	Memorial University of Newfoundland
Xiaofeng Wang	Indiana University
Mihalis Yannakakis	Columbia University
Yinyu Ye	Stanford University
Mingsheng Ying	Tsinghua University
Guochuan Zhang	Zhejiang University
Li Zhang	HP Corp.
Binhai Zhu	Montana State University

List of Subreferees

Marcos Aguilera
Susanne Albers
Helmut Alt
Arne Andersson
Arijit Bishnu
Andreas Brandstädt
Ching-Lueh Chang
Hsun-Wen Chang
Ning Chen
Xi Chen
Ming-Yang Chen
Jian-Jia Chen
Tien-Ren Chen
Kuan-Lin Chen
Hsueh-Yi Chen
Ho-Lin Chen
Shirley H.C. Cheung
Hung Chim
Kai-min Chung
Sandip Das
Vinay Deolalikar
Michael Dom
Stefan Dziembowsk
Thomas Erlebach
Jia-Hao Grant Fan
Henning Fernau
Anthony Yingjie Fu
Hiroshi Fujiwara
Jens Gramm
Jiong Guo
Falk Hüffner
Harald Hempel
Jin-Ju Hong
Chun-Hung Hsiao
Yu-Hao Huang
Jesper Jansson
Yien-Lin Jyu
Ton Kloks
Ming-Tat Ko
Dieter Kratsch

Chien-Chih Liao
Wan-Chen Lu
Daniel Mölle
Daniel Marx
Hiro-taka Ono
Mindos Siskerodir
Douglas R. Stinson
Zuowen Tan
Jörg Vogel
Yin Wang
Hsin-Wen Bertha Wei
Deshi Ye
Yunlei Zhao
Feng Zou
Hong-Yiu Lin
Guanfeng Lv
Haiko Müller
Krishnendu Mukhopadhyaya
Md. Saidur Rahman
Andreas Spillner
Susmita Sur-Kolay
Salil Vadhan
Lusheng Wang
Zhikui Wang
Sebastian Wernicke
Hai Yu
Yunhong Zhou

Sponsors

Academy of Mathematics and System Sciences,
Chinese Academy of Sciences
Department of Computer Science, City University of Hong Kong

Best Paper Nominations

Embedding Point Sets Into Plane Graphs of Small Dilation
Annette Ebbers-Baumann, Ansgar Gruene, Marek Karpinski, Rolf Klein, Christian Knauer, Andrzej Lingas

Almost Optimal Solutions for Bin Coloring Problems
Mingen Lin, Zhiyong Lin, Jinhui Xu

Complexity and Approximation of the Minimum Recombination Haplotype Configuration Problem
Lan Liu, Xi Chen, Jing Xiao,Tao Jiang

A 1.5-Approximation of the Minimal Manhattan Network Problem
Sebastian Seibert,Walter Unger

Space Efficient Algorithms for Ordered Tree Comparison
Lusheng Wang, Kaizhong Zhang

The Layered Net Surface Problems in Discrete Geometry and Medical Image Segmentation
Xiaodong Wu, Danny Chen, Kang Li, Milan Sonka

Best Student Paper Nominations

Longest Increasing Subsequences in Windows Based on Canonical Antichain Partition
Erdong Chen, Hao Yuan, Linji Yang

On the Complexity of the G-Reconstruction Problem
Zdenek Dvorak,Vit Jelinek

A Practical Algorithm for the Computation of Market Equilibrium with Logarithmic Utility Functions
Li-Sha Huang

An Improved $\tilde{O}(1.234^m)$-Time Deterministic Algorithm for SAT
Masaki Yamamoto

Improved Algorithms for Largest Cardinality 2-Interval Pattern Problem
Hao Yuan, Linji Yang, Erdong Chen

Algorithms for Local Forest Similarity
PENG, Zeshan

Table of Contents

Errata from ISAAC 2004 (LNCS 3341)

Algorithmic Problems in Wireless Ad Hoc Networks

Frances F. Yao

Department of Computer Science, City University of Hong Kong,
Hong Kong SAR., China
csfyao@cityu.edu.hk

Abstract. A wireless ad hoc network is a collection of geographically distributed radio nodes which communicate with each other without the support of fixed infrastructure. Wireless ad hoc networks have gained much attention in recent years because their potential wide applications such as environmental monitoring and emergency disaster relief. Some of the key design issues for wireless ad hoc networks include power management, network connectivity and routing. These problems require different formulations and solutions from the classical setting. In this talk, we will look at some sample problems, their mathematical modelling and solutions. It will be seen that graph theory and computational geometry can play a role in the design and analysis of ad hoc networks.

X. Deng and D. Du (Eds.): ISAAC 2005, LNCS 3827, p. 1, 2005.

Probability and Recursion

Kousha Etessami[1] and Mihalis Yannakakis[2]

[1] LFCS, School of Informatics, University of Edinburgh
[2] Department of Computer Science, Columbia University

In this talk we will discuss recent work on the modeling and algorithmic analysis of systems involving recursion and probability. There has been intense activity recently in the study of such systems [2,3,10,11,13,14,15,16,17]. The primary motivation comes from the analysis of probabilistic programs with procedures. Probability can arise either due to randomizing steps in the program, or it may reflect statistical assumptions on the behaviour of the program, under which we want to investigate its properties.

Discrete-time, finite state Markov chains have been used over the years in a broad range of applications to model the evolution of a variety of probabilistic systems. Markov Decision Processes are a useful model for control optimization problems in a sequential stochastic environment that combines probabilistic and nonprobabilistic aspects of system behavior [28,19,18]. These models have also been used extensively in particular to model probabilistic programs without procedures and to analyze their properties [7,8,24,27,33]. In the presence of recursive procedures, a natural model for probabilistic programs is *Recursive Markov Chains* (RMCs): Informally, a RMC consists of a collection of finite state component Markov chains that can call each other in a potentially recursive manner [13]. An equivalent model is *probabilistic Pushdown Automata* (pPDA) [10]. These models are essentially a succinct, finite representation of an infinite state Markov chain, which captures the global evolution of the system.

More generally, if some steps of the program/system are probabilistic while other steps are not, but rather are controllable by the designer or the environment, then such a system can be naturally modeled by a *Recursive Markov Decision Process* (RMDP) or a *Recursive Simple Stochastic Game* (RSSG)[15]. In a RMDP all the nonprobabilistic actions are controlled by the same agent (the controller or the environment), while in a RSSG, different nonprobabilistic actions are controlled by two opposing agents (eg. some by the designer and some by the environment).

Some types of recursive probabilistic models arise naturally in other contexts and have been studied earlier, some even before the advent of computer science. *Branching processes* are an important class of such processes [21], introduced first by Galton and Watson in the 19th century to study population dynamics, and generalized later on in the mid 20th century to the case of *multitype branching processes* by Kolmogorov and developed further by Sevastyanov [23,31]. They have been applied in a wide variety of contexts such as population genetics [22], models in molecular biology for RNA [30], and nuclear chain reactions [12]. Another related model is that of *stochastic context-free grammars* which have been

X. Deng and D. Du (Eds.): ISAAC 2005, LNCS 3827, pp. 2–4, 2005.

studied extensively since the 1970's especially in the Natural Language Processing community (see eg. [25]). In a certain formal sense, multitype branching processes and stochastic context-free grammars correspond to a subclass of recursive Markov chains (the class of "single-exit RMCs", where each component Markov chain has a single exit state where it can terminate and return control to the component that called it).

Recursive Markov chains, and their extension to Recursive Markov Decision Processes and Simple Stochastic Games, have a rich theory. Their analysis involves combinatorial, algebraic, and numerical aspects, with connections to a variety of areas, such as the existential theory of the reals [5,29,4], multidimensional Newton's method, matrix theory, and many others. There are connections also with several well-known open problems, such as the square root sum problem [20,32] (a 30-year old intriguing, simple problem that arises often in the numerical complexity of geometric computations, and which is known to be in PSPACE, but it is not known even whether it is in NP), and the value of simple stochastic games [6] and related games, which are in NP∩coNP, but it is not known whether they are in P.

In this talk we will survey some of this theory, the algorithmic results so far, and remaining challenges.

Acknowledgement. Work partially supported by NSF Grant CCF-04-30946.

References

1. R. Alur, M. Benedikt, K. Etessami, P. Godefroid, T. W. Reps, and M. Yannakakis. Analysis of recursive state machines. In *ACM Trans. Progr. Lang. Sys.*, 27:786-818, 2005.
2. T. Brázdil, A. Kučera, and J. Esparza. Analysis and prediction of the long-run behavior of probabilistic sequential programs with recursion. In *Proc. of FOCS'05*, 2005.
3. T. Brázdil, A. Kučera, and O. Stražovský. Decidability of temporal properties of probabilistic pushdown automata. In *Proc. of STACS'05*, 2005.
4. S. Basu, R. Pollack, and M. F. Roy. On the combinatorial and algebraic complexity of quantifier elimination. *J. ACM*, 43(6):1002–1045, 1996.
5. J. Canny. Some algebraic and geometric computations in PSPACE. In *Prof. of 20th ACM STOC*, pages 460–467, 1988.
6. A. Condon. The complexity of stochastic games. *Inf. & Comp.*, 96(2):203–224, 1992.
7. C. Courcoubetis and M. Yannakakis. The complexity of probabilistic verification. *Journal of the ACM*, 42(4):857–907, 1995.
8. C. Courcoubetis and M. Yannakakis. Markov decision processes and regular events. *IEEE Trans. on Automatic Control*, 43(10):1399–1418, 1998.
9. L. de Alfaro, M. Kwiatkowska, G. Norman, D. Parker, and R. Segala. Symbolic model checking of probabilistic processes using MTBDDs and the kronecker representation. In *Proc. of 6th TACAS*, pages 395–410, 2000.
10. J. Esparza, A. Kučera, and R. Mayr. Model checking probabilistic pushdown automata. In *Proc. of 19th IEEE LICS'04*, 2004.

11. J. Esparza, A. Kučera, and R. Mayr. Quantitative analysis of probabilistic pushdown automata: expectations and variances. *Proc. of 20th IEEE LICS*, 2005.
12. C. J. Everett and S. Ulam. Multiplicative systems, part i., ii, and iii. Technical Report 683,690,707, Los Alamos Scientific Laboratory, 1948.
13. K. Etessami and M. Yannakakis. Recursive Markov chains, stochastic grammars, and monotone systems of non-linear equations. In *Proc. of 22nd STACS'05*. Springer, 2005. (Tech. Report, U. Edinburgh, June 2004).
14. K. Etessami and M. Yannakakis. Algorithmic verification of recursive probabilistic state machines. In *Proc. 11th TACAS*, vol. 3440 of LNCS, 2005.
15. K. Etessami and M. Yannakakis. Recursive Markov Decision Processes and Recursive Stochastic Games. In *Proc. ICALP*, pp. 891-903, Springer, 2005.
16. K. Etessami and M. Yannakakis. Checking LTL Properties of Recursive Markov Chains. In *Proc. 2nd Intl. Conf. on Quantitative Evaluation of Systems*, IEEE, 2005.
17. K. Etessami and M. Yannakakis. Efficient Analysis of Classes of Recursive Markov Decision Processes and Stochastic Games, submitted.
18. E. Feinberg and A. Shwartz, editors. *Handbook of Markov Decision Processes*. Kluwer, 2002.
19. J. Filar and K. Vrieze. *Competitive Markov Decision Processes*. Springer, 1997.
20. M. R. Garey, R. L. Graham, and D. S. Johnson. Some NP-complete geometric problems. In *8th ACM Symp. on Theory of Computing*, pages 10–22, 1976.
21. T. E. Harris. *The Theory of Branching Processes*. Springer-Verlag, 1963.
22. P. Jagers. *Branching Processes with Biological Applications*. Wiley, 1975.
23. A. N. Kolmogorov and B. A. Sevastyanov. The calculation of final probabilities for branching random processes. *Dokl. Akad. Nauk SSSR*, 56:783–786, 1947. (Russian).
24. M. Kwiatkowska. Model checking for probability and time: from theory to practice. In *18th IEEE LICS*, pages 351–360, 2003.
25. C. Manning and H. Schütze. *Foundations of Statistical Natural Language Processing*. MIT Press, 1999.
26. A. Paz. *Introduction to Probabilistic Automata*. Academic Press, 1971.
27. A. Pnueli and L. D. Zuck. Probabilistic verification. *Inf. and Comp.*, 103(1):1–29, 1993.
28. M. L. Puterman. *Markov Decision Processes*. Wiley, 1994.
29. J. Renegar. On the computational complexity and geometry of the first-order theory of the reals, parts I-III. *J. Symb. Comp.*, 13(3):255–352, 1992.
30. Y. Sakakibara, M. Brown, R Hughey, I.S. Mian, K. Sjolander, R. Underwood, and D. Haussler. Stochastic context-free grammars for tRNA modeling. *Nucleic Acids Research*, 22(23):5112–5120, 1994.
31. B. A. Sevastyanov. The theory of branching processes. *Uspehi Mathemat. Nauk*, 6:47–99, 1951. (Russian).
32. P. Tiwari. A problem that is easier to solve on the unit-cost algebraic ram. *Journal of Complexity*, pages 393–397, 1992.
33. M. Vardi. Automatic verification of probabilistic concurrent finite-state programs. In *Proc. of 26th IEEE FOCS*, pages 327–338, 1985.

Embedding Point Sets into Plane Graphs of Small Dilation

Annette Ebbers-Baumann[1], Ansgar Grüne[1], Marek Karpinski[2], Rolf Klein[1], Christian Knauer[3], and Andrzej Lingas[4]

[1] University of Bonn, Institute of Computer Science I, D-53117 Bonn, Germany
{ebbers, gruene, rolf.klein}@cs.uni-bonn.de

[2] University of Bonn, Institute of Computer Science V, D-53117 Bonn, Germany
marek@cs.uni-bonn.de

[3] FU Berlin, Institute of Computer Science, D-14195 Berlin, Germany
christian.knauer@inf.fu-berlin.de

[4] Lund University, Department of Computer Science, 22100 Lund, Sweden
andrzej.lingas@cs.lth.se

Abstract. Let S be a set of points in the plane. What is the minimum possible dilation of all plane graphs that contain S? Even for a set S as simple as five points evenly placed on the circle, this question seems hard to answer; it is not even clear if there exists a lower bound > 1. In this paper we provide the first upper and lower bounds for the embedding problem.

1. Each finite point set can be embedded into the vertex set of a finite triangulation of dilation ≤ 1.1247.
2. Each embedding of a closed convex curve has dilation ≥ 1.00157.
3. Let P be the plane graph that results from intersecting n infinite families of equidistant, parallel lines in general position. Then the vertex set of P has dilation $\geq 2/\sqrt{3} \approx 1.1547$.

Keywords: Dilation, geometric network, lower bound, plane graph, spanning ratio, stretch factor.

1 Introduction

Transportation networks like railway systems can be modeled by geometric graphs: stations correspond to vertices, and the tracks between stations are represented by arcs. One measure of the performance of such a network P is given by its *vertex-to-vertex dilation*. For any two vertices, p and q, let $\pi(p, q)$ be a shortest path from p to q in P. Then,

$$\delta_P(p, q) := \frac{|\pi(p, q)|}{|pq|}$$

measures the detour one encounters in using P, in order to get from p to q, instead of traveling straight; here $|.|$ denotes the Euclidean length. The dilation of P is given by

$$\delta(P) := \sup_{p,q \text{ vertices of } P} \delta_P(p, q).$$

X. Deng and D. Du (Eds.): ISAAC 2005, LNCS 3827, pp. 5–16, 2005.

1.1 Problem Statement

Suppose we are given a set of stations, and we want to build a network connecting them whose dilation is as low as possible. In this work we are assuming that bridges cannot be used. That is, where two or more tracks cross each other, a station is required, that must also be considered in evaluating the dilation of the network.

More precisely, we are given a set S of points in the plane, and we are interested in plane graphs $P = (V, E)$ whose vertex set V contains S, such that the dilation $\delta(P)$ is as small as possible. At this point we are not concerned with the algorithmic cost of computing P, nor with its building cost in terms of the total length of all edges in E, or the size of V—only the dilation of P matters. However, to rule out degenerate solutions like the complete graph over all points in the real plane, we require that the vertex set V of P contains only a finite number of vertices in addition to S. This leads to the following definition.

Definition 1. *Let S be a set of points in the plane. Then the* dilation of S *is given by*

$$\Delta(S) = \inf \{ \ \delta(P); P = (V, E) \ \textit{plane graph} \ \& \ S \subseteq V \ \& \ V \setminus S \ \textit{finite} \ \} .$$

The challenge is in computing the dilation of a given point set. Even for a set as simple as S_5, the vertices of a regular 5-gon, $\Delta(S_5)$ is not known. It is not even clear if $\Delta(S_5) > 1$ holds.[1] Figure 1 depicts some attempts to find good embeddings for S_5.

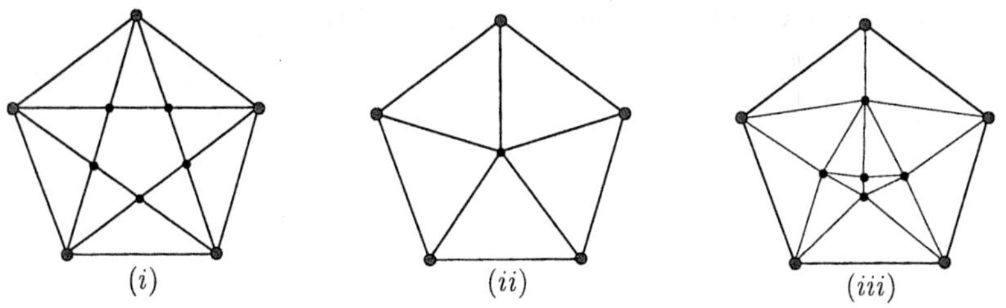

Fig. 1. Embedding five points in regular position into the vertex set of a plane graph of low dilation. (i) Constructing the complete graph results in a new 5-gon. (ii) A star yields dilation ≈ 1.05146. (iii) Using a 4-gon around an off-center point gives dilation ≈ 1.02046, as was shown by Lorenz [17].

[1] While S_5 is not contained in the vertex set of any plane graph of dilation 1, according to the characterization of dilation-free graphs given by D. Eppstein [12], there could be a sequence of plane graphs, each containing S_5, whose dilation shrinks towards 1.

1.2 Related Work

In the context of spanners, the dilation is often called the stretch factor or the spanning ratio of a graph P; see Eppstein's handbook chapter [11], Arikati et al. [3], or the forthcoming monograph [18] by Narasimhan and Smid. However, spanners are usually allowed to contain edge crossings, unlike the plane graphs considered here.

Substantial work has been done on proving upper bounds to the dilation of certain plane graphs. For example, Dobkin et al. [5] and Keil and Gutwin [16] have shown that the Delaunay triangulation of a finite point set has a dilation bounded from above by a small constant. The best upper bound known is 2.42, but a better bound of $\pi/2$ is conjectured to hold. Moreover, there are structural properties of plane graphs, like the good polygon and diamond properties, which imply that the dilation is bounded from above, see Das and Joseph [4]. This result implies that the minimum weight and the greedy triangulations also have a dilation bounded by a constant. Our approach differs from this work, in that the use of extra vertices is allowed. This will lead to an upper bound considerably smaller than $\pi/2$.

Quite recently, a related measure called geometric dilation has been introduced, see [10,8,1,6,9,7], where *all* points of the graph, vertices and interior edge points alike, are considered. The small difference in definition leads to rather different results which do not apply here. For example, plane graphs of minimum geometric dilation tend to have curved edges, whereas for the vertex-to-vertex dilation, straight edges work best.

1.3 New Results

In this paper we provide the first lower and upper bounds to the dilation of point sets, as defined in Definition 1. First, in Section 2, we prove a structural property similar in spirit to the good polygon and diamond properties mentioned above. If a plane graph P contains a face R whose diameter is a weak local maximum, so that each face in a certain neighborhood of R has a diameter at most a few percents larger than R, then the dilation of P can be bounded away from 1. We will derive the following consequence. If C denotes a closed convex curve then $\Delta(C) > 1.00157$ holds for the set of points on C, i.e., each point on curve C is considered a degree 2 vertex; see Figure 2 (1). Another consequence: If P is a plane graph whose faces have diameters bounded from above by some constant then $\delta(P) > 1.00156$ holds.

When looking for plane graphs of low dilation that can accommodate a set of given points, grids come to mind. Even the simple quadratic grid consisting of equidistant vertical and horizontal lines has a vertex-to-vertex dilation of only $\sqrt{2} \approx 1.414$, and we can force its vertex set to contain any finite number of points with rational coordinates, by choosing the cell size appropriately. How to accommodate points with real coordinates is discussed in the full paper. If we use three families of lines, as in the tiling of the plane by equilateral triangles, a smaller dilation of only $2/\sqrt{3} \approx 1.1547$ can be achieved; see Figure 3.

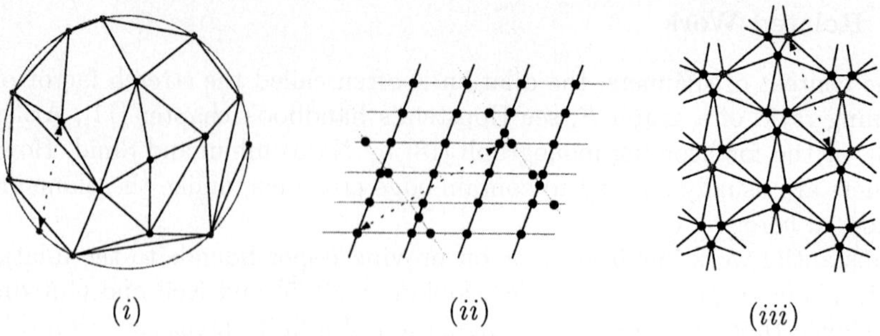

Fig. 2. Results. (i) Each embedding of a closed convex curve has dilation > 1.00157. (ii) Each such arrangement has dilation > 1.1547. (iii) Each finite point set has dilation < 1.1247.

An interesting question is if the dilation can be decreased even further by using lines of more than three different slopes. The answer is somewhat surprising, because parallel highways, a mile apart, for each orientation $2\pi i/n$, would in fact provide very low dilation to long distance traffic. But there are always vertices relatively close to each other, for which the dilation is at least $2/\sqrt{3}$ as we shall prove in Section 3.

Yet in Section 4 we introduce a way of getting below the $2/\sqrt{3}$ bound offered by the equitriangular tiling depicted in Figure 3. One can modify this tiling by replacing each vertex with a triangle, and by connecting neighboring triangles as shown in Figure 2 (iii). The resulting graph has a dilation less than 1.1247. We can scale, and slightly deform this graph, so that its vertex set contains any given finite set of points; then we cut off the unbounded part which does not host any point. These operations increase the dilation by some factor that can be made arbitrarily small. Thus we obtain that $\Delta(S) < 1.1247$ holds, for every finite point set S.

Finally, in Section 5, we address some of the questions left open and discuss future work.

2 A Lower Bound

First, we introduce some notations. Let P be a plane graph, and let R be a face of P with boundary $\partial(R)$. As usual, let

$$\mathrm{diam}(R) = \sup\{\ |ef|\ ;\ e, f \text{ vertices of } R\}$$

be the *diameter* of R; unbounded faces have infinite diameter. Now let R be a bounded face of P. For any positive number r, the r-*neighborhood of face* R is defined as the set of all faces of P that have non-empty intersection with a disk of radius r centered at the midpoint of a segment ef, where e, f are vertices on $\partial(R)$ satisfying $|ef| = \mathrm{diam}(R)$; see Figure 4. If there are more than one pair e, f of vertices of this kind we break ties arbitrarily. One should observe that the r-neighborhood of a bounded face may include unbounded faces of P.

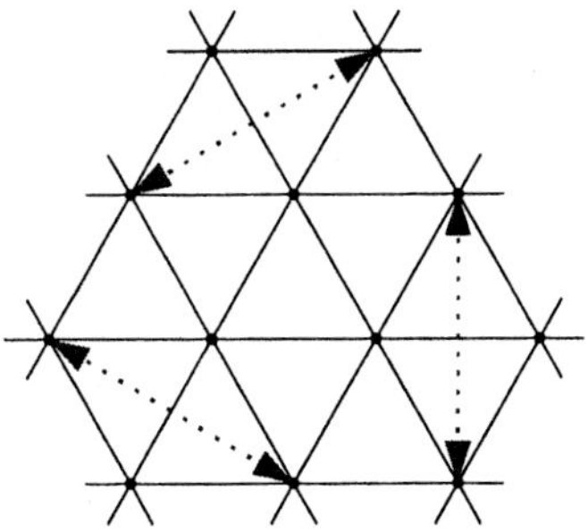

Fig. 3. The tiling by equilateral triangles is of dilation $2/\sqrt{3} \approx 1.1547$

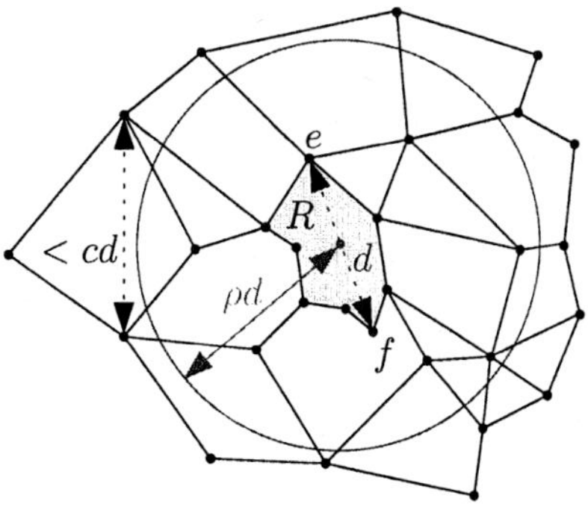

Fig. 4. No face intersected by the disk is of diameter $> cd$

The results of this section are based on the following lemma.

Lemma 1. *For each parameter $c \in [1, 1.5)$ there exist numbers $\rho > 1$ and $\delta > 1$ such that the following holds. Suppose that the plane graph P contains a bounded face R of diameter d, such that all faces in the ρd-neighborhood of R have diameter less than cd.*[2] *Then the dilation of P is at least δ.*

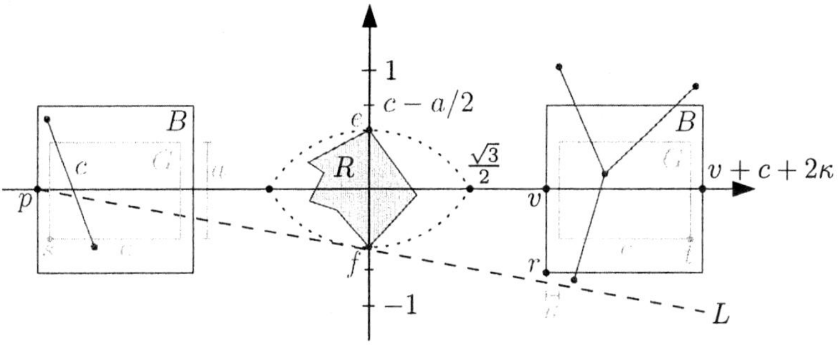

Fig. 5. Constructing a lower bound

Proof. We may assume that face R is of diameter $d = 1$. Also, we assume that ef is vertical and that its midpoint equals the origin; see Figure 5. As no vertex of $\partial(R)$ has a distance > 1 from e or from f, face R is completely contained in the lune spanned by e and f.[3]

Now we place two axis-parallel boxes of width c and height a symmetrically on the X-axis, at a distance of $v + \kappa$ to either side of the origin; these boxes are denoted by G in Figure 5. The parameters a, v, and κ will be chosen later.

[2] This implies that only bounded faces can be included in the ρd-neighborhood of R.

[3] The lune spanned by e and f equals the intersection of the two circles of radius $|ef|$ centered at e and f, respectively.

Suppose that all faces of P that intersect the disk of radius $\rho := v + c + \kappa$ about the origin have a diameter less than c.

Now, let us consider such a box, G. As its width equals c, there cannot exist two points on the left and on the right vertical side of G, respectively, that are contained in the same face R' of the graph, because this would imply $\mathrm{diam}(R') \geq c$. Thus, the vertical sides of G must be separated by P, i. e., there must be a sequence of edges, or a single edge, cutting through the upper and lower horizontal sides of G. In the first case, box G must contain a vertex of P, as shown on the right hand side in the figure. In the second case, the edge that crosses G top-down must itself be of length $< c$, because it belongs to a face intersected by the disk of radius ρ. We enclose G in the smallest axis-parallel box B which contains all line segments of length c that cross both horizontal sides of G. The outer box B is of height $2c - a$, its width exceeds the width c of G by $\kappa = \kappa(a, c)$ on either side, to include all slanted segments. The analysis of $\kappa(a, c)$ can be found in the full paper. By construction, both the upper and the lower half of B must contain a vertex of P in the second case, as shown on the left hand side of Figure 5.

Now we discuss how to choose the parameters v and a such as to guarantee a dilation of $\delta > 1$ in either possible case. First, we let $v > \sqrt{3}/2$ so that the boxes B are disjoint from the lune; consequently, every shortest path in P connecting vertices in the two boxes B has to go around the face R.

Case 1. Each of the boxes G contains a vertex of P. If a is less than $|ef| = 1$ then these vertices cause a dilation of at leas t a certain value $\delta > 1$ which depends only on a, c, v and κ.

Case 2. Each of the boxes B contains two vertices of P, one above, and one below the X-axis. Let p and r denote vertices in the upper part of the left and in the lower part of the right box B, respectively. We assume w. l. o. g. that the shortest path in P connecting them runs below vertex f, so that its length is at least $|pf| + |fr|$. If we make sure that, even at the extreme position depicted in Figure 5, vertex r lies above the line L through p and f, a dilation $\delta > 1$ is guaranteed. The equation of L is given by

$$Y = -\frac{1}{2(v + c + 2\kappa)} X - \frac{1}{2}.$$

Thus, we must ensure that

$$-(c - \frac{a}{2}) > -\frac{1}{2(v + c + 2\kappa)} v - \frac{1}{2}$$

or, equivalently,

$$a > 2c - 1 - \frac{v}{v + c + 2\kappa}$$

holds. Together with the condition $1 > a$ from Case 1 we obtain

$$3 - 2c > a + 2 - 2c > \frac{c + 2\kappa}{v + c + 2\kappa} > 0. \qquad (*)$$

Case 3. There exist at least one vertex of P in the left box G, and at least two vertices in the right box B, above and below the X-axis. Since the vertex in the box G must reside in either the upper or the lower part of the enclosing box B, Case 2 applies.

Clearly, the above conditions can be fulfilled for each given $c \in [1, 1.5)$. First, we pick $a \in (2c-2, 1)$, which guarantees $3-2c > a+2-2c > 0$. Then we choose $v > \sqrt{3}/2$ so large that the second inequality in condition (*) is satisfied. This proves Lemma 1.

It is quite straightforward to derive quantitative results from the above construction, by adjusting the values of the parameters a and v. The following numerical values for ρ and δ have been obtained using Maple.

c	1.0	1.001	1.1	1.2	1.3	1.4
δ	1.00157	1.00156	1.00043	1.000092	1.000012	1.00000056
ρ	1.923	1.925	2.46	3.9	6.9	16.5

As a first consequence, we get the following result.

Theorem 1. *Let P be a infinite graph whose faces cover the whole plane and have a diameter bounded from above by some constant. Then $\delta(P) > 1.00156$ holds for its dilation.*

Proof. By assumption, $d^* := \sup\{\text{diam}(R) : R \text{ face of } P\}$ is finite. For each $\epsilon > 0$ there exists a face R of P such that $\text{diam}(R) > (1-\epsilon)d^*$ holds. By the assumption on graph P, *all* faces R' of P, in particular those in any neighborhood of R, satisfy

$$\text{diam}(R') \leq d^* < \frac{1}{1-\epsilon}\,\text{diam}(R) \leq 1.001\,\text{diam}(R),$$

if ϵ is small enough. Thus, graph P has dilation at least 1.00156.

The second consequence of Lemma 1 is a lower bound to the Δ function.

Theorem 2. *Let C denote the set of points on a closed convex curve. Then $\Delta(C) > 1.00157$ holds for its dilation.*

Proof. (Sketch) If curve C intersects or encircles a box G, we can argue as before. If C passes between G and ef it becomes even easier to provide two points of high dilation.

For the circle one can find an embedding of dilation $(1+\epsilon)/\sin 1 \approx 1.188$ by placing a single vertex at the center and adding many equidistant radial segments.

3 A Lower Bound for Line Arrangements

A lower bound much stronger than 1.00157 can be shown for graphs that result from intersecting n families F_i of infinitely many equidistant parallel lines. Each family is defined by three parameters, its orientation α_i, the distance w_i in X-direction between consecutive lines, and the offset distance e_i from the origin to the first line in positive X-direction. We say that such families are in *general position* if the numbers w_i^{-1} are linearly independent over the rationals[4].

Theorem 3. *Given n families F_i, $2 \leq i \leq n$, each consisting of infinitely many equidistant parallel lines. Suppose that these families are in general position. Then their intersection graph P is of dilation at least $2/\sqrt{3}$.*

One should observe that this lower bound is attained by the equitriangular grid shown in Figure 3.

Proof. For $n \leq 3$ the claim can be proven quite easily without assuming general position, as shown in the full paper.

Now let $n > 3$. We shall prove the existence of a face R of P that represents so large a barrier between two vertices p and p' of P that even the Euclidean shortest path from p to p' around R is of dilation at least $2/\sqrt{3}$. In fact, we shall provide such a face and two vertices that are *symmetric* about the same center point, which greatly helps with our analysis; see Figure 6. To this end we use the general position assumption, and apply Kronecker's theorem [2] on simultaneous approximation in its following form.

Theorem 4. *(Kronecker) Let L be a line in $\mathbf{R}^n$ that passes through the origin and through some point $(y_1, \ldots, y_n)$ whose coordinates are linearly independent over the rationals. For each point $t \in \mathbf{R}^n$, and for each $\epsilon > 0$, there is an integer translate $t + m$, $m \in \mathbf{Z}^n$, of t whose $\epsilon-$neighborhood is visited by L.*

In other words, line L is dense on the torus $\mathbf{R}^n/\mathbf{Z}^n$. Proofs can be found in, e. g., Apostol [2] or Hlawka [15].

Since the families of lines are in general position, the real numbers $y_i := w_i^{-1}, 1 \leq i \leq n$, do not satisfy a linear equation with rational coefficients. Kronecker's theorem, applied to $t_i := e_i/w_i + 1/2$ and $\epsilon > 0$ yields the existence of integers m_i and of a real number x satisfying

$$|\frac{e_i}{w_i} + \frac{1}{2} + m_i - x\,\frac{1}{w_i}| < \epsilon$$

that is,

$$|e_i + \frac{1}{2}\,w_i + m_i\,w_i - x| < \epsilon\,w_i.$$

Consequently, the point $(x, 0)$ lies, for each family F_i, halfway between two neighboring lines, so that it is center of symmetry for some face R in P—up to an error that can be made arbitrarily small since the numbers w_i are fixed.

[4] This means, if $\sum_{i=1}^{n} a_i\, w_i^{-1} = 0$ holds for rational coefficients a_i then each a_i must be zero.

From now on, we assume that R is symmetric about the origin, and that its longest diagonal, d, is vertical and of length 2. Let us assume that the dilation of graph P is less than $2/\sqrt{3}$. We shall derive a contradiction by proving that there exist two families of lines that contribute to the boundary of R and have symmetric intersection vertices p, p' that cause a dilation $> 2/\sqrt{3}$ in the presence of the barrier R. Since R is symmetric, it is sufficient to provide *one* such vertex p satisfying

$$\frac{|pa| + |pb|}{2|p|} \geq 2/\sqrt{3}$$

where a and b are the endpoints of diagonal d, and $|p| = |p0|$ denotes the distance from p to the origin, as shown in Figure 6.

To this end, consider the locus E of all points where equality holds in the above inequality, see Figure 7. E satisfies the quartic equation $(X^2+Y^2-\frac{3}{2})^2 = \frac{9}{4}(1-Y^2)$. We want to show that the right part of its interior—referred to as an *ear*, due to its shape—contains a vertex p of P.

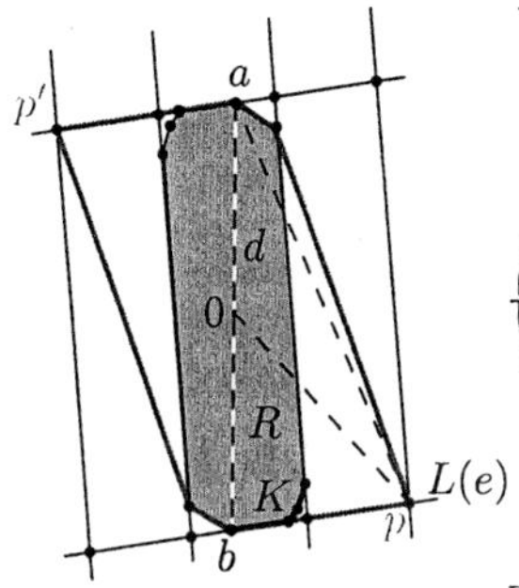

Fig. 6. A symmetric face acting as barrier

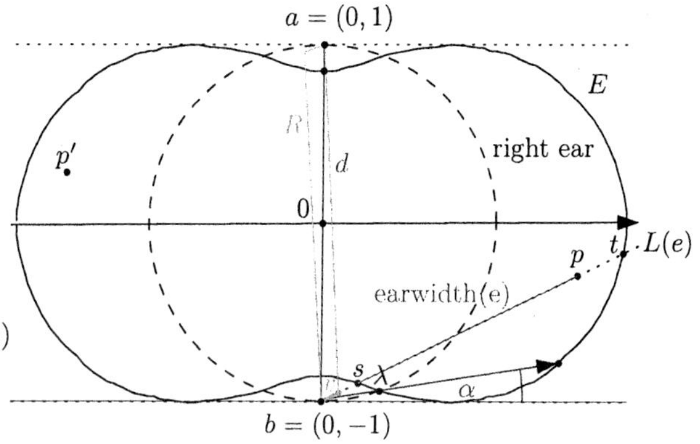

Fig. 7. The locus E of all points that cause dilation $2/\sqrt{3}$ in the presence of line segment ab. The halfline $L(e)$ extends the edge $e = bv$ to the right.

Since d is the longest diagonal of the symmetric face R, the vertices of R are contained in the circle spanned by d. On the other hand, R itself is of dilation $< 2/\sqrt{3}$, by assumption. Hence, the vertices of R must be outside of the locus curve. This leaves only the small caps at a, b for the remaining vertices of R, and $\partial(R)$ must contain two long edges.

First, assume that R is a parallelogram, as shown in Figure 7. Its shape is determined by the position of its lower right vertex v in the bottom cap. Consider the edge e from vertex b to v. Its extension beyond v, $L(e)$, intersects the right ear in a segment of length earwidth(e). In Figure 7 we have earwidth(e) $= |st|$, while the length of the intersection of $L(e)$ with the lower cap equals $|bs|$. The ratio

$|st|/|bs|$ takes on its minimum value $3.11566\ldots > 1$ at the angle $\alpha \approx 9.74°$, when $L(e)$ hits the intersection point, λ, of the locus curve with the circle. Because of

$$\text{earwidth}(e) \; > \; 3.11|bs| \; \geq \; 3.11|bv| \; > \; |bv|$$

the segment of $L(e)$ passing through the ear must contain a vertex p of the two line families bounding R. This proves Theorem 3 in case face R is a parallelogram.

It is interesting to observe that in the limiting case $\alpha = 0$, when R degenerates into its diagonal d, the largest possible dilation $2/\sqrt{3}$ is only attainable by picking p as the bottommost point of the ear. This shows why these arguments would not work for any lower bound larger than $2/\sqrt{3}$.

Now let R be a general symmetric convex polygon. We may assume that its two long edges have non-positive slope. Let K denote the convex boundary chain of R that starts at vertex b and leads to the right until it hits the rightmost long edge of R. In this situation the following holds.

Lemma 2. *There exists an edge e in chain K such that the line $L(e)$ passing through e has the following property. The intersection of $L(e)$ with the the right ear of the locus curve is at least twice as long as the vertical projection of chain K onto $L(e)$.*

The proof of Lemma 2 requires some technical effort; we skip it due to space limitations. Consider Figure 8. Let $\text{cut}(e)$ denote the length of the segment of $L(e)$ that is cut out by the extensions of the long edges of face R, and let $\text{proj}_K(e)$ be the length of the vertical projection of chain K onto $L(e)$. By translating $L(e)$ to the endpoint of K, we can see $\text{cut}(e) \leq 2\ \text{proj}_K(e)$, so Lemma 2 implies

$$\text{cut}(e) \; \leq \; 2\ \text{proj}_K(e) \; \leq \; \text{earwidth}(e).$$

This guarantees the existence of a vertex of P in the interior of the locus curve and completes the proof of Theorem 3 in the general case.

4 An Upper Bound to the Dilation of Finite Point Sets

First, we show how to modify the equitriangular grid, H, displayed in Figure 3, in order to decrease its dilation. The construction is shown in Figure 9.

We replace each vertex v of H with an equilateral triangle T that has one vertex on each of the three lines passing through v in H. The distance, a, between the vertices and the center v of T is a parameter of our construction. Next, we connect by an edge each vertex of T to the two visible vertices of its neighboring triangle T'. Afterwards, all vertices and edges of the old graph H are removed. Let $H_A = H_A(a)$ denote the resulting graph.

Theorem 5. *Within the family $H_A(a)$, the minimum dilation of* $1.1246\ldots$ *is attained for* $a \approx 0.2486$.

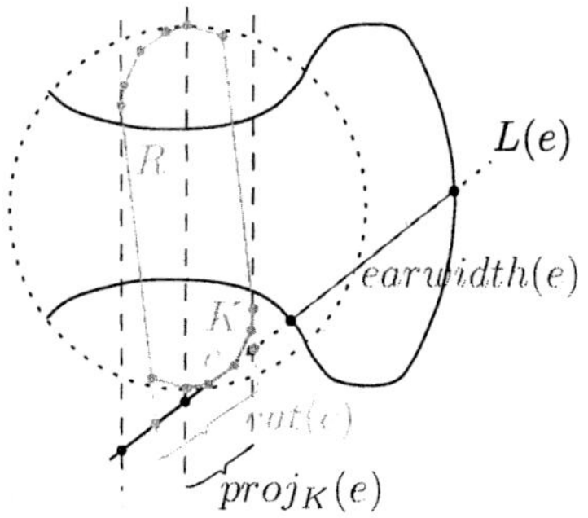

Fig. 8. Vertically projecting the convex chain K onto the line $L(e)$. The intersection of $L(e)$ with the right ear of the locus curve is of length earwidth(e).

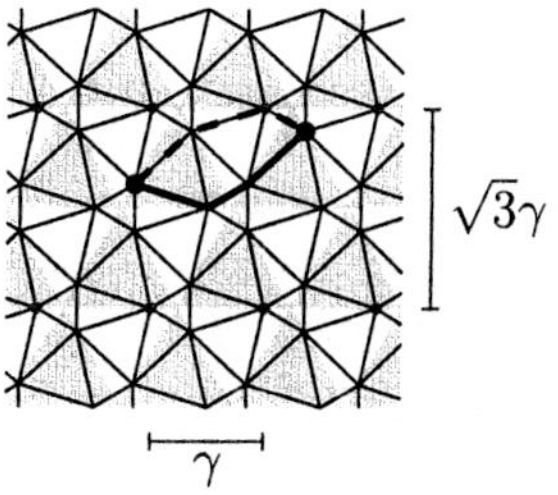

Fig. 9. The new graph H_A of dilation ≈ 1.1246

Proving this result requires considerable technical effort. For example, it is in general not true that the dilation of an unbounded graph is attained by vertices that are close to each other, as a rectangular grid of irrational aspect ratio shows. After introducing global and local coordinates for the vertices of H_A, one has to distinguish 45 shortest path types. Closer inspection leads to four functions of the integer coordinates i, j, whose maximum must be minimized by a suitable choice of parameter a.

A vertex pair causing maximum dilation is shown in Figure 9, together with two shortest paths connecting the two vertices. We obtain the following consequence of Theorem 5.

Theorem 6. *Each finite point set S is of dilation $\Delta(S) < 1.1247$.*

The proof uses a technique introduced in [8] which is also based on the approximation of reals by rationals. It allows us to scale H_A, and distort it carefully, without affecting the dilation by more than a factor arbitrary close to 1, so that the points of S can be accommodated in a finite part of the graph that contains, for any two vertices, the shortest path connecting them in the original graph H_A.

5 Conclusion

We have introduced the notion of the dilation of a set of points, and proven a non-trivial lower bound to the dilation of the points on a closed curve. The big challenge is in proving a similar lower bound for a finite set of points like, e.g., S_5. As to the arrangements of lines, we conjecture that our lower bound holds without the assumption of general position. Another interesting question is the following. What is the lowest possible dilation of a graph whose faces cover the whole plane and have bounded diameter? Our results place this value into the interval $(1.00157, 1.1247)$. Any progress on the upper bound might lead to an improvement of Theorem 6.

References

1. P. Agarwal, R. Klein, Ch. Knauer, S. Langerman, P. Morin, M. Sharir, and M. Soss. Computing the Detour and Spanning Ratio of Paths, Trees and Cycles in 2D and 3D. Manuscript, submitted for publication, 2005.
2. T. M. Apostol. Dirichlet Series in Number Theory, 2nd ed. Springer-Verlag, 1997. Pp. 148–155.
3. S. R. Arikati, D. Z. Chen, L. P. Chew, G. Das, and M. Smid. Planar Spanners and Approximate Shortest Path Queries Among Obstacles in the Plane. European Symposium on Algorithms, 514–528, 1996.
4. G. Das and D. Joseph. Which Triangulations Approximate the Complete Graph? Proc. Int. Symp. Optimal Algorithms, Springer LNCS 401, 168–192, 1989.
5. D.P. Dobkin, S.J. Friedman, and K.J. Supowit. Delaunay Graphs Are Almost as Good as Complete Graphs. *Discrete & Computational Geometry* 5:399-407 (1990).
6. A. Dumitrescu, A. Grüne, and G. Rote. On the Geometric Dilation of Curves and Point Sets. Manuscript, 2004.
7. A. Dumitrescu, A. Ebbers-Baumann, A. Grüne, R. Klein, and G. Rote. On Geometric Dilation and Halving Chords. WADS'05
8. A. Ebbers-Baumann, A. Grüne, and R. Klein. On the Geometric Dilation of Finite Point Sets. 14th International Symposium ISAAC 2003, Kyoto. In T. Ibaraki, N. Katoh, and H. Ono (Eds.) Algorithms and Computation, Proceedings, pp. 250–259, LNCS 2906, Springer-Verlag, 2003. To appear in *Algorithmica.*
9. A. Ebbers-Baumann, A. Grüne, and R. Klein. Geometric Dilation of Closed Planar Curves: New Lower Bounds. To appear in *Computational Geometry: Theory and Applications.*
10. A. Ebbers-Baumann, R. Klein, E. Langetepe, and A. Lingas. A Fast Algorithm for Approximating the Detour of a Polygonal Chain. *Computational Geometry: Theory and Applications*, 27:123–134, 2004.
11. D. Eppstein. Spanning trees and spanners. In Handbook of Computational Geometry, J.-R. Sack and J. Urrutia, editors, pp. 425-461. Elsevier, 1999.
12. D. Eppstein. The Geometry Junkyard. http://www.ics.uci.edu/~eppstein/junkyard/dilation-free/.
13. D. Eppstein and D. Hart. Shortest Paths in an Arrangement with k Line Orientations. Proceedings 10th Annual ACM-SIAM Symposium on Discrete Algorithms, pp. 310-316, 1999.
14. S. Fekete, R. Klein, and A. Nüchter. Online Searching with an Autonomous Robot. Workshop on Algorithmic Foundations of Robotics, 2004.
15. E. Hlawka. Theorie der Gleichverteilung. BI Wissenschaftsverlag.
16. J.M. Keil and C.A. Gutwin. The Delaunay Triangulation Closely Approximates the Complete Euclidean Graph. *Discrete Comput. Geom.* 7, 1992, pp. 13-28.
17. D. Lorenz. On the Dilation of Finite Point Sets. Diploma Thesis. Bonn, 2005.
18. G. Narasimhan and M. Smid. Geometric Spanner Networks. Cambridge University Press, to appear.

The Layered Net Surface Problems in Discrete Geometry and Medical Image Segmentation

Xiaodong Wu[1,*], Danny Z. Chen[2,**], Kang Li[3], and Milan Sonka[4,***]

[1] Departments of Electrical & Computer Engineering and Radiation Oncology, University of Iowa, Iowa City, Iowa 52242, USA
xiaodong-wu@uiowa.edu

[2] Department of Computer Science and Engineering, University of Notre Dame, Notre Dame, IN 46556, USA
chen@cse.nd.edu

[3] Dept. of Electrical and Computer Engineering, Carnegie Mellon University, Pittsburgh, PA 15213, USA
kangl@cmu.edu

[4] Dept. of Electrical and Computer Engineering, University of Iowa, Iowa City, IA 52242-1595, USA
sonka@engineering.uiowa.edu

Abstract. Efficient detection of multiple inter-related surfaces representing the boundaries of objects of interest in d-D images ($d \geq 3$) is important and remains challenging in many medical image analysis applications. In this paper, we study several *layered net surface (LNS)* problems captured by an interesting type of geometric graphs called *ordered multi-column graphs* in the d-D discrete space ($d \geq 3$). The LNS problems model the simultaneous detection of multiple mutually related surfaces in three or higher dimensional medical images. Although we prove that the d-D LNS problem ($d \geq 3$) on a general ordered multi-column graph is NP-hard, the (special) ordered multi-column graphs that model medical image segmentation have the self-closure structures, and admit polynomial time exact algorithms for solving the LNS problems. Our techniques also solve the related *net surface volume (NSV)* problems of computing well-shaped geometric regions of an optimal total volume in a d-D weighted voxel grid. The NSV problems find applications in medical image segmentation and data mining. Our techniques yield the first polynomial time exact algorithms for several high dimensional medical image segmentation problems. The practical efficiency and accuracy of the algorithms are showcased by experiments based on real medical data.

* This research was supported in part by a faculty start-up fund from University of Iowa.

** This research was supported in part by the National Science Foundation under Grants CCR-9988468 and CCF-0515203.

*** This research was supported in part by NIH-NHLBI research grants R01-HL063373 and R01-HL071809.

X. Deng and D. Du (Eds.): ISAAC 2005, LNCS 3827, pp. 17–27, 2005.

1 Introduction

In this paper, we study the *layered net surface* (LNS) problems and their extensions in discrete geometry. These problems arise in d-D medical image segmentation ($d \geq 3$) and other applications.

As a central problem in image analysis, image segmentation aims to define accurate boundaries for the objects of interest captured by image data. Accurate 3-D image segmentation techniques promise to improve medical diagnosis and revolutionize the current medical imaging practice. Although intensive research has been done on image segmentation in several decades, efficient and effective high dimensional image segmentation still poses one of the major challenges in image understanding. As one common practice, to identify surface representing the boundary of the sought 3-D object, 2-D image slices are more or less analyzed independently, and the 2-D results are stacked together to form the 3-D segmentation. One most successful and widely used technique is based on 2-D dynamic programming and optimal graph searching [11]. These approaches have inherent limitations – the most fundamental one stems from the lack of contextual slice-to-slice information when analyzing a sequence of consecutive 2-D images. Performing the segmentation directly on a 3-D image can produce a more consistent segmentation result, yielding 3-D surfaces for object boundaries instead of a set of individual 2-D contours. Another active and far-reaching line of research in this area rely on variational calculus and numerical methods, e.g. level set methods and deformable models [9]. Although these approaches are theoretically powerful, the interfacing between continuous formulations and discrete solutions involve numerical approximation and stability issues.

We present a novel graph-theoretic technique for the problem of simultaneous segmentation of multiple inter-related surfaces in three or higher dimensional medical images, namely, the LNS problem. This technique is practically significant since many surfaces in medical images appear in mutual relations. A number of medical imaging problems can benefit from an efficient method for simultaneous detection of multiple inter-related 3-D surfaces [11, 15, 9].

The simultaneous detection of multiple inter-related surfaces has been studied by the medical image analysis community for a long time. For the 2-D case, there are several satisfactory results [11, 1, 14]. However, little work has been done on the three and higher dimensional cases. Previous attempts [12, 3] on extending graph-search based segmentation methods for the 2-D case to identifying even a single optimal surface in 3-D medical images either made the methods computationally intractable or traded their ability to achieve global optima for computational efficiency. Motivated by this segmentation problem, Wu and Chen [13] introduced the optimal net surface problems and presented efficient polynomial time exact algorithms for them. But, the algorithms in [13] can detect only one optimal surface in 3-D. An implementation of their algorithms and experimental validation based on real 3-D medical images were presented in [7]. More recently, Li *et al.* [8] extended the approach [13, 7] to segmenting multiple inter-related surfaces in 3-D. However, their new method does not consider the very important region information (e.g., homogeneity) for the surface detection.

Modeling the simultaneous detection of multiple inter-related surfaces in high dimensional medical images, we introduce the *layered net surface* (LNS) problems on an interesting type of geometric graphs, called *ordered multi-column graphs*, embedded in the d-D discrete space for $d \geq 3$ (to be defined in Section 2). We further extend the LNS problems to a more general ordered multi-column graph (Section 5). Motivated by segmenting anatomical structures with a relatively regular geometric shape, such as the left ventricles, kidneys, livers, and lungs, we also study several *net surface volume* (NSV) problems, which aim to find well-shaped regions of an optimal "volume" in a d-D weighted voxel grid. These well-shaped geometric regions are closely related to monotonicity and convexity in d-D discrete spaces (Section 4). Our main results in this paper are summarized as follows.

- We develop an efficient algorithm for solving the LNS problem on an interesting type of ordered multi-column graphs in polynomial time, by formulating it as computing a minimum closed set in a vertex-weighted directed graph.
- We extend our LNS technique to solving the NSV problems of computing several classes of optimal well-shaped geometric regions in a d-D weighted voxel grid. These NSV problems arise in data mining [2], image segmentation [1], and data visualization. The classes of regions that we study can be viewed as generalizations of some of the pyramid structures in [2].
- We prove that the LNS problem on a general ordered multi-column graph is NP-hard. However, the (special) ordered multi-column graphs that model medical image segmentation applications have additional properties, and the LNS problem on such graphs is polynomially solvable.
- We apply our polynomial time LNS algorithms to segmenting multiple inter-related object boundaries in 3-D medical images.

We omit the proofs of the lemmas and theorems due to the page limit.

2 The Layered Net Surface (LNS) Problems

A *multi-column graph* $G = (V, E)$ embedded in the d-D discrete space is defined as follows. For a given undirected graph $B = (V_B, E_B)$ embedded in $(d-1)$-D (called the *net model*) and an integer $\kappa > 0$, G is an undirected graph in d-D *generated* by B and κ. For each vertex $v = (x_0, x_1, \ldots, x_{d-2}) \in V_B$, there is a sequence $Col(v)$ of κ vertices in G corresponding to v; $Col(v) = \{(x_0, x_1, \ldots, x_{d-2}, k) : k = 0, 1, \ldots, \kappa - 1\}$, called the *v-column* of G. We denote the vertex $(x_0, x_1, \ldots, x_{d-2}, k)$ of $Col(v)$ by v_k. If an edge $(v, u) \in E_B$, then we say that the v-column and u-column in G are *adjacent* to each other. For each vertex $v_k \in Col(v)$, v_k has edges in G to a non-empty list of consecutive vertices in every adjacent u-column $Col(u)$ of $Col(v)$, say $u_{k'}, u_{k'+1}, \ldots, u_{k'+s}$ $(s \geq 0)$; we call $(u_{k'}, u_{k'+1}, \ldots, u_{k'+s})$, in this order, the *edge interval* of v_k on $Col(u)$, denoted by $I(v_k, u)$. For an edge interval I, we denote by $Bottom(I)$ (resp., $Top(I)$) the d-th coordinate of the first (resp., last) vertex in I (e.g., $Bottom(I(v_k, u)) = k'$ and $Top(I(v_k, u)) = k' + s$ in the above example).

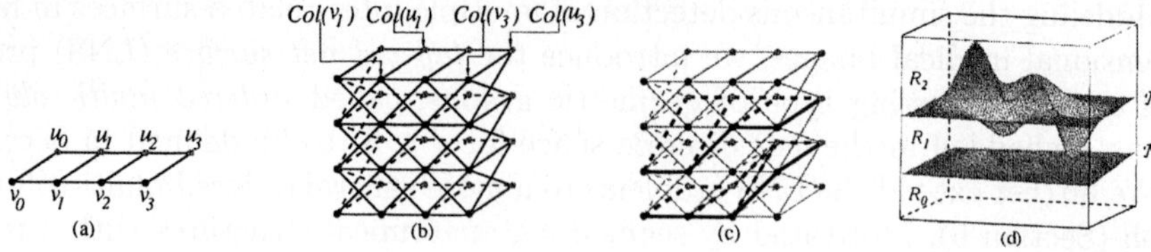

Fig. 1. (a) A 2-D net model B. (b) A 3-D properly ordered multi-column graph G generated by B and $\kappa = 4$ (the edges between $Col(u_i)$ and $Col(u_{i+1})$, $i = 0, 1, 2$, are symmetric to those between $Col(v_i)$ and $Col(v_{i+1})$, and the edges between $Col(v_j)$ and $Col(u_j)$, $j = 1, 2$, are symmetric to those between $Col(v_3)$ and $Col(u_3)$; all these edges are omitted for a better readability). (c) Two (1, 2)-separate net surfaces in G marked by heavy edges. (d) Two net surfaces divide the vertex set of G into three disjoint vertex subsets.

Two adjacent columns $Col(v)$ and $Col(u)$ in G are said to be in *proper order* if for any two vertices v_k and v_{k+1} in $Col(v)$, $Bottom(I(v_k, u)) \leq Bottom(I(v_{k+1}, u))$ and $Top(I(v_k, u)) \leq Top(I(v_{k+1}, u))$, and if the same holds for any two vertices u_k and u_{k+1} of $Col(u)$ on $Col(v)$. The corresponding edge $(v, u) \in E_B$ is called a *proper edge*. If all pairs of adjacent columns in G are in proper order, then we call G a *properly ordered* multi-column graph (briefly, a *properly ordered graph*). Figures 1(a)–1(b) show a net model and a properly ordered graph.

Note that in medical image segmentation, the boundaries of the target objects (e.g., organs) are often sufficiently "smooth". The smoothness constraint on the sought surfaces is modeled by the proper ordering of the edges in a multi-column graph G, that is, the edges connecting each vertex v_k in G to every adjacent column $Col(u)$ of $Col(v)$ form an edge interval on $Col(u)$, and such edge intervals for any two adjacent columns of G are in proper order.

A *net surface* in G (also called a *net*) is a subgraph of G defined by a function $\mathcal{N}: V_B \rightarrow \{0, 1, \ldots, \kappa - 1\}$, such that for every edge $(v, u) \in E_B$, $(v_{k'}, u_{k''})$, with $k' = \mathcal{N}(v)$ and $k'' = \mathcal{N}(u)$, is also an edge in G. For simplicity, we denote a net by its function $\mathcal{N}$. Intuitively, a net $\mathcal{N}$ in G is a special mapping of the $(d-1)$-D net model B to the d-D space, such that $\mathcal{N}$ "intersects" each v-column of G at exactly one vertex and $\mathcal{N}$ preserves all topologies of B. $\mathcal{N}$ can be viewed as a functional "surface" of B in d-D defined on the $(d-1)$-D space in which B is embedded.

Given two integers L and U, $0 < L < U$, two nets $\mathcal{N}_1$ and $\mathcal{N}_2$ of a properly ordered graph G are said to be (L, U)-*separate* if $L \leq \mathcal{N}_2(v) - \mathcal{N}_1(v) \leq U$ for every vertex $v \in V_B$. Roughly speaking, $\mathcal{N}_1$ and $\mathcal{N}_2$ do not cross each other and are within a specified range of distance (see Figure 1(c)). For a given set of $l-1$ integer parameter pairs $\{(L_i, U_i) : 0 < L_i < U_i, 1 \leq i < l\}$, $l \geq 2$, we consider l net surfaces $\mathcal{NS} = \{\mathcal{N}_1, \mathcal{N}_2, \ldots, \mathcal{N}_l\}$ in G such that $\mathcal{N}_{i+1}$ is "on top" of $\mathcal{N}_i$ (i.e., $\forall v \in V_B, \mathcal{N}_{i+1}(v) > \mathcal{N}_i(v)$), and $\mathcal{N}_i$ and $\mathcal{N}_{i+1}$ are (L_i, U_i)-separate. Then, these l net surfaces partition the vertex set V of G into $l+1$ disjoint subsets R_i, with $R_0 = \{v_k : v \in V_B, 0 \leq k \leq \mathcal{N}_1(v)\}$, $R_i = \{v_k : v \in V_B, \mathcal{N}_i(v) < k \leq \mathcal{N}_{i+1}(v)\}$ for $i = 1, 2, \ldots, l-1$, and $R_l = \{v_k : v \in V_B, \mathcal{N}_l(v) < k < \kappa\}$ (see Figure 1(d)).

Motivated by medical image segmentation [11,16,9], we assign costs to every vertex of G as follows. Each vertex $v_k \in V$ has an *on-surface cost* $b(v_k)$, which is an arbitrary real value. For each region R_i ($i = 0, 1, \ldots, l$), every vertex $v_k \subseteq V$ is assigned a real-valued *in-region cost* $c_i(v_k)$. The on-surface cost of each vertex is inversely related to the likelihood that it may appear on a desired net surface, while the in-region costs $c_i(\cdot)$ ($i = 0, 1, \ldots, l$) measure the inverse likelihood of a given vertex preserving the expected regional properties of the partition $\{R_0, R_1, \ldots, R_l\}$. Both the on-surface and in-region costs for image segmentation can be determined using low-level image features [9,11,16].

The **layered net surface (LNS)** problem seeks l net surfaces $\mathcal{NS} = \{\mathcal{N}_1, \mathcal{N}_2, \ldots, \mathcal{N}_l\}$ in G such that the total cost $\alpha(\mathcal{NS})$ induced by the l net surfaces in $\mathcal{NS}$, with

$$\alpha(\mathcal{NS}) = \sum_{i=1}^{l} b(\mathcal{N}_i) + \sum_{i=0}^{l} c_i(R_i) = \sum_{i=1}^{l} \sum_{u \in V(\mathcal{N}_i)} b(u) + \sum_{i=0}^{l} \sum_{u \in R_i} c_i(u),$$

is minimized, where $V(H)$ denotes the vertex set of a graph H.

In fact, our algorithmic framework is general enough for the cases in which each vertex has *only* an on-surface cost, *only* in-region costs, or both. We will present our approach for the case where each vertex has both the on-surface and in-region costs.

3 Algorithm for the Layered Net Surface (LNS) Problem

This section gives our polynomial time algorithm for the layered net surface problem on a d-D properly ordered graph $G = (V, E)$. We first exploit the self-closure structure of the LNS problem, and then model it as a minimum-cost closed set problem based on a nontrivial graph transformation scheme.

A *closed set* $\mathcal{C}$ in a directed graph with arbitrary vertex costs $w(\cdot)$ is a subset of vertices such that all successors of any vertex in $\mathcal{C}$ are also contained in $\mathcal{C}$ [10]. The *cost* of a closed set $\mathcal{C}$, denoted by $w(\mathcal{C})$, is the total cost of all vertices in $\mathcal{C}$. Note that a closed set can be empty (with a cost zero). The minimum-cost closed set problem seeks a closed set in the graph whose cost is minimized.

3.1 The Self-closure Property of the LNS Problem

Our algorithm for the LNS problem hinges on the following observations about the self-closure structure of any feasible LNS solution. Recall that in a set of l feasible net surfaces $\mathcal{NS} = \{\mathcal{N}_1, \mathcal{N}_2, \ldots, \mathcal{N}_l\}$ in G, $\mathcal{N}_{i+1}$ is "on top" of $\mathcal{N}_i$, for each $i = 1, 2, \ldots, l-1$.

For a vertex $v_k \in V$ (i.e., $v \in V_B$ and $0 \leq k < \kappa$) and each adjacent column $Col(u)$ of $Col(v)$ (i.e., $(v, u) \in E_B$), the *lower-eligible-neighbor* of v_k on $Col(u)$ is the vertex in $Col(u)$ with the smallest d-th coordinate that has an edge to v_k in G (i.e., the vertex in $Col(u)$ with the smallest d-th coordinate that can possibly appear together with v_k on a same feasible net surface in G).

Given the surface separation constraints, we define below the *upstream* and *downstream* vertices of any vertex in G, to help characterize the spatial relations between feasible net surfaces in G. For every vertex $v_k \in V$ and $1 \leq i < l$ (resp., $1 < i \leq l$), the *i-th upstream* (resp., *downstream*) vertex of v_k is v_{k+L_i} (resp., $v_{\max\{0,k-U_{i-1}\}}$) if $k + L_i < \kappa$ (resp., $k - L_{i-1} \geq 0$). Intuitively, if $v_k \in \mathcal{N}_i$, then the i-th upstream (resp., downstream) vertex of v_k is the vertex in $Col(v)$ with the smallest d-th coordinate that can be on $\mathcal{N}_{i+1}$ (resp., $\mathcal{N}_{i-1}$).

We say that a vertex v_k is *below* (resp., *above*) a net surface $\mathcal{N}_i$ if $\mathcal{N}_i(v) > k$ (resp., $\mathcal{N}_i(v) < k$), and denote by $LO(\mathcal{N}_i)$ the subset of all vertices of G that are on or below $\mathcal{N}_i$. For every vertex $v_k \in LO(\mathcal{N}_i)$, consider its lower-eligible-neighbor $u_{k'}$ on any adjacent $Col(u)$ of $Col(v)$. Let $r = \mathcal{N}_i(v)$ and $u_{k''}$ be the lower-eligible-neighbor of v_r on $Col(u)$ ($v_r \in Col(v)$ is on $\mathcal{N}_i$). Then by the definition of net surfaces, $k'' \leq \mathcal{N}_i(u)$. Since $k \leq \mathcal{N}_i(v)$, we have $k' \leq k''$ due to the proper ordering. Thus, $k' \leq \mathcal{N}_i(u)$, and further, $u_{k'} \in LO(\mathcal{N}_i)$. Hence, we have the following observation.

Observation 1. *For any feasible net surface $\mathcal{N}_i$ in G, if a vertex v_k is in $LO(\mathcal{N}_i)$, then every lower-eligible-neighbor of v_k is also in $LO(\mathcal{N}_i)$.*

Observation 1 characterizes the self-closure property of every set $LO(\mathcal{N}_i)$. However, our task is more involved since the l net surfaces in $\mathcal{NS}$ are inter-related. We need to further examine the closure structure between the $LO(\mathcal{N}_i)$'s.

Observation 2. *Given any set $\mathcal{NS} = \{\mathcal{N}_1, \mathcal{N}_2, \ldots, \mathcal{N}_l\}$ of l feasible net surfaces in G, the i-th upstream (resp., downstream) vertex of each vertex in $LO(\mathcal{N}_i)$ is in $LO(\mathcal{N}_{i+1})$ (resp., $LO(\mathcal{N}_{i-1})$), for every $1 \leq i < l$ (resp., $1 < i \leq l$).*

Observations 1 and 2 show an important self-closure structure of the LNS problem, which is crucial to our LNS algorithm and suggests a connection between our target problem and the minimum-cost closed set problem [10]. In our LNS approach, instead of directly searching for an optimal set of l net surfaces, $\mathcal{NS}^* = \{\mathcal{N}_1^*, \mathcal{N}_2^*, \ldots, \mathcal{N}_l^*\}$, we look for l optimal subsets of vertices in G, $LO(\mathcal{N}_1^*) \subset LO(\mathcal{N}_2^*) \subset \ldots \subset LO(\mathcal{N}_l^*)$, such that each $LO(\mathcal{N}_i^*)$ uniquely defines the net surface $\mathcal{N}_i^* \in \mathcal{NS}^*$.

3.2 The Graph Transformation Scheme

Our LNS algorithm is based on a sophisticated graph transformation scheme, which enables us to simultaneously identify $l > 1$ optimal inter-related net surfaces as a whole by computing a minimum closed set in a weighted directed graph G' that we transform from G. This section presents the construction of G', which crucially relies on the self-closure structure shown in Section 3.1.

We construct the vertex-weighted directed graph $G' = (V', E')$ from the d-D properly ordered graph $G = (V, E)$, as follows. G' contains l vertex-disjoint subgraphs $\{G_i' = (V_i', E_i') : i = 1, 2, \ldots, l\}$; each G_i' is for the search of the i-th net surface $\mathcal{N}_i$. $V' = \bigcup_{i=1}^{l} V_i'$ and $E' = \bigcup_{i=1}^{l} E_i' \cup E_s'$. The surface separation constraints between any two consecutive net surfaces $\mathcal{N}_i$ and $\mathcal{N}_{i+1}$ are enforced in G' by a subset of edges in E'_s, which connect the subgraphs G'_i and G'_{i+1}.

The construction of each subgraph $G'_i = (V'_i, E'_i)$ is similar to that in [13]. For G'_i, every vertex v_k in G corresponds to exactly one vertex $v^i_k \in V'_i$. Each column $Col(v)$ in G associates with a *chain* $v^i_{\kappa-1} \to v^i_{\kappa-2} \to \cdots \to v^i_0$ in G'_i. We then put directed edges into E'_s between G'_i and G'_{i+1}, to enforce the surface separation constraints. For each vertex v^i_k with $k < \kappa - L_i$ on the chain $Ch_i(v)$ in G'_i, a directed edge is put in E'_s from v^i_k to $v^{i+1}_{k+L_i}$ on $Ch_{i+1}(v)$ in G'_{i+1}. On the other hand, each vertex v^{i+1}_k with $k \geq L_i$ on $Ch_{i+1}(v)$ has a directed edge in E'_s to $v^i_{k'}$ on $Ch_i(v)$ with $k' = \max\{0, k - U_i\}$ (note that $v_{k'}$ in G is the $(i+1)$-th downstream vertex of v_k). Note that in this construction, each vertex v^i_k with $k \geq \kappa - L_i$ has no edge to any vertex on $Ch_{i+1}(v)$, and each vertex v^{i+1}_k with $k < L_i$ has no edge to any vertex on $Ch_i(v)$. These vertices of G' are called *deficient vertices*, whose corresponding vertices in G cannot possibly appear in any feasible solution for the LNS problem. By exploiting the geometric properties of the properly ordered graphs, all deficient vertices in G' can be pruned in linear time. We simply denote the graph thus resulted also by G'. Then, for every $v \in V_B$ and $i = 1, 2, \ldots, l$, let $\mu_i(v)$ and $\kappa_i(v)$ be the smallest and largest d-th coordinates of the vertices on the chain $Ch_i(v)$ of G'_i, respectively. We then assign a cost $w(v^i_k)$ for each vertex v^i_k in G'_i: If $k = \mu_i(v)$, then $w(v^i_k) = b(v_k) + \sum_{j=0}^{k}[c_{i-1}(v_j) - c_i(v_j)]$; otherwise, $w(v^i_k) = [b(v_k) - b(v_{k-1})] + [c_{i-1}(v_k) - c_i(v_k)]$. This completes the construction of G'.

3.3 Computing Optimal Layered Net Surfaces for the LNS Problem

The graph G' thus constructed allows us to find l optimal net surfaces in G, by computing a non-empty minimum-cost closed set in G'. Given any closed set $\mathcal{C} \neq \varnothing$ in G', we define l feasible net surfaces, $\mathcal{NS} = \{\mathcal{N}_1, \mathcal{N}_2, \ldots, \mathcal{N}_l\}$, in G, as follows. Recall that we search for each net $\mathcal{N}_i$ in the subgraph $G'_i = (V'_i, E'_i)$. Let $\mathcal{C}_i = \mathcal{C} \cap V'_i$. For each vertex $v \in V_B$, denote by $\mathcal{C}_i(v)$ the set of vertices of $\mathcal{C}_i$ on the chain $Ch_i(v)$ of G'_i. Based on the construction of G'_i, it is not hard to show that $\mathcal{C}_i(v) \neq \varnothing$. Let $r_i(v)$ be the largest d-th coordinate of the vertices in $\mathcal{C}_i(v)$. Define the function $\mathcal{N}_i$ as $\mathcal{N}_i(v) = r_i(v)$ for every $v \in V_B$. The following lemma is a key to our algorithm.

Lemma 1. *Any closed set $\mathcal{C} \neq \varnothing$ in G' specifies l feasible net surfaces in G whose total cost differs from that of $\mathcal{C}$ by a fixed value $c_l(V)$.*

Next, we argue that any l feasible net surfaces, $\mathcal{NS} = \{\mathcal{N}_1, \mathcal{N}_2, \ldots, \mathcal{N}_l\}$, in G correspond to a closed set $\mathcal{C} \neq \varnothing$ in G'. Based on the construction of G', every vertex v_k on the net $\mathcal{N}_i$ corresponds to a vertex v^i_k in G'_i (v^i_k is not a deficient vertex). We construct a closed set $\mathcal{C}_i \neq \varnothing$ in G'_i for each net $\mathcal{N}_i$, as follows. Initially, let $\mathcal{C}_i = \varnothing$. For each vertex $v \in V_B$, we add to $\mathcal{C}_i$ the subset $\mathcal{C}_i(v) = \{v^i_k \ : \ k \leq \mathcal{N}_i(v)\}$ of vertices on $Ch_i(v)$ of G'_i.

Lemma 2. *Any set $\mathcal{NS}$ of l feasible net surfaces in G defines a closed set $\mathcal{C} \neq \varnothing$ in G' whose cost differs from that of $\mathcal{NS}$ by a fixed value.*

By Lemmas 1 and 2, we compute a minimum-cost closed set $C^* \neq \varnothing$ in G', which specifies l optimal net surfaces in G. Note that G' has $O(l \cdot n)$ vertices and $O(l \cdot n \cdot \frac{m_B}{n_B})$ edges, where $n = |V|$ is the number of vertices in G, and $n_B = |V_B|$ and $m_B = |E_B|$ for the net model B. By using the minimum s-t cut algorithm in [5] to compute a minimum-cost closed set in G', we have the following result.

Theorem 1. *The LNS problem can be solved in* $O(l^2 n^2 \frac{m_B}{n_B} \log(\frac{l \cdot n \cdot n_B}{m_B}))$ *time.*

4 Algorithms for the Net Surface Volume (NSV) Problems

This section presents our algorithms for several optimal net surface volume (NSV) problems. Specifically, instead of looking for multiple inter-related net surfaces as in Section 3, for a given d-D voxel grid $\Gamma = [0..N-1]^d$ of $n = N^d$ cells, with each cell $\mathbf{x}(x_0, x_1, \ldots, x_{d-1}) \in \Gamma$ having an arbitrary real "volume" value $vol(\mathbf{x})$, we seek multiple surfaces that enclose a well-shaped region $R \subseteq \Gamma$, such that the *volume* $vol(R)$ of R, $vol(R) = \sum_{\mathbf{x} \in R} vol(\mathbf{x})$, is minimized (or maximized). Note that even the case of the NSV problem on finding an optimal simple polygon in a weighted 2-D grid is in general NP-hard [1].

We consider two classes of regions, called *weakly watershed-monotone regions* and *watershed-monotone shells*, defined as follows. For any integers $0 \leq i < d$ and $0 \leq c < N$, let $\Gamma_i(c)$ denote all voxels of Γ whose x_i-coordinate is c (note that $\Gamma_i(c)$ is orthogonal to the x_i-axis). A region R in Γ is said to be x_i-*monotone* if for any line l parallel to the x_i-axis, the intersection $R \cap l$ is either empty or a continuous segment. Further, we say that R is *watershed-monotone with respect to* $\Gamma_i(c)$ if (1) R is x_i-monotone, and (2) for any line l orthogonal to $\Gamma_i(c)$, if the intersection $R \cap l \neq \varnothing$, then $R \cap l$ intersects a voxel of $R \cap \Gamma_i(c)$. (Intuitively, the intersection of R and $\Gamma_i(c)$ is equal to the projection of R onto $\Gamma_i(c)$, and is like a "watershed" of R.) If for every $i = 0, 1, \ldots, d-1$, R is watershed-monotone to a $\Gamma_i(c_i)$ for an integer $0 \leq c_i < N$, then we say that R is *watershed-monotone.* A region $R \subseteq \Gamma$ is *weakly watershed-monotone* if R is watershed-monotone to every axis in a set of $d-1$ axes of Γ and is monotone (but need not be watershed-monotone) to the remaining axis. Clearly, watershed-monotone regions are a subclass of weakly watershed-monotone regions. Suppose R is watershed-monotone with respect to some $\Gamma_i(c_i)$, for each $i = 0, 1, \ldots, d-1$; then it is easy to see that $\cap_{i=0}^{d-1} \Gamma_i(c_i) \neq \varnothing$. A voxel in $\cap_{i=0}^{d-1} \Gamma_i(c_i)$ is called a *kernel voxel* of R. Our second region class is called the *watershed-monotone shells.* For any two watershed-monotone regions R_1 and R_2 such that R_1 and R_2 have a common kernel voxel $\mathbf{c}$ and $R_2 \subseteq R_1$, the region R in Γ bounded between R_1 and R_2, i.e., $R = R_1 - R_2$, is a *watershed-monotone shell.*

Theorem 2. *(1) The optimal weakly watershed-monotone region problem is solvable in* $O(dn^2 \log \frac{n}{d})$ *time. (2) The optimal watershed-monotone shell problem is solvable in* $O(dn^2 \log \frac{n}{d})$ *time.*

5 Algorithm for the Bipartite LNS (BLNS) Problem

In this section, we consider the layered net surface problem on a more general *ordered* multi-column graph. Recall that any two adjacent columns of a *properly* ordered multi-column graph are in proper order (Section 2). We now define the *reverse order* on two adjacent columns $Col(v)$ and $Col(u)$ in a d-D multi-column graph $G = (V, E)$ generated by a $(d-1)$-D net model $B = (V_B, E_B)$: If for any two vertices v_k and v_{k+1} in $Col(v)$, $Bottom(I(v_k, u)) \geq Bottom(I(v_{k+1}, u))$ and $Top(I(v_k, u)) \geq Top(I(v_{k+1}, u))$, and if the same holds for any two vertices u_k and u_{k+1} of $Col(u)$ on $Col(v)$, then we say that $Col(u)$ and $Col(v)$ are in *reverse order.* If every two adjacent columns in G are in either proper order or reverse order, then we call G a d-D *ordered* multi-column graph. Further, for two (L, U)-separate nets $\mathcal{N}_1$ and $\mathcal{N}_2$ in G, if two adjacent columns $Col(v)$ and $Col(u)$ are in reverse order, then $L \leq \mathcal{N}_1(v) - \mathcal{N}_2(v) \leq U$ and $L \leq \mathcal{N}_2(u) - \mathcal{N}_1(u) \leq U$. In this section, we assume that each vertex in G has only an on-surface cost.

We can prove that the LNS problem on a general d-D ordered multi-column graph ($d \geq 3$) is NP-hard, by reducing to it the minimum vertex cover problem that is known to be NP-complete [4].

Next, we consider the LNS problem on a d-D ordered multi-column graph $G = (V, E)$ with a special net model $B = (V_B, E_B)$, defined as follows. First, remove from B all reverse edges; the remaining B is a set $\mathcal{CC}$ of connected components with proper edges only. Then, contract each connected component of $\mathcal{CC}$ into a single vertex. Finally, for each (removed) reverse edge $(v, u) \in E_B$, say, v in $C' \in \mathcal{CC}$ and u in $C'' \in \mathcal{CC}$ ($C' = C''$ is possible), add an edge between the contracted vertices of C' and C''. The resulting graph is called the *p-contracted graph* of B. The **bipartite LNS** (BLNS) problem is defined on a d-D ordered multi-column graph with a net model B whose p-contracted graph is *bipartite.* Let $n = |V|$, $m_B = |E_B|$, and $n_B = |V_B|$.

Theorem 3. *The general BLNS problem can be solved in* $O(l^2 n^2 \frac{m_B}{n_B} \log(\frac{l \cdot n \cdot n_B}{m_B}))$ *time, where* l *is the number of sought net surfaces.*

6 Implementation and Experiments

To further examine the behavior and performance of our LNS algorithm, we implemented it in standard C++ templates. After the implementation, we extensively experimented with 3-D synthetic and real medical image data, and compared with a previously validated slice-by-slice 2-D segmentation approach based on graph search techniques [14]. Our LNS program was tested on an AMD Athlon MP 2000+ Dual CPU workstation running MS Windows XP.

The experiments showed that our LNS algorithm and software are computationally efficient and produce highly accurate and consistent segmentation results. The average execution time of our simultaneous 3-surface detection algorithm on images of size $200 \times 200 \times 40$ is 401.3 seconds. An accuracy assessment on images of physical phantom tubes revealed that the overall signed errors for the inner and outer diameters derived from the tube boundaries were (mean $\pm$

standard deviation) $-0.36 \pm 2.47\%$ and $-0.08 \pm 1.35\%$, respectively. Our LNS approach was tested on segmenting both the inner and outer airway wall surfaces in CT images, in which outer wall surfaces are very difficult to detect due to their blurred and discontinuous appearance and the presence of adjacent blood vessels. The CT images had a nearly isotropic resolution of $0.7 \times 0.7 \times 0.6mm^3$. The currently used 2-D dynamic programming method is unsuitable for the segmentation of the outer airway wall. Our new approach produces good segmentation results for both airway wall surfaces in a robust manner. Comparing to manual tracing on 39 randomly selected slices, our LNS technique yielded signed border positioning errors of -0.01 ± 0.15mm and 0.01 ± 0.17mm for the inner and outer wall surfaces, respectively.

References

1. T. Asano, D.Z. Chen, N. Katoh, and T. Tokuyama, Efficient algorithms for optimization-based image segmentation, *Int. Journal of Computational Geometry & Applications*, 11(2)(2001), pp. 145-166.
2. D.Z. Chen, J. Chun, N. Katoh, and T. Tokuyama, Efficient algorithms for approximating a multi-dimensional voxel terrain by a unimodal terrain, *Proc. 10th Annual Int. Computing and Combinatorics Conf.*, Jeju Island, Korea, 2004, pp. 238-248.
3. R.J. Frank, D.D. McPherson, K.B. Chandran, and E.L. Dove, Optimal surface detection in intravascular ultrasound using multi-dimensional graph search, *Computers in Cardiology*, IEEE, Los Alamitos, 1996, pp. 45-48.
4. M.R. Garey and D.S. Johnson, *Computers and Intractability: A Guide to the Theory of NP-Completeness*, Freeman, San Francisco, CA, 1979.
5. A.V. Goldberg and R.E. Tarjan, A new approach to the maximum-flow problem, *J. Assoc. Comput. Mach.*, 35(1988), pp. 921-940.
6. X. Huang, D. Metaxas, and T. Chen, MetaMorphs: Deformable shape and texture models, *Proc. IEEE Computer Society Conference on Computer Vision and Pattern Recognition (CVPR),* vol. I, June 2004, pp. 496-503.
7. K. Li, X. Wu, D.Z. Chen, and M. Sonka, Efficient optimal surface detection: Theory, implementation and experimental validation, *Proc. SPIE's Int. Symp. on Medical Imaging: Imaging Processing*, Vol. 5370, San Diego, CA, 2004, pp. 620-627.
8. K. Li, X. Wu, D.Z. Chen, and M. Sonka, Optimal Surface Segmentation in Volumetric Images — A Graph-Theoretic Approach, accepted to *IEEE Trans. on Pattern Analysis and Machine Intelligence,* 2005.
9. S. Osher and N. Paragios, Eds., *Geometric Level Set Methods in Imaging, Vision and Graphics.* Springer, 2003.
10. J.C. Picard, Maximal closure of a graph and applications to combinatorial problems, *Management Science*, 22(1976), 1268-1272.
11. M. Sonka, V. Hlavac, and R. Boyle, *Image Processing, Analysis, and Machine Vision*, 2nd edition, Brooks/Cole Publishing Company, Pacific Grove, CA, 1999.
12. D.R. Thedens, D.J. Skorton, and S.R. Fleagle, Methods of graph searching for border detection in image sequences with applications to cardiac magnetic resonance imaging, *IEEE Trans. on Medical Imaging*, 14(1)(1995), pp. 42-55.
13. X. Wu and D.Z. Chen, Optimal net surface problems with applications, *Proc. 29th International Colloquium on Automata, Languages and Programming (ICALP)*, 2002, pp. 1029-1042.

14. F. Yang, G. Holzapfel, C. Schulze-Bauer, R. Stollberger, D. Thedens, L. Bolinger, A. Stolpen, and M. Sonka, Segmentation of wall and plaque in in-vitro vascular MR image, *International Journal on Cardiovascular Imaging*, 19(5)(2003), pp. 419-428.
15. X. Zeng, L.H. Staib, R.T. Schultz, and J.S. Duncan, Segmentation and measurement of the cortex from 3-D MR images using coupled surfaces propagation, *IEEE Trans. Med. Imag.*, 18(1999), pp. 927-937.
16. S. Zhu and A. Yuille, Region competition: Unifying snakes, region growing, and Bayes/MDL for multiband images segmentation, *IEEE Trans. on Pattern Analysis and Machine Intelligence,* 18(1996), pp. 884-900.

Separability with Outliers

Sariel Har-Peled[1,*] and Vladlen Koltun[2]

[1] Department of Computer Science, University of Illinois,
201 N. Goodwin Avenue, Urbana, IL, 61801, USA
sariel@cs.uiuc.edu

[2] Computer Science Department, 353 Serra Mall,
Gates 464, Stanford University, Stanford, CA 94305, USA
vladlen@cs.stanford.edu

Abstract. We develop exact and approximate algorithms for computing optimal separators and measuring the extent to which two point sets in d-dimensional space are separated, with respect to different classes of separators and various extent measures. This class of geometric problems generalizes two widely studied problem families, namely separability and the computation of statistical estimators.

1 Introduction

Consider a family $\mathcal{F}$ of surfaces in $\mathbb{R}^d$, for $d \geq 2$, called *separators*, such that every $f \in \mathcal{F}$ partitions $\mathbb{R}^d$ into at least two connected components, some of which are labelled *inside* f, while the rest are said to be *outside* f. For instance, if $\mathcal{F}$ is a set of hyperplanes, the open halfspace lexicographically below $f \in \mathcal{F}$ is said to be inside f, while the other halfspace is outside. As another example, if $\mathcal{F}$ is a set of spheres, points inside $f \in \mathcal{F}$ are inside while the other connected component of $\mathbb{R}^d \setminus f$ is outside. Given such a family $\mathcal{F}$ and two sets of points $\mathcal{R}$ and $\mathcal{B}$ in $\mathbb{R}^d$ (said to be *red* and *blue*, respectively), such that $|\mathcal{R}| = |\mathcal{B}| = n$, the *separability* problem asks for finding a separator $f \in \mathcal{F}$, if one exists, such that all the blue (resp., red) points are inside (resp., outside) f.

The study of the separability problem is motivated in part by the fundamental classification problem in machine learning: Given two sets of objects (the *training sets*), construct a *predictor* that will facilitate a rapid classification of a new object as belonging to one of the sets. For the separability problem, the training sets are sets of points, and the computed separator is a good candidate predictor for classifying all other points in the space. Numerous statistical techniques have been developed for the classification problem, such as Support Vector Machines [17, 27]. The separability problem has also been studied from the combinatorial point of view, particularly in computational geometry, with the aim of developing algorithms with guaranteed correctness and bounds on worst-case running time. In this context it is customary to assume that the dimension d is constant.

* Work on this paper has been supported by a NSF CAREER award CCR-0132901.

X. Deng and D. Du (Eds.): ISAAC 2005, LNCS 3827, pp. 28–39, 2005.

For hyperplane and sphere separators linear-time algorithms are known [32, 34, 36]. Algorithms have been found and hardness results have been developed for a number of types of separators for $d = 2$, such as (boundaries of) strips, wedges, double-wedges [8, 29, 28], convex polygons and simple polygons [22, 24, 33]. Separability for slab, wedge, prism and cone separators in $d = 3$ has received recent attention, alongside other types of separators in 3-space, and exact algorithms with running times ranging from near-linear to near-$O(n^8)$ were presented [3, 30].

Previous works on the separability problem in computational geometry deal with finding a separator that *completely* separates the red and blue points, as described above. However, in practical applications it is likely that the point sets will be *almost*, but not completely separable, in a sense that is made precise below, due to noise, sampling and round-off errors. In such scenario the above algorithms simply report that the point sets are not separable and terminate. In this paper we develop separability algorithms that ensure robustness to inaccuracies in the input by computing the *best* separator (when the point sets are separable, this is simply a complete separator as above).

Specifically, for $p, q \in \mathbb{R}^d$ and $\mathcal{S}, \mathcal{T} \subseteq \mathbb{R}^d$ define $d(p,q) = \| p - q \|$, $d(p, \mathcal{T}) = \min_{q \in \mathcal{T}} \| p - q \|$, and $d(\mathcal{S}, \mathcal{T}) = \min_{p \in \mathcal{S}, q \in \mathcal{T}} \| p - q \|$. Consider the following measures for estimating the extent of a point set $\mathcal{S} \in \mathbb{R}^3$ with respect to a separator $f \in \mathcal{F}$.

- The combinatorial measure: $\mathcal{M}_c(f, \mathcal{S}) = |\mathcal{S}|$. (This quantity does not depend on f for a fixed $\mathcal{S}$.)
- The $\mathcal{L}_\infty$ measure: $\mathcal{M}_\infty(f, \mathcal{S}) = \max_{p \in \mathcal{S}} d(f, p)$.
- The $\mathcal{L}_1$ measure: $\mathcal{M}_1(f, \mathcal{S}) = \sum_{p \in \mathcal{S}} d(f, p)$.
- The $\mathcal{L}_2$ measure: $\mathcal{M}_2(f, \mathcal{S}) = \sum_{p \in \mathcal{S}} (d(f, p))^2$.

For $\mathcal{R}$, $\mathcal{B}$ and $f \in \mathcal{F}$ as above, let a red (resp., blue) *outlier* be defined as a point of $\mathcal{R}$ (resp., of $\mathcal{B}$) that lies inside (resp., outside) f. Let $\mathcal{R}_f \subseteq \mathcal{R}$ (resp., $\mathcal{B}_f \subseteq \mathcal{B}$) be the set of red (resp., blue) outliers and define $\mathcal{O}_f := \mathcal{R}_f \cup \mathcal{B}_f$. In this paper we develop algorithms that, given $\mathcal{R}$, $\mathcal{B}$ and $\mathcal{F}$, compute a separator $f \in \mathcal{F}$ that minimizes (exactly or approximately) one of the above extent measures of $\mathcal{O}_f$ with respect to f. That is, the separator minimizes the number of outliers, the distance of the farthest outlier from the separator, the sum of the distances of the outliers from the separator, or the sum of the squares of these distances. We call this class of problems *separability with outliers*. We concentrate in this initial study on hyperplane and sphere separators. We present exact algorithms that apply in any $d = O(1)$. Since exact algorithms quickly become impractical as the dimension increases, we present approximation algorithms for all the studied problems. Most of these are candidates for efficient implementation. Specifically, we show that:

- Let $\mathcal{F}$ be the set of hyperplanes (resp., spheres) in $\mathbb{R}^d$. The separator $f^* \in \mathcal{F}$ that minimizes $\mathcal{M}_c(f, \mathcal{O}_f)$ for $f \in \mathcal{F}$ can be computed in time $O(nk^{d+1} \log k)$ (resp., $O(nk^{d+2} \log k)$), where $k = \mathcal{M}_c(f^*, \mathcal{O}_{f^*})$. For any $\varepsilon > 0$, a separator

$f \in \mathcal{F}$, such that $\mathcal{M}_c(f, \mathcal{O}_f) \leq (1+\varepsilon)\mathcal{M}_c(f^*, \mathcal{O}_{f^*})$ can be computed in time $O\big(n(\varepsilon^{-2}\log n)^{d+1}\big)$ (resp., $O\big(n(\varepsilon^{-2}\log n)^{d+2}\big)$).

- Let $\mathcal{F}$ be the set of hyperplanes (resp., spheres) in $\mathbb{R}^d$. The separator $f^* \in \mathcal{F}$ that minimizes $\mathcal{M}_\infty(f, \mathcal{O}_f)$ for $f \in \mathcal{F}$ can be computed in time $O\big(n^{\lceil d/2 \rceil}\big)$ (resp., $O\big(n^{\lfloor d/2 \rfloor + 1}\big)$). For any $\varepsilon > 0$, a separator $f \in \mathcal{F}$, such that $\mathcal{M}_\infty(f, \mathcal{O}_f) \leq (1+\varepsilon)\mathcal{M}_\infty(f^*, \mathcal{O}_{f^*})$ can be computed in time $O\big(n/\varepsilon^{(d-1)/2}\big)$ (resp., $O\big(n/\varepsilon^{(d-1)/2} + 1/\varepsilon^{4d}\big)$).
- Let $\mathcal{F}$ be the set of hyperplanes in $\mathbb{R}^d$. The separator $f^* \in \mathcal{F}$ that minimizes $\mathcal{M}_1(f, \mathcal{O}_f)$ (resp., $\mathcal{M}_2(f, \mathcal{O}_f)$) for $f \in \mathcal{F}$ can be computed in time $O(n^d)$ (resp., $O(n^{d+1})$). For any $\varepsilon > 0$, a separator $f \in \mathcal{F}$, such that $\mathcal{M}_1(f, \mathcal{O}_f) \leq (1+\varepsilon)\mathcal{M}_1(f^*, \mathcal{O}_{f^*})$ (resp., $\mathcal{M}_2(f, \mathcal{O}_f) \leq (1+\varepsilon)\mathcal{M}_2(f^*, \mathcal{O}_{f^*})$) can be computed in time $O\big(n/\varepsilon^{(d-1)/2}\big)$ (resp., $O\big(n/\varepsilon^{d/2}\big)$).
- Let $\mathcal{F}$ be the set of spheres in $\mathbb{R}^d$. For any $\varepsilon > 0$, a separator $f \in \mathcal{F}$, such that $\mathcal{M}_1(f, \mathcal{O}_f) \leq (1+\varepsilon)\mathcal{M}_1(f^*, \mathcal{O}_{f^*})$ (resp., $\mathcal{M}_2(f, \mathcal{O}_f) \leq (1+\varepsilon)\mathcal{M}_2(f^*, \mathcal{O}_{f^*})$) can be computed in time $O\Big(\big(\frac{n}{\varepsilon}\log\frac{n}{\varepsilon}\big)^{d+1}\Big)$.

The running time for most of these algorithms can be improved by specialized techniques when the dimension is small (e.g., $d \leq 3$). We do not describe exact algorithms for finding optimal sphere separators with respect to the $\mathcal{L}_1$ and $\mathcal{L}_2$ measures, since computing such a separator requires analytic evaluation of a function that is a sum of terms, each term being an absolute value of a difference of a square root of a polynomial and a variable. Such analytic evaluation is not known to be possible.

To our knowledge, the only instance of separability with outliers that was studied before is line separability with respect to the combinatorial measure in the plane [13, 23]. The optimal line separator in this setting can be computed in time $O((n + k^2)\log k)$, where k denotes the number of outliers achieved by the optimum.

Motivated by the practical considerations expressed above, a number of previous results in computational geometry provide algorithms that are insensitive to outliers. Problems that were studied in this context include linear programming [13, 23, 31], shape fitting [26], and facility location [15].

Separability with outliers has an interesting connection to the computation of statistical estimators, which in turn is intimately related to shape fitting. Given a point set $\mathcal{S}$ and a family $\mathcal{F}$ of *estimators*, consider the problem of finding the estimator $f \in \mathcal{F}$ that minimizes one of the above extent measures of $\mathcal{S}$ with respect to f. Motivated by applications in statistical analysis and computational metrology, this class of problems has been intensively studied in recent years, specifically for hyperplane, sphere, line, and cylinder estimators [2, 4, 5, 7, 12, 14, 19, 25, 26]. It is also a special case of separability with outliers. Indeed, setting $\mathcal{R} = \mathcal{B} = \mathcal{S}$ yields precisely $\mathcal{O}_f = \mathcal{S}$ for any $f \in \mathcal{F}$, equating the computation of the optimal separator of $\mathcal{R}$ and $\mathcal{B}$ to the computation of the optimal estimator of $\mathcal{S}$. Some of the results obtained in this paper indeed use the algorithmic machinery developed for the computation of statistical estimators, appropriately extending and modifying it to handle separability with outliers.

2 The Combinatorial Measure

Throughout this paper, let $\mathcal{R}, \mathcal{B} \in \mathbb{R}^d$ be two collections of n points each.

2.1 Hyperplane Combinatorial Separators

Theorem 1. *Let $\mathcal{F}$ be the set of hyperplanes in $\mathbb{R}^d$. The separator $f^* \in \mathcal{F}$ that minimizes $\mathcal{M}_c(f, \mathcal{O}_f)$ for $f \in \mathcal{F}$ can be computed in time $O(nk_{\text{opt}}^{d+1} \log k_{\text{opt}})$, where $k_{\text{opt}} = \mathcal{M}_c(f^*, \mathcal{O}_{f^*})$.*

Proof. For $f \in \mathcal{F}$, a point $x = (x_1, x_2, \ldots, x_d) \in \mathbb{R}^d$ lies on h, if $a_1x_1 + \ldots + a_dx_d + a_{d+1} = 0$, where $a_1, \ldots, a_{d+1}$ are the coefficients of h. With a slight abuse of notation we denote $a_1x_1 + \ldots + a_dx_d + a_{d+1}$ by $f(x)$. Note that although the parametric space defining h is $d+1$ dimensional, one parameter can be eliminated by enforcing $a_{d+1} = 1$.

In the above parametric space, every point of $\mathcal{R} \cup \mathcal{B}$ defines a linear inequality. A point corresponding to a separator f violates exactly those inequalities that correspond to points in $\mathcal{O}_f$. Given the linear program formed by these inequality constraints, we are looking for a point that minimizes the number of violated constraints. Stated differently, we seek the minimum number k_{opt} of constraints that need to be removed to make the linear program feasible.

Given k, we can decide in time $O(nk^{d+1})$ whether $k_{\text{opt}} \leq k$ [31]. Performing an exponential search with $k = 2^0, 2^1, 2^2, \ldots$ results in a constant factor approximation of k_{opt}. A binary search then locates the exact optimum.

Remark: Using the improved decision procedure of Chan for $d = 2, 3$ [13], the running time in Theorem 1 can be improved to $O((n + k_{\text{opt}}^2) \log^2 n)$ for $d = 2$ and to $O(n \log^2 n + k_{\text{opt}}^{11/4} n^{1/4} \operatorname{polylog} n)$ for $d = 3$.

Theorem 2. *Let $\mathcal{F}$ be the set of hyperplanes in $\mathbb{R}^d$ and let $f^* \in \mathcal{F}$ be the separator that minimizes $\mathcal{M}_c(f, \mathcal{O}_f)$ for $f \in \mathcal{F}$. For any $\varepsilon > 0$, a separator $f \in \mathcal{F}$, such that $\mathcal{M}_c(f, \mathcal{O}_f) \leq (1+\varepsilon)\mathcal{M}_c(f^*, \mathcal{O}_{f^*})$ can be computed in time $O\big(n(\varepsilon^{-2} \log n)^{d+1}\big)$.*

Proof. We use the above reduction to the problem of estimating the number of constraints that need to be removed so that a linear program in d dimensions becomes feasible. A recent result of Aronov and Har-Peled [9] shows that this number can be approximated within a factor of $(1+\varepsilon)$ in time $O\big(n(\varepsilon^{-2} \log n)^{d+1}\big)$.

2.2 Sphere Combinatorial Separators

Theorem 3. *Let $\mathcal{F}$ be the set of spheres in $\mathbb{R}^d$. The separator $f^* \in \mathcal{F}$ that minimizes $\mathcal{M}_c(f, \mathcal{O}_f)$ for $f \in \mathcal{F}$ can be computed in time $O(nk_{\text{opt}}^{d+2} \log k_{\text{opt}})$, where $k_{\text{opt}} = \mathcal{M}_c(f^*, \mathcal{O}_{f^*})$. For any $\varepsilon > 0$, a separator $f \in \mathcal{F}$, such that $\mathcal{M}_c(f, \mathcal{O}_f) \leq (1+\varepsilon)\mathcal{M}_c(f^*, \mathcal{O}_{f^*})$ can be computed in time $O\big(n(\varepsilon^{-2} \log n)^{d+2}\big)$.*

Proof. Apply the standard "lifting transformation" [20] that maps every ball B in $\mathbb{R}^d$ to a hyperplane B^* in $\mathbb{R}^{d+1}$ and a point p in $\mathbb{R}^d$ to a point p^* on the standard paraboloid in $\mathbb{R}^{d+1}$, such that $p \in B$ (resp., $p \in \partial B, p \notin B$) if and only if p^* lies below (resp., on, above) B^*. This reduces the problem to finding the (approximately) optimal hyperplane separator in $\mathbb{R}^{d+1}$ and the results follow from Theorems 1 and 2.

Remark: Using the improved decision procedure of Chan [13], the running time in Theorem 3 can be improved to $O(n \log^2 n + k_{\text{opt}}^{11/4} n^{1/4} \operatorname{polylog} n)$ for $d = 2$.

3 The $\mathcal{L}_\infty$ Measure

3.1 Hyperplane $\mathcal{L}_\infty$-Separators

Theorem 4. *Let $\mathcal{F}$ be the set of hyperplanes in $\mathbb{R}^d$, $d \geq 3$. The separator $f^* \in \mathcal{F}$ that minimizes $\mathcal{M}_\infty(f, \mathcal{O}_f)$ for $f \in \mathcal{F}$ can be computed in time $O(n^{\lceil d/2 \rceil})$.*

Proof. We follow the approach of Chan [12] to computing the width of a point set. For a separator $f \in \mathcal{F}$, parameterize it as $f = (x_f, a_f)$, for $x_f \in \mathbb{R}^d$, $a_f \in \mathbb{R}$, such that $f = \{p \in \mathbb{R}^d \mid p \cdot x_f = a_f\}$. Consider the minimal slab S_f bounded by hyperplanes parallel to f that contains $\mathcal{O}_f$. It can be parameterized as $S_f = \{p \in \mathbb{R}^d \mid b_f \leq p \cdot x_f \leq c_f\}$, such that $b_f, c_f \in \mathbb{R}$ and $2a_f = b_f + c_f$. Since x_f, b_f and c_f are dependent, c_f can be eliminated: $S_f = \{p \in \mathbb{R}^d \mid b_f \leq p \cdot x_f \leq b_f + 1\}$. Observe that $\mathcal{M}_\infty(f, \mathcal{O}_f) = 1/(2\,\|x_f\|)$, where $1/\,\|x_f\|$ is the width of S_f. We compute the separator f^* that minimizes $1/\,\|x_f\|$. Consider the following programming problem:

$$
\begin{array}{lll}
\min & 1/\,\|x\|, & \\
\text{s.t.} & \forall p \in \mathcal{B} & x \cdot p \leq b + 1, \\
 & \forall q \in \mathcal{R} & x \cdot q \geq b, \\
 & b \in \mathbb{R}, x \in \mathbb{R}^d &
\end{array}
$$

The optimal solution vector (b^*, x^*) provides exactly the parameters b_{f^*} and x_{f^*}, respectively, of the optimal separator. The problem is a non-convex optimization problem within a convex polytope defined by $2n$ linear constraints in dimension $d+1$. It can thus be solved in time $O(n^{\lceil d/2 \rceil})$ by constructing the polytope and explicitly computing the value of the optimization function at each boundary feature [16].

Remark: The running time in Theorem 4 can be improved to $O(n^{3/2+\varepsilon})$ for any $\varepsilon > 0$ when $d = 3$, using the randomized techniques of [6], and is $O(n \log n)$ when $d = 2$. We omit the details.

Theorem 5. *Let $\mathcal{F}$ be the set of hyperplanes in $\mathbb{R}^d$ and let $f^* \in \mathcal{F}$ be the separator that minimizes $\mathcal{M}_\infty(f, \mathcal{O}_f)$ for $f \in \mathcal{F}$. For any $\varepsilon > 0$, a separator $f \in \mathcal{F}$, such that $\mathcal{M}_\infty(f, \mathcal{O}_f) \leq (1+\varepsilon)\mathcal{M}_\infty(f^*, \mathcal{O}_{f^*})$ can be computed in time $O(n/\varepsilon^{(d-1)/2})$.*

Proof. We can test in time $O(n)$ whether $\mathcal{B}$ and $\mathcal{R}$ are separable by a hyperplane f, in which case $\mathcal{M}_\infty(f, \mathcal{O}_f) = 0$ [32]. Assume therefore that this is not the case. Let $\text{conv}(\mathcal{S})$ denote the convex hull of a point set $\mathcal{S} \in \mathbb{R}^d$. Assume without loss of generality that $O \in \text{conv}(\mathcal{B}) \cap \text{conv}(\mathcal{R})$, where O denotes the origin. Let the *penetration depth* $\pi(\text{conv}(\mathcal{B}), \text{conv}(\mathcal{R}))$ denote the smallest distance by which $\text{conv}(\mathcal{B})$ can be translated so that $\text{conv}(\mathcal{B})$ and $\text{conv}(\mathcal{R})$ are interior-disjoint [6]. Denote the optimal translation vector by $t \in \mathbb{R}^d$. It is easy to see that there is a unique hyperplane H_r, that separates the interiors of $\text{conv}(\mathcal{B}+t)$ and $\text{conv}(\mathcal{R})$. Consider also $H_b = H_r - t$ and $H_m = H_r - \frac{t}{2}$. (H_m is thus the mid-hyperplane between H_r and H_b.) Observe that H_m is the optimal separator f^*. We concentrate on computing the translation t that approximately realizes the penetration depth $\pi(\text{conv}(\mathcal{B}), \text{conv}(\mathcal{R}))$, i.e., if t^* denotes the optimal translation, $\|t\| \leq (1+\varepsilon)\ \|t^*\|$ and $\text{conv}(\mathcal{B}+t) \cap \text{conv}(\mathcal{R}) = \emptyset$. Given t we can find the corresponding separator in time $O(n)$. We first observe that given a *direction* $\delta \in \mathbb{S}^{d-1}$, we can compute the shortest separating translation t_δ in direction δ in $O(n)$ time. Indeed, assume without loss of generality that $\delta = (0, \ldots, 0, 1)$ is the positive vertical direction. Consider the linear program

$$\begin{array}{lll} \min & e, & \\ \text{s.t.} & \forall b \in \mathcal{B} & (b + e\delta) \cdot f \geq g, \\ & \forall r \in \mathcal{R} & r \cdot f \leq g, \\ & e, g \in \mathbb{R}, f \in \mathbb{R}^d & \end{array}$$

The optimum in this program is the value of t_δ and can be found in time $O(n)$ [32]. Observe that if $\gamma, \delta \in \mathbb{S}^{d-1}$ are two directions at angle at most $\sqrt{\varepsilon}$ from each other, then $(1-\varepsilon)t_\delta \leq t_\gamma \leq (1+\varepsilon)t_\delta$ [6]. This implies that if $\Delta \subseteq \mathbb{S}^{d-1}$ is a $\sqrt{\varepsilon}$-net on $\mathbb{S}^{d-1}$ and t^* denotes the optimal translation, $\min_{\delta \in \Delta} \|t_\delta\| \leq (1+\varepsilon)\ \|t^*\|$. There exists such a $\sqrt{\varepsilon}$-net Δ of size $O\big(1/\varepsilon^{(d-1)/2}\big)$ [18]. Executing the above algorithm for every $\delta \in \Delta$ implies the theorem.

3.2 Sphere $\mathcal{L}_\infty$-Separators

The algorithms in this section follow the approach of Chan [12] to computing the minimum-width spherical shell.

Theorem 6. *Let $\mathcal{F}$ be the set of spheres in $\mathbb{R}^d$. The separator $f^* \in \mathcal{F}$ that minimizes $\mathcal{M}_\infty(f, \mathcal{O}_f)$ for $f \in \mathcal{F}$ can be computed in time $O\big(n^{\lfloor d/2 \rfloor + 1}\big)$.*

Proof. For a separator $f \in \mathcal{F}$, denote its center by $x_f \in \mathbb{R}^d$ and its radius by $r_f \in \mathbb{R}$. Consider the minimum-width spherical shell enclosed by two spheres centered at x_f that contains $\mathcal{O}_f$. Denote its inner and outer radii by y_f and z_f, respectively. Note that $\mathcal{M}_\infty(f, \mathcal{O}_f) = (z_f - y_f)/2$. We compute the separator f^* that minimizes $z_{f^*} - y_{f^*}$. Consider the following programming problem:

$$\begin{array}{l} \min\ z - y, \\ \text{s.t. } \forall b \in \mathcal{B}\ \|b - x\| \leq z, \end{array}$$

$$\forall r \in \mathcal{R} \qquad \|r - x\| \geq y,$$
$$y, z \in \mathbb{R}, x \in \mathbb{R}^d$$

The optimal solution vector (y^*, z^*, x^*) provides the radius $(z_{f^*} + y_{f^*})/2$ and center x_{f^*} of the optimal separator. By a change of variables, the problem is easily seen to be a non-convex optimization problem within a convex polytope defined by $2n$ linear constraints in dimension $d + 2$. It can be solved in time $O(n^{\lfloor d/2 \rfloor + 1})$ by constructing the polytope and explicitly computing the value of the optimization function at each boundary feature [16].

Theorem 7. *Let $\mathcal{F}$ be the set of spheres in $\mathbb{R}^d$ and let $f^* \in \mathcal{F}$ be the separator that minimizes $\mathcal{M}_\infty(f, \mathcal{O}_f)$ for $f \in \mathcal{F}$. For any $\varepsilon > 0$, a separator $f \in \mathcal{F}$, such that $\mathcal{M}_\infty(f, \mathcal{O}_f) \leq (1+\varepsilon)\mathcal{M}_\infty(f^*, \mathcal{O}_{f^*})$ can be computed in time $O(n/\varepsilon^{(d-1)/2} + 1/\varepsilon^{4d})$.*

Proof. Omitted due to space limitations.

Remark: In Theorems 7 and 5, more advanced arguments can be used to decouple the $1/\varepsilon$ factor in the bound from n, yielding a running time of the form $O(n + 1/\varepsilon^{O(d)})$, which can be refined for small d. We omit the details.

4 The $\mathcal{L}_1$ and $\mathcal{L}_2$ Measures

4.1 Hyperplane $\mathcal{L}_1$-Separators

Theorem 8. *Let $\mathcal{F}$ be the set of hyperplanes in $\mathbb{R}^d$. The separator $f^* \in \mathcal{F}$ that minimizes $\mathcal{M}_1(f, \mathcal{O}_f)$ for $f \in \mathcal{F}$ can be computed in time $O(n^d)$.*

Proof. Observe that the optimal separator divides $R \cup B$ into two equal sets. Our algorithm considers all such partitions of $R \cup B$. After the standard duality transformation, these partitions correspond precisely to the faces of the median level of the arrangement $\mathcal{A}$ of the hyperplanes dual to the points $R \cup B$. In fact, it can be shown that the optimal separator corresponds to a *vertex* of this level. We construct $\mathcal{A}$ and traverse the vertices of the median level using breadth-first search on the 1-skeleton of the arrangement [21]. Throughout the traversal we maintain the value of $\mathcal{M}_1(f, \mathcal{O}_f)$. Since at each step only $O(1)$ points change their relation to the separator, the value of the measure can be maintained in $O(1)$ time per step. Upon the completion of the traversal we output the minimal value of $\mathcal{M}_1(f, \mathcal{O}_f)$ encountered during the process. We omit the easy details.

Remark: The running time in Theorem 8 can be improved to $O(n^{4/3} \log n)$ when $d = 2$ using the dynamic convex hull algorithm of [10] to construct the median level in the arrangement. When $d = 3$, the running time is $O(n^{5/2+\varepsilon})$ for any $\varepsilon > 0$ using combinatorial and algorithmic results concerning median levels in three dimensions [11, 35]. Unfortunately there are no substantially sub-$O(n^d)$ bounds on the complexity of the median level of an arrangement in $\mathbb{R}^d$ for $d \geq 4$ (see [1] for the state of the art). Obtaining such a bound is an interesting open problem.

The following is an easy consequence of the techniques of Yamamoto et al [37].

Theorem 9. *Let $\mathcal{F}$ be the set of hyperplanes in $\mathbb{R}^d$ and let $f^* \in \mathcal{F}$ be the separator that minimizes $\mathcal{M}_1(f, \mathcal{O}_f)$ for $f \in \mathcal{F}$. For any $\varepsilon > 0$, a separator $f \in \mathcal{F}$, such that $\mathcal{M}_1(f, \mathcal{O}_f) \leq (1+\varepsilon)\mathcal{M}_1(f^*, \mathcal{O}_{f^*})$ can be computed in time $O(n/\varepsilon^{(d-1)/2})$.*

Proof. Parameterize $f \in \mathcal{F}$ by $x_f \in \mathbb{R}^d$, such that $f = \{p \in \mathbb{R}^d \mid p \cdot x_f = 1\}$. Let $\alpha \in \mathbb{S}^{d-1}$ be the vertical direction. For $x = (x_1, \ldots, x_d) \in \mathbb{R}^d$, define the vertical distance $d^\alpha(f,p) = \mid p \cdot x_f - 1 \mid$. Define the vertical $\mathcal{L}_1$-measure of f with respect to $S \subseteq \mathbb{R}^d$ as $\mathcal{M}_1^\alpha(f, \mathcal{S}) = \sum_{p \in \mathcal{S}} d^\alpha(f,p)$.

Lemma 1. *The separator $f^\alpha \in \mathcal{F}$ that minimizes $\mathcal{M}_1^\alpha(f, \mathcal{O}_f)$ for $f \in \mathcal{F}$ can be computed in time $O(n)$.*

Proof. Consider a point $r \in \mathcal{R}$. Observe that its contribution to $\mathcal{M}_1^\alpha(f, \mathcal{O}_f)$ is $C_r(x_f) = \max(1 - r \cdot x_f, 0)$. On the other hand, the contribution of $b \in \mathcal{B}$ to $\mathcal{M}_1^\alpha(f, \mathcal{O}_f)$ is $C_b(x_f) = \max(b \cdot x_f - 1, 0)$. We seek to minimize the function $\sum_{p \in \mathcal{B} \cup \mathcal{R}} C_p(x_f)$ over the space of separators $x_f \in \mathbb{R}^d$. We are thus searching for the minimum of the sum of piecewise linear convex functions over $\mathbb{R}^d$. This can be computed in time $O(n)$ using the pruning technique of Megiddo [32, 37].

Note that the algorithm of Lemma 1 can compute the separator that minimizes the directional distance in any direction $\gamma \in \mathbb{S}^{d-1}$ by first applying a linear transformation to the space that transforms γ to be vertical. Observe that if $\gamma, \delta \in \mathbb{S}^{d-1}$ are two directions at angle at most $\sqrt{\varepsilon}$ from each other, $(1-\varepsilon)\mathcal{M}_1^\gamma(f^\gamma, \mathcal{O}_{f^\gamma}) \leq \mathcal{M}_1^\delta(f^\delta, \mathcal{O}_{f^\delta}) \leq (1+\varepsilon)\mathcal{M}_1^\gamma(f^\gamma, \mathcal{O}_{f^\gamma})$. (We omit the simple proof.) Note also that if $\beta \in \mathbb{S}^{d-1}$ denotes the direction normal to f^*, then $\mathcal{M}_1^\beta(f^\beta, \mathcal{O}_{f^\beta}) = \mathcal{M}_1(f^*, \mathcal{O}_{f^*})$. This implies that if $\Delta \subseteq \mathbb{S}^{d-1}$ is a $\sqrt{\varepsilon}$-net on $\mathbb{S}^{d-1}$, $\min_{\delta \in \Delta} \mathcal{M}_1^\delta(f^\delta, \mathcal{O}_{f^\delta}) \leq (1+\varepsilon)\mathcal{M}_1(f^*, \mathcal{O}_{f^*})$. There exists such a $\sqrt{\varepsilon}$-net Δ of size $O(1/\varepsilon^{(d-1)/2})$ [18]. Executing the algorithm of Lemma 1 for every $\delta \in \Delta$ implies the theorem.

Remark: The algorithm of Theorem 9 immediately yields a linear (in n) ε-approximation algorithm for the computation of an $\mathcal{L}_1$ linear estimator of a point set in $\mathbb{R}^d$. To our knowledge this is the first such algorithm for this basic problem in statistics (see [37] for an exact algorithm in the plane).

4.2 Hyperplane $\mathcal{L}_2$-Separators

Theorem 10. *Let $\mathcal{F}$ be the set of hyperplanes in $\mathbb{R}^d$. The separator $f^* \in \mathcal{F}$ that minimizes $\mathcal{M}_2(f, \mathcal{O}_f)$ for $f \in \mathcal{F}$ can be computed in time $O(n^{d+1})$.*

Proof. Consider the arrangement $\mathcal{A}(\mathcal{B}^* \cup \mathcal{R}^*)$ of the set of hyperplanes dual to the points $\mathcal{B} \cup \mathcal{R}$. Every point in a specific face C of $\mathcal{A}(\mathcal{B}^* \cup \mathcal{R}^*)$ corresponds to a hyperplane f that induces a fixed partition of $\mathcal{B} \cup \mathcal{R}$ that is the same for all such

f. Thus the set of outliers O_f is fixed and the measure $\mathcal{M}_2(f, \mathcal{O}_f)$ is given by a function common to all such f. This function is a sum of $|O_f|$ quadratic functions and is thus a quadratic function. Furthermore, since the partitions induced by neighboring faces of $\mathcal{A}(\mathcal{B}^* \cup \mathcal{R}^*)$ differ by at most a constant number of points, the above measure function for a specific cell can be computed in constant time given the measure function for a neighboring cell. This leads to the following algorithm.

Construct the arrangement $\mathcal{A}(\mathcal{B}^* \cup \mathcal{R}^*)$ and traverse all its faces of all dimensions. At each step of the traversal, update in constant time the measure function and compute the point that minimizes the function. Determine in linear time [32] whether this optimum point lies in the interior of the current face. If so, compare the value at the optimum to the minimal value so far and update the minimum if necessary. Otherwise the optimal point cannot lie in the current arrangement face. When the traversal is completed the algorithm outputs the minimal value and the hyperplane that corresponds to the dual point that achieves the minimal value. The running time follows since we traverse $O(n^d)$ arrangement faces and spend $O(n)$ time in each.

Theorem 11. *Let $\mathcal{F}$ be the set of hyperplanes in $\mathbb{R}^d$ and let $f^* \in \mathcal{F}$ be the separator that minimizes $\mathcal{M}_2(f, \mathcal{O}_f)$ for $f \in \mathcal{F}$. For any $\varepsilon > 0$, a separator $f \in \mathcal{F}$, such that $\mathcal{M}_2(f, \mathcal{O}_f) \leq (1+\varepsilon)\mathcal{M}_2(f^*, \mathcal{O}_{f^*})$ can be computed in time $O(n/\varepsilon^{d/2})$.*

Proof. Here, we want to find the hyperplane that approximately minimizes the sum of square distances of the misclassified points. This can be achieved by a direct adaptation of the techniques of Theorem 9. We omit the details.

4.3 Sphere $\mathcal{L}_1$- and $\mathcal{L}_2$-Separators

Theorem 12. *Let $\mathcal{F}$ be the set of spheres in $\mathbb{R}^d$ and let $f^* \in \mathcal{F}$ be the separator that minimizes $\mathcal{M}_1(f, \mathcal{O}_f)$ for $f \in \mathcal{F}$. For any $\varepsilon > 0$, a separator $f \in \mathcal{F}$, such that $\mathcal{M}_1(f, \mathcal{O}_f) \leq (1+\varepsilon)\mathcal{M}_1(f^*, \mathcal{O}_{f^*})$ can be computed in time $O\left(\left(\frac{n}{\varepsilon} \log \frac{n}{\varepsilon}\right)^{d+1}\right)$. The same result can be achieved for the $\mathcal{M}_2$ measure.*

Proof. Throughout the proof we primarily discuss the case of the $\mathcal{M}_1$ measure. The algorithm can be naturally adapted to the $\mathcal{M}_2$ measure.

We begin by using Theorem 6 to compute an optimal $\mathcal{L}_\infty$ sphere separator f^∞ in time $O(n^{\lfloor d/2 \rfloor + 1})$ and define $r = \mathcal{M}_\infty(f^\infty, \mathcal{O}_{f^\infty})$. This provides an n-approximation to the optimal $\mathcal{L}_1$-separator. Indeed, for any sphere s in $\mathbb{R}^d$ there is a point in $\mathcal{O}_s$ at distance at least r from s. On the other hand, all points of $\mathcal{O}_{f^\infty}$ are at distance at most r from f^∞. Thus $r \leq \mathcal{M}_1(f^*, \mathcal{O}_{f^*}) \leq nr$. Define $U = nr$ and note that $\mathcal{M}_1(f^*, \mathcal{O}_{f^*}) \leq U \leq n\mathcal{M}_1(f^*, \mathcal{O}_{f^*})$.

Place around each point of $\mathcal{B} \cup \mathcal{R}$ a ball of radius $r_i = \frac{\varepsilon}{2}(U/n^2)(1+\frac{\varepsilon}{2})^i$, for $i = 0, \ldots, M$, where $M = 4\left\lceil \log_{1+\varepsilon/2}(n^2/\varepsilon) \right\rceil = O(\frac{1}{\varepsilon} \log \frac{n}{\varepsilon})$. Let $\mathcal{D}$ denote the resulting set of $O(nM) = O(\frac{n}{\varepsilon} \log \frac{n}{\varepsilon})$ balls.

Consider a sphere f. Clearly, if f fails to touch any ball of $\mathcal{D}$ around a point $p \in \mathcal{O}_f$, then $d(p, C) \geq r_M \geq Un^2$. In particular, we can ignore such spheres, since they are too expensive to be a good approximation to $\mathcal{M}_1(f^*, \mathcal{O}_{f^*})$. Furthermore, let $r_\mathcal{D}(f, p)$ denote the radius of the smallest ball in $\mathcal{D}$ that is centered at p and intersects f. Let $\alpha(f) = \sum_{p \in \mathcal{O}_f} r_\mathcal{D}(f, p)$. We have that

$$\mathcal{M}_1(f, \mathcal{O}_f) = \sum_{p \in \mathcal{O}_f} d(f, p) \leq \alpha(f) = \sum_{p \in \mathcal{O}_f} r_\mathcal{D}(f, p) \leq \sum_{p \in \mathcal{O}_f} \Big(r_0 + (1 + \frac{\varepsilon}{2}) d(f, p) \Big)$$
$$= n r_0 + (1 + \frac{\varepsilon}{2}) \sum_{p \in \mathcal{O}_f} d(f, p) \leq (1 + \varepsilon) \mathcal{M}_1(f, \mathcal{O}_f).$$

However, the value of $\alpha(f)$ is uniquely defined by the set of balls of $\mathcal{D}$ that the sphere f intersects and the points of $\mathcal{B} \cup \mathcal{R}$ that the ball defined by f contains. We now show how to enumerate all such sets efficiently. We encode a ball B in $\mathbb{R}^d$ by a cone in $\mathbb{R}^{d+1}$ with axis parallel to the x_{d+1}-axis, such that its intersection with the hyperplane $x_{d+1} = 0$ is B. Thus a ball centered at $\mathbf{x} \in \mathbb{R}^d$ with radius r is encoded by a cone with apex $(\mathbf{x}, -r)$, denoted by cone$(\mathbf{x}, -r)$. A *sphere* S centered at $\mathbf{x}$ with radius r is encoded by the point $(\mathbf{x}, r)$. Consider the set of spheres intersecting B. Clearly, this is the set of points lying above the hyperplane $x_{d+1} = 0$, above cone$(\mathbf{x}, -r)$ and below cone$(\mathbf{x}, r)$. Also, the set of spheres that contain a point $\mathbf{x} \in \mathbb{R}^d$ is the set of points lying above cone$(\mathbf{x}, 0)$. Thus, consider the arrangement of the $2|\mathcal{D}| + n$ cones induced in this way by the balls of $\mathcal{D}$ and the points of $\mathcal{B} \cup \mathcal{R}$, together with the hyperplane $x_{d+1} = 0$. Clearly, all the spheres defined by points inside a particular face in this arrangement intersect the same set of balls of $\mathcal{D}$ and enclose the same set of points of $\mathcal{B} \cup \mathcal{R}$. We can thus find an approximate $\mathcal{L}_1$ sphere separator simply by traversing the above arrangement. The complexity of the arrangement is $O((2|\mathcal{D}| + n)^{d+1}) = O\Big(\big(\frac{n}{\varepsilon} \log \frac{n}{\varepsilon}\big)^{d+1}\Big)$ and the traversal can be performed in the same asymptotic time. This implies the theorem.

5 Discussion and Open Problems

This paper is a step towards a better understanding of separability with outliers. A wide range of interesting problems remain. Can algorithms presented in this paper be improved? In particular, are there ε-approximation algorithms whose running time is near-linear in n and polynomial in $1/\varepsilon$ and d?

Can we use the ideas of this paper to handle other types of separators, such as slabs, cylinders, cones, and prisms? We can also consider the family of all convex bodies as separators.

A number of successes in designing approximation algorithms for shape fitting and extent approximation have been achieves by proving that there always exists a small *coreset* that approximately captures the extent of the point set. Can we demonstrate the existence of coresets for separability with outliers (i.e., small sets that witness the extent to which two point sets penetrating each other) or prove that they do not always exist?

It would also be interesting to consider algorithms that maintain (approximately) minimal separators as the point sets change, either as points are inserted and deleted (dynamic setting) or as points move continuously (kinetic setting)? The problem can also be considered in the streaming model of computation.

Acknowledgements

The authors wish to thank Pankaj Agarwal for suggesting the problem of separability with outliers and for initial discussions.

References

1. P. K. Agarwal, B. Aronov, T. M. Chan, and M. Sharir. On levels in arrangements of lines, segments, planes, and triangles. *Discrete Comput. Geom.*, 19:315–331, 1998.
2. P. K. Agarwal, B. Aronov, S. Har-Peled, and M. Sharir. Approximation and exact algorithms for minimum-width annuli and shells. *Discrete Comput. Geom.*, 24(4):687–705, 2000.
3. P. K. Agarwal, B. Aronov, and V. Koltun. Efficient algorithms for bichromatic separability. *ACM Transactions on Algorithms*, 2005. to appear.
4. P. K. Agarwal, B. Aronov, and M. Sharir. Line traversals of balls and smallest enclosing cylinders in three dimensions. *Discrete Comput. Geom.*, 21:373–388, 1999.
5. P. K. Agarwal, B. Aronov, and M. Sharir. Exact and approximation algorithms for minimum-width cylindrical shells. *Discrete Comput. Geom.*, 26(3):307–320, 2001.
6. P. K. Agarwal, L. J. Guibas, S. Har-Peled, A. Rabinovitch, and M. Sharir. Penetration depth of two convex polytopes in 3D. *Nordic J. Comput.*, 7(3):227–240, 2000.
7. P. K. Agarwal, S. Har-Peled, and K. R. Varadarajan. Approximating extent measures of points. *J. Assoc. Comput. Mach.*, 51:606–635, 2004.
8. E. Arkin, F. Hurtado, J. Mitchell, C. Seara, and S. Skiena. Some lower bounds on geometric separability problems. In *11th Fall Workshop on Computational Geometry*, 2001.
9. B. Aronov and S. Har-Peled. On approximating the depth and related problems. In *Proc. 16th ACM-SIAM Sympos. Discrete Algorithms*, 2005.
10. G. S. Brodal and R. Jacob. Dynamic planar convex hull. In *Proc. 43th Annu. IEEE Sympos. Found. Comput. Sci.*, pages 617–626, 2002.
11. T. M. Chan. Output-sensitive results on convex hulls, extreme points, and related problems. *Discrete Comput. Geom.*, 16:369–387, 1996.
12. T. M. Chan. Approximating the diameter, width, smallest enclosing cylinder and minimum-width annulus. *Internat. J. Comput. Geom. Appl.*, 12(2):67–85, 2002.
13. T. M. Chan. Low-dimensional linear programming with violations. In *Proc. 43th Annu. IEEE Sympos. Found. Comput. Sci.*, pages 570–579, 2002.
14. T. M. Chan. Faster core-set constructions and data stream algorithms in fixed dimensions. In *Proc. 20th Annu. ACM Sympos. Comput. Geom.*, pages 152–159, 2004.
15. M. Charikar, S. Khuller, D. M. Mount, and G. Narasimhan. Algorithms for facility location problems with outliers. In *Proc. 12th ACM-SIAM Sympos. Discrete Algorithms*, pages 642–651, 2001.

16. B. Chazelle. An optimal convex hull algorithm in any fixed dimension. *Discrete Comput. Geom.*, 10:377–409, 1993.
17. N. Cristianini and J. Shaw-Taylor. *Support Vector Machines.* Cambridge University Press, 2000.
18. R. M. Dudley. Metric entropy of some classes of sets with differentiable boundaries. *J. Approx. Theory*, 10:227–236, 1974.
19. C. A. Duncan, M. T. Goodrich, and E. A. Ramos. Efficient approximation and optimization algorithms for computational metrology. In *Proc. 8th ACM-SIAM Sympos. Discrete Algorithms*, pages 121–130, 1997.
20. H. Edelsbrunner. *Algorithms in Combinatorial Geometry*, volume 10 of *EATCS Monographs on Theoretical Computer Science.* Springer-Verlag, Heidelberg, West Germany, 1987.
21. H. Edelsbrunner, J. O'Rourke, and R. Seidel. Constructing arrangements of lines and hyperplanes with applications. *SIAM J. Comput.*, 15:341–363, 1986.
22. H. Edelsbrunner and F. P. Preparata. Minimum polygonal separation. *Inform. Comput.*, 77:218–232, 1988.
23. H. Everett, J.-M. Robert, and M. van Kreveld. An optimal algorithm for the ($\leq k$)-levels, with applications to separation and transversal problems. *Internat. J. Comput. Geom. Appl.*, 6:247–261, 1996.
24. S. P. Fekete. On the complexity of min-link red-blue separation. Manuscript, Department of Applied Mathematics, SUNY Stony Brook, NY, 1992.
25. S. Har-Peled and K. Varadarajan. High-dimensional shape fitting in linear time. In *Proc. 19th Annu. ACM Sympos. Comput. Geom.*, pages 39–47, 2003.
26. S. Har-Peled and Y. Wang. Shape fitting with outliers. *SIAM J. Comput.*, 33(2):269–285, 2004.
27. T. Hastie, R. Tibshirani, and J. Friedman. *The Elements of Statistical Learning.* Springer-Verlag, Berlin, Germany, 2001.
28. F. Hurtado, M. Mora, P. A. Ramos, and C. Seara. Two problems on separability with lines and polygonals. In *Proc. 15th European Workshop on Computational Geometry*, pages 33–35, 1999.
29. F. Hurtado, M. Noy, P. A. Ramos, and C. Seara. Separating objects in the plane with wedges and strips. *Discrete Appl. Math.*, 109:109–138, 2001.
30. F. Hurtado, C. Seara, and S. Sethia. Red-blue separability problems in 3d. In *Proc. 3rd Int. Conf. Comput. Sci. and Its Appl.*, pages 766–775, 2003.
31. J. Matoušek. On geometric optimization with few violated constraints. *Discrete Comput. Geom.*, 14:365–384, 1995.
32. N. Megiddo. Linear programming in linear time when the dimension is fixed. *J. Assoc. Comput. Mach.*, 31:114–127, 1984.
33. J. S. B. Mitchell. Approximation algorithms for geometric separation problems. Technical report, Department of Applied Mathematics, SUNY Stony Brook, NY, July 1993.
34. J. O'Rourke, S. R. Kosaraju, and N. Megiddo. Computing circular separability. *Discrete Comput. Geom.*, 1:105–113, 1986.
35. M. Sharir, S. Smorodinsky, and G. Tardos. An improved bound for k-sets in three dimensions. *Discrete Comput. Geom.*, 26:195–204, 2001.
36. V. Vapnik. *The Nature of Statistical Learning Theory.* Springer-Verlag, New York, 1996.
37. P. Yamamoto, K. Kato, K. Imai, and H. Imai. Algorithms for vertical and orthogonal L_1 linear approximation of points. In *Proc. 4th Annu. ACM Sympos. Comput. Geom.*, pages 352–361. ACM Press, 1988.

Casting an Object with a Core*

Hee-Kap Ahn[1], Sang Won Bae[1], Siu-Wing Cheng[2], and Kyung-Yong Chwa[1]

[1] Division of Computer Science,
Korea Advanced Institute of Science and Technology,
Daejon, Korea
{heekap, swbae, kychwa}@jupiter.kaist.ac.kr

[2] Dept. of Computer Science, Hong Kong University of Science and Technology,
Hong Kong, China
scheng@cs.ust.hk

Abstract. In casting, molten material is poured into the cavity of the cast and allowed to solidify. The cast has two main parts to be removed in opposite parting directions. To manufacture more complicated objects, the cast may also have a side core to be removed in a direction skewed to the parting directions. In this paper, given an object and the parting and side core directions, we give necessary and sufficient conditions to verify whether a cast can be constructed for these directions. In the case of polyhedral objects, we develop a discrete algorithm to perform the test in $O(n^3 \log n)$ time, where n is the object size. If the test result is positive, a cast with complexity $O(n^3)$ can be constructed within the same time bound. We also present an example to show that a cast may have $\Theta(n^3)$ complexity in the worst case. Thus, the complexity of our cast is worst-case optimal.

1 Introduction

Casting or injection molding [7, 12, 14] is ubiquitous in the manufacturing industry for producing consumer products. A cast can be viewed as a box with a cavity inside. Molten material (such as iron, glass or polymer) is poured into the cavity and allowed to solidify. The cast has two main parts and the hardened object is taken out by removing the two parts in opposite *parting directions.* Many common objects need a side core in additional to the two main parts in order to be manufactured. (For simplicity, we will refer to the side core as core in the rest of the paper.) For example, consider a coffee mug in Figure 1(a). The handle of the mug can only be produced using the two main parts. However, these two main parts cannot produce the cavity. Figure 1(b) shows how the coffee mug can be manufactured by incorporating a core into the cast. Cores are used widely in prevailing modes of production, and the class of castable objects may be enlarged through the use of cores [6, 12, 14, 15]. Cores naturally increase

* Work by Ahn, Bae and Chwa was supported by the Brain Korea 21 Project, The School of Information Technology, KAIST, in 2005. Work by Cheng was supported by Research Grant Council, Hong Kong, China (DAG04/05.EG21).

X. Deng and D. Du (Eds.): ISAAC 2005, LNCS 3827, pp. 40–49, 2005.

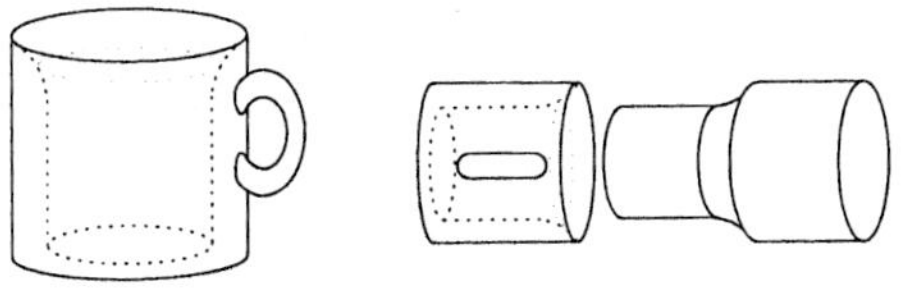

Fig. 1. (a) A coffee mug is unattainable using a 2-part cast. (b) By incorporating a core to the cast, the cavity of the coffee mug can be manufactured.

manufacturing costs and decrease production capacity [7]. Besides, the core retraction mechanism takes up much of extra space. In this paper, we deal with the case where one core is allowed.

We require that the main parts and the core should be removed without being blocked by the cast or the object. This ensures that the given object can be mass produced by re-using the same cast. This paper is concerned with the verification of the geometric feasibility, *castability*, of the cast given the parting and core directions. There has been a fair amount of work on the castability problem [1, 3, 4, 5, 9, 10, 11] for the case that there is no core. Chen, Chou and Woo [6] described a heuristic to compute a parting direction to minimize the number of cores needed. However, the parting direction returned need not be feasible. Based on the approach of Chen, Chou and Woo, Hui presented exponential time algorithms to construct a cast [8]. However, there is no guarantee that a feasible cast will be found if there is one. Ahn et al. [2] proposed a hull operator, reflex-free hull, to define cavities in polyhedron. The motivation is that the cavities limit the search space for parting and core directions.

In this paper, we give the first exact characterization of the castability of an object given the parting and core directions, assuming that the removal order of the parts and the core is immaterial. The core is often removed first in practice, so our assumption is stronger than necessary. Nevertheless, our result is the first known characterization of castability when a core is allowed. For a polyhedron of size n, we develop an $O(n^3 \log n)$-time algorithm for performing this test. The cast can be constructed within the same time bound. This paper presents, to the best of our knowledge, the first polynomial time algorithm for the problem.

2 Preliminaries

Let A be a subset of $\mathbb{R}^3$. We say that A is *open* if for any point $p \in A$, A contains some ball centered at p with positive radius. We say that A is *closed* if its complement is open. For all points $p \in \mathbb{R}^3$, p is a *boundary point* of A, if any ball centered at p with positive radius intersects both A and its complement. The *boundary of* A, denoted by $\mathrm{bd}(A)$, is the set of boundary points. The *interior* of A, $\mathrm{int}(A)$, is $A \setminus \mathrm{bd}(A)$. Note that $\mathrm{int}(A)$ must be open.

We assume that the outer shape of the cast equals a box denoted by $\mathcal{B}$. The cavity of $\mathcal{B}$ has the shape of the object $\mathcal{Q}$ to be manufactured. We assume that

$\mathcal{Q}$ is an open set so that the cast $\mathcal{B} \setminus \mathcal{Q}$ is a closed set. The box $\mathcal{B}$ is large enough so that $\mathcal{Q}$ is contained strictly in its interior. We use d_m and $-d_m$ to denote the given opposite parting directions, and d_c to denote the given core direction.

We call the main part to be removed in direction d_m the *red cast part* and denote it by $\mathcal{C}_r$. We call the other main part the *blue cast part* and denote it by $\mathcal{C}_b$. We denote the core by $\mathcal{C}_c$. We require each cast part and the core to be a connected subset of $\mathcal{B}$ such that $\mathcal{B} \setminus \mathcal{Q} = \mathcal{C}_r \cup \mathcal{C}_b \cup \mathcal{C}_c$ and these three pieces only overlap along their boundaries.

Given the object $\mathcal{Q}$ and the directions d_m and d_c, our problem is to decide if $\mathcal{Q}$ is *castable*. That is, whether $\mathcal{B}$ can be partitioned into $\mathcal{C}_r$, $\mathcal{C}_b$ and $\mathcal{C}_c$ so that they can be translated to infinity in their respective directions without colliding with $\mathcal{Q}$ and the other pieces. We assume that the order of removing the parts and the core is immaterial. In other words, if $\mathcal{Q}$ is castable, the parts and the core can be removed in any order without colliding with $\mathcal{Q}$ or the other pieces.

3 The Characterization of Castability

In this section, we develop the exact characterization of the castability of $\mathcal{Q}$ given the parting and core directions d_m and d_c. Recall that we assume $\mathcal{Q}$ to be open but we do not require $\mathcal{Q}$ to be polyhedral. We first need some visibility and monotonicity concepts.

Consider the illumination of $\mathcal{Q}$ by light sources at infinity in directions d_m and $-d_m$. We denote by $\mathcal{V}_m$ the subset of points of $\mathcal{B} \setminus \mathcal{Q}$ that do not receive light from the direction d_m or the direction $-d_m$. That is, both rays emitting from p in directions d_m and $-d_m$ intersect $\mathcal{Q}$. We use $\mathcal{V}_m^c$ to denote the points in $\mathbb{R}^3 \setminus \mathcal{Q}$ encountered when we sweep $\mathcal{V}_m$ to infinity in direction d_c. Note that $\mathcal{V}_m^c$ includes $\mathcal{V}_m$ itself. Consider the illumination of $\mathcal{Q} \cup \mathcal{V}_m^c$ by light sources at infinity in directions d_m and $-d_m$. We use $\mathcal{V}_o$ to denote those points in $\mathcal{B} \setminus (\mathcal{Q} \cup \mathcal{V}_m^c)$ that do not receive light from the direction d_m or the direction $-d_m$. That is, both rays emitting from p in directions d_m and $-d_m$ intersect $\mathcal{Q} \cup \mathcal{V}_m^c$. Then $\mathcal{V}_o^c$ denotes the points in $\mathbb{R}^3 \setminus (\mathcal{Q} \cup \mathcal{V}_m^c)$ encountered when we sweep $\mathcal{V}_o$ to infinity in direction d_c. Note that $\mathcal{V}_o^c$ includes $\mathcal{V}_o$ itself.

An object is *monotone* in direction d if for any line ℓ parallel to d, the intersection between ℓ and the interior of the object is a single interval. Notice that although $\mathcal{V}_m^c$ is constructed by sweeping $\mathcal{V}_m$ in direction d_c, the points inside $\mathcal{Q}$ are excluded. Therefore, $\mathcal{V}_m^c$ needs not be monotone in direction d_c in general. So does $\mathcal{V}_o^c$.

We will need the following lemmas. We skip the proof of the first one due to space limitation.

Lemma 1. *$\mathcal{Q} \cup \mathcal{V}_m^c \cup \mathcal{V}_o^c$ is monotone in d_m.*

Lemma 2. *Given d_m and d_c, if $\mathcal{Q}$ is castable, then $\mathcal{V}_m^c \cap \mathcal{B} \subseteq \mathcal{C}_c$ and $\mathcal{V}_o^c \cap \mathcal{B} \subseteq \mathcal{C}_c$.*

Proof. Let p be a point in $\mathcal{V}_m$. By the definition of $\mathcal{V}_m$, if we move p in direction d_m or $-d_m$ to infinity, p will hit $\mathcal{Q}$. So p cannot be a point in $\mathcal{C}_r$ or $\mathcal{C}_b$.

Thus, $\mathcal{V}_m \subseteq \mathcal{C}_c$. Since $\mathcal{Q}$ is castable, $\mathcal{C}_c$ can be translated to infinity in direction d_c without colliding with $\mathcal{Q}$ and the red and blue cast parts. This implies that $\mathcal{V}_m^c \cap \mathcal{B} \subseteq \mathcal{C}_c$. By the definition of $\mathcal{V}_o$, if we move a point $q \in \mathcal{V}_o$ in direction d_m or $-d_m$, q will hit $\mathcal{Q} \cup \mathcal{V}_m^c$. So q must be a point in $\mathcal{C}_c$, implying that $\mathcal{V}_o \subseteq \mathcal{C}_c$. Thus, the same reasoning shows that $\mathcal{V}_o^c \cap \mathcal{B} \subseteq \mathcal{C}_c$. ◻

Theorem 1. *Given d_m and d_c, $\mathcal{Q}$ is castable if and only if $\mathcal{V}_m^c \cup \mathcal{V}_o^c$ is monotone in d_c.*

Proof. If $\mathcal{Q}$ is castable, then for any point $p \in \mathcal{C}_c$, moving p to infinity in direction d_c will not hit $\mathcal{Q}$, $\mathcal{C}_r$, or $\mathcal{C}_b$. By Lemma 2, $(\mathcal{V}_m^c \cup \mathcal{V}_o^c) \cap \mathcal{B}$ is contained in $\mathcal{C}_c$. Therefore, by considering the movement of all points in $(\mathcal{V}_m^c \cup \mathcal{V}_o^c) \cap \mathcal{B}$ in direction d_c, we conclude that $\mathcal{V}_m^c \cup \mathcal{V}_o^c$ is monotone in d_c. This proves the necessity of the condition.

We prove the sufficiency by showing the construction of a cast for $\mathcal{Q}$. Ahn et al. [4] proved that an object is castable using a 2-part cast (without any side core) in parting direction d if and only if the object is monotone in direction d. Thus, Lemma 1 implies that $\mathcal{Q} \cup \mathcal{V}_m^c \cup \mathcal{V}_o^c$ is castable using a 2-part cast in direction d_m. We use the construction by Ahn et. al [4] to build $\mathcal{C}_r$ and $\mathcal{C}_b$ with the necessary modification for handling the side core. The details are as follows. Without loss of generality, we assume that d_m is the upward vertical direction, d_c makes angle of at most $\pi/2$ with d_m, and the horizontal projection of d_c aligns with the positive x-axis.

Recall that the cast is made from a rectangular axis-parallel box $\mathcal{B}$. We make $\mathcal{B}$ sufficiently large and position $\mathcal{Q}$ inside $\mathcal{B}$ so that $\mathcal{V}_m^c \cup \mathcal{V}_o^c$ intersects the interior of one vertical side facet of $\mathcal{B}$ only. Let S be that side facet of $\mathcal{B}$. Thicken S slightly to form a slab S^+. Let T be the top horizontal facet of $\mathcal{B}$. Thicken T slightly to form one slab T^+.

We move $\mathcal{Q} \cup \mathcal{V}_m^c \cup \mathcal{V}_o^c$ upward to infinity to form one swept volume. Then we subtract $\mathcal{Q} \cup \mathcal{V}_m^c \cup \mathcal{V}_o^c$ from this swept volume to form a shape $\mathcal{X}$. We can almost make $\mathcal{X} \cap \mathcal{B}$ the red cast part, but it is possible that $\mathcal{X} \cap \mathcal{B}$ is disconnected. So we add T^+ to connect the components of $\mathcal{X} \cap \mathcal{B}$ to form one red cast part $\mathcal{C}_r$. Similarly, we can almost make $(\mathcal{V}_m^c \cup \mathcal{V}_o^c) \cap \mathcal{B}$ the side core, but it may be disconnected. So we add $S^+ \setminus T^+$ to connect the components in $(\mathcal{V}_m^c \cup \mathcal{V}_o^c) \cap \mathcal{B}$ to form the side core $\mathcal{C}_c$. Lastly, we construct the blue cast part $\mathcal{C}_b$ as $\mathcal{B} \setminus (\mathcal{Q} \cup \mathcal{C}_r \cup \mathcal{C}_c)$.

We argue that any part or the side core can be removed without colliding with $\mathcal{Q}$, the other part or the side core. Since $\mathcal{V}_m^c \cup \mathcal{V}_o^c$ is monotone in direction d_c, the core $\mathcal{C}_c$ can be removed first without colliding with $\mathcal{Q}$ or the other cast parts. Consider $\mathcal{C}_r$. As $\mathcal{Q} \cup \mathcal{V}_m^c \cup \mathcal{V}_o^c$ is monotone in direction d_m, the removal of $\mathcal{C}_r$ cannot collide with $\mathcal{Q}$ or $\mathcal{C}_c$. Clearly, the removal of $\mathcal{C}_r$ cannot collide with $\mathcal{C}_b$ by construction. The argument that $\mathcal{C}_b$ can be removed first is similar. ◻

If we are given a CAD system that is equipped with visibility computation, volume sweeping, and monotonicity checking operation, the characterization in Theorem 1 can be used directly to check the castability of any object. The proof also yields the construction of the cast.

4 An Algorithm for Polyhedra

In this section, we apply Theorem 1 to check the castability of a polyhedron. The goal is to obtain a discrete algorithm whose running time depends on the combinatorial complexity of the polyhedron. To be consistent with the previous section, our object is the interior of the polyhedron and we denote it by $\mathcal{P}$. The combinatorial complexity n of $\mathcal{P}$ is the sum of the numbers of vertices, edges, and facets in $\mathrm{bd}(\mathcal{P})$. We present an $O(n^3 \log n)$-time algorithm for testing the castability of $\mathcal{P}$ given d_m and d_c. During the verification, we compute $\mathcal{V}_m^c \cup \mathcal{V}_o^c$, from which the cast $\mathcal{C}$ can be easily obtained as mentioned in the proof of Theorem 1.

Throughout this section, we assume that d_m is the upward vertical direction. We also make two assumptions about non-degeneracy. First, no facet in $\mathrm{bd}(\mathcal{P})$ is vertical. Second, the vertical projections of two polyhedron edges are either disjoint or they cross each other. These non-degeneracy assumptions simplify the presentation and they can be removed by a more detailed analysis. We call a facet of $\mathcal{P}$ an *up-facet* if its outward normal points upward, and a *down-facet* if its outward normal points downward.

Let $\mathcal{H}$ be a horizontal plane below $\mathcal{P}$. We project all facets of $\mathcal{P}$ onto $\mathcal{H}$. The projections may self-intersect and we insert vertices at the crossings. The resulting subdivision has $O(n^2)$ size and we denote it by $\mathcal{M}$. For each cell of $\mathcal{M}$, we keep the set of polyhedron facets that cover it. We can compute $\mathcal{M}$ in $O(n^2 \log n)$ time using a plane-sweep algorithm and the association of polyhedron facets to cells can also be done in $O(n^3 \log n)$ time during the plane sweep.

After computing $\mathcal{M}$, we test whether $\mathcal{V}_m^c \cup \mathcal{V}_o^c$ is monotone in d_c (see Theorem 1). We partition $\mathcal{H}$ into 2D slabs by taking vertical planes parallel to d_c through all vertices of $\mathcal{M}$. Since there are $O(n^2)$ vertices in $\mathcal{M}$ and a vertical plane parallel to d_c intersects $O(n)$ edges of $\mathcal{P}$, there are $O(n^3)$ intersections in total. So the overlay of $\mathcal{M}$ and the slabs can be computed in $O(n^3 \log n)$ time using a plane-sweep algorithm.

Consider a slab Σ on $\mathcal{H}$. From the construction, Σ contains no vertex in its interior and is partitioned into $O(n)$ regions by the edges of $\mathcal{M}$. Let d be the projection of d_c on $\mathcal{H}$. The regions in Σ are linearly ordered in direction d and we label them by $\Delta_0, \Delta_1, \ldots$ in this order. Notice that Δ_0 is unbounded in direction $-d$ and the last region is unbounded in direction d. We use ζ_i to denote the boundary edge between Δ_{i-1} and Δ_i. For each region Δ_i, we keep the set of polyhedron facets that cover it. We cannot do this straightforwardly. Otherwise, since there are $O(n^3)$ regions over all slabs and we may keep $O(n)$ polyhedron facets per region, the total time and space needed to do this would be $O(n^4)$. The key observation is that if we walk from Δ_0 along Σ in direction d and record the changes in the set of facets whenever we cross a boundary edge ζ_i, then the total number of changes in Σ is $O(n)$. Therefore, we can use a persistent search tree [13] to store the sets of polyhedron facets for all regions in Σ. This takes $O(n \log n)$ time and $O(n)$ space to build per slab. Hence, it takes a total of $O(n^3 \log n)$ time and $O(n^3)$ space.

We employ an inductive strategy for testing the monotonicity of $\mathcal{V}_m^c \cup \mathcal{V}_0^c$ in d_c within the unbounded 3D slab $\Sigma \times [\infty, -\infty]$ for each 2D slab Σ on $\mathcal{H}$. Repeating this test for all 2D slabs on $\mathcal{H}$ gives the final answer. We scan the regions in Σ in the order Δ_0, Δ_1, ... During the scanning, we incrementally grow a volume $\mathcal{V}^c$. The volume $\mathcal{V}^c$ is initially empty and $\mathcal{V}^c$ will be equal to $(\mathcal{V}_m^c \cup \mathcal{V}_o^c) \cap (\Sigma \times [\infty, -\infty])$ in the end.

We first discuss the data structures that we need to maintain during the scanning. Consider the event that we cross the boundary ζ_i and that the portion of $\mathcal{V}_m^c \cup \mathcal{V}_o^c$ encountered so far is monotone in d_c. Take the vertical strip through ζ_i. We translate this strip slightly into Δ_{i-1} (resp. Δ_i) and denote the perturbed strip by H_i^- (resp. H_i^+). Let I_i^- denote the intersection $H_i^- \cap (\mathcal{V}_m^c \cup \mathcal{V}_o^c)$ and let I_i^+ denote the intersection $H_i^+ \cap (\mathcal{V}_m^c \cup \mathcal{V}_o^c)$. Both I_i^- and I_i^+ consist of $O(n)$ trapezoids. We call the upper and lower sides of each trapezoid its *ceiling* and *floor*, respectively. The ceiling of each trapezoid τ lies on a boundary facet of $\mathcal{V}_m^c \cup \mathcal{V}_o^c$. We call this boundary facet the *ceiling-facet* of τ. This ceiling-facet may lie within a down-facet in bd($\mathcal{P}$) or it may be parallel to d_c and disjoint from bd($\mathcal{P}$). The latter kind of facets are generated by the sweeping towards d_c. Therefore, it suffices to store a polyhedron facet or a plane parallel to d_c to represent the ceiling-facet. We denote this representation by $f_u(\tau)$. Similarly, the floor of τ lies on a boundary facet of $\mathcal{V}_m^c \cup \mathcal{V}_o^c$. This boundary facet may lie within a up-facet of $\mathcal{P}$ or it may be parallel to d_c and disjoint from bd($\mathcal{P}$). We call it the *floor-facet* of τ and denote its representation by $f_\ell(\tau)$.

We are ready to describe the updating strategy when we reach a new region Δ_i. We first discuss the monotonicity test. Later, we discuss how to grow $\mathcal{V}^c$ if the test is passed. Note that there is a change in the polyhedron facets that cover Δ_{i-1} and Δ_i. There are several cases.

1. For any trapezoid $\tau \in I_i^-$, neither $f_u(\tau)$ nor $f_\ell(\tau)$ is about to vanish above ζ_i. Then some polyhedron edge e must project vertically onto ζ_i. Also, the vertical projections of the two incident polyhedron facets of e cover Δ_i but not Δ_{i-1}. Consider the projection e^- of e in direction $-d_c$ onto H_i^-. Since the monotonicity test has been passed so far, the space between two trapezoids in I_i^- is the polyhedron interior. Thus the projection e^- cannot lie between two trapezoids in I_i^-. So there are only two cases:
   (a) The projection e^- cuts across the interior of a trapezoid $\tau \in I_i^-$. In this case, we abort and report that $\mathcal{P}$ is not castable. The reason is that one polyhedron facet incident to e must block the sweeping of τ towards d_c. It follows that $\mathcal{V}_m^c \cup \mathcal{V}_o^c$ is not monotone in d_c and so $\mathcal{P}$ is not castable by Theorem 1.
   (b) The projection e^- lies above all trapezoids. The case that e^- lies below all trapezoids can be handled symmetrically. Let f be the down-facet incident to e. If we project e vertically downward, the projection either lies on some up-facet f', or a boundary facet of $\mathcal{V}_m^c \cup \mathcal{V}_o^c$ that is parallel to d_c, or lies at infinity. The last case happens when I_i^- is empty (e.g., when we cross the boundary ζ_1 between Δ_1 and Δ_0) and there is nothing

to be done for this case. We discuss the other two cases. Let e' denote this vertical downward projection of e.

  i. If e' lies on a up-facet f', then e and e' define a new trapezoid τ that lies above all trapezoids in I_i^- and that $f_u(\tau) = f$ and $f_\ell(\tau) = f'$. I_i^+ contains all trapezoids in I_i^- as well as τ. However, if the outward normal of f makes an obtuse angle with d_c, then f blocks the sweeping of τ towards d_c and we should abort and conclude as before that $\mathcal{P}$ is not castable.
  ii. If e' lies on a boundary facet of $\mathcal{V}_m^c \cup \mathcal{V}_o^c$ that is parallel to d_c, then e' actually lies on $f_u(\tau)$ where τ is the topmost trapezoid in I_i^-. Thus, we should grow τ upward and set $f_u(\tau) = f$. I_i^+ contains this updated trapezoid τ and the other trapezoids in I_i^-. There is no change in the monotonicity status.

2. For some trapezoid $\tau \in I_i^-$, $f_u(\tau)$ or $f_\ell(\tau)$ is about to vanish above ζ_i. In this case, a polyhedron edge e bounds $f_u(\tau)$ or $f_\ell(\tau)$ and e projects vertically onto ζ_i. There are two cases:
   (a) The polyhedron facets incident to e lie locally on different sides of the vertical plane through ζ_i. Let f be the incident facet of e that lies locally in direction d_c from e. In this case, the vanishing $f_u(\tau)$ or $f_\ell(\tau)$ should be replaced by f. However, if the outward normal of f makes an obtuse angle with d_c, we should abort and conclude as before that $\mathcal{P}$ is not castable.
   (b) Otherwise, both incident facets of e lie locally in direction $-d_c$ from e. There is no change in monotonicity status, but we need to perform update as follows. Let f be the vanishing $f_u(\tau)$ or $f_\ell(\tau)$ of τ. There are two cases:
      i. There are trapezoids in I_i^- that lie above and below f. Clearly, τ is one of them. Let τ' be the other trapezoid. Then $f_u(\tau')$ or $f_\ell(\tau')$ is about to vanish above ζ_i too. In this case, we should merge τ and τ' into one trapezoid. The ceiling-facet and floor-facet of this merged trapezoid are the non-vanishing ceiling-facet and floor-facet of τ and τ'. I_i^+ contains this merged trapezoid and the trapezoids in I_i^- other than τ and τ'.
      ii. All trapezoids in I_i^- lie on one side of f. Assume that τ is the topmost trapezoid in I_i^-. The other case can be handled symmetrically. Then $f = f_u(\tau)$. It means that we are about to sweep the shadow volume below f and bounded by τ into the space above Δ_i. Thus, we should set $f_u(\tau)$ to be the plane that passes through e and is parallel to d_c. I_i^+ contains this updated trapezoid τ and the other trapezoids in I_i^-.

By representing each trapezoid in I_i^- combinatorially by its ceiling-facet and floor-facet, the above description tells us how to update I_i^- combinatorially to produce I_i^+. Notice that I_i^+ will be treated as I_{i+1}^- when we are about to cross the boundary ζ_{i+1} in the future. By storing the trapezoids in I_i^- in a balanced binary search tree, the update at ζ_i can be performed in $O(\log n)$ time. Since there are $O(n)$ regions in Σ, scanning Σ takes $O(n \log n)$ time. Summing over all 2D slabs on $\mathcal{H}$ gives a total running time of $O(n^3 \log n)$.

What about growing $\mathcal{V}^c$ into the space above Δ_i? After the update, for each trapezoid $\tau \in I_i^+$, $f_u(\tau)$ and $f_\ell(\tau)$ cut $\Delta_i \times [\infty, -\infty]$ into two unbounded solid and one bounded solid B_τ. Conceptually, we can grow $\mathcal{V}^c$ by attaching B_τ for each trapezoid $\tau \in I_i^+$, but this is too consuming. Observe that if I_i^+ merely inherits a trapezoid τ from I_i^-, there is no hurry to sweep τ into the space above Δ_i. Instead, we wait until ζ_j for the smallest $j > i$ such that I_j^+ does not inherit τ from I_{j-1}^-. Then $f_u(\tau)$ and $f_\ell(\tau)$ cut $R \times [\infty, -\infty]$ into two unbounded solids and one bounded solid S_τ, where R is the area within Σ bounded by ζ_i and ζ_j. We attach S_τ to grow $\mathcal{V}^c$. By adopting this strategy, we spend $O(1)$ time to grow $\mathcal{V}^c$ when we cross a region boundary. Hence, we spend a total of $O(n^3)$ time to construct $\mathcal{V}_m^c \cup \mathcal{V}_o^c$. Once $\mathcal{V}_m^c \cup \mathcal{V}_o^c$ is available, we can construct the cast in $O(n^3)$ time as explained in the proof of Theorem 1.

Theorem 2. *Given d_m and d_c, the castability of a polyhedron with size n can be determined in $O(n^3 \log n)$ time and $O(n^3)$ space. If castable, the cast can be constructed in the same time and space bounds.*

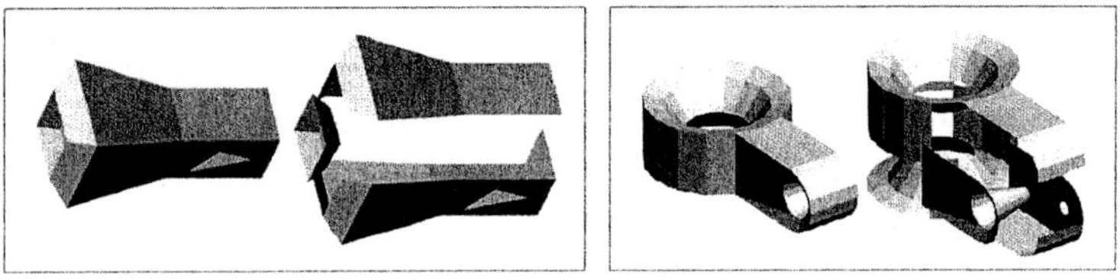

Fig. 2. The boundary of each object is partitioned into three groups in accordance with the removal directions in which the object has been verified castable

We developed a preliminary implementation of the algorithm of Theorem 2. Figure 2 shows the output of our implementation on two polyhedra: the direction d_m is the upward vertical direction and the direction d_c is the leftward direction. In the figure, the boundary of each object is partitioned into three groups depending on which cast part they belong to. For the ease of visualization, each boundary group is translated slightly in its corresponding removal direction.

5 Worst-Case Example

In this section, we present a lower bound construction showing that a castable polyhedron of size n can require a cast of $\Omega(n^3)$ size. Thus the space complexity in Theorem 2 is worst-case optimal and the time complexity of our algorithm is at most a $\log n$ factor off the worst-case optimum. Throughout this section, we assume that d_m is the upward vertical direction and d_c is the leftward direction.

Figure 3 shows our lower bound construction. The polyhedron consists of two parts: the upper part has four horizontal legs in a staircase and three slanted

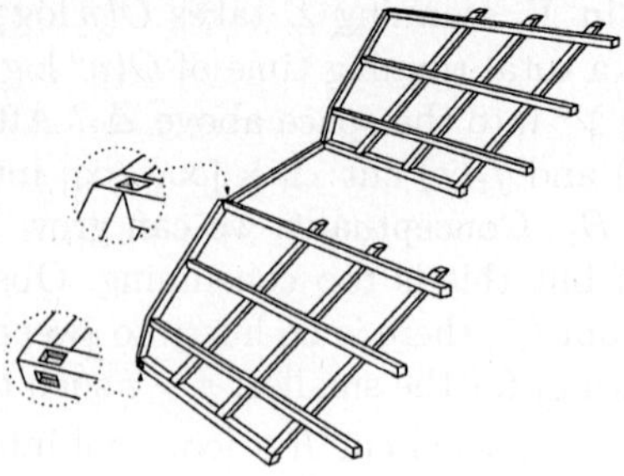

Fig. 3. The lower bound example in a perspective view

legs sitting on a horizontal leg. The lower part is an almost identical copy of the upper part, except that it has three small holes as shown in the figure. The upper hole can only be covered by the red cast part to be removed vertically upward, and the other two holes only be covered each by the core and the blue cast part. Figure 4(a) shows the front view (when we look at it from the left) and the top view of the polyhedron $\mathcal{P}$. In both projections, all three horizontal legs cross the other three slanted legs in the upper part as well as in the lower part.

Clearly, the polyhedron is castable with respect to the given d_m and d_c. We put $\Theta(n)$ horizontal legs and $\Theta(n)$ slanted legs in both the upper and the lower parts. In the upper part, each slanted leg must be in contact with both C_r and C_c. Moreover, the contacts with C_r and C_c alternate $\Theta(n)$ times along the slanted leg.

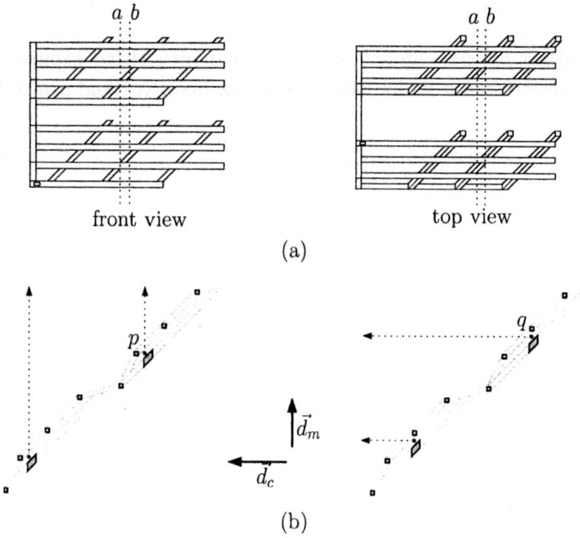

Fig. 4. (a) A top view and a side view of the lower bound construction. (b) Two cross sections along a (left) and b (right). The only way to remove p (resp. q) is translating it in d_m (resp. d_c).

As a result, the slanted legs in the upper part have a total of $\Theta(n^2)$ contacts with $\mathcal{C}_c$. These contacts sweep in direction d_c and generate $\Theta(n^2)$ swept volumes. The merging of any two such swept volumes is forbidden by the alternate appearances of the left cross-section in Figure 4(b). Each swept volume projects vertically and produces a shadow on each horizontal leg that lies below it. Thus, the total complexity of $\mathcal{C}_c$ is $\Omega(n^3)$.

References

1. H.K. Ahn, S.W. Cheng, and O. Cheong. Casting with skewed ejection direction. In *Proc. 9th Annu. International Symp. on Algorithms and Computation,*, volume 1533 of *Lecture Notes in Computer Science*, pages 139–148. Springer-Verlag, 1998.
2. H.K. Ahn, S.W. Cheng, O. Cheong, and J. Snoeyink. The reflex-free hull. *International Journal of Computational Geometry and Applications*, 14(6):453–474, 2004.
3. H.K. Ahn, O. Cheong, and R. van Oostrum. Casting a polyhedron with directional uncertainty. *Computational Geometry: Theory and Applications*, 26(2):129–141, 2003.
4. H.K. Ahn, M. de Berg, P. Bose, S.W. Cheng, D. Halperin, J. Matoušek, and O. Schwarzkopf. Separating an object from its cast. *Computer-Aided Design*, 34:547–559, 2002.
5. P. Bose and G. Toussaint. Geometric and computational aspects of gravity casting. *Computer-Aided Design*, 27(6):455–464, 1995.
6. L.L. Chen, S.Y. Chou, and T.C. Woo. Parting directions for mould and die design. *Computer-Aided Design*, 25:762–768, 1993.
7. R. Elliot. *Cast Iron Technology*. Butterworths, London, 1988.
8. K. Hui. Geometric aspects of mouldability of parts. *Computer Aided Design*, 29(3):197–208, 1997.
9. K.C. Hui and S.T. Tan. Mould design with sweep operations—a heuristic search approach. *Computer-Aided Design*, 24:81–91, 1992.
10. K. K. Kwong. *Computer-aided parting line and parting surface generation in mould design*. PhD thesis, The University of Hong Kong, Hong Kong, 1992.
11. J. Majhi, P. Gupta, and R. Janardan. Computing a flattest, undercut-free parting line for a convex polyhedron, with application to mold design. *Computational Geometry: Theory and Applications*, 13:229–252, 1999.
12. W.I. Pribble. Molds for reaction injection, structural foam and expandable styrene molding. In J.H. DuBois and W.I. Pribble, editors, *Plastics Mold Engineering Handbook*. Van Nostrand Reinhold Company, New York, 1987.
13. N. Sarnak and R.E. Tarjan. Planar point location using persistent search trees. *Communications of the ACM*, 29:669–679, 1986.
14. C.F. Walton and T.J. Opar, editors. *Iron Castings Handbook*. Iron casting society, Inc., 1981.
15. E. C. Zuppann. Castings made in sand molds. In J. G. Bralla, editor, *Handbook of Product Design for Manufacturing*, pages 5.3–5.22. McGraw-Hill, New York, 1986.

Sparse Geometric Graphs with Small Dilation*

Boris Aronov[1], Mark de Berg[2], Otfried Cheong[3], Joachim Gudmundsson[4], Herman Haverkort[2], and Antoine Vigneron[5]

[1] Department of Computer and Information Science, Polytechnic University, Brooklyn, New York, USA
http://cis.poly.edu/~aronov

[2] Department of Mathematics and Computing Science, TU Eindhoven, Eindhoven, The Netherlands
mdberg@win.tue.nl, cs.herman@haverkort.net

[3] Division of Computer Science, KAIST, Daejeon, South Korea
otfried@cs.kaist.ac.kr

[4] IMAGEN Program, National ICT Australia Ltd**, Australia
joachim.gudmundsson@nicta.com.au

[5] Department of Computer Science, National University of Singapore, Singapore
antoine@comp.nus.edu.sg

Abstract. Given a set S of n points in the plane, and an integer k such that $0 \leqslant k < n$, we show that a geometric graph with vertex set S, at most $n - 1 + k$ edges, and dilation $O(n/(k + 1))$ can be computed in time $O(n \log n)$. We also construct n-point sets for which any geometric graph with $n - 1 + k$ edges has dilation $\Omega(n/(k + 1))$; a slightly weaker statement holds if the points of S are required to be in convex position.

1 Preliminaries and Introduction

A *geometric network* is an undirected graph whose vertices are points in $\mathbb{R}^d$. Geometric networks, especially geometric networks of points in the plane, arise in many applications. Road networks, railway networks, computer networks—any collection of objects that have some connections between them can be modeled as a geometric network. A natural and widely studied type of geometric network is the *Euclidean network*, where the weight of an edge is simply the Euclidean distance between its two endpoints. Such networks for points in the plane form the topic of study of our paper.

When designing a network for a given set S of points, several criteria have to be taken into account. In particular, in many applications it is important to

* This work was supported by LG Electronics and NUS research grant R-252-000-166-112. Research by B.A. has been supported in part by NSF ITR Grant CCR-00-81964 and by a grant from US-Israel Binational Science Foundation. Part of the work was carried out while B.A. was visiting TU/e in February 2004 and in the summer of 2005. MdB was supported by the Netherlands' Organisation for Scientific Research (NWO) under project no. 639.023.301.

** National ICT Australia is funded through the Australian Government's Backing Australia's Ability initiative, in part through the Australian Research Council.

X. Deng and D. Du (Eds.): ISAAC 2005, LNCS 3827, pp. 50–59, 2005.

ensure a short connection between every pair of points in S. For this it would be ideal to have a direct connection between every pair of points; the network would then be a complete graph. In most applications, however, this is unacceptable due to the high costs. Thus the question arises: is it possible to construct a network that guarantees a reasonably short connection between every pair of points while not using too many edges? This leads to the concept of *spanners*, which we define next.

Recall that the weight of an edge $e = (u, v)$ in a Euclidean network $G = (S, E)$ on a set S of n points is the Euclidean distance between u and v, which we denote by $d(u, v)$. The *graph distance* $d_G(u, v)$ between two vertices $u, v \in S$ is the length of the shortest path in G connecting u to v. The *dilation* (or: *stretch factor*) of G, denoted $\Delta(G)$, is the maximum factor by which graph distance d_G differs from the Euclidean distance d, namely

$$\Delta(G) := \max_{\substack{u,v \in S \\ u \neq v}} \frac{d_G(u, v)}{d(u, v)}.$$

The network G is a *t-spanner* for S if $\Delta(G) \leqslant t$.

Spanners find applications in robotics, network topology design, distributed systems, design of parallel machines, and many other areas and have been a subject of considerable research [3]. Recently spanners found interesting practical applications in metric space searching [14,15] and broadcasting in communication networks [1,9,13]. The problem of constructing spanners has received considerable attention from a theoretical perspective—see the surveys [7,17].

The complete graph has dilation 1, which is optimal, but we already noted that the complete graph is generally too costly. The main challenge is therefore to design *sparse* networks that have small dilation. There are several possible measures of sparseness, for example the total weight of the edges or the maximum degree of a vertex. The measure that we will focus on is the number of edges. Thus the main question we study is this: Given a set S of n points in the plane, what is the best dilation one can achieve with a network on S that has few edges? Notice that the edges of the network are allowed to cross or overlap.

This question has already received ample attention. For example, there are several algorithms [2,12,16,18] that compute a $(1+\varepsilon)$-spanner for S, for any given parameter $\varepsilon > 0$. The number of edges in these spanners is $O(n/\varepsilon)$. Although the number of edges is linear in n, it can still be rather large both due to the dependency on ε and due to the hidden constants in the O-notation. Das and Heffernan [4] showed how to compute, for any $c > 1$, an $O(1)$–spanner with cn edges. We are interested in the case where the number of edges is close to n, not just linear in n. Any spanner must have at least $n - 1$ edges, for otherwise the graph would not be connected, and the dilation would be infinite. This leads us to define the quantity $\Delta(S, k)$:

$$\Delta(S, k) := \min_{\substack{V(G)=S \\ |E(G)|=n-1+k}} \Delta(G).$$

Thus $\Delta(S,k)$ is the minimum dilation one can achieve with a network on S that has $n-1+k$ edges. The goal of our paper is not to give an algorithm for computing a minimum-dilation network with $n-1+k$ edges for the specific input graph. (Problems of this type have been studied by several authors [6,8,10]. In general they appear quite hard.) Rather, we will study the worst-case behavior of the function $\Delta(S,k)$: what is the best dilation one can guarantee for *any* set S of n points if one is allowed to use $n-1+k$ edges? In other words, we study the quantity

$$\delta(n,k) := \sup_{\substack{S\subset\mathbb{R}^2\\|S|=n}} \Delta(S,k).$$

Keil and Gutwin [11] proved that the Delaunay triangulation of any set S of points in the plane has dilation at most $\frac{2\pi}{3\cos(\pi/6)} \approx 2.42$. Since the Delaunay triangulation has at most $3n-6$ edges (when $n \geqslant 3$), this shows that $\delta(n,2n-5) \leqslant 2.42$. We are interested in what can be achieved for smaller values of k, in particular for $0 \leqslant k < n$.

In the above definitions we have placed a perhaps unnecessary restriction that the graph may use no other vertices besides the points of S. We will also consider networks whose vertex sets strictly contain S. In particular, we define a *Steiner tree* on S as a tree T with $S \subset V(T)$. The dilation $\Delta^*(T)$ is defined analogously as

$$\Delta^*(T) := \max_{\substack{u,v\in S\\u\neq v}} \frac{d_T(u,v)}{d(u,v)}.$$

The vertices in $V(T)\setminus S$ are called *Steiner points.* We do not assume *a priori* that the network is embedded in the plane without self-intersections—the edges of T may cross or overlap. Note that T may have any number of vertices, the only restriction is on its topology.

Our results. We first show that any Steiner tree on a set S of n equally spaced points on a circle has dilation at least n/π. We prove in a similar way that $\delta(n,0) \geqslant \frac{2}{\pi}n - 1$. Eppstein [7] gave a simpler proof that $\delta(n,0) > \frac{1}{3}n$; we improve this bound by a constant factor. Our bound is tight in the sense that $\Delta(S,0) = \frac{2}{\pi}n - 1 + o(1)$ when S is a set of n equally spaced points on a circle.

We then continue with the case $0 < k < n$. Here we give an example of a set S of n points for which any network with $n-1+k$ edges has dilation at least $\frac{2}{\pi}\lfloor n/(k+1)\rfloor - 1$, proving that $\delta(n,k) \geqslant \frac{2}{\pi}\lfloor n/(k+1)\rfloor - 1$. We also prove that for points in convex position[1] the dilation can be almost as large as in the general case: there are sets of points in convex position such that any network has dilation $\Omega(n/((k+1)\log n))$.

Next we study upper bounds. We describe a $O(n\log n)$ time algorithm that computes for a given set S and parameter $0 \leqslant k < n$ a network of dilation $O(n/(k+1))$. Combined with our lower bounds, this implies that $\delta(n,k) = \Theta(n/(k+1))$. In particular, our bounds apply to the case $k = o(n)$, which was left

[1] A set of points is *in convex position* if they all lie on the boundary of their convex hull.

open by Das and Heffernan [4]. Notice that, for any constant $c \geqslant 1$, if $n \leqslant k \leqslant cn$, then we have $1 \leqslant \delta(n,k) \leqslant \delta(n,n-1)$, and thus $\delta(n,k) = \Theta(1) = \Theta(n/(k+1))$. It means that our result $\delta(n,k) = \Theta(n/(k+1))$ generalizes to the case $0 \leqslant k \leqslant cn$ for any constant $c \geqslant 1$.

Our lower bounds use rather special point sets and it may be the case that more 'regular' point sets admit networks of smaller dilation. Therefore we also study the special case where S is a $\sqrt{n} \times \sqrt{n}$ grid. We show that such a grid admits a network of dilation $O(\sqrt{n/(k+1)})$ and that this bound is asymptotically tight. We also obtain tight bounds for point sets with so-called *bounded spread.* These results and their proofs can be found in the full version of the paper.

Notation and terminology. Hereafter S will always denote a set of points in the plane. Whenever it causes no confusion we do not distinguish an edge $e = (u,v)$ in the network under consideration and the line segment uv.

2 Lower Bounds

In this section we prove lower bounds on the dilation that can be achieved with $n-1+k$ edges for $0 \leqslant k < n$.

2.1 Steiner Trees

We first show a lower bound on the dilation of any Steiner tree for S. The lower bound for this case uses the set S of n points $p_1, p_2, \ldots, p_n$ spaced equally on the unit circle, as shown in Fig. 1(a).

Theorem 1. *For any $n > 1$, there is a set S of n points in convex position such that any Steiner tree on S has dilation at least $\frac{1}{\sin(\pi/n)} \geqslant \frac{n}{\pi}$.*

Proof. Consider the set S described above and illustrated in Fig. 1(a). Let o be the center of the circle, and let T be a Steiner tree on S. First, let's assume that o does not lie on an edge of the tree.

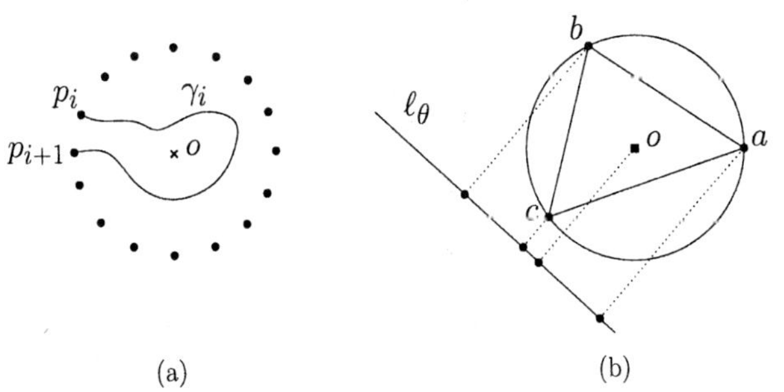

Fig. 1. (a) The homotopy class of the path γ_i. (b) Illustrating the proof of Lemma 1.

Let x and y be two points and let γ and γ' be two paths from x to y avoiding o. We call γ and γ' *(homotopy) equivalent* if γ can be deformed continuously into γ' without ever passing through the point o, that is, if γ and γ' belong to the same homotopy class in the punctured plane $\mathbb{R}^2 \setminus \{o\}$.

Let γ_i be the unique path in T from p_i to p_{i+1} (where $p_{n+1} := p_1$). We argue that there must be at least one index i for which γ_i is not equivalent to the straight segment p_ip_{i+1}, as illustrated in Fig. 1(a).

We argue by contradiction. Let Γ be the closed loop formed as the concatenation of $\gamma_1, \ldots, \gamma_n$, and let Γ' be the closed loop formed as the concatenation of the straight segments p_ip_{i+1}, for $i = 1 \ldots n$. If γ_i is equivalent to p_ip_{i+1}, for all i, then Γ and Γ' are equivalent. We now observe that, since Γ' is a simple closed loop surrounding o, it cannot be contracted to a point in the punctured plane (formally, it has winding number 1 around o). On the other hand, Γ is contained in the tree $T \not\ni o$ (viewed as a formal union[2] of its edges) and hence must be contractible in $\mathbb{R}^2 \setminus \{o\}$. Hence Γ and Γ' cannot be equivalent, a contradiction.

Consider now a path γ_i not equivalent to the segment p_ip_{i+1}. Then γ_i must "go around" o, and so its length is at least 2. The distance between p_i and p_{i+1}, on the other hand, is $2\sin(\pi/n)$, implying the theorem.

Now consider the case where o lies on an edge of T. Assume for a contradiction that there is a spanning tree T that has dilation $1/\sin(\pi/n) - \varepsilon$ for some $\varepsilon > 0$. Let o' be a point not on T at distance $\varepsilon/100$ from o. Then we can use the argument above to show that there are two consecutive points whose path in T must go around o'. By the choice of o' such a path must have dilation larger than $1/\sin(\pi/n) - \varepsilon$, a contradiction. □

2.2 The Case $k = 0$

If we require the tree to be a spanning tree without Steiner points, then the path γ_i in the above proof must not only "go around" o, but must do so using points p_j on the circle only. We can use this to improve the constant in Theorem 1, as follows. Let p_i and p_{i+1} be the consecutive points such that the path γ_i is not equivalent to the segment p_ip_{i+1}. Consider the loop formed by γ_i and $p_{i+1}p_i$. It consists of straight segments visiting some of the points of S. Let C be the convex hull of this loop. We can deal with the case where the center o lies on the boundary of C by moving it slightly, as we did in the proof of Theorem 1. Therefore we can assume that o lies in the interior of C (otherwise, the loop is contractible in the punctured plane) and there exist three vertices v_1, v_2, and v_3 of C such that $o \in \triangle v_1v_2v_3$. Since the loop visits each of these three vertices once, its length is at least the perimeter of $\triangle v_1v_2v_3$, which is at least 4 by Lemma 1 below. Therefore, we have proven

[2] Notice that T may not be properly embedded in the plane, i.e., the edges of T may cross or overlap. However, viewed not as a subset of the plane, but rather as an abstract simplicial complex, T is certainly simply connected and Γ is a closed curve contained in it and thus contractible, *within* T, to a point. Therefore it is also contractible in the punctured plane, as claimed.

Corollary 1. *For any $n > 1$ there is an n-point set S in convex position such that any spanning tree on S has dilation at least $\frac{4-2\sin(\pi/n)}{2\sin(\pi/n)} \geqslant \frac{2}{\pi}n - 1$ and thus $\delta(n,0) \geqslant \Delta(S,0) \geqslant \frac{2}{\pi}n - 1$.*

Lemma 1. *Any triangle inscribed in a unit circle and containing the circle center has perimeter at least 4.*

Proof. Let o be the circle center, and let a, b, and c be three points at distance one from o such that o is contained in the triangle $\triangle abc$. We will prove that the perimeter $p(\triangle abc)$ is at least 4.

We need the following definition: For a compact convex set C in the plane and $0 \leqslant \theta < \pi$, let $w(C,\theta)$ denote the *width of C in direction θ*. More precisely, if ℓ_θ is the line through the origin with normal vector $(\cos\theta, \sin\theta)$, then $w(C,\theta)$ is the length of the orthogonal projection of C to ℓ_θ.

The Cauchy-Crofton formula [5] allows us to express the perimeter $p(C)$ of a compact convex set C in the plane as $p(C) = \int_0^\pi w(C,\theta)d\theta$.

We apply this formula to $\triangle abc$, and consider its projection on the line ℓ_θ (for some θ). The endpoints of the projection of $\triangle abc$ are projections of two of the points a, b, and c. Without loss of generality, for a given θ, let these be a and b, and assume that c projects onto the projection of the segment ob. Then we clearly have $w(oc,\theta) \leqslant w(ob,\theta)$. Since o lies in C, we also have $w(oc,\theta) \leqslant w(oa,\theta)$ (consider the angles the segments oa, ob and oc make with the line ℓ_θ). This implies $3w(oc,\theta) \leqslant w(oa,\theta) + w(ob,\theta) + w(oc,\theta)$, and therefore

$$\begin{aligned} w(\triangle abc,\theta) &= w(oa,\theta) + w(ob,\theta) \\ &= w(oa,\theta) + w(ob,\theta) + w(oc,\theta) - w(oc,\theta) \\ &\geqslant \tfrac{2}{3}(w(oa,\theta) + w(ob,\theta) + w(oc,\theta)). \end{aligned}$$

Integrating θ from 0 to π and applying the Cauchy-Crofton formula gives $p(\triangle abc) \geqslant \frac{2}{3}(p(oa)+p(ob)+p(oc))$. Since oa, ob, and oc are segments of length 1, each has perimeter 2, and so we have $p(\triangle abc) \geqslant \frac{2}{3} \cdot 6 = 4$. □

2.3 The General Case

We now turn to the general case, and we first consider graphs with $n - 1 + k$ edges for $0 < k < n$.

Theorem 2. *For any n and any k with $0 < k < n$, there is a set S of n points such that any graph on S with $n - 1 + k$ edges has dilation at least $\frac{2-\sin(\pi/\lfloor n/(k+1)\rfloor)}{\sin(\pi/\lfloor n/(k+1)\rfloor)} \geqslant \frac{2}{\pi} \cdot \lfloor \frac{n}{k+1} \rfloor - 1$. Hence, $\delta(n,k) \geqslant \frac{2}{\pi} \cdot \lfloor \frac{n}{k+1} \rfloor - 1$.*

Proof. Our example S consists of $k+1$ copies of the set used in Theorem 1. More precisely, we choose sets S_i, for $1 \leqslant i \leqslant k+1$, each consisting of at least $\lfloor n/(k+1) \rfloor$ points. We place the points in S_i equally spaced on a unit-radius circle with center at $(2ni, 0)$, as in Fig. 2. The set S is the union of $S_1, \ldots, S_{k+1}$; we choose the sizes of the S_i such that S contains n points.

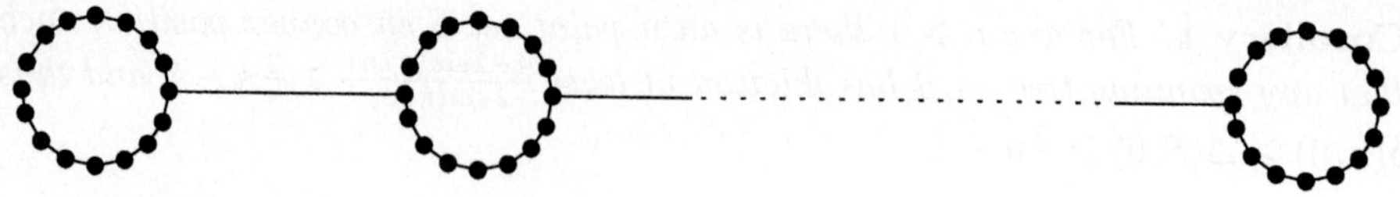

Fig. 2. Illustrating the point set S constructed in the proof of Theorem 2

Let G be a graph with vertex set S and $n-1+k$ edges. We call an edge of G *short* if its endpoints lie in the same set S_i, and *long* otherwise. Since G is connected, there are at least k long edges, and therefore at most $n-1$ short edges. Since $\sum |S_i| = n$, this implies that there is a set S_i such that the number of short edges with endpoints in S_i is at most $|S_i| - 1$. Let G' be the induced subgraph of S_i. By Corollary 1, its dilation is at least

$$\frac{2 - \sin(\pi/\lfloor n/(k+1)\rfloor)}{\sin(\pi/\lfloor n/(k+1)\rfloor)} \geqslant \frac{2}{\pi} \cdot \left\lfloor \frac{n}{k+1} \right\rfloor - 1.$$

Since any path connecting two points in S_i using a long edge has dilation at least n, this implies the claimed lower bound on the dilation of G. □

2.4 Points in Convex Position

The point set of Theorem 1 is in convex position, but works as a lower bound only for $k = 0$. In fact, by adding a single edge (the case $k = 1$) one can reduce the dilation to a constant. Now consider n points that lie on the boundary of a planar convex figure with aspect ratio at most ρ, that is, with the ratio of diameter to width at most ρ. It is not difficult to see that connecting the points along the boundary—hence, using n edges—leads to a graph with dilation $\Theta(\rho)$. However, the following theorem shows that for large aspect ratio, one cannot do much better than in the general case. The proof will be given in the full version of this paper.

Theorem 3. *For any n and k such that $0 \leqslant k < n$, there is a set S of n points in convex position such that any graph G with vertex set S and $n-1+k$ edges has dilation $\Omega(n/((k+1)\log n))$.*

3 A Constructive Upper Bound

In this section we show an upper bound on the dilation achievable with k extra edges. We make use of the fact (observed by Eppstein [7]) that a minimum spanning tree has linear dilation. More precisely, we will use the following lemma.

Lemma 2. *Let S be a set of n points in the plane, and let T be a minimum spanning tree of S. Then T has dilation at most $n-1$. In other words, $\Delta(S, 0) \leqslant \Delta(T) \leqslant n-1$ and, as it holds for all S with $|S| = n$, we have $\delta(n, 0) \leqslant n-1$.*

Proof. Let $p, q \in S$ and consider the path γ connecting p and q in T. Since T is a minimum spanning tree, any edge in γ has length at least $d(p, q)$. Since γ consists of at most $n - 1$ edges, the dilation of γ is at most $n - 1$. □

The following algorithm builds a spanner with at most $n - 1 + k$ edges:

Algorithm 1 SPARSESPANNER(S, k)

Input A set S of n points in the plane and an integer $k \geqslant 0$.
Output A graph G=(S,E).

```
1: Compute a Delaunay triangulation of S.
2: Compute a minimum spanning tree T of S.
3: if k = 0 then
4:     return T.
5: Let m ← ⌊(k + 5)/2⌋.
6: Compute m disjoint subtrees of T, each containing O(n/m) points, by removing
   m − 1 edges.
7: E ← ∅.
8: for each subtree T' do
9:     add the edges of T' to E.
10: for each pair of subtrees T' and T'' do
11:     if there is a Delaunay edge (p, q) with p ∈ T', q ∈ T'' then
12:         add the shortest such edge (p, q) to E.
13: return G = (S, E).
```

We first prove the correctness of the algorithm.

Lemma 3. *Algorithm* SPARSESPANNER *returns a graph G with at most $n-1+k$ edges and dilation bounded by $O(n/(k + 1))$.*

Proof. Lemma 2 shows that our algorithm is correct if $k = 0$, so from now on we assume that $k \geqslant 1$, and thus $m \geqslant 3$. Consider the graph G' obtained from the output graph G by contracting each subtree T' created in step 6 to a single node. G' is a planar graph with $m \geqslant 3$ vertices, without loops or multiple edges, and so it has at most $3m - 6$ edges. The total number of edges in the output graph is therefore at most

$$n - 1 - (m - 1) + 3m - 6 = n + 2m - 6 \leqslant n + k - 1.$$

Consider two points $x, y \in S$. Let $x = x_0, x_1, \ldots, x_j = y$ be a shortest path from x to y in the Delaunay graph. It has dilation [11] at most $2\pi/(3\cos(\pi/6)) = O(1)$. We claim that each edge $x_i x_{i+1}$ in the Delaunay graph has path in G with dilation $O(n/k)$. The concatenation of these paths yields a path from x to y with dilation $O(n/k)$, proving the lemma.

If x_i and x_{i+1} fall into the same subtree T', then Lemma 2 implies a dilation of $O(n/m) = O(n/k)$.

It remains to consider the case $x_i \in T'$, $x_{i+1} \in T''$, where T' and T'' are distinct subtrees of T. Let (p, q) be the edge with $p \in T'$, $q \in T''$ inserted in step 12—such an edge exists, since at least one Delaunay edge between T' and T'' has been considered, namely (x_i, x_{i+1}). Let δ' be a path from x_i to p in T', and let δ'' be a path from q to x_{i+1} in T''. Both paths have $O(n/k)$ edges. By construction, we have $d(p, q) \leqslant d(x_i, x_{i+1})$. We claim that every edge e on δ' and δ'' has length at most $d(x_i, x_{i+1})$. Indeed, if e had length larger than $d(x_i, x_{i+1})$, then we could remove it from T and insert either (x_i, x_{i+1}) or (p, q) to obtain a better spanning tree. Therefore the concatenation of δ', the edge (p, q), and δ'' is a path of $O(n/k)$ edges, each of length at most $d(x_i, x_{i+1})$, and so the dilation of this path is $O(n/k)$. □

Theorem 4. *Given a set S of n points in the plane and an integer $k \geqslant 0$, a graph G with vertex set S, $n - 1 + k$ edges, and dilation $O(n/(k + 1))$ can be constructed in $O(n \log n)$ time.*

Proof. We use algorithm SPARSESPANNER. Its correctness has been proven in Lemma 3. Steps 1 and 2 can be implemented in $O(n \log n)$ time [7]. Step 6 can be implemented in linear time as follows. Orient T by arbitrarily choosing a root node. Traverse T in postorder, keeping track of the size $|T_v|$ of the (remaining) subtree T_v rooted at the current node v. When $|T_v|$ reaches n/m, cut T_v off the main tree by removing the edge connecting v to its parent. Each of the trees rooted at the children of v has size smaller than n/m, and the maximum degree of any node of T is at most six [7,19]. Therefore, at the time T_v is cut off from the main tree, we have $n/m \leqslant |T_v| < 1 + 5(n/m)$. The argument does not apply when we reach the root, so we are left with one tree that can be arbitrarily small, but has fewer than $6n/m$ vertices. To implement steps 10–12, we scan the edges of the Delaunay triangulation, and keep the shortest edge connecting each pair of subtrees. It can be done in $O(n \log n)$ time. □

4 Conclusions and Open Problems

We have shown that for any planar n-point set S and parameter $0 \leqslant k < n$, there is a graph G with vertex set S, $n - 1 + k$ edges, and dilation at most $O(n/(k + 1))$. We also proved a lower bound of $\Omega(n/(k + 1))$ on the maximum dilation of such a graph.

Minimum dilation graphs are not well understood yet. For instance, it is not known whether a minimum dilation tree for a given point set may self-intersect [7], not even if the point set is in convex position. (On the other hand, minimum dilation paths or tours can self-overlap.) No efficient algorithm for computing the minimum dilation tree for a given point set is known. It would be interesting to either give such an algorithm, or show that the problem is NP-hard and look for algorithms that approximate the best possible dilation (instead of giving only a guarantee in terms of n and k, as we do).

References

1. K. Alzoubi, X.-Y. Li, Y. Wang, P.-J. Wan, and O. Frieder. Geometric spanners for wireless ad hoc networks. *IEEE Trans. Parallel Dist. Systems*, 14(4):408–421, 2003.
2. P. B. Callahan and S. R. Kosaraju. A decomposition of multidimensional point sets with applications to k-nearest-neighbors and n-body potential fields. *J. ACM*, 42:67–90, 1995.
3. B. Chandra, G. Das, G. Narasimhan, and J. Soares. New sparseness results on graph spanners. *Internat. J. Comput. Geom. Appl.*, 5:124–144, 1995.
4. G. Das and P. Heffernan. Constructing Degree-3 Spanners with Other Sparseness Properties. *Int. J. Found. Comput. Sci.*, 7(2):121–136, 1996.
5. M. do Carmo. *Differential Geometry of Curves and Surfaces.* Prentice Hall, Englewood Cliffs, NJ, 1976.
6. Y. Emek and D. Peleg. Approximating minimum max-stretch spanning trees on unweighted graphs (full version). In *Proc. ACM-SIAM Symp. Discrete Algo.*, pages 261–270, 2004.
7. D. Eppstein. Spanning trees and spanners. In J.-R. Sack and J. Urrutia, editors, *Handbook of Computational Geometry*, pages 425–461. Elsevier Science Publishers, Amsterdam, 2000.
8. D. Eppstein and K. Wortman. Minimum dilation stars. In *Proc. ACM Symp. Comput. Geom.*, pages 321–326, 2005.
9. A. M. Farley, A. Proskurowski, D. Zappala, and K. J. Windisch. Spanners and message distribution in networks. *Discrete Appl. Math.*, 137(2):159–171, 2004.
10. S. P. Fekete and J. Kremer. Tree spanners in planar graphs. *Discrete Appl. Math.*, 108:85–103, 2001.
11. J. M. Keil and C. A. Gutwin. Classes of graphs which approximate the complete Euclidean graph. *Discrete Comput. Geom.*, 7:13–28, 1992.
12. C. Levcopoulos and A. Lingas. There are planar graphs almost as good as the complete graphs and almost as cheap as minimum spanning trees. *Algorithmica*, 8:251–256, 1992.
13. X.-Y. Li. Applications of computational geomety in wireless ad hoc networks. In X.-Z. Cheng, X. Huang, and D.-Z. Du, editors, *Ad Hoc Wireless Networking.* Kluwer, 2003.
14. G. Navarro and R. Paredes. Practical construction of metric t-spanners. In *Proc. 5th Workshop Algorithm Eng. Exp.*, pages 69–81. SIAM Press, 2003.
15. G. Navarro, R. Paredes, and E. Chez. t-spanners as a data structure for metric space searching. In *Proc. 9th Int. Symp. String Proc. Inf. Retrieval*, volume 2476 of *Lecture Notes in Computer Science*, pages 298–309. Springer-Verlag, 2002.
16. J. S. Salowe. Constructing multidimensional spanner graphs. *Internat. J. Comput. Geom.*, 1:99–107, 1991.
17. M. Smid. Closest point problems in computational geometry. In J.-R. Sack and J. Urrutia, editors, *Handbook of Computational Geometry*, pages 877–935. Elsevier Science Publishers, Amsterdam, 2000.
18. P. M. Vaidya. A sparse graph almost as good as the complete graph on points in K dimensions. *Discrete Comput. Geom.*, 6:369–381, 1991.
19. A. C. Yao. On constructing minimum spanning trees in k-dimensional spaces and related problems. *SIAM J. Comput.*, 11, 1992.

Multiple Polyline to Polygon Matching

Mirela Tănase[1], Remco C. Veltkamp[1], and Herman Haverkort[2]

[1] Department of Computing Science, Utrecht University, The Netherlands
[2] Department of Computing Science, TU Eindhoven, The Netherlands

Abstract. We introduce a measure for computing the similarity between multiple polylines and a polygon, that can be computed in $O(km^2n^2)$ time with a straightforward dynamic programming algorithm. We then present a novel fast algorithm that runs in time $O(kmn \log mn)$. Here, m denotes the number of vertices in the polygon, and n is the total number of vertices in the k polylines that are matched against the polygon. The effectiveness of the similarity measure has been demonstrated in a part-based retrieval application with known ground-truth.

1 Introduction

The motivation for multiple polyline to polygon matching is twofold. Firstly, the matching of shapes has been done mostly by comparing them as a whole [2,8,10]. This fails when a significant part of one shape is occluded, or distorted by noise. In this paper, we address the partial shape matching problem, matching portions of two given shapes. Secondly, partial matching helps identifying similarities even when a significant portion of one shape boundary is occluded, or seriously distorted. It could also help in identifying similarities between contours of a non-rigid object in different configurations of its moving parts, like the contours of a sitting and a walking cat. Finally, partial matching helps alleviating the problem of unreliable object segmentation from images, over or undersegmentation, giving only partially correct contours.

Contribution. Firstly, we introduce a measure for computing the similarity between multiple polylines and a polygon. This similarity measure is a turning angle function-based similarity, minimized over all possible shiftings of the endpoints of the parts over the shape, and also over all independent rotations of the parts. Since we allow the parts to rotate independently, this measure could capture the similarity between contours of non-rigid objects, with parts in different relative positions. We then derive a number of non-trivial properties of the similarity measure.

Secondly, based on these properties we characterize the optimal solution that leads to a straightforward $O(km^2n^2)$-time and space dynamic programming algorithm. We then present a novel $O(kmn \log mn)$ time and space algorithm. Here, m denotes the number of vertices in the polygon, and n is the total number of vertices in the k polylines that are matched against the polygon.

Thirdly, we have experimented with a part-based retrieval application. Given a large collection of shapes and a query consisting of a set of polylines, we want to

X. Deng and D. Du (Eds.): ISAAC 2005, LNCS 3827, pp. 60–70, 2005.

retrieve those shapes in the collection that best match our query. The evaluation using a known ground-truth indicates that a part-based approach improves the global matching performance for difficult categories of shapes.

2 Related Work

Arkin et al. [2] describe a metric for comparing two whole polygons that is invariant under translation, rotation and scaling. It is based on the L_2-distance between the turning functions of the two polygons, and can be computed in $O(mn \log mn)$ time, where m is the number of vertices in one polygon and n is the number of vertices in the other.

Most partial shape matching methods are based on computing local features of the contour, and then looking for correspondences between the features of the two shapes, for example points of high curvature [1,7]. Such local features-based solutions work well when the matched subparts are almost equivalent up to a transformation such as translation, rotation or scaling, because for such subparts the sequences of local features are very similar. However, parts that we perceive as similar, may have quite different local features (different number of curvature extrema for example).

Geometric hashing [12] is a method that determines if there is a transformed subset of the query point set that matches a subset of a target point set, by building a hash table in transformation space. Also the Hausdorff distance [5] allows partial matching. It is defined for arbitrary non-empty bounded and closed sets A and B as the infimum of the distance of the points in A to B and the points in B to A. Both methods are designed for partial matching, but do not easily transform to our case of matching multiple polylines to a polygon.

Partially matching the turning angle function of two polylines under scaling, translation and rotation, can be done in time $O(m^2n^2)$ [4]. Given two matches with the same squared error, the match involving the longer part of the polylines has a lower dissimilarity. The dissimilarity measure is a function of the scale, rotation, and the shift of one polyline along the other. However, this works for only two single polylines.

Latecki et al [6], establish the best correspondence of parts in a decomposition of the matched shapes. The best correspondence between the maximal convex arcs of two simplified versions of the original shapes gives the partial similarity measure between the shapes. One drawback of this approach is that the matching is done between parts of simplified shapes at "the appropriate evolution stage". How these evolution stages are identified is not indicated in their papers, though it certainly has an effect on the quality of the matching process.

3 Polylines-to-Polygon Matching

We concentrate on the problem of matching an ordered set $\{P_1, P_2, \ldots, P_k\}$ of k polylines against a polygon P. We want to compute how close an ordered set of polylines $\{P_1, P_2, \ldots, P_k\}$ is to being part of the boundary of P in the given

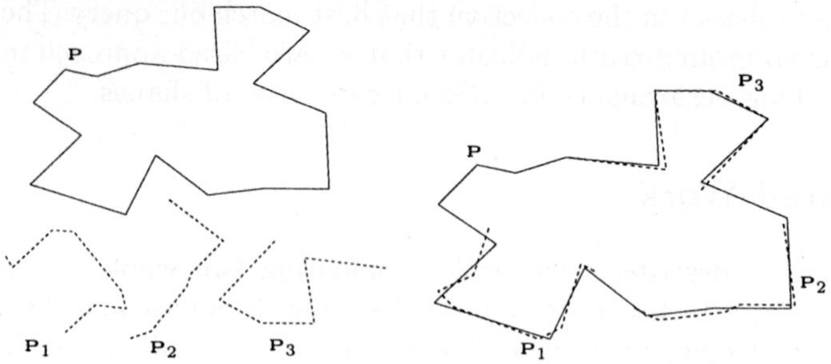

Fig. 1. Matching an ordered set $\{P_1, P_2, P_3\}$ of polylines against a polygon P

order in counter-clockwise direction around P (see figure 1). For this purpose, the polylines are rotated and shifted along the polygon P, in such a way that the pieces of the boundary of P "covered" by the k polylines are mutually disjoint except possibly at their endpoints. Note that P is a polygon and not an open polyline, only because of the intended application of part based retrieval.

3.1 Similarity Between Multiple Polylines and a Polygon

The *turning function* Θ_A of a polygon A measures the angle of the counter-clockwise tangent with respect to a reference orientation as a function of the arc-length s, measured from some reference point on the boundary of A. It is a piecewise constant function, with jumps corresponding to the vertices of A. A rotation of A by an angle θ corresponds to a shifting of Θ_A over a distance θ in the vertical direction. Moving the location of the reference point $A(0)$ over a distance $t \in [0, l_A)$ along the boundary of A corresponds to shifting Θ_A horizontally over a distance t.

Let $\Theta : [0, l] \rightarrow \mathbb{R}$ be the turning function of a polygon P of m vertices, and of perimeter length l. Since P is a closed polyline, the domain of Θ can be easily extended to the entire real line, by $\Theta(s + l) = \Theta(s) + 2\pi$. Let $\{P_1, P_2, \ldots P_k\}$ be a set of polylines, and let $\Theta_j : [0, l_j] \rightarrow \mathbb{R}$ denote the turning function of the polyline P_j of length l_j. If P_j is made of n_j segments, Θ_j is piecewise-constant with $n_j - 1$ jumps.

For simplicity of exposition, $\mathsf{f}_j(t, \theta)$ denotes the quadratic similarity between the polyline P_j and the polygon P, for a given placement (t, θ) of P_j over P: $\mathsf{f}_j(t, \theta) = \int_0^{l_j} (\Theta(s + t) - \Theta_j(s) + \theta)^2 ds$.

We assume the polylines $\{P_1, P_2, \ldots, P_k\}$ satisfy the condition $\sum_{j=1}^{k} l_j \leq l$. The similarity measure, which we denote by $\mathsf{d}(P_1, \ldots, P_k; P)$, is the square root of the sum of quadratic similarities f_j, minimized over all valid placements of $P_1, \ldots, P_k$ over P:

$$\mathsf{d}(P_1, \ldots, P_k; P) = \min_{\substack{\text{valid placements} \\ (t_1, \theta_1) \ldots (t_k, \theta_k)}} \left(\sum_{j=1}^{k} \mathsf{f}_j(t_j, \theta_j) \right)^{1/2} .$$

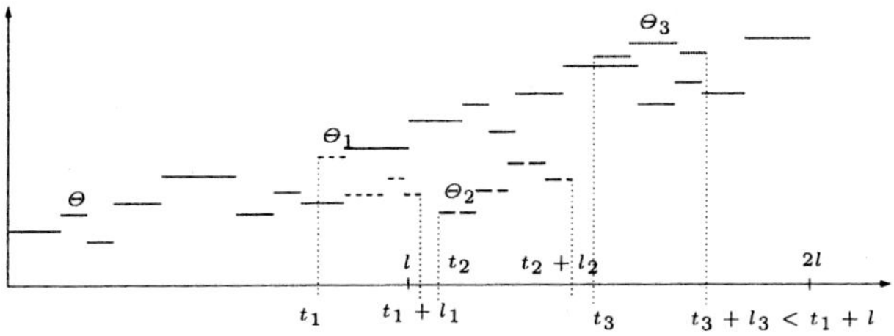

Fig. 2. To compute $\mathsf{d}(P_1, \ldots, P_k; P)$ between the polylines $P_1, \ldots, P_3$ and the polygon P, we shift the turning functions Θ_1, Θ_2, and Θ_3 horizontally and vertically over Θ

It remains to define what the valid placements are. The horizontal shifts $t_1, \ldots, t_k$ correspond to shiftings of the starting points of the polylines $P_1, \ldots, P_k$ along P. We require that the starting points of $P_1, \ldots, P_k$ are matched with points on the boundary of P in counterclockwise order around P, that is: $t_{j-1} \leq t_j$ for all $1 < j \leq k$, and $t_k \leq t_1 + l$. Furthermore, we require that the matched parts are disjoint (except possibly at their endpoints), sharpening the constraints to $t_{j-1} + l_{j-1} \leq t_j$ for all $1 < j \leq k$, and $t_k + l_k \leq t_1 + l$ (see figure 2).

The vertical shifts $\theta_1, \ldots, \theta_k$ correspond to rotations of the polylines $P_1, \ldots, P_k$ with respect to the reference orientation, and are independent of each other. Therefore, in an optimal placement the quadratic similarity between a particular polyline P_j and P depends only on the horizontal shift t_j, while the vertical shift must be optimal for the given horizontal shift. We can thus express the similarity between P_j and P for a given positioning t_j of the starting point of P_j over P as: $\mathsf{f}_j^*(t_j) = \min_{\theta \in \mathbb{R}} \mathsf{f}_j(t_j, \theta)$.

The similarity between the polylines $P_1, \ldots, P_k$ and the polygon P is thus:

$$\mathsf{d}(P_1, \ldots, P_k; P) = \min_{\substack{t_1 \in [0,l),\, t_2,\ldots,t_k \in [0,2l);\\ \forall j \in \{2,\ldots,k\}:\ t_{j-1}+l_{j-1} \leq t_j;\quad t_k+l_k \leq t_1+l}} \left(\sum_{j=1}^{k} \mathsf{f}_j^*(t_j) \right)^{1/2} . \quad (1)$$

3.2 Properties of the Similarity Function

In this section we give a few properties of $\mathsf{f}_j^*(t)$, as functions of t, that constitute the basis of the algorithms for computing $\mathsf{d}(P_1, \ldots, P_k; P)$ in sections 3.3 and 3.4. We also give a simpler formulation of the optimization problem in the definition of $\mathsf{d}(P_1, \ldots, P_k; P)$. Arkin et al. [2] have shown that for any fixed t, the function $\mathsf{f}_j(t, \theta)$ is a quadratic convex function of θ. This implies that for a given t, the optimization problem $\min_{\theta \in \mathbb{R}} \mathsf{f}_j(t, \theta)$ has a unique solution, given by the root $\theta_j^*(t)$ of the equation $\partial \mathsf{f}_j(t, \theta)/\partial\theta = 0$. As a result:

Lemma 1. *For a given positioning t of the starting point of P_j over P, the rotation that minimizes the quadratic similarity between P_j and P is given by* $\theta_j^*(t) = -\int_0^{l_j} (\Theta(s+t) - \Theta_j(s))ds/l_j$.

We now consider the properties of $\mathsf{f}_j^*(t) = \mathsf{f}_j(t, \theta_j^*(t))$, as a function of t.

Lemma 2. *The quadratic similarity $\mathsf{f}_j^*(t)$ has the following properties:*
i) it is periodic, with period l;
ii) it is piecewise quadratic, with mn_j breakpoints within any interval of length l; moreover, the parabolic pieces are concave.

For a proof of this and the following lemmas, see [11].

The following corollary indicates that to compute the minimum of the function f_j^*, we need to look only a discrete set of at most mn_j points.

Corollary 1. *The local minima of the function f_j^* are among the breakpoints between its parabolic pieces.*

We now give a simpler formulation of the optimization problem in the definition of $\mathsf{d}(P_1, \ldots, P_k; P)$. In order to simplify the restrictions on t_j in equation (1), we define: $\overline{\mathsf{f}}_j(t) := \mathsf{f}_j^*\left(t + \sum_{i=1}^{j-1} l_i\right)$. In other words, the function $\overline{\mathsf{f}}_j$ is a copy of f_j^*, but shifted to the left with $\sum_{i=1}^{j-1} l_i$. Obviously, $\overline{\mathsf{f}}_j$ has the same properties as f_j^*, that is: it is a piecewise quadratic function of t that has its local minima in at most mn_j breakpoints in any interval of length l. With this simple transformation of the set of functions f_j^*, the optimization problem defining $\mathsf{d}(P_1, \ldots, P_k; P)$ becomes:

$$\mathsf{d}(P_1, \ldots, P_k; P) = \min_{\substack{t_1 \in [0,l),\, t_2,\ldots,t_k \in [0,2l); \\ \forall j \in \{2,\ldots,k\}:\ t_{j-1} \le t_j;\quad t_k \le t_1 + l_0}} \left(\sum_{j=1}^{k} \overline{\mathsf{f}}_j(t_j) \right)^{1/2}, \quad (2)$$

where $l_0 := l - \sum_{i=1}^{k} l_i$. Notice that if $(\overline{t}_1^*, \ldots, \overline{t}_k^*)$ is a solution to the optimization problem in equation (2), then $(t_1^*, \ldots, t_k^*)$, with $t_j^* := \overline{t}_j^* + \sum_{i=1}^{j-1} l_i$, is a solution to the optimization problem in equation (1).

3.3 Characterization of an Optimal Solution

In this section we characterize the structure of an optimal solution to the optimization problem in equation (2), and give a recursive definition of this solution. This definition forms the basis of a straightforward dynamic programming solution to the problem.

Let $(\overline{t}_1^*, \ldots, \overline{t}_k^*)$ be a solution to the optimization problem in equation (2).

Lemma 3. *The values of an optimal solution $(\overline{t}_1^*, \ldots, \overline{t}_k^*)$ are found in a discrete set of points $X \subset [0, 2l)$ of the breakpoints of the functions $\overline{\mathsf{f}}_1, \ldots, \overline{\mathsf{f}}_k$, plus two copies of each breakpoint: one shifted left by l_0 and one shifted right by l_0.*

We call a point in $[0, 2l)$, which is either a breakpoint of $\overline{\mathsf{f}}_1, \ldots, \overline{\mathsf{f}}_k$, or such a breakpoint shifted left or right by l_0, a *critical point.* Since function $\overline{\mathsf{f}}_j$ has $2mn_j$ breakpoints, the total number of critical points in $[0, 2l)$ is at most $6m \sum_{i=1}^{k} n_i = 6mn$. Let $X = \{x_0, \ldots, x_{N-1}\}$ be the set of critical points in $[0, 2l)$.

With the observations above, the optimization problem we have to solve is:

$$\mathsf{d}(P_1,\ldots,P_k;P) = \min_{\substack{t_1,\ldots,t_k \in X \\ \forall j>1:\ t_{j-1}\le t_j\,;\, t_k - t_1 \le l_0}} \left(\sum_{j=1}^{k} \bar{\mathsf{f}}_j(t_j)\right)^{1/2} . \tag{3}$$

We denote:

$$D[j,a,b] = \min_{\substack{t_1,\ldots,t_j \in X \\ x_a \le t_1 \le \ldots \le t_j \le x_b}} \sum_{i=1}^{j} \bar{\mathsf{f}}_i(t_i) , \tag{4}$$

where $j \in \{1,\ldots,k\}$, $a,b \in \{0,\ldots,N-1\}$, and $a \le b$. Equation (4) describes the subproblem of matching the set $\{P_1,\ldots,P_j\}$ of j polylines to a subchain of P, starting at $P(x_a)$ and ending at $P(x_b + \sum_{i=1}^{j} l_i)$. We now show that $D[j,a,b]$ can be computed recursively. Let $(t_1^{\circledast},\ldots t_j^{\circledast})$ be an optimal solution for $D[j,a,b]$. Regarding the value of $t_j^{\circledast}$ we distinguish two cases:

- $t_j^{\circledast} = x_b$, in which case $(t_1^{\circledast},\ldots t_{j-1}^{\circledast})$ must be an optimal solution for $D[j-1,a,b]$, otherwise $(t_1^{\circledast},\ldots t_j^{\circledast})$ would not give a minimum for $D[j,a,b]$; thus in this case, $D[j,a,b] = D[j-1,a,b] + \bar{\mathsf{f}}_j(x_b)$;
- $t_j^{\circledast} \ne x_b$, in which case $(t_1^{\circledast},\ldots t_j^{\circledast})$ must be an optimal solution for $D[j,a,b-1]$; otherwise $(t_1^{\circledast},\ldots t_j^{\circledast})$ would not give a minimum for $D[j,a,b]$; thus in this case $D[j,a,b] = D[j,a,b-1]$.

We can now conclude that :

$$D[j,a,b] = \min\left(D[j-1,a,b] + \bar{\mathsf{f}}_j(x_b),\ D[j,a,b-1]\right), \mathrm{for\, j} \ge 1 \wedge \mathrm{a} \le \mathrm{b}, \tag{5}$$

where the boundary cases are $D[0,a,b] = 0$ and $D[j,a,a-1]$ has no solution.

A solution of the optimization problem (3) is then given by

$$\mathsf{d}(P_1,\ldots,P_k;P) = \min_{x_a,x_b \in X,\ x_b - x_a \le l_0} \sqrt{D[k,a,b]} . \tag{6}$$

Equations (5) and (6) lead to a straightforward dynamic programming algorithm for computing the similarity measure $\mathsf{d}(P_1,\ldots,P_k;P)$ in $O(km^2n^2)$ time.

3.4 A Fast Algorithm

The above time bound to computate of the similarity measure $\mathsf{d}(P_1,\ldots,P_k;P)$ can be improved to $O(kmn \log mn)$. The refinement of the dynamic programming algorithm is based on the following property of equation (5):

Lemma 4. *For any polyline P_j, $j \in \{1,\ldots,k\}$, and any critical point x_b, $b \in \{0,\ldots,N-1\}$, there is a critical point x_z, $0 \le z \le b$, such that:*

i) $D[j,a,b] = D[j,a,b-1]$, *for all* $a \in \{0,...,z-1\}$, *and*
ii) $D[j,a,b] = D[j-1,a,b] + \bar{\mathsf{f}}_j(x_b)$, *for all* $a \in \{z,...,b\}$.

For given j and b, we consider the function $\mathcal{D}[j,b] : \{0, N-1\} \to \mathbb{R}$, with $\mathcal{D}[j,b](a) = D[j,a,b]$. Lemma 4 expresses the fact that the values of function $\mathcal{D}[j,b]$ can be obtained from $\mathcal{D}[j,b-1]$ up to some value z, and from $\mathcal{D}[j-1,b]$ (while adding $\overline{\mathsf{f}}_j(x_b)$) from this value onwards. This property allows us to improve the time bound of the dynamic programming algorithm. Instead of computing arrays of scalars $D[j,a,b]$, we will compute arrays of functions $\mathcal{D}[j,b]$. The key to success will be to represent these functions in such a way that they can be evaluated fast and $\mathcal{D}[j,b]$ can be constructed from $\mathcal{D}[j-1,b]$ and $\mathcal{D}[j,b-1]$ fast.

Algorithm FastCompute d$(P_1,\ldots,P_k;P)$

1. Compute the set of critical points $X = \{x_0,\ldots,x_{N-1}\}$, and sort them
2. For all $j \in \{1,...,k\}$ and all $b \in \{0,...,N-1\}$, evaluate $\overline{\mathsf{f}}_j(x_b)$
3. $ZERO \leftarrow$ a function that always evaluates to zero (see Lemma 5)
4. $INFINITY \leftarrow$ a function that always evaluates to ∞ (see Lemma 5)
5. $MIN \leftarrow \infty$
6. $a \leftarrow 0$
7. **for** $j \leftarrow 1$ **to** k **do**
8. $\quad \mathcal{D}[j,-1] \leftarrow INFINITY$
9. **for** $b \leftarrow 0$ **to** $N-1$ **do**
10. $\quad \mathcal{D}[0,b] \leftarrow ZERO$
11. $\quad$ **for** $j \leftarrow 1$ **to** k **do**
12. $\quad\quad$ Construct $\mathcal{D}[j,b]$ from $\mathcal{D}[j-1,b]$ and $\mathcal{D}[j,b-1]$
13. $\quad$ **while** $x_a < x_b - l_0$ **do**
14. $\quad\quad a \leftarrow a+1$
15. $\quad$ val $\leftarrow$ evaluation of $\mathcal{D}[k,b](a)$
16. $\quad MIN \leftarrow \min(\text{val}, MIN)$
17. **return** $\sqrt{MIN}$

The running time of this algorithm depends on how the functions $\mathcal{D}[j,b]$ are represented. In order to make especially steps 12 and 15 of the above algorithm efficient, we represent the functions $\mathcal{D}[j,b]$ by means of balanced binary trees. Asano et al. [3] used an idea similar in spirit.

An efficient representation for function $\mathcal{D}[j,b]$. We now describe the tree $\mathcal{T}_{j,b}$ used for storing the function $\mathcal{D}[j,b]$. Each node ν of $\mathcal{T}_{j,b}$ is associated with an interval $[a_\nu^-, a_\nu^+]$, with $0 \le a_\nu^- \le a_\nu^+ \le N-1$. The root ρ is associated with the full domain, that is: $a_\rho^- = 0$ and $a_\rho^+ = N-1$. Each node ν with $a_\nu^- < a_\nu^+$ is an internal node that has a split value $a_\nu = \lfloor (a_\nu^- + a_\nu^+)/2 \rfloor$ associated with it. Its left and right children are associated with $[a_\nu^-, a_\nu]$ and $[a_\nu + 1, a_\nu^+]$, respectively. Each node ν with $a_\nu^- = a_\nu^+$ is a leaf of the tree, with $a_\nu = a_\nu^- = a_\nu^+$. For any index a of a critical point x_a, we will denote the leaf ν that has $a_\nu = a$ by λ_a. Note that so far, the tree looks exactly the same for each function $\mathcal{D}[j,b]$: they are balanced binary trees with N leaves, and $\log N$ height. Moreover, all trees have the same associated intervals, and split values in their corresponding nodes. With each node ν we also store a weight w_ν, such that $\mathcal{T}_{j,b}$ has the following property: $\mathcal{D}[j,b](a)$ is the sum of the weights on the path from the tree root to the leaf λ_a. Such a representation of a function $\mathcal{D}[j,b]$ is not unique. Furthermore, we store with each node ν a value m_ν which is the sum of the weights on the

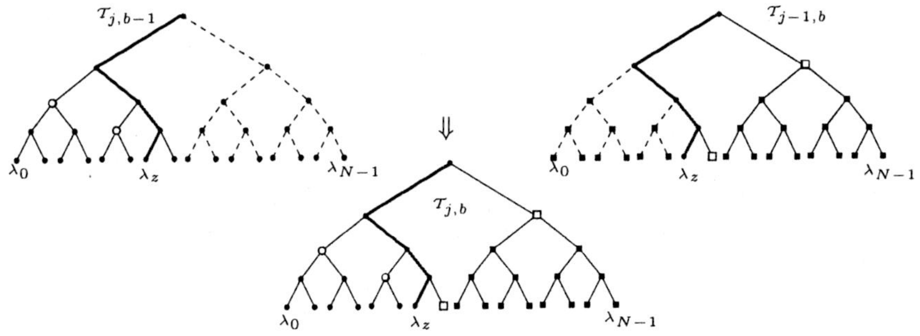

Fig. 3. The tree $\mathcal{T}_{j,b}$ is contructed from $\mathcal{T}_{j,b-1}$ and $\mathcal{T}_{j-1,b}$ by creating new nodes along the path from the root to the leaf λ_z, and adopting the subtrees to the left of the path from $\mathcal{T}_{j,b-1}$, and the subtrees to the right of the path from $\mathcal{T}_{j-1,b}$

path from the left child of ν to the leaf λ_{a_ν}, that is: the rightmost descendant of the left child of ν.

Lemma 5. *The data structure $\mathcal{T}_{j,b}$ for the representation of function $\mathcal{D}[j,b]$ can be operated on such that:*

(i) The representation of a zero-function (i.e. a function that always evaluates to zero) can be constructed in $O(N)$ time. Also the representation of a function that always evaluates to ∞ can be constructed in $O(N)$ time.
(ii) Given $\mathcal{T}_{j,b}$ of $\mathcal{D}[j,b]$, evaluating function $\mathcal{D}[j,b](a)$ takes $O(\log N)$ time.
(iii) Given $\mathcal{T}_{j-1,b}$ and $\mathcal{T}_{j,b-1}$ of the functions $\mathcal{D}[j-1,b]$ and $\mathcal{D}[j,b-1]$, respectively, a representation $\mathcal{T}_{j,b}$ of $\mathcal{D}[j,b]$ can be computed in $O(\log N)$ time.

Proof. We restrict ourselves to item (iii), the main element of the solution.

To construct $\mathcal{T}_{j,b}$ from $\mathcal{T}_{j,b-1}$ and $\mathcal{T}_{j-1,b}$ efficiently, we take the following approach. We find the sequences of left and right turns that lead from the root of the trees down to the leaf λ_z, where z is defined as in lemma 4. Note that the sequences of left and right turns are the same in the trees $\mathcal{T}_{j,b}$, $\mathcal{T}_{j,b-1}$, and $\mathcal{T}_{j-1,b}$, only the weights on the path differ. Though we do not compute z explicitly, we will show below that we are able to construct the path from the root of the tree to the leaf λ_z corresponding to z, by identifying, based on the stored weights, at each node along this path whether the path continues left or right.

Lemma 4 tells us that for each leaf left of λ_z, the total weight on the path to the root in $\mathcal{T}_{j,b}$ must be the same as the total weight on the corresponding path in $\mathcal{T}_{j,b-1}$. At λ_z itself and right of λ_z, the total weights to the root in $\mathcal{T}_{j,b}$ must equal those in $\mathcal{T}_{j-1,b}$, plus $\bar{f}_j(x_z)$. We construct the tree $\mathcal{T}_{j,b}$ with these properties as follows. We start building $\mathcal{T}_{j,b}$ by constructing a root ρ. If the path to λ_z goes into the right subtree, we adopt as left child of ρ the corresponding left child ν of the root from $\mathcal{T}_{j,b-1}$. There is no need to copy ν: we just add a pointer to it. Furthermore, we set the weight of ρ equal to the weight of the root of $\mathcal{T}_{j,b-1}$. If the path to λ_z goes into the left subtree, we adopt the right child from $\mathcal{T}_{j-1,b}$ and take the weight of ρ from there, now adding $\bar{f}_j(x_z)$.

Then we make a new root for the other subtree of the root ρ, i.e. the one that contains λ_z, and continue the construction process in that subtree. Every time we go into the left branch, we adopt the right child from $\mathcal{T}_{j-1,b}$, and every time we go into the right branch, we adopt the left child from $\mathcal{T}_{j,b-1}$ (see figure 3). For every constructed node ν, we set its weight w_ν so that the total weight of ν and its ancestors equals the total weight of the corresponding nodes in the tree from which we adopt ν's child — if the subtree adopted comes from $\mathcal{T}_{j-1,b}$, we increase w_ν by $\bar{\mathsf{f}}_j(x_z)$.

By keeping track of the accumulated weights on the path down from the root in all the trees, we can set the weight of each newly constructed node ν correctly in constant time per node. The accumulated weights together with the stored weights for the paths down to left childrens' rightmost descendants, also allow us to decide in constant time which is better: $\mathcal{D}[j,b-1](a_\nu)$ or $\mathcal{D}[j-1,b](a_\nu)+\bar{\mathsf{f}}_j(x_z)$. This will tell us if λ_z is to be found in the left or in the right subtree of ν.

The complete construction process only takes $O(1)$ time for each node on the path from ρ to λ_z. Since the trees are perfectly balanced, this path has only $O(\log N)$ nodes, so that $\mathcal{T}_{j,b}$ is constructed in time $O(\log N)$. □

Theorem 1. *The similarity* $\mathsf{d}(P_1,\ldots,P_k;P)$ *between* k *polylines* $\{P_1,\ldots,P_j\}$ *with* n *vertices in total, and a polygon* P *with* m *vertices, can be computed in* $O(kmn\log(mn))$ *time using* $O(kmn\log(mn))$ *storage.*

Proof. We use algorithm Fastcompute $\mathsf{d}(P_1,\ldots,P_k;P)$ with the data structure described above. Step 1 and 2 of the algorithm can be executed in $O(kmn + mn\log n)$ time. From lemma 5, we have that the zero-function $ZERO$ can be constructed in $O(N)$ time (line 3). Similarly, the infinity-function $INFINITY$ can be constructed in $O(N)$ time (line 4). Lemma 5 also insures that constructing $\mathcal{D}[j,b]$ from $\mathcal{D}[j-1,b]$ and $\mathcal{D}[j,b-1]$ (line 12) takes $O(\log N)$ time, and that the evaluation of $\mathcal{D}[k,b](a)$ (line 15) takes $O(\log N)$ time. Notice that no node is ever edited after it has been constructed. Thus, the total running time of the above algorithm will be dominated by $O(kN)$ executions of line 12, taking in total $O(kN\log N) = O(kmn\log(mn))$ time.

Apart from the function values of $\bar{\mathsf{f}}_j$ computed in step 2, we have to store the $ZERO$ and the $INFINITY$ function. All these require $O(kN) = O(kmn)$ storage. Notice that any of the functions constructed in step 12 requires only storing $O(\log(N)) = O(\log(mn))$ new nodes and pointers to nodes in previously computed trees, and thus we need $O(kmn\log(mn))$ for all the trees computed in step 12. So the total storage required by the algorithm is $O(kmn\log(mn))$. □

We note that the problem resembles a general edit distance type approximate string matching [9]. Global string matching under a general edit distance error model can be done by dynamic programming in $O(kN)$ time, where k and N represent the lengths of the two strings. The same time complexity can be achieved for partial string matching through a standard "assign first line to zero" trick [9]. This however does not apply here due to the condition $x_b - x_a \leq l_0$.

4 Experimental Results

Our algorithm has been implemented in C++ and is evaluated in a part-based shape retrieval application (see `http://give-lab.cs.uu.nl/Matching/Mtam/`) with the Core Experiment "CE-Shape-1" part B test set devised by the MPEG-7 group to measure the performance of similarity-based retrieval for shape descriptors. This test set consists of 1400 images: 70 shape classes, with 20 images per class. The shape descriptor selected by MPEG-7 to represent a closed contour of a 2D object or region in an image is based on the Curvature Scale Space (CSS) representation [8]. We compared our matching to the CSS method, as well as to matching the global contours with turning angle functions (GTA).

The performance of each shape descriptor was measured using the "bull's-eye" percentage: the percentage of retrieved images belonging to the same class among the top 40 matches (twice the class size). These experimental results indicate that for those classes with a low performance of the CSS matching, our approach consistently performs better. See figure 4 for two examples. The emphasis of this paper lies on the algorithmic aspects, but for a rigorous experimental evaluation, see [11]. The running time for a single query on the MPEG-7 test set of 1400 images is typically about one second on a 2 GHz PC.

CSS/GTA Query Image	Part-based Query Parts	Bull's Eye Score		
		CSS	GTA	MPP
 beetle-20	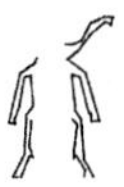	10	30	65
 ray-3	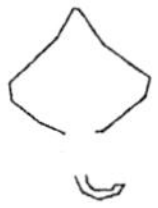	15	15	70

Fig. 4. A comparison of the Curvature Scale Space (CSS), Global Turning Angle function (GTA), and our Multiple Polyline to Polygon (MPP) matching (in %)

Acknowledgements. We thank Veli Mäkinen for the partial string matching reference, and Geert-Jan Giezeman for programming support. This research was supported by the FP6 IST projects 511572-2 PROFI and 506766 AIM@SHAPE, and the Dutch Science Foundation (NWO) project 612.061.006 MINDSHADE.

References

1. N. Ansari and E. J. Delp. Partial shape recognition: A landmark-based approach. *PAMI*, 12:470–483, 1990.
2. E. Arkin, L. Chew, D. Huttenlocher, K. Kedem, and J. Mitchell. An efficiently computable metric for comparing polygonal shapes. *PAMI*, 13:209–215, 1991.
3. T. Asano, M. de Berg, O. Cheong, H. Everett, H.J. Haverkort, N. Katoh, and A. Wolff. Optimal spanners for axis-aligned rectangles. *CGTA*, 30(1):59–77, 2005.
4. Scott D. Cohen and Leonidas J. Guibas. Partial matching of planar polylines under similarity transformations. In *Proc. SODA*, pages 777–786, 1997.

5. D. Huttenlocher, G. Klanderman, and W. Rucklidge. Comparing images using the hausdorff distance. *PAMI*, 15:850–863, 1993.
6. L. J. Latecki, R. Lakämper, and D. Wolter. Shape similarity and visual parts. In *Proc. Int. Conf. Discrete Geometry for Computer Imagery*, pages 34–51, 2003.
7. H.-C. Liu and M. D. Srinath. Partial shape classification using contour matching in distance transformation. *PAMI*, 12(11):1072–1079, 1990.
8. F. Mokhtarian, S. Abbasi, and J. Kittler. Efficient and robust retrieval by shape content through curvature scale space. In *Workshop on Image DataBases and MultiMedia Search*, pages 35–42, 1996.
9. Gonzala Navarro. A guided tour to approximate string matching. *ACM Computing Surveys*, 33(1):31–88, 2001.
10. K. Siddiqi, A. Shokoufandeh, S.J. Dickinson, and S.W. Zucker. Shock graphs and shape matching. *IJCV*, 55(1):13–32, 1999.
11. Mirela Tanase. *Shape Deomposition and Retrieval.* PhD thesis, Utrecht University, Department of Computer Science, February 2005.
12. Haim Wolfson and Isidore Rigoutsos. Geometric hashing: an overview. *IEEE Computational Science & Engineering*, pages 10–21, October-December 1997.

Minimizing a Monotone Concave Function with Laminar Covering Constraints

Mariko Sakashita[1], Kazuhisa Makino[2], and Satoru Fujishige[3]

[1] Graduate School of Informatics, Kyoto University, Kyoto, 606-8501, Japan
sakasita@amp.i.kyoto-u.ac.jp
[2] Graduate School of Information Science and Technology, University of Tokyo, Tokyo, 113-8656, Japan
makino@mist.i.u-tokyo.ac.jp
[3] Research Institute for Mathematical Sciences, Kyoto University, Kyoto, 606-8502, Japan
fujishig@kurims.kyoto-u.ac.jp

Abstract. Let V be a finite set with $|V| = n$. A family $\mathcal{F} \subseteq 2^V$ is called *laminar* if for arbitrary two sets $X, Y \in \mathcal{F}$, $X \cap Y \neq \emptyset$ implies $X \subseteq Y$ or $X \supseteq Y$. Given a laminar family $\mathcal{F}$, a demand function $d : \mathcal{F} \to \mathbb{R}_+$, and a monotone concave cost function $F : \mathbb{R}_+^V \to \mathbb{R}_+$, we consider the problem of finding a minimum-cost $x \in \mathbb{R}_+^V$ such that $x(X) \geq d(X)$ for all $X \in \mathcal{F}$. Here we do not assume that the cost function F is differentiable or even continuous. We show that the problem can be solved in $\mathrm{O}(n^2 q)$ time if F can be decomposed into monotone concave functions by the partition of V that is induced by the laminar family $\mathcal{F}$, where q is the time required for the computation of $F(x)$ for any $x \in \mathbb{R}_+^V$. We also prove that if F is given by an oracle, then it takes $\Omega(n^2 q)$ time to solve the problem, which implies that our $\mathrm{O}(n^2 q)$ time algorithm is optimal in this case. Furthermore, we propose an $\mathrm{O}(n \log^2 n)$ algorithm if F is the sum of linear cost functions with fixed setup costs. These also make improvements in complexity results for source location and edge-connectivity augmentation problems in undirected networks. Finally, we show that in general our problem requires $\Omega(2^{\frac{n}{2}} q)$ time when F is given implicitly by an oracle, and that it is NP-hard if F is given explicitly.

1 Introduction

Let V be a finite set with $|V| = n$. A family $\mathcal{F} \subseteq 2^V$ is called *laminar* if for arbitrary two sets $X, Y \in \mathcal{F}$, $X \cap Y \neq \emptyset$ implies $X \subseteq Y$ or $X \supseteq Y$. A function $F : \mathbb{R}^V \to \mathbb{R}$ is called *monotone nondecreasing* (simply *monotone*) if $F(x) \leq F(y)$ holds for arbitrary two vectors $x, y \in \mathbb{R}^V$ with $x \leq y$, and *concave* if $\alpha F(x) + (1-\alpha)F(y) \leq F(\alpha x + (1-\alpha)y)$ holds for arbitrary two vectors $x, y \in \mathbb{R}^V$ and real α with $0 \leq \alpha \leq 1$.

Given a laminar family $\mathcal{F}$, a demand function $d : \mathcal{F} \to \mathbb{R}_+$, and a monotone concave function $F : \mathbb{R}_+^V \to \mathbb{R}_+$, the problem to be considered in this paper is given as

$$
\begin{aligned}
\textbf{(P)} \qquad & \text{Minimize} \quad F(x) & & & (1.1)\\
& \text{subject to} \quad x(X) \geq d(X) & & (X \in \mathcal{F}), & (1.2)\\
& \qquad\qquad\quad x(v) \geq 0 & & (v \in V), & (1.3)
\end{aligned}
$$

X. Deng and D. Du (Eds.): ISAAC 2005, LNCS 3827, pp. 71–81, 2005.

where $\mathbb{R}_+$ denotes the set of all nonnegative reals, and $x(X) = \sum_{v \in X} x(v)$ for any $X \subseteq V$. Here we do not assume that the cost function F is differentiable or even continuous. The present problem has various applications, since laminar families represent hierarchical structures in many organizations. Moreover, the problem can be regarded as a natural generalization of the source location problem and the edge- connectivity augmentation problem in undirected networks, which do not seemingly have laminar structures. We shall show in Section 2 that they can be formulated as (**P**) by using extreme sets in given networks.

In this paper, we study the following three cases, in which the cost functions F are expressed as

$$\text{(i)}\ \ F_1(x) = \sum_{X \in \mathcal{F}} f_{\Delta X}(x[\Delta X]) \quad \text{(laminar sum)}, \tag{1.4}$$

$$\text{(ii)}\ \ F_2(x) = \sum_{v \in V} f_v(x(v)) \quad \text{(separable)}, \tag{1.5}$$

$$\text{(iii)}\ \ F_3(x) = \sum_{v \in V : x(v) > 0} (a_v x(v) + b_v) \quad \text{(fixed-cost linear)}, \tag{1.6}$$

where $\Delta X = X - \bigcup\{Y \mid Y \in \mathcal{F},\ Y \subsetneq X\}$, $x[\Delta X]$ denotes the projection of x on ΔX, $f_{\Delta X} : \mathbb{R}_+^{\Delta X} \to \mathbb{R}_+$ and $f_v : \mathbb{R}_+^{\{v\}} \to \mathbb{R}_+$ are monotone concave, and a_v and b_v are nonnegative constants. It is clear that F_2 is a special case of F_1, and F_3 is a special case of F_2 (and hence of F_1).

We consider Problem (**P**) when the cost function F is given either explicitly or implicitly. Here "implicitly" means that F is given by an oracle, i.e., we can invoke the oracle for the evaluation of $F(x)$ for any x in $\mathbb{R}_+^V$ and use the function value $F(x)$. In either case (explicitly or implicitly), we assume that $F(x)$ can be computed for any $x \in \mathbb{R}_+^V$ in $\mathrm{O}(q)$ time.

We show that if $F = F_1$, the problem can be solved in $\mathrm{O}(n^2 q)$ time, where q is the time required for the computation of $F(x)$ for each $x \in \mathbb{R}_+^V$. We also prove that the problem requires $\Omega(n^2 q)$ time, if $F (= F_2)$ is given by an oracle. This implies that our $\mathrm{O}(n^2 q)$ algorithm is *optimal* if $F (= F_1, F_2)$ is given by an oracle. Moreover, we show that the problem can be solved in $\mathrm{O}(n \log^2 n)$ and $\mathrm{O}(n(\log^2 n + q))$ time if F is explicitly and implicitly given as F_3, respectively, and the problem is intractable in general. Table 1 summarizes the complexity results obtained in this paper.

We remark that the results below remain true, even if we add the integrality condition $x \in \mathbb{Z}^V$ to the problem.

Our positive results can be applied to the source location problem and the edge-connectivity augmentation problem (see Section 2 for details). These make improve-

Table 1. Summary of the results obtained in this paper

	F_1	F_2	F_3	general
explicit	$\mathrm{O}(n^2 q)$	$\mathrm{O}(n^2 q)$	$\mathrm{O}(n \log^2 n)$	NP-hard inapproximable
implicit (oracle)	$\Theta(n^2 q)$	$\Theta(n^2 q)$	$\mathrm{O}(n(\log^2 n + q))$	$\Omega(2^{\frac{n}{2}} q)$

ments in complexity results for the problems. Our results implies that the source location problem can be solved in $O(nm + n^2(q + \log n))$ time if the cost function F is expressed as F_1, e.g., the sum of fixed setup costs and concave running costs for facilities at $v \in V$, and in $O(n(m + n \log n))$ time if the cost function F is expressed as F_3, i.e., the cost is the sum of fixed setup costs and linear running costs. We remark that the source location problem has been investigated, only when the cost function depends on the fixed setup cost of the facilities (see (2.5)). Similarly to the source location problem, our results together with the ones in [7, 8] imply that the augmentation problem can be solved in $O(nm + n^2(q + \log n))$ time if $F = F_1$, and in $O(n(m + n \log n))$ time if $F = F_3$. We remark that the augmentation problem has been investigated only when the cost function is linear (see (2.8)).

The rest of the paper is organized as follows. Section 2 presents two applications of our covering problem. Section 3 investigates the case in which F is laminar or separable, and Section 4 studies the case in which F is linear with fixed costs or general monotone concave.

Due to the space limitation, technical details are omitted.

2 Applications of Our Covering Problem

In this section we introduce two network problems as examples of our problem.

2.1 Source Location Problem in Undirected Networks

Let $\mathcal{N} = (G = (V, E), u)$ be an undirected network with a vertex set V, an edge set E, and a capacity function $u : E \to \mathbb{R}_+$. For convenience, we regard $\mathcal{N}$ as a symmetric directed graph $\hat{\mathcal{N}} = (\hat{G} = (V, \hat{E}), \hat{u})$ defined by $\hat{E} = \{(v, w), (w, v) \mid \{v, w\} \in E\}$ and $\hat{u}(v, w) = \hat{u}(w, v) = u(\{v, w\})$ for any $\{v, w\} \in E$. We also often write $u(v, w)$ instead of $u(\{v, w\})$.

A flow $\varphi : \hat{E} \to \mathbb{R}_+$ is *feasible* with a supply $x : V \to \mathbb{R}_+$ if it satisfies the following conditions:

$$\partial\varphi(v) \stackrel{\text{def}}{=} \sum_{(v,w)\in\hat{E}} \varphi(v, w) - \sum_{(w,v)\in\hat{E}} \varphi(w, v) \leq x(v) \quad (v \in V), \tag{2.1}$$

$$0 \leq \varphi(e) \leq \hat{u}(e) \qquad (e \in \hat{E}). \tag{2.2}$$

Here (2.1) means that the net out-flow value $\partial\varphi(v)$ at $v \in V$ is at most the supply at v.

Given an undirected network $\mathcal{N}$ with a demand $k > 0$ and a cost function $F : \mathbb{R}_+^V \to \mathbb{R}_+$, the source location problem is to find a minimum-cost supply x such that for each $v \in V$ there is a feasible flow φ_v such that the sum of the net in-flow value and the supply at v is at least k. The problem is rewritten as follows.

$$\begin{aligned}
&\text{Minimize} \quad F(x) \\
&\text{subject to} \ \forall\, v \in V, \ \exists \text{ a feasible flow } \varphi_v \text{ in } \mathcal{N} \text{ with a supply } x\text{:} \\
&\qquad\qquad -\partial\varphi_v(v) + x(v) \geq k, && (2.3) \\
&\qquad x(v) \geq 0 \qquad (v \in V). && (2.4)
\end{aligned}$$

Note that the flow φ_v $(v \in V)$ in (2.3) may depend on $v \in V$. From the max-flow min-cut theorem, we can show that (2.3) can be represented by a laminar covering constraint (1.2). Hence the source location problem can be formulated as **(P)** in Section 1. Since given an undirected network $\mathcal{N} = (G = (V, E), u)$ and a demand k (> 0), the family $\mathcal{F}$ of all extreme sets, as well as the deficiency $d : \mathcal{F} \to \mathbb{R}_+$, can be computed in $\mathrm{O}(n(m + n \log n))$ time [7], our results for the laminar covering problem immediately imply the ones for the source location problem (see Section 1). We remark that the source location problem has been investigated in [1,9] in a special case where the cost function is given as

$$F(x) = \sum_{v \in V : x(v) > 0} b_v. \tag{2.5}$$

Namely, the cost function depends only on the fixed setup cost of the facilities at vertices $v \in V$ with $x(v) > 0$, and is independent of the positive supply value $x(v)$.

2.2 Edge-Connectivity Augmentation in Undirected Networks

Let $\mathcal{N} = (G = (V, E), u)$ be an undirected network with a capacity function $u : E \to \mathbb{R}_+$. For a positive real k we call $\mathcal{N}$ k-edge-connected if for every two nodes $v, w \in V$ the maximum flow value between v and w is at least k. Given an undirected network $\mathcal{N}$, a positive real k, and a node-cost function $F : \mathbb{R}_+^V \to \mathbb{R}_+$, the edge-connectivity augmentation problem (e.g., [2, 3, 4, 7, 10]) is to find a set D of new edges with capacity $\mu_D : D \to \mathbb{R}_+$ for which $\mathcal{N}' = (G' = (V, E \cup D), u \oplus \mu_D)$ is k-edge-connected and $F(\partial \mu_D)$ is minimum, where $\partial \mu_D(v) = \sum_{e \in D : e \ni v} \mu_D(e)$ $(v \in V)$, and $u \oplus \mu_D$ is the direct sum of u and μ_D. From the max-flow min-cut theorem, we can see that $x = \partial \mu_D$ must satisfy $\kappa(X) + x(X) \geq k$ for any nonempty $X \subsetneq V$, which implies

$$x(X) \geq d(X) \qquad (X \in \mathcal{F}), \tag{2.6}$$

where $d(X)$ is given by

$$d(X) = \max\{k - \kappa(X), 0\} \tag{2.7}$$

and $\mathcal{F}$ is the family of all extreme sets in $\mathcal{N}$. On the other hand, it is known that any x satisfying (2.6) can create a capacity function $\mu_D : D \to \mathbb{R}_+$ for which $\mathcal{N}'$ is k-edge-connected [5, 6] and moreover, such an x of minimum $x(V)$ can be found in $\mathrm{O}(n(m+n \log n))$ time [7, 8]. Therefore, the augmentation problem can be formulated as our laminar covering problem. Our results provide an improvement over the existing ones for the augmentation problem (see Section 1). We note that the edge-connectivity augmentation problem has been studied only when the cost function F is linear, i.e.,

$$F(x) = \sum_{v \in V} a_v x(v). \tag{2.8}$$

Note that, if $a_v = \frac{1}{2}$ for all $v \in V$, then the cost $F(\partial \mu_D)$ with $\mu_D(e) = 1$ $(e \in D)$ is equal to the number of edges in D.

3 The Laminar Sum Cost Case

In this section we consider the problem whose cost function is given by F_1, i.e.,

$$(\mathbf{P}_1) \qquad \text{Minimize} \sum_{X \in \mathcal{F}} f_{\Delta X}(x[\Delta X])$$
$$\text{subject to } x(X) \geq d(X) \qquad (X \in \mathcal{F}), \tag{3.1}$$
$$x(v) \geq 0 \qquad (v \in V),$$

where $f_{\Delta X}$ is a monotone concave function on ΔX. We shall present an $\mathrm{O}(n^2 q)$ time algorithm for the problem and show the $\Omega(n^2 q)$ time bound when the cost function is given by an oracle.

3.1 Structural Properties of Optimal Solutions

This section reveals structural properties of optimal solutions of Problem $(\mathbf{P}_1)$ in (3.1), which makes it possible for us to devise a polynomial algorithm for Problem $(\mathbf{P}_1)$.

For a laminar family $\mathcal{F} = \{X_i \mid i \in I\}$ define a directed graph $T = (W, A)$ with a vertex set W and an arc set A by

$$W = \{w_i \mid i \in I \cup \{i_0\}\}$$
$$A = \{a_i = (w_i, w_j) \mid X_i \subsetneq X_j, \mathcal{F} \text{ contains no set } Y \text{ with } X_i \subsetneq Y \subsetneq X_j\}$$
$$\cup \{a_i = (w_i, w_{i_0}) \mid X_i \text{ is a maximal set in } \mathcal{F}\},$$

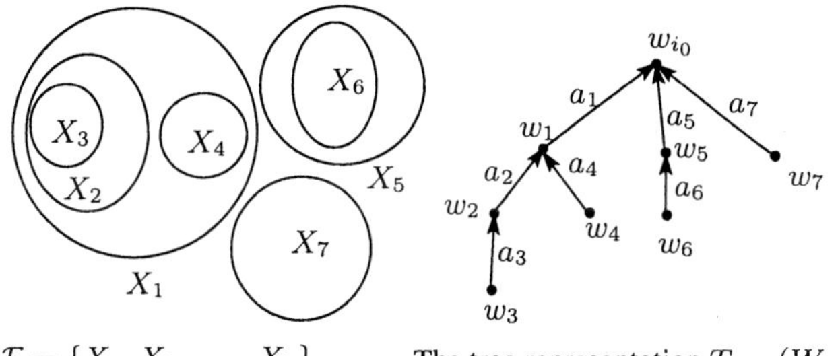

$\mathcal{F} = \{X_1, X_2, \cdots, X_7\}$ The tree representation $T = (W, A)$ of $\mathcal{F}$

Fig. 1.

where i_0 is a new index not in I. Since $\mathcal{F}$ is laminar, the graph $T = (W, A)$ is a directed tree toward the root w_{i_0} and is called the *tree representation* of $\mathcal{F}$ (see Fig. 1). For each $X_i \in \mathcal{F}$ let us define the family of the children, the incremental set and the depth by

$$\mathcal{S}(X_i) = \{X_j \mid a_j = (w_j, w_i) \in A\},$$
$$\Delta X_i = X_i \setminus \bigcup_{X_j \in \mathcal{S}(X_i)} X_j,$$
$$h(X_i) = |\{X_j \mid X_j \in \mathcal{F} \text{ with } X_j \supseteq X_i\}|.$$

For a function $d : \mathcal{F} \to \mathbb{R}$ we also define the increment Δd by $\Delta d(X) = d(X) - \sum_{Y \in \mathcal{S}(X)} d(Y)$. If $\Delta d(X) \leq 0$, we can remove constraint $x(X) \geq d(X)$ from (1.2).

Hence we assume that every set $X \in \mathcal{F}$ satisfies $\Delta d(X) > 0$. We also assume without loss of generality that $F(\mathbf{0}) = 0$.

Let $\mathcal{F}$ be a laminar family on V, and $T = (W, A)$ be the tree representation of $\mathcal{F}$. Consider the problem projected on $Y \in \mathcal{F}$ that is given as

$$
\begin{aligned}
(\mathbf{P}_Y) \qquad & \text{Minimize} \quad \sum_{X \in \mathcal{F}: X \subseteq Y} f_{\Delta X}(x[\Delta X]) \\
& \text{subject to } x(X) \geq d(X) \qquad (X \in \mathcal{F}, X \subseteq Y), \\
& \qquad\qquad\quad x(v) \geq 0 \qquad\qquad (v \in V).
\end{aligned}
\tag{3.2}
$$

We first show properties of optimal solutions of $(\mathbf{P}_Y)$, from which we derive properties of optimal solutions of $(\mathbf{P}_1)$.

Lemma 3.1. *For a minimal $Y \in \mathcal{F}$, Problem $(\mathbf{P}_Y)$ has an optimal solution $x = z_v$ for some $v \in Y$ such that*

$$
z_v(t) = \begin{cases} d(Y) \ (= \Delta d(Y)) & (t = v) \\ 0 & (t \in V \setminus \{v\}). \end{cases}
\tag{3.3}
$$

□

Lemma 3.2. *Let Y be a non-minimal set in $\mathcal{F}$. Then there exists an optimal solution x of Problem $(\mathbf{P}_Y)$ such that for some $v \in \Delta Y$*

$$
\begin{aligned}
x(t) &= \begin{cases} \Delta d(Y) & (t = v) \\ 0 & (t \in (V \setminus Y) \cup (\Delta Y \setminus \{v\})), \end{cases} \\
x(X) &= d(X) \qquad (X \in \mathcal{S}(Y)),
\end{aligned}
$$

or for some $X \in \mathcal{S}(Y)$

$$
\begin{aligned}
x(Z) &= \begin{cases} d(X) + \Delta d(Y) & (Z = X) \\ d(Z) & (Z \neq X,\ Z \in \mathcal{S}(Y)), \end{cases} \\
x(v) &= 0 \qquad (v \in (V \setminus Y) \cup \Delta Y).
\end{aligned}
$$

□

Let $W^* = \{w_i \mid X_i \in \mathcal{F}\}$. A partition $\mathcal{P} = \{P_1, \cdots, P_k\}$ of W^* is called a *path-partition* of W^* if each $P_j = \{w_{j_0}, w_{j_1}, \cdots, w_{j_{r_j}}\} \in \mathcal{P}$ forms a directed path $w_{j_0} \to w_{j_1} \to \cdots \to w_{j_{r_j}}$ in $T = (W, A)$ with $\Delta X_{j_0} \neq \emptyset$. We are now ready to describe our structure theorem.

Theorem 3.3. *Problem $(\mathbf{P}_1)$ in (3.1) has an optimal solution x^* that can be obtained from a path-partition $\mathcal{P} = \{P_1, \cdots, P_k\}$ of W^* together with $v_j \in \Delta X_{j_0}$ $(j = 1, \cdots, k)$ as follows.*

$$
x^*(t) = \begin{cases} \sum_{w_{j_i} \in P_j} \Delta d(X_i) & (t = v_j,\ j = 1, \cdots, k) \\ 0 & (t \in V \setminus \{v_j \mid j = 1, \cdots, k\}). \end{cases}
$$

□

3.2 A Polynomial Algorithm

In this section we present a polynomial algorithm for Problem $(\mathbf{P}_1)$ in (3.1). The algorithm applies dynamic programming to compute an optimal path-partition of W^*.

For any $Y \in \mathcal{F}$, we denote by w_Y the node in W corresponding to Y, and by $w_{j_0}(= w_Y), w_{j_1}, \cdots, w_{j_{h(Y)-1}}, w_{j_{h(Y)}}(= w_{i_0})$ the directed path from w_Y to the root w_{i_0}. Our dynamic programming solves the following $h(Y)$ problems for each $Y \in \mathcal{F}$.

$$(\mathbf{P}(Y,k)) \qquad \text{Minimize} \quad \sum_{X \in \mathcal{F}: X \subseteq Y} f_{\Delta X}(x[\Delta X]) \tag{3.4}$$

$$\text{subject to} \quad x(Y) \geq d(Y) + \sum_{i=1}^{k} \Delta d(X_{j_i}), \tag{3.5}$$

$$x(X) \geq d(X) \qquad (X \in \mathcal{F}, X \subsetneq Y), \tag{3.6}$$

$$x(v) \geq 0 \qquad (v \in V), \tag{3.7}$$

where $Y \in \mathcal{F}$ and $k = 0, 1, \cdots, h(Y) - 1$. Let $\alpha(Y,k)$ denote the optimal value of Problem $(\mathbf{P}(Y,k))$. By Theorem 3.3, these problems $(\mathbf{P}(Y,k))$ have optimal solutions based on a path-partition $\mathcal{P}$ of $\{w_i \mid X_i \in \mathcal{F}, X_i \subseteq Y\}$. For $P_j \in \mathcal{P}$ containing $w_Y \in W$ (that corresponds to Y), let v_j be the node in ΔX_{j_0} given in Theorem 3.3. We store v_j as $\beta(Y,k)$. It follows from Lemmas 3.1 and 3.2 that $\alpha(Y,k)$ and $\beta(Y,k)$ can be computed as follows.

For each minimal $Y \in \mathcal{F}$ (which corresponds to a leaf in T) the following z_v^k for some $v \in Y$ gives an optimal solution, due to Lemma 3.1.

$$z_v^k(t) = \begin{cases} \sum_{i=0}^{k} \Delta d(X_{j_i}) & (t = v) \\ 0 & (t \in \Delta Y \setminus \{v\}). \end{cases}$$

Hence we have

$$(\alpha(Y,k), \beta(Y,k)) = \left(\min_{v \in Y} f_Y(z_v^k), \arg\min_{v \in Y} f_Y(z_v^k) \right) \tag{3.8}$$

for $k = 0, \cdots, h(Y)-1$, where $\arg\min_{v \in Y} f_Y(z_v^k)$ denotes a vertex $v^* \in Y$ satisfying $f_Y(z_{v^*}^k) = \min_{v \in Y} f_Y(z_v^k)$.

For a non-minimal $Y \in \mathcal{F}$, Lemma 3.2 validates the following recursive formulas.

$$\alpha(Y,k) = \min \left\{ \min_{X \in \mathcal{S}(Y)} \left\{ \alpha(X, k+1) + \sum_{\substack{Z \in \mathcal{S}(Y) \\ Z \neq X}} \alpha(Z, 0) \right\}, \min_{v \in \Delta Y} \left\{ f_{\Delta Y}(z_v^k) + \sum_{X \in \mathcal{S}(Y)} \alpha(X, 0) \right\} \right\}, \tag{3.9}$$

$$\beta(Y,k) = \begin{cases} \beta(X, k+1) \text{ if } \alpha(Y,k) = \alpha(X, k+1) + \sum_{\substack{Z \in \mathcal{S}(Y) \\ Z \neq X}} \alpha(Z, 0), \\ v \qquad \text{if } \alpha(Y,k) = f_{\Delta Y}(z_v^k) + \sum_{X \in \mathcal{S}(Y)} \alpha(X, 0). \end{cases} \tag{3.10}$$

By using (3.8), (3.9), and (3.10), our algorithm first computes each α and β from the leaves toward root w_{i_0} of T. Then we obtain an optimal value $\sum_{X \in \mathcal{S}(X_{i_0})} \alpha(X, 0)$ of Problem $(\mathbf{P}_1)$ in (3.1). Next, we compute an optimal solution x^* by using β from the root toward the leaves of T.

Our algorithm is formally described as follows.

Algorithm DP
Input: A laminar family $\mathcal{F}$, a demand function d, and a cost function F as in (3.1).
Output: An optimal solution x^* for Problem $(\mathbf{P}_1)$ in (3.1).
Step 0. $\tilde{W} := W$.
Step 1. (Compute α and β) While $\tilde{W} \neq \{w_{i_0}\}$ **do**

Choose an arbitrary leaf $w \in \tilde{W}$ of $T[\tilde{W}]$ and let Y be the set in $\mathcal{F}$ corresponding to w.

/* $T[\tilde{W}]$ denotes the subtree of T induced by $\tilde{W}$. */

(1-I) Compute $\alpha(Y, k)$ and $\beta(Y, k)$ for $k = 0, \cdots, h(Y){-}1$ by using (3.8)~(3.10).

(1-II) $\tilde{W} := \tilde{W} \setminus \{w\}$.

Step 2. $\tilde{W} := W \setminus \{w_{i_0}\}$, and $x^*(v) := 0$ for all $v \in V$.
Step 3. (Compute an optimal x^*) While $\tilde{W} \neq \emptyset$ **do**

Choose an arbitrary node w of $T[\tilde{W}]$ having no leaving arc and let Y be the set in $\mathcal{F}$ corresponding to w. Let w_{j_0} be the node in W corresponding to X_{j_0} such that $\beta(Y, 0) \in \Delta X_{j_0}$ and $w_{j_0} \to w_{j_1} \to \cdots \to w_{j_l}$ $(= w)$ be a directed path in $T[\tilde{W}]$. $x^*(\beta(Y, 0)) := \sum_{i=0}^{l} \Delta d(X_{j_i})$ and $\tilde{W} := \tilde{W} \setminus \{w_{j_0}, \cdots, w_{j_l}\}$.

Step 4. Output x^* and halt. □

We now have the following theorem.

Theorem 3.4. *Algorithm* DP *computes an optimal solution for Problem* $(\mathbf{P}_1)$ *in* $O(n^2 q)$ *time.* □

3.3 The Lower Bound for the Time Complexity When F is Given by an Oracle

In this section we consider a lower bound for the time complexity of our problem when F is given by an oracle. We shall show that the oracle has to be invoked $\Omega(n^2)$ times even if we know in advance that F is given in the form of (1.5), i.e., $F = \sum_{v \in V} f_v(x(v))$. This, together with Theorem 3.4, implies that Algorithm DP is optimal if F is given by an oracle.

Suppose n is a positive even number. Let $g_0 : \mathbb{R}_+ \to \mathbb{R}_+$ be a monotone increasing and strictly concave function (e.g., $g_0(x) = \frac{-1}{x+1} + 1$ $(x \geq 0)$), and for each $i = \frac{n}{2}+1, \frac{n}{2}+2, \cdots, n$ define $g_i : \mathbb{R}_+ \to \mathbb{R}_+$ by

$$g_i(\xi) = \begin{cases} g_0(\frac{n}{2}+1) - g_0(i - \frac{n}{2}) & (\xi > 0) \\ 0 & (\xi = 0) \end{cases}.$$

Then, let $V = \{v_1, \cdots, v_n\}$ and consider a problem instance I obtained by

$$\begin{aligned} &\text{a laminar family } \mathcal{F} = \{X_i = \{v_1, \cdots, v_{\frac{n}{2}+i}\} \mid i = 0, \cdots, \tfrac{n}{2}\}, \\ &\text{a demand function } d : d(X_i) = i + 1 \quad (i = 0, 1, \cdots, \tfrac{n}{2}), \\ &\text{a cost function } F(x) = \sum_{v \in V} f_v(x(v)), \end{aligned} \tag{3.11}$$

where $f_{v_i}(\xi) = g_0(\xi)$ if $v_i \in X_0$, and $g_i(\xi)$ if $v_i \in V - X_0$.

For this problem instance, we have the following lemma.

Lemma 3.5. *The problem instance* I *defined as above requires at least* $\frac{n}{2}(\frac{n}{2}+1)$ *calls to the oracle for* F. □

This implies the following theorem.

Theorem 3.6. *If* F *is given by an oracle, then Problem (3.1) requires* $\Omega(n^2 q)$ *time.* □

We can easily see that Lemma 3.5 still holds even if each f_v is given by an oracle.

Theorem 3.7. *Let* F *be a separable monotone concave function (i.e.,* $F = \sum_{v \in V} f_v$ *with monotone concave functions* $f_v : \mathbb{R}_+ \to \mathbb{R}_+$ $(v \in V)$*). If each* f_v *is given by an oracle, then Problem (3.1) requires* $\Omega(n^2 q)$ *time.* □

Notice that Algorithm DP given in Section 3.2 is optimal, due to this theorem.

Corollary 3.8. *If the cost function* $F(= F_1$ *or* $F_2)$ *is given by an oracle, then Problem (3.1) requires* $\Theta(n^2 q)$ *time.* □

4 The FC Linear and General Cost Cases

We first consider the problem whose cost function is given as F_3, i.e., the cost function F is the sum of f_v $(v \in V)$ represented by

$$f_v(x) = \begin{cases} a_v x + b_v & (x > 0) \\ 0 & (x = 0). \end{cases}$$

Note that Problem $(\mathbf{P})$ with $F = F_3$ can be solved in $O(n^2 q)$ time by using Algorithm DP in Section 3.2. We shall show that it admits an $O(n \log^2 n)$ time algorithm. We remark that our problem requires $O(n^2 q)$ time even if F is separable.

The *lower envelope* of $f_{v_1}, \cdots, f_{v_k}$ is given as $f(x) = \min_i f_{v_i}(x)$, and note that it is piecewise linear for $x > 0$, and it is well known that the lower envelope of $f_{v_1}, \cdots, f_{v_{j+1}}$ can be computed from that of $f_{v_1}, \cdots, f_{v_j}$ and $f_{v_{j+1}}$ in $O(\log j)$ time. Hence the lower envelope of $f_{v_1}, \cdots, f_{v_k}$ can be constructed in $O(k \log k)$ time.

Our algorithm is similar to Algorithm DP in Section 3.2, but, for each $Y \in \mathcal{F}$, it constructs the lower envelope corresponding to $\alpha(Y, k)$ $(k = 0, \cdots, h(Y) - 1)$. This implicit computation of $\alpha(Y, k)$ $(k = 0, \cdots, h(Y) - 1)$ makes the algorithm faster. Although we skip the details due to the space limitations, we have the following theorem.

Theorem 4.1. *Problem* $(\mathbf{P})$ *with* $F = F_3$ *can be solved in* $O(n \log^2 n)$ *time if* F *is given explicitly, and in* $O(n(q + \log^2 n))$ *time if* F *is given implicitly by an oracle.* □

We next consider Problem $(\mathbf{P})$ whose cost function F is general monotone concave. We show that it requires $\Omega(2^{\frac{n}{2}} q)$ time to solve the problem if F is given implicitly by an oracle, and that it is NP-hard if F is given explicitly.

Let us first consider the case in which the cost function is given explicitly. We show that the problem is NP-hard, by reducing it to the following 3 SAT, which is known to be NP-hard.

3 SAT
Input: A 3-CNF $\varphi = \bigwedge_{c_j \in C} c_j$, where $c_j = (l_{j_1} \vee l_{j_2} \vee l_{j_3})$.
Question: Is φ satisfiable, i.e., does there exist an assignment $y^* \in \{0,1\}^N$ such that $\varphi(y^*)=1$? □

Here c_j is a clause containing three literals l_{j_1}, l_{j_2} and l_{j_3} in $\{y_1, \cdots, y_N, \overline{y}_1, \cdots, \overline{y}_N\}$.

Given a problem instance I of 3 SAT, we construct the corresponding instance J of Problem $(\mathbf{P})$ as follows.

Let $V = \{v_1, \cdots, v_{2N}\}$ and $\mathcal{F} = \{X_i = \{v_{2i-1}, v_{2i}\} \mid i = 1, 2, \cdots, N\}$. Let d be a demand function defined by $d(X_i) = 1$ for each $X_i \in \mathcal{F}$, and let F be a cost function given by

$$F(x)=\alpha\left(\sum_{i=1}^{N} \min\{x(v_{2i-1}), x(v_{2i})\}+\sum_{c_j \in C} \min\{x(v^{(j_1)}), x(v^{(j_2)}), x(v^{(j_3)})\}\right) \quad (4.1)$$

where $\alpha > 0$, and $v^{(j_k)} = v_{2i-1}$ if $l_{j_k} = \overline{y}_i$ and $v^{(j_k)} = v_{2i}$ if $l_{j_k} = y_i$. Note that F is a monotone concave function, and hence $\mathcal{F}$, d, and F above give a problem instance of $(\mathbf{P})$. Then we have the following lemma.

Lemma 4.2. *Let* J *denote the problem instance of* $(\mathbf{P})$ *constructed as above from a problem instance* I *of 3* SAT. *Let* $OPT(\mathrm{J})$ *denote the optimal value of* J. *Then* I *is satisfiable if and only if* $OPT(\mathrm{J}) = 0$, *and unsatisfiable if and only if* $OPT(\mathrm{J}) \geq \alpha\,(> 0)$. □

Since the value of α in (4.1) can be arbitrarily large, we have the following result.

Theorem 4.3. *It is* NP-*hard to approximate Problem* $(\mathbf{P})$ *within any constant* α. *In particular, Problem* $(\mathbf{P})$ *is* NP-*hard.*

As for implicit representation of F, we have the following theorem.

Theorem 4.4. *It takes* $\Omega(2^{\frac{n}{2}} q)$ *time to solve Problem* $(\mathbf{P})$ *if* F *is given by an oracle.* □

5 Concluding Remarks

We have considered the problem of minimizing monotone concave functions with laminar covering constraints. Our results can be summarized by Table 1 in Section 1.

In this paper we have assumed that the objective function F is monotone nondecreasing. It should be noted that this monotonicity assumption can be removed if we impose that the sum $x(V)$ be equal to a constant.

Acknowledgments

The authors would like to express their appreciation to Hiro Ito (Kyoto Univ.), Hiroshi Nagamochi (Kyoto Univ.), András Frank (Eötvös Univ.), and Tibor Jordán (Eötvös Univ.) for their valuable comments.

References

1. K. Arata, S. Iwata, K. Makino, and S. Fujishige: Locating sources to meet flow demands in undirected networks, *J. Algorithms*, **42** (2002), 54–68.
2. A. A. Benczúr and D. R. Karger: Augmenting undirected edge connectivity in $\tilde{O}(n^2)$ time, *J. Algorithms*, **37** (2000), 2–36.
3. G.-R. Cai and Y.-G. Sun: The minimum augmentation of any graph to k-edge-connected graph, *Networks*, **19** (1989), 151–172.
4. A. Frank: Augmenting graphs to meet edge-connectivity requirements, *SIAM J. Discrete Mathematics*, **5** (1992), 25-53.
5. L. Lovász: *Combinatorial Problems and Exercises*, North-Holland (1979).
6. W. Mader: A reduction method for edge-connectivity in graphs, *Ann. Discrete Mathematics*, **3** (1978), 145–164.
7. H. Nagamochi: Computing extreme sets in graphs and its applications, *Proc. of the 3rd Hungarian-Japanese Symposium on Discrete Mathematics and Its Applications* (January 21–24, 2003, Tokyo, Japan) 349–357.
8. H. Nagamochi and T. Ibaraki: Augmenting edge-connectivity over the entire range in $\tilde{O}(nm)$ time, *J. Algorithms*, **30** (1999), 253–301.
9. H. Tamura, M. Sengoku, S. Shinoda, and T. Abe: Some covering problems in location theory on flow networks, *IEICE Transactions on Fundamentals of Electronics, Communications and Computer Sciences*, **E75-A** (1992), 678–683.
10. T. Watanabe and A. Nakamura: Edge-connectivity augmentation problems, *J. Comput. System Sci.*, **35** (1987), 96–144.

Almost Optimal Solutions for Bin Coloring Problems

Mingen Lin, Zhiyong Lin, and Jinhui Xu

Department of Computer Science and Engineering,
University at Buffalo, the State University of New York,
Buffalo, NY 14260, USA
{mlin6, zlin, jinhui}@cse.buffalo.edu

Abstract. In this paper we study two interesting bin coloring problems: Minimum Bin Coloring Problem (MinBC) and Online Maximum Bin Coloring Problem (OMaxBC), motivated from several applications in networking. For the MinBC problem, we first show that it is NP-complete, and then present two near linear time approximation algorithms to achieve almost optimal solutions, i.e., no more than $OPT + 2$ and $OPT + 1$ respectively, where OPT is the optimal solution. For the OMaxBC problem, we first introduce a deterministic 2-competitive greedy algorithm, and then give lower bounds for any deterministic and randomized (against adaptive offline adversary) online algorithms. The lower bounds show that our deterministic algorithm achieves the best possible competitive ratio.

1 Introduction

Bin packing is a fundamental problem in combinatorial optimization and has been extensively studied in the past [1]. In this paper we consider two interesting variants of the packing problem, called *Minimum Bin Coloring* (MinBC) problem and *Online Maximum Bin Coloring* (OMaxBC) problem respectively. In the MinBC problem, we are given n types of unit-size items with each type containing $q_i, 1 \leq i \leq n$, items and associated with a distinct color. We are also given m bins, each with an integer capacity B such that $B = \frac{\sum_{i=1}^{n} q_i}{m}$. The objective is to minimize the maximum number of different colors in each bin. In the OMaxBC problem, all the items are given and packed in an online fashion, and the objective is to maximize the minimum number of different colors in each bin.

Our bin coloring problems are motivated from several applications in networking. In such applications, each type of items represents all the packets from a single user (or a single task) and each bin represents a burst [2, 3] or a channel. By maximizing (or minimizing) the different number of colors in the OMaxBC (or MinBC) problem, we can maximize (minimize the cost of de-assembling) the fairness for all users. Other applications of the bin coloring problem have been discussed in [4].

X. Deng and D. Du (Eds.): ISAAC 2005, LNCS 3827, pp. 82–91, 2005.

Several variants of the bin coloring problems have been studied. In [4] Krumke *et al.* considered the online version of the MinBC problem and proved an upper bound and a lower bound for the competitive ratio of a greedy-type algorithm. They also showed that some trivial algorithm has a better competitive ratio. In [5] and [6] two related bin coloring problems called *Class-Constrained Bin-Packing problem* (CCBP) and *Class-Constrained Multiple Knapsack problem* (CMKP) were studied. A dual polynomial time approximation scheme (PTAS) was presented in [5] for the CCBP problem, and a dual approximation algorithm was given in [6] for the CMKP problem. For the CMKP problem, Shachnai and Tamir showed that it is sufficient to add one more compartment to each knapsack in order to eliminate the gap between the performance of an optimal and a polynomial time algorithm. The running time of their algorithm is $O(M^2N^2)$, where M is the number of bins and N is the number of types of items.

In this paper we present the best possible solutions (under the assumption that $P \neq NP$) for both MinBC and OMaxBC problems. For the MinBC problem, we first show that it is NP-complete even for the offline version, and then present two near linear time approximation algorithms to achieve almost optimal solutions, i.e., the generated solutions are no more than $OPT+2$ and $OPT+1$ respectively, where OPT is the optimal solution. Our algorithms are based on a number of interesting observations and efficient techniques for handling various types of items. In certain way the MinBC problem can be viewed as a dual problem of the CMKP problem with the restriction that all the bins have the same size. With this restriction our algorithm for the MinBC problem improves the algorithm in [6] by a factor of $O(N)$ in the running time.

For the OMaxBC problem, we first introduce a deterministic greedy algorithm with a competitive ratio of 2, and then give lower bounds for any deterministic and randomized (against adaptive offline adversary) online algorithms. The lower bounds indicate that our deterministic algorithm achieves the best possible competitive ratio.

Our algorithms for both problems are very simple, efficient and with high quality. They can be easily implemented for practical purpose.

2 Minimum Bin Coloring Problem

In this section, we study the (offline) Minimum Bin Coloring Problem (MinBC).

Let $G_1, G_2, \cdots, G_n$ be the n types of items with each G_i containing q_i unit-size items and associated with a distinct color c_i (i.e., all the q_i items in G_i share the same color c_i), and an integer $B = \frac{\sum_{i=1}^{n} q_i}{m}$ be the capacity of each of the m bins $b_1, b_2, \cdots, b_m$. Let I_i^A be the set of items packed in bin b_i by some algorithm A, C_i^A be the number of different colors of I_i^A, and C^A be the maximum number of different colors in one of the m bins. The objective of the MinBC problem is to pack all the items in the m bins so that the maximum number of different colors in a single bin is minimized, i.e., $\min_A C^A = \min_A \max_{i=1}^{m}\{C_i^A\}$.

For simplicity of discussion, we also view each G_i as a splittable item of size q_i, but require that when G_i is split, it always yields items of integral sizes.

2.1 MinBC Is NP-Complete

In this section, we show that the (decision version of) MinBC problem is NP-complete.

First of all, it is obvious that the MinBC problem is in NP, since given a packing of the m bins, we can easily compute the maximum number of different colors in a single bin in polynomial time.

To show the NP-hardness of the MinBC problem, we reduce from the 3-partition problem, which is NP-complete. The 3-partition problem can be defined as follows. Given $3n$ integers $t_1, t_2, \ldots, t_{3n}$ and $v = \frac{1}{n}\sum_{i=1}^{3n} t_i$, find sets $S_1, \ldots, S_n$ with $|S_i| = 3$ such that for all i, $\sum_{j \in S_i} t_j = v$.

The reduction is constructed as follows. For a given instance I of the 3-partition problem, we construct an instance of the MinBC problem I' which has $3n$ distinctively colored items with size $t_1, t_2, \ldots, t_{3n}$ respectively and n bins with capacity v. We show that if I has a solution if and only if I' has a solution such that each bin has exactly 3 different colors (see details in the full paper).

Theorem 1. *The minimum bin coloring (MinBC) problem is NP-complete. For any $\epsilon > 0$, there is no approximation algorithm with approximation ratio $\frac{4}{3} - \epsilon$ for it unless $P = NP$.*

2.2 A Basic Algorithm

In this section, we present a simple algorithm and show the quality of its solutions. The main steps of our algorithm are shown in Algorithm 1.

Algorithm 1 BASICALG

1: Sort all items in a non-decreasing order of their sizes. Let the sorted order be $G_1, G_2, \cdots, G_n$.
2: Start from G_1 and the first bin b_1.
3: Pack as much of the current item G_j as possible into the current bin b_i.
4: If the current bin b_i is full, move to the next bin $b_{(i+1)\ mod\ m}$ until the current item G_j has been fully packed. Else move to the next bin $b_{(i+1)\ mod\ m}$.
5: Move to the next item.
6: Repeat steps 3 to 5 until all items are packed.

Our algorithm can be divided into two stages. In the first stage, we pack small-sized items fully into bins in a round-robin manner until some item cannot be fully packed into a single bin. In the second stage, we pack the rest of items with each item in multiple bins.

We show in Theorem 2 that even though the above algorithm is very simple, it yields near optimal solutions. Let OPT be the (maximum) number of colors in an optimal packing.

Lemma 1. *In an optimal packing, there exists a bin having at least $\lceil \frac{n}{m} \rceil$ different colors.*

Theorem 2. *The algorithm* BASICALG *yields a packing with at most $OPT + 2$ different colors in each bin.*

Proof. First, by Lemma 1 we know that if no item is packed into more than one bin, then the number of different colors in each bin is at most $\lceil \frac{n}{m} \rceil \leq OPT \leq OPT + 2$.

Second, we consider the case where at least one item is packed into more than one bin. Let G_j be the first (according to the order of items being packed) of such items and b_i be the first bin in which G_j is packed. Let the *level* of a bin be the size of all items packed in it. When the first stage of the algorithm finishes, we have the following observations.

Observation 1. *Each of the bins $b_1, b_2, \ldots, b_i$ has $\lceil \frac{j-1}{m} \rceil$ items and each of the bins $b_{i+1}, \ldots, b_m$ has $\lceil \frac{j-1}{m} \rceil - 1$ items.*

Thus each bin has no more than $\lceil \frac{j-1}{m} \rceil \leq \lceil \frac{n}{m} \rceil \leq OPT$ different colors after finishing stage 1.

Observation 2. *The levels of bins $b_i, b_{i+1}, \ldots, b_m, b_1, \ldots, b_{i-1}$ are in non-decreasing order. (The proof is omitted, see the full paper for details)*

The above observation, along with the fact that G_j has to be packed into multiple bins, implies that no item in $G_j, \ldots, G_n$ can be fully packed into a single bin in stage 2. Also in stage 2, each item is packed into consecutive bins. Thus, each bin will be packed with at most two more items in stage 2. Therefore we have the following observation.

Observation 3. *Each bin obtains at most 2 more colors in stage 2.*

Finally, putting all together, we know that when the algorithm finishes, each bin has at most $OPT + 2$ different colors. □

The above theorem gives an upper bound for the performance of Algorithm BASICALG. The following example shows that this bound is actually tight. In this example, we have 4 items a, b, c, d of size ϵN, 2 items e, f of size $(\frac{1}{3} - \epsilon)N$, 12 items $g \ldots r$ of size $\frac{1}{3}N$, 1 item x of size $(\frac{1}{3} + 2\epsilon)N$, and 2 items y, z of size $(1 - 2\epsilon)N$, where $\epsilon < \frac{1}{6}$ is a constant. All items have different colors. We also have 7 bins with capacity N. By carefully choosing N, we can make all items have integer sizes.

Our algorithm outputs a packing shown in Figure 1, where the maximum number of different colors in a single bin is 5. An optimal packing is shown in Figure 2, where the maximum number of different colors in a single bin is only 3.

Theorem 3. *The running time of Algorithm* BASICALG *is $O(m + n \log n)$.*

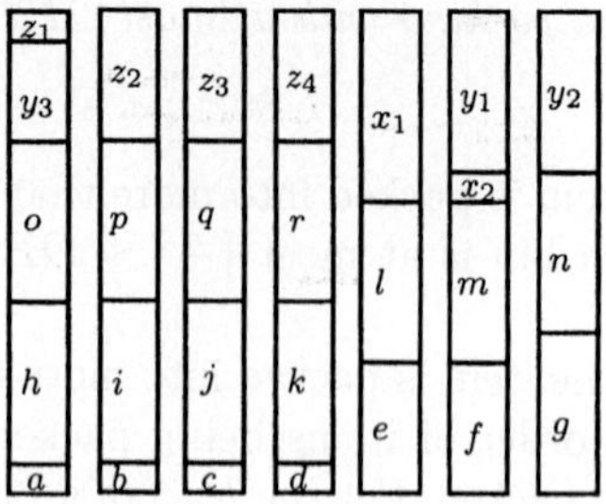

Fig. 1. The output of BASICALG where $x_1 = (\frac{1}{3}+\epsilon)N, x_2 = \epsilon N, y_1 = y_2 = \frac{1}{3}, y_3 = (\frac{1}{3} - 2\epsilon)N, z_1 = \epsilon N, z_2 = z_3 = z_4 = (\frac{1}{3} - \epsilon)N$

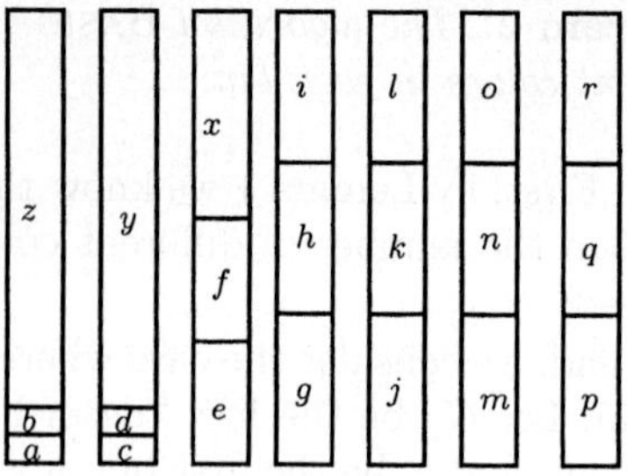

Fig. 2. An optimal packing

2.3 An Improved Algorithm

The algorithm presented in previous section yields a near optimal solution (i.e., $C^A \leq OPT + 2$). Since the MinBC problem is NP-complete, if we assume that $P \neq NP$, then the best possible (polynomial-time) solution one could hope is an algorithm with $C^A \leq (OPT+1)$ (i.e., with no more than $OPT+1$ different colors in a single bin). Thus a very interesting question is to determine whether such an algorithm exists. In this section, we give an affirmative answer by presenting such an algorithm. Our algorithm is modified from our algorithm BASICALG.

Note that in the analysis of Algorithm BASICALG, if no item is packed in more than one bin, then the resulting packing is actually optimal. Thus to improve the algorithm, we need to use different strategy to deal with those items packed in more than one bin.

Consider the case in which at least one item is packed into more than one bin. Let G_j be the first item packed in more than one bin by Algorithm BASICALG, and b_i be the first bin in which G_j is packed. Let L be the set of bins $\{b_1, b_2, \ldots, b_{i-1}\}$ and R be the set of bins $\{b_i, \ldots, b_m\}$. By Observations 1 and 3, after packing all items, each bin in R has no more than $OPT + 1$ different colors and each bin in L has no more than $OPT + 2$ different colors. Thus to achieve a better solution, we have to reduce the number of colors for bins in L. Our main idea is to move items in bins in L with $OPT+2$ colors into bins in R with OPT colors.

To do so, we modify our algorithm as follows. We define *additional color* of a bin to be the number of different colors the bin gains in stage 2 and *base* of a bin to be the level of the bin after finishing stage 1.

The correctness of the algorithm is shown as follows. Clearly if the algorithm terminates in step 2 or 3, it returns the packing generated by Algorithm BASICALG. We shall show that the algorithm will terminate in the rest of the steps.

To show that the algorithm terminates in steps 4 to 8, we shall show that (1) the index of b_{l_2} decreases after each iteration, and (2) if b_{l_2} exists, b_{r_1} exists.

Algorithm 2 IMPROVEDALG

$B_k(L) + B_k(R) \leq g - 1 < m - i + 1$ for all $k \geq 0$

1: Run Algorithm BASICALG.
2: If no item is packed into more than one bin, stop.
3: If after stage 1 of Algorithm BASICALG, there are more than $m - i + 1$ items unpacked, stop.
4: Find the last bin b_{l_2} with 2 additional colors in L.
5: Find the first bin b_{r_1} with 1 additional color in R.
6: Remove the last item packed in b_{l_2} and insert it into b_{r_1} at the position just above the base (i.e., pack it on the top of the items accumulated in stage 1). Bin b_{r_1} will overflow.
7: Continuously propagate the overflowing part of b_{r_1} to the following bins and insert it on the top of their bases. See Figure 3 .
8: Repeat steps 4 to 7 until no bin has 2 additional colors in L.

Lemma 2. *After each iteration (step 7), the index l_2 of b_{l_2} decreases and the index r_1 of b_{r_1} increases.*

Proof. Since b_{l_2} is the last bin with 2 additional colors in L, each bin in $b_{l_2+1}, \ldots,$ b_{i-1} has 1 additional color. Below we show that after each iteration (step 7), bin b_{l_2} has 1 additional color and b_{r_1} has 2 additional colors, that is, l_2 decreases after the propagation while r_1 increases. Let a and b be the two partial items packed in b_{l_2} before the propagation and a is on top of b (i.e. a is packed after b, see Figure 3). In the propagation (step 6 and 7), a is moved onto the base of bin b_{r_1} and some part c with size $|a|$ in bin b_{l_2-1} is moved into b_{l_2}. By Observation 2 and the order we pack items, we know that in stage 2 no item can be fully packed in any bin and each item is packed in consecutive bins. This implies that if b is part of an item G_k, then c has to be part of G_k, since otherwise (i.e., c contains part of another item) G_k can be fully packed in bin b_{l_2}, a contradiction. Thus, after the propagation, bin b_{l_2} has only 1 additional color, and therefore l_2 decreases in next iteration.

During each iteration, the items packed in bins between b_{l_2} and b_{r_1} do not change. By Observation 2, in stage 2, the partial item packed in bin b_{r_1} has a larger size than that of a. Thus after the iteration, b_{r_1} has 2 additional colors which implies that r_1 can only increase in the next iteration. □

Next, we show that if b_{l_2} exists, then b_{r_1} exists. Let g be the number of remaining items after finishing stage 1 of Algorithm BASICALG. When Algorithm IMPROVEDALG reaches step 4, we have $g \leq m - i + 1$. Let $B_k(L)$ (or $B_k(R)$) be the number of bins with 2 additional colors in L (or R) after the kth iteration.

Lemma 3. *After each iteration k, each bin has at most 2 additional colors and $B_k(L) + B_k(R) \leq g - 1$.*

Proof. By Observation 3, each bin has at most 2 additional colors before the first iteration. After stage 1, each item is packed in consecutive bins, so $B_0(L) + B_0(R) \leq g - 1$.

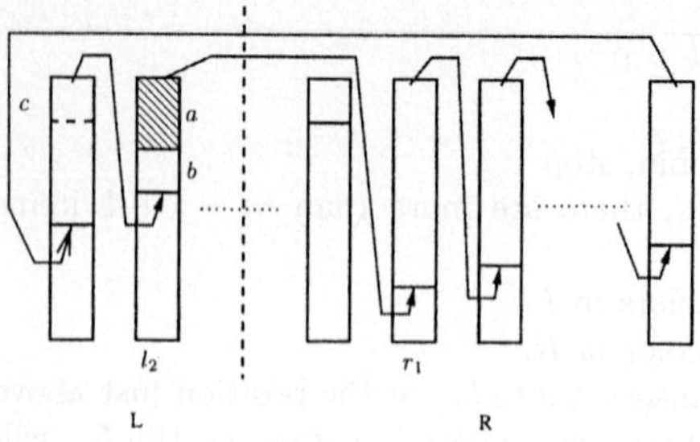

Fig. 3. The procedure of exchange items in Algorithm IMPROVEDALG

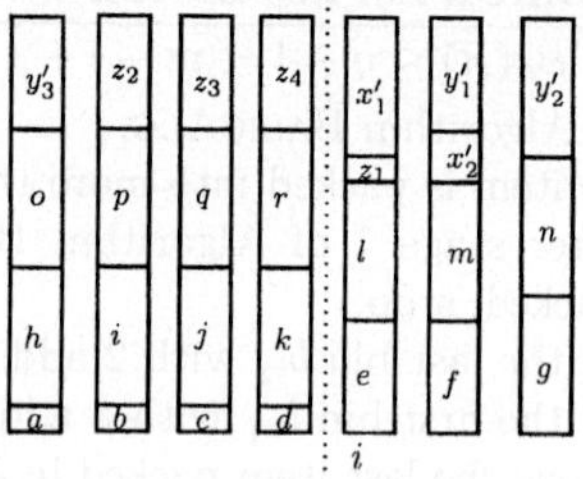

Fig. 4. The output of IMPROVEDALG where $x'_1 = \frac{1}{3}N, x'_2 = 2\epsilon N, y'_1 = (\frac{1}{3} - \epsilon)N, y'_2 = \frac{1}{3}N, y'_3 = (\frac{1}{3} - \epsilon)N$

Let X be the set of bins between b_{l_2} and b_{r_1}. Note that bins in X do not change and each of them has at most 2 additional color. Hence, we only need to show that each bin in $(L \cup R) \setminus X$ has 2 additional colors after each iteration. As shown in Lemma 2, b_{l_2} has 1 additional color and b_{r_2} has 2 additional color after each iteration. Also note that after each iteration, each item packed in bins $(L \cup R) \setminus X$ during stage 2 is still packed in consecutive bins. By the proof of Observations 3 in the proof Theorem 2 each bin in $(L \cup R) \setminus X$ has at most 2 additional colors and $B_k(L) + B_k(R) \leq g - 1$. □

From Lemma 3, we have $B_k(L) \leq g - 1 - B_k(R) < m - i + 1 - B_k(R)$. Note that $m - i + 1 - B_k(R)$ is the number of bins with 1 additional color in R after the k-th iteration . This implies that if b_{l_2} exists, then b_{r_1} must exist.

The performance of the algorithm is shown in the following theorem.

Theorem 4. *The algorithm* IMPROVEDALG *produces a packing with at most* $OPT + 1$ *different colors in each bin.*

Proof. As mentioned previously, the algorithm can terminate in any one of three cases (in step 2,3 and 8, respectively).

If the algorithm terminates in step 2, the output is an optimal packing since no item is split.

If the algorithm terminates in step 3, then after stage 1 of Algorithm BASICALG, there are more than $m - i + 1$ unpacked items. Each bin in L has $\lceil \frac{j-1}{m} \rceil$ colors, each bin in R has $\lceil \frac{j-1}{m} \rceil - 1$ colors and $|R| = m - i + 1$. There are more than $\lceil \frac{j-1}{m} \rceil (i-1) + (\lceil \frac{j-1}{m} \rceil - 1)(m - i + 1) + m - i + 1 = \lceil \frac{j-1}{m} \rceil m$ colors. By pigeon hole principle, there must have one bin with at least $\lceil \frac{j-1}{m} \rceil + 1$ different colors in any optimal solution. Thus when the algorithm terminates, each bin has at most $\lceil \frac{j-1}{m} \rceil + 2 \leq OPT + 1$ different colors.

Now we consider the case in which after stage 1, there are no more than $m - i + 1$ unpacked items. From Lemma 3, we know that when the algorithm terminates, each bin in L has one additional color and each bin in R has no more than 2 additional colors. Thus, the maximum number of different colors is $\lceil \frac{j-1}{m} \rceil + 1 \leq OPT + 1$. □

For the same instance given in subsection 2.2, our improved algorithm yields the following packing (Figure 4).

Theorem 5. *The running time of Algorithm* ImprovedAlg *is* $O(m+n\log n)$.

Proof. By Theorem 3, step 1 takes $O(m+n\log n)$ time. By carefully manipulating an linked list, we can achieve $O(m)$ running time for steps 4-8. See full paper for details. □

3 Online Maximum Bin Coloring Problem

In this section, we consider the Online Maximum Bin Coloring (OMaxBC) problem. The OMaxBC problem is defined as follows.

Online Maximum Bin Coloring Problem ($OMaxBC_{B,m}$): Given $B, m \in \mathcal{N}$ and a sequence $\delta = r_1, \cdots, r_N$, where $N = Bm$, of unit-size items, each associated with a color $c_i \in \mathcal{N}$, one is asked to pack δ into m bins $(b_1, b_2, \cdots, b_m)$ with size B so as to maximize the minimum number of different colors in a single bin. The items in δ are given in an online fashion and have to be packed online, i.e., each r_i has to be packed irrevocably before every r_k for $k > i$, and has no knowledge about r_k.

We say a bin is *open* if it is not completely filled up. Conversely a bin is *closed* if it is completely filled up. Let $A(\delta)$ denote the objective function value of the solution generated by an online algorithm A on sequence δ, and $OPT(\delta)$ denote that of an optimal offline algorithm which has full knowledge about the input sequence δ in advance. A is said to have a competitive ratio α if there exists a constant β such that $OPT(\delta) \leq \alpha \cdot A(\delta) + \beta$ for any sequence δ.

3.1 A Greedy Algorithm

In this section we introduce a greedy algorithm, called GreedyFit, and show that its competitive ratio is 2.

We denote the difference between the number of items and the number of different colors in a bin as the *repeated number* of the bin.

Algorithm 3 GreedyFit

1: For each arriving r_i, if every open bin contains color c_i, pack r_i into the open bin with the smallest repeated number (break ties arbitrarily).
2: Otherwise, pack r_i into the open bin which doesn't contain color c_i and has the minimum number of different colors (break ties arbitrarily).

After GreedyFit finishes a sequence δ of items, let C_{max} be the maximum number of different colors assigned to a bin, and b_{max} be one of the bins with C_{max} different colors. Similarly, let C_{min} be the minimum number of different colors assigned to a bin, and b_{min} be one of the bins with C_{min} different colors.

Given a bin b and an item r, if the number of different colors of b increases to i after r is packed into b, we say r is the ith *new* item of b. Otherwise r is a *repeated* item of b.

Lemma 4. *For any sequence δ, $C_{max} \le 2C_{min} + 1$.*

Proof. If $C_{max} \le C_{min} + 1$, clearly $C_{max} \le 2C_{min} + 1$. Thus we can assume that $C_{max} > C_{min} + 1$.

Consider the ith new item r_i^{max} of b_{max}, where $C_{min} + 1 < i \le C_{max}$. When r_i^{max} arrives, since GREEDYFIT puts r_i^{max} into b_{max} instead of b_{min}, there are two possibilities: either the color c_i^{max} of r_i^{max} is already in b_{min} or b_{min} is closed. If b_{min} is closed, the repeated number of b_{min} is $B - C_{min}$. Consider the moment when the last repeated item of b_{min} is processed. The number of repeated items of b_{max} is no smaller than $B - C_{min} - 1$. Hence, we have $C_{max} \le C_{min} + 1$, contradicting our assumption that $C_{max} > C_{min} + 1$. This means that when r_i^{max} is processed, b_{min} must be open and has already contained c_i^{max}. Thus, each r_i^{max} of b_{max} is already in b_{min}. We have $C_{max} - C_{min} - 1 \le C_{min}$, and the lemma follows. □

Let δ be any sequence of $N = Bm$ items. Let n be the total number of different colors in δ. For each $1 \le i \le n$, let G_i be the set of items with the ith color and q_i be the size of G_i. Define

$$q_i' = \begin{cases} q_i, & \text{if } q_i < m \\ m, & \text{otherwise} \end{cases}$$

Lemma 5. *In an optimal offline solution to the $OMaxBC_{B,m}$ problem, the minimum number of different colors in a single bin is $\lfloor \frac{\sum_{i=1}^{n} q_i'}{m} \rfloor$.*

Theorem 6. *Algorithm* GREEDYFIT *is 2-competitive for the $OMaxBC_{B,m}$ problem. In particular, $OPT(\delta) \le 2 \cdot C_{min} + 1$ for any sequence δ.*

Proof. Let b_i be the first closed bin after applying GREEDYFIT on δ. Let C_i be the number of different colors in b_i when it is closed. Then at the moment when b_i becomes closed. The followings are true.

1. b_i has $B - C_i$ repeated items. Any of the other bins has at least $B - C_i - 1$ repeated items.
2. An item r_i is repeated if and only if c_i occurs greater than m times so far.

Therefore, we have $(B - C_i - 1)(m - 1) + (B - C_i) \le Bm - \sum_{i=1}^{n} q_i'$. It implies $\sum_{i=1}^{n} q_i' \le (C_i + 1)(m - 1) + C_i \le (C_{max} + 1)(m - 1) + C_{max} = mC_{max} + m - 1$. Thus, $OPT(\delta) = \lfloor \frac{\sum_{i=1}^{n} q_i'}{m} \rfloor \le C_{max} \le 2C_{min} + 1$. □

3.2 A Lower Bound on the Competitive Ratio

In this section, we give a lower bound on the competitive ratio of any deterministic and randomized algorithm for the $OMaxBC_{B,m}$ problem.

Theorem 7. *Let $B \geq 3, m \geq 2 \in \mathcal{N}$, and A be a deterministic online algorithm for the $OMaxBC_{B,m}$ problem. Then the competitive ratio C_A of A is no less than 2, i.e., $C_A \geq 2$.*

Proof. Let h be an integer in $[2, min\{\lceil \frac{B}{2} \rceil, m\}]$. Consider the following sequence δ consisting of two sub-sequences δ_1, δ_2 of length $hm - 1$ and $(B-h)m+1$ respectively. δ_1 consists of $hm-1$ items with each having a different color. After A finishes processing δ_1, let b_{min} be the bin containing the least number of items. Let X be the set of items in b_{min}. There are two cases to consider.

- **Case 1** $0 < |X| \leq h-1$. In this case, δ_2 consists of $(B-h)m+1$ items such that all colors are in X and each color occurs at least m times. After finishing δ_2, the number of different colors in b_{min} remains the same. Thus $A(\delta) = |X|$. We have $OPT(\delta) \geq |X| + \lfloor \frac{hm-1-|X|}{m} \rfloor \geq |X| + h - 1 \geq 2|X|$
- **Case 2** $|X| = 0$. In this case, we can select a new color c different from all the colors in δ_1. δ_2 consists of $(B-h)m+1$ items of color c. It is clear that $A(\delta) = 1$ and $OPT(\delta) \geq \lfloor \frac{hm}{m} \rfloor = h \geq 2$.

Therefore, in both cases $\lim_{h\to\infty} \frac{OPT(\delta)}{A(\delta)} \geq 2$. □

Note that in the proof of Theorem 7, we construct the sequence based on the output of the algorithm. The arguments still hold when A is a randomized algorithm. Thus we have the following corollary.

Corollary 1. *Let $B \geq 3, m \geq 2 \in \mathcal{N}$. Let A be any randomized online algorithm for the $OMaxBC_{B,m}$. The competitive ratio of A is $C_A \geq 2$ against any adaptive offline adversary.*

References

1. Coffman, E.G., Garey, M.R., Johnson, D.S.: Approximation algorithms for bin packing: a survey. In: Approximation Algorithms for NP-Hard Problems. PWS Publishing Company (1997) 46–93
2. Xu, J., Qiao, C., Li, J., Xu, G.: Efficient burst scheduling algorithms in optical burst-switched networks using geometric techniques. IEEE Journal on Selected Areas in Communications **22** (2004) 1796–1811
3. Yoo, M., Qiao, C.: A high speed protocol for bursty traffic in optical networks. In: Proc. SPIE All-Opt. Commun. Syst. Volume 3230. (1997) 79–90
4. Krumke, S.O., de Paepe, W.E., Rambau, J., Stougie, L.: Online bin coloring. In: Proceedings of the 9th Annual European Symposium on Algorithms. (2001) 74–85
5. Schachnai, H., Tamir, T.: Polynomial time approximation schemes for class-constrained packing problems. In: 3rd International Workshop on Approximation Algorithms for Combinatorial Optimization. (2000) 238–249
6. Shachnai, H., Tamir, T.: On two class-constrained version of the mutiple knapsack problem. Algorithmica **29** (2001) 442–467

GEN-LARAC: A Generalized Approach to the Constrained Shortest Path Problem Under Multiple Additive Constraints*

Ying Xiao[1], Krishnaiyan Thulasiraman[1], and Guoliang Xue[2]

[1] School of Computer Science, University of Oklahoma,
200 Felgar Street, Norman, OK 73019, USA
{ying_xiao, thulsi}@ou.edu

[2] Arizona State University, Tempe, AZ 85287, USA
xue@asu.edu

Abstract. Given a network modeled as a graph G with each link associated with a cost and k weights, the Constrained Shortest Path (CSP(k)) problem asks for computing a minimum cost path from a source node s to a target node t satisfying pre-specified bounds on path weights. This problem is NP-hard. In this paper we propose a new approximation algorithm called GEN-LARAC for CSP(k) problem based on Lagrangian relaxation method. For $k = 1$, we show that the relaxed problem can be solved by a polynomial time algorithm with time complexity $O((m + n \log n)^2)$. Using this algorithm as a building block and combing it with ideas from mathematical programming, we propose an efficient algorithm for arbitrary k. We prove the convergence of our algorithm and compare it with previously known algorithms. We point out that our algorithm is also applicable to a more general class of constrained optimization problems.

1 Introduction

Recently there has been considerable interest in the design of communication protocols that deliver certain performance guarantees that are usually referred to as Quality of Service (QoS) guarantees. A problem of great interest in this context is the QoS routing problem that requires the determination of a minimum cost path from a source node to a destination node in a data network that satisfies a specified upper bound on the delay of the path. This problem is also known as the Constrained Shortest Path (CSP) problem. The CSP problem is known to be NP-hard [5]. So, in the literature, heuristic approaches and approximation algorithms have been proposed. Heuristics, in general, do not provide performance guarantees on the quality of the solution produced. On the other hand, ϵ-approximation algorithms deliver solutions within arbitrarily specified precision requirement but are not efficient in practice. References [6], [8], [13]

* The work of K. Thulasiraman has been supported by NSF ITR grant ANI-0312435
The work of G. Xue has been supported by NSF ITR grant ANI-0312635.

X. Deng and D. Du (Eds.): ISAAC 2005, LNCS 3827, pp. 92–105, 2005.

present ϵ-algorithms and contain several fundamental ideas of interest in developing such algorithms.

There are certain approximation algorithms based on mathematical programming techniques. These algorithms start with an integer linear programming (ILP) formulation and relax the integrality constraints. The relaxed problem is usually solved by Lagrangian dual (relaxation) method. The first such algorithm was reported in [7] by Handler and Zang. It is based on a geometric approach which is also called the hull approach by Mehlhorn and Ziegelmann [15]. More recently, in an independent work, Jüttner et al. [10] developed the Lagrangian Relaxation based Aggregated Cost (LARAC) algorithm which also uses the Lagrangian dual method. In contrast to the geometric method, they used an algebraic approach. In another work, Blokh and Gutin [2] defined a general class of combinatorial optimization problems of which the CSP problem is a special case. In a recent work, Xiao et al. [21] drew attention to the fact that the algorithms in [2], [7], [10] are equivalent. In view of this equivalence, we shall refer to these algorithms simply as the LARAC algorithm. Ziegelmann [24] provides a fairly complete list of references to the literature on the CSP and related problems.

The CSP(k) problem is more general than the CSP problem in that it allows more than one delay constraint. Given a communication network modeled as a graph G with each link in the graph associated with a cost and $k(\geq 1)$ weights, the CSP(k) problem asks for a minimum cost path from a source node s to a target node t satisfying multiple constraints on the path weights.

A variation of CSP(k) problem, Multi-Constrained Path (MCP) problem has also been a topic of extensive study. The difference between CSP(k) and MCP problems is that MCP problem only asks for a path satisfying all the constraints simultaneously without the requirement of minimizing the cost. For the MCP problem, a series of heuristics and approximation algorithms can be found in [4], [9],[11],[12], [16], [22], [23].

Two methods for the CSP(k) problem based on mathematical programming have been proposed by Beasley and Christofides [1], and Mehlhorn and Ziegelmann [15]. Reference [1] uses a subgradient procedure to compute the Lagrangian relaxation function of the ILP formulation. With geometrical interpretation of the algorithm of [7], the authors of [1] proposed an algorithm called hull approach which is a special case of cutting planes method [18].

In this paper we present a new approach to the CSP(k) problem using Lagrangian relaxation. We first show that for $k = 1$, an approximation solution can be computed in $O((m + n \log n)^2)$ time. Because this algorithm and LARAC are based on the same methodology and obtain the same solution, we also denote our algorithm as LARAC.

For arbitrary k, we use our LARAC algorithm as a building block and combine it with ideas from mathematical programming to achieve progressively higher values of the Lagrangian function. We present the GEN-LARAC algorithm and prove its correctness and convergence properties in Sect. 3. Simulation results comparing our algorithm with two other algorithms are presented in Sect. 4. We conclude in Sect. 5 pointing out that our approach is quite general and is applicable for the general class of combinatorial optimization problems studied in [2].

2 Problem Definition and Preliminaries

Consider a directed graph $G(N, E)$ where N is the set of nodes and E is the set of links in G. Each link (u, v) is associated with a set of $k+1$ additive non-negative integer weights $C_{uv} = (c_{uv}, w_{uv}^1, w_{uv}^2 \ldots, w_{uv}^k)$. Here c_{uv} is called the cost of link (u, v) and w_{uv}^i is called the ith delay of link (u, v). Given two nodes s and t, an s-t path in G is a directed simple path from s to t. Let P_{st} denote the set of all s-t paths in G. For path p define

$$c(p) = \sum_{(u,v)\in p} c_{uv} \text{ and } d_i(p) = \sum_{(u,v)\in p} w_{uv}^i, i = 1, 2 \ldots, k.$$

The value $c(p)$ is called the cost of path p and $d_i(p)$ is called the ith delay of path p. Given k positive integer r_1, $r_2 \ldots$, r_k, an s-t path is called feasible (resp. strictly feasible) if $d_i(p) \leq r_i$ (resp. $d_i(p) < r_i$), for all $i = 1, 2 \ldots, k$ (r_i is called the bound on the ith delay of a path).

The CSP(k) problem is to find a minimum cost feasible s-t path. An instance of the CSP(k) problem is strictly feasible if all the feasible paths are strictly feasible. Without loss of generality, we assume that the problem under consideration is always feasible. In order to guarantee strict feasibility, we do the following transformation.

For $i = 1, 2 \ldots, k$, transform the delays of link (u, v) such that the new weight vector C'_{uv} is given by $C'_{uv} = (c_{uv}, 2w_{uv}^1, 2w_{uv}^2 \ldots, 2w_{uv}^k)$.

Also transform the bounds r_i's so that the new bound vector R' is given by $R' = (2r_1 + 1, 2r_2 + 1 \ldots, 2r_k + 1)$.

In the rest of the paper, we only consider the transformed problem. Thus all link delays are even integers, and delay bounds are odd integers. We shall use symbols with capital or bold letters to represent vectors and matrices. For the simplicity of presentation, we shall use C_{uv} and R instead of C'_{uv} and R' to denote the transformed weight vector and the vector of bounds. Two immediate consequences of this transformation are stated below.

Lemma 1. *$\forall p \in P_{st}, \forall i \in \{1, 2 \ldots, k\}, d_i(p) \neq r_i$ in the transformed problem.*

Lemma 2. *A path in the original problem is feasible (resp. optimal) iff it is strictly feasible (resp. optimal) in the transformed problem.*

Starting with an ILP formulation of the CSP(k) problem and relaxing the integrality constraints we get the RELAX-CSP(k) problem below. In this formulation, for each s-t path p, we introduce a variable x_p.

$$\text{RELAX - CSP}(k):$$
$$\min \quad \sum_p c(p)x_p \tag{1}$$
$$\text{s.t.} \quad \sum_p x_p = 1 \tag{2}$$
$$\sum_p d_i(p)x_p \leq r_i, i = 1\ldots, k \tag{3}$$
$$x_p \geq 0, \forall p \in P_{st}. \tag{4}$$

The Lagrangian dual of RELAX-CSP(k) is given below.

$$\text{DUAL - RELAX-CSP}(k):$$
$$\max \quad w - \lambda_1 r_1 \cdots - \lambda_k r_k \tag{5}$$
$$\text{s.t.} \quad w - d_1(p)\lambda_1 \cdots - d_k(p)\lambda_k \leq c(p), \forall p \in P_{st} \tag{6}$$
$$\lambda_i \geq 0, i = 1\ldots, k. \tag{7}$$

In the above dual problem λ_1,λ_2...,λ_k and w are the dual variables, with w corresponding to (2) and each λ_i corresponding to the ith constraint in (3).

It follows from (6) that $w \leq c(p) + d_1(p)\lambda_1 \cdots + d_k(p)\ \lambda_k, \forall p \in P_{st}$. Since we want to maximize (5), the value of w should be as large as possible, i.e.

$$w = \min_{p \in P_{st}} \{c(p) + d_1(p)\lambda_1 \cdots + d_k(p)\lambda_k\}.$$

With the vector Λ defined as $\Lambda = (\lambda_1, \lambda_2 \ldots, \lambda_k)$, define

$$L(\Lambda) = \min_{p \in P_{st}} \{c(p) + \sum_{i=1}^{k} \lambda_i(d_i(p) - r_i)\}. \tag{8}$$

Notice that $L(\Lambda)$ is called the Lagrangian function in literature and is a continuous concave function of Λ [3].

Then DUAL-RELAX-CSP(k) can be written as follows.

$$\text{DUAL- RELAX-CSP}(k)$$
$$\max L(\Lambda), \Lambda \geq \mathbf{0}. \tag{9}$$

The Λ^* that maximizes (9) is called the maximizing multiplier and is defined as

$$\Lambda^* = \arg\max\nolimits_{\Lambda \geq 0} L(\Lambda). \tag{10}$$

Lemma 3. *If an instance of the CSP(k) problem is feasible and a path p_{opt} is an optimal path, then $\forall \Lambda \geq 0, L(\Lambda) \leq c(p_{opt})$.*

We shall use $L(\Lambda)$ as an lower bound of $c(p_{opt})$ to evaluate the approximation solution obtained by our algorithm. Given $p \in P_{st}$ and Λ, define

$$
\begin{aligned}
&C(p) \equiv (c(p), d_1(p), d_2(p) \ldots, d_k(p)),\\
&D(p) \equiv (d_1(p), d_2(p) \ldots, d_k(p)),\\
&R \equiv (r_1, r_2 \ldots, r_k),\\
&c_\Lambda(p) \equiv c(p) + d_1(p)\lambda_1 \cdots + d_k(p)\lambda_k, and\\
&d_\Lambda(p) \equiv d_1(p)\lambda_1 \cdots + d_k(p)\lambda_k.
\end{aligned}
$$

Here $c_\Lambda(p)$ and $d_\Lambda(p)$ are called the aggregated cost and the aggregated delay of path p, respectively. We shall use P_Λ to denote the set of s-t paths attaining the minimum aggregated cost w.r.t. to Λ. A path $p_\Lambda \in P_\Lambda$ is called a Λ-minimal path.

The key issue now is to search for the maximizing multipliers and termination conditions. For the case $k = 1$, we have the following theorem.

Theorem 1. *DUAL-RELAX-CSP(1) is solveable in $O((m + n \log n)^2)$ time.*

Proof. See Appendix.

Because our algorithm and LARAC are based on the same methodology and obtain the same solution, we shall also call our algorithm LARAC. In the rest of the paper, we shall discuss how to extend it for $k > 1$. In particular we develop an approach that combines the LARAC algorithm as a building block with certain techniques in mathematical programming. We shall call this new approach as GEN-LARAC.

3 GEN-LARAC for the CSP(k) Problem

3.1 Optimality Conditions

Theorem 2. *Given an instance of a feasible CSP(k) problem, a vector $\Lambda \geq 0$ maximizes $L(\Lambda)$ iff the following problem in variables u_j's is feasible.*

$$\sum_{p_j \in P_\Lambda} u_j d_i(p_j) = r_i, \forall i, \lambda_i > 0 \tag{11}$$

$$\sum_{p_j \in P_\Lambda} u_j d_i(p_j) \leq r_i, \forall i, \lambda_i = 0 \tag{12}$$

$$\sum_{p_j \in P_\Lambda} u_j = 1 \tag{13}$$

$$u_j \geq 0, \forall j, p_j \in P_\Lambda. \tag{14}$$

Proof. **Sufficiency:** Let $\boldsymbol{x} = (u_1 \ldots, u_r, 0 \ldots, 0)$ be a vector of size $|P_{st}|$, where $r = |P_\Lambda|$.

Obviously, $\boldsymbol{x}$ is a feasible solution to RELAX-CSP(k). It suffices to show that $\boldsymbol{x}$ and Λ satisfy the complementary slackness conditions.

According to (6), $\forall p \in P_{st}, w \leq c_p + d_1(p)\lambda_1 \cdots + d_k(p)\lambda_k$. Since we need to maximize (5), the optimal $w = c_p + d_1(p_\Lambda)\lambda_1 \cdots + d_k(p_\Lambda)\lambda_k, \forall p_\Lambda \in P_\Lambda$. For all other paths p, $w - c_p + d_1(p)\lambda_1 \cdots + d_k(p)\lambda_k < 0$. So $\boldsymbol{x}$ satisfies the complementary slackness conditions. By (11) and (12), Λ also satisfies complementary slackness conditions.

Necessary: Let $\boldsymbol{x}^*$ and (w, Λ) be the optimal solution to RELAX-CSP(k) and DUAL-RELAX-CSP(k), respectively. It suffices to show that we can obtain a feasible solution to (11)-(14) from $\boldsymbol{x}^*$.

We know that all the constraints in (6) corresponding to paths in $P_{st} - P_\Lambda$ are strict inequalities and $w = c_p + d_1(p_\Lambda)\lambda_1 \cdots + d_k(p_\Lambda)\lambda_k, \forall p_\Lambda \in P_\Lambda$. So, from complementary slackness conditions we get

$$x_p = 0, \forall p \in P_{st} - P_\Lambda.$$

Now let us set u_j corresponding to path p in P_Λ equal to x_p, and set all other u_j's corresponding to paths not in P_Λ equal to zero. The u_i's so elected will satisfy (11) and (12) since these are complementary conditions satisfied by (w, Λ). Since x_i's satisfy (2), u_j's satisfy (13). Thus we have identified a solution satisfying (11)-(14).

3.2 GEN-LARAC: A Coordinate Ascent Method

Our approach is based on the coordinate ascent method and proceeds as follows. Given a multiplier Λ, in each iteration we try to improve the value of $L(\Lambda)$ by updating one component of the multiplier vector. If the objective function is not differentiable, the coordinate ascent method may get stuck at a corner Λ_s not being able to make progress by changing only one component. We shall call

Algorithm 1. GEN-LARAC: A Coordinate Ascent Algorithm

Step 1: $\Lambda^0 \leftarrow (0, \ldots, 0); flag \leftarrow true; B \leftarrow 0; t \leftarrow 0$
Step 2: {Coordinate Ascent Step}
while *flag* **do**
  flag $\leftarrow$ *false*
  for $i = 1$ to k **do**
    $\gamma \leftarrow \arg\max_{\xi \geq 0} L(\lambda_1^t \ldots, \lambda_{i-1}^t, \xi, \lambda_{i+1}^t \ldots, \lambda_k^t)$
    if $\gamma \neq \lambda_i^t$ **then**
      flag $\leftarrow$ *true*
      $\lambda_j^{t+1} = \begin{cases} \gamma & \text{if } j = i; \\ \lambda_j^t & \text{if } j \neq i. \end{cases}, j = 1, 2 \ldots, k$
      $t \leftarrow t + 1$
    end if
  end for
end while
Step 3: if Λ^t is optimal then return Λ^t
Step 4: $B \leftarrow B + 1$ and stop if $B > B_{max}$ (B_{max} is the maximum number of iteration allowed)
Step 5: Compute a vector Λ^+ such that $L(\Lambda^+) > L(\Lambda^t)$.
Step 6: $t \leftarrow t + 1, \Lambda^t \leftarrow \Lambda^+$, and go to Step 2.

Λ_s pseudo optimal point which requires updates of at least two components to achieve improvement in the solution. We shall discuss how to jump to a better solution from a pseudo optimal point in Sect. 3.3. Our simulations show that the objective values attained at pseudo optimal points are usually very close to the maximum value of $L(\Lambda)$.

3.3 Verification of Optimality of Λ

In Step 3 we need to verify if a given Λ is optimal. We show that this can be accomplished by solving the following LP problem, where $P_\Lambda = \{p_1, p_2 \ldots, p_r\}$ is the set of Λ-minimal paths.

$$\max \quad 0 \tag{15}$$

$$\text{s.t.} \quad \sum_{p_j \in P_\Lambda} u_j d_i(p_j) = r_i, \forall i, \lambda_i > 0 \tag{16}$$

$$\sum_{p_j \in P_\Lambda} u_j d_i(p_j) \leq r_i, \forall i, \lambda_i = 0 \tag{17}$$

$$\sum_{p_j \in P_\Lambda} u_j = 1 \tag{18}$$

$$u_j \geq 0, \forall j, p_j \in P_\Lambda. \tag{19}$$

By Theorem 2, if the above linear program is feasible then the multiplier Λ is a maximizing multiplier.

Let $(y_1 \ldots, y_k, \delta)$ be the dual variables corresponding to the above problem. Let $Y = (y_1, y_2 \ldots, y_k)$. The dual of (15)-(19) can be written as follows

$$\min \quad RY^T + \delta \tag{20}$$

$$\text{s.t.} \quad D(p_i)Y^T + \delta \geq 0, i = 1, 2 \ldots, r \tag{21}$$

$$y_i \geq 0, \forall i, \lambda_i > 0. \tag{22}$$

Evidently the LP problem (20)-(22) is feasible. From the relationship between primal and dual problems, it follows that if the linear program (15)-(19) is infeasible, then the objective of (20) is unbounded ($-\infty$). Thus, if the optimum objective of (20)-(22) is 0, then the linear program (15)-(19) is feasible and by Theorem 2 the corresponding multiplier Λ is optimal. In summary we have the following lemma.

Lemma 4. *If (15)-(19) is infeasible, then* $\exists Y = (y_1, y_2 \ldots, y_k)$ *and* δ *satisfying (21)-(22) and* $RY^T + \delta < 0$.

The Y required in the above lemma can be identified by applying the simplex method on (20)-(22) and terminating it once the objective value becomes negative.

Let Λ be a non-optimal Lagrangian multiplier and denote $\Lambda(s, Y) = \Lambda + Y/s$ for $s > 0$.

Theorem 3. *If a multiplier $\Lambda \geq 0$ is not optimal, then*

$$\exists M > 0, \forall s > M, L(\Lambda(s, Y)) > L(\Lambda).$$

Proof. If M is big enough, $P_\Lambda \cap P_{\Lambda(s,Y)} \neq \emptyset$. Let $p_J \in P_\Lambda \cap P_{\Lambda(s,Y)}$.

$$\begin{aligned} L(\Lambda(s, Y)) &= c(p_J) + (D(p_J) - R)(\Lambda + Y/s)^T \\ &= c(p_J) + (D(p_J) - R)\Lambda^T + (D(p_J) - R)(Y/s)^T \\ &= L(\Lambda) + (D(p_J)Y^T - RY^T)/s. \end{aligned}$$

Since $D(p_J)Y^T + \delta \geq 0$ and $RY^T + \delta < 0, D(p_J)Y^T - RY^T > 0$.
Hence $L(\Lambda(s, Y)) > L(\Lambda)$.

We can find the proper value of M by binary search after computing Y. The last issue is to compute P_Λ. It can be expected that the size of P_Λ is usually very small. So we adapted the k-shortest path algorithm to compute P_Λ.

3.4 Analysis of the Algorithm

In this section, we shall discuss the convergence properties of GEN-LARAC.

Lemma 5. *If there is a strictly feasible path, then for any given τ, the set $S_\tau = \{\Lambda | L(\Lambda) \geq \tau\} \subset R^k$ is bounded.*

Proof. Let p^* be a strictly feasible path. For any $\Lambda = (\lambda_1 \cdots, \lambda_k) \in S_\tau$, we have

$$c(p^*) + \lambda_1(d_1(p^*) - r_1) \cdots + \lambda_k(d_k(p^*) - r_k) \geq L(\Lambda) \geq \tau.$$

Since $d_i(p^*) - r_i < 0$ and $\lambda_i \geq 0$ for $i = 1, 2 \ldots, k$, Λ must be bounded.

If there is only one delay constraint, i.e., $k = 1$, we have the following property [10].

Lemma 6. *A value $\lambda > 0$ maximizes the function $L(\lambda)$ if and only if there are paths p_c and p_d which are both c_λ-minimal and for which $d(p_c) \geq r$ and $d(p_d) \leq r$. (p_c and p_d can be the same; in this case $d(p_d) = d(p_c) = r$).*

By Lemma 6, we have the following lemma.

Lemma 7. *A multiplier $\Lambda \geq 0$ is pseudo optimal iff $\forall i \exists p_c^i, p_d^i \in P_\Lambda, d_i(p_c^i) \geq r_i$ and $d_i(p_d^i) \leq r_i$.*

For an n-vector $V = (v_1, v_2 \ldots, v_n)$, let $|V|_1 = |v_1| + |v_2| \cdots + |v_n|$ denote the L^1-norm.

Lemma 8. *Let Λ and H be two multipliers obtained in the same while-loop of Step 2 in Algorithm 1. Then $|L(H) - L(\Lambda)| \geq |H - \Lambda|_1$.*

Proof. Let $\Lambda = \Lambda^1, \Lambda^2 \dots, \Lambda^j = H$ be the consecutive sequence of multipliers obtained in Step 2. We first show that $|L(\Lambda^{i+1}) - L(\Lambda^i)| \geq |\Lambda^{i+1} - \Lambda^i|_1$.

Consider two cases.

Case 1: $\lambda_b^{i+1} > \lambda_b^i$. By Lemma 6 and Lemma 1, $\exists p_{\Lambda^{i+1}}$, $d_b(p_{\Lambda^{i+1}}) > r_b$.

By definition, we have:

$c(p_{\Lambda^{i+1}}) + \Lambda^{i+1} D^T(p_{\Lambda^{i+1}}) \leq c(p_{\Lambda^i}) + \Lambda^{i+1} D^T(p_{\Lambda^i})$, and

$c(p_{\Lambda^{i+1}}) + \Lambda^i D^T(p_{\Lambda^{i+1}}) \geq c(p_{\Lambda^i}) + \Lambda^i D^T(p_{\Lambda^i})$.

Then

$$\begin{aligned}
&L(\Lambda^{i+1}) - L(\Lambda^i) \\
&= c(p_{\Lambda^{i+1}}) + \Lambda^{i+1}(D(p_{\Lambda^{i+1}}) - R)^T - [c(p_{\Lambda^i}) + \Lambda^i(D(p_{\Lambda^i}) - R)^T] \\
&= c(p_{\Lambda^{i+1}}) + \Lambda^i(D(p_{\Lambda^{i+1}}) - R)^T + (\Lambda^{i+1} - \Lambda^i)(D(p_{\Lambda^{i+1}}) - R)^T \\
&\quad - [c(p_{\Lambda^i}) + \Lambda^i(D(p_{\Lambda^i}) - R)^T] \\
&\geq c(p_{\Lambda^i}) + \Lambda^i(D(p_{\Lambda^i}) - R)^T + (\Lambda^{i+1} - \Lambda^i)(D(p_{\Lambda^{i+1}}) - R)^T \\
&\quad - [c(p_{\Lambda^i}) + \Lambda^i(D(p_{\Lambda^i}) - R)^T] \\
&= (\Lambda^{i+1} - \Lambda^i)(D(p_{\Lambda^{i+1}}) - R)^T \\
&= (\lambda_b^{i+1} - \lambda_b^i)(d_b(p_{\Lambda^{i+1}}) - r_b) \geq |\lambda_b^{i+1} - \lambda_b^i|.
\end{aligned}$$

Case 2: $\lambda_b^{i+1} < \lambda_b^i$. We have $\exists p_{\Lambda^{i+1}}$, $d_b(p_{\Lambda^{i+1}}) < r_b$.

The rest of the proof is similar to Case 1.

Hence

$$\begin{aligned}
&|L(\Lambda^j) - L(\Lambda^1)| \\
&= |L(\Lambda^2) - L(\Lambda^1) + L(\Lambda^3) - L(\Lambda^2) \cdots + L(\Lambda^j) - L(\Lambda^{j-1})| \\
&= |L(\Lambda^2) - L(\Lambda^1)| + |L(\Lambda^3) - L(\Lambda^2)| \cdots + |L(\Lambda^j) - L(\Lambda^{j-1})| \\
&\geq |\Lambda^2 - \Lambda^1|_1 + |\Lambda^3 - \Lambda^2|_1 \cdots + |\Lambda^j - \Lambda^{j-1}|_1 \geq |\Lambda^j - \Lambda^1|_1.
\end{aligned}$$

The second equality holds because $L(\Lambda^1) < L(\Lambda^2) < \cdots < L(\Lambda^j)$.

Obviously, if the while-loop in Step 2 terminates in a finite number of steps, the multiplier is pseudo optimal by definition. If the while loop does not terminate in a finite number of steps (this occurs only when infinite machine precision is assumed), we have the following theorem.

Theorem 4. *Let $\{\Lambda^i\}$ be a consecutive sequence of multipliers generated in the same while-loop in Step 2 in Algorithm 1. Then the limit point of $\{L(\Lambda^i)\}$ is pseudo optimal.*

Proof. Since $L(\Lambda^1) < L(\Lambda^2) < \cdots$ and Λ^i is bounded from above, $\lim_{s\to\infty} L(\Lambda^s)$ exists and is denoted as L^*. We next show the vector $\lim_{s\to\infty} \Lambda^s$ also exists.

By Lemma 8, $\forall s, j > 0$, $|\Lambda^{s+j} - \Lambda^s|_1 \leq |L(\Lambda^{s+j}) - L(\Lambda^s)|$.

By Cauchy criterion, $\lim_{s\to\infty} \Lambda^s$ exists. We denote it as Λ^*.

We label all the paths in P_{st} as $p_1, p_2 \dots, p_N$ such that $c_{\Lambda^*}(p_1) \leq c_{\Lambda^*}(p_2) \dots \leq c_{\Lambda^*}(p_N)$. Obviously p_1 is a Λ^*-minimal path. Let

$$\theta = \min\{c_{\Lambda^*}(p_j) - c_{\Lambda^*}(p_i) | \forall p_i, p_j \in P_{st}, c_{\Lambda^*}(p_j) - c_{\Lambda^*}(p_i) > 0\}, and$$
$$\pi = max\{d_w(p_j) - d_w(p_i) | \forall p_i, p_j \in P_{st}, w = 1, 2 \dots, k\}.$$

Let M be a large number, such that $\forall t \geq M, |\Lambda^* - \Lambda^t|_1 < \theta/(2\pi)$.

Consider any component $j \in \{1, 2 \ldots, k\}$ of the multiplier after computing $\arg\max_{\xi \geq 0} L(\lambda_1 \ldots, \lambda_{j-1}, \xi, \lambda_{j+1} \ldots, \lambda_k)$ in iteration $t \geq M$.

By Lemma 6, $\exists p_c^t$ and $p_d^t \in P_{\Lambda^t}$ and $d_j(p_c^t) \geq r_j \geq d_j(p_d^t)$.

It suffices to show $P_{\Lambda^t} \subseteq P_{\Lambda^*}$. Given $p_{\Lambda^t} \in P_{\Lambda^t}$, we shall show $c_{\Lambda^*}(p_{\Lambda^t}) = c_{\Lambda^*}(p_1)$. We have

$$\begin{aligned} 0 \leq c_{\Lambda^t}(p_1) - c_{\Lambda^t}(p_{\Lambda^t}) &= [c(p_1) + d_{\Lambda^t}(p_1)] - [c(p_{\Lambda^t}) + d_{\Lambda^t}(p_{\Lambda^t})] \\ &= c(p_1) + d_{\Lambda^*}(p_1) - [c(p_{\Lambda^t}) + d_{\Lambda^*}(p_{\Lambda^t})] + (\Lambda^t - \Lambda^*)(D(p_1) - D(p_{\Lambda^t}))^T \\ &= c_{\Lambda^*}(p_1) - c_{\Lambda^*}(p_{\Lambda^t}) + (\Lambda^t - \Lambda^*)(D(p_1) - D(p_{\Lambda^t}))^T. \end{aligned}$$

Since $|(\Lambda^t - \Lambda^*)(D(p_i) - D(p_j))^T| \leq \pi |(\Lambda^t - \Lambda^*)|_1 \leq \theta/2$,

$$0 \leq c_{\Lambda^t}(p_1) - c_{\Lambda^t}(p_{\Lambda^t}) \leq c_{\Lambda^*}(p_1) - c_{\Lambda^*}(p_{\Lambda^t}) + \theta/2.$$

Then $c_{\Lambda^*}(p_{\Lambda^t}) - c_{\Lambda^*}(p_1) \leq \theta/2$, which implies that $c_{\Lambda^*}(p_{\Lambda^t}) = c_{\Lambda^*}(p_1)$. Hence, $\forall j \in \{1, 2 \ldots, k\}, \exists p_c^*, p_d^* \in P_{\Lambda^*}, d_j(p_c^*) \geq r_j \geq d_j(p_d^*)$.

4 Numerical Simulation

We use *COPT*, *OPT*, and *POPT* to denote the cost of the optimal path to the CSP(k) problem, the optimal value, and the pseudo optimal value of the Lagrangian function, respectively. In our simulation, we first verify that the objectives at pseudo optimal points are very close to the optimal objectives. We use 3 types of graphs: Power-law out-degree graph (PLO) [17], Waxman's random graph (RAN) [20], and regular graph (REG) [19]. The number of weights is 4, 8 and 12, i.e., $k = 4$, 8, and 12. Link weights are random even integers uniformly distributed between 2 and 200. We use the following two metrics to measure the quality of path p in Table 1.

$$g(p) = c(p)/POPT \text{ and } f(p) = \max_{i=1,2\ldots,k} d_i(p)/r_i.$$

By Lemma 3, $g(p)$ is the upper bound of the ratio of the cost of p and $COPT$. The $f(p)$ indicates the degree of violation of p to the constraints on its delays.

In Fig. 1, the label LARAC-REG means the results obtained by running GEN-LARAC algorithm on regular graphs. Other labels can be interpreted similarly. We only report the results on regular graph and random graph for better visibility.

We conducted extensive experiments to compare our algorithm with the Hull approach [15], the subgradient method [1], and the general-purpose LP solver CPLEX. Because the four approaches share the same objective, i.e., maximizing the Lagrangian function, they always obtain similar results. Due to space limitation we only report the number of shortest path computation which dominate the running time of all the first three algorithms. Generally, GEN-LARAC algorithm and Hull approach are faster than the subgradient methods and CPLEX (See [24] for the comparison of Hull approach and CPLEX). But GEN-LARAC

Table 1. Quality of Pseudo-optimal paths: Error = $(OPT - POPT)/OPT$ #SP = Number of invocations of shortest path algorithm

Type	k	OPT	$POPT$	**Error**	$g(p)$	$f(p)$	**#SP**	**Time(s)**
REG	4	1116.8	1109.9	0.006	1.01	1.07	15.2	0.14
REG	8	1078.3	1069.0	0.032	1.00	1.09	20.7	0.21
REG	12	1066.2	1057.8	0.008	1.00	1.08	28.2	0.32
PLO	4	401.75	382.71	0.047	1.00	1.25	9.6	0.07
PLO	8	328.95	320.77	0.025	1.01	1.15	7.2	0.04
PLO	12	368.43	342.43	0.071	1.02	1.24	18.3	0.21
RAN	4	1543.6	1531.5	0.008	1.01	1.08	13.4	0.13
RAN	8	1473.3	1456.5	0.011	1.00	1.09	20.0	0.25
RAN	12	1438.6	1423.7	0.010	1.00	1.01	17.9	0.45

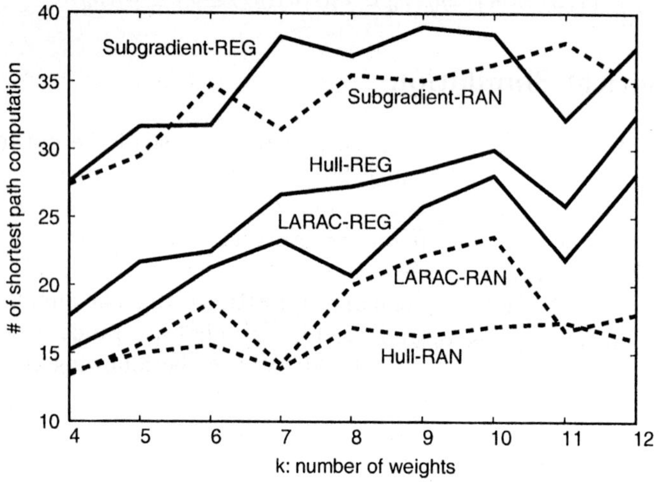

Fig. 1. The comparison of the number of shortest path computation among GEN-LARAC, Hull approach, and subgradient method. All algorithms terminate when have reached 99% of the OPT.

and Hull approach beat each other on different graphs. Figure 1 shows that on the regular graph, GEN-LARAC is the fastest. But for the random and Power-law out degree graphs, the Hull approach is the fastest. The probable reason is that the number of s-t paths is relatively small in these two types of graphs because the length (number of hops)of s-t paths is usually small even when the number of nodes is large. This will bias the results in favor of Hull approach which adds one s-t path into the linear system in each iteration [15]. So we choose the regular graph because we have a better control of the length of s-t paths.

5 Summary and Conclusion

In this paper we developed a new approach to the constrained shortest path problem involving multiple additive constraints. Our approach uses the LARAC algorithm as a building block and combines it with certain ideas from mathematical programming to design a method that progressively improves the value of the Lagrangian function until optimum is reached. The algorithm is analyzed and its convergence property has been established. Simulation results comparing our approach with two other approaches show that the new approach is quite competitive.

Since the LARAC algorithm is applicable for the general class of optimization problems (involving one additive delay constraint) studied in [2] our approach can also be extended for this class of problems to include multiple additive constraints, whenever an algorithm for the underlying optimization problem (such as Dijkstra's algorithm for the shortest path problem) is available.

References

1. J. Beasley and N. Christofides. An algorithm for the resource constrained shortest path problem. *Networks*, 19:379–394, 1989.
2. B. Blokh and G. Gutin. An approximation algorithm for combinatorial optimization problems with two parameters. *Australasian Journal of Combinatorics*, 14:157–164, 1996.
3. S. Boyd and L. Vandenberghe. *Convex Optimization.* Cambridge University Press, Cambridge, UK, 2003.
4. S. Chen and K. Nahrstedt. On finding multi-constrained path. In *ICC*, pages 874–879, 1998.
5. M. R. Garey and D. S. Johnson. *Computers and Intractability.* Freeman, San Francisco, CA, USA, 1979.
6. Ashish Goel, K. G. Ramakrishnan, Deepak Kataria, and Dimitris Logothetis. Efficient computation of delay-sensitive routes from one source to all destinations. In *INFOCOM*, pages 854–858, 2001.
7. G. Handler and I. Zang. A dual algorithm for the constrained shortest path problem. *Networks*, 10:293–310, 1980.
8. R. Hassin. Approximation schemes for the restricted shortest path problem. *Math. of Oper. Res.*, 17(1):36–42, 1992.
9. J. M. Jaffe. Algorithms for finding paths with multiple constraints. *Networks*, 14:95–116, 1984.
10. Alpár Jüttner, Balázs Szviatovszki, Ildikó Mécs, and Zsolt Rajkó. Lagrange relaxation based method for the QoS routing problem. In *INFOCOM*, pages 859–868, 2001.
11. Turgay Korkmaz and Marwan Krunz. Multi-constrained optimal path selection. In *INFOCOM*, pages 834–843, 2001.
12. Gang Liu and K. G. Ramakrishnan. A*prune: An algorithm for finding k shortest paths subject to multiple constraints. In *INFOCOM*, pages 743–749, 2001.
13. D. Lorenz and D. Raz. A simple efficient approximation scheme for the restricted shortest paths problem. *Oper. Res. Letters*, 28:213–219, 2001.

14. Nimrod Megiddo. Combinatorial optimization with rational objective functions. In *STOC '78: Proceedings of the tenth annual ACM symposium on Theory of computing*, pages 1–12, New York, NY, USA, 1978. ACM Press.
15. Kurt Mehlhorn and Mark Ziegelmann. Resource constrained shortest paths. In *ESA*, pages 326–337, 2000.
16. H. De Neve and P. Van Mieghem. Tamcra: A tunable accuracy multiple constraints routing algorithm. *Comput. Commun.*, 23:667–679, 2000.
17. C. R. Palmer and J. G. Steffan. Generating network topologies that obey power laws. In *IEEE GLOBECOM*, pages 434–438, 2000.
18. A. Schrijver. *Theory of linear and integer programming.* John Wiley, New York, 1986.
19. K. Thulasiraman and M. N. Swamy. *Graphs: Theory and algorithms.* Wiley Interscience, New York, 1992.
20. B. M. Waxman. Routing of multipoint connections. *IEEE Journal on Selected Areas in Communications*, 6(9):1617–1622, Dec. 1988.
21. Y. Xiao, K. Thulasiraman, and G. Xue. Equivalence, unification and generality of two approaches to the constrained shortest path problem with extension. In *Allerton Conference on Control, Communication and Computing, University of Illinois*, pages 905–914, 2003.
22. G. Xue, A. Sen, and R. Banka. Routing with many additive QoS constraints. In *ICC*, pages 223–227, 2003.
23. Xin Yuan. Heuristic algorithms for multiconstrained quality-of-service routing. *IEEE/ACM Trans. Netw.*, 10(2):244–256, 2002.
24. M. Ziegelmann. *Constrained shortest paths and related problems.* PhD thesis, Max-Planck-Institut fr Informatik, 2001.

Appendix: Proof to Theorem 1

Lemma 9. *If $\lambda < \lambda^*(\lambda > \lambda^*)$, then $d(p_\lambda) \geq T(d(p_\lambda) \leq T)$ for each c_λ-minimal path, where T is the path delay constraint [10].*

We next give the proof to Theorem 1.

Proof. We prove this theorem by showing an algorithm with $O((m + n\log n)^2)$ time complexity. Assume vertex 1 (n) is the source (target). Algorithm 2 computes shortest path using lexicographic order on a pair of link weights (l_{uv}, c_{uv}), $\forall(u,v) \in E$ based on parametric search [14], where $l_{uv} = c_{uv} + \lambda^* d_{uv}$ and λ^* is unknown. The algorithm is the same as Bellman-Ford algorithm except for Step 4 which needs special cares (We use Bellman-Ford algorithm here because it is easy to explain. Actually the Dijkstra's algorithm is used to get the claimed time complexity). In Algorithm 2, we need extra steps to evaluate the Boolean expression in the if statement in Step 4 since $\lambda^* \geq 0$ is unknown. If $x_v = \infty, y_v = \infty$, then the inequality holds. Assume x_v and y_v are finite (non-negative) values. We need an Oracle test to tell whether the value of $p + q\lambda^*$ is less than, equal to, or larger than 0, where $p = x_u + c_{uv} - x_v$ and $q = y_u + d_{uv} - y_v$. If $p\ q \geq 0$, it is trivial to find the sign of $p + q\lambda^*$. WLOG, assume $p\ q < 0$, i.e., $-p/q > 0$. The Oracle test is presented as Algorithm 3.

Algorithm 2. Parametric Search Based Algorithm for CSP(1) Problem

Step 1: $M_v = (x_v, y_v) \leftarrow (+\infty, +\infty)$ for $v = 2, 3 \ldots, n$ and $M_1 = (0, 0)$
Step 2: $i \leftarrow 1$
Step 3: $u \leftarrow 1$
Step 4: $\forall v, (u, v) \in E$, if $(x_v + \lambda^* y_v > x_u + \lambda^* y_u + c_{uv} + \lambda^* d_{uv})$ or
$(x_v + \lambda^* y_v = x_u + \lambda^* y_u + c_{uv} + \lambda^* d_{uv})$ and $(x_v > x_u + c_{uv})$)
$M_v \leftarrow (x_u + c_{uv}, y_u + d_{uv})$
Step 5: $u \leftarrow u + 1$ and if $u \leq n$, go to Step 4.
Step 6: $i \leftarrow i + 1$ and if $i < n$, go to Step 3.

The time complexity of the Oracle test is $O(m + n \log n)$. On the other hand, we can revise Algorithm 2 using Dijkstra's algorithm and the resulting algorithm has time complexity $O((m + n \log n)^2)$.

Next, we show how to compute the value of λ^* and $L(\lambda^*)$. Algorithm 2 computes a λ^*-minimal path p with minimal cost. Similarly, we can compute a λ^*-minimal path q with minimal delay. Then the value of λ^* is given by the following equation: $c(p) + \lambda^* d(p) = c(q) + \lambda^* d(q)$ and $L(\lambda^*) = c(p) + \lambda^*(d(p) - T)$. Notice that $d(q) < T < d(p)$.

Algorithm 3. Oracle Test with Unknown λ^*

Step 1: Let $\lambda = -p/q > 0$ and $\forall (u, v) \in E$, define link length $l_{uv} = c_{uv} + \lambda d_{uv}$
Step 2: Compute two shortest paths p_c and p_d using the lexicographic order on (l_{uv}, c_{uv}) and (l_{uv}, d_{uv}), respectively
Step 3: Obviously, $d(p_c) \geq d(p_d)$. Consider 4 cases:

1. $d(p_c) > T$ and $d(p_d) > T$: By Lemma 6 and Lemma 9, $\lambda < \lambda^*$ and thus $p + q\lambda^* < 0$ if $q < 0$ and $p + q\lambda^* > 0$ otherwise
2. $d(p_c) < T$ and $d(p_d) < T$: By Lemma 6 and Lemma 9, $\lambda > \lambda^*$ and thus $p + q\lambda^* > 0$ if $q < 0$ and $p + q\lambda^* < 0$ otherwise
3. $d(p_c) > T$ and $d(p_d) < T$: By Lemma 6, $\lambda = \lambda^*$ and thus $p + q\lambda^* = 0$
4. $d(p_c) = T$ or $d(p_d) = T$: By Lemma 1, this is impossible

Simultaneous Matchings

Khaled Elbassioni[1], Irit Katriel[2,*], Martin Kutz[1], and Meena Mahajan[3,**]

[1] Max-Plank-Institut für Informatik, Saarbrücken, Germany
{elbassio, mkutz}@mpi-sb.mpg.de
[2] BRICS ***, University of Aarhus, Åbogade 34, Århus, Denmark
irit@daimi.au.dk
[3] The Institute of Mathematical Sciences, Chennai, India
meena@imsc.res.in

Abstract. Given a bipartite graph $G = (X \dot{\cup} D, E \subseteq X \times D)$, an X-*perfect matching* is a matching in G that covers every node in X. In this paper we study the following generalisation of the X-perfect matching problem, which has applications in constraint programming: Given a bipartite graph as above and a collection $\mathcal{F} \subseteq 2^X$ of k subsets of X, find a subset $M \subseteq E$ of the edges such that for each $C \in \mathcal{F}$, the edge set $M \cap (C \times D)$ is a C-perfect matching in G (or report that no such set exists). We show that the decision problem is NP-complete and that the corresponding optimisation problem is in APX when $k = O(1)$ and even APX-complete already for $k = 2$. On the positive side, we show that a $2/(k+1)$-approximation can be found in $2^k \text{poly}(k, |X \cup D|)$ time.

1 Introduction

Matching is one of the most fundamental problems in algorithmic graph theory. The all-important notion of characterising feasibility / efficiency as polynomial-time computation came about in the context of the first efficient matching algorithm due to Edmonds [4]. Since then, an immense amount of research effort has been directed at understanding the various nuances and variants of this problem and at attacking special cases. For an overview of developments in matching theory and some recent algorithmic progress, see for instance [7,11].

In this paper we consider a generalisation of the bipartite matching problem. Suppose we are given a bipartite graph $G = (V, E)$ where the vertex set partition is $V = X \dot{\cup} D$ (so $E \subseteq X \times D$) and a collection $\mathcal{F} \subseteq 2^X$ of k *constraint sets.* A solution to the problem is a subset $M \subseteq E$ of the edges such that M is *simultaneously* a perfect matching for each constraint set in $\mathcal{F}$. More precisely, for each $C \in \mathcal{F}$, the edge set $M \cap (C \times D)$ has to be a C-perfect matching, i.e., a subgraph of G in which every vertex has degree at most 1 and every vertex of

* Supported by the Danish Research Agency (grant # 272-05-0081).

** This work was done while the fourth author was visiting the Max-Plank-Institut für Informatik, Saarbrücken, Germany.

*** Basic Research in Computer Science, funded by the Danish National Research Foundation.

X. Deng and D. Du (Eds.): ISAAC 2005, LNCS 3827, pp. 106–115, 2005.

C has degree exactly 1. Also, analogous to maximum-cardinality matchings, we may relax the perfect matching condition and ask for a largest set M such that for each $C \in \mathcal{F}$, the edge set $M \cap (C \times D)$ is a matching in G.

Why consider this generalisation of matching? Apart from purely theoretical considerations suggesting that any variant of matching is worth exploring, there is a concrete important application in constraint programming where precisely this question arises. A *constraint program* consists of a set X of variables and a set D of values. Each variable $x \in X$ has a *domain* $D(x) \subseteq D$, i.e., a set of values it can take. In addition, there is a set of *constraints* that specify which combinations of assignments of values to variables are permitted.

An extensively studied constraint is the *AllDifferent* constraint (*AllDiff*) which specifies for a given set of variables that the values assigned to them must be pairwise distinct (see, e.g., [9,10,12,14,15]). An *AllDiff* constraint can be viewed as an X-perfect matching problem in the bipartite graph that has the set X of variables on one side, the set D of values on the other side, and an edge between each variable and each value in its domain. Typically, a constraint program contains several *AllDiff* constraints, defined over possibly overlapping variable sets. This setting corresponds to the generalisation we propose.

Formally, we consider the following problems:

SIM-W-MATCH (SIMULTANEOUS WEIGHTED MATCHINGS):

Input: a bipartite graph $G = (V, E)$ with $V = X \dot{\cup} D$ and $E \subseteq X \times D$, a weight or profit $w(e)$ associated with each edge in E, and a collection of constraint sets $\mathcal{F} \subseteq 2^X$.

Feasible Solution: a set $M \subseteq E$ satisfying $\forall C \in \mathcal{F} : M \cap (C \times D)$ is a matching. The weight of this solution is $\sum_{e \in M} w(e)$.

Output: (The weight of) a maximum-weight feasible solution.

SIM-W-PERF-MATCH (SIMULTANEOUS WEIGHTED PERFECT MATCHINGS):

Input: as above

Feasible Solution: as above but only saturating (perfect) matchings allowed, i.e., a set $M \subseteq E$ satisfying $\forall C \in \mathcal{F}$: "$M \cap (C \times D)$ is a C-perfect matching."

Output: (The weight of) a maximum-weight feasible solution or a flag indicating the absence of any feasible solution.

These are the optimisation / search versions[1]; in the decision versions, an additional weight W is given as input and the answer is 'yes' if there is a feasible solution of weight at least W. When all edge weights are 1, the corresponding problems are denoted by SIM-MATCH and SIM-P-MATCH respectively. In this case, the decision version of SIM-P-MATCH does not need any additional parameter W.

We use the following notation: $n = |X|$, $d = |D|$, $m = |E|$, $k = |\mathcal{F}|$, $t = \max\{|C| : C \in \mathcal{F}\}$. We also assume, without loss of generality, that $X = \cup_{C \in \mathcal{F}} C$.

[1] Note that due to the weights, an optimal solution to the former might not saturate all sets even if such a perfect assignment exists. Thus SIM-W-PERF-MATCH is not a special case of SIM-W-MATCH.

At first sight these problems do not appear much more difficult than bipartite matching, at least when the number of constraint sets is a constant. It seems quite plausible that a modification of the Hungarian method [8] should solve this problem. However, we show in Section 2 that this is not the case; even when $k = 2$, SIM-P-MATCH is NP-hard. We also show that it remains NP-hard even if the graph is complete bipartite (i.e., $E = X \times D$) and d and t are constants; of course, in this case, k must be unbounded. Furthermore, SIM-MATCH is APX-hard, even for $k = 2$. On the positive side, SIM-W-MATCH is in APX for every constant k. These results are shown in Section 3. Finally, in Section ?? we examine the SIM-P-MATCH polytope and observe that it can have vertices that are not even half-integral.

2 NP-Completeness of SIM-P-MATCH for $k \geq 2$

The main result of this section is the following.

Theorem 1. *Determining feasibility of an instance of* SIM-P-MATCH *with k constraint sets is NP-complete for every single parameter $k \geq 2$.*

Proof. Membership in NP is straightforward. We establish NP-hardness of SIM-P-MATCH for $k = 2$ (it then follows trivially for each $k > 2$). The proof is by reduction from SET-PACKING.

SET-PACKING:
- Instance: A universe U; a collection $\mathcal{C} = \{S_1, S_2, \ldots, S_p\}$ of subsets of U.
- Decision problem: Given an integer $\ell \leq p$, is there a collection $\mathcal{C}' \subseteq \mathcal{C}$ of at least ℓ pairwise disjoint sets?
- Optimisation problem: Find a collection $\mathcal{C}' \subseteq \mathcal{C}$ of pairwise disjoint subsets such that $|\mathcal{C}'|$, the number of chosen subsets, is maximised.
- k-SET-PACKING: The restriction where every set in $\mathcal{C}$ contains at most k elements.
- SET-PACKING(r): The restriction where every vertex of U appears in at most r sets from $\mathcal{C}$.

It is known that SET-PACKING is NP-hard, and so is 3-SET-PACKING(2), the special case where the size of each set is bounded by 3 and each element occurs in at most 2 sets. See, for instance, [2].

We present the reduction from SET-PACKING to SIM-P-MATCH (with $k = 2$) in detail here because it will later serve for an APX result, too. View SIM-P-MATCH as a question of assigning values to variables (as in the constraint programming application described in Section 1). Each element a from the universe U is embodied by a single value v_a and there are ℓ variables $x_1, \ldots, x_\ell$, which belong to both constraint sets X_1 and X_2, that are fighting for these values. Between the x_i's and the v_j's we place gadgets that encode the sets in $\mathcal{C}$.

Consider the trivial situation with only singleton sets in $\mathcal{C}$. There we could simply connect each x_i to each value that occurs as such a singleton. Then a

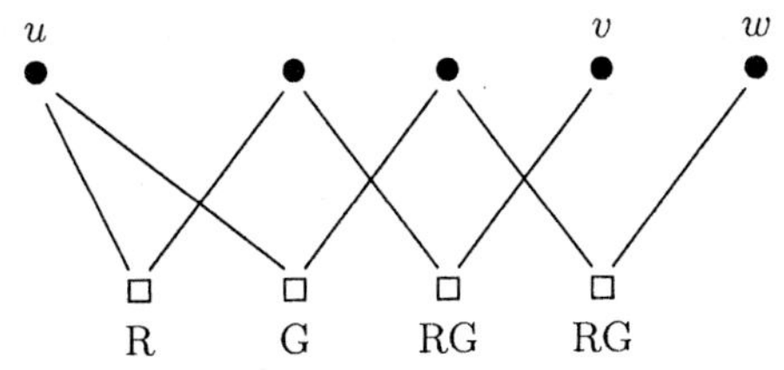

Fig. 1. The basic gadget for the reduction

"packing" of ℓ singleton sets in U would obviously give an assignment of ℓ values to the x_i's in this complete bipartite graph, and vice versa.

The difficult part is to build gadgets between the x_i and U such that a single variable occupies more than one value from U. Therefore consider the configuration in Figure 1. We have five values on the upper side and four variables on the lower, marked with letters 'R' (red) and 'G' (green) to indicate that they belong to the constraint sets X_1 and X_2, respectively (red-green colour indicating membership in both sets). If the leftmost value is assigned to some red-green variable outside the figure then the two left variables will be forced to claim the two values to their right and in turn, the remaining two variables will have to pick the values marked v and w. In other words, if a red-green variable claims the *input* value u on the left, it effectively occupies the two *output* values v and w, too. Conversely, if the value u is not required elsewhere, the four variables can all make their left-slanted connections and leave v and w untouched.

We can concatenate several such 4-variable gadgets to obtain a larger amplification. If we merge the output value v of one gadget with the input u' of another one, as shown in Figure 2, we get the effect that occupying only the input u from outside this configuration, forces the gadget variables to claim the three output values w, v', and w' that could otherwise stay untouched. (Indeed, we only get three such values and not four because the connecting value v counts no longer as an output.)

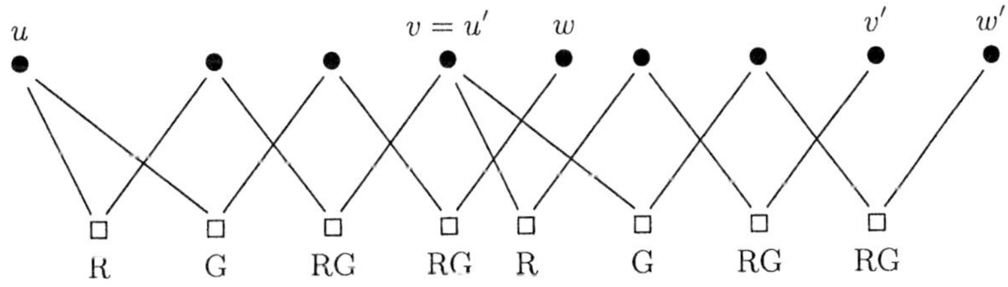

Fig. 2. Merging two 1:2-gadgets into one 1:3-gadget

For a q-element set $S = \{v_1, \ldots, v_q\} \in \mathcal{C}$, we concatenate $q-1$ gadgets and make $v_1, \ldots, v_q \in U$ their resulting output values. Then we connect each red-green variable x_i to the input value of each such set gadget. The resulting configuration has obviously the desired behaviour. We can assign values to all variables without violating the red and green constraint if and only if we can pack ℓ sets from C into U. □

Complete bipartite graphs with $d = 3$, $t = 2$. Theorem 1 states that it is NP-hard to solve instances of SIM-P-MATCH with two constraint sets X_1 and X_2. However, if the graph is a complete bipartite graph, it is straightforward to determine whether a solution exists and to find it if so: First match the vertices of X_1 with any set of distinct vertices in D. Then it remains to match the vertices of $X_2 \setminus X_1$ with vertices that were not matched with vertices from the intersection $X_1 \cap X_2$. Since the graph is complete bipartite, the existence of a solution is determined solely by the sizes of X_1, X_2, $X_1 \cap X_2$ and D.

It is therefore natural to ask whether it is always possible to solve SIM-P-MATCH on a complete bipartite graph. With a little bit of thought and inspection, one can come up with similar feasibility conditions for $k = 3, 4$. What about arbitrary k? It turns out that the problem is NP-hard if the number of constraint sets is not bounded, even if each constraint set has cardinality 2 and $d = |D| = 3$. The proof is by a reduction from 3-VERTEX-COLORING, which is known to be NP-complete (see for instance[5]).

Instance: An undirected graph $G = (V, E)$.
Decision problem: Is there a way of colouring the vertices of G, using at most 3 distinct colours, such that no two adjacent vertices get the same colour?

Proposition 1. SIM-P-MATCH *is NP-hard even when $d = 3$, $t = 2$, and the underlying graph is complete bipartite.*

Proof. Let $G = (V, A)$ be an instance of 3-VERTEX-COLORING. We construct a corresponding instance of SIM-P-MATCH as follows: let $X = V$, $D = \{1, 2, 3\}$, $E = X \times D$, and $\mathcal{F} = A$. It is straightforward to see that any feasible solution to this instance of SIM-P-MATCH is a 3-colouring of V with no monochromatic edge, and vice versa. □

3 APX-Completeness

We now examine the approximability of SIM-W-MATCH.

3.1 Membership in APX for Constant k

APX is the class of optimisation problems which have polynomial-time constant-factor approximation algorithms. That is, a maximisation problem Π is in APX if there is a constant $0 < \alpha \leq 1$ and a polynomial-time algorithm A such that for every instance x of Π we have $\alpha \cdot \text{Opt}(\Pi, x) \leq A(x) \leq \text{Opt}(\Pi, x)$, where $A(x)$ is the output of the algorithm on input x, and $\text{Opt}(\Pi, x)$ is the value of the optimum solution. Clearly, the larger the *approximation factor* α, the better the quality of approximation. For more details, see any text on approximation algorithms, such as [6,16].

Consider the following naive polynomial-time approximation algorithm for SIM-W-MATCH: Find a maximum profit matching for each constraint set $C \in \mathcal{F}$ independently and return the most profitable matching found. Clearly, an optimal solution is at most k times larger, which gives an approximation ratio of $1/k$. Hence we have:

Proposition 2. *An instance of* SIM-W-MATCH *with k constraint sets can be approximated in polynomial time within a factor of* $1/k$.

Corollary 1. *For any constant k,* SIM-W-MATCH *with k constraints is in APX.*

3.2 Improving the Approximation Factor

We can slightly improve the $1/k$ factor by considering more than one set $X_i \in \mathcal{F}$ at a time. Fix a maximum-weight feasible solution M with profit Opt. For any set $S \subseteq X$, let $f(S)$ denote the profit of the maximum weight simultaneous matching in the graph induced by $S \cup D$, and let $g(S)$ be the profit of the edges in $M \cap (S \times D)$. Clearly, for each S we have $f(S) \geq g(S)$ and further, each feasible solution on $S \cup D$ is also a solution on G, so that $\mathrm{Opt} \geq f(S)$.

First consider the case with two constraint sets X_1, X_2. We can compute $f(X_1)$ and $f(X_2)$ independently, each as an ordinary maximum-matching problem, and we can also evaluate the symmetric difference $f(X_1 \oplus X_2) := f(X_1 \setminus X_2) + f(X_2 \setminus X_1)$ as the union of two independent maximum-matching problems. Altogether we get

$$f(X_1)+f(X_2)+f(X_1 \oplus X_2) \geq g(X_1)+g(X_2)+g(X_1 \oplus X_2) = 2g(X_1 \cup X_2) = 2\,\mathrm{Opt}.$$

By averaging, the largest of the three terms on the left is at least $2/3 \cdot \mathrm{Opt}$.

For $k > 2$, we generalise the notion of symmetric differences appropriately. Define $\hat{X}_i := X_i \setminus \bigcup_{h \neq i} X_h$ and consider the reduced pairs $S_{ij} = \hat{X}_i \cup \hat{Y}_i$. As before, we can determine $f(S_{ij})$ exactly for any index pair $i \neq j$ by independent computation of maximum matchings on $\hat{X}_i$ and $\hat{X}_j$. A careful choice of coefficients yields

$$\sum_i f(X_i) + \frac{1}{k-1} \sum_{i<j} f(S_{ij}) > \sum_i g(X_i) + \frac{1}{k-1} \sum_{i<j} g(S_{ij}) \geq 2g(X) = 2\,\mathrm{Opt}$$

and again by averaging, at least one of the $f(X_i)$ or $f(S_{ij})$ is no less than $4/(3k) \cdot \mathrm{Opt}$.

Proposition 3. *For any constant k,* SIM-W-MATCH *with k constraint sets can be approximated in polynomial time within a factor of* $4/(3k)$.

We now pursue this approach to its logical conclusion. The main idea is to identify subsets S of X for which (a) $f(S)$ is upper bounded by Opt and (b) $f(S)$ can be efficiently computed. Note that for any $S \subseteq X$, (a) comes for free, since in computing $f(S)$ we consider all the original constraint sets, restricted to S. Then, for every choice of non-negative weights ξ_S, we have

$$\sum_S \xi_S f(S) \geq \sum_S \xi_S g(S) \geq F_\xi \cdot \mathrm{Opt}$$

where $F_\xi = \min_{x \in X} \sum_{S: x \in S} \xi_S$; the last inequality holds because each edge (x, d) of the optimal solution M is counted with a total weight of $\sum_{S: x \in S} \xi_S$. By averaging, the largest term $f(S)$ is at least $F_\xi \cdot \mathrm{Opt}/T_\xi$, where $T_\xi = \sum \xi_S$ is the total weight. (In proving Proposition 2, the chosen S's were precisely the X_i's, with weight 1 each, so $F_\xi = 1$ and $T_\xi = k$. In proving Proposition 3, the chosen S's were the X_i's and the S_{ij}, and $F_\xi = 2$ and $T_\xi = 3k/2$.)

Clearly, we lose nothing by considering only *maximal* subsets S for which $f(S)$ is efficiently computable. (The sets S_{ij} above were not maximal in this sense.) Below, we consider only maximal subsets. Our approach can be sketched as follows.

Let $\mathcal{R}$ denote the collection of maximal sets $S \subseteq X$ for which $f(S)$ can be efficiently computed (and is upper bounded by Opt). Also, let $\mathcal{C}$ denote the family of maximal subsets of X entirely contained in either of each X_i or $X \setminus X_i$. That is, every $C \in \mathcal{C}$ is identified with a vector $(v_1, v_2, \ldots, v_k) \in \{0,1\}^k \setminus 0^k$, such that $C = \bigcap_{i:v_i=1} X_i \setminus \bigcup_{i:v_i=0} X_i$. Clearly, $\beta = |\mathcal{C}| = 2^k - 1$. We characterise the sets in $\mathcal{R}$ and observe that $\alpha = |\mathcal{R}|$ is $O((2k)^k)$. Also, we note that for each $R \in \mathcal{R}$ and each $C \in \mathcal{C}$, C is either completely contained in or completely outside R. We now wish to compute, for each $R \in \mathcal{R}$, a weight ξ_R such that the corresponding ratio F_ξ/T_ξ is maximised. Define an $\alpha \times \beta$ 0-1 matrix $A = (a_{R,C})_{R,C}$ where $a_{R,C} = 1$ iff $C \subseteq R$. Consider the following pair of primal-dual linear programs:

$$\begin{array}{llll} \lambda_P^*(k) = \max \lambda & & \lambda_D^*(k) = \min \lambda & \\ \quad s.t. & A^T\xi \geq \lambda \mathbf{e}_\beta & \quad s.t. & Az \leq \lambda \mathbf{e}_\alpha \\ & \mathbf{e}_\alpha^T \xi = 1 & & \mathbf{e}_\beta^T z = 1 \\ & \xi \geq 0 & & z \geq 0 \end{array}$$

over $\xi \in \mathbb{R}^\alpha$ and $z \in \mathbb{R}^\beta$, where $\mathbf{e}_\alpha$ and $\mathbf{e}_\beta$ are the vectors of all ones of dimensions α and β respectively. A feasible solution ξ to the primal assigns weights (normalised so that $T_\xi = 1$) to each $R \in \mathcal{R}$ achieving an approximation factor given by the value of the corresponding objective function. We show that both the primal and the dual have feasible solutions with objective value $2/(k+1)$.

Theorem 2. *For any integer $k \geq 1$, we have $\lambda_P^*(k) = \lambda_D^*(k) = \frac{2}{k+1}$. There is a primal optimal solution $\xi \in \mathbb{R}^\alpha$ whose support has size $|\{R \in [\alpha] : \xi_R > 0\}| = k + \binom{k}{2} 2^{k-2}$.*

Due to space limitations, we skip the proof of this theorem; it can be found in the full version of this paper. Theorem 2 establishes that an approximation factor of $2/(k+1)$ is possible. To compute the running time of the approximation, note that each $f(R)$ can be computed in polynomial time, and we need to compute $f(R)$ only for those $R \in \mathcal{R}$ having non-zero ξ_R. Hence by Theorem 2, the running time is polynomial provided $k + \binom{k}{2} 2^{k-2}$ is polynomial in the input size.

Theorem 3. SIM-W-MATCH *can be approximated within a factor of* $2/(k+1)$ *by an algorithm that runs in* $2^k\text{poly}(n, m, d, k)$ *time.*

Thus, for all instances of SIM-W-MATCH with $k = O(\log N)$, where $N = \max\{n, m, d\}$, a $\frac{2}{k+1}$ approximation can be found in polynomial time. Furthermore, Theorem 2 tells us also that this factor is the best-possible via the above approach.

3.3 APX-Hardness for $k \geq 2$

Recall that completeness within APX is defined through L reductions, see for instance [16]. So an approximation scheme for an APX-complete problem translates into such a scheme for any problem in APX.

Theorem 4. *For each* $k \geq 2$, SIM-MATCH *with* k *constraint sets is APX-hard.*

Proof. We only have to modify our reduction from the proof of Theorem 1 slightly to account for the new setting. Instead of testing for a given number ℓ of variables x_i, we let $\ell = p$, the cardinality of $\mathcal{C}$. So a perfect solution would have to pack all sets into U. In order to get an approximation-preserving reduction, we have to make sure that a certain fraction of the sets can always be packed. This is achieved by restricting to 3-SET-PACKING(2), which is already APX-hard [2].

In this situation, the overall number of variables is at most $9p$ since there are p choice variables, and each gadget contributes at most 8 variables. Let M denote the number of gadget variables; then $M \leq 8p$. Since each element of U appears in at most 2 sets and since each set is of size at most 3, we can always find at least $p/4$ disjoint sets. (Just construct any maximal collection of disjoint sets. Including any one set in the collection rules out inclusion of at most 3 other sets.) Thus, if the optimal set packing has k_{opt} sets then $k_{\text{opt}} \geq p/4$.

Let s_{opt} denote the value of an optimal solution to the SIM-MATCH instance constructed. Note that s_{opt} counts variables, while k_{opt} counts sets. We claim that $s_{\text{opt}} = k_{\text{opt}} + M$. The relation "$\geq$" follows simply from assigning the k_{opt} input values of the gadgets that correspond to an optimal packing to some x_i. Then all gadget variables can be assigned values without conflict. To see "$\leq$," notice that any assignment can be transformed into one in which all gadget variables receive values, without decreasing the total number of satisfied variables. It is then easy to see that we can find a set packing with as many sets as we have x_i assigned with values. This shows the claim.

Suppose now that SIM-MATCH can be approximated within a factor of α. That is, we can find in polynomial time a feasible assignment on s variables, where s is at least αs_{opt}. Then $s = k' + M \geq \alpha(k_{opt} + M)$, so $k' \geq \alpha k_{opt} - M(1-\alpha) \geq \alpha k_{opt} - 8p(1-\alpha) \geq \alpha k_{opt} - 8(4k_{opt})(1-\alpha) = k_{opt}(33\alpha - 32)$. Thus, an α-approximation for SIM-MATCH gives a $(33\alpha - 32)$-approximation for 3-SET-PACKING(2). This gives the desired L reduction. □

Plugging the current-best known inapproximability bound of 99/100 for 3-SET-PACKING(2) from [3] into the above reduction, we learn that SIM-MATCH cannot be approximated to within a factor of $1 - 1/3300$ unless P = NP.

4 The SIMULTANEOUS MATCHINGS Polytope

Consider again instances of SIM-P-MATCH, on complete bipartite graphs, with $k = 2$. As remarked in section 2, checking feasibility in such a setting is trivial. In [1], a somewhat different aspect of this setting is considered. Assume that the set D is labelled by the set of integers $0, 1, \ldots, d-1$, and $X = \{x_1, x_2, \ldots, x_n\}$. Then every feasible solution becomes an integer vector in the n-dimensional space $\{0, 1, \ldots, d-1\}^n$. Now what is the structure of the polytope defined by the convex hull of integer vectors corresponding to feasible solutions? The authors of [1] establish the dimension of this polytope and also obtain classes of facet-defining inequalities.

We consider the variant where dimensions / variables are associated with each edge of the graph, rather than each vertex in X. Viewed as a purely graph-theoretic decision / optimisation problem, this makes eminent sense as it directly generalises the well-studied matching polytope (see for instance [11]): we wish to assign 0,1 values to each edge variable (a value of 1 for an edge corresponds to putting this edge into the solution M, 0 corresponds to omitting this edge) such that all vertices of X (or as many as possible) have an incident edge in M, and M is a feasible solution. This is easy to write as an integer program: Choose x_e for each edge e so as to

$$\text{maximise} \quad \sum_{e \in E} w_e x_e \quad \text{(for SIM-W-MATCH)}$$

$$S.T. \quad \forall x \in X: \sum_{e=(x,z): z \in D} x_e \leq 1, \qquad \forall z \in D: \sum_{e=(x,z): x \in X_1} x_e \leq 1,$$

$$\forall z \in D: \sum_{e=(x,z): x \in X_2} x_e \leq 1, \qquad \forall e: x_e \in \{0, 1\}$$

The corresponding linear program replaces the last condition above by $\forall e : x_e \in [0, 1]$. Let P_I denote the convex hull of integer solutions to the integer program, and let P_L denote the convex hull of feasible solutions to the linear program. P_I and P_L are polytopes in $\mathbb{R}^n$, with $P_I \subseteq P_L$.

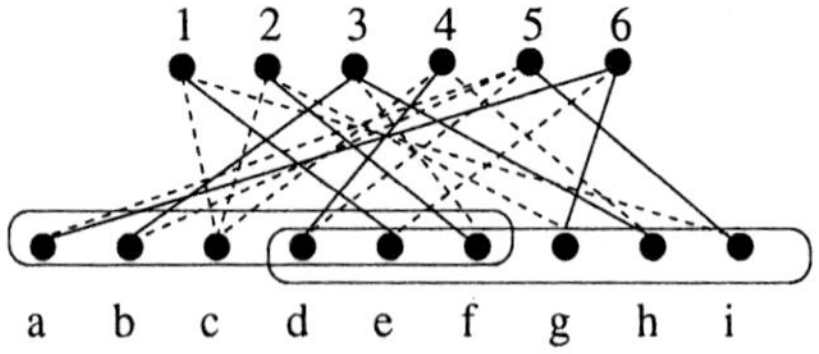

Fig. 3. A vertex of P_L that is not half-integral

The special case of the above where there is just one constraint set (either X_1 or X_2 is empty) is the bipartite matching polytope. For this polytope, it is known that every vertex is integral; i.e. $P_I = P_L$. For non-bipartite graphs, this polytope is not necessarily integral, but it is known that all vertices there are half-integral (i.e. at any extremal point of the polytope, all edge weights are from the set $\{0, 1/2, 1\}$). Unfortunately, these nice properties break down even for two constraint sets. We illustrate this with an example in Figure 3. The underlying graph is the complete bipartite graph. Assign weights of 1/3 to the edges shown by dotted lines, 2/3 to those shown with solid lines, and 0 to all other edges. This gives a feasible solution and hence a point in P_L, and it can be verified[2] that it is in fact a vertex of P_L and is outside P_I.

References

1. G. Appa, D. Magos, and I. Mourtos. On the system of two all_different predicates. *Information Processing Letters*, 94/3:99–105, 2005.
2. P. Berman and T. Fujito. Approximating independent sets in degree 3 graphs. In *WADS 1995*, volume 955 of *LNCS*, pages 449–460, 1995.
3. M. Chlebík and J. Chlebíková. Inapproximability results for bounded variants of optimization problems. In *FCT 2003*, volume 2751 of *LNCS*, pages 27–38, 2003.
4. J. Edmonds. Path, trees and flowers. *Canadian Journal of Mathematics*, pages 233–240, 1965.
5. M. Garey and D. S. Johnson. *Computers and Intractability, A Guide to the Theory of NP-Completeness*. Freeman, 1979.
6. D. S. Hochbaum, editor. *Approximation Algorithms for NP-Hard Problems*. Brooks/Cole Pub. Co., 1996.
7. M. Karpinski and W. Rytter. *Fast parallel algorithms for graph matching problems*. 1998. Oxford Lecture Series in Mathematics and its Applications 9.
8. H.W. Kuhn. The Hungarian Method for the assignment problem. *Naval Research Logistic Quarterly*, 2:83–97, 1955.
9. M. Leconte. A bounds-based reduction scheme for constraints of difference. In *Proceedings of the Constraint-96 Workshop*, pages 19–28, 1996.
10. A. Lopez-Ortiz, C.-G. Quimper, J. Tromp, and P. van Beek. A fast and simple algorithm for bounds consistency of the alldifferent constraint. In *IJCAI*, 2003.
11. L. Lovasz and M. Plummer. *Matching Theory*. North-Holland, 1986. Annals of Discrete Mathematics 29.
12. K. Mehlhorn and S. Thiel. Faster Algorithms for Bound-Consistency of the Sortedness and the AllDifferent Constraint. In *CP*, 2000.
13. PORTA. http://www.zib.de/optimization/software/porta/.
14. J.-F. Puget. A fast algorithm for the bound consistency of alldiff constraints. In *AAAI*, 1998.
15. W.J. van Hoeve. The AllDifferent Constraint: A Survey. In *Proceedings of the Sixth Annual Workshop of the ERCIM Working Group on Constraints*, 2001.
16. V. Vazirani. *Approximation Algorithms*. Springer, 2001.

[2] The software PORTA [13] was used to find this vertex.

An Optimization Problem Related to VoD Broadcasting

Tsunehiko Kameda[1], Yi Sun[1], and Luis Goddyn[2]

[1] School of Computing Science
[2] Department of Mathematics,
Simon Fraser University, Burnaby, B.C. Canada V5A 1S6

Abstract. Consider a tree T of depth 2 whose root has s child nodes and the kth child node from left has n_k child leaves. Considered as a round-robin tree, T represents a schedule in which each page assigned to a leaf under node k ($1 \le k \le s$) appears with period sn_k. By varying s, we want to maximize the total number $n = \sum_{k=1}^{s} n_k$ of pages assigned to the leaves with the following constraints: for $1 \le k \le s$, $n_k = \lfloor (m + \sum_{j=1}^{k-1} n_j)/s \rfloor$, where m is a given integer parameter. This problem arises in the optimization of a video-on-demand scheme, called *Fixed-Delay Pagoda Broadcasting*.

Due to the floor functions involved, the only known algorithm for finding the optimal s is essentially exhaustive, testing $m/2$ different potential optimal values of size $O(m)$ for s. Since computing n for a given value of s incurs time $O(s)$, the time complexity of finding the optimal s is $O(m^2)$. This paper analyzes this combinatorial optimization problem in detail and limits the search space for the optimal s down to $\kappa\sqrt{m}$ different values of size $O(\sqrt{m})$ each, where $\kappa \approx 0.9$, thus improving the time complexity down to $O(m)$.

1 Introduction

Recently, Bar-Noy et al. have formulated a combinatorial problem called the *windows scheduling problem* [1]. This problem is defined by positive integers c and $w_1, w_2, \ldots, w_n$, where c is the number of slotted channels and, for $i = 1, 2, \ldots, n$, a *window* of size w_i is associated with page i. A valid *schedule* assigns page i to slots such that it appears at least once in every window of w_i slots (not necessarily in the same channel).

Recently, there has been much interest in the broadcast-based delivery of popular videos, in order to address the scalability issue in video-on-demand (VoD). A VoD broadcasting scheme, called *Fixed-Delay Pagoda Broadcasting* (FDPB), was proposed by Pâris [5].[1]

A *channel* consists of a sequence of time slots of duration d (sec) each. The viewer initially downloads pages for md (sec) before s/he starts viewing the

[1] It has been implemented in a prototype video-on-demand (VOD) system [6]. Hollermann and Holzscherer [3] had also conceived a scheme similar to FDPB. A recent survey on VoD can be found in [4].

X. Deng and D. Du (Eds.): ISAAC 2005, LNCS 3827, pp. 116–125, 2005.

video. FDPB has the following constraints: (a) there is just one channel[2] that is divided into s subchannels, each subchannel consisting of every s^{th} slot of the channel; (b) $w_i = m+i-1$; (c) each page must be transmitted in one subchannel with a fixed period, and (d) all pages appearing in a given subchannel must be consecutive and have the same period. Our objective is to maximize the number of pages n that can be scheduled, by choosing the optimal value for s, which depends on m.

The author of [5] originally conjectured that $s = \sqrt{m}$ would be the optimal s value, but later found that it was not always the case among the examples he tested. Recently, Bar-Noy et al. [2] considered this problem and proposed a method whereby they could find the optimal value by testing $m/2$ different values of s. We analyze this dependency in detail in this paper and show that the optimal value of s is guaranteed to be one of roughly $0.9\sqrt{m}$ possible values. We thus can avoid time-consuming exhaustive search for the optimal s. The limited range for s reduced the running time of a computer program to compute the optimal value for s rather dramatically from a few hours down to a few seconds when $m = 10000$.[3]

The rest of the paper is organized as follows. Sect. 2 describes a useful tool, called the *round-robin tree*, introduced in [2]. In Sect. 3, we derive a formula for the optimal number s_{opt} of subchannels that maximizes the number of pages that can be scheduled under the constraints adopted by FDPB. We then find in Sect. 4 a small range for s in terms of m that needs to be searched, in order to find s_{opt}. Finally, in Sect. 5, we mention implications of our results, and also propose a new method of subchanneling such that a subchannel is divided into subsubchannels.

2 Preliminaries

2.1 Round-Robin Tree

The *round-robin (RR) tree* is a useful tool to represent a cyclic schedule [2]. Fig. 1 shows an example of a 2-level round-robin tree whose leaves are labeled by pages. A round-robin tree represents a schedule as follows:

1. Initially the root gets a "turn".
2. When a non-leaf node gets a turn, it passes the turn to its "next" child node. The leftmost child node gets a turn first and the order "next" means the next sibling to the right, wrapping around back to the leftmost child node.
3. When a leaf gets a turn, its associated page is scheduled and the turn goes back to the root. □

Example 1. Applying the above rules, it is seen that Fig. 1 represents the schedule, $\langle P_1, P_4, P_8, P_2, P_5, P_9, \ldots \rangle$. Note that pages P_1, P_2 and P_3 have period 9

[2] For the purpose of our discussion, this can be assumed without loss of generality. The general case is considered in the discussions section of this paper.

[3] Measured for a Java program running on a Pentium II CPU.

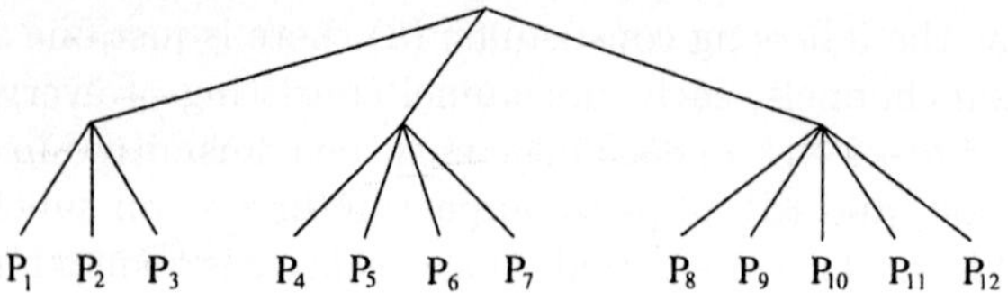

Fig. 1. A round-robin tree representation of FDPB with $s = 3$ subchannels

($= 3s$), pages $P_4, \ldots, P_7$ have period 12 ($= 4s$), and pages $P_8, \ldots, P_{12}$ have period 15 ($= 5s$). □

Lemma 1. [1] *Suppose the root of a 2-level RR tree T has s subtrees and for $k = 1, 2, \ldots, s$, the kth subtree from left has n_k child leaves. Then T represents a schedule in which each page assigned to a leaf in the kth subtree appears with period $n_k \times s$, and the minimum cycle of the schedule is given by $s \times \mathrm{LCM}(n_1, n_2, \ldots, n_k)$, where LCM stands for the least common multiple of the arguments.* □

2.2 Model

The kth subtree of an RR tree T gets one "turn" out of every s "turns". We thus consider that the given channel is divided into s *subchannels* such that subchannel k consists of the time slots t satisfying $(t \bmod s) + 1 = k$, where time slots are numbered $t = 0, 1, 2, \ldots$ [5].

Lemma 2. [1, 5] *In FDPB with parameter m, in order for the viewer to be able to view the video continuously, page i must be broadcast (scheduled) at least once in each window of size $w_i = m + i - 1$.* □

Example 2. Let us suppose $m = 9$ and choose $s = 3$. Page P_1 can be scheduled in every $w_1 = 9 + 1 - 1 = 9^{th}$ slot. Since $s = 3$, subchannel 1 consists of every 3rd slot, and P_1 needs only 1/3 of it, and P_2 and P_3 can also be scheduled in subchannel 1, Thus we create a subtree with three leaves and label them by P_1, P_2 and P_3. See Fig. 1. These three pages will each have period 9 ($= 3s$), which is adequate, since $w_2 > 9$ and $w_3 > 9$. Page P_4 must have period at most $w_4 = m + 4 - 1 = 12$ ($= 4s$). Thus, we create the second subtree (representing subchannel 2) as shown in Fig. 1. Similarly for the last subtree. In summary, by dividing a channel into three subchannels, we can now pack 12, instead of 9 (when $s = 1$), pages in a channel. □

Let n_k denote the number of leaves of the kth subtree. In order for P_1 to appear within every window of size $w_1 = m$, we must satisfy $sn_1 \leq m$ by Lemma 1. We thus get $n_1 = \lfloor m/s \rfloor$ as the maximum integer satisfying this inequality. Now that the first n_1 pages have been scheduled in subchannel 1, the next page, i.e., the $n_1 + 1^{st}$ page must have period at most $w_{n_1+1} = m + (n_1 + 1) - 1$, hence $sn_2 \leq m + n_1$, from which we get $n_2 = \lfloor (m + n_1)/s \rfloor$. In general, we have the following formula for n_k:

$$n_k = \lfloor (m + n_1 + n_2 + \cdots + n_{k-1})/s \rfloor \tag{1}$$

Let $n(m, s)$ denote the total number of pages that can be scheduled by a 2-level RR tree with s subtrees, i.e.,

$$n(m, s) = \sum_{k=1}^{s} n_k. \tag{2}$$

3 Optimization for FDPB

3.1 Problem

Fig. 2(a) plots $n(100, s)$ computed from (1) and (2) by varying s in the range $1 \le s \le 100$ (the rugged curve). It is seen that $s = 10$ maximizes $n(100, s)$.

In Fig. 2(b), $n(m, s)$ is plotted for many different values of m ($1 \le m \le 130$). For each value of m, a curve is drawn by varying s within the range $1 \le s \le m$. This can be considered as exhaustive search by which to find the optimal s that maximizes $n(m, s)$.[4] Note that the optimal value of s that maximizes $n(m, s)$ grows with m. Our main interest in this paper is to analyze the dependency of the optimal value of s on m in the hope of finding the optimal value without resorting to exhaustive search.

From what we have seen, we can use subtrees and subchannels almost synonymously. From now on, we will mainly use the term subtree. In reference to (1), for $k = 1, \cdots, s$, we define r_k ($0 \le r_k \le s - 1$) by

$$m + \sum_{j=1}^{k-1} n_j = s n_k + r_k. \tag{3}$$

We thus have for $k \ge 2$

$$\begin{aligned} n_k &= \lfloor (n_{k-1} s + r_{k-1} + n_{k-1})/s \rfloor = n_{k-1} + \lfloor (n_{k-1} + r_{k-1})/s \rfloor \\ r_k &= n_{k-1} + r_{k-1} \pmod{s}. \end{aligned} \tag{4}$$

In order to find the optimal value of s that maximizes $n(m, s)$ by differentiation, we try to approximate $n(m, s)$ by a function that doesn't contain any floor function. Let $\bar{r}$ denote the average of $\{r_k \mid k = 1, 2, \ldots, s\}$ in (3).

Lemma 3. *The total number of pages that can be assigned to the s subtrees is approximated by the following formula when s ($< m$) is sufficiently large:*

$$n(m, s) \approx (m - \bar{r}) \left((1 + \frac{1}{s})^s - 1 \right). \tag{5}$$

[4] The four dotted curves in Fig. 2(b) correspond to $m = 9, 21, 51$ and 128. (See Sect. 5.)

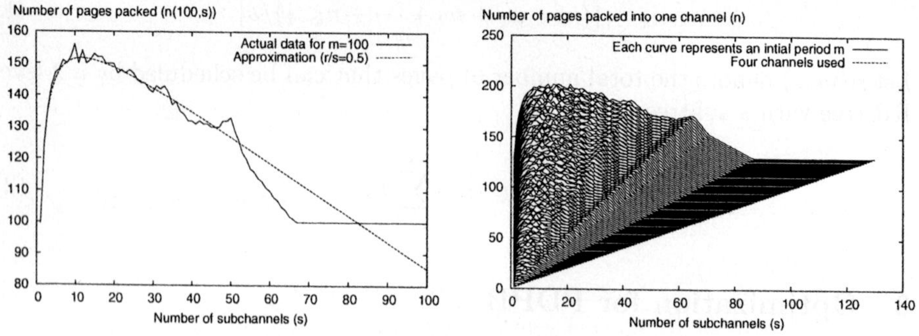

Fig. 2. (a) The rugged curve shows $n(100, s)$ for $1 \leq s \leq m = 100$. The smooth dashed curve is an approximation using (5) with $\bar{r}/s = 0.5$; (b) $n(m, s)$ for $m = 1, 2, \ldots, 130$.

Proof. Let $n_0 = m$ and rewrite (3) as follows: $r_k = \sum_{j=0}^{k-1} n_j - sn_k$, and hence $\frac{r_k}{s}(1+1/s)^{s-k} = (1+1/s)^{s-k} \sum_{j=0}^{k-1} n_j/s - (1+1/s)^{s-k} n_k$. Summing this from $k = 1$ to s, we get

$$\sum_{k=1}^{s} \frac{r_k}{s}(1+1/s)^{s-k} = \sum_{k=1}^{s}(1+1/s)^{s-k} \sum_{j=0}^{k-1} n_j/s - \sum_{k=1}^{s}(1+1/s)^{s-k} n_k.$$

We can rewrite the right hand side as follows: $\sum_{j=0}^{s-1}(n_j/s) \sum_{k=j+1}^{s}(1+1/s)^{s-k} - \sum_{k=1}^{s}(1+1/s)^{s-k} n_k = \sum_{j=0}^{s-1} n_j[(1+1/s)^{s-j} - 1] - \sum_{k=1}^{s}(1+1/s)^{s-k} n_k = m[(1+1/s)^s - 1] - n(m, s)$. We thus obtain

$$n(m, s) = m[(1+1/s)^s - 1] - \sum_{k=1}^{s} \frac{r_k}{s}(1+1/s)^{s-k}. \tag{6}$$

We notice that $n(m, 1) = n(m, m) = m$ and that $n(m, s)$ increases with s for small (relative to m) values of s. We now estimate $n(m, s)$ for large s ($< m$). For sufficiently large s, we assume that the remainders $r_1, \cdots, r_s$ are uniformly distributed in the range $[0,\ s-1]$. Substituting $r_i = \bar{r}$ into (6), we obtain (5). □

3.2 Solution

Corollary 1. *For sufficiently large s, the total number of pages $n(m, s)$ can be bounded as follows:*

$$(m - s + 1)\left((1 + \frac{1}{s})^s - 1\right) \leq n(m, s) \leq m\left((1 + \frac{1}{s})^s - 1\right)$$

Proof. Follows directly from Lemma 3 by setting $\bar{r} = s - 1$ (for the lower bound) and $\bar{r} = 0$ (for the upper bound). □

In order to find the optimal value of s that maximizes $n(m, s)$, one needs to evaluate (2) for s ranging $2 \leq s \leq m/2$ [2]. This is time-consuming when m gets large. In what follows, we show that the optimal solution s_{opt} can be found within a range containing $O(\sqrt{m})$ possible values of s. The following theorem shows how s_{opt} grows with m for large m.

Theorem 1. *The optimal number of subtrees s_{opt} grows roughly linearly with $\sqrt{m}$.*

Proof. Let $\bar{r} = a(s-1)$ in (5), where the range of parameter a is $0 < a < 1$. The optimal s satisfies the differential equation, $\frac{\partial n(m,s)}{\partial s} = 0$. Therefore, we have $(1+1/s)^s \left[ln(1+1/s) + \frac{s^2(1/s-(s+1)/s^2)}{s+1} \right] (m-a(s-1)) - a\left((1+1/s)^s - 1\right) = 0$. If s is sufficiently large, we can approximate the above equation as follows:

$$\frac{em/2 + \frac{23}{24}ea}{s^2} - (e-1)a \approx 0.$$

By solving the above quadratic equation, using the assumption $s^2 \gg a$, we obtain

$$s_{\text{opt}} \approx \sqrt{\frac{em}{2a(e-1)}}. \tag{7}$$

□

Fig. 3 plots the optimal number of subtrees, s_{opt}, computed by exhaustive search, varying s from 1 to m for each period m of the first page in the range up to 10000. It is observed that, despite the assumption $s \gg 1$ made to derive (5), practically all the data points are within an area bounded by $\sqrt{m} - 3$ and $1.54\sqrt{m}+6$. By reducing the range of search from $s \in [1, m/2]$ to $s \in [\max\{\sqrt{m}-3, 1\}, 1.54\sqrt{m}+6]$, the execution time of our search program for all values of m between 1 and 50000 went down rather drastically from more than one day to less than a minute. This range is roughly $(1.54-1)\sqrt{m} \approx 0.54\sqrt{m}$.

4 Theoretical Bounds

Although the results obtained in the previous section are quite satisfactory, there is no guarantee that nothing strange will happen beyond $m = 10000$. In order to dispel such misgivings, we shall derive theoretical upper and lower bounds in this section. They are shown in Fig. 3 as the top and bottom curves.

The formula (7) is not directly useful, since $a = a(m, s)$ is a function of s. In order to overcome this problem, let S denote a range for s such that $s_{\text{opt}} \in S$ for large m. Define $a^{up} = Sup_{s \in S}\{a(m,s)\}$ and $a^{low} = Inf_{s \in S}\{a(m,s)\}$ Then from (7), we have

$$\sqrt{\frac{em}{2a^{up}(e-1)}} \leq s_{\text{opt}} \leq \sqrt{\frac{em}{2a^{low}(e-1)}}. \tag{8}$$

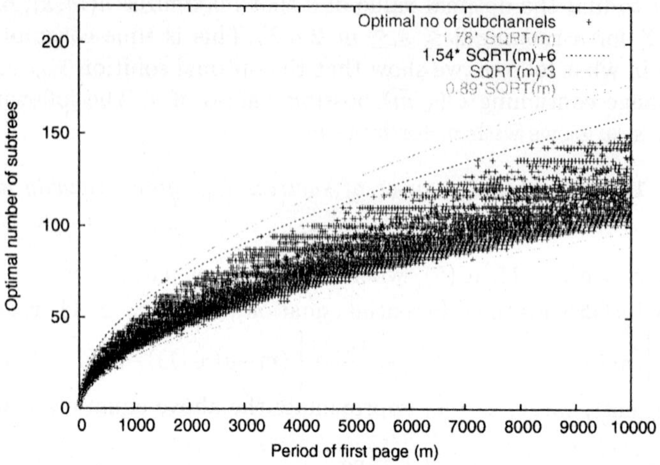

Fig. 3. The optimal number of subtrees ($s_{\rm opt}$) vs. the period (m) of the first page. The actual optima are plotted as data points.

Since $a < 1$, we obviously have $a^{up} < 1$. In the rest of this section, we shall find a lower bound on a^{low}. Fig. 4(a) plots the computation results for $a(10000,\ s)$ for s in the range $1 \leq s \leq m = 10000$. The initial part of the graph is blown up in Fig. 4(b).

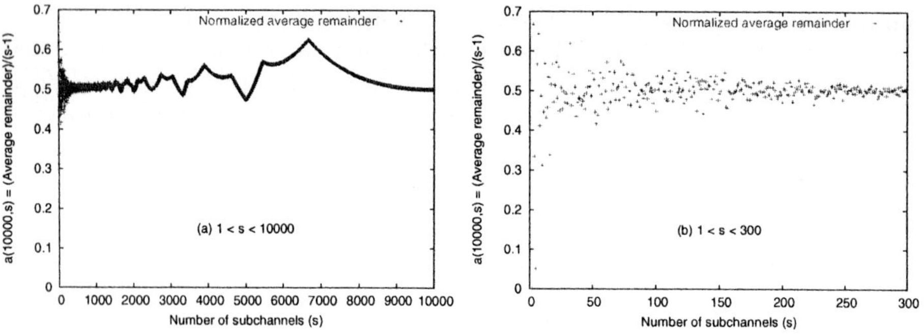

Fig. 4. (a) Quantity $a(10000, s) = \bar{r}/(s-1)$ vs. s ($1 \leq s \leq 10000$); (b) Quantity $a(10000, s) = \bar{r}/(s-1)$ vs. s ($1 \leq s \leq 300$)

In order to find a^{low} for given m, we want to investigate how the remainder changes in the vicinity of $s = \sqrt{m}$, where the optimal value for s is to be found.

Example 3. Fig. 5 illustrates the case where $m = 10000$ and the number of subtrees $s = 150$. (The optimal number of subtrees in this case is $s_{\rm opt} = 132$.) The horizontal axis represents the 150 subtrees, i.e., subtrees $k = 1$ to $k =$

150. Each number in Fig. 5 is n_k of (1) for some k, and its height represents r_k of (3). The lower horizontal line is the average of all the remainders $\{r_k \mid k = 1, 2, \ldots, s\}$, while the upper horizontal line is at height $(s-1)/2$. The dotted curves (each "dot" is represented by symbol "+") in Fig. 5 represent the difference $|r_{k+1} - r_k|$ for $k = 1, 2, \ldots, s-1$. For example, let index k be such that $n_k = 137$ and $r_k = 5$. (Of the two points in Fig. 5 that are labeled by 137, this point is the one close to the bottom.) Then by (4) we can compute $n_{k+1} = 137 + \lfloor (137+5)/150 \rfloor = 137$ and $r_{k+1} = 137 + 5 \pmod{150} = 142$, which correspond to the other point labeled by 137. The difference $r_{k+1} - r_k = 137$ is seen as a "+" just below this 137. Similarly, we can compute $n_{k+2} = 138$ and $r_{k+2} = 129$, and hence $|r_{k+2} - r_{k+1}| = 8$. This explains the position of the point labeled by 138 and the height of the next "+". □

One striking feature in Fig. 5 is the presence of quadratic curves (parabolas).We now examine the cause of the parabolas observed in the above example. For $m = 10000$ and $s > \sqrt{m} = 100$, after computing $n_1, n_2, \ldots$, suppose we reach $i-2$ such that $n_{i-2} = s-1$ and $r_{i-2} = r$ for some r. Using (4), we get $n_{i-1} = s - 1 + \lfloor (s-1+r)/s \rfloor = s, r_{i-1} = r - 1$; $n_i = s+1, r_i = r-1$; ($n_i = s+1$ appears at the bottom of the right parabola in Fig. 5. If s is chosen too large, this parabola will move to the right out of range.) $n_{i+1} = s+2, r_{i+1} = r = (r-1)+1$; $n_{i+2} = s+3, r_{i+2} = (r-1)+1+2$, and so forth, and in general,

$$r_{i+j} = (r-1) + \sum_{h=1}^{j} h = (r-1) + j(j+1)/2, \tag{9}$$

for $0 \le j \le p$, where p is the maximum number such that $(r-1) + p(p+1)/2 \le s-1$. It is seen that r_{i+j} is a quadratic function of j.

Lemma 4. *For sufficiently large m, if $s > \sqrt{m}$ then the average remainder $\bar{r} > (s-1)/4$.*

Proof. Fix the value $s > \sqrt{m}$ near $\sqrt{m}$,[5] and assume the worst case, where the remainders are concentrated near 0, i.e., the remainder in (9) has the form $r_{i+j} = j(j+1)/2$ (i.e., $r = 1$) for $0 \le j \le p$. Thus the remainders take one of the $p+1$ values, $0, 1, 3, \ldots, p(p+1)/2$, because this leads to the most skewed distribution of remainders towards 0. Note that p satisfies $(p+1)(p+2)/2 > s-1$. Then the average value of all the $p+1$ remainders is given by $R = (1/(p+1))\sum_{j=1}^{p} j(j+1)/2 = p(p+2)/6$. Using $(p+1)(p+2)/2 > s-1$, we can derive $R > (1 - 1/(p+1))(s-1)/3 > [1 - 1/(\sqrt{2(s-1)} - 1)](s-1)/3$. For $s \ge 14$, we have $[1 - 1/(\sqrt{2(s-1)} - 1)] > 3/4$, and therefore $\bar{r} > R > (s-1)/4$. □

Quadratic growth is not clearly discernible in the middle part of Fig. 5, but it is there. For example, there is a parabolic curve segment on which the points

[5] As commented earlier, if s is chosen too large, the parabola that contains a point labeled by $n_k = s$ at its bottom won't appear. In such a case, see the comments after this lemma.

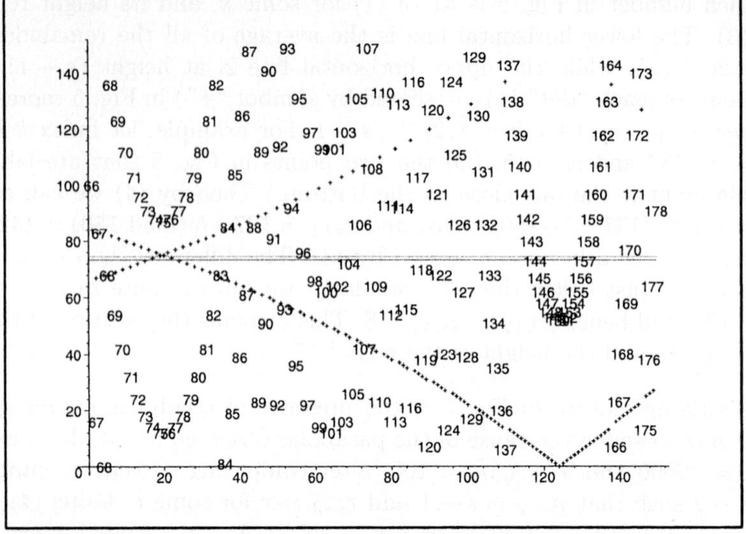

Fig. 5. Above k ($1 \le k \le s = 150$) on the horizontal axis, a point with label n_k is plotted at height r_k

labeled by 95, 96 and 97 lie. It is easy to see that in general the average of the remainders on such a parabolic curve segment is larger than $(s-1)/4$. In particular, if there are only two points on a parabolic curve segment, then their difference in height must be $> (s-1)/2$, and hence their average should be $> (s-1)/4$. There may be one or two points, e.g., the point labeled by 101 near the bottom of the graph, which do not share a parabolic curve segment with any other points. But their influence on lowering the average remainder can be compensated for by other large remainders. Note that on the left part of Fig. 5 there are parabola-like patterns, but they are not parabolas in the sense the term is used here. Two points labeled by 80, for example, form a parabolic curve segment.

Theorem 2. *For sufficiently large m, the optimal number of subtrees s_{opt} that maximizes $n(m,s)$ can be bounded as follows:*

$$\sqrt{\frac{em}{2(e-1)}} < s_{\text{opt}} < \sqrt{\frac{2em}{(e-1)}}.$$

Proof. Lemma 4 implies $a^{low} > 1/4$. Plug it and $a^{up} < 1$ into (8). □

According to the above theorem, only $\sqrt{2em/(e-1)} - \sqrt{em/2(e-1)} \approx 1.78\sqrt{m} - 0.89\sqrt{m} = 0.89\sqrt{m}$ different values of s need be tested to find s_{opt}. The bounds given by the above theorem are shown as the top and bottom curves in Fig. 3.

5 Discussions

Let us consider the general case, where we have c channels, $C_1, C_2, \ldots, C_c$. Suppose the maximum period of the initial page for channel C_1 is m_1. Then we have $m_2 = m_1 + (i_1+1) - 1$ for first page of channel C_2, if pages 1 to i_1 are packed into channel C_1. Thus by varying the parameter m in our previous analysis, we can determine how many pages can be packed into the second, third, ... channels. For example, the four dotted curves in Fig. 2(b) correspond to $m = 9, 21, 51$ and 128. The curve for $m_1 = 9$ reaches its peak 12 at $s = 3$, which implies that 12 pages can be packed into C_1 if the period of the first page is 9 and three subtrees are used. Thus page 13 is the first page to be packed in C_2, and therefore we should look at the curve for $m_2 = 9 + 13 - 1 = 21$. This curve has the peak value of 30, and thus 30 pages can be packed into C_2, and so forth.

We could use recursive subchanneling. Namely, we first divide a channel into s subchannels of equal bandwidth as before, and then further divide each of the s subchannels into subsubchannels, and so forth. In other words, instead of a 2-level RR tree, we use a RR tree with 3 or higher levels. It turns out that we are able to fit only slightly more pages into a channel for some values of m. Recursive subchanneling becomes more beneficial for larger values of m.

Although the problem addressed in this paper is rather special, we believe the approach we used could be applied to many other optimization problems that involve the floor or ceiling function.

Acknowledgement

We thank the members of the Distributed Computing Laboratory in the School of Computing Science at Simon Fraser University for stimulating discussions.

References

1. Bar-Noy, A., Ladner, R.E.: Windows scheduling problems for broadcast systems. Proc. 13th ACM-SIAM Symposium on Discrete Algorithms (2002) 433–442
2. Bar-Noy, A., R.E. Ladner, R.E., Tamir, T.: Scheduling Techniques for Media-on-Demand. Proc. 14th ACM-SIAM Symposium on Discrete Algorithms (2003) 791–800
3. Hollmann, H.D.L., Holzscherer, C.D.: Philips Tech. Rept. (1991) European Patent (1991) and US patent No. **5524271** (1995)
4. Kameda, T., Sun, T.: Survey on VoD broadcasting schemes. School of Computing Science, SFU (2003) `http://www.cs.sfu.ca/~tiko/ publications/ VODsurveyPt1.pdf`
5. Pâris, J.-F.: A fixed-delay broadcasting protocol for video-on-demand. 10th Int'l Conf. on Computer Communications and Networks (2001) 418–423
6. Thirumalai, K., Pâris, J.-F., Long, D.D.E.: Tabbycat: an inexpensive scalable server for video-on-demand. Proc. IEEE International Conference on Communications (2003) 896–900

A Min-Max Relation on Packing Feedback Vertex Sets*

Xujin Chen[1], Guoli Ding[2], Xiaodong Hu[1], and Wenan Zang[3]

[1] Institute of Applied Mathematics, Chinese Academy of Sciences,
P.O. Box 2734, Beijing 100080, China
{xchen, xdhu}@amss.ac.cn
[2] Mathematics Department, Louisiana State University,
Baton Rouge, Louisiana, USA
gxlading@cox.net
[3] Department of Mathematics, The University of Hong Kong,
Hong Kong, China
wzang@maths.hku.hk

Abstract. Let G be a graph with a nonnegative integral function w defined on $V(G)$. A family $\mathcal{F}$ of subsets of $V(G)$ (repetition is allowed) is called a *feedback vertex set packing* in G if the removal of any member of $\mathcal{F}$ from G leaves a forest, and every vertex $v \in V(G)$ is contained in at most $w(v)$ members of $\mathcal{F}$. The *weight* of a cycle C in G is the sum of $w(v)$, over all vertices v of C. In this paper we characterize all graphs with the property that, for any nonnegative integral function w, the maximum cardinality of a feedback vertex set packing is equal to the minimum weight of a cycle.

1 Introduction

We begin with a brief introduction to the theory of packing and covering. More details on this subject can be found in [6]. A *hypergraph* H is an ordered pair $(V, \mathcal{E})$, where V is a finite set and $\mathcal{E}$ is a set of subsets of V. Members of V and $\mathcal{E}$ are called *vertices* and *edges* of H, respectively. An edge is *minimal* if none of its proper subsets is an edge. A *clutter* is a hypergraph whose edges are all minimal. The *blocker* of hypergraph $H = (V, \mathcal{E})$ is the clutter $b(H) = (V, \mathcal{E}')$, where $\mathcal{E}'$ is the set of all minimal subsets $B \subseteq V$ such that $B \cap A \neq \emptyset$ for all $A \in \mathcal{E}$. We also define $b(H)^{\uparrow} = (V, \mathcal{E}'')$, where $\mathcal{E}''$ consists of all $B \subseteq V$ such that $B \cap A \neq \emptyset$ for all $A \in \mathcal{E}$. It is well known that $b(b(\mathcal{C})) = \mathcal{C} = b(b(\mathcal{C})^{\uparrow})$ holds for every clutter $\mathcal{C}$.

Let I be a set and let α be a function with domain I. Then, for any finite subset S of I, we denote by $\alpha(S)$ the sum of $\alpha(s)$, over all $s \in S$. Let $\boldsymbol{R}_+$ (resp. $\boldsymbol{Z}_+$) denote the sets of nonnegative real numbers (resp. integers). Let M be the $\mathcal{E}$-V incidence matrix of a hypergraph $H = (V, \mathcal{E})$. For any $w \in \boldsymbol{Z}_+^V$,

* Supported in part by: [1]The NSF of China under Grant No. 70221001 and 60373012, [2]NSA grant H98230-05-1-0081, NSF grant ITR-0326387, and AFOSR grant: F49620-03-1-0239-0241, and [3]The Research Grants Council of Hong Kong.

X. Deng and D. Du (Eds.): ISAAC 2005, LNCS 3827, pp. 126–135, 2005.

let $\nu_w^*(H) = \max\{\boldsymbol{x}^T\mathbf{1} : \boldsymbol{x} \in \boldsymbol{R}_+^{\mathcal{E}}, \boldsymbol{x}^T M \le w^T\}$, $\tau_w^*(H) = \min\{w^T\boldsymbol{y} : \boldsymbol{y} \in \boldsymbol{R}_+^V, M\boldsymbol{y} \ge \mathbf{1}\}$, $\nu_w(H) = \max\{\boldsymbol{x}^T\mathbf{1} : \boldsymbol{x} \in \boldsymbol{Z}_+^{\mathcal{E}}, \boldsymbol{x}^T M \le w^T\}$, $\tau_w(H) = \min\{w^T\boldsymbol{y} : \boldsymbol{y} \in \boldsymbol{Z}_+^V, M\boldsymbol{y} \ge \mathbf{1}\}$. Combinatorially, each vector $\boldsymbol{x} \in \boldsymbol{Z}_+^{\mathcal{E}}$ with $\boldsymbol{x}^T M \le w^T$ can be interpreted as a family $\mathcal{F}$ of edges (repetition is allowed) of H, for which each vertex $v \in V$ belongs at most $w(v)$ members of $\mathcal{F}$. Such a family is called a *w-packing* of H. It is clear that $\nu_w(H)$ is the maximum size of a w-packing of H. Similarly, $\tau_w(H)$ is the minimum of $w(B)$, over all edges B of $b(H)^\uparrow$. Notice from the LP Duality Theorem that

$$\nu_w(H) \le \nu_w^*(H) = \tau_w^*(H) \le \tau_w(H). \tag{1}$$

One of the fundamental problems in combinatorial optimization is to identify scenarios under which either one or two of the above inequalities holds with equality. In particular, H is *ideal* if $\tau_w^*(H) = \tau_w(H)$, for all $w \in \boldsymbol{Z}_+^V$, while H is *Mengerian* if $\nu_w^*(H) = \nu_w(H)$, for all $w \in \boldsymbol{Z}_+^V$. Obviously $b(H)$ is Mengerian iff so is $b(H)^\uparrow$. It is well known that being Mengerian is actually equivalent to $\nu_w(H) = \tau_w(H)$ for all $w \in \boldsymbol{Z}_+^V$ [3]. Thus every Mengerian hypergraph is ideal.

Our present work is a continuation of [1,2]. To clarify our motivation, we summarize the main results in [1]. For any simple graph $G = (V, E)$, let $\mathcal{C}_G = (V, \mathcal{E})$ denote the clutter in which $\mathcal{E}$ consists of $V(C)$, for all induced cycles C of G. A *Θ-graph* is a subdivision of $K_{2,3}$. A *wheel* is obtained from a cycle by adding a new vertex and making it adjacent to all vertices of the cycle. A *W-graph* is a subdivision of a wheel. An *odd ring* is a graph obtained from an odd cycle by replacing each edge $e = uv$ with *either* a cycle containing e or two triangles $uabu, vcdv$ together with two additional edges ac and bd. A subdivision of an odd ring is called an *R-graph*. Let $\mathcal{L}$ be the class of simple graphs G such that no induced subgraph of G is isomorphic to a Θ-graph, a W-graph, or an R-graph.

Theorem 1. [1] *The following are equivalent for every simple graph G:* (i) *$\mathcal{C}_G$ is Mengerian;* (ii) *$\mathcal{C}_G$ is ideal;* (iii) *$G \in \mathcal{L}$.*

Fulkerson [4] proved that a hypergraph is ideal iff its blocker is ideal. Therefore, the equivalence of (ii) and (iii) in Theorem 1 implies the following

Corollary 1. *$b(\mathcal{C}_G)^\uparrow$ is ideal if and only if $G \in \mathcal{L}$.*

At this point, Guenin [5] suggested a natural question: When is $b(\mathcal{C}_G)^\uparrow$ Mengerian? In general, the blocker of a Mengerian hypergraph does not have to be Mengerian (see [6]). However, the following theorem, our main result in this paper, says that $\mathcal{C}_G$, $b(\mathcal{C}_G)^\uparrow$, and hence $b(\mathcal{C}_G)$ are always Mengerian together.

Theorem 2. *$b(\mathcal{C}_G)^\uparrow$ is Mengerian if and only if $\mathcal{C}_G$ is.*

Let $G = (V, E)$ be a simple graph and let $w \in \boldsymbol{Z}_+^V$. A subset of V is called an *feedback vertex set* (FVS) in G if it meets every cycle in G. Since the edge set of $b(\mathcal{C}_G)^\uparrow$ is exactly the set of feedback vertex sets (FVSs) in G, we also call a w-packing of $b(\mathcal{C}_G)^\uparrow$ a *w-packing of FVSs in G*, or simply an *FVS packing*. The min-max relation in our main result can be restated as follows: if $G = (V, E)$ is a simple graph, then the maximum cardinality of a w-packing of FVSs equals the minimum weight of a cycle in G for any $w \in \boldsymbol{Z}_+^V$ iff $G \in \mathcal{L}$.

The rest of the paper is devoted to the proof of Theorem 2. In section 2, we prove some results on how Mengerian hypergraphs can be put together to get a larger Mengerian hypergraph. Then, in Section 3, we explain results from [1], which describe how graphs in $\mathcal{L}$ can be constructed from some "prime" graphs by "summing" operations. Finally, we establish Theorem 2 in Section 4 by showing that all prime graphs have the required Mengerian property.

2 Sums of Hypergraphs

The purpose of this section is to prove a few lemmas, which claim that being Mengerian is preserved under some natural summing operations.

Let $H = (V, \mathcal{E})$ be a hypergraph and $w \in \boldsymbol{Z}_+^V$. It is easy to see that $\tau_w(b(H)^\uparrow) = \min_{A \in \mathcal{E}} w(A)$. Denoting $r_w(H) = \min_{A \in \mathcal{E}} w(A)$, we have

$$b(H)^\uparrow \text{ is Mengerian iff } b(H)^\uparrow \text{ has a } w\text{-packing of size } r_w(H), \forall\, w \in \boldsymbol{Z}_+^V. \quad (1)$$

Let $H_1 = (V_1, \mathcal{E}_1)$ and $H_2 = (V_2, \mathcal{E}_2)$ be two hypergraphs. If $|V_1 \cap V_2| \in \{0, 1\}$, then $(V_1 \cup V_2, \mathcal{E}_1 \cup \mathcal{E}_2)$ is called the $|V_1 \cap V_2|$-*sum* of H_1 and H_2. If $V_1 \cap V_2 = \{x_1, x_2\}$ and for $i = 1, 2$, H_i has an edge $A_i = \{x_1, x_2, y_i\}$ that is the only edge containing y_i, then $((V_1 \cup V_2) - \{y_1, y_2\}, (\mathcal{E}_1 \cup \mathcal{E}_2) - \{A_1, A_2\})$ is called the 2-*sum* of H_1 and H_2. If $V_1 \cap V_2 = \{x_1, x_2, x_3\}$ and $A = \{x_1, x_2, x_3\}$ is an edge of H_1 and H_2, then $(V_1 \cup V_2, \mathcal{E}_1 \cup \mathcal{E}_2)$ is called the 3-*sum* of H_1 and H_2 over A. The notations given here will be used implicitly in the proofs of Lemma 1 and Lemma 2 below.

Lemma 1. *Let H be a k-sum ($k \in \{0, 1, 2\}$) of H_1 and H_2. If both $b(H_1)^\uparrow$ and $b(H_2)^\uparrow$ are Mengerian, then so is $b(H)^\uparrow$.*

Proof. The conclusion is obvious when $k \in \{0, 1\}$. We consider the case of $k = 2$. Suppose $H = (V, \mathcal{E})$. By (1), it suffices to show that $(*)$ $b(H)^\uparrow$ has a w-packing of size $r_w(H)$ for all $w \in \boldsymbol{Z}_+^V$. Suppose otherwise, $(*)$ were false for some $w \in \boldsymbol{Z}_+^V$ with $w(V)$ minimum. Let $r = r_w(H)$.

(1.1) $w(v) \le r$ for all $v \in V$.

Suppose (1.1) fails. Then $w' \in \boldsymbol{Z}_+^V$ with $w'(v) = \min\{r, w(v)\}$ for all $v \in V$ satisfies $r_{w'}(H) = r$ and $w'(V) < w(V)$, and therefore $b(H)^\uparrow$ has a w'-packing of size r, which is also a w-packing of $b(H)^\uparrow$, a contradiction. So (1.1) holds.

Let $i = 1, 2$, define $w_i \in \boldsymbol{Z}_+^V$ with $w_i(y_i) = \max\{0, r - w(x_1) - w(x_2)\}$ and $w_i(v) = w(v)$ for all $v \in V_i - \{y_i\}$. Then $r_{w_i}(H_i) \ge r$, and by (1), $b(H_i)^\uparrow$ has a w_i-packing $\mathcal{B}_i$ of size r. Choosing such $\mathcal{B}_i$ with maximum $\sum_{B \in \mathcal{B}_i} |B|$, we have
(1.2) For any $j \in \{1, 2\}$, x_j is contained in exactly $w(x_j)$ members of $\mathcal{B}_i$.

Suppose $B_1 \cap \{x_1, x_2\} = B_2 \cap \{x_1, x_2\}$ for some $B_1 \in \mathcal{B}_1$ and $B_2 \in \mathcal{B}_2$. Let χ_1, χ_2, and χ be the characteristic vectors of B_1, B_2, and $B = (B_1 \cup B_2) - \{y_1, y_2\}$, which are considered as subsets of V_1, V_2, and V, respectively. Define $w_1' =$

$w_1 - \chi_1$, $w_2' = w_2 - \chi_2$, and $w' = w - \chi$. For $i = 1, 2$, since $b(H_i)^\uparrow$ has a w_i'-packing $\mathcal{B}_i - \{B_i\}$ of size $r-1$, it follows from (1) that $r_{w_i'}(H_i) = \tau_{w_i'}(b(H_i)^\uparrow) \geq r-1$. Therefore $r_{w'}(H) \geq r-1$. Since $w'(V) < w(V)$, $b(H)^\uparrow$ has a w'-packing $\mathcal{B}'$ of size $r-1$, which yields a w-packing $\mathcal{B}' \cup \{B\}$ of $b(H)^\uparrow$. This contradiction gives

(1.3) $B_1 \cap \{x_1, x_2\} \neq B_2 \cap \{x_1, x_2\}$, for all $B_1 \in \mathcal{B}_1$ and $B_2 \in \mathcal{B}_2$.

It can be deduced from (1.2) and (1.3) that $w(x_1) + w(x_2) < r$. Recalling $w_i(A_i) = r$, we have $|B_i \cap A_i| = 1$, for all $B_i \in \mathcal{B}_i$, $i = 1, 2$, which, together with (1.2), implies a contradiction to (1.3). The lemma is proved. □

Lemma 2. *Let H is be a 3-sum of H_1 and H_2 over $A = \{x_1, x_2, x_3\}$. For $i = 1, 2$ and $1 \leq j < k \leq 3$, let H_{ijk} be obtained from H_i by adding a new vertex x_{ijk} and a new edge $A_{ijk} = \{x_{ijk}, x_j, x_k\}$. If all $b(H_{ijk})^\uparrow$ are Mengerian, then so is $b(H)^\uparrow$.*

Proof. Let $H = (V, \mathcal{E})$. As in the proof of Lemma 1, we shall prove: $(*)$ $b(H)^\uparrow$ has a w-packing of size $r_w(H)$ for all $w \in \boldsymbol{Z}_+^V$. Suppose that $(*)$ were false for a $w \in \boldsymbol{Z}_+^V$ with $w(V)$ minimum. Writing $r = r_w(H)$, we have

(2.1) $w(v) \leq r$ for all $v \in V$.

Let $1 \leq i \leq 2$, $1 \leq j < k \leq 3$, $V_{ijk} = V_i \cup \{x_{ijk}\}$, and define $w_{ijk} \in \boldsymbol{Z}_+^{V_{ijk}}$ with $w_i(x_{ijk}) = \max\{0, r - w(x_j) - w(x_k)\}$ and $w_{ijk}(v) = w(v)$ for all $v \in V_i$. Then $b(H_{ijk})^\uparrow$ has a w_{ijk}-packing $\mathcal{B}_{ijk}$ of size r. Choosing such $\mathcal{B}_{ijk}$ with $\sum_{B \in \mathcal{B}_i} |B|$ as *large* as possible, we have

(2.2) For any $1 \leq h \leq 3$, $1 \leq i \leq 2$, and $1 \leq j < k \leq 3$, x_h is contained in exactly $w(x_h)$ members of $\mathcal{B}_{ijk}$.

(2.3) $B \cap A \neq B' \cap A$, for all $B \in \mathcal{B}_{1jk}$ and $B' \in \mathcal{B}_{2j'k'}$ with $1 \leq j < k \leq 3$ and $1 \leq j' < k' \leq 3$.

(2.4) $w(x_j) + w(x_k) > r$ for all $1 \leq j < k \leq 3$.

Suppose otherwise. By symmetry, we assume $w(x_1) + w(x_2) \leq r$. Then, for $i = 1, 2$, $w_{i12}(A_{i12}) = r$, and hence no member of $\mathcal{B}_{i12}$ can contain $\{x_1, x_2\}$.

If $w(x_1) + w(x_3) > r$, then, by (2.2), some B_{i12} in $\mathcal{B}_{i12}$ $(i = 1, 2)$ contains both x_1 and x_3, which implies $B_{112} \cap A = \{x_1, x_3\} = B_{212} \cap A$ contradicting (2.3). Hence $w(x_1) + w(x_3) \leq r$, and by symmetry, $w(x_2) + w(x_3) \leq r$. Therefore $|B \cap \{x_j, x_k\}| \leq 1$ for all $B \in \mathcal{B}_{ijk}$.

Obviously $w(A) \geq r_w(H) = r$. Furthermore $w(A) > r$ as $w(A) = r$ implies a contradiction to (2.3). It follows from (2.2) that each $\mathcal{B}_{ijk}$ has an edge B_{ijk} with $|B_{ijk} \cap A| \geq 2$. Thus, by (2.3), for $1 \leq j < k \leq 3$, $\{B_{1jk} \cap A, B_{2jk} \cap A\} = \{\{x_j, x_\ell\}, \{x_k, x_\ell\}\}$ where $\ell \in \{1, 2, 3\} - \{j, k\}$. Without loss of generality, let $B_{112} \cap A = \{x_1, x_3\}$ and $B_{212} \cap A = \{x_2, x_3\}$. By (2.3), $B_{113} \cap A \neq \{x_2, x_3\}$. Thus $B_{113} = \{x_1, x_2\}$ and $B_{213} \cap A = \{x_2, x_3\}$. Now $B_{223} \cap A \in \{\{x_1, x_2\}, \{x_1, x_3\}\}$ violates (2.3), which proves (2.4).

(2.5) $|B \cap A| \leq 2$ for all $B \in \mathcal{B}_{ijk}$, where $1 \leq i \leq 2$ and $1 \leq j < k \leq 3$.

Suppose otherwise. Without loss of generality, we assume that some $\mathcal{B}_{1j_0k_0}$ has a member B_0 with $B_0 \supseteq A$. It follows from (2.3) that $|B \cap A| \leq 2$ for all

$B \in \mathcal{B}_{212} \cup \mathcal{B}_{213} \cap \mathcal{B}_{223}$. Let $1 \leq j < k \leq 3$. Since, by (2.4), $w(x_j) + w(x_k) > r$, it follows from (2.2) that $\mathcal{B}_{2jk}$ has a member B_{2jk} that contains both x_j and x_k, implying $B_{2jk} \cap A = \{x_j, x_k\}$. Therefore, by (2.3), $|B \cap A| \neq 2$ for all $B \in \mathcal{B}_{112} \cup \mathcal{B}_{113} \cup \mathcal{B}_{123}$.

Let $\{j, k, \ell\} = \{1, 2, 3\}$ with $j < k$, and let $\mathcal{B}'_{1jk}$ consist of members of $\mathcal{B}_{1jk}$ that contains x_ℓ. By (2.2), $|\mathcal{B}'_{1jk}| = w(x_\ell)$. As, by (2.4), $w_{1jk}(x_{1jk}) = 0$, we have $|B \cap \{x_j, x_k\}| \geq 1$ and hence $B \supseteq A$ for all $B \in \mathcal{B}'_{1jk}$. Consequently, $w(x_j) \geq w(x_\ell)$ and $w(x_k) \geq w(x_\ell)$. Since j, k, ℓ were chosen arbitrarily, it follows that $w(x_1) = w(x_2) = w(x_3)$, $\mathcal{B}'_{1jk} = \mathcal{B}_{1jk}$ and thus $w(x_1) = w(x_2) = w(x_3) = r$. On the other hand, since by (2.3), $|B \cap A| \leq 2$ for all $B \in \mathcal{B}_{212}$, we deduce from (2.2) that $r = |\mathcal{B}_{212}| \geq w(A)/2 = 3r/2$, a contradiction, which proves (2.5).

Finally, let $i \in \{1, 2\}$. By (2.4), $w(x_1) + w(x_2) > r$, which, together with (2.2), implies that $\mathcal{B}_{i12}$ has a member B_{i12} that contains both x_1 and x_2. Now by (2.5), we must have $B_{i12} = \{x_1, x_2\}$, contradicting (2.3) and establishing the lemma. □

3 Graphical Structures

In this section, we summarize some results form [1] that describe how graphs in $\mathcal{L}$ can be constructed from "prime" graphs.

All graphs considered are undirected, finite, and simple, unless otherwise stated. Let $G = (V, E)$ be a graph. For any $U \subseteq V$ or $U \subseteq E$, let $G \backslash U$ be the graph obtained from G by deleting U, and let $G[U]$ be the subgraph of G induced by U; when U is a single to $\{u\}$, we simply write $G \backslash u$ instead of $G \backslash \{u\}$. A *rooted graph* consists of a graph G and a specified set R of edges such that each edge of R belongs to a triangle and each triangle in G contains at most one edge from R. By *adding pendent triangles* to the rooted graph G we mean the following operation: to each edge uv in R, we introduce a new vertex t_{uv} and two new edges ut_{uv} and vt_{uv}. The readers are referred to [1] for the definitions of sums of graphs.

Lemma 3. [1] *For any graph $G \in \mathcal{L}$, at least one of the following holds.*

(i) *G is a k-sum of two smaller graphs, for $k \in \{0, 1, 2, 3\}$;*
(ii) *G is obtained from a rooted 2-connected line graph by adding pendent triangles.*

Let G be a k-sum ($k = 0, 1, 2, 3$) of graphs G_1 and G_2, then $H = \mathcal{C}_G$ is the k-sum of $H_1 = \mathcal{C}_{G_1}$ and $H_2 = \mathcal{C}_{G_2}$, and each hypergraph H_{ijk} defined in Lemma 2 is precisely $\mathcal{C}_{G_{ijk}}$, where G_{ijk} is the graph defined in the following lemma.

Lemma 4. [1] *Let $G \in \mathcal{L}$ be a k-sum of two smaller graphs. Then*

(i) *If $k \in \{0, 1, 2\}$, then G is a k-sum of two smaller graphs that belong to $\mathcal{L}$.*
(ii) *If G is a 3-sum of G_1 and G_2 over a triangle $x_1x_2x_3x_1$, then all G_{ijk} ($1 \leq i \leq 2$, $1 \leq j < k \leq 3$)) are in $\mathcal{L}$, where G_{ijk} is obtained from G_i by adding a new vertex x_{ijk} and two new edges $x_{ijk}x_j$ and $x_{ijk}x_k$.*

Two distinct edges are called *in series* if they form a minimal edge cut. Every edge is also considered as being series with itself. Being in series is an equivalence relation. Each equivalence class is called a *series family*. A series family is *nontrivial* if it has at least two edges. A graph G is *weakly even* if, for every nontrivial series family F of G with $|F|$ odd, there are two distinct edges xy and xz such that they are the only two edges of G that are incident with vertex x. A graph is *subcubic* if the degree of each vertex is at most three. A graph is *chordless* if every cycle of the graph in an induced cycle. Let K_4^- be obtained from K_4 by deleting an edge, W_4^- be obtained a wheel on five vertices by deleting a rim edge, and $K_{2,3}^+$ be obtained from $K_{2,3}$ by adding an edge between the two vertices of degree three. As usual, $L(G)$ stands for the line graph of G.

Lemma 5. [1] *Suppose $G \in \mathcal{L}$ is not a 2-sum of two smaller graphs. If G is obtained from a rooted 2-connected line graph $L(Q)$ by adding pendent triangles, where Q has no isolated vertices, then the following statements hold:* (i) *if Q has a triangle, then $G \in \{K_3, K_4^-, W_4^-, K_{2,3}^+\}$;* (ii) *$Q$ is connected, subcubic, and chordless;* (iii) *every cut edge of Q is a pendent edge; (iv) Q is weakly even.*

Lemma 6. [1] *If Q is subcubic and chord chordless, then every noncut edge belongs to a nontrivial series family.*

A path with end vertices u and v is called a *u-v path.* If a vertex v has degree three, then the subgraph formed by the three edges incident with v is called a *triad with center v*. In the next lemma, the sum of the indices is taken mod t.

Lemma 7. [1] *Suppose Q is connected and subcubic, and all its cut edges are pendent edges. If $F = \{e_1, \dots, e_t\}$ is a nontrivial series family of Q, then $Q \backslash F$ has precisely t components $Q_1, \dots, Q_t$. The indices can be renamed such that each e_i is between $V(Q_i)$ and $V(Q_{i+1})$. In addition, if $|V(Q_i)| = 2$, then the only edge in $E(Q_i)$ is a pendent edge of Q and it forms a triad with e_{i-1} and e_i; if $|V(Q_i)| > 2$, and u and v are the ends of e_{i-1} and e_i in Q_i, then $u \neq v$ and Q_i has two internally vertex-disjoint u-v paths.*

Let $G = (V, E)$ be a graph. The degree of a vertex $v \in V$ is denoted by $d_G(v)$. A *2-edge coloring of G* is an assignment of two colors to every edge in E. We say that a color is *represented* at vertex v if at least one edge incident with v is assigned that color.

Lemma 8. *Let $G = (V, E)$ be a graph and let $U \subseteq V$. Suppose $G[U]$ is bipartite and $d_G(u) \geq 2$ for all $u \in U$. Then G has a 2-edge coloring such that both colors are represented at every vertex in U.*

Let G' be a connected subgraph of G. Then the *contraction* of G' in G is obtained from $G \backslash E(G[V(G')])$ by identifying all vertices in $V(G')$. This is the same as the ordinary contraction except we also delete the resulting loops.

Lemma 9. *Let $G = (V, E)$ be subcubic, chordless, and weakly even. If $G' = (V', E')$ is obtained from G by repeatedly contracting induced cycles, and $U = (V' - V) \cup \{v \in V \cap V' : d_G(v) = 3\}$, then $G'[U]$ is bipartite.*

4 Packing Feedback Vertex Sets

The goal of this section is to prove Theorem 2, the main result of this paper. The major part of our proof consists of the following two lemmas.

Lemma 10. *Let G be obtained from a rooted 2-connected line graph $L(Q)$ by adding pendent triangles, where Q is triangle-free and satisfies* (ii)-(iv) *in Lemma 5. Let $\mathcal{C}$ be a collection of induced cycles in G, which include all triangles in G. Suppose $S \subseteq V(G)$ with $|S \cap V(C)| \geq 2$ for every $C \in \mathcal{C}$. Then S can be partitioned into R and B such that $R \cap V(C) \neq \emptyset \neq B \cap V(C)$ for every $C \in \mathcal{C}$.*

Proof. Let us call a pair (R, B) satisfying the conclusion of the lemma a *certificate* for $(G, \mathcal{C}, S)$. Suppose the lemma is false. Then we can choose a counterexample $\Omega = (G, \mathcal{C}, S)$ such that (a) $|\mathcal{C}|$ is minimized; (b) subject to (a), $t_\Omega = |C \in \mathcal{C} : |V(C)| = 3 \text{ and } V(C) \subseteq S\}|$ is minimized; (c) subject to (a) and (b), $d_\Omega = |\{v : v \in V(G) \text{ and } d_G(v) = 4\}|$ is minimized. Clearly we have
(10.1) $|\mathcal{C}| \geq 2$.

By (a)-(c), we shall define $\Omega' = (G', \mathcal{C}', S')$ such that Ω' satisfies the hypothesis of the lemma with G', $\mathcal{C}'$, S' in place of G, $\mathcal{C}$, S, respectively, and Ω' has a certificate (R', B'), from which we deduce contradiction to the assumption that Ω has no certificate.

(10.2) If $x \in V(G)$ belongs to a triangle T of G and $d_G(x) = 2$, then $x \notin S$; in particular $S \in E(Q)$.

Otherwise, $\Omega' = (G \backslash x, \mathcal{C} - \{T\}, S - \{x\})$ has a certificate (R', B'), and therefore either $(R' \cup \{x\}, B')$ or $(R', B' \cup \{x\})$ is a certificate for Ω. So (10.2) holds.

(10.3) If x is a pendent edge of Q, then $x \notin S$.
By (10.1) and (10.2), we may assume that x are contained in both a triangle T in $L(Q)$ and a pendent triangle T' in G. Since $\Omega' = (G \backslash ((V(T') \cap E(Q)) - \{x\}), \mathcal{C} - \{T, T'\}, S - \{x\})$ has a certificate (R', B'), we have $x \notin S$ as otherwise either $(R' \cup \{x\}, B')$ or $(R', B' \cup \{x\})$ is a certificate for Ω. Thus (10.3) holds.

Given $x \in E(Q)$, Q_x is obtained from Q by subdividing x with a new vertex w_x. There is a natural 1-1 correspondence between triangles in $L(Q)$ and triangles in $L(Q_x)$. Additionally, $L(Q_x)$ can be rooted the same way as $L(Q)$ was rooted. Let G_x be obtained from the rooted $L(Q_x)$ by adding pendent triangles. For every $C \in \mathcal{C}$, we define cycle C_x in G_x as follows: if C is a triangle, then C_x is a triangle in G_x that naturally corresponds to C; if C has length at least four, then $C_x = C$ when C avoids x, and $C_x = L(D_x)$ when $C = L(D)$ for cycle D through x in Q, and D_x is obtained from D by subdividing x with w_x. Set $\mathcal{C}_x = \{C_x : C \in \mathcal{C}_x\}$. An edge in Q is called *maximum* if its both ends have degree 3.

(10.4) Every maximum edge of Q belong to S.
If $x \notin S$ for some maximum edge x, then the certificate for $\Omega' = (G_x, \mathcal{C}_x, S)$ is a certificate for Ω. Hence (10.4) holds.

(10.5) $t_\Omega = 0$. That is, $|S \cap V(T)| = 2$ for all triangles T of G.

If (10.5) fails for T, then, by (10.2), T is a triad in Q with center u and contains edge $x = uv$ which is not a root edge. As, by (10.3), x is not a pendent edge, $\Omega' = (G_x, \mathcal{C}_x, (S - \{x\}) \cup \{w_x v\})$ has a certificate (R', B'). Now replacing $w_x v$ with x in (R', B') results in a certificate for Ω. This contradiction proves (10.5).

(10.6) $G = L(Q)$.

If G has a pendent triangle T, then, by (10.2) and (10.5), the certificate for $\Omega' = (G, \mathcal{C} - \{T\}, S)$ is a certificate for Ω. So we have (10.6).

By (10.5), every triad T of Q contains precisely two edges in S. Let S_T be the set of these two edges. Let $\mathcal{D}$ be the set of cycles D of Q such that $L(D) \in \mathcal{C}$. If $S_T \subseteq E(D)$ for some triad T and $D \in \mathcal{D}$, then the certificate for $\Omega' = (G, \mathcal{C} - \{L(D)\}, S)$ is a certificate for Ω. Therefore we have

(10.7) $|S_T \cap E(D)| < 2$ for all triads T of Q and all cycles $D \in \mathcal{D}$.

(10.8) No cycle in $\mathcal{D}$ contains a maximum edge.

Suppose some $D \in \mathcal{D}$ contains a maximum edge x. Then by Lemma 6 and Lemma 5(iii), x is contained in a nontrivial series family $F = \{e_1, \ldots, e_t\}$ of Q. Let components $Q_1, \ldots, Q_t$ of $Q \backslash F$ be indexed as in Lemma 7. It can be deduced from Lemma 7 and (10.3), (10.7) that $|V(Q_i)| \neq 2$ for all i. Notice that $I = \{i : 1 \leq i \leq t \text{ and } |V(Q_i)| > 2\}$ is of size at least two.

In case of $|I| = t$, (10.4) implies $F \subseteq S$. Let $Z_1 = Q \backslash V(Q_2)$ and $Z_2 = Q_2$. For $i = 1, 2$, let Q'_i be obtained from Q by contracting Z_{3-i} into a vertex z_i, and then adding a pendent edge f_i at z_i, let $G_i = L(Q'_i)$, $\mathcal{C}_i = \{C \in \mathcal{C}, V(C) \subseteq V(G_i)\} \cup \{f_i e_1 e_2 f_i\}$, and $S_i = (S \cap E(Z_i)) \cup \{e_1, e_2\}$. Each $(G_i, \mathcal{C}_i, S_i)$ has a certificate (R_i, B_i), which gives a certificate $(R_1 \cup R_2, B_1 \cup B_2)$ for Ω. In case of $|I| < t$, suppose $1 \notin I$. For every $i \in I$, let $Q'_i = Q[E(Q_i) \cup E(D)]$, $G_i = L(Q'_i)$, $\mathcal{C}_i = \{C \in \mathcal{C} : V(C) \subseteq E(Q_i) \cup \{e_{i-1}, e_i\}\} \cup \{L(D)\}$ and $S_i = S \cap E(Q_i) \cup \{e_{i-1}, e_i\}$. Then every $(G_i, \mathcal{C}_i, S_i)$, $i \in I$ has a certificate (R_i, S_i) such that for all $\{i, i+1\} \subseteq I$, if $S_i \cap S_{i+1} \neq \emptyset$, then e_i belongs to either $R_i \cap R_{i+1}$ or $B_i \cap B_{i+1}$. It follows that $(\cup_{i \in I} R_i, S - \cup_{i \in I} R_i)$ is a certificate for Ω. The contradiction establishes (10.8).

For each $D \in \mathcal{D}$, edges of Q that have precisely one end in $V(D)$ are called *connectors* of D. The combination of (10.3) and (10.7) implies

(10.9) Every $D \in \mathcal{D}$ has at least two connectors.

(10.10) Cycles in $\mathcal{D}$ are pairwise vertex-disjoint.

Suppose otherwise, D and D' are distinct cycles in $\mathcal{D}$ that share a common vertex. As the certificate (R', B') for $\Omega' = (G, \mathcal{C} - \{L(D)\}, S)$ cannot be a certificate for Ω, we may assume $S \cap E(D) \subseteq R'$, and by (10.7), all connectors of D belong to B'. Observe that D' contains at least two connector x_1, x_2 of D. For $i = 1, 2$, let y_i be the edge in $D \cap R'$ that has a common end with x_i. By (10.8), $y_1 \neq y_2$, and it can be verified that $((R' - \{y_1\}) \cup \{x_1\}, (B' - \{x_1\}) \cup \{y_1\})$ is a certificate for Ω. Hence we have (10.10).

Let $Q_{\mathcal{D}}$ be obtained from Q by contracting D, for every $D \in \mathcal{D}$, into a vertex v_D. Let $U = \{v_D : D \in \mathcal{D}\} \cup \{v \in V(Q) - \cup_{D \in \mathcal{D}} V(D) : d_Q(v) = 3\}$ and let

$Q' = Q_{\mathcal{D}}[S']$, where $S' \subseteq E(Q_{\mathcal{D}})$ is the set of edges corresponding to those in $S - \cup_{D\in\mathcal{D}} E(D)$. By Lemma 9, $Q'[U]$ is bipartite, and by (10.5), (10.7), (10.9), $d_{Q'}(u) \geq 2$ for all $u \in U$. Thus Lemma 8 guarantees a 2-edge coloring of Q' in which both colors are represented at every vertex in U. Let R' and B' be the two color classes. We view $S' = R' \cup B'$ as a subset of S. Then by (10.8), R' and B' can be easily extended to be R and B, respectively, such that (R, B) forms a certificate for Ω. The contradiction completes the proof of the lemma. □

Lemma 11. *Let G be obtained from a rooted 2-connected line graph $L(Q)$ by adding pendent triangles, where Q is triangle-free and satisfies* (ii)-(iv) *in Lemma 5. Then $b(\mathcal{C}_G)^\uparrow$ is Mengerian.*

Proof. Let $w \in \mathbf{Z}_+^V$ and $r = r_w(\mathcal{C}_G)$. By (1), we only need to show that $b(\mathcal{C}_G)^\uparrow$ has w-packing of size r. We may assume that $r \geq 2$, and $w(v) \leq r$ for all $v \in V$. Let $\mathcal{C}'$ consist of all triangles in G and $\mathcal{C}''$ consist of all other cycles in G. For any $F \subseteq V$, let $\alpha(F)$ and $\beta(F)$ be the number of cycles in $\mathcal{C}'$ and $\mathcal{C}''$, respectively, that F meets. Clearly, there is a collection $\mathcal{F}$ of subsets of V such that (a) $|\mathcal{F}| = r$; and (b) every $v \in V$ is contained in exactly $w(v)$ members of $\mathcal{F}$. We chose such an $\mathcal{F}$ such that (c) $\alpha(\mathcal{F}) = \sum_{F\in\mathcal{F}} \alpha(F)$ is maximum, and (d) subject to (c), $\beta(\mathcal{F}) = \sum_{F\in\mathcal{F}} \beta(F)$ is maximum. We prove that every member of $\mathcal{F}$ is an FVS of G, and thus $\mathcal{F}$ is a w-packing of $b(\mathcal{C}_G)^\uparrow$ of size r.

(11.1) $F \cap V(C) \neq \emptyset$, for all $F \in \mathcal{F}$ and $C \in \mathcal{C}'$.
Suppose otherwise, $F_0 \cap V(C_0) = \emptyset$ for some $F_0 \in \mathcal{F}$ and $C_0 \in \mathcal{C}'$. It follows that $|F_1 \cap V(C_0)| \geq 2$ for some $F_1 \in \mathcal{F}$. Let $F_0 \Delta F_1 = (F_0 - F_1) \cup (F_1 - F_0)$, $F_{01}^Q = (F_0 \Delta F_1) \cap E(Q)$ and $F_{01}^G = (F_0 \Delta F_1) - E(Q)$. Let $\mathcal{C}_0'$ be the set of all cycles $C \in \mathcal{C}'$ with $V(C) \cap (F_0 \cap F_1) = \emptyset$ and $|V(C) \cap F_{01}^Q| \geq 2$. For each $C \in \mathcal{C}_0'$, certain triad in Q contains all members of $V(C) \cap F_{01}^Q$. Let U be the set of the centers of all these triads. For each pendent triangle $C \in \mathcal{C}_0'$, we perform the following operations on Q. Let x, y be the two edges in $V(C) \cap F_{01}^Q$, let u be their common end, and let $z = uv$ be the other edge incident with u. We replace z with $u'v$, where u' is a new vertex. Let Q' be the resulting graph, after performing this operation over all pendent triangles $C \in \mathcal{C}_0'$. Let $Q'' = Q'[F_{01}^Q]$. By Lemma 9, $Q''[U]$ is bipartite, and by Lemma 8, Q'' has a 2-edge coloring so that both colors are represented at each vertex of U. Let R_0 and R_1 denote the two color classes. For each $z \in V(G) - E(Q)$, let T_z denote the pendent triangle of G that contains z. Let $S_0 = \{z \in F_{01}^G : |V(T_z) \cap R_0| < |V(T_z) \cap R_1|\}$ and $S_1 = F_{01}^G - S_0$. For $i = 0, 1$, let $F_i' = (F_1 \cap F_0) \cup R_i \cup S_i$. Let $\mathcal{F}' = (\mathcal{F} - \{F_0, F_1\}) \cup \{F_0', F_1'\}$. Then $\mathcal{F}'$ satisfies (a) and (b), and $\alpha(\mathcal{F}') > \alpha(\mathcal{F})$ contradicts (c), yielding (11.1).

(11.2) For any $x \in V$, if G' is a block of $G\backslash x$, then there exists a triangle-free graph Q', which satisfies (ii)-(iv) in Lemma 5, such that G' is obtained from $L(Q')$ by adding pendent triangles.

We may assume that $|V(G')| \geq 3$, and for each $z \in V(G') - E(Q)$, the pendent triangle T_z containing z is contained in G'. Let $Q_1 = Q[V(G') \cap E(Q)]$. We may assume that some $T_z\backslash z$ is not contained in any triangle of $L(Q_1)$ for otherwise $Q' = Q_1$ is as desired. Let Z be the set of all such z. Construct Q' from Q_1 by adding $|Z|$ pendent edges such that $L(Q')$ is isomorphic to $G' \backslash (V(G') - E(Q) - Z)$. It can be deduced from Lemma 5 that Q' is as desired. Thus we have (11.2).

Now we prove that each member of $\mathcal{F}$ is an FVS of G. Suppose otherwise. By (11.1) and (b), we have $F_0, F_1 \in \mathcal{F}$ and $C_0 \in \mathcal{C}''$ such that $F_0 \cap V(C_0) = \emptyset$ and $|F_1 \cap V(C_0)| \geq 2$. Suppose that $G_1, \ldots, G_k$ are all blocks of $G \backslash (F_0 \cap F_1)$. By (11.2), G_i is obtained from $L(Q_i)$ by adding pendent triangles, where Q_i is triangle-free and satisfies (ii)-(iv) in Lemma 5. Let $i \in \{1, \ldots, k\}$. Let $S_i = (F_0 \Delta F_1) \cap V(G_i)$ and let $\mathcal{C}_i$ be the set of cycles C of G_i with $|V(C) \cap S_i| \geq 2$. By (11.1), Lemma 10 applies and provides a partition (R_i, B_i) of S_i such that each cycle in $\mathcal{C}_i$ meets both R_i and B_i. By interchanging R_i with B_i if necessary, it can be assumed that if any distinct S_i and S_j have a common vertex v then either $v \in R_i \cap R_j$ or $v \in B_i \cap B_j$. Let $F_0' = (F_0 \cap F_1) \cup (R_1 \cup \cdots \cup R_k)$, $F_1' = (F_0 \cap F_1) \cup (B_1 \cup \cdots \cup B_k)$, and $\mathcal{F}' = (\mathcal{F} - \{F_0, F_1\}) \cup \{F_0', F_1'\}$. Then $\mathcal{F}'$ satisfies (a) and (b), $\alpha(\mathcal{F}') \geq \alpha(\mathcal{F})$, and $\beta(\mathcal{F}') > \beta(\mathcal{F})$, contradicting to (d). The lemma is established. □

Proof of Theorem 2. Since every Mengerian hypergarph is ideal, the "only if" part follows from Corollary 1 and Theorem 1. To establish the "if" part, we only need to show that, if $G \in \mathcal{L}$ then $b(\mathcal{C}_G)^{\uparrow}$ is Mengerian. We apply induction on $|V(G)|$. The base case $|V(G)| = 1$ is trivial, so we proceed to the induction step. By Lemma 4 and Lemma 1, 2, we may assume that G cannot be represented as a k-sum ($k = 0, 1, 2, 3$) of two smaller graphs, for otherwise we are done by induction. Then we conclude from Lemma 3 that G is obtained from a rooted 2-connected line graph $L(Q)$ by adding pendent triangles. It can be assumed that Q has no isolated vertices. If Q has a triangle, then we are done by Lemma 5(i) and (1) since for any $K = (V, E) \in \{K_3, K_4^-, W_4^-, K_{2,3}^+\}$ and $w \in \boldsymbol{Z}_+^V$, it is not hard to find a w-packing of FVSs in K of size equal to the minimum weight of a cycle in K. So we may assume that Q is a triangle-free and satisfies (ii)-(iv) in Lemma 5. Now the result follows from Lemma 11. □

Acknowledgement. The authors are indebted to Professor Bertrand Guenin for suggesting the problem and for stimulating discussion.

References

1. G. Ding and W. Zang, Packing cycles in graphs, *J. Combin. Theory Ser. B* **86** (2003), 381-407.
2. G. Ding, Z. Xu, and W. Zang, Packing cycles in graphs, II, *J. Combin. Theory Ser. B* **87** (2003), 244-253.
3. J. Edmonds and R. Giles, A min-max relation for submodular functions on graphs, *Annals of Discrete Math.*, **1** (1977), 185-204.
4. D. R. Fulkerson, Blocking and anti-blocking pairs of polyhedra, *Mathematical Programming* **1** (1971), 168-194.
5. B. Guenin, Oral communication, 2000.
6. A. Schrijver, *Combinatorial Optimization - Polyhedra and Efficiency*, Sringer-Verlag, Berlin, 2003.

Average Case Analysis for Tree Labelling Schemes

Ming-Yang Kao[1,*], Xiang-Yang Li[2,**], and WeiZhao Wang[2]

[1] Northwestern University, Evanston, IL, USA
kao@cs.northwestern.edu
[2] Illinois Institute of Technology, Chicago, IL, USA
xli@cs.iit.edu, wangwei4@iit.edu

Abstract. We study how to label the vertices of a tree in such a way that we can decide the distance of two vertices in the tree given only their labels. For trees, Gavoille *et al.* [7] proved that for any such distance labelling scheme, the maximum label length is at least $\frac{1}{8}\log^2 n - O(\log n)$ bits. They also gave a separator-based labelling scheme that has the optimal label length $\Theta(\log n \cdot \log(H_n(T)))$, where $H_n(T)$ is the height of the tree. In this paper, we present two new distance labelling schemes that not only achieve the optimal label length $\Theta(\log n \cdot \log(H_n(T)))$, but also have a much smaller expected label length under certain tree distributions. With these new schemes, we also can efficiently find the least common ancestor of any two vertices based on their labels only.

1 Introduction

For commonly used graph representations such as adjacency matrices and lists [15], one cannot determine whether or not two vertices are adjacent in the graph only based on the names of the two vertices. In contrast, Breuer and Folkman [5, 6] proposed to label the vertices in such a way that there exists a polynomial-time algorithm that can determine the adjacency of two vertices given only their labels. Such a labelling scheme is generally known as an *adjacency labelling scheme*. If the length of a label is allowed to be arbitrarily large, then one can encode any desired information. However, for a labelling scheme to be useful, the label length should be relatively short (say, polylogarithmic in the size of the graph) and yet allows one to decode the adjacency efficiently (say, time polynomial in the input label lengths). Breuer and Folkman [5, 6] proposed to use Hamming distances to label general graph. An (m, t)-*labelling scheme* labels each vertex with an m-bit label such that two vertices are adjacent if and only if their labels are at Hamming distance t or less of each other. Breuer and Folkman [6] showed that every n-vertex graph has a $(2n\Delta, 4\Delta - 4)$-labelling scheme, where Δ is the maximum vertex degree in the graph. Kannan *et al.* [14] gave adjacency labelling schemes with $O(\log n)$-bit labels for several families of graphs, including graphs of bounded degrees, graphs of bounded genuses, trees, and various intersection-based graphs such as internal graphs and c-decomposable graphs. Alstrup and Rauhe [4] improved the bound to $k \log n + O(\log^* n)$ for the family $\mathcal{A}_k$ of graphs with arboricity k and n vertices.

* Supported in part by NSF Grant IIS-0121491.
** Supported in part by NSF Grant CCR-0311174.

X. Deng and D. Du (Eds.): ISAAC 2005, LNCS 3827, pp. 136–145, 2005.

It is useful and possible to design a more general labelling scheme that also contains the distance information. A *distance labelling scheme* permits one to determine the distance between two vertices efficiently based only on their labels [7, 12]. Peleg [12] gave an $O(\log^2 n)$-bit distance labelling scheme for general trees and c-decomposable graphs. He showed [12] that for a family of n-vertices graphs with $\Omega(\exp(n^{1+\epsilon}))$ non-isomorphic graphs, any distance labelling scheme must use labels with a total length $\Omega(n^{1+\epsilon})$. Gavoille *et al.* [7] studied the bounds for the label length of the distance labelling schemes for several graph families. For general graphs, they gave a tight bound of $\Theta(n)$ bits; for planar graphs, an upper bound of $O(\sqrt{n}\log n)$ and a lower bound of $\Omega(n^{1/3})$; for bounded-degree graphs, a lower bound of $\Omega(\sqrt{n})$; and for trees, a tight bound of $\Theta(\log n \cdot \log(H_n(T)))$, where $H_n(T)$ is the height of the tree. Alstrup and Rauhe [3] built the lower-bounds of length of the label for supporting ancestor, sibling and connectivity. Recently, several distance labelling schemes considering bounded distance and weighted distance have been devised and surveyed by Gavoille and Peleg [10]. Alstrup *et al.* [2] designed a labelling scheme for a rooted tree to compute in constant time the least common ancestor from the labels of any two vertices. The labels assigned are of size $O(\log n)$ bits for a tree of n vertices. Alstrup *et al.* [1] studied labelling schemes for trees, supporting various relationships (ancestor, sibling, and connectivity) between vertices at small distance.

In this paper, we study distance labelling schemes for unweighted trees. For trees, Gavoille *et al.* [7] proved that for any distance labelling scheme, the label length is at least $\frac{1}{8}\log^2 n - O(\log n)$; they also gave a separator-based labelling scheme that has a label length $O(\log^2 n)$. Gavoille [9] improved the label scheme to $O(\log n \cdot \log(H_n(T)))$. Here, we present two new distance labelling schemes- backbone-based scheme and rake-based scheme, that not only achieve the asymptotically optimal label length $O(\log n \cdot \log(H_n(T)))$ but also have a much smaller expected label length under certain tree distributions. With these new schemes, we can also find the least common ancestor of any two vertices based on their labels only. Table 1 summarizes our main results, where k is the maximum vertex degree, $\mathbb{E}(H_n)$ is the expected height of a tree.

2 Preliminaries

Unless explicitly stated otherwise, a tree is always rooted at vertex r. The relative positions of the children are significant. The *size* of a tree T, denoted as $|T|$, is the number of the vertices in T. Given two vertices u and v in a tree T, the unique simple path between u and v in T is denoted as $\mathsf{P}(u, v, T)$, and the number of edges on $\mathsf{P}(u, v, T)$ is the *distance* between u and v, denoted as $d_T(u, v)$. The *level* of a vertex u is $d_T(u, r)$. The *height* of a tree T with n vertices, denoted as $H_n(T)$, is $\max_{u \in T} d_T(u, r)$. A vertex w is an *ancestor* of a vertex u if it is on the path $\mathsf{P}(u, r, T)$; the vertex u is then called a *descendant* of w. A vertex w is the *least common ancestor* of two vertices u, v if w has the largest level among all common ancestors of u and v. For a tree T and a vertex u, let T^u denote the subtree of T formed by u and all its descendants in T.

A *vertex labelling* for a tree T is a function L that assigns an integer $L(u, T)$ to each vertex in the tree T. A *distance calculator* is a function f that computes the distance of two vertices u,v in tree T given only their labels $L(u, T)$ and $L(v, T)$ but

Table 1. Summary of the main results of this paper

tree labelling schemes		separator-based	backbone-based	rake-based
worst case deterministic analysis		$\Theta(\log n \cdot \log(H_n(T)))$ Theorem 5	$\Theta(\log n \cdot \log(H_n(T)))$ Theorem 2	$\Theta(\log n \cdot \log(H_n(T)))$ Theorem 4
binary search tree Distribution	upper	$O(\log n \cdot \log\log n)$ Theorem 7	$O(\log n \cdot \log\log n)$ Theorem 7	$O(\log n \cdot \log\log\log n)$ Theorem 10
	lower	$\Omega(\log n \cdot \log\log n)$ Theorem 9	$\Omega(\frac{\log n \cdot \log\log n}{\log\log\log n})$ Theorem 8	$\Omega(\log n)$ Lemma 1
uniform tree distribution	upper	$O(\log^2 n)$ Theorem 5	$O(\log^2 n)$ Theorem 2	$O(\log^2 n)$ Theorem 4
	lower	$\Omega(\log^2 n)$ Theorem 13	$\Omega(\frac{\log^2 n}{\log\log n})$ Theorem 12	$\Omega(\frac{\log^2 n}{\log\log n})$ Theorem 11
distributions with $\mathbb{E}(H_n) = O(\log^\epsilon n)$	upper	$O(\log n \cdot \log\log n)$ Theorem 14	$O(\log n \cdot \log\log n)$ Theorem 14	$O(\log n \cdot \log\log n)$ Theorem 14
	lower	$\Omega(\frac{\log n \cdot \log\log n}{\log k})$ Theorem 6	$\Omega(\log n)$ Lemma 1	$\Omega(\log n)$ Lemma 1
distributions with $\mathbb{E}(H_n) = \Omega(n^\epsilon)$	Upper	$O(\log^2 n)$ Theorem 5	$O(\log^2 n)$ Theorem 2	$O(\log^2 n)$ Theorem 4
	lower	$\Omega(\log^2 n)$ Lemma 1	$\Omega(\log n)$ Lemma 1	$\Omega(\log n)$ Lemma 1

not T. A *distance labelling scheme* is a two-component tuple $\mathcal{L} = \langle L, f \rangle$ such that $f(L(u,T), L(v,T)) = d_T(u,v)$ for any pair of vertices $u, v \in T$. The *length* of a labelling scheme $\mathcal{L}$ for a tree T with n vertices, denoted as $\ell_n(\mathcal{L}, T)$, is defined as $\ell_n(\mathcal{L}, T) = \max_{u \in T} |L(u,T)|$, where $|x|$ is the number of bits in the integer x. The *length* $\ell_n(\mathcal{L})$ of a labelling scheme $\mathcal{L}$ is defined as $\ell_n(\mathcal{L}) = \max_T \ell_n(\mathcal{L}, T)$. All logarithmic functions ln in this paper are in base 2. It is easy to show that

Lemma 1. *For any tree labelling scheme $\mathcal{L}$ and tree distribution, $E(\ell_n(\mathcal{L})) \geq \log n$.*

3 Three Tree Labelling Schemes

In this section, we first present two new tree labelling schemes, namely, *the backbone-based labelling scheme* and *the rake-based labelling scheme*. We then review the separator-based labelling scheme and discuss the worst case performances of these three schemes.

3.1 Backbone-Based Labelling

Given a tree T with root r, a *backbone* $\mathcal{B}(T)$ is a path from the root r to leaf formed recursively as follows. If r has no child, then the backbone is r itself. If r has one child, say h_1, then the backbone is the path of r concatenated by $\mathcal{B}(T^{h_1})$, i.e., $\mathcal{B}(T) = r \oplus \mathcal{B}(T^{h_1})$. If r has more than one child, then the backbone is the path of r concatenated by $\mathcal{B}(T^{h_1})$ where h_1 is the child of r such that $|T^{h_1}|$ is maximum among all r's children, i.e., $\mathcal{B}(T) = r \oplus \mathcal{B}(T^{h_1})$. Here $\mathsf{P}_1 \oplus \mathsf{P}_2$ stands for the concatenatation of two paths.

Given a forest F, let $B(F) = \bigcup_{T \in F} \mathcal{B}(T)$. Define a *d-backbone* operation as first removing the edges in $\mathcal{B}(F)$ from F and then removing the resulting isolated vertices in F

Algorithm 1: Backbone-Based Vertex Labelling

for each internal vertex v_i **do**
  Assign a unique positive label $\mu(v_i, v_j)$ between 1 and W_i, where W_i is the number of v_i's child, for every vertex v_j that is v_i's child.
for $i = 0$ to $C_B(T) - 1$ **do**
  for each tree T_j in forest $\mathcal{D}^{(i)}(T)$ **do**
    Let v_j be T_j's root and $\mathcal{B}(T_j)$ be its backbone, and v_ℓ be v_j's parent if it exists.
    for every vertex $v_k \in \mathcal{B}(T_j)$ **do**
      Set $L_B(v_k, T) = L_B(v_\ell, T) \circ \langle d_T(v_k, v_j), \mu(v_\ell, v_j)\rangle$ if v_ℓ exists and set $L_B(v_k, T) = \langle d_T(v_k, v_j), 0\rangle$ otherwise. Here, the $\circ$ separates the label into *chunks*.

Algorithm 2: Backbone-Based Distance Decoder

Without loss of generality, we assume $L_B(u,T) = \mathcal{L}_0(u) \circ \cdots \circ \mathcal{L}_a(u)$ and $L_B(v,T) = \mathcal{L}_0(v) \circ \cdots \circ \mathcal{L}_b(v)$ with $a \geq b$. Here, $\mathcal{L}_i(u)$ is the $i+1$ part of the label $L_B(u,T)$.
Assume $\mathcal{L}_c(u) = \langle x, y\rangle$. For notational simplicity, we let $\mathcal{L}_c(u)[1] = x$ and $\mathcal{L}_c(u)[2] = y$.
Set $dis = 0$ and find the smallest index c such that $\mathcal{L}_c(u) \neq \mathcal{L}_c(v)$ if such c exists.
if c does not exist **then**
  $dis = dis + \mathcal{L}_i(v)[1]$ for $i = b+1$ to a.
else
  $dis = dis + \mathcal{L}_i(v)[1]$ for $i = c+1$ to a and $dis = dis + \mathcal{L}_i(u)[1]$ for $i = c+1$ to b.
  Set $dis = dis + \mathcal{L}_c(u)[1] + \mathcal{L}c(v)[1]$ if $\mathcal{L}_c(u)[2] \neq \mathcal{L}_c(v)[2]$ and set $dis = dis + |\mathcal{L}_c(u)[1] - \mathcal{L}c(v)[1]|$ otherwise.
Output $f_B(L_B(u,T), L_B(v,T)) = dis$.

Fig. 1. The Backbone-Based Distance Labelling Scheme

from F to produce a forest $\mathcal{D}(F)$. For simplicity, we denote $\mathcal{D}^{(k)}(F) = \mathcal{D}(\mathcal{D}^{(k-1)}(F))$, i.e., $\mathcal{D}^{(k)}(F)$ is the forest after k d-backbone operations on the original forest F. Let $C_B(T)$ denote the number of d-backbone operations needed to separate a tree T into isolated vertices. We have the following theorem (proof omitted):

Theorem 1. *For a tree T of n vertices, $C_B(T) \leq \log n$.*

Figure 1 presents our backbone-based labelling scheme $\mathcal{L}_B = \langle L_B, f_B\rangle$. Given a vertex u, its label $L_B(u,T)$ is a series of two element tuples separated by the " $\circ$ " symbol. We call each two element tuple a *chunk* of the label. Let $L_B(u,T) = \mathcal{L}_0(u) \circ \ldots \circ \mathcal{L}_i(u) \circ \ldots \mathcal{L}_a(u)$, where $\mathcal{L}_i(u)$ is the ith chunk of the label. Let c be the smallest index such that $\mathcal{L}_c(u) \neq \mathcal{L}_c(v)$ if it exists. Without loss of generality, assume that $\mathcal{L}_c(u) < \mathcal{L}_c(v)$. A key observation is that the vertex with label $\mathcal{L}_0(u) \circ \cdots \circ \mathcal{L}_c(u)$ is the least common ancestor of vertices with label $L_B(u,T)$ and $L_B(v,T)$.

In Algorithm 1, for every vertex v_i, when we assign child-label to v_j who is v_i's child, we assume the label length is $\log W_i$, where W_i is the number of children of v_i. However, given W_i children, when you assign a label ℓ, the label length is $\log \ell$ instead of $\log W_i$. With this observation [9], we can reduce the total tree label length by applying the following *reshuffle process*. First, we apply Algorithm 1 to obtain a label $L_B(u,T)$ for every vertex u. Initially, we mark all the internal vertices as "unprocessed" and all leaf vertices as "processed". While there is an "unprocessed" vertex , we pick one vertex v such that all of its children are processed. Without loss of generality, we assume that $v_{i_1}, v_{i_2}, \ldots, v_{i_k}$ are v's children who are not on the same backbone of v. For any vertex w in tree $T^{v_{i_j}}$, the label of $L_B(w,T)$ should contain $L_B(v,T)$ as a common prefix and the second element of $(a+1)$th chunk is also the same. Assume that $L_B(w,T) = L_B(v,T) \circ \mathcal{L}_{a+1}(w) \circ \mathcal{L}_{a+2}(w) \circ \ldots \circ \mathcal{L}_c(w)$. Define $\kappa(w) = \sum_{i=a+2}^{c} \log(\mathcal{L}_i(w)[2])$, and $\gamma(v_{i_j}) = \max_{w \in T^{v_{i_j}}} \kappa(w)$. We sort the vertices $v_{i_1}, v_{i_2}, \ldots, v_{i_k}$ according to the size of their subtrees $T^{v_{i_j}}$ in an ascending order, and let σ be the index of the sorted list, i.e., $|T^{v_{i_{\sigma(j)}}}|$ is the jth largest. Then we reassign $L_{a+1}(v)[2] = j$ to each vertex v if v is in the tree $T^{v_{i_{\sigma(j)}}}$. Observe that this reassign process does not affect the label of the first element of any chunks and the correctness is straightforward. Following Lemma reveals a property of the reshuffle process (proof omitted due to space limit).

Lemma 2. *After the reshuffle process, $\gamma(r) \leq 2\log n$ for $\mathcal{L}_B$, where r is the root.*

Notice that the reshuffle process does not depend the any specific properties of the Backbone-Based Distance Labelling Scheme. Thus, even we change the labelling scheme for the first element, as long as the label contains at most $\log n$ chunks, Lemma 2 still holds. Recall that the label of vertex u is $L_B(u,T) = \mathcal{L}_0(u) \circ \ldots \circ \mathcal{L}_k(u)$, where $\mathcal{L}_i(u)$ is tuple composed of two integers. Since $\sum_{i=1}^{k} \log(\mathcal{L}_i(u)[2]) \leq \gamma(r)$, we have

Theorem 2. *$\ell_n(\mathcal{L}_B, T)$ and the time complexity of decoding is $O(\log n \cdot \log H_n(T))$ for any tree T with n vertices.*

PROOF. From the definition of tree label length, $\ell_n(\mathcal{L}_B) = \max_T \ell_n(\mathcal{L}_B, T) \leq \log(\max\{H_n(T)\} \cdot C_B(T) + \gamma(r) \leq \log n \cdot \lceil \log(\max\{H_n(T)\} + 2 \rceil$. □

3.2 Rake-Based Scheme

In this section, we present a new tree labelling scheme based on the tree decomposition scheme by Kao [11]. A *chain* of T is a path in T such that every vertex of the given path has at most one child in T. A *tube* of T is a maximal chain of T. A *root path* of a tree is a tree path whose head is the root of that tree; similarly, a *leaf path* is one ending at a leaf. A *leaf tube* of T is a tube that is also a leaf path. Let $LT(T)$ denote the set of leaf tubes in T. Let $\mathcal{R}(T) = T - LT(T)$, i.e., the subtree of T obtained by deleting from T all its leaf tubes. The operation $\mathcal{R}$ is called the *rake operation*.

A *tube system* of a tree T is a set of tree paths $P_1, \cdots, P_m$ in T such that $T^{h_1}, \cdots, T^{h_m}$ are pairwise disjoint, where h_i is the head of P_i. We can iteratively rake T to obtain tube systems. Every rake operation produces a tube system of T until T is raked to empty. Given a tree T, let $\mathcal{R}^{(i)}(T)$ be the remaining tree after ith rake operation and $C_R(T)$ be the number of rake operations needed to make the tree empty. Similarly, we have

Theorem 3. *For any tree T of n vertices, $C_R(T) \leq \log n$.*

Based on the rake operation, we define a labelling scheme $\mathcal{L}_R = (L_R, f_R)$ as follows. For the rake-based labelling scheme defined in Algorithm 3 and Algorithm 4,similar to the backbone scheme, by assuming that $d_c(u) < d_c(v)$, a key observation

Algorithm 3: Rake-Based Vertex Labelling

1: **for** each internal vertex v_i **do**
2: Assign a unique positive ID $\mu(v_i, v_j)$ for every vertex v_j that is v_i's child, i.e., $\mu(v_i, v_a) \neq \mu(v_i, v_b)$ if v_a and v_b are v_i's children.
3: Let $C_R(T)$ be the number of rake operations needed to make T empty.
4: **for** $i = C_R(T) - 1$ down to 0 **do**
5: **for** each tube S in LT($R^i(T)$) **do**
6: Let h be the head of the tube S, i.e., the vertex with the smallest level in the tube, and let h' be the parent of h in the tree T if such h' exists.
7: **for** each vertex v_j in tube S **do**
8: Set the label of v_j as $L_R(v_j, T) = L_R(v_k, T) \circ \langle d_T(v_j, h'), \mu(h', h) \rangle$ if h' exists and set $L_R(v_j, T) = \langle d_T(v_j, r), 0 \rangle$ otherwise.
9: Apply the reshuffle process to modify the second element of the chunks of the label.

Algorithm 4: Rake-Based Distance Decoder

1: For any pair of vertices $u \neq v$, we assume $L_R(u, T) = \mathcal{L}_0(u) \circ \cdots \circ \mathcal{L}_a(u)$ and $L_R(v, T) = \mathcal{L}_0(v) \circ \cdots \circ \mathcal{L}_b(v)$ with $a \geq b$. Assume $\mathcal{L}_c(u) = \langle x, y \rangle$. For notational simplicity, we let $\mathcal{L}_c(u)[1] = x$ and $\mathcal{L}_c(u)[2] = y$.
2: Set $dis = 0$ and find the smallest index c such that $\mathcal{L}_c(u) \neq \mathcal{L}_c(v)$ if such c exists.
3: **if** c does not exist **then**
4: $dis = dis + \sum_{i=b+1}^{a} d_i(v)$.
5: **else**
6: Set $dis = dis + \sum_{i=c+1}^{a} d_i(v) + \sum_{i=c+1}^{b} d_i(u)$.
7: Set $dis = dis + \mathcal{L}_c(u)[1] + \mathcal{L}c(v)[1]$ if $\mathcal{L}_c(u)[2] \neq \mathcal{L}_c(v)[2]$ and $dis = dis + |\mathcal{L}_c(u)[1] - \mathcal{L}c(v)[1]|$ otherwise.
8: Output $f_R(L_R(u, T), L_R(v, T)) = dis$.

Fig. 2. The Rake-Based Distance Labelling Scheme

about vertex u, v's least common ancestor is that the vertex with label $\mathcal{L}_0(u) \circ \cdots \circ \mathcal{L}_c(u)$ is the least common ancestor of vertices with label $L_R(u, T)$ and $L_R(v, T)$.

From Lemma 2 and Theorem 3, $\ell_n(\mathcal{L}_R) = \max_T \ell_n(\mathcal{L}_R, T) \leq \log(H_n(T)) \cdot C_R(T) + \gamma(r) \leq \log n \cdot (\log(H_n(T)) + 2)$. We thus have

Theorem 4. *The length of $\ell_n(\mathcal{L}_R, T)$ is $O(\log n \cdot \log H_n(T))$ and the time complexity of decoding is $O(\log n \cdot \log H_n(T))$ for any tree T with n vertices.*

3.3 Separator-Based Labelling

In this section, we review a tree labelling scheme first proposed by Peleg [12] and then improved by Gavoille [9]. The key idea is to find a *separator*, i.e., a vertex here, of a tree such that the removal of the separator breaks the tree into several subtrees each with at most half of the vertices in the original tree. Iteratively remove separators of the remaining trees until all vertices are disconnected. For more details of the separator-based labelling scheme please refer to [12] and [9].

Again, a key observation here is that the vertex with label $\mathcal{L}_0(u) \circ \cdots \circ \mathcal{L}_c(u)$ is the least common ancestor of vertices with label $L_S(u,T)$ and $L_S(v,T)$. Regarding the length of the separator-based labelling scheme, we have the following two theorems (their proofs are omitted here due to space limit).

Theorem 5. *$\ell_n(\mathcal{L}_S, T)$ is $O(\log n \cdot \log(H_n(T)))$ for any tree T with n vertices.*

Theorem 6. *$\ell_n(\mathcal{L}_S, T)$ is $\Omega(\max\{\frac{\log n \cdot \log\log n}{\log k}, \log^2(H_n(T)\})$ for any tree T with n vertices and bounded degree k.*

4 Expected Label Length Under Binary Search Tree Distribution

In Section 3, we presented two tree labelling schemes which have the worst case length $\Theta(\log^2 n)$ for any binary tree. We focus on the expected label length under binary search tree distribution in this section and under uniform tree distribution in the next section.

4.1 General Upper Bound

In this subsection, we build a general but not too bad upper bound for the expected length of $\ell_n(\mathcal{L}_R, T)$ and $\ell_n(\mathcal{L}_B, T)$ when the trees are *binary search trees* with usual randomization; that is, the binary search tree is constructed in a standard fashion (n consecutive insertions) from a random permutation of $\{1, 2, \cdots, n\}$, where each permutation is equally likely. It has been proved in [13] that the expected height of a random binary search tree is $\mathbb{E}(H_n) = \alpha \log n - \beta \log\log n + O(1)$, where $\alpha \log\left(\frac{2e}{\alpha}\right) = 1, \alpha \geq 2$ and $\beta = \frac{3}{2\log\frac{\alpha}{2}}$. Numerically, $\alpha = 4.311\cdots$, and $\beta = 1.953\cdots$. With the above fact, we can give an upper bound for the expected length of both backbone-based labelling scheme and rake-based labelling scheme, and this technique can be applied to other tree randomization also. The proof is omitted due space limit.

Theorem 7. *The expected label lengths for both backbone-based scheme, rake-based scheme, and separator based scheme are at most $\log n \cdot \log\log n + \log\alpha \log n$, where α is a constant satisfying the equation $\alpha \log\left(\frac{2e}{\alpha}\right) = 1, \alpha \geq 2$.*

4.2 Lower Bound of the Expected Length for Backbone-Based Scheme and Separator-Based Scheme

Given the upper bound of expected length for backbone-based scheme, we would like to compute the lower bound for $\mathbb{E}(\ell_n(\mathcal{L}_B), T)$ and find the gap between them. Following theorem gives a lower bound for the expected length of a random binary search tree based on backbone-based scheme.

Theorem 8. *The expected label length of a random binary search tree based on the backbone-based scheme is $\Omega(\frac{\log n \cdot \log\log n}{\log\log\log n})$, i.e., $\mathbb{E}(\ell(\mathcal{L}_B), T) = \Omega(\frac{\log n \cdot \log\log n}{\log\log\log n})$.*

The proof of Theorem 8 is omitted here due to space limit. Theorem 8 gives a lower bound that is very close to the upper bound. The gap is only $\log\log\log n$, and we conjecture that the lower bound is $\Omega(\log n \cdot \log\log n)$ which is tight. Similarly, we have a lower bound of the expected length for separator-based Scheme, and it can be obtained directly from Theorem 6 since for a a binary search tree, the degree of the vertex is bounded by 3 and $\ell(\mathcal{L}_B) = \Omega(\log n \cdot \log\log n)$ from Theorem 6.

Theorem 9. *The expected label length of a random binary search tree based on the separator-based scheme is* $\Omega(\log n \cdot \log\log n)$*, i.e.,* $\mathbb{E}(\ell(\mathcal{L}_B), T) = \Omega(\log n \cdot \log\log n)$.

Theorem 9 and Theorem 7 together shows that the expected length for separator-based scheme for random binary search tree is exactly $\Theta(\log n \cdot \log\log n)$.

4.3 Upper Bound of Expected Length for Rake-Based Scheme

In this section, we give a tighter upper bound of the expected tree label length for rake-based scheme. We first present the following theorem (proof omitted here).

Theorem 10. *The expected label length of a random binary search tree for rake-based scheme is* $\log n \cdot \log\log\log n + \log\alpha \cdot \log n + o(1)$*, where* $\alpha = 2^{13} + 1$*, i.e.,* $\mathbb{E}(\ell_n(\mathcal{L}_B, T)) = \log n \cdot \log\log\log n + \log\alpha \cdot \log n + o(1)$.

Remember that for a tree with n vertices, we need at least $\log n$ bits to represent the vertices even without the requirement to recover the distance. Thus, from Theorem 10, our rake-based Scheme is almost tight. Our conjecture is that the upper bound could be improved to $O(\log n)$, which matches the lower bound. An interesting result drawn from Theorem 8 and Theorem 10 is that under the binary search tree distribution, usually the rake-based Scheme is better than backbone-based scheme. Recall that for the backbone based scheme, the length of the backbone $\mathcal{B}(T)$ is at least $\log(|T|)$. However, for rake based scheme, every rake operation decreases the height of the tree at least by 1 and most often more than 1. Thus, the last tube of the tree T, as we proved, is $O(\log\log n)$ with high probability, compared with $O(\log n)$ for the backbone. Therefore, it is natural that the rake-based scheme outperforms the backbone-based scheme.

5 Expected Label Length Under Uniform Binary Tree Distribution

In this section, we consider the binary trees with *uniform* distribution; that is every distinct binary tree with n vertices has the same probability. It is well known that there are C_n of enumeration of different binary trees with n vertex, where C_n is *Catalan Number*. Based on this fact, we have the following lower bounds for the backbone-based scheme, rake-based scheme and separator based scheme.

Theorem 11. *The expected tree label length of backbone-based scheme is* $\Omega(\frac{\log^2 n}{\log\log n})$.

Theorem 12. *The expected tree label length of rake-based scheme is* $\Omega(\frac{\log^2 n}{\log\log n})$.

Theorem 13. *The expected tree label length of separator-based scheme is* $\Theta(\log^2 n)$.

The lower bound of the expected label length of backbone-based, rake-based and separator-based are $\Omega(\frac{\log^2 n}{\log\log n})$, $\Omega(\frac{\log^2 n}{\log\log n})$ and $\Omega(\log^2 n)$ respectively. These lower bounds either are very close to or match the upper bounds $\log^2 n$, and we conjecture that the lower bounds for both the backbone-based and rake-based schemes are also $\Omega(\log^2 n)$, which is asymptotically tight.

6 Expected Label Length Under Several Other Tree Distributions

We then discuss the upper and lower bounds in a more general setting. Generally, we have the following results on the expected label length for any tree distribution:

Theorem 14. *Under any tree distribution, we have (1)* $\mathbb{E}(\ell_n(\mathcal{L}_R, T)) \leq \log \mathbb{E}(H_n(T)) \cdot \log n$*; (2)* $\mathbb{E}(\ell_n(\mathcal{L}_B, T)) \leq \log \mathbb{E}(H_n(T)) \cdot \log n$*; (3)* $\mathbb{E}(\ell_n(\mathcal{L}_S, T)) \leq \log \mathbb{E}(H_n(T)) \cdot \log n$*.*

Theorem 14 reveals an important information about the expected label length: the upper bound of expected label length relates to the expected height of the tree. For the lower bound of the expected label length, we have the following theorem.

Theorem 15. *For any degree bounded tree distribution, if the probability* $\mathbb{P}(H_n(T) \geq \mathbb{E}(H_n(T)) = \alpha$ *where* α *is some constant, then the expected length of separator-based scheme is* $\Omega(\frac{\log(n)\cdot\log(\mathbb{E}(H_n(T))}{\log k})$*, where* k *is the degree bound.*

From the previous two sections, one may observe that for bounded degree tree distribution, the label length depends on the expected tree height and size of the largest subtree. When the expected tree height is $O(n^{\epsilon})$ where ϵ is some constant, the label length for the backbone-based, rake-based and separator-based are most likely to be similar, which is close to $O(\log^2 n)$, under most distributions. When the expected tree height is $O(\log^{\epsilon} n)$, the backbone-based, rake-based and separator-based schemes can achieve a better expected label length, which is $O(\log n \cdot \log\log n)$. We also conjecture that the label length of rake-based scheme can achieve $O(\log n \cdot \log\log\log n)$ or even $O(\log n)$ under certain tree distributions, which is tight.

7 Conclusion

In this paper, we studied how to label the vertices of a tree such that we can decide, given only the labels of two vertices, their distance in the tree. Specifically, we present two new distance labelling schemes that can achieve asymptotic optimal length $O(\log n \cdot \log(H_n(T))$ and have a much smaller expected label length under certain tree distributions. In the meanwhile, we also show how to find the least common ancestor of any two vertices based on their labels only. Rake-based labelling scheme usually achieves a smaller expected label length than backbone-based and separator-based schemes for most tree distributions with average low height. A remaining future work is to close the gaps between the upper bounds and the lower bounds for various tree distributions, and to prove the conjectures listed in our full version [17]. For more details of the proof, please refer [17] also.

References

1. S. ALSTRUP, P. BILLE, AND T. RAUHE, *Labeling schemes for small distances in trees*, In Proceedings of the fourteenth annual ACM-SIAM symposium on Discrete algorithms, 2003, pp. 689–698.
2. S. ALSTRUP, C. GAVOILLE, H. KAPLAN, AND T. RAUHE, *Nearest common ancestors: A survey and a new distributed algorithm*, in SPAA'02, 2002.
3. S. ALSTRUP AND T. RAUHE, *Lower bounds for labeling schemes supporting, ancestor, sibling, and connectivity queries*, Tech. Report IT-C, nr. 10, IT University of Copenhagen, 2001.
4. ———, *Small induced universal graphs and compact implicit graph representations*, in IEEE FOCS, 2002.
5. M. A. BREUER, *Coding the vertexes of a graph*, in IEEE Transactions on Information Theory, vol. 12, April 1966, pp. 148–153.
6. M. A. BREUER AND J. FOLKMAN, *An unexpected result on coding the vertices of a graph*, in Journal of Mathematical Analysis and Applications, vol. 20, 1967, pp. 583–600.
7. S. P. C. GAVOILLE, D. PELEG AND R. RAZ, *Distance labeling in graphs*, in Proceedings of the Twelfth Annual ACM-SIAM Symposium on Discrete algorithms, 2001, pp. 210–219.
8. P. FLAJOLET AND A. ODLYZKO, *The average height of binary trees and other simple trees*, in Journal of Computer and System Sciences, vol. 25, 1982, pp. 171–213.
9. C. GAVOILLE, M. KATZ, N. KATZ, C. PAUL, AND D. PELEG, *Approximate distance labeling schemes*, in 9th Annual European Symposium on Algorithms (ESA), vol. 2161 of LNCS, 2001, pp. 476–488.
10. C. GAVOILLE AND D. PELEG, *Compact and localized distributed data structures*, Distrib. Comput., 16 (2003), pp. 111–120.
11. M.-Y. KAO, *Tree contractions and evolutionary trees*, SIAM Journal on Computing, 27 (1998), pp. 1592–1616.
12. D. PELEG, *Proximity-preserving labeling schemes and their applications*, in Proceedings of the 25th International Workshop on Graph-Theoretic Concepts in Computer Science, Springer-Verlag, 1999, pp. 30–41.
13. B. REED, *The height of a random binary search tree*, Journal of ACM, 50 (2003), pp. 306–332.
14. M. N. S. KANNAN AND S. RUDICH, *Implicit representation of graphs*, in Proceedings of the Twentieth annual ACM symposium on Theory of computing, ACM Press, 1988, pp. 334–343.
15. J. P. SPINRAD, *Efficient Graph Representations*, American Mathematical Society, June 2003.
16. D. B. WEST, *Introduction to Graph Theory*, Prentice Hall, 2nd edition ed., August 2000.
17. MING-YANG KAO, XIANG-YANG LI, AND WEIZHAO WANG, *Average Case Analysis for Tree Labelling Schemes*. Full veresion of the paper is available at `http://www.cs.iit.edu/~xli/publications-select.html`

Revisiting T. Uno and M. Yagiura's Algorithm*
(Extended Abstract)

Binh-Minh Bui Xuan, Michel Habib, and Christophe Paul

CNRS - LIRMM, 161 rue Ada, 34392 Montpellier Cedex 5, France
{buixuan, habib, paul}@lirmm.fr

Abstract. In 2000, T. Uno and M. Yagiura published an algorithm that computes all the K common intervals of two given permutations of length n in $\mathcal{O}(n+K)$ time. Our paper first presents a decomposition approach to obtain a compact encoding for common intervals of d permutations. Then, we revisit T. Uno and M. Yagiura's algorithm to yield a linear time algorithm for finding this encoding. Besides, we adapt the algorithm to obtain a linear time modular decomposition of an undirected graph, and thereby propose a formal invariant-based proof for all these algorithms.

1 Introduction

T. Uno and M. Yagiura's algorithm [23] computes all the K *common intervals* of two permutations of length n in $\mathcal{O}(n+K)$ time. Therein, each genome is regarded as a permutation on a finite set of genes, and a common interval of two genomes refers to a set of genes that are consecutive on each genome. This notion formalises the concept of a gene cluster. Afterwards, F. de Montgolfier pointed out strong relationships between *modules* of a *permutation graph* and common intervals of any of its *realiser* (made of two permutations) [12]. This allows to define the *common interval decomposition tree* for this case. From recent works, the tree turns out to own some important biological meaning [2, 19]. Particularly, common intervals help out with finding evolutionary distances between the corresponding species [4, 14, 19]. Finally, common intervals can be interpreted as pieces of each genome that have been *conserved* all along an evolutionary scenario between the involved species and their common ancestor [2].

The seminal algorithmic result on common intervals is due to T. Uno and M. Yagiura (Fig. 1). This really is a masterpiece among combinatorial algorithms as it uses a unique scan on one of the two permutations and could be seen as an application of a sweep plane paradigm as used in computational geometry [11]. However, its correctness proof is tough to understand. Later, S. Heber and J. Stoye pointed out a smaller and generating sub-family, so-called the family of *irreducible* common intervals, and succeeded in adapting T. Uno and M. Yagiura's algorithm to find all irreducible common intervals of d permutations in $\mathcal{O}(d \times n)$ time [17]. Besides, generating all the K common intervals from this sub-family is in $\mathcal{O}(K)$ time [17]. While they used T. Uno and M. Yagiura's

* Full version available at `http://www.lirmm.fr/~buixuan` as RR-LIRMM-05049

X. Deng and D. Du (Eds.): ISAAC 2005, LNCS 3827, pp. 146–155, 2005.

T. Uno and M. Yagiura's general scheme:
1. Let `Potential` be an empty list
2. **For** $i = n$ down to 1 **Do**
3. (Filter): Remove all known boundaries r in `Potential` such that for all $l \leq i$, (l, r) is not a common interval
4. (Add): Add i to the head of `Potential`
5. (Extract): While there still is some boundary r of `Potential` such that (i, r) is a common interval, output (i, r)
6. **End of for**

Fig. 1. A list `Potential` is used. It contains at each step i all boundaries $r \geq i$ such that there is some $l \leq i$ with (l, r) a common interval. Then, `Potential` is traced to output the common intervals of the form (i, r). The main difficulty of such approach relies on the linear time complexity while the idea is based on a double iteration.

scheme as a black box, they did not give further explanations for the correctness proof. Recently, A. Bergeron et al. bypassed this difficult issue and devised an alternative algorithm together with its combinatorial proof [3].

In this paper, we propose a complete invariant-based proof of T. Uno and M. Yagiura's algorithm, as well as its complexity analysis. We also show how it can easily be adapted to compute in $\mathcal{O}(n)$ time a tree representation of all common intervals of two permutations on n elements. Then, Section 3 generalises T. Uno and M. Yagiura's algorithm, and uses it as a central step for modular decomposition algorithms of undirected graphs.

2 Common Interval Decomposition

Let us denote $\mathbb{N}_n = [\![1, n]\!] = \{1, 2, \ldots, n\}$. A permutation π on a finite set V is regarded indifferently as a bijection from $\mathbb{N}_{|V|}$ to V, a total order on V, or a word in V^* without multiple occurrence. The support of a factor of π is called an *interval of* π, noted $\pi([\![l, r]\!])$ with $l, r \in \mathbb{N}_{|V|}$ its left and right boundaries. A *common interval* of two permutations on V is interval of each (see Figure 2). There could be a quadratic number of those, e.g. when the permutations are identical. The decomposition addressed in this paper is based on the seminal works on weakly partitive families [8, 22]. Let us recall some useful formalisms.

2.1 Combinatorial Decomposition Aspects

Let V be a given finite set. Two subsets of V *overlap* when none of their intersection and differences is empty. A family $\mathcal{F} \subseteq 2^V$ is *weakly partitive* if and only if $\emptyset \notin \mathcal{F}$, $\mathcal{F}$ contains the trivial subsets (singletons and V), and $\mathcal{F}$ is closed by intersection, union, and differences on overlapping subsets. It is *partitive* if weakly partitive and closed by symmetric difference on overlapping subsets [8, 22]. A weakly partitive family can have $\mathcal{O}(n!)$ members, e.g. with $\mathcal{F} = 2^V$.

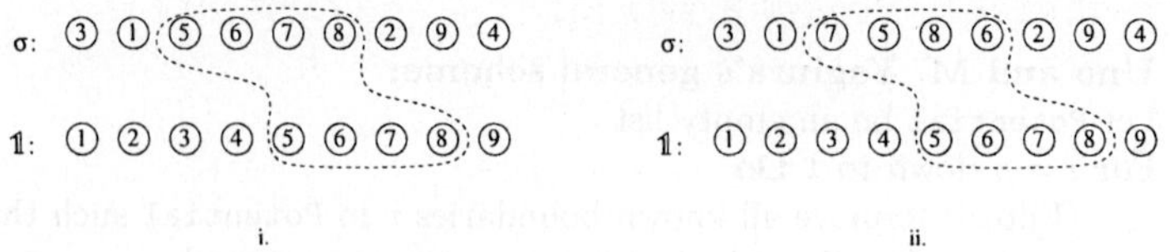

Fig. 2. In both examples, $\sigma([\![3,6]\!]) = \{5,6,7,8\} = \mathbb{1}([\![5,8]\!])$ is a common interval

Fig. 3. Common interval decomposition tree. "L" stands for *Linear* and "P" for *Prime*.

Let $\mathcal{F} \subseteq 2^V$ be weakly partitive. A member $S \in \mathcal{F}$ is *strong* when it does not overlap any other $F \in \mathcal{F}$. The subset of $\mathcal{F}$ containing all strong members of $\mathcal{F}$ is denoted $\mathcal{S}_\mathcal{F}$. The members of $\mathcal{S}_\mathcal{F}$ can be organised by inclusion order in a tree, so-called the *decomposition tree* and noted $\mathcal{T}_\mathcal{F}$. The size of $\mathcal{T}_\mathcal{F}$ is $\mathcal{O}(n)$.

Theorem 1. *[8, 22] Except for binary nodes, an internal node in $\mathcal{T}_\mathcal{F}$ satisfies one and only one of the following: (*Prime *node) no union of children belongs to $\mathcal{F}$, except for the node itself; (*Degenerate *node) all union of children belongs to $\mathcal{F}$; (*Linear *node) there is a children ordering such that a union of children belongs to $\mathcal{F}$ if and only if they are consecutive in this order.*

Roughly, the tree $\mathcal{T}_\mathcal{F}$ is a (compact) encoding of $\mathcal{F}$ from which all members of $\mathcal{F}$ can easily be generated. A permutation σ is *factorising* for $\mathcal{F}$ if and only if any strong subset $S \in \mathcal{S}_\mathcal{F}$ is an interval of σ [7]. In other words, a factorising permutation is a visit-order of the leaves of $\mathcal{T}_\mathcal{F}$ by a depth-first graph search. Though the following property is trivial, it yields a formal decomposition framework for common intervals. Fig. 3 exemplifies the common interval decomposition.

Property 1. *The family $\mathcal{CI}$ of common intervals of two permutations σ_1 and σ_2 satisfies three following properties: $\mathcal{CI}$ is weakly partitive; $\mathcal{T}_{\mathcal{CI}}$ has no* Degenerate *nodes; and both σ_1 and σ_2 are factorising.*

A common interval is *reducible* if it is union of consecutively overlapping non-trivial common intervals, and is *irreducible* when not reducible [17]. This notion can easily be generalised to any weakly partitive family. Now, it is straightforward from the definitions that the irreducible common intervals exactly are *Prime* nodes and pairs of consecutive children of *Linear* nodes of the decomposition tree. Hence, one can compute in $\mathcal{O}(n)$ time the family of irreducible common intervals from the decomposition tree and conversely. In this paper, we would rather focus on the notion of *right-strong intervals.* However, notice that both notions of irreducibility and right-strong interval merely are combinatorial tools to remove the term "K" in the raw $\mathcal{O}(n+K)$ common intervals computing time. Fortunately, both of them can be adapted in T. Uno and M. Yagiura's sweep paradigm.

2.2 Right-Strong Intervals

Let $\sigma = \sigma_1$ and σ_2 be two permutations on V. Let $\mathcal{CI}$ refer to the family of their common intervals. Then, σ is factorising for $\mathcal{CI}$. W.l.o.g., from now on, *intervals* will stand for intervals of σ. By definition, a common interval is an interval.

Definition 1 (Right-Strong Interval). *Given a factorising permutation σ for a (weakly) partitive family $\mathcal{F} \subseteq 2^V$, an interval $\sigma([\![i,j]\!]) \in \mathcal{F}$ is* right-strong *if and only if it does not overlap on its right any other interval of σ that belongs to $\mathcal{F}$, namely if and only if $i < i' \leq j < j'$ implies $\sigma([\![i',j']\!]) \notin \mathcal{F}$.*

Roughly, a right-strong interval of $\mathcal{CI}$ is a member of $\mathcal{CI}$ that does not overlap any other member of $\mathcal{CI}$ on its right in the order σ. Their number is bounded by $2 \times n$ from Corollary 1 below. To formalise their computation, let us define $\texttt{Select}(i) = \{j \mid \sigma([\![i,j]\!])$ is a right-strong interval $\}$ for all $n \geq i \geq 1$.

Definition 2 (Useless Boundary). *While inspecting σ from n down to 1, $\sigma([\![l,r]\!])$ is* visited *at step i if $i < l$,* unvisited *otherwise. Then, $r \in [\![i,n]\!]$ is* useless *w.r.t. i if none of the unvisited right-strong intervals is of the form $\sigma([\![l,r]\!])$.*

Lemma 1. *Let m_i be the maximum boundary such that $\sigma([\![i,m_i]\!]) \in \mathcal{F}$. Then, $m_i = \max \texttt{Select}(i)$ and for all $i < r < m_{i+1}$, r is useless w.r.t. i.*

Proof. If $\sigma([\![i,m_i]\!])$ overlaps $\sigma([\![i',m']\!])$ on its right, then $\sigma([\![i,m']\!]) \in \mathcal{F}$ (partitivity) and m_i is not maximum. Therefore, $m_i \in \texttt{Select}(i)$. Then, $m_i = \max \texttt{Select}(i)$ is trivial. Besides, for all $l < i+1 \leq r < m_{i+1}$, $\sigma([\![l,r]\!])$ overlaps $\sigma([\![i+1,m_{i+1}]\!])$ on its right. □

Corollary 1. $|\texttt{Select}(1)| + \ldots + |\texttt{Select}(n)| \leq 2 \times n$.

Proof. From Lemma 1, the sets $\texttt{Select}(i) \setminus \{\max \texttt{Select}(i)\}$ $(1 \leq i \leq n)$ are pairwise disjunctive and their total cardinal is bounded by n. □

2.3 Right-Strong Intervals of Two Permutations Computation

With a slight modification, i.e. by adding an one-line routine, T. Uno and M. Yagiura's algorithm computes in $\mathcal{O}(n)$ time the family of *right-strong intervals* of two permutations $\sigma = \sigma_1$ and σ_2 on V, where $n = |V|$. However, we will detail its correctness, since the original version is tough to understand. The sets $\texttt{Select}(i)$ $(n \geq i \geq 1)$ will be computed using a list `Potential`. At each step i, this list contains the right boundaries $r \geq i$ of all unvisited right-strong intervals.

`Potential` is initialised as an empty list. Each step $n \geq i \geq 1$ aims at removing from `Potential` as many useless boundaries w.r.t. i as possible. For this purpose, let $C_2(i,j)$ refer to the convex hull in σ_2 of $\sigma([\![i,j]\!])$, i.e. $C_2(i,j) = \sigma_2([\![l,r]\!])$

where $l = \min\{k \mid \sigma_2(k) \in \sigma([\![i,j]\!])\}$ and $r = \max\{k \mid \sigma_2(k) \in \sigma([\![i,j]\!])\}$. We define $\mathbb{S}_{\sigma([\![i,j]\!])} = C_2(i,j) \backslash \sigma([\![i,j]\!])$ as the *splitter set* of $\sigma([\![i,j]\!])$. Roughly, a splitter makes an interval not a common interval. Let $s(\sigma([\![i,j]\!])) = |\mathbb{S}_{\sigma([\![i,j]\!])}| = s_i(j)$. We define $\delta_i(p_j) = s_i(p_{j+1}) - s_i(p_j)$ if a member p_j of `Potential` has a successor p_{j+1}. Otherwise, $\delta_i(p_j) = +\infty$. Then, Theorem 2 below is fundamental and most results thereafter rely on it. However, from our standpoint, the theorem is easier to prove and most comprehensive when generalised to Theorem 4 in Section 3.1.

Property 2. *[23] $\sigma([\![i,j]\!])$ is a common interval if and only if $s_i(j) = 0$.*

Theorem 2. *[23] $\delta_i(p_j) < 0$ implies p_j is useless w.r.t. i.*

At each step i, assume that some Update-Detect routine provides 1. for each p_j in `Potential` a pointer to the value of $s_i(p_j)$; and 2. a list `Detected` of pointers to all p_j with $\delta_i(p_j) < 0$, and possibly to some other useless boundaries w.r.t. i. Besides, assume that the pointed $p_{j_1} < \ldots < p_{j_h}$ are organised increasingly.

Then, `Potential` is filtered twice. The first filtering (Pre-Filter) is our only addition to the original algorithm. It follows from Lemma 1, which states that it is possible to move apart some useless boundaries w.r.t. i even before considering $\sigma(i)$. Concisely, a pointer to $r_0 = \max \texttt{Select}(i+1)$ is maintained. Then, if r_0 has some predecessors in `Potential`, they are removed and r_0 receives the mark *Eaten*, which is for use in Section 2.4. The second filtering (Customised Filter) backtracks `Detected` from p_{j_h} down to p_{j_1}. Each p_{j_k} is removed from `Potential` if still there. If some removing makes the next-left boundary p' have $\delta_i(p') < 0$, p' is also removed and so on. Thus, only useless boundaries w.r.t. i are removed, and all remaining boundaries have positive δ_i. Both filtering takes linear time on the number of removed boundaries. The boundary i is then added to the head of `Potential` (Add) and the update of step i is complete. Notice that $\delta_i(i) \geq 0$.

Invariant 1. *After the update of step i, let p_{j_0} be the first member of* `Potential` *with $s_i(p_{j_0}) \neq 0$. Then,* $\texttt{Select}(i) = \{r < p_{j_0} \mid r$ *is a member of* $\texttt{Potential}\}$.

Proof. After the update, all p_j have $\delta_i(p_j) \geq 0$. If $r \in \texttt{Select}(i)$, then $s_i(r) = 0$ and $r < p_{j_0}$. Besides, $\sigma([\![i,r]\!])$ is unvisited at step i. Hence, r still is a member of `Potential`, and it is strictly before p_{j_0}. Conversely, any member $r < p_{j_0}$ of `Potential` after the update satisfies $s_i(r) = 0$. If $\sigma([\![i,r]\!])$ overlaps some $\sigma([\![i',r']\!])$ on its right, then $i < i' \leq r < r'$, $\sigma([\![i',r]\!]) \in \mathcal{CI}$, $\sigma([\![i',r']\!]) \in \mathcal{CI}$ and the Pre-Filter at step i' would remove r from `Potential` if it was still there. □

Outputting `Selected`(i) from the list `Potential` (Extract) follows from Invariant 1. Its computing time obviously is linear on the size of the output.

Turning our attention to complexity issues, Corollary 1 and the fact that each boundary is inserted exactly once in `Potential` imply the following.

Result 1. *The right-strong intervals computing time is $\mathcal{O}(n)$ if Update-Detect runs in linear time on the size of the output* `Detected` *at each iteration step i.*

T. Uno and M. Yagiura's algorithm revisited:
1. Let `Potential` be an empty list and $\texttt{Select}(n+1) = \emptyset$
2. **For** $i = n$ down to 1 **Do**
3. (Update-Detect): Collect all known useless boundaries w.r.t. i
4. (Pre-Filter): If there are some $r < r_0 (= \max \texttt{Select}(i+1))$ in `Potential`, remove them and mark r_0 as *Eaten*
5. (Customised Filter): Remove all known useless boundaries w.r.t. i
6. (Add): Add the boundary i to the head of `Potential`
7. (Extract): Find the right-most r_q in `Potential` with $s_i(r_q) = 0$ and output $\texttt{Select}(i) = \{r_1 \dots r_q\}$
8. **End of for**

Update-Detect can be as follows [23]. Let $\texttt{Potential} = [p_1(= i+1), \dots, p_l]$ at the beginning of step i. The routine updates two lists $\texttt{Min} = [Min_1, \dots, Min_s]$ and `Max`. Each $1 \leq Min_j \leq n$ is a boundary with two pointers $\texttt{first}(Min_j)$ and $\texttt{last}(Min_j)$ to two members of `Potential`. All p_j between these two members satisfy $Min_j = \min\{k \mid \sigma_2(k) \in \sigma([\![i, p_j]\!])\}$. Besides, each p_j in `Potential` has a pointer $\texttt{Min}(p_j)$ to the corresponding member of `Min`. It is analogous for `Max`. By supposing $V = [\![1, n]\!]$, computing $s_i(p_j)$ from this structure is in $\mathcal{O}(1)$ time.

Let $\texttt{Min} = [Min'_1, \dots, Min'_{s'}]$ and $\texttt{Max} = [Max'_1, \dots, Max'_{t'}]$ at the beginning of step i. Suppose inductively that $C_2(i+1, p_j) = \sigma_2([\![\texttt{Min}(p_j), \texttt{Max}(p_j)]\!])$ for all p_j and that `Min`, resp. `Max`, is strictly decreasing, resp. increasing. Notice that $\sigma_2(Min'_1) = \sigma_2(Max'_1) = \sigma(i+1)$. Now, i' with $\sigma_2(i') = \sigma(i)$ can be obtained in $\mathcal{O}(1)$ time. Then, either $i' < Min'_1$ and `Max` will be unchanged, or $Max'_1 < i'$ and `Min` unchanged. We trace `Min`, resp. `Max`, from $j = 1$ until finding the first j^* with $Min'_{j^*} \leq i' < Max'_1$, resp. $Min'_1 < i' \leq Max'_{j^*}$. Notice that $j^* > 1$ and let $p_{j_0} = \texttt{first}(Min'_{j^*-1})$, resp. $p_{j_0} = \texttt{first}(Max'_{j^*-1})$.

Lemma 2. *[23] p_j is useless w.r.t. i if $s_i(p_j) - s_{i+1}(p_j) > s_i(p_{j+1}) - s_{i+1}(p_{j+1}) \geq 0$.*

Invariant 2. *(equivalent to Lemma 2) p_j with $1 \leq j < j_0$ is useless w.r.t. i.*

W.l.o.g. $Min'_{j^*} \leq i' < Max'_1$, we set Min'_{j^*-1} to i'; point $\texttt{first}(Min'_{j^*-1})$ to p_1; and for all $1 \leq j < j_0$, point $\texttt{Min}(p_j)$ to Min'_{j^*-1}. Thus, each p_j satisfies $C_2(i, p_j) = \sigma_2([\![\texttt{Min}(p_j), \texttt{Max}(p_j)]\!])$. It is straightforward to maintain this fact until the end of step i, and the inductive hypothesis for the next step holds. Finally, `Detected` is defined as a list of pointers to $p_1 < \dots < p_{j_0-1}$. Now, the only member of `Potential` where δ_i can be negative that is not pointed by `Detected` is $p_{j_1} = \texttt{last}(Min_{j^*-1})$. Thus, if $\delta_i(p_{j_1}) < 0$, we add a pointer to p_{j_1} to the end of `Detected`. The running time is $\mathcal{O}(j_0 + j^*) = \mathcal{O}(j_0) = \mathcal{O}(|\texttt{Detected}|)$.

Result 2. *Right-strong intervals of two permutations computing time is $\mathcal{O}(n)$.*

Remark 1. Ideally, at each step i, `Potential` would contain *only* the right boundaries $r \geq i$ of all unvisited right-strong intervals. Is it true ?

2.4 Common Interval Decomposition of d Permutations

After the right-strong intervals computation, a symmetric sweep from left-to-right generates the strong common intervals. We recall that those are the nodes of the decomposition tree. Moreover, the sweep organises them by interval inclusion. Hence, constructing the tree is in $\mathcal{O}(n)$ time. Then, the labelling can use the following remarks. Since there are only *Prime* and *Linear* nodes, the strong common intervals that are marked *Eaten* by the right-strong intervals computation also have this mark in the left-strong intervals computation. Besides, a node has *Eaten* if and only if it is *Linear*. Finally, Property 2, Theorem 2, and Lemma 2 can be generalised to the case of d permutations if one replaces $C_2(i,p_j)$ with $C_j = \mathcal{S}_{\sigma([\![i,p_j]\!])} \uplus \sigma([\![i,p_j]\!]) = \cup_{h=2}^{d} C_h(i,p_j)$. Then, at each step i in the new Update-Detect, one has to maintain C_j rather than just $C_2(i,p_j)$. The hitch lays on the fact that $C_h(i,p_j)$ $(2 \leq h \leq d)$ are not pairwise disjunctive. However, as an element $\sigma(i')$ can be added to some $C_h(i'',p_j'')$ only once throughout the computation, the total maintenance can be done in $\mathcal{O}(d \times n)$ time.

Result 3. *The common interval decomposing time of d permutations is* $\mathcal{O}(d \times n)$.

3 Modular Decomposition

Let $G = (V,E)$ be a loopless simple undirected graph with $n = |V|$ and $m = |E|$. A vertex $v \in V \setminus X$ exterior to $X \neq \emptyset$ is *adjacent to* X if it is adjacent to each vertex of X, *non-adjacent to* X if non-adjacent to each vertex of X. In both cases, v is *uniform to* X. Otherwise, v is a *splitter of* X. X is a *module* if it has no splitters. Roughly, the family $\mathcal{M}$ of modules of G refers to the set of subgraphs of G that behave as one single vertex. It is well-known that $\mathcal{M}$ is partitive [8, 22], and finding efficient algorithms for computing $\mathcal{T}_\mathcal{M}$ from G has been an important challenge of the last two decades [7, 8, 9, 12, 15, 21, 22]. The *factorising permutations of* G refer to the ones of $\mathcal{M}$. Linear time algorithms for obtaining one such permutation are available for chordal graphs [18], inheritance graphs [16], and even for arbitrary graphs [15]. The decomposition approach conducted by C. Capelle is as follows. First find a factorising permutation [15], then construct the modular decomposition tree [7]. Both computations run in $\mathcal{O}(n+m)$ time even if the latter [7] is somewhat heavily fathered.

Theorem 3. *[12] Both permutations of any realiser of a permutation graph are factorising for the graph.*

Corollary 2. *The family $\mathcal{CI}$ of common intervals of two permutations is included in the family $\mathcal{M}$ of modules of the permutation graph that realises them. Besides, $\mathcal{CI}$ and $\mathcal{M}$ share the same strong subsets, namely $\mathcal{S}_{\mathcal{CI}} = \mathcal{S}_\mathcal{M}$. Finally, $\mathcal{T}_{\mathcal{CI}}$ is isomorphic to $\mathcal{T}_\mathcal{M}$, with* Degenerate *labels replaced by* Linear *ones.*

From Corollary 2, the algorithm of the previous section is an $\mathcal{O}(n)$ time modular decomposition algorithm for a permutation graph given by one realiser, yet $m = \theta(n^2)$. In this section, given a factorising permutation σ, we compute the modular decomposition tree of G. Let us first adapt Property 2 and Theorem 2.

3.1 Submodularity on the Size of the Splitter Sets

Let $\mathcal{S}_X$ refer to the splitter set of a vertex subset $X \neq \emptyset$ and $s(X) = |\mathcal{S}_X|$ count the number of its splitters. We extend $s(\emptyset) = -n$. Property 3 and Corollary 3 below are our graph versions of respectively Property 2 and Theorem 2.

Property 3. *$\sigma([\![i,j]\!])$ is a module if and only if $s_i(j) = s(\sigma([\![i,j]\!])) = 0$.*

Definition 3 (Submodularity). *(see e.g. [24]) A set function $\mu : 2^V \to \mathbb{R}$ is* submodular *when $\mu(X) + \mu(Y) \geq \mu(X \cup Y) + \mu(X \cap Y)$, for all $X, Y \subseteq V$.*

Theorem 4 (Submodularity). *The function s counting the splitters of modules of a graph is submodular.*

Proof. Since $s(\emptyset) = -n$ and $s(X) \leq n$ for $X \subseteq V$ non-empty, the only tricky issue consists of proving the submodular inequality for a pair (X, Y) of overlapping subsets of V. To do this, we first notice that $\mathcal{S}_{X\cap Y} = (\mathcal{S}_{X\cap Y} \setminus Y, \mathcal{S}_{X\cap Y} \cap Y)$. Besides, $\mathcal{S}_{X\cup Y} = (\mathcal{S}_{X\cup Y} \setminus \mathcal{S}_X, \mathcal{S}_{X\cup Y} \cap \mathcal{S}_X)$ can be reduced by definition of splitters to $\mathcal{S}_{X\cup Y} = (\mathcal{S}_{X\cup Y} \setminus \mathcal{S}_X, \mathcal{S}_X \setminus (X \cup Y))$. Similarly, $\mathcal{S}_Y = (\mathcal{S}_Y \setminus \mathcal{S}_{X\cap Y}, \mathcal{S}_{X\cap Y} \setminus Y)$. Finally, $\mathcal{S}_X = (\mathcal{S}_X \setminus Y, (\mathcal{S}_X \cap Y) \setminus \mathcal{S}_{X\cap Y}, (\mathcal{S}_X \cap Y) \cap \mathcal{S}_{X\cap Y})$ can be reduced to $\mathcal{S}_X = (\mathcal{S}_X \setminus (X \cup Y), (\mathcal{S}_X \cap Y) \setminus \mathcal{S}_{X\cap Y}, \mathcal{S}_{X\cap Y} \cap Y)$. Hence, $|\mathcal{S}_X| + |\mathcal{S}_Y| - |\mathcal{S}_{X\cup Y}| - |\mathcal{S}_{X\cap Y}| = |(\mathcal{S}_X \cap Y) \setminus \mathcal{S}_{X\cap Y}| + |\mathcal{S}_Y \setminus \mathcal{S}_{X\cap Y}| - |\mathcal{S}_{X\cup Y} \setminus \mathcal{S}_X|$.

To achieve proving the lemma, we prove that $\mathcal{S}_Y \setminus \mathcal{S}_{X\cap Y} \supseteq \mathcal{S}_{X\cup Y} \setminus \mathcal{S}_X$. Indeed, let $v \in \mathcal{S}_{X\cup Y} \setminus \mathcal{S}_X$. Then, v is exterior to X and is uniform to X. By symmetry, we suppose w.l.o.g. that $v \in N_X$. Since $X \cap Y \neq \emptyset$, there exists $w \in X \cap Y$ with $(v, w) \in E$. Now, v is a splitter of $X \cup Y$, implying $u \in Y \setminus X$ such that $(v, u) \notin E$. Hence, $v \in \mathcal{S}_Y$. It is trivial by $v \in N_X$ that $v \notin \mathcal{S}_{X\cap Y}$. □

Corollary 3. *Let $i \leq p_j < p_{j+1}$, and $\delta_i(p_j) = s_i(p_{j+1}) - s_i(p_j)$. Then, $\delta_i(p_j) < 0$ implies there is no $k \leq i$ such that $\sigma([\![k, p_j]\!])$ is a module.*

Proof. If $\delta_i(p_j) < 0$, then the submodularity on the subsets $\sigma([\![k, p_j]\!])$ and $\sigma([\![i, p_{j+1}]\!])$ for all $k \leq i$ implies that $s_k(p_j) > s_k(p_{j+1}) \geq 0$. □

3.2 Modular Decomposition Algorithm

The two latter generalisations state that Section 2.4's decomposition scheme can be used in the case of modules if one adapts the involved routines accordingly. Actually, the adaptation is straightforward, except for the Update-Detect routine and labelling the decomposition tree. For lake of space, we do not detail here the graph version of Update-Detect, which computes in $\mathcal{O}(n+m)$ global time. Now, let us show how to label the decomposition tree. First, it is well-known that a modular decomposition tree has no *Linear* nodes, and its *Degenerate* nodes are divided into *Serial* nodes – adjacency guaranteed between all children – and *Parallel* nodes – non-adjacency between children [8, 22]. Then, by analogous remarks as in Section 2.4, nodes marked *Eaten* are *Degenerate*, others are *Prime*. Besides, thanks to some *Adjacency* marks in the graph version of Update-Detect, which state the adjacency between the children of the node, we can differ the *Serial* and *Parallel* nodes in the labelling.

Result 4. *The modular decomposition is solved in $\mathcal{O}(n+m)$ time for any graph.*

There exists an $\mathcal{O}(n)$ time common interval decomposition algorithm of two permutations [20]. Unfortunately, the algorithm therein is not that simple and relies on a rather sophisticated algorithm [10]. Moreover their approach is not extended to general modular decomposition. To this aim one could use the algorithm proposed in [7]. However, this latter produces a rather heavy sequence of trees. On the other hand, our approach uses a unique paradigm for both computations of common interval and modular decomposition tree, and not only we unify the two corresponding domains but also provide very efficient algorithms.

4 Conclusion and Perspectives

We show the importance of graph layout approaches, e.g. with factorising permutations, which are based on a gateway between algorithms on permutations and those on graphs. Besides, we show strong potentials of generalising T. Uno and M. Yagiura's algorithm to the case of weakly partitive families. Thus, the use of this algorithm would be an important crux for designing future algorithms. For instance, it would be interesting to adopt the same philosophy conducted throughout our paper to other combinatorial problems such as decomposition into "inheritance-block" of an inheritance graph in $\mathcal{O}(n+m)$ time, which would yield an alternative to the algorithms proposed in [6, 16]. Another example would be the modular decomposition in $\mathcal{O}(n)$ time of a bounded tolerance graph – trapezoid graph with solely parallelograms [5, 13] – when an intersection model is provided. Then, it would be very interesting to have an $\mathcal{O}(n)$ modular decomposition time for an interval or trapezoid graph on one of its intersection model, which would give interesting links to works on gene-teams [1].

Acknowledgements. We are grateful to T. Uno for helpful discussions.

References

1. M.-P. Béal, A. Bergeron, S. Corteel, and M. Raffinot. An algorithmic view of gene teams. *Theoretical Computer Science*, 320(2-3):395–418, 2004.
2. S. Bérard, A. Bergeron, and C. Chauve. Conservation of combinatorial structures in evolution scenarios. In *International Workshop on Comparative Genomics (RECOMB04)*, volume 3388 of *LNCS*, pages 1–14, 2004.
3. A. Bergeron, C. Chauve, F. de Montgolfier, and M. Raffinot. Computing common intervals of k permutations, with applications to modular decomposition of graphs. In *13th Annual European Symposium on Algorithms (ESA05)*, 2005. to appear in LNCS.
4. A. Bergeron and J. Stoye. On the similarity of sets of permutations and its applications to genome comparison. In *9th Annual International Conference on Computing and Combinatorics (COCOON03)*, volume 2697 of *LNCS*, pages 68–79, 2003.
5. K. P. Bogart, P. C. Fishburn, G. Isaak, and L. Langley. Proper and unit tolerance graphs. *Discrete Applied Mathematics*, 60:99–117, 1995.

6. C. Capelle. Block decomposition of inheritance hierarchies. In *23rd International Workshop on Graph-Theoretic Concepts in Computer Science (WG97)*, volume 1335 of *LNCS*, pages 118–131, 1997.
7. C. Capelle, M. Habib, and F. de Montgolfier. Graph decomposition and factorizing permutations. *Discrete Mathematics and Theoretical Computer Science*, 5(1):55–70, 2002.
8. M. Chein, M. Habib, and M.C. Maurer. Partitive hypergraphs. *Discrete Mathematics*, 37(1):35–50, 1981.
9. A. Cournier and M. Habib. A new linear algorithm for modular decomposition. In *Trees in algebra and programming (CAAP 94)*, volume 787 of *LNCS*, pages 68–84, 1994.
10. E. Dahlhaus. Parallel algorithms for hierarchical clustering, and applications to split decomposition and parity graph recognition. *Journal of Algorithms*, 36(2):205–240, 2000.
11. M. de Berg, M. van Kreveld, M. Overmars, and O. Schwarzkopf. *Computational geometry*. Springer-Verlag, 1991.
12. F. de Montgolfier. *Décomposition modulaire des graphes. Théorie, extensions et algorithmes*. PhD thesis, Université Montpellier II, 2003.
13. S. Felsner. Tolerance graphs and orders. *Journal of Graph Theory*, 28(3):129–140, 1998.
14. M. Figeac and J.-S. Varré. Sorting by reversals with common intervals. In *4th International Workshop on Algorithms in Bioinformatics (WABI04)*, volume 3240 of *LNBI*, pages 26–37, 2004.
15. M. Habib, F. de Montgolfier, and C. Paul. A simple linear-time modular decomposition algorithm. In *9th Scandinavian Workshop on Algorithm Theory (SWAT04)*, volume 3111 of *LNCS*, pages 187–198, 2004.
16. M. Habib, M. Huchard, and J.P. Spinrad. A linear algorithm to decompose inheritance graphs into modules. *Algorithmica*, 13(6):573–591, 1995.
17. S. Heber and J. Stoye. Finding all common intervals of k permutations. In *12th Annual Symposium on Combinatorial Pattern Matching (CPM01)*, volume 2089 of *LNCS*, pages 207–218, 2001.
18. W.-L. Hsu and T.-M. Ma. Substitution decomposition on chordal graphs and applications. In *2nd International Symposium on Algorithms (ISA91)*, volume 557 of *LNCS*, pages 52–60, 1991.
19. G. M. Landau, L. Parida, and O. Weimann. Using pq trees for comparative genomics. In *16th Annual Symposium on Combinatorial Pattern Matching (CPM05)*, volume 3537 of *LNCS*, 2005.
20. R. M. McConnell and F. de Montgolfier. Algebraic Operations on PQ Trees and Modular Decomposition Trees. In *31st International Workshop on Graph-Theoretic Concepts in Computer Science (WG05)*, 2005. to appear in LNCS.
21. R.M. McConnell and J.P. Spinrad. Modular decomposition and transitive orientation. *Discrete Mathematics*, 201:189–241, 1999.
22. R.H. Möhring and F.J. Radermacher. Substitution decomposition for discrete structures and connections with combinatorial optimization. *Annals of Discrete Mathematics*, 19:257–356, 1984.
23. T. Uno and M. Yagiura. Fast algorithms to enumerate all common intervals of two permutations. *Algorithmica*, 26(2):290–309, 2000.
24. D. J. A. Welsh. Matroids: Fundamental concepts. In *Handbook of Combinatorics*, volume 1, pages 481–526. North-Holland, 1995.

Generating Cut Conjunctions and Bridge Avoiding Extensions in Graphs*

L. Khachiyan[1], E. Boros[2], K. Borys[2], K. Elbassioni[3], V. Gurvich[2], and K. Makino[4]

[1] Department of Computer Science, Rutgers University, 110 Frelinghuysen Road, Piscataway NJ 08854**

[2] RUTCOR, Rutgers University, 640 Bartholomew Road, Piscataway NJ 08854
{boros, kborys, gurvich}@rutcor.rutgers.edu

[3] Max-Planck-Institut für Informatik, Saarbrücken, Germany
elbassio@mpi-sb.mpg.de

[4] Division of Mathematical Science for Social Systems, Graduate School of Engineering Science, Osaka University, Toyonaka, Osaka, 560-8531, Japan
makino@sys.es.osaka-u.ac.jp

Abstract. Let $G = (V, E)$ be an undirected graph, and let $B \subseteq V \times V$ be a collection of vertex pairs. We give an incremental polynomial time algorithm to enumerate all minimal edge sets $X \subseteq E$ such that every vertex pair $(s,t) \in B$ is disconnected in $(V, E \setminus X)$, generalizing well-known efficient algorithms for enumerating all minimal s-t cuts, for a given pair $s, t \in V$ of vertices. We also present an incremental polynomial time algorithm for enumerating all minimal subsets $X \subseteq E$ such that no $(s,t) \in B$ is a bridge in $(V, X \cup B)$. These two enumeration problems are special cases of the more general cut conjunction problem in matroids: given a matroid M on ground set $S = E \cup B$, enumerate all minimal subsets $X \subseteq E$ such that no element $b \in B$ is spanned by $E \setminus X$. Unlike the above special cases, corresponding to the cycle and cocycle matroids of the graph $(V, E \cup B)$, the enumeration of cut conjunctions for vectorial matroids turns out to be NP-hard.

1 Introduction

The *cut enumeration problem* for graphs calls for listing all minimal subsets of edges whose removal disconnects two specified vertices of a given graph. This so called *two-terminal cut enumeration problem* is known to be solvable in $O(Nm+m+n)$ time and $O(n+m)$ space [10], where n and m are the numbers of vertices and edges in the input graph, and N is the number of cuts. In this paper, we study the following extension of the two-terminal cut enumeration problem:

* This research was partially supported by the National Science Foundation (Grant IIS-0118635), and by DIMACS, the National Science Foundation's Center for Discrete Mathematics and Theoretical Computer Science.

** Our friend and co-author, Leonid Khachiyan passed away with tragic suddenness, while we were working on this paper.

X. Deng and D. Du (Eds.): ISAAC 2005, LNCS 3827, pp. 156–165, 2005.

Cut Conjunctions in Graphs: *Given an undirected graph* $G = (V, E)$, *and a collection* $B = \{(s_1, t_1), \ldots, (s_k, t_k)\}$ *of* k *vertex pairs* $s_i, t_i \in V$, *enumerate all minimal edge sets* $X \subseteq E$ *such that for all* $i = 1, \ldots, k$, *vertices* s_i *and* t_i *are disconnected in* $G' = (V, E \setminus X)$.

Note that s_i and s_j or s_i and t_j or t_i and t_j may coincide for $i \neq j$. We call a minimal edge set $X \subseteq E$ for which all pairs of vertices $(s_i, t_i) \in B$ are disconnected in the subgraph $G' = (V, E \setminus X)$, a *minimal B-cut*, or simply a *cut conjunction* if B is clear from the context. Observe that each cut conjunction must indeed be the union (conjunction) of some minimal s_i-t_i cuts for $i = 1, \ldots, k$, justifying the naming of these edge sets. Let us also note that the enumeration of cut conjunctions cannot efficiently be reduced to two-terminal cut enumeration, because the majority of conjunctions of minimal s_i-t_i cuts, for $i = 1, \ldots, k$, may be non minimal B-cuts (see [1] for an example).

In what follows, we assume without any loss of generality that in the above cut conjunction problem, no pair of vertices (s_i, t_i) is connected by an edge of G, i.e. $E \cap B = \emptyset$ (since all such edges would have to belong to all cut conjunctions).

When B is the collection of all pairs of distinct vertices drawn from some vertex set $V' \subseteq V$, minimal B-cuts are known as *multiway cuts*, see e.g. [11]. The enumeration of cut conjunctions in graphs thus also includes the enumeration of multiway cuts.

It will be convenient to consider the cut conjunction problem for graphs in the context of the more general cut enumeration problem for (vectorial) matroids (see e.g. [7,12] for a thorough introduction to matroid theory).

Cut Conjunctions in Matroids: *Given a matroid* M *on ground set* S *and a set* $B \subseteq S$, *enumerate all maximal sets* $X \subseteq A = S \setminus B$ *that span no element of* B.

When M is the cycle matroid of a graph $G = (V, E \cup B)$, where $E \cap B = \emptyset$, we can let $S = E \cup B$, and then by definition, an edge set $Y \subseteq A = E$ spans $b = (s_i, t_i) \in B$ if and only if Y contains an s_i-t_i path. This means that a maximal edge set $Y \subseteq E$ spans no edge $b \in B$ if and only if $X = E \setminus Y$ is a minimal B-cut in the graph (V, E). Thus, in this special case the cut conjunction problem in matroids is equivalent with the cut conjunction problem in graphs.

Let $r : S \to \mathbb{Z}_+$ be the rank function of a matroid M on S. The *dual* matroid M^* on S is defined by the rank function $r^*(X) = r(S \setminus X) + |X| - r(S)$, see e.g. [12]. In particular, $Y \subseteq A = S \setminus B$ spans $b \in B$ in M^* if and only if $r^*(Y \cup \{b\}) = r^*(Y)$, which is equivalent to $r(X \cup B) = r(X \cup (B \setminus b)) + 1$, where as before, $X = A \setminus Y$ denotes the set complimentary to Y in A. This means that the cut conjunction problem for the dual matroid M^* is equivalent to the following enumeration problem:

Dual Formulation of Cut Conjunctions in Matroids: *Given a matroid* M *on ground set* S *and a set* $B \subseteq S$, *enumerate all minimal sets* $X \subseteq A = S \setminus B$ *such that each element* $b \in B$ *is spanned by* $X \cup (B \setminus b)$.

In particular, when M is the cycle matroid of a graph $G = (V, E)$ (and consequently, M^* is the cocyle matroid of G), the dual formulation of the cut conjunction problem for matroids leads to the following enumeration problem:

Bridge-Avoiding Extensions in Graphs: *Given an undirected graph $G = (V, E)$, and a collection of edges $B \subseteq E$, enumerate all minimal edge sets $X \subseteq E \setminus B$ such that no edge $b \in B$ is a bridge in $G' = (V, B \cup X)$.*

Let us note that in all of the mentioned problems, the output may consist of exponentially many sets, in terms of the input size. Thus, the efficiency of such enumeration algorithms customarily is measured in both the input and output sizes (see e.g., [6]). In particular, it is said that a family $\mathcal{F}$ can be enumerated in *incremental polynomial time*, if for any subfamily $\mathcal{F}' \subseteq \mathcal{F}$ the problem of finding $e \in \mathcal{F} \setminus \mathcal{F}'$ or proving that $\mathcal{F}' = \mathcal{F}$ can be solved in $poly(n, |\mathcal{F}'|)$ time, where n denotes the input size of the problem. The enumeration problem of $\mathcal{F}$ is called NP-hard, if deciding $\mathcal{F}' \neq \mathcal{F}$ for subfamilies $\mathcal{F}' \subseteq \mathcal{F}$ is NP-hard, in general. It can be shown that if the enumeration problem for $\mathcal{F}$ is NP-hard, then no algorithm can generate all elements of $\mathcal{F}$ in time $poly(n, |\mathcal{F}|)$, unless P=NP.

1.1 Our Results

We show that all of the above enumeration problems *for graphs* can be solved efficiently, i.e. in incremental polynomial time.

Theorem 1. *All cut conjunctions for a given set of vertex pairs in a graph can be enumerated in incremental polynomial time.*

Theorem 2. *All minimal bridge-avoiding extensions for a given set of edges in a graph can be enumerated in incremental polynomial time.*

In contrast, it can be shown that the more general cut conjunction problem for vectorial matroids is NP-hard:

Proposition 1. *[3] Let M be a vectorial matroid defined by a collection S of n-dimensional vectors over a field of characteristic zero or of large enough characteristic (at least $8n$), let B be a given subset of S and let $\mathcal{F}$ be the family of all maximal subsets of $A = S \setminus B$ that span no vector $b \in B$. Given a subfamily $\mathcal{X} \subseteq \mathcal{F}$, it is NP-hard to decide if $\mathcal{X} \neq \mathcal{F}$.*

In addition to indicating that the enumeration of cut conjunctions in vectorial matroids cannot be solved in incremental (or output) polynomial time, unless P=NP, the above result also implies that the dual formulation of the cut conjunction problem for vectorial matroids is similarly NP-hard. This is, because the dual of an explicitly given vectorial matroid M over a field $\mathbf{F}$ is again a vectorial matroid over the same field, for which an explicit representation can be obtained from M efficiently (see e.g., [9]).

As stated in Proposition 1, our NP-hardness result for cut conjunctions in vectorial matroids is valid for vectorial matroids over sufficiently large fields. In particular, the complexity of enumerating cut conjunctions in *binary* matroids remains open. We can only show that this problem is tractable for $|B| = 2$:

Proposition 2. *Let M be a binary matroid on ground set S and let $B = \{b_1, b_2\} \subseteq S$. All maximal subsets X of $A = S \setminus B$ which span neither b_1 nor b_2 can be enumerated in incremental polynomial time.*

Finally, it is worth mentioning that for an arbitrary B, the cut conjunction problem in binary matroids includes, as a special case, the well-known *hypergraph transversal* (also known as *hypergraph dualization*) problem [4,5]: *enumerate all maximal independent sets (equivalently, minimal transversals) for an explicitly given hypergraph* $\mathcal{H} \subseteq 2^V$. To see this inclusion, let B be the $n \times |\mathcal{H}|$ binary matrix whose columns are the characteristic vectors of the hyperedges of $\mathcal{H}$, and let I be the $n \times n$ identity matrix. Letting $M = [I, B]$ and denoting by A the columns set of I, we can identify each maximal subset of A which spans no columns of B with a maximal independent vertex set for $\mathcal{H}$. This shows that listing cut conjunctions for a binary matroid is at least as hard as listing all maximal independent sets for a hypergraph. The fastest algorithm for hypergraph dualization runs in quasi-polynomial time $poly(n) + N^{o(\log N)}$, where N is the sum of the input and output sizes, see [5].

1.2 The $X - e + Y$ Method

We prove Theorems 1, 2 by using a generic approach discussed below. Let E be a finite set and let $\pi : 2^E \to \{0, 1\}$ be a monotone Boolean function defined on the subsets of E, i.e., for which $\pi(X) \leq \pi(Y)$ whenever $X \subseteq Y$. Suppose that an efficient algorithm is available for evaluating $\pi(X)$ in $poly(|E|)$ time for any $X \subseteq E$, and that our goal is to enumerate all minimal subsets $X \subseteq S$ for which $\pi(X) = 1$. In particular, the enumeration of cut conjunctions (CC) and bridge avoiding subsets (BA) can all be embedded into this general scheme by letting $\pi_{CC}(X) = 1$ iff vertex s_i is disconnected from t_i in $(V, E \setminus X)$ for all $i = 1, \ldots, k$ and $\pi_{BA}(X) = 1$ iff no $b \in B$ is a bridge in $(V, B \cup X)$.

Returning to the general scheme, given a monotone Boolean function π, let $\mathcal{F} = \{X \mid X \subseteq E$ is a minimal set satisfying $\pi(X) = 1\}$, and let $\mathcal{G} = (\mathcal{F}, \mathcal{E})$ be the directed "supergraph" with the vertex set $\mathcal{F}$ in which two vertices $X, X' \in \mathcal{F}$ are connected by an arc $(X, X') \in \mathcal{E}$ if and only if X' can be obtained from X by the following process:

(p1) Delete an element e from X ($\pi(X \setminus e) = 0$ by the minimality of X).
(p2) Add a minimal set $Y \subseteq E \setminus X$ to restore the property $\pi((X \setminus e) \cup Y) = 1$.
(p3) Assuming a linear order on the elements of E, delete the lexicographically first minimal set $Z \subseteq X \setminus e$ for which $X' = (X \setminus (Z \cup e)) \cup Y \in \mathcal{F}$.

Note that in step **(p3)** an arbitrary fixed linear order can be used, and that the lexicographic minimization can be done efficiently, because we assume that evaluating $\pi(\cdot)$ takes $poly(|E|)$ time. Note also that in step **(p3)** we can always find a subset $Z \subseteq X \setminus e$ for which $X' = (X \setminus (Z \cup e)) \cup Y$ belongs to $\mathcal{F}$, due to the minimality of Y chosen in step **(p2)**. Since an out-neighbor of vertex X in $\mathcal{G}$ is generated for all elements $e \in X$ in step **(p1)** and all possible minimal sets Y in step **(p2)**, the resulting supergraph $\mathcal{G} = (\mathcal{F}, \mathcal{E})$ is always strongly connected.

Proposition 3. *For any monotone Boolean function $\pi : 2^E \to \{0,1\}$, and any linear ordering of E, the supergraph $\mathcal{G} = (\mathcal{F}, \mathcal{E})$ is strongly connected.*

Proof. Let $X, X' \in \mathcal{F}$ be two vertices of $\mathcal{G}$. We show by induction on $|X \setminus X'|$ that $\mathcal{G}$ contains a directed path from X to X'. If $X \setminus X' = \emptyset$ then $X \subseteq X'$, but since X' is minimal, $X = X'$ must follow. Suppose that $|X \setminus X'| > 0$ and let us show that there is an out-neighbor X'' of X such that $|X'' \setminus X'|$ is smaller than $|X \setminus X'|$. Choose an arbitrary $e \in X \setminus X'$. Since $(X \setminus e) \cup X'$ contains X' and $\pi(X') = 1$, we have $\pi((X \setminus e) \cup X') = 1$ by the monotonicity of π, and hence there is a minimal nonempty set $Y \subseteq X' \setminus X$ such that $\pi((X \setminus e) \cup Y) = 1$. Now we can lexicographically delete some elements Z of $X \setminus e$ and obtain an out-neighbor $X'' = (X \setminus (Z \cup e)) \cup Y \in \mathcal{F}$ for which $|X'' \setminus X'| \leq |X \setminus (X' \cup e)| < |X \setminus X'|$. Such a set Z exists because we have $\pi((X \setminus e) \cup (Y \setminus y)) = 0$ for all $y \in Y$ by the minimality of Y, and thus any minimal set $\tilde{X} \in \mathcal{F}$ contained in $(X \setminus e) \cup Y$ must contain Y. □

Let us remark that the number of minimal sets Y in **(p2)** may be exponential. For a given set $X \in \mathcal{F}$ and element $e \in X$, any two distinct minimal sets, Y and Y', corresponding to X and e, produce different neighbors of X in $\mathcal{G}$. Hence, every neighbor of X in $\mathcal{G}$ may be generated at most $|X|$ times.

Since the supergraph $\mathcal{G} = (\mathcal{F}, \mathcal{E})$ is always strongly connected, we can generate $\mathcal{F}$ by first computing an initial vertex $X^o \in \mathcal{F}$ and then performing a traversal (say, breadth-first search) of $\mathcal{G}$. Given our assumption that π can be evaluated in *poly*($|E|$) time, computing an initial vertex of $\mathcal{G}$ can be done in polynomial time. Steps **(p1)** and **(p3)** can also be performed in *poly*($|E|$) time. Hence we can conclude that the enumeration problem for $\mathcal{F}$ reduces to the enumeration of sets Y in step **(p2)**. In particular, due to the above remark we get the following statement:

Proposition 4. *All elements of $\mathcal{F}$ can be enumerated in incremental polynomial time whenever the enumeration problem* **(p2)** *can be done in incremental polynomial time.* □

As an illustration for the $X - e + Y$ method, consider the *path conjunction problem [2]: given an undirected graph $G = (V, E)$ and a collection of k vertex pairs $s_i, t_i \in V$, enumerate all minimal edge sets $X \subseteq E$ such that for all $i = 1, \ldots, k$, vertices s_i and t_i are connected in* (V, X). Let us define in this case $\pi(X) = 1$ if and only if every s_i is connected to t_i in (V, X). Then the minimal edge sets X for which $\pi(X) = 1$ are exactly the minimal path conjunctions. Furthermore, any such minimal path conjunction is a collection of trees $T_1, \ldots, T_l$ such that each vertex pair (s_i, t_i) belongs to a common tree T_j. Removing an edge e from X splits one of the trees into two sub-trees T'_j, T''_j such that there is at least one pair (s_i, t_i) with one vertex belonging to T'_j and the other to T''_j. Let G' be the graph obtained from G by contracting each tree of $T_1, \ldots, T'_j, T''_j, \ldots, T_l$ into a vertex, and let u and v denote the vertices corresponding to T'_j and T''_j. A minimal edge set Y restores the property that all s_i and t_i are connected in $(V, (X \setminus e) \cup Y)$ if and only if Y is a simple path from u to v in G'. Hence the $X - e + Y$ method reduces the problem to the enumeration of all u-v paths in G', which can be done via backtracking [8] incrementally efficiently. Thus by Proposition 4 enumerating all minimal path conjunctions can be done in incremental polynomial time [2].

Our proofs of Theorems 1, 2 follow this approach using the two monotone Boolean functions π_{CC}, π_{BA} defined at the beginning of this section. The remainder of the paper is organized as follows. We provide sketches for the proofs of Theorems 1 and 2 in Section 2 and 3, respectively. For a full version with all related results and proofs see [1].

2 Proof of Theorem 1

In this section we apply the $X - e + Y$ method to the Boolean function π_{CC} in order to enumerate all cut conjunctions in graphs. First, in Subsection 2.1 we state a characterization of cut conjunctions in graphs. In Subsection 2.2 we reduce the problem of enumerating all minimal sets Y in **(p2)** to the enumeration of all cut conjunctions in a graph of a simpler structure. In Subsection 2.3 we show that the latter problem can be solved efficiently by using a variant of the supergraph approach.

For a graph $G = (V, E)$, a vertex subset $U \subseteq V$, and edge subset $F \subseteq E$, let us denote by $G[U]$ the subgraph of G induced by U, by $G - U = G[V \setminus U]$ the graph obtained from G by deleting U together with all incident edges, and by $G - F = (V, E \setminus F)$ the subgraph of G obtained by deleting the edge set F. Furthermore, we write $G + U = (V \cup U, E)$, $G + F = (V, E \cup F)$, and $G - G'$ for the subgraph $G - V'$, when $G' = (V', E')$ is a subgraph of G. Finally, we write $G' + G'' = (V' \cup V'', E' \cup E'')$ for two graphs $G' = (V', E')$ and $G'' = (V'', E'')$.

2.1 Characterization of Minimal Cut Conjunctions in Graphs

Let $G = (V, E)$ be an undirected connected graph. It will be convenient to define a *cut* to be a set of edges $E(G_1, \ldots, G_l) = \bigcup_{i \neq j} \{uv \in E : u \in G_i, v \in G_j\}$ where $G_1, \ldots, G_l$ are induced subgraphs of G such that their vertex sets partition V, and G_i is connected for each $i = 1, \ldots, l$. Let $B = \{(s_1, t_1), \ldots, (s_k, t_k)\}$ be a set of distinct source-sink pairs of G. A *B-cut* is a cut $E(G_1, \ldots, G_l)$ such that, for each i, s_i and t_i do not belong to the same G_j. If the set B is clear from the context we shall call the minimal B-cut a *cut conjunction.* The following characterization of cut conjunctions follows directly from their definition.

Proposition 5. *Let $E(G_1, G_2, \ldots, G_l)$ be a B-cut. Then, $E(G_1, G_2, \ldots, G_l)$ is a minimal B-cut if and only if for every $x, y \in \{1, \ldots, l\}$ with $x \neq y$, if there is an edge of G between G_x and G_y then there must exist a source-sink pair (s_i, t_i) with exactly one vertex in G_x and the other in G_y.*

2.2 Reduction

In this section we reduce the problem of generating all minimal sets Y in **(p2)** to generating all cut conjunctions in a graph of a simpler structure.

Let F be a subset of edges of G and let $(s_i, t_i) \in B$. Suppose that s_i and t_i are in the same component of $G - F$. Then we say that (s_i, t_i) is *F-conflicting.* Let $X = E(G_1, G_2, \ldots, G_l)$ be a minimal B-cut of G and let $b \in X$. The removing b from X reconnects some two components, G_x and G_y, of $G - X$, where one endpoint of b is in G_x and the other in G_y. Thus $G - (X \setminus b)$ contains at least one $(X \setminus b)$-conflicting pair. Hence generating all minimal sets $Y \subseteq E \setminus X$ which restore the property that no s_i is connected to t_i, is equivalent to generating all minimal B'-cuts in the graph $G_x + G_y + b$ where B' is the set of $(X \setminus b)$-conflicting pairs. Let $L = G_x$ and $R = G_y$. We can always relabel the $(X \setminus b)$-conflicting pairs to guarantee that the conflicting s_i's are in L and the conflicting t_i's are in R. We denote the resulting graph by $H(X, b)$.

2.3 Enumerating Minimal Cut Conjunctions in $H(X, b)$

Let $H = H(X, b) = (V, E)$ be the graph defined at the end of Section 2.2, that is: $H = L + R + b$, where $b = v_L v_R$ is a bridge, L contains the sources $s_1, \ldots, s_k$, and R contains the sinks $t_1, \ldots, t_k$. Let $B = \{(s_1, t_1), \ldots, (s_k, t_k)\}$ and let $K = E(G_1, \ldots, G_l)$ be a cut conjunction of H, such that $K \neq \{b\}$. Without loss of generality, assume that b is in G_1. Note that every other G_j is contained either in L or in R (since G_j is connected and all paths from L to R go through b). We denote by $M = G_1$ the component containing b and call it the *root component of* K. The other components will be called *leaf components of* K. Denote the G_j's contained in L by $L_1, \ldots, L_m$ and those in R by $R_1, \ldots, R_n$. Observe that all edges of $K = E(M, L_1, \ldots, L_m, R_1, \ldots, R_n)$ lie between the root and leaf components. Hence M uniquely determines the leaf components of K. Now we define the digraph $\mathcal{H}$, the *supergraph of cut conjunctions of* H. The vertex set of $\mathcal{H}$ is the family of all cut conjunctions of H other than $\{b\}$. For each cut conjunction $K = E(M, L_1, \ldots, L_m, R_1, \ldots, R_n)$ of H we define its out-neighborhood to consist of all cut conjunctions which can be obtained from K by the following sequence of steps:

(q1) Choose a vertex v incident to $e \in K$ such that $v \notin \{v_L, v_R\}$. Depending on v we have the following three cases.
(q2-a) Suppose $v \in R_j$ is in a leaf component of K and $M+v+e$ does not contain a source-sink pair (s_i, t_i). Thus either v is not a sink, or $v = t_i$ and $s_i \notin M$. Let $W_1, \ldots, W_p$ be the components of $R_j - v$, and let $M_D = M + v + \bigcup_{u \in M}\{uv \in E\}$. Then $D = E(M_D, L_1, \ldots, L_m, R_1, \ldots, R_{j-1}, W_1, \ldots, W_p, R_{j+1}, \ldots, R_n)$ is a B-cut. Note that we have moved v from R_j to M. Removing v from R_j splits R_j into components $W_1, \ldots, W_p$.
(q2-b) Suppose $v \in R_j$ is in a leaf component of K and $M + v + e$ contains a source-sink pair (s_i, t_i). Thus $v = t_i$, $s_i \in M$ and $v_L \neq s_i$. Let $W_1, \ldots, W_p$ be the components of $R_j - t_i$ and let $U_1, \ldots, U_r$ be the components of $M - s_i$ not containing b. Denote $M_D = (M + t_i + \bigcup_{u \in M}\{ut_i \in E\}) - (s_i + U_1 + \ldots + U_r)$. Then $D = E(M_D, L_1, \ldots, L_m, s_i, U_1, \ldots, U_r, R_1, \ldots, R_{j-1}, W_1, \ldots, W_p, R_{j+1}, \ldots, R_n)$ is a B-cut. Note that we have moved t_i from R_j to M. To restore

the property that no s_i is connected to t_i, we have removed s_i from M. Removing v from R_j splits R_j into components $W_1, \ldots, W_p$, and removing s_i from M splits M into components $U_1, \ldots, U_r$ and M_D, the component containing b.
(q2-c) Suppose $v \in M - \{v_L, v_R\}$ and v is adjacent to L_j. Note that $v \notin \{t_1, \ldots, t_k\}$. Let $U_1, \ldots, U_r$ be the components of $M - v$ not containing b, and let $M_D = M - (v + U_1 + \ldots + U_r)$. Then $D = E(M_D, L_1, \ldots, L_{j-1}, L_j + v + \bigcup_{u \in L_j} uv \in E, L_{j+1}, \ldots, L_m, U_1, \ldots, U_r, R_1, \ldots, R_n)$ is a B-cut. Note that we have moved v from M to L_j splitting M into components $U_1, \ldots, U_r$ and M_D.
(q3) Let $D = E(G_1, \ldots, G_l)$ be the B-cut obtained in the previous step. Choose the lexicographically first two sets G_x and G_y such that there is an edge $e \in D$ connecting G_x and G_y and there is no $(D \setminus e)$-conflicting pair. Replace G_x and G_y in D by $G_x + G_y$. Repeat until no such edge exists, thus the B-cut is minimal. Let $K' = E(M', L'_1, \ldots, L'_{m'}, R'_1, \ldots, R'_{n'})$ be the resulting cut conjunction. Then K' is a neighbor of K in $\mathcal{H}$.

Proposition 6. *The supergraph $\mathcal{H}$ is strongly connected.*

This claim will follow from lemmas 1 and 2. To state these, let us consider two cut conjunctions K_1 and K_3, and denote by M_1, M_3 their root components. We call the vertices of M_3 *blue*, all others *green*, and denote by $\mathcal{K}$ the subgraph of $\mathcal{H}$ induced by cut conjunctions whose root components contain all blue vertices.

Lemma 1. *There exists a cut conjunction $K_2 \in \mathcal{K}$ such that there is a path from K_1 to K_2 in $\mathcal{H}$.*

Proof. Let T be an arbitrary spanning tree of M_3 containing the bridge b. For a B-cut D of H with M as its root component, we partition the edges of T into two groups. We call an edge e of T a *D-solid edge*, if $e \in M$ and e is reachable from b by using only edges of T that are in M. Otherwise e is called a *D-dashed edge*. Note that b is D-solid edge. We denote the set of D-solid edges by S_D and the set of D-dashed edges by D_D. Clearly, $|S_D| + |D_D| = |T|$.

Let $K_1 = E(M_1, L_1, \ldots, L_m, R_1, \ldots, R_n)$. We will show by induction on $|S_{K_1}|$ that there is a path from K_1 to K_2. If $|S_{K_1}| = |T|$, then $K_1 \in \mathcal{K}$. If $|S_{K_1}| < |T|$, then there exists a K_1-dashed edge vw between two blue vertices v and w such that v is in a leaf component of K_1, $w \in M_1$ and w is incident to a K_1-solid edge. Suppose that $v \in R_j$. We now show that K'_1, a neighbor of K_1, obtained by moving v from the leaf to the root component, has $|S_{K'_1}| \geq |S_{K_1}| + 1$.

Case 1: v is not a sink or $v = t_i$ and $s_i \notin M_1$. Let D be a B-cut obtained in **(q2-a)** and $M_D = M_1 + v$ be its root component. Since M_D contains both v and w, vw is a D-solid edge, so $|S_D| = |S_{K_1}| + 1$. In **(q3)** M_D can only merge with leaf components, hence $|S_{K'_1}| \geq |S_D|$.

Case 2: $v = t_i$, $s_i \in M$. Note that t_i is a blue vertex, so s_i must be green, since M_3 does not contain any source-sink pair, and in particular s_i cannot be an endpoint of b. Let D be a B-cut obtained in **(q2-b)** and M_D be its root component. Recall that $M_D = (M_1 + t_i) - (s_i + U_1 + \ldots + U_r)$, where $U_1, \ldots, U_r$ are the components of $M - s_i$ not containing b. Observe that in **(q2-b)** we did not remove any K_1-solid edge from M_1. Since s_i is a green vertex, all edges incident to s_i do not belong to T. Edges in $U_1, \ldots, U_r$ and incident to these components are also not K_1-solid, because all paths from b to $U_1, \ldots, U_r$, which use edges of T that are in M_1, must go through s_i. Thus $|S_D| = |S_{K_1}| + 1$. In **(q3)** M_D can only merge with leaf components, hence $|S_{K_1'}| \geq |S_D|$. □

Lemma 2. *For every $K_2 \in \mathcal{K}$ there is a path from K_2 to K_3 in $\mathcal{K}$.*

This lemma follows similarly to the previous one, see [1] for details.

Note finally that since $\mathcal{H}$ is strongly connected and finding an initial vertex of $\mathcal{H}$ is easy, we can enumerate all sets Y in **(p2)** in incremental polynomial time, completing the proof of Proposition 6. □

3 Proof of Theorem 2

In this section we assume that $G = (V, E)$ is a multigraph. Given $B \subseteq E$, let $\mathcal{F}$ = minimal $\{X \subseteq E \setminus B \mid \text{no } b \in B \text{ is a bridge of } (V, B \cup X)\}$. We enumerate $\mathcal{F}$ by using the $X - e + Y$ method stated in Section 1.2. Proposition 7 below implies that this can be accomplished in incremental polynomial time.

Proposition 7. *Given a set $X \in \mathcal{F}$ and an edge $e \in X$, all sets Y in* **(p2)** *can be enumerated with polynomial delay.*

Proof. Let $B' = \{b_1, \ldots, b_k\}$ be the subset of edges of B that are bridges in $(V, B \cup (X \setminus e))$. First observe that for each edge $b_i \in B'$ there is a cycle C_i in $(V, B \cup X)$ containing e and b_i. Suppose $b_i \in C_i \setminus C_j$ for some $i, j \in \{1, \ldots, k\}$. Then there is a cycle $C' \subseteq (C_i \cup C_j) \setminus e$ such that $b_i \in C'$. C' is also the cycle in $(V, B \cup (X \setminus e))$. This would contradict the definition of B'. Hence all edges of B' lie on a common cycle in $(V, B \cup X)$ containing e, and accordingly, all edges of B' belong to a common path in $(V, B \cup (X \setminus e))$.

Let $G' = (V', E')$ be the multigraph obtained from $(V, E \setminus e)$ by contracting all edges in $(B \setminus B') \cup (X \setminus e)$. Then B' is a path in G'. Furthermore, the enumeration of all sets Y in **(p2)** now reduces to the enumeration of all minimal edge sets Y' of G' for which no edge b on the path B' is a bridge in $(V', B' \cup Y')$. Thus the enumeration of cut conjunctions in cocycle matroids reduces to the special case of the same problem for multigraphs in which B is a path.

Now we argue that the latter problem is equivalent to the enumeration of all directed paths between a pair of vertices in some explicitly given directed multigraph. To see this, denote by $u_1, \ldots, u_{k+1}$ the $k+1$ vertices on the path $B' = \{b_1, \ldots, b_k\}$ in G', and assume that $b_i = u_i u_{i+1}$ for $i = 1, \ldots, k$. If no edge $b \in B'$ is a bridge in $(V', B' \cup Y')$, then for each $i = 1, \ldots, k$ there must exist a path $P \subseteq Y'$ such that P and B' are edge disjoint and the vertex set of P contains exactly two vertices u_α, u_β of B' such that $\alpha \leq i$ and $\beta \geq i+1$. Thus $Y' = P_1 \cup \ldots \cup P_s$, for some paths $P_1, \ldots, P_s$ satisfying conditions above, where no two distinct paths in Y' have a common vertex outside of B'. Denoting by u_{α_i} and u_{β_i} the intersection of the vertex set of P_i with B', we conclude $u_1 = \alpha_1 < \alpha_2 \leq \beta_1 < \alpha_3 \leq \beta_2 < \alpha_4 \leq \ldots < \alpha_s \leq \beta_{s-1} < \beta_s = u_{k+1}$.

Let us now consider the directed multigraph $\overrightarrow{G}' = (V', \overrightarrow{E}')$ obtained from the multigraph $G' = (V', E')$ by replacing the undirected path B' by the directed path $\overrightarrow{B}' = u_1 \leftarrow u_2 \leftarrow \ldots \leftarrow u_{k+1}$ and by adding two opposite arcs $u \rightarrow v$ and $v \rightarrow u$ for each edge $uv \in E' \setminus B'$. From the above discussion it follows that there exists a one to one correspondence between all minimal sets Y' and all minimal directed paths from u_1 to u_{k+1} in $\overrightarrow{G}'$. Since it is well known that all minimal directed paths between a given pair of vertices can be enumerated via backtracking [8] with polynomial delay, Proposition 7 follows. □

References

1. E. Boros, K. Borys, K. Elbassioni, V. Gurvich, L. Khachiyan, and K. Makino. Enumerating cut conjunctions in graphs and related problems. Rutcor Research Report RRR 19-2005, Rutgers University.
2. E. Boros, K. Elbassioni, V. Gurvich, L. Khachiyan, and K. Makino. Generating paths and cuts in multi-pole (di)graphs. In J. Fiala, V. Koubek, and J. Kratochvil, editors, *Mathematical Foundations of Computer Science MFCS*, volume 3153 of *Lecture Notes in Computer Science*, pages 298–309, Prague, Czech Republic, August 22-27 2004. Springer Verlag, Berlin, Heidelberg, New York.
3. E. Boros, K. Elbassioni, V. Gurvich, L. Khachiyan, and K. Makino. On the complexity of some enumeration problems for matroids. To appear in SIAM Journal on Discrete Mathematics, 2005.
4. T. Eiter and G. Gottlob. Identifyig the minimal transversals of a hypergraph and related problems. *SIAM Journal on Computing*, 24:1278–1304, 1995.
5. M. Fredman and L. Khachiyan. On the complexity of dualization of monotone disjunctive normal forms. *Journal of Algorithms*, 21:618–628, 1996.
6. E. Lawler, J. K. Lenstra, and A. H. G. Rinnooy Kan. Generating all maximal independent sets: NP-hardness and polynomial-time algorithms. *SIAM Journal on Computing*, 9:558–565, 1980.
7. J.G. Oxley. *Matroid Theory*. Oxford University Press, Oxford, New York, Tokyo, 1992.
8. R. C. Read and R. E. Tarjan. Bounds on backtrack algorithms for listing cycles, paths, and spanning trees. *Networks*, 5:237–252, 1975.
9. A. Schrijver. *Combinatorial Optimization: Polyhedra and Efficiency*, volume B. Springer Verlag, Berlin, Heidelberg, New York, 2003. page 654.
10. S. Tsukiyama, I. Shirakawa, H. Ozaki, and H. Ariyoshi. An algorithm to enumerate all cutsets of a graph in linear time per cutset. *Journal of the Association for Computing Machinery*, 27:619–632, 1980.
11. V. Vazirani. *Approximation Algorithms*. Springer Verlag, Berlin, Heidelberg, New York, 2001.
12. D.J.A. Welsh. *Matroid Theory*. Academic Press, London, 1976.

Orthogonal Drawings of Series-Parallel Graphs with Minimum Bends*

Xiao Zhou and Takao Nishizeki

Graduate School of Information Sciences, Tohoku University,
Aoba-yama 6-6-05, Sendai, 980-8579, Japan
{zhou, nishi}@ecei.tohoku.ac.jp

Abstract. In an orthogonal drawing of a planar graph G, each vertex is drawn as a point, each edge is drawn as a sequence of alternate horizontal and vertical line segments, and any two edges do not cross except at their common end. A bend is a point where an edge changes its direction. A drawing of G is called an optimal orthogonal drawing if the number of bends is minimum among all orthogonal drawings of G. In this paper we give an algorithm to find an optimal orthogonal drawing of any given series-parallel graph of the maximum degree at most three. Our algorithm takes linear time, while the previously known best algorithm takes cubic time. Furthermore, our algorithm is much simpler than the previous one. We also obtain a best possible upper bound on the number of bends in an optimal drawing.

Keywords: orthogonal drawing, bend, series-parallel graph, planar graph.

1 Introduction

Automatic graph drawings have numerous applications in VLSI circuit layouts, networks, computer architecture, circuit schematics, etc. [3, 10]. Many graph drawing styles have been introduced [1, 3, 8, 10, 14, 16]. Among them, an "orthogonal drawing" has attracted much attention due to its various applications, specially in circuit schematics, entity relationship diagrams, data flow diagrams, etc. [13, 15, 18, 19]. An *orthogonal drawing* of a planar graph G is a drawing of G such that each vertex is mapped to a point, each edge is drawn as a sequence of alternate horizontal and vertical line segments, and any two edges do not cross except at their common end. A point where an edge changes its direction in a drawing is called a *bend* of the drawing. Figure 1(a) depicts an orthogonal drawing of the planar graph in Fig. 1(b); the drawing has exactly one bend on the edge joining vertices g and t. If a planar graph G has a vertex of degree five or more, then G has no orthogonal drawing. On the other hand, if G has no vertex of degree five or more, that is, the maximum degree Δ of G is at most four, then G has an orthogonal drawing, but may need bends. If a planar graph

* This work is supported by JSPS grants.

X. Deng and D. Du (Eds.): ISAAC 2005, LNCS 3827, pp. 166–175, 2005.

represents a VLSI routing, then one may be interested in an orthogonal drawing such that the number of bends is as small as possible, because bends increase the manufacturing cost of a VLSI chip. An orthogonal drawing of a planar graph G is called an *optimal* orthogonal drawing if it has the minimum number of bends among all possible orthogonal drawings of G.

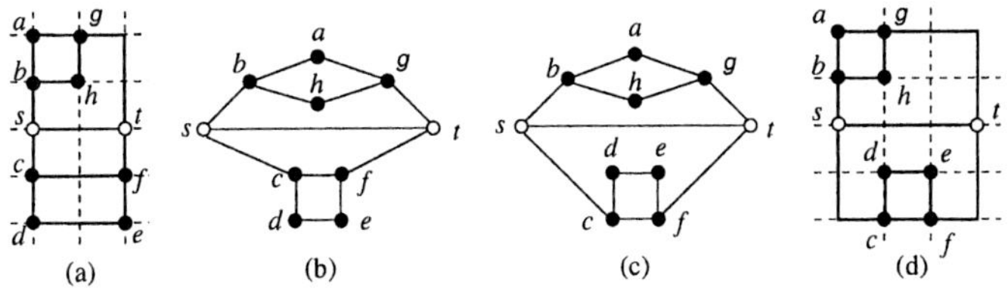

Fig. 1. (a) An optimal orthogonal drawing with one bend, (b), (c) two embeddings of the same planar graph, and (d) an orthogonal drawing with three bends

The problem of finding an optimal orthogonal drawing is one of the most famous problems in the graph drawing literature [3, 10] and has been studied both in the fixed embedding setting [6, 13, 15, 17, 19] and in the variable embedding setting [5, 7, 12]. A planar graph with a fixed embedding is called a *plane graph*. As a result in the fixed embedding, Tamassia [19] presented an algorithm to find an optimal orthogonal drawing of a plane graph G in time $O(n^2 \log n)$ where n is the number of vertices in G; he reduced the optimal drawing problem to a min-cost flow problem. Then Garg and Tamassia improved the complexity to $O(n^{7/4}\sqrt{\log n})$ [6]. As a result in the variable embedding setting, Garg and Tamassia showed that the problem is NP-complete for planar graphs of $\Delta \leq 4$ in the variable embedding setting [7]. However, Di Battista *et al.* [5] showed that the problem can be solved in polynomial time for planar graphs G of $\Delta \leq 3$. Their algorithm finds an optimal orthogonal drawing among all possible plane embeddings of G. They use the properties of "spirality," min-cost flow techniques, and a data structure, call a SPQ*R-tree that implicitely represents all the plane embeddings of G. The algorithm is complicated and takes time $O(n^5 \log n)$ for planar graph of $\Delta \leq 3$. Using the algorithm, one can find an optimal orthogonal drawing of a biconnected series-parallel simple graph of $\Delta \leq 4$ and of $\Delta \leq 3$ in time $O(n^4)$ [5] and in time $O(n^3)$ [4], respectively. Note that every series-parallel graph is planar. Series-parallel graphs arise in a variety of problems such as scheduling, electrical networks, data-flow analysis, database logic programs, and circuit layout [20]. The complexities $O(n^5 \log n)$, $O(n^4)$ and $O(n^3)$ above for the variable embedding setting are very high, and it is expected to obtain an efficient algorithm for a particular class of planar graphs of $\Delta \leq 3$ [2].

In this paper we deal with the class of series-parallel (multi)graphs of $\Delta \leq 3$, and give a simple linear algorithm to find an optimal orthogonal drawing in the variable embedding setting. The graph G in Fig. 1 is series-parallel, and has various plane embeddings; two of them are illustrated in Figs. 1(b) and (c); there is no plane embedding having an orthogonal drawing with no bend; however, the embedding in Fig. 1(b) has an orthogonal drawing with one bend as illustrated in Fig. 1(a) and hence the drawing is optimal; the embedding in Fig. 1(c) needs three bends as illustrated in Fig. 1(d); given G, our algorithm finds an optimal drawing in Fig. 1(a). Our algorithm works well even if G has multiple edges or is not biconnected, and is much simpler and faster than the algorithms for biconnected series-parallel simple graphs in [4, 5]; we use neither the min-cost flow technique nor the SPQ*R tree, but uses some structural features of series-parallel graphs, which have not been exploited in [20]. We furthermore obtain a best possible upper bound on the minimum number of bends.

The rest of the paper is organized as follows. In Section 2 we present some definitions and our main idea. In Section 3 we present an algorithm and an upper bound for biconnected series-parallel graphs. Finally Section 4 is a conclusion. We omit a linear algorithm for non-biconnected series-parallel graphs in this extended abstract, due to the page limitation.

2 Preliminaries

In this section we present some definitions and our main idea.

Let $G = (V, E)$ be an undirected graph with vertex set V and edge set E. We denote the number of vertices in G by $n(G)$ or simply by n. For a vertex $v \in V$, we denote by $G - v$ the graph obtained from G by deleting v. An edge joining vertices u and v is denoted by uv. We denote by $G - uv$ the graph obtained from G by deleting uv. We denote the degree of a vertex v in G by $d(v, G)$ or simply by $d(v)$. We denote the maximum degree of G by $\Delta(G)$ or simply by Δ. A connected graph is *biconnected* if there is no vertex whose removal results in a disconnected graph or a single-vertex graph K_1. A *plane graph* is a fixed embedding of a planar graph.

Let G be a planar graph of $\Delta \leq 3$. We denote by $bend(G)$ the number of bends of an optimal orthogonal drawing of G in the variable embedding setting. (Thus $bend(G) = 1$ for the graph G in Fig. 1.) Let D be an orthogonal drawing of G. The number of bends in D is denoted by $bend(D)$. Of course, $bend(G) \leq bend(D)$. Let $G(D)$ be a plane graph obtained from a drawing D by replacing each bend in D with a new vertex. Figures 2(a) and (b) depict $G(D)$ for the drawings D in Figs. 1(a) and (d), respectively. An angle formed by two edges e and e' incident to a vertex v in $G(D)$ is called an *angle of vertex* v if e and e' appear consecutively around v. An angle of a vertex in $G(D)$ is called an *angle of the plane graph* $G(D)$. In an orthogonal drawing, every angle is $\pi/2$, π, $3\pi/2$ or 2π. Consider a labeling l which assigns a label 1, 0, -1 or -2 to every angle of $G(D)$. Labels 1, 0, -1 and -2 correspond to angles $\pi/2$, π, $3\pi/2$ and 2π, respectively. We call l a *regular labeling* of $G(D)$ if l satisfies the following three conditions (a)–(c) [10, 19]:

(a) for each vertex v of $G(D)$,

  (a-1) if $d(v) = 1$ then the label of the angle of v is -2,

(a-2) if $d(v) = 2$ then the labels of the two angles of v total to 0, and
(a-2) if $d(v) = 3$ then the labels of the three angles of v total to 2;

(b) the sum of the labels of each inner face is 4; and
(c) the sum of the labels of the outer face is -4.

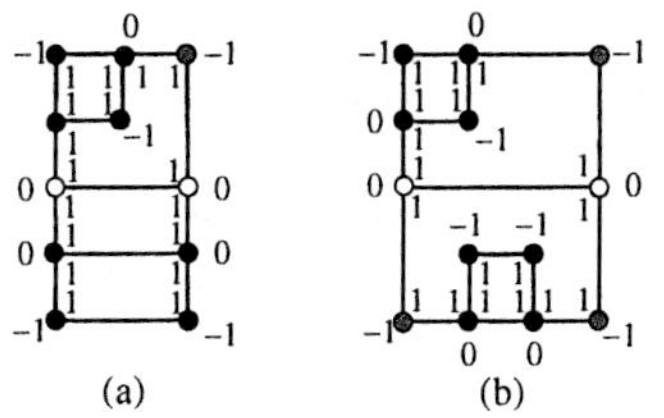

Fig. 2. Regular labelings of $G(D)$ corresponding to the drawings D in Figs. 1(a) and (d), respectively

Figures 2(a) and (b) illustrate regular labelings for the orthogonal drawings in Figs. 1(a) and (d), respectively. If D is an orthogonal drawing of G, then clearly $G(D)$ has a regular labeling. Conversely, every regular labeling of $G(D)$ corresponds to an orthogonal drawing of G [19]. An orthogonal (geometric) drawing of G can be obtained from a regular labeling of $G(D)$ in linear time, that is, in time $O(n(G) + bend(D))$ [10, 19]. Therefore, from now on, we call a regular labeling of $G(D)$ an *orthogonal drawing of a planar graph* G or simply a *drawing of* G, and obtain a regular labeling of G in place of an orthogonal (geometric) drawing of G.

A *series-parallel graph (with terminals* s *and* t*)* is recursively defined as follows:

(a) A graph G of a single edge is a series-parallel graph. The ends s and t of the edge are called the *terminals* of G.
(b) Let G_1 be a series-parallel graph with terminals s_1 and t_1, and let G_2 be a series-parallel graph with terminals s_2 and t_2.
   (i) A graph G obtained from G_1 and G_2 by indentifying vertex t_1 with vertex s_2 is a series-parallel graph, whose terminals are $s = s_1$ and $t = t_2$. Such a connection is called a *series connection.*
   (ii) A graph G obtained from G_1 and G_2 by identifying s_1 with s_2 and t_1 with t_2 is a series-parallel graph, whose terminals are $s = s_1 = s_2$ and $t = t_1 = t_2$. Such a connection is called a *parallel connection.*

For example, the graph in Fig. 1 is series-parallel.

Throughout the paper we assume that the maximum degree of a given series-parallel graph G is at most three, that is, $\Delta \leq 3$. We may assume without loss of generality that G is a simple graph, that is, G has no multiple edges, as follows. If a series-parallel multigraph G consists of exactly three multiple edges, then G has an optimal drawing of four bends; otherwise, insert a dummy vertex of degree two into an edge of each pair of multiple edges in G, and let G' be the resulting series-parallel simple graph, then an optimal drawing of the multigraph

G can be immediately obtained from an optimal drawing of the simple graph G' by replacing each dummy vertex with a bend.

A drawing D of a series-parallel graph G is *outer* if the two terminals s and t of G are drawn on the outer face of D. A drawing D is called an *optimal outer drawing* of G if D is outer and $bend(D) = bend(G)$. The graph in Fig. 1 has an optimal outer drawing as illustrated in Fig. 1(a). On the other hand, the graph in Fig. 3(a) has no optimal outer drawing for the specified terminals s and t; the no-bend drawing D in Fig. 3(b) is optimal but is not outer, because s is not on the outer face; and the drawing D° with one bend in Fig. 3(c) is outer but is not optimal.

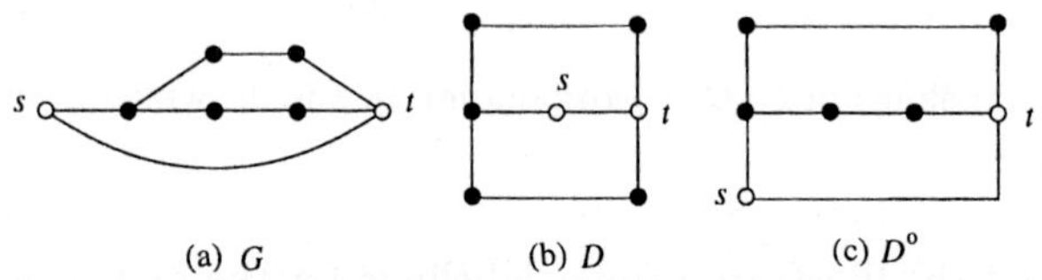

Fig. 3. (a) A biconnected series-parallel graph G, (b) an optimal drawing D, and (c) an outer drawing D°

Our main idea is to notice that a series-parallel graph G has an optimal outer drawing if G is "2-legged." We say that G is 2-*legged* if $n(G) \geq 3$ and $d(s) = d(t) = 1$ for the terminals s and t of G. The edge incident to s or t is called a *leg* of G, and the neighbor of s or t is called a *leg-vertex*. For example, the series-parallel graphs in Figs. 4(a)–(c) are 2-legged.

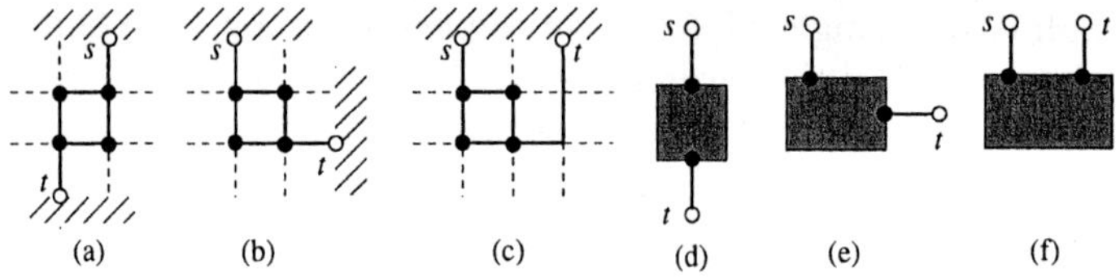

Fig. 4. (a)–(c) I-, L- and U-shaped drawings, and (d)–(f) their schematic representations

We will show in Section 3 that every 2-legged series-parallel graph G has an optimal outer drawing and the drawing has one of the three shapes, "I-shape," "L-shape" and "U-shape," defined as follows. An outer drawing D of G is *I-shaped* if D intersects neither the north side of terminal s nor the south side of terminal t after rotating the drawing and renaming the terminals if necessary, as illustrated in Fig. 4(a). D is *L-shaped* if D intersects neither the north side of s nor the east side of t after rotating the drawing and renaming the terminals if necessary, as illustrated in Fig. 4(b). D is *U-shaped* if D does not intersect the north sides of s and t after rotating the drawing and renaming the terminals if necessary, as illustrated in Fig. 4(c). In Figs. 4(a)–(c) each side is shaded. The north side and the south side of a terminal contain the horizontal line passing through the terminal, while the east side of a terminal contains the vertical line passing through the terminal. The schematic representations of I-, L-, and U-shaped drawings are depicted in Figs. 4(d), (e) and (f), respectively. D is called an *optimal X-shaped drawing*, X=I, L and U, if D is X-shaped and $bend(D) = bend(G)$.

More precisely, we will show in Section 3 that every 2-legged series-parallel graph G with $n(G) \geq 3$ has both an optimal I-shaped drawing and an optimal L-shape drawing and that G has an optimal U-shaped drawing, too, unless G is a "diamond graph," defined as follows. A *diamond graph* is either a path with three vertices or obtained from two diamond graphs by connecting them in parallel and adding two leggs.

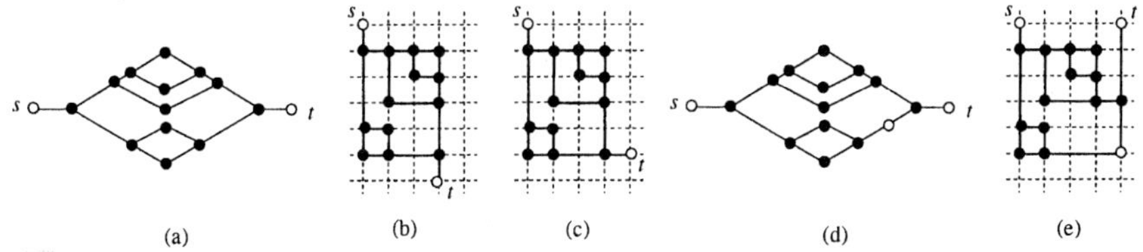

Fig. 5. (a) Diamond graph, (b) I-shaped drawing, (c) L-shaped drawing, (d) non-diamond graph, and (e) U-shaped drawing

For example, the 2-legged series-parallel graph in Fig. 5(a) is a diamond graph, and has both an optimal (no-bend) I-shaped drawing and an optimal (no-bend) L-shaped drawing as illustrated in Figs. 5(b) and (c), but does not have an optimal (no-bend) U-shaped drawing. On the other hand, the 2-legged series-parallel graph in Fig. 5(d) is obtained from the diamond graph in Fig. 5(a) simply by inserting a new vertex of degree two in an edge, and is not a diamond graph any more. It has an optimal (no-bend) U-shaped drawing, too, as illustrated in Fig. 5(e). Thus the diamond graph in Fig. 5(a) has a U-shaped drawing with one bend.

3 Optimal Drawing of Biconnected Series-Parallel Graph

In this section we give a linear algorithm to find an optimal drawing of a biconnected series-parallel graph G of $\Delta \leq 3$. We first give an algorithm for 2-legged series-parallel graphs in Subsection 3.1. Using the algorithm, we then give an algorithm for biconnected series-parallel graphs in Subsection 3.2.

3.1 2-Legged Series-Parallel Graph

We first have the following lemma on a diamond graph. (The proofs of all theorems and lemmas are omitted in this extended abstract, due to the page limitation.)

Lemma 1. *If G is a diamond graph, then*

(a) *G has both a no-bend* I*-shaped drawing D_{I} and a no-bend* L*-shaped drawing D_{L};*
(b) *D_{I} and D_{L} can be found in linear time; and*
(c) *every no-bend drawing of G is either* I*-shaped or* L*-shaped, and hence G does not have a no-bend* U*-shaped drawing.*

The proof of Lemma 1 immediately yields a linear algorithm **Diamond**$(G, D_{\mathrm{I}}, D_{\mathrm{L}})$ which recursively finds both a no-bend I-shaped drawing D_{I} and a no-bend L-shaped drawing D_{L} of a given diamond graph G.

The following lemma holds for a 2-legged series-parallel graph G which is not a diamond graph.

Lemma 2. *The following* (a) *and* (b) *hold for a 2-legged series-parallel graph G with $n(G) \geq 3$ unless G is a diamond graph:*

(a) *G has three optimal* I-, L- *and* U*-shaped drawings D_{I}, D_{L} and D_{U};*
(b) *D_{I}, D_{L} and D_{U} can be found in linear time; and*
(c) *$bend(G) \leq (n(G) - 2)/3$.*

We denote by K_n a complete graph of $n(\geq 1)$ vertices. Let G be a 2-legged series-parallel graph obtained from copies of K_2 and K_3 by connected them alternately in series, as illustrated in Fig. 6. Then $bend(G) = (n(G) - 2)/3$. Thus the bound in Lemma 2(c) is best possible.

Fig. 6. A graph attaining the bound in Lemma 2(c)

The proof of Lemma 2 immediately yields a linear algorithm **Non-Diamond** $(G, D_{\mathrm{I}}, D_{\mathrm{L}}, D_{\mathrm{U}})$ which recursively finds three optimal I-, L- and U-shaped drawings D_{I}, D_{L} and D_{U} of a given 2-legged series-parallel graph G unless G is a diamond graph. By algorithms **Diamond** and **Non-Diamond** one can find an optimal drawing of a 2-legged series-parallel graph G.

3.2 Biconnected Series-Parallel Graphs

A biconnected series-parallel graph G can be defined (without specifying terminals) as a biconnected graph which has no K_4 as a minor. For every edge uv in G, G is a series-parallel graph with terminals u and v.

A cycle C of four vertices in a graph G is called a *diamond* if two non-consecutive vertices of C have degree two in G and the other two vertices of C have degree three and are not adjacent in G. We denote by G/C the graph obtained from G by contracting C to a new single vertex v_C. (Note that $G_C = G/C$ is series-parallel if G is series-parallel. One can observe that, from every diamond graph, one can obtain a path with three vertices by repeatedly contracting a diamond.)

Noting that every biconnected series-parallel graph has a vertex of degree two, one can easily observe that the following Lemma 3 holds. (Lemma 3 is also an immediate consequence of Lemma 2.1 in [9] on general series-parallel graphs.)

Lemma 3. *Every biconnected series-parallel graph G of $\Delta \leq 3$ has, as a subgraph, one of the following three substructures* (a)–(c)*:*

(a) *a diamond C;*
(b) *two adjacent vertices u and v such that $d(u) = d(v) = 2$; and*
(c) *a complete graph K_3 of three vertices u, v and w such that $d(v) = 2$.*

Our idea is to reduce the optimal drawing problem for a biconnected series-parallel graph G to that for a smaller graph G' as in the following Lemma 4.

Lemma 4. *Let G be a biconnected series-parallel graph with $n(G) \geq 6$.*

(a) *If G has a diamond C, then $bend(G) = bend(G')$ for $G' = G/C$.*
(b) *If G has a substructure (b) in Lemma 3(b), then $bend(G) = bend(G')$ for $G' = G - uv$.*
(c) *If G has a substructure (c) in Lemma 3(c), then $bend(G) = bend(G') + 1$ for $G' = G - v - uw$.*

From the proof of Lemma 4 we have the following algorithm **Biconnected**(G, D) to find an optimal drawing D of a biconnected series-parallel graph G.

Biconnected(G, D);
begin
One may assume that $n(G) \geq 6$ (otherwise, one can easily find an optimal drawing D of G in linear time);
{ By Lemma 3 G has one of the three substructures (a)–(c) in Lemma 3. }
Case 1: G has a diamond C;
Let $G' = G/C$; { G' is a biconnected series-parallel graph. }
Biconnected(G', D');
Extend an optimal drawing D' of G' to an optimal drawing D of G simply by replacing v_C by a rectanglar drawing of C;
Case 2: G has no diamond, but has a substructure (b);
Let $G' = G - uv$;
{ G' is a 2-legged series-parallel graph with terminals u and v, and is not a diamond graph. }
Find an optimal U-shaped drawing D'_{U} of G' by **Non-Diamond**; { cf. Lemma 2 }
Extend D'_{U} to an optimal drawing D of G by drawing uv as a straight line segment; { cf. Lemma 4 }
Case 3: G has neither a diamond nor a substructure (b), but has a substructure (c);

Let $G' = G - v - uw$;
{ G' is a 2-legged series-parallel graph with terminals u and w, and is not a diamond graph. }
Find an optimal U-shaped drawing D'_U of G' by **Non-Diamond**;
Extend D'_U to an optimal drawing D of G by drawing $K_3 = uvw$ as a rectangle with one bend; { cf. Lemma 4 }
end

All substructures (a)–(c) can be found total in time $O(n)$ by a standard bookkeeping method to maintain all degrees of vertices together with all paths of length two with an intermediate vertex of degree two. One can thus observe that **Biconnected** can be executed in linear time.

We thus have the following theorem.

Theorem 1. *An optimal orthogonal drawing of a series-parallel biconnected graph G of $\Delta \leq 3$ can be found in linear time.*

4 Conclusions

In this paper, we gave a linear algorithm to find an optimal orthogonal drawing of a series-parallel graph G of $\Delta \leq 3$ in the variable embeddings setting. Our algorithm works well even if G has multiple edges or is not biconnected, and is simpler and faster than the previously known one for biconnected series-parallel simple graphs [4, 5]. One can easily extend our algorithm so that it finds an optimal orthogonal drawing of a partial 2-tree of $\Delta \leq 3$. Note that the so-called block-cutvertex graph of a partial 2-tree is a tree although the block-cutvertex graph of a series-parallel graph is a path. One can prove that $bend(G) \leq \lceil n(G)/3 \rceil$ for every biconnected series-parallel graph and $bend(G) \leq (n(G)+4)/3$ for every series-parallel graph. The bounds on $bend(G)$ are best possible.

In an *orthogonal grid drawing*, every vertex has an integer coordinate. The *size* of an orthogonal grid drawing is the sum of width and height of the minimum axis-parallel rectangle enclosing the drawing. One can prove that every biconnected series-parallel graph G of $\Delta \leq 3$ has an optimal orthogonal grid drawing of size $\leq 2N/3 + 1$, where $N = n(G) + bend(G)$.

It is left as a future work to obtain a linear algorithm for a larger class of planar graphs.

Acknowledgments. We thank Dr. Md. Saidur Rahman for fruitful discussions.

References

1. P. Bertolazzi, R. F. Cohen, G. Di Battista, R. Tamassia and I. G. Tollis, *How to draw a series-parallel digraph*, Porc. of SWAT '92, pp. 272-283, 1992.
2. F. Brandenburg, D. Eppstein, M. T. Goodrich, S. Kobourov, G. Liotta and P. Mutzel, *Selected open problems in graph drawings*, Proc. of GD '03, LNCS, 1912, pp. 515-539, 2004.

3. G. Di Battista, P. Eades, R. Tamassia and I. G. Tollis, *Graph Drawing: Algorithm for the Visualization of Graphs*, Prentice-Hall Inc., Upper Saddle River, New Jersey, 1999.
4. G. Di Battista, G. Liotta and F. Vargiu, *Spirality and optimal orthogonal drawings*, Tech. Rept. 07.94, Dipartimento di Informatica e Sistemistica, Universita' di Roma "La Sapienza," 1994.
5. G. Di Battista, G. Liotta and F. Vargiu, *Spirality and optimal orthogonal drawings*, SIAM J. Comput., 27(6), pp. 1764–1811, 1998.
6. A. Garg and R. Tamassia, *A new minimum cost flow algorithm with applications to graph drawing*, Proc. of GD '96, LNCS, 1190, pp. 201–226, 1997.
7. A. Garg and R. Tamassia, *On the computational complexity of upward and rectilinear planarity testing*, SIAM J. Comput., 31(2), pp. 601–625, 2001.
8. S. H. Hong, P. Eades and S. H. Lee, *Drawig series-parallel digraphs symmetrically*, Comput. Geom., 17, pp. 165–188, 2000.
9. M. Juvan, B. Mohar and R. Thomas, *List edge-colorings of series-parallel graphs*, The Electronic J. Combinatorics, 6(42), pp. 1–6, 1999.
10. T. Nishizeki and M. S. Rahman, *Planar Graph Drawing*, World Scientific, Singapore, 2004.
11. K. Nomura, S. Taya and S. Ueno, *On the orthogonal drawing of outerplanar graphs*, Proc. of COCOON '04, LNCS, 3106, pp. 300–308, 2004.
12. M. S. Rahman, N. Egi and T. Nishizeki, *No-bend orthogonal drawings of subdivisions of planar triconnected cubic graphs*, IEICE Trans. Inf. & Syst., E88-D(1), pp. 23-30, 2005.
13. M. S. Rahman and T. Nishizeki, *Bend-minimum orthogonal drawings of plane 3-graphs*, Proc. of WG '02, LNCS, 2573, pp. 265–276, 2002.
14. M. S. Rahman, S. Nakano and T. Nishizeki, *Rectangular grid drawings of plane graphs*, Comp. Geom. Theo. App., 10(3), pp. 203–220, 1998.
15. M. S. Rahman, S. Nakano and T. Nishizeki, *A linear algorithm for bend-optimal orthogonal drawings of triconnected cubic plane graphs*, J. of Graph Alg. App., 3(4), pp. 31-62, 1999.
16. M. S. Rahman, S. Nakano and T. Nishizeki, *Box-rectangular drawings of plane graphs*, J. of Algorithms, 37, pp. 363–398, 2000.
17. M. S. Rahman, S. Nakano and T. Nishizeki, *Orthogonal drawings of plane graphs without bends*, J. of Graph Alg. App., 7(4), pp. 335–362, 2003.
18. J. Storer, *On minimum node-cost planar embeddings*, Networks, 14, pp. 181–212, 1984.
19. R. Tamassia, *On embedding a graph in the grid with the minimum number of bends*, SIAM J. Comput., 16, pp. 421–444, 1987.
20. K. Takamizawa, T. Nishizeki and N. Saito, *Linear-time computability of combinatorial problems on series-parallel graphs*, J. Assoc. Comput. Mach., 29, pp. 623–641, 1982.

Bisecting a Four-Connected Graph with Three Resource Sets

Toshimasa Ishii[1], Kengo Iwata[2], and Hiroshi Nagamochi[3]

[1] Department of Information and Computer Sciences,
Toyohashi University of Technology,
Aichi 441-8580, Japan
ishii@ics.tut.ac.jp
[2] Mazda Motor Corporation,
Hiroshima 730-8670, Japan
[3] Department of Applied Mathematics and Physics,
Graduate School of Informatics,
Kyoto University,
Kyoto 606-8501, Japan
nag@amp.i.kyoto-u.ac.jp

Abstract. Let $G = (V, E)$ be an undirected graph with a node set V and an arc set E. G has k pairwise disjoint subsets $T_1, T_2, \ldots, T_k$ of nodes, called resource sets, where $|T_i|$ is even for each i. The partition problem with k resource sets asks to find a partition V_1 and V_2 of the node set V such that the graphs induced by V_1 and V_2 are both connected and $|V_1 \cap T_i| = |V_2 \cap T_i| = |T_i|/2$ holds for each $i = 1, 2, \ldots, k$. The problem of testing whether such a bisection exists is known to be NP-hard even in the case of $k = 1$. On the other hand, it is known that that if G is $(k+1)$-connected for $k = 1, 2$, then a bisection exists for any given resource sets, and it has been conjectured that for $k \geq 3$, a $(k+1)$-connected graph admits a bisection. In this paper, we show that for $k = 3$, the conjecture does not hold, while if G is 4-connected and has K_4 as its subgraph, then a bisection exists and it can be found in $O(|V|^3 \log |V|)$ time. Moreover, we show that for an arc-version of the problem, the $(k+1)$-edge-connectivity suffices for $k = 1, 2, 3$.

1 Introduction

In this paper, we consider the following graph partition problems: given an undirected graph $G = (V, E)$ with a set V of nodes a set E of arcs, and k pairwise disjoint sets $T_1, T_2, \ldots, T_k$ of nodes, called *resource sets*, where each $|T_i|$ is even, find a partition V_1 and V_2 of V such that the graphs induced by V_1 and V_2 are both connected and $|V_1 \cap T_i| = |V_2 \cap T_i| = |T_i|/2$ holds for each $i = 1, 2, \ldots, k$. This problem is called *the bisection problems with k resource sets*, and such a bisection is called *k-bisection* (*with respect to* $T_1, \ldots, T_k$).

Such a problem of partitioning a graph into connected subgraphs under fair-division type of constraints appears in many applications such as political districting [1,7,13], the paging system in operating systems [11] and the image

X. Deng and D. Du (Eds.): ISAAC 2005, LNCS 3827, pp. 176–185, 2005.

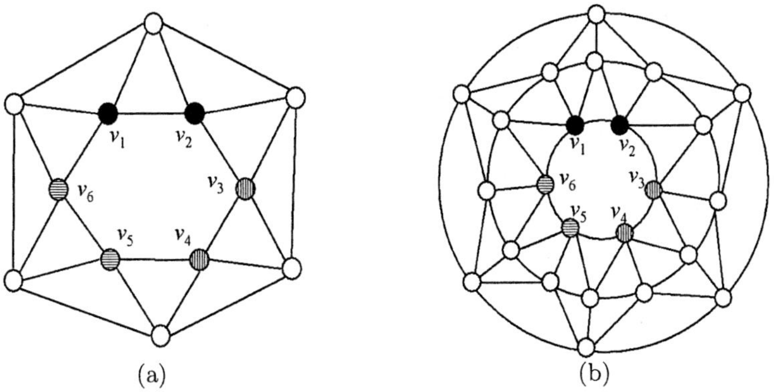

Fig. 1. Illustration of instances of 4-connected graphs which have no 3-bisection, where $T_1 = \{v_1, v_2\}$, $T_2 = \{v_3, v_4\}$, and $T_3 = \{v_5, v_6\}$ in both (a) and (b). Note that the graph (b) is also 5-connected.

processing [6]. For the political districting, a dual graph of the map which consists of regions is required to be divided into connected subgraphs, each of which represents an electoral zone, so that both the area and the number of voters in each zone is balanced over all zones.

So far, for general graphs, the problem was shown to be NP-hard even if $k = 1$ holds, since it is NP-hard to test whether a 1-bisection exists or not [3,4]. On the other hand, when $k = 1, 2$, it was shown that if a given graph is $(k+1)$-connected, then for any given resource sets, such a k-bisection exists and it can be found in linear time for $k = 1$ by Suzuki et al. [10] and by Wada and Kawaguchi [12], and in $O(|V|^2 \log |V|)$ time for $k = 2$ by Nagamochi et al. [9]. For a general $k \geq 3$, to our knowledge, any nontrivial sufficient condition for which a k-bisection exists is not known, while it was conjectured in [9] that every $(k + 1)$-connected graph admits a k-bisection.

In this paper, we consider the case of $k = 3$. We first show that there exist 4-connected graphs which have no 3-bisection, as shown in Figure 1. Indeed, in both graphs, it is not difficult to observe that for any partition V_1 and V_2 of V bisecting each T_i, V_1 or V_2 do not induce a connected graph. This indicates a negative answer to the above conjecture for $k = 3$. In particular, the graph in Figure 1(b) is also 5-connected, which shows that even 5-connected graphs may have no 3-bisection (this also indicates a negative answer to the above conjecture for $k = 4$). Instead, in this paper, we give a sufficient condition for which a 3-bisection exists; we prove that if G is 4-connected and has a complete graph K_4 of four nodes as its subgraph, then a 3-bisection exists. We also show that it can be found in $O(|V|^3 \log |V|)$ time. A key technique of the proof, which is an extension of the method by Nagamochi et al. [9], is a reduction of the problem to a geometrical problem. We first prove that every 4-connected graph containing a complete graph K^* of four nodes as its subgraph can be embedded in the 3-dimensional space $\Re^3$, in such a way that the following (i)(ii) hold: (i) the convex hull of its nodes is a trigonal pyramid corresponding to the K^*, (ii) every node not in K^* is in the convex hull of its neighbors (precise definition is given in Section 2.2). This will guarantee that, for any given plane H in $\Re^3$, each of the two subgraphs of G separated by H remains connected. Given such an embedding in $\Re^3$, we apply the so-called ham-sandwich cut algorithm, which is well known in computational geometry, to find a plane H^* that bisects T_1, T_2, and T_3 simultaneously. Consequently, the two subgraphs by the plane H^* indicates a 3-bisection. We give an algorithm for finding such a plane H^* in $O(|V|^3 \log |V|)$ time.

Moreover, we consider an arc-version of the bisection problem; given an undirected graph $G = (V, E)$ and k pairwise disjoint sets $T_1, \ldots, T_k$ of arcs, where each $|T_i|$ is even, find a partition E_1 and E_2 of E such that the graphs induced by E_1 and E_2 are both connected and $|E_1 \cap T_i| = |E_2 \cap T_i| = |T_i|/2$ holds for each $i = 1, \ldots, k$. We call such a bisection *k-bisection of the arc set.* For this problem, we show that a $(k+1)$-edge-connected graph admits a bisection for any given resource sets for $k = 1, 2, 3$.

The paper is organized as follows. Some definitions and preliminaries are described in Section 2. Section 3 describes an algorithm for finding a 3-bisection in a 4-connected graph with K_4, and Section 4 proves its correctness. In Section 5, we make some remarks on the problem for a general k and discuss the arc-version of the problem.

2 Preliminaries

Let $G = (V, E)$ stand for an undirected simple graph with a set V of *nodes* and a set E of *arcs*, where we denote $|V|$ by n and $|E|$ by m. A singleton set $\{x\}$ may be simply written as x, and "$\subset$" implies proper inclusion while "$\subseteq$" means "$\subset$" or "$=$". For a subgraph G' of G, the sets of nodes and arcs in G' are denoted by $V(G')$ and $E(G')$, respectively. For a set X of nodes in G, a node $v \in V - X$ is called a *neighbor* of X if it is adjacent to some node in X, and the set of all neighbors of X is denoted by $N_G(X)$.

For an arc $e = (u, v)$, we denote by G/e the graph obtained from G by contracting u and v into a single node (deleting any resulted self-loop), and by $G - e$ the graph obtained from G by removing e. We also say that G/e is obtained from G by contracting the arc e. A graph G is *k-connected* if and only if $|V| \geq k+1$ and the graph $G - X$ obtained from G by removing any set X of $(k-1)$ nodes remains connected. A graph G is *k-edge-connected* if and only if the graph $G - F$ obtained from G by removing any set F of $(k-1)$ arcs remains connected.

The main results of this paper are described as follows.

Theorem 1. *Let $G = (V, E)$ be a 4-connected graph which contains a complete graph with four nodes as its subgraph. Let T_1, T_2, T_3 be pairwise disjoint subsets of V such that $|T_i|$ is even for $i = 1, 2, 3$. Then G has a 3-bisection with respect to T_1, T_2, and T_3, and it can be found in $O(n^3 \log n)$ time.* ⊓

Theorem 2. *Let $G = (V, E)$ be a $(k+1)$-edge-connected graph with pairwise disjoint subsets T_i, $i = 1, \ldots, k$ of E such that each $|T_i|$ is even. If $k = 1, 2, 3$, then a k-bisection of the arc set exists.* □

In the sequel, we first give a constructive proof of Theorem 1 by reducing the problem to a geometrical problem as mentioned in Section 1. For this, we give some geometric notations in the next two subsections.

2.1 Convex Hull and Ham-Sandwich Cut

Consider the d-dimensional space $\Re^d$. For a non-zero $a \in \Re^d$ and a real $b \in \Re^1$, $H(a, b) = \{x \in \Re^d \mid \langle a \cdot x \rangle = b\}$ is called a *hyperplane*, where $\langle a \cdot x \rangle$ denotes the inner product of $a, x \in \Re^d$. Moreover, $H^+(a, b) = \{x \in \Re^d \mid \langle a \cdot x \rangle \geq b\}$ (resp., $H^-(a, b) = \{x \in \Re^d \mid \langle a \cdot x \rangle \leq b\}$) is called a *positive closed half space* (resp., *negative closed half space*) with respect to $H = H(a, b)$.

For a set $P = \{x_1, \ldots, x_k\}$ of points in $\Re^d$, a point $x' = \alpha_1 x_1 + \cdots + \alpha_k$ with $\sum_{i=1,\ldots,k} \alpha_i = 1$ and $\alpha_i \geq 0$, $i = 1, \ldots, k$ is called a *convex combination of* P, and the set of all convex combinations of P is denoted by $conv(P)$. If $P = \{x_1, x_2\}$, then $conv(P)$ is called a *segment* (connecting x_1 and x_2), denoted by $[x_1, x_2]$. A subset $S \subseteq \Re^d$ is called a *convex set* if $[x, x'] \subseteq S$ for any two points $x, x' \in S$. For a convex set $S \subseteq \Re^d$, a point $x \in S$ is called a *vertex* if there is no pair of points $x', x'' \in S - x$ such that $x \in [x', x'']$. For two vertices $x_1, x_2 \in S$, the segment $[x_1, x_2]$ is called an *edge* of S if $\alpha x' + (1 - \alpha)x'' = x \in [x_1, x_2]$ for some $0 \leq \alpha \leq 1$ implies $x', x'' \in [x_1, x_2]$. The intersection S of a finite number of closed half spaces is called a *convex polyhedron*, and is called a *convex polytope* if S is non-empty and bounded.

Given a convex polytope S in $\Re^d$, the *vertex-edge graph* $G_S = (V_S, E_S)$ is defined to be an undirected graph with node set V_S corresponding to the vertices of S and arc set E_S corresponding to those pairs of vertices x, x' for which $[x, x']$ is an edge of S. For a convex polyhedron S, a hyperplane $H(a, b)$ is called a *supporting hyperplane* of S if $H(a, b) \cap S \neq \emptyset$ and either $S \subseteq H^+(a, b)$ or $S \subseteq H^-(a, b)$. We say that a point $p \in S$ is *strictly inside* S if there is no supporting hyperplane of S containing p. If S has a point strictly inside S in $\Re^d$, S is called *full-dimensional* in $\Re^d$. The set of points strictly inside $conv(P)$ is denoted by $int(conv(P))$.

Let $P_1, \ldots, P_d$ be d sets of points in $\Re^d$. We say that a hyperplane $H = H(a, b)$ in $\Re^d$ *bisects* P_i if $|P_i \cap H^+(a, b)| \geq \lceil |P_i|/2 \rceil$ and $|P_i \cap H^-(a, b)| \geq \lceil |P_i|/2 \rceil$ hold. Thus if $|P_i|$ is odd, then any bisector H of P_i contains at least one point of P_i. If H bisects P_i for each $i = 1, \ldots, d$, then H is called a *ham-sandwich cut* with respect to $P_1, \ldots, P_d$. The following result is well-known (see, e.g., [5]).

Theorem 3. *Given d sets $P_1, \ldots, P_d$ of points in the d-dimensional space $\Re^d$, there exists a hyperplane which is a ham-sandwich cut with respect to the sets $P_1, \ldots, P_d$.* □

In [2], Chi-Yuan et al. showed that a ham-sandwich cut with respect to given sets $P_1, P_2, \ldots, P_d$ of points in $\Re^d$ with $\sum_{i=1}^{d} |P_i| = p$ can be found in $O(p^{3/2})$

time for $d = 3$, $O(p^{8/3})$ time for $d = 4$, and $O(p^{d-1-a(d)})$ time with certain small constant $a(d) > 0$ for $d \geq 5$.

2.2 Convex Embedding of a Graph

In this section, we introduce a strictly convex embedding of a graph in $\Re^d$, which was first defined by Nagamochi et al. [9].

Given a graph $G = (V, E)$, an *embedding* of G in $\Re^d$ is an mapping $f : V \rightarrow \Re^d$, where each node v is represented by a point $f(v) \in \Re^d$, and each arc $e = (u, v)$ by a segment $[f(u), f(v)]$, which may be written by $f(e)$. For two arcs $e, e' \in E$, segments $f(e)$ and $f(e')$ may cross each other. For a set $\{v_1, \ldots, v_p\} = Y \subseteq V$ of nodes, we denote by $f(Y)$ the set $\{f(v_1), \ldots, f(v_p)\}$ of points, and we denote $conv(f(Y))$ by $conv_f(Y)$.

A strictly convex embedding of a graph is defined as follows (see Figure 2).

Definition 1. [9] *Let $G = (V, E)$ be a graph without isolated nodes and let $G' = (V', E')$ be a subgraph of G. A* strictly convex embedding (*or* SC-embedding, *for short*) *of G with* boundary *G' is an embedding f of G into $\Re^d$ in such a way that*

(i) *the vertex-edge graph of the full-dimensional convex polytope $conv_f(V')$ is isomorphic to G' (such that f itself defines an isomorphism),*

(ii) *$f(v) \in int(conv_f(N_G(v)))$ holds for all nodes $v \in V - V'$,*

(iii) *the points of $\{f(v) \mid v \in V\}$ are in general position.* □

From this definition, we can see that the vertices of $conv_f(V)$ are precisely the points in the boundary $f(V')$.

The following lemma implies that given an SC-embedding of $G = (V, E)$ into $\Re^d$, each two sets of nodes obtained by bisecting $f(V)$ with an arbitrary hyperplane in $\Re^d$ induce connected graphs.

Lemma 1. [9, Lemma 4.2] *Let $G = (V, E)$ be a graph without isolated nodes and let f be an SC-embedding of G into $\Re^d$. Let $f(V_1) \subseteq H^+(a, b)$ and $f(V) \cap (H^+(a, b) - H(a, b)) \subseteq f(V_1)$ hold for some hyperplane $H = H(a, b)$ and for some $\emptyset \neq V_1 \subseteq V$. Then V_1 induces a connected graph.* □

By Theorem 3 and this lemma, if there is an SC-embedding of a given graph $G = (V, E)$ into $\Re^k$, then by bisecting the embedded graph with a hyperplane which is a ham-sandwich cut with respect to $T_1, \ldots, T_k$, we can obtain a partition V_1 and V_2 of V bisecting each T_i such that each V_j induces a connected graph, that is, a k-bisection. Based on this observation, we give an algorithm for finding a 3-bisection in the next section.

3 Algorithm for Bisecting Resource Sets

In this section, we give an algorithm, named BISECT3 for finding a 3-bisection in a 4-connected graph with K_4 in $O(n^3 \log n)$ time, which proves Theorem 1.

Algorithm BISECT3
Input: A 4-connected graph $G = (V, E)$ which has a complete graph K with 4 nodes, and three pairwise disjoint node sets T_1, T_2, and T_3 where each $|T_i|$ is even.
Output: A 3-bisection of G with respect to T_1, T_2, and T_3.
Phase 1: Find an SC-embedding f of G with boundary K into $\Re^3$.
Phase 2: By applying a ham-sandwich cut algorithm to $f(V)$ in $\Re^3$, find a plane H in $\Re^3$ which bisects T_1, T_2, and T_3. Output the bisection $\{V_1, V_2\}$ of V divided by H. □

As mentioned in Section 2.1, a ham-sandwich cut bisecting each T_i in $\Re^3$ exists, and it can be found in $O(n^{3/2})$ time. Hence, for proving the correctness of algorithm BISECT3, it suffices to show that Phase 1 can find an SC-embedding of G with boundary K into $\Re^3$ in $O(n^3 \log n)$ time. In the next section, we give a proof for this.

4 SC-Embedding of a Graph into $\Re^3$

In this section, given a 4-connected graph G which contains a complete graph with four nodes, denoted by K, we propose an algorithm, named EMBED3, for finding an SC-embedding of G with boundary K into $\Re^3$ in $O(n^3 \log n)$ time. Figure 2 shows an instance of such an SC-embedding of a 4-connected graph into $\Re^3$.

The algorithm EMBED3, which is an extension of the algorithm in $\Re^2$ given in [9], consists of two steps. First, we contract arcs in $E - E(K)$, one by one, while preserving the 4-connectivity until a complete graph G^* with 5 nodes containing K is obtained. Then we can easily obtain an SC-embedding f of G^* with boundary K into $\Re^3$; we find an embedding f' of $V(K)$ by putting them in general position (which shapes a trigonal pyramid), and we embed the node v

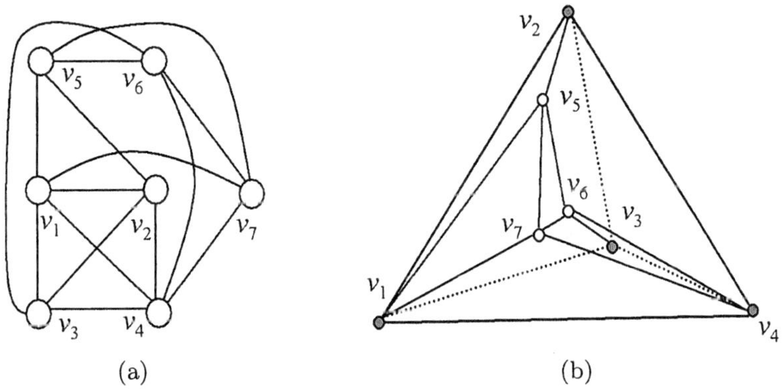

Fig. 2. Illustration of an instance of an SC-embedding; (b) shows an SC-embedding of the graph in (a) with boundary $(\{v_1, v_2, v_3, v_4\}, \bigcup_{1\le i,j\le 4}(v_i, v_j))$ into $\Re^3$

with $\{v\} = V(G^*) - V(K)$ in $int(conv_{f'}(V(K)))$. Next, by tracing the process of the contraction reversely and embedding the contracted arcs into $\Re^3$, we convert the embedding f into the one for the original graph. The outline of algorithm EMBED3 is described as follows.

Algorithm EMBED3
Input: A 4-connected graph $G = (V, E)$ which has a complete graph K with 4 nodes.
Output: An SC-embedding of G with boundary K into $\Re^3$.
Step 1: While $|V(G)| \geq 6$ holds, execute the following procedure.

Find an arc $e \in E(G) - E(K)$ such that G/e remains 4-connected, and contract the arc e. Let $G := G/e$.

/** The current graph G obtained by Step 1 is a complete graph with 5 nodes containing K. **/
Step 2: Embed G into $\Re^3$ so that its embedding is an SC-embedding f with boundary K. Next, by tracing the process of the contraction in Step 1 reversely and embedding the contracted arcs into $\Re^3$, one by one, we convert the embedding f into the one for the original graph. □

In the subsequent sections, we prove the correctness of algorithm EMBED3 by describing the details for each step. The analysis of the time complexity is omitted due to space limitation.

4.1 Correctness of Step 1

We give a proof of the following theorem for the correctness of Step 1.

Theorem 4. *Let $G = (V, E)$ be a 4-connected graph which has a complete graph K with 4 nodes. Then there exists an arc $e \in E - E(K)$ such that G/e is 4-connected.* □

We first introduce the following preparatory theorem about the contraction of arcs in 4-connected graphs.

Definition 2. *A graph G is called* uncontractible k-connected *if G is k-connected and G/e is not k-connected for any arc $e \in E(G)$.* □

Theorem 5. [8] *A graph G is uncontractible 4-connected if and only if G satisfies the following properties:*
(i) *G is 4-connected,*
(ii) *the degree of each node in $V(G)$ is exactly 4, and*
(iii) *for each arc $(u, v) \in E(G)$, there exists a node $w \in V(G) - \{u, v\}$ with $\{(u, w), (v, w)\} \subseteq E(G)$.* □

Proof of Theorem 4. Let $V_1 = V - V(K)$. We construct the new graph G^* from $G = (V, E)$, defined as follows. $V(G^*) = V_1 \cup V(K) \cup V_2$, where V_2 is a copy of V_1. An arc (u_1, u_2) belongs to $E(G^*)$ if and only if (a) $(u_1, u_2) \in E$, (b) $u_1, u_2 \in V_2$

and $u_i, i = 1, 2$ is the copy of $v_i \in V_1$ such that $(v_1, v_2) \in E$, or (c) $u_1 \in V(K)$, $u_2 \in V_2$, and u_2 is the copy of $v_2 \in V_1$ such that $(u_1, v_2) \in E$. Note that G^* is also 4-connected. Since $|N_{G^*}(v)| \geq 5$ holds for a node $v \in V(K)$, Theorem 5 implies that G^* has an arc $e \in E(G^*)$ such that G^*/e is 4-connected. Without loss of generality, let $e \in E - E(K)$ (note that $e \notin E(K)$ since $|V(K)| = 4$ and $G^* - V(K)$ is not connected).

We claim that G/e remains 4-connected, proving the theorem. Assume by contradiction that G/e would have a node set X with $|X| \leq 3$ such that $(G/e) - X$ is not connected. Then there is a component C_ℓ of $(G/e) - X$ with $V(C_\ell) \subseteq V_1$ since K is the complete graph. Also in G^*/e, $N_{G^*/e}(V(C_\ell)) \subseteq X$ holds. $V(G^*/e) - V(C_\ell) - X \neq \emptyset$ and this contradict the 4-connectivity of G^*/e. □

4.2 Correctness of Step 2

In this section, for a graph $G = (V, E)$ and a subgraph G_1 of G, we consider a situation where a graph G/e obtained from G by contracting some arc $e = (u_1, u_2)$ with $\{u_1, u_2\} - V(G_1) \neq \emptyset$ has an SC-embedding f' of G/e with boundary G_1 into $\Re^d$. For proving the correctness of Step 2, we will show by the following Lemma 2 that if $|N_G(u_i)| \geq d + 1$ holds for $i = 1, 2$, then we can find an SC-embedding of G with boundary G_1 into $\Re^d$. Since Step 1 in algorithm EMBED3 contracts arcs while preserving the 4-connectivity, it follows that the degree of every node is always at least 4 in the current graph. Also note that any arc in boundary K is not contracted through the algorithm. Hence, we can observe that the following Lemma 2 proves the correctness of Step 2.

Lemma 2. *Let $G = (V, E)$ be a graph without isolated nodes and let f' be an SC-embedding of G/e with boundary G_1 into $\Re^d$ for an arc $e = (u_1, u_2)$ with $\{u_1, u_2\} - V(G_1) \neq \emptyset$. Assume that for each node u_i, $i = 1, 2$, $|N_G(u_i)| \geq d + 1$ holds if $u_i \in V - V(G_1)$. Then there is an SC-embedding of G with boundary G_1 into $\Re^d$.* □

Before proving Lemma 2, we give some notations and one preparatory lemma for an embedding of a new point into $\Re^d$. For a convex polyhedron S in $\Re^d$, a supporting hyperplane H of S is called a *facet* of S if the dimension of $H \cap S$ is $d - 1$. It is well-known that every full-dimensional convex polyhedron can be uniquely represented by all of its facets.

Definition 3. *For a full-dimensional convex polyhedron S in $\Re^d$, let x be a vertex of S. Let $\mathcal{H}_x$ denote the family of all facets $H(a, b)$ of S containing the point x such that $S \subseteq H^+(a, b)$. Define the following polyhedron:*

$$D(x, S) = \bigcap_{H(a,b) \in \mathcal{H}_x} (H^-(a, b) - H(a, b)).$$ □

Lemma 3. *Let P be a set of points in $\Re^d$ such that $conv(P)$ is full-dimensional, and let x be a vertex of $conv(P)$. Then for a point $y \in \Re^d$, $x \in int(conv(P \cup \{y\})$ if and only if $y \in D(x, conv(P))$.* □

Proof of Lemma 2 (sketch). Let $u^* \in V(G/e)$ denote the node obtained by contracting u_1 and u_2 in G. Without loss of generality, assume $u_2 \in V - V(G_1)$ (this is possible from the assumption $\{u_1, u_2\} - V(G_1) \neq \emptyset$). Hence $|N_G(u_2)| \geq d+1$ holds. We give a constructive proof of the lemma; we show a way of finding an SC-embedding f of G with boundary G_1 into $\Re^d$. Let $f(v) := f'(v)$ for each node $v \in V(G/e) - \{u^*\} = V(G) - \{u_1, u_2\}$ and $f(u_1) := f'(u^*)$. Note that G_1 also plays the role as G' in Definition 1 (i), and that every node $v \in V(G) - V(G_1) - (N_G(u_2) \cup \{u_2\})$ satisfies $v \in int(conv_f(N_G(v)))$. We prove this lemma by showing that u_2 can be embedded so that each node $v \in \{u_2\} \cup N_G(u_2) - V(G_1)$ is strictly inside the convex hull of its neighbors. For the convexity for u_2, $u_2 \in int(conv_f(N_G(u_2)))$ must hold (the position of each node $v \in V(G) - \{u_2\}$ has been fixed, so $int(conv_f(N_G(u_2)))$ is well-defined). For the convexity for each $v \in N_G(u_2) - V(G_1)$, $u_2 \in D_v$ must hold by Lemma 3, where $D_v = D(v, conv_f(N_G(v) \cup \{v\} - \{u_2\}))$ if v is a vertex of $conv_f(N_G(v) \cup \{v\} -\{u_2\})$, $D_v = \Re^d$ otherwise. Hence it suffices to show that $int(conv_f(N_G(u_2))) \cap (\bigcap_{v \in N_G(u_2) - V(G_1)} D_v) \neq \emptyset$ (we have only to embed u_2 into this space).

Since $|N_G(u_2)| \geq d+1$ holds and the points of $\{f(v) \mid v \in N_G(u_2)\}$ are in general position, it follows that $int(conv_f(N_G(u_2))) \neq \emptyset$. Similarly, we can prove that for each v which is a vertex of $conv_f(N_G(v) \cup \{v\} - \{u_2\})$, $conv_f(N_G(v) \cup \{v\} - \{u_2\})$ is full-dimensional in $\Re^d$ and $D_v \neq \emptyset$. Moreover, since each node $v \in N_G(u_2) - \{u_1\}$ satisfies $v \in N_{G/e}(u^*)$ and $f(v) = f'(v) \in int(conv_{f'}(N_{G/e}(v)))$, it follows that $\bigcap_{v \in N_G(u_2) - \{u_1\} - V(G_1)} D_v$ contains the point $f'(u^*) = f(u_1)$ and points sufficiently close to $f(u_1)$. This and $u_1 \in N_G(u_2)$ indicate that $D^* = int(conv_f(N_G(u_2))) \cap (\bigcap_{v \in N_G(u_2) - \{u_1\} - V(G_1)} D_v) \neq \emptyset$ holds. Moreover, we have $int(conv_f(N_G(u_2))) \cap D_{u_1} \neq \emptyset$ since otherwise $u_1 \notin V(G_1)$ would hold and $f'(u^*)(= f(u_1))$ would be on a facet of $conv_{f'}(N_{G/e}(u^*))$, contradicting that f' is an SC-embedding of G/e. Therefore, it follows that $D^* \cap D_{u_1}$ contains points sufficiently close to $f(u_1)$. □

5 Remarks

For $k = 3$, the following properties hold as a corollary of Theorem 1.

Corollary 1. *Let $G = (V, E)$ be a 4-connected graph and contain pairwise disjoint subsets T_1, T_2, T_3 of V such that $|T_i|$ is even for $i = 1, 2, 3$. Then, by adding at most two extra arcs to G, a 3-bisection can be obtained. In particular, if G has K_3, then by adding at most one extra arc to G, a 3-bisection can be obtained.* □

As for a general k, we can observe from Theorem 3 and Lemma 1 that if an SC-embedding of G into $\Re^k$ exists, then G admits a k-bisection with respect to any family $\{T_1, \ldots, T_k\}$ of resource sets. Lemma 2 implies that if

(a) G has a subgraph G_1 which plays the role as G' in Definition 1 (i),
(b) $|N_G(v)| \geq k+1$ holds for each node $v \in V(G) - V(G_1)$, and
(c) we can continue contracting arcs not in $E(G_1)$ while preserving the above (b) until a graph consisting of G_1 and at most one extra node is obtained,

then an SC-embedding of G with boundary G_1 into $\Re^k$ can be found. Hence, a sufficient condition for G to satisfy the above (a)–(c) indicates one for G to have a k-bisection.

Finally, we give a proof of Theorem 2 about the arc-version of the bisection problem. This can be done by using a reduction to the node-version of the problem. Let G be a $(k+1)$-edge-connected graph. If G has exactly $k+1$ arcs, then it has exactly two nodes and it is trivial. Assume that $|E(G)| > k+1$. Let $L(G)$ denote the line graph of G, and $V_L(E') \subseteq V(L(G))$ denote the node set of $L(G)$ corresponding to an arc set $E' \subseteq E$ in G. Observe that $\{E_1, E_2\}$ is a k-bisection of the arc set with respect to $\{T_1, \ldots, T_k\}$ in G if and only if $\{V_L(E_1), V_L(E_2)\}$ is a k-bisection of the node set with respect to $\{V_L(T_1), \ldots, V_L(T_k)\}$ in $L(G)$. Moreover, if G is $(k+1)$-edge-connected, then $L(G)$ is $(k+1)$-connected and has K_{k+1}. As mentioned in Section 1 and Theorem 1, if $k = 1, 2, 3$, then $L(G)$ admits a k-bisection of the node set, and hence G also admits a k-bisection of the arc set. Finally, we remark that there exist instances that have no feasible partition unless G is $(k+1)$-edge-connected for $k = 1, 2, 3$.

References

1. B. Bozkaya, G. Laporte, E. Erkut, *A tabu search heuristic and adaptive memory procedure for political districting*, European J. Operational Research, **144** (2003), 12–26.
2. L. Chi-Yuan, J. Matoušek and W. Steiger, *Algorithms for ham-sandwich cuts*, Discrete Comput. Geom., **11** (1994), 433–452.
3. J. Chleíková, *Approximating the maximally balanced connected partition problem in graphs*, Information Processing Letters, **60** (1999), 225–230.
4. M. E. Dyer and A. M. Frieze, *On the complexity of partitioning graphs into connected subgraphs*, Discrete Applied Mathematics, **10** (1985), 139–153.
5. H. Edelsbrunner, Algorithms in combinatorial geometry, Springer-Verlag, Berlin, 1987.
6. R. C. Gonzales and P. Wintz, Digital Image Processing, Publisher Addison-Wesley, Reading, MA, 1977.
7. B. Hayes, *Machine politics*, American Scientist, **84** (1996), 522–526.
8. N. Martinov, *A recursive characterization of the 4-connected graphs*, Discrete Mathematics, **84** (1990), 105–108.
9. H. Nagamochi, T. Jordán, Y. Nakao, and T. Ibaraki, *Convex embeddings bisecting of 3-connected graphs*, Combinatorica, **22**(4) (2002), 537–554.
10. H. Suzuki, N. Takahashi, and T. Nishizeki, *A linear algorithm for bipartition of biconnected graphs*, Information Processing Letters, **33** (1990), 227–232.
11. D. C. Tsichritzis and P. A. Bernstein, Operating Systems, Academic Press, New York, 1981.
12. K. Wada and K. Kawaguchi, *Efficient algorithms for tripartitioning triconnected graphs and 3-edge-connected graphs*, Lecture Notes in Comput. Sci., 790, Springer, Graph-theoretic concepts in computer science, 1994, 132–143.
13. J. C. Williams Jr, *Political redistricting: a review*, Papers in Regional Science, **74** (1995), 12–40.

Laminar Structure of Ptolemaic Graphs and Its Applications

Ryuhei Uehara[1] and Yushi Uno[2]

[1] School of Information Science, Japan Advanced Institute of Science and Technology (JAIST), Ishikawa, Japan
uehara@jaist.ac.jp
[2] Department of Mathematics and Information Sciences, Graduate School of Science, Osaka Prefecture University, Sakai, Japan
uno@mi.s.osakafu-u.ac.jp

Abstract. Ptolemaic graphs are graphs that satisfy the Ptolemaic inequality for any four vertices. The graph class coincides with the intersection of chordal graphs and distance hereditary graphs, and it is a natural generalization of block graphs (and hence trees). In this paper, a new characterization of ptolemaic graphs is presented. It is a laminar structure of cliques, and leads us to a canonical tree representation, which gives a simple intersection model for ptolemaic graphs. The tree representation is constructed in linear time from a perfect elimination ordering obtained by the lexicographic breadth first search. Hence the recognition and the graph isomorphism for ptolemaic graphs can be solved in linear time. Using the tree representation, we also give an $O(n)$ time algorithm for the Hamiltonian cycle problem.

Keywords: algorithmic graph theory, data structure, Hamiltonian cycle, intersection model, ptolemaic graphs.

1 Introduction

Recently, many graph classes have been proposed and studied [2,10]. Among them, the class of chordal graphs is classic and widely investigated. One of the reasons is that the class has a natural intersection model and hence a concise tree representation; a graph is chordal if and only if it is the intersection graph of subtrees of a tree. The tree representation can be constructed in linear time, and it is called a clique tree since each node of the tree corresponds to a maximal clique of the chordal graph (see [17]). Another reason is that the class is characterized by a vertex ordering called a perfect elimination ordering. The ordering can also be computed in linear time, and a typical way to find it is called the lexicographic breadth first search (LBFS) introduced by Rose, Tarjan, and Lueker [16]. The LBFS is also widely investigated as a tool for recognizing several graph classes (see a comprehensive survey by Corneil [6]). Using those characterizations, many efficient algorithms have been established for chordal graphs (see, e.g., [9]).

Distance in graphs is one of the most important topics in algorithmic graph theory. The class of distance hereditary graphs was introduced by Howorka to deal with the distance property called isometric [12]. For the class, some characterizations are investigated [1,8,11], and many efficient algorithms have been proposed (see, e.g., [5,4,14]).

X. Deng and D. Du (Eds.): ISAAC 2005, LNCS 3827, pp. 186–195, 2005.

However, the recognition of distance hereditary graphs in linear time is not simple; Hammer and Maffray's algorithm [11] fails in some cases, and Damiand, Habib, and Paul's algorithm [7] requires to build a cotree in linear time (see [7, Chapter 4] for further details), where the cotree can be constructed in linear time by using recent algorithm with multisweep LBFS approach by Bretscher, Corneil, Habib, and Paul [3].

In this paper, we focus on the class of ptolemaic graphs. Ptolemaic graphs are graphs that satisfy the Ptolemaic inequality $d(x,y)d(z,w) \leq d(x,z)d(y,w) + d(x,w)d(y,z)$ for any four vertices x, y, z, w [1]. Howorka showed that the class of ptolemaic graphs coincides with the intersection of the class of chordal graphs and the class of distance hereditary graphs [13]. On the other hand, the class of ptolemaic graphs is a natural generalization of block graphs, and hence trees (see [19] for the relationships between related graph classes). However, there are relatively few known results specified to ptolemaic graphs. The reason seems that the ptolemaic graphs have no useful characterizations from the viewpoint of the algorithmic graph theory.

We propose a tree representation of ptolemaic graphs which is based on the laminar structure of cliques of a ptolemaic graph. The tree representation also gives a natural intersection model for ptolemaic graphs, which is defined over directed trees. The tree representation can be constructed in linear time. The construction algorithm can also be modified to a recognition algorithm which runs in linear time. It is worth remarking that the algorithm is quite simple, especially, much simpler than the combination of two recognition algorithms for chordal graphs and distance hereditary graphs. Moreover, the tree representation is canonical up to isomorphism. Hence, using the tree representation, we can solve the graph isomorphism problem for ptolemaic graphs in linear time.

The tree representation enables us to use the dynamic programming technique for some problems on ptolemaic graphs $G = (V, E)$. It is sure that the Hamiltonian cycle problem is one of most well known NP-hard problem, and it is still NP-hard even for a chordal graph, and that an $O(|V| + |E|)$ time algorithm is known for distance hereditary graphs [14]. Here, we show that the Hamiltonian cycle problem can be solved in $O(|V|)$ time using the technique if a ptolemaic graph is given in the tree representation.

Due to space limitation, some proofs are omitted, and can be found at `http://www.jaist.ac.jp/~uehara/pdf/ptolemaic2.pdf`.

2 Preliminaries

The *neighborhood* of a vertex v in a graph $G = (V, E)$ is the set $N_G(v) = \{u \in V \mid \{u, v\} \in E\}$, and the *degree* of a vertex v is $|N_G(v)|$ and is denoted by $\deg_G(v)$. For a subset U of V, we denote by $N_G(U)$ the set $\{v \in V \mid v \in N(u)$ for some $u \in U\}$. If no confusion can arise we will omit the index G. Given a graph $G = (V, E)$ and a subset U of V, the *induced subgraph* by U, denoted by $G[U]$, is the graph (U, E'), where $E' = \{\{u, v\} \mid u, v \in U$ and $\{u, v\} \in E\}$. Given a graph $G = (V, E)$, its *complement* is defined by $\bar{E} = \{\{u, v\} \mid \{u, v\} \notin E\}$, and is denoted by $\bar{G} = (V, \bar{E})$. A vertex set I is an *independent set* if $G[I]$ contains no edges, and then the graph $\bar{G}[I]$ is said to be a *clique*.

[1] The inequality is also known as "Ptolemy" inequality which seems to be more popular. We here use "Ptolemaic" stated by Howorka [13].

Given a graph $G = (V, E)$, a sequence of the distinct vertices $v_1, v_2, \ldots, v_l$ is a *path*, denoted by $(v_1, v_2, \ldots, v_l)$, if $\{v_j, v_{j+1}\} \in E$ for each $1 \leq j < l$. The *length* of a path is the number of edges on the path. For two vertices u and v, the *distance* of the vertices, denoted by $d(u, v)$, is the minimum length of the paths joining u and v. A *cycle* is a path beginning and ending with the same vertex. A cycle is said to be *Hamiltonian* if it visits every vertex in a graph exactly once.

An edge which joins two vertices of a cycle but is not itself an edge of the cycle is a *chord* of that cycle. A graph is *chordal* if each cycle of length at least 4 has a chord. Given a graph $G = (V, E)$, a vertex $v \in V$ is *simplicial* in G if $G[N(v)]$ is a clique in G. An ordering $v_1, \ldots, v_n$ of the vertices of V is a *perfect elimination ordering* (PEO) of G if the vertex v_i is simplicial in $G[\{v_i, v_{i+1}, \ldots, v_n\}]$ for all $i = 1, \ldots, n$. Once a vertex ordering is fixed, we denote $N(v_j) \cap \{v_{i+1}, \ldots, v_n\}$ by $N_{>i}(v_j)$. It is known that a graph is chordal iff it has a PEO (see [2]). A typical way of finding a PEO of a chordal graph in linear time is the lexicographic breadth first search (LBFS), which is introduced by Rose, Tarjan, and Lueker [16], and a comprehensive survey is presented by Corneil [6].

It is also known that a graph $G = (V, E)$ is chordal iff it is the intersection graph of subtrees of a tree T (see [2]). Let T_v denote the subtree of T corresponding to the vertex v in G. Then we can assume that each node c in T corresponds to a maximal clique C of G such that C contains v on G iff T_v contains c on T. Such a tree T is called a *clique tree* of G. From a PEO of a chordal graph G, we can construct a clique tree of G in linear time [17].

Given a graph $G = (V, E)$ and a subset U of V, an induced connected subgraph $G[U]$ is *isometric* if the distances in $G[U]$ are the same as in G. A graph G is *distance hereditary* if G is connected and every induced path in G is isometric.

A connected graph G is *ptolemaic* if for any four vertices u, v, w, x of G, $d(u, v)d(w, x) \leq d(u, w)d(v, x) + d(u, x)d(v, w)$. We will use the following characterization of ptolemaic graphs due to Howorka [13]:

Theorem 1. *The following conditions are equivalent: (1) G is ptolemaic; (2) G is distance hereditary and chordal; (3) for all distinct nondisjoint maximal cliques P, Q of G, $P \cap Q$ separates $P \setminus Q$ and $Q \setminus P$.*

Let V be a set of n vertices. Two sets X and Y are said to be *overlapping* if $X \cap Y \neq \emptyset$, $X \setminus Y \neq \emptyset$, and $Y \setminus X \neq \emptyset$. A family $\mathcal{F} \subseteq 2^V \setminus \{\{\emptyset\}\}$ is said to be *laminar* if $\mathcal{F}$ contains no overlapping sets; that is, for any pair of two distinct sets X and Y in $\mathcal{F}$ satisfy either $X \cap Y = \emptyset$, $X \subset Y$, or $Y \subset X$. Given a laminar family $\mathcal{F}$, we define *laminar digraph* $\vec{T}(\mathcal{F}) = (\mathcal{F}, \vec{E}_{\mathcal{F}})$ as follows; $\vec{E}_{\mathcal{F}}$ contains an arc (X, Y) iff $X \subset Y$ and there are no other subset Z such that $X \subset Z \subset Y$, for any sets X and Y. We denote the underlying graph of $\vec{T}(\mathcal{F})$ by $T(\mathcal{F}) = (\mathcal{F}, E_{\mathcal{F}})$. The following two lemmas for the laminar digraph are known (see, e.g., [15, Chapter 2.2]);

Lemma 1. *(1) $T(\mathcal{F})$ is a forest. (2) If a family $\mathcal{F} \subseteq 2^V$ is laminar, we have $|\mathcal{F}| \leq 2|V| - 1$.*

Hence, hereafter, we call $T(\mathcal{F})$ ($\vec{T}(\mathcal{F})$) a (directed) laminar forest. We regard each maximal (directed) tree in the laminar forest $T(\mathcal{F})$ ($\vec{T}(\mathcal{F})$) as a (directed) tree rooted at the maximal set, whose outdegree is 0 in $\vec{T}(\mathcal{F})$. We define a *label* of each node S_0 in $\vec{T}(\mathcal{F})$, denoted by $\ell(S_0)$, as follows: If S_0 is a leaf, $\ell(S_0) = S_0$. If S_0 is not a leaf and has children $S_1, S_2, \ldots, S_h$, $\ell(S_0) = S_0 \setminus (S_1 \cup S_2 \cup \cdots \cup S_h)$. Since $\mathcal{F}$ is laminar, each vertex in V appears exactly once in $\ell(S)$ for some $S \subseteq V$, and its corresponding node is uniquely determined.

3 A Tree Representation of Ptolemaic Graphs

3.1 A Tree Representation

For a ptolemaic graph $G = (V, E)$, let $\mathcal{M}(G)$ be the set of all maximal cliques, i.e., $\mathcal{M}(G) := \{M \mid M \text{ is a maximal clique in } G\}$, and $C(G)$ be the set of nonempty vertex sets defined below: $C(G) := \bigcup_{S \subseteq \mathcal{M}(G)} \{C \mid C = \cap_{M \in S} M, C \neq \emptyset\}$. Each vertex set $C \in C(G)$ is a nonempty intersection of some maximal cliques. Hence, $C(G)$ contains all maximal cliques, and each C in $C(G)$ induces a clique. We also denote by $\mathcal{L}(G)$ the set $C(G) \setminus \mathcal{M}(G)$. That is, each vertex set $L \in \mathcal{L}(G)$ is an intersection of two or more maximal cliques. The following properties are crucial.

Theorem 2. *Let $G = (V, E)$ be a ptolemaic graph. Let $\mathcal{F}$ be a family of sets in $\mathcal{L}(G)$ such that $\cup_{L \in \mathcal{F}} L \subset M$ for some maximal clique $M \in \mathcal{M}(G)$. Then $\mathcal{F}$ is laminar.*

Proof. Omitted. □

Lemma 2. *Let C_1, C_2 be any overlapping sets in $C(G)$ for a ptolemaic graph $G = (V, E)$. Then $C_1 \cap C_2$ separates $C_1 \setminus C_2$ and $C_2 \setminus C_1$.*

Proof. Omitted. □

Now we define a directed graph $\vec{T}(C(G)) = (C(G), A(G))$ for a given ptolemaic graph $G = (V, E)$ as follows: two nodes $C_1, C_2 \in C(G)$ are joined by an arc (C_1, C_2) if and only if $C_1 \subset C_2$ and there is no other C in $C(G)$ such that $C_1 \subset C \subset C_2$. We denote by $T(C(G))$ the underlying graph of $\vec{T}(C(G))$.

Theorem 3. *A graph $G = (V, E)$ is ptolemaic if and only if the graph $T(C(G))$ is a tree.*

Proof. Omitted. □

Hereafter, given a ptolemaic graph $G = (V, E)$, we call $T(C(G))$ ($\vec{T}(C(G))$) a *(directed) clique laminar tree* of G. We can naturally extend the label of a laminar forest to the directed clique laminar tree: Each node C_0 in $C(G)$ has a label $\ell(C_0) := C_0 \setminus (C_1 \cup C_2 \cup \cdots \cup C_h)$, where (C_i, C_0) is an arc on $\vec{T}(C(G))$ for $1 \leq i \leq h$. Intuitively, we additionally define the label of a maximal clique as follows; the label of a maximal clique is the set of vertices that are not contained in any other maximal cliques. We note that for each vertex in G its corresponding node in $T(C(G))$ is uniquely determined by maximal cliques. Therefore, we can define the mapping from each vertex to a vertex set in C in $T(C(G))$: We denote by $C(v)$ the clique C with $v \in \ell(C)$. When we know whether $C(v)$ is in $\mathcal{M}$ or $\mathcal{L}$, we specify it by writing $C_{\mathcal{M}}(v)$ or $C_{\mathcal{L}}(v)$. An example is given in Figure 2 (for the given ptolemaic graph (a), the clique laminar tree (b) is obtained after adding the vertices 16, 15, 14, 13, 12, 11, 10, 9, 8, and the clique laminar tree (c) is obtained after adding all vertices). In Figure 2, each single rectangle represents a non-maximal clique, each double rectangle represents a maximal clique, and each rectangle contains its label.

We also note that from $\vec{\mathcal{T}}(C(G))$ with labels, we can reconstruct the original ptolemaic graph uniquely up to isomorphism. That is, two ptolemaic graphs G_1 and G_2 are isomorphic if and only if labeled $\vec{\mathcal{T}}(C(G_1))$ is isomorphic to labeled $\vec{\mathcal{T}}(C(G_2))$.

By Theorem 3, we obtain an intersection model for ptolemaic graphs as follows:

Corollary 1. *Let $\vec{T}$ be any directed graph such that its underlying graph T is a tree. Let $\mathcal{T}$ be any set of subtrees $\vec{T_v}$ such that $\vec{T_v}$ consists of a root C and all vertices reachable from C in $\vec{T}$. Then the intersection graph over $\mathcal{T}$ is ptolemaic. On the other hand, for any ptolemaic graph, there exists such an intersection model.*

Proof. The directed clique laminar tree $\vec{\mathcal{T}}(C(G))$ is the base directed graph of the intersection model. For each $v \in V$, we define the root C such that $v \in \ell(C)$. □

3.2 A Linear Time Construction of Clique Laminar Trees

The main theorem in this section is the following:

Theorem 4. *Given a ptolemaic graph $G = (V, E)$, the directed clique laminar tree $\vec{\mathcal{T}}(C(G))$ can be constructed in $O(|V| + |E|)$ time.*

We will make the directed clique laminar tree $\vec{\mathcal{T}}(C(G))$ by separating the vertices in G into the vertex sets in $C(G) = \mathcal{M}(G) \cup \mathcal{L}(G)$.

We first compute (and fix) a PEO $v_1, v_2, \ldots, v_n$ by the LBFS. The outline of our algorithm is similar to the algorithm for constructing a clique tree for a given chordal graph due to Spinrad in [17]. For each vertex $v_n, v_{n-1}, \ldots, v_2, v_1$, we add it into the tree and update the tree. For the current vertex v_i, let $v_j := \min\{N_{>i}(v_i)\}$. Then, in Spinrad's algorithm [17], there are two cases to consider: $N_{>i}(v_i) = C(v_j)$ or $N_{>i}(v_i) \subset C(v_j)$. The first case is sipmle; just add v_i into $C(v_j)$. In the second case, Spinrad's algorithm adds a new maximal clique $C(v_i)$ that consists of $N_{>i}(v_i) \cup \{v_i\}$. However, in our algorithm, involved case analysis is required. For example, in the latter case, the algorithm have to handle three vertex sets; two maximal cliques $\{v_i\} \cup N_{>i}(v_i)$ and $C(v_j)$ together with one vertex set $N_{>i}(v_i)$ shared by them. In this case, intuitively, our algorithm makes three distinct sets C_M with $\ell(C_M) = \{v_i\}$, C_L with $\ell(C_L) = N_{>i}(v_i)$, and C with $\ell(C) = C(v_j) \setminus N_{>i}(v_i)$, and adds two arcs (C_L, C_M) and (C_L, C); this means that v_i is in $C_M = N_{>i}(v_i) \cup \{v_i\}$, C is a clique $C(v_j)$, and C_L is the vertex set shared by C_M and C. However, our algorithm has to handle more complicated cases since the set $C(v_j)$ (and hence $N_{>i}(v_i)$) can already be partitioned into some vertex sets.

In $\vec{\mathcal{T}}(C(G))$, each node C stores $\ell(C)$. Hence each vertex in G appears exactly once in the tree. To represent it, each vertex v has a pointer to the node $C(v)$ in $C(G) = \mathcal{M}(G) \cup \mathcal{L}(G)$. The detail of the algorithm is described as CliqueLaminarTree shown in Figure 1, and an example of the construction is depicted in Figure 2. In Figure 2, the left-hand graph gives a ptolemaic graph, and the right-hand tree is the clique laminar tree constructed according to the vertex ordering given in the figure. We show the correctness and a complexity analysis of the algorithm.

We will use the following property of a PEO found by the LBFS of a chordal graph:

Algorithm 1: CLIQUELAMINARTREE

Input : A ptolemaic graph $G = (V, E)$ with a PEO $v_1, v_2, \ldots, v_n$ obtained by the LBFS,
Output: A clique laminar tree T.

initialize T by the clique $C_M(v_n) := \{v_n\}$ and set the pointer from v_n to $C_M(v_n)$;
for $i := n-1$ *down to* 1 **do**
  let $v_j := \min\{N_{>i}(v_i)\}$;
  switch *condition of* $N_{>i}(v_i)$ **do**
    case *(1)* $N_{>i}(v_i) = C_M(v_j)$
      update $\ell(C_M(v_j)) := \ell(C_M(v_j)) \cup \{v_i\}$ and $|C_M(v_j)| := |C_M(v_j)| + 1$;
      set $C_M(v_i) := C_M(v_j)$;
    case *(2)* $N_{>i}(v_i) = C_L(v_j)$
      make a new maximal clique $C_M(v_i)$ with $\ell(C_M(v_i)) := \{v_i\}$ and $|C_M(v_i)| := |C_L(v_j)| + 1$;
      add an arc $(C_L(v_j), C_M(v_i))$;
    case *(3)* $N_{>i}(v_i) \subset C(v_j)$ *and* $|\ell(C(v_j))| = |C(v_j)|$
      update $\ell(C(v_j)) := \ell(C(v_j)) \setminus N_{>i}(v_i)$ and $|\ell(C(v_j))| := |\ell(C(v_j))| - |N_{>i}(v_i)|$;
      make a new vertex set $L := N_{>i}(v_i)$ with $\ell(L) := N_{>i}(v_i)$ and $|L| := |N_{>i}(v_i)|$;
      make a new maximal clique $C_M(v_i)$ with $\ell(C_M(v_i)) = \{v_i\}$ and $|C_M(v_i)| := |L| + 1$;
      add arcs $(L, C(v_j))$ and $(L, C_M(v_i))$;
    case *(4)* $N_{>i}(v_i) \subset C(v_j)$ *and* $|\ell(C(v_j))| < |C(v_j)|$
      make a new vertex set $L := N_{>i}(v_i)$ with $\ell(L) := N_{>i}(v_i) \cap \ell(C(v_j))$ and $|L| := |N_{>i}(v_i)|$;
      update $\ell(C(v_j)) := \ell(C(v_j)) \setminus L$ and $|\ell(C(v_j))| := |\ell(C(v_j))| - |L|$;
      make a new maximal clique $C_M(v_i)$ with $\ell(C_M(v_i)) = \{v_i\}$ and $|C_M(v_i)| = |L| + 1$;
      remove the arc $(L', C(v_j))$ with $L' \subset L$ and add an arc (L', L);
      add arcs $(L, C(v_j))$ and $(L, C_M(v_i))$;
    end
  end
  set the pointer from v_i to $C(v_i)$;
end
return T.

Fig. 1. A linear time algorithm for the clique laminar tree T of a ptolemaic graph $G = (V, E)$

Lemma 3. *[6, Theorem 1] Let* $v_1, v_2, \ldots, v_n$ *be a PEO found by the LBFS. Then* $i < j$ *implies* $\max\{N(v_i)\} \leq \max\{N(v_j)\}$.

We assume that Algorithm CLIQUELAMINARTREE is going to add v_i, and let $v_j := \min\{N_{>i}(v_i)\}$. We will show that all possible cases are listed, and in each case, CLIQUELAMINARTREE correctly manages the nodes in $C(G)$ and their labels in $O(\deg(v_i))$ time. The following lemma drastically decreases the number of possible cases, and simplifies the algorithm.

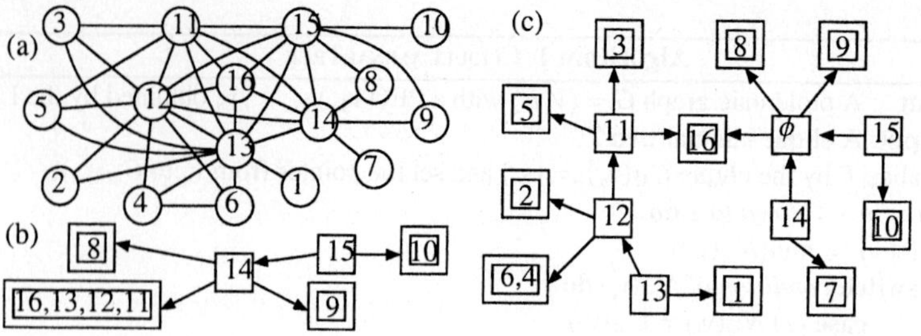

Fig. 2. A ptolemaic graph and its clique laminar tree

Lemma 4. *Let v_k be $\max\{N_{>i}(v_i)\}$. We moreover assume that the set $N_{>i}(v_i)$ has already been divided into some distinct vertex sets $L_1, L_2, \ldots, L_h$. Then, there is an ordering of the sets such that $v_k \in L_1 \subset L_2 \subset \cdots \subset L_h$.*

Proof. Omitted. □

We here describe the outline of the proof of Theorem 4, and the details can be found in Appendix. Since the graph G is chordal and the vertices are ordered in a PEO, $N_{>i}(v_i)$ induces a clique. By Lemma 4, we have three possible cases; (a) $N_{>i}(v_i) = C(v_j)$, (b) $N_{>i}(v_i) \subset C(v_j)$ and there are no vertex sets in $N_{>i}(v_i)$, and (c) $N_{>i}(v_i) \subset C(v_j)$ and there are vertex sets $L_1 \subset L_2 \subset \cdots \subset L_h \subset N_{>i}(v_i)$. In the last case, we note that $L_h \neq N_{>i}(v_i)$; otherwise, we have $v_j \in L_h$, or consequently, $L_h = C(v_j) = N_{>i}(v_i)$, which is case (a). In case (a), we have two subcases; $C(v_j)$ is a maximal clique (i.e., $N_{>i}(v_i) = C_M(v_j)$) or $C(v_j)$ is a non-maximal clique (i.e., $N_{>i}(v_i) = C_L(v_j)$).

In each case, careful analysis implies that the maintenance of the clique laminar tree can be done in $O(\deg(v_i))$ time for each v_i. Therefore the time complexity of CLIQUE-LAMINARTREE is $O(n + m)$, which completes the proof of Theorem 4.

4 Applications of Clique Laminar Trees

Theorem 5. *The recognition problem for ptolemaic graphs can be solved in linear time.*

Proof. Omitted. □

Theorem 6. *The graph isomorphism problem for ptolemaic graphs can be solved in linear time.*

Proof. Omitted. □

We note that Theorem 5 is not new. Since a graph is ptolemaic iff it is chordal and distance-hereditary [13], we have Theorem 5 by combining the results in [16,11,7,3]. We dare to state Theorem 5 to show that we can recognize if a graph is ptolemaic and then construct its clique laminar tree at the same time in linear time, and the algorithm is much simpler and more straightforward than the combination of known algorithms. (As noted in Introduction, the linear time algorithm for recognition of distance hereditary graphs is not so simple.)

Theorem 7. *The Hamiltonian cycle problem for ptolemaic graphs can be solved in $O(n)$ time.*

Due to space limitation, we describe the outline of the proof of Theorem 7.

We remind that $\vec{T}(C(G))$ takes $O(n)$ space. We first observe that if $\vec{T}(C(G))$ contains a vertex set C with $|C| = 1$, the vertex in C is a cutpoint of G, and hence G does not have a Hamiltonian cycle. This condition can be checked in $O(n)$ time over $\vec{T}(C(G))$. Hence, hereafter, we assume that G has no cutpoint, or equivalently, any vertex set C in C satisfies $|C| > 1$.

Let L be a vertex set in $C(G)$. Each vertex set L' with $(L, L') \in A(G)$ is said to be a *child* of L, and each vertex set L'' with $(L'', L) \in A(G)$ is said to be a *parent* of L. That is, a child L' and a parent L'' of L satisfy $L'' \subset L \subset L'$. We define ancestors and descendants for L as in ordinary trees. Note here that any node L in $\vec{T}(C(G))$ is an ancestor and descendant of itself. We denote by $c(L)$ and $p(L)$ the number of children of L and the number of parents of L in $\vec{T}(C(G))$, respectively. Hence $c(M) = 0$ for each maximal clique M, and $p(L) = 0$ for each minimal vertex set L.

We first consider a minimal vertex set L with $p(L) = 0$. By Lemma 2, each L in $\mathcal{L}(G)$ is a separator of G. It is easy to see that if we remove L from G, we have $c(L)$ connected components. Hence, if $|L| < c(L)$, G cannot have a Hamiltonian cycle. On the other hand, when $|L| = c(L)$, any Hamiltonian cycle uses all vertices in L to connect each connected component. This fact can be seen as follows; we first make a cycle of length $|L|$ in L, and next replace each edge by a path through the vertices in one vertex set corresponding to a child of the node L. We say that we *assign* each edge to distinct child of L. If $|L| > c(L)$, we can construct a Hamiltonian cycle with $|L| - c(L)$ edges in $G[L]$. In this case, $c(L)$ edges in L are assigned to construct a cycle, and $|L| - c(L)$ edges are left, which can be assigned in some other descendants. We then define the *margin* $m(L)$ by $|L| - c(L) = |\ell(L)| - c(L)$. That is, if $m(L) < 0$, G has no Hamiltonian cycle, and if $m(L) > 0$, we have $m(L)$ edges in L which can be assigned in some descendants. We note that a margin can be inherited only from an ancestor to an descendant.

We next define a *distribution* $\delta((C_i, C_j))$ of the margin, which is a function assigned to each arc (C_i, C_j) in $\vec{T}(C(G))$. Let $C_1, \ldots, C_k$ be the children of L. Then for $i = 1, 2, \ldots, k$ each arc (L, C_i) has a distribution $\delta((L, C_i))$ with $\sum_{i=1}^{k} \delta((L, C_i)) = m(L)$. That is, each child C_i inherits $\delta((L, C_i))$ margins from L, and some descendants of C_i will consume $\delta((L, C_i))$ margins from L. The way to compute the distribution will be discussed later.

We then consider a vertex set C with $p(C) > 0$ and $c(C) \geq 0$, that is, C is a vertex set which is not minimal. Let $P_1, P_2, \ldots, P_h$ be parents of C and $C_1, C_2, \ldots, C_k$ children of C. That is, we have $P_i \subset C \subset C_j$ for each i and j with $1 \leq i \leq h = p(C)$ and $1 \leq j \leq k = c(C)$ ($k = c(C) = 0$ when C is maximal clique). We assume that $\delta((P_i, C))$ are already defined for each P_i. In the case, we have to assign k edges in C to the children of C. Each child will replace it by the path through all vertices in the child. We also have one assigned edge from each parent, and some additional vertices from parents P_i if $\delta((P_i, C)) > 0$. Hence the margin $m(C)$ is defined by $|\ell(C)| + h + \sum_{i=1}^{h} \delta((P_i, C)) - k = |\ell(C)| + \sum_{i=1}^{h}(\delta((P_i, C)) + 1) - k$. The distribution $\delta((C, C_i))$ of the margin $m(C)$ is defined by a function with $\sum_{i=1}^{k} \delta((C, C_i)) = m(C)$.

Above discussion leads us to the following theorem:

Theorem 8. *Let $G = (V, E)$ be a ptolemaic graph. Then G has a Hamiltonian cycle if and only if there exist feasible distributions of margins such that each vertex set C in $\mathcal{C}$ satisfies $m(C) \geq 0$.*

Our linear time algorithm, say $\mathcal{A}$, runs on $T(\mathcal{C}(G))$; $\mathcal{A}$ collects the leaves in $T(\mathcal{C}(G))$, computes the margins, and repeats this process by computing the margin of C such that all neighbors of C have been processed except exactly one neighbor. The outline of the procedure for each vertex set C with parents $P_1, P_2, \ldots, P_h$ and children $C_1, C_2, \ldots, C_k$ is described as follows:

(1) When the vertex set C is a leaf of $T(\mathcal{C}(G))$, C is a maximal clique in G, and hence $\delta((P, C))$ is set to 0, where P is the unique parent of C.

(2) When C is not a leaf of $T(\mathcal{C}(G))$, let X be the only neighbor which is not processed. Without loss of generality, we assume that either $X = P_h$ or $X = C_k$. To simplify the notation, we define $h' = h - 1$ and $k' = k$ if $X = P_h$, and $h' = h$ and $k' = k - 1$ if $X = C_k$. We have three subcases, but the most complicated case is described below.

When C is not a maximal clique with $k > 0$ and $h > 0$, $\mathcal{A}$ first computes the margin $m(C) = |\ell(C)| + \sum_{i=1}^{h'}(\delta((P_i, C)) + 1) - k'$. Next, $\mathcal{A}$ distributes the margin $m(C)$ to the children $C_1, \ldots, C_{k'}$ by computing $\delta' := m(C) - \sum_{i=1}^{k'} \delta((C, C_i)) = |\ell(C)| + \sum_{i=1}^{h'}(\delta((P_i, C)) + 1) - \sum_{i=1}^{k'}(\delta((C, C_i)) + 1)$. The value δ' indicates the margin that can be consumed by X.

If X is a child C_k, $\mathcal{A}$ distributes all margins δ' to X, or sets $\delta((C, X)) = \delta'$. Thus, in this case, if $\delta' < 0$, G has no Hamiltonian cycles. When $\delta' \geq 0$, $\mathcal{A}$ will use the margin δ' when it processes the vertex set X.

On the other hand, if X is a parent P_h, the margin will be distributed from X to C. Hence, if $\delta' < 0$, the vertex C borrows margin δ' from X which will be adjusted when the vertex X is chosen by $\mathcal{A}$. Thus $\mathcal{A}$ sets $\delta((X, C)) = -\delta'$ in this case. If $\delta' \geq 0$, the margin is useless since the parent X only counts the number of its children C, and does not use their margins. Therefore, $\mathcal{A}$ does nothing.

(3) When C is the last node of the process; that is, every value of $\delta((C, C'))$ or $\delta((C', C))$ for each neighbor C' of C has been computed. In the case, $\mathcal{A}$ computes $m(C) = |\ell(C)| + \sum_{i=1}^{h}(\delta((P_i, C)) + 1) - \sum_{i=1}^{k}(\delta((C, C_i)) + 1)$. If $m(C) < 0$, C does not have enough margin. Hence G has no Hamiltonian cycle. Otherwise, every node has enough margin, and hence G has a Hamiltonian cycle.

The correctness of $\mathcal{A}$ can be proved by a simple induction for the number of nodes in $\vec{T}(\mathcal{C}(G))$ with Theorem 8. On the other hand, since $T(\mathcal{C}(G))$ contains $O(n)$ nodes, the algorithm runs in $O(n)$ time and space, which completes the proof of Theorem 7. We note that the construction of a Hamiltonian cycle can be done simultaneously in $O(n)$ time and space.

5 Concluding Remarks

In this paper, we presented new tree representations (data structures) for ptolemaic graphs. The result enables us to use the dynamic programming technique to solve some basic problems on this graph class. We presented a linear time algorithm for the Hamiltonian cycle problem, as one of such typical examples. To develop such efficient algorithms based on the dynamic programming for other problems are future works.

Acknowledgment

The authors thank Professor Hiro Ito, who pointed out a flaw in our polynomial time algorithm for finding a longest path in a ptolemaic graph stated in [18]. That motivated us to investigate the problems in this paper.

References

1. H.-J. Bandelt and H.M. Mulder. Distance-Hereditary Graphs. *J. of Combinatorial Theory, Series B*, 41:182–208, 1986.
2. A. Brandstädt, V.B. Le, and J.P. Spinrad. *Graph Classes: A Survey*. SIAM, 1999.
3. A. Bretscher, D. Corneil, M. Habib, and C. Paul. A Simple Linear Time LexBFS Cograph Recognition Algorithm. In *Graph-Theoretic Concepts in Computer Science (WG 2003)*, pages 119–130. LNCS Vol. 2880, Springer-Verlag, 2003.
4. H.J. Broersma, E. Dahlhaus, and T. Kloks. A linear time algorithm for minimum fill-in and treewidth for distance hereditary graphs. *Discrete Applied Mathematics*, 99:367–400, 2000.
5. M.-S. Chang, S.-Y. Hsieh, and G.-H. Chen. Dynamic Programming on Distance-Hereditary Graphs. In *Proceedings of 8th International Symposium on Algorithms and Computation (ISAAC '97)*, pages 344–353. LNCS Vol. 1350, Springer-Verlag, 1997.
6. D.G. Corneil. Lexicographic Breadth First Search — A Survey. In *Graph-Theoretic Concepts in Computer Science (WG 2004)*, pages 1–19. LNCS Vol. 3353, Springer-Verlag, 2004.
7. G. Damiand, M. Habib, and C. Paul. A Simple Paradigm for Graph Recognition: Application to Cographs and Distance Hereditary Graphs. *Theoretical Computer Science*, 263:99–111, 2001.
8. A. D'Atri and M. Moscarini. Distance-Hereditary Graphs, Steiner Trees, and Connected Domination. *SIAM J. on Computing*, 17(3):521–538, 1988.
9. F. Gavril. Algorithms for Minimum Coloring, Maximum Clique, Minimum Covering by Cliques, and Maximum Independent Set of a Chordal Graph. *SIAM J. on Computing*, 1(2):180–187, 1972.
10. M.C. Golumbic. *Algorithmic Graph Theory and Perfect Graphs*. Annals of Discrete Mathematics 57. Elsevier, 2nd edition, 2004.
11. P.L. Hammer and F. Maffray. Completely Separable Graphs. *Discrete Applied Mathematics*, 27:85–99, 1990.
12. E. Howorka. A Characterization of Distance-Hereditary Graphs. *Quart. J. Math. Oxford (2)*, 28:417–420, 1977.
13. E. Howorka. A Characterization of Ptolemaic Graphs. *J. of Graph Theory*, 5:323–331, 1981.
14. R.-W. Hung and M.-S. Chang. Linear-time algorithms for the Hamiltonian problems on distance-hereditary graphs. *Theoretical Computer Science*, 341:411–440, 2005.
15. B. Korte and J. Vygen. *Combinatorial Optimization*, volume 21 of *Algorithms and Combinatorics*. Springer, 2000.
16. D.J. Rose, R.E. Tarjan, and G.S. Lueker. Algorithmic Aspects of Vertex Elimination on Graphs. *SIAM J. on Computing*, 5(2):266–283, 1976.
17. J.P. Spinrad. *Efficient Graph Representations*. American Mathematical Society, 2003.
18. R. Uehara and Y. Uno. Efficient Algorithms for the Longest Path Problem. In *15th Annual International Symposium on Algorithms and Computation (ISAAC 2004)*, pages 871–883. LNCS Vol.3341, Springer-Verlag, 2004.
19. H.-G. Yeh and G.J. Chang. Centers and medians of distance-hereditary graphs. *Discrete Mathematics*, 265:297–310, 2003.

On the Complexity of the G-Reconstruction Problem

Zdeněk Dvořák and Vít Jelínek

Department of Applied Mathematics, Charles University,
Malostranské Náměstí 25, 118 00 Praha 1
{rakdver, jelinek}@kam.mff.cuni.cz

Abstract. Let G be a fixed undirected graph. The G-structure of a graph F is the hypergraph H with the same set of vertices as F and with the property that a set h is a hyperedge of H if and only if the subgraph of F induced on h is isomorphic to G. For a fixed parameter graph G, we consider the complexity of determining whether for a given hypergraph H there exists a graph F such that H is the G-structure of F. It has been proven that this problem is polynomial if G is a path with at most 4 vertices ([9], [10]). We investigate this problem for larger graphs G and show that for some G the problem is NP-complete – in fact we prove that it is NP-complete for almost all graphs G.

1 Introduction and Basic Definitions

In this paper, we study the complexity of a decision problem related to the existence of a graph with a prescribed set of induced subgraphs of a given kind.

All the graphs considered in this paper are simple undirected graphs without multiple edges or loops. If $G = (V, E)$ is a graph and h a subset of V, let $G[h]$ denote the subgraph of G induced by h, i.e., $G[h] = \left(h, E \cap \binom{h}{2}\right)$. The vertex connectivity of G is denoted by $\kappa(G)$.

A *hypergraph* $H = (V, E)$ is an ordered pair consisting of a vertex set V and a set of hyperedges $E \subseteq 2^V$. A hypergraph H is called k-*uniform* if every hyperedge of H has size k. Let G be a graph on k vertices. The G-*structure* of a graph F with vertex set V is the k-uniform hypergraph whose vertex set is V and whose edges are exactly the sets $h \subseteq V$ such that $F[h]$ is isomorphic to G. If H is the G-structure of F, we say that F is a G-*realization* of H.

The G-*reconstruction problem* for a fixed graph G is defined as follows:

Input: a k-uniform hypergraph H, where $k = |V(G)|$.
Question: does a G-realization F of the hypergraph H exist?

This question is related to the general problem of reconstructing a graph from information about its smaller parts (see [1] and [2] for examples of this kind of problems). Another motivation for this problem stems from the *Semi-Strong Perfect Graph Theorem*, which states that if two graphs have the same P_4-structure (where P_4 is a path on four vertices), then either both of them are perfect or neither of them is ([7]). Therefore, to recognize perfect graphs, it suffices to be able

X. Deng and D. Du (Eds.): ISAAC 2005, LNCS 3827, pp. 196–205, 2005.

to recognize the P_4-structures corresponding to perfect graphs. This observation motivated Chvátal to ask whether the P_4-reconstruction problem can be solved in polynomial time. This stimulated a lot of research in the area, see for example [5], [6] or [8]. The question was finally answered by Hayward et al. ([10]), who proved that there exists a polynomial time algorithm for this problem.

The P_3-reconstruction problem can also be solved in polynomial time by [9], and it is easy to see that also for all other graphs on three vertices. This led Hayward et al. ([10]) to ask whether there is a graph G such that the G-reconstruction problem is NP-complete.

In this paper, we prove that the G-reconstruction problem is NP-complete for many graphs G (in fact it is NP-complete for a sufficiently large random graph G with high probability). The smallest graph for which we are able to prove the problem to be NP-complete is $K_2 + 3K_1$ (a graph consisting of an edge and three isolated vertices). With respect to the original question of Chvátal and the results of [10] it is noteworthy that we are also able to prove that the P_n-reconstruction problem is NP-complete for all $n \geq 6$.

The paper is organized as follows: in Section 1 we prepare some technical tools. In Section 2 we prove the key theorem of this paper, which gives sufficient conditions for the G-reconstruction problem to be NP-complete. In Section 3 we show several interesting classes of graphs which satisfy these conditions. Since the problem obviously belongs to NP, we are only concerned with its NP-hardness.

2 Preliminaries

Consider the G-reconstruction problem for a fixed graph G. Throughout this paper we always assume that k denotes the number of vertices of G. Furthermore, we always assume that neither G nor its complement is a complete graph (in both of these cases, the G-reconstruction problem is trivial). Note that this implies that $k \geq 3$.

Our proof of NP-hardness proceeds by construction of hypergraphs such that prescribed subgraphs of their reconstructions must satisfy some properties. We are able to express other NP-hard problems with these properties, thus showing that it is hard to determine whether such a reconstruction exists. Let us state a few definitions that describe this idea more formally.

Definition 1. *Let $F = (V, E)$ be a graph, let $U = (u_1, \ldots, u_m)$ be an ordered m-tuple whose elements are unorderdered pairs of vertices of F. Let $X = (X_1, \ldots, X_m)$ be an ordered m-tuple of boolean values. We say that X* agrees with F on U *if X_i is true if and only if u_i is an edge of F for each $i = 1, \ldots, m$. Furthermore, if $P(x_1, \ldots, x_m)$ is an m-variate boolean predicate we say that F* satisfies P on U *if P is satisfied by the boolean values that agree with F on U.*

Definition 2. *An ordered pair $T = (H, U)$ is called a G-*gadget *for an m-variate boolean predicate P, if the following conditions hold:*

– *H is a k-uniform hypergraph.*
– *$U = (u_1, \ldots, u_m)$ is an ordered m-tuple, where $u_i \in \binom{V(H)}{2}$ are mutually distinct, but not necessarily disjoint pairs of vertices of H.*
– *For any m-tuple of boolean values $X = (X_1, \ldots, X_m)$, the predicate P is satisfied by X if and only if there exists a graph F with G-structure H such that X agrees with F on U.*

We call the pairs $u_i \in U$ as well as the vertices that belong to these pairs special. *If no ambiguity can arise, we omit G from the specification of the gadget and speak simply of a* gadget for P.

Definition 3. *Let $H_1 = (V_1, E_1)$ and $H_2 = (V_2, E_2)$ be two k-uniform hypergraphs. Let $e_1 = \{x_1, x_2\} \in \binom{V_1}{2}$ and $e_2 = \{y_1, y_2\} \in \binom{V_2}{2}$ be two pairs of vertices. We say that a hypergraph $H = (V, E)$ is obtained by* connecting H_1 and H_2 through e_1 and e_2 *if H satisfies the following conditions:*

– *V is the disjoint union of V_1 and $V_2 \setminus e_2$.*
– *Let $g : V_2 \to (V_2 \setminus e_2) \cup e_1$ be a bijection defined as follows: $g(y) = y$ for each $y \in V_2 \setminus e_2$, and $g(y_i) = x_i$ for $i \in \{1, 2\}$. Then $H = H_1 \cup \{g[h]; h \in H_2\}$, where $g[h]$ is the image of the set h under the bijection g.*

Similarly, if (H_1, U_1) and (H_2, U_2) are two G-gadgets with $U_1 = (e_1, \ldots, e_p)$ and $U_2 = (f_1, \ldots, f_q)$, then by connecting the two gadgets through e_1 and f_1 we obtain a gadget (H, U), where H is the corresponding connection of H_1 and H_2, and $U = (e_1, \ldots, e_p, f'_2, \ldots, f'_q)$, with f'_i being the image of f_i under the mapping g defined above.

Note that since $k \geq 3$, there can be no hyperedge fully contained in either e_1 or e_2. This implies that the induced subhypergraph $H[V_1]$ is equal to H_1 and $H[(V_2 \setminus e_2) \cup e_1]$ is isomorphic to H_2.

The importance of the operation introduced by Definition 3 is explained by the following lemma.

Lemma 1. *Suppose that $\kappa(G) > 2$. Let $T_1 = (H_1, U_1)$, $T_2 = (H_2, U_2)$ be G-gadgets for the predicates $P_1(x_1, \ldots, x_p)$ and $P_2(y_1, \ldots, y_q)$, respectively, where $U_1 = (e_1, \ldots, e_p)$ and $U_2 = (f_1, \ldots, f_q)$. Let $T = (H, U)$ be the gadget obtained by connecting T_1 and T_2 through e_1 and f_1, with $U = (e_1, e_2, \ldots, e_p, f'_2, \ldots, f'_q)$. Then T is a G-gadget for the following predicate P:*

$$P(x_1, \ldots, x_p, y_2, \ldots, y_q) = P_1(x_1, \ldots, x_p) \wedge P_2(x_1, y_2, \ldots, y_q).$$

Proof. Let F be a G-realization of H. Then $F[V_1]$ is a realization of H_1, thus $F[V_1]$ satisfies P_1 on U_1. Similarly, $F[V_2]$ satisfies P_2 on $(e_1, f'_2, \ldots, f'_q)$. It follows that F satisfies P on U, as required.

Conversely, let $(X_1, \ldots, X_p, Y_2, \ldots, Y_q)$ be a $(p+q-1)$-tuple of boolean values that satisfy P. Then $(X_1, \ldots, X_p)$ satisfy P_1, hence there exists a graph F_1 which is a realization of H_1 and agrees with $(X_1, \ldots, X_p)$ on U_1. Similarly, there is a realization F_2 of $H[(V_2 \setminus f_1) \cup e_1]$ that agrees with $(X_1, Y_2, \ldots, Y_q)$ on

$(e_1, f_2', \ldots, f_q')$. Since both F_1 and F_2 agree with X_1 on e_1 and the two graphs do not have any other common vertex, there is a graph F on the vertex set V such that $F[V_1] = F_1$ and $F[(V_2 \setminus f_1) \cup e_1] = F_2$. It remains to show that F is a realization of H. Clearly the only problem might arise from the k-tuples that intersect both $V_1 \setminus e_1$ and $V_2 \setminus f_1$. No such k-tuple belongs to H, thus it suffices to prove that G is not isomorphic to any graph induced by such a k-tuple. This however follows straightforwardly from the condition on the connectivity of G, since if G were isomorphic to a graph induced by some such k-tuple K, then $K \cap e_1$ would be a cut of size at most 2 in G. □

Next we are going to state several lemmas related to the existence of gadgets for various properties, concretely:

- a gadget O that forces the presence of an edge at the pair of special vertices, i.e., the one for the property $P(x) = x$,
- a gadget Z that forces the absence of an edge at the pair of special vertices, i.e., the one for the property $P(x) = \overline{x}$,
- a nonequality gadget N for the property $P(x, y) = (x \neq y)$,
- an equality gadget Q for the property $P(x, y) = (x = y)$,
- a gadget T_{SAT} for the property $P(x, y, z) = x \vee y \vee z$,
- a gadget T_{NAE} for the property $P(x, y, z) = (x \neq y) \vee (x \neq z) \vee (y \neq z)$, and
- a gadget $T_{1 \text{ in } 3}$ for the property $P(x, y, z) = (x \wedge \overline{y} \wedge \overline{z}) \vee (\overline{x} \wedge y \wedge \overline{z}) \vee (\overline{x} \wedge \overline{y} \wedge z)$.

Lemma 2. *Suppose that $\kappa(G) > 2$. Then a gadget O exists if and only if a gadget Z exists.*

Proof. We construct Z as follows: first, we take k vertices and put this k-tuple K into Z. Then we embed a fixed copy of G on these vertices, and connect a copy of the gadget O to every pair that forms an edge of this copy. We choose $U = \{u\}$, where u is any non-edge of the prescribed copy of G. By Lemma 1 this is a gadget that enforces that the subgraph of any reconstruction of Z induced by K is exactly the prescribed copy of G and thus that u is a non-edge.

By an analogical construction, we may obtain O from Z. □

Lemma 3. *Suppose that $\kappa(G) > 2$. If a gadget O exists, then also a gadget N exists.*

Proof. We take k vertices and put this k-tuple K into N. Let this hypergraph be denoted by N_0. Since G is neither complete nor edgeless, there are at least two ways how to reconstruct the N_0. We construct the sequence N_0, N_1, ..., N_p of hypergraphs satisfying the following conditions:

- $N_i \subset N_{i+1}$
- Let n_i be the number of different subgraphs of reconstructions of N_i induced by K. Then $n_i \leq 2n_{i+1}$.
- $n_i \geq 2$ for $i \neq p$, $n_p = 1$.

If we succeed, then $n_{p-1} = 2$. The graphs F_1 and F_2 obtainable as subgraphs of reconstructions of N_{p-1} induced by K must have the same number of edges (since they are isomorphic to G) and thus there exist two pairs of vertices u_1 and u_2 of K such that u_i is the edge in F_i and the nonedge in F_{3-i}, thus $(N_{p-1}, (u_1, u_2))$ is a gadget for P.

We construct the sequence as follows: Let $e_1, e_2, \ldots$ be a sequence of pairs of vertices of K, in any order. Let N_i^O and N_i^Z be hypergraphs obtained from N_{i-1} by connecting O and Z on e_i, respectively, n_i^O and n_i^Z numbers of subgraphs in reconstructions of these hypergraphs induced by K. Then $n_i^O + n_i^Z = n_{i-1}$, let n_i^X be the greater of them (any of them if they are equal) and put $N_i = N_i^X$. If $n_i^X = 1$, we let $p = i$, otherwise we continue with the construction. □

Lemma 4. *Suppose that $\kappa(G) > 2$. If a gadget N exists, then also a gadget Q exists.*

Proof. We take two copies of the gadget N and connect them using Lemma 1. □

Lemma 5. *Suppose that $\kappa(G) > 2$. If a gadget O exists, then also at least one of T_{SAT}, T_{NAE}, or $T_{1\ in\ 3}$ gadgets exists.*

Proof. We take a k-tuple of vertices K and add K to the constructed hypergraph. Let v be an arbitrary vertex in K and let as $e_1, \ldots, e_{k-1}$ be the pairs of vertices of K containing v.

Due to the connectivity condition and the fact that G is not complete we have $\delta(G) \geq 3$ and $k \geq 5$. The following cases may occur:

- There are vertices in G that have degrees $\delta(G) + 1$ and $\delta(G) + 2$. Then we attach copies of O to $e_1, \ldots, e_{\delta(G)-1}$ and copies of Z to $e_{\delta(G)}, \ldots, e_{k-4}$. The pairs e_{k-3}, e_{k-2} and e_{k-1} are the special ones. This forms a T_{SAT} gadget, as in any reconstruction at least one of the special pairs may and must be chosen to be an edge.
- There is a vertex in G that has degree $\delta(G) + 1$, no vertex of degree $\delta(G) + 2$ and $\delta(G) \leq k - 2$. Then the same construction applies and the resulting gadget is T_{NAE}.
- There is no vertex in G with degree $\delta(G) + 1$, then we attach copies of O to $e_1, \ldots, e_{\delta(G)-2}$ and copies of Z to $e_{\delta(G)-1}, \ldots, e_{k-4}$. Furthermore, we attach copies of N to e_{k-3}, e_{k-2} and e_{k-1} and take the other special pairs of these copies as special for the resulting gadget. We thus obtain $T_{\mathrm{1\ in\ 3}}$.
- $\delta(G) = k - 2$, $\Delta(G) = k - 1$, then G is a complement of a matching. If there are exactly two vertices of degree $k - 2$, we attach copies of O to all pairs of O except of a single triangle. Then we choose negations (through N) of the pairs of the triangle as special and the resulting gadget is $T_{\mathrm{1\ in\ 3}}$. Otherwise we force edges and nonedges in K using gadgets O and Z as given by a fixed copy of G except for pairs on four vertices on which two nonedges are in this fixed copy. Of these four we take any three and by taking negated pairs in this triangle as special we again obtain $T_{\mathrm{1\ in\ 3}}$.

This shows that indeed in all the cases at least one of the gadgets exists. □

Lemma 6. *Suppose that a gadget Q exists and $\kappa(G) > 2$. If a gadget $T = (H, U)$ for a predicate $P(x_1, x_2, \dots, x_m)$ exists, then also a gadget $T' = (H', U')$ for the predicate $P'(x_2, \dots, x_m) = P(x_2, x_2, x_3, \dots, x_m)$ exists.*

Proof. This lemma is similar to Lemma 1, but we must be a bit more careful since we are working only on one graph. Let u_1 and u_2 be the vertex pairs that correspond to variables x_1 and x_2 of P. The basic idea of the proof is to join the pairs u_1 and u_2 by a sufficiently long chain of gadgets Q, and use the connectivity constraints again to show that a reconstruction of such hypergraph exists whenever the predicate P' is satisfied. The details are left out due to space constraints. □

The proof of the following lemma is omitted due to space constrains.

Lemma 7. *Suppose that $\kappa(G) > 2$ and G is self-complementary (i.e., G is isomorphic to $\overline{G}$). If a gadget Q exists, then also a gadget T_{NAE} exists.*

3 NP-Completeness of the Graph Reconstruction Problem

We are now ready to prove the NP-completeness of the graph reconstruction problem modulo a few conditions on G:

Theorem 1. *If G is a graph such that $\kappa(G) > 2$ and a G-gadget O exists. Then the G-reconstruction problem is NP-complete.*

Proof. Using the previous lemmas, T_{SAT}, T_{NAE} or $T_{1 \text{ in } 3}$ exist. We proceed by reduction from one of the well-known NP-complete problems $3 - SAT$, $3 - NAE - SAT$ or 3-exact set cover ([3]), depending on which of the gadgets exists. Let T_X be this gadget.

Input of each of those problems is a CNF-formula with clauses of size exactly 3. We want to decide whether a assignment of truth values to variables exists, such that at least 1 (2 or 3, exactly 1, respectively) literal in each clause is satisfied. In the second and the third case we even do not need negations to be present in the clauses, but allowing them makes the problems only harder. Let a formula I be an instance of the problem.

We need to create an equivalent instance of G-reconstruction problem. For each clause of I, we add a copy of T_X. For each variable that occurs in n clauses, we add a star with n rays consisting of copies of Q joined by single special vertex pair p. Whether the variable is true or false will be determined by whether the edge p (and hence also its copies) is appears in the reconstruction or not. If the occurrence of a variable is negated, we attach a copy of a gadget N to the appropriate ray. For each clause C of I, we then identify the special edges of gadget T_X added for C with the ends of the rays that correspond to the appropriate occurrences of variables of C, using Lemma 1 and Lemma 6. It follows that the created instance is a gadget for formula I (where the special edges are the centers of the stars corresponding to variables, and clauses of I are interpreted according to X), and thus it is reconstructible if and only if I is satisfiable.

Since G is fixed, all the used gadgets have constant size, hence the reduction is polynomial. □

Note also that the gadget O does not exist if G is self-complementary. However this does not prevent the existence of the gadget N for this class of graphs, which justifies the following theorem:

Theorem 2. *If G is a self-complementary graph such that $\kappa(G) > 2$ and a G-gadget N exists, then the G-reconstruction problem is NP-complete.*

Proof. The proof is analogical to the proof of Theorem 1 – we may construct gadget Q using just gadget N and T_{NAE} is obtained using Lemma 7. □

We can obtain NP-completeness for some more graphs using the following trivial theorem:

Theorem 3. *If the G-reconstruction problem is NP-complete, then also the $\overline{G}$-reconstruction problem is NP-complete.*

Proof. This follows trivially from the fact that F is a G-reconstruction of a hypergraph H if and only if $\overline{F}$ is a $\overline{G}$-reconstruction of H. □

4 Classes of Graphs for Which the Problem Is Hard

We show that there are several important classes of graphs that satisfy the conditions of Theorem 1. The existence of O is the interesting part. For the classes of graphs described in this section, a gadget O exists. Therefore, provided that they (or their complements, by Theorem 3 and Lemma 2) have sufficient connectivity, the G-reconstruction for them is NP-complete. The following theorem shows that a slight variant of rigidity of a graph is sufficient to ensure existence of a gadget O.

Theorem 4. *Suppose that G is a graph on k vertices which satisfies the following conditions:*

(i) $\forall x \in V(G)$ $G - x$ has no nontrivial automorphism
(ii) $\forall x, y \in V(G)$ if $x \neq y$ then $G - x$ is not isomorphic to $G - y$

Then the following statements hold:

1. *Let H be a G-structure of a graph F which contains two hyperedges h_1 and h_2 that intersect in $k-1$ vertices, i. e. we have $h_1 = \{a, v_1, v_2, \ldots, v_{k-1}\}$ and $h_2 = \{b, v_1, v_2, \ldots, v_{k-1}\}$. Then $\{a, v_i\} \in E(F)$ if and only if $\{b, v_i\} \in E(F)$.*
2. *If G is not self-complementary, then the gadgets O and Z exist. On the other hand, if G is self-complementary, then the gadget N exists.*

Proof. First we prove part 1 of the theorem. Let f_i be the isomorphism that maps G onto $F[h_i]$ for $i \in \{1, 2\}$. For contradiction, suppose that F does not have the required property. We may assume that $\{a, v_1\} \in E(F)$ and $\{b, v_1\} \notin E(F)$. The following two cases are distinguished:

1. There is a vertex $x \in V(G)$ such that $f_1(x) = a$ and $f_2(x) = b$. Since $\{a, v_1\} \in E(F)$ and $\{b, v_1\} \notin E(F)$, we know that the preimage of v_1 under f_1 is different from the preimage of v_1 under f_2. Hence if we compose f_1 with the inverse f_2 and restrict this mapping to the graph $G - x$, we obtain a nontrivial automorphism, contrary to assumption (i) of the theorem.
2. There are two distinct vertices $x, y \in V(G)$ such that $f_1(x) = a$ and $f_2(y) = b$. But then the graphs $G - x$ and $G - y$ are both isomorphic to $F[\{v_1, \ldots, v_{k-1}\}]$, which contradicts assumption (ii) of the theorem.

To prove the second part of the theorem, we first introduce the following notation: let $\{x, y\} \in E(G)$ be an arbitrary edge of G and $\{\widehat{x}, \widehat{y}\} \in \binom{V(G)}{2} \setminus E(G)$ be an arbitrary non-edge of G. We construct a graph K by the following procedure:

1. Let V_0 be an arbitrary set of size k, let $G_0 = (V_0, E_0)$ be a fixed isomorphic embedding of G on the set V_0.
2. With each element $x_0 \in V_0$ we associate a set $S(x_0)$ of $k - 2$ new vertices denoted $\{w(x_0); w \in V(G) \setminus \{x, y\}\}$. For each vertex $y_0 \in V_0$ such that $\{x_0, y_0\}$ is an edge of E_0 we add new edges induced by the set $\{x_0, y_0\} \cup S(x_0)$ so that this set contains an isomorphic copy of G defined by the isomorphism that maps x to x_0, y to y_0 and w to $w(x_0)$ for each $w \in V(G) \setminus \{x, y\}$.
3. With each element $x_0 \in V_0$ we associate a set $\widehat{S}(x_0)$ of $k - 2$ new vertices denoted $\{\widehat{w}(x_0); w \in V(G) \setminus \{\widehat{x}, \widehat{y}\}\}$. For each vertex $y_0 \in V_0$ such that $\{x_0, y_0\}$ is not an edge of G_0 we add new edges induced by the set $\{x_0, y_0\} \cup \widehat{S}(x_0)$ so that this set contains an isomorphic copy of G defined by the isomorphism that maps $\widehat{x}$ to x_0, $\widehat{y}$ to y_0 and w to $\widehat{w}(x_0)$ for each $w \in V(G) \setminus \{\widehat{x}, \widehat{y}\}$.

Let K denote the graph on $k \cdot (2k - 3)$ vertices obtained by the previous three steps. Let H be the G-structure of K. Let K' be any graph whose G-structure is H, let $G'_0 = K'[V_0]$. Obviously, G'_0 and G_0 are both isomorphic to G, because their vertex set V_0 is a hyperedge of H. However, we can establish a stronger property:

- If G is self-complementary then $G'_0 = G_0$ or $G'_0 = \overline{G_0}$, where $\overline{G_0}$ is the complement of G_0.
- If G is not self-complementary then $G'_0 = G_0$.

This property immediately implies the second part of the theorem. To prove the property, we introduce a binary relation on $\binom{V_0}{2}$ called *friendship* defined as follows: we say that the two pairs $\{x, y_1\}$ and $\{x, y_2\}$ are friends in G_0, if they share a common vertex and if either both of them or neither of them belong to E_0. From the first part of the theorem we obtain that if two pairs of vertices

$\{x, y_1\}$ and $\{x, y_2\}$ are friends in G_0, then they are also friends in G'_0, because H contains the two hyperedges $h_i = \{x, y_i\} \cup S'(x)$ for $i \in \{1, 2\}$, where $S'(x)$ is defined as $S(x)$ if the two friends are edges of G_0, and $S'(x) = \widehat{S}(x)$ otherwise. Let $\sim$ denote the relation obtained as the transitive closure of the friendship relation in G_0. Note that $\sim$ is an equivalence on $\binom{V_0}{2}$ whose blocks are the edge sets of the connected components of G_0 and of the connected components of $\overline{G_0}$. Since friendship in G_0 implies friendship in G'_0, we have that the blocks of $\sim$ are subsets of the blocks of $\sim'$, where $\sim'$ is the closure of the friendship relation in G'_0. But the number of blocks of $\sim'$ is the number of connected components of G'_0 and its complement, which is equal to the number of blocks of $\sim$. Hence the two equivalence relations $\sim$ and $\sim'$ are equal and the two corresponding friendship relations are equal as well. Since every disconnected graph has a connected complement, we know that at least one block B of the relation $\sim$ is spanning in V_0, i. e. for each $x \in V_0$ at least one edge incident to x belongs to B. If the members of B are edges of G'_0 then all the other blocks must only contain non-edges (and vice versa), because any edge outside of B would necessarily become a friend of some edge in B by the spanning property of B. This implies that the edge set of G'_0 is either equal to B or equal to the complement of B. Clearly if G is not self-complementary, then only one of these two options is available.

This concludes the proof of the theorem. □

Modifying slightly the well-known proof of the fact that almost no random graph has a nontrivial automorphism (see for example [4]), we can show that almost all graphs satisfy the assumptions of Theorem 4, and thus we obtain the following result:

Theorem 5. *Let G be a random graph. Then with probability $1 - o(1)$ the G-reconstruction problem is NP-complete.*

We now consider several more specialized examples. The straightforward (but technical) proofs of the following theorems are left out due to the space constraints.

The previous results by Hayward et al. ([10]) showed that the P_4-reconstruction problem can be solved in a polynomial time. This makes the following result interesting:

Theorem 6. *If G is a path on k vertices with $k \geq 5$, then a gadget O and a gadget Z for G exist.*

Note that the complement of P_n satisfies $\kappa(\overline{P_n}) > 2$ whenever $n \geq 6$. Also note that a gadget Z for some graph G is at the same time a gadget O for $\overline{G}$. By the theorems 1, 3 and 6 we then have that the P_n-reconstruction problem is NP-complete for each $n \geq 6$.

The smallest graph G for that we can show that the G-reconstruction problem is NP-complete has 5 vertices, by the following result:

Theorem 7. *If G is a graph on $k \geq 4$ vertices with only one edge, then a G-gadget Z exists.*

If G is a graph with one edge and at least five vertices then $\kappa(\overline{G}) > 2$, thus the G-reconstruction problem is NP-complete.

5 Conclusion

We have proved that for many graphs G, the G-reconstruction problem is NP-complete. On the other hand, the P_4-reconstruction problem is polynomial and hence we might hope that for some other small graphs the problem could be solvable in polynomial time. However, the genericity of Theorem 1 suggests that the problem is usually hard. We are confident enough to state the following

Conjecture 1. There exists a constant k_0 such that for any graph G with $|V(G)| > k_0$, that is not edgeless or complete, the G-reconstruction problem is NP-complete.

Acknowledgements

We would like to thank Jan Kratochvíl and Pavel Valtr for introducing the problem to us and for useful discussions about it. We would also like to thank Daniel Kráľ' for careful reading of the paper and useful suggestions on improvement of the presentation.

References

1. Ulam S. M.: *A Collection of Mathematical Problems*, Wiley, New York, 1960.
2. Bondy J. A. and Hemminger R. L.: Graph Reconstruction – a survey, *J. Graph Theory* **1** (1977), 227–268.
3. M. R. Garey and D. S. Johnson. *Computers and Intractability: A Guide to the Theory of NP-Completeness.* W.H. Freeman, New York, 1979.
4. L. Babai, P. Erdos, and S. M. Selkow: Random graph isomorphism, *SIAM J. Computing*, **9** (1980), 628–635.
5. V. Chvátal, C.T. Hoàng: On the P_4-structure of perfect graphs I. Even decompositions, *J. Combin. Theory Ser. B* **39** (1985), 209–219.
6. C.T. Hoàng: On the P_4-structure of perfect graphs II. Odd decompositions, *J. Combin. Theory Ser. B* **39** (1985), 220–232.
7. B. A. Reed: A semi-strong perfect graph theorem, *J. Combin. Theory B* **43** (1987), 223–240.
8. V. Chvátal: On the P_4-structure of perfect graphs III. Partner decompositions, *J. Combin. Theory Ser. B* **43** (1987), 349–353.
9. R. Hayward: Recognizing P_3-structure: A Switching Approach, *Journal Comb. Theory (Series B)* **66** No. 2 (1996), 247–262.
10. R. Hayward, S. Hougardy and B. Reed: Polynomial time recognition of P_4-structure, *Proceedings of 13th Annual ACM-SIAM Symp. On Discrete Algorithms* (2002), 382–389.

Hybrid Voting Protocols and Hardness of Manipulation

Edith Elkind[1] and Helger Lipmaa[2,*]

[1] Department of Computer Science, University of Warwick, U.K.
[2] Cybernetica AS and University of Tartu, Estonia

Abstract. This paper addresses the problem of constructing voting protocols that are hard to manipulate. We describe a general technique for obtaining a new protocol by combining two or more base protocols, and study the resulting class of (vote-once) hybrid voting protocols, which also includes most previously known manipulation-resistant protocols. We show that for many choices of underlying base protocols, including some that are easily manipulable, their hybrids are NP-hard to manipulate, and demonstrate that this method can be used to produce manipulation-resistant protocols with unique combinations of useful features.

1 Introduction

In multiagent systems, the participants frequently have to agree on a joint plan of action, even though their individual opinions about the available alternatives may vary. Voting is a general method of reconciling these differences, and having a better understanding of what constitutes a good voting mechanism is an important step in designing better decision-making procedures. In its most general form, a voting mechanism is a mapping from a set of votes (i.e., voters' valuations for all alternatives) to an ordering of the alternatives that best represents the collective preferences. In many cases, however, the attention can be restricted to mechanisms that interpret their inputs (votes) as total orderings of the alternatives/candidates and output a single winner. A classical example here is Plurality voting, where only the top vote of each voter is taken into account, and the candidate with the largest number of top votes wins.

A fundamental problem encountered by all voting mechanisms is *manipulation*, i.e., the situation when a strategizing voter misrepresents his preferences in order to obtain a more desirable outcome. One can expect that rational agents will engage in manipulation whenever it is profitable for them to do so; as a result, the output of the voting mechanism may grossly misrepresent the actual preferences of the agents and be detrimental to the system as a whole.

It is well-known [8,11] that any nondictatorial voting mechanism for three or more candidates is susceptible to manipulation. However, while there is no information-theoretic solution to this problem, one can try to discourage potential manipulators by making manipulation infeasible. This approach is particularly attractive in multi-agent setting, when decisions have to be made in real time, and whether an agent can

* Part of this work was done when the first author was visiting Helsinki University of Technology. The authors were supported by the Finnish Defence Forces Institute for Technological Research; also, the first author was supported by the EPSRC grant GR/T07343/01 "Algorithmics of Network-sharing Games".

X. Deng and D. Du (Eds.): ISAAC 2005, LNCS 3827, pp. 206–215, 2005.

find a beneficial manipulation quickly is more important than whether such a manipulation exists in principle. It turns out that some of the voting protocols that are used in practice enjoy this property: it has been shown [1,2] that second-order Copeland and Single Transferable Vote (STV) are NP-hard to manipulate. Furthermore, in a recent paper [4], Conitzer and Sandholm showed that several protocols, including Borda, STV, Maximin and Plurality, can be modified so that manipulating them becomes computationally hard. Their method involves prepending the original protocol by a pre-round in which candidates are divided into pairs and the voters' preferences are used to determine the winner of each pair; the winners of the pre-round participate in elections conducted according to the original protocol. Different methods for pairing up the candidates and eliciting the votes give rise to different levels of complexity, such as NP-hardness, #P-hardness, or PSPACE-hardness. The advantage of this method of constructing manipulation-resistant protocols is in preserving some of the properties of the original protocol: for example, if the base protocol is Condorcet-consistent (see Sect. 6 for definition), then the modified protocol is Condorcet-consistent as well. However, for some other desirable features this is not true, and, generally, eliminating half of the candidates using a set of criteria that may be very different in spirit from those used by the original protocol, is likely to alter the outcome considerably, so that the desiderata that motivated the original protocol may no longer be attainable.

We build upon the ideas of [4] to construct a larger family of protocols that are hard to manipulate. We observe that their pre-round phase can be viewed as the first stage of the voting protocol known as Binary Cup (BC) (defined in Sect. 2). While this protocol itself is not hard to manipulate (at least, when the schedule is known in advance), the results of [4] can be interpreted as showing that combining BC with other protocols results in manipulation-resistant schemes. We generalize this idea by showing that this kind of hardness amplification is not unique to BC.

We define the class of *hybrid voting protocols* $\mathsf{Hyb}(\mathsf{X}_k, \mathsf{Y})$. In $\mathsf{Hyb}(\mathsf{X}_k, \mathsf{Y})$, after the voters have expressed their preferences, k steps of protocol X are performed to eliminate some of the candidates, and then protocol Y is run on the rest of the candidates, reusing the votes as restricted to the remaining candidates. Clearly, the protocols of [4] belong to this family, as does STV; therefore, our framework encompasses most of the known hard-to-manipulate voting mechanisms.

We show that many other hybrid protocols are NP-hard to manipulate as well. Specifically, we consider several well-known protocols, such as Plurality, Borda, STV, and Maximin, and prove that many of their hybrids are manipulation-resistant. We do this by formulating some fairly general conditions on X and Y under which the protocols of the form $\mathsf{Hyb}(\mathsf{X}_k, \mathsf{Plurality})$, $\mathsf{Hyb}(\mathsf{X}_k, \mathsf{STV})$, or $\mathsf{Hyb}(\mathsf{STV}_k, \mathsf{Y})$ are NP-hard to manipulate. Additionally, we show that a hybrid of a protocol with itself may be different from the original protocol — and much harder to manipulate. We prove that this is, indeed, the case for Borda protocol: $\mathsf{Hyb}(\mathsf{Borda}_k, \mathsf{Borda})$ is NP-hard to manipulate, while Borda itself is easily manipulable. We define a generic closure operation on protocols that makes them closed under hybridization. Interestingly, applying this operation to the easy-to-manipulate Plurality results in the hard-to-manipulate STV. We conjecture that for many other basic protocols, their closed versions are NP-hard to manipulate as well. On the flip side, we demonstrate that hybridization does not always result in hard-to-manipulate protocols: in particular, the hybrid protocols that use Plurality as their first component, are almost as easy to manipulate as their second component. Finally, we demonstrate that our techniques extend to voting protocols that allow voters to rate the candidates rather than just order them.

The value of our results is not so much in constructing specific new manipulation-resistant protocols, but rather in providing a general method for doing that, which can be used with many basic schemes. Since a hybrid inherits some of the properties of its ingredients, we get hard-to-manipulate protocols with properties not shared by the schemes from [1,2,4]. It has already been argued in [4] that it is desirable to have manipulation-resistant protocols that can be used in different real-life situations; our method fits the bill.

The use of voting and voting-related techniques is not restricted to popular elections: the ideas from this domain have been applied in rank aggregation [5,9], recommender systems [10], multiagent decision making in AI [7], etc. In many of these settings, the number of alternatives is large enough to make our results applicable, and, furthermore, the agents are both sufficiently sophisticated to attempt manipulation and may derive significant utility from doing so. Therefore, we feel that it is important to have a better understanding of what makes voting protocols hard to manipulate, as this will allow us to design more robust decision-making systems that use voting-like methods.

2 Preliminaries and Notation

We assume that there are n voters and m candidates and denote the set of all voters by $V = \{v_1, \ldots, v_n\}$ and the set of all candidates by $C = \{c_1, \ldots, c_m\}$. Most of our complexity results are in terms of m and n, i.e., unless specified otherwise, 'polynomial' always means 'polynomial in m and n'.

The set of all permutations of C is denoted by $\Pi(C)$; the preference of the ith voter is expressed by a list $\pi_i \in \Pi(C)$: the first element is the voter's most preferred candidate, etc. In particular, this means that within one voter's preference list, ties are not allowed. We write $(\ldots, c_i, \ldots, C_j, \ldots)$ to denote that a voter prefers c_i to all candidates in C_j, without specifying the ordering of candidates within C_j. For any subset $C' \subseteq C$, let $\pi|_{C'}$ be the permutation π as restricted to C' (i.e., elements not from C' are omitted). Note that $\pi|_{C'}$ corresponds to a valid preference in an election that has the candidate set C'.

A *voting protocol* is a mapping $P : \Pi(C) \times \cdots \times \Pi(C) \mapsto C$ that selects a winner $c \in C$ based on all voters' preference lists. In this paper, we consider the following common voting protocols (in all definitions that mention points, the candidate with the most points wins):

Plurality: A candidate receives 1 point for every voter that ranks it first.

Borda: For each voter, a candidate receives $m - 1$ point if it is the voter's top choice, $m - 2$ if it is the second choice, ..., 0 if it is the last.

Single Transferable Vote (STV): Winner determination proceeds in rounds. In each round, a candidate's score is the number of voters that rank it highest among the remaining candidates, and the candidate with the lowest score drops out. The last remaining candidate wins. (A vote transfers from its top remaining candidate to the next highest remaining candidate when the former drops out.)

Maximin: A candidate's score in a pairwise election is the number of voters that prefer it over the opponent. A candidate's number of points is the lowest score it gets in any pairwise election.

Binary Cup (BC): The winner determination process consists of $\lceil \log m \rceil$ rounds. In each round, the candidates are paired; if there is an odd number of candidates, one of them gets a bye. The candidate that wins the pairwise election between the two (or got a bye) advances into the next round. The schedule of the cup (i.e., which candidates face each other in each round) may be known in advance (i.e., before the votes are elicited) or it may depend on the votes.

We say that a voter v_j can *manipulate* a protocol P if there is a permutation $\pi'_j \in \Pi(C)$ such that for some values of $\pi_i \in \Pi(C), i = 1, \ldots, n$, we have $P(\pi_1, \ldots, \pi_n) = c$, $P(\pi_1, \ldots, \pi_{j-1}, \pi'_j, \pi_{j+1}, \ldots, \pi_n) = c' \neq c$, and v_j ranks c' above c. We say that v_j manipulates P *constructively* if v_j ranks c' first and *destructively* otherwise. All results in this paper are on constructive manipulation; in what follows, we omit the word 'constructive'. A voter v_j manipulates P *efficiently* if there is a polynomial time algorithm that given preference lists $\pi_1, \ldots, \pi_n$ for which such π'_j exists, can find one such π'_j.

3 Hybrid Protocols

In this section, we formally define *(vote-once) hybrid protocols*. Intuitively, a hybrid of two protocols X and Y executes several steps of X to eliminate some of the candidates, and then runs Y on the remaining set of candidates. To make this intuition precise, however, we have to define how to interpret the first protocol X as a sequence of steps. While there is no obvious way to do this for an arbitrary protocol, most well-known protocols, including the ones described in Sect. 2, admit such an interpretation. In particular, we suggest the following definitions:

- For STV, a *step* is a single stage of the protocol. That is, a step of STV consists of eliminating a candidate with the least number of first-place votes and transferring each vote for this candidate to the highest remaining candidate on that ballot.
- For Binary Cup (BC), a *step* is a single stage of the protocol as well, i.e., it consists of pairing up the candidates and eliminating the ones who lose in the pairwise comparison.
- For point-based protocols, such as Plurality, Borda, or Maximin, we first compute the scores of all candidates, order them by their scores from the lowest to the highest, and define a *step* to consist of eliminating the first (i.e., the lowest ranked) remaining candidate in this sequence. Note that the scores are not recomputed between the steps. (A similar approach can be applied to any voting protocol that can be extended to a preference aggregation rule, i.e., a function that maps votes to total orderings of the candidates. In this case, the order in which the candidates are eliminated is obtained by inverting the output of the preference aggregation rule.)

Definition 1. *A* hybrid protocol $\mathsf{Hyb}(\mathsf{X}_k, \mathsf{Y})$ *consists of two* phases. *Suppose that the voters' preference lists are described by the n-tuple $(\pi_1, \ldots, \pi_n)$. In the first phase, the* *protocol executes k steps of* $\mathsf{X}(\pi_1, \ldots, \pi_n)$*; let S be the set of candidates not eliminated in the first phase. In the second phase, the protocol applies* Y *to* $(\pi_1|_S, \ldots, \pi_n|_S)$*, i.e., the preference lists restricted to the remaining set S of candidates.*

It is straightforward to extend this definition to hybrids $\mathsf{Hyb}(\mathsf{X}^{(1)}_{k_1}, \mathsf{X}^{(2)}_{k_2}, \ldots, \mathsf{X}^{(t)}_{k_t}, \mathsf{Y})$ of three or more protocols.

4 Hardness Results

4.1 Hardness of STV-Based Hybrids

In this subsection, we show that hybrids $\mathsf{Hyb}(\mathsf{STV}_k, \mathsf{Y})$ and $\mathsf{Hyb}(\mathsf{X}_k, \mathsf{STV})$ are NP-hard to manipulate for many "reasonable" voting protocols X and Y, including the cases $\mathsf{X}, \mathsf{Y} \in \{\mathsf{Plurality}, \mathsf{Borda}, \mathsf{Maximin}, \mathsf{BC}\}$.

Theorem 1. *A hybrid of the form* $\mathsf{Hyb}(\mathsf{STV}_k, \mathsf{Y})$ *is NP-hard to manipulate as long as* Y *satisfies the following condition: Whenever there is a candidate c who receives K first-place votes and* $n - K$ *second-place votes, while all other candidates receive at most* $K - 1$ *first-place vote,* Y *declares c the winner.*

The proof can be found in the full version of the paper.

Corollary 1. *The hybrids* $\mathsf{Hyb}(\mathsf{STV}_k, \mathsf{Y})$*, where* $\mathsf{Y} \in \{\mathsf{Plurality}, \mathsf{Borda}, \mathsf{Maximin}, \mathsf{BC}, \mathsf{STV}\}$*, are NP-hard to manipulate.*

The proof of this corollary is straightforward since all these voting protocols satisfy the required property.

Theorem 2. *A hybrid of the form* $\mathsf{Hyb}(\mathsf{X}_k, \mathsf{STV})$ *is NP-hard to manipulate if* X *satisfies the following condition for some unbounded nondecreasing function* $f(\cdot)$ *and infinitely many K: Suppose that all but one voter rank some K candidates* $c_1, \ldots, c_K$ *after all other candidates, and all other candidates receive at least 2 first-place votes. Then after* $f(K)$ *steps of* X*, the set of eliminated candidates is a subset of* $\{c_1, \ldots, c_K\}$*.*

Proof (Sketch). Set $k = f(K)$. Denote the set of candidates in the construction of [2] by C'; let $C'' = \{c_1, \ldots, c_K\}$ and $C = C' \cup C''$. Modify the votes of all honest voters in that construction so that they rank C' above C''. The reduction of [2] has the property that each candidate in C' gets more than 2 first-place votes. Hence, the set of candidates eliminated in k rounds of X is a subset of C''; furthermore, the remaining candidates from C'' will be the first candidates eliminated by STV. Hence, no matter how the manipulator ranks the candidates in C'', it has no effect on the execution of the protocol, and his vote can be viewed as a vote in the original STV and vice versa. □

Corollary 2. *The hybrids of the form* $\mathsf{Hyb}(\mathsf{X}_k, \mathsf{STV})$*, where* $\mathsf{X} \in \{\mathsf{Plurality}, \mathsf{Borda}, \mathsf{Maximin}, \mathsf{BC}\}$*, are NP-hard to manipulate.*

Proof. It is easy to see that Plurality, Maximin and BC satisfy the condition of the theorem. For Borda, it is satisfied whenever the number of voters exceeds the number of candidates; in the construction of [2], the number of voters is larger than $3|C'|$, so we can set $K = |C'|$. □

4.2 Hybrids of the Form Hyb(X_k, Plurality)

In this subsection, we prove that Hyb(X_k, Plurality) is hard to manipulate whenever X satisfies Property 1, defined below. While this property might seem artificial, we show that it is possessed by at least two well-known protocols, namely, Borda and Maximin.

Property 1. For any set $G = \{g_1, \ldots, g_N\}$, any collection $S = \{s_1, \ldots, s_M\}$ of subsets of G, and any $K \leq M$, there are some k', $k' \leq M$, and $T, T > 3N$, such that it is possible to construct in polynomial time a set of $T + N(T-2) + 3N$ votes over the set of candidates $C' \cup C'' \cup \{p\}$, where $C' = \{c'_1, \ldots, c'_N\}$, $C'' = \{c''_1, \ldots, c''_M\}$, so that

1. there are T voters who rank p first;
2. for each $i = 1, \ldots, N$, there are $T-2$ voters who rank c'_i first;
3. for each $i = 1, \ldots, N$, there are 3 voters who rank all c''_j such that $g_i \in s_j$ above c'_i, and rank c'_i above all other candidates;
4. for any additional vote π, when it is tallied with all other votes, the set of candidates eliminated in the first k' rounds is a subset of C'' of size $M-K$;
5. for any subset $S' \subseteq S$, $|S'| = M-K$, one can design in polynomial time a vote $\pi_{S'}$ that, when tallied with other votes, guarantees that the set of candidates eliminated in the first k' rounds is exactly $\{c''_i \mid s_i \in S'\}$.

Theorem 3. *A hybrid of the form* Hyb(X_k, Plurality) *is NP-hard to manipulate constructively whenever X satisfies Property 1.*

Proof. We give a reduction that is based on the NP-hard problem SET COVER. Recall that SET COVER can be stated as follows: Given a ground set $G = \{g_1, \ldots, g_N\}$, a collection $S = \{s_1, \ldots, s_M\}$ of subsets of G, and an integer K, does there exist a K-cover of G, i.e., a subset S' of S, $S' = \{s_1, \ldots, s_K\}$, such that for every $g_i \in G$ there is an $s_j \in S'$ such that $g_i \in s_j$?

Construct the set of votes based on G, S, and K so that it satisfies Property 1. Let $k = k'$, and let p be the manipulator's preferred candidate. We show that the manipulator can get p elected under Hyb(X_k, Plurality) if and only if he can find a set cover for G. Indeed, after k rounds of X, all candidates in $C' \cup \{p\}$ survive, as well as exactly K candidates from C''. We show that p wins if and only if these K candidates correspond to a set cover of G. Observe that any surviving candidate from C'' has at most $3N < T$ first-place votes, so he cannot win in the last stage. Now, consider a candidate $c'_i \in C'$. Suppose that the corresponding element is not covered, i.e., all c''_j such that $g_i \in s_j$ are eliminated. Then after the end of the first phase, c'_i has $T+1$ first-place vote, while p has T first-place votes, so in this case p cannot win. On the other hand, suppose that for any $g_i \in G$ there is an $s_j \in S$ such that $g_i \in s_j$ and c''_j is not eliminated in the first phase. Then at the beginning of the second phase each $c'_i \in C'$ has $T-2$ first-place votes, while p has T first-place votes, so in this case p wins. □

Corollary 3. *The protocols* Hyb(Borda_k, Plurality) *and* Hyb(Maximin_k, Plurality) *are NP-hard to manipulate.*

Proof. Let the voters who rank p first, rank the candidates in C' above those in C'', and the voters who rank c'_i first, rank the candidates in $p \cup C'$ above those in C''. For large enough T, this guarantees that both Borda and Maximin scores of the candidates in $C' \cup \{p\}$ are much higher than those of the candidates in C'', so none of the candidates in $C' \cup \{p\}$ can be eliminated in the first phase. On the other hand, we still have enough flexibility to ensure that all candidates in C'' have the same Borda (or Maximin) score with respect to the honest voters' preferences. Then, for both protocols, the manipulator can get any $M-K$ candidates from C'' eliminated by putting them on the bottom of his vote and ranking the remaining K candidates above the candidates in $C' \cup \{p\}$. Thus, both Borda and Maximin satisfy all conditions in the statement of Theorem 3. □

Together with our results on STV and the results of [4], the constructions of this section provide a wide choice of manipulation-resistant protocols. In the next section, we add to our repertoire two more protocols that are hard to manipulate, namely, $\mathsf{Hyb}(\mathsf{Borda}_k, \mathsf{Borda})$ and $\mathsf{Hyb}(\mathsf{Maximin}_k, \mathsf{Borda})$.

5 Hybrid of a Protocol with Itself

We say that a protocol is *hybrid-proof* if a hybrid of several copies of this protocol is equivalent to the original protocol. While some protocols, such as STV or Binary Cup, have this property, for many other protocols, especially score-based ones, this is not the case, since in a hybrid protocol, the scores of all surviving candidates are recomputed in the beginning of the second phase, while in the original protocol they are computed only once. Nevertheless, any protocol can be modified to be hybrid-proof. For an arbitrary protocol X, define a *closed protocol* $\overline{\mathsf{X}}$ by $\overline{\mathsf{X}} = \mathsf{Hyb}(\mathsf{X}_1, \dots, \mathsf{X}_1)$, where the number of copies of X_1 is such that $\overline{\mathsf{X}}$ selects a single winner; a step of $\overline{\mathsf{X}}$ corresponds to a single copy of X_1. It is not hard to check that for any protocol X, the closed protocol $\overline{\mathsf{X}}$ is hybrid-proof.

Interestingly, $\mathsf{Hyb}(\mathsf{Plurality}_1, \dots, \mathsf{Plurality}_m) = \mathsf{STV}$: the vote transfer mechanism can be viewed as recomputing each candidate's Plurality score. Observe that while Plurality has particularly bad manipulation resistance properties (see, e.g., Sect. 7), STV is NP-hard to manipulate. This leads us to conjecture that for many other base protocols, the new protocols obtained in this manner are NP-hard to manipulate. Whenever this is the case, the closed protocols provide the most faithful manipulation-resistant approximation to the underlying protocols, which makes them compelling alternatives to the original protocols. This conjecture is supported by the fact that for some easy-to-manipulate protocols, a hybrid of just two copies of the protocol is NP-hard to manipulate; increasing the number of copies should make the manipulation harder, not easier. As an illustration, in the full version of the paper, we prove the following result.

Theorem 4. *The hybrids* $\mathsf{Hyb}(\mathsf{Borda}_k, \mathsf{Borda})$ *and* $\mathsf{Hyb}(\mathsf{Maximin}_k, \mathsf{Borda})$ *are NP-hard to manipulate.*

6 Properties of Voting Protocols

Voting protocols are evaluated based on various criteria, such as *Pareto-optimality* (a candidate who is ranked lower than some other candidate by every voter never wins),

Condorcet-consistency (if there is a candidate who is preferred to every other candidate by a majority of voters, this candidate wins), or *monotonicity* (with the relative order of the other candidates unchanged, ranking a candidate higher should never cause the candidate to lose, nor should ranking a candidate lower ever cause the candidate to win). In the context of this paper, a natural addition to this list is *hardness of manipulation*.

Most voting schemes that are based on pairwise comparisons, in particular, BC and Maximin, are Condorcet-consistent, while for STV, or positional methods, such as Plurality or Borda, this is not the case. One can prove that Plurality, Borda, Maximin, and BC are monotone, while STV is not. All basic voting protocols considered in this paper except BC are Pareto-optimal.

To analyze whether properties (1)–(3) are preserved under hybridization, we have to extend these definitions to multi-step protocols. We say that a multi-step protocol is *strongly Pareto-optimal* if whenever every voter ranks c_1 below c_2, c_1 is eliminated before c_2, and *strongly monotone* if ranking a candidate higher does not affect the relative order of elimination of other candidates and cannot result in him being eliminated at an earlier step; the definition of Condorcet consistency remains unchanged. It is easy to see that multi-step versions of Pareto-optimal protocols that we consider are strongly Pareto-optimal, at least for some draw resolution rules. However, not all monotone protocols are strongly monotone: for example, in Borda, moving a candidate several positions up changes other candidates' scores in a non-uniform way.

Proposition 1. *For any voting protocols* X *and* Y *and any* $k > 0$, *if both* X *and* Y *are Condorcet-consistent, so is* $\mathsf{Hyb}(\mathsf{X}_k, \mathsf{Y})$; *if* X *is strongly Pareto-optimal (strongly monotone) and* Y *is Pareto-optimal (monotone), then* $\mathsf{Hyb}(\mathsf{X}_k, \mathsf{Y})$ *is Pareto-optimal (monotone).*

We omit the proofs.

The construction proving that BC is not Pareto-optimal can be easily modified to show that any protocol of the form $\mathsf{Hyb}(\mathsf{BC}_k, \mathsf{Y})$ is not Pareto-optimal for some k, where $\mathsf{Y} \in \{\mathsf{Plurality}, \mathsf{Borda}, \mathsf{Maximin}, \mathsf{STV}\}$. Hence, prior to this work, the only Pareto-optimal mechanisms that were known to be NP-hard to manipulate were STV and the variants of the Copeland protocol that were described in [1]. Our results imply that $\mathsf{Hyb}(\mathsf{Borda}_k, \mathsf{Plurality})$, $\mathsf{Hyb}(\mathsf{Maximin}_k, \mathsf{Plurality})$, and $\mathsf{Hyb}(\mathsf{Borda}_k, \mathsf{Borda})$ also combine these two properties.

Furthermore, except for STV, all previous hard-to-manipulate protocols involved methods that use pairwise comparisons, and such methods have been criticized for relying too much on the number of victories rather than their magnitude. On the other hand, both $\mathsf{Hyb}(\mathsf{Borda}_k, \mathsf{Plurality})$ and $\mathsf{Hyb}(\mathsf{Borda}_k, \mathsf{Borda})$ are based purely on positional methods, which do not suffer from this flaw, and Maximin (and hence, hybrids of Maximin with positional methods) also takes into account the magnitude of victories.

7 Limitations and Extensions

Hybrids That Are Easy to Manipulate. Unfortunately, our method of obtaining hard-to-manipulate protocols is not universal: if the protocol used in the first phase does not provide the manipulator with sufficiently many choices, the resulting hybrid protocol is almost as easy to manipulate as its second component. In particular, this applies to Plurality protocol.

Theorem 5. *Suppose that a protocol* Y *satisfies the following property for any candidate c: Given other voters' preference profiles, the manipulator can in polynomial time find a beneficial manipulation that ranks c first or infer that no such manipulation exists. Then there is a polynomial-time algorithm that can constructively manipulate the hybrid* $\mathsf{Hyb}(\mathsf{Plurality}_k, \mathsf{Y})$ *for any k.*

Proof. For the first phase of the protocol, the only choice that the manipulator has to make is which candidate to rank first; the rest of his vote will have no effect on the elimination process. Hence, he can try all m options. Suppose that when the manipulator ranks c_i first, the set of candidates that survive the first phase is C_i. The manipulator can deduce the honest voters' preferences over C_i. If $c_i \notin C_i$, he simply has to construct a beneficial manipulation $\pi|_{C_i}$ of Y and, in his vote, rank c_i first and order the candidates in C_i as suggested by $\pi|_{C_i}$. If $c_i \in C_i$, in constructing a beneficial manipulation of Y he is restricted to orderings that rank c_i first. By our assumptions, he can find a solution to this problem in polynomial time. □

Corollary 4. *There are polynomial-time algorithms that can constructively manipulate* $\mathsf{Hyb}(\mathsf{Plurality}_k, \mathsf{Y})$*, where* $\mathsf{Y} \in \{\mathsf{Borda}, \mathsf{Maximin}, \mathsf{BC}, \mathsf{Plurality}\}$ *for any k.*

Utility-Based Voting. So far, we have focused on voting schemes that required each voter to submit a total ordering of the candidates. However, in many settings a voter may be essentially indifferent between some of the alternatives, but have a strong opinion on the relative merit of other alternatives. In this case, his preference may be better reflected by a *utility vector* $\mathbf{u} = (u_1, \ldots, u_m)$, where $0 \leq u_j \leq 1$ is the *utility* that this voter assigns to candidate c_j. To guarantee fairness, the utility vectors are normalized, i.e., we require that either $u_j = 0$ for all j or $\sum_j u_j = 1$.

The definitions of a voting protocol and manipulation can be modified in a straightforward manner. The most natural voting protocol for the utility-based framework is HighestScore, which computes the total score of each candidate, i.e., the sum of utilities assigned to this candidate by all voters, and selects the candidate with the highest total score. A hybrid of two utility-based protocols is a protocol that performs k steps of the first protocol, re-normalizes the utility vectors (restricted to the surviving candidates) and executes the second protocol on the remaining candidates. While HighestScore itself is not manipulation-resistant, the techniques we use for ordering-based protocols are applicable in this setting, too.

Theorem 6. *There is a polynomial-time algorithm that can manipulate* HighestScore*. However,* $\mathsf{Hyb}(\mathsf{HighestScore}_k, \mathsf{HighestScore})$ *is NP-hard to manipulate.*

8 Conclusions and Future Work

Our work places the results of [3,4] within a more general paradigm of hybrid voting schemes. The advantage of our approach is that it works for a wide range of protocols: while some voting procedures are inherently hard to manipulate, they may not satisfy the intuitive criteria of a given setting. On the other hand, a hybrid of two protocols retains many of their desirable properties, and sometimes may combine the best of both worlds. All of the voting protocols described in Sect. 2, as well as many others, are used in different contexts; while it would be unreasonable to expect that all of them will be replaced, say, by STV just because it is harder to manipulate, hybrids of these protocols with similar ones or even with themselves may be eventually preferred to the original protocols. Moreover, our results on utility-based voting suggest that our techniques can be useful for a wider class of problems and can be viewed as a contribution to the more general task of constructing computationally strategy-proof mechanisms.

An important issue not addressed in this paper is that of designing protocols with high average-case complexity. However, even asking this question properly, i.e., coming up with a natural distribution of voter's preferences with respect to which the average-case hardness is computed is itself a difficult task: clearly, in most scenarios one cannot expect preferences to be uniformly distributed. Initial results in this direction can be found in [6]; however, this topic should be explored further.

References

1. John J. Bartholdi, III, Craig A. Tovey, and James B. Orlin. The Computational Difficulty of Manipulating an Election. *Social Choice and Welfare*, 6:227–241, 1989.
2. John J. Bartholdi, III and James B. Orlin. Single Transferable Vote Resists Strategic Voting. *Social Choice and Welfare*, 8(4):341–354, 1991.
3. Vincent Conitzer and Tuomas Sandholm. Complexity of Manipulating Elections with Few Candidates. In *Proceedings of the Eighteenth National Conference on Artificial Intelligence and Fourteenth Conference on Innovative Applications of Artificial Intelligence*, pages 314–319, Edmonton, Alberta, Canada, July 28 — August 1 2002. AAAI Press.
4. Vincent Conitzer and Tuomas Sandholm. Universal Voting Protocol Tweaks to Make Manipulation Hard. In *Proceedings of the International Joint Conference on Artificial Intelligence (IJCAI)*, pages 781–788, Acapulco, Mexico, August 9–15 2003.
5. Cynthia Dwork, Ravi Kumar, Moni Naor, and D. Sivakumar, Rank aggregation methods for the Web, In *Proc. 10th International World-Wide Web Conference (WWW)*, pages 613–622.
6. Edith Elkind and Helger Lipmaa. Small Coalitions Cannot Manipulate Voting. In *Proceedings of Financial Cryptography and Data Security - Ninth International Conference*, Roseau, The Commonwealth Of Dominica, February 28–March 3, 2005.
7. Eithan Ephrati and Jeffrey S. Rosenschein. Multi-Agent Planning as a Dynamic Search for Social Consensus. In *Proceedings of the International Joint Conference on Artificial Intelligence (IJCAI)*, 1993.
8. Allan F. Gibbard. Manipulation of Voting Schemes: A General Result. *Econometrica*, 41:597–601, 1973.
9. Ronald Fagin, Ravi Kumar, Mohammad Mahdian, D. Sivakumar, Erik Vee. Comparing and Aggregating Rankings with Ties. In *Proceedings of 23rd ACM Symposium on Principles of Database Systems (PODS)*, pages 47–58.
10. David M. Pennock, Eric Horvitz, and C. Lee Giles. Social Choice Theory and Recommender Systems: Analysis of the Axiomatic Foundations of Collaborative Filtering. In *Proceedings of the Seventeenth National Conference on Artificial Intelligence*, July 2000.
11. Mark A. Satterthwaite. *The Existence of Strategy-Proof Voting Procedures: A Topic in Social Choice Theory*. PhD thesis, University of Wisconsin, Madison, 1973.

On the Complexity of Rocchio's Similarity-Based Relevance Feedback Algorithm

Zhixiang Chen[1] and Bin Fu[2,3]

[1] Department of Computer Science, University of Texas-Pan American, Edinburg, TX 78541, USA
chen@cs.panam.edu

[2] Department of Computer Science, University of New Orleans, New Orleans, LA 70148, USA
fu@cs.uno.edu

[3] Research Institute for Children, 200 Henry Clay Avenue, New Orleans, LA 80118, USA

Abstract. In this paper, we prove for the first time that the learning complexity of Rocchio's algorithm is $O(d+d^2(\log d+\log n))$ over the discretized vector space $\{0,\ldots,n-1\}^d$, when the inner product similarity measure is used. The upper bound on the learning complexity for searching for documents represented by a monotone linear classifier $(\mathbf{q},0)$ over $\{0,\ldots,n-1\}^d$ can be improved to $O(d+2k(n-1)(\log d+\log(n-1)))$, where k is the number of nonzero components in $\mathbf{q}$. An $\Omega(\binom{d}{2}\log n)$ lower bound on the learning complexity is also obtained for Rocchio's algorithm over $\{0,\ldots,n-1\}^d$. In practice, Rocchio's algorithm often uses fixed query updating factors. When this is the case, the lower bound is strengthened to $2^{\Omega(d)}$ over the binary vector space $\{0,1\}^d$. In general, if the query updating factors are bounded by $O(n^c)$ for some constant $c\geq 0$, an $\Omega(n^{d-1-c}/(n-1))$ lower bound is obtained over $\{0,\ldots,n-1\}^d$.

1 Introduction

Information retrieval has a long history of research on relevance feedback (see, for example, [1,8,14,15,17,19]), and becomes a necessary part of our daily life due to the popularity of the Web. It is regarded as the most popular query reformation strategy [1,17]. The central idea of relevance feedback is to improve search performance for a particular query by modifying the query step by step, based on the user's judgments of the relevance or irrelevance of some of the documents retrieved. In his popular textbook [19], van Vijsbergen describes the relevance feedback as a fixed error correction procedure and relates it to the linear classification problem. When the inner product similarity is used, relevance feedback is just a Perceptron-like learning algorithm. Wong, Yao and Bollmann [20] studied the linear structure in information retrieval. They designed a very nice gradient descent procedure to compute the coefficients of a linear function and analyzed its performance. In order to update the query vector adaptively, their gradient descent procedure must know the user preference which is in

X. Deng and D. Du (Eds.): ISAAC 2005, LNCS 3827, pp. 216–225, 2005.

practice the unknown target to be learned by an information retrieval system. In [3,4], multiplicative adaptive algorithms are devised for user preference retrieval with provable, efficient performance.

There are many different variants of relevance feedback in information retrieval. However, in this paper we only study Rocchio's similarity-based relevance feedback algorithm [15,17]. In spite of its popularity in various applications, there is little rigorous analysis of its complexity as a learning algorithm in literature. As a first step towards formal analysis of Rocchio's similarity-based relevance feedback algorithm, the work in [7] establishes a linear lower bound on the learning complexity for the algorithm in searching for documents represented by a monotone linear classifier $x_{i_1} \vee x_{i_2} \vee \cdots \vee x_{i_k}$ over the binary vector space $\{0,1\}^d$, when any of the four typical similarity measures (inner product, dice coefficient, cosine coefficient, and Jaccard coefficient) listed in [17] is used. The linear lower bound obtained in [7] is independent of the query updating factors and the classification threshold that are used by the algorithm. A number of challenging problems regarding further analysis of the algorithm remain open [7].

In practice, a fixed query updating factor and a fixed classification threshold are often used in Rocchio's similarity-based relevance feedback algorithm [1,17]. Using a fixed query updating factor has many merits, such as simplicity and efficiency. As another example, the popular Winnow algorithm [11] uses a fixed updating factor. When this is the case, the work in [5] strengthened the linear lower bound in [7] to a quadratic lower for the algorithm in searching for documents represented by $x_{i_1} \vee x_{i_2} \vee \cdots \vee x_{i_k}$ over $\{0,1\}^d$.

In this paper, we prove for the first time that the learning complexity of Rocchio's algorithm in searching for documents represented by a linear classifier is $O(d + d^2(\log d + \log n))$ over the discretized vector space $\{0,\ldots,n-1\}^d$, when the inner product similarity measure is used. The upper bound on learning complexity for documents represented by a monotone linear classifier $(\mathbf{q}, 0)$ over $\{0,\ldots,n-1\}^d$ can be improved to $O(d + 2k(n-1)(\log d + \log(n-1)))$, where k is the number of nonzero components in $\mathbf{q}$. An $\Omega((\frac{d}{2})\log n)$ lower bound on the learning complexity is also obtained for Rocchio's algorithm over $\{0,\ldots,n-1\}^d$. In practice, Rocchio's algorithm often uses fixed query updating factors. When this is the case, the lower bound is strengthened to $2^{\Omega(d)}$ over $\{0,1\}^d$. In general, if the query updating factors are bounded by $O(n^c)$ for some constant $c \geq 0$, an $\Omega(n^{d-1-c}/(n-1))$ lower bound is obtained over $\{0,\ldots,n-1\}^d$.

2 Rocchio's Similarity-Based Relevance Feedback Algorithm

Let $\mathcal{R}$ be the set of all real values. Let d and n be two integers with $d \geq 1$ and $n \geq 2$. In the binary vector space model in information retrieval [17,18], a collection of d features (or terms) $T_1, T_2, \ldots, T_d$ are used to represent documents and queries. Each document $\mathbf{x}$ is represented as a vector $v_{\mathbf{x}} = (x_1, x_2, \ldots, x_d)$ such that for any i, $1 \leq i \leq d$, the i-th component of $v_{\mathbf{x}}$ is one if the i-th feature T_i appears in $\mathbf{x}$ or zero otherwise. Each query $\mathbf{q}$ is represented by a vector

$v_{\mathbf{q}} = (q_1, q_2, \ldots, q_d)$ such that for any i, $1 \leq i \leq d$, the i-th component of $v_{\mathbf{q}} \in \mathcal{R}$ is a real value used to determine the relevance (or weight) of the i-th feature T_i. Because of the unique vector representations of documents and queries, for convenience we simply use $\mathbf{x}$ and $\mathbf{q}$ to stand for their vector representations $v_{\mathbf{x}}$ and $v_{\mathbf{q}}$, respectively. If term frequencies are used to index a document $\mathbf{x}$, then $\mathbf{x}$ is a vector in the discretized vector space $\{0, \ldots, n-1\}^d$. Note that fractional term frequency/inverse document frequency vectors can be converted into vectors in $\{0, \ldots, n-1\}^d$. A similarity measure, called similarity for short, in general is a function m from $R^n \times R^n$ to non-negative real values. A similarity m is used to determine the *relevance closeness* of documents to the search query and to rank documents according to such closeness. Among a variety of similarity measures, vector inner product similarity is commonly used [17,18,1]. To simply presentation, we will focus on this similarity throughout this paper.

Definition 1. *A linear classifier over the d-dimensional discretized vector space* $\{0, \ldots, n-1\}^d$ *is a pair* $(\mathbf{q}, \phi)$*, where* $\mathbf{q} \in R^d$ *is the query/weight vector and* $\phi \in \mathcal{R}$ *is a threshold. The classifier classifies any documents* $\mathbf{x} \in \{0, \ldots, n-1\}^d$ *as relevant if* $\mathbf{q} \cdot \mathbf{x} > \phi$ *or irrelevant otherwise.*

Definition 2. *An adaptive learning algorithm A for learning a unknown target linear classifier* $(\mathbf{q}, \phi)$ *over the n-dimensional discretized vector space* $\{0, \ldots, n-1\}^d$ *from examples is a game played between the algorithm A and the user in a step by step fashion, where the query/weight vector* $\mathbf{q}$ *and the threshold* ϕ *are unknown to the algorithm A. At any step* $t \geq 1$*, A gives a linear classifier* $(\mathbf{q}_t, \phi)$ *as a hypothesis to the target linear classifier to the user, where* $\mathbf{q}_t \in R^d$ *and* $\phi \in \mathcal{R}$*. If the hypothesis is equivalent to the target, then the user says "yes" to conclude the learning process. Otherwise, the user presents a counterexample* $\mathbf{x}_t \in \{0, \ldots, n-1\}^d$ *such that the target classifier and the hypothesis classifier differ at* $\mathbf{x}_t$*. In this case, we say that the algorithm A makes a mistake. At step* $t+1$*, the algorithm A constructs a new hypothetical linear classifier* $(\mathbf{q}_{t+1}, \phi_{t+1})$ *to the user based on the received counterexamples* $\mathbf{x}_1, \ldots, \mathbf{x}_t$*. The learning complexity (or the mistake bound) of the algorithm A is in the worst case the maximum number of counterexamples that it may receive from the user in order to learn some classifier over* $\{0, \ldots, n-1\}^d$*.*

Definition 3. *Rocchio's similarity-based relevance feedback algorithm is an adaptive learning algorithm for learning any linear classifier* $(\mathbf{q}, \phi)$ *over the d-dimensional discretized vector space* $\{0, \ldots, n-1\}^d$ *from examples. Let* $(\mathbf{q}_1, \phi_1)$ *be the initial hypothesis. Assume that at the beginning of step* $t > 1$ *the algorithm has received a sequence of counterexamples* $\mathbf{x}_1, \ldots, \mathbf{x}_{t-1}$*, then the algorithm uses the following modified query vector* $\mathbf{q}_t$ *for its next classification:*

$$\mathbf{q}_t = \alpha_{t_0}\mathbf{q}_1 + \sum_{j=1}^{t-1} \alpha_{t_j}\mathbf{x}_j, \tag{1}$$

where $\alpha_{t_j} \in \mathcal{R}$*, for* $j = 0, \ldots, t-1$*, are called query updating factors.*

Please note that our definition above is a generalized version of Rocchio's original algorithm.

3 An Upper Bound for Documents Represented by a Linear Classifier

In this section, we will prove an upper on the learning complexity of Rocchio's similarity-based algorithm in searching for documents represented by a linear classifier over the discretized vector space $\{0, \ldots, n-1\}^d$. We first prove Lemma 1 using linear independence to allow Rocchio's algorithm to simulate any adaptive learning algorithm for learning linear classifiers. Utilizing linear independence to derive upper bounds on learning complexity can be found, for examples, in [2,6,10].

Lemma 1. *Let A be any given adaptive learning algorithm for learning linear classifiers over the d-dimensional discretized vector space* $\{0, \ldots, n-1\}^d$. *Let* $l(A)$ *and* $t(A)$ *denote respectively the learning complexity and the time complexity of the algorithm A. Then, the learning complexity of Rocchio's similarity-based relevance feedback in searching for documents represented by a linear classifier over* $\{0, \ldots, n-1\}^d$ *is at most* $d + l(A)$, *and it time complexity is* $O(d^2 \log^2 n(d + l(A)) + t(A))$.

Proof. By Definition 1, the algorithm A works in a step by step fashion. At step $t \geq 1$. A computes a hypothesis linear classifier $(\mathbf{q}_t, \phi_t)$. We design the following procedure to allow Rocchio's algorithm to simulate the algorithm A:

```
For t = 1, do
    Call A to generate the hypothesis (q_1, θ_1).
    Set q*_1 = 0 and present the hypothesis (q*_1, φ_1) to the user.
    If the user answers "yes", then stop.
    Otherwise a counterexample x_1 is received as relevance feedback.
For t > 1, do
    If x_{t-1} is also a counterexample to A's current hypothesis linear
    classifier (q_{t-1}, φ_{t-1}), then call A to generate a new hypothesis (q_t, φ_t)
    using x_1, ..., x_{t-1}.
    If x_{t-1} is not a counterexample to A's current hypothesis linear
    classifier (q_{t-1}, φ_{t-1}), then simply let q_t = q_{t-1} and φ_t = φ_{t-1}.
    Compute q*_t as the projection of q_t onto the linear
    space defined by {x_1, ..., x_{t-1}}.
    Let Rocchio's algorithm present the new hypothesis (q*_t, φ_t) to the user.
    If the user answers "yes", then stop.
    Otherwise a counterexample x_t is received as relevance feedback.
    Repeat the process for t > 1.
```

We note that Rocchio's algorithm in the above simulation procedure uses a zero initial query vector. For any $t > 1$, the algorithm uses a query vector $\mathbf{q}_t^* = \sum_{i=1}^{t-1} \alpha_i \mathbf{x}_i$ for some $\alpha_i \in \mathcal{R}$, because $\mathbf{q}_t^*$ is the projection of $\mathbf{q}_t$ onto the linear space defined by $\{\mathbf{x}_1, \ldots, \mathbf{x}_{t-1}\}$. This means that the query vector is updated following expression (1). Also, since $\mathbf{q}_t^*$ is the projection of $\mathbf{q}_t$ onto the linear space defined by by $\{\mathbf{x}_1, \ldots, \mathbf{x}_{t-1}\}$, we have $\mathbf{q}_t = \mathbf{q}_t^* + \mathbf{r}_t$ for some vector $\mathbf{r}_t$ such that $\mathbf{r}_t \cdot \mathbf{x}_i = \mathbf{0}$ for $i = 1, \ldots, t-1$.

For any $t > 1$, if $\mathbf{x}_t$ is linearly dependent on $\mathbf{x}_1, \ldots, \mathbf{x}_{t-1}$, i.e., $\mathbf{x}_t = \sum_{i=1}^{t-1} \beta_i \mathbf{x}_i$ for some $\beta_i \in \mathcal{R}$, then we have

$$\begin{aligned}
\mathbf{q}_t \cdot \mathbf{x}_t &= (\mathbf{q}_t^* + \mathbf{r}_t) \cdot \sum_{i=1}^{t-1} \beta_i \mathbf{x}_i \\
&= \mathbf{q}_t^* \cdot \sum_{i=1}^{t-1} \beta_i \mathbf{x}_i + \sum_{i=1}^{t-1} \beta_i \mathbf{r}_t \cdot \mathbf{x}_i \\
&= \mathbf{q}_t^* \cdot \sum_{i=1}^{t-1} \beta_i \mathbf{x}_i = \mathbf{q}_t^* \cdot \mathbf{x}_t
\end{aligned}$$

Hence, the algorithm A has the same classification on $\mathbf{x}_t$ as Rocchio's algorithm does. In other words, if both A and Rocchio's algorithm have different classifications on $\mathbf{x}_t$, then $\mathbf{x}_t$ must be linearly independent of $\mathbf{x}_1, \ldots, \mathbf{x}_{t-1}$. Note that $\mathbf{x}_t$ is a counterexample for the hypothesis linear classifier $(\mathbf{q}_t^*, \phi_t)$ of Rocchio's algorithm. The above analysis implies that if $\mathbf{x}_t$ is not a counterexample for the hypothesis linear classifier $(\mathbf{q}_t, \phi_t)$ of the algorithm A, then $\mathbf{x}_t$ must be linearly independent of $\mathbf{x}_1, \ldots, \mathbf{x}_{t-1}$. Since there are at most d many linearly independent vectors over the vector space $\{0, \ldots, n-1\}^d$, this can only happen at most d times. This follows that the learning complexity of Rocchio's algorithm is at most $d + l(A)$. For each $t > 1$, the projection $\mathbf{q}_t^*$ can be computed using standard matrix operations in $O(d^2 \log^2 n)$ time, and the above simulation procedure runs at most $d + l(A)$ iterations. Therefore, the time complexity of Rocchio's algorithm is $O(d^2 \log^2 n(d + l(A)) + t(A))$. □

Theorem 1. *The learning complexity of Rocchio's similarity-based relevance feedback algorithm in searching for documents represented by a linear classifier over the d-dimensional discretized vector space* $\{0, \ldots, n-1\}^d$ *is* $O(d + d^2(\log d + \log n))$. *Moreover, the time complexity of achieving this upper bound is polynomial in* d *and* $\log n$.

Proof. It is proved by Maass and Turán in [12] that linear classifiers over the d-dimensional vector space $\{0, \ldots, n-1\}^d$ can be learned by an adaptive learning algorithm from examples with learning complexity $O(d^2(\log d + \log n))$. Their algorithm uses linear classifiers over $\{0, \ldots, n-1\}^d$ as hypotheses, and its time complexity is polynomial in d and $\log n$. Therefore, the above theorem follows from Lemma 1 via simulating the learning algorithm by Maass and Turán [12]. □

4 Upper Bound for Documents Represented by a Monotone Linear Classifier

In this section, we consider the learning complexity of Rocchio's similarity-based relevance feedback algorithm in searching for documents represented by a monotone linear classifier $(\mathbf{q}, 0)$ over the discretized vector space $\{0, \ldots, n-1\}^d$, where $q_i \geq 0$, $1 \leq i \leq d$. For a monotone linear classifier $(\mathbf{q}, 0)$, any $\mathbf{x} \in \{0, \ldots, n-1\}^d$ is classified as relevant if

$$\mathbf{q} \cdot \mathbf{x} = q_1 x_1 + q_2 x_2 + \cdots + q_d x_d > 0, \tag{2}$$

or irrelevant otherwise. The efficient learnability of monotone linear classifiers has been extensively studied in machine learning (for example, [11]).

Theorem 2. *The learning complexity of Rocchio's similarity-based relevance feedback algorithm in searching for documents represented by a monotone linear classifier* $(\mathbf{q}, 0)$ *over the discretized vector space* $\{0, n-1\}^d$ *is at most* $d + 2k(n-1)(\log d + \log(n-1))$. *Here,* k *is the number of nonzero components of* $\mathbf{q}$.

We postpone the proof to the end of this section, but first give the following corollary, which gives the bound of the well-known algorithm Winnow1 [11]:

Corollary 1. *The learning complexity of Rocchio's similarity-based relevance feedback algorithm in searching for documents represented by a monotone linear classifier* $(\mathbf{q}, 0)$ *over the binary vector space* $\{0, 1\}^d$ *is at most* $d + 2k \log d$. *Here,* k *is the number of nonzero components of* $\mathbf{q}$.

We now extend the multiplicative query updating technique developed by Littlestone [11] for learning monotone linear classifiers over the boolean vector space $\{0, 1\}^d$ to the discretized vector space $\{0, \ldots, n-1\}^d$ to search for documents represented by a monotone linear classifier $(\mathbf{q}, 0)$ satisfying expression (2). The algorithm MAL is given in the following. We observe that the algorithm MAL differs from the algorithm Winnow1 [11] at the promotion step as follows: MAL uses $\alpha^{\frac{x_i}{n-1}} q_{i,t}$ to update $q_{i,t+1}$, while the algorithm Winnow1 uses $\alpha q_{i,t}$. The performance of the algorithm MAL, when it is used to search for documents represented by a monotone linear classifier $(\mathbf{q}, 0)$ satisfying expression (2), is given in the following two lemmas.

Lemma 2. *Let* u *denote the total number of promotions that the algorithm MAL needs to search for documents represented by a monotone the linear classifier* $(\mathbf{q}, 0)$. *If* $\alpha > n - 1$, *then,*

$$u \leq k \frac{\log \phi}{\log \frac{\alpha}{n-1}}.$$

Here, k *is the number of nonzero components* $q_i > 0$.

Lemma 3. *Let* T *denote the learning complexity of the algorithm MAL in searching for documents represented a monotone linear classifier* $(\mathbf{q}, 0)$ *over the discretized vector space* $\{0, \ldots, n-1\}^d$. *let* k *denote the number of nonzero components of* $\mathbf{q}$. *Suppose* $\alpha > n - 1$. *Then,*

$$T \leq \frac{d(n-1)}{\phi} + k\alpha \frac{\log \phi}{\log \frac{\alpha}{n-1}}. \tag{3}$$

Proof of Theorem 2. It follows directly from the above Lemma 3 and Lemma 1 in the previous section with the choices of $\phi = d(n-1)$ and $\alpha = 2(n-1)$. □

```
Algorithm MAL:
(i) Inputs:
        q_1 = 1, the initial query vector
        α: the query updating factor
        φ ≥ 0, the classification threshold
(ii) Set t = 1.
(iii) Classify and rank documents with the linear classifier (q_t, φ).
(iv) While (the user judged the relevance of a document x) do {
        for i = 1, ..., d, do {
            /* q_t = (q_{1,t}, ..., q_{d,t}), x = (x_1, ..., x_d) */
            if (x_i ≠ 0) {
                if (x is relevant)   /* promotion */
                        set q_{i,t+1} = α^{x_i/(n-1)} q_{i,t}
                else    /* demotion */
                        set q_{i,t+1} = 0
            } else
                set q_{i,t+1} = q_{i,t}
        }
    }
(v) If no documents were judged in the k-th step, then stop.
 O therwise, let t = t + 1 and go to step (iv).
/* The end of the algorithm MAL */
```

5 Lower Bounds

Maass and Turán [12] have derived the following lower bound on the number of different linear classifiers over the discretized vector space $\{0, \ldots, n-1\}^d$:

Proposition 1. *(Maass and Turán [12]) The number of different linear classifiers over the discretized vector space* $\{0, \ldots, n-1\}^d$ *is at least* $n^{\binom{d}{2}}(n-1)^d$.

Based on the above lower bound on the number of linear classifiers and the binary decision tree technique devised by Littlestone [11], they obtained an $\Omega(\binom{d}{2} \log n)$ for any adaptive learning algorithm for learning linear classifier over $\{0, \ldots, n-1\}^d$. This implies the following corollary:

Corollary 2. *The learning complexity of Rocchio's algorithm in searching for documents represented by a linear classifier over the discretized vector space* $\{0, \ldots, n-1\}^d$ *is at least* $\Omega(\binom{d}{2} \log n)$. *In particular, in the binary vector space* $\{0, 1\}^d$, *the lower bound is* $\Omega(d^2)$.

Remark 1. The above lower bound does not apply to the case of searching for documents represented by a monotone linear classifier over $\{0,\ldots,n-1\}^d$, because there are fewer monotone linear classifiers over $\{0,\ldots,n-1\}^d$ than general linear classifiers, so that the $\Omega(\binom{d}{2}\log n)$ lower bound on the number of linear classifiers in general does not hold for monotone linear classifiers. In particular, there are at most $\binom{d}{k}$ monotone disjunctions $x_{i_1} \vee x_{i_2} \vee \cdots \vee x_{i_k}$ over the binary vector space $\{0,1\}^d$. We can only derive an $\Omega(k \log d)$ lower bound for searching for documents represented by a monotone disjunction of k binary relevant features. When k is a constant, this lower bound becomes $\Omega(\log d)$. In [7], an $\Omega(n)$ lower bound is obtained for monotone disjunctions of k relevant features over $\{0,1\}^d$.

In practice, a fixed query updating factor α us used for Rocchio's algorithm. At any step $t \geq 1$, for $1 \leq i \leq d$, the i-component of the query vector $\mathbf{q}_{t+1}$ is updated with respect to the counterexample example $\mathbf{x}_t = (x_{1,t},\ldots,x_{d,t})$ as follows: $q_{i,t+1} = q_{i,t} + \alpha x_{i,t}$ if $\mathbf{x}_t$ is relevance, otherwise $q_{i,t+1} = q_{i,t} - \alpha x_{i_t}$. Since in general the classification threshold can be reviewed as an additional variable, it can be updated as $\phi_{t+1} = \phi_t + \alpha$ if $\mathbf{x}_t$ is relevance, otherwise $\phi_{t+1} = \phi_t - \alpha$. Following the approach for deriving a lower bound for *k-bound* learning algorithm in [12], we have the following theorem:

Theorem 3. *If a fixed query updating factor is used, then the learning complexity of Rocchio's similarity-based relevance feedback algorithm in searching for documents represented by a linear classifier over the Boolean vector space* $\{0,1\}^d$ *is at least* $2^{\Omega(d)}$.

Proof. Note that at any step $t \geq 1$, the counterexample example $\mathbf{x}_t = (x_{1,t},\ldots,x_{d,t})$ to the query vector $\mathbf{q}_t$ is a binary vector in $\{0,1\}^d$. As analyzed above, for $1 \leq i \leq d$, the i-component $q_{i,t+1}$ of the query vector $\mathbf{q}_{t+1}$ can be updated as either $q_{i,t}$, or $q_{i,t} + \alpha$, or $q_{i,t} - \alpha$, depending on whether $\mathbf{x}_t$ is relevant or not and whether x_i is zero or one. By iteration from expression (1), $q_{i,t+1}$ is obtained from $q_{i,1}$ with t many operations, each of which is one of three types $+0$, $+\alpha$, and $-\alpha$. The order of these operations do not affect the value of $q_{i,t+1}$. The value of $q_{i,t+1}$ is determined by the number of each type of operations involved. Thus, there are at most t^3 many possible values for $q_{i,t+1}$. This means there are at most t^{3d} many possible choices for $\mathbf{q}_{t+1}$. Similarly, ϕ_{t+1} can be updated as either $\phi_t + \alpha$ or $\phi_t - \alpha$, implying there are at most t^2 many possible values for ϕ_{t+1}. Therefore, at step t, Rocchio's algorithm with a fixed query updating factor can generate at most $t^{3(d+1)}$ many possible hypotheses. By Proposition 1, in order to search for document sets represented by any linear classifier over the binary vector space $\{0,1\}^d$, we must have

$$t^{3(d+1)} \geq 2^{\binom{d}{2}}.$$

Hence, $t \geq 2^{\Omega(d)}$. □

Remark 2. As commented in Remark 1, the lower bound in Theorem 3 does not apply to the case of searching for documents represented by a monotone linear classifier over $\{0,\ldots,n-1\}^d$, because there are fewer monotone linear classifiers over $\{0,\ldots,n-1\}^d$ than general linear classifiers, so that the $\Omega(\binom{d}{2}\log n)$ lower bound on the number of linear classifiers in general does not hold for monotone linear classifiers. In [5], an $\Omega(k(n-k))$ lower bound is obtained for monotone disjunctions of k relevant features over $\{0,1\}^d$, when a constant query updating factor is used.

Unfortunately, the proof of Theorem 3 cannot be generalized to the discretized vector space $\{0,\ldots,n-1\}^d$. In this case, $q_{i,k+1}$ is obtained from $q_{i,1}$ with k many operations, each of which is one of three types of operations $+0$, $+\alpha x_{i,k}$ and $-\alpha x_{i,k}$. Since $x_{i,k}$ in general may not be 1 or 0, the value of $q_{i,k+1}$ is determined by not only the number of each type of operations involved, but also the order of these operations. However, we have the following lower bound:

Theorem 4. *Suppose that the query updating factors used by Rocchio's similarity-based algorithm are bounded by $O(n^c)$ from some constant $c \geq 0$ during its process of searching for documents represented by a linear classifier over the discretized vector space $\{0,\ldots,n-1\}^d$. Then, learning complexity of Rocchio's algorithm is at least $\Omega(n^{d-1-c}/(n-1))$.*

Proof. It is proved in Hampson and Volper [9] that there are linear classifiers $(\mathbf{q},0)$ over $\{0,\ldots,n-1\}^d$ such that $q_i = \theta(n^{d-1})$ for some $1 \leq i \leq d$. At each step $k \geq 1$, Rocchio's algorithm can update its query vector by a magnitude of at most $O(n^c(n-1)$. Hence, the algorithm needs at least $\Omega(n^{d-1-c}/(n-1))$ steps to to learn $q_i = \theta(n^{d-1})$. □

Remark 3. The exponential lower bounds obtained in Theorems 3 and 4 do not contradict to the $O(d + d^2(\log d + \log n))$ upper bound obtained in Theorem 1 and the $O(d + 2k(n-1)(\log d + \log(n-1)))$ upper bound obtained in Theorem 2. In Theorems 3 and 4, the query updating factors are either fixed or bounded by $O(n^c)$, while there are no such requirements in Theorems 1 and 2. In reality, when computing the projection of a query vector onto a linear subspace spanned by a list of counterexamples in the proof of Lemma 1, exponential query updating factors may occur.

6 Concluding Remarks

It would be very interesting to analyze the average-case learning complexity of Rocchio's algorithm. We feel that this problem is very challenging, because any nontrivial average case complexity analysis will reply on realistic models of document distribution, index term distribution, and the user preference distribution as well. We feel that it is not easy to model those distributions nor to analyze the complexity under those distributions. The probabilistic corpus model proposed in [13] may shed some light on this problem.

References

1. R. Baeza-Yates and B. Ribeiro-Neto, eds. Modern Information Retrieval, Addison-Wesley, 1999.
2. N. Bshouty, Z. Chen, S. Decatur and S. Homer, On the learnability of Z_N-DNF formulas, Proceedings of COLT'95, 198-205, 1995.
3. Z. Chen, Multiplicative adaptive algorithms for user preference retrieval, Proceedings of COCOON'01, 540-549, 2001.
4. Chen Z, Multiplicative adaptive user preference retrieval and its applications to Web search. In: Zhang G, Kandel A, Lin T, and Yao Y, ed. Computational Web Intelligence: Intelligent Technology for Web Applications. World Scientific, 303-328, 2004.
5. Z. Chen and B. Fu, A quadratic lower bound for Rocchio's similarity-based relevance feedback algorithm, Proceedings of COCOON'05, 955-964, 2005.
6. Z. Chen and S. Homer, Learning counting functions with queries, Theoretical Computer Science, 180(1-2):155-168, 1997.
7. Z. Chen and B. Zhu, Some formal analysis of Rocchio's similarity-based relevance feedback algorithm. Information Retrieval **5**: 61-86, 2002. (The preliminary version of this paper appeared in Proceedings of ISAAC'00, LNCS 1969, pp 108-119, 2000.)
8. W. Frakes and R. Baeza-Yates, eds. Information Retrieval: Data Structures and Algorithms. Prentice Hall, 1992.
9. S. Hampson and D. Volper, Representing and learning boolean functions of multivalued features, IEEE Trans. on Systems, Man, and Cybernetics, 20(1):67-80, 1990.
10. J. Kivinen, M. Warmuth and P. Auer, The perceptron algorithm vs. Winnow: linear vs. logarithmic mistake bounds when few input variables are relevant, Artificial Intelligence, (1-2):325-343, 1997.
11. N. Littlestone, Learning quickly when irrelevant attributes abound: A new linear-threshold algorithm, Machine Learning **2**:285-318, 1988.
12. W. Maass and G. Turán, How fast can a threshold gate learn? Computational Learning Theory and Natural Learning Systems **1**:381-414, 1994.
13. C. Papadimitriou, P. Raghavan and H. Tamaki, Latent semantic indexing: A probabilistic analysis, Journal of Computer and System Science **61**(2):217-235, 2000.
14. V. Raghavan and S. Wong, A critical analysis of the vector space model for information retrieval, Journal of the American Society for Information Science **37**(5):279-287, 1986.
15. J. Rocchio, Relevance feedback in information retrieval. In: Salton G, ed. The Smart Retrieval System - Experiments in Automatic Document Processing. Prentice-Hall, Englewood Cliffs, NJ, 313-323, 1971.
16. F. Rosenblatt, (1958) The perceptron: A probabilistic model for information storage and organization in the brain, Psychological Review **65**(6):386-407, 1958.
17. G. Salton, eds. Automatic Text Processing: The Transformation, Analysis, and Retrieval of Information by Computer, Addison-Wesley, 1989.
18. Salton G, Wong S and Yang C (1975) A vector space model for automatic indexing. Comm. of ACM **18**(11):613–620.
19. CJ van Vijsbergen, *Information Retrieval*, Butterworths, 1979.
20. S. Wong, Y. Yao and P. Bollmann, Linear structures in information retrieval, Proceedings of the 1988 ACM-SIGIR Conference on Information Retrieval, 219-232, 1988.

Correlation Clustering and Consensus Clustering

Paola Bonizzoni[1], Gianluca Della Vedova[2], Riccardo Dondi[1,*], and Tao Jiang[3,4,**]

[1] Dipartimento di Informatica, Sistemistica e Comunicazione, Università degli Studi di Milano-Bicocca Milano - Italy
[2] Dipartimento di Statistica, Università degli Studi di Milano-Bicocca Milano - Italy
[3] Department of Computer Science and Engineering University of California - Riverside - USA
[4] Center for Advanced Study, Tsinghua University, Beijing, China
bonizzoni@disco.unimib.it, gianluca.dellavedova@unimib.it
riccardo.dondi@unimib.it, jiang@cs.ucr.edu

Abstract. The Correlation Clustering problem has been introduced recently [5] as a model for clustering data when a binary relationship between data points is known. More precisely, for each pair of points we have two scores measuring respectively the similarity and dissimilarity of the two points, and we would like to compute an optimal partition where the value of a partition is obtained by summing up scores of pairs involving points from a same cluster and scores of pairs involving points from different clusters. A closely related problem is Consensus Clustering, where we are given a set of partitions and we would like to obtain a partition that best summarizes the input partitions. The latter problem is a restricted case of Correlation Clustering. In this paper we prove that Min Consensus Clustering is APX-hard even for three input partitions, answering an open question, while Max Consensus Clustering admits a PTAS on instances with a bounded number of input partitions. We exhibit a combinatorial and practical $\frac{4}{5}$-approximation algorithm based on a greedy technique for Max Consensus Clustering on three partitions. Moreover, we prove that a PTAS exists for Max Correlation Clustering when the maximum ratio between two scores is at most a constant.

1 Introduction

The problem of analyzing a set of points in order to isolate subsets of points that are closely related is known as *clustering*. Clustering is an important problem in computer science due to its broad applications in areas such as datamining, machine learning, and bioinformatics. An example application taken from [5] is to cluster a set of documents into topics without having a clear notion of topic, but with a measure of similarity (or dissimilarity) between pairs of documents. We construct a graph G whose vertices are the documents. Each edge e is labeled with two weights, where the first weight $a(e)$ measures the similarity between the documents incident on e and the second weight $b(e)$ measures the dissimilarity between the documents incident on e. Given a partition

* Corresponding author.

** Tao Jiang research is supported in part by NSF grant CCR-0309902, National Key Project for Basic Research (973) grant 2002CB512801, and a fellowship from the Center for Advanced Study, Tsinghua University.

X. Deng and D. Du (Eds.): ISAAC 2005, LNCS 3827, pp. 226–235, 2005.

π of the vertex set of G, the value of π can be formally defined as the summation of $a(e)$ for each edge internal to a cluster and $b(e)$ for each edge whose endpoints are in two different clusters. The MAX CORRELATION CLUSTERING problem asks for a partition π of vertices of G of maximum value. An interesting property of this approach is that the number of clusters to be obtained is not predetermined *a priori*, as in the case of the k-median or k-center problems, and it depends uniquely on the instance.

There are in fact several versions of the problem. For instance, both the maximization and minimization versions of the problem have been studied in the literature and shown to have different approximation properties, as we will see in this paper. Note that the minimization version asks for a partition π of vertices of G whose value is defined by summing the weight $a(e)$ for each edge e whose endpoints are in two different clusters and weight $b(e)$ for each edge internal to a cluster. Moreover, in the case of arbitrary weights (i.e. $a(e) = b(e) = 0$ is allowed) the minimization version of the problem, *i.e.* MIN CORRELATION CLUSTERING, has an $O(\log n)$-approximation algorithm [6,7,8], while MAX CORRELATION CLUSTERING is APX-hard and has a 0.7664-approximation algorithm [6], improved to 0.7666 in [13]. *Unweighted* versions of CORRELATION CLUSTERING (that is all scores $a(e)$ are either 0 or 1 and $b(e) = 1 - a(e)$) have also been considered before, proving that the unweighted MIN CORRELATION CLUSTERING is APX-hard [8,6] via a reduction from minimum multicut, while admitting a 4-approximation algorithm [6] and a randomized 3-approximation algorithm [3]. The unweighted MAX CORRELATION CLUSTERING admits a probabilistic PTAS [5]. Advances on variants of CORRELATION CLUSTERING have been recently achieved in [3]. In this paper, we will describe a PTAS for MAX CORRELATION CLUSTERING on input graphs where the ratio between the largest and smallest weights is bounded by a constant. Such a restriction clearly implies that all weights are strictly larger than zero, and hence the result is applicable only to complete graphs. The PTAS is based on the smooth polynomial programming technique introduced in [2] and exploits the natural denseness of the problem.

Another variant of CORRELATION CLUSTERING is the CONSENSUS CLUSTERING problem. Here, we are given a set of partitions and we want to compute the *median* partition, *i.e.* a partition having the total maximum similarity (or minimum distance) to the input partitions. CONSENSUS CLUSTERING has been studied extensively in the literature [12,14], and its NP-hardness over general graphs is well-known. More attention has been given to the problem recently because of its application in bioinformatics, in particular, microarray data analysis. It is observed in [10,9] that microarray experiments provide measures of gene expression levels, and clustering genes with similar expression levels could provide information useful for the construction of genetic networks. Since different experimental conditions may result in significantly different expression data (thus partitions of genes), it is often useful to compute the consensus of the partitions given by a collection of gene expression data. It is easy to see that CONSENSUS CLUSTERING is actually a special case of (weighted) CORRELATION CLUSTERING. Consider an instance of CONSENSUS CLUSTERING. For each pair of elements of the universe, x_1 and x_2, define an edge $e = (x_1, x_2)$ with weight $a(e)$ equal to the number of input partitions containing x_1 and x_2 in the same set (*i.e.* they are *co-clustered*) and weight $b(e)$ equal to the number of input partitions containing x_1 and x_2 in different sets (*i.e.* they are *not co-clustered*). Then, solving CONSENSUS CLUSTERING on the instance would be equivalent to solving CORRELATION CLUSTERING. A number of heuristics have been proposed for CONSENSUS CLUSTERING, which are based on cutting-plane [11] and simulated annealing [9] techniques. In the latter paper, it was observed that the problem is trivially solvable for instances of at most two partitions, while an open question, as recently recalled in [3], is the complexity of the problem (minimization and maximization versions) for k input partitions, for any constant $k > 2$.

In this paper, we settle the open question by showing that MIN CONSENSUS CLUSTERING is MAX SNP-hard on instances of three input partitions, which is the strongest inapproximability results we could hope for. On the positive side, we exhibit a combinatorial and practical $\frac{4}{5}$-approximation algorithm based on a greedy technique for MAX CONSENSUS CLUSTERING on three partitions. To our knowledge, this is the first combinatorial approximation algorithm for CONSENSUS CLUSTERING, except for the trivial approximation algorithm that picks the best among the input partitions (which has approximation factor $\frac{3}{4}$ when applied to instances consisting of three partitions). The approximation factor of 0.8 achieved by our algorithm improves on the approximation factor obtained by applying the best algorithm for CORRELATION CLUSTERING based on semi-definite programming. Moreover, we show that MAX CONSENSUS CLUSTERING on instances of a bounded number of partitions admits a PTAS. This is achieved by first showing that MAX CORRELATION CLUSTERING on instances with bounded weights has a PTAS, by using the smooth polynomial programming technique.

The proof are given in the full version of the paper.

2 Preliminary Definitions

In this section, we introduce some basic notations and definitions that we will need later. Let π be a partition and r_π be the characteristic vector associated with π defined as $r_\pi(i,j) = 1$ if (i,j) are co-clustered in π, and $r_\pi(i,j) = 0$ if (i,j) are not co-clustered in π. Closely related is the notion of *correlation graph*, that is a weighted graph G such that each edge $e = (i,j)$ is labeled with two weights $a(e)$ and $b(e)$, which are respectively the similarity and the distance between i and j. In this paper we will study two problems whose instance is a correlation graph $G = (V,E)$ and the output is a partition π of V. For the MIN CORRELATION CLUSTERING we want to find a partition minimizing the sum $\sum_{e=(i,j)} (r_\pi(i,j)b(e) + (1 - r_\pi(i,j))a(e))$, while for MAX CORRELATION CLUSTERING we want to maximize the sum $\sum_{e=(i,j)} (r_\pi(i,j)a(e) + (1 - r_\pi(i,j))b(e))$.

For studying our versions of CONSENSUS CLUSTERING we need two notions of distance and similarity between two partitions of a universe set U. The *symmetric difference distance* of π_1, π_2, denoted by $d(\pi_1, \pi_2)$ is the number of pairs of elements co-clustered in exactly one of $\{\pi_1, \pi_2\}$, while the *similarity measure*, denoted by $s(\pi_1, \pi_2)$, is the number of pairs of elements co-clustered in both partitions plus the number of pairs of elements not co-clustered in both partitions π_1 and π_2. Given two elements i, j of the universe set U and a set $\Pi = \{\pi_1, \ldots, \pi_k\}$ of partitions of U, we denote by $s_\Pi(i,j)$ (or simply $s(i,j)$ whenever Π is known from the context) the number of partitions of Π in which i,j are co-clustered. We denote by $d_\Pi(i,j)$ (or simply $d(i,j)$ whenever Π is known from the context) the number of partitions of Π in which i,j

are not co-clustered. Clearly, for each pair (i,j), $d_\Pi(i,j)+s_\Pi(i,j)=k$. Now we formally introduce the other two problems studied in the paper, both have a set $\Pi = \{\pi_1,\pi_2,...,\pi_k\}$ of partitions over universe U as instances and the output is a partition π of U. For the MIN CONSENSUS CLUSTERING we want to minimize $\sum_{i=1}^{k} d(\pi_i,\pi)$, which is equivalent to minimize $\sum_{(i<j)}(r_\pi(i,j)d_\Pi(i,j)+(1-r_\pi(i,j))s_\Pi(i,j))$, defined as the *cost* $c(\pi)$ of a solution π. For MAX CONSENSUS CLUSTERING we want to maximize $\sum_{i=1}^{k} s(\pi_i,\pi)$, which is equivalent to maximize $\sum_{(i<j)}(r_\pi(i,j)s_\Pi(i,j)+(1-r_\pi(i,j))d_\Pi(i,j))$, defined as the *similarity value* $v(\pi)$ of a solution π.

Consider the problem restricted to three partitions, *3 Consensus Clustering*, 3*CC*. A fundamental notion used in the paper is that of 2-*component* of an instance I of 3*CC* over universe U. A 2-component is a maximal subset X of U such that each pair of elements of X is co-clustered in at least two input partitions and $|X| \geq 2$. It is easy to see that it is possible to compute efficiently 2-components of an instance I of 3*CC*.

An important tool used in the reduction and in the approximation algorithm of Section 4 is the *component graph* constructed from the 2-components of an instance I of 3CC. Let $C = \{C_1, C_2, \ldots, C_m\}$ be the set of 2-components of I, then the component graph associated with I is the graph $G_c = (C, E_c)$ where $(C_i, C_j) \in E_c$ iff $C_i \cap C_j \neq \emptyset$.

3 Inapproximability of Consensus Clustering

In this section, we show the inapproximability of MIN CONSENSUS CLUSTERING within a certain constant factor. We consider a restricted case of MIN CONSENSUS CLUSTERING where each instance consists of exactly three partitions and no pair of elements is co-clustered in all three partitions: we call this case 3-RESTRICTED CONSENSUS CLUSTERING. We will prove that MIN 3-RESTRICTED CONSENSUS CLUSTERING, in short *MR3CC*, is MAX SNP-hard via an L-reduction from the MAX INDEPENDENT SET problem on cubic graphs, where we are asked for the largest subset of vertices that are not connected by any edge [1] (see [4] for details on L-reductions).

The proof of MAX SNP-hardness of MR3CC consists of two separate reductions: (1) an L-reduction from the MAX INDEPENDENT SET problem on cubic graphs to the problem of finding a maximum independent set on an artificial class of graphs, called *gadget graphs*, or *G-graphs* in short, (2) an L-reduction from the Max Independent Set problem on G-graphs to MR3CC. The latter reduction is based on two main steps: proving that **(a)** a G-graph associated with a cubic graph is the component graph of an instance of MR3CC, **(b)** the size of an independent set of a G-graph is related by a constant to the number of co-clustered pairs in a feasible solution of MR3CC that is constructed from 2-components or vertices of the G-graph.

The first reduction. Given a cubic graph $G = (V,E)$, we will associate with G a G-graph $\mathcal{G}$ by constructing for each vertex of G a vertex gadget and for each edge $(v_i, v_j) \in E$, an edge gadget connecting the two vertex gadgets associated with v_i and v_j. For each vertex $v_i \in V$, the vertex gadget VG_i is the graph represented in Fig.1.

Since, in a cubic graph, a vertex v_i may be adjacent to three edges, the vertex gadget VG_i has three vertices, $c_{i_1}, c_{i_4}, c_{i_{12}}$, called *docking vertices*, each one connecting VG_i to an edge gadget associated with an edge adjacent to v_i. We denote with $VG(\mathcal{G})$ the set of vertex gadgets associated with the vertex set of graph $\mathcal{G}$. Given two adjacent vertices

$v_i, v_j \in V$ and the corresponding vertex gadgets VG_i and VG_j, respectively, there is an edge gadget $EG_{i,j}$ associated with the edge $(v_i, v_j) \in E$. The edge gadget $EG_{i,j}$ is the graph of 6 vertices joining VG_i, VG_j in two of their docking vertices, see Fig. 2.

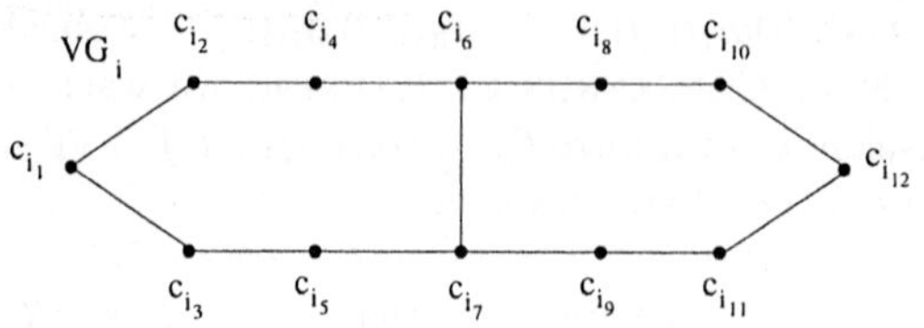

Fig. 1. A vertex gadget VG_i

We will say that two vertex gadgets are *independent* if there is no edge gadget between them; otherwise they are *adjacent*. Please notice that each vertex gadget VG_i has a unique maximum independent set containing 6 vertices ($\{c_{i_1}, c_{i_4}, c_{i_5}, c_{i_8}, c_{i_9}, c_{i_{12}}\}$) and that all the docking vertices of VG_i are part of this set. We denote this independent set of a vertex gadget as *type 1*. There is an independent set of VG_i having cardinality 5 with no docking vertices, for example ($\{c_{i_2}, c_{i_3}, c_{i_6}, c_{i_{10}}, c_{i_{11}}\}$). We denote this independent set of a vertex gadget as *type 2*. Observe that no two adjacent vertex gadgets have an independent set of type 1 in a maximum independent set of $\mathcal{G}$. We call 1,2-*solution* an independent set of $\mathcal{G}$ where all vertex gadgets have an independent set of type 1 or type 2 and for each pair VG_i, VG_j of vertex gadgets connected by an edge gadget $EG_{i,j}$, at least one of VG_i, VG_j is of type 2.

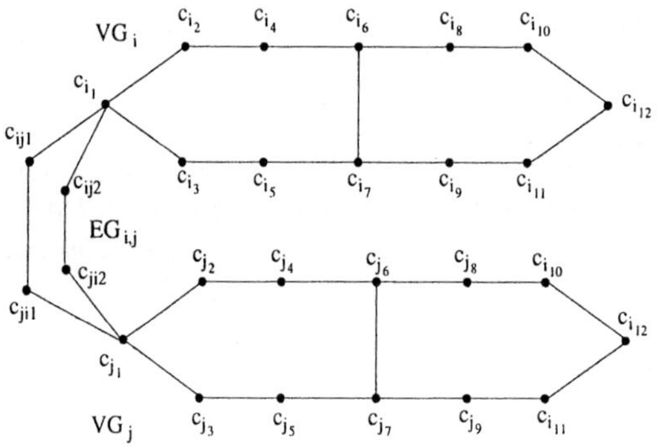

Fig. 2. Vertex gadgets VG_i, VG_j and edge gadget EG_{ij}

The following property derives from the fact that an edge gadget has an independent set of size at most 2 not including docking vertices, i.e. vertices of the vertex gadgets. Given a G-graph $\mathcal{G}$ associated with $G = (V, E)$ and given an independent set I of $\mathcal{G}$ such that I is not a 1,2-solution, then (in polynomial time) it is possible to compute a type 1,2-solution larger than I. The following theorem shows the reduction from independent set on cubic graphs to the same problem on G-graphs.

Theorem 1. *Let $G = (V, E)$ be a cubic graph with $|E| = m$, $|V| = n$, and let $\mathcal{G}$ be the G-graph obtained from G by our reduction. Then G has an independent set of size h if and only if $\mathcal{G}$ has an independent set of size at least $6h + 5(n - h) + 2m$.*

For each cubic graph $G = (V, E)$, $|E| = \frac{3}{2}|V|$ and there exists an independent set of size $|V|/4$, hence the reduction is actually an L-reduction, and consequently computing the maximum independent set is MAX SNP-hard also for G-graphs.

The second reduction. The first basic step is to build from a G-graph associated with a cubic graph an instance of MR3CC whose component graph is the given G-graph.

Step (a). In the following, we first associate with each vertex and edge gadget a set of 2-components, and then we construct three partitions having exactly such 2-components. Consider a vertex gadget VG_i and the corresponding 2-components as represented in Fig. 3. Associated with VG_i there are a universe set $U = \{i_1, i_2, ..., i_{35}\}$ and 3 partitions over U whose 2-components are the ones of Fig. 3: $\pi_1(VG_i) = (\{c_{i_1} \cup c_{i_2} \cup c_{i_4}\}, \{c_{i_5} \cup c_{i_7}\}, \{c_{i_8}\}, \{c_{i_{11}} \cup c_{i_{12}}\}, \{i_8\}, \{i_9\}, \{i_{17}\}, \{i_{26}\}, \{i_{27}\}, \{i_{28}\}, \{i_{29}\})$, $\pi_2(VG_i) = (\{c_{i_1} \cup c_{i_3}\}, \{c_{i_4} \cup c_{i_6} \cup c_{i_7} \cup c_{i_9}\}, \{c_{i_{10}} \cup c_{i_{12}}\}, \{i_5\}, \{i_6\}, \{i_{15}\}, \{i_{16}\}, \{i_{23}\}, \{i_{24}\}, \{i_{31}\}, \{i_{32}\})$, $\pi_3(VG_i) = (\{c_{i_2}\}, \{c_{i_3} \cup c_{i_5}\}, \{c_{i_6} \cup c_{i_8} \cup c_{i_{10}}\}, \{c_{i_9} \cup c_{i_{11}}\}, \{i_3\}, \{i_4\}, \{i_{11}\}, \{i_{12}\}, \{i_{21}\}, \{i_{34}\}, \{i_{35}\})$.

It is easy to verify that each set c_{i_z} is a 2-component. Now let $c_{i_1}, c_{i_4}, c_{i_{12}}$ be the docking vertices of vertex VG_i. Note that each of them shares two elements with some other 2-components of the same vertex gadget, while two elements, called *private*, are not shared with any other 2-components of the vertex gadgets. Denoting by $d_1(EG_{i,j})$, $d_2(EG_{i,j})$ the two docking vertices of the edge gadget $EG_{i,j}$, and by $p_1(d_k(EG_{i,j}))$, $p_2(d_k(EG_{i,j}))$ the two private elements of the docking vertex $d_k(EG_{i,j})$, we can then describe the 2-components associated with $EG_{i,j}$. These 2-components are clearly the two docking vertices and $c_{ij1} = \{p_1(d_1(EG_{i,j})), h_{i,j,1}, e_{i,j,1}\}$, $c_{ij2} = \{p_2(d_1(EG_{i,j})), e_{i,j,2}\}$, $c_{ji1} = \{p_1(d_2(EG_{i,j})), h_{i,j,2}, e_{i,j,1}\}$, $c_{ji2} = \{p_2(d_2(EG_{i,j})), e_{i,j,2}\}$, where $e_{i,j,1}$, $e_{i,j,2}$, $h_{i,j,1}$ and $h_{i,j,2}$ are new elements associated with $EG_{i,j}$.

Now we show how to construct the three partitions π_1, π_2, π_3 associated with a G-graph $\mathcal{G}$. Note that for each pair VG_i, VG_j of vertex gadgets, the three partitions associated with VG_i are over a universe set that is disjoint from the one of the three partitions of VG_j. Consequently, we can construct three partitions associated with all vertex gadgets as the union of the partitions for each single gadget. Formally, for each $i \in \{1,2,3\}$, then $\pi_i(VG(\mathcal{G})) = \bigcup_{vg \in VG(\mathcal{G})} \pi_i(vg)$. Furthermore, other sets "producing" the 2-components associated with edge gadgets are appropriately added to the three partitions $\pi_1(VG(\mathcal{G}))$, $\pi_2(VG(\mathcal{G}))$ and $\pi_3(VG(\mathcal{G}))$ to get π_1, π_2, π_3. Then we can prove the following lemma.

Lemma 2. *Let $\mathcal{G}$ be a G-graph. Then there exists a set I of three partitions π_1, π_2, π_3 computed by a polynomial algorithm such that $\mathcal{G}$ is the component graph of I. Then I is an instance of* MR3CC *associated with $\mathcal{G}$.*

Step (b). The second basic step of the reduction from the Max Independent Set problem on a G-graph $\mathcal{G}$ to MR3CC, consists in relating the cost of a feasible solution of the

$c_{i_1} = \{i_1, i_2, i_3, i_4\}$ $\quad c_{i_2} = \{i_1, i_5, i_6, i_7\}$
$c_{i_3} = \{i_2, i_8, i_9, i_{10}\}$ $\quad c_{i_4} = \{i_7, i_{11}, i_{12}, i_{13}\}$
$c_{i_5} = \{i_{10}, i_{14}, i_{15}, i_{16}\}$ $\quad c_{i_6} = \{i_{13}, i_{17}, i_{18}, i_{19}\}$
$c_{i_7} = \{i_{14}, i_{19}, i_{20}, i_{21}\}$ $\quad c_{i_8} = \{i_{18}, i_{22}, i_{23}, i_{24}\}$
$c_{i_9} = \{i_{20}, i_{26}, i_{27}, i_{25}\}$ $\quad c_{i_{10}} = \{i_{22}, i_{28}, i_{29}, i_{30}\}$
$c_{i_{11}} = \{i_{25}, i_{31}, i_{32}, i_{33}\}$ $\quad c_{i_{12}} = \{i_{30}, i_{33}, i_{34}, i_{35}\}$

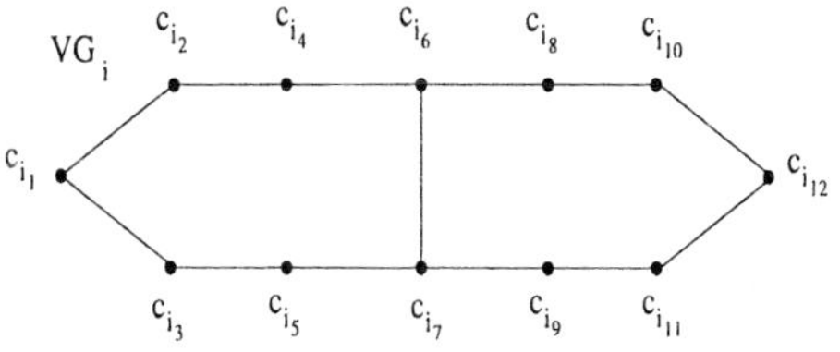

Fig. 3. A vertex gadget and its 2-components

instance of MR3CC associated with $\mathcal{G}$ to the 2-components or vertices of $\mathcal{G}$. To this aim it is east to see that, given an arbitrary solution π of MR3CC, there exists a solution π^* of cost at most equal to the one of π, but such that each set in π^* is a subset of a 2-component of the component graph associated with the instance: we denote such type of solution as *special solution*.

The following theorem is the basic result used in our second L-reduction.

Theorem 3. *There is an independent set I of a G-graph $\mathcal{G}= (V,E)$ of size at most $6k+5(n-k)+2m$ if and only if there is a special solution π of the associated instance of* MR3CC *with $41k+40(n-k)+3m$ co-clustered pairs, where n, m and k are respectively the number of vertex gadgets, the number of edge gadgets and k is the number of mutually independent vertex gadgets.*

Given a special solution π^* with at least $41h+40(n-h)+3m$ pairs, we can find in polynomial time a set of h independent vertex gadgets in the G-graph.

Lemma 4. *Let π be a special solution of* MR3CC, *it is possible to compute in polynomial time a special solution π^* that has at least the same number of co-clustered pairs of π and contains* 41 *pairs for each vertex gadget in a set of h independent vertex gadgets,* 40 *pairs for each of the remaining $n-h$ vertex gadgets and* 3 *pairs for each edge gadget.*

Since the cost of a feasible solution of MR3CC is related by a constant to the number of pairs in the solution, Theorem 3 provides a L-reduction, thus leading to the APX-hardness of MR3CC.

4 A $\frac{4}{5}$-Approximation Algorithm for Max Consensus Clustering

In this section, we present a $\frac{4}{5}$-approximation algorithms for MAX CONSENSUS CLUSTERING on instances of 3 partitions (*Max-3CC*) based on a combinatorial approach. The algorithm constructs a partition by selecting 2-components (*i.e.* co-clustered pairs), using a greedy technique, from a component graph built from the input instance. The trivial approximation algorithm that picks the best of the input partitions has approximation factor $\frac{3}{4}$. In what follows, given a set X, we denote with $P(X)$ the set of all pairs of elements over X.

Lemma 5. *Given an instance $\Pi = \{\pi_1, \pi_2, \pi_3\}$ of Max-3CC, let A, B be 2-components of Π, then all elements in $A \cap B$ are co-clustered in three partitions. Moreover, there is at most another 2-component C with elements in $A \cap B$ and it must be $A \cap B = A \cap C = B \cap C$.*

Let $X_1, \ldots, X_n$ be the 2-components of the instance. If 2-components X_i and X_j are not disjoint, let $A_h = X_i \cap X_j$. Denote with $\mathcal{A} = \{A_1, \ldots, A_k\}$ the set of non-empty intersections of pairs of 2-components. Note that, if $|A_i| \geq 2$, then each pair of elements $(x,y) \in A_i$ is co-clustered in three input partitions.

Let Π be an instance of Max-3CC and let (i,j) be a pair of elements of U. Recall that $s_\Pi(i,j)$ denotes the number of input partitions in which i and j are co-clustered, while $d_\Pi(i,j)$ denotes the number of input partitions in which i and j are not co-clustered.

Algorithm 1: Greedy-CC

Data: an instance Π over universe U

$\pi \leftarrow \emptyset$; $G \leftarrow$ the component graph associated with Π;
while $|V(G)| > 0$ **do**
  X is a maximum 2-components in G; $\pi \leftarrow \pi \cup \{X\}$;
  remove all elements of X from U and update Π;
  $G \leftarrow$ the component graph associated with Π;
end
foreach *$u \in U$, u is not in a set of π* **do**
  add $\{u\}$ to π;
end

Result: π

Define the optimal weight of (i,j), denoted by $w_o(i,j)$, as the maximum of $s_\Pi(i,j)$ and $d_\Pi(i,j)$. Given a solution π, we define the weight of (i,j) in π, denoted by $w_\pi(i,j)$ as $s_\Pi(i,j)$ if (i,j) is co-clustered in π, $d_\Pi(i,j)$ otherwise. Thus the similarity value $v(\pi)$ of a solution π can be expressed simply as $\sum_{(i<j)} w_\pi(i,j)$. Let *opt* be an optimal solution of Π. Then: $w_{opt}(p) \leq w_o(p)$ for each $p \in P(U)$. Let P' be a set of pairs, the *similarity value* of P' in a solution π is $v(P',\pi) = \sum_{p \in P'} w_\pi(p)$.

Let π be the solution returned by Algorithm 1. Since every set in the solution π is a subset of a 2-component, then all elements co-clustered in π are co-clustered in at least two partitions of the input and thus $w_\pi(p) = w_{opt}(p)$ for each pair p of such pairs.

Moreover, observe that, since Greedy-CC constructs a partition π from subsets of 2-components, if a pair p is co-clustered in the instance in less than two partitions, it follows that p is not co-clustered in π and thus $w_\pi(p) = w_o(p)$. Similarly, it is easy to see that for each pair p co-clustered in three partitions, it follows that $w_\pi(p) = w_o(p) = 3$.

Consequently, $w_\pi(p) = w_o(p)$ except for some pairs $p = (a,b)$, such that $a,b \in X$, where X is a 2-component of graph G, and $w_\pi(p) = 1 < w_o(p) = 2$, i.e. p is "not optimal". Let us denote by $P_{\pi,loss}$ the set of such (non optimal) pairs. Then, we show that the number of pairs in $P_{\pi,loss}$ is limited and we can bound from above the similarity value of such pairs as follows.

Let us denote by $P_{\pi,o}$ the set of pairs contained in 2-components of G such that $w_\pi(p) = w_o(p) = 2$, by $P_{\pi,1}$ the subset of pairs p that are not in 2-components and such that $w_\pi(p) = w_o(p) = 2$ (p is not co-clustered in π). Moreover, denote by P_{A_i} the set of pairs with elements in a certain A_i and $P_A = \cup_i P_{A_i}$. Hence $v(\mathcal{A})$ is the similarity value of the pairs in P_A, that is $v(\mathcal{A}) = \sum_{A_i \in \mathcal{A}} \frac{3|A_i|(|A_i|-1)}{2}$. The following basic inequality is used to prove the approximation factor:

$$v(P_{\pi,o},\pi) + v(\mathcal{A}) + v(P_{\pi,1},\pi) \geq 3v(P_{\pi,loss},\pi). \quad (1)$$

The detailed analysis used to prove the above equation is based on properties of 2-components. Since for each p in $P_{\pi,loss}$ $w_\pi(p) = 1 < w_o(p) = 2$, it follows that $v(P_{\pi,loss},\pi) \geq \frac{1}{2} v(P_{\pi,loss},Opt)$. Hence it is easy to see that the following theorem holds.

Theorem 6. *Algorithm* Greedy-CC *achieves an approximation factor of* $\frac{4}{5}$.

5 The PTAS

In this section, we will present a polynomial-time approximation scheme for MAX CORRELATION CLUSTERING problem when the ratio between the maximum and the minimum weights is upper-bounded by a constant. We recall that the value of a partition π is $\sum_{i<j} a(i,j)r_\pi(i,j) + \sum_{i<j} b(i,j)(1-r_\pi(i,j))$, where $r_\pi(i,j) = 1$ if and only if the elements i and j are co-clustered in π. Since we are interested only in instances where the ratio $\max_{i,j}\{a(i,j),b(i,j)\}/\min_{i,j}\{a(i,j),b(i,j)\}$ is at most a constant, it is not restrictive to assume that the $\min_{i,j}\{a(i,j),b(i,j)\} = 1$ and $\max_{i,j}\{a(i,j),b(i,j)\}$ is equal to a certain constant w_{max}.

The scheme is based on the smooth polynomial programming technique of [2]. We will briefly recall the relevant material from this paper, rephrased to take into account maximization problems instead of minimization. A *c-smooth polynomial integer program* (or PIP) is a problem of the form *maximize* $p_0(x_1,\dots,x_n)$, subject to constraints $l_j \le p_j(x_1,\dots,x_n) \le u_j$, $x_i \in \{0,1\}$ *for* $i = \{1,\dots,n\}$ where each p_j is an n-variate polynomial of maximum degree d, and coefficients of each degree-ℓ monomial (term) are at most $c \cdot n^{d-\ell}$. Let *opt* denote the optimal value of a PIP. The fundamental result that we will use, Theorem 1.10 of [2], asserts that, for each $\delta > 0$, there exists an approximation algorithm that, in time $O(n^{\frac{1}{\delta^2}})$, computes a 0/1 assignment $< y_1,\dots,y_n >$ to the variables $< x_1,\dots,x_n >$ of a c-smooth PIP whose value is at least $opt - \delta n^d$. Moreover the assignment $< y_1,\dots,y_n >$ satisfies each linear constraint within an *additive* error of $O(\delta\sqrt{n\log n})$. Notice that $< y_1,\dots,y_n >$ is not necessarily a feasible solution.

Our goal is therefore to write the restriction of CORRELATION CLUSTERING where the number of sets in the computed partition is at most $8w_{max}/\varepsilon$ (that is a constant) as a quadratic PIP. The simplest formulation uses variables $x_{i,k}$ whose value is 1 if and only if the i-th element is in the k-th set of the partition, exploiting the fact that the quantity $\sum_k x_{i,k}x_{j,k}$ is equal to 1 if and only if the i-th and the j-th element are in the same set of the partition. Formally, we want to maximize $\sum_{i,j}(a_{i,j}\sum_k x_{i,k}x_{j,k} + b_{i,j}(1-\sum_k x_{i,k}x_{j,k}))$ subject to the constraints $\sum_k nx_{i,k} = n$ and $x_{i,k} \in \{0,1\}$.

Notice that it is trivial to compute the partition associated to each 0/1 assignment of $x_{i,k}$. It has already been shown in [5] that the optimal value is at least $\Theta(n^2)$, hence an approximation algorithm with additive error δn^2 is a PTAS. Also notice that only $O(n)$ variables appear in the above quadratic PIP, therefore the additive error is at most δn^2. The solution computed by the algorithm consists of at most $8w_{max}/\varepsilon$ sets, we still have to show that it is not too far from the optimum when the number of sets of the solution is unrestricted. In fact it is possible to prove that there is at most another εn^2 additional error, which leads to an overall an *additive* error of $2\varepsilon n^2$ of the optimum of CORRELATION CLUSTERING.

Remember that the assignment computed by applying Theorem 1.10 of [2] may violate each linear constraint at most by an additive error of $O(\varepsilon\sqrt{n\log n})$. Actually it is possible to show that no linear constraint is violated in our formulation. In fact for each linear constraint and for each 0/1 assignment, both the left-hand side and the right-hand side are multiple of n. Since $n \ge \varepsilon\sqrt{n\log n}$ for sufficiently large values of n, the 0/1 assignment must satisfy exactly all linear constraint.

The idea of the PTAS above can be applied also to the problem of MAX CONSENSUS CLUSTERING where there are at most a constant number of input partitions. In fact, given an instance of MAX CONSENSUS CLUSTERING consisting of k input partitions, let $a(i,j)$, $b(i,j)$ be respectively the number of input partitions where the i-th and the j-th elements are co-clustered (respectively are not co-clustered). Clearly $a(i,j)+b(i,j)=k$ for all i,j, which implies that the optimal value is at least $kn^2/8$, moreover $0 \leq a(i,j), b(i,j) \leq k$. Since k is at most a constant, even though $a(i,j)$ or $b(i,j)$ can be equal to zero, similarly to the case of CORRELATION CLUSTERING we can restrict ourselves to looking for a partition with a constant number of sets. Applying Theorem 1.10 of [2], gives a 0/1 assignment to the variables $x_{i,k}$ with an additive error εn^2, which immediately leads to a PTAS.

References

1. P. Alimonti and V. Kann. Some APX-completeness results for cubic graphs. *Theoretical Computer Science*, 237(1–2):123–134, 2000.
2. S. Arora, D. Karger, and M. Karpinski. Polynomial time approximation schemes for dense instances of $\mathcal{NP}$-hard problems. *Journal of Computer and System Sciences*, 58:193–210, 2000.
3. N. Ailon, M. Charikar, A. Newman. Aggregating Inconsistent Information: Ranking and Clustering. In *Proc. 37th Symposium on Theory of Computing (STOC 2005)*, pages 684 – 693, 2005
4. G. Ausiello, P. Crescenzi, V. Gambosi, G. Kann, A. Marchetti-Spaccamela, and M. Protasi. *Complexity and Approximation: Combinatorial optimization problems and their approximability properties*. Springer-Verlag, 1999.
5. N. Bansal, A. Blum, and S. Chawla. Correlation clustering. In *Machine Learning* 56(1-3): 89–113 (2004).
6. M. Charikar, V. Guruswami, and A. Wirth. Clustering with qualitative information. In *Proc. 44th Symp. Foundations of Computer Science (FOCS)*, pages 524–533, 2003.
7. E. D. Demaine and N. Immorlica. Correlation clustering with partial information. In *6th Workshop on Approximation Algorithms for Combinatorial Optimization Problems (APPROX)*, pages 1–13, 2003.
8. D. Emanuel and A. Fiat. Correlation clustering – minimizing disagreements on arbitrary weighted graphs. In *Proc. 11th European Symp. on Algorithms (ESA)*, pages 208–220, 2003.
9. V. Filkov and S. Skiena. Integrating microarray data by consensus clustering. In *Proc. 15th International Conference on Tools with Artificial Intelligence (ICTAI)*, pages 418–425, 2003.
10. V. Filkov and S. Skiena. Heterogeneous data integration with the consensus clustering formalism. In *Data Integration in the Life Sciences, First Intern. Workshop, (DILS)*, pages 110–123, 2004.
11. M. Grtschel and Y. Wakabayashi. A cutting plane algorithm for a clustering problem. *Mathematical Programming*, 45:52–96, 1989.
12. M. Krivanek and J. Moravek. Hard problems in hierarchical-tree clustering. *Acta Informatica*, 23:311–323, 1986.
13. C. Swamy. Correlation clustering: maximizing agreements via semidefinite programming. In *Proc. 15th Symp. on Discrete Algorithms (SODA)*, pages 526–527, 2004.
14. Y. Wakabayashi. The complexity of computing medians of relations. *Resenhas*, 3(3):323–349, 1998.

An Approximation Algorithm for Scheduling Malleable Tasks Under General Precedence Constraints

Klaus Jansen[1,*] and Hu Zhang[2,**]

[1] Institute of Computer Science and Applied Mathematics, University of Kiel, Olshausenstraße 40, D-24098 Kiel, Germany
kj@informatik.uni-kiel.de

[2] Department of Computing and Software, McMaster University, 1280 Main Street West, Hamilton, ON L8S 4K1, Canada
zhanghu@mcmaster.ca

Abstract. In this paper we study the problem of scheduling malleable tasks with precedence constraints. We are given m identical processors and n tasks. For each task the processing time is a function of the number of processors allotted to it. In addition, the tasks must be processed according to the precedence constraints. The goal is to minimize the makespan (maximum completion time) of the resulting schedule. The best previous approximation algorithm (that works in two phases) by Lepére et al. [18] has a ratio $3 + \sqrt{5} \approx 5.236$. We develop an improved approximation algorithm with a ratio at most $100/43 + 100(\sqrt{4349} - 7)/2451 \approx 4.730598$. We also show that our resulting ratio is asymptotically tight.

1 Introduction

In recent development of information technology, traditional super computers have been gradually replaced by systems with large number of standard units. All units have similar structure with certain processing ability in such a system [3]. To manage these resources, algorithms with satisfactory performance guarantee are needed. However, classical scheduling algorithms usually are not able to play such a role, mostly due to the large amount of communications between

* Research supported in part by the EU Thematic Network APPOL II, Approximation and Online Algorithms for Optimization Problems, IST-2001-32007, and by the EU Project CRESCCO, Critical Resource Sharing for Cooperation in Complex Systems, IST-2001-33135.

** This work was performed in part when this author was studying at the University of Kiel, Germany. Research supported in part by the DFG Graduiertenkolleg 357, Effiziente Algorithmen und Mehrskalenmethoden, by the EU Thematic Network APPOL II, Approximation and Online Algorithms for Optimization Problems, IST-2001-32007, by the EU Project CRESCCO, Critical Resource Sharing for Cooperation in Complex Systems, IST-2001-33135, by an MITACS grant of Canada, and by the NSERC Discovery Grant DG 5-48923.

X. Deng and D. Du (Eds.): ISAAC 2005, LNCS 3827, pp. 236–245, 2005.

units. There have been many models for this problem [2,9,14,22,24]. Among them scheduling *malleable tasks* [24] is an important and promising model. In this model, the processing time of a malleable task depends on the number of processors allotted to it. The influence of communications between processors allotted to the same task, synchronization and other overhead are included in the processing time.

We assume that the malleable tasks are linked by *precedence constraints*, which are determined in advance by the data flow among tasks. Let $G = (V, E)$ be a directed graph, where $V = \{1, \ldots, n\}$ represents the set of malleable tasks, and $E \subseteq V \times V$ represents the set of precedence constraints among the tasks. If there is an arc $(i, j) \in E$, then task J_j can not be processed before the completion of processing of task J_i. Task J_i is called a predecessor of J_j, while J_j a successor of J_i. Each task J_j can be processed on any number $l \in \{1, \ldots, m\}$ of identical processors, and the corresponding processing time is $p_j(l)$. The goal of the problem is to find a feasible schedule minimizing the makespan $C_{\max}$ (maximum completion time).

According to the usual behaviour of parallel tasks in practice, Blayo et al. [1] proposed a realistic model of malleable tasks under the following *monotonous penalty assumptions*:

Assumption 1: The processing time $p(l)$ of a malleable task J is non-increasing in the number l of the processors allotted, i.e., $p(l) \leq p(l')$, for $l \geq l'$;

Assumption 2: The work $W(l) = w(p(l)) = lp(l)$ of a malleable task J is non-decreasing in the number l of the processors allotted, i.e., $W(l) \leq W(l')$ for $l \leq l'$.

Assumption 1 indicates that if more processors are allotted to a malleable task, then the task can not run slower. Furthermore, Assumption 2 implies that the increase of processors allotted leads to an increasing amount of communication, synchronization and scheduling overhead.

In a schedule each task J_j has two associated values: the starting time τ_j and the number of processors l_j allotted to task J_j. A task J_j is called *active* during the time interval from its starting time τ_j to its completion time $C_j = \tau_j + p_j(l_j)$. A schedule is *feasible* if at any time t, the number of active processors does not exceed the total number of processors $\sum_{j:t\in[\tau_j,C_j]} l_j \leq m$ and if the precedence constraints $\tau_i + p_i(l_i) \leq \tau_j$ are fulfilled for all $i \in \Gamma^-(j)$, where $\Gamma^-(j)$ is the set of predecessors of task J_j.

Related Work: The problem of scheduling independent malleable tasks (without precedence constraints) is strongly $\mathcal{NP}$-hard even for only 5 processors [4]. Approximation algorithms for the problem of scheduling independent malleable tasks with a ratio 2 was addressed in [7,19]. This was improved to $\sqrt{3}$ by Mounié et al. [20], and further to 3/2 [21]. For the case of fixed m, Jansen and Porkolab proposed a PTAS [11]. If $p(l) \leq 1$, for arbitrary m an AFPTAS was addressed by Jansen [10]. Du and Leung [4] showed that scheduling malleable tasks with precedence constraints is strongly $\mathcal{NP}$-hard for $m = 3$. Furthermore, there is no polynomial time algorithm with approximation ratio less than 4/3, unless $\mathcal{P} = \mathcal{NP}$ [16]. If the precedence graph is a tree, a 4-approximation algorithm was developed in [17]. The idea of the two-phase algorithms was proposed initially in [17] and further used in [18] to obtain the previous best known approximation algorithm for general precedence constraints with a ratio $3 + \sqrt{5} \approx 5.236$. In addition, in [18] the ratio was improved to $(3 + \sqrt{5})/2 \approx 2.618$ if the precedence graph is a tree. More details on the problem of scheduling independent or precedence constrained malleable tasks can be found in [5].

In [23] the discrete *time-cost tradeoff problem* is studied. An instance of the discrete time-cost tradeoff problem is a *project* given by a finite set J of *activities* with a *partial order* $(J, \prec)$ on the set of activities. All activities have to be executed in accordance with the precedence constraints given by the partial order. Each activity $J_j \in J$ has a set of feasible *durations* $\{d_{j_1}, \ldots, d_{j_{k(j)}}\}$ sorted in a non-decreasing order, and has a non-increasing non-negative *cost* function $c_j : \mathbb{R}_+ \to \mathbb{R}_+ \cup \infty$, where $c_j(x_j)$ is the amount paid to run J_j with duration x_j. Skutella [23] showed that for a given deadline L (or budget $B - P$) and a fixed rounding parameter $\rho \in (0, 1)$, there exists an algorithm that computes a realization x for this problem such that $\hat{c}(x) \leq (B - P)/(1 - \rho)$ and $C_{\max} \leq L/\rho$, where $B - P$ is the optimal cost for the linear relaxation with a deadline L (respectively T is the optimal project length for the linear relaxation with a budget $B - P$). This problem is equivalent to the allotment problem of scheduling malleable tasks with precedence constraints with unbounded number of processors. The algorithm in [23] was used in [18] for solving the allotment problem approximately. We will also borrow ideas from [23] to develop our approximation algorithm.

Our Contribution: In this paper, we develop an improved approximation algorithm for scheduling malleable tasks with general precedence constraints. Our algorithm is based on the two-phase approximation algorithm in [17,18]. In the first phase we solve an *allotment problem* approximately. For a given set of malleable tasks, the goal of the allotment problem is to find an allotment $\alpha : V \to \{1, \ldots, m\}$ deciding the numbers of processors allotted to execute the tasks such that the maximum between both opposite criteria of critical path length and average work (total work divided by total number of processors) is minimized. In the second phase a variant of a list scheduling algorithm is employed to generate a new allotment and to schedule all tasks according to the precedence constraints. In [18], the allotment problem is formulated as a bicriteria version of the discrete time-cost tradeoff problem, which can be approximately solved by the algorithm in [23] together with a binary search strategy. The time-cost tradeoff problem corresponds to the scheduling problem with an unbounded number of processors and bounded total work. A rounding parameter $\rho = 1/2$ is fixed in their algorithm. We borrow the ideas of the algorithm in [23]. In our strategy, we construct a "reduced" instance of the time-cost tradeoff problem and further transform it into instances of the linear time-cost tradeoff problem with only two durations. Here we include two new constraints for the linear program such that the binary search procedure can be avoided, and a fractional solution to the allotment problem is obtained. In the rounding procedure, we do not fix the parameter $\rho = 1/2$. Instead, we leave it as an unspecified additional parameter for the second phase. In the second phase, the allotment parameter $\mu \in \{1, \ldots, \lfloor (m+1)/2 \rfloor\}$ is used to generate a new allotment such that no task is allotted a number of processors more than μ. The variant of the list scheduling algorithm runs with the resulting restricted allotment. Furthermore, we carefully study the rounding technique. We notice that a certain amount of work is not involved in the rounding procedure. By counting this term carefully, we obtain some new bounds depending on ρ for the total work of the rounded solution. Together with a rounding parameter μ employed in the second phase, we develop a min-max nonlinear program, whose optimal objective value is an upper bound on the approximation ratio by choosing appropriate values of ρ and μ. In fact this min-max nonlinear program measures the gap between the optimal fractional allotment solution and the length of the generated schedule. Next we analyze the nonlinear program to obtain optimum values for the parameters ρ and μ. Using $\rho = 0.43$ and $\mu = (93m - \sqrt{4349m^2 - 4300m})/100$ we obtain an improved approximation algorithm with a ratio at most $100/43 + 100(\sqrt{4349} - 7)/2451 \approx 4.730598$. The values of the above bound for small m's are listed and they are better than those of the algorithm in [18]. In addition, we show that asymptotically the best ratio is 4.730577 when $m \to \infty$. This indicates that our result is asymptotically tight. Recently, a 3.291919-approximation algorithm is proposed for the scheduling problem with an additional assumption that the work function is convex in processing time [13,25]. It is worth noting that their model is a special case of the model we study in this paper. Due to the limit of space we do not give proofs of all results in this version. We refer our full version [12,25] for details.

2 Approximation Algorithm

Our algorithm is based on the two-phase approximation algorithm in [18]. In the first phase of their algorithm, a $(2, 2)$-approximate solution to the bicriteria version of the discrete time-cost tradeoff problem is computed with a fixed rounding parameter $\rho = 1/2$. However, in the solution the case of small project length with large work can happen as there is no bound on the number of processors available. Thus a binary search procedure is conducted to find a solution such that the maximum of the corresponding project length and the average work (the total work divided by m) is minimized. We also use the idea of the discrete time-cost tradeoff problem. Different from their strategy, by including some additional constraints we formulate a linear program relaxation for the allotment problem to avoid the binary search procedure. In our algorithm we do not fix $\rho = 1/2$ as in [18]. In fact, we set ρ as a parameter in the second phase in order to enlarge the set of feasible solution and to find a better solution.

In the initialization step, we compute the values of the rounding parameter ρ and the allotment parameter μ depending on the input m (See Section 3 for the formulae).

In the first phase, we develop a linear program based on the approximation algorithm for discrete time-cost tradeoff problem in [23]. By rounding its fractional solution with the parameter $\rho \in [0,1]$ we are able to obtain a feasible allotment α' such that each task J_j is allotted l'_j number of processors.

We borrow the ideas of the approximation algorithm for the discrete time-cost tradeoff problem in [23] but not directly use it. For any malleable task J_j, its processing time $p_j(l)$ and work $W_j(l) = lp_j(l)$ on l processors fulfil Assumption 1 and 2. We also denote by $w(\cdot)$ the work function in processing time, i.e., $w_j(p_j(l)) = W_j(l)$. Denote by x_j the fractional duration of the allotment problem (or the processing time). Based on the ideas of the linear relaxation of the discrete time-cost tradeoff problem in [23], we need to solve the following linear program:

$$
\begin{array}{lll}
\min C = \max\{L, W/m\} & & \\
\text{s.t. } 0 \le C_j \le L, & \text{for all } j; & \\
\quad C_j + x_k \le C_k, & \text{for all } j \text{ and } k \in \Gamma^+(j); & \\
\quad x_j \le p_j(1), & \text{for all } j; & \\
\quad x_{j_i} \le x_j, & \text{for all } j \text{ and } i = 1,\ldots,m; & (1)\\
\quad 0 \le x_{j_i} \le p_j(i), & \text{for all } j \text{ and } i = 2,\ldots,m; & \\
\quad x_{j_1} = p_j(m), & \text{for all } j; & \\
\quad \hat{w}_j(x_j) = \sum_{i=1}^m \bar{w}_{j_i}(x_{j_i}), & \text{for all } j; & \\
\quad P = \sum_{j=1}^n p_j(1); & & \\
\quad \sum_{j=1}^n \hat{w}_j(x_j) + P \le W, & &
\end{array}
$$

where the "virtual" work function $\bar{w}_j(x_{j_m}) = 0$, and for all j and $i = 1,\ldots,m-1$:

$$\bar{w}_{j_i}(x_{j_i}) = [W_j(i+1) - W_j(i)](p_j(i) - x_{j_i})/p_j(i). \qquad (2)$$

Clearly, if $x_{j_i} = 0$ for all $i < l$ and $x_{j_i} = p_j(i)$ for all $i \ge l$, then $\hat{w}_j(x_j) = lp_j(l) - p(1)$ and $x_j = \max_i x_{j_i} = p_j(l)$. In (1), the two more constraints come from the following facts: In any schedule, any critical path length should be bounded by the makespan. In addition, the makespan is an upper bound on the average work (total work divided by m).

We apply the rounding technique similar to [23] for the fractional solution to (1). For any solution $x^*_{j_i} \in [0, p_j(i)]$, if $x^*_{j_i} < \rho p_j(i)$, we round it to $x_{j_i} = 0$; otherwise we round it to $x_{j_i} = p_j(i)$, where $\rho \in (0,1)$ is a rounding parameter to be determined later. With the rounded solution $\max_{1 \le i \le m} x_{j_i} \in \{p_j(m), \ldots, p_j(1)\}$ we are able to identify an l'_j such that $p_j(l'_j)$ equals to the rounded solution. This gives an allotment α' where each job J_j is allotted a number l'_j processors.

In the second phase, with the resulting allotment α' and the pre-computed allotment parameter μ, the algorithm generates a new allotment α and runs **LIST**, a variant of the list scheduling algorithm, in Table 1 (as proposed in [8,18]) and a feasible scheduling is delivered for the instance.

3 Analysis of the Approximation Algorithm

We shall show the approximation ratio of our algorithm. Denote by L, W, $C_{\max}$ and by L', W', $C'_{\max}$ the critical path lengths, the total works and the makespans

Table 1. Algorithm **LIST**

LIST (J, m, α', μ)
initialization: allot $l_j = \min\{l'_j, \mu\}$ processors to task J_j, for $j \in \{1, \dots, n\}$;
$SCHEDULED = \emptyset$;
if $SCHEDULED \neq J$ **then**
  $READY = \{J_j | \Gamma^-(j) \subseteq SCHEDULED\}$;
  compute the earliest possible starting time under α for all tasks in $READY$;
  schedule the task $J_j \in READY$ with the smallest earliest starting time;
  $SCHEDULED = SCHEDULED \cup \{J_j\}$;
end

of the final schedule delivered by our algorithm and the schedule corresponding to the allotment α' generated in the first phase, respectively. Furthermore, we denote by $C^*_{\max}$ the optimal objective value of (1), and L^*, W^* the (fractional) optimal critical path length and the (fractional) optimal total work in (1). It is worth noting that here $W^* = P + \sum_{j=1}^n \hat{w}_j(x^*_j)$ and $W' = \sum_{j=1}^n l'_j p_j(l'_j)$. We now estimate the bounds of L' and W'. According to the rounding in the first phase of our algorithm described in Section 2, if in the optimal solution of (1) $x^*_{j_i} < \rho p_j(i)$, it is rounded to $x_{j_i} = 0$. In this case its processing time is not increasing while the work may increase. According to the definition of the "virtual" work (2), $\bar{w}(x_{j_i}) = W_j(i+1) - W_j(i)$ and $\bar{w}(x^*_{j_i}) = [W_j(i+1) - W_j(i)][p_j(i) - x^*_{j_i}]/p_j(i) \geq [W_j(i+1) - W_j(i)](1-\rho)$. In the other case, when $x^*_{j_i} \geq \rho p_j(i)$, it is rounded to $x_{j_i} = p_j(i)$. Now the processing time may increase so the work is not increasing. So we have $x_{j_i} \leq x^*_{j_i}/\rho$. Combining these two cases we have that the following bounds:

$$L' \leq L^*/\rho \text{ and } (W' - P) \leq (W^* - P)/(1-\rho). \tag{3}$$

Denote by OPT the overall optimal makespan (over all feasible schedules with integral number of processors allotted to all tasks). It is obvious that

$$\max\{L^*, W^*/m\} \leq C^*_{\max} \leq OPT. \tag{4}$$

We define a piecewise work function $w_j(x)$ in fractional processing time x_j for task J_j based on (2) as follows: For $x_j \in [p_j(m), p_j(1)]$, if $l = \min\{k | k \in \{1, \dots, m\}, p_j(k) = x_j\}$, then $w_j(x_j) = l p_j(l)$. If $x_j \in (p_j(l+1), p_j(l))$ and $l = 1, \dots, m-1$, then

$$w_j(x_j) = \frac{p_j(l) p_j(l+1)}{p_j(l) - p_j(l+1)} - \frac{(l+1)p_j(l+1) - l p_j(l)}{p_j(l) - p_j(l+1)} x_j, \tag{5}$$

In allotments α and α', a task J_j is allotted l_j and l'_j processors, and their processing times are $p_j(l_j)$ and $p_j(l'_j)$, respectively. In the optimal (fractional) solution to (1), each task J_j has a fractional processing time x^*_j. We define the fractional number of processors allotted as $l^*_j = w_j(x^*_j)/x^*_j$.

Lemma 1. *For any malleable task J_j, if $p_j(l+1) \le x_j^* \le p_j(l)$ for some $l \in \{1,\ldots,m-1\}$, then $l \le l_j^* \le l+1$.*

Lemma 1 shows that l_j^* is well defined. We notice that l_j^* is just a notation and we only have knowledge that the real fractional number of processors corresponding to x_j^* should be in the interval $[l, l+1]$ if $p_j(l+1) \le x_j^* \le p_j(l)$. The notation l_j^* here fulfils this property and is convenient for our following analysis.

Same as in [18], in the final schedule, the time interval $[0, C_{\max}]$ consists of three types of time slots. In the first type of time slots, at most $\mu - 1$ processors are busy. In the second type of time slots, at least μ and at most $m-\mu$ processors are busy. In the third type at least $m - \mu + 1$ processors are busy. Denote the sets of the three types time slots by T_1, T_2 and T_3, and $|T_i|$ the overall lengths for $i \in \{1,2,3\}$. In the case that $\mu = (m+1)/2$ and m odd, $T_2 = \emptyset$. In other cases all three types of time slots may exist. Then we have the following bound on $|T_1|$ and $|T_2|$:

Lemma 2. $\rho|T_1| + \min\{\mu/m, \rho\}|T_2| \le C^*_{\max}$.

Proof. We construct a "heavy" directed path $\mathcal{P}$ in the final schedule, similar to [18,13,25]. For any job J_j in $T_1 \cap \mathcal{P}$, the processing time of the fractional solution to (1) increases by at most a factor $1/\rho$ according to (3). The processing time does not change in the second phase as in α' the job J_j is only allotted a number $l'_j \le \mu$ of processors such that it can be in the time slot of T_1. Therefore for such kind of jobs we have $p_j(l_j) = p_j(l'_j) \le x_j^*/\rho$. For any job J_j in $T_2 \cap \mathcal{P}$, there are two cases. In the first case, in α' a job J_j is allotted $l'_j \le \mu$ processors. This is same as the case before and we also have $p_j(l_j) \le x_j^*/\rho$. In the second case, in α' a job J_j is allotted $l'_j > \mu$ processors, and $l_j = \mu$. Then there are two subcases according to the solution to (1). In the first subcase, in the fractional solution to (1) there are $l_j^* \ge \mu$ processors allotted. Since μ is an integer, we have $l_j^* \ge \lfloor l_j^* \rfloor \ge \mu \ge l_j$. Then $l_j p_j(l_j) = W_j(l_j) \le W_j(\mu) \le W_j(\lfloor l_j^* \rfloor) \le w_j(x_j^*) = l_j^* x_j^* \le W_j(\lceil l_j^* \rceil)$ due to Assumption 2 and the definition of l_j^*. Because $l_j^* \le m$, and $w_j(x_j^*) = x_j^* l_j^* \ge p_j(l_j)l_j = W_j(l_j)$, it holds that $p_j(l_j) \le x_j^* l_j^*/l_j \le x_j^* m/\mu$. In the second subcase, in the fractional solution there are $l_j^* < \mu$ processors allotted to J_j. Then in the rounding procedure of the first phase the processing time must be rounded down from x_j^* to $p_j(l'_j)$ as only in this way the assumption that $l'_j > \mu$ of this case can be satisfied. Then in the second phase, J_j is allotted μ processors and from Assumption 2, $p_j(l_j)l_j \le p_j(l'_j)l'_j$. Since there are at most m processors allotted to J_j in α', we have $p_j(l_j) \le p_j(l'_j)l'_j/l_j \le p_j(l'_j)m/\mu \le x_j^* m/\mu$. Therefore for any job J_j in $T_2 \cap \mathcal{P}$, $p_j(l_j) \le x_j^* \max\{1/\rho, m/\mu\}$. With the construction of the directed path $\mathcal{P}$, it covers all time slots in $T_1 \cup T_2$ in the final schedule. In addition, in the schedule resulted from the fractional solution to (1), the jobs processed in T_1 in the final schedule contribute a total length of at least $\rho|T_1|$ to $L^*(\mathcal{P})$. In addition, the tasks processed in T_2 contribute a total length of at least $|T_2|\min\{\rho, \mu/m\}$ to $L^*(\mathcal{P})$. Since the critical path $L^*(\mathcal{P})$ is not more than the makespan $C^*_{\max}$ according to (4), we have proved the claimed inequality.

Furthermore, we consider the "fixed" cost $P = \sum_{j=1}^n p_j(1)$. Let us consider a schedule with makespan exactly the amount of P. The schedule is constructed by scheduling all tasks on the same processor. In this schedule for each task the processing time is $p_j(1)$ (the largest processing time) and obviously the makespan is equal to or larger than any feasible schedule. Thus we have

$$|T_1| + |T_2| + |T_3| \leq P. \tag{6}$$

For each task J_j, the number l of processors allotted in α is not more than the number l' of processors allotted in α'. According to Assumption 2, its work does not increase. Therefore $W \leq W'$. The total work W^* of the optimal solution to (1) is the sum of the "fixed" cost and the remaining unrounded part, i.e., $W^* = P + \sum_{j=1}^{n} \hat{w}_j(x_j^*)$. According to the bound for the rounded solution (3), we have that

$$W \leq W' \leq (W^* - P)/(1-\rho) + P = W^*/(1-\rho) - \rho P/(1-\rho). \tag{7}$$

Define the normalized overall length of the i-th type of time slots by $x_i = |T_i|/C^*_{\max}$ for $i = 1, 2, 3$, and $x_P = P/C^*_{\max}$. Thus we are able to obtain a min-max nonlinear program as follows where its optimal value is an upper bound on the approximation ratio r:

Lemma 3. *The optimal approximation ratio of our algorithm is bounded by the optimal objective value of the following min-max nonlinear program:*

$$\begin{aligned} \min_{\mu,\rho} \max_{x_1,x_2} \; & \frac{x_1(m-\mu)(1-\rho) + x_2(1-\rho)(m-2\mu+1) + m}{(m-\mu)(1-\rho)+1} \\ s.t.\; & \rho x_1 + \min\{\rho, \mu/m\} x_2 \leq 1; \\ & x_1 + x_2(\mu(1-\rho)+\rho) \leq m; \\ & x_1, x_2 \geq 0; \\ & \rho \in (0,1); \\ & \mu \in \{1, \ldots, \lfloor (m+1)/2 \rfloor\}. \end{aligned} \tag{8}$$

To solve (8) a complicated case study is needed. Furthermore, we can show [12,25] that to obtain the optimal ρ and μ we need to find roots to a polynomial with degree 6. In general it is impossible to solve it analytically. So in our algorithm, we fix $\rho = 0.43$ and $\mu = (93m - \sqrt{4349m^2 - 4300m})/100$. Then we have the following theorem:

Theorem 1. *There exists an algorithm for the problem of scheduling malleable tasks under precedence constraints with an approximation ratio*

$$r \leq \begin{cases} 2, & \text{if } m = 2; \\ \dfrac{100}{43} + \dfrac{100}{43} \dfrac{(43m-100)(57\sqrt{4349m^2-4300m} - 399m - 4300)}{139707m^2 - 174021m - 184900}, & \text{otherwise.} \end{cases}$$

Corollary 1. *For all $m \in \mathbb{N}$ and $m \geq 2$, the approximation ratio r is at most* $100/43 + 100(\sqrt{4349} - 7)/2451 \approx 4.730598$. *Furthermore, when $m \to \infty$, the upper bound in Theorem 1 tends to the above value.*

The corresponding bounds depending on m for $m = 2, \ldots, 33$ are listed as follows:

m	$\mu(m)$	$\rho(m)$	$r(m)$	m	$\mu(m)$	$\rho(m)$	$r(m)$	m	$\mu(m)$	$\rho(m)$	$r(m)$	m	$\mu(m)$	$\rho(m)$	$r(m)$
2	1	0.500	2.0000	10	3	0.430	3.9078	18	6	0.430	4.3249	26	8	0.430	4.4281
3	2	0.430	2.7551	11	4	0.430	4.0639	19	6	0.430	4.3083	27	8	0.430	4.4113
4	2	0.430	3.1080	12	4	0.430	4.0656	20	6	0.430	4.2938	28	8	0.430	4.3961
5	2	0.430	3.3125	13	4	0.430	4.0669	21	6	0.430	4.2880	29	8	0.430	4.4663
6	2	0.430	3.4458	14	4	0.430	4.1739	22	7	0.430	4.3857	30	9	0.430	4.4593
7	3	0.430	3.7507	15	5	0.430	4.2173	23	7	0.430	4.3685	31	9	0.430	4.4433
8	3	0.430	3.7995	16	5	0.430	4.2065	24	7	0.430	4.3531	32	9	0.430	4.4287
9	3	0.430	3.8356	17	5	0.430	4.1973	25	7	0.430	4.3897	33	10	0.430	4.4995

This is to compare with the result in [18], and gives an improvement of all m.

Furthermore, we can also analyze the asymptotic behaviour of the parameters and the ratio. The numerical results for $m = 2, \ldots, 33$ are listed below:

m	$\mu(m)$	$\rho(m)$	$r(m)$	m	$\mu(m)$	$\rho(m)$	$r(m)$	m	$\mu(m)$	$\rho(m)$	$r(m)$	m	$\mu(m)$	$\rho(m)$	$r(m)$
2	1	0.500	2.0000	10	3	0.420	3.8867	18	5	0.401	4.2579	26	8	0.477	4.3918
3	2	0.618	2.6180	11	4	0.500	4.0000	19	6	0.484	4.2630	27	8	0.469	4.3747
4	2	0.581	2.9610	12	4	0.500	4.0000	20	6	0.467	4.2520	28	8	0.440	4.3822
5	2	0.562	3.1717	13	4	0.462	4.0198	21	6	0.429	4.2837	29	8	0.414	4.4139
6	2	0.445	3.4139	14	4	0.408	4.1196	22	7	0.480	4.3466	30	9	0.475	4.4250
7	3	0.523	3.6617	15	5	0.490	4.1653	23	7	0.481	4.3276	31	9	0.456	4.4142
8	3	0.519	3.7104	16	5	0.491	4.1526	24	7	0.451	4.3257	32	9	0.431	4.4268
9	3	0.500	3.7500	17	5	0.441	4.1787	25	7	0.420	4.3581	33	9	0.409	4.4571

We can also show that when $m \to \infty$, $\rho^* \to 0.430991$, $\mu^* \to 0.270875m$ and the approximation ratio $r \to 4.730577$ (see [12,25]). In this way we conjecture that there exists a 4.730577-approximation algorithm for the problem of scheduling malleable tasks with precedence constraints. However, our algorithm already has a ratio asymptotically tight.

References

1. E. Blayo, L. Debrue, G. Mounié and D. Trystram, Dynamic load balancing for ocean circulation with adaptive meshing, *Proceedings of the 5th European Conference on Parallel Computing*, Euro-Par 1999, LNCS 1685, 303-312.
2. D. E. Culler, R. Karp, D. Patterson, A. Sahay, E.Santos, K. Schauser, R. Subramonian and T. von Eicken, LogP: A practical model of parallel computation, *Communications of the ACM*, 39 (11), 1996, 78-85.
3. D. E. Culler, J. P. Singh and A. Gupta, *Parallel computer architecture: A hardware/software approach*, Morgan Kaufmann Publishers, San Francisco, 1999.
4. J. Du and J. Leung, Complexity of scheduling parallel task systems, *SIAM Journal on Discrete Mathematics*, 2 (1989), 473-487.
5. P.-F. Dutot, G. Mounié and D. Trystram, Scheduling parallel tasks – approximation algorithms, *in Handbook of Scheduling: Algorithms, Models, and Performance Analysis, J. Y.-T. Leung (Eds.)*, CRC Press, Boca Raton, 2004.

6. D. R. Fulkerson, A network flow computation for project cost curves, *Management Science*, 7 (1961) 167-178.
7. M. Garey and R. Graham, Bounds for multiprocessor scheduling with resource constraints, *SIAM Journal on Computing*, 4 (1975), 187-200.
8. R. L. Graham, Bounds for certain multiprocessing anomalies, *Bell System Technical Journal*, 45 (1966), 1563-1581.
9. J. J. Hwang, Y. C. Chow, F. D. Anger and C. Y. Lee, Scheduling precedence graphs in systems with interprocessor communication times, *SIAM Journal on Computing*, 18(2), 1989, 244-257.
10. K. Jansen, Scheduling malleable parallel tasks: an asymptotic fully polynomial-time approximation scheme, *Proceedings of the 10th European Symposium on Algorithms*, ESA 2002, LNCS 2461, 562-573.
11. K. Jansen and L. Porkolab, Linear-time approximation schemes for scheduling malleable parallel tasks, *Proceedings of the 10th Annual ACM-SIAM Symposium on Discrete Algorithms*, SODA 1999, 490-498.
12. K. Jansen and H. Zhang, Improved approximation algorithms for scheduling malleable tasks with precedence constraints, Technical Report, http://www.informatik.uni-kiel.de/~hzh/malle.ps.
13. K. Jansen and H. Zhang, Scheduling malleable tasks with precedence constraints, *Proceedings of the 17th ACM Symposium on Parallelism in Algorithms and Architectures*, SPAA 2005, 86-95.
14. T. Kalinowski, I. Kort and D. Trystram, List scheduling of general task graphs under LogP, *Parallel Computing*, 26(9), 2000, 1109-1128.
15. J. E. Kelley, Critical path planning and scheduling: mathematical bases, *Operations Research*, 9 (1961) 296-320.
16. J. K. Lenstra and A. H. G. Rinnooy Kan, Complexity of scheduling under precedence constraints, *Operations Research* 26 (1978), 22-35.
17. R. Lepère, G. Mounié and D. Trystram, An approximation algorithm for scheduling trees of malleable tasks, *European Journal of Operational Research*, 142(2), 242-249 (2002).
18. R. Lepère, D. Trystram and G. J. Woeginger, Approximation algorithms for scheduling malleable tasks under precedence constraints, *International Journal of Foundations of Computer Science*, 13(4): 613-627 (2002).
19. W. Ludwig and P. Tiwari, Scheduling malleable and nonmalleable parallel tasks, *Proceedings of the 5th ACM-SIAM Symposium on Discrete Algorithms*, SODA 1994, 167-176.
20. G. Mounié, C. Rapine and D. Trystram, Efficient approximation algorithms for scheduling malleable tasks, *Proceedings of the 11th Annual ACM Symposium on Parallel Algorithms and Architectures*, SPAA 1999, 23-32.
21. G. Mounié, C. Rapine and D. Trystram, A 3/2-dual approximation algorithm for scheduling independent monotonic malleable tasks, *manuscript*.
22. G. N. S. Prasanna and B. R. Musicus, Generalised multiprocessor scheduling using optimal control, *Proceedings of the 3rd Annual ACM Symposium on Parallel Algorithms and Architectures*, SPAA 1991, 216-228.
23. M. Skutella, Approximation algorithms for the discrete time-cost tradeoff problem, *Mathematics of Operations Research*, 23 (1998), 909-929.
24. J. Turel, J. Wolf and P. Yu, Approximate algorithms for scheduling parallelizable tasks, *Proceedings of the 4th Annual Symposium on Parallel Algorithms and Architectures*, SPAA 1992, 323-332.
25. H. Zhang, Approximation Algorithms for Min-Max Resource Sharing and Malleable Tasks Scheduling, *Ph.D. Thesis*, University of Kiel, Germany, 2004.

A 1.5-Approximation of the Minimal Manhattan Network Problem

Sebastian Seibert[1,*] and Walter Unger[2,**]

[1] ETH Zürich, Department Informatik, ETH Zentrum, CH-8092 Zürich
sseibert@inf.ethz.ch
[2] RWTH Aachen, Lehrstuhl für Informatik I, D-52056 Aachen
quax@cs.rwth-aachen.de

Abstract. Given a set of points in the plane, the Minimal Manhattan Network Problem asks for an axis-parallel network that connects every pair of points by a shortest path under L_1-norm (Manhattan metric). The goal is to minimize the overall length of the network.

We present an approximation algorithm that provides a solution of length at most 1.5 times the optimum. Previously, the best known algorithm has given only a 2-approximation.

1 Introduction

Constructing connection networks of cheapest possible cost is an elementary task in network design. Here, we consider the *Minimal Manhattan Network Problem* which is defined as follows.

We are given a finite set of n points in the plane, and as distance measure we use the L_1-norm, also called Manhattan metric. A *Manhattan path* in the plane is a path that consists solely of axis-parallel line segments. A shortest Manhattan path connecting two points $p_1 = (x_1, y_1)$ and $p_2 = (x_2, y_2)$ will always be a staircase, and its length is the L_1-distance between p_1 and p_2, i.e. $|x_1 - x_2| + |y_1 - y_2|$. The task is to construct a network that connects each pair of points by a minimal Manhattan path. The goal is to minimize the overall length of the network.

Clearly there always exists a Manhattan network. For example, a complete grid on the given point set will always do. That is, we take the set of x- and y-coordinates $\{x_1, \ldots, x_n\}$, and $\{y_1, \ldots, y_n\}$ respectively, with minimal and maximal elements $x_{min}, x_{max}, y_{min}, y_{max}$. Now the complete grid consist of the vertical lines $\{(x_i, y) \mid y_{min} \le y \le y_{max}\}$, and of the horizontal lines $\{(x, y_j) \mid x_{min} \le x \le x_{max}\}$, where $1 \le i, j \le n$. However, this grid can be more expensive than the optimum at a rate of $\Theta(n)$ as pointed out already in [8].

Obviously, this problem has applications in VLSI design and related area. Note that the given problem can be seen as a special case of finding a spanner.

* Partially supported by SNF 200021109252/1: Solution quality and efficiency in discrete optimization.

** Partially supported by EU 6th Framework Programme, contract 001907 (DELIS).

X. Deng and D. Du (Eds.): ISAAC 2005, LNCS 3827, pp. 246–255, 2005.

The problem of finding cheap spanners has numerous applications, and it has been widely studied, especially metric subproblems [1,4,5,7,9].

Under any norm L_b, the problem to find a k-spanner in the plane is to find a network that connects each pair of the given point set by a path of length at most k times the distance of the points pair under the given norm. Here we are asking for a 1-spanner, a problem that under L_b-norm, $b \geq 2$, becomes trivial (the complete graph of direct connections in the plane is the minimal 1-spanner). Under L_1-norm however, there is a multitude of possible solutions, and very different minimal ones can exist.

Though it is not known whether the Manhattan network problem is $\mathcal{NP}$-hard, only approximative solutions have been found so far. Gudmundsson, Levcopoulos, and Narasimhan [8] gave the first approximation algorithms with ratio 4 and running time $\mathcal{O}(n^3)$, and a fast ($\mathcal{O}(n \log n)$) algorithm with ratio 8. Later, Kato, Imai, and Asano [10] gave a supposed 2-approximation algorithm with running time $\mathcal{O}(n^3)$, however, the proof turned out to be incomplete, as pointed out in [2,3]. Building on some ideas from [10], Benkert, Shirabe, Widmann, and Wolff [2,3] gave a 3-approximation algorithm with running time $\mathcal{O}(n \log n)$. Also in [2], exact solutions in exponential time were studied. Just recently Chepoi, Nouioua, and Vaxes presented a new 2-approximation algorithm [6].

In this paper, we devise a completely new approach and obtain a 1.5-approximation algorithm with running time $\mathcal{O}(n^3)$.

The remainder of the paper is structured as follows: In Section 2 we introduce some basic definition. The algorithm and its analysis are presented in Section 3.

2 Notation and Preliminary Remarks

As mentioned before, the complete grid on the given point set can be a very inefficient solution. However, it is easy to see that w.l.o.g., we may restrict to those solutions only which are part of this grid. In other words, suppose we have a solution that uses a line segment which is not part of a line on which at least one of the given points lies. Then such a line segment can be moved until it aligns with at least one of the given points, without increasing the overall costs.

A central though rather easy observation lies in the fact that we have to look at only some pairs of points in order to establish an admissible solution of our problem. Each pair of points p, q from the given set P spans an axis-parallel rectangle $R(p, q)$ having p and q as its corners.

First, we note that $R(p, q)$ may degenerate into a line. But then, this straight line, connecting p and q, will necessarily be part of every admissible solution. Consequently, these degenerate cases are not problems in themselves. They might influence other parts of the solution but only in the same way as placing other line segments will do. In the following, we will picture the problem in general without two or more points on the same line, and where necessary we will remark how these cases will be included.

We call $R(p, q)$ a *critical rectangle* if it contains no other point besides p and q. Now we may focus on those rectangles only. Since if there is another point r in $R(p,q)$, then already the shortest paths from p to r and from r to q together form a shortest path from p to q.

Remark 1. A set of axis parallel lines solves the Minimal Manhattan Network Problem on input point set P iff every pair $p,q \in P$ of points that form a critical rectangle $R(p,q)$ is connected by a shortest L_1-path.

Now let's have a look at the following basic situation which will be the central element of all considerations. Let a and b be two vertical neighboring points, i.e. in the set of y-coordinates, there exists no other point between the y-coordinates of a and b. We consider two segments of the horizontal lines where a and b lie, namely the upper and lower part of a grid cell, see Figure 1. Clearly, any solution (that uses only grid lines) uses one of the two line segments drawn bold.

Fig. 1. line segments between points a,b

The construction of a complete solution will essentially consist of making, for all such pairs of segments the choice which one to use. Obviously, choices on neighboring lines are not independent of each other. In general, we have two basic situations. We call a *block* a sequence of neighboring grid cells. Here, we describe only the placement of horizontal line segments, i.e. we look at a vertical sequence of cells. The placement of vertical line segments is completely analogue.

Lemma 1. *In a vertical block where on k consecutive horizontal lines separating the block cells, points are lying alternatingly left and right of the block, $\lfloor \frac{k}{2} \rfloor$ line segments are necessary and sufficient to connect those points.*

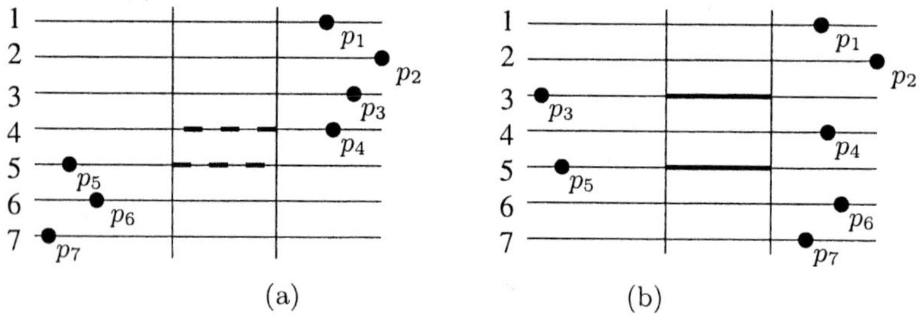

Fig. 2. Segments in a block

Proof. The situation is shown in Figure 2 for the cases of two (a) respectively five (b) alternating points. Note that for an odd number of points there is only one possible way to use the minimal number of segments.

As remarked above, we need at least one of two line segments for each consecutive pair of points. This immediately implies that $\lfloor \frac{k}{2} \rfloor$ is a lower bound on the number of used segments.

On the other hand, by choosing every other line segment, all requirements of using one out of two consecutive lines are fulfilled. And we can connect all points in question by using additionally only vertical line segments (as well as horizontal ones outside the block under consideration). □

3 The Approximation Algorithm

In this section, we will describe an algorithm that in a first phase basically just constructs sets of segments as described in Lemma 2, independently for the vertical and horizontal orientation. Consequently, the resulting set of line segments is not more expensive than an optimal solution but no solution in itself. The second phase of the algorithm will overcome this at the price of a moderate cost increase.

Algorithm 1.

Input: A set P of points in the plane.
1. Compute a set H of horizontal line segments.
2. Compute a set V of vertical line segments.
3. Chose the cheaper of the sets V, H and compute a solution S as an augmentation of that set H respectively V, by adding some locally optimal parts.
Output: S

Before showing how to implement this, let us first demonstrate why this is an 1.5-approximation algorithm.

Lemma 2. *For the sets H, V, S, constructed in Algorithm 1, the following holds.*

(a) Each admissible solution needs horizontal line segments of a total length at least the same as H.

(b) The analogue holds for V.

(c) S can be constructed using additional line segments of a total weight not more than an optimal solution.

As a consequence of (a) and (b), $H \cup V$ costs at most as much as an optimal solution. Together with (c) this gives the desired approximation ratio, as explained in more detail at the end of the section.

It suffices to describe Steps 1 and 3, since Step 2 is a complete analogue of Step 1. First, let us focus on Step 1, the construction of H.

Here, the essential point is that for each critical rectangle R, horizontal line segments are added inside R that cover the whole width of R. Conversely, each added line segment h will be necessary according to Lemma 1. This will immediately proof Lemma 2 (a).

As we will see, H, and analogously V, will have some other properties additionally that will be needed to proof Lemma 2 (c).

The Construction of H: To construct H, a sweep is performed over the point set P, from left to right. For each point q newly reached by this sweep, we describe how the horizontal line segments are continued, discontinued, or started to the right of q (up to the horizontal position of the next point).

To this end, we have to distinguish three basic cases, depending on the position of the next *vertical* neighbors, p and r, of q. Each case can split into sub-cases depending on what line segments exist to the left of q. For the moment, let us assume that there are no two points on the same grid line. We will deal with this exception later on.

As a first case assume that p and r both are to the left of q, as seen in Figure 3. Please note that grid lines 2, 3, and 4 are supposed to be neighbors. However, grid lines 1 and 5 need not be neighbors to 2 and 4, respectively. Rather on Line 1 lies the next point s above q among those that are to the right of q, and analogously, on Line 5 lies t, the next point below q, among those that are to the right of q.

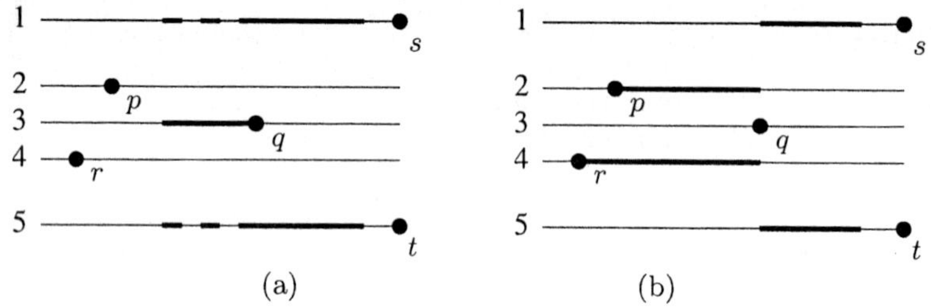

Fig. 3. New point to the right of its neighbors

Now, regardless of whether one or two horizontal segments on Lines 2 to 4 arriving at q are part of H, these are not continued beyond q. Rather, new segments are started (or potentially continued) on lines 1 and 5. These segments are immediately justified by the necessity to connect q as well as p and r with s and t. The creation of these lines may then influence the construction of H in that above Line 1, or below Line 5 switches may occur. These will be explained later on.

For the moment just note that if to the right of q there are no more points above or below, i.e. if s or t do not exist, then the corresponding segment of Line 1, respectively 5, will simply be not constructed.

Now assume that p and q are both to the right of q.

If two of the three lines 2-4 are used left of q (Figure 4 (a)) this can be the case only because they are needed to connect points left of q (on lines 1 and 5) to p, q, r. Then these segments will be continued toward p and q.

If just one or even no line out of 2-4 is used left of q, there are two possible continuations to the right of q, shown dashed versus dotted in Figure 4 (b).

If there are points like s and t left of q above and below it, there must be line segments in H to connect s with p and t with r. The choice of the dashed line

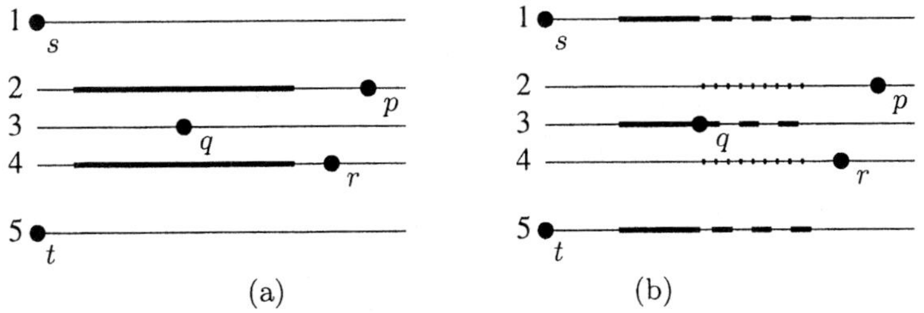

Fig. 4. New point to the left of its neighbors

segment means just to continue existing line segments, while the dotted solution might result in switches below and above q. We choose the latter one only if it saves line segments overall.

If no points exist to the left and above q, Line 1 does not exist, and we have to compare dashed and dotted solutions (two segments each) only with respect to a switch that might occur below Line 5. The case that there are no points to the left and below q is dealt with symmetrically. If no points are to the left of q at all, then q is the leftmost point of P, and we just start a single segment on Line 3 to the right of q.

Finally, assume that w.l.o.g. p is to the right of q, and r is to the left, see Figure 5.

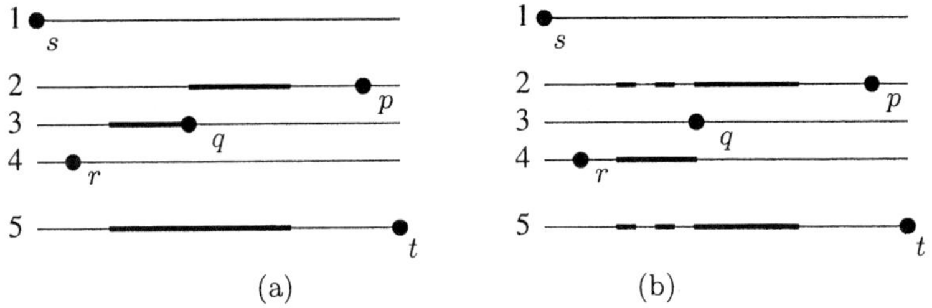

Fig. 5. New point with mixed neighbors

If there are any of the points right of q below it, we assume the next lower point t to be on Line 5. Consequently, we include in H a segment of Line 5 from q to the right. This segment is needed independent of q since r needs to be connected to t. (Note that it wouldn't be placed better on Line 4 since there is no point to the right and above to which such a placement could help connect r because those points will always be connected to r via q.) If there are no points like t, obviously, we don't need any segments to the right of q lower than Line 3.

It remains to connect q to p to which end we chose a segment from Line 2 since that might help save segments via a switch above. Also, we calculate if the segment on Line 5 (if existent) causes some switch below.

Now we have to state an important feature of the line segments constructed so far. Each line segment that was newly begun (i.e. not a continuation of a segment to the left) either originated from the point currently looked at, or it was started on a line on which there is a point to the right of it. Also, line segments from the left are never discontinued if there is a point more to the right on their line. In other words, line segments are either started from a point rightwards, or they are started left of a point, in which case they always continue at least up to that point. Recalling that V is constructed analogously top-down, we can summarize the properties of H and V as follows.

Remark 2. (a) Each line segment in H and V has a point from P on it.
(b) If p and q are on vertically consecutive lines, and p left of q, there is in H a line segment from p rightwards and one from q leftwards that either overlap or end at points above each other. (One may be empty if the other covers the whole distance.)
(c) If p and q are on horizontally consecutive lines, and p above q, there is in V a line segment from p downwards and one from q upwards that either overlap or end at points beside each other. (One may be empty if the other covers the whole distance.)

When we show now how switches are done, we will make sure that the above features of H and V never are violated.

A switch: In the above considerations, we have shown how a point influences the construction of line segments in its immediate vicinity. Now we consider its further influence. Due to symmetry, we only need to study the situation below q, see Figure 6.

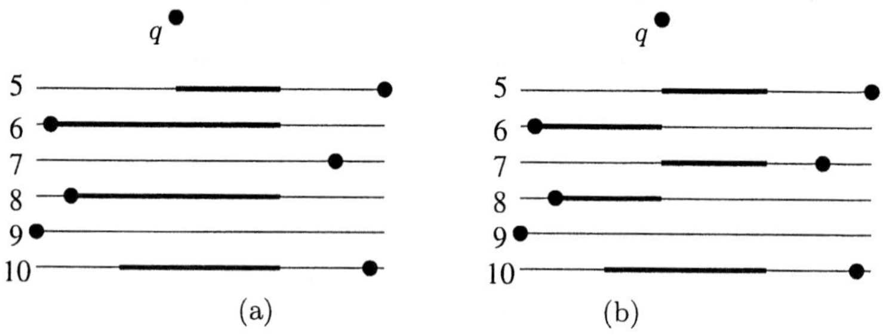

Fig. 6. A switch

Assume we have started a new segment on Line 5, and that on the lines below, all segments are simply continued rightwards, see Figure 6 (a). Then, in the alternating points situation from Line 5 to 10, we would have more lines than necessary (see Lemma 1). Consequently, we have to switch lines as shown in Figure 6 (b). That is, we start new segments on those lines where there is a point to the right.

Multiple points on the same line: We remark that two or more points on the same line rather simplify the problem to solve in that they create a mandatory line segment between them, i.e. one that needs to be included in every admissible solution. Consequently, we just obtain a few more cases that can be dealt with similarly or even simpler than those above. Due to lack of space we defer these to the full paper.

The construction of solution S: Assume that V is the cheaper one of sets V and H. Following Remark 1, we will have to construct shortest paths for critical rectangles only, in order to obtain a solution S by augmenting V.

Let us look at points p and q forming a critical rectangle $R(p, q)$. From Remark 2 (b), we can deduce that there always exist in H line segments between p as shown bold in Figure 7 (a), where s may be identical with p and t with q. To see this, just follow the vertical order of points from p down to q. All of them except p and q need to be outside of $R(p, q)$ since that rectangle is critical. Consequently there exist a point s left of $R(p, q)$ (or $s = p$) whose successor t is right of $R(p, q)$ (or $t = q$). Now the shown pattern exists immediately by Remark 2 (b).

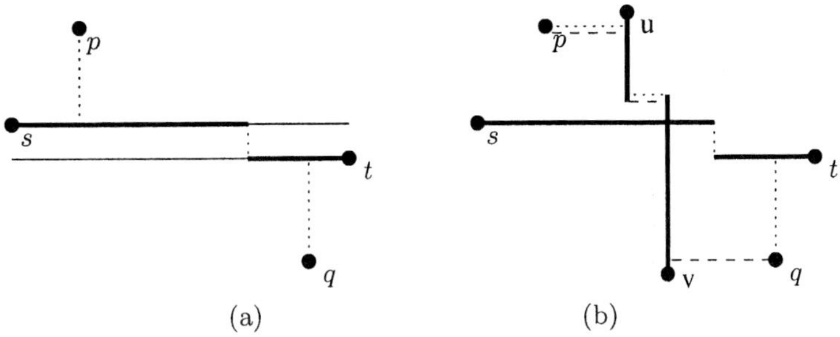

Fig. 7. Critical rectangles with line segments from $H \cup V$

Please note that the dotted staircase from p to q may consist of fewer parts if $s = p$ or $t = q$, or if the jump in the connection from s to t is outside $R(p, q)$. But the staircase has never more parts, that is, there is always a horizontal connection through $R(p, q)$ consisting of at most two parts.

One fact we have proven already by this is that we can always obtain an admissible solution by adding vertical lines only.

When we apply the same argument to V, we obtain in general a situation as shown in Figure 7 (b). In order to obtain a solution, one has to add some additional line segments. Two possible solutions are shown by dotted respectively dashed lines.

Now if the depicted points were the only ones to be considered, we could generate a solution by moving and doubling lines from H, see Figure 8 (a).

However, in general several critical rectangles may overlap. That means we would have, between v and t, a set of points that all need to be connected to

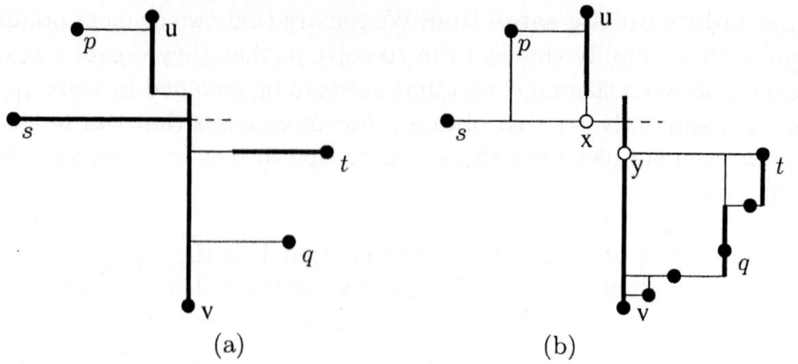

Fig. 8. Constructing a solution from H

points left of v and above t. But these points form an anti-diagonal, i.e. they lay top-right and bottom-left of each other. (Otherwise they would not all be part of a critical rectangle with a point in the upper left part.) And they need only be connected to a point y which is the closest up and left of them on a V-segment, see Figure 8 (b). Note that x and y are helper points, not elements of P.

Now we have a sub-problem that can be solved optimally by a dynamic programming approach: Connect a set of points that are bottom-left respectively top-right of each other to a point that is top-left of all of them, including connections between those points themselves. Note that parts of V may be existing already as part of the solution, as drawn bold in Figure 8 (b).

The important point is that solving all sub-problems of this type costs all in all not more than the optimal solution, and V (respectively H) is chosen to be the cheaper one of H, V, i.e. costs at most half of an optimal solution. This gives the desire result.

Theorem 1. *Algorithm 1 generates in time $\mathcal{O}(|P|^3)$ a solution S with length at most 1.5 times the optimum.* □

To conclude this section, we describe more precisely how the dynamic programming approach is used here. However, for a complete exposition we have to defer to the full paper due to space restrictions.

The problem to solve is to connect a set of points $p_1, \ldots, p_k$ to each other and to one specified point $z_{1,k}$. We assume that the x and y coordinates each are ordered, say $x_1 \leq x_2 \leq \ldots \leq x_k$ and $y_1 \leq y_2 \leq \ldots \leq y_k$ to match the above picture. (All other cases are dealt with analogously). The specified point for any subsequence of these is the left upper limit, i.e. $z_{i,j} = (x_i, y_j)$ for $1 \leq i \leq j \leq k$.

Now we can obtain an optimal solution by dynamic programming.

First we note that for singleton subproblems, nothing is to be done since $z_{i,i} = p_i$, i.e. the solutions for that subproblem has cost zero.

For any larger subsequence $p_i, \ldots, p_j$, we can compare all possibly optimal solutions. For each subdivision of the set into $p_i, \ldots, p_l$ and $p_{l+1}, \ldots, p_j$, a solution is obtained by connecting $z_{i,j}$ with both $z_{i,l}$ and $z_{l+1,j}$, as well as p_l with

p_{l+1}, together with the previously computed solutions of the two sub-problems. Now for $p_i, \ldots, p_j$ the cheapest of all computed solutions is taken.

The optimality comes from the fact that we consider after all every possibility to connect $z_{i,k}$ with all points in a tree-like manner. Since every admissible solution needs to contain such a tree (or else some connection would be missing), we include in the search an optimal solution.

Additionally, we have to remark that in our construction, we need to connect the point $z_{1,k}$ with its counterpart to the upper left. Part of this is done by using the segments from V, but a horizontal line segment is needed to connect these. However since it is created in a part where there are no line segments above or below it that where part of the sub-problem solutions, and we know that some segment is needed here, we can account for this when we estimate that the cost of the sub-problems together with this segment are not more expensive than an optimal solution.

Finally we estimate the running time of the algorithm. The sweeps in Steps 1 and 2 can be implemented in time $\mathcal{O}(|P|^2)$ while the dynamic programming in step three is cubic in the number of involved points, that is $\mathcal{O}(|P|^3)$ in the worst case.

References

1. I. Althöfer, G. Das, D. Dobkin, D. Joseph, J. Soares, On Sparse Spanners of Weighted Graphs, *Discrete Comput. Geoam.* 9 (1993), 81–100.
2. M. Benkert, T. Shirabe, A. Wolff, The Minimum Manhattan Network Problem—Approximations and Exact Solution. In *Proc. 20th European Workshop on Computational Geometry (EWCG'04)*, 209-212.
3. M. Benkert, F. Widmann, A. Wolff, The Minimum Manhattan Network Problem: A Fast Factor-3 Approximation. In *Proc. 8th Japanese Conf. on Discrete and Computational Geometry (JCDCG'04)*, 85–86.
4. B. Chandra, G. Das, G. Narasimhan, J. Soares, New Sparseness Resuls on Graph Spanners. *Internat. J. Comput. Geom. Appl.* 5 (1995), 125–144.
5. D. Chen, G. Das, M. Smid, Lower bounds for computing geometric spanners and approximate shortest paths. *Discrete Applied Math.* 110 (2001), 151–167.
6. V. Chepoi, K. Nouioua, Y. Vaxes, A rounding algorithm for approximating minimum Manhattan networks. *8th International Workshop on Approximation Algorithms for Combinatorial Optimization Problems, APPROX 2005* Springer Lect. Notes Comp. Sci. 3624 (2005)
7. G. Das, G. Narasimhan, A Fast Algorithm for Constructing Sparse Euclidian Spanners. *Internat. J. Comput. Geom. Appl.* 7 (1997), 297–315.
8. J. Gudmundsson, C. Levcopoulos, G. Narasimhan, Approximating a Minimum Manhattan Network. *Nordic J. Computing* 8 (2001), 219–232.
9. J. Gudmundsson, C. Levcopoulos, G. Narasimhan, Fast Greedy Algorithms for Constructing Sparse Geometric Spanners. *SIAM J. Computing* 31 (2002), 1479–1500.
10. R. Kato, K. Imai, T. Asano, An Improved Algorithm for the Minimum Manhattan Network Problem. In P. Bose., P. Morin (eds.) *Algorithms and Computation, 13th International Symposium, Proc. ISAAC '02* Springer Lect. Notes Comp. Sci. 2518 (2002), 344–356.

Hardness and Approximation of Octilinear Steiner Trees

Matthias Müller-Hannemann[1] and Anna Schulze[2]

[1] Technische Universität Darmstadt, Department of Computer Science, Hochschulstraße 10, 64289 Darmstadt, Germany
muellerh@algo.informatik.tu-darmstadt.de
http://www.algo.informatik.tu-darmstadt.de

[2] Zentrum für Angewandte Informatik Köln, Weyertal 80, 50931 Köln, Germany
schulze@zpr.uni-koeln.de

Abstract. Given a point set K of terminals in the plane, the octilinear Steiner tree problem is to find a shortest tree that interconnects all terminals and edges run either in horizontal, vertical, or $\pm 45°$ diagonal direction. This problem is fundamental for the novel octilinear routing paradigm in VLSI design, the so-called X-architecture.

As the related rectilinear and the Euclidian Steiner tree problem are well-known to be NP-hard, the same was widely believed for the octilinear Steiner tree problem but left open for quite some time. In this paper, we prove the NP-completeness of the decision version of the octilinear Steiner tree problem.

We also show how to reduce the octilinear Steiner tree problem to the Steiner tree problem in graphs of polynomial size with the following approximation guarantee. We construct a graph of size $O(\frac{n^2}{\varepsilon^2})$ which contains a $(1+\varepsilon)$–approximation of a minimum octilinear Steiner tree for every $\varepsilon > 0$ and $n = |K|$. Hence, we can apply any α-approximation algorithm for the Steiner tree problem in graphs (the currently best known bound is $\alpha \approx 1.55$) and achieve an $(\alpha+\varepsilon)$- approximation bound for the octilinear Steiner tree problem. This approximation guarantee also holds for the more difficult case where the Steiner tree has to avoid blockages (obstacles bounded by octilinear polygons).

Keywords: octilinear Steiner trees, NP-completeness, VLSI design, approximation algorithms, blockages.

1 Introduction

Background and motivation. In recent years there has been strong and growing interest in a new routing paradigm in VLSI design: octilinear routing, the so-called X-architecture [1]. In addition to vertical and horizontal wires, octilinear routing allows wiring in 45- and 135-degree directions. Compared to traditional and state-of-the-art rectilinear (Manhattan) routing, such a technology promises clear advantages in wire length and via reduction. As a consequence a significant chip performance improvement and power reduction can be obtained (with estimations being in the range of 10% to 20% improvement) [2, 3]. To

X. Deng and D. Du (Eds.): ISAAC 2005, LNCS 3827, pp. 256–265, 2005.

enable such a technology, novel algorithmic approaches for the construction of octilinear Steiner trees are needed. An *octilinear Steiner tree* is a tree that interconnects a set of points (*terminals*) in the plane with minimum length such that every line segment uses one of the four given orientations.

Even more general routing architectures are obtained if a fixed set of uniformly oriented directions is allowed. For an integral parameter $\lambda \geq 2$, consecutive orientations are separated by a fixed angle of π/λ. A *λ-geometry* is a routing environment in which every line segment uses one of the given orientations. Manhattan routing can then be seen as the special case $\lambda = 2$ and the X-architecture as the case $\lambda = 4$. A Steiner minimum tree in a λ-geometry is called λ-SMT.

In this paper we focus on the octilinear case (although most of our results can be generalized to arbitrary $\lambda \geq 2$). We study approximation algorithms for the octilinear Steiner tree problem with and without obstacles. The rectilinear and the Euclidean Steiner tree problem have been shown to be NP-hard in [4] and [5], respectively. It is widely believed that the Steiner tree problem is NP-hard for every fixed λ (although this question seems to be open [6]). Here, we present the proof that the octilinear Steiner tree problem is indeed NP-hard.

Blockages. In VLSI design routing is often restricted by the presence of blockages (or *obstacles*) which exclude certain areas for possible interconnections. Throughout this paper, an *obstacle* is a connected region in the plane bounded by a simple polygon. For a given set of obstacles $\mathcal{O}$ we require that the obstacles be disjoint, except for possibly a finite number of common points. If all boundary edges of an obstacle are rectilinear, we call such an obstacle a *rectilinear obstacle.* Analogously, if all obstacle edges lie within the 4-geometry, such an obstacle is called *octilinear obstacle.* In practice, obstacles are caused by preplaced macros or other circuits and can be assumed to be rectilinear.

Previous work. It is fairly easy to see that the approximation schemes of Arora [7] and Mitchell [8] are also applicable to the octilinear Steiner tree problem (without obstacles). Rao and Smith [9] even improved the running time of a $(1+\varepsilon)$–approximation to $O(n \log n)$. Unfortunately, the hidden constants of the asymptotic running time grow exponentially depending on ε. Hence, in spite of its theoretical importance, the practical value of these approximation schemes might be limited. Heuristics have been proposed by Kahng et al. [10] and Zhu et al. [11]. Exact approaches to the octilinear Steiner tree problem have been developed by Nielsen, Winter and Zachariasen [12] and Coulston [6]. Nielsen et al. report the exact solution to a large instance with 10000 terminals within two days of computation time. However, we are not aware of exact approaches or approximations in the presences of obstacles.

Transformation to Steiner tree problem in graphs. For rectilinear Steiner tree problems for point sets in the plane the most successful approaches are based on transformations to the related Steiner tree problem in graphs. Given a connected graph $G = (V, E)$, a length function ℓ, and a set of terminals $S \subseteq V$, a *Steiner tree* is a tree of G containing all vertices of S. A Steiner tree T is a

Steiner minimum tree of G if the length of T is minimum among all Steiner trees. The best available approximation guarantee for the Steiner problem in general graphs is $\alpha = 1 + \frac{\ln 3}{2} \approx 1.55$, obtained by Robins and Zelikovsky [13].

Given a finite point set K in the plane, the so-called *Hanan grid* [14] is obtained by constructing a vertical and a horizontal line through each point of K. The importance of the Hanan grid lies in the fact that it contains a rectilinear Steiner minimum tree. An implementation by Althaus, Polzin and Daneshmand [15] is the currently strongest available exact approach for both the Steiner tree problem in graphs and the rectilinear Steiner tree problem.

Du and Hwang [16] generalized the Hanan grid construction to λ-geometries. They define grids $G_k(K)$ recursively in the following way. For an instance with point set K, $G_0(K) = K$. The grid $G_1(K)$ is constructed by taking λ (infinite) lines with orientations $\pi/\lambda, 2\pi/\lambda, \ldots, (\lambda - 1)\pi/\lambda, \pi$ for each point of K. The k-th grid $G_k(K)$ for $k > 1$ is constructed from the $(k-1)$-th grid by adding for each intersection point x of lines in $G_{k-1}(K)$ additional lines through x with orientations $\pi/\lambda, 2\pi/\lambda, \ldots, (\lambda - 1)\pi/\lambda, \pi$. Lee and Shen [17] showed that for every instance of the Steiner tree problem in a λ-geometry with $\lambda \in \mathbb{N}_{\geq 2}$, there is a minimum λ-Steiner tree which is contained in $G_{n-2}(K)$. This result has been strengthened for octilinear Steiner trees by Lin and Xue [18]. They showed that a minimum octilinear Steiner tree is already contained in the grid $G_{(\lceil 2n/3 \rceil - 1)}(K)$. Unfortunately, the graph $G_k(K)$ has $O(n^{2^k})$ vertices and edges. Hence, in general an optimal solution requires an exponentially large graph.

It is therefore an interesting open question which approximation guarantee for the octilinear (or λ–) Steiner tree problem can be achieved if one works with a graph $G_k(K)$ for some fixed constant k. Some partial answers to this question are obvious. Since $G_1(K)$ contains a shortest path between any pair of terminals it also contains the solution obtained from the minimum spanning tree heuristic to approximate the Steiner minimum tree. Therefore, its performance guarantee cannot be worse than the Steiner ratio. The *Steiner ratio* is the smallest upper bound on the ratio between the length of a minimum spanning tree and the length of a Steiner minimum tree. The Steiner ratio in the octilinear case is $\frac{4}{2+\sqrt{2}}$ [19, 20]. This implies that $G_1(K)$ contains a solution which is not more than about 17.15% above the minimum. In this paper, we show how to modify $G_1(K)$ so that we can derive stronger approximation guarantees. For any $k \in \mathbb{N}$ we construct a graph of size $O(k^2 n^2)$, which contains a $(1 + \frac{1}{k})$–approximation.

Our contribution. We summarize the main results of this paper:

- We establish the NP-completeness of the decision version of the octilinear Steiner tree problem.
- For a given set of n terminals in the plane and for every $\varepsilon > 0$ we construct a graph of size $O(\frac{n^2}{\varepsilon^2})$ which contains a $(1+\varepsilon)$–approximation of a minimum octilinear Steiner tree.
- If α denotes the approximation guarantee of an algorithm for the Steiner tree problem in graphs, then we achieve an $(\alpha+\varepsilon)$–approximation guarantee for the octilinear Steiner tree problem with or without blockages.

Overview. The remaining part of the paper is organized as follows. In Section 2 we state some basic definitions and facts about octilinear Steiner trees. Afterwards, we sketch our NP-completeness proof. Then, in Section 4, we derive our approximation for the case without blockages. Finally, we briefly point out why the same method also works in the presence of blockages. Several proofs had to be omitted due to space restrictions. For details we refer to a preprint of the full paper available at `http://www.algo.informatik.tu-darmstadt.de/muellerh/`.

2 Basic Definitions and Facts for Octilinear Steiner Trees

In this section we recall some basic definitions and known facts about octilinear Steiner trees which will be used in the later analysis of our approach, see for example [17].

Property 1. The number of Steiner points for a Steiner tree on n terminals is at most $n-2$.

Property 2. The degree of any Steiner point is either three or four. There exists a Steiner minimum tree such that every degree-4 Steiner point is adjacent to four terminals which form a cross.

Property 3. There exists an octilinear Steiner minimum tree T_{opt} such that the three angles around a degree-3 Steiner point are $\frac{\pi}{2}, \frac{3\pi}{4}, \frac{3\pi}{4}$ (in some order).

A Steiner tree is a *full* Steiner tree if all its terminals are leaves. Any Steiner tree can be decomposed into its full components.

Property 4 ([21]). Given a set of terminals K such that every Steiner minimum tree is a full Steiner tree, there is a Steiner minimum tree T_{opt} such that all but at most one edge are straight edges. The latter one may bend once.

3 NP-Completeness of the Octilinear Steiner Tree Problem

In this section we sketch the proof of the fact that the decision version of the octilinear Steiner tree problem is NP-complete. We have the following decision problem:

Problem: Octilinear Steiner tree decision problem
Instance: A set K of terminals with integral coordinates in the plane and a number $L \in \mathbb{N}$.
Task: Is there an octilinear Steiner tree T with $l(T) \leq L$?

At a first glance, it might not even be clear whether the decision version of the octilinear Steiner tree problem belongs to the class NP, since the distance between terminals and/or Steiner points may be irrational. In sharp contrast to the Euclidian version where this question is still open, we can prove membership in NP for the octilinear case.

Lemma 1. *The decision version of the octilinear Steiner tree problem belongs to the class NP.*

Proof. To show membership in NP, we need a certificate which a nondeterministic algorithm can guess and we can verify in polynomial time. In this case, an appropriate certificate is the topology of an optimal tree T. (In this context, *topology* means a graph which includes the terminals and Steiner points as vertices and specifies the connections between these vertices as edges.)

From the topology of a tree, one can compute an optimal realization in the plane in linear time [22]. Using our assumption that all terminals have integral coordinates, it is easy to see that all Steiner points have rational coordinates (namely by traversing the tree from its leaves). Moreover, the encoding length of each Steiner point does not become too large. If s denotes the maximum number of bits to store the coordinates of some input terminal, then the encoding length of each Steiner point is upper bounded by $O(s + |K|)$. This follows from the fact that all line segments are either horizontal, vertical or have slopes ± 1. Hence, we can express the length of each tree edge e by $\ell(e) = a_e + b_e \cdot \sqrt{2}$, where a_e and b_e are rational numbers of polynomial size with respect to the input. The length of the Steiner tree can be evaluated as $\ell(T) = a + b\sqrt{2}$, where $a = \sum_{e \in T} a_e$ and $b = \sum_{e \in T} b_e$. Hence, $\ell(T) \leq L$ if and only if $2 \leq \left(\frac{L-a}{b}\right)^2$. □

Theorem 1. *The decision version of the octilinear Steiner tree problem is NP-complete.*

To prove this theorem, we basically use the same idea of a reduction as Garey, Graham and Johnson [5] provided in their hardness proof for the Euclidean Steiner tree problem. The main difference lies in the proof that this reduction is correct. In the remainder of this section, we briefly sketch the construction and point out the technical differences to the Euclidean case in the proof of its correctness. A detailed formal description and the full proof are deferred to the journal version of this extended abstract due to space restrictions. The problem we reduce to the octilinear Steiner tree decision problem is that of EXACT COVER BY 3-SETS:

Problem: Exact cover by 3-sets
Instance: A family $\mathcal{F} = \{F_1, F_2, \ldots, F_t\}$ of 3-element subsets of a set F of $3n$ elements. Without loss of generality let $F = \{1, 2, \ldots, 3n\}$.
Task: Is there a subfamily $\mathcal{F}' \subseteq \mathcal{F}$ such that distinct elements of $\mathcal{F}'$ are disjoint and $\bigcup_{F_i \in \mathcal{F}'} F_i = F$?

The 3-dimensional matching problem shown to be NP-complete in [23] is a special case of the problem EXACT COVER BY 3-SETS. Therefore, EXACT COVER BY 3-SETS is also NP-complete.

The main difficulty in the NP-completeness proof is due to the problem that it is hard to argue about the optimality of some Steiner tree unless we have very few terminals or very restricted locations for them. Hence, a reduction requires gadgets of very small size. As gadgets have to be combined with each other, we

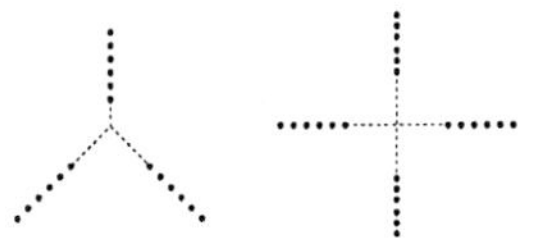

Fig. 1. Basic gadgets: terminals are dots; possible locations of Steiner points which are not excluded by probes are displayed with dashed lines

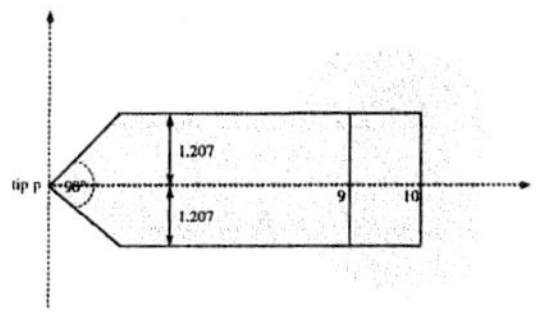

Fig. 2. The region of a probe P with tip p is the shaded area

would like to have that the possible configurations of optimal Steiner trees can easily be enumerated for each subset of terminals contained in the gadget. The overall configuration is composed by gadgets of rows of terminals which meet in "triangles" or "squares", see Fig. 1. Gadgets of these types are connected by long rows of terminals. Hence, they are placed far enough from each other, so that they do not directly mutually interact. In contrast, adjacent terminals of the same row are relatively near to each other, they have a distance of at most 1/10. This has the important effect that there is a spanning tree such that each edge has length at most 1. For a tree T denote by $m(T)$ the maximum length of some edge in T. From this property, we can conclude that also every edge in an optimum Steiner tree T_{opt} cannot be longer than 1, that is $m(T_{opt}) \leq 1$. The composition of gadgets is the same as in the proof for the Euclidian case [5].

A fundamental idea in the proof is to restrict the possible locations of Steiner points to certain so-called *active regions*. To exclude other regions for Steiner points, we use a discrete version of the concept of so-called *probes* which are regions with a geometric shape as in Fig. 2. The central node of a probe is called the *tip* of the probe. We say that a probe P is *valid* if it is rotated by an integral multiple of $\pi/4$ around its tip. In the Euclidean case, probes can be rotated by any angle and have a slightly different shape.

Lemma 2. *If p is a Steiner point of an octilinear Steiner minimum tree with $m(T_{opt}) \leq 1$, then there must be at least one terminal located inside every valid probe P with tip p.*

These properties will enable us to conclude that any two terminals which are at most 1/10 apart from each other must be connected by an edge in any minimum Steiner tree. Hence, the whole combinatorial difficulty lies in the problem how to connect these components to a tree for the overall configuration of terminals.

Given a configuration of terminals which encode an instance $\mathcal{F}$ of EXACT COVER BY 3-SETS, one has to show that an optimal Steiner tree for this configuration does not exceed a value $L = L(\mathcal{F})$ if and only if there is an exact cover. The value of L depends only on parameters t and n of $\mathcal{F}$. Again one can argue along the lines of the proof in [5]. However, compared to the Euclidian case, optimal subtrees inside active regions have different lengths. Therefore, it is crucial to note that certain inequalities about the relative lengths of such subtrees remain valid.

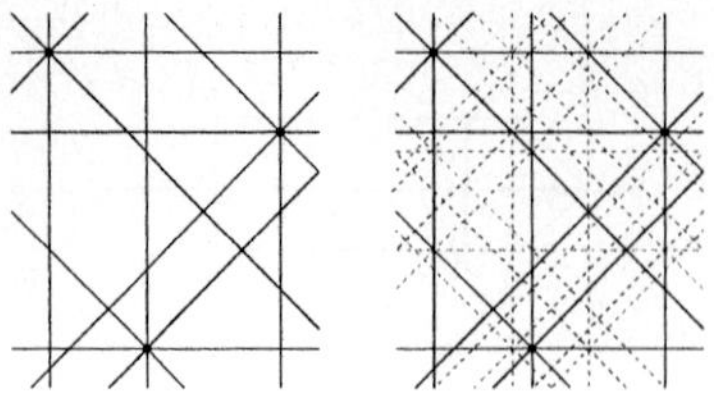

Fig. 3. Example of the graph G_1 for a set of three terminals (left) and its refinement G_1^k with $k = 2$ (right)

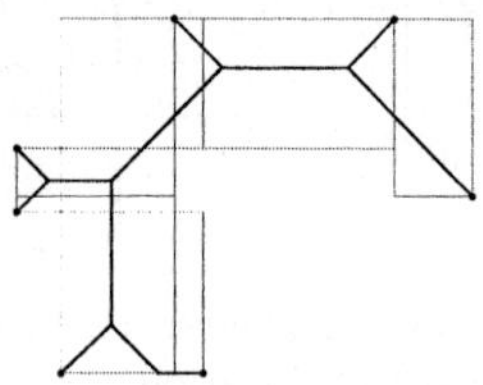

Fig. 4. Example of the covering of an octilinear Steiner tree by rectangles with edges in G_1^k

4 Error Bounds for Graph-Based Approximations

In this section we show how to improve upon the approximation guarantee obtained for $G_1(K)$ for octilinear Steiner trees. To this end we construct a graph $G_1^k(K)$ which is parameterized by some constant k.

Recall that the graph G_1 is the graph induced by four lines (vertical, horizontal, and both main diagonals) through each terminal. The idea is to refine G_1 by superimposing $O(k)$ additional lines. This is done as follows. Given a set of terminals K with $|K| = n$, let $BB(K)$ denote the *bounding box* of this point set, that is, the smallest axis-parallel rectangle which includes all terminals. We subdivide each side of $BB(K)$ equidistantly with k points into $k + 1$ segments and add for each subdivision point additional lines in all four feasible orientations of the octilinear geometry. See Fig. 3 for a small example. Since we have $O(n + k)$ lines in each feasible direction, we get $O((n + k)^2)$ intersection points of these lines. Hence, the induced graph G_1^k has $O((n + k)^2)$ many vertices and edges. For the bounding box $BB(K)$ with side lengths bb_x and bb_y, denote by $bb := \max\{bb_x, bb_y\}$ its maximum side length.

We next define how to *cover a Steiner tree T by a set of axis-parallel rectangles* as follows (the rectangles may overlap). For each Steiner point s of T, the set $\mathcal{R}$ contains a smallest rectangle including s with horizontal and vertical edges from G_1^k. In the degenerate case that s lies on a vertex or an edge of G_1^k we add no rectangle. We also add a smallest enclosing rectangle for each point p where an edge of T bends. Degenerate cases are handled as with Steiner points. For each straight-line segment of T not covered by previous rectangles we independently add to $\mathcal{R}$ a smallest enclosing rectangle bounded by vertical and horizontal edges from G_1^k. Thus, we finally have the following partition of the Steiner tree: $T = \cup_{R \in \mathcal{R}}(T \cap R)$. See Fig. 4 for an example.

For a given tree T, we construct an approximating Steiner tree T_{app} with edges in G_1^k as follows. For each rectangle $R \in \mathcal{R}$ let S_R be the set of intersection points of T_{opt} with the boundary of R. We connect the point set S_R in the shortest possible way by (portions of) edges in G_1^k, yielding a tree T_R. From the union of all these trees T_R we eliminate in a postprocessing step the longest edge of each cycle which may occur and all leaves and incident edges of the resulting tree which are not terminals. We thereby obtain our approximation T_{app}. The following technical lemma shows that we can bound for each rectangle R included in $\mathcal{R}$ the length $\ell(T_{app} \cap R)$ of $T_{app} \cap R$ in terms of the length $\ell(T \cap R)$ of $T \cap R$.

Lemma 3. *For each $R \in \mathcal{R}$, the following bound holds:*

$$\ell(T_{app} \cap R) - \ell(T \cap R) \leq (4 - \sqrt{2})\frac{bb}{k+1}.$$

Lemma 4. *The graph G_1^k contains an octilinear Steiner tree which is at most a factor of*

$$1 + \frac{(2n-3)(4-\sqrt{2})}{k+1}$$

longer than the optimal one.

Proof. Let K be a set of points in the plane with $|K| = n$ and T_{opt} be some octilinear Steiner minimum tree for K. We have to show that there is some octilinear Steiner tree T_{app} within G_1^k which approximates T_{opt} sufficiently well.

We cover T_{opt} by a set $\mathcal{R}$ of axis-parallel rectangles as described above. Let us assume that T_{opt} is composed of $k \geq 1$ full Steiner trees with $n_1, n_2, \ldots, n_k \geq 2$ vertices each. Then $\sum_{i=1}^{k} n_i = n+k-1$. Each full component may have at most $s_i \leq n_i - 2$ Steiner points. Hence, the total number of Steiner points satisfies $\sum_{i=1}^{k} s_i \leq n-k-1$. If m_i denotes the number of edges in the i-th full component, we have $m_i = n_i + s_i - 1$ for a total of $m = \sum_{i=1}^{k} m_i \leq 2n-2-k$ edges in T_{opt}.

The cover $\mathcal{R}$ of T_{opt} by rectangles contains at most one rectangle per Steiner point, one rectangle for each edge and at most two additional rectangles per bending edge (one for the bending point and one for the second part of the edge). By Property 4, we may assume that each full component has at most one bending edge. Thus $|\mathcal{R}| \leq \sum_{i=1}^{k} s_i + \sum_{i=1}^{k} m_i + 2k \leq 3n-3$.

Next we analyze the length $\ell(T_{app})$ of T_{app} in comparison to the optimal length $\ell(T_{opt})$. All edges of T_{opt} which are incident to a terminal are represented in G_1^k. Hence, for all corresponding rectangles $\ell(T_{opt} \cap \mathcal{R}) = \ell(T_{app} \cap \mathcal{R})$. Clearly, there are at least n edges incident to terminals. This implies, that for at most $2n-3$ rectangles of $\mathcal{R}$ there will be a difference between $\ell(T_{app} \cap \mathcal{R})$ and $\ell(T_{opt} \cap \mathcal{R})$. Thus, we have

$$\frac{\ell(T_{app})}{\ell(T_{opt})} = \frac{\ell(T_{opt}) + \sum_{R \in \mathcal{R}}(\ell(T_{app} \cap \mathcal{R}) - \ell(T_{opt} \cap \mathcal{R}))}{\ell(T_{opt})}$$
$$\leq \frac{\ell(T_{opt}) + (2n-3) \cdot \max_{R \in \mathcal{R}}\{\ell(T_{app} \cap \mathcal{R}) - \ell(T_{opt} \cap \mathcal{R})\}}{\ell(T_{opt})}.$$

Hence, it suffices to show that

$$\max_{R \in \mathcal{R}}\{\ell(T_{app} \cap \mathcal{R}) - \ell(T_{opt} \cap \mathcal{R})\} \leq (4 - \sqrt{2}) \cdot \frac{\ell(T_{opt})}{k+1}.$$

This relation follows from Lemma 3 and the observation that $\ell(T_{opt}) \geq bb$, since every Steiner tree must connect the terminals which define the bounding box $BB(K)$. □

Theorem 2. *For a given set of n terminals in the plane and for every $\varepsilon > 0$ there is a graph of size $O(\frac{n^2}{\varepsilon^2})$ which contains a $(1 + \varepsilon)$-approximation of a minimum octilinear Steiner tree.*

Proof. The approximation guarantee follows directly from Lemma 4 if we choose $k := \frac{(4-\sqrt{2})2n}{\varepsilon}$. With such a choice of k, the graph has the claimed size. □

Blockages. Let us now sketch the necessary modifications in the presence of obstacles. Let K be a set of points (terminals) in the plane and O be a set of octilinear (or rectilinear) obstacles. Denote by V_O the set of obstacle vertices. Let $n = |K| + |V_O|$. Analogously to the definition of $G_1(K)$, we now define a graph $G(K, O)$ which is induced by the set L of lines in all feasible directions in 4-geometry going through terminals or obstacle vertices. For a given parameter k, we refine $G(K, O)$ by adding lines. For any two parallel lines in L which are neighbored (i.e., no third line with the same orientation lies between them) we add k additional lines with the same orientation between them and place them equidistantly. In total, we have $O(nk)$ lines. From the resulting induced graph, we erase all vertices and their incident edges which lie strictly inside some obstacle. The latter guarantees that every Steiner tree in this graph corresponds to a tree in the plane which avoids all obstacles.

Theorem 3. *Let α denote the approximation guarantee for an algorithm solving the Steiner tree problem in graphs. Given a terminal set K, a set of octilinear obstacles O, and some $\varepsilon > 0$, there is an $(\alpha + \varepsilon)$-approximation of the octilinear Steiner tree problem with obstacles which have to be avoided.*

The proof of this theorem follows basically the same ideas as that for the case without obstacles. There is one essential difference, however. In the presence of obstacles, edges between terminals and/or Steiner points may be forced to bend several times. But if such an edge bends, then all but at most two of its straight segments will lie on $G(K, O)$. This observation implies that the number of rectangles in a cover of some optimal Steiner tree on which we have to find an approximative solution is upper bounded by $6n - 11$ which suffices for an analogous result as in Lemma 4.

References

1. http://www.xinitiative.org (2005)
2. Teig, S.L.: The X architecture: not your father's diagonal wiring. In: SLIP '02: Proceedings of the 2002 international workshop on System-level interconnect prediction, ACM Press (2002) 33–37
3. Paluszewski, M., Winter, P., Zachariasen, M.: A new paradigm for general architecture routing. Proceedings of the 14th ACM Great Lakes Symposium on VLSI (GLSVLSI) (2004) 202–207
4. Garey, M., Johnson, D.: The rectilinear Steiner tree problem is NP-complete. SIAM Journal on Applied Mathematics **32** (1977) 826–834
5. Garey, M., Graham, R., Johnson, D.: The complexity of computing Steiner minimal trees. SIAM Journal on Applied Mathematics **32** (1977) 835–859

6. Coulston, C.: Constructing exact octagonal Steiner minimal trees. In: ACM Great Lakes Symposium on VLSI. (2003) 1–6
7. Arora, S.: Polynomial time approximation schemes for the Euclidean traveling salesman and other geometric problems. Journal of the ACM **45** (1998) 753–782
8. Mitchell, J.: Guillotine subdivisions approximate polygonal subdivisions: A simple polynomial-time approximation scheme for geometric TSP, k-MST, and related problems. SIAM Journal on Computing **28** (1999) 1298–1309
9. Rao, S., Smith, W.: Approximating geometric graphs via "spanners" and "banyans". Proceedings of the 30th ACM Symposium on Theory of Computing (1998) 540–550
10. Kahng, A., Măndoiu, I., Zelikovsky, A.: Highly scalable algorithms for rectilinear and octilinear Steiner trees. Proceedings 2003 Asia and South Pacific Design Automation Conference (ASP-DAC) (2003) 827–833
11. Zhu, Q., Zhou, H., Jing, T., Hong, X., Yang, Y.: Efficient octilinear Steiner tree construction based on spanning graphs. Proceedings 2004 Asia and South Pacific Design Automation Conference (ASP-DAC) (2004) 687–690
12. Nielsen, B., Winter, P., Zachariasen, M.: An exact algorithm for the uniformly-oriented Steiner tree problem. In: 10th Annual European Symposium on Algorithms (ESA 2002). Volume 2461 of Lecture Notes in Computer Science. Springer (2002) 760–772
13. Robins, G., Zelikovsky, A.: Improved Steiner tree approximation in graphs. Proceedings of the 11th Annual ACM-SIAM Symposium on Discrete Algorithms (2000) 770–779
14. Hanan, M.: On Steiner's problem with rectilinear distance. SIAM Journal on Applied Mathematics **14** (1966) 255–265
15. Althaus, E., Polzin, T., Daneshmand, S.: Improving linear programming approaches for the Steiner tree problem. Research Report MPI-I-2003-1-004, Max-Planck-Institut für Informatik, Saarbrücken, Germany (2003)
16. Du, D.Z., Hwang, F.: Reducing the Steiner problem in a normed space. SIAM Journal on Computing **21** (1992) 1001–1007
17. Lee, D., Shen, C.F.: The Steiner minimal tree problem in the λ-geometry plane. In: Proceedings 7th International Symposium on Algorithms and Computations (ISAAC 1996). Volume 1178 of Lecture Notes in Computer Science., Springer (1996) 247–255
18. Lin, G.H., Xue, G.: Reducing the Steiner problem in four uniform orientations. Networks **35** (2000) 287–301
19. Koh, C.: Steiner problem in octilinear routing model. Master thesis, National University of Singapore (1995)
20. Shen, C.: The λ-geometry Steiner minimal tree problem and visualization. PhD thesis, Northwestern University, Evanston, IL, USA (1997)
21. Brazil, M., Thomas, D., Winter, P.: Minimum networks in uniform orientation metrics. SIAM Journal on Computing **30** (2000) 1579–1593
22. Brazil, M., Thomas, D., Weng, J., Zachariasen, M.: Canonical forms and algorithms for Steiner trees in uniform orientation metrics. Technical Report TR-02/22, DIKU, Department of Computer Science, Copenhagen, Denmark (2002). To appear in Algorithmica.
23. Karp, R.: Reducibility among combinatorial problems. In Miller, R., Thatcher, J., eds.: Complexity of Computer Computations. Plenum Press, New York (1972) 85–104

Dense Subgraph Problems with Output-Density Conditions

Akiko Suzuki and Takeshi Tokuyama

Graduate School of Information Sciences,
Tohoku University, Sendai, 980-8579, Japan
{akiko, tokuyama}@dais.is.tohoku.ac.jp

Abstract. We consider the dense subgraph problem that extracts a subgraph with a prescribed number of vertices that has the maximum number of edges (total edge weight in the weighted case) in a given graph. We give approximation algorithms with improved theoretical approximation ratios—assuming that the density of the optimal output subgraph is high, where density is the ratio of number of edges (or sum of edge weights) to the number of edges in the clique on the same number of vertices. Moreover, we investigate the case where the input graph is bipartite, and design a pseudo-polynomial time approximation scheme that can become a PTAS even if the size of the optimal output graph is comparatively small. This is a significant improvement in a theoretical sense, since no constant-ratio approximation algorithm was known previously if the output graph has $o(n)$ vertices.

1 Introduction

We consider the weighted *dense subgraph problem* (often called the maximum dispersion problem or dense k-subgraph problem) defined as follows:

> Consider a weighted graph $G = (V, E)$, where $|V| = n$, and each edge e has a nonnegative weight $0 \le w(e) \le 1$. Given a natural numbers $k \le n$, find a subgraph $H = (X, F)$ of G such that $|X| = k$ and $w(F) = \sum_{e \in F} w(e)$ is maximized.

Its bipartite version is as follows:

> Consider a weighted bipartite graph $G = (U, V, E)$, where $|U| = m$, $|V| = n$, and each edge e has a nonnegative weight $0 \le w(e) \le 1$. Given two natural numbers $m' \le m$ and $n' \le n$, find a subgraph $H = (X, Y, F)$ of G such that $|X| = m'$, $|Y| = n'$, and $w(F) = \sum_{e \in F} w(e)$ is maximized.

We note the condition $0 \le w(e) \le 1$ is given since it is convenient for presenting our theoretical results, although we can define each problem without it. We say *unweighted dense subgraph problem* if $w(e) = 1$ for each edge. We define the density Δ of the output subgraph H to be $\Delta = \frac{2w(F)}{k(k-1)}$ for the non-bipartite

X. Deng and D. Du (Eds.): ISAAC 2005, LNCS 3827, pp. 266–276, 2005.

case and $\Delta = \frac{w(F)}{m'n'}$ for the bipartite case. In other words, density is the ratio of the weight sum to the number of edges in a clique (or a bipartite clique) of the same size. We mainly consider the case where the density of the optimal output subgraph is high (e.g., $\Delta = \Omega(1)$), and aim to design efficient approximation algorithms.

CLIQUE and dense subgraph with a density condition. We first show a motivating scenario to convince readers that the problem is interesting even for the special case where $\Delta = 1$ and the graph is unweighted. Suppose that we want to find a clique of size k in a graph G, suspecting that such a clique exists. Suppose that such a clique indeed exists. Since CLIQUE is NP-complete, we want to have an approximate solution. One possibility is to find a clique of a size $k' < k$ such that the approximation ratio k/k' is small. Unfortunately, it is very difficult to obtain a clique of a large size, since it is known to be hard to attain the approximation ratio $n^{1-\epsilon}$ unless CO-NP = NP, and the current best ratio is $O(n/\log^2 n)$ [4]. Another possible solution is to find a dense subgraph with k vertices. Again, unfortunately, the current known approximation ratio for the general dense subgraph problem is high. We show in this paper that the approximation ratio is significantly improved utilizing the fact that $\Delta = 1$, and hence this formulation of clique approximation has a nice theoretical guarantee.

Previous work. The densest subgraph problem that finds a subgraph with the maximum average degree without any size constraint of the subgraph can be solved in polynomial time [9]. However, the unweighted dense subgraph problem (where k is given) is NP-hard since the max-clique problem is reduced to it.

Therefore, several approximation algorithms have been proposed [1,3,5,7,10] in the literature for the dense subgraph problem. We use the convention that an algorithm has an approximation ratio $r > 1$ for a maximization problem if its objective value is at least r^{-1} times the optimal value. In practice, a greedy algorithm removing the smallest weighted-degree vertex one by one often works well; however, its approximation ratio is $2n/k$ (ignoring smaller terms) if $k < n/3$, and it is asymptotically tight [3]. One nice feature of this algorithm is that it gives a constant approximation ratio if $k = \Omega(n)$; however, the linear dependency of the ratio in n is not satisfactory from the theoretical point of view. The currently best algorithm has an approximation ratio that is slightly better than $n^{1/3}$, and it is conjectured that there exists some constant ϵ such that n^{ϵ} approximation is hard [5,7,2,10]. As for the lower bound, Feige [6] showed that it is R3SAT-hard to approximate within a ratio better than some constant when $k = \Omega(n)$.

Unweighted dense subgraph problems with some additional density conditions on the input/output graph have been also considered. In particular, when the optimal subgraph has $\Omega(n^2)$ edges (thus, $\Delta = \Omega(1)$ and $k = \Omega(n)$), PTAS algorithms are known [1,5]. However, in many applications of the dense subgraph problem, it is desired to solve the problem with only a density assumption on the optimal output graph. An n^{ϵ}-approximation algorithm with a time complexity $O(n^{1/\epsilon})$ is known for the case $\Delta = \Omega(1)$ under an additional condition that the average degree of the input graph is $\Omega(k)$ [7]. It is also claimed in [7] that if the optimal subgraph is a clique, there is an $n^{O((1/\epsilon)\log(n/k))}$ time algorithm to have an $(1+\epsilon)$-approximation solution: This time complexity is polynomial only if $k = \Omega(n)$. As far as the authors know, no constant-ratio algorithm is known for $k = o(n)$ even for the unweighted problem. It is an interesting question whether we can relax the requirement $k = \Omega(n)$, and this is one of our motivations of this research.

Next, let us consider the bipartite dense subgraph problem. Although the bipartite dense subgraph problem looks easier than the general dense subgraph problem, it has not been revealed how much easier it is. Indeed, it is not much easier. The problem is NP-hard, since the NP-hard edge-maximizing bipartite clique problem is reduced to this problem. Note that bipartite clique problem is polynomial-time solvable if the criterion is to maximize the number of vertices. (See [8] for the complexities of bipartite clique problems.)

Our contribution. For the bipartite dense subgraph problem, let $p = m/m'$, and $q = n/n'$. We have an algorithm with time complexity $O(mn2^{O(\Delta^{-1}\epsilon^{-2}\log p \log q)})$ which outputs a subgraph $H_0 = (X_0, Y_0, F_0)$ such that $|X_0| = m'$, $|Y_0| = n'$ and its weighted density Δ_0 satisfies $(1+\epsilon)\Delta_0 \geq \Delta$ for any positive $\epsilon < 1$. The time complexity implies that we have a PTAS if either (1) Δ and $\min(p,q)$ are constants, (2) both p and q are constants and $\Delta = \Omega(\log^{-1} n)$, or (3) Δ is a constant and $\max(p,q) = 2^{O(\sqrt{\log mn})}$. We also give a polynomial-time approximation algorithm with an approximation ratio $(\min(p,q))^{\epsilon}$ for any constant $\epsilon > 0$ only with the density condition $\Delta = \Omega(1)$. These results imply that Δ is the principal parameter to control the computational complexity.

As direct consequences of the bipartite problem, we have the following results for the non-bipartite dense subgraph problem: We give an algorithm with an approximation ratio $(n/k)^{\epsilon}$ if the optimal solution has $\Omega(k^2)$ edges (or $w(F) = \Omega(k^2)$ for the weighted problem). This improves the previous n^{ϵ} approximation ratio of [7] when k is large, and also the additional condition on the average degree of the input graph is removed. Moreover, we give a $(2+\epsilon)$-approximation algorithm for any $\epsilon \leq 1$ that runs in $O(n^2 2^{O(\Delta^{-1}\epsilon^{-2}\log^2(n/k))})$ time. This implies that we have a $(2+\epsilon)$-approximation polynomial-time algorithm if $\Delta = \Omega(1)$ and $n/k < 2^{\sqrt{\log n}}$. As far as the authors know, this is the first constant-ratio polynomial-time algorithm for the dense subgraph problem that works for some $k = o(n)$.

Technically, we apply the framework of Arora *et al.* [1] solving combinatorial problems on a dense input graph by using quadratic integer programming (QIP) and sampling. For our problem, the QIP is bilinear, and hence the error due to the sampling can be more precisely analyzed. This enables to replace the density condition of the input graph with that of the output graph. Our algorithms are randomized. However, we can derandomize them as follows: We can use any pseudo random generator with pairwise independence [11] for graph decomposition given in Section 2. For the random sampling algorithms given in Sections 3 and 4, we can use a derandomization technique given in [1] using an expander.

2 Reduction to the Bipartite Problem

We show that we can reduce the general dense subgraph problem to the bipartite case by increasing the approximation factor by 2. We have a graph $G = (V, E)$ and want to find a dense k-vertex subgraph $H = (X, F)$. We first randomly divide V into two sets U' and V', where each vertex of V is assigned to U' with probability 0.5. Thus, we have a bipartite graph $G' = (U', V', E')$. Consider what happens on the optimal subgraph H. Let $F' = E' \cap F$. Then, we have a bipartite subgraph $W = (U' \cap X, V' \cap X, F')$ of G'. By using the argument of randomized max-cut [11], the expected value of $w(F')$ is $w(F)/2$. Thus, with a nonnegligible probability, $(2 + o(1))w(F') \geq w(F)$, and it is easy to see that the density condition inherits. Next, we find an approximation solution W' for the bipartite dense subgraph problem that has an approximation ratio r. Here, although we may try all combinations of m' and n' satisfying $m' + n' = k$, it suffices to consider the case where $n' \approx k/2$ (we omit details in this version). Then, the subgraph in G induced by the vertex set of W' gives an $(2 + o(1))r$ approximation solution of the original problem. Hence, we have our results on the dense subgraph problem from those on the bipartite dense subgraph problem.

3 Quasi-Polynomial Time Approximation Scheme for the Bipartite Problem

We give an algorithm named ABDense (Approximate Bipartite-Dense) for computing a dense subgraph with a given combination (m', n') of numbers of vertices in a bipartite graph. In a weighted graph, the weighted degree of a vertex is the sum of weights of edges incident to it. We need to give parameters $\gamma < 1$ and I (the number of iterations of random sampling) to implement the algorithm. The following is a pseudo-code of the algorithm:

Algorithm *ABDense(G, m', n')*
(∗ Output is a subgraph H with density D ∗)
1. $D = 0$;
2. **for** $i \leftarrow 1$ **to** I;
3. **do**
4. Randomly select a subset $X_0 \subset U$ of size $\gamma m'$;
5. Let Y_1 be the set of vertices with n' largest weighted-degrees in the subgraph induced by $X_0 \cup V$ in G;
6. Let X_1 be the set of vertices with m' largest weighted-degrees in the subgraph induced by $U \cup Y_1$ in G;
7. **if** density D_1 of the graph H_1 induced by $X_1 \cup Y_1$ is larger than D;
8. **then** $D = D_1$, $H = H_1$:
9. **return** H;

The algorithm itself is a familiar one. If $\gamma = 1$, it is a common naive algorithm for this problem that is a core of many heuristics (e.g., multi-start local search or evolutional methods), and its deterministic version is utilized as a constituent of a hybrid algorithm of [7] with an $n^{1/3}$ approximation ratio.

Our contribution is a precise analysis of this algorithm. The success probability that $D > (1+\epsilon)^{-1}\Delta$ depends on the parameters γ and I. We analyze the performance of the algorithm to give suitable choice of γ and I, and show the following result: Recall that $p = m/m'$, $q = n/n'$, and $\Delta = \frac{w(F)}{m'n'}$.

Theorem 1. *For any $0 < \epsilon < 1$ ABDense computes a $(1+\epsilon)$-approximation solution for the bipartite dense subgraph problem in $O(mn2^{O(\Delta^{-1}\epsilon^{-2}\log p \log q)})$ randomized expected time.*

3.1 Framework of the Analysis

We first design another algorithm that is easier to analyze, and then simplify the algorithm to obtain an analysis for ABDense. The algorithm is a two-step sampling algorithm following a framework given by Arora *et al.* [1] for solving the dense subgraph problem: The algorithm first selects a sample set Γ of size γm and then search in its power set to find a subset of size $\gamma m'$ that leads to an approximate solution with the desired theoretical quality.

We can formulate the problem into a quadratic integer programming problem. We write $U = \{u_1, u_2, \dots, u_m\}$, $V = \{v_1, v_2, \dots, v_n\}$. We introduce a matrix $W = (w_{i,j})_{1\le i\le m, 1\le j\le n}$ indicating the graph structure of G. For the unweighted case, $w_{i,j}$ is binary, and $w_{i,j} = 1$ if and only if $(u_i, v_j) \in E$. For the weighted problem, we regard $w_{i,j}$ as the weight of the edge (u_i, v_j). It is straightforward to see that the bipartite dense subgraph problem is equivalent to the following **QIP**.

$$\textbf{QIP:}\quad \text{Maximize } {}^t\mathbf{x}W\mathbf{y}, \quad \text{subject to}$$
$$\sum_{1\le i\le m} x_i = m', \quad \sum_{1\le j\le n} y_j = n', \text{ and } \; x_i, y_j \in \{0,1\}.$$

Here, ${}^t\mathbf{x}W\mathbf{y} = \sum_{1\le i\le m}\sum_{1\le j\le n} w_{i,j}x_iy_j$ is the matrix product (we consider $\mathbf{x}$ and $\mathbf{y}$ as column vectors and ${}^t\mathbf{x}$ is the transpose of $\mathbf{x}$). A very useful feature is that the objective function is bilinear in x_i and y_j.

Let $(\mathbf{x}^{opt}, \mathbf{y}^{opt})$ be an optimal solution of **QIP** and $z^{\text{opt}} = {}^t\mathbf{x}^{opt}W\mathbf{y}^{opt}$. For a given $\mathbf{x}$, we define an n-dimensional vector $\mathbf{a}(\mathbf{x}) = {}^t\mathbf{x}W$. Then, $z^{\text{opt}} = \mathbf{a}(\mathbf{x}^{opt}) \cdot \mathbf{y}^{opt}$, where $\cdot$ is the inner product operation. Thus, if $\mathbf{a} = \mathbf{a}(\mathbf{x}^{opt})$ is known, it suffices to solve the following problem:

$$\textbf{IP-y(a):}\quad \text{Maximize } \mathbf{a}\cdot\mathbf{y} \quad \text{subject to } \sum_{1\le j\le n} y_j = n' \quad \text{and} \quad y_j \in \{0,1\}.$$

Symmetrically, we have the following **IP-x**:

$$\textbf{IP-x(b):}\quad \text{Maximize } \mathbf{b}\cdot\mathbf{x} \quad \text{subject to } \sum_{1\le i\le m} x_i = m' \quad \text{and} \quad x_i \in \{0,1\}.$$

Both of **IP-y(a)** and **IP-x(b)** can be solved by greedy algorithms. Indeed, given $\mathbf{a} = (a_1, a_2, \dots, a_n)$ (resp. $\mathbf{b} = (b_1, b_2, \dots, b_m)$), we consider the largest

n' (resp. m') entries (breaking tie arbitrary) of $\mathbf{a}$ (resp. $\mathbf{b}$), and let $J(\mathbf{a}) \subset \{1, 2, \ldots, n\}$ (resp. $J(\mathbf{b}) \subset \{1, 2, \ldots, m\}$) be the set of corresponding indices.

We define the binary vector $\mathbf{y}(\mathbf{a}) = (y_1(\mathbf{a}), y_2(\mathbf{a}), \ldots, y_n(\mathbf{a}))$ such that $y_j(\mathbf{a}) = 1$ if and only if $j \in J(\mathbf{a})$. Similarly, we define a binary vector $\mathbf{x}(\mathbf{b})$ such that $x_i(\mathbf{b}) = 1$ if and only if $i \in J(\mathbf{b})$. Then, it is easy to see that the vectors $\mathbf{y}(\mathbf{a})$ and $\mathbf{x}(\mathbf{b})$ are optimal solutions for **IP-y(a)** and **IP-x(b)**, respectively.

Now, we can design an algorithm to find a feasible solution for **QIP**. Start with any nonnegative vector $\mathbf{a}^*$ and solve **IP-y(a*)**. Next, using the output $\mathbf{y}^1 = \mathbf{y}(\mathbf{a}^*)$ of **IP-y(a*)**, we compute the vector $\mathbf{b}^* = \mathbf{b}(\mathbf{y}^1) = W\mathbf{y}^1$, and solve **IP-x (b*)** to have an output vector $\mathbf{x}^1 = \mathbf{x}(\mathbf{b}^*)$. The following lemma is straightforward:

Lemma 1. *The pair* $(\mathbf{x}^1, \mathbf{y}^1)$ *is a feasible solution of* **QIP**. *Moreover, let* $(\mathbf{x}^0, \mathbf{y}^0)$ *be any feasible solution of* **QIP**, *and let* $(\mathbf{x}^1, \mathbf{y}^1)$ *be the vector obtained by applying the above procedure to* $\mathbf{a}(\mathbf{x}^0)$. *Then,* ${}^t\mathbf{x}^1 W \mathbf{y}^1 \geq {}^t\mathbf{x}^0 W \mathbf{y}^0$.

Let $z^1 = {}^t\mathbf{x}^1 W \mathbf{y}^1$ be the objective function value associated with $(\mathbf{x}^1, \mathbf{y}^1)$. We compare z^1 to z^{opt}. We first claim that if $\mathbf{a}^*$ is a good approximation of $\mathbf{a}^{opt} = \mathbf{a}(\mathbf{x}^{opt})$, then $z^{\mathrm{opt}} - z^1$ is small; in other words, $(\mathbf{x}^1, \mathbf{y}^1)$ gives a good approximation solution for **QIP**. Next, we give a method to obtain an $\mathbf{a}^*$ that approximates $\mathbf{a}^{opt}$.

Lemma 2. *Let* $J = J(\mathbf{a}^{opt})$ *and* $J^* = J(\mathbf{a}^*)$. *Then,* $\sum_{j \in J^*} a_j^* \geq \sum_{j \in J} a_j^*$ *and* $z^1 \geq z^* = \sum_{j \in J^*} a_j^{opt}$.

Proof. The first formula is straightforward from the definition of J^*. For the second formula, $z^* = \sum_{j \in J^*} a_j^{opt} = {}^t\mathbf{x}^{opt} W \mathbf{y}^1$ is the objective function value of a feasible solution $(\mathbf{x}^{opt}, \mathbf{y}^1)$ of **QIP**. However, if we fix $\mathbf{y}^1$, $\mathbf{x}^1$ is the best possible assignment of x values. Thus, this solution cannot be better than $(\mathbf{x}^1, \mathbf{y}^1)$.

We define $F_1(\mathbf{a}^*) = \sum_{j \in J} a_j^{opt} - \sum_{j \in J} a_j^*$ and $F_2(\mathbf{a}^*) = \sum_{j \in J^*} a_j^* - \sum_{j \in J^*} a_j^{opt}$.

Lemma 3. $z^{\mathrm{opt}} - z^1 \leq F_1(\mathbf{a}^*) + F_2(\mathbf{a}^*)$.

Proof. From the previous lemma, $z^{\mathrm{opt}} - z^1 \leq \sum_{j \in J} a_j^{opt} - \sum_{j \in J^*} a_j^{opt} \leq \sum_{j \in J} a_j^{opt} - \sum_{j \in J^*} a_j^{opt} + \sum_{j \in J^*} a_j^* - \sum_{j \in J} a_j^* = F_1(\mathbf{a}^*) + F_2(\mathbf{a}^*)$.

Thus, in order to obtain an approximate solution whose objective function value is at least $(1-\epsilon)z^{\mathrm{opt}}$, it suffices to find $\mathbf{a}^*$ such that $F_1(\mathbf{a}^*) + F_2(\mathbf{a}^*) \leq \epsilon z^{\mathrm{opt}}$.

Now, we consider a random sample $\Gamma \subset U$ of size $|\Gamma| = \gamma m$. We identify U and the index set $\{1, 2, \ldots, m\}$, thus we regard $\Gamma \subset \{1, 2, \ldots, m\}$. Let $h_j = \sum_{i \in \Gamma} w_{i,j} x_i^{opt}$, for each $j = 1, 2, \ldots, n$. We define $\mathbf{a}(\Gamma) = (a_1(\Gamma), a_2(\Gamma), \ldots, a_n(\Gamma))$ by $a_j(\Gamma) = \gamma^{-1} h_j$. Since $\sum_{1 \leq i \leq m} w_{i,j} x_i^{opt} = a_j^{opt}$, the expected values of h_j and $a_j(\Gamma)$ are represented by $\mathbf{E}(h_j) = \gamma a_j^{opt}$ and $\mathbf{E}(a_j(\Gamma)) = a_j^{opt}$, respectively.

This $\mathbf{a}(\Gamma)$ is our candidate for $\mathbf{a}^*$, and we estimate $F_1(\mathbf{a}(\Gamma)) + F_2(\mathbf{a}(\Gamma))$. Note that we are not ready to claim that we can compute $\mathbf{a}(\Gamma)$ at this stage, since we do not know $\mathbf{x}^{opt}$.

3.2 Analysis of $F_1(\mathbf{a}(\Gamma)) + F_2(\mathbf{a}(\Gamma))$

Let us start the detailed analysis. Since the corresponding terms in F_1 and F_2 cancels out for indices in $J \cap J^*$, we can remove them from the estimation. Thus, the worst case occurs when $J \cap J^* = \emptyset$, and we assume this situation without loss of generality in our analysis. We denote the expected value $\mathbf{E}(h_j)$ by μ_j for $j = 1, 2, \ldots n$.

Lemma 4. *$Pr[h_j > \mu_j + \delta] < e^{-\delta^2/(2\mu_j+\delta)}$ and $Pr[h_j < \mu_j - \delta] < e^{-\delta^2/2\mu_j}$ for each $1 \leq j \leq n$. Here, $e = 2.718\ldots$ is the base of natural logarithm.*

Proof. We take each element of U in the sample with probability γ, and the element u_i contributes by $w_{i,j}x_i^{opt}$ to h_j if it is selected. Thus, for each fixed j, we define a random variable $Z_{i,j}$ that takes $w_{i,j}x_i^{opt}$ with probability γ and 0 with probability $1-\gamma$ for each $i = 1, 2, \ldots, m$. Note that if $w_{i,j}x_i^{opt} = 0$, $Z_{i,j} \equiv 0$. This gives a series of independent variables for each fixed j, and $h_j = \sum_{i=1}^m Z_{i,j}$. Thus, we apply Chernoff's bounds to obtain the lemma.

Corollary 1. *If $\mu_j \leq f$, $Pr[h_j < \mu_j - rf] < e^{-r^2 f/2}$ and $Pr[h_j > \mu_j + rf] < e^{-r^2 f/(2+r)}$ for any $r > 0$.*

We utilize the following combinatorial fact (proof is omitted):

Lemma 5. *Let $S = \{n_1, n_2, \ldots n_k\}$ be a set of k real numbers satisfying that $\sum_{1 \leq i \leq k} n_i \geq N$ for a given number N. Then there exists a natural number ℓ such that at least $2^{-\ell+1}k$ entries of S exceed $\ell N/3k$.*

Lemma 6. *If $\gamma = \frac{cn' \ln q}{z^{\mathrm{opt}}\epsilon^2} = \frac{c \ln q}{\Delta m' \epsilon^2}$ for a sufficiently large constant c, $F_i(\mathbf{a}(\Gamma)) < \epsilon z^{\mathrm{opt}}/2$ with a probability at least 0.9 for each of $i = 1, 2$.*

Proof. We only prove for F_2. Suppose that $F_2(\mathbf{a}(\Gamma)) \geq \epsilon z^{\mathrm{opt}}/2$. We apply Corollary 1 for $f = \gamma z^{\mathrm{opt}}/n' = c\epsilon^{-2} \ln q$ that is the average of γa_j^{opt} over $j \in J$. Recall that we can assume $J^* \cap J = \emptyset$. J is the set of indices for the n' largest elements of a_j^{opt}. Thus, $\mu_j \leq f$ if $j \notin J$, and accordingly $\mu_j \leq f$ for $j \in J^*$. Thus, $Pr[h_j > \mu_j + r\epsilon f] < e^{-cr^2 \ln q/(2+r\epsilon)} < q^{-cr^2/(2+r)}$. We define $P(r) = q^{-cr^2/(2+r)}$. Thus, the expected number of indices $j \in \{1, 2, \ldots, n\}$ such that $h_j - \mu_j > r\epsilon f$ is bounded by $nP(r)$. Hence, from Markov's inequality, the probability that there are L such indices is bounded by $nP(r)/L$.

We use Lemma 5 by setting $n_j = h_j - \mu_j$ for $j \in J^*$, $k = n'$ and $N = \gamma\epsilon z^{opt}/2 = cn'\epsilon^{-1} \ln q/2$. (We ignore all $j \notin J^*$.) If $F_2(\mathbf{a}(\Gamma)) \geq \epsilon z^{\mathrm{opt}}/2$, $\sum_{j \in J^*} n_j \geq N$. Thus, there exists an ℓ such that at least $2^{-\ell+1}n'$ entries of S exceeds $\ell N/3n' = \frac{\ell cn'\epsilon^{-1} \ln q}{6n'} \geq \frac{c\ell\epsilon^{-1} \ln q}{6} = \frac{\ell\epsilon f}{6}$. From the argument above, the probability that there are $2^{-\ell+1}n'$ such indices is bounded by $nP(\ell/6)/(2^{-\ell+1}n') = qP(\ell/6)2^{\ell-1}$. It is routine to show $qP(\ell/6)2^{\ell-1} < e^{-3\ell}$ if c is sufficiently large and $q > e$. Thus, the probability that there exists at least one such ℓ is bounded by $\sum_{\ell=1}^{\log n'} e^{-3\ell} < 1/10$. Thus, the probability that $F_2(\mathbf{a}(\Gamma)) < \epsilon z^{\mathrm{opt}}/2$ is at most 0.1.

Corollary 2. *If* $\gamma \geq \frac{cn' \ln q}{z^{\mathrm{opt}} \epsilon^2}$ *for a sufficiently large constant c,* $F_1(\mathbf{a}(\Gamma)) + F_2(\mathbf{a}(\Gamma)) < \epsilon z^{\mathrm{opt}}$ *with a probability at least* 0.8.

We consider a sample Γ of size γm suggested in the above corollary. Let $Z(\Gamma)$ be the subset of Γ defined by $Z(\Gamma) = \{i \in \Gamma | x_i^{opt} = 1\}$. $h_j = \sum_{i \in Z(\Gamma)} w_{i,j}$ and $a_j(\Gamma) = \gamma^{-1} h_j$ are computed from $Z(\Gamma)$, thus we obtain a $(1-\epsilon)$ approximation solution with probability 0.8 if we can correctly guess $Z(\Gamma)$.

We can apply a version of exhaustive search to find $Z(\Gamma)$. However, the number of all subsets of Γ is $2^{O(\Delta^{-1}\epsilon^{-2} p \ln q)}$. Thus, if we exhaustively search all subsets of Γ to find $Z(\Gamma)$, the computation time is exponential in p. Fortunately, we do not need to examine all subsets, since the expected value of the size of $Z(\Gamma)$ is $\gamma m' = p^{-1}|\Gamma|$, and the following lemma is obtained from Chernoff's bounds.

Lemma 7. *The probability that* $||Z(\Gamma)| - \gamma m'| > 3\sqrt{\gamma m'}$ *is at most* $2e^{-2}$.

From the above lemma and Corollary 2, we have the following:

Corollary 3. *With probability* $0.8 - 2e^{-2}$, *we have a sample* Γ *such that* $||Z(\Gamma)| - \gamma m'| < 3\sqrt{\gamma m'}$ *and* $Z(\Gamma)$ *gives an* $(1+\epsilon)$*-approximation solution for* **QIP**.

The number of subsets of Γ whose cardinality is at most $r = \gamma m' + 3\sqrt{\gamma m'}$ is $O(((p+1)e)^r)$ from the Stirling's formula. Since $r = O(\Delta^{-1}\epsilon^{-2} \ln q)$, we have the following:

Theorem 2. *We can compute an* $(1+\epsilon)$*-approximation solution for the bipartite dense subgraph problem in* $O(mn2^{O(\Delta^{-1}\epsilon^{-2} \ln p \ln q)})$ *time with high probability.*

3.3 Simplifying the Algorithm

In the above algorithm, we take a sample Γ, and exhaustively search all its subsets of size approximately $\gamma m'$ to find $Z(\Gamma)$. However, it is easy to see that there exists a subset of size exactly $\lfloor \gamma m' \rfloor$ to give an approximation algorithm, if we allow to increase the error ratio ϵ slightly to $(1 + 3p^{-0.5})\epsilon$. Now, instead of two-step sampling, we can randomly select a subset X_0 of size $\gamma m'$ directly from U, and apply **IP-y** and **IP-x**; thus, we obtain the algorithm ABDense. Our analysis given in the previous subsection works to give the same time complexity for ABDense, where the number I of iterations is $O(2^{O(\Delta^{-1}\epsilon^{-2} \ln p \ln q)})$.

4 Approximation Algorithm Without Size Restriction

Now, let us consider the case where both p and q are large. By symmetry, we assume $p \geq q$ without loss of generality. If we could naively set $\epsilon = \sqrt{\log q}$ in the complexity of our quasi-polynomial time algorithm, it would imply a polynomial time algorithm with an $O(\sqrt{\log q})$ approximation ratio. Unfortunately, the analysis only holds under the condition $\epsilon < 1$. However, we can prove the following theorem:

Theorem 3. *For any fixed* $\epsilon > 0$, *we have a* q^ϵ*-approximation polynomial-time algorithm for the weighted bipartite dense subgraph problem if* $\Delta = \Omega(1)$.

We assume q is larger than a sufficiently large constant, since otherwise we have already given a PTAS. For technical reasons, we prove $2q^\epsilon$ approximation ratio, since it is easy to remove the factor 2 by decreasing ϵ slightly. The algorithm is the same as the one in Section 3.1 except the sample size. Here, we take a sample Γ of size $\tilde{\gamma}m = \frac{cmn'}{z^{\mathrm{opt}}}$, where c is a constant dependent on ϵ. Using the analysis given in the previous section, the time complexity becomes $O(mn2^{O(\ln p\Delta^{-1})}) = O(mnp^{O(\Delta^{-1})})$, which is polynomial if $\Delta = \Omega(1)$. Therefore, it suffices to estimate the approximation ratio.

Lemma 8. *If we take* c *sufficiently large, and take a random sample* Γ *such that* $|\Gamma| = \tilde{\gamma}m$, $F_1(\mathbf{a}(\Gamma)) < z^{\mathrm{opt}}/2$ *with probability at least* 0.9.

Proof. The proof is analogous to Lemma 6. The reason that we can save a $\ln q$ factor in the sample size is that J is independent of choice of Γ, since a^{opt} does not rely on sampling. Thus, instead of the sum of the worst n' errors of random variables (among n candidates) in Lemma 6, we only need to estimate the sum of errors of known n' random variables. We omit routine details in this version.

Claim. If we choose the constant c sufficiently large, $F_2(\mathbf{a}(\Gamma)) \leq (q^\epsilon - 1)z^*$ with a large probability (say, ≥ 0.9).

Theorem 3 follows from the above claim. Recall that $z^* = \sum_{j \in J^*} a_j^{\mathrm{opt}}$. The formula $F_2(\mathbf{a}(\Gamma)) \leq (q^\epsilon - 1)z^*$ implies that either $z^{\mathrm{opt}} \leq 2q^\epsilon z^*$ or $F_2(\mathbf{a}(\Gamma)) \leq \frac{1-q^{-\epsilon}}{2} z^{\mathrm{opt}}$. In the first case, we have $2q^\epsilon$-approximation since the objective function value of our solution is at least z^*. In the second case, with probability 0.8, $z^* \geq z^{\mathrm{opt}} - (F_1(\mathbf{a}(\Gamma)) + F_2(\mathbf{a}(\Gamma))) \geq \frac{q^{-\epsilon}}{2} z^{\mathrm{opt}}$, and hence we also attain $2q^\epsilon$-approximation.

Now, let us prove the claim. The following lemma follows from Chernoff's bounds for large deviation cases:

Lemma 9. *Let* $\tau_j = \max\{1, \mu_j/\mu\}$, *and we assume* $\theta > e^2 \approx 7.39$. *Then, we have the following: If* $\mu_j > \mu$, $Pr[h_j > (1+\theta)\mu_j] < (\theta+1)^{-\theta r^{-1} c\tau_j/2}$. *If* $\mu_j < \mu$, $Pr[h_j > (1+\theta)\mu] < (\theta+1)^{-\theta r^{-1} c\tau_j/2}$. *Here,* $\mu = r^{-1}\tilde{\gamma} z^{\mathrm{opt}}/n' = r^{-1}c$ *is the average of* μ_j *for* $j \in J^*$.

The following lemma is a refinement of Lemma 5 (proof is omitted).

Lemma 10. *Let* $S = \{n_1, n_2, \ldots n_k\}$ *and* $T = \{m_1, m_2, \ldots m_k\}$ *be sets of positive numbers. If* $\sum_{1 \leq j \leq k} m_j = M$ *and* $\sum_{1 \leq j \leq k} n_j > NM$, *then there exists a natural number* ℓ *and* $I_\ell \subset \{1, 2, \ldots, k\}$ *such that* $\sum_{j \in I_\ell} m_j > 2^{-\ell+1}M$ *and* $n_j \geq \frac{\ell N m_j}{3}$ *for each* $j \in I_\ell$.

We apply Lemma 10 setting $m_j = \tau_j$, $n_j = h_j - \mu_j$ for $j \in J^*$ and $N = \frac{(r-1)\mu}{2}$, where we ignore indices outside J^*. We remark that $n' \leq M = \sum_{j \in J^*} \tau_j < 2n'$. Because $\sum_{j \in J^*} h_j = r \sum_{j \in J^*} \mu_j = rn'\mu$, we have $\sum_{j \in J^*}(h_j - \mu_j) = (r-1)n'\mu > (r-1)M\mu/2 = NM$. Thus, there exists a natural number ℓ and a set I_ℓ such that $\sum_{j \in I_\ell} \tau_j > 2^{-\ell+1}n'$ and $n_j \geq \frac{\ell N \tau_j}{3}$ for each $j \in I_\ell$. Thus, $h_j - \mu_j \geq \frac{\ell(r-1)\tau_j \mu}{6}$ for each $j \in I_\ell$. From Lemma 9, the probability P that $h_j - \mu_j \geq \frac{\ell(r-1)\tau_j\mu}{6}$ is bounded by $(\ell(r-1)\tau_j)^{-c\ell\tau_j(r-1)/12r} < (\ell r)^{-c\ell\tau_j/13}$, if r is sufficiently large.

Now, suppose that the claim is false and $r > q^\epsilon$, and set the constant $c = 26\epsilon^{-1}$. Then, the probability P is $(\ell q)^{-2\ell\tau_j} < q^{-2\ell\tau_j}$. By using Markov's inequality, we can show that the probability of the existence of I_ℓ for a fixed ℓ is at most $nP \leq \frac{n}{2^{-\ell+1}n'} q^{-2\ell} < q^{-\ell}$. Thus, the probability of existence of such a pair (ℓ, I_ℓ) is bounded by $\sum_{\ell=1}^{\log q} q^{-\ell}$, which is very small if q is large. Therefore, we have a contradiction. Thus, we have proven the claim, and hence Theorem 3.

5 Concluding Remarks

It is important to seek for more practical algorithms with approximation ratios as good as those given in this paper. Our algorithms are not practically efficient, but we may have good solution by taking a smaller number of instances combined with heuristics so that the process finishes within practical computation time.

Finally, it would be nice to remove the factor of 2 caused by the max-cut.

Acknowledgement. We thank an anonymous referee for suggesting derandomization of algorithms. This research is supported by JSPS Grant-in-Aid for Scientific Research on priority areas of New Horizons in Computing.

References

1. S. Arora, D. Karger, and M. Karpinski, Polynomial Time Approximation Schemes for Dense Instances of NP-hard Problems, *Proc. STOC'95* (1995), pp. 284-293.
2. Y. Asahiro, R. Hassin, and K. Iwama, Complexity of Finding Dense Subgraphs, *Discrete Applied Math.* **121** (2002), pp. 15-26.
3. Y. Asahiro, K. Iwama, H. Tamaki, and T. Tokuyama, Greedily Finding Dense Subgraphs, *Journal of Algorithms* **34** (2000), pp. 203-221.
4. G. Ausiello, P. Crescenzi, G. Gambosi, V. Kann, A. Marchetti-Spaccamela, M. Protasi, *Complexity and Approximation, Combinatorial Optimization Problems and Their Approximability Properties*, Springer-Verlag (1999).
5. A. Czygrinow, Maximum Dispersion Problem in Dense Graphs, *Operation Research Letters* **27** (2000), pp. 223-227.
6. U. Feige, Relations between Average Case Complexity and Approximation Complexity, *Proc. STOC 02* (2002), pp. 534-543.
7. U. Feige, G. Kortsarz, D. Peleg, The Dense k-Subgraph Problem, *Algorithmica* **29** (2001), pp. 410-421.
8. D. S. Hochbaum, Approximating clique and biclique problems, *J. Algorithms* **29** (1998), pp. 174-200.

9. G. Gallo, M. D. Grigoriadis, R. E. Tarjan, A Fast Parametric Maximum Flow Algorithm and Applications, *SIAM J. Comput.* **18** (1989), pp. 30-55.
10. G. Kortsarz, D. Peleg, On Choosing a Dense Subgraph, *Proc. FOCS93* (1993), pp. 692-701.
11. R. Motwani, P. Raghavan, *Randomized Algorithms*, Cambridge University Press (1995).

A Complete Characterization of Tolerable Adversary Structures for Secure Point-to-Point Transmissions Without Feedback

Yvo Desmedt[1,3,⋆], Yongge Wang[2,⋆⋆], and Mike Burmester[3,⋆⋆⋆]

[1] University College London, UK
y.desmedt@cs.ucl.ac.uk
[2] UNC Charlotte, USA
yonwang@uncc.edu
[3] Florida State University, USA
burmester@cs.fsu.edu

Abstract. Problems of unconditionally secure communication have been studied extensively in network models. Dolev-Dwork-Waarts-Yung considered the Byzantine threats model, in which the adversary can only take over a number of nodes bounded by a threshold. They studied two cases:

1. all communication links (edges in the graph) are two-way communication,
2. all communication links are one-way communication, and there is no feedback.

The node sets that the adversary can take over was generalized by Hirt-Maurer to an adversary structure. At PODC 2002, Kumar-Goundan-Srinathan-Rangan generalized Dolev-Dwork-Waarts-Yung's first scenario to the case of a general adversary structure. In this paper we generalize Dolev-Dwork-Waarts-Yung's second scenario to the case of a general adversary structure. As in Dolev-Dwork-Waarts-Yung, our work relies on the use of secret sharing.

Keywords: Network security, Byzantine threats, Secret Sharing, adversary structure, unconditional security.

1 Introduction

Originally work on secure distributed computation (see, e.g., [1,3]) assumed that the parties were connected by a complete network of reliable and private point-to-point communication channels, with an adversary that could control up to k (Byzantine) nodes.

In practical networked environments it is often the case that two parties are not connected with a private and authenticated channel. They then need to use intermediate nodes to help them to carry out secure communication and secure multiparty computations. In this case, it is important to design secure communication protocols to achieve

⋆ A part of this work has been funded by CCR-0209092. The author is BT Professor of Information Security.

⋆⋆ This work was supported, in part, by funds provided by The University of North Carolina at Charlotte.

⋆⋆⋆ A part of this work has been funded by CCR-0209092.

X. Deng and D. Du (Eds.): ISAAC 2005, LNCS 3827, pp. 277–287, 2005.

participant cooperation in the presence of faults. Dolev, Dwork, Waarts, and Yung [6] initiated the study of reducing the requirements for network connectivity in secure multiparty computation by providing protocols that achieve private and reliable communication. In the case of k Byzantine faults, they studied the cases:

1. all communication links (edges in the graph) are two-way communication. Reliable and private communication is achievable if and only if the communication network is $2k + 1$ connected.
2. all communication links are one-way communication, and there is no feedback. Then $3k + 1$ connectivity is necessary and sufficient for reliable and private communications.

In 2002, Desmedt and Wang [4] extended this to the case when there are feedback channels.

The Byzantine faults model typically addresses threat scenarios in which the faults are *independent*. This assumption is unrealistic when dealing with computer viruses, such as the ILOVEYOU [11] virus and the Internet virus/worm [7] that only spread to Windows, respectively Unix. A hacker who can exploit a weakness in one platform, can with almost the same ease attack many computers, if not all, on that same platform. There is therefore a need to design a model in which nodes on the same platform can be distinguished. One such model, the *adversary structure* model, was proposed by Hirt and Maurer [10] for secure multiparty computation (for an earlier version see [9]). In this model the adversary is characterized by a structure (or family) of subsets of the set of parties, which are the sets that the adversary can corrupt. Hirt and Maurer gave necessary and sufficient conditions for secure multiparty computation in this adversary model. Similar to the classical results for multiparty computation, Hirt and Maurer assumed that communication networks are complete.

As in Dolev, Dwork, Waarts, and Yung [6], in this paper, we study the problem of reducing the requirements for network connectivity in secure multiparty computation in the sense of Hirt and Maurer [10] under the adversary structure model. We give necessary and sufficient conditions on the communication network, with respect to a given adversary structure $\mathcal{Z}$, such that we have reliable and private point-to-point communication. It should be noted that results for not necessarily complete bi-direction networks have been obtained by Kumar-Goundan-Srinathan-Rangan in [13]. They used an induction argument to prove the sufficient condition. There are two essential differences between our results and the results in [13].

- The secure message transmission protocols in [13] only apply to networks in which all links are two-way. That is, one can send messages in two directions. Our protocols work for networks that are not necessarily complete and the links are one-way. That is, one may send message from one direction though not the other direction.
- Our protocols for secure transmissions are slightly more efficient than these in [13] due to the induction processes in [13]. Specifically, our protocols will use *minimal* $\mathcal{Z}$-connected path-sets. We shall find a bound on the number of paths in such path-sets, and show that it is sharp. Although we focus on point-to-point networks, our results can easily be extended to broadcast networks.

The organization of this paper has as follows. In Section 2, we describe our model and give definitions. In Section 3 we find necessary and sufficient conditions for secure communication in our adversary model and propose secure transmission protocols. In Section 4 we propose bounds.

2 Model and Background

2.1 Threshold Secret Sharing Schemes

Let $\mathbf{F}$ be a finite field, and let k, n be integers such that $0 \leq k < n$. A $(k+1)$-out-of-n *secret sharing scheme* is a probabilistic function S: $\mathbf{F} \rightarrow \mathbf{F}^n$ with the property that: for any $M \in \mathbf{F}$ and $\mathrm{S}(M) = (v_1, \ldots, v_n)$, no information about M can be inferred from any k entries of $(v_1, \ldots, v_n)$, whereas M can be recovered from any $k+1$ entries of $(v_1, \ldots, v_n)$.

2.2 The Adversary

We employ an unconditional setting. That is, the adversary has unlimited resources. The adversary is characterized by an adversary structure $\mathcal{Z}$ [10] that consists of all sets of parties that the adversary can corrupt. Formally, let $\mathcal{P}$ be a party set. A subset $\Gamma_{\mathcal{P}}$ of the power set $2^{\mathcal{P}}$ of $\mathcal{P}$ is called an access structure on $\mathcal{P}$ [12]. It is monotone if and only if $\emptyset \notin \Gamma_{\mathcal{P}}$ and supersets of elements $\Gamma_{\mathcal{P}}$ also belong to $\Gamma_{\mathcal{P}}$, i.e., we require that if $A \in \Gamma_{\mathcal{P}}$ and $A \subseteq A' \subseteq \mathcal{P}$, then $A' \in \Gamma_{\mathcal{P}}$. Let $\mathcal{Z}_{\mathcal{P}} = \Gamma_{\mathcal{P}}$. We call $\mathcal{Z}_{\mathcal{P}} \subset 2^{\mathcal{P}}$, or simply $\mathcal{Z}$ when there is no confusion, an adversary structure on $\mathcal{P}$ if its complement, i.e., $\mathcal{Z}_{\mathcal{P}}^c = 2^{\mathcal{P}} \setminus \mathcal{Z}_{\mathcal{P}}$ is a monotone access structure.

If $\mathcal{Z}_1$ and $\mathcal{Z}_2$ are adversary structures for $\mathcal{P}$, then $\mathcal{Z}_1 + \mathcal{Z}_2 = \{Z_1 \cup Z_2 : Z_1 \in \mathcal{Z}_1, Z_2 \in \mathcal{Z}_2\}$ is also an adversary structure for $\mathcal{P}$. $2\mathcal{Z}$ and $3\mathcal{Z}$ are the adversary structures $\mathcal{Z} + \mathcal{Z}$ and $\mathcal{Z} + \mathcal{Z} + \mathcal{Z}$ respectively. Finally, a set of parties $Z \in \mathcal{Z}$ is *maximal* if: $Z' \supset Z \Rightarrow Z' \notin \mathcal{Z}$. In the remaining part of the paper, we will use $2\mathcal{Z}$ and $3\mathcal{Z}$ to denote the adversary structures $\mathcal{Z} + \mathcal{Z}$ and $\mathcal{Z} + \mathcal{Z} + \mathcal{Z}$ respectively.

We consider two kinds of adversary: a *passive* and an *active* adversary. A passive adversary (or gossiper adversary) can only read the traffic (the variables in the view) of the parties in Z. An active adversary has unlimited computational power and can read the traffic of Z and also control the parties in Z. Both kinds of adversary are assumed to know the complete protocol specification, message space, and the complete structure of the graph. In this paper, we shall not consider dynamic adversaries who can change the parties they corrupt from round to round, but only static adversaries. That is, the adversary selects which set of parties $Z \in \mathcal{Z}$ to corrupt, before the start of the protocol.

2.3 The Communication Network

We model the communication network by a directed graph $G = G(V, E)$ whose nodes are the parties and whose edges are point-to-point reliable and private communication channels.

2.4 Message Transmission Protocols

Let π be a message transmission protocol, let A be the sender and B the receiver, and let $\mathcal{Z}$ be an adversary structure. The sender A selects a message M^A drawn from a message space $\mathcal{M}$ with a certain probability distribution. At the beginning of the protocol, the adversary flips coins and chooses a set $Z \in \mathcal{Z}$ of nodes to corrupt. At the end of protocol π, the receiver B outputs a message $M^B \in \mathcal{M}$. We will assume that the message space $\mathcal{M}$ is a subset of a finite field $\mathbf{F}$ (our results easily extend to message spaces of tuples over $\mathbf{F}$).

For any message transmission protocol we denote by adv the adversary's view of the execution of the protocol and by $adv(M, r)$ the view when $M^A = M$ and when the sequence of coin flips used by the adversary is r.

Definition 1. *Let π be transmission protocol, let M^A the message selected by A and M^B the message output by B, let $\mathcal{Z}$ be an adversary structure.*

1. *We say that π is $\mathcal{Z}$-*reliable *if B outputs $M^B = M^A$ with probability* 1. *(The probability is taken over the choices of M^A and the coin flips of all nodes.)*
2. *We say that π is perfectly $\mathcal{Z}$-*private *if for any two messages M_0, M_1, and for any coin tosses r, we have* $\Pr[adv(M_0, r) = c] = \Pr[adv(M_1, r) = c]$. *The probabilities are taken over the coin flips of the honest parties.*
3. *We say that π is perfectly $\mathcal{Z}$-*secure *if it is $\mathcal{Z}$-*reliable *and perfectly $\mathcal{Z}$-*private.

2.5 Connectivity

Definition 2. *Let $G(V, E)$ be a directed graph, A, B be nodes in $G(V, E)$, and $\mathcal{Z}$ be a an adversary structure on $V \setminus \{A, B\}$.*

- *A, B are $\mathcal{Z}$-*separable *in G, if there is a set $Z \in \mathcal{Z}$ such that all paths from A to B go through at least one node in Z. We say that Z separates A and B.*
- *A, B are $(\mathcal{Z} + 1)$-*connected *if they are not $\mathcal{Z}$-separable in G.*

Observe that if $(A, B) \in E$, then A, B are $(\mathcal{Z}+1)$-*connected* for any $\mathcal{Z}$ on $V \setminus \{A, B\}$. The following result will be needed in our later discussions.

Theorem 1. *Let $G = G(V, E)$ be a directed graph, A, B be nodes in G, and $\mathcal{Z}_1, \mathcal{Z}_2$ be adversary structures on $V \setminus \{A, B\}$. Then A, B are $(\mathcal{Z}_1 + \mathcal{Z}_2 + 1)$-connected if, and only if: for all sets $Z_1 \in \mathcal{Z}_1$ there is a set S_{Z_1} of paths between A and B such that,*

- *the paths in S_{Z_1} are free from nodes of Z_1,*
- *for every $Z_2 \in \mathcal{Z}_2$ there is at least one path in S_{Z_1} that is free from nodes of Z_2.*

Proof. First consider the case when A, B are $(\mathcal{Z}_1 + \mathcal{Z}_2 + 1)$-connected. We prove that the conditions are satisfied. For any set $Z_1 \in \mathcal{Z}_1$, let S_{Z_1} be the set of all paths from A to B free from nodes of Z_1. Assume that there is a set $Z_2 \in \mathcal{Z}_2$ such that all paths of S_{Z_1} go through Z_2. Then $Z_1 \cup Z_2$ separates A, B in G. That is, A, B are $(\mathcal{Z}_1 + \mathcal{Z}_2)$-separable in G. This is a contradiction.

For the converse observe that the conditions on the paths S_{Z_1} make it impossible to have a set $Z_2 \in \mathcal{Z}_2$ such that $Z = Z_1 \cup Z_2$ separates A, B. Indeed if there where such a set Z separating A, B then there would be no path in S_{Z_1} free of Z_1 and Z_2.

3 Secure Message Transmissions

We start by discussing message transmissions that tolerate passive adversaries. First we briefly describe the intuition behind our protocols by observing that, whereas in the Byzantine threats model the sender and receiver use vertex disjoint paths, for general adversary structures this is not necessarily the case.

Theorem 2. *Let $G = G(V, E)$ be a directed graph, A, B be two nodes in G, and $\mathcal{Z}$ be an adversary structure on $V \setminus \{A, B\}$. Suppose that the adversary is passive.*

1. *We have polynomial time (in the graph size) $\mathcal{Z}$-reliable message transmission from A to B if, and only if, A, B are $(\{\emptyset\} + 1)$-connected in G.*
2. *We have polynomial time (in the graph size) perfectly $\mathcal{Z}$-secure message transmission from A to B if, and only if, A, B are $(\mathcal{Z} + 1)$-connected in G.*

Proof. Result 1 is straightforward. For Result 2, first observe that if A, B are $\mathcal{Z}$-separable in G, then there is a set $Z \in \mathcal{Z}$ such that all paths from A go through Z. Therefore $\mathcal{Z}$-private message transmission from A to B is impossible. Next suppose that A, B are $(\mathcal{Z} + 1)$-connected in G. In the following we describe a polynomial time (in the graph size) protocol for A to securely send a message M^A to B (our protocol is similar to a protocol in [8, Lemma 2]).

1. Let $G'(V'E')$ be the maximum subgraph of $G(V, E)$ such that for each node $v \in V'$, there is a direct path from v to B.
2. For each edge $(u, v) \in E'$, u chooses a random message (group elements chosen according to the uniform distribution) $r_{u,v}$ and sends it to the node v.
3. Every node computes the sum of messages it has received and substracts the sum of messages it has sent out. If the node is the actual sender A, then it adds to this total the messsage M^A. Call this sum the "final result" for this node.
4. Each final result from Step 3, except for the final result held by the actual receiver B, is propagated by the nodes openly to the receiver B through a series of transmissions. The sum of all final results, including the final result held by the receiver B, is the message M^B.

Using a similar proof to that in [8, Lemma 2], we can show that the above protocol is a $\mathcal{Z}$-secure message transmission from A to B if A, B are $(\mathcal{Z} + 1)$-connected in G.

Next we consider secure message transmission protocols for active adversaries.

Theorem 3. *Let $G = G(V, E)$ be a directed graph, A, B be nodes in G, and $\mathcal{Z}$ be an adversary structure on $V \setminus \{A, B\}$. We have $\mathcal{Z}$-reliable message transmission from A to B if, and only if, A, B are $(2\mathcal{Z} + 1)$-connected in G.*

Proof. First assume that A, B are $(2\mathcal{Z} + 1)$-connected in G, and let S be the set of all directed paths from A to B. The paths in S will be used by the sender A to transmit messages to B. Let M^A be the message that A wants to send to B. For each path $p \in S$, A sends M^A to B over p. At the end of the protocol, B receives M_p^B through each path $p \in S$. Then by using Theorem 1, B finds a node set $Z_1 \in \mathcal{Z}$ whose path set S_{Z_1} is such that the same message is received on all its paths. Let M^B be this message. It is sufficient to show that $M^B = M^A$. Suppose that the adversary selects $Z_2 \in \mathcal{Z}$. We have:

- If $Z_1 \cap Z_2 = \emptyset$ then $M^A = M^B$.
- If $Z_1 \cap Z_2 \neq \emptyset$ then by Theorem 1, since A, B are $(2\mathcal{Z}+1)$-connected, there will be a path $p_0 \in S_{Z_1}$ free from nodes of Z_2. On this path $M^B_{p_0} = M^A$. Since B receives the same message from all paths in S_{Z_1}, we must have $M^A = M^B_{p_0} = M^B$.

It follows that B can reliably recover the message M^A.

Next assume that A, B are not $(2\mathcal{Z} + 1)$-connected in G. That is, there are sets $Z_1, Z_2 \in \mathcal{Z}$ such that all directed paths from A to B passes through some nodes in $Z_1 \cup Z_2$. Let M_0 be the message that A transmits. The adversary will attempt to maintain a simulation of the possible behavior of A by executing the message transmission protocol for some other message M_1. The strategy of the adversary is to flip a coin and then, depending on the outcome, decide which set of Z_1 or Z_2 to control (our model is not dynamic, so the selection is done before the protocol starts). Let Z_b be the chosen set. In each execution step of the transmission protocol, the adversary causes each node in Z_b to follow the protocol as if the transmitted message were M_1. Since B does not know whether $b = 1$ or $b = 2$, with probability better than 1/2, at the end of the protocol B cannot decide whether A has transmitted M_0 or M_1 with probability better than 1/2.

Theorem 4. *Let $G = G(V, E)$ be a directed graph, A, B be nodes in G, and $\mathcal{Z}$ be an adversary structure on $V \setminus \{A, B\}$. If there are no directed paths from B to A, then we have perfectly $\mathcal{Z}$-secure message transmission from A to B if and only if, A, B are $(3\mathcal{Z} + 1)$-connected in G.*

Proof. First we show that the condition is sufficient. We shall describe a message transmission protocol π that is perfectly $\mathcal{Z}$-secure. Let $\mathcal{Z} = \{Z_1, \ldots, Z_t\}$ be the adversary structure (to make our protocol more efficient, we could take a list of maximal adversary sets) and let M^A be the message that A wants to send to B. The sender A first uses a t-out-of-t secret sharing scheme to get shares $(s^A_1, \ldots, s^A_t)$ of the message M^A. For each $i \leq t$, A reliably sends s^A_i to B via the reduced graph $G_{V\setminus Z_i} = G_{V\setminus Z_i}(V\setminus Z_i, E_{V\setminus Z_i})$, where $E_{V\setminus Z_i} = E \setminus \{(u, v) : u \in Z_i \text{ or } v \in Z_i\}$. For each $i \leq t$, B reliably receives s^B_i on the reduced graph $G_{V\setminus Z_i}$, and hence recovers the message M^B from the shares $(s^B_1, \ldots, s^B_t)$. Now assume that the adversary controls all nodes in Z_{i_0}. Then the adversary will learn no information about M^A from $s^A_{i_0}$. Therefore the transmission protocol π is perfectly $\mathcal{Z}$-private. It remains to show that π is $\mathcal{Z}$-reliable. Since A, B are $(3\mathcal{Z}+1)$-connected, it is straightforward to see that for each $Z_i \in \mathcal{Z}$, the reduced graph $G_{V\setminus Z_i}$ is $(2\mathcal{Z} + 1)$-connected. From Theorem 3 we get that B receives reliably all the shares $(s^A_1, \ldots, s^A_t)$. It follows that the protocol is perfectly $\mathcal{Z}$-secure.

Next we show that the condition is necessary. Assume that A, B are not $(3\mathcal{Z} + 1)$-connected. Then there are sets $Z_1, Z_2, Z_3 \in \mathcal{Z}$ such that all directed paths from A to B pass through some nodes in $Z_1 \cup Z_2 \cup Z_3$. Let M_0 be the message that A transmits. The adversary will attempt to maintain a simulation of the possible behavior of A by executing the transmission protocol π for some other message M_1. The strategy of the adversary is to flip coins and then, depending on the outcome, decide which set of Z_1, Z_2 or Z_3 to control. Let Z_b be the chosen set. In each execution step of the transmission protocol, the adversary causes each node in Z_b to follow the protocol π as if the protocol were transmitting the message M_1. At the end of the protocol, the view of B could be divided into three parts $view_{Z_1}, view_{Z_2}$, and $view_{Z_3}$, where $view_{Z_i}$ $(i = 1, 2, 3)$

consists of all the information that the nodes in Z_i have learned. Since neither A nor B knows whether $b = 1$, $b = 2$ or $b = 3$, and since π is a perfectly private message transmission protocol, M^A cannot be recovered from any single $view_{Z_i}$. Since π is a reliable message transmission protocol and the adversary controls one of Z_1, Z_2, or Z_3, B should be able to recover the secret message from two of these three views. That is, if we regard $(view_{Z_1}, view_{Z_2}, view_{Z_3})$ as shares of M^A in a 2-out-of-3 secret sharing scheme, then this scheme should be able to detect and simultaneously correct one error. However 2-out-of-3 secret sharing schemes can only detect one error, and cannot correct any errors (see, e.g., [14]). So we get a contradiction. This proves that there is no perfectly secure message transmission protocol from A to B when A, B are only $3\mathcal{Z}$-connected.

4 Bounds and Other Properties

4.1 Introduction

A *threshold adversary structure* $\mathcal{Z}$ is a structure whose maximal sets Z have cardinality bounded by a threshold k. An example of such a structure is the classical Byzantine faults structure. Evidently if one restricts oneself to threshold adversary structures, then there is a description of the adversary structure which is logarithmic in the number of elements in it. In this context, it should be noted that the "polynomial" algorithms for multiparty computation in Hirt and Maurer [10] are polynomial in the size of the adversary access structure which is generally exponential in the size of the network. Hirt and Maurer do not study the problem for restricted access structures.

For a threshold adversary structure $\mathcal{Z}$ we get $\mathcal{Z}$-tolerable communication in both the passive and active case [5,6] via path-sets S for which,

- the paths are vertex-disjoint, and so
- the number of paths is polynomially bounded (in the size of the graph).

The goal of this section is to show that both properties are false in the general adversary case, when limiting the adversary structures. Our results do *not* rely on unproven assumptions.

We shall use a family of specific adversary structures $\mathcal{Z}$ that was informally introduced in [2]. This structure consists of sets Z of nodes of a colored graph that have at most k colors (e.g. computers running the same/different operating system are respectively colored with the same/different color). In other words, the adversary can corrupt any set of nodes provided it has no more than k colors. Evidently the number of nodes in Z can be much larger than k (if many nodes have the same color). We now define this structure formally.

Definition 3. *A colored graph is a tuple $G = G(V, E, C, f)$, with V the node set, E the edge set, C the color set, and f a map from V onto C. The structure*

$$\mathcal{Z}_{C,k} = \{Z \mid Z \subset V \text{ and } |f(Z)| \leq k\}.$$

is called a k-color adversary structure.

Note that if all nodes of the colored graph have different colors then we have the classical Byzantine faults structure, with no colors.

4.2 Bounds

To study the properties of path-sets that we shall use for $\mathcal{Z}$-tolerable communication we define the following. (This definition can also be used for general adversary structures, not based on colors.)

Definition 4. *Let $G(V,E)$ be a directed graph, A, B be nodes in G, S be a set of simple paths in G between A and B, and G_S be the graph obtained by removing all nodes and edges of G not in S. Let $\mathcal{Z}$ be an adversary structure. We say that S is a minimal $(\mathcal{Z}+1)$-connected path-set from A to B in G, if*

1. *A and B are $(\mathcal{Z}+1)$-connected in G_S, and*
2. *for each path $p \in S$, A and B are $\mathcal{Z}$-separable in $G_{S\setminus\{p\}}$.*

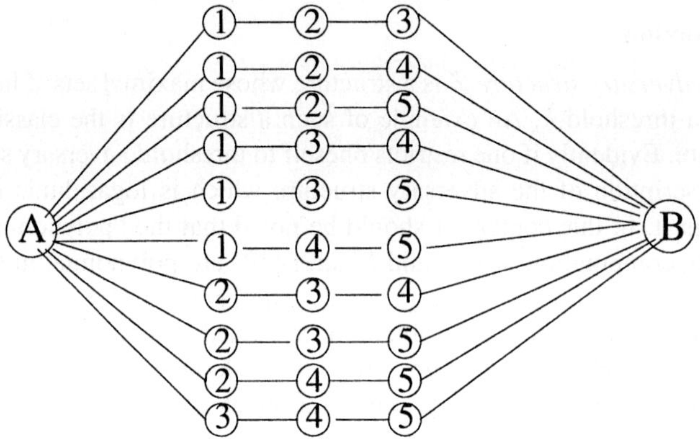

Fig. 1. A minimal $(\mathcal{Z}_{C,k}+1)$-connected path-set of a graph with 5 colors for a 2-color adversary structure ($c=5$, $k=2$)

Lemma 1. *Let $G = G(V,E,C,f)$ be a colored graph, S be a minimal $(\mathcal{Z}_{C,k}+1)$-connected path-set between A and B, and c be the number of colors in the graph. Then $|S| \le \binom{c}{k}$. This bound is tight.*

Proof. We apply Theorem 1 with $\mathcal{Z}_1 = \{\emptyset\}$ and $\mathcal{Z}_2 = \mathcal{Z}_{C,k}$, and construct a minimal path-set, starting with $S = \emptyset$. For every maximal $Z_2 \in \mathcal{Z}_2$, find a path free from nodes of Z_2, label it Z_2, and add it to S. If there is already such a path in S, then just add the extra label Z_2 to this path.

Note that non-maximal sets do not add extra paths. Indeed, if after running this construction one takes a set $Z_2' \subset Z_2$, then the path with label Z_2 will also receive the Z_2' label. Since all the different sets $Z_2 \in \mathcal{Z}_2$ have been considered, the set S will be $(\mathcal{Z}_{C,k}+1)$-connected (by Theorem 1).

Now note that a maximal set Z_2 must have k colors and Z_2 must contain all nodes with these colors. So there are $\binom{c}{k}$ maximal sets in $\mathcal{Z}_2$. Combining this with the previous argument we get $|S| \le \binom{c}{k}$.

To show that the bound is tight we describe an appropriate colored graph $G = G(V, E, C, f)$ and a minimal path-set S. Figure 1 illustrates our proof for $c = 5$ and $k = 2$.

Let $c \geq k+1$. We construct V and E as follows. Start with $V = \{A, B\}$ and $E = \emptyset$. Then add $\binom{c}{k}$ node-disjoint paths connecting A to B as follows:

For i from 1 till $\binom{c}{k}$ do:

Step 1 Select a not yet selected subset $C' \subset C$ of k colors.
Step 2 Add $c - k$ new nodes $v_{i,1}, \ldots, v_{i,c-k}$ to V.
Step 3 Color these $c - k$ new nodes with the $C \setminus C'$ different colors, in such a way that each gets a different color.
Step 4 Add the edges $(A, v_{i,1})$, $(v_{i,1}, v_{i,2})$, $\ldots$, $(v_{i,c-k-1}, v_{i,c-k})$, $(v_{i,c-k}, B)$ to E.

end-for-loop.

It is easy to verify that we get a minimal $(\mathcal{Z}_{C,k} + 1)$-connected path-set from A to B. Indeed, if one picks a maximal element from $\mathcal{Z}_{C,k}$ and removes all its nodes (which is equivalent to picking any k colors and removing all nodes with these colors) one is left with just one path. So, A and B are $(\mathcal{Z}_{C,k} + 1)$-connected. This also implies that if one removes a single path from this path-set then A, B become $\mathcal{Z}_{C,k}$-separable.

Remark 1. Though the minimal path-set in Lemma 1 is super-polynomial in k, it is still polynomial in the graph size. The construction in the proof of Lemma 1 will be used to prove Theorem 5.

Theorem 5. *There are directed graphs $G(V, E)$ and adversary structures $\mathcal{Z}$ for which the number of paths in a minimal $(\mathcal{Z} + 1)$-connected path-set from A to B is super-polynomial in the size of the graph.*

Proof. We modify the colored graph $G(V, E, C, f)$ constructed in Lemma 1 to get a graph $G'(V', E')$ as follows. Take $|C| = 2k + 1$ (so $c - k = k + 1$) and construct G' with $|E'| = |E|$ and $|V'| = (k + 1)^2 + 2$. Start from G. For each path in this graph reorder linearly the nodes accordingly to their colors (if the colors are labeled $1, 2, \ldots$ put the nodes with the smallest color label the closest to A). Make the graph directed, i.e. the paths go from A to B. Write the nodes and their colors in a $\binom{c}{k} \times (k+1)$ matrix $T = [(v_{i,j}, f(v_{i,j})]$.

Now we explain how V' is constructed. In each column of T collapse the nodes with the same color: that is, if $f(v_{i,j}) = f(v_{i',j})$ then $v'_{i,j} = v'_{i',j}$. From elementary combinatorics we get that the number of different nodes that remain in each column is $k + 1$. So, we have a total of $(k + 1)^2 + 2$ different nodes including A and B. It is now easy to verify that the size of this path-set is super-polynomial.

Observe that the paths of the minimal path-set in Lemma 1 are *not* node-disjoint.

Remark 2. Though the number of paths in a minimal $(\mathcal{Z}+1)$-connected path-set for the directed graph $G(V, E)$ in Theorem 5 is super-polynomial in the graph size, one may still be able to design polynomial time algorithms (in the size of the graph) for perfect secure communication. Indeed, for the case of privacy against a passive adversary, we have an algorithm that runs in polynomial time in the graph size, which is more efficient than in polynomial time in the graph and adversary size (see Theorem 2). It is an **open** problem whether there exists a graph such that all perfect secure communication protocols are super-polynomial in the graph size.

More details for colored graphs are given in the full version of this paper.

4.3 Other Results for Color Adversary Structures

Since color adversary structures are of interest on their own (to model platform dependent attacks) [2], we prove the following result:

Theorem 6. *Let $G = G(V, E, C, f)$ be a colored graph which is $(\mathcal{Z}_{C,k}+1)$-connected. If the number of colors is minimal then the paths in a minimal path-set are node-disjoint and each path is monochrome (all nodes on one path have the same color).*

Proof. Using Lemma 1 we see that the number of colors must be at least $k + 1$. We now prove that we can have $k+1$ colors and that there is a minimal path-set with $k+1$ monochrome paths. Apply Theorem 1 with $\mathcal{Z}_1 = \{\emptyset\}$ and $\mathcal{Z}_2 = \mathcal{Z}_{C,k}$ to construct a minimal path-set. Start with $S = \emptyset$. Let Z_2 be a maximal subset of $\mathcal{Z}_2$. Obviously Z_2 contains nodes of k different colors. Moreover, due to its maximality, it contains all nodes that have these colors. Then by Theorem 1 there must be at least one path which is free from the nodes of Z_2, that is, free from the k colors of Z_2. Since there are only $k + 1$ colors, this path must be monochrome. Add one of these paths to S. Continue with a new maximal $Z_2' \in \mathcal{Z}_2$ with at least one color different from Z_2 until all k-color sets are exhausted. In this way we get a minimal path-set consisting of $k+1$ paths, one for each color. Clearly these paths are disjoint.

References

1. M. Ben-Or, S. Goldwasser, and A. Wigderson. Completeness theorems for non-cryptographic fault-tolerant distributed computing. In: *Proc. ACM STOC '88*, pages 1–10.
2. M. Burmester and Y. G. Desmedt. Is hierarchical public-key certification the next target for hackers? *Communications of the ACM*, 47(8), pp. 68–74, August 2004.
3. D. Chaum, C. Crépeau, and I. Damgård. Multiparty unconditionally secure protocols. In *Proc. ACM STOC*, pp. 11–19, May 2–4, 1988.
4. Y. Desmedt and Y. Wang. Perfectly secure message transmission revisited. In: *Proc. Eurocrypt '02*, Lecture Notes in Computer Science 2332, Springer-Verlag, pp. 502–517, 2002.
5. D. Dolev. The Byzantine generals strike again. *Journal of Algorithms*, 3, pp. 14–30, 1982.
6. D. Dolev, C. Dwork, O. Waarts, and M. Yung. Perfectly secure message transmission. *Journal of the ACM*, 40(1), pp. 17–47, January 1993.
7. M. W. Eichin and J. A. Rochlis. With microscope and tweezers: an analysis of the Internet virus of November 1988. In *Proc. IEEE Sym. on Security and Privacy*, pp. 326–343, 1989.
8. M. Franklin and M. Yung. Secure hypergraphs: privacy from partial broadcast. In: *Proc. ACM STOC '95*, pages 36–44, ACM Press, 1995.
9. M. Hirt and U. Maurer. Complete Characterization of Adversaries Tolerable in Secure Multi-Party Computation. In *Proc. of the 16th ACM PODC*, pp. 25–34, August, 1997.

10. M. Hirt and U. Maurer. Player Simulation and General Adversary Structures in Perfect Multiparty Computation. Journal of Cryptology 13(1): 31-60 (2000)
11. 'ILOVEYOU' computer bug bites hard, spreads fast. `http://www.cnn.com/2000/TECH/computing/05/04/iloveyou.01/index.html`, May 4, 2000.
12. M. Ito, A. Saito and T. Nishizeki. Secret sharing schemes realizing general access structures. *Proc. IEEE Global Telecommunications Conf., Globecom'87*, pp. 99–102, 1987.
13. M. Kumar, P. Goundan, K. Srinathan, and C. Rangan. On perfectly secure communication over arbitrary networks. Proc. of ACM PODC 2002, pages 193–202.
14. F. J. MacWilliams and N. J. A. Sloane. *The theory of error-correcting codes*. North-Holland Publishing Company, 1978.

Network Game with Attacker and Protector Entities*

Marios Mavronicolas[1], Vicky Papadopoulou[1], Anna Philippou[1], and Paul Spirakis[2]

[1] Department of Computer Science, University of Cyprus, Nicosia CY-1678, Cyprus
{mavronic, viki, annap}@ucy.ac.cy

[2] Department of Computer Engineering and Informatics, University of Patras, 265 00 Patras, Greece, & Research and Academic Computer Technology Institute, 261 10 Patras, Greece
spirakis@cti.gr

Abstract. Consider an information network with harmful procedures called *attackers* (e.g., viruses); each attacker uses a probability distribution to choose a node of the network to damage. Opponent to the attackers is the *system protector* scanning and cleaning from attackers some part of the network (e.g., an edge or a path), which it chooses independently using another probability distribution. Each attacker wishes to maximize the probability of escaping its cleaning by the system protector; towards a conflicting objective, the system protector aims at maximizing the expected number of cleaned attackers.

We model this network scenario as a non-cooperative strategic game on graphs. We focus on the special case where the protector chooses a single edge. We are interested in the associated *Nash equilibria,* where no network entity can unilaterally improve its local objective. We obtain the following results:

- No instance of the game possesses a pure Nash equilibrium.
- Every mixed Nash equilibrium enjoys a graph-theoretic structure, which enables a (typically exponential) algorithm to compute it.
- We coin a natural subclass of mixed Nash equilibria, which we call *matching Nash equilibria,* for this game on graphs. Matching Nash equilibria are defined using structural parameters of graphs, such as independent sets and matchings.
  - We derive a characterization of graphs possessing matching Nash equilibria. The characterization enables a linear time algorithm to compute a matching Nash equilibrium on any such graph with a given independent set and vertex cover.
  - Bipartite graphs are shown to satisfy the characterization. So, using a polynomial-time algorithm to compute a perfect matching in a bipartite graph, we obtain, as our main result, an efficient graph-theoretic algorithm to compute a matching Nash equilibrium on any instance of the game with a bipartite graph.

* This work was partially supported by the IST Programs of the European Union under contract numbers IST-2001-33116 (FLAGS) and IST-2004-001907 (DELIS).

X. Deng and D. Du (Eds.): ISAAC 2005, LNCS 3827, pp. 288–297, 2005.

1 Introduction

Motivation and Framework. Consider an information network represented by an undirected graph. The nodes of the network are insecure and vulnerable to infection. A *system protector* (e.g., antivirus software) is available in the system; however, its capabilities are limited. The system protector can guarantee safety only to a small part of the network, such as a path or even a single edge, which it may choose using a probability distribution. A collection of *attackers* (e.g., viruses or Trojan horses) are also present in the network. Each attacker chooses (via a separate probability distribution) a node of the network; the node is harmed unless it is covered by the system protector. Apparently, the attackers and the system protector have conflicting objectives. The system protector seeks to protect the network as much as possible, while the attackers wish to avoid being caught by the network protector so that they be able to damage the network. Thus, the system protector seeks to maximize the expected number of attackers it catches, while each attacker seeks to maximize the probability it escapes from the system protector.

Naturally, we model this scenario as a strategic game with two kinds of players: the *vertex players* representing the attackers, and the *edge player* representing the system protector. The Individual Cost of each player is the quantity to be maximized by the corresponding entity. We are interested in the *Nash equilibria* [5,6] associated with this game, where no player can unilaterally improve its Individual Cost by switching to a more advantageous probability distribution. We focus on the simplest case where the edge player chooses a singe edge.

Summary of Results. Our results are summarized as follows:

- We prove that no instance of the game has a pure Nash equilibrium (pure NE) (Theorem 1).
- We then proceed to study mixed Nash equilibria (mixed NE). We provide a graph-theoretic characterization of mixed NE (Theorem 2). Roughly speaking, the characterization yields that the support of the edge player and the vertex players are an edge cover and a vertex cover of the graph and an induced subgraph of the graph, respectively. Given the supports, the characterization provides a system of equalities and inequalities to be satisfied by the probabilities of the players. Unfortunately, this characterization only implies an exponential time algorithm for the general case.
- We introduce *matching* Nash equilibria, which are a natural subclass of mixed Nash equilibria with a graph-theoretic definition (Definition 1). Roughly speaking, the supports of vertex players in a matching Nash equilibrium form together an independent set of the graph, while each vertex in the supports of the vertex players is incident to only one edge from the support of the edge player.
- We provide a characterization of graphs admitting a *matching* Nash equilibrium (Theorem 3). We prove that a *matching* Nash equilibrium can be computed in linear time for any graph satisfying the characterization once a *suitable* independent set is given for the graph.

– We finally consider bipartite graphs for which we show that they satisfy the characterization of *matching* Nash equilibria; hence, they always have one (theorem 5). More importantly, we prove that a *matching* Nash equilibrium can be computed in polynomial time for bipartite graphs (6).

Due to space limits, some proofs are omitted; we include them in the full version of the paper [3].

Significance. Our work joins the booming area of *Algorithmic Game Theory.* Our work is the *first*[1] to model realistic scenarios about infected networks as a strategic game and study its associated Nash equilibria. Our results contribute towards answering the general question of Papadimitriou [7] about the complexity of Nash equilibria for our special game. Our results highlight a fruitful interaction between *Game Theory* and *Graph Theory.* We believe that our *matching* Nash equilibria (and extensions of them) will find further applications in other network games and establish themselves as a candidate Nash equilibrium for polynomial time computation in other settings as well. Towards this direction, in a subsequent work of ours [4], we compute polynomial time Nash equilibria for the problem studied here in some practical families of graphs, such as regular graphs, graphs with perfect matching and trees. In the same work, we study Nash equilibria for a generalized variation of the problem considered here.

2 Framework

Throughout, we consider an undirected graph $G(V, E)$, with $|V(G)| = n$ and $|E(G)| = m$. Given a set of vertices $X \subseteq V$, the graph $G\backslash X$ is obtained by removing from G all vertices of X and their incident edges. A graph H, is an *induced* subgraph of G, if $V(H) \subseteq V(G)$ and $(u, v) \in E(H)$, whenever $(u, v) \in E(G)$. Denote $\Delta(G)$ the maximum degree of the graph G. For any vertex $v \in V(G)$, denote $Neigh(v) = \{u : (u, v) \in E(G)\}$, the set of neighboring vertices of v. For a set of vertices $X \subseteq V$, denote $Neigh_G(X) = \{u \notin X : (u, v) \in E(G)$ for some $v \in X\}$. For all above properties of a graph G, when no confusion raises, we omit G.

2.1 The Model

An information network is represented as an undirected graph $G(V, E)$. The vertices represent the network hosts and the edges represent the communication links. We define a non-cooperative game $\Pi(G) = \langle \mathcal{N}, \{S_i\}_{i \in \mathcal{N}}, \{\mathsf{IC}\}_{i \in \mathcal{N}} \rangle$ as follows:

– The set of players is $\mathcal{N} = \mathcal{N}_{vp} \cup \mathcal{N}_{ep}$, where $\mathcal{N}_{vp}$ is a finite set of *vertex* players vp_i, $i \geq 1$, and $\mathcal{N}_{ep}$ is a singleton set of an *edge* player ep. Denote $\nu = |\mathcal{N}_{ep}|$.

[1] To the best of our knowledge, [1] is a single exception. It considers inoculation strategies for victims of viruses and establishes connections with variants of the Graph Partition problem.

– The strategy set S_i of each player vp_i, $i \in \mathcal{N}_{vp}$, is V; the strategy set S_{ep} of the player *ep* is E. Thus, the strategy set $\mathcal{S}$ of the game is $\left(\times_{i \in N_{vp}} S_i \right) \times S_{ep} = V^{|\mathcal{N}_{vp}|} \times E$.
– Take any *strategy profile* $\mathbf{s} = \langle s_1, \ldots, s_{|\mathcal{N}_{vp}|}, s_{ep} \rangle \in \mathcal{S}$, also called a *configuration.*
  • The *Individual Cost* of vertex player vp_i is a function $\mathsf{IC}_i : \mathcal{S} \rightarrow \{0,1\}$ such that $\mathsf{IC}_i(\mathbf{s}) = \begin{cases} 0, & s_i \in s_{ep} \\ 1, & s_i \notin s_{ep} \end{cases}$; intuitively, vp_i receives 1 if it is not caught by the edge player, and 0 otherwise.
  • The *Individual Cost* of the edge player *ep* is a function $\mathsf{IC}_{ep} : \mathcal{S} \rightarrow \mathbb{N}$ such that $\mathsf{IC}_{ep}(\mathbf{s}) = |\{s_i : s_i \in s_{ep}\}|$.

The configuration $\mathbf{s}$ is a *pure Nash equilibrium* [5,6] (abbreviated as *pure NE*) if for each player $i \in \mathcal{N}$, it minimizes IC_i over all configurations $\mathbf{t}$ that differ from $\mathbf{s}$ only with respect to the strategy of player i.

A *mixed strategy* for player $i \in \mathcal{N}$ is a probability distribution over its strategy set S_i; thus, a mixed strategy for a vertex player (resp., edge player) is a probability distribution over vertices (resp., over edges) of G. A *mixed strategy profile* $\mathbf{s}$ is a collection of mixed strategies, one for each player. Denote $P_\mathbf{s}(ep, e)$ the probability that edge player *ep* chooses edge $e \in E(G)$ in $\mathbf{s}$; denote $P_\mathbf{s}(vp_i, v)$ the probability that vertex player vp_i chooses vertex $v \in V$ in $\mathbf{s}$. Note $\sum_{v \in V} P_\mathbf{s}(vp_i, v) = 1$ for each vertex player vp_i; similarly, $\sum_{e \in E} P_\mathbf{s}(ep, e) = 1$. Denote $P_\mathbf{s}(vp, v) = \sum_{i \in \mathcal{N}_{vp}} P_\mathbf{s}(vp_i, v)$ the probability that vertex v is chosen by some vertex player in $\mathbf{s}$.

The *support* of a player $i \in \mathcal{N}$ in the configuration $\mathbf{s}$, denoted $D_\mathbf{s}(i)$, is the set of pure strategies in its strategy set to which i assigns strictly positive probability in $\mathbf{s}$. Denote $D_\mathbf{s}(vp) = \bigcup_{i \in \mathcal{N}_{vp}} D_\mathbf{s}(i)$; so, $D_\mathbf{s}(vp)$ contains all pure strategies (that is, vertices) to which some vertex player assigns strictly positive probability. Let also $ENeigh_\mathbf{s}(v) = \{(u,v)E : (u,v) \in D_\mathbf{s}(ep)\}$; that is $ENeigh_\mathbf{s}(v)$ contains all edges incident to v that are included in the support of the edge player in $\mathbf{s}$.

A mixed strategic profile $\mathbf{s}$ induces an *Expected Individual Cost* IC_i for each player $i \in \mathcal{N}$, which is the expectation, according to $\mathbf{s}$, of its corresponding Individual Cost (defined previously for pure strategy profiles). The mixed strategy profile $\mathbf{s}$ is a *mixed Nash equilibrium* [5,6] (abbreviated as mixed NE) if for each player $i \in \mathcal{N}$, it maximizes IC_i over all configurations $\mathbf{t}$ that differ from $\mathbf{s}$ only with respect to the mixed strategy of player i.

For the rest of this section, fix a mixed strategy profile $\mathbf{s}$. For each vertex $v \in V$, denote $Hit(v)$ the event that the edge player hits vertex v. So, the probability (according to $\mathbf{s}$) of $Hit(v)$ is $P_\mathbf{s}(Hit(v)) = \sum_{e \in ENeigh_\mathbf{s}(v)} P_\mathbf{s}(ep, e)$. Define the minimum hitting probability $P_\mathbf{s}$ as $\min_v P_\mathbf{s}(Hit(v))$. For each vertex $v \in V$, denote $m_\mathbf{s}(v)$ the expected number of vertex players choosing v (according to $\mathbf{s}$). For each edge $e = (u,v) \in E$, denote $m_\mathbf{s}(e)$ the expected number of vertex players choosing either u or v; so, $m_\mathbf{s}(e) = m_\mathbf{s}(u) + m_\mathbf{s}(v)$. It is easy to see that for each vertex $v \in V$, $m_\mathbf{s}(v) = \sum_{i \in \mathcal{N}_{vp}} P_\mathbf{s}(vp_i, v)$. Define the maximum expected number of vertex players choosing e in $\mathbf{s}$ as $\max_e m_{\mathbf{s}}(e)$. We proceed to calculate the Expected Individual Cost. Clearly, for the vertex player $vp_i \in \mathcal{N}_{vp}$,

$$\begin{aligned}\mathsf{IC}_i(\mathbf{s}) &= \sum_{v \in V(G)} P_{\mathbf{s}}(vp_i, v) \cdot (1 - P_{\mathbf{s}}(Hit(v)) \\ &= \sum_{v \in V(G)} P_{\mathbf{s}}(vp_i, v) \cdot (1 - \sum_{e \in ENeigh_{\mathbf{s}}(v)} P_{\mathbf{s}}(ep, e))\end{aligned} \tag{1}$$

For the edge player ep,

$$\begin{aligned}\mathsf{IC}_{ep}(\mathbf{s}) &= \sum_{e=(u,v) \in E(G)} P_{\mathbf{s}}(ep, e) \cdot (m_{\mathbf{s}}(u) + m_{\mathbf{s}}(v)) \\ &= \sum_{e=(u,v) \in E(G)} P_{\mathbf{s}}(ep, e) \cdot (\sum_{i \in N_{vp}} P_{\mathbf{s}}(vp_i, u) + P_{\mathbf{s}}(v_i, v))\end{aligned} \tag{2}$$

2.2 Background from Graph Theory

Throughout this section, we consider the (undirected) graph $G = G(V, E)$. $G(V, E)$ is *bipartite* if its vertex set V can be partitioned as $V = V_1 \cup V_2$ such that each edge $e = (u, v) \in E$ has one of its vertices in V_1 and the other in V_2. Such a graph is often referred to as a V_1, V_2-bigraph. Fix a set of vertices $S \subseteq V$. The graph G is an *S-expander* if for every set $X \subseteq S$, $|X| \leq |Neigh_G(X)|$.

A set $M \subseteq E$ is a *matching* of G if no two edges in M share a vertex. Given a matching M, say that set $S \subseteq V$ is *matched in* M if for every vertex $v \in S$, there is an edge $(v, u) \in\in M$. A *vertex cover* of G is a set $V' \subseteq V$ such that for every edge $(u, v) \in E$ either $u \in V'$ or $v \in V'$. An *edge cover* of G is a set $E' \subseteq E$ such that for every vertex $v \in V$, there is an edge $(v, u) \in E'$. Say that an edge $(u, v) \in E$ (resp., a vertex $v \in V$) is *covered* by the vertex cover V' (resp., the edge cover E') if either $u \in V'$ or $v \in V'$ (resp., if there is an edge $(u, v) \in E'$). A set $IS \subseteq V$ is an *independent set* of G if for all vertices $u, v \in IS$, $(u, v) \notin E$. Clearly, $IS \subseteq V$ is an independent set of G if and only if the set $VC = V \backslash IS$ is a vertex cover of G. We will use the following consequence of Hall's Theorem [2, Chapter 6] on the marriage problem.

Proposition 1 (Hall's Theorem). *A graph G has a matching M in which the vertex set $X \subseteq V$ is matched if and only if for each for each subset $S \subseteq X$, $|Neigh(S)| \geq |S|$.*

3 Nash Equilibria

Theorem 1. *If G contains more than one edges, then $\Pi(G)$ has no pure Nash equilibrium.*

Proof. Consider any graph G with at least two edges and any configuration $\mathbf{s}$ of $\Pi(G)$. Let e the edge selected by the edge player in $\mathbf{s}$. Since G contains more than one edges, there exists an $e' \in E$ not selected by the edge player in $\mathbf{s}$, such that e and e' contain at least one different endpoint, assume u. If there is at least one v.p. located on e, it will prefer to go to u so that not to get arrested by the edge player and gain more. Thus, this case can not be a pure NE for the vertex players. Otherwise, the edge e contains no vertex player. But in this case, the edge player would like to change current action and select another edge, where there is at least one vertex player, so that to gain more. Thus, again this case can not be a pure NE, for the edge player this time. Since always in any case, one of the two kinds of players is not satisfied by $\mathbf{s}$, $\mathbf{s}$ is not a pure NE. □

Theorem 2 (Characterization of Mixed NE). *A mixed strategy profile* $\mathbf{s}$ *is a Nash equilibrium for any* $\Pi(G)$ *if and only if:*

1. $D_{\mathbf{s}}(ep)$ *is an* edge cover *of* G *and* $D_{\mathbf{s}}(vp)$ *is a* vertex cover *of the graph obtained by* $D_{\mathbf{s}}(ep)$.
2. *The probability distribution of the edge player over* E, *is such that, (a)* $P_{\mathbf{s}}(Hit(v)) = P_{\mathbf{s}}(Hit(u)) = \min_v P_{\mathbf{s}}(Hit(v)),\ \forall\ u, v \in D_{\mathbf{s}}(vp)$ *and (b)* $\sum_{e \in D_{\mathbf{s}}(ep)} P_{\mathbf{s}}(ep, e) = 1$.
3. *The probability distributions of the vertex players over* V *are such that, (a)* $m_{\mathbf{s}}(e_1) = m_{\mathbf{s}}(e_2) = \max_e m_{\mathbf{s}}(e),\ \forall\ e_1 = (u_1, v_1), e_2 = (u_2, v_2) \in D_{\mathbf{s}}(ep)$ *and (b)* $\sum_{v \in V(D_{\mathbf{s}}(ep))} m_{\mathbf{s}}(v) = \nu$.

4 Matching Nash Equilibria

The obvious difficulty of solving the system of Theorem 2 directs us in trying to investigate the existence of some polynomially computable, solutions of the system, corresponding to mixed NE of the game. We introduce a class of such configurations, called *mathing*. We prove that they can lead to mixed NE, we investigate their existence and their polynomial time computation.

Definition 1. *A* **matching configuration** $\mathbf{s}$ *of* $\Pi(G)$ *satisfies:* **(1)** $D_{\mathbf{s}}(vp)$ *is an independent set of* G *and* **(2)** *each vertex* v *of* $D_{\mathbf{s}}(vp)$ *is incident to only one edge of* $D_{\mathbf{s}}(ep)$.

Lemma 1. *For any graph* G, *if in* $\Pi_E(G)$ *there exists a* matching *configuration which additionally satisfies condition* **1** *of Theorem 2, then there exists probability distributions for the vertex players and the edge player such that the resulting configuration is a mixed Nash equilibrium for* $\Pi_E(G)$. *These distributions can be computed in polynomial time.*

Proof. Consider any configuration $\mathbf{s}$ as stated by the lemma (assuming that there exists one) and the following probability distributions of the vertex players and the edge player on $\mathbf{s}$:

$$\forall e \in D_{\mathbf{s}}(ep),\ P_{\mathbf{s}}(ep, e) := 1/|D_{\mathbf{s}}(ep)|,\ \forall e' \in E,\ e' \notin D_{\mathbf{s}}(ep),\ P_{\mathbf{s}}(ep, e') := 0 \tag{3}$$

$$\begin{aligned} &\forall\ i \in \mathcal{N}_{vp},\ \forall\ v \in D_{\mathbf{s}}(vp),\ P_{\mathbf{s}}(vp_i, v) := \tfrac{1}{|D_{\mathbf{s}}(vp)|}, \\ &\forall u \in V,\ u \notin D_{\mathbf{s}}(vp),\ P_{\mathbf{s}}(vp_i, u) := 0 \end{aligned} \tag{4}$$

Proposition 2.

$$\forall\, v \in D_{\mathbf{s}}(vp),\ \ m_{\mathbf{s}}(v) = \frac{\nu}{|D_{\mathbf{s}}(vp)|} \ \ \textit{and}\ \forall\, u \in V,\ u \notin D_{\mathbf{s}}(vp),\ m_{\mathbf{s}}(u) = 0$$

We show that s satisfies all conditions of Theorem 2, thus it is a mixed NE. **2.(a)**: $P_{\mathbf{s}}(Hit(v)) = \frac{1}{|D_{\mathbf{s}}(ep)|},\ \forall\ v \in D_{\mathbf{s}}(vp)$, by condition (2) of the definition of a *matching* configuration and equation (3) above. **3.(a)**: $m_{\mathbf{s}}(e_1) = m_{\mathbf{s}}(v_1) + m_{\mathbf{s}}(u_1) = 0 + \frac{\nu}{|D_{\mathbf{s}}(vp)|} = \frac{\nu}{|D_{\mathbf{s}}(vp)|},\ \forall\ e_1 = (u_1, v_1) \in D_{\mathbf{s}}(ep)$, because D_{ep} is an edge cover of G (by assumption), D_{vp} is an independent set of G (condition (1) of the definition of a *matching* configuration) and recalling Proposition 2 above. **3.(c)**: Since $D_{evp}(\mathbf{s})$ is an edge cover of G (by assumption) and by Proposition 2, we have $\sum_{v \in V(D_{\mathbf{s}}(ep))} m_{\mathbf{s}}(v) = \sum_{v \in V} \frac{\nu}{|D_{\mathbf{s}}(vp)|} = |D_{\mathbf{s}}(vp)| \cdot \frac{\nu}{|D_{\mathbf{s}}(vp)|} = \nu$. The rest of the conditions, can be easily shown to be fulfilled in s; see [3]. □

Definition 2. *A* matching *configuration which additionally satisfies condition* 1 *of Theorem 2 is called a* **matching mixed NE.**

We proceed to characterize graphs that admit *matching* Nash equilibria.

Theorem 3. *For any graph G, $\Pi(G)$ contains a* matching *mixed Nash equilibrium if and only if the vertices of the graph G can be partitioned into two sets IS, VC ($VC \cup IS = V$ and $VC \cap IS = \emptyset$), such that IS is an independent set of G (equivalently, VC is a vertex cover of the graph) and G is a VC-expander graph.*

Proof. *We first prove that if G has an independent set IS and the graph G is a VC-expander graph, where $VC = V \backslash IS$, then $\Pi_E(G)$ contains a* matching *mixed NE.* By the definition of a VC-expander graph, it holds that $|Neigh(VC')| \geq VC'$, for all $VC' \subseteq VC$. Thus, by the Marriage's Theorem 1, G has a matching M such that each vertex $u \in VC$ is matched into $V \backslash VC$ in M; that is there exists an edge $e = (u, v) \in M$, where $v \in V \backslash VC = IS$. Partition IS into two sets IS_1, IS_2, where set IS_1 consists of vertices $v \in IS$ for which there exists an $e = (u, v) \in M$ and $u \in VC$. Let IS_2 the remaining vertices of the set, i.e. $IS_2 = \{v \in IS : \forall\, u \in VC,\ (u, v) \notin M\}$.

Now, recall that there is no edge between any two vertices of set IS, since it is independent set, by assumption. Henceforth, since G is a connected graph, $\forall\ u \in IS_2 \subseteq IS$, there exists $e = (u, v) \in E$ and moreover $v \in V \backslash IS = VC$. Now, construct set $M_1 \subseteq E$ consisting of all those edges. That is, initially set $M := \emptyset$ and then for each $v \in IS_2$, add one edge $(u, v) \in E$ in M_1. Note that, by the construction of the set M_1, each edge of it is incident to only one vertex of IS_2. Next, construct the following configuration s of $\Pi_E(G)$: Set $D_{\mathbf{s}}(vp) := IS$ and $D_{\mathbf{s}}(ep) := M \cup M_1$.

We first show that that $\mathbf{s}$ is a *matching* configuration. Condition (1) of a matching configuration is fulfilled because $D_{\mathbf{s}}(vp)(= IS)$ is an independent set. We show that condition (2) of a *matching* configuration is fulfilled. Each vertex of set IS belongs either to IS_1 or to IS_2. By definition, each vertex of IS_1 is incident to only one edge of M and each vertex of IS_2 is incident to no edge in M. Moreover, by the construction of set M_1, each vertex of IS_2 is incident to exactly one edge of M_1. Thus, each vertex $v \in D_{\mathbf{s}}(vp)(= IS)$ is incident to only one edge of $D_{\mathbf{s}}(ep)(= M \cup M_1)$, i.e. condition (2) holds as well. Henceforth, $\mathbf{s}$ is a *matching* configuration.

We next show that condition **1** of Theorem 2 is satisfied by $\mathbf{s}$. We first show that $D_{\mathbf{s}}(ep)$ is an edge cover of G. This is true because (i) set $M \subseteq D_{\mathbf{s}}(ep)$ covers all vertices of set VC and IS_1, by its construction and (ii) set $M_1 \subseteq D_{\mathbf{s}}(ep)$ covers all vertices of set IS_2, which are the remaining vertices of G not covered by set M, also by its construction. We next show that $D_{\mathbf{s}}(vp)$ is a vertex cover of the subgraph of G obtained by set $D_{\mathbf{s}}(ep)$. By the definition of sets IS_1, $IS_2 \subseteq IS$, any edge $e \in M$ is covered by a vertex of set IS_1 and each edge $e \in M_1$ is covered by a vertex of set IS_2. Since $D_{\mathbf{s}}(ep) = M \cup M_1$, we get that all edges of the set are covered by $D_{\mathbf{s}}(vp) = IS_1 \cup IS_2$. This result combined with the above observation on $D_{\mathbf{s}}(ep)$ concludes that condition **1** of Theorem 2 is satisfied by $\mathbf{s}$. Henceforth, by lemma 1, it can lead to a *matching* mixed NE of $\Pi_E(G)$.

We proceed to show that if G contains a matching *mixed NE, assume* $\mathbf{s}$, *then G has an independent set IS and the graph G is a VC-expander graph, where $VC = V \backslash IS$.* Define sets $IS = D_{\mathbf{s}}(vp)$ and $VC = V \backslash IS$. We show that these sets satisfy the above requirements for G. Note first that, set IS is an independent of G since $D_{\mathbf{s}}(vp)$ is an independent set of G by condition (1) of the definition of a *matching* configuration.

We next show G contains a matching M such that each vertex of VC is matched into $V \backslash VC$ in M. Since $D_{\mathbf{s}}(ep)$ is an edge cover of G (condition **1** of a mixed NE of Theorem 2), for each $v \in VC$, there exists an edge $(u, v) \in D_{\mathbf{s}}(ep)$. Note that for edge (u, v), it holds that $v \in IS$, since otherwise IS would not be a vertex cover of $D_{\mathbf{s}}(ep)$ (Condition **1** of a mixed NE). Now, construct a set $M \subseteq E$ consisting of all those edges. That is, That is, initially set $M := \emptyset$ and then for each $v \in VC$, add one edge $(u, v) \in D_{\mathbf{s}}(ep)$ in M. By the construction of set M and condition (2) of a *matching* mixed NE, we get that M is a matching of G and that each vertex of VC is matched into $V \backslash VC$ in M. Thus, by the Marriage's Theorem 1, we get that $|Neigh(VC')| \geq |VC'|$, for all $VC' \subseteq VC$ and so G is a VC-expander and condition (2) of a matching configuration also holds in $\mathbf{s}$.

□

4.1 A Polynomial Time Algorithm

The previous Theorems and Lemmas enables us to develop a polynomial time algorithm for finding *matching* mixed NE for any $\Pi(G)$, where G is a graph satisfying the requirements of Theorem 3. The Algorithm is described in pseudocode in Figure 1.

Theorem 4. *Algorithm A computes a* matching *mixed Nash equilibrium for $\Pi(G)$ and needs linear time $O(m)$.*

Algorithm $A(\Pi(G), IS, VC)$

INPUT: A game $\Pi(G)$ and a partition of $V(G)$ into sets IS, $VC = V \backslash IS$, such that IS is an independent set of G and G is a VC-expander graph.
OUTPUT: A mixed NE s for $\Pi(G)$.

1. Compute a set $M \subseteq E$, as follows:
   (a) *Initialization*: Set $M := \emptyset$, $Matched := \emptyset$ (currently matched vertices in M), $Unmatched := VC$ (currently unmatched vertices in M vertices of VC), $Unused := IS$, $i := 1$, $G_i := G$ and $M_1 := \emptyset$.
   (b) While $Unmatched \neq \emptyset$ Do:
      i. Consider a $u \in Unmatched$.
      ii. Find a $v \in Unused$ such that $(u, v) \in E_i$. Set $M := M \cup (u, v)$, $Unused := Unused \backslash \{v\}$.
      iii. *Prepare next iteration*: Set $i := i + 1$, $Matched := Matched \cup \{u\}$, $Unmatched := Unmatched \backslash \{u\}$, $G_i := G_{i-1} \backslash u \backslash v$.
2. Partition set IS into two sets IS_1, IS_2 as follows: $IS_1 := \{u \in IS : \exists (u, v) \in M\}$ and $IS_2 := IS \backslash IS_1$. Note that $IS_2 := \{u \in IS : \nexists (u, v) \in M, v \in VC\}$. Compute a set M_1 as follows: $\forall u \in IS_2$, set $M_1 := M_1 \cup (u, v)$, for any $(u, v) \in E$, $v \in VC$.
3. Define a s with the following support: $D_{\mathbf{s}}(vp) := IS$, $D_{\mathbf{s}}(ep) := M \cup M_1$.
4. Determine the probabilities distributions of the vertex players and the edge player of configuration s′ using equations (3) and (4) of Lemma 1.

Fig. 1. Algorithm A

5 Bipartite Graphs

Lemma 2. *In any bipartite graph G there exists a matching M and a vertex cover VC such that* (1) *every edge in M contains exactly one vertex of VC and* (2) *every vertex in VC is contained in exactly one edge of M.*

Proof. Let X, Y the bipartition of the bipartite graph G. Consider any minimum vertex cover of the graph G, VC. We are going to construct a matching M of G so that conditions (1) and (2) of the Lemma hold. Let R the vertices of VC contained in set X, i.e. $R = VC \cap X$ and T the vertices of VC contained in set Y, i.e. $T = VC \cap Y$. Note that $VC = R \cup T$. Let H and H' the subgraphs of G induced by $R \cup (Y - T)$ and $T \cup (X - R)$, respectively. We are going to show that G contains a matching in M as required by the Lemma.

Since $R \cup T$ is a vertex cover, G has no edge from $Y - T$ to $X - R$. We show that for each $S \subseteq R$, $|Neigh_H(S)| \subseteq |Y - T|$. If $|Neigh_H(S)| < |S|$, then we can substitute $Neigh_H(S)$ for S in VC to obtain a smallest vertex cover (∗1). This is true because (i) $Neigh_H(S)$ covers all edges incident to S that are not covered by T and (ii) since G is a bipartite graph there are no edges between the vertices of set S, so that a possible substitute of set S do not need to cover any such edge.

Thus, $|Neigh_H(S)| \geq |S|$, for all $S \subseteq R$. By Hall's Theorem (Theorem 1), H has a matching M_H such that each vertex of R is matched in M_H. Using similar arguments for set T, we can prove that for each $S' \subseteq T$, $|Neigh_{H'}(S')| \geq |S'|$. Henceforth, H' has a matching $M_{H'}$ such that each vertex of T is matched in $M_{H'}$. Now define $M = M_H \cup M_{H'}$. Since each H, H' is an induced subgraph of G and the two subgraphs have disjoint sets of vertices, we get that M is matching of G and that each vertex of $VC = R \cup T$ is matched in M. This result combined with the fact that M is a matching of G concludes that condition (2) of the Lemma holds.

We proceed to prove condition (1). That is, to show that every edge of M contains exactly one vertex of VC. Observe first that by the construction of set M, every edge of M contains at least one vertex of VC. Moreover, note that only one of the endpoints of the edge is contained in M. This is true because by the construction of set M each edge of set M matches either (i) a vertex of set $R \subseteq X$ to a vertex of set $(Y - T) \subseteq Y$ or (ii) a vertex of set $T \subseteq Y$ to a vertex of set $(X - R) \subseteq X$. So, for any case exactly one of the two endpoints of the edge is not contained in VC. □

Lemma 3. *Any X, Y-bigraph graph G can be partitioned into two sets IS, VC ($IS \cup VC = V$ and $IS \cap VC = \emptyset$) such that VC is a vertex cover of G (equivalently, IS is an independent set of G) and G is a VC-expander graph.*

Lemma 3 and Theorem 3 finally imply:

Theorem 5. *Any $\Pi(G)$ for which G is a connected bipartite graph, contains a* matching *mixed Nash equilibrium.*

Theorem 6. *For any $\Pi(G)$, for which G is a bipartite graph, a* matching *mixed Nash equilibrium of $\Pi(G)$ can be computed in polynomial time,* $\max\{O(m\sqrt{n}), O(n^{2.5}/\sqrt{\log n})\}$*, using Algorithm A.*

References

1. Aspnes, J., Chang, K., A. Yampolskiy: Inoculation Strategies for Victims of Viruses and the Sum-of-Squares Problem. Proceedings of the 16th Annual ACM-SIAM Symposium on Discrete Algorithms (2005) 43-52
2. Asratian, A. S., Tristan, D., Häggkvist, M. J.: Bipartite Graphs and Their Applications. Cambridge Tracts in Mathematics, **131** (1998)
3. M. Mavronicolas, V. Papadopoulou, A. Philippou and P. Spirakis: A Network Game with Attacker and Protector Entities. *TR-05-13*, Univ. of Cyprus, July (2005)
4. Mavronicolas, M., Papadopoulou, V., Philippou, A., Spirakis, P.: A Graph-Theoretic Network Security Game. Accepted in 1st Workshop on Internet and Network Economics, August (2005)
5. Nash, J. F.: Equilibrium Points in n-Person Games. Proceedings of the National Acanemy of Sciences of the United States of America **36** (1950) 48-49
6. Nash, J. F.: Noncooperative Games. Annals of Mathematics **54**(2) (1951) 286-295
7. C. H. Papadimitriou: Algorithms, Games, and the Internet. Proceedings of the 33rd Annual ACM Symposium on Theory of Computing (2001) 749-753

SkipTree: A Scalable Range-Queryable Distributed Data Structure for Multidimensional Data

Saeed Alaei, Mohammad Toossi, and Mohammad Ghodsi

Computer Engineering Department,
Sharif University of Technology, Tehran, Iran

Abstract. This paper presents the SkipTree, a new balanced, distributed data structure for storing data with multidimensional keys in a peer-to-peer network. The SkipTree supports range queries as well as single point queries which are routed in $O(\log n)$ hops. SkipTree is fully decentralized with each node being connected to $O(\log n)$ other nodes. The memory usage for maintaining the links at each node is $O(\log n \log\log n)$ on average and $O(\log^2 n)$ in the worst case. Load balance is also guaranteed to be within a constant factor.

1 Introduction and Related Work

Over the past few years, there has been a trend to move from centralized server based network architectures toward decentralized and distributed architectures and peer to peer networks. The term *Scalable Distributed Data Structure*(SDDS) first introduced by Litwin et al. in LH* [9] refers to this class of data structures. Litwin et al. modified the original hash-based LH* structure to support range queries in RP*[8, 9]. Based on the previous work of distributed data structures, RP* and Distributed Random Tree (DRT) [6], new data structures based on either hashing or key comparison have been proposed like Chord[13], Viceroy[10], Koorde[4], Pastry[12], and P-Grid [2]. Most existing peer-to-peer overlays require $\Theta(\log n)$ links per node in order to achieve $O(\log n)$ hops for routing. Viceroy and Koorde which are based on DHTs are the remarkable exceptions in that they achieve $O(\log n)$ hops with only $O(1)$ links per node at the cost of restricted or no load balancing.

Typically, those systems which are based on DHTs and hashing lack range-query, locality properties and control over distribution of keys due to hashing. In contrast, those which are based on key comparison, although requiring more complicated load balancing techniques, do better in those respects. P-Grid is one of the systems based on key comparison which uses a distributed binary tree to partition a single dimensional space with network nodes representing the leaves of the tree and each node having a link to some node in every sibling subtree along the path from the root to that node. Other systems like P-Tree have been proposed that provide range queries in single dimensional space.

SkipNet [3] on which our new system relies heavily, is another system for single dimensional spaces based on an extension to skip lists which supports range queries in single dimensional case too.

X. Deng and D. Du (Eds.): ISAAC 2005, LNCS 3827, pp. 298–307, 2005.

RAQ [11] is also another solution for the multidimensional case which incorporates a distributed partition tree structure to partition the space. Its network requires $O(h)$ links at each node and routes in $O(h)$ hops where h is the height of the partition tree which can be of $O(n)$ for an unbalanced tree. Although it has been shown [1] that even for such unbalanced trees the number of messages required to resolve a query still remains of $O(\log n)$ on average if the links are chosen randomly, the number of links that a node should maintain and the memory requirement at each node for storing information about the path from that node to the root still remain of $O(h)$ which is as bad as $O(n)$ for unbalanced trees.

In this paper we propose a new efficient scalable distributed data structure called the *SkipTree* for storage of keys in multidimensional spaces. Our system uses a distributed partition tree to partition the space into smaller regions with each network node being a leaf node of that tree and responsible for one of the regions. In contrast to similar tree-based solutions the partition tree here is used only to define an ordering between the regions. The routing mechanism and link maintenance is similar to that of SkipNet and independent of the shape of the partition tree, so in general our system does not need to balance the partition tree. It maintains a SkipNet by the leaves of the tree in which the sequence of nodes in the SkipNet is the same sequence defined by the leaves of the partition tree from left to right. Handling a single key query is almost similar to that of an ordinary SkipNet while range queries are quiet different due to the multidimensional nature of the SkipTree. From another point of view, our system can be seen as an extension to the SkipNet for the multidimensional spaces.

In section 2 we explain the basic structure of the SkipTree including the structure of the partition tree, its associated SkipNet and the additional information that needs to be stored in each node. In section 3, single and range queries are explained. In section 4, the procedure for joining and leaving the network is described. In section 5, some techniques for load balancing in SkipTrees are discussed. In section 6 we modify the SkipTree structure to reduce the amount of information that needs to be stored in each node about the partition tree and finally section 7 concludes the paper.

2 Basic SkipTree Structure

The distributed data structure used in the SkipTree consists of a *Partition Tree* whose leaves also form a SkipNet.

We assume that each data element has a key which is a point in our k-dimensional search space. This space is split into n regions corresponding to the n network nodes. Let $S(v)$ denote the region assigned to node v. v is the node responsible for every data element whose key is in $S(v)$. We extend the definition of $S(v)$ to also denote the region assigned to the internal nodes of the tree. A sample partition-tree is depicted in Figure 1.

For network node u, which corresponds to a leaf in the partition tree, we call the path connecting the root of the tree to u the *Principal Path* of node u. We

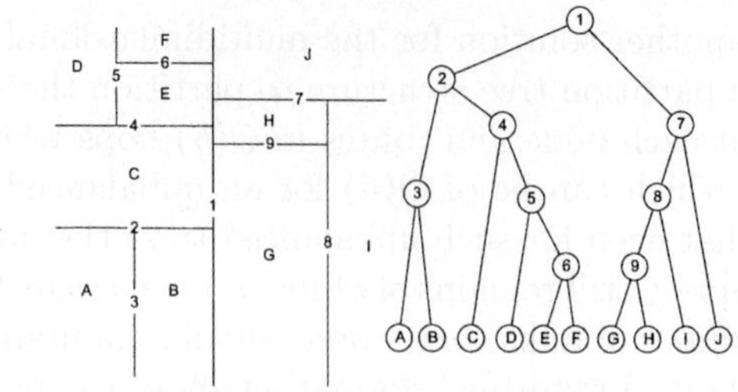

Fig. 1. A sample two dimensional partition tree and its corresponding space partitioning. Each internal node in the partition tree, labelled with a number, divides a region using the line labelled with the same number. Each leaf of the partition tree is a network node responsible for the region labelled with the same letter.

refer to the hyperplane equations assigned to the nodes of the principal path of node u (including information about on which side of those hyperplanes u resides) as the *Characteristic Plane Equations* of u or *CPE* of u for short. Every leaf node in the SkipTree stores its own CPE as well as the CPE of its links. Using theses CPE information, every node like u can locally identify if a given point belongs to a node to the left or the right of u or to the left or right of any of its links in the partition tree. The latter is useful in routing queries as explained in section 3. Hereafter, whenever we refer to a plane, we actually refer to a hyperplane of $k-1$ dimensions in a k-dimensional space.

We link the network nodes in the SkipTree together as shown in Figure 2 by forming a SkipNet among the leaves of the partition tree described before.

Finally, we note that a real number p_v is assigned to each node v. p_v is randomly generated when v joins the SkipTree so that $p_a < p_v < p_b$ where a and b are v's left and right leaf in the tree. This number is used in subsection 3.2 to handle range queries more efficiently.

3 Handling Queries

3.1 Single Point Query

The routing algorithm for single point queries is essentially the same algorithm used in the SkipNet, that is every node receiving the query along the path, sends it through its farthest link which does not point past the destination node as shown in Figure 3. The distance to the destination node is at least halved at each hop. This implies that the query reaches the destination after at most $\log_2 n$ hops. However, because SkipNet uses a probabilistic method for selecting and maintaining links in the network, it guarantees routing in $O(\log n)$ hops w.h.p. A formal proof of this can be found in [3].

For the above procedure to be effective we must be able to compare points against nodes to identify whether the node containing a given point lies before or after another node in the sequence. To do so, a node compares the point against the planes in its own CPE in the order they appear in its principal path starting

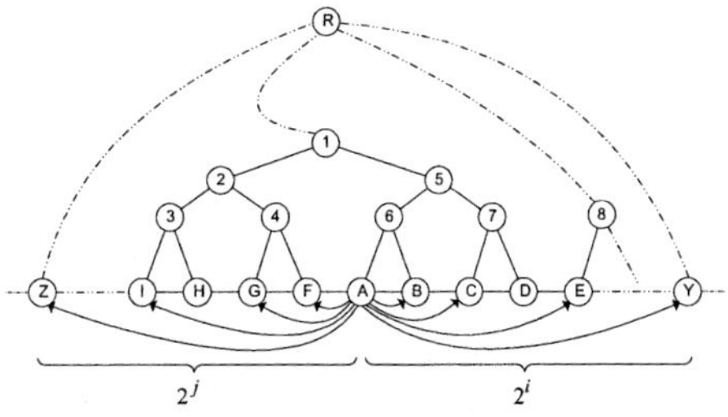

Fig. 2. The links maintained by node A in the ideal SkipTree. The target nodes are independent of the tree structure. The tree only helps us to put an ordering on the nodes. The ith link in each direction skips over $2^{i-1} - 1$ nodes in that direction.

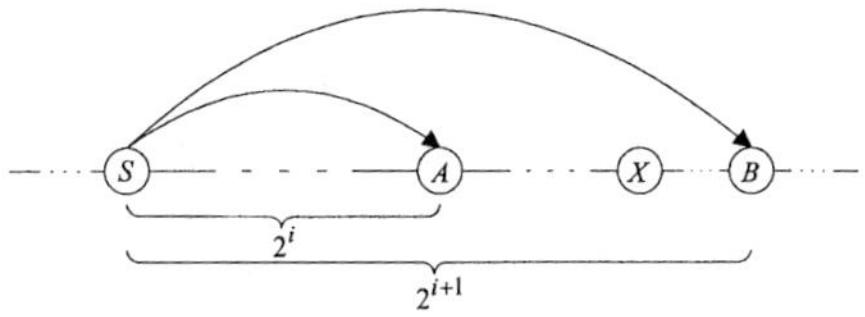

Fig. 3. A point query is routed through the farthest link which does not point past the destination node. Here, S receives a query targeting node X, so it routes the query to A. The distance to the destination node is at least halved at each hop.

from the root until it finds the first plane where the current node and the point lie on different sides of the plane. This is where the point is contained in a region belonging to a sibling subtree. If that subtree is a left (right) subtree, all of its nodes as well as the node containing the point must also be to the left (right) of the current node. The above procedure leads to $O(min(h \log n, n))$ memory usage at each node for storing the CPE. We will modify the tree structure in section 6 to improve this.

3.2 Range Query

A range query in the SkipTree is a 3-tuple of the form (R, fs, ls) where R is the query range in the multidimensional space and only the network nodes whose sequence numbers reside in the interval $[fs, ls]$ are searched. A normal range query takes the form $(R, -\infty, +\infty)$, so that all of the nodes are searched regardless of their sequence numbers. Note that the region defined by R can be of any shape as long as every node can locally identify whether R intersects with a given hypercube.

When a node S receives a range query (R, fs, ls) it sends the query to each of its links whenever there is node which intersects with the region R between that link and the next link. Assume that A and B shown in Figure 4 are two nodes corresponding to some two consecutive links maintained by S. S sends a copy of the query to A if there is any node between A and B which intersects R. Every such node, if any, must reside in one of the crosshatched subtrees illustrated in

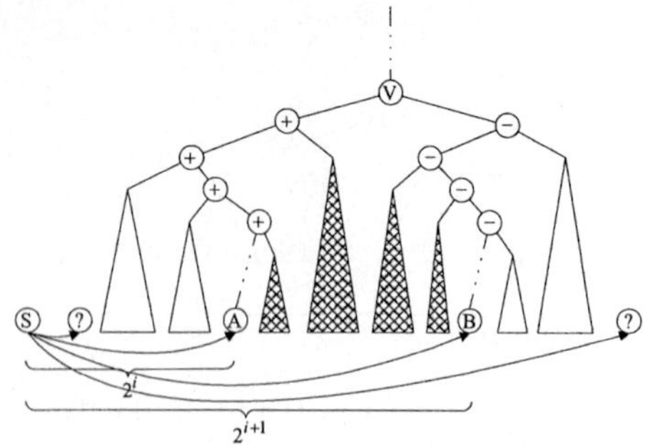

Fig. 4. A range query is propagated through each of the links maintained by S whenever there is node which intersects R between that link and the next link. Here, a copy of the query is propagated to A if any of the nodes between A and B intersects with R.

the figure. In fact, such a node must be to the right of the nodes marked with $+$ and to the left of the node marked with $-$ and because S has all of CPEs corresponding to its links, it also has access to the plane equations corresponding to the internal nodes marked with a $+$ or $-$ sign. So, it can easily identify from those equations the regions in the multidimensional space associated with each of the subtrees between A and B and from that it can determine whether there is any subtree between A and B whose region intersects with R and if there is such a subtree, it must also contain a node whose region intersects with R. Note that the fs and ls fields of the query are modified appropriately before a copy of the query is sent through a link. The reason is to restrict the sequence of nodes to be searched to prevent duplicate queries. For example in Figure 4, suppose that a copy of the form (R, fs, ls) is to be sent from S to A. Also assume that $A.seq$ and $B.seq$ are the sequence number of A and B respectively. Then S computes the interval $[fs', ls']$ as the intersection of $[fs, ls]$ and $[A.seq, B.seq]$ and it sends the query (R, fs', ls') to A. This will ensure that no nodes in the network receives the query more than once.

4 Node Join and Departure

4.1 Joins

To join the SkipTree, a new node v has to be able to contact an existing node u.

The node v first inserts itself in the partition tree by splitting $S(u)$ using a new plane P to two regions. One region is then assigned to v while u retains control of the other region. Also, v copies its CPE from u and appends P to both CPEs. The plane P can be arbitrarily chosen as our load balancing protol will gradually change the partitioning to a more balanced configuration.

After updating the Partition Tree, v establishes its network links by joining the SkipNet. Node sequence numbers are used here to define a total ordering among the nodes. The SkipNet join algorithm is described in [3] and involves only $O(\log n)$ steps w.h.p.

To complete the join, u transfers the data items which are no longer in its assigned region to v.

4.2 Departures

Similar to node joins, when the node v is leaving the SkipTree, it has to follow three steps.

The first step is to update the Partition Tree. Suppose that the last plane in CPE_v, called P, splits its parent region into regions $S(v)$ and R. To update the Partition Tree, node v sends a special range query to the nodes in R and instructs them to remove the plane P from their CPE. This will effectively remove v from the partition tree.

Next step is to transfer the data items, v can simply find the node responsible for each item using a single point query and transfer the item accordingly. However, a more efficient method is to collect all possible target regions and evaluate the queries locally.

In the last step, v has to remove itself from the SkipNet. As [3] points out, this can be reduced to removing $O(1)$ links by using background repair processes similar to Chord and Pastry.

5 Load Balancing

Many distributed lookup protocols use hashing to distribute keys uniformly in the search space and achieve some degree of load balance. Hashing cannot be used in the SkipTree as it makes range queries impossible. As a result, a load balancing mechanism is necessary to deal with the nonuniform key distribution.

Our load balancing protocol is derived from the *Item Balancing* technique in [5]. Load balancing is achieved using a randomized algorithm that requires a node to be able to contact random nodes in the network.

Let l_i, the load on node i, be the number of data items stored on i and α be a constant number so that $\alpha > 1$. We will prove that the SkipTree's load will be balanced w.h.p. if each node performs a minimum number of *load balancing tests* as per system *half-life* [7].

Load Balancing Test. In a load balancing test, node i asks a randomly chosen node j for l_j. If $l_j \geq \alpha l_i$ or $l_i \geq \alpha l_j$, i performs a *load balancing operation.*

Load Balancing Operation. Assume w.l.o.g that $l_i < l_j$. First, node i normally leaves the SkipTree using the algorithm given in subsection 4.2. Then, i joins the network once again at node j and selects a hyperplane for the newly created internal node in the partition tree in a way that the number of data elements is halved at both sides of the hyperplane. This makes both l_i and l_j to become equal to half the old value of l_j.

Theorem 1. *If each node performs $\Omega(\log n)$ load balancing operations per half-life as well as whenever its own load doubles, then the above protocol has the following properties where N is the total number of stored data items.*

– *With high probability, the load of all nodes is between* $\frac{N}{8\alpha n}$ *and* $\frac{16\alpha N}{n}$.
– *The amortized number of items moved due to load balancing is* $O(1)$ *per insertion or deletion, and* $O(N/n)$ *per node insertion or deletion.*

The proof of this theorem using potential functions can be found in [5].

6 Memory Optimization

In this section we enforce some constraints on the plane equations that a node stores, so that for a SkipTree of height h only $O(\log h)$ of the plane equations of any CPE will be needed. The constraints that we enforce are the following:

– The planes must be perpendicular to a principal axis. So, in a k-dimensional space of $(x_1, x_2, \cdots, x_k)$ it must take the form of $x_i = c$ for some $1 \leq i \leq k$ and some value of c.
– If the search space is k-dimensional, we precisely define the form of the plane equation that may be assigned to an internal node depending on the depth of that node. We first introduce the following notation:
  d_A: for a node A in the SkipTree, the depth of A is represented by d_A and is defined to be the length of the principal path corresponding to A plus one as illustrated in Figure 5.
  l_A: for every node A in the SkipTree, the level of A is indicated by l_A where $l_A = \lceil \log_2 (\frac{d_A}{k} + 1) \rceil$ as illustrated in Figure 5.
  d'_A: for a node A in the SkipTree, the relative depth of A is represented by d'_A and is defined as $d'_A = d_A - k(2^{l_A - 1} - 1)$ as illustrated in Figure 5.
  s_A: for a node A in the SkipTree, the section number of A is represented by s_A where $s_A = \lceil \frac{d'_A}{k} \rceil$.
  We are now ready to state the last constraint:
  If A is an internal node, the plane equation assigned to A must be of the form $x_{s_A} = c$ for an arbitrary value of c, that is for any given i, all of the nodes

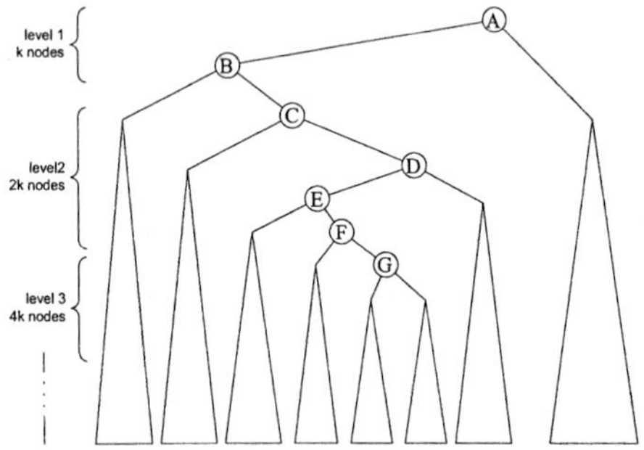

Fig. 5. A sample SkipTree for a two dimensional space. Nodes A to G have depths 1 to 7 respectively. A and B are on level 1; C, D, E and F are on level 2 and G is on level 3. The relative depth are: $d'_A = 1$, $d'_B = 2$, $d'_C = 1$, $d'_D = 2$, $d'_E = 3$, $d'_F = 4$, $d'_G = 1$.

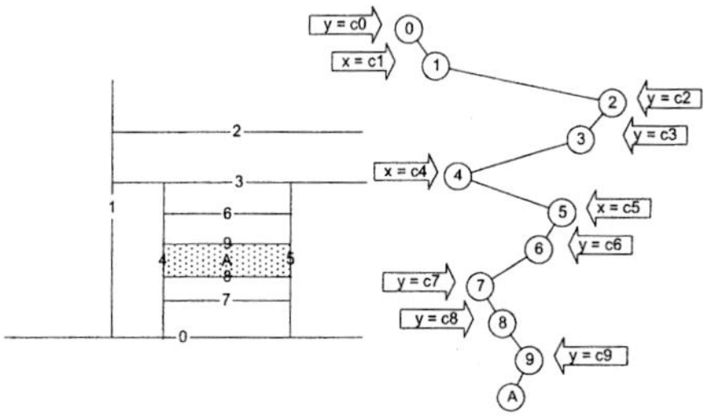

Fig. 6. The left is a sample partitioning of a 2-dimensional space under the memory optimization constraints, from the view point of node A and the right is the principal path of node A. The plane equations assigned to the internal nodes are shown in the arrows.

whose section numbers are i are assigned plane equations of the form $x_i = c$. A typical 2-dimensional space partitioned under the above constraints and its associated tree are shown in Figure 6.

Lemma 1. *In any principal path of length h nodes are partitioned to at most $k\lceil \log_2 (\frac{h}{k} + 1)\rceil$ different sections.*

Proof: Since we defined the level of a node at depth d to be $\lceil log_2(\frac{d}{k} + 1)\rceil$, nodes in any principal cannot be partitioned to more than $\lceil log_2(\frac{h}{k} + 1)\rceil$ levels. Nodes at each level are further partitioned to k sections so there can be at most $k\lceil log_2(\frac{h}{k} + 1)\rceil$ sections in any principal path.

Lemma 2. *For any leaf node A in a SkipTree, A needs to store only two plane equations for each section of its principal path. we call the sequence of these pairs of plane equations that node A stores, the* Reduced-Characteristic Plane Equations *of node A or for short the* RCPE *of node A.*

Proof: All of the planes on the same section partition the space based on the value of the same field x_i. So for each of the sections, A needs to store an inequality of the form $a \leq x_i < b$. Therefore an RCPE can be stored as an ordered sequence of inequalities of the form $a \leq x_i < b$, one for each section in the principal path. When a node like A receives a point query it finds the first inequality in the RCPE sequence that does not hold for the queried point. Then the first constraint we introduced on the beginning of this section ensures that the destination node which is responsible for the queried point will be to the left of the current node if the point is to the left of the interval represented by the first unsatisfied inequality and the destination node will be to the right of the current node otherwise. The situation with range queries is quite similar. The sequence of inequalities in the RCPE for the node A in Figure 6 is shown bellow: level=1, section=1 : $c_0 \leq y < +\infty$; level=1, section=2 : $c_1 \leq x < +\infty$; level=2, section=1 : $c_0 \leq y < c_3$; level=2, section=2 : $c_4 \leq x < c_5$; level=3, section=1 : $c_8 \leq y < c_9$.

6.1 Node Join and Departure

Joining mechanism is the same as before except that a new node must obey the constraints mentioned earlier. However, when leaving, the situation is a little different since deleting a node may cause an internal node and its associated plane to be deleted which in turn may invalidate the memory optimization constraints. If this is the case we can swap the node to be deleted with a lower node in the tree which can be deleted without causing any problem. and then we can delete the node.

6.2 Complexity

The memory requirement of any node A for storing its RCPE as well as the RCPE of its links as described earlier is of $O(\log h \log n)$ where h is the height of the tree which is a major improvement over the $O(min(h \log n, n))$ memory requirement in the default case.

7 Conclusion and Future Work

In this paper we introduced the *SkipTree* which is designed to handle point and range queries over a multidimensional space in a distributed environment. Our data structure maintains $O(\log n)$ links at each node and guarantees an upper bound of $O(\log n)$ messages w.h.p for point queries and also guarantees range queries with depth of $O(\log n)$ message w.h.p. Besides, using the memory optimization of section 6, each node needs only to store the RCPE of itself and its links that requires $O(\log h \log n)$. We also adapted some load balancing techniques and a memory optimization technique to improve our data structure. Another important areas which needs further investigation is fault tolerance in presence of node failures.

References

1. K. Aberer. Scalable data access in p2p systems using unbalanced search trees. In *Proceedings of Workshop on Distributed Data and Structures(WDAS-2002)*, 2002.
2. K. Aberer, P. Cudr-Mauroux, A. Datta, Z. Despotovic, M. Hauswirth, M. Punceva, and R. Schmidt. P-grid: a self-organizing structured p2p system. *SIGMOD Rec.*, 32(3):29–33, 2003.
3. N. HARVEY, M. JONES, S. SAROIU, M. THEIMER, and A. WOLMAN. Skipnet: A scalable overlay network with practical locality properties, 2003.
4. M. Kaashoek and D. Karger. Koorde: A simple degree-optimal distributed hash table, 2003.
5. D. R. Karger and M. Ruhl. Simple efficient load balancing algorithms for peer-to-peer systems. In *SPAA '04: Proceedings of the sixteenth annual ACM symposium on Parallelism in algorithms and architectures*, pages 36–43. ACM Press, 2004.
6. B. Kroll and P. Widmayer. Distributing a search tree among a growing number of processors. In *SIGMOD '94: Proceedings of the 1994 ACM SIGMOD international conference on Management of data*, pages 265–276. ACM Press, 1994.

7. D. Liben-Nowell, H. Balakrishnan, and D. Karger. Analysis of the evolution of peer-to-peer systems. In *PODC '02: Proceedings of the twenty-first annual symposium on Principles of distributed computing*, pages 233–242. ACM Press, 2002.
8. W. Litwin, M.-A. Neimat, and D. A. Schneider. Rp*: A family of order preserving scalable distributed data struc tures. In J. B. Bocca, M. Jarke, and C. Zaniolo, editors, *VLDB'94, Proceedings of 20th International Conference on Very Large Data Bases*, sep 1994.
9. W. Litwin, M.-A. Neimat, and D. A. Schneider. Lh* – a scalable, distributed data structure. *ACM Trans. Database Syst.*, 21(4):480–525, 1996.
10. D. Malkhi, M. Naor, and D. Ratajczak. Viceroy: a scalable and dynamic emulation of the butterfly. In *PODC '02: Proceedings of the twenty-first annual symposium on Principles of distributed computing*, pages 183–192. ACM Press, 2002.
11. H. Nazerzadeh and M. Ghodsi. RAQ: A range-queriable distributed data structure (extended version). In *Proceeding of Sofsem 2005, 31st Annual Conference on Current Trends in Theory and Practice of Informatics, LNCS 3381, pp. 264-272*, February 2005.
12. A. I. T. Rowstron and P. Druschel. Pastry: Scalable, decentralized object location, and routing for large-scale peer-to-peer systems. In *Middleware 2001: Proceedings of the IFIP/ACM International Conference on Distributed Systems Platforms Heidelberg*, pages 329–350. Springer-Verlag, 2001.
13. I. Stoica, R. Morris, D. Karger, F. Kaashoek, and H. Balakrishnan. Chord: A scalable Peer-To-Peer lookup service for internet applications. pages 149–160.

The Phase Matrix

Peter Høyer*

Department of Computer Science, University of Calgary, Canada
hoyer@cpsc.ucalgary.ca

Abstract. Reducing the error of quantum algorithms is often achieved by applying a primitive called amplitude amplification. Its use leads in many instances to quantum algorithms that are quadratically faster than any classical algorithm. Amplitude amplification is controlled by choosing two complex numbers ϕ_s and ϕ_t of unit norm, called phase factors. If the phases are well-chosen, amplitude amplification reduces the error of quantum algorithms, if not, it may increase the error. We give an analysis of amplitude amplification with a emphasis on the influence of the phase factors on the error of quantum algorithms. We introduce a so-called phase matrix and use it to give a straightforward and novel analysis of amplitude amplification processes. We show that we may always pick identical phase factors $\phi_s = \phi_t$ with argument in the range $\frac{\pi}{3} \leq \arg(\phi_s) \leq \pi$. We also show that identical phase factors $\phi_s = \phi_t$ with $-\frac{\pi}{2} < \arg(\phi_s) < \frac{\pi}{2}$ never leads to an increase in the error, generalizing a recent result of Lov Grover who shows that amplitude amplification becomes a quantum analogue of classical repetition if we pick phase factors $\phi_s = \phi_t$ with $\arg(\phi_s) = \frac{\pi}{3}$.

Keywords: Quantum Computing. Algorithms. Amplitude Amplification. Randomized Algorithms.

1 Introduction to Quantum Searching

A decade ago, Lov Grover [5, 6] discovered a quantum mechanical algorithm that has since become one of the most famous quantum algorithms. The algorithm solves one of the most basic problems in algorithmics, that of searching an unstructured search space. It is based intrinsically on quantum mechanical effects and demonstrates beautifully a fundamental relationship between quantum mechanics and computations. The algorithm is very simple and runs quadratically faster than any classical algorithm in terms of query complexity. His algorithm is likely the most studied quantum algorithm ever, seconded only by the quantum Fourier transform. One of the early work related to Grover's algorithm is a generalization called *amplitude amplification* [2], which in rough terms shows that the benefits of Grover's algorithm carry over to arbitrary search processes and settings [2, 3]. Amplitude amplification is possibly an even simpler concept than

* Supported by the Canadian Institute for Advanced Research (CIAR), Canada's Natural Sciences and Engineering Research Council (NSERC), and The Mathematics of Information Technology and Complex Systems (MITACS).

X. Deng and D. Du (Eds.): ISAAC 2005, LNCS 3827, pp. 308–317, 2005.

Grover's algorithm. It allows to talk about arbitrary processes and algorithms, and helps devise simple yet optimal solutions to problems that can be related to searching. Amplitude amplification is sometimes referred to as *quantum searching* when applied specifically to search problems.

Grover's algorithm [5] and more generally amplitude amplification [2] are limited by the unitarity condition of quantum mechanics. In popular terms, the unitarity condition implies that if we run some algorithm and it solves some problem, then if we keep on running the algorithm (as opposed to stopping it and outputting the result found at that time), eventually the solution will be lost. The reason is that unitarity implies reversibility, and it thus seems tempting to conclude that there thus can be no fixed-point or one-way quantum search algorithms. However, the unitary limitation applies only if the following three conditions are satisfied: (1) we keep running the *same* process, (2) that process is *unitary*, and (3) we run it on a closed and *finite* quantum system. The remedy is to invalidate one of more of these conditions, which in earlier work [1, 3, 9] is done by introducing measurements, thus voiding (2), or embedding the quantum system into a much larger system, thus effectively voiding (3). Though these proposals certainly are remedies, they come with several downsides: they are somewhat technical, add to the overall complexity of the algorithm, are harder to understand that the original algorithm, are rather classical fixes to a quantum mechanical obstacle, and complicate applications.

In recent work, Grover [7] has suggested an idea that aims at voiding (1). When applying amplitude amplification, we pick two complex numbers of unit norm $\phi_s, \phi_t \in \mathbb{C}^*$, which are called phase factors. Grover [7] observes that if we pick the phase factor $\phi_s = \phi_t = e^{\imath\pi/3}$, then the amplitude amplification process automatically slows down as the error decreases. This naturally leads to the question if other phase factors have similar properties, and more generally, what influence does the phase factors have on the amplitude amplification process.

We first give a novel analysis of amplitude amplification with emphasis on the choice of phases. Our analysis uses only basic linear algebra, yet still it allows us to prove new fundamental properties of amplitude amplification, properties that have not been identified using more involved methods of analysis. As our main analytical tool, we introduce a so-called phase matrix of dimension 2×2. We use it to explain why $e^{\imath\pi/3}$ is an appropriate phase factor. We also show that any nontrivial phase factor with positive real guarantees a reduction in the error in quantum searching. That is, pick any complex number $\phi \in \mathbb{C}^*$ of unit norm with $0 < \mathrm{Re}(\phi) < 1$, and amplitude amplification with ϕ reduces the overall error. This is (at least to this author) a somewhat surprisingly strong result given that amplitude amplification is a very well-studied and often used primitive. Our analysis yields several other corollaries as well.

2 Amplitude Amplification

Amplitude amplification [2, 3] is a technique for manipulating the amplitudes of quantum states. One of its uses is to boost the success probability of quantum algorithms. To keep this paper self-contained, we give a concise description of amplitude amplification with emphasis on error reduction. Further details and other applications can be found in e.g. [3].

Consider that we are given, or have developed, some quantum algorithm for solving some problem. We assume the algorithm does not use any measurements, so we can describe the algorithm by a unitary operator A. We may run algorithm A on some initial state $|s\rangle$, producing a final state $|\Psi\rangle = \mathsf{A}|s\rangle$. Suppose that our aim is in producing some target state $|t\rangle$ that represents a solution to the problem. The success probability of algorithm A is the squared overlap $|\langle t|\Psi\rangle|^2$ of the state $|\Psi\rangle$ produced by A with the target state $|t\rangle$. Let angle $0 \leq \theta < \pi/2$ be such that $|\langle t|\Psi\rangle|^2 = \sin^2\theta$. Then θ is an angle that represents the success probability of A. The closer θ is to $\pi/2$, the higher success probability. We write[1] $|\Psi\rangle = \sin\theta|t\rangle + \cos\theta|t^\perp\rangle$, where $|t^\perp\rangle$ is the normalized projection of $|\Psi\rangle$ onto the subspace complementary to the subspace spanned by $|t\rangle$.

Let ϵ be such that $|\langle t|\Psi\rangle|^2 = \sin^2\theta = 1-\epsilon$, so that ϵ is the error probability of A. We want ϵ to be small. If ϵ is large, we may want to apply some boosting method. A standard classical boosting technique is by repetition. If we run A, say, three times, each time on the initial state $|s\rangle$, then the probability of all three runs failing is $\epsilon' = \epsilon^3$. Amplitude amplification offers a quantum alternative to classical repetition. It works as follows. Pick any two complex numbers of unit norm $\phi_s, \phi_t \in \mathbb{C}^*$, which we refer to as phase factors. We define two operators $\mathsf{R}_s = \mathsf{I} - (1-\phi_s)|s\rangle\langle s|$ and $\mathsf{R}_t = \mathsf{I} - (1-\phi_t)|t\rangle\langle t|$, where I is the identity operator. These two operators are pseudo-reflections, where the first acts nontrivially on the ray spanned by $|s\rangle$, and the second nontrivially on the ray spanned by $|t\rangle$. Set $\mathsf{Q} = \mathsf{Q}(\phi_s,\phi_t) = -\mathsf{A}\mathsf{R}_s\mathsf{A}^{-1}\mathsf{R}_t$. Applying operator Q is amplitude amplification.

Operator Q acts invariantly on the two-dimensional subspace spanned by the orthogonal states $|t\rangle$ and $|t^\perp\rangle$, and with respect to the ordered basis $(|t\rangle, |t^\perp\rangle)$, has the following matrix representation [3],

$$\mathsf{Q} = \begin{bmatrix} \phi_t((1-\phi_s)\sin^2\theta - 1) & (1-\phi_s)\cos\theta\sin\theta \\ \phi_t(1-\phi_s)\sin\theta\cos\theta & -(1-\phi_s)\sin^2\theta - \phi_s \end{bmatrix}. \tag{1}$$

Analysis on amplitude amplification can be conducted via an analysis of this matrix, and is thus a basic exercise which surprisingly previously has eluded a full understanding of the rôle of the phase factors, including the phase factor of $e^{\imath\pi/3}$. One aim of this work is to provide such an understanding.

3 The Phase Matrix

Returning to the general setting, consider we apply algorithm A on the initial state $|s\rangle$, hereby producing $|\Psi\rangle = \mathsf{A}|s\rangle$. The probability that a measurement $\mathcal{M}$ of state $\mathsf{A}|\Psi\rangle$ does not yield the outcome $|t\rangle$, is ϵ. If we repeat this experiment three times independently, we may reduce the error of never measuring $|t\rangle$ to ϵ^3.

[1] For simplicity, we ignore a possible global phase factor, which has no consequences in this work.

If we use amplitude amplification instead of classical boosting, we apply Q on $\mathsf{A}|s\rangle$, unitarily producing the state $\mathsf{AR}_s\mathsf{A}^{-1}\mathsf{R}_t\mathsf{A}|s\rangle$, using a total number of three applications of A and its inverse. It is well-established that if we pick phase factors $\phi_s = \phi_t = -1$, we may achieve a quadratic improvement over classical repetition in many settings [5, 1]. The question is what happens for other choices of ϕ_s and ϕ_t.

We first write $|\Psi\rangle = \mathsf{A}|s\rangle$ with respect to the ordered basis $(|t\rangle, |t^\perp\rangle)$ as a column vector, $\mathsf{A}|s\rangle = [\sin\theta, \cos\theta]^T$. Applying Q on $\mathsf{A}|s\rangle$ produces the state $\mathsf{QA}|s\rangle$ represented by

$$\begin{bmatrix} \phi_t((1-\phi_s)\sin^2\theta - 1) & (1-\phi_s)\cos\theta\sin\theta \\ \phi_t(1-\phi_s)\sin\theta\cos\theta & -(1-\phi_s)\sin^2\theta - \phi_s \end{bmatrix} \begin{bmatrix} \sin\theta \\ \cos\theta \end{bmatrix}.$$

Our cardinal step is to isolate the influence of the phase factors ϕ_s and ϕ_t, rewriting $\mathsf{QA}|s\rangle$ as

$$\begin{bmatrix} \sin\theta & 0 \\ 0 & \cos\theta \end{bmatrix} \begin{bmatrix} -\phi_s\phi_t & 1-\phi_s-\phi_t \\ -1+\phi_t-\phi_s\phi_t & -\phi_s \end{bmatrix} \begin{bmatrix} \sin^2\theta \\ \cos^2\theta \end{bmatrix}.$$

We now define the *phase matrix* P as

$$\mathsf{P} = \mathsf{P}(\phi_s, \phi_t) = \begin{bmatrix} -\phi_s\phi_t & 1-\phi_s-\phi_t \\ -1+\phi_t-\phi_s\phi_t & -\phi_s \end{bmatrix}. \tag{2}$$

This matrix allows us to give a simple intuitive study of the choices of phases, expressed independently of angle θ and thus also of the error probability ϵ. We may summarize the action of the amplitude amplification operator Q as follows.

$$\mathsf{Q} : \begin{bmatrix} \sin\theta \\ \cos\theta \end{bmatrix} \longmapsto \begin{bmatrix} \alpha\sin\theta \\ \beta\cos\theta \end{bmatrix} \quad \text{where} \quad \begin{bmatrix} \alpha \\ \beta \end{bmatrix} = \mathsf{P}\begin{bmatrix} \sin^2\theta \\ \cos^2\theta \end{bmatrix} = \mathsf{P}\begin{bmatrix} 1-\epsilon \\ \epsilon \end{bmatrix}. \tag{3}$$

Let ϵ' be the probability that a measurement $\mathcal{M}$ of state $\mathsf{QA}|\Psi\rangle$ does not yield the outcome $|t\rangle$. Then $\frac{\epsilon'}{\epsilon} = \beta^2$ is the relative change in error by applying amplitude amplification operator Q. We want β to be small.

In the next section, we use the above interpretation of amplitude amplification to give an analysis of the choices of phases.

4 Analysis of the Phase Matrix

Our analysis of the phase matrix P given by Eq. 2 begins with a consideration of the entry P_{21}. Recall that the phase matrix P is written with respect to the ordered basis $(|t\rangle, |t^\perp\rangle)$. The entry P_{21} thus captures how much amplitude may potentially be moved out of the solution subspace spanned by $|t\rangle$ to the error subspace spanned by $|t^\perp\rangle$. If entry P_{21} is zero, any amplitude that is moved into the solution subspace, stays in the solution subspace. Conversely, if entry P_{21} is non-zero, some amplitude might escape from the solution subspace.

If we insist on picking phase factors so that P_{21} is zero, then we effectively eliminate interference between the two subspaces, and the quantum algorithm becomes classical of nature. Indeed this is our interpretation of the main result of Grover [7] who shows that if we pick $\phi_s = \phi_t = e^{\imath\pi/3}$, then if the error of algorithm $\mathsf{A}|s\rangle$ is ϵ, the error of algorithm $\mathsf{QA}|s\rangle$ is ϵ^3. It achieves the exact same error reduction as classical repetition. It is different from classical repetition in that it it does not use intermediate measurements or additional storage space (e.g., an ancilla used for counting purposes). It is a truly in-place quantum mechanical version of classical repetition. It is fairly easy to see that this phase factor is unique up to conjugation, and thus does play a pivotal rôle in amplitude amplification.

Proposition 1. *Entry* P_{21} *of the phase matrix is zero iff* $\phi_s = \phi_t = e^{\pm\imath\pi/3}$.

With this uplifting simple explanation of Grover's result, we now study the phase matrix in more detail. Entry P_{21} is the sum of three unit numbers. We say that three unit numbers $e^{\imath x}, e^{\imath y}, e^{\imath z}$ lie *within a half circle* if there exists a unit number w so that $e^{\imath x}w, e^{\imath y}w, e^{\imath z}w$ all have non-negative real part. We say they lie *strictly* within a half circle if there exists a unit number w so that $e^{\imath x}w, e^{\imath y}w, e^{\imath z}w$ all have nonzero and positive real part. The three vectors do not lie within a half circle if and only if every hyperplane strictly separates them. The sum of three unit vectors is related to whether they lie within a half circle.

Proposition 2. *The absolute value of the sum of three unit numbers is greater than one if they lie strictly within a half circle, is one if two of the numbers are each others negation, and is less than one otherwise.*

Proof. Consider the sum of three unit vectors $e^{\imath x}, e^{\imath y}, e^{\imath z}$. Without loss of generality, we may assume that $0 \leq z - y \leq \pi$. Fix y and z and consider the function $f(x) = |e^{\imath x} + e^{\imath y} + e^{\imath z}|$. The function is strictly monotonically increasing in the interval $[\frac{z+y}{2} - \pi, \frac{z+y}{2}]$, and it is one when $x = z - \pi$. Similarly, the function strictly monotonically decreasing in the interval $[\frac{z+y}{2} - 2\pi, \frac{z+y}{2} - \pi]$ and it is one when $x = y - \pi$. In summary, $f(x)$ is bigger than 1 if and only if the three vectors lie strictly within a half circle, the function is one if and only if $x = z - \pi$ or $x = y - \pi$, and it is less than one in all other cases. □

We multiply entry P_{21} with $-\overline{\phi_t}$, apply Proposition 2 on $-1, \phi_s, \overline{\phi_t}$, and obtain the following characterization of $|\mathsf{P}_{21}|$.

Lemma 1 (Contribution from solution subspace to error subspace).

$$|\mathsf{P}_{21}| = \begin{cases} > 1 & \text{if } -1, \phi_s, \overline{\phi_t} \text{ lie strictly within a half circle} \\ = 1 & \text{if } \phi_s = 1,\ \phi_t = 1, \text{ or } \phi_s\phi_t = -1 \\ = 0 & \text{if } \phi_s = \phi_t = e^{\pm\imath\pi/3} \\ < 1 & \text{otherwise.} \end{cases}$$

To exhibit the significance of Lemma 1, reconsider Equation 3. We are interested in having $|\beta| < 1$ so that the error ϵ' of $\mathsf{QA}|s\rangle$ is smaller than the error ϵ of $\mathsf{A}|s\rangle$, the algorithm without amplitude amplification. The phase matrix P is multiplied on the right by the column vector $[\sin^2\theta, \cos^2\theta]^T$. Since entry $|\mathsf{P}_{22}|$ is of unit norm, if $|\mathsf{P}_{21}| < 1$ then $|\beta| < 1$ and hence $\epsilon' < \epsilon$.

Theorem 1 (Strictly decreasing error). *If* $-1, \phi_s, \overline{\phi_t}$ *do not lie within a half circle,* $\epsilon' < \epsilon$.

The above theorem gives a sufficient condition for strict error reduction in amplitude amplification processes. Prior to Grover's recent work on the phase factor $e^{\imath\pi/3}$, no such condition was known. Our general condition is that $-1, \phi_s, \overline{\phi_t}$ do not lie close together.

5 Error Reduction with Phase Matching

A desirable property of the phase matrix P is to have $\arg(\mathsf{P}_{21}) = \arg(-\mathsf{P}_{22})$. If the two entries P_{21} and P_{22} point in opposite directions as vectors, the error introduced by the term $\mathsf{P}_{21}\sin^2\theta$ partly cancels with the error introduced by the other term $\mathsf{P}_{22}\cos^2\theta$. For fixed $|\mathsf{P}_{21}|$, having $\arg(\mathsf{P}_{21}) = \arg(-\mathsf{P}_{22})$ minimizes ϵ' (ϵ' is defined in Section 3 above). Our next lemma shows that the only nontrivial choices of phases for which $\arg(\mathsf{P}_{21}) = \arg(-\mathsf{P}_{22})$ is when $\phi_s = \phi_t$.

Lemma 2 (Phase matching). $\arg(\mathsf{P}_{21}) = \arg(-\mathsf{P}_{22})$ *if and only if* $\phi_s = \phi_t$ *and* $\mathrm{Re}(\phi_s) < \frac{1}{2}$.

Proof. Suppose $\arg(\mathsf{P}_{21}) = \arg(-\mathsf{P}_{22})$. Then $\arg(1 - \phi_t + \phi_s\phi_t) = \arg(-\phi_s)$, and by adding $-\phi_s$ on the left hand side, $\arg((1-\phi_t)(1-\phi_s)) = \arg(-\phi_s)$. Multiplying through by $\overline{\phi_s}$, this implies that $\arg((1-\overline{\phi_s})(1-\phi_t)) = 0$, which is satisfied only if $\phi_s = \phi_t$, $\phi_s = 1$, or $\phi_t = 1$. Conversely, if $\phi_s = \phi_t$ and $\mathrm{Re}(\phi_s) < \frac{1}{2}$, then $\arg(\mathsf{P}_{21}) = \arg(-\mathsf{P}_{22})$, while in the other cases, $\arg(\mathsf{P}_{21}) = \arg(\mathsf{P}_{22})$. □

Note that for any number $0 \le p \le 3$, there always exists a phase factor $\phi \in \mathbb{C}^*$ so that if we set $\phi = \phi_s = \phi_t$, then $|-1 + \phi_t - \phi_s\phi_t| = p$. This implies that for any choice of phases $\tilde{\phi}_s, \tilde{\phi}_t \in \mathbb{C}^*$, there exists a phase factor $\phi \in \mathbb{C}^*$ so that the error if applying $\mathsf{Q}(\phi,\phi)$ is no larger than if applying $\mathsf{Q}(\tilde{\phi}_s, \tilde{\phi}_t)$. That is, picking identical phases is never suboptimal in terms of error reduction.

Picking identical phases is already known to have other advantageous properties. In work on Grover's algorithm, Long, Li, Zhang, and Niu [9], consider the choices of phases for which amplitude amplification may yield exact algorithms and they derive the phase matching condition $\phi_s = \phi_t$ (the term 'exact quantum algorithm' is explained in the next paragraph). Together with a later paper by Long, Xiao, and Song [8], they give a general and throughly analysis of the conditions we must put on ϕ_s and ϕ_t for obtaining exact algorithms. One instance of their scenario requires the *phase matching* condition $\phi_s = \phi_t$. Phase matching is particular appealing as then the relationship between ϕ_s and ϕ_t does not depend on ϵ, it helps simplifying the analysis, and it might be easier to implement just one phase factor instead of working with two.

Suppose we are given a classical randomized algorithm that succeeds in solving some problem with some probability $1-\epsilon$. Then we may repeat the algorithm several times, hereby reducing the error probability. With only polynomially many repetitions, we can reduce the error to being exponentially small. However, there are no known schemes that would allow us to reduce the error to zero with only polynomial overhead for arbitrary processes. There are in short no general derandomization schemes. In contrast, amplitude amplification can be used to obtain exact quantum algorithms in some settings. We say a quantum algorithm is *exact* if its error probability is zero. Exact quantum algorithms for decision problems that run in polynomial time comprise the quantum analogue $\mathcal{EQP}$ of the classical complexity class $\mathcal{P}$.

Under phase matching, the phase matrix simplifies to

$$\mathsf{P} = \mathsf{P}(\phi, \phi) = \begin{bmatrix} -\phi^2 & 1 - 2\phi \\ -1 + \phi - \phi^2 & -\phi \end{bmatrix}. \tag{4}$$

Theorem 2 gives a necessary and sufficient condition for having $\epsilon' < \epsilon$. It generalizes Theorem 1 under phase matching to include necessary conditions, and states that $\epsilon' < \epsilon$ if and only if $\mathrm{Re}(\phi) > -\frac{\epsilon}{1-\epsilon}$ for nontrivial ϕ. The right hand side, $-\frac{\epsilon}{1-\epsilon}$, may be arbitrarily close to 0, and thus if the expression $\mathrm{Re}(\phi) > -\frac{\epsilon}{1-\epsilon}$ is to be true for all nonzero ϵ, we effectively require that $\mathrm{Re}(\phi) \geq 0$. Expressed in terms of trigonometric functions, $\epsilon' < \epsilon$ if and only if $\cos(\varphi) > -\tan^2(\theta)$ where $\phi = e^{\imath\varphi}$.

Theorem 2 (Error reduction with phase matching). *Suppose that* $\phi = \phi_s = \phi_t$. *Then*

$$\epsilon' = \begin{cases} = 0 & \textit{if } \mathrm{Re}(\phi) = \frac{1}{2}(1 - \frac{\epsilon}{1-\epsilon}) \\ < \epsilon & \textit{if } \mathrm{Re}(\phi) > -\frac{\epsilon}{1-\epsilon} \textit{ and } \phi \neq 1 \\ = \epsilon & \textit{if } \mathrm{Re}(\phi) = -\frac{\epsilon}{1-\epsilon} \\ > \epsilon & \textit{if } \mathrm{Re}(\phi) < -\frac{\epsilon}{1-\epsilon}. \end{cases}$$

Proof. First note that $\epsilon' < \epsilon$ if and only if the absolute value of $\mathsf{P}_{21}(1-\epsilon)+\mathsf{P}_{22}\epsilon$ is less than one. If $\phi = \phi_s = \phi_t$ then $\mathsf{P}_{21} = \phi(1 - 2\mathrm{Re}(\phi))$ and $\mathsf{P}_{22} = -\phi$, and thus $|\mathsf{P}_{21}(1-\epsilon)+\mathsf{P}_{22}\epsilon| = |(1-2\mathrm{Re}(\phi))(1-\epsilon)-\epsilon|$. If $0 < \mathrm{Re}(\phi) < 1$, then $|1-2\mathrm{Re}(\phi)| < 1$ and hence $\epsilon' < \epsilon$. Now, suppose $\mathrm{Re}(\phi) \leq 0$. Then $(1 - 2\mathrm{Re}(\phi))(1 - \epsilon) - \epsilon < 1$ if and only if $\mathrm{Re}(\phi) > -\frac{\epsilon}{1-\epsilon}$.

Similarly, if $\mathrm{Re}(\phi) < -\frac{\epsilon}{1-\epsilon}$ then $\epsilon' > \epsilon$. Finally, $(1 - 2\mathrm{Re}(\phi))(1 - \epsilon) - \epsilon = 0$ if and only if $\mathrm{Re}(\phi) = \frac{1}{2}(1 - \frac{\epsilon}{1-\epsilon})$. □

It follows that any phase factor with positive real ensures strict error reduction.

Corollary 1 (Strictly decreasing error). *If* $\phi_s = \phi_t \neq 1$ *and* $0 \leq \mathrm{Re}(\phi_s) < 1$, *then* $\epsilon' < \epsilon$.

It ϵ is known [1], we can obtain an exact quantum algorithm by choosing the phase factor ϕ so that $\mathsf{P}_{21}(1 - \epsilon) + \mathsf{P}_{22}\epsilon = 0$.

Corollary 2 (Exact algorithm). *If* $\phi_s = \phi_t$ *and* $\mathrm{Re}(\phi_s) = \frac{1}{2}(1 - \frac{\epsilon}{1-\epsilon})$, *then* $\epsilon' = 0$.

Note that the condition $\mathrm{Re}(\phi_s) = \frac{1}{2}(1 - \frac{\epsilon}{1-\epsilon})$ can be satisfied if and only if $\epsilon \leq \frac{3}{4}$. That is, we can achieve $\epsilon' = 0$ by a one-round amplitude amplification process if and only if the original error ϵ is at most $\frac{3}{4}$.

The next theorem states that phase matching with a phase factor $\phi \in \mathbb{C}^*$ with $\mathrm{Re}(\phi) \leq \frac{1}{2}$ is never suboptimal in terms of error reduction. This implies that when applying amplitude amplification (as considered in this paper) we may restrict our attention to picking a single phase factor ϕ with $-1 \leq \mathrm{Re}(\phi) \leq \frac{1}{2}$, as opposed to considering all possible choices of pairs $(\phi_s, \phi_t) \in \mathbb{C}^* \times \mathbb{C}^*$.

Theorem 3 (Phase matching with $\mathrm{Re}(\phi) \leq \frac{1}{2}$). *For all phase factors $\phi_s, \phi_t \in \mathbb{C}^*$ there exists a phase factor $\phi \in \mathbb{C}^*$ with $\mathrm{Re}(\phi) \leq \frac{1}{2}$ so that the error of $\mathsf{Q}(\phi, \phi)\mathsf{A}|s\rangle$ is no larger than the error of $\mathsf{Q}(\phi_s, \phi_t)\mathsf{A}|s\rangle$. Phase factor ϕ depends only on ϕ_s and ϕ_t, and not on A, $|s\rangle$, and $|t\rangle$.*

Proof. Let ϵ denote the error probability of $\mathsf{Q}(\phi_s, \phi_t)\mathsf{A}|s\rangle$. By the comments succeeding Lemma 2, we can without loss of generality assume that $\phi = \phi_s = \phi_t$ and thus use the simplified matrix given by Eq. 4. If we pick ϕ so that $\mathrm{Re}(\phi) = \frac{1}{2}$, then $|(1 - 2\mathrm{Re}(\phi))(1-\epsilon) - \epsilon| = \epsilon$ whereas if $\mathrm{Re}(\phi) > \frac{1}{2}$ then $|(1 - 2\mathrm{Re}(\phi))(1-\epsilon) - \epsilon| > \epsilon$. It follows that using ϕ with $\mathrm{Re}(\phi) > \frac{1}{2}$ is never better than using ϕ with $\mathrm{Re}(\phi) = \frac{1}{2}$. □

We mention that picking conjugate phases aligns the entries P_{11} and P_{12}.

Proposition 3 (Conjugate phase matching). $\arg(\mathsf{P}_{11}) = \arg(\mathsf{P}_{12})$ *if and only if (1) $\phi_s = \overline{\phi_t}$ and $\mathrm{Re}(\phi_s) > \frac{1}{2}$, (2) $\phi_s = 1$, or (3) $\phi_t = 1$.*

6 Computational Considerations

Thus far we have primarily been concerned with error reduction in amplitude amplification and mostly ignored the computational costs. Amplitude amplification maps operator A to operator $\mathsf{A}_1 = \mathsf{QA}$,

$$\mathsf{A} \mapsto \mathsf{A}_1 = \mathsf{A}\mathsf{R}_s\mathsf{A}^{-1}\mathsf{R}_t\mathsf{A} \tag{5}$$

at the cost of in total three applications of A and its inverse, and one application of each of R_s and R_t. We may repeat the amplification process on the thus formed algorithm A_1, mapping

$$\mathsf{A}_1 \mapsto \mathsf{A}_2 = \mathsf{A}_1\mathsf{R}_s\mathsf{A}_1^{-1}\mathsf{R}_t\mathsf{A}_1.$$

Applying the mapping k times utilizes $\frac{1}{2}(3^k + 1)$ applications of A, $\frac{1}{2}(3^k - 1)$ applications of the inverse A^{-1}, and $\frac{1}{2}(3^k - 1)$ applications of each of the pseudo-reflections R_s and R_t. That is, we use $\Theta(K)$ applications of each of the four operators $\mathsf{A}, \mathsf{A}^{-1}, \mathsf{R}_s, \mathsf{R}_t$, where $K = 3^k$.

The error ϵ' after k recursive applications of the mapping given by Eq. 5 depends on the choice of phase factors. If we apply Q with phase factor $e^{\imath\pi/3}$, then $\epsilon' = \epsilon^K$. This error is the exact same error as we could achieve by K classical repetitions, as exemplified in Section 3. Amplitude amplification with phase factor $e^{\imath\pi/3}$ is a genuinely quantum mechanical version of classical repetition. Note that $\mathrm{Re}(e^{\imath\pi/3}) = \frac{1}{2}$. By Theorem 3, applying amplitude amplification with a phase factor ϕ having $\mathrm{Re}(\phi) > \frac{1}{2}$ is suboptimal in terms of error reduction.

For matching phases ϕ with $\mathrm{Re}(\phi) < \frac{1}{2}$, the error reduction β^2 is given by the square of $(1-2\mathrm{Re}(\phi))(1-\epsilon)-\epsilon$, where β is defined by Eq. 3. The factor of error reduction by three classical repetitions is $\frac{\epsilon^3}{\epsilon} = \epsilon^2$. Amplitude amplification is thus superior to classical boosting if and only if $-\epsilon < (1-2\mathrm{Re}(\phi))(1-\epsilon)-\epsilon < \epsilon$ which is valid if and only if $\frac{1}{2}\frac{1-3\epsilon}{1-\epsilon} < \mathrm{Re}(\phi) < \frac{1}{2}$. The left-hand side $\frac{1}{2}\frac{1-3\epsilon}{1-\epsilon}$ is less than -1 if $\epsilon > \frac{3}{5}$, in which case the error reduction β^2 in amplitude amplification $\mathsf{Q}(-1,-1)$ with phase factor $\phi = -1$ is better than the classical error reduction of ϵ^2 obtained by three repetitions.

Theorem 4 (Amplitude amplification works well for large error). *For $\epsilon > \frac{3}{5}$, amplitude amplification with phase factor $\phi = -1$ is superior to classical repetition.*

Buhrman, Cleve, de Wolf, and Zalka show in [4] that any quantum algorithm for improving the error from a constant, say $\frac{1}{2}$, to δ requires $\Omega(\log\frac{1}{\delta})$ applications of A. In particular, for small ϵ, amplitude amplification requires asymptotically the same number of applications of A as does classical repetition. The above theorem implies that in the range $1 > \epsilon > \frac{3}{5}$, standard amplitude amplification with phase factor $\phi = -1$ is superior to classical boosting in terms of error reduction.

7 Concluding Remarks

Amplitude amplification is one of the most used and well-studied primitives in quantum algorithmics. One of its obstacles is that applying standard amplitude amplification (with phase factor -1) can sometimes be harmful. If the given quantum algorithm A has small error, then applying amplitude amplification may produce an algorithm QA that has larger error than A itself. It is in most cases not desirable to use computational efforts in creating worse algorithms.

In recent work [7], Grover shows that applying amplitude amplification with phase factor $e^{\imath\pi/3}$ is never harmful. If the error of algorithm A is ϵ, then amplitude amplification produces an algorithm QA with error ϵ^3, which is never larger than ϵ. The limitation of the phase factor $e^{\imath\pi/3}$ is that the exact same error reduction can be achieved using the same computational efforts by classical repetition.

Applying amplitude amplification as a subroutine, as is done in many quantum algorithms, can be a challenge: Phase factor -1 might be harmful, phase factor $e^{\imath\pi/3}$ leads to no better error reduction than classical. Possibly one may consequently decide to choose phase factors ϕ_s, ϕ_t somewhere in between in lack of better.

In this paper, we provide a set of tools and results on picking phase factors. There is no unifying answer on phase picking—the preferable choice depends on the application at hand. We give a novel and simple, yet rigorous, analysis of the interplay between phase factors and error reduction. We show that one may always limit oneself to consider matching phases $\phi = \phi_s = \phi_t$ with argument in the range $\frac{\pi}{3} \leq \arg(\phi) \leq \pi$. We show that any phase with $\mathrm{Re}(\phi) > 0$ reduces the error for all $0 < \epsilon < 1$. The optimal choice in terms of error reduction is to pick $\phi = -1$ if $\epsilon \geq \frac{3}{4}$ and otherwise to pick ϕ so that $\mathrm{Re}(\phi) = \frac{1}{2}\frac{1-2\epsilon}{1-\epsilon}$.

Grover's work has already lead to new applications based on the pivotal phase factor $e^{\imath\pi/3}$ [10, 11]. Our set of analytical tools and results may initiate other results related to amplitude amplification and quantum error reduction in more general settings.

References

1. M. Boyer, G. Brassard, P. Høyer, and A. Tapp. Tight bounds on quantum searching. *Fortschr. Phys.*, **46**(4–5):493–505, 1998.
2. G. Brassard and P. Høyer. An exact quantum polynomial-time algorithm for Simon's problem. In *Proc. 5th Israeli Symp. Theory of Comput. and Systems*, pp. 12–23, 1997.
3. G. Brassard, P. Høyer, M. Mosca, and A. Tapp. Quantum amplitude amplification and estimation. In: Quantum Computation and Quantum Information: A Millennium Volume. *AMS Contemp. Math.*, **305**:53–74, 2002.
4. H. Buhrman, R. Cleve, R. de Wolf, and Ch. Zalka. Bounds for small-error and zero-error quantum algorithms. In *Proc. 40th IEEE Symp. Found. Comput. Sci.*, pp. 358–368, 1999.
5. L. K. Grover. A fast quantum mechanical algorithm for database search. In *Proc. 28th ACM Symp. Theory Comput.*, pp. 212–219, 1996.
6. L. K. Grover. Quantum mechanics helps in searching for a needle in a haystack. *Phys. Rev. Lett.*, **79**(2):325–328, 1997.
7. L. K. Grover. A different kind of quantum search. quant-ph/0503205, May 2005.
8. G.-L. Long, X. Li, and Y. Sun. Phase matching condition for quantum search with a generalized initial state. *Phys. Lett. A*, **294**(3-4):143-152, 2002.
9. G.-L. Long, Y. S. Li, W. L. Zhang, and L. Niu. Phase matching in quantum searching. *Phys. Lett. A*, **262**(1):27–34, 1999.
10. T. Tulsi, L. K. Grover, and A. Patel. A new algorithm for directed quantum search. quant-ph/0505007, May 2005.
11. L. Xiao and J. Jones. An NMR implementation of Grover's fixed-point quantum search algorithm. quant-ph/0504054, April 2005.

ISB-Tree: A New Indexing Scheme with Efficient Expected Behaviour*

Alexis Kaporis[1], Christos Makris[1], George Mavritsakis[1], Spyros Sioutas[1], Athanasios Tsakalidis[1,2], Kostas Tsichlas[1], and Christos Zaroliagis[1,2]

[1] Dept of Computer Eng and Informatics, University of Patras, 26500 Patras, Greece
[2] Computer Technology Institute, N. Kazantzaki Str, Patras University Campus, 26500 Patras, Greece
{kaporis, makri, mayritsa, sioutas, tsak, tsihlas, zaro}@ceid.upatras.gr

Abstract. We present the interpolation search tree (ISB-tree), a new cache-aware indexing scheme that supports update operations (insertions and deletions) in $O(1)$ worst-case (w.c.) block transfers and search operations in $O(\log_B \log n)$ expected block transfers, where B represents the disk block size and n denotes the number of stored elements. The expected search bound holds with high probability for a large class of (unknown) input distributions. The w.c. search bound of our indexing scheme is $O(\log_B n)$ block transfers. Our update and expected search bounds constitute a considerable improvement over the $O(\log_B n)$ w.c. block transfer bounds for search and update operations achieved by the B-tree and its numerous variants. This is also suggested by a set of preliminary experiments we have carried out. Our indexing scheme is based on an externalization of a main memory data structure based on interpolation search.

1 Introduction

More than three decades after its invention, B-tree [3, 4] and its variants remain the ubiquitous (external memory) data structure for indexing and organizing large data sets with numerous applications, especially in database systems. Its popularity is mainly due to the stable and guaranteed performance for search and update (insertion and deletion) operations, which both cost $O(\log_B n)$ block transfers in the worst-case, with B and n representing the number of elements in a disk block and the number of stored elements, respectively. The most heavily used application is the efficient answering of one-dimensional range search queries using $O(\log_B n + r)$ block transfers, where $R = rB$ is the number of elements reported. In such a query, the elements in a range $[z_1, z_2]$ can be found by first searching the B-tree for z_1 and then performing an in-order traversal

* This work was partially supported by the IST Programme (6th FP) of EC under contracts IST-2001-33058 (PANDA) and IST-2002-001907 (integrated project DELIS), and by the Action PYTHAGORAS of the Operational Programme for Educational & Vocational Training II, with matching funds from the European Social Fund and the Greek Ministry of Education.

X. Deng and D. Du (Eds.): ISAAC 2005, LNCS 3827, pp. 318–327, 2005.

in the tree from z_1 to z_2. These bounds hold for any cache-aware disk-access model, that is, a model that accounts memory transfers in disk blocks, as these transfers are the dominating operation w.r.t. time. In this paper, we consider one of the most known and widely used such models, namely the two-level memory hierarchy model introduced in [1]. In this model, the memory hierarchy consists of an internal (main) memory and an arbitrarily large external memory (disk) partitioned into blocks of size B. The data from the external to the main memory and vice versa are transferred in blocks (one block at a time).

A vast number of variants of the B-tree have been proposed since its appearance — B^+-trees [4] and B^*-trees [4, 11] are some representative examples; see the excellent survey by Vitter [20] for an extended accounting of these and other variants and their applications — in order to improve its performance in practice for various applications, to make it parallel for use in multi-disk environments [18], to tune it for concurrency and recovery purposes [19], to extend it to cover other than the original field [6], etc. However, to the best of our knowledge, the aforementioned search and update bounds of B-tree and its variants remained untouched all these years. The same applies to the one-dimensional range search query bound, although some variants (with B^+-tree being the most popular) offer a slightly different procedure, since the leaves are linked together and hence allow for sequential access. Regarding the update operation, it should be noted that it consists of three consecutive phases: a search phase (to locate the element), a value-updating phase (to replace the element's key with its new value), and a rebalancing phase (to restore the B-tree structure). Excluding the first phase (search operation), the dominating phase of an update operation is the rebalancing one, since the value-updating phase takes typically $O(1)$ block transfers (and/or time). In the case of B-tree and its variants, the rebalancing phase requires $O(\log_B n)$ block transfers in the worst-case. This implies that the update operation takes $O(\log_B n)$ block transfers, even in the case where the update position (block within which the update will take place) is given. Note that there are certain applications (see e.g., [13]) which justify the exclusion of the search phase in an update operation: once the requested element has been found, then the next element to be searched is located "near by" and hence a new search is redundant.

In this work, we present a new indexing scheme, called *ISB-tree* (Interpolation Search B-tree), that supports search operations in $O(\log_B \log n)$ expected block transfers *with high probability* (w.h.p.) for a large class of input distributions (including both uniform and non-uniform classes) that are explained below, and update operations in $O(1)$ block transfers, provided that the update position is given. The search bound implies that a one-dimensional range search query can be supported in $O(\log_B \log n + r)$ expected block transfers w.h.p.. The worst-case block transfers for the search operation in our indexing scheme are $O(\log_B n)$.

To achieve our expected search bound we consider a rather general scenario of *μ-random* insertions and *random* deletions, where μ is a so-called *smooth* probability density [2, 16]. An insertion is μ-random if the key to be inserted is drawn randomly with density function μ; a deletion is random if every key present in the data structure is equally likely to be deleted (see [12]). Informally, a distribution defined over an interval I is *smooth* if the probability density over any subinterval of I does not exceed a specific bound, however small this subinterval is (i.e., the distribution does not contain sharp peaks). Smooth distributions are a superset of uniform, bounded, and several non-uniform distributions (e.g., the class of regular distributions introduced by Willard [21]).

Our indexing scheme is a two-level data structure. The upper level is an externalization of the static interpolation search tree presented in [9]. The lower level is a forest of buckets, each of which is implemented by a new variant of the classical B-tree, the *Lazy B-tree*, which is introduced in [10]. The lazy B-tree supports a search operation in $O(\log_B n)$ block transfers and an update operation in $O(1)$ block transfers, provided that the update position is given. However, a straightforward combination of the above structures does not necessarily lead to better bounds, since: (i) the number of elements within a bucket may grow arbitrarily large, as insertions are performed; and (ii) we strive for creating a *robust indexing scheme*, that is, a data structure that works correctly *without* apriori knowledge of the particular smooth distribution μ. To overcome these problems, we employ the combinatorial game of bins and balls introduced in [9] that allows to upper bound the number of elements in a bucket, and to approximate an unknown distribution by an almost uniform one.

To the best of our knowledge, this is the first work that uses the dynamic interpolation search paradigm in the framework of indexing data in external memory. External data structures related to our approach are those based on hashing [11, 15, 20]. The main representatives of external memory hashing methods are: extendible hashing [5], linear hashing [14], and external perfect hashing [7]. These hashing schemes and their variants need $O(1)$ expected block transfers for answering search queries, but they share various disadvantages when compared to our structure: (i) they do not support range queries; (ii) their expected case analysis usually assumes *uniform* input distributions (or input distributions that produce uniform hash key values); and (iii) they exhibit poor worst case performance.

The remainder of the paper is organized as follows. In Section 2, we discuss preliminary notions and results that are used throughout the paper. The main result of this paper, the *ISB-tree*, with the complexity analysis of its operations is discussed in Section 3. Section 4 provides an experimental evaluation of our theoretical findings. We conclude in Section 5. Due to space limitations the full details of the paper can be found in [10].

2 Preliminaries

The *B-tree* is a $\Theta(B)$-ary tree (with the root possibly having smaller degree) built on top of $\Theta(n/B)$ leaves. The degree of internal nodes, as well as the number of elements in a leaf, is typically kept in the range $[B/2, B]$ such that a node or leaf can be stored in one disk block. All leaves are on the same level and the tree has height $O(\log_B n)$. This guarantees that a search operation can be accomplished within $O(\log_B n)$ block transfers. An insertion is performed in $O(\log_B n)$ block transfers by first searching down the tree for the relevant leaf l. If there is room for the new element in l, then we simply store it there. Otherwise, we split l into two leaves l' and l'' of approximately the same size and insert the new element in the relevant leaf. The split of l results in the insertion of a new routing element in the parent of l, and thus the split may propagate up the tree. Propagation of splits can often be avoided by sharing some of the (routing) elements of the full node with a non-full sibling. A new (degree 2) root is produced when the root splits and the height of the tree grows by one. Similarly, a deletion can be performed in $O(\log_B n)$ block transfers by first searching down the tree for the relevant leaf l and then removing the deleted element. If this results in l containing too few elements, then we either fuse it with one of its siblings (corresponding to deleting l and inserting its elements in a sibling), or we perform a share operation by moving elements from a sibling to l. Fuse operations may also propagate up the tree and eventually result in the height of the tree decreasing by one.

One of the first works, in the context of internal memory data structures, that investigated non-uniform distributions regarding insertions in an update sequence was that of Willard [21], who introduced the so-called *regular* distributions. A probability density μ is regular if there are constants b_1, b_2, b_3, b_4 such that $\mu(x) = 0$ for $x < b_1$ or $x > b_2$, and $\mu(x) \geq b_3 > 0$ and $|\mu'(x)| \leq b_4$ for $b_1 \leq x \leq b_2$. This has been further pursued by Mehlhorn and Tsakalidis [16], who introduced the *smooth* input distributions, a notion that was further generalized and refined in [2]. Given two functions f_1 and f_2, a density function $\mu = \mu[a, b](x)$ is (f_1, f_2)*-smooth* [2] if there exists a constant β, such that for all c_1, c_2, c_3, $a \leq c_1 < c_2 < c_3 \leq b$, and all integers n, it holds that

$$\int_{c_2 - \frac{c_3 - c_1}{f_1(n)}}^{c_2} \mu[c_1, c_3](x)dx \leq \frac{\beta \cdot f_2(n)}{n}$$

where $\mu[c_1, c_3](x) = 0$ for $x < c_1$ or $x > c_3$, and $\mu[c_1, c_3](x) = \mu(x)/p$ for $c_1 \leq x \leq c_3$ where $p = \int_{c_1}^{c_3} \mu(x)dx$. Intuitively, function f_1 partitions an arbitrary subinterval $[c_1, c_3] \subseteq [a, b]$ into f_1 equal parts, each of length $\frac{c_3 - c_1}{f_1} = O(\frac{1}{f_1})$; that is, f_1 measures how fine is the partitioning of an arbitrary subinterval. Function f_2 guarantees that no part, of the f_1 possible, gets more probability mass than $\frac{\beta \cdot f_2}{n}$; that is, f_2 measures the sparseness of any subinterval $[c_2 - \frac{c_3 - c_1}{f_1}, c_2] \subseteq [c_1, c_3]$. The class of (f_1, f_2)-smooth distributions (for appropriate choices of f_1 and f_2) is a superset of both regular and uniform classes of distributions, as well as of several non-uniform classes [2, 9]. Actually, *any* probability distribution is $(f_1, \Theta(n))$-smooth, for a suitable choice of β.

The *static interpolation search tree* [9] is a static, explicit, and refined version of the search trees used in [2, 16]. A static interpolation search tree containing n elements can be fully characterized by three nondecreasing functions $H(n)$, $R(n)$ and $I(n)$, where $H(n)$ denotes the height of the tree, $R(n)$ denotes the out-degree of the root, and $I(n)$ denotes how fine is the partition of the set of elements and is defined by $I(n) = n \cdot g(H(n))$, where $g(n)$ should satisfy $\sum_{i=1}^{\infty} g(i) = \Theta(1)$.

To guarantee the height of $H(n)$, it should hold that $n/R(n) = H^{-1}(H(n)-1)$. The children of the root have $n' = \Theta(n/R(n))$ leaves. Their height will be $H(n') = H(n)-1$, their out-degree is $R(n') = \Theta(H^{-1}(H(n)-1)/H^{-1}(H(n)-2))$, and $I(n') = n' \cdot g(H(n'))$. In general, consider an internal node v at depth i and assume that n_i leaves are stored in the subtree rooted at v. Then we have that $R(n_i) = \Theta(H^{-1}(H(n)-i+1)/H^{-1}(H(n)-i))$, and $I(n_i) = n_i \cdot g(H(n)-i)$. The node v is associated with an array REP$[1..R(n_i)]$ of sample elements, one sample element for each of its subtrees, and an ID$[1..I(n_i)]$ array that stores a set of sample elements approximating the inverse distribution function. By using the ID array, we can interpolate the REP array to determine the subtree in which the search procedure will continue. In particular, the ID array for node v is an array ID$[1..m]$, where $m = I(n_i)$, with ID$[i] = j$ iff REP$[j] < \ell + i(u-\ell)/m \leq$ REP$[j+1]$, where ℓ and u are the minimum and the maximum element, resp., stored in the subtree rooted at v. Let x be the element we seek. To interpolate REP, compute the index $j = $ ID$[\lfloor(m(x-\ell)/(u-\ell))\rfloor]$, and then search the REP array from REP$[j+1]$ until the appropriate subtree is located. For each node we explicitly maintain parent and child pointers. The required pointer information can be easily incorporated in the construction of the static interpolation search tree. Throughout the paper, we say that an event E occurs *with high probability* if $\Pr[E] = 1 - o(1)$.

3 The ISB-Tree

The ISB-tree is a two-level data structure. The lower level is a set of buckets each of which contains a subset of the stored elements and is represented by a unique representative. The representatives of the buckets are stored in the upper level structure.

The upper level data structure is an external version of the static interpolation search tree (SIST) described in [9] (see also Section 2), with parameters $R(s) = s^\delta$, $I(s) = s/(\log\log s)^{1+\epsilon}$, where $\epsilon > 0$, $\delta = 1 - \frac{1}{B}$, and s is the number of stored elements in the tree. The specific choice of δ guarantees the desirable $O(\log_B \log s)$ height of the upper level structure. For each node that stores more than $B^{1+\frac{1}{B-1}}$ elements in its subtree, we represent its REP and ID arrays as static external sorted arrays; otherwise, we store all the elements in a constant number of disk blocks. In particular, let v be a node and n_v be the number of stored elements in its subtree, with $n_v \geq B^{1+\frac{1}{B-1}}$. Node v is associated with two external arrays EREP$_v$ and EID$_v$ that represent the REP$_v$ and ID$_v$ arrays of the original SIST structure. The EID$_v$ array uses $O(\frac{I(n_v)}{B})$ contiguous blocks, the EREP$_v$ array uses $O(\frac{R(n_v)}{B})$ contiguous blocks, while an arbitrary element of the arrays can be accessed with $O(1)$ block transfers, given its index. Moreover, the choice of the parameter $B^{1+\frac{1}{B-1}}$ guarantees that each of the EREP$_v$ and EID$_v$ arrays contains at least B elements, and hence we do not waste space (in terms of underfull blocks) in the external memory representation.

On the other hand, the lower level is a set of ρ buckets. Let S_0 be the set of elements to be stored where the elements take values in $[a, b]$. Each bucket

$\mathcal{B}_i$, $1 \leq i \leq \rho$, stores a subset of elements and is represented by the element $rep(i) = \max\{x : x \in \mathcal{B}_i\}$. The set of elements stored in the buckets constitute an ordered collection $\mathcal{B}_1, \ldots, \mathcal{B}_\rho$ such that $\max\{x : x \in \mathcal{B}_i\} < \min\{y : y \in \mathcal{B}_{i+1}\}$ for all $1 \leq i < \rho - 1$. In other words, $\mathcal{B}_i = \{x : x \in (rep(i-1), rep(i)]\}$, for $2 \leq i \leq \rho$, and $\mathcal{B}_1 = \{x : x \in [rep(0), rep(1)]\}$, where $rep(0) = a$ and $rep(\rho) = b$.

The elements of each $\mathcal{B}_i$ are stored in a *Lazy B-tree*, which is a new variant of the classical B-tree. Due to space limitations, the details of the Lazy B-tree are discussed in the extended version of the paper [10]. The following theorem summarizes the properties of a Lazy B-tree.

Theorem 1. *A Lazy B-Tree supports search operations with $O(\log_B n)$ worst-case block transfers and update operations with $O(1)$ worst-case block transfers, provided that the update position is given.*

The ISB-tree is maintained by incrementally performing global reconstructions [17]. Let S_0 be the set of stored elements at the latest reconstruction, and assume that $S_0 = \{x_1, \ldots, x_{n_0}\}$ in sorted order. The reconstruction is performed as follows. We partition S_0 into two sets S_1 and S_2, where $S_1 = \{x_{i \cdot \ln n_0} : i = 1, \ldots, \frac{n_0}{\ln n_0} - 1\} \cup \{b\}$, and $S_2 = S_0 - S_1$. The i-th element of S_1 is the representative $rep(i)$ of the i-th bucket $\mathcal{B}_i$, where $1 \leq i \leq \rho$ and $\rho = |S_1| = \frac{n_0}{\ln n_0}$. The representatives $rep(i)$, $1 \leq i \leq \rho$, are stored in the external SIST (note that there is no need to store $rep(0)$). An element $x \in S_2$ is stored in $\mathcal{B}_i$, iff $rep(i-1) < x \leq rep(i)$, for $i \in \{2, \ldots, \frac{n_0}{\ln n_0}\}$; otherwise ($x \leq rep(1)$), x is stored in $\mathcal{B}_1$. The same condition holds for every new element inserted in the structure. In order to insert/delete an element, given the position (block) of the update, we simply have to insert/delete the element to/from the *Lazy B-tree* storing the elements of the corresponding bucket. Each time the number of updates exceeds cn_0, where c is a predefined constant, the whole data structure is reconstructed. Let n be the number of stored elements at this time. After the reconstruction, the number of buckets is equal to $\lceil \frac{n}{\ln n} \rceil$.

The search procedure for locating a query element x can be decomposed into two phases: (i) the traversal of internal nodes of the external SIST locating a bucket $\mathcal{B}_i$, and (ii) the search for x in the *Lazy B-tree* storing the elements of $\mathcal{B}_i$. Phase (i) starts from the root of the external SIST. It checks the external arrays on the root and by interpolating it decides into which child the search procedure will continue. More specifically, let v be a node in the search path for query element x, n_v be the number of leaves in its subtree, and let l_v and u_v be the minimum and the maximum element, resp., stored in the subtree rooted at v. As we have already mentioned, node v is associated with two external arrays EREP_v and EID_v that implement the REP_v and ID_v arrays of the SIST definition. To interpolate, we compute the value $i = \lfloor \frac{x - l_v}{u_v - l_v} R(n_v) \rfloor$ and find the index $j = \mathrm{EID}_v[i]$, by retrieving the $\lceil \frac{i}{B} \rceil$-th block of the EID_v array. We then scan the blocks of the EREP_v array, starting from its $\lceil \frac{j}{B} \rceil$-th block, until locating an index l such that $\mathrm{EREP}_v[l] \leq x < \mathrm{EREP}_v[l+1]$. If the l-th son of v is not a bucket, then we continue recursively in the same manner in the l-th son of v, until we locate the representative of a bucket $\mathcal{B}_i$. In this case, the search procedure is concluded by entering phase (ii) and by searching further in the *Lazy B-tree* of the bucket $\mathcal{B}_i$.

In the following, we will analyze the bounds of the search and update operations. Our result holds for the very broad class of $(n/(\log\log n)^{1+\epsilon}, n^{1-\delta})$-smooth densities, where $\delta = 1 - \frac{1}{B}$ and includes the uniform, regular, bounded as well as several non-uniform distributions [2, 9], and is stated by the following theorem.

Theorem 2. *Suppose that the upper level of the ISB-tree is an external static interpolation search tree with parameters* $R(s_0) = s_0^{\delta}$, $I(s_0) = s_0/(\log\log s_0)^{1+\epsilon}$, *where* $\epsilon > 0$, $\delta = 1 - \frac{1}{B}$, $s_0 = \frac{n_0}{\ln n_0}$ *and* n_0 *is the number of elements in the latest reconstruction, and that the lower level is implemented as a forest of Lazy B-trees. Then, the ISB-tree supports search operations in* $O(\log_B \log n)$ *expected block transfers with high probability, where* n *denotes the current number of elements, and update operations in* $O(1)$ *worst-case block transfers, if the update position is given. The worst-case update bound is* $O(\log_B n)$ *block transfers, and the structure occupies* $O(n/B)$ *blocks.*

Proof. (sketch) As we have already mentioned, the search operation in the ISB-tree can be decomposed into two basic steps: (i) the traversal of internal nodes of the external SIST, and (ii) the search for x in the *Lazy B-tree* in the bucket that we located from step (i).

We can prove (in a way similar to that in the proof of [9–Theorem 1]) that the expected number of block transfers for step (i) is $O(h)$ w.h.p, where h is the height of the external SIST. The main point of the proof is that the expected number of blocks in the EREP_v array, which we need to linearly scan when interpolating at a node v, is $O(1)$ w.h.p. Since in our case, the height of the tree is $h = O(\log_B \log s_0)$, where s_0 is the number of stored elements at the latest reconstruction and $s_0 = O(n)$, we get that the expected number of block transfers for step (i) is $O(\log_B \log n)$ w.h.p.

Regarding the complexity of step (ii), we can use the same combinatorial game of balls and bins introduced in [9] and prove (similarly to [9–Theorem 6]) that w.h.p. the expected number of elements in each bucket is $O(\log n)$, when elements are μ-randomly inserted and randomly deleted, and μ is an unknown smooth probability density. Since we store the elements of each bucket in a lazy B-tree, we get from Theorem 1 that the block transfers of step (ii) are also $O(\log_B \log n)$. Consequently, the total expected complexity of the search procedure is bounded by $O(\log_B \log n)$ block transfers w.h.p.

Let us now consider the update bound. Between reconstructions the block transfers for an update are clearly $O(1)$, since we only have to update the appropriate Lazy B-tree which can be done in $O(1)$ block transfers (cf. Theorem 1). The reconstructions can be easily handled by using the technique of global rebuilding [13]. With this technique the linear work spent during a global reconstruction of the upper level structure may be spread out on the updates in such a way that a rebuilding cost of $O(1)$ block transfers is spent at each update.

Finally, the worst-case search complexity of $O(\log_B n)$ block transfers can be achieved by using two data structures, an ISB-tree and a Lazy B-Tree, and hence storing each element twice. A search for a query element is performed by searching simultaneously both structures and terminating the search when locating for the first time the sought element. The worst-case update and space complexity remain asymptotically unaffected and so the theorem is proven. □

4 Experimental Evaluation

In this section, we investigate the practical merits of the ISB-tree. We have conducted an experimental study making the customary assumption that the page size is 4096 bytes, the length of each key is 8 bytes, and the length of each pointer is 4 bytes. Consequently, each block contains $B = 341$ elements. We considered data sets of size $n_0 \in [10^6, 10^{12}]$ elements generated by a variety of smooth distributions, namely uniform, regular, normal and Gaussian. We compared the implementation of the ISB-tree with that of a B-tree on the same data sets (implementations were carried out in C++). Our main concern was to measure the performance, in simulated block transfers (I/Os), of the search and update operations. The experimental results regarding the search operations are reported in Fig. 1. The sequence σ of search operations had length equal to its corresponding data set and the reported values are averages over the whole sequence. Our experiments revealed that the expected number of block transfers in the ISB-tree remains constant even for gigantic data sets (Terabytes - TB). Moreover, for data sets larger than 100 GB, the expected number of block transfers is reduced by a factor ranging from 1/3 (for normal and Gaussian distributions) to 1/2 (for uniform and regular distributions) compared with the B-tree. This behaviour is justified by the time complexity of the search operation and by the fact that for data sets up to 1 TB and block size of 341 elements, the ISB-tree is a two level structure, where the first level (SIST structure) consists of only one node equipped with the appropriate EID and EREP arrays, while the second level (lazy B-tree) consists of only one block of elements. Thus, we need 2 block transfers in the first level (one for each array) and 1 block transfer in the second level. Our experiments also show that for uniform and regular distributions, the position of EREP array (which has been located by its corresponding entry in EID) points in almost all cases to the correct subset within which the search has to be continued in the second level. For the case of normal and Gaussian distributions, we often had to move to the immediately next block and this adds one additional block transfer to the search operation. Naturally, for small data sets (smaller than 10 MB), our data structure becomes less efficient than B-trees, due to the overhead of the two-level structure.

As a final remark, we note that there are applications with uniform key sizes larger than 8 bytes, resulting in a smaller value of B. The main example of such applications involve manipulation of strings. In this case, the size of the block may be as small as 2. Consequently, we expect that in such cases the ISB-tree will have a much better performance.

Regarding the number of block transfers required for rebalancing after an update operation to the data structure, we again considered the above values of $n_0 \in [10^6, 10^{12}]$ for our initial data sets upon which we performed update

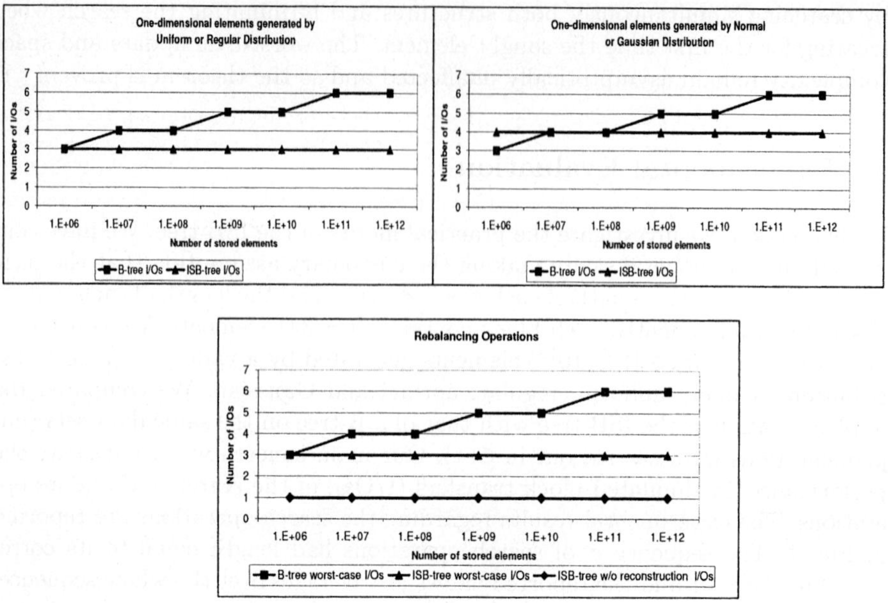

Fig. 1. Search performance for uniform and regular distributions (upper left) and normal and Gaussian distributions (upper right). Block transfers of rebalancing operations after an update (bottom).

sequences of length $n_0/2$ and $2n_0$. The data structure is reconstructed every n_0 operations (i.e., we chose $c = 1$). Our experimental results are reported in Fig. 1. The values represent averages over the smaller update sequence (where no reconstruction occurs) and the larger one (where a reconstruction indeed occurs). We have observed that both numbers of rebalancing operations are independent of the distribution.

5 Conclusions

We presented a new indexing scheme, the ISB-tree, that supports update operations in $O(1)$ worst-case block transfers and search operations in $O(\log_B \log n)$ expected block transfers w.h.p. for a large class of input distributions. The ISB-tree is innovative in the sense that it shoots down for the first time the optimal $O(\log_B n)$ block transfer bound of B-tree and its variants when the updates are drawn from a large class of input distributions. Its analysis is based on the traditional I/O model of [1], but we feel that it can also be implemented in the cache-oblivious model [8] with the same complexities.

Acknowledgements. We thank Y. Theodoridis for several interesting discussions.

References

1. A. Aggarwal and J.S. Vitter. The Input/Output Complexity of Sorting and Related Problems. *Communications of the ACM*, 31(9):1116-1127, 1988.
2. A. Andersson and C. Mattson. Dynamic Interpolation Search in $o(\log\log n)$ Time. *In Proc. 14th International Colloquium on Automata, Languages and Programming (ICALP).* LNCS 700, pp. 15-27, 1993.
3. R. Bayer and E. McCreight. Organization of large ordered indexes. *Acta Informatica*, 1:173-189, 1972.
4. D. Comer. The Ubiquitous B-tree. *ACM Computing Surveys*, 11(2):121-137, 1979.
5. R. Fagin, J. Nievergelt, N. Pippinger, H.R. Strong. Extendible Hashing-A fast access method for dynamic files. *ACM Transactions on Database Systems*, 4(3):315-344, 1979.
6. P. Ferragina and R. Grossi. The String B-tree: A New Data Structure for String Search in External Memory and Its Applications. *Journal of the ACM*, 46(2):236-280, 1999.
7. E. Fox, Q. Chen, A. Daoud. Practical Minimal Perfect Hash Functions for Large Databases. *Communications of the ACM*, 35(5):105-121, 1992.
8. M. Frigo, C.E. Leiserson, H. Prokop, and S. Ramachandran. Cache-oblivious algorithms. *In Proc. 40th IEEE Symp. on Foundations of Computer Science (FOCS)*, pp. 285-297, 1999.
9. A. Kaporis, C. Makris, S. Sioutas, A. Tsakalidis, K. Tsichlas, and C. Zaroliagis: Improved Bounds for Finger Search on a RAM. *In 11th Ann. European Symp. on Algorithms (ESA), LNCS* Vol. 2832, pp. 325-336. Full version as Tech. Rep. TR-2003/07/01, Computer Technology Institute, Patras, July 2003.
10. A. Kaporis, C. Makris, G. Mavritsakis, S. Sioutas, A. Tsakalidis, K. Tsichlas and C. Zaroliagis. ISB-Tree: A New Indexing Scheme with Efficient Expected Behaviour. Computer Technology Institute Tech. Report TR 2005/09/02, September 2005.
11. D.E. Knuth. *Sorting and Searching*, Vol. 3 of *The Art of Computer Programming*, Addison-Wesley, 1973.
12. D.E. Knuth. Deletions that preserve randomness. *IEEE Trans. Softw. Eng.*, 3:351-359, 1977.
13. C. Levcopoulos and M.H. Overmars: Balanced Search Tree with $O(1)$ Worst-case Update Time. *Acta Informatica*, 26:269-277, 1988.
14. W. Litwin. Linear Hashing: A new tool for files and tables addressing. In *In Proc. Int. Conf. on Very Large Databases (VLDB)*, 6:212-223, 1980.
15. Y. Manolopoulos, Y. Theodoridis, V. Tsotras. *Advanced Database Indexing.* Kluwer Academic Publishers, 2000.
16. K. Mehlhorn and A. Tsakalidis. Dynamic Interpolation Search. *Journal of the ACM*, 40(3):621-634, 1993.
17. M. Overmars, J. van Leeuwen. Worst Case Optimal Insertion and Deletion Methods for Decomposable Searching Problems. *Information Processing Letters*, 12(4):168-173.
18. B. Seeger and P.A. Larson. Multi-Disk B-trees. *In Proc. SIGMOD Conference*, pp. 436-445, 1991.
19. V. Srinivasan and M.J. Carey. Performance of B+ Tree Concurrency Algorithms. *VLDB Journal*, 2(4):361-406, 1993.
20. J.S. Vitter. External memory algorithms and data structures: dealing with massive data. *ACM Computing Surveys*, 33(2):209-271, 2001.
21. D.E. Willard. Searching Unindexed and Nonuniformly Generated Files in $\log\log N$ Time. *SIAM Journal of Computing*, 14(4):1013-1029, 1985.

External Data Structures for Shortest Path Queries on Planar Digraphs

Lars Arge[1,*] and Laura Toma[2]

[1] Duke University, Durham, NC 27708 USA
large@cs.duke.edu
[2] Bowdoin College, Brunswick, ME 04011 USA
ltoma@bowdoin.edu

Abstract. In this paper we present space-query trade-offs for external memory data structures that answer shortest path queries on planar directed graphs. For any $S = \Omega(N^{1+\epsilon})$ and $S = O(N^2/B)$, our main result is a family of structures that use S space and answer queries in $O(\frac{N^2}{SB})$ I/Os, thus obtaining optimal space-query product $O(N^2/B)$. An S space structure can be constructed in $O(\sqrt{S} \cdot \text{sort}(N))$ I/Os, where sort(N) is the number of I/Os needed to sort N elements, B is the disk block size, and N is the size of the graph.

1 Introduction

Let $G = (V, E)$ be a directed graph (digraph) with real edge weights. If G has no negative-weight cycles, the shortest path $\delta(s,t)$ from vertex s to vertex t is the minimum length path from s to t in G, where the length of a path is defined as the sum of the weights of its edges. The length of the shortest path $\delta(s,t)$ is called the *distance* from s to t in G. Shortest path computation is a fundamental and well-studied problem that appears in a diverse set of applications. In recent years, an increasing number of these applications involve massive graphs. Massive planar graph problems and in particular shortest paths computation arise frequently in Geographic Information System (GIS), where datasets such as the ones acquired by missions like NASA Earth Observing System (EOS) or Space Radar Topography Mission (SRTM) are on the order of terabytes. Environmental researchers often need to compute shortest paths for instance when planning and assessing the impact of new development, or modeling the communication between areas of conservation for endangered species. When working with such massive graphs that do not fit in the main memory of even state-of-the-art machines, transfer of data between main memory and external memory (such as disk), rather than internal computation time, is often the performance bottleneck. In such cases it is important to consider algorithms that minimize Input-Output (or simply I/O) communication.

* Supported in part by the National Science Foundation through RI grant EIA–9972879, CAREER grant CCR–9984099, ITR grant EIA–0112849, and U.S.–Germany Cooperative Research Program grant INT–0129182.

X. Deng and D. Du (Eds.): ISAAC 2005, LNCS 3827, pp. 328–338, 2005.

The most commonly studied shortest path problems are the *single-source-shortest-path* (SSSP) problem and the *all-pair-shortest-path* (APSP) problem, where the goal is to find the shortest paths from a source vertex s to all other vertices in G, and between all pairs of vertices in G, respectively. Several authors have considered I/O-efficient algorithms for these problems. In this paper we study another variant of the problem, namely the design of I/O-efficient data structures for answering shortest path queries on *planar* directed graphs (digraphs that can be embedded in the plane such that no edges intersect). In particular, we study the space-query trade-off for such structures; using $O(N^2)$ space we can obviously design a structure that can answer a shortest path distance query in $O(1)$ I/Os, simply by storing the shortest paths between every pair of vertices in G. At the other extreme, we can design an $O(N)$ space structure by simply running an SSSP algorithm to answer a query. Since we are interested in massive graphs, we are of course interested in data structures that use close to linear space but answer queries more efficiently that by computing SSSP on-the-fly. In this paper we develop a family of structures with a trade-off between space use and the number of I/Os needed to answer a query. Although this problem has been extensively studied in internal memory [7, 13, 12, 10, 9], this is the first result of its type in external memory.

1.1 I/O-Model and Related Work

We will be working in the standard two-level I/O model [1], where M is the number of vertices that can fit into internal memory, and B is the number of vertices that can fit into a disk block, with[1] $M < N$ and $1 \leq B \leq M/2$. An *I/O* is the operation of transferring a block of data between main memory and disk, and the complexity of an algorithm is measured in terms of the number of disk blocks and I/Os it uses to solve a problem.

In the I/O-model, the minimal number of I/Os needed to read N input elements (the "linear bound") is obviously $\text{scan}(N) = N/B$. The number of I/Os needed to sort N elements is $\text{sort}(N) = \Theta(\frac{N}{B}\log_{M/B} N/B)$ [1]. For realistic values of N, B, and M, $\text{scan}(N) < \text{sort}(N) \ll N$, and the difference in running time between an algorithm performing N I/Os and one performing $\text{scan}(N)$ or $\text{sort}(N)$ I/Os can be very significant.

On general digraphs the best known algorithm for SSSP, as well the best algorithms for the simpler BFS and DFS problems, use $\Omega(|V|)$ I/Os. More precisely, their I/O-complexity is $O(\min\{(|V| + |E|/B) \cdot \log|V| + \text{sort}(|E|), |V| + \frac{|V|}{M}\frac{|E|}{B}\})$ [8, 11, 17] . However, improved algorithms have been developed for *planar* digraphs [5, 6, 3]. On such graphs, SSSP and BFS can be solved in $O(\text{sort}(N))$ I/Os [5], and DFS in $O(\text{sort}(N)\log N/M)$ I/Os [6]; all these algorithms are based on I/O-efficient reductions [2, 4, 5, 6] and on an $O(\text{sort}(N))$ I/O planar graph separator algorithm [19].

[1] The planar separator algorithm [19] makes the stronger but realistic assumption that $M > B^2 \lg^2 B$; we make this assumption indirectly as we rely on planar separators.

The only known I/O-efficient external data structure for answering shortest path queries is a structure for planar digraphs in [16]. The structure uses $O(N\sqrt{N})$ space and answers shortest path distance queries in $O(\sqrt{N}/B)$ I/Os (and can report the shortest path with additional $O(K/B)$ I/Os, where K is the number of edges on the path). Note that the space-query product is $O(N^2/B)$. The structure in [16] is based on an internal memory data structure obtained independently by Arikati *et al* [7] and Djidjev [12] using planar separators and ideas due to Frederickson [14, 15]. In internal memory, this structure has been generalized to obtain a family of structures, such that a structure using $S \in [N, N^2]$ space can answer shortest path distance queries in $O(N^2/S)$ time [12, 10]. Note that the space-query product is $O(N^2)$. Improved results have been obtained for values of S larger then $N^{4/3}$ [12, 10], as well as for special classes of graphs [13, 9]. Similar space-query results and improvements have not been obtained in external memory; the $O(N\sqrt{N})$ space use of the structure by Hutchinson *et al* [16] probably means that it is mostly of theoretical interest if N is large. Ideas from previous work do not extend in external memory to small space. Finding space-query trade-offs for close to linear S is harder and it is precisely the small values of S that are interesting in external memory.

1.2 Our Results

In this paper we obtain the first space-query trade-offs for external data structures for answering shortest path queries on planar digraphs. Our main result is a family of structures that can answer shortest path distance queries in $O(\frac{N^2}{SB})$ I/Os using $S = \Omega(N^{1+\epsilon})$ (and $S = O(N^2/B)$) space, for any $\epsilon > 0$. Note that, similarly to the internal memory results, the space-query product is $O(N^2/B)$. An S space structure can be constructed in $O(\sqrt{S} \cdot \text{sort}(N))$ I/Os. For values of $S = o(N^{1+\epsilon})$, we show that we can still achieve a trade-off but at the cost of an increased space-query product. More precisely, we show that for any $S \in [N\frac{\log^2 N}{\log\log N}, \frac{N^2}{B}]$, there exists a data structure of size S that can answer distance queries in $O(\frac{N^2}{SB} \cdot \frac{\log N}{\log(S/(N\log N))})$ I/Os. Our structures can be extended to answer shortest path queries with an extra $O(K/B)$ I/Os, where K is the number of edges on the path; for brevity we only consider distance queries.

Our results use ideas similar to the ones in the previously developed internal structures, i.e. planar separators, but with non-trivial external memory modifications to make the structure efficient for small S. In Sec. 2 we review planar separators, and in Sec. 4 we present our new structure. It relies on a structure for computing distances to the boundary of a planar graph presented in Sec. 3.

2 Preliminaries

A $f(N)$-*separator* of an N-vertex graph $G = (V, E)$ is a subset V_S of the vertices V of size $f(N)$, such that the removal of V_S partitions G into two subgraphs G_1 and G_2 of size at most $2N/3$. Lipton and Tarjan [18] showed that any planar

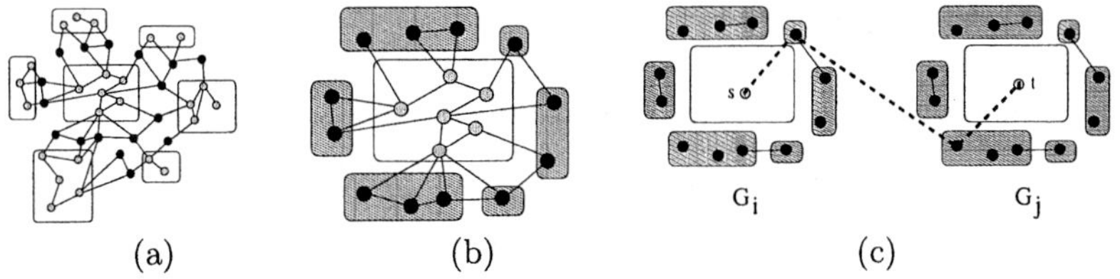

Fig. 1. (a) Partition of G into clusters G_i (boxed) and separators vertices V_S (black). (b) One cluster G_i in the partition and its adjacent boundary sets. (c) The shortest path from s to t is is $\delta(s,t) = \min_{v \in \partial G_i, w \in \partial G_j}\{\delta_{\overline{G_i}}(s,v) + \delta(v,w) + \delta_{\overline{G_j}}(w,t)\}$.

graph has an $O(\sqrt{V})$-separator. Using this result recursively, Frederickson [15] showed that for any parameter $R \leq N$ there exists a subset V_S of $\Theta(N/\sqrt{R})$ vertices, such that the removal of V_S partitions G into $\Theta(N/R)$ subgraphs G_i of size $O(R)$, where (the vertices in) each G_i is (are) adjacent to $O(\sqrt{R})$ vertices of V_S. We call such a partitioning an *R-partition.* The vertices in V_S are called the *separator vertices* and each of the graphs G_i a *cluster.* The set of separator vertices adjacent to G_i are called the *boundary vertices* ∂G_i (or simply the *boundary*) of G_i. We use $\overline{G_i}$ to denote the graph consisting of G_i, ∂G_i and the subset of edges of E connecting G_i and ∂G_i (Fig. 1(a)). The set of separator vertices can be partitioned into maximal subsets so that the vertices in each subset are adjacent to the same set of clusters G_i. These sets are called the *boundary sets* of the partition (Fig. 1(b)). If the graph has bounded degree, which can be ensured for planar graphs using a simple transformation, there exists an R-partition with only $O(N/R)$ boundary sets [15]. It is shown in [19] how to compute such an R-partition in $O(\text{sort}(N))$ I/Os, provided that $M > B^2 \log^2 B$.

All the known internal memory shortest path data structures for planar graphs exploit R-partitions [12, 7, 10]. Consider an R-partition of a planar digraph, and let $\delta_{\overline{G_i}}(s,t)$ denote the length of the shortest path from vertex s to vertex t in $\overline{G_i}$. The shortest path from $s \in G_i$ to $t \in G_j$ must go through the boundaries ∂G_i and ∂G_j of G_i and G_j, and thus we can compute the distance from s to t as $\delta(s,t) = \min_{v \in \partial G_i, w \in \partial G_j}\{\delta_{\overline{G_i}}(s,v) + \delta(v,w) + \delta_{\overline{G_j}}(w,t)\}$ (Fig. 1(c)). The basic idea is to store the distance from a set of vertices of G to the separator vertices V_S in order to be able to efficiently evaluate this formula for given s and t. The size of this set of vertices is a function of the available space S: For small values of S ($S \in [N, N^{3/2}]$) we store only the distances between separator vertices; to answer a query we basically need to solve a SSSP problem in the two clusters $\overline{G}_i$ and $\overline{G}_j$. For large values of S ($S \in [N^{3/2}, N^2]$) we store the distances from all vertices in G to all separator vertices, as well as a shortest path data structure [7, 12] for each cluster; answering a query reduces to using the stored distances if s, t are in different clusters, or querying the cluster, otherwise.

For large values of S ($S \in [N^{2/3}, N^2]$) we can adapt the above strategy to external memory using the known external shortest path data structure [16]; we obtain a structure that answers distance queries in $O(\frac{N^2}{SB})$ I/Os, i.e. has the desired $O(N^2/B)$ space-query product (details in the full version of this paper). However, for small values of S ($S \in [N, N^{3/2}]$), which are the ones we are normally interested in when handling massive graphs, a similar adaption leads to an $O(\text{sort}(\frac{N^2}{S}))$ query bound, mainly because of the need to solve SSSP problems in two clusters. For small values of S, this is no better than running SSSP from scratch. Note the difference between answering distance queries in internal and external memory: In internal memory small values of S are easy to handle because the clusters in the R-division are small enough for it to be efficient to compute SSSP in linear time in the cluster on-the-fly. In external memory the problem is easy if S is large because we can store enough additional information using S, and becomes harder as S gets smaller. However it is precisely the small values of S that are interesting external memory.

In this paper we show how to use the R-partition in a novel way in order to obtain a small space data structure. One of the main ingredients in our solution is a structure for answering *all-boundary-shortest-path queries*, that is, for finding shortest paths lengths from a vertex s in a cluster G_i to the vertices on the boundary ∂G_i. We describe such a structure below.

3 All-Boundary-Shortest-Path Structure

Assume we are given a cluster H of size N and its boundary ∂H such that $|\partial H| \leq c \cdot \sqrt{N}$, for some constant $c \geq 1$. As usual, we denote $H \cup \partial H$ as $\overline{H}$ and let $\delta_{\overline{H}}(s,t)$ denote the length of the shortest path from s to t in $\overline{H}$. Let $\delta_{\overline{H}}(s, \partial H)$ denote the list of distances from s to the vertices in ∂H, sorted by the id of the vertices in ∂H. Similarly, let $\delta_{\overline{H}}(\partial H, s)$ denote the list of distances from the vertices in ∂H to s, sorted by the id of the vertices in ∂H.

This section describes an I/O-efficient data structure for all-boundary-shortest-path queries, that is, for finding the shortest paths $\delta_{\overline{H}}(s, \partial H)$ and $\delta_{\overline{H}}(\partial H, s)$ between a vertex s in the cluster and the vertices on its boundary. Our structure improves the straightforward $O(\text{sort}(N))$ I/Os bound obtained by running SSSP in $\overline{H}$. More precisely, we prove the following.

Lemma 1. *Given an N-vertex cluster H and its $O(\sqrt{N})$-vertex boundary ∂H we can construct a data structure using $O(N \log N)$ space such that $\delta_{\overline{H}}(s, \partial H)$ or $\delta_{\overline{H}}(\partial H, s)$ can be computed in $O(N/B)$ I/Os for any s in H. The structure can be constructed in $O(\sqrt{N} \cdot \text{sort}(N))$ I/Os.*

Our all-boundary-shortest-path data structure is constructed as follows[2]: we first compute an $N/2$-partition for H, that is, a partition of H using a set V_S of $O(\sqrt{N})$ separator vertices into $O(1)$ clusters, each of which contains at most $N/2$ vertices and is adjacent to at most $\sqrt{N}/2$ separators. As usual define the boundary ∂H_i of a cluster H_i to be the set of vertices in $\partial H \cup V_S$ that are adjacent to vertices in H_i. We then recursively construct an all-boundary-shortest-path data structure for each cluster H_i and its boundary (to do so we first process

[2] We only discuss how to compute $\delta_{\overline{H}}(s, \partial H)$. Computing $\delta_{\overline{H}}(\partial H, s)$ can be done similarly.

each H_i in turn such that its boundary has at most $c \cdot \sqrt{|H_i|}$ vertices; details in the full paper). For each separator or boundary vertex $u \in V_S \cup \partial H$ we compute the shortest paths in $\overline{H}$ from u to all vertices v in ∂H. We store these distances ordered by the vertex id of $v \in \partial H$ and secondarily by vertex id of u; thus the list of distances from ∂H_i to a vertex $v \in \partial H$ can be retrieved by scanning this list in $|V_S \cup \partial H|/B = \sqrt{N}/B$ I/Os.

Query: Consider an all-boundary-shortest-path query $\delta_{\overline{H}}(s, \partial H)$. The interesting case is when $s \in H_i$ (if s is a separator vertex we simply return the list of $O(\sqrt{N})$ pre-computed distances from s to ∂H). Let w be an arbitrary vertex in ∂H; we compute $\delta(s, w)$ as the shortest way to get from s to a boundary vertex v of H_i in $\overline{H_i}$ and from v to w in $\overline{H}$: that is, $\delta(s,w) = \min_{v \in \partial H_i}\{\delta_{\overline{H_i}}(s,v) + \delta(v,w)\}$. It can be shown that $\delta(s,w)$ is indeed the shortest path from s to w in $\overline{H}$. To compute $\delta_{\overline{H}}(s, \partial H)$ we first find the all-boundary-shortest-paths $\delta_{\overline{H_i}}(s, \partial H_i)$ using the recursive data structure for the cluster $\overline{H_i}$ containing s; let L be the list of shortest paths returned, $L = \{\delta_{\overline{H_i}}(s,v) | v \in \partial H_i\}$, sorted by the id of the vertex $v \in \partial H_i$. For every vertex $w \in \partial H$, we compute $\delta(s,w)$ by scanning in parallel the list L and the list of distances from $v \in V_S \cup \partial H$ to w (stored in the structure) and compute the minimum sum $\delta(s,v) + \delta(v,w)$. This takes $O(\sqrt{N}/B)$ I/Os for every $w \in \partial H$, or $O(\sqrt{N} \cdot \sqrt{N}/B) = O(N/B)$ I/Os in total. Thus, the number of I/Os to answer an all-boundary-shortest-path query $\delta_{\overline{H}}(s, \partial H)$ is given by the recurrence $Q(N) = O(N/B) + Q(N/2)$ with solution $Q(N) = O(N/B)$ I/Os.

Space: Storing the shortest paths in $\overline{H}$ from $u \in V_S \cup \partial H$ to all vertices v in ∂H uses $O(\sqrt{N} \cdot \sqrt{N}) = O(N)$ space. Thus the total space is given by $S(N) \leq O(N) + 2S(N/2)$ with solution $S(N) = O(N \log N)$.

Construction: Processing each cluster H_i such that its boundary has at most $c \cdot \sqrt{|H_i|}$ vertices can be done as in the R-partition algorithm [19]; this uses $O(\text{sort}(N))$ I/Os in total for all clusters. Computing the shortest paths in $\overline{H}$ from any vertex $u \in V_S \cup \partial H$ to all vertices v in ∂H can be done in $O(\sqrt{N} \cdot \text{sort}(N))$ I/Os. Thus the total pre-processing time is given by the recurrence $P(N) \leq O(\sqrt{N} \cdot \text{sort}(N)) + 2P(N/2)$ with solution $P(N) = O(\sqrt{N} \cdot \text{sort}(N))$ I/Os. This concludes the proof of Lemma 1.

4 Shortest Path Data Structure

This section describes the main result of this paper, a data structure for shortest path queries. As all related work, our structure is based on R-divisions and pre-computing shortest paths between separator vertices, but it combines these ideas in a novel way that avoids computation of SSSP inside the cluster and obtains an optimal space-time trade-off.

Let R be a parameter $B \leq R \leq N/2$ and G a bounded-degree[3] planar digraph. Our shortest path data structure for G is constructed as follows:

1. Step 1: Compute an R-partition of G.
2. Step 2: Compute and store the distances between the separator vertices V_S, such that the list of distances between a vertex in ∂G_i and the $O(\sqrt{R})$ vertices on the boundary of any cluster ∂G_j can be retrieved sorted by vertex id using $O(\sqrt{R}/B)$ I/Os.
3. Step 3: Recursively construct a shortest path structure for each cluster G_i.
4. Step 4: We do the following for each cluster G_i
   (a) If $M < R \leq N^{2/3}$, we compute and store the distances between all vertices in G_i and vertices in ∂G_i, such that for any $s \in G_i$, $\delta_{\overline{G_i}}(s, \partial G_i)$ and $\delta_{\overline{G_i}}(\partial G_i, s)$ can be retrieved sorted by vertex id using $O(\sqrt{R}/B)$ I/Os.
   (b) If $R > N^{2/3}$, we construct an all-boundary-shortest-path data structure (Lemma 1) for each $\overline{G_i}$.

Below we show how to answer distance queries using the data structure and analyze its space usage and construction time.

Answering distance queries: To find the distance $\delta(s,t)$ between two query vertices s and t we consider the following 4 cases:

(1) If both s and t are separator vertices then we know that $\delta(s,t)$ is stored explicitly (Step 2) and we can thus answer a query in $O(1)$ I/Os.

(2) If s is a separator vertex and t is in cluster G_i (or the symmetrical case) then it can be shown that $\delta(s,t) = \min_{v \in \partial G_i}\{\delta(s,v) + \delta_{\overline{G_i}}(v,t)\}$. To answer the query we first obtain the list of distances from s to ∂G_i (sorted by vertex id) using $O(\sqrt{R}/B)$ I/Os (they are stored explicitly in Step 2). Then we obtain the list $\delta_{\overline{G_i}}(\partial G_i, t)$ of distances from vertices on ∂G_i to t (sorted by vertex id) in $O(R/B)$ I/Os as follows: If $R \leq M$ we load $\overline{G_i}$ in memory using $O(R/B)$ I/Os and compute the distances on the fly. If $M < R \leq N^{2/3}$ the distances $\delta_{\overline{G_i}}(\partial G_i, t)$ are stored explicitly in the data structure (Step 4 (a)) and we can retrieve them in $O(\sqrt{R}/B)$ I/Os; If $R > N^{2/3}$ we obtain $\delta_{\overline{G_i}}(\partial G_i, t)$ from the all-boundary-shortest-path data structure for $\overline{G_i}$ (Step 4(b)) using $O(R/B)$ I/Os (Lemma 1). Since the two obtained lists of distances (from s to ∂G_i and from ∂G_i to t) are both sorted by the vertex id of v we can scan them together to compute the min sum $\delta(s,v) + \delta(v,t)$; thus we answer the query using $O(R/B)$ I/Os in total.

(3) If s is in cluster G_i and t is in a different cluster G_j then it can be shown that $\delta(s,t) = \min_{v \in \partial G_i, w \in \partial G_j}\{\delta_{\overline{G_i}}(s,v) + \delta(v,w) + \delta_{\overline{G_j}}(w,t)\}$. We obtain the lists of distances $\delta_{\overline{G}_i}(s, \partial G_i)$ sorted by vertex id of $v \in \partial G_i$ and $\delta_{\overline{G}_j}(\partial G_j, t)$ sorted by vertex id of $w \in \partial G_j$ using $O(R/B)$ I/Os as in the previous case. We compute $\delta(s,t)$ as follows: scan the list $\delta_{\overline{G}_i}(s, \partial G_i)$ and read the distance from s to the next vertex v on ∂G_i. For each such distance we scan the distances $\delta(v, \partial G_j)$ and $\delta_{\overline{G}_j}(\partial G_j, t)$. Since these lists are sorted by id of vertex in ∂G_j, we

[3] Any graph can easily be transformed into a graph with each vertex having degree at most 3 [15].

can compute $\delta_{\overline{G_i}}(s,v)+\delta(v,w)+\delta_{\overline{G_j}}(w,t)$ for each vertex $w \in \partial G_j$ by scanning the lists in parallel. Thus, for every vertex in ∂G_i we scan two lists of size $\sqrt{R}$ each. Throughout this step we keep a running minimum of the smallest distance encountered so far. This takes in total $\sqrt{R} \cdot \sqrt{R}/B = O(R/B)$ I/Os.

(4) If s and t are in the same cluster G_i then it can be shown that $\delta(s,t) = \min\{\delta_{G_i}(s,t), \min_{v,w \in \partial G_i}\{\delta_{\overline{G_i}}(s,v) + \delta(v,w) + \delta_{\overline{G_i}}(w,t)\}\}$. To answer a query in this case we compute $\delta_{G_i}(s,t)$ using the recursive structure for G_i (Step 3) and the other term in the minimum in $O(R/B)$ I/Os as above. Computing the overall minimum thus takes $O(R/B)$ I/Os. Overall the number of I/Os used to answer a query is given by the recurrence $Q(N) = O(R/B)+Q(R)$ with solution $Q(N) = O(R/B)$.

Construction. An R-partition of G (Step 1) can be computed in $O(\text{sort}(N))$ [19]. The shortest paths in G between the $O(N/\sqrt{R})$ separator vertices (Step 2) can be computed in $O(N/\sqrt{R} \cdot \text{sort}(N))$ I/Os using the $O(\text{sort}(N))$ I/O SSSP algorithm [5] for each separator vertex. Let L be the list containing the shortest paths between the separator vertices: $L = \{\delta(u,v) | u,v \in V_S\}$. We need to store L such that for any separator vertex u, the distances between u and the boundary of any cluster G_j can be retrieved from the list in $\text{scan}(|\partial G_j|) = O(\sqrt{R}/B)$ I/Os. To do so we first tag each separator vertex with the indices of the clusters it is adjacent to. Then by sorting and scanning, we merge this list with L; this gives an augmented list that contains, for every pair of separator vertices (u,v), the distance $\delta(u,v)$ and the indices of the clusters that contain u and v respectively on their boundary; we then scan the augmented list and, for each pair (u,v), for every i such that $u \in \partial G_i$ and for every j such that $v \in \partial G_j$, we output $(u,v,\delta(u,v),i,j)$. Because we assume G to have bounded degree, each separator vertex is adjacent to $O(1)$ clusters; therefore every pair (u,v) appears in the list $O(1)$ times and the size of the list remains $O(|L|) = O(N^2/R)$. Finally, we sort this list by (i,j) and, within the same cluster, by vertex id. Overall we use $O(\text{sort}(N^2/R))$ I/Os. It can be seen that the distances between a vertex $u \in \partial G_i$ and all vertices in ∂G_j are adjacent in the constructed list and can thus be retrieved in $O(\sqrt{R}/B)$ I/Os. Also, the list L_{ij} of $O(R)$ shortest path distances from ∂G_i to ∂G_j can be retrieved in $O(R/B)$ I/Os. Overall the preprocessing required by Step 2 of our data structure uses $O(N/\sqrt{R} \cdot \text{sort}(N) + \text{sort}(N^2/R))$ I/Os.

Finally, the shortest paths between the boundary vertices of a cluster and all vertices in the cluster (Step 4 (a)) can be computed in $O(\sqrt{R} \cdot \text{sort}(R))$ I/Os for each cluster using the $O(\text{sort}(N))$ I/O SSSP algorithm [5] for each boundary vertex. Processing a cluster into an all-boundary-shortest-path data structure (Step 4 (b)) takes $O(\sqrt{R} \cdot \text{sort}(R))$ I/Os by Lemma 1. Thus the preprocessing required by Step 4 of our data structure uses $O(\frac{N}{R} \cdot \sqrt{R} \cdot \text{sort}(R)) = O(N/\sqrt{R} \cdot \text{sort}(R))$ I/Os.

Overall the total pre-processing time $P(N)$ is given by the following recurrence: $P(N) = O(\text{sort}(N) + N/\sqrt{R} \cdot \text{sort}(N) + \text{sort}(N^2/R) + N/\sqrt{R} \cdot \text{sort}(R)) + N/R \cdot P(R) = O(N/\sqrt{R} \cdot \text{sort}(N)) + N/R \cdot P(R)$ with solution $P(N) = O(N/\sqrt{R} \cdot \text{sort}(N))$ I/Os.

Space. Storing the shortest paths between all $O(N/\sqrt{R})$ separator vertices in the partition (Step 2) uses $O(N^2/R)$ space. If $M < R \leq N^{2/3}$ (Step 4(a)) we use $N/R \cdot R\sqrt{R} = O(N\sqrt{R})$ space in total to store the distances between each vertex and all vertices on the boundary of its cluster, which is $O(N^2/R)$. If $R > N^{2/3}$ (Step 4(b)) we use $O(R \log R)$ space on the all-boundary-shortest-path structures (Lemma 1) in each of the N/R clusters, for a total of $O(N \log R)$ space. The total space used by the data structure is thus given by the recurrence $S(N) = O(N^2/R) + O(N \log R) + N/R \cdot S(R)$ with solution $S(N) = O(\max\{N^2/R, N \log R\} \cdot \log_{N/R} N)$.

Overall we have proved the following result:

Theorem 1. *Given a planar digraph G and $R \in [B, N/2]$, a data structure of size $O(\max\{N \log R, N^2/R\} \cdot \log_{N/R} N)$ can be constructed in $O(\frac{N}{\sqrt{R}} \cdot sort(N))$ I/Os such that distance queries can be answered in $O(R/B)$ I/Os.*

Consider the effect of R on the space and query time of the data structure. Note that $\max\{N \log R, N^2/R\}$ is N^2/R for any $R < N/\log N$ and $N \log R$ otherwise. Choosing $R = B$, we obtain a data structure that uses $\Theta(N^2/B)$ space and answers distance queries in $O(1)$ I/Os. This corresponds to building a B-partition and storing APSP between the $N/\sqrt{B}$ separators. As R increases, the space used by the structure decreases and the query time increases. At the other extreme, choosing $R = \frac{N}{\log N}$ we obtain a data structure of size $\Theta(N \frac{\log^2 N}{\log \log N})$ that answers queries in $O(\frac{N}{B \log N})$ I/Os. A simple calculation shows that if $R > \frac{N}{\log N}$ then the space used by the data structure is $N \log R \cdot \log_{N/R} N = \Omega(N \frac{\log^2 N}{\log \log N})$. This means that when increasing R beyond $N/\log N$ both the space of the data structure and the query time go up and we can no longer trade query time for space. Thus, the space used by our structure is lower bounded by $\Omega(N \frac{\log^2 N}{\log \log N})$. One interesting question is whether one can use less then $o(N \frac{\log^2 N}{\log \log N})$ space and answer distance queries faster than $O(\text{sort}(N))$ I/Os.

To express the query time directly in terms of the space S used by the data structure we let R such that $S = O(\frac{N^2}{R} \cdot \log_{N/R} N)$, or $\frac{1}{R} = \frac{1}{N} \frac{S}{N \log N} \log \frac{S}{N \log N}$. Substituting in Theorem 1 above, we obtain:

Lemma 2. *Given a planar digraph G and $S \in [N \frac{\log^2 N}{\log \log N}, \frac{N^2}{B}]$, a data structure of size S can be constructed in $O(\sqrt{\frac{S}{\log N}} \cdot \log \frac{S}{N \log N} \cdot sort(N))$ I/Os such that distance queries can be answered in $O(\frac{N^2}{SB} \cdot \frac{\log N}{\log(S/(N \log N))})$ I/Os*

The bounds in Lemma 2 simplify if S is such that $S/N = \Omega(N^\epsilon)$ for some $\epsilon > 0$. In this case note that $\frac{\log N}{\log(N^\epsilon/\log N)} = \frac{1}{\epsilon} = O(1)$. Thus we obtain our main result stated in Section 1.2:

Theorem 2. *Given a planar digraph G and $S = \Omega(N^{1+\epsilon})$ for some $0 < \epsilon \leq 1$, a data structure of size $O(S)$ can be constructed in $O(\sqrt{S} \cdot sort(N))$ I/Os such that distance queries can be answered in $O(\frac{N^2}{SB})$ I/Os.*

The space-query product of our data structure is $\frac{N^2}{B} \cdot \frac{\log N}{\log(S/(N \log N))}$. For $S = \Omega(N^{1+\epsilon})$ this is $\frac{N^2}{B}$. As S drops below $N^{1+\epsilon}$ the space-query product increases, up to a maximum of $\frac{N^2}{B} \cdot \frac{\log N}{\log \log N}$, when $S = N \frac{\log^2 N}{\log \log N}$.

References

1. A. Aggarwal and J. S. Vitter. The Input/Output complexity of sorting and related problems. *Communications of the ACM*, 31(9):1116–1127, 1988.
2. L. Arge, G. S. Brodal, and L. Toma. On external memory MST, SSSP and multi-way planar graph separation. *Journal of Algorithms*, 53(2):186–2006, 2004.
3. L. Arge, U. Meyer, and L. Toma. External memory algorithms for diameter and all-pairs shortest-paths on sparse graphs. In *Proc. International Colloquium on Automata, Languages, and Programming*, pages 146–157, 2004.
4. L. Arge, U. Meyer, L. Toma, and N. Zeh. On external-memory planar depth first search. *Journal of Graph Algorithms*, 7(2):105–129, 2003.
5. L. Arge, L. Toma, and N. Zeh. I/O-efficient topological sorting of planar DAGs. In *Proc. ACM Symp. on Parallel Algorithms and Architectures*, 2003.
6. L. Arge and N. Zeh. I/O-efficient strong connectivity and depth-first search for directed planar graphs. In *Proc. IEEE Symp. on Foundations of Computer Science*, pages 261–270, 2003.
7. S. Arikati, D. Chen, L. Chew, G. Das, M. Smid, and C. Zaroliagis. Planar spanners and approximate shortest path queries among obstacles in the plane. In *Proc. European Symp. on Algorithms, LNCS 1136*, pages 514–528. Springer, 1996.
8. A. Buchsbaum, M. Goldwasser, S. Venkatasubramanian, and J. Westbrook. On external memory graph traversal. In *Proc. ACM-SIAM Symp. on Discrete Algorithms*, pages 859–860, 2000.
9. S. Chaudhuri and C. Zaroliagis. Shortest path queries in digraphs of small treewidth. In *Proc. International Colloquium on Automata, Languages, and Programming, LNCS 944*, pages 244–255. Springer, 1995.
10. D. Chen and J. Xu. Shortest path queries in planar graphs. In *Proc. ACM Symp. on Theory of Computation*, pages 469–478. ACM Press, 2000.
11. Y. Chiang, M. Goodrich, E. Grove, R. Tamassia, D. Vengroff, and J. S. Vitter. External-memory graph algorithms. In *Proc. ACM-SIAM Symp. on Discrete Algorithms*, pages 139–149, 1995.
12. H. Djidjev. Efficient algorithms for shortest path queries in planar digraphs. In *Proc. Graph-Theoretic Concepts in Comp. Science*, pages 151–165. Springer, 1996.
13. H. Djidjev, G. Pantziou, and C. Zaroliagis. Computing shortest paths and distances in planar graphs. In *Proc. International Colloquium on Automata, Languages, and Programming, LNCS 510*, pages 327–338. Springer, 1991.
14. G. Frederickson. Data structures for on-line updating of minimum spanning trees, with applications. *SIAM J. Comput.*, 14(4):781–798, 1985.
15. G. Frederickson. Fast algorithms for shortest paths in planar graphs, with applications. *SIAM Journal on Computing*, 16:1004–1022, 1987.
16. D. Hutchinson, A. Maheshwari, and N. Zeh. An external-memory data structure for shortest path queries. In *Proc. Annual Combinatorics and Computing Conference, LNCS 1627*, pages 51–60, 1999.

17. V. Kumar and E. Schwabe. Improved algorithms and data structures for solving graph problems in external memory. In *Proc. IEEE Symp. on Parallel and Distributed Processing*, pages 169–177, 1996.
18. R. Lipton and R. Tarjan. A separator theorem for planar graphs. *SIAM Journal of Applied Math.*, 36:177–189, 1979.
19. A. Maheshwari and N. Zeh. I/O-optimal algorithms for planar graphs using separators. In *Proc. ACM-SIAM Symp. on Discrete Algorithms*, pages 372–381, 2002.

Improved Approximate String Matching Using Compressed Suffix Data Structures

Tak-Wah Lam[1], Wing-Kin Sung[2], and Swee-Seong Wong[2,*]

[1] Department of Computer Science, The University of HongKong, HongKong
twlam@cs.hku.hk
[2] School of Computing, National University of Singapore, Singapore
{ksung, wongss}@comp.nus.edu.sg

Abstract. Approximate string matching is about finding a given string pattern in a text by allowing some degree of errors. In this paper we present a space efficient data structure to solve the 1-mismatch and 1-difference problems. Given a text T of length n over a fixed alphabet A, we can preprocess T and give an $O(n\sqrt{\log n})$-bit space data structure so that, for any query pattern P of length m, we can find all 1-mismatch (or 1-difference) occurrences of P in $O(m \log\log n + occ)$ time, where occ is the number of occurrences. This is the fastest known query time given that the space of the data structure is $o(n \log^2 n)$ bits.

The space of our data structure can be further reduced to $O(n)$ if we can afford a slow down factor of $\log^\epsilon n$, for $0 < \epsilon \leq 1$. Furthermore, our solution can be generalized to solve the k-mismatch (and the k-difference) problem in $O(|A|^k m^k (k + \log\log n) + occ)$ and $O(\log^\epsilon n(|A|^k m^k (k + \log\log n) + occ))$ query time using an $O(n\sqrt{\log n})$-bit and an $O(n)$-bit indexing data structures, respectively.

1 Introduction

Consider a text T of length n and a pattern P of length m, both strings over a constant size alphabet A. The approximate string matching problem is to find all approximate occurrences of P in T. Depending on the definition of "error", this problem has two variations: (1) The k-difference problem is to find all occurrences of P in T that have edit distance at most k from P (edit distance is the minimum number of character insertions, deletions and replacements to convert one string to another); and (2) The k-mismatch problem is to find all occurrences of P in T that have Hamming distance at most k from P (Hamming distance is the minimum number of character replacements to convert one string to another). Both k-difference and k-mismatch problems are well-studied and they found applications in many areas including computational biology, text retrieval, multimedia data retrieval, pattern recognition, signal processing, handwriting recognition, etc.

Recently, people are interested in the offline approximate matching problem, in which, we can preprocess the text T and build some indexing data structure so that any pattern query can be answered in a shorter time. Jokinen and Ukkonen [6] were the first to treat the approximate offline matching problem. Since then, many different approaches have been proposed. Some techniques are fast on the average [8, 9]. However,

* Supported by Institute for Infocomm Research, a member of A*STAR, Singapore.

X. Deng and D. Du (Eds.): ISAAC 2005, LNCS 3827, pp. 339–348, 2005.

they incur a query time complexity depending on n, i.e., in the worst case, they are inefficient even if the pattern is very short and k is as small as one. The first solution with query time complexity independent of n is proposed by Ukkonen [13]. When $k = 1$ (that is, 1-mismatch or 1-difference problem), Cobbs [3] gave an indexing data structure using $O(n \log n)$ bits space and having $O(m^2 + occ)$ query time. Later, Amir et al [1] proposed an $O(n \log^3 n)$-bit indexing data structure with $O(m \log n \log \log n + occ)$ query time. Then, Buchsbaum et al [2] proposed another indexing data structure which uses $O(n \log^2 n)$ bits space so that every query can be solved in $O(m \log \log n + occ)$ time. Cole *et. al.* [4] further improved the query time. They gave an $O(n \log^2 n)$-bit data structure so that both the 1-mismatch and the 1-difference problems can be solved in $O(m + \log n \log \log n + occ)$ time, respectively. Recently, in order to target large text indexing like in genomic sequences, Trinh *et. al.* [12] improves upon the space-efficiency. They proposed two data structures of size $O(n \log n)$ bits and $O(n)$ bits with query time $O(m \log n + occ)$ and $O(m \log^2 n + occ \log n)$, respectively.

Some of the above results can be generalized for $k > 1$. Cobbs's $O(n \log n)$-bit indexing data structure can answer both k-mismatch and k-difference queries in $O(m^{k+2}|A|^k + occ)$ time [3]. Cole *et. al.* [4] proposed an $O(n \frac{(c_3 \log n)^k}{k!} \log n)$-bit indexing data structure with query times of $O(\frac{(c_1 \log n)^k \log \log n}{k!} + m + occ)$ and $O(\frac{(c_2 \log n)^k \log \log n}{k!} + m + 3^k \cdot occ)$ for the k-mismatch and k-difference problems, respectively, where c_1, c_2, c_3 are constants with $c_2 > c_1$. Trinh *et. al.* [12] gave $O(n \log n)$-bit and $O(n)$-bit data structures that can answer a k-mismatch (or a k-difference) query in $O(|A|^k m^k \log n + occ)$ time and $O(|A|^k m^k \log^2 n + occ \log n)$ time, respectively.

All previous data structures for supporting the 1-mismatch (or 1-difference) query either require a space of $\Omega(n \log^2 n)$ bits or $\Omega(m \log n + occ)$ time. It is an open problem whether there exists an $O(n \log n)$ or even $o(n \log n)$-bit data structure so that every 1-mismatch (or 1-difference) query can be answered in $o(m \log n + occ)$ time. In this paper, we resolve this open problem in the affirmative by presenting a data structure which uses $O(n\sqrt{\log n})$ bit space while every 1-mismatch (or 1-difference) query can be answered in $O(m \log \log n + occ)$ time.

Our result can be further extended in two ways. First, we show that the space of the data structure can be reduced to $O(n)$ bits if we accept a slow down factor of $\log^\epsilon n$ for the query time where $0 < \epsilon \leq 1$. Second, the data structure can be extended to solve the k-mismatch (or the k-difference) problem for $k \geq 1$. Our solution can solve the k-mismatch (or the k-difference) problem in $O(|A|^k m^k (k + \log \log n) + occ)$ or $O(\log^\epsilon n\ (|A|^k m^k (k + \log \log n) + occ))$ query time, when the text is encoded using $O(n\sqrt{\log n})$-bit or $O(n)$-bit indexing data structures, respectively.

2 Preliminaries

2.1 Suffix Array, Inverse Suffix Array, and Ψ Function

Let $T[0..n] = t_0 t_1 \cdots t_{n-1}$ be a text of length n over an alphabet A, appended with a special symbol t_n = '\$' that is not in A and is smaller than any other symbol in A. The j-th suffix of T is defined as $T[j..n] = t_j \cdots t_n$ and is denoted by T_j.

The *suffix array* $SA[0..n]$ of T is an array of integers so that $T_{SA[i]}$ is lexicographically smaller than $T_{SA[j]}$ if and only if $i < j$. Note that $SA[0] = n$. The *inverse suffix array* of T is denoted as $SA^{-1}[0..n]$, that is, $SA^{-1}[i]$ equals the number of suffixes which are lexicographically smaller than T_i.

In this paper, an interval $[st..ed]$ is called the range of the suffix array of T corresponding to a string P if $[st..ed]$ is the largest interval such that P is a prefix of every suffix T_j for $j = SA[st], SA[st+1], \dots, SA[ed]$. We write $[st..ed] = range(T, P)$.

A concept related to the suffix array is the $\Psi[0..n]$ [5], which is defined as follows:

$$\Psi[i] = SA^{-1}[SA[i] + 1].$$

Let t_{SA} and t_{Ψ} be the access time of each entry on SA and Ψ respectively. In this paper, we need a data structure $\mathcal{D}$ which supports, for any i, the following operations.

- reports $SA[i]$ in t_{SA} time,
- reports $SA^{-1}[i]$ in t_{SA} time,
- reports $\Psi[i]$ in t_{Ψ} time, and
- reports *substring(i,l)* $= T[SA[i]..SA[i] + l - 1]$ in $O(lt_{\Psi})$ time for some length l.

Lemmas 1 and 3 give two implementations of the data structure $\mathcal{D}$.

Lemma 1. *[5] The data structure $\mathcal{D}$ can be implemented in $O(n \log |A|)$ bits so that $t_{SA} = O(\log^{\epsilon} n)$ and $t_{\Psi} = O(1)$, where $1 \geq \epsilon > 0$.*

Below lemma is needed for the second implementation of the data structure $\mathcal{D}$.

Lemma 2. *Let $X_1, \dots, X_\ell$ be ℓ subsets of $\{0, \dots, n-1\}$ such that $|X_j| = n/\ell$, $1 \leq j \leq \ell$. Then $\{\Psi^j[z] | z \in X_j\}$ for all j, can be stored in an $O(n\ell \log |A| + |A|^{\ell} \log n)$-bit data structure such that $\Psi^j[z]$, for any $z \in X_j$, can be accessed in $O(1)$ time.*

Proof. The result follows from data structures described in Rao's paper [10]. □

Lemma 3. *The data structure $\mathcal{D}$ can be implemented in $O(n\sqrt{\log n} \log |A|)$ bits so that $t_{SA} = O(1)$ and $t_{\Psi} = O(1)$.*

Proof. Building the $O(n \log |A|)$-bit data structure in Lemma 1, the Ψ function can be accessed in $O(1)$ time. Below, we describe $O(n\sqrt{\log n} \log |A|)$-bit data structures so that both SA and SA^{-1} can be computed in $O(1)$ time.

For the access of SA value, recall that Rao [10] gives an implementation of the compressed suffix array that reports $SA[i]$ in $O(1)$ time using $O(n\sqrt{\log n})$ bits for binary text string (refer to Theorem 4 in [10]). For text on a fixed finite alphabet A, Rao's idea can be generalized so that $SA[i]$ can be accessed in constant time using an $O(n\sqrt{\log n} \log |A|)$-bit data structure.

For the access of SA^{-1} value, we need the following data structure. Let $\ell = \sqrt{\log n}$. First, we store $SA^{-1}[x\ell]$ for all $0 \leq x \leq \lfloor n/\ell \rfloor$, which requires $O(n\sqrt{\log n})$ bits. Then, we need a data structure so that $\Psi^j[z]$ can be accessed in $O(1)$ time for any $0 \leq x \leq \lfloor n/\ell \rfloor$ and any $1 \leq j \leq \ell$. By Lemma 2, such data structure can be stored in $O(n\sqrt{\log n} \log |A| + |A|^{\sqrt{\log n}} \log n) = O(n\sqrt{\log n} \log |A|)$ bits.

Now we show how to access $SA^{-1}[i]$ given i in constant time. Let $y = \lfloor i/\ell \rfloor$, $k' = i - y\ell$, and $z' = SA^{-1}[y\ell]$. We claim that $SA^{-1}[i] = \Psi^{k'}[z']$ and $k' \leq \ell$. Then, using the data structures above, $SA^{-1}[i]$ can be computed in $O(1)$ time.

Note that $y\ell \leq i \leq (y+1)\ell$ and thus, $k' = i - y\ell \leq \ell$. It is then easy to verify that $SA[\Psi^{k'}[z']] = SA[z'] + k'$. Since $SA[z'] = y\ell$, we have $SA[\Psi^{k'}[z']] = y\ell + k' = i$. Thus, the claim follows. □

2.2 Suffix Tree

A suffix tree for the text T is an edge-labeled rooted directed tree with exactly $n+1$ leaves numbered 0 to n. Each edge is labeled with a non-empty substring of T such that no two outgoing edges from a node have labels with the same first character. For every node v, its path label $plabel(v)$ is constructed by concatenating the edge labels, in order, from the root to the node. Note that the path label of every leaf i, is a suffix of T that starts at position i.

We assumed that the suffixes of the leaves in the suffix tree are lexicographically ordered so that the collection of leaf nodes from left to right will form the suffix array denoted by $SA[0..n]$. For our approach, we require a suffix tree that support the following operations:

$label(u, v)$: returns the label on the edge joining node u to v in $O(xt_{SA})$ time where x is the length of the edge label of (u, v).

$plen(v)$: returns the length of the path label $plabel(v)$ in $O(t_{SA})$ time .

$leftmost(v)$: returns the SA index of the leftmost leaf in the subtree rooted at node v in $O(1)$ time.

$rightmost(v)$: returns the SA index of the rightmost leaf in the subtree rooted at node v in $O(1)$ time.

$slink(v)$: returns a node u if there is a suffix link from node v to node u in $O(t_{\Psi})$ time.

$child(v, c)$: returns a child w of the node v if c is a prefix character to string $label(v, w)$ in $O(|A|t_{SA})$ time.

Lemma 4. *A suffix tree with the above properties can be implemented using (1) $O(n \log |A|)$ bits for $t_{SA} = O(\log^{\epsilon} n)$ and $t_{\Psi} = O(1)$, or (2) $O(n\sqrt{\log n} \log |A|)$ bits for $t_{SA} = O(1)$ and $t_{\Psi} = O(1)$.*

Proof. We refer to Sadakane's paper [11] on compressed suffix tree (CST) implementation that uses data structure $\mathcal{D}$ and $O(n)$ bits for the balanced parentheses representation of the suffix tree [7]. The space complexities follow from Lemmas 1 and 3. □

The following result on LCP query is also available.

Lemma 5. *[11] Given SA indexes i and j, the length of the longest common prefix (LCP) between suffixes at positions $SA[i]$ and $SA[j]$, denoted by $|lcp(i,j)|$, can be computed in $O(t_{SA})$ time using additional $O(n)$ bits data structure. The lowest common ancestor (LCA) node between any two nodes in the suffix tree can also be computed in $O(t_{SA})$ time.*

2.3 Other Data Structures

Given a suffix tree ST built from the text T, and a query pattern P of length m, we define the following terminologies and data structures:

Definition 1. *Given a node x in ST, let x_{le} and x_{ri} denote indexes of SA corresponding to the leftmost and rightmost leaf nodes in the subtree spanned by x.*

Based on the above definition, for any node x in ST, we have $[x_{le}..x_{ri}] = range(T, plabel(x))$.

Definition 2. *Arrays $F_{st}[1..m]$ and $F_{ed}[1..m]$ are such that $[F_{st}[i]..F_{ed}[i]] = range(T, P[i..m])$ for $1 \leq i \leq m$. We also define $F_{st}[j] = 0$ and $F_{ed}[j] = n$ for $j > m$.*

Lemma 6. *$F_{st}[1..m]$ and $F_{ed}[1..m]$ can be constructed in $O(mt_{\Psi} + m|A|t_{SA})$ time.*

Proof. This can be done using the suffix links in ST in $O(mt_{\Psi})$ time, given that $range(T, P[1..m])$ can be obtained by traversing the suffix tree in $O(m|A|t_{SA})$ time. □

Furthermore, the following lemma is needed to support exact pattern search over a subtree in the suffix tree.

Lemma 7. *Given a pattern P, let x be a node such that $[x_{le}..x_{ri}] = range(T, P)$. For any position i in T, P is a prefix of $T[i..n]$ if and only if $x_{le} \leq SA^{-1}[i] \leq x_{ri}$. Also, $SA^{-1}[SA[x_{le}] + |P|] < SA^{-1}[SA[x_{le} + 1] + |P|] < \ldots < SA^{-1}[SA[x_{ri}] + |P|]$.*

2.4 Heavy Path Decomposition

We introduce a standard technique to partition $O(n)$ nodes of a tree into $O(\log n)$ levels. The heavy path decomposition scheme is as such: Given a suffix tree ST, we assign a level to every node in ST. The root is assigned $level$ 1. If a node v has $level\ i$, we assign $level\ i$ to the single child node of v, that has the largest subtree (in terms of number of nodes) among all the other child nodes of v. The other child nodes of v are assigned $level\ i + 1$. Edges joining 2 nodes with the same $level$ are denoted as *core edges* and the rest of edges that join nodes at $level\ i$ to nodes at $level\ i + 1$ are denoted as *side edges*. An internal node will have exactly one outgoing *core edge* and the rest of the outgoing edges are *side edges*. We also denote a node with an incoming core edge as a *core node* and otherwise, a *side node*. The root, is by default, a *side node*. There are $O(\log n)$ levels. The followings are also observed:

Lemma 8. *There are $O(\log n)$ side edges on the path from the root to any node in the suffix tree.*

Lemma 9. *Consider any two distinct side edges e_1 and e_2 with end nodes v_1 and v_2 respectively. If both v_1 and v_2 have $level\ i$, then the two subtrees rooted at v_1 and v_2 are disjoint. In other words, the subtrees rooted at any two distinct side nodes of the same level are disjoint.*

Lemma 10. *Given any side node v, that begins a core path (where all edges in the path are core edges), we can find the leaf node u that terminates the core path in $O(1)$ time using additional $O(n)$ bits data structure.*

Proof. Let u_i be the i-th node in the suffix tree according to the prefix order, and $t = O(n)$ be the number of nodes in the suffix tree. Let $X[1..t]$ be an array such that X[i]='(' if u_i is an internal side node and $X[j] =')'$ if u_j is a core leaf node. It can easily check that the parentheses in $X[1..t]$ is balance. More importantly, for every pair of parentheses (i, j) in X, u_i and u_j form the start and the end of some core path. By the data structure for balanced parentheses [7], the lemma follows. □

3 1-Approximate String Matching Problem

3.1 The Data Structure for 1-Approximate Matching

Our 1-approximate matching data structure is basically the suffix tree ST of the text T (see Section 2.2), together with two other data structures. First, for every side node v (see Section 2.4), let u be the parent node of v, we maintain a set $\Gamma_v = \{SA^{-1}[SA[i] + plen(u) + 1] \mid i = 1(\text{mod } \log^2 n) \text{ and } v_{le} \leq i \leq v_{ri}\}$.

Second, for every core leaf node u (whose SA index is k), let v be the start of the corresponding core path, we maintain 2 lists of SA indexes, $H_u^l = \{i \mid i = 1(\text{mod } \log^2 n), i \leq k \text{ and } |lcp(k,i)| \geq plen(v)\}$ and $H_u^r = \{i \mid i = 1(\text{mod } \log^2 n), i > k \text{ and } |lcp(k,i)| \geq plen(v)\}$. The values in H_u^l and H_u^r are ordered by increasing longest common prefix length $|lcp(k,i)|$.

Lemma 11. *We can store Γ_v for all side nodes v using $O(n)$ bit space. In addition, we can answer any range query in Γ_v using $O(\log\log n)$ time.*

Proof. By Lemma 9, all subtrees rooted at different side nodes, of the same level-ℓ, are disjoint. Hence, the total size of Γ_v for all level-ℓ side nodes v is at most $n/\log^2 n$. Since there are $O(\log n)$ levels, the total size of Γ_v for all side nodes v is $O(n/\log n)$. We store Γ_v, for every side node v, using the *y-fast trie* [14] data structure. The size of the data structure is $O(|\Gamma_v|\log n)$ bits and it allows efficient range query in $O(\log\log n)$ time. Since the total size of all Γ_v is $O(n/\log n)$, the lemma follows. □

Lemma 12. *We can store H_u^l and H_u^r for all core leaf nodes u (whose SA index is k) using $O(n)$ bit space. In addition, for any range $[x..y]$, we can report the values $i^l \in H_u^l$ and $i^r \in H_u^r$ such that $x \leq lcp(i^l, k) \leq y$ and $x \leq lcp(i^r, k) \leq y$ using $O(\log\log n)$ time.*

Proof. There are at most $n/\log^2 n$ selected leaf nodes and each leaf node is reachable from at most $\log n$ different core paths. Hence, the total size of H_u^l and H_u^r for all core paths is $O(n/\log n)$. Each H_u^l and H_u^r is stored using the *y-fast trie* [14] data structure. The size of the data structures is $O((|H_u^l| + |H_u^r|)\log n)$ bits. Since the total size of all H_u^l and H_u^r is $O(n/\log n)$, the lemma follows. □

By Lemmas 4, 11 and 12,

Lemma 13. *The 1-approximate matching data structure can be stored in* $O(n \log |A|)$ *and* $O(n\sqrt{\log n} \log |A|)$ *bits for* $t_{SA} = O(\log^{\epsilon} n)$ *and* $t_{SA} = O(1)$ *respectively. In both cases,* $t_{\psi} = O(1)$.

Below is the key lemma for our algorithm.

Lemma 14. *Consider (1) a node* u *in* ST *such that* $P_1 = plabel(u)$, *(2) a character* c, *and (3) another string* P_2 *with* $[st..ed] = range(T, P_2)$. *Let* $v = child(u, c)$. *Then, all occurrences of* P_1cP_2 *can be computed in* $O(t_{SA}(\log\log n + occ))$ *time where occ is the total number of occurrences of* P_1cP_2 *in* T.

Proof. Note that $[v_{le}..v_{ri}] = range(T, P_1c)$. If $P = P_1cP_2$ occurs in T, as P_1c is a prefix of P, P must occur at position $SA[i]$ for some $v_{le} \leq i \leq v_{ri}$. By Lemma 7, we can verify if P occurs at position $SA[i]$ by checking if $st \leq SA^{-1}[SA[i] + |P_1| + 1] \leq ed$. Hence, the occurrence of P can be found by performing the above checking for all $i \in [v_{le}..v_{ri}]$. Moreover, by Lemma 7, $SA^{-1}[SA[i] + |P_1| + 1]$ is increasing for $i \in [v_{le}..v_{ri}]$. Note that, for any i, $SA^{-1}[SA[i] + |P_1| + 1]$ can be retrieved in $O(t_{SA})$ time. Hence, one occurrence of P can be found in $O(t_{SA} \log(v_{ri} - v_{le}))$ time by binary search.

If v is a side node, recall that we associate a set Γ_v to it, where $\Gamma_v = \{SA^{-1}[SA[i] + |P_1| + 1] \mid i = 1 (\text{mod } \log^2 n) \text{ and } i \in [v_{le}..v_{ri}]\}$. There are 3 cases.

- Case 1: Γ_v is empty. This means that the number of leaves in the subtree of v $(v_{ri} - v_{le} + 1)$ is $< \log^2 n$. Using the method we just discussed, one occurrence of P can be found in $O(t_{SA} \log(v_{ri} - v_{le})) = O(t_{SA} \log\log n)$ time.
- Case 2: Γ_v is non-empty and, by Lemma 11, we find some i such that $st \leq SA^{-1}[SA[i] + |P_1| + 1] \leq ed$. Since any range query of *y-fast trie* takes $O(\log\log n)$ time, the second case follows.
- Case 3: Γ_v is non-empty and, by Lemma 11, we cannot find any i such that $st \leq SA^{-1}[SA[i] + |P_1| + 1] \leq ed$. In this case, using $O(\log\log n)$ time, we apply *y-fast trie* to find a and b such that $SA^{-1}[SA[a] + |P_1| + 1] \in \Gamma_v$ is just smaller than st and $SA^{-1}[SA[b] + |P_1| + 1] \in \Gamma_v$ is just bigger than ed. Note that $b - a \leq \log^2 n$. Then, using the method described at the beginning of the proof, we can found one occurrence of P in $O(t_{SA} \log(b - a)) = O(t_{SA} \log\log n)$ time.

If v is a core node, let CP be the core path containing v. Since the side node lies on the path from the root node to v and would have been uncovered from traversing the suffix tree to obtain P_1, here we assume that the side node that begins the core path CP is known. We obtain the terminating leaf node x (whose SA index is k) of CP by Lemma 10 using $O(1)$ time. Next, we search for the node $r \in CP$ whose path label is of length $|P_1| + q + 1$ where $q = |lcp(SA^{-1}[SA[k] + |P_1| + 1], st)|$. By Lemma 5, q is computed in $O(t_{SA})$ time. There are 3 cases.

- Case 1: H_x is empty. This means that the number of leaves hanging from the core path is $< \log^2 n$. We can find one occurrence of P in $O(t_{SA} \log(v_{ri} - v_{le})) = O(t_{SA} \log\log n)$ time.

- Case 2: $q \geq |P_2|$. This means that leaf node x corresponds to a suffix with P as its prefix. We have recovered one occurrence of P.
- Case 3: $q < |P_2|$. First, we would like to find $j^l \in H_x^l$ and $j^r \in H_x^r$ such that $|lcp(j^l - \log^2 n, k)| \leq |P_1| + q + 1 \leq |lcp(j^l, k)|$ and $|lcp(j^r, k)| \leq |P_1| + q + 1 \leq |lcp(j^r + \log^2 n, k)|$ respectively. This can be computed in $O(\log\log n)$ time given Lemma 12. Next, using binary search, we locate i^l and i^r within the range of $j^l - \log^2 n ... j^l$ and $j^r ... j^r + \log^2 n$ such that $|lcp(i^l, k)| = |P_1| + q + 1$ and $|lcp(i^r, k)| = |P_1| + q + 1$ respectively. If both j^l and j^r are not found, the binary search is performed for $k - \log^2 n \leq i^l \leq k + \log^2 n$. The binary search takes $O(t_{SA} \log\log n)$ time. Given i^l or i^r whichever one is found, we can recover node r by performing a LCA on the leaf node at i^l or i^r with x in $O(t_{SA})$ time. If P_2 is not completely matched after node r, we can continue to search the outgoing side edges from node r as described above (case where v is a side node) using additional $O(t_{SA} \log\log n)$ time. Overall, the time taken is still $O(t_{SA} \log\log n)$.

Once we confirm that P occurs in position $SA[i]$, the remaining occurrences of P can be found by performing, for entries i' to the left and to the right of i, the above checking (that is, $st \leq SA^{-1}[SA[i'] + |P_1| + 1] \leq ed$), until we reach the boundary of $[v_{le}..v_{ri}]$ or a false case occurs. The time required is $O(t_{SA}(occ + 2))$. □

Here, we define the procedure $TreeSearch(u, c, [st..ed])$ to be the routine which finds all the occurrences of $P_1 c P_2$ where $P_1 = plabel(u)$ and $[st..ed] = range(T, P_2)$. By Lemma 14, this procedure runs in $O(t_{SA}(\log\log n + occ))$ time.

3.2 The 1-Approximate Matching Algorithm

The algorithm traverses the suffix tree from the root to find the pattern P character by character. Then, for every position i, it introduces an "error" at that position and checks for occurrences by calling $TreeSearch$. The details of the algorithm is stated in Figure 1.

Lemma 15. *Given the indexing data structure in Section 3.1, we can locate all* 1-*mismatch (or* 1-*difference) occurrences of a length-*m *pattern* P *in* T*, using* $O(t_\Psi m + t_{SA}(|A| m \log\log n + occ))$ *time.*

Proof. By Lemma 6, Step 1 takes $O(m t_\Psi + m|A| t_{SA})$ time. Step 2 takes $O(1)$ time.

When we traverse down the suffix tree to a node u (with $plabel(u) = P[1..i-1]$), we will execute Steps 3(a-c) for the node u. By Lemma 14, Steps 3(a-c) in total takes $O(t_{SA}(|A| m \log\log n + occ))$ time.

For Step 3(d), it takes $O(t_{SA}|A|m)$ time. The lemma follows. □

By Lemmas 13 and 15, we get the following 2 theorems:

Theorem 1. *Given an* $O(n\sqrt{\log n} \log|A|)$*-bit indexing data structure, the 1-mismatch or 1-difference problem can locate all* 1*-approximate occurrences of a length-*m *pattern* P *in* T*, using* $O(|A| m \log\log n + occ)$ *time.*

Theorem 2. *Given an* $O(n \log|A|)$*-bit indexing data structure, the 1-mismatch or 1-difference problem can locate all* 1*-approximate occurrences of a length-*m *pattern* P *in* T*, using* $O(\log^\epsilon n(|A| m \log\log n + occ))$ *time, where* $0 < \epsilon \leq 1$.

Algorithm *1-approximate match*

1. Construct $F_{st}[1..m]$ and $F_{ed}[1..m]$ such that $[F_{st}[i]..F_{ed}[i]] = range(T, P[i..m])$.
2. $u =$ root node, $i = 1$.
3. Repeat
   /* Note: we maintain the invariant that $plabel(u) = P[1..i-1]$. */
   (a) Deletion at i (find occurrences of $P[1..i-1]P[i+1..m]$)
      If $P[i] \neq P[i+1]$
         report the occurrences found by $TreeSearch(u, P[i+1], [F_{st}[i+2]..F_{ed}[i+2]])$.
   (b) Substitution at i (find occurrences of $P[1..i-1]cP[i+1..m]$ for all $c \in A - \{P[i]\}$)
      For $c \in A - \{P[i]\}$,
         report the occurrences found by $TreeSearch(u, c, [F_{st}[i+1]..F_{ed}[i+1]])$.
   (c) Insertion at i (find occurrences of $P[1..i-1]cP[i..m]$ for all $c \in A - \{P[i]\}$)
      For $c \in A - \{P[i]\}$,
         report the occurrences found by $TreeSearch(u, c, [F_{st}[i]..F_{ed}[i]])$.
   (d) No insertion, deletion, and substitution at i
      Let $v = child(u, P[i])$, $E = label(u, v)$.
      If $P[i..i+|E|-1] = E$
         $u = v, i = i + |E|$
      Else
         Find the smallest $j > i$ such that $P[j] \neq E[j-i+1]$.
         Report all the occurrences of P so that the error is at j.
         Terminate and return.

Fig. 1. Algorithm for 1-mismatch and 1-difference

3.3 The k-Approximate Matching Problem with $k \geq 1$

Theorem 3. *After preprocessing the text T of length n and obtain an $O(n\sqrt{\log n}\log|A|)$ or $O(n\log|A|)$ bits data structure, the k-mismatch or k-difference problem can locate all approximate occurrences of a length-m pattern P in T, using $O(|A|^k m^k(k + \log\log n) + occ)$ or $O(\log^\epsilon n(|A|^k m^k(k + \log\log n) + occ))$ time respectively, where $0 < \epsilon \leq 1$, $|A|$ is the fixed alphabet size and occ is the number of approximate occurrences of P in T.*

Proof. We give a sketch of the proof here. The k-approximate matching problem can be solved using dynamic programming on the suffix tree in $O(|A|^k m^k mk)$ time. The factor $|A|^k m^k$ is the number of different paths down the suffix tree that are to be traversed, which is bounded by the number of edit traces [13]. By merging our data structure in Section 3.1 to compute only for the kth error, with dynamic programming over the suffix tree, we can obtain the stated complexities. □

4 Conclusion

In this paper, we give an $O(n\sqrt{\log n})$-bit data structure which can answer 1-mismatch (and 1-difference) query in $O(m\log\log n + occ)$ time. If we can afford a slow down factor of $\log^\epsilon n$ for the query time, the size of the data structure can be further reduced to $O(n)$ bits. We also generalize our solution to solve k-mismatch and k-difference problems.

Acknowledgments

The authors would like to thank Limsoon Wong for his helpful comments. This work is supported in part by the NUS Academic Research Grant R-252-000-119-112.

References

1. A. Amir, D. Keselman, G. M. Landau, M. Lewenstein, N. Lewenstein, and M. Rodeh. Text indexing and dictionary matching with one error. *Journal of Algorithms*, 37(2):309–325, 2000.
2. A. L. Buchsbaum, M. T. Goodrich, and J. R. Westbrook. Range searching over tree cross products. In *Proceedings of the 8th Annual European Symposium on Algorithms*, pages 120–131, 2000.
3. A. L. Cobbs. Fast approximate matching using suffix trees. In *Proceedings of the 6th Annual Symposium on Combinatorial Pattern Matching*, pages 41–54, July 1995.
4. R. Cole, L-A. Gottlieb, and M. Lewenstein. Dictionary matcing and indexing with errors and don't cares. In *Proceedings of the 36th Annual ACM Symposium on Theory of Computing*, pages 91–100, 2004.
5. R. Grossi and J. S. Vitter. Compressed suffix arrays and suffix trees with applications to text indexing and string matching. In *Proceedings of the 32nd ACM Symposium on Theory of Computing*, pages 397–406, 2000.
6. P. Jokinen and E. Ukkonen. Two algorithms for approximate string matching in static texts. In *Proceedings of the 16th International Symposium on Mathematical Foundations of Computer Science*, pages 240–248, September 1991.
7. J. I. Munro, V. Raman, and S. S. Rao. Space efficient suffix trees. *Journal of Algorithms*, 39:205–222, 2001.
8. G. Navarro and R.Baeza-Yates. A hybrid indexing method for approximate string matching. *Journal of Discrete Algorithms*, 1(1):205–239, 2000.
9. G. Navarro, E. Sutinen, J. Tanninen, and J. Tarhio. Indexing text with approximate q-grams. In *Proceedings of the 11th Annual Symposium on Combinatorial Pattern Matching*, pages 350–365, 2000.
10. S. S. Rao. Time-space trade-offs for compressed suffix arrays. *Information Processing Letters*, 82:307–311, 2002.
11. K. Sadakane. Compressed suffix trees with full functionality. *Theory of Computing Systems*, Accepted.
12. H. N. D. Trinh, W. K. Hon, T. W. Lam, and W. K. Sung. Approximate string matching using compressed suffix arrays. In *Proceedings of the 15th Annual Symposium on Combinatorial Pattern Matching*, pages 434–444, 2004.
13. E. Ukkonen. Approximate string-matching over suffix trees. In *Proceedings of the 4th Annual Symposium on Combinatorial Pattern Matching*, pages 228–242, 1993.
14. D. E. Willard. Log-logarithmic worst-case range queries are possible in space $\theta(n)$. *Information Processing Letters*, 17:81–84, August 1983.

Monitoring Continuous Band-Join Queries over Dynamic Data*

Pankaj K. Agarwal, Junyi Xie, Jun Yang, and Hai Yu

Department of Computer Science, Duke University,
Durham, NC 27708-0129, USA
{pankaj, junyi, junyang, fishhai}@cs.duke.edu

Abstract. A *continuous query* is a standing query over a dynamic data set whose query result needs to be constantly updated as new data arrive. We consider the problem of constructing a data structure on a set of continuous *band-join* queries over two data sets R and S, where each band-join query asks for reporting the set $\{(r,s) \in R \times S \mid a \le r - s \le b\}$ for some parameters a and b, so that given a data update in R or S, one can quickly identify the subset of continuous queries whose results are affected by the update, and compute changes to these results.

We present the first nontrivial data structure for this problem that simultaneously achieves subquadratic space and sublinear query time. This is achieved by first decomposing the original problem into two independent subproblems, and then carefully designing data structures suitable for each case, by exploiting the particular structure in each subproblem.

A key step in the above construction is a data structure whose performance increases with the degree of *clusteredness* of the band-joins being indexed. We believe that this structure is of independent interest and should have broad impact in practice. We present the details in [1].

1 Introduction

In contrast to traditional queries, where each query is executed once on a given set of items, a *continuous query* is a standing query over a set of items that, once issued by the user, needs to keep generating new results (or changes to old results) subject to the same query condition, as new items continue to arrive in a stream. Continuous query processing has recently attracted much interest from the database community because of its wide range of traditional and emerging applications, especially in data streaming settings; see [3, 4, 6, 13] and the references therein.

One of the main challenges in continuous query processing is how to monitor a *large* number of continuous queries over dynamically changing data. For each

* Research by P.A. and H.Y. is supported by NSF under grants CCR-00-86013, EIA-01-31905, CCR-02-04118, and DEB-04-25465, by ARO grants W911NF-04-1-0278 and DAAD19-03-1-0352, and by a grant from the U.S.–Israel Binational Science Foundation. Research by J.X. and J.Y. is supported by NSF CAREER award under grant IIS-0238386.

X. Deng and D. Du (Eds.): ISAAC 2005, LNCS 3827, pp. 349–359, 2005.

incoming data update, one needs to identify the subset of continuous queries whose results are affected by the data update, and compute changes to these results. If there are many continuous queries, then a brute-force approach that processes each of them in turn will be too inefficient to meet the response-time requirement of most target applications, and faster techniques are needed.

An interesting aspect of continuous query processing is the interchangeable roles played by queries and data: Continuous queries can be treated as data, while each data update can be treated as a query requesting the subset of continuous queries affected by the update. Thus, to preprocess simple continuous queries, it is natural to apply indexing and query processing techniques which were traditionally intended for preprocessing data. For example, one can simply use R-trees to monitor a large number of continuous rectangular range queries over a dynamic point set in $\mathbb{R}^2$. However, for more complex continuous queries, especially those involving *joins* between multiple dynamic data sets, designing both space- and query-efficient data structures turns out to be interesting.

Problem Statement. Let $R, S \subseteq \mathbb{R}$ be two different data sets. A *band-join* query on R and S, denoted by $J(a, b)$ for some parameters $a, b \in \mathbb{R}$, asks for reporting the set $J(a, b) = \{(r, s) \in R \times S \mid a \leq r - s \leq b\}$. Band-join queries naturally arise in many continuous query applications, and form the basis of more complex join queries [5, 6, 9]. In this paper we are interested in monitoring a set of continuous band-join queries when data are inserted or deleted from R or S. More precisely, let $\mathcal{Q} = \{Q_1, \cdots, Q_n\}$ be a set of continuous band-join queries on R and S, where $Q_i = J(a_i, b_i)$ and $a_i \leq b_i$, $a_i, b_i \in \mathbb{R}$. We are interested in supporting the following two functions when an element s is inserted into or deleted from S (the case where R is updated is entirely symmetric):

(Maintaining affected joins). Report all pairs (r, Q_i), where $r \in R$ and $Q_i \in \mathcal{Q}$, such that $a_i \leq r - s \leq b_i$. Each reported pair (r, Q_i) corresponds to an element $(r, s) \in R \times S$ to be inserted into or deleted from the result of Q_i.

(Reporting affected joins). Report all $Q_i \in \mathcal{Q}$ such that $a_i \leq r - s \leq b_i$ for some $r \in R$. They are the subset of queries in $\mathcal{Q}$ whose results have been affected by the insertion or deletion of s. This functionality is useful in settings in which it suffices to detect and report that Q_i has been affected, while the actual result of Q_i may be computed later on demand.

These two functionalities can be reduced to the following abstract problem. We say that an interval $\gamma \subset \mathbb{R}$ is *stabbed* by a point $x \in \mathbb{R}$ if $x \in \gamma$. Let $X = \{x_1, \ldots, x_m\}$ be a set of m points in $\mathbb{R}$, and $I = \{[a_1, b_1], \ldots, [a_n, b_n]\}$ be a set of n (closed) intervals in $\mathbb{R}$. Consider in the following queries on X and I: **(Q1)** Given a displacement $\Delta x \in \mathbb{R}$, report the set of all point-interval pairs (x, γ), where $x \in X$ and $\gamma \in I$, so that $x + \Delta x$ stabs γ. **(Q2)** Given a displacement $\Delta x \in \mathbb{R}$, report the set of all intervals in I that are stabbed by at least one point in $X + \Delta x$.

We are also interested in the following update operations: **(U1)** Add/remove a point in X. **(U2)** Add/remove an interval in I.

In the context of monitoring continuous band-joins, note that if we set $X_R = R$ and $I_{\mathcal{Q}} = \{[a_i, b_i] \mid Q_i \in \mathcal{Q}\}$, then maintaining affected joins corresponds to a (Q1) query on X_R and $I_{\mathcal{Q}}$ with $\Delta x = -s$, and reporting affected joins corresponds to a (Q2) query on X_R and $I_{\mathcal{Q}}$ with the same Δx. Also note that when an element is inserted into or deleted from R, or when a query is added into or deleted from $\mathcal{Q}$, we need to update the data structure for X_R and $I_{\mathcal{Q}}$; this corresponds to the operations (U1) and (U2).

Our Results. As the problem of monitoring continuous band-joins is equivalent to (Q1) and (Q2) queries, we shall focus our attention on these two queries from now on. For either query, a straightforward data structure with $O(m+n)$ size and $O(m+n)$ query time has been known and actually used in practice for some time in the database community [5]. However, the query time of this approach is quite unsatisfactory. In the other extreme, a data structure with $O(mn)$ size and $O(\log(mn))$ query time is not hard to design either (see Lemma 1). But in this case, the size of the structure becomes problematic.

A natural open problem is whether there exists a data structure that fits between the above two extremes. We answer this question in the affirmative. In Section 2, we present the first nontrivial data structure for (Q1) and (Q2) that simultaneously achieves $o(mn)$ size and $o(m+n)$ query time. This demonstrates an intriguing tradeoff between the space and query time for both queries. Our construction proceeds by decomposing the original problem into two independent subproblems, and then carefully designing data structures suitable for each case, by exploiting the particular structure in each subproblem.

We should point out that the above structure is mostly of theoretical interest. However, in solving one of the subproblems, we have devised a data structure whose performance increases with the degree of *clusteredness* of the input intervals in I. Because in practice the intervals are often naturally clustered, we believe that this structure is of independent interest and can lead to a practically more efficient approach. We further comment on this structure in Section 2.2; details are given in [1].

Related Work. Since dynamic data is ubiquitous in the real world, there has been a lot of work on designing efficient dynamic data structures for answering various queries; see, for example, the monograph by Overmars [15] and the survey by Chiang and Tamassia [7]. Motivated by more recent applications such as sensor networks, there has also been much research on maintaining various synopsis structures of the data in the data stream model [14], such as ε-approximations [2, 16], histograms [12] and so on, so that various statistical queries can be answered quickly. However, most of the above work only deals with non-continuous queries (i.e., each query is executed once on the current data set). Recently, motivated by a wide range of emerging applications, there has be a flurry of activity on studying continuous queries in the database community (see e.g., [3, 4]). But to the best of our knowledge, the problem of monitoring continuous queries has not yet received much attention from algorithms research community so far.

2 The Construction

The main result of this section is a data structure of size $o(mn)$ that answers (Q1) and (Q2) queries in $o(m+n)$ time. We first decompose the original problem into two subproblems, and then solve the two subproblems respectively (Sections 2.1 and 2.2). By putting pieces together and choosing appropriate parameters, we obtain the desired bounds on the size and query time (Section 2.3).

Let $r \in \mathbb{N}$ be a parameter to be fixed later. We first partition I into $O(n/r)$ groups in the following manner (see Figure 1). We sort the $2n$ endpoints of I and pick every $2r$-th endpoint. Thus a set of n/r points are picked. Let I_0 be the subset of intervals in I that are stabbed by any of these picked points. Note that each of the remaining intervals in $I \setminus I_0$ lies entirely between two consecutive picked points. We let $I_i \subset I$ denote the subset of intervals lying entirely between the $(i-1)$-th and the i-th picked points, for $1 \le i \le n/r+1$, where the 0-th and $(n/r+1)$-th picked points are defined to be $-\infty$ and $+\infty$. In this way, the set I is partitioned into a collection of disjoint subsets $I_0, I_1, \cdots, I_{n/r+1}$, where $|I_i| \le r$ for all $i \ge 1$.

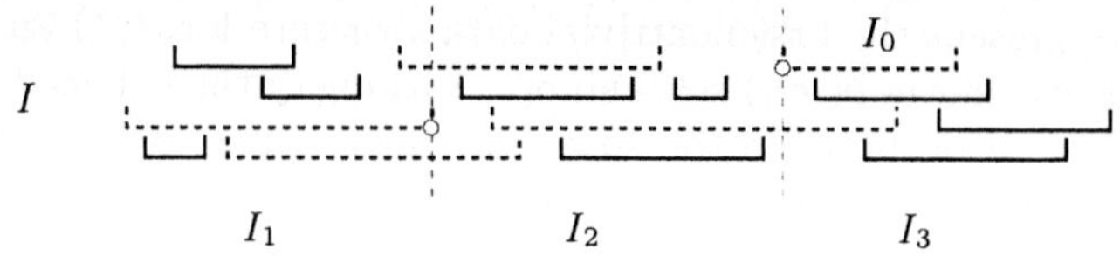

Fig. 1. The partition of I. Dashed Intervals belong to I_0, and solid intervals belong to $I_1, I_2, \ldots$, respectively.

Next, assume that the elements in $X = \{x_1, x_2, \cdots, x_m\}$ are in sorted order. We also partition X into m/r subsets $X_1, X_2, \cdots$, each of size at most r, where $X_i = \{x_{(i-1)r+1}, x_{(i-1)r+2}, \ldots, x_{ir}\}$, for $1 \le i \le m/r$.

The overall data structure $\mathcal{D}(X, I)$ for answering (Q1) and (Q2) queries on X and I consists of two separate structures: $\mathcal{D}(X, I_0)$ for answering the same queries on X and I_0, and $\mathcal{D}(X, I \setminus I_0)$ for answering the same queries on X and $I \setminus I_0$. The structure $\mathcal{D}(X, I \setminus I_0)$ further consists of a collection of data structures $\mathcal{D}(X_i, I_j)$ on X_i and I_j, for every $i \ge 1$ and $j \ge 1$. We now describe each of these components.

2.1 Structure $\mathcal{D}(X, I \setminus I_0)$

The structure $\mathcal{D}(X, I \setminus I_0)$ consists of a collection of data structures $\mathcal{D}(X_i, I_j)$ for each $i, j \ge 1$. We first describe a general data structure of quadratic size and logarithmic query time for (Q1) and (Q2) queries, and then apply it to each $\mathcal{D}(X_i, I_j)$. In the subsequent discussions, we assume that the reader is familiar with interval trees [8].

Clearly, an interval $[a_i, b_i] \in I$ is stabbed by $x_j + \Delta x$ if and only if $\Delta x \in [a_i - x_j, b_i - x_j]$. Therefore, to answer (Q1), we build an interval tree on the set of intervals $I - X = \{[a_i - x_j, b_i - x_j] \mid [a_i, b_i] \in I, x_j \in X\}$. The size of the tree is bounded by $O(|I - X|) = O(mn)$, and the preprocessing time is $O(mn \log(mn))$. Given a query $\Delta x \in \mathbb{R}$, we find the set of intervals in $I - X$ stabbed by Δx. For each such interval $[a_i - x_j, b_i - x_j]$, we report the pair $(x_j, [a_i, b_i]) \in X \times I$. The query time is bounded by $O(\log(mn) + k)$, where k is the output size. The data structure can be made dynamic by using a dynamic interval tree [7]. Note that in our context, insertion or deletion of a point in X or an interval in I involves inserting or deleting n or m intervals, respectively, in the interval tree.

Similarly, to answer (Q2), observe that an interval $[a_i, b_i] \in I$ is stabbed by any point of $X + \Delta x$ if and only if $\Delta x \in \bigcup_{x_j \in X}[a_i - x_j, b_i - x_j]$. The set $\bigcup_{x_j \in X}[a_i - x_j, b_i - x_j]$ can be written as the union of a set I_i of at most m mutually *disjoint* intervals. We build an interval tree on intervals in $\bigcup_{i=1}^{n} I_i$. Given a query $\Delta x \in \mathbb{R}$, we find the set of intervals in $\bigcup_{i=1}^{n} I_i$ that are stabbed by Δx. For each such interval γ, if $\gamma \in I_i$, we report the interval $[a_i, b_i] \in I$. Note that each interval in I is reported at most once, and hence the query time is $O(\log(mn) + k)$, where k is the number of intervals reported. The data structure can also be made dynamic easily.

Lemma 1. *A data structure of size $O(mn)$ can be constructed in $O(mn \log(mn))$ time so that (Q1) and (Q2) can be answered in $O(\log(mn) + k)$ time, where k is the output size. Each (U1) operation takes $O(n \log(mn))$ time, and each (U2) operation takes $O(m \log(mn))$ time.*

We use Lemma 1 to preprocess each pair X_i and I_j $(i, j \geq 1)$ into a data structure $\mathcal{D}(X_i, I_j)$ of size $O(r^2)$ so that (Q1) and (Q2) on X_i and I_j can be answered in $O(\log r + k)$ time, where k is the output size. Given a displacement Δx, we can find all stabbing pairs in $(X + \Delta x) \times (I \setminus I_0)$ by scanning $X_1, X_2, \ldots$ and $I_1, I_2, \ldots$, and querying at most $O((m + n)/r)$ data structures $\mathcal{D}(X_i, I_j)$ as follows. For convenience, we denote the leftmost point in X_i by $l(X_i)$, and the rightmost point in X_i by $r(X_i)$. Similarly, we denote the leftmost endpoint in I_i by $l(I_i)$ and the rightmost endpoint in I_i by $r(I_i)$. Notice that $r(X_i) \leq l(X_{i+1})$ and $r(I_i) \leq l(I_{i+1})$. A (Q1) or (Q2) query on X and $I \setminus I_0$ can be answered in a way similar to the merge procedure of the merge-sort algorithm. Initially we let $Y = X_1$ and $J = I_1$. Suppose at a certain stage $Y = X_i$ and $J = I_j$. We first check whether $[l(Y) + \Delta x, r(Y) + \Delta x] \cap [l(J), r(J)] \neq \emptyset$. If so, there is a potential stabbing between a point in Y and an interval in J, and we use $\mathcal{D}(Y, J)$ to report all such stabbings; otherwise Y does not stab J, and we report nothing. Then, if $r(Y) + \Delta x > r(J)$, we let $J = I_{j+1}$; otherwise, we let $Y = X_{i+1}$. The procedure is repeated until all X_i and I_j are processed. Clearly, the above procedure probes at most $O((m + n)/r)$ data structures $\mathcal{D}(X_i, I_j)$, each requiring $O(\log r)$ time plus the time for reporting. Therefore the total time spent is $O(\frac{m+n}{r} \cdot \log r + k)$.

There is one problem: If we rely on Lemma 1 to construct $\mathcal{D}(X_i, I_j)$ for each $i, j \geq 1$, we need $O(r^2)$ space for each $\mathcal{D}(X_i, I_j)$, leading to an overall data structure of size $O(mn)$. To fix this problem, we store all $\mathcal{D}(X_i, I_j)$ compactly, using a method from Dumitrescu and Steiger [10]. The observation is that, for the data structure described in Lemma 1 on a point set Y and an interval set J, the combinatorial description of this structure is fully determined by the ordering of all the endpoints of $J - Y$. Using this combinatorial description of the structure and the values of J and Y, (Q1) and (Q2) can be answered as before. If the sizes of Y and J are at most r, then the number of possible different orderings of the endpoints of $J - Y$ is $r^{O(r)}$ [11]. For each possible ordering, we construct a complete combinatorial description of the data structure with respect to this ordering. For each pair of X_i and I_j, we determine the ordering of the endpoints of $I_j - X_i$, and map this pair to the combinatorial description of $\mathcal{D}(X_i, I_j)$ according to this ordering. Therefore, the total space needed is $O(mn/r^2 + r^{O(r)} \cdot r^2) = O(mn/r^2 + r^{O(r)})$.

Lemma 2. *The data structure $\mathcal{D}(X, I \setminus I_0)$ has size $O(mn/r^2 + r^{O(r)})$, and answers (Q1) or (Q2) queries on X and $I \setminus I_0$ in $O(\frac{m+n}{r} \cdot \log r + k)$ time.*

2.2 Structure $\mathcal{D}(X, I_0)$

Given a set I of n intervals, a *stabbing set* of I is a set $P \subseteq \mathbb{R}$ of points so that each interval in I is stabbed by at least one point of P. We denote by $\tau(I)$ the size of the smallest stabbing set of I. It is well known that a stabbing set of I of size $\tau(I)$ can be computed in $O(n \log n)$ time by the following greedy algorithm: first sort the intervals in I by their left endpoints; then examine each of them from left to right, and find a maximal number of leftmost intervals so that they have a nonempty common intersection, stab these intervals by an arbitrary point from their common intersection, and repeat this step for the remaining intervals.

We first present a general data structure for answering (Q1) and (Q2) queries on X and I whose size and query time depend on the parameter $\tau = \tau(I)$. Then $\mathcal{D}(X, I_0)$ is constructed by simply applying this data structure on X and I_0. We treat (Q1) and (Q2) separately; in particular, the data structure for (Q1) is simpler than that for (Q2).

(Q1) queries. The data structure relies on the following observation.

Lemma 3. *A data structure of size $O(n)$ can be built in $O(n \log n)$ time so that the stabbing query on a set I of n intervals (i.e., reporting the subset of intervals in I that are stabbed by a query point x) can be answered in $O(\log \tau + k)$ time, where k is the number of reported intervals.*

Proof. We first compute a stabbing set $P = \{p_1, p_2, \ldots, p_\tau\}$ of size τ using the greedy algorithm described above. In fact, this algorithm also returns $\mathcal{I} = \{I_i \mid 1 \le i \le \tau\}$, a partition of I into τ subsets, so that every interval in I_i is stabbed by p_i. Let $l_i = \min_{[a_j, b_j] \in I_i} a_j$ and $r_i = \max_{[a_j, b_j] \in I_i} b_j$; clearly we have $l_i \le p_i \le r_i$. Let $\mathcal{H} = \{[l_i, r_i] \mid 1 \le i \le \tau\}$ be the *covering interval set* of I with respect to P; intuitively, each *covering interval* $[l_i, r_i]$ in $\mathcal{H}$ is the union of all intervals in I_i. We construct an interval tree [8] $\mathcal{T}$ on the set $\mathcal{H}$. In addition, for each I_i, we store two copies of I_i: (i) I_i^l, which sorts the intervals in I_i in increasing order of their left endpoints, and (ii) I_i^r, which sorts the intervals in I_i in decreasing order of their right endpoints. The size of the data structure is $O(\tau + n) = O(n)$, and the preprocessing time is $O(n \log n)$.

Given a query point x, the stabbing query is answered as follows. With the interval tree $\mathcal{T}$, we find in $O(\log \tau + k')$ time the k' covering intervals in $\mathcal{H}$ that are stabbed by x. If a covering interval $[l_i, r_i] \in \mathcal{H}$ contains x, we compare p_i and x. If $x \leq p_i$, then we scan I_i^l and report each interval until we encounter an interval whose left endpoint lies to the right of x. Symmetrically, if $x > p_i$, then we scan I_i^r and report each interval until we encounter an interval whose right endpoint lies to the left of x. The correctness of this procedure is easy to verify. Note that if $[l_i, r_i]$ is stabbed by x, then at least one interval in I_i is stabbed by x, and therefore $k' \leq k$, where k is the number of reported intervals. Hence, the total query time can be bounded by $O(\log \tau + k)$. □

Returning to our original problem, note that a (Q1) query is equivalent to a stabbing query on $I - X$ with the query point Δx, as shown in Section 2.1. Thus naturally we want to apply Lemma 3. Observe that if $P = \{p_1, p_2, \ldots, p_\tau\}$ is a stabbing set for I, then $P - X = \{p_i - x_j \mid p_i \in P, x_j \in X\}$ is a stabbing set of size at most $m\tau$ for $I - X$, and if $\mathcal{I} = \{I_1, \ldots, I_\tau\}$ is a partition of I with respect to P, then $\mathcal{I} - X = \{I_i - x_j \mid 1 \leq i \leq \tau, 1 \leq j \leq m\}$ is a partition of $I - X$ with respect to $P - X$. If we simply apply Lemma 3 directly, we would get a structure of size $O(|I - X|) = O(mn)$ for (Q1) queries. Hence, we need to be more careful. Observe that for any $1 \leq i \leq \tau$ and $1 \leq j \leq m$, I_i and $I_i - x_j$ are congruent in the sense that the sequence $I_i - x_j$ is obtained from I_i by translating each interval by $-x_j$. Thus, it suffices to store I_i. In more detail, we proceed as follows.

Let $\mathcal{H}$ be the covering interval set of I with respect to P, as defined in the proof of Lemma 3. We preprocess the intervals in $\mathcal{H} - X$ for stabbing queries, using an interval tree, as in Lemma 3. In addition, we store the sequences I_i^l and I_i^r for $1 \leq i \leq \tau$. Let Δx be a query displacement. We first report all intervals in $\mathcal{H} - X$ stabbed by Δx. Suppose an interval $[l_i - x_j, r_i - x_j]$ is stabbed by Δx. We then wish to report all intervals in $I_i - x_j$ that contain Δx, which is equivalent to reporting the intervals in I_i stabbed by $x_j + \Delta x$. This reporting can be done by the procedure described in Lemma 3. The overall query time is $O(\log(m\tau) + k)$, where k is the output size. The data structure uses $O(m\tau + n)$ storage, and can be constructed in $O(n \log n + m\tau \log(m\tau))$ time.

(Q2) queries. Similar bounds for (Q2) can also be obtained. The major difference is that, we have to make sure each stabbed interval is reported only once. This is achieved by using the notion of *(i, j)-intervals* introduced below. Again, we first compute a stabbing set $P = \{p_1, p_2, \ldots, p_\tau\}$, the corresponding partition $\mathcal{I} = \{I_i \mid 1 \leq i \leq \tau\}$ of I, and the covering interval set $\mathcal{H} = \{[l_i, r_i] \mid 1 \leq i \leq \tau\}$. For simplicity, let us assume $-\infty = x_0 < x_1 < x_2 < \cdots < x_m < x_{m+1} = +\infty$.

Definition 1. *Given $1 \leq i \leq \tau$ and $0 \leq j \leq m$, an* (i, j)-interval *is a maximal interval $\gamma \subset [p_i - x_{j+1}, p_i - x_j)$ so that for any $\Delta x \in \gamma$, $(X + \Delta x) \cap [l_i, r_i] \neq \emptyset$.*

We remark that there may be more than one (but at most two) (i, j)-interval for a fixed i and j. Intuitively, each (i, j)-interval corresponds to a contiguous range of displacement values Δx for which some interval in I_i is stabbed by some point in $X+\Delta x$. Furthermore, as illustrated in Figure 2 (a), given a displacement value Δx in an (i,j)-interval, the stabbed intervals in I_i are precisely those stabbed by either $x_j+\Delta x$ or $x_{j+1}+\Delta x$. This is because p_i, the stabbing point of I_i, lies in between $x_j+\Delta x$ and $x_{j+1}+\Delta x$.

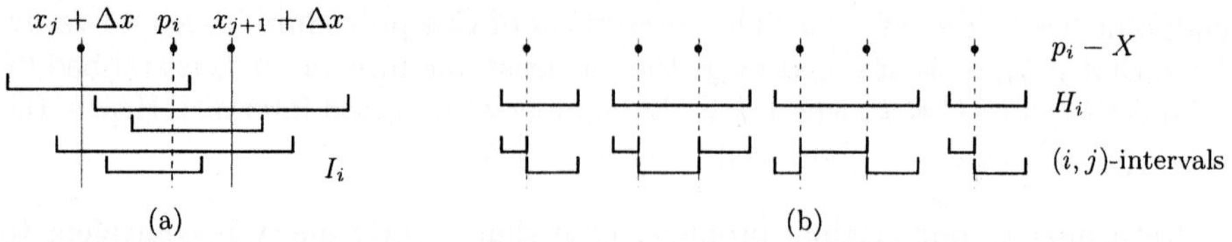

Fig. 2. (a) For any displacement Δx in an (i,j)-interval, an interval of I_i is stabbed by $X+\Delta x$ if and only if it is stabbed by either $x_j+\Delta x$ or $x_{j+1}+\Delta x$. (b) Computing the set of all (i,j)-intervals for a fixed i.

Lemma 4. *Let Π be the set of all (i,j)-intervals. Then $|\Pi| = O(m\tau)$, and Π can be computed in $O(m\log m + m\tau)$ time.*

Proof. For a fixed i, the set of all (i,j)-intervals, for $0 \le j \le m$, can be found as follows. In order for $X+\Delta x$ to stab $[l_i, r_i]$, Δx has to lie within the range $H_i = \bigcup_{x_j\in X}[l_i - x_j, r_i - x_j]$. Regard H_i as the union of a set of mutually disjoint intervals. We throw the m points from $p_i - X$ into H_i. These points further divide H_i into a set of atomic intervals, each of which is a maximal interval whose interior does not contain any point from $p_i - X$ (see Figure 2 (b)). It can be easily verified that each such atomic interval is an (i,j)-interval, if it lies between $p_i - x_{j+1}$ and $p_i - x_j$. Clearly, there are $O(m)$ such atomic intervals, all of which can be computed in $O(m)$ time, once X is sorted. Thus, the total time for computing all these intervals, for $i = 1,\cdots,\tau$, is $O(m\log m + m\tau)$. □

We preprocess Π into an interval tree, and store the sequence I_i^l and I_i^r for each $I_i \in \mathcal{I}$. By Lemma 4, the size of the structure is $O(m\tau + n)$. To answer (Q2) for a displacement Δx, we first report in $O(\log(m\tau) + k')$ time all k' intervals of Π that are stabbed by Δx. Suppose an (i,j)-interval of Π is reported. We scan the sequence I_i^l and report all of its intervals whose left endpoints lie to the left of $x_j+\Delta x$. We also report all intervals in the sequence I_i^r whose right endpoints lie to the right of $x_{j+1}+\Delta x$ but left endpoints lie to the right of $x_j+\Delta x$ (as otherwise they would have already been reported when we scan I_i^l); see Figure 2 (a).

Clearly, the overall query time is $O(\log(m\tau)+k'+k)$, where k' is the number of intervals in Π stabbed by Δx, and k is the output size. Note that for a fixed i, an (i,j_1)-interval is disjoint from an (i,j_2)-interval if $j_1 \ne j_2$. Therefore at most one (i,j)-interval is reported for any fixed i. Moreover, for each reported (i,j)-interval, at least one interval of I_i is stabbed by $X+\Delta x$. This implies that $k' \le k$. Hence the overall query time is in fact $O(\log(m\tau)+k)$.

We thus have the following lemma.

Lemma 5. *A data structure of size $O(m\tau + n)$ can be constructed in total time $O(m\tau \log(m\tau) + n \log n)$ so that (Q1) and (Q2) can be answered in $O(\log(m\tau) + k)$ time, where k is the output size.*

Note that $\tau(I_0) = O(n/r)$. Applying Lemma 5 on X and I_0, we thus obtain:

Lemma 6. *The data structure $\mathcal{D}(X, I_0)$ has size $O(mn/r + n)$, and answers (Q1) or (Q2) queries on X and I_0 in $O(\log(mn/r) + k)$ time.*

Remark. As shown by Lemma 5 above, the size and query time size and query time of this data structure depend on the parameter $\tau(I)$ of the input I. We call this an *input-sensitive* data structure. For small values of $\tau(I)$, i.e., when input intervals are highly clustered, this data structure almost achieves both linear size and logarithmic query time.

Recall that in (Q1) and (Q2) queries, the input intervals come from the continuous band-joins being indexed. These intervals naturally reflect users' interests on the data, and it is plausible to assume that in practice most of the users' interests would gather around a small number of "hotspots." In other words, in practice the intervals involved in continuous band-joins often admit a small stabbing set.[1] To exploit this pleasant property further, an alternative is to apply Lemma 5 to the entire set of intervals to be indexed (as opposed to just I_0). The result, described in detail in [1], is a dynamic practical structure whose performance increases with the degree of clusteredness of the input intervals. We also show in [1] how to dynamize that structure to support (U1) and (U2) operations, which is nontrivial because the smallest stabbing set of a set of intervals can completely change after every constant number of update operations.

2.3 Putting It Together

Overall, a query on X and I is answered by processing the query on X and I_0 as well as on X and $I \setminus I_0$ as described in the previous two sections. By Lemmas 2 and 6, the total size of the data structure is bounded by

$$S = O(mn/r + n + mn/r^2 + r^{O(r)}) = O(mn/r + r^{O(r)}),$$

and the query time is bounded by

$$T = O(\log(mn/r) + (m/r + n/r) \log r + k) = O((m/r + n/r) \log r + k).$$

Finally, we choose $r = c \log(mn)/\log\log(mn)$, where $c > 0$ is a sufficiently small constant, so that $S = O(mn \log\log(mn)/\log(mn)) = o(mn)$ and $T =$

[1] For example, in practice most band-join queries are only interested in the absolute difference of the data (i.e, query of the form $J(-a, a) = \{(r, s) \in R \times S \mid |r - s| \leq a\}$), or the one-sided difference of the data (i.e., query of the form $J(-\infty, a)$ or $J(a, +\infty)$). These query intervals can be all stabbed by one of the three points: 0, $-\infty$ and $+\infty$.

$O((m+n)\log^2\log(mn)/\log(mn)+k) = o(m+n)+O(k)$, as desired. We also note that the total preprocessing time can be bounded by

$$O(m\log m+n\log n+(mn/r)\log(mn/r)+mn\log r+r^{O(r)}) = O(mn\log\log(mn)).$$

Theorem 1. *A data structure of size $o(mn)$ can be constructed in total time $O(mn\log\log(mn))$ so that (Q1) and (Q2) can be answered in $o(m+n)+O(k)$ time, where k is the output size.*

Remark. (1) The data structure supports each (U1) or (U2) operation in near-linear time. We omit the details here. Note that although the near-linear bound appears horrible, we show in [1] that it is essentially the best possible.

(2) Although we have been careful in the construction, the size of the data structure is only slightly subquadratic, and the query time is only slightly sub-linear. We leave finding the best tradeoff between the space and query time for (Q1) and (Q2) queries as an interesting open problem.

References

1. P. K. Agarwal, J. Xie, J. Yang, and H. Yu. Monitoring continuous band-join queries over dynamic data. Technical report, Department of Computer Science, Duke University, Durham, North Carolina, USA, Sept. 2005. Available at `http://www.cs.duke.edu/dbgroup/papers/2005-axyy-joinidx.pdf`.
2. A. Bagchi, A. Chaudhary, D. Eppstein, and M. Goodrich. Deterministic sampling and range counting in geometry data streams. In *Proc. 20th ACM Sympos. Comput. Geom.*, pages 144–151, 2004.
3. D. Carney et al. Monitoring streams: A new class of data management applications. In *Proc. 28th Intl. Conf. on Very Large Data Bases*, pages 215–226, 2002.
4. S. Chandrasekaran and M. J. Franklin. Streaming queries over streaming data. In *Proc. 28th Intl. Conf. on Very Large Data Bases*, pages 203–214, 2002.
5. S. Chandrasekaran and M. J. Franklin. PSoup: a system for streaming queries over streaming data. *The VLDB Journal*, 12(2):140–156, 2003.
6. J. Chen, D. J. DeWitt, F. Tian, and Y. Wang. NiagraCQ: A scalable continuous query system for internet databases. In *Proc. 19th ACM SIGMOD Intl. Conf. on Management of Data*, pages 379–390, 2000.
7. Y.-J. Chiang and R. Tamassia. Dynamic algorithms in computational geometry. *Computational Geometry: Theory & Applications*, 80(9):1412–1434, 1992.
8. M. de Berg, M. van Kreveld, M. Overmars, and O. Schwarzkopf. *Computational Geometry: Algorithms and Applications.* Springer-Verlag, 2nd edition, 2000.
9. D. DeWitt, J. Naughton, and D. Schneider. An evaluation of nonequijoin algorithms. In *Proc. 17th Intl. Conf. on Very Large Data Bases*, pages 443–452, 1991.
10. A. Dumitrescu and W. Steiger. Space-time tradeoffs for some ranking and searching queries. *Inform. Process. Lett.*, 79(5):237–241, 2001.
11. M. Fredman. How good is the information theory bound on sorting? *Theoret. Comput. Sci.*, 1:355–361, 1976.
12. S. Guha, N. Koudas, and K. Shim. Data streams and histograms. In *Proc. 33rd ACM Sympos. Theory of Computing*, pages 471–475, 2001.
13. L. Liu, C. Pu, and W. Tang. Continual queries for internet scale event-driven information delivery. *IEEE Trans. on Knowledge and Data Engineering*, 11(4):610–628, 1999.

14. S. Muthukrishnan. Data streams: algorithms and applications. Available at `http://www.cs.rutgers.edu/~muthu`.
15. M. H. Overmars. *The Design of Dynamic Data Structures.* Lecture Notes in Computer Science. Springer-Verlag, NY, 1987.
16. S. Suri, C. Toth, and Y. Zhou. Range counting over multidimensional data streams. In *Proc. 20th ACM Sympos. Comput. Geom.*, pages 160–169, 2004.

Approximate Colored Range Queries*

Ying Kit Lai, Chung Keung Poon, and Benyun Shi

Dept. of Computer Science, City U. of Hong Kong, China

Abstract. In this paper, we formulate a class of colored range query problems to model the multi-dimensional range queries in the presence of categorical information. By applying appropriate sketching techniques on our framework, we obtained efficient data structures that provide approximate solutions to these problems. In addition, the framework can be employed to attack other related problems by finding the appropriate summary structures.

1 Introduction

Range query problems are concerned with the storage and maintenance of a set of multi-dimensional data points so that given any query region, certain information about the data points lying within the region can be answered efficiently. Typical information of interest includes the set of points, the number of them, the sum, maximum and minimum of their associated values, etc. The corresponding problems are referred to as the range report, count, sum, max and min queries respectively.

In many applications, the data points are associated with categorical attributes and we are interested in information about the categories of the points lying within a query region, instead of the individual points themselves. Such problems can be abstracted as *colored range query* problems in which points in the given dataset are labeled with colors and we ask for information about the colors of points within a query region. As pointed out by Agarwal et al. [1], colored range queries are highly prevalent in database queries. Applications of such queries can also be found in document retrieval [13] and in indexing multi-dimensional strings [9].

To give a concrete example and to motivate our definition of a whole class of colored range queries, consider a database of sales records, each of which consisting of the following attributes: time, branch location (x, y-coordinates), product ID and sales. Suppose we are interested in information about sales grouped by product. Then we can model the database as a set of points in a 3-dimensional space (defined by the time and branch attributes) where each point is associated with a color (the product ID) as well as a value (the sales).

A specific query to the database can be to ask for the number of different products sold during a period of time within a region of branch locations. In

* The work described in this paper was fully supported by a grant from the Research Grants Council of the Hong Kong Special Administrative Region, China [Project No. CityU 1198/03E].

X. Deng and D. Du (Eds.): ISAAC 2005, LNCS 3827, pp. 360–369, 2005.

other words, we want to compute the number of distinct colors for the points lying within the corresponding query region. This is called the *colored range counting* problem. Another possible query is to ask for the best-selling products, i.e., we group the individual points in the region by product ID, sum the sales of each group and then find those groups that contribute most for the total sales of all products. This is a query that has not been formulated and studied before.

1.1 Problem Formulation

Generalizing the above sample queries, we can say that a general class of queries is to aggregate the points by colors (where the aggregation can be a simple count of points or a sum of values associated with points having that color) and then extract certain information about the colors. To capture such queries, we formalize the following class of colored range query problems. We will assume that our dataset consists of n points lying over a d-dimensional space and each point is associated with a color as well as a numeric value. We also assume there are m distinct colors in the universe of colors where m is large. Otherwise, one can construct an ordinary data structure for each color and solve the colored range queries by querying the m structures one by one.

Conceptually, one can view the set of all possible colors as a *colored* dimension. After the aggregation by colors, we have a set of points (colors) along the colored dimension, each associated with a value. Now, consider the set $\mathcal{F}$ of possible functions on the colored dimension. One may be interested in reporting the set or number of colors. These we denote by report and count respectively. One can also ask for the total, maximum or minimum value associated with the colors. These are denoted as sum, max and min respectively. We denote by heavy the query that asks for the colors with values above a certain threshold. Thus, we define $\mathcal{F} = \{\mathsf{sum}, \mathsf{count}, \mathsf{max}, \mathsf{min}, \mathsf{report}, \mathsf{heavy}\}$. Then for any $f \in \mathcal{F}$, a *colored range-f query* on a query region $R_1 \times R_2 \times \cdots \times R_d$ (where each R_i is an interval in dimension i) is to apply f on the set of colors within the query region. Depending on whether the value associated with a color is: (1) a count of points having that color, or (2) a sum of values associated with points having that color, we called the corresponding query *unweighted* or *weighted* respectively.

Note that for $f = \mathsf{sum}$, the weighted and unweighted colored range-f queries are just ordinary range sum and range count queries (without colors) respectively. Nevertheless, this gives an alternative formulation of these ordinary range query problems. Also, both the weighted and unweighted colored range-report queries are equivalent to the colored range reporting problem. For $f = \mathsf{count}$, both the weighted and unweighted queries are equivalent to the colored range counting problem we mentioned earlier.

1.2 Previous Work

Except for the colored range reporting and counting problems, none of the problems we defined above have been investigated before. As one will see below, many of these problems are difficult and it is not clear if existing techniques for range query problems can yield efficient data structures for them.

The colored range reporting problem is probably the most studied among all the known colored range query problems and there are rather efficient solutions. To our knowledge, Janardan and Lopez [12] were the first to investigate the static case for 1- and 2-dimensional points. Gupta et al [11] devised a *chaining* technique that transforms a set of 1-dimensional colored points into a set of 2-dimensional points (without colors). Using this technique, they obtained dynamic structures for colored range reporting up to 2 dimensions and a static structure for 3 dimensions. The structures of [12,11] generally require $O(n \cdot polylog(n))$ space and $O(polylog(n) + k)$ query time (where k is the number of distinct colors in the result set) and $O(polylog(n))$ update time. Recently Agarwal et al. [1] studied the variant of the problem where the data points are lying on an integer grid. In terms of techniques, they adapted the intuition and ideas from [12,11] and extended them to work for higher dimensions. Motivated by a number of document retrieval problems, Muthukrishnan [13] studied yet another variant where the points are stored in a 1-dimensional array. Nanopoulos et al. [14] studied the problem in the context of large, disk resident data sets. They proposed a multi-tree index to solve the problem but did not provide any analytical bound for their method.

For the colored range counting problem, much less is known. In the 1-dimensional case, Gupta et al. [11] obtained several efficient static and dynamic data structures with $O(n \cdot polylog(n))$ space and $O(polylog(n))$ query/update time. Using their chaining technique (mentioned above) and the best-known results on 2-dimensional range counting [15,6], one can actually improve their bounds, see Table 1. For 2 dimensions, their approach is to first obtain a persistent 1-dimensional data structure and then apply a standard line-sweeping approach. This results in static 2-dimensional structures with high (quadratic and cubic) storage requirement. Moreover, the approach does not readily yield a dynamic 2-dimensional structure.

There are other variants of colored queries which are outside the scope of this paper, . This includes the various colored intersection problems studied in [12,11] and the *colored significant-presence* queries proposed in [4]. The colored significant-presence queries is to report only those colors with at least a certain fraction of their points fallen into the query range. It looks similar but is in fact different from our colored range-heavy queries.

Table 1. Summary of results on colored range counting

Problem	Space	Query Time	Update Time	Reference
exact dynamic 1-d	$n \log n$	$\log^2 n$	$\log^2 n$	Gupta et al. [11]
	$n \log n$	$\log n$	$\log n$	Gupta et al. [11] + Willard [15]
	n	$\log^2 n$	$\log^2 n$	Gupta et al. [11] + Chazelle [6]
approx. dynamic 1-d	n	$\log^2 n$	$\log^2 n$	this paper
exact **static** 2-d	$n^2 \log^2 n$	$\log n$		Gupta et al. [11] + Willard [15]
approx. dynamic 2-d	$n \log n$	$\log^3 n$	$\log^3 n$	this paper

1.3 Our Contribution

In this paper, we propose a general methodology for constructing randomized data structures that can produce approximate answers. Our approach significantly differs from previous techniques for colored range queries in that sketching techniques are employed. For most of the problems, our method yields efficient solutions. In particular, our colored range counting structure has much smaller space than previous structures for 2-dimensions or above, as shown in Table 1.

We now illustrate our methodology using 1-dimensional range queries. A common technique for (1-dimensional) range query is to build a balanced binary search tree T where each leaf represents a data point and each internal node represents a group of data points, i.e., those represented by the leaves in the subtree of that internal node. Denote by $S(u)$ the set of points represented by node u. It is well-known that given any query range (an interval), one can always represent the set of points within the query range as a disjoint union of $S(u)$'s for at most $O(\log n)$ u's, with no more than two such u's in each level of the tree T. This union of sets can be obtained by traversing T with the left and right boundaries of the query interval in $O(\log n)$ time. Thus, if we store with each node u certain summary information about $S(u)$, a range query may be solvable in $O(\log n)$ time. For example, for range sum queries, the relevant information about $S(u)$ would be the sum of values among all points in $S(u)$.

Unfortunately, the success of this approach very much depends on the additivity of the summary information. Consider the colored range counting problem in which we are to find the number of distinct colors in a query range. Suppose we store the number of distinct colors in $S(u)$ for each tree node u. However, the number of distinct colors in the query range cannot be computed by simply adding the count for each $S(u)$ for the $O(\log n)$ u's that partitions the query region. Similarly, problems such as colored range minimum cannot be solved by associating with each node u in T the minimum aggregate value among the colors of points in $S(u)$.

To overcome the problem, our approach is to store an appropriate sketch with each node u to summarize the relevant information about $S(u)$. Specifically, the sketch should possess the following properties: (1) the size of a sketch is small, (2) the sketches are additive, and (3) an approximate answer can be obtained from the sketch. Property (1) ensures that the time and space of the data structure are within good bounds. Property (2) allows us to generate the required sketch for any query region on-the-fly (i.e., during the query time). By the structure of T, we only need to add $O(\log n)$ sketches to form the desired sketch. Finally, property (3) is essential if the sketch is to be useful for our problem at all.

Using this approach, we obtain a data structure for the 1-dimensional colored range counting problem with $O(hn)$ space and $O(h \log n)$ query and update times where $h = O(\frac{1}{\epsilon^2} \log n)$ is the size of an appropriate sketch. The data structure is randomized in the sense that with probability $1 - n^{-\Omega(1)}$, it produces an approximate answer accurate to within ϵ factor of error for *all* possible queries. We can further reduce the storage to $O(n)$ without increasing the asymptotic complexities of queries and updates (treating ϵ as a fixed constant). Applying standard techniques [3], we obtain multi-dimensional colored range query structures with a blow-up factor of $\log n$ in storage and operation time per dimension. More specifically, we utilize the additivity of the appropriate sketches to obtain a sketch that summaries the d-dimensional query region and then obtain the desired information from the sketch. In general, our data structures require $O(dn \log^{d-1} n)$ space and support queries and updates in $O(d \log^{d+1} n)$ time.

Using the same methodology but with different sketches, we obtain data structures with similar performance bounds for many other colored range query problems we defined here.

2 Preliminaries

2.1 Notations and Sketches

We will represent information about colors using vectors. Thus, we will be considering vectors and their norms. For a vector $\boldsymbol{a} = (a_1, a_2, \ldots, a_m)$, we denote by $||\boldsymbol{a}||_p$ its l_p-norm, i.e., $||\boldsymbol{a}||_p = (\sum_{i=1}^{m} |a_i|^p)^{1/p}$. When $p = 0$, $||\boldsymbol{a}||_0$ is defined as the number of non-zero components in $\boldsymbol{a}$.

The sketch of a vector $\boldsymbol{a}$ is the projection of $\boldsymbol{a}$ onto a set of random basis vectors. The purpose of a sketch is to estimate certain quantities about $\boldsymbol{a}$ to within certain error with high probability. The number, h, of required random basis vectors, and hence the size of the sketch, depends on the two parameters, ϵ and δ, that controls the error and failure probability respectively; and may also depend on m. Usually, we have $h \ll m$ so that storage reduction is achieved. Another important property of sketches is that additivity holds for many types of sketches, i.e., the sketch of $\boldsymbol{a} \pm \boldsymbol{b}$ can be computed by adding/subtracting the sketch of $\boldsymbol{a}$ with that of $\boldsymbol{b}$. Below we describe two sketches that we will employ, namely the *Count-Min* or *CM sketch* and the *l_0-sketch.* They both possess properties (1) and (2) mentioned in section 1.3.

2.2 Count-Min Sketch

When the components of a vector $\boldsymbol{a}$ are all non-negative, its *CM sketch* ([8]), denoted $CM(\boldsymbol{a})$, allows us to estimate any component, a_i, of $\boldsymbol{a}$ by specifying its index i. We call this a *point estimate* on $\boldsymbol{a}$. A certain collection of CM sketches, denoted $CCM(\boldsymbol{a})$, allows us to report the set of indices i such that a_i is larger than a certain threshold fraction ϕ of $||\boldsymbol{a}||_1$. We call this *heavy hitters estimate.*

Point Estimates. We first describe the construction of $CM(\boldsymbol{a})$ and the algorithm for point estimate. The sketch is essentially a two-dimensional array of counts with width $w = \lceil \frac{e}{\epsilon} \rceil$ and depth $d = \lceil \ln \frac{1}{\delta} \rceil$, given the sketch parameters of ϵ and δ. The size of the sketch is thus $h = wd = O(\frac{1}{\epsilon} \log \frac{1}{\delta})$. We represent the counts in the sketch by $count(1,1), ..., count(d,w)$.

Given $\boldsymbol{a}$, we choose d hash functions $h_1, \ldots, h_d$ ($h_i : \{1, \ldots, m\} \rightarrow \{1, \ldots, w\}$) uniformly at random from a family of pairwise independent hash functions ([5]). We initialize all the counts in the sketch to 0. Then for each $j \in \{1, \ldots, d\}$, we hash the components of $\boldsymbol{a}$ into different cells of row j according to h_j. That is, for each $i \in \{1, \ldots, m\}$, the value a_i is added to the cell $count(j, h_j(i))$. Besides the sketch, we also store d hash functions which takes negligible storage. Any component a_i of $\boldsymbol{a}$ is estimated as $\hat{a}_i = \min_{j=1}^{d}\{count(j, h_j(i))\}$.

Theorem 1. *[8] For any vector* $\boldsymbol{a} = (a_1, \ldots, a_m)$ *and any pair of parameters* $\epsilon, \delta > 0$, *the CM sketch of* $\boldsymbol{a}$, $CM(\boldsymbol{a})$, *has size* $h = O(\frac{1}{\epsilon}\log\frac{1}{\delta})$ *and supports point estimates in* $g = O(\log\frac{1}{\delta})$ *time. The estimate* $\hat{a}_i$ *of* a_i *satisfies: (1)* $a_i \leq \hat{a}_i$ *and (2) with probability* $1 - \delta$, $\hat{a}_i \leq a_i + \epsilon||\boldsymbol{a}||_1$. *Further, the sketch can be updated (for changes in a component of* $\boldsymbol{a}$*) in* $O(g)$ *and constructed in* $O(gm + h)$ *time.*

Heavy Hitters Estimates. We first explain the notion of *dyadic intervals*. For convenience, we assume m is a power of 2. A *dyadic interval* $I_{j,k}$ on the universe $\{1, \ldots, m\}$ is an interval of the form $[k2^j, (k+1)2^j - 1]$, for $j \in \{1, \ldots, \log m\}$ and $k \in \{0, \ldots, m/2^j - 1\}$. The parameter j of a dyadic interval is its *resolution level* from the *finest*: $I_{0,i} = \{i\}$ where $i \in \{1, \ldots, m\}$ to the *coarsest* $I_{\log m, 0} = \{1, \ldots, m\}$. There are $\log m$ resolution levels and $2m - 1$ dyadic intervals altogether, organized in a binary tree-like structure.

To support heavy hitter estimates, the collection $CCM(\boldsymbol{a})$ consists of $\log m$ CM sketches, each of size $dw = \log(\frac{\log m}{\delta\phi}) \times \frac{e}{\epsilon} = O(\frac{1}{\epsilon}\log(\frac{\log m}{\delta\phi}))$. The first sketch is just $CM(\boldsymbol{a})$. The second sketch is the CM sketch on a vector with $m/2$ components, each being the sum of a_i's over all i in a dyadic interval $I_{1,k}$ where $0 \leq k \leq m/2 - 1$. The third up to the $\log m$-th sketch are CM sketches of $\boldsymbol{a}$ at a progressively coarser and coarser resolution.

To find the heavy hitters, we start from the sketch of the coarsest resolution level and check for the dyadic interval(s) with (estimated) weights exceeding the threshold $(\phi + \epsilon)||\boldsymbol{a}||_1$. We then explore their children dyadic intervals at the next finer level of resolution. Repeating this until we arrive at dyadic intervals of length 1, we can locate all the components a_i such that the estimated a_i exceeds $(\phi+\epsilon)||\boldsymbol{a}||_1$. Note that in each sketch, at most $2/\phi$ dyadic intervals are examined. As for updates, since each point in the universe $\{1, \ldots, m\}$ is a member of $\log m$ sketches, each of the sketches are updated when an update on a point arrives.

Theorem 2. *[8] For any* $\boldsymbol{a} = (a_1, \ldots, a_m)$ *and* $\epsilon, \delta, \phi > 0$, *the collection of sketches* $CCM(\boldsymbol{a})$ *has size* $h = O(\frac{1}{\epsilon}\log m \log(\frac{\log m}{\delta\phi}))$. *In* $O(\epsilon h/\phi)$ *time, it can report all the components with weight at least* $(\phi + \epsilon)||\boldsymbol{a}||_1$, *and with probability* $1 - \delta$, *it reports no component with weight less than* $\phi||\boldsymbol{a}||_1$. *The sketch can be updated in* $g = O(\log m \log(\frac{\log m}{\delta\phi}))$ *time and constructed in* $O(gm + h)$ *time.*

2.3 l_0-Sketch

When the magnitude of each component of $\boldsymbol{a}$ is bounded from above by some value U, we can compute its l_0-sketch ([7]), denoted $L_0(\boldsymbol{a})$, which allows us to approximate, $||a||_0$, the number of non-zero entries in $\boldsymbol{a}$.

Theorem 3. *[7] For any* $\boldsymbol{a} = (a_1, \ldots, a_m)$ *such that* $|a_i| \leq U$ *for every* i, *and any* $\epsilon, \delta > 0$, *the* l_0*-sktech of* $\boldsymbol{a}$, $L_0(\boldsymbol{a})$, *has size* $h = O(\frac{1}{\epsilon^2}\log(\frac{1}{\delta}))$. *Using the sketch, the* l_0*-norm,* $||\boldsymbol{a}||_0$, *can be approximated to within a factor of* $1 \pm \epsilon$ *with* *probability* $1 - \delta$ *in* $O(h)$ *time. Further, it can be updated in* $O(h)$ *time and constructed in* $O(hm)$ *time.*

3 Colored Range Counting

Construction and Storage. To solve for the 1-dimensional case, we construct a balanced binary search tree T to store the points. For each node u in T, we apply the l_0-sketch to summarize the at most m distinct colors present in $S(u)$. More precisely, we define $\boldsymbol{a} = (a_1, a_2, \ldots, a_m)$ to be the color frequency vector so that a_i represents the number of occurrences of color i in a region. In the region, we would like to know the number of non-zero components of the vector $\boldsymbol{a}$, or formally the l_0-norm of $\boldsymbol{a}$. In general m can be very large and hence storing an $\boldsymbol{a}$ explicitly for each node u will be very space-consuming. Therefore, we will store the l_0-sketch of $\boldsymbol{a}$ so that we only need $h = O(\frac{1}{\epsilon^2} \log(\frac{1}{\delta}))$ space per sketch. (The value of ϵ and δ will be determined later.) Hence in total we need $O(hn)$ space.

The l_0-sketch of a leaf u (representing one data point) can be constructed in $O(h)$ time. By additivity, the l_0-sketch of an internal node v can be computed by adding that of its two children. This takes again $O(h)$ time. Hence, the construction of the whole data structure requires $O(hn)$ time.

Querying. To handle a query $[x, y]$, we search T for x and y in order to identify the set V of internal nodes the union of whose leaves is the set of points lying within $[x, y]$. Note that $|V| = O(\log n)$. By additivity, the l_0-sketch of any query region can be computed by adding the corresponding $O(\log n)$ many l_0-sketches. Thus, a query takes $O(h \log n)$ time. We ignore the query time for the resulting sketch here as the time for the additions clearly dominates.

Updating. When inserting or deleting a point x without changing the structure of T, the update can be done straightforwardly. Suppose the point has color i and value μ. Then for each node v along the path from the root to x, its sketch has to be updated. Let $\boldsymbol{a}$ be the color frequency vector associated with v. Then inserting (deleting) the point will cause a_i to increase (decrease) by μ. Let $\boldsymbol{b} = (b_1, b_2, \ldots, b_m)$ such that $b_i = \mu$ and $b_j = 0$ for all $j \neq i$. Then $L_0(\boldsymbol{a})$ should be updated to $L_0(\boldsymbol{a}) \pm L_0(\boldsymbol{b})$ by additivity and this takes $O(h)$ time. The total update time along the path is $O(h \log n)$.

When the update causes the structure of T to change, we need to modify the sketches of all the affected nodes. For each such node, its sketch can be re-computed by adding the sketches of all its children (in the new tree). For example, if we use a red-black tree [2,10] for T, an update requires at most 3 rotations and propagating the change from the points of rotation to the root by adding sketches takes $O(h \log n)$ time in the worst case. (For higher dimensions, the update time can be made worst case using standard techniques [16].)

Storage Reduction. We can reduce the space requirement of the data structure to be independent of ϵ and δ. The idea is to store the sketch only for nodes v in T with sufficient number of leaves. More precisely, let F be the set of internal nodes $v \in T$ such that v has fewer than $O(h)$ leaves while parent of v has at least $O(h)$ leaves. We call the nodes in F the *frontier nodes*. Clearly, there are $O(n/h)$

of them. We will only store the l_0-sketch for nodes above the frontier F. Then the storage requirement is reduced to $O(n)$. When querying for a range $[x, y]$, we search for x and y in T as before. When the search reaches a node v in F, we construct the l_0-sketch for the portion of the subtree rooted at v that lies within the query region on the fly. This is done by traversing the subtree of T rooted at v to construct the color frequency vector $\boldsymbol{a}$ for only the points lying within the query region. This takes $O(h)$ time by definition of F. Then we construct $L_0(\boldsymbol{a})$ using $O(h^2)$ time. The total query time now becomes $O(\log n + h^2)$.

Choosing δ. Now we consider the value of δ which controls the failure probability. We will generate all the $O(n)$ sketches using the same set of basis vectors. Moreover, for each query region, we will also generate the corresponding sketch on the fly. In total, there are at most $O(n^2)$ query regions. We want to have a random basis such that with high probability (i.e., $1 - \delta'$ where $\delta' = n^{-\Omega(1)})$), all the $O(n^2)$ sketches give good enough approximations. Clearly, $\delta' = O(n^2\delta)$. Thus, we set $\delta = n^{-\Omega(1)}$ and so $h = O(\log n)$ for constant ϵ. In general for d dimensions, number of query regions become $O(n^{2d})$ and thus $h = O(d \log n)$.

Since the size of sketches now depends on n, we may have to change the size of the sketches as n changes. However, it can be shown that by buidling the structure with sketch size h doubled and rebuilding the whole structure when n becomes too large, one can achieve an amortized update bound of $O(\log^2 n)$.

4 Colored Range-Min Queries

Construction and Storage. For the 1-dimensional case, we consider the base tree T. For each node u in T, we will store the CM sketch of the color vector $\boldsymbol{a} = (a_1, a_2, \ldots, a_m)$ associated with u. Each a_i represents the sum or count of the individual point values with color i under the subtree $T(u)$.

However, instead of estimating individual component, we use $CM(\boldsymbol{a})$ to estimate the minimum component of $\boldsymbol{a}$. We do this by finding the minimum value in each row of $CM(\boldsymbol{a})$ and then taking the minimum among the $\lceil \ln \frac{1}{\delta} \rceil$ rows. In other words, we find the minimum value within the whole sketch $CM(\boldsymbol{a})$.

There is a small technical issue. It is possible that some cells in $CM(\boldsymbol{a})$ are not hashed into by any index $i \in \{1, \ldots, m\}$. Then our method will return a zero value (the initial sketch cell value). To identify and exclude those cells, we slightly modify the sketch construction. First, we initialize the whole table to some negative values. Then, when hashing the components of $\boldsymbol{a}$ into the table, the first component hashed into a cell will overwrite the initial negative value while subsequent components will add to the existing value. When computing the minimum cell value, we only consider those non-negative cells. The following lemmas prove that the estimated minimum is accurate with high probability.

Lemma 1. *[8] For any* $i \in \{1, \ldots, m\}$, $j \in \{1, \ldots, \lceil \ln \frac{1}{\delta} \rceil\}$, $E[count(j, h_j(i)) - a_i] \leq \frac{\epsilon}{e} ||a||_1$.

Lemma 2. *Let* $a_t = \min_{i=1}^{m}\{a_i\}$. *For all* j, *define* $m_j = \min\{count(j, i') | count(j, i') \geq 0\}$. *Then* $E[m_j - a_t] \leq \frac{\epsilon}{e} ||\boldsymbol{a}||_1$.

Lemma 3. *Let* $a_t = \min_{i=1}^{m}\{a_i\}$. *Define* $cmm = \min_j\{m_j\}$. *Then* $cmm \geq a_t$ *and* $cmm \leq a_t + \epsilon||\boldsymbol{a}||_1$ *with probability* $1-\delta$.

We need $h = O(\frac{1}{\epsilon}\log\frac{1}{\delta})$ space to store a CM sketch and thus for all node u of tree T, we need $O(hn)$ space where n is the total number of points. The CM sketch of a leaf can be constructed in $O(h)$ time. By additivity, the sketch of an internal node can be constructed by adding those of its two children using $O(h)$ time. The construction of the whole data structure thus requires $O(hn)$ time.

Querying. Like the query in colored range counting, we first identify the set V of $O(\log n)$ internal nodes corresponding to the query range. Then, by the additivity, we compute the sketch of the query range by adding the corresponding $O(\log n)$ sketches. So, it takes $O(h \log n)$ time for a query.

Updating. In the case where structure of T remains unchanged, the update procedure is just similar to Section 3 and the only difference is the update time. Note that we need only $O(g) = O(\log\frac{1}{\delta})$ time to update a node in T with a single value. So the total time we need for updating will be $O(g \log n)$. In case where rotations occur due to changes in the structure of T (implemented as a red-black tree), the time required to update the sketch of an affected node by adding those of its children is $O(h)$. Since the number of affected nodes is at most $O(\log n)$, an update requires $O(h \log n)$ time in the worst case.

Storage Reduction and choosing δ. If we just store the CM sketches for nodes above the *frontier* F as defined in Section 3, $h = O(\frac{1}{\epsilon}\log\frac{1}{\delta})$ in this case, we can reduce the storage requirement to $O(n)$ while the query time increases to $O(h\log n + hg)$ and update time remains to be $O(h \log n)$. Choosing $\delta = n^{-\Omega(1)}$, the asymptotic complexity of query and update times are both $O(\log^2 n)$.

5 Heavy Colors

Construction and Storage. Our main result in this section is the solution to the approximate colored range-heavy problem, i.e. $f = \mathsf{heavy}$. We will focus on the unweighted case. (The weighted case is very similar.) Again, we associate with a node u in T the vector $\boldsymbol{a} = (a_1, \ldots, a_m)$ such that a_i is the number of occurrences of color i in the region (a_i is the sum of values with color i when weighted). We urther assume all components of $\boldsymbol{a}$ to be non-negative. Our problem then is to report those colors with $a_i \geq \phi||\boldsymbol{a}||_1$, where $\phi \in [0,1]$. We store for each node u the collection of sketches $CCM(\boldsymbol{a})$ for the vector $\boldsymbol{a}$. Each of the sketches represents a level of dyadic intervals (as described in Section 2.2). In this setting, we require only $h = O(\frac{1}{\epsilon}\log(m)\log(\frac{\log m}{\delta\phi}))$ space per node. Thus, the total space needed is $O(hn)$ for the whole tree T.

Using similar idea as in the previous section, the construction time of sketches for the whole tree takes $O(hn)$ time. Querying takes $O(h\log n + \epsilon h/\phi) = O(h \log n)$ time according to Theorem 2 and since ϕ is constant. Updating takes $O(h \log n)$ time. Furthermore, the storage can be reduced to $O(n)$ while the query time is increased to $O(h\log n + h^2)$ and update time remains to be $O(h \log n)$.

6 Conclusion

In this paper, we have formulated a new class of colored range problems and proposed a general approach of using sketches to solve many of them approximately. We believe that there are two benefits in adopting this approach. First, advancements in the research on sketches and summary structures with the required properties immediately improve the solutions for the corresponding colored range query problems. Second, by finding the appropriate summary structures, the method can be employed to solve other unsolved problems. Also, our approach can be applied to other indexing structures, e.g. R-tree and its variants.

References

1. P.K. Agarwal, S. Govindarajan, and S. Muthukrishnan. Range searching in categorical data: colored range searching on grid. In *ESA'02*, pages 17–28, 2002.
2. R. Bayer. Symmetric binary B-trees: data structures and maintenance algorithms. *Acta Informatica*, 1:290–306, 1972.
3. J.L. Bentley. Multidimensional divide-and-conquer. *Comm. ACM*, 23(4):214–228, April 1980.
4. M. Berg and H.J. Haverkort. Significant-presence range queries in categorical data. In *WADS'03*, *LNCS 2748*, pages 462–473, 2003.
5. J. L. Carter and M. N. Wegman. Universal classes of hash functions. *J. Computer and System Sciences*, 18:143–154, 1979.
6. B. Chazelle. A functional approach to data structures and its use in multidimensional searching. *SIAM J. Computing*, 17(3):427–462, June 1988.
7. G. Cormode, M. Datar, P. Indyk, and S. Muthukrishnan. Comparing data streams using Hamming norms (how to zero in). In *Proc. of 28th Int'l Conf. on Very Large Data Bases*, pages 335–345, 2002.
8. G. Cormode and S. Muthukrishnan. An improved data stream summary: the count-min sketch and its applications. In *6th Latin American Theoretical Informatics (LATIN)*, *LNCS 2976*, pages 29–38, 2004.
9. P. Ferragina, N. Koudas, D. Srivastava, and S. Muthukrishnan. Two-dimensional substring indexing. In *Proc. 20th ACM Symp. on Principles of Database Systems*, pages 282–288, 2001.
10. L.J. Guibas and R. Sedgewick. A dichromatic framework for balanced trees. In *FOCS'78*, pages 8–21, 1978.
11. P. Gupta, R. Janardan, and M. Smid. Further results on generalized intersection searching problems: counting, reporting, and dynamization. *J. Algorithms*, 19:282–317, 1995.
12. R. Janardan and M. Lopez. Generalized intersection searching problems. *Int'l J. Computational Geometry and Applications*, 3(1):39–69, 1993.
13. S. Muthukrishnan. Efficient algorithms for document retrieval problems. In *SODA'02*, pages 657–666, 2002.
14. A. Nanopoulos and P. Bozanis. Categorical range queries in large databases. In *SSTD'03*, *LNCS 2750*, pages 122–139, 2003.
15. D.E. Willard. New data structures for orthogonal queries. *SIAM J. Computing*, 14(1):232–253, February 1985.
16. D.E. Willard and G.S. Lueker. Adding range restriction capability to dynamic data structures. *J. ACM*, 32(3):597–617, July 1985.

Complexity and Approximation of the Minimum Recombination Haplotype Configuration Problem

Lan Liu[1], Xi Chen[3], Jing Xiao[3], and Tao Jiang[1,2]

[1] Department of Computer Science and Engineering, University of California, Riverside, CA 92521, USA
[2] Center for Advanced Study, Tsinghua University, Beijing, China
{lliu, jiang}@cs.ucr.edu
[3] Department of Computer Science and Technology, Tsinghua University, Beijing 100084, P.R. China
{xichen00, xiaojing00}@mails.tsinghua.edu.cn

Abstract. We study the complexity and approximation of the problem of reconstructing haplotypes from genotypes on pedigrees under the Mendelian Law of Inheritance and the minimum recombinant principle (MRHC). First, we show that MRHC for simple pedigrees where each member has at most one mate and at most one child (*i.e.* binary-tree pedigrees) is NP-hard. Second, we present some approximation results for the MRHC problem, which are the first approximation results in the literature to the best of our knowledge. We prove that MRHC on two-locus pedigrees or binary-tree pedigrees with missing data cannot be approximated (the formal definition is given in section 1.2) unless P=NP. Next we show that MRHC on two-locus pedigrees without missing data cannot be approximated within any constant ratio under the Unique Games Conjecture and can be approximated within ratio $O(\sqrt{\log(n)})$. Our L-reduction for the approximation hardness gives a simple alternative proof that MRHC on two-locus pedigrees is NP-hard, which is much easier to understand than the original proof. We also show that MRHC for tree pedigrees without missing data cannot be approximated within any constant ratio under the Unique Games Conjecture, too. Finally, we explore the hardness and approximation of MRHC on pedigrees where each member has a bounded number of children and mates mirroring real pedigrees.

Keywords: Haplotyping, pedigree, recombinant, SNP, complexity, approximation, L-reduction, positive result, negative result, bounded number, children, mates.

1 Introduction and Definitions

The secret mechanism behind phenotypic variation and inheritance has intrigued the study of genetic markers. With the discovery of genetic markers such as microsatellite DNA sequences and Single Nucleotide Polymorphisms (SNPs), it is now possible to provide a unique genetic map to track the variation and inheritance of genetic markers. The international HapMap project launched in October 2002, aims to discover the haplotype structure of human beings and examine the common haplotypes among populations [17].

X. Deng and D. Du (Eds.): ISAAC 2005, LNCS 3827, pp. 370–379, 2005.

Homologous recombination, the combination of genetic material between chromosome pairs during meiosis, is essential in diploid organisms such as humans [7]. Unfortunately, the diploid structure of humans makes it very expensive to collect haplotype data directly to display the recombination events. In a large-scale sequencing project, genotype data instead of haplotype data are collected. However, haplotype data are required in many genetic marker applications, such as linkage disequilibrium analysis and disease association mapping to name a few [12,13]. Therefore, combinatorial algorithms and statistical methods to reconstruct haplotypes from genotypes (*i.e.* the haplotype phasing or inference problem) are urgently needed.

The input data for this problem can be SNP fragments from an individual, genotype data in a population or genotype data in a family [8,9,10,11,15]. There are many combinatorial [1,2,14,16] and statistical ways [11,19] of tackling the phasing problem. They are usually quite computationally demanding.

Some of the commonly used combinatorial methods [1,2,14,16] take advantage of the availability of pedigree data. In other words, given a pedigree and the genotype information, they reconstruct a haplotype configuration for each individual in the pedigree by trying to solve the Minimum Recombinant Haplotype Configuration (MRHC) problem [1]. During the process of reconstruction, the minimum recombinant criterion is used as the objective function. Because this objective attempts to reduce the number of candidate haplotype configurations, it naturally preserves common haplotype structures.

All the existing methods to the MRHC problem are time and space consuming for realistic applications. For example, a Pentium IV computer with 256MB RAM is used to solve MRHC on an input pedigree with 29 members and 51 SNP markers. An effective combinatorial algorithm ILP takes about 5 hours to find an exact solution, whereas a well-known statistical approach SimWalk2 takes even more than 6 days to find a haplotype configuration with the maximum likelihood [21]. While over 5 millions of SNPs have been identified in the public database dbSNP [17], there is a great need for efficient algorithms that could scale up to the whole genome level. This difficulty motivates us to analyze the hardness and approximability of MRHC problems from a theoretical point of view.

1.1 Formal Definition of the MRHC Problem

In this subsection, we give a formal definition of the MRHC problem as well as the issue of pedigree representation and biological background. We follow the conventions in [1].

Definition 1. *A pedigree graph is a connected directed acyclic graph (DAG) $G=\{V, E\}$, where $V= M \cup F \cup N$, M represents the male nodes, F represents the female nodes, N represents the matting nodes, and $E= \{ (u, v): u \in M \cup F$ and $v \in N$ or $u \in N$ and $v \in M \cup F\}$. $M \cup F$ is called individual nodes. The in-degree of each individual node is at most one. The in-degree of a mating node must be two, with one edge starting from a male (called the father) node and the other edge from a female node (called the mother) and the out-degree of a mating node must be larger than zero.*

In a pedigree, the individual nodes outgoing from a mating node are called the *children*. The individual nodes with zero in-degree are called the *founders*. The induced subgraph by the father, the mother and one child adjacent to the same mating node is called a *family trio*. If there are two node-disjoint paths between two mating nodes in the pedigree graph, this pedigree has a *mating loop*. A pedigree without mating loops is called a *tree pedigree*. A pedigree where each member has at most one mate and at most one child looks like a binary tree, so this kind of pedigree is called a *binary-tree pedigree*. Fig. 1 demonstrates an example pedigree drawn in both the formal and conventional ways. In the conventional way, the mating nodes are omitted. For convenience, we use conventional drawings of pedigrees throughout this paper.

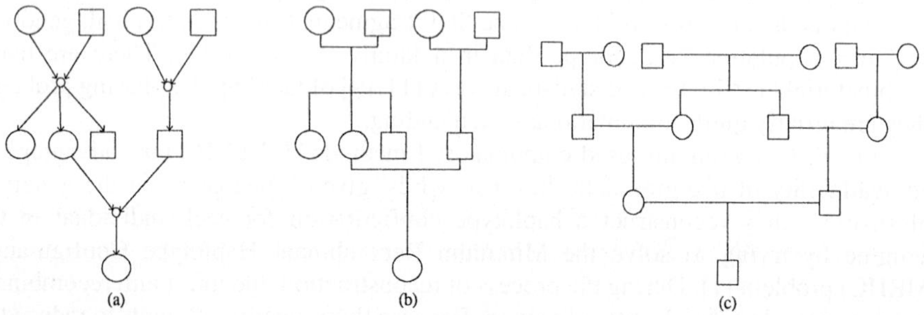

Fig. 1. (a) A pedigree drawn in the formal way. (b) The pedigree drawn in the conventional way. (c) A pedigree with a mating loop.

A *genetic marker* is a short non-redundant discriminative DNA sequence that can be used to trace inheritance. Some common genetic markers are microsatellite DNA sequences or SNP data. Each polymorphism state of a genetic marker is called an *allele*. Different kinds of markers have different numbers of alleles. For instance, a microsatellite marker has multiple possible alleles occurring at a locus, which is called *multi-allelic*. An SNP marker commonly has only two possible alleles occurring at a locus, which is called *bi-allelic*. We will mostly be interested in bi-allelic markers because they are becoming the most popular markers in practice. Bi-alleles can be in exactly one of the two alternative states, such as 1 or 2. If an allele is missing at some locus, it is denoted as a "*".

In diploid organisms, because chromosomes come in pairs, at each locus there is a pair of alleles, which is referred to the *genotype* of this locus. If these alleles are the same, the genotype at this locus is *homozygous*; otherwise, the genotype is *heterozygous*. The alleles on the same chromosome form a *haplotype*. Each individual has a pair of haplotypes.

If there is no genetic mutation in a meiosis process, the child inherits one haplotype from the mother and the other one from the farther. This is the well-known *Mendelian law of inheritance*. The haplotype inherited from the mother is called the *maternal haplotype* while the one from the father is called the *paternal haplotype*. Given a pair of haplotypes of an individual, if it is known which one was inherited from his (or her) father and which was from his (or her) mother, the haplotypes and the inheritance information together are called a *haplotype configuration* (*i.e.* a *configuration* in short); otherwise, the haplotypes without inheritance information form a *haplotype grouping* (*i.e.* a *grouping* in short).

Usually, an entire haplotype of the mother's (or father's) haplotype pair is passed onto the child during meiosis. However, crossover between the haplotype pair might occur, where the haplotype pair gets shuffled and one of the mixed haplotypes is passed onto the child. This crossover is called a recombinant.

A PS (or phase) value represents the paternal or maternal information about the alleles at a locus. The PS value can take the values 0 or 1, where 1 means that the allele with the smaller identification number is from the mother and the allele with the larger identification number is from the father, and 0 otherwise. Thus, the reconstruction of haplotype configuration for an input pedigree can be viewed as assigning PS values to each locus of every member of the pedigree.

Now, the MRHC problem is defined as follows:

Definition 2 (MRHC [1]). *Given a pedigree and genotype information for each member of the pedigree, find a haplotype configuration of the pedigree that obeys the Mendelian law of inheritance and requires the minimum number of recombinants.*

1.2 Variants of MRHC and Some Related Problems

We give the definitions of the variants of MRHC and list the related problems that are going to be discussed later in the paper.

Definition 3. *MRHC(k, j) is defined the same as MRHC except that each member in the pedigree has at most k mates and at most j children with each mate. Binary-tree-MRHC is defined as MRHC on a binary-tree pedigree. Binary-tree -MRHC* is defined the same as binary-tree-MRHC except it is allowed to have missing alleles. 2-locus-MRHC is MRHC on a two-locus pedigree without missing data. 2-locus-MRHC* is defined the same as 2-locus-MRHC except it is allowed to have missing data. Tree-MRHC is MRHC on a pedigree without mating loops or missing data.*

In order to discuss the hardness and approximation of the variants, we are going to make use of some related problems or properties, such as the Min UnCut [5] (*i.e.* 2-Linear-Equations Mod 2 [4]), Min UnCut(*k*) (the same definition as Min UnCut except that each variable occurs at most *k* times), Min 2CNF Deletion [4, 5] problems, consistency and satisfiablility property. The Min UnCut and Min 2CNF Deletion problems are known to be NP-hard [5]. We will show that the Min UnCut(*k*) problem is also NP-hard in this paper.

For any NP-hard minimization (or maximization) problem, if there is some polynomial time algorithm to give a solution with the objective value no more (or less, respectively) than $f(n)\cdot$OPT (or OPT/$f(n)$, respectively), where $f(n)$ can be any function of the input size n, the problem can be approximated within ratio $f(n)$; otherwise, the problem cannot be approximated.

1.3 Previous Complexity Results on MRHC

Qian and Bechmann proposed a ruled-based algorithm to reconstruct haplotype configurations based on six rules [16]. Their algorithm is a heuristic without theoretical analysis. Li and Jiang first proved that MRHC on two-locus pedigree is NP-hard [1]. Doi, Li and Jiang further proved that MRHC on tree pedigrees is also NP-hard in the general case [2], even though MRHC can be solved by dynamic programming algorithms when the number of members or loci in the input pedigree is bounded by a constant. However, the NP-hardness proof requires pedigrees containing individuals with an unbounded number of mates or children. It was left as an open question if the proof can be improved to work for tree pedigrees where every individual has a bounded number of mates and children.

Consistency checking of the Mendelian law of inheritance (*i.e.* the Mendelian law checking problem) is closely related to the MRHC problem. The purpose of Mendelian law checking is to determine whether the given genotype data obey the classic Mendelian law of inheritance. Mendelian law checking usually needs to be done ahead of phasing haplotype configurations. Aceto *et al.* showed that the Mendelian law checking problem is NP-hard in general, although checking the consistency on pedigrees with bi-allelic data or with no mating loops [3] can be done in polynomial time.

In this paper, we consider a simple variant of MRHC, which involves pedigrees with members that has at most one mate and one child (*i.e.* binary-tree-MRHC). It is an open question if binary-tree MRHC is NP-hard. A polynomial-time algorithm for it, if exists, could be useful for solving the general-case MRHC problem. Another important question is whether a good approximation algorithm exists for MRHC. Here, in terms of computing the minimum-recombinant haplotype, the accuracy is sacrificed to improve the efficiency. Previously, there is no known polynomial-time approximation algorithm for MRHC with guaranteed ratio.

Table 1. The known hardness results of the Mendelian law checking and MRHC problems

Pedigree / Problem	Loop?	Multi-allelic?	Unbounded number of loci?	Unbouned number of members?	Hardness
Mendelian law checking	Yes	Yes			NP-hard [3]
		No			P [3]
	No				P [3]
MRHC	Yes	No	No	Yes	NP-hard [1]
	No	No	No	Yes	P [2]
	No	No	Yes	No	P [2]
	No	No	Yes	Yes	NP-hard [2]

1.4 Our Results

We will consider pedigrees with bi-allelic genotype data throughout this paper. First, we reduce ≠3SAT to the binary-tree-MRHC problem and show that this problem is NP-hard, which answers an open question in [2]. Second, we study the approximability of MRHC on pedigree data with the following restrictions: (*I*) 2-locus genotype data with missing alleles, (*II*) binary tree pedigrees with missing alleles, (*III*) 2-locus genotype data without missing alleles, and (*IV*) tree pedigrees without missing alleles. These four restricted cases of MRHC are NP-hard problems shown either in the literature [1,2] or in this paper. We demonstrate that for MRHC in the former two cases *I* and *II* cannot be approximated unless P = NP. We also prove that it is NP-hard to approximate problems *III* and *IV* within any constant ratio under the Unique Games

Conjecture [4]. Moreover, we show that problem *III* can be approximated with ratio $O(\sqrt{\log(n)})$ in polynomial time by reducing it to the Min 2CNF Deletion problem, Finally, we discuss the approximation of MRHC on pedigrees where each member has a bounded number of children and mates, mirroring pedigrees in real applications.

1.5 Organization of the Paper

The paper is organized as follows. We briefly give definitions of the MRHC problem and other closely related problems, introduce the related biological background in section 1. We prove binary-tree-MRHC is NP-hard and state the approximatability of MRHC on pedigrees with missing data in section 2. We show the approximation lower bound of MRHC on pedigrees without missing data and the approximation upper bound of 2-locus-MRHC in section 3. In section 4, we tentatively explore the approximation hardness of MRHC on the pedigrees where each member has a bounded number of mates and children. We organize our hardness results and conclude this paper with a few remarks in section 5. Due to space limitations, the proofs are omitted in the main text and are given in the full version [22].

2 Approximation of MRHC on Pedigrees with Missing Data

In this section, we prove the hardness of approximating MRHC on pedigree data with missing alleles. Two variants are considered.

Lemma 1. *If it is NP-hard to decide whether OPT(R)=0 for a minimization problem R, R cannot be approximated unless P=NP.*

2.1 Hardness and Approximation of Binary-Tree-MHRC(*)

Theorem 2. *Binary-tree-MRHC is NP-hard.*

Theorem 3. *It is NP-hard to decide whether OPT(binary-tree-MRHC*)=0.*

Corollary 4. *Binary-tree- MRHC* cannot be approximated unless P=NP.*

2.2 Approximation of 2-Loop-MHRC*

Theorem 5. *It is NP-hard to decide whether OPT(2-locus-MRHC*)=0*

Corollary 6. *2-locus-MRHC* cannot be approximated unless P=NP.*

3 Approximation of MRHC on Pedigrees Without Missing Data

In this section, we consider the approximability of the same variants of MRHC without missing data. In order to show the negative result, we need to use some gap-introducing reduction (or gap-preserving reduction) for MRHC. We will use the concept of *L-reduction* proposed by Papadimitriou and Yannakakis [18].

3.1 Approximation of Tree-MRHC

Lemma 7. *There is an L-reduction from Min UnCut to tree-MRHC that transforms a set of Boolean constraints* φ *to a tree pedigree* ξ *such that:*

(i) $OPT_{Min\ UnCut}(\varphi) = OPT_{tree\text{-}MRHC}(\xi)$*, and*

(ii) Given a haplotype solution for ξ *with k recombinants, we can construct a solution for* φ *with at most k unsatisfied clauses.*

Theorem 8. *It is NP-hard to approximate tree-MRHC within any constant ratio under the Unique Games Conjecture* [4].

3.2 Approximation of 2-Locus-MRHC

We will present a lower bound and an upper bound on the approximation ratio for the 2-locus-MRHC problem.

3.2.1 Negative Result for Approximating 2-Locus-MRHC

Lemma 9. *There is a polynomial-time L-reduction from Min UnCut to 2-locus-MRHC that transforms a Boolean constraints set* φ *to a pedigree* ξ *such that*

(i) $OPT_{Min\ UnCut}(\varphi) = OPT_{2\text{-}locus\text{-}MRHC}(\xi)$*, and*

(ii) Given any haplotype solution for ξ *with k recombinants, we can find in polynomial time a truth assignment for* φ *with at most k unsatisfied constraints.*

Theorem 10. *It is NP-hard to approximate 2-locus-MRHC within any constant ratio under the Unique Games Conjecture* [4].

3.2.2 Positive Result for Approximating 2-Locus-MRHC

We first would like to reduce an instance of 2-locus-MRHC so that each member of the pedigree can be described by one Boolean variable. Since only two loci are involved, there are three types of members in a pedigree: (*I*) both loci are homozygous, (*II*) one locus is homozygous, and (*III*) both loci are heterozygous. A type *I* (or *II*) member has a fixed haplotype grouping. A type *III* member has a variable haplotype grouping.

Agarwal and Charikar recently presented a randomized polynomial-time $O(\sqrt{\log(n)})$ approximation algorithm for the Min 2CNF Deletion problem [5], where n is the number of variables in the input 2CNF constraints.

Theorem 11. *There is a randomized polynomial-time* $O(\sqrt{\log(n)})$ *approximation algorithm for 2-locus-MRHC, where n is the number of members in the input pedigree.*

Observe that the results in this section show that, in terms of approximability, the 2-locus-MRHC problem is easier than the Min 2CNF Deletion problem and harder than the Min UnCut problem. Also, Lemma 9 presents an alternative proof that 2-locus-MRHC is NP-hard, which is much easier to understand than the original proof in [1].

4 Approximation of MRHC(k, j)

The proof of Lemma 7 uses a pedigree that contains members with a variable number of children, although every member in the pedigree has only one mate. Can we get the same hardness result for tree-MRHC if we bound the number of mates instead of the number of children? In addition, the pedigrees in the proofs of Theorem 5 and Lemma 9 contain members with a variable number of children or mates. Another question is whether MRHC on two-locus pedigrees with a bounded number of children and mates leads to the same hardness result. In this section, we discuss the approximation of MRHC on pedigrees with bounded number of children and mates. For the convenience of comparison, we state strengthened versions of the previous theorems in the order they appear in this paper. We use u to present an integer variable.

First, we refine Theorem 5. The hardness result in this theorem holds for 2-locus-MRHC(u,1), because some member might appear in every clause gadget and every member has at most one child in the proof of Theorem 5.

Theorem 12. *2-locus-MRHC*(4,1) cannot be approximated unless P=NP.*

Next, let us look at Lemma 7. This lemma actually works for tree-MRHC(1,u). It is natural to consider tree-MRHC on pedigrees where members have a bounded number of children with each mate. In order to decrease the number of children and mates in the pedigree, we need a bounded version of Min UnCut like the one for Max 3SAT.

In fact, there is an L-reduction from Min UnCut to Min UnCut(15) that transforms a Boolean constraints set φ to another Boolean constraints set ψ such that

(*i*) $\mathrm{OPT}_{\mathrm{UnCut}}(\varphi) = \mathrm{OPT}_{\mathrm{UnCut(15)}}(\psi)$, and

(*ii*) Given any truth assignment for ψ with k unsatisfied constraints, we can find in polynomial time a truth assignment for φ with at most k unsatisified constraints.

This L-reduction from Min UnCut to Min UnCut(15) can be constructed using the same idea as the L-reduction that transforms Max 3SAT to Max 3SAT(29) in [6] with just a few minor modifications. The details of this L-reduction are omitted here. Based on the property of this L-reduction, we know that it is NP-hard to approximate Min UnCut(15) within any constant ratio under the Unique Games Conjecture [4].

Theorem 13. *It is NP-hard to approximate tree-MRHC(u,1) within any constant ratio under the Unique Games Conjecture [4].*

Finally, we consider Lemma 9. The hardness result actually holds for 2-locus-MRHC(u, u), because neither the number of mates nor the number of children for a member is bounded by any constant.

Theorem 14. *It is NP-hard to approximate 2-locus-MRHC(16,15) within any constant ratio under the Unique Games Conjecture [4].*

5 Discussion and Conclusion

The results presented in this paper are organized in Table 2. First, we showed that binary-tree-MRHC is NP-hard. Binary-tree-MRHC is a simplest variant of MRHC because one mate and one child are the minimum requirement to express the inheritance of human beings. Second, we showed some approximability results concerning the MRHC problem. With the presence of missing data, it is NP-hard to tell if an instance of 2-locus-MRHC* and binary-tree-MRHC* requires any recombinant.

This gives an interesting contrast to the results in [1] where the problem of finding a zero-recombinant haplotype solution for MRHC was shown to be solvable in polynomial time. This result also implies that 2-locus-MRHC* and binary-tree-MRHC* is not approximable in polynomial time. Without the presence of missing data, 2-locus-MRHC can be approximated with the ratio $O(\sqrt{\log(n)})$. In addition it is NP-hard to approximate 2-locus-MRHC and tree-MRHC within any constant ratio under the Unique Games Conjecture [4]. Our final results concern the inapproximability of MRHC on pedigrees where each member has a bounded number of mates and/or a bounded number of children with each mate.

Table 2. Our hardness and approximation results for MRHC with bi-alleles

	Loop ?	Missing data?	Unbounded Number of loci?	Unbounded number of members?	Hardness	Lower bound of approx. ratio	Assumption	The lower bound holds for	Upper bound of approx. ratio
Binary-tree-MRHC	No	No	Yes	Yes	NP				
2-locus-MRHC*	Yes	Yes	No	Yes		Any f(n)	P≠ NP	2-locus-MRHC* (4,1)	
Binary-tree-MRHC*	No	Yes	Yes	Yes		Any f(n)	P≠ NP	Binary-tree-MRHC*	
2-locus-MRHC	Yes	No	No	Yes		Any constant	P≠ NP, the Unique Games Conjecture	2-locus-MRHC (16,15)	$O(\sqrt{\log(n)})$
Tree-MRHC	No	No	Yes	Yes		Any constant	P≠ NP, the Unique Games Conjecture	Tree-MRHC(1,u) Tree-MRHC(u,1)	

Acknowledgement

We would like to thank Dr. Neal Young and Dr. Marek Chrobak for their valuable suggestions and discussion. This research is supported in part by NSF grant CCR-0309902, National KeyProject for Basic Research (973) grant 2002CB512801, and a fellowship from the Center for Advanced Study, Tsinghua University.

References

1. J. Li and T. Jiang: Efficient rule-based haplotyping algorithm for pedigree data. Proc. of the 7th Annual Conference on Research in Computational Molecular Biology (RECOMB'03), pages 197-206, 2003.
2. K. Doi, J. Li and T. Jiang: Minimum recombinant haplotype configuration on tree pedigrees. Proc. of the 3rd Annual Workshop on Algorithms in Bioinformatics (WABI'03), pages 339-353, 2003.
3. L. Aceto *et al.*: The complexity of checking consistency of pedigree information and related problems. J. Comp. Sci. Tech., 19(1): 42–59, 2004.
4. S. Khot: On the power of 2-Prover 1-Round Games. Proc. of the 34th ACM Symposium on Theory of Computing (STOC'02), pages 767-775, 2002.
5. A. Agarwal, M. Charikar: $O(\sqrt{\log(n)})$ approximation algorithms for min UnCut, min 2CNF deletion, and directed cut problems. Proc. STOC'05, pages 573-581, 2005.
6. G. Ausiello *et al.*: Complexity and approximation: combinatorial optimization problems and their approximability properties, pages 276-279, Springer, 1999.
7. L. Jorde: Where we are hot, they are not. Science, Volume 308, pages 60-62, 2005.

8. L. Li, J.H. Kim, and M. S. Waterman: Haplotype reconstruction from SNP alignment. Proc. RECOMB'03, pages 207-216, 2003.
9. R. Lippert, *et al.*: Algorithmic strategies for the single nucleotide polymorphism haplotype assembly problem. Briefings in Bioinformatics 3(1): 23-31, 2002.
10. E. Eskin E. Halperin, and R.M. Karp: Large scale reconstruction of haplotypes from genotype data. Proc. RECOMB'03, pages 104-113, 2003.
11. L. Excoffier and M. Slatkin: Maximum–likelihood estimation of molecular haplotype frequencies in a diploid population. Mol. Biol. Evol., 12: 921-927, 1995.
12. H. Seltman, K. Roeder, and B. Devlin: Transmission/disequilibrium test meets measured haplotype analysis: family-based association analysis guided by evolution of haplotypes. Am. J . Hum. Genet., 68(5): 1250-1263, 2001.
13. S. Zhang *et al.*: Transmission/ disequilibrium test based on haplotype sharing for tightly linked markers. Am. J. Hum. Genet., 73(3): 556-579, 2003.
14. J. Li and T. Jiang: An exact solution for finding minimum recombinant haplotype configurations on pedigrees with missing data by integer linear programming. Proc. RECOMB'04, pages 20-29, 2004.
15. J.R. O'Connell: Zero-recombinant haplotyping: applications to fine mapping using SNPs. Genet. Epidemiol., 19 Suppl. 1: S64-70, 2000.
16. D. Qian and L. Beckmann: Minimum-recombinant haplotyping in pedigrees. Am. J. Hum. Genet., 70(6): 1434-1445, 2002.
17. The International HapMap Consortium: The International HapMap Project. Nature, Volume 426, pages 789-796, December 2003.
18. C.H. Papadimitriou and M. Yannakakis: Optimization, Approximation, and Complexity Classes. J. Comp. System Sci., pages 425-440, 1991.
19. M. Stephens, N. J. Smith, and P. Donnelly: A new statistical method for haplotype reconstruction from population data. Am. J. Hum. Genet., 68(4):978–989, 2001.
20. T.J. Schaefer: The complexity of satisfiability problems. Proc. of the 10th STOC, pages 216-226, 1978.
21. J. Li and T. Jiang: Computing the Minimum Recombinant Haplotype Configuration from incomplete genotype data on a pedigree by integer linear programming. Proc. RECOMB'04, pages 20-29, 2004.
22. Available at http://www.cs.ucr.edu/~lliu/App_MRHC.pdf.

Space Efficient Algorithms for Ordered Tree Comparison

Lusheng Wang[1] and Kaizhong Zhang[2]

[1] Department of Computer Science, City University of HongKong, HongKong
cswangl@cityu.edu.hk

[2] Dept. of Computer Science, University of Western Ontario,
London, Ont. N6A 5B7, Canada
kzhang@csd.uwo.ca

Abstract. In this paper we present techniques to significantly improve the space complexity of several ordered tree comparison algorithms without sacrificing the corresponding time complexity. We present new algorithms for computing the constrained ordered tree edit distance and the alignment of (ordered) trees. The techniques can also be applied to other related problems.

Keywords: Space efficient algorithms, Constrained tree edit distance, Alignment of trees.

1 Introduction

Ordered labeled trees are trees whose nodes are labeled and in which the left to right order among siblings is significant. Comparing such trees with distance and/or similarity measures has applications in several diverse areas such as computational molecular biology [8,5], computer vision, pattern recognition, programming compilation and natural language processing.

Algorithms have been developed for ordered labeled tree comparisons [10,9,14,5,12]. The degree-one edit distance was introduced by Selkow [9] in which insertions and deletions are restricted to the leaves of the trees. The degree-two edit distance was introduced by Zhang et al. [13,16] in which insertions and deletions are restricted to tree nodes with zero or one child. Zhang [12] introduced the constrained edit distance which is a bit more general than degree-two edit distance. The (general) edit distance between ordered labeled trees was introduced by Tai [10]. His algorithm has been improved by several authors [14,6,3]. The alignment of trees was introduced by Jiang et al. in [5].

Given two ordered trees T_1 and T_2, the degree-one edit distance, the degree-two edit distance, and the constrained edit distance can all be computed in $(|T_1||T_2|)$ time and space [9,12,13], where $|\cdot|$ is the number of nodes in the tree. Richter [7] presented an algorithm for the constrained edit distance with $O(|T_1||T_2|deg(T_1)deg(T_2))$ time and $O(|T_1|dep(T_2)deg(T_2))$ space, where $deg(\cdot)$ is the degree of the tree and $dep(\cdot)$ is the depth of the tree. For small degree and low depth trees, this is a space improvement. According to a recent survey

X. Deng and D. Du (Eds.): ISAAC 2005, LNCS 3827, pp. 380–391, 2005.

on tree edit distance by Bille [1], the algorithms of Zhang [12] and Richter [7] are currently the best for the constrained tree edit distance. In this paper, we present an algorithm for the constrained tree edit distance with $O(|T_1||T_2|)$ time and $O(\log(|T_1|)|T_2|)$ space. The techniques used can also be used for the degree-one and the degree-two edit distance to achieve the same time and space complexities.

For alignment of trees, the algorithm in [5] runs in $O(|T_1|T_2|(deg(T_1) + deg(T_2))^2)$ time and needs $O(|T_1||T_2|(deg(T_1) + deg(T_2))$ space. Recently, there is a strike to reduce the space [11]. The space required for the new algorithm is $O(\log(|T_1|)|T_2|(deg(T_1) + deg(T_2))deg(T_1))$. However, the running time is increased to $O(|T_1|^2|T_2|(deg(T_1) + deg(T_2))^2)$. In this paper, we proposed a new algorithm that keeps the same space complexity and reduces the run timep by a factor of $O(|T_1|)$.

2 Constrained Edit Distance Between Ordered Trees

The nodes in an ordered tree of size n are numbered from 1 to n according to the postorder, where the left siblings are ordered before the right siblings. Given an ordered labeled tree T, the ith node of tree T is represented as $t[i]$, the subtree of T rooted at node $t[i]$ is represented as $T[i]$, and the subforest obtained by deleting $t[i]$ from $T[i]$ is represented as $F[i]$. The parent of node $t[i]$ in T is denoted as $t[p(i)]$. The number of nodes in a tree T is denoted as $|T|$.

Let θ be the empty tree, and λ a inserted space. $\gamma(a, \lambda)$, $\gamma(\lambda, a)$, and $\gamma(a, b)$ denote the cost of deleting a, inserting a, and substituting a with b, respectively.

Consider two trees T_1 and T_2 to compare. Suppose that the degrees of node $t_1[i]$ and node $t_2[j]$ are m_i and n_j, respectively. Denote the children of $t_1[i]$ from left to right as $t_1[i_1], ..., t_1[i_{m_i}]$ and children of $t_2[j]$ from left to right as $t_2[j_1], ..., t_2[j_{n_j}]$.

The constrained edit distance metric is based on a constrained edit mapping allowed between two trees. We will first give the definition of constrained editing mapping and then use it to define the constrained editing distance metric.

Formally we define a triple (M, T_1, T_2) to be a constrained edit mapping from T_1 to T_2, where M is any set of pairs of integers (i, j) satisfying:

(1) $1 \leq i \leq\mid T_1 \mid, 1 \leq j \leq\mid T_2 \mid$;
(2) For any pair of (i_1, j_1) and (i_2, j_2) in M,
  (a) $i_1 = i_2$ iff $j_1 = j_2$ (one-to-one)
  (b) $t_1[i_1]$ is to the left of $t_1[i_2]$ iff $t_2[j_1]$ is to the left of $t_2[j_2]$ (sibling order preserved);
  (c) $t_1[i_1]$ is an ancestor of $t_1[i_2]$ iff $t_2[j_1]$ is an ancestor of $t_2[j_2]$ (ancestor order preserved);
(3) For any triple (i_1, j_1), (i_2, j_2) and (i_3, j_3) in M, let $lca()$ represent least common ancestor function,
$t_1[lca(i_1, i_2)]$ is a proper ancestor of $t_1[i_3]$ iff $t_2[lca(j_1, j_2)]$ is a proper ancestor of $t_2[j_3]$.

This definition adds constraint (3) to the definition of the edit mapping [14]. This is why we call it a constrained edit mapping. The intuitive idea behind this definition is that two separate subtrees of T_1 should be mapped to two subtrees of T_2 and vice versa.

We will use M instead of (M, T_1, T_2) if there is no confusion. Let M be a constrained editing mapping from T_1 to T_2. We can define the cost of M:

$$\gamma(M) = \sum_{(i,j)\in M} \gamma(t_1[i], t_2[j]) + \sum_{i \notin M} \gamma(t_1[i], \lambda) + \sum_{j \notin M} \gamma(\lambda, t_2[j])$$

We can now define the constrained edit distance between T_1 and T_2 as:

$$D(T_1, T_2) = \min_M \{\gamma(M) \mid M \text{is a constrained edit mapping between } \mathrm{T_1 to T_2}\}$$

2.1 A Simple Algorithm

We now present an algorithm for computing the constrained edit distance between two ordered trees. This algorithm is given in [12] and is the basis of our new space efficient algorithm.

The algorithm in [12] for the constrained edit distance between two ordered trees is given in Figure 1. While we do not show the details of the correctness of the algorithm, we now explain the ideas of the algorithm.

The computation of $D(T_1[i], T_2[j])$ has several cases. In one case, $T_1[i]$ is matched to a child subtree of $T_2[j]$. In a symmetric case, $T_2[j]$ is matched to a child subtree of $T_1[i]$. In the last case, $t_1[i]$ is matched with $t_2[j]$ and therefore $F_1[i]$ is matched with $F_2[j]$ which means that we need to know $D(F_1[i], F_2[j])$.

Similarly, the computation of $D(F_1[i], F_2[j])$ has several cases. In the first case, $F_1[i]$ is matched to $F_2[j_k]$, $1 \leq k \leq n_j$, a subforest of a child of $F_2[j]$. In the second case, $F_2[j]$ is matched to $F_1[i_k]$, $1 \leq k \leq m_i$, a subforest of a child of $F_1[i]$. In the third case, there is a matching between $T_1[i_1], ..., T_1[i_{m_i}]$ and $T_2[j_1], ..., T_2[j_{n_j}]$. We uae $E(m_i, n_j)$ to denote the optimal matching between $T_1[i_1], ..., T_1[i_{m_i}]$ and $T_2[j_1], ..., T_2[j_{n_j}]$. $E(m_i, n_j)$ can be computed using the sequence edit distance algorithm treating each subtree as an unit. The recurrence equations are shown in Figure 1. All the values can be computed in a bottom up fashion.

The time and space complexity for computing $E(m_i, n_j)$ is obviously $O(m_i \times n_j)$. The complexity of computing $D(T_1[i], T_2[j])$ is bounded by $O(m_i + n_j)$. The complexity of computing $D(F_1[i], F_2[j])$ is bounded by the $O(m_i + n_j)$.

Hence for any pair i and j, the complexity of computing $D(T_1[i], T_2[j])$ and $D(F_1[i], F_2[j])$ is bounded by $O(m_i \times n_j)$. Therefore the complexity of the algorithm is

$$\sum_{i=1}^{|T_1|} \sum_{j=1}^{|T_2|} O(m_i \times n_j) = O(\sum_{i=1}^{|T_1|} m_i \times \sum_{j=1}^{|T_2|} n_j) = O(|T_1| \times |T_2|)$$

$$E(s,t) = \min \begin{cases} E(s,t-1) + D(\theta, T_2[j_t]) \\ E(s-1,t) + D(T_1[i_s], \theta) \\ E(s-1,t-1) + D(T_1[i_s], T_2[j_t]), \end{cases}$$

$$D(F_1[i], F_2[j]) = \min \begin{cases} D(F_1[i], \theta) + \min_{1 \le s \le m_i} \{D(F_1[i_s], F_2[j]) - D(F_1[i_s], \theta)\} \\ D(\theta, F_2[j]) + \min_{1 \le t \le n_j} \{D(F_1[i], F_2[j_t]) - D(\theta, F_2[j_t])\} \\ E(n_i, n_j), \end{cases}$$

$$D(T_1[i], T_2[j]) = \min \begin{cases} D(T_1[i], \theta) + \min_{1 \le s \le m_i} \{D(T_1[i_s], T_2[j]) - D(T_1[i_s], \theta)\} \\ D(\theta, T_2[j]) + \min_{1 \le t \le n_j} \{D(T_1[i], T_2[j_t]) - D(\theta, T_2[j_t])\} \\ D(F_1[i], F_2[j]) + \gamma(t_1[i] \to t_2[j]). \end{cases}$$

Fig. 1. The equations for the simple algorithm

2.2 Reducing the Space Complexity

In practice, the space required for the simple algorithm maybe a bottleneck. In this section, we propose a method to modify Algorithm 1 so that the space required is reduced to $O(\log(|T_1|) \cdot |T_2|)$.

The basic idea is straightforward. In order to compute the constrained edit distance, some computed values are no longer useful after they were used. We simple release the space that is no longer useful. To achieve our goal, we need to be more careful about the computational order of nodes of tree T_1, not just an arbitrary postorder. We also modify the computation of node $t_1[i]$ slightly. When the computation for node $t_1[i_1]$ is completed, we immediately start the computation of $E[i,j]$ without waiting for the computation of $t_1[i_2]$.

Here we give a more restricted order for the nodes in T_1. Consider a node $t_1[i]$ in T_1. Let $t_1[i_1]$, $t_1[i_2]$, ..., $t_1[i_{m_i}]$ be the children of $t_1[i]$ in T_1 and $|T_1[i_k]|$ be the number of nodes in the subtree $T_1[i_k]$. The *computational order* of the children of $t_1[i]$ is that the child with largest number of nodes, which we refer to as the favorable child, would be the first one and then from left to right for the rest of children. This order is applied to all the nodes and if $t_1[u]$ and $t_1[v]$ are siblings in T_1 and $t_1[u]$ is ordered before $t_1[v]$ then all nodes in $T_1[u]$ are ordered before any node in $T_1[v]$. Figure 2 gives an example of the new order.

The modified algorithm will use this new order for T_1 and the same post order as in the simple algorithm for T_2. For each node $t_1[i]$, we will compute $D(F_1[i], F_2[j])$ and $D(T_1[i], T_2[j])$ for all nodes $t_2[j]$ in T_2. Now suppose that $t_1[p(i)]$ is the parent of $t_1[i]$, then we can immediately use $D(F_1[i], F_2[j])$ and $D(T_1[i], T_2[j])$ for the computation of $D(F_1[p(i)], F_2[j])$ and $D(T_1[p(i)], T_2[j])$. There are two cases. One case is that $t_1[i]$ is the favorable child of $t_1[p(i)]$ and it is not the leftmost child of $t_1[p(i)]$. In this case, we need to keep the values of $D(F_1[i], F_2[j])$ and $D(T_1[i], T_2[j])$ since the computation of $E(p(i), j)$ is from left to right. The other case is that $t_1[i]$ is not the favorable child or it is the favorable child and also the leftmost child. In this case, we can use it immediately for the computation of $E(p(i), j)$. Therefore for each node $t_1[i]$ in T_1 such that at least its favorable child's computation is completed, we at most will need to keep two sets of values: the values for its favorable child and the partial computation of

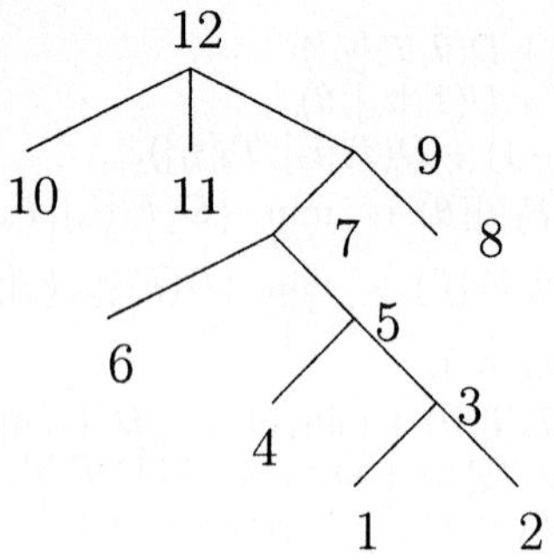

Fig. 2. An example of the new order

$E(m_i, n_j)$. Notice that for any node $t_1[i]$ in T_1, if the computation of its favorable child is not completed, we do not need to store any value and if the computation of all its children have been completed, then in next step the computation for node $t_1[i]$ is completed and we can release the space used.

By using such a new order for T_1 and the above straightforward idea we now have an algorithm with much less space. Assuming the initial values are set, the modified algorithm is given in Figure 3 as Algorithm 2.

The following lemma shows that the space complexity of Algorithm 2 is reduced to $O(\log(|T_1|)|T_2|)$.

Lemma 1. *The time and space complexities of Algorithm 2 are* $O(|T_1||T_2|)$ *and* $O(\log(|T_1|)|T_2|)$.

Proof. From the algorithm, it is clear that for each i of T_1, the number of steps is bounded by $O(\sum_{j=1}^{|T_2|} n_j)$. Therefore the time complexity of Algorithm 2 is $O(\sum_{i=1}^{|T_1|}\sum_{j=1}^{|T_2|} n_j) = O(|T_1||T_2|)$.

For the space complexity, let us consider the status of a node in T_1 with respect to the space requirement. We say a node $t_1[i]$ in T_1 is active if the computation of its favorable child is completed and the computation for node $t_1[i]$ itself has not been completed. Therefore a node is inactive if the computation of its favorable child has not been completed or if the computation for node i is completed.

If a node is inactive because its computation is completed and this node is the favorable child of its parent, then we cannot immediately release the space used since these values will be used for the computation of its parent. However in this situation, we can contribute the space requirement of the favorable child to its parent, which is active, when counting the space requirement. Therefore, we can assume that if a node is inactive, then we can release the space used for that node. This means that we only have to check maximum space used by active nodes during the execution of the algorithm.

For any active node $t_1[p]$, assuming that its favorable child is $t_1[p_f]$, then we need $O(|T_2|)$ to store $D(F_1[p_f], F_2[j])$ and $D(T_1[p_f], T_2[j])$ for $1 \le j \le |T_2|$, and

Input: T_1 and T_2
Output: $D(T_1, T_2[j])$, where $1 \leq j \leq |T_2|$

for $i = 1$ **to** $|T_1|$ (the new order)
  for $j = 1$ **to** $|T_2|$

$$D(F_1[i], F_2[j]) = \min \begin{cases} D(\theta, F_2[j]) + \min_{1 \leq t \leq n_j} \{D(F_1[i], F_2[j_t]) - D(\theta, F_2[j_t])\} \\ F_i \\ E_i(n_j); \end{cases}$$

$$D(T_1[i], T_2[j]) = \min \begin{cases} D(\theta, T_2[j]) + \min_{1 \leq t \leq n_j} \{D(T_1[i], T_2[j_t]) - D(\theta, T_2[j_t])\} \\ G_i \\ D(F_1[i], F_2[j]) + \gamma(t_1[i], t_2[j]); \end{cases}$$

Let $t_1[p(i)]$ be the parent of $t_1[i]$ and $t_1[p_f]$ be the favorable child of $t_1[p(i)]$
if $t_1[i]$ is the favorable child of $t_1[p(i)]$ **then**
  $E_p(0) = 0$;
  $F_p = D(F_1[p(i)], \theta) + D(F_1[i], F_2[j]) - D(F_1[i], \theta)$;
  $G_p = D(T_1[p(i)], \theta) + D(T_1[i], T_2[j]) - D(T_1[i], \theta)$;
  for $t = 1$ **to** n_j
    $E_p(t) = E(t-1) + D(\theta, T_2[j_t])$;
if $t_1[i]$ is not the favorable child of $t_1[p(i)]$ or $t_1[i]$ is the leftmost child of $t_1[p(i)]$ **then**
  $F_p = \min\{F_p, D(F_1[p(i)], \theta) + D(F_1[i], F_2[j]) - D(F_1[i], \theta)\}$;
  $G_p = \min\{G_p, D(T_1[p(i)], \theta) + D(T_1[i], T_2[j]) - D(T_1[i], \theta)\}$;
  for $t = 0$ **to** n_j
    $E_p^0(t) = E_p(t)$;
  $E_p(0) = E_p^0(0) + D(T_1[i], \theta)$;
  for $t = 1$ **to** n_j

$$E_p(t) = \min \begin{cases} E_p(t-1) + D(\theta, T_2[j_t]) \\ E_p^0(t) + D(T_1[i], \theta) \\ E_p^0(t-1) + D(T_1[i], T_2[j_t]); \end{cases}$$

if $t_1[i]$ is the left sibling of the favorable child $t_1[p_f]$ of $t_1[p(i)]$ **then**
  $F_p = \min\{F_p, D(F_1[p(i)], \theta) + D(F_1[p_f], F_2[j]) - D(F_1[p_f], \theta)\}$;
  $G_p = \min\{G_p, D(T_1[p(i)], \theta) + D(T_1[p_f], T_2[j]) - D(T_1[p_f], \theta)\}$;
  for $t = 0$ **to** n_j
    $E_p^0(t) = E_p(t)$;
  $E_p(0) = E_p^0(0) + D(T_1[p_f], \theta)$;
  for $t = 1$ **to** n_j

$$E_p(t) = \min \begin{cases} E_p(t-1) + D(\theta, T_2[j_t]) \\ E_p^0(t) + D(T_1[p_f], \theta) \\ E_p^0(t-1) + D(T_1[p_f], T_2[j_t]). \end{cases}$$

Fig. 3. A more space efficient algorithm, Algorithm 2

$O(\sum_{j=1}^{|T_2|} n_j) = O(|T_2|)$ to store $E_p(t)$ where $1 \leq t \leq n_j$ for all j in T_2. Therefore for each active node, the space required is $O(|T_2|)$. Because of the new order, if two nodes are active at the same time, then one has to be the ancestor of the other. Therefore all active nodes of T_1 are on a path in T_1. Let $t_1[s]$ and $t_1[t]$ be two neighboring active nodes on this path and $t_1[s]$ is an ancestor of $t_1[t]$, then

$|T_1[s]| \geq 2|T_1[t]|$ since the computation of $t_1[s_f]$, the favorable child of $t_1[s]$ with maximum size, is already completed and therefore $t_1[t]$ is not a descendant of $t_1[s_f]$.

This means that there are at most $O(\log(|T_1|))$ active nodes. Therefore the space complexity is $O(\log(|T_1|)|T_2|)$.

2.3 Finding the Optimal Constrained Edit Mapping Between Two Trees

In previous section we show that $D(T_1, T_2)$ can be computed in $O(|T_1||T_2|)$ time and $O(\log(|T_1|)|T_2|)$ space. However in real application the optimal mapping that achieves $D(T_1, T_2)$ maybe required.

Finding the optimal mapping in $O(\log(|T_1|)|T_2|)$ space is in fact a more difficult task. This is similar to the situation of computing edit distance between two sequences. Computing the edit distance in linear space is easy, but finding the optimal edit script in linear space is more involved. Hirschberg [4] presented a clever way to do this.

In this section we will present a method that can find the optimal mapping between two ordered trees in $O(|T_1||T_2|)$ time and $O(\log(|T_1|)|T_2|)$ space. When the input is two sequences (trees such that each node only has one child), then our method produces the optimal edit script for sequence edit distance in $O(|T_1||T_2|)$ time and $O(|T_2|)$ space.

The main idea is as the follows. Given T_1 and T_2, there is a unique node, called key node, $t_1[k]$ in T_1 such that in $O(|T_1||T_2|)$ time and $O(\log(|T_1|)|T_2|)$ space we can determine, in the optimal mapping, what subtree and subforest in T_2 that $T_1[k]$ and $F_1[k]$ would match to produce $D(T_1, T_2)$. With this information, we then decompose the optimal mapping into several components mapping such that for each component the subtree or subforest of T_1 involved has a size less than or equal to half of $|T_1$. This means that in the next step if we repeat the same process for each component then the total cost for all the components is $O(0.5|T_1||T_2|)$ time using $O(\log(|T_1|)|T_2|)$ space. Therefore, repeating this process at most $\log(|T_1|)$ times, we can compute the optimal mapping in $O(|T_1||T_2|)$ time and $O(\log(|T_1|)|T_2|)$ space.

Given T_1, the unique key node $t_1[k]$, with children $t_1[k_1], t_2[k_2], ..., t_1[k_{m_k}]$, is a node satisfying the following properties.

- $|T_1[k]| > 0.5|T_1|$,
- $|T_1[k_s]| \leq 0.5|T_1|$ for $1 \leq s \leq m_k$.

With a small modification of algorithm 2, when computing $D(T_1, T_2)$, we can also compute two integers t_1 and t_2 for subtree $T_1[k]$ and two integers f_1 and f_2 for subforest $F_1[k]$.

We now define the meaning of t_1 and t_2. If in the optimal mapping $T_1[k]$ is deleted, then $t_1 = t_2 = 0$. If in the optimal mapping subtree $T_1[k]$ is matched with subtree $T_2[l]$ such that in this mapping $T_1[k_s]$ is matching with $T_2[l]$ and the rest of $T_1[k]$ is deleted, then $t_1 = k_s$ and $t_2 = l$. If in the optimal mapping node $t_1[k]$ is matched with node $t_2[l]$, then $t_1 = k$ and $t_2 = l$. If in the optimal mapping $t_1[k]$ is deleted and $F_1[k]$ is matched with $F_2[l]$, then $t_1 = k + 1$ and $t_2 = l$.

f_1 and f_2 are defined similarly. Note that f_1 and f_2 are only meaningful when $t_1 = k$ or $t_1 = k+1$, Otherwise $f_1 = f_2 = -1$. If in the optimal mapping $F_1[k]$ is deleted, then $f_1 = f_2 = 0$. If in the optimal mapping $F_1[k]$ is matched with $F_2[l]$ such that in this mapping $F_1[k_s]$ is matching with $F_2[l]$ and the rest of $F_1[k]$ are deleted, then $f_1 = k_s$ and $f_2 = l$. If in the optimal mapping $F_1[k]$ is matched with $F_2[l]$ and $D(F_1[k], F_2[l]) = E(m_k, n_l)$, then $f_1 = k$ and $f_2 = l$.

We now give the formulae for computing t_1, t_2, f_1, and f_2. We use $t\_T(i,j)$ to represent (t_1, t_2) in the optimal mapping of $D(T_1[i], T_2[j])$. We use $f\_T(i,j)$ to represent (f_1, f_2) in the optimal mapping of $D(T_1[i], T_2[j])$. We use $t\_F(i,j)$ to represent (t_1, t_2) in the optimal mapping of $D(F_1[i], F_2[j])$. We use $f\_F(i,j)$ to represent (f_1, f_2) in the optimal mapping of $D(F_1[i], F_2[j])$.

Lemma 2. *Let $t_1[k]$ be the key node of T_1, then*

$$f\_F(k,j) = \begin{cases} (k,j) & \text{if } D(F_1[k], F_2[j]) = E(m_k, n_j) \\ (k_s, j) & \text{if } D(F_1[k], F_2[j]) = D(F_1[k], \theta) + D(F_1[k_s], F_2[j]) - D(F_1[k_s], \theta) \\ f\_F(k, j_t) & \text{if } D(F_1[k], F_2[j]) = D(\theta, F_2[j]) + D(F_1[k], F_2[j_t]) - D(\theta, F_2[j_t]), \end{cases}$$

$$t\_T(k,j) = \begin{cases} (k,j) & \text{if } D(T_1[k], T_2[j]) = D(F_1[k], F_2[j]) + \gamma(t_1[k], t_2[j]) \\ (k_s, j) & \text{if } D(T_1[k], T_2[j]) = D(T_1[k], \theta) + D(T_1[k_s], T_2[j]) - D(T_1[k_s], \theta) \\ t\_T(k, j_t) & \text{if } D(T_1[k], T_2[j]) = D(\theta, T_2[j]) + D(T_1[k], T_2[j_t]) - D(\theta, T_2[j_t]), \end{cases}$$

$$f\_T(k,j) = \begin{cases} f\_F(k,j) & \text{if } D(T_1[k], T_2[j]) = D(F_1[k], F_2[j]) + \gamma(t_1[k], t_2[j]) \\ (-1,-1) & \text{if } D(T_1[k], T_2[j]) = D(T_1[k], \theta) + D(T_1[k_s], T_2[j]) - D(T_1[k_s], \theta) \\ f\_T(k, j_t) & \text{if } D(T_1[k], T_2[j]) = D(\theta, T_2[j]) + D(T_1[k], T_2[j_t]) - D(\theta, T_2[j_t]). \end{cases}$$

Lemma 3. *Let $t_1[i]$ be the parent of $t_1[k]$. Let x be t or f in the following formula for $x\_T(i,j)$.*

$$t\_F(i,j) = \begin{cases} (0,0) & \text{if } D(F_1[i], F_2[j]) = E(m_i, n_j) \text{ and } T_1[k] \text{ is deleted} \\ t\_T(k, j_t) & \text{if } D(F_1[i], F_2[j]) = E(m_i, n_j) \text{ and } T_1[k] \text{ is matched to } T_2[j_t] \\ (0,0) & \text{if } D(F_1[i], F_2[j]) = D(F_1[i], \theta) + D(F_1[i_s], F_2[j]) - D(F_1[i_s], \theta) \\ & \quad i_s \neq k \\ (k+1, j) & \text{if } D(F_1[i], F_2[j]) = D(F_1[i], \theta) + D(F_1[k], F_2[j]) - D(F_1[k], \theta) \\ t\_F(i, j_t) & \text{if } D(F_1[i], F_2[j]) = D(\theta, F_2[j]) + D(F_1[i], F_2[j_t]) - D(\theta, F_2[j_t]), \end{cases}$$

$$f\_F(i,j) = \begin{cases} (0,0) & \text{if } D(F_1[i], F_2[j]) = E(m_i, n_j) \text{ and } T_1[k] \text{ is deleted} \\ f\_T(k, j_t) & \text{if } D(F_1[i], F_2[j]) = E(m_i, n_j) \text{ and } T_1[k] \text{ is matched to } T_2[j_t] \\ (0,0) & \text{if } D(F_1[i], F_2[j]) = D(F_1[i], \theta) + D(F_1[i_s], F_2[j]) - D(F_1[i_s], \theta) \\ & \quad i_s \neq k \\ f\_F(k, j) & \text{if } D(F_1[i], F_2[j]) = D(F_1[i], \theta) + D(F_1[k], F_2[j]) - D(F_1[k], \theta) \\ f\_F(i, j_t) & \text{if } D(F_1[i], F_2[j]) = D(\theta, F_2[j]) + D(F_1[i], F_2[j_t]) - D(\theta, F_2[j_t]), \end{cases}$$

$$x\_T(i,j) = \begin{cases} x\_F(i,j) & \text{if } D(T_1[i], T_2[j]) = D(F_1[i], F_2[j]) + \gamma(t_1[i], t_2[j]) \\ (0,0) & \text{if } D(T_1[i], T_2[j]) = D(T_1[i], \theta) + D(T_1[i_s], T_2[j]) - D(T_1[i_s], \theta) \\ & \quad i_s \neq k \\ x\_T(k,j) & \text{if } D(T_1[i], T_2[j]) = D(T_1[i], \theta) + D(T_1[k], T_2[j]) - D(T_1[k], \theta) \\ x\_T(i, j_t) & \text{if } D(T_1[i], T_2[j]) = D(\theta, T_2[j]) + D(T_1[i], T_2[j_t]) - D(\theta, T_2[j_t]). \end{cases}$$

Lemma 4. *Let $t_1[i]$ be a proper ancestor of $t_1[p(k)]$ and $t_1[i_k]$ be the child of $t_1[i]$ which is an ancestor of $t_1[k]$. Let x be t or f in the following formulae for $x\_F(i,j)$ and $x\_T(i,j)$.*

$$x\_F(i,j)=\begin{cases}(0,0) & \text{if } D(F_1[i],F_2[j]) = E(m_i,n_j) \text{ and } T_1[i_k] \text{ is deleted}\\ x\_T(i_k,j_t) & \text{if } D(F_1[i],F_2[j]) = E(m_i,n_j) \text{ and } T_1[i_k] \text{ is matched to } T_2[j_t]\\ (0,0) & \text{if } D(F_1[i],F_2[j]) = D(F_1[i],\theta)+D(F_1[i_s],F_2[j])-D(F_1[i_s],\theta)\\ & \quad s \neq k\\ x\_F(i_k,j) & \text{if } D(F_1[i],F_2[j]) = D(F_1[i],\theta)+D(F_1[i_k],F_2[j])-D(F_1[i_k],\theta)\\ x\_F(i,j_t) & \text{if } D(F_1[i],F_2[j]) = D(\theta,F_2[j])+D(F_1[i],F_2[j_t])-D(\theta,F_2[j_t]),\end{cases}$$

$$x\_T(i,j)=\begin{cases}x\_F(i,j) & \text{if } D(T_1[i],T_2[j]) = D(F_1[i],F_2[j]) + \gamma(t_1[i],t_2[j])\\ (0,0) & \text{if } D(T_1[i],T_2[j]) = D(T_1[i],\theta) + D(T_1[i_s],T_2[j]) - D(T_1[i_s],\theta)\\ & \quad s \neq k\\ x\_T(i_k,j) & \text{if } D(T_1[i],T_2[j]) = D(T_1[i],\theta) + D(T_1[i_k],T_2[j]) - D(T_1[i_k],\theta)\\ x\_T(i,j_t) & \text{if } D(T_1[i],T_2[j]) = D(\theta,T_2[j]) + D(T_1[i],T_2[j_t]) - D(\theta,T_2[j_t]).\end{cases}$$

From these formulae, it is clear that we can computer $t\_T(|T_1|,|T_2|)$ and $f\_T(|T_1|,|T_2|)$ with a small modification of Algorithm 2.

We now consider how to use $t\_T(|T_1|,|T_2|) = (t_1,t_2)$ and $f\_T(|T_1|,|T_2|) = (f_1,f_2)$ to decompose the optimal mapping into smaller component mappings.

case 1: $t_1 = t_2 = 0$. In this case, since $T_1[k]$ is deleted, it is not in the optimal mapping for $D(T_1,T_2)$. Therefore the optimal mapping for $D(T_1,T_2)$ would be the optimal mapping for $D(T_1/T_1[k],T_2)$ where $T_1/T_1[k]$ is T_1 with $T_1[k]$ deleted. Note that $|T_1/T_1[k]| < 0.5|T_1|$.

case 2: $t_1 = k_s$ and $t_2 = l$. In this case, $T_1[k]$ is matched with $T_2[l]$ and in this matching, $T_1[k_s]$ is matched with $T_2[l]$ and the rest of $T_1[k]$ is deleted. Therefore the optimal mapping for $D(T_1,T_2)$ has two component mappings. One component is the optimal mapping for $D(T_1/F_1[k],T_2/F_2[l])$ with additional condition that $t_1[k]$ has to match with $t_2[l]$ though (k,l) is not in the optimal mapping of $D(T_1,T_2)$ since $t_1[k]$ is deleted. The other component is the optimal mapping of $D(T_1[k_s],T_2[l]$. Note that $|T_1/F_1[k]| \leq 0.5|T_1|$ and $|T_1[k_s]| \leq 0.5|T_1|$.

Computing $D(T_1/F_1[k],T_2/F_2[l])$ with condition that some of the leaves of the two trees are forced to match would not be more difficult then computing $D(T_1/F_1[k],T_2/F_2[l])$ without any condition. In fact the condition will force the ancestors of these matched leaves of the two trees to match making the computation less expensive.

case 3: $t_1 = k$ and $t_2 = l$, $f_1 = f_2 = 0$. In case 3, case 4 and case 5, $t_1 = k$ and $t_2 = l$. This means that in the optimal mapping for $D(T_1,T_2)$, $t_1[k]$ is matched with $t_2[l]$. Therefore one component mapping (for case 3-5) is the optimal mapping for $D(T_1/F_1[k],T_2/F_2[l])$ with additional condition that $t_1[k]$ matched with $t_2[l]$. Note that $|T_1/T_1[k]| < 0.5|T_1|$.

In this case, since $f_1 = f_2 = 0$ there is no additional component.

case 4: $t_1 = k$ and $t_2 = l$, $f_1 = k_s$ and $f_2 = l$. In this case, one component mapping is the same as that in case 3 and the other component mapping is the optimal mapping for $D(F_1[k_s],F_2[l])$. Note that $|F_1[k_s]| \leq 0.5|T_1|$).

case 5: $t_1 = k$ and $t_2 = l$, $f_1 = k$ and $f_2 = l$. In this case, one component mapping is the same as that in case 3 and the other component mapping is the optimal mapping for $D(F_1[k], F_2[l])$. Since in this case, $D(F_1[k], F_2[l]) = E(m_k, n_l)$, the optimal mapping for $D(F_1[k], F_2[l])$ can be further decomposed into the optimal mappings of the matching subtree pairs of $E(m_k, n_l)$. Note that $|T_1[k_s]| \leq 0.5|T_1|)$ for $1 \leq s \leq m_k$.

case 6-8: $t_1 = k + 1$ and $t_2 = l$. This means that in the optimal mapping for $D(T_1, T_2)$, $t_1[k]$ is deleted and $F_1[k]$ is matched with $F_2[l]$. Therefore one component mapping (for case 6-8) is the optimal mapping for $D(T_1/F_1[k], T_2/F_2[l])$ with additional condition that $t_2[l]$ matched with a proper ancestor of $t_1[k]$. Note that $|T_1 - T_1[k]| < 0.5|T_1|$.

There is no additional component for case 6. The other component for case 7 is exactly the same as that of case 4. The other components for case 8 are exactly the same as that of case 5.

In all the above cases, with t_1, t_2, f_1, and f_2, we can determine components needed except matching subtree pairs of $E(m_k, n_l)$ for case 5 and case 8. Finding the matching subtree pairs of $E(m_k, n_l)$ can be done in $O(|F_1[k]||F_2[l]|)$ time and $O(\log(|F_1[k]|)|F_2[l]|)$ space. First we find $T_1[k_s]$ such that $\sum_{i=1}^{s-1} |T_1[k_i]| \leq 0.5|F_1[k]|$ and $\sum_{i=1}^{s} |T_1[k_i]| > 0.5|F_1[k]|$. We then determine $T_2[l_t]$, in $O(|F_1[k]||F_2[l]|)$ time and $O(\log(|F_1[k]|)|F_2[l]|)$ space, such that either $T_1[k_s]$ matched with $T_2[l_t]$ in $E(m_k, n_l)$ or $T_1[k_s]$ is deleted, $T_1[k_1], ..., T_1[k_{s-1}]$ are matched with $T_2[l_1], ..., T_2[l_t]$, and $T_1[k_{s+1}], ..., T_1[k_{m_k}]$ are matched with $T_2[l_{t+1}], ..., T_2[l_{n_l}]$ in $E(m_k, n_l)$. In both cases, since $\sum_{i=1}^{s-1} |T_1[k_i]| \leq 0.5|F_1[k]$ and $\sum_{i=s+1}^{m_k} |T_1[k_i]| \leq 0.5|F_1[k]$, we can repeat and get all the subtree matching pairs for $E(m_k, n_l)$ in $O(|F_1[k]||F_2[l]|)$ time and $O(\log(|F_1[k]|)|F_2[l]|)$.

Therefore using $c|T_1||T_2|$ time for some constant c and $O(\log(|T_1|)|T_2|)$ space we can decompose the optimal mapping for $D(T_1, T_2)$ in to some components such that for each component the involved subtree or subforest of T_1 has a size less or equal to $0.5|T_1|$.

In the next step we will use the same algorithm for all the components and the total time is bounded by $c0.5|T_1||T_2|$.

The above analysis means that for our algorithm the space complexity is bounded by $O(\log(|T_1|)|T_2|)$ and the time complexity is bounded by $c|T_1||T_2| + c\frac{1}{2}|T_1||T_2| + c\frac{1}{4}|T_1||T_2| + ... \leq 2c|T_1|T_2| = O(|T_1||T_2|)$.

Theorem 1. *Given two ordered trees T_1 and T_2, the constrained edit distance and the optimal constrained mapping between them can be computed in $O(|T_1||T_2|)$ time and $O(\log(|T_1|)|T_2|)$ space.*

Proof. *Immediately from the above analysis.*

3 Alignment of Trees

The alignment of tree is another measure for comparison of ordered trees [5]. The definition of *alignment of trees* is as follows: Inserting a node u into T means that for some node v (could be a leaf) in T, we make u the parent of the consecutive subsequence of the children of v (if any) and then v the parent of u. We also allow to directly add/insert a node as a child of a leaf in the tree. Given two trees T_1 and T_2, an alignment of the two trees can be obtained by first inserting nodes labeled with spaces into T_1 and T_2 such that the two resulting tree T_1' and T_2' have the same structure, (i.e., they are identical if labels are ignored,) and then overlaying T_1' and T_2'. A score is defined for each pair of labels. The value of an alignment is the total score of all the opposing labels in the alignment. The problem here is to find an alignment with the optimal (maximum or minimum) value.

Theorem 2. *There exists an algorithm for alignment of trees that runs in* $O(|T_1||T_2|(deg(T_1) + deg(T_2))^2)$ *time and requires* $O(\log(|T_1|)|T_2|(deg(T_1)+deg(T_2))deg(T_1))$ *space.*

The basic ideas of the algorithm are similar to that for constrained edit distance. We first use the method in [11] to compute the cost of an optimal alignment. The remaining task is to get the alignment.

Let q be the node in T_1 such that $|T_1[q]| \geq 0.5|T_1|$ and for any child q_i of node q, $|T_1[q_i]| < 0.5|T_1|$. q is refereed to as a *cutting* point in T_1. By definition, there always exists a cutting point for any T_1.

By the definition of the alignment, node q is either aligned with a node $T_2[j]$ or aligned with an inserted node (labeled with empty). We can modify the algorithm for computing the cost of an optimal alignment so that when the cost of an optimal alignment is computed, we immediately know the configuration of node q in the alignment. Thus, we can decompose the alignment of the whole two tree T_1 and T_2 into two parts (1) the alignment of subtree $T_1[q]$ with a subtree or subforest of T_2, and (2) the alignment of the remaining parts of T_1 and T_2.

Repeating the process, we can get the alignment. We can show that the time required is still $O(|T_1||T_2|(deg(T_1) + deg(T_2))^2)$.

4 Conclusion

We have presented space efficient algorithms for the computation of constrained edit distance and tree alignment for ordered rooted trees. The techniques can also be applied to other tree comparison problems such as degree-one and degree-two edit distance between ordered trees.

Acknowledgements. This work is supported by a grant from City University of Hong Kong (Project No. 7001696) and the Natural Sciences and Engineering Research Council of Canada under Grant No. OGP0046373.

References

1. P. Bille, "A survey on tree edit distance and related problems", *Theoretical Computer Science*, no. 337, pp. 217-239, 2005.
2. Y. C. Cheng and S. Y. Lu, "Waveform correlation by tree matching", *IEEE Trans. PAMI*, vol. 7, pp.299-305, 1985

3. S. Dulucq and H. Touzet, "Decomposition algorithm for tree editing distance", *Journal of Discrete Algorithms*, 2004.
4. D.S. Hirschberg, "A linear space algorithm for computing maximal common subsequences", *Communications of the ACM*, no. 18, pp. 341-343, 1975.
5. T. Jiang, L. Wang and K. Zhang, 'Alignment of trees - an alternative to tree edit', *Theoretical Computer Science* vol. 143, no. 1, pp. 137-148, 1995.
6. P. Klein, "Computing the edit-distance between unrooted ordered trees", *Proceedings of 6th European Symposium on Algorithms*, pp. 91-102, 1998.
7. T. Richter, "A new measure of the distance between ordered trees and its applications", *Technical Report 85166-cs*, Department of Computer Science, University of Bonn, 1997.
8. B. Shapiro and K. Zhang, 'Comparing multiple RNA secondary structures using tree comparisons', *Comput. Appl. Biosci* vol. 6, no.4, pp.309-318, 1990
9. S.M. Selkow, "The tree-to-tree editing problem", *Information Processing Letters*, vol. 6, pp.184-186, 1977.
10. K.C. Tai, "The tree-to-tree correction problem", *J. ACM*, vol. 26, pp.422-433, 1979.
11. L. Wang and J. Zhao, "Parametric alignment of ordered trees", *Bioinformatics*, vol. 19, pp. 2237-2245, 2003.
12. K. Zhang, "Algorithms for the constrained editing distance between ordered labeled trees and related problems". *Pattern Recognition*, vol.28, no. 3, pp. 463-474, 1995.
13. K. Zhang, "Efficient parallel algorithms for tree editing problems", *Proceedings of the Seventh Symposium on Combinatorial Pattern Matching*, Laguna Beach, California, June 1996, *Springer-Verlag's Lecture Notes in Computer Science 1075*, pp 361-372.
14. K. Zhang and D. Shasha, 'Simple fast algorithms for the editing distance between trees and related problems', *SIAM J. Computing* vol. 18, no. 6, pp.1245-1262, 1989
15. K. Zhang, L. Wang, and B. Ma, 'Computing similarity between RNA structures', *Proceedings of the Tenth Symposium on Combinatorial Pattern Matching. LNCS 1645*, pp. 281-293, 1999.
16. K. Zhang, J.T.L. Wang and D. Shasha, 'On the editing distance between undirected acyclic graphs', *International Journal of Foundations of Computer Science*, vol. 7, no. 1, pp. 43-57, 1996.

A 1.75-Approximation Algorithm for Unsigned Translocation Distance

Yun Cui[1], Lusheng Wang[2], and Daming Zhu[1]

[1] School of Computer Science and Technology,
Shandong University, PR China
yuncui@cityu.edu.hk
dmzhu@sdu.edu.cn

[2] Department of Computer Science,
City University of HongKong, HongKong
cswangl@cityu.edu.hk

Abstract. The translocation operation is one of the popular operations for genome rearrangement. In this paper, we present a 1.75-approximation algorithm for computing unsigned translocation distance which improves upon the best known 2-approximation algorithm [1].

Keywords: Unsigned translocation distance, Approximation algorithm.

1 Introduction

A *chromosome* $X = x_1, x_2, \ldots, x_p$ is a sequence of genes, where each gene x_i is represented by an integer. A gene x_i has a direction. When the direction of every gene is known, we use a signed integer to indicate the direction. When the directions of genes are unknown, we use unsigned integers to represent the genes. Throughout this paper, each x_i in a *signed chromosome* is a signed integer, and each x_i in an *unsigned chromosome* is an unsigned integer. A *signed genome* is a set of signed chromosomes and an *unsigned* genome is a set of unsigned chromosomes.

For two unsigned chromosomes $X = x_1, x_2, \ldots, x_m$ and $Y = y_1, y_2, \ldots, y_n$ in a genome, a *translocation* swaps the segments in the chromosomes and generates two new chromosomes. A prefix-prefix translocation $\rho_{pp}(X, Y, i, j)$ generates two new chromosomes: $x_1, \ldots, x_{i-1}, y_j, \ldots, y_n$ and $y_1, \ldots, y_{j-1}, x_i, \ldots, x_m$. A prefix-suffix translocation $\rho_{ps}(X, Y, i, j)$ generates two new chromosomes: $x_1, \ldots, x_{i-1}, y_{j-1}, \ldots, y_1$ and $x_m, \ldots, x_i, y_j, \ldots, y_n$.

For two signed chromosomes $X = x_1, x_2, \ldots, x_m$ and $Y = y_1, y_2, \ldots, y_n$ in a genome, a prefix-prefix translocation $\rho_{pp}(X, Y, i, j)$ generates two new chromosomes: $x_1, \ldots, x_{i-1}, y_j, \ldots, y_n$ and $y_1, \ldots, y_{j-1}, x_i, \ldots, x_m$. A prefix-suffix translocation $\rho_{ps}(X, Y, i, j)$ generates two new chromosomes: $x_1, \ldots, x_{i-1}, -y_{j-1}, \ldots, -y_1$ and $-x_m, \ldots, -x_i, y_j, \ldots, y_n$.

The *translocation distance* between two (signed/unsigned) genomes is the minimum number of translocations used to transform one genome into the other.

X. Deng and D. Du (Eds.): ISAAC 2005, LNCS 3827, pp. 392–401, 2005.

Hannenhalli designed the first $O(n^3)$ algorithm [2] for computing translocation distance for signed genomes. The time complexity was improved to $O(n^2)$ in [3]. In [5], an error originated in [2] was fixed. The problem of computing translocation distance for unsigned genomes was recently proved to be NP-hard [4]. Kececioglu and Ravi gave a ratio-2 approximation algorithm for the translocation distance for unsigned genomes [1].

In this paper, we present a ratio-1.75 approximation algorithm for computing the translocation distance of unsigned genomes which improves upon the best known 2-approximation algorithm [1]. Our algorithm uses the maximum match method to find a cycle decomposition that contains enough number of 2-cycles (cycle containing exactly two black edges). By doing this, we give each unsigned gene a sign and the problem becomes the computation of translocation distance for signed genomes. Thus, we can use the algorithm in [3, 5] for signed genomes to finally get an approximation solution.

2 Signed and Unsigned Translocation

The basic idea of our approximation algorithm for unsigned genomes is to carefully assign a sign to each gene in the genomes and use the algorithm for signed genomes to compute the translocation distance. The approximation ratio purely depends on the quality of the sign assignment of each gene.

First, let us introduce the computation method for signed genomes.

2.1 Signed Translocation

Given signed genomes $\boldsymbol{A}$ and $\boldsymbol{B}$, the *breakpoint graph* $G_s(\boldsymbol{A},\boldsymbol{B})$ can be obtained as follows: for every chromosome $X = x_1, x_2, \ldots, x_n$ of $\boldsymbol{A}$, replace each x_i with an ordered pair $(l(x_i), r(x_i))$ of vertices. If x_i is positive, $(l(x_i),\ r(x_i)) = (x_i^t,\ x_i^h)$; if x_i is negative, $(l(x_i),\ r(x_i)) = (x_i^h,\ x_i^t)$. The vertices $r(x_i)$ and $l(x_{i+1})$ are *neighbors* in $\boldsymbol{A}$. The neighbors in $\boldsymbol{B}$ are defined analogously. For two vertices u and v, if they are neighbors in $\boldsymbol{A}$, then we use a black edge to connect them; if they are neighbors in $\boldsymbol{B}$, then we use a grey edge to connect them.

Every vertex in $G_s(\boldsymbol{A},\boldsymbol{B})$ is incident with at most one black and one grey edge. Therefore, $G_s(\boldsymbol{A},\boldsymbol{B})$ can be uniquely decomposed into *cycles*. A cycle containing exactly i black (grey) edges is called an *i-cycle*. A cycle is *long* if it is not a 1-cycle.

Let $X = x_1, x_2, \ldots, x_p$ be a chromosome in $\boldsymbol{A}$. A *subpermutation* (*SP*) is an interval $x_i, x_{i+1}, \ldots, x_{i+l}$ in X containing at least three genes such that there is another interval of the same length $y_j,\ y_{j+1}, \ldots, y_{j+l}$ in a chromosome Y of $\boldsymbol{B}$ satisfying $\{|x_i|, |x_{i+1}|, \ldots, |x_{i+l}|\} = \{|y_j|, |y_{j+1}|, \ldots, |y_{j+l}|\}$, $x_i = y_j$, $x_{i+l} = y_{j+l}$ and $x_i, x_{i+1}, \ldots, x_{i+l-1}, x_{i+l} \neq y_j, y_{j+1}, \ldots, y_{j+l-1}, y_{j+l}$. Here x_i and x_{i+l} are the two *ending* genes of the *SP*. A *minimal subpermutation* (*minSP*) is a *SP* not containing any other *SP*. By the definition of *SP*, we have

Lemma 1. *Let* $I = r(x_i), l(x_{i+1}), r(x_{i+1}), \ldots, l(x_{j-1}), r(x_{j-1}), l(x_j)$ *denote a* SP *in* $G_s(\boldsymbol{A}, \boldsymbol{B})$, *then the grey edge* $(r(x_i), l(x_j))$ *is not in* $G_s(\boldsymbol{A}, \boldsymbol{B})$. *Moreover, the two (ending) genes* x_i *and* x_j *cannot be neighbors in* $\boldsymbol{B}$.

The translocation distance for signed genomes is closely related to the number of cycles and the number of $minSP$'s. If all $minSP$'s in $G_s(\boldsymbol{A}, \boldsymbol{B})$ are in a SP, say, I, and the total number of $minSP$'s is even, then call I an *even-isolation*. Clearly there is at most one even-isolation in $G_s(\boldsymbol{A}, \boldsymbol{B})$.

Let n be the number of genes in the two genomes and N the number of chromosomes in the genomes. c denotes the total number of cycles in the breakpoint graph and s denotes the number of $minSP$'s. f is the *remaining index* which is defined as follows: (1) $f = 1$ if s is odd; (2) $f = 2$ if there is an even-isolation; (3) $f = 0$ otherwise. Lemma 2 gives the formula to compute the translocation distance $d_s(\boldsymbol{A}, \boldsymbol{B})$ for the two signed genomes $\boldsymbol{A}$ and $\boldsymbol{B}$.

Lemma 2. *[2]*

$$d_s(\boldsymbol{A}, \boldsymbol{B}) = n - N - c + s + f. \tag{1}$$

2.2 Unsigned Translocation

Consider unsigned genomes $\boldsymbol{A}$ and $\boldsymbol{B}$. For every chromosome $X = x_1, x_2, \ldots, x_n$ of $\boldsymbol{A}$, x_i and x_{i+1} are *neighbors* in $\boldsymbol{A}$. The neighbors in $\boldsymbol{B}$ are defined analogously. To define the *breakpoint graph* $G(\boldsymbol{A}, \boldsymbol{B})$, we use a vertex to represent a gene. Two vertices are connected with a black edge if they are neighbors in $\boldsymbol{A}$ and two vertices are connected with a grey edge if they are neighbors in $\boldsymbol{B}$.

Note that every vertex is incident either with one black and one grey edge, or with two black and two grey edges. Therefore, the cycle decompositions for $G(\boldsymbol{A}, \boldsymbol{B})$ are not unique. Once we have a cycle decomposition for the breakpoint graph of two unsigned genomes, we actually assign a sign to each gene in the genomes. Thus, one way to compute the translocation distance for two unsigned genomes is to (1) try all possible ways to get cycle decomposition (thus we can get a sign for each gene), and (2) compute the translocation distance for signed genomes and select the minimum value among all possible cycle decompositions.

3 The Approximation Algorithm

If we can give a good approximation of the cycle decomposition of the unsigned case, we can get a good approximation solution for the unsigned translocation distance. Our main idea of the approximation algorithm is to give a cycle decomposition of $G(\boldsymbol{A}, \boldsymbol{B})$ that contains the maximum number of 1-cycles and a sufficient number of 2-cycles.

Why the ratio could be better than 2?
Now, we give an intuitive explanation that if we keep the maximum number of 1-cycles and maximum number of 2-cycles in assigning signs to genes, then the best performance ratio we can expect is 1.5.

Suppose that we ignore the effect of s and f in formula (1). That is, we assume that $s = 0$ and $f = 0$ in the optimal cycle decomposition. Then $d_s(\boldsymbol{A}, \boldsymbol{B}) = n - N - c$. Let c_i^* be the number of i-cycles in the optimal cycle decomposition. Then

$$d_s(\boldsymbol{A}, \boldsymbol{B}) = n - N - c = n - N - c_1^* - c_2^* - \sum_{i \geq 3} c_i^*. \quad (2)$$

$n - N$ is the number of black edges in the breakpoint graph. We further assume that $c_1^* = 0$, $c_2^* = 0$ and all black edges are in 3-cycles in the optimal cycle decomposition. In this case, $d_s(\boldsymbol{A}, \boldsymbol{B}) = n - N - \frac{n-N}{3} = \frac{2}{3}(n - N)$. If in the approximation solution, we do not care about i-cycles for $i \geq 3$, the distance for the approximation solution could be $n - N$. Thus, the ratio becomes $\frac{3}{2}$. In our approximation algorithm, we cannot get the maximum number of 2-cycles, but we get a large number of 2-cycles. Besides, we have to design sophisticated ways to deal with the other two parameters s and f in the analysis.

The cycle decomposition algorithm

Given unsigned genomes $\boldsymbol{A}$ and $\boldsymbol{B}$, a cycle decomposition of $G(\boldsymbol{A}, \boldsymbol{B})$ can be computed in the following three steps.

Step 1: Decomposition of 1-cycles

If two vertices are joined by a black edge and a grey edge in $G(\boldsymbol{A}, \boldsymbol{B})$, then assign proper signs to the two vertices to obtain the 1-cycle containing the black edge and the grey edge. Thus, if two genes are neighbors in both genomes, the corresponding 1-cycle is kept in the cycle decomposition.

Step 2: Decomposition of 2-cycles

From $G(\boldsymbol{A}, \boldsymbol{B})$, we define a new graph, called *match graph*, $F_{\boldsymbol{AB}}$ as follows: (1) For every black edge in $G(\boldsymbol{A}, \boldsymbol{B})$ with at least one end not assigned a sign in Step 1, we create a vertex of $F_{\boldsymbol{AB}}$. (2) For every two vertices of $F_{\boldsymbol{AB}}$ (representing two black edges in $G(\boldsymbol{A}, \boldsymbol{B})$), we create an edge connecting them in $F_{\boldsymbol{AB}}$ if the two black edges in $G(\boldsymbol{A}, \boldsymbol{B})$ can form a 2-cycle. $F_{\boldsymbol{AB}}$ can be constructed in $O(n^2)$ time where n is the number of genes.

Let M denote a maximum match of $F_{\boldsymbol{AB}}$. $|M|$ is the size of the match. A maximum match of any graph can be found in $O(|V|\,|E|^{\frac{1}{2}})$ time, where $|V|$ is the number of vertices and $|E|$ is the number of edges [11]. Since $F_{\boldsymbol{AB}}$ contains at most n vertices and $O(n)$ edges, M can be found in $O(n^{\frac{3}{2}})$ time. Every edge in M represents a 2-cycle of $G(\boldsymbol{A}, \boldsymbol{B})$. By the construction, two 2-cycles in M cannot share any black edge of $G(\boldsymbol{A}, \boldsymbol{B})$. However, they may share a grey edge in $G(\boldsymbol{A}, \boldsymbol{B})$. In that case, the two 2-cycles cannot be kept in the cycle decomposition simultaneously. A 2-cycle in M is *isolated* if it does not share any grey edge with any other 2-cycles in M. Otherwise, the 2-cycle is *related*. Since a 2-cycle has two grey edges, it is related to at most two 2-cycles.

A *related component* U consists of related cycles $C_1, C_2, \ldots, C_k$, where C_i is related to C_{i-1} $(2 \leq i \leq k)$, and every 2-cycle in U is not related to any 2-cycle not in U. A related component involves at most two chromosomes, and can be one of the four types shown in Figure 1.

In our cycle decomposition, we keep all the isolated 2-cycles and alternatively select 2-cycles from every related component. Assume that a maximum match M of $F_{\boldsymbol{AB}}$ contains z isolated 2-cycles. In our cycle decomposition approach, we can keep at least $\lceil\frac{|M|-z}{2}\rceil+z$, i.e., $\lceil\frac{|M|+z}{2}\rceil$ 2-cycles in Step 2.

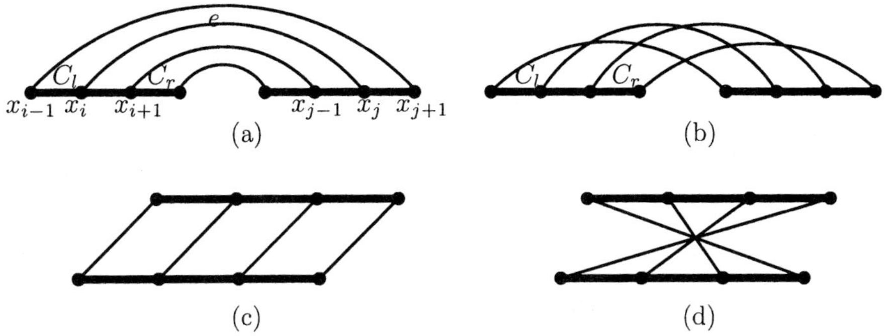

Fig. 1. The four cases of related components including three 2-cycles

Step 3: Decomposition of other long cycles

After the decomposition of 2-cycles, the other long cycles can be arbitrarily selected from the remaining graph.

The long cycles created in Step 2 are called *selected* cycles and the cycles created in Step 3 are called *arbitrary* cycles.

Our approximation algorithm for unsigned translocation problem is as follows:

Algorithm 1:

Input: $G(\boldsymbol{A},\boldsymbol{B})$

1. Compute the cycle decomposition of $G(\boldsymbol{A},\boldsymbol{B})$ as described before. Denote the resulting graph as $G_s^A(\boldsymbol{A},\boldsymbol{B})$.

2. Solve the signed case using the standard algorithm.

Let n be the number of genes in the given genomes. $G(\boldsymbol{A},\boldsymbol{B})$ and $F_{\boldsymbol{AB}}$ can be constructed in $O(n^2)$ time. A maximum match of $F_{\boldsymbol{AB}}$ can be found in $O(n^{\frac{3}{2}})$ time. The algorithm in [3] requires $O(n^2)$ time to compute an optimal sequence of translocations for signed case. Thus, the total time required for our approximation algorithm is $O(n^2)$.

A *minSP* $I=r(x_i),x_{i+1},\ldots,x_{j-1},l(x_j)$ *contains* a cycles C if all vertices of C are in $\{r(x_i),l(x_{i+1}),r(x_{i+1})\ldots,l(x_{j-1}),r(x_{j-1}),l(x_j)\}$. A cycle C is *outside* I if no vertex of C is in $\{r(x_i),l(x_{i+1}),r(x_{i+1})\ldots,l(x_{j-1}),r(x_{j-1}),l(x_j)\}$.

Lemma 3. *If a minSP contains a selected related 2-cycle in* $G_s^A(\boldsymbol{A},\boldsymbol{B})$, *then this minSP contains at least one arbitrary cycle.*

4 Analysis of the Performance Ratio

In this section, we will show that the performance ratio of the algorithm is 1.75. We use several new bounds in our analysis.

Suppose that each of the given genomes has n genes and N chromosomes. Let $d(\boldsymbol{A}, \boldsymbol{B})$ denote the (optimal) translocation distance between two unsigned genomes $\boldsymbol{A}$ and $\boldsymbol{B}$, and $G_s^{opt}(\boldsymbol{A}, \boldsymbol{B})$ the breakpoint graph of an optimal cycle decomposition.

4.1 1-Cycles

In this subsection, we will show that Step 1 in the cycle decomposition algorithm always leads to a good approximation solution.

Lemma 4. *We modify $G_s^{opt}(\boldsymbol{A}, \boldsymbol{B})$ as follows: if two vertices in $G(\boldsymbol{A}, \boldsymbol{B})$ are connected by a black edge and a grey edge in $G(\boldsymbol{A}, \boldsymbol{B})$, then we re-assign the signs of the two genes to obtain a 1-cycle. Assume that the resulting breakpoint graph has c' cycles and s' minSP's. We have $d(\boldsymbol{A}, \boldsymbol{B}) \geq n - N - c' + s' + f^o$, where f^o is the remaining index for $G_s^{opt}(\boldsymbol{A}, \boldsymbol{B})$.*

4.2 A Lower Bound

In this subsection, we give a lower bound for $d(\boldsymbol{A}, \boldsymbol{B})$. This lower bound will be used as the starting point of our analysis.

Note that every *minSP* contains at least one long cycle. A *simple minSP* (*S*-MSP) is a *minSP* containing *one* 2-cycle as its *unique* long cycle. By definition, a simple *minSP* is a segment of genes in a chromosome containing 1-cycle(s) in the middle of the segment and a 2-cycle containing the two black edges at the two ends of the segments. The two grey edges in the 2-cycle must be "twisted" since by Lemma 1 $(r(x_i), l(x_j))$ cannot be a grey edge for the two ending genes x_i and x_j. The whole analysis of the approximation algorithm depends heavily on the special treatment of simple *minSP*'s.

Given unsigned genomes $\boldsymbol{A}$ and $\boldsymbol{B}$, a *candidate simple minSP* (*CS*-MSP for short) is defined as an interval $I_c = x_i, x_{i+1}, \ldots, x_{i+l-1}, x_{i+l}$ containing at least *four* genes in a chromosome of $\boldsymbol{A}$ such that there is another interval of the same length $y_j, y_{j+1}, \ldots, y_{j+l}$ in a chromosome Y of $\boldsymbol{B}$ satisfying $x_i = y_j$, $x_{i+l} = y_{j+l}$ and $x_{i+k} = y_{j+l-k}$ $(1 \leq k \leq l-1)$. Any *CS*-MSP can be turned into a *S*-MSP by assigning proper signs to all genes in it. For convenience, we also call the unique 2-cycle in the *S*-MSP, the *unique 2-cycle* in the *CS*-MSP.

Given signed genomes $\boldsymbol{A}$ and $\boldsymbol{B}$, let $I_s = x_i, x_{i+1}, \ldots, x_{j-1}, x_j$ be a *S*-MSP in $G_s(\boldsymbol{A}, \boldsymbol{B})$. A cycle $C = r(x_{i-1}), l(x_i), \ldots, l(x_{j+1}), r(x_j), \ldots, r(x_{i-1})$ in $G_s(\boldsymbol{A}, \boldsymbol{B})$ containing the two black edges $(r(x_{i-1}), l(x_i))$ and $(r(x_j), l(x_{j+1}))$ on the left and right of I_s is called a *removable cycle.* (See Figure 2 (a).) If there is a removable cycle C for I_s, then I_s is called a *removable simple minSP* (*RS*-MSP for short).

Lemma 5. *Given a RS-MSP* $I_s = x_i, x_{i+1}, \ldots, x_{j-1}, x_j$, *if we change the signs of* x_i *and* x_j, *then we have*

(*a*) $I_1 = -x_i, x_{i+1}, \ldots, x_{j-1}, -x_j$ *is no longer a minSP;*

(*b*) *the number of cycles in the new breakpoint graph remains the same and the number of minSP's is not increased.*

(*c*) *every black edge in the CS-MSP is in a long cycle containing a black edge that is either* $(r(x_{i-1}), l(x_i))$, *or* $(r(x_j), l(x_{j+1}))$.

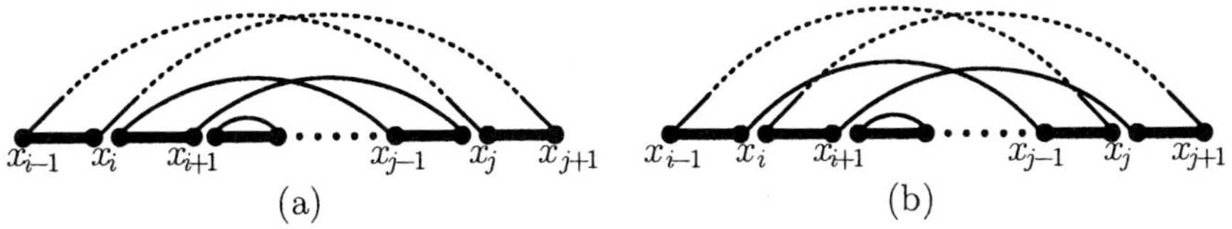

Fig. 2. The breakpoint graphs before and after a *RS*-MSP is destroyed

The lower bound of $d(\boldsymbol{A}, \boldsymbol{B})$ we are going to develop is based on the modification of CS-MSP's in an optimal cycle decomposition $G_s^{opt}(\boldsymbol{A}, \boldsymbol{B})$.

Modifying an optimal cycle decomposition $G_s^{opt}(\boldsymbol{A}, \boldsymbol{B})$

Let I_c be a CS-MSP and $l(I_c)$ and $r(I_c)$ denote the leftmost and rightmost genes of I_c. The modification method is as follows:

ModificationMethod:
Input: $G_s^{opt}(\boldsymbol{A}, \boldsymbol{B})$
1. **For** every chromosome X of $\boldsymbol{A}$,
2. Obtain possible 1-cycles as described in Step 1 of cycle decomposition.
3. **For** every chromosome X of $\boldsymbol{A}$,
4. Process each CS-MSP I_c in X from left to right:
5. Assign proper signs to $l(I_c)$ and $r(I_c)$ to turn I_c into a S-MSP I_s.
6. If I_s is a RS-MSP, then remove it by changing the signs of both $l(I_s)$ and $r(I_s)$.

Theorem 1. c^* *and* s^* *denote the number of cycles and number of minSP's in the new breakpoint graph after ModificationMethod. We have* $d(\boldsymbol{A}, \boldsymbol{B}) \geq n - N - c^* + s^* + f^o$, *where* f^o *is the remaining index for* $G_s^{opt}(\boldsymbol{A}, \boldsymbol{B})$.

4.3 A Key Inequality

Given unsigned genomes $\boldsymbol{A}$ and $\boldsymbol{B}$, let s_c denote the number of CS-MSP's in $G(\boldsymbol{A}, \boldsymbol{B})$. Let c_i^* denote the number of i-cycles and s_e^* the number of S-MSP's in the new breakpoint graph after applying ModificationMethod.

Theorem 2. $\sum_{i\geq 2}(i-1)c_i^* \geq 2(s_c - s_e^*)$.

For unsigned genomes $\boldsymbol{A}$ and $\boldsymbol{B}$, let $G_s^*(\boldsymbol{A},\boldsymbol{B})$ be the breakpoint graph produced by running ModificationMethod on $G_s^{opt}(\boldsymbol{A},\boldsymbol{B})$. $G_s^A(\boldsymbol{A},\boldsymbol{B})$ is the breakpoint graph produced by Algorithm 1. f is the remaining index for $G_s^A(\boldsymbol{A},\boldsymbol{B})$. We use $d^A(\boldsymbol{A},\boldsymbol{B})$ to represent the translocation distance obtained by Algorithm 1. Let $d(\boldsymbol{A},\boldsymbol{B})$ be the (optimal) translocation distance between the two unsigned genomes. Now, we are ready to show the performance ratio.

4.4 The Performance Ratio When $f = 0$

Assume that $G_s^A(\boldsymbol{A},\boldsymbol{B})$ contains z isolated 2-cycles. Let $z^{(o)}$ denote the number of isolated 2-cycles outside all $minSP$'s. Consider the $minSP$'s containing only isolated 2-cycles and 1-cycles. Let $s^{(s)}$ denote the number of (simple) $minSP$'s containing only one isolated 2-cycle. $s^{(m)}$ denotes the number of $minSP$'s containing at least two isolated 2-cycles without any selected related 2-cycle or arbitrary 2-cycle. Let $c_i^{(o)}$ be the number of arbitrary i-cycles ($i \geq 2$) outside all $minSP$'s in $G_s^A(\boldsymbol{A},\boldsymbol{B})$.

Theorem 3. *If f=0, then $d^A(\boldsymbol{A},\boldsymbol{B}) \leq \frac{7}{4}d(\boldsymbol{A},\boldsymbol{B})$. That is, the performance ratio of Algorithm 1 is 1.75 if $f = 0$.*

Proof. By definition, $2s^{(m)} + s^{(s)} \leq z - z^{(o)}$. Thus, we have

$$s^{(m)} \leq \frac{z - z^{(o)} - s^{(s)}}{2}. \tag{3}$$

Suppose that $G_s^A(\boldsymbol{A},\boldsymbol{B})$ has s $minSP$'s. c_i ($i \geq 1$) denotes the number of i-cycles in $G_s^A(\boldsymbol{A},\boldsymbol{B})$. Similarly, c_i^* denotes the number of i-cycles in $G_s^*(\boldsymbol{A},\boldsymbol{B})$. By Lemma 3, a $minSP$ contains (at least) an isolated 2-cycle or an arbitrary cycle. Thus, there are $s - s^{(m)} - s^{(s)}$ $minSP$'s, each containing at least one arbitrary cycle. Since there are at least $\lceil\frac{|M|+z}{2}\rceil$ selected 2-cycles created in Step 2 of the cycle decomposition algorithm, the number of arbitrary cycles in $minSP$'s is less than or equal to $\sum_{i\geq 2} c_i - (\frac{|M|}{2} + \frac{z}{2}) - \sum_{i\geq 2} c_i^{(o)}$. We have

$$s - s^{(m)} - s^{(s)} \leq \sum_{i\geq 2} c_i - (\frac{|M|}{2} + \frac{z}{2}) - \sum_{i\geq 2} c_i^{(o)}. \tag{4}$$

Combining (3) and (4), we have

$$s \leq \sum_{i\geq 2} c_i - \sum_{i\geq 2} c_i^{(o)} - \frac{|M|}{2} - \frac{z^{(o)}}{2} + \frac{s^{(s)}}{2}. \tag{5}$$

By Lemma 2,

$$d^A(\boldsymbol{A},\boldsymbol{B}) = n - N - c_1 - c_2 - \sum_{i\geq 3} c_i + s + f. \tag{6}$$

From Theorem 1, we have

$$d(\boldsymbol{A},\boldsymbol{B}) \geq n - N - c_1^* - c_2^* - \sum_{i\geq 3} c_i^* + s^* + f^o. \tag{7}$$

Let $\triangle = \frac{7}{4}d(\boldsymbol{A},\boldsymbol{B}) - d^A(\boldsymbol{A},\boldsymbol{B})$. Since $G_s^*(\boldsymbol{A},\boldsymbol{B})$ and $G_s^A(\boldsymbol{A},\boldsymbol{B})$ contain all possible 1-cycles, $c_1 = c_1^*$. Since a cycle decomposition of $G(\boldsymbol{A},\boldsymbol{B})$ contains at most $|M|$ 2-cycles, then $c_2^* \leq |M|$. From (7) and (6), we have

$$\begin{aligned}\triangle &= \frac{7}{4}d(\boldsymbol{A},\boldsymbol{B}) - d^A(\boldsymbol{A},\boldsymbol{B}) \\ &\geq \frac{7}{4}(n - N - c_1^* - c_2^* - \sum_{i\geq 3} c_i^* + s^* + f^o) - (n - N - c_1 - \sum_{i\geq 2} c_i + s + f) \\ &= \frac{3}{4}(n - N - c_1^*) - \frac{5}{4}c_2^* - \frac{|M|}{2} - \frac{7}{4}\sum_{i\geq 3} c_i^* + \frac{7}{4}s^* + \frac{7}{4}f^o + \sum_{i\geq 2} c_i - s - f. \end{aligned} \tag{8}$$

From (8) and (5), we have

$$\begin{aligned}\triangle \geq &\frac{1}{4}(n - N - c_1^* - c_2^* - \sum_{i\geq 3} c_i^*) + \frac{1}{2}(n - N - c_1^* - 2c_2^* - 3\sum_{i\geq 3} c_i^*) - \frac{|M|}{2} \\ &+ \frac{7}{4}s^* + \frac{7}{4}f^o + \sum_{i\geq 2} c_i^{(o)} + \frac{|M|}{2} + \frac{z^{(o)}}{2} - \frac{s^{(s)}}{2} - f. \end{aligned} \tag{9}$$

Since there are $n - N$ black edges in $G_s^*(\boldsymbol{A},\boldsymbol{B})$ and each black edge is in a cycle, we have $n - N == \sum_{i\geq 1} ic_i^*$. That is,

$$n - N - c_1^* - c_2^* - \sum_{i\geq 3} c_i^* = \sum_{i\geq 2}(i-1)c_i^*. \tag{10}$$

From (9) and (10), we can immediately obtain

$$\begin{aligned}\triangle \geq &\frac{1}{4}(\sum_{i\geq 2}(i-1)c_i^* + 2s^* - 2s^{(s)}) + \frac{1}{2}\sum_{i\geq 4}(i-3)c_i^* + \frac{5}{4}s^* + \frac{7}{4}f^o \\ &+ \sum_{i\geq 2} c_i^{(o)} + \frac{z^{(o)}}{2} - f. \end{aligned} \tag{11}$$

From Theorem 2, $\sum_{i\geq 2}(i-1)c_i^* \geq 2(s_c - s_e^*)$. Moreover, by definitions, $s_c \geq s^{(s)}$ and $s^* \geq s_e^*$. Thus, we have

$$\sum_{i\geq 2}(i-1)c_i^* + 2s^* - 2s^{(s)} \geq 0. \tag{12}$$

From (12), (11) becomes

$$\triangle \geq \frac{1}{2}\sum_{i\geq 4}(i-3)c_i^* + \frac{5}{4}s^* + \frac{7}{4}f^o + \sum_{i\geq 2} c_i^{(o)} + \frac{z^{(o)}}{2} - f. \tag{13}$$

From the fact that all variables in (13) are non-negative, we can immediately conclude that $\triangle \geq 0$ when $f = 0$. □

Corollary 1. *There exists an polynomial time algorithm with ratio-*$(1.75 + \epsilon)$ *for computing the translocation distance for unsigned genomes.*

Proof. For any constant d, if the translocation distance between $\boldsymbol{A}$ and $\boldsymbol{B}$ is less than d, we can give an optimal solution in polynomial time. The ratio of Algorithm 1 could be large than 1.75 when $f = 1$ or $f = 2$. However, if the distance between $\boldsymbol{A}$ and $\boldsymbol{B}$ is large enough, the ratio is at most $1.75 + \epsilon$.

With a more precise analysis, we can show that

Theorem 4. *The performance ratio of Algorithm 1 is* 1.75.

The details will be given in the full paper.

Acknowledgements. The work is supported by a grant from the Research Grants Council of the Hong Kong Special Administrative Region, China [Project No. CityU 1070/02E].

References

1. J. Kececioglu and R. Ravi. Of mice and men: Algorithms for evolutionary distances between genomes with translocation. In *6th ACM-SIAM Symposium on Discrete Algorithms*, 604-613, 1995.
2. Sridhar Hannenhalli. Polynomial-time Algorithm for Computing Translocation Distance between Genomes. *CPM'95*, 162-176, 1995.
3. Lusheng Wang, Daming Zhu, Xiaowen Liu, and Shaohan Ma. An $O(n^2)$ algorithm for signed translocation, *Journal of Computer and System Sciences*, **70**, 284-299, 2005.
4. Daming Zhu and Lusheng Wang. On the Complexity of Unsigned Translocation Distance. *submited to theoretical computer science.*
5. Anne Bergeron, Julia Mixtacki, Jens Stoye, On sorting by translocation, *RECOMB'05*, 615-629, 2005.
6. S. Hannenhalli and P. A. Pevzner. Transforming cabbage into turnip (polynomial algorithm for sorting signed permutations by reversals). *STOC'95*, 178-189, 1995.
7. H. Kaplan, R. Shamir, and R. E. Tarjan. Faster and simpler algorithm for sorting signed permutations by reversals. *SIAM Journal on Computing*, **29**(3), 880-892, 2000.
8. A.Caprara. Sorting by reversals is difficult. *Proceedings of the 1st Annual International Conference on Research Computational Molecular Biology*, 84-93, 1999.
9. V. Bafna and P. Pevzner. Genome rearrangements and sorting by reversals. *SIAM Journal on Computing*, **25**(2), 272-289, 1996.
10. D.A. Christie. A 3/2 Approximation Algorithm for Sorting by Reversals. *Proceedings of the 9th Annual ACM-SIAM Symposium on Discrete Algorithms*,244-252, 1998.
11. L. Lovász and M. D. Plummer. *Annals of Discrete Mathematics (29): Matching Theory.* North-Holland, Amsterdam, 1986

Fast Algorithms for Computing the Tripartition-Based Distance Between Phylogenetic Networks

Nguyen Bao Nguyen, C. Thach Nguyen, and Wing-Kin Sung

National University of Singapore, 3 Science Drive 2, Singapore 117543
{baonguyen, thachnguyen, dcsswk}@nus.edu.sg

Abstract. Consider two phylogenetic networks N and N' of size n. The tripartition-based distance finds the proportion of tripartitions which are not shared by N and N'. This distance is proposed by Moret et al (2004) and is a generalization of Robinson-Foulds distance, which is orginally used to compare two phylogenetic trees. This paper gives an $O(\min\{kn \log n, n \log n + hn\})$-time algorithm to compute this distance, where h is the number of hybrid nodes in N and N' while k is the maximum number of hybrid nodes among all biconnected components in N and N'. Note that $k << h << n$ in a phylogenetic network. In addition, we propose algorithms for comparing galled-trees, which are an important, biological meaningful special case of phylogenetic network. We give an $O(n)$-time algorithm for comparing two galled-trees. We also give an $O(n + kh)$-time algorithm for comparing a galled-tree with another general network, where h and k are the number of hybrid nodes in the latter network and its biggest biconnected component respectively.

1 Introduction

Phylogenetic trees are traditionally used to describe evolutionary relationships among a set of objects. However, evolutionary events such as horizontal gene transfer or hybrid speciation (often referred to as *recombination events*) which suggest convergence between objects cannot be adequately represented in a single tree structure. In the famous Science paper [3] by Doolittle, he also pointed out that the phylogenetic tree is inadequated to represent the "true" evolution history. To solve the shortcoming, phylogenetic networks were introduced. A *phylogenetic network* is a distinctly leaf-labeled directed acyclic graph where the in-degree and out-degree of all nodes are bounded above by 2. The nodes with in-degree 2 are called hybrid nodes and they are used to model the recombination events. Fig. 1(a) shows an example of a phylogenetic network.

Recently, a lot of works have been proposed to reconstruct phylogenetic networks [1,4,5,6,8,9,10,11,15,16,17,19]. To access the topological accuracy of different construction methods, two measurements were proposed for comparing networks. They are Maximum Agreement Phylogenetic Subnetwork (MASN)[2,13], and Tripartition-based distance[14]. This paper focuses on the latter measurement.

X. Deng and D. Du (Eds.): ISAAC 2005, LNCS 3827, pp. 402–411, 2005.

The tripartition-based distance is a generalization of the Robinson-Foulds measure [18], which is a well-known method for comparing phylogenetic trees. Given two phylogenetic networks N and N' which have the same leaf set, the tripartition-based distance $tri(N, N')$ computes the proportion of tripartitions (defined in Section 2.3) which are not shared by N and N'. The tripartition-based distance is shown to be a distance metric[14]. More importantly, when both N and N' are trees, $tri(N, N')$ equals the Robinson-Foulds distance between them.

In this paper, we compute $tri(N, N')$ in $O(\min\{kn \log n, n \log n + hn\})$ time where $n = \max\{V(N), V(N')\}$, h is the maximum number of hybrid nodes in N and N', and k is the maximum number of hybrid nodes among all the biconnected components in N and N'. As the number of hybrid nodes in a network is relatively rare (recombination events do not happen frequently), $k << h << n$. Thus, in practice, the running time of our algorithm achieves $O(n \log n)$.

We also consider comparing an important, biologically motivated special case of phylogenetic networks, known as galled-trees. A galled-tree [2,5,10,11,12,13,16,19] (also referred to in the literature as a *a level-1 network* [2,12], a *gt-network* [16], or a *topology with independent recombination events* [19]) is a phylogenetic network in which all cycles in the underlying undirected graph are node-disjoint (see the network N in Fig. 2 for an example). When both N and N' are galled-trees, we show that $tri(N, N')$ can be computed in $O(n)$ time. If only N is known to be a galled-tree, $tri(N, N')$ can be computed in $O(n + kh)$ where h and k are the number of hybrid nodes in N' and in the biggest biconnected component of N' respectively.

Our improvement is stemmed from a novel labeling technique which labels the nodes of the networks to facilite efficient identification of common tripartitions between two networks.

The rest of the paper is organized as follows. We first present the preliminaries in Section 2. Section 3 details the results for computing $tri(N, N')$ when at least one of N and N' is a galled-tree. Finally, we present the result for computing the tripartition-based distance for two general networks in Section 4.

2 Preliminaries

2.1 Phylogenetic Network

A *phylogenetic tree* is a binary, rooted, unordered tree whose leaves are distinctly labeled. A *phylogenetic network* is a generalization of a phylogenetic tree formally defined as a rooted, connected, directed acyclic graph in which: (1) exactly one node has indegree 0 (the *root*), and all other nodes have indegree 1 or 2; (2) all nodes with indegree 2 (referred to as *hybrid nodes*) have outdegree 1, and all other nodes have outdegree 0 or 2; and (3) all nodes with outdegree 0 (the *leaves*) are distinctly labeled.

For each hybrid node h, there are several nodes, called *split nodes* of h, from which there are two disjoint paths, called *merge paths*, to h. Two merge paths of

h starting from the same split node u form a simple cycle, called the *recombinant cycle* of u. For example, in Fig. 2, the root of N' and v_2 are split nodes of v_5.

For any phylogenetic network N, let $\mathcal{U}(N)$ be the undirected graph obtained from N by replacing each directed edge by an undirected edge. The level of N [2] is the maximum number of hybrid nodes among all biconnected components of $\mathcal{U}(N)$. A *galled-tree* is a network of level 1.

In the following, we will use $V(N)$, $E(N)$, $L(N)$, $h(N)$ and $k(N)$ to denote the set of nodes, edges, leaves, the number of hybrid node and the level of the network N.

2.2 Component Tree of a Network

We decompose $\mathcal{U}(N)$ into biconnected components (or simply components). In each biconnected component C in $\mathcal{U}(N)$, there is a node being an ancestor of all the others, called the component's sub-root and denoted by $r(C)$. For a node $x \in C$, a child u of x is called its *internal child* if $u \in C$; otherwise, u is x's *external child. Internal descendants* and *external descendants* of a node are similarly defined.

Given a component C, its *reduced component* C^r is obtained by contracting all nodes which have one external and one internal children (by "contracting" a node, we mean deleting it and letting all its children become its parent's children). The *reduced network* N^r is obtained by replacing each component of N by its reduced component. Fig. 1(a) and (c) show a network N and its reduced network N^r.

If we consider every biconnected component in N as a node, the resulting graph is a tree and we denoted it as biconnected component tree $\mathcal{T}(N)$ (see Fig. 1(b) for an example). One property of $\mathcal{T}(N)$ is that edges from one component to another have the sub-roots of the latter as their heads.

We will use the term *children* to indicate both children of a node in a network N as well as children of a component in $\mathcal{T}(N)$. *Parents*, *ancestors* and *descendants* will also be used in this way. Thus, children, parents, descendants and ancestors of a node are nodes in N whereas those of a component are components in $\mathcal{T}(N)$.

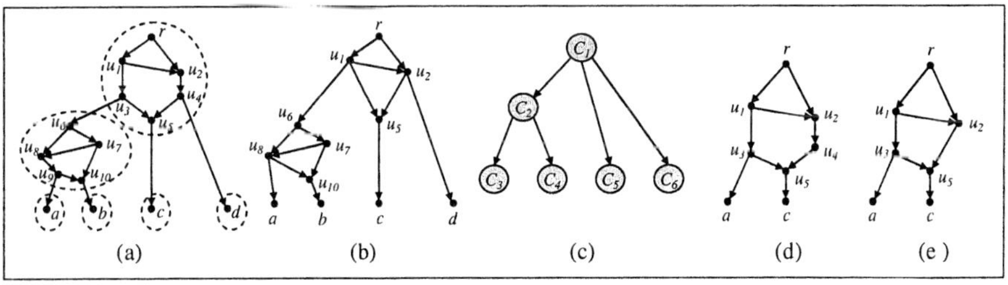

Fig. 1. (a) A network N. (b) The reduced network N^r. (c) The component tree $\mathcal{T}(N)$. (d) $N^*_{\{a,c\}}$. (e) $N_{\{a,c\}}$.

The following lemma states an important property of the reduced components.

Lemma 1. *Each reduced component contains $O(t)$ nodes and $O(t)$ internal edges where t is the number of its hybrid nodes.*

Proof. A node in a reduced component belongs to one of 4 types A, B, C and D whose internal in and out degrees are 0 and 2, 1 and 2, 2 and 1 and 2 and 0 respectively. Let a, b, c, d be the numbers of nodes belong to each type respectively. We have $c + d = t$ and $2a + 2b + c = b + 2c + 2d$. This implies $a + b \leq 2a + b = c + 2d \leq 2(c + d) = 2t$. Thus $a + b + c + d \leq 3t$. □

2.3 Tripartition-Based Measure

Consider a phylogenetic network N leaf-labeled by S. For a node $u \in V(N)$, an ancestor v of u is called its *strict ancestor* if all paths from the root of N to u contain v. Otherwise, v is called a *non-strict ancestor* of u. The tripartition of u is $(A(u), B(u), C(u))$ where $A(u) = \{s \in S | u$ is a strict ancestor of $s\}$; $B(u) = \{s \in S | u$ is a non-strict ancestor of $s\}$; and $C(u) = \{s \in S | u$ is not an ancestor of $s\}$. We also denote $A(u) \cup B(u)$ as $D(u)$.

Given two networks N and N' having the same leaf set, a node u in one network is *unmatched* if and only if there is no node v in the other network such that $A(u) = A(v)$, $B(u) = B(v)$ and $C(u) = C(v)$. An edge (u, v) is unmatched if and only if v is unmatched. The tripartition-based distance $tri(N, N')$ between N and N' is defined by:

$$\left(\frac{|\{e \in E(N) | e \text{ is unmatched}\}|}{|E(N)|} + \frac{|\{e \in E(N') | e \text{ is unmatched }\}|}{|E(N')|}\right)/2$$

Fig. 2 shows an example of how to compute tripartition of all nodes in N and N'. All the nodes are unmatched. Hence, all the edges whose heads are of these nodes are unmatched and $tri(N, N') = (7/10 + 7/10)/2 = 0.7$.

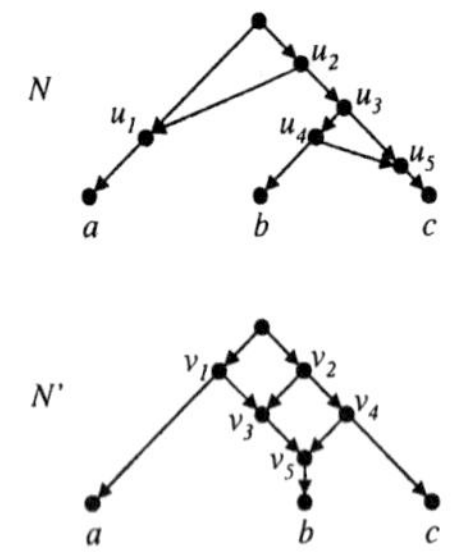

Node	A(u)	B(u)	C(u)
u1	{a}	∅	{b,c}
u2	{b,c}	{a}	∅
u3	{b,c}	∅	{a}
u4	{b}	{c}	{a}
u5	{c}	∅	{a,b}

Node	A(u)	B(u)	C(u)
v1	{a}	{b}	{c}
v2	{c}	{b}	{a}
v3	∅	{b}	{a,c}
v4	{c}	{b}	{a}
v5	{b}	∅	{a,c}

Fig. 2. Two networks N and N' whose tripartition-based distance $tri(N, N') = 0.7$

3 Comparing a Galled-Tree and a General Network

Given a galled-tree N and a general network N' having the same leaf set, this section describes an algorithm to identify their unmatched nodes. Once all such nodes are identified, $tri(N, N')$ can be readily computed. The algorithm processes in 3 steps as in Fig. 3.

Algorithm *UnmatchedNode*
Input: A galled-tree N and a general network N' of the same leaf set
Output: The unmatched nodes in the two networks
1 Label the nodes of N and N' such that two nodes $u \in V(N)$ and $u' \in V(N')$ have the same label if and only if they induce the same tripartition.
2 Sort the nodes of both networks by their labels.
3 Compare the sorted lists of labels to identify unmatched nodes in the two networks.
End *UnmatchedNode*

Fig. 3. Algorithm to identify all unmatched nodes in two phylogenetic networks

The first step of Fig. 3 is achieved in two phases. Phase 1 reindexes the leaves by numbers from 1 to $|L(N)|$ so that for each node u of the galled-tree N, both $D(u)$ and $B(u)$ form sub-intervals of $[1..|L(N)|]$. Phase 2 labels each non-leaf node u by a 6-tuple of integers $(m_d(u),\ M_d(u),\ n_d(u),\ m_b(u),\ M_b(u),\ n_b(u))$ whose meaning is:

- $m_d(u), M_d(u), n_d(u)$ indicate the minimum leaf index, maximum leaf index and the number of leaves, respectively, in $D(u)$.
- $m_b(u), M_b(u), n_b(u)$ indicate the minimum leaf index, maximum leaf index and the number of leaves, respectively, in $B(u)$.

The labeling satisfies the following property.

Lemma 2. *For* $u \in V(N)$ *and* $u' \in V(N')$, $(m_d(u),\ M_d(u),\ n_d(u)) = (m_b(v),\ M_b(v),\ n_b(v))$ *if and only if* u *and* v *induce the same tripartition.*

Given the labeling, Steps 2 and 3 can identify all unmatched nodes. Below, we detail the reindexing and the labeling.

3.1 Reindexing the Leaves of a Galled-Tree

We now describe how to index the leaves of a galled-tree so that for each internal node u, both $D(u)$ and $B(u)$ are sub-intervals of $[1, l]$. First, we state a property of the biconnected components of a galled-tree.

Lemma 3. *Each biconnected component* C *of a galled-tree consists of either a single node or the (only) recombinant cycle of its sub-root.*

Algorithm *Reindex*
Input: A galled-tree N of l leaves and an integer i
Output: A new indexing of N's leaves so that for each node $u \in V(N)$, both $D(u)$ and $B(u)$ are sub-intervals of $[i, i+l-1]$.
1 **if** N consists of a single leaf **then**
1.1 reindex the leaf as i
elseif the root of N is a tree node **then**
1.2 let N_l and N_r be the left and right subnetworks attached to the left and right children of the root of N
1.3 $Reindex(N_l,\, i)$
1.4 $Reindex(N_r,\, i + |L(N_l)|)$
elseif the root of N is a split node **then**
1.5 let $N_1, N_2, \ldots N_x$ be the list of subnetworks attached to the recombinant cycle of the root of N in counter clockwise order
1.6 **for** $j = 1$ to x **do**
1.6.1 $Reindex(N_j,\, i + |L(N_1)| + |L(N_2)| + \cdots |L(N_{j-1})|)$
endfor
endif
End *Reindex*

Fig. 4. Algorithm to reindexing the leaves of a galled-tree

Based on the lemma, we design the reindexing algorithm as in Fig. 4. The correctness and time complexity of this algorithm is stated above.

Lemma 4. *Reindex(N, 1) runs in $O(|E(N)|)$ time and it reindexes the leaves of a galled-tree N such that, for each $u \in V(N)$, both $D(u)$ and $B(u)$ are sub-intervals of* $[1, |L(N)|]$.

3.2 Labeling the Nodes of a Network

Given a network N, we go bottom up on $\mathcal{T}(N)$ and label the nodes in each component visited. To reduce the time of labeling the nodes in each component, we divide the process into two steps. First, all the nodes in the reduced component are labeled. Then the remaining nodes are labeled based on the labeled nodes.

For every $x \in V(C^r)$, let $Ex(x)$ and $In(x)$ be the set of external children and internal descendants of x, respectively, i.e., $Ex(x) = \{v | (x, v) \in E(N^r)$ and $v \notin V(C^r)\}$ and $In(x) = \{v | v \in V(C^r)$ and v is a descendant of $x\}$. In addition, let $e(x) = \sum_{v \in Ex(x)} n_d(v)$ and $H(x) = \{v | v \in V(C^r)$ and x is a non-strict ancestor of $v\}$.

The following lemmas help us label the nodes of N. Lemma 5 and Lemma 6 compute the labels of nodes in a reduced component C^r whereas Lemma 7 computes the labels of the other nodes.

Lemma 5. *For each node $x \in V(C^r)$ whose children are u and v, we have: $n_d(x) = e(x) + \sum_{w \in In(x)} e(w)$, $m_d(x) = \min\{m_d(u), m_d(v)\}$, and $M_d(x) = \max\{M_d(u), M_d(v)\}$.*

Lemma 6. *For each node $x \in V(C^r)$, $n_b(x) = \sum_{v \in H(x)} e(v)$, $m_b(x) = \min_{v \in H(x)} m_d(v)$ and $M_b(x) = \max_{v \in H(x)} M_d(v)$.*

Lemma 7. *For each node x having an internal child u and an external child v.*

$$\begin{aligned}
n_d(x) &= n_d(u) + n_d(v) \\
m_d(x) &= \min\{m_d(u), m_d(v)\} \\
M_d(x) &= \max\{M_d(u), M_d(v)\} \\
n_b(x) &= n_b(u) \text{ if } u \text{ is a tree node or } n_d(u) \text{ if } u \text{ is a hybrid node} \\
m_b(x) &= m_b(u) \text{ if } u \text{ is a tree node or } m_d(u) \text{ if } u \text{ is a hybrid node} \\
M_b(x) &= M_b(u) \text{ if } u \text{ is a tree node or } M_d(u) \text{ if } u \text{ is a hybrid node}
\end{aligned}$$

Lemma 8. *The tripartition distance between a galled-tree N and a general network N' can be computed in $O(|E(N)| + |E(N')| + k(N')h(N'))$.*

Proof. (Sketch) The time needed to compute $e(x)$, $In(x)$ and $H(x)$ for all $x \in V(N^r)$ is $O(|E(N)| + \sum_i h(C_i)^2) = O(|E(N)| + k(N)h(N))$ where C_i for $i = 1, 2, \ldots$ is a non-singleton biconnected component of N and $h(C_i)$ is the number of hybrid nodes in C_i. Then, by Lemmas 5 and 6, the labels of all $x \in V(N^r)$ can be computed in $O(|V(N)| + k(N)h(N))$. Finally, the labels of other nodes are computed by Lemma 7 using $O(|E(N)|)$ time. The lemma then follows. □

Corollary 1. *The tripartition distance between two galled-tree N and N' can be computed in $O(|E(N)| + |E(N')|)$.*

4 Comparing Two General Networks

4.1 The Subnetwork Induced by a Set of Leaves

We first define the subnetwork induced by a set of leaves, which will play an important role in comparing two general networks.

Given a network N and a set of leaves $X = \{l_1, l_2, \ldots l_t\}$, we denote $\mathcal{T}_X(N)$ be a subtree of $\mathcal{T}(N)$ induced by X, which is a tree such that (1) whose nodes are X and their lowest common ancestors in $\mathcal{T}(N)$; and (2) whose edges preserve the ancestor-descendant relationship of $\mathcal{T}(N)$.

Let $E'_X = \{(u, v) | u \in C_1, v \in C_2, (C_1, C_2) \subset V(\mathcal{T}_X(N))$ and there exists a path from u to v which does not pass through any nodes belonging to some components in $V(\mathcal{T}_X(N))\}$. Let N^*_X be a subnetwork of N such that (1) whose node set is $\bigcup_{C \in V(\mathcal{T}_X(N))} V(C)$ and (2) whose edge set is $E'_X \cup \bigcup_{C \in V(\mathcal{T}_X(N))} E(C)$. We denote N_X be the subnetwork of N induced by X, which is a network formed by contracting all nodes in N^*_X whose in-degree and out-degree are equal to 1. Fig. 1(d) and (e) show example of $N^*_{\{a,c\}}$ and $N_{\{a,c\}}$ where N is the network in Fig. 1(a).

Lemma 9. $|E(N_X)| = O(\min\{h(N) + |X|, k(N)|X|\})$.

Lemma 10. *Given t disjoint leaf sets $X_1, X_2, \ldots X_t$ such that $\bigcup X_i = L(N)$, the subnetworks $N_{X_1}, N_{X_2}, \ldots N_{X_t}$ can be computed in total $O(\sum |N_{X_i}|)$ time.*

4.2 *DB*-labeling

We also use *UnmatchedNode* (Fig. 3) to identify all unmatched nodes of two general networks N and N'. However, the non-leaf nodes of the networks are labeled in a different way. Each of them is assigned a *DB-label*, which is a pair of integers $(d(u), b(u))$ such that (1) $d(u) = d(v)$ if and only if $D(u) = D(v)$; (2) $b(u) = b(v)$ if and only if $B(u) = B(v)$; and (3) $d(u) = b(v)$ if and only if $D(u) = B(v)$. Furthermore, $d(u) = 0$ if and only if $D(u) = \emptyset$ and $b(v) = 0$ if and only if $B(v) = \emptyset$.

It is clear that two nodes have the same *DB*-label if and only if they induce the same tripartition. The above labeling is called a *DB-labeling* of N and N'.

Algorithm *DBlabeling*
Input: two networks N and N' of the same leaf set S
Output: the *DB*-labeling of N and N'

1 Consider the singleton sets $X_1, X_2, \ldots X_{|S|}$, each containing a distinct leaf in S. For each i, find the *DB*-labeling for N_{X_i} and N'_{X_i}.
2 Repeat the following for $\log |S|$ rounds: Let $X_1, X_2, \ldots$ be the sets of leaves considered in last round. Pair up X_i's and let $X_{2i-1} = X_{2i-1} \cup X_{2i}$. Delete all X_{2i}'s and rename X_{2i-1}'s as X_i's. For each i, compute the *DB*-labeling of N_{X_i} and N'_{X_i} based on the result of last round.

End *DBlabeling*

Fig. 5. Algorithm to compute the *DB*-labeling of two networks of the same leaf set

We compute the *DB*-labeling of N and N' incrementally in a way similar to [7] as in Fig. 5. In step 2, given the *DB*-labeling of N_X and N_Y, we find a *DB*-labeling of $N_{X \cup Y}$ by the following relabeling procedure. This procedure utilizes the concept of Z-stamp of a node u where Z is a set of leaves, which is a pair of integers $(d_Z(u), b_Z(u))$ such that (1) $d_Z(u) = d_Z(v)$ if and only if $D(u) \cap Z = D(v) \cap Z$; and (2) $d_Z(u) = b_Z(v)$ if and only if $D(u) \cap Z = B(v) \cap Z$; and (3) $b_Z(u) = b_Z(v)$ if and only if $B(u) \cap Z = B(v) \cap Z$.

Relabeling procedure:

1. For each $u \in V(N_{X \cup Y})$, compute its X-stamp and Y-stamp as follow. If $u \in V(N_X)$, by Lemma 11, the X-stamp of u equals its label in N_X. Otherwise, its X-stamp is computed by Lemmas 12 and 13. The Y-stamps are calculated similarly.

2. Sort all the pairs $(d_X(u), d_Y(u))$ and $(b_X(u), b_Y(u))$ together and replace each pair by a number such that two pairs are replaced by the same number if and only if they are identical.

Lemma 11. *Let the tripartition of u in N_X be $(A_X(u), B_X(u), C_X(u))$. We have $A_X(u) = A(u) \cap X$, $B_X(u) = B(u) \cap X$ and $C_X(u) = C(u) \cap X$.*

Lemma 12. *Let u be a node in a component C which is in $V(\mathcal{T}_X(N))$. If u is in $V(N_{X\cup Y})$ but not in $V(N_X)$ then it must have an internal child v and an external child w. Furthermore, $d_X(w) = 0$, $d_X(u) = d_X(v)$ and $b_X(u) = d_X(v)$ if v is a hybrid node and $b_X(v)$ otherwise.*

Lemma 13. *Each component C in $V(\mathcal{T}_{X\cup Y}(N)) - V(\mathcal{T}_X(N))$ has at most one child C' in $\mathcal{T}_{X\cup Y}(N)$ such that $D(r(C')) \neq \emptyset$. If no such C' exists, $D_X(u) = \emptyset$ for all $u \in C$. Otherwise, for each $u \in V(C) \cap V(N_{X\cup Y})$, $d_X(u) = d_X(r(C'))$ if u is an ancestor of $r(C')$ and 0 otherwise and $b_X(u) = d_X(r(C'))$ if u is a non-strict ancestor of $r(C')$ and 0 otherwise.*

From the above two lemmas, we can compute the X-stamps and Y- stamps of nodes in $N_{X\cup Y}$ and $N'_{X\cup Y}$ separately. To achieve good running time, for each node in a reduced component, we store the sets of its ancestors and non-strict ancestors in that reduced component. These sets for all nodes in N can be pre-calculated in total $O(k(N)h(N))$ time.

Lemma 14. *The sets of ancestors and non-strict ancestors of any node $u \in V(C) \cap V(N_{X\cup Y})$ can be computed in $O(|V(C) \cap V(N_{X\cup Y}))$.*

Let $k = \max\{k(N), k(N')\}$, $h = \max\{h(N), h(N')\}$ and $n = \max\{|V(N)|, |V(N')|\}$. The following lemmas state the time complexity of comparing two general networks.

Lemma 15. *Given the DB-labeling of N_X, N'_X, N_Y and N'_Y, the above procedure computes DB-labeling of $N_{X\cup Y}$ and $N'_{X\cup Y}$ using $O(\min\{h + |X \cup Y|, k|X \cup Y|\})$ time.*

Lemma 16. *The tripartition-based distance between two general networks N and N' can be computed in $O(\min\{hn + n\log n, kn\log n\})$ time.*

References

1. D. Bryant and V. Moulton. NeighborNet: an agglomerative method for the construction of planar phylogenetic networks. In *Proc. of the 2nd Workshop on Algorithms in Bioinformatics* (WABI 2002), volume 2452 of *LNCS*, pages 375–391. Springer, 2002.
2. C. Choy, J. Jansson, K. Sadakane, and W.-K. Sung. Computing the maximum agreement of phylogenetic networks. *Theoretical Computer Science*, 335(1):93–107, 2005.
3. W. F. Doolittle. Phylogenetic classification and the universal tree. *Science*, 284:2124–2128, 1999.

4. D. Gusfield and V. Bansal. A fundamental decomposition theory for phylogenetic networks and incompatible characters. In *Proc. of the 9th Annual International Conf. on Research in Computational Molecular Biology* (RECOMB 2005), pages 217–232, 2005.
5. D. Gusfield, S. Eddhu, and C. Langley. Efficient reconstruction of phylogenetic networks with constrained recombination. In *Proc. of the Computational Systems Bioinformatics Conference* (CSB2003), pages 363–374, 2003.
6. J. Hein. Reconstructing evolution of sequences subject to recombination using parsimony. *Mathematical Biosciences*, 98(2):185–200, 1990.
7. Wing-Kai Hon, Ming-Yang Kao, Tak Wah Lam, Wing-Kin Sung, and Siu-Ming Yiu. Non-shared edges and nearest neighbor interchanges revisited. *Inf. Process. Lett.*, 91(3):129–134, 2004.
8. D. H. Huson, T. Dezulian, T. Klöpper, and M. Steel. Phylogenetic super-networks from partial trees. In *Proc. of the 4th Workshop on Algorithms in Bioinformatics* (WABI 2004), pages 388–399, 2004.
9. D. H. Huson, T. Klopper, P. J. Lockhart, and M. A. Steel. Reconstruction of reticulate networks from gene trees. In *Proc. of the 9th Annual International Conf. on Research in Computational Molecular Biology* (RECOMB 2005), pages 233–249, 2005.
10. T. N. D. Huynh, J. Jansson, N. B. Nguyen, and W. K. Sung. Constructing a smallest refining galled phylogenetic network. In *Proc. of the 9th Annual International Conf. on Research in Computational Molecular Biology* (RECOMB 2005), pages 265–280, 2005.
11. J. Jansson, N. B. Nguyen, and W. K. Sung. Algorithms for combining rooted triplets into a galled phylogenetic network. In *Proc. of the 16th Annual ACM-SIAM Symposium on Discrete Algorithms* (SODA 2005), pages 349–358, 2005.
12. J. Jansson and W.-K. Sung. Inferring a level-1 phylogenetic network from a dense set of rooted triplets. In *Proc. of the 10th International Computing and Combinatorics Conference* (COCOON 2004), 2004.
13. J. Jansson and W. K. Sung. The maximum agreement of two nested phylogenetic networks. In *Proc. of the 15th Annual International Symposium on Algorithms and Computation* (ISAAC 2004), pages 581–593, 2004.
14. B. M. E. Moret, L. Nakhleh, T. Warnow, C. R. Linder, A. Tholse, A. Padolina, J. Sun, and R. Timme. Phylogenetic networks: Modeling, reconstructibility, and accuracy. *IEEE Transactions on Computational Biology and Bioinformatics*, 1(1):1–12, 2004.
15. L. Nakhleh, J. Sun, T. Warnow, C. R. Linder, B. M. E. Moret, and A. Tholse. Towards the development of computational tools for evaluating phylogenetic reconstruction methods. In *Proc. of the 8th Pacific Symposium on Biocomputing* (PSB 2003), pages 315–326, 2003.
16. L. Nakhleh, T. Warnow, and C. R. Linder. Reconstructing reticulate evolution in species – theory and practice. In *Proc. of the 8th Annual International Conf. on Research in Computational Molecular Biology* (RECOMB 2004), pages 337–346, 2004.
17. D. Posada and K. A. Crandall. Intraspecific gene genealogies: trees grafting into networks. *TRENDS in Ecology & Evolution*, 16(1):37–45, 2001.
18. D. F. Robinson and L. R. Foulds. Comparison of phylogenetic trees. *Mathematical Biosciences*, 53:131–147, 1981.
19. L. Wang, K. Zhang, and L. Zhang. Perfect phylogenetic networks with recombination. *Journal of Computational Biology*, 8(1):69–78, 2001.

Improved Algorithms for Largest Cardinality 2-Interval Pattern Problem

Hao Yuan, Linji Yang, and Erdong Chen

Department of Computer Science and Engineering,
Shanghai Jiao Tong University,
200030 Shanghai, P.R. China
{hyuan, ljyang, edchen}@cs.sjtu.edu.cn

Abstract. The 2-INTERVAL PATTERN problem is to find the largest constrained pattern in a set of 2-intervals. The constrained pattern is a subset of the given 2-intervals such that any pair of them are R-comparable, where model $R \subseteq \{<, \sqsubset, \between\}$. The problem stems from the study of general representation of RNA secondary structures. In this paper, we give three improved algorithms for different models. Firstly, an $O(n \log n + \mathcal{L})$ algorithm is proposed for the case $R = \{\between\}$, where $\mathcal{L} = O(dn) = O(n^2)$ is the total length of all 2-intervals (density d is the maximum number of 2-intervals over any point). This improves previous $O(n^2 \log n)$ algorithm. Secondly, we use dynamic programming techniques to obtain an $O(n \log n + dn)$ algorithm for the case $R = \{<, \sqsubset\}$, which improves previous $O(n^2)$ result. Finally, we present another $O(n \log n + \mathcal{L})$ algorithm for the case $R = \{\sqsubset, \between\}$ with disjoint support(interval ground set), which improves previous $O(n^2\sqrt{n})$ upper bound.

1 Introduction

In the area of prediction and analysis of RNA secondary structures, arc-annotated sequence focuses on the very detailed description of the structure itself — the sequence of the bases and the bonds between the bases [1]. However, using arc-annotated sequence to further predict other homogeneous RNA structures is proved sometimes hard. Derived from arc-annotated sequence, 2-intervals representation considers only the bonds between the bases and the patterns of the bonds, such as knots, hairpin structures and pseudoknots [2]. Thus, it has become a well macroscopic describer of RNA secondary structures.

A 2-interval is two disjoint intervals on a line. Two disjoint 2-intervals can be defined in the relations of precedence($<$), nest($\sqsubset$) or cross($\between$). A constrained pattern is a set of 2-intervals such that any pair of them are R-comparable, where $R \subseteq \{<, \sqsubset, \between\}$. 2-INTERVAL PATTERN problem introduced by Vialette [2] is to find the largest constrained pattern in a set of 2-intervals, and it is closely related to the problem of PATTERN MATCHING OVER SET OF 2-INTERVALS [2,3] and LONGEST ARC-PRESERVING COMMON SUBSEQUENCE [1,4,5].

The R-comparable relations of 2-intervals can be formulated in different graph classes [6], and some graph-theoretic algorithms have been used to solve the 2-INTERVAL PATTERN problem efficiently [2,7]. In the paper of Blin et al.[7], they

X. Deng and D. Du (Eds.): ISAAC 2005, LNCS 3827, pp. 412–421, 2005.

almost complete the NP-Completeness results for 2-INTERVAL PATTERN problems under three different types of support models (unlimited, unitary, disjoint) classified by Vialette[2]. Recently, Crochemore et al. studied the approximation algorithms for 2-INTERVAL PATTERN problem [8]. In our paper, we give several algorithms to improve the time complexity of finding optimal solutions for some models to $O(n \log n + \mathcal{L})$, which is worst-case quadratic.

The rest of this paper is organized as follows. In Section 2, we define some basic terminologies for 2-INTERVAL PATTERN problem. In Section 3, 4 and 5, we will give improved algorithms for $R = \{ \between \}$, $R = \{ <, \sqsubset \}$ and $R = \{ \sqsubset, \between \}$ respectively. Finally, conclusions are made in Section 6. Due to space limit, some obvious proofs are omitted.

2 Preliminaries

First, we'll review the terminologies used in [2,7]. Let $I = [a, b]$ be an interval $(a \leq b)$, define $l(I) = a$ and $r(I) = b$. A 2-interval is the union of two disjoint intervals I and J, denoted by $D = (I, J)$ such that $I < J$, where the strict precedence order $<$ means I is strictly to the left of J, i.e. $r(I) < l(J)$. The left interval I and right interval J of D are denoted by Left(D) and Right(D) respectively.

For any two 2-intervals $D_1 = (I_1, J_1)$ and $D_2 = (I_2, J_2)$, we say they are disjoint if and only if $(I_1 \cup J_1) \cap (I_2 \cup J_2) = \emptyset$. Any pair of disjoint 2-intervals must satisfy one of the following three relations:

PRECEDE	$D_1 < D_2$	$\Leftrightarrow$	$I_1 < J_1 < I_2 < J_2$;	transitive
NEST	$D_1 \sqsubset D_2$	$\Leftrightarrow$	$I_2 < I_1 < J_1 < J_2$;	transitive
CROSS	$D_1 \between D_2$	$\Leftrightarrow$	$I_1 < I_2 < J_1 < J_2$;	not symmetric

D_1 and D_2 are called τ-comparable if $D_1 \tau D_2$ for some $\tau \in \{<, \sqsubset, \between\}$. A 2-interval set is called R-comparable if and only if for any two elements of it, there exists a relation $\tau \in R$ such that they are τ-comparable. Let $\mathscr{D} = \{D_1, D_2, \ldots, D_n\}$ denote a set of n 2-intervals. The support (or interval ground set) of $\mathscr{D}$ is denoted by Support($\mathscr{D}$) $= \bigcup \{I_i, J_i \mid D_i = (I_i, J_i)\}$. Let X denote the set of interesting coordinates $\bigcup_{D \in \mathscr{D}} \{r(\text{Left}(D)), l(\text{Right}(D))\}$. Assume that the elements of $X = \{x_1, x_2, \ldots, x_{|X|}\}$ are sorted, i.e. $x_1 < x_2 < \cdots < x_{|X|}$. Since $|X| \leq 2n$, the sorting process takes $O(n \log n)$ time.

Given a 2-interval set $\mathscr{D}$ and a model $R \subseteq \{ <, \sqsubset, \between \}$, the 2-INTERVAL PATTERN problem is to find the largest cardinality subset $\mathscr{D}' \subseteq \mathscr{D}$, so that $\mathscr{D}'$ is R-comparable. In [2], Vialette classified the problem into three types:

- UNITARY: All the intervals in Support($\mathscr{D}$) are of the same size;
- DISJOINT: The elements of Support($\mathscr{D}$) are disjoint, and have equal size;
- UNLIMITED: No restriction on the support of $\mathscr{D}$.

In our paper, we only concern the cases of DISJOINT and UNLIMITED.

To better illustrate our improvements, we define a parameter $\mathcal{L}$, which means the total length of 2-intervals. The length of a 2-interval D is defined to be

Length$(D) = k_2 - k_1$, where $x_{k_1} = r(\text{Left}(D))$ and $x_{k_2} = l(\text{Right}(D))$. The density of $\mathscr{D}$, denoted by d, is the maximum number of 2-intervals over any point. Formally, $d = \max_{x \in X} \left| \{ D = (I, J) \in \mathscr{D} \mid r(I) \le x < l(J) \} \right|$. It is easy to see $\mathcal{L} \le dn \le n^2$. The following table summarizes the results of our work.

Table 1. Summarized results for different models of 2-INTERVAL PATTERN problem. The improved results of this paper are marked by $\star$.

<table>
<tr><th rowspan="2">MODEL</th><th colspan="2">SUPPORT</th></tr>
<tr><th>DISJOINT</th><th>UNLIMITED</th></tr>
<tr><td>$\{<, \sqsubset, \between\}$</td><td>$O(n\sqrt{n})$ [9]</td><td>APX-Hard [10]</td></tr>
<tr><td>$\{<, \between\}$</td><td>?</td><td>NP-Complete [7]</td></tr>
<tr><td>$\{\sqsubset, \between\}$</td><td>$O(n \log n + \mathcal{L})$ $\star$</td><td>NP-Complete [2]</td></tr>
<tr><td>$\{<, \sqsubset\}$</td><td colspan="2">$O(n \log n + dn)$ $\star$</td></tr>
<tr><td>$\{\between\}$</td><td colspan="2">$O(n \log n + \mathcal{L})$ $\star$</td></tr>
<tr><td>$\{<\}$</td><td colspan="2">$O(n \log n)$ [2]</td></tr>
<tr><td>$\{\sqsubset\}$</td><td colspan="2">$O(n \log n)$ [7]</td></tr>
</table>

3 Improved Algorithm for $\{\between\}$-Structured Pattern

In this section, we will give an $O(n \log n + \mathcal{L})$ algorithm for the model $R = \{\between\}$. This improves the $O(n^2 \log n)$ algorithm given in [2].

Our algorithm is based on a sweep-line method. Let $\mathscr{D}[x]$ be the set of 2-intervals crossing a vertical line whose horizontal coordinate is x, that is $\mathscr{D}[x] = \{ D_i \in \mathscr{D} \mid r(\text{Left}(D_i)) \le x \text{ and } l(\text{Right}(D_i)) > x \}$. For convenience, set $x_0 = -\infty$ and let $\mathscr{D}^{(k)}$ denote $\mathscr{D}[x_k]$ for $0 \le k \le |X|$, hence $\mathscr{D}^{(0)} = \emptyset$. When the vertical line sweeps from left to right (Fig. 1) passing an interesting point $x_k \in X$, some 2-intervals may be added to $\mathscr{D}[x]$, and some may be removed. Let $P^{(k)}$ and $Q^{(k)}$ be the differences between $\mathscr{D}^{(k-1)}$ and $\mathscr{D}^{(k)}$, we have

$$P^{(k)} = \mathscr{D}^{(k)} \setminus \mathscr{D}^{(k-1)} = \{ D \in \mathscr{D} \mid r(\text{Left}(D)) = x_k \},$$
$$Q^{(k)} = \mathscr{D}^{(k-1)} \setminus \mathscr{D}^{(k)} = \{ D \in \mathscr{D} \mid l(\text{Right}(D)) = x_k \}.$$

Theorem 1. *Let $\omega(\mathscr{D})$ be the cardinality of the largest $\{\between\}$-comparable subset of $\mathscr{D}$, then $\omega(\mathscr{D}) = \max_{1 \le k \le |X|} \omega(\mathscr{D}^{(k)})$.*

Proof. Observe that for any two 2-intervals D_i and D_j which are $\between$-comparable, we have $r(\text{Left}(D_j)) < l(\text{Right}(D_i))$. Thus for any subset $\mathscr{D}' \subseteq \mathscr{D}$ which is $\{\between\}$-comparable, we have $\max_{D'_j \in \mathscr{D}'} r(\text{Left}(D'_j)) < \min_{D'_i \in \mathscr{D}'} l(\text{Right}(D'_i))$. This implies that $\mathscr{D}'$ is also a subset of $\mathscr{D}^{(k)}$, where $x_k = \max_{D' \in \mathscr{D}'} r(\text{Left}(D'))$. Hence, $\omega(\mathscr{D})$ must be equal to $\omega(\mathscr{D}^{(k)})$ for a specific k. $\square$

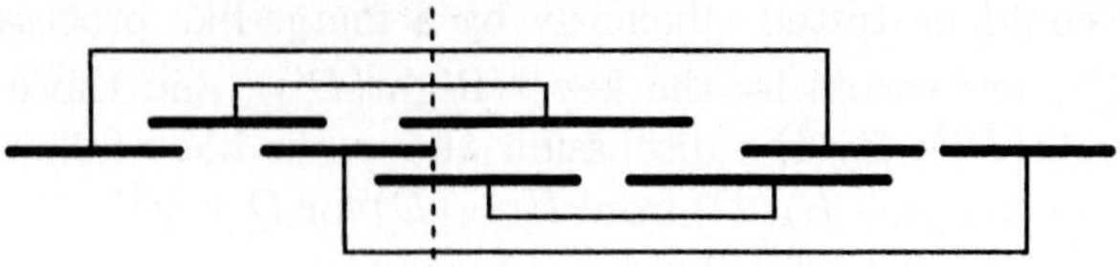

Fig. 1. A sweep-line sweeps from left to right

By Theorem 1, we can get $\omega(\mathscr{D})$ by computing $\omega(\mathscr{D}^{(k)})$ for each k. Vialette shows that $\omega(\mathscr{D}^{(k)})$ can be computed by finding a maximum independent set of corresponding trapezoid graphs in $O(n \log n)$ time [11], so his algorithm works in $O(n^2 \log n)$ total time. To improve the complexity, we utilize the dynamic structure of $\mathscr{D}^{(k)}$ by discovering the relationship between $\omega(\mathscr{D}^{(k-1)})$ and $\omega(\mathscr{D}^{(k)})$.

Definition 1. *The height of D under $\mathscr{D}^{(k)}$, denoted by $H_k(D)$, is the cardinality of the largest $\{\between\}$-comparable subset of $\mathscr{D}^{(k)}$, whose maximal element must be D under the relation $\between$. That is $H_k(D) = \omega\left(\{D' \in \mathscr{D}^{(k)} \mid D' \between D\}\right) + 1$ for $D \in \mathscr{D}^{(k)}$, and $H_k(D) = 0$ for $D \notin \mathscr{D}^{(k)}$.*

Obviously, we have $\omega(\mathscr{D}^{(k)}) = \max_{D \in \mathscr{D}^{(k)}} H_k(D)$. If we can compute $H_k(D)$ for each k and D efficiently, then we can get a better upper bound. To achieve this, we first exam the the relationship between $H_k(D)$ and $H_{k-1}(D)$, and show that every non-zero $H_k(D)$ can be computed efficiently.

Lemma 1. *For any k, the $\between$ relation is transitive in $\mathscr{D}^{(k)}$.*

Theorem 2. *For $D \in \mathscr{D}^{(k-1)}$, after the sweep-line goes from x_{k-1} to x_k, the height of D decreases by at most one, or remains the same, i.e. $H_{k-1}(D) - 1 \leq H_k(D) \leq H_{k-1}(D)$.*

Proof. First, we can see that $H_k(D) = h > 1$ if and only if there exists a $D' \in \mathscr{D}^{(k)}$ so that $H_k(D') = h - 1$ and $D' \between D$ (due to Lemma 1). Since no 2-interval in $P^{(k)}$ can cross D, the height of D will not increase. Let $\{D_1, D_2, \ldots, D_{h-1}, D\}$ be a $\{\between\}$-comparable subset of $\mathscr{D}^{(k-1)}$, where $h = H_{k-1}(D)$. Because no pair of 2-intervals in $Q^{(k)}$ are $\between$-comparable, at least $h-1$ elements from the previous set will be preserved in $\mathscr{D}^{(k)}$, hence $H_k(D) \geq h - 1 = H_{k-1}(D) - 1$. □

Let $C_h^{(k)} = \left\{D \in \mathscr{D}^{(k)} \mid H_k(D) = h\right\}$. The key issue is to test efficiently for a 2-interval $D \in C_h^{(k-1)}$ whether there exists a $D' \in C_{h-1}^{(k)}$ such that $D' \between D$. In other words, we have to check whether the following set is empty

$$\left\{D' \in C_{h-1}^{(k)} \;\middle|\; r(\text{Left}(D')) < l(\text{Left}(D)) \text{ and } r(\text{Right}(D')) < l(\text{Right}(D))\right\}.$$

Alternatively, it is to test whether

$$\min\left\{r(\text{Left}(D')) \;\middle|\; D' \in C_{h-1}^{(k)} \text{ and } r(\text{Right}(D')) < l(\text{Right}(D))\right\} < l(\text{Left}(D)).$$

This property could be tested efficiently by a merge-like process, when the 2-intervals in $C_{h-1}^{(k)}$ are sorted by the key $r(\text{Right}(D'))$, and those in $C_h^{(k-1)}$ are sorted by $l(\text{Right}(D))$. By the discussion above, we have following procedure FALLDOWN(k) to compute $H_k(D)$ from $H_{k-1}(D)$ for $D \in \mathscr{D}^{(k-1)}$ in $O(|\mathscr{D}^{(k-1)}| + |\mathscr{D}^{(k)}|)$ time. For a fixed k that $0 < k \leq |X|$, the procedure partitions $\mathscr{D}^{(k)}$ into $|\mathscr{D}^{(k)}|$ 2-interval set $C_1^{(k)}, C_2^{(k)}, \ldots, C_{|\mathscr{D}^{(k)}|}^{(k)}$.

Procedure 3.1 FALLDOWN(k)

Function: Compute $H_k(D)$ for every $D \in \mathscr{D}^{(k-1)}$ from $H_{k-1}(D)$.
Notation: Let $C_h^{(k)}[i]$ denote the i_{th} element in sorted $C_h^{(k)}$.

Sort the 2-intervals in each $C_h^{(k-1)}$ according to the key $l(\text{Right}(D))$.
$C_1^{(k)} \leftarrow C_1^{(k-1)} \setminus Q^{(k)}$, sort it by the key $r(\text{Right}(D))$
for $h \leftarrow 2$ to $|\mathscr{D}^{(k-1)}|$ **do**
  $C_h^{(k)} \leftarrow C_h^{(k-1)}$; $j \leftarrow 1$, $i \leftarrow 1$; LEFTMOST $\leftarrow +\infty$;
  while $j \leq |C_h^{(k-1)}|$ **do**
    while $r(\text{Right}(C_{h-1}^{(k)}[i])) < l(\text{Right}(C_h^{(k-1)}[j]))$ **do**
      LEFTMOST $\leftarrow \min\left\{\text{LEFTMOST}, r(\text{Left}(C_{h-1}^{(k)}[i]))\right\}$
      $i \leftarrow i + 1$
    end while
    if $LEFTMOST < l(\text{Left}(C_h^{(k-1)}[j]))$ **then**
      Do nothing. // the height of $C_h^{(k-1)}[j]$ remains the same
    else
      $C_{h-1}^{(k)} \leftarrow C_{h-1}^{(k)} \cup \left\{C_h^{(k-1)}[j]\right\}$ // the height of $C_h^{(k-1)}[j]$ decreases by one
      $C_h^{(k)} \leftarrow C_h^{(k)} \setminus C_h^{(k-1)}[j]$
    end if
    $j \leftarrow j + 1$
  end while
  Sort the 2-intervals in $C_h^{(k)}$ by the key $r(\text{Right}(D))$
end for

FALLDOWN(k) over all k can be implemented in $O\left(n \log n + \sum |\mathscr{D}^{(k)}|\right)$ time.

Next, we will give a procedure to compute $H_k(D)$ for $D \in P^{(k)}$. From the fact that there is not any pair of 2-intervals in $P^{(k)}$ that are $\between$-comparable, the height of $D_j \in P^{(k)}$ under $\mathscr{D}^{(k)}$ can be calculated by $H_k(D_j) = \max_{D_i \between D_j} H_k(D_i) + 1$ if there is at least one $D_i \in \mathscr{D}^{(k)} \setminus P^{(k)}$ which cross D_j. Otherwise, $H_k(D_j) = 1$.

Definition 2. *For each* $0 < h \leq |\mathscr{D}^{(k)}|$, *define the boundary with respect to* y,

$$B_h^{(k)}(y) = \min_{D \in C_h^{(k)}} \left\{ r(\mathit{Left}(D)) \;\middle|\; r(\mathit{Right}(D)) < y \right\}. \tag{1}$$

Lemma 2. *For any* $0 < h_1 < h_2 \le |\mathscr{D}^{(k)}|$, *we have* $B_{h_1}^{(k)}(y) < B_{h_2}^{(k)}(y)$

Proof. For a fixed k and y, let D_{h_2} denote the 2-interval from $C_{h_2}^{(k)}$ which minimizes $B_{h_2}^{(k)}$ in Equation (1). Since $h_2 > h_1$, there must be at leat one $D_{h_1} \in C_{h_1}^{(k)}$ that $D_{h_1} \between D_{h_2}$. Since $r(\text{Right}(D_{h_1})) < r(\text{Right}(D_{h_2})) < y$, therefore, we have $B_{h_1}^{(k)}(y) \le r(\text{Left}(D_{h_1})) < r(\text{Left}(D_{h_2})) = B_{h_2}^{(k)}(y)$. □

Procedure 3.2 JumpUp(k)

Calculate the Heights of $P^{(k)}$ under $\mathscr{D}^{(k)}$

1: set $B_h^{(k)} \leftarrow \infty$ for each $1 \le h \le |\mathscr{D}^{(k)}|$ and set $B_0^{(k)} \leftarrow -\infty$

2: sort 2-intervals in $\mathscr{D}^{(k)}$ according to key(D), where

$$\text{key}(D) = \begin{cases} r(\text{Right}(D)), & \text{if } D \in \mathscr{D}^{(k)} \setminus P^{(k)} \\ l(\text{Right}(D)), & \text{if } D \in P^{(k)} \end{cases}$$

If there is a tie, let the elements in $P^{(k)}$ go first.

3: **for** $i \leftarrow 1$ to $|\mathscr{D}^{(k)}|$ **do**

4: let D be i_{th} element in $\mathscr{D}^{(k)}$

5: **if** $D \in P^{(k)}$ **then**

6: find the largest h such that $B_h^{(k)} < l(\text{Left}(D))$

7: $H_k(D) \leftarrow h + 1$

8: **else**

9: let $h \leftarrow H_k(D)$, then set $B_h^{(k)} \leftarrow \min\{B_h^{(k)}, r(\text{Left}(D))\}$

10: **end if**

11: **end for**

12: **for** each $0 < h \le |\mathscr{D}^{(k)}|$ **do**

13: $C_h^{(k)} \leftarrow C_h^{(k)} \cup \left\{H_k(D) = h \mid D \in P^{(k)}\right\}$

14: **end for**

The procedure JumpUp(k) is used for computing the heights of 2-intervals in $P^{(k)}$ under $\mathscr{D}^{(k)}$. The sorting process in Line 2 and the merge process in Lines 12-14 can be done in $O\left(|\mathscr{D}^{(k)}|\right)$ time, if we first sort each $P^{(k)}$ in a global stage. Line 6 can be implemented by binary search (Lemma 2) in $O\left(\log|\mathscr{D}^{(k)}|\right)$ time. So the total complexity for all JumpUp(k) is $\sum O\left(|\mathscr{D}^{(k)}| + |P^{(k)}|\log|\mathscr{D}^{(k)}|\right)$ $= O\left(\sum|\mathscr{D}^{(k)}| + \sum|P^{(k)}|\log n\right) = O(\sum|\mathscr{D}^{(k)}| + n\log n)$.

Combining procedures FallDown(k) and JumpUp(k), we finally obtain algorithm 3.3. It is easy to see that the space complexity is $O(n)$. By the equation $\sum|\mathscr{D}^{(k)}| = \mathcal{L}$, we have the total time complexity $O\left(n\log n + \mathcal{L}\right)$.

Proposition 1. *The* $\{\between\}$-Structured 2-Interval Pattern *problem can be solved in* $O(n\log n + \mathcal{L})$ *time.*

For the case of disjoint support, it can be transformed to the problem of finding a maximum clique in circle graphs [6]. Each 2-interval maps to a vertex, and two vertices are adjacent if and only if their corresponding 2-interval are $\between$-comparable. By the result of Masuda et al. [12], we have

Algorithm 3.3 $\{\between\}$-Structured 2-Interval Pattern

sort the 2-intervals and their endpoints
calculate $P^{(k)}$ and $Q^{(k)}$ for each $1 \leq k \leq |X|$
set $C_h^{(0)} = \emptyset$ for all $1 \leq h \leq n$
for $k \leftarrow 1$ to $|X|$ **do**
    call Procedure FallDown(k)
    call Procedure JumpUp(k)
    $\omega(\mathscr{D}^{(k)})$ equals to the largest h such that $C_h^{(k)} \neq \emptyset$, or zero if $\mathscr{D}^{(k)} = \emptyset$
end for
return $\max \omega(\mathscr{D}^{(k)})$

Proposition 2. *The* Disjoint Support $\{\between\}$-Structured 2-Interval Pattern *problem can be solved in $O(n \log n + \min\{m, dn\})$ time, where m is the number of 2-interval pairs that are $\between$-comparable.*

4 Improved Algorithm for $\{<, \sqsubset\}$-Structured Pattern

In this section, we will give an $O(n \log n + dn)$ algorithm for $\{<, \sqsubset\}$-Structured 2-Interval Pattern problem, and it is $O(n \log n + \mathcal{L})$ for the case that no 2-intervals share the same rightmost endpoint. Our algorithm improves the $O(n^2)$ upper bound given in [2] since $d = O(n)$, and it is similar to the maximum weighted independent set algorithm for circle graphs which is proposed recently by Valiente [13].

Definition 3. *Let $\mathscr{D}[x_1, x_2]$ represent the 2-intervals lie in $[x_1, x_2]$, where $x_1 < x_2$, i.e. $\mathscr{D}[x_1, x_2] = \{D \in \mathscr{D} \mid x_1 \leq \mathfrak{l}(Left(D)) < \mathfrak{r}(Right(D)) \leq x_2\}$.*

Definition 4. *Let $\alpha(\mathscr{D})$ be the cardinality of largest $\{<, \sqsubset\}$-comparable subset of $\mathscr{D}$. Take $\alpha[x_1, x_2]$ instead of $\alpha(\mathscr{D}[x_1, x_2])$ for short. Let $\alpha(D)$ denote the cardinality of largest $\{<, \sqsubset\}$-comparable subset of $\mathscr{D}[\mathfrak{l}(Left(D)), \mathfrak{r}(Right(D))]$), with the constraint that D must be in that largest cardinality subset.*

Lemma 3. $\alpha(D) = 1 + \alpha[\mathfrak{r}(Left(D)) + 1, \mathfrak{l}(Right(D)) - 1]$.

Lemma 4. *Let $\alpha'[x_1, x_2] = \max\{\alpha[x_1, \mathfrak{l}(Left(D)) - 1] + \alpha(D)\}$, where $D \in \mathscr{D}[x_1, x_2]$ and $\mathfrak{r}(Right(D)) = x_2$. We have $\alpha[x_1, x_2] = \max\{\alpha[x_1, x_2 - 1], \alpha'[x_1, x_2]\}$.*

Combining the two Lemmas above, we have a dynamic programming algorithm working in $O(n^2)$ time. To achieve a tighter bound, we can first compute $\alpha(D)$ for every $D \in \mathscr{D}$ in the order of their lengths (see Lemma 5). After that, $\alpha(\mathscr{D}) = \alpha[x_1, x_{|X|}]$ can be calculated by Lemma 4 in $O(|X|)$ time.

Lemma 5. *For an 2-interval $D_i \in \mathscr{D}$, if we know the values of $\alpha(D')$ for all $D' \sqsubset D_i$ that $Length(D') < Length(D_i)$, then we can compute $\alpha(D_i)$ in $O(Length(D_i) + m_i)$ time, where m_i is the number of 2-intervals nested in D_i.*

The total complexity is $O(n \log n + \mathcal{L} + \sum m_i + |X|)$. It is not too difficult to see that $\sum m_i \leq dn$. If no 2-intervals share the same rightmost endpoint, then it is easy to see that $m_i \leq \text{Length}(D_i)$, hence we have $\sum m_i \leq \mathcal{L}$ in this case.

Proposition 3. *The* $\{<, \sqsubset\}$*-STRUCTURED 2-INTERVAL PATTERN problem can be solved in* $O(n \log n + dn)$ *time. If no 2-intervals share the same rightmost endpoint, then the complexity can be improved to* $O(n \log n + \mathcal{L})$.

5 Improved Algorithm for $\{\sqsubset, \between\}$-Structured Pattern

In this section, we will give an $O(n \log n + \mathcal{L})$ algorithm for the model $R = \{\sqsubset, \between\}$ with disjoint support. This improves the $O(n^2\sqrt{n})$ algorithm given by Blin et al. [7].

In the case with disjoint support, define $\imath(D)$ and $\tau(D)$ to be the left and right endpoints respectively for a 2-interval $D \in \mathscr{D}$, i.e. $\imath(D) = \imath(\text{Left}(D)) = \tau(\text{Left}(D))$ and $\tau(D) = \imath(\text{Right}(D)) = \tau(\text{Right}(D))$.

Lemma 6. *Let* $\mathscr{D}'$ *be a* $\{\sqsubset, \between\}$*-comparable subset of* $\mathscr{D}$*, then*

$$\max_{D'_j \in \mathscr{D}'} \imath(D'_j) < \min_{D'_i \in \mathscr{D}'} \tau(D'_i).$$

Define $\varphi(\mathscr{D})$ to be the largest largest $\{\sqsubset, \between\}$-comparable subset of $\mathscr{D}$ with disjoint support. Similar to section 3, we have $\varphi(\mathscr{D}) = \max \varphi(\mathscr{D}^{(k)})$. Blin et al. calculate each $\varphi(\mathscr{D}^{(k)})$ by finding the maximum cardinality matching in the corresponding bipartite graphs. We still apply this idea, but the complexity is improved by discovering the dynamic structure of the bipartite graphs.

Definition 5. *For a fixed* $0 < k \leq |X|$*, let* $G^{(k)} = (U^{(k)}, V^{(k)}, E^{(k)})$ *be a bipartite graph corresponding to* $\mathscr{D}^{(k)}$ *defined as follows:* $U^{(k)} = \{x \in X \mid x \leq x_k\}$*, and* $V^{(k)} = \{x \in X \mid x > x_k\}$*, the edges* $E^{(k)} = \{\langle \imath(D), \tau(D) \rangle \mid D \in \mathscr{D}^{(k)}\}$.

Obviously, a maximum matching of $G^{(k)}$ corresponds to a maximum cardinality $\{\sqsubset, \between\}$-comparable subset of $\mathscr{D}^{(k)}$. Let $M^{(k)}$ be a maximum matching in $G^{(k)}$.

Theorem 3. *The cardinality of maximum matching in* $G^{(k)}$ *and* $G^{(k-1)}$ *differs at most one, i.e.* $|M^{(k-1)}| - 1 \leq |M^{(k)}| \leq |M^{(k-1)}| + 1$.

Proof. Define the differences of $E^{(k)}$ and $E^{(k-1)}$ by

$$F_+^{(k)} = E^{(k)} \setminus E^{(k-1)} = \{\langle x_k, x_j \rangle \mid x_j = \tau(D) \text{ for } D \in \mathscr{D} \text{ where } \imath(D) = x_k\},$$
$$F_-^{(k)} = E^{(k-1)} \setminus E^{(k)} = \{\langle x_i, x_k \rangle \mid x_i = \imath(D) \text{ for } D \in \mathscr{D} \text{ where } \tau(D) = x_k\}.$$

Let $M = M^{(k-1)} \setminus \{\langle x_i, x_k \rangle\}$ if there is an edge $\langle x_i, x_k \rangle \in F_-^{(k)}$, otherwise $M = M^{(k-1)}$. Obviously, M is a matching of $G^{(k)}$, so $|M^{(k-1)}| - 1 \leq |M| \leq |M^{(k)}|$. The second part of inequalities can be proved in the same way. Let $M' = M^{(k)} \setminus \{\langle x_k, x_j \rangle\}$ if there is an edge $\langle x_k, x_j \rangle \in F_+^{(k)}$, otherwise $M' = M^{(k)}$. Now M' is a matching of $G^{(k-1)}$, so $|M^{(k)}| - 1 \leq |M'| \leq |M^{(k-1)}|$. □

Theorem 4 (Hopcroft and Karp[14]). *Let M_1 and M_2 be two matchings, if $|M_1| = s$, $|M_2| = r$ and $r > s$, then there $M_1 \oplus M_2$ contains at least $r - s$ vertex disjoint augmenting paths relative to M_1. Where the operation $\oplus$ means*

$$M_1 \oplus M_2 = \{e \mid e \in M_1 \text{ and } e \notin M_2\} \cup \{e \mid e \notin M_1 \text{ and } e \in M_2\}.$$

By Theorem 3 and Theorem 4, we have Algorithm 5.1 to find the maximum matching of $G^{(k)}$ for each $0 < k \leq |X|$ efficiently. Since the number of edges is at most $|\mathscr{D}|$, so the space complexity is $O(n)$. It is easy to see that the time complexity from Line 4 to Line 9 is $O(\sum |E^{(k)}|) = O(\sum |\mathscr{D}^{(k)}|) = O(\mathcal{L})$. Thus, our algorithm is worst-case quadratic, which improves the previous best known upper bound $O(n^2\sqrt{n})$.

Algorithm 5.1 DISJOINT SUPPORT $\{\sqsubset, \between\}$-STRUCTURED 2-INTERVAL PATTERN

1: sort the endpoints of $\mathscr{D}$
2: calculate $F_+^{(k)}$ and $F_-^{(k)}$ for every $0 < k \leq |X|$
3: $E \leftarrow \emptyset$, $M \leftarrow \emptyset$
4: **for** $k \leftarrow 1$ to $|X|$ **do**
5: $\quad E \leftarrow E \setminus F_-^{(k)}$, $M \leftarrow M \setminus F_-^{(k)}$
6: $\quad$ **if** there exist an augmenting path $P \in E$ relative to M, **then** $M \leftarrow M \oplus P$
7: $\quad E \leftarrow E \cup F_+^{(k)}$
8: $\quad$ **if** there exist an augmenting path $P \in E$ relative to M, **then** $M \leftarrow M \oplus P$
9: **end for**

Proposition 4. *The* DISJOINT SUPPORT $\{\sqsubset, \between\}$-STRUCTURED 2-INTERVAL PATTERN *problem can be solved in $O(n \log n + \mathcal{L})$ time.*

6 Conclusions

In this paper, we give several improved algorithms for different models of 2-INTERVAL PATTERN problem. The case of $R = \{\between\}$ and $R = \{\sqsubset, \between\}$ with disjoint support are solved both in $O(n \log n + \mathcal{L})$ time by sweep-line based method. An $O(n \log n + dn)$ algorithm is given to solve the case $R = \{<, \sqsubset\}$, which is similar to previous known maximum independent set algorithm for circle graphs. All of our algorithms require only linear space.

Acknowledgements

Thanks to Hong Zhu, Binhai Zhu, Yunfeng Tao and the reviewers for their helpful comments. Thanks to Prof. Yong Yu[1] for his support on this work.

---

[1] The advisor of the undergraduate program in Department of Computer Science and Engineering, Shanghai Jiao Tong University.

References

1. Patricia A. Evans. Finding common subsequences with arcs and pseudoknots. In Maxime Crochemore and Mike Paterson, editors, *Combinatorial Pattern Matching, 10th Annual Symposium, CPM 99, Proceedings*, pages 270–280. Springer, 1999.
2. Stéphan Vialette. On the computational complexity of 2-interval pattern matching problems. *Theoretical Computer Science*, 312(2-3):223–249, 2004.
3. Jens Gramm. A polynomial-time algorithm for the matching of crossing contact-map patterns. In *Algorithms in Bioinformatics, 4th International Workshop, WABI 2004, Proceedings*, pages 38–49. Springer, 2004.
4. Jochen Alber, Jens Gramm, Jiong Guo, and Rolf Niedermeier. Computing the similarity of two sequences with nested arc annotations. *Theoretical Computer Science*, 312(2-3):337–358, 2004.
5. Tao Jiang, Guohui Lin, Bin Ma, and Kaizhong Zhang. The longest common subsequence problem for arc-annotated sequences. *J. Discrete Algorithms*, 2(2):257–270, 2004.
6. M.C. Golumbic. *Algorithmic graph theory and perfect graphs.* Academic Press, New York, NY, 1980.
7. Guillaume Blin, Guillaume Fertin, and Stéphane Vialette. New results for the 2-interval pattern problem. In *Combinatorial Pattern Matching, 15th Annual Symposium, CPM 2004, Proceedings*, pages 311–322. Springer, 2004.
8. Maxime Crochemore, Danny Hermelin, Gad M. Landau, and Stephane Vialette. Approximating the 2-Interval Pattern problem. To be appeared in 13th Annual European Symposium on Algorithms, ESA 2005.
9. S. Micali and V.V. Vazirani. An $O(\sqrt{|V|}|E|)$ algorithm for finding maximum matching in general graphs. In *Proceedings of the 21st Annual Symposium on Foundation of Computer Science*, pages 17–27. IEEE, 1980.
10. Reuven Bar-Yehuda, Magnús M. Halldórsson, Joseph Naor, Hadas Shachnai, and Irina Shapira. Scheduling split intervals. In *Proceedings of the 13th Annual ACM-SIAM Symposium on Discrete Algorithms*, pages 732–741, 2002.
11. Stefan Felsner, Rudolf Müller, and Lorenz Wernisch. Trapezoid graphs and generalizations, geometry and algorithms. *Discrete Applied Mathematics*, 74(1):13–32, 1997.
12. Sumio Masuda, Kazuo Nakajima, Toshinobu Kashiwabara, and Toshio Fujisawa. Efficient algorithms for finding maximum cliques of an overlap graph. *Networks*, 20:157–171, 1990.
13. Gabriel Valiente. A new simple algorithm for the maximum-weight independent set problem on circle graphs. In *Algorithms and Computation, 14th International Symposium, ISAAC 2003, Proceedings*, pages 129–137. Springer, 2003.
14. John E. Hopcroft and Richard M. Karp. An $n^{5/2}$ algorithm for maximum matchings in bipartite graphs. *SIAM Journal on Computing*, (4), 1973.

Preemptive Semi-online Scheduling on Parallel Machines with Inexact Partial Information*

Yong He and Yiwei Jiang

Department of Mathematics, State Key Lab of CAD & CG, Zhejiang University, Hangzhou 310027, P.R. China
jywzju@163.com

Abstract. In semi-online scheduling problems, we always assume that some partial additional information is exactly known in advance. This may not be true in some application. This paper considers semi-online problems on identical machines with inexact partial information. Three problems are considered, where we know in advance that the optimal value, or the largest job size are in given intervals, respectively, while their exact values are unknown. We give both lower bounds of the problems and competitive ratios of algorithms as functions of a so-called disturbance parameter $r \in [1, \infty)$. We establish that for which r the inexact partial information is useful to improve the performance of a semi-online algorithm with respect to its pure online problem. Optimal preemptive semi-online algorithms are then obtained.

1 Introduction

In scheduling theory, a problem is called *offline* if we have full information on the job data before constructing a schedule. If jobs arrive one by one over list and are required to be scheduled irrevocably on machines as soon as they are given, without any knowledge of the jobs that will arrive later, the problem is called *online*. If some partial additional information about the jobs is available in advance, and we cannot rearrange any job that has been assigned to machines, then the problem is called *semi-online*. Different partial information produces different semi-online problems. Algorithms for (semi-) online problems are called *(semi-) online algorithms*. Naturally, one wishes to achieve improvement of the performance of a semi-online algorithm with respect to its corresponding online problem by exploiting additional information. Though it is a relatively new area, various papers and a number of results on semi-online algorithms for scheduling problems have been published in the last decade. This paper will also consider the design and analysis of algorithms for semi-online scheduling.

We use the *competitive ratio* to measure the the performance of an (semi-) online algorithm. For an algorithm A and a job sequence $\mathcal{J}$ and let $c^A(\mathcal{J})$ (or in short c^A) denote the objective function value produced by A and let $c^*(\mathcal{J})$ (or in short c^*) denote the optimal value in an offline version. Then the competitive ratio of A is defined as the smallest number C such that for any $\mathcal{J}$,

* Research supported by the TRAPOYT of China and NSFC (10271110, 60021201).

X. Deng and D. Du (Eds.): ISAAC 2005, LNCS 3827, pp. 422–432, 2005.

$c^*(\mathcal{J}) \le Cc^A(\mathcal{J})$. If A is randomized, we use $E(c^A(\mathcal{J}))$ instead of $c^A(\mathcal{J})$ in the above definition. An (semi-) online problem has a *deterministic* or *randomized lower bound* ρ if no (semi-) online deterministic or randomized algorithm has a competitive ratio of smaller than ρ. An (semi-) online algorithm is called *optimal* if its competitive ratio matches the lower bound. With these definitions, we say that an information is *useful* if it can admit an optimal semi-online algorithm with a competitive ratio smaller than that of an optimal online algorithm or the lower bound of the pure online problem.

In semi-online research, it is crucial to determine which type of partial information, and in how much extent, can improve the performance of a semi-online algorithm. To illustrate our idea and method, we consider the classical parallel machine scheduling in this paper. It can be described as follows. We are given a sequence $\mathcal{J}$ of independent jobs with positive sizes $p_1, p_2, \ldots, p_n$, which must be scheduled onto m uniform machines $M_1, M_2, \cdots, M_m$. We identify the jobs with their sizes. Machine M_i has speed $s_i \ge 1$. Without loss of generality, we assume $1 = s_1 \le s_2 \le \cdots \le s_m = s$. If job p_j is completely assigned to machine M_i, then p_j/s_i time units are required to process this job. Machines are available at time zero. Our aim is to minimize the maximum machine completion time $C_{\max}$, called *makespan*.

We consider preemptive algorithms in this paper, hence each job may be cut into a few pieces. These pieces are to be assigned to possibly different machines, in non-overlapping time slots. It is not required that a machine is busy all the time until the last piece of a job being processed on the machine is completed. A period of time when a machine is not processing a job, while it has not yet completed all the pieces of the jobs assigned to it (i.e., it is assigned to process a job at a later time), is called *idle time*. Although it is not common to design algorithms that introduce idle time during the assignment procedure, it may be beneficial to keep room for jobs that arrive later in the online and semi-online problems, for example, see Refs. [5],[11], [12]. Using the three-field notation for scheduling problems [10], we denote our problem as $Q|pmpt|C_{\max}$ for the general case of $s \ge 1$ (uniform machines) and $P|pmpt|C_{\max}$ for the special case of $s = 1$ (identical machines).

The study of preemptive online algorithms for parallel machine scheduling problems date back to Chen et al. [2], who presented an optimal algorithm with a competitive ratio of $\frac{m^m}{m^m-(m-1)^m} \to \frac{e}{e-1}$ for $Pm|pmpt, online|C_{\max}$, where $e \approx 2.7183$. Wen and Du [16] and Epstein et al. [6] independently presented an optimal algorithm with a competitive ratio of $\frac{(s+1)^2}{s^2+s+1}$ for the problem $Q2|pmpt, online|C_{\max}$. Those results are extended to a class of m uniform machines with the non-decreasing speed ratios in [3]. All the algorithms for the above preemptive online problems are deterministic and their competitive ratios also match their respective randomized lower bounds. Hence, randomization does not help for these problems. Ebenlendr and Sgall [8] presented an algorithm with a competitive ratio of 4 for the problem $Qm|pmpt, online|C_{\max}$, and a randomized algorithm with a competitive ratio of e. A randomized lower bound of 2 was given in [7].

Several *basic semi-online* variants have been studied so far. Among others, Epstein [4] considered the information that the optimal value c^* is known in advance (denoted by *opt*), which is also called the *online bin stretching problem* [1]. She presented an optimal algorithm with a competitive ratio of 1 for $Q2|pmpt, online|C_{\max}$. Ebenlendr and Sgall [8] further presented an optimal algorithm with the same competitive ratio for $Qm|pmpt, online|C_{\max}$. It is remarkable that by combining this optimal semi-online algorithm with a doubling strategy for guessing the optimal value, Ebenlendr and Sgall obtained a deterministic algorithm with a competitive ratio of 4 for the problem $Qm|pmpt, online|C_{\max}$, and a randomized algorithm with a competitive ratio of e. It shows that the research of semi-online algorithms is also of theoretical significance, besides their important applications. Seiden et al. [15] considered the information that jobs arrive in an order of decreasing job sizes (denoted by *decr*). They proposed an optimal algorithm with a competitive ratio of $\max\limits_{k=0,\cdots,m} \frac{m^2+2mk}{m^2+k^2+2k} \rightarrow \frac{\sqrt{3}+1}{2}$ for $Pm|pmpt, decr|C_{\max}$. Furthermore, they showed that the algorithm still works with the same competitive ratio for $Pm|pmpt, max|C_{\max}$, where *max* denotes the semi-online version that the largest job size $p_{\max} = \max_{j=1,\cdots,n} p_j$ is known in advance. Epstein and Favrholdt [5] considered $Q2|pmpt, decr|C_{\max}$. They proposed an optimal algorithm with a competitive ratio of $\frac{3(s+1)}{3s+2}$ if $s \in [1,3]$, and $\frac{2s(s+1)}{2s^2+s+1}$ if $s \in [3,+\infty)$. He and Jiang [11] considered $Q2|pmpt, max|C_{\max}$. They proposed an optimal algorithm with a competitive ratio of $\frac{2s^2+3s+1}{2s^2+2s+1}$. It follows that the usefulness of the information *max* is weaker than that of *decr* for uniform machine case with regard to competitive analysis, which is unlike that for identical machine case. Note that randomization does not help for all above semi-online problems, too.

In all the above considered semi-online problems, we assume that the known partial information is exact. However, this assumption may not be true in some application. Instead, we may know some partial information in advance, but this information is sensitive and not accurate, or with uncertainty. That is, we only know some *disturbed partial information* in advance. We would like to see whether it is still useful, and how to design an algorithm based on this inexact information accordingly. In this paper, we propose to introduce this concept in the context of semi-online scheduling. Two variants regarding the basic semi-online versions *opt*, *max* will be studied. For the first variant, *dist opt*, we know in advance that there exist some $p > 0$ and $r \geq 1$ such that $c^* \in [p, rp]$. For the second one, denoted by *dist max*, we know in advance that there exist some $p > 0$ and $r \geq 1$ such that $p_{\max} \in [p, rp]$. We call r the *disturbance parameter*. Three problems, $Pm|pmpt, dist\ opt|C_{\max}$, $Q2|pmpt, dist\ opt|C_{\max}$ and $Q2|pmpt, dist\ max|C_{\max}$, will be considered. We will present their respective optimal semi-online algorithms.

The competitive ratio is a function of the r for each problem, by which we can see in what extent the disturbed information is useful. For example, since the competitive ratio (i.e., lower bound) of the first problem $Pm|pmpt, dist\ opt|C_{\max}$ is $\frac{m^m}{m^m-(m-1)^m}$ when $r \geq (\frac{m}{m-1})^{m-1}$, we conclude that the disturbed

information becomes useless. When $1 \le r < (\frac{m}{m-1})^{m-1}$, the disturbed information is still useful. Further, we can also see how the disturbance parameter affects the performance guarantee from the obtained parametric competitive ratios. The results are summarized in Table 1.

Table 1. The obtained results in this paper

the problem	the competitive ratio of the optimal algorithm	the interval of r where the information is useless
$Pm\|pmpt, dist\ opt\|C_{\max}$ $(q = \frac{m-1}{m})$	$\frac{mr}{mr(1-q^k)+m-k}$ if $r \in [(\frac{1}{q})^{k-1}, (\frac{1}{q})^k)$, $k = 1, 2, \cdots, m-1$; $\frac{1}{1-(\frac{1}{q})^m}$ if $r \in [(\frac{1}{q})^{m-1}, +\infty)$	$r \in [(\frac{1}{q})^{m-1}, +\infty)$
$Q2\|pmpt, dist\ opt\|C_{\max}$	$\frac{r(s+1)}{rs+1}$ if $r \in [1, s+1]$; $\frac{(s+1)^2}{s^2+s+1}$ if $r \in [s+1, +\infty)$	$r \in [s+1, +\infty)$
$Q2\|pmpt, dist\ max\|C_{\max}$	$\frac{(s+1)(1+s+rs)}{1+2s+s^2+rs^2}$ if $r \in [1, s+1]$; $\frac{(s+1)^2}{s^2+s+1}$ if $r \in [s+1, +\infty)$	$r \in [s+1, +\infty)$

The paper is organized as follows. Section 2 presents some preliminary results and lower bounds of the considered problems. Sections 3-5 present respective optimal semi-online algorithms for $Pm|pmpt, dist\ opt|C_{\max}$ and $Q2|pmpt, dist\ opt|C_{\max}$ and $Q2|pmpt, dist\ max|C_{\max}$, respectively.

2 Preliminary and Lower Bounds

Let L_j^* be the makespan of an optimal schedule for the first j jobs. Thus $c^* = L_n^*$. Denote $S_j = \sum_{i=1}^{j} p_i$ and $p_j^{\max} = \max\{p_i | i = 1, \cdots, j\}$. The following result is well-known.

Lemma 1. (1) ([14]) *For any instance of the problem $Pm|pmpt|C_{\max}$, we have $c^* = \max\{\frac{S_n}{m}, p_n^{\max}\}$.* (2) ([9],[13]) *For any instance of the problem $Q2|pmpt|C_{\max}$, we have $c^* = \max\{\frac{S_n}{s+1}, \frac{p_n^{\max}}{s}\}$.*

Define $q = \frac{m-1}{m}$, $\alpha_k(r) = \frac{mr}{mr(1-q^k)+m-k}$ for $k = 1, 2, \cdots, m-1$ and $\alpha_m(r) = \frac{1}{1-q^m}$. To simplify presentation, we will drop the dependence on r and always write α_k instead of $\alpha_k(r)$ for $k = 1, 2, \cdots, m$. Define

$$\alpha = \begin{cases} \alpha_k, \text{ if } (\frac{1}{q})^{k-1} \le r < (\frac{1}{q})^k, \ k = 1, 2, \cdots, m-1, \\ \alpha_m, \text{ if } r > (\frac{1}{q})^{m-1}. \end{cases}$$

Theorem 1. *Any deterministic or randomized preemptive semi-online algorithm A for the problem $Pm|pmpt, dist\ opt|C_{\max}$ has a competitive ratio α.*

Proof. First we assume that r satisfy $(1/q)^{k-1} \le r < (1/q)^k$ for some $k = 1, 2, \cdots, m-1$. Let A be a randomized algorithm with a competitive ratio of

C. Consider the sequence $p_1 = (m-1)(q^{k-1}r-1)$, $p_2 = \cdots = p_m = 1$ and $p_{m+i} = q^{k-i}r$, $i = 1, 2, \cdots, k$. It is clear that the total size is mr for any k, and thus $c^* = r \in [1, r]$.

Let $X_i, i = 1, 2, \cdots, m$, be the last time when at least i jobs are running (after scheduling all $m+k$ jobs). Note that X_i is a random variable. For $i = 1, \cdots, m$, as long as i jobs are running, at least one of the first $m+k+1-i$ jobs is running. Therefore, the average makespan of the first $m+k+1-i$ jobs is at least $E[X_i]$. Since by Lemma 1(1) the optimal schedule for these $m+k+1-i$ jobs has makespan $q^{i-1}r$ if $1 \le i \le k$ and 1 if $k+1 \le i \le m$, we have

$$E[X_i] \le \begin{cases} C \cdot q^{i-1}r, & \text{if } 1 \le i \le k, \\ C, & \text{if } k+1 \le i \le m. \end{cases}$$

So, we have

$$\sum_{i=1}^{m} E[X_i] \le C(\sum_{i=1}^{k} q^{i-1}r+m-k) = C(\frac{1-q^k}{1-q}r+m-k) = C((1-q^k)mr+m-k).$$

Combining it with $\sum_{i=1}^{m} X_i \ge \sum_{i=1}^{m+k} p_i = mr$, we obtain $C \ge \frac{mr}{(1-q^k)mr+m-k} = \alpha_k$.

For any $r \ge (1/q)^{m-1}$, it suffices to consider the sequence $p_1 = \cdots = p_m = 1$, and $p_{m+i} = (1/q)^i, i = 1, \cdots, m-1$. Then a similar argument as above can reach the goal. □

Theorem 2. *Any semi-online algorithm A for $Q2|pmpt, dist\ opt|C_{\max}$ has a competitive ratio of at least* $\begin{cases} \frac{r(s+1)}{rs+1}, & \text{if } 1 \le r < s+1, \\ \frac{(s+1)^2}{s^2+s+1}, & \text{if } r \ge s+1. \end{cases}$

Theorem 3. *Any semi-online algorithm A for $Q2|pmpt, dist\ max|C_{\max}$ has a competitive ratio of at least* $\begin{cases} \frac{(1+s)(1+s+rs)}{1+2s+s^2+rs^2}, & \text{if } 1 \le r < s+1, \\ \frac{1+2s+s^2}{1+s+s^2}, & \text{if } r \ge s+1. \end{cases}$

3 Optimal Algorithm for $Pm|pmpt, dist\ opt|C_{\max}$

This section considers the preemptive scheduling problem on m identical machines, where we know in advance that the optimal value is in the interval $[p, rp]$. By normalization, we assume $p = 1$, resulting in $c^* \in [1, r]$.

Since the lower bound for $Pm|pmpt, dist\ opt|C_{\max}$ is the same as the competitive ratio of the optimal algorithm for the problem $Pm|pmpt, on-line|C_{\max}$ when $r \ge (1/q)^{m-1}$ [2], we focus on the case $1 \le r < (1/q)^{m-1}$ in the remainder of this section. We will present a semi-online algorithm $A1$ that is optimal for every $1 \le r < (1/q)^{m-1}$. The algorithm consists of $m-1$ procedures which are identical except for the value of a parameter. We first define the procedure.

Procedure $A0(a)$

0. Let $j = 0$.
1. **If** no new job arrives, stop. Otherwise, set $j = j + 1$.
2. Compute L_j^* according to Lemma 1(1).
   2.1 **If** $p_j \le a\max\{1, L_j^*\} - L_{j-1}^1$, schedule p_j on machine M_1.
   2.2 **If** $p_j > a\max\{1, L_j^*\} - L_{j-1}^1$, compute $l_j = \max\{1 \le i \le m : L_{j-1}^i + p_j \ge a\max\{1, L_j^*\}\}$. Schedule $a\max\{1, L_j^*\} - L_{j-1}^{l_j}$ part of p_j on M_{l_j}, and the leftover of p_j, if any, on M_{l_j+1}.
3. Re-index the machines with respect to their current loads such that $L_j^1 \ge L_j^2 \ge \cdots \ge L_j^m$, and return 1.

Algorithm $A1$

If $(1/q)^{k-1} \le r < (1/q)^k$ for some $k = 1, \cdots, m-1$, let $a = \alpha_k$ and run $A0(\alpha_k)$.

Remember that in this section the moment j represents the moment right after the j-th job has been scheduled and the machines have been re-indexed in non-decreasing order of their current loads. Hence, if we say "right after assigning job p_j", it means that machines are not yet re-indexed in non-decreasing order of their current loads.

In the remainder of this section, let p_{j_i} be the first job (or a part of it) processed on the i-th machine at moment j_i.

Lemma 2. *$L_{j_i}^1 \ge \cdots \ge L_{j_i}^{i-1} \ge \alpha_k$ at moment j_i for every $2 \le i \le m$.*

Proof. We prove the result by induction. For $i = 2$, we claim that p_{j_2} must be assigned by step 2.2. Otherwise, we have $p_{j_2} \le \alpha_k\max\{1, L_{j_2}^*\} - L_{j_2-1}^1$, and p_{j_2} is assigned to machine M_1 by the algorithm rule, a contradiction. Hence, the algorithm assigns the portion $\alpha_k\max\{1, L_{j_2}^*\} - L_{j_2-1}^1$ to machine M_1 (as $l_{j_2} = 1$) and the leftover to machine M_2. Right after assigning p_{j_2}, we have

$$L_{j_2}^1 = L_{j_2-1}^1 + (\alpha_k\max\{1, L_{j_2}^*\} - L_{j_2-1}^1) = \alpha_k\max\{1, L_{j_2}^*\} \ge \alpha_k,$$

so we conclude that $L_{j_2}^1 \ge \alpha_k$ still holds at moment j_2.

Assume that the result is true when $i = t$, that is, we have $L_{j_t}^1 \ge \cdots \ge L_{j_t}^{t-1} \ge \alpha_k$ at moment j_t. Next we consider the case $i = t + 1$. Since for any job $p_j, j_t < j < j_{t+1}$, no new machine starts to process jobs by the definition of $p_{j_{t+1}}$. Then all these jobs are assigned to machine M_1, and we can conclude that $L_j^1 \ge \cdots \ge L_j^{t-1} \ge \alpha_k$ holds at any moment $j_t < j < j_{t+1}$. Now we focus on the moment j_{t+1}. We can also conclude that $p_{j_{t+1}}$ is scheduled by step 2.2, and $p_{j_{t+1}}$ is the first job processed on the $t+1$-th machine. Hence, we see that $l_j = t$, i.e., we have

$$L_{j_{t+1}}^t = L_{j_{t+1}-1}^t + (\alpha_k\max\{1, L_{j_{t+1}}^*\} - L_{j_{t+1}-1}^t) = \alpha_k\max\{1, L_{j_{t+1}}^*\} \ge \alpha_k,$$

right after assigning $p_{j_{t+1}}$. Moreover, the loads of the first $t-1$ machines are unchanged at that time. Hence, we have one more machine with load at least α_k. It implies that $L_{j_{t+1}}^1 \ge L_{j_{t+1}}^2 \ge \cdots \ge L_{j_{t+1}}^t \ge \alpha_k$ at moment j_{t+1}. Therefore, the result is true for $i = t + 1$. The lemma follows. □

The next two lemmas show that the algorithm $A1$ is well defined, and it maintains the following invariants at any moment $j \geq 1$:

(I1) $L_j^1 \leq \alpha_k \max\{1, L_j^*\}$;

(I2) for any $1 \leq t \leq k-1$, $\sum_{i=1}^{t} L_j^i \geq (1-q^t)S_j\alpha_k$.

Lemma 3. *If* (I1) *and* (I2) *are fulfilled at moment* $j-1$, *then the assignment of job* p_j *is feasible, and* (I1) *is still fulfilled at moment* j.

Lemma 4. *If* (I1) *and* (I2) *are fulfilled at moment* $j-1$, *then* (I2) *is still fulfilled at moment* j.

Theorem 4. *The competitive ratio of* $A1$ *is at most* α_k *when* $(1/q)^{k-1} \leq r < (1/q)^k$ *for some* $k = 1, \cdots, m-1$, *and thus it is optimal for any* $1 \leq r < (1/q)^{m-1}$.

Proof. It is easy to verify that both (I1) and (I2) are fulfilled at moment 1. Combining it with Lemmas 3 and 4, we conclude that the assignment of all jobs is feasible, and both (I1) and (I2) are fulfilled at any moment. Together with the condition $L_n^* = c^* \in [1, r]$, (I1) implies that $\frac{c^{A1}}{c^*} = \frac{L_n^{A1}}{\max\{1, L_n^*\}} \leq \alpha_k$. Moreover, algorithm $A1$ is optimal by Theorem 1. □

4 Optimal Algorithm for $Q2|pmpt, dist\ opt|C_{\max}$

This section considers the preemptive scheduling problem on two uniform machines, where we know in advance that the optimal value is in the interval $[p, rp]$. By normalization, we assume $p = 1$, resulting in $c^* \in [1, r]$, and $1 = s_1 \leq s_2 = s$.

Since the lower bound for $Q2|pmpt, disturbed\ opt|C_{\max}$ is the same as the competitive ratio of the optimal algorithm for the problem $Q2|pmpt, on-line|C_{\max}$ when $r \geq s+1$ [6, 16], we focus on the case $1 \leq r < s+1$ in the remainder of this section. We will present an semi-online algorithm $A2$ which is optimal for any $1 \leq r < s+1$.

Let $\beta = \frac{r(s+1)}{rs+1}$ and $t = \min\{j | S_j > \beta\}$.

Algorithm $A2$**:**

0. $j = 0$.
1. **If** no new job arrives, stop. Otherwise, set $j = j+1$.
2. **If** $j < t$, schedule p_j on M_1 completely, and return 1.
3. **If** $j = t$, we do
   3.1 **If** $p_t < \beta + (s-1)L_{t-1}^1$, schedule the portion $\beta - L_{t-1}^1$ (denoted by p_t^1) of p_t on M_1 at time L_{t-1}^1, and the leftover (if exists, denoted by p_t^2) on M_2 at time zero. Return 1.
   3.2 **If** $\beta + (s-1)L_{t-1}^1 \leq p_t < (\beta - L_{t-1}^1)s$, schedule the portion $\frac{(\beta - L_{t-1}^1)s - p_t}{s-1}$ (denoted by p_t^1) of p_t on M_1 at time L_{t-1}^1 and the leftover (denoted by p_t^2) on M_2 at time L_t^1. Return 1.

3.3 **If** $(\beta - L^1_{t-1})s \le p_t < s\beta$, schedule p_t on M_2 completely at time $\beta - \frac{p_t}{s}$. Return 1.
3.4 **If** $p_t \ge s\beta$, schedule p_t on M_2 completely at time zero. Return 1.
4. **If** $j > t$, we do
4.1 **If** p_t is scheduled by step 3.1, schedule job p_j on M_2. Return 1.
4.2 **If** p_t is scheduled by step 3.2 or 3.3, schedule p_j in the idle time on M_2, once the idle time is fully occupied, scheduled the leftover (if exists) on M_1 until time β. Once exceed, scheduled the leftover (if exists) on M_2. Return 1.
4.3 **If** p_t is scheduled by step 3.4, schedule it on M_1 until time β. Once exceeded, scheduled the leftover (if exists) on M_2. Return 1.

Remark 1. Note that before we run algorithm $A2$, we do not need to know the value of t in advance. In fact, in the description of the algorithm, the expressions $t > j$, $t < j$ and $t = j$ only represent the situations that the current total job size S_j is larger/smaller than, and equal to β, respectively, which can be checked online. Hence, our algorithm runs in a semi-online way. It is also valid for the later algorithm.

Remark 2. The following four figures show the different situations with respect to the assignment of p_t. It is clear that in Figures 2 and 3 idle time is introduced on M_2 when assigning the job p_t, later jobs will be scheduled during the idle time window and no more idle time will be introduced.

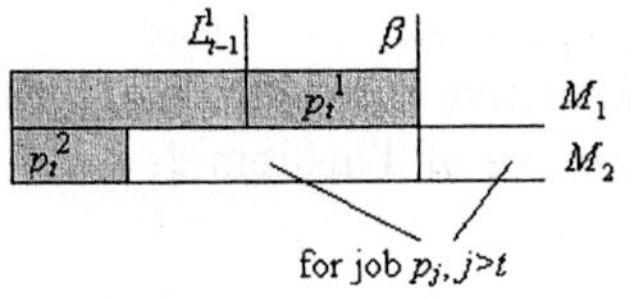

Fig. 1. Step 3.1

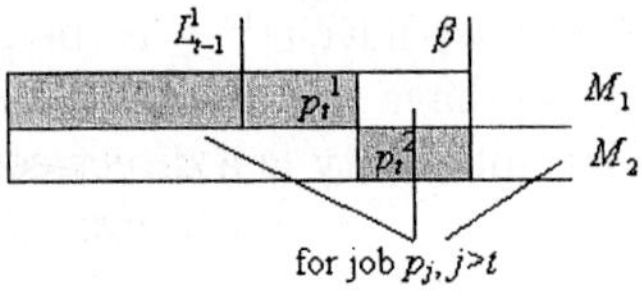

Fig. 2. Step 3.2

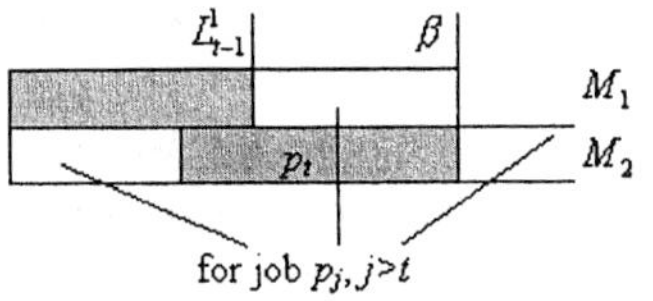

Fig. 3. Step 3.3

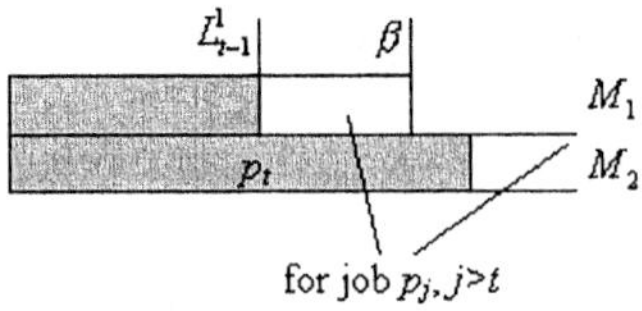

Fig. 4. Step 3.4

Lemma 5. *The algorithm A2 is feasible.*

Lemma 6. *The makespan yielded by algorithm H2 satisfies one of the two following conditions: (i)* $c^{A2} \le \beta$*; (ii)* $c^{A2} \le \max\{(S_n - \beta)/s, p_n^{\max}/s\}$.

Proof. If $n < t$, i.e., all the jobs have a total size not greater than β. Hence, we obtain that $c^{A2} = S_n \leq \beta$. Hence, we obtain the result.

If $n = t$, we see that $c^{A2} = \beta$ if p_t is scheduled by one of steps 3.1, 3.2 and 3.3, and $c^{A2} = p_t/s \leq p_t^{\max}/s$ if p_t is assigned by step 3.4. The result is true.

If $n > t$, by the algorithm, we still have $c^{A2} = L_t^{A2} = \beta$ if the idle time slots before time β are not fully occupied (see Figures 1-4). Now we consider the case that the idle time slots before time β have been fully occupied. If $c^{A2} = L_t^{A2}$, then either $c^{A2} = \beta$ or $c^{A2} = p_t$. Hence, the result is true trivially by the above argument for the case $n = t$. Then we assume that $c^{A2} > L_t^{A2} \geq \beta$. We deduce that the load of machine M_1 must be β because no more jobs (or parts of jobs) are scheduled on M_1 by the rule of step 4 starting from the time β. Therefore, we can obtain that the makespan of $A2$ is $L_n^2 = (S_n - L_n^1)/s = (S_n - \beta)/s$, which is the desired result. The proof is complete. □

Theorem 5. *The algorithm A2 has a competitive ratio of β, thus it is optimal for any $1 \leq r < s+1$ and $s \geq 1$.*

Proof. Recall that $c^* \in [1, r]$. By Lemma 6, we have $c^{A2} \leq \beta$ or $c^{A2} \leq \max\{(S_n - \beta)/s, p_n^{\max}/s\}$. If $c^{A2} \leq \beta$, then from $c^* \geq 1$, we have $c^{A2}/c^* \leq \beta$. If $c^{A2} \leq (S_n - \beta)/s$, then since $c^* \geq \frac{S_n}{s+1}$ due to Lemma 1 and $c^* \leq r$, we have $S_n \leq r(s+1)$. Hence, we have

$$\frac{c^{A2}}{c^*} \leq \frac{(S_n - \beta)/s}{S_n/(s+1)} = \frac{s+1}{s}(1 - \frac{\beta}{S_n}) \leq \frac{s+1}{s}(1 - \frac{\beta}{r(s+1)}) = \beta.$$

If $c^{A2} \leq p_n^{\max}/s$, we have $c^{A2} \leq c^*$ because Lemma 1 states that $c^* \geq p_n^{\max}/s$.

Therefore, we have shown that the competitive ratio of the algorithm $A2$ is at most β. Its optimality is a direct consequence of Theorem 2. □

5 Optimal Algorithm for $Q2|pmpt, dist\ max|C_{\max}$

This section considers the preemptive problem on two uniform machines, where we know in advance that the largest job size is in the interval $[p, rp]$. By normalization, we assume that $p = 1$, and $1 = s_1 \leq s_2 = s$.

Since the lower bound for $Q2|pmpt, dist\ max|C_{\max}$ is the same as the competitive ratio of the optimal algorithm for the problem $Q2|pmpt, on-line|C_{\max}$ when $r \geq s+1$ [6, 16], we focus on the case $1 \leq r < s+1$ in the remainder of this section. We will present a semi-online algorithm $A2$ which is optimal for every $1 \leq r < s+1$.

Define $\gamma = \frac{(s+1)(1+s+rs)}{1+2s+s^2+rs^2}$, $b_j = \gamma \max\{\frac{1}{s}, L_j^*\}$ and $t = \min\{j | S_j > b_j\}$.

Algorithm $A3$**:**

0. $j = 0$.
1. **If** no new job arrives, stop. Otherwise, set $j = j + 1$
2. **If** $j < t$, schedule p_j on M_1 completely. Return 1.

3. **If** $j = t$, we do
   3.1 **If** $p_t < (b_j - L^1_{t-1})s$, schedule the portion $\frac{(b_j - L^1_{t-1})s - p_t}{s-1}$ (denoted by p^1_t) of p_t on M_1 at time L^1_{t-1} and the leftover (denoted by p^2_t) on M_2 at time L^1_t. Return 1.
   3.2 **If** $p_t > (b_j - L^1_{t-1})s$, schedule p_t on M_2 completely at time $b_j - \frac{p_t}{s}$. Return 1.
4. **If** $j > t$, schedule the portion $p^1_j = (b_j - L^2_{j-1})s$ of p_j on M_2 at time L^2_{j-1}, and the leftover (if exists) in the idle time on M_2. Once the idle time is fully occupied, scheduled the leftover (if exists) on M_1. Return 1.

Remark 3. As the descriptions in precious section, we can obtain the following two figures which show the different situations with respect to the assignment of p_t. It is clear that in both figures idle time is introduced on M_2 when assigning the job p_t, subsequent arriving jobs will be scheduled to filling up the idle time period and no more idle time will be introduced.

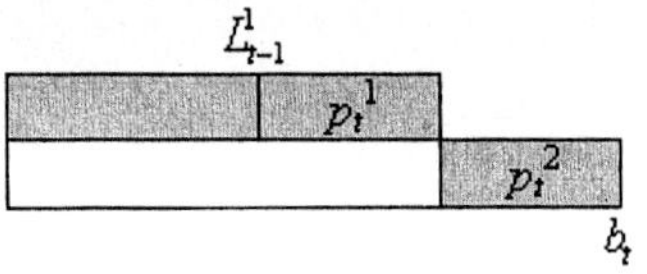

Fig. 5. Step 3.1

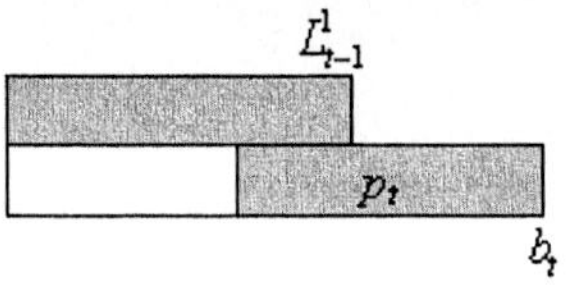

Fig. 6. Step 3.2

Theorem 6. *The competitive ratio of A3 is γ, and thus it is optimal.*

References

1. Y. Azar, O. Regev, Online bin stretching, *Theoretical Computer Science*, 268, 2001, 17-41.
2. B. Chen, A. van Vliet, G. Woeginger, An optimal agorithm for preemptive on-line scheduling, *Operations Research Letters*, 18, 1995, 127-131.
3. L. Epstein, Optimal preemptive on-line scheduling on uniform processors with non-decreasing speed ratios, *Operations Research Letters*, 29, 2001, 93-98.
4. L. Epstein, Bin stretching revisited, *Acta Informatica*, 39, 2003, 97-117.
5. L. Epstein, L. Favrholdt, Optimal preemptive semi-online scheduling to minimize makespan on two related machines, *Operations Research Letters*, 30, 2002, 269-275.
6. L. Epstein, J. Noga, S. Seiden, J. Sgall, G. Woeginger, Randomized on-line scheduling on two uniform machines, *Journal of Scheduling*, 4, 2001, 71-92.
7. L. Epstein, J. Sgall, A lower bound for on-line scheduling on uniformly related machines, *Operations Research Letters*, 26, 2000, 17-22.
8. T. Ebenlendr, J. Sgall, Optimal and online preemptive scheduling on uniformly related machines, *Lecture Notes in Computer Science* 2996, pages 199-210. Springer, 2004.
9. T. Gonzalez, S. Sahni, Preemptive scheduling of uniform processor systems, *Journal of the Association for Computing Machinery*, 25, 1978, 92-101.

10. R. L. Graham, E. L. Lawler, J. K. Lenstra, A. H. G. Rinnooy Kan, Optimization and approximation in deterministic sequencing and scheduling: a survey, *Annals of Operations Research*, 5, 1979, 287-326.
11. Y. He, Y.W. Jiang, Optimal algorithms for semi-online preemptive scheduling problems on two uniform machines, *Acta Informatica*, 40, 2004, 367-383.
12. Y. He, Y.W. Jiang, Optimal semi-online preemptive algorithms for machine covering on two uniform machines, *Theoretical Computer Science*, 339, 2005, 293-314.
13. E. C. Horvath, S. Lam, R. Sethi, A level algorithm for preemptive scheduling, *Journal of the Association for Computing Machinery*, 24, 1977, 32-43.
14. R. McNaughton, Scheduling with deadlines and loss functions, *Management Sciences*, 6, 1959, 1-12.
15. S. Seiden, J. Sgall, G. Woeginger, Semi-online scheduling with decreasing job sizes, *Operations Research Letters*, 27, 2000, 215-221.
16. J. Wen, D. Du, Preemptive on-line scheduling for two uniform processors, *Operations Research Letters*, 23, 1998, 113-116.

On-Line Computation and Maximum-Weighted Hereditary Subgraph Problems

Marc Demange[1], Bernard Kouakou[2], and Éric Soutif[3]

[1] ESSEC, SID department, Avenue Bernard HIRSH, BP 105, F-95021 Cergy Pontoise Cedex, France
demange@essec.fr

[2] CERMSEM, Université Paris 1, 106–112 bd de l'Hôpital, F-75013 Paris, France
kouakou@univ-paris1.fr

[3] CEDRIC, CNAM, 292 rue Saint-Martin, F-75003 Paris, France
soutif@cnam.fr

Abstract. In this paper, we study the on-line version of maximum-weighted hereditary subgraph problems. In our on-line model, the final instance-graph (which has n vertices) is revealed in t clusters, $2 \leq t \leq n$. We first focus on the on-line version of the following problem: finding a maximum-weighted subgraph satisfying some hereditary property. Then, we deal with the particular case of the independent set. For all these problems, we are interested in two types of results: the competitivity ratio guaranteed by the on-line algorithm and hardness results that account for the difficulty of the problems and for the quality of algorithms developed to solve them.

Keywords: on-line algorithm, hereditary property, independent set, competitivity ratio.

1 Introduction

On-line computation aims to solve combinatorial problems with the constraint that the instance is not a priori completely known before one begins to solve it. In other words, the data-set is revealed step-by-step and one has, at the end of each step, to irrevocably decide on the final solution dealing with this step. On-line algorithms deal with a large class of problems subjected to time constraints (one must take decisions before knowing all data). An on-line problem is defined by:

- a combinatorial optimisation problem,
- a set of rules R describing how the final instance will be revealed,
- a set of rules R' defining what kind of decisions are allowed.

1.1 Maximum Weighted Hereditary Subgraph Problems

In this paper, we deal with on-line versions of the problem of finding, in a weighted graph G, a maximum weighted subgraph satisfying a non-trivial hereditary property π.

X. Deng and D. Du (Eds.): ISAAC 2005, LNCS 3827, pp. 433–442, 2005.

Definition 1. *Let us consider a graph-property π from the set of graphs to $\{0,1\}$. π is said to be satisfied for a graph G if and only if $\pi(G) = 1$. Property π is hereditary if, whenever it is satisfied by a graph, it is also satisfied by each of its induced subgraph; it is non-trivial if it is true for an infinite number of graphs and false for an infinite number of graphs. If $G = (V, E)$ is a graph, then a set of vertices $V' \subset V$ is said to satisfy π if and only if the induced subgraph $G[V']$ satisfies π.*

Since π is assumed to be non trivial, then for every n there exists a graph of order at least n satisfying π; which implies that an isolate vertex satisfies π.

Definition 2. *Given a graph $G = (V, E)$, the maximum hereditary subgraph problem, HG, consists of finding a subgraph of maximum order of G satisfying a given non-trivial hereditary property π.*

WHG denotes the weighted version of HG: the input is a weighted graph (each vertex x has a weight $w(x)$) and one looks for a maximum-weighted subgraph satisfying π. Some on-line versions of HG have been studied in [3]. This paper is devoted to extend the results to the weighted case.

1.2 On-Line Models and Competitive Analysis

Let $LWHG$ denote the on-line version of WHG. By using the framework described in [7], it is defined by (WHG, R, R'). Given a problem, we can pass from an on-line version to another by changing only the rules R and R'. For $LWHG$ we consider the same model as in [3]:

R: the graph G is revealed in t steps. At each step, a weighted subgraph $G_i = (V_i, E_i)$ (called cluster) of G is revealed together with the edges linking the vertices of G_i and already revealed vertices. Consequently at step i, the graph already revealed is $G[V_1, \cdots, V_i]$, the subgraph of the whole instance induced by $V1, \cdots, V_i$.
R': at each step, one irrevocably decides which new vertices are introduced in the solution.

The quality of an on-line algorithm is expressed by the competitive ratio defined below.

Let us consider a maximization on-line problem P, an instance σ of P and an algorithm **LA** for solving P. Furthermore, we consider a function $c_{\mathbf{LA}}(n)$ (n denotes the order of the instance). Let $\mathbf{LA}(\sigma)$ denote the value of the solution of P computed by the algorithm LA and $\beta(\sigma)$ the value of the solution returned by an optimal algorithm knowing beforehand the final instance. The algorithm **LA** is said to guarantee the competitivity ratio $c_{\mathbf{LA}}(n)$ if, for any instance σ of P:

$c_{\mathbf{LA}}(n) \times \beta(\sigma) \leq \mathbf{LA}(\sigma)$ ($c_{\mathbf{LA}}(n) \times \mathbf{LA}(\sigma) \leq \beta(\sigma)$ for a minimization problem).
LA is also said to be $c_{\mathbf{LA}}(n)$-competitive.

In what follows, we supposed that the number of vertices and the total weight of the final graph are known at the beginning of the on-line process. In the opposite

case, it is easy to see that no interesting result can be found; only the trivial ratio $\frac{W_{min}}{W(G)-W_{min}}$ can be guaranteed (where $W(G)$ and W_{min} denote respectively the total and the minimum weight of the final graph G).

1.3 Approximation Ratio

In the case of approximation, given a polynomial-time algorithm **A** computing, for every instance σ of an $NP-hard$ (off-line) minimization problem (the maximization case is defined in a similar manner), a feasible solution, **A** is said to guarantee an approximation ratio of ρ_A if, for every instance σ of order n, the approximation value $\mathbf{A}(\sigma)$ satisfies: $\rho_A(n) \times \beta(\sigma) \leq \mathbf{A}(\sigma)$ where $\beta(\sigma)$ denotes the value of the optimal solution of the instance.

For $LWHG(t)$ (t *steps with* $t << n$), we are interested in links between off-line and on-line algorithms. A central question in this work is how it is possible to exploit (polynomial-time) algorithms in order to devise (polynomial-time) on-line algorithm and moreover, is it possible to transfer performance guarantees from off-line to on-line framework?

For this study, we suppose we are given a $\rho(n)$-approximation algorithm FWA for WHG such that the following hypotheses **H** hold:

1. The considered approximation algorithm **FWA** solving WHG is supposed to guarantee ratio of the form $\frac{f(n)}{n}$, where f is an increasing function in n beyond k.
2. Algorithm **FWA** satisfies $w(\mathbf{FWA}(G)) \geq \frac{w(V)}{n}$, $\forall G$ (the ratio $1/n$ is always guaranteed, i.e $f(n) \geq 1$);

These hypotheses are not restrictive (see also [3]); in particular item 1 is satisfied for every known approximation ratios for HG and WHG: both problems can be polynomially approximated within $O(\log n/n)$ and this ratio can be improved to $O(\log^2 n/n)$ for maximum independent set [4, 2]. Condition 2 is also natural since the ratio $\frac{w(V)}{n}$ can always be guaranteed.

In section 2, we show that $LWHG(t)$ reduces to WHG. More precisely, we show that a (polynomial-time) $f(n)/n$-approximation algorithm leads to a (polynomial) on-line algorithm with competitive ratio $c_{\mathbf{LWA}}(G) \geq \frac{1-f(n)^{1/t}}{1-f(n)} \frac{f(n)}{n}$. For $t = 2$, it leads to a result of [3] which was obtained by adapting to $LWHG(2)$ the methodology used for $LHG(t)$. The proof of our result is totally different while the on-line algorithm is quite similar. At each step, it computes a solution of the problem restricted to the new cluster and decides either to include the whole solution performed if its value is sufficiently good or to reject it. We call such an algorithm a *threshold algorithm.*

In section 3, we perform lower and upper bounds for a class of algorithms including threshold ones. The main result is shown in the case where π is either *clique* or *independent set.* Finally, section 4 deals with on-line independent set problem for which we propose an hardness result that bounds below the competitive ratio of any algorithm (not only threshold ones). It points out that, for this problem, threshold algorithms are almost optimal.

1.4 Notations

We will consider only simple graphs $G = (V, E)$, $n = |V|$ denotes the order of G. Every edge in G will be denoted either by $(i, j) = (j, i)$ or simply by $ij = ji$. $\bar{G} = (V, \bar{E})$, $\bar{E} = \{(i, j), i \neq j, ij \notin E\}$ denotes the complement of G. If $V' \subset V$, $G[V']$ is the subgraph of G induced by V'. An independent set of G is a set of vertices which are mutually not linked by an edge: $V' \subset V$, $\forall (i, j) \in V' \times V', ij \notin E$, in other words, $G[V']$ has no edge. A clique of G is a set of vertices such that $G[V']$ is a complete graph or, equivalently, V' is an independent set of $\bar{G}$. $w(G)$ denotes the sum of the weights of the vertices of G. The on-line version of any hereditary weighted graph problem in which the final instance is revealed in t steps will be denoted by $LWHG(t)$. $LWIS$ denotes the on-line version of the weighted independent set problem.

2 Competitive Ratio for *LWHG*

Recall that the total number of vertices and the total weight are supposed to be known in advance. This section is devoted to prove the following result:

Theorem 1. *Suppose that WHG admits a polynomial-time $\rho(n)$−approximation algorithm* **FWA** *with $\rho(n) = \frac{f(n)}{n}$. Then, under the hypotheses H, there exists an on-line polynomial-time algorithm* **LWA**, *for $LWHG$, achieving for all graph G of order n a competitive ratio of:*

$$c_{\mathbf{LWA}}(G) \geq \frac{1 - f(n)^{1/t}}{1 - f(n)} \rho(n).$$

Moreover, for $\epsilon > 0$ and n large enough: $c_{\mathbf{LWA}}(G) \geq (1 - \epsilon)\frac{f(n)^{1/t}}{n}$.

Proof. Let us consider **LWA**, the following algorithm which receives the input-graph G in t subgraphs $G_1, G_2, \ldots, G_t$ of respective order $n_1, n_2, \ldots, n_t$, and returns the solution $\mathbf{LWA}(G)$ for the $LWHG$ problem :

Algorithm 1

1. $i \leftarrow 1$;
2. $W \leftarrow w(G)$;
3. $r \leftarrow n$;
4. while $w(\mathbf{FWA}(G_i)) < \frac{W - w(G_i)}{r - n_i} f(r - n_i)^{1/t}$ and $i < t$ do
5. $\quad W \leftarrow W - w(G_i)$;
6. $\quad r \leftarrow r - n_i$;
7. $\quad i \leftarrow i + 1$;
8. end while
9. return $\mathbf{LWA}(G) \leftarrow \mathbf{FWA}(G_i)$;

Remark 1. This is a "threshold-algorithm" using the threshold $\frac{W-w(G_i)}{r-n_i} f(r-n_i)^{1/t}$. It is polynomial since algorithm **FWA** is polynomial-time.

For $i \leq t-1$, we denote $R_i = G[V_{i+1} \cup \cdots \cup V_t]$, and $r_i = n_{i+1} + \cdots + n_t$. Let us denote by $k \leq t$ the value of i at the end of the while loop. We will consider two cases whether $k < t$ or $k = t$. In the first case, the following statements hold:

$$w(\mathbf{FWA}(G_i)) < \frac{w(R_i)}{r_i} f(r_i)^{1/t} \qquad \forall i < k \tag{1}$$

$$w(\mathbf{FWA}(G_k)) \geq \frac{w(R_k)}{r_k} f(r_k)^{1/t} \geq \frac{w(R_k)}{r_k} \tag{2}$$

In the second case, only relation 1 holds.
We first point out the following result:

Lemma 1.

$$\forall i < k,\ w(\boldsymbol{FWA}(G_i)) \leq f(n)^{\frac{k-i}{t}} w(\boldsymbol{FWA}(G_k))$$

Proof. (of lemma 1)
Let us first consider that $k < t$. For $i < k$, since $w(R_i) = w(G_{i+1}) + \cdots + w(G_k) + w(R_k)$, relation 1 becomes

$$w(\mathbf{FWA}(G_i)) < \frac{w(G_{i+1}) + \cdots + w(G_k) + w(R_k)}{n_{i+1} + \cdots + n_k + r_k} f(r_i)^{1/t}.$$

By using item 1 of **H**, we deduce:
$w(\mathbf{FWA}(G_i)) \leq \frac{n_{i+1} w(\mathbf{FWA}(G_{i+1})) + \cdots + n_k w(\mathbf{FWA}(G_k)) + r_k w(\mathbf{FWA}(G_k))}{n_{i+1} + \cdots + n_k + r_k} f(n)^{1/t}$
which implies $w(\mathbf{FWA}(G_i)) \leq f(n)^{1/t} \sup_{i+1 \leq q \leq k} w(\mathbf{FWA}(G_q))$.

Then, a simple backward induction concludes the result. The case $k = t$ is simmilarly solved by putting $w(R_t) = 0$ and $r_t = 0$, which concludes the proof of lemma 1.

Let us continue the demonstration of the theorem:
Case 1: algorithm stops with $k < t$. Heredity implies that $\beta(G) \leq \beta(G_1) + \cdots + \beta(G_k) + \beta(R_k)$, moreover, $\beta(R_k) \leq w(R_k) \leq \frac{r_k}{f(r_k)^{1/t}} w(\mathbf{FWA}(G_k))$, which implies:
$\beta(G) \leq \rho(n_1)^{-1} w(\mathbf{FWA}(G_1)) + \cdots + \rho(n_k)^{-1} w(\mathbf{FWA}(G_k)) + \frac{r_k}{f(r_k)^{1/t}} w(\mathbf{FWA}(G_k))$
By using the increasing of f and the decreasing of $\rho(n) = f(n)/n$ (item 1 of hypothesis H), we deduce:
$\beta(G) \leq \rho(n)^{-1} \left(w(\mathbf{FWA}(G_1)) + \cdots + w(\mathbf{FWA}(G_k)) + f(n)^{\frac{t-1}{t}} w(\mathbf{FWA}(G_k)) \right)$
Then lemma 1 implies

$$\beta(G) \leq \rho(n)^{-1} (f(n)^{\frac{t-2}{t}} + \cdots + f(n)^{\frac{1}{t}} + 1 + f(n)^{\frac{t-1}{t}}) w(\mathbf{FWA}(G_k))$$

and finally

$$\beta(G) \leq \rho(n)^{-1} \frac{1 - f(n)}{1 - f(n)^{\frac{1}{t}}} w(\mathbf{FWA}(G_k))$$

The related ratio is:

$$\frac{w(S)}{\beta(G)} \geq \rho(n)\frac{1-f(n)^{\frac{1}{t}}}{1-f(n)}.$$

Case 2: the algorithm runs until the tth iteration.
By using similar arguments to the first case, we successively get:
$\beta(G) \leq \beta(G_1) + \cdots + \beta(G_t)$
$\beta(G) \leq \rho(n_1)^{-1} w(\mathbf{FWA}(G_1)) + \cdots + \rho(n_t)^{-1} w(\mathbf{FWA}(G_t))$
$\beta(G) \leq \rho(n)^{-1}((w(\mathbf{FWA}(G_1)) + \cdots + w(\mathbf{FWA}(G_t)))$
$\beta(G) \leq \rho(n)^{-1}(f(n)^{\frac{t-1}{t}} + \cdots + f(n)^{\frac{1}{t}} + 1)w(\mathbf{FWA}(G_t))$ so $\frac{w(S)}{\beta(G)} \geq \rho(n)\frac{1-f(n)^{\frac{1}{t}}}{1-f(n)}$
Since f is supposed to infinitly increase, the asymptotic equivalent immediately follows.

We deduce from approximation results for WHG and WIS:

Corollary 1. *For fixed values of t,*

i *$LWHG(t)$ admits a polynomial $O((\log n)^{1/t}/n)$-competitive algorithm;*
ii *$LWIS(t)$ admits a polynomial $O((\log n)^{2/t}/n)$-competitive algorithm.*

If we consider not only polynomial-time algorithms, the result also holds with $\rho(n) = 1$:

Corollary 2. *For fixed values of t, $LWHG(t)$ (and also $LWIS(t)$) admits a $O(n^{1/t-1})$-competitive algorithm.*

3 Limit of Threshold-Algorithms (Hardness Result)

In this section, we first suppose that π is either a *clique* or an *independent set.*

Let us consider an approximation algorithm **FWA** for WHG satisfying the set of hypotheses **H**. In [1], for problem $LWHG(2)$, a lower bound of $2\sqrt{1+\mu}\frac{\sqrt{f(n)}}{n}$ for the competitive ratio of threshold-algorithm is devised by assuming the following hypothesis $H'(\mu)$:

$H'(\mu)$**:** There exists G_1, a graph of order n_1 such that $\beta(G_1)\rho(n_1) \leq \mathbf{FWA}(G_1) < (1+\mu)\beta(G_1)\rho(n_1)$.

If μ is close to 0, $H'(\mu)$ means that the approximation ratio of the algorithm **FWA** cannot be significantly improved.

We show that this result can be generalized for any $t \geq 2$; moreover, it shows that the analysis performed in the proof of Theorem 1 is asymptocally tight.

Proposition 1. *If π is either clique or independent set and if **FWA** satisfies $H'(\mu)$ for a given approximation ratio $\rho(n) = f(n)/n$ and $\mu > 0$, then the corresponding threshold-algorithm (algorithm 1, section 2) cannot guarantee the competitive ratio* $t(1+\mu)^{1-\frac{1}{t}}\frac{f(n)^{\frac{1}{t}}}{n}$.
The result holds even if the sequence of the weights is known at the beginning of the on-line process and if clusters are of the same size.

Proof. FWA satisfies $H'(\mu)$. So there exists a graph G_1 of order n_1 such that $\beta(G_1)\rho(n_1) \leq \mathbf{FWA}(G_1) < (1+\mu)\beta(G_1)\rho(n_1)$. We consider $t-1$ real numbers $x_1, x_2, \dots, x_t$ strictly positive such that:

$$\mathbf{FWA}(G_1) < x_t < \cdots < x_2 < x_1 < (1+\mu)\beta(G_1)\rho(n_1).$$

Set $w_k = \frac{nx_k}{t}(1+\mu)^{\frac{k-t-1}{t}} f(n)^{\frac{1-k}{t}}, \forall k \geq 2$.

We apply the algorithm **LWA** to a graph of $n = tn_1$ vertices and of total weight $W = w(G_1) + w_2 + \cdots + w_t$, and we suppose that G_1 is revealed at the first step. Then we apply the following strategy to reveal a final graph having n vertices and whose total weight is W so that the algorithm cannot return a good solution (actually the worst one).

1. For $i < t$, if the algorithm selects some vertices in G_i, we reveal, for $k = 1 \cdots t-i$, clusters G_{i+k} so that : for the *independent set* problem, G_{i+k} is an independent set of order n_1; every vertex of G_{i+k} is of weight $\frac{w_{i+k}}{n_1}$ and is linked to all vertices in $G[V_1 \cdots V_i]$, i.e vertices revealed at steps $1, \cdots, i$. (For the *clique* problem, G_{i+k} will be a clique with no link with $G[V_1 \cdots V_i]$).
2. For $i < t$, if the algorithm does not select vertices in G_i, we reveal G_{i+1} a graph of order n_1 so that: for the independent set problem, G_{i+1} is a clique with vertices of weight $\frac{w_{i+1}}{n_1}$ and is linked to all vertices in $G[V_1 \cdots V_i]$. (With a similar construction one can deal with the case of the *clique* problem).

We then distinguish the following three cases (recall that vertices of the final solution returned by algorithm LWA are all in a same cluster G_i):
i)We first consider the case where the solution returned by **LWA** contains vertices of G_1.
ii) Then, we deal with the case where the first selected vertices belong to G_i, with $1 < i < t$. In this case, the result returned by **LWA** is $\mathbf{FWA}(G_i)$ since rule 1 is applied to i.
iii) Finally, we study the case where the algorithm does not select vertices in the subgraphs $G_1, G_2, \dots, G_{t-1}$ (so the algorithm returns $\mathbf{FWA}(G_t)$). Rule 2 is applied at step $t-1$.

In three cases one can establish that $\frac{\mathbf{LWA}(G)}{\beta(G)} < t(1+\mu)^{1-\frac{1}{t}} \frac{f(n)^{\frac{1}{t}}}{n}$.

We have thus shown that there exists a graph having n vertices such that $c_{\mathbf{LWA}}(n) < t(1+\mu)^{1-\frac{1}{t}} \frac{f(n)^{\frac{1}{t}}}{n}$.

Corollary 3. *Under hypotheses H and $H'(\mu)$, for any ϵ and for n large enough, we have:*

$$(1-\epsilon)\frac{f(n)^{\frac{1}{t}}}{n} < c_{LWA}(n) < t(1+\mu)\frac{f(n)^{\frac{1}{t}}}{n}$$

This bound is asymptotically tight since the ratio between the upper and the lower bounds is $t\frac{(1+\mu)}{1-\epsilon}$.

4 On-Line Maximum-Weighted Independent Set Problem (*LWIS*)

Theorem 1 holds for maximum independent set problem. The lower bound devised for threshold algorithms also applies to this problem. Roughly speaking, those results show that the only way to improve the competitive ratio of algorithm 1 (*LWA*) for *LWIS* is to improve the performance guarantee of the off-line algorithm **FWA** used by **LWA**. Netherless, this method is limited by the bound $O(n^{1/t-i})$ obtained by using an optimal algorithm as **LWA**.. This raises the following question: is it possible to get a best competitive ratio by using another type of algorithms? In this section, we bring a negative answer to this question.

Theorem 2. *Let* **LWA** *be an on-line algorithm solving* $LWIS$ *for* $t \geq 2$ *(the graph is revealed in* t *clusters). Assume that the weights of clusters are known by the algorithm as soon as the game starts, then its competitivity ratio* $c_{\boldsymbol{LWA}}$ *satisfies for every* n*:* $c_{\boldsymbol{LWA}}(n) \leq t\frac{n^{\frac{1}{t}}}{n}$.

Proof. Let **LWA** be an on-line algorithm solving $LWIS$; consider $t \geq 2$ and $n = kt$, $k \geq 2$. Set $w_i = k^{1-\frac{i-1}{t}}$; the total weight of the final graph is then $W = \frac{k-1}{k^{\frac{1}{t}}-1} + q$ (if we suppose $k > 1$, i.e. $n > 2t - 2$).

We apply algorithm **LWA** to a graph of order n and of total weight W. A first cluster consisting of a clique of $q+1$ vertices of weight 1, is revealed. Then, we apply the following strategy.

1. If at step $i < t$, **LWA** has not selected any vertex yet; a clique G_{i+1} of k vertices is revealed. Each vertex of the clique is of weight $\frac{1}{k^{\frac{i}{t}}}$ and is not linked to vertices already revealed.
2. If **LWA** selects some vertices at the step i, then clusters $G_{i+1}, G_{i+2}, \ldots, G_t$ are independent sets of size k. Their vertices are respectively of weights $\frac{1}{k^{\frac{i}{t}}}, \ldots, \frac{1}{k^{\frac{t-1}{t}}}$, and are (all) linked to an already selected vertex.

We distinguish two cases.

Case 1: rule 2 has never been used, namely the algorithm has only taken some vertices in the last cluster which is a clique; so $w(\mathbf{LWA}(G)) = \frac{1}{k^{\frac{t-1}{t}}}$.

Now $\alpha(G) \geq \alpha(G_1) = 1$, so $\frac{w(\mathbf{LWA}(G))}{\alpha(G)} \leq k^{\frac{1-t}{t}}$.

Case 2: rule 2 has been used at the step i, $i < t$. It is clear that only one vertex of weight $\frac{1}{k^{\frac{i-1}{t}}}$ has been selected (in G_i) since only cliques have been revealed until the ith step and all the next vertices are linked to an already selected vertex: $w(\mathbf{LWA}(G)) = \frac{1}{k^{\frac{i-1}{t}}}$.

Moreover, $\alpha(G) \geq \alpha(G_{i+1}) = \frac{k^{\frac{i}{t}}}{k}$, So $\frac{w(\mathbf{LWA}(G))}{\alpha(G)} \leq k^{\frac{1-t}{t}}$.

In conclusion, we get (since $n = kt$): $c_{\mathbf{LWA}}(n) \leq k^{\frac{1-t}{t}} \leq (\frac{n}{t})^{\frac{1-t}{t}}$

So, $c_{\mathbf{LWA}}(n) \leq t\frac{n^{\frac{1}{t}}}{n}$.

5 Concluding Remarks

This work points out that results for LGH and $LWHG(2)$ can be generalized to the weighted class $LWHG$. In the off-line case, HG and WHG are equivalently solved by polynomial-time algorithms ([2, 7]). The situation is rather different in on-line framework.

Indeed, the comparison between theorem 1 and result of [3] brings to the fore a gap between competitive ratio of $LHG(t)$ and $LWHG(t)$, which is asymptotically: $\frac{\rho_{LWIS}}{\rho_{LIS}} \sim f(n)^{\frac{1}{t}-\frac{1}{2}}\sqrt{t}$. More generally, weights induce many questions in on-line framework. In this work, we focused on an on-line model in which weights are revealed on-line together with vertices. But one can also study either a model where the sequence of the weights is revealed at the beginning while the graph is revealed on-line or a model where the graph is known in advance and weight are on-line.

Such models being particular cases of the ones we deal with in this paper, the competitive analysis performed in Theorem 1 remains valid in both cases. Let us now consider this question from the hardness results point of view. By revisiting the proof of theorem 2, let us point out of that weights are known beforehand. So this hardness result also holds for a model in which weights are removed from the on-line process.

Let us now consider a dual situation where the graph is fixed and weights are given on-line. The following proposition shows that the same hardness result raises for this model:

Proposition 2. *Let* **LWA** *be an on-line algorithm solving the problem $LWIS$ for $t \geq 2$ (the set of weights is completely revealed in t steps); the total weight of each cluster is known at the beginning of the process. Then, there exists a graph of order n for which the competitivity ratio $c_{\mathbf{LWA}}$ satisfies: $c_{\mathbf{LWA}}(n) \leq t\frac{n^{\frac{1}{t}}}{n}$.*

Proof. We set $n = tk$ where $k \geq 2$, $k \in N$. $w_i = k^{1-\frac{i-1}{t}}$, $n_i = k$, $\forall i \geq 1$. So the total weight of the final graph is $W = k\frac{k-1}{k^{\frac{1}{t}}-1}$.
This time, we apply the algorithm **LWA** to a complete graph of order n and of weight W. The first cluster contains $(q+1)$ weights, each of them being 1. Then we use the following strategy.

1. If at step $i < t$, **LWA** has not selected any vertex yet, we reveal a set of k identical weights $\frac{w_{i+1}}{k} = \frac{1}{k^{\frac{i}{t}}}$, in order to form the cluster G_{i+1} of total weight $w_{i+1} = k^{1-\frac{i}{t}}$.
2. If **LWA** selects vertices at step i, then the next clusters, $G_{i+1}, G_{i+2}, \ldots, G_t$ (each of them has k vertices) are such that:
for each of the clusters, $k-1$ vertices have a null weight and only one vertex supports the weight of the whole cluster.

We conclude by using the same arguments as in Theorem 2.

Let us finally underline a main difference between Theorems 1 and 2. The first one is only valid for a class of algorithms, nevertheless, it gives some informations not only about general algorithms but also about polynomial-time ones

while the second one does not allow us to take complexity considerations into account. Theorem 1 brings to the fore that a threshold algorithm parametrized by an exact off-line algorithm is almost optimal among on-line algorithms. An interesting question is to know if the same holds for polynomial time on-line algorithms. Theorem 2 induces that any improvement dealing with the approximation of WHG would immediately induce an improvement of the competitive ratio that can be guaranteed in polynomial-time. What about the converse? In [1] a reduction from WIS to $LWIS(2)$ is proposed, with improvement of the ratio allowing to show that any improvement of $LWIS(2)$'s competitive ratio would imply an improvement of WIS's approximation ratio. A consequence is an hardness result for polynomial-time algorithms. A generalization of this result to $LWIS(t)$ or, more generally, the design of such reductions from off-line to on-line seems to be a fruitful research area.

References

1. Demange, M.: Reduction off-line to on-line: an example and its applications. Yugoslav Journal of Operations Research **13**(1), 3-24, 2003
2. Demange, M., Paschos, V. Th.: Improved approximations for weighted and unweighted graph problems. Theory of Computing Systems. To appear
3. Demange, M., Paradon, X., Paschos, V. Th.: On-line maximun-order induced hereditary subgraph problems. International Transaction in Operational Research, **12**(2), 185-201, 2005
4. Halldòrsson M.M.: Approximations of weighted independent set and heditary subset problems. Proc. 5th Ann. Int. Conf. on Computing and Combinatories, Lecture Notes in Computer Science, Springer-Verlag, 261-270, 1999
5. Crescenzi P., Silvestri R., Trevisan L.:, To weight or not to weight: where is the question? In Proc. 4th Israel Symposium on Theory and Computing and Systems, IEEE Computer Society, 68-77, 1996
6. Garey M.R., Johnson. D.S:, Computers and intractability. A guide to the theory of NP-completness. CA.Freeman, San Francisco, 1979
7. Paradon X.:, Algorithmique on-line. Thèse de doctorat, Université Paris Dauphine, 2000 (in french)

A Novel Adaptive Learning Algorithm for Stock Market Prediction

Lean Yu[1,2], Shouyang Wang[1,2,3], and Kin Keung Lai[3,4]

[1] Institute of Systems Science, Academy of Mathematics and Systems Science, Chinese Academy of Sciences, Beijing 100080, China
[2] School of Management, Graduate University of Chinese Academy of Sciences, Chinese Academy of Sciences, Beijing 100039, China
{yulean, sywang}@amss.ac.cn
[3] College of Business Administration, Hunan University, Changsha 410082, China
[4] Department of Management Sciences, City University of Hong Kong, Tat Chee Avenue, Kowloon, Hong Kong
mskklai@cityu.edu.hk

Abstract. In this study, a novel adaptive learning algorithm for feed-forward network based on optimized instantaneous learning rates is proposed to predict stock market movements. In this new algorithm, the optimized adaptive learning rates are used to adjust the weight changes dynamically. For illustration and testing purposes the proposed algorithm is applied to two main stock price indices: S&P 500 and Nikkei 225. The experimental results reveal that the proposed algorithm provides a promising alternative to stock market prediction.

1 Introduction

Stock market prediction is regarded as a challenging task because of high fluctuation and irregularity. There have many studies using artificial neural networks (ANNs) in this area. Back-propagation neural network (BPNN) is the most popular class of ANNs which have been widely applied to time series prediction. The basic learning rule of BPNN is based on the gradient descent optimization method and the chain rule, as initially proposed by Werbos [1] in the 1970s. Since the basic learning rule is based on the gradient descent method, which is known for its slowness and its frequent confinement to local minima [2], many improved BP algorithms are developed such as variable step size, adaptive learning [3-4] and others [5-6]. Generally, these algorithms have an improved convergence property, but most of these methods do not use the optimized instantaneous learning rates. In their studies, the learning rate is set to a fixed value when learning. However, it is critical to determine a proper fixed learning rate for the applications of the BPNN. If the learning rate is large, learning may occur quickly, but it may also become unstable and even will not learn at all. To ensure stable learning, the learning rate must be sufficiently small, but with a small learning rate the BPNN may be lead to a long learning time. Also, just how small the learning rate should be is unclear. In addition, for different structures of BPNN and for different applications, the best fixed learning rates are different.

X. Deng and D. Du (Eds.): ISAAC 2005, LNCS 3827, pp. 443 – 452, 2005.

There are other ways to accelerate the network learning using second-order gradient based nonlinear optimization methods, such as the conjugate gradient algorithm [6] and Levenberg-Marquardt algorithm [7]. The crucial drawbacks of these methods, however, are that in many applications computational demands are so large that their effective use in many practical problems is not viable.

A common problem with the all above mentioned methods is a non-optimal choice of the learning rate even with the adaptive change of the learning rate. A solution is to derive optimal learning rate formulae for BPNN and then allow an adaptive change at each iteration step during the learning process. The resulting algorithm will eliminate the need for a search for the proper fixed learning rate and provide fast convergence.

Due to the highly nonlinearity of neural networks, it is difficult to obtain the optimum learning rate. In this paper, a new method based on matrix and optimization techniques is proposed to derive the optimal learning rate and construct an adaptive learning algorithm. To test the efficiency of the proposed algorithm, two important stock indices, S&P500 and Nikkei225, are used. The rest of this work is organized as follows. In Section 2, the proposed adaptive learning algorithm with optimal learning rate is presented. In order to testing the proposed algorithm, Section 3 gives an experiment and reports the results. Finally, the conclusions are made in Section 4.

2 The Proposed Adaptive Learning Algorithm

Consider a three-layer BPNN, which has p nodes in the input layer, q nodes in the hidden layer and k nodes in the output layer. Mathematically, the basic structure of the BPNN model is described by

$$Y(t+1)=\begin{bmatrix} y_1(t+1)\\ y_2(t+1)\\ \cdots \\ y_k(t+1)\end{bmatrix}=\begin{bmatrix} f_2[\sum_{i=1}^{q} f_1(\sum_{j=1}^{p} w_{ij}(t)x_j(t)+w_{i0}(t))v_{1i}(t)+v_{10}(t)]\\ f_2[\sum_{i=1}^{q} f_1(\sum_{j=1}^{p} w_{ij}(t)x_j(t)+w_{i0}(t))v_{2i}(t)+v_{20}(t)]\\ \cdots \\ f_2[\sum_{i=1}^{q} f_1(\sum_{j=1}^{p} w_{ij}(t)x_j(t)+w_{i0}(t))v_{ki}(t)+v_{k0}(t)]\end{bmatrix}$$

$$=\begin{bmatrix} f_2[\sum_{i=0}^{q} f_1(\sum_{j=0}^{p} w_{ij}(t)x_j(t))v_{1i}(t)]\\ f_2[\sum_{i=0}^{q} f_1(\sum_{j=0}^{p} w_{ij}(t)x_j(t))v_{2i}(t)]\\ \cdots \\ f_2[\sum_{i=0}^{q} f_1(\sum_{j=0}^{p} w_{ij}(t)x_j(t))v_{ki}(t)]\end{bmatrix}=\begin{bmatrix} f_2[V_1^T F_1(W(t)X(t))]\\ f_2[V_2^T F_1(W(t)X(t))]\\ \cdots \\ f_2[V_k^T F_1(W(t)X(t))]\end{bmatrix}$$

$$=F_2[V^T(t)F_1(W(t)X(t))] \tag{1}$$

where $x_j(t)$, $j = 1, 2, \ldots, p$, are the inputs of the BPNN; y is the output of the BPNN; $w_{ij}(t)$, $i = 1, \ldots, q, j = 1, \ldots, p$, are the weights from the input layer to the hidden layer; $w_{i0}(t)$, $i = 1, \ldots, q$, are the biases of the hidden nodes; $v_{ij}(t)$, $i = 1, \ldots, q, j = 1, \ldots, k$, are the weights from the hidden layer to the output layer; $v_{i0}(t)$ is the bias of the output node; t is a time factor; f_1 is the activation function of the nodes for the hidden layer and f_2 is the activation function of the nodes for the output layer. Generally, the activation function for nonlinear nodes is assumed to be a symmetric hyperbolic tangent

function, i.e., $f_1(x) = \tanh(u_0^{-1}x)$, and its derivative is $f_1'(x) = u_0^{-1}[1 - f_1^2(x)]$, $f_1''(x) = -2u_0^{-1} f_1(x)[1 - f_1^2(x)]$, where u_0 is the shape factor of the activation function. Specially, some notations in Equation (1) are defined as follows:

$$X = (x_0, x_1, \cdots, x_p)^T \in R^{(p+1)\times 1},\quad Y = (y_0, y_1, \cdots, y_k)^T \in R^{k\times 1},$$

$$W = \begin{pmatrix} w_{10} & w_{11} & \cdots & w_{1p} \\ w_{20} & w_{21} & \cdots & w_{2p} \\ \cdots & \cdots & \cdots & \cdots \\ w_{q0} & w_{q1} & \cdots & w_{qp} \end{pmatrix} = (W_0, W_1, \cdots, W_p) \in R^{q\times(p+1)},\quad V = \begin{pmatrix} v_{10} & v_{20} & \cdots & v_{k0} \\ v_{11} & v_{21} & \cdots & v_{k1} \\ \cdots & \cdots & \cdots & \cdots \\ v_{1q} & v_{2q} & \cdots & v_{kq} \end{pmatrix} = (V_1, V_2, \cdots, V_k) \in R^{(q+1)\times k},$$

$$F_1(W(t)X(t)) = \left(F_1(net_0^H(t)) \quad F_1(net_1^H(t)) \quad \cdots \quad F_1(net_q^H(t))\right)^T \in R^{(q+1)\times 1},$$

$$net_i^H(t) = \sum_{j=0}^{p} w_{ij}(t)x_j(t),\ i = 0, 1, \ldots, q.$$

Usually, by estimating model parameter vector (W, V) via BPNN training and learning, we can realize the corresponding tasks such as function approximation, system identification or prediction. In fact, the model parameter vector (W, V) can be obtained by iteratively minimizing a cost function $E(X: W, V)$. In general, $E(X: W, V)$ is a sum of the error squares cost function with k output nodes and N training pairs or patterns, that is,

$$E(X:W,V) = \frac{1}{2}\sum_{j=1}^{N}\sum_{i=1}^{p} e_{ij}^2 = \frac{1}{2}\sum_{j=1}^{N} e_j^T e_j = \frac{1}{2}\sum_{j=1}^{N}[y_j - y_j(X:W,V)]^T[y_j - y_j(X:W,V)] \tag{2}$$

where y_j is the jth actual value and $y_j(X: W,V)$ is the jth estimated value.

Given the time factor t, Equation (2) can be rewritten as

$$E(t) = \frac{1}{2}\sum_{j=1}^{N}\sum_{i=1}^{k} e_{ij}^2(t) = \frac{1}{2}\sum_{j=1}^{N} e_j^T(t)e_j(t) = \frac{1}{2}\sum_{j=1}^{N}[y_j - y_j(t)]^T[y_j - y_j(t)] \tag{3}$$

where $e_j(t) = [e_{1j}(t) \quad e_{2j}(t) \quad \cdots \quad e_{kj}(t)]^T \in R^{k\times 1},\ j = 1,2,\cdots,N.$

By applying the steepest descent method to the error cost function $E(t)$ (i.e., Equation (3)), we can obtain the gradient of $E(t)$ with respect to V and W, respectively.

$$\nabla_V E(t) = \frac{\partial E(t)}{\partial V(t)} = \sum_{j=1}^{N}\sum_{i=1}^{k} e_{ij}(t)\frac{\partial e_{ij}(t)}{\partial V(t)} = -\sum_{j=1}^{N}\sum_{i=1}^{k} e_{ij}(t)\frac{\partial y_{ij}(t)}{\partial V(t)}$$
$$= -\sum_{j=1}^{N} e_j(t)F_{2(j)}'[V^T F_{1(j)}(WX)]' = -\sum_{j=1}^{N} F_{1(j)}(WX)e_j^T(t)F_{2(j)}' = -\sum_{j=1}^{N} F_{1(j)}e_j^T(t)F_{2(j)}'$$

$$\nabla_W E(t) = \frac{\partial E(t)}{\partial W(t)} = \sum_{j=1}^{N}\sum_{i=1}^{k} e_{ij}(t)\frac{\partial e_{ij}(t)}{\partial W(t)} = -\sum_{j=1}^{N}\sum_{i=1}^{k} e_{ij}(t)\frac{\partial y_{ij}(t)}{\partial W(t)}$$
$$= -\sum_{j=1}^{N} e_j(t)F_{2(j)}'[V^T F_{1(j)}(WX)]' = -\sum_{j=1}^{N} F_{1(j)}'\bar{V}F_{2(j)}'e_j(t)x_j^T(t) = -\sum_{j=1}^{N} \bar{F}_{1(j)}'\bar{V}F_{2(j)}'e_j x_j^T$$

So, the updated formulae of weights are given by, respectively

$$\Delta V = -\eta\nabla_V E(t) = \eta\sum_{j=1}^{N} F_{1(j)}e_j^T F_{2(j)}' \tag{4}$$

$$\Delta W = -\eta \nabla_W E(t) = \eta \sum_{j=1}^{N} \overline{F}'_{1(j)} \overline{V} F'_{2(j)} e_j x_j^T \tag{5}$$

where η is the learning rate;

$$\overline{F}'_{1(j)} = \text{diag}[f'_{1(1)} \quad f'_{1(2)} \quad \cdots \quad f'_{1(q)}] \in R^{q \times q},\ f'_{1(i)} = f'_1(net_i^H) = \frac{\partial f_1(net_i^H)}{\partial net_i^H}, i = 1,2,\cdots,q,$$

$$\overline{v}_i = [v_{i1} \quad \cdots \quad v_{iq}]^T \in R^{q \times 1}, i = 1,2,\cdots,q, F'_2 = diag[f'_{2(1)} \quad f'_{2(2)} \quad \cdots \quad f'_{2(k)}] \in R^{k \times k},$$

$$\overline{V} = \begin{bmatrix} v_{11} & v_{21} & \cdots & v_{k1} \\ v_{12} & v_{22} & \cdots & v_{k2} \\ \cdots & \cdots & \cdots & \cdots \\ v_{1q} & v_{2q} & \cdots & v_{kq} \end{bmatrix} = [\overline{v}_1 \ \overline{v}_2 \ \cdots \ \overline{v}_k] \in R^{q \times p},\ f'_{2(i)} = f'_2[v_i^T F_1(WX)] = \frac{\partial f_2[v_i^T F_1(WX)]}{\partial [v_i^T F_1(WX)]}, i = 1,2,\cdots k.$$

To derive the optimal learning rate, let Δ be an increment operator and consider the general error equation:

$$\Delta e(t+1) = e(t+1) - e(t) = y - y(t+1) - y + y(t) = -\Delta y(t+1) \tag{6}$$

where $\Delta y(t+1) \equiv 0$. This means there is no change in the pattern during the neural networks' learning procedure and the change of output of neural networks is

$$\Delta y(t+1) = \eta \xi(t) e(t) \tag{7}$$

where $\xi(t) = \mathbf{F}'_2[(\mathbf{F}_1^T \mathbf{F}_1) \otimes \mathbf{I}_{k^2} + (\mathbf{X}^T \mathbf{X}) \times (\overline{V} \mathbf{F}'_1 \mathbf{F}'_1 \overline{V})] \mathbf{F}'_2$ with

$$\mathbf{F}'_2 = \begin{bmatrix} F'_{2(1)} & 0 & \cdots & 0 \\ 0 & F'_{2(2)} & \cdots & 0 \\ \cdots & \cdots & \cdots & \cdots \\ 0 & 0 & \cdots & F'_{2(N)} \end{bmatrix},\quad \mathbf{F}_1^T \mathbf{F}_1 = \begin{bmatrix} F_{1(1)}^T F_{1(1)} & F_{1(1)}^T F_{1(2)} & \cdots & F_{1(1)}^T F_{1(N)} \\ F_{1(2)}^T F_{1(1)} & F_{1(2)}^T F_{1(2)} & \cdots & F_{1(2)}^T F_{1(N)} \\ \cdots & \cdots & \cdots & \cdots \\ F_{1(N)}^T F_{1(1)} & F_{1(N)}^T F_{1(2)} & \cdots & F_{1(N)}^T F_{1(N)} \end{bmatrix},$$

$$\mathbf{X}^T \mathbf{X} = \begin{bmatrix} x_1^T x_1 & x_1^T x_2 & \cdots & x_1^T x_N \\ x_2^T x_1 & x_2^T x_2 & \cdots & x_2^T x_N \\ \cdots & \cdots & \cdots & \cdots \\ x_N^T x_1 & x_N^T x_2 & \cdots & x_N^T x_N \end{bmatrix},\quad \mathbf{F}'_1 \mathbf{F}'_1 = \begin{bmatrix} \overline{F}'_{1(1)} \overline{F}'_{1(1)} & \overline{F}'_{1(1)} \overline{F}'_{1(2)} & \cdots & \overline{F}'_{1(1)} \overline{F}'_{1(N)} \\ \overline{F}'_{1(2)} \overline{F}'_{1(1)} & \overline{F}'_{1(2)} \overline{F}'_{1(2)} & \cdots & \overline{F}'_{1(2)} \overline{F}'_{1(N)} \\ \cdots & \cdots & \cdots & \cdots \\ \overline{F}'_{1(N)} \overline{F}'_{1(1)} & \overline{F}'_{1(N)} \overline{F}'_{1(2)} & \cdots & \overline{F}'_{1(N)} \overline{F}'_{1(N)} \end{bmatrix}.$$

Here, $\otimes$ indicates a direct product, and $\times$ indicates a cross product.

In order to prove Equation (7), a lemma must be introduced.

Lemma: The total time derivative of the BPNN single output $v^T F_1(WX)$ is given by

$$\frac{d[v^T F_1(WX)]}{dt} = F_1(WX)\frac{dv}{dt} + \overline{v}^T \overline{F}'_1(WX)\frac{dW}{dt}X = \frac{dv^T}{dt}F_1(WX) + \overline{v}^T \overline{F}'_1(WX)\frac{dW}{dt}X$$

Proof: Derivation of $\frac{d[v^T F_1(WX)]}{dt}$ is as follows:

$$\frac{d[v^T F_1(WX)]}{dt} = \frac{d\left[\sum_{i=0}^{q} v_i f_1\left(\sum_{j=0}^{p} w_{ij} x_j\right)\right]}{dt}$$

$$= \sum_{i=0}^{q} \frac{\partial\left[\sum_{i=0}^{q} v_i f_1\left(\sum_{j=0}^{p} w_{ij} x_j\right)\right]}{\partial v_i}\frac{dv_i}{dt} + \sum_{i=0}^{q}\sum_{j=0}^{p} \frac{\partial\left[\sum_{i=0}^{q} v_i f_1\left(\sum_{j=0}^{p} w_{ij} x_j\right)\right]}{\partial w_{ij}}\frac{dw_{ij}}{dt}$$

$$= \sum_{i=0}^{q} f_1\left(\sum_{j=0}^{p} w_{ij} x_j\right)\frac{dv_i}{dt} + \sum_{i=0}^{q}\sum_{j=0}^{p} v_i f_1'\left(\sum_{j=0}^{p} w_{ij} x_j\right) x_j \frac{dw_{ij}}{dt}$$

$$= f_1(net_0)\frac{dv_0}{dt} + f_1(net_1)\frac{dv_1}{dt} + \cdots + f_1(net_q)\frac{dv_q}{dt}$$

$$+ v_0 f_1'(net_0)\left[x_0\frac{dw_{00}}{dt} + x_1\frac{dw_{01}}{dt} + \cdots + x_p\frac{dw_{0p}}{dt}\right]$$

$$+ v_1 f_1'(net_1)\left[x_0\frac{dw_{10}}{dt} + x_1\frac{dw_{11}}{dt} + \cdots + x_p\frac{dw_{1p}}{dt}\right]$$

$$+ v_q f_1'(net_q)\left[x_0\frac{dw_{q0}}{dt} + x_1\frac{dw_{q1}}{dt} + \cdots + x_p\frac{dw_{qp}}{dt}\right]$$

$$= \left[f_1(net_0) \quad f_1(net_1) \qquad f_1(net_q)\right]\left[\frac{dv_0}{dt} \quad \frac{dv_1}{dt} \quad \cdots \quad \frac{dv_q}{dt}\right]^T$$

$$+ \left[v_1 \quad v_2 \quad \cdots \quad v_q\right]\begin{bmatrix} f'(net_1) & \cdots & 0 \\ \cdots & \cdots & \cdots \\ 0 & \cdots & f'(net_q)\end{bmatrix} \quad \begin{pmatrix}\text{due to } f(net_0)\equiv 1, \\ f'(net_0)=0\end{pmatrix}$$

$$\times\begin{bmatrix} \frac{dw_{00}}{dt} & \frac{dw_{01}}{dt} & \cdots & \frac{dw_{0p}}{dt} \\ \frac{dw_{10}}{dt} & \frac{dw_{11}}{dt} & \cdots & \frac{dw_{1p}}{dt} \\ \cdots & \cdots & \cdots & \cdots \\ \frac{dw_{q0}}{dt} & \frac{dw_{q1}}{dt} & \cdots & \frac{dw_{qp}}{dt}\end{bmatrix}\begin{bmatrix} x_0 \\ x_1 \\ \cdots \\ x_p\end{bmatrix}$$

$$= F_1(WX)\frac{dv}{dt} + \bar{v}^T\overline{F_1'}(WX)\frac{dW}{dt}X = \frac{dv}{dt}^T F_1(WX) + \bar{v}^T\overline{F_1'}(WX)\frac{dW}{dt}X \qquad \square$$

In the following, we start to prove Equation (7). Let us consider the change of output of BPNN for the *m*th pattern first. The above Lemma together with Equations (4) and (5) gives

$$\Delta y_m(t+1) = \Delta F_2\left[V^T F_1(Wx_m)\right] = \begin{bmatrix}\Delta f_{2(1),m} \\ \Delta f_{2(2),m} \\ \cdots \\ \Delta f_{2(k),m}\end{bmatrix} = \begin{bmatrix} f'_{2(1),m}\cdot(F_{1,m}^T\Delta v_1 + v_1^T\overline{F'_{1,m}}\Delta Wx_m) \\ f'_{2(2),m}\cdot(F_{1,m}^T\Delta v_2 + v_2^T\overline{F'_{1,m}}\Delta Wx_m) \\ \cdots \\ f'_{2(k),m}\cdot(F_{1,m}^T\Delta v_k + v_k^T\overline{F'_{1,m}}\Delta Wx_m)\end{bmatrix}$$

$$= \begin{bmatrix} f'_{2(1),m}\cdot(F_{1,m}^T\sum_{j=1}^{N}\Delta v_{1j} + \bar{v}_1^T\overline{F'_{1,m}}\Delta Wx_m) \\ f'_{2(2),m}\cdot(F_{1,m}^T\sum_{j=1}^{N}\Delta v_{2j} + \bar{v}_2^T\overline{F'_{1,m}}\Delta Wx_m) \\ \cdots \\ f'_{2(k),m}\cdot(F_{1,m}^T\sum_{j=1}^{N}\Delta v_{kj} + \bar{v}_k^T\overline{F'_{1,m}}\Delta Wx_m)\end{bmatrix}$$

$$= \begin{bmatrix} f'_{2(1),m} \cdot \left[F_{1,m}^T \left(\eta \sum_{j=1}^N F_{1j} e_{1j} f'_{2(1),j}\right) + v_1^T \overline{F'_{1,m}} \left(\eta \sum_{j=1}^N \overline{F'_{1j}} \overline{V} \overline{F'_{2j}} e_j x_j^T\right) x_m\right] \\ f'_{2(2),m} \cdot \left[F_{1,m}^T \left(\eta \sum_{j=1}^N F_{1j} e_{2j} f'_{2(2),j}\right) + v_2^T \overline{F'_{1,m}} \left(\eta \sum_{j=1}^N \overline{F'_{1j}} \overline{V} \overline{F'_{2j}} e_j x_j^T\right) x_m\right] \\ \cdots \\ f'_{2(k),m} \cdot \left[F_{1,m}^T \left(\eta \sum_{j=1}^N F_{1j} e_{kj} f'_{2(k),j}\right) + v_k^T \overline{F'_{1,m}} \left(\eta \sum_{j=1}^N \overline{F'_{1j}} \overline{V} \overline{F'_{2j}} e_j x_j^T\right) x_m\right] \end{bmatrix}$$

$$= \eta \sum_{j=1}^N \begin{bmatrix} f'_{2(1),m} \cdot (F_{1,m}^T F_{1j} e_{1j} f'_{2(1),j} + \overline{v}_1^T \overline{F'_{1,m}}\, \overline{F'_{1j}} \overline{V} \overline{F'_{2j}} e_j x_j^T x_m) \\ f'_{2(2),m} \cdot (F_{1,m}^T F_{1j} e_{2j} f'_{2(2),j} + \overline{v}_2^T \overline{F'_{1,m}}\, \overline{F'_{1j}} \overline{V} \overline{F'_{2j}} e_j x_j^T x_m) \\ \cdots \\ f'_{2(k),m} \cdot (F_{1,m}^T F_{1j} e_{kj} f'_{2(k),j} + \overline{v}_k^T \overline{F'_{1,m}}\, \overline{F'_{1j}} \overline{V} \overline{F'_{2j}} e_j x_j^T x_m) \end{bmatrix}$$

$$= \eta \sum_{j=1}^N \begin{bmatrix} f'_{2(1),m} & 0 & \cdots & 0 \\ 0 & f'_{2(2),m} & \cdots & 0 \\ \cdots & \cdots & \cdots & \cdots \\ 0 & 0 & \cdots & f'_{2(k),m} \end{bmatrix} \begin{bmatrix} F_{1,m}^T F_{1j} e_{1j} f'_{2(1),j} + \overline{v}_1^T \overline{F'_{1,m}}\, \overline{F'_{1j}} \overline{V} \overline{F'_{2j}} e_j x_j^T x_m \\ F_{1,m}^T F_{1j} e_{2j} f'_{2(2),j} + \overline{v}_2^T \overline{F'_{1,m}}\, \overline{F'_{1j}} \overline{V} \overline{F'_{2j}} e_j x_j^T x_m \\ \cdots \\ F_{1,m}^T F_{1j} e_{kj} f'_{2(k),j} + \overline{v}_k^T \overline{F'_{1,m}}\, \overline{F'_{1j}} \overline{V} \overline{F'_{2j}} e_j x_j^T x_m \end{bmatrix}$$

$$= \eta \sum_{j=1}^N F'_{2,m} \cdot (F'_{2j} e_j F_{1,m}^T F_{1j} + \overline{V}^T F'_{1,m} F'_{1j} \overline{V} \overline{F'_{2j}} e_j x_j^T x_m)$$

$$= \eta \sum_{j=1}^N F'_{2,m} \cdot (F_{1,m}^T F_{1j} I_{k^2} + \overline{V}^T F'_{1,m} F'_{1j} \overline{V} x_j^T x_m) \cdot F'_{2j} e_j$$

So, the total change caused by all patterns is

$$\Delta y(t+1) = \begin{bmatrix} \Delta y_1(t+1) \\ \Delta y_2(t+1) \\ \cdots \\ \Delta y_m(t+1) \\ \cdots \\ \Delta y_N(t+1) \end{bmatrix} = \begin{bmatrix} \eta \sum_{j=1}^N F'_{2,1} \cdot (F_{1,1}^T F_{1j} I_{k^2} + \overline{V}^T F'_{1,1} F'_{1j} \overline{V} x_j^T x_1) \cdot F'_{2j} e_j \\ \eta \sum_{j=1}^N F'_{2,2} \cdot (F_{1,2}^T F_{1j} I_{k^2} + \overline{V}^T F'_{1,2} F'_{1j} \overline{V} x_j^T x_2) \cdot F'_{2j} e_j \\ \cdots \\ \eta \sum_{j=1}^N F'_{2,m} \cdot (F_{1,m}^T F_{1j} I_{k^2} + \overline{V}^T F'_{1,m} F'_{1j} \overline{V} x_j^T x_m) \cdot F'_{2j} e_j \\ \cdots \\ \eta \sum_{j=1}^N F'_{2,N} \cdot (F_{1,N}^T F_{1j} I_{k^2} + \overline{V}^T F'_{1,N} F'_{1j} \overline{V} x_j^T x_N) \cdot F'_{2j} e_j \end{bmatrix}$$

$$= \eta \begin{bmatrix} F'_{2,1} & 0 & \cdots & 0 \\ 0 & F'_{2,2} & \cdots & 0 \\ \cdots & \cdots & \cdots & \cdots \\ 0 & 0 & \cdots & F'_{2,N} \end{bmatrix}$$

$$\times \begin{bmatrix} (F_{1,1}^T F_{11} I_{k^2} + \overline{V}^T F'_{1,1} F'_{11} \overline{V} x_1^T x_1) & (F_{1,1}^T F_{12} I_{k^2} + \overline{V}^T F'_{1,1} F'_{12} \overline{V} x_1^T x_2) & \cdots & (F_{1,1}^T F_{1N} I_{k^2} + \overline{V}^T F'_{1,1} F'_{1N} \overline{V} x_1^T x_N) \\ (F_{1,2}^T F_{11} I_{k^2} + \overline{V}^T F'_{1,2} F'_{11} \overline{V} x_2^T x_1) & (F_{1,2}^T F_{12} I_{k^2} + \overline{V}^T F'_{1,2} F'_{12} \overline{V} x_2^T x_2) & \cdots & (F_{1,2}^T F_{1N} I_{k^2} + \overline{V}^T F'_{1,2} F'_{1N} \overline{V} x_2^T x_N) \\ \cdots & \cdots & \cdots & \cdots \\ (F_{1,N}^T F_{11} I_{k^2} + \overline{V}^T F'_{1,N} F'_{11} \overline{V} x_N^T x_1) & (F_{1,N}^T F_{12} I_{k^2} + \overline{V}^T F'_{1,N} F'_{12} \overline{V} x_N^T x_2) & \cdots & (F_{1,N}^T F_{1N} I_{k^2} + \overline{V}^T F'_{1,N} F'_{1N} \overline{V} x_N^T x_N) \end{bmatrix}$$

$$\times \begin{bmatrix} F'_{21} e_1 \\ F'_{22} e_2 \\ \cdots \\ F'_{2N} e_N \end{bmatrix} = \eta \begin{bmatrix} F'_{21} & 0 & \cdots & 0 \\ 0 & F'_{22} & \cdots & 0 \\ \cdots & \cdots & \cdots & \cdots \\ 0 & 0 & \cdots & F'_{2N} \end{bmatrix} \left\{ \begin{bmatrix} F_{11}^T F_{11} I_{k^2} & F_{11}^T F_{12} I_{k^2} & \cdots & F_{11}^T F_{1N} I_{k^2} \\ F_{12}^T F_{11} I_{k^2} & F_{12}^T F_{12} I_{k^2} & \cdots & F_{12}^T F_{1N} I_{k^2} \\ \cdots & \cdots & \cdots & \cdots \\ F_{1N}^T F_{11} I_{k^2} & F_{1N}^T F_{12} I_{k^2} & \cdots & F_{1N}^T F_{1N} I_{k^2} \end{bmatrix} \right.$$

$$+\begin{bmatrix} x_1^T x_1 & x_1^T x_2 & \cdots & x_1^T x_N \\ x_2^T x_1 & x_2^T x_2 & \cdots & x_2^T x_N \\ \cdots & \cdots & \cdots & \cdots \\ x_N^T x_1 & x_N^T x_2 & \cdots & x_N^T x_N \end{bmatrix} \times \left(\overline{V}^T \begin{bmatrix} \overline{F}_{11}'\overline{F}_{11}' & \overline{F}_{11}'\overline{F}_{12}' & \cdots & \overline{F}_{11}'\overline{F}_{1N}' \\ \overline{F}_{12}'\overline{F}_{11}' & \overline{F}_{12}'\overline{F}_{12}' & \cdots & \overline{F}_{12}'\overline{F}_{1N}' \\ \cdots & \cdots & \cdots & \cdots \\ \overline{F}_{1N}'\overline{F}_{11}' & \overline{F}_{1N}'\overline{F}_{12}' & \cdots & \overline{F}_{1N}'\overline{F}_{1N}' \end{bmatrix} \overline{V} \right)\Bigg\}$$

$$\times \begin{bmatrix} F_{21}' & 0 & \cdots & 0 \\ 0 & F_{22}' & \cdots & 0 \\ \cdots & \cdots & \cdots & \cdots \\ 0 & 0 & \cdots & F_{2N}' \end{bmatrix} \begin{bmatrix} e_1 \\ e_2 \\ \cdots \\ e_N \end{bmatrix}$$

$$= \eta \mathbf{F}_2'[(\mathbf{F}_1^T \mathbf{F}_1) \otimes \mathbf{I}_{k^2} + (\mathbf{X}^T \mathbf{X}) \times (\overline{V}\mathbf{F}_1'\mathbf{F}_1'\overline{V})]\mathbf{F}_2' e(t) = \eta \xi(t) e(t) \qquad \square$$

Substituting (7) into (6), we obtain

$$e(t+1) = e(t) - \eta \xi(t) e(t) \tag{8}$$

The objective here is to derive an optimal learning rate η. That is, at iteration t, an optimal value of the learning rate, $\eta^*(t)$, which minimizes $E(t+1)$ is obtained. Define the cost function:

$$E(t+1) = \frac{1}{2} e^T(t+1) e(t+1) \tag{9}$$

Using Equation (8), Equation (9) may be written as

$$E(t+1) = 0.5[e(t) - \eta \xi(t) e(t)]^T [e(t) - \eta \xi(t) e(t)] \tag{10}$$

which gives the error e, at iteration t+1, as a function of the learning rate η, which minimizes $E(t+1)$. Now we use the first and second order conditions

$$\left.\frac{dE(t+1)}{d\eta}\right|_{\eta=\eta^*(t)} = -\frac{1}{2}[\xi(t)e(t)]^T [e(t) - \eta^*(t)\xi(t)e(t)] - \frac{1}{2}[e(t) - \eta^*(t)\xi(t)e(t)]^T \xi(t)e(t) = 0$$

$$\left.\frac{d^2 E(t+1)}{d\eta^2}\right|_{\eta=\eta^*(t)} = e^T(t)\xi^T(t)\xi(t)e(t) > 0$$

Since $\xi(t)$ is positively defined, the second condition is met and the optimum value of the learning rate is found to be

$$\eta^*(t) = \frac{e^T(t)\xi^T(t)e(t)}{e^T(t)\xi^T(t)\xi(t)e(t)} \tag{11}$$

Finally, the increments of the BP neural network parameters, by using the optimal learning rate, are obtained by replacing the η^* given by Equation (11) to Equations (4) and (5), which yield

$$\Delta V = -\eta \nabla_V E(t) = \frac{e^T(t)\xi^T(t)e(t)}{e^T(t)\xi^T(t)\xi(t)e(t)} \sum_{j=1}^{N} F_{1(j)} e_j^T F_{2(j)}' \tag{12}$$

$$\Delta W = -\eta \nabla_W E(t) = \frac{e^T(t)\xi^T(t)e(t)}{e^T(t)\xi^T(t)\xi(t)e(t)} \sum_{j=1}^{N} \overline{F}'_{1(j)} \overline{V} F'_{2(j)} e_j x_j^T \tag{13}$$

Using the new weight update formulae with optimal learning rates, a new learning algorithm is generated. To verify the effectiveness of the proposed adaptive learning model, two major stock indices (S&P500 and Nikkei225) are used as testing targets. A detailed experiment is presented in the following.

3 Empirical Study

3.1 Data Description

In the experiments, two stock indices, S&P500 and Nikkei225, are daily and are obtained from *Datastream*. The entire data set covers the period from January 1 2000 to December 31 2004. The data sets are divided into two periods: the first period covers from January 1 2000 to December 31 2003 while the second period is from January 1 2004 to December 31 2004. The first period, which is assigned to in-sample estimation, is used to network learning and training. The second period is reserved for out-of-sample evaluation. For brevity, the original data are not listed in the paper, and detailed data can be obtained from the sources.

To examine the forecasting performance, the root mean squared error (*RMSE*) and directional change statistics (D_{stat}) of stock index movement are employed in this study. The directional change statistics (D_{stat}) can be expressed as

$$D_{stat} = \sum_{t=1}^{N} a_t \Big/ N \tag{14}$$

where a_t =1 if $(x_{t+1} - x_t)(\hat{x}_{t+1} - x_t) \geq 0$, and a_t =0 otherwise.

3.2 Experiment Results

When the data are prepared, we begin to train BPNN model. In these experiments, we prepare 5 years' daily data. We use the first 4 years' daily data to train and validate the network, and use the last one years' data to test the prediction performance. For comparison, the standard three-layer BP neural network is used as benchmark model. This study varies the number of nodes in the hidden layer and stopping criteria for training. In this study, 5, 10, 20 hidden nodes for each stopping criteria because the BP network does not have a general rule for determining the optimal number of hidden nodes. The study uses 500, 1000, 2000 and 4000 learning epochs for the stopping criteria of BPNN. For standard BPNN model, the learning rate is set to 0.25. The hidden nodes use the sigmoid transfer function and the output node uses the linear transfer function. The study allows 5 input nodes in terms of the results of auto-regression testing. The comparison of experiment results are reported in Table 1.

Table 1. The prediction performance comparison of various BPNN models

Stock indices	Training epochs	Number of hidden nodes	RMSE Standard BPNN	RMSE Adaptive BPNN	$D_{stat}(\%)$ Standard BPNN	$D_{stat}(\%)$ Adaptive BPNN
S&P 500	500	5	35.3315	24.5342	50.36	61.02
		10	31.3562	20.3236	52.34	63.38
		20	30.2489	18.2457	51.58	64.47
	1000	5	27.5533	22.3589	52.63	62.38
		10	20.3658	15.3547	55.68	66.71
		20	21.8145	14.9631	56.74	68.36
	2000	5	14.5741	8.7423	54.35	70.25
		10	***7.4984***	***1.5843***	***57.88***	***75.24***
		20	9.3657	3.5138	57.37	72.35
	4000	5	11.2547	9.8566	52.24	68.78
		10	8.7534	5.4237	55.68	69.14
		20	8.1254	5.5243	58.75	70.25
Nikkei 225	500	5	100.3541	70.1124	49.63	59.44
		10	89.6472	54.3589	51.58	63.38
		20	70.5428	41.2547	52.63	62.38
	1000	5	81.5477	50.3584	50.36	58.76
		10	51.8545	39.6874	52.05	61.02
		20	37.5426	21.2387	52.63	64.47
	2000	5	40.3376	14.3541	53.54	66.71
		10	22.5474	8.4579	***55.68***	70.25
		20	***16.3785***	***5.4763***	55.41	***72.39***
	4000	5	35.4754	20.2378	52.24	65.65
		10	36.3687	13.3782	54.35	72.35
		20	21.3523	7.8524	52.63	68.36

As can be seen from Table 1, for different stock indices, the neural network architecture is different. For S&P 500, the best prediction performance for the testing data is produced when the number of hidden neurons is 10 and the training epochs are 2000 for both standard BPNN and adaptive BPNN. For the best standard BPNN, the RMSE of the testing data is 7.4984 and D_{stat} of the testing data is 57.88, while for the best adaptive BPNN, the RMSE is 1.5843 and the D_{stat} is 75.24. For Nikkei 225, the best prediction performance for the testing data is generally produced when the number of hidden nodes is 20 and the training epochs are 2000. Interestingly, we find that the number of hidden nodes for the case of Nikkei 225 is larger than that of S&P 500. The main reason may be that Nikkei 225 has high volatility relative to the S&P 500.

Focusing on the RMSE indicator, we can find (1) the performance of the proposed adaptive BPNN model is much better than that of the standard BPNN model in both S&P 500 and Nikkei 225. (2) Generally speaking, the prediction performance improves with the increase of training epochs and hidden nodes. (3) Usually, the few training epochs and hidden nodes can not lead to a good forecasting result.

Focusing on D_{stat} of Table 1, we find the proposed adaptive BPNN model performs much better than the standard BPNN models in all testing cases. These results indi-

cate the feasibility of the adaptive BPNN model in stock index forecasting. Furthermore, from the business practitioners' point of view, D_{stat} is more important than RMSE because the former is an important decision criterion. With reference to Table 1, the differences between the different models are very significant. For example, for the S&P 500 test case, the D_{stat} for the best standard BPNN model is only 57.88%; while for the proposed adaptive BPNN forecasting model, D_{stat} reaches 75.24%, which is much higher than the standard BPNN model, implying that the adaptive BPNN is an efficient algorithm for stock market prediction.

4 Conclusions

In this study, an adaptive BP learning algorithms with optimal learning rate is first proposed. And then this exploratory research examines the potential of using an adaptive BPNN model to predict two main international stock indices, S&P 500 and Nikkei 225. Our empirical results suggest that the adaptive BPNN model may provide better forecasts than the standard BPNN model. The comparative evaluation is based on a variety of statistics such as *RMSE* and D_{stat}. For two stock indices included in our empirical investigation, the adaptive BPNN model outperforms the standard BPNN model in terms of *RMSE* and D_{stat}. Furthermore, our experimental analyses reveal that the *RMSE* and D_{stat} for two stock indices using the proposed adaptive BPNN model are significantly better than those obtained using the standard BPNN model. This implies that the proposed adaptive BPNN model can be used as a feasible solution for stock market prediction.

References

1. Widrow, B., Lehr, M.A.: 30 Years of Adaptive Neural Networks: Perception, Madaline, and Backprpagation. Proceedings of the IEEE Neural Networks I: Theory & Modeling (Special issue) 78 (1990) 1415-1442
2. Rumelhart, D.E., McClelland, J.L.: Parallel Distributed Processing. MIT Press, Cambridge, MA 1986
3. Tollenaere, T.: SuperSAB: Fast Adaptive Back Propagation with Good Scaling Properties. Neural Networks 3 (1990) 561-573
4. Park, D.C., El-Sharkawi, M.A., Marks II, R.J.: An Adaptive Training Neural Network. IEEE Transactions on Neural Networks 2 (1991) 334-345
5. Jacobs, R.A.: Increase Rates of Convergence through Learning Rate Adaptation. Neural Networks 1 (1988) 295-307
6. Brent, R.P.: Fast Training Algorithms for Multilayer Neural Nets. IEEE Transactions on Neural Networks 2 (1991) 346-35
7. Hagan, M.T., Menhaj, M.: Training Feedforward Networks with Marquardt Algorithm. IEEE Transactions on Neural Networks 5 (1994) 989-993

Uniformization of Discrete Data

Lei Yang*

Tsinghua University, Beijing 100084, China
yanglei96@mails.tsinghua.edu.cn

Abstract. Some kind of discrete data sets can be practically transformed into uniform by the related distribution function. By addressing the sparsity of data which measures the discreteness, this paper demonstrates that the sparsity decides the uniformity of the transformed data, and that could be a good reason to explain both the success of the bucket sort in PennySort 2003 and the failure for the same algorithm with the data modified. So the sparsity provides a good criterion to predict whether the algorithm works or not.

1 Introduction

Uniform data is favored in general. In theory any data pattern may be recreated out of uniform, and algorithms such as bucket sort and Hash Tables[1] have been affectively used with uniform data. As the standard method, any continuous data may be transformed into uniform by the distribution function (named as *d-transformation* in this paper). In PennySort 2003 [2, ?] a revised bucket sort algorithm achieved high performance[4]. The original PennySort data were found to be not uniform and the direct application of bucket sort was prohibited. But by a roughly-estimated distribution function the PennySort data were successfully transformed into uniform, enabling the bucket sort algorithm to work. The powerful ability of the *d*-transformation was well shown by the success of the bucket sort algorithm in PennySort.

But the algorithm failed in our further experiments where the data were modified into discrete by cutting some of the key bits away. It seems that the *d*-transformation handles only continuous data, but strictly the unmodified PennySort data were also discrete. Then what made the difference? In fact the continuous cases exist only in theory and all practical data are discrete, how can they be treated as continuous and then transformed into uniform?

The hint is that the failure lied not on the discreteness but on the *degree* of the discreteness. By examining the distribution $D(x) = \Pr(\boldsymbol{X} \leq x)$, [1] this paper presents the concept of *sparsity* to measure the discreteness of a data set $\boldsymbol{X}$. The

* Supported by the Chinese National Key Foundation Research & Development Plan (2004CB318108), Natural Science Foundation (60223004, 60321002, 60303005) and the Key Project of Chinese Ministry of Education (No.104236).

[1] We take $\Pr(\boldsymbol{X} \leq x)$ as the proportion of data no greater than x. Treat each record as a random variable X then it equals to the probability $\Pr(X \leq x)$. See appendix.

X. Deng and D. Du (Eds.): ISAAC 2005, LNCS 3827, pp. 453–462, 2005.

d-transformation transforms $\boldsymbol{X}$ into $d\boldsymbol{X} = D(\boldsymbol{X})$ and the derived distribution $\Pr(d\boldsymbol{X} \leq x)$ is calculated where $d\boldsymbol{X} = D(\boldsymbol{X})$. [2] The mathematics states that the sparsity ϵ decides the uniformity of the transformed data set, and experiments show that the algorithm in [4] failed only with small sparsity. The failure could not be avoided by more precisely estimated distribution because the sparsity is an internal property. Though fortunately the sparsity of the unmodified PennySort data is so small that such risk could be ignored, we must still be aware that the d-transformation works only with small sparsity.

The paper is organized as below. Section 2 clarifies the concept of *continuous* and *uniform*, and then presents *sparsity* to measure discreteness. In Sect.3 the distribution of the transformed data set $\Pr(d\boldsymbol{X} \leq z)$ is calculated to show that the uniformity of $d\boldsymbol{X}$ is decided by the sparsity of $\boldsymbol{X}$. Section 4 presents the experiments to demonstrate how bucket sort algorithm fails with large sparsity. In Sect.5 a brief summarization is presented together with some discussions.

2 Preparations

As in PennySort we are just interested in real data set $\boldsymbol{X} \subset \mathbb{R}$ with records independent to each other but following identical distributions. The distribution $D(x) = \Pr(\boldsymbol{X} \leq x) \in [0,1]$ is defined as the proportion of data no greater than $x \in \boldsymbol{X}$, and the concepts of *continuous* is defined as below[5]:

Definition 1 (Continuous). *A set* $\boldsymbol{X}$ *is continuous if there exists the density function* $p(x) \geq 0$ *defined on* $\boldsymbol{X}$ *that* $D(x) = \int_{-\infty}^{x} p(t)dt$.

A data set is discrete when not continuous. Obviously $d\boldsymbol{X} \subset [0,1]$ where $d\boldsymbol{X} = D(\boldsymbol{X}) = \{D(x)|x \in \boldsymbol{X}\}$ is the transformed set and also the range of $D(x)$. With continuity we get:

Lemma 1. $d\boldsymbol{X} = [0,1]$ *if* $\boldsymbol{X}$ *is continuous.*

The unit interval $[0,1]$ is of special interest in this paper, implying a simplified definition of *uniform*:

Definition 2 (Uniform). *The continuous* $\boldsymbol{Z}$ *with distribution function* $D(z)$ *is called uniform on* $[0,1]$ *when* $D(z) = z, \forall z \in [0,1]$.

By Lemma 1, $\boldsymbol{X}$ is discrete when $d\boldsymbol{X} \neq [0,1]$. Since no irrational number can be the ratio of two finite integers, all finite sets are in theory discrete. But in PennySort the discreteness is weak enough to be ignored. To measure how discrete the data set is we introduce the concept of ϵ-dense:

Definition 3 (ϵ-Dense). *Given* $\epsilon \geq 0$. *A set* $\boldsymbol{Z} \subset [0,1]$ *is called* ϵ-dense *in* $[0,1]$ *if* $\boldsymbol{Z} \cap (w - \epsilon - \delta, w + \delta) \neq \emptyset$ *for all* $w \in (\epsilon, 1)$ *and any* $\delta > 0$.

[2] In this paper we abbreviate the distribution $D(\cdot)$ into the d-operator, for example, $dx = D(x)$. Here $d\boldsymbol{X} = D(\boldsymbol{X})$ is the transformed set. In appendix we also have $dX = D(X)$ where X is a random variable and dX is the transformed variable by the distribution function $D(\cdot)$.

Note that if $\boldsymbol{Z}$ is ϵ_0-dense in $[0,1]$, it is ϵ-dense for all $\epsilon > \epsilon_0$. So the definition below is valid:

Definition 4 (Sparsity). *The smallest* $\epsilon \geq 0$ *for* $d\boldsymbol{X}$ *to be* ϵ*-dense in* $[0,1]$ *is called the sparsity of* $\boldsymbol{X}$.

Sparsity instead of density is named since it decreases as the set gets denser. This concept measures how dense $d\boldsymbol{X}$ is or how discrete $\boldsymbol{X}$ is. The sparsity is 0 for all continuous $\boldsymbol{X}$, while with a positive sparsity $\boldsymbol{X}$ must be discrete.

For example when $\boldsymbol{X} = \{1, 2, \cdots, N\}$ we have $D(x) = \frac{\lfloor x \rfloor}{N}, x \in [0, N]$. As the range of $D(x)$ the transformed set is $d\boldsymbol{X} = \{D(x)|x \in \boldsymbol{X}\} = \{0, \frac{1}{N}, \frac{2}{N}, \cdots, \frac{N-1}{N}, 1\}$. The sparsity is calculated to be $\frac{1}{N}$ which diminishes to zero as N goes to infinity.

3 Transformation by Distribution

Now we are to transform data set $\boldsymbol{X}$ using the distribution function $D(x)$ as we did in PennySort to see whether the transformed set is uniform or how uniform it is. For simplicity we abbreviate the transformation into a new term:

Definition 5 (d-Transformation). *For data set* $\boldsymbol{X}$ *with distribution* $D(x)$*, the transformation from each record* x *to* $dx = D(x)$ *is called the d-transformation on* $\boldsymbol{X}$ *which transforms* $\boldsymbol{X}$ *into* $d\boldsymbol{X} = D(\boldsymbol{X})$.

To apply the bucket sort algorithm in PennySort, a uniform $d\boldsymbol{X}$ is expected by the d-transformation. Specially when $\boldsymbol{X}$ is already uniform on $[0,1]$, Definition 2 tells us that the distribution is identical on $[0,1]$ and $d\boldsymbol{X} = \boldsymbol{X}$, both are uniform. The Lemma 2 below shows that $d\boldsymbol{X}$ is still uniform when $d\boldsymbol{X} = [0,1]$:

Lemma 2. *Given data set* $\boldsymbol{X} \subset \mathbb{R}$. *For all* $x \in \boldsymbol{X}$ *we have*

$$D_{d\boldsymbol{X}}(dx) = dx \tag{1}$$

where $dx = D(x)$ *and the function* $D_{d\boldsymbol{X}}(z)$ *is the distribution of the d-transformed set* $d\boldsymbol{X} \subset [0,1]$*, namely, the proportion of data in* $d\boldsymbol{X}$ *no greater than* z.

Lemma 2 is just another form of (2.3C) on pp.8 in [7]. It shows that $d\boldsymbol{X}$ is uniform once for any $z \in [0,1]$ exists $x \in \boldsymbol{X}$ that $dx = z$, that is, $d\boldsymbol{X} = [0,1]$. By Lemma 1 we get:

Theorem 1. *The d-transformation transforms* $\boldsymbol{X}$ *into uniform if* $d\boldsymbol{X} = [0,1]$. *Especially the d-transformation transforms continuous data into uniform.*

Practically the original PennySort data were regarded as satisfying $d\boldsymbol{X} = [0,1]$ that they were transformed into uniform and the bucket sort algorithm worked well. But as a finite subset of $[0,1]$ the data are in fact discrete. In our further experiments such approximation resulted in failure where less key bits were kept. How such approximation affects the transformation? Theorem 2 calculates the exact distribution of $d\boldsymbol{X}$ without assuming it to cover $[0,1]$ [3]:

[3] Readers unfamiliar with the notion *sup* may simply take it as *max*. See the appendix.

Theorem 2. *Given data set* $\boldsymbol{X} \subset \mathbb{R}$*. For all* $x \in \boldsymbol{X}$ *we have*

$$D_{d\boldsymbol{X}}(dx) = \sup\{dx|x \in \boldsymbol{X} : dx \leq z\} \ . \tag{2}$$

Theorem 2 states that the distribution of $d\boldsymbol{X}$ on $z \in [0,1]$ is the *greatest* number no greater than z in $d\boldsymbol{X}$. It may be strictly smaller than z when $z \notin d\boldsymbol{X}$ and hense declares a nonuniform $d\boldsymbol{X}$. For the original PennySort data the probability for any single key to occur is about $\epsilon_{(10)} = (1/95)^{10} \approx 10^{-20}$, and such $x \in \boldsymbol{X}$ that $dx \leq z$ can always be found above $z - \epsilon_{(10)}$. So $\Pr(d\boldsymbol{X} \leq z) = \sup\{dx|x \in \boldsymbol{X} : dx \leq z\}$ is rather close to z which implies almost uniform $d\boldsymbol{X}$ by Def.2. While with less key-bits the key probability gets much greater, for example, $\epsilon_{(1)} = 1/95 \approx 10^{-2}$ when with only one byte and $\Pr(d\boldsymbol{X} \leq z)$ may be about 10^{-2} smaller than z. Theorem 3 below shows that about 1% records might crowd in one bucket which should contain only $\frac{1}{b}$ records in average where $b = 300$ in [4]. Once a bucket exceeded the memory capacity, the algorithm failed. Intuitively more key bits generate data of smaller sparsity, and formally we have:

Theorem 3. *Given data set* $\boldsymbol{X} \subset \mathbb{R}$ *with sparsity* ϵ*,* $\forall z \in [0,1]$ *we have*

$$z - \epsilon \leq \Pr(d\boldsymbol{X} \leq z) \leq z \tag{3}$$

Then the uniformity of the transformed data is decided by the *sparsity* which measures how discrete $\boldsymbol{X}$ is. A parallel result is $\Pr(d\boldsymbol{X} \in [s, s+\delta]) \leq \delta + \epsilon$. In the algorithm in [4] each bucket was defined by two boundaries x_L and x_U. The bucket probabilities would be $\Pr(x_L < X \leq x_U) = \frac{1}{b}$ where $b = 300$ in [4] had the transformed data been uniform. While considering the discreteness these probabilities became $\frac{1}{b} + \epsilon$. For the original PennySort data, the sparsity was as low as 10^{-20} and the additional ϵ might be completely ignored. But when leaving only one byte as the key, the sparsity increased to about $\frac{1}{95}$ which influenced the probability greatly. Chances were good for much more records than average to crowd in just several buckets, and the algorithm failed once any bucket could no longer be held in memory.

4 Transforming PennySort Data

PennySort is one of the sort benchmarks proposed by Jim Gray, the 1998 Turing Award winner, to follow the world computer developement [2, ?]. In 2003 a revised external bucket sort algorithm won in the PennySort Indy group[4]. Traditionally the bucket sort distributes data into some predefined *buckets* that the records in one bucket are smaller than those in the next. Then sort the records in each bucket and join them in order, we get the sorted file. Bucket sort is rather efficient except for that the memory must be large enough to hold the greatest bucket and to sort all its records internally. The exact number of records in a bucket won't be known until all records have been distributed. Once any bucket exceeds the memory capacity, the algorithm fails.

So bucket sort works best with uniform data. PennySort records keys are made up of 10 random bytes ranging from 32 to 126. It is found that for 7

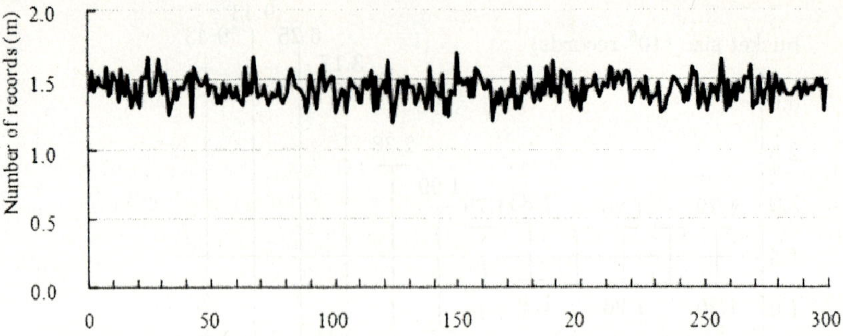

Fig. 1. Bucket sizes in [4] (*433M records, 70K samples, 300 buckets*)

out the 10 bytes (byte $2, 3, 4, 6, 7, 8, 10$) the values appear evenly with constant frequency $1/95 = 1.05\%$. For the other three bytes (byte $1, 5, 9$) peaks are found: 2% records take their values on 32, 1.3% on 33, and only 1.03% elsewhere. The peaks prohibit the direct application of bucket sort.

To overcome this, [4] tried to transform the PennySort data into uniform by the distribution function (the d-transformation) and to generate buckets of approximate sizes. By sampling a rather small fragment of the data to estimate the distribution, that was successfully done. Figure 1 shows that the maximal bucket was only 18% greater than average.

But the results changed with some modifications on data. In our further experiments we devised more data by cutting some PennySort key bits away (unused bits were manually set into zero). Like in [4] we still took 70K records as the sample set to calculate the boundaries of the 300 buckets (estimating the distribution function) and then distributed the whole data set (433M records where 1K = 1024 and 1M = 1024K) into each bucket. The average bucket size kept constant: 433M/300 =1.44M. Figure 2 shows that the maximum/minimum bucket sizes remained almost unchanged with keys of more than 14 bits but grew/dropped rapidly of less bits. For example, when with the 12-bit keys more than 2M records crowded in the largest bucket, breaking down the sorting system designed in [4] because it is too large.

The *sparsity* explains the failure. For finite set like in PennySort, the sparsity is calculated to be the highest probability of a single key to occur. With the full 10-byte keys, the sparsity of the original PennySort data is $\epsilon = (2.00\%)^3 \times (1.03\%)^7 \approx 10^{-19}$ (slightly greater than $(\frac{1}{95})^{10} \approx 10^{-20}$ when taking the bytes as uniform in the last section). Far less than the bucket probability $\frac{1}{300}$, this sparsity was overwhelmed and the transformed data might be securely treated as uniform. But with the 12-bit keys, the sparsity exceeded the bucket probability: $2.00\% \times 2^4 \times 1.05\% = 3.37 \times 10^{-3}$. The term $\frac{1}{b} + \epsilon$ doubled the bucket probability for such a large ϵ and chances for data to fall into some buckets became much larger. Figure 3 calculates the sparsities, showing that the maximum bucket sizes relate closely to the sparsity compared with the bucket probability $\frac{1}{b}$.

Figure 3 tells that bucket sort works only with PennySort keys of 14 bits or more. To exclude the influence of the the roughly estimated distribution, exper-

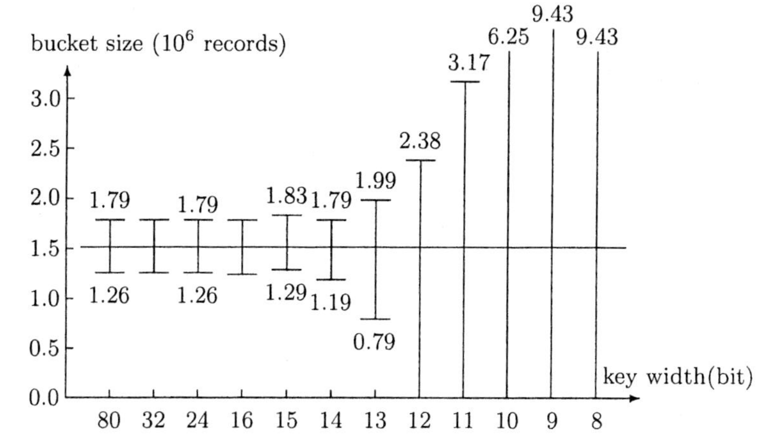

Fig. 2. Min/max bucket size v.s. key width (*433M records, 70K samples, 300 buckets)*

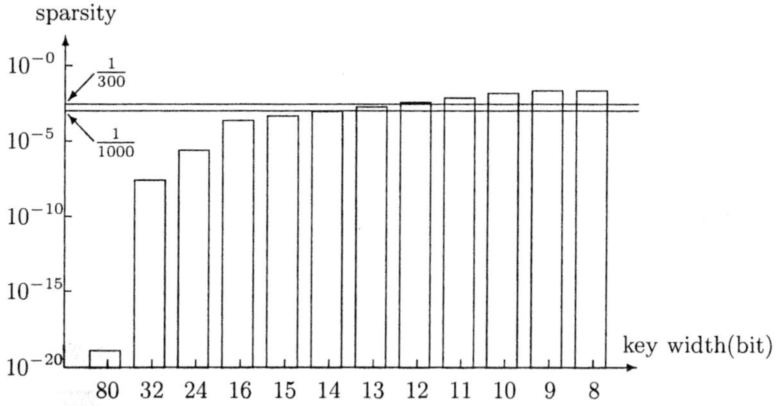

Fig. 3. Sparsities of the PennySort data v.s. key width

iments were repeated with a larger sample of 1M records, see Fig.4. With more precisely calculated distribution the maximal bucket size got smaller for keys of more than 12 bits. But for the 12-bit keys, the largest bucket still contained 2.38×10^6 records, no less than in Fig.2. Actually the distribution helps only the bucket probability $\frac{1}{b}$. Given large ϵ, the upper bound $\frac{1}{b} + \epsilon$ kept large even with precisely calculated distribution. The algorithm could not be saved.

In our last experiment we take 1000 buckets (still 1M samples to calcuate distribution), see Fig.5. With 15 or more key bits where sparsities were less than $\frac{1}{1000}$ the buckets generated were of approximate sizes around average: 433M/1000 $= 4.54 \times 10^5$. While with less key bits that the sparsities were large relative to $\frac{1}{1000}$, the largest buckets were much more crowded than average(for example, 75% more records than average were contained in the largest bucket for 14-bit keys). Actually the bucket probability $\frac{1}{b}$ acted as a threshold because bucket that doubled the average size could seldom be hold in memory in practice.

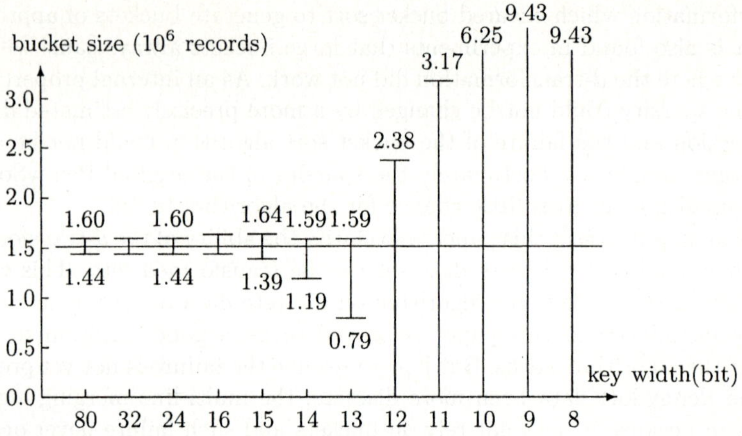

Fig. 4. Min/max bucket size v.s. key width (*433M records, 1M samples, 300 buckets*)

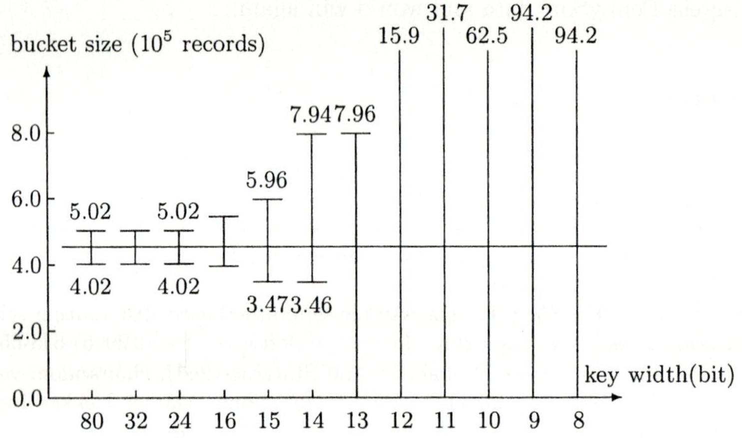

Fig. 5. Min/max bucket size v.s. key width (*433M records, 1M samples, 1000 buckets*)

5 Conclusions and Discussions

By the distribution function, some discrete data may be transformed into *practically* uniform so that algorithms like bucket sort can be applied. While for some other data the *d*-transformation does not work and bucket sort fails. By analyzing the behavior of the transformed data in theory, this paper shows that the uniformity is decided by the *sparsity* which measures how discrete the data are before transformation. Discrete data can be transformed into uniform if the sparsity is small enough. Otherwise, the *d*-transformation does not work.

In spite of the success in 2003 PennySort Indy, the algorithm in [4] failed when some key bits were cut away. The failure was imputed to the discreteness because less key bits produced more discrete data. The discreteness disabled the

d-transformation which ensured bucket sort to generate buckets of approximate sizes. It is also found in experiments that large buckets always came with large sparsity where the d-transformation did not work. As an internal property of the data, the sparsity could not be changed by a more precisely estimated distribution function and the failure of the bucket sort algorithm could not be avoided by a larger sample set. Fortunately the sparsity of the original PennySort data was so small that it gives little chance for the algorithm to fail.

The main purpose of this paper is to clarify the ability of the d-transformation to see how uniform a discrete data set can be transformed into. This explains the failure of the bucket sort algorithm on discrete data as an application. The sparsity introduced in this paper is proved to be a good criterion to predict whether the algorithm works. But how to resolve the failure is not yet presented. Had the PennySort data been more discrete, the multi-line merging algorithm might win because it does not rely on buckets and such failure never occurs[8]. For data of large sparsity, can they be transformed into more uniform than d-transformation? Further work is needed so that bucket sort might be applied to more discrete PennySort data and would win again.

References

1. Knuth, D. E.: The Art of Computer Programming, vol3: Sorting and Searching. Addison-Wesley Inc., Reading, Mass.(1973) 506-549
2. Sort Benchmark homepage. http://research.microsoft.com/barc/SortBenchmark/.
3. Gray, J., Coates, J., Nyberg C.: Performance/price sort and PennySort. Technical Report, MS-TR-98-45, Microsoft Research (1998)
4. Yang, L., Huang H., Song T.: The sample-seperator based distributing scheme of the external bucket sort algorithm. Journal of Software, **16**(5)(2005) 643-651
5. Giri, N. C.: Introduction to Probability and Statistics(2ed), ch3:random variables, probability distributions and characteristic functions. Marcel Dekker Inc., New York(1993) 55-135
6. Wade, W. R.: An Introduction to Analysis(2ed), ch1:the real number system. Prentice-Hall International Inc., Upper Saddle River, NJ(2000) 1-33
7. Evans, M., Hastings, N., Peacock, B.: Statistical Distributions(3ed), ch2:terms and symbols. John Wiley & Sons, Inc., New York(2000) 3-17
8. Shi, Y., Zhang, L., Liu, P.: THSORT: A single-processor parallel sorting algorithm. Journal of software, **14**(2)(2003) 159-165

A The Mathematic Deductions

Given data set $\boldsymbol{X} \in \mathbb{R}$, we may take each record as a random variable before observation. Under the assumption of independence and identical distributions, the probability distribution of each data record $D_X(x) = \Pr(X \leq x) \in [0, 1]$ is approached by the statistical distribution $D(x) \in [0, 1]$. So all results on random variables below may be applied directly to data sets. Like the previous notions $d\boldsymbol{X}$ and $D_{d\boldsymbol{X}}(z)$ ($\boldsymbol{X}$ in bold face) we also have $dX = D(X)$, the transformed

variable by the distribution function and $D_{dX}(z) = \Pr(dX \le z)$, the probability distribution function of dX.

In Sect.2 we mention the *supreme* of set $\boldsymbol{S}$ defined as its lowest upper bound denoted as $\sup \boldsymbol{S}$. As a fundamental property of the real numbers, any bounded set has its unique supreme in $\mathbb{R}$ unless it is empty[6].

Lemma 3. *For any subset $\boldsymbol{S} \subset \mathbb{R}$, we can always find an ascendent sequence $x_1 \le x_2 \le \cdots (x_n \in \boldsymbol{S}, n = 1, 2, \cdots)$ such that $\boldsymbol{S} = \bigcup_{n=1,2,\cdots} \{x \in \boldsymbol{S} | x \le x_n\}$.*

Proof. If $\boldsymbol{S}$ is not bounded in $\mathbb{R}$ then $\forall n \in \mathbb{N}$ we may find some $x_n \in \boldsymbol{S}$ that $x_n \ge n$. Given any $x \in \boldsymbol{S}$ exists $n \in \mathbb{N}$ that $x \le n \le x_n$.

When $\boldsymbol{S}$ is bounded we can always find a finite $\sup \boldsymbol{S} = \sup \{x | x \in \boldsymbol{S}\} \in \mathbb{R}$. If $\sup \boldsymbol{S} \in \boldsymbol{S}$ we simply take $x_n = \sup \boldsymbol{S}, n = 1, 2, \cdots$ and $\{x \in \boldsymbol{S} | x \le x_n\} = \{x \in \boldsymbol{S} | x \le \sup \boldsymbol{S}\} = \boldsymbol{S}$.

Otherwise when $\sup \boldsymbol{S} \notin \boldsymbol{S}$ we have $\forall n > 0, \exists x_n \in \boldsymbol{S}$ that $x_n \ge \sup \boldsymbol{S} - \frac{1}{n}$. For any $y \in \boldsymbol{S}$ we may find some $n \in \mathbb{N}$ that $\frac{1}{n} \le \sup \boldsymbol{S} - y$ because $\sup \boldsymbol{S} \notin \boldsymbol{S}$, implying $y \le \sup \boldsymbol{S} - \frac{1}{n} \le x_n$.

An ascendent sequence may be created by taking the maximums. □

Lemma 4 (2'). *Given variable $X \in \mathbb{R}$, for all $x \in \boldsymbol{X}$ we have*

$$\Pr(dX \le dx) = \Pr(X \le x) = dx \text{ where } dx = D_X(x) \ .$$

Proof. Note that $y \le x \Rightarrow dy \le dx$ we get $\{y \in \boldsymbol{X} | dy \le dx\} = \{y \in \boldsymbol{X} | y \le x\} \cup \{y \in \boldsymbol{X} | (y > x) \wedge (dy = dx)\}$. Let $\boldsymbol{S} = \{y \in \boldsymbol{X} | (y > x) \wedge (dy = dx)\}$ and we are to prove $\Pr(X \in \boldsymbol{S}) = 0$.

By Lemma 3 we may find an ascendent sequence $x_1 \le x_2 \le \cdots (x_n \in \boldsymbol{S}, n = 1, 2, \cdots)$ such that $\boldsymbol{S} = \bigcup_{n=1,2,\cdots} \{y \in \boldsymbol{S} | x < y \le x_n\}$. Then

$$\Pr(X \in \boldsymbol{S}) = \Pr\left(\bigcup_{n=1,2,\cdots} \{y \in \boldsymbol{S} | x < y \le x_n\}\right) \tag{4}$$

$$\le \Pr\left(\bigcup_{n=1,2,\cdots} \{y \in \mathbb{R} | x < y \le x_n\}\right) \tag{5}$$

$$= \lim_{n \to \infty} \Pr(x < X \le x_n) \tag{6}$$

$$= \lim_{n \to \infty} (dx_n - dx) \tag{7}$$

$$= 0 \text{ since } x_n \in \boldsymbol{S} \ . \tag{8}$$

□

Theorem 4 (2'). *Given variable $X \in \mathbb{R}$, for all $z \in [0, 1]$ we have:*

$$D_{dX}(z) = \Pr(dX \le z) = \sup \{dx | x \in \boldsymbol{X} : dx \le z\} \ .$$

Proof. By Lemma 2 we have $\forall x \in \boldsymbol{X}$:

$$dx \le z \Rightarrow dx = \Pr(dX \le dx) \le \Pr(dX \le z) \tag{9}$$
$$\Rightarrow \sup\{dx | x \in \boldsymbol{X} : dx \le z\} \le \Pr(dX \le z) \ . \tag{10}$$

On the other hand let $\boldsymbol{S} = \{x \in \boldsymbol{X} | dx \le z\}$ we may find an ascendent sequence $x_1 \le x_2 \le \cdots$ where $x_n \in \boldsymbol{S}$ and $\boldsymbol{S} = \bigcup_{n=1,2,\cdots} \{x \in \boldsymbol{S} | x \le x_n\}$. The *countable additivity of probability* ensures that

$$\Pr(dX \le z) = \Pr(X \in \boldsymbol{S}) \tag{11}$$
$$= \Pr\left(\bigcup_{n=1,2,\cdots} \{x \in \boldsymbol{S} | x \le x_n\}\right) \tag{12}$$
$$\le \Pr\left(\bigcup_{n=1,2,\cdots} \{x \in \mathbb{R} | x \le x_n\}\right) \tag{13}$$
$$= \sup\{dx_n | n = 1, 2, \cdots\} \tag{14}$$
$$\le \sup\{dx | x \in \boldsymbol{X} : dx \le z\} \ . \tag{15}$$

□

Theorem 5 (3'). *Given* $\boldsymbol{X} \in \mathbb{R}$ *with sparsity* ϵ. *For all* $z \in [\epsilon, 1]$ *we have:*

$$z - \epsilon \le \Pr(dX \le z) \le z \ .$$

Proof. The sparsity implies that X is ϵ-dense in $[0, 1]$.

By Theorem 2 we have $D_{dX}(z) = \sup\{dx | x \in \boldsymbol{X} : dx \le z\} \le z$.

On the other hand $\forall \delta > 0, \exists x \in \boldsymbol{X}$ that $dx \ge z - \epsilon - \delta$ because $d\boldsymbol{X}$ is ϵ-dense in $[0, 1]$, ensuring $\sup\{dx | x \in \boldsymbol{X} : dx \le z\} \ge z - \epsilon$. □

A Practical Algorithm for the Computation of Market Equilibrium with Logarithmic Utility Functions
(Extended Abstract)

Li-Sha Huang[1,*]

State Key Laboratory of Intelligent Technology and Systems,
Dept. of Computer Science and Technology,
Tsinghua Univ., Beijing, 10084, China

Abstract. We develop an algorithm for computing the equilibrium price in the Fisher's exchange market model with logarithmic utility functions. The algorithm is proved to converge to the equilibrium price in finite time and performs better than existing polynomial-time algorithms in experimental tests.

1 Introduction

The paper studies the computation of equilibria in exchange markets. The problem has been an active research area in the economists' society since Arrow and Debreu published their classical paper [1] on the existence of market equilibria in 1954. Recently, the problem is attracting the computer scientists' attention after Deng, Papadimitriou and Safra's study [2] from the algorithmic complexity point of view.

Within three years, various approaches are developed to solve the problem for different settings. The most successful approaches include [3,4,5,6,7]. Codenotti, Pemmaraju and Varadarajan have a comprehensive survey [8] for this topic.

The paper studies the algorithm of computing the equilibrium in the Fisher's model with logarithmic utility functions. The logarithmic utility function, which was first studied by Chen *et al* [9], has the form

$$u_i(x_i) = \sum_j \alpha_{ij} \log(\beta_{ij} + x_{ij}).$$

The logarithmic utility function has an interesting property that the equilibrium point of the market is rational, whenever the input data are rational. Up till now, only linear function and logarithmic function are known to possess the property. Assume the number of agents is n and the number of commodities is m. Chen *et al* [9] propose a polynomial-time algorithm for the Fisher's model, when either

* Supported by Natural Science Foundation of China (No.60135010,60321002) and the Chinese National Key Foundation Research and Development Plan (2004CB318108).

X. Deng and D. Du (Eds.): ISAAC 2005, LNCS 3827, pp. 463–472, 2005.

m or n is bounded. In fact, the general model can be solved in polynomial time by convex programming [3,4,10] or ellipsoid methods [7]. Deng *et al* [10] analyze the time complexity and conclude that the interior-point algorithm can reach the exact equilibrium point in $O((n+m)n^{1.5}m^{3.5}L)$ time.

In the paper, we propose a primal-dual type algorithm for computing the equilibrium price in the Fisher's model with logarithmic utility functions. We prove that the algorithm will converge in finite steps. Although its time complexity is unknown, the algorithm is easy to implement and performs well in experiments. The algorithm can also be applied as a black box to solve the general model.

2 Preliminaries

2.1 Model and Definitions

Let $|m|$ denote the set $\{1, 2, ..., m\}$. Let $\mathbb{R}^n_+$ denote the set of nonnegative vectors in n-dimensional Euclidean spaces and $\mathbb{R}^n_{++}$ denote the set of positive vectors in $\mathbb{R}^n_+$.

Assume there are n agents and m kinds of commodities in the market and each agent is both a seller and a buyer. The agents are endowed with bundles of commodities initially, denoted by n nonnegative vectors $\{e_i \in \mathbb{R}^m_+ \mid i \in |n|\}$. Without loss of generality, assume that the total amount of each commodity is normalized to 1, i.e., $\sum_{i=1}^{n} e_{ij} = 1, (\forall j \in |m|)$.

Each agent has a utility function to express his utility for bundles of commodities. In the paper, we assume that the utility functions are logarithmic functions. The utility function of agent i is

$$u_i(x_i) = u_i(x_{i1}, x_{i2}, ..., x_{im}) = \sum_{j=1}^{m} \alpha_{ij} \ln(\beta_{ij} + x_{ij})$$

where $x_i \in \mathbb{R}^m_+$ is the bundle of commodities distributed to agent i and α_{ij} and β_{ij} are non-negative constants. W.l.o.g, we assume that for any agent i, there exists a commodity j such that $\alpha_{ij} > 0$ and for any commodity j, there exists an agent i such that $\alpha_{ij} > 0$. This assumption guarantees that the price p is strictly positive, i.e., $p \subset \mathbb{R}^m_{++}$.

Given a certain price p, the agents exchange their commodities. Each agent tries to maximize his utility under the budget constraint. An equilibrium of the market is defined as follows.

Definition 1. *An equilibrium in an exchange economy is a price $\bar{p} \in \mathbb{R}^m_+$ and distributions of commodities $\{\bar{x}_i \in \mathbb{R}^m_+, i \in |n|\}$, such that*

$$\begin{aligned} \bar{x}_i &\in argmax\{u_i(x_i)|x_i \geq 0, \langle x_i, \bar{p}\rangle \leq \langle e_i, \bar{p}\rangle\}, \forall i \in |n| \\ \sum_{i=1}^{n} \bar{x}_{ij} &\leq 1, \forall j \in |m| \end{aligned}$$

A special case of the above exchange market is the Fisher's model. In the model, the initial endowments of agents are money and the commodities are held by the market. Each agent buys goods from the market to maximize his utility under the budget constraint. Assume agent i's money is $w_i \in \mathbb{R}_+$, then the equilibrium in the Fisher's model is defined as follows.

Definition 2. *An equilibrium in the Fisher's model is a price vector* $\bar{p} \in \mathbb{R}^m_+$ *and bundles of commodities* $\{\bar{x}_i \in \mathbb{R}^m_+, i \in |n|\}$, *such that*

$$\begin{aligned} \bar{x}_i &\in \mathit{argmax}\left\{u_i(x_i) | x_i \geq 0, \langle x_i, \bar{p}\rangle \leq w_i\right\}, \forall i \in |n| \\ \sum_{i=1}^{n} \bar{x}_{ij} &\leq 1, \forall j \in |m| \\ \sum_{j=1}^{m} p_j &= \langle p, \textstyle\sum_{i=1}^{n} \bar{x}_i\rangle \end{aligned}$$

Usually, computing the equilibrium in the Fisher's model is much easier than the general model.

2.2 Equilibrium Conditions

In the subsection, we propose sufficient and necessary equilibrium conditions for the Fisher's model with logarithmic utility functions. First, we write the utility function $u_i(x_i)$ in a more convenient form by substituting x_{ij} with y_{ij}/p_j, where y_{ij} is the amount of money agent i spends on commodity j and p_j is the price of the commodity j.

Given a price $p = (p_1, ..., p_m)^T \in \mathbb{R}^m_+$, consider the following optimization problem for agent i:

$$\begin{aligned} \max &\sum_{j=1}^{m} \alpha_{ij} \ln(\beta_{ij} + y_{ij}/p_j) \\ s.t. &\sum_{j=1}^{m} y_{ij} \leq w_i \\ & y_{ij} \geq 0, \forall j \end{aligned} \tag{2.1}$$

Since the utility function is strictly concave, the solution of Problem (2.1) is unique. By the KKT theorem, we have the following lemma:

Lemma 1. $y_i \in \mathbb{R}^m_+$ *is the optimum of Problem (2.1) if and only if there exists an index set* $\Gamma_i \subseteq |m|$ *and* $\lambda_i \in \mathbb{R}_{++}$ *such that*

$$\lambda_i = (w_i + \sum_{l \in \Gamma_i} \beta_{il} p_l) / (\sum_{l \in \Gamma_i} \alpha_{il})$$

$$y_{ij} = \begin{cases} \alpha_{ij}\lambda_i - \beta_{ij} p_j & \mathit{for}\ j \in \Gamma_i \\ 0 & \mathit{for}\ j \notin \Gamma_i \end{cases}$$

For an instance of our problem, a *configuration* is a couple (Γ, p), where $\Gamma = \{\Gamma_1, \Gamma_2, ..., \Gamma_n\}$ are collection of n subsets of $|m|$ and $p \in \mathbb{R}^m_+$ is the price.

For a configuration ($\Gamma = \{\Gamma_1, ..., \Gamma_n\}, p$), we introduce the following functions:

$$\lambda_i(\Gamma, p) = (w_i + \sum_{l \in \Gamma_i} \beta_{il} p_l) / (\sum_{l \in \Gamma_i} \alpha_{il}) \tag{2.2}$$

$$y_{ij}(\Gamma, p) = \begin{cases} \alpha_{ij}\lambda_i(\Gamma, p) - \beta_{ij} p_j & \text{for } j \in \Gamma_i \\ 0 & \text{for } j \notin \Gamma_i \end{cases} \tag{2.3}$$

$$F_{ij}(\Gamma, p) = \frac{\beta_{ij}}{\alpha_{ij}} p_j - \lambda_i(\Gamma, p) \tag{2.4}$$

$$D_j(\Gamma, p) = p_j - \sum_{i=1}^{n} y_{ij}(\Gamma, p) \tag{2.5}$$

Intuitively, λ_i is the incremental value that the trader i would gain from another dollar of wealth at current spend. The function F_{ij} can be viewed as the difference between λ_i and the incremental value of good j. D_j represents the excess demands of good j. With these notations, Lemma 1 can be restated as the follow lemma:

Lemma 2. *Given a configuration* (Γ, p), $\{y_{ij}(\Gamma, p)\}$ *maximize every agent's utility if and only if:*

$$\begin{cases} F_{ij}(\Gamma, p) \le 0\,, & \forall j \in \Gamma_i, \quad \forall i \in |n| \\ F_{ij}(\Gamma, p) \ge 0\,, & \forall j \notin \Gamma_i, \quad \forall i \in |n| \end{cases}$$

In order to make the market clear, we further require that

$$D_j(\Gamma, p) = p_j - \sum_{i=1}^{n} y_{ij}(\Gamma, p) = 0, \quad \forall j \in |m|$$

Combining the preceding discussions, we have proved the sufficient and necessary conditions for a configuration (Γ, p) to be a market equilibrium:

Theorem 1. *A configuration* (Γ, p) *is an equilibrium in the Fisher's model with logarithmic utility functions if and only if*

$$F_{ij}(\Gamma, p) \le 0\,, \quad \forall j \in \Gamma_i, \quad \forall i \in |n| \tag{2.6}$$

$$F_{ij}(\Gamma, p) \ge 0\,, \quad \forall j \notin \Gamma_i, \quad \forall i \in |n| \tag{2.7}$$

$$D_j(\Gamma, p) = 0\,, \quad \forall 1 \le j \le m \tag{2.8}$$

Remark 1. The advantage of the substitution of y_{ij}/p_j to x_{ij} is that the functions $\lambda_i(\Gamma, p)$, $y_{ij}(\Gamma, p)$, $F_{ij}(\Gamma, p)$ and $D_j(\Gamma, p)$ are all linear to p. The linearity can help us to build a primal-dual type algorithm.

3 The Primal-Dual Algorithm

Given a configuration (Γ, p), we define two subsets of $|m| = \{1, ..., m\}$:

$$TIGHT(\Gamma, p) = \{j \in |m|, \text{s.t. } D_j(\Gamma, p) = 0\}$$
$$ACTIVE(\Gamma, p) = \{j \in |m|, \text{s.t. } D_j(\Gamma, p) < 0\}$$

Due to Theorem 1, finding an equilibrium is equivalent to finding a configuration (Γ, p) such that:

$$\begin{aligned}&TIGHT(\Gamma,p) = |m|;\\&ACTIVE(\Gamma,p) = \emptyset;\\&F_{ij}(\Gamma,p) < 0, \quad \forall j \in \Gamma_i, \quad \forall i \in |n|;\\&F_{ij}(\Gamma,p) \geq 0, \quad \forall j \notin \Gamma_i, \quad \forall i \in |n|.\end{aligned}$$

Lemma 3. *For any instance of the Fisher's model with logarithmic utility functions, there exists a configuration* $(\Gamma = \{\Gamma_1, ..., \Gamma_n\}, p)$ *such that*

$$\begin{aligned}&\Gamma_i = |m|, \quad \forall i \in |n|;\\&TIGHT(\Gamma,p) = \emptyset;\\&ACTIVE(\Gamma,p) = |m|;\\&F_{ij}(\Gamma,p) < 0, \quad \forall i,j\end{aligned}$$

Proof. Assume $\Gamma = \{\Gamma_i = |m|, i \in |n|\}$. For any i, we have

$$\lim_{p\to 0} \lambda_i(\Gamma,p) = w_i / \sum_{l\in\Gamma_i} \alpha_{il} > 0$$

And for any j,

$$\begin{aligned}\lim_{p\to 0} D_j(\Gamma,p) &= -\sum_{i=1}^{n} \lim_{p\to 0} y_{ij}(\Gamma,p)\\&= -\sum_{i=1}^{n} \alpha_{ij} \lim_{p\to 0} \lambda_i(\Gamma,p) < 0\end{aligned}$$

Hence there exists a price p small enough to guarantee that $F_{i,j}(\Gamma,p) < 0$ for all i,j and $D_j(\Gamma,p) < 0$ for all j. □

Lemma 3 guarantees that the algorithm can starts from a very low price p such that all commodities are active and $\Gamma = (|m|, |m|, ..., |m|)$. After that, the algorithm increases the price linearly. The increase is parameterized by $p(t) = p_0 + rt$, where $r \in \mathbb{R}^m$ is the *velocity* of the price. We require that the velocity r should preserve the TIGHT set. The requirement is satisfied by solving a linear system that will be explained later.

Since all the functions $\lambda_i(\Gamma,p)$, $y_{ij}(\Gamma,p)$, $F_{ij}(\Gamma,p)$, and $D_j(\Gamma,p)$ are linear functions of p, their partial derivatives to p are quantities that only depend on Γ:

$$\frac{\partial \lambda_i}{\partial p_k} = \begin{cases} \beta_{ik} / \sum_{l\in\Gamma_i} \alpha_{il} & \text{for } k \in \Gamma_i \\ 0 & \text{for } k \notin \Gamma_i \end{cases} \tag{3.1}$$

$$\frac{\partial y_{ij}}{\partial p_k} = \begin{cases} \alpha_{ij} \frac{\partial \lambda_i}{\partial p_k} - \delta_{jk}\beta_{ij} & \text{for } j \in \Gamma_i \\ 0 & \text{for } j \notin \Gamma_i \end{cases} \tag{3.2}$$

$$\frac{\partial F_{ij}}{\partial p_k} = \frac{\beta_{ij}}{\alpha_{ij}} \delta_{jk} - \frac{\partial \lambda_i}{\partial p_k} \tag{3.3}$$

$$\frac{\partial D_j}{\partial p_k} = \delta_{jk} - \sum_{i=1}^{n} \frac{\partial y_{ij}}{\partial p_k} \tag{3.4}$$

Here $\delta_{jk} = 1$ if $j = k$, otherwise $\delta_{jk} = 0$.

Let $p(t) = p_0 + rt$, where $r \in \mathbb{R}^m$ is the velocity of the price. Let $D_j(t)$ denote $D_j(\Gamma, p(t))$. Note that $\frac{dp_k}{dt} = r_k$, then

$$\frac{dD_j}{dt} = \sum_{k=1}^{m} \frac{\partial D_j}{\partial p_k} r_k \tag{3.5}$$

In the algorithm, we require the velocity r to preserve the set TIGHT. Therefore, the velocity r should satisfy the following linear equations:

$$\frac{dD_j}{dt} = \sum_{k \in TIGHT} \frac{\partial D_j}{\partial p_k} r_k + \sum_{l \in ACTIVE} \frac{\partial D_j}{\partial p_l} r_l = 0, \quad \forall j \in TIGHT$$

Without loss of generality, assume that $TIGHT = \{1, ..., N\}$ and $ACTIVE = \{N+1, ..., m\}$. We set $r_k = 1$ for $k \in ACTIVE$.

Lemma 4. *The linear equation*

$$\frac{dD_j}{dt} = \sum_{k=1}^{N} \frac{\partial D_j}{\partial p_k} r_k + \sum_{l=N+1}^{m} \frac{\partial D_j}{\partial p_l} = 0 \qquad (1 \le j \le N) \tag{3.6}$$

always admits a solution. Moreover, the solution r satisfies that $\sum_{j=1}^{m} r_j > 0$.

Proof. The matrix on the l.h.s of the equation is non-degenerated since it is diagonal dominated. Therefore, the linear system (3.6) always has a unique solution.

Note that $\frac{dD_j(\Gamma, p(t))}{dt} > 0$ for any $j \in ACTIVE$ and $\sum_{j=1}^{m} y_{ij}(\Gamma, p) = w_i$. Then

$$\begin{aligned}
\sum_{j=1}^{m} p_j(t) &= \sum_{j=1}^{m} D_j(\Gamma, p(t)) + \sum_{i=1}^{n} \sum_{j=1}^{m} y_{ij}(\Gamma, p(t)) \\
&= \sum_{j \in ACTIVE} D_j(\Gamma, p(t)) + \sum_{i=1}^{n} w_i \qquad \Rightarrow \\
\sum_{j=1}^{m} r_j &= \sum_{j \in ACTIVE} \frac{dD_j(\Gamma, p(t))}{dt} > 0
\end{aligned}$$

□

Lemma 4 shows that we can always find a velocity of the price which preserves the set TIGHT and the total price is monotonous increasing.

Let $F_{ij}(t) = F_{ij}(\Gamma, p(t))$. Given the velocity of the price, we can calculate the *velocities* of $\{F_{ij}(t), \forall i, j\}$:

$$\frac{dF_{ij}}{dt} = \sum_{k=1}^{m} \frac{\partial F_{ij}}{\partial p_k} r_k \tag{3.7}$$

When the prices $p(t)$ linearly increase with the parameter t, three events may happen:

Event 1. A commodity j leaves Γ_i, i.e., $F_{ij}(0) < 0$ and $F_{ij}(\Delta t_1) = 0$ for some $\Delta t_1 > 0$;
Event 2. A commodity j enters Γ_i, i.e., $F_{ij}(0) \geq 0$ and $F_{ij}(\Delta t_2) = 0$ for some $\Delta t_2 > 0$.
Event 3. A commodity j moves from $ACTIVE$ to $TIGHT$, i.e., $D_j(0) < 0$ and $D_j(\Delta t_3) = 0$ for some $\Delta t_3 > 0$.

Notice that $F_{ij}(t)$ and $D_j(t)$ are all linear functions with respect to t. Hence, it is easy to determine which event happens first during the increase of the price.

When one of the three events happens, the algorithm updates the configuration and recompute the velocity (solves the linear system (3.6)). The algorithm repeats the process iteratively until all commodities are tight.

Let a^+ denote $\max\{a, 0\}$. According to the above discussion, we have the pseudo-codes for the algorithm.

$p_j = \epsilon$ for some $\epsilon > 0$, $\forall j$; $\Gamma_i \leftarrow |m|$ for all i;
$ACTIVE \leftarrow |m|$; $TIGHT \leftarrow \emptyset$.
While $\#ACTIVE > 0$ **do**
Compute $\lambda_i(\Gamma, p)$, $y_{ij}(\Gamma, p)$, $F_{ij}(\Gamma, p)$ and $D_j(\Gamma, p)$ according to (2.2)-(2.5).
Compute $\frac{\partial D_j}{\partial p_k}$ and $\frac{\partial y_{ij}}{\partial p_k}$ according to (3.2) and (3.4).
Set $r_j = 1$ for $j \in ACTIVE$
compute r_k for $k \in TIGHT$ by solving the equations (3.6).
Compute $\dot{F}_{ij} = \frac{dF_{ij}}{dt}$ and $\dot{D}_j = \frac{dD_j}{dt}$ according to (3.5) and (3.7).
Let $\Delta t_1 = \min\left\{(-F_{ij}/\dot{F}_{ij})^+ \mid \forall i \text{ and } \forall j \in \Gamma_i\right\}$
$\Delta t_2 = \min\left\{(-F_{ij}/\dot{F}_{ij})^+ \mid \forall i \text{ and } \forall j \notin \Gamma_i\right\}$
$\Delta t_3 = \min_{j \in ACTIVE}\left\{(-D_j/\dot{D}_j)^+\right\}$ and $\Delta t = \min\{\Delta t_1, \Delta t_2, \Delta t_3\}$.
If $\Delta t = \Delta t_i$, handle Event i.
End While

Algorithm 1: The Primal-Dual Algorithm

4 Convergence and Performance Test

4.1 Convergence of the Algorithm

The algorithm adjusts the price in the space $\mathbb{R}^m_+$. It starts from the point near the origin and raises the price step by step until it reaches the plane $\sum_{j=1}^{m} p_j = \sum_{i=1}^{n} w_i$. During the adjustment process, the algorithm may encounter three types of events (see Sec. 3). Note that the algorithm keeps the invariants:

- **Invariant 1** $D_j < 0$ for $j \in ACTIVE$ and $D_j = 0$ for $j \in TIGHT$.
- **Invariant 2** $F_{ij} < 0$ for $j \in \Gamma_i$ and $F_{ij} \geq 0$ for $j \notin \Gamma_i$.

Due to **Invariant 1**, the third event (a commodity becomes tight) happens exactly m times. For the first and second events, we consider an arbitrary $\Gamma = \{\Gamma_1, ..., \Gamma_n\}$, a collection of n subsets of $|m|$, which naturally corresponds to a polyhedron in the price space:

$$\begin{cases} \frac{\beta_{ij}}{\alpha_{ij}} p_j < (w_i + \sum_{l \in \Gamma_i} \beta_{il} p_l)/(\sum_{l \in \Gamma_i} \alpha_{il}), & \forall i \in |n|, \quad \forall j \in \Gamma_i \\ \frac{\beta_{ij}}{\alpha_{ij}} p_j \geq (w_i + \sum_{l \in \Gamma_i} \beta_{il} p_l)/(\sum_{l \in \Gamma_i} \alpha_{il}), & \forall i \in |n|, \quad \forall j \notin \Gamma_i \end{cases}$$

Let $P(\Gamma_i, j)$ denote the plane $\frac{\beta_{ij}}{\alpha_{ij}} p_j = (w_i + \sum_{l \in \Gamma_i} \beta_{il} p_l)/(\sum_{l \in \Gamma_i} \alpha_{il})$ in the price space. The cone $\left\{p \in \mathbb{R}_+^m \mid \sum_{j=1}^m p_j \leq \sum_{i=1}^n w_i\right\}$ can be divide into finite number of polyhedrons, whose boundary are such planes.

When the first or second event happens, the price crosses a plane and moves from one polyhedron to another. We can project the path drawn by the price in the adjustment process to the line $l \subset \mathbb{R}_+^m$, where l is parameterized by the equation $l(t) = \mathbf{1} \cdot t$. The projected path starts from a point near the origin and ends at $l(\sum_{i=1}^{n} w_i/m)$. By Lemma 4, the total price increases monotonously, thus the projected path never turns back. Hence, the algorithm will converge to an equilibrium point in finite number of steps.

The algorithm and its convergence also builds a constructive proof of the existence of equilibrium in the Fisher's setting with logarithmic utilities.

4.2 Performance Test

Although the speed of convergence of this algorithm is unknown, it does well in practice. The algorithm has been implemented on a PC and tested against markets of various sizes. We randomly generate markets with n agents and m commodities, for $n = 50$, $m = 20, 40, 60, 80, 120, 160$ and $m = 50$, $n = 20, 40, 60, 80, 120, 160$. For each pair of n and m, we generate 5 instances and record the average running time. The running time is measured by the number of *steps*, where a step is a single loop with an events happens in Algorithm 1. Note that the time complexity of each step is $O(m^3)$, which is spent on solving the linear equation (3.6).

The results are plotted in Fig 4.2 and Fig 4.2. We can see from the figure that the number of steps is almost a perfect linear function with respect to the number of agents or the number of commodities.

The algorithm can also be utilized as a black box to solve general models. In order to do so, we adopt the *welfare adjustment scheme.* Given a certain price, the scheme turns a general market model to a Fisher's model and computes the Fisher's equilibrium price by calling a solver of the Fisher's model. The procedure naturally yields a map from the space of price to itself whose fixed points are the equilibrium points in the general model. We may expect that the simple iteration can converge to one of the fixed points. The details of the welfare adjustment scheme can be found in [8]. We test the scheme on the markets with hundreds

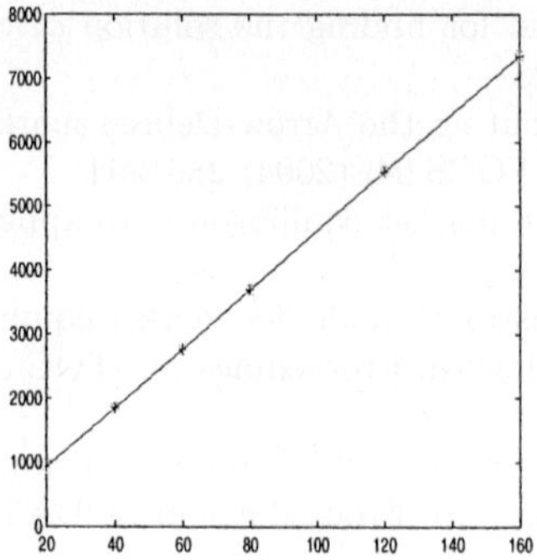

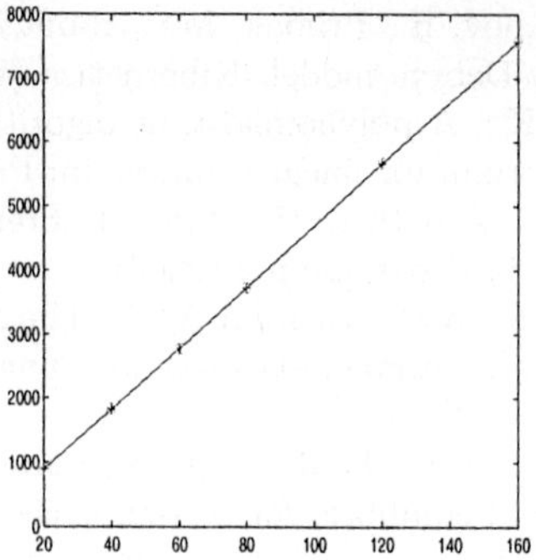

Fig. 1. The X-axis is the number of commodities. The Y-axis is the number of running steps. The number of agents is fixed to 50.

Fig. 2. The X-axis is the number of agents. The Y-axis is the number of running steps. The number of commodities is fixed to 50.

of agents and commodities and usually reach the equilibrium price within ten oracle invocations.

5 Conclusion

In this paper, we propose a primal-dual type algorithm to solve the equilibrium problem in the Fisher's market model with logarithmic utility functions. The logarithmic function is interesting because it is non-homogenous and always admits rational equilibrium point, whenever the input data are rational. With a tricky transformation of the utility function, the equilibrium conditions become linear functions of the price, so that the algorithm can increase the price linearly and predict all events during the increase. The algorithm is proved to converge in finite steps and performs well in practice. From the experimental tests, we may conjecture that the algorithm can reach the equilibrium point in $O(m^4n)$ time, which is better than existing techniques (convex programming or ellipsoid algorithms).

Acknowledgment

The author thanks an anonymous referee for pointing out a few errors and some valuable comments.

References

1. Arrow, K.J., Debreu, G.: Existence of an equilibrium for a competitive economy. Econometrica **22** (1954) 265–290
2. Deng, X., Papadimitriou, C., Safra, S.: On the complexity of price equilibria. Journal of Computer and System Sciences **67** (2003) 311–324

3. Nenakhov, E., Primak, M.: About one algorithm for finding the solution of the Arrow-Debreu model. Kibernetica (1983) 127–128
4. Jain, K.: A polynomial time algorithm for computing the Arrow-Debreu market equilibrium for linear utilities. In: Proceeding of FOCS'04. (2004) 286–294
5. Ye, Y.: A path to the Arrow-Debreu competitive market equilibrium. to appear in Math. Programming (2004)
6. Devanur, N.R., Vazirani, V.V.: The spending constraint model for market equilibrium: algorithmic, existence and uniqueness results. In: Proceedings of STOC'04. (2004) 519–528
7. Codenotti, B., Pemmaraju, S., Varadarajan, K.: On the polynomial time computation of equilibria for certain exchange economies. In: Proceedings of SODA'05. (2005) 72–81
8. Codenotti, B., Pemmaraju, S., Varadarajan, K.: Algorithms column, the computation of market equilibria. ACM SIGACT News **35** (2004)
9. Chen, N., Deng, X., Sun, X., Yao, A.C.: Fisher equilibrium price with a class of concave utility functions. In: Proceedings of ESA'04. (2004) 169–179
10. Deng, X., Huang, L.S., Ye, Y.: Computation of the Arrow-Debreu equilibrium for a class of non-homogenous utility functions. (Manuscript)

Boosting Spectral Partitioning by Sampling and Iteration*

Joachim Giesen and Dieter Mitsche

Institute for Theoretical Computer Science, ETH Zürich, CH-8092 Zürich
{giesen, dmitsche}@inf.ethz.ch

Abstract. A partition of a set of n items is a grouping of the items into k disjoint classes of equal size. Any partition can be modeled as a graph: the items become the vertices of the graph and two vertices are connected by an edge if and only if the associated items belong to the same class. In a planted partition model a graph that models the planted partition is obscured by random noise, i.e., edges within a class can get removed and edges between classes can get inserted at random. We study the task to reconstruct the planted partition from this graph whose complexity can be controlled by the number k of classes if the noise level is fixed. The best bounds on k where the classes can be reconstructed correctly almost surely are achieved by spectral algorithms. We show that a combination of random sampling and iterating the spectral approach can boost its performance in the sense that the number of classes that can be reconstructed correctly asymptotically almost surely can be as large as $k = c\sqrt{n}/\log\log n$ for some constant c. This extends the range of k for which such guarantees can be given for any efficient algorithm.

1 Introduction

The partition reconstruction problem that we study in this paper is related to the k-partition problem. In the latter problem the task is to partition the vertices of a given graph into k equally sized classes such that the number of edges between the classes is minimized. This problem is NP-hard already for $k = 2$, i.e., in the graph bisection case [5]. This worst case complexity need not necessarily show up in specialized but from an application point of view (especially clustering) meaningful graph families - especially families of random graphs, see for example [2,1,3] and the references therein. The random graph families typically assume a given partition of the vertices of the graph (planted partition modeled as cliques), which is obscured by random noise (randomly removed or inserted edges). The goal becomes to assess the ability of a partitioning algorithm to reconstruct the planted partition. This ability can be measured by the probability that the algorithm can reconstruct the planted partition.

The best studied random graph family for the partition reconstruction problem is the following: an edge in the graph appears with probability p if its two

* Partly supported by the Swiss National Science Foundation under the grant "Nonlinear manifold learning".

X. Deng and D. Du (Eds.): ISAAC 2005, LNCS 3827, pp. 473–482, 2005.

incident vertices belong to the same planted class and with probability $q < p$ otherwise, independently from all other edges. The probabilities p and q control the density / sparsity of the graph and can depend on the number n of vertices. In general there is a trade-off between the sparsity of the graph and the number k of classes that can be reconstructed, i.e., the sparser the graph the less classes can be reconstructed correctly and vice versa the larger k the denser the graph has to be in order to reconstruct the classes correctly. Here we assume that p and q are fixed, i.e., we deal with dense graphs only. That leaves only k as a free parameter. We believe that this should be enough to assess the power of most partitioning algorithms.

In this model the most powerful efficient (polynomial in n) algorithms known so far are the algorithm of Shamir and Tsur [10], which builds on ideas of Condon and Karp [2], and the algorithm of McSherry [9]. Both algorithms can with high probability reconstruct correctly up to $k = O(\sqrt{n/\log n})$ planted classes. The algorithm of McSherry falls in the category of spectral clustering algorithms that make use of the eigenvalues and eigenvectors of the adjacency matrix of the input graph. In [6] we design an efficient spectral algorithm for which we cannot show that it reconstructs the planted partition with high probability even for quite small values for k, but we can prove that the relative number of misclassifications for $k = o(\sqrt{n})$ goes to zero with high probability. In [7] we design an algorithm, which with high probability can reconstruct up to $k = c\sqrt{n}$ partitions, where c is a small constant. However this algorithm needs super-polynomial time in n.

Here we design an efficient spectral algorithm and prove that it asymptotically almost surely reconstructs a planted k-partition for $k \leq c\sqrt{n}/\log\log n$ for some constant c. At the core of this algorithm is a spectral algorithm with weaker guarantees. This algorithm uses a small random sub-matrix of the adjacency matrix of the input graph to correctly reconstruct a large fraction of the planted classes. Iterating this algorithm allows us to reconstruct the remaining classes. Doing this in the naive way would allow us to correctly reconstruct up to $k = O(\sqrt{n/\log n})$ classes. In order to boost the power of the algorithm we prune from the small random sub-matrix the algorithm is currently working with all entries that correspond to classes that were already reconstructed in earlier iterations. This means that the number of entries in this matrix that correspond to a not yet reconstructed class is increasing relatively to the size of the matrix. This makes the reconstruction problem easier and thus the relative fraction of not yet reconstructed classes that the algorithm reconstructs increases in every iteration.

2 Planted Partitions

In this section we introduce the planted partition reconstruction problem, see [3] for a motivation of this problem and the underlying model. We first define the $A(\varphi, p, q)$ distribution, see also McSherry [9].

$A(\varphi, p, q)$ distribution. Given a surjective function $\varphi : \{1, \ldots, n\} \to \{1, \ldots, k\}$ and probabilities $p, q \in (0, 1)$ with $p > q$. The $A(\varphi, p, q)$ distribution is a

distribution on the set of $n \times n$ symmetric, 0-1 matrices with zero trace. Let $\hat{A} = (\hat{a}_{ij})$ be a matrix drawn from this distribution. It is $\hat{a}_{ij} = 0$ if $i = j$ and for $i \neq j$,

$$\begin{aligned}
P(\hat{a}_{ij} = 1) &= p && \text{if } \varphi(i) = \varphi(j) \\
P(\hat{a}_{ij} = 0) &= 1 - p && \text{if } \varphi(i) = \varphi(j) \\
P(\hat{a}_{ij} = 1) &= q && \text{if } \varphi(i) \neq \varphi(j) \\
P(\hat{a}_{ij} = 0) &= 1 - q && \text{if } \varphi(i) \neq \varphi(j),
\end{aligned}$$

independently. The *matrix of expectations* $A = (a_{ij})$ corresponding to the $A(\varphi, p, q)$ distribution is given as

$$\begin{aligned}
a_{ij} &= 0 \;\; \text{if } i = j \\
a_{ij} &= p \;\; \text{if } \varphi(i) = \varphi(j) \text{ and } i \neq j \\
a_{ij} &= q \;\; \text{if } \varphi(i) \neq \varphi(j)
\end{aligned}$$

Lemma 1 (Füredi and Komlós [4], Vu [12], Krivelevich and Vu [8]). *Let $\hat{A}$ be a matrix drawn from the $A(\varphi, p, q)$ distribution and A be the matrix of expectations corresponding to this distribution. Let $c = \min\{p(1-p), q(1-q)\}$ and assume that $c^2 \gg (\log n)^6/n$. Then*

$$|A - \hat{A}| \leq \sqrt{n}$$

with probability at least $1 - 2e^{-c^2 n/8}$. Here $|\cdot|$ denotes the L_2 matrix norm, i.e., $|B| = \max_{|x|=1} |Bx|$.

Planted partition reconstruction problem. Given a matrix $\hat{A}$ drawn from from the $A(\varphi, p, q)$ distribution. Assume that all classes $C_\ell := \varphi^{-1}(l), l \in \{1, \ldots, k\}$ have the same size n/k. Then the function φ is called a *partition function.* The planted partition reconstruction problem asks to reconstruct φ up to a permutation of $\{1, \ldots, k\}$ only from $\hat{A}$ (up to permutations of $\{1, \ldots, k\}$).

3 Spectral Properties

Any real symmetric $n \times n$ matrix has n real eigenvalues and $\mathbb{R}^n$ has a corresponding eigenbasis. Here we are concerned with two types of real symmetric matrices. First, any matrix $\hat{A}$ drawn from an $A(\varphi, p, q)$ distribution. Second, the matrix A of expectations corresponding to the distribution $A(\varphi, p, q)$.

We want to denote the eigenvalues of $\hat{A}$ by $\hat{\lambda}_1 \geq \hat{\lambda}_2 \geq \ldots \geq \hat{\lambda}_n$ and the vectors of a corresponding orthonormal eigenbasis of $\mathbb{R}^n$ by $v_1, \ldots, v_n$, i.e., it is $\hat{A}v_i = \hat{\lambda}_i v_i$, $v_i^T v_j = 0$ if $i \neq j$ and $v_i^T v_i = 1$, and the $v_1, \ldots, v_n$ span the whole $\mathbb{R}^n$.

For the sake of analysis we want to assume here without loss of generality that the matrix A of expectations has a block diagonal structure, i.e., the elements in the i-th class have indices from $\frac{n}{k}(i-1)+1$ to $\frac{n}{k}i$ in $\{1, \ldots, n\}$. It is easy to verify that the eigenvalues $\lambda_1 \geq \ldots \geq \lambda_n$ of A are $(\frac{n}{k}-1)p + (n - \frac{n}{k})q$, $\frac{n}{k}(p-q) - p$ and $-p$ with corresponding multiplicities 1, $k-1$ and $n-k$, respectively. A possible

orthonormal basis of the eigenspace corresponding to the k largest eigenvalues of A is u_i, $i = 1, \ldots, k$, whose j-th coordinates are given as follows,

$$u_{ij} = \begin{cases} \sqrt{\frac{k}{n}}, & j \in \{\frac{n}{k}(i-1)+1, \ldots, \frac{n}{k}i\} \\ 0, & \text{else.} \end{cases}$$

Spectral separation. The *spectral separation* $\delta_k(A)$ of the eigenspace of the matrix A of expectations corresponding to its k largest eigenvalues from its complement is defined as the difference between the k-th and the $(k+1)$-th eigenvalue, i.e., it is $\delta_k(A) = \frac{n}{k}(p-q)$.

Projection matrix. The matrix $\hat{P}$ that projects any vector in $\mathbb{R}^n$ to the eigenspace corresponding to the k largest eigenvalues of a matrix $\hat{A}$ drawn from the distribution $A(\varphi, p, q)$, i.e., the projection onto the space spanned by the vectors $v_1, \ldots, v_k$, is given as

$$\hat{P} = \sum_{i=1}^{k} v_i v_i^T.$$

The matrix P that projects any vector in $\mathbb{R}^n$ to the eigenspace corresponding to the k largest eigenvalues of the matrix A of expectations can be characterized even more explicitly. Its entries are given as

$$p_{ij} = \begin{cases} \frac{k}{n}, & \varphi(i) = \varphi(j) \\ 0, & \varphi(i) \neq \varphi(j) \end{cases}$$

In [7] we prove the following two lemmas.

Lemma 2. *All the k largest eigenvalues of $\hat{A}$ are larger than $\sqrt{n}$ and all the $n-k$ smallest eigenvalues of $\hat{A}$ are smaller than $\sqrt{n}$ with probability at least $1 - 2e^{-c^2 n/8}$ provided that n is sufficiently large and $k < \frac{p-q}{4}\sqrt{n}$.*

Lemma 3. *With probability at least $1 - 2e^{-c^2 n/8}$,*

$$q \le \frac{k}{k-1}\frac{\hat{\lambda}_1}{n} + \frac{k}{k-1}\frac{1}{\sqrt{n}} \quad \text{and} \quad q \ge \frac{k}{k-1}\frac{\hat{\lambda}_1}{n} - \frac{k}{k-1}\left(\frac{1}{\sqrt{n}} + \frac{1}{k}\right)$$

and with the same probability

$$p \le \frac{k\hat{\lambda}_2 + \frac{k}{k-1}\hat{\lambda}_1}{n-k} + \frac{k\sqrt{n}}{(n-k)(k-1)} + \frac{k\sqrt{n}}{n-k} \quad \text{and}$$

$$p \ge \frac{k\hat{\lambda}_2 + \frac{k}{k-1}\hat{\lambda}_1}{n-k} - \frac{k\sqrt{n}}{(n-k)(k-1)} - \frac{n}{(n-k)(k-1)} - \frac{k\sqrt{n}}{n-k}.$$

Theorem 1 (Stewart [11]). *Let $\hat{P}$ and P be the projection matrices as defined above. It holds*

$$|P - \hat{P}| \le \frac{2|A - \hat{A}|}{\delta_k(A) - 2|A - \hat{A}|}$$

if $\delta_k(A) > 4|A - \hat{A}|$ where $|\cdot|$ is the L_2 matrix norm.

4 An Iterative Spectral Algorithm

Now we have all prerequisites at hand that we need to describe our spectral algorithm to solve the planted partition reconstruction problem.

SpectralReconstruct($\hat{A}$)

$\hat{k}, k' :=$ number of eigenvalues $\hat{\lambda}_i$ of $\hat{A}$ that are larger than $\sqrt{n}$.
$\hat{p} := \frac{\hat{k}}{\hat{k}-1}\frac{\hat{\lambda}_1}{n}$
$\hat{q} := \frac{\hat{k}\hat{\lambda}_2 + \frac{\hat{k}}{\hat{k}-1}\hat{\lambda}_1}{n-\hat{k}}$
$m := n/c\log\log n$
Randomly partition $\{1,\ldots,n\}$ into $c\log\log n$ equal size subsets I_i.
$i := 1;\quad \mathcal{C} := \emptyset$
while $i < c\log\log n - 1$ **do**
  $\hat{A}_i :=$ restriction of $\hat{A}$ to I_i.
  $\hat{P}_i :=$ projector onto the space spanned by the k' largest eigenvectors of $\hat{A}_i$.
  $m_i := |I_i|$
  for $j := 1$ **to** m_i **do**
    for $l := 1$ **to** m_i **do**
      $(c_j)_l := \begin{cases} 1 & : \ (\hat{P}_i)_{lj} \geq \frac{\hat{k}}{2m} \\ 0 & : \ \text{otherwise} \end{cases}$
    end for
    if $0.89\frac{m}{\hat{k}} \leq |c_j|^2 \leq 1.11\frac{m}{\hat{k}}$ **do**
      mark the element of rank j in I_i.
    end if
  end for
  while $\{l \in I_i : l \text{ marked}\} \neq \emptyset$ **do**
    choose arbitrary $l \in \{l' \in I_i : l' \text{ marked}\}$ and unmark l.
    $C_1 := \left\{ l' \in I_i : l' \text{ marked}, {c_{\mathrm{rk}_i(l)}}^T c_{\mathrm{rk}_i(l')} \geq 0.79\frac{m}{\hat{k}} \right\}$
    if $0.89\frac{m}{\hat{k}} \leq |C_1| \leq 1.11\frac{m}{\hat{k}}$ **do**
      $C_2 := \left\{ l \in I_{i+1} : \sum_{l' \in C_1} \hat{a}_{ll'} \geq \frac{3}{4}|C_1|\hat{p} + \frac{1}{4}|C_1|\hat{q} \right\}$
      if $|C_2| \geq \frac{3}{4}\frac{m}{\hat{k}}$ **do**
        $C_3 := \left\{ l \in I_{i+2} : \sum_{l' \in C_2} \hat{a}_{ll'} \geq \frac{3}{4}|C_2|\hat{p} + \frac{1}{4}|C_2|\hat{q} \right\}$
        $C_4 := \left\{ l \in I_{i+3} : \sum_{l' \in C_3} \hat{a}_{ll'} \geq \frac{3}{4}|C_3|\hat{p} + \frac{1}{4}|C_3|\hat{q} \right\}$
        $C_1 := \left\{ l \in I_i \quad : \sum_{l' \in C_3} \hat{a}_{ll'} \geq \frac{3}{4}|C_3|\hat{p} + \frac{1}{4}|C_3|\hat{q} \right\}$
        $C_2 := \left\{ l \in I_{i+1} : \sum_{l' \in C_4} \hat{a}_{ll'} \geq \frac{3}{4}|C_4|\hat{p} + \frac{1}{4}|C_4|\hat{q} \right\}$
        $C := C_1 \cup C_2 \cup C_3 \cup C_4$
        for $j := 1$ **to** $c\log\log n$ **do**
          if $j \notin \{i, i+1, i+2, i+3\}$ **do**
            $C := C \cup \left\{ l \in I_j : \sum_{l' \in C} \hat{a}_{ll'} \geq \frac{3}{4}|C|p + \frac{1}{4}|C|q \right\}$
          end if
          $I_j := I_j \setminus (I_j \cap C)$
        end for
        $\mathcal{C} := \mathcal{C} \cup \{C\};\quad k' := k' - 1$

```
37         end if
38       end if
39     end while
40     i := i + 4
41   end while
42   return C
```

In a nutshell the algorithm SPECTRALRECONSTRUCT works as follows: in lines 1 to 6 we do some pre-processing. The main loop of the algorithm is enclosed by lines 7 and 41. In lines 8 to 18 we compute binary vectors c_j of size m_i that potentially are close in Hamming distance to characteristic vectors of planted classes C_ℓ restricted to the index set I_i. Note that very likely these vectors are not exactly characteristic vectors of planted classes. To account for this we have to work with four index sets $I_i, \ldots, I_{i+3}$. In lines 19 to 39 we use the vectors c_j to reconstruct whole classes, i.e., not just the restriction to $I_i \cup \ldots \cup I_{i+3}$. Removing reconstructed elements from not yet processed index sets I_j in line 34 boosts the performance of the algorithm as a potential source of inter class noise between an already reconstructed and a not yet reconstructed class gets removed. That is, in subsequent iterations we expect to get relatively more vectors that are close to characteristic vectors of restricted planted classes. Our subsequent analysis shows that this is indeed the case.

The running time of the algorithm is determined by $c \log\log n$ times the time to compute the k' largest eigenvectors of an $m \times m$ matrix. In the following we give a more detailed description of the algorithm.

PREPROCESSING: In line 1 we estimate the number of classes k by the number of eigenvalues of $\hat{A}$ larger than $\sqrt{n}$, which by Lemma 2 is a.a.s. correct. In lines 2 and 3 we estimate the parameters p and q of the $A(\varphi, p, q)$ distribution from which the matrix $\hat{A}$ is drawn. By Lemma 3 these estimates almost surely converge to the true values when n goes to infinity, provided $k = o(\sqrt{n})$. In line 5 we randomly partition the index set $\{1, \ldots, n\}$ into $c \log\log n$ sets of equal size. We choose the constant c such that m and $c \log\log n$ are integers and $c \log\log n$ is divisible by 4. In line 6 we initialize the $\mathcal{C}$ that is going to contain the reconstructed classes.

MAIN LOOP: In line 9 we compute the projector $\hat{P}_i$ onto the space spanned by the eigenvectors that correspond to the k' largest eigenvalues of $\hat{A}_i$. In the variable k' we store the number of classes that still have to be reconstructed. In lines 11 to 18 we compute for every column of $\hat{P}_i$ a binary vector c_j. If we did this for the projector derived from the unperturbed adjacency matrix A_i then this would be the characteristic vectors of planted classes C_ℓ restricted to the index set I_i. Thus if the noise is small most of the vectors should be close to characteristic vectors. In lines 15 to 17 we discard c_j if it does not have the right size in order to be close to a characteristic vector. In the loop enclosed by lines 19 and 39 the actual reconstruction takes place. In lines 20 and 21 we use the binary vectors that belong to marked elements in I_i to compute a first rough reconstruction C_1 of some class C_ℓ restricted to I_i. In order to do

so we have to use the rank function $\text{rk}_i(\cdot)$, which gives the rank of an element in I_i, to map the elements of I_i into $\{1, \ldots, m_i\}$, i.e., the set that contains the column indices of $\hat{P}_i$. The intuition behind the computations in line 21 is that if $c_{\text{rk}_i(l)}$ and $c_{\text{rk}_i(l')}$ are close in Hamming distance to the characteristic vector of the same class C_ℓ restricted to I_i then their dot product should be large. Note that very likely C_1 does not contain all elements from $C_\ell \cap I_i$ and may contain elements from other classes than C_ℓ. If C_1 contains roughly as many elements as we expect to be in $C_\ell \cap I_i$ then it likely contains many elements from C_ℓ and few elements from other classes. In this case we compute in line 23 a set C_2 that contains the elements from I_{i+1} such that every column of the matrix $\hat{A}$ with index in $C_2 \cap C_\ell$ has many entries that are '1' at row indices in C_1. We will show that with high probability C_2 contains only elements from C_ℓ, but maybe not all of them. If C_2 is large enough then we can use it in line 23 to compute (as we computed C_2 from C_1) a set $C_3 \subset I_{i+2}$, for which we can show that with high probability $C_3 = C_\ell \cap I_{i+2}$. Similarly we compute C_4 and re-compute C_1 from C_3 and re-compute C_2 from C_4. We will show that with high probability it also holds that $C_1 = C_\ell \cap I_i$, $C_2 = C_\ell \cap I_{i+1}$ and $C_4 = C_\ell \cap I_{i+3}$. The reason for computing four sets is that we need some probabilistic independence in the analysis of the algorithm. In practice it probably is sufficient to re-compute C_1 from C_2 and afterwards C_2 from C_1. Similarly we will show that we compute with high probability the set $C_\ell \cap I_j$ in line 32 and put it into the reconstruction C of C_ℓ. In line 34 we remove all elements in C from the sets I_j. In line 36 we add C to the set of reconstructed classes.

5 Analysis of the Algorithm

In the following we use the notation as in the algorithm SPECTRALRECONSTRUCT and let $C_{\ell i} = C_\ell \cap I_i$ for I_i as computed in line 5 of the algorithm. We say that an event occurs asymptotically almost surely, or short a.a.s., if the probability that it holds goes to 1, as n tends to infinity. Instead of the spectral matrix norm $|\cdot|$ we will often use the related Frobenius norm $|\cdot|_F$, which is bounded by the spectral norm times the square root of the rank of the matrix. The Frobenius norm, which is the square root of the sum of the squared entries of the matrix, makes it easier to deal with individual entries of matrices.

Here is a short outline of the proof, which is by induction on the number of iterations of the algorithm. The induction is anchored in Lemmas 6 and 7 and the induction step is proved in Theorem 2 and Corollary 2. But at first we state two technical lemmas, whose proofs are almost identical to proofs of similar statements in [7], and derive a simple corollary from them.

Lemma 4. *The size of $C_{\ell i}$ is a.a.s. contained in the interval*

$$\left[(1-\delta)\frac{m_i}{k}, (1+\delta)\frac{m_i}{k}\right] \quad \text{with} \quad \delta := 3(km_i^{-3/4})^{1/3}.$$

Lemma 5. *The spectral separation $\delta_k(A_1)$ is a.a.s. at least $(1-\delta)\frac{m_1}{k}(p-q)$, where δ as in Lemma 4.*

Corollary 1. *If $k < c'\sqrt{m_1}$ then a.a.s. $|P_1 - \hat{P}_1|^2 < \left((1-\delta)\frac{p-q}{2c'} - 1\right)^{-1} =: \varepsilon$, where δ as in Lemma 4.*

Proof. Follows immediately from Stewart's theorem and the theorem of Füredi and Komlós and Lemma 5.

In the following we will use δ from Lemma 4 and ε from Corollary 1. Note that we have that δ goes to 0 and ε goes to $1/\left(\frac{p-q}{c'} - 1\right)$ as n (and thus also m_1) goes to infinity.

Dangerous element. Let $x \in C_{\ell i}$ and $r_i(x)$ be the rank of x in I_i. The element x is called *dangerous* if the $r_i(x)$'th column in the matrix $P_i - \hat{P}_i$ has more than $\frac{1}{10}\frac{m}{k}$ entries whose absolute value is at least $\frac{1}{3}\frac{k}{m}$.

Safe class. A class C_ℓ is called *safe* with respect to I_i if it satisfies the following two conditions:

(1) At most $\frac{1}{10}\frac{m}{k}$ elements of $C_{\ell i}$ are dangerous.
(2) At most $\frac{1}{10}\frac{m}{k}$ columns of $\hat{P}_i$ whose index is the rank $r_i(\cdot)$ of an element in $I_i \setminus C_\ell$ have at least $\frac{1}{10}\frac{m}{k}$ entries of value at least $\frac{1}{3}\frac{k}{m}$ at row indices that are the ranks of elements in $C_{\ell i}$.

Lemma 6. *If $k < c'\sqrt{m}$ then a.a.s. there are at least $(1-3600\varepsilon)k$ classes C_ℓ that are safe with respect to I_1, provided that c' is small enough such that $3600\varepsilon \ll 1$.*

Proof. The event whose probability we want to lower bound here can also be stated as: there are at most $3600\varepsilon k$ classes C_ℓ that are not safe with respect to I_1. Here it is easier to work with the latter formulation of the event. For a class to be not safe either condition (1) or condition (2) in the definition of a safe class has to be violated.

For condition (1), observe that for every dangerous element $x \in C_{\ell 1}$ the contribution of the $r_1(x)$'th column to the Frobenius norm $|P_1 - \hat{P}_1|_F^2$ is at least $(\frac{k}{3m})^2\frac{m}{10k} = \frac{k}{90m}$. Hence, for a class C_ℓ to violate condition (1), the contribution to $|P_1 - \hat{P}_1|_F^2$ is at least $\frac{k}{90m}\frac{m}{10k} = \frac{1}{900}$. Since we have chosen c' sufficiently small Corollary 1 gives $|P_1 - \hat{P}_1| < \varepsilon < 1$ a.a.s. This implies that a.a.s. $|P_1 - \hat{P}_1|^2 \leq |P_1 - \hat{P}_1|$. Using that $|P_1 - \hat{P}_1|_F^2 \leq \text{rk}(P_1 - \hat{P}_1)|P_1 - \hat{P}_1|^2$ and given that the rank $\text{rk}(P_1 - \hat{P}_1)$ is at most $2k$ with the probability stated in the theorem of Füredi and Komlós, we have for the Frobenius norm $|P_1 - \hat{P}_1|_F^2 \leq 2\varepsilon k$ a.a.s. Thus we obtain for the number y of classes violating condition (1), $\frac{1}{900}y \leq 2\varepsilon k$ and equivalently $y \leq 1800\varepsilon k$ a.a.s.

For condition (2), let C_ℓ be a class that violates condition (2). That is, there are at least $\frac{m}{10k}$ elements $x \in I_1 \setminus C_\ell$ such that the columns with index rank $r_1(x)$ each have at least $\frac{m}{10k}$ entries of value at least $\frac{k}{3m}$ at row indices that are the ranks of elements in $C_{\ell 1}$. Hence the contribution of these columns to

the Frobenius norm $|P_1 - \hat{P}_1|_F^2$ is at least $(\frac{k}{3m})^2 \frac{m}{10k}\frac{m}{10k} = 1/900$. If we denote by y the number of classes violating condition (2), we get using as above that $|P_1 - \hat{P}_1|_F^2 \le 2\varepsilon k$ a.a.s. that $\frac{1}{900}y \le 2\varepsilon k$ and equivalently $y \le 1800\varepsilon k$ a.a.s.

In combination we get that a.a.s. there are at most $3600\varepsilon k$ classes which are not safe with respect to I_1. This proves the statement of the lemma.

Lemma 7. *A.a.s., the following holds, provided $k < c'\sqrt{m}$:*

(1) Every class C_ℓ that is safe with respect to I_1 is reconstructed correctly.
(2) Every class C_ℓ that is not safe with respect to I_1 is either reconstructed correctly or none of its elements gets removed from any index set I_i.

Proof. Is left for the full version of the paper.

Lemma 7 basically states that after the first iteration of the outer while-loop of the algorithm all classes C_ℓ that are safe with respect to I_1, but probably even more classes, are reconstructed correctly. Thus we get from Lemma 6 that after the first iteration of the outer while-loop a.a.s. at least $(1-3600\varepsilon)k$ classes are reconstructed correctly, or equivalently a.a.s. at most $3600\varepsilon k$ remain to be reconstructed.

Theorem 2. *A.a.s., after the j'th iteration of the outer while loop of the algorithm* SpectralReconstruct *there remain at most*

$$(3600\varepsilon')^{2(1.5^j-1)}k \quad \textit{with} \quad \varepsilon' = \frac{2\sqrt{m(1+\delta)}}{\frac{m}{k}(p-q)(1-\delta) - 2\sqrt{m(1+\delta)}}$$

classes to be reconstructed and all other classes have been reconstructed correctly, provided $k < c'\sqrt{m}$ and c' sufficiently small.

Proof. Is left for the full version of the paper.

Note that in the statement of Theorem 2 the boosting reflects itself in the double exponential dependence of the decrease factor on j. Without boosting we would only have a single exponential dependence on j.

Corollary 2. *A.a.s., the algorithm* SpectralReconstruct *reconstructs all k classes correctly, provided $k \le c'\frac{\sqrt{n}}{\log\log n}$ for a small enough constant c'.*

Proof. By Theorem 2, we have that after the $(\frac{c}{4}\log\log n)$'th iteration of the outer while-loop, a.a.s., the number of not yet reconstructed classes is at most $(3600\varepsilon')^{2(1.5^{\frac{c}{4}\log\log n}-1)}k$. If c' is small enough then we have $3600\varepsilon' \ll 1$ and

$$\lim_{n\to\infty}(3600\varepsilon')^{2(1.5^{\frac{c}{4}\log\log n}-1)}k = 0.$$

Hence, for n sufficiently large, $(3600\varepsilon')^{2(1.5^{\frac{c}{4}\log\log n}-1)}k$ is strictly less than 1. Since the number of not yet reconstructed classes is an integer, it has to be 0 for n large enough. Thus a.a.s. all classes are reconstructed correctly.

6 Conclusion

The simple technique of sampling and iteration that we used here to boost spectral partitioning should also be applicable to other partitioning and clustering algorithms, e.g., [9,10]. It would be interesting to see if one could also boost their performance with this technique.

References

1. B. Bollobas and A.D. Scott. Max cut for random graphs with a planted partition. *Combinatorics, Probability and Computing*, 13:451–474, 2004.
2. A. Condon and R. Karp. Algorithms for graph partitioning on the planted partition model. *Random Structures and Algorithms 8*, 2:116–140, 1999.
3. A. Frieze and C. McDiarmid. Algorithmic theory of random graphs. *Random Structures and Algorithms*, 10:5–42, 1997.
4. Z. Füredi and J. Komlós. The eigenvalues of random symmetric matrices. *Combinatorica I*, 3:233–241, 1981.
5. M. R. Garey, D. S. Johnson, and L. Stockmeyer. Some simplified NP-complete graph problems. *Theoretical Computer Science*, 1:237–267, 1976.
6. J. Giesen and D. Mitsche. Bounding the misclassification error in spectral partitioning in the planted partition model. *Proceedings of the 31st International Workshop on Graph-Theoretic Concepts in Computer Science*, 2005.
7. J. Giesen and D. Mitsche. Reconstructing many partitions using spectral techniques. *Proceedings of the 15th International Symposium on Fundamentals of Computation Theory*, 2005.
8. M. Krivelevich and V. H. Vu. On the concentration of eigenvalues of random symmetric matrices. *Microsoft Technical Report*, 60, 2000.
9. F. McSherry. Spectral partitioning of random graphs. *Proceedings of 42nd IEEE Symosium on Foundations of Computer Science*, pages 529–537, 2001.
10. R. Shamir and D. Tsur. Improved algorithms for the random cluster graph model. *Proceedings 7th Scandinavian Workshop on Algorithm Theory*, pages 230–259, 2002.
11. G. Stewart and J. Sun. *Matrix perturbation theory.* Academic Press, Boston, 1990.
12. V. H. Vu. Spectral norm of random matrices. In *STOC '05: Proceedings of the thirty-seventh annual ACM symposium on Theory of computing*, pages 423–430, New York, NY, USA, 2005. ACM Press.

Smoothed Analysis of Binary Search Trees*

Bodo Manthey** and Rüdiger Reischuk

Universität zu Lübeck, Institut für Theoretische Informatik,
Ratzeburger Allee 160, 23538 Lübeck, Germany
{manthey, reischuk}@tcs.uni-luebeck.de

Abstract. Binary search trees are one of the most fundamental data structures. While the height of such a tree may be linear in the worst case, the average height with respect to the uniform distribution is only logarithmic. The exact value is one of the best studied problems in average-case complexity.

We investigate what happens in between by analysing the smoothed height of binary search trees: Randomly perturb a given (adversarial) sequence and then take the expected height of the binary search tree generated by the resulting sequence. As perturbation models, we consider partial permutations, partial alterations, and partial deletions.

On the one hand, we prove tight lower and upper bounds of roughly $\Theta(\sqrt{n})$ for the expected height of binary search trees under partial permutations and partial alterations. This means that worst-case instances are rare and disappear under slight perturbations. On the other hand, we examine how much a perturbation can increase the height of a binary search tree, i.e. how much worse well balanced instances can become.

1 Introduction

To explain the discrepancy between average-case and worst-case behaviour of the simplex algorithm, Spielman and Teng introduced the notion of *smoothed analysis* [5]. Smoothed analysis interpolates between average-case and worst-case analysis: Instead of taking the worst-case instance or, as in average-case analysis, choosing an instance completely at random, we analyse the complexity of (worst-case) objects subject to slight random perturbations, i.e. the expected complexity in a small neighbourhood of (worst-case) instances. Smoothed analysis takes into account that a typical instance is not necessarily a random instance and that worst-case instances are often artificial and rarely occur in practice.

Let C be some complexity measure. The worst-case complexity is $\max_x C(x)$, and the average-case complexity is $\mathbb{E}_{x \sim \Delta} C(x)$, where $\mathbb{E}$ denotes the expectation with respect to some probability distribution Δ. The smoothed complexity is defined as $\max_x \mathbb{E}_{y \sim \Delta(x,p)} C(y)$. Here, x is chosen by an adversary and y is randomly chosen according to some probability distribution $\Delta(x, p)$ that depends

* A full version of this work with all proofs and experimental data is available as Report 05-063 of the Electronic Colloquium on Computational Complexity (ECCC).

** Supported by DFG research grant RE 672/3.

X. Deng and D. Du (Eds.): ISAAC 2005, LNCS 3827, pp. 483–492, 2005.

on x and a parameter p. The distribution $\Delta(x,p)$ should favour instances in the vicinity of x, i.e. $\Delta(x,p)$ should put almost all weight on the neighbourhood of x, where "neighbourhood" has to be defined appropriately depending on the problem considered. The smoothing parameter p denotes how strong x is perturbed, i.e. we can view it as a parameter for the size of the neighbourhood. Intuitively, for $p = 0$, smoothed complexity becomes worst-case complexity, while for large p, smoothed complexity becomes average-case complexity.

The smoothed complexity of continuous problems seems to be well understood. There are, however, only few results about smoothed analysis of discrete problems. For such problems, even the term "neighbourhood" is often not well defined. Thus, special care is needed when defining perturbation models for discrete problems. Perturbation models should reflect "natural" perturbations, and the probability distribution for an instance x should be concentrated around x, particularly for small values of the smoothing parameter p.

Here, we will conduct a smoothed analysis of an ordering problem, namely the *smoothed height of binary search trees.* Binary search trees are one of the most fundamental data structures and, as such, building blocks for many advanced data structures. The main criteria of the "quality" of a binary search tree is its height. Unfortunately, the height is equal to the number of elements in the worst case, i.e. for totally unbalanced trees generated by an ordered sequence of elements. On the other hand, if a binary search tree is chosen at random, then the expected height is only logarithmic in the number of elements. Thus, there is a huge discrepancy between the worst-case and the average-case behaviour of binary search trees.

We analyse what happens in between: An adversarial sequence will be perturbed randomly and then the height of the binary search tree generated by the sequence thus obtained is measured. Thus, our instances are neither adversarial nor completely random.

The height of a binary search tree obtained from a sequence of elements depends only on the ordering of the elements. Therefore, one should use a perturbation model that slightly perturbs the order of the elements of the sequence. We consider the perturbation models *partial permutations*, *partial alterations*, and *partial deletions*. For all three, we show tight lower and upper bounds. As a by-product, we obtain tight bounds for the smoothed number of left-to-right maxima, which is the number of new maxima seen when scanning a sequence from the left to the right. This improves a result by Banderier et al. [1].

In smoothed analysis one analyses how fragile worst-case instances are. We suggest examining also the dual property: Given a good instance, how much can the complexity increase by slightly perturbing the instance? In other words, how stable are best-case instances? We show that there are best-case instances that indeed are not stable, i.e. there are sequences that yield trees of logarithmic height, but slightly perturbing them yields trees of polynomial height.

Existing Results. Spielman and Teng introduced smoothed analysis as a hybrid of average-case and worst-case complexity [5]. Since then, smoothed analysis has been applied to a variety of fields [4].

Banderier, Beier, and Mehlhorn [1] applied smoothed analysis to ordering problems. In particular, they analysed the number of left-to-right maxima of a sequence. Here the worst case is the sequence $1, 2, \ldots, n$, which yields n left-to-right maxima. On average we expect $\sum_{i=1}^{n} 1/i \approx \ln n$ left-to-right maxima. Banderier et al. used the perturbation model of *partial permutations*, where each element of the sequence is independently selected with a probability of $p \in [0,1]$ and then a random permutation on the selected elements is performed. They proved that the number of left-to-right maxima under partial permutations is $O(\sqrt{(n/p)\log n})$ in expectation for $0 < p < 1$. Furthermore, they showed a lower bound of $\Omega(\sqrt{n/p})$ for $0 < p \leq 1/2$.

Given a sequence $\sigma = (\sigma_1, \sigma_2, \ldots, \sigma_n)$ of n distinct elements from any ordered set, we obtain a binary search tree $T(\sigma)$ by iteratively inserting $\sigma_1, \sigma_2, \ldots, \sigma_n$ into the initially empty tree (this is formally described in Section 2). The study of binary search trees is one of the most fundamental problems in computer science since they are the building blocks for a large variety of data structures.

The worst-case height of a binary search tree is obviously n: just take the sequence $\sigma = (1, 2, \ldots, n)$. (We define the length of a path as the number of vertices.) The expected height of the binary search tree obtained from a random permutation (with all permutations being equally likely) has been the subject of a considerable amount of research in the past, culminating in Reed's result [3] that the expectation of the height is $\alpha \ln n + \beta \ln(\ln n) + O(1)$ with $\alpha \approx 4.31107$ being the larger root of $\alpha \ln(2e/\alpha) = 1$ and $\beta = \frac{3}{2\ln(\alpha/2)} \approx 1.953$. Drmota [2] and Reed [3] proved independently of each other that the variance of the height is $O(1)$.

Although the worst-case and average-case height of binary search trees are very well understood, nothing is known in between, i.e. when the sequences are not completely random, but the randomness is limited.

New Results. We consider the height of binary search trees subject to slight random perturbations, i.e. the expected height under limited randomness.

We consider three perturbation models, which will formally be defined in Section 3. *Partial permutations*, introduced by Banderier et al. [1], rearrange some elements, i.e. they randomly permute a small subset of the elements of the sequence. The other two perturbation models are new. *Partial alterations* do not move elements, but replace some elements by new elements chosen at random. Thus, they change the rank of the elements. *Partial deletions* remove some of the elements of the sequence without replacement, i.e. they shorten the input. This model turns out to be useful for analysing the other two models.

We prove matching lower and upper bounds for the expected height of binary search trees under all three perturbation models (Section 5). More precisely: For all $p \in (0,1)$ and all sequences of length n, the expectation of the height of a binary search tree obtained via p-partial permutation is at most $6.7 \cdot (1-p) \cdot \sqrt{n/p}$ for sufficiently large n. On the other hand, the expected height of a binary search tree obtained from the sorted sequence via p-partial permutation is at least $0.8 \cdot (1-p) \cdot \sqrt{n/p}$. This lower bound matches the upper bound up to a constant factor.

For the number of left-to-right maxima under partial permutations, we are able to prove an even better upper bound of $3.6 \cdot (1-p) \cdot \sqrt{n/p}$ for all sufficiently large n and a lower bound of $0.4 \cdot (1-p) \cdot \sqrt{n/p}$ (Section 4).

All these bounds hold for partial alterations as well.

For partial deletions, we obtain $(1-p) \cdot n$ both as lower and upper bound.

In smoothed analysis one analyses how fragile worst case instances are. We suggest examining also the dual property: Given a good instance, how much can the complexity increase if the instance is perturbed slightly?

The main reason for considering partial deletions is that we can bound the expected height under partial alterations and permutations by the expected height under partial deletions (Section 6). The converse holds as well, we only have to blow up the sequences quadratically.

We exploit this when considering the stability of the perturbation models in Section 7: We prove that partial deletions and, thus, partial permutations and partial alterations as well can cause best-case instances to become much worse. More precisely: There are sequences of length n that yield trees of height $O(\log n)$, but the expected height of the tree obtained after smoothing the sequence is $n^{\Omega(1)}$.

2 Preliminaries

For any $n \in \mathbb{N}$, let $[n] = \{1, 2, \ldots, n\}$ and $[n - \frac{1}{2}] = \{\frac{1}{2}, \frac{3}{2}, \ldots, n - \frac{1}{2}\}$.

Let $\sigma = (\sigma_1, \ldots, \sigma_n) \in S^n$ for some ordered set S. We call σ a **sequence** of length n. Usually, we assume that all elements of σ are distinct. In most cases, σ will simply be a permutation of $[n]$. We denote the sorted sequence $(1, 2, \ldots, n)$ by σ^n_{sort}. When considering partial alterations, we define $\sigma^n_{\text{sort}} = (\frac{1}{2}, \frac{3}{2}, \ldots, n - \frac{1}{2})$ instead (this will be clear from the context).

Let $\sigma = (\sigma_1, \ldots, \sigma_n)$ be a sequence. We obtain a **binary search tree $T(\sigma)$** from σ by iteratively inserting the elements $\sigma_1, \sigma_2, \ldots, \sigma_n$ into the initially empty tree as follows: The root of $T(\sigma)$ is the first element σ_1 of σ. Let $\sigma_< = \sigma_{\{i|\sigma_i < \sigma_1\}}$ be σ restricted to elements smaller than σ_1. The left subtree of the root σ_1 of $T(\sigma)$ is obtained inductively from $\sigma_<$. Analogously, let $\sigma_> = \sigma_{\{i|\sigma_i > \sigma_1\}}$ be σ restricted to elements greater than σ_1. The right subtree of σ_1 of $T(\sigma)$ is obtained inductively from $\sigma_>$. Figure 1 shows an example. We denote the height of $T(\sigma)$, i.e. the number of nodes on the longest path from the root to a leaf, by **height(σ)**.

The element σ_i is called a **left-to-right maximum** of σ if $\sigma_i > \sigma_j$ for all $j \in [i-1]$. Let **ltr(σ)** denote the number of left-to-right maxima of σ. We have $\text{ltr}(\sigma) \leq \text{height}(\sigma)$ since the number of left-to-right maxima of a sequence is equal to the length of the right-most path in the tree $T(\sigma)$.

3 Perturbation Models for Permutations

Since we deal with ordering problems, we need perturbation models that slightly change a given permutation of elements. There seem to be two natural

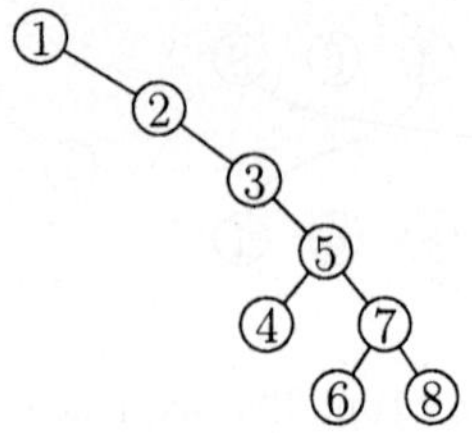

Fig. 1. The tree $T(\sigma)$ obtained from $\sigma = (1, 2, 3, 5, 7, 4, 6, 8)$. We have height$(\sigma) = 6$

possibilities: Either *change the positions* of some elements, or *change the elements* themselves.

Partial permutations implement the first option: A subset of the elements is randomly chosen, and then these elements are randomly permuted.

The second possibility is realised by partial alterations. Again, a subset of the elements is chosen randomly. These elements are then replaced by random elements.

The third model, partial deletions, also starts by randomly choosing a subset of the elements. These elements are then removed without replacement.

For all three models, we obtain the random subset as follows. Let σ be a sequence of length n and $p \in [0, 1]$ be a probability. Every element of σ is marked independently of the others with probability p.

By **height-perm$_p(\sigma)$**, **height-alter$_p(\sigma)$**, and **height-del$_p(\sigma)$** we denote the expected height of the binary search tree $T(\sigma')$, where σ' is the sequence obtained from σ by performing a p-partial permutation, alteration, and deletion, respectively (all three models will be defined formally in the following). Analogously, by **ltr-perm$_p(\sigma)$**, **ltr-alter$_p(\sigma)$**, and **ltr-del$_p(\sigma)$** we denote the expected number of left-to-right maxima of the sequence σ' obtained from σ via p-partial permutation, alteration, and deletion, respectively.

Partial Permutations. The notion of ***p*-partial permutations** was introduced by Banderier et al. [1]. Given a random subset M_p^n of $[n]$, the elements at positions in M_p^n are permuted according to a permutation drawn uniformly at random: Let $\sigma = (\sigma_1, \dots, \sigma_n)$. Then the sequence $\sigma' = \Pi(\sigma, M_p^n)$ is a random variable with the following properties:

- Π chooses a permutation π of M_p^n uniformly at random and
- sets $\sigma'_{\pi(i)} = \sigma_i$ for all $i \in M_p^n$ and $\sigma'_i = \sigma_i$ for all $i \notin M_p^n$.

Figure 2 shows an example of a partial permutation.

By varying p, we can interpolate between the average and the worst case: for $p = 0$, no element is marked and $\sigma' = \sigma$, while for $p = 1$, σ' is a random permutation of the elements of σ with all permutations being equally likely.

Partial permutations are a suitable perturbation model since the distribution of $\Pi(\sigma, M_p^n)$ favours sequences close to σ. To show this, we have to introduce a metric on sequences. Let σ and τ be two sequences of length n. We assume that

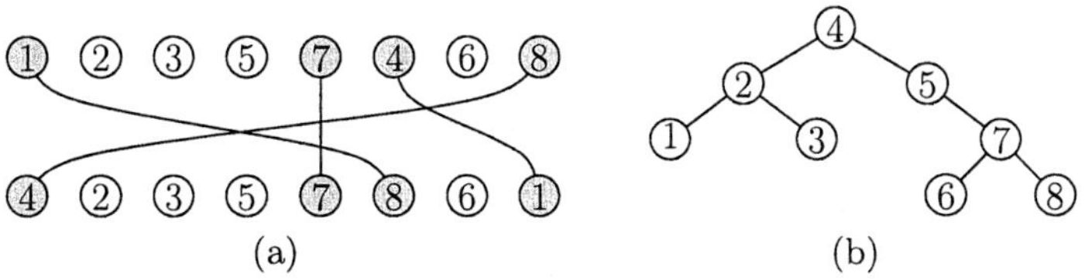

Fig. 2. A partial permutation. (a) Top: The sequence $\sigma = (1, 2, 3, 5, 7, 4, 6, 8)$; Figure 1 shows $T(\sigma)$. The first, fifth, sixth, and eighth element is (randomly) marked, thus $M_p^n = \{1, 5, 6, 8\}$. Bottom: The marked elements are randomly permuted. The result is the sequence $\sigma' = \Pi(\sigma, \mu)$, in this case $\sigma' = (4, 2, 3, 5, 7, 8, 6, 1)$. (b) $T(\sigma')$ with $\text{height}(\sigma') = 4$.

both are permutations of $[n]$ and define the distance $d(\sigma, \tau)$ between σ and τ as $d(\sigma, \tau) = |\{i \mid \sigma_i \neq \tau_i\}|$, thus d is a metric.

The distribution of $\Pi(\sigma, M_p^n)$ is symmetric around σ with respect to d. Furthermore, the probability that $\Pi(\sigma, M_p^n)$ equals some τ decreases exponentially with $d(\sigma, \tau)$. Thus, the distribution of $\Pi(\sigma, M_p^n)$ is highly concentrated around σ.

Partial Alterations. Let us now introduce ***p*-partial alterations**. For this perturbation model, we restrict the sequences of length n to be permutations of $[n - \frac{1}{2}] = \{\frac{1}{2}, \frac{3}{2}, \ldots, n - \frac{1}{2}\}$.

Every element at a position in M_p^n is replaced by a real number drawn uniformly and independently at random from $[0, n)$ to obtain a sequence σ'. All elements in σ' are distinct with probability one.

Instead of considering permutations of $[n - \frac{1}{2}]$, we could also consider permutations of $[n]$ and draw the random values from $[\frac{1}{2}, n + \frac{1}{2})$. This would not change the results. Another possibility would be to consider permutations of $[n]$ and draw the random values from $[0, n + 1)$. This would not change the results by much either. However, for technical reasons, we consider partial alterations as introduced above.

Like partial permutations, partial alterations interpolate between the worst case ($p = 0$) and the average case ($p = 1$). Partial alterations are somewhat easier to analyse: The majority of results on the average-case height of binary search trees is actually not obtained by considering random permutations. Instead, the binary search trees are grown from a sequence of n random variables that are uniformly and independently drawn from $[0, 1)$. This corresponds to partial alterations for $p = 1$. There is no difference between partial permutations and partial alterations for $p = 1$. This appears to hold for all p in the sense that the lower and upper bounds obtained for partial permutations and partial alterations are equal for all p.

Partial Deletions. As the third perturbation model, we introduce ***p*-partial deletions**: Again, we have a random marking M_p^n. Then we remove all marked elements.

Partial deletions do not really perturb a sequence: any ordered sequence remains ordered even if elements are deleted. The reason for considering partial deletions is that they are easy to analyse when considering the stability of

perturbation models (Section 7). The results obtained for partial deletions then carry over to partial permutations and partial alterations since the expected heights with respect to these three models are closely related (Section 6).

4 Tight Bounds for the Number of Left-to-Right Maxima

Partial Permutations. The main idea for proving the following theorem is to estimate the probability that one of the k largest elements of σ is among the first k elements, which would bound the number of left-to-right maxima by $2k$.

Theorem 1. *Let $p \in (0,1)$. Then for all sufficiently large n and for all sequences σ of length n,*

$$\text{ltr-perm}_p(\sigma) \leq 3.6 \cdot (1-p) \cdot \sqrt{n/p}\,.$$

The following lemma is an improvement of the lower bound proof for the number of left-to-right maxima under partial permutations presented by Banderier et al. [1]. We obtain a lower bound with a much larger constant that holds for all $p \in (0,1)$; the lower bound provided by Banderier et al. holds only for $p \leq 1/2$.

The idea of the proof is as follows. Let $K_c = c\sqrt{n/p}$ and let $\sigma = (n - K_c + 1, \ldots, n, 1, \ldots, n - K_c)$. The probability that none of the first K_c elements of σ, which are also the K_c largest elements of σ, is moved further to the front is bounded from below by $\exp(-c^2/\alpha)$ for any fixed $\alpha > 1$. In such a case, all unmarked elements of the first K_c elements are left-to-right maxima.

Lemma 1. *Let $p \in (0,1)$, $\alpha > 1$, and $c > 0$. For all sufficiently large n, there exists a sequence σ of length n with* $\text{ltr-perm}_p(\sigma) \geq \exp(-c^2\alpha) \cdot c \cdot (1-p) \cdot \sqrt{n/p}$.

We obtain the strongest lower bound from Lemma 1 by choosing α close to 1 and $c = 1/\sqrt{2\alpha}$. This yields the following theorem.

Theorem 2. *For all $p \in (0,1)$ and all sufficiently large n, there exists a sequence σ of length n with*

$$\text{ltr-perm}_p(\sigma) \geq 0.4 \cdot (1-p) \cdot \sqrt{n/p}\,.$$

Theorem 2 also yields the same lower bound for $\text{height-perm}_p(\sigma)$ since the number of left-to-right maxima of a sequence is a lower bound for the height of the binary search tree obtained from that sequence. We can, however, prove a stronger lower bound for the smoothed height of binary search trees (Theorem 4).

A consequence of Lemma 1 is that there is no constant c such that the number of left-to-right maxima is at most $c \cdot (1-p) \cdot \sqrt{n/p}$ with high probability, i.e. with a probability of at least $1 - n^{-\Omega(1)}$. Thus, the bounds proved for the expected tree height or the number of left-to-right maxima cannot be generalised to bounds that hold with high probability. However, we can prove that with high probability, the height under partial permutations is $O(\sqrt{(n/p) \cdot \log n})$. Clearly, this bound holds for the number of left-to-right maxima as well.

Partial Alterations. Similar to the results for partial permutations, we obtain an upper bound of $3.6 \cdot (1-p) \cdot \sqrt{n/p}$ and a lower bound of $0.4 \cdot (1-p) \cdot \sqrt{n/p}$.

Again, we cannot achieve a bound of $O((1-p) \cdot \sqrt{n/p})$ for the number of left-to-right maxima that holds with high probability, but we can show that the height after partial alteration is $O(\sqrt{(n/p) \cdot \log n})$ with high probability.

5 Tight Bounds for the Height of Binary Search Trees

Partial Permutations. The following theorems is one of the main results of this work. The idea for proving it is as follows: We divide the sequence into blocks $B_1, B_2, \ldots$, where B_d is of size $cd^2\sqrt{n/p}$ for some $c > 0$. Each block B_d is further divided into d^4 parts $A_d^1, \ldots, A_d^{d^4}$, each consisting of $cd^{-2}\sqrt{n/p}$ elements. Assume that on every root-to-leaf path in the tree obtained from the perturbed sequence, there are elements of at most two such A_d^i for every d. Then the height can be bounded from above by $\sum_{d=1}^{\infty} 2 \cdot cd^{-2}\sqrt{n/p} = (c\pi^2/3)\sqrt{n/p}$.

The probability for such an event is roughly $O(\exp(-c^2)^2/(1-\exp(-c^2)))$. We obtain the upper bound claimed in the theorem mainly by carefully applying this bound and by exploiting the fact that only a fraction of $(1-p)$ of the elements are unmarked. Marked elements contribute at most $O(\log n)$ to the expected height of the tree.

Theorem 3. *Let $p \in (0,1)$. Then for all sufficiently large n and all sequences σ of length n,*

$$\text{height-perm}_p(\sigma) \leq 6.7 \cdot (1-p) \cdot \sqrt{n/p}\,.$$

We can also prove the following bound for the tree height: With probability $1 - n^{-\Omega(1)}$, the height is at most $O(\sqrt{(n/p) \cdot \log n})$. More precisely: The probability that the height after partial permutation is at most $c \cdot \sqrt{(n/p) \cdot \log n}$ is at least $1 - n^{-(c/3)^2/\alpha + 0.5}$ for sufficiently large n and arbitrary $\alpha > 1$.

As a counterpart to Theorem 3, we prove the following lower bound. Interestingly, the lower bound is obtained for the sorted sequence, which is not the worst case for the expected number of left-to-right maxima under partial permutation; the expected number of left-to-right maxima of the sequence obtained by partially permuting the sorted sequence is only logarithmic [1].

Theorem 4. *Let $p \in (0,1)$. Then for all sufficiently large $n \in \mathbb{N}$,*

$$\text{height-perm}_p(\sigma_{\text{sort}}^n) \geq 0.8 \cdot (1-p) \cdot \sqrt{n/p}\,.$$

Partial Alterations. The results proved for partial permutation can be carried over to partial alterations. This means particularly that we obtain the same upper bound of $\text{height-alter}_p(\sigma) \leq 6.7 \cdot (1-p) \cdot \sqrt{n/p}$ for all $p \in (0,1)$ and all sequences σ with elements from $[n - \frac{1}{2}]$ for sufficiently large n.

Furthermore, we obtain the same upper bound of $n^{-(c/3)^2/\alpha+0.5}$ on the probability that the height after partial alteration is greater than $c \cdot \sqrt{(n/p) \cdot \log n}$

Finally, we have $\text{height-alter}_p(\sigma_{\text{sort}}^n) \geq 0.8 \cdot (1-p) \cdot \sqrt{n/p}$ for all $p \in (0,1)$ and sufficiently large n.

6 Partial Deletions Versus Permutations and Alterations

For partial deletions, we easily obtain $\text{height-del}_p(\sigma) \leq (1-p)\cdot n$ for all sequences σ of length n and all $p \in [0,1]$ as an upper bound and $\text{height-del}_p(\sigma^n_{\text{sort}}) = (1-p)\cdot n$ as a lower bound.

Partial deletions are in some sense the worst of the three models: Trees are usually expected to be higher under partial deletions than under partial permutations or alterations, even though they contain fewer elements. The expected height under partial deletions yields upper bounds (up to an additional $O(\log n)$ term) for the expected height under partial permutations and alterations. The same holds for the number of left-to-right maxima.

Lemma 2. *For all sequences σ of length n and $p \in [0,1]$,* $\text{height-perm}_p(\sigma) \leq \text{height-del}_p(\sigma) + O(\log n)$. *If σ is a permutation of $[n-\frac{1}{2}]$, then* $\text{height-alter}_p(\sigma) \leq \text{height-del}_p(\sigma) + O(\log n)$.

The converse is not true, but we can bound the expected height under partial deletions by the expected height under partial permutations or alterations by padding the sequences considered. The following lemma holds also for partial alterations if the sequence σ is a permutation of $[n-\frac{1}{2}]$.

Lemma 3. *Let $p \in (0,1)$ be fixed and let σ be a sequence of length n with* $\text{height}(\sigma) = d$ *and* $\text{height-del}_p(\sigma) = d'$. *Then there exists a sequence $\tilde{\sigma}$ of length $O(n^2)$ with* $\text{height}(\tilde{\sigma}) = d + O(\log n)$ *and* $\text{height-perm}_p(\tilde{\sigma}) \in \Omega(d')$.

7 The (In-)Stability of Perturbations

Having shown that worst-case instances become much better when smoothed, we now provide a family of best-case instances for which smoothing results in an exponential increase in height. We consider the following family of sequences: $\sigma^{(1)} = (1)$ and $\sigma^{(k+1)} = (2^k, \sigma^{(k)}, 2^k + \sigma^{(k)})$, where $c + \sigma = (c+\sigma_1, \ldots, c+\sigma_n)$ for a sequence σ of length n. For instance, $\sigma^{(3)} = (4,2,1,3,6,5,7)$. Let $n = 2^k - 1$. Then $\sigma^{(k)}$ contains the numbers $1, 2, \ldots, n$, and we have $\text{height}(\sigma^{(k)}) = \text{ltr}(\sigma^{(k)}) = k \in \Theta(\log n)$.

Deleting the first element of $\sigma^{(k)}$ roughly doubles the number of left-to-right maxima in the resulting sequence. This is the idea behind the following theorem.

Theorem 5. *For all $p \in (0,1)$ and all $k \in \mathbb{N}$,* $\text{ltr-del}_p(\sigma^{(k)}) = \frac{1-p}{p}\cdot((1+p)^k - 1)$.

Since the number of left-to-right maxima is a lower bound for the height of a binary search tree, we obtain $\text{height-del}_p(\sigma^{(k)}) \geq \frac{1-p}{p}\cdot((1+p)^k - 1)$.

We conclude that there are some best-case instances that are quite fragile under partial deletions: From logarithmic height they "jump" via smoothing to a height of $\Omega(n^{\log(1+p)})$. (We have $\frac{1-p}{p}\cdot((1+p)^k - 1) \in \Theta(n^{\log(1+p)})$.)

We can transfer this result to partial permutations due to Lemma 3. The result holds also for partial alterations. This means that there are sequences that yield trees of height $O(\log n)$, but perturbing them with partial permutations or partial alterations yields trees of height $\Omega(n^\delta)$ for some fixed $\delta > 0$.

8 Conclusions

We have analysed the height of binary search trees obtained from perturbed sequences and obtained asymptotically tight bounds of roughly $\Theta(\sqrt{n})$ for the height under partial permutations and alterations. This stands in contrast to both the worst-case and the average-case height of n and $\Theta(\log n)$, respectively.

Interestingly, the sorted sequence turns out to be the worst-case for the smoothed height of binary search trees in the sense that the lower bounds are obtained for σ^n_{sort}. This is in contrast to the fact that the expected number of left-to-right maxima of σ^n_{sort} under p-partial permutations is roughly $O(\log n)$ [1]. We believe that for binary search trees, σ^n_{sort} is indeed the worst case.

We performed experiments to estimate the constants in the bounds for the height of binary search trees. The results led to the conjecture that the expected height of trees obtained by performing a partial permutation on σ^n_{sort} is $(\gamma+o(1))\cdot(1-p)\cdot\sqrt{n/p}$ for some $\gamma \approx 1.8$ and for all $p \in (0,1)$. Proving this conjecture would immediately improve our lower bound. Provided that the sorted sequence is indeed the worst case, this conjecture would also improve the upper bound for binary search trees and left-to-right maxima.

The bounds obtained in this work for partial permutations and partial alterations are equal. We suspect that this is always true for binary search trees.

Finally, we are interested in generalising these results to other problems based on permutations, like sorting (Quicksort under partial permutations has already been investigated by Banderier et al. [1]), routing, and other algorithms and data structures. Hopefully, this will shed some light on the discrepancy between the worst-case and average-case complexity of these problems.

References

1. Cyril Banderier, René Beier, and Kurt Mehlhorn. Smoothed analysis of three combinatorial problems. In Branislav Rovan and Peter Vojtáš, editors, *Proc. of the 28th Int. Symp. on Mathematical Foundations of Computer Science (MFCS)*, volume 2747 of *Lecture Notes in Computer Science*, pages 198–207. Springer, 2003.
2. Michael Drmota. An analytic approach to the height of binary search trees II. *Journal of the ACM*, 50(3):333–374, 2003.
3. Bruce Reed. The height of a random binary search tree. *Journal of the ACM*, 50(3):306–332, 2003.
4. Daniel A. Spielman. The smoothed analysis of algorithms. In Maciej Liśkiewicz and Rüdiger Reischuk, editors, *Proc. of the 15th Int. Symp. on Fundamentals of Computation Theory (FCT)*, volume 3623 of *Lecture Notes in Computer Science*, pages 17–18. Springer, 2005.
5. Daniel A. Spielman and Shang-Hua Teng. Smoothed analysis of algorithms: Why the simplex algorithm usually takes polynomial time. *Journal of the ACM*, 51(3):385–463, 2004.

Simple and Efficient Greedy Algorithms for Hamilton Cycles in Random Intersection Graphs*

C. Raptopoulos[1,2] and P. Spirakis[1,2]

[1] Computer Technology Institute, P.O. Box 1122, 26110 Patras, Greece
spirakis@cti.gr, raptopox@ceid.upatras.gr
[2] University of Patras, 26500 Patras, Greece

Abstract. In this work we consider the problem of finding Hamilton Cycles in graphs derived from the uniform random intersection graphs model $G_{n,m,p}$. In particular, (a) for the case $m = n^{\alpha}, \alpha > 1$ we give a result that allows us to apply (with the same probability of success) any algorithm that finds a Hamilton cycle with high probability in a $G_{n,k}$ graph (i.e. a graph chosen equiprobably form the space of all graphs with k edges), (b) we give an **expected polynomial time** algorithm for the case $p =$ constant and $m \leq \alpha\sqrt{\frac{n}{\log n}}$ for some constant α, and (c) we show that the greedy approach still works well even in the case $m = o(\frac{n}{\log n})$ and p just above the connectivity threshold of $G_{n,m,p}$ (found in [21]) by giving a greedy algorithm that finds a Hamilton cycle in those ranges of m, p with high probability.

1 Introduction

Random graphs, introduced by P. Erdös and A. Rényi, still continue to attract a huge amount of research and interest in the communities of Theoretical Computer Science, Graph Theory and Discrete Mathematics.

There exist various models of random graphs. The most famous is the $G_{n,p}$ random graph, a sample space whose points are graphs produced by randomly sampling the edges of a graph on n vertices independently, with the same probability p. Other models have also been quite a lot investigated: $G_{n,r}$ (the "random regular graphs", produced by randomly and equiprobably sampling a graph from all regular graphs of n vertices and vertex degree r) and $G_{n,k}$ (produced by randomly and equiprobably selecting an element of the class of graphs on n vertices having k edges). For an excellent survey of these models, see [1, 3].

In this work we address the problem of finding Hamilton cycles in the Uniform Random Intersection Graphs Model $G_{n,m,p}$, which was introduced by

* This work has been partially supported by the IST Programme of the European Union under contract number 001907 (DELIS) and by the GSRT PENED 2003 ALGO.D.E.S. Project.

X. Deng and D. Du (Eds.): ISAAC 2005, LNCS 3827, pp. 493–504, 2005.

M. Karoński, E.R. Sheinerman and K.B. Singer-Cohen [13] and K.B. Singer-Cohen [21]. Also, Godehardt and Jaworski [11] considered similar models. The formal definition of this model is given below.

Definition 1 (Uniform random intersection graph). *Let us consider a universe $M = \{1, 2, \ldots, m\}$ of elements and a set of vertices $V = \{v_1, v_2, \ldots, v_n\}$. If we assign independently to each vertex v_j, $j = 1, 2, \ldots, n$, a subset S_{v_j} of M by choosing each element $i \in M$ independently with probability p and put an edge between two vertices v_{j_1}, v_{j_2} if and only if $S_{v_{j_1}} \cap S_{v_{j_2}} \neq \emptyset$, then the resulting graph is an instance of the uniform random intersection graph $G_{n,m,p}$. In this model we also denote by L_l the set of vertices that have chosen label $l \in M$.*

The representation matrix *of a random intersection graph $G_{n,m,p}$ is a matrix R that has n rows and m columns corresponding to the vertices and the labels of the graph respectively. Element R_{ij} equals 1 if the i-th vertex has selected the j-th label (for some prespecified ordering of the vertices and labels) and 0 otherwise.*

Importance and Motivation. First of all, we note that (as proved in [14]) any graph is a random intersection graph. Thus, the $G_{n,m,p}$ model is very general. Furthermore, for some ranges of the parameters m, p ($m = n^{\alpha}, \alpha > 6$) the spaces $G_{n,m,p}$ and $G_{n,p}$ are equivalent (as proved by Fill, Sheinerman and Singer-Cohen [10], showing that in this range the total variation distance between the graph random variables has limit 0).

Second, random intersection graphs may model real-life applications more accurately (compared to the $G_{n,p}$ case). This is because in many cases the independence of edges is not well-justified. In fact, objects that are closer (like moving hosts in mobile networks or sensors in smart dust networks) are more probable to interact with each other. Even epidemiological phenomena (like spread of disease) tend to be more accurately captured by this "proximity-sensitive" random intersection graphs model. Other applications may include oblivious resource sharing in a distributed setting, interactions of mobile agents traversing the web etc.

Other Related Work. The question of how close $G_{n,m,p}$ and $G_{n,\hat{p}}$ are for various values of m, p (the value of $\hat{p}$ is also given as a function of m, p) has been studied by Fill, Sheinerman and Singer-Cohen in [10]. In [17] the authors introduce two variations of the uniform random intersection graphs model, i.e. the general and the regular random intersection graph models, and address the problem of finding large independent sets of vertices in these models. Also, geometric proximity between randomly placed objects is nicely captured by the model of random geometric graphs (see e.g. [6, 8, 19]) and important variations (like random scaled sector graphs, [7]). Other extensions of random graph models (such as random regular graphs) and several important combinatorial properties (connectivity, expansion, existence of a giant connected component) are performed in [16, 18]. Thresholds for the existence of Hamilton cycles in random intersection graphs were given by members of our team in [9].

An expected polynomial time algorithm for finding a Hamilton path in a $G_{n,p}$ graph with $p \geq \frac{1}{2}$ was developed in [12]. In [22] Thomason gave a linear expected

time algorithm for the Hamilton cycle problem in the case where $p > 0$ is any constant probability. Also the authors in [4] propose a polynomial time algorithm that finds a Hamiltonian cycle in a random graph with high probability. Other $\mathcal{NP}$-hard problems for which expected polynomial time algorithms exist when the input follows some specific distribution are the colouring of random graphs [5], the random knapsack [2] etc.

Our Contribution.

1. We first make a reduction of the $G_{n,m,p}$ model to the model $G_{n,k}$ (i.e. the model where we randomly and equiprobably select an element of the class of graphs on n vertices and k edges) in the case where $m = n^\alpha, \alpha > 1$. In particular, for this range of m and p we give a way to apply any result (algorithmic and combinatoric) concerning increasing properties of $G_{n,k}$ graphs to the $G_{n,m,p}$ graphs.
2. We give a polynomial expected time algorithm for finding Hamilton cycles in a $G_{n,m,p}$ graph, in the case where p is constant and the number of labels is at most $\alpha\sqrt{\frac{n}{\log n}}$ for some constant α. The algorithm uses first a randomized greedy algorithm and if it fails it then uses an exhaustive algorithm to solve the problem.
3. Finally we show that the greedy approach gives an algorithm that succeeds with high probability in the case $m = o(\frac{n}{\log n})$ and p is not constant. We give a polynomial time randomized algorithm that finds with high probability a Hamilton cycle in the case where p is just above the connectivity threshold for these graphs. This algorithm also serves as a way to find a quite tight bound on the probability p that ensures that with high probability the uniform random intersection graph has a Hamilton cycle.

2 A Reduction to the $G_{n,k}$ Model

Let $G_{n,1,p}$ be a random intersection graph with only one label. It is obvious from the definition of the random intersection graphs model that this graph will either contain a single clique, or it will be the empty graph. We note that given that the $G_{n,1,p}$ graph has a clique of size k, then this clique is equiprobably any of the $\binom{n}{k}$ cliques of size k. Let now $\hat{p}$ denote the probability that the $G_{n,1,p}$ graph has at least one edge. Then, for $np \to 0$ we get

$$\hat{p} \stackrel{\text{def}}{=} P\{G_{n,1,p} \text{ is non-empty}\} = 1 - (1-p)^n - np(1-p)^{n-1} \sim \frac{1}{2}n^2p^2.$$

Now, let us return to the case of m labels. We construct a graph H in the following way:

1. Initially H is the empty graph with n vertices.
2. In each step $i = 1, 2, \ldots, m$ we independently add a single edge to H with probability $\frac{1}{2}n^2(1+\epsilon)^2p^2$. This edge can be any of the $\binom{n}{2}$ possible edges of H with equal probability. Also, with probability $1 - \frac{1}{2}n^2(1+\epsilon)^2p^2$ we do nothing.

We note that the graph H constructed above is a multigraph, since a single edge may appear more than once. Furthermore, given that H has exactly k edges, then the graph H is with equal probability any of the multigraphs with k edges. The mean number of all edges of H is $m\frac{1}{2}n^2(1+\epsilon)^2p^2$. By using Chernoff bounds we get for any constant β in $(0,1)$ that

$$P\{\#\text{ of all edges of } H \leq (1-\beta)m\frac{1}{2}n^2(1+\epsilon)^2p^2\ \} \leq e^{-\beta^2 mn^2(1+\epsilon)^2p^2/4}.$$

So, if we choose $mn^2(1+\epsilon)^2p^2 \to \infty$, then with high probability H will have at least $(1-\beta)m\frac{1}{2}n^2(1+\epsilon)^2p^2$ edges (counting multiple eges as many times as their multiplicity).

Let now $e(H)$ be the number of different edges of H (i.e. we count multiple edges as one). Then, by a slightly modified coupon collector's problem (see Appendix in the full version of this paper [20]), we can see that $e(H)$ must satisfy the inequality $e(H) \geq \frac{1-\beta}{1+\gamma}mn^2(1+\epsilon)^2p^2$, with probability at least $1-e^{-\beta^2 mn^2(1+\epsilon)^2p^2/4} - \frac{1}{\gamma^2\binom{n}{2}} \to 1$, where γ is any positive constant.

We note here that given that $e(H) = k$, the graph H is with equal probability any graph with n vertices and k edges, hence it is distributed like $G_{n,k}$. We can therefore prove the following:

Theorem 1. *Let Q be an increasing property on the number of edges and suppose that for some integer k we have*

$$P\{G_{n,k}\ \textit{has property}\ Q\} = 1 - g(n)$$

where $g(n) = o(1)$. Let also m, p, ϵ be such that $m = n^\alpha$, $\alpha > 1$, $np \to 0$, ϵ is a (small) positive constant and

$$k \leq \frac{1-\beta}{1+\gamma}mn^2(1+\epsilon)^2p^2 \tag{1}$$

where β, γ are any two small positive constants. Then

$$P\{H\ \textit{has property}\ Q\} \geq 1 - g(n) - e^{-\beta^2 mn^2(1+\epsilon)^2p^2/4} - \frac{1}{\gamma^2\binom{n}{2}} \to 1$$

We now use Theorem 1 for the case where Q is the property "existence of a Hamilton cycle". We can prove the following

Corollary 1. *If $m = n^\alpha$, $\alpha > 1$, $p = \sqrt{\frac{\log n}{nm}}$ and ϵ is a small constant* [1], *then*

$$P\{G_{n,m,(1+2\epsilon)p}\ \textit{has a Hamilton cycle}\} \geq 1 - g(n) - e^{-\beta^2 mn^2(1+\epsilon)^2p^2/4} - \frac{1}{\gamma^2\binom{n}{2}} \to 1.$$

[1] In fact the constant ϵ can be as small as to ensure that the constant quantity $\frac{1-\beta}{1+\gamma}(1+\epsilon)^2$ is as close to 1 as possible (but greater than 1). Since β, γ are small (and can be controlled to become as small as possible), we can get a quite small value for ϵ.

Proof. See full paper [20]. □

We finally note that, as shown in [21], the value $p = \sqrt{\frac{\log n}{nm}}$ is exactly the connectivity threshold of a random intersection graph with $m = n^{\alpha}$, $\alpha > 1$.

3 Hamilton Cycles in $G_{n,m,p}$ with Constant p

In this section we provide an expected polynomial time algorithm for Hamilton Cycles in random intersection graphs with constant p and $m \leq \alpha\sqrt{\frac{n}{\log n}}$, where $\alpha < \frac{\beta p}{\sqrt{2}}$ and $\beta \in (0,1)$ are positve constants. By a straightforward analysis (see Appendix in the full version of this paper [20]) we can show that even in this restricted range of values of p, m the problem of finding a hamilton cycle in $G_{n,m,p}$ is non-trivial. Before presenting the algorithm we give some preliminary results.

It is easy to see that if the $G_{n,m,p}$ has a Hamilton cycle, then there exists a sequence

$$HC := l_1 \rightarrow v_1 \rightarrow l_2 \rightarrow \cdots \rightarrow l_k \rightarrow v_k \rightarrow l_{k+1} (= l_1) \qquad (2)$$

that satisfies the following four conditions:

1. $l_i \in M$ and $v_i \in V$,
2. $\bigcup_{i=1}^{k} L_{l_i} = V$,
3. $l_i, l_{i+1} \in S_{v_i}$, $i = 1, 2, \ldots, k$ and
4. $v_i \neq v_j$, for any $i \neq j$.

We note here that given an HC sequence of the form (2) that satisfies the four constraints above we can construct a Hamilton cycle in time $\Theta(n \cdot m)$ (we give such a construction in the algorithm that follows this section).

Suppose now that our graph has a Hamilton cycle hence there is a sequence HC of the form (2) satisfying the four constraints above. We will refer to the integer k as the *size* of the sequence HC. Let x denote the number of different labels used by HC and let HC_{min} be a sequence of the form (2) that satisfies the four conditions, uses exactly the labels used by HC and has minimum length. If we denote by k_{min} the size of HC_{min} then it is obvious that $x \leq k_{min}$. We can also prove the following:

Lemma 1. *If k_{min} and x are defined as above, then $k_{min} \leq 1 + \frac{x(x-1)}{2}$.*

Proof. See full paper [20]. □

Suppose now that we are interested in the number of different sequences HC_{min} that use x specific labels. Because of Lemma 1 we know that their size must be at most $1 + \frac{x(x-1)}{2}$. We now suppose that we have to fill in $1 + \frac{x(x-1)}{2}$ label-cells by choosing for each one any of x different labels and $1+\frac{x(x-1)}{2}$ vertex-cells by chosing for each any of n different vertices. In that way, sequences with size less than the maximum will correspond to some sequence of size $1 + \frac{x(x-1)}{2}$

that has at least one label that repeats itself in the following label-cell. We easily see then that the total number of ways to fill in these cells is an upper bound to the number of different HC_{min} sequences.

In view of the above, the number of sequences of the form (2) that an exhaustive algorithm that searches for a Hamilton cycle needs to check is at most

$$\sum_{x=1}^{m} \binom{m}{x} (x \cdot n)^{1+\frac{x(x-1)}{2}} \le \sum_{x=1}^{m} \frac{m^x}{x!} (x \cdot n)^{\frac{x^2}{2}} \le \exp\left\{m^2 \log n\right\} = n^{m^2}. \quad (3)$$

where in the last inequality we used the upper bound on m.

3.1 Expected Polynomial Time Algorithm for Constant p

In this section we use the previous results in order to present an algorithm for finding Hamilton Cycles in a $G_{n,m,p}$ graph with constant p and $m \le \alpha\sqrt{\frac{n}{\log n}}$, where $\alpha < \frac{\beta p}{\sqrt{2}}$ and $\beta \in (0,1)$ are positve constants.

The algorithm first uses a greedy algorithm (steps 1-10) to form a sequence HC sequence of the form (2) that uses all the labels and if this fails (which happens with some small probability of failure) it runs the exhaustive algorithm (step 12) implied in the previous section. Both the randomized algorithm and the exhaustive algorithm try to find a sequence HC of the form (2) that satisfies all four constraints. If either of them succeeds in finding one, then they run the following procedure that constructs a Hamilton cycle.

`procedure` CONSTRUCT_HAM(HC)

1. let $HC := l_1 \to v_1 \to l_2 \to \cdots \to l_k \to v_k \to l_{k+1}(= l_1)$;
2. $i = 1$; $A = V \backslash \bigcup_{i=1}^{k} \{v_i\}$;
3. `while` $i \le k$ `do`
4. let D_i be any ordered list of $L_{l_i} \cap A$;
5. $A = A \backslash \{D_i\}$; $i = i + 1$;
6. `output` $D_1 \to v_1 \to D_2 \to \cdots D_k \to v_k$;

We now show that if the HC sequence satisfies all four constraints, then the output of procedure CONSTRUCT_HAM(HC) is a Hamilton cycle. First, we note that since HC satisfies the 2nd constraint, all vertices are contained at most once in the output. Second, because of the definition of the sets D_i, step 5 of the algorithm and the 4th condition, each vertex is contained exacty once in the output. Third, because of the definition of the sets D_i and the 3rd condition, every two consecutive vertices of the output have at least one label in common, hence they are connected. Finally, we close the Hamilton cycle by noting that v_k has at least one label in common with any vertex in D_1 (i.e. label l_1).

The expected polnomial time algorithm is shown below. We denote by L_y the set of vertices having chosen label $y \in M$.

`Algorithm I:`
`Input:` The representation matrix $R_{n,m,p}$ of a graph $G_{n,m,p}$.
`Output:` A Hamilton cycle of the graph corresponding to $R_{n,m,p}$.

1. let $l_1, l_2, \ldots, l_m$ be a *random* ordering of the labels;
2. consider the sequence $l_1, l_2, \ldots, l_m, l_{m+1} = l_1$;
3. `if` $\bigcup_{i=1}^{m} L_{l_i} \neq V$ `then`
4.   `output` "The graph has no Hamilton Cycle"; `exit`;
5. $i = 1$; $A = V$; HC = empty list;
6. `while` $i < m + 1$ `do`
7.   `if` $L_{l_i} \cap L_{l_{i+1}} \cap A = \emptyset$ `then goto L1`;
8.   `select` a *random* vertex $v_i \in L_{l_i} \cap L_{l_{i+1}} \cap A$;
9.   `set` $HC = HC \rightarrow \{l_i\} \rightarrow \{v_i\}$;
10.   `set` $A = A \backslash \{v_i\}$; $i = i + 1$;
11. `goto L2`
12. `L1: if` there is a sequence of the form (2) and size $k \leq 1 + \frac{m(m-1)}{2}$ that satisfies all four costraints `then`
13.   let HC be such a sequence; `goto L2`;
14. `else`
15.   `output` "The graph has no Hamilton Cycle"; `exit`;
16. `L2:` CONSTRUCT_HAM(HC);

3.2 Analysis of Algorithm I

We will first find an upper bound for the probability of failure of the greedy part of the algorithm (lines 1-10). Notice that the greedy algorithm can fail to output a solution only because of step 7. That is, the algorithm fails only in the case where two consecutive labels in the sequence $l_1, l_2, \ldots, l_m, l_{m+1} = l_1$ do not have a common vertex that is also not used to "connect" two previous consecutive labels.

Let us now denote by X_{ij} the number of verices having chosen both labels $i, j \in M$. Obviously, X_{ij} is a random variable that is binomially distributed with parameters n and p^2. So, $E[X_{ij}] = np^2$. But from Chernoff bounds we have that for any constant β in $(0, 1)$

$$P\{X_{ij} \leq (1 - \beta)np^2\} \leq e^{-\beta^2 p^2 n/2}$$

and by applying Boole's inequality

$$P\{\exists i, j \in M : X_{ij} \leq (1 - \beta)np^2\} \leq \binom{m}{2} e^{-\beta^2 p^2 n/2} \overset{def}{=} \phi.$$

We have thus proven that with probability at least $1 - \binom{m}{2} e^{-\beta^2 p^2 n/2}$ every pair of labels has $\Theta(n)$ vertices in common. Bearing in mind that the vertices used by the greedy algorithm are exactly $m = o(n)$, every pair of labels is left with at least $\Theta(n) - o(n) = \Theta(n)$ vertices in common. So, the probability that

the greedy algorithm fails is at most $\binom{m}{2}e^{-\beta^2 p^2 n/2}$. Moreover, it is easy to see that the running time of the greedy algorithm is $O(n \cdot m)$ in the worst case.

If the greedy algorithm fails (which means that it does not output a Hamilton cycle nor does it output "The graph has no Hamilton Cycle") then we run the exhaustive part of the algorithm (included in step 12). In order to check if a particular sequence of the form (2) and of size at most $1 + \frac{x(x-1)}{2}$ satisfies the four constraints we need $O(n \cdot m)$ time in the worst case.

Finally, the running time time of the procedure CONSTRUCT_HAM is $O(n \cdot k)$ in the worst case, where k is the size of the HC sequence. If the HC sequence is the one produced by the greedy algorithm, then $k = m$, otherwise $k \leq 1 + \frac{m(m-1)}{2}$. Therefore, by using inequality (3), it is easy to see that

$$E[\text{running time of Algorithm I}] \leq (1 - \phi)O(n \cdot m) + \phi n^{m^2} O(n \cdot m)$$

which becomes $O(n \cdot m)$ for $m \leq \alpha\sqrt{\frac{n}{\log n}}$, where $\alpha < \frac{\beta p}{\sqrt{2}}$ is a constant. It follows then that Algorithm I runs in polynomial expected time.

4 A Second Greedy Algorithm for Smaller *p*

In this section we will see that the greedy approach still works quite well for smaller p, in the case where m is at most $o(\frac{n}{\log n})$. More specifically, we will present an algorithm that with high probability finds a Hamilton cycle in a $G_{n,m,p}$ graph, in the case where $m = o(\frac{n}{\log n})$ and p is just above the connectivity threshold of $G_{n,m,p}$, i.e. $p = \frac{\log n + h(n)}{m}$, for any function $h(n)$ that goes to ∞ as slowly as possible (see [21]).

Suppose that we partition the set of vertices V into two sets V_1, V_2 of $\frac{n}{2}$ vertices each. We then use only vertices in V_1 to make an HL sequence of the form

$$HL := l_1 \to v_1 \to l_2 \to \cdots \to l_k \tag{4}$$

that is as long as possible and satisfies the following conditions:

1. $l_i \in M$ and $v_i \in V_1$,
2. $l_i, l_{i+1} \in S_{v_i}$, $i = 1, 2, \ldots, k-1$,
3. $v_i \neq v_j$, for any $i \neq j$ and
4. $l_x \neq l_y$, for any $x \neq y$.

We use the following procedure in order to construct a sufficiently large HL sequence:

`procedure` MAKE_HL(V_1, M)

1. `set` $F_V = V_1$; `set` $F_M = M$;
2. `select a` *random* label $l \in M$; `set` $F_M = F_M \backslash \{l\}$; `set` $HL = l$;
3. `L1: if` $|L_l \cap F_V| < \log n$ `then goto Lo;`

```
4. else
5.     let v_1,...,v_{log n} ∈ L_l ∩ F_V;
6.     if ∃l', v : l' ∈ F_M, v ∈ {v_1,...,v_{log n}} and l' ∈ S_v then
7.         set F_M = F_M\{l'}; F_V = F_V\{v_1,...,v_{log n}};
8.         set HL = HL → v → l'; set l = l'; goto L1;
9.     else goto Lo;
10. Lo: output HL
```

If we assume that we can perform a random selection of an element in a set of n items in constant time, the above procedure runs obviously in polynomial time. Indeed, the loop between steps 3 to 9 is executed at most $\frac{n}{2\log n}$ times because in step 7 we reduce the set F_V by $\log n$ vertices. Furthermore, the most expensive steps of the loop are the if-condition of step 3 which can take at most $n^2/4$ time steps (because F_V and F_M have at most $n/2$ vertices each) and the if-condition of step 6 which can take at most $m^2 \log n$ time steps (because F_M has at most m labels and each set S_v has at most m labels). Hence, taking into consideration that $m = o(\frac{n}{\log n})$, we have proved the following:

Lemma 2. *Procedure MAKE_HL terminates in at most $O(\frac{n^3}{\log n})$ time steps.*

As we will now see, procedure MAKE_HL produces with high probability an HL sequence of the form (4) that has the following additional properties:

1. The size of HL is at least $m - \frac{m}{\log n}$ and
2. $\bigcup_{i=1}^{k} l_i = V$.

We will prove these in that order. First we note that steps 6-9 of the procedure can be executed at most m times, since each time step 7 is executed, the set F_M loses one label. Hence, whenever step 3 is executed, the set F_V will always consist of at least $|V_1| - m\log n = \frac{n}{2} - o(n)$ vertices. Also, it is easy to see that the label l selected in step 6 of the procedure chooses each vertex in F_V independently with probability p. So, $|L_l|$ is stochastically greater than a random variable $X \sim \mathcal{B}(|V_1| - m\log n, p)$. Clearly, $E[X] = \mu = (|V_1| - m\log n)p \geq (\frac{n}{2} - m\log n)\frac{\log n}{m} = \Theta(\frac{n\log n}{2m})$.By Chernoff bounds and Boole's inequality we then have, for any constant β,

$$P\{\text{an execution of step 3 finds } |L_l \cap F_V| \leq (1-\beta)\mu\} \leq me^{-\beta^2\mu/2} \to 0.$$

We have thus proved that with probability at least $1 - \exp\left\{-\frac{\beta^2 n\log n}{6m}\right\} \to 1$, the procedure never stops due to the if-condition of step 3.

The only other way to end the procedure is by step 6 (and as we said before the number of executions of step 7 is at most m). Suppose then that at some point of the execution of the procedure the HL sequence has been extended so as to contain $m - k$ labels (i.e. the test of step 6 has been passed $m-k-1$ times), leaving exactly k "free" labels in F_M. It is true that these labels continue to

select each member of F_V independently with probability p. This means that the probability that none of the free labels contains some of the vertices $v_1, \ldots, v_{\log n}$ of step 5 is exactly $(1-p)^{k \log n}$. Now, by Boole's inequality, for $k = \frac{m}{\log n}$, the probability that some of the $m-k-1$ first tests of step 6 fails is at most

$$m(1-p)^{k \log n} \leq \exp\{\log m - pm\} \leq \exp\{\log m - \log n\} \to 0.$$

So, with probability $1 - \exp\{\log m - \log n\} \to 1$ the HL sequence of procedure MAKE_HL(V_1, M) has size at least $m - \frac{m}{\log n}$.

Given now that the size of the HL sequence is at least $m - \frac{m}{\log n}$ we can see that the mean number of vertices not covered by (i.e. not contained in) any label of HL is at most

$$\begin{aligned} E[\#\text{ vertices not covered by } HL] &\leq n(1-p)^{m-\frac{m}{\log n}} \\ &\leq \exp\left\{\log n - mp + \frac{mp}{\log n}\right\} \to 0 \end{aligned}$$

for the values of m, p we have assumed. So, by Markov's inequality, the HL sequence of procedure MAKE_HL(V_1, M) covers all the vertices in V with probability at least $1 - \exp\left\{-h(n) + 1 + \frac{h(n)}{\log n}\right\} \to 1$. We have thus proved the following:

Theorem 2. *With probability at least*

$$1 - \exp\left\{-\frac{\beta^2 n \log n}{6m}\right\} - \exp\{\log m - \log n\} - \exp\left\{-h(n) + 1 + \frac{h(n)}{\log n}\right\} \to 1$$

procedure MAKE_HL(V_1, M) constructs an HL sequence of the form (4) that covers all the vertices of the graph.

We now notice that if we manage to "close" the sequence HL in order to get a sequence HC of the form (2), then we can run CONSTRUCT_HAM(HC) to get a Hamlilton cycle. This is where we use the set of vertices V_2. Suppose that we run MAKE_HL(V_2, M) but instead of choosing a random label to begin with at step 2, we select label l_1 of the HL sequence constructed by MAKE_HL(V_1, M). By symmetry, a theorem similar to theorem 2 is also true in this case. This means that all the vertices of V will be covered by the HL' sequence constructed by MAKE_HL(V_2, M). Also, by Chernoff bounds we can see that if l_f is the last label of HL, then

$$P\{|L_l \cap V_2| \leq (1-\beta)\frac{np}{2}\} \leq e^{-\beta^2 np/4} \to 0$$

for any constant β. Hence, label l has almost definitely some vertices in V_2.

Suppose then that at some point of the execution of MAKE_HL(V_2, M) HL' has the form

$$HL' := l_1 \to v_1' \to \cdots \to l_k'$$

and that $L_{l_f} \cap L_{l_k'} \cap F_V \neq \emptyset$ (because of theorem 2 such a point always exists with high probability). Then, if $v \in L_{l_f} \cap L_{l_k'} \cap F_V$, we can stop the procedure and set

$$HC := HL \to v \to l_k' \to \cdots \to v_1' \to l_1.$$

Then it is easy to see that the sequence HC is of the form (2) and satisfies all four conditions specified in section 3 so that we may run CONSTRUCT_HAM(HC) to get a Hamilton cycle.

Consequently, we have shown a polynomial (due to Lemma 2) greedy way of finding a Hamilton cycle in a $G_{n,m,p}$ graph with probability at least $1 - 2\exp\left\{-h(n) + 1 + \frac{h(n)}{\log n}\right\} \to 1$.

5 Conclusions and Further Work

We studied the problem of finding Hamilton cycles in uniform random intersection graphs. The discovery of efficient algorithms for other interesting graph objects (e.g. long paths, giant components, dominating sets etc) is a subject of our future work.

References

1. N. Alon and J.H. Spencer, *"The Probabilistic Method"*, Second Edition, John Wiley & Sons, Inc, 2000.
2. Rene Beier and Berthold Vöcking, *"Random Knapsack in Expected Polynomial Time"*, in the Proc. of the 35th Annual ACM Symposium on Theory of Computing, ACM Press, 2003, pp. 232-241.
3. B. Bollobás, *"Random Graphs"*, Second Edition, Cambridge University Press, 2001.
4. B. Bollobás, T.I. Fenner and A.M. Frieze, *"An algorithm for Finding Hamilton Paths and Cycles in Random Graphs"*, Combinatorica 7, 327-341.
5. Amin Coja-Oghlan and Anusch Taraz, *"Colouring Random Graphs in Expected Polynomial Time"*, in the Proc. of the 20th Symposium of Theoretical Computer Science (STACS 2003), Springer-Verlag, 2003, pp. 487-498.
6. J. Díaz, M.D. Penrose, J.Petit and M. Serna, *"Approximating Layout Problems on Random Geometric Graphs"*, Journal of Algorithms, 39:78-116, 2001.
7. J. Díaz, J.Petit and M. Serna, *"A Random Graph Model for Optical Networks of Sensors"*, in the 1st International Workshop on Efficient and Experimental Algorithms, (WEA), 2003. Also in the IEEE Transactions on Mobile Computing Journal, 2(3):186-196, 2003.
8. J. Díaz, J.Petit and M. Serna, *"Random Geometric Problems on* $[0,1]^2$*"*, in J. Rolim, M. Luby and M. Serna, editors, Randomization and Approximation Techniques in Computer Science, volume 1518 of Lecture Notes in Computer Science, pages 294-306, Springer Verlag, Berlin, 1998.

9. H. Efthymiou and P. Spirakis, *"On the Existence of Hamiltonian Cycles in Random Intersection Graphs"*, in the Proc. of the 32nd International Colloquium on Automata, Languages and Programming (ICALP), Lecture Notes in Computer Science, Vol. 3580 (Springer Verlag), pp. 690-701, 2005.
10. J.A. Fill, E.R. Sheinerman and K.B Singer-Cohen, *"Random Intersection Graphs when $m = \omega(n)$: An Equivalence Theorem Relating the Evolution of the $G(n,m,p)$ and $G(n,p)$ models"*, http://citeseer.nj.nec.com/fill98random.html
11. E. Godehardt and J. Jaworski, *"Two models of Random Intersection Graphs for Classification"*, Studies in Classification, Data Analysis and Knowledge Organisation, Opitz O., Schwaiger M, (Eds), Springer Verlag, Berlin, Heidelberg, New York (2002), 67-82.
12. Y. Gurevich and S. Shelah, *"Expected Computation Time for Hamiltonian Path Problem"*, SIAM Journal on Computing, 16:486-502, 1987.
13. M. Karoński, E.R. Scheinerman and K.B. Singer- Cohen, *"On Random Intersection Graphs: The Subgraph Problem"*, Combinatorics, Probability and Computing journal (1999) 8, 131-159.
14. E. Marczewski, *"Sur deux propriétés des classes d' ensembles"*, Fund. Math. 33, 303- 307 (1945).
15. R. Motwani and P. Raghavan, *"Randomized Algorithms"*, Cambridge University Press, 1995.
16. S. Nikoletseas, K. Palem, P. Spirakis and M. Yung, *"Short Vertex Disjoint Paths and Multiconnectivity in Random Graphs: Reliable Network Computing"*, in the Proceedings of the 21st International Colloquium on Automata, Languages and Programming (ICALP), Lecture Notes in Computer Science Vol. 820 (Springer Verlag), pp. 508-519, 1994. Also, in the Special Issue on Randomized Computing of the International Journal of Foundations of Computer Science (IJFCS), Vol. 11 No. 2 (2000), pp. 247-262, World Scientific Publishing Company, 2000
17. S. Nikoletseas, C. Raptopoulos and P. Spirakis, *"The Existence and Efficient Construction of Large Independent Sets in General Random Intersection Graphs"*, in the Proceedings of the 31st International Colloquium on Automata, Languages and Programming (ICALP), Lecture Notes in Computer Science (Springer Verlag), 2004.
18. S. Nikoletseas and P. Spirakis, *"Expander Properties in Random Regular Graphs with Edge Faults"*, in the Proceedings of the 12th Annual Symposium on Theoretical Aspects of Computer Science (STACS), Lecture Notes in Computer Science Vol. 900 (Springer Verlag), pp. 421-432,1995
19. M. Penrose, *"Random Geometric Graphs"*, Oxford Studies in Probability, 2003.
20. C. Raptopoulos and P. Spirakis, *"Simple and Efficient Greedy Algorithms for Hamilton Cycles in Random Intersection Graphs"*, accepted paper in the 16th International Symposium on Algorithms and Computation (ISAAC), 2005, http://students.ceid.upatras.gr/~raptopox/fullpaper.ps.
21. K.B. Singer-Cohen, *"Random Intersection Graphs"*, PhD thesis, John Hopkins University, 1995.
22. A. Thomason, *"A simple linear expected time algorithm for the Hamilton cycle problem"*, Discrete Math., 75 (1989), pp. 373–379.

Counting Distinct Items over Update Streams

Sumit Ganguly

Indian Institute of Technology, Kanpur
sganguly@cse.iitk.ac.in

Abstract. We present two novel algorithms for tracking the number of distinct items over high speed data streams consisting of insertion and deletion operations that improves on the space and time complexity of existing algorithms.

1 Introduction

Data streaming applications occur naturally in many established and emerging application areas, such as database systems, network monitoring, sensor networks etc.. These applications are characterized by rapidly arriving voluminous data and require algorithms that, (a) have low time complexity of processing each stream update, and, (b) have sub-linear space complexity. In this paper, we consider a class of data streaming applications that correspond to the *update* model of streaming, that is, records arriving over a stream correspond to both insertion and deletion of data (e.g., establishment and termination of a network connection). In particular, consider the problem of estimating the number of distinct items that have arrived over an update stream. Queries that count the number of distinct items in a stream(s) satisfying given predicates play a significant role in decision-support applications. For example, consider the following network monitoring query: find the number of distinct connections that have been routed through router R_1 and R_2 but not through router R_3.

We view *update data streams* as a sequence of arrivals of the form (i, v), where, i is the identity of an item, assumed to be from the domain $\mathcal{D} = \{0, 1, \ldots, N-1\}$, and v is the change in frequency of i. That is, $v > 0$ specifies v insertions of the item i and $v < 0$ correspondingly specifies v deletions of i. The frequency of an item i, denoted by f_i, is defined as $f_i = \sum_{(i,v)\in\mathcal{S}} v$. We assume that items cannot have negative frequency. The metric F_0 of a stream is defined as the number of distinct items i in the stream whose frequency f_i is positive (i.e., the zeroth frequency moment $F_0 = \sum_{f_i>0} f_i^0$). A canonical problem in the class of distinct item queries is to estimate F_0. The problem of estimating the size of distinct item queries over stream(s) is a straightforward generalization of the problem of estimating F_0 [8]. We therefore restrict ourselves to the problem of estimating F_0.

Previous work. The append only model of streaming data refers to streams without deletion operations. . The problem of estimating F_0 has received considerable attention for append-only data streams [7,1,2,4,5,10,11,12,3,13]. The algorithm

X. Deng and D. Du (Eds.): ISAAC 2005, LNCS 3827, pp. 505–514, 2005.

of [3] is the most time and space efficient among the algorithms that are applicable to append-only streams only. [13] presents a lower bound of $\Omega(\frac{1}{\epsilon^2})$ for estimating F_0 to within an accuracy factor of $1 \pm \epsilon$. (We use $\epsilon > 0$ and $0 < \delta < 1$ to denote the accuracy and confidence parameters, respectively) The algorithms of [10,11,12,3] use the technique of distinct samples[10,11,12,3], which is a simple and elegant technique for estimating F_0.

Algorithms for estimating F_0 over streams with deletion operations [1,2,8,9] use the technique of hashing introduced in [7]. The algorithms of [1,2,8] use space $O(\frac{1}{\epsilon^2} \cdot \log N \log m \cdot \log \frac{1}{\delta})$ bits (where, m is the sum of frequencies of the items in stream) and are the most space-economical among the algorithms for estimating F_0 over update streams. However, they require time $O(\frac{1}{\epsilon^2} \log \frac{1}{\delta})$ operations for processing each stream update. The algorithm of [9] is the most time-efficient among algorithms that estimate F_0 over update streams and requires time $O(\log N \cdot (\log \frac{1}{\epsilon} + \log \frac{1}{\delta}))$ to process each stream update. However, it uses an additional factor of $O(\log N \log \frac{1}{\delta})$ in space, as compared to [1,2,8], that is, its space complexity is $O(\frac{1}{\epsilon^2} \log^2 N \log m \log^2 \frac{1}{\delta})$ bits. This algorithm designs an updatable distinct sample, which is a data structure that can be used for both insertions and deletions of items and effectively yields a distinct sample [10,11,12,3].

Our Contributions. We present two novel algorithms for estimating F_0 over update streams that improves upon the time and space complexity of existing algorithms. Our first algorithm has time complexity $O(\log \frac{1}{\epsilon} + \log \frac{1}{\delta})$, which is a logarithmic factor smaller than the time complexity of the most time efficient algorithm currently known for this problem [9]. Its space complexity is $O(\frac{1}{\epsilon^2}(\log m + \log N) \log N(\log \frac{1}{\delta} + \log \frac{1}{\epsilon}) \log \frac{1}{\delta})$ bits, which is lower than that of [9] by a logarithmic factor. The improvement is a consequence of a more efficient design of the updatable distinct samples structure. Our second algorithm for estimating F_0 presents a time versus space tradeoff vis-a-vis our first algorithm. It takes $O\left((\log \frac{1}{\epsilon})(\log \frac{1}{\delta})\right)$ time to process each stream update and requires space $O(\frac{1}{\epsilon^2}(\log N + \log m) \log N \log \frac{1}{\delta})$ bits, thus, using a logarithmic factor less space and an extra logarithmic factor in time, in comparison with our first algorithm.

Organization. Sections 2 and 3 present the first and second algorithms respectively.

2 Algorithm I

The distinct samples data structure [10,11,12] was used to estimate F_0 over append-only or sliding window data streams. In this section, we generalize the structure so that it can be used to estimate F_0 for streams with deletion operations. We begin with the TESTSINGLETON data structure.

2.1 TESTSINGLETON Data Structure

A TESTSINGLETON data structure supports two operations, namely, (a) an update operation corresponding to insertion and deletion operations over the

stream, called TESTSINGLETONUPDATE, and, (b) a procedure TESTCARD that tests whether F_0 is either 0, 1 or greater than 1, and accordingly returns EMPTY, SINGLETON or COLLISION, respectively. Further, if TESTCARD returns SINGLETON, then the unique member i comprising the singleton set and its frequency f_i is also returned.

A straightforward counting argument shows that a TESTSINGLETON structure must use $\Omega(\log N)$ bits. We keep three counters m, U and V (all initialized to zero) with the following properties.

$$m = \sum_{i \in \text{ stream}} f_i, \quad U = \sum_{i \in \text{ stream}} f_i \cdot i, \quad \text{and} \quad V = \sum_{i \in \text{stream}} f_i \cdot i^2 .$$

The counters are easily maintained in the face of stream updates as shown below.

TESTSINGLETONUPDATE(i, v): $m := m + v$; $U := U + v \cdot i$; $V := V + v \cdot i^2$;

The space required is $O(\log m + \log N)$ bits and the time required for the update operation is $O(1)$. The operation TESTCARD is given below.

TESTCARD: **if** $(m = 0)$ **return** EMPTY
else if $(U^2 = m \cdot V)$ **then return** (SINGLETON, $\frac{U}{m}$, m)
else return COLLISION

The correctness of this test is proved below.

Theorem 1. *$F_0 = 1$ if and only if $m > 0$ and $U^2 = m \cdot V$. If $F_0 = 1$, then, the unique item in the stream has identity $\frac{U}{m}$ and has frequency m.*

Proof. Let the stream consist of a single item i with frequency $f_i > 0$. Then, $m = f_i > 0$, $U = m \cdot i$ and $V = m \cdot i^2$. Therefore, $U^2 = m \cdot V$, $m > 0$ and $i = \frac{U}{m}$.

To prove the converse, let x be a random variable such that for $0 \le i \le N-1$, $x = i$ with probability $\frac{f_i}{m}$. Then, $\mathrm{E}[x] = \sum_{i=0}^{N-1} \frac{f_i}{m} \cdot i$ and $\mathrm{E}[x^2] = \sum_{i=1}^{N-1} \frac{f_i}{m} \cdot i^2$. Since $m > 0$, the condition, $U^2 = mV$ is equivalent to the condition, $(\mathrm{E}[x])^2 = \mathrm{E}[x^2]$, or that, $\mathrm{Var}[x] = 0$. Therefore, the set of items in the stream is either empty or consists of a singleton item. Since, $m > 0$, the stream is not empty. Therefore, $F_0 = 1$. □

2.2 *K*-Set Structure

A k-set structure, for parameter k, is a dictionary data structure that supports insertion and deletion of items and an operation RETRSET, that either returns the set of items S present in the dictionary or returns NIL. Further, if the number of items $|S|$ in the dictionary is at most k, then RETRSET returns S (i.e., does not return NIL in this case).

Classic dictionary structures including heaps, binary search trees, red-black trees, AVL trees etc., satisfy the requirements of a k-set data structure for all values of k. However, these structures require space $\Omega(|S|)$, whereas, we require a design whose space complexity is close to the lower bound of $\Omega(k \log \frac{N}{k})$ bits.

A randomized k-set data structure takes a confidence parameter δ such that if $|S| \leq k$, then RETRSET returns S with probability $1-\delta$.

We now present a design for a k-set structure. Our structure is a two dimensional array $H[R \times B]$ and can be thought of as consisting of $R = \lceil \log \frac{k}{\delta} \rceil$ hash tables, each consisting of B buckets indexed from 0 through $B-1$, where, $B = 2k$. Each bucket $H[r, b]$ is a TESTSINGLETON data structure. The rth hash table uses a randomly chosen pair-wise independent hash function $h_r : \{0, 1, \ldots, N-1\} \rightarrow \{0, 1, \ldots, B-1\}$; the random bits used by each of the R hash tables are independent. Further, we maintain a variable m that tracks the total sum of the frequencies, that is, $m = \sum_{i \in S} f_i$. The structure is initialized so that all counters are initially zeros. For every stream update of the form (i, v), and for every hash table index r, we hash the item to its bucket number $h_r(i)$ and propagate the update to the corresponding TESTSINGLETON structure in that bucket.

KSETUPDATE(i, v): $H[r, h_r(i)]$.TESTSINGLETONUPDATE(i, v), for $1 \leq r \leq R$.

The space used by the k-set structure is $O(k(\log m + \log N) \log \frac{k}{\delta})$ bits and the time complexity of processing each stream update is $O(\log \frac{k}{\delta})$.

The RETRSET operation works as follows. We iterate over each of the B buckets of each of the R hash tables checking whether the bucket $H[r, b]$ is a singleton or not, for $r = 1, \ldots, R$, $b = 0, \ldots, B-1$. If the bucket is a singleton, then the item and its frequency is retrieved using the TESTCARD subroutine of the TESTSINGLETON structure. In this manner, we retrieve the union $\hat{S}$ of all the distinct items and their frequencies. Next, we preform a verification step, that is, the frequencies of the (distinct) retrieved items are added and compared with m. If they match, then $\hat{S}$ is returned, otherwise, NIL is returned. The correctness of the RETRSET procedure is proved below.

Theorem 2. *If* RETRSET $\neq$ NIL, *then* RETRSET $= S$. *Further, if* $|S| \leq k$, *then* $\Pr\{\text{RETRSET} = S\} \geq 1 - \delta$.

Proof. If a bucket $H[r, b]$ is singleton, then, the TESTSINGLETON structure correctly retrieves the item and its frequency. Therefore, the set $\hat{S}$ of retrieved items is always a subset of S and $\sum_{i \in \hat{S}} f_i \leq m$. RETRSET returns non-NIL only if $\sum_{i \in \hat{S}} f_i = m$, which implies that $\hat{S} = S$. Now suppose that $|S| = s \leq k$. Fix an item $i \in S$ and a hash table index r. Let C_i be the number of items from $S - \{i\}$ that collide with i. Then, $\mathrm{E}[C_i] = \sum_{j \in S, j \neq i} \Pr\{h_r(i) = h_r(j)\} = \frac{s-1}{B}$, by pairwise independence of h_r. By Markov's inequality, $\Pr\{C_i = 0\} \geq 1 - \frac{s-1}{B} > \frac{1}{2}$, since, $s \leq k$ and $B \geq 2k$. Therefore, the probability that i is not returned from any of the R hash tables is less than $\frac{1}{2^R}$. Thus, the total error probability is at most $\frac{k}{2^R} \leq \delta$. □

2.3 Updatable Distinct Samples Structure

A distinct sample of a stream with sampling probability p is a set D of items such that each of the F_0 distinct items in the stream has an equal and independent chance of p of being included in D. An *updatable distinct sample* is

a data structure that enables the extraction of a distinct sample over streams with insertions and deletion operations, thereby enabling distinct sampling based estimation algorithms to be used for update streams.

Let $F = GF(2^d)$ be a field such that $|F| \geq N^2$, where, N is the size of the domain of the items in the stream. An updatable distinct sample, with *capacity parameter* s, can be implemented using an array $D[1, \ldots, \log|F|]$, where, each entry of the array, namely, $D[r]$, is a k-set structure with $k = s$. A d-wise independent random hash function $h : F \rightarrow F$ is chosen (d is a parameter) and is used to define the function level : $F \rightarrow \{1, \ldots, \log|F|\}$ function as follows. $\text{level}(i) = 1$ if $h(i) = 0$; otherwise, $\text{level}(i) = \text{lsb}(h(i))$. Here, $\text{lsb}(a)$ denotes the least significant bit position of a. Corresponding to a stream update of the form (i, v), the data structure is updated by invoking the update operation of the s-set structure $D[\text{level}(i)]$.

$\text{DISTSAMPUPDATE}(i, v)$: $D[\text{LEVEL}(i)].\text{KSETUPDATE}(i, v)$

The *current level* l_{curr} of the updatable distinct sample is the lowest value of $1 \leq l \leq \log N$ such that $D[l].\text{RETRSET}$ is non-NIL. The distinct sample is defined as $D[l_{\text{curr}}].\text{RETRSET}$. The space required to store an updatable distinct sample with capacity parameter s is $O(s \cdot (\log m + \log N) \cdot \log N \cdot \log \frac{s}{\delta})$ and the time required to process each stream update is $O(\log \frac{s}{\delta} + \log \frac{1}{\epsilon})$.[1]

F_0 can be estimated for append-only streams using distinct samples of size $s = O(\frac{1}{\epsilon^2} \log \frac{1}{\delta})$ items and using a degree of independence of $d = O(\log \frac{1}{\epsilon})$ for the hash family [3]. By using updatable distinct samples in place of distinct samples, we obtain an algorithm for estimating F_0 over update streams, that processes each stream update in time $O(\log \frac{1}{\epsilon} + \log \frac{1}{\delta})$ and uses space $O(\frac{1}{\epsilon^2} \cdot (\log m + \log N) \cdot \log N \cdot (\log \frac{1}{\delta} + \log \frac{1}{\epsilon}) \cdot \log \frac{1}{\delta})$ bits.

3 Algorithm II

In this section, we present our second algorithm[2] for estimating F_0.

The basic data structure is a two-dimensional table $T[L \times K]$, where, $L = \log\ |F|$ (F is a field containing $\{0, 1, \ldots, N-1\}$ and such that $|F| \geq N^2$) and $K = \frac{720}{\epsilon^2}$. We keep two random hash functions namely, $h : F \rightarrow F$ and $g : F \rightarrow \{0, 1, \ldots, K-1\}$ that are d-wise independent, where, $d = O(\log \frac{1}{\epsilon})$. h is used to define a randomized level function as in Section 2.3, that is, $\text{level}(i) = 1$ if $h(i) = 0$, else $\text{level}(i) = \text{lsb}(h(i))$. Corresponding to each stream update of the form (i, v), the data structure is updated as follows.

$\text{ALGO2UPDATE}(i, v)$: $T[\text{level}(i), g(i)].\text{TESTSINGLETONUPDATE}(i, v)$

That is, the item i is first hashed using the hash function h to obtain its level, say l. Next, the item is hashed to a bucket index $0 \leq g(i) \leq K-1$ and the

---

[1] In comparison, the updatable distinct samples structure of [9] requires space $O(s \cdot \log m \cdot \log^2 N \cdot \log \frac{s}{\delta})$ bits and time $O(\log N \cdot \log \frac{s}{\delta} + \log \frac{1}{\epsilon})$ to process each stream update.

[2] We do not optimize the constants used in the algorithm.

TESTSINGLETON structure at the entry $T[l, g(i)]$ is updated accordingly. We keep $R = 96 \cdot \log \frac{2}{\delta}$ independent copies of the basic data structure. Finally, we also keep a k-set structure, where, $k = \frac{126}{\epsilon^2}$, as explained in Section 2.2, and maintain it in the presence of stream updates. The space used by the data structure is $O(\frac{1}{\epsilon^2} \log N(\log m + \log N) \log \frac{1}{\delta})$ bits and the time complexity of processing each stream update is $O(\log \frac{1}{\epsilon} \cdot \log \frac{1}{\delta})$ operations.

Each copy of the basic data structure yields an estimate $\hat{F}_0$ for F_0 as follows. We find the lowest level l such that the number X of the singleton buckets at this level exceeds $\frac{12}{\epsilon^2}$ and the number Z of the non-empty buckets exceeds $\frac{7K}{8}$. If there is no such level, then, we return $\hat{F}_0 = \perp$. Otherwise, we (numerically) solve the equation $X = \frac{\hat{F}_0}{2^l}\left(1 - \frac{1}{K}\right)^{\frac{\hat{F}_0}{2^l} - 1}$ and return the smaller of the two possible roots of this equation as $\hat{F}_0$. The median of the observations, denoted $\hat{F}_0^{\text{median}}$ from the copies is computed (ignoring those copies which have returned $\perp$). If the median value is at least $\frac{126}{\epsilon^2}$, then, the median value is returned, otherwise, an estimate for F_0 is obtained from the parallel k-set structure and returned.

The procedure is different from past work [1,2,3,8] in that, in previous work, the count of the number of singleton buckets (or, a similar measure, such as the number of non-empty buckets) was taken across *independent observations.* In our case, the sum is taken across the buckets in the same hash table, and hence, the observations are not even pair-wise independent. We are not aware of Chernoff-like tail inequalities that are applicable in such a scenario[3]. Theorem 3 states the correctness property of the algorithm. Its proof relies on Lemma 1, which is the main technical lemma of this section.

Theorem 3. *Suppose* $\epsilon \leq \frac{1}{8}$, $K = \frac{720}{\epsilon^2}$, $d = O(\log \frac{1}{\epsilon})$ *and* $R = 96 \log \frac{2}{\delta}$. *Then, the estimate returned by Algorithm II is within* $(1 \pm 3\epsilon)F_0$ *with probability at least* $1 - 2\delta$.

Proof. Suppose $\hat{F}_0^{\text{median}} \geq (1 + \frac{3}{4})\frac{72}{\epsilon^2} = \frac{126}{\epsilon^2}$. Then, by Lemma 1, this observation happens with probability at most δ, provided, $F_0 < \frac{72}{\epsilon^2}$. In this case, with probability $1 - \delta$, the algorithm returns the estimate from the k-set structure, where, $k = \frac{126}{\epsilon^2}$, which is exactly correct with probability $1 - \delta$. Otherwise, by Lemma 1, the algorithm returns $\hat{F}_0^{\text{median}}$, which is within $(1 \perp 3\epsilon)F_0$ with probability $1 - \delta$. □

Lemma 1. *Suppose* $\epsilon \leq \frac{1}{8}$, $K = \frac{720}{\epsilon^2}$, $F_0 \geq \frac{72}{\epsilon^2}$, $d = O(\log \frac{1}{\epsilon})$ *and* $R = 96 \cdot \log \frac{2}{\delta}$. *Then,* $\Pr\left\{|\hat{F}_0^{\text{median}} - F_0| \leq 3\epsilon F_0\right\} > 1 - \delta$.

Analysis. Fix a level l. Consider the set of items in the stream that hash to level l (i.e., level$(i) = l$). For each such item, let p denote the probability that an item i is placed in a given bucket, that is, $p = \frac{1}{K}$. Corresponding to bucket

[3] [6] presents tail inequalities for sums of negatively dependent boolean variables that assume that the hash function that distributes the balls to the buckets is *fully independent.* The techniques of [6] do not apply to the scenario when hash functions have limited dependence.

numbered i at level l, $1 \le i \le K$, we define an indicator variable x_i that is 1 if exactly one ball has mapped to this bucket and is 0 otherwise. Let X denote the number of singleton buckets at this level, that is, $X = \sum_{i=1}^{K} x_i$. If the random hash functions are chosen from a d-wise independent hash family, then, we denote the corresponding probability function by $\Pr_d\{\cdot\}$, and the corresponding expectations by $\mathrm{E}_d[\cdot]$. If the random hash function is chosen from a fully independent hash family, then, we denote the corresponding probability function by $\Pr\{\cdot\}$ and the corresponding expectations as $\mathrm{E}[\cdot]$. Suppose that the number of items that have hashed to level l is n, and let $\Pr\{x_i = 1\}$ be denoted by q, then, $q = \frac{n}{K}(1-\frac{1}{K})^{n-1}$, and $\mathrm{E}[X] = Kq$. The following lemma can be used to obtain bounds for $\mathrm{E}_d[X]$.

Lemma 2 ([3,8]). *If h is $d = O(\log \frac{1}{\epsilon})$-wise independent and $q \le \frac{1}{4}$, then $|\Pr_d\{x_i = 1\} - q| \le \epsilon^4 q$. Therefore, $|\mathrm{E}_d[X] - Kq| \le \epsilon^4 Kq$.* □

The majority of the analysis is devoted to showing that $X \in (1 \pm \epsilon)\mathrm{E}[X]$ (with probability at least $\frac{7}{8}$). Lemma 3 justifies this line of argument by showing that if X is close to $\mathrm{E}[X]$, then, $\hat{F}_0(X)$, calculated as the smaller of the two roots of the equation $X = \frac{\hat{F}_0}{2^l}(1 - \frac{1}{K})^{\frac{\hat{F}_0}{2^l}-1})$ is also close to F_0. We denote $f(x) = \frac{x}{2^l}(1-\frac{1}{K})^{\frac{x}{2^l}-1}$, so that, $f(F_0) = \mathrm{E}[X]$. Lemma 3 also shows that the gap between the two roots is substantial.

Lemma 3. *Let $x = F_0$ and $\epsilon \le \frac{1}{8}$. If $\frac{x}{2^l} \le \frac{5K}{24}$, $|f(x) - f(y)| \le \epsilon f(x)$ then either $|y - x| \le 3\epsilon x$ or $y > 12x$.*

Proof. Let $0 < \gamma \le 1$ and $x = F_0$. $f(x + \gamma x) = f(x)(1+\gamma)\left(1 - \frac{1}{K}\right)^{x\gamma/2^l}$. Therefore, $|f(x+\gamma x) - f(x)| = f(x)|((1+\gamma)\left(1 - \frac{1}{K}\right)^{x\gamma/2^l} - 1)| \ge f(x)((1+\gamma)(1-\frac{\gamma}{4})-1) \ge f(x)\frac{14\gamma}{24}$. Since, $|f(x+\gamma x) - f(x)| \le \epsilon f(x)$, therefore, $\epsilon f(x) \ge \frac{14}{24}|\gamma| f(x)$, or that, $\gamma \le \frac{24}{14}\epsilon$. If $-1 < \gamma < 0$, then a similar analysis holds.

Another solution to $f(x + \gamma x) = (1+\epsilon) f(x)$ occurs when $\gamma \ge \gamma'$, where, γ' satisfies the equation $\frac{1+\gamma'}{1+\epsilon} = e^{\frac{5\gamma'}{24}}$. For $\epsilon \le \frac{1}{8}$, $\gamma' > 11$. □

We now present an overview of the analysis, leading to a proof of Lemma 3. The main article of interest is the following lemma that bounds the variance of X. Let n denote the number of items that have mapped to the level l.

Lemma 4. *Suppose that $n \le \frac{K}{4}$, $\epsilon \le \frac{1}{4}$ and $d \ge \max(3 + e, 2 \log K)$. Then, $\mathrm{Var}_d[X] \le 2Kq + 2\epsilon^4 K^2 q^2 + K^{-2}$.*

Proof. $\Pr\{x_i = 1\}$ can be calculated to be $\frac{n(K-1)^{n-1}}{K^n}$. For a fixed pair of distinct buckets, i and j, let $\phi = \phi(i,j)$ denote the event that after a random experiment, buckets i and j are singleton. Therefore, $\Pr\{\phi\} = \Pr\{x_i = 1 \text{ and } x_j = 1\} = \frac{n(n-1)(K-2)^{n-2}}{K^n} = n(n-1)p^2(1-2p)^{n-2} < \Pr\{x_i = 1\}\Pr\{x_j = 1\}$. The expression $\Pr\{\phi\} = n(n-1)p^2(1-2p)^{n-2}$ can be viewed as a polynomial in $p = \frac{1}{K}$. Using the principle of inclusion and exclusion, an expression can be obtained for $\Pr_d\{\phi\}$ that is identical to the terms of this polynomial from degree 2 through d

(inclusive). Let T_d denote the coefficient of p^d in the polynomial and let S_d denote the partial sum from $s = 0$ through $d-2$, that is, until the term for p^d. Since, the polynomial expression $\Pr\{\phi\}$ is an alternating sum, $|\Pr\{\phi\} - S_d| \le |T_d|$ and $|\Pr_d\{\phi\} - S_d| \le |T_d|$. Therefore, using triangle inequality and the assumptions that $p = \frac{1}{K}$ and $n \le \frac{K}{4}$,

$$|\Pr_d\{\phi\} - \Pr\{\phi\}| \le 2|T_d| = 2^{d-1}n(n-1)p^d\binom{n-2}{d-2} < K^{-4} \ .$$

Further, $\Pr\{\phi\} < q^2$. Therefore, $\mathrm{E}_d[x_i x_j] < q^2 + K^{-4}$.

$$\begin{aligned}\mathrm{Var}_d[X] &= \mathrm{E}_d[X^2] - (\mathrm{E}_d[X])^2 = \mathrm{E}_d[X] + 2\sum_{i<j}\mathrm{E}_d[x_i x_j] - (\mathrm{E}_d[X])^2 \\ &\le Kq(1+\epsilon^4) + 2\binom{K}{2}(q^2 + K^{-4}) - (Kq)^2(1-2\epsilon^4) \\ &\le 2Kq + 2\epsilon^4K^2q^2 + 2K^{-2} \ . \qquad \square\end{aligned}$$

Using Chebychev's inequality and Lemma 4, the following lemma can be shown.

Lemma 5. *If $\epsilon \le \frac{1}{4}$, $\mathrm{E}_d[X] \ge \frac{8}{\epsilon^2}$, $n \le \frac{K}{4}$ and $d \ge 4\log K$, then,*

$$\Pr_d\{|X - \mathrm{E}_d[X]| \ge \epsilon\mathrm{E}_d[X]\} \le \tfrac{1}{8} \ . \qquad \square$$

Our proof obligation is now two-fold, namely, (a) if a level satisfying the above conditions on X and Z are found, and $F_0 \ge \frac{72}{\epsilon^2}$, then, $\Pr_d\{|X - \mathrm{E}[X]| > \epsilon\mathrm{E}[X]\} < \frac{1}{8}$, and, (b) if $F_0 \ge \frac{72}{\epsilon^2}$, then, there is a level l such that $X \ge \frac{12}{\epsilon^2}$ and $Z \ge \frac{7K}{8}$. We consider part (a) first. Lemma 5 assumes that $\mathrm{E}[X] \ge \frac{8}{\epsilon^2}$ and $n \le \frac{K}{4}$. Assuming that we have found a level satisfying $X \ge \frac{12}{\epsilon^2}$, Lemmas 6 shows that $\mathrm{E}[X] \ge \frac{8}{\epsilon^2}$.

Lemma 6. *If $X \ge \frac{12}{\epsilon^2}$ and $K \ge \frac{512}{\epsilon^2}$ then $\mathrm{E}[X] \ge \frac{8}{\epsilon^2}$.*

Proof. Clearly, $n \ge X \ge \frac{12}{\epsilon^2}$. For $\frac{12}{\epsilon^2} \le n \le \frac{K}{4}$, $Kq = n(1-\frac{1}{K})^{n-1}$ is an increasing function of n. Therefore, in this range of n, $Kq \ge \frac{12}{\epsilon^2}(1 - \frac{12/\epsilon^2}{512/\epsilon^2}) > \frac{8}{\epsilon^2}$. $\square$

We now show in Lemma 7 that if we have observed $X \ge \frac{12}{\epsilon^2}$, then, $n \in (1\pm\epsilon)\mathrm{E}_d[n]$ with reasonable probability.

Lemma 7. *If $\epsilon \le \frac{1}{4}$, $X \ge \frac{12}{\epsilon^2}$ and $d \ge 2$, then, $\Pr_d\{|n - \mathrm{E}_d[n]| \ge \epsilon\mathrm{E}_d[n]\} \le \frac{1}{8}$.*

Proof. Since, $X \ge \frac{12}{\epsilon^2}$, $n \ge \frac{12}{\epsilon^2}$. Assuming $d \ge 2$, $\mathrm{Var}[n] \le \mathrm{E}[n]$. If $\mathrm{E}[n] \ge \frac{8}{\epsilon^2}$, then, $\Pr\{|n - \mathrm{E}_d[n]| \ge \epsilon\mathrm{E}_d[n]\} < \frac{\mathrm{Var}[n]}{(\epsilon\mathrm{E}_d[n])^2} \le \frac{1}{\epsilon^2\mathrm{E}_d[n]} = \frac{1}{8}$. Otherwise, $\Pr\left\{n \ge \frac{12}{\epsilon^2}\right\} \le \frac{\mathrm{E}_d[n]}{(12/\epsilon^2 - \mathrm{E}_d[n])^2} \le \frac{8/\epsilon^2}{16/\epsilon^4} \le \frac{1}{32}$. $\square$

We now return to the proof obligation of $n \le \frac{K}{4}$. This follows from the observation that Z, the number of non-empty buckets, is at least $\frac{7K}{8}$, since, intuitively, as n becomes a larger proportion of K, the number of empty buckets observed

must drop. Let y_i be an indicator variable that is 1 if bucket i is non-empty and is 0 otherwise. Let $Y = \sum_{i=0}^{K-1} y_i$ denote the number of non-empty buckets. Then, $r = \Pr\{y_i = 1\} = 1 - \left(1 - \frac{1}{K}\right)^n$ and $\mathrm{E}[Y] = Kr$. Further, using the arguments above, $|\Pr_d\{y_i = 1\} - r| \le \epsilon^4 r$.

Lemma 8. *Suppose* $\epsilon \le \frac{1}{4}$, $n \ge \frac{K}{4}$ *and* $K \ge \frac{512}{\epsilon^2}$, *then,* $\Pr_d\left\{Y \le \frac{K}{8}\right\} \le \frac{1}{8}$.

Proof. Since $n \ge \frac{K}{4}$, $r \ge 1 - (1 - \frac{1}{K})^{\frac{K}{4}} \ge \frac{3}{4}$. Arguing similarly as in Lemma 4, $\mathrm{Var}_d[Y_d] \le 2Kr + 2\epsilon^4 K^2 r^2 + 2K^{-2}$. Further, if $n \ge \frac{K}{4}$, then, $\mathrm{E}[Y] = Kr \ge \frac{3K}{4}$. The lemma now follows by applying Chebychev's inequality. □

Adding error probabilities in Lemmas 5, 7 and 8, we obtain that if $X \ge \frac{12}{\epsilon^2}$ and $Z \ge \frac{7K}{8}$, then, $\mathrm{E}[X] \ge \frac{8}{\epsilon^2}$, $X \in (1 \pm \epsilon)\mathrm{E}[X]$ and $n \le \frac{K}{4}$, with probability at least $1 - \frac{1}{8} - \frac{1}{8} - \frac{1}{8} = \frac{5}{8}$. We now consider the second part of the proof obligation, namely, to show that there is a reasonable probability of finding a level l such that $X \ge \frac{12}{\epsilon^2}$ and $Z \ge \frac{7K}{8}$, provided, $F_0 \ge \frac{72}{\epsilon^2}$.

Lemma 9. *Suppose* $\epsilon \le \frac{1}{8}$, $K = \frac{720}{\epsilon^2}$, $F_0 \ge \frac{72}{\epsilon^2}$ *and* $l = \lceil \log \frac{F_0}{36/\epsilon^2} \rceil$. *Then,* $Z \ge \frac{7K}{8}$ *and* $\frac{12}{\epsilon^2} \le X \le \frac{81}{\epsilon^2}$ *is satisfied with probability* $\frac{7}{8}$.

Proof. The expected number of items that map to level l is $\mathrm{E}_d[n] = \frac{F_0}{2^l}$ which lies in the range of $\frac{36}{\epsilon^2}$ and $\frac{72}{\epsilon^2}$. Arguing as in Lemma 7, it follows that $\Pr_d\{|n - \mathrm{E}_d[n]| \ge \epsilon n\} < \frac{1}{24}$. Therefore, we have $\frac{36}{\epsilon^2}(1-\epsilon) \le n \le \frac{72}{\epsilon^2}(1+\epsilon)$, with probability at least $\frac{23}{24}$. Therefore $Z \ge K - \frac{81}{\epsilon^2} = \frac{720}{\epsilon^2} - \frac{81}{\epsilon^2} > \frac{7K}{8}$. Further, $\mathrm{E}[X] = Kq = n(1 - \frac{1}{K})^{n-1} \ge n(1 - \frac{n}{K}) > \frac{30}{\epsilon^2}$ and $X \le n \le \frac{72(1+\epsilon)}{\epsilon^2} = \frac{81}{\epsilon^2}$. Therefore, by Chebychev's inequality, it follows that,

$$\Pr_d\left\{X \le \frac{12}{\epsilon^2}\right\} \le \frac{\mathrm{Var}_d[X]}{(\mathrm{E}_d[X] - \frac{12}{\epsilon^2})^2} \le \frac{2Kq + 2\epsilon^4 K^2 q^2 + K^{-2}}{(Kq(1-\epsilon^4) - \frac{12}{\epsilon^2})^2} \le \frac{1}{16} .$$

Total error probability is $\frac{1}{16} + \frac{1}{24} = \frac{1}{8}$. □

Proof (Of Lemma 1). If $F_0 \ge \frac{72}{\epsilon^2}$, then, by Lemma 9, there exists a level with probability at least $\frac{7}{8}$ such that $\frac{12}{\epsilon^2} \le X \le \frac{81}{\epsilon^2}$ and $Z \ge \frac{7K}{8}$. In this case, the algorithm returns a non-$\perp$ value as the estimate of F_0. Since we keep keep $96 \cdot \log \frac{2}{\delta}$ independent copies, by Chernoff bounds, the probability that the number of copies that returns a non-$\perp$ value is at least $60 \cdot \log \frac{2}{\delta}$ is at least $1 - \frac{\delta}{8}$. By previous arguments, for each such estimate, the probability that $X \in (1 \pm \epsilon)\mathrm{E}[X]$ is at least $1 - \frac{3}{8} = \frac{5}{8}$. By Lemma 3, in each of these cases, $\hat{F}_0 \in (1 \pm 3\epsilon)F_0$. Therefore, by a standard application of Chernoff's bounds of returning the median of $60 \log \frac{2}{\delta}$ estimates $\hat{F}_0$, the error probability is reduced to $\frac{\delta}{2}$. Adding the error probabilities $\frac{\delta}{8} + \frac{\delta}{2} < \delta$, we obtain the statement of the lemma. □

Acknowlegement

The author thanks Barna Saha for comments on the paper.

References

1. Noga Alon, Yossi Matias, and Mario Szegedy. "The Space Complexity of Approximating the Frequency Moments". In *Proceedings of the 28th Annual ACM Symposium on the Theory of Computing STOC, 1996*, pages 20–29, Philadelphia, Pennsylvania, May 1996.
2. Noga Alon, Yossi Matias, and Mario Szegedy. "The space complexity of approximating frequency moments". *Journal of Computer Systems and Sciences*, 58(1):137–147, 1998.
3. Ziv Bar-Yossef, T.S. Jayram, Ravi Kumar, D. Sivakumar, and Luca Trevisan. "Counting distinct elements in a data stream". In *Proceedings of the 6th International Workshop on Randomization and Approximation Techniques in Computer Science, RANDOM 2002*, Cambridge, MA, 2002.
4. Andrei Z. Broder, Moses Charikar, Alan M. Frieze, and Michael Mitzenmacher. "Min-wise independent permutations (Extended Abstract)". In *Proceedings of the 30th Annual ACM Symposium on the Theory of Computing STOC, 1998*, pages 327–336, Dallas, Texas, May 1998.
5. Andrei Z. Broder, Moses Charikar, Alan M. Frieze, and Michael Mitzenmacher. "Min-wise independent permutations". *Journal of Computer Systems and Sciences*, 60(3):630–659, 2000.
6. Devdatt Dubhashi, Volker Priebe, and Desh Ranjan. "Negative Dependence through the FKG Inequality". Basic Research in Computer Science, Report Series, BRICSRS-96-27.
7. Philippe Flajolet and G.N. Martin. "Probabilistic Counting Algorithms for Database Applications". *Journal of Computer Systems and Sciences*, 31(2):182–209, 1985.
8. Sumit Ganguly, Minos Garofalakis, and Rajeev Rastogi. "Processing Set Expressions over Continuous Update Streams". In *Proceedings of the 2003 ACM SIGMOD International Conference on Management of Data*, San Diego, CA, 2003.
9. Sumit Ganguly, Minos Garofalakis, Rajeev Rastogi, and Krishna Sabnani. "Streaming Algorithms for Robust, Real-Time Detection of DDoS Attacks". Bell Laboratories Technical Memorandum, 2004.
10. Philip B. Gibbons. "Distinct Sampling for Highly-accurate Answers to Distinct Values Queries and Event Reports". In *Proceedings of the 27th International Conference on Very Large Data Bases*, Roma, Italy, September 2001.
11. Philip B. Gibbons and Srikant Tirthapura. "Estimating simple functions on the union of data streams". In *Proceedings of the 13th Annual ACM Symposium on Parallel Algorithms and Architectures, SPAA 2001*, pages 281–291, Heraklion, Crete, Greece, July 2001.
12. Philip B. Gibbons and Srikant Tirthapura. "Distributed streams algorithms for sliding windows". In *Proceedings of the 14th Annual ACM Symposium on Parallel Algorithms and Architectures, SPAA 2002*, pages 63–72, Winnipeg, Manitoba, Canada, August 2002.
13. Piotr Indyk and David Woodruff. "Tight Lower Bounds for the Distinct Elements Problem". In *Proceedings of the 35th ACM Symposium on Theory of Computing (STOC), 2003*, San Diego, CA, 2003.

Randomized Algorithm for the Sum Selection Problem*

Tien-Ching Lin** and D.T. Lee**

Department of Computer Science and Information Engineering,
National Taiwan University, Taipei, Taiwan
{kero, dtlee}@iis.sinica.edu.tw

Abstract. Given a sequence of n real numbers $A = a_1, a_2, \ldots, a_n$ and a positive integer k, the SUM SELECTION PROBLEM is to find the segment $A(i, j) = a_i, a_{i+1}, \ldots, a_j$ such that the rank of the sum $s(i, j) = \sum_{t=i}^{j} a_t$ is k over all $\frac{n(n-1)}{2}$ segments. We will give a randomized algorithm for this problem that runs in expected $O(n \log n)$ time. Applying this algorithm we can obtain an algorithm for the k MAXIMUM SUMS PROBLEM, i.e., the problem of enumerating the k largest sum segments, that runs in expected $O(n \log n + k)$ time. The previously best known algorithm for the k MAXIMUM SUMS PROBLEM runs in $O(n \log^2 n + k)$ time in the worst case.

Keywords: computational geometry, randomized algorithm, random sampling, order-statistic tree, k maximum sums problem, sum selection problem, maximum sum problem, maximum sum subarray problem.

1 Introduction

Given a sequence A of real numbers $a_1, a_2, \ldots, a_n$, the MAXIMUM SUM PROBLEM is to find the segment $A(i, j) = a_i, a_{i+1}, \ldots, a_j$ whose sum $s(i, j) = \sum_{t=i}^{j} a_t$ is the maximum among all possible $1 \le i \le j \le n$. This problem was first introduced by Bentley [1,2] and can be easily solved in $O(n)$ time [2,3].

Given an $m \times n$ matrix of real numbers (assuming that $m \le n$), the MAXIMUM SUM SUBARRAY PROBLEM is to find the submatrix, the sum of whose entries is the maximum among all $O(m^2 n^2)$ submatries. The problem can be solved in $O(m^2 n)$ time [2,3,4]. Tamaki and Tokuyama [5] gave the first sub-cubic time algorithm for this problem and Takaoka [6] later gave a simplified algorithm achieving sub-cubic time as well. Many parallel algorithms under different parallel models of computation were also obtained [4,7,8,9].

* Research supported in part by the National Science Council under the Grants NSC-92-3112-B-001-018-Y, NSC-92-3112-B-001-021-Y, NSC-92-2218-E-001-001, NSC 93-2422-H-001-0001, NSC 93-2213-E-001-013 and NSC 93-2752-E-002-005-PAE, and by the Taiwan Information Security Center (TWISC), National Science Council under the Grants NSC 94-3114-P-001-001-Y and NSC 94-3114-P-011-001.

** Also with Institute of Information Science, Academia Sinica, Nankang, Taipei 115, Taiwan.

X. Deng and D. Du (Eds.): ISAAC 2005, LNCS 3827, pp. 515–523, 2005.

The MAXIMUM SUM PROBLEM can find many applications in pattern recognition, image processing and data mining [10,11]. A natural generalization of the above MAXIMUM SUM PROBLEM is to find the k segments such that their sums are the k largest over all $\frac{n(n-1)}{2}$ segments. Bae and Takaoka [12] presented an $O(kn)$ time algorithm for this problem. Bengtsson and Chen [13] recently gave an algorithm that runs in $O(n \log^2 n + k)$ time in the worst case. In this paper we will give an expected $O(n \log n + k)$ time algorithm based on a randomized algorithm which finds in expected $O(n \log n)$ time the segment whose sum is the kth smallest, for any given positive integer $1 \leq k \leq \frac{n(n-1)}{2}$. The latter problem is referred to as the SUM SELECTION PROBLEM. The randomized algorithm is based on a random sampling technique given in [14,15].

The rest of the paper is organized as follows. Section 2 gives three subroutines for solving the SUM SELECTION PROBLEM. Section 3 gives a randomized algorithm for the SUM SELECTION PROBLEM and and a randomized algorithm of the K MAXIMUM SUMS PROBLEM. Section 4 gives some conclusion.

2 Subroutines for Sum Selection Problem

We define the *rank* $r(x, P)$ of an element x in a set P to be the number of elements in P no greater than x, i.e. $r(x, P) = |\{y | y \in P, y \leq x\}|$. Given a sequence A of real numbers $a_1, a_2, \ldots, a_n$, and a positive integer $1 \leq k \leq \frac{n(n-1)}{2}$, the SUM SELECTION PROBLEM is to find the segment $A(i, j)$ over all $\frac{n(n-1)}{2}$ segments such that the rank of the sum $s(i, j) = \sum_{t=i}^{j} a_t$ in the set of possible subsequence sums is k. We will give a randomized algorithm for this problem that runs in expected $O(n \log n)$ time. Our randomized algorithm for the SUM SELECTION PROBLEM is based on three subroutines. In this section we will consider these three subproblems first.

For ease of notation, we assume each sum $s(i, j) = \sum_{t=i}^{j} a_t$ is associated with the indices i and j of the segment, and only the sum will be produced rather than the actual segment itself. For convenience we shall without confusion use the segment $A(i, j)$ and its corresponding sum $s(i, j)$ interchangeably.

We define the set $P = \{x_0, x_1, \ldots, x_n\}$, where $x_i = \sum_{t=1}^{i} a_t$, $i = 1, 2, \ldots, n$ and $x_0 = 0$ is the prefix sum $A(1, i)$ of the sequence $A(1, n)$. Thus, $s(i, j)$ of $A(i, j)$ is then equal to $x_j - x_{i-1}$. Let $P_j = \{x_0, x_1, \ldots, x_j\}$. We will maintain an order-statistic tree $T(P_j)$ on P_j. An *order-statistic tree* [16] is simply a balanced binary search tree with additional information, $size[z]$, stored in each node z of the tree, containing the total number of nodes in the subtree rooted at z. $size[z]$ is defined as $size[z] = size[left[z]] + size[right[z]] + 1$, where $left[z]$ and $right[z]$, denote respectively the left and right children of node z, and $size[z] = 1$, if z is a leaf node. The order-statistic tree $T(P_j)$ allows the rank of an element x to be determined in $O(\log n)$ time, i.e. we can find the rank $r(x, P_j) = |\{y | y \in P_j, y \leq x\}|$ for any x not necessarily in P_j in $O(\log n)$ time, retrieve an element in $T(P_j)$ with a given rank in $O(\log n)$ time and maintain both insertion and deletion operations in $T(P_j)$ in $O(\log n)$ time.

The first subproblem, called *reporting version* of SUM RANGE QUERY PROBLEM, is considered as follows: Given a sequence A of n real numbers $a_1, a_2, \ldots, a_n$ and two real numbers s_ℓ, s_r with $s_\ell \leq s_r$, find all segments $A(i,j)$, $1 \leq i \leq j \leq n$, among all $\frac{n(n-1)}{2}$ segments such that their sums $s(i,j)$ satisfy $s_\ell \leq s(i,j) \leq s_r$. To solve this subproblem, it suffices to iterate on each j finding all $x_i \in P_{j-1}$ such that $s_\ell \leq x_j - x_i \leq s_r$. At each iteration j, we can maintain $T(P_{j-1})$ dynamically such that we can find all the numbers $x_i \in [x_j - s_r, x_j - s_\ell]$ by binary search in $O(\log n + h_j)$ time, where h_j is the total number of segments $s(i,j)$, for all $i < j$ such that $s_\ell \leq s(i,j) \leq s_r$ and then we can insert x_j into $T(P_{j-1})$ to obtain $T(P_j)$ in $O(\log n)$ time.

Lemma 1. *The reporting version of the* SUM RANGE QUERY PROBLEM *can be solved in $O(n)$ space and $O(n \log n + h)$ time, where h is the output size.*

The second subproblem, called *counting version* of SUM RANGE QUERY PROBLEM, is defined as before, except that we want to find the *number* of segments that satisfy the range query.

To solve this subproblem, it suffices to iterate on each j counting the total number of elements x_i in P_{j-1} such that $s_\ell \leq x_j - x_i \leq s_r$. At each iteration j, we can make a rank query to the order-statistic tree $T(P_{j-1})$ to count the total number, say α_j, of elements $x_i \in P_{j-1}$ such that $x_i < x_j - s_r$ and make another rank query to count the total number, say β_j, of elements $x_i \in P_{j-1}$ such that $x_i \leq x_j - s_\ell$. After these two queries we insert x_j into $T(P_{j-1})$ to obtain $T(P_j)$. We get $t_j = \beta_j - \alpha_j$ to be the total number of elements in P_{j-1} lying in $[x_j - s_r, x_j - s_\ell]$. Hence $t = t_1 + t_2 + \ldots + t_n$ is the total number of segments such that their sums are between s_ℓ and s_r.

Lemma 2. *The counting version of the* SUM RANGE QUERY PROBLEM *can be solved in $O(n)$ space and $O(n \log n)$ time.*

We now address the third subproblem called, RANDOM SAMPLING SUM SELECTION PROBLEM which is defined as follows: Given a sequence A of n real numbers $a_1, a_2, \ldots, a_n$ and two real numbers s_ℓ, s_r with $s_\ell \leq s_r$, randomly generate n segments among all $\frac{n(n-1)}{2}$ segments such that their sums are between s_ℓ and s_r. Let $N \geq n$ be the total number segments among all $\frac{n(n-1)}{2}$ segments such that their sums are between s_ℓ and s_r. Note that $N = t_1 + t_2 + \ldots + t_n$ can be obtained by running the counting algorithm for SUM RANGE QUERY PROBLEM, where t_j is the total number of segments $s(i,j), i < j$, such that $s_\ell \leq s(i,j) \leq s_r$. We first select allowing duplicates, n random integers $R = \{r_1, r_2, \ldots, r_n\}$ distributed uniformly in the range from 1 to N. Since $N = O(n^2)$ we can sort them by radix sort and rename them such that $r_1 \leq r_2 \leq \ldots \leq r_n$ in $O(n)$ time. Let $\tau_j = t_1 + t_2 + \ldots + t_j$. For each j there exist $r_c, r_{c+1}, \ldots, r_{c+d} \in R$ such that $\tau_{j-1} < r_c \leq r_{c+1} \leq \ldots \leq r_{c+d} \leq \tau_j$. We would like to find segments $\overline{s}_c, \overline{s}_{c+1}, \ldots, \overline{s}_{c+d}$ with a one-to-one correspondence respectively to $r_c, r_{c+1}, \ldots, r_{c+d}$. To do so, we first make a rank query to the order-statistic tree $T(P_{j-1})$ to count the total number α_j of elements $x_i \in P_{j-1}$ such that $x_i < x_j - s_r$. We then retrieve the element $x_{q_i} \in P_{j-1}$ so that x_{q_i} has a rank equal

to $(\alpha_j + r_{c+i} - r_{j-1})$ in P_{j-1}, and let $\bar{s}_{c+i} = s(q_i, j)$ for each $i = 0, 1, ..., d$. We thus can obtain the set of random sampling sums or segments, $\overline{S} = \{\bar{s}_1, \bar{s}_2, \ldots, \bar{s}_n\}$.

Lemma 3. *The* RANDOM SAMPLING SUM SELECTION PROBLEM *can be solved in $O(n)$ space and $O(n \log n)$ time.*

3 Algorithm for Sum Selection Problem and k Maximum Sums Problem

In this section we consider the SUM SELECTION PROBLEM and then use it to solve the k MAXIMUM SUMS PROBLEM. Given a sequence A of real numbers $a_1, a_2, \ldots, a_n$, and a positive integer k, the SUM SELECTION PROBLEM is to find the segment $A(i,j)$ over all $\frac{n(n-1)}{2}$ segments such that the rank of the sum $s(i,j), 1 \leq i \leq j \leq n$ is k. We will give a randomized algorithm for this problem by random sampling technique [14,15].

We shall consider a more general version, called SUM SELECTION RANGE QUERY PROBLEM defined as follows. Given an interval $[s_l, s_r]$ which contains N segments whose sums are between s_l and s_r, we would like to find the kth smallest segment among these N segments in the interval. Let s^* denote the sum of the kth smallest segment in the interval. Note that the SUM SELECTION PROBLEM is just a special case of this problem such that $N = \frac{n(n-1)}{2}$ and $[s_\ell, s_r] = (-\infty, \infty)$.

The randomized algorithm for the SUM SELECTION RANGE QUERY PROBLEM will contract the interval $[s_l, s_r]$ into a smaller subinterval $[s_{l'}, s_{r'}]$ such that it also contains s^* and the new subinterval $[s_{l'}, s_{r'}]$ contains at most $O(N/\sqrt{n})$ segments. It will repeat to contract the interval several times until the interval $[s_{l'}, s_{r'}]$ contains not only s^* but also at most $O(n)$ segments. It then outputs all the segments in $[s_{l'}, s_{r'}]$ by the reporting algorithm of the SUM RANGE QUERY PROBLEM and find the solution segment with an appropriate *rank* and whose sum is s^* by using any standard selection algorithm.

Our randomized algorithm for the SUM SELECTION RANGE QUERY PROBLEM runs as follows: We first use the random sampling subroutine to select out of a total of N segments, n randomly independent segments $\overline{S} = \{\bar{s}_1, \bar{s}_2, \ldots, \bar{s}_n\}$ whose sums lie in the interval $[s_l, s_r]$ in $O(n \log n)$ time.

Whenever we select a random segment in $[s_l, s_r]$, it has the probability $\frac{k}{N}$ such that it is smaller than s^*. Consider such an event as a "success" in performing n independent Bernoulli trials, each with a probability of $p = \frac{k}{N}$. Let m be a random number denoting the total number of successes in the n independent Bernoulli trials. It is easy to see that random variable m will obey binomial distribution with probability density function

$$b(n, m, p) = \binom{n}{m} p^m (1-p)^{n-m}.$$

The expected value of m is $\mu = np = n\frac{k}{N}$ and the standard deviation is $\sigma = \sqrt{np(1-p)} \leq \frac{\sqrt{n}}{2}$. Hence we expect that the wth smallest element in $\overline{S}$,

where $w = \lfloor np \rfloor = \lfloor n\frac{k}{N} \rfloor$ should be a good approximation for the kth smallest segment s^*. Let $l' = \max\{1, \lfloor n\frac{k}{N} - t\frac{\sqrt{n}}{2} \rfloor\}$ and $r' = \min\{n, \lceil n\frac{k}{N} + t\frac{\sqrt{n}}{2} \rceil\}$, for some constant t to be determined later. After random sampling, we can find the l'^{th} smallest element $s_{\ell'}$ and the r'^{th} smallest element $s_{r'}$ in $\overline{S}$ by any standard selection algorithm in $O(\overline{S})$ time to obtain the subinterval $[s_{\ell'}, s_{r'}]$.

The key step of the randomized algorithm is as follows. We check the following two conditions by the counting algorithm of Sum Range Query Problem in $O(n \log n)$ time:

(1) The sum s^* of the kth smallest segment lies in the subinterval $[s_{\ell'}, s_{r'}]$.
(2) The subinterval $[s_{\ell'}, s_{r'}]$ contains at most $t^2N/((t-1)\sqrt{n})$ $(< 2tN/\sqrt{n})$ segments.

Let k_1 and k_2 be the total number of segments lying in $[s_\ell, s_{\ell'})$ and $[s_\ell, s_{r'}]$ respectively. Note that s^* lies in the subinterval $[s_{\ell'}, s_{r'}]$ if and only if $k_1 < k$ and $k_2 \geq k$. If both of these conditions hold, we replace the current interval $[s_\ell, s_r]$ by the subinterval $[s_{\ell'}, s_{r'}]$ and let $k' = k - k_1$. If either (1) or (2) is violated, we repeat the algorithm from scratch again until both (1) and (2) are satisfied: i.e. We need to select n random independent segments with replacement in the interval $[s_\ell, s_r]$ by running the random sampling algorithm again to obtain a new subinterval $[s_{\ell'}, s_{r'}]$ and then check the above two conditions (1) and (2) for the new subinterval $[s_{\ell'}, s_{r'}]$.

Since the algorithm of Sum Selection Problem starts with $N = \frac{n(n-1)}{2}$ segments in the initial interval $[s_\ell, s_r] = (-\infty, \infty)$, after the first successful random sampling which satisfies conditions (1) and (2) we have an interval $[s_{\ell'}, s_{r'}]$ which contains the k'^{th} smallest segment with sum s^* and has $O(n^2/\sqrt{n}) = O(n^{\frac{3}{2}})$ segments and after the second successful random sampling which satisfies conditions (1) and (2) we have an interval $[s_{\ell''}, s_{r''}]$ which contains the k''^{th} smallest segment with sum s^* and has $O(n^{\frac{3}{2}}/\sqrt{n}) = O(n)$ segments. Then, we can enumerate all segments in the interval $[s_{\ell''}, s_{r''}]$ in $O(n \log n + n) = O(n \log n)$ time by the reporting algorithm of Sum Range Query Problem and select the k''^{th} smallest segment with sum s^* from those segments by using any standard selection algorithm in $O(n)$ time.

We now show that with a high probability, the kth smallest segment with sum s^* lies in the subinterval $[s_{\ell'}, s_{r'}]$ and the subinterval $[s_{\ell'}, s_{r'}]$ contains at most $t^2N/((t-1)\sqrt{n})$ segments.

Let us introduce the famous Chernoff's bound in probability theory.

Lemma 4 (Chernoff's bound [17,18]). *Let $X_1, X_2, \cdots, X_n$ be independent random variables, each attaining value 1 with probability p and value 0 with probability $1-p$. Let $X = X_1 + X_2 + \cdots + X_n$ and $\mu = np$ be the expectation of X. Then for any $t \geq 0$ we have*

(1) $Pr[X \leq \mu - t\sqrt{n}] \leq e^{-2t^2}$.
(2) $Pr[X \geq \mu + t\sqrt{n}] \leq e^{-2t^2}$.

(3) $Pr[X \leq (1-t)\mu] \leq e^{-t^2\mu/2}$.
(4) $Pr[X \geq (1+t)\mu] \leq (\frac{e^t}{(1+t)^{(1+t)}})^{\mu}$.

Lemma 5. *For a random choice of n independent segments with replacement among the N segments in the interval* $[s_\ell, s_r]$*, the probability that the subinterval* $[s_{\ell'}, s_{r'}]$ *contains at least* $t^2N/((t-1)\sqrt{n})$ *segments is at most* $e^{-\sqrt{n}/(2(t-1))}$*, where* ℓ' *and* r' *are as defined earlier .*

Proof. Assume that a random sampling $\overline{S}$ in the algorithm of the Sum Selection Range Query Problem obtains a subinterval $[s_{\ell'}, s_{r'}]$ containing more than $t^2N/((t-1)\sqrt{n})$ segments for a random choice of n independent segments with replacement among the N segments in the interval $[s_\ell, s_r]$. Hence, whenever we select a random segment s_i in $[s_\ell, s_r]$, it has probability larger than $\frac{t^2N/((t-1)\sqrt{n})}{N} = \frac{t^2}{(t-1)\sqrt{n}}$ such that s_i lies in $[s_{\ell'}, s_{r'}]$. We again think such an event as a "success", each with a probability of success equal to $p \geq t^2/((t-1)\sqrt{n})$. Let X_i be the random variable, attaining value 1 with probability $p \geq t^2/((t-1)\sqrt{n})$ if the ith selected segment falls in $[s_{\ell'}, s_{r'}]$ and value 0 with probability $1-p$ if the ith selected segment falls outside $[s_{\ell'}, s_{r'}]$. Let $X = X_1 + X_2 + \cdots + X_n$ be the total number of selected segments falling in $[s_{\ell'}, s_{r'}]$. The expectation of the random experiment is $\mu = np \geq nt^2/((t-1)\sqrt{n}) = t^2\sqrt{n}/(t-1)$. Note that $s_{\ell'}$ and $s_{r'}$ are the l'^{th} and r'^{th} smallest elements in the random sampling $\overline{S}$ respectively. It means that the random sampling $\overline{S}$ contains exactly $r' - \ell'$ ($\leq t\sqrt{n}$) successful sample segments lying in $[s_{\ell'}, s_{r'}]$. By the Chernoff bound, we have $Pr[X \leq t\sqrt{n}] \leq Pr[X \leq (1-\frac{1}{t})\mu] \leq e^{-\mu/2t^2} \leq e^{-\sqrt{n}/2(t-1)}$.

Lemma 6. *For a random choice of n independent segments with replacement among the N segments in the interval* $[s_\ell, s_r]$*, the probability that the* k^{th} *smallest segment with sum* s^* *not lying in the subinterval* $[s_{\ell'}, s_{r'}]$ *is at most* $2e^{-t^2/2}$*, where* ℓ' *and* r' *are defined as earlier.*

Proof. Let X_i be the random variable, attaining value 1 with probability $p = \frac{k}{N}$ if the ith sample segment is smaller than or equal to s^* and value 0 with probability $1-p$ if the ith sample segment is larger than s^*. If the r'^{th} smallest segment $s_{r'}$ in $\overline{S}$ smaller than s^*, it means that at least r' among the n randomly sample segments fall before s^*. Let $X = X_1 + X_2 + \cdots + X_n$ be the total number of sample segments falling before s^* and $\mu = np$. By the Chernoff bound, we have $Pr[X \geq r'] = Pr[X \geq \mu + t\frac{\sqrt{n}}{2}] \leq e^{-t^2/2}$. Similarly, By the Chernoff bound, we have $Pr[X \leq l'] = Pr[X \leq \mu - t\frac{\sqrt{n}}{2}] \leq e^{-t^2/2}$.

Now, we can choose t large enough such that $2e^{-t^2/2} \leq \frac{1}{4}$ and then we can find some large enough positive integer n_0 such that $e^{-\sqrt{n}/(2(t-1))} \leq \frac{1}{4}$ for all $n \geq n_0$. For example, we can choose $t = 3$ and $n_0 = 31$ respectively. Therefore, if the size n of the input sequence is larger than or equals to n_0, we just need to repeat the key step at most twice on the average in the randomized algorithm

of the SUM SELECTION RANGE QUERY PROBLEM. Otherwise we can solve the SUM SELECTION RANGE QUERY PROBLEM directly by brute force.

We thus conclude with the following theorem.

Theorem 1. *The* SUM SELECTION PROBLEM *can be solved in $O(n)$ space and expected $O(n \log n)$ time.*

Theorem 2. *The k* MAXIMUM SUMS PROBLEM *can be solved in $O(n)$ space and expected $O(n \log n + k)$ time.*

Proof. Let $\ell = \frac{n(n-1)}{2} - k + 1$ and $r = \frac{n(n-1)}{2}$. We can run the randomized algorithm of the SUM SELECTION PROBLEM to obtain the ℓth smallest segment s_ℓ and r^{th} smallest segment s_r respectively in expected $O(n \log n)$ time and then we can enumerate them by the algorithm for the *reporting version* of the SUM RANGE QUERY PROBLEM in the interval $[s_\ell, s_r]$ in $O(n \log n + k)$ time.

Remark 1. Combining the algorithm for the SUM SELECTION PROBLEM and the algorithm for the *reporting version* of the SUM RANGE QUERY PROBLEM not only allows us to find the first k largest sum segments in expected $O(n \log n + k)$ time but also allows us to find all segments the ranks of whose sum, are between k_1 and k_2 in expected $O(n \log n + (k_2 - k_1))$ time, where k_1 and k_2 are two positive integers such that $1 \leq k_1 \leq k_2 \leq \frac{n(n-1)}{2}$.

We now consider the K MAXIMUM SUMS PROBLEM for higher dimensional case. Bengtsson and Chen [13] use one-dimensional algorithm of K MAXIMUM SUMS PROBLEM as a subroutine to solve the problem for higher dimensional case. We will follow their idea but use our one-dimensional randomized algorithm of K MAXIMUM SUMS PROBLEM as a subroutine. We consider the two-dimensional k MAXIMUM SUMS PROBLEM as an example. Given an $m \times n$ matrix $A = (a_{i,j})$ of real numbers (assuming that $m \leq n$), the objective is to find the k submatrices such that their sums are the k largest values over all $O(m^2 n^2)$ submatrices. The pseudo-code of our two-dimensional randomized algorithm of k MAXIMUM SUMS PROBLEM is as follows:

1. Compute a new matrix $B = (b_{i,j})$, where $b_{i,j} = \sum_{t=1}^{i} a_{t,j}$, in $O(mn)$ time;
2. For each i and j, $1 \leq i \leq j \leq m$
   (a) Create a sequence $A_{i,j} = a_1, a_2, \ldots, a_n$, where $a_t = b_{j,t} - b_{i-1,t}$;
   (b) Solve the k MAXIMUM SUMS PROBLEM on $A_{i,j}$ by one-dimensional algorithm and output the set $S_{i,j}$ of the k segments such that their sums are the k largest values over all segments in $A_{i,j}$;
3. Select the k largest elements in $\bigcup_{1 \leq i \leq j \leq m} S_{i,j}$ by any standard selection algorithm in $O(m^2 k)$ time;

Theorem 3. *The two-dimensional k* MAXIMUM SUMS PROBLEM *of an $m \times n$ matrix can be solved in expected $O(m^2(n \log n + k))$ time.*

We can further apply the Bengtsson and Chen's d-dimensional algorithm to solve the d-dimensional k MAXIMUM SUMS PROBLEM in $O(n^{2d-2}(n \log n + k))$ time. It improves the previous algorithm [13] that runs in $O(n^{2d-2}(n \log^2 n + k))$ time in the worst case.

4 Conclusion

In the paper we have presented a randomized algorithm for the SUM SELECTION PROBLEM that runs in expected $O(n \log n)$ time. We have applied it to give a more efficient algorithm for the K MAXIMUM SUMS PROBLEM that runs in expected $O(n \log n + k)$ time. It is better than the previously best known result for the problem, but whether one can find a deterministic algorithm for the problem that runs within the same time bound is still open.

References

1. Bentley, J. Programming perals: algorithm design techniques. *Commun. ACM*, 27, 9:865-873, 1984.
2. Bentley, J. Programming perals: algorithm design techniques. *Commun. ACM*, 27, 11:1087-1092, 1984.
3. Gries, D. A note on the standard strategy for developing loop invariants and loops. *Science of Computer Programming*, 2:207-214, 1982.
4. Smith, D. Applications of a strategy for designing divide-and-conquer algorithms. *Science of Computer Programming*, 8:213-229, 1987.
5. Tamaki, H., Tokuyama, T. Algorithms for the maximum subarray problem based on matrix multiplication. *Proceedings of the ninth annual ACM-SIAM symposium on Discrete algorithms*, 446-452, 1998.
6. Takaoka, T. Efficient algorithms for the maximum dubarray problem by fistance matrix multiplication. *Proceedings of the 2002 australian theory symposium*, 189-198, 2002.
7. Alk, S., Guenther, G. Application of broadcasting with selective reduction to the maximal sum subsegment problem. *International journal of high speed computating*, 3:107-119, 1991.
8. Qiu, K., Alk, S. Parallel maximum sum algorithms on interconnection networks. *Technical Report No. 99-431, Jodrey School of Computer Science, Acadia University, Canada*, 1999.
9. Perumalla, K., Deo, N. Parallel algorithms for maximum subsequence and maximum subarray. *Parallel Processing Letters*, 5:367-373, 1995.
10. Fukuda, T., Morimoto, Y., Morishita, S. Tokuyama, T. Mining association rules between sets of items in large databases. *Proceedings of the 1996 ACM SIGMOD international conference on management of data*, 13-23, 1996.
11. Agrawal, R., Imielinski, T. Swami, A. Data mining using two-dimensional optimized association rules: scheme, algorithms, and visualization. *Proceedings of the 1993 ACM SIGMOD international conference on management of data*, 207-216, 1993.
12. Bae, S. E., Takaoka, T. Algorithms for the problem of k maximum sums and a VLSI algorithm for the k maximum subarrays problem. *2004 International Symposium on Parallel Architectures, Algorithms and Networks*, 247-253, 2004.
13. Bengtsson, F., Chen, J. Efficient Algorithms for k Maximum Sums. *Algorithms and Computation, 15th International Symposium, ISAAC 2004*, 137-148.
14. Michael H. Dillencourt, David M. Mount, and Nathan S. Netanyahu. A Randomized Algorithm for slope selection. *International Journal of Computational Geometry and Applications*, vol. 2, No. 1:1–27, 1992.

15. J. Matoušek. Randomized optimal algorithm for slope selection. *Information Processing Letters*, 39(4):183–187, 1991.
16. Thomas. H. Cormen, Charles. E. Leiserson. and Ronald L. Rivest. Introdution to Algorithms, MIT Press.
17. Noga Alon, Joel H. Spencer and Paul Erdős. The Probabilistic Method, Wiley-Interscience Series.
18. Rajeev Motwani, Probhakar Raghavan. Randomized Algorithms, Cambridge.

An Improved Interval Routing Scheme for Almost All Networks Based on Dominating Cliques
(Extended Abstract)

Martin Nehéz[1] and Daniel Olejár[2]

[1] Department of Computer Science and Engineering,
FEE CTU Prague, Karlovo náměstí 13,
121 35 Praha 2, Czech Republic
nehez@fel.cvut.cz
[2] Department of Computer Science,
FMPI, Comenius University in Bratislava,
Mlynská dolina,
842 48 Bratislava, Slovak Republic

Abstract. Motivated by the peer-to-peer content sharing systems in large-scale networks, we will study interval routing schemes in Erdös-Rényi random graphs. C. Gavoille and D. Peleg [13] posed an open question of whether almost all networks support a shortest-path interval routing scheme with 1 interval. In this paper, we answer this question partially by proving that in almost all networks, there is an interval routing scheme with 1 interval up to additive stretch 2. Our proof is based on the properties of dominating cliques in random graphs.

Keywords: Interval routing, random graphs, dominating cliques, additive stretch routing scheme.

1 Introduction

Routing in networks is a fundamental operation in distributed computing. However, the effectiveness of routing can deteriorate rapidly especially in large networks such as Internet in case of heavy traffic. Hence, the question of effectiveness of routing strategies is of great significance. Several universal routing techniques are known. We will focus of one of them in this paper - interval routing.

The interval routing is space-efficient routing method which historically has been implemented in INMOS C104 router [22] and recently, in Freenet peer-to-peer (shortly P2P) content sharing system [3]. The latter one uses the idea of interval routing for retrieving files from local datastores according to keys. This application has been successfully implemented in decentralized large-scale distributed systems based on the P2P paradigm. An interesting question arises in this context that what is an authentic model of typical large-scale decentralized networks. Based on the several previous works, we will focus on the networks modelled by random graphs [4,9,13,19].

X. Deng and D. Du (Eds.): ISAAC 2005, LNCS 3827, pp. 524–532, 2005.

1.1 Model of Random Graphs

We consider a point-to-point communication network modeled by a simple connected graph $G = (V, E)$, where V is the set of nodes (or processors or routers) and E is a set of edges (or bidirectional communication links). To describe almost all networks we will use the random graph model defined as follows. Let p, $0 \leq p \leq 1$, be a *probability of an edge*. The *(probabilistic) model of random graphs* $\mathbb{G}(n,p)$ consists of all graphs with n-node set V such that each graph has at most $\binom{n}{2}$ edges being inserted independently with probability p. Equivalently, if G is a graph with node set V and it has $|E(G)|$ edges, then:

$$\Pr[G] = p^{|E(G)|}(1-p)^{\binom{n}{2}-|E(G)|} ,$$

where Pr is a probability measure defined on $\mathbb{G}(n,p)$. This model is also called *Erdös-Rényi random graph model* [1,16].

Let A be any set of graphs from $\mathbb{G}(n,p)$ with a property Q. We say that *almost all graphs (networks)* have the property Q iff:

$$\Pr[A] \to 1 \quad as \quad n \to \infty .$$

1.2 Interval Routing

Given an n-node graph $G = (V, E)$, an interval routing is based on a suitable labeling scheme for nodes and edges in G. A node label is an element of the set $\{1, \ldots, n\}$ and an arc label (i.e., outgoing edge label) is a cyclic interval $[a, b]$ with $a, b \in \{1, \ldots, n\}$, where $[a, b] = \{a, a+1, \ldots, b\}$ for $a \leq b$ and $[a, b] = \{a, a+1, \ldots, n, 1, \ldots, b\}$ for $a > b$. Formally, a pair $(\mathcal{L}, \mathcal{I})$ is an *interval labeling scheme* on G (shortly ILS), if:

1. $\mathcal{L} : V \to \{1, \ldots, |V|\}$ is one-to-one mapping (labeling) of V,
2. $\mathcal{I}$ is an arc-labeling $\mathcal{I} : E \to 2^{\mathcal{L}(V)}$, such that:
   - for every $v \in V$, $\{ \mathcal{I}(v,u) \mid (v,u) \in E \} \cup \mathcal{L}(v) = \{1, \ldots, |V|\}$;
   - for every distinct arcs (v,u), (v,w) of E, $\mathcal{I}(v,u) \cap \mathcal{I}(v,w) = \emptyset$.

If the edge labeling assigns at most k intervals per arc and the routing strategy guarantees that the messages always arrive at their destinations, then the ILS is said to be a *k interval routing scheme* (k-IRS).

We will deal with *additive stretched routing schemes*, rather than multiplicative stretched ones. A path of a graph is *δ-stretched* if the length of the path is at most the length of a shortest path between its extremities plus δ. A δ-stretched k-IRS, denoted by (k, δ)-IRS, is a k-IRS for which all the routes are δ-stretched paths. A $(k, 0)$-IRS is also called a *shortest-path* k-IRS. Fig. 1 depicts two interval routing schemes on a 6-cycle. For more precise formulations and other details, see [11,12,26]. The problem of determining whether a graph supports a shortest path 1-IRS is NP-complete [6].

If G is a connected graph, then we denote by $\mathrm{IRN}_\delta(G)$ the smallest integer k such that G supports a (k, δ)-IRS. The number $\mathrm{IRN}_0(G)$ is also called the *compactness* of G.

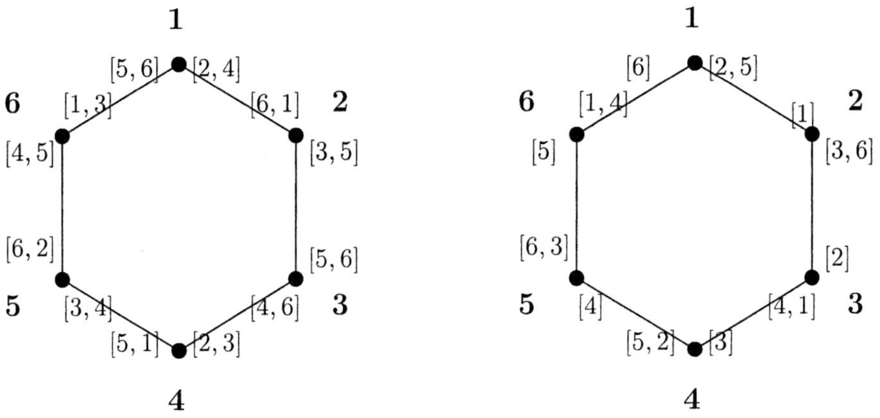

Fig. 1. A shortest-path 1-IRS (left side) and a 2-stretched 1-IRS (right side)

1.3 Previous Works and Our Result

Flammini, van Leeuwen, and Marchetti-Spaccamela [7] proved a non-constant lower bound on $(k,0)$-IRS on $\mathbb{G}(n,p)$. They pointed out that, with high probability, a random graph from $\mathbb{G}(n,p)$ requires $\Omega(n^{1-1/\Theta(\sqrt{\log n})})$ intervals per outgoing edge for some specific value of p, namely for $p = 1/n^{1-1/\Theta(\sqrt{\log n})}$. On the other hand, Gavoille and Peleg [13] proved that almost all graphs (that is a fraction of $1-o(1)$ of all n-node graphs, or equivalently the graphs of $\mathbb{G}(n,p)$ for $p = 0.5$ and with high probability) support a $(2,0)$-IRS. Actually, they constructed a routing scheme such that every node has at most $O(\log^3 n)$ outgoing edges with 2 intervals, all the other ones having 1 interval. They also posed an open question of whether almost all graphs support $(1,0)$-IRS.

Our main result is a refinement of the result from Gavoille and Peleg [13]. We prove that almost all graphs support a $(1,2)$-IRS.

Theorem 1. *For every fixed $p > 1/2$, a random graph $G \in \mathbb{G}(n,p)$ satisfies $IRN_2(G) = 1$ with probability $1 - O\left((\log n)^{-3}\right)$.*

The rest of this paper contains the proof of the Theorem 1. The main idea is based on the properties of dominating cliques in graphs. Conditions for the existence of dominating cliques in random graphs are analyzed in Section 2. A construction of a $(1,2)$-IRS for graphs that contain dominating cliques is described in Section 3. By combining of these two results we prove Theorem 1.

2 Dominating Cliques in Random Graphs

To obtain our main result it is necessary to study properties of dominating cliques in random graphs. Given a graph $G = (V,E)$, a set $S \subseteq V$ is said to be a *dominating set* of G if each node $v \in V$ is either in S or is adjacent to a node

in S. The *domination number* $\gamma(G)$ is the minimum cardinality of a dominating set of G.

There are several alternative definitions of the dominating set [15]. The following one is important for the our purpose. Given two nodes $u, v \in V$, let $d_G(u,v)$ denote the distance between u and v in G. Let $\Gamma(u) = \{v \in V \mid d_G(u,v) \leq 1\}$ denote a *ball* of radius 1 centered at u. For every subset $S \subseteq V$, let $\Gamma(S) = \cup_{u \in S}\Gamma(u)$. A subset S is said to be a *dominating set* of G if $\Gamma(S) = V$.

A *clique* in G is a maximal set of mutually adjacent nodes of G, i.e., it is a maximal complete subgraph of G. The *clique number*, denoted $cl(G)$, is the number of nodes of clique of G. If a subgraph induced by a dominating set is a clique in G then the induced subgraph is called a *dominating clique* in G. Dominating sets and cliques are basic structures in graphs that have been investigated very intensively. To determine whether the domination number of a graph is at most r is an NP-complete problem [10]. The maximum-clique problem is one of the first shown to be NP-hard [18]. A well-known celebrated result of B. Bollobás, P. Erdös et al. is a proof that the clique number in random graphs is bounded by a very tight bounds [1,2,17,21,24,25]. The domination number of a random graph have been studied by B. Wieland and A. P. Godbole in [28].

2.1 Results for Random Graphs

For $r > 1$, let S be a r-node subset of an n-node graph G. Let A denote the event that "S is a dominating clique of $G \in \mathbb{G}(n,p)$". Let in_r be the associated 0-1 (indicator) random variable on $\mathbb{G}(n,p)$ defined as follows: $in_r = 1$ if G contains a dominating clique S and $in_r = 0$, otherwise. Let X_r be a random variable that denotes the number of r-node dominating cliques. More precisely, $X_r = \sum in_r$ where the summation ranges over all sets S. The following lemma expresses the expectation of X_r.

Lemma 1. *For the expectation $E(X_r)$ of the random variable X_r*

$$E(X_r) = \binom{n}{r} p^{\binom{r}{2}} (1 - p^r - (1-p)^r)^{n-r} . \tag{1}$$

Proof. The linearity of the expectation leads to

$$E(X_r) = \sum E(in_r) = \sum in_r \cdot \Pr[A] ,$$

over all r-node sets S. The nodes of the S can be chosen in $\binom{n}{r}$ ways. Since S is a complete subgraph, every of its r nodes has to be joined with the remaining $r-1$ nodes of S. Hence, the probability of this fact is $p^{\binom{r}{2}}$. The last term in (1) expresses the probability that S is a clique spanning a dominating set of $G \in \mathbb{G}(n,p)$. More precisely, let v be an arbitrary but fixed node, $v \notin V(S)$; v is said to be a "good" node (i.e., the node which does not spoil the "cliqueness" and the "domination" of S), if it cannot be joined neither with all nodes of S

nor with none of them. It follows that v has to be joined at least with one and at most with $r-1$ of nodes of S. Therefore,

$$\Pr[\ v \textit{ is a "good" node with respect to } S\ ] = \sum_{0<j<r} \binom{r}{j} p^j (1-p)^{r-j} =$$

$$= \left[\sum_{j=0}^{r} \binom{r}{j} p^j (1-p)^{r-j} \right] - p^r - (1-p)^r = 1 - p^r - (1-p)^r .$$

All of $(n-r)$ nodes from $V(G) \setminus V(S)$ must be "good". Hence, the proof is complete. □

The estimation of the variance $Var(X_r)$ seems to be more difficult problem. However, we can use the fact that the clique number in random graphs is bounded within the tight interval. This assumption leads to the simplification of the enumeration of the variance. Therefore, we introduce the following notations.

Let us denote $1/p$ by b. Let

$$r_0 = \log_b n - 2\log_b \log_b n + \log_b 2 + \log_b \log_b e , \tag{2}$$

$$r_1 = 2\log_b n - 2\log_b \log_b n + 2\log_b e + 1 - 2\log_b 2 . \tag{3}$$

J. G. Kalbfleisch and D. W. Matula [17,21] proved that a random graph from $\mathbb{G}(n,p)$ does not contain cliques of the order greater than $\lceil r_1 \rceil$ and less or equal than $\lfloor r_0 \rfloor$. (See also [2,24,25].) D. Olejár and E. Toman [24] used the bounds (2) and (3) to obtain an estimation of the number of cliques in random graphs. To obtain an estimation of the $Var(X_r)$ we will apply a similar approach.

Lemma 2. *Let p be fixed, $0 < p < 1$ and $\lfloor r_0 \rfloor \le r \le \lceil r_1 \rceil$. Let*

$$\beta = \min\{\ 2/3,\ \ -2\log_b(1-p)\ \} .$$

Then:

$$Var(X_r) = E(X_r)^2 \cdot O\left(\frac{(\log n)^3}{n^\beta} \right) . \tag{4}$$

The following claim expresses the number of the dominating cliques in random graphs.

Lemma 3. *Let p, r and β be as before, and*

$$X_r = \binom{n}{r} p^{\binom{r}{2}} (1 - p^r - (1-p)^r)^{n-r} \times \left\{ 1 + O\left(\frac{(\log n)^3}{n^{\beta/2}} \right) \right\} . \tag{5}$$

With probability $1 - O\left((\log n)^{-3}\right)$, a random graph from $\mathbb{G}(n,p)$ contains X_r dominating cliques on r nodes.

Proof. It follows from the Chebyshev's inequality [16]: if $Var(X)$ exists, then:

$$\Pr[\ |X - E(X)| \geq t\ ] \geq \frac{Var(X)}{t^2}\ , \qquad t > 0\ .$$

Letting $t = E(X_r) \cdot (\log n)^3 \cdot n^{-\beta/2}$ and using lemma 2, we obtain the assertion of lemma 3. □

For $r > 1$, let Y_r be the random variable on $\mathbb{G}(n,p)$ which denotes the number of r-node cliques. According to [24],

$$Y_r = \binom{n}{r} p^{\binom{r}{2}} (1 - p^r)^{n-r} \times \left\{ 1 + O\left(\frac{(\log n)^3}{\sqrt{n}} \right) \right\} \tag{6}$$

with probability $1 - O\left((\log n)^{-3}\right)$.

The ratio X_r/Y_r expresses the relative number of dominating cliques to all cliques of $\mathbb{G}(n,p)$ and it attains the value within the interval $[0, 1]$. By analysis of cases whether X_r/Y_r tends to 1, we obtain the main result of this section.

Lemma 4. *For every fixed $p > 1/2$, a random graph from $\mathbb{G}(n,p)$ contains a dominating clique with probability* $1 - O\left((\log n)^{-3}\right)$.

3 Interval Routing in Dominating Cliques

In this section, we will construct an $(1,2)$-IRS for graphs that contains dominating cliques. According to Lemma 4, it completes the proof of the Theorem 1.

Lemma 5. *Let a graph G contains a nonempty dominating clique. Then:*

$$IRN_2(G) = 1\ .$$

Sketch of the proof. Let $G = (V, E)$ be an n-node graph that contains a nonempty dominating clique $S \subseteq V$ in G. Assume S to be maximal. Assume $S \subseteq V$ with $|S| = 1$. Then G is a complete bipartite graph $K_{1,n-1}$. Hence, G is a tree and $\mathrm{IRS}_0(G) = 1$, as was stated in [27]. The assertion trivially holds.

Let $|S| = \nu \geq 2$. We will describe a linear-time algorithm that constructs an interval routing scheme $\mathcal{R} = (\mathcal{L}, \mathcal{I})$ on G. The motivation for this algorithm comes from the fact that each n-node complete graph K_n supports an 1-IRS. The general idea of our construction is based on the partition of the node-set V into ν mutually disjoint subsets $X_1, \ldots, X_\nu$ such that each X_i contains exactly one node from S. Consequently, the all nodes from the partition $\cup_{i=1}^{\nu} X_i$ will attain labels from the set $\{1, \ldots, n\}$ such that $\mathcal{L}(X_i)$ forms one particular interval of the node label set $\{1, \ldots, n\}$, where

$$\mathcal{L}(X) = \cup_{v \in X} \mathcal{L}(v)\ .$$

We will proceed by induction.

First, select an arbitrary node $x \in S$ and let X_1 contain x and all nodes from $V \setminus S$ that are adjacent to x. Set the label of x to be 1 and label any other node $u \in X_1$ by a distinct number from $\{2, \ldots, |X_1|\}$ such that for all $u, w \in X_1 : \mathcal{L}(u) \neq \mathcal{L}(w)$ whenever $u \neq w$.

Assume that $X_1, \ldots, X_m$ are mutually disjoint sets such that each of which contains exactly one node from S and $2 \leq m < \nu$. It holds that all nodes from the union $\cup_{i=1}^{m} X_i$ are labeled by the numbers from the set $\{1, \ldots, l\}$, where $l = | \cup_{i=1}^{m} X_i |$. Select an arbitrary node $x^* \in (S \setminus (\cup_{i=1}^{m} X_i))$ and construct X_{m+1} such that it contains the node x^* and all the nodes from $V \setminus [(\cup_{i=1}^{m} X_i) \cup S]$ that are adjacent to x^*. Label each node from X_{m+1} by a distinct number from $\{l+1, \ldots, l + |X_{m+1}|\}$ such that for all $u, w \in X_{m+1}$ it holds $\mathcal{L}(u) \neq \mathcal{L}(w)$ whenever $u \neq w$.

The node labeling construction is done if $m = \nu$. Observe that for $m = \nu$ it holds that $\cup_{i=1}^{m} X_i = V$ and each X_i, X_j are mutually disjoint $(1 \leq i, j \leq \nu)$. The properties of the node labeling algorithm and the fact that S is dominating set yield that the function $\mathcal{L}$ is one-to-one mapping. It implies that each X_i corresponds to one unique interval of labels from $\{1, \ldots, n\}$. For each i such that $1 \leq i \leq \nu$, let us denote the interval of the labels from $\{1, \ldots, n\}$ by I_i iff it corresponds to the labels for all nodes from X_i. Formally,

$$I_i = \cup_{v \in X_i} \mathcal{L}(v) \ , \quad for \ \ i = 1, \ldots, \nu \ .$$

Note that $\cup_{i=1}^{\nu} I_i = \{1, \ldots, n\}$ and each I_i, I_j are mutually disjoint $(1 \leq i, j \leq \nu)$.

The construction of the arc-labeling follows accordingly from the node labeling. We will distinguish two cases.

Case 1. Let $v \in S$. For each $x \in S$ such that $x \neq v$ set the label of the outgoing edge (v, x) to be interval I_j corresponding to the set X_j such that $x \in X_j$. (Note that there exists exactly one such j.) It means that the messages from v and addressed for the all nodes from X_j will be sent via the arc (v, x) with the label I_j. For each $w \in (V \setminus S)$ with label $\mathcal{L}(w)$ such that v and w are adjacent, let us set the label of the outgoing edge (v, w) to be the one-member interval that contains only label $\mathcal{L}(v)$.

Case 2. Let $v \in (V \setminus S)$ and let $\mathcal{L}(v)$ denote its label. Thus, there exists a unique index j such that $v \in X_j$ and the intersection $S \cap X_j$ contains exactly one node, say x. Hence, the nodes v and x are adjacent. Let the label of the outgoing edge (v, x) be the interval $[\mathcal{L}(v) + 1, \mathcal{L}(v) - 1]$. It is the cyclic interval that contains all labels from the set $\{1, \ldots, n\}$ except the number $\mathcal{L}(v)$. It means that all messages sent from v will be routed via the arc (v, x). The correctness of such a routing scheme follows from the fact that x is the node of the dominating set which is also the complete subgraph. It means that x is adjacent to each node from S and each node from $V \setminus S$ is adjacent to any node from S, since it is a dominating set.

We have obtained the interval routing scheme $\mathcal{R}$. According to $\mathcal{R}$, the length of the longest routing path between two arbitrary nodes of G is at most 3, since the length of the shortest path between them may be 1. Hence, $\mathcal{R}$ is δ-stretched path for $\delta = 2$. The correctness of $\mathcal{R}$ follows from the previous description. Thus, $\mathcal{R}$ induces an $(1, 2)$-IRS for G and consequently, $\mathrm{IRN}_2(G) = 1$. □

4 Conclusions

We have proved that almost all networks support interval routing scheme with 1 interval and additive stretch at most 2. The summary of the results for random graphs is listed in Table 1. Our result (the 3rd line of the table) is a refinement of the result of C. Gavoille and D. Peleg [13] (the 2nd line). These both results hold also for the Kolmogorov model of almost all networks [4] which is closely related to $\mathbb{G}(n, 1/2)$ [5]. Clearly, the question of C. Gavoille and D. Peleg about the existence $(1, 0)$-IRS for random graphs remains open.

Table 1. Summary of results for random graphs $\mathbb{G}(n, p)$

Probability p	$\mathrm{IRN}_\delta(G)$	δ	Reference
$1/n^{1-1/\Theta(\sqrt{\log n})}$	$\Omega(n^{1-1/\Theta(\sqrt{\log n})})$	0	[7]
constant	≤ 2	0	[13]
$p > 1/2$	1	2	**Theorem 1**

Another interesting problem is to extend our result to more realistic networks. It is known that real-world networks (e.g., small world networks, internet-like networks) have several characteristics different from random graphs [20]. However, the main idea of our routing scheme could be applied if we were able to decompose large real-world networks into subnetworks (e.g., clusters) with dominating cliques.

Acknowledgement. The authors thank prof. Pavel Tvrdík for valuable comments and fruitful discussion.

This research has been supported by MŠMT under research program MSM 6840770014.

References

1. B. Bollobás: *Random Graphs (2nd edition)*, Cambridge Studies in Advanced Mathmatics 73, 2001.
2. B. Bollobás, P. Erdös: *Cliques in random graphs*, Math. Proc. Cam. Phil. Soc. (1976), 80, pp. 419–427.
3. L. Bononi: *A Perspective on P2P Paradigm and Services*, Slide courtesy of A. Montresor, URL: `http://www.cs.unibo.it/people/faculty/bononi//AdI2004/AdI11.pdf`
4. H. Buhrman, J.-H. Hoepman, P. M. B. Vitányi: *Space-efficient Routing Tables for Almost All Networks and the Incompressibility Method*, SIAM Journal on Computing, 28(4), 1999, pp. 1414–1432.
5. H. Buhrman, M. Li, J. Tromp, P. M. B. Vitányi: *Kolmogorov Random Graphs and the Incompressibility Method*, SIAM Journal on Computing, 29(2), 1999, pp. 590–599.

6. T. Eilam, S. Moran, S. Zaks: *The Complexity of the Characterization of Networks Supporting Shortest-Path Interval Routing*, In Proc. 4th Int. Colloquium on Struct. Information & Communication Complexity SIROCCO'97, (D. Krizanc and P. Widmayer, eds.), Carleton Scientific, 1997, pp. 99–111.
7. M. Flammini, J. van Leeuwen, A. Marchetti-Spaccamela: *The complexity of interval routing on random graphs*, The Computer Journal, 41(1), 1998, pp. 16–25.
8. P. Fraigniaud, C. Gavoille: *Interval Routing Schemes*, Algorithmica 21(2), 1998, pp. 155–182.
9. A. J. Ganesh, A.-M. Kermarrec, L. Massoulié: *Peer-to-Peer Membership Management for Gossip-Based Protocols*, IEEE Trans. Computers, 52(2), 2003, pp. 139–149.
10. M. R. Garey, D.S. Johnson: *Computers and Intractability*, Freeman, New York, 1979.
11. C. Gavoille: *A survey on interval routing*, Theoretical Computer Science, 245(2), 2000, pp. 217–253.
12. C. Gavoille, M. Nehéz: *Interval routing in reliability networks*, Theoretical Computer Science, 333(3), 2005, pp. 415–432.
13. C. Gavoille, D. Peleg: *The compactness of interval routing for almost all graphs*, SIAM Journal on Computing, 31(3), 2001, pp. 706–721.
14. C. Gavoille, D. Peleg, A. Raspaud, E. Sopena: *Small k-Dominating Sets in Planar Graphs with Applications*, In Proc. 27th Int. Workshop Graph-Theoretic Concepts in Comp. Sci. WG 2001, LNCS 2204, Springer-Verlag, 2001, pp. 201–216.
15. J. L. Gross, J. Yellen: *Handbook of Graph Theory*, CRC Press, 2003.
16. S. Janson, T. Luczak, A. Rucinski: *Random Graphs*, John Wiley & Sons, New York, 2000.
17. J. G. Kalbfleisch: *Complete subgraphs of random hypergraphs and bipartite graphs*, In Proc. 3rd Southeastern Conf. of Combinatorics, Graph Theory and Computing, Florida tlantic University, 1972, pp. 297–304.
18. R. M. Karp: *Reducibility among combinatorial problems*, In Complexity of Computer Computation, (R. E. Miller and J. W. Thatcher, eds.), Plenum Press, 1972, 24, pp. 85–103.
19. A.-M. Kermarrec, L. Massoulié, A. J. Ganesh: *Probabilistic Reliable Dissemination in Large-Scale Systems*, IEEE Trans. Parallel Distrib. Syst. 14(3), 2003, pp. 248–258.
20. D. Krioukov, K. R. Fall, X. Yang: *Compact Routing on Internet-like Graphs*, INFOCOM 2004 (full version published in CoRR cond-mat/0308288, 2003).
21. D. W. Matula: *The largest clique size in a random graph*, Technical report CS 7608, Dept. of Comp. Sci. Southern Methodist University, Dallas, 1976.
22. D. May, P. Thompson: *Transputers and Routers: Components for concurrent machines*, INMOS Ltd. 1990.
23. M. Nehéz: *The compactness lower bound of shortest-path interval routing on $n \times n$ tori with random faulty links*, Technical Report 582, KAM-DIMATIA Series, Charles University, Praha, 2002.
24. D. Olejár, E. Toman: *On the Order and the Number of Cliques in a Random Graph*, Math. Slovaca, 47(5), 1997, pp. 499–510.
25. E. M. Palmer: *Graphical Evolution*, John Wiley & Sons, Inc., New York, 1985.
26. P. Ružička: *On efficiency of path systems induced by routing and communication schemes*, Computing and Informatics, 20(2), 2001, pp. 181–205.
27. J. van Leeuwen, R. B. Tan: *Interval routing*, The Computer Journal, 30(4), 1987, pp. 298–307.
28. B. Wieland, A. P. Godbole: *On the Domination Number of a Random Graph*, Electronic Journal of Combinatorics, 8(1), #R37, 2001.

Basic Computations in Wireless Networks*

Ioannis Caragiannis[1], Clemente Galdi[1,2], and Christos Kaklamanis[1]

[1] Research Academic Computer Technology Institute,
Department of Computer Engineering and Informatics,
University of Patras, 26500, Rio Greece
[2] Dipartimento di Informatica ed Applicazioni "R.M. Capocelli",
Universitá di Salerno, 84081, Baronissi (SA), Italy

Abstract. In this paper we address the problem of estimating the number of stations in a wireless network. Under the assumption that each station can detect collisions, we show that it is possible to estimate the number stations in the network within a factor 2 from the correct value in time $O(\log n \log \log n)$. We further show that if no station can detect collisions, the same task can be accomplished within a factor of 3 in time $O(\log^2 n)$ and maximum energy $O(\log n)$ per node, with high probability. Finally, we present an algorithm that computes the minimum value held by the stations in the wireless network in time $O(\log^2 n)$.

1 Introduction

In the last years wireless networks have attracted a lot of attention of the scientific community. Such networks essentially constitute large scale dynamic distributed systems in which each node has a low computational power and limited lifetime. An extreme example of wireless networks are sensor networks in which, thousands of low-cost, independent entities are deployed in an area and their task is to self-organize a network in order to accomplish a specific task. Each node is equipped with sensing devices and, depending on the specific task, the sensors may be identical or there may be different kinds of nodes. The communication among the nodes is guaranteed by means of radio or laser transmitters/receivers.

One of the assumptions that facilitate the design of algorithms in wireless networks is the knowledge of the number (or an upper bound on the number) of stations in the network. However, especially in highly dynamic networks this assumption seems to be strong.

Network models. The radio network consists of n *stations*, devices running on batteries and with low computational capabilities. The stations communicate by means of a shared channel. The stations are anonymous, in the sense that they do not have, or it is not feasible to retrieve any ID or serial number.

Each station has a local clock and all the clocks are synchronized. Although such devices may have limited computational capabilities, this is not a strong assumption because of the existence of various protocols for clock synchronization

* This work was partially supported by the European Union under IST FET Integrated Project 015964 AEOLUS.

X. Deng and D. Du (Eds.): ISAAC 2005, LNCS 3827, pp. 533–542, 2005.

(e.g., [4]). During each time step, each station can send and/or receive a message. We assume a *reliable single-hop* network, that is, whenever exactly one station transmits a message, all the other stations receive it. If two or more stations send a message in the same time slot, a *collision* occurs. We distinguish three different models, depending on the information obtained by a station whenever a collision occurs: (a) *CD*: All the stations can detect collisions; (b) *strong-noCD*: The stations that send a message can detect a collision while the other stations cannot distinguish between channel noise and a collision; (c) *weak-noCD*: No station can detect a collision.

As usual, the time needed by an algorithm to execute a task is one performance measure we will consider. Furthermore, since stations have limited lifetime, energy saving is an important issue to consider. We assume that each station can be in two different states: *active/awake* or *inactive/sleep*. When a station is active, it can send and or receive messages and execute computations and we assume consumes energy 1 for each time step. When a station is in a sleep state it neither sends/receives messages nor executes computations. A station, before switching to the sleep state, sets a timer. Whenever the timer expires, the station goes back to the active state. In the sleep state we assume that the station consumes no energy.

Previous work. Various computation problems have been studied in the literature in the above models. In the case of reliable radio networks, specific solutions have been designed for computing the function sum ([1]), sorting and ranking ([3]), and for solving the problem of leader election [9,12] and network initialization, i.e., the problem of assigning unique identifiers to stations in anonymous networks, [8]. More generally, the authors in [11] present an energy efficient algorithm to simulate parallel algorithms on mesh-like wireless networks. In [6], it is proved that any algorithm designed for a single-hop network in the strong-noCD model can be simulated in the weak-noCD model, with no slowdown. However, for this solution, a $O(n)$ preprocessing time is needed. Notice that both the above solution assume the knowledge of the number of stations in the network.

All the algorithms above can be divided essentially in two (or three) different classes depending on whether or not the number of stations (or an upper bound for it) is known. It is immediate that the second scenario is much more realistic. On the other hand, designing algorithms in such a scenario is a much more complicated task. In [5] the authors provide an algorithm that computes an approximation of the number of stations in the network in the strong-noCD model, in time $O(\log^{2+\epsilon} n)$ where each station spends energy $O((\log \log n)^{\epsilon})$, for any constant $\epsilon > 0$. Note that in [5], the energy is assumed to be proportional to the number of messages transmitted/listened and not to their total size. However, several messages in that protocol may have size of $\Omega(\log n)$ bits. The algorithm presented in [5] assumes no collision detection capability and outputs, with high probability, an estimation n_0 of the number of stations that satisfies the following: $n/2^6 \leq n_0 \leq 2n$. The authors in [6] claim that similar techniques can be used also in the weak-noCD model.

Our results. In this paper we present algorithms that compute the number of stations in the network in the CD and the weak-noCD models. Let n be the number of stations in the network. We show that if each station can detect collisions, in time $O(\log n \log\log n)$ the algorithm outputs an estimation m of the number of stations such that $m/2 < n < 2m$, where m denotes the output of the algorithm. In the weak-noCD model, we present an algorithm that accomplishes the same task in time $O(\log^2 n)$, maximum energy $O(\log n)$ per station and outputs an estimation m such that $m/3 < n < 3m$. Finally we show that, under the assumption that the approximate number of station is known, there exists an algorithm that computes the minimum in the weak-noCD model in time $O(\log^2 n)$. Our algorithms are randomized and the results hold with high probability (i.e., probability $1 - O(n^{-c})$ for some positive constant c). Our protocols for estimating the number of stations use only bit messages. Due to lack of space, we only outline the proofs here. Formal proofs will appear in the final version of the paper.

2 Estimating the Number of Stations

In this section we present two algorithms to estimate the number of stations in the network. We first start by describing the algorithm in the CD model. Then, we present an algorithm in the weak-noCD model.

2.1 Estimating the Number of Stations in the CD Model

In this section we present an algorithm that computes an approximation of the number of stations based on the assumption that, whenever a collision occurs, each station in the network detects it.

The basic idea of the algorithm is the following. Assume m is an estimation of the number of stations in the network. In each *phase* of the algorithm, each station may select uniformly at random and independently from the other stations, an integer r between 1 and m. At this point, the algorithm consists of m rounds. During round r, all the stations that hold a value equal to r send a message on the channel. Since stations can detect collisions, each station counts the number of "empty" slots, i.e., the number of rounds in which no station sent a message. This number is a random variable occurring in the well-known occupancy problem where k balls are randomly thrown into b bins and is known to be sharply concentrated around its expected value as the following theorem states.

Theorem 1 ([7]). *Let $r = k/b$, and let Z be the number of empty bins when k balls are thrown randomly into b bins. For $\lambda > 0$ it holds that:*

$$\mu = E(Z) = b\left(1 - \frac{1}{b}\right)^k \simeq be^{-r} \qquad \Pr\left[|Z - \mu| \geq \lambda\right] \leq 2\exp\left(-\frac{\lambda^2}{2k}\right)$$

By Theorem 1, if the estimation is close to the actual number of stations, the number of empty slots will be around m/e. So, if the number of empty bins is

much higher (resp. much smaller) than m/e, the estimation m is increased (resp. decreased) and a new phase is executed. The above solution clearly needs $\Omega(n)$ rounds to terminate.

In order to avoid this unnecessary waste of time, we would like to use an algorithm in which each phase is executed by a polylogarithmic number of stations. Indeed, in this case, the number of stations is big enough to guarantee a good estimation of the actual number of stations in the network and, at the same time, the number of rounds needed to execute the algorithm is low. Notice that, from the point of view of energy consumption, this strategy also leads to low consumption since, in each phase, most of the stations will be in a sleep state. Unfortunately, the above strategy seems to require the knowledge of the number of stations.

One issue to address is the function we use to compute the new estimation of the number of stations. A first idea could be to simply double (or reduce by a half) the current estimation in order to obtain the new one. It is immediate that using this strategy, $O(\log n)$ phases will be enough in order to compute a good approximation of the number of sensors. On the other hand it is possible to reduce this number phases by choosing a function that reaches n more quickly, say in $O(\log \log n)$ phases. The problem with this second type of function is that if we compute an estimation m that is much higher than the actual value n, we may end up spending a huge number of rounds before realizing that the estimation is wrong.

What we are going to use is somehow derived by the strategy above. During each phase, a station decides to sleep or to execute the algorithm with a probability that is (roughly) inversely proportional to the current estimation of m. The reason of such relation is based on the idea that, whenever m grows, the number of stations executing each phase decreases. Using this relation between the number of active stations and the estimation, we guarantee that, with high probability each step in the algorithm is correct. On the other hand, the number of time slots in each phase is, of course, a function that is increasing as m increases. However, this function is polylogarithmic in m and this guarantees that every phase does not take too much time. This also allows us to use a "quick" way to compute a very rough estimation of n, starting from which we refine the estimation using a binary search.

We are now ready to define algorithm CountCD. We will use a simple building block, the procedure CountEmpty. This procedure takes as input the current estimation m of the number of sensors and works as follows. Let $f(m) = \alpha \log m$, for sufficiently large α. When invoked, a station switches to the sleep mode for $f(m)$ time slots with probability $1 - f(m)/m$, or stands awake and participates in the current phase with probability $f(m)/m$. If the station is awake, it selects an integer r between 1 and $f(m)$ and, during round r, it sends a message on the channel. Based on the number of empty slots, each node participating in the current phase decides whether the current estimation is higher than, lower than, or very close to the actual number of stations. At the end of the phase, all the stations wake up and the station that first transmitted successfully during

the current phase announces whether the current estimation is higher than, lower than, or very close to the actual number of stations. In case no station successfully sent a message during a given phase, we distinguish between two cases. If the participating stations have decided that the current estimation is lower than the actual number of stations then they all send a "lower" message at the last round. If a collision appears, then this is realized by all stations as a "lower" message. Otherwise, if the participating stations have decided that the current estimation is higher than the actual number of stations then no station sends any message and this silence is realized as a "higher" message.

As stated above, we divide algorithm CountCD into two epochs which we call QuickStart and SlowEnd. In the first epoch, starting with an estimation $m = 2$, each station runs the procedure CountEmpty with input the current estimation, sets the current estimation to m^2 and repeats until the call of CountEmpty on input the current estimation returns that the current estimation is higher than or very close to the number of stations. If an estimation very close to the number of stations is found, then the algorithm CountCD terminates. Otherwise, let m be the estimation which CountEmpty found to be higher than the actual number of stations. Then, the second epoch begins and executes a binary search in the set $\{i, i+1, \ldots, 2i\}$ where i is such that $\sqrt{m} = 2^i \leq n \leq 2^{2i} = m$ (i.e., $i = O(\log n)$), based on the outcomes of CountEmpty. Assuming that procedure CountEmpty correctly decides whether the current estimation is much higher, much lower, or very close to the actual number of stations, algorithm CountCD runs in $O(\log n \log\log n)$ time (i.e., $O(\log\log n)$ calls of CountEmpty in each of the two epochs QuickStart and SlowEnd).

It is clear that one of the crucial points in the algorithm is the correct estimation of the range in which the number of empty slots should belong to in order for the current estimation of the number of stations to be accepted by CountEmpty. The next lemma provides bounds on the number of empty slots in a phase depending on whether the estimation is much higher than, much lower than, or very close to the actual number of stations. Intuitively, the number of stations that are awake in this round will be much higher, much lower, or very close to $f(m)$ and the number of empty slots will be much higher, much lower, or very close to $f(m)/e$, respectively. The proof uses Theorem 1 and Chernoff bounds.

Lemma 1. *Consider a phase of algorithm* CountCD, *let m be the current estimation of the number of stations in the network and denote by O the number of empty time slots. There exist positive constants α_1, α_2 and α_3 such that the following hold:*

- *If $m \geq 2n$, then* $\Pr[O < 0.55 f(m)] < 2n^{-\alpha_1}$.
- *If $m \leq n/2$, then* $\Pr[O > 0.2 f(m)] < 2n^{-\alpha_2}$.
- *If $n/\sqrt{2} \leq m \leq n\sqrt{2}$, then* $\Pr[O < 0.2 f(m)$ *or* $O > 0.55 f(m)] < 2n^{-\alpha_3}$.

By Lemma 1, procedure CountEmpty suffices to compare the number of empty slots O in each phase with the lower and upper thresholds $0.2f(m)$ and $0.55f(m)$. If O is higher than the upper threshold or lower than the lower threshold, then

the current estimation is almost surely larger than $2n$ or smaller than $n/2$, while if O is between the two thresholds, then the current estimation is almost surely correct.

We can thus prove the following statement.

Theorem 2. *Let $n > 2$ be the number of stations in the network. With high probability, algorithm* CountCD *runs in time $O(\log n \log\log n)$ and outputs an estimation m of the number of stations such that $n/2 < m < 2n$.*

2.2 Estimating the Number of Stations Without Collision Detections

The algorithm presented in the previous section exploits the collision detection capability in order to verify whether the current estimation is smaller or bigger than the actual number of stations. As stated above, during each round, each station can determine whether or not a message was sent. More specifically, when the estimation is smaller than the actual number of stations, the number of rounds in which no station sends a message is small. Conversely, when the current estimation is bigger than the number of stations, the number of rounds in which no station sends a message is high.

It could be tempting to use similar arguments in the weak no-CD model. In this model we may count the number of time slots in which a message was successfully sent. In order to use this idea, we need the following counterpart of Theorem 1 for the number of bins containing exactly one ball in the classical balls-to-bins process.

Theorem 3. *Let $r = k/b$, and let Z be the number of bins containing exactly one ball when k balls are thrown randomly into b bins. For $\lambda > 0$ it holds that:*

$$\mu = E(Z) = k\left(1 - \frac{1}{b}\right)^{k-1} \simeq ke^{-r} \qquad \Pr[|Z - \mu| \geq \lambda] \leq 2\exp\left(-\frac{\lambda^2}{2k}\right)$$

It turns out that in the case of no collision detection, it is not possible to use the QuickStart algorithm to obtain a rough estimation of the number of stations. Consider the case in which the current estimation is smaller than the number of stations. In this case, the number of collision is high and the number of slots containing exactly one message is small. Conversely, consider the case in which the estimation is bigger than the number of stations. In this case, the number of empty bins will be high and, again, the number of rounds in which a message is successfully sent is small. Since in the weak no-CD model it is not possible to distinguish between a collision and an empty slot, an algorithm cannot distinguish between the two cases.

On the other hand, whenever the current estimation is close to the number of stations, the number of rounds in which a message is successfully sent should be around $f(m)/e$. For this reason, starting from an estimation $m = 2$, whenever m is wrong, we can double it until we reach a value close to the actual number of stations. We are left to determine the thresholds to be used in the algorithm

for deciding whether or not the current estimation should be accepted. We call CountSingle the procedure that checks whether the current estimation is close to the number of stations or not. Again, each station wakes up at the end of each phase in order to realize whether or not the algorithm should continue. The station that first transmitted successfully in the current phase will announce the result at the end of the phase. Silence is realized as a "continue" message.

As described so far, we have outlined the algorithm CountnoCD which works in the strong-noCD mode. The main problem in the weak-noCD model is that the station that first transmitted successfully during a phase does not know it (since it cannot detect whether its message was successfully sent) in order to announce the result at the end of the phase. To overcome this and extend our protocol in the weak-noCD model, we can use messages of two bits in each round. The first bit is used as above while the second bit is used to encode the binary representation of the round of the first successful transmission. After i successful transmissions in the current phase (for $i \geq 1$), the second bit of the message of any transmitting node is set to the i-th least significant bit of the round in which the first node successfully transmitted.

Using Theorem 3 and Chernoff bounds we can prove the following lemma.

Lemma 2. *Consider a phase of algorithm* CountnoCD, *let m be the current estimation of the number of stations in the network and denote by O the number of time slots with successful transmissions. There are positive constants β_1 and β_2 such that the following hold:*

– *If $m \leq n/3$ or $m \geq 3n$, then* $\Pr\left[O > \frac{f(m)}{4}\right] < 2n^{-\beta_1}$.
– *If $2n/3 \leq m \leq 4n/3$, then* $\Pr\left[O < \frac{f(m)}{4}\right] < 2n^{-\beta_3}$.

By Lemma 2, procedure CountSingle suffices to compare the number O of time slots with successful transmissions in each phase with the threshold $f(m)/4$. If O is lower than the threshold, then the current estimation is almost surely wrong, while if O is higher than the threshold, then the current estimation is almost surely correct.

We can also show the following technical lemma which can be used to compute an upper bound of $O(\log n)$ on the energy of each node (i.e., total number of rounds at which a node is awake). The same bound holds for algorithm CountCD as well.

Lemma 3. *Consider the sequence of independent random variables X_i for $i = 1, ..., k$, such that $X_i \in \{0, i\}$ with $\Pr[X_i = i] = i/2^i$. Let $X = \sum_{i=1}^{k} X_i$. It holds that $\Pr[X > 6k] < 4^{-k}$.*

We can thus prove the following statement for algorithm CountnoCD.

Theorem 4. *Let $n > 2$ be the number of stations in the network. With high probability, the algorithm* CountnoCD *runs in time $O(\log^2 n)$, requires maximum energy $O(\log n)$ per node and outputs an estimation m of the number of stations such that $n/3 < m < 3n$.*

3 Computing the Minimum

We now turn to the problem of computing the minimum in the weak-noCD model. Assume each station possesses a value chosen with unknown distribution from an unknown range. We wish to design an algorithm that outputs the minimum value in the network. We assume that an upper bound on the number of stations in the network is known, otherwise we can use the algorithm presented in the previous section to estimate its value.

Let us assume that the stations hold different values. A simple idea is the following: each station randomly selects a round between 1 and n during which it sends its value on the channel. Whenever a value is successfully transmitted, all the station that hold a value bigger than the one sent, switch to the sleep state. The remaining stations continue the algorithm until a single station is alive. It is immediate that, whenever the probability distribution of the values is unknown, such an algorithm requires $O(n^2)$.

Ideally, the above algorithm tries to divide into two sets the stations so that all the station in the first set, i.e., the ones with value bigger than the received one, switch to the sleep state, while the station in the second set continue the execution. If, in each phase, we could guarantee that the set of awake stations is a constant fraction of the corresponding set in the previous phase, we could reduce the number of phases to $O(\log n)$.

To solve this problem, we will use the oversampling technique introduced in [10] that can be described as follows: from a set of n values, select equiprobably ps samples. Let $x_1, \ldots, x_{ps}$ be the sorted sequence. Consider the subsequence $x_s, x_{2s}, \ldots, x_{(p-1)s}$ and assign to the set $j = 2, \ldots, p-1$ all the values in the range $(x_{(j-1)s}, x_{js})$. All the values less than x_s will be assigned to the set with index 1 and, similarly, all the values greater than $x_{(p-1)s}$ will be assigned to the set p.

The next theorem states that the size of the sets constructed using the above technique is approximately the same with high probability:

Theorem 5 ([2]). *Let n be the number of values, let p be the number of sets*[1]*, and let s be the oversampling ratio. Then, for any $\alpha \geq 1 + 1/s$, the probability that a set contains more than $\alpha n/p$ values is at most $ne^{-(1-1/\alpha)^2 \alpha s/2}$*

Given the above theorem, we can divide the set of values held by the stations into two subsets, of approximately the same size with high probability, by using $s - O(\log n)$. As discussed above, this reduces the number of phases to $O(\log n)$.

One of the hidden assumption of Theorem 5 is the fact that the values are distinct. In order to meet this requirement, the station will randomly select in the first phase a value r in the range $\{1, \ldots, n^2\}$. A station holding a value m will broadcast the pair (m, r). We write $(m_1, r_1) < (m_2, r_2)$ iff $m_1 < m_2$ or ($m_1 = m_2$ and $r_1 < r_2$).

[1] Notice that in [10] the value p represents the number of processors in a parallel machine. However, this restriction is immaterial as in the proof of the theorem p is used as a parameter in the selection probability.

In order to reduce the number of rounds within each phase, we need to reduce the number of stations that send a message. As in the previous sections, we will use self-selection during each phase. More specifically, let c be a constant. During phase i, for $i = 0, \ldots, \log n - \log\log n$, executes phase i with probability $2^i \log n/n$. Each awake station selects a round in the range $\{1...12ce\log n\}$ in which it will send its value on the channel. At the end of phase i, on average, each station has received $12c\log n$ values. By Theorem 5, with $\alpha = 3/4$ and $p = 2$, the probability that the number of awake stations in the subsequent phase is at least $3n/4$ is at most n^{-c}.

The algorithm works as follows: All stations are initially awake. Each station runs $3\log n - \log\log n$ phases. Each phase has $\log n$ rounds. For $i = 0, ..., \log n - \log\log n - 1$, each station which is awake at the beginning of phase i, participates to the algorithm with probability $\frac{2^i}{n}\log n$. If it participates to phase i, it equiprobably selects one of the rounds of phase i to transmit its value. For $i = \log n - \log\log n, ..., 3\log n - \log\log n$, each station which is awake at the beginning of phase i equiprobably selects one of the rounds of phase i to transmit its value. When a station hears a value smaller than its value, it becomes sleeping. The minimum value is the last transmitted value.

We will show that the algorithm correctly computes the minimum value with high probability. Let n_i be the number of awake stations at the beginning of phase i. If $n_i \leq n/2^{i+1}$, then, it is certainly $n_{i+1} \leq n/2^{i+1}$. Assume that $n/2^{i+1} < n_i \leq n/2^i$. Then, the number of successfully transmitted values in phase i is at least $\alpha\log n$ for some positive constant α. This follows by considering stations transmitting within the phase as balls and rounds as bins; then the number of successful transmissions within the phase is the number of bins receiving exactly one ball in the corresponding balls-to-bins game. The probability that more than half of the n_i stations that are awake at the beginning of phase i will still be awake at the beginning of phase $i+1$ is the probability that the transmitted values will be larger than the values stored in the $n_i/2$ stations holding the smallest values, i.e., at most $n^{-\alpha}$.

Now consider a phase of the last $2\log n$ phases. Each of the stations that are awake at the beginning of phase $\log n - \log\log n$ has constant probability (say β) of neither transmitting successfully nor sleeping during each of the next phases. So, the probability that some node is still awake after the last phase is at most $\log n \cdot \beta^{2\log n} \leq n^{-c}$ for some constant c.

Theorem 6. *There exists an algorithm that computes the minimum in the weak-noCD model in time $O(\log^2 n)$, with high probability.*

4 Conclusions

In this paper we have presented an algorithm for estimating the number of stations in an anonymous wireless network. The algorithm is based on the assumptions the station can detect collisions. The algorithm presented runs in time $O(\log n \log\log n)$, its expected energy consumption is $O(\log\log n)$ and it outputs and estimation m such that $n/2 \leq m \leq 2n$. Furthermore we have shown that a

similar technique can be used to compute in time $O(\log^2 n)$ and maximum energy $O(\log n)$ an estimation of the number of stations in the weak_noCD model. In this case the estimation m computed by the algorithm is such that $n/3 \leq m \leq 3n$. Finally we have shown that in the weak-noCD model it is possible to compute the minimum in time $O(\log^2 n)$.

References

1. R. S. Bhuvaneswaran, J. L. Bordim, J. Cui, and K. Nakano. Fundamental Protocols for Wireless Sensor Networks. In *Proc. of the 15th International Parallel and Distributed Processing Symposium (IPDPS '01)*, 2001.
2. G.E. Blelloch, C.E. Leiserson, B.M. Maggs, G.C. Plaxton, S.J. Smith and M. Zagha. An Experimental Analysis of Parallel Sorting Algorithms. *Theory of Computing Systems*, 31(2), pp. 135-167, 1998.
3. J. L. Bordim, K. Nakano, and H. Shen. Sorting on Single-Channel Wireless Sensor Networks. In *Proc. of the International Symposium on Parallel Architectures, Algorithms, and Networks (I-SPAN '02)*, pp 153-158, 2002.
4. J. Elson and D. Estrin. Time Synchronization for Wireless Station Networks. In *Proc. of the 15th International Parallel and Distributed Processing Symposium (IPDPS '01)*, Workshop on Parallel and Distributed Computing Issues in Wireless and Mobile Computing, 2001.
5. T. Jurdzinski, M. Kutylowski, J. Zatopianski. Energy-Efficient Size Approximation of Radio Networks with No Collision Detection. In *Proc. of the 8th Annual International Conference on Computing and Combinatorics (COCOON '02)*, LNCS 2387, Springer, pp. 279-289, 2002.
6. T. Jurdzinski , M. Kutylowski , J. Zatopianski. Weak Communication in Radio Networks. In *Proc. of the 8th International Euro-Par Conference (EuroPar '02)*, LNCS 2400, Springer, pp. 965-972, 2002.
7. R. Motwani and P. Raghavan. Randomized Algorithms. *Cambridge University Press*, 1995.
8. K. Nakano, S. Olariu. Energy-Efficient Initialization Protocols for Radio Networks with No Collision Detection. In *Proc. of the 2000 International Conference on Parallel Processing (ICPP '00)*, pp. 263-270, 2000.
9. K. Nakano, S. Olariu. Randomized Leader Election Protocols in Radio Networks with No Collision Detection. In *Proc. of the 11th International Conference on Algorithms and Computation (ISAAC '00)*, LNCS 1969, Springer, pp. 362-373, 2000.
10. J.H. Reif and L.G. Valiant. A Logarithmic Time Sort for Linear Size Networks. *Journal of the ACM*, 34(1), pp. 60-75, 1987.
11. M. Singh, V. Prasanna, J. Rolim, and C. Raghavendra. Collaborative and Distributed Computation in Mesh-Like Wireless Sensor Arrays. In *Proc. of the 8th IFIP-TC6 International Conference on Personal Wireless Communications (PWC '03)*, LNCS 2775, Springer, pp. 1-11, 2003.
12. D. E. Willard. Log-logarithmic Selection Resolution Protocols in a Multiple Access Channel. *SIAM Journal on Computing*, 15, pp. 468-477, 1986.

A Simple Optimal Randomized Algorithm for Sorting on the PDM*

Sanguthevar Rajasekaran[1] and Sandeep Sen[2]

[1] Department of CSE, University of Connecticut
rajasek@engr.uconn.edu
[2] Department of CSE, IIT Kharagpur
ssen@cse.iitkgp.ernet.in

Abstract. The Parallel Disks Model (PDM) has been proposed to alleviate the I/O bottleneck that arises in the processing of massive data sets. Sorting has been extensively studied on the PDM model due to the fundamental nature of the problem. Several randomized algorithms are known for sorting. Most of the prior algorithms suffer from undue complications in memory layouts, implementation, or lack of tight analysis. In this paper we present a simple randomized algorithm that sorts in optimal time **with high probablity** and has all the desirable features for practical implementation.

1 Introduction

When the amount of data an application has to deal with is enormous, out-of-core computing techniques have to be invoked. In this case, the I/O bottleneck has to be dealt with. The PDM has been proposed to alleviate this I/O bottleneck. In a PDM, there is a (sequential or parallel) computer that has access to $D(\geq 1)$ disks. In one I/O operation, it is assumed that a block of size B can be fetched into the main memory. One typically assumes that the main memory has size M where M is a (small) constant multiple of DB[1].

Efficient algorithms have been devised for the PDM for numerous fundamental problems. In the analysis of these algorithms, typically, the number of I/O operations needed are optimized. Since local computations take much less time than the time needed for the I/O operations, these analyzes are reasonable. Since sorting is a fundamental and highly ubiquitous problem, a lot of effort has been spent on developing sorting algorithms for the PDM. It has been shown by Aggarwal and Vitter [2] that $\Omega\left(\frac{N}{DB}\frac{\log(N/B)}{\log(M/B)}\right)$[2] I/O operations are needed to sort N keys (residing in D disks) when the block size is B. Here M is the size of the internal memory. Many asymptotically optimal algorithms have been devised

* This research been supported in part by the NSF Grants CCR-9912395 and ITR-0326155.

[1] This is the hardest case as we will show later.

[2] In this paper we use log to denote logarithms to the base 2 and ln to denote logarithms to the base e.

X. Deng and D. Du (Eds.): ISAAC 2005, LNCS 3827, pp. 543–552, 2005.

as well (see e.g., Arge [3], Nodine and Vitter [16], and Vitter and Hutchinson [22]). The LMM sort of Rajasekaran [18] is optimal when N, B, and M are polynomially related and is a generalization of Batcher's odd-even merge sort [7], Thompson and Kung's s^2-way merge sort [21], and Leighton's columnsort [14].

Notation. We say the amount of resource (like time, space, etc.) used by a randomized algorithm is $\widetilde{O}(f(N))$ if the amount of resource used is no more than $c\alpha f(N)$ with probability $\geq (1 - N^{-\alpha})$ for any $N \geq n_0$, where c and n_0 are constants and α is a constant ≥ 1. We could also define the asymptotic functions $\widetilde{\Theta}(.)$, $\widetilde{o}(.)$, etc. in a similar manner.

2 Prior Algorithms and Our Result

Most of the previous PDM sorting algorithms can be categorized under two families - ones based on bucketsort and the others on mergesort. The first kind is based on distribution sort [23,22,15] where keys are classified into buckets depending on their values and this is repeated recursively within each bucket till each bucket reaches a managable size (corresponding to the base case). The randomized versions of distribution sort (like quicksort) are often simpler than their deterministic counterparts. The basic idea is sampling and due to Frazer and McKellar. Given a sequence X of n keys to sort: 1) a random sample of s keys are picked from X; 2) these sample keys are sorted to get the sequence $l_1, l_2, \ldots, l_s$; 3) X is partitioned into $s+1$ parts $X_0, X_1, \ldots, X_s$ using the sample keys as splitters. In particular, $X_0 = \{q \in X : q \leq l_1\}$, $X_i = \{q \in X : l_i < q \leq l_{i+1}\}$ for $1 \leq i \leq (s-1)$, and $X_s = \{q \in X : q > l_s\}$; and 4) the parts $X_0, X_1, \ldots, X_s$ are sorted recursively and independently.

The second kind of sorting algorithms on the PDM are based on R-way merging for some suitable value of R that minimizes the number of passes through the data for the given size of internal memory [1,6,11,16,18].

The primary dificulty in both the approaches is exploiting parallelism in writing (in the case of distribution sort) or reading (in mergesort). That is, how do we come up with *balanced* read/write schedules across the D disks when the data is aritrarily distributed at the beginning.

In this context the algorithm of Barve, Grove, and Vitter [6] deserves special mention. It uses a value of $R = M/B$. This algorithm stripes the runs across the disks such that for each run the first block is stored in a random disk and the other blocks are stored in a cyclic fashion starting from the random disk. They only analyze the expected performance of the algorithm (and no high probability bounds have been derived). Their algorithm called *Simple Randomized Mergesort* (SRM), has an optimal expected performance only when the internal memory size M is $\Omega(BD \log D)$. However, the standard assumption on M is that $M = O(DB)$.

This problem has been redressed by the algorithm of Hutchinson, Sanders and Vitter [13]. By using locally randomized scheduling like Fully Random (FR) or Random Cycling (RC) for writing blocks to the disks in parallel, the expected

number of write steps can be bounded by its optimal value. The core of the analysis can be found in Sanders, Egner and Korst [20] and Vitter and Hutchinson [22], who used asymptotic queueing theoretic analysis to bound the expected number of writes in a batched arrival queueing system with a bounded buffer. A batch corresponds to a memory-load of keys that we are trying to classify into buckets and the bounded buffer is a part of the memory. The FR schedule is more complicated to implement and is not read-optimal for $M = o(BD \log D)$. The RC scheduling resulted in optimal distributed sort (RCD) and optimal mergesort (RCM) via duality [13]. These have been shown to be very practical [11]. However, these algorithms have been shown to be optimal only in expectation and no high probability bounds have been derived.

Recently a third approach has been pioneered by [10], where the focus is to maximize the problem size for a fixed number of passes. Being inherently oblivious in nature, they are able to achieve the desired parallelism more easily (see [8,9] for more details). Unfortunately, the number of passes made by these algorithms deteriorate rapidly from their optimal values with increasing problem sizes. The problem sizes have been increased further in [19].

In this paper we present a simple randomized algorithm for sorting on the PDM that makes only $\widetilde{O}\left(\frac{\log(N/M)}{\log(M/B)}\right)$ passes through the data. Note that this bound holds **with high probability** for any value of N unlike the previous randomized algorithms for which only expected bounds have been proved. In our analysis we rely only on standard tools that do not rely on asymptotic convergence. In addition, we are able to adhere to desirable properties like striping and simplicity. The underlying approach in our algorithm is to first generate a random permutation and subsequently sort the random permutation using a simple mergesort. One can easily argue that these are symmetric problems and we show how to solve them optimally. In the process we introduce some fundamentally novel ideas that could find independent applications.

3 Integer Sorting and Random Permutation

Often, the keys to be sorted are integers in some range $[1, R]$. Numerous sequential and parallel algorithms have been devised for sorting integers. Several efficient out-of-core algorithms have been devised by Arge, Ferragina, Grossi, and Vitter [4] for sorting strings. In [19], two integer sorting algorithms are given. These algorithms have optimal expected performance, the expectation being over the space of all possible inputs. In particular, these algorithms have optimal performance on an overwhelming fraction of all possible inputs. We summarize their results here and more details can be found in [19]. The following Theorem is proven in [19]:

Theorem 1. *N random integers in the range $[1, R]$ (for any R) can be sorted in an expected $(1+\mu)\frac{\log(N/M)}{\log(M/B)}+1$ passes through the data, where μ is a constant < 1 and is B is $\Omega(\log N)$. In fact, this bound holds for a large fraction ($\geq 1 - N^{-\alpha}$ for any fixed $\alpha \geq 1$) of all possible inputs.*

The proof of the above Theorem is based on two algorithms. The first algorithm (called IntegerSort) sorts N random keys where each key is an integer in the range $[1, \Omega(M/B)]$. This algorithm takes $O(1)$ passes through the data. The second algorithm (called RadixSort) employs forward radix sorting. In each stage of sorting, the keys are sorted with respect to some number of their MSBs. Keys that have the same value with respect to all the bits that have been processed up to some stage are said to form a *bucket* in that stage. A description of the algorithm follows. In this algorithm, δ is any constant > 0.

Algorithm RadixSort

for $i := 1$ **to** $(1+\delta)\frac{\log(N/M)}{\log(M/B)}$ **do**

1. Employ IntegerSort to sort the keys with respect to their ith most significant $\log(M/B)$ bits.
2. Now the size of each bucket is $\leq M$. Read and sort the buckets.

We show how to randomly permute N given keys such that each permutation is equally likely. We employ RadixSort for this purpose. The idea is to assign a random label with each key in the range $[1, N^{1+\beta}]$ (for any fixed $0 < \beta < 1$) and sort the keys with respect to their labels. This can be done in $(1+\mu)\frac{\log(N/M)}{\log(M/B)} + 1$ passes through the data with probability $\geq 1 - N^{-\alpha}$ for any fixed $\alpha \geq 1$. Here μ is a constant < 1. For many applications, this permutation may suffice. But we can ensure that each permutation is equally likely with one more pass through the data.

When each key gets a random label in the range $[1, N^{1+\beta}]$, the labels may not be unique. The maximum number of times any label is repeated is $\widetilde{O}(1)$ from the observation that the number of keys falling in a bucket is binomially distributed with mean $1/n$ and applying Chernoff bounds (equation 1 in Appendix A). We have to randomly permute keys with equal labels which can be done in one more pass through the data as follows. We think of the sequence of N input keys as $S_1, S_2, \ldots, S_{N/DB}$ where each S_i $(1 \leq i \leq N/(DB))$ is a subsequence of length DB. Note that keys with the same label can only span two such subsequences. We bring in DB keys at a time into the main memory. We assume a main memory of size $2DB$. There will be two subsequences at any time in the main memory. Required permutations of keys with equal labels are done and DB keys are shipped out to the disks. The above process is repeated until all the keys are processed.

Remark: With more care we can eliminate this extra pass by combining it with the last stage of radix sort.

Thus we get the following:

Theorem 2. *We can permute N keys in $O(\frac{\log(N/M)}{\log(M/B)})$ passes through the data with probability $\geq 1 - N^{-\alpha}$ for any fixed $\alpha \geq 1$, where μ is a constant < 1, provided $B = \Omega(\log N)$.*

Remark: In the above Theorem, we assume that $B = \Omega(\log N)$. This assumption is made in [19] as well. This is a very benign assumption that readily holds

in practice. However, we later show how to eliminate this assumption without sacrificing the asymptotic performance.

4 Randomized Sorting

In this section we present a randomized sorting algorithm that sorts N given keys in $\widetilde{O}\left(\frac{\log(N/M)}{\log(M/B)}\right)$ passes through the data. Our algorithm employs the permutation algorithm from Section 3.

4.1 First Attempt

We start with a simple version of the algorithm (called sf RSort1). Unfortunately, RSort1 may not run in an optimal number of I/O's. This algorithm is modified in the next subsection to achieve optimality. RSort1 helps one understand the basic ideas behind the optimal algorithm. In the following algorithm, $R = M/B$.

Algorithm RSort1

1. Randomly permute the input N keys using the algorithm of Theorem 2.
2. In one pass through the data form runs of length $M = DB$ each.
3. **for** $i := 1$ **to** $\frac{\log(N/M)}{\log R}$ **do**

   while there are more runs **do**

   Merge the next R runs as follows.

   Start by bringing in two blocks from each run. Assume that $R = D = M/B$ and the main memory is of size $2DB$. Merge the runs to ship M keys out to the disks. This run becomes an input run for the next iteration. If one of the runs becomes empty before M output keys are formed, start all over again. (We will show that the probability of this happening is negligible).

   From here on, maintain the invariant that for each run we have two leading blocks in the memory. Form BD output keys and ship them. Repeat this until the R runs are merged.

Analysis. Let a *phase* refer to one run of step 3 of RSort1. In the following discussion ignore the number of parallel I/Os needed to do one scan through the data while merging the runs.

Consider the problem of merging any R runs in some phase of RSort1. Consider some point in time when there are $2D$ blocks in the main memory with 2 blocks per run. We merge these blocks to form M output keys. Note that the M output keys are such that each key is equally likely to have come out of the R runs. Thus the expected number of keys that come out any particular run is B. Using Chernoff bounds, this number lies in the interval $[(1-\epsilon)B,\ (1+\epsilon)B]$ with probability $\geq 2[1-\exp(-\epsilon^2 B)/3]$, ϵ being any constant > 0. In other words,

each run gets consumed at the rate of at least $(1-\epsilon)$ blocks per $(1+\epsilon)$ blocks brought in (from each run). For $B \geq \log N$, this holds with high probability. Note that the probability of ϵ being very close to 1 is very low and hence the event of two blocks getting consumed before M output keys are formed is very low.

In summary, with high probability, it takes at most $1/(1-\epsilon)$ scans through a run before it gets consumed completely. As a result, RSort1 makes $\widetilde{O}\left(\frac{\log(N/M)}{\log(M/B)}\right)$ scans through the input.

Even though RSort1 makes an optimal number of scans through the input, each scan may take more than an optimal number of I/Os. This can be seen as follows. At the beginning of the algorithm, the runs are striped in a cyclic fashion. Let the runs be $R_1, R_2, \ldots, R_q$. The first block of run i will be in disk $(i-1) \bmod D+1$; the second block of run i will be in disk $i \bmod D+1$; and so on (for $1 \leq i \leq q$). If whenever blocks are accessed from different runs these blocks come from different disks, then it will mean that the number of I/O operations is optimal as well. For instance if each run gets consumed at the rate of one block per block brought in, then this will hold.

However, the runs get consumed at different rates. For instance, there could come a time when we need a block from each run and all of these blocks are in the same disk. An occupancy analysis similar to the one in [6] will imply that the expected number of I/O operations in the worst case could be nonoptimal unless M/B is $\Omega(D \log D)$.

In the next subsection we modify RSort1 to make it optimal and still retain the simplicity.

4.2 A Second Algorithm: Periodic Resetting

The key ideas to make RSort1 optimal are: 1) Let $Q_1, Q_2, \ldots, Q_R$ be the runs to be merged at some point in time. Let a *stage* refer to the step of bringing in required keys, merging the $2DB$ keys in memory and forming M output keys. We keep the R runs such that the leading blocks for the runs are in successive disks (or very nearly so); and 2) When there are many blocks in every run, the above property may be difficult to maintain since as time progresses, the leading blocks deviate more and more from the expected disk locations. We periodically rearrange the leading M keys of each run so that the above property is reinstated after the rearrangement. Again we assume that $R = M/B$.

In the description that follows, we use M to denote DB and we assume that the actual internal memory has size $2DB$.

Algorithm RSort2

1. Permute the input N keys using the algorithm of Theorem 2.
2. In one pass through the data form runs of length $M = DB$ each.
3. **for** $i := 1$ **to** $\frac{\log(N/M)}{\log R}$ **do**

while there are more runs **do**
Merge the next R runs as follows.

Begin by bringing in $\frac{2M}{RB} = 2$ blocks from each run. Merge the runs to ship M keys out to the disks to be used as an input run for the next iteration. If one of the runs becomes empty before M output keys are formed, start all over again. The probability of this happening is low.

Maintain the property that there are exactly $\frac{2M}{RB} = 2$ leading blocks per run. Form BD output keys and ship them. Call the step of bringing in enough keys to have $2M/(RB)$ blocks per run, merging them and outputting M keys as a *stage* of the algorithm. After every (M/B) stages perform a rearrangement of runs. In particular, read the leading M keys of each run and write them back so that the leading blocks of the runs are in successive disks. Use separate areas in the disks for the purpose of rewriting.

Repeat the above step until the R runs are merged.

Theorem 3. RSort2 *takes* $\widetilde{O}\left(\frac{\log(N/M)}{\log(M/B)}\right)$ *read passes through the data provided* $B = \Omega(\sqrt{M \log N})$.

Proof. Let a *phase* of the algorithm refer to one run of step 3 and let a *stage* of the algorithm refer to bringing in enough keys per run (i.e. $2M/R$ keys per run), merging the runs, and shipping M keys out to the disks.

It suffices to prove that each phase of the algorithm takes $\widetilde{O}(N/DB)$ I/Os.

In each stage of the algorithm the expected number of blocks consumed from each run is 1. If the starting block of a run is i then after q stages, the leading block of this run is expected to be in disk $(i + q - 1) \bmod D + 1$. Assume that the leading block of each run continues to be within one disk of its expected disk. Consider the task of bringing into main memory at most K leading blocks of each run. How many I/Os will be needed? It is easy to see that in the worst case, $3 \cdot K$ I/Os will suffice as each disk can have at most 3 of the leading blocks. Thus when $K = 1$, three I/O's suffice.

We can actually obtain a stronger result:

Lemma 1. *Consider D disks and R runs with $R \leq D$. The runs are striped in the usual way. Let the leading block of each run stray away from its expected disk by at most q disks. The problem of bringing in K blocks from each run can be accomplished in at most $K + 2q$ I/Os.*

Proof. Assume that $R = D$ since this corresponds to the worst case. Also assume that $K \leq D$ for simplicity (though the result is general). Consider any disk d. From out of the blocks we want to fetch, how many blcoks will reside in d? If the starting blocks of the runs are equidistant, then K blocks will reside in d. The starting blocks of runs can be at most q disks away from their expected starting disks. The starting disks of some of the runs could be to the right of

their expected disks and the starting blocks of some others could be to the left of their starting disks. The number of blocks in d (excluding those K runs that are expected to have a block each in d) from out of the first kind of runs is at most q and the number of blocks from the second kind is at most q. (We assume that there is sufficient buffer, i.e. $(K+2q)DB$ for this purpose.)

When we perform R stages, the expected number of keys coming out of each run is M. This number will stray away from its expected value by at most $\sqrt{\alpha M \ln N}$ with probability $\geq (1-N^{-\alpha})$. Thus the leading block of each run will stray away by at most one disk provided $B \geq \sqrt{\alpha M \ln N}$. This condition is readily satisfied in practice. In this case, each stage of the algorithm can be completed in three I/Os with high probability.

Also, all the rearrangements of keys take an additional $(1+\nu)\log(N/M)/\log(M/B)$ read passes and the same number of write passes, where ν is any constant > 0.

In summary, the number of read passes taken by the algorithm is $4(1+\nu)\frac{\log(N/M)}{\log(M/B)} + (1+\mu)\frac{\log(N/M)}{\log(M/B)} + 2$ with probability $\geq (1-N^{-\alpha})$ for any fixed $\alpha \geq 1$. Here ν is any constant > 0 and μ is a constant < 1.

4.3 Relaxing the Constraint on B

RSort2 assumes that $B = \Omega(\sqrt{M \log N})$. This assumption can be relaxed with the following idea. Employ a value of $R = (M/B)^\epsilon$ for any constant $1 > \epsilon > 0$.

Note that when $R = (M/B)^\epsilon$, we have $R < D$. If $Q_1, Q_2, \ldots, Q_R$ are the runs, we stripe Q_1 starting from disk 1, Q_2 starting from disk $1 + D/R$, Q_3 starting from $1+2D/R$, and so on. (Assume w.l.o.g. that D is an integral multiple of R). In other words, the leading blocks of the runs are D/R disks apart. Therefore, even if the leading blocks of the runs stray by D/R blocks each stage can be performed in three I/Os.

The resultant algorithm RSort is the same as RSort2 except that rearrangements are done every $(M/B)^\epsilon$ stages and we use a value of $R = (M/B)^\epsilon$.

When $(M/B)^\epsilon$ stages are performed, the expected number of keys consumed from each run is M. The actual value for any run can stray away from its expected value by $\sqrt{\alpha M \ln N}$, with probability $\geq (1-N^{-\alpha})$. We want this number to be $\leq DB/R$. This happens when $\sqrt{\alpha M \ln N} \leq M^{1-\epsilon}B^\epsilon$. This implies that $B \geq M^{(\epsilon-1/2)/\epsilon}(\alpha \ln N)^{1/(2\epsilon)}$.

When $\epsilon = 1/2$, the above condition becomes: $B \geq \alpha \ln N$. This is a benign condition and readily holds in practice. For a value of $\epsilon = \frac{1}{2} - \delta$ (for any fixed $\delta > 0$), the above condition becomes, $B \geq 1$, provided $N \leq e^{M^\delta/\alpha}$. For either choice of ϵ the number of passes made by the algorithm is no more than $8(1+\nu)\frac{\log(N/M)}{\log(M/B)} + (1+\mu)\frac{\log(N/M)}{\log(M/B)} + 2$ with probability $\geq (1-N^{-\alpha})$ for any constant $\alpha \geq 1$. Here ν is any constant > 0 and μ is a constant < 1.

Thus we get the following Theorem:

Theorem 4. RSort *takes* $\widetilde{O}\left(\frac{\log(N/M)}{\log(M/B)}\right)$ *read passes through the data without considering the time for random permutation.*

Observation 1: Note that the above analysis is very conservative and in practice the underlying constants will be smaller.

Observation 2: When B is large, we can decrease the number of read passes made by RSort as follows. Let $B = M^{\beta}$. When $\beta = 3/4$, the number of read passes is $2(1+\nu)\frac{\log(N/M)}{\log(M/B)} + (1+\mu)\frac{\log(N/M)}{\log(M/B)} + 2$.

Observation 3: The algorithms IntegerSort and RadixSort assume that $B = \Omega(\ln N)$. As a consequence, the algorithm of Theorem 2 also makes this assumption. We can relax this constraint in exactly the same manner as in Section 4.3.

Observation 4: In practice, the internal memory could be larger than a constant multiple of BD. We prove the following:

Theorem 5. RSort *can be modified for the case of* $M = 2qDB$ *to run in* $\widetilde{O}\left(\frac{\log(N/(qBD))}{\log(qD)}\right)$ *read passes through the data.*

5 Conclusions

In this paper we have presented optimal randomized sorting algorithms for the PDM model. These algorithms overcome the shortcomings present in prior randomized sorting algorithms on the PDM. These algorithms and the accompanying analyses are quite simple. They have the potential of performing well in practice.

References

1. A. Aggarwal and G. Plaxton, Optimal parallel sorting in multi-level storage, *Proc of the ACM-SIAM SODA 1994*, pp. 659 – 668.
2. A. Aggarwal and J. S. Vitter, The Input/Output Complexity of Sorting and Related Problems, *Communications of the ACM* 31(9), 1988, pp. 1116-1127.
3. L. Arge, The Buffer Tree: A New Technique for Optimal I/O-Algorithms, *Proc. 4th International Workshop on Algorithms and Data Structures (WADS)*, 1995, pp. 334-345.
4. L. Arge, P. Ferragina, R. Grossi, and J. S. Vitter, On Sorting Strings in External Memory, *Proc. ACM Symposium on Theory of Computing*, 1995.
5. L. Arge, M. Knudsen, and K. Larsen, A General Lower Bound on the I/O-Complexity of Comparison-based Algorithms, *Proc. Third Workshop on Algorithms and Data Structures (WADS)*, 1993.
6. R. Barve, E. F. Grove, and J. S. Vitter, Simple Randomized Mergesort on Parallel Disks, *Parallel Computing* 23(4-5), 1997, pp. 601-631.
7. K. Batcher, Sorting Networks and their Applications, *Proc. AFIPS Spring Joint Computing Conference* 32, 1968, pp. 307-314.
8. G. Chaudhry and T. H. Cormen, Getting More From Out-of-Core Columnsort, *Proc. 4th Workshop on Algorithm Engineering and Experiments (ALENEX)*, 2002, pp. 143-154.

9. G. Chaudhry, T. H. Cormen, and E. A. Hamon, Parallel Out-of-Core Sorting: The Third Way, to appear in *Cluster Computing*.
10. G. Chaudhry, T. H. Cormen, and L. F. Wisniewski, Columnsort Lives! An Efficient Out-of-Core Sorting Program, *Proc. 13th Annual ACM Symposium on Parallel Algorithms and Architectures*, 2001, pp. 169-178.
11. R. Dementiev and P. Sanders, Asynchronous Parallel Disk Sorting, *Proc. ACM Symposium on Parallel Algorithms and Architectures*, 2003, pp. 138-148.
12. E. Horowitz, S. Sahni, and S. Rajasekaran, *Computer Algorithms*, W. H. Freeman Press, 1998.
13. D. Hutchinson, P. Sanders and J. Vitter, Duality between prefetching and queued writing with parallel disks, *9th European Symposium on Algorithms 2001*, LNCS 2161, pp. 62 – 73.
14. T. Leighton, Tight Bounds on the Complexity of Parallel Sorting, *IEEE Transactions on Computers* C34(4), 1985, pp. 344-354.
15. M. Nodine and J. Vitter, Deterministic distribution sort in shared and distributed memory multirocessors, *Proc. of the ACM SPAA 1993*, pp. 120 – 129.
16. M. H. Nodine and J. S. Vitter, Greed Sort: Optimal Deterministic Sorting on Parallel Disks, *Journal of the ACM* 42(4), 1995, pp. 919-933.
17. S. Rajasekaran, Sorting and selection on interconnection networks, *DIMACS Series in Discrete Mathematics and Theoretical Computer Science 21*, 1995, pp. 275-296.
18. S. Rajasekaran, A Framework for Simple Sorting Algorithms on Parallel Disk Systems, *Theory of Computing Systems*, 34(2), 2001, pp. 101-114.
19. S. Rajasekaran and S. Sen, PDM Sorting Algorithms That Take A Small Number Of Passes, manuscript, 2004.
20. P. Sanders, S. Enger and J. Korst, Fast concurrent access to parallel disks, *Proc of the ACM-SIAM SODA 2000*, pp. 849 – 858.
21. C.D. Thompson and H.T. Kung, Sorting on a Mesh Connected Parallel Computer, *Communications of the ACM* 20(4), 1977, pp. 263-271.
22. J. S. Vitter and D. A. Hutchinson, Distribution Sort with Randomized Cycling, *Proc. 12th Annual SIAM/ACM Symposium on Discrete Algorithms*, 2001.
23. J. S. Vitter and E. A. M. Shriver, Algorithms for Parallel Memory I: Two-Level Memories, *Algorithmica* 12(2-3), 1994, pp. 110-147.

Appendix A: Chernoff Bounds

If a random vraiable X is the sum of n iid Bernoulli trials with a success probability of p in each trial, the following equations give us concentration bounds of deviation of X from the expected value of np. The first equation is more useful for large deviations whereas the the other two are useful for small deviations from a large expected value.

$$Prob(X \geq m) \leq \left(\frac{np}{m}\right)^m e^{m-np} \tag{1}$$

$$Prob(X \leq (1-\epsilon)pn) \leq exp(-\epsilon^2 np/2) \tag{2}$$

$$Prob(X \geq (1+\epsilon)np) \leq exp(-\epsilon^2 np/3) \tag{3}$$

for all $0 < \epsilon < 1$.

Efficient Parallel Algorithms for Constructing a k-Tree Center and a k-Tree Core of a Tree Network

Yan Wang, Deqiang Wang, Wei Liu, and Baoyu Tian

Dalian Maritime University,
Dalian 116026, China
{wangyannihao_199, collionliu}@163.com
dqwang@dlmu.edu.cn
tianbaob@newmail.dlmu.edu.cn

Abstract. In this paper, we propose two efficient parallel algorithms for constructing a k-tree center and a k-tree core of a tree network, respectively. Both algorithms take $O(\log n)$ time using $O(n)$ work on the EREW PRAM. Our algorithms improve the algorithms previously proposed by Wang (IEEE Trans. Par. Dist. Sys. 1998) and Peng et al. (J. Algorithms 1993).

1 Introduction

Optimally locating a facility in a network is an important problem in the fields of transportation and communication. Due to the variety of facility kinds and different criteria for optimality, many location problems have been defined and studied [3,8,7,12]. These location problems usually have important applications in transportation and communication and thus have received much attention from researchers in the fields. The criteria for optimality extensively studied in the literature are the minimum eccentricity criterion in which the distance to the farthest vertex from the facility is minimized and the minimum distancesum criterion in which the total distance from the facility to the vertices is minimized.

Traditionally, network location theory has been concerned with the optimal location of a single-point facility. Slater [10] extended the network location theory to include a facility that is not merely a single point but a path. This extended theory has practical applications in proving a stretch of road in a highway and in establishing a high-speed transmission line in a communication network. Slater's work was confined to tree networks. A path in a tree network with the minimum eccentricity is defined as a center; a path in a tree network with the minimum distancesum is defined as a core. In Ref. [6], two kinds of facilities are considered: paths and trees; and four optimization criteria are considered: minimum eccentricity, minimum distancesum, maximum eccentricity, and maximum distancemum. In total, there are eight different problems. In this paper, we only consider the minimum eccentricity and minimum distancesum on trees.

X. Deng and D. Du (Eds.): ISAAC 2005, LNCS 3827, pp. 553–562, 2005.

In Ref. [8], Peng et al. first presented an algorithm for constructing a k-tree core on tree network, which has time complexity of $O(kn)$, where n is the number of vertices in the tree network. That is a greedy-type procedure that first finds a core of the tree and then adds to it k-2 paths obtaining a k-tree core. Efficient algorithm for finding a k-tree core of a tree network has an application in the placement of the k copies of a data object in the tree network [8]. But Peng et al.'s algorithm performs in a step-by-step refining manner. Thus, it is hard to parallelize their algorithms. So, Wang gave an efficient parallel algorithm that was proposed for finding a k-tree core of a tree network [13]. The proposed algorithm performs on the EREW PRAM in $O(\log n \log^* n)$ time using $O(n)$ work.

In this paper, two efficient parallel algorithms are proposed for constructing a k-tree center and a k-tree core of a tree network. Our algorithms perform on the EREW PRAM in $O(\log n)$ time using $O(n)$ work. They are more efficient than the two algorithms previously proposed in Ref. [8] and simpler than the algorithm proposed in Ref. [13].

The rest of the paper is organized as follows: Section 2 introduces the notations and preliminary results. The algorithms on a k-tree center and a k-tree core can be found in Section 3 and Section 4, respectively. Section 5 concludes this paper.

2 Notations and Preliminary Results

Let $T = (V, E)$ denote the tree network under consideration, where V is the vertex set and E is the edge set. Let $n = |V|$. The n vertices in V are labelled with 1, 2, ..., n, respectively. Denote $label(v)$ as the label of the vertex $v \in V$. The tree network is undirected. Each edge $e \in E$ has an arbitrary positive length $w(e)$. A leaf of T is a vertex with degree one. Let m be the number of leaves of T. For any two vertices a and b in V, the distance between a and b, denoted by $d(a, b)$, is the length of the unique path connecting a and b. The distance from a vertex v to a subtree X is defined as

$$d(v, X) = \min_{u \in V(X)} \{d(v, u)\},$$

where $V(X)$ is the vertex set of X. The distancesum and eccentricity of a subtree X of T, denoted by $Sum(X)$ and $Ecc(X)$, defined as follows, respectively,

$$Sum(X) = \sum_{v \in V} d(v, X), Ecc(X) = \max_{v \in V}\{d(v, X)\}.$$

The center of T is any vertex of T whose eccentricity is minimized. Handler and Mirchandani showed that the center of T can be determined in linear time by the following technique [3]: First, select any vertex v of T and find a vertex u that is farthest from v. Then, find a vertex z that is farthest from u. The path from u to z is called a diameter path of T. The mid-point of any diameter path of T is the unique center.

A k-tree center of T is a minimum eccentricity subtree of T, which contains exactly k leaves. A k-tree core of T is a minimum distancesum subtree of T, which contains exactly k leaves.

In this paper, we apply the Euler-tour technique [11] and tree contraction [5,9], which are two of the major parallel techniques. We assume that the data structure representing T is an adjacency list, to which Euler-tour technique and tree contraction can be applied efficiently to construct the process of the algorithms that we will propose in following section.

The following results are needed for the algorithm we shall propose in the next section. All the results are derived from the two techniques mentioned above.

Lemma 1 ([7]). *The center of a tree can be computed in $O(\log n)$ time using $O(n)$ work on the EREW PRAM.*

Lemma 2 ([4]). *An undirected tree can be oriented into a rooted tree with a specified root in $O(\log n)$ time using $O(n)$ work on the EREW PRAM.*

3 Constructing a k-Tree Center

Let c be the center of T. For easy discussion, we assume that c is a vertex of T (In case c is a point on an edge (u, v) of T, we can simply introduce a new vertex at c and split the edge (u, v) into two edges (u, c) and (c, v)). Throughout the remainder of this section, we assume that T is rooted at c. Denote T_v, $p(v)$, $size(v)$ and $depth(v)$ as the subtree rooted at v, the parent of v, the number of vertices contained in the subtree rooted at v and the distance from the root to v, respectively.

In Ref. [13], Wang defined the terms "beat" and "importance" as follows:

Let l_1 and l_2 be two leaves of a subtree T_v. Say that l_1 *beats* l_2 at the vertex v if and only if either

(1) $d(v, l_1) > d(v, l_2)$ or

(2) $d(v, l_1) = d(v, l_2)$ and $label(l_1) > label(l_2)$.

If no leaf in T_v beats l at v, then we say that the leaf l of T_v dominates T_v. Denote $DomiLeaf(v)$ as the leaf dominating T_v. Let l be a leaf in T and the path $P_{c,l}$ be given as $< u_1, u_2, \ldots, u_t >$, where $u_1 = c$ and $u_t = l$. Clearly, there exists a vertex $u_q (1 \le q \le t)$, such that the subtrees T_{u_q}, $T_{u_{q+1}}$, $\ldots$, T_{u_t} are dominated by l, but the other subtrees T_{u_1}, T_{u_2}, $\ldots$, $T_{u_{q-1}}$ are not. Denote $DomiEnd(l)$ as the vertex u_q. The dominated path of l, denoted by $DomiPath(l)$, is the path from $DomiEnd(l)$ to l. Clearly, $DomiPath(l_1) \bigcap Domipath(l_2) = \emptyset$, for any two leaves l_1 and l_2 in T.

The *importance* of a leaf l is defined as:

$$R(l) = \begin{cases} d(x, l) & \text{if } l \text{ dominates } T, \\ d(p(x), l) & \text{otherwise,} \end{cases}$$

where $x = DomiEnd(l)$. Denote $Rank(l)$ as the number of leaves in T that are more important than or as important as l. Denote $b_i (1 \le i \le m)$ as the ith

most important leaf in T. Clearly, $DomiEnd(b_1) = p(DomiEnd(b_2)) = c$ and the path from b_1 to b_2 is a diameter path of T.

As an illustrative example, let us consider the tree network depicted in Fig.1. In the figure, any edge $e \in E$ of the tree network has $w(e) = 1$. The dominated paths of the leaves of the network are depicted in Fig.2. For each leaf u_i in the network, we list the values of $DomiEnd(u_i)$, $R(u_i)$, and $Rank(u_i)$in Table 1.

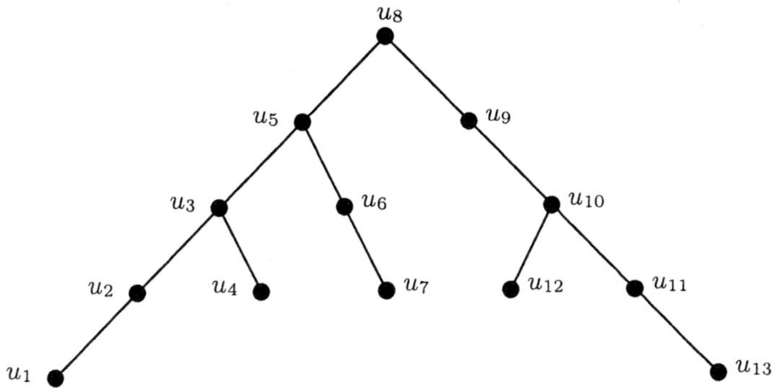

Fig. 1. A tree network

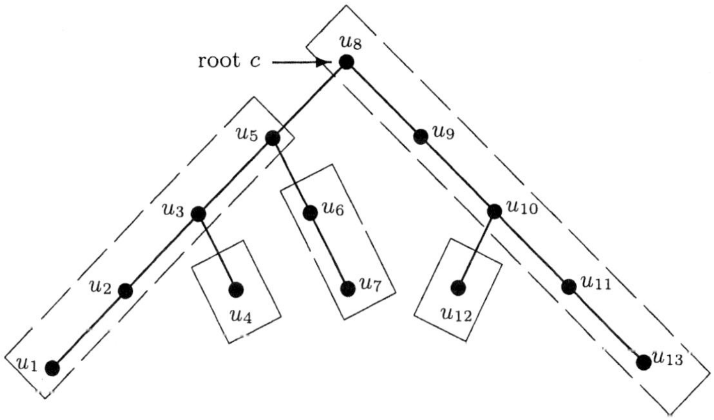

Fig. 2. k-tree center: the dominated paths

Let $T(i)$ be a subtree of T with i leaves, and let $T(1) = DomiPath(b_1)$. Clearly, $T(m)$ is just T. $T(i)(2 \leq i \leq m)$ is constructed as follows:

$$T(i) = T \setminus \bigcup_{x=i+1}^{m} DomiPath(b_x)(2 \leq i \leq m).$$

Namely, we apple the upward tree accumulation technique (see Ref. [9]) to these $DomiPath(b_x)$ $(i+1 \leq x \leq m)$ of T.

Table 1. k-tree center: $DomiEnd(u_i)$, $R(u_i)$ and $Rank(u_i)$

u_i	u_1	u_4	u_7	u_{12}	u_{13}
$DomiEnd(u_i)$	u_5	u_4	u_6	u_{12}	u_8
$R(u_i)$	4	1	2	1	4
$Rank(u_i)$	2	5	3	4	1

Lemma 3. *$T(k)$ is a subtree of T with k leaves ($2 \le k \le m$), and,*

$$Ecc(T(k)) = \begin{cases} R(b_{k+1}) & 2 \le k < m, \\ 0 & k = m. \end{cases}$$

Proof. Clearly, $T(k)$ is a subtree of T with k leaves. Since $T(m) = T$, we have $Ecc(T(m)) = 0$. In the following, we show that $Ecc(T(k)) = R(b_{k+1})$ for $2 \le k < m$. Assume $2 \le k < m$. Let p be the farthest distance vertex in T from $T(k)$. Let q be the vertex on $T(k)$ that is closest to p and w be second vertex on the path from q to p. Let $x = DomiLeaf(w)$. Note that x is a leaf in $\{b_{k+1}, b_{k+2}, \ldots, b_m\}$. Since T_q is not dominated by x, we have $d(q,x) = R(x) \le R(b_{k+1})$. Thus, we have $Ecc(T(k)) = d(q,p) \le d(q,x) \le R(b_{k+1})$. On the other hand, we have $Ecc(T(k)) \ge d(b_{k+1}, T(k)) \ge R(b_{k+1})$. Combining these two statements, we can conclude that $Ecc(T(k)) = R(b_{k+1})$. The lemma holds. □

Theorem 1. *The subtree $T(k)$ is a k-tree center of $T(2 \le k \le m)$.*

Proof. Clearly, the theorem holds for $k = m$. Assume that $T(k)$ is a k-tree center, we show that $T(k-1)$ is a $(k-1)$-tree center. Note that the set of leaves in $T(k)$ is $\{b_1, b_2, \ldots, b_k\}$. Let x be any positive number smaller than $R(b_k)$. For every $i(1 \le i \le k)$, let u_i be the points on the path from c to b_i and $d(u_i, b_i) = x$. From the Ref. [13], we know that any subtree of T that contains the root c and k points $u_i(i = 1, 2, \ldots, k)$, has at least k leaves. Let Y be a $(k-1)$-tree center of T. To complete the proof, in the following, we show that $Ecc(Y) > Ecc(T(k-1))$. Since Y has only $(k-1)$ leaves and Y contains the root c, at least one of $u_i(i = 1, 2, \ldots, k)$, is not contained in Y. Thus, at least one of $b_i(i = 1, 2, \ldots, k)$, is at a distance not smaller than x from Y. Thus, we have $Ecc(Y) > x$. Recall that x is any positive number smaller than $R(b_k)$. Therefore, we can conclude that $Ecc(Y) \ge R(b_k)$. By Lemma 3, we have $Ecc(T(k-1)) = R(b_k) \le Ecc(Y)$. The theorem holds. □

From the construction of $T(k)$ and Theorem 1, we can easily obtain two corollaries as follows:

Corollary 1. *Every $(k-1)$-tree center is contained in a k-tree center ($2 \le k \le m$) of an unweighted tree network .*

Corollary 2. *The vertex $p(DomiEnd(b_i))$ is contained in $T(i-1)(2 \le i \le m)$.*

Now, we are ready to propose our algorithm for constructing a k-tree center of T as follows:

Algorithm 1 k-Tree Center($T = (V, E)$)
```
Input: a tree T = (V,E)
Output: a k-tree center of T
Begin
Step 1: Identify an endpoint c as the center of T. And then,
        orient T into a rooted tree with root c.
Step 2: For each internal vertex v in T, compute DomiLeaf(v).
Step 3: For each leaf l in T, determine DomiEnd(l), R(l) and
        Rank(l).
Step 4: Find the dominated Path of each leaf l of T.
Step 5: Compute T(k), which is a k-tree center of T .
```

$$T(k) \longleftarrow T \backslash \bigcup_{x=k+1}^{m} DomiPath(b_x).$$

```
Step6: return T(k).
End
```

The correctness of the above algorithm is ensured by Lemma 3 and Theorem 1. Now, we discuss the parallel running time of the algorithm as follows: By Lemmas 1 and 2, step 1 take $O(\log n)$ time using $O(n)$ work. Step 2, 3 and 4 can be done in $O(\log n)$ time using $O(n)$ work. It was showed by Cole and Vishkin that the prefix computation of a linked list can be computed in $O(\log n)$ time using $O(n)$ work on the EREW PRAM [2]. Let $\odot$ denote a binary associative operator. Give an ordered data set $\{a_0, a_1, \ldots, a_{t-1}\}, t > 0$, computing $ps_j = a_0 \odot a_1 \odot \cdots \odot a_j$ for $0 \leq j < t$ is referred to as the prefix computation. Step 5 can be performed in $O(\log n)$ time using $O(n)$ work. Therefore, all steps in the algorithm of constructing a k-tree center can be implemented in $O(\log n)$ time using $O(n)$ work. We have the following theorem.

Theorem 2. *A k-tree center of a tree network can be computed in $O(\log n)$ time using $O(n)$ work on the EREW PRAM.*

4 Constructing a k-Tree Core

Let r be an endpoint of any diameter path of T. In this section, we assume that T is rooted at r. For each vertex v in T, the terms of the $p(v)$, T_v, $depth(v)$, and $size(v)$ are uniform to the above section, respectively.

Let $P_{v,l}$ be the path from a vertex v to a leaf l of T_v. The distance saving of this path is defined as [13]

$$Save(P_{v,l}) = Sum(v) - Sum(P_{v,l}) = \sum_{u \in V} d(u,v) - \sum_{u \in V} d(u, P_{v,l}),$$

$$MaxSave(v) = \max\{Save(P_{v,l}) \mid l \text{ is a leaf of } T_v\}.$$

The algorithm we shall propose for constructing a k-tree core is similar to the algorithm we have proposed for constructing a k-tree center. Here, we quote several definitions in Ref. [13] as follows:

Let l_1 and l_2 be two leaves of a subtree T_v . We say that l_1 beats l_2 at the vertex v if and only if either

(1) $Save(P_{v,l_1}) > Save(P_{v,l_2})$ or
(2) $Save(P_{v,l_1}) = Save(P_{v,l_2})$ and $label(l_1) > label(l_2)$.

The terms "*dominate*","*DomiLeaf*(v)", "*DomiEnd*(l)" and "*DomiPath*(l)" are redefined correspondingly[13].

The *importance* of a leaf l of T is as follows:

$$R(l) = \begin{cases} Save(P_{x,l}) & \text{if } l \text{ dominates } T, \\ w(p(x), x) \times size(x) + Save(P_{x,l}) & \text{otherwise,} \end{cases}$$

where $x = DomiEnd(l)$. Say that the leaf l_1 is more important than another leaf l_2 if and only if either

(1)$R(l_1) > R(l_2)$ or
(2)$R(l_1) = R(l_2)$and $label(l_1) > label(l_2)$.

The *rank* of a leaf l in T defined as the same as the section 3, denote $Rank(l)$. Throughout this section, denote a_i as the leaf with rank i in $T(1 \leq i \leq m-1)$. Clearly, a_1 is the leaf dominates T and, thus, we have $DomiEnd(a_1) = r$.

Lemma 4 ([13]). *If a leaf l_1 beats another leaf l_2 at some vertex, we have $R(l_1) \geq R(l_2)$ and $Rank(l_1) < Rank(l_2)$.*

Let

$$F_i = T \setminus \bigcup_{x=i+1}^{m-1} DomiPath(a_x)(1 \leq i \leq m-1).$$

That is, F_i is a subtree that using upward tree accumulation technique [9] to these $DomiPath(a_x)(i+1 \leq x \leq m-1)$ of T.

Lemma 5. *The vertex $p(DomiEnd(a_i))$ is contained in $F_{i-1}(2 \leq i \leq m-1)$*

Proof. Let $x = DomiEnd(a_i)$. By definition, the subtree $T_{p(x)}$ is not dominated by a_i. Let y be the leaf that dominates $T_{p(x)}$. Since y beats a_i at $p(x)$, by Lemma 4, we have $Rank(y) < i$. Thus, the path $DomiPath(y)$ is included in $F_{i-1}(2 \leq i \leq m-1)$. Since $p(x)$ is a vertex on $DomiPath(y)$, it is contained in $F_{i-1}(2 \leq i \leq m-1)$. The lemma holds. □

From Lemma 5, we can easily conclude that F_i is a subtree of T with $i+1$ endpoints(including the root r). The two subtrees F_{i-1} and F_i differ only in the path from $p(DomiEnd(a_i))$ to a_i. Thus, we have

$$\begin{aligned} Sum(F_i) &= Sum(F_{i-1}) - Save(P_{p(DomiEnd(a_i)),a_i}) \\ &= Sum(F_{i-1}) - R(a_i). \end{aligned} \quad (1)$$

Since a_1 dominates T, among all paths from r to the leaves, $DomiPath(a_1)$ is the one with the maximum distance saving. We obtain the following lemma.

Lemma 6. *The subtree F_1 is a 2-tree core of T.*

Lemma 7 ([8]). *For any k-tree core $S \neq T$, there exists a $(k+1)$-tree core S^* such that $S \subset S^*$.*

Theorem 3. *The subtree F_{k-1} is a k-tree core of $T(2 \leq k \leq m)$.*

Proof. We prove this theorem by induction on k. By Lemma 6, the base case $k = 2$ is established. Suppose, by induction, that the theorem is true for all values less than k: we will show that the theorem holds for k as well.

By the induction hypothesis, F_{k-2} is a $(k-1)$-tree core of T. For each vertex $a_i(k-1 \leq i \leq m-1)$, denote $Near(a_i)$ as the vertex in F_{k-2} that is closest to a_i. Let x be the leaf in the set $\{a_{k-1}, a_k, \dots, a_{m-1}\}$ satisfying

$$Save(P_{Near(x),x}) = \max_{k-1 \leq i \leq m-1} \{Save(P_{Near(a_i),a_i})\}$$

By Lemma 7, we can obtain a k-tree core from F_{k-2} by adding one of the paths $P_{Near(a_i),a_i} (i = k-1, k, \dots, m-1)$. Thus, we can conclude that the subtree $F^* = F_{k-2} \cup P_{Near(x),x}$ is a k-tree core of T with $Sum(F^*) = Sum(F_{k-2}) - Save(P_{Near(x),x})$. To show that F_{k-1} is a k-tree core of T, in the following, we show that $Sum(F_{k-1}) \leq Sum(F^*)$.

Let $y = Near(x)$ and q be the second vertex on the path from y to x. Since q is not contained in F_{k-2}, all leaves of T_q are in the set $\{a_{k-1}, a_k, \dots, a_{m-1}\}$. Let $z = DomiLeaf(q)$. Note that $q = DomiEnd(z)$. Since the subtree T_z is not dominated by z. We have

$$\begin{aligned} Sum(F^*) &= Sum(F_{k-2}) - Save(P_{y,x}) \\ &\geq Sum(F_{k-2}) - Save(P_{y,z}) \\ &\geq Sum(F_{k-2}) - R(z). \end{aligned}$$

Since z is a leaf in $\{a_{k-1}, a_k, \dots, a_{m-1}\}$, we have $R(a_{k-1}) \geq R(z)$. From equation (1), we have $Sum(F_{k-1}) = Sum(F_{k-2}) - R(a_{k-1})$. Thus, we have $Sum(F_{k-1}) \leq Sum(F^*)$. The theorem holds. □

Now, we are ready to propose our algorithm for constructing a k-tree core of T,which is as follows:

Algorithm 2 k-Tree Core $T = (V, E)$
```
Input: a tree T = (V, E)
Output: a k-tree core of T
Begin
Step 1: Identify an endpoint r of a two-core of T. And then,
        orient T into a rooted tree with root r.
Step 2: For each vertex v in T, compute size(v) and MaxSave(v).
Step 3: For each vertex v in T, determine DomiLeaf(v).
Step 4: For each leaf l in T, compute DomiEnd(l), R(l) and
        Rank(l).
```

Step 5: For each leaf l in T, find the $DomiPath(l)$.
Step 6: Compute F_{k-1}, which is a k-tree core of T,

$$F_{k-1} \longleftarrow T \backslash \bigcup_{x=k}^{m-1} DomiPath(a_x).$$

Step 7: Return F_{k-1}.
End

The correctness of the above algorithm is ensured by Lemma 7 and Theorem 3. The parallel running time of the algorithm is discussed as follows: By Lemma 1, Step 1 takes $O(\log n)$time using $O(n)$ work. From the Ref. [4], we know the computation of $size(v)$ takes $O(\log n)$ time using $O(n)$ work, using tree contraction, $MaxSave(v)$ can be computed in $O(\log n)$ time using $O(n)$ work. The computation of $Domileaf(v)$ and $DomiPath(l)$ is similar to that of $size(v)$. So, Step 2 and 3 can be done in $O(\log n)$ time using $O(n)$ work. Step 4 and 5 can be done in $O(\log n)$ time using $O(n)$ work. It was showed by Cole and Vishkin [2] that the prefix computation of a linked list can be computed in $O(\log n)$ time using $O(n)$ work on the EREW PRAM. Step 6 can be implemented in $O(\log n)$ time using $O(n)$ work. So, the parallel running time of the algorithm is $O(\log n)$ time using $O(n)$ work on EREW PRAM.

5 Conclusion

Eight different problems are considered by Minieka [6]. In this paper, parallel algorithms are proposed for only two of the problems, which are minimum eccentricity and minimum distancesum. A k-tree center of T is a minimum eccentricity subtree of T with exactly k leaves. A k-tree core of T is a minimum distancesum subtree of T with exactly k leaves. As mentioned in the Introduction, the k-tree core problem has an application in a distributed database system [8]. basing on the algorithms in Ref. [13], we proposed the algorithms on a k-tree center and a k-tree core of a tree network in this paper. We improved the methods of constructing a k-tree center and a k-tree core of a tree network. Our algorithms are more effective and simpler than the algorithm proposed by Wang [13], and they perform on the EREW PRAM in $O(\log n)$ time using $O(n)$ work.

References

1. Cole, R.: An Optimally Efficient Selection Algorithm. Information Processing Letters **26**(1988) 295–299
2. Cole, R., Vishkin, U.: Approximate Parallel Scheduling, Part I: The Basic Technique with Applications to Optimal Parallel List Ranking in Logarithmic Time. SIAM J. Computing **17** (1983) 128–142
3. Handler, G.Y., Mirchandani, P.: Location on Network. Cambridge,Mass:MIT Press (1979)

4. Jaja, J.: An Introduction to Parallel Algorithms. Addison Wesley (1992)
5. Miller, G.L., Reif, J.: Parallel Tree Contraction and Its Applications. Proc. 26th Ann. IEEE Symp. Foundations of Computer Science (1985) 478–489
6. Minieka, E., and Patel, N.H.: On finding the core of a tree with a specified length. Journal of Algorithms **4**(1983) 345–352
7. Peng, S., Wang, T.: The Optimal Location of a Structured Facility in a Tree Network. Parallel Algorithms and Applications **2** (1994) 43–60
8. Peng, S., Stephens, A.B., Yesha, Y.: Algorithms for a Core and k-tree Core of a Tree. J. Algorithms **15**(1993) 143–159
9. Sevilgen, F.E., Aluru, S., Futamura, F.: Parallel algorithms for tree accumulations. J. Parallel Distrib Comput **65**(2005) 85–93
10. Slater, P.J.: Locating central paths in a network. Transportation Science **16**(1982) 1–18
11. Tarjan, R.E., Vishkin, U.: Finding Biconnected Components and Computing Tree Functions in Logarithmic Parallel Time. SIAM J. Computing **14**(1985) 861–874
12. Tansel, B.C., Francis, R.L., Lowe, T.J.: Location on Networks: a Survey. Management Science **29**(1983) 482–511
13. Wang, B.F.: Finding a k-tree core and a k-tree center of a tree network in parallel. IEEE Transactions on Parallel and Distributed Systems **9**(1998) 186–191

A Tight Bound on the Number of Mobile Servers to Guarantee the Mutual Transferability Among Dominating Configurations*

Satoshi Fujita

Department of Information Engineering,
Graduate School of Engineering, Hiroshima University
fujita@se.hiroshima-u.ac.jp

Abstract. In this paper, we propose a new framework to provide continuous services to users by a collection of mobile servers distributed over an interconnection network. We model those mobile servers as a subset of host computers, and assume that a user host can receive the service if at least one adjacent host computer (including itself) plays the role of a server; i.e., we assume that the service could not be routed via the interconnection network. The main results obtained in this paper are summarized as follows: For the class of trees with n hosts, $\lceil (n+1)/2 \rceil$ mobile servers are necessary and sufficient to realize continuous services by the mobile servers, and for the class of Hamiltonian graphs with n hosts, $\lceil (n+1)/3 \rceil$ mobile servers are necessary and sufficient.

1 Introduction

In recent years, it emerges an increasingly strong requirement for high quality services provided over a large-scale interconnection network, such as mobile cellular phone systems and miscellaneous contents delivery systems. In those systems, on-line services should be provided to the users in a transparent manner; i.e., it is strongly required to provide a *common* service to *all* users at *any time*. This motivates the study of a server allocation problem in computer networks; i.e., the problem of finding an allocation of servers to the hosts that is "good" in terms of the latency of contents delivery, minimum bandwidth of contents delivery paths, and the maximum number of clients associated to each server. Many of such metrics could be treated as a constraint to be satisfied by considering a logical network derived from the original physical network; e.g., by logically connecting any two hosts whose round trip time is smaller than a predetermined threshold, we have a logical network in which at least one end vertex of each edge must be allocated a server to guarantee the contents delivery within a given latency, and a similar logical network could also be constructed to guarantee the minimum communication bandwidth of the contents delivery paths.

In such logical networks, a given constraint could be naturally represented by using the notion of **dominating set**. Given a network $G = (V, E)$ with vertex

* This research was partially supported by the Grant-in-Aid for Scientific Research.

X. Deng and D. Du (Eds.): ISAAC 2005, LNCS 3827, pp. 563–572, 2005.

set V and edge set E, a dominating set for G is defined as a subset of vertices such that for any vertex $u \in V$, either u is contained in the subset or at least one neighbor of that is contained in the subset. The notion of dominating set has been extensively studied in the literature during the past three decades, from various aspects including graph theoretic characterization [2,5,6,13,15,21], computational complexity of finding a dominating set with a minimum cardinality [10,14,17], and polynomial time algorithms for a special class of graphs such as interval graphs and perfect graphs [1,3,4,16,18]. It has also been investigated from a practical point of view, and it is pointed out by many researchers that the notion of dominating set is closely related with the resource allocation problem in networks [8], and the design of efficient routing schemes for ad hoc wireless networks [7,11,19,20].

In this paper, we will consider a model in which servers allocated to hosts are allowed to move to other hosts distributed over the network. In recent years, several platforms that allow such a mobility of servers have been implemented based on the technique of mobile agents, and those systems are expected to realize a transparent service to the users. One of the most important problems for such systems with mobile servers is how to realize *continuous services* to the users, provided that the frequent spatial transitions of those mobile servers. More concretely, when a server moves to other host, those client hosts that were assigned to the server must be reassigned to other server to realize a continuous service subject to such a spatial transition; i.e., the client hosts must be dominated by at least two servers to allow such mobility (a formal definition of the term "continuous service" will be given later).

This paper proposes a new framework to realize such a continuous service by a set of mobile servers. We model those mobile servers as a subset of host computers, i.e., a dominating set for the given network, and model the spatial transition of servers by a transition between two dominating configurations. Given such a model of spatial transition, we will consider the following theoretical problem in this paper: *Given two dominating configurations A and B, can we transfer configuration A to B by keeping continuous services to the users?* The main results obtained here could be summarized as follows: 1) For the classes of tree networks with n hosts, $\lceil (n+1)/2 \rceil$ mobile servers are necessary and sufficient to realize mutual transfers among dominating configurations, and 2) for the classes of Hamiltonian networks with n hosts, $\lceil (n+1)/3 \rceil$ mobile servers are necessary and sufficient. Readers should note that, to the authors' best knowledge, this is the first work to investigate the mutual transferability among dominating configurations, although the relation between the notion of dominating set and the server allocation problem has frequently been pointed out in the literature.

The remainder of this paper is organized as follows. In Section 2, we introduce several necessary notations that will be used throughout of the paper, that includes the definition of transferability between two dominating sets. Section 3 overviews an outline of our contribution. The proof of each theorem will be given in Section 4. Finally Section 5 concludes the paper with future problems.

2 Preliminaries

Let $G = (V(G), E(G))$ be an undirected graph with vertex set $V(G)$ and edge set $E(G)$. A dominating set for G is a subset U of $V(G)$ such that for any vertex $u \in V(G)$, either $u \in U$ or there exists a vertex $v \in U$ such that $\{u, v\} \in E(G)$. In this paper, by technical reasons, we assume that dominating set is a multiset; i.e., it can contain each vertex in $V(G)$ several times. Let $\mathcal{D}(G)$ denote an (infinite) set of all dominating (multi)sets for G. A dominating set is said to be minimal if the removal of any vertex from that violates the condition of domination (by definition, any minimal dominating set cannot be a multiset). The **domination number** $\gamma(G)$ of G is the size of a minimum dominating set for G, and the **upper domination number** $\Gamma(G)$ of G is the size of a minimal dominating set for G with a maximum cardinality [12].

For any $S_1, S_2 \in \mathcal{D}(G)$, we say that S_1 is *single-step transferable* to S_2, and denote it as $S_1 \to S_2$, if there are two vertices u and v in $V(G)$ such that $S_1 - \{u\} = S_2 - \{v\}$ and $\{u, v\} \in E(G)$. Note that a single-step transfer from S_1 to S_2 is realized by moving the "role of dominating vertex" from $u \in S_1$ to its neighbor $v \in S_2$, where each vertex can own more than one roles, since each dominating set is assumed to be a multiset. For example, in a ring network consisting of four vertices $\{a, b, c, d\}$, a dominating configuration with vertices $\{a, c\}$ is transferred to a dominating configuration with vertices $\{b, c\}$ in a single-step by moving the role of dominating vertex from a to its neighbor b. A transitive closure of the relation of single-step transferability naturally defines the notion of transferability, that will be denoted as $S_1 \overset{*}{\to} S_2$, in what follows. Note that every subset of vertices appearing in a transfer from S_1 to S_2 must be a dominating set for G. A set $\mathcal{D}' \subseteq \mathcal{D}(G)$ is said to be **mutually transferable** if it holds $S_1 \overset{*}{\to} S_2$ for any $S_1, S_2 \in \mathcal{D}'$, where a sequence of single-step transfers from S_1 to S_2 can contain a subset not in $\mathcal{D}'$, although all subsets in it must be an element in $\mathcal{D}(G)$.

3 Main Theorems

The first theorem gives a tight bound for the class of trees (proofs of all theorems will be given in the next section).

Theorem 1 (Trees). *For any tree T with n vertices, the set of dominating sets for T consisting of $k \geq \lfloor n/2 \rfloor$ vertices is mutually transferable, and there is a tree T with n vertices such that $\gamma(T) = \lfloor n/2 \rfloor$.*

Next, we provide a lower bound on the number of dominating vertices that is necessary to guarantee the mutual transferability among dominating sets, for all graphs contained in a class of Hamiltonian graphs with n vertices.

Theorem 2 (Lower Bound). *For any $r \geq 2$ and $n \geq 1$, there is a Hamiltonian r-regular graph G with more than n vertices such that the set of dominating sets for G with cardinality at least $\lceil (n+1)/3 \rceil - 1$ is not mutually transferable.*

It is worth noting that for any Hamiltonian graph G consisting of n vertices, $\gamma(G) \leq \lceil (n+1)/3 \rceil - 1$, since it contains a ring of size n as a subgraph. It is in contrast to the case of trees, since the theorem claims that there is a Hamiltonian r-regular graph G such that $\lceil (n+1)/3 \rceil - 1$ dominating vertices are not sufficient to guarantee the mutual transferability among dominating configurations, while $\lceil (n+1)/3 \rceil - 1$ vertices are sufficient to dominate it. By combining Theorem 2 with the following theorem, we could derive the tightness of $\lceil (n+1)/3 \rceil$ bound for the class of Hamiltonian graphs with n vertices.

Theorem 3 (Hamiltonian Graphs). *For any Hamiltonian graph G with n vertices, the set of dominating sets for G consisting of $k \geq \lceil (n+1)/3 \rceil$ vertices is mutually transferable.*

4 Proofs

4.1 Theorem 1

Let T be a tree with at least two vertices. Let u be a leaf vertex in T and v be the unique neighbor of u. Any dominating set for T containing u is (single-step) transferable to a dominating set that contains v instead of u, and this transformation allows us to reduce the problem of dominating a tree with n vertices by a set with $k = \lfloor n/2 \rfloor$ dominating vertices to the problem of dominating a tree with at most $n-2$ vertices by a set with $k-1 = \lfloor (n-2)/2 \rfloor$ dominating vertices.

By repeatedly applying this operation, we will have a situation in which either: 1) a tree consisting of at most three vertices is dominated by one vertex, or 2) a tree consisting of n' vertices is dominated by (at least) n' vertices, and in each of the cases, dominating configurations corresponding to the case are trivially mutually transferable. Note that a sequence of such reduction steps is characterized by a sequence of vertices, that are leaves in the corresponding reduced trees. Since any dominating configuration with k vertices can be transferred to such normalized configurations with the same sequence of vertices, we can conclude that the set of dominating sets for T with k vertices is mutually transferable via those normalized configurations. Hence the theorem follows.

4.2 Theorem 2

The given claim immediately holds for $r = 2$ since in ring networks consisting of $3m$ vertices, two dominating sets with $\lceil (3m+1)/3 \rceil - 1 = m$ vertices are not mutually transferable if $m \geq 2$. For $r = 3$, we may consider the following graph $G_1 = (V_1, E_1)$ consisting of 24 vertices, where

$$V_1 \stackrel{\text{def}}{=} \{1, 2, \ldots, 24\}, \quad \text{and}$$

$$E_1 \stackrel{\text{def}}{=} \{(i, i+1) \mid 1 \leq i \leq 11, 13 \leq i \leq 23\}$$
$$\cup \{(12, 1), (24, 13)\} \cup \{(i, i+12) \mid 1 \leq i \leq 12\}.$$

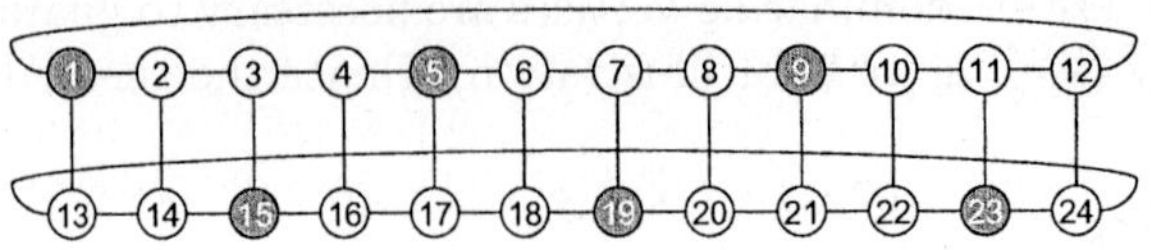

(a) A minimum dominating set for G_1.

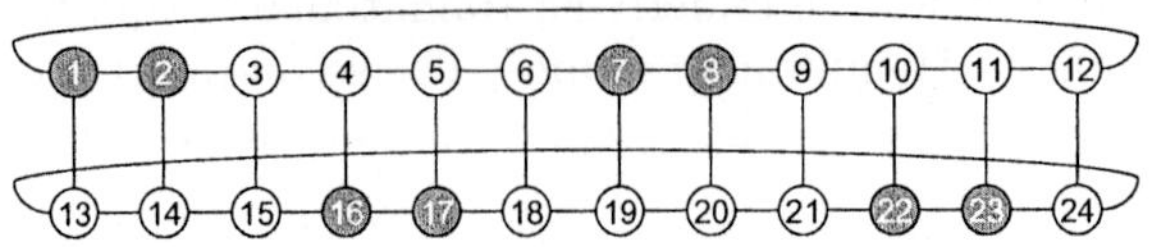

(b) A dominating set for G_1.

Fig. 1. Two dominating sets for graph G_1 (dominating vertices are painted gray)

Note that G_1 is Hamiltonian, cubic, and the domination number of G_1 is six (e.g., $\{1, 5, 9, 15, 19, 23\}$ is a minimum dominating set for G_1; see Figure 1 (a) for illustration). Now let us consider a subset of vertices $S = \{1, 2, 7, 8, 16, 17, 22, 23\}$. It is obvious that S is a dominating set for G_1 and *any* vertex in S cannot move the role of dominating vertex to its neighbor without violating the condition of domination, since each vertex in S "privately" dominates two vertices. For example, as is shown in Figure 1 (b), in the dominating set S for G_1, vertex 2 dominates vertices 3 and 14 and those two vertices are not dominated by the other vertices. Thus, in order to realize a mutual transfer among dominating sets, nine ($= \lceil (24 + 1)/3 \rceil$) vertices are necessary. The above construction can be directly extended to larger cubic graphs consisting of $24x$ vertices for all $x \geq 1$; i.e., we can show that $8x + 1$ ($= \lceil (24x + 1)/3 \rceil$) dominating vertices are necessary to guarantee the mutual transferability among dominating sets.

An extension to larger r's can be easily realized as well, i.e., we can show that there is a Hamiltonian r-regular graph consisting of $3(r - 1)x$ vertices for $x \geq 2$,

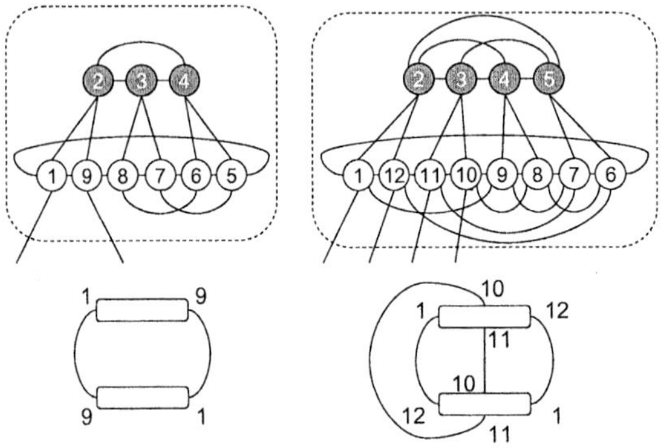

Fig. 2. Extension to larger r's (the upper figures represent the basic component and the lower figures represent how to connect of those components; the label associated with connecting edges is the label of terminal vertices in each component)

such that $(r-1)x+1$ dominating vertices are necessary to guarantee the mutual transferability (see Figure 2 for illustration). Hence the theorem follows.

4.3 Theorem 3

The proof of Theorem 3 consists of two parts. In the first part, we show that the claim holds if we restrict the underlying Hamiltonian graph to rings (note that a ring is a "simplest" Hamiltonian graph). The second part gives a transfer of a dominating set for a Hamiltonian graph to a dominating set for a Hamiltonian cycle contained in it.

Lemma 1 (Rings). *For ring R_n with n vertices, the set of dominating sets for R_n consisting of $k \geq \lceil (n+1)/3 \rceil$ vertices is mutually transferable.*

The proof of this lemma can be found in our previous paper [9].

Preprocessing for Reduction. Let $G = (V, E)$ be a Hamiltonian graph with n vertices, and R be a Hamiltonian cycle in it. In what follows, edges contained in R will be referred to as *ring edges* and the other edges in G will be referred to as *chord edges*. Let $S \subseteq V$ be a dominating set for G with at least $\lceil (n+1)/3 \rceil$ vertices. In the following, we will transfer S to a dominating configuration for R by consecutively removing chord edges and by moving the role of dominating vertices accordingly.

In the first step of the transfer, we apply the following rule until it could not be applied to the resultant graph:

Rule 1: If the removal of a chord edge does not violate the condition of domination for its end vertices, then remove it.

Let G' be the resultant graph. Figure 3 (a) illustrates the resultant graph. Note that S is a dominating set for G', and graph G' contains at most $n - \lceil (n+1)/3 \rceil -$

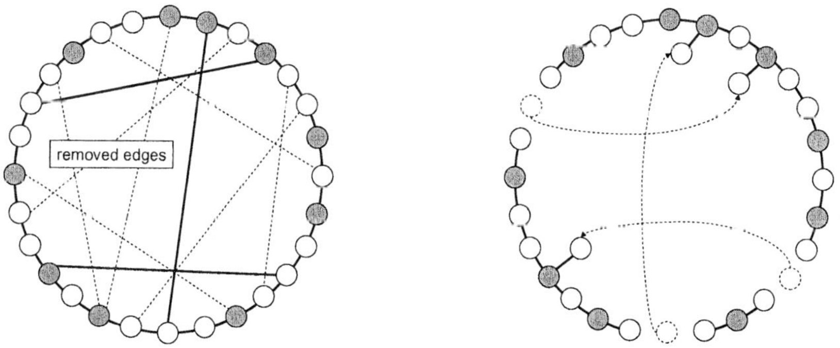

(a) An example of graph G'. (b) An example of graph G''.

Fig. 3. Explanation of the proof of Theorem 4

2 (= $|V - S| - 2$) chord edges, since there are at most $|V - S|$ vertices to be dominated by vertices in S, and at least two of them have already been dominated via ring edges. In addition, for any chord edge in G', exactly one of the end vertices must be a member of S and the other vertex must be connected with exactly one chord edge (otherwise, Rule 1 can be applied to remove a chord edge).

As the next step, we consider a subgraph G'' of G' that is obtained by removing all ring edges incident to the vertices dominated via chord edges. Figure 3 (b) shows an example of the resultant graph. By construction, G'' is a forest of trees such that every leaf is a member in $V - S$ and every vertex with degree more than two is a member in S (in what follows, we call such a vertex "branch" vertex). Since $|S| \geq \lceil (n+1)/3 \rceil$ is assumed, in at lease one of the resultant trees, the number of dominating vertices exceeds one third of the number of vertices. Let T be one of such trees and S_T ($\subseteq S$) be the set of dominating vertices contained in T.

In the following, we will show that in graph G'', S_T can be transferred to a dominating configuration for T in which at least one leaf is a dominating one, by using the fact that $|S_T|$ is greater than one third of the number of vertices in T. Note that the proof of the above claim completes the proof of the theorem since it implies that at least one chord edge can always be removed from G' and the same argument holds for the resultant graph as long as there remains a chord edge in it; i.e., we could transfer the given configuration S for G to a dominating configuration for R (note that in the sequence of reductions, we will have to replace G'' with a new subgraph after removing a chord edge from G').

Transfer of S_T. Tree T contains exactly two leaf vertices dominated via ring edges. Let $u_1, u_2, \ldots, u_m$ be the sequence of vertices on the path connecting those two leaf vertices, i.e., u_1 and u_m are vertices dominated via ring edges and are connected with vertices dominated via chord edges in G'.

If T contains no branch vertices, i.e., it is a linear path, then the claim obviously holds since we can immediately transfer S_T to a configuration in which either u_1 or u_m is a dominating vertex. Thus the following remark holds.

Remark 1. We may assume that T contains at least one branch.

Let u_i be the first branch in T; i.e., vertices $u_1, u_2, \ldots, u_{i-1}$ form a linear path connecting to u_i. Note that $u_i \in S$. Here, we may assume $i = 2$, without loss of generality, by the following two reasons:

- If $i \neq 3j + 2$ for any $j \geq 0$, then we can transfer S_T in such a way that u_1 is a dominating vertex.
- If $i = 3j + 2$ for $j \geq 1$, then we could reduce T to a smaller tree by removing vertices $u_1, u_2, \ldots, u_{i-2}$ without violating the condition on the ratio of dominating vertices, since those $i - 2$ vertices should be dominated by $(i-2)/3$ vertices.

Remark 2. We may assume that vertex u_2 is the first branch in T.

In addition, if T contains exactly one branch vertex, by the same reason to above, 1) we can transfer S_T to a configuration in which u_m is a dominating

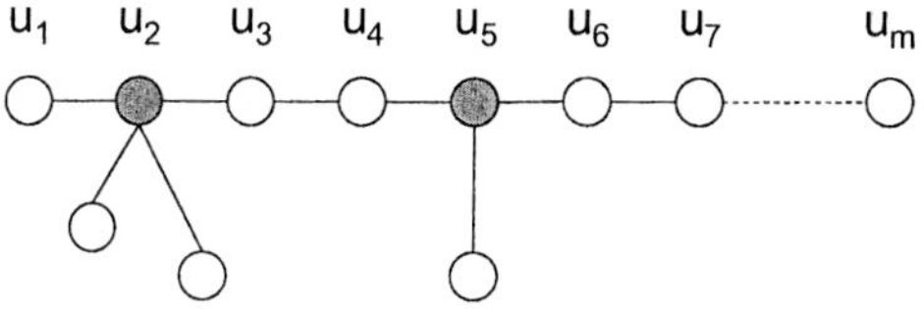

Fig. 4. Example of tree T

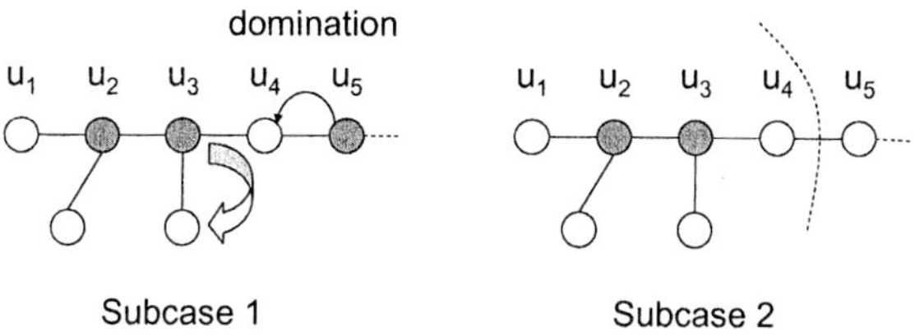

Fig. 5. Case 1

vertex, or 2) we can reduce T to a star-shaped tree with at least four vertices centered at u_2 that is dominated by at least two vertices, i.e., we can transfer S_T to a configuration in which at least one leaf vertex is a dominating one. Thus in the following, we assume T contains at least two branch vertices. Let u_j be the next branch to u_2.

Remark 3. We may assume that T contains its second branch u_j ($\neq u_2$).

Figure 4 shows an example of tree T. In the following, we show that for any $j \geq 3$, we can reduce T to a smaller tree or we can transfer S_T to a configuration with a dominating leaf, by examining the following four cases separately:

Case 1 (When $j = 3$): If u_3 is connected with two or more leaves, then we can reduce T by cutting edge $\{u_2, u_3\}$, and could apply the same argument to the remaining tree containing u_3, since u_2 dominates three vertices including itself. If u_3 is connected with exactly one leaf, on the other hand, we may consider the following two subcases separately (see Figure 5 for illustration): 1) if u_4 is commonly dominated by other vertex (i.e., u_4 or u_5), then u_3 can move the role of dominating vertex to its neighboring leaf (see Figure 5 (a)); and 2) if u_4 is privately dominated by u_3 (i.e., $u_5 \not\in S_T$), then we could reduce T by cutting edge $\{u_4, u_5\}$ and could apply the same argument to the remaining tree containing u_5, since vertices u_2 and u_3 dominate at least six vertices and the domination of u_4 by u_5 in T allows u_3 to move the role of dominating vertex to a leaf (see Figure 5 (b)).

Case 2 (When $j = 4$ or 5): We may assume that no vertices from u_3 to u_{j-1} are contained in S_T, since otherwise, it could move the role of dominating vertex to any leaf vertex in T. Then, by a similar reason to Case 1, we can reduce T by

cutting edge $\{u_{j-2}, u_{j-1}\}$, and could apply the same argument to the remaining tree containing u_{j-1}.

Case 3 (When $j = 3k$ or $3k+1$ for $k \geq 2$): Since $\lceil \frac{j-2}{3} \rceil = k$, the path connecting u_1 and u_{j-2} must be dominated by at least k vertices in S_T (including u_2, of course). In the following, we assume that the number of such dominating vertices is exactly equal to k, since otherwise, we could move the role of dominating vertex to any leaf in T. In addition, since $j - 2 \not\equiv 0 \pmod 3$, we may assume $u_{j-1} \in S_T$, without loss of generality (even if S_T does not contain u_{j-1}, we can easily transform it to a dominating configuration containing u_{j-1}). Let T' be the tree containing u_2 that is obtained by removing edge $\{u_{j-1}, u_j\}$ from T. By the above assumptions, in tree T', at least j vertices are dominated by $k = \lfloor \frac{j}{3} \rfloor$ vertices in S_T. Thus we could apply the same argument to Case 1 by identifying u_{j-1} with vertex u_2 in the proof of Case 1.

Case 4 (When $j = 3k+2$ for $k \geq 2$): By the same reason to Case 3, we may assume that the path connecting u_1 and u_{j-2} is dominated by exactly k $(= (j-2)/3)$ vertices. Let T'' be the tree containing u_2 that is obtained by removing edge $\{u_{j-2}, u_{j-1}\}$ from T. By the above assumptions, in tree T'', at least $j-1$ vertices are dominated by $k = (j-2)/3$ vertices in S_T. Thus we could apply the same argument to Case 2 by identifying u_{j-3} with vertex u_2 in the proof of the case for $j = 5$.

Hence the theorem follows.

5 Concluding Remarks

In this paper, we proposed a new framework to provide continuous services to users by a collection of mobile servers, proved several tight bounds on the number of mobile servers to guarantee the mutual transferability among dominating configurations.

There remain several interesting open problems listed below:

- How can we extend the discussion to general graphs ?
- How can we reduce the number of single-step transfers connecting two dominating configurations ? It could be asymptotically bounded as $O(n)$, but we did not find an exact value of the coefficient.
- Is it possible to construct a distributed scheme that could be executed autonomously with no global information about the overall configuration ?

References

1. M. J. Atallah, G. K. Manacher, and J. Urrutia. Finding a minimum independent dominating set in a permutation graph. *Discrete Appl. Math.*, 21:177–183, 1988.
2. D. W. Bange, A. E. Barkauskas, and P. T. Slater. Efficient dominating sets in graphs. In R. D. Ringeisen and F. S. Roberts, editors, *Applications of Discrete Mathematics*, pages 189–199. SIAM, 1988.

3. A. A. Bertossi. On the domatic number of interval graphs. *Information Processing Letters*, 28(6):275–280, August 1988.
4. G. J. Chang, C. P. Rangan, and S. R. Coorg. Weighted independent perfect domination on cocomparability graphs. Technical Report 93-24, DIMACS, April 1993.
5. E. J. Cockayne and S. T. Hedetniemi. Optimal domination in graphs. *IEEE Trans. Circuit and Systems*, CAS-22:855–857, 1975.
6. E. J. Cockayne and S. T. Hedetniemi. Towards a theory of domination in graphs. *Networks*, 7:247–261, 1977.
7. F. Dai and J. Wu. An Extended Localized Algorithm for Connected Dominating Set Formation in Ad Hoc Wireless Networks. *IEEE Transactions on Parallel and Distributed Systems*, 53(10):1343–1354, 2004.
8. S. Fujita, M. Yamashita, and T. Kameda. A Study on r-Configurations – A Resource Assignment Problem on Graphs. *SIAM J. Discrete Math.* 13(2): 227-254 (2000).
9. S. Fujita, Y. Liang. How to Provide Continuous Services by Mobile Servers in Communication Networks. In *Proc. of PDCAT 2004*, LNCS 3320, pp. 326–329 (2004).
10. M. R. Garey and D. S. Johnson. *Computers and Intractability: A Guide to the Theory of NP-Completeness.* W.H. Freeman and Company, San Francisco, 1979.
11. S. Guha and S. Khuller. Approximation Algorithms for Connected Dominating Sets. In *Proc. European Symposium on Algorithms*, 179-193, 1996.
12. T. W. Haynes, S. T. Hedetniemi, and P. J. Slater. *Fundamentals of Domination in Graphs.* Marcel Dekker, Inc., 1998.
13. T. W. Haynes, S. T. Hedetniemi, and P. J. Slater. *Domination in Graphs: Advanced Topics.* Marcel Dekker, Inc., 1998.
14. R. W. Irving. On approximating the minimum independent dominating set. *Information Processing Letters*, 37:197–200, 1991.
15. M. Livingston and Q. F. Stout. Perfect dominating sets. *Congressus Numerantium*, 79:187–203, 1990.
16. T. L. Lu, P. H. Ho, and G. J. Chang. The domatic number problem in interval graphs. *SIAM J. Disc. Math.*, 3:531–536, 1990.
17. L. R. Matheson and R. E. Tarjan. Dominating sets in planar graphs. Technical Report TR-461-94, Dept. of Computer Science, Princeton University, May 1994.
18. A. Srinivasa Rao and C. P. Rangan. Linear algorithm for domatic number problem on interval graphs. *Information Processing Letters*, 33(1):29–33, October 1989.
19. J. Wu and H. Li. Domination and Its Applications in Ad Hoc Wireless Networks with Unidirectional Links. In *Proc. of International Conference on Parallel Processing*, pages 189–200, 2000.
20. J. Wu. Extended Dominating-Set-Based Routing in Ad Hoc Wireless Networks with Unidirectional Links. *IEEE Transactions on Parallel and Distributed Computing*, 22:327–340, 2002.
21. C. C. Yen and R.C.T. Lee. The weighted perfect domination problem. *Information Processing Letters*, 35:295–299, 1990.

Bounding the Number of Minimal Dominating Sets: A Measure and Conquer Approach

Fedor V. Fomin[1,*], Fabrizio Grandoni[2,**], Artem V. Pyatkin[1,***], and Alexey A. Stepanov[1]

[1] Department of Informatics, University of Bergen, N-5020 Bergen, Norway
{fomin, Artem.Pyatkin}@ii.uib.no, ljosha@ljosha.org
[2] Dipartimento di Informatica, Università di Roma "La Sapienza",
Via Salaria 113, 00198 Roma, Italy
grandoni@di.uniroma1.it

Abstract. We show that the number of minimal dominating sets in a graph on n vertices is at most 1.7697^n, thus improving on the trivial $\mathcal{O}(2^n/\sqrt{n})$ bound. Our result makes use of the measure and conquer technique from exact algorithms, and can be easily turned into an $\mathcal{O}(1.7697^n)$ listing algorithm.

Based on this result, we derive an $\mathcal{O}(2.8805^n)$ algorithm for the domatic number problem, and an $\mathcal{O}(1.5780^n)$ algorithm for the minimum-weight dominating set problem. Both algorithms improve over the previous algorithms.

Keywords: exact (exponential) algorithms, minimum dominating set, minimum set cover, domatic number, weighted dominating set.

1 Introduction

One of the typical questions in graph theory is: how many subgraphs satisfying a given property can a graph on n vertices contain? For example, the number of perfect matchings in a simple k-regular bipartite graph on $2n$ vertices is always between $n!(k/n)^n$ and $(k!)^{n/k}$. (The first inequality was known as van der Waerden Conjecture [22] and was proved in 1980 by Egorychev [6] and the second is due to Bregman [2].) Another example is the famous Moon and Moser [18] theorem stating that every graph on n vertices has at most $3^{n/3}$ maximal cliques (independent sets). Such combinatorial bounds are of interests not only on their own but also because they are used for algorithm design as well. Lawler [17] used Moon-Moser bound on the number of maximal independent sets to construct an $\mathcal{O}((1+\sqrt[3]{3})^n)$ time graph coloring algorithm which was the fastest coloring algorithm for 25 years. Recently Byskov and Eppstein [3] obtain an $\mathcal{O}(2.1020^n)$

* Supported by Norges forskningsråd projects 160778/V30 and 162731/V00.

** Supported by EC Project DELIS, and project WEBMINDS of the Italian Ministry of University and Research (MIUR).

*** Additional support by grants of the Russian Foundation for Basic Research (project code 05-01-00395) and INTAS (project code 04–77–7173).

X. Deng and D. Du (Eds.): ISAAC 2005, LNCS 3827, pp. 573–582, 2005.

time coloring algorithm which is, again, based on a 1.7724^n combinatorial upper bound on the number of maximal bipartite subgraphs in a graph.

The dominating set problem is a classic NP-complete graph optimization problem which fits into the broader class of domination and covering problems. Hundreds of papers have been written on them; see e. g. the survey [14] by Haynes et al. However, despite the importance of minimum dominating set problem, nothing better than the trivial $\mathcal{O}(2^n/\sqrt{n})$ bound was known on the number of minimal dominating sets in a graph.

Our interest is motivated by the design of fast exponential-time algorithms for hard problems. The story of such algorithms dates back to the sixties and seventies. In 1962 Held and Karp presented an $\mathcal{O}(n^2\, 2^n)$ time algorithm for the travelling salesman problem which is still the fastest one known [15]. In 1977 Tarjan and Trojanowski [21] gave an $\mathcal{O}(2^{n/3})$ algorithm for maximum independent set problem. The last decade has seen a growing interest in fast exponential-time algorithms for NP-hard problems. Examples of recently developed fast exponential algorithms are algorithms for maximum independent set [1], satisfiability [4, 16], coloring [7], treewidth [10], and many others. For a good overview of the field we refer to the recent survey written by Woeginger [23].

Previous results. Although minimum dominating set is a natural and very interesting problem concerning the design and analysis of exponential-time algorithms, no exact algorithm for it faster than $2^n \cdot n^{\mathcal{O}(1)}$ had been known until very recently. In 2004 several different sets of authors obtained algorithms breaking the trivial "2^n-barrier". The algorithm of Fomin et al. [11] runs in time $\mathcal{O}(1.9379^n)$. The algorithm of Randerath and Schiermeyer [19] uses a very nice and cute idea (including matching techniques) to restrict the search space. The most time consuming part of their algorithm enumerates all subsets of nodes of cardinality at most $n/3$, thus the overall running time is $\mathcal{O}^*(1.8999^n)$. Grandoni [12, 13], described a $\mathcal{O}(1.8019^n)$ algorithm and finally, Fomin et al. [9] reduced the running time to $\mathcal{O}(1.5137^n)$. All the mentioned results work only in the unweighted case, and cannot be used to list all the minimal dominating sets. The best algorithm for the weighted case prior to this paper is the trivial $\mathcal{O}(2^n n^{\mathcal{O}(1)})$ one.

There are not so many known exact algorithms for the domatic number. Applying an algorithm similar to Lawler's dynamic programming algorithm [17] to the domatic number problem one obtains an $3^n \cdot n^{\mathcal{O}(1)}$ algorithm. Nothing better was known for this problem. For three domatic number problem, which is a special case of the domatic number problem, very recently Reige and Rothe succeed to break the 3^n barrier with an $\mathcal{O}(2.9416^n)$ algorithm [20].

Our results. In this paper we show that the number of minimal dominating sets in a graph on n vertices is at most 1.7697^n. Our result is inspired by the *measure and conquer* technique [9] from exact algorithms, which works as follows. The running time of exponential recursive algorithms is usually bounded by measuring the progress made by the algorithm at each branching step. Though these algorithms may be rather complicated, the measures used in their analysis are often trivial. For example in graph problems the progress is usually measured

in terms of number of nodes removed. The idea behind measure and conquer is to chose the measure more carefully: a good choice can lead to a tremendous improvement of the running time bounds (for a fixed algorithm). One of the main contributions of this paper is showing that the same basic idea can be successfully applied to derive stronger combinatorial bounds. In particular, the inductive proof of Theorem 1 is based on the way we choose the measure of the problem.

Our combinatorial result is algorithmic in spirit, and can be easily turned into an algorithm listing all minimal dominating sets in time $\mathcal{O}(1.7697^n)$. Based on the listing algorithm, we derive an $\mathcal{O}(1.5780^n)$ algorithm for the minimum-weight dominating set problem, and an $\mathcal{O}(2.8805^n)$ algorithm for the domatic number. Both algorithms improve on previous best trivial bounds. Note that our algorithm for the domatic number is even faster than the (non-trivial) algorithm of Reige and Rothe [20] for the three domatic number problem, which is a special case.

2 Definitions and Preliminaries

Let $G = (V, E)$ be a graph. A set $D \subseteq V$ is called a *dominating set* for G if every vertex of G is either in D, or adjacent to some node in D. A dominating set is *minimal* if all its proper subsets are not dominating. We define $\mathbf{DOM}(G)$ to be the number of minimal dominating sets in a graph G. The *domination number* $\gamma(G)$ of a graph G is the cardinality of a smallest dominating set of G. The *Minimum Dominating Set* problem (MDS) asks to determine $\gamma(G)$. The *domatic number* $\mathbf{DN}(G)$ of a graph G is the maximum k such that the vertex set $V(G)$ can be split into k pairwise nonintersecting dominating sets. Since any dominating set contains a minimal dominating set, the domatic number $\mathbf{DN}(G)$ is the maximum number of pairwise nonintersecting minimal dominating sets in G.

In the *Minimum Set Cover* problem (MSC) we are given a universe $\mathcal{U}$ of elements and a collection $\mathcal{S}$ of (non-empty) subsets of $\mathcal{U}$. The aim is to determine the minimum cardinality of a subset $\mathcal{S}^* \subseteq \mathcal{S}$ which *covers* $\mathcal{U}$, i. e. such that

$$\cup_{S \in \mathcal{S}^*} S = \mathcal{U}.$$

The *frequency* of $u \in \mathcal{U}$ is the number of subsets $S \in \mathcal{S}$ in which u is contained.

A covering is *minimal* if it contains no smaller covering. We denote by $\mathbf{COV}(\mathcal{U}, \mathcal{S})$ the number of minimal coverings in $(\mathcal{U}, \mathcal{S})$.

The problem of finding $\mathbf{DOM}(G)$ can be naturally reduced to finding $\mathbf{COV}(\mathcal{U}, \mathcal{S})$ by imposing $\mathcal{U} = V$ and $\mathcal{S} = \{N[v] \mid v \in V\}$. Note that $N[v] = \{v\} \cup \{u \mid uv \in E\}$ is the set of nodes dominated by v. Thus D is a dominating set of G if and only if $\{N[v] | \ v \in D\}$ is a set cover of $(\mathcal{U}, \mathcal{S})$. So, each minimal set cover of $(\mathcal{U}, \mathcal{S})$ corresponds to a minimal dominating set of G.

The following properties of minimal coverings are easy to verify.

Proposition 1. *Let $\mathcal{S}^*$ be a minimal covering of $(\mathcal{U}, \mathcal{S})$. Then the following statements hold.*

- *For every subset $S \in \mathcal{S}^*$ at least one of the elements $u \in S$ is covered only by S;*
- *If $\mathcal{S}^*$ contains two subsets S_1 and S_2 such that $S_1 \setminus S_2 = \{u_1\}$ and $S_2 \setminus S_1 = \{u_2\}$ then no other subset in $\mathcal{S}^*$ may contain u_1 or u_2.*

In the next section we prove an upper bound on $\mathbf{DOM}(G)$. Based on this result we show how to compute the domatic number of a graph in time $\mathcal{O}(2.8805^n)$.

3 Listing Minimal Dominating Sets

Here we prove the following

Theorem 1. *For any graph G on n vertices, $\mathbf{DOM}(G) < 1.7697^n$.*

Proof. Since every MDS problem can be reduced to MSC problem, we will prove an upper bound for $\mathbf{COV}(\mathcal{U}, \mathcal{S})$ first.

Consider an arbitrary instance of the MSC problem with a universe $\mathcal{U}$ of elements and a collection $\mathcal{S}$ of (non-empty) subsets of $\mathcal{U}$. Denote by s_i the number of subsets of cardinality i for $i = 1, 2, 3$ and by s_4 the number of the subsets of cardinality at least 4 in S. We use the following measure $k(\mathcal{U}, \mathcal{S})$ of $(\mathcal{U}, \mathcal{S})$:

$$k(\mathcal{U}, \mathcal{S}) = |\mathcal{U}| + \sum_{i=1}^{4} \varepsilon_i s_i,$$

where the values of $0 < \varepsilon_1 < \varepsilon_2 < \varepsilon_3 < \varepsilon_4$ will be defined later. We refer to the value $k = k(\mathcal{U}, \mathcal{S})$ as to the *size* of the MSC problem $(\mathcal{U}, \mathcal{S})$.

Let $\mathbf{COV}(k)$ be the maximum value of $\mathbf{COV}(\mathcal{U}, \mathcal{S})$ among all MSC problems of size at most k. Let $d_2 = \min\{\varepsilon_1, \varepsilon_2 - \varepsilon_1\}, d_3 = \min\{d_2, \varepsilon_3 - \varepsilon_2\}$, and $d_4 = \min\{d_3, \varepsilon_4 - \varepsilon_3\}$. We need the following

Lemma 1. $\mathbf{COV}(k) \le \alpha^k$, *where α satisfies the following inequalities:*

$$\alpha^k \ge \max \begin{cases} r\alpha^{k-r\varepsilon_1-1}, \ r \ge 2 \\ \alpha^{k-\varepsilon_4} + \alpha^{k-5-\varepsilon_4} \\ \alpha^{k-\varepsilon_4} + \alpha^{k-4-\varepsilon_4-4d_4} \\ \alpha^{k-1-\varepsilon_1-\varepsilon_2} + \alpha^{k-2-\varepsilon_1-\varepsilon_2-d_3} \\ 2\alpha^{k-2-2\varepsilon_2} \\ 2\alpha^{k-2-2\varepsilon_2-d_3} + \alpha^{k-3-2\varepsilon_2-2d_3} \\ \alpha^{k-\varepsilon_3} + \alpha^{k-3-\varepsilon_3-6d_3} \\ \alpha^{k-\varepsilon_2} + \alpha^{k-2-\varepsilon_1-\varepsilon_2-3d_2} \\ \alpha^{k-\varepsilon_2} + \alpha^{k-2-2\varepsilon_2-2d_2} \\ 3\alpha^{k-2-3\varepsilon_2-2d_2} + 3\alpha^{k-3-6\varepsilon_2} + \alpha^{k-4-6\varepsilon_2} \\ \alpha^{k-\varepsilon_2} + 2\alpha^{k-2-4\varepsilon_2-3d_2} \end{cases} \quad (1)$$

Proof. We use induction on k. Clearly, $\mathbf{COV}(0) = 1$. Suppose that $\mathbf{COV}(l) \le \alpha^l$ for every $l < k$. Let $\mathcal{S}$ be a set of subsets of $\mathcal{U}$ such that the MSC problem $(\mathcal{U}, \mathcal{S})$ is of size k. We consider different cases.

Case 0. *There is an element $u \in \mathcal{U}$ of frequency one.* Since u must be covered by the only set S containing it, we have that every minimal cover contains S. So,

$$\mathbf{COV}(\mathcal{U}, \mathcal{S}) \le \mathbf{COV}(\mathcal{U} \setminus \{u\}, \mathcal{S} \setminus S) \le \alpha^{k-1-\varepsilon_1}.$$

Case 1. *$\mathcal{U}$ has an element u belonging only to subsets of cardinality one.* Let $S_1 = S_2 = \cdots = S_r = \{u\}$, where $r \ge 2$ be all the subsets containing u. Then by Proposition 1, every minimal covering should contain exactly one subset from $S_1, \ldots, S_r$. Thus

$$\mathbf{COV}(\mathcal{U}, \mathcal{S}) \le r \cdot \mathbf{COV}(k - r\varepsilon_1 - 1) \le r\alpha^{k-r\varepsilon_1-1}.$$

Case 2. *$\mathcal{S}$ has a subset with $r \ge 5$ elements.* Let $S = \{u_1, u_2, \ldots, u_r\}$ be such a subset. Every minimal set cover either contains S, or does not. The number of minimal set covers that do not contain S is at most $\mathbf{COV}(\mathcal{U}, \mathcal{S} \setminus S)$. Clearly, the number of minimal set covers containing S is at most $\mathbf{COV}(\mathcal{U} \setminus \{u_1, u_2, \ldots, u_r\}, \mathcal{S}')$. Here $\mathcal{S}'$ consists of all nonempty subsets $S' \setminus \{u_1, u_2, \ldots, u_r\}$ where $S' \in \mathcal{S}$. Note that $S \notin \mathcal{S}'$. Thus

$$\begin{aligned}\mathbf{COV}(\mathcal{U}, \mathcal{S}) &\le \mathbf{COV}(\mathcal{U}, \mathcal{S} \setminus S) + \mathbf{COV}(\mathcal{U} \setminus \{u_1, u_2, \ldots, u_r\}, \mathcal{S}') \\ &\le \mathbf{COV}(k - \varepsilon_4) + \mathbf{COV}(k - 5 - \varepsilon_4) \le \alpha^{k-\varepsilon_4} + \alpha^{k-5-\varepsilon_4}.\end{aligned}$$

So, we may suppose that all subsets contain at most four elements and the minimum frequency of the elements is two.

Case 3. *$\mathcal{S}$ has a subset of cardinality four.* Let $S = \{u_1, u_2, u_3, u_4\}$ be such a subset. Again, the number of minimal set covers that do not contain S is at most $\mathbf{COV}(\mathcal{U}, \mathcal{S} \setminus S)$ and the number of minimal set covers containing S is at most $\mathbf{COV}(\mathcal{U} \setminus \{u_1, u_2, u_3, u_4\}, \mathcal{S}')$. Since there are no elements of frequency one and all subsets have cardinality at most four, removal of every element u_1, u_2, u_3, u_4 from $\mathcal{U}$ reduces the size of the problem by at least $d_4 + 1$. Thus

$$\begin{aligned}\mathbf{COV}(\mathcal{U}, \mathcal{S}) &\le \mathbf{COV}(\mathcal{U}, \mathcal{S} \setminus S) + \mathbf{COV}(\mathcal{U} \setminus \{u_1, u_2, u_3, u_4\}, \mathcal{S}') \\ &\le \mathbf{COV}(k - \varepsilon_4) + \mathbf{COV}(k - \varepsilon_4 - 4(d_4 + 1)) \le \alpha^{k-\varepsilon_4} + \alpha^{k-4-\varepsilon_4-4d_4}.\end{aligned}$$

Now we may suppose that all subsets contain at most three elements.

Case 4. *There is $u \in \mathcal{U}$ of frequency two.* Denote by S_1 and S_2 the subsets containing u. Let $|S_2| \ge |S_1|$. Since the condition of Case 1 does not hold, we have that S_2 is of cardinality at least two. There are three subcases.

Subcase 4A. $|S_1| = 1$. Then every minimal cover contains exactly one of these subsets. If it contains S_1, then we remove u, S_1, and S_2. Otherwise, we remove also the other element of the subset S_2 from $\mathcal{U}$ and thus, in addition, reduce the size of the problem by at least $d_3 + 1$. So we have

$$\begin{aligned}\mathbf{COV}(\mathcal{U}, \mathcal{S}) &\le \mathbf{COV}(k - 1 - \varepsilon_1 - \varepsilon_2) + \mathbf{COV}(k - 2 - \varepsilon_1 - \varepsilon_2 - d_3) \\ &\le \alpha^{k-1-\varepsilon_1-\varepsilon_2} + \alpha^{k-2-\varepsilon_1-\varepsilon_2-d_3}.\end{aligned}$$

Subcase 4B. $|S_1| \geq 2$ and $S_1 \subseteq S_2$. Again, every minimal cover contains exactly one of these subsets and we clearly have

$$\mathbf{COV}(\mathcal{U}, \mathcal{S}) \leq 2\mathbf{COV}(k - 2 - 2\varepsilon_2) \leq 2\alpha^{k-2-2\varepsilon_2}.$$

Subcase 4C. $|S_1| \geq 2$ and $S_1 \not\subseteq S_2$. Now every minimal cover contains either exactly one of these subsets, or both of them. If the first alternative happens (there are two possibilities for that), then we remove two subsets, two elements and reduce the cardinality of at least one other subset. Otherwise, we remove two subsets, three elements and decrease either the cardinalities of at least two other subsets by one or the cardinality of one subset by two (anyway, reducing the size of the instance by at least $2d_3$). Hence,

$$\begin{aligned}\mathbf{COV}(\mathcal{U}, \mathcal{S}) &\leq 2\mathbf{COV}(k - 2 - 2\varepsilon_2 - d_3) + \mathbf{COV}(k - 3 - 2\varepsilon_2 - 2d_3) \\ &\leq 2\alpha^{k-2-2\varepsilon_2-d_3} + \alpha^{k-3-2\varepsilon_2-2d_3}.\end{aligned}$$

Now we assume that the minimum frequency of the elements is three.

Case 5. *$\mathcal{S}$ has a subset of cardinality three.* This case is analyzed similar to Case 3, but since now all elements have frequency at least three and all subsets have cardinality at most three, removing each of the elements decrease the cardinalities of at least two other subsets, reducing the size of the instance by $1 + 2d_3$. Therefore

$$\begin{aligned}\mathbf{COV}(\mathcal{U}, \mathcal{S}) &\leq \mathbf{COV}(k - \varepsilon_3) + \mathbf{COV}(k - \varepsilon_3 - 3(1 + 2d_3)) \\ &\leq \alpha^{k-\varepsilon_3} + \alpha^{k-3-\varepsilon_3-6d_3}.\end{aligned}$$

Now we may suppose that all subsets contain either one or two elements.

Case 6. *There are $S, S' \in \mathcal{S}$ such that $S' \subset S$.* Since we are not in Case 1, we have that $|S| \geq 2$. By Proposition 1, every minimal covering containing S does not contain S'. Thus for $|S'| = 1$ we have

$$\begin{aligned}\mathbf{COV}(\mathcal{U}, \mathcal{S}) &\leq \mathbf{COV}(k - \varepsilon_2) + \mathbf{COV}(k - 2 - \varepsilon_1 - \varepsilon_2 - 3d_2) \\ &\leq \alpha^{k-\varepsilon_2} + \alpha^{k-2-\varepsilon_1-\varepsilon_2-3d_2}\end{aligned}$$

and for $|S'| > 1$, we have

$$\begin{aligned}\mathbf{COV}(\mathcal{U}, \mathcal{S}) &\leq \mathbf{COV}(k - \varepsilon_2) + \mathbf{COV}(k - 2 - 2\varepsilon_2 - 2d_2) \\ &\leq \alpha^{k-\varepsilon_2} + \alpha^{k-2-2\varepsilon_2-2d_2}.\end{aligned}$$

(Here we use the fact that by Case 4 the minimum frequency of the elements is three.)

Now we assume that all subsets are of cardinality two.

Case 7. *There is $u \in \mathcal{U}$ of frequency three.* Let $S_i = \{u, u_i\}$, $i = 1, 2, 3$ be the subsets containing u. Then

— There are at most $3 \cdot \mathbf{COV}(k - 2 - 3\varepsilon_2 - 2d_2)$ minimal covers of $\mathcal{S}$ containing exactly one of these subsets. In each of the three cases we remove three subsets S_1, S_2, S_3, two elements (u and one of u_i), and reduce the cardinalities of at least two other subsets.

— There are at most $3 \cdot \mathbf{COV}(k - 3 - 6\varepsilon_2)$ minimal covers containing exactly two of these subsets. Indeed, if S_1 and S_2 are in a minimal cover then by Proposition 1 no other subset containing u_1 or u_2 may lie in this cover. Since the frequencies of u_1 and u_2 are at least three (Case 4) and at most one subset may contain u_1 and u_2 together (Case 6) we can remove at least three other subsets containing u_1 or u_2.
— There are at most $\mathbf{COV}(k - 4 - 6\varepsilon_2)$ minimal covers containing all three of these subsets.

Therefore,

$$\mathbf{COV}(\mathcal{U}, \mathcal{S}) \leq 3\alpha^{k-2-3\varepsilon_2-2d_2} + 3\alpha^{k-3-6\varepsilon_2} + \alpha^{k-4-6\varepsilon_2}.$$

Case 8. *$\mathcal{S}$ does not satisfy any of the conditions from Cases* 1–7. Let $S = \{u, v\}$ be a subset of $\mathcal{S}$. Denote by $\mathcal{S}_u$ and $\mathcal{S}_v$ all other subsets containing u and v respectively. Since the minimum frequency of the elements is four (Case 7), $|\mathcal{S}_u| \geq 3$ and $|\mathcal{S}_v| \geq 3$. By Proposition 1, if $\mathcal{S}^*$ is a minimal cover containing S, then $\mathcal{S}^* \cap \mathcal{S}_u = \emptyset$ or $\mathcal{S}^* \cap \mathcal{S}_v = \emptyset$. Thus we have at most $\mathbf{COV}(k - \varepsilon_2)$ minimal covers that do not contain S and at most $2 \cdot \mathbf{COV}(k - 2 - 4\varepsilon_2 - 3d_2)$ covers containing S. Then

$$\mathbf{COV}(\mathcal{U}, \mathcal{S}) \leq \alpha^{k-\varepsilon_2} + 2\alpha^{k-2-4\varepsilon_2-3d_2}.$$

Summarizing Cases 1–7 (recurrence of Case 0 is trivial), we obtain the inequalities (1). This completes the proof of Lemma 1. □

For any graph G on n vertices the size of the corresponding instance of MSC is at most $|\mathcal{U}| + \varepsilon_4|\mathcal{S}| = (1 + \varepsilon_4)n$. Thus the estimation of $\mathbf{COV}(k)$ boils up to choosing the weights ε_i, $i = 1, \ldots, 4$ and α, minimizing $\alpha^{1+\varepsilon_4}$. This optimization problem is interesting in its own and we refer to Eppstein's work [8] on quasi-convex programming for general treatment of such problems. We numerically obtained the following values of the variables: $\varepsilon_1 = 2.9645, \varepsilon_2 = 3.5218, \varepsilon_3 = 3.9279, \varepsilon_4 = 4.1401$, and $\alpha < 1.117446$. Therefore, $\mathbf{DOM}(G) \leq \mathbf{COV}((1 + \varepsilon_4)n) < 1.117446^{5.1401n} < 1.7697^n$. This completes the proof of Theorem 1. □

Using standard methods one can easily transform the proof of the Theorem 1 into an algorithm listing all minimal dominating sets.

Corollary 1. *There is an algorithm for listing all minimal dominating sets in an n vertex graph G in time $\mathcal{O}(1.7697^n)$.*

In the *Minimum Weighted Dominating Set* problem each vertex v of the graph has weight $w(v)$ and we search for the dominating set D of minimum weight $w(D) = \sum_{v \in D} w(v)$. Clearly, the Corollary 1 allows us to solve this problem in time $\mathcal{O}(1.7697^n)$. Note, however, that the running time can be greatly reduced by exploiting the fact that, if all the subsets have cardinality at most two, Minimum Weighted Set Cover can be solved in polynomial time via reduction to the minimum-weight perfect matching problem (see [5]). This way we obtain the following theorem (we omit the proof details in this extended abstract).

Theorem 2. *There is an algorithm computing a minimum-weight dominating set in time $O(1.5780^n)$.*

4 Algorithm for the Domatic Number

The results of the previous section can be used to compute the domatic number of a graph $G = (V, E)$. Our algorithm has similarities with the classical algorithm computing the chromatic number due to Lawler [17] (see also [7]), but the analysis of our algorithm is based on Lemma 1.

For every set $X \subseteq V$ denote by $\mathbf{DN}(G|X)$ the maximum number of pairwise nonintersecting subsets of X such that each of these subsets is a minimal dominating set in G. Clearly, $\mathbf{DN}(G|V) = \mathbf{DN}(G)$. Note that if X is not dominating, then $\mathbf{DN}(G|X) = 0$

We use an array A, indexed by the 2^n subsets of V, for which we compute the numbers $\mathbf{DN}(G|X)$ for all subsets $X \subseteq V$. We initialize this array by assigning 0 to all $A[X]$. Then we run through the subsets X of V in an order such that all proper subsets of each set X are visited before X. To compute $A[X]$, we run through all minimal dominating sets $D \subseteq X$ of G, and put

$$A[X] = \max\{A[X \setminus D] + 1 \mid D \subseteq X \text{ and } D \text{ is a minimal dominating set in } G\}.$$

Finally, after running through all subsets, we return the value in $A[V]$ as the domatic number of G.

Theorem 3. *The domatic number of a graph G on n vertices can be computed in time $\mathcal{O}(2.8805^n)$*

Proof. The correctness of the algorithm DN can be shown by an easy induction. Let X be a subset of V. Suppose that after running the algorithm, for every proper subset S of X the value $A[S]$ is equal to $\mathbf{DN}(G|S)$. Note that $A[\emptyset] = 0$. If X contains no dominating subsets (i. e. X is not dominating), then we have that $A[X] = \mathbf{DN}(G|X) = 0$. Otherwise, $\mathbf{DN}(G|X)$ is equal to $\max\{\mathbf{DN}(G|(X \setminus D)) + 1\}$, where maximum is taken over all minimal dominating sets $D \subset X$, and thus the value $A[X]$ computed by the algorithm is equal to $\mathbf{DN}(G|X)$.

For a set $X \subseteq V$, let $\mathbf{DOM}(G|X)$ be the number of minimal dominating sets of G which are subsets of X. To estimate the running time of the algorithm, let us bound first $\mathbf{DOM}(G|X)$. We use the following reduction to the MSC problem. Let $\mathcal{U} = V$ and $\mathcal{S} = \{N[v] \mid v \in X\}$. Then, $\mathbf{DOM}(G|X) = \mathbf{COV}(\mathcal{U}, \mathcal{S})$. Note that the size of this problem is at most $|\mathcal{U}| + \varepsilon_4 \cdot |\mathcal{S}| = n + \varepsilon_4 \cdot |X|$. By Lemma 1, $\mathbf{COV}(\mathcal{U}, \mathcal{S}) \le \mathbf{COV}(n + \varepsilon_4 \cdot |X|) \le \alpha^{n+\varepsilon_4 \cdot |X|}$, where α and $\varepsilon_i, i = 1, \ldots, 4$ must satisfy (1). As in Theorem 1, one also can list in time $\mathcal{O}(\alpha^{n+\varepsilon_4 \cdot |X|})$ (and polynomial space) all minimal dominating sets contained in X. The main loop of the algorithm generating all minimal dominating sets contained in X can be performed in time $\mathcal{O}(\alpha^{n+\varepsilon_4 \cdot |X|})$ and the running time of the algorithm can be bounded by

$$\mathcal{O}(\sum_{i=0}^{n} \binom{n}{i} \alpha^{n+\varepsilon_4 \cdot i}) = \mathcal{O}(\alpha^n (1 + \alpha^{\varepsilon_4})^n).$$

We numerically found the following values : $\varepsilon_1 = 3.3512, \varepsilon_2 = 4.0202, \varepsilon_3 = 4.4664, \varepsilon_4 = 4.7164$, and $\alpha < 1.105579$. Then $\mathcal{O}(\alpha^n(1+\alpha^{\varepsilon_4})^n) = \mathcal{O}(2.8805^n)$ □

5 Conclusions and Open Problems

We have shown that the number of minimal dominating sets in a graph on n vertices is at most 1.7697^n. Dieter Kratsch (private communication) found graphs ($n/6$ disjoint copies of the octahedron) containing $15^{n/6} \approx 1.5704^n$ minimal dominating sets. We conjecture that Kratsch's graphs are the graphs with the maximal number of minimal dominating sets. This suggests the possibility that minimal dominating sets can be listed even faster.

As an algorithmic application of our combinatorial bound based on measure and conquer technique, we obtained a faster exponential algorithm to compute the domatic number of a graph. We also obtained an algorithm computing a minimum-weight dominating set in time $O(1.5780^n)$. An interesting open question is if the analysis of our algorithms can be refined, possibly via a further refined measure of the size of the set cover.

Acknowledgement. We are grateful to Dieter Kratsch for bringing our attention to the problem and for many fruitful discussions.

References

1. R. Beigel, *Finding maximum independent sets in sparse and general graphs*, in Proceedings of the 10th ACM-SIAM Symposium on Discrete Algorithms (SODA'1999), ACM and SIAM, 1999, pp. 856–857.
2. L. M. Brègman, *Certain properties of nonnegative matrices and their permanents*, Doklady Akademii Nauk BSSR, 211 (1973), pp. 27–30.
3. M. Byskov and D. Eppstein, *An algorithm for enumerating maximal bipartite subgraphs*, manuscript, (2004).
4. E. Dantsin, A. Goerdt, E. A. Hirsch, R. Kannan, J. Kleinberg, C. Papadimitriou, P. Raghavan, and U. Schöning, *A deterministic* $(2-2/(k+1))^n$ *algorithm for k-SAT based on local search*, Theoretical Computer Science, 289 (2002), pp. 69–83.
5. J. Edmonds and E. L. Johnson, *Matching: A well-solved class of integer linear programs*, in Combinatorial Structures and their Applications, Gordon and Breach, New York, 1970, pp. 89–92.
6. G. P. Egorychev, *Proof of the van der Waerden conjecture for permanents*, Sibirsk. Mat. Zh., 22 (1981), pp. 65–71, 225.
7. D. Eppstein, *Small maximal independent sets and faster exact graph coloring*, Journal of Graph Algorithms and Applications, 7 (2003), pp. 131–140.
8. D. Eppstein, *Quasiconvex analysis of backtracking algorithms*, in Proceedings of the 15th ACM-SIAM Symposium on Discrete Algorithms (SODA'2004), ACM and SIAM, 2004, pp. 781–790.
9. F. V. Fomin, F. Grandoni, and D. Kratsch, *Measure and conquer: Domination – a case study*, in Proceedings of the 32nd International Colloquium on Automata, Languages and Programming (ICALP 2005), LNCS, Springer, Berlin, 2005 (to appear).

10. F. V. Fomin, D. Kratsch, and I. Todinca, *Exact algorithms for treewidth and minimum fill-in*, in Proceedings of the 31st International Colloquium on Automata, Languages and Programming (ICALP 2004), vol. 3142 of LNCS, Springer, Berlin, 2004, pp. 568–580.
11. F. V. Fomin, D. Kratsch, and G. J. Woeginger, *Exact (exponential) algorithms for the dominating set problem*, in Proceedings of the 30th Workshop on Graph Theoretic Concepts in Computer Science (WG 2004), vol. 3353 of LNCS, Springer, Berlin, 2004, pp. 245–256.
12. F. Grandoni, *Exact algorithms for hard graph problems*, PhD thesis, Università di Roma "Tor Vergata", Roma, Italy, (March, 2004).
13. ——, *A note on the complexity of minimum dominating set*, Journal of Discrete Algorithms, (to appear).
14. T. W. Haynes and S. T. Hedetniemi, eds., *Domination in graphs*, Marcel Dekker Inc., New York, 1998.
15. M. Held and R. M. Karp, *A dynamic programming approach to sequencing problems*, Journal of SIAM, 10 (1962), pp. 196–210.
16. J. Iwama and S. Tamaki, *Improved upper bounds for 3-sat*, in 15th ACM-SIAM Symposium on Discrete Algorithms (SODA 2004), ACM and SIAM, 2004, p. 328.
17. E. L. Lawler, *A note on the complexity of the chromatic number problem*, Information Processing Lett., 5 (1976), pp. 66–67.
18. J. W. Moon and L. Moser, *On cliques in graphs*, Israel Journal of Mathematics, 3 (1965), pp. 23–28.
19. B. Randerath and I. Schiermeyer, *Exact algorithms for MINIMUM DOMINATING SET*, Technical Report zaik-469, Zentrum für Angewandte Informatik Köln, Germany, 2004.
20. T. Reige and J. Rothe, *An exact* 2.9416^n *algorithm for the three domatic number problem*, in Proceedings of the 30th International Symposium on Mathematical Foundations of Computer Science (MFCS 2005), LNCS, Springer, Berlin, 2005, p. to appear.
21. R. E. Tarjan and A. E. Trojanowski, *Finding a maximum independent set*, SIAM Journal on Computing, 6 (1977), pp. 537–546.
22. B. van der Waerden, *Problem 45*, Jahresber. Deutsch. Math.-Verein., 35 (1926), p. 117.
23. G. Woeginger, *Exact algorithms for NP-hard problems: A survey*, in Combinatorial Optimization - Eureka, you shrink!, vol. 2570 of LNCS, Springer-Verlag, Berlin, 2003, pp. 185–207.

Collective Tree Spanners in Graphs with Bounded Genus, Chordality, Tree-Width, or Clique-Width

Feodor F. Dragan and Chenyu Yan

Department of Computer Science, Kent State University, Kent, OH 44242
{dragan, cyan}@cs.kent.edu

Abstract. In this paper we study collective additive tree spanners for special families of graphs including planar graphs, graphs with bounded genus, graphs with bounded tree-width, graphs with bounded clique-width, and graphs with bounded chordality. We say that a graph $G = (V, E)$ *admits a system of* μ *collective additive tree* r*-spanners* if there is a system $\mathcal{T}(G)$ of at most μ spanning trees of G such that for any two vertices x, y of G a spanning tree $T \in \mathcal{T}(G)$ exists such that $d_T(x,y) \leq d_G(x,y)+r$. We describe a general method for constructing a "small" system of collective additive tree r-spanners with small values of r for "well" decomposable graphs, and as a byproduct show (among other results) that any weighted planar graph admits a system of $O(\sqrt{n})$ collective additive tree 0–spanners, any weighted graph with tree-width at most $k-1$ admits a system of $k \log_2 n$ collective additive tree 0–spanners, any weighted graph with clique-width at most k admits a system of $k \log_{3/2} n$ collective additive tree (2w)–spanners, and any weighted graph with size of largest induced cycle at most c admits a system of $\log_2 n$ collective additive tree $(2\lfloor c/2 \rfloor \mathrm{w})$–spanners and a system of $4 \log_2 n$ collective additive tree $(2(\lfloor c/3 \rfloor + 1)\mathrm{w})$–spanners (here, w is the maximum edge weight in G). The latter result is refined for weighted weakly chordal graphs: any such graph admits a system of $4 \log_2 n$ collective additive tree (2w)-spanners. Furthermore, based on this collection of trees, we derive a compact and efficient routing scheme for those families of graphs.

1 Introduction

Many combinatorial and algorithmic problems are concerned with the distance d_G on the vertices of a possibly weighted graph $G = (V, E)$. Approximating d_G by a simpler distance (in particular, by tree–distance d_T) is useful in many areas such as communication networks, data analysis, motion planning, image processing, network design, and phylogenetic analysis. Given a graph $G = (V, E)$, a spanning subgraph H is called a *spanner* if H provides a "good" approximation of the distances in G. More formally, for $t \geq 1$, H is called a *multiplicative* t*–spanner* of G [22] if $d_H(u,v) \leq t \cdot d_G(u,v)$ for all $u, v \in V$. If $r \geq 0$ and $d_H(u,v) \leq d_G(u,v)+r$ for all $u, v \in V$, then H is called an *additive* r*–spanner* of G [21]. The parameters t and r are called, respectively, the *multiplicative* and the *additive stretch factors*. When H is a tree one has a *multiplicative tree t-spanner* [4] and an *additive tree* r*-spanner* [23] of G, respectively.

X. Deng and D. Du (Eds.): ISAAC 2005, LNCS 3827, pp. 583–592, 2005.

In this paper, we continue the approach taken in [6, 10, 11, 18] of studying *collective tree spanners*. We say that a graph $G = (V, E)$ *admits a system of* μ *collective additive tree* r*-spanners* if there is a system $\mathcal{T}(G)$ of at most μ spanning trees of G such that for any two vertices x, y of G a spanning tree $T \in \mathcal{T}(G)$ exists such that $d_T(x, y) \leq d_G(x, y) + r$ (a multiplicative variant of this notion can be defined analogously). Clearly, if G admits a system of μ collective additive tree r-spanners, then G admits an additive r-spanner with at most μ $(n-1)$ edges (take the union of all those trees), and if $\mu = 1$ then G admits an additive tree r-spanner. In particular, we examine the problem of finding "small" systems of collective additive tree r-spanners for small values of r on special classes of graphs such as *planar graphs, graphs with bounded genus, graphs with bounded tree-width, graphs with bounded clique-width, and graphs with bounded chordality.*

Previously, collective tree spanners of particular classes of graphs were considered in [6, 10, 11, 18]. Paper [11] showed that any unweighted chordal graph, chordal bipartite graph or cocomparability graph admits a system of at most $\log_2 n$ collective additive tree 2–spanners. These results were complemented by lower bounds, which say that any system of collective additive tree 1–spanners must have $\Omega(\sqrt{n})$ spanning trees for some chordal graphs and $\Omega(n)$ spanning trees for some chordal bipartite graphs and some cocomparability graphs. Furthermore, it was shown that any unweighted c-chordal graph admits a system of at most $\log_2 n$ collective additive tree $(2\lfloor c/2 \rfloor)$–spanners and any unweighted circular-arc graph admits a system of two collective additive tree 2–spanners. Paper [10] showed that any unweighted AT-free graph (graph without asteroidal triples) admits a system of two collective additive tree 2-spanners and any unweighted graph having a dominating shortest path admits a system of two collective additive tree 3-spanners and a system of five collective additive tree 2-spanners. In paper [6], it was shown that no system of constant number of collective additive tree 1-spanners can exist for unit interval graphs, no system of constant number of collective additive tree r-spanners can exist for chordal graphs and $r \leq 3$, and no system of constant number of collective additive tree r-spanners can exist for weakly chordal graphs and any constant r. On the other hand, [6] proved that any unweighted interval graph of diameter D admits an easily constructable system of $2\log(D-1)+4$ collective additive tree 1-spanners, and any unweighted House-Hole-Domino free graph with n vertices admits an easily constructable system of at most $2\log_2 n$ collective additive tree 2-spanners. Only paper [18] has investigated (so far) collective (multiplicative) tree spanners in the *weighted graphs* (they were called *tree covers* there). It was shown that any weighted n–vertex planar graph admits a system of $O(\sqrt{n})$ collective multiplicative tree 1-spanners (equivalently, additive tree 0-spanners) and a system of at most $2\log_{3/2} n$ collective multiplicative tree 3–spanners.

One of the motivations to introduce this new concept steams from the problem of designing compact and efficient routing schemes in graphs. In [13, 25], a shortest path routing labeling scheme for trees of arbitrary degree and diameter is described that assigns each vertex of an n-vertex tree a $O(\log^2 n/\log\log n)$-

bit label. Given the label of a source vertex and the label of a destination, it is possible to compute in constant time, based solely on these two labels, the neighbor of the source that heads in the direction of the destination. Clearly, if an n-vertex graph G admits a system of μ collective additive tree r-spanners, then G admits a routing labeling scheme of deviation (i.e., additive stretch) r with addresses and routing tables of size $O(\mu \log^2 n/ \log\log n)$ bits per vertex. Once computed by the sender in μ time (by choosing for a given destination an appropriate tree from the collection to perform routing), headers of messages never change, and the routing decision is made in constant time per vertex (for details see [10, 11]).

Our results. In this paper we generalize and refine the method of [11] for constructing a "small" system of collective additive tree r-spanners with small values of r to weighted and larger families of "well" decomposable graphs. We define a large class of graphs, called (α, γ, r)*–decomposable*, and show that any weighted (α, γ, r)–decomposable graph G with n vertices admits a system of at most $\gamma \log_{1/\alpha} n$ collective additive tree $2r$–spanners. Then, we show that all weighted planar graphs are $(2/3, \sqrt{6n}, 0)$–decomposable, all weighted graphs with genus at most g are $(2/3, O(\sqrt{gn}), 0)$–decomposable, all weighted graphs with tree-width at most $k-1$ are $(1/2, k, 0)$–decomposable, all weighted graphs with clique-width at most k are $(2/3, k, \mathsf{w})$–decomposable, all weighted graphs with size of largest induced cycle at most c are $(1/2, 1, \lfloor c/2 \rfloor \mathsf{w})$-decomposable, $(1/2, 6, \lfloor (c+2)/3 \rfloor \mathsf{w})$-decomposable and $(1/2, 4, (\lfloor c/3 \rfloor + 1)\mathsf{w})$-decomposable, and all weighted weakly chordal graphs are $(1/2, 4, \mathsf{w})$-decomposable. Here and in what follows, w denotes the maximum edge weight in G, i.e., $\mathsf{w} := max\{w(e) : e \in E(G)\}$.

As a consequence, we obtain that any weighted planar graph admits a system of $O(\sqrt{n})$ collective additive tree 0–spanners, any weighted graph with genus at most g admits a system of $O(\sqrt{gn})$ collective additive tree 0–spanners, any weighted graph with tree-width at most $k-1$ admits a system of $k \log_2 n$ collective additive tree 0–spanners, any weighted graph with clique-width at most k admits a system of $k \log_{3/2} n$ collective additive tree $(2\mathsf{w})$–spanners, any weighted graph with size of largest induced cycle at most c admits a system of $\log_2 n$ ($6 \log_2 n$ and $4 \log_2 n$) collective additive tree $(2\lfloor c/2 \rfloor \mathsf{w})$–spanners (respectively, $(2\lfloor (c+2)/3 \rfloor \mathsf{w})$–spanners and $(2(\lfloor c/3 \rfloor + 1)\mathsf{w})$–spanners), and any weighted weakly chordal graph admits a system of $4 \log_2 n$ collective additive tree $(2\mathsf{w})$-spanners. Furthermore, based on this collection of trees, we derive compact and efficient routing schemes for those families of graphs.

Basic notions and notation. All graphs occurring in this paper are connected, finite, undirected, loopless and without multiple edges. Our graphs can have (non-negative) weights on edges, $w(e)$, $e \in E$, unless otherwise is specified. In a weighted graph $G = (V, E)$ the *distance* $d_G(u, v)$ between the vertices u and v is the length of a shortest path connecting u and v. If the graph is unweighted then, for convenience, each edge has weight 1.

The (*open*) *neighborhood* of a vertex u in G is $N(u) = \{v \in V : uv \in E\}$ and the *closed neighborhood* is $N[u] = N(u) \cup \{u\}$. Define the layers of G with respect to a vertex u as follows: $L_i(u) = \{x \in V : x$ can be connected to u by a path

with i edges but not by a path with $i-1$ edges$\}$, $i = 0, 1, 2, \ldots$. In a path $P = (v_0, v_1, \ldots, v_l)$ between vertices v_0 and v_l of G, vertices $v_1, \ldots, v_{l-1}$ are called *inner vertices.* Let r be a non-negative real number. A set $D \subseteq V$ is called an *r-dominating set* for a set $S \subseteq V$ of a graph G if $d_G(v, D) \leq r$ holds for any $v \in S$.

A tree-decomposition [24] of a graph G is a tree T whose nodes, called *bags*, are subsets of $V(G)$ such that: 1) $\bigcup_{X \in V(T)} X = V(G)$; 2) for all $\{u, v\} \in E(G)$, there exists $X \in V(T)$ such that $u, v \in X$; and 3) for all $X, Y, Z \in V(T)$, if Y is on the path from X to Z in T then $X \cap Z \subseteq Y$. The *width* of a tree-decomposition is one less than the maximum cardinality of a bag. Among all the tree-decompositions of G, let T be the one with minimum *width*. The width of T is called the *tree-width* of the graph G and is denoted by $tw(G)$. We say that *G has bounded tree-width* if $tw(G)$ is bounded by a constant. It is known that the tree-width of an outerplanar graph and of a series-parallel graph is at most 2 (see, e.g., [19]).

A related notion to tree-width is *clique-width.* Based on the following operations on vertex-labeled graphs, namely (1) creation of a vertex labeled by integer l, (2) disjoint union, (3) join between all vertices with label i and all vertices with label j for $i \neq j$, and (4) relabeling all vertices of label i by label j, the notion of *clique-width* $cw(G)$ of a graph G is defined in [12] as the minimum number of labels which are necessary to generate G by using these operations. Clique-width is a complexity measure on graphs somewhat similar to tree-width, but more powerful since every set of graphs with bounded tree-width has bounded clique-width [7] but not conversely (cliques have clique-width 2 but unbounded tree-width). It is well-known that the clique-width of a cograph is at most 2 and the clique-width of a distance-hereditary graph is at most 3 (see [17]).

The *chordality* of a graph G is the size of the largest (in the number of edges) induced cycle of G. Define *c-chordal graphs* as the graphs with chordality at most c. Then, the well-known chordal graphs are exactly the 3-chordal graphs. An induced cycle of G of size at least 5 is called a *hole.* The complement of a hole is called an *anti-hole.* A graph G is *weakly chordal* if it has neither holes nor anti-holes as induced subgraphs. Clearly, weakly chordal graphs and their complements are 4-chordal. A *cograph* is a graph having no induced paths on 4 vertices (P_4s).

The *genus* of a graph G is the smallest integer g such that G embeds in a surface of genus g without edge crossings. Planar graphs can be embedded on a sphere, hence $g = 0$ for them. A planar graph is *outerplanar* if all its vertices belong to its outerface.

2 (α, γ, r)-Decomposable Graphs and Their Collective Tree Spanners

Let α be a positive real number smaller than 1, γ be a positive integer and r be a non-negative real number. We say that an n-vertex graph $G = (V, E)$ is *(α, γ, r)-decomposable* if there is a separator $S \subseteq V$, such that the following three conditions hold:

Balanced Separator condition - the removal of S from G leaves no connected component with more than αn vertices;

Bounded r-Dominating Set condition - there exists a set $D \subseteq V$ such that $|D| \leq \gamma$ and for any vertex $u \in S$, $d_G(u, D) \leq r$ (we say that D *r-dominates* S);

Hereditary Family condition - each connected component of the graph, obtained from G by removing vertices of S, is also an (α, γ, r)–decomposable graph.

Note that, by definition, any graph having an r-dominating set of size at most γ is (α, γ, r)–decomposable, for any positive $\alpha < 1$.

Using these three conditions, one can construct for any (α, γ, r)-decomposable graph G a *(rooted) balanced decomposition tree* $\mathcal{BT}(G)$ as follows. If G has an r-dominating set of size at most γ, then $\mathcal{BT}(G)$ is a one node tree. Otherwise, find a balanced separator S with bounded r-dominating set in G, which exists according to the first and second conditions. Let $G_1, G_2, \ldots, G_p$ be the connected components of the graph $G \setminus S$ obtained from G by removing vertices of S. For each graph G_i $(i = 1, \ldots, p)$, which is (α, γ, r)-decomposable by the Hereditary Family condition, construct a balanced decomposition tree $\mathcal{BT}(G_i)$ recursively, and build $\mathcal{BT}(G)$ by taking S to be the root and connecting the root of each tree $\mathcal{BT}(G_i)$ as a child of S. Clearly, the nodes of $\mathcal{BT}(G)$ represent a partition of the vertex set V of G into *clusters* $S_1, S_2, \ldots, S_q$, each of them having in G an r-dominating set of size at most γ. For a node X of $\mathcal{BT}(G)$, denote by $G(\downarrow X)$ the (connected) subgraph of G induced by vertices $\cup\{Y : Y$ is a descendent of X in $\mathcal{BT}\}$ (here we assume that X is a descendent of itself).

It is easy to see that a balanced decomposition tree $\mathcal{BT}(G)$ of a graph G with n vertices and m edges has depth at most $\log_{1/\alpha} n$, which is $O(\log_2 n)$ is α is a constant. Moreover, assuming that a special balanced separator (mentioned above) can be found in polynomial, say $p(n)$, time, the tree $\mathcal{BT}(G)$ can be constructed in $O((p(n) + m) \log_{1/\alpha} n)$ total time.

Consider now two arbitrary vertices x and y of an (α, γ, r)-decomposable graph G and let $S(x)$ and $S(y)$ be the nodes of $\mathcal{BT}(G)$ containing x and y, respectively. Let also $NCA_{\mathcal{BT}(G)}(S(x), S(y))$ be the nearest common ancestor of nodes $S(x)$ and $S(y)$ in $\mathcal{BT}(G)$ and $(X_0, X_1, \ldots, X_t)$ be the path of $\mathcal{BT}(G)$ connecting the root X_0 of $\mathcal{BT}(G)$ with $NCA_{\mathcal{BT}(G)}(S(x), S(y)) = X_t$ (in other words, $X_0, X_1, \ldots, X_t$ are the common ancestors of $S(x)$ and $S(y)$). Clearly, any path $P^G_{x,y}$, connecting vertices x and y in G, contains a vertex from $X_0 \cup X_1 \cup \cdots \cup X_t$. Let $SP^G_{x,y}$ be a shortest path of G connecting vertices x and y, and let X_i be the node of the path $(X_0, X_1, \ldots, X_t)$ with the smallest index such that $SP^G_{x,y} \cap X_i \neq \emptyset$ in G. Then, it is easy to show that $d_G(x, y) = d_{G'}(x, y)$, where $G' := G(\downarrow X_i)$.

Let D_i be an r-dominating set of X_i in $G' = G(\downarrow X_i)$ of size at most γ. For the graph G', consider a set of $|D_i|$ *Shortest-Path-trees (SP-trees)* $\mathcal{T}(D_i)$, each rooted at a (different) vertex from D_i. Then, there is a tree $T' \in \mathcal{T}(D_i)$ which has the following distance property with respect to those vertices x and y.

Lemma 1. *For vertices $x, y \in G(\downarrow X_i)$, there exits a tree $T' \in \mathcal{T}(D_i)$ such that $d_{T'}(x, y) \leq d_G(x, y) + 2r$.*

Let now $B_1^i, \ldots, B_{p_i}^i$ be the nodes on depth i of the tree $\mathcal{BT}(G)$ and let $D_1^i, \ldots, D_{p_i}^i$ be the corresponding r-dominating sets. For each subgraph $G_j^i := G(\downarrow B_j^i)$ of G ($i = 0, 1, ..., depth(\mathcal{BT}(G)$, $j = 1, 2, \ldots, p_i$), denote by $\mathcal{T}_j^i$ the set of SP-trees of graph G_j^i rooted at the vertices of D_j^i. Thus, for each G_j^i, we construct at most γ Shortest-Path-trees. We call them *local subtrees.* Lemma 1 implies

Theorem 1. *Let G be an (α, γ, r)-decomposable graph, $\mathcal{BT}(G)$ be its balanced decomposition tree and $\mathcal{LT}(G) = \{T \in \mathcal{T}_j^i : i = 0, 1, \ldots, depth(\mathcal{BT}(G)), j = 1, 2, \ldots, p_i\}$ be its set of local subtrees. Then, for any two vertices x and y of G, there exists a local subtree $T' \in \mathcal{T}_{j'}^{i'} \subseteq \mathcal{LT}(G)$ such that $d_{T'}(x, y) \le d_G(x, y) + 2r$.*

This theorem implies two import results for the class of (α, γ, r)-decomposable graphs. Let G be an (α, γ, r)-decomposable graph with n vertices and m edges, $\mathcal{BT}(G)$ be its balanced decomposition tree and $\mathcal{LT}(G)$ be the family of its local subtrees (defined above). Consider a graph H obtained by taking the union of all local subtrees of G (by putting all of them together), i.e.,

$$H := \bigcup\{T : T \in \mathcal{T}_j^i \subseteq \mathcal{LT}(G)\} = (V, \cup\{E(T) : T \in \mathcal{T}_j^i \subseteq \mathcal{LT}(G)\}).$$

Clearly, H is a spanning subgraph of G and for any two vertices x and y of G, $d_H(x, y) \le d_G(x, y) + 2r$ holds. Also, since for any level i ($i = 0, 1, \ldots, depth(\mathcal{BT}(G))$) of balanced decomposition tree $\mathcal{BT}(G)$, the corresponding graphs $G_1^i, \ldots, G_{p_i}^i$ are pairwise vertex-disjoint and $|\mathcal{T}_j^i| \le \gamma$ ($j = 1, 2, \ldots, p_i$), the union $\bigcup\{T : T \in \mathcal{T}_j^i, j = 1, 2, \ldots, p_i\}$ has at most $\gamma(n-1)$ edges. Therefore, H has at most $\gamma(n-1)\log_{1/\alpha} n$ edges in total. Thus, we have proven the following result.

Theorem 2. *Any (α, γ, r)-decomposable graph G with n vertices admits an additive $2r$-spanner with at most $\gamma(n-1)\log_{1/\alpha} n$ edges.*

Let $\mathcal{T}_j^i := \{T_j^i(1), T_j^i(2), \ldots, T_j^i(\gamma-1), T_j^i(\gamma)\}$ be the set of SP-trees of graph G_j^i rooted at the vertices of D_j^i. Here, if $p := |D_j^i| < \gamma$ then we can set $T_j^i(k) := T_j^i(p)$ for any k, $p+1 \le k \le \gamma$. By arbitrarily extending each forest $\{T_1^i(q), T_2^i(q), \ldots, T_{p_i}^i(q)\}$ ($q \in \{1, \ldots, \gamma\}$) to a spanning tree $T^i(q)$ of the graph G, for each level i ($i = 0, 1, \ldots, depth(\mathcal{BT}(G))$) of the decomposition tree $\mathcal{BT}(G)$, we can construct at most γ spanning trees of G. Totally, this will result into at most γ $depth(\mathcal{BT}(G))$ spanning trees $\mathcal{T}(G) := \{T^i(q) : i = 0, 1, \ldots, depth(\mathcal{BT}(G)), q = 1, \ldots, \gamma\}$ of the original graph G. Thus, from Theorem 1, we have the following.

Theorem 3. *Any (α, γ, r)-decomposable graph G with n vertices admits a system $\mathcal{T}(G)$ of at most $\gamma \log_{1/\alpha} n$ collective additive tree $2r$-spanners.*

Corollary 1. *Every (α, γ, r)-decomposable graph G with n vertices admits a routing labeling scheme of deviation $2r$ with addresses and routing tables of size $O(\gamma \log_{1/\alpha} n \log^2 n / \log\log n)$ bits per vertex. Once computed by the sender in $\gamma \log_{1/\alpha} n$ time, headers never change, and the routing decision is made in constant time per vertex.*

3 Particular Classes of (α, γ, r)-Decomposable Graphs

Graphs having balanced separators of bounded size. Here we consider graphs that have balanced separators of bounded size.

To see that planar graphs are $(2/3, \sqrt{6n}, 0)$-decomposable, we recall the following Separator Theorem for planar graphs from [20] (see also [8]) : *The vertices of any n-vertex planar graph G can be partitioned in $O(n)$ time into three sets A, B, C, such that no edge joins a vertex in A with a vertex in B, neither A nor B has more than 2/3n vertices, and C contains no more than $\sqrt{6n}$ vertices.* Obviously, every connected component of $G \setminus C$ is still a planar graph. The Separator Theorem for planar graphs was extended in [9, 14] to bounded genus graphs: *a graph G with genus at most g admits a separator C of size $O(\sqrt{gn})$ such that any connected component of $G \setminus C$ contains at most $2n/3$ vertices.* Evidently, each connected component of $G \setminus C$ has genus bounded by g, too. Hence, the following results follow.

Theorem 4. *Every n–vertex planar graph is $(2/3, \sqrt{6n}, 0)$-decomposable. Every n–vertex graph with genus at most g is $(2/3, O(\sqrt{gn}), 0)$-decomposable.*

There is another extension of the Separator Theorem for planar graphs, namely, to the graphs with an excluded minor [2]. A graph H is a *minor* of a graph G if H can be obtained from a subgraph of G by contracting edges. By an H-minor one means a minor of G isomorphic to H. Thus the Pontryagin-Kuratowski-Wagner Theorem asserts that planar graphs are those without K_5 and $K_{3,3}$ minors. The following result was proven in [2]: *Let G be an n-vertex graph and H be an h-vertex graph. If G has no H-minor, then the vertices of G can be partitioned into three sets A, B, C, such that no edge joins a vertex in A with a vertex in B, neither A nor B has more than 2/3n vertices, and C contains no more than $\sqrt{h^3 n}$ vertices. Furthermore A, B, C can be found in $O(\sqrt{hn}(n+m))$ time, where m is the number of edges in G.*

Since induced subgraphs of an H-minor free graph are H-minor free, we conclude.

Theorem 5. *Let G be an n-vertex graph and H be an h-vertex graph. If G has no H-minor, then G is $(2/3, \sqrt{h^3 n}, 0)$-decomposable.*

Note that, any shortest path routing labeling scheme in n-vertex planar graphs requires at least $\Omega(\sqrt{n})$-bit labels [1]. Hence, by Corollary 1, there must exist n-vertex planar graphs, for which any system of collective additive tree 0-spanners needs to have at least $\Omega(\sqrt{n} \log\log n / \log^2 n)$ trees.

Now we turn to graphs with bounded tree-width. The following theorem is true (proof is omitted).

Theorem 6. *Every graph with tree-width at most k is $(1/2, k+1, 0)$-decomposable.*

Table 1 summarizes the results on collective additive tree spanners of graphs having balanced separators of bounded size. The results are obtained by combining Theorem 3 with Theorems 4, 5 and 6. Note that, for planar graphs, the

number of trees in the collection is at most $O(\sqrt{n})$ (not $\sqrt{6n}\log_{3/2} n$ as would follow from Theorem 3). This number can be obtained by solving the recurrent formula $\mu(n) = \sqrt{6n} + \mu(2/3n)$. Similar argument works for other two families of graphs.

Table 1. Collective additive tree spanners of n-vertex graphs having balanced separators of bounded size

Graph class	Number of trees in the collection, μ	additive stretch factor, r
Planar graphs	$O(\sqrt{n})$	0
Graphs with genus g	$O(\sqrt{gn})$	0
Graphs without an h-vertex minor	$O(\sqrt{h^3 n})$	0
Graphs with tree-width $k-1$	$k \log_2 n$	0

We conclude this subsection with a lower bound, which follows from a result in [6]. Recall that all outerplanar graphs have tree-width at most 2.

Observation 1. it No system of constant number of collective additive tree r-spanners can exist for outerplanar graphs, for any constant $r \geq 0$.

Graphs with bounded clique-width. Here we will prove that each graph with clique-width at most k is $(2/3, k, \mathsf{w})$-decomposable. Recall that w denotes the maximum edge weight in a graph G, i.e., $\mathsf{w} := max\{w(e) : e \in E(G)\}$.

Theorem 7. *Every graph with clique-width at most k is* $(2/3, k, \mathsf{w})$*-decomposable.*

Proof. It was shown in [3] that the vertex set V of any graph $G = (V, E)$ with n vertices and clique-width $cw(G)$ at most k can be partitioned (in polynomial time) into two subsets A and $B := V \setminus A$ such that both A and B have no more than $2/3n$ vertices and A can be represented as the disjoint union of at most k subsets $A_1, \dots, A_k$ (i.e., $A = A_1 \dot\cup \dots \dot\cup A_k$), where each A_i ($i \in \{1, \dots, k\}$) has the property that any vertex from B is either adjacent to all $v \subset A_i$ or to no vertex in A_i. Using this, we form a balanced separator S of G as follows. Initially set $S := \emptyset$, and in each subset A_i, arbitrarily choose a vertex v_i. Then, if $N(v_i) \cap B \neq \emptyset$, put v_i and $N(v_i) \cap B$ into S. Since for any edge $ab \in E$ with $a \in A$ and $b \in B$, vertex b must belong to S, we conclude that S is a separator of G, separating $A \setminus S$ from $B \setminus S$. Moreover, each connected component of $G \setminus S$ lies entirely either in A or in B and therefore has at most $2/3n$ vertices. By construction of S, any vertex $u \in B \cap S$ is adjacent to a vertex from $A' := A \cap S$. As $|A'| \leq k$ and w is an upper bound on any edge weight, we deduce that A' w-dominates S in G. Thus, S is a balanced separator of G and is w-dominated by a set A' of cardinality at most k. To finish the proof, it remains to recall that induced subgraphs of a graph with clique-width at most k have clique-width at most k, too (see, e.g., [7]), and therefore, by induction, the connected components of $G \setminus S$ induce $(2/3, k, \mathsf{w})$-decomposable graphs.

Combining Theorem 7 with the results of Section 2, we obtain

Corollary 2. *Any graph with n vertices and clique-width at most k admits a system of at most $k \log_{3/2} n$ collective additive tree $2\mathsf{w}$-spanners, and such a system of spanning trees can be found in polynomial time.*

To complement the above result, we give the following lower bound.

Observation 2. it There are (infinitely many) unweighted n-vertex graphs with clique-width at most 2 for which any system of collective additive tree 1-spanners will need to have at least $\Omega(n)$ spanning trees.

Graphs with bounded chordality. Here we consider the class of c-chordal graphs and its subclasses. Proofs of all results of this subsection are omitted.

Theorem 8. *Every n-vertex c-chordal graph is $(1/2, 1, \lfloor c/2 \rfloor \mathsf{w})$-decomposable, $(1/2, 4, (\lfloor c/3 \rfloor + 1)\mathsf{w})$-decomposable and $(1/2, 6, \lfloor (c+2)/3 \rfloor \mathsf{w})$-decomposable.*

Corollary 3. *Every n-vertex c-chordal graph admits a system of at most $\log_2 n$ collective additive tree $(2\lfloor c/2 \rfloor \mathsf{w})$-spanners, a system of at most $4 \log_2 n$ collective additive tree $(2(\lfloor c/3 \rfloor + 1)\mathsf{w})$-spanners and a system of at most $6 \log_2 n$ collective additive tree $(2(\lfloor (c+2)/3 \rfloor \mathsf{w})$-spanners. Moreover, such systems of spanning trees can be constructed in polynomial time.*

These results can be refined for 4-chordal graphs and weakly chordal graphs.

Theorem 9. *Every 4-chordal graph is $(1/2, 6, \mathsf{w})$-decomposable. Every 4-chordal graph not containing $\overline{C}_6$ as an induced subgraph is $(1/2, 4, \mathsf{w})$-decomposable.*

Corollary 4. *Any n-vertex 4-chordal graph admits a system of at most $6 \log_2 n$ collective additive tree $2\mathsf{w}$-spanners. Any n-vertex 4-chordal graph not containing $\overline{C}_6$ as an induced subgraph (in particular, any weakly chordal graph) admits a system of at most $4 \log_2 n$ collective additive tree $2\mathsf{w}$-spanners. Moreover, such systems of spanning trees can be constructed in polynomial time.*

Note that the class of weakly chordal graphs properly contains such known classes of graphs as interval graphs, chordal graphs, chordal bipartite graphs, permutation graphs, trapezoid graphs, House-Hole-Domino–free graphs, distance-hereditary graphs and many others. Hence, the results of this subsection generalize some known results from [6, 11]. We recall also that, as it was shown in [6], no system of constant number of collective additive tree r-spanners can exist for unweighted weakly chordal graphs for any constant $r \geq 0$.

Corollary 5. *Any n-vertex 4-chordal graph admits an additive $2\mathsf{w}$-spanner with at most $O(n \log n)$ edges. Moreover, such a sparse spanner can be constructed in polynomial time.*

The last result improves and generalizes the known results from [5, 11, 22] on sparse spanners of unweighted chordal graphs.

In full version, we discuss also implication of these results to designing compact routing labeling schemes for graphs under consideration.

References

1. I. Abraham, C. Gavoille, and D. Malkhi, Compact routing for graphs excluding a fixed minor, Proc. of *DISC 2005, LNCS 3724,* pp. 442-456.
2. N. Alon, P. Seymour, and R. Thomas, A Separator Theorem for Graphs with an Excluded Minor and its Applications, Proc. of *STOC 1990, ACM*, pp. 293–299.
3. R. Borie, J.L. Johnson, V. Raghavan, and J.P. Spinrad, Robust polynomial time algorithms on clique-width k graphs, *Manuscript*, 2002.
4. L. Cai and D.G. Corneil, Tree spanners, *SIAM J. Discr. Math.,* 8 (1995), 359–387.
5. V. D. Chepoi, F. F. Dragan, and C. Yan, Additive Spanners for k-Chordal Graphs, Proc. of *CIAC 2003, LNCS 2653*, pp. 96-107.
6. D.G. Corneil, F.F. Dragan, E. Köhler, and C. Yan, Collective tree 1-spanners for interval graphs, Proc. of *WG 2005, LNCS*, to appear.
7. B. Courcelle and S. Olariu, Upper bounds to the clique-width of graphs, *Discrete Appl. Math.*, 101 (2000), 77-114.
8. H.N. Djidjev, On the problem of partitioning planar graphs, *SIAM, J. Alg. Disc. Meth.*, 3 (1982), 229-240.
9. H.N. Djidjev, A separator theorem for graphs of fixed genus, *Serdica*, 11 (1985), 319-329.
10. F.F. Dragan, C. Yan and D.G. Corneil, Collective tree spanners and routing in AT-free related graphs, Proc. of *WG 2004, LNCS 3353,* pp. 68-80.
11. F.F. Dragan, C. Yan and I. Lomonosov, Collective tree spanners of graphs, Proc. of *SWAT 2004, LNCS 3111,* pp. 64-76.
12. J. Engelfriet and G. Rozenberg, Node replacement graph grammars, *Handbook of Graph Grammars and Computing by Graph Transformation, Foundations*, Vol. I, World Scientific, Singapore, 1997, pp. 1-94.
13. P. Fraigniaud and C. Gavoille, Routing in Trees, Proc. of *ICALP 2001, LNCS 2076*, pp. 757–772.
14. J.R. Gilbert, J.P. Hutchinson, and R.E. Tarjan, A separator theorem for graphs of bounded genus, *Journal of Algorithms*, 5 (1984), 391–407.
15. J.R. Gilbert, D.J. Rose, and A. Edenbrandt, A separator theorem for chordal graphs. *SIAM J. Alg. Discrete Meth.*, 5 (1984), 306–313.
16. M.C. Golumbic, *Algorithmic Graph Theory and Perfect Graphs*, Academic Press, New York, 1980.
17. M.C. Golumbic and U. Rotics, On the Clique-Width of Perfect Graph Classes, Proc. of *WG 1999, LNCS 1665*, pp. 135-147.
18. A. Gupta, A. Kumar and R. Rastogi, Traveling with a Pez Dispenser (or, Routing Issues in MPLS), *SIAM J. Comput.*, 34 (2005), pp. 453-474.
19. T. Kloks, Treewidth: Computations and Approximations, *Lecture Notes in Computer Science* 842, Springer, Berlin, 1994.
20. R. J. Lipton and R. E. Tarjan, A separator theorem for planar graphs, *SIAM J. Appl. Math.*, 36 (1979), 346-358.
21. A.L. Liestman and T. Shermer, Additive graph spanners, *Networks,* 23 (1993), 343–364.
22. D. Peleg, and A.A. Schäffer, Graph Spanners, *J. Graph Theory*, 13 (1989), 99-116.
23. E. Prisner, D. Kratsch, H.-O. Le, H. Müller, and D. Wagner, Additive tree spanners, *SIAM J. Discrete Math.*, 17 (2003), 332–340.
24. N. Robertson and P. D. Seymour, Graph minors II: Algorithmic aspects of tree-width, *Journal of Algorithms,* 7 (1986), 309-322.
25. M. Thorup and U. Zwick, Compact routing schemes, Proc. of *SPAA 2001, ACM,* pp. 1–10.

Sampling Unlabeled Biconnected Planar Graphs

Manuel Bodirsky[1], Clemens Gröpl[2,*], and Mihyun Kang[1,**]

[1] Humboldt-Universität zu Berlin, Institut für Informatik,
Unter den Linden 6, D-10099 Berlin, Germany
{bodirsky, kang}@informatik.hu-berlin.de
[2] Freie Universität Berlin, Institut für Informatik,
Takustraße 9, D-14195 Berlin, Germany
groepl@inf.fu-berlin.de

Abstract. We present an expected polynomial time algorithm to generate a 2-connected *unlabeled* planar graph uniformly at random. To do this we first derive recurrence formulas to count the exact number of *rooted* 2-connected planar graphs, based on a decomposition along the connectivity structure. For 3-connected planar graphs we use the fact that they have a unique embedding on the sphere. Special care has to be taken for rooted graphs that have a *sense-reversing* or a *pole-exchanging* automorphism. We prove a bijection between such symmetric objects and certain *colored networks*. These colored networks can again be decomposed along their connectivity structure. All the numbers can be evaluated in polynomial time by dynamic programming. To generate 2-connected unlabeled planar graphs *without a root* uniformly at random we apply *rejection sampling* and obtain an expected polynomial time algorithm.

1 Introduction

While there is an exact recursive counting formula and an efficient sampling algorithm for *rooted maps* [21], for *labeled and unlabeled outerplanar* graphs [8], and for *labeled planar graphs* [7], there is no such counting formula and sampling algorithm known for *unlabeled* planar graphs, i.e., graphs that can be embedded in the plane, considered up to isomorphism. Such a counting formula and sampling algorithm would be useful to verify statements about the *random planar graph*, which recently attracted attention [2,9,10,15,18,20], mainly in the labeled setting due to the lack of techniques for unlabeled planar structures. It is well known that almost all graphs have a small automorphism group. However, this is not the case for planar graphs, even if they are 2-connected: Bender, Gao, and Wormald [2] showed that almost all 2-connected planar graphs have a large automorphism group. Thus the difference between the labeled and the unlabeled setting is essential.

* Supported by the German Federal Ministry of Education and Research within the Berlin Center for Genome Based Bioinformatics (BMBF grant no. 031 2705A).
** Supported by the Deutsche Forschungsgemeinschaft (DFG Pr 296/7-3).

X. Deng and D. Du (Eds.): ISAAC 2005, LNCS 3827, pp. 593–603, 2005.

All approaches to count planar graphs are based on decompositions along the k-connected components of a graph [2,4,7,16,26]. A graph is decomposed into components, components into blocks, and blocks into bricks, which are essentially the 3-connected parts of the graph. Three-connected graphs have a unique embedding on the sphere, and thus can be further decomposed by geometric arguments (as they correspond to isomorphism types of the edge graphs of convex polyhedra [22]). The asymptotic number [3] and sampling procedures [5,13] for 3-connected planar graphs are known.

In this paper we present an algorithm that generates *two-connected* graphs on m edges uniformly at random in expected polynomial time in m. Such graphs have in general many automorphisms and also might have many embeddings on the sphere. It is easy to modify the algorithm e.g. to sample graphs with a given number of edges *and* vertices.

Strategy. To count and sample unlabeled 2-connected planar graphs, we first have to *root* them. Here, the *root* of a graph is a distinguished directed edge, and rooted planar graphs are counted up to isomorphisms that map the root of one graph to the root of the other graph. We also call such a rooted 2-connected planar graph a *(planar) network*. Clearly, generating a random rooted planar graph and then simply ignoring the root edge does not yield the uniform distribution, since unlabeled graphs might correspond to different numbers of rooted graphs. But this imbalance can be compensated by *rejection sampling*, i.e., the sampling procedure is restarted with a probability that is inverse proportional to the size of the orbit of the root. In this way we can sample unlabeled 2-connected planar graphs in expected polynomial time.

We *decompose* networks along their connectivity structure. Our approach is related to the one described in [7] for labeled planar graphs, but for unlabeled structures several new techniques are necessary. A classical theorem of Whitney (see e. g. [12]) says that rooted 3-connected planar graphs, i.e., 3-connected networks, can have either one or two embeddings in the plane where the root edge is embedded on the outer face. Such embedded three-connected networks are called *c-nets*. If both embeddings of a 3-connected graph are isomorphic, we say that it has a *sense-reversing automorphism* or it is *symmetric*. To count symmetric c-nets we present a new bijective correspondence to *colored networks* (defined below), and a decomposition of these objects. We also have to consider rooted graphs with an automorphism that reverts the direction of the root; such a graph is called *pole-symmetric*. We present a decomposition of pole-symmetric networks, and finally also that of pole-symmetric c-nets with a sense-reversing automorphism. We exploit the fact that the dual of a pole-symmetric c-net is a c-net with a sense-reversing automorphism.

As a final step we use a deterministic polynomial time sampling algorithm for c-nets described in [5]. A faster algorithm to sample c-nets uniformly at random in expected polynomial running time can be found in [1]. However, we need the new algorithm, since it can be adapted to generate c-nets with a certain specified number of edges on the outer face, which we need in the generation algorithm for unlabeled 2-connected planar graphs.

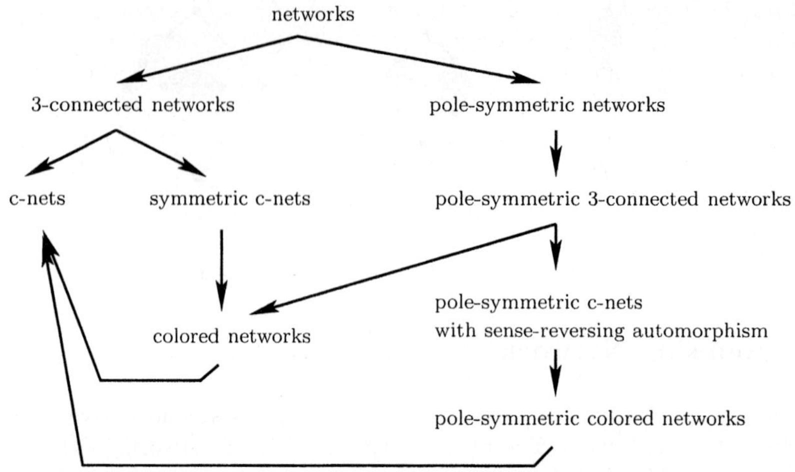

Fig. 1. Dependencies of the sections and concepts in this paper

2 Graph Decompositions

In this section we introduce concepts, which we need for a decomposition of graphs. A *(planar) network* N is a simple connected graph with distinguished vertices $s \neq t$ and a directed edge $e = st$ such that if we insert the edge e into N, then the resulting multi-graph N^* is 2-connected and planar. The edge $e = st$ is called the *root* of the network, and the vertices s and t are called the *poles* of the network; if s and t were already connected by an edge in N, we introduce a multi-edge in N^*. We say that a network N is k-connected iff N^* is k-connected.

A k-point set K of a graph G such that $G-K$ becomes disconnected is called a *k-cut* of G. A single point that forms a 1-cut is called a *cut-vertex.* Every 2-cut $\{k_1, k_2\}$ of a network N induces a partition of the edge set, and each of the partition classes is again a network with the poles k_1 and k_2 . These networks are called *subnetworks* of N, and they are called *proper* if they contain at least two edges. Every network is either an *s-, p-, or h-network*; this appeared in various slightly different forms in the literature [23,24,25].

Theorem 1. *A network N with more than two edges is of precisely one of the following types:*

s: There is a cut-vertex in N that separates s and t.
p: The vertices s and t are adjacent, or $\{s, t\}$ is a 2-cut of N.
h: The vertices s and t are not adjacent, and N is built from a uniquely determined 3-connected network H by replacing edges of H with other networks.

For examples of the three types of networks see Figure 2. As already mentioned in the introduction, a 3-connected network might have one or two embeddings where the root lies on the outer face. A 3-connected network embedded in one of these ways is called a *c-net.*

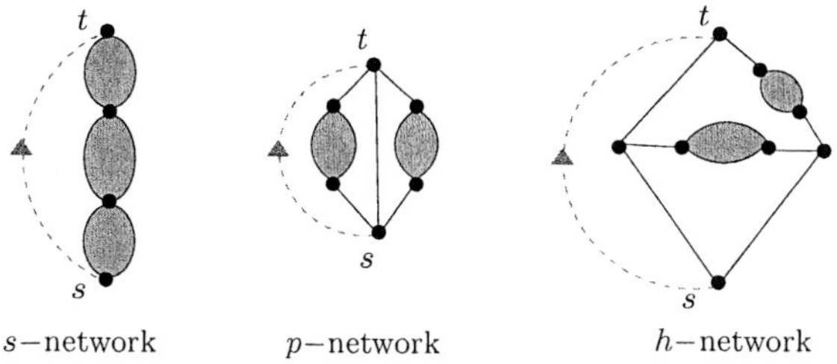

Fig. 2. Examples of the three types of networks

3 Counting Networks

In this section we present a decomposition of networks and derive recurrence formulas to count them. Let $n(m)$ be the number of networks with m edges. According to Theorem 1 we have $n(m) = s(m) + p(m) + h(m)$, where $s(m)$, $p(m)$, and $h(m)$ counts the number of s-, p-, and h-networks with m edges, respectively. The generation of these objects can be considered as a reversed decomposition, where the decisions are made with the right probabilities that can be computed using the counting formulas.

s-networks. Note that each s-network has a unique cut vertex v that is closest to the pole s (in this paper, *closest* is meant with respect to the length of the shortest connecting path). Then s and v determine a subnetwork, which is a p- or an h-network, and the remaining network with the poles v and t. Thus we have:

$$s(m) = \sum_{j=1} (p(j) + h(j))n(m-j) .$$

p-networks. Let $p_l(m)$ denote the number of p-networks where the number of edges of the largest network that replaces an edge of the core is bounded by l. The index k in the formula below denotes the number of networks of order l that replace an edge in the core.

$$p(m) = p_m(m)$$

$$p_l(m) = \sum_{k=0}^{\lfloor m/l \rfloor} \binom{s(l) + h(l) + k - 1}{k} p_{l-1}(m - kl) .$$

h-networks. Let N be an h-network. Theorem 1 asserts that there is a unique rooted 3-connected network H, such that we can derive N from H by replacing edges of H with subnetworks. We call H the *core* of N and denote $H = \text{core}(N)$. We call N *symmetric* if it has a *sense-reversing automorphism* φ, i. e., $\varphi \neq \text{id}$, but $\varphi(s) = s$ and $\varphi(t) = t$, and *asymmetric* otherwise. If N is asymmetric, one can uniquely order the edges of core(N): The idea is to label the vertices of the core according to their occurrence in a depth first search traversal of the core, beginning with the root edge and visiting the neighbors of a vertex in clockwise order with respect to one of the (at most two) possible embeddings

of the core. The edges are then labeled by the vertex labels obtained from the depth first search traversal. Then we lexicographically compare the sequence of these edge labels in the order they were visited by the depth first search. If the core is asymmetric, one of the sequences is smaller than the other; thus we can distinguish between the two embeddings. If the network has a symmetric core, both edge sequences are the same unless we have inserted two different subnetworks into a pair of core edges, in which case we can again distinguish between the two embeddings.

If H is symmetric, we order the edges of the core in the following way. We start with the edges uv where $u = \varphi(u)$ and $v = \varphi(v)$ according to the traversal; We call such edges *blue*. Then we list the edges uv where $u = \varphi(v)$ and $v = \varphi(u)$ according to the traversal; We call such edges *red*. We continue with the edges that are not fixed by the nontrivial automorphism φ, and order them according to the above traversal. Edges and their images, which we call *corresponding edges*, are ordered arbitrarily.

To count the number of symmetric and asymmetric h-networks we repeatedly replace subnetworks in the above order. It is not difficult to fomulate corresponding recursive counting formulas; for details we refer to the full version of the paper [6]. We are finally left with the problems (a) to count and sample networks *with a pole-exchanging automorphism* – see Section 6, (b) to count and sample *3-connected symmetric* networks – see Section 4, and (c) to sample *3-connected asymmetric* networks. For this last task, we apply rejection sampling. That is, we first generate an arbitrary 3-connected network. We then check whether it has such a symmetry, which can be done in linear time [17]. If yes, we restart the algorithm. If no, we output the asymmetric network. Since almost all 3-connected networks do not have a sense-reversing automorphism (see [3] for a much stronger result), the expected number of restarts is constant, and we obtain an expected polynomial time algorithm.

4 Symmetric c-Nets

This section contains one of the main ideas to deal with symmetries when counting unlabeled planar graphs. We want to count 3-connected planar networks with a distinguished directed edge, up to isomorphisms that fix this edge. There might be one or two embeddings where the root lies at the outer face. Embedded 3-connected networks are called *c-nets*. As mentioned in the introduction, counting formulas and sampling procedures for c-nets are known. If a network has a nontrivial automorphism that fixes the root edge, we call this automorphism *sense-reversing*, and say that the network is *symmetric*. Clearly, if we can compute the number of symmetric 3-connected networks, then we can also compute the number of asymmetric 3-connected networks.

Let H be a symmetric 3-connected planar network, and φ its nontrivial sense-reversing automorphism. A vertex v of H is called *blue* if $\varphi(v) = v$, and *red* if v is connected to $\varphi(v)$ by an edge. The edge $v\varphi(v)$ is also called red. An edge uv of a colored network is blue if both u and v are blue. (Red and blue edges

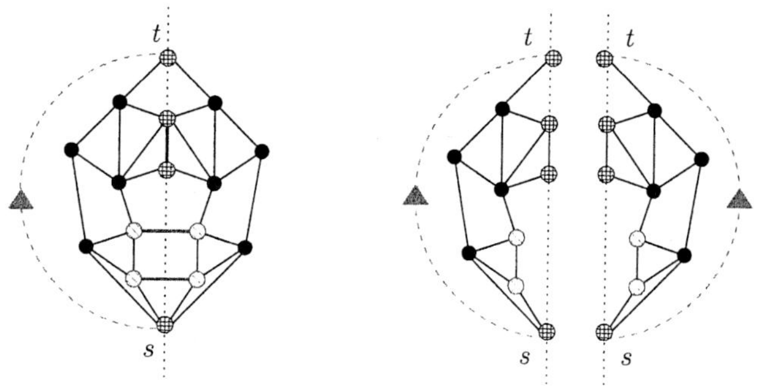

Fig. 3. Decomposition of a symmetric h-network

were already defined in Section 3.) Thus a vertex or an edge is either blue, red, or uncolored, and the poles and the root are blue. We can think of H as being embedded in the plane in such a way that φ corresponds to a reflection, the blue vertices being aligned on the reflection axis, and the red vertices having an edge crossing this axis perpendicularly (see Fig. 3, left part). Our arguments, however, do not rely on such a representation.

If we remove from H the blue vertices and their incident edges, and also remove the red edges (that is, we cut H along the symmetry axis), then the resulting graph has exactly two connected components (see Fig. 3). The graphs induced by these components and the blue vertices are isomorphic and will be called H_1 and H_2. We claim that H_1^* is 2-connected, and hence H_1 is a network rooted at s and t: Suppose there is a cut-vertex in H_1^*. Then this cut-vertex together with the corresponding cut-vertex in H_2^* is a 2-cut in H^*, contradicting the 3-connectivity of H^*. Now we extract some more properties of the graphs H_1 and H_2 and define *colored networks.* They are defined in such a way that we can recursively decompose them, and that we can establish a bijection between symmetric h-networks and certain colored networks.

Definition 1. *A* colored network *is a network N, where some vertices are colored* red *and* blue, *satisfying the following.*

(P1) *N^* has a plane embedding s.t. all colored vertices and the poles lie on the outer face.*
(P2) *N and every proper subnetwork of N contain at least one colored vertex.*
(P3) *No subnetwork of N has two blue poles.*

Then the bijection to symmetric 3-connected networks is as follows. The proof can be found in the long version of the paper [6].

Theorem 2. *For all m, b, r there is a bijection between the following two sets of objects:*

(i) *colored networks with $(m + b - r)/2$ edges and blue poles, where b is the number of blue edges and r the number of red vertices, and*

(ii) 3-connected networks with m edges having a nontrivial automorphism that fixes $b + r$ edges, and point-wise fixes the root and b other edges.

5 Counting Colored Networks

The recurrences to count the number of colored networks follow very much the decomposition that we had in Section 3, but we have to control the possible colors of the poles. Another difficulty is that in the recursive decomposition we might or might not have a blue cut-vertex in the colored network without the root edge. However, we can handle this with the help of appropriately chosen counting functions. The details and counting formulas can be found in the full version of the paper [6]. We briefly comment on all the types of networks.

Colored s-networks. Let u be the cut-vertex in S that is closest to s. If at least one of the poles s, t is blue, then S can not have a blue cut-vertex (in particular, u is not blue). The cut-vertex u induces a colored p- or h-network with poles s, u, and a remaining part with poles u, t, which is an arbitrary colored network that has no blue cut-vertex.

Colored p-networks. Due to property (P1 – P2) all the colored vertices of a colored p-network must lie in one of its parts, and the remaining networks must consist of a single edge. If at least one of the poles is blue, the colored part has no blue cut-vertex.

Colored h-networks. There is a unique embedding of the core of a colored h-network H into the plane where the root edge and the core edges replaced by colored networks lie on the outer face. To decompose H, we need to control the number of edges on the outer face. If an edge uv on the outer face of the core is not an edge in H, $\{u, v\}$ is a 2-cut in H and determines a subnetwork S. Due to property (P3) it is not possible that both u, v are blue. If either u or v is blue, then $\{u, v\}$ induces a colored network with no blue cut-vertex. It might be the case that all colored vertices lie in S. Then the remaining network after the replacement of S is 3-connected with a specified number of edges on the outer face. The number of such graphs is counted in [5].

6 Pole-Symmetric Networks

We saw in Section 3 that in a symmetric h-network with a sense-reversing automorphism φ a red edge uv of the core (i.e., $\varphi(u) = v$ and $\varphi(v) = u$) can only be replaced by a *pole-symmetric* subnetwork, that is, a subnetwork with an automorphism ψ that exchanges s and t. Such networks are further decomposed in this section.

Pole-symmetric s-networks. Here we split off the same p- or h-network at both poles simultaneously. What remains is either again a pole-symmetric network, or an edge, or a vertex.

Pole-symmetric *p*-networks. There may be several pole- symmetric s- or h-networks between s and t, and s and t may or may not be adjacent.

Pole-symmetric *h*-networks. We want to control the number of pole-symmetric h-networks with and without a sense-reversing automorphism φ satisfying $\varphi(s) = s$ and $\varphi(t) = t$. In the case where we do not have a sense-reversing automorphism, we order the edges of the core of H in such a way that blue edges uv where $\psi(u) = u$ and $\psi(v) = v$ come first, followed by the red edges uv where $\psi(u) = v$ and $\psi(v) = u$. Finally we have the uncolored edges, ordered in such a way that corresponding uncolored edges with respect to the pole-symmetry are consecutive – but we do not care about their order.

In the case that we have a sense-reversing automorphism φ, we order the edges of the core in such a way that we start with the blue edges with respect to φ, and then the blue edges with respect to ψ. Next we list the red edges with respect to φ and then the red edges with respect to ψ. Finally, we list corresponding edges with respect to φ consecutively, which are followed by the two corresponding edges with respect to ψ, respectively. Similarly as in Section 3 it is now possible to formulate recurrences for these functions, and a sampling procedure; again we have to refer to the full version of the paper [6] for details.

7 Pole-Symmetric c-Nets with a Sense-Reversing Automorphism

To compute the number of pole-symmetric networks with a sense-reversing automorphism, we again use colored networks, but impose the additional constraint that the colored network has a pole-exchanging automorphism. Along the lines of Theorem 2 we have a bijection between these pole-symmetric colored networks and pole-symmetric networks with a sense-reversing automorphism. The decomposition of pole-symmetric colored networks is a straightforward combination of the ideas in Section 4 and 6.

When we remove the last colored subnetwork in a pole-symmetric colored h-network, we have an embedded 3-connected pole-symmetric network with l edges on the outer face. The dual of such an object is an embedded 3-connected network with a sense-reversing automorphism where the s-pole has degree l (blue edges correspond to red edges and vice versa). It is possible to modify the decomposition of colored networks in Section 4 to control also this parameter.

8 Conclusion

We presented a decomposition strategy for unlabeled 2-connected planar graphs along the connectivity structure. In order to count these objects we need a *unique* decomposition. Thus we used the well-known concept of a *root* and *planar networks*. For 3-connected networks, however, we also had to control whether there is a sense-reversing automorphism or not, which in turn required to control networks that have a pole-exchanging automorphism, and networks that have

both a sense-reversing and a pole-exchanging automorphism. For this purpose we introduced the concept of *colored* networks, and proved several bijections. The decomposition together with the counting formulas can be used for a polynomial time sampling procedure for planar networks.

Theorem 3. *There is an algorithm that generates an unlabeled 2-connected planar graph with m edges uniformly at random in expected polynomial time.*

Proof. The algorithm first generates a planar network N with m edges, using the above decomposition and the values of the counting formulas that can be computed efficiently using dynamic programming. Note that the representation size of all the numbers in this paper is linear, since we deal with unlabeled structures. We use at most six-dimensional tables (in Section 6). The summation there runs over one parameter, and within the sum we have to perform a multiplication with large numbers, which can be done in quadratic time. Hence, the overall running time for the computation of the values is within $O(m^9)$.

With these values we can make the correct probabilistic decisions in a recursive construction of a planar network according to the presented decomposition – this method is standard and known as the *recursive method* for sampling [19,11,14]. Then the algorithm computes the number of orbits o in the automorphism group of the unrooted graph, which can be done in linear time, see e.g. [17], and outputs the graph with probability $1/o$. Since the number of edges in a planar graph is linear, the expected number of restarts is also linear. Thus the overall expected running time is in $O(m^9)$. If we do not charge for the costs for computing the values in the table and the partial sums of the formulas, e.g. because we performed a precomputation step, the generation can be done in cubic time. □

The counting formulas resulting from the ideas presented in this paper can easily be extended to graphs with a specified number of vertices. It is also easy to adapt the enumeration and the sampling procedure for multi-graphs with parallel edges and/or self-loops. The recursive formulas have a form that allows to formulate them with equations between the corresponding generating functions. It is sometimes possible to solve these equations and obtain closed formulas or asymptotic estimates from the solutions. However, due to the large number of parameters needed in the decomposition, it will not be easy to handle these equations. In the simpler case of labeled planar graphs the equations recently lead to asymptotic expressions for the number of labeled planar graphs [16].

References

1. C. Banderier, P. Flajolet, G. Schaeffer, and M. Soria. Random maps, coalescing saddles, singularity analysis, and Airy phenomena. *Random Structures and Algorithms*, 19:194–246, 2001.
2. A. Bender, Z. Gao, and N. Wormald. The number of labeled 2-connected planar graphs. *Electronic Journal of Combinatorics*, 9(43), 2002.

3. E. A. Bender and N. Wormald. Almost all convex polyhedra are asymmetric. *Can. J. Math.*, 27(5):854–871, 1985.
4. M. Bodirsky, O. Giménez, M. Kang, and M. Noy. On the number of series-parallel and outerplanar graphs. In *the Proceedings of European Conference on Combinatorics, Graph Theory, and Applications (EuroComb 2005)*, DMTCS Proceedings Series Volume AE, pages 383 – 388, 2005.
5. M. Bodirsky, C. Gröpl, D. Johannsen, and M. Kang. A direct decomposition of 3-connected planar graphs. In *Proceedings of the 17th Annual International Conference on Formal Power Series and Algebraic Combinatorics (FPSAC05), Taormina*, 2005.
6. M. Bodirsky, C. Gröpl, and M. Kang. Sampling unlabeled biconnected planar graphs. Full version, available at `http://www.informatik.hu-berlin.de/Forschung_Lehre/algorithmen/en/forschung/planar/`.
7. M. Bodirsky, C. Gröpl, and M. Kang. Generating labeled planar graphs uniformly at random. In *Thirtieth International Colloquium on Automata, Languages and Programming (ICALP'03)*, pages 1095–1107, 2003.
8. M. Bodirsky and M. Kang. Generating outerplanar graphs uniformly at random. Accepted for publication in Combinatorics, Probability and Computing. Presented at the 1st workshop on Algorithms for Listing, Counting, and Enumeration (ALICE 03), 2003.
9. N. Bonichon, C. Gavoille, and N. Hanusse. An information-theoretic upper bound of planar graphs using triangulation. In *In 20th Annual Symposium on Theoretical Aspects of Computer Science (STACS)*, 2003.
10. A. Denise, M. Vasconcellos, and D. Welsh. The random planar graph. *Congressus Numerantium*, 113:61–79, 1996.
11. A. Denise and P. Zimmermann. Uniform random generation of decomposable structures using floating-point arithmetic. *Theoretical Computer Science*, 218:233–248, 1999.
12. R. Diestel. *Graph Theory.* Springer–Verlag, New York, 1997.
13. D. P. Eric Fusy and G. Schaeffer. Dissections and trees: applications to optimal mesh encoding and random sampling. In *Proceedings of the sixteenth annual ACM-SIAM symposium on Discrete algorithms (SODA05)*, pages 690 – 699, 2005.
14. P. Flajolet, P. Zimmerman, and B. Van Cutsem. A calculus for the random generation of labelled combinatorial structures. *Theoretical Computer Science*, 132(1-2):1–35, 1994.
15. S. Gerke and C. McDiarmid. On the number of edges in random planar graphs. *Comb. Prob. and Computing*, 13:358–402, 2004.
16. O. Giménez and M. Noy. Asymptotic enumeration and limit laws of planar graphs. preprint.
17. J. E. Hopcroft and J. K. Wong. Linear time algorithm for isomorphism of planar graphs. *Annual ACM Symposium on Theory of Computing*, pages 172–184, 1974.
18. C. McDiarmid, A. Steger, and D. Welsh. Random planar graphs. *Journal of Combinatorial Theory, Series B*, 93:187–205, 2005.
19. A. Nijenhuis and H. Wilf. *Combinatorial algorithms.* Academic Press Inc., 1979.
20. D. Osthus, H. J. Prömel, and A. Taraz. On random planar graphs, the number of planar graphs and their triangulations. *J. Combinatorial Theory, Series B*, pages 119–143, 2003.
21. G. Schaeffer. Random sampling of large planar maps and convex polyhedra. In *Proc. of the Thirty-first Annual ACM Symposium on the Theory of Computing (STOC'99)*, pages 760–769, Atlanta, Georgia, May 1999.

22. E. Steinitz. Polyeder und Raumeinteilungen. *Encyclopädie der mathematischen Wissenschaften*, Band III(9), 1922.
23. B. A. Trakhtenbrot. Towards a theory of non-repeating contact schemes. *Trudi Mat. Inst. Akad. Nauk SSSR*, 51:226–269, 1958. [In Russian].
24. W. T. Tutte. *Graph Theory.* Cambridge University Press, 1984.
25. T. Walsh. Counting labelled three-connected and homeomorphically irreducible two-connected graphs. *J. Combin. Theory*, 32:1–11, 1982.
26. T. Walsh and V. A. Liskovets. Ten steps to counting planar graphs. In *18th South-eastern International Conference on Combinatorics, Graph Theory, and Computing, Congr. Numer.*, volume 60, pages 269–277, 1987.

Configurations with Few Crossings in Topological Graphs

Christian Knauer[1], Étienne Schramm[2,⋆], Andreas Spillner[3], and Alexander Wolff[2,⋆]

[1] Institute of Computer Science, Freie Universität Berlin
christian.knauer@inf.fu-berlin.de
[2] Fakultät für Informatik, Universität Karlsruhe, P.O. Box 6980, D-76128 Karlsruhe
i11www.ira.uka.de/algo/group
[3] Institute of Computer Science, Friedrich-Schiller-Universität Jena
spillner@minet.uni-jena.de

Abstract. In this paper we study the problem of computing subgraphs of a certain configuration in a given topological graph G such that the number of crossings in the subgraph is minimum. The configurations that we consider are spanning trees, s–t paths, cycles, matchings, and κ-factors for $\kappa \in \{1,2\}$. We show that it is NP-hard to approximate the minimum number of crossings for these configurations within a factor of $k^{1-\varepsilon}$ for any $\varepsilon > 0$, where k is the number of crossings in G. We then show that the problems are fixed-parameter tractable if we use the number of crossings in the given graph as the parameter. Finally we present a simple but effective heuristic for spanning trees.

1 Introduction

An undirected graph $G(V,E)$ that is embedded in the plane such that no two edges share an unbounded number of points is called a *topological graph.* If all edges are straight-line embedded, then G is called a *geometric* graph. A *crossing* $\{e,e'\}$ is a pair of edges in G such that $e \cap e' \not\subseteq V$. We call $\mu_{ee'} = |(e \cap e') \setminus V|$ the *multiplicity* of the crossing $\{e,e'\}$. Note that $\mu \equiv 1$ for geometric graphs. Let $X \subseteq \binom{E}{2}$ be the set of pairs of crossings in E. Note that c edges intersecting in a single non-endpoint give rise to $\binom{c}{2}$ crossings. We will use n, m, and k as shorthand for the cardinalities of V, E, and X, respectively. We define the *weighted number of crossings* of G as $\sum_{\{e,e'\}\in X} \mu_{ee'}$.

In this paper we study the problem of computing subgraphs of a certain configuration in a given topological graph such that the weighted number of crossings in the subgraph is minimum. The configurations that we consider are spanning trees, s–t paths, cycles, matchings, and κ-factors, i.e. subgraphs in which every node $v \in V$ has degree κ, for $\kappa \in \{1,2\}$. In the version of matching that we consider the number M of desired matching edges is part of the input. We will refer to this version as *M-matching.*

⋆ Supported by grant WO 758/4-2 of the German Science Foundation (DFG).

X. Deng and D. Du (Eds.): ISAAC 2005, LNCS 3827, pp. 604–613, 2005.

Algorithms that find subgraphs with few crossings have applications in VLSI design and pattern recognition [3]. For example, a set of processors (nodes) on a chip and a number of possible wire connections (edges) between the processors induce a topological graph G. A spanning tree in G with few crossings connects all processors to each other and can help to find a wire layout that uses few layers, thus reducing the chip's cost.

There is also a connection to matching with geometric objects. Rendl and Woeginger [7] have investigated the following problem. Given a set of $2n$ points in the plane they want to decide whether there is a perfect matching (i.e. a 1-factor) where the matched points are connected by axis-parallel line segments. They give an $O(n \log n)$-time algorithm for this problem and show that the problem becomes NP-hard if the line segments are not allowed to cross.

Kratochvíl et al. [5] have shown that for topological graphs it is NP-hard to decide whether they contain a crossing-free subgraph for any of the configurations mentioned above. Later Jansen and Woeginger [3] have shown that for spanning trees, 1- and 2-factors the same even holds in geometric graphs with just two different edge lengths or with just two different edge slopes.

These results do not rule out the existence of efficient constant-factor approximation algorithms. However, as we will show in Section 2, such algorithms do not exist unless $\mathcal{P} = \mathcal{NP}$. In Section 3 we complement these findings by a simple polynomial-time factor-$(k - c)$ approximation for any constant integer c. This result being far from satisfactory, we turn our attention to other possible ways to attack the problems: in Section 4 we show that the problems under consideration are fixed-parameter tractable with k being the parameter. While there are simple algorithms that show tractability, it is not at all obvious how to improve them. It was a special challenge to beat the 2^k-term in the running time of the simple fixed-parameter algorithm for deciding the existence of a crossing-free spanning tree. We also give optimization algorithms.

Finally, in Section 5 we give a simple heuristic for computing spanning trees with few crossings. Due to our findings in Section 2 our heuristic is unlikely to have a constant approximation factor. However, it performs amazingly well, both on random examples and on real-world instances. We use a mixed-integer programming (MIP) formulation (see [4]) as baseline for our evaluation.

Our fixed-parameter algorithm and the MIP formulation for the 1-factor problem can both be used to solve the above-mentioned problem of Rendl and Woeginger [7]. In our MIP formulation the numbers of variables and constraints depends only linearly on k. This makes the MIP formulation superior to the FPT algorithm for large values of k. Neither of our exact methods exploits the geometry of the embedded graph. Thus they also work if we are given an abstract graph G and a set X of crossings, and we view a crossing simply as a set of two edges not supposed to be in the solution at the same time.

In the whole paper we assume that the set of crossings X in the input graph has already been computed. Depending on the type of curves representing the graph edges this can be done using standard algorithms [2]. Whenever we want to stress that X is given, we use the notation $G(V, E, X)$.

2 Hardness of Approximation

For each of the configurations mentioned in the introduction we now show that it is hard to approximate the problem of finding subgraphs of that configuration with the minimum number of crossings in a given geometric graph G. The reductions are simple and most of them follow the same idea.

We begin with the problem of finding a spanning tree with as few crossings as possible. We already know [3] that the problem of deciding whether or not G has a crossing-free spanning tree is NP-hard. Our reduction employs this result directly. Given a graph G with k crossings and a positive integer d, we build a new graph G' by arranging k^d copies of G along a horizontal line and by then connecting consecutive copies by a single edge as in Figure 1. The new graph G' has k^{d+1} crossings. Now if G has a crossing-free spanning tree then G' has a crossing-free spanning tree. Otherwise every spanning tree in G' has at least k^d crossings. Let $\phi(G)$ be 1 plus the minimum number of crossings in a spanning tree of G. Since we can choose d arbitrarily large we have the following theorem. All theorems in this section hold for any $\varepsilon \in (0, 1]$.

Theorem 1. *It is NP-hard to approximate $\phi(G)$ within a factor of $k^{1-\varepsilon}$.*

We also consider a kind of dual optimization problem: Find a crossing-free spanning forest in G with as few trees as possible. Let $\phi'(G)$ denote the minimum number of trees in a spanning forest of G. Since there is a spanning forest with one tree if and only if there is a crossing-free spanning tree in G, we immediately have the following theorem.

Theorem 2. *It is NP-hard to approximate $\phi'(G)$ within a factor of $k^{1-\varepsilon}$.*

Next let us briefly consider the problems of finding M-matchings, 1- and 2-factors in G with as few crossings as possible. Again we already know that the related decision problems are NP-hard [3]. Let $\eta(G)$ denote 1 plus the minimum number of crossings in a subgraph of G of the desired configuration. Arguing along the same lines as for spanning trees we obtain the following theorem. Note that in this reduction we do *not* connect the copies of G in G'.

Theorem 3. *It is NP-hard to approximate $\eta(G)$ within a factor of $k^{1-\varepsilon}$.*

Now we turn to the problem of finding a path between two given vertices s and t (an s–t path for short) with as few crossings as possible. We can show that the problem of deciding whether or not a given geometric graph has a crossing-free s–t path is NP-hard by a simple adaption of the reduction from planar 3SAT presented in [5] for topological graphs. Figure 2 reproduces Figure 7 from [5]. Every clause is represented by a triple of edges between two vertices. Each edge in such a triple represents a variable occurring in the corresponding clause. It is shown in [5] that it is possible to draw the edges in such a way that occurrences of variables that cannot be set true simultaneously correspond to edges that intersect. Thus there is a satisfying truth assignment if and only if there is a crossing-free s–t path in the constructed graph. It is not hard to see that we can

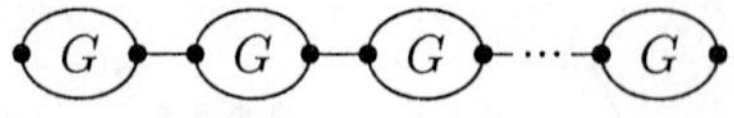

Fig. 1. Graph G': k^d copies of G

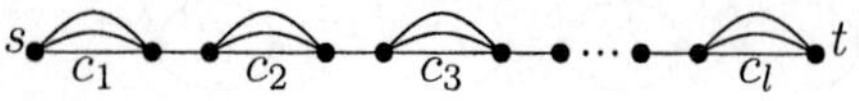

Fig. 2. The basis of the reduction in [5]

substitute a sequence of straight line segments for the drawing of every edge in the topological graph.

Now we apply the same trick as in the case of spanning trees to turn the NP-hardness of the decision problem into a hardness-of-approximation result. This is indicated in Figure 3. If there is a crossing-free s–t path in G then there is a crossing-free s_1–t_{k^d} path in G'. If there is at least one crossing in every s–t path in G then there are at least k^d crossings in every s_1–t_{k^d} path in G'. Let $\gamma(G)$ denote 1 plus the minimum number of crossings in a s–t path in G. Then we have the following theorem.

Theorem 4. *It is NP-hard to approximate $\gamma(G)$ within a factor of $k^{1-\varepsilon}$.*

In [5] it is shown that it is even possible to draw the edges of the graph in Figure 2 in such a way that in addition to the crossings which ensure the consistency of the chosen truth setting we can make the edges in every clause pairwise intersecting. Thus we can choose at most one edge in every clause gadget and the constructed graph does not contain a crossing-free cycle. Now we connect vertices s and t by an extra sequence of edges as indicated in Figure 4 and easily obtain the following corollary.

Corollary 1. *It is NP-hard to decide whether or not a geometric graph contains a crossing-free cycle.*

Unfortunately it seems impossible to apply the same trick again to obtain the hardness of approximation, since there may be cycles that do not pass through vertices s and t and that have only few crossings. Thus we have to punish the usage of a crossing in forming a cycle in the graph. To achieve this goal for every crossing in the constructed graph we make the sequences of straight line edges cross many times. This is indicated in Figure 5. By choosing the number of bends large enough we obtain the following theorem where $\zeta(G)$ denotes 1 plus the minimum number of crossings in a cycle in G.

Theorem 5. *It is NP-hard to approximate $\zeta(G)$ within a factor of $k^{1-\varepsilon}$.*

3 Approximation Algorithms

After this long list of negative results on approximability let us now give a positive remark. Trivially, any spanning tree in a geometric graph with k crossings $(k+1)$-approximates $\phi(G)$. However, with just a little more effort we can compute a factor-k approximation by checking whether all edges in $\bigcup X$ are cut edges. If yes, then *every* spanning tree of G has k crossings (and hence is optimal). Otherwise we can compute a spanning tree that avoids one of the non-cut

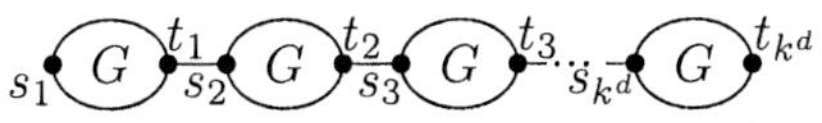

Fig. 3. The graph G'

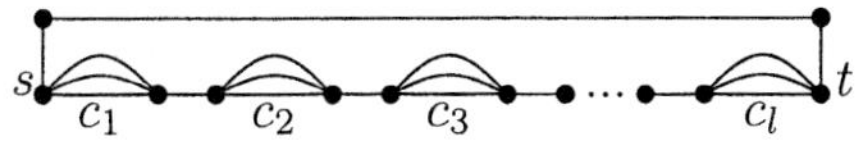

Fig. 4. Adding edges between s and t

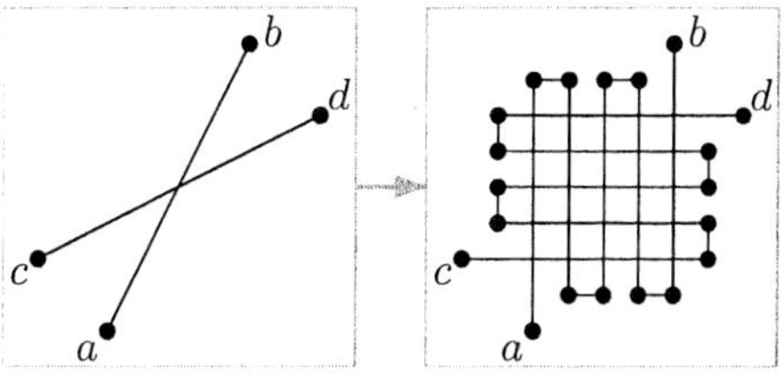

Fig. 5. Punishing the usage of crossings

edges, which yields a factor-k approximation. Along the same lines we obtain factor-$(k-c)$ approximations for every constant $c > 0$ in polynomial time.

Theorem 6. *For every constant integer $c \in (0, k)$ there is a polynomial-time factor-$(k-c)$ approximation for $\phi(G)$.*

4 Fixed-Parameter Algorithms

In this section we present fixed-parameter algorithms using the total number k of crossings as the parameter. The intuition behind the concept of fixed-parameter algorithms [1] is to find a quantity associated with the input such that the problem can be solved efficiently if this quantity is small. The number k suggests itself naturally since on the one hand the problems under consideration become trivial if $k = 0$ and on the other hand the reductions in Section 2 employ graphs with many crossings.

4.1 A Simple General Approach

We assume that the input graph G has a subgraph of the desired configuration and we only try to find one with the minimum weighted number of crossings. For example, when looking for spanning trees we assume that the input graph is connected. We set $E_X = \bigcup X$. Thus E_X contains exactly those edges that participate in a crossing. Note that $|E_X| \leq 2k$. Now we can proceed as follows:

1. Form the crossing-free graph G' by removing all edges in E_X from G.
2. For all crossing-free subsets $H \subset E_X$ check whether the graph $G' \cup H$ has a subgraph of the desired configuration.

The graph G' can be constructed in $O(m)$ time. Let $\text{check}_{\mathcal{C}}(n, m)$ be the time needed for checking whether $G' \cup H$ has a subgraph of configuration $\mathcal{C}$. Since $\text{check}_{\mathcal{C}}(n, m) = \text{poly}(n, m)$ for all the configurations we consider, the two-step procedure shows that the corresponding decision problems can be solved in $O(m + \text{check}_{\mathcal{C}}(n, m)\, 4^k)$ time and thus are all fixed-parameter tractable. However, it is easy to do better.

Observation 1. *To check the existence of a crossing-free configuration in G it suffices to go through all maximal (w.r.t. the subgraph relation) crossing-free subgraphs of G and check whether one of them has a subgraph of the desired configuration.*

By induction on k we get that E_X has at most 2^k maximal crossing-free subsets H. We perform step 2 only on these.

Theorem 7. *Given a topological graph $G(V, E, X)$ and a configuration $\mathcal{C}$, we can decide in $O(m+\text{check}_{\mathcal{C}}(n, m)\, 2^k)$ time whether G has a crossing-free subgraph of configuration $\mathcal{C}$.*

If the desired configuration $\mathcal{C}$ is an M-matching we have $\text{check}_{\mathcal{C}}(n, m) \in O(\sqrt{n}m)$ [6, 9]. Note that 1-factors are only a special kind of M-matching. For 2-factors we can employ the graph transformation of Tutte [8] and obtain $\text{check}_{\mathcal{C}}(n, m) \in O(n^4)$.

Observe that in step 2 the only interesting connected components of G' are those that contain an endpoint of an edge from E_X. However, there are at most $4k$ such connected components. For the configurations spanning tree, s–t path and cycle this observation yields a reduction to a *problem kernel* [1], i.e. to a problem whose size depends only on the parameter k, but not on the size of the input. Now it is clear that for these configurations generating and checking a subset H can be done in $O(k)$ time.

Corollary 2. *Given a topological graph $G(V, E, X)$, we can decide in $O(m + k2^k)$ time whether G has a crossing-free spanning tree, s–t path or cycle.*

Finally we want to present a simple approach to deal with the corresponding optimization problems, i.e. the problem of finding a desired configuration with minimum weighted number of crossings. For every subset X' of the set of crossings X we do the following.

1. Compute the graph $G'' = G' \cup \bigcup X'$.
2. Compute the subset of crossings X'' from X that do not share an edge with any crossing in X'.
3. Decide whether there is a crossing-free subset H of $\bigcup X''$ such that $G'' \cup H$ contains a desired configuration.

We filter out those subsets of crossings X' for which we get a positive answer in step 3. Among the filtered out subsets we can easily keep track of the one which yields a minimum weighted number of crossings. Suppose in step 3 we use a decision algorithm running in $O(\text{check}_{\mathcal{C}}(n, m)\beta^k)$ time.

Theorem 8. *Given a topological graph $G(V, E, X)$, we can compute in $O(m + \sum_{j=0}^{k} \binom{k}{j} \text{check}_{\mathcal{C}}(n, m)\beta^{k-j}) = O(m + \text{check}_{\mathcal{C}}(n, m)(1 + \beta)^k)$ time a subgraph of configuration $\mathcal{C}$ in G with minimum weighted number of crossings.*

4.2 Spanning Trees

In this section we want to improve the 2^k-term in the running time of the simple decision algorithm. It will turn out that we barely achieve this goal. We will get the exponential term in the running time down to 1.9999996^k. While this improvement seems marginal, the fact that we managed to beat the trivial algorithm is of theoretical interest. Moreover, we think that our methods can be applied in a wider scenario. A similar approach for s–t paths and cycles yields 1.733^k, see the long version of this article [4].

Imagine the process of selecting the edges for set H as a search tree. Branchings in the tree correspond to possible choices during the selection process. By selecting edges in E_X to be in H or not to be in H we reduce the number of crossings from which we can still select edges. The leaves of the search tree correspond to particular choices of H. Let $T(k)$ denote the maximum number of leaves in the search tree for input graphs with k crossings. Note that $T(k)$ also bounds the number of interior nodes of the search tree.

First we will see that it is rather easy to speed up the algorithm as long as the crossings in G are not pairwise disjoint. Let e be an edge such that exactly z crossings $c_1 = \{e, f_1\}, \ldots, c_z = \{e, f_z\}$ in X share edge e and $2 \leq z \leq k$. If we select edge e to be in H then none of the edges $f_1, \ldots, f_z$ can be in H. If we select edge e not to be in H then we can select edges $f_1, \ldots, f_z$ as if crossings $c_1, \ldots, c_z$ would not exist. Thus for both choices there are only $k - z$ crossings left from which we still can select edges. This leads to the recurrence $T(k) \leq 2T(k - z)$ which solves to $T(k) \in O(2^{k/z})$.

It remains to consider the case that the crossings in X are pairwise disjoint. Up to now we concentrated on the set E_X. It was only after selecting an edge e to be in H that we took a look at the connected components of G' that are possibly connected by e. Now we also take the connected components of G' into consideration to guide the selection process. Observe that any connected component C (in order to make G' connected) must be connected by at least one edge from E_X to the rest of G'. This puts some restriction on which crossing-free subsets of the edges in E_X with an endpoint in C need to be checked. After introducing some notation, Lemma 1 will make this more precise.

We can assume that for every crossing $c \in X$ none of the two edges in c connects vertices in the same connected component of G', since such an edge cannot help to make G' connected and thus need not be selected to be in H.

We define the degree of a connected component of G' as the number of edges in E_X with one endpoint in this component. Now consider some connected component C of G'. Let d denote the degree of C and $E(C)$ the set of edges in E_X incident to a vertex of C. Let $X(C)$ denote the set of crossings contained in $E(C)$. We set $x = |X(C)|$, $R = \bigcup X(C)$ and $S = E(C) \setminus R$. Then S contains the $d - 2x$ edges in $E(C)$ that do not cross any other edge in $E(C)$. To connect C with the rest of G' we select subsets T of $E(C)$ such that T contains exactly one edge from each crossing in $X(C)$ and a subset of the edges in S.

Lemma 1. *It suffices to check* $2^{d-x} - 1$ *subsets of* $E(C)$.

For the proof of Lemma 1 refer to [4]. The result leads to the recurrence $T(k) \leq (2^{d-x} - 1)T(k - (d - x))$ which solves to $T(k) \in O((2^{d-x} - 1)^{k/(d-x)})$. Of course, this solution is of little use if we cannot guarantee the existence of a component C with appropriately bounded degree d. It turns out that we can do that if there is a constant α $(0 < \alpha < 1)$ such that G' has more than αk connected components: Let δ denote the minimum degree of a component of G'. Then we have $4k \geq 2|E_X| > \delta\alpha k$ and hence $4/\alpha > \delta$. It is desirable to choose α as large as possible. We will use $\alpha = 0.211$ for reasons that will become clear soon. Then we can guarantee the existence of a component with degree at most 18 and obtain $T(k) \in O(\beta_{18}^k)$, where $\beta_i = \sqrt[i]{2^i - 1} < 2$. For each of the at most $T(k)$ nodes and leaves of the search tree, we need $O(k)$ time.

It remains to treat the case that the crossings are pairwise disjoint and G' has at most αk connected components. Set $l = \lfloor \alpha k \rfloor$. Observe that we can make G' connected without crossing edges iff there are l crossings in X such that G' becomes connected by using one edge from each of these l crossings. Thus we simply check every subset of l crossings from X, which can be done in $O(\binom{k}{l}2^l k)$ time. Using $\alpha = 0.211$, this is in $O(k1.985^k)$.

Theorem 9. *Given a topological graph $G(V, E, X)$, we can decide in $O(m + k\beta_{18}^k)$ time whether G has a crossing-free spanning tree ($\beta_{18} < 1.9999996$).*

If we are willing to resort to a randomized algorithm we can improve the result of Theorem 9. We are looking for a new way to treat the case that the crossings are pairwise disjoint and G' has at most αk connected components. Observe that if G has a connected crossing-free spanning subgraph at all, then there are at least $2^{(1-\alpha)k}$ maximal crossing-free subsets $H \subseteq E_X$ such that $G' \cup H$ is connected. This can be seen as follows: Suppose there is a crossing-free subset $F \subseteq E_X$ that makes G' connected. Then we can choose such an F with $|F| < \alpha k$. Let $X_F = \{c \in X \mid c \cap F = \emptyset\}$. Since the elements of X are pairwise disjoint we have $|X_F| > (1 - \alpha)k$. If we select one edge from each crossing in X_F and add these edges to F the resulting set of edges is still crossing-free. There are at least $2^{(1-\alpha)k}$ possible ways to select edges. Thus there are at least $2^{(1-\alpha)k}$ maximal crossing-free subsets $H \subseteq E_X$ such that $G' \cup H$ is connected.

This suggests the following randomized algorithm: From each element of X we randomly select one edge and check if the resulting crossing-free graph is connected. If the given geometric graph G has a crossing-free connected spanning subgraph, the probability of success is at least $2^{(1-\alpha)k}/2^k = 2^{-\alpha k}$. Thus $O(2^{\alpha k})$ iterations suffice to guarantee a probability of success greater than 1/2. The running time is in $O(k2^{\alpha k})$. Setting $\alpha = 4/5$ yields the following theorem.

Theorem 10. *There is a Monte Carlo algorithm with one sided error, probability of success greater than 1/2 and running time in $O(m + k\beta_4^k)$ which tests whether a given topological graph has a crossing-free spanning tree ($\beta_4 < 1.968$).*

For the dual optimization problem of finding a spanning forest of G consisting of as few trees as possible we can proceed similarly as in the algorithm for the decision problem above. Thus we can solve the dual optimization problem in $O(m + k\beta_{18}^k)$ time.

5 Heuristic

Due to our inapproximability results in Section 2, we cannot hope to find a constant-factor approximation for the number of crossings in any of the configurations we consider. Instead, we now describe a simple heuristic for computing spanning trees with few crossings in geometric graphs. Our heuristic uses a set of rules that simplify the input graph without changing the number of crossings of an optimal spanning tree. Initially all edges are *active*. During the process, edges can be deleted or *selected*. The solution will consist of the edges that are selected during the process. The heuristic applies the rules to the input graph until no more rule can be applied. Then a heuristic decision is taken. We decided to delete the edge e that maximizes $A(e) + 3S(e)$, where $A(e)$ and $S(e)$ are the numbers of active and selected edges that e crosses, respectively.

We now specify the rules. They are only applied to active edges. Connected components refer to the graph induced by all nodes and the selected edges.

1. If an edge has no crossings with other edges, it is selected.
2. If an edge is a cut edge, it is selected.
3. If both endpoints of an edge belong to the same connected component, then this edge is deleted.
4. If two edges e_1 and e_2 connect the same connected components, and if every edge crossed by e_1 is also crossed by e_2, then e_2 is deleted.

Our heuristic always finds a spanning tree since rule 2 makes sure that no cut edge is deleted. A brute-force implementation runs in $O(nm^3)$ time.

We have implemented the heuristic (except for rule 4) in C++ using the LEDA graph library. It can be tested via a Java applet at `http://i11www.ira.uka.de/few_crossings`. To compute optimal solutions at least for small graphs, we also implemented the MIP formulation described in [4]. We used the MIP solver *Xpress-Optimizer* (2004) by Dash Optimization with the C++ interface of the BCL library. Both heuristic and MIP were run on an AMD Athlon machine with 2.6 GHz and 512 MB RAM under Linux-2.4.20.

We generated random graphs with 20 nodes and $24, 26, \ldots, 80$ edges as follows. Edges were drawn randomly until the desired graph size was obtained. The edge set was discarded if it was not connected. Vertex coordinates were chosen uniformly from the unit square. To these graphs we applied our heuristic and the MIP solver. Figure 6 shows spanning trees of a random graph with 16 vertices and 26 edges. Figure 7 shows the average number of crossings of the spanning trees found by the heuristic and the MIP solver, as well as the number of crossings in the input graph. For each data point, we generated 30 graphs. The average was taken only over those which were solved by the MIP solver within three hours (at least 27 of the 30 graphs per data point).

As real-world data we used three graphs whose vertices correspond to airports and whose edges correspond to direct flight connections in either direction. The flight-connection graphs had each very few high-degree nodes and many leaves. We used the Mercator projection for planarization and then embedded the edges straight-line. The results are given in Table 1.

Table 1. Number of crossings of spanning trees in airline graphs

Data set				Heuristic		MIP	
	nodes	edges	crossings	crossings	time [sec.]	crossings	time [sec.]
Lufthansa Europe	68	283	1760	66	0.2	66	304.1
Air Canada	77	276	1020	83	0.1	83	379.7
Lufthansa World	163	696	8684	128	1.8	121	59.4

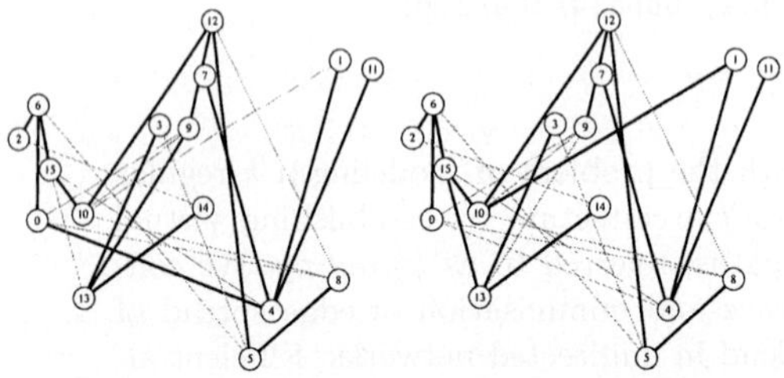

Heuristic: 5 crossings MIP: 4 crossings

Fig. 6. Solutions for a 16-node random graph

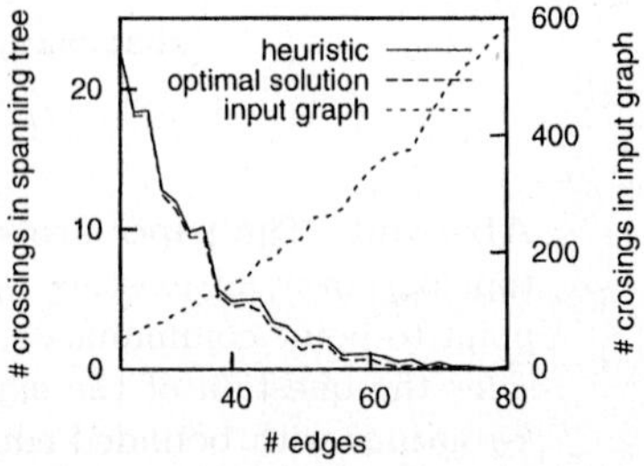

Fig. 7. Performance of heuristic and MIP on 20-node random graphs

Given our inapproximability results in Section 2 we were surprised to see how well our simple heuristic performs: in 77 % of the random graphs and in two of the three real-world instances the heuristic performed optimally. For random graphs it used at most five edge crossings above optimal.

Acknowledgments. We thank Jiong Guo, Georg Kliewer, and Markus Völker.

References

1. R. G. Downey and M. R. Fellows. *Parameterized Complexity.* Springer, 1999.
2. D. Halperin. Arrangements. In *Handbook of Discrete and Computational Geometry*, chapter 24, pages 529–562. CRC Press, 2004.
3. K. Jansen and G. J. Woeginger. The complexity of detecting crossingfree configurations in the plane. *BIT*, 33:580–595, 1993.
4. C. Knauer, É. Schramm, A. Spillner, and A. Wolff. Configurations with few crossings in topological graphs. Techical Report 2005-24, Universität Karlsruhe, Sept. 2005. `http://www.ubka.uni-karlsruhe.de/cgi-bin/psview?document=/ira/2005/24`.
5. J. Kratochvíl, A. Lubiw, and J. Nešetřil. Noncrossing subgraphs in topological layouts. *SIAM J. Disc. Math.*, 4(2):223–244, 1991.
6. S. Micali and V. V. Vazirani. An $O(\sqrt{|V|}|E|)$ algorithm for finding maximum matching in general graphs. In *Proc. IEEE Symp. Found. Comp. Sci.*, pp. 17–27, 1980.
7. F. Rendl and G. Woeginger. Reconstructing sets of orthogonal line segments in the plane. *Discrete Mathematics*, 119:167–174, 1993.
8. W. T. Tutte. A short proof of the factor theorem for finite graphs. *Canad. J. Math.*, 6:347–352, 1954.
9. V. V. Vazirani. A theory of alternating paths and blossoms for proving correctness of the $O(|V|^{1/2}|E|)$ general graph matching algorithm. *Combinatorica*, 14:71–91, 1994.

On Bounded Load Routings for Modeling k-Regular Connection Topologies

Adrian Kosowski[1], Michał Małafiejski[1,*], and Paweł Żyliński[2,**]

[1] Gdańsk University of Technology, Dept. of Algorithms and System Modeling
[2] University of Gdańsk, Institute of Mathematics
{kosowski, mima, buba}@sphere.pl

Abstract. The paper deals with the problem of modeling a k-regular topology over an existing network architecture by establishing virtual point-to-point communication paths, referred to as *k-routing*. We consider the question of the existence and minimisation of edge spread of k-routings with bounded edge load in undirected networks. Efficient algorithms are presented for determining minimal k-routings with edge load 1 and for certain cases with edge load 2. On the negative side, the problems of finding a 6-routing with load 2 and of minimising a 2-routing with load 2 are proven to be NP-hard (though the latter is approximable within 7/6). The results imply the NP-hardness of the well known all-to-all routing problem for bounded edge load.

1 Introduction

Numerous problems related to communication in distributed systems may be discussed in terms of routing in computer networks. The topology of the network is modeled in the form of a graph whose vertices correspond to nodes, while edges — to direct physical connections between nodes.

In the static routing model [3,9,11,13] which is the main topic of interest of this paper, it is assumed that all pairs of vertices forming the routed instance are initially known and all paths are determined by the routing algorithm at the same time. Static routing may be considered as a special case of the problem of embedding a virtual network topology in an existing physical network. Such virtual connections are often established for a significant timespan.

The load of the network, and consequently the efficiency of communication, is influenced by the choice of paths in the routing, and sometimes also other parameters of the considered paths. The most commonly used criterion for the assessment of the quality of a routing is the so called *edge load*, i.e. the maximum number of paths of the routing containing the same edge. We also give some attention to criteria applied in optical networks with wave division multiplexing (WDM), where each fibre-optic link forms a number of optical channels used for simultaneous communication.

* Supported by the TASK grant.
** Supported by the KBN Grant No. 4 T11C 047 25.

X. Deng and D. Du (Eds.): ISAAC 2005, LNCS 3827, pp. 614–623, 2005.

Throughout the rest of the paper we consider a special case of virtual network topology design in which the routed instance is not fixed, but may be chosen as any *k-regular* instance on the set of all nodes (i.e. each node has to be connected by paths to exactly k other nodes). The routing obtained for such an instance is required to be legitimate with respect to the stated criteria.

Problem Definition and Motivation. The physical architecture of the network is given in the form of an undirected graph $G = (V, E)$, where V denotes the set of nodes, while E represents the set of connections between them (we denote: $n = |V|, m = |E|$). A sequence of edges $P = (e_1, e_2, \ldots, e_l) \in E^l$, such that $e_i = \{v_i, v_{i+1}\}$ for some two vertices $v_i \in V, v_{i+1} \in V$, is called a *path* of length $l = |P|$ in G, with endpoints v_1 and v_{l+1}. The symbol $P_{\{u,v\}}$ is used to denote any path in G with endpoints $u \in V, v \in V$. A pair of paths P_1 and P_2 is called *conflicting* if there exists an edge $e \in E$ such that $e \in P_1$ and $e \in P_2$.

A virtual network topology for network G is defined as a multigraph $H = (V, I)$, with the same set of vertices V as G, and a multiset of edges I such that each edge of I represents a single required virtual connection between a pair of nodes. In further considerations H will usually be assumed to be a simple graph. A *routing* R of instance H in network G is any set of paths of the following form: $R = \{P_{\{u,v\}} : \{u, v\} \in I\}$. For use in further considerations, we define the following parameters for any routing R:

- *edge load* $\pi(R)$, given by the formula: $\pi(R) = \max_{e \in E} |\{P \in R : e \in P\}|$,
- *dilation* $\mathtt{d}(R)$, defined as the length of the longest path in routing R: $\mathtt{d}(R) = \max_{P \in R} |P|$,
- *edge spread* $\mathtt{m}(R)$, defined as the total number of edges of G used in R: $\mathtt{m}(R) = |\{e \in E : \exists_{P \in R}\ e \in P\}|$,
- *number of wavelengths* $\mathtt{w}(R)$ used in the optimal wavelength assignment for routing R in WDM network G ($\mathtt{w}(R)$ is the chromatic number of graph $(R, C(R))$, where $C(R)$ denotes the set of pairs of conflicting paths in R [6]).

Using the introduced terminology, the simplest version of the considered k-routing problem with bounded edge load may be posed as follows.

The l-loaded k-routing problem [l-LOADED k-ROUTING]

Input: A connected graph G representing a network topology.
Problem: Find an instance H for G such that H is a k-regular simple graph, and a routing R of instance H for which $\pi(R) \leq l$, or decide that such an instance does not exist.

The above problem models the process of establishing a communication scheme among nodes in which every node communicates with exactly k other nodes, with bounded communication load on links. In distributed computing, the problem is practically feasible for nodes considered indistinguishable from the point of view of computation. It is worth noting some special cases of extremal values of k. The case when $k = 1$ corresponds to computation by nodes coupled in

pairs, when $k = 2$ — computations in cycles, $k = 3$ — computations in 3-regular structures, etc. Other values of k may also be applied in practice, either to increase computation speed, or more typically for increased system reliability and fault detection. Notice that by allowing the value of parameter k to be directly related to the number of vertices n of G, we may consider the intensively studied problem of establishing links between all pairs of nodes (so called *all-to-all routing*, or routing of the *gossiping instance*) in terms of k-routing, or more precisely — as $(n-1)$-routing.

When considering solutions to the l-LOADED k-ROUTING problem for some graph G we adopt a natural additional optimisation criterium in the form of minimisation of the edge spread of the sought routing. Minimising edge spread decreases the total load of the system resources and operation cost, whilst guaranteeing that the mean path length in the routing is not excessively long.

The Minimum l-loaded k-routing Problem [MIN l-LOADED k-ROUTING]

Input: A connected graph G representing a network topology.
Problem: Find a solution R to the l-LOADED k-ROUTING problem for G such that the value of edge spread $\mathtt{m}(R)$ is the minimum possible.

Previous results. The problems of routing with bounded edge load π and with a bounded number of wavelengths $\mathtt{w}$ have been studied in a variety of contexts [5,6,9,13]. In the case when both the graph G and the instance H are explicitly given at input, both routing problems are known to be *NP*-hard [3].

Intensive research has been carried out into the problem of routing the all-to-all instance ($(n-1)$-routing). In the general case, the problem of finding an all-to-all routing minimising π is known to be approximable within an $O(\log n)$-factor [4]. Exact minimum values of π and $\mathtt{w}$ can be found in polynomial time for certain special classes of networks G, such as trees of rings [2], complete $2D$-grids, tori and hypercubes [5]. In the general case, both the complexity of minimising π and of $\mathtt{w}$ remained open (a problem posed in e.g. [3,4,6,7] and many other).

Our contribution and outline of the paper. In Subsection 3.1 we provide an efficient $O(m^3 \log m)$ algorithm for solving the 1-LOADED k-ROUTING problem in any network G. For $k = 1$, the 1-LOADED 1-ROUTING problem can be solved in linear time. In Subsection 3.2 we use the results from Subsection 3.1 to formulate strong sufficient conditions for the existence of a 2-loaded k-routing. We prove that the 2-LOADED k-ROUTING problem can be solved in $O(mn)$ time if the minimum degree $\delta(G) \geq k$. However, if $\delta(G) < k$, a 2-loaded k-routing need not always exist, and we state that the problem of determining its existence is *NP*-complete for any constant $k \geq 6$. We use this result to prove that determining the existence of 2-loaded all-to-all routing in a general network is *NP*-complete. In Section 4 we discuss the MIN l-LOADED k-ROUTING problem, showing it to be easy if the edge load bound is $l = 1$, and hard for $l = 2$. We additionally give a 7/6 approximation algorithm for the *NP*-hard problem of MIN 2-LOADED

2-ROUTING. Finally, in Section 5 we briefly summarise our results in the form of tables and consider their implications for routing in WDM networks.

2 Simple Properties of k-Routings

The existence of a k-routing with bounded load in a graph G has to be regarded as a feature of the topology of G itself, and not of the potentially routed traffic. Necessary conditions may be formulated by analysing specific subgraphs of G.

Odd and even factors of a graph. Let F_v be a set of nonnegative integers defined for each vertex v. An *F-factor* in G is a set of edges such that the number of edges of this set incident to vertex v belongs to F_v. A *k-odd* (*k-even*, respectively to the parity of k) *factor* is defined as an F-factor such that each set F_v consists of all odd (even) numbers from the range $[k, \deg(v)]$, where $\deg(v)$ is the degree of vertex v in graph G (i.e. the number of edges incident to v).

Using results of Tutte [14] and Gabow [10], the problem of determining a k-odd (k-even) factor with the minimum number of edges may be solved efficiently by reduction to a minimum weighted perfect matching problem.

Proposition 1. *The problem of finding a k-odd (k-even) factor in graph G with the minimum possible number of edges can be solved in $O(m^3 \log m)$ time.*

Necessary conditions for the existence of a k-routing. Let us consider a connected graph G and any k-routing R in G. Observe that R can always be correctly defined provided there exists a k-regular instance H spanning the vertex set of G. We thus have the following necessary and sufficient condition for the existence of a k-routing in G.

Proposition 2. *For any connected graph G of n vertices there exists a k-routing in G if and only if there exists a k-regular graph of n vertices (i.e. $n > k$ and $n \cdot k$ is even).*

Next, consider a k-routing R with load equal to 1. Let G' denote the graph induced by all the edges from routing R. Each vertex v is adjacent to k edges of G' (corresponding to paths ending at v) and an even number of other edges of G' (paths crossing v). Consequently we have the following necessary condition.

Proposition 3. *If there exists a solution to the* 1*-loaded k-routing problem in G then G has a k-odd (k-even) factor.*

The converse of Proposition 3 is also true; this fact is proven as Theorem 2.

3 The l-Loaded k-Routing Problem

The l-LOADED k-ROUTING problem constitutes the main topic of interest of this paper. Polynomial time algorithms are discussed in Subsections 3.1 and 3.2, while completeness results are given in Subsection 3.3.

3.1 Polynomial-Time Algorithms for 1-Loaded k-Routing

An $O(m+n)$ algorithm for 1-loaded 1-routing

Lemma 1. *For any tree T with an even number of vertices there exists a solution R to the* 1-LOADED 1-ROUTING *which can be determined in linear time, and additionally* $\mathtt{d}(R) \leq 2$.

Proof. Consider a tree T of order n with an arbitrarily chosen root vertex r. By performing bottom-up search in T we can arrange the vertices of T in a sequence S according to non-increasing distance from r. Let v be the first non-leaf vertex of S. All the children of v are obviously leaves. Depending on the number of children of v, which is obviously positive, we add paths to form the 1-routing R as follows: if v has an even number of children we connect them in pairs by paths of length 2; in the opposite case we pair all children of v except one using paths of length 2 and connect the remaining child to v by a path of length 1. After adding the appropriate paths to R, all used edges and vertices are removed from T and the process continues for the next vertices from sequence S. It is easy to verify that the constructed routing R fulfills the assumptions of the lemma. □

By finding a spanning tree of connected graph G in time $O(m+n)$, and applying the algorithm stated in Lemma 1 we can construct a 1-loaded k-routing in G with dilation at most 2.

Theorem 1. *For any graph G the* 1-LOADED 1-ROUTING *problem can be solved in $O(n+m)$ time, additionally preserving the condition* $\mathtt{d}(R) \leq 2$.

An $O(m^3 \log m)$ algorithm for 1-loaded k-routing

We will now prove that finding a k-routing with load 1 in graph G is equivalent to finding a k-odd (k-even) factor in graph G.

Theorem 2. *For any graph G the* 1-LOADED k-ROUTING *problem can be solved in $O(m^3 \log m)$ time, additionally preserving the condition* $\mathtt{d}(R) \leq 2$. *The problem has a solution if and only if graph G has a k-odd (k-even) factor.*

Proof. Without loss of generality let us assume that k is odd. If G does not admit a k-odd factor then by Proposition 3 the 1-LOADED k-ROUTING problem has no solution for G. Otherwise, let M denote any edge-minimal k-odd factor in G; by Proposition 1 M can be determined in $O(m^3 \log m)$ time.

The sought 1-loaded k-routing R with $\mathtt{d}(R) \leq 2$ will be constructed by iterative inclusion of paths. For any vertex $v \in V$ we call an edge e, adjacent to v in M, *inaccessible* if v is the center of a path of length 2 belonging to R and containing e; all other edges of M adjacent to v are called *accessible*. The number of inaccessible and accessible edges for vertex v is denoted $\deg^+(v)$ and $\deg^-(v)$ respectively. Before the start of the construction of R for every vertex v we have $\deg^+(v) = \deg_M(v) \geq k$ and the value $\deg^-(v)$ is always even. Routing R may easily be determined by stating for each vertex which pairs of inaccessible edges form paths in R.

The construction of R relies upon the iterated choice of edges $e_1 = \{v, u_1\}$, $e_2 = \{v, u_2\}$ ($e_1, e_2 \in M$) in such a way as to fulfill the following assumptions:

1. Edges e_1 and e_2 are accessible for both their end-vertices.
2. For vertices v, u_1, u_2 we have $\deg^+(v) \geq k+2$ and $\deg^+(u_1) = \deg^+(u_2) = k$
3. No path of the form $P_{\{u_1,u_2\}}$ belongs to R.

The path $P = (e_1, e_2)$ is then added to R, the accessible sets for vertices are updated, and the process is repeated. The construction of R is considered complete when for each vertex v we have $\deg^+(v) = k$. Finally, by adding all edges of M not belonging to any path of R as 1-edge paths to the routing R, we obtain a correct 1-loaded k-routing with dilation bounded by 2.

It remains to be shown that the presented construction of routing R can always be performed. By the minimality of factor M we obtain that at every stage of the construction of routing R, the subgraph F of graph M induced by the set of vertices $U = \{v \in V : \deg^+(v) > k\}$ has to be a forest, since otherwise F would have to contain a cycle whose removal from M would lead to a smaller k-odd factor in G, a contradiction. Let T be any tree belonging to forest F and let v be an arbitrary leaf of T. Since $v \in U$ we have $\deg^+(v) > k$ and consequently $\deg^+(v) \geq k+2$. Because $\deg_F(v) \leq 1$, the set of vertices N adjacent to v along an edge accessible for v and belonging to $V \setminus U$ consists of at least $k+1$ vertices. Let $u_1 \in N$ be arbitrarily chosen. We have $\deg^+(u_1) = k$ and the edge $\{u_1, v\}$ is accessible for both its ends; hence from the bound on cardinality of set N, $|N| \geq k+1$, we immediately obtain that there must exist a vertex $u_2 \in N$, $u_2 \neq u_1$ such that $P_{\{u_1,u_2\}}$ does not belong to R. It is easy to show that the path $P = (\{u_1, v\}, \{u_2, v\})$ fulfills assumptions 1, 2 and 3 of the considered construction and may be added to routing R, which completes the proof. □

3.2 Polynomial-Time Algorithms for 2-Loaded k-Routing

An $O(m+n)$ algorithm for 2-loaded 2-routing

Let us recall that a *spider* is a tree of at least 3 vertices with one distinguished vertex known as the *center* or *corpus* of the spider, such that all other vertices are of degree not more than 2 and at a distance of at most 2 from the center.

Corollary 1. *For every spider, there exists a 2-loaded 2-routing with dilation at most 3.*

It is easy to see that for any given graph G, by treating a spanning tree T of G as rooted at an arbitrary vertex and traversing it in the bottom-up manner, tree T can be disconnected into a family of spiders. Thus graph G has a spanning forest with each connected component in the form of a spider and by Corollary 1 we obtain the following theorem.

Theorem 3. *Given a connected graph G of order at least 3 there exists an $O(m+n)$-time algorithm for finding a 2-loaded 2-routing R in G, additionally fulfilling the condition* $\mathrm{d}(R) \leq 3$.

An $O(mn)$ algorithm for 2-loaded k-routing when $\delta(G) \geq k$
The class of graphs with $\delta(G) \geq k$ is the widest known class which admits a 2-loaded k-routing (provided any k-routing at all exists in the considered graph).

Theorem 4. *Let G be a graph with minimum degree $\delta(G) \geq k$. Then there exists an $O(nm)$-time algorithm for the* 2-LOADED k-ROUTING *problem in G, and such a routing always exists provided $n > k$ and $n \cdot k$ is even.*

Proof (sketch). Let us consider a minimal (not necessarily minimum) subgraph $H \subseteq G$ such that $\delta(H) \geq k$. Naturally, the set of vertices $U = \{v \in V : \deg_H(v) > k\}$ forms an independent set in V. Using a technique similar to that applied in the proof of Theorem 2, it is possible to find a routing R_H in H such that all vertices from $V \setminus U$ are connected by exactly one path to each of k other vertices, whereas all vertices from U are connected by exactly one path to each of either $k-1$, or k other vertices from $V \setminus U$; moreover, the load of routing R_H is 1. It now suffices to find a routing R_U with load 1 in G which connects vertices connected by $k-1$ paths in R_H into pairs. Such a routing may always be found by an easy modification of Theorem 1. The sought 2-loaded k-routing R may be given as the sum of the sets of paths $R_H \cup R_U$. □

3.3 Hardness Results for 2-Loaded k-Routing and Related Problems

In the general case, for some values of k the complexity status of 2-LOADED k-ROUTING remains open for graphs with $\delta(G) < k$, even when $k = 3, 4$ or 5. However we give a partial answer to the considered question, showing that the 2-LOADED 6-ROUTING problem is NP-hard in 4-regular graphs.

Hardness of 2-loaded 6-routing and its implications
The proof of the NP-hardness of the problem of 2-loaded 6-routing in 4-regular graphs proceeds by reduction from the problem of $3D$-matching, restricted to subcubic instances (details were made available to the reviewers).

Theorem 5. *The problem of determining whether there exists a* 6*-routing with load* 2 *in a* 4*-regular graph is NP-complete.*

The 2-loaded 6-routing problem in 4 regular graphs is a good starting point for proofs of NP-hardness of other problems related to k-routing. For instance, by an easy reduction from this problem we may write the following statement.

Corollary 2. *The problem of determining whether there exists a k-routing with load* 2 *in a given graph is NP-complete for any fixed value of parameter $k \geq 6$.*

Completeness results for the all-to-all routing problem
The problem of laying out paths between all pairs of nodes to form a complete virtual topology is referred to as all-to-all routing, and in the terminology of this paper, as $(n-1)$-routing. Deciding whether a given physical network G admits an l-loaded all-to-all routing is highly relevant for both WDM and ATM

networks [3,4,6,7]. We give an answer to the most fundamental question, that the problem of deciding whether a given network admits a 2-loaded all-to-all routing is NP-complete.

Theorem 6. *The problem of determining whether a network G admits an all-to-all routing with $\pi \leq 2$ is NP-complete.*

Proof (sketch). The proof proceeds by reduction from the problem of 2-loaded 6-routing in 4-regular graphs (see Theorem 5). Consider an instance of the 2-loaded 6-routing problem, a 4-regular graph G of $n \geq 16$ vertices. We construct the graph G^* as the product $G^* = G \times K_{n-7}$ (i.e. G^* is a copy of graph G connected with a copy of the complete graph K_{n-7} by all possible edges). The proof of the observation that there exists a reduction from 2-loaded 6-routing in G to 2-loaded all-to-all routing in G^* is constructive and non-trivial; details were made available to the reviewers.

4 The Minimum l-Loaded k-Routing Problem

Quite naturally, the minimisation version of the l-LOADED k-ROUTING problem with respect to edge spread value $\mathtt{m}(R)$ is noticeably harder than the original version. Although the problem is easy for $l = 1$ or $k = 1$ (Subsection 4.1) it turns out to be APX-hard for $l = 2$ and any fixed value $k \geq 2$ (Subsection 4.2).

4.1 Positive Results for Min l-Loaded k-Routing with $l = 1$ or $k = 1$

First, let us consider the MIN 1-LOADED k-ROUTING problem. By the proof of Theorem 2, finding a minimum cardinality k-odd (k-even) factor guarantees that the constructed routing uses the minimum number of edges and consequently is the optimal solution to the MIN 1-LOADED k-ROUTING, and can be determined in $O(m^3 \log m)$ time.

The case of MIN l-LOADED 1-ROUTING is even easier to analyse, since for any graph G the solution to the MIN 1-LOADED 1-ROUTING is the optimal solution to MIN l-LOADED 1-ROUTING, regardless of l. The MIN 1-LOADED 1-ROUTING problem may in turn be treated as a special case of MIN 1-LOADED k-ROUTING, and solved in $O(m^3 \log m)$ time.

4.2 Completeness and Approximability of Min 2-Loaded 2-Routing

An equivalent characterisation of the MIN 2-LOADED 2-ROUTING problem may be obtained from the following lemma (the proof was provided for inspection by the reviewers).

Lemma 2. *Let p be the number of spiders in a maximum decomposition of graph G into spiders and let p_3 be the cardinality of a maximum P_3-matching in G (i.e. the number of vertices in a packing of vertex-disjoint paths of order 3 in G). Then $p = p_3$ and solutions to these problems are equivalent in the sense that one can be constructed from the other.*

By the above Lemma and the proof of Theorem 3 it easy to observe that the Min 2-loaded 2-routing problem is equivalent to the problem of finding the maximum number of vertex-disjoint paths of order 3 in a network topology. However in [12] the authors proved that the maximum P_3-matching problem is APX-hard even for subcubic graphs, thus we get the following theorem.

Theorem 7. *The* Min 2-loaded 2-routing *problem is APX-hard even for subcubic graphs.*

The theorem may easily be generalised to prove the APX-hardness of Min 2-loaded k-routing, for any $k \geq 2$.

Approximation algorithms for Min 2-loaded 2-routing
For general graphs, the best known approximation algorithm for the maximum P_3-matching problem achieves 3/2-ratio [1]. As the number of edges in any spider decomposition of a graph is equal to $n - p$, where p is the number of spiders, then by Lemma 2, it follows that any approximation algorithm to the maximum P_3-matching problem can be applied to the Min 2-loaded 2-routing problem.

Lemma 3. *Let A be an r-approximation algorithm for the maximum P_3-matching problem. Then there exists a $\frac{3r-1}{2r}$-approximation for the* Min 2-loaded 2-routing *problem.*

Proof. Let p and a be any optimal solution and any r-approximation for the maximum P_3-matching problem, respectively. As $p/a \leq r$ and $\frac{2}{3}n \leq n-p \leq n-a$, then

$$\frac{n-a}{n-p} \leq \frac{n-\frac{p}{r}}{n-p} \leq 1 + \frac{p\frac{r-1}{r}}{n-p} \leq \frac{3r-1}{2r}.$$ □

Theorem 8. *There is a polynomial-time 7/6-approximation algorithm for the* Min 2-loaded 2-routing *problem.*

5 Final Remarks

A summary of the major results presented in this paper for the l-loaded k-routing problem and its minimisation version is given in Tables 1 and 2.

The problem of k-routing with a bounded number of wavelengths
The question of minimising the parameter $\mathtt{w}$ rather than π in routings is usually discussed in the context of all-optical networks. In the context of the aspects of k-routing, for load $\pi = 1$ and load $\pi = 2$, described in previous sections, the values of $\mathtt{w}$ and π were always equal (note that for general instances of routing this need not hold, and the ratio $\mathtt{w}/\pi$ may be arbitrarily large). In particular, we can easily formulate Theorem 6 using the parameter $\mathtt{w}$ (the same proof holds), thus closing the open problem of the complexity of all-to-all routing in the optical version [4,6].

Theorem 9. *Determining whether a WDM all-optical undirected network G admits an all-to-all routing with $\mathtt{w} \leq 2$ is NP-complete.*

Table 1. Complexity of l-LOADED k-ROUTING for $l = 1$ and $l = 2$

bound	$k = 1$	$k = 2$	$k = 3, 4, 5$	$k \geq 6$	$k = n - 1$
$l = 1$	$O(m + n)$	$O(m^3 \log m)$			
$l = 2$		$O(m + n)$	open	NPH; $O(mn)$ for $\delta \geq k$	

Table 2. Complexity and approximability of MIN l-LOADED k-ROUTING for $l = 1$ and $l = 2$ (minimisation with respect to edge spread m)

bound	$k = 1$	$k = 2$	$k \geq 3$
$l = 1$	$O(m^3 \log m)$		
$l = 2$		$APXH$; 7/6-approx.	$APXH$

References

1. K.M.J. De Bontridder, B.V. Haldórsson, M.M. Haldórsson, C.A.J. Hurkens, J.K. Lenstra, R. Ravi, L. Stougie, Approximation algorithms for the test cover problem, *Math. Programming* **98** (2003), 477–491.
2. B. Beauquier, All-to-all communication in some wavelength-routed all-optical networks. *Networks*, 33 (1999), 179–187.
3. B. Beauquier, J.C. Bermond, L. Gargano, P. Hell, S. Pèrennes and U. Vaccaro, Graph problems arising from wavelength routing in all-optical networks. *Proc. WOCS'97*, 1997, Geneve, Switzerland.
4. B. Beauquier, S. Pèrennes, and M. Syska, Efficient Access to Optical Bandwidth Routing and Grooming in WDM Networks: state-of-the-art survey, CRESCCO report IST-2001-33135, Universite de Nice-Sophia Antipolis, 2002, France.
5. J.C. Bermond, L. Gargano, S. Perennes, A. A. Rescigno, and U. Vaccaro, Efficient collective communication in optical networks. *Proc. ICALP'96*, LNCS 1099 (1996), 574–585.
6. J. Białogrodzki, Path Coloring and Routing in Graphs. In: *Graph Colorings*, Kubale M. ed., Contemporary Mathematics series 352, AMS (2004), Providence, Rhode Island, USA, 139–152.
7. S. Choplin, L. Narayanan, and J. Opatrny, Two-Hop Virtual Path Layout in Tori. *Proc. SIROCCO'04*, LNCS 3104 (2004), 69–78.
8. F. R. K. Chung, E. G. Coman, M. I. Reiman, and B. E. Simon, The forwarding index of communication networks. *IEEE Trans. on Inf. Theory*, 33 (1987), 224–232.
9. T. Erlebach, and K. Jansen, The complexity of path coloring and call scheduling. *Theoret. Comp. Sci.* 255 (2001), Elsevier Science, 33–50.
10. H.N.Gabow, Data structures for weighted matching and nearest common ancestors with linking. *Proc. 1st ACM-SIAM SODA* (1990) 434–443.
11. P.E. Green, *Fiber-optic networks.* Prentice-Hall (1992), New Jersey, USA.
12. M. Małafiejski, P. Żyliński, Weakly cooperative guards in grids. *Proc. ICCSA'05*, LNCS 3480 (2005), 647–656.
13. R. Raghavan, and E. Upfal, Efficient routing in all-optical networks. *Proc. 26th ACM STOC* (1994), 134–143.
14. W.T. Tutte, A contribution to the theory of chromatic polynomials. *Canad. J. Math* 6 (1954), 80–91.

On the Complexity of Global Constraint Satisfaction*
(Extended Abstract)

Cristina Bazgan[1] and Marek Karpinski[2]

[1] LAMSADE, Université Paris-Dauphine, Paris
bazgan@lamsade.dauphine.fr
[2] Department of Computer Science, University of Bonn, Bonn
marek@cs.uni-bonn.de

Abstract. We study the computational complexity of decision and optimization problems that may be expressed as boolean contraint satisfaction problem with the global cardinality constraints. In this paper we establish a characterization theorem for the decision problems and derive some new approximation hardness results for the corresponding global optimization problems.

1 Introduction

Constraints of the global nature arise naturally in some optimization problems. For example, MIN BISECTION can be viewed as MIN CUT with the restriction that the two sets of vertices that determine the cut must be of equal size. It is known that MIN CUT is polynomial while MIN BISECTION is *NP*-hard. MIN BISECTION, MAX BISECTION and other optimization problems can be written as boolean constraint satisfaction problems where a feasible solution is a balanced assignment (where the number of variables set to 1 is the same as the number of variables set to 0). It was an increased interest in global optimization problems recently, cf. [10, 7, 16].

In this paper we study the complexity of decision and optimization problems of the balanced versions of boolean constraint satisfaction problems depending on the type of constraints. Schaefer [24] established a dichotomy theorem for the boolean constraint satisfaction problems distinguishing six polynomial time solvable cases. For the decision versions we show that if the set of constraints contains only equations of width 2 or it contains only conjunctions of literals, then the balanced version is polynomial time solvable and otherwise it is *NP*-complete.

Creignou [3] and Khanna and Sudan [18] established a dichotomy theorem for maximization versions of boolean constraint satisfaction problems that classify

* The first author was partially supported by DAAD grant number A/04/20431 and PROCOPE Project 333587. The second author was partially supported by a DFG grant, IST grant 14036, and PROCOPE Project 333587. The full version of this paper can be found under the URL address http://www.lamsade.dauphine.fr/notes.html

X. Deng and D. Du (Eds.): ISAAC 2005, LNCS 3827, pp. 624–633, 2005.

the problems into polynomially solvable or APX-hard. The balanced versions of these problems where also studied. Sviridenko [25] proved that the balanced version of MAX SAT is $1/(1-\frac{1}{e})$-approximable. For the balanced version of MAX 2SAT, Blaser and Manthey [2] established a 1.514-approximation factor and Hofmeister [12] a 4/3-approximation factor. Lower bound were also studied for these problems. Holmerin [13] showed that the balanced version of MAX E4-0H-LIN2 (see for the definition Section 2) cannot be approximated within 1.0957 in polynomial time, unless $P=NP$. Also Holmerin and Khot [14] showed that balanced version of MAX E3-0H-LIN2 is hard to approximate within $\frac{4}{3}-\varepsilon$ and in [15] they improved their result showing that this problem is hard to approximate within $2-\varepsilon$, for any $\varepsilon > 0$, if $NP \not\subseteq \cap_{\delta>0} DTIME(2^{n^\delta})$, thus obtaining the best possible inapproximability factor result for this problem. We prove in this paper that all the cases that were considered by Creignou [3] and Khanna and Sudan [18] in the dichotomy theorem become APX-hard and also that most of the trivial maximization constraint satisfaction problems have their balanced version APX-hard.

Khanna, Sudan and Trevisan [19] established a classification theorem for minimization versions of boolean constraint satisfaction problems. The complexity of approximation of MIN BISECTION was for long time widely open. Feige and Krautghamer [6] established an approximation algorithm for this problem within $O(\log^2 n)$ approximation factor. This result has been recently improved to $O(\log^{1.5} n)$ by the recent result of Arora, Rao and Vazirani [1]. Very recently, Khot [22] established that under the assumption that $NP \not\subseteq \cap_{\delta>0} BTIME(2^{n^\delta})$, for $BTIME$ denoting randomized polynomial time, MIN BISECTION has no polynomial time approximation scheme. Under the assumption that refuting SAT formulas is hard to approximate on average, Feige [5] proved also that MIN BISECTION is hard to approximate below $\frac{4}{3}$.

Holmerin [13] studied the hardness of approximating some generalizations of MIN BISECTION. In particular he showed that the balanced version of MIN E4-1H-LIN2 is not $(2-\varepsilon)$-approximable for any $\varepsilon > 0$, unless $P=NP$. We prove several inapproximability result for balanced minimization problems. In particular, using the inapproximability result for DENSEST k SUBGRAPH established by Khot [22], we prove that the balanced version of MIN MONOTONE-E2SAT has no polynomial time approximation scheme, if $NP \not\subseteq \cap_{\delta>0} BTIME(2^{n^\delta})$.

The paper is organized as follows: in Section 2 we introduce some preliminary notation and definitions, and Section 3 contains our results on decision problems. In Sections 4 and 5 we present our results concerning maximization and minimization optimization problems.

2 Preliminaries

We refer a general reader to [19, 21, 20, 4] for a background on the boolean constraint satisfaction problems.

A constraint is a boolean function $f : \{0,1\}^k \to \{0,1\}$. A constraint application is a pair $< f, (i_1, \ldots, i_r) >$ where r is the arity of f and the $i_\ell \in [n]$

indicate to which r of the n boolean variables a given constraint is applied. This constraint application will be denoted in the following by $f(x_{i_1}, \ldots, x_{i_r})$.

Let $\mathcal{F} = \{f_1, \ldots, f_t\}$ be a finite collection of boolean functions. An $\mathcal{F}$-set of constraints on n boolean variables $x_1, \ldots, x_n$ is a collection of constraint applications $\{f_j(x_{j_1}, \ldots, x_{j_{r_j}})\}_{j=1}^m$ for some integer m, where $f_j \in \mathcal{F}$ and r_j is the arity of f_j. We say that an assignment satisfies an $\mathcal{F}$-set of constraints if it satisfies every constraint in the collection.

The satisfiability problem CSP($\mathcal{F}$) consists of deciding whether there exists an assignment that satisfies a given $\mathcal{F}$-set of constraints. kCSP($\mathcal{F}$) (respectively, EkCSP($\mathcal{F}$)) is the variant of CSP($\mathcal{F}$) where each boolean function f_j is a function of at most (respectively, exactly) k variables, for $j \leq t$. The problems MAX (MIN) CSP($\mathcal{F}$) consist of finding a boolean assignment that maximizes (minimizes) the number of constraints that are satisfied. MAX (MIN) kCSP($\mathcal{F}$) (respectively, MAX (MIN) EkCSP($\mathcal{F}$)) are variants of MAX (MIN) CSP($\mathcal{F}$) where each constraint depends on at most (respectively, exactly) k literals.

Given a problem A, the BALANCED version of A is the problem A with a new set of feasible solutions being assignments where the number of variables set to *true* (denoted by 1) is the same as the number of variables set to *false* (denoted by 0). Such assignments will be called *balanced* assignments.

We consider also a generalization of this problem. Given a problem A, the α-BALANCED version of A, $0 < \alpha < 1$, is the problem A with a new set of feasible solutions being assignments with the number of *true* variables being an α ratio of the total number of variables. Such assignments will be called in the following α-balanced.

In this paper we study the complexity of decision and optimization problems related to BALANCED CSP($\mathcal{F}$) depending on the type of constraints defined by a class $\mathcal{F}$.

kSAT (respectively, EkSAT) is the version of SAT where each clause is of size at most (respectively, exactly) k. MONOTONE-EkSAT is the variant of the EkSAT problem where either all clauses contain only positive literals or all clauses contain only negative literals.

In this paper we use the notation AND instead of DNF for the problem of deciding whether a set of conjunctions of literals has a satisfiable assignment. MONOTONE-EkAND is the variant of the EkAND problem where either all conjunctions contain only positive literals or they contain only negative literals.

The input of the EkLIN2 problem is a set of equations of the type mod 2 and we have to decide if there is a satisfiable assignment. In Ek-bH-LIN2, $b \in \{0, 1\}$, the input consists of a set of equations of the type $x_{i_1} \oplus \ldots \oplus x_{i_k} = b$ on n boolean variables $x_1, \ldots, x_n$ and the problem is to decide if there is an assignment satisfying all equations.

MAX kSAT is the problem of *constructing* for a given set of clauses an assignment satisfying a maximum number of clauses. MAX kAND, MAX EkLIN2, MIN kSAT, MIN kAND, MIN EkLIN2 are defined in a similar way.

We will use basic notation of [24]. We refer to [17] for the precise definition of an E-reduction.

3 Complexity of Decision Problems

The decision complexity of boolean constraint satisfaction problems is well established. In particular, Schaefer [24] established the following remarkable dichotomy theorem:

Theorem 1 (Dichotomy Theorem for CSP($\mathcal{F}$) [24]). *Given an $\mathcal{F}$-set of constraints, the problem* CSP($\mathcal{F}$) *is polynomial time computable if $\mathcal{F}$ satisfies one of the conditions below, and* CSP($\mathcal{F}$) *is NP-complete otherwise.*

1. *Every function in $\mathcal{F}$ is 0-valid.*
2. *Every function in $\mathcal{F}$ is 1-valid.*
3. *Every function in $\mathcal{F}$ is weakly positive.*
4. *Every function in $\mathcal{F}$ is weakly negative.*
5. *Every function in $\mathcal{F}$ is affine.*
6. *Every function in $\mathcal{F}$ is bijunctive.*

Motivated by the above result, we aim at formulating analogous result for balanced problems. Firstly we show that for any $\mathcal{F}$-set of constraints, BALANCED CSP($\mathcal{F}$) is at least as difficult as CSP($\mathcal{F}$).

Lemma 1. *If* CSP($\mathcal{F}$) *is NP-complete, then* BALANCED CSP($\mathcal{F}$) *is also NP-complete.*

We turn now to a polynomial time case. We formulate our result in slightly more general setting of the α-balanced problems.

Theorem 2. *For any $0 < \alpha < 1$, α-*BALANCED E2-LIN2 *is solvable in polynomial time.*

Proof. Let us consider first $\alpha = \frac{1}{2}$. Given an instance I of BALANCED E2-LIN2 on n variables and m equations, we construct some equivalence classes on the set of literals by considering the equations one after another as follows. Given an equation $x_i \oplus x_j = 0$ ($x_i \oplus x_j{=}1$), we distinguish the following cases.

- If literals $x_i, \bar{x}_i, x_j, \bar{x}_j$ do not appear in a class, then we construct a new class and we put together x_i and x_j (x_i and $\bar{x}_j$ respectively).
- If either x_i or $\bar{x}_i$ appears in a class C_k and $x_j, \bar{x}_j$ do not appear in a class, then
  - if $x_i \in C_k$ then we introduce x_j ($\bar{x}_j$ respectively) in C_k.
  - if $\bar{x}_i \in C_k$ then we introduce $\bar{x}_j$ (x_j respectively) in C_k.
- If literals x_i or $\bar{x}_i$ and x_j or $\bar{x}_j$ appear in the same class C_k then I is not satisfiable if $\{x_i, \bar{x}_j\} \subseteq C_k$ or $\{\bar{x}_i, x_j\} \subseteq C_k$ ($\{x_i, x_j\} \subseteq C_k$ or $\{\bar{x}_i, \bar{x}_j\} \subseteq C_k$ respectively).
- If either x_i or $\bar{x}_i$ appears in a class C_k and either x_j or $\bar{x}_j$ appears in a class C_ℓ then
  - if $x_i \in C_k$ and $x_j \in C_\ell$ then we put together the literals of both classes C_k and C_ℓ (we put together the literals of the class C_k with the negated literals of the class C_ℓ).

- if $x_i \in C_k$ and $\bar{x}_j \in C_\ell$ then we put together the literals of the class C_k with the negated literals of the class C_ℓ (we put together the literals of both classes C_k and C_ℓ).

Suppose that at the end we obtain t equivalence classes $C_1, \dots, C_t$. Denote by a_{2i-1} and a_{2i} the number of literals that appear positive and respectively negative in C_i. BALANCED E2-LIN2 on I consists of deciding if there exists a partition of these $2t$ integers in two equal size sets P and N such that P and N contain exactly one of a_{2i-1}, a_{2i} for $i = 1, \dots, t$. This problem in solvable in polynomial time by dynamic programming [8]. If such a partition P, N exists then the following assignment is balanced and satisfies I:

– if $a_{2i-1} \in P$ then we assign to the positive variables of C_i the value 1 and to the negated variables of C_i the value 0.
– if $a_{2i-1} \in N$ then we assign to the positive variables of C_i the value 0 and to the negated variables of C_i the value 1.

If $\alpha \neq \frac{1}{2}$ then as below we construct equivalence classes $C_1, \dots, C_t$ and compute integers $a_1, \dots, a_{2t}$. We add two other integers $a_{2t+1} = n|1 - 2\alpha|$, $a_{2t+2} = 0$ and solve the above partition problem on this new instance. □

The above result contrast interestingly with Theorem 3.

Theorem 3. *For any $k \geq 3$, $b \in \{0, 1\}$,* BALANCED Ek-bH-LIN2 *is NP-complete.*

MONOTONE-2SAT is a trivial problem. In contrast to this, we show that α-BALANCED MONOTONE-E2SAT is, in fact, *NP*-hard.

Theorem 4. α-BALANCED MONOTONE-E2SAT *is NP-complete, for any $\alpha > 0$.*

Proof. We reduce α-CLIQUE (cf. [8]) to α-BALANCED MONOTONE-E2SAT. An instance of α-CLIQUE has an input a graph on n vertices and we have to decide if it contains a clique of size at least αn. The reduction is as follows: given a graph $G = (V, E)$ on n vertices, we construct an instance I on n boolean variables $x_1, \dots, x_n$, one for each vertex of G. For any $i, j \in V$ such that $(i, j) \notin E$, we add the clause $\bar{x}_i \vee \bar{x}_j$. It is clear that if C is a clique in G of size αn, then the assignment $x_i = 1$ if $i \in C$ and $x_i = 0$ if $i \notin C$ satisfies each clause of I since for each $(i, j) \notin E$, x_i or x_j is false. Conversely, if an α-balanced assignment satisfies I, then the set $C = \{i : x_i = 1\}$ is a clique of size αn. Since α-CLIQUE is *NP*-hard [8], α-BALANCED MONOTONE-E2SAT is *NP*-hard as well. □

Theorem 5. BALANCED MONOTONE-EkSAT *is NP-complete for any $k \geq 3$.*

Since BALANCED AND is trivial we can formulate the following

Theorem 6 (Characterization Theorem for BALANCED CSP($\mathcal{F}$)**).** *Given an $\mathcal{F}$-set of constraints, the problem α-*BALANCED CSP($\mathcal{F}$) *is polynomial time solvable (if every function in $\mathcal{F}$ is affine with width 2 or if every function in $\mathcal{F}$ is a conjunction of literals), otherwise it is NP-complete.*

4 Approximation of Global Maximum Constraint Satisfaction

We state first the following known classification theorem of MAX CSP($\mathcal{F}$) (cf. [3, 18]).

Theorem 7 (Characterization Theorem for MAX CSP($\mathcal{F}$) **[3, 18]).** MAX CSP($\mathcal{F}$) *is either polynomial time computable or is APX-complete. Moreover, it is in P if and only if $\mathcal{F}$ is either 0-valid or 1-valid or 2-monotone.*

Some upper bounds have been established for these balanced versions of MAX CSP($\mathcal{F}$). α-BALANCED MAX SAT was proven to be $1/(1-\frac{1}{e})$-approximable ([25]). α-BALANCED MAX 2SAT was proven to be 1.514-approximable ([2]) and BALANCED MAX 2SAT was proven to be 4/3-approximable ([12]).

We state first the following direct lemma:

Lemma 2. MAX CSP($\mathcal{F}$) *is E-reducible to* BALANCED MAX CSP($\mathcal{F}$).

The following theorem shows that the three polynomial cases for MAX CSP($\mathcal{F}$) became difficult for the balanced version.

Theorem 8. BALANCED MAX MONOTONE-EkSAT *is APX-hard, for $k \geq 2$.*

A particular case of the following problem is equivalent to a BALANCED MAX CSP($\mathcal{F}$) problem for some particular $\mathcal{F}$ as it will be proved later.

We introduce now a new problem.

DENSEST k SUBGRAPH
Input: A graph $G = (V, E)$ on n vertices where n is even.
Output: A subset $S \subseteq V$ of size k that maximize the number of edges with both extremities in S.

The hardness of the approximation of DENSEST k SUBGRAPH remained open for long time. Recently, Khot [22] was able to establish such a result using a special PCP technique.

Theorem 9 ([22]). DENSEST k SUBGRAPH *has no polynomial time approximation scheme if NP $\not\subseteq \cap_{\delta>0}$ BTIME($2^{n^{\delta}}$).*

More precisely, Khot [22] has proved the previous result for DENSEST k SUBGRAPH when $k = cn$ for c a constant.

Proposition 1. DENSEST k SUBGRAPH *is E-reducible to* DENSEST $\frac{n}{2}$ SUBGRAPH.

Proposition 2. BALANCED MAX MONOTONE-E2AND *is E-equivalent to* DENSEST $\frac{n}{2}$ SUBGRAPH.

Theorem 10. BALANCED MAX MONOTONE-EkAND, $k \geq 2$, *has no polynomial time approximation scheme if NP $\not\subseteq \cap_{\delta>0}$ BTIME($2^{n^{\delta}}$).*

Proof (sketch). We can E-reduce Balanced Max Monotone-EkAND to Balanced Max Monotone-E$(k+1)$AND, for $k \geq 2$, and thus using Propositions 1, 2 and Theorem 9 (Khot's result [22]) the result follows. □

We consider in the following the balanced version of affine constraints.

Max E2-1H-Lin2, that is Max Cut, is known to be *APX*-hard [23] and Balanced Max E2-1H-Lin2 that is Max Bisection is known to be *APX*-hard [23, 11]. Each instance of Max E2-0H-Lin2 is satisfied by the trivial assignment 0. We show a relation between the complexity of Balanced Max Monotone-E2AND and Balanced Max E2-0H-Lin2 (or Balanced Max UnCut).

Proposition 3. Balanced Max Monotone-E2AND *is E-reducible to* Balanced Max E2-0H-Lin2.

Thus we establish an inapproximability result for Balanced Max UnCut.

Theorem 11. Balanced Max UnCut *has no polynomial time approximation scheme if NP $\not\subseteq \cap_{\delta>0}$ BTIME($2^{n^{\delta}}$).*

Proof. The result is a consequence of Propositions 2, 3 and Theorem 9. □

When k is odd, Max Ek-bH-Lin2 is trivial since the assignment b for all variables satisfies all equations. When k is even, Max Ek-0H-Lin2 is also trivial since the assignment 0 for all variables satisfies all equations. For $k \geq 4$ even, Max Ek-1H-Lin2 is not know to be hard to approximate.

Theorem 12. Balanced Max Ek-bH-Lin2 *is APX-hard, for $k \geq 3$, $b \in \{0, 1\}$.*

Balanced Max Ek-bH-Lin2 was studied for particular cases of k and $b = 0$. More precisely, Holmerin [13] proved that Balanced Max E4-0H-Lin2 cannot be approximated within 1.0957 in polynomial time, unless *P=NP*. Also Holmerin and Khot showed in [14] that Balanced Max E3-0H-Lin2 is hard to approximate within $\frac{4}{3} - \varepsilon$ and in [15] they improved their result showing that Balanced Max E3-0H-Lin2 is hard to approximate within $2 - \varepsilon$ if *NP* $\not\subseteq \cap_{\delta>0}$ *DTIME*$(2^{n^{\delta}})$, thus obtaining the best possible inapproximability bound result for this problem (under this assumption).

Theorem 13 (Characterization Theorem for Balanced Max CSP($\mathcal{F}$)**).** Balanced Max CSP($\mathcal{F}$) *is APX-hard.*

5 Approximation of Global Minimum Constraint Satisfaction

A classification theorem for Min CSP($\mathcal{F}$) was formulated in [19].

We can show directly, like for the decision and maximization constraint satisfaction problems, that the balanced version of a minimization problem is at least as hard as an underlying problem.

Lemma 3. MIN CSP($\mathcal{F}$) *is E-reducible to* BALANCED MIN CSP($\mathcal{F}$).

MIN MONOTONE-EkSAT for $k \geq 2$ are trivial problems. For the balanced situation we formulate

Proposition 4. BALANCED MAX MONOTONE-E2AND *is E-reducible to* BALANCED MIN MONOTONE-E2SAT.

We derive now

Theorem 14. BALANCED MIN MONOTONE-E2SAT *has no polynomial time approximation scheme if NP* $\not\subseteq \cap_{\delta>0}$ *BTIME*(2^{n^δ}).

Proof. The result is a consequence of Proposition 4 and Theorem 10. □

We first show that a hardness approximation result for BALANCED MIN MONOTONE-E2SAT implies a hardness approximation result for MIN BISECTION.

Proposition 5. BALANCED MIN MONOTONE-E2SAT *is E-reducible to* MIN BISECTION.

Proof. Given an instance I of BALANCED MIN MONOTONE-E2SAT on n variables $x_1, \ldots, x_n$ and m clauses, we construct an instance I' of MIN BISECTION on $n+2$ variables $x_1, \ldots, x_n$ and two new variables y and z and $3m$ equations as follows : for each clause $x_1 \vee x_2$ we add 3 equations $x_1 \oplus x_2 = 1, x_1 \oplus z = 1, x_2 \oplus z = 1$. We have $opt(I') \leq 2opt(I)$ since the assignment satisfying $opt(I)$ clauses in I and $z = 0$ and $y = 1$ satisfies $2opt(I)$ equations in I'. Given a balanced assignment v for I' satisfying val' equations, we can consider $z = 0$. If $y = 1$ then, the restriction of v on x variables is balanced and satisfies $\frac{val'}{2}$ clauses. If $y = 0$ then the restriction of v on x variables satisfies $\frac{val'}{2}$ clauses in I but is not balanced. Observe that the balanced assignment obtained by changing the value of an x variable from 1 to 0 satisfies at most $\frac{val'}{2}$ clauses. □

We establish now an E-reduction between BALANCED MAX UNCUT and MIN BISECTION.

Proposition 6. BALANCED MAX E2-0H-LIN2 *is E-reducible to* BALANCED MIN E2-1H-LIN2.

We formulate now

Theorem 15. BALANCED MIN MONOTONE-EkSAT, $k \geq 2$, *has no polynomial time approximation scheme if NP* $\not\subseteq \cap_{\delta>0}$ *BTIME*(2^{n^δ}).

Proof (sketch). We can E-reduce BALANCED MIN MONOTONE-EkSAT to BALANCED MIN MONOTONE-E$(k+1)$SAT, for $k \geq 2$, and thus using Theorem 14, the result follows. □

Min E2-0H-Lin2 is Min UnCut that is known to be *APX*-hard by [9] and thus Balanced Min E2-0H-Lin2 is Balanced Min UnCut is also *APX*-hard. Min E2-1H-Lin2 that is Min Cut is polynomial solvable. Balanced Min E2-1H-Lin2 is Min Bisection for which the hardness of approximation was proved very recently [22]. For $k \geq 3$, Min Ek-1H-Lin2 is trivial since the assignment 0 for all variables satisfies no equation. When k is odd, Min Ek-0H-Lin2 is also trivial since the assignment 1 for all variables satisfies no equation and when $k \geq 4$ is even, it is not known if Min Ek-0H-Lin2 is hard to approximate.

Theorem 16 ([14]). Balanced Min E3-bH-Lin2, $b \in \{0,1\}$, *is NP-hard to approximate within any constant factor.*

The proof of the above result uses a PCP technique. We can prove without using directly a PCP method a somewhat weaker result

Theorem 17. Balanced Min Ek-bH-Lin2, $b \in \{0,1\}$, *is APX-hard for every* $k \geq 3$.

References

1. S. Arora, S. Rao and U. Vazirani, *Expander flows and a $\sqrt{\log n}$-approximation to sparsest cut*, Proceedings of the 36th Annual ACM Symposium on Theory of Computing (STOC 2004), 222 – 231.
2. M. Blaser and B. Manthey, *Improved Approximation Algorithms for Max 2Sat with Cardinality Constraint*, The 13th Annual International Symposium on Algorithms and Computation (ISAAC 2002), LNCS 2518, 2002, 187–198.
3. N. Creignou, *A dichotomy theorem for maximum generalized satisfiability problems*, Journal of Computer and System Sciences, 51(3), 1995, 511–522.
4. N. Creignou, S. Khanna and M. Sudan, *Complexity Classifications of Boolean Constraint Satisfaction Problems*, SIAM Monographs on Discrete Mathematics and Applications, 2001.
5. U. Feige, *Relations between average case complexity and approximation complexity*, Proceedings of the 34th Annual ACM Symposium on Theory of Computing (STOC 2002), 534–543.
6. U. Feige and R. Krauthgamer, *A polylogarithmic approximation of the Minimum Bisection*, Proceedings of the 41st Annual Symposium on Foundations of Computer Science, 2000, 105–115.
7. U. Feige and M. Langberg, *Approximation Algorithms for Maximization Problems arising in Graph Partitioning*, Journal of Algorithms 41 (2001), 174–211.
8. M. R. Garey and D. S. Johnson, *Computers and intractability. A guide to the theory of NP-completeness*, Freeman, C.A. San Francisco (1979).
9. N. Garg, V. Vazirani and M. Yannakakis, *Approximate max-flow min-(multi)cut theorems and their applications*, Proceedings of the 25th Annual ACM Symposium on Theory of Computing (STOC 1993), 698–707.
10. E. Halperin, U. Zwick, *A unified framework for obtaining improved approximation algorithms for maximum graph bisection problems*, Proceedings of the 8th Conference on Integer Programming and Combinatorial Optimization (IPCO 2001), 210–225.

11. J. Håstad, *Some optimal inapproximability results*, Proceedings of the 29th Annual ACM Symposium on Theory of Computing (STOC 1997), 1-10.
12. T. Hofmeister, *An Approximation Algorithm for Max 2Sat with Cardinality Constraint*, Proceedings of the 11th Annual European Symposium on Algorithms, 301–312, 2003.
13. J. Holmerin, *PCP with Global Constraints – Balanced Homogeneous Linear Equations*, Manuscript, 2002.
14. J. Holmerin and S. Khot, *A strong inapproximability result for a generalization of Minimum Bisection*, Proceedings of the 18th IEEE Conference on Computational Complexity, 371–378, 2003.
15. J. Holmerin and S. Khot, *A new PCP Outer Verifier with Applications to Homogeneous Linear Equations and Max-Bisection*, Proceedings of the 36th Annual ACM Symposium on Theory of Computing (STOC 2004), 11–17.
16. G. Jger, A. Srivastav, *Improved Approximation Algorithms for Maximum Graph Partitioning Problems*, Proceedings of the 24th International Conference Foundations of Software Technology and Theoretical Computer Science (FSTTCS 2004), 348–359.
17. S. Khanna, R. Motwani, M. Sudan and U. Vazirani, *On syntactic versus computational views of approximability*, Proceedings of the 35th Annual IEEE Annual Symposium on Foundations of Computer Science, 819–830, 1994, Also published in SIAM Journal on Computing, 28(1), 1999, 164–191.
18. S. Khanna and M. Sudan, *The optimization complexity of constraint satisfaction problems*, Technical note STAN-CS-TN-96-29, Stanford University, CA, 1996.
19. S. Khanna, M. Sudan and L. Trevisan, *Constraint Satisfaction: the Approximability of Minimization Problems*, Proceedings of 12th IEEE Computational Complexity, 1997, 282–296.
20. S. Khanna, M. Sudan, L. Trevisan and D. P. Williamson, *The Approximability of Constraint Satisfaction Problems*, SIAM Journal of Computing, 30(6), 1863-1920 (2001).
21. S. Khanna, M. Sudan and D. Williamson, *A Complete Classification of the Approximability of Maximization Problems Derived from Boolean Constraint Satisfaction*, Proceedings of 29th ACM Symposium on Theory of Computing (STOC 1997), 11–20.
22. S. Khot, *Ruling Out PTAS for Graph Min-Bisection, Densest Subgraph and Bipartite Clique*, Proceedings of the 45th Annual IEEE Annual Symposium on Foundations of Computer Science (FOCS 2004), 136–145.
23. C. H. Papadimitriou and M. Yannakakis, *Optimization, approximation, and complexity classes*, Journal of Computing and System Science, 43 (1991), 425-440.
24. T. Schaefer, *The complexity of satisfiability problems*, In Conference Record of the 10th Annual ACM Symposium on Theory and Computing (STOC 1978), 216–226.
25. M. I. Sviridenko, *Best possible approximation algorithm for MAX SAT with cardinality constraint*, Algorithmica, 30(3), 2001, 398–405.

Polynomial Space Suffices for Deciding Nash Equilibria Properties for Extensive Games with Large Trees*,**

Carme Àlvarez, Joaquim Gabarró, and Maria Serna

Universitat Politècnica de Catalunya,
Jordi Girona, 1-3, Barcelona 08034, Spain
{alvarez, gabarro, mjserna}@lsi.upc.edu

Abstract. We study the computational complexity of deciding the existence of a Pure Nash Equilibrium or a subgame perfect Nash equilibrium with a given payoff and other related problems in finite multi-player extensive games with perfect information. We propose three ways of representing a game with different degrees of succinctness for the components of the game. We show that when the number of moves of each player is large and the player function and the utilities are represented succinctly the considered problems are PSPACE-complete. In contraposition, when the game is described extensively by means of its associated tree all the problems are decidable in polynomial time.

1 Introduction

Many situations involving interactions between independent agents can be modeled and analyzed by game theory. Game theory has been successfully applied to economics and computer science, among other disciplines. In recent times a lot of attention has been devoted to the computational aspects of game theory and specially to the relationship between Internet and games [15]. The design and analysis of efficient algorithms for computing Nash equilibria are the principal aims of the algorithmic game theory. The fundamental question posed by Papadimitriou [15] about the complexity of computing a Nash equilibrium in two players strategic games has initiated a line of research towards understanding the complexity of computing a pure or a mixed Nash equilibrium [5,6,4,9,2,7,1,3,16]. Recall that strategic games are one shot games as the players play simultaneously and only once.

Extensive games provide a natural way to model interactions between agents that involve sequential decisions. The game can be represented by a tree of decision nodes, and actions are represented as arcs in the tree. In a *perfect*

* Work partially supported by IST program of the EU under contract IST-2004-015964 (AEOLUS) and by Spanish CICYT under grant TIC2002-04498-C05-03 (Tracer).

** Due to lack of space proofs are omitted, we refer the interested reader to the extended version of the paper report LSI-05-39-R available at `http://www.lsi.upc.edu`.

X. Deng and D. Du (Eds.): ISAAC 2005, LNCS 3827, pp. 634–643, 2005.

Non-uniform representation	IsPN & IsSPN k-players ($k \geq 1$)	IsPNOut & IsSPNOut k-players ($k \geq 2$)	PNGrant & SPNGrant k-players ($k \geq 2$)
implicit	coNP-complete	PSPACE-complete	PSPACE-complete
general	coNP-complete	PSPACE-complete	PSPACE-complete
explicit	P	P	P

Fig. 1. Summary of the results of the paper

information game, each player has complete information of the game's history. In an *imperfect information game*, agents are uncertain about which node they are currently in. Most of the algorithmics and complexity results on extensive games deal with imperfect information games whose tree is given as input [17,10]. In terms of Game Theory two notions of equilibria play a fundamental role in the analysis of extensive games: the Nash equilibrium and subgame perfect Nash equilibrium. The fundamental question of the existence of a subgame perfect Nash equilibria, and therefore of a Nash equilibria, has been answered positively since it is known that any finite extensive game with perfect information has a subgame perfect Nash equilibrium [11]. Furthermore, one Nash equilibrium can be compute using *backward induction* [13]. This is not the case for imperfect information games.

Backward induction appears to have two limitations. It cannot be used to compute a Nash equilibrium with a required payoff, for example. Furthermore, it computes a strategy profile which needs space proportional to the size of the tree of the game. When one thinks on traditional games that involve sequential decisions, like Chess, Geography and other games considered in the computational complexity community [14,8], the tree of the game is exponentially large with respect to the size of the description of the game, thus backward induction will require exponential space. This apparent limitations lead us to study the complexity of deciding the existence of a Nash equilibrium with particular properties for the case of extensive games with perfect information when the game trees might have large size with respect to the number of moves of each player, i.e. the number of sequential decisions that can be taken by all the players.

In this paper we initiate a work towards understanding fully the components that play a fundamental role in the complexity of problems on extensive games. We note that games computationally meaningful should have a large number of players, a large number of actions for each player, or a large number of moves. Furthermore the set of actions and the payoff functions can be given in some explicit or implicit way, using more or less space. Particular attention must be paid to the representability of the data sets appearing in the definition of a game, the player strategies and the game outcomes. Following the systematic of [1] for strategic games, we consider three natural ways of describing a game as the input to a program. Each of them captures a different level of succinctness, depending on the description explicit or implicit of some of its components.

We study the complexity of the following decisional problems. The IsPN problem asks whether a given strategy profile is a pure Nash equilibrium. The IsPNOut problem determines whether a given history is the outcome of a Nash

equilibrium. Finally, the PNGRANT problem asks whether there is a pure Nash equilibrium in which the first player gets a payoff u. The same three questions can be addressed replacing the term pure Nash equilibrium by subgame perfect Nash equilibrium, giving raise to the problems IsSPN, IsSPNOUT and SPNGRANT. Our results, summarized in Figure 1, show that for games represented explicitly all the problems can be solved in polynomial time. When some of the components of the game are represented implicitly (by means of Turing machines) the IsPN and the IsSPN become coNP-complete, while the other four problems become PSPACE-complete. Furthermore, the hardness results hold also for games with two players.

The main difficulty in finding algorithms for deciding equilibria properties is that when the strategy profile is not given as a part of the input then it requires space proportional to the size of the tree of the game in order to be represented which might require exponential space in the number of sequential moves. To go below exponential space we provide recursive characterizations of the existence of a pure Nash equilibrium, or a subgame perfect Nash equilibrium, with a guaranteed payoff for the first player in terms of histories instead of strategy profiles. Those characterizations are crucial from the complexity point of view since a history requires only polynomial space in the number of sequential moves (depth of the game tree) and the length of the actions.

The hardness results follows from adequate reductions so that a quantified boolean formula is described as a game in which the particular equilibria we are looking for appear only when the quantified boolean formula is satisfiable.

Finally, we want to point out the differences with the case of strategic games obtained in [1]. For strategic games, the succinctness of the representation of the set of actions is relevant since the PNE problem that asks for the existence of a pure Nash equilibrium in a given strategic game is polynomial time solvable when the representation is explicit, NP-complete when the game is given in general form, and Σ_2^p-complete when the game is given in implicit form. The number of players is also relevant, for a constant number of players $k \geq 2$ the PNE is polynomial time solvable when the game is given in general form. Contrasting with this, in the extensive games we obtain the same complexity results when the game is given in general form or in implicit form, even for 2-players games.

2 Extensive Games with Perfect Information

We start introducing some operations and notation on languages and functions. Given a language $L \subseteq \Sigma^*$ and a word $a \in \Sigma^*$ we define the languages $aL = \{aw \mid w \in L\}$ and $a^{-1}L = \{w \mid aw \in L\}$. Given a function $\tau : L \to B$ and a symbol $a \in \Sigma$ we define the functions $a\tau : aL \to B$, where $a\tau(aw) = \tau(w)$, and $a^{-1}\tau : a^{-1}L \to B$, where $a^{-1}\tau(w) = \tau(aw)$. The previous operations can be extended to prefix formed by words in the usual way, $(az)L = a(zL)$ and $(za)^{-1}L = z^{-1}(a^{-1}L)$ and similarly for functions. Given a collection of pairwise disjoint languages $(L_1, \ldots, L_k)$ and a set of functions $\tau_1, \ldots, \tau_k$ where, for each $1 \leq j \leq k$, $\tau_j : L_j \to B$, their *union* is the function $\tau = (\tau_1, \ldots, \tau_k)$ defined on

$\cup_{i=1}^{k} L_i$ as $\tau(w) = \tau_i(w)$ for $w \in L_i$. Given a function $\tau = (\tau_1, \ldots, \tau_k)$, for any i, $1 \le i \le k$, τ_{-i} is the function $(\tau_1, \ldots, \tau_{i-1}, \tau_{i+1}, \ldots, \tau_k)$. For an element $b \in B$, the function τ_b is defined on $\{\lambda\}$ as $\tau_b(\lambda) = b$.

Now we state the definition of finite extensive games with perfect information as it is given in [13]. We have expressed some conditions using standard Computer Science notation for alphabets, words, and concatenation.

Definition 1. *A* finite extensive game with perfect information *is a structure* $\Gamma = (N, (A_i)_{i\in N}, H, Z, P, (u_i)_{i\in N})$ *where*

- $N = \{1, \ldots n\}$ *is the set of players. For each player* $i \in N$, A_i *is a finite set of actions.*
- H *is a finite set of words defined on the alphabet* $A_1 \cup \cdots \cup A_n$. *Each element of* H *is called a* history. H *satisfies the following properties: The empty word* λ *is a member of* H *and if* $h \in H$ *all the prefixes of* h *belong to* H.
- $Z \subseteq H$ *is formed by the terminal histories. A history* $h \in H$ *is* terminal *if* h *is not a proper prefix of another history.*
- P *is the* player function. P *assigns to each nonterminal history,* $h \in H \setminus Z$, *a player.* $P(h)$ *is the player who is allowed to take an action after the history* h. *Thus, for any* $h \in H \setminus Z$, *if for some action* a *we have that* $h \cdot a \in H$ *then* $a \in A_{P(h)}$.
- *For each player* $i \in N$, *a utility function* u_i *assigning to each terminal history a rational value.*

The simplest extensive game is an *empty game* denoted by $\Gamma_\lambda(v_1, \ldots, v_n)$. This game has n players and $H = Z = \{\lambda\}$, therefore Γ_λ has only one history that is terminal, and, for any $1 \le i \le n$, $u_i(\lambda) = v_i$. Observe that no player plays, but player i gets immediately the payoff $u_i(\lambda)$. The *length of a game*, $\mathsf{length}(\Gamma)$, is the length of the longest history in the game. Thus $\mathsf{length}(\Gamma_\lambda(v_1, \ldots, v_n)) = 0$.

An alternative definition of game is done by means of a rooted tree. The *tree of a game* is a rooted tree with labeled nodes and arcs. We assume that the arcs of the tree are directed from parent to child. We require all the labels in the outgoing arcs of a node to be different. Such a tree defines a game $\Gamma = (N, (A_i)_{i\in N}, H, Z, P, (u_i)_{i\in N})$ as follows. N is the set of the labels of the internal nodes. For $i \in N$, A_i is the set formed with the labels of the outgoing arcs of a node labeled i. The word obtained by the concatenation of the labels in the path from the root to a node u is the history associated to u. H is the set of histories associated to the tree nodes and Z is the set of histories associated to the leaves. The player function assigns to a non terminal history the label of its associated node. The leaves of the tree hold a table storing one rational for each player, thus defining the utility of the associated terminal history for each player. The tree associated to an empty game is a leaf holding the utility values associated to the history λ. In the following we formalize the notion of subgame.

Definition 2. *Given an extensive game* $\Gamma = (N, (A_i)_{i\in N}, H, Z, P, (u_i)_{i\in N})$, *the* subgame *of* Γ after history h *is the extensive game*

$$h^{-1}\Gamma = (N, (A_i)_{i\in N}, h^{-1}H, h^{-1}Z, h^{-1}P, (h^{-1}u_i)_{i\in N}).$$

The following result follows directly from the definitions.

Lemma 1. *Given $h \in H \setminus Z$ and an action a such that $ha \in H$ it holds $(ha)^{-1}\Gamma = a^{-1}(h^{-1}\Gamma)$.*

Our second operation is the composition of games. The intented meaning of the composition is "player i starts the game and chooses an action, from a set of ℓ actions, each selection makes the players continue playing a particular game among a set of ℓ games".

Definition 3. *Given a collection of ℓ games $(\Gamma_1, \ldots, \Gamma_\ell)$ where, for any $1 \leq \alpha \leq \ell$, $\Gamma_\alpha = (N, (A_i)_{i\in N}, H^\alpha, Z^\alpha, P^\alpha, (u_i^\alpha)_{i\in N})$, a player $i \in N$, and ℓ different actions from A_i, $a_{i_1}, \ldots a_{i_\ell}$, the* composite game *$\Gamma(i, a_{i_1}\Gamma_1 \ldots a_{i_\ell}\Gamma_\ell)$ is the game $(N, (A_i)_{i\in N}, H, Z, P, (u_i)_{i\in N})$, where $H = \{\lambda\} \cup a_{i_1}H^1 \cup \cdots \cup a_{i_\ell}H^\ell$, $Z = a_{i_1}Z^1 \cup \cdots \cup a_{i_\ell}Z^\ell$, $P = (\tau_i, a_{i_1}P^1, \ldots, a_{i_\ell}P^\ell)$, and for every player $j \in N$, $u_j = (a_{i_1}u_j^1, \ldots, a_{i_\ell}u_j^\ell)$,*

Over the tree of the game, the composition takes the tree of the component games, creates a new root with label i and join the new root to the root of game Γ_α with an arc labeled a_{i_α}. The composition operation allows us to decompose games with positive length recursively from subgames. Given a non-terminal history h, let $\mathsf{next}(h) = \{a \in A_{P(h)} \mid ha \in H\}$. Note that, an extensive game with perfect information Γ with $length(\Gamma) > 0$ can be factorized as $\Gamma = (i, a_{i_1}\Gamma_1, \ldots, a_{i_\ell}\Gamma_\ell)$ where $\mathsf{next}(\lambda) = \{a_{i_1}, \ldots, a_{i_\ell}\}$, $i = P(\lambda)$, and, for any $1 \leq \alpha \leq \ell$, $\Gamma_\alpha = a_{i_\alpha}^{-1}\Gamma$.

Now we turn our attention to how the players play. Let Γ be an extensive game with perfect information, for any player $i \in N$, define $H_i = \{h \in H \mid P(h) = i\}$. Notice that the languages in the collection $(H_i)_{i\in N}$ are pairwise disjoint. A *strategy of player* $i \in N$ in Γ is a function $s_i : H_i \to A_i$. A *strategy profile* $s = (s_1, \ldots, s_n)$ is the union of a set of n strategies, one for each player. Note that over the tree of the game a strategy profile is a selection of an outgoing arc for every internal node. When $\Gamma = \Gamma_\lambda(v_1, \ldots, v_n)$ the players can only play the *empty strategy* ($H_i = \emptyset$) and thus the game has a unique strategy profile, the *empty strategy profile.*

For a player $i \in N$, the *adversary team* is represented by $-i$ meaning all the other players. Given a strategy profile s the set of strategies for the team $-i$ is represented by s_{-i}, formally $s_{-i} = (s_1, \ldots, s_{i-1}, s_{i+1}, \ldots, s_n)$.

For each *strategy profile* $s = (s_1, \ldots s_n)$ the *outcome* of s, $O(s)$, is the terminal history that results when each player $i \in N$ follows the precepts of s_i. That is, $O(s) = a_1 \cdots a_k$ where $a_1 = s_{P(\lambda)}(\lambda)$ and, for any $1 \leq i < k$, $a_{i+1} = s_{P(a_1 \ldots a_i)}(a_1 \ldots a_i)$. Looking at the tree of the game, the outcome of a strategy profile is the history associated to the unique leaf accessible from the root following the arcs selected by the strategy profile. Note that different strategy profiles s and s' might determine the same outcome.

Finally, we define the *utility for player* i of a strategy profile s as the utility of the outcome determined by s, that is $u_i(s) = u_i(O(s))$.

Let us consider strategies in front of subgames. Let Γ be a game with positive length and let $s = (s_1, \ldots, s_n)$ be a strategy profile. The strategy profile induced

by s in the game after history h, $h^{-1}\Gamma$, is the function $h^{-1}s$. Assuming that Γ factorizes as $(i, a_{i_1}\Gamma_1, \ldots, a_{i_\ell}\Gamma_\ell)$, for any $1 \leq \alpha \leq \ell$, the strategy profile profile induced by s on Γ_α is $a_{i_\alpha}^{-1}s$.

On the reverse side we can define strategies by extension of strategies on subgames. Assume that $\Gamma = (i, a_{i_1}\Gamma_1, \ldots, a_{i_\ell}\Gamma_\ell)$. Given β, $1 \leq \beta \leq \ell$, and for any α, $1 \leq \alpha \leq \ell$, a strategy profile s^α. The *composite strategy profile* is $s = (\tau_{a_{i_\beta}}, a_{i_1}s^1, \ldots, a_{i_\ell}s^\ell)$.

We have given the definitions for strategy profiles, they can be easily extended to player strategies, replacing s by s_i, or to strategy profiles for the adversary teams, replacing s by s_{-i}. Finally, we state the definitions of Pure Nash equilibrium and subgame perfect Nash equilibrium as it is done in Game Theory.

Definition 4. *A strategy profile s^* in Γ is a* Nash equilibrium *if for every player $i \in N$ we have $u_i(s^*_{-i}, s^*_i) \geq u_i(s^*_{-i}, s_i)$, for every strategy s_i of player i. A strategy profile s^* in Γ is a* subgame perfect Nash equilibrium *if for every history h, the strategy profile $h^{-1}s^*$ is a Nash equilibrium in $h^{-1}\Gamma$.*

Note that every subgame perfect Nash equilibrium is also a Nash equilibrium but the reverse is not always true. Furthermore, for every history h, the strategy profile $h^{-1}s^*$ is a subgame perfect Nash equilibrium in $h^{-1}\Gamma$. Observe that, for an empty game, the empty strategy profile is a subgame perfect Nash equilibrium, and thus a Nash equilibrium.

3 Problems on Games and Their Input Description

As we have pointed out any finite extensive game has a strategy profile that is a subgame perfect Nash equilibrium [11], and therefore also a pure Nash equilibrium. Therefore, we are interested in deciding the existence of Nash equilibria or subgame perfect Nash equilibria that verify additional properties. We consider the following problems.

Is pure Nash equilibrium? (IsPN). Given an extensive game Γ and a strategy profile s, decide whether s is a Nash equilibrium of Γ.

Is pure Nash equilibrium outcome? (IsPNOut). Given an extensive game Γ and a history h, decide whether h is the outcome of a Nash equilibrium of Γ.

Pure Nash equilibrium with guarantee (PNGrant). Given an extensive game Γ and a positive number u, decide whether Γ has a Nash equilibrium s such that $u_{P(\lambda)}(s) = u$.

Is subgame perfect Nash equilibrium? (IsSPN). Given an extensive game Γ and a strategy profile s, decide whether s is a subgame perfect Nash equilibrium of Γ.

Is subgame perfect Nash equilibrium outcome? (IsSPNOut). Given an extensive game Γ and an history h, decide whether h is the outcome of a subgame perfect Nash equilibrium of Γ.

Subgame perfect Nash equilibrium with guarantee (SPNGrant). Given an extensive game Γ and a positive number u, decide whether Γ has a Nash equilibrium s such that $u_{P(\lambda)}(s) = u$.

The first decision to take is about the representation of a game as an input to a program. Some of the components of a game can be given with different degrees of succinctness. For example the player and the utility functions can be given tabulated or by means of Turing machines. In the following definitions P and M denote the player and utility TMs and t denotes the computation time bound. The *player machine* P implements a player function according to the following convention: Given an input sequence h, P computes in time at most t an integer i, whenever $P(h) \in \{1, \ldots, n\}$ it is interpreted as the player function on history h; $P(h) = 0$ indicates that $h \in Z$; otherwise P indicates that $h \notin H$. The *utility* machine M on input (h, i) outputs in at most t steps $u_i(h)$. We consider three natural ways of giving a game, according to the succinctness of actions, player function, and utilities. The criteria are similar to those proposed for strategic games in [1]. In the following definitions t is the computation time bound for any TM appearing in the description.

Extensive games in implicit form. The tuple $\Gamma = \langle 1^n, 1^m, 1^h, P, M, 1^t \rangle$ describes a game with n players, such that for any i, $1 \leq i \leq n$, the set of actions of player i is Σ^m. The set H of histories is a subset of Σ^{mh}.

Extensive games in general form. $\Gamma = \langle 1^n, A_1, \ldots, A_n, 1^h, P, M, 1^t \rangle$. describes a n-players game, in which, for any i, $1 \leq i \leq n$, the set of actions of player i is A_i, which is given explicitly. The set H of histories is a subset of $(\cup_{1 \leq i \leq n} A_i)^h$.

Extensive games in explicit form. The tuple $\Gamma = \langle 1^n, T \rangle$ represents a game with n players where T is the game tree.

Observe that an explicit description of a strategy requires space proportional to the size of the tree associated to a game. Therefore, since strategies can be given either implicitly or explicitly we take the following natural convention. When a game Γ is described in a succinct (implicit or general) form, the strategies will be given implicitly. A strategy is represented implicitly by a TM S, *the global strategy* TM so that on any non terminal history $h \in H \setminus Z$ outputs the action to be undertaken by the player that is allowed to play, formally $S(h) \in A_{P(h)}$. In the case of games given in explicit form, the strategy will be also represented in explicit form. This is a natural decision, having a strategy given in explicit form as input will allow us to compute a explicit form of the game in polynomial time. For sake of simplicity whenever we want to represent a game and a strategy we will incorporate a unique time bound that will be used in all the TMs involved in the input's description.

In the following sections we analyze the computational complexity of this set of six problems depending on the form in which the input game is given.

4 A Characterization of the Outcomes of Nash Equilibria

We provide first a characterization of the existence of a Pure Nash equilibrium with a particular outcome. This result will be crucial in the classification of problems because it reduces the computation space from exponential to polynomial.

Our recursive characterization will allow us to place the problem in PSPACE. We start with one technical definition.

Definition 5 (**(i,u)-blocking**). *Let Γ be an extensive game, i a player, and u a value. We say that Γ is* (i,u)-blocking *if* $\mathsf{length}(\Gamma) = 0$ *and* $u_i(\lambda) \leq u$, *or if* $\mathsf{length}(\Gamma) > 0$ *and either*
- *$P(\lambda) \neq i$ and there is $a \in \mathsf{next}(\lambda)$ such that $a^{-1}\Gamma$ is (i,u)-blocking, or*
- *$P(\lambda) = i$ and, for all $a \in \mathsf{next}(\lambda)$, the game $a^{-1}\Gamma$ is (i,u)-blocking.*

Our first result is a characterization of the previous property in terms of an adversarial team strategy s_{-i} that keeps player i benefit at most u.

Lemma 2. *Given an extensive game Γ, a player i and value u, Γ is (i,u)-blocking if and only if there is a strategy profile s_{-i} such that, for any s_i it holds $u_i(s_{-i}, s_i) \leq u$.*

Our second definition is the key to capture, through a recursive definition, the concept of outcome of a Nash equilibrium as it is shown in the next result.

Definition 6 (**Γ-stable**). *Let Γ be an extensive game and h be a terminal history in Γ. We say that h is Γ-stable if $|h| = 0$ or, otherwise, assuming that $\Gamma = (i, a_{i_1}\Gamma_1, \dots, a_{i_\ell}\Gamma_\ell)$ and that $h = a_{i_1}h'$,*
- *h' is Γ_1-stable and*
- *for all $1 < \alpha \leq \ell$, Γ_α is $(i, u_i(h))$-blocking.*

Theorem 1. *Let Γ be an extensive game with perfect information. A terminal history h is the outcome of some Nash equilibrium iff h is Γ-stable.*

The previous result is proved by induction on the length of the game. The proof that the outcome of a Nash equilibrium is indeed Γ-stable follows from the definitions of composition and Nash equilibrium. For the reverse side we show how to extend the Nash equilibrium of a subtree to a Nash equilibrium of the composition, without changing the outcome, using the definition of Γ-stable and Lemma 2.

Our PSPACE-harness results follow from reductions from the *Quantified boolean formula* problem (QBF). We define two extensive 2-player games $\Gamma(\Phi)$ and $\Gamma'(\Phi)$ associated to a boolean formula $\Phi \equiv \exists\, x_1 \;\forall\, x_2 \dots Qx_n \; F(x_1, \dots, x_n)$. In $\Gamma(\Phi)$ we set $A_1 = A_2 = \{0,1\}$. Player 1 controls all the variables with an existential quantifier and player 2 controls the other set of variables, they alternate assigning a truth value to the variables in order, note that $Z = \{0,1\}^n$. The utility function is given by $u_1(h) = F(h)$ and $u_2(h) = 1 - u_1(h)$. In $\Gamma'(\Phi)$ we set $A_1 = \{0,1\}$ and $A_2 = \{A, B, 0, 1\}$. Let $\Gamma_1 = \Gamma_\lambda(1,0)$ and set $\Gamma'(\Phi) = (2, A\Gamma_1, B\Gamma(\Phi))$.

Theorem 2. *Let Φ be a quantified boolean formula on n variables and let $\Gamma = \Gamma(\Phi)$ and $\Gamma' = \Gamma'(\Phi)$ (defined above). The following statements are equivalent*
- *Φ is satisfiable.*
- *Γ has a subgame perfect Nash equilibrium in which the first player gets 1.*

- *Γ has a Nash equilibrium in which the first player gets 1.*
- *A is the outcome of some subgame perfect Nash equilibrium of Γ'.*
- *A is the outcome of some Nash equilibrium of Γ'.*

The next theorem summarizes our complexity results.

Theorem 3. *For games given in explicit form, the* IsPN, *the* PNOut *and the* PNGrant *problems belongs to* P. *For games given in general or implicit form, the* IsPN *problem is* coNP*-complete while the* PNOut *and the* PNGrant *problems become* PSPACE*-complete.*

The PSPACE hardness follows from Theorem 2 and from the fact that the QBF problem is PSPACE-complete [8]. PSPACE membership follows from Theorem 1 that allows us to design a recursive algorithm so that the number of successive recursive calls is bounded by $\mathsf{length}(\Gamma)$ and for each recursive call we need to store only one action.

5 A Characterization of the Outcomes of Subgame Perfect Nash Equilibria

An alternative characterization of subgame perfect Nash is given in [12]. Consider the *one-deviation property:* no player i can increase their payoff by changing their action at the start of any subgame in which player i is the first mover, given the other player's strategies and the rest of player i strategy.

Lemma 3 (Proposition 438.1 of [12]). *A strategy profile in a finite extensive game with perfect information is a subgame perfect Nash equilibrium iff it satisfies the one-deviation property.*

For our complexity results we need a further refinement of the above property to transfer the existence of a strategy profile to the existence of a terminal history that is the outcome of some subgame perfect Nash equilibrium. Consider the following property defined recursively.

Definition 7 (Γ-fixable). *Let Γ be an extensive game and h be a terminal history in Γ. We say that h is Γ-fixable if $|h| = 0$ or, otherwise, assuming that $\Gamma = \Gamma(i, a_{i_1}\Gamma_1, \ldots, a_{i_\ell}\Gamma_\ell)$ and that $h = a_{i_1}h'$,*
- *h' is Γ_1-fixable and*
- *for all $1 < \alpha \leq \ell$, there is $h^\alpha \subset a_{i_\alpha}^{-1}Z$ such that $u_i(a_{i_\alpha}h^\alpha) \leq u_i(h)$ and h^α is Γ_α-fixable.*

Now we can state the characterization of the outcomes of a subgame perfect Nash equilibrium.

Theorem 4. *Let Γ be an extensive game with perfect information. A terminal history h is the outcome of some subgame perfect Nash equilibrium of Γ iff h is Γ-fixable.*

The next theorem summarizes our complexity results. Their proof uses arguments similar to those of the proof of Theorem 3.

Theorem 5. *When the game is given in explicit form, the* IsSPN, *the* SPNOut *and the* SPNGrant *problems belongs to* P. *When the game is given in general or implicit form the* IsSPN *problem is* coNP*-complete while the* SPNOut *and the* SPNGrant *problems become* PSPACE*-complete.*

References

1. C. Àlvarez, J. Gabarro, and M. Serna. Pure Nash equilibrium in strategic games with a large number of actions. In *MFCS 2005*, pages 95–106, 2005.
2. V. Conitzer and T. Sandholm. Complexity results about Nash equilibria. In *IJCAI 2003*, pages 765–771, 2003.
3. K. Daskalakis and C. Papadimitriou. The complexity of games on highly regular graphs. In *ESA 05*, LNCS to appear, 2005.
4. A. Fabrikant, C. Papadimitriou, and K. Talwar. The complexity of pure Nash equilibria. In *STOC 2004*, pages 604–612, 2004.
5. D. Fotakis, S. Kontogiannis, E. Koutsoupias, M. Mavronicolas, and P. Spirakis. The structure and complexity of Nash equilibria for a selfish routing game. In *ICALP 2002*, pages 123–134, 2002.
6. D. Fotakis, S. Kontogiannis, and P. Spirakis. Selfish unsplittable flows. In *ICALP 2004*, pages 593–605, 2004.
7. M. Gairing, T. Lcking, M. Mavronicolas, B. Monien, and M. Rode. Nash equilibria in discrete routing games with convex latency functions. In *ICALP 2004*, pages 645–657, 2004.
8. Garey and Johnson. *Computers and intractability. A guide to the NP-completeness.* Freeman, 1979.
9. G. Gottlob, G. Greco, and F. Scarcello. Pure Nash equilibria: Hard and easy games. In *Theoretical Aspects of Rationality and Knowledge.*, pages 215–230, 2003.
10. D. Koller, N. Megiddo, and B. Stengel. Efficient computation of equilibria for extensive two-person games. *Games and Economic Behavior*, 14:247–259, 1996.
11. H.W. Kuhn. Extensive games and the problem of information. In H.W. Kuhn and A.W. Tucker, editors, *Contribution to the theory of games, Volume II*, number 28 in Annals of Mathematics Studies, pages 193–216. Princeton University press, 1953.
12. M.J. Osborne. *A Introductions to Game Theory.* Oxford University Press, 2004.
13. M.J. Osborne and A. Rubinstein. *A Course in Game Theory.* MIT Press, 1994.
14. C. Papadimitriou. *Computational Complexity.* Addison-Wesley, 1994.
15. C. Papadimitriou. Algorithms, games and the Internet. In *STOC 2001*, pages 4–8, 2001.
16. G.R. Schoenebeck and S. Vadham. The complexity of Nash equilibria in concisely represented games. Technical Report 52, Electronic Colloquium on Computational Complexity, 2005.
17. B. Stengel. Computational complexity of correlated equilibria for extensive games. Technical Report LSE-CDAM-2001-03, CDAM, 2001.

An Improved $\widetilde{\mathcal{O}}(1.234^m)$-Time Deterministic Algorithm for SAT

Masaki Yamamoto

Tokyo Institute of Technology
masaki.yamamoto@is.titech.ac.jp

Abstract. We improve an upper bound by Hirsch on a deterministic algorithm for solving general CNF satisfiability problem. With more detail analysis of Hirsch's algorithm, we give some improvements, by which we can prove an upper bound $\widetilde{\mathcal{O}}(1.234^m)$ w.r.t. the number m of input clauses, which improves Hirsch's bound $\widetilde{\mathcal{O}}(1.239^m)$.

1 Introduction

During the past decades, many algorithms for Boolean satisfiability problems have been proposed, and some of them were proved to improve the nontrivial worst-case upper bounds for the problems. Such worst-case analysis was initiated by Monien and Speckenmeyer [6] for k-SAT. Given a formula F in conjunctive normal form (in short, a CNF formula), the problem of deciding whether F is satisfiable is called *SAT*; for CNF formulas consisting of clauses at most k-literals (in short, k-CNF formulas), the problem is called *k-SAT*. Since the work of Monien and Speckenmeyer, k-SAT problems, in particular 3-SAT, have been studied intensively, and various interesting algorithms have been proposed. On the other hand, not so much improvements have been done for the general SAT problem, the satisfiability problem for general CNF formulas. Note that CNF formulas with no clause size restriction are useful for expressing some combinatorial problems such as graph problems. In this paper, we focus on such general CNF formulas and propose an improved algorithm for the general SAT. Throughout this paper, we denote by n and m, the number of variables and the number of clauses, respectively.

We briefly summarize a history of improving the bounds for SAT: randomized algorithms and deterministic algorithms, respectively. We below give only the exponential parts of the bounds, omitting polynomial factors.

1.1 Randomized Algorithms for SAT

The first nontrivial upper bound was given by Pudlák [7]: $2^{n-0.5\sqrt{n}}$. A slightly better bound was given by Dantsin et al. [1]: $2^{n-0.712\sqrt{n}}$. These are bounds with respect to the number of variables. Schuler gave an algorithm with respect to the number of variables and clauses [8]: $2^{n(1-1/\log 2m)}$. This bound is better than $2^{n-c\sqrt{n}}$ for any constant c when $m = o(2^n)$.

X. Deng and D. Du (Eds.): ISAAC 2005, LNCS 3827, pp. 644–653, 2005.

1.2 Deterministic Algorithms for SAT

The first nontrivial upper bound was given by Dantsin et al. [1]: $2^{n(1-2\sqrt{1/n \log m})}$. A better algorithm was given by Dantsin and Wolpert [2]: $2^{n(1-1/\log 2m)}$. This was obtained by derandomizing Schuler's algorithm [8]. These bounds are asymptotically 2^n as m gets large. Hirsch gave an algorithm of running time: $2^{0.30897m} \approx 1.239^m$. This works better than the above algorithms when $m < n/0.30897$.

In this paper, we improve on Hirsch's algorithm [5], and obtain a better algorithm with 1.234^m running time, which has not been improved for a few years. The basic approach of our algorithm is the same as Hirsch's; in fact, our algorithm is almost the same. By more careful analysis, we guarantees that this almost the same algorithm indeed achieves the desired upper bound. The advantage of our analysis is to guarantee some paths of the recursion tree of an execution reaching a trivial formula while the recursion tree by Hirsch's analysis is of depth only two. See Fig. 3.

2 Preliminaries

We give some basic notions and notations, and briefly review how we analyze the running time of splitting algorithms. Then we present Hirsch's algorithm [5], which we will improve in the next section.

2.1 Basic Notions and Notations

Let X be a finite set of Boolean variables. A *literal* is a variable $x \in X$ or its negation $\bar{x}$. A *clause* is a disjunction of literals. We alternatively regard a clause as a set of literals. The empty clause is interpreted as `false`. The size of a clause C (denoted by $|C|$) is the number of literals in C. A k-clause is a clause of size k. A k^+-clause and a k^--clause are clauses of size at least k and at most k, respectively. A *conjunctive normal form* (CNF for short) formula is a conjunction of clauses. Again, we alternatively regard a CNF formula as a set of clauses. The empty formula is interpreted as `true`. The size of a CNF formula F (denoted by $|F|$) is the number of clauses in F. A *truth assignment* (*assignment* for short) to X is a mapping from X to $\{$`true`, `false`$\}$. We denote `true` by 1 and `false` by 0. A clause is said to be *satisfied* by an assignment if at least one literal in the clause is assigned 1 by the assignment. A formula is said to be *satisfied* by an assignment if every clause in the formula is satisfied by the assignment. A formula is said to be *satisfiable* if there exists an assignment by which the formula is satisfied, otherwise, *unsatisfiable*. Given a CNF formula F, SAT is a problem of deciding whether F is satisfiable.

Let F be a CNF formula, and l be a literal in F. We denote by $F|_{l=1}$, a formula obtained from F by an assignment $l = 1$, that is, a formula transformed from F by eliminating clauses which contain l and by eliminating $\bar{l}$. Formula $F|_{l=0}$ is defined similarly. For any literal l, if it occurs positively (i.e., as l) occurs i times

and occurs negatively (i.e., as $\bar{l}$) j times in F, then we say that l is a *(i,j)-literal.* It is equivalent to saying that $\bar{l}$ is a *(j,i)-literal.* We sometimes call a (i,j)-literal a *i-literal* for short. Also by, e.g., "(i^+,j^-)-literal", we mean a (i',j')-literal for any $i' \geq i$ and $j' \leq j$.

As many heuristic algorithms for SAT, our algorithm (as well as Hirsch's algorithm) is based on "splitting" [4,3]: Given a CNF formula F. Choose an appropriate literal l (depending on the heuristics) in F, and split F on l, that is, obtain two sub-formulas $F|_{l=1}$ and $F|_{l=0}$. Then, we have that F is satisfiable iff one of $F|_{l=1}$ and $F|_{l=0}$ is satisfiable. Hence, the splitting algorithm can decide whether F is satisfiable by recursively calling this procedure above.

Sub-formulas produced by splitting can be simplified by simple transformation rules: "pure literal elimination", and "unit clause propagation" [4,3]. The pure literal elimination is to assign $l = 1$ for any *pure literal*, a literal l such that $\bar{l}$ never occurs in F. The unit clause propagation is to assign $l = 1$ if F has a clause consisting only this literal l. Another standard rule for simplifying formulas is "resolution". For any literal, let C and C' be clauses such that $l \in C$ and $\bar{l} \in C'$ [1]. Then, we call $(C \cup C')\backslash\{l,\bar{l}\}$ the *resolvent* by l of C' and C. Given a formula F and a literal l in F, the resolution on l is the following procedure: (1) add to F all resolvents by l, and (2) eliminate from F all clauses containing l or $\bar{l}$. Note that all these three operations do not change the satisfiability of a given formula F; that is, letting F' be a formula obtained by one of these operation to F, we have F is satisfiable iff F' is satisfiable [4].

The pure literal elimination and unit clause propagation always decrease the number of variables as well as the formula size, i.e., the number of clauses of a formula. On the other hand, the resolution does not necessarily decreases formula size while the number of variables does decrease. There are, however, we can guarantee some formula size reduction; when a resolution is made on some 1-literal, then the number of clauses gets decreased. Such a resolution is called *1-literal resolution.*

Hirsch's algorithm makes use of one more simplification rule, which is based on the *black-and-white literal principle* [?]. For a given formula F, suppose that all $(2,3)$-literals appear with some $(3,2)$-literals in F's clauses; that is, every clause containing a $(2,3)$-literal also has some $(3,2)$-literal. Then we can simply assign `false` to all $(2,3)$-literals, which does not change the satisfiability of F. This is because by this assignment, any clause containing $(2,3)$-literal is satisfied by some co-existing $(3,2)$-literal that is the negation of some $(2,3)$-literal and hence is assigned `true`. We call this simplification *$(2,3)$-literal elimination by the black-and-white literal principle.* Though the case where this simplification is applicable is rare, it removes one special case, which helps us to design a splitting algorithm.

2.2 Analysis of the Running Time of Splitting Algorithms

An execution of a splitting algorithm can be considered as a *branching tree*, whose nodes are labelled with formulas. That is, given a formula F and a variable x

[1] In the standard definition of resolution, it is also required that l is the only literal such that $l \in C$ and $\bar{l} \in C'$. But here we may remove this restriction.

which an execution of the splitting algorithm is split on. Then, in the branching tree, a node labelled with F has two children labelled with $F|_{x=1}$ and $F|_{x=0}$, respectively. We abuse a formula as a node labelled with the formula. Note that the running time of a splitting algorithm is a polynomial of input length times the number of leaves of the branching tree. Let F be a formula labelling a node of a branching tree, and $F_1, \cdots, F_s$ be its children. A *branching vector* of a node is an s-tuple $(t_1, \cdots, t_s)$, where t_i is a positive number bounded by $|F| - |F_i|$. The *characteristic polynomial* of a branching vector $\boldsymbol{t}$ is defined by $h_{\boldsymbol{t}}(x) = 1 - \sum_{i=1}^{s} x^{-t_i}$. Then, the equation $h_{\boldsymbol{t}} = 0$ has exactly one positive root, which means that it is the largest root of the equation. We denote this root by $\tau(\boldsymbol{t})$, and call it the *branching number* at the node of F. The branching number of a branching tree is the largest branching numbers among all its nodes, denoted by $\tau_{\max}$. The following lemma proved by Kullmann and Luckhardt allows us to estimate the number of leaves in a branching tree. (See [5] for the details.)

Lemma 1. Let $\tau_{\max}$ be the branching number of a branching tree representing an execution of a splitting algorithm which has taken as input a CNF formula F with m clauses. Then the number of leaves in the branching tree does not exceed $(\tau_{\max})^m$.

2.3 Hirsch's Algorithm

We now review Hirsch's algorithm [5], which we denote HIRSCH($\cdot$) in this paper. Hirsch's algorithm (see below for the description[2]) makes two types of splits. For both splits, split literals are chosen among all possible ones so that a branching number satisfies some specified bounds[3]. For each splitting (i.e., after fixing split literal value(s)), the simplification explained above is made on formulas. For any formula F, let REDUCE(F) be a formula obtained by applying to F one of the following operations until no further simplification is made: pure literal elimination, unit clause propagation, resolution such that the number of clauses doesn't increase, and $(2,3)$-literal elimination by the black-and-white literal principle.

Function HIRSCH(F)

If $F = \emptyset$ (meaning F is satisfiable), then return `true`.
If $\emptyset \in F$ (meaning F is unsatisfiable), then return `false`.

(S1) Splitting into two sub-problems.
For any literal l in F, consider the following split.

$$F_1 = \text{REDUCE}(F|_{l=1}) \quad \text{and} \quad F_0 = \text{REDUCE}(F|_{l=0}).$$

[2] For our later explanation, we state a description slightly different from the original one; but it is not so hard to check that they are equivalent.

[3] Intuitively, splitting is better if more clauses are removed with smaller branching variables, and the optimal one should be chosen. But for our target upper bound, we only have to choose one satisfying the specified bounds.

If there exists some literal l with branching number $\tau(|F|-|F_1|, |F|-|F_0|) \leq \tau(3,4)$, then execute this split, i.e., call HIRSCH(F_1) and HIRSCH(F_0) and return HIRSCH(F_1) $\vee$ HIRSCH(F_0).

(S2) (If the condition of (S1) fails, then) Splitting into four sub-problems. For any literal l in F, and any literals l' and l'' in $F|_{l=1}$ and $F|_{l=0}$ respectively, consider the following split.

$$F_{11} = \mathrm{REDUCE}(F|_{l=1,l'=1}), \quad F_{10} = \mathrm{REDUCE}(F|_{l=1,l'=0}), \quad \text{and}$$
$$F_{01} = \mathrm{REDUCE}(F|_{l=0,l''=1}), \quad F_{01} = \mathrm{REDUCE}(F|_{l=0,l''=0}).$$

Then, as it is shown in [5], there must be some split satisfying $\tau(|F|-|F_{11}|, |F|-|F_{10}|, |F|-|F_{01}|, |F|-|F_{00}|) \leq \tau(6,7,6,7)$. Choose one of such splits, call HIRSCH(F_{11}), HIRSCH(F_{10}), HIRSCH(F_{01}), and HIRSCH(F_{00}), and return `true` iff one of these calls yields `true`.

Note that $\tau(6,7,6,7) < \tau(3,4)$, and in [5], it is shown that branching number $\leq \tau(6,7,6,7)$ can be achieved at each split in the recursive execution HIRSCH(F) for any formula F. This proves the following bound.

Theorem 1. (Hirsch [5]) Given a CNF formula F with m clauses as input. The algorithm HIRSCH(F) decides whether F is satisfiable in time $\tau(6,7,6,7)^m = 2^{0.30897m} \approx 1.239^m$.

3 Improving on Hirsch's Algorithm

In this section, we give an algorithm and prove that it runs in time 1.234^m. Our algorithm is almost the same as Hirsch's, with some very minor modification on the function HIRSCH($\cdot$). What is important is that by careful analysis we prove that this almost the same algorithm indeed achieves the desired upper bound. Let us define some terminologies to give an overview of our analysis.

Definition 1. Let F be a CNF formula, and let l and l' be literals of F such that a clause of F contains l and l' both. We say that l and l' are *coincident* (alternatively say that l is *coincident* with l') if there is another clause in F containing l and l' both.

Definition 2. Let F be a CNF formula. We say that F is *normal* if there are only $(3^+, 3^-)$-literals in F, and there is no pair of coincident literals in F.

Consider any recursive execution of HIRSCH($\cdot$). Suppose that a formula F input to HIRSCH(F) consists of only $(3,3)$-literals. If there exists a pair (l, l') of coincident literals in F, we can proceed into step (S1) (not into (S2)), where we split F on l and resolve $F|_{l=1}$ on l' into a formula with size one fewer. Thus, the branching number at F is $\tau(3,4)$. Otherwise, i.e., if there exists no pair of coincident literals in F, that is, F is normal with no 2^--literals, then we proceed into step (S2), where we split F on an arbitrary literal l. and then we choose

an appropriate literal for each of $F_1 = F|_{l=1}$ and $F_0 = F|_{l=0}$, which has the branching number $\tau(3,4)$. Since each assignment $l = 1$ and $l = 0$ eliminates three clauses, we can obtain F has the branching number $\tau(6,7,6,7)$. This analysis of the worst case is exactly Hirsch's, and the recursion tree corresponding to the analysis is of depth two. See the tree enclosed with the dotted line in Fig. 3. We will show that in the worst case, for each root node of F_1 and F_0 there exists at least one path from the root to a leaf such that every node of the path has the branching number $\tau(3,4)$. See the recursion tree below in Fig 3. The following lemma guarantees such a bound.

Lemma 2. Given a CNF formula H with h clauses. Suppose that H is normal, and contains at least one 2^--literal. Then, one of the followings is satisfied: (a) H has a 1^--literal, (b) H has one of the following branching numbers: $\tau(4,4)$ and $\tau(3,5)$, and (c) we have sub-formulas H' with $|H'| = h - 3$ and H'' with $|H''| = h - 4$ (meaning H has the branching number $\tau(3,4)$) such that H' and H'' are two children of H and at least one of H' and H'' is normal with a 2^--literal.

This lemma works in our algorithm as follows. Consider the previous formulas F_1 and F_0, which are normal with a 2^--literal. Then, we can apply the lemma to F_1 and F_0, and split each of F_1 and F_0 as (a), (b), or (c) of the lemma. The precise algorithm of the splitting is stated in the proof of the lemma. (We denote our algorithm by HIRSCH$'(\cdot)$.) According to the lemma, if H only has the branching number $\tau(3,4)$, then at least one of the two children has the branching number $\tau(3,4)$. Furthermore, this recursively repeats to a leaf (meaning a trivial formula), or until a node with the branching number $(3,5)$ or $(4,4)$ (or even better) found. We call such a path $(3,4)$-*path.* It is not hard to see that the worst case of possible branching trees is a tree as shown below: a node with the branching number $\tau(3,4)$ continues to grow along the branch where four clauses are eliminated.

We call such a tree structure in Fig. 3 *the worst case branching tree.* In the branching tree, a black node represents a formula F at which the worst case branching tree resume, and a white node represents a formula of an inner node of the worst case branching tree. The number which an edge has in the figure is the number of eliminated clauses in the transformation from a parent node to its child node.

The proof of the lemma

We first assume that there is no 1^--literal in H. (We have excluded (a).) Thus, we only have the following types of literals in H: $(3,3)$-literals, $(2,3)$-literals, $(3,2)$-literals, and $(2,2)$-literals.

We first consider the case that there exists a $(2,3)$-literal in H. Let x be a $(2,3)$-literal. We further consider the following sub-cases:

(1) : x occurs with a $(3,3)$-literal in any clause
(2) : x occurs with a $(2,2^+)$-literal in a 2-clause
(3) : x occurs with only $(2,2^+)$-literals in a 3^+-clause

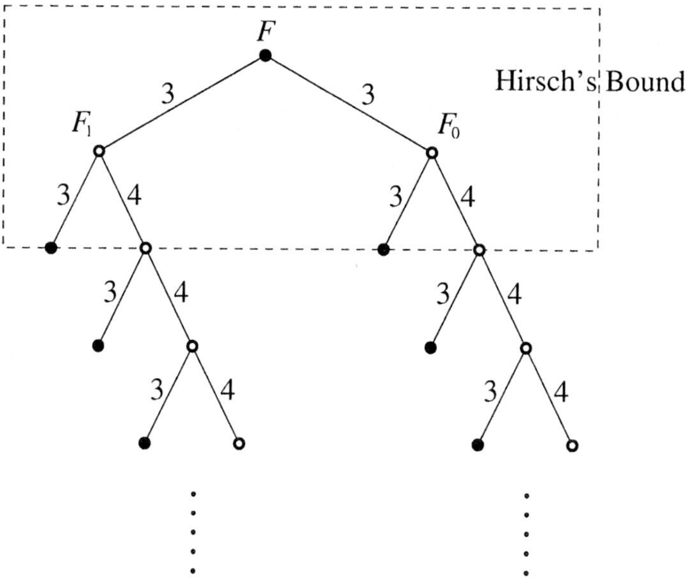

Fig. 1. The branching tree in the worst case

Let y be a $(3,3)$-literal of (1). Branching on y, the assignment $y = 1$ makes x become a 1-literal (since exactly one occurrence of x is eliminated because of no coincidence). Thus, that decreases at least one *additional* clause by the resolution on x, and this branch $y = 1$ totally decreases at least four clauses. Let $H'' = \mathrm{res}_x(H|_{y=1})$ where $\mathrm{res}_w(W)$ is a formula derived from W by the resolution on w. On the other hand, the assignment $y = 0$ eliminates three clauses containing $\bar{y}$, that results in $H' = H|_{y=0}$. It is clear that H' is normal (since H' is obtained from H just by eliminating literals and clauses). Suppose (on the contrary to (c)) that H' (as well as H'') has no 2^--literals, which means that H' consists of only $(3,3)$-literals. On this assumption, clauses of H containing $\bar{y}$ must have contained a 1-literal since any literal but $(1,1)$-literals, $(1,0)$-literals, and $(0,1)$-literals becomes a 2^--literal in H' (because of no coincidence). This is a contradiction to our assumption (which is there is no 1-literal in H). Thus, H' is normal with a 2^--literal. (This case satisfies (c).)

Let y be a $(2,2^+)$-literal of (2), i.e., the 2-clause C is $x \vee y$. Branching on x, the assignment $x = 1$ makes y become a 1-literal. Thus, this branch $x = 1$ totally decreases at least three clauses. (Let $H' = \mathrm{res}_y(H|_{x=1})$.) On the other hand, the assignment $x = 0$ makes C become a unit clause, i.e., $C = (y)$, forcing $y = 1$. Thus, the assignment $x = 0$ and $y = 1$ eliminates at least four clauses (three clauses containing $\bar{x}$ and C itself), that results in $H'' = H|_{x=0,y=1}$. It is clear that H'' is normal. If the other occurrence of y doesn't occur with $\bar{x}$ in H, then the assignment $x = 0$ and $y = 1$ eliminates five clauses ($|H''| = h - 5$), therefore, H has the branching number $\tau(3,5)$. Otherwise (i.e., the other occurrence of y occurs with $\bar{x}$ in H), the assignment $x = 0$ and $y = 1$ eliminates exactly four

clauses. Suppose (again on the contrary to (c)) that H'' (as well as H') has no 2^--literals. On this assumption, clauses of H containing $\bar{x}$ must have contained a 1-literal. This is a contradiction to our assumption. Thus, H'' is normal with a 2^--literal. (This case satisfies (c).)

Let y and z be those $(2, 2^+)$-literals of (3), i.e., the 3^+-clause C is $x \vee y \vee z \vee \cdots$. Branching on x, the assignment $x = 1$ makes y and z become 1-literals. Thus, that decrease at least additional two clauses by each resolution on y and z (because of no coincidence), and this branch $x = 1$ totally decreases at least four clauses. (Let $H'' = \text{res}_{y,z}(H|_{x=1})$.) On the other hand, the assignment $x = 0$ eliminates three clauses containing $\bar{x}$, that results in $H' = H|_{x=0}$. By the same argument as the previous, we can conclude that H'' is normal with a 2^--literal. (This case satisfies (c).)

If there is no such $(2, 3)$-literal x as (1), (2), or (3) above, that means, each $(2, 3)$-literal occurs with some $(3, 2)$-literal, then we can assign true to all of $(3, 2)$-literals with no effect on satisfiability of H. (This is by the black-and-white literals principle.)

We next consider the case that there is no $(2, 3)$-literal in H, which means that there are only $(2, 2)$-literals and $(3, 3)$-literals in H. Let x be a $(2, 2)$-literal. If x occurs with a $(3, 3)$-literal in any clause, it is the same case as (1). Otherwise, that is, every $(2, 2)$-literal occurs with only $(2, 2)$-literals, let C_1, C_2 be clauses containing x, and let D_1, D_2 be clauses containing $\bar{x}$. We consider the following sub-cases:

(4) : for one of C_i, $|C_i| \geq 3$ and for one of D_i, $|D_i| \geq 3$
(5) : $|C_1| = |C_2| = 2$ and $|D_1| = |D_2| = 2$
(6) : either $|C_1| = |C_2| = 2$ or $|D_1| = |D_2| = 2$

For (4), let $|C_1| \geq 3$ and $|D_1| \geq 3$, i.e., $C_1 = (x \vee y \vee z \vee \cdots)$ for some literals y, z, and $D_1 = (\bar{x} \vee y' \vee z' \vee \cdots)$ for some literals y', z'. Branching on x, the assignment $x = 1$ makes y and z become 1-literals. Thus, this branch $x = 1$ totally decreases at least four clauses. The assignment $x = 0$ is the same as $x = 1$. Therefore, H has the branching number $\tau(4, 4)$.

For (5), let $C_1 = (x \vee y)$, $C_2 = (x \vee z)$ for some literals y, z, and $D_1 = (\bar{x} \vee y')$, $D_1 = (\bar{x} \vee z')$ for some literals y', z'. Branching on x, the assignment $x = 1$ makes D_1 and D_2 become unit clauses, i.e., $D_1 = (y')$ and $D_2 = (z')$, forcing $y' = 1$ and $z' = 1$. Thus, this branch $x = 1$ totally decreases at least four clauses. The assignment $x = 0$ is the same as $x = 1$. Therefore, H has the branching number $\tau(4, 4)$.

For (6), let $|C_1| = |C_2| = 2$, i.e., $C_1 = (x \vee y)$ and $C_2 = (x \vee z)$ for some literal y, z. Note that for at least one of D_1 and D_2, the size is at least three. Branching on x, the assignment $x = 1$ makes y and z become 1-literals. That only guarantees to decrease at least one additional clause by the resolution on y or z because the other occurrences of y and z could be in the same clause. Thus, this branch $x = 1$ totally decreases at least three clauses. On the other hand, the assignment $x = 0$ makes C_1 and C_2 become unit clauses, i.e., $C_1 = (y)$ and $C_2 = (z)$, forcing $y = 1$ and $z = 1$. If one of the other occurrences of y and z doesn't occur with $\bar{x}$, this branch $x = 0$ totally decreases at least five clauses.

Then, H has the branching number $\tau(3,5)$. Otherwise, i.e., the other occurrences of y and z both occur with $\bar{x}$, it only guarantees that this branch $x = 0$ totally decreases at least four clauses. However, there should be another (2, 2)-literal w occurring with $\bar{x}$ (since $|D_1| \geq 3$ or $|D_2| \geq 3$). The literal w becomes a 1-literal by $x = 0$ and $y = z = 1$. Thus, the assignment $x = 0$ decreases at least one more additional clause by the resolution on w, and this branch $x = 0$ totally decreases at least five clauses. Therefore, H has the branching number $\tau(3,5)$. ■

Remark 1. We should apply REDUCE(·) where the resolution which doesn't decrease the number of clauses is prohibited, to the transformation between formulas on the (3, 4)-*path*. This is because otherwise, the resolution could spoil the coincidence of formulas while the number of clauses does not decrease by that resolution.

Lemma 3. The number of leaves in the worst case branching tree shown in Fig. 3 whose root is a formula with size m, is at most $O(1.234^m)$.

Proof. Let $T(m)$ be the number of leaves in the sub-tree whose root is a black node with m clauses. Let $S(m)$ be the number of leaves in the sub-tree whose root is a white node with m clauses. Then, we have recurrent equations:

$$\begin{aligned} T(m) &= 2S(m-3) \\ S(m-3) &= T(m-6) + S(m-7) \end{aligned}$$

From two equations above, we obtain $T(m) = O(1.234^m)$. (This comes from the following calculation: We first obtain $S(m) = S(m-4) + 2S(m-6)$ from the equations, and then the characteristic polynomial $1 - 1/x^4 - 2/x^6 = 0$ is derived from the recurrent equation. This polynomial corresponds to the branching number $\tau(4,6,6) \approx 1.234$, which means $S(m) = O(1.234^m)$.) ■

Theorem 2. Given a CNF formula F with m clauses. The upper bound of the running time of HIRSCH$'(F)$ is $\widetilde{O}(1.234^m)$.

4 Conclusion

We have shown the bound 1.234^m for m clauses by improving on the case of formulas consisting of only (3, 3)-clauses. We have so far obtained the branching number $\tau(3,4)$ except for such case. Thus, if we also obtained the same branching number for that case, we could improve our bound to $\tau(3,4)^m \approx 1.221^m$.

Acknowledgement

I thank Prof. Osamu Watanabe for all his support: giving valuable comments, discussing elaborately, and improving this paper. I also thank Prof. Kazuhisa Makino for discussing about what this paper concerns.

References

1. Dantsin E., Hirsch E. A., and Wolpert A., "Algorithms for SAT based on search in Hamming balls", Proc. of the 21st Annual Symposium on Theoretical Aspects of Computer Science (STACS04), 141-151, 2004.
2. Dantsin E. and Wolpert A., "Derandomization of Schuler's algorithms for SAT", Proc. of the 7th International Conference on Theory and Applications of Satisfiability Testing (SAT04), 69-75, 2004.
3. Davis M., Logemann G., and Loveland D., "'A machine program for theorem-proving", Comm. ACM(5), 394-397, 1962.
4. Davis M., and Putnam H., "A computing procedure for quantification theory", J. of ACM(7), 201-215, 1960.
5. Hirsch E. A., "New Worst-Case Upper Bounds for SAT", J. of Automated Reasoning, 24, 397-420, 2000. (It is also in Proc. of the 9th Annual ACM-SIAM Symposium on Discrete Algorithms (SODA98), 521-530, 1998.)
6. Monien B, and Speckenmeyer E., "Solving satisfiability in less than 2^n steps", Discrete Appl. Math.(10), 287-295, 1985.
7. Pudlák P., "Satisfiability - algorithm and logic", Proc. of the 23rd International Symposium on Mathematical Foundations of Computer Science (MFCS98), 129-141, 1998.
8. Schuler R., "An algorithm for the satisfiability problem of formulas in conjunctive normal form", J. of Algorithms, 54 40-44, 2005.

Solving Minimum Weight Exact Satisfiability in Time $O(2^{0.2441n})$

Stefan Porschen

Institut für Informatik, Universität zu Köln, D-50969 Köln, Germany
porschen@informatik.uni-koeln.de

Abstract. We show that the NP-hard optimization problem minimum weight exact satisfiability for a CNF formula over n propositional variables equipped with arbitrary real-valued weights can be solved in time $O(2^{0.2441n})$. To the best of our knowledge, the algorithm presented here is the first handling the weighted XSAT optimization version in non-trivial worst case time.

Keywords: minimum weight exact satisfiability, branching tree, minimum perfect matching, exact algorithm, NP-completeness.

1 Introduction and Notation

Recently, the exact satisfiability problem (XSAT, also called 1-in-SAT) attracted much attention regarding the decision and the counting versions [3,2,4,7]. XSAT gets as input a conjunctive normal form (CNF) formula C over Boolean variables $x \in \{0,1\}$ and asks whether there exists a truth assignment setting exactly one literal in each clause of C to 1. The first breakthrough-result by Monien, Speckenmeyer, and Vornberger [6], regarding the decision variant XSAT, dates back to 1981 and provides an algorithm of $O(2^{0.2441n})$ time deciding XSAT for a CNF formula over n propositional variables. Until 2003 this bound has been the best known, then, based on the techniques in [6], it has been slightly improved to $O(2^{0.2325n})$ by Byskov et al. in [2]. Also Dahllöf et al. in 2004 [4], presented an XSAT decision algorithm using a different methodology but having worst case time $O(2^{0.2519n})$.

In this paper we address the optimization problem minimum weight XSAT (MINW-XSAT). Given a CNF formula whose variables are equipped with arbitrary real-valued weights, MINW-SAT searches for an XSAT model of minimum weight. We show that MINW-XSAT can be solved in worst case time $O(2^{0.2441n})$. Our algorithm essentially uses the branching strategy provided in [6], and in addition it benefits from appropriate simplification steps preserving the minimum weight XSAT status of each intermediate weighted formula.

Let us fix some terminology. A *literal* is a propositional variable $x \in \{0,1\}$ or its negation $\overline{x} := \neg x$ (negated variable). The *complement* of a literal l is $\overline{l}$. A *clause* c is the disjunction of different literals and is represented as a literal set. A CNF formula C is a conjunction of different clauses and is represented as a clause set. Throughout we use the term *formula* meaning a clause set as

X. Deng and D. Du (Eds.): ISAAC 2005, LNCS 3827, pp. 654–664, 2005.

defined. For a given formula C, clause c, by $V(C), V(c)$ we denote the set of variables contained in C, c, respectively. Similarly, given a literal l, $V(l)$ denotes the underlying variable. $V_+(C)$ (resp. $V_-(C)$) denotes the set of all variables occuring unnegated (resp. negated) in C. We distinguish between the length $\|C\|$ of a formula C and the number $|C|$ of its clauses. Let CNF denote the set of all formulas and let CNF_+ denote the set of *positive monotone* formulas, i.e., each clause contains only variables, no negated variables. $C \in \text{CNF}_+$ is called a *matching formula* if each $x \in V(C)$ occurs at most twice in C. For $C \in \text{CNF}_+$, a clause $c \in C$ is called a *2-3-clause* if c contains a variable x that occurs at least three times in C and all other variables in c occur at least twice in C. A formula $C \in \text{CNF}_+$ containing a 2-3-clause is called a *2-3-formula.* $C \in \text{CNF}_+$ is called a *1-3-formula* if each clause $c \in C$ that contains a variable occuring at least three times in C also contains a unique variable in C. Observe that an arbitrary formula $C \in \text{CNF}_+$ either is a matching formula, or a 2-3-formula, or a 1-3-formula.

The exact satisfiability problem (XSAT) asks in its *decision* version, whether there is a truth assignment $t : V(C) \to \{0, 1\}$ assigning exactly one literal in each clause of C to 1, such a truth assignment is called XSAT *model* or *x-model.* XSAT is known to be NP-complete [8]. In the *search* version one has to decide whether $C \in$ XSAT and in positive case one has to find a x-model t for C. The empty set also is a formula: $\varnothing \in$ CNF which is exactly satisfiable. However, a formula C containing the empty clause cannot be exactly satisfiable. An optimization variant of XSAT naturally appears when weights are assigned to the variables: Given $C \in$ CNF and $w : V(C) \to \mathbb{R}$, problem MINW-XSAT asks whether $C \in$ XSAT and in the positive case one has to find a *minimum x-model* for C, i.e., a model t of the least weight among all x-models of C. The weight of a x-model t is defined by $w(t) = \sum_{x \in t^{-1}(1)} w(x) = \sum_{x \in V(C)} w(x)t(x)$. Observe that MINW-XSAT is NP-hard.

Organisation of the paper: Section 2 describes the global structure of our branching algorithm, followed by an explanation of the simplification steps in Section 3. In Section 4 the formulas corresponding to the leaves of the branching tree are shown to be polynomial-time solvable. In Section 5 we analyse the branching strategy and show that our algorithm runs in $O(2^{0.2441n})$ worst case time.

2 Structure of the Algorithm

The main method of the algorithm is *branching*, that means, at a given state, we take a variable x of the current formula and obtain two branches by setting it to 0 resp. to 1. So, a binary search tree, the *branching tree* is generated in a depth first search manner. Its root node corresponds to the input formula equipped with the input variable weight function, and all other nodes correspond to those weighted formulas that are calculated by branching at a variable of its parent node formula and simplifying afterwards. The leaf nodes of the branching tree correspond to weighted matching formulas. As we shall see in Section 4, we can

compute a minimum weight XSAT solution in polynomial time for a matching formula. Clearly, each leaf defines a unique path to the root in the branching tree. Suppose that each simplification step performed at every weighted node formula preserves its minimum weight exact satisfiability status in the sense that we can obtain a minimum weight solution for the non-simplified formula if we have a minimum weight XSAT solution for the simplified one. Hence traversing the path bottom up to the root setting variables according to the current path and performing the reverse simplification steps, we obtain at the root, with respect to the current path, a global minimum weight XSAT solution. Doing so for each leaf when expanding the tree, we obviously get a global minimum weight XSAT solution for the input formula. Storing the branching and simplification information of the current path in a stack we are able to obtain the correct inverse transformation in constant time when traversing bottom up along that path to the root node. Thus a global solution, if existing, is found if we have traversed each root-leaf path twice, once top-down and once bottom-up. Next we state our algorithm constructed as a procedure recursively calling itself until the global solution is found, if it exists.

Algorithm MINW-XSAT($C, w, sol, currsol, S$):
Input: $C \in \mathrm{CNF}, w : V(C) \to \mathbb{R}$
Output: Minimum weight XSAT model sol for (C, w), or **nil**

begin
(1) **if** C contains the empty clause **then return nil**
(2) SIMPLIFY($C, w, currsol, S$)
(3) **if** C is a matching formula **then**
(4) MINPERFMATCH($C, w, currsol, S$)
(5) RETURNTOROOT($C, w, currsol, S$)
(6) $S \leftarrow$ empty stack
(7) **if** $sol\_weight > currsol\_weight$ **then** $sol \leftarrow currsol$
end if
(8) **if** C is a 2-3-Formula **then** BRANCHING-MSV($C, w, currsol, S$)
(9) **if** C is a 1-3-Formula **then** BRANCHING($C, w, currsol, S$)
end

Any internal statement in each of the subprocedures of Algorithm MINW-XSAT is followed by a recursive call of itself. The algorithm uses two parameters $currsol, sol$ for storing the current local solution, resp. the current global solution. Initially, i.e., before the first call of MINW-XSAT is performed, both parameters are assumed to be set to **nil**. Moreover the weights of the corresponding solutions, i.e., $currsol\_weight$ and sol_weight are initially assumed to be set to ∞. Further, a stack S is used for storing the history of the path from the root to the current node in the branching tree: For a current node of the branching tree, this history consists of the simplifications made in each predecessor node during Procedure SIMPLIFY (cf. Section 3) and also by the variable assignments determining that branching path. Branching at a node in form $x \leftarrow 0$ and $x \leftarrow 1$ yields two subproblem instances by evaluating the current formula accordingly.

For sequentially treating the resulting subformulas in the recursive procedure, the current stack S is passed to each of them as parameter. Branching operations are executed in Procedures BRANCHING-MSV, line (8), and BRANCHING in line (9) of the algorithm as long as the current formula is no matching formula (cf. Section 5).

If the current formula is a matching formula, then a leaf of the tree is reached (line (4)). During Procedure MINPERFMATCH a minimum weight exact solution is explicitly computed (cf. Section 4). Afterwards in Procedure RETURNTOROOT, the operations stored in the stack are performed on the found local solution inversely and in reversed order yielding a minimum solution with respect to the current branching tree path.

3 Simplifying Transformations

Now we present the simplifying transformations performed on a current formula in Procedure SIMPLIFY of Algorithm MINW-XSAT. These transformations are invertible and preserve the minimum weight XSAT status of weighted formulas in the sense that from a minimum weight XSAT solution of the image formula one can compute in polynomial time a minimum weight XSAT solution of the original formula. This set of transformations ensures that a current formula particularly satisfies the conditions needed for BRANCHING-MSV as given in [6]. Without explicitly mentioning it is understood that all transformations are put on the stack. Before beginning, we state a useful tool. Let $\mathrm{X}(C)$ denote the set of all (total) x-models of C. Similarly, given $w : V(C) \to \mathbb{R}$, let $\mathrm{X}_{\min}(C, w) \subseteq \mathrm{X}(C)$ denote the set of all minimum weight x-models of C with respect to w.

Lemma 1. *For $C, C' \in$ CNF and arbitrary real-valued weight functions w, w' defined on $V(C)$ resp. $V(C')$, assume that there exists a bijection*

$$F : \mathrm{X}(C) \ni t \mapsto t' := F(t) \in \mathrm{X}(C')$$

such that $(*)$: $w(t) = w'(t') + \alpha$, where $\alpha \in \mathbb{R}$ is a constant independent of t and t'. Then $F_{\min} := F|\mathrm{X}_{\min}(C, w)$ is a bijection between $\mathrm{X}_{\min}(C, w)$ and $\mathrm{X}_{\min}(C', w')$; and we have $|\mathrm{X}_{\min}(C, w)| = |\mathrm{X}_{\min}(C', w')|$.

Proof. Let $t \in \mathrm{X}_{\min}(C, w)$ and assume that $t' := F_{\min}(t) \notin \mathrm{X}_{\min}(C', w')$. Then there is a model $t'_0 \in \mathrm{X}(C')$ with $w'(t'_0) < w'(t')$. Let $t_0 := F^{-1}(t'_0)$ be the corresponding x-model of C. Applying $(*)$ twice we obtain $w(t_0) = w'(t'_0) + \alpha < w'(t') + \alpha = w(t)$, contradicting the assumption that t is minimum. Hence $F_{\min}(t) \in \mathrm{X}_{\min}(C', w')$ holds for each $t \in \mathrm{X}_{\min}(C, w)$.

Finally, let $t' \in \mathrm{X}_{\min}(C', w')$ and assume $t := F^{-1}(t') \notin \mathrm{X}_{\min}(C, w))$. Then there is a model $t_0 \in \mathrm{X}(C')$ with $w(t_0) < w(t)$. Let $t'_0 := F(t_0)$ be the corresponding x-model of C'. As above, by $(*)$, we derive $w'(t'_0) = w(t_0) - \alpha < w(t) - \alpha = w'(t')$, contradicting the assumption that t' is minimum. Hence $F^{-1}(t') \in \mathrm{X}_{\min}(C, w)$ holds for each $t' \in \mathrm{X}_{\min}(C', w')$. Thus, F^{-1} restricted to $\mathrm{X}_{\min}(C', w')$ equals $F^{-1}_{\min}$ from which the assertion follows. □

Some easy steps:

Step 1: Initially the input formula is duplicate-free. If an intermediate formula contains a clause c repeatedly, all except for one occurences of c can obviously be removed, independent of weights.
Step 2: Similarly, if a clause contains the same literal twice, all except for one can be set to 0, since all have the same weight.
Step 3: If C contains a unit clause, then the literal must be set to true in each XSAT model of the current formula, hence unit clauses can be removed from C.
Step 4: This step is to remove 2-clauses from a current formula. In the following, we call a formula *cp-free* if none of its clauses contains a complemented pair of variables.

Lemma 2. *For $C \in$ CNF cp-free and $w : V(C) \to \mathbb{R}$, let $c \in C$ be a 2-clause. Then there is a pair (C', w'), $w' : V(C') \to \mathbb{R}$, such that C' does not contain c and each $t' \in X_{\min}(C', w')$ uniquely determines an element $t \in X_{\min}(C, w)$.*

Proof. For (C, w) with $C \in$ CNF cp-free and $w : V(C) \to \mathbb{R}$, let $c = \{l_1, l_2\} \in C$ and $V(c) = \{x_1, x_2\}$, where $x_i = V(l_i)$. Case 1: $V(c) \cap V(C - \{c\}) = \varnothing$. Then we set $C' := C - \{c\}$, hence $V(C') = V(C) - V(c)$, and define w' as the restriction $w' := w|V(C')$. Because c can be treated independently, this case is clear. Case 2: $V(c) \cap V(C - \{c\}) \neq \varnothing$. Now, we define C' as the formula that is obtained from C by first removing c and second replacing each occurence of l_2 ($\bar{l}_2$) by $\bar{l}_1$ (l_1). It is easy to see that (**): $C \in$ XSAT iff $C' \in$ XSAT and $V(C') = V(C) - \{x_2\}$. For defining w' we distinguish two cases. We either have (i) $c = \{x_1, \bar{x}_2\}$ (resp. $c = \{\bar{x}_1, x_2\}$), or (ii) $c = \{x_1, x_2\}$ (resp. $c = \{\bar{x}_1, \bar{x}_2\}$).

In subcase (i), obviously holds: (a) each truth assignment t x-satisfying C hence c necessarily fulfills $t(x_1) = t(x_2)$. We define w' as $w'(x_1) := w(x_1) + w(x_2)$ and $w'(x) = w(x)$ for each $x \in V(C') - \{x_1\}$. Define the map $F : X(C) \to X(C')$ as follows: for each $t \in X(C)$, let $F(t) := t' = t|V(C')$, and for each $t' \in X(C')$, we set $F^{-1}(t') := t$ uniquely defined by the extension to $V(C)$ by setting $t(x_2) := t(x_1)$. Clearly, F is well defined because of (**) and is a bijection, because for given t the value $F(t)$ is uniquely determined. Indeed suppose there exists $t_1 \in X(C)$ such that $F(t_1) = F(t)$. Then t, t_1 can be different at $t(x_2)$ only because all other values are the same, especially $t(x_1) = t_1(x_1)$, and by (a) we obtain $t(x_2) = t(x_1) = t_1(x_1) = t_1(x_2)$. The converse direction is obvious.

Now we prove that relation (*) in Lemma 1 holds true for F, from which the proof for subcase (i) follows immediately. So, let $t \in X(C)$, then by $t(x_1) = t(x_2)$:

$$\begin{aligned} w(t) &= [w(x_1) + w(x_2)]t(x_1) + \sum_{x \in V(C) - \{x_1, x_2\}} w(x)t(x) \\ &= w'(x_1)t'(x_1) + \sum_{x \in V(C') - \{x_1\}} w'(x)t'(x) \quad = \quad w'(t') \end{aligned}$$

Subcase (ii) proceeds completely analogously taking into account that each truth assignment t x-satisfying c necessarily fulfills $t(x_1) = 1 - t(x_2)$. □

The next steps are used to recursively obtain a positive monotone formula, i.e., to get rid of negated literals:

Step 5: If a clause contains more than one complemented pairs, then it can never be exactly satisfiable, hence a formula containing such a clause has 0 x-models. However, clauses containing exactly one complemented pair can be removed from the formula such that the minimum x-model sets are in bijection, as stated in the following lemma which can be proven applying Lemma 1:

Lemma 3. *For $C \in$ CNF with weight function $w : V(C) \to \mathbb{R}$, let $c \in C$ contain exactly one complemented pair: $x, \overline{x} \in c$. Let C_c be the formula obtained from C by removing c and assigning all literals to 0 that occur in $c' := c - \{x, \overline{x}\}$ and finally removing all duplicate clauses. Let w_c be the restriction of w to $V(C_c) = V(C) - V(c')$. Then: $|\mathrm{X}_{\min}(C, w)| = |\mathrm{X}_{\min}(C_c, w_c)|$.*

Step 6: The transformation in the next lemma removes literals exclusively occuring negated; its correctness also can easily be proven using Lemma 1.

Lemma 4. *For a cp-free formula $C \in$ CNF with weight function $w : V(C) \to \mathbb{R}$, let $x \in V(C)$ be a variable occuring negated, only, in C. Let C_x be the formula obtained from C by replacing each occurence of $\overline{x}$ by x and let $w_x : V(C) \to \mathbb{R}$ be defined as w except for $w_x(x) := -w(x)$. Then: $|\mathrm{X}_{\min}(C, w)| = |\mathrm{X}_{\min}(C_x, w_x)|$.*

Step 7: Next we state a transformation called *simple resolution* which in a different form is used in [6]. Let $C(l) := \{c \in C : l \in c\}$, for a literal l.

Lemma 5. *Let $C \in$ CNF be a cp-free formula and $w : V(C) \to \mathbb{R}$ be an arbitrary weight function. Let $c_i = \{x\} \cup u, c_j = \{\overline{x}\} \cup v \in C$ where $x \in V(C)$ and u, v are literal sets. Let C_{ij} be the formula obtained from C as follows:*

(1) Replace every clause $c \in C(x)$ by the clause $c - \{x\} \cup v$,
(2) replace every clause $c \in C(\overline{x})$ by the clause $c - \{\overline{x}\} \cup u$,
(3) set all literals in $u \cap v$ to 0,
(4) remove all duplicate clauses from the current clause set.

Let $w_{ij} := V(C_{ij}) \to \mathbb{R}$ be the weight function defined as follows: for each $y \in V(C_{ij}) - V(u \oplus v)$, set $w_{ij}(y) := w(y)$, and
(1') if $V_+(u \oplus v) \cap V_-(u \oplus v) = \{z\}$, then set $\forall y \in V(u \oplus v) - \{z\} : w_{ij}(y) := w(y)$ and

$$w_{ij}(z) := \begin{cases} w(z) + w(x), & \text{if } \overline{z} \in u, z \in v \\ w(z) - w(x), & \text{else} \end{cases}$$

(2') if $V_+(u \oplus v) \cap V_-(u \oplus v) = \varnothing$, then set $\forall y \in V(v - u) : w_{ij}(y) := w(y)$ and $\forall y \in V_+(u - v) : w_{ij}(y) := w(y) - w(x)$ and $\forall y \in V_-(u - v) : w_{ij}(y) := w(y) + w(x)$.
Then we have $V(C_{ij}) = V(C) - \{x\} - V(u \cap v)$, $|C_{ij}| \leq |C| - 1$ and: $|\mathrm{X}_{\min}(C, w)| = |\mathrm{X}_{\min}(C_{ij}, w_{ij})|$.

Proof. Since C is assumed to be cp-free and clauses are duplicate-free, neither u nor v can contain x or $\overline{x}$. It follows that, because of (1), (2), (3), c_i and c_j are transformed into the same clause $u \oplus v$ (denoting the symmetric difference),

hence x disappears from the variable set and, because of (4), we also have $|C_{ij}| \leq |C| - 1$. Hence, we obtain $V(C_{ij}) = V(C) - \{x\} - V(u \cap v)$, because no other variable can be removed during step (4).

Consider the map $F : \mathrm{X}(C) \ni t \mapsto F(t) := t|V(C_{ij}) \in \mathrm{X}(C_{ij})$ where $t := F^{-1}(t')$ is defined as the extension of t' to $V(C)$ by setting all literals in $u \cap v$ to 0 and $t(x) = 0$, if t' sets *all* literals in v to 0; and $t(x) = 1$ otherwise. Obviously, both F and F^{-1} are one-to-one and F, in fact, is a bijection of x-model spaces [7].

It remains to verify that F above satisfies relation $(*)$ of Lemma 1 w.r.t. the weight functions w and w_{ij}. To that end, assume $C \in$ XSAT and let $t \in \mathrm{X}(C)$, $t' := F(t)$. Because $\forall y \in V(C_{ij}) - V(u \oplus v) : w_{ij}(y) = w(y)$ and $\forall y \in V(C_{ij}) : t(y) = t'(y)$, we obtain from $w(t) = \sum_{y \in V(C)} w(y)t(y)$

$$w(t) = w(x)t(x) + \sum_{y \in V(u \oplus v)} w(y)t'(y) + \underbrace{\sum_{y \in V(C_{ij}) - V(u \oplus v)} w_{ij}(y)t'(y)}_{=:\hat{w}_{ij}(t')} + \alpha_1$$

where $\alpha_1 := \sum_{y \in V_-(u \cap v)} w(y) \in \mathbb{R}$ is a constant.

Clearly, any x-model of C can assign exactly one literal in $u \oplus v$ to 1, independent of the truth value of x. Hence, if $u \oplus v$ contains more than one complemented pair then C and C_{ij} cannot be exactly satisfiable. Thus, it remains to distinguish the two cases $|V_+(u \oplus v) \cap V_-(u \oplus v)| = 1$ or 0. In the first case, assume that z is the only variable in the intersection, then w_{ij} is uniquely defined by (1'), since C is assumed to be cp-free. If (a) $z \in u$ and $\overline{z} \in v$, then the truth values of z and x must be related as $t(x) = 1 - t(z) = 1 - t'(z)$ iff $C \in$ XSAT. Due to (1') and $\forall y \in V(u \oplus v) - \{z\} : w(y)t(y) = w_{ij}(y)t'(y)$ we have

$$w(t) = \alpha_1 + w(x) + w_{ij}(z)t'(z) + \hat{w}_{ij}(t') + \sum_{y \in V(u \oplus v) - \{z\}} w_{ij}(y)t'(y)$$

which means $w(t) = w_{ij}(t') + \alpha$, hence $(*)$ with $\alpha := \alpha_1 + w(x) \in \mathbb{R}$. Subcase (b), i.e., $z \in v$ and $\overline{z} \in u$ is equivalent to $t(x) = t(z) = t'(z)$ and by similar calculations one obtains relation $(*)$ where $\alpha = \alpha_1$. In the remaining case $V_+(u \oplus v) \cap V_-(u \oplus v) = \emptyset$, defining $w_{uv} := \sum_{y \in V(u \oplus v)} w(y)t'(y)$, first observe that

$$\begin{aligned} w_{uv} &= \sum_{y \in V(v \quad u)} w_{ij}(y)t'(y) + \sum_{y \in V_-(u-v)} w(y)t'(y) + \sum_{y \in V_+(u-v)} w(y)t'(y) \\ &= \sum_{y \in V(u \oplus v)} w_{ij}(y)t'(y) + w(x) \left[\sum_{y \in V_+(u-v)} t'(y) - \sum_{y \in V_-(u-v)} t'(y) \right] \end{aligned}$$

thus $w(t) = w(x)t(x) + w_{ij}(t') + w(x) \left[\sum_{y \in V_+(u-v)} t'(y) - \sum_{y \in V_-(u-v)} t'(y) \right] + \alpha_1$. Now, defining $|V_-(u - v)| =: p = const$, first assume $t(x) = 0$, then exactly one literal in $u - v$ must be set to 1 by t all other literals in $u \oplus v$ are set to 0. If the literal set to 1 belongs to a variable in $V_+(u - v)$ we have $w(t) =$

$w_{ij}(t') + \alpha_1 + w(x)[1 - p]$ and if it belongs to a variable in $V_-(u - v)$ then $w(t) = w_{ij}(t') + \alpha_1 + w(x)[-(p-1)]$ holds. Finally, for $t(x) = 1$ we have that all literals in $u - v$ are set to 0, hence $w(t) = w(x) + w_{ij}(t') - w(x)p + \alpha_1$. Thus we obtain relation (*) $w(t) = w_{ij}(t') + \alpha$ with $\alpha = \alpha_1 + w(x)[1 - p] \in \mathbb{R}$ in each subcase, completing the proof. □

Finally, we have two simplification steps regarding certain monotone clauses:

Step 8: Let $C \in \text{CNF}_+$ contain two clauses c, c' such that $c \subset c'$, then defining C' by setting all variables in $c' - c$ to 0 obviously yields an XSAT-equivalent formula. Clearly each XSAT model of C uniquely determines an XSAT model of C' and vice versa. Restricting the corresponding bijection between X(C) and X(C') to the minimum XSAT model spaces obviouly satisfies (*) of Lemma 1.
Step 9: Let $C \in \text{CNF}_+$ contain $c = u \cup x, c' = u \cup v$ where $x \in V(C)$ is a variable not contained in the subclause v. Then defining C' as the formula obtained from C by setting $x \leftarrow v$, obviously yields an XSAT equivalent formula C' with $V(C') = V(C) - \{x\}$. Each x-model of C determines a unique x-model of C' by restricting to all variables except for x. Vice versa, $t' \in$ X(C') yields a unique $t \in$ X(C) since the extension to x is determined by the assignments of t' to the variables $y \in v$: if and only if exactly one of them is set to 1, we must have $t(x) = 1$. Moreover, let w' be defined as w on $V(C') - v$ and set $w'(y) := w(y) - w(x)$, for all $y \in v$. As in the proof of Lemma 5, we obtain by Lemma 1 that the transformation $(C, w) \rightarrow (C', w')$ is bijective and preserves the minimum weight XSAT status.

We are finished by easily verifying that all steps presented in this section can be performed in polynomial time.

4 Treating Matching Formulas

This section describes how a minimum weight XSAT solution is computed in polynomial time for a matching formula $C \in \text{CNF}_+$ in Procedure MINPERF-MATCH of Algorithm MINW-XSAT. Actually this is implemented by an appropriate reduction to the minimum weight perfect matching (MIN-PM) problem for an edge weighted graph $G = (V, E)$. MIN-PM asks for a perfect matching of minimum weight. Recall that a perfect matching is a subset $P \subseteq E$ of pairwise non-adjacent edges in G such that *every* vertex of G is incident to (exactly) one edge in P. We construct a certain graph corresponding to C, called the *matching graph* G_M, which is a modification of the intersection graph G_C of C. Recall that the intersection graph of a set system has a vertex for each set and two vertices are joined by an edge if their sets have non-empty intersection.

G_M is constructed depending on whether there are unique variables in C or not:

1.) If there exists no clause in C containing a unique variable then $G_M := G_C$. Label each edge of G_C by the variable with the smallest weight occuring in the intersection of the corresponding clauses.
2.) If there exists a clause c in C that contains a unique variable, then construct two copies of G_C and join both copies by introducing an additional edge between

each two vertices in either copy that contain at least one unique variable (both vertices clearly correspond to one and the same clause). Label each additional edge by the (unique) variable of the smallest weight in that clause.

It is not hard to see that a minimum weight perfect matching in G directly corresponds to a minimum XSAT model of C, where the matching edges exactly define the variables that have to be set to 1. Clearly, the matching graph can be constructed in $O(|C|^2 \cdot |V(C)|)$ time. For $C \in$ CNF, we have $\sum_{c\in C} |c| = ||C|| = \sum_{x\in V(C)} \omega(x)$, where $\omega(x)$ is the number of occurences of x in C regardless whether negated or unnegated. Clearly, a matching formula satisfies $w(x) \leq 2$, for each x, implying $|C| \leq ||C|| \leq 2|V(C)|$. Given $G = (V, E), w : E \to \mathbb{R}$, MIN-PM can be solved in $O(|V|^2 \cdot |E|)$ time [1], thus we obtain:

Proposition 1. *For a matching formula C, and $w : V(C) \to \mathbb{R}$, we can solve minimum weight* XSAT *in $O(n^3)$ time, where $n := |V(C)|$.* □

5 The Branching Operations

The Procedure BRANCHING-MSV in Algorithm MINW-XSAT is, for a 2-3-formula, based upon the techniques provided by Monien, Speckenmeyer, and Vornberger (MSV) in [6]. There, it has been proven that the usual XSAT decision problem for arbitrary $C \in$ CNF with n variables can be solved in time $O(2^{0.2441n})$, when the current formula is simplified in the same manner as ensured by Procedure SIMPLIFY. Recall that iteratively branching at a variable means to search the whole space of feasible XSAT solutions independently of their weights. Hence, we can rely on the result by Monien et al. in [6] if and only if the current formula has the structure that is required in their branching analysis. To that end, the formula must be a 2-3-formula C. In the recursive algorithm proposed in [6] a leaf of the branching tree is reached when the current formula is a 1-3-formula. The decision whether such a formula is in XSAT can be made in polynomial time which then also yields the decision for the input formula, as all transformations on the path from the root are XSAT-equivalent. Unfortunately, in the weighted case 1-3-formulas cannot be treated within polynomial time, here a leaf formula must be a matching formula. So, in addition to Procedure BRANCHING-MSV treating only 2-3-formulas, we need a branching procedure that also treats 1-3-formulas, namely Procedure BRANCHING in Algorithm MINW-XSAT. This procedure takes a variable x that occurs at least three times in the current formula and performes a branching step at x. We claim that the worst case recurrence relations occuring during Procedure BRANCHING, are already covered by those recurrence relations produced by Procedure BRANCHING-MSV in [6]. This set of critical recurrences determined by BRANCHING-MSV in [6] contains, among others, the following:

$$T(n) \leq T(n-4) + T(n-5) + 1, \quad T(n) \leq T(n-6) + T(n-3) + 1$$

Here $T(n)$ denotes the number of recursive calls the underlying algorithm performs on a formula of n variables until a leaf of the branching tree is reached. The

two summands on each right hand side correspond to the subproblems defined by setting a variable to 0 resp. to 1.

So let $C \in \text{CNF}_+$ be a 1-3-formula, then by Procedure SIMPLIFY we know that each clause has length at least three, and by definition each clause containing a variable x occurring ≥ 3 times also contains a unique variable. Clearly, the worst case is given when such a variable occurs exactly three times restricting the number of clauses on which x has an impact. First assume that all clauses containing x have length exactly three:

$$c_1 = x \vee e_1 \vee a_1, \quad c_2 = x \vee e_2 \vee a_2, \quad c_3 = x \vee e_3 \vee a_3$$

where e_i are the pairwise distinct unique variables and a_i $(i = 1, 2, 3)$ are variables some or all of which may coincide. The subproblem defined by $x \leftarrow 1$ forces $e_1 \leftarrow e_2 \leftarrow \cdots \leftarrow a_3 \leftarrow 0$ thus eliminating at least 5 variables. And the subproblem defined by $x \leftarrow 0$ forces $e_i \leftarrow a_i$, for $i = 1, 2, 3$, thus eliminating at least 4 variables, i.e., recursively this case leads to the recurrence $T(n) \leq T(n-4) + T(n-5) + 1$ which is covered by the set above. Now suppose there is at least one clause containing x of length four:

$$c_1 = x \vee e_1 \vee a_1 \vee b_1, \quad c_2 = x \vee e_2 \vee a_2 \vee b_2, \quad c_3 = x \vee e_3 \vee a_3 \vee b_3$$

here the e_i and a_i are as above and the b_i $(i = 1, 2, 3)$ are additional variables some or all of which may coincide resp. at most 2 of which may be the constant 0 (i.e. they are "empty"). Again the subproblem defined by $x \leftarrow 1$ eliminates at least 6 variables, and the subproblem defined by $x \leftarrow 0$ eliminates at least 3 variables, which is ensured by Procedure SIMPLIFY (Steps 8, 9) that is performed in the next recursive call. Hence we obtain the recurrence $T(n-6)+T(n-3)+1$ that also is covered by the set above. Clearly, all other cases perform better, hence we are justified to conclude from [6], that all the branching operations and thus Algorithm MINW-XSAT have a worst case time bounded by $O(2^{0.2441n})$:

Theorem 1. *Algorithm* MINW-XSAT *solves minimum weight* XSAT, *for* $C \in \text{CNF}, w : V(C) \to \mathbb{R}$ *in time* $O(2^{0.2441n})$. □

References

1. D. Applegate and W. Cook, Solving large-scale matching problems, in: D. S. Johnson, C. C. McGeoch (Eds.), Algorithms for Network Flows and Matching Theory, American Mathematical Society, pp. 557-576, 1993.
2. J. M. Byskov, B. Ammitzboll Madsen and B. Skjernaa, New Algorithms for Exact Satisfiability, BRICS, Technical Report RS-03-30, University of Aarhus, 2003.
3. V. Dahllöf, P. Jonsson, An Algorithm for Counting Maximum Weighted Independent Sets and its Applications, in: Proceedings of the 13th ACM-SIAM Symposium on Discrete Algorithms, pp. 292-298, 2002.
4. V. Dahllöf, P. Jonsson, and R. Beigel, Algorithms for four variants of the exact satisfiability problem, Theoretical Comp. Sci. 320 (2004) 373-394.
5. M. R. Garey and D. S. Johnson, Computers and Intractability: A Guide to the Theory of NP-Completeness, W. H. Freeman and Company, San Francisco, 1979.

6. B. Monien, E. Speckenmeyer and O. Vornberger, Upper Bounds for Covering Problems, Methods of Operations Research 43 (1981) 419-431.
7. S. Porschen, On Some Weighted Satisfiability and Graph Problems, in: "P. Vojtas, et al. (Eds.), Proceedings of the 31st Conference on Current Trends in Theory and Practice of Informatics", Lecture Notes in Computer Science, Vol. 3381, pp. 278-287, Springer-Verlag, Berlin, 2005.
8. T. J. Schaefer, The complexity of satisfiability problems, in: Proceedings of the 10th ACM Symposium on Theory of Computing, pp. 216-226, 1978.

Efficient Algorithms for Finding a Longest Common Increasing Subsequence

Wun-Tat Chan[1,*], Yong Zhang[1,3], Stanley P.Y. Fung[2], Deshi Ye[1], and Hong Zhu[3]

[1] Department of Computer Science, University of Hong Kong, Hong Kong
{wtchan, yzhang, yedeshi}@cs.hku.hk
[2] Department of Computer Science, City University of Hong Kong, Hong Kong
pyfung@cityu.edu.hk
[3] Department of Computer Science and Engineering, Fudan University, China
hzhu@fudan.edu.cn

Abstract. We study the problem of finding a longest common increasing subsequence (LCIS) of multiple sequences of numbers. The LCIS problem is a fundamental issue in various application areas, including the whole genome alignment and pattern recognition. In this paper we give an efficient algorithm to find the LCIS of two sequences in $O(\min(r \log \ell, n\ell + r) \log\log n + n \log n)$ time where n is the length of each sequence and r is the total number of ordered pairs of positions at which the two sequences match and ℓ is the length of the LCIS. For m sequences where $m \geq 3$, we find the LCIS in $O(\min(mr^2, mr \log \ell \log^m r) + mn \log n)$ time where r is the total number of m-tuples of positions at which the m sequences match. The previous results find the LCIS of two sequences in $O(n^2)$ and $O(n\ell \log n)$ time. Our algorithm is faster when r is relatively small, e.g., for $r < \min(n^2/\log\log n, n\ell)$.

1 Introduction

Given m sequences of numbers, the *longest common increasing subsequence (LCIS)* is the longest sequence that is an increasing sequence and also a subsequence of each of the m sequences. The LCIS problem is a fundamental issue in different application areas including the whole genome alignment [3,7] and pattern recognition [2,9]. In the system for aligning multiple genome sequences [3], one basic step is to extract the longest possible set of MUMs (maximal unique matches) that occur in the same order in the genomic sequences. One approach to implement this step is to label each MUM on different sequences by the position it appears in the first sequence. They form multiple sequences of labels. Then we find the MUMs corresponding to the LCIS of these sequences of labels. In matching two images [2], one method is to consider for each image the pixel intensity values and label each pixel by the rank of its intensity value among all pixels. Then the two images form two rank matrices, which can also be considered as two permutations of $\{1, 2, \ldots, n\}$, where n is tho number of pixels in each

* This research was supported by Hong Kong RGC Grant HKU-5172/03E.

X. Deng and D. Du (Eds.): ISAAC 2005, LNCS 3827, pp. 665–674, 2005.

image. The similarity of the two images can be measured by the similarity of the two permutations, which is the length of the LCIS of the two permutations.

The study of the LCIS problem originated from two classical subsequence problems, the *longest common subsequence (LCS)* and the *longest increasing subsequence (LIS)*. The LCS problem for two sequences of n elements can be solved easily in $O(n^2)$ time, by applying a simple dynamic programming technique. However, the only $o(n^2)$ result known so far was that by Masek and Paterson [10] which runs in $O(n^2/\log n)$ time. For some special cases, Hunt and Szymanski [4] gave a faster algorithm that runs in $O((r+n)\log n)$ time, where r is the total number of ordered pairs of positions at which the two sequences match. For finding the LIS of a sequence of n numbers, Schensted [11] and Knuth [6] gave an $O(n \log n)$ time algorithm. If the input sequence is a permutation of $\{1, 2, \ldots, n\}$, both the algorithms from Hunt and Szymanski [4], and Bespamyatnikh and Segal [1] run in $O(n \log\log n)$ time. The LCIS problem has not been studied until recently that Yang, Huang and Chao gave an $O(n^2)$ time algorithm [14] to find the LCIS of two sequences of n numbers. If the length of the LCIS, ℓ, is small, Katriel and Kutz [5] gave a faster algorithm which runs in $O(n\ell \log n)$ time.

Our results: Consider m sequences where each sequence consists of n numbers. Let r be the total number of m-tuples of positions at which the m sequences match. In this paper we give a general approach in solving the LCIS problem for m sequences. For $m = 2$, we give two efficient implementations which run in $O(r \log \ell \log\log n + n \log n)$ and $O((n\ell + r) \log\log n + n \log n)$ time. For $r < \min(n^2/\log\log n, n\ell)$, our algorithm has a better time complexity than the previous algorithms. For $m \geq 3$, we give a straightforward implementation which runs in $O(mr^2 + mn \log n)$ time. In addition, we show that it is possible to improve the running time in some cases, by applying a dynamic data structure for the orthogonal range query problem [13], to $O(mr \log \ell \log^m r + mn \log n)$. Table 1 gives a summary of the results. Similar to the proof of LCS problem of arbitrary number of sequences is NP-complete [8], we can prove that the LCIS problem of arbitrary number of sequences is also NP-complete and the proof will be given in the full paper. Consider our algorithm for general m, it may run in exponential time as r may be as large as n^m. However, for the special case that the number of matches between the m sequences is relatively small, e.g., when the m sequences are permutations of $\{1, 2, \ldots, n\}$, we have $r = n$ and our algorithms run in $O(n \log n \log\log n)$ time for $m = 2$ and $O(\min\{mn^2, mn \log^{m+1} n\})$ time for $m > 2$.

Remark: We can show that the algorithms proposed in this paper can be adopted to solve a similar problem which is to find the longest common non-decreasing subsequence, with the same time complexities. The details will be given in the full paper.

The rest of the paper is organized as follows. Section 2 gives some basic definitions. Sections 3 and 4 present the algorithms to find the LCIS for two sequences and m sequences, respectively.

Table 1. Previous and present results of the LCIS problem for m sequences of n numbers, where ℓ is the length of the LCIS and r is the total number of m-tuples of positions at which the m sequences match

	Previous results	Results of this paper
$m = 2$	$O(n^2)$	$O(r \log \ell \log \log n + n \log n)$
	$O(n\ell \log n)$	$O((n\ell + r) \log \log n + n \log n)$
$m \geq 3$	nil	$O(mr^2 + mn \log n)$
		$O(mr \log \ell \log^m r + mn \log n)$

2 Preliminary

In this section we give some basic notations and assumption which are necessary for further discussion in the paper. Given a sequence $A = a_1 a_2 \ldots a_n$ of numbers, we say that A is an increasing sequence if $a_1 < a_2 < \cdots < a_n$. A sequence $S = s_1 s_2 \ldots s_\ell$ is a subsequence of A if for some integers $1 \leq i_1 < i_2 < \cdots < i_\ell \leq n$, $s_j = a_{i_j}$ for $1 \leq j \leq \ell$. Given m sequences, $A^1, A^2, \ldots, A^m$, a sequence S is a *common increasing subsequence (CIS)* of the m sequences if S is an increasing sequence and S is a subsequence of A^i for all $1 \leq i \leq m$. The *longest common increasing subsequence (LCIS)* of $A^1, A^2, \ldots, A^m$ is the longest sequence among all CIS of $A^1, A^2, \ldots, A^m$.

Let σ be the number of distinct numbers among the m sequences and $\sigma \leq n$. We can see that σ is also an upper bound on the length of the LCIS. Without loss of generality, we can assume that all m sequences are constructed from the alphabet set $\{1, 2, \ldots, \sigma\}$ because we can always map those σ distinct numbers (in ascending order) to $\{1, 2, \ldots, \sigma\}$.

3 LCIS of Two Sequences of Numbers

Let $A = a_1 a_2 \ldots a_n$ and $B = b_1 b_2 \ldots b_n$ be the two input sequences of n numbers. We need some definitions. For $1 \leq i \leq n$, denote A_i the prefix of A with the first i numbers, i.e., $a_1 a_2 \ldots a_i$. Similarly, we have $B_i = b_1 b_2 \ldots b_i$. We say that (i, j) is a *match* of A and B if $a_i = b_j$, and a_i (or b_j) is called the *match value.* We say that a match (i', j') *dominates* another match (i, j) or that (i', j') is a *dominating match* of (i, j) if $a_{i'} < a_i$ and $i' < i$ and $j' < j$. It is a necessary condition for both $a_{i'}$ and a_i (or $b_{j'}$ and b_i) to appear in the LCIS. For every match (i, j), define the *rank* of the match, denoted by $R(i, j)$, to be the length of the LCIS of A_i and B_j such that a_i (and b_j) is the last element of the LCIS. If the length of the LCIS of A_i and B_j corresponding to match (i, j) is k which is greater than 1, then there must be an LCIS of $A_{i'}$ and $B_{j'}$ corresponding to a match (i', j') with length $k - 1$ and $i' < i$, $j' < j$ and $a_{i'} = b_{j'} < a_i$. Therefore, $R(i, j)$ can be defined by the following recurrence equation.

$$R(i,j) = \begin{cases} 0 & \text{if } (i,j) \text{ is not a match,} \\ 1 & \text{if } (i,j) \text{ is a match and thereis no match } (i',j') \text{ that} \\ & \text{dominates } (i,j), \\ 1+\max\{R(i',j') \mid (i',j') \text{ is a match that dominates } (i,j) \\ & \text{otherwise.} \end{cases} \quad (1)$$

It is easy to see that the length of the LCIS of A and B is $\ell = \max\{R(i,j)\}$. First we give a straightforward algorithm to compute ℓ (see Algorithm LCIS() below). Then, based on the same framework we give an efficient implementation that runs in $O(\min(r \log \ell, n\ell + r) \log\log n + n \log n)$ time. In order to locate all the matches, we can first sort A and B individually and identify the matches in a sequential search on the sorted A and B. In this sequential search we can actually list out the matches in ascending order of their values. This process takes $O(n \log n + r)$ time. Then, starting from the match with the smallest values to the match with the largest values, it takes $O(r)$ time for each match to compute its rank by checking all matches for the dominating matches. This straightforward implementation of Algorithm LCIS() which to compute the length of the LCIS takes $O(r^2 + n \log n)$ times. Note that r can be as large as n^2, which implies a time complexity of $O(n^4)$. In addition, to retrieve the LCIS we can record for each match (i,j) its dominating match (i',j') with the largest rank. Therefore, in the subsequent discussion for efficient implementation we will focus on computing the ranks of the matches only.

Algorithm 1 LCIS(): Find the length of the LCIS of two sequences, A and B.

Assume that we have all matches between A and B in ascending order of their values
for each match (i,j), in ascending order of their values **do**
  Find the match (i',j') that dominates (i,j) and with the largest rank
  Set $R(i,j) = R(i',j') + 1$
end for
Output $\max\{R(i,j) \mid (i,j) \text{ is a match}\}$

3.1 Efficient Implementations

In this section we give two efficient implementations of Algorithm LCIS(), which run in $O(r \log \ell \log\log n + n \log n)$ and $O((n\ell + r) \log\log n + n \log n)$ time. By comparing the values of $r \log \ell$ and $n\ell$ before applying which implementation, we actually have an $O(\min(r \log \ell, n\ell + r) \log\log n + n \log n)$ time algorithm.

Both implementations speed up the step for finding the largest rank dominating match of a given match, which are based on a property described in the following. Consider the execution of Algorithm LCIS(). At any time, let M be the set of matches whose ranks have been computed so far. There are at most ℓ (disjoint) partitions of M, namely $M^1, M^2, \ldots, M^\ell$, where M^k contains the matches with rank k. Note that some of the partitions may be empty. Since the ranks of the matches are computed in ascending order of their values, when the

rank of a match is to be computed, the ranks of all dominating matches of this match (if exist) should have been computed and those matches should appear in M. The ℓ partitions of M possess a kind of monotone property that helps locating the largest rank dominating match of a given match efficiently. The property is that if no dominating match of a given match exists in a partition, say M^k, then no partition corresponding to rank larger than k contains dominating match of the given match, and this property is shown in the following lemma.

Lemma 1. *Given a match t, if there is no match in M^k dominating t, there is no match in M^x dominating t for all $x \geq k$.*

Proof. If there is a match t' in M^x dominating t, i.e., t' is with rank x, then there is match t'' dominating t' with rank $x-1$, i.e., t'' in M^{x-1}, which also dominates t. The argument can be extended to have a match in M^k that dominates t, which is a contradiction. □

In view of Lemma 1, the search for the largest rank dominating match of a given match t can be carried out in a binary search on $M^1, M^2, \ldots, M^\ell$. First, we determine if there is a dominating match of t in $M^{\ell/2}$ (assuming that ℓ is a power of 2). If there is one in $M^{\ell/2}$, continue the binary search in $M^{\ell/2}, \ldots, M^\ell$. Otherwise, continue the binary search in $M^1, \ldots, M^{\ell/2-1}$. The search space can be reduced by half each time and the search terminates when there is only one partition left in the search space.

The framework of the efficient implementations is as follows. We initialize the ℓ partitions to empty sets. For each of the matches t in ascending order of their values, we compute the rank of t by finding the largest rank dominating match of t, using the binary search as described above. Suppose the largest rank among the dominating matches is k. The rank of t is then set to $k+1$ and t is inserted to M^{k+1}.

In order to achieve the desired running time, we need an efficient method to determine if a dominating match of a given match exists in a partition, which exploits more properties of the partitions. Suppose that the ranks of all matches of values no more than v have been computed and those matches are then in M. A match (i, j) in a partition is called a *critical match* if there is no match (i', j') of the same partition with $i' \leq i$ and $j' \leq j$. For each partition, we only consider the critical matches, and those non-critical matches can be removed. The following lemma shows that to find in M the dominating match of a given match with value larger than v it is sufficient to consider just the critical matches.

Lemma 2. *Suppose that M contains all matches of values no more than v. Given a match t of value larger than v, for any partition M^k, if there is a match in M^k dominating t, then there is also a critical match in M^k dominating t.*

Proof. Suppose that $t = (i, j)$ and there is a match (i', j') in M^k dominating t but (i', j') is not a critical match. Then $i' < i$ and $j' < j$ and there must be a critical match (i'', j'') in M^k with $i'' \leq i'$ and $j'' \leq j'$. It implies that (i'', j'') dominates t. □

For each partition, to determine if there is a match in the partition that dominates a given match (i, j), it is sufficient to verify if the critical match (i', j') of the partition with the maximum j' that $j' < j$ (or equivalently the maximum i' that $i' < i$) dominates (i, j) or not. This property is shown in the following lemma.

Lemma 3. *Suppose that M contains all matches of values no more than v. Given a match (i, j) of value larger than v, for any partition M^k, if there is a match in M^k dominating (i, j), then the critical match (i', j') in M^k with the maximum j' that $j' < j$ also dominates (i, j).*

Proof. We prove by contradiction. Suppose that there is a match (i^*, j^*) in M^k dominating (i, j) but the critical match (i', j') in M^k with the maximum j' that $j' < j$ does not dominate (i, j). Then we have $i^* < i \leq i'$ and $j^* \leq j' < j$ that implies (i', j') is not critical, which is a contradiction. □

In order to maintain only the critical matches in a partition, before we insert a new match (i, j) of rank k into M^k, we should remove in M^k those matches (i', j') that will become non-critical, i.e., $i \leq i'$ and $j \leq j'$. See Figure 1 for an example. The following lemma shows which particular critical match of the partition we should verify to see if the match will become non-critical or not.

Lemma 4. *Given a match (i, j), if there is any critical match (i', j') in M^k with $i \leq i'$ and $j \leq j'$, then the critical match (i^*, j^*) in M^k with the minimum j^* that $j^* \geq j$ also satisfies the property that $i \leq i^*$ and $j \leq j^*$.*

Proof. We prove by contradiction. Suppose that $i \leq i'$ and $j \leq j'$ and the critical match (i^*, j^*) with the minimum j^* that $j^* \geq j$ has $i^* < i$. We can see from $i^* < i \leq i'$ and $j \leq j^* \leq j'$ that (i', j') is not a critical match, which is a contradiction. □

For each of M^k for $1 \leq k \leq \ell$, we employ the data structure *van Emde Boas (vEB) tree* [12] to store the critical matches in M^k. A vEB tree is a data structure that stores a set of integers, and supports all the operations insert, delete, and search in $O(\log \log n)$ time where n is the largest integer to be inserted in the vEB tree. In addition, the search operation can return for a given number its nearest numbers, i.e., for a given number j, the largest number $j' < j$ and the smallest number $j' \geq j$. In our implementation each critical match (i, j) is indexed by the value j in a vEB tree. We define four operations specific to our problem, in which all of them run in $O(\log \log n)$ time.

- $FindRight(k, (i, j))$ returns the match (i', j') in the vEB tree corresponding to M^k with the minimum j' that $j' \geq j$;
- $FindLeft(k, (i, j))$ returns the match (i', j') in the vEB tree corresponding to M^k with the maximum j' that $j' < j$;
- $Insert(k, (i, j))$ inserts the match (i, j) in the vEB tree corresponding to M^k; and
- $Delete(k, (i, j)$ deletes the match (i, j) from the vEB tree corresponding to M^k.

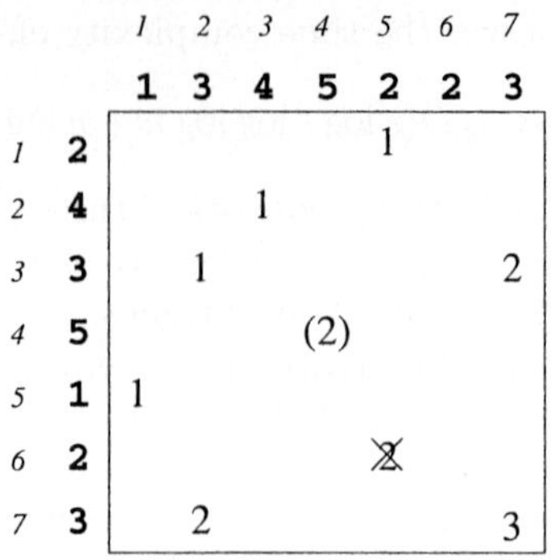

Fig. 1. With two sequences 2435123 and 1345223, the ranks of the critical matches, before and after match $(4,4)$ is considered, are shown. Match $(4,4)$ finds two dominating matches $(2,3)$ and $(3,2)$, which are with the largest rank 1. Then match $(4,4)$ has rank 2 and match $(6,5)$ becomes non-critical.

In the following two sections we use the above the operations to give two implementations of the Algorithm LCIS(), which runs in $O(r \log \ell \log \log n + n \log n)$ and $O((n\ell + r) \log \log n + n \log n)$ time. Then we have the following theorem.

Theorem 1. *Our algorithm takes* $O(\min(r \log \ell, n\ell + r) \log \log n + n \log n)$ *time to find the LCIS of two sequences of numbers.*

Implementation I. In the first implementation of LCIS() we begin with ℓ empty vEB trees, namely T^k for $1 \leq k \leq \ell$, to store the set of critical matches in M^k for $1 \leq k \leq \ell$, respectively. Then all matches between A and B are identified. The matches are arranged in ascending order of their values. For the matches (i, j) with the same value, they are arranged in descending order of j and then i. For each of the matches (i, j) in this order, we find the largest rank dominating match by a binary search on the ℓ vEB trees as described before. To determine if there is a dominating match in T^k, we check if $FindLeft(k, (i, j))$ returns a match (i', j') with $i' < i$ and $j' < j$. After the rank of a match (i, j) is computed, we remove every match (i', j') in $T^{R(i,j)}$ that becomes non-critical, i.e., $i \leq i'$ and $j \leq j'$. Then (i, j) is inserted to $T^{R(i,j)}$. Match (i, j) must be a new critical match in $M^{R(i,j)}$, which can be proved by the following lemma.

Lemma 5. *In Algorithm LCIS-1(), when* (i, j) *is inserted to* $T^{R(i,j)}$, (i, j) *must be a critical match of* $M^{R(i,j)}$.

Proof. Let $R(i, j) = k$. Suppose on the contrary that there is a match (i', j') in T^k with $i' \leq i$ and $j' \leq j$. If the value of (i', j') is less than that of (i, j), then the rank of (i, j) should be at least $k + 1$, which is a contradiction. If the value of (i', j') equals the value of (i, j), because of the order we consider the matches with the same value, either $j' > j$ or $i' > i$, which is also a contradiction. Finally, match value of (i', j') cannot be larger than that of (i, j) because we consider the matches in ascending order of their values. □

The following lemma proves the time complexity of LCIS-1().

Lemma 6. *LCIS-1() runs in* $O(r \log \ell \log\log n + n \log n)$ *time.*

Proof. The running time of LCIS-1() consists of three parts. (1) It takes $O(n \log n + r)$ time to sort the two sequences A and B and identify the matches between A and B in ascending order of their values; (2) It takes $O(\log \ell \log\log n)$ time to find the largest rank dominating match of a given match in the binary search because the function $FindLeft$ is called $O(\log \ell)$ times. Hence it takes $O(r \log \ell \log\log n)$ time to find the ranks of all matches; (3) Each match can be inserted and deleted at most once from a vEB tree. Each of these operations takes $O(\log\log n)$ time, and hence it takes $O(r \log\log n)$ time for all matches. Altogether, it takes $O(r \log \ell \log\log n + n \log n)$ time. □

Implementation II. In this section we give an implementation runs in $O((n\ell + r) \log\log n + n \log n)$ time. In Implementation II we do not use the binary search to find the ranks of individual matches. Rather, we use a sequential search to find the ranks of a group of matches. Later we show that at most n sequential searches are needed.

Similar to Implementation I, we begin with ℓ empty vEB trees, T^k for $1 \leq k \leq \ell$, to store the set of critical matches in M^k for $1 \leq k \leq \ell$, respectively. All matches between A and B are identified and arranged in ascending order of their values, so that we can find the ranks of the matches in this order. For the set of matches with the same value, say H, we divide H into groups such that a group consists of the matches (i, j) in H with the same value of j. Suppose a group consists of the matches $(i_1, j), (i_2, j), \ldots, (i_x, j)$ where $i_1 < i_2 < \cdots < i_x$. For any two of these matches (i_a, j) and (i_b, j) with $a < b$, we have $R(i_a, j) \leq R(i_b, j)$ because any dominating match of (i_a, j) is also a dominating match of (i_b, j). Therefore, we can apply a sequential search on T^1 to T^ℓ to determine the ranks of all the matches $(i_1, j), (i_2, j), \ldots, (i_x, j)$. Similarly, a different sequential search can be applied to each of all the other groups of H. After the ranks of the matches with the same value are computed, those matches are inserted to the corresponding vEB trees.

The following lemma proves the time complexity of LCIS-2().

Lemma 7. *LCIS-2() runs in* $O((n\ell + r) \log\log n + n \log n)$ *time.*

Proof. The running time of LCIS-2() consists of three parts. Two of them are similar to that of Implementation I. (1) It takes $O(n \log n + r)$ time to sort the two sequences and identify all matches in ascending order of their values; (2) Each match can be inserted and deleted at most once to and from a vEB tree, respectively. Hence it takes $O(r \log\log n)$ time for these operations of all matches.

(3) For the remaining part, we analyze the time taken for the sequential searches for the ranks of all matches. Denote the set of matches with the same value v by H_v. Recall that for each H_v, we divide it into groups such that a group consists of the matches (i, j) in H_v with the same value of j. Let d_v

be the number of groups in H_v. We have $\sum_{1\leq v\leq \ell} d_v = n$ because matches (i, j) of different groups should either be with different match values or with different value of j. As we use a sequential search for each group, totally we need n sequential searches. For each sequential search we call $FindLeft$ at most $\ell + g$ times where g is the number of matches in the group. Summing up for all sequential searches, it takes $O((n\ell + r)\log\log n)$ time. Altogether, LCIS-2() takes $O((n\ell + r)\log\log n + n\log n)$ time. □

4 LCIS of m Sequences of Numbers

In this section we give algorithms to find the LCIS of m sequences of numbers, especially for $m \geq 3$. We first generalize the Algorithm LCIS() for two sequences to m sequences. Let the m sequences be $A^1, A^2, \ldots, A^m$ where each of them has n elements. Suppose $A^k = a_1^k a_2^k \ldots a_n^k$. Again we can assume that the alphabet is $\{1, 2, \ldots, \sigma\}$ where $\sigma \leq n$. Let A_i^k denote the prefix of A^k with i elements, i.e., $a_1^k \ldots a_i^k$. We generalize the notations *match* and *dominating match* and *rank* as follows. A match is an m-tuple, $(i_1, \ldots, i_m)$ where $a_{i_1}^1 = a_{i_2}^2 = \ldots = a_{i_m}^m$. A match $(\delta_1, \ldots, \delta_m)$ is called a dominating match of another match $(\theta_1, \ldots, \theta_2)$ if $a_{\delta_1}^1 < a_{\theta_1}^1$ and $\delta_i < \theta_i$ for all $1 \leq i \leq m$. The rank of a match $\Delta = (\delta_1, \ldots, \delta_m)$, denoted by $R(\Delta)$, is the length of the LCIS of $A_{\delta_1}^1, A_{\delta_2}^2, \ldots, A_{\delta_m}^m$ such that $a_{\delta_1}^1$ is the last element of the LCIS. $R(\Delta)$ can be defined similar to Equation 1.

We generalize the Algorithm LCIS() to find the rank of each match. It takes $O(r + mn\log n)$ time to sort the m sequences and identify all the matches. It takes $O(m)$ time to determine if a match dominates another match because a match is an m-tuple which contains m integers. For each match, it takes $O(mr)$ time to compute the rank of the match. Hence it takes $O(mr^2 + mn\log n)$ time to compute the ranks of all matches and also the length of the LCIS of m sequences.

It is possible to improve the running time with the binary search approach, by using data structures for *multidimensional orthogonal range queries* as described below. Given a set of N points in d-dimensional Euclidean space, each associated with a numerical value, design a data structure for the points that supports efficient insertion, deletion, and query of the form: return a point that falls within a hyper-rectangular region $[x_1, y_1] \times [x_2, y_2] \times \ldots \times [x_d, y_d]$. There is a data structure for the above problem that supports each update and query in $O(log^d N)$ time, using space $O(N log^{d-1} N)$ [13].

In our problem, a match is considered as a m-dimensional point. The task of finding a dominating match in a partition is in fact a multidimensional orthogonal range query. To determine if there is a dominating match of a match $(i_1, \ldots, i_m)$ in a partition, we could check if there is any match (or point) of the partition that falls within the hyper-rectangular region $[1, i_1 - 1] \times [1, i_2 - 1] \times \ldots \times [1, i_m - 1]$. Then, adopting the binary search approach in Section 3.1, we can find the LCIS in $O(mr\log\ell\log^m r + mn\log n)$ time. Note that this time complexity is better than $O(mr^2 + mn\log n)$ when r is relatively small, e.g., when r is a polynomial of m and n.

Acknowledgement. We thank Chung-Keung Poon for introducing to us the orthogonal range query problem and the anonymous referees for their detailed reports.

References

1. S. Bespamyatnikh and M. Segal. Enumerating longest increasing subsequences and patience sorting. *Inf. Process. Lett.*, 76(1-2):7–11, 2000.
2. D. N. Bhat. An evolutionary measure for image matching. In *ICPR '98: Proc. of the 14th International Conference on Pattern Recognition-Volume 1*, pages 850–852. IEEE Computer Society, 1998.
3. A. L. Delcher, S. Kasif, R. D. Fleischmann, J. Peterson, O. White, and S. L. Salzberg. Alignment of whole genomes. *Nucleic Acids Res.*, 27:2369–2376, 1999.
4. J. W. Hunt and T. G. Szymanski. A fast algorithm for computing longest common subsequences. *Commun. ACM*, 20(5):350–353, 1977.
5. I. Katriel and M. Kutz. A faster algorithm for computing a longest common increasing subsequence. Research Report MPI-I-2005-1-007, Max-Planck-Institut für Informatik, Stuhlsatzenhausweg 85, 66123 Saarbrücken, Germany, March 2005.
6. D. E. Knuth. *Sorting and Searching, The Art of Computer Programming*, volume 3. Addison-Wesley, Reading, MA, 1973.
7. S. Kurtz, A. Phillippy, A. Delcher, M. Smoot, M. Shumway, C. Antonescu, and S. Salzberg. Versatile and open software for comparing large genomes. *Genome Biology*, 5(2), 2004.
8. D. Maier. The complexity of some problems on subsequences and supersequences. *J. ACM*, 25(2):322–336, 1978.
9. A. Marcolino, V. Ramos, M. Ramalho, and J. R. Caldas Pinto. Line and word matching in old documents. In *Proc. of SIARP'2000 - 5th IberoAmerican Symposium on Pattern Recognition*, pages 123–135, 2000.
10. W. J. Masek and M. Paterson. A faster algorithm computing string edit distances. *J. Comput. Syst. Sci.*, 20(1):18–31, 1980.
11. C. Schensted. Longest increasing and decreasing subsequences. *Canad. J. Math.*, 13:179–191, 1961.
12. P. van Emde Boas. Preserving order in a forest in less than logarithmic time. In *Proc. of the 16th Symposium on Foundations of Computer Science (FOCS)*, pages 75–84, 1975.
13. D. E. Willard and G. S. Lueker. Adding range restriction capability to dynamic data structures. *J. ACM*, 32(3):597–617, 1985.
14. I.-H. Yang, C.-P. Huang, and K.-M. Chao. A fast algorithm for computing a longest common increasing subsequence. *Inf. Process. Lett.*, 93(5):249–253, 2005.

Decision Making Based on Approximate and Smoothed Pareto Curves*

Heiner Ackermann, Alantha Newman, Heiko Röglin, and Berthold Vöcking

Department of Computer Science – RWTH Aachen
{ackermann, alantha, roeglin, voecking}@cs.rwth-aachen.de

Abstract. We consider bicriteria optimization problems and investigate the relationship between two standard approaches to solving them: (i) computing the *Pareto curve* and (ii) the so-called *decision maker's approach* in which both criteria are combined into a single (usually non-linear) objective function. Previous work by Papadimitriou and Yannakakis showed how to efficiently approximate the Pareto curve for problems like SHORTEST PATH, SPANNING TREE, and PERFECT MATCHING. We wish to determine for which classes of combined objective functions the approximate Pareto curve also yields an approximate solution to the decision maker's problem. We show that an FPTAS for the Pareto curve also gives an FPTAS for the decision maker's problem if the combined objective function is growth bounded like a quasi-polynomial function. If these functions, however, show exponential growth then the decision maker's problem is NP-hard to approximate within any factor. In order to bypass these limitations of approximate decision making, we turn our attention to Pareto curves in the probabilistic framework of smoothed analysis. We show that in a smoothed model, we can efficiently generate the (complete and exact) Pareto curve with a small failure probability if there exists an algorithm for generating the Pareto curve whose worst case running time is pseudopolynomial. This way, we can solve the decision maker's problem w.r.t. any non-decreasing objective function for randomly perturbed instances of, e.g., SHORTEST PATH, SPANNING TREE, and PERFECT MATCHING.

1 Introduction

We study *bicriteria optimization* problems, in which there are two criteria, say cost and weight, that we are interested in optimizing. In particular, we consider bicriteria SPANNING TREE, SHORTEST PATH and PERFECT MATCHING problems. For such problems with more than one objective, it is not immediately clear how to define an optimal solution. However, there are two common approaches to bicriteria optimization problems.

The first approach is to generate the set of *Pareto optimal* solutions, also known as the *Pareto set*. A solution S^* is Pareto optimal if there exists no other

* This work was supported in part by the EU within the 6th Framework Programme under contract 001907 (DELIS) and by DFG grant Vo889/2-1.

X. Deng and D. Du (Eds.): ISAAC 2005, LNCS 3827, pp. 675–684, 2005.

solution S that *dominates* S^*, i.e. has cost and weight less or equal to the cost and weight of S^* and at least one inequality is strict. The set of cost/weight combinations of the Pareto optimal solutions is called the *Pareto curve*. Often it is sufficient to know only one solution for each possible cost/weight combination. Thus we assume that the Pareto set is reduced and does not contain two solutions with equal cost and equal weight. Under this assumption there is a one-to-one mapping between the elements in the reduced Pareto set and the points on the Pareto curve.

The second approach is to compute a solution that minimizes some *non-decreasing* function $f : \mathbb{R}_+^2 \to \mathbb{R}_+$. This approach is often used in the field of *decision making*, in which a decision maker is not interested in the whole Pareto set but in a single solution with certain properties. For example, given a graph $G = (V, E)$ with cost $c(e)$ and weight $w(e)$ on each edge, one could be interested in finding an s-t-path P that minimizes the value $(\sum_{e \in P} w(e))^2 + (\sum_{e \in P} c(e))^2$. For a given function $f : \mathbb{R}_+^2 \to \mathbb{R}_+$ and a bicriteria optimization problem Π we will denote by f-Π the problem of minimizing f over all solutions of Π.

Note that these two approaches are actually related: for any *non-decreasing* function f, there is a solution that minimizes f that is also Pareto optimal. A function $f : \mathbb{R}_+^2 \to \mathbb{R}_+$ is non-decreasing if for any $x_1, x_2, y_1, y_2 \in \mathbb{R}_+$ where $x_1 \le x_2$ and $y_1 \le y_2$: $f(x_1, y_1) \le f(x_2, y_2)$. Thus, if for a particular bicriteria optimization problem, we can find the Pareto set efficiently and it has polynomial size, then we can efficiently find a solution that minimizes any given non-decreasing function. It is known, however, that there are instances of SPANNING TREE, SHORTEST PATH and PERFECT MATCHING problems such that even the reduced Pareto set is exponentially large [6]. Moreover, while efficient (i.e. polynomial in the size of the Pareto set) algorithms are known for a few standard bicriteria optimization problems such as the SHORTEST PATH problem [7,17], it is not known how to generate the Pareto set efficiently for other well-studied bicriteria optimization problems such as the SPANNING TREE and the PERFECT MATCHING problem.

There has been a long history of *approximating* the Pareto set starting with the pioneering work of Hansen [7] on the SHORTEST PATH problem. We say a solution S is ε-approximated by another solution S' if $c(S')/c(S) \le 1 + \varepsilon$ and $w(S')/w(S) \le 1 + \varepsilon$ where $c(S)$ and $w(S)$ denote the total cost and weight of a solution S. We say that $\mathcal{P}_\varepsilon$ is an ε-approximation of a Pareto set $\mathcal{P}$ if for any solution $S \in \mathcal{P}$ there is a solution $S' \in \mathcal{P}_\varepsilon$ that ε-approximates it. Papadimitriou and Yannakakis showed that for any Pareto set $\mathcal{P}$, there is an ε-approximation of $\mathcal{P}$ with polynomially many points [13] (w.r.t. the input size and $1/\varepsilon$). Furthermore they gave necessary and sufficient conditions under which there is an FPTAS to generate $\mathcal{P}_\varepsilon$. Vassilvitskii and Yannakakis [16] showed how to compute ε-approximate Pareto curves of almost minimal size.

1.1 Previous Work

There exists a vast body of literature that focuses on f-Π problems. For instance it is well known that, if f is a concave function, an optimal solution of the f-Π

problem can be found on the border of the convex hull of the solutions [9]. For some problems there are algorithms generating this set of solutions. In particular, for the SPANNING TREE Problem it is known that there are only polynomially many solutions on the border of the convex hull [5], and efficient algorithms for enumerating them exist [1]. Thus, there are polynomial-time algorithms for solving f-SPANNING TREE if f is concave. Katoh has described how one can use f-SPANNING TREE problems with concave objective functions to solve many other problems in combinatorial optimization [10]. For instance, a well studied application is the MINIMUM COST RELIABILITY SPANNING TREE Problem, where one is interested in finding a spanning tree minimizing the ratio of cost to reliability. This approach, however, is limited to optimizing the ratio of these two criteria. It is also known how to solve the f-SHORTEST PATH problem for concave objective functions f in polynomial time [8]. Tsaggouris and Zaroliagis [15] investigated the NON-ADDITIVE SHORTEST PATH Problem (NASP), which is to find a path P minimizing $f_c(c(P)) + f_w(w(P))$, for some convex functions f_c and f_w. This problem arises as core problem in different applications, e.g., in the context of computing traffic equilibria. Tsaggouris and Zaroliagis developed exact algorithms with exponential running time using a Lagrangian relaxation and the so called *Extended Hull Algorithm* to solve NASP.

We consider bicriteria optimization problems in the smoothed analysis framework of Spielman and Teng [14]. Spielman and Teng consider a semi-random input model where an adversary specifies an input which is then randomly perturbed. Input instances occurring in practice usually possess a certain structure but usually also have small random influences. Thus, one can hope that semi-random input models are more realistic than worst case and average case input models since the adversary can specify an arbitrary input with a certain structure that is subsequently only slightly perturbed. Since the seminal work of Spielman and Teng explaining the efficiency of the Simplex method in practical applications [14], many other problems have been considered in the framework of smoothed analysis. Of particular relevance to the results in this paper are the results of Beier and Vöcking [3,4]. First, they showed that the expected number of Pareto optimal solutions of any bicriteria optimization problem with two linear objective functions is polynomial if the coefficients in the objective functions are randomly perturbed [3]. Then they gave a complete characterization which linear binary optimization problems have polynomial smoothed complexity, namely they showed that a linear binary optimization problem has polynomial smoothed complexity if and only if there exists an algorithm whose running time is pseudopolynomially bounded in the perturbed coefficients [4]. The only way to apply their framework to multicriteria optimization is by moving all but one of the criteria from the objective function to the constraints.

1.2 Our Results

We study the complexity of the bicriteria optimization problems f-SHORTEST PATH, f-SPANNING TREE and f-PERFECT MATCHING under different classes of functions f. Our study begins with an analysis showing that these problems are

NP-hard even under seemingly harmless objective functions of the form *Minimize* $(\sum_{e\in S} c(e))^a + (\sum_{e\in S} w(e))^b$, where a, b are arbitrary natural numbers with $a \geq 2$ or $b \geq 2$. Thus, we focus on the approximability of these problems. It is not surprising that an FPTAS to approximate the Pareto curve of a problem Π can be transformed into an FPTAS for f-Π for any polynomial function f. Somewhat surprisingly, we show that this transformation works, in fact, for a larger class of functions, namely for *quasi-polynomial* functions or, more generally, for non-decreasing functions whose first derivative is bounded from above like the first derivative of a quasi-polynomial function. Additionally, we show that the restriction to quasi-polynomial growth is crucial.

In order to bypass the limitations of approximate decision making seen above, we turn our attention to Pareto curves in the probabilistic framework of smoothed analysis. We show that in a smoothed model, we can efficiently generate the (complete and exact) Pareto curve of Π with a small failure probability if there exists an algorithm for generating the Pareto curve whose worst case running time is pseudopolynomial (w.r.t. costs and weights). Previously, it was known that the number of Pareto optimal solutions is polynomially bounded if the input numbers are randomly perturbed [3]. This result, however, left open the question of how to generate the set of Pareto-optimal solutions efficiently (except for the SHORTEST PATH problem). The key result in the smoothed analysis presented in this paper is that typically the smallest gap (in cost and weight) between neighboring solutions on the Pareto curve is bounded by $n^{-O(1)}$ from below. This result enables us to generate the complete Pareto curve by taking into account only a logarithmic number of bits of each input number. This way, an algorithm with pseudopolynomial worst-case complexity for generating the Pareto curve can be turned into an algorithm with polynomial smoothed complexity.

It can easily be seen that, for any bicriteria problem Π, a pseudopolynomial algorithm for the exact and single objective version of Π (e.g. an algorithm for answering the question "Does there exist a spanning tree with costs exactly C?") can be turned into an algorithm with pseudopolynomial worst-case complexity for generating the Pareto curve. Therefore, in the smoothed model, there exists a polynomial-time algorithm for enumerating the Pareto curve of Π with small failure probability if there exists a pseudopolynomial algorithm for the exact and single objective version of Π. Furthermore, given the exact Pareto curve for a problem Π, one can solve f-Π exactly. Thus, in our smoothed model, we can, for example, find spanning trees that minimize functions that are hard to approximate within any factor in the worst case.

2 Approximating Bicriteria Optimization Problems

In this section, we consider bicriteria optimization problems in which the goal is to minimize a single objective function that takes two criteria as inputs. We consider functions of the form $f(x, y)$ where x represents the total cost of a solution and y represents the total weight of a solution. In Section 2.1, we present NP-hardness and inapproximability results for the f-SPANNING TREE, f-SHORTEST

PATH, and f-PERFECT MATCHING problems for general classes of functions. In Section 2.2, we show that we can give an FPTAS for any f-Π problem for a large class of quasi-polynomially bounded non-decreasing functions f if there is an FPTAS for generating an ε-approximate Pareto curve for Π. Papadimitriou and Yannakakis showed how to construct such an FPTAS for approximating the Pareto curve of Π given an exact pseudopolynomial algorithm for the problem [13]. For the exact s-t-PATH problem, dynamic programming yields a pseudopolynomial algorithm [17]. For the exact SPANNING TREE problem, Barahona and Pulleyblank gave a pseudopolynomial algorithm [2]. For the exact MATCHING problem, there is a fully polynomial RNC scheme [12,11]. Thus, for any quasi-polynomially bounded non-decreasing objective function, these problems have an FPTAS.

2.1 Some Hardness Results

In this section we present NP-hardness results for the bicriteria f-SPANNING TREE, f-SHORTEST PATH and f-PERFECT MATCHING problems in which the goal is to find a feasible solution S that minimizes an objective function in the form $f(x, y) = x^a + y^b$, where $x = c(S)$, $y = w(S)$, and $a, b \in \mathbb{N}$ are constants with $a \geq 2$ or $b \geq 2$. Note that the NP-hardness of such functions when $a = b$ follows quite directly from a simple reduction from PARTITION. When a and b differ, one can modify this reduction slightly by scaling the weights.

Lemma 1. *Let $f(x, y) = x^a + y^b$ with $a, b \in \mathbb{N}$ and $a \geq 2$ or $b \geq 2$. Then the f-SPANNING TREE, f-SHORTEST PATH, and f-PERFECT MATCHING problems are NP-hard.*

We will now have a closer look at exponential functions $f(x, y) = 2^{x^\delta} + 2^{y^\delta}$ for some $\delta > 0$. In the following, we assume that there is an oracle, which given two solutions S_1 and S_2, decides in constant time whether $f(c(S_1), w(S_1))$ is larger than $f(c(S_2), w(S_2))$ or vice versa. We show that even in this model of computation there is no polynomial time approximation algorithm with polynomial approximation ratio, unless $P = NP$. (The proofs of Lemma 1 and Lemma 2 can be found in a full version of this paper.)

Lemma 2. *Let $f(x, y) = 2^{x^\delta} + 2^{y^\delta}$ with $\delta > 0$. There is no approximation algorithm for the f-SPANNING TREE, f-SHORTEST PATH, and f-PERFECT MATCHING problem with polynomial running time and approximation ratio less than 2^{B^d} for any constant $d > 0$ and $B = \sum_{e \in E} c(e) + w(e)$, unless $P = NP$.*

2.2 An FPTAS for a Large Class of Functions

In this section we present a sufficient condition for the objective function f under which there is an FPTAS for the f-SPANNING TREE, the f-SHORTEST PATH and the f-PERFECT MATCHING problem. In fact, our result is not restricted to these problems but applies to every bicriteria optimization problem Π with an FPTAS for approximating the Pareto curve.

We begin by introducing a restricted class of functions f.

Definition 3. *We call a non-decreasing function* $f : \mathbb{R}_+^2 \to \mathbb{R}_+$ quasi-polynomially bounded *if there exist constants* $c > 0$ *and* $d > 0$ *such that for every* $x, y \in \mathbb{R}_+$

$$\frac{\partial f(x,y)}{\partial x} \cdot \frac{1}{f(x,y)} \leq \frac{c \cdot \ln^d x \cdot \ln^d y}{x}$$

and

$$\frac{\partial f(x,y)}{\partial y} \cdot \frac{1}{f(x,y)} \leq \frac{c \cdot \ln^d x \cdot \ln^d y}{y}.$$

Observe that every non-decreasing polynomial is quasi-polynomially bounded. Furthermore the sum of so-called quasi-polynomial functions of the form $f(x,y) = x^{\mathrm{polylog}(x)} + y^{\mathrm{polylog}(y)}$ is also quasi-polynomially bounded, whereas the sum of exponential functions $f(x,y) = 2^{x^\delta} + 2^{y^\delta}$ is not quasi-polynomially bounded. We are now ready to state our main theorem for this section.

Theorem 4. *There exists an FPTAS for any f-Π problem in which f is monotone and quasi-polynomially bounded if there exists an* FPTAS *for approximating the Pareto curve of Π.*

Proof (Sketch). Our goal is to find a solution for the f-Π problem in question with value no more than $(1+\varepsilon)$ times optimal. The FPTAS for the f-Π problem of relevance is quite simple. It uses the FPTAS for approximating the Pareto curve to generate an ε'-approximate Pareto curve $\mathcal{P}_{\varepsilon'}$ and tests which solution in $\mathcal{P}_{\varepsilon'}$ has the lowest f-value. Recall that the number of points in $\mathcal{P}_{\varepsilon'}$ is polynomial in the size of the input and $1/\varepsilon'$ [13]. The only question to be settled is how small ε' has to be chosen to obtain an ε-approximation for f-Π by this approach. Moreover, we have to show that $1/\varepsilon'$ is polynomially bounded in $1/\varepsilon$ and the input size since then, an ε'-approximate Pareto curve contains only polynomially many solutions and, thus, our approach runs in polynomial time.

Let S^* denote an optimal solution to the f-Π problem. Since f is non-decreasing we can w.l.o.g. assume S^* to be Pareto optimal. We denote by C^* the cost and by W^* the weight of S^*. We know that an ε'-approximate Pareto curve contains a solution S' with cost C' and weight W' such that $C' < (1+\varepsilon')C^*$ and $W' \leq (1+\varepsilon')W^*$. We have to choose $\varepsilon' > 0$ such that $f(C', W') \leq (1+\varepsilon) f(C^*, W^*)$ holds, in fact, we will choose ε' such that

$$f((1+\varepsilon') \cdot C^*, (1+\varepsilon') \cdot W^*) \leq (1+\varepsilon) \cdot f(C^*, W^*). \tag{1}$$

A technical calculation shows that choosing

$$\varepsilon' = \frac{\varepsilon^2}{c 2^{d+4} \cdot \ln^{d+1} C \cdot \ln^{d+1} W},$$

where C denotes sum of all costs $c(e)$ and W denotes the sum of all weights $w(e)$, satisfies (1). Observe that $1/\varepsilon'$ is polynomially bounded in $1/\varepsilon$ and $\ln C^*$ and $\ln W^*$, i.e. the input size. □

Observe that Theorem 4 is almost tight since for every $\delta > 0$ we can construct a function f for which the quotients of the partial derivatives and $f(x, y)$ are lower bounded by $\delta/x^{1-\delta}$ respectively by $\delta/y^{1-\delta}$ and for which the f-Π problem does not posses an FPTAS, namely $f(x, y) = 2^{x^\delta} + 2^{y^\delta}$.

3 Smoothed Analysis of Bicriteria Problems

In the previous section we have shown that f-Π problems are NP-hard even for simple polynomial objective functions, and we have also shown that it is even hard to approximate them for rapidly increasing objective functions, if Π is either the bicriteria SPANNING TREE, SHORTEST PATH or PERFECT MATCHING problem. In this section we will analyze f-Π problems in a probabilistic input model rather than from a worst-case viewpoint. In this model, we show that, for every $p > 0$ for which $1/p$ is polynomial in the input size, the f-Π problem can be solved in polynomial time for *every* non-decreasing objective function with probability $1-p$, if there exists a pseudopolynomial time algorithm for generating the Pareto set of Π. It is known that for the bicriteria graph problems we deal with the expected size of the Pareto set in the considered probabilistic input model is polynomially bounded [3]. Thus, if we had an algorithm for generating the set of Pareto optimal solutions whose running time is bounded polynomially in the input size and the number of Pareto optimal solutions then we could, for any non-decreasing objective function f, devise an algorithm for the f-Π problem that is efficient on semi-random inputs.

For a few problems, e.g. the SHORTEST PATH [17,7] problem, efficient (w.r.t. the input size and the size of the Pareto set) algorithms for generating the Pareto set are known. But it is still unknown whether such an algorithm exists for the SPANNING TREE or the PERFECT MATCHING problem, whereas it is known that there exist for, e.g., the SPANNING TREE and the PERFECT MATCHING problem pseudopolynomial time algorithms (w.r.t. cost and weight) for generating the reduced Pareto set. This follows since the exact versions of the single objective versions of these problems, i.e. the question, "Is there a spanning tree/perfect matching with cost exactly c?", can be solved in pseudopolynomial time (w.r.t to the costs) [2,12,11]. We will show how such pseudopolynomial time algorithms can be turned into algorithms for efficiently generating the Pareto set of semi-random inputs.

3.1 Probabilistic Input Model

Usually, the input model considered in smoothed analysis consists of two stages: First an adversary chooses an input instance then this input is randomly perturbed in the second stage. For the bicriteria graph problems considered in this paper, the input given by the adversary is a graph $G = (V, E, w, c)$ with weights $w : E \to \mathbb{R}_+$ and costs $c : E \to \mathbb{R}_+$ and in the second stage these weights and costs are perturbed by adding independent random variables to them.

We can replace this two-step model by a one-step model where the adversary is only allowed to specify a graph $G = (V, E)$ and, for each edge $e \in E$, two

probability distributions, namely one for $c(e)$ and one for $w(e)$. The costs and weights are then independently drawn according to the given probability distributions. Of course, the adversary is not allowed to specify arbitrary distributions since this would include deterministic inputs as a special case. We place two restrictions upon the distributions concerning the expected value and the maximal density. To be more precise, for each weight and each cost, the adversary is only allowed to specify a distribution which can be described by a piecewise continuous density function $f : \mathbb{R}_+ \rightarrow \mathbb{R}_+$ with expected value at most 1 and maximal density at most ϕ, i.e. $\sup_{x \in \mathbb{R}_+} f(x) = \phi$, for a given $\phi \geq 1$.

Observe that restricting the expected value to be at most 1 is without loss of generality, since we are only interested in the Pareto set which is not affected by scaling weights and costs. The parameter ϕ can be seen as a parameter specifying how close the analysis is to a worst case analysis. The larger ϕ the more concentrated the probability distribution can be. Thus, the larger ϕ, the more influence the adversary has. We will call inputs created by this probabilistic input model *ϕ-perturbed inputs.*

Note that the costs and weights are irrational with probability 1 since they are chosen according to continuous probability distributions. We ignore their contribution to the input length and assume that the bits of these coefficients can be accessed by asking an oracle in time $O(1)$ per bit. Thus, in our case only the representation of the graph $G = (V, E)$ determines the input length. In the following let m denote the number of edges, i.e. $m = |E|$.

We assume that there do not exist two different solutions S and S' with either $w(S) = w(S')$ or $c(S) = c(S')$. We can assume this without loss of generality since in our probabilistic input model two such solutions exist only with probability 0.

3.2 Generating the Pareto Set

In this section we will show how a pseudopolynomial time algorithm $\mathcal{A}$ for generating the Pareto set can be turned into a polynomial time algorithm which succeeds with probability at least $1 - p$ on semi-random inputs for any given $p > 0$ where $1/p$ is polynomial in the input size. In order to apply $\mathcal{A}$ efficiently it is necessary to round the costs and weights, such that they are only polynomially large after the rounding, i.e., such that the length of their representation if only logarithmic. Let $\lfloor c \rfloor_b$ and $\lfloor w \rfloor_b$ denote the costs and weights rounded down to the b-th bit after the decimal point. We denote by $\mathcal{P}$ the Pareto set of the ϕ-perturbed input $G = (V, E, w, c)$ and by $\mathcal{P}_b$ the Pareto set of the rounded ϕ-perturbed input $G = (V, E, \lfloor w \rfloor_b, \lfloor c \rfloor_b)$.

Theorem 5. *For $b = \Theta\left(\log\left(\frac{m\phi}{p}\right)\right)$ it holds that $\mathcal{P} \subseteq \mathcal{P}_b$ with probability at least $1 - p$.*

This means, we can round the coefficients after only a logarithmic number of bits and use the pseudopolynomial time algorithm, which runs on the rounded input in polynomial time, to obtain $\mathcal{P}_b$. With probability at least $1 - p$ the set

$\mathcal{P}_b$ contains all Pareto optimal solutions from $\mathcal{P}$ but it can contain solutions which are not Pareto optimal w.r.t. to w and c. By removing these superfluous solutions we obtain with probability at least $1-p$ the set $\mathcal{P}$.

Corollary 6. *There exists an algorithm for generating the Pareto set of Π on ϕ-perturbed inputs with failure probability at most p and running time* $\text{poly}(m, \phi, 1/p)$ *if there exists a pseudopolynomial time algorithm for generating the reduced Pareto set of Π.*

In this extended abstract we will only try to give intuition why Theorem 5 is valid. Details of the proof can be found in a full version of this paper. From the definition of a Pareto optimal solution, it follows that the optimal solution S of a constrained problem, i.e. the weight-minimal solution among all solutions fulfilling a cost constraint $c(S) \leq t$, is always a Pareto optimal solution. This is because if there were a solution S' that dominates S, then S' would also be a better solution to the constrained problem. We will show that, for every $S \in \mathcal{P}$, with sufficiently large probability we can find a threshold t such that S is the optimal solution to the constrained problem $\min \lfloor w \rfloor_b(S)$ w.r.t. $\lfloor c \rfloor_b(S) \leq t$, i.e. with sufficiently large probability every $S \in \mathcal{P}$ is Pareto optimal w.r.t. the rounded coefficients.

In the proof we will, for an appropriate z, consider z many constrained problems each with weights $\lfloor w \rfloor_b$ and costs $\lfloor c \rfloor_b$. The thresholds we consider are $t_i = i \cdot \varepsilon$, for $i \in [z] := \{1, 2, \ldots, z\}$, for an appropriately chosen ε. By Δ_{min} we will denote the minimal cost difference between two different Pareto optimal solutions, i.e.

$$\Delta_{\text{min}} = \min_{\substack{S_1, S_2 \in \mathcal{P} \\ S_1 \neq S_2}} |c(S_1) - c(S_2)|.$$

If Δ_{min} is larger than ε, then $\mathcal{P}$ consists only of solutions to constrained problems of the form $\min w(T)$, w.r.t. $c(t) \leq t_i$, since, if $\varepsilon < \Delta_{\text{min}}$ we do not miss a Pareto optimal solution by our choice of thresholds. Based on results by Beier and Vöcking [4] we will prove that, for each $i \in [z]$, the solution $S^{(i)}$ to the constrained problem $\min w(S)$ w.r.t. $c(S) \leq t_i$ is the same as the solution $S_b^{(i)}$ to the constrained problem $\min \lfloor w \rfloor_b(S)$ w.r.t. $\lfloor c \rfloor_b(S) \leq i \cdot \varepsilon$ with sufficiently large probability. Thus, if $\varepsilon < \Delta_{\text{min}}$ and $S^{(i)} = S_b^{(i)}$ for all $i \in [z]$, then $\mathcal{P} \subseteq \mathcal{P}_b$.

We do not know how to determine Δ_{min} in polynomial time but we can show a lower bound ε for Δ_{min} that holds with a certain probability. Based on this lower bound, we can appropriately choose ε. We must choose z sufficiently large so that $c(S) \leq z \cdot \varepsilon$ holds with sufficiently high probability for every solution S. Thus, our analysis fails only if one of the following three failure events occurs:

$\mathcal{F}_1$: Δ_{min} is smaller than the chosen ε.
$\mathcal{F}_2$: For one $i \in [z]$ the solution $S^{(i)}$ to $\min w(S)$ w.r.t. $c(S) \leq t_i$ does not equal the solution $S_b^{(i)}$ to $\min \lfloor w \rfloor_b(S)$ w.r.t. $\lfloor c \rfloor_b(S) \leq i \cdot \varepsilon$.
$\mathcal{F}_3$: There exists a solution S with $c(S) > z \cdot \varepsilon$.

For appropriate values of z, ε and b we can show that these events are unlikely, yielding Theorem 5.

References

1. Pankaj K. Agarwal, David Eppstein, Leonidas J. Guibas, and Monika Rauch Henzinger. Parametric and kinetic minimum spanning trees. In *IEEE Symposium on Foundations of Computer Science*, pages 596–605, 1998.
2. F. Barahona and W.R. Pulleyblank. Exact arborescences, matchings and cycles. *Discrete Applied Mathematics*, 16:91–99, 1987.
3. R. Beier and B. Vöcking. Random Knapsack in Expected Polynomial Time. In *Journal of Computer and System Sciences*, volume 69(3), pages 306–329, 2004.
4. R. Beier and B. Vöcking. Typical Properties of Winners and Losers in Discrete Optimization. In *Proc. of the 36th Annual ACM Symposium on Theory of Computing (STOC-2004)*, pages 343–352, 2004.
5. Tamal K. Dey. Improved bounds on planar k-sets and k-levels. In *IEEE Symposium on Foundations of Computer Science*, pages 165–161, 1997.
6. Matthias Ehrgott. *Multicriteria Optimization.* Lecture Notes in Economics and Mathematical Systems Vol. 491. Springer-Verlag, 2000.
7. P. Hansen. Bicriterion path problems. In *Multiple Criteria Decision Making: Theory and Applications*, volume 177 of *Lecture Notes in Economics and Mathematical Systems*, pages 109 – 127, 1980.
8. Mordechai I. Henig. The shortest path problem with two objective functions. *European Journal of Operational Research*, 25(2):281–291, 1986.
9. Reiner Horst and Hoang Tuy. *Global Optimization.* Springer-Verlag, 1990.
10. Naoki Katoh. Bicriteria network optimization problems. *IEICE Transactions Fundamentals of Electronics, Communications and Computer Sciences*, E75-A:321–329, 1992.
11. K. Mulmuley, U.V. Vazirani, and V.V. Vazirani. Matching is as easy as matrix inversion. *Combinatorica*, 7(1):105–114, 1987.
12. C.H. Papadimitriou and M. Yannakakis. The complexity of restricted spanning tree problems. *Journal of the ACM*, 29(2):285–309, 1982.
13. Christos H. Papadimitriou and Mihalis Yannakakis. On the approximability of trade-offs and optimal access of web sources. In *FOCS '00: Proceedings of the 41st Annual Symposium on Foundations of Computer Science*, pages 86–92. IEEE Computer Society, 2000.
14. D. A. Spielman and S.-H. Teng. Smoothed Analysis of Algorithms: Why The Simplex Algorithm Usually Takes Polynomial Time. In *Journal of the ACM*, volume 51(3), pages 385–463, 2004.
15. George Tsaggouris and Christos Zaroliagis. Non-additive shortest paths. In *Algorithms - ESA 2004*, Lecture Notes of Computer Sciene Vol. 3221, pages 822–834, 2004.
16. Sergei Vassilvitskii and Mihalis Yannakakis. Efficiently computing succinct tradeoff curves. In *ICALP*, pages 1201–1213, 2004.
17. Arthur Warburton. Approximation of Pareto optima in multiple-objective, shortest-path problems. *Operations Research*, 35(1):70–78, 1987.

Computing Optimal Solutions for the MIN 3-SET COVERING Problem*

Federico Della Croce[1] and Vangelis Th. Paschos[2]

[1] D.A.I., Politecnico di Torino, Italy
federico.dellacroce@polito.it
[2] LAMSADE, CNRS UMR 7024 and Université Paris-Dauphine, France
paschos@lamsade.dauphine.fr

Abstract. We consider MIN SET COVERING when the subsets are constrained to have maximum cardinality three. We propose an exact algorithm whose worst case complexity is bounded above by $O^*(1.4492^n)$.

1 Introduction and Preliminaries

In MIN SET COVERING, we are given a universe U of elements and a collection $\mathcal{S}$ of (non-empty) subsets of U. The aim is to determine a minimum cardinality sub-collection $\mathcal{S}' \subseteq \mathcal{S}$ which covers U, i.e., $\cup_{S \in \mathcal{S}'} S = U$ (we assume that $\mathcal{S}$ covers U). The frequency f_i of $u_i \in U$ is the number of subsets $S_j \in \mathcal{S}$ in which u_i is contained. The cardinality d_j of $S_j \in \mathcal{S}$ is the number of elements $u_i \in U$ that S_j contains. We say that S_j *hits* S_k if both S_j and S_k contain an element u_i and that S_j *double-hits* S_k if both S_j and S_k contain at least two element u_i, u_l. Finally, we denote by n the size (cardinality) of $\mathcal{S}$ and by m the size of U. In what follows, we restrict ourselves to MIN SET COVERING-instances such that:

1. no element $u_i \in U$ has frequency $f_i = 1$;
2. no set $S_i \in \mathcal{S}$ is subset of another set $S_j \in \mathcal{S}$.
3. no pair of elements u_i, u_j exists such that every subset $S_i \in \mathcal{S}$ containing u_i contains also u_j.

Indeed, if item 1 is not verified, then the set containing u_i belongs to any feasible cover of U. On the other hand, if item 2 is not verified, then S_i can be replaced by S_j in any solution containing S_i and the resulting cover will not be worse than the one containing S_i. Finally, if item 3 is not verified, then element u_j can be ignored as every sub-collection $\mathcal{S}'$ covering u_i will necessarily cover also u_j. So, for any instance of MIN SET COVERING, a preprocessing of data, obviously performed in polynomial time, leads to instances where all items 1, 2 and 3 are verified.

Let $T(\cdot)$ be a super-polynomial and $p(\cdot)$ be a polynomial, both on integers. In what follows, using notations in [1], for an integer n, we express running-time bounds of the form $p(n).T(n)$ as $O^*(T(n))$, the asterisk meaning that we

* Research performed while the first author was in visit at the LAMSADE on a research position funded by the CNRS.

X. Deng and D. Du (Eds.): ISAAC 2005, LNCS 3827, pp. 685–692, 2005.

ignore polynomial factors. We denote by $T(n)$ the worst case time required to exactly solve the MIN SET COVERING problem with n subsets. We recall (see, for instance, [2]) that, if it is possible to bound above $T(n)$ by a recurrence expression of the type $T(n) \leqslant \sum T(n - r_i) + O(p(n))$, we have $\sum T(n - r_i) + O(p(n)) = O^*(\alpha(r_1, r_2, \ldots)^n)$ where $\alpha(r_1, r_2, \ldots)$ is the largest zero of the function $f(x) = 1 - \sum x^{-r_i}$.

There exist to our knowledge few results on worst-case complexity of exact algorithms for MIN SET COVERING or for cardinality-constrained versions of it. Let us note that an exhaustive algorithm computes any solution for MIN SET COVERING in $O(2^n)$. For MIN SET COVERING the most recent non-trivial result is the one of [3] (that has improved the result of [4]) deriving a bound (requiring exponential space) of $O^*(1.2301^{(m+n)})$. We consider here, the most notorious cardinality-constrained version of MIN SET COVERING, the MIN 3-SET COVERING, namely, MIN SET COVERING where $d_j \leqslant 3$ for all $S_j \in \mathcal{S}$. It is well known that MIN 3-SET COVERING is **NP**-hard, while MIN 2-SET COVERING (where any set has cardinality at most 2) is polynomially solvable by matching techniques ([5,6]). Our purpose is to devise an exact (optimal) algorithm with provably improved worst-case complexity for MIN 3-SET COVERING. In what follows, we propose a search tree-based algorithm with running time $O^*(1.4492^n)$ (notice, for instance, that the bound of [3] for $f_i = 2$, $u_i \in U$, and $d_j = 3$, for any $S_j \in S$ corresponds to $O^*(1.2301^{(5n/2)}) \approx O^*(1.6782^n)$).

Consider, the following algorithm, denoted by `SOLVE-3-SET-COVERING`:

- repeat until possible:
  1. for any unassigned subset S_j test if the preprocessing induced by items 1, 2 and 3 reduces the size of the instance;
  2. select for branching the unassigned subset whose branching induces the minimum worst-case complexity (in case of tie select the subset with smallest index).

2 The Result

The following straightforward lemma holds, inducing some useful domination conditions for the solutions of MIN SET COVERING.

Lemma 1. *There exists at least one optimal solution of* MIN SET COVERING *where*

1. *for any subset S_j with $d_j = 2$ containing elements u_i, u_p, if S_j double-hits S_k, then S_j is excluded from S' (in case also $d_k = 2$, then it is immaterial to exclude either S_j or S_k);*
2. *for any subset S_j with $d_j = 2$ containing elements u_i, u_p, if S_j is included in $\mathcal{S}'$, then all subsets S_k hitting S_j are excluded from $\mathcal{S}'$;*
3. *for any subset S_j with $d_j = 3$ containing elements u_i, u_p, u_q, where S_j double-hits another subset S_k with $d_k = 3$ on u_i and u_p, if S_j is included in $\mathcal{S}'$ then S_k must be excluded from $\mathcal{S}'$ and viceversa;*

4. *for any subset S_j with $d_j = 3$ containing elements u_i, u_p, u_q, if S_j is included in S', then either all subsets S_k hitting S_j on element u_i are excluded from S', or all subsets S_k hitting S_j on elements u_p and u_q are excluded from S'.*

Proof. 1. Notice that the configuration implied by item 1 cannot occur thanks to the first hypothesis (item 2 in Section 1) on the form of the MIN SET COVERING-instances dealt.

2. Assume, without loss of generality, that S_j hits S_k on u_i and S_l on u_p. Suppose by contradiction that the optimal solution S' includes S_j and S_k. Then, it cannot include also S_l or else it would not be optimal as a better cover would be obtained by excluding S_j from S'. On the other hand, suppose that S' includes S_j, S_k but does not include S_l. Then, an equivalent optimal solution can be derived by swapping S_j with S_l.

For items 3 and 4, the same kind of analysis as for item 2 holds.

Proposition 1. *Algorithm SOLVE-3-SET-COVERING optimally solves* MIN 3-SET COVERING *within time $O^*(1.4492^n)$.*

Proof. The algorithm either detects by means of item 1 a subset S_j to be immediately included in (excluded from) S' or an element u_i to be ignored (correspondingly reducing the degree of several subsets), or applies a branching on subset S_j, where the following exhaustive relevant branching cases may occur.

1. $d_j = 2$: then no double-hitting occurs to S_j or else, due to Lemma 1, S_j can be excluded from s' without branching. The following subcases occur.
   (a) S_j contains elements u_i, u_k with $f_i = f_k = 2$ where S_j hits S_l on u_i and S_m on u_k. Due to Lemma 1, if S_j is included in S', then both S_l and S_m must be excluded from S'; alternatively, S_j is excluded from S' and, correspondingly, both S_l and S_m must be included in S' to cover elements u_i, u_k. This can be seen as a binary branching where, in both cases, three subsets (S_j, S_l, S_m) are fixed. Then, $T(n) \leqslant 2T(n-3) + O(p(n))$, where the term $T(n-3)$ measures the time for solving the same problem with $n-3$ subsets. Correspondingly, we have $T(n) = O^*(\alpha^n)$, where α is the largest real root of the equation $\alpha^3 = 2$, i.e., $\alpha \approx 1.2599$, implying a time complexity of $O^*(1.2599^n)$.
   (b) S_j contains elements u_i, u_k with $f_i = 2$ and $f_k \geqslant 3$, where S_j hits S_l on u_i and S_m, S_p on u_k. Due to Lemma 1, if S_j is included in S', then S_l, S_m, S_p must be excluded from S'; alternatively, S_j is excluded from S' and, correspondingly, S_l must be included in S' to cover element u_i. This can be seen as a binary branching where either 2 subsets (S_j, S_l), or 4 subsets (S_j, S_l, S_m, S_p) are fixed; hence, $T(n) \leqslant T(n-2) + T(n-4) + O(p(n))$. This results in a time-complexity of $O^*(1.2721^n)$.
   (c) S_j contains elements u_i, u_k with $f_i \geqslant 3$ and $f_k \geqslant 3$ where S_j hits S_l, S_m on u_i and S_p, S_q on u_k. Due to Lemma 1, if S_j is included in S', then S_l, S_m, S_p, S_q must be excluded from S'; alternatively, S_j is excluded from S'. This can be seen as a binary branching where either 1 subset (S_j), or 5 subsets $(S_j, S_l, S_m, S_p, S_q)$ are fixed; hence, $T(n) \leqslant T(n-1) + T(n-5) + O(p(n))$. This results in a complexity of $O^*(1.3248^n)$.

2. $d_j = 3$ (that is, there does not exist $S_k \in S$ such that $d_k = 2$) with S_j double-hitting one or more subsets. Notice that if S_j double-hits S_k on elements u_i, u_l, then $f_i \geqslant 3$ and $f_l \geqslant 3$ due to the preprocessing step 1 of the algorithm. The following exhaustive subcases may occur.
   (a) S_j double-hits at least 3 subsets S_k, S_l, S_m. Due to Lemma 1, if S_j is included in $\mathcal{S}'$ then S_k, S_l, S_m must be excluded from $\mathcal{S}'$; alternatively, S_j is excluded from $\mathcal{S}'$. This can be seen as a binary branching where either 1 subset (S_j) is fixed, or 4 subsets (S_j, S_k, S_l, S_m) are fixed and hence, $T(n) \leqslant T(n-1) + T(n-4) + O(p(n))$. This results in a time-complexity of $O^*(1.3803^n)$.
   (b) S_j double-hits 2 subsets S_k, S_l and hits at least one more subset S_m (we assume that S_j hits S_m on element u_i). Due to Lemma 1, if S_j is included in $\mathcal{S}'$, then S_k, S_l must be excluded from $\mathcal{S}'$ and a further branching on subset S_m with $d_m = 2$ can be applied (as u_i is already covered by S_j) where, in the worst-case, subcase 1c holds; alternatively, S_j is excluded from $\mathcal{S}'$; this can be seen as a binary branching where either 1 subset (S_j) is fixed, or 3 subsets (S_j, S_k, S_l) are fixed and a branching of type 1c on subset S_m with $n' = n-3$ variables holds. Then $T(n) \leqslant T(n-1)+T(n'-1) + T(n'-5) + O(p(n)) = T(n-1) + T(n-4) + T(n-8) + O(p(n))$. This results in a time-complexity of $O^*(1.4271^n)$.
   (c) S_j contains elements u_i, u_k, u_l and double-hits one subset S_k on elements u_i, u_k. The following exhaustive subcases must be considered.
      i. $f_i \geqslant 3$, $f_k \geqslant 3$, $f_l = 2$ with u_i contained at least by S_j, S_k, S_m, u_k contained at least by S_j, S_k, S_p and u_l contained by S_j, S_q. A composite branching can be devised:
         – either S_j and S_q are included in $\mathcal{S}'$ and, by Lemma 1, S_k, S_m, S_p must be excluded from $\mathcal{S}'$,
         – or S_j is included in $\mathcal{S}'$ and S_q is excluded from $\mathcal{S}'$ and, correspondingly, S_k must be excluded from $\mathcal{S}'$,
         – or S_j is excluded from $\mathcal{S}'$ and, correspondingly, S_q must be included in $\mathcal{S}'$.

         Then, $T(n) \leqslant T(n-2)+T(n-3)+T(n-5)+O(p(n))$. This results in a time-complexity of $O^*(1.4292^n)$.
      ii. $f_i = 3$, $f_j = 3$, $f_l \geqslant 3$ with u_i contained by S_j, S_k, S_m, u_k contained by S_j, S_k, S_p and u_l contained at least by S_j, S_q, S_r. A composite branching can be devised:
         – either S_j and S_k are excluded from $\mathcal{S}'$ and S_m, S_p must be included in $\mathcal{S}'$ (to cover u_i and u_k),
         – or S_j is included in $\mathcal{S}'$ and (due to Lemma 1) either S_k, S_q, S_r are excluded from $\mathcal{S}'$ or S_k, S_m, S_p are excluded from $\mathcal{S}'$,
         – or S_k is included in $\mathcal{S}'$ and (due to Lemma 1) either S_j, S_m, S_p are excluded from $\mathcal{S}'$ or S_j, S_λ, S_μ are excluded from $\mathcal{S}'$, where S_λ, S_μ are the subsets hitting S_k on another element (recall $f_k = 3$) u_v.

         This would induce $T(n) \leqslant 5T(n'-4)+O(p(n))$, but in all subcases a consequent branching on an unassigned subset (any of those hitting a subset just included in $\mathcal{S}'$) having (therefore) cardinality 2 holds

where, in the worst case, subcase 1c holds. Then, $T(n) \leqslant 5T(n-5)+5T(n-9)+O(p(n))$. This results in a time-complexity of $O^*(1.4389^n)$.

iii. $f_i = 3$, $f_j \geqslant 4$, $f_l \geqslant 3$, with u_i contained by S_j, S_k, S_m, u_k contained at least by S_j, S_k, S_p, S_q and u_l contained at least by S_j, S_r, S_u. A composite branching can be devised:
   - either S_j and S_k are excluded from $\mathcal{S}'$ and (to cover u_i) S_m must be included in $\mathcal{S}'$,
   - or S_j is included in $\mathcal{S}'$ and (due to Lemma 1) either S_k, S_p, S_q are excluded from $\mathcal{S}'$ or S_k, S_m, S_r, S_u are excluded from $\mathcal{S}'$,
   - or S_k is included in $\mathcal{S}'$ and (due to Lemma 1) either S_j, S_p, S_q are excluded from $\mathcal{S}'$ or $S_j, S_m, S_\lambda, S_\mu$ are excluded from $\mathcal{S}'$, where S_λ, S_μ are the subsets hitting S_k on another element (recall $f_k = 3$) u_v.

   This would induce $T(n) \leqslant T(n-3) + 2T(n-4) + 2T(n-5) + O(p(n))$, but in all subcases a consequent branching on an unassigned subset (any of those hitting a subset just included in $\mathcal{S}'$) having (therefore) cardinality 2 holds where, in the worst case, subcase 1c holds. Then, $T(n) \leqslant T(n-4) + 2T(n-5) + 2T(n-6) + T(n-8) + 2T(n-9) + 2T(n-10) + O(p(n))$. This results in a time-complexity of $O^*(1.4331^n)$.

iv. $f_i \geqslant 4$, $f_j \geqslant 4$, $f_l \geqslant 3$, with u_i contained at least by S_j, S_k, S_m and S_p, u_k contained at least by S_j, S_k, S_q and S_r and u_l contained at least by S_j, S_u and S_v. A composite branching on subset S_j can be devised (due to Lemma 1):
   - S_j is included in $\mathcal{S}'$, S_k, S_m, S_p are excluded from $\mathcal{S}'$ and a further branching on subset S_q can be applied with $d_q = 2$ (as u_k is already covered by S_j),
   - or S_j is included in $\mathcal{S}'$, S_k, S_q, S_r, S_q, S_v are excluded from $\mathcal{S}'$ and a further branching on subset S_m can be applied with $d_m = 2$ (as u_i is already covered by S_j),
   - or S_j is excluded from $\mathcal{S}'$.

   This can be seen as a composite branch where either 1 or 4 or 6 subsets have been included in or excluded from S' that is $T(n) \leqslant T(n-1) + T(n-4) + T(n-6) + O(p(n))$, where however, in the latter two branches a consequent branching on an unassigned subset having cardinality 2 holds where, in the worst case, subcase 1c holds. Then, $T(n) \leqslant T(n-1) + T(n-5) + T(n-7) + T(n-9) + T(n-11 + O(p(n))$. This results in a time-complexity of $O^*(1.4343^n)$.

3. $d_j = 3$ and no double-hitting occurs to S_j (nor to any other subset) that contains elements u_i, u_k, u_l. The following subcases occur.
   (a) $f_i = f_k = f_l = 2$ with u_i contained by S_j, S_k, u_k contained by S_j, S_l and u_l contained by S_j, S_m. A binary branching on S_j can be devised: either S_j is excluded from $\mathcal{S}'$ and then (to cover u_i, u_k, u_l) S_k, S_l, S_m must be included in $\mathcal{S}'$, or S_j is included in $\mathcal{S}'$. This would induce $T(n) \leqslant T(n-1)+T(n-4)+O(p(n))$, but in all subcases, a consequent branching on an unassigned subset (any of those hitting a subset just included in $\mathcal{S}'$)

having (therefore) cardinality 2 holds where, in the worst case, subcase 1c holds. Then, $T(n) \leqslant T(n-2)+T(n-5)+T(n-6)+T(n-9)+O(p(n))$. This results in a time-complexity of $O^*(1.3515^n)$.

(b) $f_i = f_k = 2$, $f_l \geqslant 3$ with u_i contained by S_j, S_k, u_k contained by S_j, S_l and u_l contained at least by S_j, S_m, S_p. A composite branching on S_j can be devised:
   - either S_j is excluded from $\mathcal{S}'$ and then (to cover u_i, u_k) S_k, S_l must be included in $\mathcal{S}'$,
   - or S_j is included in $\mathcal{S}'$ and S_k, S_l are excluded from $\mathcal{S}'$,
   - or S_j is included in $\mathcal{S}'$ and S_m, S_p are excluded from $\mathcal{S}'$.

   This would induce $T(n) \leqslant 3T(n-3) + O(p(n))$, but in all subcases, a consequent branching on an unassigned subset (any of those hitting a subset just included in $\mathcal{S}'$) having (therefore) cardinality 2 holds where, in the worst case, subcase 1c holds. Then, $T(n) \leqslant 3T(n-4) + 3T(n-8) + O(p(n))$. This results in a time-complexity of $O^*(1.3954^n)$.

(c) $f_i = 2$, $f_k \geqslant 3$, $f_l \geqslant 3$, where u_i is contained by S_j and S_k, u_k is contained by S_j, S_l and S_m, and u_l is contained at least by S_j, S_p and S_q. A composite branching on S_j can be devised: either S_j is excluded from $\mathcal{S}'$ and then (to cover u_i) S_k must be included in $\mathcal{S}'$, or S_j is included in $\mathcal{S}'$ and S_k, S_l, S_m are excluded from $\mathcal{S}'$, S_j is included in $\mathcal{S}'$ and S_p, S_q are excluded from $\mathcal{S}'$. This would induce execution time expressed as $T(n) \leqslant T(n-2) + T(n-3) + T(n-4) + O(p(n))$; but, in all subcases, a consequent branching on an unassigned subset (any of those hitting a subset just included in $\mathcal{S}'$) having (therefore) cardinality 2 is possible where, at worst, sub-case 1c holds. In this case, $T(n) \leqslant T(n-3)+T(n-4) + T(n-5) + T(n-7) + T(n-8) + T(n-9) + O(p(n))$. This results in a time-complexity of $O^*(1.4066^n)$.

(d) $f_i = 3$, $f_k \geqslant 3$, $f_l \geqslant 3$ with u_i contained by S_j, S_k, S_l, u_k contained by S_j, S_m, S_p and u_l contained at least by S_j, S_q, S_r. Also, both S_k and S_l have degree 3 and all elements contained by S_k or S_l have frequency at least 3 or else subcase 3c would hold either on subset S_k or on subset S_l. A composite branching can be devised:
   - either S_j is included in $\mathcal{S}'$ and then either S_k, S_l are excluded from $\mathcal{S}'$, or S_m, S_p, S_q and S_r are excluded from $\mathcal{S}'$,
   - or S_j is excluded from $\mathcal{S}'$, S_k is included in $\mathcal{S}'$ and there are at least 5 other subsets hitting S_k and, hence, either two of these subsets are excluded from $\mathcal{S}'$ or three of these subsets are excluded from $\mathcal{S}'$,
   - or S_j, S_k are excluded from $\mathcal{S}'$, S_l is included in $\mathcal{S}'$ (to cover u_i) and there are at least 4 other subsets hitting S_k and, hence, either two of these subsets are excluded from $\mathcal{S}'$, or the other two of these subsets are excluded from $\mathcal{S}'$.

   This would induce $T(n) \leqslant T(n-3) + T(n-4) + 4T(n-5) + O(p(n))$, but in all subcases, a consequent branching on an unassigned subset (any of those hitting a subset just included in $\mathcal{S}'$) having (therefore) cardinality 2 holds where, in the worst case, subcase 1c holds. Then,

$T(n) \leqslant T(n-4)+T(n-5)+4T(n-6)+T(n-8)+T(n-9)+4T(n-10)+O(p(n))$. This results in a time-complexity of $O^*(1.4492^n)$.

(e) $f_i \geqslant 4$, $f_k \geqslant 4$, $f_l \geqslant 4$, u_i is contained by S_j, S_k, S_l, S_m, u_k is contained by S_j, S_p, S_q, S_r and u_l is contained at least by S_j, S_t, S_u, S_v. A composite branching on S_j can be devised:
   - either S_j is excluded from $\mathcal{S}'$,
   - or S_j is included in $\mathcal{S}'$, S_k, S_l, S_m are excluded from $\mathcal{S}'$ and a further branching on subset S_p can be applied with $d_p = 2$ (as u_k is already covered by S_j),
   - or S_j is included in $\mathcal{S}'$, $S_p, S_q, S_r, S_t, S_u, S_w$ are excluded from $\mathcal{S}'$ and a further branching on subset S_m can be applied with $d_m = 2$ (as u_i is already covered by S_j).

   This can be seen as a composite branch where either 1 or 4 or 7 subsets have been included in or excluded from S' that is $T(n) \leqslant T(n-1)+T(n-4)+T(n-7)+O(p(n))$, where however, in the latter two branches a consequent branching on an unassigned subset having cardinality 2 holds where, in the worst case, subcase 1c holds. Then, $T(n) \leqslant T(n-1)+T(n-5)+T(n-8)+T(n-9)+T(n-12)+O(p(n))$. This results in a time-complexity of $O^*(1.4176^n)$.

Putting things together the global worst case complexity is $O^*(1.4492^n)$.

3 Conclusion

We have presented an analysis for the worst-case complexity of a tree-search based algorithm, which, based upon simple domination properties for the optimal solutions of MIN 3-SET COVERING leads to an $O^*(1.4492^n)$ time bound. We notice that a straightforward (improvable) analysis along the lines of the above one, leads to an $O^*(1.1679^n)$ time bound for minimum vertex covering in graphs of maximum degree 3. This bound, even though dominated by the ones in [7,8], $O^*(1.1252^n)$ and $O^*(1.152^n)$, respectively, already dominates the one in [9].

References

1. Wœginger, G.J.: Exact algorithms for NP-hard problems: a survey. In Juenger, M., Reinelt, G., Rinaldi, G., eds.: Combinatorial Optimization - Eureka! You shrink! Volume 2570 of Lecture Notes in Computer Science. Springer-Verlag (2003) 185–207
2. Eppstein, D.: Improved algorithms for 3-coloring, 3-edge-coloring, and constraint satisfaction. In: Proc. Symposium on Discrete Algorithms, SODA. (2001) 329–337
3. Fomin, F., Grandoni, F., Kratsch, D.: Measure and conquer: domination – a case study. Reports in Informatics 294, Department of Informatics, University of Bergen (2005) To appear in the Proceedings of ICALP'05.
4. Grandoni, F.: A note on the complexity of minimum dominating set. J. Discr. Algorithms (2005) To appear.

5. Berge, C.: Graphs and hypergraphs. North Holland, Amsterdam (1973)
6. Garey, M.R., Johnson, D.S.: Computers and intractability. A guide to the theory of NP-completeness. W. H. Freeman, San Francisco (1979)
7. Beigel, R.: Finding maximum independent sets in sparse and general graphs. In: Proc. Symposium on Discrete Algorithms, SODA. (1999) 856–857
8. Chen, J., Liu, L., Jia, W.: Improvement on vertex cover for low-degree graphs. Networks **35** (2000) 253–259
9. Chen, J., Kanj, I., Jia, W.: Vertex cover: further observations and further improvements. J. Algorithms **41** (2001) 280–301

Efficient Algorithms for the Weighted 2-Center Problem in a Cactus Graph

Boaz Ben-Moshe, Binay Bhattacharya*, and Qiaosheng Shi

School of Computing Science, Simon Fraser University,
Burnaby B.C., Canada. V5A 1S6
{benmoshe, binay, qshi1}@cs.sfu.ca

Abstract. In this paper, we provide efficient algorithms for solving the weighted center problems in a cactus graph. In particular, an $O(n \log n)$ time algorithm is proposed that finds the weighted 1-center in a cactus graph, where n is the number of vertices in the graph. For the weighted 2-center problem, an $O(n \log^3 n)$ time algorithm is devised for its continuous version and showed that its discrete version is solvable in $O(n \log^2 n)$ time. No such algorithm was previously known. The obnoxious center problem in a cactus graph can now be solved in $O(n \log^3 n)$. This improves the previous result of $O(cn)$ where c is the number of distinct vertex weights used in the graph [8]. In the worst case c is $O(n)$.

1 Introduction

The *p-center problem* is to determine a set C of p facilities in a graph G such that the maximum weighted distance $S(C, G)$ from a vertex of the graph to its closest facility in C, called as the service cost of C, is minimized. This problem has received a strong interest since its formulation by Hakimi in 1964 [2].

The problems in tree networks are well studied [3,5,6]. In the 1-center problem, the service cost function $S(C, G)$ is convex on every simple path of G. Based on this observation, Kariv and Hakimi [3] designed an $O(n \log n)$ algorithm which is improved to $O(n)$ by Megiddo [5] with a "trimming" technique, where n is the number of vertices in G. For the p-center problem in trees, most known algorithms [6] are based on parametric search technique. However, this technique is hard to generalize for other classes of graphs. Tamir [7] gave an $O(m^p n^p \log^2 n)$ algorithm for p-center problem in general graphs with m edges.

Although most of the reported works on the center problems are for trees or for general graphs, more and more attention is being paid for the classes of graphs that are between these two extremes, viz. cactus graphs [4,8], partial k-trees. In this paper we consider cactus graphs. In [4], Lan et al. designed a linear time algorithm to solve the unweighted 1-center problem in a cactus graph. In [8], Zmazek et al. proposed an algorithm that finds the obnoxious center of a cactus graph in $O(cn)$ time. Here c is the number of distinct vertex weights. However, no algorithm was known for solving the weighted p-center

* Research of the second author was partially supported by MITACS and NSERC.

X. Deng and D. Du (Eds.): ISAAC 2005, LNCS 3827, pp. 693–703, 2005.

problems (even when $p = 1, 2$) in cactus graphs. Our main results in this paper are proposed algorithms, first of their kinds, for the weighted 1-center and 2-center problem in cactus graphs and an improved bound on the obnoxious center problem in cactus graphs. The results are summarized in Table 1.

Table 1. Complexity bounds for the algorithms presented in this paper

Problems	Previous result	Our result
weighted continuous/discrete 1-center	$O(n^2 \log^2 n)$ [7]	$O(n \log n)$
weighted obnoxious center	$O(cn)$ [8]	$O(n \log^3 n)$
weighted continuous 2-center	$O(n^4 \log^2 n)$ [7]	$O(n \log^3 n)$
weighted discrete 2-center	$O(n^4 \log^2 n)$ [7]	$O(n \log^2 n)$

The basic technique used in our algorithms is the divide-and-conquer technique. We explore the properties of the weighted 1-center and the split edges (defined in Sect.4) of the weighted 2-center in a cactus graph. In our algorithm for the weighted 2-center problem, one important ingredient is to get, in the query mode, the service cost of any point in sublinear time. For this purpose, we have proposed a two-level tree decomposition structure over a cactus graph. This structure can easily be extended to general partial k-trees, k fixed.

The paper is organized as follows. Section 2 begins with definitions and problem formulation. The well-known tree structure of a cactus graph is also reviewed in this section. Section 3 provides an $O(n \log n)$-time algorithm to solve the weighted 1-center problem. Our algorithms for the weighted 2-center problem are presented in Sect.4. Section 5 gives a brief summary and future outlook.

2 Definitions and Problem Formulation

Let $G = (V, E, w, l)$ be a simple network with vertex set V, $|V| = n$, and edge set E, $|E| = m$, where each vertex $v \in V$ is associated with positive weight $w(v)$ and each edge $e \in E$ is associated with positive length $l(e)$. Let $A(G)$ denote the continuum set of points on the edges of G. For $V' \subseteq V$, let $G(V')$ denote the induced subgraph with vertex set V'. For a subgraph G' of G, let $V(G'), E(G'), A(G')$ denote the vertex set, the edge set and the continuum set of points on the edges of G', respectively. $P_{u,v}$ denotes the *shortest path* in G from u to v, $u, v \in V$. $d_{u,v}$ denotes the length of $P_{u,v}$. We similarly define $P_{a,v}$ and $d_{a,v}$ between a point a and a vertex v, $a \in A(G), v \in V$.

Definition 1. *A* cactus *is a connected graph in which two cycles have at most one vertex in common.*

In order to facilitate the overview of the algorithms devised for the center problems in cactus graphs, we start with the well-known tree structure over a cactus graph [1]. The vertex set V is partitioned into three different subsets. A *C-vertex* is a vertex of degree 2 which is included in exactly one cycle. A

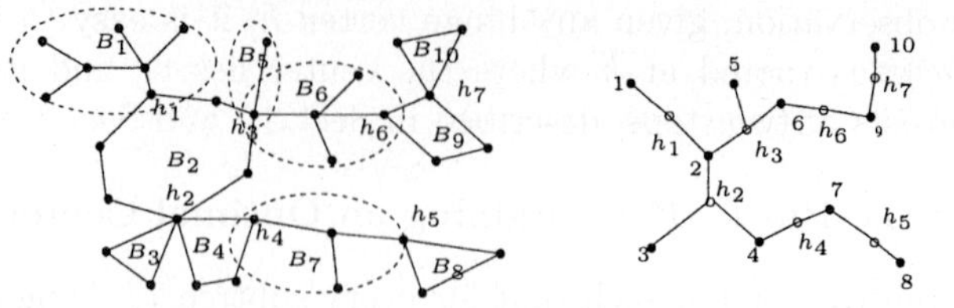

(a) Cactus graph G (b) Tree structure T_G over G

Fig. 1. A cactus graph and its corresponding tree structure

G-vertex is a vertex not included in any cycle. The remaining vertices, if any, will be referred to as *H-vertices* or *hinges*. Consider Fig.1(a) as a reference. We use the dotted ellipses to emphasize the grafts.

It's not difficult to see that a cactus consists of blocks, which are either a cycle or a graft, and these blocks are glued with H-vertices. So, we can use a tree $T_G = (V_G, E_G)$ to represent the important structure over G, where each element in V_G represents a block or a hinge vertex in G. See Fig.1(b). Let B_b denote the block represented by a block node $b \in V_G$. There is an edge between a block node b and a hinge node h if $h \in V(B_b)$.

Let $S(X, G')$ denote the *maximum weighted distance from a set* $X : \{\alpha_1, \dots, \alpha_p\}$ *to a sub-cactus* G', that is,

$$S(X, G') = \max_{v \in V(G')} \{w(v) \cdot d_{X,v}\}, \text{ where } d_{X,v} = \min_{j=1,\dots,p} d_{\alpha_j, v}.$$

The weighted 1-center problem: The weighted 1-center problem in a cactus graph G is to locate a point $\alpha_G^* \in A(G)$ such that $S(\alpha_G^*, G)$ is minimized. Let γ_G denote the *weighted radius* $S(\alpha_G^*, G)$ of G. When the weights of all the vertices in G are negative, it is called the *obnoxious center problem.*

The weighted 2-center problem: The weighted 2-center problem in the cactus graph G seeks to find a set $C : \{\alpha_1, \alpha_2\} \subset A(G)$ that minimizes $S(C, G)$. When the two centers are restricted to be vertices of G, we call it *discrete weighted* 2*-center problem.*

3 An Algorithm for the Weighted 1-Center Problem

We know that the service cost function $S(\alpha, G)$ is convex on every simple path in G if G is a tree. Unfortunately, this convexity property does not hold in cactus graphs (even in circles) [3]. However, we obtain a similar property of the weighted 1-center of a cactus, which provides the basis of our divide-and-conquer algorithm for the weighted 1-center problem: Let $G_1, \dots, G_k$ be the connected components hanging from a hinge vertex h. If the maximum value in $\{S(h, G_1) \dots, S(h, G_k)\}$ is attained at more than one component then h itself is the 1-center. On the other hand, if the maximum is attained in a unique G_i, then the center must lie in G_i.

With this observation, given any hinge vertex h, it is easy to find (in linear time) a sub-cactus rooted at h where the center lies or find its center. Our algorithm consists of two steps, described in Sect.3.1 and Sect.3.2.

3.1 Locate the Block B^* Containing an Optimal Center α_G^*

Let o be the centroid of T_G, such that for every subtree T'_G hanging from o has no more than half the nodes of T_G.

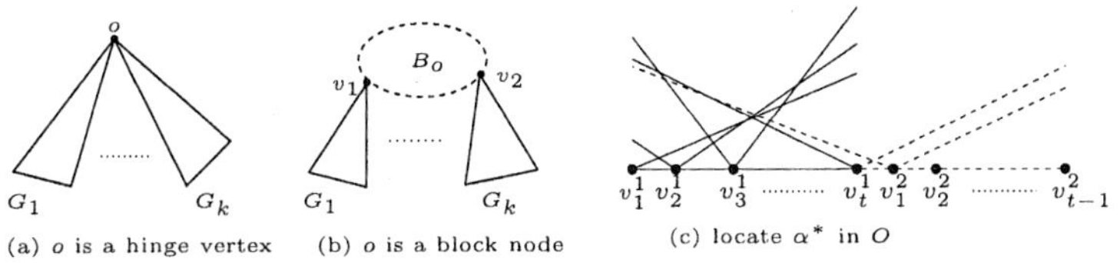

Fig. 2. Locate the sub-cactus where the center lies and locate α^* in a cycle block O

Case1: o is a hinge node. See Fig.2(a). If there are two subgraphs G_i, G_j adjacent to o, such that $S(o, G_i) = S(o, G_j) \geq S(o, G_l)$ for every other subgraph G_l adjacent to o, then o itself is the optimal 1-center α_G^*. Suppose that the subgraph G_i with largest value of $S(o, G_i)$ is unique. Then, the center lies in G_i. Since $G_j, 1 \leq j \leq k$ are pairwise disjoint, it follows that we can evaluate all the $S(o, G_j), 1 \leq j \leq k$ in linear time.
Case 2: o is a block node. See Fig.2(b). If there are two subgraphs G_i, G_j adjacent to B_o, such that $S(v_i, G_i) = S(v_j, G_j) \geq S(v_l, G_l)$ for every other subgraph G_l adjacent to B_o, then the center should be in the block B_o (So B_o is B^*). Suppose that the subgraph G_i with largest value of $S(v_i, G_i)$ is unique. Now, the center lies either in block B_o or in the sub-cactus G_i. Compare $S(v_i, G_i)$ with $S(v_i, G \setminus G_i)$. If $S(v_i, G_i) < S(v_i, G \setminus G_i)$, then the center certainly lies in block B_o (So B_o is B^*). Similarly, if $S(v_i, G_i) > S(v_i, G \setminus G_i)$, then the center lies in G_i. The remaining case is when $S(v_i, G_i) = S(v_i, G \setminus G_i)$. In this case, the hinge vertex v_i itself is the center α_G^*. All of these steps require $O(n)$ time.

Now we have the following cases: Either α_G^* is found and the algorithm terminates; B^* is found then locate α_G^* with the algorithm in Sect.3.2; Or a sub-cactus G_i containing α_G^* is obtained then the process is repeated on G_i.

Lemma 1. *It takes $O(n \log n)$ time to get B^*.*

3.2 Determining α_G^* in B^*

The second step is to locate α_G^* in B^*. If B^* is a graft, $S(x, G)$ is convex on every simple path of B^*. Therefore the algorithm in [3] can be used to determine α_G^* with the center restricted to lie in B^*. This can be done in $O(n \log n)$.

Suppose B^* is a cycle block. Observe that locating α_G^* in the cycle block B^* is equivalent to locating 1-center in a cycle. We now concentrate on solving

the weighted 1-center problem in a cycle O containing t vertices. Let α^* denote the 1-center of O. We notice that there is one edge not used by α^* to cover the vertices of O, called *cut-edge*. It follows that the 1-center on the path constructed by removing the cut-edge from O is also the 1-center in O. So our idea is that: consider each edge as a cut-edge and compute the 1-center on the resulting path.

Our algorithm is described as follows. Consider Fig.2(c) as reference. The vertices on the cycle are indexed as $v_1, v_2, \ldots, v_t$ in the counterclockwise order and the edge link vertices $v_{i-1}v_i$ is indexed as $e_{i-1}, 1 < i \leq t$ ($e_t : v_t v_1$). We put the $2t-1$ vertices $\{v_1^1 = v_1, v_2^1 = v_2, \ldots, v_1^2 = v_1, \ldots, v_{t-1}^2 = v_{t-1}\}$ on a real line. Let $Pos(v_i)$ denote the position of v_i where $Pos(v_1^1) = 0$, $Pos(v_i^1) = Pos(v_{i-1}^1) + l(e_i)$ $(1 < i \leq t)$, $Pos(v_1^2) = Pos(v_t^1) + l(e_t)$, and so on.

The path constructed by removing e_i from O is called the i-th path, which is the path from v_{i+1}^1 to v_i^2 in Fig.2(c). Let V^i be the vertex set of i-th path. The service cost function $f^i(x)$ on i-th path is defined as $f^i(x) = \max_{v \in V^i} w(v)$ $|x - Pos(v)|$, where x is a point (real number) on i-th path. Let x^i denote the 1-center of the i-th path. It is inefficient to separately compute functions $f^i(x), 1 \leq i \leq t$.

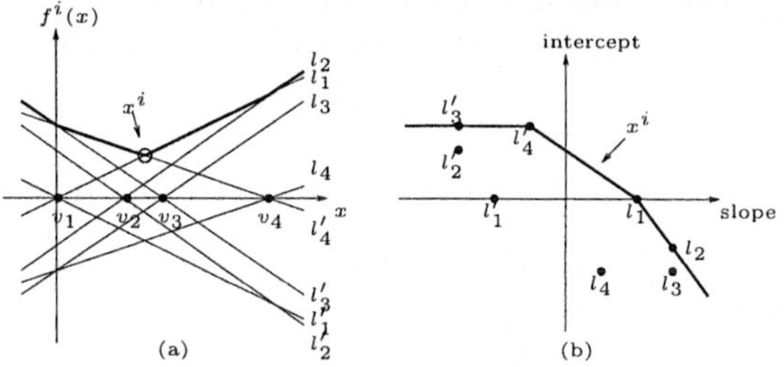

Fig. 3. $f^i(x)$ and its dual

Refer to Fig.3, $f^i(x)$ is the upper envelope of $2t$ linear functions. It can be computed by the convex hull of points in the dual plane. Also, the optimal solution x^i is represented by the edge of the convex hull in the dual plane that links one vertex whose primal line has negative slope with its adjacent one whose primal line has positive slope. Observe that $(i+1)$-th path is constructed from the i-path by simply removing v_{i+1}^1 and inserting v_{i+1}^2. The resulting updating of the convex hull can be implemented in amortized constant time by carefully storing the history of the updating during the insertion step. Thus all the optimal solutions with different cut edges can be computed in $O(n \log n)$ time.

The discrete version of the weighted 1-center problem is very similar. Summing up, we have the following theorem.

Theorem 1. *The continous/discrete weighted 1-center problem in a cactus graph can be solved in $O(n \log n)$ time and linear space.*

4 An Algorithm for the Weighted 2-Center Problem

The algorithm for the weighted 2-center problem in a cactus also consists of two steps. The first step is to locate the block B^* where an optimal split-edge set lies, called as *split-block*. The second step computes an optimal split-edge set R^*.

Let $C = \{\alpha_1, \alpha_2\} \subset A(G)$. Let $V_i \subseteq V$ be the set of vertices closest to $\alpha_i \in C, i = 1, 2$. The edges whose endpoints belong to different subgraphs $G(V_i), i = 1, 2$ are called *split edges*. Thus, locating two centers in a graph is equivalent to finding a set of split edges whose removal defines two connected components such that the maximum cost of the 1-center of these components is equal to the optimal 2-center cost of the entire graph. A property on the number of split edges in a 2-center solution of a cactus graph is shown in Lemma 2.

Lemma 2. *The split edges corresponding to one 2-center solution in a cactus graph G lie in one block B_i. If B_i is graft, the number of split edge is one. Otherwise (B_i is cycle), it is two.*

Let R denote the set of split edges for the 2-center problem. Let G_R^1, G_R^2 denote the two subgraphs corresponding to the split-edge set R. Let $\phi(R) = \max\{\gamma_{G_R^1}, \gamma_{G_R^2}\}$ denote *the service cost function of G for the split-edge set R*. A split-edge set R^* is called *an optimal split-edge set* of G if it satisfies $\phi(R^*) = \min_{R \subseteq B_i, i=1,\ldots,t} \phi(R)$, where t is the number of blocks in G.

4.1 Locate the Block B^* That Contains an Optimal Split-Edge Set

We focus on exploring the property of the split-edge set of the weighted 2-center in a cactus graph.

Lemma 3. *(Refer to Fig.2(a).) Given a hinge vertex $o \in V$, let $G_1, \ldots, G_k$ be the sub-cacti hanging from o. We can either identify the component G_i that contains an optimal split-edge set or identify an optimal split-edge set and this process takes $O(n \log n)$ time.*

Lemma 4. *(Refer to Fig.2(b).) Given a block B_o, let $G_1, \ldots, G_k$ be the sub-cacti hanging from B_o. We can either identify the component G_i that contains an optimal split-edge set, determine the block B^* or identify an optimal split-edge set, and this process takes $O(n \log n)$ time.*

The proofs of Lemma 3 and Lemma 4 are very similar to the arguments used in Sect.3.1. Using these two lemmas, we can recursively search an optimal split-edge set in a component which has at most half of the blocks as that in the previous component. Thus after an $O(n \log n \cdot \log |V_G|)$ process, either we already have an optimal split-edge set or know the split block B^*.

4.2 Computing R^* in B^*

When B^* is a graft, R^* contains exactly one edge. We can locate an optimal split edge recursively using the centroid decomposition of B^*. Each recursive

step takes $O(n \log n)$ time (Lemma 3 also works for G-vertex). Therefore, in this case it takes $O(n \log n \cdot \log |B^*|)$ time to obtain R^*.

For the other case, an optimal split-edge set contains two edges. Let $e_i = (v_i, v_{i+1}), i = 1, \ldots, t$ be the edges of B^* in counterclockwise order. Let $[e_i, e_j]$ denote the set of all the edges in the counterclockwise order from e_i to e_j in B^*. Similarly, let $[v_i, v_j]$ denote the set of vertices in the counterclockwise order from v_i to v_j in B^*.

If one edge in an optimal split edge is known, we can locate the other one in $O(n \log^2 n)$ time since it is equivalent to locating one optimal split edge on a graft: here, the graft is a path. With respect to one split edge e_i, the edge $e'_i \in E(B^*)$ is called the *match-edge* of e_i if $\phi(\{e_i, e'_i\}) = \min_{e_k \in E(B^*)} \phi(\{e_i, e_k\})$. We cannot afford to separately find the match-edge of each edge in B^*. However, the following simple observation is helpful: Assume that e'_i is the match-edge of $e_i, i = 1, \ldots, t$, and $e_j \in (e_i, e'_i)$. The match-edge e'_j of e_j should be in $[e'_i, e_i]$.

Our algorithm to locate R^* in B^* proceeds as follows. Start from e_1. After the match-edge e'_i of e_i is found, the first edge $e_j \in [e'_i, e_1]$, which satisfies $\phi(\{e_{i+1}, e_j\}) < \phi(\{e_{i+1}, e_{j+1}\})$, is the match-edge e'_{i+1} of e_{i+1} with the convexity property. Thus, the running time is determined by the complexity of computation of the service cost for a given split-edge set R, shown in next section.

4.3 Compute $\phi(R : \{e_i, e_j\})$ in Amortized $O(\log^3 n)$ Time

Let G'_k denote the subgraph hanging from B^* rooted at v_k. Assume that $S(v_{i_1}, G'_{i_1}) \geq S(v_{i_2}, G'_{i_2}) \geq S(v_k, G'_k), 1 \leq k \leq t$ and $k \neq i_1, i_2$. Thus the subgraphs G'_{i_1} and G'_{i_2} are the first two deepest components attached to B^*.

Lemma 5. *The two centers for the given split-edge set $R \in E(B^*)$ lie either in the split block B^* or in G'_{i_1}, or G'_{i_2}.*

Proof. Suppose that one of two centers, say α_1, lies in one sub-cactus G'_k with $k \neq i_1, i_2$. We can use the vertex v_k as the center instead, without increasing the service cost. □

Suppose that $v_{i_1} \in V(G^1_R)$. We can compute $\phi(R)$ by computing (I) the local optimal center α'_1 (resp. α'_2) in $[v_{i+1}, v_j]$ (resp. $[v_{j+1}, v_i]$) of G^1_R (resp. G^2_R), (II) the local optimal center α''_1 (resp. α''_2) in the G'_{i_1} (resp. G'_{i_2}) of G^1_R (resp. G^2_R if $v_{i_2} \in V(G^2_R)$, otherwise α''_2 is undefined).

In order to compute all such local optimal centers α'_1, α'_2 in B^*, we are able to use a similar algorithm used to locate 1-center in a cycle. As we have seen already, these local optimal centers can be computed in $O(n \log n)$ time.

Since computing α''_2 is similar to compute α''_1, we concentrate on computing α''_1 only. Let $G' = G^1_R \setminus G'_{i_1}$. G' is dynamic as R changes, see Fig.4. Observe that $S(x, G^1_R), x \in A(G'_{i_1})$ can be computed in $O(\log^2 n)$ since $S(x, G^1_R) = \max\{S(x, G'), S(x, G'_{i_1})\}$ and $S(x, G'_{i_1})$ can be computed in $O(\log^2 n)$ (see Lemma 11). Let β_1 denote the center of G'_{i_1}, and B denotes the block where β_1 lies.

Lemma 6. *(Refer to Fig.4(a)(b).) The local center α_1'' lies in one of the blocks that the shortest path $P_{v_{i_1},\beta_1}$ goes through.*

Proof. Suppose that α_1'' lies in some block B' that $P_{v_{i_1},\beta_1}$ doesn't go through. Let h denote the closest vertex to α_1'' in the blocks that $P_{v_{i_1},\beta_1}$ goes through. We can see that the service cost $S(h, G_R^1)$ is less than the service cost $S(\alpha_1'', G_R^1)$. □

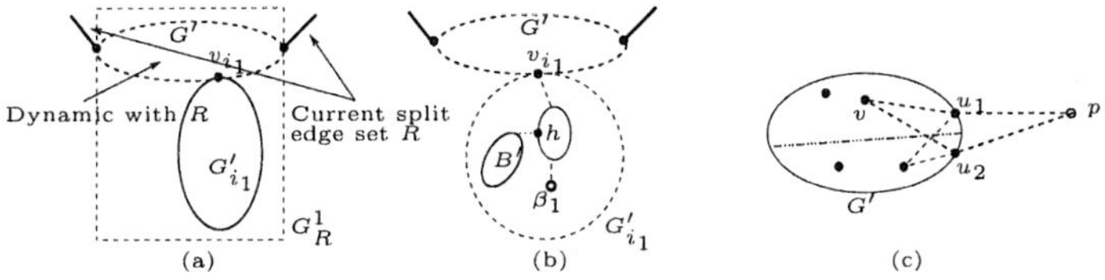

Fig. 4. (a)(b): Example with a split edge set R; (c): 2-separator between p and G'

Forcing the convexity of $S(x, G'_{i_1})$ on every edge of the block path: The service cost function $S(x, G)$ in a cactus graph may not be convex even on an edge. Fortunately, for a cactus graph we can make the service cost function convex on each edge of the *block path*, which contains all the blocks that $P_{v_{i_1},\beta_1}$ goes through, by adding extra vertices. It is done as follows. If one block on the path is a graft, then, clearly, the service cost function is convex on each edge of this block. Otherwise, for every vertex v in this block, find the *match-point* p_v in the block such that $d^c_{v,p_v} = d^{cc}_{v,p_v}$ where d^{cc}_{v,p_v} is the counterclockwise distance from v to p_v. We assign weight zero to these added vertices. In this way, the distance function of points on each edge in this block is monotone. As a consequence of this, the service cost function on the edge is convex.

Ordering the edges in the block path: Let $u_1 = v_{i_1}, u_2, \ldots, u_k$ denote the list of hinge vertices on the block path. Let $B_{u_1}, \ldots, B_{u_k} = B$ denote the sequence of the blocks that lie on the block path. The *equal-point* v' of a vertex v in a cycle B_{u_i} is a point on B_{u_i} such that $d^c_{u_i,v'} = d^{cc}_{u_i,v'}$. We can see that the number of vertices created is at most n. This allows the ordering of the edges of a block by their distances from $u_i, i = 1, \ldots, k$.

In the following we assume that G'_{i_1} contains the match-points and equal-points of the vertices in the cycle blocks on the block path. The algorithm to compute α_1'' consists of two parts. First we determine the block containing α_1'', and then determine α_1'' within this block. Observe that the weighted farthest vertex v'_j in G'_{i_1} to u_j must lie below u_j, otherwise, β_1 can't be 1-center of G'_{i_1}.

Lemma 7. *For any j, $1 \leq j \leq k$, if the farthest (weighted) vertex to u_j in G_R^1 comes from G', then α_1'' lies above u_j; Otherwise, the farthest (weighted) vertex to u_j in G_R^1 lies below u_j, then α_1'' lies below u_j.*

Using Lemma 7, we can locate the block that contains α_1'' in $O(\log n)$ steps. In each step, we need to compute $S(u_i, G_R^1), i \in [1, k]$. We can precompute all of

the $S(u_i, G'_{i_1})$ in $O(n \log^2 n)$ time. Since $S(u_i, G^1_R) = \max\{S(u_i, G'), S(u_i, G'_{i_1})\}$, $S(u_i, G^1_R)$ can be computed in $O(\log n)$ time. Hence, with $n \log^2 n$ preprocessing time, the block containing α''_1, denoted by $B_{u_i^*}$, can be found in $O(\log^2 n)$ time.

Lemma 8. *If the local optimal center q of G'_{i_1} on an edge e has a larger service cost to G'_{i_1} than a point p on another edge closer to v_{i_1}, then α''_1 can't lie on e.*

Proof. $S(x, G^1_R) = \max\{S(x, G'), S(x, G'_{i_1})\}$. Since the local minimum service cost to G'_{i_1} in e is $> S(p, G'_{i_1})$ and p is closer to v_{i_1} than any point in e, the service cost $S(p, G^1_R)$ is always less than the service cost of any point in e. □

The following lemma allows us to efficiently identify α''_1 in $B_{u_i^*}$.

Lemma 9. *Given an edge e in $B_{u_i^*}$ whose minimum service cost to G'_{i_1} is less than the minimum service cost to G'_{i_1} of all the points of the edges of $B_{u_i^*}$ above it. If the service cost of G^1_R for the local center on e is determined by some vertex in G', then all the edge segments of $B_{u_i^*}$ below e cannot contain α''_1. Otherwise, all the edge segments of $B_{u_i^*}$ above e cannot contain α''_1.*

The computation of local optimal center of each edge in $B_{u_1}, \ldots, B_{u_k}$ can be done in $O(n \log^3 n)$ with the convexity of $S(x, G'_{i_1})$ on each edge. For the discrete case, similar and somewhat simpler method can be applied, in which α''_1 can be computed in amortized $O(\log^2 n)$ time for a given split-edge set. Therefore,

Theorem 2. *The continuous weighted 2-center problem in a cactus graph can be solved in $O(n \log^3 n)$ time complexity. The discrete weighted 2-center problem in a cactus graph can be solved in $O(n \log^2 n)$ time complexity.*

4.4 A Brief Description of the Two-Level Tree Decomposition

One of most important property of trees is the existence of a 1-separator between two disjoint subtrees in each of them. Partial k-trees is more general class of graphs for which similar property is available. Such property is represented by a tree decomposition with bounded treewidth k, which can be found in linear time for fixed k by H.L. Bodlaender.

Cactus graphs are partial 2-trees. Assume that the tree decomposition TD of G is known. Given a subgraph G' represented by a subtree TD' of TD, there is a 2-separator in G' between G' and a point outside G'. With this property, we preprocess the local information of G' to be able to get the service cost of any point outside G' to cover all the clients in G' quickly.

Refer to Fig.4(c). Given any point p connecting to vertices in G' by 2-separator $\{u_1, u_2\}$, the service cost needed to cover $v \in V(G')$ is either $w(v) \cdot (d_{v,u_1} + d_{u_1,p})$ or $w(v) \cdot (d_{v,u_2} + d_{u_2,p})$. So we sort all the vertices in G' based on their distance difference to the 2-separator, and build a balanced binary-search tree over the sorted sequence. Hence, all the vertices v in G' whose shortest path to p pass through u_1 (or u_2) can be found in $O(\log |G'|)$ time. *More importantly, by this data structure, the service cost of p to cover all the clients in G' can be computed in $O(\log |G'|)$ if the distances to the 2-separator from p are given.*

Next, we add another tree decomposition over TD such that the height of the new tree $\check{TD}$ is logarithmic. There are several methods to achieve it, such as centroid tree decomposition, top-tree decomposition. After the two-level tree decomposition of G, for each node in TD, there are $O(\log n)$ subtrees of $\check{TD}$ containing all the other nodes in TD. Thus, we get the following lemma:

Lemma 10. *The complete data structure in a cactus graph can be computed in $O(n \log^2 n)$ time and $O(n \log n)$ space.*

Since distance queries in partial k-trees can be answered in constant time after almost linear-time preprocessing for fixed k (by S. Chaudhuri and C.D. Zaroliagis), the following lemma is easy to obtain.

Lemma 11. *The service cost of a point in a cactus graph can be answered in $O(\log^2 n)$ by the two-level tree decomposition data structure.*

5 Conclusion and Future Work

In this paper we have studied the center problems in a tree-like graph: cactus, and proposed the first sub-quadratic time algorithms to solve the weighted 1,2-center problems in a cactus graph. To obtain them, the properties that allow us to use the divide-and-conquer technique have been discovered. Since the service cost function on an edge is not convex in a cactus graph, a simple mechanism has been suggested that forces convexity on an edge in a cactus graph. Using this property and the two-level tree decomposition structure proposed in Sect.4.4, the local minima on an edge can be computed in amortized $O(\log^3 n)$ time. The obnoxious center problem in a cactus graph can now be solved in $O(n \log^3 n)$. This improves the previous result of $O(cn)$ where c is the number of distinct vertex weights used in the graph [8]. In the worst case c is $O(n)$.

Theorem 3. *The obnoxious center problem in a cactus graph can be solved in $O(n \log^3 n)$ time.*

Many issues are still left open. For instance, it would be interesting to find out whether there exists an optimal linear-time algorithm for the weighted 1-center problem in a cactus graph, whether the two-level tree decomposition combined with the split method is helpful to solve the weighted p-center problem for $p > 2$, and whether the parametric search can be applied in the weighted p-center problem in a cactus graph for any p. Another challenging work is to find efficient algorithms to solve the weighted 2-center problem in partial k-trees.

References

1. R.E. Burkard, J. Krarup, "A linear algorithm for the pos/neg-weighted 1-median problem on cactus", *Comput.* 60 (1998) 498–509.
2. S.L. Hakimi, "Optimum location of switching centers and the absolute centers and medians of a graph", *Oper. Res.* 12 (1964) 450–459.

3. O. Kariv and S.L. Hakimi, "An algorithmic approach to network location problems, Part I. The p-centers", *SIAM J. Appl. Math.* 37 (1979) 513–538.
4. Y.-F. Lan, Y.-L. Wang, H. Suzuki, "A linear-time algorithm for solving the center problem on weighted cactus graphs", *Inform. Process. Lett.* 71 (1999) 205–212.
5. N. Megiddo, "Linear-time algorithms for linear programming in R^3 and related problems", *SIAM J. Comput.* 12:4 (1983) 759–776.
6. N. Megiddo, A. Tamir, "New results on the complexity of p-center problems", *SIAM J. Comput.* 12:4 (1983) 751–758.
7. A. Tamir, "Improved complexity bounds for center location problems on networks by using dynamic data structures", *SIAM J. Disc. Math.* 1:3 (1988) 377–396.
8. B. Zmazek, J. Žerovnik, "The obnoxious center problem on weighted cactus graphs", *Disc. Appl. Math.* 136 (2004) 377–386.

Algorithms for Local Forest Similarity

Zeshan Peng

Computer Science Department, The University of Hong Kong, Hong Kong
zspeng@cs.hku.hk

Abstract. An *ordered labeled tree* is a rooted tree, where the left-to-right order among *siblings* is significant and each node associates a label. A *forest* is a sequence of ordered labeled trees. Suppose F is a forest. A *substructure* is a connected subgraph of F. Define a *subforest* as a sequence of substructures of F such that their roots are siblings. The *local forest similarity* problem for forests F and G is to find two most similar subforests of F and G. We answer an open problem in Paper [3] and provide efficient algorithms on it. Comparing with previous results, our algorithms achieve better performance.

1 Introduction

Trees are important data structures to represent hierarchical data such as RNA secondary structures [9] and XML documents [11]. The problem of calculating the similarity between trees is studied deeply [12]. It is a global measure which cannot solve the local similarity problem directly since they do not explore the exponential local cases. Recently the study of the *Local Forest Similarity* (LFS) problem, which is motivated mainly by locating the functional and structural regions in RNA secondary structures, attracts much attention [3,2,10]. In a nutshell, the LFS problem is to find two most similar *subforests* of two input forests.

In this paper, we refer to trees as *ordered labeled trees.* A *forest* is a sequence of trees. There are variant existing subforest definitions. A *substructure* is a connected subgraph of a forest [10]. Suppose $S_1, \cdots, S_s$ are substructures in forest F. $\langle S_1, \cdots, S_s\rangle$ is a *closed subforest* if they are successive subtrees [3,2]. The parent of a closed subforest $\langle S_1, \cdots, S_s\rangle$ is defined as the parent of the root of S_1, if it has. $\langle S_1, \cdots, S_s\rangle$ is a *gapped subforest* if it can be obtained from a closed subforest by excluding some closed subforests and any two of the excluded closed subforests do not share the same parent [3]. Jansson *et al.*[3] provided an open problem: "Another interesting task is to further generalize our local gapped subforest alignment problem by allowing exclusions of more than one closed subforest sharing the same parent node." We define $\langle S_1, \cdots, S_s\rangle$ to be a *general subforest* if the roots of $S_1, \cdots, S_s$ are *siblings*, which fulfills the task exactly.

There are two main distance measures for the similarity of forests. One is the *forest edit distance(fed)* [12] and the other is the *forest alignment distance* (*fad*) [5]. From the aspect of the similarity score, *fed* is always better than the *fad* [4]. We adopt *fed* as the similarity measure in this paper.

X. Deng and D. Du (Eds.): ISAAC 2005, LNCS 3827, pp. 704–713, 2005.

Suppose F is a forest. The *degree* of node u is the number of children of u and the *depth* of node u is the number of ancestors of u. The degree of F, $dg(F)$, is defined to be the maximal degree of nodes in F and the depth of F, $dp(F)$, is defined to be the maximal depth of nodes in F. Let $L(F)$ be the set of all leaves in F and $|F|$ be the number of nodes in F. We provide algorithms on the LFS problem for two most similar substructures, closed subforests and general subforests with *fed*. Note that the number of closed subforests in forest F is $\Omega(|F|dg(F))$ and that of substructures or general subforests may be an exponential size. The result comparison of the LFS problem is listed in Table. 1 which shows that our algorithms achieve better performance.

Table 1. Time complexities of the LFS problem for variant subforests

substructures with *fed*(Sec. 3.1)	$O(\|F\| \min\{\|L(F)\|, dp(F)\}\|G\| \min\{\|L(G)\|, dp(G)\})$
general subforests with *fed*(Sec. 3.1)	$O(\|F\| \min\{\|L(F)\|, dp(F)\}\|G\| \min\{\|L(G)\|, dp(G)\})$
closed subforests with *fed*(Sec. 3.2)	$O(\|F\|\|G\|\|L(F)\|\|L(G)\|)$
closed subforests with *fad* [3]	$O(\|F\|\|G\|(dg(F)+dg(G))^2)$
gapped subforests with *fad* [3]	$O(\|F\|\|G\|(dg(F)+dg(G))dg(F)dg(G))$
maximum substructures within d with *fed* [10]	$O(\|F\| \min\{\|L(F)\|, dp(F)\}$ $\|G\| \min\{\|L(G)\|, dp(G)\}d^2)$

Motivation: A main motivation of the LFS problem is to automatically discover functional/structural regions (*motifs*) in RNA secondary structures. Fig. 1 shows an example of the RNA GI:2347024 structure, where (a) is a segment of the RNA sequence, (b) is its secondary structure and (c) is the forest representation. The Hammerhead motif of this RNA is represented by bold in (a), (b) and (c). If edit operations are applied on RNA secondary structures such that all bonded pairs are treated as units, then the similarity of RNA secondary structures can be obtained by comparing their corresponding forests [9,6,7,2]. The algorithm in [10] is not flexible because of the difficulty to estimate the value d, especially for unknown RNA structures. Substructures, closed subforests and gapped subforests cannot represent general motifs well such as Hammerhead motifs [8,1]. Most general motifs can be represented by general subforests [3,2,1,8]. The LFS problem for two most similar general subforests fulfills the task for the general case [3].

The organization of this paper is as follows. Section 2 gives the formal definitions and notations used in our paper. Section 3 presents our algorithms on the LFS problem and analyzes their time and space complexities.

2 Definitions and Notations

Consider nodes in a rooted tree T whose root is denoted as $r(T)$. Denote the parent of node v as $p(v)$. Nodes u and v are *siblings* if $p(u)=p(v)$. Denote all siblings of node u as $A(u)$. T is said to be an *ordered labeled tree* if the left-to-right order among siblings is significant and each node associates a label.

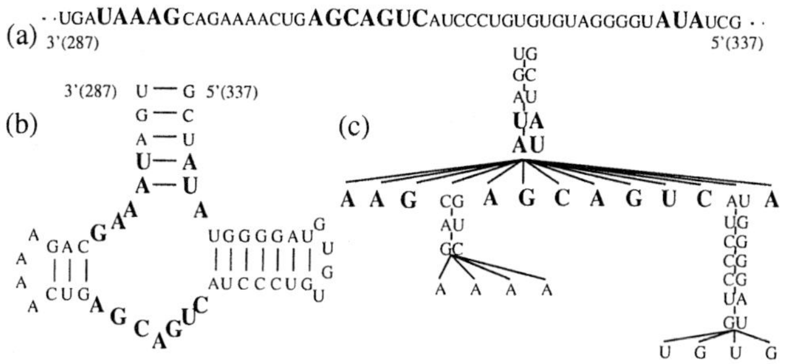

Fig. 1. RNA primary structure, RNA secondary structure and its forest representation

In the later discussion, we refer to trees as ordered labeled trees. A *forest* is a sequence of trees $F=\langle T_1,\cdots,T_t\rangle$. $\{r(T_1),\cdots,r(T_t)\}$ are said to be *siblings.* Apply a post-order numbering on forest F. Denote $f(i)$ (or i if no confusion) as the ith post-order traveling node. Denote the *subtree* rooted at node $i \in F$ as $F[i]$. Let $m(i)$ be the smallest numbered node in $F[i]$. The label of node i is denoted as $label(i)$. Node i is at the left (right resp.) of node j if $i<j$ ($i>j$ resp.). Denote the left-most sibling of i_1 as $b(i_1)$ and the right-most sibling of i_1 as $e(i_1)$. If node $i \in F[j]$, then node j is an ancestor of i, denoted as $j \in anc(i)$. Denote $i{:}j$ as the set of nodes whose numbers are from i to j inclusively. $i{:}j$ is empty if $j<i$.

Two *edit operations* are defined [12] as follows. (1)*Relabeling* node i means changing its label. (2)*Deleting* node i means making its children (if it has) as the children of $p(i)$ (if it exists) as well as removing node i and the edge between i and $p(i)$. An *edit mapping* between forests F and G, denoted as (M, F, G) (or M for short), is a set of node pairs (i, j), where node $i \in F$ and node $j \in G$, such that for any two pairs $(i_1, j_1), (i_2, j_2) \in M$, the following three properties are satisfied: (1)$i_1=i_2$ if and only if $j_1=j_2$; (2)$i_1 \in anc(i_2)$ if and only if $j_1 \in anc(j_2)$; (3)i_1 is at the left of i_2 if and only if j_1 is at the left of j_2. For any $(i, j) \in M$, we say node i is *linked* with node j. Any subset $M' \subseteq M$ is an edit mapping since it agrees the above three properties. For any edit mapping (M, F, G), define its *left-linked set* as $M_F=\{i|(i,j) \in M\}$ and its *right-linked set* as $M_G=\{j|(i,j) \in M\}$; define its *left-unlinked set* as $R_F=F \backslash M_F$ and its *right-unlinked set* as $R_G=G \backslash M_G$. An edit mapping (M,F,G) uniquely determines a sequence of deleting and relabeling operations on forests F and G such that their resulting forests are equal.

Given a fixed alphabet Σ and a space $- \notin \Sigma$, define a distance function as $\gamma{:}(\Sigma_- \times \Sigma_-)\backslash(-,-) \rightarrow \Re$, where $\Sigma_-=\Sigma \cup \{-\}$ and $\Re$ is the real number set. Without loss of generality we assume that $\gamma(a,a)<0$ and $\gamma(a,b)>0$ if $a \neq b$. For any element $(i, j) \in (M, F, G)$, the distance between node i and node j, $\gamma(i, j)$, is defined as $\gamma(i,j) = \gamma(label(i), label(j))$. Define the cost of an edit mapping (M, F, G) as $\delta(M)=\delta(M,F,G)=\sum_{(i,j)\in M}\gamma(i,j)+\sum_{i \in R_F}\gamma(i,-)+\sum_{j \in R_G}\gamma(-,j)$. Define $\delta(\emptyset) = 0$ for an empty set $\emptyset$. An optimal edit mapping M between forests F and G is an edit mapping with minimum cost, called *forest edit distance (fed)*, denoted as $\delta(F, G) = \delta(M)$.

For any node subset S of F, the *restricted subforest* of F on S, $F\|_S$, is defined as the forest obtained from F by deleting all nodes not in S. A *substructure* is defined as a connected subgraph of F. Suppose $S_1, \cdots, S_s$ are substructures in F. $\langle S_1, \cdots, S_s \rangle$ is said to be a *general subforest* if their roots are siblings. The parent of the root of S_1 (if it has) is said to be the parent of the general subforest. $\langle S_1, \cdots, S_s \rangle$ is said to be a *closed subforest* if $S_1, \cdots, S_s$ are successive subtrees. $\langle S_1, \cdots, S_s \rangle$ is said to be a *gapped subforest* if $S_1, \cdots, S_s$ are obtained from a closed subforest by excluding some closed subforests and any two of excluded closed subforests do not share the same parent [3]. Fig. 2 shows subforest examples of forest F in Fig. 1(c), where $F_1=F\|_S$ is a restricted subforest on set {3,5,6,9,13,22,31}, F_2 is a substructure, F_3 is a closed subforest, F_4 is a gapped subforest and F_5 is a general subforest. Note that F_5 is not a gapped subforest. The *Local Forest Similarity* (LFS) problem for forests F and G is to find a subforest F' of F and a subforest G' of G such that their *fed* is minimum among all subforest pairs of F and G.

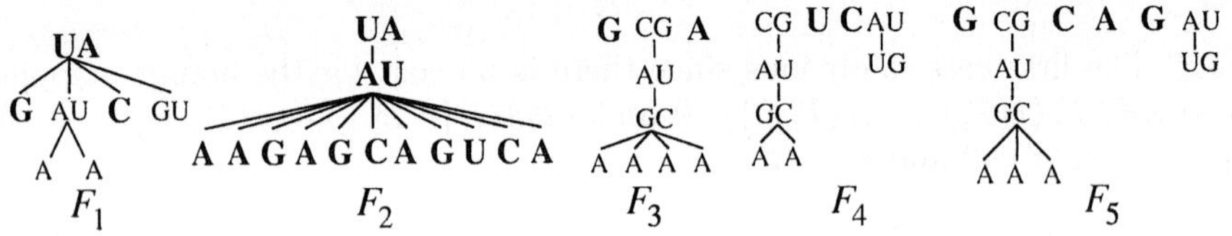

Fig. 2. Examples of variant subforests

3 Algorithms for the LFS Problem

Suppose $F=\langle T_1, \cdots, T_{x_1} \rangle$ and $G=\langle P_1, \cdots, P_{x_2} \rangle$ are two input forests. We provide efficient algorithms returning two most similar substructures (closed subforests and general subforests) of forests F and G.

Create two dummy nodes labeled by $\diamond$ as the roots of F and G, denoted as $r(F)$ and $r(G)$, such that $T_1, \cdots, T_{x_1}$ and $P_1, \cdots, P_{x_2}$ are the left-to-right children of $r(F)$ and $r(G)$ respectively. Let $\gamma(\diamond, \diamond)=0$. Define the key nodes in forest F as $K(F)=\{r(F) \cup r(T_1) \cup \cdots \cup r(T_{x_1}) \cup i \mid i \in F$ has a left sibling$\}$. Suppose $K(F), K(G)$ and $L(F), L(G)$ are calculated and stored as arrays in their post-orders. $m(i)$ for any node i in F and G is also calculated. All these as a preprocessing step of the following algorithms can be done in linear time.

3.1 Substructures and General Subforests

In this section, we provide two algorithms. One returns two most similar substructures (MSSs) of F and G. The other returns two most similar general subforests (MSGSs) of F and G.

A substructure rooted at node i can be obtained from $F[i]$ by removing some subtrees in $F[i]$. We define another edit operation: *cutting* node $i \in F$ means

removing the whole subtree $F[i]$ from forest F. We also say it is a cut-node. Two cut-nodes are said to be *consistent* if there is no ancestor relationship between them. Let C be a set of cut-nodes in F. C is said to be consistent if any pair of its cut-nodes is consistent. Let $\mathcal{S}(F)$ be all possible consistent sets of cut-nodes in F. Note that the size of $\mathcal{S}(F)$ may be exponential. Denote the remaining subforest after cutting all nodes in some $C \in \mathcal{S}(F)$ as *subf*(F, C). We have Fact 1.

Fact 1. *For* $C \in \mathcal{S}(F[i])$, *subf*$(F[i], C)$ *is a substructure rooted at node* i *if* $C \neq \{i\}$.

To get two MSSs of forests F and G, we must find two MSSs of subtrees $F[i_1]$ and $G[j_1]$ for all $1 \leq i_1 \leq |F|$ and $1 \leq j_1 \leq |G|$. From Fact 1, it equals to get $C_1 \in \mathcal{S}(F[i_1])$ and $C_2 \in \mathcal{S}(G[j_1])$ such that $\delta(subf(F[i_1], C_1), subf(G[j_1], C_2))$ is minimal over all combinations $\mathcal{S}(F[i_1]) \times \mathcal{S}(G[j_1])$ for all $1 \leq i_1 \leq |F|$ and $1 \leq j_1 \leq |G|$. Denote $\Upsilon(F, G)$ as the minimum *fed* between *subf*(F, C_1) and *subf*(G, C_2) over all $C_1 \in \mathcal{S}(F)$ and $C_2 \in \mathcal{S}(G)$. Then the objective is to find $\Upsilon(F[i_1], G[j_1])$ for all $1 \leq i_1 \leq |F|$ and $1 \leq j_1 \leq |G|$. Lemma 1 tells how to calculate some special cases.

Lemma 1. $\Upsilon(\emptyset, \emptyset) = 0$; $\Upsilon(F, \emptyset) = 0$; $\Upsilon(\emptyset, G) = 0$.

Proof. The first case is obvious since there is no cost for the empty mapping. Since *subf*$(F, \{r(T_1), \cdots, r(T_{x_1})\}) = \emptyset$ and *subf*$(G, \{r(P_1), \cdots, r(P_{x_2})\}) = \emptyset$. We get that $\Upsilon(F, \emptyset) = 0$ and $\Upsilon(\emptyset, G) = 0$. □

Consider how to calculate $\Upsilon(F[i_1], G[j_1])$ for subtrees $F[i_1]$ and $G[j_1]$. For any nodes $i \in F[i_1]$ and $j \in G[j_1]$, $F\|_{m(i_1):i}$ and $G\|_{m(j_1):j}$ are two restricted subforests. Then the key is to calculate $\Upsilon(F\|_{m(i_1):i}, G\|_{m(j_1):j})$ since $\Upsilon(F[i_1], G[j_1]) = \Upsilon(F\|_{m(i_1):i_1}, G\|_{m(j_1):j_1})$. Consider node $f(i)$. It may be deleted, or relabeled or cut. Note that if $f(i)$ is cut during the calculation, then the whole subtree $F[i]$ is removed and $F\|_{m(i_1):i}$ becomes a new restricted subforest $F\|_{m(i_1):m(i)-1}$. $f(i)$ is cut if and only if $\Upsilon(F\|_{m(i_1):i}, G\|_{m(j_1):j}) = \Upsilon(F\|_{m(i_1):m(i)-1}, G\|_{m(j_1):j})$, which means a cut operation has zero cost. Intuitively, this means subtree $F[i]$ is dissimilar with any subtree in $G[j_1]$ and the whole subtree $F[i]$ should be removed for the minimal cost. The deletion and relabeling cases are studied deeply in [12]. Node $g(j)$ follows the same cases respectively. To summarize the above analysis, Lemma 2 shows how to calculate $\Upsilon(F\|_{m(i_1):i}, G\|_{m(j_1):j})$.

Lemma 2. $\Upsilon(F\|_{m(i_1):i}, G\|_{m(j_1):j}) = \min$

$$\begin{cases} \Upsilon(F\|_{m(i_1):m(i)-1}, G\|_{m(j_1):j}); \\ \Upsilon(F\|_{m(i_1):i}, G\|_{m(j_1):m(j)-1}); \\ \Upsilon(F\|_{m(i_1):m(i)-1}, G\|_{m(j_1):m(j)-1}); \\ \Upsilon(F\|_{m(i_1):i-1}, G\|_{m(j_1):j}) + \gamma(f(i), -); \\ \Upsilon(F\|_{m(i_1):i}, G\|_{m(j_1):j-1}) + \gamma(-, g(j)); \\ \Upsilon(F\|_{m(i_1):m(i)-1}, G\|_{m(j_1):m(j)-1}) + \Upsilon(F\|_{m(i):i-1}, G\|_{m(j):j-1}) + \gamma(f(i), g(j)); \end{cases}$$

Proof. Suppose $C_1^o \in \mathcal{S}(F\|_{m(i_1):i})$ and $C_2^o \in \mathcal{S}(G\|_{m(j_1):j})$ are two optimal consistent cut-node sets such that $\delta(subf(F\|_{m(i_1):i}, C_1^o), subf(G\|_{m(j_1):j}, C_2^o))$ is minimum. Suppose M is an optimal edit mapping between *subf*$(F\|_{m(i_1):i}, C_1^o)$ and *subf*$(G\|_{m(j_1):j}, C_2^o)$. Consider the nodes $f(i), g(j)$ and the sets C_1^o, C_2^o.

- $f(i) \in C_1^o$: then $f(i)$ is a cut-node and the whole subtree $F[i]$ is removed. In this case, we get $\Upsilon(F\|_{m(i_1):i}, G\|_{m(j_1):j}) = \Upsilon(F\|_{m(i_1):m(i)-1}, G\|_{m(j_1):j})$;
- $g(j) \in C_2^o$: then $g(j)$ is a cut-node and the whole subtree $G[j]$ is removed. In this case, we get $\Upsilon(F\|_{m(i_1):i}, G\|_{m(j_1):j}) = \Upsilon(F\|_{m(i_1):i}, G\|_{m(j_1):m(j)-1})$;
- $f(i) \in C_1^o$ and $g(j) \in C_2^o$: then both $F[i]$ and $G[j]$ are removed. In this case, we get $\Upsilon(F\|_{m(i_1):i}, G\|_{m(j_1):j}) = \Upsilon(F\|_{m(i_1):m(i)-1}, G\|_{m(j_1):m(j)-1})$;
- $f(i) \notin C_1^o$ and $g(j) \notin C_2^o$: then node $f(i)$ is either deleted or linked with some node in $G\|_{m(j_1):j}$. So is for node $g(j)$. Consider $f(i)$, $g(j)$ and M.
  - $f(i) \notin M_F$, which means $f(i)$ is deleted from $F\|_{m(i_1):i}$. In this case, we get $\Upsilon(F\|_{m(i_1):i}, G\|_{m(j_1):j}) = \Upsilon(F\|_{m(i_1):i-1}, G\|_{m(j_1):j}) + \gamma(f(i), -)$;
  - $g(j) \notin M_G$, which means $g(j)$ is deleted from $G\|_{m(j_1):j}$. In this case, we get $\Upsilon(F\|_{m(i_1):i}, G\|_{m(j_1):j}) = \Upsilon(F\|_{m(i_1):i}, G\|_{m(j_1):j-1}) + \gamma(-, g(j))$;
  - $f(i) \in M_F$ and $g(j) \in M_G$: then $(f(i), g(j)) \in M$. In this case, we get $\Upsilon(F\|_{m(i_1):i}, G\|_{m(j_1):j}) = \Upsilon(F\|_{m(i_1):m(i)-1}, G\|_{m(j_1):m(j)-1}) + \Upsilon(F\|_{m(i):i-1}, G\|_{m(j):j-1}) + \gamma(f(i), g(j))$.

All cases are considered and finally we get $\Upsilon(F\|_{m(i_1):i}, G\|_{m(j_1):j})$ as the minimum of the above cases. □

In Lemma 2, from the definitions of Υ and m, it is easy to get that $\Upsilon(F\|_{m(i_1):i}, G\|_{m(j_1):j}) \le \Upsilon(F\|_{m(i_1):m(i)-1}, G\|_{m(j_1):m(j)-1}) + \Upsilon(F[i], G[j])$ and $\Upsilon(F[i], F[j]) \le \Upsilon(F\|_{m(i):i-1}, G\|_{m(j):j-1}) + \gamma(f(i), g(j))$. If $m(i) = m(i_1)$ and $m(j) = m(j_1)$, then $\Upsilon(F\|_{m(i_1):m(i)-1}, G\|_{m(j_1):m(j)-1}) = \Upsilon(\emptyset, \emptyset)$. So we have Theorem 1.

Theorem 1. *If* $m(i) = m(i_1)$ *and* $m(j) = m(j_1)$, *we have* **case 1:** $\Upsilon(F\|_{m(i_1):i}, G\|_{m(j_1):j}) = \Upsilon(F[i], G[j]) =$

$$\min \begin{cases} \Upsilon(F\|_{m(i_1):m(i)-1}, G\|_{m(j_1):j}); \\ \Upsilon(F\|_{m(i_1):i}, G\|_{m(j_1):m(j)-1}); \\ \Upsilon(F\|_{m(i_1):m(i)-1}, G\|_{m(j_1):m(j)-1}); \\ \Upsilon(F\|_{m(i_1):i-1}, G\|_{m(j_1):j}) + \gamma(f(i), -); \\ \Upsilon(F\|_{m(i_1):i}, G\|_{m(j_1):j-1}) + \gamma(-, g(j)); \\ \Upsilon(F\|_{m(i_1):i-1}, G\|_{m(j_1):j-1}) + \gamma(f(i), g(j)); \end{cases}$$

Otherwise, we have **case 2:** $\Upsilon(F\|_{m(i_1):i}, G\|_{m(j_1):j}) =$

$$\min \begin{cases} \Upsilon(F\|_{m(i_1):m(i)-1}, G\|_{m(j_1):j}); \\ \Upsilon(F\|_{m(i_1):i}, G\|_{m(j_1):m(j)-1}); \\ \Upsilon(F\|_{m(i_1):m(i)-1}, G\|_{m(j_1):m(j)-1}); \\ \Upsilon(F\|_{m(i_1):i-1}, G\|_{m(j_1):j}) + \gamma(f(i), -); \\ \Upsilon(F\|_{m(i_1):i}, G\|_{m(j_1):j-1}) + \gamma(-, g(j)); \\ \Upsilon(F\|_{m(i_1):m(i)-1}, G\|_{m(j_1):m(j)-1}) + \Upsilon(F[i], G[j]); \end{cases}$$

Applying Theorem 1, Algorithm 1 calculates $\Upsilon(F[i_1], G[j_1])$ for all $i_1 \in K(F)$ and $j_1 \in K(G)$. Create matrices with size of $|F[i_1]| \times |G[j_1]|$ for all $i_1 \in K(F)$ and $j_1 \in K(G)$. Store the values $\Upsilon(F[i], G[j])$ for all $m(i_1) \le i \le i_1$ and $m(j_1) \le j \le j_1$ in Algorithm 1. Lemma 3 follows.

Lemma 3. *For any nodes* $i_1 \in K(F)$ *and* $j_1 \in K(G)$, *and any nodes* $m(i_1) \le i \le i_1$ *and* $m(j_1) \le j \le j_1$, $\Upsilon(F\|_{m(i_1):i}, G\|_{m(j_1):j})$ *can be obtained in constant time.*

Main Loop
Input: two forests F and G;
1: **for** $x_1 := 1, \cdots, |K(F)|$ **do**
2: **for** $x_2 := 1, \cdots, |K(G)|$ **do**
3: $i_1 = K(F)[x_1]$; $j_1 = K(G)[x_2]$; Calculate $\Upsilon(F[i_1], G[j_1])$;
Tree-Tree Procedure $\Upsilon(F[i_1], G[j_1])$:
1: $\Upsilon(\emptyset, \emptyset) = 0$;
2: **for** $i := m(i_1), \cdots, i_1$ **do** $\Upsilon(F\|_{m(i_1):i}, \emptyset) = 0$;
3: **for** $j := m(j_1), \cdots, j_1$ **do** $\Upsilon(\emptyset, G\|_{m(j_1):j}) = 0$;
4: **for** $i := m(i_1), \cdots, i_1$ **do**
5: **for** $j := m(j_1), \cdots, j_1$ **do**
6: **calculate** $\Upsilon(F\|_{m(i_1):i}, G\|_{m(j_1):j})$ according to Theorem 1.

Algorithm 1: Substructure Similarity Algorithm

Proof. To prove this statement, it equals to prove that in both case 1 and base 2 of Theorem 1, $\Upsilon(F\|_{m(i_1):i}, G\|_{m(j_1):j})$ in Algorithm 1 can be calculated in constant time. In Algorithm 1, indices i and j are increased successively (see Step 4 and 5 in Algorithm 1). $\Upsilon(F\|_{m(i_1):i}, G\|_{m(j_1):j})$ can be obtained in constant time in case 1 since all values required for its calculation are known before its calculation. Now we prove that it can be obtained in constant time in case 2 too.

The calculation of $\Upsilon(F\|_{m(i_1):i}, G\|_{m(j_1):j})$ depends on values (1)$\Upsilon(F\|_{m(i_1):i}, G\|_{m(j_1):j-1})$, (2)$\Upsilon(F\|_{m(i_1):m(i)-1}, G\|_{m(j_1):m(j)-1})$, (3)$\Upsilon(F\|_{m(i_1):i-1}, G\|_{m(j_1):j})$, (4)$\Upsilon(F\|_{m(i_1):m(i)-1}, G\|_{m(j_1):j})$, (5)$\Upsilon(F\|_{m(i_1):i-1}, G\|_{m(j_1):j-1})$, (6)$\Upsilon(F\|_{m(i_1):i}, G\|_{m(j_1):m(j)-1})$, (7)$\Upsilon(F[i], G[j])$. Indices i and j are increased one by one and $i \leq i_1$ and $j \leq j_1$ (see Step 4 and 5 in Algorithm 1). Then the first six values are known before. Now we prove that (7) is also known before.

Note that $K(F)$ and $K(G)$ are stored in their post-orders, and $i_1 \in K(F)$ and $j_1 \in K(G)$. If $i \in K(F)$ and $j \notin K(G)$, then there is a node $j' \in K(G)$ on the path from j to j_1 such that $\Upsilon(F[i], G[j]) = \Upsilon(F[i], G\|_{m(j'):j})$ which is calculated in previous loops. Note that if $m(j') = m(j_1)$, then $j' = j_1$. If $i \notin K(F)$ and $j \in K(G)$, then there is a node $i' \in K(F)$ on the path from i to i_1 such that $\Upsilon(F[i], G[j]) = \Upsilon(F\|_{m(i'):i}, G[j])$ which is calculated in previous loops. Note that if $m(i') = m(i_1)$, then $i' = i_1$. If $i \notin K(F)$ and $j \notin K(G)$, then there is a node $i' \in K(F)$ on the path from i to i_1 and a node $j' \in K(G)$ on the path from j to j_1 such that $\Upsilon(F[i], G[j]) = \Upsilon(F\|_{m(i'):i}, G\|_{m(j'):j})$ which is calculated in previous loops. So for all cases, $\Upsilon(F[i], G[j])$ is known before the calculation of $\Upsilon(F\|_{m(i_1):i}, G\|_{m(j_1):j})$. The final minimizing can be done in constant time. So we get the proof. □

Given forests F and G in Algorithm 1, Lemma 3 shows that $\Upsilon(F\|_{m(i_1):i}, G\|_{m(j_1):j})$ is calculated for all $i_1 \in K(F)$, $j_1 \in K(G)$ and $m(i_1) \leq i \leq i_1$ and $m(j_1) \leq j \leq j_1$. Theorem 2 shows Algorithm 1 calculates $\Upsilon(F[i], G[j])$ for all $i \in F$ and $j \in G$ efficiently, and thereby obtains two most similar substructures of F and G.

Theorem 2. *We can find two* MSSs *of forests F and G in $O(|F||G|)$-space and $O(|F||G| \min\{|L(F)|, dp(F)\} \min\{|L(G)|, dp(G)\})$-time.*

Proof. To prove its correctness, it equals to prove that $\Upsilon(F[i], G[j])$ is calculated in Algorithm 1 for all $i \in F$ and $j \in G$. For any $i \in F$ and $j \in G$, consider the relationships between i, j and $K(F), K(G)$:

- $i \in K(F)$ and $j \in K(G)$: in this case, $\Upsilon(F[i], G[j])$ is calculated correctly at Step 3 in the main loop of Algorithm 1.
- $i \in K(F)$ and $j \notin K(G)$: in this case, there exists $j_1 \in anc(j)$ such that $j_1 \in K(G)$ and $\Upsilon(F[i], G[j]) = \Upsilon(F[i], G\|_{m(j_1):j})$ which is calculated correctly at Step 3 in the main loop of Algorithm 1.
- $i \notin K(F)$ and $j \in K(G)$: in this case, there exists $i_1 \in anc(i)$ such that $i_1 \in K(F)$ and $\Upsilon(F[i], G[j]) = \Upsilon(F\|_{m(i_1):i}, G[j])$ which is calculated correctly at Step 3 in the main loop of Algorithm 1.
- $i \notin K(F)$ and $j \notin K(G)$: in this case, there exist $i_1 \in anc(i)$ and $j_1 \in anc(j)$ such that $i_1 \in K(F)$, $j_1 \in K(G)$ and $\Upsilon(F[i], G[j]) = \Upsilon(F\|_{m(i_1):i}, G\|_{m(j_1):j})$ which are calculated at Step 3 in the main loop of Algorithm 1.

Thus in all cases, $\Upsilon(F[i], G[j])$ is calculated in Algorithm 1 correctly for all $i \in F$ and $j \in G$. $\min\{\Upsilon(F[i], G[j]) \mid i \in F, j \in G\}$ is the minimum *fed* among all substructure pairs of F and G. We can backtrack these two most similar substructures correctly in the same complexity.

In [12], it is proved that $|K(F)| \le \min\{|L(F)|, dp(F)\}$. Combining Theorem 1 and Lemma 3, it is not difficult to get that the time complexity of Algorithm 1 is $O(|F||G| \min\{|L(F)|, dp(F)\} \min\{|L(G)|, dp(G)\})$ and its space complexity is $O(|F||G|)$. So we get the proof. □

Now we provide an efficient algorithm for two MSGSs of F and G. Suppose the closed subforests $F' = \langle S_1, \cdots, S_s \rangle$ and $G' = \langle S'_1, \cdots, S'_{s'} \rangle$ are two non-empty most similar general subforests of F and G, where $S_1, \cdots, S_s$ are substructures of F whose roots are $i_1, \cdots, i_s$ respectively, and $S'_1, \cdots, S'_{s'}$ are substructures of G whose roots are $j_1, \cdots, j_{s'}$ respectively. Then $i_1, \cdots, i_s$ are left-to-right siblings in F and $j_1, \cdots, j_{s'}$ are left-to-right siblings in G. Consider $F\|_{m(b(i_1)):e(i_1)}$ and $G\|_{m(b(j_1)):e(j_1)}$. It is clear that F' and G' are also two most similar general subforests of $F\|_{m(b(i_1)):e(i_1)}$ and $G\|_{m(b(j_1)):e(j_1)}$. This claim comes from cutting all nodes $A(i_1) \backslash \{i_1, \cdots, i_s\}$ in $F\|_{m(b(i_1)):e(i_1)}$ and all nodes $A(j_1) \backslash \{j_1, \cdots, j_{s'}\}$ in $G\|_{m(b(j_1)):e(j_1)}$. So $\delta(F', G') = \Upsilon(F\|_{m(b(i_1)):e(i_1)}, G\|_{m(b(j_1)):e(j_1)})$.

Since $b(i_1)$ and $b(j_1)$ have no left siblings, then $b(i_1) \notin K(F)$ and $b(j_1) \notin K(G)$. There exist $p_i \in anc(i_1)$ and $p_j \in anc(j_1)$ such that $p_i \in K(F)$ and $m(b(i_1)) = m(p_i)$, and $p_j \in K(G)$ and $m(b(j_1)) = m(p_j)$. From Theorem 2, $\Upsilon(F\|_{m(i_1):i}, G\|_{m(j_1):j})$ and $\Upsilon(F[p_i], G[p_j])$ are calculated in Algorithm 1 for all $i_1 \in F, m(i_1) \le i \le i_1$ and $j_1 \in G, m(j_1) \le j \le j_1$. So value $\Upsilon(F\|_{m(b(i_1)):e(i_1)}, G\|_{m(b(j_1)):e(j_1)}) = \Upsilon(F\|_{m(p_i):e(i_1)}, G\|_{m(p_j):e(j_1)})$ is calculated. We can find the minimum score of $\Upsilon(F\|_{m(i_1):i_1-1}, G\|_{m(j_1):j_1-1})$ over all $i_1 \in F$ and $j_1 \in G$ in Algorithm 1. Two most similar general subforests can be found by backtracking the calculation of $F[i_1]$ and $G[j_1]$. It is obvious that it takes $O(|F||G| \min\{|L(F)|, dp(F)\} \min\{|L(G)|, dp(G)\})$ time. During the calculation, we are only asked to store the score of $\Upsilon(F[i_1], G[j_1])$ and $\Upsilon(F\|_{m(i_1):i_1-1}, G\|_{m(j_1):j_1-1})$ for all $i_1 \in F$ and $j_1 \in G$. So its space complexity is $O(|F||G|)$. Finally we have Theorem 3.

Theorem 3. *We can find two* MSGSs *of forests* F *and* G *in* $O(|F||G| \min\{|L(F)|, dp(F)\} \min\{|L(G)|, dp(G)\})$ *time and* $O(|F||G|)$ *space.*

3.2 Most Similar Closed Subforests (MSCSs)

Denote $i \cdot\cdot j$ as the set of siblings which are at the right of i and at the left of j inclusively. If i, j are not siblings or i is a right sibling of j, then $i \cdot\cdot j$ is empty. Denote $F[i \cdot\cdot j]$ as all subtrees rooted at $i \cdot\cdot j$. Then any *closed subforest* of F can be represented as $F[i \cdot\cdot j]$ [3,2]. In this section we provide a simple algorithm finding two most similar closed subforests (MSCSs) of F and G with *fed* measure.

To find two MSCSs of F and G, the search space in F is $O(|F|dg(F))$ and that in G is $O(|G|dg(G))$. Suppose $F[i_1 \cdot\cdot i_2]$ and $G[j_1 \cdot\cdot j_2]$ are two MSCSs of F and G. $F[i_1 \cdot\cdot i_2]{=}F\|_{m(i_1):i_2}$ and $G[j_1 \cdot\cdot j_2]{=}G\|_{m(j_1):j_2}$. It is not difficult to get that the algorithm provided in Paper [12] and Algorithm 1 in Section 3.1 do not work for such two MSCSs since the values of $\delta(F\|_{m(i_1):i_2}, G\|_{m(j_1):j_2})$ are not calculated if $i_1{\in}K(F), i_1{<}i_2{<}p(i_1)$ or $j_1{\in}K(G), j_1{<}j_2{<}p(j_1)$. For any nodes $i_1 \in F$ and $j_1 \in G$, and any nodes $i \in F[i_1]$ and $j \in G[j_1]$, $F\|_{m(i_1):i}$ and $G\|_{m(j_1):j}$ are two restricted subforests. Lemma 4 tells how to calculate $\delta(F\|_{m(i_1):i}, G\|_{m(j_1):j})$ efficiently. Paper [12] provides its detailed proof.

Lemma 4. *If* $m(i_1) = m(i)$ *and* $m(j_1) = m(j)$ *then:*
$\delta(F\|_{m(i_1):i}, G\|_{m(j_1):j}) = \delta(F[i], G[j]) =$

$$\min \begin{cases} \delta(F\|_{m(i_1):i-1}, G\|_{m(j_1):j}) + \gamma(f(i), -); \\ \delta(F\|_{m(i_1):i}, G\|_{m(j_1):j-1}) + \gamma(-, g(j)); \\ \delta(F\|_{m(i_1):i-1}, G\|_{m(j_1):j-1}) + \gamma(f(i), g(j)); \end{cases}$$

Otherwise: $\delta(F\|_{m(i_1):i}, G\|_{m(j_1):j}) =$

$$\min \begin{cases} \delta(F\|_{m(i_1):i-1}, G\|_{m(j_1):j}) + \gamma(f(i), -); \\ \delta(F\|_{m(i_1):i}, G\|_{m(j_1):j-1}) + \gamma(-, g(j)); \\ \delta(F\|_{m(i_1):m(i)-1}, G\|_{m(j_1):m(j)-1}) + \delta(F[i], G[j]); \end{cases}$$

Applying Lemma 4, we provide Algorithm 2 finding two MSCSs of F and G.

Main Loop
Input: Two forests F and G.

```
for x1 := |L(F)|, ···, 1 do
    for x2 := |L(G)|, ···, 1 do
        l1 = L(F)[x1];    l2 = L(G)[x2];    RTreeED(l1, l2);
```

Tree-Tree Procedure $RTreeED(l_1, l_2)$:

```
δ(∅, ∅) = 0
for i := l1, ···, |F| do     δ(F||l1:i, ∅) = δ(F||l1:i−1, ∅) + γ(f(i), −);
for j := l2, ···, |G| do     δ(∅, G||l2:j) = δ(∅, G||l2:j−1) + γ(−, g(j));
for i := l1, ···, |F| do
    for j := l2, ···, |G| do
        calculate δ(F||l1:i, G||l2:j) according to Lemma 4.
```

Algorithm 2: Closed Subforest Similarity Algorithm

Theorem 4. *Given two forests F and G, we can find two* MSCSs *with* fed *of F and G in $O(|F||G||L(F)||L(G)|)$ time and $O(|F||G|)$ space.*

Proof. Algorithm 2 calculates the values of $\delta(F\|_{l_1:i}, G\|_{l_2:j})$ for all leaves $l_1 \in L(F), i \in l_1{:}|F|$ and $l_2 \in L(G), j \in l_2{:}|G|$ from Lemma 4. To prove the correctness, we must prove that Algorithm 2 explores the search space $O(|F|dg(F)|G|dg(G))$.

Suppose $F[i_1 \cdot \cdot i_2] = F\|_{m(i_1):i_2}$ and $G[j_1 \cdot \cdot j_2] = G\|_{m(j_1):j_2}$ are two non-empty MSCSs of F and G. Since $m(i_1) \in L(F)$ and $m(j_1) \in L(G)$, $m(i_1) \leq i_2 \leq |F|$ and $m(j_1) \leq j_2 \leq |G|$, then $\delta(F[i_1 \cdot\cdot i_2], G[j_1 \cdot\cdot j_2]) = \delta(F\|_{m(i_1):i_2}, G\|_{m(j_1):j_2})$ is calculated in Algorithm 2. So we can find two most similar closed subforests by backtracking the closed subforest pair with a minimum score of $\min\{\delta(F\|_{m(i_1):i_2}, G\|_{m(j_1):j_2}) \mid i_1, i_2$ are siblings in F, j_1, j_2 are siblings in $G\}$.

RTreeED(l_1,l_2) in Algorithm 2 calculates the values of $\delta(F\|_{l_1:i}, G\|_{l_2:j})$ for all $i \in l_1{:}|F|$ and $j \in l_2{:}|G|$. $\delta(F\|_{l_1:i}, G\|_{l_2:j})$ can be calculated in constant time since all values for its calculation are obtained if we store all subtree to subtree scores using $O(|F||G|)$ space. It takes $O(|F|dg(F)|G|dg(G))$ searching time since each node in F has at most $\Omega(dg(F))$ siblings. So Algorithm 2 runs in $O(|F||G||L(F)||L(G)|)$ time and $O(|F||G|)$ space. □

References

1. S. Griffiths-Jones. The microRNA Registry. *Nucl. Acids Res.*, 32, D109-11, 2004.
2. Matthias Höchsmann, Thomas Töller, Robert Giegerich, and Stefan Kurtz. Local similarity in RNA secondary structures. In *Proceedings of the IEEE computer society conference on Bioinformatics*, pp.159–168, 2003.
3. Jesper Jansson, Ngo Trung Hieu, and Wing-Kin Sung. Local gapped subforest alignment and its application in finding RNA structural motifs. In *Proceedings of the international symposium on algorithms and computation*, pp.569–580, 2004.
4. Tao Jiang, Guohui Lin, Bin Ma, and Kaizhong Zhang. A general edit distance between RNA structures. *Journal of Molecular Biology*, 9(2):371–388, 2002.
5. Tao Jiang, Lusheng Wang, and Kaizhong Zhang. Alignment of trees - an alternative to tree edit. *Theor. Comput. Sci.*, 143:137–148, 1995.
6. Guo-Hui Lin, Bin Ma, and Kaizhong Zhang. Edit distance between two RNA structures. In *Proceedings of the international conference on computational biology*, pp.211–220, 2001.
7. Bin Ma, Lusheng Wang, and Kaizhong Zhang. Computing similarity between RNA structures. *Theor. Comput. Sci.*, 276:111–132, 2002.
8. Motifs Database. `http://subviral.med.uottawa.ca/cgi-bin/motifs.cgi`.
9. Bruce A. Shapiro and Kaizhong Zhang. Comparing multiple RNA secondary structures using tree comparisons. *Comput. Appl. Biosci.*, 6(4):309–318, 1990.
10. Jason Tsong-Li Wang, Bruce A. Shapiro, Dennis Shasha, Kaizhong Zhang, and Kathleen M. Currey. An algorithm for finding the largest approximately common substructures of two trees. *IEEE Trans. Pattern Anal. Mach. Intell.*, 20(8):889–895, 1998.
11. Extensible markup language (XML) 1.0 (third edition), W3C recommendation. `http://www.w3.org/TR/REC-xml/`, 2004.
12. Kaizhong Zhang and Dennis Shasha. Simple fast algorithms for the editing distance between trees and related problems. *SIAM J. Comput.*, 18(6):1245–1262, 1989.

Fast Algorithms for Finding Disjoint Subsequences with Extremal Densities

Anders Bergkvist[1] and Peter Damaschke[2,*]

[1] Department of Molecular Biology, Göteborg University, P.O. Box 462, 40530 Göteborg, Sweden
abk@molbio.gu.se

[2] School of Computer Science and Engineering, Chalmers University, 41296 Göteborg, Sweden
ptr@cs.chalmers.se

Abstract. We derive fast algorithms for the problem of finding, on the real line, a prescribed number of intervals of maximum total length that contain at most some prescribed number of points from a given point set. Basically this is a typical dynamic programming problem, however, for input sizes much bigger than the two parameters we can improve the obvious time bound by selecting a restricted set of candidate intervals that are sufficient to build some optimal solution. As a byproduct, the same idea improves an algorithm for a similar subsequence problem recently brought up by Chen, Lu and Tang at IWBRA 2005. The problems are motivated by the search for significant patterns in certain biological data. While the algorithmic idea for the asymptotic worst-case bound is rather evident, we also consider further heuristics to save even more time in typical instances. One of them, described in this paper, leads to an apparently open problem of computational geometry flavour (where we are seeking a subquadratic algorithm) which might be interesting in itself.

1 Problem Statement and Results

Finding large empty regions (big holes) in data sets is relevant for data mining and statistical inference. The problem in its most general form is given by a space X, a family $\mathcal{F}$ of subsets of X, a size function (satisfying some usual axioms) that assigns a positive real number to every set in $\mathcal{F}$, and a finite set $D \subset X$ of n data points. The goal is to find the largest set in $\mathcal{F}$ disjoint to D. In a commonly considered case, X is a finite-dimensional real vector space, $\mathcal{F}$ is the family of axis-parallel boxes, and the size of a box is its volume.

It is quite natural to generalize the problem in two directions: One may be interested in several disjoint big holes in X, and one may tolerate a limited number of data points in these holes. Therefore we introduce two parameters s, p and look for s disjoint sets with maximum total size which may contain up to p data points. This parameterized problem is challenging already in two dimensions,

* Contact author.

X. Deng and D. Du (Eds.): ISAAC 2005, LNCS 3827, pp. 714–723, 2005.

i.e., for axis-parallel rectangles in the plane. It might be NP-complete, we leave this as an open question. In the present paper we study the one-dimensional case where $\mathcal{F}$ is the set of sub-intervals of a finite-length interval X, and the size of an interval is its length. This one-dimensional case is already interesting for some applications, see Section 2. While this problem is in principle easy to solve by dynamic programming, the question of algorithms being as fast as possible for given n, s, p turns out to be nontrivial.

Note that an optimal solution always consists of s open intervals (excluding their endpoints) with data points as ends. (The two outermost intervals of a solution may reach an end of X.) This gives us a more convenient formal description. In a sequence of real numbers $x_0, x_1, x_2, \ldots, x_n$ called *items*, we call any set of consecutive items $x_i, \ldots, x_j$ $(i \leq j)$ an *interval of length* $x_i + \ldots + x_j$. Applied to our "big holes" problem, x_i (for $0 < i < n$) is obviously the distance of the ith and $(i+1)$st data point in their natural ordering in X, while x_0 and x_n is the distance of the first and last data point to the left and right end, respectively, of X. Now our problem can be stated as follows:

Disjoint Intervals of Maximum Length (DIMaxL)

Given a sequence $x_0, x_1, x_2, \ldots, x_n$ of real numbers, and integers $s \geq 1$ and $p \geq 0$, find s pairwise disjoint intervals with a total of $s+p$ items and maximum total length.

A batch of remarks is in order here. If $x_i \geq 0$ for all i, an optimal solution always uses exactly $s + p$ items, otherwise we could add another item to some interval. Hence, the problem would not change if we wrote "at most $s+p$ items" in the specification. In our version with exactly $s+p$ items, the x_i can be arbitary real numbers, and an optimal solution may have to take some negative items. Also, a problem instance does not change if any constant is added to all x_i. On the other hand, an optimal solution may have chains of, say, t incident intervals. We may split such chains arbitrarily in t other intervals without changing s, p, and the total length.

In some lucky cases, a trivial solution exists and can be verified in $O(n)$ time: If the $s + p$ largest items form (at most) s intervals, this is obviously an optimal solution. In particular, DIMaxL with $p = 0$ is a trivial problem. However, in general the $s + p$ largest items are more scattered, and then the optimum is no longer obvious: Parts of an optimal solution may consist of medium-length items that are clustered, and then we have to find the best among many candidates.

We also mention that DIMaxL can be rephrased as a special case of the Weighted Independent Set problem in interval graphs, where every vertex belongs to at most p consecutive cliques, the solution has to consist of s vertices, and the weights are the interval lengths in a given interval model. The problem is well studied for general weights and interval graphs (and other intersection graphs as well, see [2] for pointers to the vast literature), but to our best knowledge, efficient algorithms for our specific restrictions are not known. Thus we will not further use the relationship to interval graphs.

Organisation of the paper and results: In Section 2 we give several motivations, in Section 3 we develop our algorithms for DIMaxL. A straightforward dynamic prgramming algorithm solves DIMaxL in $O(spn)$ time. By some preprocessing that selects possible intervals for an optimal solution, without missing any candidate, we can run dynamic programming on the smaller candidate set and achieve a time bound $O(pn + s^2p^3)$. Since $s, p \ll n$ in statistically relevant cases, this alternative algorithm is then an improvement. We do not expect that we can completely get rid of the p factor, too, however this question motivates a heuristic and another (open) computational problem that looks challenging. We shall discuss it in Section 4. This and other heuristics considered in the full paper do not further improve our worst-case bound but are apparently advantageous in many instances.

Recently we became aware of a similar numerical subsequence problem: Define the density of an interval $x_i, \ldots, x_j$ to be the ratio $(x_i + \ldots + x_j)/(j - i + 1)$. Let us call the problem introduced and attacked in [7]

Disjoint Intervals of Maximum Density (DIMaxD)
Given a sequence $x_1, x_2, \ldots, x_n$ of real numbers, and integers $k \geq 1$ and $l \geq 1$, find k disjoint intervals, each with *at least* l items, so as to maximize the sum of their densities.

Chen, Lu and Tang [7] gave algorithms running in time $O(kln)$, $O(n)$, and $O(n + l^2)$ for the general case, $k = 2$, and $k = 3$, respectively. They explicitly asked if the general case can be solved in $o(kln)$ time. In Section 3 of our paper we give an affirmative answer. As a byproduct of the idea used above, we solve DIMaxD in $O(ln + k^2l^2)$ time. Actually, this scheme is applicable to any similar problem where one has to find a set of disjoint subsequences that optimizes some objective function. The main property we need from the problem is that the candidate intervals are of limited length.

2 Bioinformatics Motivations and Related Work

Finding big holes in data sets is well investigated in algorithm theory and in practice [1, 3, 6, 10, 13, 15]. But apparently very little has been done for the more general, parameterized version introduced in Section 1.

To motivate DIMaxL, suppose that certain objects are characterized by pairs of real coordinates x, y, where y is hard to measure, and that we have a random sample of data points (x, y). When we get a new measured value x_0, we would like to predict (more modestly: to restrict the range of) y_0. Let us consider a set D of data points with x around the measured x_0, e.g., within some user-defined radius. Under mild smoothness assumptions on the underlying distribution of the (x, y), we can conclude that, with high probability, y_0 is not in any of the big holes in this set. The approach can be readily extended to more dimensions. An "orthogonal" way to use holes for range predictions is to partition the data set into slices parallel to the x-axis and to compute, in each slice, the big holes in the projection to the x-axis. For a measured x_0 we would discard such values

of y where x_0 is in a big hole in the slice of y. This way we discard the true y_0 with some limited error probability only.

In a more purified setting, suppose that we sample n data points (real numbers) from an unknown probability distribution on a finite-length real interval X, and we want to "learn" a subset of X, as large as possible, where we do not expect future data points, up to some fraction (error probability) which is roughly p/n. If we simply took the p largest intervals between the data points, we would get a very scattered subset that reflects more the random locations of sampled data points than intrinsic big holes with low probability in the underlying distribution. Therefore we introduce a defragmentation parameter s. From another point of view, we might be interested in a prescribed number s of non-overlapping holes, but we should tolerate, say, p data points in these holes, since the number of data points in a low-probability interval follows a Poisson distribution. If, just as an example, the expected value of data points in some interval is 1, we get no or one data point with the same probability. Moreover, p can also account for outliers due to measurement errors. We will not further go into the discussion of statistical aspects, as the paper is focused on the algorithmic questions. We just make the point that $p > 0$ is motivated, and parameters s and p have to be chosen small compared to n.

Specifically, we were led to the problem by an ongoing project where we are trying to predict backbone torsion angles in proteins from measured nuclear magnetic resonance (NMR) chemical shifts. (We refer to [5, 8, 18, 19] for more background, but this is not needed to understand the problem studied here.) In a nutshell, it is well-known that chemical shifts are exquisitely sensitive indicators of local molecular structure. Certain torsion angles are particularly important for protein structures and they are correlated to these NMR chemical shifts. The distribution of chemical shifts versus torsion angles can be extracted from public databases. Existing technology allows protein-wide measurement and assignment of chemical shifts to residues at reasonable costs. This suggests the usage of measured chemical shifts for predicting torsion angle restraints at each residue, which can then be input in any of the existing computer programs that complete the structure prediction. Restraints should have small ranges of likely torsion angle values. At the same time, a large fraction of computed restraints should really contain the true angles. Both crtiteria are crucial to efficient and correct structure reconstruction. The process of discarding unlikely torsion angle intervals must run automatically, since *very many* restraints are to be computed. Thus it is desirable to have *fast* and easy-to-implement algorithms. Even moderate constant factors in the running time can be interesting.

The problem of [7] we called DIMaxD originates from several other pattern recognition applications in molecular biology, such as locating GC-rich regions and CpG islands in DNA, see [7] for more hints to the literature. Detection of special or unusual subsequences in biological sequences (DNA, proteins) has inspired various similar problems. To name a further example: Given a sequence of scores, find non-overlapping contiguous subsequences with maximum total scores [17]. However this problem is quite different from ours, in that the number

of items is not limited. Thus it requires different techniques. Finally we mention that sparse dynamic programming algorithms have been successfully applied to string distance computations [11, 12, 14] which are, of course, also relevant in molecular biology.

3 Selection as Preprocessing for Faster Dynamic Programming

First of all, there is a simple dynamic programming algorithm for DIMaxL. Since it is straightforward, we omit the proof.

Proposition 1. *DIMaxL can be solved in $O(spn)$ time.*

However, this is not the most efficient way if parameters s, p are small compared to n. Intuitively, we should be able to restrict the search to the largest sums of at most $p+1$ consecutive x_i. It is not possible to give a minimum length of candidate *items* that may appear in an optimal solution, since it may be necessary to insert a very short item that links two long intervals. (Examples are easy to construct.) But we can restrict the *candidate intervals* instead:

For $q = 1, \dots, p+1$ let $I(q)$ be the set of intervals with q items. We will show that some optimal solution consists only of intervals among the $r(q)$ largest from each $I(q)$, where $r(q)$ is some bound that solely depends on q and parameters s, p but not on the input size n. Moreover, $r(q)$ is polynomial in s, p:

Lemma 1. *There exists an optimal solution S where all vertices from $I(q)$ in S are among the $r(q) = (s-2)q + p + (s+p)/q + 1$ largest intervals in $I(q)$.*

Proof. Consider an optimal solution (set of disjoint intervals) S. For any $v \in S \cap I(q)$ and $u \in I(q) \setminus S$ longer than v, vertex u is *blocked*, i.e., adjacent to some of the $s-1$ vertices in $S \setminus \{v\}$, since otherwise $(S \setminus \{v\}) \cup \{u\}$ would be a better solution. For $j = 1, \dots, p+1$ let s_j denote the number of intervals in $S \cap I(j)$. Clearly, $\sum_j s_j = s$ and $\sum_j js_j = s + p$. Every interval in $(S \setminus \{v\}) \cap I(j)$ can block at most $q+j-1$ intervals in $I(q)$. Hence the number of intervals blocked by $S \setminus \{v\}$ in $I(q)$ is at most $(\sum_j s_j(q+j-1)) - (2q-1)$, the last term is subtracted because $v \in I(q)$, but factor s_q is used in the sum. Since at most $s_q - 1$ intervals in $S \cap I(q)$ stand before v, the rank of v in $I(q)$, sorted in non-increasing order of lengths, is at most $s_q - 2q + 1 + \sum_j s_j(q+j-1) = s_q - 2q + 1 + (q-1)s + (s+p) = (s-2)q + p + s_q + 1$. Observation $s_q \le (s+p)/q$ gives the result. □

Due to Lemma 1, only the $r(q)$ top intervals in each $I(q)$ need to be considered as candidates for an optimal solution. Bounding $\sum_q r(q)$ yields, after a little calculation:

Corollary 1. *The number of candidate intervals is at most $(1/2 + o(1))sp^2$.*

Now we use these findings to develop a more efficient algorithm for $s, p \ll n$. One idea would be to run the algorithm from Proposition 1 on the set of items occuring in the candidate intervals. However, we found that the following, different dynamic programming scheme that works with whole candidate intervals is faster.

Algorithm and analysis:

(1) Extract the $r(q)$ candidate intervals from every set $I(q)$. In more detail: Compute all sums of q consecutive items in an obvious way in $O(n)$ time, and then find the $r(q)$ largest sums. It is well-known that the r largest elements in an (unsorted) set of n numbers can be extracted via a Selection algorithm in $O(n)$ time, independent of r. (For an overview of results on Selection cf. [9, 16].) Altogether, this phase needs $O(pn)$ time.

(2) Sort all $O(sp^2)$ candidate intervals by their rightmost items. This costs $O(sp^2(\log s + \log p))$ time, or alternatively $O(n + sp^2)$ time if bucketsort is used.

(3) Run dynamic programming on the sorted list of candidate intervals as follows. For increasing i being the right end of some candidate interval, and for each $t = 1, \ldots, s$ and $q = 1, \ldots, p$, we compute the optimal solution with at most t intervals, $t + q$ items, and rightmost item before or at i. Note that, for any two solutions S and S' with the same t, q but with different i and i' $(i < i')$, solution S is worse than S', or we can discard the latter one. In other words, it suffices to maintain a set of solutions in which, for any fixed t, q, the lengths are monotone increasing in i.

In every step of dynamic programming we take the next candidate interval *next* from the list and append it to solutions whose rightmost item is to the left of *next*. By the monotonicity property, for every t, q it suffices to append *next* to only one solution: that with i immediately preceding the left end of *next*. Then add 1 to t, add the number of new items minus 1 to q, and discard new solutions that violate the monotonicity, by comparison to the solution with the previous end-item and same paramters t, q. It is straightforward to fill in the details. We spend $O(sp)$ time on every new candidate interval, hence $O(s^2p^3)$ time is needed for all.

Summing up the time bounds we obtain the following main result.

Theorem 1. *DIMaxL can be solved in* $O(pn + s^2p^3)$ *time.*

This is an improvement upon Proposition 1 provided that $sp^2 = o(n)$. Most noticeably, we have saved a factor s in the $O(n)$ term. The next obvious question is whether we can get rid of factor p, too. Phase (1) of our algorithm processes all $I(q)$, $q = 1, \ldots, p + 1$ in isolation, although it always works on the same input sequence. This situation suggests to do the selection somehow for all q simultaneously.

Another open question is whether the polynomial term in s, p can be reduced, so that we still save time if the parameters are comparable to $n^{1/3}$. Observe in the proof of Lemma 1 that the worst case appears if there are big gaps between the intervals of S. In more "compact" solutions S, nearby intervals in S block the same candidates. This indicates that fewer candidates are sufficient to find the best of these compact solutions, and the best "spread-out" solution might be faster to construct as well, as the large gaps might simplify the dynamic programming. However it is not clear whether these observations can give an improvement. Our full paper proposes some heuristics which, in many instances, throw out further candidates by simple rules, so that the optimization phase

is faster. Finally, an approach that is completely different from left-to-right dynamic programming and, e.g., considers items in decreasing-length order, might also succeed.

Some data structure issues:

Theorem 1 holds for any $O(n)$ time Seletion algorithm used inside. Note that a Selection algorithm with large hidden constant factor or with complicated internal administration would destroy the effect of having saved factor s, or make things even worse. However we can keep the auxiliary data structures simple and the constant factor in the Selection procedure small:

Recall that $r(q) \approx sp^2/2$ is small compared to n in our case. To select the $r(q)$ largest elements in $I(q)$, we may construct a heap (see any textbook on data structures) with the largest sums on top. This is easily done in $O(n)$ time. Then, we first remove the root, which splits the heap in two disjoint heaps. After $j-1$ steps we have j heaps, and the next largest sum is the maximum at their j roots. Using another small heap for the sums at the current roots, which supports insert and delete-max operations, we manage Selection in $O(n + r(q) \log r(q))$ time, with a small hidden factor.

Another idea is to sample k random elements and to determine a pivot element with a rank slightly above $r(q)k/n$ in the sample. Its expected rank in $I(q)$ is then slightly above $r(q)$. Now, simply compare each member of $I(q)$ to the pivot. If the selected set is smaller than $r(q)$, repeat the process similarly on the remaining set. But in general it will be larger than $r(q)$ by some percentage. Sample size k must give a compromise between the time for sampling/pivot selection and the deviation of the pivot rank from $r(q)$. A good choice is $k = \sqrt{n}$.

Solving DIMaxD faster:

We apply the same principal idea to DIMaxD and improve the earlier $O(kln)$ time bound of [7] for arbitrary parameters. A simple lemma from [7] states:

Lemma 2. *There exists an optimal solution where each interval has less than $2l$ items.*

The following result beats the $O(kln)$ bound. Note that $kl < n$, since otherwise the problem would not make sense.

Theorem 2. *DIMaxD can be solved in $O(ln + k^2l^2)$ time.*

Proof. By Lemma 2 it is enough to consider candidate segments with fewer than $2l$ items, these are $O(ln)$ segments. Due to the bound $2l$, each candidate segment intersects only $O(l)$ other candidate segments. We say that the candidate segment with rth largest density has rank r, ties are broken arbitrarily. Now assume that we have an optimal solution U containing a segment s of rank larger than $O(kl)$ (with an appropriate constant factor). The other $k-1$ segments in U intersect only $O(kl)$ other candidate segments. Hence there exists a segment s' which is no worse than s and disjoint to the $k-1$ other segments in U. Substitute s with s' in U. Iterating this argument, it follows the existence of an optimal solution where all segments have a rank $O(kl)$.

Select the $O(kl)$ candidate segments in $O(ln)$ time. The candidate segments have only $O(kl)$ different endpoints rather than n. Hence, a dynamic programming algorithm as in [7] but applied to this restricted set costs only $O(k^2l^2)$ time. □

4 Acceleration Heuristics for DIMaxL

We come back to the question of possible improvement of the $O(pn)$ term in Theorem 1. Note that Corollary 1 bounds the *number* of candidate intervals, but still we have all freedom regarding the way to compute this set. Instead of doing Selection on all q separately, we may also exploit "locality" and determine the longest intervals in subsequences of $O(p)$ items, for all q simultaneously, and then pick the best from all these subproblems.

More precisely, we cover the input sequence by overlapping intervals which we call *windows*. The ith window is $x_{(i-1)(p+1)+1}, \ldots, x_{(i+1)(p+1)}$ (except the last window which may have up to $3p$ items, depending on n). Obviously, every interval of $q \leq p+1$ items is entirely in one (or two) of these windows.

Proposition 2. *Assume that we can compute the longest interval from $I(q)$, for all $q = 1, \ldots, p+1$, within each window in $T(p)$ time. Then we could also solve DIMaxL in $O(nT(p)/p + s^2p^3)$ time.*

Proof. Recall that Phase (1) of our DIMaxL algorithm has to find the $r(q)$ longest intervals in each $I(q)$. Lemma 1 gives $r(q) = O(sp)$. Once we have the longest intervals in all $O(n/p)$ windows, we can select, for each q, the $r(q)$ longest of them in $O(n/p)$ time, thus we need $O(n)$ time for all q. At this stage we know, for each q, that all candidates are in the $O(sp)$ windows that contain these $r(q)$ top intervals, all other windows can be disregarded since already the maximum from $I(q)$ therein is too bad. Thus we collect the $O(sp^2)$ intervals of $I(q)$ from the candidate windows and do the final selection in $O(sp^2)$ time. This takes $O(sp^3)$ time for all q. The rest works as in Theorem 1. □

Trivially we have $p \leq T(p) = O(p^2)$. Note that any bound $T(p) = o(p^2)$ would speed up the candidate selection. Here we cannot deliver such a subquadratic worst-case bound nor prove an $\Omega(p^2)$ lower bound, however we give a heuristic that should be faster in almost all practical cases, as it needs $O(p^2)$ time only if the input exhibits some regularities that are unlikely in real data. First we give a nicer formulation of the essence of the problem. In the following, m is the window size.

MAXIMUM CONSECUTIVE SUMS (MaxCS)

Given a sequence $x_1, \ldots, x_m$, find for each $q = 1, \ldots, m$ the longest interval with q items.

Prefix and suffix sums $u_j := \sum_{i=1}^{j} x_i$ and $v_k := \sum_{i=m-k+1}^{m} x_i$ are computable in $O(m)$ time. Since $j + k = m - q$ holds for any prefix, suffix and enclosed interval, substituting $m - q$ with q gives the following problem with the same complexity:

Lowest Midpoints (LMP)

Given two sequences $u_1, \ldots, u_m$ and $v_1, \ldots, v_m$, find for each $q = 1, \ldots, m$ the minimum of $u_j + v_k$ with $j + k = q$.

Our naming LMP comes from a geometric interpretation: Given red and blue points in the plane with coordinates (j, u_j) and (k, v_k) respectively, find for each q the lowest midpoint, with abscissa $q/2$, of a red and a blue point.

This geometric view suggests a simple heuristic: Sweep a line with fixed slope upwards and maintain the set M of all midpoints of all red-blue pairs of points below this line. For every abscissa $q/2$ appearing in M, clearly, we only have to check the red-blue pairs where at least one partner is below the line. For every $q/2$ not appearing in M we finally compute the lowest midpoint naively, by checking all $O(m)$ red-blue pairs. This easily yields:

Proposition 3. *If, for some line, we have b points below the line, and their red-blue pairs have midpoints with a different abscissa values, then the given instance of LMP is solvable in $O(ab + (m - a)m)$ time.*

Our b and a increase during sweeping. If we are most lucky, the first $b = \Theta(\sqrt{m})$ points in sweeping direction generate nearly all abscissa values. In this case we need only $O(m^{3/2})$ computations. Of course, the set of abscissa values of the first b points may have periodicities for all b, so that many pairwise averages are equal. The worst case is arithmetic sequences. Then we are stuck with $a = O(b)$, and the time is still $O(n^2)$. We argue that such bad cases should be rare in irregular real data. Moreover, we may try different slopes and hope that one of them breaks such regularities, since the order of points is changed. Still, it would be very nice to develop an algorithm with worst-case time $T(m) = o(m^2)$. We hardly believe that LMP belongs to the class of 3SUM-complete problems where the trivial $O(m^2)$ bound is hard to beat [4].

5 Conclusions

We gave a scheme for fast algorithms for problems appearing naturally in the analysis of biological data and other linear data sets, where disjoint subsequences are sought that optimize the sum of certain values (lengths, densities). We developed a fast algorithm for a problem called DIMaxL. Some additional heuristics not mentioned in this extended abstract are useful but could not further reduce the worst-case time bound. We have to leave this and other open problems (complexity of two-dimensional analogues, the lowest-midpoints problem, impact of our heuristics) for further theoretical and experimental research.

Acknowledgments. The first author has been supported by the Knut and Alice Wallenberg Foundation, Carl Tryggers Foundation, and Assar Gabrielsson Foundation. The second author received support from the Swedish Research Council (Vetenskapsrådet), project "Algorithms for searching and inference in genetics", grant no. 621-2002-4574. He would also like to thank Ferdinando Cicalese for constructive discussions during a stay at the University of Bielefeld.

References

1. A. Aggarwal, S. Suri: Fast algorithms for computing the largest empty rectangle, *Symp. on Comput. Geometry 1987*, 278-290
2. G. Agnarsson, P. Damaschke, M.M. Halldórsson: Powers of geometric intersection graphs and dispersion algorithms, *Discrete Applied Mathematics* 132 (2003), Special Issue *Stability in Graphs and Related Topics* (eds. V. Lozin, D. de Werra), 3-16. Preliminary version in: Proc. of SWAT 2002, *LNCS* 2368, 140-149
3. M.J. Attalah, G.N. Fredrickson: A note on finding a maximum empty rectangle, *Discrete Applied Math.* 13 (1986), 87-91
4. I. Baran, E. Demaine, M. Patrascu: Subquadratic algorithms for 3SUM, *9th WADS 2005*, *LNCS* 3608, 409-421
5. R.D. Beger, P.H. Bolton: Protein ϕ and ψ dihedral restraints determined from multidimensional hypersurface correlations of backbone chemical shifts and their use in the determination of protein tertiary structures, *J. of Biomol. NMR* 10 (1997), 129-142
6. B. Chazelle, L.R.S. Drysdale, D.T. Lee: Computing the largest empty rectangle, *SIAM J. Comp.* 15 (1986), 550-555
7. Y.H. Chen, H.I. Lu, C.Y. Tang: Disjoint segments with maximum density, *International Workshop on Bioinformatics Research and Applications IWBRA 2005* (within *ICCS 2005*), *LNCS* 3515, 845-850
8. G. Cornilescu, F. Delaglio, A. Bax: Protein backbone angle restraints from searching a database for chemical shift and sequence homology, manuscript 1998, `http://spin.niddk.nih.gov/bax/software/TALOS`
9. D. Dor: Selection algorithms, PhD thesis, Tel-Aviv Univ. 1995
10. J. Edmonds, J. Gryz, D. Liang, R.J. Miller: Mining for empty rectangles in large data sets, *Theoretical Computer Science* 296 (2003), 435-452
11. D. Eppstein, Z. Galil. R. Giancarlo, G.F. Italiano: Sparse dynamic programming I: Linear cost functions, *J. of the ACM* 39 (1992), 519-545
12. Z. Galil, K. Park: Dynamic programming with convexity, concavity and sparsity, *Theor. Computer Science* 92 (1992), 49-76
13. B. Liu, L.P. Ku, W. Hsu: Discovering interesting holes in data, *15th IJCAI'1997*, 930-935
14. V. Mäkinen, G. Navarro, E. Ukkonen: Algorithms for transposition invariant string matching, *20th STACS 2003*, *LNCS* 2607, 191-202
15. M. Orlowski: A new algorithm for the largest empty rectangle problem, *Algorithmica* 5 (1990), 65-73
16. M.S. Paterson: Progress in selection, *5th SWAT'96*, *LNCS* 1097, 368-379
17. W.L. Ruzzo, M. Tompa: A linear time algorithm for finding all maximal scoring subsequences, *7th Int. Conf. Intelligent Systems for Molecular Biology 1999*, AAAI, 234-241
18. Y. Wang, O. Jardetzky: Probability-based protein secondary structure identification using combined NMR chemical-shift data, *Protein Science* 11 (2002), 852-861
19. X.P. Xu, D.A. Case: Probing multiple effects on ^{15}N, $^{13}C^{\alpha}$, $^{13}C^{\beta}$ and $^{13}C'$ chemical shifts in peptides using density functional theory, *Biopolymers* 65 (2002), 408-423

A Polynomial Space and Polynomial Delay Algorithm for Enumeration of Maximal Motifs in a Sequence

Hiroki Arimura[1,*] and Takeaki Uno[2]

[1] Hokkaido University, Kita 14-jo, Nishi 9-chome, Sapporo 060-0814, Japan
arim@ist.hokudai.ac.jp
[2] National Institute of Informatics, Tokyo 101–8430, Japan
uno@nii.jp

Abstract. In this paper, we consider the problem of enumerating all maximal motifs in an input string for the class of repeated motifs with wild cards. A maximal motif is such a representative motif that is not properly contained in any larger motifs with the same location lists. Although the enumeration problem for maximal motifs with wild cards has been studied in (Parida et al., CPM'01), (Pisanti et al.,MFCS'03) and (Pelfrene et al., CPM'03), its output-polynomial time computability is still open. The main result of this paper is a polynomial space polynomial delay algorithm for the maximal motif enumeration problem for the repeated motifs with wild cards. This algorithm enumerates all maximal motifs in an input string of length n with $O(n^3)$ time per motif with $O(n^2)$ space and $O(n^3)$ delay. The key of the algorithm is depth-first search on a tree-shaped search route over all maximal motifs based on a technique called prefix-preserving closure extension. We also show an exponential lowerbound and a succinctness result on the number of maximal motifs, which indicate the limit of a straightforward approach.

1 Introduction

Pattern discovery is to find all patterns within a class of combinatorial patterns that appear in an input data satisfying a specified constraint, and it is a central task in computational biology, temporal sequence analysis, sequence and text mining [3]. We consider the pattern discovery problem for the class of patterns with wild cards, which are strings consisting of constant symbols (called *solid letters*) drawn from an alphabet and variables '$\circ$' (called *wild cards*) that matches any symbol [9,13]. For instance, $\mathtt{B} \circ \mathtt{AB}$ and $\mathtt{B} \circ \mathtt{AB} \circ \circ \mathtt{B}$ are examples of patterns. Given a positive integer θ called *quorum* and an input string s, a frequent motif (or *motif*, for short) in s is a pattern that appears at least θ times in s.

Frequent motif discovery has a drawback that a huge number of motifs are often generated from an input string without conveying any useful information.

* This work is done during the first author's visit in LIRIS, University Claude-Bernard Lyon 1, France.

X. Deng and D. Du (Eds.): ISAAC 2005, LNCS 3827, pp. 724–737, 2005.

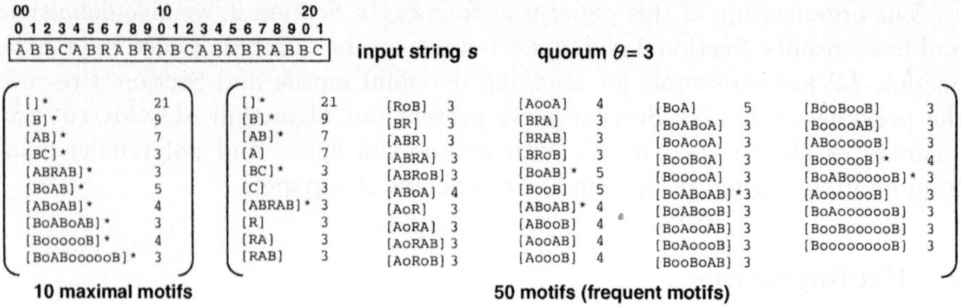

Fig. 1. Examples of maximal motifs (left) and motifs (right) for an input string s and a quorum θ, where * indicates a maximal motif and the number associated to each motif indicates its frequency. These 10 maximal motifs are representatives containing the whole information on the occurrences of all motifs in s.

To overcome this problem, we focus on discovery of *maximal motifs* [9,12,13]. The semantics of a pattern x is given by the location list $\mathcal{L}(x)$ consisting of the positions in an input string s at which the pattern occurs. A motif is said to be *maximal* if it is not properly contained by other motifs with the equivalent location lists allowing position shift. For example, we show in Fig. 1 all maximal motifs and all motifs on input sting $s =$ ABBCABRABRABCABABRABBC for quorum $\theta = 3$. Then, the pattern $x_1 = \mathtt{R} \circ \mathtt{B}$ is a motif having location list $\mathcal{L}(x_1) = \{6, 9, 17\}$ in s but not a maximal one since there is another motif $x_2 =$ ABRAB which contains x_1 and whose location list $\mathcal{L}(x_2) = \{4, 7, 15\}$ is obtained from the location list $\mathcal{L}(x_1)$ by shifting leftward with two. In this example, we can also observe that there are only 10 maximal motifs among 50 motifs. In general, the number of maximal motifs can be exponentially smaller than the number of motifs. while the former is exponential in the input size.

In this paper, we study the problem of enumerating all maximal motifs in an input string of length n. In particular, from practical viewpoint, we are interested in those algorithms that have small space and delay complexities independent from the output size in addition to polynomial amortized time per motif. However, the output-polynomial time computability for maximal motif discovery with polynomial space and delay is still open.

We first show an exponential lowerbound and a succinctness result on the number of maximal motifs, which show the limit of a straightforward approach. By examining the previous approaches [9,13,12], we present a simple output-polynomial time algorithm for maximal motif enumeration by breadth-first search, which possibly requires exponential space and delay. Then, we present an efficient algorithm that enumerates all maximal motifs in an input string of length n with $O(n^3)$ time per motif with $O(n^2)$ space and $O(n^3)$ delay. A key of the algorithm is depth-first search on a *tree-shaped search route* for all maximal motifs build by *prefix-preserving closure extension*, which enable us to enumerate all maximal motifs without storing discovered motifs for duplication and maximality tests. To the best of our knowledge, this is the first result on a polynomial space polynomial delay algorithm for maximal pattern discovery for sequences.

The organization of this paper is as follows. In Section 2, we give definitions and basic results. Section 3 gives lowerbounds on the number of maximal motifs. Section 2.2 prepares tools for studying maximal motifs and Section 4 reviews the previous results. In Section 5, we present our algorithm MaxMotif that enumerates all maximal motifs with polynomial space and polynomial delay from an input string. In Section 6, we conclude this paper.

2 Preliminaries

2.1 Maximal Motifs

We briefly introduce basic definitions and results on maximal pattern enumeration according to [13,12]. For definitions not found here, see text books on string algorithms, e.g., [6,8]. Given an alphabet Δ, a *string* of length $n \geq 0$ is a consecutive sequence of letters $s = a[0] \cdots a[n-1] \in \Delta^*$, where $a[i] \in \Delta$, $0 \leq i \leq n-1$. For every $0 \leq i \leq j \leq n-1$, $s[i..j]$ denotes the substring $a_i a_{i+1} \cdots a_j$. Δ^* denotes the set of all possibly empty strings over Δ, and ε denotes the empty string. If $s = uvw$ for some $u, v, w \in \Delta^*$, then we say that u is a *prefix* and w is a *suffix* of s. For a set $S \subseteq \Delta^*$ of strings, we denote by $|S|$ the cardinality of S and by $||S|| = \sum_{s \in S} |s|$ the total length of S.

Let Σ be an alphabet of *solid characters* (or *constant letters*). Let $\circ \notin \Sigma$ be a distinguished letter not belonging to Σ, called the *wild card* (or *don't care*). A wild card $\circ$ matches any solid character $c \in \Sigma$ and also matches $\circ$ itself. An *input string* is a string $s = s[1] \cdots s[n] \in \Sigma^*$ consisting of solid characters of length $n \geq 0$.

Definition 1 (pattern [9,13,12]). *A* pattern *over Σ is a string x in $\Sigma(\Sigma \cup \{\circ\})^*\Sigma$ that starts and ends with a solid character, or an empty string ε.*

We denote the class of patterns by $\mathcal{P} = \{\varepsilon\} \cup \Sigma \cup (\Sigma \cdot (\Sigma \cup \{\circ\})^* \cdot \Sigma)$. For example, ABC and B $\circ$ C are patterns, but $\circ$BC and $\circ\circ$B$\circ$ are not. Note that ε is a pattern in our definition. We define a binary relation $\preceq$ over letters and patterns, called the *specificity relation*.[1] For letters $a, b \in \Sigma \cup \{\circ\}$, we define $a \preceq b$ if either $a = b$ or $a = \circ$ holds. For patterns x and y, We say that x *occurs at position* p *in* y if there exists some index $0 \leq p \leq |y| - |x|$ such that for every index $0 \leq i \leq |x| - 1$, $x[i] \preceq y[p+i]$ holds. Then, we also say that p is an *occurrence* of x in y, and that x *matches* the substring $y[p..p+|x|-1]$.

Example 1. Pattern $x =$ B $\circ$ D occurs in pattern $y =$ AB $\circ$ DA at position 1, and x occurs three times in string $s =$ EABCDABEDBCDE at positions 2, 6 and 9.

We extend binary relation $\preceq$ from letters to patterns as follows. Let x and y be patterns in $\mathcal{P}$. If x occurs at some position p in y, then we define $x \preceq y$ and say that either x is *contained by* y or y is *more specific to* x. For any pattern x,

---

[1] The binary relation $\preceq$ is also called the *generalization relation* or the *subsumption relation* in artificial intelligence and data mining.

we define $\varepsilon \preceq x$. If $x \preceq y$ but $y \not\preceq x$, then we define $x \prec y$ and say that either x is *properly contained by* y or y is *properly more specific to* x. We can see that if $x \preceq y$ and $y \preceq x$ hold, then x and y are identical each other. Furthermore, $\preceq$ is a partial order over $\mathcal{P}$. A maximal motif y is a *successor* of maximal motif x (within $\mathcal{M}$) if $x \prec y$ and there is no maximal motif z such that $x \prec z \prec y$.

Definition 3 (location list [9,13,12]). *For an input string $s \in \Sigma^*$ of length $n \geq 0$, the* location list *of pattern x is the set $\mathcal{L}(x) \subseteq \{0, \ldots, n-1\}$ of all the positions in s at which x occurs. The* frequency *of x on s is $|\mathcal{L}(x)|$.*

Example 2. The location list of pattern $x = \mathtt{B} \circ \mathtt{D}$ in the input string $s =$ EA BCD A BED BCD E is $\mathcal{L}(x) = \{2, 6, 9\}$.

A *quorum* (or *minimum frequency threshold*) is any positive number $\theta \geq 1$. Let $\theta \geq 1$ be a quorum. We say that pattern x is a *θ-motif* (or *motif*, for short) in s if $|\mathcal{L}(x)| \geq \theta$ holds [9,13,12]. Let $\mathcal{L}$ be any location list and d be any integer. Then, we define the *shift of $\mathcal{L}$ with displacement d* by $\mathcal{L} + d = \{\, \ell + d \mid \ell \in \mathcal{L} \,\}$. We write $\mathcal{L} - x$ to represent the set $\mathcal{L} + y$ with $y = -x$.

Definition 5 (Parida *et al* [9]). *Let $\theta \geq 1$ be a quorum. A motif x is* maximal *in s if for any motif y that properly contains x, there is no integer d such that $\mathcal{L}(y) = \mathcal{L}(x) + d$.*

In other words, θ-motif x is maximal in s iff there exists no θ-motif in s properly containing x that is equivalent to x under shift-invariance. Let θ be a quorum. We denote by $\mathcal{F}$ and $\mathcal{M}$ the sets of all (frequent) motifs and all maximal motifs, respectively. Clearly, $\mathcal{M} \subseteq \mathcal{F} \subseteq \mathcal{P}$ for any s and θ.

Lemma 1 ([9,12,13]). *Let $\theta \geq 1$ and $x, y \in \mathcal{P}$ be any motifs. If $x \preceq y$ then $\mathcal{L}(x) \supseteq \mathcal{L}(y) + d$ for some integer $d \geq 0$. The converse does not holds in general.*

Example 3. Let $s = \mathtt{EABCDABEDABCDE}$ over $\Sigma = \{\mathtt{A}, \mathtt{B}, \mathtt{C}, \mathtt{D}, \mathtt{E}\}$ be an input string. Consider motifs $x = \mathtt{AB} \circ \mathtt{D}$ with location list $\mathcal{L}(x) = \{1, 5, 8\}$, $y = \mathtt{B} \circ \mathtt{D}$ with $\mathcal{L}(y) = \{2, 6, 9\}$, and $z = \mathtt{D}$ with $\mathcal{L}(z) = \{4, 8, 11\}$. We can see that $z \prec y \prec z$ holds and x, y, z are equivalent each other. For instance, $\mathcal{L}(z) = \mathcal{L}(x) + d$ with the displacement $d = 3$. Then, x is maximal in s, but y and z are not.

Now, we state our problem as follows.

Definition 6. *The* maximal motif enumeration problem *is, given an input string s of length n and a quorum $\theta \geq 1$, to enumerate all maximal motifs in s without repetition.*

2.2 Merge and Closure

In this subsection, we define merge and closure operations which are originally introduced in [1,3,12,13].

An *infinite string* is a function from integers to symbols in $\Sigma \cup \{\circ\}$. For a finite string $x \in (\Sigma \cup \{\circ\})^*$, the *infinite or expanded version* of x is an infinite

string $\lfloor x \rfloor$ defined by $\lfloor x \rfloor[i] = x[i]$ for $0 \leq i \leq |x| - 1$ and $\lfloor x \rfloor[i] = \circ$ otherwise. For an infinite string x, its *finite string version* (or *trimmed* version), denoted by $\lceil x \rceil$, is the longest substring of x that starts and ends with a solid character in Σ, i.e., $\lceil x \rceil \in \mathcal{P}$, if it exists and *varepsilon* otherwise. By definition, $\lceil \lfloor x \rfloor \rceil = x$ for any $x \in \mathcal{P}$. Let d be an integer called a displacement. For an infinite string x, the infinite string $(x + d)$ is defined by $(x + d)[i] = x[i - d]$ for every i. For a finite string x, $(x + d) = \lfloor x \rfloor + d$. Then, $(x + d)$ is called the *shift of x by d*.

Example 4. Given a finite string s = EABCDABED, its infinite version is $\lfloor s \rfloor = \cdots \circ \circ \downarrow \mathtt{EABCDABED} \circ \circ \cdots$, where $\downarrow$ indicates the origin $i = 0$. Then, $(\lfloor s \rfloor + 2) = \cdots \circ \circ \mathtt{EA} \downarrow \mathtt{BCDABED} \circ \circ \cdots$, and its finite version is $\lceil (\lfloor s \rfloor + 2) \rceil = \downarrow \mathtt{EABCDABED} = \mathtt{s}$.

Merge of infinite and finite strings. Next, we define the merge operator $\oplus$. For letters $a, b \in \Sigma$, we define $a \oplus a = a$ and $a \oplus \circ = \circ \oplus a = a \oplus b = \circ$ if $a \neq b$. For infinite strings α, β, the *merge* of α and β, denoted by $\alpha \oplus \beta$, is the infinite string such that $(\alpha \oplus \beta)[i] = \alpha[i] \oplus \beta[i]$ for every integer i. For finite strings $x, y \in \mathcal{P}$, the *merge* of α and β, denoted by $x \oplus y$, is the finite string $x \oplus y = \lceil \lfloor x \rfloor \oplus \lfloor y \rfloor \rceil \in \mathcal{P}$. Note that the operator $\oplus$ is associative and commutative. For a location list $\mathcal{L} = \{d_1, \ldots, d_{|\mathcal{L}|}\}$, The *merge* of $\mathcal{L}$ ([3,12]) is the pattern $\bigoplus \mathcal{L} \in \mathcal{P}$ defined by

$$\bigoplus \mathcal{L} = \lceil (\lfloor s \rfloor + d_1) \oplus \cdots \oplus (\lfloor s \rfloor + d_{|\mathcal{L}|}) \rceil.$$

Lemma 2. *Let $\mathcal{L}, \mathcal{L}'$ be any location lists.*

1. *If $\mathcal{L} \supseteq \mathcal{L}'$ then $\bigoplus \mathcal{L} \preceq \bigoplus \mathcal{L}'$.*
2. *$\bigoplus \mathcal{L} = \bigoplus (\mathcal{L} + d)$ for any integer d.*

Lemma 3. *Let θ be a quorum and $\mathcal{L}$ be any location list such that $|\mathcal{L}| \geq \theta$. Then, $\bigoplus \mathcal{L}$ is a maximal motif.*

Definition 7 (closure operation [12,13]). *Given a pattern x and an input string s, the maximal motif $Clo(x) = \bigoplus \mathcal{L}(x)$ is called the* closure of x on s.

The above definition is a generalization of the closure operation [11,16] from sets to motifs, and is introduced by [12,13].

Lemma 4. *The closure $Clo(x)$ of a pattern x is unique and computable in $O(mn)$ time from x and $\mathcal{L}(x)$, where $n = |s|$ and $m = |\mathcal{L}(x)| \leq n$.*

Lemma 5 (properties of closure). *Let x, y be any patterns occurring in s and X, Y be any location lists.*

1. *$x \preceq Clo(x)$.*
2. *$Clo(x) = Clo(Clo(x))$.*
3. *If $x \preceq y$ then $Clo(x) \preceq Clo(y)$.*

Theorem 1 (characterization of maximal motifs [12]). *Let θ a quorum and x be a motif pattern in an input string s. Then, the following (i)–(iii) are equivalent:*

(i) x *is a maximal motif.*
(ii) $x = \bigoplus \mathcal{L}$ *and* $|\mathcal{L}| \geq \theta$ *for some* $\mathcal{L} \subseteq \{0, \ldots, |s| - 1\}$.
(iii) $x = Clo(x)$.

A maximal motif x is the unique maximal element with respect to $\preceq$ in the equivalence class $\{\, y \mid \mathcal{L}(x) = \mathcal{L}(y) + d$ for some integer $d \,\}$.

Lemma 6 ([12]). *Let* $\theta \geq 1$ *be a quorum and* $x, y \in \mathcal{M}$ *be maximal motifs.*

1. *Then,* $x \preceq y$ *iff* $\mathcal{L}(x) \supseteq \mathcal{L}(y) + d$ *for some integer* d.
2. *Then,* $x = y$ *iff* $\mathcal{L}(x) = \mathcal{L}(y) + d$ *for some integer* d.

Proof. (1) The only-if direction is obvious from Lemma 1. The if direction follows from Theorem ?? and Property 1 of Lemma 5. □

Example 5. Let s = ABBCABRABRABRABCABABRABBC be an input string. Let x = B∘∘∘∘A be a pattern with location list $\mathcal{L}(x) = \{2, 5, 8\}$. First, we compute the alignment of infinite strings $\mathcal{S} = \{(\lfloor s \rfloor - 2), (\lfloor s \rfloor - 5), (\lfloor s \rfloor - 8)\}$ as follows:

$$\begin{array}{ll} \circ\circ\circ\circ\circ\circ\mathtt{AB}{\downarrow}\underline{\mathtt{B}}\mathtt{C}\underline{\mathtt{AB}}\mathtt{R}\underline{\mathtt{AB}}\mathtt{RABCABABRABBC} & = s - 2 \\ \circ\circ\circ\mathtt{ABBCA}{\downarrow}\underline{\mathtt{B}}\mathtt{R}\underline{\mathtt{AB}}\mathtt{R}\underline{\mathtt{AB}}\mathtt{CABABRABBC}\circ\circ\circ & = s - 5 \\ \mathtt{ABBCABRA}{\downarrow}\underline{\mathtt{B}}\mathtt{R}\underline{\mathtt{AB}}\mathtt{C}\underline{\mathtt{AB}}\mathtt{ABRABBC}\circ\circ\circ\circ\circ\circ & = s - 8 \end{array}$$

where the underlines indicate the common letters. Then, we compute the merge $\bigoplus \mathcal{S} = (\lfloor s \rfloor - 2) \oplus (\lfloor s \rfloor - 5) \oplus (\lfloor s \rfloor - 8)$ of the infinite strings in $\mathcal{S}$ as follows:

$$\circ\circ\circ\circ\circ\circ\circ\circ{\downarrow}\mathtt{B}\circ\mathtt{AB}\circ\mathtt{AB}\circ\circ\circ\circ\circ\circ\circ\circ\circ\circ\circ\circ\circ \quad = \bigoplus \mathcal{S}$$

Finally, we get the closure $Clo(x)$ = B∘AB∘AB by taking its infinite version. □

2.3 Enumeration Algorithms

We introduce terminology for enumeration algorithms according to [7,15]. An *enumeration algorithm* for an enumeration problem Π is an algorithm $\mathcal{A}$ that receives an *instance* I and outputs all *solutions* S in the answer set $\mathcal{S}(I)$ into a write-only output stream O without duplicates. Let $N = ||I||$, $M = |\mathcal{S}(I)|$ be the input and the output sizes on I, and $T_{\mathcal{A}}$ be the total running time of $\mathcal{A}$ for computing all solutions on I. Then, $\mathcal{A}$ is of *output-polynomial* (P-OUTPUT) if $T_{\mathcal{A}}$ is bounded by a polynomial $q(N, M)$. $\mathcal{A}$ is of *polynomial enumeration time* (P-ENUM) if the *amortized time* for each solution $x \in \mathcal{S}$ is bounded by a polynomial $p(N)$ in N, i.e., $T_{\mathcal{A}} = O(M \cdot p(N))$. $\mathcal{A}$ is of *polynomial delay* (P-DELAY) if the *delay*, which is the maximum computation time between two consecutive outputs, is bounded by a polynomial $p(N)$ in the input size N. $\mathcal{A}$ is of *polynomial space* (P-SPACE) if the maximum size of its working space, except the size of output stream O, is bounded by a polynomial $p(N)$. By definition, P-OUTPUT is weakest and P-DELAY is strongest among P-OUTPUT, P-ENUM, and P-DELAY.

3 Lower Bounds for the Number of Maximum Motifs

We show the following lower bound of the number of maximal motifs in a given sequence, which justifies output-sensitive algorithms for the maximal motif enumeration problem. The upper bound of $|\mathcal{M}|$ is obviously $2^{O(n)}$.

Theorem 2 (exponential lowerbound of maximal motifs). *There is an infinite series of input strings $s_0, s_1, s_2, \ldots$, such that for every $i = 0, 1, 2, \ldots$, the number $|\mathcal{M}|$ of maximal motifs in s_i is bounded below by $2^{\Omega(n)}$, that is, exponential in $n = |s_i|$.*

The following theorem says that the number of motifs can be exponentially larger than the number of maximal motifs.

Theorem 3 (succinctness of maximal motifs). *There is an infinite series of input strings $s_0, s_1, s_2, \ldots$, such that for every $i \geq 0$ with quorum $\theta = \frac{1}{2}n$, the number $F = |\mathcal{F}|$ of motifs in s_i is exponential (more precisely $2^{\Omega(n)}$) in the input size n, while the number $M = |\mathcal{M}|$ of maximal motifs in s_i is linear in n, where $n = |s_i|$.*

See the full paper [2] for the proofs of the above theorems. From Theorem 3, we know that a straightforward algorithm for $\mathcal{M}$ based on enumeration of motifs does not work efficiently. This is also true for most real world datasets. Fig. 1 shows an example, where there are only 10 maximal motifs among 50 motifs in a string of length 21.

4 Previous Approaches for Maximal Motif Enumeration

We give a brief review on possible approaches for output-sensitive computation of $\mathcal{M}$ and summarize the previous results.

4.1 Previous Approaches

A most straightforward method of generating maximal motifs is to use frequent pattern generation. We enumerate all motifs in s, classify them into equivalence classes according to their location lists, and find the maximal motifs for each equivalence class. This method requires $O(|\mathcal{F}|)$ time and $O(||\mathcal{F}||)$ memory. Since $O(|\mathcal{F}|)$ can be exponentially larger than $|\mathcal{M}|$, we cannot obtain any output-sensitive algorithm in time and memory in this way.

Another possible method is to use the *basis* for maximal motifs [10,12,13]. Parida *et al* [9] introduced the use of the basis for maximal motif enumeration. A *basis* for $\mathcal{M}$ is a subset $\mathcal{B} \subseteq \mathcal{M}$ of motifs such that $\mathcal{M}$ can be generated by finite applications of an operation, e.g., $\oplus$, over $\mathcal{M}$. Presently, the basis $\mathcal{B}_I$ of *irredundant motifs* [9], the basis $\mathcal{B}_T$ of *tiling motifs* [13], and the basis $\mathcal{B}_P$ of *primitive motifs* [12] have been proposed. A maximal motif $x \in \mathcal{M}$ is *tiling* if for any maximal motifs $y_1, \ldots, y_k \in \mathcal{M}$ and any integers $d_1, \ldots, d_k$ with $x \preceq y_i$,if

Algorithm MaxBasis(θ: quorum, s: input string, $\mathcal{B}$: basis)
$\mathcal{M}^0 := \mathcal{B}$; $i := 0$
while ($\Delta \neq \emptyset$) **do begin**
    $\Delta := \emptyset$;
    foreach $y \in \mathcal{M}^i$ and $d \in \{0, \ldots, n-1\}$ **do**
        if $y \oplus (s+d) \notin (\bigcup_{k=0}^{i} \mathcal{M}^k \cup \Delta)$ **then** $\Delta = \Delta \cup \{y \oplus (s+d)\}$; output x;
    $\mathcal{M}_{i+1} := \Delta$; $i := i + 1$;
end

Fig. 2. An polynomial time enumeration algorithm for generating $\mathcal{M}$ from $\mathcal{B}$ based on breadth-first search. This algorithm does not have polynomial space or polynomial delay.

$\mathcal{L}(x) = \bigcup_i \mathcal{L}(y_i)$ then $x = y_i$ for some i. Pisanti *et al.* [14] describe a simple algorithm for computing $\mathcal{M}$ from $\mathcal{B}_T$ in $O(|\mathcal{M}|^2 \cdot n)$ total time and $O(||\mathcal{M}||)$ space. However, the total time is not linear in $|\mathcal{M}|$. Thus, it is of P-OUTPUT but not of P-ENUM.

4.2 An Improved Algorithm for Generating $\mathcal{M}$ from a Basis $\mathcal{B}_T$

We can improve Pisanti *et al.*'s method for $\mathcal{M}$ adopting an idea used in [12] for generation of $\mathcal{B}_T$ from s. The next lemma is essential for our algorithm.

Lemma 9. *Any maximal motif $x \in \mathcal{M}$ satisfies either (i) $x \in \mathcal{B}_T$, or (ii) there exist some $y \in \mathcal{M}$ and some integer d such that $x \prec y$ and $x = y \oplus (s+d)$.*

Fig. 2 shows our algorithm MaxBasis that computes $\mathcal{M}$ from $\mathcal{B}_T$. We use a trie to store $\mathcal{M}^i$ and Δ for $O(|x|)$ membership of pattern x. Thus, we can implement MaxBasis to run in $O(|\mathcal{M}| \cdot n^2)$ total computation time.

Theorem 5 (generation of maxmal motifs from the basis). *Given a quorum $\theta \geq 1$, an input string s of length n, and the basis $\mathcal{B}_T$ of tiling motifs, the algorithm* MaxBasis *in Fig. 2 enumerates all maximal motifs of $\mathcal{M}$ from $\mathcal{B}$ in $O(n^2)$ amortized time per motif with $O(||\mathcal{M}||)$ space.*

Since its space complexity and delay are $O(||\mathcal{M}||)$ and $O(|\mathcal{M}| \cdot n^2)$, respectively, MaxBasis is neither a polynomial space or polynomial delay even given a basis $\mathcal{B}_T$ as input. Note that it is still open whether the basis $\mathcal{B}_T$ (or $\mathcal{B}_P$) is output-polynomial time computable from s since the total running time of the algorithms in [12] and [14] are only bounded by $O(n^\theta \sum_{i=1}^{\theta} |\mathcal{B}_T^i|)$ or $n^{O(\theta)}$, where $\mathcal{B}_T^i$ is the basis for quorum $i \geq 1$. Hence, it seems difficult to obtain output-polynomial time algorithm for $\mathcal{B}_T$ and thus $\mathcal{M}$ in this approach. [2]

[2] Parida *et al.* [10] presented an output-polynomial time algorithm for the class of *flexible motifs*, and claimed that they also presented a similar algorithm for maximal motifs with wild cards in [9]. Since these algorithms seem to depend on an unproved conjecture in [9], however, we did not include them. At least, the algorithm in [10] requires the space and the delay proportional to the output size $|\mathcal{M}|$. Thus, it is not polynomial space and polynomial delay.

5 A Polynomial Space Polynomial Delay Algorithm Using Depth-First Search

In this section, we present an efficient depth-first search algorithm MaxMotif that, given a quorum $\theta \geq 1$ and an input string s of length n, enumerates all maximal motifs x in s in $O(|\mathcal{L}(x)| \cdot n^2)$ delay and $O(|\mathcal{L}(x)| \cdot |x|)$ space. In what follows, we fix input string s of length $n \geq 1$ and $1 \leq \theta \leq n$. Unlike MaxBasis in the previous section, MaxMotif uses depth-first search over $\mathcal{M}$ to avoid the use of extra storage for keeping all discovered motifs. In the following sections, we explain the details of the algorithm.

5.1 Building Tree-Shaped Search Route for Maximal Motifs

We first build a tree-shaped search route $\mathcal{T} = (\mathcal{V}, \mathcal{P}, \perp)$ for traversing all maximal patterns (Fig. 3). The node set $\mathcal{V} = \mathcal{M}$ consists of all maximal motifs of $\mathcal{M}$, $\mathcal{P}$ is the set of reverse edges defined later, and $\perp = Clo(\varepsilon)$ is the root called the *root motif*. If s contains at least two solid letters then $\perp = \varepsilon$, otherwise $\perp = a$ for the only letter a in s.

Lemma 10. *$\perp = Clo(\varepsilon)$ is the unique shortest maximal motif in s.*

Proof. Since $\mathcal{L}(\perp) = \{0, \ldots, n-1\}$ is the largest location list on s, it follows from Lemma 6 that $Clo(\varepsilon) \preceq x$ for any maximal motif x. □

Next, we define the set $\mathcal{P}$ of reverse edges from a child to its parent as follows. Given a maximal motif x, the *core index* of x, denoted by $core\_i(x)$, is the smallest index $0 \leq \ell \leq |x|-1$ such that $\mathcal{L}(x) = \mathcal{L}(y)$ for the prefix $y = x[0..\ell]$. Then, we assign the unique parent to each non-root maximal motifs.

Definition 8 (parent of maximal motif). *Let y be a maximal motif such that $y \neq \perp$. If $\ell = core\_i(y)$ is the core index of y, then the* parent *of y, denoted by $\mathcal{P}(y)$, is the pattern $\mathcal{P}(y) = Clo(\lceil y[0..\ell-1] \rceil)$.* [3]

Lemma 11. *For every maximal motif y such that $y \neq \perp$, $\mathcal{P}(y)$ is always exists, unique, and maximal. Furthermore, $\mathcal{P}(y) \prec y$ holds.*

Proof. Let $p = \lceil y[0..\ell-1] \rceil$. If $y \neq \perp$, then $\ell - 1 = core\_i(y) - 1 \geq -1$ and p is always defined. If ℓ is the core index of y, then $\mathcal{L}(p) \supset \mathcal{L}(y) \neq \emptyset$, and thus $\mathcal{P}(y)$ is defined and maximal. Furthermore, if x, y are maximal then Lemma 2 and Theorem 6 imply that $x = Clo(p) \prec Clo(y) = y$. □

Theorem 6. *$\mathcal{T} = (\mathcal{V}, \mathcal{P}, \perp)$ is a spanning tree for all maximal motifs in $\mathcal{M}$.*

Proof. From Lemma 11, all maximal motifs y but $\perp$ have the unique parent $\mathcal{P}(y)$ such that $\mathcal{P}(y) \prec y$. Since the relation $\preceq$ is *acyclic* on $\mathcal{M}$, i.e., there is no infinite decreasing chain of maximal motifs of $\mathcal{M}$, the result follows. □

The remaining task is to show how to enumerate all children y of a given parent motif x without using extra space. This is not an easy task since we have only reverse edges. We discuss this issue in the next subsection.

[3] In the definition, $y[0..\ell-1] \in \Sigma(\Sigma \cup \{\circ\})^* \cup \{\varepsilon\}$ may not be a proper pattern. Thus, we use $\lceil y[0..\ell-1] \rceil$ instead of $y[0..\ell-1]$ to remove the trailing $\circ$'s.

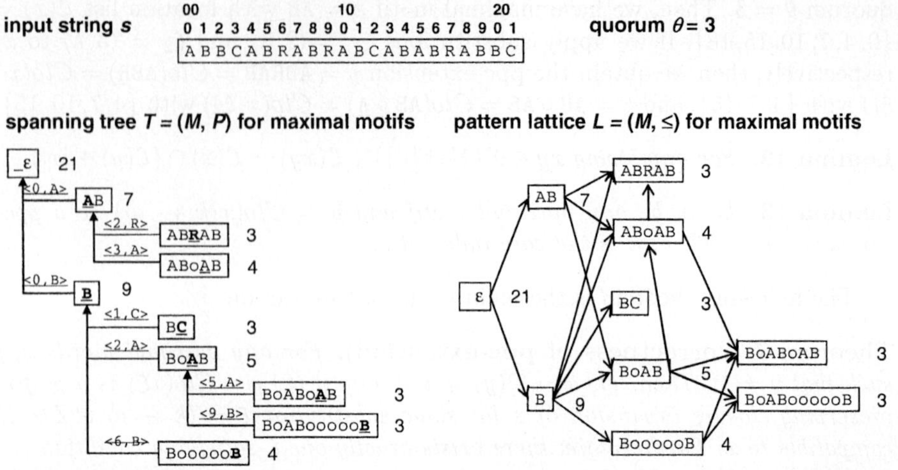

Fig. 3. The spanning tree $\mathcal{T} = (\mathcal{M}, \mathcal{P})$ (left) and the pattern lattice $\mathcal{L} = (\mathcal{M}, \preceq)$ (right) for maximal motifs of $\mathcal{M}$ on quorum $\theta = 3$ and input string s (top). Each box represents maximal motif x in $\mathcal{M}$ and the number right to the box indicates its frequency $|\mathcal{L}(x)|$. Each arrow indicates ordering, $\mathcal{P}$ or $\preceq$, of a tree/lattice. (Sec. 5.1). An arrow in the tree $\mathcal{T}$ indicates the ppc-extension with seed $\langle k, c\rangle$. The newly introduced letter c is written in bold face. (Sec. 5.2). There are 10 maximal motifs among 49 motifs in s.

5.2 Prefix-Preserving Closure Extension

We introduce the prefix-preserving closure extension defined as follows. A *substitution* for motif x is a pair $\xi = \langle k \leftarrow c\rangle \in \mathsf{Z} \times \Sigma$ of integer k and solid letter c. If $\lfloor x \rfloor[k] = \circ$, ξ is *compatible* to x. The *application* of ξ to x, denoted by $x\xi = x\langle k \leftarrow c\rangle$, is the motif $\lceil y \rceil$, where y is the infinite string such that for every integer i, $y[i] = c$ if $i = k$ and $y[i] = \lfloor x \rfloor[i]$ otherwise. For example, if $x = \mathtt{BA} \circ \mathtt{B}$, then $x\langle -1 \leftarrow \mathtt{C}\rangle = \mathtt{CBA} \circ \mathtt{B}$, $x\langle 2 \leftarrow \mathtt{C}\rangle = \mathtt{BACB}$, and $x\langle 6 \leftarrow \mathtt{C}\rangle = \mathtt{BA} \circ \mathtt{B} \circ \circ \mathtt{C}$.

Definition 9 (ppc-extension). *For any maximal motifs x, y such that $y \neq \bot$, a motif y is a* prefix-preserving closure expansion *(or a* ppc-extension*) of x if the following (i)–(iii) hold:*

(i) $y = Clo(x\langle k \leftarrow c\rangle)$ for some substitution, called the seed, *$\xi = \langle k \leftarrow c\rangle \in \mathsf{Z} \times \Sigma$ compatible to x, that is, y is obtained by first substituting c at index i and then taking its closure,*

(ii) the index k satisfies $k > core\_i(x)$, and

(iii) $x[0..k-1] = y[0..k-1]$, that is, the prefix of length $i-1$ is preserved, where $x[0..k-1]$ is the string $\lfloor x \rfloor[0..k-1]$ obtained from x by padding trailing $\circ$'s if necessary.

Example 6. In Fig. 3, we show an example of the spanning tree for $\mathcal{M}$ generated by the ppc-extension for an input string s = ABBCABRABRABCABABRABBC and

quorum $\theta = 3$. Then, we have maximal motif $x = \mathtt{AB}$ with location list $\mathcal{L}(x) = \{0, 4, 7, 10, 15, 18\}$. If we apply substitutions $\xi_1 = \langle 2, \mathtt{R} \rangle$ and $\xi_2 = \langle 3, \mathtt{A} \rangle$ to x, respectively, then we obtain the ppc-extension $y = \mathtt{ABRAB} = Clo(\mathtt{AB\underline{R}}) = Clo(x \cdot \xi_1)$ with $\{4, 7, 15\}$, and $z = \mathtt{AB} \circ \mathtt{AB} = Clo(\mathtt{AB} \circ \underline{\mathtt{A}}) = Clo(x \cdot \xi_2)$ with $\{4, 7, 10, 15\}$.

Lemma 12. *For any string $xy \in \Sigma(\Sigma \cup \{\circ\})^*$, $\mathcal{L}(xy) = \mathcal{L}(x) \cap (\mathcal{L}(y) + |x|)$.*

Lemma 13. *Let x be any maximal motif and $y = Clo(x\langle k \leftarrow c \rangle)$ be a ppc-extension of x. Then, k is the core index of y.*

The following theorem is the main result of this section.

Theorem 7 (correctness of ppc-extension). *For any maximal motifs x, y such that $y \neq \perp$. Then, (1) $x = \mathcal{P}(y)$ if and only if (2) $y = Clo(x\xi)$ is a prefix-preserving closure expansion of x for some substitution $\xi = \langle k \leftarrow c \rangle \in \mathsf{Z} \times \Sigma$ compatible to x. Furthermore, there exists exactly one ξ satisfying condition (2) for each y.*

Proof. We give a sketch of the proof. Please see the full paper [2] for the details. (1) to (2): Suppose $x = \mathcal{P}(y)$, and thus $x = Clo(y[0..\ell - 1])$ for the core index ℓ of y. First, we can show that $x[0..\ell - 1] = y[0..\ell - 1]$ holds. Otherwise, we can construct a more specific motif $y' = x[0..\ell - 1]y[\ell..|y| - 1]$ such that $y \prec y'$ but $\mathcal{L}(y) = \mathcal{L}(y')$ using Lemma 12. Next, if we take $\xi = \langle \ell \leftarrow y[\ell] \rangle$, then we can show that $y = Clo(x\xi)$ and ξ satisfies the condition of ppc-extension. This proves this direction. (2) to (1): This direction is easy. Suppose that $y = Clo(x\langle k \leftarrow c \rangle)$. Then, it follows from Lemma 13 that $\mathcal{P}(y) = Clo(y[0..k - 1])$. If $y[0..k - 1] = x[0..k - 1]$, then $Clo(y[0..k - 1]) = Clo(x[0..k - 1])$. Since x is maximal, we have $k - 1 \geq core\_i(x)$, and thus $Clo(x[0..k - 1]) = x$. Hence, we have $\mathcal{P}(y) = x$. By condition (iii) of ppc-extension, we can show that choice of ξ is unique. □

5.3 A Polynomial Space Polynomial Delay Algorithm

Based on Theorem 7, we present in Fig. 4 our algorithm MaxMotif that enumerates all maximal motif in a given input string by the depth-first search over $\mathcal{M}$ applying the ppc-extension to each maximal motifs.

A straightforward implementation of the procedure Expand in Fig. 4 requires $O(|\mathcal{L}(x)| \cdot n)$ time for each of $n \cdot |\Sigma|$ possible children at line 4 to line 9 even when none of them satisfies the quorum θ. This only yields an algorithm with $O(|\Sigma| \cdot |\mathcal{L}(x)| \cdot n^2) = O(|\Sigma| n^3)$ time and delay. Then, we have the following theorem.

Theorem 8. *Given a quorum $\theta \geq 1$ and an input string s of length n, the algorithm* MaxMotif *in Fig. 4 enumerates all maximal motifs x of $\mathcal{M}$ in $O(mn^2)$ amortized time per motif with $O(\ell m)$ space and $O(mn^2)$ delay, where $\ell = |x|$ and $m = |\mathcal{L}(x)|$.*

Proof. By Theorem 6 and Theorem 7, we see that the algorithm MaxMotif visits all maximal motifs on the spanning tree $\mathcal{T}$ starting from the root $\perp$. Since

---

Algorithm MAXMOTIF(θ: quorum, s: input string)
0 $\bot = Clo(\varepsilon)$; //the root motif $\bot$.
1 **call** EXPAND($\bot, -1, \theta, s$); //$core\_i(\bot) = -1$.

Procedure EXPAND(x: motif, ℓ: core index, θ: quorum, s: input string)
2 **if** $|\mathcal{L}(x)| < \theta$ **then return**;
3 **output** x;
4 **for** $k := \ell + 1$ **to** $|s|$ **do** //$core\_i(x) = \ell$ is ensured.
5   **foreach** $c \in \Sigma$ **do begin**
6     $y = Clo(x\langle k \leftarrow c\rangle)$; //ppt-extension.
7   **if** $\lfloor x \rfloor[0..k-1] = \lfloor y \rfloor[0..k-1]$ **then**
8   **call** EXPAND(y, k, s, θ);
9   **end for**

Fig. 4. A polynomial space polynomial delay enumeration algorithm for $\mathcal{M}$

$\mathcal{T}$ is a tree and any maximal motif appears in $\mathcal{T}$, the algorithm enumerate $\mathcal{M}$ without duplicates. Let $W(x)$ be the work that EXPAND spends for each parent maximal motif x except the recursive call. Since the closure $Clo(x)$ at line 6 takes $O(|\mathcal{L}(x)| \cdot n)$ time, $W(x) = O(|\Sigma| \cdot |\mathcal{L}(x)| \cdot n^2)$ time and this gives the delay per maximal motif. Note that we have to charge to the parent x the work for all children since x may have no maximal children. To reduce computation time further, we replace line 5 to line 9 by the following procedure OCCURRENCEDELIVER [4], which improves the work to $W(x) = O(|\mathcal{L}(x)| \cdot n^2)$, and thus gives the delay $D = W(x) \cdot |x|$ time since the depth of $\mathcal{T}$ is $\Theta(|x|)$.

OCCURRENCEDELIVER($x, \mathcal{L}(x), k$) $\equiv$
1 Initialize all $\mathcal{L}(c) := \emptyset$ for all $c \in \Sigma'$, and then $\Sigma' := \emptyset$;
2 **foreach** $d \in \mathcal{L}(x)$ **do**
3   $\mathcal{L}(c) := \mathcal{L}(c) \cup \{d\}$ and $\Sigma' = \Sigma' \cup \{c\}$, where $c := s[d+k]$;
5 **foreach** $c \in \Sigma'$ **do**
6   Compute $y = Clo(x\langle k \leftarrow c\rangle)$ using $\mathcal{L}(c)$

By applying the technique by Uno [15] that transforms any tree-search enumeration algorithm with work time $W(x)$ at each node into a $W(x)$ delay algorithm, we finally obtain an algorithm with delay $O(|\mathcal{L}(x)| \cdot n^2)$. □

In summary, MAXMOTIF computes all maximal motifs in $O(n^3)$ time per motif with $O(n^2)$ space and $O(n^3)$ delay in the input size n.

Corollary 2. *The maximal motif enumeration problem is solvable in polynomial space and polynomial delay in the input size* $n = |s|$.

[4] Occurrence deliver is introduced for closed itemsets enumeration in Uno *et al.* [16].

6 Conclusion

In this paper, we presented a polynomial space polynomial delay algorithm for enumerating all maximal motifs in an input string for the class of motifs with wild cards. By the use of depth-first search based on the prefix-preserving expansion, the algorithm enumerate all motifs without explicitly storing and checking the motifs enumerated so far. This drastically improves the space and the delay complexities compared with the previous algorithms with breadth-first search. As future research, we plan to empirical evaluation of MaxMotif algorithm on the real world datasets such as biological datasets.

In data mining, maximal motifs for *sets* are called *closed itemsets* [11]. There are a number of closed pattern discovery algorithms for itemsets [5,11], while only a few algorithms are known for sequences and trees [3,16,17]. Thus, it is an interesting research direction to extend the result of this paper for motif discovery in trees and graphs. Extension of the result for classes of unions of motifs [4] and basis of tiling motifs [13] is another research direction.

References

1. A. Apostolico and L. Parida, Compression and the wheel of fortune, In *Proc. the 2003 Data Compression Conference (DCC'03)*, IEEE, 2003.
2. H. Arimura, T. Uno, A polynomial space polynomial delay algorithm for enumeration of maximal motifs in a sequence, Technical Report Series A, TCS-TR-A-05-6, Division of Computer Science, Hokkaido Univeristy, July 2005. `http://www-alg.ist.hokudai.ac.jp/tra.html`
3. H. Arimura, T. Uno, An output-polynomial time algorithm for mining frequent closed attribute trees, In *Proc. ILP'05*, LNAI 3625, 1–19, August 2005.
4. H. Arimura, T. Shinohara, S. Otsuki, Finding minimal generalizations for unions of pattern languages and its application to inductive inference from positive data, In *STACS'94*, LNCS 775, Springer-Verlag, 649–660, 1994.
5. E. Boros V. Gurvich, L. Khachiyan, K. Makino, The complexity of generating maximal frequent and minimal infrequent sets, In *Proc. STACS '02*, LNCS, 133-141, 2002.
6. M. Crochemore and W. Rytter, *Jewels of Stringology*, World Scientific, 2002.
7. L. A. Goldberg, Polynomial space polynomial delay algorithms for listing families of graphs, In *Proc. the 25th STOC*, ACM, 218–225, 1993.
8. D. Gusfield, Algorithms on strings, trees, and sequences, Cambridge, 1997.
9. L. Parida, I. Rigoutsos, A. Floratos, D. Platt, and Y. Gao, Pattern discovery on character sets and real-valued data: linear bound on irredundant motifs and effcient polynomial time algorithm, In *Proc. the 11th SIAM Symposium on Discrete Algorithms (SODA'00)*, 297–308, 2000.
10. L. Parida, I. Rigoutsos, D. E. Platt, An Output-Sensitive Flexible Pattern Discovery Algorithm. In *Proc. CPM'01*, LNCS 2089, 131–142, 2001.
11. N. Pasquier, Y. Bastide, R. Taouil, L. Lakhal, Discovering Frequent Closed Itemsets for Association Rules, In *Proc. ICDT'99*, 398–416, 1999.
12. J. Pelfrêne, S. Abdeddaïm, and J. Alexandre, Extending Approximate Patterns, In *Proc. CPM'03*, LNCS 2676, 328–347, 2003.

13. N. Pisanti, M. Crochemore, R. Grossi, and M.-F. Sagot, A basis of tiling motifs for generating repeated patterns and its complexity for higher quorum, In *Proc. MFCS'03*, LNCS 2747, 622–631, 2003.
14. N. Pisanti, M. Crochemore, R. Grossi, and M.-F. Sagot, A comparative study of bases for motif inference, In *String Algorithmics*, KCL publications, 2004.
15. T. Uno, Two general methods to reduce delay and change of enumeration algorithms, NII Technical Report, NII-2003-004E, April 2003.
16. T. Uno, T. Asai, Y. Uchida, H. Arimura, An efficient algorithm for enumerating closed patterns in transaction databases, In *Proc. DS'04*, LNAI 3245, 16-30, 2004.
17. X. Yan, J. Han, CloseGraph: Mining Closed Frequent Graph Patterns In *Proc. SIGKDD'03*, 2003.

5-th Phylogenetic Root Construction for Strictly Chordal Graphs

William Kennedy* and Guohui Lin**

Algorithmic Research Group and Bioinformatics Research Group,
Department of Computing Science, University of Alberta,
Edmonton, Alberta T6G 2E8, Canada
{kennedy, ghlin}@cs.ualberta.ca

Abstract. Reconstruction of an evolutionary history for a set of organisms is an important research subject in computational biology. One approach motivated by graph theory constructs a relationship graph based on pairwise evolutionary closeness. The approach builds a tree representation equivalent to this graph such that leaves, corresponding to the organisms, are within a specified distance of k in the tree if connected in the relationship graph. This problem, the k-th phylogenetic root construction, has known linear time algorithms for $k \leq 4$. However, the computational complexity is unknown when $k \geq 5$. We present a polynomial time algorithm for strictly chordal relationship graphs when $k = 5$.

Keywords: Computational biology, phylogeny reconstruction, phylogenetic root, Steiner root, chordal, strictly chordal.

1 Introduction

A phylogeny is the development and history through evolution of a set of organisms or evolutionary units. A phylogenetic tree is a visual representation with the leaves labeled by the evolutionary units and distances in the tree representing evolutionary closeness; reconstruction of such trees is a fundamental question in computational biology. One approach, based on graph theory, uses an input graph representing the known relationships of the units; the vertices are labeled by the units and connected if their relative closeness is greater than some prespecified threshold. The approach then constructs a tree, if it exists, where the unit-labeled leaves are within a given path distance in the tree if and only if the evolutionary units' vertices are connected in the input graph. This problem is a variation of the well-studied graph root and graph power problem.

The k-th power of a graph $G = (V, E)$, denoted G^k, is another graph on the same vertex set with an edge between $x, y \in V$ if and only if a path in G of length at most k exists between x, y. Given a graph, it is efficient to compute a k-th power, but the reverse direction, finding a k-th root, is generally more

* Supported partially by NSERC.
** Supported by NSERC and CFI.

X. Deng and D. Du (Eds.): ISAAC 2005, LNCS 3827, pp. 738–747, 2005.

complex. For example, finding a square root of graph is NP-complete [9]. A few polynomial time algorithms exist for computing if a graph has a k-th root where this root is a tree [3,8]. For a graph G, a *k-th leaf root* is a tree T where the vertex set of G corresponds to the leaves of T and the k-th power T^k induced on the leaf set of T is isomorphic to G. The internal vertices in T are *Steiner points.* For a Steiner point the number of Steiner points it is adjacent to in T is its *Steiner degree.* All strictly chordal graphs are leaf powers, and finding the k-th leaf root of a strictly chordal graph for $k \geq 4$ has a linear time solution [4].

The *k-root phylogenetic tree* problem is a restriction of the k-th leaf root with, in addition, all Steiner points necessarily having degree at least 3; this derived from the notion of an internal point in the tree representing a genetic split from a common ancestor. The problem of finding a k-root phylogenetic tree has a linear time solution for $k \leq 4$ [6]. [5,7] have produced algorithms for the $k = 5$ case that runs in linear time if the critical clique graph $CC(G)$ is a tree.

The *k-root Steiner* problem is a relaxation of the k-th leaf root with vertices of G now represented by both the leaves and some internal vertices of the tree. [6] used this problem as an important intermediary step for constructing the 3rd and 4th root phylogenetic tree; we employ an analogous approach in this paper. The main contribution of this paper will be to produce an algorithm that decides if a given strictly chordal graph has a 5-root phylogeny tree, and if such a tree exists, constructs this tree.

2 Preliminaries

A forest is a simple acyclic graph; a tree is a connected forest. We will assume that a tree is undirected. A graph is *chordal* [1] if it contains no induced cycle of length four or more. Chordal graphs can be recognized in linear time [1]. All graphs that have a leaf power representation are chordal graphs [6].

Let $G = (V, E)$ be a graph. A *clique* is a set of pairwise adjacent vertices. A clique is *maximal* if it is not properly contained in any other clique. A *homogeneous clique* is a clique S such that either $|S| = 1$ or for all $v_1, v_2 \in S$ and $w \in V \setminus S$, $v_1 w \in E$ if and only if $v_2 w \in E$. A *critical clique* is a homogeneous clique that is not a proper subset of any other homogeneous clique [6]. The *critical clique cardinality* of a maximal clique K, denoted $cccard(K)$, is the number of critical cliques it contains. We define a maximal clique to be *large* if it has critical clique cardinality three or more. A critical clique is *internal* if contained in at least two maximal cliques and *external* otherwise.

The set of critical cliques of a graph is a partition of the vertex set of the graph. Two critical cliques C_1 and C_2 are adjacent if the vertices of C_1 are adjacent to the vertices of C_2. A critical clique graph $CC(G)$ with nodes of $CC(G)$ corresponding to the critical cliques of G such that two nodes are adjacent if and only if the critical cliques that they represent are adjacent [6]. To avoid confusion we will refer the vertices in $CC(G)$ as nodes.

For a vertex set V and a *hyperedge* set $\mathcal{E} = \{E_i | i = 0, 1, 2, ..., n\}$ such that $E_i \subseteq V$, the pair $\mathcal{H} = (V, \mathcal{E})$ is a *hypergraph* [1]. A *twig* in a hypergraph $\mathcal{H}(V, \mathcal{E})$,

denoted E_t, is a hyperedge such that there exists another hyperedge, E_b called a branch, with the property that $E_t \cap (\cup_{E \in \mathcal{E}-E_t} E) = E_t \cap E_b$. A hypergraph is a *hypertree* if there exists an ordering $(E_1, E_2, \ldots, E_m)$ of the hyperedges such that E_i is a twig in $\mathcal{H}(V, \mathcal{E}_i)$ for $1 \leq i \leq m$ given that $\mathcal{E}_i = (E_1, E_2, \ldots, E_i)$. This definition of hypertree corresponds to [4,2] and differs from the definition in [1]. Given a set of hyperedges $\mathcal{E}' \subseteq \mathcal{E}$, We say the intersection I of $\mathcal{E}'$ is *maximal* [4] if no other hyperedges exist which contain all the vertices of I. We say that an intersection is *strict* [4] if for every pair of edges, $E', E'' \in \mathcal{E}$, $E' \cap E'' = I$ and $E^* \cap I = \emptyset$ for all $E^* \in \mathcal{E} - \mathcal{E}'$. A hypertree is *strict* [4] if all its intersections are strict. Define a *clique hypergraph* of a graph $G = (V, E)$ with vertex set V and hyperedges set as the maximal cliques of G. A graph is *strictly chordal* if it is chordal and its clique hypergraph is a strict hypertree. There exists a linear time algorithm to recognize strictly chordal graphs [4].

A k-root Steiner tree T of a graph G is S*-restricted* for a set of critical cliques S if T has no degree 2 Steiner points and critical cliques in S are internal in T.

Lemma 1. *[7] For a graph G, let S be the critical cliques of size 1 in G. If no (k-2)-root Steiner tree T is associated with G such that T is an S-restricted Steiner tree, then no k-root phylogenetic tree is equivalent to G.*

We now give a brief overview of the strategy we employ to produce a 5-root phylogenetic tree from a graph G. Starting with G, check if G is strictly chordal, and if yes, we build the critical clique graph $CC(G)$. Using $CC(G)$, we produce the set of tree chordal graphs $\mathcal{T}$ by removing edges in large maximal cliques (see Sect. 2.1). For each tree chordal graph $T_i \in \mathcal{T}$, we apply $Alg(T_i)$, a modification of the algorithm for producing a 5-root phylogenetic tree for tree chordal graphs given by Lin et al. [7]. If $Alg(T_i)$ fails to produce a 3-root Steiner tree we will find no phylogenetic tree by Lemma 2. If for all $T_i \in \mathcal{T}$, $Alg(T_i)$ produces a Steiner tree S_i we then continue to consider the edges removed from the large maximal cliques. We combine each 3-root Steiner tree S_i until either we come to a contradiction or we have a valid S-restricted Steiner tree where S is the critical cliques of size 1 in G. If we construct such a Steiner tree, by Lemma 1 we are always able to produce a corresponding 5-root phylogenetic tree.

Before we present the algorithms, we first discuss the tree chordal algorithm presented by Lin et al. in [7] (Sect. 2.1) and discuss the structure of large maximal cliques (Sect. 2.2). We will present the algorithm in three progressively less restrictive parts in Sects. 3.1 – 3.2. Due to space constraints, we do not provide proofs of all lemmas and theorems. The interested read may refer to the full paper available at URL http://www.cs.ualberta.ca/research/techreports/2005.php.

2.1 Tree Chordal Graphs

A graph G is *tree chordal* [7] if $CC(G)$ is a tree. [7] showed a polynomial time algorithm to construct a 5-root phylogenetic tree from a tree chordal graph. Starting with $CC(G)$, this algorithm produces an S-restricted 3-root Steiner tree where the set S contains external nodes of size 1 in $CC(G)$; from this

Steiner tree a 5-root phylogenetic tree is easily produced. The non-trivial cases for this algorithm are internal nodes of size 1 and external nodes of size 2 and 3.

Our algorithm for 5-PRP on strictly chordal graphs uses the algorithm for 5-PRP on tree chordal graphs and the following observation. Observe that we can produce a set of tree chordal graphs $\mathcal{T}$ from a strictly chordal graph in the following way. Using the critical clique graph $CC(G)$, remove the edges from large maximal cliques to produce $CC(\mathcal{T})$, which is a set of trees. To create $\mathcal{T}$, re-substitute each critical clique in for the node that it represents in $CC(\mathcal{T})$.

Let G be a strictly chordal graph with an S-restricted 3rd Steiner root tree T, where S is the set of all size 1 critical cliques in G. Let $\mathcal{T}$ be a forest of tree chordal graphs decomposed from G. Let c be a critical clique contained in a large maximal clique in G and contained in a tree chordal graph T_i, decomposed into at least two nodes in $CC(T_i)$.

We now consider the construction of an S-restricted 3rd Steiner root tree T_i' for a T_i in $\mathcal{T}$ as an intermediary step in the process of the 5-PRP algorithm for strictly chordal graphs, therefore, we will allow a critical clique such as c to be adjacent to a degree 2 Steiner point or c to have a single size 1 leaf representative. These inconsistencies are allowable as long as they do not exist in the final S-restricted 3rd Steiner root tree T for G. If c has a single size 1 leaf representative r_c in T_i' then r_c must be internal in T; as we will see in Sect. 3, this is always able to be done. For the degree 2 Steiner point, we note that c must have a single representative r_c as it is adjacent to another critical clique in T_i. r_c will need to be adjacent to an additional Steiner point in T, as the degree must be at least 3 in T and if another representative is adjacent to this Steiner point then it will be indistinguishable from r_c.

Therefore, for the following S-restricted 3rd Steiner root tree constructions we allow a slight relaxation with critical cliques such as c to be adjacent to a degree 2 Steiner point or to have a single size 1 leaf representative.

Corollary 3.1 of [7] can easily be adapted to show that no size 1 leaf exists in trivial tree chordal graphs. Lemma 3.2 of [7] shows that a critical clique has diameter at most 2 in the Steiner tree. Finally, Lemma 3.9 of [7] shows that leaf nodes of size 2 or 3 are represented by a single vertex adjacent to a neighbors representative and Lemma 3.10 shows that leaf nodes of size at least 4 are represented by two representatives adjacent to a common Steiner point.

A critical clique c is *constrained* in K if c has two or more representatives in the Steiner tree T, c has a single representative adjacent to a Steiner point with Steiner degree 1, or c has a single representative adjacent to the representative of another critical clique. As shown in Sect. 2.2, at most one critical clique can be constrained in a large maximal clique of a strictly chordal graph with an S-restricted 3rd Steiner root tree. Therefore, the algorithm aims to produce constrained critical cliques only when necessary.

We first deal with the case when $CC(G)$ contains less than 3 nodes - trivial tree chordal graphs. A tree chordal graph T is *trivial* when $CC(T)$ is a single node. No connected graph will have a $CC(G)$ with two nodes; as the two adjacent critical cliques would be one large critical clique. Trivial tree chordal graphs T

can arise in two ways, when decomposed from a strictly chordal graph: T was part of only large maximal cliques in G, or T was part of a large maximal clique and decomposed into a tree chordal graph of exactly two critical cliques. Therefore, we describe an algorithm to handle trivial tree chordal graphs.

If a tree chordal graph was part of only large maximal cliques in G, the corresponding S-restricted 3rd Steiner root tree to the T_i will be a single representative. For the second case, we present the following algorithm $TrivAlg(T_i)$.

1. if both c_1 and c_2 are internal critical cliques in G then represent c_1 and c_2 by two single representatives connected by a path of two Steiner points; or,
2. if only one is an internal critical clique, assume c_1, then:
   a. if $|c_1| = 1$ and $1 < |c_2| < 4$ then represent c_1 and c_2 by two single adjacent representatives;
   b. if $|c_1| > 1$ and $1 < |c_2| < 4$ then represent c_1 by r'_{c_1} and r''_{c_1}, c_2 by r_{c_2}, and create path $r'_{c_1} - r''_{c_1} - r_{c_2}$;
   c. if $|c_2| > 3$ then represent c_1 by a single representative, represent c_2 by two representatives of sizes $\lceil |c_2|/2 \rceil$ and $\lfloor |c_2|/2 \rfloor$, and make all adjacent to a common Steiner point; or,
   d. otherwise ($|c_2| = 1$) no S-restricted 3rd Steiner root tree exists.

The trees produced by $TrivAlg(T_i)$ satisfy the condition of being an S-restricted 3rd Steiner root tree for T_i with respect to relaxation given above. Notice that all configurations for this critical clique are constrained. The choice for critical cliques represented by a path of representatives are not chosen in this case as all adjacent critical cliques will need to have single representatives and, as we will show, one possibility will force one maximal cliques contained critical cliques to have single representatives. Therefore, we choose the less restrictive case. This leaves two options for the S-restricted 3rd Steiner root tree: (1) the option presented in the algorithm and (2) letting c_1 and c_2 be adjacent, with no Steiner points. For (1) c_1 and c_2 must be contained in two additional maximal cliques each in G, one adjacent to c_1 or c_2 (if $|c_1| = 1$ or $|c_2| = 1$) and the other adjacent to the Steiner points adjacent to c_1 and c_2. For (2) that c_1 and c_2 must also be contained in two additional maximal cliques each. The difference is all maximal cliques adjacent to c_1 and c_2 will have to all adjacent critical cliques in $CC(G)$ as unconstrained for (2), whereas only one maximal clique needs all contained critical cliques as unconstrained for (1) (See Lemma 7). Cases 2a and 2b satisfy Lemma 3.9 of [7], where case 2a must be contained in at least two additional maximal cliques in G and case 2b is only contained in at least one. Case 2c is ideal with no restriction place on the maximal cliques adjacent to c_1. For Case 2d no S-restricted 3rd Steiner root tree exist for G, as c_2's representative will always be external and have size 1.

We now describe and justify a modification of the tree chordal algorithm of Lin et al. [7] to minimize constrained critical cliques. Given a tree chordal graph $T_i \in \mathcal{T}$ decomposed from a graph G, set S corresponding to nodes in T_i of size 1 in $CC(G)$, and a set R corresponding to nodes of $CC(T_i)$ contained in maximal cliques of size three or more in $CC(G)$ produce an S-restricted 3rd Steiner root tree as follows. Denote this modified algorithm $ALG(G)$.

- If T_i was part of only large maximal cliques, return a single representative.
- If T_i was part of a large maximal clique and decomposed into a tree chordal graph of exactly two critical cliques, return tree as in $TrivAlg(T_i)$.
- Produce tree chordal graph $T_i^\star$ as follows:
  - size two and three external nodes contained in R, change size to four;
  - size one external nodes contained in R adjacent to degree-2 size-2 node in $CC(T_i)$, change size to four;
  - remaining size one external nodes contained in R, change size to two;
  - size one internal nodes contained in R which are not adjacent to an external node not contained in R, change size to two.
- Call the tree chordal algorithm with the modified tree $T_i^\star$.
- return no if the tree chordal algorithm fails, or return the S-restricted 3rd Steiner root tree.

Lemma 2. *Given a strictly chordal graph G decomposed into a forest of tree chordal graphs $\mathcal{T}$ and set S corresponding to nodes in G of size 1, if $ALG(T)$ fails to produce a valid S-restricted 3rd Steiner root tree for any $T_i \in \mathcal{T}$ then no S-restricted 3rd Steiner root tree exists for G.*

By Lemmas 1 and 2, if $ALG(T)$ fails for any tree chordal graph, we can return no, as no S-restricted 3rd Steiner root tree exits and therefore no 5th phylogenetic root tree will exist. We now enumerate the possibilities of a critical clique returned by $ALG(T)$.

Lemma 3. *Given a tree chordal graph G with at least two critical cliques, $ALG(G)$ leaves the representatives of any critical cliques in the 3rd Steiner root tree T in exactly one of the follow states:*

c1: Representatives adjacent to a Steiner point of Steiner degree one; nearest representative of another critical clique is at distance of three with:
- *a: a single representative, or*
- *b: two representatives,*

c2: One representative adjacent to a degree two Steiner point, with:
- *a: nearest representative of another critical clique is at distance of three, or*
- *b: nearest representative of another critical clique is at distance of two,*

c3: One representative at distance of one to another leaf critical clique and a Steiner point, other critical cliques are at a distance of three,

c4: One representative adjacent to one a representative of another critical clique.

c5: Two adjacent representatives; one adjacent to another leaf's representative.

2.2 Structure of Large Maximal Cliques

The following lemma, Lemma 4, is an example of structure that is a potential problem for construction of an S-restricted 3rd Steiner root tree; the following section shows why this poses a problem and how it becomes unnecessary in the construction of an S-restricted 3rd Steiner root tree.

Lemma 4. *[4] Let G be a graph with a 3rd Steiner root T. Assume there exist in G three maximal cliques K_1, K_2, K_3 such that $K_1 \cap K_2 = I_1 \neq \emptyset$, $K_2 \cap K_3 = I_3 \neq \emptyset$, and $K_1 \cap K_3 = \emptyset$. Let $I_2 = K_2 - I_1 - I_3$. If $I_1 = \{u_1, u_1'\}$, $I_3 = \{u_3, u_3'\}$, and $|I_2| > 0$, then u_1-u_1'-u_3'-u_3 is a path in T and every representative for a critical clique in I_2 is adjacent to either u_1' or u_3'.*

The critical cliques contained in I_1 and I_3 are as $c5$ and the critical cliques contained in I_2 are as $c4$, implying all critical cliques in I_1 and $I_2$are constrained. We show we can maintain all distance constraints while changing this maximal clique K_2 to have all critical cliques as unconstrained.

Lemma 5. *[4] Let G be a graph with a 3rd Steiner root T, then each maximal cliques with critical clique cardinality of 3 or more either has exactly two critical cliques each with two representatives as in Lemma 4, or has at most one internal critical clique with two are more representatives in T.*

Lemma 6. *Let K be a maximal clique represented by the situation of Lemma 4 in a 3rd Steiner root T, then there exists an equivalent representation with a central Steiner point adjacent to the representatives of its critical cliques in K.*

The following corollary follows easily from Lemmas 5 and 6.

Corollary 1. *Let G be a graph with a 3rd Steiner root T. Then there exists a representation in which all maximal cliques with critical clique cardinality of 3 or more have at most one internal critical clique with two or more representatives.*

3 5PRP on Strictly Chordal Graphs

This section deals with the combination of the Steiner trees returned by $ALG(G)$ and progresses from the most trivial case to the complete case: the solution of the 5-root phylogeny problem on strictly chordal graphs.

3.1 Structural Restriction 1

A *small leaf* is an external critical clique of size 1 in a maximal clique of critical clique cardinality at least three. We remind the reader that a constrained critical clique in a maximal clique K is a critical clique with either two or more representatives in the Steiner tree T, a single representative adjacent to a Steiner point with Steiner degree 1 in T, or a single representative adjacent to the representative of another critical clique in T. In the following section we will assume that the input graph G contains no small leaves and large maximal cliques K contain at most one critical clique which is constrained. These simplifying assumptions imply that at most one critical clique in a large maximal clique will be as $c1a$, $c1b$, $c3$, $c4$, or $c5$. Case $c4$ is a restrictive case as the following lemma shows.

Lemma 7. *Let G be a graph with an S-restricted 3rd Steiner root tree T and a maximal clique K. If K has an internal critical clique as in $c4$ then the critical clique must be part of at least two maximal cliques with critical clique cardinality three or more with other critical cliques unconstrained.*

The structure of a critical clique in case $c1a$, is a single representative adjacent to a Steiner point with Steiner degree one; as such, if the corresponding critical clique c has size 1 then we must increase the degree of both this representative and the Steiner point. It follows that c must be part of at least three maximal cliques; the following operation shows how to increase the degree of both the Steiner point and the representative of c.

Definition 1 (Operation 1). *Let c be a critical clique part of at least three maximal cliques $K_1, K_2, ..., K_n$. If K_1 is in c1a or c2b and K_2 has all critical cliques unconstrained then assign c a single representative and let the Steiner point adjacent to c in K_1 be adjacent to the Steiner point from K_2 such that all its critical cliques are at distance exactly three from c. For $K_3, ..., K_n$, c now corresponds to c2a and is unconstrained.*

If a critical clique needs to have Operation 1 performed, check that there exists a maximal clique containing it that has all critical cliques unconstrained. If no maximal clique exists, we check if an adjacent critical clique is as $c1a$ or $c1b$, and apply Operation 1. In a similar fashion continue searching for a resolvable path through maximal cliques. Note that such a search is a depth first search through the tree, and in the worst case, and has a linear runtime. Notice that the choice made to change a path by Operation 1 will never affect another path as the search will assign a single representative for a critical clique, and this critical clique will be now unconstrained. Therefore, we pick the first resolvable path.

Theorem 1. *Let G be a connected strictly chordal graph G such that G contains no small leaves and large maximal cliques contain at most one constrained critical clique, there exists a $O(|V|^3)$ time algorithm to recognize whether G has 5th phylogenetic root tree T, and if so, return such a T.*

3.2 Structural Restriction 2

In following section, we assume that the input graph G contains no small leaves. A strictly chordal graph may have a large maximal clique having more than one constrained critical clique; if all except one cannot be modified to be unconstrained, then the algorithm returns no, by Corollary 1. If a critical clique in a 3-root Steiner tree is as $c4$, Lemma 7 forces the structure for all maximal cliques it is contained in; $c3$ and $c5$ are similarly restrictive.

Lemma 8. *Let G be a graph with an S-restricted 3rd Steiner root tree T with $S = \emptyset$ and a maximal clique K. If K has a critical clique c as in c3 or c5 of Lemma 3 then any other maximal cliques with critical clique cardinality three or more containing C will have all critical cliques as unconstrained.*

Thus, given a representative as in cases $c3$, $c4$, or $c5$, we can immediately decide if the maximal clique can be recombined. As $c2a$ and $c2b$ both have a single representative adjacent to a Steiner point of degree at least three, we now deal with the cases $c1a$ and $c1b$.

Lemma 9. *Let G be a graph with an S-restricted 3rd Steiner root tree T and c be a critical clique, where c is part of maximal cliques $K_1, K_2, ..., K_n$ and K_1 is as in $c1a$ or $c1b$, then at least one maximal clique must have all critical cliques other than c unconstrained.*

Theorem 2. *Let G be a connected strictly chordal graph, G contains no small leaves, there exists a $O(|V|^3)$ time algorithm to recognize whether G has a 5th phylogenetic root tree T, and if so, return such a T.*

3.3 No Restrictions

Lemma 10. *Given a strictly chordal graph G and a corresponding S-restricted 3rd Steiner root tree T, if there exists a small leaf l in a maximal clique K, then:*

1. *l is internal in T,*
2. *each critical clique $c \in K \setminus l$ has all adjacent critical cliques not in K at a distance of at least 2 in T,*
3. *at least one critical clique $c \in K \setminus l$ has all adjacent critical cliques not in K at a distance of 3 in T, and*
4. *every critical clique in K has a single or 2 adjacent representatives.*

By this lemma, a critical clique c in a maximal clique containing a small leaf can be as $c1a$, $c1b$, $c2a$ or $c2b$. $c4$ is impossible as the critical clique is adjacent to another critical clique failing to satisfy condition 2. $c3$ is impossible as the small leaf would have to be at distance exactly three from the single representative, but then it would be a leaf in T, failing to satisfy condition 1. For $c5$, a small leaf would be distance three from the degree two representative and a leaf in T.

We now introduce two operations to change a critical clique to satisfy condition 3. The algorithm applies these operations if no suitable critical clique exists to satisfy condition 3 of Lemma 10. Notice that only one of these operations can apply to a set of maximal cliques. In addition, the search as done for Operation 1 can be applied to these operations.

Definition 2 (Operation 2a). *Given a critical clique, c, which is part of at least three maximal cliques, $K_1, K_2, ..., K_n$. If at most one of $K_2, ..., K_n$ was part of a decomposed tree chordal graph with c as in $c1a$, $c2b$ $c2a$, or $c2b$ and all critical cliques in the remainder are unconstrained. Then give all critical cliques c a single representative and let the Steiner point adjacent to c be adjacent to the Steiner points from $K_2, ..., K_n$ such that each remaining critical clique is at distance exactly three from c.*

Definition 3 (Operation 2b). *Given a critical clique, $|c| \geq 2$, which is part of exactly two large maximal cliques K_1 and K_2. If all critical cliques in K_1 and K_2 other than c are unconstrained then create two Steiner points, p_1 and p_2; let all critical cliques in K_1 other than c be adjacent to p_1, all critical cliques in K_2 other than c be adjacent to p_2, and give c two adjacent representatives where one is adjacent to p_1 and the other to p_2.*

Lemma 11. *Given a graph with an S-restricted 3rd Steiner root tree and a large maximal clique K containing a small leaf l, then one critical clique $C \in K \setminus l$ can have Operation 2a applied, Operation 2b applied, or is as c2a.*

Lemma 12. *Given a strictly chordal graph G and a corresponding S-restricted 3rd Steiner root tree T, if there exists a small leaf l in a maximal clique K, then:*

1. *if $\text{cccard}(K) = 3$ and there exists exactly one critical clique $c \in K \setminus l$ with two adjacent representative, then all other critical cliques have all adjacent critical cliques not in K at a distance of exactly 3 in T,*
2. *if $\text{cccard}(K) = 3$ and no critical clique $c \in K \setminus l$ has two adjacent representative, then all critical cliques having all adjacent critical cliques not in K at a distance of exactly 3 in T, and*
3. *if $\text{cccard}(K) \geq 4$ then there exists a critical clique $c \in K \setminus l$ with Operation 2a applicable or c is as c2a.*

Theorem 3. *Let G be a strictly chordal graph. Then there exists a $O(|V|^3)$ time algorithm to recognize whether G has a 5th phylogenetic root tree T, and if so, return such a T.*

4 Concluding Remarks

The general complexity for the k-th root phylogeny problem, for $k \geq 5$, is still unknown. Our future work will be towards development of an algorithm for all graphs when $k = 5$. Of additional interest is the development of approximation algorithms for the k-th phylogenetic root problem.

References

1. A. Brandstadt, V. B. Le, and J. P. Spinrad. *Graph Classes: A Survey.* SIAM, Monographs on Discrete Mathematics and Applications, SIAM, Philadelphia, 1999.
2. R. Haenni and N. Lehmann. Efficient hypertree construction. Technical Report 99-3, Institute of Informatics, University of Fribourg, 1999.
3. P. E. Kearney and D. G. Corneil. Tree powers. *Journal of Algorithms*, 29(1):111 – 131, 1998.
4. W. Kennedy, G. Lin, and G. Yan. Strictly chordal graphs are leaf powers. *Journal of Discrete Algorithms*, 2005.
5. H. Kong and G. Yan. Algorithm for phylogenetic 5-root problem. unpublished, 2003.
6. G. Lin, T. Jiang, and P. E. Kearney. Phylogenetic k-root and steiner k-root. In *ISAAC*, volume 1969, pages 539–551, 2000.
7. G. Lin, W. Kennedy, H. Kong, and G. Yan. The 5-root phylogeny construction for tree chordal graphs. submitted to Discrete Applied Mathematics, 2005.
8. Y. L. Lin and S. S. Skiena. Algorithms for square roots of graphs. *SIAM Journal on Discrete Mathematics*, 8:99 – 118, 1995.
9. R. Motwani and M. Sudan. Computing roots of graphs is hard. *Discrete Applied Mathematics*, 54:81 – 88, 1994.

Recursion Theoretic Operators for Function Complexity Classes

Kenya Ueno

Department of Computer Science,
Graduate School of Information Science and Technology,
the University of Tokyo
kenya@is.s.u-tokyo.ac.jp

Abstract. We characterize the gap between time and space complexity of functions by operators and completeness. First, we introduce a new notion of operators for function complexity classes based on recursive function theory and construct an operator which generates $FPSPACE$ from FP. Then, we introduce new function classes composed of functions whose output lengths are bounded by the input length plus some constant. We characterize FP and $FPSPACE$ by using these classes and operators. Finally, we define a new notion of completeness for $FPSPACE$ and show a $FPSPACE$-complete function.

1 Introduction

Recursive function theory expresses the computable function class as the smallest class containing some initial functions and closed under operators that generate new functions. Similarly, it is known that many function complexity classes are characterized by recursion theoretic scheme [1,2]. In 1953, Grzegorczyk [3] introduced the hierarchical classes $\mathcal{E}^n$ by composition and an operator called bounded recursion. Later, Ritchie [4] showed that $\mathcal{E}^2$ is equal to the linear space computable function class in 1963. In 1964, Cobham [5] characterized FP by composition and bounded recursion on notation. In 1972, Thompson [6] characterized $FPSPACE$ by composition and bounded recursion.

Whereas these studies expanded various fields [7] and have many interesting applications, there are no attempts to study the relation among function complexity classes from a recursion theoretic viewpoint. Existent studies have only concerned about how to characterize function complexity classes simpler way or fewer initial functions. However, this direction is nonsense when we set our goal separation of complexity classes. It is the relation among them that is the most important thing for separation. From this perspective, we extend the notion of operators to operators that act on general function complexity classes.

Operators for complexity classes are useful to clarify the relation among them [8,9,10]. Toda's theorem $PH \subset P^{PP}$ [11] makes use of some properties of operators. There are many operators studied like $\exists$ and $\forall$, which are the operators that generates NP and $coNP$ from P respectively. Many other operators that express the gaps between P and complexity classes like RP, BPP, PP and $\oplus P$

X. Deng and D. Du (Eds.): ISAAC 2005, LNCS 3827, pp. 748–756, 2005.

have been also studied. Then, what about the case of P and $PSPACE$? Is there any appropriate one that simply expresses the gap between P and $PSPACE$? In fact, it is hard to construct it because operators like $\exists$ and $\forall$ are defined according to the number of accepting configurations. But, in the case of function complexity classes, we can say that there exists an appropriate one.

Function complexity classes related with NP are studied widely [12]. However, there are few studies about $FPSPACE$. In this paper, we clarify the structure of $FPSPACE$. The importance of this study is that we show "functions can do what languages cannot do". This is one of the few examples which clearly indicate merits of studies on function complexity classes compared with language complexity classes.

We construct an operator which exactly expresses the gap between FP and $FPSPACE$. We introduce new operators $Comp^*$ and $BRec$ and show $FPSPACE = Comp^*(BRec(FP))$. This result can be generalized. For example, this operator generates the linear space computable function class from the linear time computable function class as $FSPACE(n) = Comp^*(BRec(FTIME(n)))$.

Moreover, we introduce an operator $RecN$ and new function complexity classes $AGTIME$ and $AGSPACE$. We show the structures of FP and $FPSPACE$ by showing $FP = Comp^*(RecN(AGTIME))$ and $FPSPACE = Comp^*(RecN(AGSPACE))$. The case of FP is proven by simulating polynomially time bounded DTMs by using a function called "step", which receives a configuration and returns the next configuration. To prove the case of $FPSPACE$, we introduce a new function called "next" and substitute it for the step function. It simulates partial works of DTMs by manipulating configurations within limited space.

The next function contains a further implication. We introduce completeness for $FPSPACE$, which is a new notion while there is a study about completeness for polynomial space counting classes [13], and show that this function is $FPSPACE$-complete under FP Turing reductions. It is also $FSPACE(n)$-complete under $FTIME(n)$ Turing reductions. Thus, we can say that the next function exactly expresses the gap between time and space.

2 Function Complexity Classes

We assume that the readers are familiar with the notion of Turing machines, resource bounded computations and the definitions of complexity classes like P and $PSPACE$. A configuration is representation of the whole state of a Turing machine at a certain time.

Functions discussed in this paper are total functions from a natural number to a natural number. If we say that a Turing machine computes a function, it means that natural numbers are represented by strings of the binary notation and it converts a string into a string.

First, we define the following function complexity classes.

Definition 1.

- $FTIME(t(n))$ = *the class of functions computable in* $c \cdot t(n)$ *time for some constant* c *by DTMs*
- $FSPACE(s(n))$ = *the class of functions computable in* $c \cdot s(n)$ *space for some constant* c *by DTMs*

Output lengths of functions in $FSPACE(s(n))$ are bounded by $c \cdot s(n)$ for some constant c.

Then, FP and $FPSPACE$ are defined as follows.

Definition 2.

- $FP = \bigcup_{c \geq 1} FTIME(n^c)$
- $FPSPACE = \bigcup_{c \geq 1} FSPACE(n^c)$

Output lengths of functions in $FPSPACE$ are also bounded by polynomial to the input length.

These function complexity classes are closely related with language complexity classes.

Theorem 1. $P = PSPACE \Leftrightarrow FP = FPSPACE$

This theorem is proven by the prefix searching method, which appears in [14] and [15]. This theorem implies that there is little difference whether we study languages or functions for separating complexity classes. So, we may choose an appropriate one according to the case.

Furthermore, we introduce new function classes, which are composed of functions whose output lengths are bounded by the input length plus some constant. We call them additive growth functions.

Definition 3.

- $AGTIME$ = *the class of functions computable in* $n + c$ *time by DTMs for some constant* c *where* n *is the input length*
- $AGSPACE$ = *the class of functions computable in* $n + c$ *space by DTMs for some constant* c *where* n *is the input length*

The output lengths of functions contained in $AGSPACE$ are bounded by $n + c$ for some constant c.

3 Recursion Theoretic Operators

Now, we define the operators based on recursive function theory. Let C be an arbitrary function complexity class and we define operators by applying them to C. We write multiple argument function as $f(\boldsymbol{x}) = f(x_1, \cdots, x_n)$ for ease. Then, the operator *Comp* is defined as follows.

Definition 4.

$$Comp(C) = C \cup \{f \mid f(\boldsymbol{x}) = g(h_1(\boldsymbol{x}), \cdots, h_n(\boldsymbol{x})),\ g, h_1, \cdots, h_n \in C\}$$

In addition, we define $Comp^*(C)$ as the smallest class that contains C and is closed under $Comp$.

Bounded recursion is recursion which restricts output values of functions by a certain function. The operator $BRec$ is defined as follows. ($BRec^*$ is defined similarly as the case of $Comp^*$)

Definition 5.

$$\begin{aligned} BRec(C) = C \cup \{f \mid\ & f(0, \boldsymbol{y}) = g(\boldsymbol{y}), \\ & f(x, \boldsymbol{y}) = h(x, \boldsymbol{y}, f(x-1, \boldsymbol{y})), \\ & \forall x, \forall \boldsymbol{y}, f(x, \boldsymbol{y}) \le k(x, \boldsymbol{y}), \\ & g, h, k \in C\} \end{aligned}$$

Recursion on notation is recursion that decreases length instead of value whenever it carries out recursion. The operator $RecN$ is defined as follows.

Definition 6.

$$\begin{aligned} RecN(C) = C \cup \{f \mid\ & f(0, \boldsymbol{y}) = g(\boldsymbol{y}), \\ & f(x, \boldsymbol{y}) = h(x, \boldsymbol{y}, f(\lfloor \frac{x}{2} \rfloor, \boldsymbol{y})), \\ & g, h \in C\} \end{aligned}$$

4 Closure Properties

Before proceeding the main result, we describe some basic properties related with operators defined above. First, we see the closure properties related with $Comp$.

Proposition 1. *$FTIME(n)$, $FSPACE(n)$, FP and $FPSPACE$ are closed under $Comp$.*

Next, we show the closure properties related with $BRec$.

Lemma 1. *If $s(n) \ge n$, $BRec(FSPACE(s(n))) \subset FSPACE(s(2 \cdot\ s(n)))$.*

Proof. We consider a function $f \in BRec(FSPACE(s(n)))$, which is obtained by applying $BRec$ to $g, h, k \in FSPACE(s(n))$. The program computing f is described as follows.

```
input x, y
z = g(y);
for(i = 1 to x){
z = h(i, y, z);
}
output z
```

Space required for i is $|x|$ and the input length of z is bounded by $s(|x-1|+|y|)$ because it is bounded by k. Thus, the following inequality holds because $s(n) \geq n$.

$$s'(|x|+|y|) = s(|x|+|y|+s(|x-1|+|y|)) + |x| \leq 2 \cdot s(2 \cdot s(|x|+|y|))$$

Here, $s'(n)$ is the space complexity of this program. □

By this lemma, the following proposition holds.

Proposition 2. *$FSPACE(n)$, $FPSPACE$ are closed under $BRec$.*

5 Time, Space and Bounded Recursion

In this section, we show that $BRec$ characterizes the relation between time and space. First, we prove the following lemma.

Lemma 2. *If $s(n) \geq n$, $FSPACE(s(n)) \subset Comp^*(BRec(FTIME(s(n))))$.*

Proof. Let f be an arbitrary function computable in $k \cdot s(n)$ space for some constant k. We define the following functions defined for the Turing machine computing f.

1. $init(x)$ is a function that returns the initial configuration on input x
2. $step(c)$ is a function that returns the configuration when when it changes to the next state from configuration c
3. $out(c)$ is a function that returns the output of configuration c

These functions are all computable in linear time. Thus, they are also in $FTIME(s(n))$ on assumption of $s(n) \geq n$.

Then, there exists a function $T \in FTIME(s(n))$ that satisfies $\forall x$, $|T(x)| \geq k \cdot s(|x|)$. There also exists a function $S \in FTIME(s(n))$ that satisfies $\forall t, x$, $|S(t,x)| \geq |T(x)| + k'$ for any constant k' that depends only on f. This function is defined in order to represent all configurations in $|S(t,x)|$ space.

Applying $BRec$ to these functions, we construct the following function $F \in BRec(FTIME(s(n)))$.

$$\begin{aligned} F(0,x) &= init(x) \\ F(t,x) &= step(F(t-1,x)) \\ \forall t, \forall x, F(t,x) &\leq S(t,x) \end{aligned}$$

Remark that $k \cdot s(x)$ space bounded computations are simulated by $2^{k \cdot s(x)}$ time bounded computations. Thus, the computation of f is simulated as $f(x) = out(F(T(x),x))$ because recursion is carried out $2^{|T(x)|}$ times, which is larger than $2^{k \cdot s(x)}$. □

Thus, we can obtain an operator that express the gap between FP and $FPSPACE$.

Theorem 2. $FPSPACE = Comp^*(BRec(FP)) = Comp^*(BRec^*(FP))$

Proof. $Comp^*(BRec^*(FP)) \subset FPSPACE$ because $FPSPACE$ is closed under $Comp$ and $BRec$. Moreover, $FPSPACE \subset Comp^*(BRec(FP))$ by Lemma 2. □

Corollary 1. *FP is closed under $BRec \Leftrightarrow P = PSPACE$*

As we can see from the proof, it does not seem to hold $Comp^*(BRec(C)) = FPSPACE$ where C is any function complexity class smaller than FP. To include $FPSPACE$ by $Comp^*(BRec(C))$, it is necessary for C to be stronger than FP. So, we can say that it is an operator that exactly expresses the gap between FP and $FPSPACE$.

By the same way, this operator expresses the gap between $FTIME(n)$ and $FSPACE(n)$.

Theorem 3.

$$FSPACE(n) = Comp^*(BRec(FTIME(n))) = Comp^*(BRec^*(FTIME(n)))$$

6 Structure of Polynomially Bounded Functions

In this section, we show that a combination of the operator $RecN$ and additive growth functions expresses polynomially bounded functions.

Lemma 3. *There is a function $p \in Comp^*(RecN(AGTIME))$ that satisfies $\forall x,\ |p(x)| > c \cdot |x|^k$ for any constants c and k.*

Proof. To prove the existence of such a function p, we consider the function $cat(x, y) \in AGTIME$ that returns concatenation of the inputs x, y. By the definition, $|cat(x, y)| = |x| + |y|$ holds. By applying $RecN$ to this function, we construct the following function $\sigma_1 \in RecN(AGTIME)$.

$$\sigma_1(0) = 0,\ \sigma_1(x) = cat(x, \sigma_1(\lfloor \frac{x}{2} \rfloor))$$

This function satisfies the following equation.

$$|\sigma_1(x)| = |x| + (|x| - 1) + (|x| - 2) + \cdots + 1 = \frac{|x| \cdot (|x| + 1)}{2}$$

Moreover, if we define $\sigma_2(x) = cat(\sigma_1(x), \sigma_1(x))$, then we obtain the function σ_2 that satisfies $|\sigma_2(x)| > |x| \cdot (|x| + 1)$. Then, we construct the function σ_3 that satisfies $\sigma_3(x) > |x|^k$ for any constant k by applying $Comp$ to σ_2 enough times as $\sigma_3(x) = \sigma_2(\cdots(\sigma_2(x)))$. Consequently, we can obtain the function $p \in Comp^*(RecN(AGTIME))$ as $p(x) = cat(\sigma_3(x), cat(\cdots cat(\sigma_3(x), \sigma_3(x))))$. □

Now, we show the theorem by using this lemma.

Theorem 4. $FP = Comp^*(RecN(AGTIME))$

Proof. First, we show $Comp^*(RecN(AGTIME)) \subset FP$. Since FP is closed under $Comp$, it is sufficient to show $RecN(AGTIME) \subset FP$. To show this, we consider a function $f \in (RecN(AGTIME))$, which is obtained by applying $RecN$ to $g, h \in AGTIME$. The program computing f is described as follows.

input $x, \boldsymbol{y}$
$z = g(\boldsymbol{y})$;
for($i = 1$ to $|x|$){
x_i = from first bit to $(|x| - i)$ th bit of x
$z = h(x_i, \boldsymbol{y}, z)$;
}
output z

In this program, the length of z increases at most the sum of the length of $x, \boldsymbol{y}$ plus some constant at each repetition because $h \in AG$. So, the length of z is bounded by the polynomial to the sum of the length of $x, \boldsymbol{y}$ during execution of the program. Consequently, this program can compute f in polynomial time.

Next, we prove $FP \subset Comp^*(RecN(AGTIME))$. To simulate the computation of $f \in FP$, we use the three functions *init*, *step* and *out* described in Lemma 2 defined for the Turing machine computing f. It is easy to see that these functions are in $AGTIME$ by using special representations of configurations.

By applying the operator $RecN$ to *init* and *step*, we obtain the following function $F \in RecN(AGTIME)$.

$$F(0, x) = init(x)$$
$$F(t, x) = step(F(\lfloor \frac{t}{2} \rfloor, x))$$

Whenever it carries out recursion once, the length of t decreases one. So, the length of t corresponds to the time of the computation. We can construct the time bounding function $p \in Comp^*(RecN(AGTIME))$ for f by Lemma 3. Namely, $f(x)$ is computable in $|p(x)|$ time. Then, the computation of f can be simulated as $f(x) = out(F(p(x), x))$. □

This result can be extended to other function complexity classes related to polynomial like $FPSPACE$. To achieve this extension, we define the following function.

Definition 7. *$next(e, c)$ is the function that returns the configuration when the computation begins from c and the Turing machine's head comes the next position for the first time from the position at the time of c, for the Turing machine represented by e. (if there is not such a configuration, then return the final configuration, in the other case undefined)*

We describe the work of the next function below. It may simulate only one step of a Turing machine's transition. On the other hand, it may simulate exponentially many steps of a Turing machine's transition by a zigzag path. In the

figure, the area of oblique lines means the computational space used till the time which corresponds to the configuration after $s(n)$-th call of the next function. So, $s(n)$ space bounded computations can be simulated by calling the next function at most $s(n)$ times. Thus, polynomial space computations are simulated by calling the next function polynomial times.

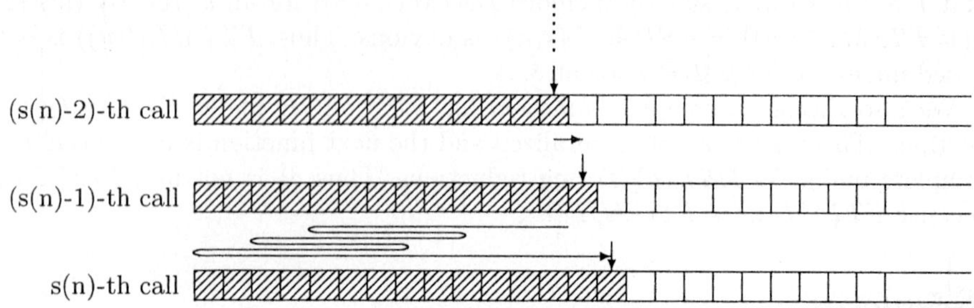

The next function partially simulates the works of deterministic Turing machines. This is done within limited space.

Proposition 3. $next \in AGSPACE$

By using this function, we can obtain the following theorem. The proof is similar to the case of Theorem 4, but we should substitute the next function for the step function.

Theorem 5. $FPSPACE = Comp^*(RecN(AGSPACE))$

7 FPSPACE-Completeness

We denote FP^f as the class of functions computable in polynomial time by oracle DTMs which are given f as oracle. Then, we define $FPSPACE$-completeness as follows.

Definition 8. *If* $FPSPACE \subset FP^f \wedge f \in FPSPACE$, *then* f *is* $FPSPACE$-*complete.*

It is easy to see $f \in FP \Leftrightarrow FP = FPSPACE$ for any $FPSPACE$-complete function f. We can prove that all characteristic functions of $PSPACE$-complete languages are $FPSPACE$-complete by the prefix searching method. Moreover, we show the following theorem.

Theorem 6. *next is* $FPSPACE$-*complete.*

Proof. From the property of *next*, polynomial space bounded computations can be simulated by calling the *next* oracle polynomial times. □

Corollary 2. $next \in FP \Leftrightarrow P = PSPACE$.

One interesting point is that the next function is $FPSPACE$-complete and at the same time it is in $AGSPACE$.

8 Concluding Remarks

We proved $FSPACE(s(n)) \subset Comp^{*}(BRec(FTIME(s(n)))$ and showed the operator which expresses the gap between FP and $FPSPACE$. Although whether $PSPACE$ strictly includes P is still an open problem, it is known that $DSPACE(s(n))$ strictly includes $DTIME(s(n))$ for all s [16]. By this result, $FTIME(s(n)) \neq FSPACE(s(n))$ is obvious. Thus, $FTIME(t(n))$ is not closed under $Comp$ or $BRec$ for all t.

We also showed a complete function for $FPSPACE$ under FP Turing reductions. This result can be generalized and the next function is $FSPACE(n)$-complete under $FTIME(n)$ Turing reductions. Thus, it is not in $FTIME(n)$ because $FTIME(n) \neq FSPACE(n)$,

References

1. Rose, H.E.: Subrecursion: Functions and Hierarchies. Claredon Press (1984)
2. Clote, P., Kranakis, E.: Boolean Functions and Computation Models. Springer (2002)
3. Grzegorczyk, A.: Some classes of recursive functions. Rozprawy Mathematyczne **4** (1953)
4. Ritchie, R.W.: Classes of predictably computable functions. Transactions of the American Mathematical Society **106** (1963) 139–173
5. Cobham, A.: The intrinsic computational difficulty of functions. In: Proceedings of the 1964 Congress for Logic, Methodology, and the Philosophy of Science, North-Holland (1964) 24–30
6. Thompson, D.B.: Subrecursiveness: Machine-independent notions of computability in restricted time and storage. Mathematical Systems Theory **6** (1972) 3–15
7. Bellantoni, S., Cook, S.: A new recursion-theoretic characterization of the polytime functions. Computational Complexity **2** (1992) 97–110
8. Ogiwara, M., Hemachandra, L.A.: A complexity theory for feasible closure properties. In: Structure in Complexity Theory Conference. (1991) 16–29
9. Vollmer, H., Wagner, K.: Classes of counting functions and complexity theoretic operators. Technical report, Universitat Wurzburg (1991)
10. Zachos, S., Pagourtzis, A.: Combinatory complexity: Operators on complexity classes. In: Proceedings of 4th Panhellenic Logic Symposium. (2003)
11. Toda, S.: PP is as hard as the polynomial-time hierarchy. SIAM Journal on Computing **20** (1991) 865–877
12. Selman, A.L.: Much ado about functions. In: Proceedings, Eleventh Annual IEEE Conference on Computational Complexity. (1996) 198–212
13. Ladner, R.E.: Polynomial space counting problems. SIAM Journal on Computing **18** (1989) 1087–1097
14. Balcázar, J., Díaz, J., Gabarró, J.: Structural Complexity I. 2nd edn. Springer (1994)
15. Selman, A.L., Xu, M.R., Book, R.V.: Positive relativizations of complexity classes. SIAM Journal on Computing **12** (1983) 565–579
16. Hopcroft, J.E., Paul, W., Valiant, L.G.: On time versus space. Journal of the ACM **24** (1977) 332–337

From Balls and Bins to Points and Vertices*

Ralf Klasing[1], Zvi Lotker[2], Alfredo Navarra[3], and Stephane Perennes[4]

[1] LaBRI - Université Bordeaux 1, 351 cours de la Liberation,
33405 Talence cedex, France
Ralf.Klasing@labri.fr

[2] Centrum voor Wiskunde en Informatica Kruislaan 413,
NL-1098 SJ Amsterdam, Netherlands
lotker@cwi.nl

[3] Computer Science Department, University of L'Aquila, Italy
navarra@di.univaq.it

[4] MASCOTTE project, I3S-CNRS/INRIA/Univ. Nice–Sophia Antipolis, France
Stephane.Perennes@sophia.inria.fr

Abstract. Given a graph $G = (V, E)$ with $|V| = n$, we consider the following problem. Place n points on the vertices of G independently and uniformly at random. Once the points are placed, relocate them using a bijection from the points to the vertices that minimizes the maximum distance between the random place of the points and their target vertices.

We look for an upper bound on this maximum relocation distance that holds with high probability (over the initial placements of the points).

For general graphs, we prove the $\#P$-hardness of the problem and that the maximum relocation distance is $O(\sqrt{n})$ with high probability. We also present a Fully Polynomial Randomized Approximation Scheme when the input graph admits a polynomial-size family of witness cuts while for trees we provide a 2-approximation algorithm.

1 Introduction

Given a set of n uniform random points inside a given square $D \subseteq \mathbb{R}^d$ and n points of a square grid covering D, an interesting question is the "cost" of ordering the random points P on the grid vertices. A natural cost function is the measure of the distance that the random points have to move in order to achieve the grid order. Among all the possible bijections $f : P \rightarrow Grid$, we are interested in minimizing the maximum distance between P and $f(P)$, i.e. $\min_f \max_{1 \le i \le n} ||p_i - f(p_i)||_1$ with $p_i \in P$. In [1,2,3], the relation between two basic, fundamental structures like Uniform Random points and d-dimensional Grid points was studied. Those papers show that the expected minimax grid matching distance is $\Theta(\log(n)^{3/4})$ for $d = 2$ and $\Theta(\log(n)^{1/d})$ for $d > 2$. In a more general setting, we are interested in the *Points and Vertices* problem for arbitrary graphs $G = (V, E)$ with $|V| = n$ which can be described as follows:

* The research was partially funded by the European project IST FET AEOLUS and by the European COST Action 293, "Graphs and Algorithms in Communication Networks" (GRAAL).

X. Deng and D. Du (Eds.): ISAAC 2005, LNCS 3827, pp. 757–766, 2005.

1. Throw n points independently and randomly onto the n vertices of G.
2. Remap the points on G such that the load of each vertex is exactly 1, minimizing the maximal distance that any point has to move (on G).

The Points and Vertices problem may be viewed as an extension of the classical *Balls into Bins* problem, where m balls are thrown (independently and uniformly at random) into n bins, by adding graph-structural properties to the bins. The bins become vertices and there is an edge between two vertices if they are "close" enough (see e.g., [4,5,6,7] for a formal definition of the Balls into Bins problem and some of its variations). Usually, in the Balls into Bins problem the aim is to find out the distribution of the most loaded bins. In the Points and Vertices problem, instead, we are interested in the accumulation of several vertices, not only one.

The interest in the Points and Vertices problem arises from the fact that it captures in a natural way the "distance" between the *randomness* of throwing points (independently and uniformly at random) onto the vertices of G, and the *order* of the points being evenly balanced on G. In fact, our problem can be considered as the opposite of the "Discrepancy" (see for instance [8]).

Besides the pure theoretical interest, the *Points and Vertices* problem has applications in several fields. E.g., in the field of robot deployment as well as in sensor networks, one of the main problems is how to organize a huge number of randomly spread devices. The goal is usually to obtain a nearly equidistant formation so as to maximize the coverage of interesting areas [9,10]. In the field of computer graphics, the mapping of points onto cells (pixels) of a regular grid is a well-studied topic [11]. Another application in which our study can be applied concerns Geometric Pattern Matching problems [12]. In fact, we can derive good bounds on the number of edges of the bipartite graph. For more general topologies, instead, we can consider the token distribution [13,14] and load balancing problems [6,15]. The general case is constituted by a set of k tokens that must be assigned to n processors connected by a general graph. Our problem appears when the tokens are arriving randomly uniformly and when the cost is the maximum distance that some token has to travel.

Our Results. We formalize the *Points and Vertices* problem by defining a random variable $\rho(G, \omega)$ for the remapping distance on G. We relate the behavior of $\rho(G, \omega)$ to the graph expansion properties and study the complexity of computing essential parameters of $\rho(G, \omega)$. This distance turns out to be somewhat difficult to capture since it is related to global phenomena on G. We study $\rho(G, \omega)$ for general graphs and trees. (Note that results for classical topologies like paths and grids can be found in [1,2,3].) More specifically, we obtain:

1. $\#P$-hardness for the general case.
2. A Fully Polynomial Randomized Approximation Scheme (FPRAS) when the graph admits a polynomial-size family of witness cuts.
3. $\rho(G, \omega) = O(\sqrt{n})$ with high probability (w.h.p.) for any connected graph G.
4. A greedy algorithm $\mathcal{A}$ that remaps the points on any tree T with remapping distance $\rho_{\mathcal{A}}(T, \omega) \leq 2\rho(T, \omega)$.

The paper is organized as follows. Section 2 provides a formal definition of the *Points and Vertices* problem. Section 3 contains some general observations from which we derive the related computational hardness results. In Section 4, the greedy algorithm on arbitrary trees that computes the remapping distance up to a factor of 2 is presented. Finally, Section 5 gives some conclusive remarks.

2 Formalizing the Points and Vertices Problem

We study how far is a random structure from a regular one assuming that two structures are close if there exists a "short" bijection from one to the other.

Actually, we are interested in bounding the maximum distance performed by the movement of the points randomly and uniformly distributed over the vertices of a graph $G = (V, E)$ to the vertices V by moving the points over the edges E in such a way that the final setting is given by one point for each vertex.

Definition 1. *Given a metric space with metric d and $\rho \in I\!R^+$, a one-to-one mapping $f : A \to B$ is called* mapping with stretch ρ from a set A to a set B *if $d(x, f(x)) \leq \rho$ for all $x \in A$.*

Definition 2. *Given a metric space with metric d and two sets A and B, we define $\delta(A, B)$ as the minimum $\rho \in I\!R^+$ such that there exists a one-to-one mapping with stretch ρ from A to B.*

Let $G = (V, E)$ be a graph with $n = |V(G)|$ vertices. In what follows, $\Omega = V(G)^n$ is the probabilistic space associated to uniform independent choices of n points over the nodes $V(G)$. The events will either be considered as (indexed) sets or as positive integral weight functions on the ground set $V(G)$ with the adequate measure.

On graphs, we use the usual distance metric (assuming edges with positive length) and, unless differently specified, the edges have length 1.

Problem 1. Given a graph G with $n = |V(G)|$ *vertices* and a random set $P(G, \omega)$, $\omega \in \Omega$ of n *points* lying on the vertices of G, the aim is to study the random variable $\rho(G, \omega) = \delta(P(\omega), V(G))$.[1]

Problem 1 can be generalized as follows:

Problem 2. Given a graph G, a set of locations $L \subseteq V(G)$ and a random set $P(L, \omega)$ of *points* chosen according to a distribution[2] F, with $\omega \in \Omega$ and $|P| = |L|$, the aim is to study the random variable $\rho(L, \omega) = \delta(P(L, \omega), L)$.

In what follows, for any graph G and any $\rho \in \mathbb{R}^+$, we will denote by $\mu(G, \rho)$ the probability that there exists a stretch ρ one-to-one mapping from $P(G, \omega)$ to $V(G)$, and we define $\rho(G, \omega) = \min\{\rho \in \mathbb{R}^+ | \mu(G, \rho) = 1 - o(1)\}$. For instance, $\rho(G, \omega) = \sqrt{n}$ means that there exists a function $f \in o(1)$ such that $\mu(G, \sqrt{n}) > 1 - f$. Whenever it will not be ambiguous, we omit the parameters G and ω.

---

[1] Abusing notation, with ρ we represent both a real positive number and a function.

[2] In the rest of the paper, we will assume such a distribution to be the Uniform one unless specified differently.

3 Hardness Results

We will often replace our process by a Poisson process with intensity 1, since the points and vertices process is simply the Poisson process conditioned by the fact that the total number of points is $|V|$.[3] Note that the Poisson process will always fail when the number of points is not $|V|$. It follows that, denoting by $\mu_{Poisson}(G,\rho)$ the probability of finding a stretch ρ one-to-one mapping for the Poisson model, we have $\mu(G,\rho) \sim \mu_{Poisson}(G,\rho)\sqrt{2\pi|V|}$.

Perfect matching and Duality. The Points and Vertices problem can also be stated in terms of perfect matchings. Given a set of random points P, we build the following auxiliary bipartite graph. On one side of the graph we take as vertices the random points and on the other side the original vertices. We then connect any random point to the vertices at distance at most ρ. A stretch ρ mapping from P to $V(G)$ is exactly a perfect matching in the auxiliary graph. It follows that for any fixed event ω, $\rho(G,\omega)$ can be computed in polynomial time, moreover duality can be used to prove bounds on $\rho(G,\omega)$. In order to apply the Koenig-Hall lemma (see [16]) to the associated bipartite graph, we need the following notation. For any set $X \subseteq V(G)$ and any event $\omega \in \Omega$, we denote by $\eta(X,\omega)$ the number of random points that lie inside X. For any set $X \subseteq V(G)$, let $\Gamma^\rho(X) = \{v \in V | d(X,v) \leq \rho\}$ and $\partial^\rho(X) = \Gamma^\rho(X) \setminus X$. The Koenig-Hall lemma can then be expressed as follows:

Lemma 1. $\rho(G,\omega) = \min\{\rho \in I\!R^+ \mid \forall X \subseteq V(G) : |\eta(X,\omega) - |X|| \leq |\partial^\rho(X)|\}$.

For a given ω, we will say that X is a *bad ρ-cut* whenever $|\eta(X,\omega) - |X|| > |\partial^\rho(X)|$. The lemma implies that the graph expansion properties are strongly related to the distribution of $\rho(G,\omega)$. The random variable $\eta(X,\omega)$ will "usually" be distributed almost like the sum of $|X|$ independent Poisson variables with intensity 1[4]. So, $\eta(X,\omega)$ will be concentrated around its mean $|X|$ in a normal way, $\Pr(|\eta(X,\omega) - |X|| \geq t\sqrt{|X|}) \sim \frac{e^{-t^2}}{\sqrt{2}t}$.

It follows that, given a fixed $t > 0$, whenever there exists a set X such that $|X| \leq \frac{n}{2}, |\partial^\rho(X)| \leq t\sqrt{|X|}$, the probability for X to be a bad ρ-cut will be non-vanishing (around e^{-t^2}).

Isoperimetric properties may also lead to some upper bounds, but these will usually not be tight, indeed by the first moment method it follows:

$$\mu(G,\rho) = \Pr(\forall X \subseteq V, X \text{ is not a bad } \rho\text{-cut}) \leq \sum_{X \subseteq V(G)} \Pr(X \text{ is not a bad } \rho\text{-cut}).$$

Notice that such a bound is usually weak since when there exists a bad cut it is likely to happen that the event induces a very high number of bad cuts. Moreover, the bound is not easy to estimate since among the $2^{|V(G)|}$ cuts some are much more likely to be bad cuts than others (e.g., in the 2-dimensional grid a disk is much more likely to be bad than a random set of vertices).

[3] The intensity of a Poisson process represents the mean of the number of events occurring per time unit.

[4] This is not true when $|X|$ is too small or too close to n.

Computational Issues. Our problem consists in computing the number of points in a polytope defined by an exponential number of constraints but that admits a polynomial time separation oracle (namely the perfect matching algorithm). Let the vector $(x_1, x_2, \ldots, x_n)$ with $\sum_{i=1}^{n} x_i = n$ represent the event with i points at vertex v_i, then the polytope F of feasible events for $\rho = 1$ is the set satisfying the linear constraints[5]: $\{(x_1, ..., x_n) : \forall X \subseteq V(G), \ |\sum_{v_i \in X} x_i - |X|| \leq |\partial(X)|\}$ and we wish to compute $\sum_{x \in F} \ell(x)$ where $\ell(x)$ is a discrete measure derived from Ω (e.g., $\Pr(x_i = k) \sim \frac{1}{k!}$).

This suggests connections with $\#P$ counting problems or volume estimation and with $\#P$ problems for which the decision problem is in P: matchings, Eulerian cycles and in particular reliability estimation problems. With the next theorem we prove that Problem 2 is $\#P$-hard by reducing it to the problem of counting the number of matchings in a graph.

Theorem 1. *Problem 2 is #P-hard.*

Proof. Let us assume that it is possible to compute $\mu(G, 1)$ for any graph $G = (V, E)$ in polynomial time. Let $G' = (V', E')$ be the graph obtained from G by replacing each edge $e \in E$ by a path of length two (note that $|V'| = |V| + |E|$ and $|E'| = 2|E|$). We set as locations L the nodes corresponding to the original vertices of G, that is, $|L| = |V|$. Let F be a distribution of random points obtained by choosing $\frac{|V|}{2}$ vertices of G' and placing 2 points in each one. In order to obtain the number of matchings in G, it is then sufficient to multiply the probability to have a bijection between the thrown points and L with $\binom{|V'|}{\frac{|V|}{2}}$.□

Our sample space is extremely simple, and we can check if $\rho(G, \omega) \leq \rho$ in polynomial time. So, for any fixed graph G, it is "usually" easy to compute a $(1 + \varepsilon)$-approximation of $\mu(G, \rho)$ (resp. $1 - \mu(G, \rho)$) using the Monte Carlo method. It is efficient only as long as one can observe successful (resp. failing) events. Indeed, as noticed by Karp and Luby [17] if an event has probability p, a Monte Carlo estimation with $O(\frac{\log n}{\epsilon^2 p})$ samples is a $(1 + \varepsilon)$-approximation of p with probability $\frac{1}{n}$. Since our goal is not to approximate $\mu(G, \rho)$ when it is close to zero (since then we would consider $\rho' > \rho$), we are left with the problem of computing an approximation of $1 - \mu(G, \rho)$ when $\mu(G, \rho)$ is close to 1.

FPRAS to estimate $1 - \mu(G, \rho)$ when there is a small set of witness cuts. We say that a family F of cuts is a family of *Witness Cuts*, whenever the probability that some cut $C \in F$ is a bad ρ-cut, conditioned on the fact that some bad ρ-cut exists is almost 1.

In the case we have a polynomial-size family of witness cuts, following [18], we can evaluate the probability that an event violates a cut of the family, conditioned on the fact that the event is bad. Then, we can estimate the probability of a conjunction of "simple" events like in the case of DNF formulas [17]. We refer to Vazirani [19] for a detailed comprehensive presentation.

[5] The next inequality produces two linear constraints.

Let $C_w, w \in W$ be a set of witness cuts, A_w be the event *Cut C_w fails* (i.e. A_w is true when the cut fails), and p_w be the probability that this happens. By hypothesis, we have that $(1 - \mu(\rho)) \sim \Pr(A_1 \vee A_2 \vee \ldots \vee A_w)$.

Let $c(\omega)$ denote the number of cuts violated by an event ω. We have $E[c(\omega)] = E[c(\omega) \mid \omega \text{ fails}]Pr(\omega \text{ fails})$, and $E[c(\omega)]$ is simply $\sum_{w \in W} p_w$. It follows that computing $(1 - \mu(\rho))$ reduces to computing $E[c(\omega) \mid \omega \text{ fails}]$.
Consider the following sampling process:

(1) Choose ω with probability $\frac{p_w}{\sum_{v \in W} p_v}$.
(2) Pick uniformly an event ω failing for A_w.
(3) Output ω with weight $\frac{1}{c(\omega)}$.

This process samples the space of true events, moreover each true event is sampled with uniform probability. In order to get a sample space with measure m we need in the worst case $|W|m$ steps. If now we want to estimate, using T Monte Carlo trials, the value of $c(\omega)$ under the failed condition, we simply need to count $\sum_{1,2,\ldots,T} c(\omega)\frac{1}{c(\omega)} = T$ and to divide by the sample measure $\sum_{t=1,2,\ldots,T} \frac{1}{c(\omega)}$.

So, using a sample with T elements we have

$$(1 - \mu(\rho)) \sim \sum_{w \in W} p_w \frac{\sum_{t=1,2,\ldots,T} \frac{1}{c(\omega)}}{T}$$

In order to show that our algorithm is polynomial, we simply need to check that p_w can be estimated and that the space $\Omega \mid C_w$ fails can be sampled.

In the case of *i.i.d.* points, this is straightforward, p_w is obtained via a closed formula and sampling $\Omega \mid C_w$ fails simply means conditioning on the event $\eta(C_w)$ whose distribution is also known.

Graphs with no polynomial set of witness cuts. Unfortunately, there exist graphs on which in order to solve the points and vertices problem, we have to consider an exponential number of cuts. In the example below, for any polynomial family of cuts, most of the events will satisfy all the cut inequalities while still violating some random[6] cut inequality.

Let us consider the following graph G. We start from a clique with k vertices and add ℓ "leaves" that are connected to all the clique nodes. The diameter of G is 2, so $\mu(2) = 1$. Let us study $\mu(1)$ with the Poisson paradigm. For any set X of leaves, we have $|\Gamma(X)| = |X| + k$, and a cut fails if $\eta(X, \omega) > |X| + k$ or $\eta(X, \omega) < |X| - k$. So, only two cuts induce the failure, but they are random, that is, the set of leaves with at least 1 point or the set of leaves with 0 points.

Since the probability for a vertex to receive p points is $\frac{1}{p!e}$, we find about $\frac{\ell}{e}$ leaves with 0 points and about $\frac{\ell}{e}$ extra points in the set of leaves with 1 or more points. So the set of vertices with 1 or more points is a bad cut with high probability as soon as $k < \frac{\ell}{e}(1 - \epsilon)$.

6 In the Kolmogorov acceptation.

Taking for instance $k = n^{1-\epsilon}$ and $\ell = n-k$ it follows that $\mu(1)$ is exponentially small. If we consider now a fixed cut X, it follows that $|\partial(X)| \geq k = n^{1-\epsilon}$ and the probability that $\eta(X,\omega)$ deviates from $|X|$ by more than $t\sqrt{n}$ is exponentially small. Consequently, in this graph, the probability that no matching exists is exponentially larger than the probability for a cut to fail. This means that cuts are not correlated and that the failure probability is induced by an exponential number of cuts. Notice that to get an example with $\mu(1) \sim 1$, we can choose an appropriate $k \sim \frac{\ell}{e}$.

Consequences for general graphs. Let P_n be a path with n vertices. It is well-known that $\rho(P_n,\omega) = \sqrt{n}$ (see for instance [3]). From this example, we derive a general result for arbitrary graphs.

Intuitively, paths look like the graphs with the worst possible ρ. We can motivate this intuition as follows. Since for any graph G, G^3 contains a Hamiltonian path [20], we conclude that for any graph $\Pr(\rho(G,\omega) \geq 3k\sqrt{n}) \leq e^{-k^2}$ and so $\mu(G, \sqrt{n \log n}) \sim 1$.

Theorem 2. *For any graph G, $\rho(G,\omega) = O(\sqrt{n})$.*

The following example shows that for some graph G_0, $\rho(G_0,\omega) >> \sqrt{D} >> 1$ (where D is the diameter of G_0), i.e. $\Pr(\rho(G_0,\omega) \geq \sqrt{D})$ is small. Consider two complete graphs with n nodes connected with a path of length ℓ, with $\ell \leq \sqrt{n/4}$. If the number of points in one of the complete graphs deviates by more than ℓ (this happen with finite probability), $\rho(G_0,\omega)$ is larger than ℓ, so we have $D = \sqrt{n}$ and $\rho(G_0,\omega) = \Theta(D) = \Theta(\sqrt{n})$ with large probability. Notice that we can replace the complete graphs by binary trees to get a bounded-degree example.

4 Trees

Previous results for paths and grids can be found in [1,2,3]. In this section, we consider tree topologies and we show that $\mu(\rho)$ is quite well described by a few cut inequalities. Hence, we describe a greedy algorithm that for a given tree T and a set of points $P(T,\omega)$ evaluates up to a factor of 2 the value $\rho(T,\omega)$.

A greedy approximation algorithm. We use a labelling process, each node v receives a family of labels. Label $+\ell$ (resp. $-\ell$) means that one point (resp. vertex) at distance ℓ from v in the subtree rooted at v need to be assigned an image (resp. a pre-image) outside the subtree.

To each leaf we associate a label -1 if there are no points inside it, 0 if there is 1 point, $p-1$ times $+1$ if there are p points. Then for each subtree whose vertices are already labelled except for the root, we compute the number of positive labels minus the number of negative labels. Let us call s such a number. If $s > 0$ we label the root with the smallest $s-1$ positive numbers contained in the previous labels increased by 1 and a $+1$ for each point contained in it. If $s < 0$, let s' be the number of points contained in the root. If $s' > |s|$ then we

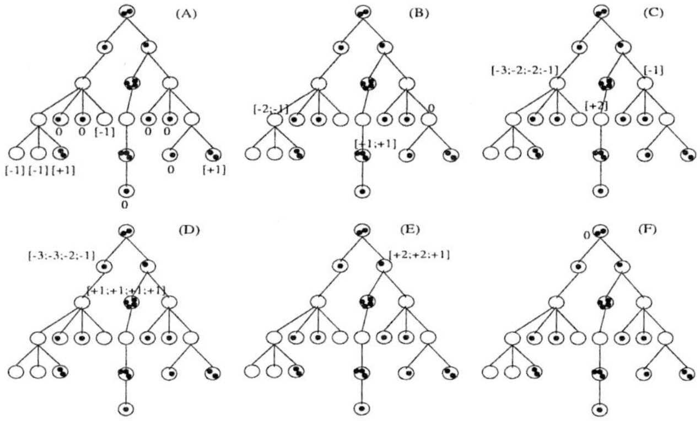

Fig. 1. An example of the labelling process. The maximum absolute value obtained is $M = 3$.

label the root with $s' + s - 1$ times $+1$ (hence with 0 if $s' + s - 1 = 0$); if $s' < |s|$ then with the biggest $|s| - s' - 1$ negative numbers contained in the previous labels decreased by 1 and a -1; if $s' = |s|$ just with a -1. We can then continue the process until the whole tree is labelled. Since we are considering a number of points equal to the number of vertices, the last vertex will be labelled by just a 0, see Figure 1.

Let $m(v)$ be the biggest absolute value appearing as a label for a node v, it is possible to prove by induction that any matching will have to use a path with length at least $m(v)$ going through v. This property is due to the fact that the algorithm always pushes up the smallest possible set of "ordered" labels (according to the positive cone order $u > v$ when $u - v$ is a positive vector). It follows that if M is the biggest absolute value of a label, then $\rho \geq M$.

Now, remark that we can easily find a matching with stretch $2M$ by associating positive labels with negative labels.

Analysis of the algorithm. In order to compute the probability of finding a matching between random points and the tree vertices, we would normally apply the Hall theorem to every vertex-subset of the tree. The greedy algorithm tells us that we can actually reduce our attention to specific subsets obtained that correspond to edge-cuts. There are $2(n - 1)$ such subsets, reducing the number of witness cuts from an exponential to a linear number.

Definition 3. *For a given tree T, $T' < T$ if T' is one of the two subtrees obtained by removing one edge of T.*

Lemma 2. *Given a tree $T = (V, E)$, $T' < T$ and stretch ρ, it is possible to compute in polynomial time the probability that T' induces a bad cut for ρ.*

Proof. Using standard binomial coefficient evaluation, we can compute $\Pr[\eta(T', \omega) > |\Gamma^\rho(T')|] = \sum_{i=|\Gamma^\rho(T')|+1}^{|V|} \binom{|V|}{i} \left(\frac{|T'|}{|V|}\right)^i \left(1 - \frac{|T'|}{|V|}\right)^{|V|-i}$; and do the same for $\Pr[\eta(T', \omega) < |\Gamma^\rho(T')|]$. □

Theorem 3. *Given a tree $T = (V, E)$ and any stretch ρ, it is possible to approximate $1 - \mu(T, \rho)$ within $2|V|$.*

Proof. From the previously described labelling scheme, $1 - \mu(\rho) \le \Pr(\exists T' < T$ such that $\eta(T', \omega) \ge |\Gamma^{2\rho}(T')|)$ and $\sum_{T'<T} Pr(\eta(T', \omega) \ge |\Gamma^{2\rho}(T')|) \le 2(|V| - 1) \max_{T'<T} Pr(\eta(T', \omega) \ge |\Gamma^{2\rho}(T')|)$.

Moreover, $\max_{T'<T} \Pr(\eta(T', \omega) > |\Gamma^{2\rho}(T')|) \le 1 - \mu(2\rho) \le 1 - \mu(\rho)$. It follows that $1 - \mu(\rho) \le 2(|V| - 1) \max_{T'<T} \Pr(\eta(T', \omega) \ge |\Gamma^{2\rho}(T')|) \le (1 - \mu(\rho))2(|V| - 1)$. Since by Lemma 2 such probabilities can be computed in polynomial time, the claim holds. □

The previous proof can be interpreted as follows. If there exists a bad cut then there exists a bad cut that is defined by a subtree obtained by removing one edge. This means that we can use n witness cuts and get a good estimation of the probability of failing using simple cut considerations. Since on the line there is only one witness cut (the half line) we wonder if the same happens for trees. Consider a subdivided star with k branches of length k, we see that for $\rho = \sqrt{k}$ no cut is likely to be bad. Now, consider an event, with high probability any branch will contain $k + \Theta(\sqrt{k \log k})$ points, and then each branch is an independent Poisson process conditioned on its number of points. Let $C_i, i = 1, 2, \ldots, k$ be the set containing the $k/2$ points of branch i that are at distance at least $k/2$ from the central node. Since we have k branches, one of them will deviate by about $\sqrt{m \log k}$ where m is its mean. So for some C_i, $|\eta(C_i, \omega) - |C_i|| = \Theta(\sqrt{k \log k})$ and we need $\rho = \Theta(\sqrt{k \log k})$ to get $\mu(\rho) \sim \frac{1}{2}$, and $\mu(t\sqrt{k}) \le e^{-t^2}k$. On this graph we find about $k = \sqrt{n}$ "independent" cuts and ρ is chosen such that the probability of each of these cuts to be bad is less than $\frac{1}{\sqrt{n}}$.

5 Conclusion

We have introduced the *Points and Vertices* problem for a graph G, which captures in a natural way the "distance" between the *randomness* of throwing points (independently and uniformly at random) onto the vertices of G, and the *order* of the points being evenly balanced on G. We have derived several results on the problem with exact balancing of the points. Besides the pure theoretical interest, the *Points and Vertices* problem comes out to be of relevant interest in several fields motivating further investigation.

References

1. Leighton, F.T., Shor, P.W.: Tight bounds for minimax grid matching with applications to the average case analysis of algorithms. Combinatorica **9** (1989) 161–187
2. Shor, P.W.: The average-case analysis of some on-line algorithms for bin packing. Combinatorica **6** (1986) 179–200
3. Shor, P.W., Yukich, J.E.: Minimax grid matching and empirical measures. The Annals of Probability **19** (1991) 1338–1348

4. Cole, R., Frieze, A.M., Maggs, B.M., Mitzenmacher, M., Richa, A.W., Sitaraman, R.K., Upfal, E.: On balls and bins with deletions. In: Proc. of the 2nd Int. Workshop on Randomization and Approximation Techniques in Computer Science (RANDOM). (1998) 145–158
5. Drinea, E., Frieze, A., Mitzenmacher, M.: Balls and bins models with feedback. In: Proc. of the 13th Annual ACM-SIAM Symposium on Discrete Algorithms (SODA). (2002) 308–315
6. Iwama, K., Kawachi, A.: Approximated two choices in randomized load balancing. In: Proc. of the 15th Annual Int. Symposium on Algorithms and Computation (ISAAC). Volume 3341 of LNCS. (2004) 545–557
7. Raab, M., Steger, A.: "Balls into bins" - a simple and tight analysis. In: Proc. of the 2nd Int. Workshop on Randomization and Approximation Techniques in Computer Science (RANDOM). (1998) 159–170
8. Chazelle, B.: The Discrepancy Method Randomness and Complexity. Cambridge University Press (2002)
9. Dudenhoeffer, D.D., Jones, M.P.: A formation behavior for large-scale micro-robot force deployment. In: Proc. of the 32nd Conference on Winter Simulation. (2000) 972–982
10. Gobriel, S., Melhem, R., Mosse, D.: A unified interference/collision analysis for power-aware adhoc networks. In: Proc. of the 23rd Conference of the IEEE Communications Society. (2004)
11. Banez, J.M.D., Hurtado, F., Lopez, M.A., Sellares, J.A.: Optimal point set projections onto regular grids. In: Proc. of the 14th Annual Int. Symposium on Algorithms and Computation (ISAAC). Volume 2906 of LNCS. (2003) 270–279
12. Indyk, P., Motwani, R., Venkatasubramanian, S.: Geometric matching under noise: combinatorial bounds and algorithms. In: Proc. of the 10th Annual ACM-SIAM Symposium on Discrete Algorithms. (1999) 457–465
13. Meyer, F., Oesterdiekhoff, B., Wanka, R.: Strongly adaptive token distribution. Algorithmica **15** (1993) 413–427
14. Peleg, D., Upfal, E.: The token distribution problem. SIAM J. Comput. **18** (1989) 229–243
15. Mitzenmacher, M., Prabhakar, B., Shah., D.: Load balancing with memory. In: Proc. of the 43rd Annual IEEE Symposium on Foundations of Computer Science (FOCS). (2002) 799–808
16. Diestel, R.: Graph Theory. Springer-Verlag, New York (2nd edition, 2000)
17. Karp, R.M., Luby, M.G.: Monte carlo algorithms for planar multiterminal network reliability problems. Journal of Complexity **1** (1985) 45–64
18. Karger, D.R.: A randomized fully polynomial time approximation scheme for the all-terminal network reliability problem. SIAM Journal on Computing **29** (2000) 492–514
19. Vazirani, V.: Approximation Algorithms. Springer (2001)
20. Karaganis, J.J.: On the cube of a graph. Canad. Math. Bull. **11** (1969) 295–296

Simulating Undirected st-Connectivity Algorithms on Uniform JAGs and NNJAGs*

Pinyan Lu[1], Jialin Zhang[2], Chung Keung Poon[3], and Jin-Yi Cai[4]

[1] State Key Laboratory of Intelligent Technology and Systems, Department of Computer Science and Technology, Tsinghua University, Beijing, China
lpy@mails.tsinghua.edu.cn

[2] Department of Physics, Tsinghua University, Beijing, China
zhanggl02@mails.tsinghua.edu.cn

[3] Department of Computer Science, City University of Hong Kong, Hong Kong, China
ckpoon@cs.cityu.edu.hk

[4] Computer Sciences Department, University of Wisconsin, Madison, WI 53706, USA; and Tsinghua University, Beijing, China
jyc@cs.wisc.edu

Abstract. In a breakthrough result, Reingold [17] showed that the Undirected *st*-Connectivity problem can be solved in $O(\log n)$ space. The next major challenge in this direction is whether one can extend it to directed graphs, and thereby lowering the deterministic space complexity of $\mathcal{RL}$ or $\mathcal{NL}$. In this paper, we show that Reingold's algorithm, the $O(\log^{4/3} n)$-space algorithm by Armoni et al. [3] and the $O(\log^{3/2} n)$-space algorithm by Nisan et al. [14] can all be carried out on the RAM-NNJAG model [15] (a uniform version of the NNJAG model [16]). As there is a tight $\Omega(\log^2 n)$ space lower bound for the Directed *st*-Connectivity problem on the RAM-NNJAG model implied by [8], our result gives an obstruction to generalizing Reingold's algorithm to the directed case.

1 Introduction

The *st*-Connectivity problem is one of the most widely studied problems in computer science. It is a fundamental problem with many applications and yet is simple to state: Given a directed graph G together with two vertices s and t, the (directed) *st*-connectivity problem STCON is to determine if there is a directed path from s to t. In the special case when the graph G is undirected, we denote the problem by USTCON.

STCON is important in complexity theory as it is complete for the complexity class Non-deterministic Logspace $\mathcal{NL}$ under Deterministic Logspace reductions. Thus determining its (deterministic) space complexity is to answer the question

* The work described in this paper was fully supported by a grant from the Research Grants Council of the Hong Kong Special Administrative Region, China [CityU 1071/02E], an NSF Grant, USA (CCR-0208013) and the Natural Science Foundation of China (60223004, 60321002).

X. Deng and D. Du (Eds.): ISAAC 2005, LNCS 3827, pp. 767–776, 2005.

whether nondeterminism helps in space bounded computation—a space analog of the "$\mathcal{NP} = \mathcal{P}$?" question. The special case USTCON is also important for the dual reasons of being the core problem in many applications as well as being complete for the complexity class Symmetric Logspace $\mathcal{SL}$ [11].

To solve STCON, a natural approach is to perform a Depth-First Search or Breadth-First Search from node s trying to discover node t. This requires $\Omega(n)$ bits of storage in the worst case. Currently, the best known space upper bound is $O(\log^2 n)$ using Savitch's algorithm [19]. Proving any non-trivial ($\omega(\log n)$) space lower bound on a general Turing machine is beyond the reach of current techniques, let alone proving that Savitch's algorithm is optimal. Thus Cook and Rackoff [6] proposed a model called "Jumping Automata for Graphs" (JAG), in order to abstract away certain inessential features of existing st-connectivity algorithms, and to focus on its essential feature of moving from vertices to vertices. Briefly, the JAG model can only examine the input graph by a set of pebbles that traverse the graph from s. Moreover, it can only tell which pebbles are on the same nodes but cannot see the node names. Other than that, a JAG is unrestricted. Although Savitch's algorithm needs to cycle through all nodes in the graph, which is generally impossible for a JAG, Cook and Rackoff adapted the algorithm to run on a JAG in $O(\log^2 n)$ space. Moreover, they proved a space lower bound of $\Omega(\log^2 n/\log\log n)$ for directed STCON on this model. The lower bound was then extended to a randomized JAG (i.e., a JAG that can flip random coins to determine its moves) by Berman and Simon [5].

Later, Immerman [10] and Szelepcsényi [20] discovered a surprising nondeterministic logspace algorithm for st-nonconnectivity which seems to require node names in an essential way and is not known to be implementable on a nondeterministic JAG (i.e., a JAG that can make nondeterministic moves). Then Poon [16] proposed the NNJAG model which is more natural as it can see the names of the pebbled nodes. Further, he showed how to implement the Immerman-Szelepcsényi algorithm on a non-deterministic NNJAG while extending the lower bounds of Cook and Rackoff, and Berman and Simon to the deterministic and randomized NNJAG model. The lower bound is further improved to $\Omega(\log^2 n)$ for a randomized NNJAG by Edmonds et al. [8]. Other major newer algorithms for STCON, including the time-space tradeoff [4] by Barnes et. al., and the randomized STCON algorithm [9] by Gopalan et. al., can all be implemented on a deterministic or randomized NNJAG as appropriate. Thus the NNJAG model seems to be general enough to capture all existing STCON algorithms.

For USTCON, Aleliunas et al. [1] showed that a random walk from any node s will hit all other nodes in its connected component in expected $O(nm)$ steps, m being the number of edges in the graph. This puts USTCON in Randomized Logspace $\mathcal{RL}$ and also implies the existence of a polynomial length Universal Traversal Sequence (UTS), i.e., a sequence of edge labels following which one can reach every node in a connected component from any starting node in that component. A deterministic JAG can trivially simulate such UTS using one pebble and $O(mn)$ states as it is a non-uniform computation device. A randomized JAG can also easily simulate a random walk in the same $O(\log n)$ space.

Restricting our attention to uniform deterministic computation, we witnessed a decrease in the space complexity of USTCON from $O(\log^2 n)$ of Savitch [19] and Nisan [13] to $O(\log^{3/2} n)$ by the NSW algorithm [14] and to $O(\log^{4/3} n)$ by the ATWZ algorithm [3]. Finally, Reingold [17] settled the deterministic space complexity of USTCON by discovering a (deterministic) $O(\log n)$ space algorithm. It follows that $\mathcal{SL} = \mathcal{L}$. Given the success of NNJAG in simulating algorithms for STCON, it is natural to ask if it can simulate those algorithms for USTCON as well. To our knowledge, no one has carefully investigated this question.

In this paper, we show that all the three algorithms: the NSW algorithm, the ATWZ algorithm and Reingold's algorithm can be implemented on a RAM-NNJAG which is the uniform counterpart of an NNJAG. Note that it is important to consider the simulation on a uniform model since a non-uniform one can follow a UTS to visit all nodes in the connected component of s and so USTCON in logspace becomes trivial.

As mentioned, a central focus in this area is the deterministic space complexity of STCON. One possible direction is to extend Reingold's algorithm to the directed case. In fact, Dinur et. al. [7] has extended Reingold's algorithm to directed graphs which are bi-regular. However, the feasibility of these three algorithms on a RAM-NNJAG model implies an obstruction to generalizing them to the directed case: Since Edmonds et. al. [8] proved a tight space lower bound of $\Omega(\log^2 n)$ for solving STCON on the NNJAG model (and hence the RAM-NNJAG model), our result implies that Reingold's algorithm does not immediately apply to directed graphs and, if we are to make progress on improving the space complexity for STCON, some new algorithmic techniques need to be developed.

In the next section, we will introduce the JAG and related models. Section 3 is an overview of the simulations followed by the details in Section 4 and 5.

2 The JAG and Related Models

A JAG as introduced in [6] is a non-uniform model. It consists of a sequence of automata $\{J_1, J_2, \ldots\}$ where the n-th automaton $J = J_n$ consists of a set of p distinguishable pebbles numbered 1 to p, a set of q states and a transition function δ. In general, p and q will be functions of n and the transition function δ also depends on n.

The input to J is a triple (G, s, t) where G is an n-node graph containing nodes s and t. For every node in G, its out-edges are labelled by consecutive integers starting from 0. The nodes in G are also labelled from 0 up to $n-1$. We define the *instantaneous description* (id) of J as the pair (Q, Π) where Q is the current state and Π is a mapping of pebbles to nodes specifying the current location of each pebble in the graph. When J is in id (Q, Π), the transition function δ determines the next move for J based on: (1) the state Q, and (2) which pebbles are on the same node and which are not, according to Π (i.e., for each pair of pebbles P and P', whether $\Pi(P) = \Pi(P')$). A move is either a *walk* or a *jump*. A *walk* (P, i, Q') consists of moving the pebble P along the edge labelled i that comes out of the node $\Pi(P)$ and then assuming state Q'. (If there

is no such edge, the pebble just remains on the same node.) A *jump* (P, P', Q') consists of moving pebble P to the node $\Pi(P')$ and then assuming state Q'. The automaton J is initialized to have state Q_0 and with pebbles $P_1, \ldots, P_{p-1}$ on node s and pebble P_p on node t (which makes node t distinguishable from the rest). It is said to *accept* an input (G, s, t) if it enters an accepting state on this input. It solves STCON for n-node graphs if for every input (G, s, t) where G is an n-node directed graph, it accepts the input iff there is a directed path from s to t in G. The definition for USTCON is similar.

It is easy to see that an id of a JAG can be specified using $p \log n + \log q$ bits by any ordinary computation model such as a Turing machine or a Random Access Machine. Thus we heuristically define this quantity as the space used by a JAG. The time used is the number of moves it made.

An NNJAG [16] is similar to a JAG except that the transition function depends on Q and Π. In other words, in NNJAG the transition function δ can use the actual names $\Pi(P_i)$. Due to its ability to see and work with node names, we need not put a pebble on t to make it distinguishable from the others. So we can put all pebbles on s initially and the definition would guarantee that every node that has ever been pebbled are reachable from s.

The RAM-JAG [15] consists of a finite state control together with p pebbles and a number of $O(\log n)$-bit registers which in total require $O(\log q)$ bits of storage. Its storage is defined as $(p \log n + \log q)$ bits. It can perform the usual RAM operations on the registers and also three special instructions: *walk*, *jump* and *compare*. The instructions *walk(P,j)* and *jump(P,P')* are the same as that in a JAG. The instruction *compare(P,P',R)* checks whether pebbles P and P' are on the same node and stores the result (T or F) in a register R. A RAM-NNJAG is similar to a RAM-JAG except that the instruction *compare*(P, P', R) is replaced by *copy*(P, R) which copies the name of the node pebbled by P to the register R.

It is straightforward to show that a RAM-JAG (resp. RAM-NNJAG) can be simulated by an ordinary JAG (resp. NNJAG) with $O(p)$ pebbles and $q^{O(1)}$ states. Thus, any lower bound on the JAG (resp. NNJAG) carries over to the RAM-JAG (resp. RAM-NNJAG) model.

3 Overview of the Simulations

At the highest level, all three algorithms are to (conceptually) construct a sequence of graphs $G_1, G_2, \ldots, G_\ell$ from the input graph $G = G_0$ such that connectivity between s and t in G is the same as that between two nodes s' and t' in G_ℓ.

For the NSW algorithm, G_k is obtained by choosing at most $|G_{k-1}|/f$ representative nodes in G_{k-1} and putting in edges so that two nodes u and v in G_{k-1} are connected if and only if their representatives in G_k are connected. Setting $f = 2^{\sqrt{\log n}}$ and $\ell = \sqrt{\log n}$, G_ℓ contains at most a constant number of nodes and hence st-connectivity can be trivially determined, say, by a DFS or BFS from the representative of s in G_ℓ. The ATWZ algorithm is similar except that it chooses $f = 2^{\log^{2/3} n}$ and $\ell = \log^{1/3} n$.

For Reingold's algorithm, G_k is obtained from G_{k-1} by performing a zigzag product with an expander graph H, followed by a graph powering operation. The number of nodes actually increases by a constant factor but the degree remains constant while the graph becomes more expanding. After $\ell = O(\log n)$ levels, the diameter is guaranteed to be $O(\log n)$. Thus st-connectivity is solved by exhausting all the (polynomially many) paths of length $L = O(\log n)$ on G_ℓ from one of the nodes corresponding to s in G_ℓ.

Due to the space limitation, one cannot store all the graphs. Instead, these algorithms just find out if an arbitrary pair of nodes u, v are connected by an edge in G_k (or which node is connected to a node u via its x-th edge) when necessary; and they achieve this by recursively asking for the appropriate edges and nodes in G_{k-1}. To carry out this on-demand scheme on a RAM-JAG/NNJAG, we show how an edge traversal in G_k can be effected by traversing appropriate edges in G_{k-1}. Specifically, we take the following approach:

1. Design a way to *store* a node of G_k in a RAM-JAG/NNJAG.
2. Design a procedure that, given a node u in G_k stored in a RAM-JAG/NNJAG and an edge label x, simulates the traversal of the x-th edge emanating from u in G_k such that the node thus reached is stored.

A natural idea to store a node u is to have a pebble on u. This is sufficient for the NSW and ATWZ algorithm since nodes in G_k are also nodes in the original graph G. In contrast, Reingold's algorithm blows up the number of nodes by a constant factor for each level. So we need a more generalized concept of storing a node and traversing an edge in G_k (to be given in next section).

4 Simulating Reingold's Algorithm

4.1 Reingold's Algorithm

In Reingold's algorithm, a preliminary step transforms the input graph G into a D-regular non-bipartite graph for some well chosen constant D. This is done in Logspace by replacing each node with a cycle and adding self-loops if necessary. Furthermore, it assumes that the graph G is specified by a *rotation map* $Rot_G : [n] \times [D] \to [n] \times [D]$ such that $Rot_G(u, a) = (v, b)$ if the a-th edge incident to u is $e = \{u, v\}$ which leads to v and this edge is the b-th edge incident to v.

Given an n-node, D-regular graph G and a D-node, d-regular graph H, the zig-zag product $G ⓩ H$ ([18]) is a graph with vertex set $[n] \times [D]$ such that every vertex has d^2 edges labelled by $(x, y) \in [d] \times [d]$. Its *rotation map* $Rot_{G ⓩ H}$ is defined as:

$$Rot_{G ⓩ H}((u, a), (x, y)) = ((v, b), (y', x')),$$

where $Rot_H(a, x) = (a', x')$, $Rot_G(u, a') = (v, b')$, and $Rot_H(b', y) = (b, y')$, see Figure 1 for an illustration. Note that in reverse, $Rot_{G ⓩ H}((v, b), (y', x')) = ((u, a), (x, y))$.

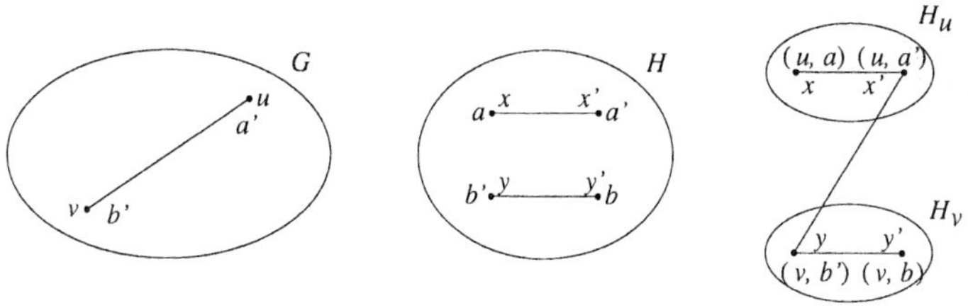

Fig. 1. Zigzag graph product: (Left) An edge in G, (Middle) Two edges in H, (Right) A path of 3 edges in the cross product of G and H. It corresponds to the edge labelled by (x, y) from node (u, a) in $G \circledz H$.

In [18] a remarkable property concerning the spectral gap (i.e., the difference between 1 and its second largest eigenvalue of the normalized adjacency matrix) is proved which has the following direct consequence [17]:

$$1 - \lambda(G \circledz H) \geq \frac{1}{2}(1 - \lambda(H)^2)(1 - \lambda(G))$$

where $\lambda(G)$ denotes the second largest eigenvalue of the normalized adjacency matrix of graph G. Essentially this shows that if H is chosen to be a good expander (i.e., a graph with a small second largest eigenvalue), then $G \circledz H$ will have a spectral gap not much smaller than that of G. Meanwhile a powering of $G \circledz H$ will increase the spectral gap of $G \circledz H$. Reingold chose $D = d^{16}$ and an appropriate H with D vertices. Then $(G \circledz H)^8$ is D-regular again, and has an increased spectral gap than G.

The main part of Reingold's algorithm is to repeatedly apply in turn the zig-zag product and graph powering to the input graph: For $k > 0$, define $G_k = (G_{k-1} \circledz H)^8$ where $\circledz$ is the zig-zag operator and H is a fixed D-node, d-regular expander graph. Thus, G_k will be a d^{16}-regular graph of size $D|G_{k-1}|$. As $D = d^{16}$, G_k is D-regular again. For example, one can use the construction by Alon and Roichman [2] which gives a d^{16}-node, d-regular expander graph H for some constant d, with $\lambda(H) \leq 1/2$.

The effect of one step of the above transformation from G_{k-1} to G_k is to turn the (connected components of the) previous graph G_{k-1} into a more expanding one in G_k, measured in terms of the spectral gap. At the same time, the number of nodes of the transformed graph increases by a factor of D while its degree remains to be $D = d^{16}$. Each node u in $G = G_0$ corresponds to D^k nodes in G_k, for $k > 0$. They are connected among themselves (in G_k) and we denote them by $(u, a_0, \ldots, a_{k-1})$ where $(a_0, \ldots, a_{k-1}) \in [D]^k$. The transformation ensures that two nodes u and v in G are connected iff $(u, a_0, \ldots, a_{k-1})$ and $(v, b_0, \ldots, b_{k-1})$ are connected in G_k, for each $(a_0, \ldots, a_{k-1})$ and $(b_0, \ldots, b_{k-1})$ in $[D]^k$.

Reingold showed that after $\ell = O(\log n)$ steps, (any connected component of) G_ℓ has a spectral gap greater or equal to $1/2$. That means any pair of nodes in the same connected component in G_ℓ are joined by a path of length $L = O(\log n)$. Thus, on G_ℓ, we simply enumerate all possible paths of length L from, say, $(s, 1, 1, \ldots, 1)$ and see if we can reach $(t, 1, 1, \ldots, 1)$.

4.2 Simulation on a RAM-JAG

In this section, we will actually describe our simulation on a RAM-JAG as we do not need the power of an NNJAG to see the node names.

Since a node in G_k is of the form $u_k = (u, a_0, a_1, \ldots, a_{k-1})$ where u is a node in G and $(a_0, a_1, \ldots, a_{k-1}) \in [D]^k$, we *store* a node u_k by having a pebble P on the node u in G and storing the values $a_0, \ldots, a_{k-1}$ in k registers, $A_0, \ldots, A_{k-1}$, each of size $\log D$ bits. Note that the RAM-JAG initially has pebbles on node s in G. Therefore it is easy to store node $(s, 1, \ldots, 1)$ in G_ℓ by initializing ℓ registers appropriately.

Next, consider the traversal of an edge labelled a_k from a node u_k in G_k. Let $Rot_{G_k}(u_k, a_k) = (v_k, b_k)$. We will prove by induction on $k \geq 0$ that if u_k and a_k are stored by the RAM-JAG, it can traverse the a_k-th edge of u_k so that v_k and b_k are stored.

For $k = 0$, assume that the RAM-JAG stores u_k in pebble P (i.e., pebble P is on node u_k) and a_k is stored in a register A. Then the RAM-JAG can easily move pebble P from u_k to v_k by walking along the edge labelled a_k in G_k, i.e., the original input graph G. To compute the reverse label b_k, we try all possible edges from v_k and see which one brings us back to node u_k. That is, we first mark the node u_k with an extra pebble P'. Then we move P from node v_k along its first edge and see if it meets P'. If not, we jump P to P' (so P is at u_k again) and walk along the a_k-th edge so that P arrives at v_k again. Then we try the second edge of v_k and so on. In this way, the RAM-JAG can compute and store b_k in a register and have node v_k pebbled.

For $k > 0$, recall that $G_k = (G_{k-1} ⓩ H)^8$. Therefore, an edge in G_k corresponds to a path of length eight in $G_{k-1} ⓩ H$. Thus, we write a_k as a sequence of 8 edge labels, $(x_{k,1}, y_{k,1})$, $(x_{k,2}, y_{k,2})$, $\ldots$, $(x_{k,8}, y_{k,8})$ in $G_{k-1} ⓩ H$. This in turn can be viewed as a path of 8 edges in G_{k-1} beginning from node u_{k-1} but with edge labels "permuted" by the expander H. Suppose $u_k = (u_{k-1}, a_{k-1})$ and $Rot_H(a_{k-1}, x_{k,1}) = (a'_{k-1}, x'_{k-1})$. Then the first edge label in G_{k-1} to follow is a'_{k-1}. Note that the RAM-JAG can figure out (a'_{k-1}, x'_{k-1}) without traversing G_{k-1} since H is fixed. Let $Rot_{G_{k-1}}(u_{k-1}, a'_{k-1}) = (v_{k-1}, b'_{k-1})$. Since u_k stored in the RAM-JAG implies u_{k-1} is also stored, we can assume (by induction hypothesis) that the RAM-JAG can traverse the a'_{k-1}-th edge of u_{k-1} in G_{k-1} so that v_{k-1} and b'_{k-1} is stored. Suppose $Rot_H(b'_{k-1}, y_{k,1}) = (b_{k-1}, y'_{k,1})$. Then again, the RAM-JAG can compute b_{k-1} and $y'_{k,1}$ from b'_{k-1} and $y_{k,1}$ using the fixed structure of H only. In terms of $G_{k-1} ⓩ H$, the first edge leads to the node (v_{k-1}, b_{k-1}). Note that since the reverse edge b_{k-1} is known, one can traverse the next edge $(x_{k,2}, y_{k,2})$ in $G_{k-1} ⓩ H$. Thus, the RAM-JAG can repeat the traversal for the remaining seven edges, $(x_{k,2}, y_{k,2}), \ldots, (x_{k,8}, y_{k,8})$ in order to complete the traversal of one edge in G_k. Finally, observe that the 8 reverse edge labels on the path of length 8 in $G_{k-1} ⓩ H$ arranged in the order $(y'_{k,8}, x'_{k,8}), \ldots, (y'_{k,1}, x'_{k,1})$ is nothing but the reverse edge label, b_k, for the single edge just traversed in G_k. Thus, the RAM-JAG is also able to store b_k.

Thus our induction statement is proved and it follows that the traversals of the D^L paths in G_ℓ can be carried out on a RAM-JAG.

Now consider the storage per level. Storing the 8 reverse edge labels takes $O(\log D)$ space. To store a node $u_k = (u, a_0, \ldots, a_{k-1})$ in G_k, note that the storage for $u_{k-1} = (u, a_0, \ldots, a_{k-2})$ can be charged to level $k-1$ or below. So we just charge $O(\log D)$ bits for storing a_{k-1} in level k. Hence, the total storage for all the levels is $O(\log n)$.

5 Simulating the NSW and ATWZ Algorithms

5.1 The NSW and ATWZ Algorithms

The core of the algorithms is the *shrink* procedure which takes a graph G_{k-1} as input and return G_k. It computes the set of representatives of G_{k-1}, i.e., the node set of G_k, as follows. First, it computes a set L_k of landmark nodes which includes s and t together with some nodes drawn from a pairwise independent space. This can be implemented as $L_k = \{s,t\} \cup \{u \in G_{k-1} | 1 \leq ua_k + b_k \leq q/(6f)\}$ for some pair (a_k, b_k) drawn from a field F_q of size polynomial in n.

Then for any u in G_{k-1}, we find a neighbourhood, $N(u)$, of u. In NSW, we follow a short universal traversal sequence (UTS) of length $2^{O(\log^2 f)}$ from u and define $N(u)$ as the set of all nodes encountered in the UTS as well as their immediate neighbours. In ATWZ, we run a number of pseudo-random walks of length $2^{O(\log^2 f)}$ from u to approximate the average number of such walks hitting each node v or its immediate neighbors. Any node v with approximate average at least $1/n$ is put into $N(u)$.

With $N(v)$, we define the representative, $rep_{L_k}(u)$, of u as follows. If $N(u)$ contains the whole component of u, then u is not represented in G_k (component too small). Otherwise, u is either represented by the minimum v in $N(u) \cap L_k$ or by itself in case $N(u) \cap L_k$ is empty. (We will always treat s and t as the minimum nodes in L_k. This ensures that they are always represented by themselves unless their components are too small.) It was shown in [14] that $|G_{k-1}|/|G_k| \geq f$ for a constant fraction of (a_k, b_k)'s, i.e., the size reduction is guaranteed if we go over the polynomially many (a_k, b_k)'s. Finally, for all $(u, v) \in G_{k-1}$, we have $(rep_{L_k}(u), rep_{L_k}(v)) \in G_k$.

5.2 Simulation on a RAM-NNJAG

Since a RAM-NNJAG cannot cycle through all the nodes in G_k, it cannot immediately see if the choice of (a_k, b_k) achieves the required size reduction factor f. However, a RAM-NNJAG can try all possible sequences of $(a_1, b_1), (a_2, b_2), \ldots (a_\ell, b_\ell)$. At the end, at least one choice of the sequence will be good enough.

Now fix a sequence of $(a_1, b_1), \ldots, (a_\ell, b_\ell)$. We will store a node u by having a pebble on it. We will (conceptually) assign labels to edges in G_k such that the x-th edge of node u will lead to its x-th smallest neighbour in G_k. Suppose node u in G_k is stored and we are to traverse its x-th edge. In other words, we need to find the x-th smallest neighbours of u in G_k, To this end, we compute all those nodes u' in G_{k-1} represented by u in G_k (to be described in next paragraph). Then we compute all nodes v' which is an immediate neigbour of some u', followed

by their representatives, $rep_{L_k}(v')$, using a forward UTS. Finally, we choose the x-th smallest representative.

The most difficult step is to compute those u' in G_{k-1} represented by u in G_k For NSW, we make use of a reversible UTS. The idea is to convert the graph G_{k-1} into one with edges symmetrically labelled, i.e if the ith neighbor of vertex v is the vertex u, then the ith neighbor of vertex u is also the vertex v. Suppose this is done and suppose u is the i-th node along a short UTS from u' and let the sequence of edge labels in the UTS be $\sigma_1\sigma_2\cdots\sigma_i$. Then the i-th reverse UTS, $\sigma_i\cdots\sigma_2\sigma_1$, starting from u will bring us to u'. Thus to find all u' represented by u, we try, for each possible i, to walk from u using the i-th reverse UTS to reach a candidate node u'. Then we walk from u' using the (forward) UTS to verify if u is the minimum node in L_k. For ATWZ, the idea is similar except that we try for each pseudorandom walk and for each i, the i-th reverse pseudorandom walk from u to arrive at a candidate u'. Note that checking for $v \in L_k$ is easily done by checking if $1 \leq a_k v + b_k \leq q/(6f)$. Note also that when $u = s$ and in the process of discovering those u' represented by u, we hit t, then we can stop and answer: "s,t connected". Thus if $N(s)$ contains the whole component of s and we have not discovered t, we can stop and answer: "s,t not connected". More detail will be given in our technical report [12].

To convert the graph into a symmetrically labelled one, we expand a degree-d node v into a cycle of d nodes $v_0,\ldots,v_{d-1}$ if d is even; or $d+1$ nodes $v_0,\ldots,v_d$ if d is odd. The edges on the cycles are labelled such that the edge (v_i, v_{i+1}) (where for the vertex indices, we take modulo d if d is even; or modulo $d+1$ if d is odd) is labelled with 0 (or 1) if i is even (or odd respectively). For arbitrary edge (u,v) in the original graph, if u is the k-th neighbor of v and v is the l-th neighbor of u, then connect v_k and u_l using an edge labelled 2 at both ends. Finally for each node v of odd degree, add a self-loop to v_d with edge label 2.

As for the storage, storing the sequence $(a_1,b_1),\ldots,(a_\ell,b_\ell)$ requires $O(\ell \times \log n)$ space. For the NSW algorithm, each of the ℓ levels requires $O(1)$ pebbles and $O(\log^2 f)$ bits to generate and follow the short UTS. Thus $O(\log^{3/2} n)$ space is needed. For the ATWZ algorithm, each level requires $O(\log n)$ space and overall it also needs $O(\log^2 f + \ell \log n) = O(\log^{4/3} n)$ space to generate the pseudorandom walks. Hence in total $O(\log^{4/3} n)$ space is needed.

Acknowledgments

We would like to thank Xiaotie Deng, Andrew Yao, Frances Yao and Ming Li for several discussions. We especially thank Steve Cook for helpful comments on a previous draft of the paper.

References

1. R. Aleliunas, R. M. Karp, R. J. Lipton, L. Lovász, and C. Rackoff. Random walks, universal traversal sequences, and the complexity of maze problems. In *20th Annual Symposium on Foundations of Computer Science*, pages 218–223, San Juan, Puerto Rico, October 1979. IEEE.

2. N. Alon and Y. Roichman. Random Cayley graphs and and expanders. *Random Structures and Algorithms*, 5(2):271–284, 1994.
3. R. Armoni, A. Ta-Shma, A. Wigderson, and S. Zhou. An $O(\log(n)^{4/3})$ space algorithm for (s,t) connectivity in undirected graphs. *Journal of the ACM*, 47(2):294–311, 2000.
4. G. Barnes, J. F. Buss, W. L. Ruzzo, and B. Schieber. A sublinear space, polynomial time algorithm for directed s-t connectivity. *SIAM Journal on Computing*, 27(5):1273–1282, 1998.
5. P. Berman and J. Simon. Lower bounds on graph threading by probabilistic machines. In *24th Annual Symposium on Foundations of Computer Science*, pages 304–311, Tucson, AZ, November 1983. IEEE.
6. S. A. Cook and C. W. Rackoff. Space lower bounds for maze threadability on restricted machines. *SIAM Journal on Computing*, 9(3):636–652, August 1980.
7. I. Dinur, O. Reingold, L. Trevisan, and S. Vadhan. Finding paths in nonreversible markov chains. Technical Report TR05-022, Electronic Colloquium on Computational Complexity, 2005.
8. J. Edmonds, C. K. Poon, and D. Achlioptas. Tight lower bounds for st-connectivity on the NNJAG model. *SIAM Journal on Computing*, 28(6):2257–2284, 1999.
9. P. Gopalan, R. Lipton, and A. Mehta. Randomized time-space tradeoffs for directed graph connectivity. In *FSTTCS'03*, pages 208–216, 2003. LNCS 2914.
10. N. Immerman. Nondeterministic space is closed under complementation. *SIAM Journal on Computing*, 17(5):935–938, October 1988.
11. H. R. Lewis and C. H. Papadimitriou. Symmetric space-bounded computation. *Theoretical Computer Science*, 19(2):161–187, August 1982.
12. P. Lu, J. Zhang, C.K. Poon, and J. Cai. Simulating undirected st-oonnectivity algorithms on uniform JAGs and NNJAGs. Technical report, 2005.
13. N. Nisan. $RL \subseteq SC$. In *Proceedings of the Twenty Fourth Annual ACM Symposium on Theory of Computing*, pages 619–623, Victoria, B.C., Canada, May 1992.
14. N. Nisan, E. Szemerédi, and A. Wigderson. Undirected connectivity in $O(\log^{1.5} n)$ space. In *33rd Annual Symposium on Foundations of Computer Science*, Pittsburgh, PA, October 1992. IEEE.
15. C. K. Poon. *On the Complexity of the ST-Connectivity Problem.* PhD thesis, University of Toronto, 1996.
16. C. K. Poon. A space lower bound for st-connectivity on node-named JAGs. *Theoretical Computer Science*, 237(1-2):327–345, 2000.
17. O. Reingold. Undirected st-connectivity in log-space. In *Proceedings of the Thirty Seventh Annual ACM Symposium on Theory of Computing*, pages 376–385, 2005.
18. O. Reingold, S. Vadhan, and A. Wigderson. Entropy waves, the zig-zag graph product, and new constant-degree expanders. *Annals of Mathematics*, 155(1), 2001.
19. W. J. Savitch. Relationships between nondeterministic and deterministic tape complexities. *Journal of Computer and System Sciences*, 4(2):177–192, 1970.
20. R. Szelepcsényi. The method of forcing for nondeterministic automata. *Bulletin of the European Association for Theoretical Computer Science*, 33:96–100, October 1987.

Upper Bounds on the Computational Power of an Optical Model of Computation

Damien Woods

Boole Centre for Research in Informatics and School of Mathematics,
University College Cork, Ireland
d.woods@bcri.ucc.ie

Abstract. We present upper bounds on the computational power of an optical model of computation called the $\mathcal{C}_2$-CSM. We show that $\mathcal{C}_2$-CSM time is no more powerful than sequential space, thus giving one of the two inclusions that are necessary to show that the model verifies the parallel computation thesis. Furthermore we show that $\mathcal{C}_2$-CSMs that simultaneously use polynomial space and polylogarithmic time decide no more than the class NC.

1 Introduction

The computational model we study is relatively new and is called the continuous space machine (CSM) [11, 12, 13, 20, 21, 22, 23]. The original definition of the model was by Naughton [11, 12]. The CSM is inspired by classical Fourier optics and uses complex-valued images, arranged in a grid structure, for data storage. The program also resides in images. The CSM has the ability to perform Fourier transformation, complex conjugation, multiplication, addition, thresholding and resizing of images. It has simple control flow operations and is deterministic. We analyse the model in terms of seven complexity measures inspired by real-world resources.

A rather general variant of the model was previously shown [23] to decide the membership problem for all recursively enumerable languages, and as such is unreasonable in terms of implementation. Also, the growth in resource usage was shown for each CSM operation, which in some cases was unreasonably large [21]. This work motivated the definition of the $\mathcal{C}_2$-CSM, a more realistic and restricted CSM.

Recently [22] we have given lower bounds on the computational power of the $\mathcal{C}_2$-CSM by showing that it is at least as powerful as models that verify the parallel computation thesis. This thesis [4, 6] states that parallel time corresponds, within a polynomial, to sequential space for reasonable parallel models. See, for example, [18, 14, 8, 7] for details. Furthermore we have shown that $\mathcal{C}_2$-CSMs that simultaneously use polynomial SPACE and polylogarithmic TIME accept at least the class NC [22].

Here we present the other of the two inclusions that are necessary in order to verify the parallel computation thesis; we show that $\mathcal{C}_2$-CSMs computing in

X. Deng and D. Du (Eds.): ISAAC 2005, LNCS 3827, pp. 777–788, 2005.

TIME $T(n)$ accept at most the languages accepted by deterministic Turing machines in $O(T^2(n))$ space: $\mathcal{C}_2$-CSM-TIME$(T(n)) \subseteq$ DSPACE$(O(T^2(n)))$. Also we show that $\mathcal{C}_2$-CSMs that simultaneously use polynomial SPACE and polylogarithmic TIME accept at most the class NC: $\mathcal{C}_2$-CSM-SPACE, TIME$(n^{O(1)}, \log^{O(1)} n)$ $\subseteq$ NC. These inclusions are established via $\mathcal{C}_2$-CSM simulation by a polynomial sized, log depth, logspace uniform circuit.

2 The CSM

We begin by informally describing the model, this brief overview is not intended to be complete more details are to be found in [23, 20].

A complex-valued image (or simply, image) is a function $f : [0,1) \times [0,1) \to \mathbb{C}$, where $[0,1)$ is the half-open real unit interval. We let $\mathcal{I}$ denote the set of complex-valued images. Let $\mathbb{N}^+ = \{1,2,3,\ldots\}$, $\mathbb{N} = \mathbb{N}^+ \cup \{0\}$, and for a given CSM M let $\mathcal{N}$ be a countable set of images that encode M's addresses. Additionally, for a given M there is an *address encoding function* $\mathfrak{E} : \mathbb{N} \to \mathcal{N}$ such that $\mathfrak{E}$ is Turing machine decidable, under some *reasonable* representation of images as words. An address is an element of $\mathbb{N} \times \mathbb{N}$.

Definition 1 (CSM). *A CSM is a quintuple $M = (\mathfrak{E}, L, I, P, O)$, where*

$\mathfrak{E} : \mathbb{N} \to \mathcal{N}$ *is the address encoding function*
$L = ((s_\xi, s_\eta), (a_\xi, a_\eta), (b_\xi, b_\eta))$ *are the addresses: sta, a, and b,*
$I = ((\iota_{1_\xi}, \iota_{1_\eta}), \ldots, (\iota_{k_\xi}, \iota_{k_\eta}))$ *are the addresses of the k input images,*
$P = \{(\zeta_1, p_{1_\xi}, p_{1_\eta}), \ldots, (\zeta_r, p_{r_\xi}, p_{r_\eta})\}$ *are the r programming symbols and their addresses where* $\zeta_j \in (\{h, v, *, \cdot, +, \rho, st, ld, br, hlt\} \cup \mathcal{N}) \subset \mathcal{I}$,
$O = ((o_{1_\xi}, o_{1_\eta}), \ldots, (o_{l_\xi}, o_{l_\eta}))$ *are the addresses of the l output images.*

Each address is an element from $\{0, 1, \ldots, \Xi - 1\} \times \{0, 1, \ldots, \mathcal{Y} - 1\}$ *where* $\Xi, \mathcal{Y} \in \mathbb{N}^+$. *Addresses a and b are distinct.*

Addresses whose contents are not specified by P in a CSM definition are assumed to contain the constant image $f(x,y) = 0$. We interpret this definition to mean that M is (initially) defined on a grid of images bounded by the constants Ξ and $\mathcal{Y}$, in the horizontal and vertical directions respectively.

In our grid notation the first and second elements of an address tuple refer to the horizontal and vertical axes of the grid respectively, and image $(0,0)$ is at the lower left-hand corner of the grid. The images have the same orientation as the grid. Figure 1 gives the CSM operations in this grid notation. Configurations are defined in a straightforward way as a tuple $\langle c, e \rangle$ where c is an address called the control and e represents the grid contents. For a more thourough introduction see [20, 23].

Next we define some CSM complexity measures. All resource bounding functions map from $\mathbb{N}$ into $\mathbb{N}$ and are assumed to have the usual properties [1].

Definition 2. *The* TIME *complexity of a CSM M is the number of configurations in the computation sequence of M, beginning with the initial configuration and ending with the first final configuration.*

[h]	: replace image in a with its horizontal 1D Fourier transform (FT).
[v]	: replace image in a with its vertical 1D FT.
[$*$]	: replace image in a with its complex conjugate.
[$\cdot$]	: multiply (point by point) the two images in a and b. Store result in a.
[$+$]	: perform a complex (point by point) addition of a and b. Store result in a.
[ρ | z_{l} | z_{u}]	: $z_{\mathrm{l}}, z_{\mathrm{u}} \in \mathcal{I}$; filter the image in a by amplitude using z_{l} and z_{u} as lower and upper amplitude threshold images, respectively.
[st | ξ_1 | ξ_2 | η_1 | η_2]	: $\xi_1, \xi_2, \eta_1, \eta_2 \in \mathbb{N}$; copy the image in a into the rectangle of images whose bottom left-hand corner address is (ξ_1, η_1) and whose top right-hand corner address is (ξ_2, η_2).
[ld | ξ_1 | ξ_2 | η_1 | η_2]	: $\xi_1, \xi_2, \eta_1, \eta_2 \in \mathbb{N}$; copy into a the rectangle of images whose bottom left-hand corner address is (ξ_1, η_1) and top right-hand corner address is (ξ_2, η_2).
[br | ξ | η]	: $\xi, \eta \in \mathbb{N}$; unconditionally branch to the image at address (ξ, η).
[hlt]	: halt.

Fig. 1. The set of CSM operations, given in our grid notation

Definition 3. *The* GRID *complexity of a CSM M is the minimum number of images, arranged in a rectangular grid, for M to compute correctly on all inputs.*

Let $S : \mathcal{I} \times (\mathbb{N} \times \mathbb{N}) \rightarrow \mathcal{I}$, where $S(f(x,y), (\Phi, \Psi))$ is a raster image, with $\Phi\Psi$ constant-valued pixels arranged in Φ columns and Ψ rows, that approximates $f(x,y)$. If we choose a reasonable and realistic S then the details of S are not important.

Definition 4. *The* SPATIALRES *complexity of a CSM M is the minimum $\Phi\Psi$ such that if each image $f(x,y)$ in the computation of M is replaced with $S(f(x,y), (\Phi, \Psi))$ then M computes correctly on all inputs.*

Definition 5. *The* DYRANGE *complexity of a CSM M is the ceiling of the maximum of all the amplitude values stored in all of M's images during M's computation.*

In earlier treatments [21, 23] we defined the complexity measures AMPLRES, PHASERES and FREQ. AMPLRES and PHASERES are measures of the cardinality of discrete amplitude and phase values of the complex numbers in the range of CSM images. In the present work AMPLRES and PHASERES both have constant value of 2 which means that all images are of the form $f : [0,1) \times [0,1) \rightarrow \{0, \pm\frac{1}{2}, \pm 1, \pm\frac{3}{2}, \ldots\}$. Furthermore we are studying a restricted CSM to which FREQ does not apply.

Often we wish to make analogies between space on some well-known model and CSM 'space-like' resources. Thus we define the following convenient term.

Definition 6. *The* SPACE *complexity of a CSM M is the product of all of M's complexity measures except* TIME.

We have defined the complexity of a computation (sequence of configurations) for each measure. We extend this definition to the complexity of a (possibly non-final) configuration in the obvious way. Also, we sometimes talk about the complexity of an image, this is simply the complexity of the configuration that the image is in.

In [21, 20] we defined the $\mathcal{C}_2$-CSM, a restricted and more realistic class of CSM.

Definition 7 **($\mathcal{C}_2$-CSM).** *A $\mathcal{C}_2$-CSM is a CSM whose computation* TIME *is defined for $t \in \{1, 2, \ldots, T(n)\}$ and has the following restrictions:*

- *For all* TIME *t both* AMPLRES *and* PHASERES *have constant value of 2.*
- *For all* TIME *t each of* SPATIALRES, GRID *and* DYRANGE *is $O(2^t)$ and* SPACE *is redefined to be the product of all complexity measures except* TIME *and* FREQ.
- *Operations h and v compute the discrete FT (DFT) in the horizontal and vertical directions respectively.*
- *Given some* reasonable *binary word representation of the set of addresses $\mathcal{N}$, the address encoding function $\mathfrak{E} : \mathbb{N} \to \mathcal{N}$ is decidable by a logspace Turing machine.*

3 Circuits and Representation

In this work we are using logspace uniform circuits over the complete basis $\wedge$, $\vee$ and $\neg$ with the usual complexity measures of size and depth [1, 15]. For convenience we frequently write "uniform" instead of "logspace uniform". Given a circuit c_n, its encoding $\bar{c}_n$ is a string of 4-tuples, where each tuple encodes a single gate and is of the form $(g, b, g_l, g_r) \in (\{0,1\}^+, \{\wedge, \vee, \neg\}, \{0,1\}^+ \cup \varnothing, \{0,1\}^+ \cup \varnothing)$. The tuple specifies the gate label g, the operation b that the gate computes, and the inputs g_l and g_r. For $\neg$ gates exactly one of g_l or g_r is the null symbol $\varnothing$.

Let U-SIZE,DEPTH(size(n), depth(n)) be the class of languages recognised by logspace uniform bounded fan-in circuits of size size(n) and depth depth(n), respectively. The equality NSPACE$(S^{O(1)}(n))$ = U-SIZE, DEPTH$(2^{n^{O(1)}}, S^{O(1)}(n))$ is well-known [3, 8]. Circuit depth is a measure of parallel time, hence logspace uniform circuits verify the parallel computation thesis. Specifically we are interested in the inclusion [1, 8]

$$\text{U-SIZE, DEPTH}(2^{S(n)}, S(n)) \subseteq \text{DSPACE}(O(S(n))). \qquad (1)$$

Suppose we are simulating a $\mathcal{C}_2$-CSM M that has TIME, GRID, SPATIALRES, DYRANGE and SPACE of $T(n)$, $G(n)$, $R_S(n)$, $R_D(n)$ and $S(n)$ respectively. From Definition 7, $S(n) = G(n)R_S(n)R_D(n)4 \leqslant c_G 2^{T(n)} c_{R_S} 2^{T(n)} c_{R_D} 2^{T(n)}$ for constants c_G, c_{R_S}, c_{R_D}. To simplify things (by removing the constants) we redefine the the TIME of M by increasing it by a constant, $T'(n) = \lceil \log(c 2^{T(n)}) \rceil$ where $c = \max(c_G, c_{R_S}, c_{R_D})$. In the sequel we write $T'(n)$ as $T(n)$. SPACE is now bounded above by $2^{T(n)} 2^{T(n)} 2^{T(n)}$, specifically each of GRID, SPATIALRES and DYRANGE is now bounded above by $2^{T(n)}$. Next let $G(n) = G_x(n)G_y(n)$ and $R_S(n) = R_{S_x}(n)R_{S_y}(n)$, where $G_x(n)$ and $G_y(n)$ are the number of grid images in the

horizontal and vertical directions respectively, and where $R_{S_x}(n)$ and $R_{S_y}(n)$ are the number of image pixels in the horizontal and vertical directions respectively. Define $G_x(n) = G_y(n) = R_{S_x}(n) = R_{S_y}(n) = 2^{T(n)}$. Thus $G(n) = R_S(n) = 2^{2T(n)}$ and $R_D(n) = 2^{T(n)}$ and we get our final upper bound on M's SPACE: $S(n) \leqslant 2^{2T(n)}2^{2T(n)}2^{T(n)} = 2^{5T(n)}$. These adjustments do not affect M's computation, we have simply defined M to be more TIME and SPACE inefficient.

M's grid of images are represented by a single binary word $\mathcal{G}$. The word $\mathcal{G}$ is composed of $G(n)$ *image subwords* of equal length such that for each image i in M there is an image subword $\mathcal{G}_i$, and vice versa. Specifically $\mathcal{G} = \mathcal{G}_0\mathcal{G}_1 \cdots \mathcal{G}_{G(n)-1}$. We order M's images first horizontally and then vertically; beginning with the lower leftmost grid image, proceeding left to right and then bottom to top.

Next we show how each pixel in image i is represented in the image subword $\mathcal{G}_i$. This representation scheme is analogous to the previous representation of images as subwords. The image subword $\mathcal{G}_i$ is composed of $R_S(n)$ *pixel subwords* of equal length, $\mathcal{G}_i = \mathcal{G}_i[0]\mathcal{G}_i[1] \cdots \mathcal{G}_i[R_S(n)-1]$. For each pixel j in image i there is a pixel subword $\mathcal{G}_i[j]$ and vice versa. Analogous to the image ordering, we order the pixels in each image first by the horizontal and then by the vertical direction, beginning with the lower leftmost pixel.

Given DYRANGE of $R_D(n)$ it follows directly that the value (or range) of each pixel in a $\mathcal{C}_2$-CSM configuration is from the set $\{0, \pm\frac{1}{2}, \pm 1, \pm\frac{3}{2}, \ldots, \pm R_D(n)\}$. To represent this set as a set of binary words we use the 2's complement binary representation of integers, with a slight modification: the binary sequence is shifted by one place to take care of the halves.

At this point we have defined the entire structure of the word $\mathcal{G}$ representing M's grid. The length of $\mathcal{G}$ is $|\mathcal{G}| = R_S(n)G(n)\lceil \log(4R_D(n)+1)\rceil$. Hence if M's SPACE complexity is $O(S(n))$ then $|\mathcal{G}|$ is also $O(S(n))$. We substitute to define $|\mathcal{G}|$ in terms of TIME, $|\mathcal{G}| = 2^{2T(n)}2^{2T(n)}\lceil \log(4 \cdot 2^{T(n)}+1)\rceil$. Specifically the length of each pixel subword is $|\mathcal{G}_i[j]| = \lceil \log(4 \cdot 2^{T(n)}+1)\rceil$ and the length of each image subword is $|\mathcal{G}_i| = 2^{2T(n)}\lceil \log(4 \cdot 2^{T(n)}+1)\rceil$. These expressions are useful for giving bounds on circuit complexity in terms of TIME $T(n)$.

A $\mathcal{C}_2$-CSM configuration $\langle c, e\rangle$ is represented as a binary word ctrl$\mathcal{G}$, which we write as $(\text{ctrl}, \mathcal{G})$, where ctrl is of length $2T(n)$ and $\mathcal{G}$ is as given above. (We interpret the 'instruction pointer' ctrl as a number that indexes the location of the next $\mathcal{C}_2$-CSM operation).

4 Circuit Simulation of $\mathcal{C}_2$-CSM

To simulate the $T(n)$ computation steps of the $\mathcal{C}_2$-CSM M we will design a logspace uniform circuit c_M that is of size $S^{O(1)}(n)$ and depth $O(T^2(n))$. We assume that M is a language deciding $\mathcal{C}_2$-CSM [23, 20], and that the input word w is of length n. M is simulated by the circuit c_M in the following way. At the first step of the simulation the circuit c_M is presented with the input word $(\text{ctrl}_{\text{sta}}, \mathcal{G}_{\text{sta}})$ that represents M's initial configuration (including M's input). The circuit c_M has $T(n)$ identical layers numbered 0 (the input layer) to $T(n)-1$

(the output layer). A layer contains a circuit c_{op} for each $\mathcal{C}_2$-CSM operation *op*, where the circuit encoding function $1^n \rightarrow \overline{c}_{op}$ is computable by a transducer Turing machine in workspace $\log \text{size}(\overline{c}_{op})$.

4.1 Circuits Computing $+$, $\cdot$, $*$, ρ, h and v

We begin by simulating the $+$ operation. Addition of two nonnegative integers written in binary is computable by an unbounded fan-in circuit of size $O(m^2)$ and constant depth and so is an AC^0 problem (e.g. use the well-known carry look-ahead algorithm, for example see [8]). Hence this problem is also in NC^1. Krapchenko [9], and Ladner and Fischer [10] give tighter NC^1 adders that have depth $O(\log m)$ and lower the size bound to $O(m)$.

Theorem 1 (circuit simulation of $+$). *The $\mathcal{C}_2$-CSM operation $+$ is simulated by a logspace uniform circuit $c_{a:=a+b}$ of size $O(2^{2T(n)}T(n))$ and depth $O(\log T(n))$.*

Proof. Circuit: To add two pixel words we use Ladner and Fischer's NC^1 addition algorithm [10]. Extending this algorithm to work for the 2's complement binary representation is straightforward. Recall that the addition operation adds images a and b in a parallel point by point fashion and places the result in a. The circuit $c_{a:=a+b}$ has one adder subcircuit for each pixel j in a. Under the representation scheme described above, pixel word $\mathcal{G}_a[j]$ is added to pixel word $\mathcal{G}_b[j]$, resulting in a new word $\mathcal{G}_{a'}[j]$. There are $R_S = 2^{2T(n)}$ pixels in each of a and b thus the circuit $c_{a:=a+b}$ consists of $2^{2T(n)}$ parallel adders. Each adder has depth $O(\log p)$ and size $O(p)$ where $p = \lceil \log(4 \cdot 2^{T(n)} + 1) \rceil$ is pixel word length. Since each adder has size $O(T(n))$, the circuit has size $O(2^{2T(n)}T(n))$. The circuit has depth $O(\log p) = O(\log T(n))$.

Uniformity: We show that $1^n \rightarrow \overline{c}_{a:=a+b}$ is computable by a transducer that uses at most log workspace. At any computation step the transducer will have at most the encoding of the current gate and a constant number of counters on its worktapes. The circuit has $O(2^{2T(n)}T(n))$ gates, thus each gate label has length $O(T(n))$ and is computable in space $\log |\overline{c}_{a:=a+b}| = O(T(n))$. There are counters for the index of the current gate and for the current adder circuit. Each adder circuit is logspace uniform and hence a constant number of [length $O(T(n))$] counters, is sufficient to construct each one. All gates and counters are computable in space $\log |\overline{c}_{a:=a+b}|$. □

As with addition there are NC^1 circuits for multiplication of binary numbers [8], for example Schönage and Strassen's circuit [16], which uses the DFT, has size $O(m \log m \log \log m)$ and depth $O(\log m)$. Unlike addition, Furst, Saxe and Sipser [5] showed that multiplication is not in AC^0, by showing that parity is not in AC^0 [2]. The proof of the following theorem is identical to the previous one except that we use a polynomial sized NC^1 multiplication circuit as opposed to the linear sized NC^1 adder used above.

Theorem 2 (circuit simulation of $\cdot$). *The $\mathcal{C}_2$-CSM operation $\cdot$ is simulated by a logspace uniform circuit $c_{a:=a\cdot b}$ of size $O(2^{2T(n)}T^2(n))$ and depth $O(\log T(n))$.*

Each pixel in a C_2-CSM is rational valued, hence the complex conjugation operation $*$ is the identity and is simulated by the identity circuit $(\neg\neg a)$.

Theorem 3 (circuit simulation of $*$). *The C_2-CSM operation $*$ is simulated by a logspace uniform circuit $c_{a:=a^*}$ of size $O(2^{2T(n)})$ and constant depth.*

Theorem 4 (circuit simulation of ρ). *The C_2-CSM operation ρ is simulated by a logspace uniform circuit $c_{a:=\rho(a,z_{\mathrm{l}},z_{\mathrm{u}})}$ of size $O(2^{2T(n)}T^2(n))$ and depth $O(\log T(n))$.*

Proof. Circuit: First we build a circuit $c_{u>v}$ that tests if the number represented by one pixel word is greater than another. It is straightforward to give constant size and depth circuits that tell if two bits $b_1, b' \in \{0,1\}$ are equal and if one is greater than the other: The circuits $c_{b\equiv b'}$ and $c_{b>b'}$ respectively compute the $\equiv$ and $>$ expressions $b \equiv b' = (b \wedge b') \vee (\neg b \wedge \neg b')$ and $b > b' = b \wedge \neg b'$. On pixel words u and v, we define the Boolean expression $u > v = \bigvee_{m=0}^{|u-1|}((u_m > v_m) \wedge (\bigwedge_{k=0}^{m-1}(u_k \equiv v_k)))$. We build $c_{u>v}$ as follows. For each m the circuit computing $\bigwedge_{k=0}^{m-1}(u_k \equiv v_k)$ is realised as a $O(m)$ size, $O(\log m)$ depth tree. There are $|u|$ such trees, the root of each has a $\wedge$ test with the constant depth circuit for $u_m > v_m$. We take the OR of these ANDs using a $\log|u|$ depth OR tree. The circuit $c_{u>v}$ has size $O(|u|^2)$ and depth $O(\log|u|)$.

Using a similar construction we build the circuit $c_{u<v}$, this has the same complexity as $c_{u>v}$. Moreover these circuits can be extended to work on the 2's complement representation with only a multiplicative constant increase in size and additive constant increase in depth. We combine these circuits to create the pixel thresholding circuit $c_{v:=\rho(v,l,u)}$ that evaluates to l if $c_{v<l} \equiv 1$, to u if $c_{v>u} \equiv 1$, and to v otherwise.

Recall that the operation $\rho(a, z_{\mathrm{l}}, z_{\mathrm{u}})$ thresholds image a in a parallel point by point fashion and places the result in a. The circuit $c_{a:=\rho(a,z_{\mathrm{l}},z_{\mathrm{u}})}$ has one pixel-thresholding subcircuit for each pixel j in a. Pixel word $\mathcal{G}_a[j]$ is thresholded below by pixel word $\mathcal{G}_{z_{\mathrm{l}}}[j]$ and above by pixel word $\mathcal{G}_{z_{\mathrm{u}}}[j]$ resulting in a new word $\mathcal{G}_{a'}[j]$. The circuit $c_{a:=\rho(a,z_{\mathrm{l}},z_{\mathrm{u}})}$ consists of $2^{2T(n)}$ parallel pixel thresholding subcircuits. Each pixel thresholding subcircuit has $O(\log p)$ depth and size $O(p^2)$ where $p = \lceil \log(4 \cdot 2^{T(n)} + 1) \rceil$ is pixel word length. Since there are $2^{2T(n)}$ pixel thresholding subcircuits (each has size $O(T^2(n))$) the circuit has size $O(2^{2T(n)}T^2(n))$. The entire circuit has depth $O(\log p) = O\left(\log \lceil \log(4 \cdot 2^{T(n)} + 1) \rceil\right) = O(\log T(n))$.

Uniformity: By stepping through the construction and applying the arguments given in Theorem 1 it is seen that the length of each gate label is $O(T(n))$ and a constant number of variables is sufficient to construct the encoding. □

Next we give simulations of the DFT operations via logspace uniform fast Fourier transform (FFT) circuits. On input length m the FFT circuit has size bounded above by $2m \log m$ and depth bounded above by $2 \log m$ [15].

Theorem 5 (circuit simulation of h and v). *The C_2-CSM horizontal (respectively, vertical) DFT operation h (respectively, v) is simulated by a logspace uniform circuit $c_{a:=h(a)}$ of size $O(2^{2T(n)}T(n))$ and depth $O(T(n))$.*

Proof. For each row of pixel words in image word a, the circuit $c_{a:=h(a)}$ has a single FFT subcircuit. The output is an image word a' such that each row in a' is the DFT of the same row in a. It is straightforward to verify the size, depth and uniformity. The circuit $c_{a:=v(a)}$ that simulates the C_2-CSM vertical DFT operation v is constructed similarly to $c_{a:=h(a)}$, except we replace the word "row" with "column". □

4.2 Circuits Computing *ld* and *st*

The simulations of ld and st are much more involved than the previous constructions. We briefly sketch the simulations, details are to be found in [20].

Theorem 6 (circuit simulation of ld by $c_{a:=[(\xi_1',\eta_1'),(\xi_2',\eta_2')]}$). *The C_2-CSM operation $ld\,(\xi_1', \eta_1', \xi_2', \eta_1')$ is simulated by the uniform circuit $c_{a:=[(\xi_1',\eta_1'),(\xi_2',\eta_2')]}$ of size $O(2^{6T(n)}T(n))$ and depth $O(T(n))$. This circuit takes as input the image words ξ_1', η_1', ξ_2' and η_2', and outputs an image word a.*

Proof (sketch). The address image words ξ_1', η_1', ξ_2', η_2' are decoded into four binary number words by four circuits that respectively compute $\mathfrak{E}^{-1}_{2^{T(n)}}(\xi_1') = \xi_1$, $\mathfrak{E}^{-1}_{2^{T(n)}}(\eta_1') = \eta_1$, $\mathfrak{E}^{-1}_{2^{T(n)}}(\xi_2') = \xi_2$, $\mathfrak{E}^{-1}_{2^{T(n)}}(\eta_2') = \eta_2$, where $\mathfrak{E}^{-1}_{2^{T(n)}}$ is a function that encodes the inverse of the logspace computable address encoding function $\mathfrak{E}$ from Definition 1.

In the next step we wish to select the 'rectangle' of image words that are to be loaded to image word a. Suppose that the rectangle contains more than one image, then naïvely copying the entire rectangle to the image word a would cause problems. (The rectangle to be loaded may contain up to $2^{4T(n)} = R_{\mathrm{S}}(n)G(n)$ pixel words whereas the image word a should contain exactly $2^{2T(n)}$ pixel words). However, observe that SPATIALRES is bounded above by $2^{2T(n)}$, thus the rectangle to be loaded contains at most $2^{2T(n)}$ regions of distinct value (otherwise image a would contain $> 2^{2T(n)}$ pixels after execution of ld). Hence we have to select only $2^{2T(n)}$ representative pixel words out of the total of $2^{2T(n)}(\xi_2 - \xi_1 + 1)(\eta_2 - \eta_1 + 1) \leqslant 2^{4T(n)}$ pixels. We choose the pixel with the lowest index in each (possibly) distinct area. For each pixel word i in image word a, the pixel j to be loaded is defined as $j = \mathrm{col}(i) + \mathrm{row}(i)R_{\mathrm{S}_x}G_x$ where $\mathrm{col}(i) = (i \mod R_{\mathrm{S}_x})(\xi_2 - \xi_1 + 1) + R_{\mathrm{S}_x}\xi_1$ and $\mathrm{row}(i) = \lfloor \frac{i}{R_{\mathrm{S}_x}} \rfloor (\eta_2 - \eta_1 + 1) + R_{\mathrm{S}_y}\eta_1$ and as usual $R_{\mathrm{S}_y} = R_{\mathrm{S}_y} = G_x = 2^{T(n)}$.

For each i, a circuit computes the binary number j, the index of the pixel we want to 'load', j is then passed to another circuit which selects pixel word j from the grid word $\mathcal{G}$. The output of the ith circuit represents the ith pixel in image a after a ld operation. □

So far each simulated operation affects only image a. The simulation of st differs in that a *rectangle* of images defined by the coordinates (ξ_1, η_1) and (ξ_2, η_2) is affected.

Theorem 7 (circuit simulation of st by $c_{[(\xi_1',\eta_1'),(\xi_2',\eta_2')]:=a}$). *The C_2-CSM operation $st\,(\xi_1', \eta_1', \xi_2', \eta_1')$ is simulated by a logspace uniform circuit*

$c_{[(\xi'_1,\eta'_1),(\xi'_2,\eta'_2)]:=a}$ *of size* $O(2^{12T(n)}T^6(n))$ *and depth* $O(T(n))$. *This circuit takes as input the image words* ξ'_1, η'_1, ξ'_2 *and* η'_2. *It outputs a word of length* $|\mathcal{G}|$ *that contains the rectangle (defined by* $(\xi'_1,\eta'_1),(\xi'_2,\eta'_2)$*) of image words to be stored and zeros at all other positions.*

Proof (sketch). Let i be the index of a pixel word in image word a. For each i, the index j of each pixel word that will be stored to, satisfies $j = (\text{col}(i) + u) + (\text{row}(i) + v)R_{S_x}G_x$ for $0 \leqslant u \leqslant \xi_2 - \xi_1$ and $0 \leqslant v \leqslant \eta_2 - \eta_1$, where $R_{S_x} = G_x = 2^{T(n)}$ and col(i) and row(i) were given in the proof of Theorem 6.

The address image words ξ'_1, η'_1, ξ'_2, η'_2 are decoded into the binary numbers ξ_1, η_1, ξ_2, η_2 (as in Theorem 6). Then for each $i \in \{0, \ldots, 2^{2T(n)}\}$ there is a circuit $c_{st\text{Pixel}(i)}$. The circuit $c_{st\text{Pixel}(i)}$ consists of $2^{4T(n)}$ subcircuits, each tests $j = j'$ for a unique pixel word j' in $\mathcal{G}$. Essentially $c_{st\text{Pixel}(i)}$ generates a 'mask' word m, $|m| = 2^{4T(n)}$. For the given i, mask m has the property that $m_{j'} = 1$ if and only if j' is the index of a pixel word that is to be overwritten with pixel word i from image word a.

Next, the circuit $c_{[(\xi'_1,\eta'_1),(\xi'_2,\eta'_2)]:=a}$ uses i subcircuits as follows. For each pixel word i in image word a: Subcircuit i ANDs the kth symbol in pixel word i with each of the $2^{4T(n)}$ outputs of $c_{st\text{Pixel}(i)}$. At this point we have i grid words; the ith grid word is 0 everywhere except for the 'rectangular' part of the grid that pixel i is stored to. These i grid words are ORed using an OR tree, giving the final output grid word as defined in the theorem statement. This final output word is a mask that contains the 'stored rectangle' and all other pixel words contain only zeros. □

4.3 Control Flow and Main Results

C_2-CSM control flow is straightforward to simulate. Recall from Section 3 that the binary word ctrl represents the C_2-CSM control (or instruction pointer). Simulating *br* involves finding a new value for ctrl from the *br* parameters.

Theorem 8 (circuit simulation of *br* by $c_{br\,(\xi',\eta')}$). *The* C_2-CSM *branch operation br* (ξ',η') *is simulated by a logspace uniform circuit* $c_{br\,(\xi',\eta')}$ *of size* $O(2^{2T(n)}T(n))$ *and depth* $O(T(n))$.

Proof. The circuit $c_{br\,(\xi',\eta')}$ decodes its address image word parameters into the binary numbers ξ and η (as in Theorem 6) which are then translated into an image word index i by evaluating $\xi + \eta G_x = i$. Index i points to the image word that encodes the next operation to be executed. □

Let $C_{(i)}$ be an arbitrary C_2-CSM configuration and let $\vdash_M$ be a relation on configurations that defines the operational semantics of C_2-CSM M [23, 20]. The configuration $C_{(i)}$ is encoded as $(\text{ctrl}_{(i)}, \mathcal{G}_{(i)})$ as described in Section 3.

Theorem 9 (circuit simulation of $C_{(i)} \vdash_M C_{(i+1)}$ by c_{step}). *Let* M *be a* C_2-CSM. *The uniform circuit* c_{step} *simulates* $C_{(i)} \vdash_M C_{(i+1)}$ *and is of size* $O(2^{12T(n)}T^6(n))$ *and depth* $O(T(n))$.

Proof (sketch). The circuit c_{step} computes $(\text{ctrl}_{(i)}, \mathcal{G}_{(i)}) \rightarrow (\text{ctrl}_{(i+1)}, \mathcal{G}_{(i+1)})$. A control flow simulating circuit updates $\text{ctrl}_{(i)}$ using either $c_{br\,(\xi',\eta')}$ or a circuit that simulates sequential control flow by incrementing $\text{ctrl}_{(i)}$ by one of $\{0,1,3,5\}$ depending on the current operation. Another circuit updates $\mathcal{G}_{(i)}$ by making use of the circuits that were given earlier and special mask words to simulate whatever $\mathcal{C}_2$-CSM operation is pointed to by $\text{ctrl}_{(i)}$. □

Next we give the resource use for our circuit simulation of a $\mathcal{C}_2$-CSM.

Theorem 10 (circuit simulation of M by c_M). *Let M be a $\mathcal{C}_2$-CSM that computes for* TIME *$T(n)$. The logspace uniform circuit c_M simulates M and is of size $O(2^{12T(n)}T^7(n))$ and depth $O(T^2(n))$.*

Proof. Circuit c_M is the composition of $T(n)$ instances of c_{step} from Theorem 9. The circuit is given the initial configuration $(\text{ctrl}_{\text{sta}}, \mathcal{G}_{\text{sta}})$ of M as input. After $O(T^2(n))$ parallel timesteps c_M outputs a word representing the final configuration of M. □

The size bound in the previous theorem seems quite high, however one should keep in mind that M has SPACE of $S(n) = O(2^{3T(n)})$. (This was the original SPACE bound on M before we redefined SPACE to suit our simulations).

If M is a language deciding $\mathcal{C}_2$-CSM [20, 23] we augment c_M so that it ORs the bits of the relevant output image word, thus computing a $\{0,1\}$-valued function. The resulting circuit has only a constant factor overhead in the size and depth of c_M. From this we state:

Corollary 1.
$\mathcal{C}_2\text{-CSM-TIME}(T(n)) \subseteq \text{U-SIZE, DEPTH}(O(2^{12T(n)}T^7(n)),\ O(T^2(n)))$

Let $T(n), S(n) = \Omega(\log n)$. From the inclusion given by Equation (1) we state:

Corollary 2. $\mathcal{C}_2\text{-CSM-TIME}(T(n)) \subseteq \text{DSPACE}(O(T^2(n)))$

Combining the above result with the converse inclusion (given in [20, 22]) gives a relationship between nondeterministic (sequential) space, $\mathcal{C}_2$-CSM TIME and deterministic space.

Corollary 3. $\text{NSPACE}(S(n)) \subseteq \mathcal{C}_2\text{-CSM-TIME}(O(S(n) + \log n)^4) \subseteq \text{DSPACE}(O(S(n) + \log n)^8)$

To summarise, the $\mathcal{C}_2$-CSM satisfies the parallel computation thesis:

Corollary 4. $\text{NSPACE}(S^{O(1)}(n)) = \mathcal{C}_2\text{-CSM-TIME}(S^{O(1)}(n))$

This establishes a link between space bounded sequential computation and TIME bounded $\mathcal{C}_2$-CSM computation, e.g. $\mathcal{C}_2\text{-CSM-TIME}(n^{O(1)}) = \mathsf{PSPACE}$.

The thesis relates parallel time to sequential space, however in our simulations we explicitly gave *all* resource bounds. As a final result we show that the class of $\mathcal{C}_2$-CSMs that simultaneously use polynomial SPACE and polylogarithmic TIME decide at most the languages in NC. Let $\mathcal{C}_2\text{-CSM-SPACE, TIME}(S(n), T(n))$ be the class of languages decided by $\mathcal{C}_2$-CSMs that use SPACE $S(n)$ and TIME $T(n)$.

For uniform circuits it is known [8] that U-SIZE, DEPTH$(n^{O(1)}, \log^{O(1)} n) = \mathsf{NC}$. From the resource overheads in our simulations:

$$\mathcal{C}_2\text{-CSM-SPACE, TIME}(2^{O(T(n))}, T(n)) \subseteq \text{U-SIZE, DEPTH}(2^{O(T(n))}, T^{O(1)}(n)).$$

For the case of $T(n) = \log^{O(1)} n$ we have our final result.

Corollary 5. $\mathcal{C}_2$-CSM-SPACE, TIME$(n^{O(1)}, \log^{O(1)} n) \subseteq \mathsf{NC}$

In [20, 22] it was shown that the converse inclusion also holds. Thus $\mathcal{C}_2$-CSMs that simultaneously use both polynomial SPACE and polylogarithmic TIME exactly characterise NC.

5 Discussion

We have given upper bounds on $\mathcal{C}_2$-CSM power in terms of uniform circuits. Combining this with previously shown lower bounds [22]; the $\mathcal{C}_2$-CSM verifies the parallel computation thesis and $\mathcal{C}_2$-CSMs with polynomial SPACE and polylogarithmic TIME decide exactly the languages in NC.

Our simulations could probably be improved to get a tighter relationship between $\mathcal{C}_2$-CSM TIME and sequential space. For example, the size bounds on the circuit simulation of *ld* and *st* could be improved with the aim of reducing the degree of the polynomials in Corollary 3, maybe even to a quadratic. Any improvement beyond that would be difficult, since it would imply an improvement to the quadratic bound in Savitch's theorem. Such improvements would enable us define a tighter bound on simultaneous resource usage between the $\mathcal{C}_2$-CSM and (say) uniform circuits, in the hope of exactly characterising NC^k by varying a parameter k to the model.

By how much can we generalise the $\mathcal{C}_2$-CSM definition and still preserve the upper bounds presented here? For example, in analogy with Simon's result for vector machines [17], we believe it should be possible to remove the $O(2^{3t})$ SPACE restriction from the $\mathcal{C}_2$-CSM definition.

Our results are the first to establish a general relationship (upper and lower bounds) between optically inspired computation and standard complexity classes. On a related note, our results show that the kind of optics modelled by the $\mathcal{C}_2$-CSM is simulated in reasonable time on any of the parallel architectures that verify the parallel computation thesis.

References

1. J. L. Balcázar, J. Díaz, and J. Gabarró. *Structural complexity, vols I and II.* EATCS Monographs on Theoretical Computer Science. Springer, Berlin, 1988.
2. R. B. Boppana and M. Sipser. *The complexity of finite functions.* Volume A of van Leeuwen [19], 1990.
3. A. Borodin. On relating time and space to size and depth. *SIAM Journal on Computing*, 6(4):733–744, Dec. 1977.

4. A. K. Chandra and L. J. Stockmeyer. Alternation. In *FOCS 1976*, pages 98–108, Houston, Texas, Oct. IEEE.
5. M. L. Furst, J. B. Saxe, and M. Sipser. Parity, circuits and the polynomial-time hierarchy. *Mathematical Systems Theory*, 17(1):13–27, 1984.
6. L. M. Goldschlager. *Synchronous parallel computation.* PhD thesis, University of Toronto, Computer Science Department, Dec. 1977.
7. R. Greenlaw, H. J. Hoover, and W. L. Ruzzo. *Limits to parallel computation: P-completeness theory.* Oxford university Press, Oxford, 1995.
8. R. M. Karp and V. Ramachandran. *Parallel algorithms for shared memory machines.* Volume A of van Leeuwen [19], 1990.
9. V. Krapchenko. Asymptotic estimation of addition time of a parallel adder. *Syst. Theory Res.*, 19:105–222, 1970.
10. R. E. Ladner and M. J. Fischer. Parallel prefix computation. *Journal of the ACM*, 27(4):831–838, 1980.
11. T. J. Naughton. Continuous-space model of computation is Turing universal. In *Critical Technologies for the Future of Computing*, Proc. SPIE vol. 4109, pages 121–128, Aug. 2000.
12. T. J. Naughton. A model of computation for Fourier optical processors. In *Optics in Computing 2000*, Proc. SPIE vol. 4089, pages 24–34, Quebec, June 2000.
13. T. J. Naughton and D. Woods. On the computational power of a continuous-space optical model of computation. In *Machines, Computations and Universality: 3rd Int. Conference*, volume 2055 of *LNCS*, pages 288–299, Chişinău, 2001. Springer.
14. I. Parberry. *Parallel complexity theory.* Wiley, 1987.
15. J. E. Savage. *Models of computation: Exploring the power of computing.* Addison Wesley, 1998.
16. A. Schönhage and V. Strassen. Schnelle multiplikation grosser zahlen. *Computing*, 7(3-4):281–292, 1971.
17. J. Simon. On feasible numbers. In *Proc. 9th Annual ACM Symposium on Theory of Computing*, pages 195–207. ACM, 1977.
18. P. van Emde Boas. *Machine models and simulations.* Volume A of van Leeuwen [19], 1990.
19. J. van Leeuwen, editor. *Handbook of Theoretical Computer Science*, volume A. Elsevier, Amsterdam, 1990.
20. D. Woods. *Computational complexity of an optical model of computation.* PhD thesis, National University of Ireland, Maynooth, 2005.
21. D. Woods and J. P. Gibson. Complexity of continuous space machine operations. In *New Computational Paradigms, First Conference on Computability in Europe*, volume 3526 of *LNCS*, pages 540–551, Amsterdam, June 2005. Springer.
22. D. Woods and J. P. Gibson. Lower bounds on the computational power of an optical model of computation. In *Fourth International Conference on Unconventional Computation*, volume 3699 of *LNCS*, pages 237–250, Sevilla, Oct. 2005. Springer.
23. D. Woods and T. J. Naughton. An optical model of computation. *Theoretical Computer Science*, 334(1–3):227–258, Apr. 2005.

Complexity of the Min-Max (Regret) Versions of Cut Problems*

Hassene Aissi, Cristina Bazgan, and Daniel Vanderpooten

LAMSADE, Université Paris-Dauphine, France
{aissi, bazgan, vdp}@lamsade.dauphine.fr

Abstract. This paper investigates the complexity of the min-max and min-max regret versions of the $s-t$ min cut and min cut problems. Even if the underlying problems are closely related and both polynomial, we show that the complexity of their min-max and min-max regret versions, for a constant number of scenarios, are quite contrasted since they are respectively strongly *NP*-hard and polynomial. Thus, we exhibit the first polynomial problem, $s-t$ min cut, whose min-max (regret) versions are strongly *NP*-hard. Also, min cut is one of the few polynomial problems whose min-max (regret) versions remain polynomial. However, these versions become strongly *NP*-hard for a non constant number of scenarios. In the interval data case, min-max versions are trivially polynomial. Moreover, for min-max regret versions, we obtain the same contrasted result as for a constant number of scenarios: min-max regret $s-t$ cut is strongly *NP*-hard whereas min-max regret cut is polynomial.

Keywords: min-max, min-max regret, complexity, min cut, $s-t$ min cut.

1 Introduction

The definition of an instance of a combinatorial optimization problem requires to specify parameters, in particular objective function coefficients, which may be uncertain or imprecise. Uncertainty/imprecision can be structured through the concept of *scenario* which corresponds to an assignment of plausible values to model parameters. Each scenario s can be represented as a vector in $\mathbb{R}^m$ where m is the number of relevant numerical parameters. Kouvelis and Yu [7] proposed the min-max and min-max regret criteria, stemming from decision theory, to construct solutions hedging against parameters variations. In min-max optimization, the aim is to find a solution having the best worst case value across all scenarios. In min-max regret problem, it is required to find a feasible solution minimizing the maximum deviation, over all possible scenarios, of the value of the solution from the optimal value of the corresponding scenario. Two

* This work has been partially funded by grant CNRS/CGRI-FNRS number 18227. The second author was partially supported by the ACI Sécurité Informatique grant-TADORNE project 2004.

X. Deng and D. Du (Eds.): ISAAC 2005, LNCS 3827, pp. 789–798, 2005.

natural ways of describing the set of all possible scenarios S have been considered in the literature. In the *discrete scenario case*, S is described explicitly by the list of all vectors $s \in S$. In this case, we distinguish situations where the number of scenarios is constant from those where the number of scenarios is non constant. In the *interval data case*, each numerical parameter can take any value between a lower and upper bound, independently of the values of the other parameters. Thus, in this case, S is the cartesian product of the intervals of uncertainty for the parameters.

Complexity of the min-max (regret) versions has been studied extensively during the last decade. In the discrete scenario case, this complexity was investigated for several combinatorial optimization problems in [7]. In general, these versions are shown to be harder than the classical versions. For a constant number of scenarios, pseudo-polynomial algorithms, based on dynamic programming, are given in [7] for the min-max (regret) versions of shortest path, knapsack and minimum spanning tree for grid graphs. The latter result is extended to general graphs in [1]. However, up to now, no polynomial problem was known to have min-max (regret) versions which are strongly *NP*-hard. When the number of scenarios is not constant, these versions usually become strongly *NP*-hard, even if the underlying problem is polynomial. In the interval data case, extensive research has been devoted for studying the complexity of min-max regret versions of various optimization problems including shortest path [5], minimum spanning tree [4,5] and assignment [2].

We investigate in this paper the complexity of min-max (regret) versions of two closely related polynomial problems, min cut and $s-t$ min cut. Quite interestingly, for a constant number of scenarios, the complexity status of these problems is widely contrasted. More precisely, min-max (regret) versions of min cut are polynomial whereas min-max (regret) versions of $s-t$ min cut are strongly *NP*-hard even for two scenarios. We also prove that for a non constant number of scenarios, min-max (regret) min cut become strongly *NP*-hard.

In the interval data case, min-max versions are trivially polynomial. Moreover, for min-max regret versions, we obtain the same contrasted result as for a constant number of scenarios: min-max regret $s-t$ cut is strongly *NP*-hard whereas min-max regret cut is polynomial.

After presenting preliminary concepts (Section 2), we investigate the complexity of min-max (regret) versions of min cut and $s-t$ min cut in the discrete scenario case (Section 3), and in the interval data case (Section 4).

2 Preliminaries

Let us consider an instance of a 0-1 minimization problem Q with a linear objective function defined as:

$$\begin{cases} \min \sum_{i=1}^{m} c_i x_i & c_i \in \mathbb{N} \\ x \in X \subset \{0,1\}^m \end{cases}$$

This class encompasses a large variety of classical combinatorial problems, some of which are polynomial-time solvable (shortest path problem, minimum spanning tree, ...) and others are *NP*-difficult (knapsack, set covering, ...).

In the discrete scenario case, the min-max (regret) version associated to Q has as input a finite set of scenarios S where each scenario $s \in S$ is represented by a vector $(c_1^s, \ldots, c_m^s)$. In the interval data case, each coefficient c_i can take any value in the interval $[\underline{c}_i, \overline{c}_i]$. In this case, the scenario set S is the cartesian product of the intervals $[\underline{c}_i, \overline{c}_i]$, $i = 1, \ldots, m$.

We denote by $val(x, s) = \sum_{i=1}^{m} c_i^s x_i$ the value of solution $x \in X$ under scenario $s \in S$, by x_s^* an optimal solution under scenario s, and by $val_s^* = val(x_s^*, s)$ the optimal value under the scenario s.

The min-max optimization problem corresponding to Q, denoted by Min-Max Q, consists of finding a solution x having the best worst case value across all scenarios, which can be stated as:

$$\min_{x \in X} \max_{s \in S} val(x, s)$$

This version is denoted by Discrete Min-Max Q in the discrete scenario case, and by Interval Min-Max Q in the interval data case.

Given a solution $x \in X$, its *regret*, $R(x, s)$, under scenario $s \in S$ is defined as $R(x, s) = val(x, s) - val_s^*$. The *maximum regret* $R_{max}(x)$ of solution x is then defined as $R_{max}(x) = \max_{s \in S} R(x, s)$.

The min-max regret optimization problem corresponding to Q, denoted by Min-Max Regret Q, consists of finding a solution x minimizing the maximum regret $R_{max}(x)$ which can be stated as:

$$\min_{x \in X} R_{max}(x) = \min_{x \in X} \max_{s \in S} \{val(x, s) - val_s^*\}$$

This version is denoted by Discrete Min-Max Regret Q in the discrete scenario case, and by Interval Min-Max Regret Q in the interval data case.

In the interval data case, for a solution $x \in X$, we denote by $c^-(x)$ the worst scenario associated to x, where $c_i^-(x) = \overline{c}_i$ if $x_i = 1$ and $c_i^-(x) = \underline{c}_i$ if $x_i = 0$, $i = 1, \ldots, m$. Then we can establish easily that $R_{max}(x) = R(x, c^-(x))$, as shown e.g. in [9] in the specific context of the minimum spanning tree problem.

In this paper, we focus on the min-max (regret) versions of the two following cut problems:

Min Cut
Input: A connected graph $G = (V, E)$ with weight w_{ij} associated with each edge $(i, j) \in E$.
Output: A cut in G, that is a partition of V into two sets, of minimum value.

$s - t$ Min Cut
Input: A connected graph $G = (V, E)$ with weight w_{ij} associated with each edge $(i, j) \in E$, and two specified vertices $s, t \in V$.
Output: An $s - t$ cut in G, that is a partition of V into two sets V_1 and V_2, with $s \in V_1$ and $t \in V_2$, of minimum value.

In order to prove our complexity results we use the two following problems proved strongly NP-hard in [6].

MIN BISECTION
Input: A graph $G = (V, E)$ with an even number of vertices.
Output: A bisection in G, that is a partition of V into two equal cardinality sets, of minimum value.

$s - t$ MIN BISECTION
Input: A graph $G = (V, E)$ with an even number of vertices, and two specified vertices $s, t \in V$.
Output: An $s - t$ bisection in G, that is a partition of $V = V_1 \cup V_2$ such that $s \in V_1$, $t \in V_2$, and $|V_1| = |V_2|$, of minimum value.

3 Discrete Scenario Case

We show in this section the first polynomial-time solvable problem, $s - t$ MIN CUT, which becomes strongly NP-hard when considering its min-max or min-max regret version.

Min-max cut was proved polynomially solvable for a constant number of scenarios [3]. We show that min-max regret cut also remains polynomial for a constant number of scenarios. When the number of scenarios is not constant, min-max (regret) versions become strongly NP-hard.

3.1 $s - t$ Min Cut

In order to prove these results, we construct polynomial reductions from the decision version of MIN BISECTION.

Theorem 1. DISCRETE MIN-MAX (REGRET) $s - t$ CUT *are strongly NP-hard even for two scenarios.*

Proof. Consider an instance $G = (V, E)$ of MIN BISECTION with $|V| = 2n$, and a positive integer v. We construct an instance $\widetilde{G} = (\widetilde{V}, \widetilde{E})$ of DISCRETE MIN-MAX $s - t$ CUT with the scenario set $S = \{s_1, s_2\}$. The node set is $\widetilde{V} = V \cup \{s, t\}$ where s and t correspond to a source and a sink respectively. The edge set $\widetilde{E} = E \cup \{(s, i) : i \in V\} \cup \{(i, t) : i \in V\}$. Edge weights in scenarios s_1 and s_2 are assigned for each edge $(i, j) \in \widetilde{E}$ as follows:

$$w_{ij}^1 = \begin{cases} 1 & \text{if } (i,j) \in E \\ n^2 + 1 & \text{if } i = s \text{ or } j = s \\ 0 & \text{if } i = t \text{ or } j = t \end{cases} \quad \text{and} \quad w_{ij}^2 = \begin{cases} 1 & \text{if } (i,j) \in E \\ 0 & \text{if } i = s \text{ or } j = s \\ n^2 + 1 & \text{if } i = t \text{ or } j = t \end{cases}$$

We claim that there exists a bisection C in G of value at most v if and only if there exists an $s - t$ cut $\widetilde{C}$ in $\widetilde{G}$ with $\max\{val(\widetilde{C}, s_1), val(\widetilde{C}, s_2)\} \leq v + (n^2 + 1)n$.
$\Rightarrow$ Consider a bisection $C = (V_1, V_2)$ of value $x \leq v$. We construct an $s - t$ cut $\widetilde{C} = (\widetilde{V}_1, \widetilde{V}_2)$ where $\widetilde{V}_1 = V_1 \cup \{s\}$, and $\widetilde{V}_2 = V_2 \cup \{t\}$. Consequently, we have $val(\widetilde{C}, s_1) = val(\widetilde{C}, s_2) = x + (n^2 + 1)n \leq v + (n^2 + 1)n$.

$\Leftarrow$ Consider now an s–t cut $\widetilde{C} = (\widetilde{V}_1, \widetilde{V}_2)$ verifying $\max\{val(\widetilde{C}, s_1), val(\widetilde{C}, s_2)\} \leq v + (n^2+1)n$. Let $V_1 = \widetilde{V}_1 \setminus \{s\}$ and $V_2 = \widetilde{V}_2 \setminus \{t\}$. We have by construction $val(\widetilde{C}, s_1) = y + |V_2|(n^2+1)$ and $val(\widetilde{C}, s_2) = y + |V_1|(n^2+1)$, where y is the number of edges from E that have one endpoint in V_1 and one endpoint in V_2. Suppose that $|V_1| = n + z$ and $|V_2| = n - z$, $z \geq 0$. Then $val(\widetilde{C}, s_1) = y + (n^2+1)n - z(n^2+1)$, $val(\widetilde{C}, s_2) = y + (n^2+1)n + z(n^2+1)$ and $\max\{val(\widetilde{C}, s_1), val(\widetilde{C}, s_2)\} = y + (n^2+1)n + z(n^2+1) \leq v + (n^2+1)n$. Since $v \leq n^2$ we have $z = 0$ and thus $|V_1| = |V_2| = n$ and $y \leq v$.

In order to prove the result for the min-max regret version, we use exactly the same graph $\widetilde{G} = (\widetilde{V}, \widetilde{E})$. Let C_i^* denote the optimal solution in scenario s_i, $i = 1, 2$. We have $C_1^* = (\widetilde{V} \setminus \{t\}, \{t\})$ and $C_2^* = (\{s\}, \widetilde{V} \setminus \{s\})$ with $val(C_1^*, s_1) = val(C_2^*, s_2) = 0$. Therefore, there exists a bisection in G of value at most v if and only if there exists an $s-t$ cut $\widetilde{C}$ in $\widetilde{G}$ with $R_{max}(\widetilde{C}) \leq v + (n^2+1)n$. □

3.2 Min Cut

Armon and Zwick [3] constructed a polynomial-time algorithm for Discrete Min-Max Cut, in the case of a constant number of scenarios, based essentially on the result of Nagamochi, Nishimura and Ibaraki [8] for computing all α-approximate cuts in time $O(m^2 n + mn^{2\alpha})$. A cut C in a graph G is called an α-approximate cut if $val(C) \leq \alpha\, opt$, where opt is the value of a minimum cut in G.

Theorem 2 ([3]). Discrete Min-Max Cut *is solvable in polynomial time for a constant number of scenarios.*

In a graph on n vertices and m edges and with k scenarios, Armon and Zwick's algorithm [3] constructs an optimal solution in $O(mn^{2k})$.

We show in the following that this algorithm can be modified in order to obtain a polynomial-time algorithm for Discrete Min-Max Regret Cut.

Theorem 3. Discrete Min-Max Regret Cut *is solvable in polynomial time for a constant number of scenarios.*

Proof. Consider an instance I of the problem given by graph $G = (V, E)$ on n vertices and m edges and a set of k scenarios S such that each edge $(i, j) \in E$ has a weight w_{ij}^s in scenario s. We construct, as before, an instance I' of Min Cut on the same graph, where $w'_{ij} = \sum_{s \in S} w_{ij}^s$. The algorithm consists firstly of computing all k-approximate cuts and secondly of choosing among these cuts one with a minimum maximum regret.

We prove now the correctness of the algorithm. Let C^* be an optimal min-max regret cut in G. We show that for any cut C of G, we have $val'(C^*) \leq k val'(C)$, where $val'(C)$ is the value of cut C in I'. In fact,

$$val'(C^*) = \sum_{s \in S} val(C^*, s) = \sum_{s \in S} (val(C^*, s) - val_s^*) + \sum_{s \in S} val_s^* \leq$$

$$k \max_{s \in S}\{val(C^*, s) - val_s^*\} + \sum_{s \in S} val_s^* \leq k \max_{s \in S}\{val(C, s) - val_s^*)\} + \sum_{s \in S} val_s^* \leq$$

$$k\sum_{s\in S}(val(C,s)-val_s^*)+\sum_{s\in S}val_s^* = k\sum_{s\in S}val(C,s)-(k-1)\sum_{s\in S}val_s^* \leq kval'(C)$$

In particular, if C is a minimum cut in I', we obtain $val'(C^*) \leq kopt(I')$. Thus all optimal solutions to DISCRETE MIN-MAX REGRET CUT are among the k-approximate cuts in I'.

The running time of the algorithm is $O(mn^{2k})$. □

The algorithms described above to solve DISCRETE MIN-MAX (REGRET) CUT are exponential in k. We prove in the following that when k is not constant, both problems become strongly NP-hard.

Theorem 4. DISCRETE MIN-MAX (REGRET) CUT *are strongly NP-hard for a non constant number of scenarios.*

Proof. We use a reduction from MIN BISECTION. Consider an instance $G = (V, E)$ of MIN BISECTION with $V = \{1, \ldots, 2n\}$, and a positive integer v. We construct an instance $\widetilde{G} = (\widetilde{V}, \widetilde{E})$ of DISCRETE MIN-MAX CUT with a scenario set S of size $2n$. The node set is $\widetilde{V} = V \cup \{1', \ldots, 2n'\} \cup \{1'', \ldots, 2n''\}$. The edge set $\widetilde{E} = E \cup \{(i', j') : i, j = 1, \ldots, 2n\} \cup \{(i, i'), (i', i'') : i = 1, \ldots, 2n\}$. Scenario set S corresponds to nodes of G. The weights of the edges in any scenario $s_i \in S$ are defined as follows: $w^i_{hj} = 1$ for all $(h, j) \in E$; $w^i_{i'j'} = n^2$, for $j = 1, \ldots, 2n$; $w^i_{h'j'} = 0$ for $h \neq i$ and $j \neq i$; $w^i_{ii'} = w^i_{i'i''} = n^3 + n^2 + 1$; $w^i_{jj'} = w^i_{j'j''} = 0$, for $j \neq i$.

We claim that there exists a bisection C in G of value at most v if and only if there is a cut $\widetilde{C}$ in $\widetilde{G}$ with $\max_{s\in S} val(\widetilde{C}, s) \leq n^3 + v$.

$\Rightarrow$ Consider a bisection $C = (V_1, V_2)$ in G of value $x \leq v$. We construct a cut $\widetilde{C} = (\widetilde{V}_1, \widetilde{V}_2)$ in $\widetilde{G}$ where $\widetilde{V}_1 = V_1 \cup \{i', i'' : i \in V_1\}$ and $\widetilde{V}_2 = V_2 \cup \{i', i'' : i \in V_2\}$. For any scenario $s \in S$, we have $val(\widetilde{C}, s) \leq n^3 + v$ and thus $\max_{s\in S} val(\widetilde{C}, s) \leq n^3 + v$.

$\Leftarrow$ Consider now a cut $\widetilde{C} = (\widetilde{V}_1, \widetilde{V}_2)$ in $\widetilde{G}$ such that $\max_{s\in S} val(\widetilde{C}, s) \leq n^3 + v$. Cut $\widetilde{C}$ does not contain any edge (i, i') or (i', i'') for some $i = 1, \ldots, 2n$, since otherwise, we have $\max_{s\in S} val(\widetilde{C}, s) \geq n^3 + n^2 + 1 > n^3 + v$. Denote by V_i, for $i = 1, 2$, the restriction of $\widetilde{V}_i$ to the vertices of V. Suppose now that $|V_1| < |V_2|$, then for any scenario s_i such that $i \in V_1$ we have $val(\widetilde{C}, s_i) \geq (n + 1)n^2 + 1 > n^3 + v$. Thus, we have necessarily $|V_1| = |V_2|$ and the value of the bisection (V_1, V_2) is at most v.

In order to prove the result for the min-max regret version, we use exactly the same graph $\widetilde{G} = (\widetilde{V}, \widetilde{E})$. Notice that, for any scenario $s_i \in S$, cut $C_i^* = (\{j''\}, \widetilde{V} \setminus \{j''\})$ for some $j \neq i$ is a minimum cut in scenario s_i, with value 0. Therefore, there exists a bisection in G of value at most v if and only if there exists a cut $\widetilde{C}$ in $\widetilde{G}$ with $R_{max}(\widetilde{C}) \leq n^3 + v$. □

Observe that in the previous proof we used the same graph $\widetilde{G}$ both for the min-max and min-max regret versions. Actually, a slightly simpler proof can be obtained, for the min-max part, considering only the subgraph of $\widetilde{G}$ induced by $\widetilde{V} \setminus \{1'', \ldots, 2n''\}$. Vertex subset $\{1'', \ldots, 2n''\}$ is necessary, for the min-max regret part, to ensure the existence of minimum cuts of value 0 for each scenario.

4 Interval Data Case

We first state the polynomiality of the min-max cut problems (Section 4.1), then we establish the strong *NP*-hardness of INTERVAL MIN-MAX REGRET $s-t$ CUT (Section 4.2) and the polynomiality of INTERVAL MIN-MAX REGRET CUT (Section 4.2).

4.1 Min-Max Versions

In the interval data case, the min-max version of a minimization problem corresponds to solving this problem in the worst-case scenario defined by the upper bounds of all intervals. Therefore, a minimization problem and its min-max version have the same complexity. INTERVAL MIN-MAX $s-t$ CUT and INTERVAL MIN-MAX CUT are thus polynomial-time solvable.

4.2 Min-Max Regret Versions

When the number $u \leq m$ of uncertain/imprecise parameters, corresponding to non-degenerate intervals, is small enough, then the problem becomes polynomial. More precisely, as shown by Averbakh and Lebedev [5] for general networks problems solvable in polynomial time, if u is fixed or bounded by the logarithm of a polynomial function of m, then the min-max regret version is also solvable in polynomial time (based on the fact that an optimal solution for the min-max regret version corresponds to one of the optimal solutions for the 2^u extreme scenarios, where extreme scenarios have values on each edge corresponding to either the lower or upper bound of its interval). This clearly applies to the $s-t$ min cut and min cut problems.

$s-t$ min cut
We show now that INTERVAL MIN-MAX REGRET $s-t$ CUT is strongly *NP*-hard. For this purpose, we construct a reduction from the decision version of $s-t$ MIN BISECTION.

Theorem 5. INTERVAL MIN-MAX REGRET $s-t$ CUT *is strongly NP-hard.*

Proof. Consider $G = (V, E)$ an instance of $s-t$ MIN BISECTION with $|V| = 2n$, where $V = \{s = 1, \ldots, t = 2n\}$. We construct from G an instance $\widetilde{G} = (\widetilde{V}, \widetilde{E})$ of INTERVAL MIN-MAX REGRET $s-t$ CUT as illustrated in Figure 1. The vertex set is $\widetilde{V} = V \cup \{1', \ldots, 2n'\} \cup \{1'', \ldots, 2n''\} \cup \{1''', \ldots, 2n'''\} \cup \{\widetilde{s}, 2n+1\}$, and $\widetilde{t} = t$.

The edge set is $\widetilde{E} = E \cup \{(i', i''), (i'', i''') : i = 1, \ldots, 2n\} \cup \{(i, i'') : i = 2, \ldots, 2n-1\} \cup \{(2n+1, i') : i = 1, \ldots, 2n\} \cup \{(i''', t) : i = 1, \ldots, 2n\} \cup \{(\widetilde{s}, 2n+1), (\widetilde{s}, s)\}$.

Let p and q verifying, respectively, $p > n^2$ and $q > 4n(p+1)^2$. The weights are defined as follows :

- $\underline{w}_{ij} = \overline{w}_{ij} = 1$, for all $(i,j) \in E$;
- $\underline{w}_{i'i''} = \begin{cases} q \text{ for } i = 1 \\ 0 \text{ otherwise} \end{cases}$ and $\overline{w}_{i'i''} = \begin{cases} q & \text{for } i = 1 \\ p^2 + p & \text{otherwise} \end{cases}$
- $\underline{w}_{i''i'''} = \overline{w}_{i''i'''} = \begin{cases} p^2 + np & \text{for } i = 1 \\ p^2 & \text{for } i = 2, \ldots, 2n-1 \\ q & \text{for } i = 2n \end{cases}$
- $\underline{w}_{ii''} = \overline{w}_{ii''} = q$, for $i = 2, \ldots, 2n-1$;
- $\underline{w}_{(2n+1)i'} = \begin{cases} 0 & \text{for } i = 1 \\ 2p & \text{otherwise} \end{cases}$ and $\overline{w}_{(2n+1)i'} = q$, for $i = 1, \ldots, 2n$;
- $\underline{w}_{i'''t} = \overline{w}_{i'''t} = q$, for $i = 1, \ldots, 2n$;
- $\underline{w}_{\widetilde{s}(2n+1)} = 2np$ and $\overline{w}_{\widetilde{s}(2n+1)} = q$;
- $\underline{w}_{\widetilde{s}s} = 0$ and $\overline{w}_{\widetilde{s}s} = q$.

Clearly this transformation can be obtained in polynomial time.

We first establish the following property.

For any $\widetilde{s} - \widetilde{t}$ cut $\widetilde{C} = (\widetilde{V}_1, \widetilde{V}_2)$ in $\widetilde{G}$ not including any edge $(i,j) \in \widetilde{E}$ with $\overline{w}_{ij} = q$, a minimum $\widetilde{s} - \widetilde{t}$ cut $C^*_{w^-(\widetilde{C})}$ in, $w^-(\widetilde{C})$, the worst scenario associated to $\widetilde{C}$, has value $val(C^*_{w^-(\widetilde{C})}, w^-(\widetilde{C})) = 2p \min\{n, |V_2|\}$, where $V_2 = \widetilde{V}_2 \cap V$.

Indeed, consider such a cut $\widetilde{C} = (\widetilde{V}_1, \widetilde{V}_2)$ with $\widetilde{s} \in \widetilde{V}_1$, $\widetilde{t} \in \widetilde{V}_2$ and denote $V_1 = \widetilde{V}_1 \cap V$. Clearly, vertices $2n+1$, $1''$ and i', $i = 1, \ldots, 2n$ belong to $\widetilde{V}_1$. Also, vertices $2n''$ and i''', $i = 1, \ldots, 2n$ belong to $\widetilde{V}_2$. Moreover, i and i'' belong to the same part, $\widetilde{V}_1$ or $\widetilde{V}_2$. It follows that

$$val(\widetilde{C}, w^-(\widetilde{C})) = x + (n + |V_2|)p + 2np^2 \tag{1}$$

where x denotes the number of edges that have one endpoint in V_1 and one endpoint in V_2.

By construction, $C^*_{w^-(\widetilde{C})}$ necessarily cuts edge $(\widetilde{s}, s)$. Furthermore, there exist two cases:

1. If $|V_2| \leq n$ then $C^*_{w^-(\widetilde{C})} = (\widetilde{V}_1^*, \widetilde{V}^* \setminus \widetilde{V}_1^*)$, where $\widetilde{V}_1^* = \{\widetilde{s}, 2n+1\} \cup \{i' : i'' \in \widetilde{V}_1, i \neq 1\}$ and thus $val(C^*_{w^-(\widetilde{C})}, w^-(\widetilde{C})) = 2|V_2|p$.
2. If $|V_2| > n$ then $C^*_{w^-(\widetilde{C})} = (\{\widetilde{s}\}, \widetilde{V} \setminus \{\widetilde{s}\})$ and thus $val(C^*_{w^-(\widetilde{C})}, w^-(\widetilde{C})) = 2np$.

We claim that there exists an $s - t$ bisection $C = (V_1, V_2)$ of value no more than v if and only if there exists an $\widetilde{s} - \widetilde{t}$ cut $\widetilde{C} = (\widetilde{V}_1, \widetilde{V}_2)$ in $\widetilde{G}$ with $R_{max}(\widetilde{C}) \leq v + 2np^2$.

$\Rightarrow$ Consider an $s - t$ bisection $C = (V_1, V_2)$ in G of value $x \leq v$. We construct an $\widetilde{s} - \widetilde{t}$ cut $\widetilde{C}$ in $\widetilde{G}$ deduced from C as follows: $\widetilde{V}_1 = \{\widetilde{s}, 2n+1\} \cup \{1', \ldots, 2n'\} \cup V_1 \cup \{i'' : i \in V_1\}$ and $\widetilde{V}_2 = \{1''', \ldots, 2n'''\} \cup V_2 \cup \{i'' : i \in V_2\}$. It is easy to verify that $val(\widetilde{C}, w^-(\widetilde{C})) = x + 2n(p + p^2)$ and using the previous result, we have $R_{max}(\widetilde{C}) = x + 2np^2 \leq v + 2np^2$.

$\Leftarrow$ Consider an $\widetilde{s} - \widetilde{t}$ cut $\widetilde{C}$ in $\widetilde{G}$ with $R_{max}(\widetilde{C}) \leq v + 2np^2$. Cut $\widetilde{C}$ does not cut any edge $(i,j) \in \widetilde{E}$ such that $\overline{w}_{ij} = q$, since otherwise, $val(\widetilde{C}, w^-(\widetilde{C})) \geq q$, and, since

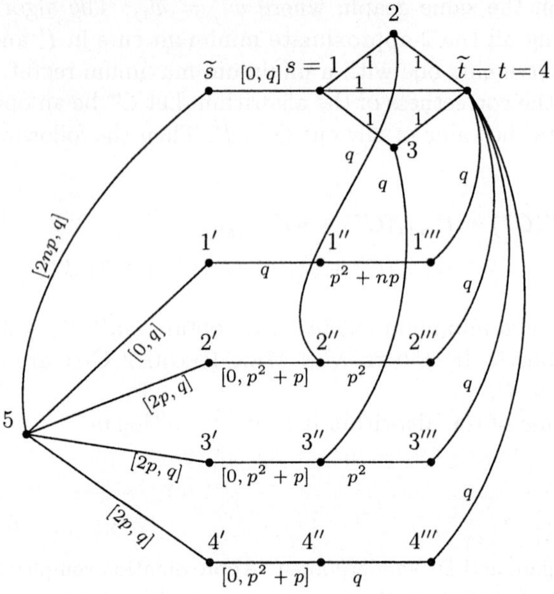

Fig. 1. INTERVAL MIN-MAX REGRET $s - t$ CUT instance resulting from $s - t$ MIN BISECTION instance

a minimum $\widetilde{s}-\widetilde{t}$ cut $C^*_{w^-(\widetilde{C})}$ in $w^-(\widetilde{C})$, does not cut any edge $(i,j) \in \widetilde{E}$ such that $\overline{w}_{ij} = q$, we have, using (1), $val(C^*_{w^-(\widetilde{C})}, w^-(\widetilde{C})) \leq n^2+3np+2np^2 < 4np+2np^2$ and consequently, we have $R_{max}(\widetilde{C}) > 2np^2 + v$.

Thus $val(\widetilde{C}, w^-(\widetilde{C})) = y + 2np^2 + np + p|V_2|$ where y is the value of the cut induced by $\widetilde{C}$ in E. It follows that

$$R_{max}(\widetilde{C}) = \begin{cases} y + (n - |V_2|)p + 2np^2 \text{ if } |V_2| \leq n \\ y + (|V_2| - n)p + 2np^2 \text{ if } |V_2| > n \end{cases}$$

Consequently, since $R_{max}(\widetilde{C}) \leq v + 2np^2$, and $p > n^2 \geq v$, we have $|V_1| = n = |V_2|$ and $y \leq v$. □

Min cut
We prove in this section that the min-max regret version of min-cut problem is polynomial in the interval data case.

Theorem 6. INTERVAL MIN-MAX REGRET CUT *is solvable in polynomial time in the interval data case.*

Proof. Consider an instance I of INTERVAL MIN-MAX REGRET CUT given by graph $G = (V, E)$ on n vertices and m edges. The weight w_{ij} of each edge $(i,j) \in E$ can take any value in the interval $[\underline{w}_{ij}, \overline{w}_{ij}]$. We construct an instance

I' of MIN CUT on the same graph, where $w'_{ij} = \overline{w}_{ij}$. The algorithm consists firstly of computing all the 2-approximate minimum cuts in I' and secondly of choosing among these cuts one with a minimum maximum regret.

We prove now the correctness of the algorithm. Let C^* be an optimal cut in I and $val'(C)$ denote the value of any cut C in I'. Then the following inequalities hold:

$$\begin{aligned} val'(C^*) &= R_{max}(C^*) + val^*_{w^-(C^*)} \\ &\leq R_{max}(C) + val(C, w^-(C^*)) \leq 2val'(C) \end{aligned}$$

In particular, if C is a minimum cut in I', we obtain $val'(C^*) \leq 2opt(I')$. Thus all optimal solutions to INTERVAL MIN-MAX REGRET CUT are among the 2-approximate cuts in I'.

The running time of the algorithm is $O(mn^5 + n^6 \log m)$. □

References

1. H. Aissi, C. Bazgan, and D. Vanderpooten. Approximation complexity of min-max (regret) versions of shortest path, spanning tree, and knapsack. In *Proceedings of the 13th Annual European Symposium on Algorithms (ESA 2005), Mallorca, Spain*, 2005. to appear.
2. H. Aissi, C. Bazgan, and D. Vanderpooten. Complexity of the min-max and min-max regret assignment problem. *Operations Research Letters*, 33:634–640, 2005.
3. A. Armon and U. Zwick. Multicriteria global minimum cuts. In *Proceedings of the 15th International Symposium on Algorithms and Complexity (ISAAC 2004), Hong Kong, China*, LNCS 3341, pages 65–76. Springer-Verlag, 2004.
4. I. D. Aron and P. Van Hentenryck. On the complexity of the robust spanning tree with interval data. *Operations Research Letters*, 32:36–40, 2004.
5. I. Averbakh and V. Lebedev. Interval data min-max regret network optimization problems. *Discrete Applied Mathematics*, 138:289–301, 2004.
6. M. Garey and D. Johnson. *Computers and intractability: a guide to the theory of NP-completeness.* Freeman, San Francisco, 1979.
7. P. Kouvelis and G. Yu. *Robust Discrete Optimization and its Applications.* Kluwer Academic Publishers, Boston, 1997.
8. H. Nagamochi, K. Nishimura, and T. Ibaraki. Computing all small cuts in an undirected network. *SIAM Journal on Discrete Mathematics*, 10(3):469–481, 1997.
9. H. Yaman, O. E. Karaşan, and M. C. Pinar. The robust spanning tree problem with interval data. *Operations Research Letters*, 29:31–40, 2001.

Improved Algorithms for the k Maximum-Sums Problems

Chih-Huai Cheng[1], Kuan-Yu Chen[1], Wen-Chin Tien[1], and Kun-Mao Chao[1,2,*]

[1] Department of Computer Science and Information Engineering
[2] Graduate Institute of Networking and Multimedia,
National Taiwan University, Taipei, Taiwan 106
kmchao@csie.ntu.edu.tw

Abstract. Given a sequence of n real numbers and an integer k, $1 \leq k \leq \frac{1}{2}n(n-1)$, the k maximum-sum segments problem is to locate the k segments whose sums are the k largest among all possible segment sums. Recently, Bengtsson and Chen gave an $O(\min\{k+n\log^2 n, n\sqrt{k}\})$-time algorithm for this problem. In this paper, we propose an $O(n + k\log(\min\{n,k\}))$-time algorithm for the same problem which is superior to Bengtsson and Chen's when k is $o(n\log n)$. We also give the first optimal algorithm for delivering the k maximum-sum segments in non-decreasing order if $k \leq n$. Then we develop an $O(n^{2d-1}+k\log\min\{n,k\})$-time algorithm for the d-dimensional version of the problem, where $d > 1$ and each dimension, without loss of generality, is of the same size n. This improves the best previously known $O(n^{2d-1}C)$-time algorithm, also by Bengtsson and Chen, where $C = \min\{k + n\log^2 n, n\sqrt{k}\}$. It should be pointed out that, given a two-dimensional array of size $m \times n$, our algorithm for finding the k maximum-sum subarrays is the first one achieving cubic time provided that k is $O(\frac{m^2 n}{\log n})$.

1 Introduction

The maximum-sum subarray problem was first surveyed by Bentley in his "Programming Pearls" column of CACM [3]. The one-dimensional case is also called the maximum subsequence problem and is well known linear-time solvable using Kadane's algorithm [3]. In the two-dimensional case, the task is to find a subarray such that the sum of its elements is maximized. The maximum subsequence (subarray) problem is widely used in pattern recognition [8], image processing [7], biological sequence analysis [6,9,10,11], data mining [7], and many other applications.

Computing the k largest sums over all possible segments is a natural extension of the maximum-sum segment problem. This extension has been considered from two perspectives, one of which allows the segments to overlap, while the other disallows. Linear-time algorithms for finding all the non-overlapping maximal segments were given in [4,12]. In this paper, we focus on finding the k maximum-sum

* Corresponding author.

X. Deng and D. Du (Eds.): ISAAC 2005, LNCS 3827, pp. 799–808, 2005.

segments whose overlapping is allowed. We will use the terms "k maximum segments" and "k maximum-sum segments" interchangeably. A naïve approach is to choose the k largest from the sums of all possible contiguous subsequences which requires $O(n^2)$ time. Bae and Takaoka [1] presented an $O(kn)$-time algorithm for the k maximum segment problem. Recently, an improvement to $O(\min\{k + n\log^2 n, n\sqrt{k}\})$ was given by Bengtsson and Chen [2]. In this paper, we propose an $O(n + k\log(\min\{n,k\}))$-time algorithm which is superior to Bengtsson and Chen's when k is $o(n\log n)$. It is not difficult to see that a lower bound of the k maximum segment problem is $\Omega(n+k)$. There is still a gap between the trivial lower bound and our method. However, if the k maximum segments are requested in non-decreasing order, we give an $\Omega(n + k\log k)$-time lower bound for the $k \leq n$ case. A simple variant of our algorithm can deliver the k maximum segments in non-decreasing order in $O(n + k\log k)$ time, which is optimal if $k \leq n$.

To avoid misunderstanding, we will use the term "subarray" instead of "segment" in the multiple-dimensional cases. In the two-dimensional case, we are given an $m \times n$ array of real numbers. The fastest algorithm for the maximum subarray problem stayed at $O(m^2n)$ time for a long period of time. In 1998, the first subcubic-time algorithm was proposed by Tamaki and Tokuyama [14]. The time complexity of the latest algorithm for the maximum subarray problem presented by Takaoka [13] is $O(m^2n(\log\log m/\log m)^{1/2})$. Clearly, it is still close to $O(m^2n)$. Our goal for the two-dimensional case is to find the k maximum-sum subarrays in the array. Bae and Takaoka [1] gave an $O(m^2nk)$-time algorithm for this problem. Bengtsson and Chen [2] presented an improved algorithm in $O(\min\{m^2C, m^2n^2\})$ time, where $C = \min\{k + n\log^2 n, n\sqrt{k}\}$. We propose an $O(m^2n + k\log(\min\{n,k\}))$-time algorithm, which is superior to the previous results for every value of k. Notice that our algorithm is the first cubic-time algorithm for the k maximum subarray problem when k is $O(\frac{m^2n}{\log n})$. For the d-dimensional case, the best previously known algorithm, by Bengtsson and Chen [2], runs in $O(n^{2d-1}C)$ time, where $C = \min\{k + n\log^2 n, n\sqrt{k}\}$. We propose an improved $O(n^{2d-1} + k\log\min\{n,k\})$-time algorithm.

The rest of the paper is organized as follows. Section 2 gives a formal definition of the k maximum segment (subarray) problem. In Section 3, we give the algorithm for the k maximum segment problem based on an iterative partial-table building approach and discuss the issue of reporting the k maximum segments in non-decreasing order. We extend the results to the multiple-dimensional cases in Section 4. Finally, we close the paper by mentioning a few open problems.

2 Problem Definitions and Notations

Given a sequence of n real numbers $A[1 \ldots n]$, a segment is simply a contiguous subsequence of that sequence. Let P denote the prefix-sum array of A where $P[i] = a_1 + a_2 + \ldots + a_i$ for $i = 1, \ldots, n$. Let $A[i \ldots j]$ denote the segment $<a_i, a_{i+1}, \ldots, a_j>$. Let $S(i,j)$ be the sum of $A[i \ldots j]$, i.e. $S(i,j) = a_i + a_{i+1} + \ldots + a_j$. It is easy to see that $S(i,j) = P[j] - P[i-1]$. Let $[i,j]$ denote the set $\{i, i+1, \ldots, j\}$ for $i \leq j$.

Problem 1: k Maximum-Sum Segments
Input: a sequence of n real numbers $A = <a_1, a_2, \ldots, a_n>$ and an integer k, $1 \leq k \leq \frac{1}{2}n(n-1)$
Output: k maximum-sum segments such that the sums of these segments are the k largest among all possible segment sums

We also consider the k maximum-sum problem in higher dimensions. In particular, the two-dimensional problem is sometimes referred to as the k maximum-sum subarray problem.

Problem 2: k d-Dimensional Maximum-Sum Subarrays
Input: a d-dimensional array of real numbers and a positive integer k
Output: k d-dimensional subarrays such that the sums of these subarrays are the k largest among all possible subarray sums

3 The k Maximum-Sum Segment Problem

Finding the k largest elements of a sequence is essential in the construction of our algorithm for the k maximum-sum segment problem. Thus, we first describe how to find the k largest elements in Lemma 1.

Lemma 1. *Given a sequence of numbers, the k maximum (or minimum) elements can be found in linear time.*

Proof. According to [5], the kth maximum element of a sequence can be found in linear time. Suppose a_i denotes the kth maximum element. To obtain the k maximum elements, we simply compare a_i with all the elements of the sequence. We first output those elements whose values are greater than a_i. Then, we append additional elements equal to a_i to the output so that k elements are yielded. □

A naïve quadratic-time solution to the k maximum segment problem is to build a table of size $n \times n$, storing all the possible segments. By Lemma 1, the k maximum segments can be retrieved from the table in $O(n^2)$ time. To speed up, we introduce a partial-table building method for the k maximum segment problem.

3.1 An Iterative Partial-Table Building Approach

Instead of building the entire table at once, we adopt an iterative strategy, in the sense that in each iteration we build only a partial table. Before introducing our main algorithm, let us define some notations first.

Definition 1. *Let $R_{i,j}$ denote the segment ending at index i such that $R_{i,j}$ is the jth largest among those segments that end at i. That is, $R_{i,j} = A[p \ldots i]$ where $S(p,i)$ is the jth largest among $S(q,i)$ for all $q \in [1,i]$.*

Definition 2. *Let $T_{i,j}$ denote the set of segments $R_{i,1}, R_{i,2}, \ldots, R_{i,j}$. In other words, $T_{i,j}$ contains all the j largest segments ending at index i.*

The naïve approach, described at the beginning of this section, compares all the segments in $T_{1,n}, T_{2,n}, \ldots, T_{n,n}$, each of which contains at most n segments. However, we know that if $R_{i,j}$ is not one of the k maximum segments of A, then neither are $R_{i,j+1}, R_{i,j+2}, \ldots, R_{i,n}$ since each of them has a smaller sum than $R_{i,j}$ by definition. Furthermore, given an integer ℓ, there are only two possible cases for every index as follows.

1. For some index i, if not all the segments in $T_{i,\ell}$ belong to the k largest segments retrieved from $T_{1,\ell}, T_{2,\ell}, \ldots, T_{n,\ell}$, then we need not consider segments $R_{i,\ell+1}, R_{i,\ell+2}, \ldots, R_{i,n}$ anymore.
2. Conversely, for some index i', if all the segments in $T_{i',\ell}$ belong to the k largest segments retrieved from $T_{1,\ell}, T_{2,\ell}, \ldots, T_{n,\ell}$, then we need to consider $R_{i',\ell+1}, R_{i',\ell+2}, \ldots, R_{i',n}$ since they are still candidates for the k maximum segments of A.

In conclusion, by comparing segments in $T_{1,\ell}, T_{2,\ell}, \ldots, T_{n,\ell}$, we can bypass the indices in case 1, and only have to consider the indices in case 2 since they may contribute more segments. We call those indices in case 2 the "qualified right ends".

Algorithm 1 KMaxSums

$Q \leftarrow \{1, 2, \ldots n\}$, $m \leftarrow n$, $K \leftarrow \phi$;
repeat
  find ℓ such that $m \times (\ell - 1) < 2k \leq m \times \ell$;
  if $\ell > n$ **then**
    $\ell \leftarrow n$;
  end if
  for all $i \in Q$ **do**
    compute $T_{i,\ell}$;
  end for
  $K \leftarrow$ the k largest segments from $K \cup \bigcup_{i \in Q} T_{i,\ell}$;
  $Q \leftarrow \{i \mid T_{i,\ell} \subseteq K \ \forall\, i \in Q\}$;
  $m \leftarrow |Q|$;
until $m = 0$ or $\ell = n$
output the segments in K;

The pseudo code for the problem of k maximum-sum segments is given in Algorithm 1. Let Q denote the list of m distinct qualified right ends, which are initially the n positions of A and $Q[1], Q[2], \ldots, Q[m]$ refer to the m right ends in Q, respectively. Let m be the number of qualified right ends, assigned n in the beginning. We let K, initially empty, denote a list of candidate segments for the k maximum segments. The algorithm repeats the following procedure. In each iteration, we choose $\ell = 2\left\lceil \frac{k}{m} \right\rceil$ and then compute $T_{Q[1],\ell}, T_{Q[2],\ell}, \ldots, T_{Q[m],\ell}$ which contain around $2k$ segments. We next retrieve the k largest segments from the

set of segments, obtained by incorporating segments in $T_{Q[1],\ell}, T_{Q[2],\ell}, \ldots, T_{Q[m],\ell}$ with the segments in K. It should be noted here that the k largest segments retrieved in this manner are not certainly the k maximum-sum segments of A because we only consider the ℓ largest segments ending at $Q[1], Q[2], \ldots, Q[m]$. Since only k largest segments are retrieved, there would be at most half qualified right ends left over for the next iteration. (Imagining k balls are thrown into an $m \times \ell$ table, $m \times \ell \cong 2k$, we know that there would be at most half rows full of balls.) Meanwhile, the value of ℓ will at least double in the next iteration. We set Q the qualified right ends, set m the size of Q, and then restart the procedure. The procedure will terminate either when the number of qualified right ends decreases to zero or ℓ increases to n.

Lemma 2. *Algorithm KMaxSums terminates in at most* $\log n + 1$ *iterations.*

Proof. Suppose algorithm KMaxSums terminates in the ith iteration, and the values of ℓ in each iteration are denoted by $\ell_1, \ell_2, \ldots, \ell_i$, respectively. Our goal is to prove that $i \leq \log n + 1$. For any two consecutive iterations j and $j+1$, suppose m_j and m_{j+1} are the corresponding values of m in the jth iteration and the $j+1^{th}$ iteration. We have $m_j \times (\ell_j - 1) < 2k \leq m_j \times \ell_j$ and $m_{j+1} \times (\ell_{j+1} - 1) < 2k \leq m_{j+1} \times \ell_{j+1}$. By definition, m_{j+1} is the number of qualified right ends with all their ℓ_j largest segments being retrieved, so it is clear that $m_{j+1} \leq \lfloor k/\ell_j \rfloor$. This yields $2k \leq m_{j+1} \times \ell_{j+1} \leq \lfloor k/\ell_j \rfloor \times \ell_{j+1} \leq \frac{k}{\ell_j} \times \ell_{j+1}$. We conclude that $\ell_{j+1} > 2\ell_j$.

Next we show by induction that $\ell_i \geq 2^{i-1}$. The basis holds since $\ell_1 \geq 2^0 = 1$. For any j we know $\ell_{j+1} \geq 2\ell_j$, so by inductive hypothesis, $\ell_j \geq 2^{j-1}$, we can deduce that $\ell_{j+1} \geq 2^j$. Thus, $\ell_i \geq 2^{i-1}$ by induction. Moreover, since ℓ_i is at most n, it follows that $n \geq \ell_i \geq 2^{i-1}$, which leads to $i \leq \log n + 1$. □

Lemma 3. *Given* p *distinct increasing indices* $i_1, i_2, \ldots, i_p$, $1 \leq p \leq n$, *and a positive integer* q, $1 \leq q \leq n$, *we can compute* $T_{i_1,q}, T_{i_2,q}, \ldots, T_{i_p,q}$ *in* $O(n + pq)$ *time.*

Proof. The input sequence A is partitioned into $p+1$ contiguous subsequences, $A[1 \ldots i_1]$, $A[i_1+1 \ldots i_2]$, $\ldots$, $A[i_{p-1}+1 \ldots i_p]$, $A[i_p+1 \ldots n]$. Let ℓ_j denote the length of the jth contiguous subsequence, i.e. ℓ_j is the length of $A[i_{j-1} \ldots i_j]$. Notice that $\ell_1 + \ell_2 + \ldots + \ell_p$ is less than or equal to n.

To compute $T_{i_j,q}$, it suffices to find the q minimum values in the prefix-sum array $P[1 \ldots i_j]$ for all $j \in [1,p]$. So, in the first step we find the q minimum values among $P[1 \ldots i_1]$, which can be done in $O(\ell_1)$ time by Lemma 1. Let Q record these q minimum values. In the second step, the q minimum values among $P[1 \ldots i_2]$ is found by retrieving the q minimum values from Q and $P[i_1+1 \ldots i_2]$. This requires $O(\ell_2+q)$ time by Lemma 1. Proceeding in this manner, we compute the $T_{i_j,q}$ for each index i_j in $O(\ell_j+q)$ time. Therefore, the total running time is $\ell_1+(\ell_2+q)+(\ell_3+q)+\ldots+(\ell_j+q)+\ldots+(\ell_p+q) = (\ell_1+\ell_2+\ldots+\ell_p)+(p-1)q = O(n+pq)$. □

Lemma 4. *Algorithm KMaxSums runs in $O((n + k)\log n)$ time.*

Proof. Since $m \times (\ell - 1) < 2k \leq m \times \ell \Rightarrow 2k \leq m \times \ell < 2k + m \leq 2k + n$, we can derive that $m \times \ell = O(n+k)$. By Lemma 3, the time required for computing the l largest segments ending at m qualified ends is $O(n+m\times\ell) = O(n+k)$. Retrieving k largest segments from $O(k)$ segments takes only $O(k)$ time by Lemma 1. So, it takes $O(n+k)$ time in each iteration. Since there are at most $\log n + 1$ iterations by Lemma 2, we conclude that the total time is $O((n + k)\log n)$. □

If $k \geq n$, we can write the time complexity of KMaxSums as $O(k \log n)$. If $k < n$, we can write the time complexity as $O(n \log n)$. In what follows, we show that in the $k < n$ case, we can further reduce the running time to $O(n + k \log k)$, which leads to Theorem 1.

Theorem 1. *The problem of finding the k maximum-sum segments can be solved in $O(n + k \log(min\{n, k\}))$ time.*

As we will see, the $O(n)$ term in $O(n + k \log k)$ comes from the time needed to compress the input sequence, and the $O(k \log k)$ term comes from the time of executing k iterations, each of which costs $O(\log k)$ time.

3.2 Improving on the Time Complexity in the $k < n$ Case

The strategy is to compress the input sequence A into a sequence of size at most $2k$ by preprocessing A in $O(n)$ time. We can find k distinct positions containing all the right ends of the k maximum segments. Similarly, we find k distinct positions containing all the left ends. To find the k maximum segments, we only have to consider these $2k$ positions. Specifically, let $r_1, r_2, \ldots, r_k$ denote the right ends of the k largest segments retrieved from $R_{1,1}, R_{2,1}, \ldots, R_{n,1}$.

Lemma 5. *The k maximum-sum segments of A must end at the indices in $r_1, r_2, \ldots r_k$.*

Proof. Assume that there exists one of the k maximum-sum segments, say S_i, whose right end, i, is not in $r_1, r_2 \ldots r_k$. By definition of $R_{i,1}$, we know S_i cannot have a larger sum than $R_{i,1}$. Moreover, $R_{i,1}$ has a smaller sum than $R_{r_1,1}, R_{r_2,1} \ldots, R_{r_k,1}$ since $R_{i,1}$ is not the k largest by assumption. So, we have deduced that there are at least k segments having a larger sum than S_i. This contradicts to the assumption. □

Similarly, we find the largest segments starting at each index of A. To do so, we scan the suffix sum array in the reverse order and keep the minimum value on the fly. Let us use a similar notation to refer to these segments, say $L_{1,1}, L_{2,1}, \ldots, L_{n,1}$. Let $l_1, l_2, \ldots, l_k$ denote the left ends of the k largest segments retrieved from $L_{1,1}, L_{2,1}, \ldots, L_{n,1}$. It can be shown in the same way that the k maximum segments of A must start at $l_1, l_2, \ldots, l_k$.

We construct a compressed sequence recording the prefix sums of $l_1, l_2, \ldots, l_k$ and $r_1, r_2 \ldots r_k$. Algorithm KMaxSums takes this compressed sequence as input and runs in $O(k \log k)$ time. Since we use $O(n)$ time to compress A, the total time is $O(n + k \log k)$.

3.3 Finding k Maximum-Sum Segments in Order

The problem of k maximum segments has a trivial lower bound $\Omega(n+k)$. Notice that the k maximum segments are not sorted. Lemma 6 states that if we want to output k maximum segments in non-decreasing order, the lower bound becomes $O(n + k \log k)$ when k is no more than n.

Lemma 6. *When $k \leq n$, it requires $\Omega(n+k \log k)$ time to output the k maximum segments in non-decreasing order.*

Proof. If there exists an $o(n + k \log k)$-time algorithm for computing the k maximum segments in non-decreasing order, we show that sorting k random numbers can be done in $o(k \log k)$ time. Assume to the contrary that there exists an algorithm Λ computing the k sorted maximum segments in $o(n + k \log k)$ time. Given k random numbers, we obtain a new sequence of length $2k - 1$ as follows. For each two consecutive random numbers, we augment a negative number whose absolute value is larger than them. That way the k maximum segments in this new sequence are all atomic elements. The output of Λ on this new sequence is equivalent to the k sorted random numbers. The running time of Λ is $o((2k-1) + (2k-1)\log(2k-1)) = o(k \log k)$. However, the lower bound of sorting k random numbers is well-known to be $O(k \log k)$ which contradicts to our assumption. □

Corollary 1. *A simple variant of algorithm $KMaxSums$ yields an optimal solution to the problem of finding the k sorted maximum segments when $k \leq n$.*

Proof. We first run algorithm $KMaxSums$ to find the k maximum segments in $O(n + k \log k)$ time. We next sort the k maximum segments by sum, which requires $O(k \log k)$ time. □

It is not difficult to see that when $k > n$, algorithm $KMaxSums$ also leads to an $O(k \log k)$-time solution to the problem of finding the k sorted maximum segments. However, we do not know if $O(k \log k)$ is the actual lower bound of this sorted problem when $k > n$.

4 Multiple-Dimensional Cases

It is helpful to introduce the two-dimensional case before extending the results to the multiple-dimensional cases. Recall the definition of the k maximum-sum subarray problem in d dimensions. Its goal is to find k d-dimensional subarrays such that the sums of those subarrays are the k largest among all possible d-dimensional subarray sums.

4.1 Two-Dimensional Case

The input sequence is replaced by a two-dimensional array $X = [x_{ij}]_{1 \leq i \leq m, 1 \leq j \leq n}$. We define $X[p \ldots q, r \ldots s]$ as the subarray expanded by the four corners (p, r),

(p, s), (q, r) and (q, s).The idea is to transform the input array $X[1 \dots m, 1 \dots n]$ into a pile of one-dimensional sequences. The original problem is then reduced to finding the k largest segments from these one-dimensional sequences. Using similar techniques presented in the previous sections, we can solve the k two-dimensional maximum subarrays problem. Let us show the transformation in details. Given two indices i and j where $1 \leq i \leq j \leq m$, we convert the subarray $X[i \dots j, 1 \dots n]$ into a new sequence $X_{i,j}[1 \dots n]$ such that $X_{i,j}[q] = \sum_{p=i}^{j} x_{pq}$ for $q = 1, \dots, n$. Clearly, each segment $X_{i,j}[p \dots q]$ corresponds to the subarray $X[i \dots j, p \dots q]$ respectively. The maximum-sum subarray problem is equivalent to finding k maximum segments from the $O(m^2)$ converted sequences, $X_{i,j}$ for $1 \leq i \leq j \leq m$.

Given an integer ℓ, observe that each converted sequence's ℓ maximum segments are the ℓ local maxima with respect to the k maximum subarray problem. Obviously, the k maximum subarrays of X are the k global maxima. A naïve approach is to find every converted sequence's k maximum segments, and the k maximum subarrays are the k largest among the $O(m^2k)$ local maxima. Instead of finding $O(m^2k)$ local maxima at once, we adopt the same trick to speed up the computation. That is, in each iteration we compute only $2k$ local maxima and eliminate half of them.

The pseudo code for the k maximum-sum subarray problem is given in Algorithm 2. Let n_s denote the number of "qualified sequences", which will be defined later, and n_s is initialized as $O(m^2)$. In each iteration, we use algorithm KMax-Sums to find ℓ maximum segments, $\ell = 2\lceil \frac{k}{n_s} \rceil$, from n_s converted sequences and then retrieve the k largest which are the candidates of the k maximum subarrays. Clearly, if the ℓth largest local maximum is not one of the candidates, neither is the $(\ell + 1)^{th}$ largest local maximum. We call the sequences whose ℓ^{th} largest local maximum belongs to the k candidates the "qualified sequences", and the rest the unqualified sequences. Only the qualified sequences need to be considered in the next iteration. The algorithm terminates when all the sequences are unqualified.

Now we turn to the time complexity analysis. In line 1, we compute the prefix sums for each column of X in $O(mn)$ time. It takes $O(m^2n)$ time to transform input array into sequences in lines 2, 3 and 4. Recall that at most half of the qualified sequences are left over after each round. We know that finding the ℓ maximum segments takes $O(n + \ell \log(\min\{n, \ell\}))$ time by Theorem 1. Below, we discuss it in two possible cases. When $k \leq n$, the number of qualified sequences is reduced to k in the second iteration. So, the entire while-loop takes $O(m^2 \times (n + 1 \log 1) + k) + O(k \times (n + 2 \log 2) + k) + O(\frac{k}{2} \times (n + 4 \log 4) + k) + \dots + O(1 \times (n + k \log k) + k) = O(m^2 n + k \log k)$ time. When $k > n$, $O(m^2 \times (n + \ell \log \ell) + k) + O(\frac{m^2}{2} \times (n + 2\ell \log 2\ell) + k) + O(\frac{m^2}{4} \times (n + 4\ell \log 4\ell) + k) + \dots + O(\frac{m^2}{\min\{m^2, k\}} \times (n + (\ell \min\{m^2, k\}) \log(\ell \min\{m^2, k\}) + k) = O(m^2 n + k \log n)$ where $\ell = 2\lceil \frac{k}{m^2} \rceil$. Therefore, we have the following theorem.

Theorem 2. *Algorithm KMaxSums2D finds the k maximum-sum subarrays in $O(m^2 n + k \log(\min\{n, k\}))$ time.*

Algorithm 2 KMaxSums2D

$Q \leftarrow \{(i,j) \mid 1 \le i \le j \le m\}$, $n_s \leftarrow m(m-1)/2$, $K \leftarrow \phi$;
Compute a new array, $Y = [y_{ij}]$ of order $m \times n$, where $y_{ij} = \sum_{h=0}^{j} x_{ih}$
for each i and j, $1 \le i \le j \le m$ **do**
  Compute sequence $X_{i,j}[1 \ldots n]$ such that $X_{i,j}[h] = y_{jh} - y_{ih}$ for $h = 1, 2, \ldots, n$
end for
repeat
  find ℓ such that $n_s \times (\ell - 1) < 2k \le n_s \times \ell$;
  if $\ell > n(n-1)/2$ **then**
    $\ell \leftarrow n(n-1)/2$;
  end if
  for all $(i,j) \in Q$ **do**
    $L_{i,j} \leftarrow \ell$ maximum segments of $X_{i,j}$ computed by KMaxSums;
  end for
  $K \leftarrow$ the k largest segments from $K \cup \bigcup_{(i,j) \in Q} L_{i,j}$;
  $Q \leftarrow \{(i,j) \mid L_{i,j} \subseteq K \ \forall \ (i,j) \in Q\}$;
  $n_s \leftarrow |Q|$;
until $n_s = 0$ or $\ell = n(n-1)/2$
output the segments in K;

4.2 Higher-Dimensional Cases

Without loss of generality, we assume each dimension is of equal size n. Given a d-dimensional array of real numbers, we wish to find k d-dimensional subarrays with maximum sums. Similar techniques in the two-dimensional case are used here. We reduce the higher-dimensional k maximum subarray problem to several k maximum segment problems. We have to transform the d-dimensional input array into sequences first. We store $(d-1)$-dimensional values to each element of a converted sequence. Because there are $O(n^2)$ combinations in every dimension, we transform the d-dimensional input array into $O(n^{2d-2})$ converted sequences. A similar analysis to the two-dimensional case yields the following theorem.

Theorem 3. *The k d-dimensional maximum-sum subarrays can be found in $O(n^{2d-1} + k \log \min\{n, k\})$ time.*

5 Conclusions

We close this paper by mentioning a few open problems. First, is there an algorithm running in $o(k \log k)$ time for finding the k maximum segments in non-decreasing order when $k > n$? Second, it would be interesting to find a tight lower bound for the multiple-dimensional k maximum subarray problem.

Acknowledgements

Chih-Huai Cheng, Kuan-Yu Chen, Wen-Chin Tien and Kun-Mao Chao were supported in part by NSC grants 92-2213-E-002-059 and 93-2213-E-002-029 from the National Science Council, Taiwan.

References

1. S.E. Bae and T. Takaoka, Algorithms for the Problem of k Maximum Sums and a VLSI Algorithm for the k Maximum Subarrays Problem, *Proceedings of the 7th International Symposium on Parallel Architectures, Algorithms and Networks*, 247–253, 2004.
2. F. Bengtsson and J. Chen, Efficient Algorithms for k Maximum Sums, *Proceedings of the 15th International Symposium on Algorithms And Computation, LNCS* 3341, 137–148, 2004.
3. J. Bentley, Programming Pearls: Algorithm Design Techniques, *Communications of the ACM*, 865–871, 1984.
4. K.-Y. Chen and K.-M Chao, On the Range Maximum-Sum Segment Query Problem, *Proceedings of the 15th International Symposium on Algorithms And Computation, LNCS* 3341, 294–305, 2004.
5. T.H. Cormen, C.E. Leiserson, R.L. Rivest and C. Stein, Introduction to Algorithms, *The MIT Press: 2nd Edition*, 185–196, 1999.
6. T.-H. Fan, S. Lee, H.-I Lu, T.-S. Tsou, T.-C. Wang, and A. Yao, An Optimal Algorithm for Maximum-Sum Segment and Its Application in Bioinformatics. *Proceedings of the Eighth International Conference on Implementation and Application of Automata*, LNCS 2759, 251–257, 2003.
7. T. Fukuda, Y. Morimoto, S. Morishita, and T. Tokuyama, Data Mining Using Two-Dimensional Optimized Association Rules: Scheme, Algorithms, and Visualization, *Proceedings of the 1996 ACM SIGMOD International Conference on Management of Data*, 13–23,1996.
8. U. Grenander, Pattern Analysis, *Springer-Verlag, New York*, 1978.
9. X. Huang, An Algorithm for Identifying Regions of a DNA Sequence that Satisfy a Content Requirement. *Computer Applications in the Biosciences*, 10: 219–225, 1994.
10. Y.-L. Lin, X. Huang, T. Jiang, and K.-M. Chao, MAVG: Locating Non-Overlapping Maximum Average Segments in a Given Sequence, *Bioinformatics*, 19: 151–152, 2003.
11. Y.-L. Lin, T. Jiang, and K.-M. Chao, Efficient Algorithms for Locating the Length-constrained Heaviest Segments with Applications to Biomolecular Sequence Analysis. *Journal of Computer and System Sciences*, 65: 570–586, 2002.
12. W.L. Ruzzo and M. Tompa, A Linear Time Algorithm for Finding All Maximal Scoring Subsequences, *Proceedings of the 7th International Conference on Intelligent Systems for Molecular Biology*, 234–241, 1999.
13. T. Takaoka, Efficient Algorithms for the Maximum Subarray Problem by Distance Matrix Multiplication, *Electronic Notes in Theoretical Computer Science*, 61: 1–10, 2002.
14. T. Tamaki and T. Tokuyama, Algorithms for the Maximum Subarray Problem Based on Matrix Multiplication, *Proceedings of the Ninth Annual ACM-SIAM Symposium on Discrete Algorithms*, 446–452, 1998.

Network Load Games*

Ioannis Caragiannis[1], Clemente Galdi[1,2], and Christos Kaklamanis[1]

[1] Research Academic Computer Technology Institute,
Department of Computer Engineering and Informatics,
University of Patras, 26500, Rio Greece
[2] Dipartimento di Informatica ed Applicazioni "R.M. Capocelli",
Universitá di Salerno, 84081, Baronissi (SA), Italy

Abstract. We study *network load games*, a class of routing games in networks which generalize selfish routing games on networks consisting of parallel links. In these games, each user aims to route some traffic from a source to a destination so that the maximum load she experiences in the links of the network she occupies is minimum given the routing decisions of other users. We present results related to the existence, complexity, and price of anarchy of Pure Nash Equilibria for several network load games. As corollaries, we present interesting new statements related to the complexity of computing equilibria for selfish routing games in networks of restricted parallel links.

1 Introduction

We study algorithmic questions related to a particular class of games in networks. A *game* with $n \geq 2$ *players* or *users* is a finite set of *actions* or *strategies* $\mathbf{S}_i$ for each user and a *payoff function* u_i (or, alternatively, a cost function c_i) defined over the users and the set $\mathbf{S}_1 \times \mathbf{S}_2 \times ... \times \mathbf{S}_n$. The elements of $\mathbf{S}_1 \times \mathbf{S}_2 \times ... \times \mathbf{S}_n$ are called *states.* A *Pure Nash Equilibrium* is a state $\mathbf{s} = \langle P_1, ..., P_n \rangle$ such that $u_i(P_1, ..., P_i, ..., P_n) \geq u_i(P_1, ...P_i', ..., P_n)$ for any $P_i' \in \mathbf{S}_i$. In general, a game does not have a Pure Nash Equilibrium.

In a *load game* the input is a set of n users, a set E of m resources, and the action sets are $\mathbf{S}_i \subseteq 2^E$. There are also given a load function L mapping $E \times \{1, ..., n\}$ to positive real numbers denoting how much a user increases the load of a resource it uses and a cost function ℓ mapping $E \times R$ to positive real numbers ($\ell_e(x)$ is non-decreasing in x). The payoffs/costs are computed as follows. Let $\mathbf{s} = (P_1, ..., P_n)$ be a state. Then

$$c_i(\mathbf{s}) = -u_i(\mathbf{s}) = \max_{e \in P_i} \left\{ \ell_e \left(\sum_{j: e \in P_j} L_e(j) \right) \right\}.$$

Intuitively, each user chooses a set of resources and the cost incurred by user i is the maximum load of the resources used by i.

* This work was partially supported by the European Union under IST FET Integrated Project 015964 AEOLUS.

X. Deng and D. Du (Eds.): ISAAC 2005, LNCS 3827, pp. 809–818, 2005.

In *network load games*, the resources correspond to links of a directed network and the action sets correspond to paths in the network. Links are related in the sense that they have (possibly different) bandwidths and the users have (possibly different) traffic weights. Then the load function for edge e and user i is defined as $L_e(i) = w_i/b_e$ where w_i is the weight of user i and b_e is the bandwidth of edge e. Hence, there is given a directed network $G = (V, E)$, a source-destination pair s_i, t_i for each user i. The subsets of E available as actions to the user i is the set of all simple directed paths from s_i to t_i. In a state $\mathbf{s} = (P_1, ..., P_n)$, the payoff/cost for user i is defined as

$$c_i(\mathbf{s}) = -u_i(\mathbf{s}) = \max_{e \in P_i} \left\{ \frac{1}{b_e} \sum_{j:e \in P_j} w_j \right\}.$$

The *social cost* of a state $\mathbf{s} = (P_1, ..., P_n)$ is defined as the maximum load over all edges of the network, i.e., $SC(\mathbf{s}) = \max_{e \in E} \left\{ \frac{1}{b_e} \sum_{j:e \in P_j} w_j \right\}$.

The *price of anarchy* (or *coordination ratio* [11]) for a network load game is defined as the ratio of the worst social cost over all Pure Nash Equilibria over the optimal social cost. Intuitively, it gives a measure for the degradation of performance due to the selfish behavior of the users.

In their full generality, network load games are defined on multicommodity networks (i.e., the source-destination pairs may be different) while we are also interested in single-commodity networks or multicommodity networks with either one source or one destination node. Given a single-commodity network, we call *stretch* the ratio of the length of the longest simple path over the length of the shortest path from s to t. An important class of networks is that of layered networks; in these networks, all s-t paths have the same length (i.e., the stretch is 1). The simplest single-commodity network is a network of m parallel links connecting the source to the destination. In the generalization of restricted parallel links studied in [7] each user has a permissible set of links corresponding to her set of strategies. This game can be thought of as a network load game on a multicommodity network with one source and many destinations as it has been observed in [10]. Another combinatorial problem to which this game is related is the task scheduling on unrelated machines.

In the following we survey the results of the literature which are related to load games; these include results on the well-studied congestion games. We consider only on the existence, complexity, and price of anarchy of Pure Nash Equilibria. The extensive literature on mixed equilibria (their existence is guaranteed by the famous theorem of Nash [13]) is not discussed here.

The network load game on m parallel links (also known as selfish routing on parallel links and also studied in the context of task scheduling on related parallel machines) is the mostly studied game in the literature starting with the work of Papadimitriou and Koutsoupias in [11]. For this game, polynomial-time algorithms for computing Pure Nash Equilibria are known in the most general case of links with different bandwidths and users with different traffics while the problems of computing Pure Nash Equilibria of best of worst social cost are NP-

hard [5]. The nashification technique presented in [4] shows that, starting from any assignment of users to links, a Pure Nash Equilibrium of not larger social cost can be computed in polynomial time. Using a polynomial-time approximation scheme (PTAS) for task scheduling on related parallel machines [8], a PTAS for computing a Pure Nash Equilibrium of best social cost (i.e., a Pure Nash Equilibrium of cost at most $1+\epsilon$ times the best social cost, for any constant $\epsilon > 0$) is obtained. Concerning the price of anarchy of Pure Nash Equilibria in such games, there is a tight bound of $2 - 1/m$ [11].

In the restricted links model, there is a polynomial-time 2-approximation algorithm for computing a Pure Nash Equilibrium of best social cost in the case of identical links (and users with different traffic) due to [7]. In the same paper, tight bounds on the price of anarchy of Pure Nash Equilibria for these games are presented. The problem of computing an assignment of best social cost is a special case of the single-source unsplittable flow problem for which constant approximations are presented in [9,10]. Unfortunately, the solutions computed by these algorithms are not Pure Nash Equilibria in general.

Network load games generalize the games on (restricted) parallel links. Another generalization is the classes of *congestion games* and *network congestion games* which have received significant attention in the literature. The selfish routing games on (restricted) parallel links are special cases of congestion games as well. The definition of congestion games is similar to that of load games; the main difference being that the payoff of each user is defined as the sum of the loads on the links (or resources) used by the user (as opposed to the maximum which is assumed in load games). As we will see, this subtle difference implies that load games are essentially different than congestion games with respect to the existence, complexity, and price of anarchy of Pure Nash Equilibria. Formally, in congestion games there is a set E of resources, a set of n users with action sets $\mathbf{S}_i \subseteq 2^E$, and a *delay function* mapping $E \times \{1, ..., n\}$ to the integers ($d_e(j)$ is non-decreasing on j). Given a state $\mathbf{s} = (P_1, ..., P_n)$, let $f_{\mathbf{s}}(e) = |\{i : e \in P_i\}|$. Then, the payoff/cost of user i is defined as $c_i(\mathbf{s}) = -u_i(\mathbf{s}) = \sum_{e \in P_i} d_e(f_{\mathbf{s}}(e))$. In network congestion games, the set E corresponds to the links of a directed network $G = (V, E)$, each user has a source-destination pair $s_i, t_i \in V$ and her strategies are all simple directed paths from s_i to t_i in G. *Linear network congestion games* have linear delay functions which is equivalent to assuming that links have bandwidths b_e and the payoffs are computed as $c_i(\mathbf{s}) = -u_i(\mathbf{s}) = \sum_{e \in P_i} \frac{f_{\mathbf{s}}(e)}{b_e}$. In *weighted linear network congestion games* each user i may have a positive weight w_i and her payoff/cost is defined as $c_i(\mathbf{s}) = -u_i(\mathbf{s}) = \sum_{e \in P_i} \frac{\sum_{i:e \in P_i} w_i}{b_e}$.

Rosenthal [14] has shown that any congestion game has a Pure Nash Equilibrium. His proof is based on the definition of a potential function associated to the states of the game whose local minima correspond to Pure Nash Equilibria. Since that paper, the use of potential function arguments is the main tool for proving the existence of Pure Nash Equilibria. An interesting characterization of games with respect to the potential functions they admit is presented in [12]. Interestingly, generalizations of congestion games on single-commodity networks may not admit Pure Nash Equilibria as it is shown in [6]. Pure Nash Equilib-

ria of best social cost can be computed efficiently in weighted linear network congestion games on single-commodity layered networks [6].

Fabrikant et al. in [3] study the complexity of computing any Pure Nash Equilibrium in congestion games. They show that, in general, the problem of computing a Pure Nash Equilibrium is PLS-complete (i.e., as hard as computing any object whose existence is guaranteed by a polynomially-computable potential function). For symmetric network congestion games (i.e., congestion games with users having the same set of strategies), a Pure Nash Equilibrium can be computed by a polynomial-time algorithm which minimizes Rosenthal's potential function through a reduction to min-cost flow [3].

In this paper, we present new results for the existence, complexity, and price of anarchy of Pure Nash Equilibria in network load games. In particular:

- In Section 2, we show that any load game has a Pure Nash Equilibrium. This is very interesting since the class of load games is especially broad. Also, this result stands in contrast with a result in [6] for generalized versions of network congestion games which do not always have Pure Nash Equilibria. We also study the relation of network load games to congestion games with respect to the potential functions they admit.
- We study the complexity of the problem of computing the best Pure Nash Equilibrium in network load games. The NP-completeness of the problem in single-commodity networks follows by the NP-completeness of the problem in the network of m parallel links. We use the Nashification technique of [4] in the case of single-commodity networks with identical links and users with different traffics, and we obtain a polynomial time approximation scheme for computing the best social cost. In the case of users with identical traffic, we show that Pure Nash Equilibria of best social cost can be computed in polynomial time in networks with either a single source or a single destination using a reduction to min-cost flow (the proof can be thought of as the minimization of a potential function). These networks include networks of restricted parallel links [7] as a special case. Concerning Pure Nash Equilibria of worst social cost, we show that the problem of computing them is inapproximable in single-commodity networks with stretch strictly larger than 1 and NP-hard in single-commodity layered networks and networks of restricted parallel links. Here, we exploit the intuitive relations of Pure Nash Equilibria in network load games to maximal matchings and longest paths in graphs. All these results are presented in Section 3.
- We also present tight bounds of $\Theta(\sqrt{\alpha m})$ on the price of anarchy in single-commodity networks with m identical links and stretch $\alpha \geq 1$. Due to our related inapproximability results, our upper bound also yield almost optimal approximations to the worst social cost of Pure Nash Equilibria in single-commodity networks with stretch strictly larger than 1. We also show a higher lower bound on the price of anarchy for single-commodity networks with arbitrary links. These results are presented in Section 4.

We conclude with open problems in Section 5. Due to lack of space, all the proofs but one have been omitted.

2 Existence of Pure Nash Equilibria

Although the definition of load games is quite general, we show that any such game has a Pure Nash Equilibrium. This comes in contrast with generalized versions of congestion games which may not have a Pure Nash Equilibrium [6]. The proof uses an appropriately defined potential function over the states of the game.

Theorem 1. *Any load game has a Pure Nash Equilibrium.*

In the following, we focus on network load games. We first attempt a characterization of these games with respect to the potential functions they admit. A potential function Φ defined over the states $\mathbf{s}_1$ and $\mathbf{s}_2$ of a game is called an exact potential if for any two states differing in the strategy of user i, it is $\Phi(\mathbf{s}_1) - \Phi(\mathbf{s}_2) = c_i(\mathbf{s}_1) - c_i(\mathbf{s}_2) = u_i(\mathbf{s}_2) - u_i(\mathbf{s}_1)$. By extending this definition, a potential function Φ defined over the states $\mathbf{s}_1$ and $\mathbf{s}_2$ of a game is called a ξ-potential for a positive vector ξ defined over the users if for any two states differing in the strategy of user i, it is $\Phi(\mathbf{s}_1) - \Phi(\mathbf{s}_2) = \xi_i(u_i(\mathbf{s}_1) - u_i(\mathbf{s}_2))$. Monterer and Shapley have proved in [12] that every finite potential game (i.e., a game admitting an exact potential function) is isomorphic to a congestion game. Weighted network congestion games in single-commodity layered networks are known to admit a ξ-potential [6]. Clearly, network load games with identical users on networks of parallel links are congestion games. Furthermore, it can be easily seen that load games with two identical users on networks with identical links admit an exact potential while load games with two users with different weights on networks with identical links admit a ξ-potential. The following lemma states that these are the only similarities load games share with congestion games with respect to the potential functions they admit.

Lemma 1. *There exists a load game with 3 identical users on a layered network with identical links and a load game with 2 users on a layered network with different links that do not admit a ξ-potential.*

3 Complexity of Computing Pure Nash Equilibria

Computing Pure Nash Equilibria of best social cost. We first consider single-commodity networks with identical links and show how to extend the nashification technique for selfish routing on parallel links [4] in this case. In this way, we obtain a polynomial time approximation scheme (PTAS) for computing a Pure Nash Equilibrium with best social cost.

Theorem 2. *There is a PTAS for computing a Pure Nash Equilibrium of best social cost in single-commodity networks with identical links.*

The NP-hardness of the problem of computing a Pure Nash Equilibrium of best social cost in single-commodity networks with arbitrary users follows by the NP-hardness of the problem on the network of m parallel links. The proof

assumes identical links. We can prove the following theorem which essentially shows that what makes the parallel links model hard is the existence of arbitrary users. The result is more general and also applies to selfish routing games on restricted parallel links [7].

Theorem 3. *A Pure Nash Equilibrium with the best social cost in networks with either one source or one destination and with identical users can be computed in polynomial time.*

Proof. Consider a network load game with n users with identical traffic on a network $G = (V, E)$ with a single source s and k destination nodes $t_1, ..., t_k$ and m links (the proof for networks with one destination is very similar). Denote by n_i the number of users wishing to route their traffic from s to t_i.

Consider all pairs (i, e) of the cartesian product $\{1, 2, ..., n\} \times E$ and sort them in non-decreasing order with respect to the value of the quantity i/b_e. Let $L \leq nm$ be the number of different values of i/b_e, and let $r(i, e)$ be such that i/b_e is the $r(i, e)$-th smallest among the L different values.

The proof uses a reduction to min-cost flow. We construct a network N having the same set of nodes as G by replacing each link of the original network by n parallel directed edges. Each edge has a unit capacity. The cost of the i-th edge corresponding to the link e is equal to $(m+1)^{r(i,e)-1}$. The min-cost flow problem is defined by $f_s = n$, $f_{t_i} = -n_i$, for each destination node t_i, and $f_u = 0$ for any other node. Intuitively, this flow problem asks for pushing a total amount n units of flows from node s so that amounts of n_i units of flow reach the sink nodes t_i satisfying the capacity constraints so that the total cost of the edges carrying flow is minimized. We use a min-cost flow algorithm to to compute an optimal solution f for this problem. Note that, although the cost function on the edges is exponential, min-cost flow algorithms work in polynomial time and perform a polynomial number of operations (i.e., comparisons, additions, multiplications, etc.) which do not depend on the cost function. In our case, the costs are representable with a polynomial number of bits and each of the operations mentioned above can be implemented with a polynomial number of bit operations. Overall, both the space required and the running time are polynomial (see [1] for an extensive overview of min-cost flow algorithms).

Using the optimal solution to the above min-cost flow we construct an assignment for the original network load game as follows. We decompose the flow in N into n disjoint paths. Each of these paths corresponds to the strategy of the user in the obvious way. If a path uses some of the parallel edges corresponding to link e, the corresponding user 's path in G uses e.

Observe that if i flow paths use some of the parallel edges between two nodes u and v in the flow solution f, then these paths should traverse the first i edges, otherwise the solution would not be optimal.

We first show that the assignment produced in this way is a Pure Nash Equilibrium. Assume that this is not the case and that the maximum load in the path assigned to some user j by the flow algorithm is ℓ while user j has an incentive to change her strategy p_j and use another path p'_j of maximum load $\ell' < \ell$. Denote by i_e the number of users using link e in the assignment produced

by the flow algorithm. Then ℓ corresponds to the load of some link e' which is used by i' users and it is $i'/b_{e'} = \ell$ while ℓ' corresponds to some link e'' which is used by $i_{e''}$ users and it is $\frac{i''+1}{b_{e''}} = \ell'$. This also implies that $r(i_{e''}, e'') < r(i_{e'}, e')$. Without loss of generality, we may assume that for each link $e \in p_j$, the flow path corresponding to path p_j traverses the i_e-th parallel edge corresponding to the link e of G. If this is not the case and some other flow path uses the i_e-th parallel edge, we can trivially exchange the edges used by the two flows without affecting the cost of f. Now, consider the flow f' which is obtained by f by changing the route of the flow path corresponding to user j and route it through the i_e-th parallel edge corresponding to link e for each $e \in p'_j \cap p_j$ and through the $i_e + 1$-th parallel edge corresponding to link e for each $e \in p'_j \backslash p_j$ (while no flow is routed through the i_e-th parallel edge corresponding to the link e for each $e \in p_j \backslash p'_j$). The cost of the new flow f' is

$$\begin{aligned} COST(f') &= COST(f) - \sum_{e \in p_j \backslash p'_j} (m+1)^{r(i_e, e)-1} + \sum_{e \in p'_j \backslash p_j} (m+1)^{r(i_e+1, e)-1} \\ &\leq COST(f) - (m+1)^{r(i_{e'}, e')-1} + (m-1)(m+1)^{r(i_{e''}, e'')-1} \\ &< COST(f) \end{aligned}$$

which contradicts the fact that f is a flow of minimum cost.

We now show that the Pure Nash Equilibrium defined by the optimal solution f of the flow problem has optimal social cost. Again, denote by ℓ the social cost of this assignment which corresponds to some link e' used by i' users so that $i'/b_{e'} = \ell$. Assume that there was another assignment with maximum load $\ell' < \ell$ corresponding to a pair (i'', e'') meaning that link e'' is assigned to i'' users and has load $i''/b_{e''} = \ell'$. Clearly, $r(i'', e'') \leq r(i', e') - 1$. Using i'_e to denote the number of users using link e in the second assignment, we obtain that the cost of the flow solution f' defined by this assignment is at most

$$\begin{aligned} COST(f') &\leq \sum_{e \in E} \sum_{j=1}^{i'_e} (m+1)^{r(j,e)-1} \leq \sum_{e \in E} \sum_{j=0}^{r(i'', e'')-1} (m+1)^j \\ &= (m+1)^{r(i'', e'')} - 1 < (m+1)^{r(i', e')-1} \leq COST(f) \end{aligned}$$

which again contradicts the fact that f is a flow of minimum cost. □

For networks of parallel links with identical users, we make the following observation.

Lemma 2. *The social cost of all Pure Nash Equilibria in a network of parallel links with identical users is the same.*

Computing Pure Nash Equilibria of worst social cost. Lemma 2 trivially yields that computing the worst PNE is in P. Clearly, in single-commodity networks, Pure Nash Equilibria may have different social costs.

In the following, we show that computing the worst social cost in single-commodity networks is inherently more difficult than in the case of parallel links. The proof uses an approximation-preserving reduction from Longest Directed Path in Hamiltonian Graphs [2].

Theorem 4. *For any constants $\delta, \epsilon > 0$, there is no polynomial-time algorithm which approximates the worst social cost within $O((am)^{1/2-\epsilon})$ in single-commodity networks with m identical links and stretch $\alpha > 1+\delta$, unless P=NP. Also, there is no polynomial-time algorithm which approximates the worst social cost within $o(\sqrt{am}/\log^2 m)$, unless the Exponential Time Hypothesis fails.*

In the case of networks with arbitrary links we can show a stronger result by slightly modifying the construction in the proof of Theorem 4.

Corollary 1. *For any constants $\delta, \epsilon > 0$, there is no polynomial-time algorithm which approximates the worst social cost within $O(m^{1-\epsilon})$ in single-commodity networks with m arbitrary links and stretch $\alpha > 1+\delta$, unless P=NP. Also, there is no polynomial-time algorithm which approximates the worst social cost within $o(m/\log^2 m)$, unless the Exponential Time Hypothesis fails.*

The proofs of the above two statements make use of the fact that the stretch is strictly larger than 1. In the case of single-commodity layered networks, we can still show a negative result. The proof is based on a reduction from Minimum Maximal Bipartite Matching [16].

Theorem 5. *Computing a Pure Nash Equilibrium of worst social cost in single-commodity layered networks with identical links and users with identical traffic is NP-hard.*

The case of restricted parallel links. Next, we consider the case of restricted parallel links studied in [7], a special case of network load games on multicommodity networks. We consider the case of users with identical traffic. Recall that Theorem 3 yields a polynomial-time algorithm for computing a Pure Nash Equilibrium of the best social cost. The following theorem states that computing Pure Nash Equilibria of worst social cost becomes difficult in restricted parallel links with identical users (as opposed to parallel links). The proof uses a reduction from Minimum Maximal Bipartite Matching [16].

Theorem 6. *Computing the worst Pure Nash Equilibrium for users of identical traffic in restricted parallel identical links is NP-hard.*

4 Bounds on the Price of Anarchy

The results presented in the following establish a tight bound of $\Theta(\sqrt{\alpha m})$ on the price of anarchy of Pure Nash Equilibria on network load games on single-commodity networks with identical links. The upper bound is stated in Theorem 7 while the lower bound is stated in Theorem 8. In particular, Theorem 7 also

implies that any Pure Nash Equilibrium (including the one of the best social cost) gives an almost optimal approximation to the worst social cost. Assuming that the Exponential Time Hypothesis (that Satisfiability has no subexponential-time algorithms) holds, the corresponding approximation ratio is optimal within polylogarithmic factors. Our constructions in the proof of Theorem 8 are generalizations of the construction yielding the Braess Paradox in other selfish routing games (see e.g., [15]).

Theorem 7. *The price of anarchy of network load games in single-commodity networks with m identical links and stretch α is at most $O(\sqrt{\alpha m})$.*

Theorem 8. *For any integer $m \geq 9$ and α such that $1 \leq \alpha \leq m-1$, there exists a network load game on a single-commodity network with at most m identical links and stretch at most α such that the price of anarchy is $\Omega(\sqrt{\alpha m})$.*

By slightly modifying the construction in the proof of Theorem 8, we obtain an even worse lower bound on the price of anarchy in network load games on single-commodity networks with arbitrary links (and identical users).

Theorem 9. *For any integer $m \geq 7$, there exists a network load game with identical users on a single-commodity layered network of at most m arbitrary links with price of anarchy $\Omega(m)$.*

5 Open Problems

Our work reveals some interesting open questions:

- We have not provided any polynomial-time algorithm for computing any Pure Nash Equilibrium in single-commodity networks with arbitrary links and users with arbitrary traffic. The nashfication technique of [4] for arbitrary users and arbitrary parallel links does not seem to apply in this case. Furthermore, the following question is very challenging. Is there a constant approximation algorithm or even a PTAS for computating of a Pure Nash Equilibrium of best social?
- What is approximability of computing a Pure Nash Equilibrium of worst social cost in single-commodity layered networks? Although we have proved that the problem is NP-hard, the only known upper bound on its approximability is the $O(\sqrt{m})$ implied by Theorem 7 for single-commodity layered networks with identical links.
- Is there a polynomial-time algorithm for computing the best social cost in networks with either a single source or a single destination with arbitrary links when the number of different traffic weights of the users is constant? Note that all possible values of the potential function are representable with a polynomial number of bits in this case as well but the reduction of Theorem 3 does not seem to extend in this case. Also, even in the simplest case of restricted parallel arbitrary links and users of arbitrary traffic, computing any Pure Nash Equilibrium in polynomial time is still open.

– Is there a load game or even a network load game which is PLS-complete? A characterization of these games like the one presented in [3] for congestion games would be very interesting.

References

1. R.K. Ahuja, T.L. Magnati, and J.B. Orlin. Network flows, Theory, Algorithms, and Applications. *Prentice Hall*, 1993.
2. A. Björklund, T. Husfeldt, and S. Khanna. Approximating longest directed paths and cycles. In *Proceedings of the 31st International Colloquium on Automata, Languages, and Programming (ICALP '04)*, LNCS 3142, Springer, pp. 222-233, 2004.
3. A. Fabrikant, C. Papadimitriou, and K. Talwar. The complexity of pure nash equilibria. In *Proceedings of the 36th Annual ACM Symposium on Theory of Computing (STOC '04)*, pp. 604-612, 2004.
4. R. Feldmann, M. Gairing, T. Lücking, B. Monien, and M. Rode. Nashification and the coordination ratio for a selfish routing game. In *Proceedings of the 30th International Colloquium on Automata, Language, and Programming (ICALP '03)*, LNCS 2719, Springer, pp. 514-526, 2003.
5. D. Fotakis, S. Kontogiannis, E. Koutsoupias, M. Mavronicolas, and P. Spirakis. The structure and complexity of nash equilibria for a selfish routing game. In *Proceedings of the 29th International Colloquium on Automata, Language, and Programming (ICALP '02)*, LNCS 2380, Springer, pp. 123-134, 2002.
6. D. Fotakis, S. Kontogiannis, and P. Spirakis. Selfish unsplittable flows. In *Proceedings of the 31st International Colloquium on Automata, Language, and Programming (ICALP '04)*, LNCS 3142, Springer, pp. 593-605, 2004.
7. M. Gairing, T. Lücking, M. Mavronicolas, and B. Monien. Computing Nash equilibria for restricted parallel links. In *Proceedings of the 36th Annual ACM Symposium on Theory of Computing (STOC '04)*, pp. 613-622, 2004.
8. D.S. Hochbaum and D. Shmoys. A polynomial approximation scheme for scheduling on uniform processors: using the dual approximation approach. *SIAM Journal on Computing*, 17(3), pp. 539-551, 1988.
9. J. Kleinberg. Single-source unsplittable flow. In *Proceedings of the 37th Annual Symposium on Foundations of Computer Science (FOCS '97)*, pp. 68-77, 1996.
10. S. Kolliopoulos and C. Stein. Approximation algorithms for single-source unsplittable flow. *SIAM Journal on Computing*, 31, pp. 919-946, 2002.
11. E. Koutsoupias and C. Papadimitriou. Worst-case equilibria. In *Proceedings of the 16th International Symposium on Theoretical Aspects of Computer Science (STACS '99)*, LNCS 1563, Springer, pp. 404-413, 1999.
12. D. Monterer and L. S. Shapley. Potential games. *Games and Economic Behavior*, Vol. 14, pp. 124-143, 1996.
13. J. F. Nash. Non-cooperative games. *Annals of Mathematics*, 54(2), pp. 286-295, 1951.
14. R. W. Rosenthal. A class of games possessing pure-strategy Nash equilibria. *International Journal of Game Theory*, Vol. 2, pp. 65-67, 1973.
15. T. Roughgarden and E. Tardos. How bad is selfish routing? *Journal of the ACM*, 49(2), pp. 236-259, 2002.
16. M. Yannakakis and F. Gavril. Edge dominating sets in graphs. *SIAM Journal on Applied Mathematics*, Vol. 38(3), pp. 364-372, 1980.

Minimum Entropy Coloring

Jean Cardinal[1], Samuel Fiorini[2], and Gwenaël Joret[3]

[1] Computer Science Dept., CP 212
[2] Mathematics Dept., CP 216
[3] Computer Science Dept., CP 212, Aspirant du F.N.R.S.,
Université Libre de Bruxelles,
Boulevard du Triomphe,
B-1050, Brussels, Belgium
{jcardin, sfiorini, gjoret}@ulb.ac.be

Abstract. We study an information-theoretic variant of the graph coloring problem in which the objective function to minimize is the entropy of the coloring. The minimum entropy of a coloring is called the chromatic entropy and was shown by Alon and Orlitsky to play a fundamental role in the problem of coding with side information. In this paper, we consider the minimum entropy coloring problem from a computational point of view. We first prove that this problem is NP-hard on interval graphs. We then show that it is NP-hard to find a coloring whose entropy is within $(\frac{1}{7} - \epsilon) \log n$ of the chromatic entropy for any $\epsilon > 0$, where n is the number of vertices of the graph. A simple polynomial case is also identified. It is known that the traditional graph entropy is a lower bound for the chromatic entropy. We prove that this bound can be arbitrarily bad, even for chordal graphs. Finally, we consider the minimum number of colors required to achieve minimum entropy and prove a Brooks-type theorem.

1 Introduction

The standard minimum graph coloring problem asks to color the vertices of a given graph with a minimum number of colors so that no two adjacent vertices have the same color. The minimum number of colors in a coloring of G is the *chromatic number* of G, denoted by $\chi(G)$. Numerous variants of this problem have been studied, with different objective functions and constraints [15]. An example of such alternative graph coloring problem is the *optimum cost chromatic partition problem* [14], in which the cost of a color grows linearly with the size of the color class. In the problem we consider, the cost is a concave function of the size of the color class.

The problem is actually defined on specific vertex-weighted graphs called *probabilistic graphs*. A probabilistic graph is a graph equipped with a probability distribution on its vertices. Let (G, P) be a probabilistic graph, and let X be any random variable over the vertex set $V(G)$ of G with distribution P. We define the *entropy* $H(\phi)$ of a coloring ϕ as the entropy of the random variable $\phi(X)$. In other words, the entropy of ϕ is the sum over all colors i of $-c_i \log c_i$, where

X. Deng and D. Du (Eds.): ISAAC 2005, LNCS 3827, pp. 819–828, 2005.

$c_i = \sum_{x:\phi(x)=i} P(x)$ is the probability that X has color i. The *chromatic entropy* $H_\chi(G, P)$ of the probabilistic graph (G, P) is the minimum entropy of any of its colorings. We consider the problem of finding a minimum entropy coloring of a probabilistic graph.

The notion of chromatic entropy was first proposed in an information-theoretic context by Alon and Orlistky [3]. They considered the problem of *(zero-error) coding with side information*, in which a random variable X must be transmitted to a receiver having already some partial information about X. Witsenhausen [20] showed how this transmission scenario could be encoded in a *characteristic graph* G, the set of vertices of which is the set of possible values of X. Alon and Orlitsky [3] proved that the minimum achievable rate for coding with side information is between $H_\chi(G, P)$ and $H_\chi(G, P) + 1$ where P the probability distribution of X.

Given a minimum entropy coloring of (G, P), a Huffman code computed from the color probabilities will provide a suitable code with average length at most $H_\chi(G, P) + 1$. So minimum entropy colorings directly yield good codes for the problem of coding with side information. Heuristic algorithms for practical coding with side information based on minimum entropy coloring have been proposed by Effros et al. [21].

Minimum entropy coloring also applies to the compression of digital image partitions created by segmentation algorithms [1,2].

While the problem has received attention in the information theory and data compression community, it has not yet been studied thoroughly from a computational and combinatorial point of view. Our contribution aims at filling this gap. Preliminary results have already been presented in [6]. Note that another combinatorial optimization problem with an entropy-like objective function has been recently studied by Halperin and Karp [13].

We first give in section 2 some useful lemmas concerning the structure of minimum entropy colorings, and introduce the definition of maximal color-feasible sequences.

In section 3, we consider the computational complexity of the minimum entropy coloring problem. We show that the problem is NP-hard even if the input graph G is an interval graph, a class of graphs on which many classical NP-hard problems become polynomial. We also study the approximability of the problem. Since the chromatic entropy takes value in the interval $[0, \log n]$, where n is the number of vertices of the graph, it is natural to consider additive approximations, i.e. approximations within an additive term. This translates to a multiplicative factor if we consider $2^{H(\phi)}$ instead of $H(\phi)$ as objective function. We show that, unless P=NP, there is no polynomial algorithm finding a coloring of entropy at most $H_\chi(G, P) + (1/7 - \epsilon) \log n$ for any $\epsilon > 0$. This result holds even if P is the uniform distribution. We end the section by giving a simple polynomial case, namely when the input graph G satisfies $\alpha(G) \leq 2$.

Alon and Orlitsky showed that the chromatic entropy was bounded from below by a well-known quantity called graph entropy [18], also known as *Körner entropy.* They left open the question of how tight a lower bound the Körner

entropy is. In section 4, we first note that the ratio between those two quantities is unbounded and then prove that the difference between them can be made arbitrarily large, even if the graph is chordal and the probability distribution is uniform.

Finally, we prove in section 5 that, if P is uniform, a Brooks-type theorem holds for the minimum number $\chi_H(G, P)$ of colors required to achieve minimum entropy: $\chi_H(G, P)$ is at most the maximum degree of G, provided G is connected, and different from an odd cycle or a complete graph.

2 Preliminaries

Consider a probabilistic graph (G, P), where G is a graph and P a probability distribution defined on $V(G)$. For simplicity, we denote by $P(S)$ the sum $\sum_{x \in S} P(x)$, where $S \subseteq V(G)$. We see colors of a coloring ϕ of G as positive integers. We also use $\phi^{-1}(i)$ for the set of vertices colored with color i. As above, let c_i be the probability mass of the i-th color class. Hence we have $c_i = P(\phi^{-1}(i)) = Pr[\phi(X) = i]$, where $X \sim P(x)$ is a random vertex with distribution P. The *color sequence* of ϕ with respect to P is the infinite vector $c = (c_i)$.

A sequence c is said to be *color-feasible* for a given probabilistic graph (G, P) if there exists a coloring ϕ of G having c as color sequence. Most of the time, we will restrict to nonincreasing color sequences, that is, color sequences c such that $c_i \geq c_{i+1}$ for all i. This can be easily achieved for a given color sequence by renaming the colors. Note that color sequences define discrete probability distributions on $\mathbb{N}^+$. The entropy of a coloring is the entropy of the discrete random variable having its color sequence as distribution. In other words, we have $H(\phi) = H(c)$ whenever c is the color sequence of ϕ, where (with a slight abuse of terminology) $H(c)$ is the *entropy of color sequence c*.

The following lemma is of fundamental importance for the remaining proofs and was noted by Alon and Orlitsky [3]. The proof is straightforward and only relies on the concavity of the function $p \mapsto -p \log p$.

Lemma 1 (Alon and Orlitsky [3]). *Let c be a nonincreasing color sequence, let i, j be two indices such that $i < j$ and let α a real number such that $0 < \alpha \leq c_j$. Then we have $H(c) > H(c_1, \ldots, c_{i-1}, c_i + \alpha, c_{i+1}, \ldots, c_{j-1}, c_j - \alpha, c_{j+1}, c_{j+2}, \ldots)$.*

We now examine the consequences of this lemma. We say that a color sequence c *dominates* another color sequence d if $\sum_{i=1}^{j} c_i \geq \sum_{i=1}^{j} d_i$ holds for all j. We denote this by $c \succeq d$. The partial order $\succeq$ is known as the *dominance ordering*. It is often restricted to nonincreasing color sequences in order to avoid unwanted incomparabilities. A nonincreasing color sequence is said to be *maximal color-feasible* when it is not dominated by any other nonincreasing color sequence of the considered probabilistic graph. The next lemma indicates that color sequences of minimum entropy colorings are always maximal color-feasible. (Proof omitted due to lack of space.)

Lemma 2. *Let* c *and* d *be two distinct nonincreasing rational color sequences such that* $c \succeq d$*. Then we have* $H(c) < H(d)$*.*

A similar property was observed for other coloring problems, in particular by de Werra et al. for minimum cost edge colorings [7,8]. A further consequence of Lemma 1 is that any minimum entropy coloring can be constructed by iteratively removing maximal stable sets, i.e., subsets of pairwise nonadjacent vertices that are inclusionwise maximal. (Proof omitted due to lack of space.)

Lemma 3. *Assume that* $P(x) > 0$ *holds for all vertices of a probabilistic graph* (G, P)*. Let* ϕ *be a minimum entropy coloring of* G *with respect to* P*. If the color sequence of* ϕ *is nonincreasing, then the* i*-th color class of* ϕ *is a maximal stable set in the subgraph of* G *induced by the vertices with colors* $j \geq i$*.*

3 Complexity and Approximability

We study in this section the complexity of the minimum entropy coloring problem and its approximability. We first note that the minimum entropy coloring problem has already been shown to be NP-hard on planar graphs with the uniform distribution [6].

An *interval graph* is the intersection graph of a set of intervals on the real line: vertices correspond to intervals and two distinct vertices are adjacent if the corresponding intervals overlap. Our first result shows that finding a minimum entropy coloring of a probabilistic interval graph is NP-hard. Since the numerators and denominators of the probabilities that are used in our reduction are polynomial in the size of the input, the proof also shows that NP-hardness holds in the strong sense.

Theorem 1. *Finding a minimum entropy coloring of a probabilistic interval graph is strongly NP-hard.*

Proof. Our reduction is from the problem of deciding if a circular arc graph G is k-colorable, which is NP-complete [10]. A *circular arc graph* is the intersection graph of arcs on a circle: vertices correspond to arcs and two distinct vertices are adjacent whenever the corresponding arcs intersect each other. Given a circular arc graph G, one can construct a circular representation for G in polynomial time [19]. Without loss of generality, we assume that all arcs in the representation are open (i.e., with the endpoints removed). The basic idea of the proof is to start with a circular-arc graph and cut it open somewhere to obtain an interval graph. The same idea is used in [17], where it is proved that finding a minimum sum coloring of an interval graph is NP-hard.

Let y be an arbitrary point on the circle that is not the endpoint of any of the arcs in the considered representation of G. Let k' be the number of arcs in which y is included. If $k' > k$, then G is not k-colorable. If $k' < k$, we add to the representation $k - k'$ sufficiently small arcs that intersect only arcs including y. This clearly does not increase the chromatic number of G above k. Thus, it can be assumed that y is contained in exactly k arcs.

Denote $a_1, \ldots, a_k$ the arcs that contain y. By splitting each arc a_i into two parts l_i and r_i at point y we obtain an interval representation of some interval graph G'. As is easily checked, G is k-colorable if and only if there is a k-coloring of G' in which l_j and r_j receive the same color for $1 \le j \le k$.

Using an algorithm designed for chordal graphs [11], we list in linear time all maximal cliques of G' (as interval graphs are chordal). For each such clique K, we do the following. If $|K| > k$ then we reject the input as this implies that G is not k-colorable. Now $|K| \le k$. By the Helly property for intervals, there exists a point z of the real line contained in the intervals of K and in no other. We extend the clique K by adding $k - |K|$ sufficiently small intervals in the interval representation around z. This is done in such a way that the new intervals intersect only intervals corresponding to vertices of K. As before, this operation does not increase the chromatic number of G' above k. Let H denote the resulting interval graph. By construction, all maximal cliques of H are also maximum.

Let $\mathcal{K}$ denote the set of maximum cliques of H and let $C = |\mathcal{K}|$. Consider the auxiliary bipartite graph B having $V(H)$ and $\mathcal{K}$ as color classes in which $x \in V(H)$ is adjacent to $K \in \mathcal{K}$ whenever $x \in K$. We define a probability distribution P on the vertices of H as follows:

$$P(x) = \lambda \deg_B(x) + \begin{cases} \lambda j & \text{if } x = l_j \text{ or } r_j, \\ 0 & \text{otherwise,} \end{cases}$$

where $\lambda > 0$ is chosen such that the sum of $P(x)$ over all vertices x of H equals 1, and $\deg_B(x)$ denotes the degree of x in the auxiliary graph B.

We claim that G is k-colorable if and only if the sequence $c^* = \lambda(2k+C, 2(k-1)+C, \ldots, 2+C, 0, \ldots)$ is color-feasible for (H, P). First assume that G is k-colorable. Then there exists a k-coloring ϕ of (H, P) assigning the same color to l_j and r_j for all j. Let c denote the color sequence of ϕ. We assume that c is nonincreasing. For every color i used in ϕ, we have

$$c_i = \sum_{x \in \phi^{-1}(i)} P(x) \tag{1}$$

$$= 2\lambda(k - i + 1) + \lambda \sum_{x \in \phi^{-1}(i)} \deg_B(x) \tag{2}$$

$$= \lambda(2(k - i + 1) + C). \tag{3}$$

The second equality holds because c is nonincreasing. Because ϕ is proper, no two vertices of H with color i are contained in the same maximum clique. Moreover, if some maximum clique is disjoint from the i-th color class then ϕ cannot possibly be a k-coloring, a contradiction. It follows that every maximum clique of H contains exactly one vertex of H with color i. The third equality follows.

Now assume that $c^* = \lambda(2k+C, 2(k-1)+C, \ldots, 2+C, 0, \ldots)$ is color-feasible and let ψ denote a coloring of (H, P) that has c^* as color sequence. Let S be a stable set of H. No two vertices of S are contained in the same clique. For the auxiliary graph B, this means that no $K \in \mathcal{K}$ is adjacent to two distinct

elements of S. It follows that the sum $\sum_{x \in S} \deg_B(x)$ is at most C. Hence the total probability mass of S in (H, P) is at most $\lambda(2k + C)$, with equality if and only if S contains both l_k and r_k. In particular, the first color class of ψ contains both l_k and r_k. By iterating this argument, we conclude that the i-th color class of ψ contains both l_{k-i+1} and r_{k-i+1} for all i, which means that G is k-colorable.

Let c^* be defined as above. The arguments used in the last paragraph show that c^* is color-feasible for (H, P) if and only if it is the unique maximal color-feasible sequence of (H, P). From Lemma 2, we then infer that c^* is color-feasible for (H, P) if and only if every minimum entropy coloring of (H, P) has c^* as color sequence. We conclude that finding a minimum entropy coloring of a probabilistic interval graph is (strongly) NP-hard. □

We now consider the approximability of the minimum entropy coloring problem.

Theorem 2. *Let c be a real such that $0 < c \leq 1$ and assume that there exists a polynomial time algorithm that finds a coloring ϕ of a graph G such that the entropy of ϕ with respect to the uniform distribution U on the vertices of G is at most $H_\chi(G, U) + (c - \epsilon) \log n$ for some positive real ϵ. Then there exists a polynomial time algorithm coloring G with at most $n^{c-\epsilon/2}\chi(G)$ colors.*

Proof. Suppose that A is an algorithm filling the assumptions of the claim. Without loss of generality, we can assume $\epsilon < c$. Let $n = |G|$ be the order of G and $\chi = \chi(G)$ be the chromatic number of G. We claim that some color class in the coloring ϕ found by the algorithm contains at least $n^{1-c+\epsilon}/\chi$ vertices. In order to show this, list the color classes of ϕ in nonincreasing cardinalities as S_1, S_2, ..., S_k. Letting $H(\phi)$ denote the entropy of ϕ with respect to the uniform distribution U, we have

$$-\log \frac{|S_1|}{n} \leq H(\phi) \leq H_\chi(G, U) + (c - \epsilon) \log n \leq \log \chi + (c - \epsilon) \log n.$$

The first inequality follows from the relation $\sum_i -P(i) \log P(i) \geq -\log P_{max}$ for a probability distribution P whose maximum is P_{max}. The middle one holds by hypothesis and the last one comes from the fact that the entropy of a minimum cardinality coloring is at most $\log \chi$. Hence the size of S_1 is at least $n^{1-c+\epsilon}/\chi$, so our claim holds.

Let A' denote the polynomial time algorithm that uses A as a subroutine to find in any graph G with n vertices and chromatic number χ a stable set of size at least $n^{1-c+\epsilon}/\chi$. Now we iteratively use A' to color any graph by coloring with the same color all the vertices in the stable set output by A' and removing these vertices from the graph. Let $G_0 = G$, $G_1 = G_0 - A'(G_0)$, $G_2 = G_1 - A'(G_1)$, ..., $G_\ell = G_{\ell-1} - A'(G_{\ell-1})$ be the sequence of graphs considered, and let $t = (\chi n^{c-\epsilon})/(\chi n^{c-\epsilon} - 1)$. For each i between 1 and ℓ, we have

$$|G_i| \leq |G_{i-1}| - \frac{|G_{i-1}|^{1-c+\epsilon}}{\chi(G_{i-1})} \leq |G_{i-1}| - \frac{|G_{i-1}|}{\chi n^{c-\epsilon}} = \frac{|G_{i-1}|}{t}.$$

It follows that $|G_i| \leq n/t^i$ for all i. Because G_ℓ is nonempty, we have $n/t^\ell \geq 1$ and hence $\ell \leq \log_t n = \ln n / \ln t$. The number of colors in the resulting coloring of G equals $\ell + 1$. By what precedes, we have

$$\ell + 1 \leq \frac{\ln n}{\ln t} + 1 \leq \frac{\ln n}{(t-1) - \frac{(t-1)^2}{2}} + 1 = (n^{c-\epsilon}\chi - 1)\frac{\ln n}{1 - \frac{1}{2(\chi n^{c-\epsilon} - 1)}} + 1.$$

In the second inequality we used that $\ln(x+1) \geq x - \frac{x^2}{2}$ for $x \geq 0$. Because the case $\chi = 1$ is trivial, we can assume that $\chi \geq 2$. It follows that

$$\ell + 1 \leq 2(n^{c-\epsilon}\chi - 1)\ln n + 1 = 2n^{c-\epsilon}\ln n \cdot \chi - 2\ln n + 1.$$

If n is large enough, that is, greater or equal to some constant depending on ϵ, we find $\ell + 1 \leq n^{c-\epsilon/2}\chi$. Consequently, we can in polynomial time find a coloring of a graph G with at most $n^{c-\epsilon/2}\chi$ colors. (Indeed, if n is small we use a brute force algorithm to color the graph exactly and we can easily detect if $\chi = 1$ in polynomial time.) □

It is known that the existence of a polynomial time algorithm coloring any graph G with at most $n^{1-\epsilon}\chi(G)$ colors for some positive real ϵ implies ZPP = NP [9]. Moreover, if the number of colors used by such an algorithm is bounded by $n^{1/7-\epsilon}\chi(G)$ then it implies P = NP [4]. Combining these results with Theorem 2 leads to the following corollaries.

Corollary 1. *Let ϵ be any positive real. There is no polynomial time algorithm that finds a coloring ϕ of a graph G such that the entropy of ϕ with respect to the uniform distribution U on the vertices of G is at most $H_\chi(G, U) + (1 - \epsilon)\log n$, unless ZPP = NP.*

Corollary 2. *Let ϵ be any positive real. There is no polynomial time algorithm that finds a coloring ϕ of a graph G such that the entropy of ϕ with respect to the uniform distribution U on the vertices of G is at most $H_\chi(G, U) + (1/7 - \epsilon)\log n$, unless P = NP.*

We end this section by identifying an easy polynomial case for the minimum entropy coloring problem. (Proof omitted due to lack of space.)

Theorem 3. *There exists a polynomial algorithm for the minimum entropy coloring problem restricted to graphs G satisfying $\alpha(G) \leq 2$.*

4 Bounds and Körner Entropy

We first give a definition of a previously known quantity that is often referred to as graph entropy. Following [3] and to avoid ambiguities, we call it Körner entropy.

The *Körner entropy* $H_\kappa(G, P)$ of a probabilistic graph (G, P) is defined by $H_\kappa(G, P) = \min_{a \in \text{STAB}(G), a > 0} - \sum_{x \in V(G)} P(x) \log a_x$, where STAB$(G)$ is the

stable set polytope of G, defined in $\mathbb{R}^{V(G)}$ as the convex hull of the characteristic vectors of the stable sets of G. The Körner entropy has a number of applications, the most prominent of which being the problem of sorting with partial information studied by Kahn and Kim in their celebrated paper [16].

We also let $\alpha(G, P)$ be the maximum weight $P(S)$ of a stable set S of (G, P).

Lemma 4. *For any probabilistic graph (G, P), we have*

$$-\log \alpha(G, P) \leq H_\kappa(G, P) \leq H_\chi(G, P) \leq \log \chi(G).$$

Proof. The last inequality comes from the fact that in the worst case, the distribution of the colors in a minimum entropy coloring is uniform, hence its entropy is at most $\log \chi(G)$. The second inequality is proved in [3]. This can be done by remarking that the optimization problem defining the Körner entropy is a relaxation of the minimum entropy coloring problem.

The first inequality is derived as follows. Let $a \in \mathrm{STAB}(G)$. A stable set has weight at most $\alpha(G, P)$, so we have $\sum_{x \in V(G)} P(x) a_x \leq \alpha(G, P)$. Combining this with the concavity of $x \mapsto \log(x)$ yields

$$-\sum_{x \in V(G)} P(x) \log a_x \geq -\log \sum_{x \in V(G)} P(x) a_x \geq -\log \alpha(G, P).$$

□

Although Lemma 4 holds for any probabilistic graph, the bounds on the chromatic entropy can be computed in polynomial time only for certain classes of probabilistic graphs. In particular, when G is a perfect graph, the two lower bounds can be computed (to any fixed accuracy) in polynomial time [12]. The chromatic number can also be computed in polynomial time on these graphs.

The question of the quality of the Körner entropy as lower bound for the chromatic entropy was raised by Alon and Orlistky [3]. We clarify this point in the following lemmas. (Proof omitted due to lack of space.)

Lemma 5. *Let (G, P) be a probabilistic graph. Then the ratio $H_\chi(G, P)/H_\kappa(G, P)$ can be arbitrarily large.*

We known that it makes sense to look for an approximation of $H_\chi(G, P)$ within an additive term. We thus consider this case in the next lemma.

Lemma 6. *Let (G, P) be a probabilistic graph. Then the quantity $H_\chi(G, P) - H_\kappa(G, P)$ can be arbitrarily large, even if G is chordal and P is the uniform distribution.*

In order to prove this result, we define a graph $G_k(n)$ $(n \geq 2, k \geq 1)$ inductively on k. The graph $G_1(n)$ is the single vertex graph K_1, and for $k \geq 2$ the graph $G_k(n)$ is obtained as follows:

- start with the complete graph $K_{n^{k-1}}$ on n^{k-1} vertices,
- partition its vertex set $V(K_{n^{k-1}})$ in n sets $V_1, V_2, \ldots, V_n$ of equal sizes,

– for each set V_i ($1 \leq i \leq k$) add a disjoint copy G^i of $G_{k-1}(n)$ and make it completely adjacent to vertices of V_i (in other words, all possible edges between vertices of G^i and vertices in V_i are added).

It can easily be checked that $G_k(n)$ is chordal. We study in the next two lemmas the behavior of $H_\kappa(G_k(n), U)$ and $H_\chi(G_k(n), U)$ when k is fixed and n goes to infinity. (Proofs are omitted due to lack of space.)

Lemma 7. $H_\kappa(G_k(n), U) \leq \frac{(k-1)}{2} \log n + o(1)$.

Lemma 8. $H_\chi(G_k(n), U) \geq \log k + \frac{(k-1)}{2} \log n - o(1)$.

Lemma 6 follows from Lemma 7 and Lemma 8.

5 Number of Colors

We consider in this section the number of colors used in a minimum entropy coloring. We denote by $\chi_H(G, P)$ the minimum number of colors in a minimum entropy coloring of the probabilistic graph (G, P). We assume here that we have $P(x) > 0$ for all vertices x.

Brooks [5] showed a classical result stating that if G is a connected graph different from a complete graph or an odd cycle, then $\chi(G) \leq \Delta(G)$, where $\Delta(G)$ is the maximum degree of a vertex in G. In the next theorem we prove that this statement is also true for $\chi_H(G, P)$ when P is the uniform distribution. (The proof is omitted due to lack of space.)

Theorem 4. *If G is a connected graph different from a complete graph or an odd cycle, then $\chi_H(G, U) \leq \Delta(G)$, where U is the uniform distribution over $V(G)$.*

Acknowledgements

We thank Stefan Langerman and Gilles Van Assche for their valuable comments on an earlier version of this paper.

References

1. M. Accame, F.G.B. De Natale, and F. Granelli. Efficient labeling procedures for image partition encoding. *Signal Processing*, 80(6):1127–1131, June 2000.
2. S. Agarwal and S. Belongie. On the non-optimality of four color coding of image partitions. In *Proc. IEEE Int. Conf. Image Processing*, 2002.
3. N. Alon and A. Orlitsky. Source coding and graph entropies. *IEEE Trans. Inform. Theory*, 42(5):1329–1339, September 1996.
4. M. Bellare, O. Goldreich, and M. Sudan. Free bits, PCPs and non-approximability—towards tight results. In *36th Annual Symposium on Foundations of Computer Science (Milwaukee, WI, 1995)*, pages 422–431. IEEE Comput. Soc. Press, Los Alamitos, CA, 1995.

5. R. L. Brooks. On colouring the nodes of a network. *Proc. Cambridge Philos. Soc.*, 37:194–197, 1941.
6. J. Cardinal, S. Fiorini, and G. Van Assche. On minimum entropy graph colorings. In *Proc. IEEE Int. Symposium on Information Theory*, page 43, 2004.
7. D. de Werra, F. Glover, and E. A. Silver. A chromatic scheduling model with costs. *IIE Trans.*, 27:181–189, 1995.
8. D. de Werra, A. Hertz, D. Kobler, and N. V. R. Mahadev. Feasible edge colorings of trees with cardinality constraints. *Discrete Math.*, 222(1-3):61–72, 2000.
9. U. Feige and J. Kilian. Zero knowledge and the chromatic number. *J. Comput. Syst. Sci.*, 57(2):187–199, 1998.
10. M. R. Garey, D. S. Johnson, G. L. Miller, and C. H. Papadimitriou. The complexity of coloring circular arcs and chords. *SIAM J. Algebraic Discrete Methods*, 1(2):216–227, 1980.
11. M. C. Golumbic. *Algorithmic graph theory and perfect graphs.* Academic Press, 2004.
12. M. Grötschel, L. Lovász, and A. Schrijver. *Geometric algorithms and combinatorial optimization*, volume 2 of *Algorithms and Combinatorics.* Springer-Verlag, Berlin, second edition, 1993.
13. E. Halperin and R. M. Karp. The minimum-entropy set cover problem. In *Automata, languages and programming*, volume 3142 of *Lecture Notes in Comput. Sci.*, pages 733–744. Springer, Berlin, 2004.
14. K. Jansen. Approximation results for the optimum cost chromatic partition problem. *Journal of Algorithms*, 34:54–89, 2000.
15. T.R. Jensen and B. Toft. *Graph Coloring Problems.* Wiley Interscience, 1995.
16. J. Kahn and J.H. Kim. Entropy and sorting. *J. Comput. Syst. Sci.*, 51(3):390–399, 1995.
17. D. Marx. A short proof of the NP-completeness of minimum sum interval coloring. *Oper. Res. Lett.*, 33(4):382–384, 2005.
18. G. Simonyi. Perfect graphs and graph entropy. An updated survey. In *Perfect graphs*, Wiley-Intersci. Ser. Discrete Math. Optim., pages 293–328. Wiley, Chichester, 2001.
19. A. Tucker. An efficient test for circular-arc graphs. *SIAM J. Comput.*, 9(1):1–24, 1980.
20. H. S. Witsenhausen. The zero-error side information problem and chromatic numbers. *IEEE Trans. Inform. Theory*, 22(5):592–593, 1976.
21. Q. Zhao and M. Effros. Low complexity code design for lossless and near-lossless side information source codes. In *Proc. IEEE Data Compression Conf.*, 2003.

Algorithms for Max Hamming Exact Satisfiability

Vilhelm Dahllöf*

Dept. of Computer and Information Science,
Linköping University,
SE-581 83 Linköping, Sweden
vilda@ida.liu.se

Abstract. We here study MAX HAMMING XSAT, i.e., the problem of finding two XSAT models at maximum Hamming distance. By using a recent XSAT solver as an auxiliary function, an $O(2^n)$ time algorithm can be constructed, where n is the number of variables. This upper time bound can be further improved to $O(1.8348^n)$ by introducing a new kind of branching, more directly suited for finding models at maximum Hamming distance. The techniques presented here are likely to be of practical use as well as of theoretical value, proving that there are non-trivial algorithms for maximum Hamming distance problems.

1 Introduction

Most previous algorithms for optimization problems have contented themselves with producing *one* best or good-enough solution. However, often there is an actual need for *several solutions* that are at a maximum (or at least great) Hamming distance. For instance, when scheduling a group of people one typically wants to present substantially different alternatives to choose between. Somewhat surprisingly, the MAX HAMMING CSP problem has only recently become an area of academic research. The first paper (to the best of our knowledge) by Rossi *et al.* [3] came in 2002. In their paper they present some results on the hardness of approximating the problem for CSPs on Boolean domains. Angelsmark and Thapper [1] have presented exact and randomized algorithms for the general finite domain problem as well as dedicated algorithms for MAX HAMMING SAT. Hebrard *et al.* [7] consider a broader range of problems, including finding solutions that are similar. They also test some heuristic methods. The so far best exact algorithm for MAX HAMMING SAT by Angelsmark and Thapper [1] runs in $O(4^n)$ time (where n is the number of variables) and polynomial space.

In this paper we will consider MAX HAMMING XSAT. The XSAT problem asks for an assignment to the variables such that exactly one literal be true in each clause. XSAT is NP-complete as shown by Schaefer [12]. The problem is well studied, and many exact algorithms have been presented, e.g. [6,8,11,5,2]. The so far best algorithm by Byskov *et al.* [2] have a running time in $O(1.1749^n)$

* The research is supported by CUGS – National Graduate School in Computer Science, Sweden.

X. Deng and D. Du (Eds.): ISAAC 2005, LNCS 3827, pp. 829–838, 2005.

and uses polynomial space. XSAT can be used to model for instance the graph colourability problem (since every vertex must have exactly one colour, for an example see [10]). Furthermore, there is a close connection between XSAT and more general cardinality constraints (see [4]). MAX HAMMING XSAT is not efficiently approximable (see [3]) and so exact algorithms are of real-world interest.

We will present two polynomial space algorithms P and Q. Previous algorithms for maximum Hamming problems have relied on an external solver for the base problem. P is also such an algorithm, however, there is a novelty: by using a polynomial time test, many unnecessary calls to the solver can be avoided. Thereby the running time is improved substantially. Q represents something totally new in this area, because it works directly on the inherent structure of the MAX HAMMING XSAT problem. More precisely, a new kind of DPLL branching is introduced. Apart from the immediate interest of the MAX HAMMING XSAT problem itself, we hope that the ideas presented here will also be applicable for other problems such as MAX HAMMING SCHEDULING, MAX HAMMING CLIQUE and the like.

For the sake of conciseness, we phrase the algorithms in such a way that they answer the question "what is the maximum Hamming distance between any two models?". However, it is trivial to see how they can be modified to actually produce two such models.

In what follows we first give some preliminaries and then in Section 3 we present P and Q. In Section 4 some conclusions and possible future research directions are given. At http://arxiv.org/abs/cs.DS/0509038 an extended version of this paper can be found.

2 Preliminaries

A *propositional variable* (or *variable* for short) has either the value *true* or *false*. A *literal* is a variable p or its negation $\bar{p}$. We say that the literals p and $\bar{p}$ are *derived* from the variable p. When *flipping* p ($\bar{p}$) one gets $\bar{p}$ (p). The literal p is *true* iff it is derived from the variable p which has the value *true* and $\bar{p}$ is *true* iff it is derived from the variable p which has the value *false*. A *clause* is a number of literals connected by logical or ($\vee$). The *length* of a clause x, denoted $|x|$, is the number of literals in it. We will sometimes need a sub-clause notation in this way: $(a \vee b \vee C)$, such that $C = c_0 \vee \ldots \vee c_n$ is a disjunction of one or more literals. In the following, literals will be indicated by lower-case letters and sub-clauses by upper-case letters. A *formula* is a set of clauses. For a formula F, $Var(F)$ denotes the set of variables appearing in a clause of F. The *degree of* c, denoted $\delta(c)$, is the number of appearances of the variable c, that is, the number of clauses that contain either c or $\bar{c}$. If $\delta(c) = 1$ we call c a *singleton*. From a formula one gets the *formula graph* by letting the variables form the vertices and every pair of variables occuring together in a clause is joined by an edge. Hence, graph concepts such as "connected components" can be used for formulae.

An *x-model* is an assignment to the variables of a formula F such that there is exactly one true literal in every clause. The problem of determining whether

F allows an x-model is called XSAT. A literal that exactly satisfies a clause is called a *satisfactor*.

We now reach two central definitions: The *Hamming distance* between two assignments is the number of assignments to the individual variables that disagree. MAX HAMMING XSAT is the problem of determining for a formula F the maximum Hamming distance between any two x-models of F.

Substitution of a by δ in the formula F is denoted $F(a/\delta)$; the notation $F(a/\delta; b/\gamma)$ indicates repeated substitution: $F(a/\delta)(b/\gamma)$ (first a is replaced and then b). $F(B/false)$ means that every literal of B is replaced by $false$.

We will deal with variants of the XSAT problem, and in order not to clutter the algorithms with trivialities, we shall assume that the substitution performs a little more than just a syntactical replacement, namely propagation in the following sense: Given a formula F, assume that there are three clauses $x = (a \vee b \vee c), y = (b \vee f \vee g \vee h)$ and $z = (\bar{c} \vee d \vee e)$ in a formula. When substituting true for a ($F(a/true)$), b and c must both be replaced by false (because in the context of XSAT exactly one literal must be true in each clause). This means that y will become $(false \vee f \vee g \vee h)$ which can be simplified to $(f \vee g \vee h)$ and that z will become $(true \vee d \vee e)$ which implies that d and e are false, and so on. Other trivial simplifications are also made. For instance, the occurrence of both a and $\bar{a}$ in a clause is replaced by true. This process goes on until no more simplifications can be done. If the substitution discovers that the formula is x-unsatisfiable (for instance if there is a clause $(true \vee a \vee \bar{a})$) the unsatisfiable formula $\{()\}$ is returned.

When analyzing the running time of the algorithms, we will encounter recurrences of the form $T(n) \le \sum_{i=1}^{k} T(n - r_i) + \text{poly}(n)$. They satisfy $T(n) \in O(\tau(r_1, \ldots, r_k)^n)$ where $\tau(r_1, \ldots, r_k)$ is the largest, real-valued root of the function

$$f(x) = 1 - \sum_{i=1}^{k} x^{-r_i} \tag{1}$$

see [9]. Since this bound does not depend on the polynomial factor poly(n), we ignore all polynomial-time calculations. Let $R = \sum_{i=1}^{k} r_i$ and then note that due to the nature of the function $f(x) = 1 - \sum_{i=1}^{k} x^{-r_i}$, the smallest possible real-valued root (and hence the best running time) will appear when each r_i is as close to R/k as possible, i.e., when the decrease of size of the instance is balanced through the branches. Say for instance that $R = 4, k = 2$. Then $\tau(1,3) = \tau(3,1) \approx 1.4656$ and $\tau(2,2) \approx 1.4142$. We will refer to this as *the balanced branching effect.* We will use the shorthand notation $\tau(r^k \ldots)$ for $\tau(\underbrace{r, r \ldots r}_{k}, \ldots)$, e.g., $\tau(5^2, 3^3)$ for $\tau(5, 5, 3, 3, 3)$.

3 Exact Algorithms for MAX HAMMING XSAT

In what follows we present the two poly-space algorithms P and Q for MAX HAMMING XSAT and prove that they run in $O(2^n)$ and $O(1.8348^n)$ time respectively.

Though the running time of P is slightly inferior to the running time of Q, there are good reasons to present both algorithms: P resembles previous algorithms and gives a hint on how they can be improved, and it is easy to implement given an external XSAT solver. Furthermore, if one is content with getting two models that have at least the Hamming distance d, for some constant d, then P will have a provably better upper time bound than Q.

As a convention, when we present a clause $(a \vee \ldots)$, it is intended to cover all dual cases as well, i.e., $(\bar{a} \vee \ldots)$.

3.1 Using an External XSAT Solver

One solution to the MAX HAMMING XSAT problem is this algorithm which bears resemblance to the $O(4^n)$ time MAX HAMMING SAT algorithm by Angelsmark and Thapper [1]. If the formula is x-unsatisfiable $\perp$ is returned. The answer 0 of course indicates that there is only one model.

algorithm $P(F)$
$ans := \perp$
for $k := 0$ to n **do**
  for every subset $X \subseteq Var(F)$ of size k **do**
    Let C be the set of clauses containing any literal derived from X
    Let C' be a copy of C where every literal derived from X is flipped
    if all clauses of C contain either 0 or 2 literals from X **then**
      if $F \cup C'$ is x-satifiable **then** $ans := k$
return ans

Before stating the correctness of P we need an auxiliary lemma. The proof is trivial.

Lemma 1. *Assume that M and M' are x-models for F and that X is the subset of variables assigned different values. Then each clause of F contains either zero or two literals derived from X.*

Theorem 1. *$P(F)$ decides* MAX HAMMING XSAT *for F*

Proof. For completeness: Assume there are two models M and M' at maximum hamming distance k and that the differing variables are collected in X. The clauses containing zero literals from X remain the same under both models, the interesting case is a clause $(a \vee b \vee C)$ where a and b are from X (by Lemma 1 this is the only possible case). Assume w.l.o.g. that a is true and b is false under M and the opposite holds for M'. Then the clause $(\bar{a} \vee \bar{b} \vee C)$ is x-satisfied under both models.

For soundness: Assume we have a model M for $F \cup C'$. Then it is possible to form another model M' by assigning all variables of X the opposite values.

We can now start examining the running time of P. Let an *allowed subset* S of variables in a formula F be a subset such that each clause of F contains either 0 or two members of S. The following lemma establishes an upper bound for the number of allowed subsets. The proof is trivial.

Lemma 2. *For any formula F the number N of allowed subsets is in $O(7^{n/4}) \subseteq O(1.6266^n)$.*

Theorem 2. *$P(F)$ runs in polynomial space and time $O(2^n)$.*

Proof. Clearly P uses polynomial space. Furthermore, the running time is $O(2^n + N \cdot C^n)$, where N is the constant of Lemma 2 and C is a constant such that XSAT is solvable in polynomial space and time $O(C^n)$. The currently best value for C is 1.1749 (by Byskov *et al.*, [2]) and so the upper time bound is $O(2^n + 1.6266^n 1.1749^n) \subseteq O(2^n + 1.9111^n) = O(2^n)$.

3.2 Using Branching

We will now move on to another poly-space algorithm Q with a provably better running time than P. It is a DPLL-style algorithm relying on the fact that under two models M and M', any variable a has either the same or opposite value. If a is true under both models, then all variables occuring in a clause $w = (a \vee \ldots)$ can be removed (because only one literal is true). If a is false under both models it can be removed. If a has different values then by Lemma 1 there is exactly one more variable a' in w that has different values and we need to examine all possible cases of a'. During the branching some simplifications of the formula are made, for instance, superfluous singletons are removed. We need to store information about removals of variables due to simplifications and therefore the following is needed: To every variable a we associate two (possibly empty) sets of variables: $sing(a)$ and $dual(a)$. We also need a marker $sat(a)$. As a consequence of the simplifications, in the leaves of the recursion tree a kind of generalized models are found, that summarize several models. For now, we hide the details in the helper algorithm Gen_H which we will come back to after the presentation of the main algorithm. The reason for doing so, is that we first need to see how the simplifications work.

Another technicality: like P, Q may return $\perp$ if F is unsatisfiable. Therefore we define $\perp < 0$ and $\perp + 1 = \perp$; furthermore, $\max_{\perp}(\perp, Z)$ returns Z, even if $Z = \perp$. Before $Q'(F)$ is used, all sets $dual(a)$ and $sing(a)$ are assumed to be empty, and every marker $sat(a)$ assumed to be unassigned. During the execution of Q', if there is a clause $(a \ldots)$ where a is a singleton assigned a satisfactor, then $sat(a) := true$, in the dual case where the clause looks like $(\bar{a} \ldots)$, $sat(a) := false$. This allows us to find out the role of a in a model.

For clarity of presentation we will first present a simplified algorithm Q'. Later an optimization to improve the running time will be added.

algorithm $Q'(F)$
As long as there is a clause $(a_1 \vee a_2 \ldots)$ where a_1 and a_2 are singletons, remove a_2 and let $sing(a_1) := sing(a_1) \cup \{a_2\} \cup sing(a_2)$.
As long as there is a clause $(a \vee b)$, assume w.l.o.g. that b is a non-singleton (otherwise pick a) and let $F := F(a/\bar{b})$ and let $dual(b) := dual(b) \cup dual(a) \cup \{a\}$. If a singleton was created, goto the previous line.
if $F = \{()\}$ **then return** $\perp$
elsif $F = \{\}$ **then return** $Gen_H(F)$

elsif F is not connected **then** assume the components are $F_1 \dots F_k$ and return $\sum_{i=1}^{k} Q'(F_i)$
else
  Pick a longest clause $w = (a_1 \vee a_2 \dots a_k)$ and assume w.l.o.g. that a_1 is a non-singleton. Now do the following:
  $ans_{true} := Q'(F(a_1/true))$
  $ans_{false} = Q'(F(a_1/false))$
  if $ans_{true} = \bot$ **or** $ans_{false} = \bot$ **then return** $\max_{\bot}(ans_{true}, ans_{false})$
  else
    for $i = 2$ to k **do**
      Let $ans_i := Q'(F(a_1/\bar{a}_i))$
    return $\max_{\bot}(ans_{true}, ans_{false}, (ans_2 + 1) \dots, (ans_k + 1))$

We are now ready to take a closer look at how the result of the simplifications are handled by Gen_H. Note that during the execution of Q', every removed variable is kept in exactly one set $sing(a)$ or $dual(a)$, for (possibly) different variables a. This motivates the following definition:

A *generalized assignment* is a partial assignment, such that every unassigned variable is contained in exactly one set $sing(a)$ or $dual(a)$ (i.e., for all the sets $sing(a_1), dual(a_1), sing(a_2) \dots$, every intersection is empty). We say that a variable a' is *transitively linked* to the variable a if either 1) $a' \in sing(a) \cup dual(a)$ or 2) a is transitively linked to a member of $sing(a) \cup dual(a)$.

We will also need the two following auxiliary algorithms. Intuitively, $Fix(a_1)$ corresponds to the maximum number of variables transitively linked to a_1 that can have different values under a model M where a_1 is a satisfactor and a model M' where a_1 is not a satisfactor. The recursive algorithm $di(a_1)$ calculates the maximum number of variables, transitively linked to a_1 that can be assigned different values while a_1 is a non-satisfactor. In the recursive calls, it might be that the argument is a satisfactor. The variable k is assumed to be initialized to 0.

algorithm $Fix(a_1)$
$fix := 0$
if $sing(a) \neq \varnothing$ **then**
  Let $\{a_1, a_2 \dots a_m\} := \{a_1\} \cup sing(a_1)$
  $sing(a_1) := \varnothing$
  $fix := \max(Fix(a_1), Fix(a_2) \dots Fix(a_m))$
elsif $dual(a) \neq \varnothing$ **then**
  Let $\{a_1, a_2 \dots a_m\} := \{a_1\} \cup dual(a_1)$
  $dual(a_1) := \varnothing$
  $fix := \sum(Fix(a_1), Fix(a_2) \dots Fix(a_m))$
else
  $fix := 1$
return fix

algorithm $di(a_1)$
if $sing(a_1) \neq \varnothing$ **and** a_1 is a satisfactor **then**
  $k := k + Fix(a_1)$
else
  for each member $b_i \in dual(a_1) \cup sing(a_1)$ **do**
    assign b_i a value according to a_1; $k := k + di(b_i)$
return k

We are now ready to present $Gen_H(F)$. Although F is an empty formula, it is assumed that from it, every variable assigned a value during the execution of Q' can be reached.

algorithm $Gen_H(F)$
$k := 0$
for every variable a_1 assigned a value **do**
  if $sing(a_1) = \varnothing$ and $dual(a_1) = \varnothing$ **then** do nothing
  elsif $sing(a_1) = \{a_2 \ldots a_m\}$ and a_1 is a satisfactor **then**
    Pick two members a' and a'' from $\{a_2 \ldots a_m\}$, there are $\binom{m}{2}$ choices. Try all and for each choice calculate $k_1 := Fix(a') + Fix(a'') + \sum di(a_i)$ such that $a_i \in \{a_1 \ldots a_m\} \setminus \{a' \cup a''\}$. The maximum k_1 found is added to k.
  else
    $k := k + di(a_1)$
return k

We are now ready to state the correctness of Q':

Theorem 3. *$Q'(F)$ decides* MAX HAMMING XSAT *for F*

Proof. We inspect the lines of Q':
Lines 2–5: Let us start by looking at Lines 2 and 3 to see that they do not alter the x-satisfiability of F and that they indeed produce a generalized assignment. As for Line 2, it is clear that removing all singletons but one does not alter the x-satisfiability. It is also clear that every removed singleton will be in one and only one set *sing*. Concerning Line 3, the clause $(a \vee b)$ implies that a and b have opposite values, hence $F := F(a/\bar{b})$ does not alter the x-satisfiability. By the previous line, one of a and b is a non-singleton and so every variable removed by this line is found in exactly one set *dual*. The formula $\{()\}$ is unsatisfiable and thus $\perp$ is returned. When it comes to Gen_H, we need to justify the calculation of the maximum Hamming distance for a generalized assignment. Consider two models M and M' at maximum Hamming distance, contained in the generalized assignment at hand. Clearly, all variables that are assigned a fixed value and have empty sets *sing* and *dual* will have the same value under both models. Furthermore, whenever there is a situation with a satisfactor a_1 having a non-empty set $sing(a_1)$, one of $a_1 \ldots a_m$ will be a satisfactor under M and one under M'. When a_i is a satisfactor under M, $Fix(a_1)$ is the largest number of variables

transitively linked to it that can get assigned one value under M and another value under M'. Also, even though a variable is not a satisfactor itself, it may well be that one variable transitively linked to it is. As in a general assignment every variable is either assigned a fixed value or transitively linked to such a variable, the distance between M and M' can be found as the sum of the values calculated for the assigned variables.

Line 6: If F is not connected every model for one component can be combined with any model for another component in order to form a model for F.

Lines 7–15: Assume there are two models M and M' at maximum Hamming distance k. If a_1 is true under both models then the formula where all other literals of w are set to false is x-satisfiable and the recursive call will return k (assuming that the algorithm is correct for smaller input). Similarly for Line 10. If both Lines 9 and 10 returned an integer, we know that there are models under which a_1 is false and models under which a_1 is true. Thus M and M' may assign different values to a_1. Assume this is the case. By Lemma 1 we know that M and M' differ in exactly one more variable in w. Assume w.l.o.g. that a_2 is that literal. Then we know that a_1 and a_2 have different values and that the other literals of w are false.

As for the running time of Q', the handling of clauses of length 4 will cause an unnecessarily bad upper time bound. The problem is that in Line 10 only one variable is removed. However, a clause of length 3 is created which can be exploited. Hence we replace Line 10 in Q' by the following, thereby obtaining the algorithm Q. The correctness is easily seen, because it is the same kind of branching we have already justified.

1 **if** $|W| \neq 4$ **then** $ans_{false} = Q(F(a_1/false))$
2 **else**
3 let $W = (a_1 \vee a_2 \vee a_3 \vee a_4)$ and assume that a_2 is a non-singleton
4 $ans^1_f := Q(F(a_1/false; a_2/true)$; $ans^2_f := Q(F(a_1/false; a_2/false)$
5 **if** $ans^1_f = \perp$ **or** $ans^2_f = \perp$ **then** $ans_{false} := \max_\perp(ans^1_f, ans^2_f)$
6 **else**
7 $ans^3_f := Q(F(a_1/false; a_2/\bar{a}_3)$; $ans^4_f := Q(F(a_1/false; a_2/\bar{a}_4)$
8 $ans_{false} := \max_\perp(ans^1_f, ans^2_f, (ans^3_f + 1), (ans^4_f + 1))$

Theorem 4. *$Q(F)$ runs in polynomial space and time $O(1.8348^n)$*

Proof. Let $T(n)$ be the running time for $Q(F)$. The analysis will proceed by examining what the running time would be if Q always encountered the same case. It is clear that the worst case will decide an overall upper time bound for Q. We inspect the lines of Q:

Line 1–5: All these lines are polynomial time computable.

Line 6: This line does not increase the running time as clearly, $\sum_{i=1}^{k} T(n_i) \leq T(n)$ when $n = \sum_{i=1}^{k} n_i$.

Lines 7–: It is clear that the worst clause length will decide an overall upper time bound for Q. Note that if there are variables left in F, then there will be at least two clauses left and one of the cases below must be applicable.

1. $|w| \geq 5$. Already a rough analysis suffices here: In the call $Q(F(a_1/true))$ a_1 as well as all the other variables in w get a fixed value and hence $|w|$ variables are removed. The next call only removes one variable, namely a_1. In every of the other $|w|-1$ calls $|w|-1$ variables are removed. Hence, the running time will be in $O(\tau(|w|, 1, (|w|-1)^{|w|-1})^n)$ and the worst case is $O(\tau(5, 1, 4^4)^n) \subseteq O(1.7921^n)$.
2. $|w| = 4$ For a better readability, assume $w = (a \vee b \vee c \vee d)$. As a and b are not singletons there are clauses $a \in y$ and $b \in z$. There are several possibilities for y and z, but due to the balanced branching effect, we may disregard cases where $a \in w$ but $\bar{a} \in y$ etc.
   (a) $y = (a \vee e \vee f \vee g)$, $z = (b \vee h \vee i \vee j)$. The call $Q(F(a/true))$ removes 7 variables – all variables of w and y. The call $Q(F(a/false; b/true))$ removes 7 variables – all variables of w and z. The call $Q(F(a/false; b/false))$ removes 3 variables, because the clause $w = (c \vee d)$ will in the next recursive step be simplified. The call $Q(F(a/false; b/\bar{c}))$ removes 3 variables, because the clause $w = (c \vee \bar{c} \vee d)$ implies $d =$ false, which will be effectuated by the substitution operation. The call $Q(F(a/false; b/\bar{d}))$ removes 3 variables for the same reasons. The call $Q(F(a/\bar{b}))$ removes 3 variables – c and d must be false. Similarly for the remaining two calls. Hence, the running time is in $O(\tau(7^2, 3^6)^n) \subseteq O(1.8348^n)$.
   If $|z| = 3$, then regardless of y we get cases better than the above case:
   (b) $z = (b \vee e \vee f)$. Counting removed variables as previously we get that this case runs in time $O(\tau(6, 4^4, 3^3)^n) \subseteq O(1.7605^n)$.
   (c) $z = (a \vee b \vee e)$ or $z = (b \vee c \vee e)$ or $z = (b \vee d \vee e)$. All these cases run in time $O(\tau(5^2, 4^6)^n) \subseteq O(1.6393^n)$.
   If $|y| = 3$, then regardless of y we get cases better than the so far worst:
   (d) $y = (a \vee e \vee f)$. This case runs in time $O(\tau(6, 5, 4^3, 3^3)^n) \subseteq O(1.7888^n)$.
   (e) $y = (a \vee b \vee e)$. Already examined.
   (f) $y = (a \vee c \vee e)$ or $y = (a \vee d \vee e)$. These cases run in time $O(\tau(5^2, 4^5, 3)^n) \subseteq O(1.6749^n)$.
   If y shares more than one variable with w, then regardless of z we get cases better than the so far worst:
   (g) $y = (a \vee b \vee c \vee e)$ or $y = (a \vee c \vee d \vee e)$ or $y = (a \vee b \vee e \vee f)$. These cases run in time $O(\tau(5^2, 4^6)) \subseteq O(1.6393^n)$, $O(\tau(5^4, 4^4)) \subseteq O(1.5971^n)$ and $O(\tau(6^2, 5, 4, 3^4)) \subseteq O(1.7416^n)$, respectively.
   (h) $y = (a \vee c \vee e \vee f)$ or $y = (a \vee d \vee e \vee f)$. These cases run in time $O(\tau(6, 5^2, 4, 3^4))^n) \subseteq O(1.7549^n)$.
   If z shares more than one variable with w, then regardless of y we get cases better than the so far worst:
   (i) $z = (a \vee b \ldots)$. Already examined.
   (j) $z = (b \vee c \vee d \vee e)$. This case runs in $O(\tau(5^2, 4^6)) \subseteq O(1.6393^n)$.

(k) $z = (b \vee c \vee e \vee f)$ or $z = (b \vee d \vee e \vee f)$. These cases run in time $O(\tau(6^2, 5, 4, 3^4))^n) \subseteq O(1.7416^n)$.

3. $|w| = 3$. We know that there is another clause y such that $|y| = 3$ and $a \in y$ and $y \neq w$. Hence we have a running time in $O(\tau(4, 3, 2^2)) \subseteq O(1.7107^n)$.

4 Conclusions

We have presented two non-trivial, exact, poly-space algorithms for MAX HAMMING XSAT and provided interesting upper bounds on their running time. Both algorithms point out new interesting research directions. Using P as a template when constructing an algorithm for a max Hamming problem, the goal is to analyze the instance at hand to see which calls to the external solver that are superfluous. Q indicates that it is possible to take direct advantage of the inherent structure of the problem itself.

References

1. O. Angelsmark and J. Thapper. Algorithms for the maximum hamming distance problem. In *Boi Faltings, Adrian Petcu, François Fages, Francesca Rossi (Eds.): Recent Advances in Constraints, Joint ERCIM/CoLogNet International Workshop on Constraint Solving and Constraint Logic Programming*, LNCS 3419, pages 128–141, 2004.
2. J. M. Byskov, B. A. Madsen, and B. Skjernaa. New algorithms for exact satisfiability. *Theoretical Computer Science*, 332(1–3):515–541, 2005.
3. P. Crescenzi and G. Rossi. On the hamming distance of constraint satisfaction problems. *Theoretical Computer Science*, 288(1):85–100, 2002.
4. V. Dahllöf. Applications of general exact satisfiability in propositional logic modelling. In *Proceedings 11th International Conference on Logic for Programming, Artificial Intelligence and Reasoning (LPAR-2004)*, pages 95–109, 2004.
5. V. Dahllöf, P. Jonsson, and R. Beigel. Algorithms for four variants of the exact satisfiability problem. *Theoretical Computer Science*, 320(2–3):373–394, 2004.
6. L. Drori and D. Peleg. Faster solutions for some NP-hard problems. *Theoretical Computer Science*, 287:473–499, 2002.
7. E. Hebrard, B. Hnich, B. O'Sullivan, and T. Walsh. Finding diverse and similar solutions in constraint programming. In *Proceedings 20th International Conference on AI (AAAI-2005)*, pages 372–377, 2005.
8. E. Hirsch and A. Kulikov. A $2^{n/6.15}$-time algorithm for X3SAT.
9. O. Kullmann. New methods for 3-SAT decision and worst-case analysis. *Theoretical Computer Science*, 223:1–72, 1999.
10. L. Liu and M. Truszczyński. Local-search techniques for propositional logic extended with cardinality constraints. In *Proceedings 9th International Conference on Principles and Practice of Constraint Programming (CP-2003)*, pages 495–509, 2003.
11. S. Porschen, B. Randerath, and E. Speckenmeyer. X3SAT is decidable in time $O(2^{n/5})$. In *Proceedings 5th International Symposium on Theory and Applications of SAT*, pages 231–235, 2002.
12. T. J. Schaefer. The complexity of satisfiability problems. In *Proceedings 10th Annual ACM Symposium on the Theory of Computing (STOC-1978)*, pages 216–226, 1978.

Counting Stable Strategies in Random Evolutionary Games*

Spyros Kontogiannis[1,2] and Paul Spirakis[2]

[1] Computer Science Department, University of Ioannina,
45110 Ioannina, Greece
kontog@cs.uoi.gr

[2] Research Academic Computer Technology Institute, N. Kazantzakis Str.,
University Campus, 26500 Rio-Patra, Greece
{kontog, spirakis}@cti.gr

Abstract. In this paper we study the notion of the Evolutionary Stable Strategies (ESS) in evolutionary games and we demonstrate their qualitative difference from the Nash Equilibria, by showing that a random evolutionary game has on average *exponentially less* number of ESS than the number of Nash Equilibria in the underlying symmetric 2-person game with random payoffs.

1 Introduction

Game theory is the study of *interactive* decision making, in the sense that those involved in the decisions are affected not only by their own choices, but also by the decisions of others. This study is guided by two principles: (1) The choices of players are affected by well-defined (fixed) *preferences* over outcomes of decisions. (2) Players act *strategically*, ie, they take into account the interaction between their choices and the ways other players act. The dominant aspect of game theory is the belief that players are *rational* and this rationality is common knowledge. This common knowledge of rationality gives hope to equilibrium play: Players use their equilibrium strategies because of what would happen if they had not.

The point of departure for **evolutionary game theory** is the view that the players are not always rational. In evolutionary games, "good" strategies *emerge* from a trial-and-error learning process, in which players discover that some strategies perform better than others, or decide to play in a different way in hope of getting a better payoff. The players may do very little reasoning during this process. Instead, they simply take actions by rules of thumb, social norms, analogies for similar situations, or by other (possibly more complex) *methods* for converting stimuli into actions. Thus, in evolutionary games we may say that the players are "programmed" to adopt some strategies. Typically, the evolution process deals with an infinite population of players. As time proceeds, many small games are conducted (eg, among pairs of players that "happen" to meet). One

* This work was partially supported by the 6th Framework Programme under contract 001907 (DELIS).

X. Deng and D. Du (Eds.): ISAAC 2005, LNCS 3827, pp. 839–848, 2005.

then expects that strategies with higher payoffs will spread within the population (by learning, copying successful strategies, or even by infection).

Indeed, evolutionary games in large populations of players create a *dynamic process*, where the frequencies of the strategies played (by population members) *change in time* because of the learning or selection forces guiding the players' strategic choices. Clearly, the rate of changes depends on the current strategy mix in the population. Such dynamics can be described by stochastic or deterministic models. The subject of evolutionary game theory is *exactly* the study of these dynamics. An excellent presentation of the evolutionary game dynamics can be found in [7]. For a more thorough study the reader is referred to [3].

Not surprisingly, evolutionary game processes that converge to stable states have usually the property that those states are also self-confirming equilibria (eg, Nash equilibria). This is one of the most robust results in evolutionary game theory, the "folk theorem", that stability implies Nash equilibrium. In fact, one of the main approaches in the study of evolutionary games is the concept of **Evolutionary Stable Strategies** (ESS), which are nothing more than Nash Equilibria together with an additional stability property. This additional property is interpreted as ensuring that if an ESS is established in a population, and if a small proportion of the population adopts some *mutant* behavior, then the process of selection (or learning) will eliminate the latter. Once an ESS is established in a population, it should therefore be able to withstand the pressures of mutation and selection.

Related Work and Our Contribution. Evolutionary game theory is quite interesting on its own, and Evolutionary Stable Strategies (ESS) are the most popular notion of stability in these games. Consequently, knowing the number of ESS that an evolutionary game may end up in, is a very important issue since it demonstrates the inherent complexity of the game. On the other hand, evolutionary game theory may be seen as a methodology for either computing (approximate) solutions, or estimating the convergence times to such solutions, for hard combinatorial problems.

For example, [5] exploited the popular notion of replicator dynamics of evolutionary games, to prove that for an evolutionary game whose underlying strategic game is a single–commodity selfish routing game, the convergence time to an ε–approximate Nash Equilibrium is polynomial in $1/\varepsilon$ and logarithmic in the maximum-to-optimal latency ratio. A weakened version of this result was also shown for the more general case of evolutionary games for which the underlying strategic game is a multi–commodity network congestion game. Some years earlier ([1]) the notion of replicator dynamics had been used as a heuristic for the **NP**–hard Max Weight Clique problem. Recently ([4]) the problem of deciding the existence of an ESS in an arbitrary evolutionary game has been proved to be both **NP**–hard and **coNP**–hard. On the other hand, the existence of a regular ESS is an **NP**–complete problem.

Concerning the number of ESS in an evolutionary game, to our knowledge, there is not much in the literature. The only interesting work that we could find is [2], which gives some (worst-case) exponential upper and lower bounds on the maxi-

mum number of ESS in evolutionary games whose underlying payoff matrix is symmetric. On the other hand, some very interesting results have appeared concerning the expected number of Nash equilibria in strategic games. In particular, [9] provides a (computationally feasible) formula for the mean number of Nash Equilibria in a random game where all the players have the same strategy set and the entries of their payoff matrices are independent (uniform) random variables. This formula is used in [10] in order to prove that the expected number of Nash equilibria in a random 2-person (strategic) game tends to $\exp(0.281644N + \mathcal{O}(\log N))$ as N (the number of pure strategies of a player) tends to infinity. Although it is trivial to show that each ESS indicates a (symmetric) Nash Equilibrium for the underlying strategic game, it was not known (until the present work) whether there is a qualitative difference between the set of ESS in an evolutionary game and the set of Nash Equilibria of the underlying strategic game. Indeed, [2] gives some evidence that the notion of ESS may be of the same hardness as that of Nash Equilibria by giving exponential (worst-case) bounds on their number. On the other hand, in this work we show that the number of ESS in a random evolutionary game is *exponentially* smaller than the number of Nash Equilibria in the underlying random symmetric 2-person game. We prove this by exploiting a quite interesting (necessary and sufficient) condition for a strategy being an ESS of an evolutionary game, given that the underlying symmetric strategies profile is a Nash equilibrium ([6]). Our approach is based on constructing sufficiently many (independent of each other) certificates for an arbitrary strategy (such that the underlying profile is a symmetric Nash Equilibrium) being an ESS, for which the joint probability of being true is very small.

The paper is organized as follows: In section 2 we give the main definitions and notation of non-cooperative game theory, we introduce the symmetric 2-player games, the evolutionary games and the evolutionary stable strategies. In section 3 we present our result on the number of ESS in a random 2-player game.

2 Preliminaries

Non-cooperative Games and Equilibria. We restrict our view in this paper to the class of *finite games in strategic form.* More precisely, let $I = [n]$[1] be a set of **players**, where n is a positive integer. For each player $i \in I$, let S_i be her (finite) set of allowable actions, called the **action set** of player i. The deterministic choice of a specific action $s_i \in S_i$ by a player $i \in I$, is called a **pure strategy** for this player. Thus the action set of a player can also be seen as her **pure strategy set.** A vector $\mathbf{s} = (s_1, \ldots, s_n) \in \times_{i \in I} S_i$, where $s_i \in S_i$ is the pure strategy adopted by player $i \in I$, is called a **pure strategies profile** or a **configuration** of the players. The space of all the pure strategies profiles in the game is thus the Cartesian product $S = \times_{i \in I} S_i$ of the players' pure strategies sets (usually called the **configuration space**).

For any configuration $\mathbf{s} \in S$ and any player $i \in I$, let $\pi_i(\mathbf{s})$ be a real number indicating the payoff to player i upon the adoption of this configuration by all the players of the game. The finite collection of the real numbers $\{\pi_i(\mathbf{s}) : \mathbf{s} \in S\}$

[1] For any integer $k \in \mathbb{N}$, $[k] \equiv \{1, 2, \ldots, k\}$.

defines player i's pure strategies payoff function. Let $\pi(\mathbf{s}) = (\pi_i(\mathbf{s}))_{i\in I}$ be the vector function of all the players' payoffs. Thus, a game in strategic form is simply described by a triplet $\Gamma = (I, S, \pi)$ where I is the set of players, S is the configuration space of the players, and π is the vector function of all the players' payoffs. Let $\mathcal{P}_k \equiv \{\mathbf{z} \in [0,1]^k : \sum_{i\in[k]} z_i = 1\}$ denote the $(k-1)$-dimensional simplex. For any player $i \in I$, any point $\mathbf{x_i} = (x_{i,1}, x_{i,2}, \ldots, x_{i,m_i}) \in \mathcal{P}_{m_i}$ (where $m_i = |S_i|$) is called a **mixed strategy** for her. A player that adopts the mixed strategy $\mathbf{x_i}$ is assumed to determine her own choice of an action *randomly* (ie, using $\mathbf{x_i}$) and independently of all the other players. $\mathcal{P}_{m_i}$ is the **mixed strategies set** of player i. A **mixed strategies profile** is a vector $\mathbf{x} = (\mathbf{x_1}, \ldots, \mathbf{x_n})$ whose components are themselves mixed strategies of the players, ie, $\forall i \in I,\ \mathbf{x_i} \in \mathcal{P}_{m_i}$. We call the Cartesian product $\Delta = \times_{i\in I}\mathcal{P}_{m_i} \subseteq \mathbb{R}^m$ the **mixed strategies space** of the game $(m = m_1 + \cdots + m_n)$.

When the players adopt a mixed strategies profile $\mathbf{x} \in \Delta$, we can compute what is the *average* payoff, u_i, that player i gets (for $\mathbf{x}$) in the usual way: $u_i(\mathbf{x}) \equiv \sum_{\mathbf{s}\in S} P(\mathbf{x},\mathbf{s}) \cdot \pi_i(\mathbf{s})$, where, $P(\mathbf{x},\mathbf{s}) \equiv \prod_{i\in I} x_i(s_i)$ is the occurrence probability of configuration $\mathbf{s} \in S$ wrt the mixed profile $\mathbf{x} \in \Delta$. This (extended) function $u_i : \Delta \mapsto \mathbb{R}$ is called the **(mixed) payoff function** for player i.

Let us indicate by $(\mathbf{x_i}, \mathbf{y_{-i}})$ a mixed strategies profile where player i adopts the mixed strategy $\mathbf{x_i} \in \mathcal{P}_{m_i}$, and all other players adopt the mixed strategies that are determined by the profile $\mathbf{y} \in \Delta$. This notation is particularly convenient when a single player i considers a unilateral "deviation" $\mathbf{x_i} \in \mathcal{P}_{m_i}$ from a given profile $\mathbf{y} \in \Delta$. A **best response** of player i to a mixed strategies profile $\mathbf{y} \in \Delta$ is any element of the set $B_i(\mathbf{y}) \equiv \arg\max_{\mathbf{x_i}\in\mathcal{P}_{m_i}} \{u_i(\mathbf{x_i}, \mathbf{y_{-i}})\}$. A **Nash Equilibrium** (NE in short) is any profile $\mathbf{y} \in \Delta$ having the property that $\forall i \in I,\ \mathbf{y_i} \in B_i(\mathbf{y})$. The nice thing about NE is that each finite strategic game $\Gamma = (I, S, \pi)$ has at least one NE [11].

The subclass of symmetric 2-player games provides the basic setting for much of the evolutionary game theory. Indeed, many of the important insights can be gained already in this (special) case.

Definition 1. *A finite strategic game $\Gamma = (I, S, \pi)$ is a* **2-player game** *when $I = \{1, 2\}$. It is called a* **symmetric** *2-player game if in addition, $S_1 = S_2$ and $\forall(s_1, s_2) \in S, \pi_1(s_1, s_2) = \pi_2(s_2, s_1)$.*

Note that in the case of a symmetric 2-player strategic game, the payoff functions of Γ can be represented by two $|S_1| \times |S_2|$ real matrices Π_1, Π_2 such that $\Pi_1[s_i, s_j] = \pi_1(s_i, s_j)$, $\forall(s_1, s_2) \in S$ and $\Pi_1 = \Pi_2^T = \Pi$. For any mixed strategies profile $\mathbf{x} = (\mathbf{x_1}, \mathbf{x_2}) \in \Delta$, the expected payoff of player 1 for this profile is $u_1(\mathbf{x}) = \mathbf{x_1}^T \Pi \mathbf{x_2}$ while the payoff of player 2 is $u_2(\mathbf{x}) = \mathbf{x_1}^T \Pi_2 \mathbf{x_2} = \mathbf{x_2}^T \Pi \mathbf{x_1}$. Thus we can fully describe a symmetric 2-player game by (S, Π), where S is the common action set and Π is the payoff matrix of the row player.

Two useful notions, especially in the case of 2-player games, are the support and the extended support. In a 2-player strategic game $\Gamma = (\{1, 2\}, S, \pi)$, the support of a mixed strategy $\mathbf{x_1} \in \mathcal{P}_{m_1}$ ($\mathbf{x_2} \in \mathcal{P}_{m_2}$) is the set of allowable actions of player 1 (player 2) that have *non-zero* probability in $\mathbf{x_1}$ ($\mathbf{x_2}$). Formally, $\forall i \in \{1, 2\}, supp(\mathbf{x_i}) \equiv \{j \in S_i : x_i(j) > 0\}$. The **extended support** of a mixed

strategy $\mathbf{x_2} \in \mathcal{P}_{m_2}$ of player 2 is the set of *pure best responses* of player 1 to $\mathbf{x_2}$. That is, $extsupp(\mathbf{x_2}) \equiv \{j \in S_1 : u_1(j, \mathbf{x_2}) = \max_{\mathbf{x_1} \in \mathcal{P}_{m_1}}\{u_1(\mathbf{x_1}, \mathbf{x_2})\}\}$. Similarly, the **extended support** of a mixed strategy $\mathbf{x_1} \in \mathcal{P}_{m_1}$ of player 1, is the set of *pure best responses* of player 2 to $\mathbf{x_1}$. That is, $extsupp(\mathbf{x_1}) \equiv \{j \in S_2 : u_2(\mathbf{x_1}, j) = \max_{\mathbf{x_2} \in \mathcal{P}_{m_2}}\{u_2(\mathbf{x_1}, \mathbf{x_2})\}\}$. The following lemma is a direct consequence of the definition of NE (a proof is provided in the full version of the paper):

Lemma 1. *If* $(\mathbf{x_1}, \mathbf{x_2}) \in \Delta$ *is a NE of a 2-player strategic game, then* $supp(\mathbf{x_1}) \subseteq extsupp(\mathbf{x_2})$ *and* $supp(\mathbf{x_2}) \subseteq extsupp(\mathbf{x_1})$.

Due to their connection to the notion of ESS, a class of NE of particular interest is that of symmetric Nash equilibria. A strategy pair $(\mathbf{x_1}, \mathbf{x_2}) \in \mathcal{P}_{m_i} \times \mathcal{P}_{m_i}$ for a symmetric 2-player game $\Gamma = (S, \Pi)$ is a **symmetric Nash Equilibrium** (SNE in short), if and only if (a) $(\mathbf{x_1}, \mathbf{x_2})$ is a NE for Γ, and (b) $\mathbf{x_1} = \mathbf{x_2}$. Not all NE of a symmetric 2-player game need be symmetric. However it is known that each finite symmetric 2-player game has at least one SNE [11].

When we wish to argue about the vast majority of symmetric 2-player games, one way is to assume that the real numbers in the set $\{\Pi[i,j] : (i,j) \in S\}$ are independently drawn from a probability distribution F. For example, F can be the uniform distribution in an interval $[a, b]$ for some $a, b \in \mathbb{R} : a < b$. Then, a typical symmetric 2-player game Γ is just an instance of the implied random experiment that is described in the following definition.

Definition 2. *A symmetric 2-player game* $\Gamma = (S, \Pi)$ *is an instance of a (symmetric 2-player)* random game *wrt the probability distribution* F*, if and only if* $\forall i, j \in S$*, the real number* $\Pi[i,j]$ *is an independently and identically distributed random variable drawn from* F*.*

Evolutionary Stable Strategies. From now on we shall only consider symmetric 2-player strategic games. So, fix such a game $\Gamma = (S, \Pi)$, for which the mixed strategies space is $\Delta = \mathcal{P}_n \times \mathcal{P}_n$. Suppose that all the individuals of an infinite population are programmed to play the same (either pure or mixed) *incumbent* strategy $\mathbf{x} \in \mathcal{P}_n$, whenever they are involved in Γ. Suppose also that a small group of *invaders* appears in the population. Let $\varepsilon \in (0,1)$ be the share of invaders in the post–entry population. Assume that all the invaders are programmed to play the (pure or mixed) strategy $\mathbf{y} \in \mathcal{P}_n$ whenever they are involved in Γ. Pairs of individuals in this *dimorphic* post–entry population are now repeatedly drawn at random to play always the same symmetric 2-player game Γ against each other. If an individual is chosen to participate, the probability that her (random) opponent will play strategy $\mathbf{x}$ is $1-\varepsilon$, while that of playing strategy $\mathbf{y}$ is ε. This is equivalent to saying that the opponent is an individual who plays the *mixed* strategy $\mathbf{z} = (1-\varepsilon)\mathbf{x} + \varepsilon\mathbf{y}$. The post–entry payoff to the incumbent strategy $\mathbf{x}$ is then $u(\mathbf{x}, \mathbf{z})$ and that of the invading strategy $\mathbf{y}$ is just $u(\mathbf{y}, \mathbf{z})$ ($u = u_1 = u_2$). Intuitively, evolutionary forces will select *against* the invader if $u(\mathbf{x}, \mathbf{z}) > u(\mathbf{y}, \mathbf{z})$. The most popular notion of stability in evolutionary games is the Evolutionary Stable Strategy (ESS). A strategy $\mathbf{x}$ is **evolutionary stable**

(ESS in short) if for any strategy $\mathbf{y} \neq \mathbf{x}$ there exists a barrier $\bar{\varepsilon} = \bar{\varepsilon}(\mathbf{y}) \in (0,1)$ such that $\forall 0 < \varepsilon \leqslant \bar{\varepsilon}$, $u(\mathbf{x},\mathbf{z}) > u(\mathbf{y},\mathbf{z})$ where $\mathbf{z} = (1-\varepsilon)\mathbf{x} + \varepsilon\mathbf{y}$. An ESS $\mathbf{x}$ is called **regular** if it holds that its support matches its extended support, ie, $supp(\mathbf{x}) = extsupp(\mathbf{x})$. One can easily prove the following characterization of ESS, which sometimes appears as an alternative definition (cf. [7]):

Proposition 1. *Let $\mathbf{x} \in \mathcal{P}_n$ be a (mixed in general) strategy profile that is adopted by the whole population. The following sentences are equivalent:*

(i) $\mathbf{x}$ *is an evolutionary stable strategy.*
(ii) $\mathbf{x}$ *satisfies the following properties,* $\forall \mathbf{y} \in \Delta \setminus \{\mathbf{x}\}$*:*
[P1] $u(\mathbf{y},\mathbf{x}) \leqslant u(\mathbf{x},\mathbf{x})$, and **[P2]** If $u(\mathbf{y},\mathbf{x}) = u(\mathbf{x},\mathbf{x})$ then $u(\mathbf{y},\mathbf{y}) < u(\mathbf{x},\mathbf{y})$.

Observe that the last proposition implies that $(\mathbf{x},\mathbf{x})$ has to be a SNE of the underlying symmetric 2-player strategic game Γ (due to **[P1]**) and $\mathbf{x}$ has to be strictly better than any invading strategy $\mathbf{y} \in \Delta \setminus \{\mathbf{x}\}$, against $\mathbf{y}$ itself, in case that $\mathbf{y}$ is a best-response strategy against $\mathbf{x}$ in Γ (due to **[P2]**).

A mixed strategy $\mathbf{x} \in \Delta$ is **completely mixed** if and only if $supp(\mathbf{x}) = S$. It is well (Haigh 1975, [6]) that if a completely mixed strategy $\mathbf{x} \in \mathcal{P}_n$ is an ESS, then it is the *unique* ESS of the evolutionary game.

3 Mean Number of ESS in Random Evolutionary Games

In this section we study the expected number of ESS in a generic evolutionary game with an action set $S = [n]$ and a payoff matrix which is an $n \times n$ matrix U whose entries are **iid** r.v.s drawn from a probability distribution F. For any non-negative vector $\mathbf{x} \in \mathcal{P}_k$ for some $k \in \mathbb{N}$, let $Y_{\mathbf{x}} \equiv \mathcal{P}_k \setminus \{\mathbf{x}\}$. The following statement, proved by Haigh, is a necessary and sufficient condition for a mixed strategy $\mathbf{s} \in \mathcal{P}_n$ being an ESS, given that $(\mathbf{s},\mathbf{s})$ is a SNE of $\Gamma = (S,U)$.

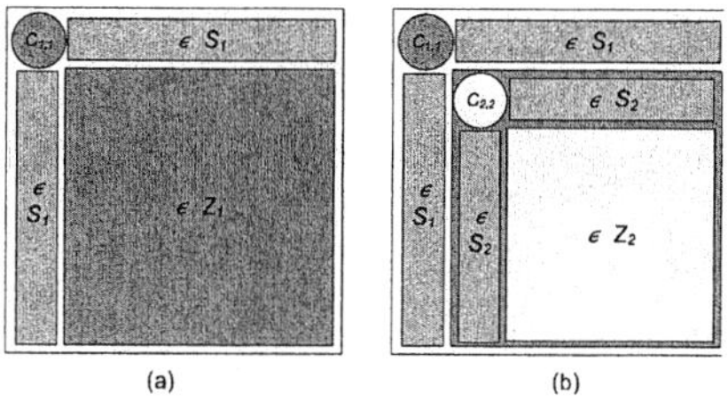

Fig. 1. The partition of C

Lemma 2 (Haigh 1975, [6]). *Let $(\mathbf{s},\mathbf{s}) \in \mathcal{P}_n \times \mathcal{P}_n$ be a SNE for the symmetric game $\Gamma = (S,U)$ and let $M = extsupp(\mathbf{s})$. Let also $\mathbf{x}$ be the projection of $\mathbf{s}$ on M, and C the submatrix of U consisting of the rows and columns indicated by M. Then $\mathbf{s}$ is an ESS if and only if $\forall \mathbf{y} \in Y_{\mathbf{x}}$, $(\mathbf{y}-\mathbf{x})^T C(\mathbf{y}-\mathbf{x}) < 0$.*

The following lemma also holds (a simple consequence of the definition of ESS):

Lemma 3. *Let $\mathbf{s} \in \mathcal{P}_n$ be an ESS of an evolutionary game whose underlying (symmetric) 2-player game is $\Gamma = (S,U)$. Then $(\mathbf{s},\mathbf{s})$ is a SNE for Γ.*

Combining lemmas 2 and 3 we observe that suffices to examine only SNE of $\Gamma = (S,U)$ in order to find out if this game possesses an ESS. In the sequel we focus our attention to random evolutionary games for which the payoff matrix has entries that are independent r.v.s which are symmetric about their mean:

Definition 3. *A random variable X is called* **symmetric about the mean** *if and only if* $\forall a \geqslant 0, \mathbb{P}\{\mu \leqslant X \leqslant \mu + a\} = \mathbb{P}\{\mu - a \leqslant X \leqslant \mu\}$, *where* $\mu = \mathbb{E}\{X\}$.

The following lemma will be useful in our investigation (a proof is in the full version of the paper):

Lemma 4. *Let X, Y be two independent, continuous, symmetric about the mean r.v.s, such that $\mu = \mathbb{E}\{X\} = \mathbb{E}\{Y\}$. Then $\mathbb{P}\{X \geqslant Y\} = \frac{1}{2}$.*

We now show our main theorem:

Theorem 1. *Let $\Gamma = (S, U)$ be an instance of a random symmetric 2-player game whose payoff entries are* **iid** *r.v.s drawn uniformly from $[0, A]$, for some constant $A > 0$. Then $\mathbb{E}\{\#ESS\} = o(\mathbb{E}\{\#SNE\})$.*

Proof. Consider an arbitrary SNE $(\mathbf{s}, \mathbf{s}) \in \mathcal{P}_n \times \mathcal{P}_n$, and assume wlog that $extsupp(\mathbf{s}) = [m]$ for some $1 \leqslant m \leqslant n$ (by reordering the action set S). Assume also that $s_1 \geqslant s_2 \geqslant \cdots \geqslant s_r > 0 = s_{r+1} = \cdots = s_m$ for some $1 \leqslant r \leqslant m$ (ie, its support is $supp(\mathbf{s}) = [r]$). Let $\mathbf{x} = \mathbf{s}|_{[m]} \equiv (s_1, \ldots, s_m) \in \mathcal{P}_m$ be the projection of $\mathbf{s}$ to its extended support. Let also $C = U|_{[m],[m]}$ be the submatrix of U consisting of its first m rows and columns. By lemmas 2 and 3 we know that a necessary condition for $\mathbf{s}$ being an ESS is that C is negative definite, ie, $\forall \mathbf{y} \in Y_{\mathbf{x}}$, $(\mathbf{y} - \mathbf{x})^T C (\mathbf{y} - \mathbf{x}) < 0$. We shall prove that this is highly unlikely to hold for any mixed strategy $\mathbf{s}$ with support of size $r \gg 1$. Set $\varepsilon = s_r > 0$. Consider the following collection of vectors from $Y_{\mathbf{x}}$: $\forall 1 \leqslant k \leqslant r^* \equiv \min\{r, \frac{m}{2}\}$, $\mathbf{y^k} = \left(x_1, \ldots, x_{k-1}, x_k - \varepsilon, x_{k+1} + \frac{\varepsilon}{m-k}, \ldots, x_m + \frac{\varepsilon}{m-k}\right)$ and $\mathbf{z^k} = \mathbf{y^k} - \mathbf{x} = \left(0, \ldots, 0, -\varepsilon, \frac{\varepsilon}{m-k}, \ldots, \frac{\varepsilon}{m-k}\right)$. Then we have: $\forall 1 \leqslant k \leqslant r^*$,

$$(\mathbf{z^k})^T C \mathbf{z} = \varepsilon^2 \cdot C_{k,k} - \frac{\varepsilon^2}{m-k} \sum_{j=k+1}^{m} [C_{k,j} + C_{j,k}] + \frac{\varepsilon^2}{(m-k)^2} \sum_{k+1 \leqslant i,j \leqslant m} C_{i,j} \quad (1)$$

By lemma 2 we know that a necessary condition for the mixed strategy $\mathbf{s}$ (for which we already assumed that it is such that $(\mathbf{s}, \mathbf{s})$ is a SNE for Γ) to be an ESS is the following: $\forall k \in [r^*], (\mathbf{y^k} - \mathbf{x})^T C (\mathbf{y^k} - \mathbf{x}) = (\mathbf{z^k})^T C \mathbf{z} < 0 \overset{/*\ \varepsilon > 0\ */}{\Longrightarrow}$

$$\Rightarrow \forall k \in [r^*], \tfrac{1}{m-k} \textstyle\sum_{j=k+1}^{m} [C_{k,j} + C_{j,k}] > C_{k,k} + \tfrac{1}{(m-k)^2} \sum_{k+1 \leqslant i,j \leqslant m} C_{i,j} \quad (2)$$

Consider now the collection of events $E = \left\{\mathcal{E}_k \equiv \mathbb{I}_{\{S_k > C_{k,k} + Z_k\}}\right\}_{1 \leqslant k \leqslant r^*}$ where $S_k \equiv \frac{1}{m-k} \sum_{j=k+1}^{m} [C_{k,j} + C_{j,k}]$ and $Z_k \equiv \frac{1}{(m-k)^2} \sum_{k+1 \leqslant i,j \leqslant m} C_{i,j}$. We know that $\mathbf{s}$ is an ESS only if $\bigcap_{1 \leqslant k \leqslant r^*} \mathcal{E}_k$ holds true. Applying repeatedly the Bayes rule we get:

$$\mathbb{P}\{\cap_{1 \leqslant k \leqslant r^*} \mathcal{E}_k\} = \mathbb{P}\{\mathcal{E}_{r^*}\} \mathbb{P}\{\mathcal{E}_{r^*-1} \mid \mathcal{E}_{r^*}\} \cdots \mathbb{P}\left\{\mathcal{E}_1 \mid \cap_{j=2}^{r^*} \mathcal{E}_j\right\} \quad (3)$$

where $r^* = \min\{r, \frac{m}{2}\}$. In order to prove that this probability is very small, we proceed as follows: We calculate first that for any $1 \leqslant k \leqslant r^*$, the probability of $\mathcal{E}_k$ being true (independently of all other events).

Lemma 5. $\forall 1 \leqslant k \leqslant r^*$, $\mathbb{P}\{\mathcal{E}_k\} = \frac{1}{2}$.

Proof. Fix arbitrary $1 \leqslant k \leqslant r^*$. Observe that: (a) Z_k does not include in this sum any of $C_{k,j}, C_{j,k}, C_{k,k}$. Thus, Z_k is *independent* of S_k and $C_{k,k}$. (b) S_k is not affected at all by the value of $C_{k,k}$. Therefore, S_k is also *independent* of $C_{k,k}$. Let $R_k \equiv C_{k,k} + Z_k$. By our previous remarks, we get the following claim:

Claim. R_k is a r.v. independent of S_k.

By linearity of expectation and assuming that any r.v. that is distributed according to F has an expectation μ, we have that $\mathbb{E}\{R_k\} = \mathbb{E}\{C_{k,k}\} + \frac{1}{(m-k)^2}\sum_{k+1 \leqslant i,j \leqslant m} \mathbb{E}\{C_{i,j}\} = \mu + \frac{1}{(m-k)^2} \cdot (m-k)^2 \cdot \mu = 2\mu$. Similarly, $\mathbb{E}\{S_k\} = \frac{1}{m-k}\sum_{k+1 \leqslant j \leqslant m}[\mathbb{E}\{C_{k,j}\} + \mathbb{E}\{C_{k,j}\}] = \frac{1}{m-k} \cdot (m-k) \cdot 2\mu = 2\mu$. That is, we deduce that

Claim. $\mathbb{E}\{R_k\} = \mathbb{E}\{S_k\}$.

Notice also that the following claim holds, whose proof is straightforward:

Claim. Let $X_1, \ldots, X_t$ be **iid** *uniform* r.v.s drawn from $[0, A]$. Then $X = \sum_{j=1}^{t} X_i$ is a symmetric r.v. about its expectation $\mathbb{E}\{X\} = \frac{tA}{2}$, in the interval $[0, tA]$.

Since $\{C_{k,j}, C_{j,k}\}_{k+1 \leqslant j \leqslant m}$ is a collection of $2(m-k)$ **iid** uniform r.v.s on $[0, A]$, $(m-k) \cdot S_k$ is symmetric r.v. about its expectation $2(m-k)\mu$ in $[0, 2(m-k)A]$, or equivalently, S_k is a symmetric r.v. about its expectation 2μ in $[0, 2A]$. Similarly, $\{C_{i,j}\}_{k+1 \leqslant i,j \leqslant m}$ is a collection of $(m-k)^2$ **iid** uniform r.v.s on $[0, A]$ and thus, $(m-k)^2 Z_k$ is a symmetric r.v. on $[0, (m-k)^2 A]$, or equivalently, Z_k is a symmetric r.v. (around $\mathbb{E}\{Z_k\} = \mu$) on $[0, A]$. So, $R_k = Z_k + C_{k,k}$ is also a symmetric r.v. (about its expectation 2μ) on $[0, 2A]$. Thus, we conclude that:

Claim. R_k and S_k have the same expectation 2μ, they are independent of each other, and they are both symmetric r.v.s in the interval $[0, 2A]$.

The following exploits the symmetry of the r.v.s that we compare in each certificate (of $\mathbf{s}$ being an ESS) $\mathcal{E}_k$:

Claim. $\mathbb{P}\{\mathcal{E}_k\} = \mathbb{P}\{S_k > R_k\} = \frac{1}{2}$.

Proof. $\mathcal{E}_k$ compares the values of the (independent) r.v.s S_k and R_k. But these two r.v.s have the same expectation and they are symmetric about their (common) mean. By applying lemma 4 we get that $\mathbb{P}\{\mathcal{E}_k\} = \frac{1}{2}$.

This completes the proof of lemma 5.

Recall that we are interested in $\mathbb{P}\{\mathcal{E}_k | \cap_{k+1 \leqslant j \leqslant r^*} \mathcal{E}_j\}$, $\forall 1 \leqslant k \leqslant r^* - 1$ (see equation (3)). Our goal is to determine an upper bound on the occurrence probability of each event $\mathcal{E}_k$, despite its dependence on its nested events $\{\mathcal{E}_j\}_{k+1 \leqslant j \leqslant r^*}$. Observe that we reveal the outcome of the conditional events involved in eq. (3) *sequentially* starting from $\mathcal{E}_{r^*}$, then $\mathcal{E}_{r^*-1} | \mathcal{E}_{r^*}$, etc, up to $\mathcal{E}_1 | \cap_{2 \leqslant j \leqslant r^*} \mathcal{E}_j$.

Lemma 6. *The probability of a SNE* s *being an ESS is* $\mathbb{P}\left\{\bigcap_{1\leqslant k\leqslant r^*}\mathcal{E}_k\right\} \leqslant \left(\frac{1+\delta'}{2}\right)^{r^*} + 2\exp\left(-\frac{\delta^2\cdot(m-r^*)^2 A}{4+(4/3)\delta}\right)$, *where* $\delta' = \delta + \frac{r^*-1}{m-1} - \delta\cdot\frac{r^*-1}{m-1}$ *and* $\delta > 0$.

Proof. $\forall k \in [r^*-1]$, $\hat{Z}_k$ denotes the sum of r.v.s from C involved in Z_k, (ie, $\hat{Z}_k = (m-k)^2 Z_k$); $\hat{S}_k$ is the sum of the r.v.s in S_k, (ie, $\hat{S}_k = (m-k)S_k$). Then,

$$\forall 1 \leqslant k \leqslant r^*-1,\ \hat{Z}_k = C_{k+1,k+1} + \hat{Z}_{k+1} + \hat{S}_{k+1} \tag{4}$$

Observe now that we can get a new set of inequalities, exploiting the fact that we check for the truth of an event $\mathcal{E}_k$ conditioned on the truth of *all* its nested events: For $\mathcal{E}_{r^*}$, $\hat{S}_{r^*} > (m-r^*)C_{r^*,r^*} + \frac{\hat{Z}_{r^*}}{m-r^*}$, for $\mathcal{E}_{r^*-1}|\mathcal{E}_{r^*}$, $\hat{S}_{r^*-1} > (m-r^*+1)C_{r^*-1,r^*+1} + \frac{\hat{Z}_{r^*-1}}{m-r^*+1} \overset{/*\ (4),\ \mathcal{E}_{r^*}\ */}{\Longrightarrow} \hat{S}_{r^*-1} > (m-r^*+1)C_{r^*-1,r^*-1} + C_{r^*} + \frac{\hat{Z}_{r^*}}{m-r^*}$, …, for $\mathcal{E}_1|\bigcap_{2\leqslant j\leqslant r^*}\mathcal{E}_j$, $\hat{S}_1 > (m-1)C_{1,1} + \frac{\hat{Z}_1}{m-1} \overset{/*\ (4),\ \mathcal{E}_2,\ldots,\mathcal{E}_{r^*}\ */}{\Longrightarrow} \hat{S}_1 > (m-1)C_{1,1} + \sum_{j=2}^{r^*} C_j + \frac{\hat{Z}_{r^*}}{m-r^*}$. From this we define a new necessary condition:

$$\forall 1 \leqslant k \leqslant r^*,\ S_k > C_{k,k} + \frac{m-r^*}{m-k} Z_{r^*} \tag{5}$$

So, we consider a new collection of events $\{\mathcal{E}'_k\}_{1\leqslant k\leqslant r^*}$ (described by (5)), each of which involves a r.v. (S_k) that is twice the average of $2(m-k)$ **iid** uniform r.v.s, a unique uniform r.v. ($C_{k,k}$) of expectation $A/2$ that is not considered at all by the other events, and a last term that is handled as a *constant* that is asymptotically equal to the expected value $A/2$ of a uniform r.v. in $[0, A]$ (since $m - r^* = \mathcal{O}(m)$ and Z_{r^*}, the average value of $\mathcal{O}(m^2)$ **iid** r.v.s, tends rapidly to their common expectation $A/2$, provided $A \ll m$). Observe that these events are *independent*, since each of them compares the values of two (unique) independent and symmetric r.v.s, of (asymptotically) the same expectation. Using an argument similar to that of lemma 5, it is not hard to prove that the probability of s being an ESS is upper bounded by the value claimed in the statement of the lemma. The complete presentation of this proof is provided in the full version of the paper.

As a direct consequence of lemma 6, we get the following corollary:

Corollary 1. *The probability of a SNE* (s, s) *indicating an ESS* s *is upper bounded by* $2\cdot\left(\frac{\sqrt{3+\delta}}{2}\right)^m = \mathcal{O}(\sqrt{m})\cdot\left(\frac{\sqrt{3}}{2}\right)^m$, *if we set* $r^* = \frac{m}{2}$ *and* $\delta = \mathcal{O}\left(\frac{1}{\sqrt{m}}\right)$.

It is worth mentioning that for the random 2-player games considered in [10] (where each entry of the payoff matrices is uniformly distributed in the unit sphere) almost all the NE have supports whose sizes are approximately $0.315915n$. The scaling of the unit interval $[0, 1]$ to $[0, A]$ for any $A > 0$ does not affect this result, since a NE is determined by linear constraints wrt the support (each support is defined as a system of *linear* inequalities for a 2-player game, cf. [8]) and we can scale by A each inequality, so long as $A > 0$.

Let now $q = \#NE$ be the number of NE in our game, that is of course a random variable. For each NE $\mathbf{x}$ of the game, let $I(\mathbf{x}) = \mathbb{I}_{\{\mathbf{x} \text{ is ESS}\}}$ be the corresponding indicator variable of $\mathbf{x}$ being also an ESS. Since an ESS implies our event $E \equiv \cap_{1 \leqslant k \leqslant r^*} \mathcal{E}_k$, we have that $\mathbb{E}\{I(\mathbf{x})\} = \mathbb{P}\{I(\mathbf{x}) = 1\} = \mathbb{P}\{\mathbf{x} \text{ is an ESS}\} \leqslant \mathbb{P}\{E\} \Rightarrow \mathbb{E}\{I(\mathbf{x})\} = \mathcal{O}(\sqrt{m}) \cdot \left(\frac{\sqrt{3}}{2}\right)^m \Rightarrow \mathbb{E}\{\#ESS\} = \mathbb{E}\{\sum_{\mathbf{x}=NE} I(\mathbf{x})\} = \mathcal{O}(\sqrt{m}) \cdot \left(\frac{\sqrt{3}}{2}\right)^m \cdot \mathbb{E}\{\#NE\}$ (by Wald's inequality for a random sum of random variables). So, we have established that, since almost all NE have support $0.315915n \leqslant r \leqslant m$, $\mathbb{E}\{\#ESS\} = \mathbb{E}\{\#NE\} \cdot \mathcal{O}(\sqrt{n}) \cdot \left(\frac{\sqrt{3}}{2}\right)^{0.315915n}$, which proves our main theorem.

Remark 1. If we also adopt the numerical analysis of [10] on the expected number of NE in such a game, according to which the expected number of NE is $\exp(0.281644n + \mathcal{O}(\log n))$, then we will come to the conclusion that $\mathbb{E}\{\#ESS\} = \exp(0.281644n + \mathcal{O}(\log n)) \cdot \mathcal{O}(\sqrt{n}) \cdot \left(\frac{\sqrt{3}}{2}\right)^{0.315915n} = \exp(0.137803n + \mathcal{O}(\log n))$. This is still exponential, but also exponentially smaller than the expected number of NE.

Acknowledgments. The authors wish to thank an anonymous referee for some nice remarks on their structural argument, in an earlier version of the paper.

References

1. Bomze I.M., Pelillo M., Stix V. Approximating the maximum weight clique using replicator dynamics. *IEEE Transactions on Neural Networks*, 11(6):1228–1241, November 2000.
2. Broom M. Bounds on the number of esss of a matrix game. *Mathematical Biosciences*, 167(2):163–175, October 2000.
3. Cressman R. *Evolutionary dynamics and extensive form games.* MIT Press, 2003.
4. Etessami K., Lochbihler A. The computational complexity of evolutionary stable strategies. Technical Report 55, Electronic Colloquium on Computational Complexity (ECCC), 2004. ISSN 1433-8092.
5. Fischer S., Vöcking B. On the evolution of selfish routing. In *Proc. of the 12th European Symposium on Algorithms (ESA '04)*, pages 323–334. Springer, 2004.
6. Haigh J. Game theory and evolution. *Advances in Applied Probability*, 7:8–11, 1975.
7. Hofbauer J., Sigmund K. Evolutionary game dynamics. *Bulletin of the American Mathematical Society*, 40(4):479–519, 2003.
8. Koutsoupias E., Papadimitriou C. Worst-case equilibria. In *Proc. of the 16th Annual Symposium on Theoretical Aspects of Computer Science (STACS '99)*, pages 404–413. Springer-Verlag, 1999.
9. McLennan A. The expected numer of nash equilibria of a normal form game. *Econometrica*, 73(1):141–174, January 2005.
10. McLennan A., Berg J. The asymptotic expected number of nash equilibria of two player normal form games. To appear in the *Games and Economic Behavior*, 2005.
11. Nash J. F. Noncooperative games. *Annals of Mathematics*, 54:289–295, 1951.

Exact and Approximation Algorithms for Computing the Dilation Spectrum of Paths, Trees, and Cycles

Rolf Klein[1,*], Christian Knauer[2], Giri Narasimhan[3], and Michiel Smid[4,**]

[1] Institute of Computer Science I, Universität Bonn, Römerstraße 164, D-53117 Bonn, Germany
rolf.klein@uni-bonn.de

[2] Freie Universität Berlin, Institute of Computer Science, Berlin, Germany
christian.knauer@inf.fu-berlin.de

[3] School of Computer Science, Florida, International University, ECS389, University Park, Miami, FL 33199, USA
giri@cs.fiu.edu

[4] School of Computer Science, Carleton University, 1125 Colonel By Drive, Ottawa, Ontario K1S 5B6, Canada
michiel@scs.carleton.ca

Abstract. Let G be a graph embedded in Euclidean space. For any two vertices of G their *dilation* denotes the length of a shortest connecting path in G, divided by their Euclidean distance. In this paper we study the spectrum of the dilation, over all pairs of vertices of G. For paths, trees, and cycles in 2D we present $O(n^{3/2+\epsilon})$ randomized algorithms that compute, for a given value $\kappa \geq 1$, the exact number of vertex pairs of dilation $> \kappa$. Then we present deterministic algorithms that approximate the number of vertex pairs of dilation $> \kappa$ up to an $1 + \eta$ factor. They run in time $O(n \log^2 n)$ for chains and cycles, and in time $O(n \log^3 n)$ for trees, in any constant dimension.

1 Introduction

Let S be a set of n points in $\mathbb{R}^d$, where $d \geq 1$ is a small constant, and let G be an undirected connected graph having the points of S as its vertices. The length of any edge (p, q) of G is defined as the Euclidean distance $|pq|$ between the two vertices p and q. Such graphs are called *Euclidean graphs*. Let $d_G(x, y)$ denote the length of a shortest path in G that connects x and y. Then the *dilation* between x and y in G is defined as

$$\delta_G(x, y) = \frac{d_G(x, y)}{|xy|}.$$

Euclidean graphs are frequently used for modeling traffic or transportation networks. In order to measure their performance, the *dilation* of G [13] has been

* supported by DFG Kl 655/14.
** supported by NSERC.

X. Deng and D. Du (Eds.): ISAAC 2005, LNCS 3827, pp. 849–858, 2005.

used, which is defined as the maximum dilation over all pairs of vertices in G, i.e.,

$$\sigma(G) = \max_{p \neq q \in V} \delta_G(p,q). \tag{1}$$

This value is also called the stretch factor, the spanning ratio, or the distortion [10] of G. A lot of work has been done on the construction of good spanners, i. e., of sparse graphs of low dilation that connect a given vertex set and enjoy other desirable properties; see the handbook chapter [5] or the forthcoming monograph [12]. How to compute the dilation of a given Euclidean graph has first been addressed in [11]. They gave an $O(n \log n)$ algorithm for approximating, up to an $1-\epsilon$ factor, the dilation of paths, trees, and cycles in Euclidean space of constant dimension. In [1] exact randomized algorithms were given that run in time $O(n \log n)$ for paths in the plane, and in time $O(n \log^2 n)$ for computing the dilation of a plane tree or cycle. In 3D, time $O(n^{4/3+\epsilon})$ is sufficient for either type of graph; see also [2, 9]. For general graphs, nothing better seems to be known than running Dijkstra's algorithm for each vertex of G, which leads to an $O(mn + n^2 \log n)$ algorithm [4]; here n denotes the number of vertices, while m is the number of edges of G. In the recent paper [6], the interesting question has been addressed how to decrease the dilation of a graph as much as possible by inserting another edge.

In this paper we are studying the vertex-to-vertex dilation of graphs from a different perspective. Computing the dilation, as defined in (1), simply points to the pair of vertices for which the dilation is maximized. In real networks, one may wish to tolerate some number of pairs of vertices with high dilations, if the network provides good connections to the majority of vertex pairs. Therefore, we are interested in computing (exactly or approximately) the *dilation spectrum* of a graph G. That is, for a given threshold $\kappa > 1$ we are interested in the number

$$\pi_G(\kappa) := |\{ (p,q) \in S^2;\ \delta_G(p,q) > \kappa \}| \tag{2}$$

of all vertex pairs whose dilation exceeds κ. This distribution of the dilation could also be helpful in understanding structural properties of a given geometric graph.

Clearly, the cost $O(mn + n^2 \log n)$ of running Dijkstra's algorithm from each vertex of G is an upper bound on the time complexity of computing the dilation spectrum of G. For some classes of graphs, better running times can be obtained. For example, it has been shown in [7] that the distances in G between all pairs of vertices of a plane geometric graph can be computed in $O(n^2)$ total time. The same upper bound holds for the dilation spectrum.

This paper is organized as follows. In Section 2 we provide randomized algorithms for polygonal paths, trees, and cycles in the plane, that allow $\pi_G(\kappa)$ to be computed in time $O(n^{3/2+\epsilon})$. To this end, we first use a geometric transformation scheme introduced in [1], in order to reduce the problem of computing $\pi_G(\kappa)$ to a counting problem, and then apply suitable range counting techniques. Then, in Section 3, faster algorithms will be presented for *approximating* the dilation spectrum. More precisely, for given reals $\kappa \geq 1$ and $\epsilon > 0$ we shall compute a number M satisfying

$$\pi_G(\kappa\ (1+\epsilon)^2) \leq M \leq \pi_G(\kappa) \tag{3}$$

that approximates the number of vertex pairs of dilation $> \kappa$. The run time of these deterministic algorithms is in $O(n \log^2 n)$ for paths and cycles, and in $O(n \log^3 n)$ for trees. Our approach is based on the well-separated pair decomposition [3]; hence, it works in any constant dimension. Finally, in Section 4 we mention some open problems.

2 Computing the Exact Dilation Spectrum

In this section we derive exact algorithms for computing the dilation distribution $\pi_G(\kappa)$ of certain types of plane graphs, given a threshold $\kappa > 1$.

2.1 Paths

We start with a randomized algorithm for computing $\pi_P(\kappa)$ for a simple polygonal path P in the plane. First, we describe a reduction that rephrases the problem of computing $\pi_P(\kappa)$ as a counting problem in three-dimensional space. Then we apply range counting algorithms to solve that problem.

Reduction to range counting. We shall consider a slightly more generally version of our problem. Namely, let A, B be two disjoint vertex sets of G; then we are going to compute

$$\pi_G(\kappa, A, B) = |\{\ (x, y) \in A \times B \mid \delta_G(x, y) > \kappa\ \}|,$$

the number of vertex pairs in $A \times B$ whose dilation exceeds κ.

Assume that some orientation $<_P$ of P is given, and let p_0 be the first vertex of P. For a vertex $p \in A$, we define the *weight* of p to be $\omega(p) = d_P(p_0, p)/\kappa$. Let $\check{C}$ denote the cone $z = \sqrt{x^2 + y^2}$ in $\mathbb{R}^3$. We map each vertex $p = (p_x, p_y) \in A$ to the cone $C_p = \check{C} + (p_x, p_y, \omega(p))$. If we regard C_p as the graph of a bivariate function then for any point $x \in \mathbb{R}^2$, $C_p(x) = |xp| + \omega(p)$. Set $\mathcal{C}(A) = \{C_p \mid p \in A\}$. We map a vertex $q = (q_x, q_y) \in B$ to the point $\hat{q} = (q_x, q_y, \omega(q)) \in \mathbb{R}^3$. Let $\hat{B} = \{\hat{q} \mid q \in B\}$.

Lemma 1. *For any vertex pair $(p, q) \in A \times B$ such that $p <_P q$, we have that $\delta(q, p) \le \kappa$ if and only if $\hat{q}$ lies below C_p, i.e., $\omega(q) \le C_p(q)$.*

Proof.

$$\begin{aligned}
\delta(p, q) \le \kappa &\iff \frac{d_P(p, q)}{|qp|} \le \kappa \iff \frac{d_P(p_0, q) - d_P(p_0, p)}{|qp|} \le \kappa \\
&\iff \frac{d_P(p_0, q)}{\kappa} \le |qp| + \frac{d_P(p_0, p)}{\kappa} \\
&\iff \omega(q) \le |qp| + \omega(p) \iff \omega(q) \le C_p(q).
\end{aligned}$$

If there is a vertex pair $(p, q) \in A \times B$ with $\delta(p, q) > \kappa$ we cannot be sure that p that lies before q on P. Hence, we must also consider the symmetric situation with the orientation of P reversed.

Counting points above cones. In the previous section we have seen that the problem of computing $\pi(\kappa, A, B)$ amounts to count the number of point-cone pairs $(p, C) \in \hat{B} \times \mathcal{C}(A)$ such that p lies above C.

Suppose we are given a set P of n points and a set $\mathcal{C}$ of m cones in $\mathbb{R}^3$ whose axes are vertical and whose apices are their bottommost points. We describe a randomized algorithm to compute $\underline{\mu}(P, \mathcal{C})$, the number of point-cone pairs $(p, C) \in P \times \mathcal{C}$ such that p lies above C.

We fix a sufficiently large constant r, choose a random sample $\mathcal{R}$ of $O(r)$ cones in $\mathcal{C}$, and compute the vertical decomposition $\mathcal{A}_{\|}$ of the arrangement $\mathcal{A}$ of the cones in $\mathcal{R}$. As is known (c.f. [14], Theorem 8.21), $\mathcal{A}_{\|}$ has $O(r^3 \log^4 r)$ cells. For each cell $\Delta \in \mathcal{A}_{\|}$, let $P_\Delta = \{p \in P \mid p \in \Delta\}$, let $\mathcal{C}_{\not\Delta} \subseteq \mathcal{C}$ be the set of cones crossing Δ, and let $\mathcal{C}_{\underline{\Delta}} \subseteq \mathcal{C}$ be the set of cones which lie completely below Δ. Clearly $\underline{\mu}(P, \mathcal{C}) = \sum_{\Delta \in \mathcal{A}_{\|}} |P_\Delta||\mathcal{C}_{\underline{\Delta}}| + \underline{\mu}(P_\Delta, \mathcal{C}_{\not\Delta})$.

Set $n_\Delta = |P_\Delta|$ and $m_\Delta = |\mathcal{C}_{\not\Delta}|$. Obviously, $\sum_\Delta n_\Delta = n$ and by the theory of random sampling [8], $m_\Delta \leq m/r$ with high probability, for all Δ. If this condition is not satisfied for the sample $\mathcal{R}$, we pick a new random sample. The expected number of trials until we get a 'good' sample is constant. If m_Δ or n_Δ is less than a prespecified constant, then we use a naive procedure to determine $\underline{\mu}(P_\Delta, \mathcal{C}_\Delta)$. Otherwise, we recursively compute $\underline{\mu}(P_\Delta, \mathcal{C}_\Delta)$. For $m, n > 0$, let $T(n, m)$ denote the expected running time of the algorithm on a set of n points and a set of m cones. We get the following probabilistic recurrence:

$$T(n, m) = \sum_{\Delta \in \mathcal{A}_{\|}} T\left(n_\Delta, \frac{m}{r}\right) + O(m + n),$$

where $\sum_\Delta n_\Delta \leq n$. The solution to the above recurrence is, for any $\epsilon > 0$,

$$T(n, m) = O(m^{3+\epsilon} + n \log m).$$

To improve the running time of the algorithm we can perform the following dualization step: Let $\hat{C}$ denote the cone $z = -\sqrt{x^2 + y^2}$ in $\mathbb{R}^3$. If we map each point $p = (p_x, p_y, p_z) \in P$ to the cone $\delta(p) = C_p = \hat{C} + (p_x, p_y, p_z)$ and each cone $C = \check{C} + (p_x, p_y, p_z)$ to the point $\delta(C_p) = p_C = (p_x, p_y, p_z)$ we have that p lies above C if and only if p_C lies below C_p, in other words $\underline{\mu}(P, \mathcal{C}) = \bar{\mu}(\delta(\mathcal{C}), \delta(P))$.

We can use this to tune our algorithm as follows. The recursion proceeds as earlier except that when $m_\Delta > n_\Delta^3$, we switch the roles of P_Δ and $\mathcal{C}_\Delta$ using the duality transformation δ, and compute $\underline{\mu}(P_\Delta, \mathcal{C}_\Delta) = \bar{\mu}(\delta(\mathcal{C}_\Delta), \delta(P_\Delta))$ in $T(m_\Delta, n_\Delta) = O(n_\Delta^{3+\epsilon} + m_\Delta \log n_\Delta) = O(m_\Delta^{1+\epsilon})$ time with the algorithm just described. With these modifications, the recurrence becomes

$$T(n, m) = \begin{cases} \sum_{\Delta \in \mathcal{A}_{\|}} T\left(n_\Delta, \frac{m}{r}\right) + O(m + n) & \text{if } m \leq n^3 \\ O(m^{1+\epsilon}) & \text{if } m > n^3, \end{cases}$$

where $\sum_\Delta n_\Delta \leq n$. It can be shown that, for any $\epsilon > 0$,

$$T(n, m) = O((mn)^{3/4+\epsilon} + m^{1+\epsilon} + n^{1+\epsilon}).$$

Thus, we obtain the following result.

Theorem 1. *Let P be a polygonal path on n vertices in $\mathbb{R}^2$, and let $\kappa \geq 1$. Then we can compute $\pi_P(\kappa)$, i.e., the number of vertex-pairs of P that attain at least dilation κ, in $O(n^{3/2+\epsilon})$ randomized expected time for any $\epsilon > 0$.*

Note that we can use the same approach to report all pairs of vertices for which the dilation is larger than κ, in additional time that is proportional to the size of the output.

The same result can be obtained for trees, using a decomposition technique that will be presented in Section 3.4 for computing the approximate dilation spectrum.

2.2 Cycles

Let us now consider the case in which P is a simple closed polygonal curve. This case is more difficult because there are two paths between any two points x, y of P. Let $P[x, y]$ denote the portion of P from x to y in clockwise direction, and let $d_P(x, y)$ denote its length. Moreover, for a vertex p of P, let $\nu(p)$ denote the last vertex on P in clockwise direction such that $d_P(p, \nu(p))$ does not exceed half the perimeter of P.

Suppose we are given four vertices $t_1, t_2, b_1 = v(t_1), b_2 = v(t_2)$ of P in clockwise order, and we want to compute

$$\pi(t_1, t_2, b_1, b_2) := \pi_P(\kappa, P[t_1, t_2], P[b_1, b_2]).$$

First observe that $d_P(b_1, b_2) \leq |P|/2$ holds. Let m, n be the number of edges in $P[b_1, b_2]$ and $P[t_1, t_2]$, respectively. If $\min\{m, n\} = 1$, then we compute $\pi(t_1, t_2, b_1, b_2)$ in $O(m + n)$ time, by brute force. Otherwise, suppose that $n \geq m$ holds. Let t be the median vertex of $P[t_1, t_2]$, and let $b = \nu(t)$. Clearly, $b \in P[b_1, b_2]$ and

$$\pi(t_1, t_2, b_1, b_2) = \pi(t_1, t, b, b_2) + \pi(t, t_2, b_1, b) + \pi(t_1, t, b_1, b) + \pi(t, t_2, b, b_2).$$

The quantities $\pi(t_1, t, b_1, b)$ and $\pi(t, t_2, b, b_2)$ are computed recursively. Since $P[t, t_2]$ and $P[b_1, b]$ lie in $P[t, \nu(t)]$, we can compute $\pi(t, t_2, b_1, b)$ according to Theorem 1 in $O((n + m)^{3/2+\epsilon})$ randomized expected time for any $\epsilon > 0$. Almost the same argument applies to $\pi(t_1, t, b, b_2)$.

Let m_1 be the number of edges in $P[b_1, b]$. Then $P[b, b_2]$ contains at most $m - m_1 + 1$ edges. Let $T(n, m)$ denote the maximum expected time of computing $\pi(t_1, t_2, b_1, b_2)$. Then we obtain the following recurrence for any $\epsilon > 0$:

$$T(n, m) \leq T(n/2, m_1) + T(n/2, m - m_1 + 1) + O((n + m)^{3/2+\epsilon}), \quad \text{if } n \geq m,$$

with a symmetric inequality for $m \geq n$, and $T(n, 1) = O(n), T(1, m) = O(m)$. The solution to the above recurrence is

$$T(n, m) = O((n + m)^{3/2+\epsilon}).$$

To compute $\pi_P(\kappa)$, we choose a vertex $v \in P$ and let $P_1 = P[v, \nu(v)]$ and $P_2 = P[\nu(v), \nu(\nu(v))]$. In

$$\pi_P(\kappa) = \pi_{P_1}(\kappa) + \pi_{P_2}(\kappa) + \pi(v, \nu(v), \nu(v), \nu(\nu(v)))$$

the values π_{P_1}, π_{P_2} can be computed in $O(n^{3/2+\epsilon})$ expected time using Theorem 1, and $\pi(v, \nu(v), \nu(v), v)$ can be computed within the same time by the recursion just described. We obtain

Theorem 2. *Let P be a closed polygonal path on n vertices in $\mathbb{R}^2$, and let $\kappa \geq 1$. Then we can compute $\pi_P(\kappa)$ in $O(n^{3/2+\epsilon})$ randomized expected time.*

3 Computing the Approximate Dilation Spectrum

Now we set out to give faster algorithms for computing an approximation of the dilation spectrum. Our reduction uses the well-separated pair decomposition, thus adding to the list of applications of this powerful method.

3.1 Well-Separated Pairs

We briefly review this decomposition and some of its relevant properties. Let d be a fixed dimension, and let s denote a fixed constant, called the *separation constant.* Two point sets A, B in $\mathbb{R}^d$ are *well separated* with respect to s if they can be circumscribed by two disjoint d-dimensional balls of some radius r that are at least $s \cdot r$ apart. Then the distance between any two points of the same set is at most $1 + 4/s$ times the distance between any two points of different sets, while point pairs of different sets differ in distance by at most a factor $2/s$; these basic WSPD properties will be used in the sequel.

Given a set S of n points, a *well-separated pair decomposition (WSPD)* consists of a sequence $\{A_1, B_1\}, \{A_2, B_2\}, \ldots, \{A_k, B_k\}$ of well-separated pairs of subsets of S such that, for any two points $p \neq q$ of S, there is a unique index i such that $p \in A_i$ and $q \in B_i$ holds, or vice versa.

Callahan and Kosaraju [3] showed that a WSPD of size $k = O(n)$ always exists, and that it can be efficiently computed in time $O(n \log n)$ using a split tree. We are going to use a modified version of their result where each pair $\{A_i, B_i\}$ contains a singleton set and size k is allowed to be in $O(n \log n)$.

3.2 A General Algorithm

Given a real number $\kappa > 1$ and a geometric graph G, we show how to compute the number $\rho_G(\kappa)$ of all pairs of vertices in the graph for which the dilation is at most κ.[1] Now consider a WSPD

$$\{A_1, B_1\}, \{A_2, B_2\}, \ldots, \{A_k, B_k\}$$

[1] Clearly, we can obtain the number $\pi_G(\kappa)$ of vertex pairs whose dilation exceeds κ by subtracting $\rho_G(\kappa)$ from $\binom{n}{2}$.

for the set of vertices of G. We may assume that each A_i is a singleton set, $\{a_i\}$ and that $k = O(n \log n)$. We know that all geometric distances between a_i and a point of B_i are roughly equal. Let the distance between a_i and any representative point in B_i be denoted by D_i. Hence, if we compute for each i, $1 \le i \le k$, all points $b_i \in B_i$ whose distance $d_G(a_i, b_i)$ in G is at most $\kappa(1+\epsilon) \cdot D_i$, then we would have effectively computed all pairs of points in the pair $\{A_i, B_i\}$ for which their approximate dilation is at most κ. This observation leads to the general algorithm, $\mathcal{A}$, presented below in Figure 3.2; it could be easily modified

General Algorithm $\mathcal{A}$
Input: A geometric graph G on a set S of points in $\mathbb{R}^d$ and a constant $\epsilon > 0$.
Output: The number of pairs of vertices of G of dilation at most $\kappa(1+\epsilon)$.
Step 1: Using separation constant $s = 4/\epsilon$, compute a WSPD

$$\{A_1, B_1\}, \{A_2, B_2\}, \ldots, \{A_k, B_k\}$$

for the set S, with the added condition that $|A_i| = 1$, for each $i = 1, \ldots, k$.
Step 2: For each i, $1 \le i \le k$, let $A_i = \{a_i\}$, and D_i denote the length $|a_i b|$, where b is an arbitrary element of B_i. For $i = 1, \ldots, k$. compute n_i, the number of points $b_i \in B_i$, such that $d_G(a_i, b_i) \le \kappa(1+\epsilon) \cdot D_i$.
Step 3: Report $N = \sum_{i=1}^{k} n_i$.

to output the pairs of points of dilation $\le \kappa(1+\epsilon)$.

The following lemma compares N, the output of Algorithm $\mathcal{A}$, with the true value of $\rho_G(\kappa)$.

Lemma 2. *The number of pairs of vertices in G with stretch factor at most κ is at most equal to N. The number of pairs of vertices in G with stretch factor at most $\kappa(1+\epsilon)^2$ is at least equal to N. In other words,*

$$\rho_G(\kappa) \le N \le \rho_G(\kappa(1+\epsilon)^2)$$

Proof. In step 2 of algorithm $\mathcal{A}$, D_i is equal to $|a_i b|$, where b is an arbitrary element of B_i. Step 2 of the algorithm counts all pairs of points $b_i \in B_i$ for which $|a_i b_i| \le \kappa(1+\epsilon) \cdot D_i$.

If $d_G(a_i, b_i) \le \kappa |a_i b_i|$, then by the WSPD properties, $d_G(a_i, b_i) \le \kappa(1+\epsilon) \cdot D_i$, and thus the pair (a_i, b_i) is counted in step 2 of the algorithm, proving that $\rho_G(\kappa) \le N$.

If $d_G(a_i, b_i) \le \kappa(1+\epsilon)D_i$, then it is counted in step 2 of the algorithm. But then, by the WSPD property, $d_G(a_i, b_i) \le \kappa(1+\epsilon)^2 \cdot |a_i b_i|$, implying that this pair has a dilation of at most $\kappa(1+\epsilon)^2$. Thus, $N \le \rho_G(\kappa(1+\epsilon)^2)$.

This completes the proof.

3.3 Paths

Let the graph G be a simple path $(p_0, p_1, \ldots, p_{n-1})$ on the points of the set S, and let $\epsilon > 0$ be a constant.

Following our general algorithm $\mathcal{A}$ of Section 3.2, we start by computing a split tree T and a corresponding WSPD

$$\{A_1, B_1\}, \{A_2, B_2\}, \ldots, \{A_m, B_m\}$$

for S, with separation constant $s = 4/\epsilon$, where $|A_i| = 1$, for $1 \leq i \leq m$, and $m = O(n \log n)$. This can be done in time $O(n \log n)$.

By a simple postorder traversal of T, we store with each node u of T the sorted list of indices of all points stored at the leaves of the subtree of u. Note that this involves merging two sorted sublists at each node of the tree. Each set B_i is represented by a node v_i in the split tree T such that their subtrees contain exactly the points of B_i in their leaves. Let the sorted list of the points for node v_i be denoted by V_i. Also since $|A_i| = 1$, it is represented by a leaf node u_i in the split tree.

Step 2 of our algorithm is implemented as follows. First, we traverse the path, and compute for each vertex p_j, $0 < j < n$, the distance $d_G(p_0, p_j)$. Using this information, we can compute for any $0 \leq j < k < n$, the distance $d_G(p_j, p_k)$ in constant time, as the difference between $d_G(p_0, p_k)$ and $d_G(p_0, p_j)$.

Then, for $1 \leq i \leq m$, use binary search to identify the two sublists of points in V_i that lie before and after point a_i. It is easy to observe that the two sublists are effectively sorted by their distance from a_i. Finally use two binary searches to search the two sublists separately to identify the number of points $b_i \in V_i$ such that $d_G(a_i, b_i) \leq \kappa(1+\epsilon)D_i$.

It is easy to see that this correctly implements Step 2. Since it involves $2m = O(n \log n)$ binary searches, the running time of the entire algorithm is bounded by $O(n \log^2 n)$.

Theorem 3. *Let S be a set of n points in $\mathbb{R}^d$, let G be a simple path on the points of S, and let ϵ be a positive constant. In $O(n \log^2 n)$ time, we can compute a number N that lies between $\rho_G(\kappa)$ and $\rho_G(\kappa(1+\epsilon)^2)$.*

Due to space limitations we can only mention that the same result can be obtained for cycles.

3.4 Trees

It is well known that in any tree G having n vertices, there is a vertex v, whose removal gives two graphs G'_1 and G'_2, each having at most $2n/3$ vertices. Moreover, such a vertex v can be found in $O(n)$ time. Each of the two graphs G'_i is a forest of trees. We will call v a *centroid vertex* of G. Each of the graphs $G_1 := G'_1 \cup \{v\}$ and $G_2 := G'_2 \cup \{v\}$ is connected and, hence, a tree again.

Let S be a set of n points in $\mathbb{R}^d$, and let G be an arbitrary tree having the points of S as its vertices. We will identify the vertices of G with the points of S. Let $\epsilon > 0$ be a constant. The following recursive algorithm, which is inspired by the general algorithm $\mathcal{A}$, solves the problem when the input graph G is a tree.

Step 1: Compute a centroid vertex v of G, and the corresponding decomposition into trees G_1 and G_2. Note that v is a vertex of both these trees. Traverse each

tree in preorder, and store with each vertex p the distance $d_G(p, v)$ between p and the centroid vertex v.

Step 2: Use the same algorithm to recursively solve the problem on graph G_1 and on graph G_2.

Step 3: Let $s := 4/\epsilon$. Compute a split tree T and a corresponding WSPD

$$\{A_1, B_1\}, \{A_2, B_2\}, \ldots, \{A_m, B_m\}$$

for the points of S, with separation constant s, having size $m = O(n \log n)$.

Step 4: For each node u of the split tree, denote by S_u the set of points of S that are stored in the subtree of u. Traverse T in postorder, and compute for each of its nodes u the sorted sequence of nodes, S_u^1, consisting of nodes in G_1 sorted according to their distance from the centroid v. Similarly, compute the sorted sequence of nodes, S_u^2, consisting of nodes in G_2 sorted according to their distance from the centroid v.

Step 5: For each i, $1 \leq i \leq m$, we do the following. Consider the pair $\{A_i, B_i\}$ in our WSPD, with $A_i = \{a_i\}$, and the node v_i in the split tree such that $B_i = S_{v_i}$. Let the two sets computed for v_i be $S_{v_i}^1$ and $S_{v_i}^2$. If $a_i \in G_1$, then use binary search to identify the number of points $b_i \in S_{v_i}^2$ such that $d_G(a_i, b_i) \leq \kappa(1+\epsilon)D_i$. Otherwise the search is performed in $S_{v_i}^2$.

The correctness of the above algorithm is proved by induction and results from the fact that Step 2 counts all pairs of nodes that involve a_i and another node in G_1 (assuming that $a_i \in G_1$), while Step 5 counts all pairs of nodes that involve a_i and another node in G_2. Note that the distance in G from point $a_i \in G_1$ to point $b_i \in G_2$ is given by adding their distances to the centroid vertex v.

To prove the time complexity, let $T(n)$ denote the running time of our algorithm on an input tree having n vertices. Then, $T(n) = O(n \log^2 n) + T(n') + T(n'')$, where n' and n'' are positive integers such that $n' \leq 2n/3$, $n'' \leq 2n/3$, and $n' + n'' = n + 1$. The $O(n \log^2 n)$ term comes about because the binary search spends $O(\log n)$ time on each of the $O(n \log n)$ well-separated pairs. The above recurrence relation solves to $T(n) = O(n \log^3 n)$.

Theorem 4. *Let S be a set of n points in $\mathbb{R}^d$, let G be a tree on the points of S, and let ϵ be a positive constant. In $O(n \log^3 n)$ time, we can compute a number N such that it lies between $\rho_G(\kappa)$ and $\rho_G(\kappa(1+\epsilon)^2)$.*

4 Open Problems

Besides the obvious questions about possible improvement or generalization of our algorithms, there seem to be some interesting structural problems one may want to study. To what extend does the dilation spectrum of a graph, that is, the function $\pi_G(\kappa)$, reflect the graph's structure? And, what types of functions $\pi_G(\kappa)$ can occur?

Acknowledgement. The authors would like to thank the organizers and all participants of the Korean workshop 2004 at Schloß Dagstuhl for establishing the nice and stimulating atmosphere that gave rise to this work.

References

1. P. Agarwal, R. Klein, Ch. Knauer, S. Langerman, P. Morin, M. Sharir, and M. Soss, *Computing the detour and spanning ratio of paths, trees and cycles in 2D and 3D*, to appear in Discrete and Computational Geometry.
2. P. Agarwal, R. Klein, Ch. Knauer, and M. Sharir, *Computing the detour of polygonal curves*, Technical Report B 02-03, Freie Universität Berlin, Fachbereich Mathematik und Informatik, 2002.
3. P. B. Callahan and S. R. Kosaraju, *A decomposition of multidimensional point sets with applications to k-nearest-neighbors and n-body potential fields*, J. ACM, 42 (1995), pp. 67–90.
4. T. H. Cormen, C. E. Leiserson, and R. L. Rivest, *Introduction to Algorithms*, MIT Press, Cambridge, MA, 1990.
5. D. Eppstein, *Spanning trees and spanners*, in Handbook of Computational Geometry, J.-R. Sack and J. Urrutia, eds., Elsevier Science, Amsterdam, 1999, pp. 425–461.
6. M. Farshi, P. Giannopoulos, and J. Gudmundsson, *Finding the best shortcut in a geometric network*, 21st Ann. ACM Symp. Comput. Geom. (2005), pp. 327–335.
7. G. N. Frederickson, *Fast algorithms for shortest paths in planar graphs, with applications*, SIAM J. Comput., 16 (1987), pp. 1004–1022.
8. D. Haussler and E. Welzl, *Epsilon-nets and simplex range queries*, Discrete Comput. Geom. 2 (1987), pp. 127–151.
9. S. Langerman, P. Morin, and M. Soss, *Computing the maximum detour and spanning ratio of planar chains, trees and cycles*, In *Proceedings of the 19th International Symposium on Theoretical Aspects of Computer Science (STACS 2002)*, volume 2285 of *LNCS*, pages 250–261. Springer-Verlag, 2002.
10. N. Linial, E. London, and Y. Rabinovich, *The geometry of graphs and some of its algorithmic applications*, Combinatorica, 15 (1995), pp. 215–245.
11. G. Narasimhan and M. Smid, *Approximating the stretch factor of Euclidean Graphs*, SIAM J. Comput., 30(3) (2000), pp. 978–989.
12. G. Narasimhan and M. Smid, *Geometric Spanner Networks*, Cambridge University Press, to appear.
13. D. Peleg and A. Schäffer, *Graph spanners*, J. Graph Theory, 13 (1989), pp. 99–116.
14. M. Sharir and P. Agarwal, *Davenport-Schinzel sequences and their geometric applications*, Cambridge University Press, New York, 1995.
15. M. Smid, *Closest-point problems in computational geometry*, in Handbook of Computational Geometry, J.-R. Sack and J. Urrutia, eds., Elsevier Science, Amsterdam, 1999, pp. 877–935.

On the Computation of Colored Domino Tilings of Simple and Non-simple Orthogonal Polygons

Chris Worman and Boting Yang

Department of Computer Science, University of Regina

Abstract. We explore the complexity of computing tilings of orthogonal polygons using colored dominoes. A colored domino is a rotatable 2×1 rectangle that is partitioned into two unit squares, which are called faces, each of which is assigned a color. In a colored domino tiling of an orthogonal polygon P, a set of dominoes completely covers P such that no dominoes overlap and so that adjacent faces have the same color. We describe an $O(n)$ time algorithm for computing a colored domino tiling of a simple orthogonal polygon, where n is the number of dominoes used in the tiling. We also show that deciding whether or not a non-simple orthogonal polygon can be tiled with colored dominoes is NP-complete.

1 Introduction

We study a computational tiling problem where the tiles model the commonly used domino game piece. We consider tiling orthogonal polygons with so-called *colored dominoes*, which are rotateable 2×1 rectangles that are partitioned into two colored faces. The colors on the domino faces are used to restrict how the dominoes can be positioned: when two domino faces are positioned next to each other in a tiling, their colors must coincide. Colored dominoes are similar to *Wang tiles* [1, 11, 7, 17], which are unit squares with colored sides. Wang tiles find application in DNA computing [1, 11] and image processing [5, 9].

Many variants of domino tiling problems have been studied, but most of the results have focused on "colorless" dominoes, which are simply 2×1 rotateable rectangles (see, for example, [12, 6, 16, 14, 15, 8]). Results concerning colored domino tilings have only arisen relatively recently [18, 3, 19]. In the colored domino tiling problems studied in [18] and [3], a multiset of dominoes is provided, and in tilings the multiplicity of the dominoes cannot exceed those in the multiset. In [18], the multiplicity of the provided dominoes equals that of the multiplicity of the dominoes used in the tiling. An algorithm is described for computing colored domino tilings of so-called "paths" or "cycles". This algorithm runs in time linear in the number of dominoes used in the tiling. The authors of [18] also consider a colored domino tiling problem where some dominoes have already been positioned on the polygon, and we are asked to decide if the tiling can be completed. This problem is shown to be NP-complete. In [3], Biedl studies two variants of a domino tiling problem. In the first problem, known as *EXACT DOMINO TILING*, the multiplicity of dominoes in the provided set is equal to multiplicity of the dominoes used in the tiling. Biedl shows

X. Deng and D. Du (Eds.): ISAAC 2005, LNCS 3827, pp. 859–868, 2005.

that *EXACT DOMINO TILING* is NP-complete, even for a very restricted class of polygons known as "caterpillars". It is also shown that *EXACT DOMINO TILING* remains NP-complete when the dominoes are restricted to three colors. In the second domino tiling problem studied in [3], which is called *PARTIAL DOMINO TILING*, not all the dominoes are used in the tiling. It is shown that *PARTIAL DOMINO TILING* is NP-complete, and remains so even for so-called "paths" or "cycles". It is also shown that *PARTIAL DOMINO TILING* is NP-complete when the dominoes are restricted to three colors.

In the problems that we study, the set of dominoes used is not a multiset, and each domino has two colors and can be used an unlimited number of times. Thus in this kind of colored domino tiling problem, the set of dominoes provides a set of possible "types" of dominoes that can be used in the tiling. In a previous work we demonstrated that for simple layout polygons that can be tiled with colored dominoes, two colors are always sufficient [19]. We also showed that for tileable non-simple layout polygons, four colors are always sufficient and sometimes necessary.

In the current paper we describe an $O(n)$ time algorithm for computing a colored domino tiling of a simple orthogonal polygon, where n is the number of dominoes used in the tiling[1]. We also show that deciding whether or not a non-simple orthogonal polygon can be tiled with colored dominoes is NP-complete. Due to space constraints, many details have been omitted in this version of the paper.

2 Definitions

A *colored domino* is a rotatable 2×1 rectangle that has been partitioned into two colored unit squares, which we refer to as the *faces* of the colored domino. Since all dominoes considered herein will be colored, we will often refer to "colored dominoes" simply as "dominoes". All polygons that we consider are orthogonal polygons with integer coordinates. We adopt terminology similar to that found in [18, 3, 19], and refer to such polygons as *layout polygons*. We refer to the set of unit squares induced by the integer grid within a layout polygon P as the *squares* in P, which are denoted by $\rho(P)$. A *tiling* is a placement of dominoes from a set D of dominoes onto a layout polygon P such that each square in $\rho(P)$ is covered by exactly one domino face, adjacent faces of dominoes have the same color, and the union of all the dominoes in the tiling is P. We say that $p, q \in \rho(P)$ are *matched* under a tiling if p and q are covered by the same domino in the tiling. A tiling of a layout polygon P therefore describes a perfect matching of $\rho(P)$.

We are concerned with tilings that disallow *twin* dominoes, which are monochromatic colored dominoes. The reason for excluding twin dominoes is necessitated from the problem statement: if we allow twin dominoes then color becomes irrelevant and the problem reduces to that of partitioning a polygon into 2×1 rectangles. Motivated by this restriction, we define D_k as follows: D_1 is a singleton containing a solitary twin domino, and for $k \geq 2$, we define D_k to be the set of $\binom{k}{2}$ non-twin dominoes over k colors.

[1] This algorithm was presented as an extended abstract in [19].

Definition 1. k-tileable: *A layout polygon P is* k-tileable *if and only if there exists a tiling of P using dominoes from D_k.*

Extending this notion slightly, we define a *k-tiling* to be a tiling that uses dominoes from D_k. We study the following tiling decision problem:

Definition 2. TWINLESS TILING:
INSTANCE: *A layout polygon P.*
DECISION: *Is P k-tileable for $k \geq 2$?*

In [19] it was shown that a simple layout polygon is k-tileable for $k \geq 2$ if and only if it is 2-tileable. Thus for simple layout polygons the decision associated with the *TWINLESS TILING* problem can be restated as "Is P 2-tileable?".

Next we introduce the main tools that we use to analyze colored domino tilings.

Definition 3. Twin-Forcing Arrangement: *Let P be a layout polygon with a perfect matching M of $\rho(P)$. A* twin-forcing arrangement *is a set of four squares $p, q, r, s \in \rho(P)$ contained in a 2×2 subrectangle of P, such that p and q are adjacent and matched with each other, while r and s are adjacent, but r is not matched with s.*

One can easily show that a twin-forcing arrangement requires that p and q must be covered with a twin domino. Next we introduce the concepts of *leaves* and *corners* in layout polygons.

Definition 4. Leaf: *Let L be a 2×1 subrectangle of P. Then L is a* leaf *of P if L contains a square p such that p is adjacent to exactly one square in $\rho(P)$.*

Definition 5. Corner: *Let C be a 2×2 subrectangle of P. Then C is a* corner *of P if C contains a square p such that p is adjacent to only two other squares in $\rho(P)$.*

The following two Lemmas about leaves and corners will be essential to our algorithm for simple layout polygons.

Lemma 1. *Let P be a k-tileable layout polygon for $k \geq 2$ that contains a leaf L. In any k-tiling of P there is only one domino that covers L.*

Proof. (omitted)

Lemma 2. *Let P be a k-tileable layout polygon for $k \geq 2$ that contains a corner C. In any k-tiling of P there are only two dominoes that cover C.*

Proof. (omitted)

A key observation about leaves and corners is their existence in simple layout polygons.

Lemma 3. *If P is a simple layout polygon such that $|\rho(P)| \geq 2$, then P contains a leaf or a corner.*

Proof. (omitted)

3 Tiling Simple Layout Polygons

In this Section we describe an algorithm for the *TWINLESS TILING* problem for simple layout polygons. We note that our algorithm not only decides whether or not a simple layout polygon is k-tileable for $k \geq 2$, it also computes such a tiling if one exists. Our algorithm operates on the following decomposition of the layout polygon.

Definition 6. Leaf-Corner Decomposition: A leaf-corner decomposition *of a layout polygon P, denoted $\mathcal{L}(P)$, is defined recursively as follows: if P does not contain a leaf or a corner, then $\mathcal{L}(P) := \emptyset$. Otherwise, let X be a leaf or corner of P, and let $P_1, P_2, ..., P_l$ be the simple layout polygons obtained by removing X from P. Then we define $\mathcal{L}(P) := \{X\} \bigcup \mathcal{L}(P_1) \bigcup \mathcal{L}(P_2) \bigcup ... \bigcup \mathcal{L}(P_l)$.*

We pause here to consider an important note concerning the members of a leaf-corner decomposition $\mathcal{L}(P)$. A member of $\mathcal{L}(P)$ is not necessarily a leaf or a corner of P, although each member of $\mathcal{L}(P)$ is a leaf or a corner of some subpolygon of P. For convenience., we will refer to the members of $\mathcal{L}(P)$ as leaves or corners. We can justify this naming convention by observing that Lemma 1 and Lemma 2 hold for the members of $\mathcal{L}(P)$:

Lemma 4. *Let P be a simple layout polygon that is k-tileable for $k \geq 2$, and let $\mathcal{L}(P)$ be a leaf-corner decomposition of P. If $L \in \mathcal{L}(P)$ is a 1×2 rectangle then in any k-tiling of P for $k \geq 2$, there is only one domino that covers L. Futhermore, if $C \in \mathcal{L}(P)$ is a 2×2 rectangle then in any k-tiling of P, for $k \geq 2$, there are only two dominoes that cover C.*

Proof. (omitted)

We say that a leaf-corner decomposition $\mathcal{L}(P)$ *covers* P if and only if the union of all the members in $\mathcal{L}(P)$ is P.

Lemma 5. *Given a simple layout polygon P and a leaf-corner decomposition $\mathcal{L}(P)$, if $\mathcal{L}(P)$ does not cover P then P is not k-tileable for $k \geq 2$.*

Proof. (omitted)

Due to the following result, the main objective of our algorithm is to compute a perfect matching of $\rho(P)$ that does not contain any twin-forcing arrangements.

Lemma 6. *Let P be a simple layout polygon. P is k-tileable with matching M, for $k \geq 2$, if and only if M is a perfect matching of $\rho(P)$ that does not contain a twin-forcing arrangement.*

Proof. (omitted)

Let L be a leaf, and C be a corner, from a leaf-corner decomposition $\mathcal{L}(P)$ of a simple layout polygon P. According to Lemma 4, the squares in L must be matched, and the squares contained C may be matched one of two ways, in any

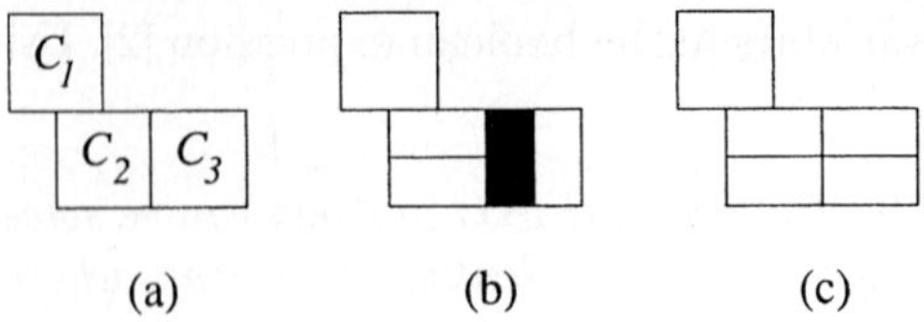

Fig. 1. A matching that forces another matching. (a) Three label corners. (b) Matchings of C_2 and C_3 that cause of twin-forcing arrangement. (c) A matching of C_2 that forces a matching of C_3.

k-tiling of P, for $k \geq 2$. Suppose we are given a perfect matching of $\rho(P)$ that is associated with a k-tiling of P for $k \geq 2$, and let p be the highest and leftmost square in C. If p is matched with its right neighbor, then we say that C is matched "horizontally", and deem C matched "vertically" otherwise. The key to our algorithm is noticing that the matching of a leaf or a corner may "force" neighboring leaves and corners to be matched a certain way if we are to avoid twin-forcing arrangements. To illustrate this, consider the corners C_1, C_2, and C_3 from the leaf-corner decomposition of a layout polygon depicted in Figure 1(a). Suppose the corner C_2 is matched as in Figure 1(b). We can see that C_1 can be matched in either of two ways without causing a twin-forcing arrangement. This is not the case for C_3. If C_3 is matched "vertically" (see Figure 1(b)) then a twin-forcing arrangements arises. This forces C_3 to be matched "horizontally" as in Figure 1(c).

We can generalize these observations in the following Lemma. We say that two members X and Y of a leaf-corner decomposition are *adjacent* if there exists two squares $p, q \in \rho(P)$, such that $p \in X$, $q \in Y$, and p and q are adjacent.

Lemma 7. *Let P be a simple layout polygon with a perfect matching M of $\rho(P)$, and a leaf-corner decomposition $\mathcal{L}(P)$ that covers P. $\mathcal{L}(P)$ satisfies the following rules with respect to M if and only if M does not contain a twinforcing arrangement:*

1. *For each edge e of each leaf $L \in \mathcal{L}(P)$, there is at most one member of $\mathcal{L}(P)$ that both intersects e and is adjacent to L.*
2. *If $C_i, C_j \in \mathcal{L}(P)$ are two corners that share an entire edge, then C_i and C_j are both matched horizontally or they are both matched vertically.*
3. *If $L, C_i \in \mathcal{L}(P)$ are a leaf and a corner such that L shares an entire vertical (resp. horizontal) edge with C_i, then C_i is matched vertically (resp. horizontally).*
4. *If $X, Y \in \mathcal{L}(P)$ are both adjacent to a corner $C_i \in \mathcal{L}(P)$, such that both X and Y intersect the same vertical (resp. horizontal) edge of C_i, then C_i is matched horizontally (resp. vertically).*

Proof. (omitted)

Our algorithm encodes rules (1)-(4) to create a boolean expresion which is an instance of the 2SAT problem, which is known to be solveable in $O(n)$ time, where

n is the number of variables in the boolean expression [2]. Using this idea, we can show the following:

Theorem 1. *The TWINLESS TILING problem can be solved in $O(|\rho(P)|)$ time on simple layout polygons, or equivalently, $O(n)$ time, where n is the number of dominoes needed in the tiling.*

Proof. (omitted)

4 Tiling Non-simple Layout Polygons

In this Section we show that the *TWINLESS TILING* problem is NP-complete for non-simple layout polygons, i.e. layout polygons that contain holes. This problem is in NP since we can "guess" a tiling and verify that this tiling is a k-tiling for $k \geq 2$ in non-deterministic polynomial time.

Our reduction is from a boolean satisfiability problem called *PLANAR 3,4SAT*. Let ϕ be a boolean expression in conjunctive normal form. We define the *inclusion graph* of ϕ, denoted $G(\phi)$, to be the graph whose vertices are the variables and clauses of ϕ. An edge (v_i, c_j) exists in $G(\phi)$ if and only if the variable v_i appears negated or unnegated in clause c_j.

Definition 7. *PLANAR 3,4SAT*
INSTANCE: *A boolean expression ϕ in conjunctive normal form such that: (i) Each clause contains exactly 3 literals, (ii) Each variable appears negated or unnegated at most 4 times in ϕ, and (iii) The inclusion graph $G(\phi)$ is planar.*
DECISION: *Is ϕ satisfiable?*

The PLANAR 3,4SAT problem is known to be NP-complete (see [10] and [13]). Our reduction will construct a layout polygon P that is k-tileable for $k \geq 2$ if and only if ϕ is satisfiable. To accomplish this, we will use a planar embedding of $G(\phi)$ as a template for P. Specifically, we will use a *planar orthogonal grid drawing* of $G(\phi)$, which we denote as $\mathcal{D}(\phi)$, which has the following properties: (i) Each edge of $G(\phi)$ is drawn as a chain of orthogonal line segments, (ii) Each bend in the drawing of an edge lies at an integer valued coordinate, (iii) Each node of $G(\phi)$ is drawn as a point that is located at an integer valued coordinate, and (iv) All edges are disjoint except when their endpoints meet at a node. We maintain the graph theoretic terminology when refering to $\mathcal{D}(\phi)$. For example, when we say "an edge e in $\mathcal{D}(\phi)$" we are referring to the chain of orthogonal line segments that constitute the drawing of the edge e in $G(\phi)$. Similarily, when we say "the vertex v in $\mathcal{D}(\phi)$" we are referring to the point in the integer plane associated with the vertex v in $G(\phi)$.

As already noted, $\mathcal{D}(\phi)$ will be used as a template for P. So that our reduction can be computed in polynomial time, we must ensure that the edges in $\mathcal{D}(\phi)$ are not too long. We can guarantee that each edge in $\mathcal{D}(\phi)$ has a length in $O(n^2)$ if $\mathcal{D}(\phi)$ can occupies $O(n^2)$ space, where n is the number of variables in ϕ. Such a planar orthogonal grid drawing can be computed in $O(n)$ (see, for example, [4]).

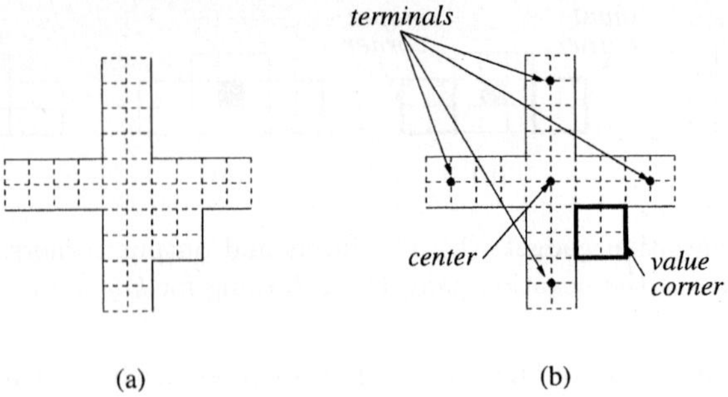

Fig. 2. (a) The variable gadget. (b) The terminals, center, and value corner of a variable gadget.

We note here that as part of our reduction, $\mathcal{D}(\phi)$ will also be scaled by a constant in order to ensure that P can be constructed correctly. The exact value of this constant will be discussed in what follows.

To construct P, we will overlay various gadgets over the nodes and edges in $\mathcal{D}(\phi)$. These gadgets, which are layout polygons, will correspond to different aspects of ϕ. We employ gadgets that represent variables, negation, and clauses. We also make use of a special gadget called a *spacer* that is used to ensure proper connnections between variable gadgets and clause gadgets.

Variable Gadget: A variable gadget represents a particular variable from ϕ. The tiling of a variable gadget will correspond to the variable being assigned a particular truth value. The variable gadget is depicted in Figure 2(a). Figure 2(b) identifies important aspects of the variable gadget. The *terminals* of a variable gadget will be used to attach a variable gadget to the appropriate clause gadgets. The *center* will be used to position a variable gadget on the appropriate node in $\mathcal{D}(\phi)$. Specifically, if a variable gadget is representing a variable v_i, then the variable gadget is positioned in the plane so its center is located at the node associated with v_i in $\mathcal{D}(\phi)$. The *value corner* is used to "set" the truth value of variable. Specifically, the orientation (i.e. horizontal or vertical) of the matching of the value corner with respect to a k-tiling for $k \geq 2$ will determine the truth value of the variable, where horizontal corresponds to "true", and vertical corresponds to "false". Although we have omitted the details in this version of the paper, we can show the following:

Lemma 8. *In a k-tiling of a variable gadget for $k \geq 2$, the value corner of a variable gadget is matched horizontally (resp. vertically) if and only if all squares in the variable gadget are also matched horizontally (resp. vertically).*

Negation Gadget: The "truth" value of a variable gadget will be transmitted to a clause gadget according to the orientation of the tiling of the variable gadget. When a variable appears negated in a clause, we must "flip" this orientation. The gadget that accomplishes this is called the negation gadget and is depicted in Figure 3(a). Figure 3(b) identifies two important regions of the negation gadget called the *input corner* and the *output corner*. Figure 3(c) and (d) illustrates the

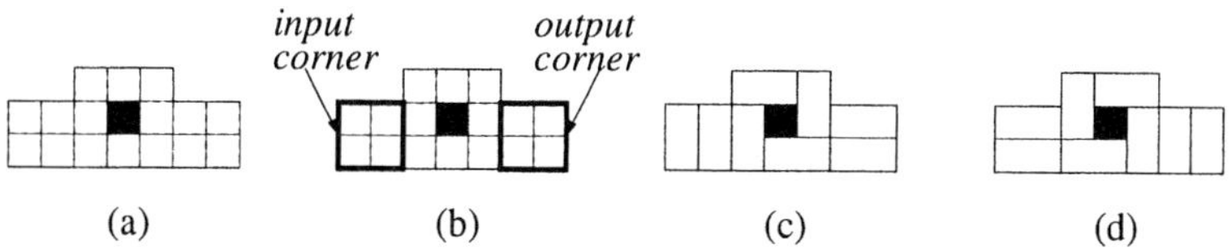

Fig. 3. (a) The negation gadget. (b) The input and output corners. (c)(d) The two possible matchings of the negation gadget in a k-tiling for $k \geq 2$.

fact that the orientation of the matching associated with a k-tiling for $k \geq 2$ of the input corner is always opposite to that of the output corner. Thus we have demonstrated the following:

Lemma 9. *In any k-tiling of a negation gadget for $k \geq 2$, if the input corner of the negation gadget is matched horizontally (resp. vertically), then the output corner is matched vertically (resp. horizontally).*

Clause Gadget: A clause gadget is used to represent a clause from ϕ (see Figure 4(a)). Figure 4(b) identifies important aspects of the clause gadget. The *terminals* in a clause gadget are used to connect a clause gadget to the appropriate variable gadgets. The *center* of the clause-gadget is used to position a clause gadget on the appropriate vertex in $\mathcal{D}(\phi)$. Specifically, if a clause gadget is representing a clause c_i, then the clause gadget is positioned in the plane so that the center is located at the vertex associated with c_i. The clause gadget is then rotated if necessary so that each of the terminals intersect a line segment that is incident with the node associated with c_i.

The clause gadget is constructed such that it is k-tileable for $k \geq 2$ if and only if at least one of the variable gadgets that is connected to it is tiled horizontally, i.e. set to true. The details of this have been omitted in this version of the paper, but we can show the following:

Lemma 10. *In any k-tiling for $k \geq 2$ of a clause gadget, at least one of the 2×2 squares that surround the terminals must be tiled horizontally, i.e. in the "true" orientation.*

The final gadget that we require is called the *spacer* gadget, which is used to ensure a proper connection between variable and clause gadgets (see Figure 5).

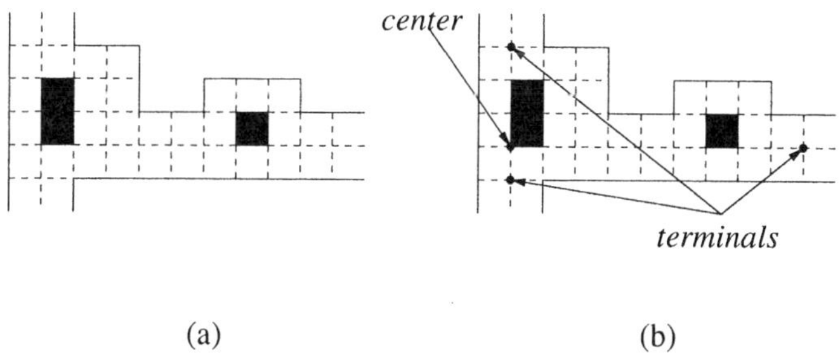

Fig. 4. (a) The clause gadget. (b) The terminals and center of the clause gadget.

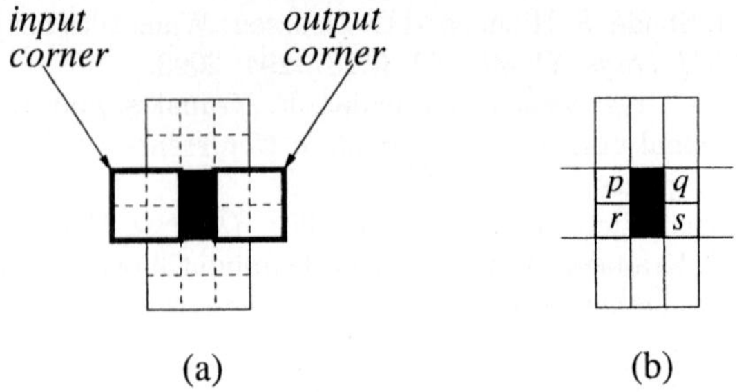

Fig. 5. (a) The spacer gadget with the input and output corners identified. (b) The forced matchings of the spacer gadget.

We can show that the spacer gadget is an identity gadget in the sense that the input and output corners are always matched with the same orientation in any k-tiling for $k \geq 2$. Also, the distance between the input and ouput corners is odd, and hence the spacer gadget is used to adjust the parity of connections between variable and clause gadgets. This is needed when we make connections between variable and clause gadgets on line segments of odd length in $\mathcal{D}(\phi)$. The details concerning the spacer gadget have been omitted in this version of the paper.

Lemma 11. *In any k-tiling of a spacer gadget for $k \geq 2$, the input and output corners are either both matched horizontally or both vertically.*

From our construction we can see that P is k-tileable if and only if ϕ is satisfiable. Thus we have the following:

Theorem 2. *The TWINLESS TILING problem is NP-complete for non-simple layout polygons.*

Acknowledgements

We would like to thank Therese Biedl for useful comments regarding the results presented in Section 3.

References

1. L. Adleman, J. Kari, L. Kari, and D. Reishus. On the decidability of self-assembly of infinite ribbons. *Proceedings of FOCS 2002, IEEE Symposium on Foundations of Computer Science*, pages 530–537, 2002.
2. B. Aspvall, M. F. Plass, and R. E. Tarjan. A linear-time algorithm for testing the truth of certain quantified boolean formulas. *Inf. Process. Lett.*, 8(3):121–123, 1979.
3. T. Biedl. The complexity of domino tiling. In *Canadian Conference on Computational Geometry (CCCG'05)*, pages 187–190, 2005.
4. T. Biedl and G. Kant. A better heuristic for orthogonal graph drawings. *Comput. Geom. Theory Appl.*, 9(3):159–180, 1998.

5. M. F. Cohen, J. Shade, S. Hiller, and O. Deussen. Wang tiles for image and texture generation. *ACM Trans. Graph.*, 22(3):287–294, 2003.
6. G. Csizmadia, J. Czyzowicz, L. Gasieniec, F. Kranakis, and J. Urrutia. Domino tilings of orthogonal polygons. In *Canadian Conference on Computational Geometry (CCCG'99)*, pages 154–157, 1999.
7. K. Culik. An aperiodic tiling of 13 wang tiles. *Discrete Math*, 160:245–251, 1996.
8. J. Czyzowicz, E. Kranakis, and J. Urrutia. Domino tilings and two-by-two squares. In *Canadian Conference on Computational Geometry (CCCG'97)*, 1997.
9. S. Hiller, O. Deussen, and A. Keller. Tiled blue noise samples. In *VMV '01: Proceedings of the Vision Modeling and Visualization Conference 2001*, pages 265–272. Aka GmbH, 2001.
10. K. Jansen and H. Müller. The minimum broadcast time problem for several processor networks. *Theoretical Computer Science*, 147(1–2):69–85, 1995.
11. J. Kari. Infinite snake tiling problems. In M. Ito and M. Toyama, editors, *Developments in Language Theory, 6th International Conference, DLT 2002, Kyoto, Japan, September 18-21, 2002, Revised Papers*, volume 2450 of *Lecture Notes in Computer Science*, pages 67–77. Springer, 2003.
12. C. Kenyon and R. Kenyon. Tiling a polygon with rectangles. In *33rd Fundamentals of Computer Science (FOCS)*, pages 610–619, 1992.
13. D. Lichtenstein. Planar formulae and their uses. *SIAM J. Comput.*, 11(2):329–343, 1982.
14. M. L. N. Elkies, G. Kuperberg and J. Propp. Alternating sign matrices and domino tilings, part 1. *Journal of Algebraic Combinatorics 1*, pages 111–132, 1992.
15. M. L. N. Elkies, G. Kuperberg and J. Propp. Alternating sign matrices and domino tilings, part 2. *Journal of Algebraic Combinatorics 1*, pages 219–234, 1992.
16. J. Propp. A reciprocity theorem for domino tilings. *The Electronic Journal of Combinatorics*, 8, 2001.
17. H. Wang. Proving theorems by pattern recognition. ii. *Bell Systems Technical Journal*, (40):1–41, 1961.
18. C. Worman and M. Watson. Tiling layouts with dominoes. In *Proceedings of the 16th Canadian Conference on Computational Geometry (CCCG'04)*, pages 86–90, 2004.
19. C. Worman and B. Yang. On the computation and chromatic number of colored domino tilings. In *Canadian Conference on Computational Geometry (CCCG'05)*, 2005.

Optimal Paths for Mutually Visible Agents

Joel Fenwick and V. Estivill-Castro

Institute for Integrated and Intelligent Systems,
Griffith University, Australia
j.fenwick@student.gu.edu.au, v.estivill-castro@griffith.edu.au

Abstract. We present linear-time algorithms for a pair of robots to travel inside a simple polygon on paths of total minimum length while maintaining visibility with one another. We show that the optimal paths for this mutually visible constraint are almost always each agent's shortest path. The this may not happen only on a sub-case of when the line of visibility of the source points crosses the line of visibility of the target points. We also show that the travel schedule is computable, but that it also suffers from a pathological case.

Keywords: Polygon, Shortest Path, Teams of Robots, Line of Visibility.

1 Introduction

Recently, tasks performed by groups of robots have sparked significant research in robotics and multi-agent systems because many potential benefits arise (better resource utilization, due to specialization and load balancing, and robustness as failure of one agent does not necessarily imply the task is impossible by the team). The focus has been on teams of robots performing task related to mobility and in particular to exploration of terrains and map construction [1,7,11,13,12]. Settings vary from situations where the robots have previous knowledge of the physical layout of their environment to settings where such knowledge is distributed and requires fusion [10]. Some of these research efforts consider constraints, like maintaining line of sight or proximity to enable communication [9].

However, little has been explored from the perspective of the theoretical possibilities of some of these tasks in abstractions of the problem. There is extensive research in shortest paths problems [8]. For example, if one robot is to navigate a large room for which it has a map that can be modelled as a simple polygon with n vertices, the task of finding the shortest path from a source point $\boldsymbol{s}$ to a target $\boldsymbol{t}$ can be achieved in $O(n)$ time (first triangulate the polygon in $O(n)$ time [3], and then find the shortest path in the resulting planar graph [5]). If the room is more complicated and needs to be modelled as a polygon with holes, then the complexity of the problem blows out to exponential time since there may be an exponential number of shortest paths. However, if the homotopy type is given [6], the problem is back to polynomial time and much research has been recently dedicated to this variant of the problem [2,4]. If there are multiple agents, and agent a_i is to travel from point $\boldsymbol{s}_i$ to point $\boldsymbol{t}_i$, then we can minimize the total

X. Deng and D. Du (Eds.): ISAAC 2005, LNCS 3827, pp. 869–881, 2005.

distance travelled by computing the shortest path for each robot (unless there is some other constraint). We study here what we believe is a reasonable constraint that one may require of a pair of robots. Namely, we require the robots to start in a configuration where they are in line of sight and maintain that visibility condition until they reach their destinations. We show that this constraint involves a series of interesting results. Some of these are positive, in particular, we show settings where two robots can maintain mutual visibility and travel each on their shortest path. But, we also show in Section 4 that small changes in the assumptions result in negative results. With this in mind Section 2 presents the necessary definitions in full detail.

2 The Moving Points Case

We establish necessary definitions to present the problem formally.

Definition 1. *A sequence* $\langle \boldsymbol{x}_0, \ldots, \boldsymbol{x}_{n-1} \rangle$ *of different 2D points such that only consecutive line segments* $\overline{\boldsymbol{x}_{i-1 \bmod n}\boldsymbol{x}_i}$ *and* $\overline{\boldsymbol{x}_i\boldsymbol{x}_{i+1 \bmod n}}$ *intersect and they do so only at* $\boldsymbol{x}_i$ *(for* $i = 0, \ldots, n-1$*) defines a* simple polygon. *The polygon consists of both the interior and boundary. The line segment between two points* $\boldsymbol{u}$ *and* $\boldsymbol{v}$ *is* $\overline{\boldsymbol{uv}}$ *and has Euclidean length len(*$\overline{\boldsymbol{uv}}$*). The line through* $\boldsymbol{u}$,$\boldsymbol{v}$ *is line(*$\boldsymbol{u}, \boldsymbol{v}$*).*

A natural restriction explored in this paper is that agents must maintain a line of sight. This is a reasonable abstraction for agents that communicate via infrared signals, gestures, and even other wireless methods.

Definition 2. *Two points in a polygon are* mutually visible *if the line segment joining the points does not intersect the exterior of the polygon. In this case, the line segment is named the* sightline *for those two points.*

It is customary to model a robot as a point [8] (but see Section 4 for the impact of changing this assumption). Note that the shortest path between two points $\boldsymbol{s}$ and $\boldsymbol{t}$ in a polygon is composed of a sequence of line segments starting at $\boldsymbol{s}$ and ending at $\boldsymbol{t}$ and joining at vertices of the polygon [8]. For the purposes of this discussion a path containing three or more Co-linear vertices (of the polygon) in sequence will not be simplified to a single line segment.

Definition 3. *A* mutually visible path problem*(MVPP) is given by a simple polygon* P *and four points (*$\boldsymbol{s}_1$, $\boldsymbol{s}_2$, $\boldsymbol{t}_1$ *and* $\boldsymbol{t}_2$*). The points* $\boldsymbol{s}_1$ *and* $\boldsymbol{s}_2$ *are mutually visible, and so are* $\boldsymbol{t}_1$ *and* $\boldsymbol{t}_2$. *Two agents* a_1 *and* a_2 *are modelled as points capable of moving at constant speed (this constant is the same for both agents) or waiting (with acceleration being instantaneous). Find paths* π_1 *and* π_2 *for the agents and sequences of pauses* w_1 *and* w_2 *such that agents moving along the paths and pausing according to the sequences are mutually visible at all times.*

This abstraction allows us to explore fundamental questions. The first question is if such MVPPs always have a feasible solution. For this issue we need to introduce some notation. In an MVPP, the *start sightline* is $S = \overline{\boldsymbol{s}_1\boldsymbol{s}_2}$ while the *target sightline* is $T = \overline{\boldsymbol{t}_1\boldsymbol{t}_2}$. Together they will be referred as *the sightlines* of

the MVPP. For an MVPP, let Π_1 and Π_2 be the shortest paths between (s_1,t_1) and (s_2,t_2) respectively. We denote the number of line segments in path Π_i by $|\Pi_i|$.

Definition 4. *Two sightlines S and T are said to* cross *if $S \cap T = \{c_p\}$ and $c_p \notin \{s_1, s_2, t_1, t_2\}$. The point c_p will be called the crossing point. This definition is extended to apply to paths in the obvious way.*

If $\|S \cap T\| > 1$ or c_p is an endpoint, we say that the sightlines do not cross, but intersect. We first make the observation that, if the sightlines S and T cross, the MVPP does have a solution. Consider Fig. 1 (the dotted lines are the sightlines, the dashed lines are shortest paths and the thicker lines are part of the polygon boundary). When we discuss optimal solutions (shortest in Euclidean metric), this special case will be important. One feasible solution is simply for agent a_1 to move to the crossing point c_p along the starting sightline while agent a_2 is stationary. Then agent a_1 is stationary while agent a_2 moves to the crossing point of the MVPP's sightlines also along the starting sightline. Note that we allow the agents to occupy the same point. From here, agent a_1 travels to its target along the target sightline while a_2 is stationary. The last step in the schedule is for agent a_2 to move to its target along the target sightline while a_1 is stationary. This example also illustrates the type of feasible solutions for an MVPP. In the case where the sightlines do not cross we not only establish the existence of solutions, but we can characterize optimal solutions in the following sense.

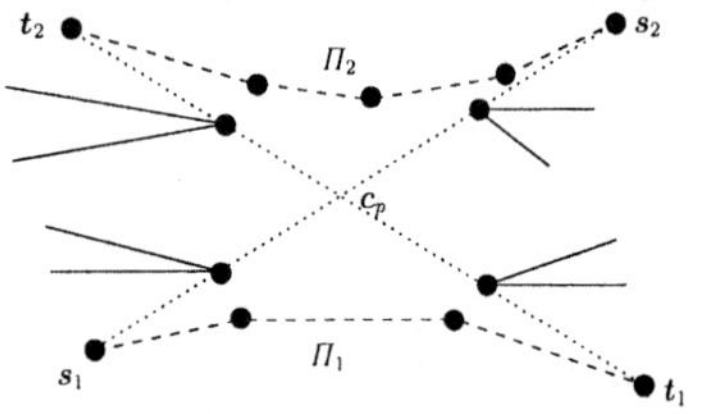

Fig. 1. A MVPP where the sightlines cross

Definition 5. *In an MVPP, a solution (π_1, π_2, w_1, w_2) is optimal if the total Euclidean length of the paths π_1 and π_2 is minimal among feasible solutions of the MVPP.*

The main results of this paper are the following theorems:

Theorem 1. *If the sightlines of an MVPP do not cross, then the shortest paths Π_1 and Π_2 between (s_1, t_1) and (s_2, t_2) respectively are a solution, and therefore they are an optimal solution. Moreover, computing the sequence of pauses for both agents requires linear time.*

Theorem 2. *If the sightlines of an MVPP cross, then either: the shortest paths Π_1 and Π_2 are the paths of an optimal solution, or paths π_1' and π_2' (which are the paths for an optimal solution) can be produced from Π_1 and Π_2 with a combined length as close as desired to Π_1 and Π_2. Moreover, distinguishing the two cases above is computable in linear time.*

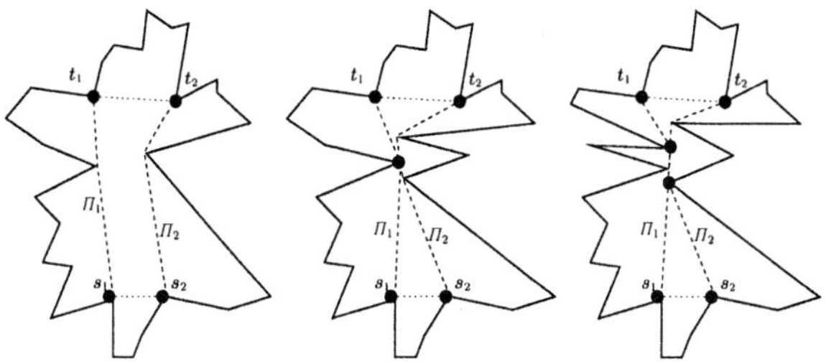

Fig. 2. No obstacle can be in the interior of the region bounded by Π_1, Π_2, S, T

We treat the crossing and non-crossing cases separately.

2.1 Non-crossing Case

We prove Theorem 1 using the following two lemmas.

Lemma 1. *Consider an MVPP where the sightlines do not cross and* Π_1, Π_2 *do not cross (other than at a vertex on both paths). If vertices* $\boldsymbol{u} \in \Pi_1, \boldsymbol{v} \in \Pi_2$ *are mutually visible then:*

1. *- If* $\boldsymbol{u}$ *and* $\boldsymbol{v}$ *are not final vertices, then they can see their successors* $\boldsymbol{u}_{+1}$ *and* $\boldsymbol{v}_{+1}$ *respectively.*
2. *- The interior of the region R bounded by the shortest paths and the sightlines (ie the region bounded by* Π_1,Π_2,S *and* T*) is free of obstacles.*
3. *- The quadrilateral* $\boldsymbol{v}$,$\boldsymbol{v}_{+1}$,$\boldsymbol{u}_{+1}$,$\boldsymbol{u}$ *is free of obstacles.*

Proof. Paths are composed of unbroken line segments so no obstacle can block vision from a vertex to its successor. The second assertion is illustrated in Fig 2. No obstacle can cut any link in the path (assertion 1), no obstacle can cut either of the sightlines and there are no holes in the polygons. Hence the interior of the region must be free of obstacles. Since the quadrilateral in assertion 3 is a subset of R, it must be free of obstacles. □

Lemma 2. *Consider an MVPP where the sightlines do not cross and a vertex* $\boldsymbol{u}$ *on one shortest path can see vertices* $\boldsymbol{v}$ *and* $\boldsymbol{v}_{+1}$ *on the other shortest path. Then,* $\boldsymbol{u}$ *can see every point on the path between* $\boldsymbol{v}$ *and* $\boldsymbol{v}_{+1}$.

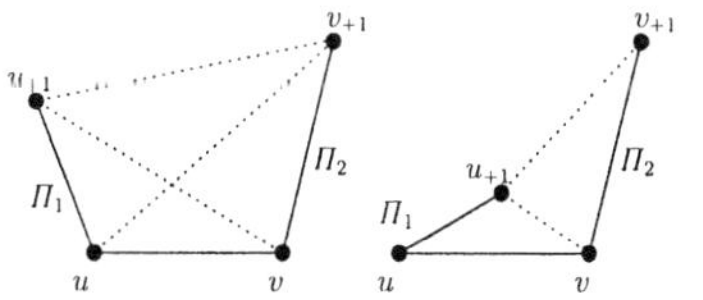

Fig. 3. Two successive points on Π_1 and Π_2 where $\boldsymbol{u}$ and $\boldsymbol{v}$ are mutually visible. The solid lines are known to not have obstacles obstructing them.

Proof. Proof by contradiction: Suppose $\exists$ point $\boldsymbol{p}$ on the path between $\boldsymbol{v}$ and $\boldsymbol{v}_{+1}$ which is not visible to $\boldsymbol{u}$. The region bounded by $\overline{\boldsymbol{uv}}$, $\overline{\boldsymbol{uv}_{+1}}$ and the path must be free of obstacles (proof of Lemma 1) so the path must be non-convex. But we know that the path must be convex since it has minimal length and has no obstacles affecting it on one side. □

We now prove Theorem 1. Suppose that the agents have successfully moved to $\boldsymbol{u} \in \Pi_1$ and $\boldsymbol{v} \in \Pi_2$ respectively and that $\boldsymbol{u}_{+1} \in \Pi_1$ and $\boldsymbol{v}_{+1} \in \Pi_2$ are the next vertices on each path. The quadrilateral defined by $\boldsymbol{u}, \boldsymbol{u}_{+1}, \boldsymbol{v}_{+1}, \boldsymbol{v}$ must be free of obstacles (see Figure 3, and Lemma 1). Thus, it must be possible to advance one of the agents to its next point. Note it is not always possible to advance both agents simultaneously since if the quadrilateral has a reflex vertex vision may be interrupted.

The pause sequences can be updated to take this movement into account and the process repeated. If the paths cross (in the same sense as defined previously for sightlines) then an extra vertex must be introduced at the intersection point (allowing us to apply Lemma 1). If one path has fewer segments than the other, the other agent can move to its target without pausing (Lemma 2). So we have paths and a sequence of pauses which solves the problem. The solution is optimal since the paths used are of minimal length. □

The claim regarding linear-time computability is left till Section 3.

2.2 Crossing Case

This section proves Theorem 2. While we have shown that Fig. 1 has a solution, it illustrates the difficulties of the optimal solution being the shortest paths. Note that if one agent moves without the other moving as well, then the sightlines will intersect the exterior of the polygon. That is, sightlines crossing as shown in Fig. 1 imply that neither agent can start its sequence with a pause. For the rest of this section, assume that we have an MVPP where the sightlines cross. We also make the following simplifying observation. Consider an MVPP where the lines cross, and assume that the first line segment in $\Pi_{i\in\{1,2\}}$ is on the start sightline. Then the MVPP is equivalent to another MVPP where $\boldsymbol{s}_i$ is replaced by the endpoint of the first line segment of Π_i. This is simply because agent a_i can move along the start sightline to the endpoint of the first line segment in Π_i without losing site of the other agent. A similar reduction holds for the last segment of Π_i. Thus, in what follows we assume the sightlines cross and each of the shortest paths coincides with the sightlines only at each of their sources and each of their targets.

In order to identify conditions under which an optimal solution still consists of the shortest paths we start with the following lemma.

Lemma 3. *Consider an MVPP where the sightlines cross. Then, each shortest path $\Pi_{i\in\{1,2\}}$ must be convex chain, and for all points $\boldsymbol{p} \in \Pi_1$ and all points $\boldsymbol{q} \in \Pi_2$, either the line segment $\overline{\boldsymbol{pq}}$ is a sightline or it is not a sightline and $\overline{\boldsymbol{pq}} \cap (\Pi_1 \cup \Pi_2) = \{\boldsymbol{p}, \boldsymbol{q}\}$ (that is, the obstacle blocking vision must be a section of the boundary that is not part of the shortest paths).*

Proof. First we show Π_1 is convex and by symmetry Π_2 is convex. But, as in Lemma 2, this is easy to see from properties of shortest paths, the only way a shortest path changes curvature is if there are boundary vertices on both sides of the path. Since the sightlines ensure that there are no polygon vertices on one side of the path, the path must be convex (Fig. 1 helps visualise this argument).

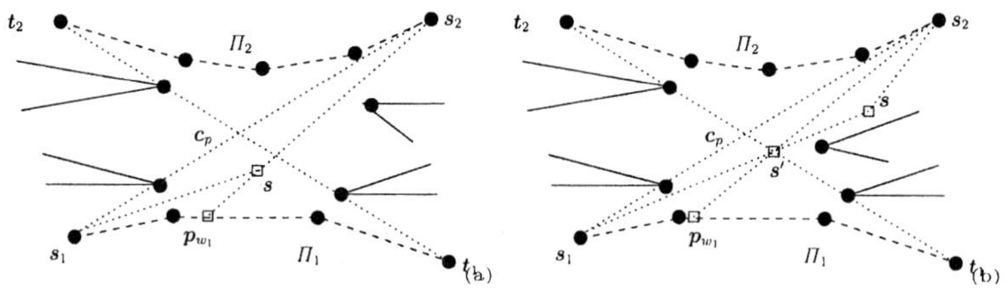

Fig. 4. Agents can start along their shortest path if both see an additional point s

The second claim now follows from the first. Consider the area bounded by the line segments $\overline{s_1 c_p}$ and $\overline{c_p t_1}$ and Π_1. This area does not intersect the exterior of the polygon and similarly for the corresponding area for Π_2. Thus, any possible visibility problems must come from some other part of the polygon. □

We will show that if at least one of the agents can get started (and can land at its target), then the shortest paths are also solution and thus such solution is optimal. We start by describing the conditions under which they can get started.

Lemma 4. *In an MVPP where sightlines cross, if there is a point $s \notin line(s_1, s_2)$ such that s is visible to both s_1 and s_2, then the agents can start a feasible solution along the shortest paths Π_1 and Π_2.*

Proof. Without loss of generality, assume s is below the start sightline (refer to Fig. 4). If s is above the start sightline reverse the roles of s_1 and s_2. Then, the interior of the triangle s_1, s_2, s is empty of obstacles. Consider two cases:

1.- If s is below or on the target sightline, then the segment $\overline{s_2 s}$ can be extended until Π_1 (refer to Fig. 4 (a)). Let p_{w_1} be the intersection of Π_1 and the line through s_2 and s. Then, clearly agent a_1 can start moving until p_{w_1} while agent a_2 waits at s_2 since the interior of the area bounded by the start sightline, the sightline $\overline{s_2 p_{w_1}}$ and Π_1 is empty of obstacles. Note that after a_1 reaches p_{w1}, a_2 can move to the intersection of line(p_{w1}, c_p) and Π_2.
2.- The point s is above the target sightline (refer to Fig. 4 (b)). Then, consider the crossing point s' of the sightline $\overline{s_1 s}$ and the target sightline. Using s' this case reduces to the one before. □

A symmetric proof provides the following lemma.

Lemma 5. *Consider an MVPP where sightlines cross. If there is a point $t \notin \overline{t_1 t_2}$ such that t is visible to both t_1 and t_2, then the agents can terminate a feasible solution along the shortest paths Π_1 and Π_2.*

Now, we only have one more case. But its treatment remains delicate. This is the case where the start and target sightlines cross and no agent can start with a pause. In this case, the set of points visible to both s_1 and s_2 is the maximal unbroken segment of line(s_1, s_2). Because both agents must move simultaneously,

an optimal solution would coincide with the shortest paths if there is a point s'_1 in the first line segment of Π_1, and a point s'_2 in the second line segment of Π_2 such that len($\overline{s'_1 s_1}$) =len($\overline{s'_2 s_2}$) and the sightline $\overline{s'_1 s'_2}$ does not cross the exterior of the polygon. Moreover, it must be the case that for all d with $0 \leq d < len(\overline{s'_1 s_1})$, the sightline between $s^d{}_1 = s_1 + d(s'_1 - s_1)/len(\overline{s'_1 s_1})$ and $s^d{}_2 = s_2 + d(s'_2 - s_2)/len(\overline{s'_1 s_1})$ does not cross the exterior of the polygon. We show next that this condition can be tested in linear time. For convenience of the argument consider the situation after a translation and a rotation that makes $s_2 = (0,0)$ and s_1 to have its x coordinate 0 and y coordinate $s_{1_y} = -len(\overline{s_1 s_2})$. Fig 5 illustrates this setting.

With the length $L = -s_{1_y}$ of the starting line known we can find an expression for the coordinates of the points the agents travel in. Namely,

$$s^d{}_1 = (d\sin(\pi - \alpha), -(L + d\cos(\pi - \alpha))) = (d\sin\alpha, -L + d\cos\alpha)$$

and $s^d{}_2 = (-d\sin(\pi - \beta), d\cos(\pi - \beta)) = (-d\sin\beta, -d\cos\beta)$. Therefore, we can express the sightline parametrically since $\overline{s^d{}_2 s^d{}_1} = (-d\sin\beta + \lambda(d\sin\alpha + d\sin\beta), -d\cos\beta + \lambda(d\cos\alpha - L + d\cos\beta))$, with $0 \leq \lambda \leq 1$. We are interested in the point z where the sightline of the agents intersects the start sightline. This point has $x = 0$, and this gives $\lambda_z = \sin\beta/(\sin\alpha + \sin\beta)$. With this value we obtain the following derivation.

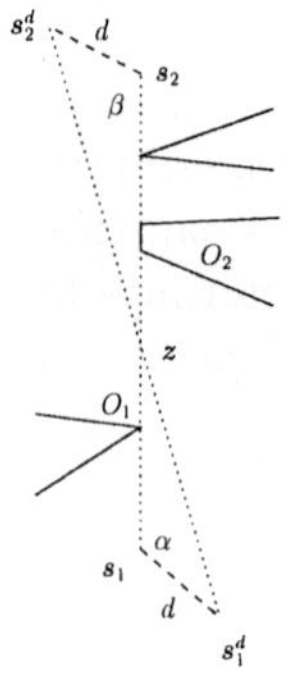

Fig. 5. Both agents must move

$$\begin{aligned} z_y &= -d\cos\beta + \frac{\sin\beta}{\sin\alpha + \sin\beta}(d\cos\alpha - L + d\cos\beta) \\ &= \frac{-d\cos\beta(\sin\alpha + \sin\beta) + \sin\beta(d\cos\alpha - L + d\cos\beta)}{\sin\alpha + \sin\beta} \\ &= \frac{-d\sin\alpha\cos\beta - d\sin\beta\cos\beta + d\sin\beta\cos\alpha - L\sin\beta + d\sin\beta\cos\beta}{\sin\alpha + \sin\beta} \\ &= \frac{d(-\sin\alpha\cos\beta + \sin\beta\cos\alpha) - L\sin\beta}{\sin\alpha + \sin\beta} = \frac{d\sin(\beta - \alpha) - L\sin\beta}{\sin\alpha + \sin\beta} \end{aligned}$$

We are now in a position to analyse the change in z_y as the agents travel. This gives $\partial z_y/\partial d = \frac{\sin(\beta-\alpha)}{\sin\alpha + \sin\beta}$ and therefore, we find that that $\partial z_y/\partial d = 0$ implies $\sin(\beta - \alpha) = 0$. We see that in this case $\beta = \alpha$ (since $\alpha, \beta \in (0, \pi)$). In the case where $\partial z_y/\partial d > 0$, $\sin(\beta - \alpha) > 0$ or $\beta > \alpha$ (ie the intersection point z moves closer to s_2) Conversely, it goes closer to s_1 when $\partial z_y/\partial d < 0$. This is $\sin(\beta - \alpha) < 0$ or $\beta < \alpha$.

With respect to Fig. 5, we say that a vertex v of the polygon and on the start sightline *touches left* if neither of the polygon edges incident on v intersect the interior of the half-plane to the right of the start sightline (the obstacle O_1 illustrates this). Similarly, we say *touches right* and an example is the vertices of the obstacle O_2. Now, traverse the boundary of the polygon and find the closest vertex v_r to s_1 that touches right and the closest vertex v_l to s_2 that touches left. Any sightlines for a successful solution must pass through the interval $\overline{v_r v_l}$

on the start sightline. We know this interval must exist since the target sightline must pass through it. Moreover, we can verify this in $O(n)$ time.

Lemma 6. *In an MVPP where the sightlines cross and all points visible to both* $\boldsymbol{s}_1$ *and* $\boldsymbol{s}_2$ *are in* $line(\boldsymbol{s}_1, \boldsymbol{s}_2)$ *it is possible to verify in linear time if the shortest paths can start an optimal solution.*

The proof of this lemma is placed in the appendix but these results lead to the following definition.

Definition 6. *We say that a feasible solution can* start *on shortest paths if there is a point* $s \notin line(\boldsymbol{s}_1, \boldsymbol{s}_2)$ *such that* $\boldsymbol{s}$ *is visible to both* $\boldsymbol{s}_1$ *and* $\boldsymbol{s}_2$*, or if* $\boldsymbol{z}_y$ *is between* $\boldsymbol{v}_R$ *and* $\boldsymbol{u}_L$ *(as defined above).*

A similar definition with can also be tested in linear time can be made for when a feasible solution can *terminate* on the shortest paths. We show here that once we can start and terminate, then there is obstacle free space so the agents can alternate between one moving and the other pausing, always making progress. First, let us argue that we can always move one agent while the other waits.

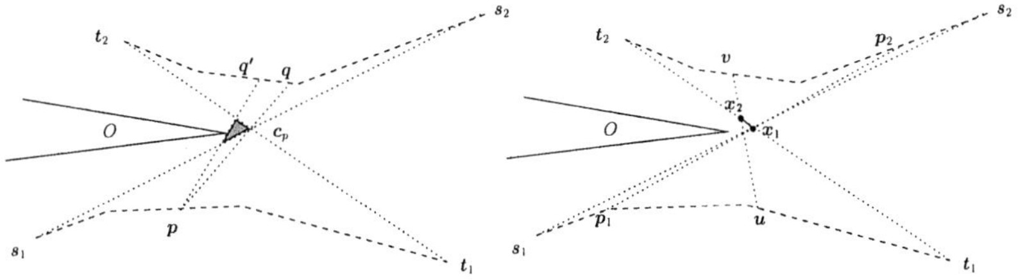

Fig. 6. (a) One crab step made by a_2. (b) A sightline through $\overline{\boldsymbol{x}_1\boldsymbol{x}_2}$.

Lemma 7. *In an MVPP where the sightlines cross: Once the agents start, there is no need to move them simultaneously unless this is necessary to terminate.*

Proof. Consider a situation such as in Figure 6(a), in this case $\overline{\boldsymbol{pq}}$ is a sightline. Agent a_2 moves to $\boldsymbol{q}'$ where further progress is prevented by obstacle O. Now the region between $\boldsymbol{pq}'$ and the start and end sightlines (the shaded region in the figure) must be free of obstacles (Lemma 3) so a_1 must be able to make some progress while a_2 remains at $\boldsymbol{q}'$. □

Lemma 8. *In an MVPP with crossing sightlines, once the agents start, there are points* $\boldsymbol{p}_1 \in \Pi_1$ *and* $\boldsymbol{p}_2 \in \Pi_2$ *such that* $\overline{\boldsymbol{p}_1, \boldsymbol{p}_2}$ *is a sightline and there are points* $\boldsymbol{x}_1$ *and* $\boldsymbol{x}_2$ *in the polygon such that:*

1. $\overline{\boldsymbol{x}_1\boldsymbol{x}_2}$ *intersects* $\overline{\boldsymbol{p}_1\boldsymbol{p}_2}$ *at a single point (ie* $\overline{\boldsymbol{x}_1\boldsymbol{x}_2} \cap \overline{\boldsymbol{p}_1\boldsymbol{p}_2} = \{\boldsymbol{p}\}$ *for some point* $\boldsymbol{p}$ *in the polygon)*

2. *$\forall u \in \Pi_1$ past $\mathbf{p}_1$ and $\forall v \in \Pi_2$ past $\mathbf{p}_2$ such that $\overline{\boldsymbol{uv}} \cap \overline{\boldsymbol{x}_1\boldsymbol{x}_2}$ is a singleton, then $\overline{\boldsymbol{uv}}$ is a sightline. (Note that if $\boldsymbol{u} = \mathbf{t}_1$ and $\boldsymbol{v} = \mathbf{t}_2$, we have a sightline,but in that case $\overline{\boldsymbol{uv}} \cap \overline{\boldsymbol{x}_1\boldsymbol{x}_2} = \overline{\boldsymbol{x}_1\boldsymbol{x}_2}$)*

Using the points $\boldsymbol{x}_1$ and $\boldsymbol{x}_2$ in the lemma above, we can move agents alternately (switching to the other agent when the sightline touches $\boldsymbol{x}_1$ or $\boldsymbol{x}_2$).

Lemma 9. *For all $d > 0$, the alternating process advances both agents until they are both at less that d from their targets.*

The ability to start and finish combined with the ability to reach a point arbitrarily close to the target gives us the required pause sequences. In cases where Π_i are not part of an optimal solution π_i' can be produced by starting and finishing at slightly steeper angles than Π_i. This concludes Theorem 2. The detail of this proofs as well as the proofs of the last two lemmas are in the appendix.

3 Complexity

Deciding if the shortest paths are an optimal solution does not require polygon triangulation in the case that the sightlines do not cross. So this check is constant time in this case. To produce the schedule would require linear time ($O(|\Pi_1| + |\Pi_2|)$) after the production of the shortest paths because the proof of Theorem 1 ensures we always advance at least one vertex of a shortest path. In theory, producing the shortest paths is $O(n)$ by triangulation, so overall producing the schedule takes linear time.

However in the case where the sightlines cross, we need to find if the boundary of the polygon creates a situation such as in Fig 4. Thus, any algorithm here will require $\Omega(n)$ time. But after computation of the shortest-paths in $O(n)$ time, the alternation algorithms is $O(n + \|w_1\|)$ time where the length $\|w_1\|$ does not depend on n. Depending on the model of computation we use, we may have several bounds for the computational time that do not depend on n (an alternative we may wish to explore is if k bits are allowed for each coordinate of the vertices of the polygon). In any case, $\|w_1\| = O(\max_i \|\overline{\boldsymbol{s}_i\boldsymbol{t}_i}\|/m)$.

4 Final Remarks

What effect do changes in the problem definitions have? Polygons with non-zero genus would significantly complicate matters since we could no longer assume that the area bounded by the individual shortest paths and the sightlines was free of obstacles. If the polygon is not simple, then the case where the sightlines cross (eg if two vertices coincide at the intersection point) will admit Π_1,Π_2 as a solution only in a much more restricted set of problems. On the other hand allowing the agents to move at varying speeds would increase the number of problems which are solvable.

One significant difference occurs if the problem definition is modified to allow the agents to have non-zero radius. Simply reducing to the point case by

"thickening" the polygon boundary is insufficient. For example consider the case where the agent's position is constrained by its radius but vision is defined as the line segments joining the center points being unobstructed. In this case there are simple problems which have no feasible solution (see Figure 7).

Even in the point case, what happens if more agents are added to the problem? The team of agents would maintain a visibility graph but the constraints imposed on such a graph could vary. The graph may merely need to be connected, it may need to maintain a given topology or be allowed to change as long as it remains connected. Alternatively one could insist on the graph being complete. Many other variants are possible. This paper demonstrates that small modifications to the assumptions shift the problem from practical polynomial time, to completely intractable.

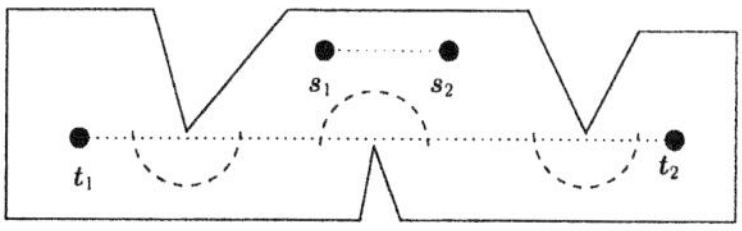

Fig. 7. Non-zero radius problem with no solution. There are no paths which allow both agents to be at their targets at the same time. (Dashed lines show exclusion radius around the obstacles).

References

1. T Bailey. Constrained initialisation for bearing-only SLAM. In *IEEE International Conference on Robotics and Automation*, volume 2, pages 1966–1971, Taipei, Taiwan, September 14-19 2003.
2. S. Bespamyatnikh. Computing homotopic shortest paths in the plane. *Journal of Algorithms*, 49(2):284–303, November 2003.
3. B. Chazelle. A theorem on polygon cutting with applications. In *23th Annu. IEEE Sympos. Found. Comput. Sci*, pages 339–34, 1982.
4. S. Efrat, S.G. Kobourov, and A. Lubiw. Computing homotopic shortest paths efficiently. In *Proceedings of the 10th Annual European Symposium on Algorithms*, pages 411–423, September 17-21 2002.
5. L. Guibas, J. Hershberger, D. Leven, M. Sharir, and R.E. Tarjan. Linear-time algorithms for visibility and shortest path problems inside triangulated simple polygons. *Algorithmica*, 2:209–233, 1987.
6. J. Hershberger and J. Snoeyink. Computing minimum length paths of a given homotopy class. *Computational Geometry: Theory and Applications*, 4(2):63–97, June 1994.
7. Andrew Howard. Multi-robot mapping using manifold representations. In *Proceedings of the 2004 IEEE International Conference on Robotics and Automation*, volume 4, pages 4198–4203, New Orleans, LA, April 2004.
8. J.S.B Mitchell. Geometric shortest paths and network optimization. In J.-R. Sack, editor, *Handbook of Computational Geometry*, pages 633–701, Amsterdam, 2000. Elsevier, North-Holland.
9. E. Nettleton, S. Thrun, H. Durrant-Whyte, and S. Sukkarieh. Decentralised SLAM with low-bandwidth communication for teams of vehicles. In *Intl. Conf. on Field and Service FSR 2003*, Lake Yamanaka, Japan, July 2003.

10. M. Rosencrantz, G. Gordon, and S. Thrun. Decentralized sensor fusion with distributed particle filters. In *Proceedings of Conf. Uncertainty in Artificial Intelligence*, Acapulco, Mexico, 2003.
11. J. Sullivan, S. Waydo, and M. Campbell. Using stream functions for complex behavior and path generation. In *AIAA Guidance, Navigation, and Control Conference*, Austin, Texas, August 11-14 2003. American Institute of Aeronautics and Astronautics. AIAA-2003-5800.
12. S. Williams, G. Dissanayake, and H.F. Durrant-Whyte. An efficient approach to the simultaneous localisation and mapping problem. In *Proceedings of the IEEE International Conference on Robotics and Automation*, volume 1, pages 406–411, Washington, DC, May 2002.
13. S.B. Williams, G. Dissanayake, and H. Durrant-Whyte. Towards multi-vehicle simultaneous localisation and mapping. In *IEEE International Conference on Robotics and Automation*, volume 3, pages 2743–274, Washington DC, May 2002.

Appendix

Proof (of Lemma 6). Traverse the boundary to find $\overline{\boldsymbol{v}_r\boldsymbol{v}_l}$ and also to find *min_dist*, the minimum perpendicular distance from vertices of the polygon which do not lie on the start sightline to the start sightline. *min_dist* is the smallest of at most n positive real numbers, so it is positive. Let $d = min\_dist$.

Now, if $\beta = \alpha$ compute $\boldsymbol{z} = (0, \lim_{d\to 0^+}(\boldsymbol{z}_y)$ by Equation (1). That is, $\boldsymbol{z}_y = -L\sin\beta/\sin(\beta+\alpha)$. If $\boldsymbol{z}$ lies now between $\boldsymbol{v}_l$ and $\boldsymbol{v}_r$ we know that $\boldsymbol{z}$ will be stationary as the agents move and thus they can start.

If $\beta \neq \alpha$, test for which one is larger. In the case $\beta > \alpha$, we see if there is space between $\boldsymbol{z}$ and $\boldsymbol{v}_r$ (namely $\boldsymbol{z}_y$ is smaller than the y coordinate of $\boldsymbol{v}_r$). In this case the agents can start. If $\boldsymbol{v}_r$ has a y-coordinate smaller than $\boldsymbol{z}_y$, the optimal solution does not use the shortest paths.

Finally, the case $\beta < \alpha$ is a test for space between $\boldsymbol{z}$ and $\boldsymbol{v}_r$. No space implies the shortest paths cannot be the optimal solution, while if the y-coordinate of $\boldsymbol{v}_l$ is below $\boldsymbol{z}_y$ then the agents can start. A situation where an obstacle cuts the sightline cannot arise since d is smallest among all distances from a polygon vertex not in the starting sightline to the starting sightline. □

Proof (of Lemma 8). Figure 6(b) illustrates the symbols used in this proof. Definition 6 describes the two ways an agent can start. First suppose that agents can start because there is an $\boldsymbol{s}' \in \overline{\boldsymbol{t}_1\boldsymbol{t}_2}$ visible to $\boldsymbol{s}_1$ and $\boldsymbol{s}_2$ and $\boldsymbol{s}' \notin line(\boldsymbol{s}_1, \boldsymbol{s}_2)$ (from the proof of Lemma 4: if we have an $\boldsymbol{s}$ we can always choose such an $\boldsymbol{s}'$). Recall that we can chose the first pause of a_2 as $\boldsymbol{p}_2 = \Pi_2 \cap line(\boldsymbol{s}_1, \boldsymbol{s}')$ and then agent a_1 moves to $\boldsymbol{p}_1 = line(\boldsymbol{c}_p, \boldsymbol{p}_2) \cap \Pi_1$. Thus, $\boldsymbol{p}_1$ and $\boldsymbol{p}_2$ are as required (ie $\overline{\boldsymbol{p}_1\boldsymbol{p}_2}$ is a sightline). Now chose $\boldsymbol{x}_1 = \boldsymbol{c}_p$ and $\boldsymbol{x}_2 = \boldsymbol{s}'$. Then, the first condition is satisfied since $\overline{\boldsymbol{x}_1\boldsymbol{x}_2} \cap \overline{\boldsymbol{p}_1\boldsymbol{p}_2} = \boldsymbol{x}_1$. The second condition is fulfilled since the wedge $\boldsymbol{s}_1$, $\boldsymbol{x}_2$, $\boldsymbol{t}_1$ and the wedge $\boldsymbol{s}_2$, $\boldsymbol{x}_1$, $\boldsymbol{t}_2$ are free of obstacles.

The second way of starting is as in item 2 in Def. 6. Although $\boldsymbol{c}_p$ may be a vertex of the polygon there can not be obstacles touching $\boldsymbol{c}_p$ on both sides (in this case the polygon would not be simple). If necessary we reverse the identifiers 1 and 2 so that there is no obstacle touching $\boldsymbol{c}_p$ on the left (ie above

the starting sightline). Let $\boldsymbol{p}_1 \in \Pi_1$ and $\boldsymbol{p}_2 \in \Pi_2$ be the first pauses of both agents that moved simultaneously. Clearly $\overline{\boldsymbol{p}_1\boldsymbol{p}_2}$ is a sightline as required. Now let $\boldsymbol{x}_1 = \boldsymbol{c}_p$. Assume that $\overline{\boldsymbol{p}_1\boldsymbol{p}_2}$ intersects $\overline{\boldsymbol{t}_1\boldsymbol{t}_2}$ above the starting sightline or at $\boldsymbol{c}_p$ (otherwise, instead of fixing $\boldsymbol{p}_2$ the argument follows fixing $\boldsymbol{p}_1$). If point $\boldsymbol{p'}_2$ traces along Π_2 from $\boldsymbol{p}_2$ while keeping a_1 stationary at $\boldsymbol{p}_1$ then either $\boldsymbol{p'}_2$ reaches $\boldsymbol{t}_2$ or further progress is prevented by an obstacle. Let $\boldsymbol{x}_2 = \overline{\boldsymbol{p'}_2\boldsymbol{p}_1} \cap \overline{\boldsymbol{t}_1\boldsymbol{t}_2}$. Now we have $\overline{\boldsymbol{x}_1\boldsymbol{x}_2} \cap \overline{\boldsymbol{p}_1\boldsymbol{p}_2} = \boldsymbol{x}_1$. Also, the wedge $\boldsymbol{p}_1$, $\boldsymbol{x}_2$, $\boldsymbol{t}_1$ is free of obstacles as well as the wedge $\boldsymbol{t}_2$, $\boldsymbol{x}_1$, $\boldsymbol{p}_2$ so we have the second condition as well. □

Proof (of Lemma 9). Without loss of generality, consider a single step of the alternating process where a_1 is stationary at $\boldsymbol{p}$ and a_2 moves as illustrated by Fig. 8. Let $\boldsymbol{x}_1$, $\boldsymbol{x}_2$ be as in Lemma 8. Consider a line L parallel to line$(\boldsymbol{s}_1, \boldsymbol{t}_1)$ through $\boldsymbol{x}_2$. Let $\theta > 0$ be the angle L makes with $\overline{\boldsymbol{x}_1\boldsymbol{x}_2}$. Note that if $\boldsymbol{x}_1$ is below L (as in the figure), use $-\theta$ instead. We let ψ be the angle between line$(\boldsymbol{p}, \boldsymbol{x}_1)$ and $\overline{\boldsymbol{s}_1\boldsymbol{t}_1}$. We are interested in bounding Δ, the displacement on $\overline{\boldsymbol{s}_2\boldsymbol{t}_2}$ made by a_2. Bounding this from below proves the result since we know progress on the shortest paths Π_i is monotonic with respect to the segment $\overline{\boldsymbol{s}_i\boldsymbol{t}_i}$. It is not hard to see that Δ is bounded by M, the length of the segment from $\boldsymbol{x}_2$ to the intersection of L and line$(\boldsymbol{p}, \boldsymbol{x}_1)$. If we let m =len$(\overline{\boldsymbol{x}_1\boldsymbol{x}_2})$, we find that $M(\psi) = \frac{\sin(\theta+\psi)m}{\sin\psi}$. Now, $M(\psi)$ is continuous, differentiable and non-zero for permissible range of angles. The function $M(\psi)$ has no stationary points in the interval $(0, \pi)$ since $0 = \partial M/\partial\psi = (\sin\psi\cos(\theta+\psi) + \sin(\theta+\psi)\cos\psi) / \sin^2\psi$ implies $\sin\psi\cos(\theta+\psi) = \sin(\theta+\psi)\cos\psi$ or simply $\tan\psi = \tan(\theta+\psi)$.

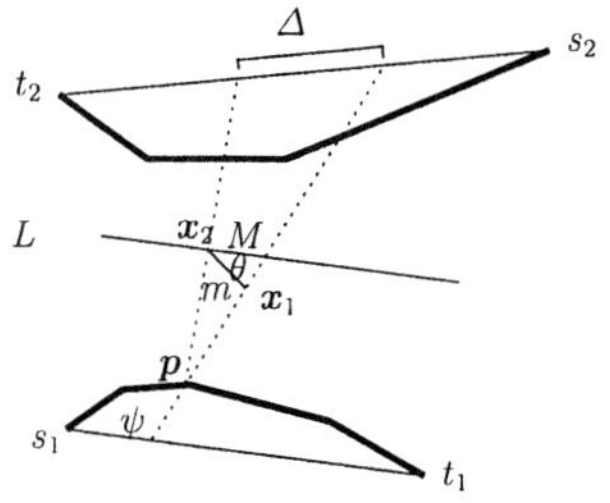

Fig. 8. Δ is bounded below by a positive constant

Choose a point $\boldsymbol{p}_d$ and two angles b_1, b_2 as $\boldsymbol{p}_d \in \Pi_1$ and at distance d from t_1. Let b_1 be the angle between $\overline{\boldsymbol{s}_1\boldsymbol{s}_2}$ and $\overline{\boldsymbol{s}_1\boldsymbol{t}_1}$. Let b_2 be the angle between line$(\boldsymbol{p}_d, \boldsymbol{x}_1)$ and $\overline{\boldsymbol{s}_1\boldsymbol{t}_1}$ (measured on the same side as b_1). This means that $b_1 < \psi < b_2$, $0 < b_1 < \pi$ and $0 < b_2 < \pi$ so $\Delta \geq M$ is bounded below by $k = \min(M(b_1), M(b_2)) > 0$. Since the amount of progress for each agent up to and including the desired position is at least k they are guaranteed to reach the desired point. Because we can find a $d > 0$ so that we know how to terminate simultaneously, the process is to alternate until both agents are within d of their targets and then to terminate simultaneously. This proves the first claim of Theorem 2.

In order to show the second claim we need to show how to construct a solution which has length as close as desired to the length of Π_1 and Π_2. The conditions under which Π_1, Π_2 are not part of an optimal solution are described in Lemma 4. However if we are willing to leave Π_1,Π_2 then $\boldsymbol{z}$ can be moved by modifying the angles which the agents move at initially. So pick some small $d > 0$ and angles α', β' such that $\alpha' \leq \alpha, \beta' \leq \beta$ and $Z_y(\alpha', \beta', D)$ lies between

$\boldsymbol{v}_l$ and $\boldsymbol{v}_r$ $\forall D < d$. Obviously d will also need to be small enough that the agents can actually travel distance d without leaving the region between the sightlines. Call the points the agents reach after this initial step $\boldsymbol{q}_1$, $\boldsymbol{q}_2$. If there is no problem starting but rather in finishing let $\boldsymbol{q}_i = \boldsymbol{s}_i (i \in \{1, 2\})$. A similar process applied to the target points can produce $\boldsymbol{r}_1$, $\boldsymbol{r}_2$ if required (if not let $\boldsymbol{r}_1$,$\boldsymbol{r}_2$ be the respective target points).

Let $\pi'_{i \in \{1,2\}}$ be the shortest path from $\boldsymbol{q}_i$ to $\boldsymbol{r}_i$. These paths will be convex and hence progress along them will result in monotonic progress relative to $\overline{\boldsymbol{s}_i \boldsymbol{t}_i}$. Notice also that these paths will not leave the region between the sightlines. We can now use the reasoning from Lemma 7 and Lemma 8 to show that the agents following such paths will reach $\boldsymbol{r}_1$, $\boldsymbol{r}_2$. So the starting and finishing diversions and π'_i are a feasible solution. The only thing left is to prove they approach Π_i. This fact follows simply from the fact that the end points of Π_i and π'_i can be made as close as desired by reducing d. □

Stacking and Bundling Two Convex Polygons*

Hee-Kap Ahn and Otfried Cheong

Division of Computer Science, Korea Advanced Institute of Science and Technology, Daejeon, Korea
{heekap, otfried}@tclab.kaist.ac.kr

Abstract. Given two compact convex sets C_1 and C_2 in the plane, we consider the problem of finding a placement φC_1 of C_1 that minimizes the area of the convex hull of $\varphi C_1 \cup C_2$. We first consider the case where φC_1 and C_2 are allowed to intersect (as in "stacking" two flat objects in a convex box), and then add the restriction that their interior has to remain disjoint (as when "bundling" two convex objects together into a tight bundle). In both cases, we consider both the case where we are allowed to reorient C_1, and where the orientation is fixed. In the case without reorientations, we achieve exact near-linear time algorithms, in the case with reorientations we compute a $(1+\varepsilon)$-approximation in time $O(\varepsilon^{-1/2}\log n + \varepsilon^{-3/2}\log \varepsilon^{-1/2})$, if two sets are convex polygons with n vertices in total.

1 Introduction

We consider the problem of stacking two flat objects into a box. The objects will lie on top of each other, and our goal is to design a box that is as small as possible, while (for simplicity) restricting it to be convex. This problem can be modelled as follows: let C_1 and C_2 be planar compact convex sets, find the rigid motion φ such that the area of the convex hull of $\varphi C_1 \cup C_2$ is minimized. (Note that considering convex sets only is no restriction, as the target function does not change when we replace a set by its convex hull.)

We give an ε-approximation algorithm for this problem, which computes a rigid motion φ^{app} such that the area of the convex hull of $\varphi^{\text{app}} C_1 \cup C_2$ is at most $1+\varepsilon$ times the optimally achievable area. The running time is $O(\varepsilon^{-1/2} T_C + \varepsilon^{-3/2}\log \varepsilon^{-1/2})$, where T_C is the time needed to find a point extreme in a given direction, or the intersection of a given line with C_1 and C_2.

We also consider a related problem, where we wish to minimize the same function, but with the restriction that C_1 and C_2 remain disjoint. This problem arises in various applications, consider for instance the problem of minimizing the cross section of a wire bundle consisting of two subsets of wires.

There has been a fair amount of work on the problem of maximizing the overlap (or, equivalently, minimizing the symmetric difference) of two shapes in the context of shape matching [2,3,4,5,6,7,11] under translations or rigid motions. Surprisingly, little is known about the problem of minimizing the convex hull

* This research was supported by LG Electronics.

X. Deng and D. Du (Eds.): ISAAC 2005, LNCS 3827, pp. 882–891, 2005.

of two shapes. Recently, Ahn et al. [2] gave an approximation algorithm to find the line ℓ such that the convex hull of a given convex set C and a reflected copy of C along ℓ is minimized. Their solution, however, does not generalize to our more general problem. With application in the clothing industry, Milenkovic [10] studied a problem of packing a set of polygons into another polygon (container) without overlapping. Sugihara et al. [12] recently considered a two-dimensional disk packing problem of finding the smallest enclosing circle containing a set of disks (the cross section of a wire bundle), and gave a "shake-and-shrink" algorithm that shakes the disks and shrinks the enclosing circle step-by-step.

2 Preliminaries

For a compact set S in the plane, let $d(S)$ and $w(S)$ denote the diameter and width of S (the width is the minimum distance between two parallel lines enclosing S). Also, let $|S|$ denote the area of *the convex hull of* S. Note that $d(S) = d(\text{conv}(S))$, $w(S) = w(\text{conv}(S))$, and (by definition) $|S| = |\text{conv}(S)|$.

We denote the interior of a set $A \subseteq \mathbb{R}^2$ by $\text{int}(A)$ and its closure by $\text{cl}(A)$.

In all problems considered in this paper, our goal is to minimize $|\varphi C_1 \cup C_2|$, where C_1 and C_2 are given convex sets. We distinguish between the case where φ ranges over all possible translations, and where it can be any rigid motion. In addition, we distinguish the "box" case, where φC_1 and C_2 are allowed to intersect (and always will in the optimal solution), and the "bundle" case, where φC_1 and C_2 must be disjoint.

3 Stacking Polygons Without Reorientation

Let P and Q be convex polygons in the plane with n vertices in total. For a vector $r \in \mathbb{R}^2$, let $P+r$ denote the translation of P by r, that is, $P+r = \{p+r \mid p \in P\}$. We consider the function $\omega(r) := |(P+r) \cup Q|$.

Lemma 1. *Let P and Q be convex polygons in the plane. Then the function $r \mapsto \omega(r)$ is convex (that is, the volume above the graph of the function is convex).*

Proof. Since a function $\mathbb{R}^2 \to \mathbb{R}$ is convex if any cross section along a line is convex, it suffices to prove the latter fact. Without loss of generality, we restrict ourselves to horizontal lines. For $t \in \mathbb{R}$, let $P_t := P+(t,0)$, that is, the translation of P by t along the x-axis. We will show that the function $t \mapsto \omega(t) := |P_t \cup Q|$ is convex. We do so by showing that the derivative $\omega'(t)$ is non-decreasing.

Consider the convex polygon $C_t := \text{conv}(P_t \cup Q)$. It consists of edges of P_t (type 0 edges), edges of Q (type 2 edges), and type 1 edges connecting one vertex of P_t with one vertex of Q. As t increases, the polygon P_t moves to the right, while Q remains stationary. We consider how the area of C_t changes. As long as the combinatorial structure of C_t remains the same, the change in $|C_t|$ can be expressed as the sum of changes incurred by the individual edges: moving edges on the right side of C_t add area to C_t, while moving edges on the left side of C_t remove area from C_t. For a compact set S, let $h(S)$ denote the "height" of S,

that is, the length of the projection of S onto the y-axis. For a type 0 edge e, the area swept over when t increases by δ is $\delta h(e)$. For a type 1 edge e, the area swept over is $\delta h(e)/2$ (since one endpoint moves by δ, the area swept is a triangle). Type 2 edges are stationary, and do not contribute to the change in $|C_t|$.

It follows that as long as $C_{t+\delta}$ and C_t have the same combinatorial structure, we have

$$\omega(t+\delta)-\omega(t) = |C_{t+\delta}|-|C_t| = \sum_{e\in R_t^0} \delta h(e)+\sum_{e\in R_t^1} \frac{\delta h(e)}{2}-\sum_{e\in L_t^0} \delta h(e)-\sum_{e\in L_t^1} \frac{\delta h(e)}{2},$$

where R_t^i is the set of right edges of C_t of type i, and L_t^i is the set of left edges of C_t of type i. Taking the limit for $\delta \to 0$, we get

$$2\omega'(t) = 2\sum_{e\in R_t^0} h(e) + \sum_{e\in R_t^1} h(e) - 2\sum_{e\in L_t^0} h(e) - \sum_{e\in L_t^1} h(e),$$

Now note that $\sum h(e) = h(C)$, when the sum is taken over all left edges or over all right edges. It follows that

$$\sum_{e\in R_t^1} h(e) = h(C) - \sum_{e\in R_t^0} h(e) - \sum_{e\in R_t^2} h(e),$$

$$\sum_{e\in L_t^1} h(e) = h(C) - \sum_{e\in L_t^0} h(e) - \sum_{e\in L_t^2} h(e).$$

Substituting into the previous equality we obtain

$$2\omega'(t) = \sum_{e\in R_t^0} h(e) - \sum_{e\in R_t^2} h(e) - \sum_{e\in L_t^0} h(e) + \sum_{e\in L_t^2} h(e).$$

Consider now a left edge e of P_t. This edge appears as a type 0 edge of C_t if and only if no vertex of Q lies on the left of the supporting line ℓ_e of e. Since P_t moves horizontally rightwards, there is a unique *edge event* for e where the moving ℓ_e *touches* the first vertex v_e of the stationary Q. Let $t(e)$ be the "time" of this event, that is, the value of t such that $v_e \in \ell_e$. Clearly, $e \in L_t^0$ if $t < t(e)$, and $e \notin L_t^0$ if $t > t(e)$ (and either can be true at $t = t(e)$). See Fig. 1. This implies that the function $t \mapsto \sum_{e\in L_t^0} h(e)$ is non-increasing. Similarly, a right edge e of P_t is in R_t^0 if and only if no vertex of Q lies to the right of ℓ_e. This implies that there is a "time" $t(e)$ such that $e \notin R_t^0$ if $t < t(e)$ and $e \in R_t^0$ if $t > t(e)$, and it follows that $t \mapsto \sum_{e\in R_t^0} h(e)$ is non-decreasing. Analogously, we can show that $t \mapsto \sum_{e\in L_t^2} h(e)$ is non-decreasing, while $t \mapsto \sum_{e\in R_t^2} h(e)$ is non-increasing. It follows that $\omega'(t)$ is non-decreasing, and so $\omega(t)$ is convex, proving the lemma. ◻

The proof above can be turned into an algorithm to compute the optimal placement of P along a line.

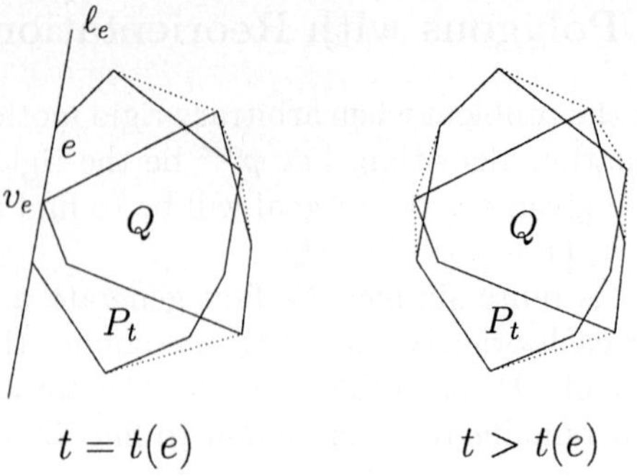

Fig. 1. The unique *edge event* at time $t = t(e)$ for a left edge e

Lemma 2. *Let P and Q be convex polygons with n vertices in total, and let ℓ be a line. Then we can compute $\min_{r \in \ell} \omega(r)$ in time $O(n)$.*

Proof. We choose a coordinate-system such that ℓ is the x-axis, and end up with the setting of the proof of Lemma 1. Our task is to find a value $t \in \mathbb{R}$ such that $\omega'(t - \varepsilon) \leqslant 0$ and $\omega'(t + \varepsilon) \geqslant 0$ for ε small enough. As we saw in Lemma 1, $\omega'(t)$ can be expressed as a linear combination of $h(e)$, over all edges of P_t and Q, and changes only whenever a vertex of Q lies on the supporting line of P_t, and vice versa.

There are $O(n)$ such values of t, and they can be computed in linear time (observing that the vertex of Q hitting a line ℓ_e first must have a tangent parallel to ℓ_e, and so it suffices to merge the lists of slopes of P and Q).

If we were willing to spend $O(n \log n)$ time, we could just sort all these values of t, and scan them in order while maintaining the value of $\omega'(t)$. We can do better than this (at least asymptotically) by using a decimation approach that repeatedly finds the median from a list of candidate values. The details are similar to the algorithm described by Ahn et al. [2], and we omit them in this extended abstract. □

We now have all the tools we need to find the optimal placement.

Theorem 3. *Let P and Q be convex polygons with n vertices in total. Then we can compute a translation $r \in \mathbb{R}^2$ minimizing $|(P + r) \cup Q|$ in time $O(n \log n)$.*

Proof. De Berg et al. [7] gave an $O(n \log n)$ time algorithm to compute the translation r maximizing $|(P + r) \cap Q|$. Their algorithm made use of a subroutine to find the best translation restricted to a line, and exploits the unimodularity of the function $r \mapsto |(P+r) \cap Q|$. Since we have established the analogous properties for our function $r \mapsto |(P+r) \cup Q|$ in Lemmas 1 and 2, their technique can be applied to our problem as well. We omit the details in this extended abstract. □

4 Stacking Polygons with Reorientations

We now consider the problem when arbitrary rigid motions of P are allowed, and give an approximation algorithm. Let φ^{opt} be the rigid motion that minimizes $|\varphi^{\mathrm{opt}}P \cup Q|$. For a given $\varepsilon > 0$, our goal will be to find a rigid motion φ^{app} such that $|\varphi^{\mathrm{app}}P \cup Q| \leqslant (1+\varepsilon)|\varphi^{\mathrm{opt}}P \cup Q|$.

Our algorithm is quite simple: We first generate a set D_ε of $O(1/\varepsilon)$ orientations of P. For each orientation, we then compute the optimal translation of (the rotated copy of) P, using Theorem 3. The total running time is clearly $O((1/\varepsilon)n \log n)$, it remains to describe how to find D_ε and to prove the approximation bound.

The difficulty is that we cannot simply sample orientations uniformly. This works when P and Q are rather round and "fat", but the sampling resolution would have to too fine when they are long and skinny. Fortunately, in the latter case we can prove that the diameters of P and Q must be nearly aligned, and it suffices to sample very finely around the orientation that achieves alignment.

We first show a lower bound on $|C_1 \cup C_2|$ for two convex sets C_1 and C_2 based on their diameter and width.

Lemma 4. *Let C_1 and C_2 be convex sets in the plane. Then*

$$|C_1 \cup C_2| \geqslant \frac{1}{2} \cdot \max\{d(C_1), d(C_2)\} \cdot \max\{w(C_1), w(C_2)\}$$

Proof. Let pq be a diameter of C_1, let R be a rectangle circumscribed to C_1 with two sides parallel to pq such that C_1 touches all four sides of R at points p, q, r and s, and let w be the side of R orthogonal to pq.

Assume that $d(C_1) \geqslant d(C_2)$. Clearly, $\mathrm{conv}(p, q, r, s)$ is contained in $\mathrm{conv}(C_1 \cup C_2)$ and consists of two triangles with a common base pq. Since $w \geqslant w(C_1)$, it has area at least $d(C_1) \cdot w(C_1)/2$. Now let $p'q'$ be a line segment in C_2 which has length $w(C_2)$ and is orthogonal to pq. Note that there always exists such a segment: in fact, C_2 contains a segment of length $w(C_2)$ of every direction [9, pg. 12, ex.4(a)]. Then $\mathrm{conv}(p, q, p', q')$ is contained in $\mathrm{conv}(C_1 \cup C_2)$, and it has area at least $d(C_1) \cdot w(C_2)/2$. Therefore,

$$|C_1 \cup C_2| \geqslant d(C_1) \cdot \frac{1}{2} \cdot \max\{w(C_1), w(C_2)\},$$

the lemma follows. ◻

We now prove that diameters need to be nearly aligned for long and skinny objects.

Lemma 5. *Let C_1 and C_2 be convex sets in the plane, let φ^{opt} be the rigid motion minimizing $|\varphi^{\mathrm{opt}}C_1 \cup C_2|$, and let ϑ be the smaller angle between two diameters of $\varphi^{\mathrm{opt}}C_1$ and C_2. Then*

$$\sin\vartheta \leqslant \frac{2 \cdot \max\{w(C_1), w(C_2)\}}{\min\{d(C_1), d(C_2)\}}$$

Proof. Let pq be a diameter of $\varphi^{\text{opt}}C_1$ and let $p'q'$ be a diameter of C_2. Assume that two diameters pq and $p'q'$ make an angle $\vartheta \in [0, \pi/2]$, as in Fig. 2. If two diameters intersect at a point x, the convex hull consists of two triangles with a common base pq and has area

$$|\text{conv}(p,q,p',q')| \geqslant \frac{1}{2} \cdot d(C_1) \cdot (|p'x| + |xq'|) \sin\vartheta = \frac{1}{2} \cdot d(C_1) \cdot d(C_2) \cdot \sin\vartheta.$$

If two diameters do not intersect, we can always translate one of them until they intersect while the area function is non-increasing. Since $\text{conv}(p,q,p',q') \subset \text{conv}(\varphi^{\text{opt}}C_1 \cup C_2)$, we have $|\varphi^{\text{opt}}C_1 \cup C_2| \geqslant \frac{1}{2} \cdot d(C_1) \cdot d(C_2) \cdot \sin\vartheta$.

Consider now rectangles R_i circumscribed to C_i, with sides $d(C_i)$ and $w(C_i)$, for $i = 1, 2$. There is a rigid motion φ such that $\varphi R_1 \cup R_2$ fits in a rectangles with sides $\max\{d(C_1), d(C_2)\}$ and $\max\{w(C_1), w(C_2)\}$, and so

$$|\varphi^{\text{opt}}C_1 \cup C_2| \leqslant |\varphi C_1 \cup C_2| \leqslant |\varphi R_1 \cup R_2| \leqslant \max\{d(C_1), d(C_2)\} \cdot \max\{w(C_1), w(C_2)\}.$$

Combining the upper and lower bounds for $|\varphi^{\text{opt}}C_1 \cup C_2|$ proves the lemma. □

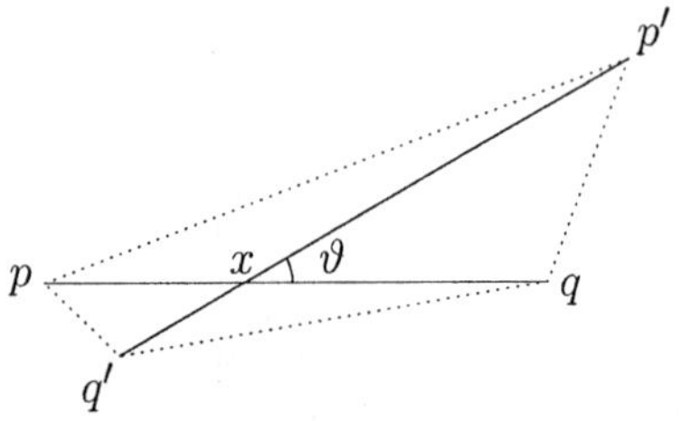

Fig. 2. The convex hull of two diameters has area at least $\frac{1}{2} \cdot d(C_1) \cdot d(C_2) \cdot \sin\vartheta$

We will now prove that sampling orientations works. Let $\text{peri}(C)$ denote the perimeter of a convex set C; we will make use of the inequality that for a convex set C, $\text{peri}(C) \leqslant \pi d(C)$ [13, pg. 257, ex.7.17a]. We also need the following lemma proven by Ahn et al. [2].

Lemma 6. *([2]) Let C be a convex set, let $r > 0$, and let C' be the set of points at distance at most r from C (in other words, C' is the Minkowski sum of C and a disk of radius r). Then $|C'| = |C| + r\,\text{peri}(C) + \pi r^2$.*

Lemma 7. *Let C_1, C_2 be convex sets, let $\varepsilon > 0$, and let ρ be a rotation of angle*

$$\delta \leqslant \frac{\varepsilon}{24} \frac{\max\{w(C_1), w(C_2)\}}{\min\{d(C_1), d(C_2)\}}.$$

around a point in C_1. Then $|\rho C_1 \cup C_2| \leqslant (1+\varepsilon)|C_1 \cup C_2|$.

Proof. Without loss of generality, we assume that $d(C_1) \leqslant d(C_2)$. Let $Q = \text{conv}(C_1 \cup C_2)$ and let $Q' = \text{conv}(\rho C_1 \cup C_2)$. Note that any point q in $Q' \setminus Q$

is at distance at most $\delta d(C_1)$ from the boundary of Q. Let T denote the set of points that are at distance at most $\delta d(C_1)$ from the boundary of Q. Then we have $(Q \cup T) \supset Q'$. By Lemma 6, the area of T is

$$\begin{aligned}
|T| &\leqslant \delta d(C_1) \cdot \mathrm{peri}(Q) + \pi(\delta d(C_1))^2 \\
&\leqslant \delta d(C_1) \cdot \pi d(Q) + \pi(\delta d(C_1))^2 \\
&\leqslant \delta d(C_1) \cdot 2\pi d(C_2) + \pi\delta^2 d(C_1) \cdot d(C_2) \\
&\leqslant \delta d(C_1) \cdot 2\pi d(C_2) + \delta(\pi^2/2) d(C_1) \cdot d(C_2) \\
&= \delta d(C_1) \cdot d(C_2) \cdot (2\pi + \pi^2/2) \\
&\leqslant 12\delta d(C_1) \cdot d(C_2) \\
&\leqslant \frac{12\varepsilon}{24} \frac{d(C_1) \cdot d(C_2)}{\min\{d(C_1), d(C_2)\}} \max\{w(C_1), w(C_2)\} \\
&= \frac{\varepsilon}{2} \max\{d(C_1), d(C_2)\} \max\{w(C_1), w(C_2)\} \\
&\leqslant \varepsilon |Q|.
\end{aligned}$$

(The last inequality follows from Lemma 4.) □

We can now state the algorithmic result.

Lemma 8. *Let P and Q be convex polygons with n vertices in total, and let $\varepsilon > 0$. Then we can compute a rigid motion φ^{app} such that $|\varphi^{\mathrm{app}} P \cup Q| \leqslant (1+\varepsilon)|\varphi^{\mathrm{opt}} P \cup Q|$ in time $O((1/\varepsilon) n \log n)$.*

Proof. In linear time, we compute width and diameter of both polygons. We then sample orientations at interval $\frac{\varepsilon}{24} \frac{\max\{w(P), w(Q)\}}{\min\{d(P), d(Q)\}}$, but omitting all directions where the computed diameters make an angle ϑ with $\sin\vartheta > \frac{2 \cdot \max\{w(P), w(Q)\}}{\min\{d(P), d(Q)\}}$. Clearly we have sampled $O(1/\varepsilon)$ directions. For each of these, we then compute the translation minimizing $|\varphi P \cup Q|$ using Theorem 3. The whole procedure takes time $O((1/\varepsilon) n \log n)$, and the approximation bound follows from Lemmas 5 and 7. □

The time bound in Lemma 8 can be improved by a classical idea: we first replace the input by approximations with $O(1/\sqrt{\varepsilon})$ vertices (that is, the complexity of the problem is now independent of n). We then apply Lemma 8 to the approximations. This will also allow us to solve the problem for more general convex sets (not neccssarily polygons).

Following Agarwal et al. [1], we say that a convex set $C' \subset C$ is an ε-kernel of C if and only if for all $u \in \mathcal{U}$, $(1-\varepsilon)\,\mathrm{dwidth}(u, C) \leqslant \mathrm{dwidth}(u, C')$, where $\mathcal{U}$ is the set of unit vectors in the plane and $\mathrm{dwidth}(u, C)$ denotes the directional width of C in direction u, that is

$$\mathrm{dwidth}(u, C) := \max_{x \in C} \langle x, u \rangle - \min_{x \in C} \langle x, u \rangle.$$

There is a constant $c > 0$ such that if C' is an ε/c-kernel of C, then $|C \backslash C'| \leqslant \varepsilon |C|$.

Based on Dudley's constructive proof from 1974 [8], Ahn et al. [2] gave an algorithm that computes inner and outer approximations to a given convex set C. The algorithm requires only two operations on the set C, namely (a) given a query direction $u \in \mathcal{U}$, find an extreme point in direction u; and (b) given a line ℓ, find the line segment $\ell \cap C$. Let T_C denote the time needed to answer any of these queries.

Lemma 9. *([2]) Given a planar convex set C and $\varepsilon > 0$, one can construct in time $O(T_C/\sqrt{\varepsilon})$ a convex polygon C' with $O(1/\sqrt{\varepsilon})$ vertices such that $C \subset C'$ and C is an ε-kernel of C'. If C is a convex n-gon whose vertices are given in a sorted array, then we can compute C' in time $O(\log n/\sqrt{\varepsilon})$.*

We can now give the main result of this section.

Theorem 10. *Given two convex sets C_1 and C_2 in the plane and $\varepsilon > 0$, we can compute in time $O((T_{C_1} + T_{C_2})\varepsilon^{-1/2} + (\varepsilon^{-3/2})\log(\varepsilon^{-1/2}))$ a rigid motion φ^{app} such that $|\varphi^{\mathrm{app}} C_1 \cup C_2| \leqslant (1+\varepsilon)\min_\varphi |\varphi C_1 \cup C_2|$, where the minimum is over all rigid motions.*

Proof. Let $\varepsilon' := \varepsilon/(3c)$. We construct the outer ε'-approximation of Lemma 9 to C_1 and C_2 in time $O((T_{C_1} + T_{C_2})\varepsilon^{-1/2})$. Let P_i denote the outer approximation of C_i. We then apply Lemma 8 to compute a rigid motion φ^{app} such that $|\varphi^{\mathrm{app}} P_1 \cup P_2| \leqslant (1+\varepsilon/3)\min_\varphi |\varphi P_1 \cup P_2|$. This takes time $O((\varepsilon^{-3/2})\log(\varepsilon^{-1/2}))$, and it remains to prove the approximation bound.

By Lemma 9, C_i is an ε'-kernel of P_i, for $i = 1, 2$. This implies that $\mathrm{conv}(\varphi^{\mathrm{opt}} C_1 \cup C_2)$ is an ε'-kernel of $\mathrm{conv}(\varphi^{\mathrm{opt}} P_1 \cup P_2)$, where φ^{opt} is the rigid motion minimizing $|\varphi^{\mathrm{opt}} C_1 \cup C_2|$. This implies that

$$|\varphi^{\mathrm{opt}} P_1 \cup P_2| \leqslant (1 + \frac{\varepsilon}{3})|\varphi^{\mathrm{opt}} C_1 \cup C_2|,$$

and so we get

$$\begin{aligned}|\varphi^{\mathrm{app}} C_1 \cup C_2| &\leqslant |\varphi^{\mathrm{app}} P_1 \cup P_2| \leqslant (1 + \frac{\varepsilon}{3})\min_\varphi |\varphi P_1 \cup P_2| \\ &\leqslant (1 + \frac{\varepsilon}{3})|\varphi^{\mathrm{opt}} P_1 \cup P_2| \leqslant (1 + \frac{\varepsilon}{3})^2|\varphi^{\mathrm{opt}} C_1 \cup C_2| \\ &\leqslant (1+\varepsilon)|\varphi^{\mathrm{opt}} C_1 \cup C_2|,\end{aligned}$$

proving the approximation bound. □

5 Bundling Polygons with Translations

We now consider the problem of bundling two convex polygons, that is, we wish to minimize $|\varphi P \cup Q|$ under the restriction that the interiors of φP and Q are disjoint.

Again we first consider the case of translations only, so we are looking for $r \in \mathbb{R}^2$ such that $|(P + r) \cup Q|$ is minimal under the restriction that $\mathrm{int}(P + r) \cap \mathrm{int}(Q) = \emptyset$. Clearly, in the optimal solution $P + r$ and Q are touching, and so it suffices to "slide" P around Q, keeping their boundaries in contact.

Theorem 11. *Let P and Q be convex polygons with n vertices in total. Then we can compute a translation $r \in \mathbb{R}^2$ minimizing $|(P+r) \cup Q|$ under the restriction $\mathrm{int}(P+r) \cap \mathrm{int}(Q) = \emptyset$ in time $O(n)$.*

Proof. We "slide" P around Q, that is we translate P such that it describes a complete loop around Q while always remaining in contact with Q. We note that the combinatorial structure of $\mathrm{conv}(P \cup Q)$ changes only when a line supporting an edge of Q becomes tangent to P, or vice versa. Since each vertex of P describes a convex path around Q, there are only $O(n)$ such events. By merging the lists of slopes of both polygons, we can precompute all such events (in order of their occurrence) in linear time. Finally, it suffices to simulate the entire sliding process. We omit the details in this extended abstract. □

6 Bundling Polygons with Rigid Motions

We proceed very similarly to the stacking case: we prove that sampling directions is sufficient, restrict our attention to directions where the two diameters are aligned (depending on the skinniness of the objects), and finally improve the running time by using the core set approximation of Lemma 9. One may conjecture that the optimal solution is obtained when two edges are in contact. This is, however, not always the case (See Fig. 3).

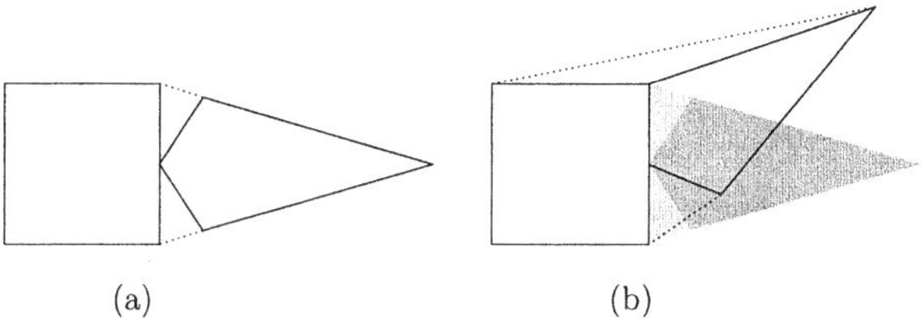

Fig. 3. (a) The optimal solution with edge-vertex contact. (b) By rotating the right polygon slightly in counterclockwise orientation, the area of the convex hull has been increased.

Lemma 12. *Let C_1 and C_2 be convex sets in the plane, let φ^{opt} be a rigid motion minimizing $|\varphi^{\mathrm{opt}} C_1 \cup C_2|$ while keeping their interiors disjoint, and let ϑ be the smaller angle between two diameters of $\varphi^{\mathrm{opt}} C_1$ and C_2. Then,*

$$\sin \vartheta \leqslant \frac{4 \cdot \max\{w(C_1), w(C_2)\}}{\min\{d(C_1), d(C_2)\}}$$

Proof. As in Lemma 5, we first argue that $|\varphi^{\mathrm{opt}} C_1 \cup C_2| \geqslant \frac{1}{2} \cdot d(C_1) \cdot d(C_2) \cdot \sin \vartheta$ (the restriction can only increase the optimum). We then again consider the rectangles R_i with sides $d(C_i)$ and $w(C_i)$ circumscribed to C_i, for $i = 1, 2$. There is a rigid motion φ that aligns them such that $\varphi R_1 \cup R_2$ fits in a rectangle with sides $2 \cdot \max\{d(C_1), d(C_2)\}$ and $\max\{w(C_1), w(C_2)\}$, implying the bound. □

Theorem 13. *Given two convex sets C_1 and C_2 in the plane and $\alpha > 0$, we can compute in time $O((T_{C_1} + T_{C_2})\alpha^{-1/2} + (\alpha^{-3/2}))$ a rigid motion φ^{app} such that the area of $\mathrm{conv}(\varphi^{\mathrm{app}} C_1 \cup C_2)$ is at most $1 + \varepsilon$ times the minimum over all rigid motions.*

We omit the proof in this extended abstract.

References

1. P. K. Agarwal, S. Har-Peled, K. R. Varadarajan. Approximating Extent Measures of Points. *Journal of the ACM* **51** (2004) 606–635.
2. H.-K. Ahn, P. Brass, O. Cheong, H.-S. Na, C.-S. Shin, and A. Vigneron. Inscribing an axially symmetric polygon and other approximation algorithms for planar convex sets. To appear in *Comput. Geom. Theory Appl.*
3. H.-K. Ahn, O. Cheong, C.-D. Park, C.-S. Shin, and A. Vigneron. Maximizing the overlap of two planar convex sets under rigid motions, *Proc. 21st Annu. Symp. Comput. geometry*, (2005) 356–363.
4. H. Alt, J. Blömer, M. Godau, and H. Wagener. Approximation of convex polygons, *Proc. 17th Internat. Colloq. Automata Lang. Program.*, Lecture Notes Comput. Sci. **443**, p. 703–716, Springer-Verlag 1990.
5. H. Alt, U. Fuchs, G. Rote, and G. Weber. Matching convex shapes with respect to the symmetric difference. *Algorithmica*, 21:89–103, 1998.
6. M. de Berg, S. Cabello, P. Giannopoulos, C. Knauer, R. van Oostrum, and R. C. Veltkamp. Maximizing the area of overlap of two unions of disks under rigid motion. *Proc. Scandinavian Workshop on Algorithm Theory*, 138–149, 2004.
7. M. de Berg, O. Cheong, O. Devillers, M. van Kreveld, and M. Teillaud. Computing the maximum overlap of two convex polygons under translations. *Theo. Comp. Sci.*, 31:613–628, 1998.
8. R. M. Dudley. Metric entropy of some classes of sets with differentiable boundaries, *J. Approximation Theory* **10** (1974) 227–236; Erratum in *J. Approx. Theory* **26** (1979) 192–193.
9. J. Matoušek. *Lectures on Discrete Geometry.* Springer-Verlag, New York, 2002
10. V. J. Milenkovic.Rotational polygon containment and minimum enclosure. *Proc. 14th Annu. Symp. Comput. geometry*, 1–8, 1998
11. D. M. Mount, R. Silverman, and A. Y. Wu. On the area of overlap of translated polygons. *Computer Vision and Image Understanding: CVIU*, 64(1):53–61, 1996.
12. K. Sugihara, M. Sawai, H. Sano, D.-S. Kim and D. Kim. Disk Packing for the Estimation of the Size of a Wire Bundle, *Japan Journal of Industrial and Applied Mathematics*, 21(3):259–278, 2004.
13. I. M. Yaglom and V. G. Boltyanskii. *Convex figures.* Holt, Rinehart and Winston, New York, 1961.

Algorithms for Range-Aggregate Query Problems Involving Geometric Aggregation Operations*

Prosenjit Gupta

Algorithms and Computation Theory Laboratory, International Institute of Information Technology, Gachibowli, Hyderabad 500019, India
pgupta@iiit.net

Abstract. We consider *variations* of the standard orthogonal range searching motivated by applications in database querying and VLSI layout processing. In a generic instance of such a problem, called a *range-aggregate query* problem we wish to preprocess a set S of geometric objects such that given a query orthogonal range q, a certain intersection or proximity query on the objects of S intersected by q can be answered efficiently. Efficient solutions are provided for point enclosure queries, 1-d interval intersection, 2-d orthogonal segment intersection and 1- and 2-d closest pair problems in this framework. Although range-aggregate queries have been widely investigated in the past for aggregation functions like average, count, min, max, sum etc. we consider geometric aggregation operations in this paper.

1 Introduction

1.1 Range-Aggregate Query Problems

Range searching is a fairly well-studied problem in Computational Geometry [1]. In such a problem, we are given a set S of n geometric objects. The goal is to efficiently report or count the intersections of the given set of objects with a given query range q. Since we are required to perform queries on the geometric data set several times, it is worthwhile to arrange the information into a data structure to facilitate searching.

In this paper, we consider a class of problems called *range-aggregate query* problems [16] which deal with some composite queries involving range searching, where we need to do more than just a simple range reporting or counting. In a generic instance of a range-aggregate query problem, we wish to preprocess a set of geometric objects S, such that given a query range q, a certain aggregation function that operates on the objects of $S' = S \cap q$ can be computed efficiently. As an example consider the *range-aggregate closest pair* problem studied in [11]:

* This research is supported in part by grants SR/S3/EECE/22/2004 and DST/INT/US/NSF-RPO-0155/04 from the Department of Science and Technology, Government of India.

X. Deng and D. Du (Eds.): ISAAC 2005, LNCS 3827, pp. 892–901, 2005.

"Preprocess a set S of points in $\mathbb{R}^2$ such that given a query rectangle q, the closest pair of points in $S \cap q$ can be reported efficiently". An R-tree based solution is provided for the problem in [11]. Aggregation functions are divided into three classes [12,6]: distributive, algebraic and holistic. Distributive aggregates (e.g. count, max, min, sum) can be computed by partitioning the input into disjoint sets, aggregating each set individually and then obtaining the final result by further aggregation of the partial results. Algebraic aggregates (e.g. average) can be expressed as a function of distributive aggregates. Holistic aggregates (e.g. median) cannot be computed by dividing the input into parts. Aggregation functions may be geometric in nature like intersection, convex hull etc. [12].

1.2 Applications

In on-line analytical processing (OLAP), geographic information systems (GIS) and other applications range aggregate queries play an important role in summarizing information [16] and hence large number of algorithms and storage schemes have been proposed. However most of these work consider functions like count, sum, min, max, average etc. [5,7,16,18,17,19]. Other than the work of [11] on the range-aggregate closest-pair problem, little or no work has been done on range-aggregate queries with geometric aggregation functions.

In this paper, we consider some range-aggregate query problems with geometric aggregation functions. We mention some applications where such problems can be useful. One of the most time-consuming steps in the map overlay processing is *line-breaking*, which we can abstract as the pairwise segment intersection problem [8]. If an user is interested in an overlay operation in a particular window of interest, a range-aggregate segment intersection query is useful. In a VLSI layout editing environment [13], geometric queries commonly arise. However, the user often zooms to a part of the layout and is interested in queries with respect to the portion of the layout on the screen. The 2-d range-aggregate point enclosure query and the 2-d range-aggregate interval intersection query are two fundamental operations which can be useful in this context. VLSI design rules are often based on the so-called *lambda* (λ) based design rules made popular by Mead and Conway [9]. Design rule checking (DRC) is the process of checking if the layout satisfies the given set of rules. One problem of interest to the designer is to check whether certain features are apart at least by a required separation. To check for violations in a part of the circuit, we can check if any two points in a query range violate the minimum separation rule. This can be answered using the range-aggregate closest-pair query.

1.3 Potential Approaches and Pitfalls

A range-aggregate query SR can be thought of as the composition of two queries: the range query R and another query S. We can solve the range-aggregate query problems using standard approaches. However, neither the technique of applying a simple range search to solve problem R and then applying the rest of the query S to the answer nor applying S followed by filtering using the query range R

Table 1. Summary of results for range-aggregate query problems on S and T; the query is an orthogonal range; k is the output size.

Underlying space	*Objects in S*	*Objects in T*	*Query*	*Space*	*Query time*
$\mathbb{R}^1$	points	intervals	point enclosures	$n \log n$	$\log n + k$
$\mathbb{R}^d$ $d \geq 2$	points	hyper rectangles	point enclosures	$n \log^d n$	$\log^d n + k$
$\mathbb{R}^1$	intervals	intervals	intersections	$n \log n$	$\log n + k$
$\mathbb{R}^2$	orthogonal segments	orthogonal segments	intersections	$n \log^2 n$	$\log^2 n + k$
$\mathbb{R}^1$	points	-	closest pair	n	$O(1)$
$\mathbb{R}^2$	points	-	closest pair	$n^2 \log^3 n$	$\log^3 n$

yields efficient solutions. This is also experimentally validated in [11] for the range-aggregate closest pair problem.

As as example, let us consider the 2-dimensional range-aggregate orthogonal segment intersection problem. Let S be a set of n orthogonal line segments in $\mathbb{R}^2$. We wish to preprocess S into a data structure such that given a query rectangle $q = [a, b] \times [c, d]$, we can efficiently report the pairs $(s_1, s_2), s_i \in S$ where s_1 intersects s_2 and the intersection point of s_1 and s_2 lies inside q.

One simple solution would be to preprocess S such that given q, the t segments of $S' \subseteq S$ that intersect q are retrieved. We can then run the $O(t \log t + k)$ algorithm for finding pairwise segment intersections [3] on S'. Thus the overall solution takes $O(\log n + t \log t + k)$ time, where k is the actual output size. If t is large, this solution is inefficient.

An alternative solution will be to find all pairs of intersections in S in $O(n \log n + t)$ time, using the algorithm of [3] (where t is the number of pairs of segments that intersect) and preprocess the intersection points for 2-dimensional orthogonal range searching which takes $O(t)$ space and $O(\log t + k) = O(\log n + k)$ time, where k is the actual output size. If t is large, this solution is inefficient in terms of space. The challenge is to develop low-space, low-query-time solutions which are also output-sensitive.

1.4 Our Contributions

From the above discussions, it is clear that different techniques are required to solve the range-aggregate query problems efficiently. In this paper, we provide efficient output-sensitive solutions for several range-aggregate query problems. We solve the 1-d range-aggregate point enclosure problem in Section 2, the 1-d range-aggregate interval intersection problem in Section 3, the range-aggregate point enclosure problem for $d \geq 2$ in Section 4 and the 2-d range-aggregate orthogonal segment intersection problem in Section 5. We revisit the range-aggregate closest-pair problem of [11] in Section 6. To the best of our knowledge, [11] is the only earlier work on this class of problems which considers geometric aggregation functions. Ours is the first systematic study of some fundamental intersection

and proximity problems in this framework. Our results are summarized in Table 1. Due to lack of space, we omit some proofs and many details.

2 1-d Range-Aggregate Point Enclosure

In this section we consider the 1-d range-aggregate point enclosure problem. Here the goal is to preprocess a set of points and intervals on the real line to efficiently report all point-interval incidences that lie within a query interval.

Problem 1. Preprocess a set S of points and a set T of intervals on the x-axis, with $|S| + |T| = n$ such that given a query interval $q = [a, b]$, all pairs $(s, t), s \in S, t \in T$ satisfying $s \in t \cap q$ can be reported efficiently.

We sort the points in S in non-decreasing order and remove any point in S which is not covered by any interval in T. We store the reduced set S' in the leaves of a binary search tree BT. We search in BT with the intervals $t \in T$ to find the points in S that intersect t. If no such points exist for an interval $t \in T$, we remove it from T. Let $T' \subseteq T$ be the reduced set. We preprocess the intervals in T' into a segment tree ST. The intervals in T' partition the x-axis it into $2|T'| + 1$ *elementary intervals* (some of which may be empty). We build a *segment tree* ST which stores these elementary intervals at the leaves. Let v be any node of ST. We associate with v an x-interval $Int(v)$, which is the union of the elementary intervals stored at the leaves in v's subtree. We say that an interval $i \in T$ is *allocated* to a node $v \in ST$ iff $Int(v) \neq \emptyset$ and i covers $Int(v)$ but not $Int(parent(v))$. Let $I(v)$ be the set of intervals allocated to v. For any node v, let $ANC(v)$ denote the nearest ancestor u of v such that $I(u) \neq \emptyset$.

For each point $s \in S'$, if s lies at the boundary of two elementary intervals, we let $leftleaf(s)$ (respectively $rightleaf(s)$) denote the leaf in ST which corresponds to the elementary interval to the left (respectively right) of s. If s lies in the interior of an elementary interval i, we let $leftleaf(s)$ and $rightleaf(s)$ both point to i.

Given a query interval $q = [a, b]$, first we search in BT to locate a' (respectively b'), the points with the smallest (respectively largest) x-coordinate greater than or equal to (respectively less than or equal to) a (respectively b). Then for each leaf ℓ of BT in the interval $[a', b']$, we locate the leaves $\ell_1 = leftleaf(\ell)$ and $\ell_2 = rightleaf(\ell)$ in ST. In ST, we trace the path $\Pi(\ell_1)$ from the node ℓ_1 to the root using the $ANC(v)$ pointers and report for each node v on the path, pairs (p, i) for all $i \in I(v)$ where p is the point corresponding to leaf ℓ_1 of BT. If $\ell_2 \neq \ell_1$, we trace the path $\Pi(\ell_2)$ in ST from ℓ_2 to the root, reporting point-interval incidences as above.

Theorem 1. *A set S of points and a set T of intervals on the x-axis where $|S| + |T| = n$, can be preprocessed into a data structure of size $O(n \log n)$ such that given a query interval $q = [a, b]$, all pairs $(s, t), s \in S, t \in T$ such that $s \in t \cap q$ can be reported in time $O(\log n + k)$ where k is the output size.*

3 1-d Range-Aggregate Interval Intersection

In this section, we consider the 1-d *range-aggregate interval intersection* problem. Here the goal is to preprocess a set S of intervals in the real line to efficiently report pairwise intersections of intervals in S such that the intersections overlap with the query interval q.

Problem 2. Preprocess a set S of n intervals on the x-axis, such that given a query interval $q = [a, b]$, all pairs $(s, t), s \in S, t \in S$ satisfying $s \cap t \cap q \neq \emptyset$ can be reported efficiently.

Lemma 1. *A pair of intervals (s, t) of S satisfies $s \cap t \cap q \neq \emptyset$ iff*
(i) An endpoint of s is in $t \cap q$ or
(ii) An endpoint of t is in $s \cap q$ or
(iii) $q \subseteq (s \cap t)$

To report interval pairs satisfying conditions (i) or (ii) of Lemma 1, we simply preprocess the intervals of S and the endpoints of the intervals into an instance $D1$ of the data structure of Theorem 1. Given q, we query $D1$ and for every pair (i, p) found, we report (i, p') where p' is the interval one of whose endpoints is p. To report interval pairs satisfying condition (iii) of Lemma 1, we map each interval $i = [c, d]$ into the point $\mathcal{F}(i) = (c, d) \in \mathbb{R}^2$. We preprocess such points into a data structure $D2$ for 2-d quadrant searching. This can be implemented using a priority search tree [10]. We map the interval $q = [a, b]$ into the northwest quadrant $NW(q)$ of the point $(a, b) \in \mathbb{R}^2$. We query $D2$ with $NW(q)$ and store the result in a temporary list $L(q)$. We report for each pair of points (p_1, p_2), $p_1 \in L(q)$, $p_2 \in L(q)$, the interval pair (p'_1, p'_2) where p'_1 (respectively p'_2) is the interval corresponding to p_1 (respectively p_2). Since $|L(q)| \leq \min\{n, k\}$, the overhead only contributes a constant factor to the space and time bounds.

Theorem 2. *A set S of n intervals on the x-axis can be preprocessed into a data structure of size $O(n \log n)$ such that given a query interval $q = [a, b]$, all pairs $(s, t), s \in S, t \in S$ such that $s \cap t \cap q \neq \emptyset$ can be reported in time $O(\log n + k)$ where k is the output size.*

4 Range-Aggregate Point Enclosure for $d \geq 2$

First, we consider the 2-d range-aggregate point enclosure problem. Here the goal is to preprocess a set S of points and a set T of orthogonal rectangles, all in $\mathbb{R}^2$ to report all point-rectangle incidences inside a query orthogonal rectangle.

Problem 3. Preprocess a set S of points and a set T of axes-parallel rectangles in $\mathbb{R}^2$ with $|S| + |T| = n$, such that that given a query rectangle $q = [a, b] \times [c, d]$, all pairs $(s, t), s \in S, t \in T$ satisfying $s \in (t \cap q)$ can be reported efficiently.

Lemma 2. *For a query rectangle q, a point s and a rectangle t satisfy $s \in (t \cap q)$ iff*
(i) $s_x \in t_x \cap q_x$ and
(ii) $s_y \in t_y \cap q_y$

where s_x, t_x and q_x are the x-projections of s, t and q respectively and where s_y, t_y and q_y are the y-projections of s, t and q respectively.

The problem is decomposable along the dimensions. We construct a two-level data structure where the outer structure DS is an instance of the data structure of Theorem 1 for 1-d range-aggregate point enclosure problem built on the x-projections of the points in S and the rectangles in T. DS consists of a binary search tree $BT(S)$ and a segment tree $ST(S)$. During preprocessing, we map the the x-coordinates of the points in $BT(S)$ to elementary intervals in $ST(S)$ and create the $leftleaf$ and $rightleaf$ pointers. At each internal node v of $ST(S)$, let $S(v)$ denote the set of points from $BT(S)$ lying in the elementary intervals at the leaves of the subtree rooted at v. For node v, recall that $I(v)$ is the set of intervals (in this case x-projections of rectangles) allocated to v. At each such node v, we create an instance $D(v)$ of the data structure of Theorem 1 built on the y-coordinates of the points in $S(v)$ and the y-projections of the rectangles in T, whose x-projections are in $I(v)$.

Given the query rectangle $q = [a, b] \times [c, d]$, we query DS with $q' = [a, b]$ and identify the canonical nodes v such that q' covers $Int(v)$ but q' does not cover $Int(parent(v))$. At each such node v, we query $D(v)$ with $q'' = [c, d]$. We report the union of all the answers retrieved.

Lemma 3. *Given a set S of points and a set T of axes-parallel rectangles in $\mathbb{R}^2$, and a query rectangle q, the query algorithm reports a pair $(s, t), s \in S, t \in T$ iff $s \in (t \cap q)$.*

Proof. ($\Leftarrow$) Let $s \in (t \cap q)$. Hence $s_x \in q_x$. During preprocessing, s is mapped to the appropriate elementary interval corresponding to a leaf ℓ_s of the outer segment tree $ST(S)$ and is in each set $S(v)$ for all nodes v in the path $\Pi(\ell_s)$ from ℓ_s to the root of $ST(S)$. Note that $s_x \in i$ for each $i \in I(v)$ for all nodes v on $\Pi(\ell_s)$. Since $s_x \in t_x$, by construction of the segment tree, t_x must correspond to some such interval i stored at a node u on $\Pi(\ell_s)$. Since $D(u)$ is built with y-coordinates of points in $S(u)$ (which includes s) and y-projections of rectangles in T whose x-projections are allocated to u (which includes t), and $s_y \in t_y \cap q_y$, from the correctness of the structure of Theorem 1, it follows that the pair (s, t) is reported while querying $D(u)$.
($\Rightarrow$) Let the query algorithm report a pair $(s, t), s \in S, t \in T$. It must be reported while querying $D(u)$ for some node u in $ST(S)$. Hence $s_y \in t_y \cap q_y$. Since $D(u)$ is built with y-coordinates of points in $S(u)$ and y-projections of rectangles in T whose x-projections are in $I(u)$, $S(u)$ must include s and $I(u)$ must include t_x. Hence $s_x \in t_x$. Since $D(u)$ is selected as a canonical node, $Int(u) \subseteq q_x$. Since $s_x \in Int(u)$, we conclude that $s_x \in q_x$. Since $s_x \in t_x$ and $s_x \in q_x$, $s_x \in t_x \cap q_x$. Since $s_y \in t_y \cap q_y$ also, we conclude that $s \in t \cap q$.

Since the problem is decomposable along the dimensions, we can extend the result to dimensions $d > 2$. We conclude:

Theorem 3. *A set S of points and a set T of axes-parallel hyper-rectangles in $\mathbb{R}^d$ for $d \geq 2$ with $|S| + |T| = n$, can be preprocessed into a data structure of*

size $O(n \log^d n)$ such that given a query hyper-rectangle $q = [a_1, b_1] \times [a_2, b_2] \times \ldots \times [a_d, b_d]$, all pairs $(s,t), s \in S, t \in T$ such that $s \in (t \cap q)$ can be reported in time $O(\log^d n + k)$ where k is the output size.

5 Range-Aggregate Orthogonal Segment Intersection

In this section, we consider the 2-d range-aggregate orthogonal segment intersection problem. Here the goal is to preprocess a set S of orthogonal segments in $\mathbb{R}^2$ to report all intersections inside a query orthogonal range q.

Problem 4. Preprocess a set S of axes-parallel segments in $\mathbb{R}^2$ such that that given a query rectangle $q = [a, b] \times [c, d]$, all pairs $(s_1, s_2), s_1 \in S, s_2 \in S$ such that s_1 intersects s_2 and the intersection point lies in q, can be reported efficiently.

First we consider reporting the horizontal-vertical intersections:

Problem 5. Preprocess a set H of horizontal and a set V of vertical segments all in $\mathbb{R}^2$ such that that given a query rectangle $q = [a, b] \times [c, d]$, all pairs $(s_1, s_2), s_1 \in H, s_2 \in V$ such that s_1 intersects s_2 and the intersection point lies in q, can be reported efficiently.

Our outer structure is an instance DS of the data structure of Theorem 1. DS is built on a set of 1-dimensional points and a set of 1-dimensional intervals. The set of points is the set of x-projections of the segments in V and the set of intervals is the set of x-projections of the horizontal segments H. DS consists of a segment tree $ST(S)$ and a binary search tree $BT(S)$. At each internal node v of $ST(S)$, let $S(v)$ denote the set of points from $BT(S)$ lying in the elementary intervals at the leaves of the subtree rooted at v. For node v, recall that $I(v)$ is the set of intervals (in this case x-projections of horizontal segments) allocated to v. At each such node v, we create an instance $D(v)$ of the data structure of Theorem 1. $D(v)$ is built on a set of intervals and points. The set of intervals that $D(v)$ is built on are the y-projections of the vertical segments whose x-projections are in $S(v)$. The set of points that $D(v)$ is built on are the y-projections of the horizontal segments in H, whose x-projections are intervals allocated to v.

Given the query rectangle $q = [a, b] \times [c, d]$, we query DS with $q' = [a, b]$ and identify the canonical nodes v such that q' covers $Int(v)$ but q' does not cover $Int(parent(v))$. At each such node v, we query $D(v)$ with $q'' = [c, d]$. The query retrieves pairs of the form (i, p). For any such pair (i, p), we report (i', p'), where i' is the vertical segment whose y-projection is the interval i and p' is the horizontal segment whose y-projection is the point p. We report the union of all the answers retrieved.

Lemma 4. *The query algorithm reports a pair $(s_1, s_2), s_1 \in H, s_2 \in V$ iff $s_1 \cap s_2 \in q$.*

Proof. ($\Leftarrow$) Let $s_1 \cap s_2 \in q$. Then the x-projection of the vertical segment s_2 lies in the x-projection of q. Hence the x-projection of s_2 is in $S(v)$ for a canonical node v in $ST(S)$ as identified while querying with the x-projection of q. This

implies that the interval set that $D(v)$ is built on will include the y-projection of s_2. Since the x-projection of the horizontal segment s_1 must also include the x-projection of s_2, by construction of the segment tree $ST(S)$, $I(v)$ must include the x-projection of s_1. Then the point set that $D(v)$ is built on will include the y-projection of s_1. By the correctness of the structure of Theorem 1, the pair (s_1, s_2) will be reported.
($\Rightarrow$) Let the pair $(s_1, s_2), s_1 \in H, s_2 \in V$ be reported. From the correctness of the structure of Theorem 1, the y-projection of s_1 must be in the y-projection of s_2 and also in the y-projection of q. If the pair is reported while querying $D(v)$ at node v, $I(v)$ must include the x-projection of s_1 and $S(v)$ must include the x-projection of s_2. Hence the x-projection of s_2 must be in the x-projection of s_1. Since v is a canonical node identified by querying $ST(S)$ with the x-projection of q, the x-projection of s_2 must also be in the x-projection of q. Hence $s_1 \cap s_2 \in q$.

The case of intersections amongst horizontal segments is symmetrical to that amongst vertical segments. We describe the former here. Formally the problem can be stated as:

Problem 6. Preprocess a set H of horizontal segments in $\mathbb{R}^2$ such that that given a query rectangle $q = [a, b] \times [c, d]$, all pairs (s_1, s_2), $s_1 \in H, s_2 \in H$ satisfying $s_1 \cap s_2 \cap q \neq \emptyset$ can be reported efficiently.

Lemma 5. *Let s_1 and s_2 be horizontal segments. Then given a rectangle $q = [a, b] \times [c, d]$, $s_1 \cap s_2 \cap q \neq \emptyset$ iff*
(i) s_1 has an endpoint in $s_2 \cap q$ or
(ii) s_2 has an endpoint in $s_1 \cap q$ or
(iii) The left edge of q intersects $s_1 \cap s_2$

We first build a data structure D_1 to cover cases (i) and (ii) of Lemma 5. We preprocess the x-projections of the segments into a segment tree ST. At each internal node v of ST, the set $S(v)$ includes the endpoints of the segments which define the endpoints of the elementary intervals at the leaves of the subtree rooted at v. We store the y-coordinates of the intervals stored in $I(v)$ in the leaves of a binary search tree $BT(v)$. During preprocessing, we map the y-coordinates of the points in $S(v)$ to the leaves of $BT(v)$. Each leaf node ℓ is associated with a list $YL(\ell)$ of such points.

Given the query rectangle $q = [a, b] \times [c, d]$, we query ST with $q' = [a, b]$ and identify the canonical nodes v such that q' covers $Int(v)$ but q' does not cover $Int(parent(v))$. At each such node v, we query $BT(v)$ with $q'' = [c, d]$. For each leaf node ℓ of $BT(v)$ retrieved by the query, we report (s_1, s_2) where s_1 is the segment whose y-coordinate is stored in ℓ and s_2 is the segment whose endpoint is in $L(\ell)$. We report the union of all the answers retrieved.

Next we build a data structure D_2 to cover case (iii) of Lemma 5. This is similar to D_1 with the exception of the secondary structure $BT(v)$. In this case, we store the y-coordinates of the intervals stored in $I(v)$ in the leaves of a binary search tree $BT(v)$. Each leaf node ℓ is associated with a list $YL(\ell)$ of such y-coordinates. Given the query rectangle $q = [a, b] \times [c, d]$, we query ST with a

and identify the canonical nodes v such that $a \in Int(v)$. At each such node v, we query $BT(v)$ with $q'' = [c,d]$. For each leaf node ℓ of $BT(v)$ retrieved by the query, we report each pair (s_1, s_2) where s_1 and s_2 are segments whose y-coordinates are stored in $L(\ell)$. We report the union of all the answers retrieved.

Theorem 4. *A set S of n orthogonal segments in $I\!R^2$ can be preprocessed into a data structure of size $O(n\log^2 n)$ such that given a query rectangle $q = [a,b] \times [c,d]$, all pairs (s_1, s_2), $s_1 \in S, s_2 \in S$ and $s_1 \cap s_2 \cap q \neq \emptyset$ can be reported in time $O(\log^2 n + k)$, where k is the output size.*

6 Proximity in a Range

Geometric proximity problems arise in numerous applications and have been widely studied in computational geometry. The closest-pair problem involves finding a pair of points in a set such that the distance between them is minimal. Work on the closest-pair and some related problems are surveyed in [14]. The range-aggregate version was considered in [11] and a R-tree based solution was given. We assume that all distances are euclidean distances.

To solve the 1-dimensional range-aggregate closest pair problem, we associate with each point $s \in S$ a real number $g(s)$ being the distance between s and the point immediately to its right. For the rightmost point r let $g(r) = \infty$. We preprocess the points into an instance D of the data structure of [4] for 1-dimensional range minimum query to find a point in a query range with minimum $g(s)$.

Theorem 5. *A set S of n points in $I\!R^1$ can be preprocessed into a data structure of size $O(n)$ such that given a query interval $q = [a,b]$, the closest pair of points in $S \cap q$ can be reported in time $O(1)$.*

To solve the 2-dimensional problem, we first find all pairwise distances between the points in S. We create $O(n^2)$ tuples $(a_x, a_y, b_x, b_y, d(a,b))$ where $a = (a_x, a_y)$ and $b = (b_x, b_y)$ are points in S and $d(a,b)$ is the euclidean distance between a and b. We preprocess these tuples into a data structure D for the 4-dimensional range minimum query of [4]. An instance of D built on a set of n points occupies $O(n\log^3 n)$ space and can be queried in time $O(\log^3 n)$. Given q, we query D to find a pair (a,b) with the minimum value of $d(a,b)$.

Theorem 6. *A set S of n points in $I\!R^2$ can be preprocessed into a data structure of size $O(n^2\log^3 n)$ such that given a query rectangle $q = [a,b] \times [c,d]$, the closest pair of points in $S \cap q$ can be reported in time $O(\log^3 n)$.*

References

1. P.K. Agarwal, and J. Erickson. Geometric range searching and its relatives. In B. Chazelle, J. E. Goodman, and R. Pollack, editors, *Advances in Discrete and Computational Geometry*, Contemporary Mathematics, 23, 1999, 1–56, American Mathematical Society Press.

2. M. de Berg, M. van Kreveld, M. Overmars, and O. Schwarzkopf. *Computational Geometry: Algorithms and Applications*, Springer Verlag, 2000.
3. I.J. Balaban. An optimal algorithm for finding segment intersections. *Proceedings, 11th Annual ACM Symposium on Computational Geometry*, 1995, 211–219.
4. H.N. Gabow, J.L. Bentley, and R.E. Tarjan. Scaling and related techniques for geometry problems. *Proceedings, ACM Symposium on Theory of Computing*, 1984, 135–142.
5. S. Govindarajan, P.K. Agarwal, and L. Arge. CRB-Tree: An efficient indexing scheme for Range Aggregate Queries. *Proceedings, International Conference on Database Theory*, 2003, 143–157.
6. J. Gray, S. Chaudhuri, A. Bosworth, A. Layman, D. Recichart, M. VenkatRao, F. Pellow, and H. Pirahesh. Data Cube: A relational aggregation operator generalizing Group-By, Cross-Tab, and Sub-Totals. *Data Mining and Knowledge Discovery*, 1(1), 1997, pp. 29–53.
7. S. Hong, B. Song, and S. Lee. Efficient execution of range-aggregate queries in data warehouse environments. *Proceedings, ICCM 01*, Springer Verlag LNCS, Vol. 2224, 2001, 299–310.
8. R. Jampani, R. Thonangi, and P. Gupta. Overlaying multiple maps efficiently. *Proceedings, CIT 04*, Springer Verlag LNCS, Vol. 3356, 2004, 263–272.
9. C.A. Mead, and L.A. Conway. *Introduction to VLSI Systems*, Addison Wesley, USA, 1980.
10. E.M. McCreight. Priority search trees, *SIAM Journal of Computing*, 14(2), 1985, 257–276.
11. J. Shan, D. Zhang, and B. Salzberg. On spatial-range closest-pair query. *Proceedings, Symposium on Spatial and Temporal Databases*, Springer Verlag LNCS, Vol. 2750, 2003, 252–269.
12. S. Shekhar, and S. Chawla. *Spatial Databases: A Tour*, Prentice Hall, 2002.
13. N. Sherwani. *Algorithms for VLSI Physical Design Automation*, Kluwer Academic, 1998.
14. M. Smid. Closest point problems in computational geometry. *Handbook of Computational Geometry*, J. Sack and J. Urrutia editors, Elsevier, 2000, 877–935.
15. T.G. Szymanski, and C.J. van Wyk. Layout analysis and verification, in *Physical Design Automation of VLSI Systems*, B. Preas and M. Lorenzetti eds., Benjamin/Cummins, 1988, pp. 347–407.
16. Y. Tao and D. Papadias. Range aggregate processing in spatial databases. *IEEE Transactions on Knowledge and Data Engineering*, 16(12), 2004, 1555–1570.
17. D. Zhang, V.J. Tsotras. Improving min/max aggregation over spatial objects. *VLDB Journal* 14(2), 2005, 170–181.
18. D. Zhang, A. Markowetz, V. Tsotras, D. Gunopulos, and B. Seeger. Efficient computation of temporal aggregates with range restrictions. *Proceedings, Symposium on Principles of Database Systems*, 2001.
19. D. Zhang, V.J. Tsotras, D. Gunopulos. Efficient aggregation over objects with extent. *Proceedings, Symposium on Principles of Database Systems*, 2002, 121–132.

A $(2 - c\frac{1}{\sqrt{N}})$–Approximation Algorithm for the Stable Marriage Problem

Kazuo Iwama[1,⋆], Shuichi Miyazaki[2,⋆⋆], and Naoya Yamauchi[1]

[1] Graduate School of Informatics, Kyoto University
[2] Academic Center for Computing and Media Studies, Kyoto University
{iwama, nyamauchi}@kuis.kyoto-u.ac.jp, shuichi@media.kyoto-u.ac.jp

Abstract. We consider the problem of finding a stable matching of maximum size when both ties and unacceptable partners are allowed in preference lists. This problem is NP-hard and the current best known approximation algorithm achieves the approximation ratio $2 - c\frac{\log N}{N}$, where c is an arbitrary positive constant and N is the number of men in an input. In this paper, we improve the ratio to $2 - c\frac{1}{\sqrt{N}}$, where c is a constant such that $c \leq \frac{1}{4\sqrt{6}}$.

1 Introduction

An instance of the stable marriage problem consists of N men, N women, and each person's preference list. A preference list is a totally ordered list including all members of the opposite sex depending on his/her preference. For a matching M between men and women, a pair of a man m and a woman w is called a *blocking pair* if m prefers w to his current partner and w prefers m to her current partner. A matching with no blocking pair is called *stable*. Gale and Shapley showed that every instance admits at least one stable matching, and proposed a polynomial-time algorithm to find one, which is known as the Gale-Shapley algorithm [9]. There are several examples of using the stable marriage problem in assignment systems. Probably, one of the most famous applications is to assign medical students to hospitals based on preference lists of both sides, which is known as NRMP in the U.S. [11, 28], CaRMS in Canada [6], SPA in Scotland [17, 18], and JRMP in Japan [23]. Another application is to assign students to schools in Singapore [32].

Considering such applications, it is unrealistic to require each participant to submit a preference list *strictly* ordering *all* members of the opposite side. One natural extension is to allow an *incomplete list*, namely, one may drop persons from the list whom he/she does not want to be matched with. In this case, a stable matching may not be a perfect matching, but all stable matchings for a

---

⋆ Supported in part by Scientific Research Grant, Ministry of Japan, 16300002, 16092101 and 16092215.

⋆⋆ Supported in part by Scientific Research Grant, Ministry of Japan, 15700010, 17700015 and 16092215.

X. Deng and D. Du (Eds.): ISAAC 2005, LNCS 3827, pp. 902–914, 2005.

fixed instance are of the same size [10], and a slight modification of the Gale-Shapley algorithm can find one in polynomial time. In fact, a lot of well-known applications allow agents to submit incomplete lists. Another natural extension is to allow *ties* in the list [11, 16]. Again, it is easy to find a stable matching by a modified Gale-Shapley algorithm.

However, if we allow *both* extensions, one instance can admit stable matchings of different sizes, and in a worst case, a smallest stable matching is of size only a half of a largest one. It is known that the size of a stable matching we obtain is essentially based on the way of tie-breaking (see [29] for example). If a matching system does not allow ties, participants may have to break ties independently by themselves, which may reduce the size of a resulting stable matching to a half the optimal. Hence, it is important to allow not only incomplete lists but also ties at the same time. Unfortunately, the problem of finding a maximum stable matching, which we call *MAX SMTI* (MAXimum Stable Marriage with Ties and Incomplete lists), is NP-hard [21, 29], and the current best polynomial-time approximation algorithm achieves an approximation ratio $2 - c\frac{\log N}{N}$, where c is an arbitrary positive constant [22].

Simply speaking, the algorithm in [22] starts from an arbitrary initial stable matching, and at each iteration, successively improves the size of a stable matching. While $|M|$, the size of the current solution, is at most $\frac{OPT}{2} + c \log N$, where OPT is the size of an optimal solution, it can increase the size by at least one. Hence, the size of a stable matching is eventually greater than $\frac{OPT}{2} + c \log N$. To increase the size of a matching, it first finds a subset H of M, and removes H from M. It then adds a set K of new edges to the matching $M - H$, where K is computed from H. $|K|$ is closely related to the approximation ratio; if *all* edges of K satisfy some good property (say, Ψ), we can guarantee that the size of a stable matching we finally obtain is at least $\frac{OPT}{2} + O(|K|)$. In [22], it was proved that there exists a good H that produces K such that $|K| = 2|H|$ and all edges of K satisfy Ψ, but it was not known how to find it efficiently. Hence, we had to rely on an exhaustive search, which allows $|H|$ (and hence $|K|$) to be at most $O(\log N)$ for polynomial-time computation. So, the obvious problem was how to find this good subset H efficiently instead of the naive exhaustive search.

Our Contribution. In this paper, we modify the algorithm so that it does not need exhaustive search, improving the approximation ratio to $2-c\frac{1}{\sqrt{N}}$ ($c \leq \frac{1}{4\sqrt{6}}$). To achieve this improvement, we relax the condition mentioned above. If K contains a subset K' such that all edges in K' satisfy Ψ, then we can obtain a stable matching of size at least $\frac{OPT}{2} + O(|K'|)$. With this relaxation, we show that $|K'|$ can be set as large as $O(\sqrt{|M|})$. To prove the correctness of this relaxed condition, we use the fact that a minimum vertex cover for bipartite graphs can be found in polynomial time. Also, for finding K that contains large K', we exploit a polynomial-time algorithm for finding a maximum matching in bipartite graphs, and multiple use of the Gale-Shapley algorithm.

Related Results. As mentioned above, MAX SMTI was proved to be NP-hard [21] and then to be APX-hard [12]. Subsequently, it was shown that MAX

SMTI cannot be approximated within 21/19 unless P≠NP [14]. For restricted inputs, there are a couple of approximation algorithms with factor better than two [13, 14]. Other than SMTI, research on stable matchings has been quite intensive recently, which includes studies on strong stability [20, 26], rank-maximal matchings [19], Pareto optimal matchings [1], popular matchings [2], and others [7, 3, 8].

There are several optimization problems that resemble MAX SMTI, where designing a 2-approximation algorithm is trivial but obtaining a $(2-\epsilon)$-approximation algorithm for a positive constant ϵ is extremely hard, such as Minimum Vertex Cover (MIN VC for short) and Minimum Maximal Matching (MIN MM for short). As is the case with MAX SMTI, there are a lot of approximability results for these problems by restricting instances. For example, MIN VC is approximable within 7/6 if the maximum degree of an input graph is bounded by 3 [5], or within $2/(1+\epsilon)$ if every vertex has degree at least $\epsilon|V|$ [25]. For MIN MM, there is a $(2-1/d)$-approximation algorithm for regular graphs with degree d [33], and PTAS for planar graphs [31]. For general inputs, $(2-o(1))$-approximation algorithms are presented for MIN VC, namely, $2-\frac{\log\log|V|}{2\log|V|}$ and $2-(1-o(1))\frac{2\ln\ln|V|}{\ln|V|}$ [30, 4, 15]. Very recently, it is improved to $2-\Theta(\frac{1}{\sqrt{\log|V|}})$ [24].

2 Preliminaries

2.1 Notations and Definitions

In this section, we formally define MAX SMTI and approximation ratios. An instance I of MAX SMTI comprises of N men, N women and each person's preference list that may be incomplete and may include ties. We usually use m, m_i, etc for men, and w, w_i, etc for women. If a man m writes a woman w in his list, we say that w is *acceptable* to m. If m strictly prefers w_i to w_j in I, we write $w_i \succ_m w_j$. If w_i and w_j are tied in m's list, we write $w_i =_m w_j$. The statement $w_i \succeq_m w_j$ is true if and only if $w_i \succ_m w_j$ or $w_i =_m w_j$. We use similar notations for women's preference lists.

A matching M is a set of pairs (m, w) such that m is acceptable to w and vice versa, and each person appears at most once in M. For convenience, we sometimes use the notation $m \in M$, which means that m is matched in M, namely, $\exists w\ (m, w) \in M$. If $m \notin M$, we say that m is *single* in M. We sometimes call a pair $(m, w) \in M$ *an edge* of M.

If a man m is matched with a woman w in M, we write $M(m) = w$ and $M(w) = m$. We say that m and w form a *blocking pair* for M (or simply, (m, w) *blocks* M) if the following three conditions are met: (i) $M(m) \neq w$ but m and w are acceptable to each other. (ii) $w \succ_m M(m)$ or m is single in M. (iii) $m \succ_w M(w)$ or w is single in M. A matching M is called *stable* if there is no blocking pair for M. MAX SMTI is the problem of finding a largest stable matching.

A goodness measure of an approximation algorithm T of a maximization problem is defined as usual: the *approximation ratio* of T is $\max\{opt(x)/T(x)\}$ over all instances x of size N, where $opt(x)$ and $T(x)$ are the size of the optimal and the algorithm's solution, respectively. Throughout this paper, we denote the optimal cost by OPT.

2.2 Algorithm LocalSearch(I)

In this section, we briefly review the algorithm LocalSearch presented in [22]. It takes a MAX SMTI instance I and outputs a stable matching for I. It consists of two subroutines, called Increase and Stabilize. Increase takes a stable matching M for I and outputs a (not necessarily stable) matching M' such that $|M'| > |M|$ and M' satisfies the following property Π:

Π: For any blocking pair (m, w) for M', at least one of m and w is single in M'.

Increase may fail to find such a matching. In that case, it returns error. Stabilize takes a matching M' with property Π, and outputs a stable matching of size at least $|M'|$. Stabilize never fails. The whole description of LocalSearch is as follows:

Algorithm LocalSearch(I)

1: M = arbitrary stable matching for I.
  /* This can be done in polynomial time by arbitrary tie-breaking
   and applying the Gale-Shapley algorithm. */
2: while (true)
3: { M' = Increase (M).
   If Increase returns an error, exit while loop and output M.
4: M = Stabilize (M'). }

It was shown in [22] that while $|M| \le \frac{OPT}{2} + c \log N$, Increase never fails, and hence we finally obtain a stable matching of size more than $\frac{OPT}{2} + c \log N$. This gives an upper bound on the approximation ratio of LocalSearch.

3 Improved Algorithm

To obtain a better upper bound, we improve the subroutine Increase, and show that it never fails if $|M| < \frac{OPT}{2} + c'\sqrt{|M|}$. Here, c' is a constant such that $c' \le \frac{1}{8\sqrt{6}}$. So, LocalSearch can obtain a stable matching of size at least $\frac{OPT}{2} + c'\sqrt{|M|}$.

Before presenting Increase, we give a couple of procedures used in Increase. The following procedure MaxMatching takes a stable matching M, and outputs a matching that matches a subset of women in M with a subset of men single in M.

Procedure MaxMatching(M)

1: Construct a bipartite graph $G = (U, V, E)$ in the following way.
  U is the set of women in M.
  V is the set of men single in M.
  Give an edge between $w \in U$ and $m \in V$ if and only if $M(w) =_w m$.

2: Find a maximum matching in G, and output it.

Next, we give the procedure MultipleGS. It takes a stable matching M and its subset $S \subseteq M$. It outputs several matchings, each of which matches a subset of men in S with a subset of women single in M. Fix a constant c' ($c' \le \frac{1}{8\sqrt{6}}$), and define $B = 4c'\sqrt{|M|}$. (For the Gale-Shapley algorithm, see [11] for example.)

Procedure MultipleGS(M, S)

1: Let X be the set of men in S.

2: Let Y be the set of women single in M.

3: For each man $m \in X$, do the following.

4: { Delete all women not in Y from m's preference list.

5: Break all ties in m's preference list in an arbitrary way.
  /* Call the resulting lists *modified lists.* */ }

6: For each woman $w \in Y$, do the following.

7: { Delete all men not in X from w's preference list.

8: Break all ties in w's preference list in an arbitrary way. }

9: For $k = 1$ to $2B - 1$, do the following.

10: { For each man $m \in X$, delete, from m's preference list, all women whose position (with respect to his modified list) is greater than k.

11: Apply men-propose Gale-Shapley algorithm to X and Y, and let the resulting matching be Q_k. }

12: Partition X into $|X|/2B$ sets $X_1, X_2, \cdots, X_{|X|/2B}$ arbitrarily, each of which is of size $2B$.

13: For each X_i, do the following.

14: { For each man $m \in X_i$, delete, from m's preference list, all women whose position (with respect to his modified list) is greater than $2B$.

15: Apply men-propose Gale-Shapley algorithm to X_i and Y, and let the resulting matching be Q_{2B-1+i}. }

16: Output $Q_1, \cdots, Q_{2B-1+|S|/2B}$. /* Note that $|S| = |X|$ */

Finally, we give Increase. Fig. 1 shows an example of the execution of Increase up to the 8th line.

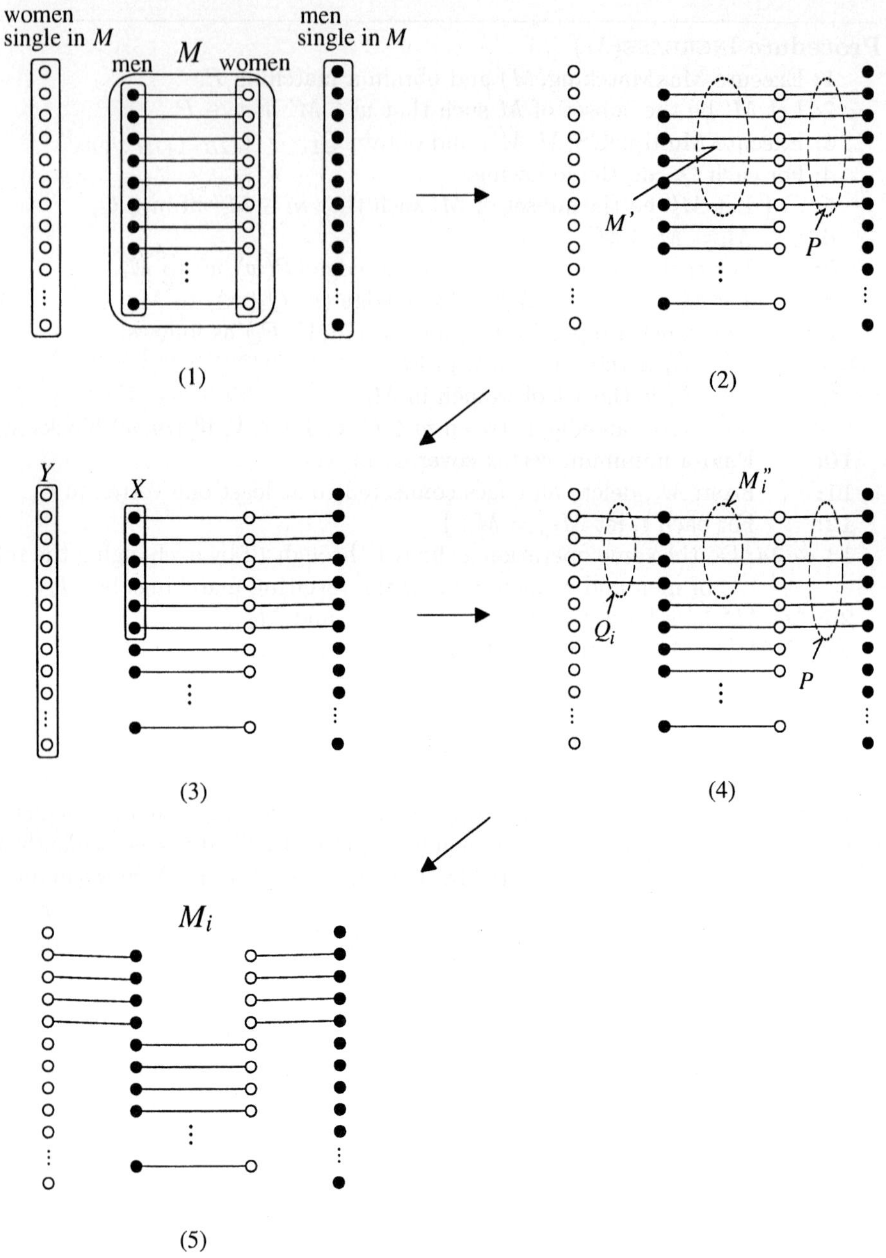

Fig. 1. An example of computation by INCREASE

Procedure INCREASE(M)

1: Execute MaxMatching(M) and obtain a matching P.
2: Let M' be the subset of M such that $w \in M'$ iff $w \in P$.
3: Execute MultipleGS(M, M') and obtain $Q_1, \cdots, Q_{2B-1+|M'|/2B}$.
4: For each Q_i, do the following.
5: { Let M_i'' be the subset of M' such that $m \in M_i''$ iff $m \in Q_i$.
6: $M_i = M - M_i''$.
7: For each woman $w \in M_i''$, add an edge $(P(w), w)$ to M_i.
8: For each man $m \in M_i''$, add an edge $(m, Q_i(m))$ to M_i.
9: Construct a bipartite graph $G_i = (U_i, V_i, E_i)$ as follows.
U_i is the set of men in M_i.
V_i is the set of women in M_i.
Give an edge between $m \in U_i$ and $w \in V_i$ iff (m, w) blocks M_i.
10: Find a minimum vertex cover C_i for G_i.
11: From M_i, delete all edges connected to at least one vertex in C_i.
12: For each i, let $M_{1,i} = M_i$. }
13 $\sim$ 24: Do the same operation as lines 1 through 12 by exchanging the role of men and women, and let the resulting matchings be $M_{2,i}$.
25: Let M^* be a largest matching of all $M_{1,i}$ and $M_{2,i}$.
26: If $|M^*| > |M|$, output M^*. Otherwise, output error.

Here, let us roughly explain an execution of the algorithm. INCREASE executes computation on all possible i. An example of computation on fixed i is depicted in Fig. 1. Vertices corresponding to men are denoted by filled circles, while those corresponding to women are denoted by unfilled circles. Fig. 1 (1) shows an input matching M. M is given to Procedure MaxMatching, and it returns P (Fig. 1 (2)). This P defines M', and then, M and M' are given to Procedure MultipleGS. Fig. 1 (3) shows X and Y used in Procedure MultipleGS. MultipleGS returns several matchings. Fix one matching Q_i (Fig. 1 (4)). Then, M_i'' is defined by this Q_i. Next, INCREASE increases the size of M (lines 6 through 8) by removing edges $(m, w) \in M_i''$ from M, and by matching m and w to single woman and man, respectively, according to P and Q_i. The resulting matching M_i is shown in Fig. 1 (5). However, M_i may break the property Π. At lines 9 through 11, it removes some edges of M_i so that the matching satisfies Π. This decreases the size of matching, but in the next section, we will show that there is a good i such that the size does not decrease too much.

4 Correctness Proof

It is not hard to see that INCREASE runs in time polynomial in N. (Note that both maximum matchings and minimum vertex covers for bipartite graphs can be computed in polynomial time [27].) Also, observe that all $M_{1,i}$ at the end of

INCREASE satisfy Π. For, if there are edges $e_1 = (m, w)$ and $e_2 = (m', w')$ in M_i such that m and w' form a blocking pair, then at least one of e_1 and e_2 is removed from M_i at line 11 of INCREASE. Similarly, all $M_{2,i}$ satisfy Π. Hence, the output of INCREASE satisfies Property Π. It remains to show that $|M^*| > |M|$ if $|M| < \frac{OPT}{2} + c'\sqrt{|M|}$. To this end, it suffices to show that there are $a \in \{1, 2\}$ and i^* such that M_{a,i^*} at the end of INCREASE satisfies $|M_{a,i^*}| > |M|$.

First, let us fix an optimal solution M_{opt}, a largest stable matching for I, and let M be an input for INCREASE, a stable matching for I. Let us define the following bipartite graph $G_{M_{opt},M}$: Each vertex of $G_{M_{opt},M}$ corresponds to a person in I. There is an edge between vertices m and w if and only if $M_{opt}(m) = w$ or $M(m) = w$. If both $M_{opt}(m) = w$ and $M(m) = w$ hold, we give two edges between m and w; hence $G_{M_{opt},M}$ is a multigraph. An edge (m, w) associated with $M_{opt}(m) = w$ is called an *OPT-edge*. Similarly, an edge associated with $M(m) = w$ is called an *M-edge*. Observe that the degree of each vertex is at most two, and hence each connected component of $G_{M_{opt},M}$ is a simple path, a cycle, or an isolated vertex. Partition M-edges of $G_{M_{opt},M}$ into good edges and bad ones. If an edge is in the path of length three starting from and ending with OPT-edges, then it is called *good*. Otherwise, it is *bad*. The following lemmas are proved in [22].

Lemma 1. *[22] If (m, w) is a good edge of M, then (i) $w \succeq_m M_{opt}(m)$, (ii) $m \succeq_w M_{opt}(w)$, and (iii) either $w =_m M_{opt}(m)$ or $m =_w M_{opt}(w)$.*

Lemma 2. *[22] Let t be a positive integer. If $|M| < \frac{|M_{opt}|}{2} + t$, then the number of bad edges in $G_{M_{opt},M}$ is less than $4t$.*

Since we assume that $|M| < \frac{|M_{opt}|}{2} + c'\sqrt{|M|}$, the number of bad edges in M is less than $B(= 4c'\sqrt{|M|})$.

Let (m, w) be a good edge. Then, by Lemma 1 (iii), either $w =_m M_{opt}(m)$ or $m =_w M_{opt}(w)$. Without loss of generality, we assume that at least half of good edges of M satisfy $m =_w M_{opt}(w)$, and prove that INCREASE finds a desirable i^* during the execution of 1st through 12th lines. Otherwise, it finds i^* during the execution of lines 13 through 24, whose proof is similarly obtained and hence is omitted in this paper.

Lemma 3. *Let P be the matching obtained at line 1 of* INCREASE. *Consider a woman $w \in P$ such that (m, w) is a good edge of M. Then $P(w) \succeq_w M_{opt}(w)$.*

Proof. By the construction of the bipartite graph G in MaxMatching, $P(w) =_w M(w)$, and by Lemma 1 (ii), $M(w) \succeq_w M_{opt}(w)$. Hence, $P(w) \succeq_w M_{opt}(w)$. □

Lemma 4. $|P| \geq \frac{|M|-B}{2}$.

Proof. By Lemma 2, the number of good edges in M is at least $|M| - B$. Recall that we assume that at least half of these good edges (m, w) satisfy $m =_w M_{opt}(w)$. Hence, this number of such w has an edge to $M_{opt}(w)$ in the bipartite graph G constructed in MaxMatching. Observe that $M_{opt}(w) \neq M_{opt}(w')$ if

$w \neq w'$. So a maximum matching in G is of size at least $\frac{|M|-B}{2}$, which implies $|P| \geq \frac{|M|-B}{2}$. □

Lemma 5. *There exists an* i^* *such that* Q_{i^*} *at line 3 of* INCREASE *contains at least* B *men* m *such that* $Q_{i^*}(m) \succeq_m M_{opt}(m)$.

Proof. Consider the execution of MultipleGS(M, M'). Let us call preference lists of men after MultipleGS(M, M') executes lines 4 and 5 *modified lists.*

Partition X into $A_0, A_1, \cdots, A_{2B}$ in the following way. A_0 is the set of men m such that $(m, M(m))$ is a bad edge of M. Since there are less than B bad edges, $|A_0| < B$. Next, note that if $(m, M(m)) \in M'$ is a good edge of M, then $M_{opt}(m)$ is single in M, namely, $M_{opt}(m) \in Y$. This implies that $M_{opt}(m)$ remains in m's modified list. For each i $(1 \leq i \leq 2B-1)$, let A_i be the set of men m such that $M_{opt}(m)$ is at the ith position of m's modified list. Finally, define $A_{2B} = X - (A_0 \cup A_1 \cup \cdots \cup A_{2B-1})$. We consider the following three cases: (1) $|A_1| \geq 2B$. (2) There is an i $(2 \leq i \leq 2B-1)$ such that $|A_i| \geq 3B$. (3) $|A_1| < 2B$ and $|A_i| < 3B$ for all i $(2 \leq i \leq 2B-1)$.

Case (1). We show that i^*, which we want to prove the existence, is 1. We first show that $|Q_1| \geq 2B$. Consider the execution of lines 10 and 11 for $k = 1$. Call preference lists of men after MultipleGS(M, M') executes line 10 *short lists.* Consider a man $m \in A_1$. There is only one woman in his short list, and she is $M_{opt}(m)$ by the definition of A_1. Note that $M_{opt}(m) \neq M_{opt}(m')$ for $m \neq m'$. So, there are at least $|A_1| \geq 2B$ different women on the short lists of men in X. During the execution of the Gale-Shapley algorithm, every man proposes to a woman at the top of his list. So, at least $2B$ women receive a proposal. Once a woman receives a proposal, she is matched at the end of the algorithm. So, $|Q_1| \geq 2B$.

Now, consider an arbitrary man $m \in Q_1 \cap (A_1 \cup A_2 \cup \cdots \cup A_{2B})$. $M_{opt}(m)$ is in his modified list as mentioned above, and m is matched with his first choice woman in the modified list. This means that $Q_1(m) \succeq_m M_{opt}(m)$. So, every man in $Q_1 \cap (A_1 \cup A_2 \cup \cdots \cup A_{2B})$ satisfies the condition of this lemma. Since $|A_0| < B$, $|Q_1 \cap (A_1 \cup A_2 \cup \cdots \cup A_{2B})| \geq |Q_1| - |A_0| > B$ as required.

Case (2). Consider the execution of MultipleGS at lines 10 and 11 for $k = i$ such that $|A_i| \geq 3B$. Define a short list similarly as (1). For $0 \leq j \leq 2B$ and $1 \leq \ell \leq i$, let $a_{j,\ell}$ be the number of men in A_j who is matched in Q_i with a woman of his ℓth choice with respect to his short list. Define

$$S_1 = \sum_{1 \leq j \leq i, 1 \leq \ell \leq j} a_{j,\ell} + \sum_{i < j \leq 2B, 1 \leq \ell \leq i} a_{j,\ell} \quad \text{and} \quad S_2 = \sum_{0 \leq j < \ell \leq i} a_{j,\ell}.$$

Then, the number of men who satisfy our condition, namely, who are matched with a woman at least as good as his partner in M_{opt}, is at least S_1, and the size of matching Q_i is $|Q_i| = \sum_{0 \leq j \leq 2B, 1 \leq \ell \leq i} a_{j,\ell} = S_1 + S_2$. Note that, for each j $(1 \leq j \leq i)$, $\sum_{1 \leq \ell < j} a_{j,\ell}$ men in A_j are matched with a woman strictly better than his partner in M_{opt} (with respect to modified list). So, remaining

$|A_j| - (\sum_{1\le\ell<j} a_{j,\ell})$ men made a proposal to their jth choice, namely, their partner in M_{opt}. Hence, the number of men who made a proposal to his partner in M_{opt} is at least $\sum_{1\le j\le i}(|A_j| - (\sum_{1\le\ell<j} a_{j,\ell}))$. It then results that this number of different women receive a proposal during the Gale-Shapley algorithm, and hence the matching size is at least this number (for the same reason as given in case (1)). Namely, $|Q_i| \ge \sum_{1\le j\le i}(|A_j| - (\sum_{1\le\ell<j} a_{j,\ell}))$. Because $S_2 \le \sum_{0\le j<i} |A_j|$,

$$S_1 = |Q_i| - S_2 \ge |A_i| - |A_0| - (\sum_{1\le j\le i, 1\le\ell<j} a_{j,\ell}). \tag{1}$$

Also, the following inequality is obvious.

$$S_1 \ge \sum_{1\le j\le i, 1\le\ell<j} a_{j,\ell}. \tag{2}$$

By adding (1) and (2), we obtain $S_1 \ge (|A_i| - |A_0|)/2 \ge B$. This completes the proof.

Case (3). Consider the execution of MultipleGS at lines 12 through 15. Observe that $|X| = |M'| = |P| \ge \frac{|M|-B}{2}$ (the last inequality comes from Lemma 4). Recall that $B = 4c'\sqrt{|M|}$ and $c' \le \frac{1}{8\sqrt{6}}$. So, $6B^2 - 3B < |X|/2$.

Next, by the condition of this case, we have that

$$\sum_{0\le i\le 2B-1} |A_i| < B + 2B + 3B(2B-2) = 6B^2 - 3B < \frac{|X|}{2}.$$

Hence, when we partition $|X|$ into $|X|/2B$ sets, each of which contains $2B$ men at line 12, there is at least one X_i such that $|X_i \cap A_{2B}| \ge B$. Each man in $X_i \cap A_{2B}$ has his partner in M_{opt} at position greater than or equal to $2B$ in his modified list. This means that each man in $X_i \cap A_{2B}$ has a modified list with length at least $2B$. Because preference lists are cut into length $2B$ at the 14th line, if a man in $X_i \cap A_{2B}$ gets a partner in Q_{2B-1+i}, she is at least as good as his partner in M_{opt}.

It remains to show that all men in $X_i \cap A_{2B}$ are matched in Q_{2B-1+i}: In the execution of the Gale-Shapley algorithm at line 15, each man in X_i has a preference list of length $2B$. Suppose that there is a man $m \in X_i \cap A_{2B}$ who is single in Q_{2B-1+i}. Then, m was rejected by all $2B$ women on his list. At each time a woman rejected m, she was matched. She never becomes single again and hence these $2B$ women are matched at the end of the Gale-Shapley algorithm. This contradicts the assumption that m is single in Q_{2B-1+i} since there are only $2B$ men. □

Now, let us consider the matching M_{i^*} after INCREASE executes line 8. Note that $|M_{i^*}| = |M| - |M''_{i^*}| + 2|M''_{i^*}| = |M| + |M''_{i^*}|$. We then count the number of edges removed at line 11. There are six types of edges in M_{i^*}: (1) Good edges originally existed in M. (2) Bad edges originally existed in M. (3) Edges $(m, w) \in P$ added at line 7 such that $P(w) \succeq_w M_{opt}(w)$. (4) Edges in P added

at line 7, not of type (3). (5) Edges $(m, w) \in Q_{i^*}$ added at line 8 such that $Q_{i^*}(m) \succeq_m M_{opt}(m)$. (6) Edges in Q_{i^*} added at line 8, not of type (5).

For $\ell = 1$ through 6, let D_ℓ (resp., E_ℓ) denote the set of men (resp., women) who are matched in M_{i^*} by an edge of type (ℓ) above.

Lemma 6. *(m, w) is not a blocking pair for M_{i^*} in each of the following cases: (i) $m \in D_1 \cup D_2 \cup D_3 \cup D_4$. (ii) $m \in D_5$ and $w \in E_1 \cup E_3$. (iii) $m \in D_5 \cup D_6$ and $w \in E_5 \cup E_6$.*

Proof. (i) If $m \in D_1 \cup D_2$, m is matched with the same woman as in M, namely, $M_{i^*}(m) = M(m)$. If $m \in D_3 \cup D_4$, m is single in M. If $w \in E_1 \cup E_2$, $M_{i^*}(w) = M(w)$. If $w \in E_3 \cup E_4$, $M_{i^*}(w) =_w M(w)$ by the construction of P. If $w \in E_5 \cup E_6$, w is single in M. In any combination of m and w, it is easy to see that if (m, w) blocks M_{i^*} then it also blocks M, which contradicts the stability of M.

(ii) By the definition of D_5, $M_{i^*}(m) \succeq_m M_{opt}(m)$. If $w \in E_1$, $M_{i^*}(w) \succeq_w M_{opt}(w)$ by Lemma 1 (ii). If $w \in E_3$, $M_{i^*}(w) \succeq_w M_{opt}(w)$ by the definition of E_3. So, if (m, w) blocks M_{i^*}, it also blocks M_{opt}, which contradicts the stability of M_{opt}.

(iii) Recall that Q_{i^*} is obtained by breaking ties in preference lists, and applying the Gale-Shapley algorithm to $D_5 \cup D_6$ and $E_5 \cup E_6$. So there can never be a blocking pair between $D_5 \cup D_6$ and $E_5 \cup E_6$ by the nature of the Gale-Shapley algorithm. □

By this lemma, if there is a blocking pair (m, w) for M_{i^*}, then $m \in D_6$ or $w \in E_2 \cup E_4$. Consider the graph G_{i^*} constructed at line 9. By the above observation, the set of vertices corresponding to $D_6 \cup E_2 \cup E_4$ is a vertex cover for G_{i^*}; so $|C_{i^*}| \leq |D_6| + |E_2| + |E_4|$.

If a woman w is in E_4, then $(M(w), w)$ is a bad edge of M, for if $(M(w), w)$ is good, w must be in E_3 by Lemma 3. Also, by the definition of E_2, if a woman w is in E_2, then $(M(w), w)$ is a bad edge. Hence, $|E_2| + |E_4| \leq$ (the number of bad edges of M) $< B$ by Lemma 2. By the definition of D_5 and D_6, $|D_5| + |D_6| = |M''_{i^*}|$ and by Lemma 5, $|D_5| \geq B$. So, $|D_6| \leq |M''_{i^*}| - B$ and hence $|C_{i^*}| < |M''_{i^*}|$. For each vertex in C_{i^*}, at most one edge is removed from M_{i^*} at line 11. Hence the size of M_{i^*} after removing edges at line 11 is at least $|M| + |M''_{i^*}| - |C_{i^*}| > |M|$. This completes the correctness proof.

Theorem 1. *If M, an input for* Increase, *satisfies $|M| < \frac{OPT}{2} + c'\sqrt{|M|}$, then* Increase *does not fail. Here c' ($c' \leq \frac{1}{8\sqrt{6}}$) is a constant fixed for the procedure* MultipleGS.

Corollary 1. *The approximation ratio of* LocalSearch *is at most $2 - c\frac{1}{\sqrt{N}}$, where c is a constant such that $c \leq \frac{1}{4\sqrt{6}}$.*

Proof. By Theorem 1, LocalSearch finds a solution M such that $|M| \geq \frac{OPT}{2} + c'\sqrt{|M|}$. By multiplying both sides by $\frac{2}{|M|}$, we have $\frac{OPT}{|M|} \leq 2 - 2c'\frac{1}{\sqrt{|M|}} \leq 2 - 2c'\frac{1}{\sqrt{N}}$. The last inequality follows from the fact that $N \geq \sqrt{|M|}$. □

References

1. D. J. Abraham, K. Cechlárová, D. F. Manlove and K. Mehlhorn, "Pareto optimality in house allocation problems," *Proc. ISAAC 2004*, LNCS 3341, pp. 3–15, 2004.
2. D. J. Abraham, R. W. Irving, T. Kavitha and K. Mehlhorn, "Popular matchings," *Proc. SODA 2005*, pp. 424–432, 2005.
3. V. Bansal, A. Agrawal and V. Malhotra, "Stable marriages with multiple partners: efficient search for an optimal solution," *Proc. ICALP 2003*, LNCS 2719, pp. 527–542, 2003.
4. R. Bar-Yehuda and S. Even, "A local-ratio theorem for approximating the weighted vertex cover problem," In Analysis and Design of Algorithms for Combinatorial Problems, volume 25 of Annals of Disc. Math., Elsevier Science Publishing Company, Amsterdam, pp. 27–46, 1985.
5. P. Berman and T. Fujito, "On the approximation properties of independent set problem in degree 3 graphs," *Proc. WADS 95*, LNCS 955, pp. 449–460, 1995.
6. Canadian Resident Matching Service (CaRMS), `http://www.carms.ca/`
7. K. Cechlárová, "On the complexity of exchange-stable roommates," *Discrete Applied Mathematics*, Vol. 116, pp. 279–287, 2002.
8. T. Fleiner, "A fixed-point approach to stable matchings and some applications," *Mathematics of Operations Research*, Vol. 28, Issue 1, pp. 103–126, 2003.
9. D. Gale and L. S. Shapley, "College admissions and the stability of marriage," *Amer. Math. Monthly*, Vol. 69, pp. 9–15, 1962.
10. D. Gale and M. Sotomayor, "Some remarks on the stable matching problem," *Discrete Applied Mathematics*, Vol. 11, pp. 223–232, 1985.
11. D. Gusfield and R. Irving, "The Stable Marriage Problem: Structure and Algorithms," MIT Press, Boston, MA, 1989.
12. M. M. Halldórsson, R. Irving, K. Iwama, D. Manlove, S. Miyazaki, Y. Morita and S. Scott, "Approximability results for stable marriage problems with ties," *Theoretical Computer Science*, Vol. 306, pp. 431–447, 2003.
13. M. M. Halldórsson, K. Iwama, S. Miyazaki and H. Yanagisawa, "Randomized approximation of the stable marriage problem," *Theoretical Computer Science*, Vol. 325, No. 3, pp. 439–465, 2004.
14. M. M. Halldórsson, K. Iwama, S. Miyazaki and H. Yanagisawa, "Improved approximation of the stable marriage problem," *Proc. ESA 2003*, LNCS 2832, pp. 266–277, 2003.
15. E. Halperin, "Improved approximation algorithms for the vertex cover problem in graphs and hypergraphs," *Proc. SODA 2000*, pp. 329–337, 2000.
16. R. Irving, "Stable marriage and indifference," *Discrete Applied Mathematics*, Vol. 48, pp. 261–272, 1994.
17. R. Irving, "Matching medical students to pairs of hospitals: a new variation on an old theme," *Proc. ESA 98*, LNCS 1461, pp. 381–392, 1998.
18. R. W. Irving, D. F. Manlove and S. Scott, "The hospital/residents problem with ties," *Proc. SWAT 2000*, LNCS 1851, pp. 259–271, 2000.
19. R. Irving, T. Kavitha, K. Mehlhorn, D. Michail, and K. Paluch, "Rank-maximal matchings," *Proc. SODA 2004*, pp. 68–75, 2004.
20. R.W. Irving, D.F. Manlove, S. Scott, "Strong stability in the hospitals/residents problem," *Proc. STACS 2003*, LNCS 2607, pp. 439–450, 2003.
21. K. Iwama, D. F. Manlove, S. Miyazaki, and Y. Morita, "Stable marriage with incomplete lists and ties," *Proc. ICALP 99*, LNCS 1644, pp. 443–452, 1999.

22. K. Iwama, S. Miyazaki and K. Okamoto, "A $(2 - c\log N/N)$-approximation algorithm for the stable marriage problem, *Proc. SWAT 2004*, LNCS 3111, pp. 349–361, 2004.
23. Japanese Resident Matching Program (JRMP), `http://www.jrmp.jp/`
24. G. Karakostas, "A better approximation ratio for the Vertex Cover problem," ECCC Report, TR04-084, 2004.
25. M. Karpinski, "Polynomial time approximation schemes for some dense instances of NP-hard optimization problems," *Proc. RANDOM 97*, LNCS 1269, pp. 1–14, 1997.
26. T. Kavitha, K. Mehlhorn, D. Michail, and K. Paluch, "Strongly stable matchings in time $O(nm)$ and extension to the hospitals-residents problem," *Proc. STACS 2004*, pp. 222–233, 2004.
27. E. L. Lawler, "Combinatorial Optimization: Networks and Matroids," HOLT, RINEHART AND WINSTON, 1976.
28. T. Le, V. Bhushan, and C. Amin, "First aid for the match, second edition," McGraw-Hill, Medical Publishing Division, 2001.
29. D. F. Manlove, R. W. Irving, K. Iwama, S. Miyazaki, Y. Morita, "Hard variants of stable marriage," *Theoretical Computer Science*, Vol. 276, Issue 1-2, pp. 261–279, 2002.
30. B. Monien and E. Speckenmeyer, "Ramsey numbers and an approximation algorithm for the vertex cover problem," *Acta Inf.*, Vol. 22, pp. 115–123, 1985.
31. H. Nagamochi, Y. Nishida and T. Ibaraki, "Approximability of the minimum maximal matching problem in planar graphs," Institute of Electronics, Information and Communication Engineering, Transactions on Fundamentals, vol.E86-A, pp. 3251-3258, 2003.
32. C.P. Teo, J.V. Sethuraman and W.P. Tan, "Gale-Shapley stable marriage problem revisited: Strategic issues and applications," *Proc. IPCO 99*, pp. 429–438, 1999.
33. M. Zito,"Randomized techniques in combinatorial algorithmics," PhD thesis, Dept. of Computer Science, University of Warwick, 1999.

Approximating the Traffic Grooming Problem*
(Extended Abstract)

Michele Flammini[1], Luca Moscardelli[1],
Mordechai Shalom[2], and Shmuel Zaks[2]

[1] Dipartmento di Informatica,
Universita degli Studi dell'Aquila, L'Aquila, Italy
{flammini, moscardelli}@di.univaq.it
[2] Department of Computer Science, Technion, Haifa, Israel
{cmshalom, zaks}@cs.technion.ac.il

Abstract. The problem of grooming is central in studies of optical networks. In graph-theoretic terms, this can be viewed as assigning colors to the lightpaths so that at most g of them (g being the *grooming factor*) can share one edge. The cost of a coloring is the number of optical switches (ADMs); each lightpath uses two ADM's, one at each endpoint, and in case g lightpaths of the same wavelength enter through the same edge to one node, they can all use the same ADM (thus saving $g-1$ ADMs). The goal is to minimize the total number of ADMs. This problem was shown to be NP-complete for $g = 1$ and for a general g. Exact solutions are known for some specific cases, and approximation algorithms for certain topologies exist for $g = 1$. We present an approximation algorithm for this problem. For every value of g the running time of the algorithm is polynomial in the input size, and its approximation ratio for a wide variety of network topologies - including the ring topology - is shown to be $2 \ln g + o(\ln g)$. This is the first approximation algorithm for the grooming problem with a general grooming factor g.

Keywords: Wavelength Assignment, Wavelength Division Multiplexing(WDM), Optical Networks,Add-Drop Multiplexer(ADM), Traffic Grooming.

1 Introduction

1.1 Background

Optical wavelength-division multiplexing (WDM) is today the most promising technology, that enables us to deal with the enormous growth of traffic in communication networks, like the Internet. A communication between a pair of nodes is done via a *lightpath*, which is assigned a certain wavelength. In graph-theoretic terms, a lightpath is a simple path in the network, with a color assigned to it.

* This research was supported in part by the EU COST 293 (GRAAL) research fund.

X. Deng and D. Du (Eds.): ISAAC 2005, LNCS 3827, pp. 915–924, 2005.

Most of the studies in optical networks dealt with the issue of assigning colors to lightpaths, so that every two lightpaths that share an edge get different colors.

When the various parameters comprising the switching mechanism in these networks became clearer, the focus of studies shifted, and today a large portion of the studies concentrates with the total hardware cost. The key point here is that each lightpath uses two ADM's, one at each endpoint. If two adjacent lightpaths are assigned the same wavelength, then they can use the same ADM. An ADM may be shared by at most two lightpaths. The total cost considered is the total number of ADMs. Lightpaths sharing ADM's in a common endpoint can be thought as concatenated, so that they form longer paths or cycles. These paths/cycles do not use any edge $e \in E$ twice, for otherwise they cannot use the same wavelength which is a necessary condition to share ADM's.

Moreover, in studying the hardware cost, the issue of *grooming* became central. This problem stems from the fact that the network usually supports traffic that is at rates which are lower than the full wavelength capacity, and therefore the network operator has to be able to put together (= groom) low-capacity demands into the high capacity fibers. In graph-theoretic terms, this can viewed as assigning colors to the lightpaths so that at most g of them (g being the *grooming factor*) can share one edge. In terms of ADMs, each lightpath uses two ADM's, one at each endpoint, and in case g lightpaths of the same wavelength enter through the same edge to one node, they can all use the same ADM (thus saving $g-1$ ADMs). The goal is to minimize the total number of ADMs. Note that the above coloring problem is simply the case of $g = 1$.

We note that we deal with the *single hop* problem, where a connection is carried along one wavelength. A nice review on traffic grooming problems can be found in [1].

1.2 Previous Work

The problem of minimizing the number of ADMs for the case $g = 1$ was introduced in [2] for ring topology. The problem was shown to be NP-complete for ring networks in [3]. An approximation algorithm for the ring topology with approximation ratio of $3/2$ was presented in [4], and was improved in [5, 6] to $10/7 + \epsilon$ and $10/7$ respectively. For a general topology [3] describes an algorithm with approximation ratio of $8/5$. The same problem was studied in [7], and an algorithm with approximation ratio $3/2 + \epsilon$ was presented.

The notion of traffic grooming ($g > 1$) was introduced in [8] for the ring topology. The problem was shown to be NP-complete in [9] for ring networks and a general g. The uniform all-to-all traffic case, in which there is the same demand between each pair of nodes, is studied in [9, 10] for various values of g; an optimal construction for the uniform all-to-all problem, for the case $g = 2$ in a path network was given in [11].

The hardness results of [3, 9] are for $g = 1$ and for general g, respectively. NP-completeness results for ring and path networks are shown in [12] for general values of g (in the strong sense) and for any fixed value of g.

1.3 Our Contribution

We present an approximation algorithm for the general instance of the traffic grooming problem, namely general topology and general set of requests. The approximation ratio of our algorithm is $2\ln g + o(\ln g)$ in ring networks, with arbitrary set of requests. The ring topology is the most widely studied topology due to its implementation in SONET networks. Therefore and for matter of presentation, our discussion deals only with ring topologies. The extensions are briefly discussed in Section 5. Note that the approximation ratio of any algorithm for this problem is between 1 and $2g$. To the best of our knowledge this is the first approximation algorithm for the grooming problem with a general grooming factor g. In Section 2 we describe the problem and make some preliminary observations. The algorithm presented in Section 3, and analyzed in Section 4. We conclude in Section 5 with possible extensions of this result and some open problems. Some proofs are sketched or omitted in this Extended Abstract.

2 Problem Definition and Basic Observations

An instance of the *traffic grooming problem* is a triple (G, P, g) where $G = (V, E)$ is an undirected graph, P is a set of simple paths in G and g is a positive integer, namely the grooming factor.

Given such an instance we define the following:

Definition 1. *Given a subset $Q \subseteq P$ and an edge $e \in E$, Q_e is the set of paths from Q using edge e. $l_Q(e)$ is the number of these paths, or in networking terminology, the load induced on the edge e by the paths in Q. L_Q is the maximum load induced by the paths in Q on any edge of G. When $Q = P$, we will omit the indices and simply write $l(e)$ and L instead of $l_P(e)$ and L_P respectively. Formally,*

$$\forall Q \subseteq P, \forall e \in E:$$

$$Q_e \stackrel{def}{=} \{p \in Q | e \in p\}$$

$$l_Q(e) \stackrel{def}{=} |Q_e|$$

$$L_Q \stackrel{def}{=} \max_{e \in E} l_Q(e)$$

Definition 2. *A coloring (or wavelength assignment) of (G, P) is a function $w : P \mapsto \mathbb{N}^+ = \{1, 2, ...\}$. We extend the definition of w on any subset Q of P as $w(Q) = \cup_{p \in Q} w(p)$. For a coloring w, a color λ and any $Q \subseteq P$, Q^w_λ is the subset of paths from Q colored λ by w and $Q^w_{e,\lambda}$ is the set of paths from Q, using edge e and colored λ by w. Formally,*

$$Q^w_\lambda \stackrel{def}{=} w^{-1}(\lambda) \cap Q = \{p \in Q | w(p) = \lambda\}$$

$$Q^w_{e,\lambda} \stackrel{def}{=} Q_e \cap Q^w_\lambda.$$

Definition 3. *A proper coloring (or wavelength assignment) w of (G, P, g) is a coloring of P, in which for any edge e at most g paths using e are colored with the same color. Formally, $\forall \lambda \in \mathbb{N}^+, L_{P_\lambda^w} \leq g$.*

Definition 4. *A coloring w is a W-coloring of $Q \subseteq P$, if it colors the paths of Q using exactly W colors. Formally, if $|w(Q)| = W$. A set Q is W-colorable if there exists a proper W-coloring for it.*

For a W-coloring of P, we will assume w.l.o.g. that $w(P) = 1, 2, ..., W$.

Observe that a set $Q \subseteq P$ is 1-colorable iff $L_Q \leq g$.

Now we define the cost function $\#ADM$, under the assumption that G is a cycle.

Definition 5. *For a coloring w of P, a subset $Q \subseteq P$ and a node $v \in V$, Q_v is the subset of paths from Q having an endpoint in v. $Q_{v,\lambda}^w$ is the subset of paths from Q_v colored λ by w. $\#ADM_\lambda^w(v)$ is the number of ADM's operating at wavelength λ at node v.*

For each pair $v \in V, \lambda \in \{1, 2, ..., W\}$ we need one ADM operating at wavelength λ in node v iff there is at least one path colored λ among the paths having an endpoint at v. Formally,

$$
\begin{aligned}
Q_v &\stackrel{def}{=} \{p \in Q | v \text{ is an endpoint of } p\} \\
touches(Q, v) &\stackrel{def}{=} \begin{cases} 0 \text{ if } Q_v = \emptyset \\ 1 \text{ otherwise} \end{cases} \\
endpoints(Q) &\stackrel{def}{=} \sum_{v \in V} touches(Q, v) \\
Q_{v,\lambda}^w &\stackrel{def}{=} Q_v \cap Q_\lambda^w \\
\#ADM_\lambda^w(v) &\stackrel{def}{=} touches(P_\lambda^w, v) \\
\#ADM_\lambda^w(Q) &\stackrel{def}{=} endpoints(Q_\lambda^w) \\
\#ADM_\lambda^w &\stackrel{def}{=} \#ADM_\lambda^w(P) \\
\#ADM^w &\stackrel{def}{=} \sum_\lambda \#ADM_\lambda^w
\end{aligned}
$$

Definition 6. *For any subset $Q \subseteq P$ and any subset $U \subseteq V$, Q_U is the set of paths in Q having at least one endpoint in U. Formally,*

$$Q_U \stackrel{def}{=} \bigcup_{u \in U} Q_u.$$

The traffic grooming problem is the optimization problem of finding a proper coloring w of (G, P, g) minimizing $\#ADM^w$.

Observe that *endpoints* and consequently $\#ADM_\lambda^w$ are monotone non decreasing functions. Formally, if $R \subseteq Q \subseteq P$ then

$$
\begin{aligned}
endpoints(R) &\leq endpoints(Q) \\
\#ADM_\lambda^w(R) &\leq \#ADM_\lambda^w(Q).
\end{aligned}
$$

3 Algorithm GROOMBYSC(k)

Given an instance (G, P, g) of the traffic grooming problem, our algorithm has a parameter k which depends only on g. The value of k will be determined in the analysis (see Section 4).

The algorithm has three phases. During phase 1 it computes 1-colorable sets and their corresponding weights. It considers subsets of the paths P, of size at most $k \cdot g$. Whenever a 1-colorable set is found, it is added to the list of relevant sets, together with its corresponding weight. In phase 2 it finds a set cover of P using subsets calculated in phase 1. It uses the GREEDYSC approximation algorithm for the minimum weight set cover problem presented in [13]. In phase 3 it transforms the set cover into a partition by eliminating intersections, then colors the paths according the partition. Each set in the partition is colored with one color.

1. Phase 1- Prepare the input for GREEDY:
   - $S \leftarrow \emptyset$
   - For each $U \subseteq V$, such that $|U| \leq k$ {
     - For each $Q \subseteq P_U$, such that $|Q| \leq k \cdot g$ {
       - If Q is 1-colorable then {
         - $S \leftarrow S \cup \{Q\}$
         - $weight[Q] = endpoints(Q)$ // weight[] is an associative
         - // array containing a weight for each set
       - }
     - }
   - }
2. Phase 2- Run GREEDYSC:
   - $SC \leftarrow GREEDYSC(S, weight)$.// Assume w.l.o.g $SC = \{S_1, S_2, ..., S_W\}$
3. Phase 3- Transform the Set Cover SC into a Partition $PART$:
   - $PART \leftarrow \emptyset$
   - For $i = 1$ to W $\{PART_i \leftarrow S_i\}$
   - As long as there are two intersecting sets $PART_i, PART_j$ {
     - $PART_i \leftarrow PART_i \setminus PART_j$
   - }
   - For $\lambda = 1$ to W{
     - $PART \leftarrow PART \cup \{PART_\lambda\}$
     - For each $p \in PART_\lambda \{w(p) = \lambda\}$
   - }

4 Analysis

4.1 Correctness

Claim. w calculated by the algorithm is a coloring.

Proof. During phase 1, each path $p \in P$ is included at least in one set $Q \in S$. This is because the set $\{p\}$ is considered during the loop and it is clearly found to be 1-colorable. As SC is calculated in phase 2 a set cover of these sets, p is an element of at least one set $S_i \in SC$. During phase 3 intersections are eliminated, therefore p is an element of exactly one set of $PART$. Therefore each p is assigned exactly one value $w(p)$ during phase 3. □

Lemma 1. *w calculated by the algorithm is a proper coloring.*

Proof. For every color $\lambda \in \{1, 2, ..., W\}$ the set of paths colored λ is exactly $PART_\lambda$. It suffices to show that the sets $PART_\lambda$ are 1-colorable.

A subset of an x-colorable set is x-colorable. By the code of phase 3 $PART_\lambda \subseteq S_\lambda$. By phase 1, S_λ is 1-colorable, therefore $PART_\lambda$ is 1-colorable. □

4.2 Running Time

Claim. The running time of $GROOMBYSC(k)$ is polynomial in $n = |P|$ and $m = |E|$, for any given g and for all instances (G, P, g).

Proof. We will show that the running time of each one of the three phases is $poly(n, m)$.

- **Phase 1:**
  The number of subsets of P considered during the first phase is $O(n^{g \cdot k})$ since their sizes are at most $g \cdot k$. To check whether a set is 1-colorable takes $O(g \cdot k \cdot m)$ time. To calculate $endpoints(Q)$ can be done in $O(g \cdot k \log |Q|) = O(g \cdot k \cdot \log m)$ time.
  For any constant g, k is determined as a function of g only. Then $g \cdot k$ is a constant. Therefore the running time of phase 1 is polynomial in n and m for any given g.
- **Phase 2:**
  The number of the sets in S is at most $n^{g \cdot k}$. The running time of GREEDYSC is polynomial in $|S|$ and $|P|$, namely $poly(n^{g \cdot k}, n) = poly(n)$.
- **Phase 3:**
  The running time of phase 3 is polynomial in the size of the cover which is in turn polynomial in n. □

4.3 Approximation Ratio

Lemma 2. *Let $H_n = 1 + \frac{1}{2} + ... + \frac{1}{n}$ be the n-th harmonic number. $GROOMBYSC$ (k) is a $H_{g \cdot k}(1 + \frac{2g}{k})$ approximation algorithm for the traffic grooming problem in ring networks.*

Proof. Recall that in the Minimum Weight Set Cover problem, each subset S_i has an associated weight, $weight[S_i]$. The weight of a cover is the sum of the individual weights of its sets.

Let w be the coloring returned by $GROOMBYSC(k)$ and w^* an optimal coloring. We will use the shortcut $\#ADM^*$ for $\#ADM^{w^*}$.

On one hand

$$\#ADM^w = \sum_\lambda \#ADM^w_\lambda = \sum_\lambda endpoints(PART_\lambda)$$
$$\leq \sum_\lambda endpoints(S_\lambda) = \sum_\lambda weight[S_\lambda] = weight(SC). \quad (1)$$

On the other hand GREEDYSC is an H_f-approximation algorithm, where f is the maximum cardinality of the sets in the input. In our case $f = g \cdot k$. In other words if SC^* is a minimum weight set cover on the set S, we have

$$weight(SC) \leq H_{g\cdot k}\, weight(SC^*). \quad (2)$$

Clearly if $\overline{SC}$ is an arbitrary set cover of S, by definition

$$weight(SC^*) \leq weight(\overline{SC}). \quad (3)$$

Combining the inequalities (1), (2) and (3) we get

$$\#ADM^w \leq H_{g\cdot k}\, weight(\overline{SC})$$

for any set cover $\overline{SC}$ of S.

In the following claim we will show the existence of a set cover $\overline{SC}$ satisfying $weight(\overline{SC}) \leq \#ADM^* \left(1 + \frac{2g}{k}\right)$, which implies

$$\#ADM^w \leq \#ADM^*\, H_{g\cdot k} \left(1 + \frac{2g}{k}\right). \qquad \square$$

Claim. There exists a set cover $\overline{SC}$ of S, such that $weight(\overline{SC}) \leq \#ADM^* \left(1 + \frac{2g}{k}\right)$.

Proof. Let $w^*(P) = \{1, 2, ..., W^*\}$ and $1 \leq \lambda \leq W^*$. Consider the set V^*_λ of nodes v such that $ADM^*_\lambda(v) = 1$, namely having an ADM operating at wavelength λ at node v. We divide V^*_λ into sets of k nodes starting from an arbitrary node and going clockwise along the cycle (see Figure 1). Let $V_{\lambda,j}$ be the subsets of nodes obtained in this way. Let

$$\#ADM^*_\lambda = |V^*_\lambda| = kq_\lambda + r_\lambda \quad (4)$$

where $r_\lambda = |V^*_\lambda| \mod k$ and $0 \leq r_\lambda < k$.

Clearly $\forall 1 \leq j \leq q_\lambda, |V_{\lambda,j}| = k$, and in case $r_\lambda > 0$ we have $|V_{\lambda,q_\lambda+1}| < k$. In both cases $|V_{\lambda,j}| \leq k$. Therefore, each $V_{\lambda,j}$ is considered in the outer loop of phase 1 of the algorithm, and hence, is added to S.

For $V_{\lambda,j}$ we define $\overline{S}_{\lambda,j}$ to be the set of paths in $P^{w^*}_\lambda$ having their counter-clockwise endpoint in $V_{\lambda,j}$. As $V_{\lambda,j}$ has at most k nodes, and every node may be the clockwise endpoint of at most g paths from a 1−colorable set, we have

$|\overline{S}_{\lambda,j}| \leq g \cdot k$. Therefore, $\overline{S}_{\lambda,j}$ is considered by the algorithm in the inner loop of phase 1. Being 1-colorable it should be added to S, thus $\overline{S}_{\lambda,j} \in S$.

Every $p \in P_\lambda^{w^*}$ has its both endpoints in the sets $V_{\lambda,j}$. In particular, it has its clockwise endpoint in $V_{\lambda,j}$ for a certain j, thus it is an element of some $\overline{S}_{\lambda,j}$. Therefore $\overline{SC}_\lambda \stackrel{def}{=} \cup_j \{\overline{S}_{\lambda,j}\}$ is a cover of $P_\lambda^{w^*}$. Considering all colors $1 \leq \lambda \leq W^*$ we conclude that $\overline{SC} \stackrel{def}{=} \cup_{\lambda=1}^{W*} \overline{SC}_\lambda$ is a cover of P.

Therefore $\overline{SC}$ is a cover of P with sets from S. It remains to show that its weight has the claimed property.

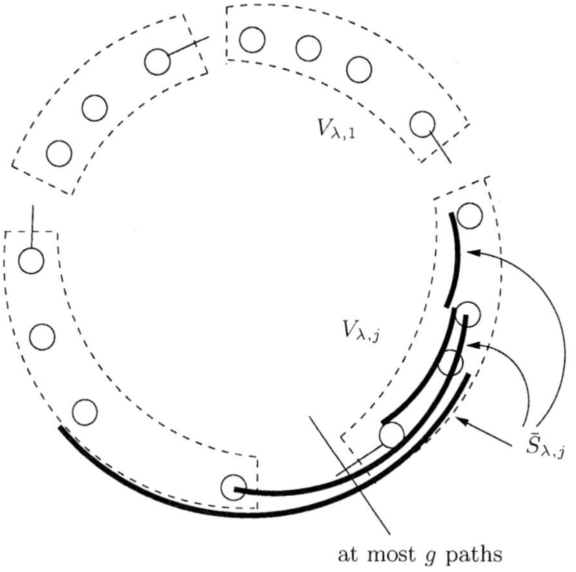

Fig. 1. The sets $V_{\lambda,j}$ and $\overline{S}_{\lambda,j}$ $(k = 4)$

Summing up equation (4) over all possible values of λ we obtain $\#ADM^* = k\sum_\lambda q_\lambda + \sum_\lambda r_\lambda$, which implies:

$$\sum_\lambda q_\lambda \leq \frac{\#ADM^*}{k} \tag{5}$$

We claim that $\forall j \leq q_\lambda, weight[\overline{S}_{\lambda,j}] = endpoints(\overline{S}_{\lambda,j}) \leq k + g$. This is because:

- The endpoints of the paths with both endpoints in $\overline{S}_{\lambda,j}$ are in $V_{\lambda,j}$ and $|V_{\lambda,j}| = k$.
- The number of paths having only the clockwise endpoint in set $V_{\lambda,j}$ is at most g. This follows from the observation that these paths should use the unique edge in the clockwise cut of $V_{\lambda,j}$. As the set $\overline{S}_{\lambda,j}$ is 1-colorable, the number of these paths is at most g.

For the set $j = q_\lambda + 1$ (which exists only if $r_\lambda > 0$) the above bound becomes $weight(\overline{S}_{\lambda,q_\lambda+1}) \leq r_\lambda + g \cdot q_\lambda$. This is because:

- The endpoints of the paths with both endpoints in $\overline{S}_{\lambda,q_\lambda+1}$ are in $V_{\lambda,q_\lambda+1}$ and $|V_{\lambda,q_\lambda+1}| = r_\lambda$.
- By the same argument as before, the paths having only the clockwise endpoint in $V_{\lambda,q_\lambda+1}$ are at most g in number. When $q_\lambda \geq 1$, $g \leq g \cdot q_\lambda$ and we are done. Otherwise $q_\lambda = 0$ meaning that $V_{\lambda,1}$ is the unique set. Then the number of paths having exactly one endpoint in the set is zero.

Summing up for all $1 \leq j \leq q_\lambda + 1$ we get:

$$weight(\overline{SC}_\lambda) \leq \sum_{j=1}^{q_\lambda}(k+g)+r_\lambda+g\cdot q_\lambda = (k+g)q_\lambda+r_\lambda+g\cdot q_\lambda = kq_\lambda+r_\lambda+2g\cdot q_\lambda$$

Summing up for all λ and recalling (4) and (5) we get:

$$\begin{aligned}weight(\overline{SC}) &= \sum_\lambda weight(\overline{SC}_\lambda) \leq \sum_\lambda (kq_\lambda + r_\lambda + 2g \cdot q_\lambda) = \#ADM^* + 2g\sum_\lambda q_\lambda \\ &\leq \#ADM^* + 2g\frac{\#ADM^*}{k} = \left(1 + \frac{2g}{k}\right)\#ADM^*. \end{aligned}$$

□

Theorem 1. *There is a $2\ln g + o(\ln g)$-approximation algorithm for the traffic grooming problem in ring networks.*

Proof. The approximation ratio ρ of GROOMBYSC(k) is at most $H_{g\cdot k}\left(1+\frac{2g}{k}\right)$.
We substitute $k = g \ln g$ and get:

$$\begin{aligned}\rho &\leq H_{g^2 \ln g}\left(1 + \frac{2}{\ln g}\right) \leq (1 + \ln(g^2 \ln g))\left(1 + \frac{2}{\ln g}\right) \\ &= (1 + 2\ln g + \ln\ln g)\left(1 + \frac{2}{\ln g}\right) = 2\ln g + o(\ln g)\end{aligned}$$

□

5 Discussion and Open Problems

We presented an approximation algorithm for ring networks, whose approximation ratio is $2\ln g + o(\ln g)$. Note that the approximation ratio of any algorithm for this problem is between 1 and $2g$.

Our algorithm can be used in arbitrary networks. In some topologies the analysis will yield a similar result. For this, note that the only point in the analysis that used the fact that the topology is a ring is where we considered the unique edge between the blocks of an optimal solution. Therefore a similar analysis follows for any topology and set of demands in which any solution can be partitioned in a similar way. This clearly includes all graphs which consists of blocks $B_0, B_1, ..., B_b$ whose sizes are bounded by $\alpha \leq k$ (k is the parameter used in our analysis) and at most $\beta = O(1)$ edges connecting consecutive blocks B_i and $B_{i+1 \bmod b}$.

We mention few open problems which arise from this study.

- Improve the analysis of algorithm $GROOMBYSC(k)$.
- Find an algorithm with a better performance guarantee.
- Analyze algorithm $GROOMBYSC(k)$ for general topology and set of requests.

References

1. K. Zhu and B. Mukherjee. A review of traffic grooming in wdm optical networks: Architecture and challenges. *Optical Networks Magazine*, 4(2):55–64, March-April 2003.
2. O. Gerstel, P. Lin, and G. Sasaki. Wavelength assignment in a wdm ring to minimize cost of embedded sonet rings. In *INFOCOM'98, Seventeenth Annual Joint Conference of the IEEE Computer and Communications Societies*, pages 69–77, 1998.
3. T. Eilam, S. Moran, and S. Zaks. Lightpath arrangement in survivable rings to minimize the switching cost. *IEEE Journal of Selected Area on Communications*, 20(1):172–182, Jan 2002.
4. G. Călinescu and P-J. Wan. Traffic partition in wdm/sonet rings to minimize sonet adms. *Journal of Combinatorial Optimization*, 6(4):425–453, 2002.
5. M. Shalom and S. Zaks. A $10/7 + \epsilon$ approximation scheme for minimizing the number of adms in sonet rings. In *First Annual International Conference on Broadband Networks, San-José, California, USA*, October 2004.
6. L. Epstein and A. Levin. Better bounds for minimizing sonet adms. In *2nd Workshop on Approximation and Online Algorithms, Bergen, Norway*, September 2004.
7. G. Călinescu, Ophir Frieder, and Peng-Jun Wan. Minimizing electronic line terminals for automatic ring protection in general wdm optical networks. *IEEE Journal of Selected Area on Communications*, 20(1):183–189, Jan 2002.
8. O. Gerstel, R. Ramaswami, and G. Sasaki. Cost effective traffic grooming in wdm rings. In *INFOCOM'98, Seventeenth Annual Joint Conference of the IEEE Computer and Communications Societies*, 1998.
9. A. L. Chiu and E. H. Modiano. Traffic grooming algorithms for reducing electronic multiplexing costs in wdm ring networks. *Journal of Lightwave Technology*, 18(1):2–12, January 2000.
10. J-C. Bermond and D. Coudert. Traffic grooming in unidirectional WDM ring networks using design theory. In *IEEE ICC*, Anchorage, Alaska, May 2003.
11. Jean-Claude Bérmond, Laurent Braud, and David Coudert. Traffic grooming on the path. In *12 th Colloqium on Structural Information and Communication Complexity, Le Mont Saint-Michel, France*, May 2005.
12. M. Shalom, W. Unger, and S. Zaks. On the complexity of the traffic grooming problem. *In preparation*, 2005.
13. V. Chvátal. A greedy heuristic for the set covering problem. *Mathematics of Operation Research*, 4:233–235, 1979.

Scheduling to Minimize Makespan with Time-Dependent Processing Times

L.Y. Kang[1,2,⋆], T.C.E. Cheng[2], C.T. Ng[2], and M. Zhao[1]

[1] Department of Mathematics, Shanghai University, Shanghai 200444, China
lykang@staff.shu.edu.cn
[2] Department of Logistics, The Hong Kong Polytechnic University

Abstract. In this paper we study the scheduling problem of minimizing makespan on identical parallel machines with time-dependent processing times. We first consider the problem with time-dependent processing times on two identical machines to minimize makespan, which is NP-hard. We give a fully polynomial-time approximation scheme for the problem. Furthermore, we generalize the result to the case with m machines.

Keywords: makespan; fully polynomial approximation scheme; parallel machines scheduling.

1 Introduction

In real-life applications, many systems exhibit dynamic behaviors characterized by a set of dynamic parameters. This fact is commonly recognized in control theory, system engineering and many other areas. Based on some scheduling problems with dynamic parameters considered by Gupta et al. [13], Gupta and Gupta [12] introduced an interesting scheduling model in which the processing time of a task is a polynomial function of its starting time. From a modelling perspective, however, the makespan scheduling problem with quadratic time-dependent processing times is already very intricate. For this reason, most subsequent research along this line has concentrated on problem with linear time-dependent processing times[2,3,4,5,6,7,8,9,16]. Generally, there are two groups of models to describe this kind of scheduling processes. The first group is devoted to the problems in which the job processing times are characterized by non-decreasing functions of their starting times, and the second group concerns the problems in which the job processing times are nonincreasing functions of their processing times. In this paper, we study the latter group of problems. In this standard form, the processing time of a job is given by $f_i(t) = a_i - b_i t$, where t is the job start time, a_i is the 'normal processing time', which is the length of time required to complete the job if it is scheduled first ($t = 0$), and $b_i > 0$ is the job-dependent decreasing rate, which determines the job's (actual) processing time at $t > 0$.

Application examples of the non-decreasing model of job processing time are quite intuitively different from their non-increasing counterparts. The

⋆ Corresponding author.

X. Deng and D. Du (Eds.): ISAAC 2005, LNCS 3827, pp. 925–933, 2005.

latter model can be used to describe the process by which aerial threats are to be recognized by a radar station. In this case, a radar station has detected some objects approaching it. The time required to recognize the objects decreases as the objects get closer. Thus, the later the objects are detected, the smaller is the time for their recognition. Another example refers to the so-called learning effect. Assume that a worker has to assemble a large number of similar products. The time required by the worker to assemble one product depends on his knowledge, skills, organization of his working place and others. The worker learns how to produce over time. After some time, he is better skilled, his working place is better organized and his knowledge is increased. As a result of his learning, the time required to assemble subsequent products decreases.

We adopt the three-field notation of Graham et al. [11] for describing traditional scheduling problem, $\alpha/\beta/\gamma$, to denote our problems. The single machine model with $p_j = a_j - b_j t$ has been suggested by Ho, Leung and Wei [14]. They show that the sequence in nonincreasing order of a_j/b_j is optimal for the problem of minimizing maximum lateness and the jobs have a common due date. Chen [10] gives an $O(n^2)$ algorithm for the problem where the objective is to minimize number of tardy jobs. Cheng et al. [6] study the problems of scheduling a set of jobs on a single or multiple machine without idle times where the processing time of a job is piecewise linear nonincreasing function of its time. They prove that the problem $P2|p_j = a_j - bt|C_{max}$ is NP-hard and the problem $P|p_j = a_j - bt|C_{max}$ is strongly NP-hard.

In this paper, we study the following m identical machines scheduling problem. There are n independent jobs that need to be processed by m identical machines. The machine is continuously available for processing from time zero onwards and it can handle at most one job at a time. Each job J_j has basic processing time a_j $(j = 1, \ldots, n)$. The actual processing time p_j of J_j depends on its start time t, in the following way $p_j = a_j - b_j t$, where $b_j > 0$ is the decreasing rate of J_j. It is assumed that the decreasing rate b_j satisfies the following condition: $b_j(\sum_{i=1}^{n} a_i - a_j) < a_j$. The condition ensures that all job-processing times are positive in non-delay schedules. Without loss of generality, we assume that a_j is integral. The problem is to schedule the jobs to minimize makespan. We shall denote this problem by $Pm|p_j = a_j - b_j t|C_{max}$. This paper considers only cases where non-delay schedules are optimal.

The remainder of this paper is organized as follows. In Section 2 we give a fully polynomial approximation scheme for two identical parallel machines and prove its correctness and complexity. We generalize the result to the case with m machines in Section 3.

2 A Fully Polynomial Time Approximation Scheme for $P2|p_j = a_j - b_j t|C_{max}$

An algorithm $\mathcal{A}$ is a *ρ-approximation* algorithm for a minimization problem if it produces a solution which is at most ρ times the optimal one, in time that is polynomial in the input size. A family of algorithm $\{\mathcal{A}_\epsilon\}$ is called a *fully*

polynomial time approximation scheme (FPTAS) if, for each $\epsilon > 0$, the algorithm $\mathcal{A}_\epsilon$ is a $(1+\epsilon)$-approximation algorithm running in time that is polynomial in the input size and $\frac{1}{\epsilon}$. In the sequel, we assume $0 < \epsilon \le 1$.

Ho et al [14] showed that the nonincreasing order of a_j/b_j is optimal for $1|p_j = a_j - b_j t|C_{max}$. So the following lemma follows.

Lemma 1. *There exists an optimal job sequence for problem $P2|p_j = a_j - b_j t|C_{max}$ such that in each machine jobs are sequenced in nonincreasing order of a_j/b_j.*

Let us index all jobs so that $a_1/b_1 \ge a_2/b_2 \ge \ldots \ge a_n/b_n$. We introduce 0-1 variables $x_j, j = 1, 2, \ldots, n$, where $x_j = 1$ if job J_j is executed on machine 1 and $x_j = 0$ if job J_j is executed on machine 2. Let X be the set of all 0-1 vectors $x = (x_1, x_2, \ldots, x_n)$. Set

$$\begin{aligned} F_0(x) &= 0, \ G_0(x) = 0; \\ F_j(x) &= F_{j-1}(x) + x_j(a_j - b_j F_{j-1}(x)); \\ G_j(x) &= G_{j-1}(x) + (1 - x_j)(a_j - b_j G_{j-1}(x)); \\ Q(x) &= \max(F_n(x), G_n(x)). \end{aligned}$$

Thus the problem $P2|p_i = a_i - b_j t|C_{max}$ reduces to the following problem

$$\text{Minimize } Q(x) \text{ for } x \in X.$$

We introduce procedure *Partition* (A, H, ν) presented in [15] where $A \subseteq X_j$, H is a nonnegative function on X_j, and ν is a number. This procedure partitions A into disjoint subsets $A_1^H, A_2^H, \ldots, A_{r_H}^H$ such that $|H(x) - H(x')| \le \nu \min\{H(x), H(x')\}$ for any x, x' from the same subset $A_l^H, l = 1, \ldots, r_H$. The following description gives details of Partition (A, H, ν).

Procedure Partition (A, H, ν)

Arrange vector $x \in A$ in order $x^{(1)}, x^{(2)}, \ldots, x^{(|A|)}$, where $0 \le H(x^{(1)}) \le H(x^{(2)}) \le \ldots \le H(x^{(|A|)})$. Assign vectors $x^{(1)}, x^{(2)}, \ldots, x^{(i_1)}$ to set A_1^H until you find i_1 such that $H(x^{(i_1)}) \le (1+\nu)H(x^{(1)})$ and $H(x^{(i_1+1)}) > (1+\nu)H(x^{(1)})$. If such an i_1 does not exist, then take $A_1^H = A$ and stop.

Assign $x^{(i_1+1)}, x^{(i_1+2)}, \ldots, x^{(i_2)}$ to set A_2^H until you find i_2 such that $H(x^{(i_2)}) \le (1+\nu)H(x^{(i_1+1)})$ and $H(x^{(i_2+1)}) > (1+\nu)(x^{(i_1+1)})$. If such an i_2 does not exist, then take $A_2^H = A - A_1^H$ and stop.

Continue the above construction until $x^{(|A|)}$ is included in $A_{r_H}^H$, for some r_H.

The main properties of *Partition* were given in [15].

Proposition 1. $|H(x) - H(x')| \le \nu min\{H(x), H(x')\}$ *for any* $x, x' \in A_l^H, l = 1, \ldots, r_H$.

Proposition 2. $r_H \le log H(x^{|A|})/\nu + 2$ *for* $0 < \nu \le 1$ *and* $1 \le H(x^{|A|})$.

In the following we give a fully polynomial approximation scheme for $P2|p_j = a_j - b_j t|C_{max}$.

Algorithm $\mathcal{A}_\epsilon$

Step 1. Number jobs so that $a_1/b_1 \geq a_2/b_2 \geq \ldots \geq a_n/b_n$. Set $Y_0 = \{(0,0,\ldots,0)\}$ and $j = 1$.

Step 2. For set Y_{j-1}, generate set Y_j' by adding 0 and 1 in position j of each vector for Y_{j-1}, i.e., $Y_j' = Y_{j-1} \cup \{x + (0,\ldots,0,x_j = 1,0,\ldots)|x \in Y_{j-1}\}$. Calculate the following for any $x \in Y_j'$.

$$F_j(x) = F_{j-1}(x) + x_j(a_j - b_j F_{j-1}(x))$$
$$G_j(x) = G_{j-1}(x) + (1 - x_j)(a_j - b_j G_{j-1}(x)).$$

If $j = n$ then set $Y_n = Y_n'$ and go to Step 3.

If $j < n$ then set $\nu = \frac{\epsilon}{2(n+1)}$ and perform the following computations.

Call partition (Y_j', F_j, ν) to partition set Y_j' into disjoint subsets $Y_1^{F_j}, Y_2^{F_j}, \ldots, Y_{r_f}^{F_j}$.

Call partition (Y_j', G_j, ν) to partition set Y_j' into disjoint subsets $Y_1^{G_j}, Y_2^{G_j}, \ldots, Y_{r_g}^{G_j}$.

Divided set Y_j' into disjoint subsets $Y_{ab} = Y_a^{F_j} \cap Y_b^{G_j}, a = 1,\ldots,r_f, b = 1,\ldots,r_g$. In each non-empty subset Y_{ab}, choose a vector $x^{(ab)}$ such that

$$F_j(x^{(ab)}) = \min\{F_j(x)|x \in Y_{ab}\}.$$

Set $Y_j := \{x^{(ab)}|a = 1,\ldots,r_f, b = 1,\ldots,r_g \text{ and } Y_a^{F_j} \cap Y_b^{G_j} \neq \emptyset\}; j := j+1$, go to Step 2.

Step 3. Select vector $x^0 \in Y_n$ such that $Q(x^0) = \min\{Q(x)|x \in Y_n\} = \min\{\max(F_n(x), G_n(x))|x \in Y_n\}$.

Let $x^* = (x_1^*, \ldots, x_n^*)$ be an optimal solution to the problem $P2|p_j = a_j - b_j t|C_{max}$. Let $L = log\ max\{n, 1/\epsilon, a_{max}\}$. We have the following theorem.

Theorem 1. *Algorithm $\mathcal{A}_\epsilon$ find $x^0 \in X$ such that $Q(x^0) \leq (1+\epsilon)Q(x^*)$ in $O(n^3L^3/\epsilon^2)$.*

Proof. Suppose that $(x_1^*, \ldots, x_j^*, 0, \ldots, 0) \in Y_{ab} \subseteq Y_j'$ for some j and a, b. Algorithm $\mathcal{A}_\epsilon$ may not choose $(x_1^*, \ldots, x_j^*, 0, \ldots, 0)$ for further construction, however, for a vector $x^{(ab)}$ chosen instead of it, we have

$$|F_j(x^*) - F_j(x^{(ab)})| \leq \nu F_j(x^*)$$

and

$$|G_j(x^*) - G_j(x^{(ab)})| \leq \nu G_j(x^*).$$

Set $\nu_1 = \nu$. Consider vectors $(x_1^*, \ldots, x_j^*, x_{j+1}^*, 0, \ldots, 0)$ and $\tilde{x}^{(ab)} = (x_1^{(ab)}, \ldots, x_j^{(ab)}, x_{j+1}^*, 0, \ldots, 0)$. We have

$$\begin{aligned}
&|F_{j+1}(x^*) - F_{j+1}(\tilde{x}^{(ab)})| \\
&= |F_j(x^*) + x_{j+1}^*(a_{j+1} - b_{j+1}F_j(x^*)) - (F_j(x^{(ab)}) + x_{j+1}^*(a_{j+1} - b_{j+1}F_j(x^{(ab)})))|
\end{aligned}$$

$$
\begin{aligned}
&= |(1 - b_{j+1}x^*_{j+1})(F_j(x^*) - F_j(x^{(ab)}))| \\
&\le (1 - b_{j+1}x^*_{j+1})\nu F_j(x^*) \\
&\le \nu_1(F_j(x^*) + x^*_{j+1}(a_{j+1} - b_{j+1}F_j(x^*))) \\
&= \nu_1 F_{j+1}(x^*). \qquad (1)
\end{aligned}
$$

Similarly,

$$|G_{j+1}(x^*) - G_{j+1}(\tilde{x}^{(ab)})| \le \nu_1 G_{j+1}(x^*). \qquad (2)$$

Consequently,

$$
\begin{aligned}
F_{j+1}(\tilde{x}^{(ab)}) &\le (1 + \nu_1)F_{j+1}(x^*), \\
G_{j+1}(\tilde{x}^{(ab)}) &\le (1 + \nu_1)G_{j+1}(x^*).
\end{aligned}
$$

Assume that $\tilde{x}^{(ab)} \in Y_{de} \subseteq Y'_{j+1}$ and Algorithm A_ϵ chooses $x^{(de)} \in Y_{de}$ instead of $\tilde{x}^{(ab)}$ in the $(j+1)$st iteration, we have

$$
\begin{aligned}
|F_{j+1}(\tilde{x}^{(ab)}) - F_{j+1}(x^{(de)})| &\le \nu F_{j+1}(\tilde{x}^{(ab)}) \\
&\le \nu(1 + \nu_1)F_{j+1}(x^*). \qquad (3)
\end{aligned}
$$

and

$$|G_{j+1}(\tilde{x}^{(ab)}) - G_{j+1}(x^{(de)})| \le \nu(1 + \nu_1)G_{j+1}(x^*). \qquad (4)$$

From (1) and (3), we obtain

$$
\begin{aligned}
&|F_{j+1}(x^*) - F_{j+1}(x^{(de)})| \\
&\le |F_{j+1}(x^*) - F_{j+1}(\tilde{x}^{(ab)})| + |F_{j+1}(\tilde{x}^{(ab)}) - F_{j+1}(x^{(de)})| \\
&\le (\nu + (1 + \nu)\nu_1)F_{j+1}(x^*) \qquad (5)
\end{aligned}
$$

From (2) and (4), we obtain

$$|G_{j+1}(x^*) - G_{j+1}(x^{(de)})| \le (\nu + (1 + \nu)\nu_1)G_{j+1}(x^*). \qquad (6)$$

Set $\nu_l = \nu + (1 + \nu)\nu_{l-1}, l = 2, \ldots, n - j$. From (5) and (6) we obtain

$$|F_{j+1}(x^*) - F_{j+1}(x^{(de)})| \le \nu_2 F_{j+1}(x^*),$$

and

$$|G_{j+1}(x^*) - G_{j+1}(x^{(de)})| \le \nu_2 G_{j+1}(x^*).$$

By repeating the above argument for $j + 2, \ldots, n$. We show that there exists $x' \in Y_n$ such that

$$|F_n(x') - F_n(x^*)| \le \nu_{n-j+1} F_n(x^*),$$

and

$$|G_n(x') - G_n(x^*)| \leq \nu_{n-j+1} G_n(x^*).$$

Since

$$\begin{aligned} \nu_{n-j+1} &\leq \nu \sum_{j=0}^{n} (1+\nu)^j \\ &= (1+\nu)^{n+1} - 1 \\ &= \sum_{j=1}^{n+1} \frac{(n+1)(n)\dots(n-j+2)}{j!(n+1)^j} (\frac{\epsilon}{2})^j \\ &\leq \sum_{j=1}^{n+1} \frac{1}{j!} \frac{\epsilon}{2} \leq \epsilon, \end{aligned}$$

we have

$$|F_n(x') - F_n(x^*)| \leq \epsilon F_n(x^*), \tag{7}$$

and

$$|G_n(x') - G_n(x^*)| \leq \epsilon G_n(x^*). \tag{8}$$

By (7) and (8), we have

$$|\max(F_n(x'), G_n(x')) - \max(F_n(x^*), G_n(x^*))| \leq \epsilon \max(F_n(x^*), G_n(x^*)). \tag{9}$$

Then, in Step 3, vector x^0 will be chosen such that

$$\begin{aligned} &|\max(F_n(x^0), G_n(x^0)) - \max(F_n(x^*), G_n(x^*))| \\ &\leq |\max(F_n(x'), G_n(x')) - \max(F_n(x^*), G_n(x^*))| \\ &\leq \epsilon \max(F_n(x^*), G_n(x^*)). \end{aligned}$$

Then

$$\max(F_n(x^0), G_n(x^0)) \leq (1+\epsilon)\max(F_n(x^*), G_n(x^*)).$$

So

$$Q(x^0) \leq (1+\epsilon) Q(x^*).$$

We now establish the time complexity of Algorithm $\mathcal{A}_\epsilon$. Step 2 requires $O(|Y_j'| log |Y_j'|)$ time to complete. By Algorithm $\mathcal{A}_\epsilon$, we have $|Y_{j+1}'| \leq 2|Y_j| \leq 2r_f r_g$. By Proposition 3.2, $r_f \leq 2(n+1)log(na_{max})/\epsilon + 2 \leq 4(n+1)L/\epsilon + 2$ and $r_g \leq 2(n+1)log(na_{max}))/\epsilon + 2 \leq 4(n+1)L/\epsilon + 2$. Thus $|Y_j'| = O(n^2 L^2/\epsilon^2)$, and $|Y_j'| log |Y_j'| = O(n^2 L^3/\epsilon^2)$. Then Algorithm $\mathcal{A}_\epsilon$ runs in $O(n^3 L^3/\epsilon^2)$. □

3 A Fully Polynomial Approximation Algorithm Scheme for the Problem with m Machines

In this section, we will generalize the above result to the case with m identical parallel machines. We introduce variable $x_j, j = 1, \ldots, n$, where $x_j = k$ if job J_j is executed on machine k, $k \in \{1, \ldots, m\}$. Let X be the set of all vector $x = (x_1, \ldots, x_n)$ with $x_j = k, j = 1, \ldots, n; k = 1, \ldots, m$. Set

$$\begin{aligned} F_0^i(x) &= 0, i = 1, \ldots, m; \\ F_j^k(x) &= F_{j-1}^k(x) + a_j - b_j F_{j-1}^k(x) \text{ for } x_j = k; \\ F_j^i(x) &= F_{j-1}^i(x) \text{ for } x_j = k, i \neq k; \\ Q(x) &= \max_{j=1}^m F_n^j(x). \end{aligned}$$

The problem $Pm|p_j = a_j - b_j t|C_{max}$ reduces to the following problem

$$\text{minimize } Q(x) \text{ for all } x \in X.$$

Algorithm $\mathcal{A}_\epsilon{}^m$

Step 1. Number jobs so that $a_1/b_1 \geq a_2/b_2 \geq \ldots \geq a_n/b_n$. Set $Y_0 = \{(0, \ldots, 0)\}$, $j = 1$, and $F_0^i = 0, i = 1, \ldots, m$.

Step 2. For set Y_{j-1}, generate set Y_j' by adding $k, k = 1, \ldots, m$ in position j of each vector for Y_{j-1}. Calculate the following for any $x \in Y_j'$, assume $x_j = k$.

$$\begin{aligned} F_j^k(x) &= F_{j-1}^k(x) + a_j - b_j F_{j-1}^k(x) \\ F_j^i(x) &= F_{j-1}^i(x), i \neq k. \end{aligned}$$

If $j = n$ then $Y_n = Y_n'$ and go to Step 3.

If $j < n$ then set $\nu = \epsilon/2(n+1)$ and perform the following computations.

Call partition $(Y_j', F_j^i, \nu)(i = 1, \ldots, m)$ to partition set Y_j' into disjoint subsets $Y_1^{F_j^i}, Y_2^{F_j^i} \ldots, Y_{r_{f_i}}^{F_j^i}$. Divided set Y_j' into disjoint subsets $Y_{a_1 \ldots a_m} = Y_{a_1}^{F_j^1} \cap \ldots \cap Y_{a_m}^{F_j^m}$, $a_1 = 1, 2, \ldots, r_{f_1}; \ldots; a_m = 1, 2, \ldots, r_{f_m}$. In each non-empty subset $Y_{a_1 \ldots a_m}$, choose a vector $x^{(a_1 \ldots a_m)}$ such that

$$F_j^1(x^{(a_1 \ldots a_m)}) = \min\{F_j^1(x) | x \in Y_{a_1 \ldots a_m}\}.$$

Set $Y_j := \{x^{(a_1 \ldots a_m)} | a_1 = 1, 2, \ldots, r_{f_1}; \ldots; a_m = 1, 2, \ldots, r_{f_m}$ and $Y_{a_1}^{F_j^1} \cap \ldots \cap Y_{a_m}^{F_j^m} \neq \emptyset\}$, $j := j + 1$, go to Step 2.

Step 3. Select vector $x^0 \in Y_n$ such that $Q(x^0) = \min\{Q(x) | x \in Y_n\}$.

Let $x^* = (x_1^*, \ldots, x_n^*)$ be an optimal solution to the problem $Pm|p_j = a_j - b_j t|C_{max}$. Let $L = log\ max\{n, 1/\epsilon, a_{max}\}$. As the similar argument in Theorem 2.1, we have the following result.

Theorem 2. *Algorithm $\mathcal{A}_\epsilon{}^m$ finds $x^0 \in X$ for $Pm|p_j = a_j - b_j t|C_{max}$ such that $Q(x^0) \leq (1 + \epsilon) Q(x^*)$ in $O(n^{m+1} L^{m+1} / \epsilon^m)$.*

4 Conclusion

This paper studies the problem of scheduling jobs with time-dependent processing times on m identical parallel machines. Our objective is to minimize makespan. We first give a fully polynomial time approximation scheme for two machines, then generalize the result to the the case with m machines.

Future research may focus on scheduling of time-dependent processing times for more general decreasing types. It will also be interesting to investigate the 'mirror' problem in which the job processing times are nondecreasing function of their start times.

Acknowledgments. This research was support in part by Hong Kong Polytechnic University under Grant Number G-YW81. L.Y. Kang was also support by the National Natural Science Foundation of China.

References

1. M. Behzad, G. Chartrand, and C. Wall, On minimal regular digraphs with given girth, *Fund. Math.*, **69** (1970) 227-231
2. A. Bachman and A. Janiak, Minimizing maximum lateness under linear deterioration, *European Journal of Operational Research*, **126** (2000), 557-566.
3. A. Bachman, A. Janiak and M.Y. Kovalyov, Minimizing the total weighted completion time of deteriorating jobs, *Information Processing Letters*, **81** (2002), 81-84.
4. S. Browne and U. Yechiali, Scheduling deteriorating jobs on a single processor, *Operations Research*, **38** (1990), 495-498.
5. T.C.E. Cheng, L. Kang and CT Ng, Due-date assignment and single machine scheduling with deteriorating jobs, *Journal of Operational Research Society*, **55** (2004), 198-203.
6. T.C.E. Cheng, Q. Ding, M.Y. Kovalyov, A. Bachman and A. Janiak, Scheduling jobs with piecewise linear decreasing processing times, *Naval Research Logistics*, **50** (2003), 531-554.
7. T.C.E. Cheng, Q. Ding and B.M.T. Lin, A concise survey of scheduling with time-dependent processing times, *European Journal of Operational Research*, **152** (2004), 1-13.
8. T.C.E. Cheng and Q. Ding, Single machine scheduling with deadlines and increasing rates of processing times, *Acta Informatica*, **36** (2000), 673-692.
9. Z.L. Chen, Parallel machine scheduling with time-dependent processing times, *Discrete Applied Mathematics*, 70(1996), 81-93.
10. Z.L. Chen, A note on single-processor scheduling with time dependent execution time, *Operational Research Letter*, **17** (1995), 127-129.
11. R.L. Graham, E.L. Lawler, J.K. Lenstra, and A.H.G. Rinnooy Kan, Optimization and approximation in deterministic sequencing and scheduling: A survey, *Ann Discrete Math*, **5** (1979), 287-326.
12. J.N.D. Gupta and S.K. Gupta, Single facility scheduling with nonlinear processing times, *Computers and Industrial Engineering*, **14** (1988), 387-394.
13. S.K. Gupta, A.S. Kunnathur and K. Dandapani, Optimal repayment polices for multiple loans, *OMEGA*, **15** (1987), 323-330.

14. K.I.J. Ho, J.Y.T. Leung and W.D. Wei, Complexity of scheduling tasks with time dependent execution time, *Information Processing Letter*, *48* (1993), 315-320.
15. M.Y. Kovalyov and W. Kubiak, A fully polynomial approximation scheme for minimizing makespan of deteriorating jobs, *Journal of Heuristics*, **3** (1998), 287-297.
16. G. Mosheiov, Multi-machines scheduling with linear deterioration, *INFOR*, **36** (1998), 205-214.

On Complexity and Approximability of the Labeled Maximum/Perfect Matching Problems

Jérôme Monnot

CNRS, LAMSADE, Université Paris-Dauphine, Place du Maréchal de Lattre de, Tassigny, 75775, Paris Cedex 16, France
monnot@lamsade.dauphine.fr

Abstract. In this paper, we deal with both the complexity and the approximability of the labeled perfect matching problem in bipartite graphs. Given a simple graph $G = (V, E)$ with n vertices with a color (or label) function $L : E \to \{c_1, \dots, c_q\}$, the labeled maximum matching problem consists in finding a maximum matching on G that uses a minimum or a maximum number of colors.

Keywords: labeled matching; colored matching; bipartite graphs; **NP**-complete; approximate algorithms.

1 Introduction

The maximum matching problem is one of the most known combinatorial optimization problem and arises in several applications such as images analysis, artificial intelligence or scheduling. A matching M on a graph $G = (V, E)$ on n vertices is a subset of edges that are pairwise non adjacent; M is said maximum if the size of the matching is maximum among the matchings of G and perfect if it covers the vertex set of G (that is $|M| = \frac{n}{2}$). In the labeled maximum matching problem (Labeled MM in short), we are given a simple graph $G = (V, E)$ on $|V| = n$ vertices with a color (or label) function $L : E \to \{c_1, \dots, c_q\}$ on the edge set of G. For $i = 1, \dots, q$, we denote by $L^{-1}(\{c_i\}) \subseteq E$ the set of edges of color c_i. The goal of Labeled $Min\ MM$ (resp., $Max\ MM$) is to find a maximum matching on G using a minimum (resp., a maximum) number of colors. An equivalent formulation of Labeled $Min\ MM$ could be the following: if $G[\mathcal{C}]$ and m^* denote the subgraph induced by the edges of colors from $\mathcal{C} \subseteq \{c_1, \dots, c_q\}$ and the size of the maximum matching of G respectively, then Labeled $Min\ MM$ aims at finding a subset $\mathcal{C}$ of minimum size such that $G[\mathcal{C}]$ contains a matching of size m^*. The restriction of Labeled MM to the case where each color occurs at most r times in $I = (G, L)$ (i.e., $|L^{-1}(\{c_i\})| \leq r$ for $i = 1, \dots, q$) will be denoted by Labeled MM_r. For the particular case where we deal with perfect matchings instead of maximum matchings, the labeled maximum matching problem is called the labeled perfect matching problem and denoted by Labeled PM.

The Labeled $Min\ MM$ problem has some relationship with the timetable problem, since a solution may be seen as a matching between classes and teachers that satisfies additional restrictions (for instance, a color corresponds to a school

X. Deng and D. Du (Eds.): ISAAC 2005, LNCS 3827, pp. 934–943, 2005.

where we assume that a professor may teach in several schools). An inspector would like to assess all teachers during one lecture of each one of them and it would be desirable that (s)he visits not twice the same class. Hence the lectures to be attended would form a maximum matching. For convenience the inspector would like these lectures to take place in the smallest possible number of schools. Then clearly the inspector has to construct a maximum matching meeting a minimum number of colors in the graph associated with the lectures.

2 Previous Related Works and Generalization

Labeled problems have been mainly studied, from a complexity and an approximability point of view, when Π is polynomial, [7,8,9,12,19,23,24]. For example, the first labeled problem introduced in the literature is the LABELED minimum spanning tree problem, which has several applications in communication network design. This problem is **NP**-hard and many complexity and approximability results have been proposed in [7,9,12,19,23,24]. On the other hand, the LABELED maximum spanning tree problem has been shown polynomial in [7]. More recently, the LABELED path and the LABELED cycle problems have been studied in [8]; in particular, the authors proved that the LABELED minimum path problem is **NP**-hard and provided some exact and approximation algorithms. Note that the **NP**-completeness also appears in [11] since the LABELED minimum path problem is a special case of the red-blue set cover problem. To our knowledge, the LABELED minimum (or maximum) matching problem has not been studied yet in the literature. However, the restricted perfect matching problem [17] is very closed to the LABELED perfect matching. This latter problem aims at determining, given a graph $G = (V, E)$, a partition $E_1, \ldots, E_k$ of E and k positive integers $r_1, \ldots, r_k$, whether there exists a perfect matching M on G satisfying for all $j = 1, \ldots, k$ the restriction $|M \cap E_j| \leq r_j$. The restricted perfect matching problem is proved to be **NP**-complete in [17], even if (i) $|E_j| \leq 2$, (ii) $r_j = 1$, and (iii) G is a bipartite graph. On the other hand, it is shown in [25] that the restricted perfect matching problem is polynomial when G is a complete bipartite graph and $k = 2$; some others results of this problem can be found in [13]. A perfect matching M only verifying condition (ii) (that is to say $|M \cap E_i| \leq 1$) is called good in [10]. Thus, we deduce that the LABELED maximum perfect matching problem is **NP**-hard in bipartite graph since the value of an optimal solution $opt(I) = \frac{n}{2}$ iff G contains a good matching.

Most of the labelled problems can be embedded in the following framework. Let Π be a **NPO** problem accepting simple graphs $G = (V, E)$ as instances, edge-subsets $E' \subseteq E$ verifying a given polynomial-time decidable property $Pred$ as solutions, and the solutions cardinality as objective function; the labeled problem associated to Π, denoted by LABELED Π, seeks, given an instance $I = (G, L)$ where $G = (V, E)$ is a simple graph and L is a mapping from E to $\{c_1, \ldots, c_q\}$, in finding a subset E' verifying $Pred$ that optimizes the size of the set $L(E') = \{L(e) : e \in E'\}$. Note that two versions of LABELED Π may be considered

according to the optimization goal: LABELED *Min Π* that consists of minimizing $|L(E')|$ and LABELED *Max Π* that consists of maximizing $|L(E')|$. Roughly speaking, the mapping L corresponds to assigning a color (or a label) to each edge and the goal of LABELED *Min Π* (resp., *Max Π*) is to find an edge subset using the fewest (resp., the most) number of colors. If a given **NPO** problem Π is **NP**-hard, then the associated labeled problem LABELED Π is clearly **NP**-hard (consider a distinct color per edge). For instance, the LABELED Longest path problem or the LABELED maximum induced matching problem are both **NP**-hard. Moreover, if the decision problem associated to Π is **NP**-complete, (the decision problem aims at deciding if a graph G contains an edge subset verifying *Pred*), then LABELED *Min Π* can not be approximated within performance ratio better than $2-\varepsilon$ for all $\varepsilon > 0$ unless **P**=**NP**, even if the graph is complete. Indeed, if we color the edges from $G = (V, E)$ with a single color and then we complete the graph, adding a new color per edge, then it is **NP**-complete to decide between $opt(I) = 1$ and $opt(I) \geq 2$, where $opt(I)$ is the value of an optimal solution. Notably, it is the case of the LABELED traveling salesman problem (LABELED *TSP* in short) or the LABELED minimum partition problem into paths of length k for any $k \geq 2$. Note that the problem consisting in deciding whether $opt(I) = n$ in colored complete graphs K_n for LABELED *Max TSP*, has been studied. For instance in [1,14,16], some conditions (mainly using probabilistic methods) were mentioned for a colored complete graph K_n to contain a hamiltonian cycle using n colors.

In this paper, we go into the investigation of the complexity and the approximability of labeled matching problems in bipartite graphs. More precisely, we deal with 2 extreme classes of 2-regular or $\frac{n}{2}$-regular bipartite graphs. For both cases, we obtain hardness results. For these graphs, observe that a maximum matching is a perfect matching; thus, in these graphs LABELED *MM* and LABELED *PM* are the same problem. In section 3, we analyze both the complexity and the approximability of the LABELED minimum perfect matching problem and the LABELED maximum perfect matching problem in 2-regular bipartite graphs. Finally, section 4 focuses on the case of complete bipartite graphs $K_{n,n}$.

Now, we introduce some terminology and notations that will be used in the paper. We denote by $opt(I)$ and $apx(I)$ the value of an optimal and an approximate solution, respectively. We say that an algorithm $\mathcal{A}$ is an ε-approximation of LABELED *Min PM* with $\varepsilon \geq 1$ (resp., *Max PM* with $\varepsilon < 1$) if $apx(I) \leq \varepsilon \times opt(I)$ (resp., $apx(I) \geq \varepsilon \times opt(I)$) for any instance $I = (G, L)$, for more details see for instance [4].

3 The 2-Regular Bipartite Case

In this section, we deal with a particular class of graphs that consist in a collection of pairwise disjoints cycles of even length; note that such graphs are 2-regular bipartite graphs.

Theorem 1. LABELED *Min PM_r is **APX**-complete in 2-regular bipartite graphs for any $r \geq 2$.*

Proof. Observe that any solution of LABELED $Min\ PM_r$ is an r-approximation. The rest of the proof will be done via an approximation preserving reduction from the minimum balanced satisfiability problem with clauses of size at most r, MIN BALANCED r-SAT for short. An instance $I = (\mathcal{C}, X)$ of MIN BALANCED r-SAT consists of a collection $\mathcal{C} = (C_1, \dots, C_m)$ of clauses over the set $X = \{x_1, \dots, x_n\}$ of boolean variables, such that each clause C_j has at most r literals and each variable appears positively as many times as negatively; let B_i denote this number for any $i = 1, \dots, n$. The goal is to find a truth assignment f satisfying a minimum number of clauses. MIN BALANCED 2-SAT where $2 \leq B_i \leq 3$ has been shown **APX**-complete by the way of an L-reduction from MAX BALANCED 2-SAT where $B_i = 3$, [6,18].

We only prove the case $r = 2$. Let $I = (\mathcal{C}, X)$ be an instance of MIN BALANCED 2-SAT on m clauses $\mathcal{C} = \{C_1, \dots, C_m\}$ and n variables $X = \{x_1, \dots, x_n\}$ such that each variable x_i has either 2 occurrences positive and 2 occurrences negative, or 3 occurrences positive and 3 occurrences negative. We build the instance $I' = (H, L)$ of LABELED $Min\ PM_2$ where H is a collection of pairwise disjoints cycles $\{H(x_1), \dots, H(x_n)\}$ and L colors edges of H with colors $c_1, \dots, c_j, \dots, c_m$, by applying the following process:

- For each variable x_i, create $2B_i$-long cycle $H(x_i) = \{e_{i,1}, \dots, e_{i,k}, \dots, e_{i,2B_i}\}$.
- Color the edges of $H(x_i)$ as follows: if x_i appears positively in clauses $C_{j_1}, \dots, C_{j_{B_i}}$ and negatively in clauses $C_{j'_1}, \dots, C_{j'_{B_i}}$, then set $L(e_{i,2k}) = c_{j_k}$ and $L(e_{i,2k-1}) = c_{j'_k}$ for $k = 1, \dots, B_i$.

Clearly, H is made of n disjoint cycles and is painted with m colors. Moreover, each color appears at most twice.

Let f^* be an optimal truth assignment on I satisfying m^* clauses and consider the perfect matching $M = \cup_{i=1}^{n} M_i$ where $M_i = \{e_{i,2k} | k = 1, \dots, B_i\}$ if $f(x_i) = true$, $M_i = \{e_{i,2k-1} | k = 1, \dots, B_i\}$ otherwise; M uses exactly m^* colors and thus:

$$opt(I') \leq m^* \tag{1}$$

Conversely, let M' be a perfect matching on H using $apx(I') = m'$ colors; if one sets $f'(x_i) = true$ if $e_{i,2} \in M'$, $f'(x_i) = false$ otherwise, we can easily observe that the truth assignment f' satisfies m' clauses.

$$apx(I) = m' \tag{2}$$

Hence, using inequalities (1) and (2) the result follows.

Trivially, the problem becomes obvious when each color is used exactly once. We now show that we have a 2-approximation in 2-regular bipartite graphs, showing that the restriction of LABELED $Min\ PM$ to 2-regular bipartite graphs is as hard as approximate as MINSAT.

Theorem 2. *There exists an approximation preserving reduction from* LABELED $Min\ PM$ *in 2-regular bipartite graphs to* MINSAT *of expansion* $c(\varepsilon) = \varepsilon$.

Proof. The result comes from the reciprocal of the previous transformation. Let $I = (G, L)$ be an instance of LABELED *Min PM* where $G = (V, E)$ is a collection $\{H_1, \ldots, H_n\}$ of disjoint cycles of even length and $L(E) = \{c_1, \ldots, c_m\}$ defines the label set, we describe every cycle H_i as the union of two matchings M_i and $\overline{M_i}$. We construct an instance $I' = (\mathcal{C}, X)$ of MINSAT where $\mathcal{C} = \{C_1, \ldots, C_m\}$ is a set of m clauses and $X = \{x_1, \ldots, x_n\}$ is a set of n variables, as follows. The clause set $\mathcal{C}$ is in one to one correspondence with the color set $L(E)$ and the variable set X is in one to one correspondence with the connected components of G; a literal x_i (resp., $\overline{x_i}$) appears in C_j iff $c_j \in L(M_i)$ (resp., $c_j \in L(\overline{M_i})$). We easily deduce that any truth assignment f on I' that satisfies k clauses can be converted into a perfect matching M_f on I that uses k colors.

Using the 2-approximation of MINSAT [20] and the Theorem 2, we deduce:

Corollary 1. LABELED *Min PM in 2-regular bipartite graphs is 2-approximable.*

Dealing with LABELED *Max* PM_r, the result of [17] shows that computing a *good matching* is **NP**-hard even if the graph is bipartite and each color appears at most twice; a good matching M is a perfect matching using $|M|$ colors. Thus, we deduce from this result that LABELED *Max* PM_r is **NP**-hard for any $r \geq 2$. We strengthen this result using a reduction from MAX BALANCED 2-SAT.

Theorem 3. LABELED *Max* PM_r *is* ***APX***-*complete in 2-regular bipartite graphs for any* $r \geq 2$.

In the same way, there exists an approximation preserving reduction from LABELED *Max PM* in 2-regular bipartite graphs to MAXSAT of expansion $c(\varepsilon) = \varepsilon$. Thus, using the approximate result for MAXSAT [3], we obtain

Corollary 2. LABELED *Max PM in 2-regular bipartite graphs is 0.7846-approximable.*

4 The Complete Bipartite Case

When considering complete bipartite graphs, we obtain several results:

Theorem 4. LABELED *Min* PM_r *is* ***APX***-*complete in bipartite complete graphs* $K_{n,n}$ *for any* $r \geq 6$.

Proof. We give an approximation preserving L-reduction (cf. Papadimitriou & Yannakakis [21]) from the restriction MINSC$_3$ of the set cover problem, MINSC for short. Given a family $\mathcal{S} = \{S_1, \ldots, S_{n_0}\}$ of subsets and a ground set $X = \{x_1, \ldots, x_{m_0}\}$ (we assume that $\cup_{i=1}^{n_0} S_i = X$), a set cover of X is a sub-family $\mathcal{S}' = \{S_{f(1)}, \ldots, S_{f(p)}\} \subseteq \mathcal{S}$ such that $\cup_{i=1}^{p} S_{f(i)} = X$; MINSC is the problem of determining a minimum-size set cover $\mathcal{S}^* = \{S_{f^*(1)}, \ldots, S_{f^*(q)}\}$ of X. Its restriction MINSC$_3$ to instances where each set is of size at most 3 and each

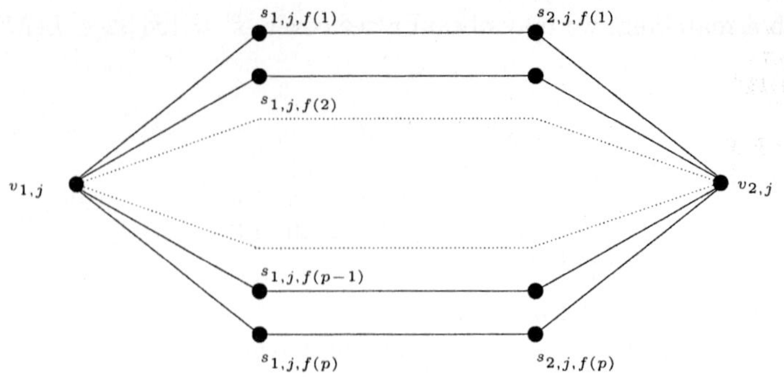

Fig. 1. The gadget $H(x_j)$

element x_j appears in at most 3 and at least 2 different sets has been proved **APX**-complete in [21].

Given an instance $I_0 = (\mathcal{S}, X)$ of MINSC, its characteristic graph $G_{I_0} = (L_0, R_0; E_{I_0})$ is a bipartite graph with a left set $L_0 = \{l_1, \dots, l_{n_0}\}$ that represents the members of the family $\mathcal{S}$ and a right set $R_0 = \{r_1, \dots, r_{m_0}\}$ that represents the elements of the ground set X; the edge-set E_{I_0} of the characteristic graph is defined by $E_{I_0} = \{[l_i, r_j] : x_j \in S_i\}$. Note that G_{I_0} is of maximum degree 3 iff I_0 is an instance of MINSC$_3$. From I_0 an instance of MINSC$_3$, we construct the instance $I = (K_{n,n}, L)$ of LABELED *Min PM*$_6$. First, we start from a bipartite graph having m_0 connected components $H(x_j)$ and $n_0 + m_0$ colors $\{c_1, \dots, c_{n_0+m_0}\}$, described as follows:

- For each element $x_j \in X$, we build a gadget $H(x_j)$ that consists of a bipartite graph of $2(d_{G_{I_0}}(r_j)+1)$ vertices and $3d_{G_{I_0}}(r_j)$ edges, where $d_{G_{I_0}}(r_j)$ denotes the degree of vertex $r_j \in R_0$ in G_{I_0}. The graph $H(x_j)$ is illustrated in Figure 1.
- Assume that vertices $\{l_{f(1)}, \dots, l_{f(p)}\}$ are the neighbors of r_j in G_{I_0}, then color $H(x_j)$ as follows: for any $k = 1, \dots, p$, $L(v_{1,j}, s_{1,j,f(k)}) = L(v_{2,j}, s_{2,j,f(k)}) = c_{f(k)}$ and $L(s_{1,j,f(k)}, s_{2,j,f(k)}) = c_{n_0+j}$.
- We complete $H = \cup_{x_j \in X} H(x_j)$ into $K_{n,n}$, by adding a new color per edge.

Clearly, $K_{n,n}$ is complete bipartite and has $2n = 2\sum_{r_j \in R_0}(d_{G_{I_0}}(r_j) + 1) = 2|E_{I_0}| + 2m_0$ vertices. Moreover, each color is used at most 6 times.

Let $\mathcal{S}^*$ be an optimal set cover on I_0. From $\mathcal{S}^*$, we can easily construct a perfect matching M^* on I using exactly $|\mathcal{S}^*| + m_0$ colors (since we assume that each element appears in at least 2 sets) and thus:

$$opt_{\text{LABELED } Min\ PM_6}(I) \le opt_{\text{MINSC}_3}(I_0) + m_0 \tag{3}$$

Conversely, we can show that any perfect matching M may be polynomially transformed into a perfect matching M" with value $apx(I)$, using the edges of H and verifying:

Property 1. $apx(I) \le |L(M)|$

From such a matching, we may obtain a set cover $\mathcal{S}" = \{S_k | c_k \in L(M")\}$ on I_0 verifying:

$$|\mathcal{S}"| = apx(I) - m_0 \tag{4}$$

Using (3) and (4), we deduce $opt_{\text{LABELED } Min\ PM_6}(I) = opt_{\text{MINSC}_3}(I_0) + m_0$ and $|\mathcal{S}"| - opt_{\text{MINSC}_3}(I_0) \leq |L(M))| - opt_{\text{LABELED } Min\ PM_6}(I)$. Finally, since $opt_{\text{MINSC}_3}(I_0) \geq \frac{m_0}{3}$ the result follows.

Applying the same kind of proof from the vertex cover problem (MINVC in short) in cubic graphs [2], we obtain a stronger result.

Theorem 5. LABELED *Min* PM_r *is* ***APX****-complete in bipartite complete graphs* $K_{n,n}$ *for any* $r \geq 3$.

Proof. Starting from a cubic graph $G = (V, E)$ instance of MINVC, we associate to each edge $e = [x, y] \in E$ a 4-long cycle $\{a_{1,e}, a_{2,e}, a_{3,e}, a_{4,e}\}$ together with a coloration L given by: $L(a_{1,e}) = c_x$, $L(a_{2,e}) = c_y$ and $L(a_{3,e}) = L(a_{4,e}) = c_e$. We complete this graph into a complete bipartite graph, adding a new color per edge.

Unfortunately, we can not apply the proof of Theorem 2 since in this latter, on the one hand, we have some cycles of size 6 and, on the other hand, a color may occur in different gadgets. One open question concerns the complexity of LABELED *Min* PM_2 in bipartite complete graphs. Moreover, from Theorem 4, we can also obtain a stronger inapproximability result concerning the general problem LABELED *Min PM*: one can not compute in polynomial-time an approximate solution of LABELED *Min PM* that uses less than $(1/2 - \varepsilon)\ln(opt_{\text{LABELED } Max\ PM}(I))$ colors in complete bipartite graphs.

Corollary 3. *For any* $\varepsilon > 0$, LABELED *Min PM is not* $(\frac{1}{2} - \varepsilon) \times \ln(n)$ *approximable in complete bipartite graphs* $K_{n,n}$, *unless* $\textbf{NP} \subset \textbf{DTIME}(n^{\log\log n})$.

Proof. First, we apply the construction made in Theorem 4, except that $I_0 = (\mathcal{S}, X)$ is an instance of MINSC such that the number of elements m_0 is strictly larger than the number of sets n_0. From I_0, we construct n_0 instances $I'_1, \ldots, I'_{n_0}$ of LABELED *Min PM* where $I'_i = (H, L_i)$. The colors $L_i(E)$ are the same as $L(E)$, except that we replace colors $c_{n_0+1}, \ldots, c_{n_0+m_0}$ by c_i. Finally, as previously, we complete each instance I'_i into a complete bipartite graph $K_{n,n}$ by adding a new color by edge.

Let $\mathcal{S}^*$ be an optimal set cover on I_0 and assume that $S_i \in \mathcal{S}^*$, we consider the instance I_i of LABELED *Min PM*. From $\mathcal{S}^*$, we can easily construct a perfect matching M_i^* of I_i that uses exactly $|\mathcal{S}^*|$ colors. Conversely, let M_i be a perfect matching on I_i; by construction, the subset $\mathcal{S}' = \{S_k : c_k \in L(M_i)\}$ of $\mathcal{S}$ is a set cover of X using $|L(M_i)|$ sets. Finally, let A be an approximate algorithm for LABELED *Min PM*, we compute n_0 perfect matchings M_i, applying A on instances I_i. Thus, if we pick the matching that uses the minimum number of colors, then we can polynomially construct a set cover on I_0 of cardinality this number of colors.

Since $n_0 \leq m_0 - 1$, the size n of a perfect matching of $K_{n,n}$ verifies: $n = |E_{I_0}| + m_0 \leq n_0 \times m_0 + m_0 \leq m_0(m_0 - 1) + m_0 = m_0^2$. Hence, from any algorithm A solving LABELED *Min PM* within a performance ratio $\rho_A(I) \leq \frac{1}{2} \times \ln(n)$, we can deduce an algorithm for MINSC that guarantees the performance ratio $\frac{1}{2}\ln(n) \leq \frac{1}{2}\ln(m_0^2) = \ln(m_0)$. Since the negative result of [15] holds when $n_0 \leq m_0 - 1$, i.e., MINSC is not $(1-\varepsilon) \times \ln(m_0)$ approximable for any $\varepsilon > 0$, unless $\mathbf{NP} \subset \mathbf{DTIME}(n^{loglogn})$, we obtain a contradiction.

On the other hand, dealing with LABELED *Max PM*$_r$ in $K_{n,n}$, the result of [10] shows that the case $r = 2$ is polynomial, whereas it becomes **NP**-hard when $r = \Omega(n^2)$. Indeed, it is proved in [10] that, on the one hand, we can compute a good matching in $K_{n,n}$ within polynomial-time when each color appears at most twice and, on the other hand, there always exists a good matching in such a graph if $n \geq 3$. An interesting question is to decide the complexity and the approximability of LABELED *Max PM*$_r$ when r is a constant greater than 2.

4.1 Approximation Algorithm for LABELED *Min PM*$_r$

Let us consider the greedy algorithm for LABELED *Min PM*$_r$ in complete bipartite graphs that iteratively picks the color that induces the maximum-size matching in the current graph and delete the corresponding vertices. Formally, if $L(G')$ denotes the colors that are still available in the graph G' at a given iteration and if $G'[c]$ (resp., $G'[V']$) denotes the subgraph of G' that is induced by the edges of color c (resp., by the vertices V'), then the greedy algorithm consists of the following process:

Greedy

1 Set $C' = \emptyset$, $V' = V$ and $G' = G$;
2 While $V' \neq \emptyset$ do
  2.1 For all $c \in L(G')$, compute a maximum matching M_c in $G'[c]$;
  2.2 Select a color c^* maximizing $|M_c|$;
  2.3 $C' \leftarrow C' \cup \{c^*\}$, $V' \leftarrow V' \setminus V(M_{c^*})$ and $G' = G[V']$;
3 output C';

Theorem 6. *Greedy is an $\frac{H_r + r}{2}$-approximation of* LABELED *Min PM*$_r$ *in complete bipartite graphs where H_r is the r-th harmonic number $H_r = \sum_{i=1}^{r} \frac{1}{i}$, and this ratio is tight.*

Proof. Let $I = (G, L)$ be an instance of LABELED *Min PM*$_r$. We denote by C'_i for $i = 1, \ldots, r$ be the set of colors of the approximate solution which appears exactly i times in C' and by p_i its cardinality (thus, $\forall c \in C'_i$ we have $|M_c| = i$ in $G'[c]$); finally, let M_i denote the matching with colors C'_i. If $apx(I) = |C'|$, then we have:

$$apx(I) = \sum_{i=1}^{r} p_i \tag{5}$$

Let $\mathcal{C}^*$ be an optimal solution corresponding to the perfect matching M^* of size $opt(I) = |\mathcal{C}^*|$; we denote by E_i the set of edges of M^* that belong to $G[\cup_{k=1}^{i} V(M_k)]$, the subgraph induced by $\cup_{k=1}^{i} V(M_k)$ and we set $q_i = |E_i \setminus E_{i-1}|$ (where we assume that $E_0 = \emptyset$). For any $i = 1, \ldots, r-1$, we get:

$$opt(I) \geq \frac{1}{i} \sum_{k=1}^{i} q_k \tag{6}$$

Indeed, $\sum_{k=1}^{i} q_k = |E_i|$ and by construction, each color appears at most i times in $G[\cup_{k=1}^{i} V(M_k)]$.

We also have the following inequality for any $i = 1, \ldots, r-1$:

$$opt(I) \geq \frac{1}{r} \left(2 \sum_{k=1}^{i} k \times p_k - \sum_{k=1}^{i} q_k \right) \tag{7}$$

Since M^* is a perfect matching, the quantity $2\sum_{k=1}^{i} k \times p_k - \sum_{k=1}^{i} q_k$ counts the edges of M^* of which at least one endpoint belongs to $G[\cup_{k=1}^{i} V(M_k)]$. Because each color appears on at most r edges, the result follows.

Finally, since $\sum_{k=1}^{r} k \times p_k$ is the size of a perfect matching of G, the following inequality holds:

$$opt(I) \geq \frac{1}{r} \sum_{k=1}^{r} k \times p_k \tag{8}$$

Using equality (5) and adding inequalities (6) with coefficient $\alpha_i = \frac{1}{2(i+1)}$ for $i = 1, \ldots, r-1$, inequalities (7) with coefficient $\beta_i = \frac{r}{2i(i+1)}$ for $i = 1, \ldots, r-1$ and inequality (8), we obtain:

$$apx(I) \leq \left(\frac{H_r + r}{2} \right) opt(I) \tag{9}$$

The proof of the tightness is omitted.

We conjecture that LABELED *Min PM* is not $O(n^\varepsilon)$-approximable in complete bipartite graphs. Thus, a challenge will be to give better approximate algorithms or to improve the lower bound.

References

1. M. Albert, A. Frieze and B. Reed. Multicoloured Hamilton cycles. *Electron. J. Combin.*, 2, 1995 (Research Paper 10).
2. P. Alimonti and V. Kann. Some APX-completeness results for cubic graphs. *Theoretical Computer Science* 237:123-134, 2000.
3. T. Asano and D. P. Williamson Improved Approximation Algorithms for MAX SAT. *Journal of Algorithms* 42(1):173-202, 2002.
4. G. Ausiello, P. Crescenzi, G. Gambosi, V. Kann, A. Marchetti-Spaccamela and M. Protasi. Complexity and Approximation (Combinatorial Optimization Problems and Their Approximability Properties). *Springer, Berlin*, 1999.

5. P. Berman, M. Karpinski, and A. D. Scott. Approximation Hardness of Short Symmetric Instances of MAX-3SAT. *ECCC TR-03-049*, 2003.
6. P. Berman and M. Karpinski. On Some Tighter Inapproximability Results. *ECCC TR-05-029*, 1998.
7. H. Broersma and X. Li. Spanning trees with many or few colors in edgecolored graphs. *Discussiones Mathematicae Graph Theory*, 17(2):259-269, 1997.
8. H. Broersma, X. Li, G. J. Woeginger and S. Zhang. Paths and cycles in colored graphs. *Australasian J. Combin.*, 31:, 2005.
9. T. Brüggemann, J. Monnot, and G. J. Woeginger. Local search for the minimum label spanning tree problem with bounded color classes. *Operations Research Letters* 31(3):195-201, 2003.
10. K. Cameron. Coloured matchings in bipartite graphs. *Discrete Mathematics*, 169:205-209, 1997.
11. R. D. Carr, S. Doddi, G. Konjevod and M. V. Marathe. On the red-blue set cover problem. *SODA*, 345-353, 2000.
12. R-S. Chang and S-J. Leu. The minimum labeling spanning trees. *Information Processing Letters*, 63:277-282, 1997.
13. M. C. Costa, D. de Werra, C. Picouleau and B. Ries. Bicolored matchings in some classes of graphs. *technical report*, 2004.
14. P. Erdös, J. Nešetřil and V. Rödl. On some problems related to partitions of edges of a graph. *Graphs and other combinatorial topics, Teubner, Leipzig*, 54-63, 1983.
15. U. Feige. A threshold of for approximating set cover. *J. ACM*, 45:634-652, 1998.
16. A. Frieze and B. Reed. Polychromatic Hamilton cycles. *Discrete Math.* 118:69-74, 1993.
17. A. Itai, M. Rodeh, and S. Tanimoto. Some matching problems in bipartite graphs. *J. ACM*, 25(4):517-525, 1978.
18. M. Karpinski. Personnal communication. 2005.
19. S. O. Krumke and H-C. Wirth. On the minimum label spanning tree problem. *Information Processing Letters*, 66:81-85, 1998.
20. M. V. Marathe, S. S. Ravi. On Approximation Algorithms for the Minimum Satisfiability Problem. *Information Processing Letters*, 58(1):23-29 1996.
21. C.H. Papadimitriou and M. Yannakakis. Optimization, approximation, and complexity classes. *J. of Comp. and Sys. Sci.*, 43:425-440, 1991.
22. R. Raz and S. Safra. A sub-constant error-probability low-degree test, and subconstant error-probability PCP characterization of NP. *Proc. 29th Ann. ACM Symp. on Theory of Comp., ACM*, 475-484, 1997.
23. Y. Wan, G. Chen, and Y. Xu. A note on the minimum label spanning tree. *Information Processing Letters*, 84;99-101, 2002.
24. Y. Xiong, B. Golden and E. Wasil. Worst-case behavior of the MVCA heuristic for the minimum labeling spanning tree problem. *Operations Research Letters* 33(1):77-80, 2005.
25. T. Yi, K. G. Murty and C. Spera. Matchings in colored bipartite networks. *Discrete Applied Mathematics* 121(1-3):261-277, 2002.

Finding a Weight-Constrained Maximum-Density Subtree in a Tree

Sun-Yuan Hsieh and Ting-Yu Chou

Department of Computer Science and Information Engineering, National Cheng Kung University, No. 1, Ta-Hsueh Road, Tainan 701, Taiwan
hsiehsy@mail.ncku.edu.tw, janus@algorithm.csie.ncku.edu.tw

Abstract. Given a tree $T = (V, E)$ of n vertices such that each node v is associated with a *value-weight pair* (val_v, w_v), where *value* val_v is a real number and *weight* w_v is a non-negative integer, the *density* of T is defined as $\frac{\Sigma_{v\in V} val_v}{\Sigma_{v\in V} w_v}$. A *subtree* of T is a connected subgraph (V', E') of T, where $V' \subseteq V$ and $E' \subseteq E$. Given two integers w_{min} and w_{max}, the *weight-constrained maximum-density subtree problem* on T is to find a maximum-density subtree $T' = (V', E')$ satisfying $w_{min} \leq \Sigma_{v\in V'} w_v \leq w_{max}$. In this paper, we first present an $O(w_{max}n)$-time algorithm to find a weight-constrained maximum-density path in a tree, and then present an $O(w_{max}{}^2 n)$-time algorithm to find a weight-constrained maximum-density subtree in a tree. Finally, given a node subset $S \subset V$, we also present an $O(w_{max}{}^2 n)$-time algorithm to find a weight-constrained maximum-density subtree of T which covers all the nodes in S.

1 Introduction

Given a sequence S of n number pairs (a_i, w_i) with $w_i > 0$ for $i = 1, \ldots, n$, and two weight bounds $w_{\min}$ and $w_{\max}$, the *maximum-density segment problem* is to find a consecutive subsequence $S(i, j)$ of S such that the density of $S(i, j)$, i.e., $\frac{a_i+a_{i+1}+\cdots+a_j}{w_i+w_{i+1}+\cdots+w_j}$ is maximum over all other subsequence of S satisfying $w_{min} \leq w_i + w_{i+1} + \cdots + w_j \leq w_{max}$. This problem arises from the investigation of non-uniformity of nucleotide composition within genomic sequences, which was first revealed through thermal melting and gradient centrifugation experiments [4,9]. Researchers observed that the compositional heterogeneity is highly correlated to the GC content of the genomic sequences [10,12], and this motivates finding GC-rich segments. For the maximum-density segment problem, Goldwasser, Kao, and Lu [2] proposed an $O(n \log(w_{\max} - w_{\min} + 1))$-time algorithm. Chung and Lu [1] gave an $O(n)$-time algorithm based on bypassing the complicated preprocessing step required in [2]. More research related to the non-general case, i.e., $w_i = 1$ for all indices i, can be found in [2,3,5,6,7,10,11] and the references therein.

There are some invariants related to the maximum-density segment problem. Given a tree with weight and length on each edge, Wu, Chao, and Tang [14] presented an efficient algorithms for locating a maximum-weight path in a tree

X. Deng and D. Du (Eds.): ISAAC 2005, LNCS 3827, pp. 944–953, 2005.

whose length is at most a given upper bound. Their algorithm runs in $O(n \log^2 n)$ time and can be reduced to $O(n \log n)$ if the edge lengths are all integers in the range $[1, 2, \ldots, O(n)]$, where n is the number of nodes of the input tree.

Quite recently, Lin, Kuo, and Chao [8] studied the problem of finding a length-constrained maximum-density path in a rooted tree with each edge e is associated with a weight $w(e)$. They defined the density of a path with edges $e_1, e_2, \ldots, e_k$ as $\frac{\sum_{i=1}^{k} w(e_i)}{k}$, and presented an $O(nL)$-time algorithm to find a maximum-density path of length at least L.

In this paper, we assume that each node v of a tree $T = (V, E)$ is associated with a *value-weight pair* (val_v, w_v), where *value* val_v is a real number and *weight* w_v is a non-negative integer. The *density* of T, denoted by $d(T)$, is defined as $\frac{\Sigma_{v \in V} val_v}{\Sigma_{v \in V} w_v}$. The *weight-sum of* T, denoted by $w(T)$, is $\Sigma_{v \in V} w_v$. We consider the following three problems.

Definition 1 (The weight-constrained maximum-density path problem). Given a tree T of n nodes associated with value-weight pairs, and two bounded integers $w_{\min}$ and $w_{\max}$, find a maximum-density path P in T such that $w_{\min} \le w(P) \le w_{\max}$.

Definition 2 (The weight-constrained maximum-density subtree problem). Given a tree T of n nodes associated with value-weight pairs, and two bounded integers $w_{\min}$ and $w_{\max}$, find a maximum-density subtree T' in T such that $w_{\min} \le w(T') \le w_{\max}$.

Definition 3 (The weight-constrained maximum-density Steiner tree problem). Given a tree T of n nodes associated with value-weight pairs, two bounded integers $w_{\min}$ and $w_{\max}$, and a set of *terminals* $S \subset V(T)$, find a maximum-density subtree T' such that $S \subseteq V(T')$ and $w_{\min} \le w(T') \le w_{\max}$.

In this paper, we develop pseudo-polynomial time algorithms for the above problems. We first show that the weight-constrained maximum-density path problem on trees can be solved in $O(w_{max} n)$ time. Based on this result, we also show that both the weight-constrained maximum-density subtree problem and the weight-constrained maximum-density Steiner tree problem on trees can be solved in $O({w_{max}}^2 n)$ time.

2 Preliminaries

A graph G is a pair (V, E), where V and E are the vertex set and the edge set, respectively. We also use $V(G)$ and $E(G)$ to denote its vertex set and edge set, respectively. A path is a simple graph whose vertices can be ordered so that two vertices are adjacent if an only if they are consecutive in the list. A *subgraph* of $G = (V, E)$ is a graph (V', E') such that $V' \subseteq V$ and $E' \subseteq E$. An *induced subgraph* is an edge-preserving subgraph, that is, (V', E') is an induced subgraph of (V, E) iff $V' \subseteq V$ and $E' = \{(x, y) \in E | \ x, y \in V'\}$. For graph-theoretic terminologies and notations not mentioned here, readers should refer to [13].

A *tree* is a connected graph and has no cycles. A rooted tree is a tree with one node chosen as the root. For a node v in a rooted tree T with the root r, the *parent* of v, denoted by $par(v)$, is its neighbor on the unique path from v to r; its *children*, denoted by $child_T(v)$, are its other neighbors; and its *descendants* are the nodes u such that the unique path from u to r contains v. The *leaves* are the nodes with no children. In contrast to the unrooted case, by the degree of a node v, denoted by $deg(v)$, we mean the number of children of v in the given rooted tree. For a node v in T, the *level* of v, denoted by $level(v)$, is the number of edges between v and the root.

For a tree T, a *subtree* of T is a connected subgraph (V', E') of T, where $V' \subseteq V$ and $E' \subseteq E$. For a rooted tree T, the *descendent-subtree rooted at* v is the subtree induced by all the descendants of v, rooted at v, which is denoted by T_v. Note that for a rooted tree T, every descendent-subtree rooted at v is a subtree rooted at v; the converse is not always true. For a given node v in a rooted tree, the paths starting from v can be classified into two types: One is to stretch downward its children only, called *downward paths*, and the other is to include at least its parent, called *upward paths*. In the following sections, the input instance of the problem is restricted to be the rooted tree. If it is not the case, the tree can be easily transformed into a rooted one by selecting an arbitrary node as the root.

3 Finding a Weight-Constrained Maximum-Density Path

For a node v in a rooted tree T, we use $D_v^1[i]$ and $D_v^2[i]$ to denote the highest value and the second-best value of those downward paths starting from node v of weight i, respectively. We aim at computing $D_v^1[i]$ and $D_v^2[i]$ for each node v in T by bottom-up dynamic programming, where $i = w_v, w_v + 1, \ldots, w_{\max}$. The function "max" always selects the maximum value of the downward paths. If there exist at least two maximum-value downward paths, the function "2nd-max" selects the maximum value; otherwise, it selects the second-best value. If v is a leaf, then $D_v^1[i]$ equals val_v if $i = w_v$, and equals $-\infty$ for otherwise. Moreover, we define $D_v^2[i]$ to be $-\infty$ for all $0 \leq i \leq w_{max}$.[1]

Suppose that the internal node v has m children $v_1, v_2, \ldots, v_m$. Then, we have

$$D_v^1[i] = \max\{D_{v_j}^1[i - w_v] + val_v |\ 1 \leq j \leq m\}, \quad (1)$$

and

$$D_v^2[i] = \text{2nd-max}\{D_{v_j}^1[i - w_v] + val_v |\ 1 \leq j \leq m\}. \quad (2)$$

For each internal node v, we use $d_v^1[i]$ to record the *contributor of* $D_v^1[i]$, which is a child of v providing the value to $D_v^1[i]$, i.e., $d_v^1[i] = v_k$ if $D_v^1[i] = D_{v_k}^1[i - w_v] + val_v$. We also use $d_v^2[i]$ to record the *contributor of* $D_v^2[i]$, which

[1] For convenience, throughout this paper, we assume the value of an entry equals $-\infty$ if the value of such an entry is not assigned.

is another child of v (different from the first contributor) providing the value to $D_v^2[i]$ if such a child exists. Let $d_v^1[i]$(or $d_v^2[i]$)=NULL if the corresponding contributor does not exist. Note that if there exist at least two downward paths with the same highest value, then the node recorded in $d_v^2[i]$ is different with that recorded in $d_v^1[i]$.

Lemma 1. *The downward entries D_v^1 and D_v^2, together with d_v^1 and d_v^2, for all nodes v in a rooted tree T can be computed in $O(w_{\max}n)$ time.*

Proof. Without loss of generality, we consider an internal node v has m children. For a fixed weight i, it takes $O(m)$ time to determine $D_v^1[i]$ and $D_v^2[i]$, together with $d_v^1[i]$ and $d_v^2[i]$. Thus it takes $O(w_{\max}m)$ time to computed the downward entries of node v. By a bottom-up evaluation of T, the downward entries of all internal nodes v in tree T can be computed in $O(w_{\max}n)$ time.

The computation of upward entries is similar to that of downward entries. Let $U_v[i]$ denote the highest value of those upward paths starting from node v of weight i. For $i = w_v, w_v + 1, \ldots, w_{\max}$, we compute $U_v[i]$ for each node v in tree T by a top-down dynamic programming. Initially, $U_r[i] = val_r$ if $i = w_r$, and $U_r[i] = -\infty$ if $i \neq w_r$.

Let $par(v)$ be the parent of node v. Then, we have

$$U_v[i] = \max \begin{cases} U_{par(v)}[i - w_v] + val_v, & \\ D^1_{par(v)}[i - w_v] + val_v & \text{if } v \text{ isn't the contributor of} \\ & D^1_{par(v)}[i - w_v], \\ D^2_{par(v)}[i - w_v] + val_v & \text{if } D^2_{par(v)}[i - w_v] \text{ exists, and } v \text{ is} \\ & \text{the contributor of } D^1_{par(v)}[i - w_v]. \end{cases} \quad (3)$$

For each non-root node v, we also use $u_v[i]$ to record the *contributor of* $U_v[i]$, which is defined as follows:

$$u_v[i] = \begin{cases} par(par(v)) & \text{if } U_v[i] = U_{par(v)}[i - w_v] + val_v, \\ d^1_{par(v)}[i - w_v] & \text{if } U_v[i] = D^1_{par(v)}[i - w_v] + val_v, \\ d^2_{par(v)}[i - w_v] & \text{if } U_v[i] = D^2_{par(v)}[i - w_v] + val_v. \end{cases} \quad (4)$$

Note that $u_v[i]$ could be a child of $par(v)$ or the parent of $par(v)$.

Lemma 2. *The upward entries U_v and u_v for all nodes v in a rooted tree T can be computed in $O(w_{\max}n)$ time.*

Proof. By Lemma 1, the downward entries of D_v^1 and D_v^2, together with d_v^1 and d_v^2, for all nodes v in T can be computed in $O(w_{\max}n)$ time. Therefore, according to Equations 3 and 4, $U_v[i]$ and $u_v[i]$ can be determined in $O(1)$ time for a fixed i. Thus U_v and u_v for all nodes v can be computed in $O(w_{\max}n)$ time.

Now we present an algorithm shown in Figure 1 for computing the density of a maximum-density path under the weight-constraint.

Algorithm MDP($T, r, w_{\min}, w_{\max}$)
Input: A tree T rooted at r with n nodes, and two weight bounds $w_{\min}$ and $w_{\max}$.
Output: The density of a weight-constrained maximum-density path in T.

1: Compute level(v)'s for all $v \in V(T)$
2: **for all** $v \in V(T)$ in decreasing order of level(v) **do**
3:   **if** v is a leaf **then**
4:     $D^1_v[w_v] \leftarrow val_v$
5:   **else**
6:     **for** $i \leftarrow w_v$ **to** $w_{\max}$ **do**
7:       $D^1_v[i] \leftarrow \max\{D^1_{v_j}[i - w_v] + val_v \mid v_j \text{ is a child of } v\}$
8:       $D^2_v[i] \leftarrow \text{2nd-max}\{D^1_{v_j}[i - w_v] + val_v \mid v_j \text{ is a child of } v\}$
9:       compte $d^1_v[i]$ and $d^2_v[i]$
10:     **end for**
11:   **end if**
12: **end for**
13: **for all** $v \in V(T)$ in increasing order of level(v) **do**
14:   **if** v is the root **then**
15:     $U_v[w_v] \leftarrow val_v$
16:   **else**
17:     **for** $i \leftarrow w_v$ **to** $w_{\max}$ **do**
18:       **if** v isn't the contributor of $D^1_{par(v)}[i - w_v]$ **then**
19:         $U_v[i] \leftarrow \max\{U_{par(v)}[i - w_v] + val_v, D^1_{par(v)}[i - w_v] + val_v\}$
20:         compute $u_v[i]$ according to Equation 4
21:       **else**
22:         $U_v[i] \leftarrow \max\{U_{par(v)}[i - w_v] + val_v, D^2_{par(v)}[i - w_v] + val_v\}$
23:         compute $u_v[i]$ according to Equation 4
24:       **end if**
25:     **end for**
26:   **end if**
27: **end for**
28: **return** $\max_{v \in V(T)} \{ \max_{\substack{w_{\min} \le i \le w_{\max} \\ i \neq 0}} \{ \frac{D^1_v[i]}{i}, \frac{U_v[i]}{i} \} \}$

Fig. 1. Computing the density of a weight-constrained maximum-density path on T

Lemma 3. *The density of a weight-constrained maximum-density path in a tree can be found in $O(w_{\max} n)$ time.*

Proof. Given a tree, we first transform it into a rooted tree T. It is not difficult to see that the density of a weight-constrained maximum-density path can be computed by running Algorithm MDP on T. According to Lemmas 1 and 2, all the U, D^1, D^2, u, d^1, d^2 entries can be computed in $O(w_{\max} n)$ time. Since the number of nodes of T is $O(n)$ and i is an integer between $w_{\min}$ and w_{max}, the density formula shown in Line 28 of Algorithm MDP can be computed in $O(w_{\max} n)$ time. Thus the result holds.

Based on the information obtained after executing Algorithm MDP, we can also actually find a maximum-density path in $O(n)$ time.

Theorem 1. *The weight-constrained maximum-density path problem on trees can be solved in $O(w_{\max} n)$ time.*

Proof. By Lemma 3 and an $O(n)$-time Algorithm RECONMDP, the result holds.

4 Finding a Weight-Constrained Maximum-Density Subtree

In this section, we first prove that the decision version of the weight-constrained maximum-density subtree problem is NP-complete. We then present an algorithm to find a weight-constrained maximum-density subtree in a tree in $O(w_{\max}{}^2 n)$ time.

4.1 The Decision Version is NP-Complete

We formally define the decision version of the weight-constrained maximum-density subtree problem as follows.

Definition 4 (decision version). *Given a tree T of n nodes in which each node v is associated with a value-weight pair (val_v, w_v), two integers $w_{\min}$ and $w_{\max}$, and a target-density d, the problem is to determine whether there exists a subtree T' with density no less than d and $w_{\min} \leq w(T') \leq w_{\max}$.*

A *star* is a tree consisting of one node (*central node*) adjacent to all the others (leaves). The n-node star is the complete bipartite graph $K_{1,n}$.

Theorem 2. *The decision version of the weight-constrained maximum-density subtree problem is NP-complete.*

Proof. First, the problem can be easily verified to be in NP.

Next, we give a polynomial-time reduction from a well-known NP-complete problem SUBSET-SUM to the problem. The *SUBSET-SUM problem* is defined as follows: Given a finite set $S \subset \mathbf{N}$ of n numbers $x_1, x_2, \ldots, x_n$ and a *target number* $t \in \mathbf{N}$, we want to determine whether there is a subset $S' \subseteq S$ whose elements sum to t.

The reduction converts the given set S to an n-node star $\mathcal{T}$ as follows:

1. Create a leaf v_i of $\mathcal{T}$ for each element x_i in S, and also create a central node c.
2. Assign a weight-value pair $(val_{v_i}, w_{v_i}) = (x_i, x_i)$ for each leaf v_i, and assign $(0, 0)$ for the central node c.
3. Let both $w_{\min}$ and $w_{\max}$ be the target number t, and let $d = 1$.

Clearly, this reduction can be done in polynomial time.

Claim. There is a subset S' of S whose elements sum to t if and only if the n-node star $\mathcal{T}$ has a subtree $\mathcal{T}'$ satisfying $d(\mathcal{T}') \geq d$ and $w_{\min} \leq w(\mathcal{T}') \leq w_{\max}$.

PROOF OF THE CLAIM: Suppose that $S' = \{x_{j_1}, x_{j_2} \ldots, x_{j_q}\}$ is a subset such that $\sum_{i=1,2,\ldots,q} x_{j_i} = t$. Then, the subtree of $\mathcal{T}$, which is formed by corresponding nodes of S' and central node c obviously has value t and weight t. Therefore, the density of the above subtree equals $1 (\geq d = 1)$.
Now suppose that the n-node star $\mathcal{T}$ has a subtree $\mathcal{T}'$ of density at least 1 and $w_{\min} = t \leq w(\mathcal{T}') \leq t = w_{\max}$. Therefore, $w(\mathcal{T}') = t$. Since the value and weight of each node of $\mathcal{T}$ are the same by the reduction, we have $d(\mathcal{T}') = \frac{\sum_{v \in V(\mathcal{T}')} val_v}{w(\mathcal{T}')} = \frac{t}{t} = 1$. Therefore, the corresponding elements of the leaves of $\mathcal{T}'$ form a subset S' of S whose elements sums to t. □

By Claim 4.1, we have a polynomial-time reduction from the SUBSET-SUM problem to the decision version of the weight-constrained maximum-density subtree problem. Therefore, the problem is NP-complete.

4.2 A Polynomial-Time Algorithm for Finding a Weight-Constrained Maximum-Density Subtree

Here we present an $O({w_{\max}}^2 n)$-time algorithm using the dynamic-programming technique to solve the weight-constrained maximum-density subtree problem on trees.

We first show that the problem has an optimal substructure property. Recall that T_v is a descendent-subtree of T rooted at v, and $deg(v)$ is the number of children of a node v in a rooted tree T. For an internal node $v \in V(T)$ with children $v_1, v_2, \ldots, v_{deg(v)}$, let $C_j^v[i]$ be the maximum total value of subtrees rooted at $v_1, v_2, \ldots, v_j$, whose weights sum to i. Also let $B_v[i]$ be the maximum-value of a subtree of T_v rooted at v, with weight i. We have the following result.

Lemma 4 (optimal substructure). *Suppose that T' is a weight-constrained maximum-density subtree of a rooted tree T, and r' is the root of T'. Let $child_{T'}(r') = \{v_1, v_2, \ldots, v_q\} (\subseteq child_T(r'))$. Then, $T' - T'_{v_j}$ is a maximum-density subtree of $T - T_{v_j}$, rooted at r', with the weight $w(T' - T'_{v_j})$, where $1 \leq j \leq q$.*

For a leaf v of T, we compute $B_v[i] = val_v$ if $i = w_v$; and $B_v[i] = -\infty$ for otherwise. For an internal node v with children $v_1, v_2, \ldots, v_{deg(v)}$, we first compute $C_j^v[0] = 0$ and $C_j^v[i]$ for $i = 1, \ldots, w_{\max} - w_v$ and $j = 1, 2, \ldots, deg(v)$, based on Lemmas ?? and 4 as follows:

$$C_j^v[i] = \begin{cases} \max\limits_{0 \leq w \leq i} \{C_{j-1}^v[i-w] + B_{v_j}[w]\} & \text{if } 1 < j \leq deg(v), \\ B_{v_1}[i] & \text{if } j = 1. \end{cases} \tag{5}$$

After computing $C_{deg(v)}^v[w_{\max} - w_v]$, we obtain the maximum value of the subtrees rooted at all the children of v, whose weights from 0 to $w_{\max} - w_v$. We then compute $B_v[i]$ where $i = 0, 1, \ldots, w_{\max}$ as follows:

$$B_v[i] = \begin{cases} C_{deg(v)}^v[i - w_v] + val_v & \text{if } w_v \leq i \leq w_{max}, \\ 0 & \text{if } i = 0, \\ -\infty & \text{otherwise.} \end{cases} \tag{6}$$

We also record the following information for constructing a target tree.

$$c_j^v[i] = \begin{cases} k & \text{if } C_j^v[i] = C_{j-1}^v[i-k] + B_j[k], \\ i & \text{if } j = 1 \text{ and } C_1^v \text{ exists.} \\ \text{NULL} & \text{otherwise.} \end{cases} \quad (7)$$

Lemma 5. *All the entries $B_v[i]$, $C_j^v[i]$ ($c_j^v[i]$) can be computed in $O({w_{\max}}^2 n)$ time, where n is the number of nodes of the given tree T.*

Proof. For fixed i and j, $C_j^v[i]$ can be determined using $O(i)$ comparisons. Since $0 \le i \le w_{\max} - w_v$, it takes $O({w_{\max}}^2)$ time to calculate $C_j^v[i]$. Therefore, computing $C_j^v[i]$'s for all internal nodes takes $O({w_{\max}}^2 n)$ time. According to Equation 7, all the $c_j^v[i]$'s can be determined in $O({w_{\max}}^2 n)$ time. By Equation 6, we can computed $B_v[i]$ from $C_{deg(v)}^v[i - w_v]$ in $O(w_{\max})$ time, for all $0 \le i \le w_{\max}$. Therefore, computing $B_v[i]$'s for all nodes v takes $O(w_{\max} n)$ time. Thus the result holds.

An algorithm for computing the density of a weight-constrained maximum-density subtree in a rooted tree is presented in Figure 2.

Lemma 6. *The density of a weight-constrained maximum-density subtree in a rooted tree can be computed in $O({w_{\max}}^2 n)$ time.*

Proof. The correctness of Algorithm MDT follows from Equations 5–7. The density of a weight-constrained maximum-density subtree can be computed using the formula described in Line 22. We next analyze the time complexity of the algorithm. Computing $level(v)$'s for all nodes v in T can be carried out in $O(n)$ time using the breath-first-search of T. Based on the counting sort to sort n integers of range $[0, n]$, the decreasing order of all the levels of T can be determined in $O(n)$ time. Moreover, according to Lemma 5, Lines 2–21 can be implemented to run in $O({w_{\max}}^2 n)$ time. In Line 22, computing the desired density takes $O((w_{\max} - w_{\min})n)$ time. Therefore, the result holds.

After executing Algorithm MDT on T, we can also actually find a weight-constrained maximum-density subtree in a rooted tree T can be found in $O(n)$ time.

Theorem 3. *The weight-constrained maximum-density subtree problem on trees can be solved in $O({w_{\max}}^2 n)$ time.*

Proof. We first transform a tree T into a rooted one. According to Lemmas 6, we can solve the problem in $O({w_{\max}}^2 n)$ time.

5 Finding a Weight-Constrained Maximum-Density Steiner Tree

In this section, we consider the weight-constrained maximum-density Steiner tree problem. Given a set of terminals $S \subset V$ of a rooted tree $T = (V, E)$, a

Algorithm MDT($T, r, w_{\min}, w_{\max}$)
Input: A tree T rooted at r with n nodes, and two weight bounds $w_{\min}$ and $w_{\max}$.
Output: The density of a weight-constrained maximum-density subtree in T.

```
compute level(v)'s for all v ∈ V(T)
for all v ∈ V(T) in decreasing order of level(v) do
  if v is a leaf node then
    B_v[w_v] = val_v
  else
    for j ← 1 to deg(v) do
      for i ← 0 to w_max − w_v do
        if j = 1 then
          C^v_1[i] ← B_{v_1}[i]
          compute c^v_1[i] according to Equation 7
        else
          C^v_j[i] ← max_{0≤w≤i} {C^v_{j−1}[i − w] + B_{v_j}[w]}
          compute c^v_j[i] according to Equation 7
        end if
      end for
    end for
    for i ← w_v to w_max do
      B_v[i] ← C^v_{deg(v)}[i − w_v] + val_v
    end for
  end if
end for
return max_{v∈V(T)} { max_{w_min≤i≤w_max, i≠0} { B_v[i] / i } }
```

Fig. 2. An algorithm for computing the density of a weight-constrained maximum-density subtree in a rooted tree T

set C_S is a *connecting set of* S if $S \subseteq C_S$ and the subgraph (i.e., subtree) of T induced by C_S is connected. The set C_S is further said to be *minimal* if $C_S - v$ is not a connecting set of S for any node $v \in C_S$. Note that if the subtree of T induced by S is connected, then $C_S = S$. For a connecting set C_S of a rooted tree $T = (V, E)$, *contracting* C_S *to a supernode* s in T is to obtain another tree $T' = (V - C_S + s, E')$, where E' is obtained by deleting $\{(x, y) | \ x \in C_S \text{ or } y \in C_S\}$ from E and adding one new edge (x, s) for each edge (x, y), where $x \notin C_S$ and $y \in C_S$. Assume that $Q = \{(x, y) | \ x \notin C_S \text{ and } y \in C_S\}$. There are totally $|Q|$ created edges.

The key concept of our algorithm is first to root the given tree on an arbitrary node v in terminal set S. We then find a minimal connecting set C_S, and contract it to a super-node s. This can be implemented in $O(n)$ time. Note that s is the root of the resulting tree T'. The value-weight pair associated with s is $(val_s, w_s) = (\sum_{v \in C_S} val_v, \sum_{v \in C_S} w_v)$. By executing Algorithm MDT on T', we can find a weight-constrained maximum-density subtree rooted at s, denoted by

T^s. A weight-constrained maximum-density Steiner tree is thus the subtree of T induced by $V(T^s) - s + \mathcal{C}_S$.

Theorem 4. *The weight-constrained maximum-density Steiner tree problem on trees can be solved in* $O({w_{\max}}^2 n)$ *time.*

References

1. K. M. Chung and H. I. Lu, "An optimal algorithm for the maximum-density segment problem," *SIAM Journal on Computing*, 34(2):373–387, 2004.
2. M. H. Goldwasser, M. Y. Kao, and H. I. Lu, "Fast algorithms for finding maximum-density segments of a sequence with applications to bioinformatics," *Proceedings of the Second International Workshop of Algorithms in Bioinformatics*, Lecture Notes in Computer Science 2452, pp. 157–171, Rome, Italy, 2002, Springer-Verlag.
3. X. Huang, "An algorithm for identifying regions of a DNA sequence that satisfy a content requirement," *Computer Applications in the Biosciences*, 10(3):219–225, 1994.
4. R. B. Inman, "A denaturation map of the 1 phage DNA molecule determined by electron microscopy," *Journal of Molecular Biology*, 18:464–476, 1966.
5. S. K. Kim, "Linear-time algorithm for finding a maximum-density segment of a sequence," *Information Processing Letters*, 86(6):339–342, 2003.
6. Y. L. Lin, X. Huang, T. Jiang, and K. M. Chao, "MAVG: locating non-overlapping maximum average segments in a given sequence," *Bioinformatics*, 19(1):151–152, 2003.
7. Y. L. Lin, T. Jiang, and K. M. Chao, "Algorithms for locating the length-constrained heaviest segments, with aplications to biomolecular sequences analysis," *Journal of Computer and System Sciences*, 65(3):570–586, 2002.
8. R. R. Lin, W. H. Kuo, and K. M. Chao, "Finding a length-constrained maximum-density path in a tree," *Journal of Combinatorial Optimization*, 9(2):147–156, 2005.
9. G. Macaya, J. P. Thiery, and G. Bernardi, "An approach to the organization of eukaryotic genomes at a macromolecular level," *Journal of Molecular Biology*, 108:237-254, 1976.
10. A. Nekrutenko and W. H. Li, "Assesment of compositional heterogeneity within and between eukaryotic genomes," *Genome Research*, 10:1986-1995, 2000.
11. P. Rice, I. Longden, and A. Bleasby, "EMBOSS: The European molecular biology open software suite," *Trends in Genetics*, 16(6):276-277, 2000.
12. N. Stojanovic, L. Florea, C. Riemer, D. Gumucio, J. Slightom, M. Goodman, W. Miller, and R. Hardison, "Comparison of five methods for finding conserved sequences in multiple alignments of gene regulatory regions," *Nucleic Acids Research*, 27:3899-3910, 1999.
13. D. B. West, *Introduction to Graph Theory*, 2nd ed, Prentice Hall, Upper Saddle River, NJ, 2001.
14. B. Y. Wu, K. M. Chao, and C. Y. Tang, "An efficient algorithm for the length-constrained heaviest path problem on a tree," *Information Processing Letters*, 69:63-67, 1999.

Finding Two Disjoint Paths in a Network with Normalized α^+-MIN-SUM Objective Function

Bing Yang[1], S.Q. Zheng[1], and Enyue Lu[2]

[1] Department of Computer Science, University of Texas at Dallas, Richardson, TX 75083-0688, USA
binyang@cisco.com, sizheng@utdallas.edu

[2] Mathematics and Computer Science Department, Salisbury University, Salisbury, MD 21801, USA
ealu@salisbury.edu

Abstract. Given a number α with $0 < \alpha < 1$, a network $G = (V, E)$ and two nodes s and t in G, we consider the problem of finding two disjoint paths P_1 and P_2 from s to t such that $length(P1) \leq length(P_2)$ and $length(P_1) + \alpha \cdot length(P_2)$ is minimized. The paths may be node-disjoint or edge-disjoint, and the network may be directed or undirected. This problem has applications in reliable communication. We prove an approximation ratio $\frac{1+\alpha}{2\alpha}$ for all four versions of this problem, and also show that this ratio cannot be improved for the two directed versions unless P = NP. We also present Integer Linear Programming formulations for all four versions of this problem. For a special case of this problem, we give a polynomial-time algorithm for finding optimal solutions.

1 Introduction

A reliable telecommunication network, which is modeled by a graph $G = (V, E)$, is designed in such a way that multiple connections exist between every pair of communicating nodes. Usually, paths are selected according to an objective function. Each edge e in G is assigned a nonnegative length $l(e)$, which reflects the resource and/or performance associated with the edge, such as cost, distance, latency, etc. We use $l(P)$ to denote the length of path P. To avoid single point of failure, the paths may be node-disjoint or edge disjoint, and the network may be directed or undirected. Thus, a problem of finding disjoint paths has four versions, namely

- node-disjoint paths in directed graphs (ND-D multiple-path problem);
- edge-disjoint paths in directed graphs (ED-D multiple-path problem);
- node-disjoint paths in undirected graphs (ND-UD multiple-path problem);
- edge-disjoint paths in undirected graphs (ED-UD multiple-path problem).

Various problems of finding optimized disjoint paths between two nodes s and t in G have been investigated. Ford and Fulkson gave a polynomial-time algorithm for finding two paths with minimum total length (called *MIN-SUM 2-Path Problem*), using the algorithm of finding minimum weighted network flows

X. Deng and D. Du (Eds.): ISAAC 2005, LNCS 3827, pp. 954–963, 2005.

[3] for all four versions. Suurballe and Tarjan provided different treatment, and presented algorithms that are more efficient [11] [12]. Li *et al.* proved that all four versions of the problem of finding two disjoint paths such that the length of the longer path is minimized (called the *MIN-MAX 2-Path Problem*) are strongly NP-complete [4]. They also considered a generalized MIN-SUM problem (which we call the *G-MIN-SUM k-Path Problem*) assuming that each edge is associated with k different lengths. The objective of this problem is to find k disjoint paths such that the total length of the paths is minimized, where the jth edge-length is associated with the jth path. They showed that all four versions of the G-MIN-SUM k-path problem are strongly NP-complete even for $k = 2$ [5].

In [7], we considered the problem of finding two disjoint paths such that the length of the shorter path is minimized (*MIN-MIN 2-Path Problem*). We showed that all four versions of the MIN-MIN 2-path problem are strongly NP-complete[7]. In the same paper, we also showed there does not exist any polynomial-time approximation algorithm with a constant approximation ratio for any of these four versions of the MIN-MIN 2-path problem unless $P = NP$. In this paper we consider a generalized weighted 2-path problem. Let P_1 and P_2 be two disjoint paths from s to t in a given graph G, and α a non-negative value. Define

$$L_\alpha(P_1, P_2) = l(P_1) + \alpha \cdot l(P_2).$$

Our objective is to find two disjoint paths P_1 and P_2 such that $L_\alpha(P_1, P_2)$ is minimized. We call this problem the *α-MIN -SUM 2-Path Problem.* According to the relative values of P_1 and P_2, this problem can be treated as having two versions. One is to minimize

$$L_{\alpha^-}(P_1, P_2) = \max\{l(P_1), l(P_2)\} + \alpha \cdot \min\{l(P_1), l(P_2)\},$$

and the other is to minimize

$$L_{\alpha^+}(P_1, P_2) = \min\{l(P_1), l(P_2)\} + \alpha \cdot \max\{l(P_1), l(P_2)\}.$$

We name the former as the *α^--MIN-SUM 2-path problem*, and the latter as the *α^+-MIN-SUM 2-path problem*. It is clear that if $\alpha = 0$, the α^--MIN-SUM 2-path problem, and the α^+-MIN-SUM 2-path problem degenerates to the MIN-MAX 2-path problem and MIN-MIN 2-path problem, respectively, which are NP-complete for all four versions. But if $\alpha = 1$, both degenerate to the MIN-SUM 2-path problem, which is polynomial-time solvable. Investigation on the α-MIN-SUM 2-path problem is of theoretical interest: what are the computational complexities of the α-MIN-SUM 2-path problem with $0 < \alpha < 1$? Since

$$l(P_1) + \alpha \cdot l(P_2) = \alpha \cdot \left(\frac{1}{\alpha} \cdot l(P_1) + l(P_2) \right),$$

an α-MIN-SUM problem with $\alpha \geq 1$ can be transformed into a problem with $\alpha \leq 1$. Hence, it is sufficient to only consider the problem with $0 < \alpha \leq 1$. Hence, it is sufficient to only consider the problem with $0 < \alpha \leq 1$. We call

the α^--MIN-SUM 2-path problem (resp. α^+-MIN-SUM 2-path problem) with $0 \leq \alpha \leq 1$ the *normalized α^--MIN-SUM 2-path problem* (resp. *normalized α^+-MIN-SUM 2-path problem*). The relationship among these 2-path problems is shown in Figure 1.

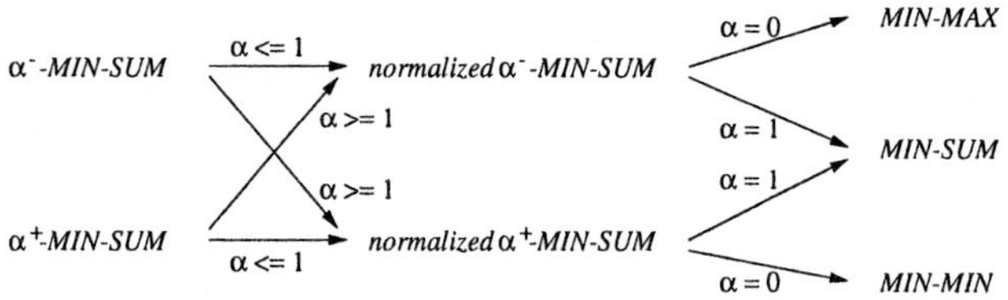

Fig. 1. Relations among various 2-path problems

In [4], Li *et al.* showed that the ED-D and ND-D versions of the normalized α^+-MIN-SUM are NP-complete. In [8], we showed that all four versions of the normalized α^--MIN-SUM 2-path problem are NP-complete. We also showed that an approximation ratio $\frac{2}{1+\alpha}$ can be achieved for all four versions of the normalized α^--MIN-SUM 2-path problem by using MIN-SUM 2-path solutions to approximate normalized α^--MIN-SUM solutions, and proved that for directed graphs there does not exist any polynomial-time approximation algorithm guaranteeing an approximation ratio smaller than $\frac{2}{1+\alpha}$ unless $P = NP$.

In this paper, we show that an approximation ratio $\frac{1+\alpha}{2\alpha}$ can be achieved for all four versions of the normalized α^+-MIN-SUM 2-path problem, and prove that for directed graphs there does not exist any polynomial-time approximation algorithm guaranteeing an smaller approximation ratio. We also present Integer Linear Programming formulations for finding optimal solutions for all four versions of the normalized α^+-MIN-SUM 2-path problem. For special cases of the ND-D and ED-D versions on acyclic graphs, we provide a polynomial-time algorithm for finding optimal solutions.

2 Approximation Analysis

The ND-D and ED-D versions of the normalized α^+-MIN-SUM 2-path problem have been proved to be NP-complete by Li *et al.* in [4]. We conjecture that the ND-UD and ED-UD versions of the normalized α^+-MIN-SUM 2-path problem are also NP-complete, but we have not been able to provide a proof.

In this section, we consider the problem of finding approximation solutions of the normalized a^+-MIN-SUM 2-path problem. We say that an instance of the normalized α^+-MIN-SUM 2-path problem is feasible if for which a feasible solution exists. We say that there exists an approximation algorithm with approximation ratio for the normalized α^+-MIN-SUM 2-path problem $\mathbf{\Pi}$ if there exists a polynomial-time algorithm $\mathcal{A}$ such that for the feasible instance space I of $\mathbf{\Pi}$

$$\frac{l(P_1) + \alpha \cdot l(P_2)}{l(P_1^*) + \alpha \cdot l(P_2^*)} \le E, \tag{1}$$

where (P_1, P_2) is the solution produced by algorithm $\mathcal{A}$, (P_1^*, P_2^*) is an optimal solution, and E is an constant. We have the following results.

Lemma 1. *For any* $0 < \beta < \gamma < 1$, *let* (P_1^β, P_2^β) *and* (P_1^γ, P_2^γ) *denote any optimal solution for the normalized* α^+*-MIN-SUM 2-path problem with* $\alpha = \beta$ *and* $\alpha = \gamma$, *respectively. Then,* $\frac{l(P_1^\gamma)+\beta \cdot l(P_2^\gamma)}{l(P_1^\beta)+\beta \cdot l(P_2^\beta)} \le \frac{\gamma(1+\beta)}{\beta(1+\gamma)}$.

Theorem 1. *The optimal solution for the MIN-SUM 2-path problem corresponding to the normalized* α^+*-MIN-SUM 2-path problem is an approximation solution to the normalized* α^+*-MIN-SUM 2-path problem with an approximation ratio* $\frac{1+\alpha}{2\alpha}$.

Theorem 2. *For the two directed versions of the normalized* α^+*-MIN-SUM 2-path problem, if there exists a polynomial-time approximation algorithm with an approximation ratio smaller than* $\frac{1+\alpha}{2\alpha}$, *then* $P = NP$.

Due to page limit, we omit formal proofs in the entire paper.

3 Integer Programming Solutions

The approximation ratio $\frac{1+\alpha}{2\alpha}$, which cannot be improved for ND-D and ED-G versions of the normalized α^+-MIN-SUM 2-path problem assuming $P \neq NP$, can be very large when α becomes small. When the input graph is not so large, and there is a need to find an optimal solution, integer linear programming (ILP) may be useful. In this section, we formulate all versions of the normalized α^+-MIN-SUM 2-path problem as ILP problems. Though these ILP problems, which belong to the class of minimum cost unsplittable flow problems (defined by Kleinberg [13]), are also NP-complete in general, some of their instances may be practically solvable using commercially available software packages with speed-up heuristics (e.g. [14][15]).

3.1 Edge-Disjoint Paths in Directed Graphs

We use network flow approach to solve the normalized α^+-MIN-SUM 2-path problem. For easy presentation, we use c_i to denote the length of edge e_i; i.e. $c_i = l(e_i)$ in this section.

Linear Programming Formulation of MIN-SUM 2-Path Problem. Some of network path problems are solvable using linear programming (LP), and the MIN-SUM 2-path problem is one of such problems. This is a classical formulation [9]. Let $P = (e_{j_1} \cdots e_{j_k})$ be a path from s to t in G. The cost of P is:

$$c(P) = \sum_{i=1}^{k} c_{j_i}$$

For a directed graph G of n nodes and m edges, we define its $n \times m$ node-edge incidence matrix $A = [a_{ij}]$ by

$$a_{ij} = \begin{cases} +1 & \text{if directed edge (arc) } e_j \text{ is incident out of node } i \\ -1 & \text{if directed edge (arc) } e_j \text{ is incident into node } i \\ 0 & \text{otherwise} \end{cases}$$

Let $\mathbf{c}$ and $\mathbf{f}$ be the vectors of the cost and the flow on the edges. Then an LP formulation for the MIN-SUM 2-path problem is given as follows:

$$\begin{cases} \min \ z = \mathbf{c}^T \mathbf{f} \\ \mathbf{Af} = \begin{bmatrix} +2 \\ -2 \\ 0 \\ 0 \\ 0 \\ 0 \end{bmatrix} \\ 0 \le f_i, 1 \le i \le m \end{cases} \tag{2}$$

Integer Linear Programming Formulation for the Normalized α^+-MIN-SUM 2-Path Problem with $\alpha = \frac{1}{2}$. We intend to use a similar method as Equation (2) to solve the normalized α^+-MIN-SUM 2-path problem. We introduce the following constraints as the first requirement that ensures the flows on two edge-disjoint paths to be 1 and 2, respectively:

$$\begin{cases} Af = \begin{bmatrix} +3 \\ -3 \\ 0 \\ \vdots \\ 0 \end{bmatrix} \\ f_i \le 2 \\ f_i \in Z, 1 \le i \le m \end{cases} \tag{3}$$

Z is the set of non-negative integers. In addition, we need to prevent the cases of path splicing and path splitting. We introduce a column vector $\mathbf{g}$ to characterize the number of inputs on nodes, and add following constraints:

$$\begin{cases} Ag = \begin{bmatrix} +2 \\ -2 \\ 0 \\ \vdots \\ 0 \end{bmatrix} \\ g_i \le 1 \\ g_i \in Z, 1 \le i \le m \end{cases} \tag{4}$$

Then, minimizing the total cost of $\min\{P_1, P_2\} + \frac{1}{2}\max\{P_1, P_2\}$ is enforced by using flow ratio $1:2$. However, as $\mathbf{g}$ defines two disjoint paths, we want to further

enforce the paths given by **f** are exactly the same as those given by **g**. It can be easily done by adding

$$\begin{cases} f_i - g_i \ge 0 \\ 2g_i - f_i \ge 0 \end{cases}$$

Combining all above constraints, we have an ILP formulation

$$\begin{cases} \min\ z = \mathbf{c}^T\mathbf{f} \\ \mathbf{Af} = \begin{bmatrix} +3 \\ -3 \\ 0 \\ \vdots \\ 0 \end{bmatrix}, \quad \mathbf{Ag} = \begin{bmatrix} +2 \\ -2 \\ 0 \\ \vdots \\ 0 \end{bmatrix} \\ f_i - g_i \ge 0,\ 2g_i - f_i \ge 0,\ g_i \le 1 \\ f_i, g_i \in Z, 1 \le i \le m \end{cases} \tag{5}$$

Theorem 3. *For $\alpha = \frac{1}{2}$, we can solve the normalized ED-D α^+-MIN-SUM 2-path problem by solving the corresponding ILP problem formulated by Equation (5). The two disjoint paths P_1^* and P_2^* consist of the edges with flow 2 and edges with flow 1, respectively, in the optimal solution of the ILP. Its actual optimal value is $\frac{1}{2}$ of the optimal value of the corresponding ILP solution.*

Integer Linear Programming Formulation for the Normalized α^+-MIN-SUM 2-Path Problem with $\alpha = \frac{1}{M}$. Now we consider a more general case that $\alpha = \frac{1}{M}$, with M being an integer greater than 1. We use the same technique as previous section to tackle the problem: let first path have flow value M and the second path have flow value 1. Then, we have the following ILP formulation:

$$\begin{cases} \min\ z = \mathbf{c}^T\mathbf{f} \\ \mathbf{Af} = \begin{bmatrix} +(M+1) \\ -(M+1) \\ 0 \\ \vdots \\ 0 \end{bmatrix}, \quad \mathbf{Ag} = \begin{bmatrix} +2 \\ -2 \\ 0 \\ \vdots \\ 0 \end{bmatrix} \\ f_i - g_i \ge 0,\ Mg_i - f_i \ge 0,\ g_i \le 1 \\ f_i, g_i \in Z, 1 \le i \le m \end{cases} \tag{6}$$

Theorem 4. *For $\alpha = \frac{1}{M}$, we can solve the normalized ED-D α^+-Min-Sum 2-path problem by solving the corresponding ILP problem formulated by Equation (6). Its actual optimal value is $\frac{1}{M}$ of the optimal value of the corresponding ILP solution. The two paths can be constructed from the solution of Equation (6) by the following operations:*

(i) *Remove all the edges with flow 0 from the graph.*
(ii) *Find the shortest path from s to t in the remaining graph, which is P_1^*.*
(iii) *The second path P_2^* consists of all the remaining edges.*

Integer Linear Programming Formulation for the Normalized α^+-MIN-SUM 2-Path Problem with $\alpha = \frac{N}{M}$. Now we consider the situations that α is a rational number; i.e. $\alpha = \frac{N}{M}$ with N and M being integers and $N < M$. The idea is to find two disjoint paths such that the first carries flow M, and the second carries flow N. This can be reduced to the problem with $\alpha = \frac{1}{M}$ as follows: (a) Let $X = N - 1$ and $Y = M - (N - 1)$, and find 2 disjoint paths with flows Y and 1; and (b) add flow X to both paths. The corresponding ILP formulation is:

$$\begin{cases} \min\ z = X\mathbf{c}^T\mathbf{g} + \mathbf{c}^T\mathbf{f} \\ \mathbf{Af} = \begin{bmatrix} +(Y+1) \\ -(Y+1) \\ 0 \\ \vdots \\ 0 \end{bmatrix}, \quad \mathbf{Ag} = \begin{bmatrix} +2 \\ -2 \\ 0 \\ \vdots \\ 0 \end{bmatrix} \\ f_i - g_i \geq 0,\ Yg_i - f_i \geq 0,\ g_i \leq 1 \\ f_i, g_i \in Z, 1 \leq i \leq m \end{cases} \tag{7}$$

Theorem 5. *If Equation (7) has an optimal solution $z^* = \min z$, then the normalized ED-D α^+-MIN-SUM 2-path problem with $\alpha = \frac{1}{M}$ has an optimal solution with total length $\frac{1}{M}z^*$, and the two paths can be constructed from the solution of Equation (6) by the following operations:*

(i) *Remove all the edges with flow 0 from the graph.*
(ii) *Find the shortest path from s to t in the remaining graph, which is P_1^*.*
(iii) *The second path P_2^* consists of all the remaining edges.*

Integer Linear Programming for the normalized α^+-MIN-SUM 2-Path Problem with Irrational α. For the case with α being an irrational number, we could not use our previous technique. For all practical purposes, however, we do not need to worry about this for the following two reasons:

(i) All the numbers represented in computer systems are rational numbers.
(ii) It is easy to use rational numbers to approximate irrational numbers.

3.2 ND-D, ED-UD and ND-UD Versions

We can reduce the ND-D version of the normalized α^+-MIN-SUM 2-path problem to the ED-D version using the well-known node-splitting technique.

Given a directed graph G, we obtain a directed graph G' as follows: replace each node v (except s and t) by two nodes v' and v'' such that all edges terminating at v are now terminating at v', and all edges originating from v are now originating from v''; in addition, there add exactly one edge from v' to v'' (see Figure 2). Clearly, the problem of finding node-disjoint paths in G is equivalent to the problem of finding edge-disjoint paths in G'. Then we can apply the methods discussed in previous section by assigning cost 0 and capacity ∞ to all newly introduced edges (due to node-splitting).

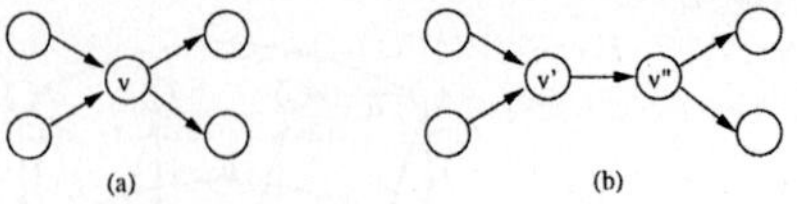

Fig. 2. (a) Before node split. (b) After node split.

The ED-UD and ND-UD versions of the normalized α^+-MIN-SUM 2-path problem can be easily reduced to the ED-D and ND-D versions by converting an undirected graph G to a directed graph G', with each undirected edge in G replaced by two directed edges with opposite directions.

4 Polynomial-Time Algorithm for Acyclic Directed Case

For the special case of the normalized α^+-MIN-SUM 2-path problem on acyclic directed graphs, there exists a polynomial-time algorithm to find optimal solutions. The algorithm is obtained by extending Perl and Shiloach's algorithm for finding two disjoint paths between two sources and two destinations on acyclic directed graphs [9].

For the acyclic ND-D version, we adopt a technique used in a pseudo-polynomial time algorithm of [4]. Given an acyclic directed graph $G = (V, E)$ and source s and destination t, we can relabel nodes with number 1 to $|V|$ to ensure that any edge $u \to v$ in E satisfies $u < v$, $s = 1$ and $t = |V|$ [9] (it is assumed that $s \to t \notin E$; otherwise we can add a node u and replace $s \to t$ by $s \to u$ and $u \to t$). After the relabeling, we can transform graph G to an acyclic directed graph $\overline{G} = (\overline{V}, \overline{E})$, whose nodes are arranged as a $|V| \times |V|$ array, as follows:

$$\overline{V} = \{\langle u, v\rangle | u, v \in V, \text{ and } u \neq v \text{ unless } u = v = s \text{ or } u = v = t\}$$
$$\overline{E} = \{\langle u, v\rangle \to \langle u, w\rangle | v \to w \in E \text{ and } v \leq u\}$$
$$\cup\{\langle v, u\rangle \to \langle w, u\rangle | v \to w \in E \text{ and } v \leq u\}$$

Assign the lengths to edges in $\overline{G}$ as follows:
$l(\langle u, v\rangle \to \langle u, w\rangle) = \alpha \cdot l(v \to w)$, and
$l(\langle v, u\rangle \to \langle w, u\rangle) = l(v \to w)$.

Figure 3 shows an example of G and its corresponding $\overline{G}$.

We call x and y of a node $\langle x, y\rangle$ in $\overline{G}$ the first and second label of the node, respectively. Given a path $\overline{P} = \langle u_1, v_1\rangle \to \langle u_2, v_2\rangle \to \cdots \to \langle u_m, v_m\rangle$ from $\langle 1, 1\rangle$ to $\langle |V|, |V|\rangle$ in $\overline{G}$, let $H(\overline{P})$ and $V(\overline{P})$ be the set of horizontal and vertical edges in $\overline{P}$ respectively. Define $V_1(\overline{P}) = (u_{i,1}, u_{i,2}, \cdots, u_{i,k})$ (resp. $V_2(\overline{P}) = (v_{j,1}, v_{j,2}, \cdots, v_{j,k'})$) as the sequence of distinct first (resp. second) labels of nodes in $V(\overline{P})$ (resp. $H(\overline{P})$). By a straightforward extension of the results of [9], we know that there exist two node-disjoint paths $P_1 = u_{i,1} \to u_{i,2} \to \cdots \to u_{i,k}$ and $P_2 = v_{j,1} \to v_{j,2} \to \cdots \to v_{j,k'}$ from s to t (from 1 to $|V|$ after relabeling) in G if and only if there exists a path $\overline{P}$ from $\langle 1, 1\rangle$ to $\langle |V|, |V|\rangle$ such that $V_1(\overline{P}) = u_{i,1} \to u_{i,2} \to \cdots \to u_{i,k}$ and $V_2(\overline{P}) = v_{j,1} \to v_{j,2} \to \cdots \to v_{j,k'}$. That is, P_1 and P_2 correspond to the vertical edges and horizontal edges in

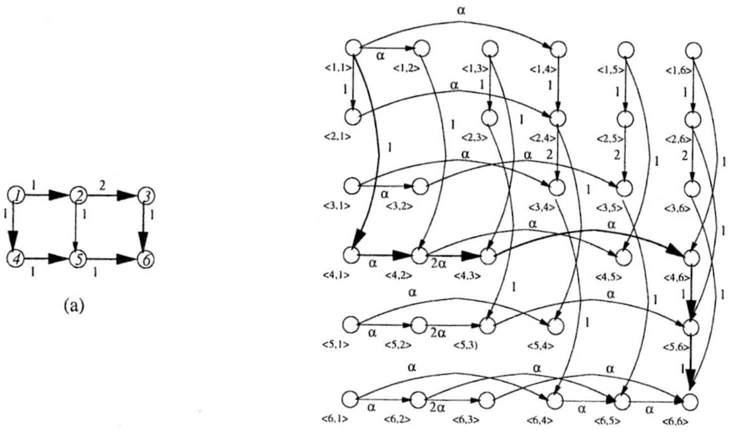

(a) (b)

Fig. 3. (a) An acyclic graph G. (b) Graph $\overline{G}$ of (a)

$\overline{P}$, respectively. In the example of Figure 3, there are two node-disjoint paths $P_1 = 1 \rightarrow 4 \rightarrow 5 \rightarrow 6$ and $P_2 = 1 \rightarrow 2 \rightarrow 3 \rightarrow 6$ in G of (a). The corresponding path in $\overline{G}$ is $\overline{P} = \langle 1,1\rangle \rightarrow \langle 4,1\rangle \rightarrow \langle 4,2\rangle \rightarrow \langle 4,3\rangle \rightarrow \langle 4,6\rangle \rightarrow \langle 5,6\rangle \rightarrow \langle 6,6\rangle$, as shown in (b).

Let $\overline{P}^*$ be a shortest path from $\langle 1,1\rangle$ to $\langle |V|,|V|\rangle$ in $\overline{G}$. By the edge length assignment of $\overline{G}$, the two paths P_1 and P_2 corresponding to $V_1(\overline{P}^*) = (u_{i,1}, u_{i,2}, \cdots, u_{i,k})$ and $V_2(\overline{P}^*) = (v_{j,1}, v_{j,2}, \cdots, v_{j,k'})$, respectively, form an optimal node-disjoint solution, and furthermore, $l(P_1) \leq l(P_2)$. Based on these discussions, we have the following 3-step algorithm:

1. Transform the given acyclic directed graph G into its corresponding graph $\overline{G}$, with all its edges assigned lengths as defined above.
2. Use Dijkstra's shortest-path algorithm to find a shortest path $\overline{P}$ from $\langle s,s\rangle$ to $\langle t,t\rangle$ in $\overline{G}$. If such a path does not exist, then two node-disjoint paths do not exist in G.
3. The sets of vertical edges and the horizontal edges in $\overline{P}$ correspond to two node-disjoint paths from s to t in G, which is a optimal solution.

Clearly, step 1) can be done in $O(|V|\cdot|E|)$ time, step 2) can be done in $O(|V|^4)$ time, and step 3) can be done in $O(|V|)$.

For the acyclic ED-D version, we directly apply a technique of [4] that transforms the acyclic edge-disjoint case to the acyclic node-disjoint case, and then apply the algorithm presented in the previous section. First, we transform the given graph $G = (V,E)$ to a corresponding directed line-graph $\overline{G}$ [1], which contains $O(|E|)$ nodes and $O(|V|)$ edges. This can be done in $O(|E|)$ time. Then, finding two node-disjoint paths in $\overline{G}$ can be done in $O(|E|^4)$ time. For details, refer to [4].

5 Concluding Remarks

A few problems remain open. Are the ED-UD and ND-UD versions of the normalized α^+-MIN-SUM 2-path problem also NP-complete? If the answer is affirmative, is the approximation ratio $\frac{1+\alpha}{2\alpha}$ also best possible? For what other special graphs polynomial-time algorithms exist?

References

1. F. Harary, *Graph Theory*, Addison-Wesley, 1972.
2. S. Even, *Graph Algorithms*, Computer Science (1979).
3. L.R. Ford and D.R. Fulkerson, *Flows in Networks*, Princeton, 1962.
4. C. L. Li, S. T. McCormick, and D. Simchi-Levi, "The Complexity of Finding Two Disjoint Paths with Min-Max Objective Function", *Discrete Appl. Math.* 26(1) (1990), 105-115.
5. C. L. Li, S. T. McCormick, and D. Simchi-Levi, "Finding Disjoint Paths with Different Path-Costs: Complexity and Algorithms", *Networks*, 22 (1992), 653-667.
6. S. Fortune, J. Hopcroft, and J. Wyllie, "The directed subgraph homeomorphism problem", *Theoret. Comput. Sci.*, 10 (1980), 111-121.
7. B. Yang, S.Q. Zheng and S. Katukam, "Finding Two Disjoint Paths in a Network with Min-Min Objective Function", *Proceedings of the 15th IASTED International Conference on Parallel and Distributed Computing and Systems*, (2003), 75-80.
8. B. Yang, S.Q. Zheng and E. Lu, "Finding Two Disjoint Paths in a Network with Normalized α^--Min-Sum Objective Function", Manuscript.
9. Y. Perl and Y. Shiloach, "Finding Two Disjoint Paths between Two Pairs of Vertices in a Graph", *J. ACM*, 25(1) (1978), 1-9.
10. Y. Shiloach, A polynomial solution to the undirected two paths problem. J. ACM 27(3) (1980) 445-456.
11. J. W. Suurballe, "Disjoint Paths in a Network", *Networks* 4 (1974), 125-145.
12. J. W. Suurballe and R. E. Tarjan, "A Quick Method for Finding Shortest Pairs of Disjoint Paths", *Networks* 14 (1984), 325-336.
13. J. M. Kleinberg, "Single-source unsplittable flow.", *Proceedings of the 37th Annual Symposium on foundations of Computer Science* (1996) 68-77.
14. F. Fallah, S. Liao, S. Devadas, "Solving Covering Problems Using LPR-Based Lower Bounds", *IEEE Transactions on Very Large Scale Integration (VLSI) Systems* 8(1) (2000), 9-17.
15. F. Fallah, S. Devadas, K. Keutzer, "Functional Vector Generation for HDL Models Using Linear Programming and Boolean Satisfiability", *IEEE Transactions on computer Aided Design of Integrated Circuits and Systems*, 20(8) (2001), 994-1002.

Sensitivity Analysis of Minimum Spanning Trees in Sub-inverse-Ackermann Time

Seth Pettie

Max Planck Institut für Informatik

Abstract. We present a deterministic algorithm for computing the *sensitivity* of a minimum spanning tree or shortest path tree in $O(m \log \alpha(m,n))$ time, where α is the inverse-Ackermann function. This improves upon a long standing bound of $O(m\alpha(m,n))$ established by Tarjan. Our algorithms are based on an efficient *split-findmin* data structure, which maintains a collection of sequences of weighted elements that may be split into smaller subsequences. As far as we are aware, our split-findmin algorithm is the first with superlinear but sub-inverse-Ackermann complexity.

1 Introduction

Split-findmin is a little known but key data structure in modern graph optimization algorithms. It was originally designed for use in the weighted matching and undirected all-pairs shortest path algorithms of Gabow and Tarjan [9, 12] and has since been rediscovered as a critical component of the hierarchy-based shortest path algorithms of Thorup [33], Hagerup [15], Pettie-Ramachandran [28], and Pettie [24, 22, 23]. In this paper we apply split-findmin to the problem of performing *sensitivity analysis* on minimum spanning trees (MST) and shortest path trees. The MST sensitivity analysis problem is, given a graph G and minimum spanning tree $T = \mathrm{MST}(G)$, to decide how much each individual edge weight can be perturbed without invalidating the identity $T = \mathrm{MST}(G)$.

A twenty year old result of Tarjan [32] shows that MST sensitivity analysis can be solved in $O(m\alpha(m,n))$ time, where m and n are the number of edges and vertices and α the inverse-Ackermann function. Furthermore, he showed that single-source shortest path sensitivity analysis can be reduced to MST sensitivity analysis in linear time. Tarjan's algorithm has not seen any unqualified improvements, though Dixon et al. [7] did present two MST sensitivity algorithms, one running in *expected* linear time and another which is deterministic and provably optimal, but whose complexity is only known to be $O(m\alpha(m,n))$.

In this paper we present a simpler and faster MST sensitivity analysis algorithm running in time $O(m \log \alpha(m,n))$. Given the notoriously slow growth of the inverse-Ackermann function, an improvement on the order of $\alpha/\log\alpha$ is unlikely to have a devastating real-world impact. Although our algorithm is substantially simpler than that of [7] and may very well be empirically faster than the competition, its real significance has little to do with practical issues, nor does it have much to do with the sensitivity problem as such. As one may observe

X. Deng and D. Du (Eds.): ISAAC 2005, LNCS 3827, pp. 964–973, 2005.

in Figure 1, the MST sensitivity analysis problem is related, via a tangled web of reductions, to many fundamental algorithmic and data structuring problems. Among those depicted in Figure 1, the two most important unsolved problems are *set maxima* and the *minimum spanning tree* problem. MST sensitivity can be expressed as a set maxima problem. In this paper we show that MST sensitivity is reducible to *both* the minimum spanning tree problem itself and the split-findmin data structure. These connections suggest that it is impossible to solve such important optimization problems as minimum spanning trees and single-source shortest paths without first understanding *why* a manifestly simpler problem like MST sensitivity still eludes us. We view the MST sensitivity analysis problem as an ideal test bed for experimenting with possible approaches to solving its more difficult, high profile cousins.

Organization. In Section 1.1 we define all the MST related problems and data structures mentioned in Figure 1. In Section 2 we define the split-findmin data structure and give a new algorithm for MST sensitivity analysis. In Section 3 we present a faster split-findmin data structure.

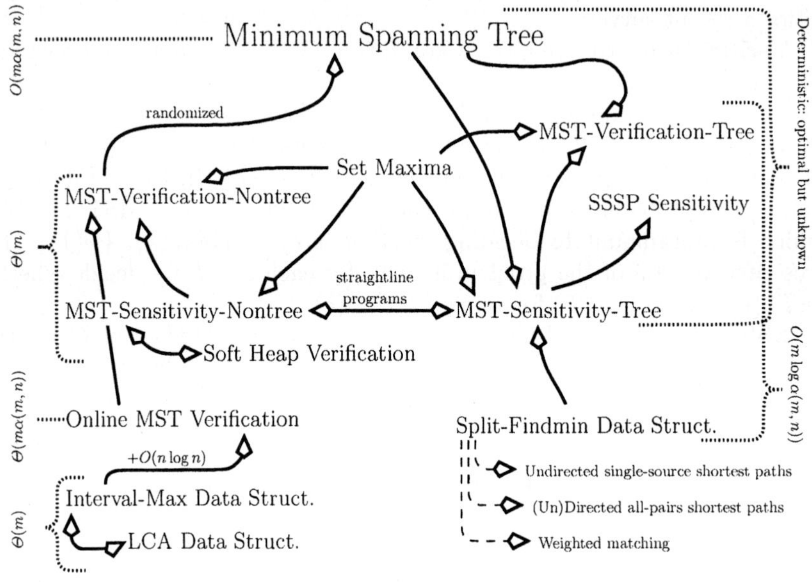

Fig. 1. The incestuous family of minimum spanning tree related problems. Here $\mathcal{A} \longrightarrow \mathcal{B}$ means that problem $\mathcal{B}$ can be solved by an algorithm that uses standard linear-time routines (graph contraction, least common ancestor computations, etc.) plus calls to an algorithm for problem $\mathcal{A}$. The dashed arrows emanating from *split-findmin* are only meant to illustrate other applications of the data structure. Some arrows are labeled with properties of the reduction. For instance, the reduction [16] from the minimum spanning tree problem to MST verification of non-tree edges is randomized.

1.1 The Problems

Minimum Spanning Tree. Given a connected undirected graph $G = (V, E, w)$, find the spanning tree $T \subseteq E$ minimizing $w(T) = \sum_{e \in T} w(e)$. For simplicity, assume throughout that edge weights are distinct. In finding and verifying MSTs there are two useful properties to keep in mind. *Cut property:* The lightest edge crossing the cut $(V', V \backslash V')$ is in MST(G), for any $V' \subset V$. *Cycle property:* The heaviest edge on any cycle is not in MST(G). The best bound on the deterministic complexity of this problem is $O(m\alpha(m, n))$, due to Chazelle [5]. Pettie and Ramachandran [26] gave a deterministic, provably optimal algorithm whose running time is asymptotic to the decision-tree complexity of the minimum spanning tree problem itself. Karger et al. [16] presented a randomized MST algorithm running in expected linear time. Since the only non-trivial component of this algorithm is a minimum spanning tree verification procedure, Figure 1 depicts this result as a randomized reduction from MST to MST verification. Pettie and Ramachandran [27] presented alternative MST algorithms that run in expected linear time but use far fewer random bits. See Graham and Hell [14] for a survey on the early history of the MST problem and Pettie [23–Chapter 7] for a recent survey.

MST Verification. We are given a graph $G = (V, E, w)$ and a (not necessarily minimum) spanning tree $T \subset E$. For $e \notin T$ let $C(e) \cup \{e\}$ be the unique cycle in $T \cup \{e\}$ and for $e \in T$ let $C^{-1}(e) = \{f \notin T \,:\, e \in C(f)\}$. That is, $C(e)$ is the path in T connecting the endpoints of e and $C^{-1}(e)$ is the set of non-tree edges crossing the cut defined by $T \backslash \{e\}$. In the non-tree-edge half of the problem we decide for each $e \notin T$ whether $e \in$ MST$(T \cup \{e\})$, which is tantamount to deciding whether $w(e) < \max_{f \in C(e)}\{w(f)\}$. The tree-edge version of the problem is dual: for each $e \in T$ we decide whether $w(T \backslash \{e\} \cup \{f\}) < w(T)$ for some $f \in C^{-1}(e)$.

MST/SSSP Sensitivity Analysis. We are given a weighted graph G and tree $T =$ MST(G). The sensitivity analysis problem is to decide how much each individual edge weight can be altered without invalidating the identity $T =$ MST(G). We must compute for each edge e:

$$\text{sens}(e) = \begin{cases} \max_{f \in C(e)}\{w(f)\} & \text{for } e \notin T \\ \\ \min_{f \in C^{-1}(e)}\{w(f)\} & \text{for } e \in T \end{cases}$$

where $\min \emptyset - \infty$. By the cut and cycle properties it follows that a non-tree edge e can be increased in weight arbitrarily or reduced by less than $w(e) - \text{sens}(e)$. Similarly, if e is a tree-edge it can be reduced arbitrarily or increased by less than $\text{sens}(e) - w(e)$.

Komlós [19] demonstrated that verification and sensitivity analysis of *non-tree* edges requires a linear number of comparisons. Linear *time* implementations of Komlós's algorithm were provided by Dixon et al. [7], King [18], and Buchsbaum et al. [3]. In earlier work Tarjan [30] gave a verification/sensitivity analysis algorithm for non-tree edges that runs in time

$O(m\alpha(m,n))$ and showed, furthermore, that it could be transformed [32] into a verification/sensitivity analysis algorithm for tree edges with identical complexity.

Online MST Verification. Given a weighted tree T we must preprocess it in some way so as to answer online queries of the form: for $e \notin T$, is $e \in \mathrm{MST}(T \cup \{e\})$? The query edges e are not known in advance. Pettie [21] proved that any data structure answering m queries must take $\Omega(m\alpha(m,n))$ time, where n is the size of the tree. This bound is tight [4, 1].

Soft Heap Verification. Pettie and Ramachandran [27] consider a *generic* soft heap (in contrast to Chazelle's concrete data structure [6]) to be any data structure that supports the priority queue operations insert, meld, delete, and findmin, without the obligation that findmin queries be answered correctly. Given a transcript of priority queue operations, including their arguments and outputs, some elements may *bear witness* to the fact that a particular findmin query was answered incorrectly. These elements are said to be corrupted. The soft heap verification problem is to determine the corrupted elements and more generally, to compute the minimum amount that every element needs to be increased such that all findmin queries are answered correctly. It was observed in [27] that there are mutual reductions between soft heap verification and MST sensitivity (non-tree edges), and consequently, that soft heap verification can be solved in linear time.

Online Interval-Max. The problem is to preprocess a sequence $(e_1, \ldots, e_n)$ of elements from a total order such that for any two indices $\ell < r$, $\max_{\ell \le i \le r}\{e_i\}$ can be reported quickly. It is known [10, 19, 2] that answering interval-max or -min queries is exactly the problem of answering least common ancestor (LCA) queries and that with linear preprocessing both types of queries can be handled in constant time. Katriel et al. [17] also proved that with an additional $O(n \log n)$ time preprocessing, the Online MST Verification problem can be reduced to Online Interval-Max.

Set Maxima. The input is a set system (or hypergraph) $(\chi, \mathcal{S})$ where χ is a set of n weighted elements and $\mathcal{S} = \{S_1, \ldots, S_m\}$ is a collection of m subsets of χ. The problem is to compute $\{\max S_1, \ldots, \max S_m\}$ by comparing elements of χ. Goddard et al. [13] gave a *randomized* algorithm for set maxima that performs $O(\min\{n \log(2\lceil m/n \rceil), n \log n\})$ comparisons, which is optimal. Although the dependence on randomness can be reduced [25], no one has yet to produce a non-trivial deterministic algorithm. The current bound of $\min\{n \log n, \sum_{i=1}^{m} |S_i|, n + m2^m\}$ comes from applying one of three trivial algorithms. All instances of MST verification and sensitivity analysis are reducible to set maxima. In the non-tree-edge version of these problems n and m refer to the number of vertices and edges, respectively, and in their tree-edge versions the roles of n and m are reversed.

Split-Findmin. A precise definition of this structure appears in Section 2. Split-findmin may be regarded as a weighted version of split-find [20], which is itself a time-reversed version of union-find [29]. On a pointer machine union-find, split-find, and split-findmin all have $\Omega(n + m\alpha(m,n))$ lower bounds [31, 20], where m is the number of operations and n the size of the

structure. The same lower bound applies to union-find [8] in the cell-probe and RAM models. Split-find, on the other hand, admits a trivial linear time algorithm in the RAM model; see Gabow and Tarjan for the technique [11]. The results of this paper establish that the *comparison* complexity of split-findmin is $O(n+m\log\alpha(m,n))$ and that on a RAM there is a data structure with the same running time.

2 Sensitivity Analysis and Split-Findmin

The Split-Findmin structure maintains a set of sequences of weighted elements. It supports the following operations:

init$(e_1, e_2, \ldots, e_n)$: Initialize the sequence set $\mathcal{S} := \{(e_1, e_2, \ldots, e_n)\}$ with $\kappa(e_i) := \infty$ for all i. $S(e_i)$ denotes the unique sequence in $\mathcal{S}$ containing e_i.

split(e_i) : Let $S(e_i) = (e_j, \ldots, e_{i-1}, e_i, \ldots, e_k)$.
Set $\mathcal{S} := \mathcal{S} \backslash S(e_i) \cup \{(e_j, \ldots, e_{i-1}), (e_i, \ldots, e_k)\}$.

findmin(e) : Return $\min_{f \in S(e)}\{\kappa(f)\}$.

decreasekey(e, w) : Set $\kappa(e) := \min\{\kappa(e), w\}$.

In Section 3 we give a data structure that maintains the minimum element in each sequence at all times. It executes m operations in $O(m \log\alpha(m,n) + n)$ time.

2.1 Sensitivity Analysis in Sub-inverse-Ackermann Time

In this subsection we calculate the sensitivity of tree edges. We create a split-findmin structure where the initial sequence consists of a list of the vertices in some preorder, w.r.t. an arbitrary root vertex. In general the sequences will correspond to vertices or subtrees of the MST. We maintain the invariant that $\kappa(v)$ corresponds to the minimum weight edge incident to v crossing the cut $(S(v), V \backslash S(v))$. If r is the root of the subtree corresponding to $S(v)$ then the sensitivity of the edge $(r, \text{parent}(r))$ can be calculated from the minimum among $S(r)$.

Step 1. Root the spanning tree at an arbitrary vertex; the ancestor relation is w.r.t. this orientation. For each non-tree edge (u, v), unless v is an ancestor of u or the reverse, replace (u, v) with $(u, \text{lca}(u, v))$ and $(v, \text{lca}(u, v))$, where the new edges inherit the weight of the old. If we have introduced multiple edges between the same endpoints we discard all but the lightest.

Step 2. init$(u_1, \ldots, u_n)$, where u_i is the vertex with pre-order number i. Note that for any subtree, the pre-order numbers of vertices in that subtree form an unbroken interval.

Step 3. 3.1 For $i := 1$ to n
3.2 If $i > 1$, sens$(u_i, \text{parent}(u_i)) := \text{findmin}(u_i)$
3.3 Let $u_{c_1}, \ldots, u_{c_\ell}$ be the children of u_i
3.4 For $j := 1$ to ℓ,
3.5 split(u_{c_j})
3.6 For all non-tree edges (u_k, u_i) where $k > i$
3.7 decreasekey$(u_k, w(u_k, u_i))$

The following lemma is used to prove that correct sens-values are assigned in step 3.2.

Lemma 1. *Let $(u_j, \ldots, u_i, \ldots)$ be the sequence in the split-findmin structure containing u_i, after an arbitrary number of iterations of steps 3.1–3.7. Then this sequence contains exactly those vertices in the subtree rooted at u_j and:*

$$\kappa(u_i) = \min\{w(u_i, u_k) \ : \ k < j \text{ and } (u_i, u_k) \in E\}$$

where $\min \emptyset = \infty$. *Furthermore, just before the ith iteration $i = j$.*

Proof. By induction on the ancestry of the tree. The lemma clearly holds for $i = 1$, where u_1 is the root of the tree. For $i > 1$ the sequence containing i is, by the induction hypothesis, $(u_i, u_{c_1}, \ldots, u_{c_2}, \ldots, u_{c_\ell}, \ldots)$. We only need to show that the combination of the splits in Step 3.5 and the decreasekeys in Step 3.7 ensure that the induction hypothesis holds for iterations $u_{c_1}, u_{c_2}, u_{c_3}, \ldots, u_{c_\ell}$ as well. After performing split$(u_{c_1}), \ldots,$ split(u_{c_ℓ}) the sequences containing $u_{c_1}, u_{c_2}, \ldots, u_{c_\ell}$ clearly correspond to their respective subtrees. Let u_{c_j} be any child of u_i and u_k be any vertex in the subtree of u_{c_j}. Before Step 3.7 we know, by the induction hypothesis, that $\kappa(u_k) = \min\{w(u_k, u_\nu) \ : \ \nu < i \text{ and } (u_k, u_\nu) \in E\}$. To finish the induction we must show that after Step 3.7 $\kappa(u_k)$ is correct w.r.t. its new sequence beginning with u_{c_j}. That is, we must consider all edges (u_k, u_ν) with $\nu < c_j$ rather than $\nu < i$. Since the graph is simple and all edges connect nodes to their ancestors, the only edge that could affect $\kappa(u_k)$ is (u_k, u_i). Performing decreasekey$(u_k, w(u_k, u_i))$ restores the invariant w.r.t. u_k. Since the ith iteration of Step 3.1 only performs splits and decreasekeys on elements in the subtree of u_i, all iterations in the interval $i+1, \ldots, c_j - 1$ do not have any influence on u_{c_j}'s sequence.

Theorem 1. *The sensitivity of a minimum spanning tree or single-source shortest path tree can be computed in $O(m \log \alpha(m, n))$ time, where m is the number of edges, n the number of vertices, and α the inverse-Ackermann function.*

Proof. Correctness. Clearly Step 1 at most doubles the number of edges and does not affect the sensitivity of any MST edge. In iteration i, sens$(u_i, \text{parent}(u_i))$ is set to findmin(u_i), which, by Lemma 1, is precisely the minimum weight of any edge whose fundamental cycle includes $(u_i, \text{parent}(u_i))$. *Running time.* Step 1 requires a a least common ancestor computation, which takes linear time. Step 2 computes a pre-order numbering in $O(n)$ time. After Step 1 the number of non-tree edges is at most $2(m - n + 1)$. In Step 3 each non-tree edge induces one decreasekey and each tree vertex induces one findmin and one split. By Theorem 2 the total cost of all split-findmin operations is $O(m \log \alpha(m, n))$.

3 A Faster Split-Findmin Structure

In this section we present a relatively simple split-findmin data structure that runs in $O(n + m \log \alpha(m,n))$ time, where n is the length of the initial sequence and m the number of operations. Our structure borrows many ideas from Gabow's [9] original split-findmin data structure.

The analysis makes use of Ackermann's function and its inverses:

$$\begin{aligned} A(1,j) &= 2^j && \text{for } j \geq 1 \\ A(i,1) &= 2 && \text{for } i > 1 \\ A(i+1,j+1) &= A(i+1,j) \cdot A(i, A(i+1,j)) && \text{for } i,j \geq 1 \end{aligned}$$

$$\lambda_i(n) = \min\{j \,:\, A(i,j) > n\} \quad \text{and} \quad \alpha(m,n) = \min\{i \,:\, A(i, \left\lceil \frac{2n+m}{n} \right\rceil) > n\}$$

The definition of split-findmin from Section 2 says that all κ-values are initially set to ∞. Here we consider a different version where κ-values are given. The asymptotic complexity of these two versions is the same, of course. However in this section we pay particular attention to the constant factors involved.

Lemma 2 shows that any split-findmin solver can be systematically transformed into another with substantially cheaper splits and incrementally more expensive decreasekeys.

Lemma 2. *If there is a split-findmin structure that requires $O(i)$ time and $2i+1$ comparisons per decreasekey, and $O(in\lambda_i(n))$ time and $3in\lambda_i(n)$ comparisons for all other operations, then there is also a split-findmin structure with parameters $O(i+1), 2i+3, O((i+1)n\lambda_{i+1}(n))$, and $3(i+1)n\lambda_{i+1}(n)$.*

Proof. Let $\mathcal{SF}_i$ and $\mathcal{SF}_{i+1}$ be the assumed and derived data structures. At any moment in its execution $\mathcal{SF}_{i+1}$ treats each sequence of length n' as the concatenation of at most $2(\lambda_{i+1}(n')-1)$ *plateaus* and at most 2 singleton elements, where a level j plateau is partitioned into less than $A(i+1,j+1)/A(i+1,j) = A(i, A(i+1,j))$ *blocks* of size exactly $A(i+1,j)$. In each sequence the plateaus are arranged in a bitonic order, with at most two plateaus per level. At initialization $\mathcal{SF}_{i+1}$ scans the whole sequence, partitioning it into at most $\lambda_{i+1}(n)-1$ plateaus and at most one singleton. Each plateau is managed by $\mathcal{SF}_i$ as a separate instance of split-findmin, where elements of $\mathcal{SF}_i$ correspond to plateau blocks and the key of an element is the minimum among the keys of its corresponding block. Each plateau keeps a pointer to the sequence that contains it. Every block and sequence keeps a pointer to its minimum element. Answering findmin queries clearly requires no comparisons. To execute a decreasekey(e,w) we spend one comparison updating $\kappa(e) := \min\{\kappa(e), w\}$ and another updating the sequence minimum. If e is not a singleton then it is contained in some block b. We finish by calling decreasekey(b,w), where the decreasekey function is supplied by $\mathcal{SF}_i$. If $\mathcal{SF}_i$ makes $2i+1$ comparisons then $\mathcal{SF}_{i+1}$ makes $2i+3$, as promised.

Consider a split operation that divides a level j block b in plateau p. Using the split operation given by $\mathcal{SF}_i$, we split p just before and after the element

corresponding to b. Let b_0 and b_1 be the constituent elements of b to the left and right of the splitting point. We partition b_0 and b_1 into blocks and plateaus (necessarily of levels less than j) just as in the initialization procedure. Notice that to retain the bitonic order of plateaus we scan b_0 from left to right and b_1 from right to left. One of the two new sequences inherits the minimum element from the original sequence. We find the minimum of the other sequence by taking the minimum over each of its plateaus—this uses $\mathcal{SF}_i$'s findmin operation—and the at most two singleton elements.

The comparisons performed in split operations can be divided into (a) those used to find block minima, (b) those used to find sequence minima, and (c) those performed by $\mathcal{SF}_i$. During the execution of the data structure each element appears in at most $\lambda_{i+1}(n) - 1$ blocks. Thus, the number of comparisons in (a) is $\sum_{j\geq 1}^{\lambda_{i+1}(n)-1}(n - n/A(i+1,j))$, which is less than $n(\lambda_{i+1}(n) - 1.5)$. For (b) the number is $n(2\lambda_{i+1}(n) - 1)$ since in any sequence there are at most $2(\lambda_{i+1}(n) - 1)$ plateaus and 2 singletons. For (c), notice that every element corresponding to a block of size $A(i+1,j)$ appears in an instance of $\mathcal{SF}_i$ with less than $A(i+1,j+1)/A(i+1,j) = A(i, A(i+1,j))$ elements. Thus the number contributed by (c) is:

$$\sum_{1\leq j<\lambda_{i+1}(n)} \frac{3in\lambda_i(A(i,A(i+1,j)) - 1)}{A(i+1,j)} = 3in(\lambda_{i+1}(n) - 1)$$

Summing up (a)–(c), the number of comparisons performed outside of decreasekeys is less than $3(i+1)n\lambda_{i+1}(n)$.

We can apply Lemma 2 to a simple split-findmin algorithm that supports decreasekeys in constant time and splits in $O(\log n)$ time. We omit the proof of Lemma 3.

Lemma 3. *There is a split-findmin structure such that decreasekeys require $O(1)$ time and 3 comparisons and other operations require $O(n \log n)$ time in total and less than $3n \log n - 2n$ comparisons.*

Theorem 2. *There is a split-findmin structure that performs $O(m \log \alpha(m,n))$ comparisons. On a pointer machine it runs in $O((m+n)\alpha(m,n))$ time and on a random access machine it runs in $O(n + m \log \alpha(m,n))$ time, where n is the length of the original sequence and m the number of decreasekeys.*

Proof. In conjunction, Lemmas 2 and 3 prove that $\mathcal{SF}_\alpha$ runs in $O(m\alpha(m,n) + \alpha(m,n)n\lambda_{\alpha(m,n)}(n))$ time, which is $O((m+n)\alpha(m,n))$ since $\lambda_{\alpha(m,n)}(n) = O(1 + m/n)$. We reduce the number of comparisons in two ways, then improve the running time. Suppose we are performing a decreasekey on some element e. In every $\mathcal{SF}_i$ e is represented in at most one element and sequence. Let E_i and S_i be the element and sequence containing e in $\mathcal{SF}_i$, i.e., E_i and S_i correspond to a set of elements that include e. Let E_0 be the block containing e in $\mathcal{SF}_1$. If we assume for simplicity that none of $E_\alpha, E_{\alpha-1}, \ldots, E_1$ correspond to singletons, then:

$$\{e\} = E_\alpha \subseteq E_{\alpha-1} \subseteq \cdots \subseteq E_1 \subseteq E_0 \subseteq S_1 \subseteq S_2 \subseteq \cdots \subseteq S_{\alpha-1} \subseteq S_\alpha = S(e)$$

Thus a decreasekey on e can only affect some *prefix* of the the min-pointers in $E_\alpha, \ldots, E_1, E_0, S_1, \ldots, S_\alpha$. Using a binary search this prefix can be determined and updated in $O(\alpha)$ time but with only $\lceil \log(2\alpha + 2) \rceil$ comparisons. We have reduced the number of comparisons to $O(n\alpha(m,n) + m \log \alpha(m,n))$. To get rid of the $n\alpha(m,n)$ term we introduce another structure $\mathcal{SF}_i^*$. Upon initialization $\mathcal{SF}_i^*$ divides the full sequence into blocks of size i. At any point each sequence consists of a subsequence of unbroken blocks and possibly two partial blocks, one at each end. The unbroken blocks are handled by $\mathcal{SF}_i$, where each is treated as a single element. $\mathcal{SF}_i^*$ keeps the keys of each block in sorted order. Findmins are easy to handle, as are splits, which take $O(i(n/i)\lambda_i(n/i)) = O(n\lambda_i(n))$ comparisons and $O(in + n\lambda_i(n))$ time. The routine for decreasekeys takes $O(i)$ time and $O(\log i)$ comparisons. If element e lies in block b then decreasekey(e, w) will call $\mathcal{SF}_i$'s decreasekey(b, w) then update the sorted order of block b.

Once we have bounded the number of element comparisons we can derive a RAM-based data structure with the same running time. The idea is to precompute a lookup table whose entries contain the decision-trees for each possible operation, as applied to sequences of length $\log \log n$. A simple counting argument shows the table can be computed in $o(n)$ time. This data structure can be composed with $\mathcal{SF}_2$ to yield a split-findmin data structure for sequences of length n. See [7, 3, 28] for a more leisurely description of this technique.

References

1. A. Alon and B. Schieber. Optimal preprocessing for answering on-line product queries. Technical Report TR-71/87, Institute of Computer Science, Tel Aviv University, 1987.
2. M. A. Bender and M. Farach-Colton. The LCA problem revisited. In *Proceedings 4th Latin American Symp. on Theoretical Informatics (LATIN), LNCS Vol. 1776*, pages 88–94, 2000.
3. A. L. Buchsbaum, H. Kaplan, A. Rogers, and J. R. Westbrook. Linear-time pointer-machine algorithms for LCAs, MST verification, and dominators. In *Proc. 30th ACM Symposium on Theory of Computing (STOC)*, pages 279–288, 1998.
4. B. Chazelle. Computing on a free tree via complexity-preserving mappings. *Algorithmica*, 2(3):337–361, 1987.
5. B. Chazelle. A minimum spanning tree algorithm with inverse-Ackermann type complexity. *J. ACM*, 47(6):1028–1047, 2000.
6. B. Chazelle. The soft heap: an approximate priority queue with optimal error rate. *J. ACM*, 47(6):1012–1027, 2000.
7. B. Dixon, M. Rauch, and R. E. Tarjan. Verification and sensitivity analysis of minimum spanning trees in linear time. *SIAM J. Comput.*, 21(6):1184–1192, 1992.
8. M. L. Fredman and M. Saks. The cell probe complexity of dynamic data structures. In *Proc. 21st ACM Symposium on Theory of Computing*, pages 345–354, 1989.
9. H. N. Gabow. A scaling algorithm for weighted matching on general graphs. In *Proc. 26th Symp. on Foundations of Computer Science (FOCS)*, pages 90–100, 1985.
10. H. N. Gabow, J. L. Bentley, and R. E. Tarjan. Scaling and related techniques for geometry problems. In *Proceedings of the 16th Annual ACM Symposium on Theory of Computing*, pages 135–143, 1984.

11. H. N. Gabow and R. E. Tarjan. A linear-time algorithm for a special case of disjoint set union. *J. Comput. Syst. Sci.*, 30(2):209–221, 1985.
12. H. N. Gabow and R. E. Tarjan. Faster scaling algorithms for general graph-matching problems. *J. ACM*, 38(4):815–853, 1991.
13. W. Goddard, C. Kenyon, V. King, and L. Schulman. Optimal randomized algorithms for local sorting and set-maxima. *SIAM J. Comput.*, 22(2):272–283, 1993.
14. R. L. Graham and P. Hell. On the history of the minimum spanning tree problem. *Ann. Hist. Comput.*, 7(1):43–57, 1985.
15. T. Hagerup. Improved shortest paths on the word RAM. In *Proc. 27th Int'l Colloq. on Automata, Languages, and Programming (ICALP), LNCS vol. 1853*, pages 61–72, 2000.
16. D. R. Karger, P. N. Klein, and R. E. Tarjan. A randomized linear-time algorithm for finding minimum spanning trees. *J. ACM*, 42:321–329, 1995.
17. I. Katriel, P. Sanders, and J. L. Träff. A practical minimum spanning tree algorithm using the cycle property. In *Proc. 11th Annual European Symposium on Algorithms, LNCS Vol. 2832*, pages 679–690, 2003.
18. V. King. A simpler minimum spanning tree verification algorithm. *Algorithmica*, 18(2):263–270, 1997.
19. J. Komlós. Linear verification for spanning trees. *Combinatorica*, 5(1):57–65, 1985.
20. H. LaPoutré. Lower bounds for the union-find and the split-find problem on pointer machines. *J. Comput. Syst. Sci.*, 52:87–99, 1996.
21. S. Pettie. An inverse-Ackermann type lower bound for online minimum spanning tree verification. *Combinatorica.* to appear.
22. S. Pettie. On the comparison-addition complexity of all-pairs shortest paths. In *Proc. 13th Int'l Symp. Algorithms and Computation (ISAAC)*, pages 32–43, 2002.
23. S. Pettie. *On the Shortest Path and Minimum Spanning Tree Problems.* Ph.D. thesis, The University of Texas at Austin, 2003. Department of Computer Sciences Technical Report TR-03-35.
24. S. Pettie. A new approach to all-pairs shortest paths on real-weighted graphs. *Theoretical Computer Science*, 312(1):47–74, 2004.
25. S. Pettie and V. Ramachandran. Minimizing randomness in minimum spanning tree, parallel connectivity and set maxima algorithms. In *Proc. 13th Ann. ACM-SIAM Symp. on Discrete Algorithms (SODA)*, pages 713–722, 2002.
26. S. Pettie and V. Ramachandran. An optimal minimum spanning tree algorithm. *J. ACM*, 49(1):16–34, 2002.
27. S. Pettie and V. Ramachandran. New randomized minimum spanning tree algorithms using exponentially fewer random bits. manuscript, submitted, 2005.
28. S. Pettie and V. Ramachandran. A shortest path algorithm for real-weighted undirected graphs. *SIAM J. Comput.*, 34(6):1398–1431, 2005.
29. R. E. Tarjan. Efficiency of a good but not linear set merging algorithm. *J. ACM*, 22:215–225, 1975.
30. R. E. Tarjan. Applications of path compression on balanced trees. *J. ACM*, 26(4):690–715, 1979.
31. R. E. Tarjan. A class of algorithms which require nonlinear time to maintain disjoint sets. *J. Comput. Syst. Sci.*, 18(2):110–127, 1979.
32. R. E. Tarjan. Sensitivity analysis of minimum spanning trees and shortest path problems. *Info. Proc. Lett.*, 14(1):30–33, 1982. See Corrigendum, IPL **23**(4):219.
33. M. Thorup. Undirected single-source shortest paths with positive integer weights in linear time. *J. ACM*, 46(3):362–394, 1999.

Approximation Algorithms for Layered Multicast Scheduling

Qingbo Cai and Vincenzo Liberatore

Case Western Reserve University, Cleveland, Ohio 44106, USA
{qxc6, vxl11}@cwru.edu

Abstract. Layered multicast is a scalable solution to data dissemination in heterogeneous networks such as the Internet. In this paper, we study the scheduling problem arising in the layered multicast context. Our goal is to generate a multicast schedule for two different objectives, i.e., to minimize the weighted sum (the L_1 objective) of the per-layer average waiting time, and to minimize the maximum approximation ratio (the L_∞ objective) of the subschedules on individual layers. Compared to the previous work on multicast scheduling, this paper addresses the data popularity and the interaction among layers simultaneously. We present a simple randomized algorithm for both objectives of the layered multicast scheduling problem. For the L_1 objective, we provide a deterministic 2-approximation algorithm for the general multi-layer cases. For the L_∞ objective, we present an algorithm for the two-layer case which is 1.6875-approximation ignoring an additive constant.

1 Introduction

Multicast is a scalable solution to data dissemination from a source node to multiple destination nodes. However, in heterogeneous networks such as the Internet, there is no single transmission rate that can simultaneously enable efficient bandwidth utilization on heterogeneous network connections. This problem can be addressed with *layered multicast* communication [15,8], where the communication channel is divided into several logical layers and each receiver subscribes to a set of layers according to its available bandwidth.

In layered multicast, the transmission rate increases exponentially from the lower layers to upper ones. Layers are *cumulative* in the sense that receivers subscribe to all layers that have bandwidth less than a certain threshold. The performance of layered multicast critically depends on how data items are scheduled for transmission. The corresponding layered multicast scheduling problem is an intricate generalization of the non-layered multicast scheduling problem due to the interaction among layers.

Non-Layered Multicast Scheduling (NLMS). The NLMS problem has been the focus of much previous work. The input to NLMS is a set of M equal size data items each of which is associated with an access probability p_i, and the capacity of the communication channel W. The output is a schedule S specifying the data

X. Deng and D. Du (Eds.): ISAAC 2005, LNCS 3827, pp. 974–983, 2005.

items to be transmitted at each time-slot. The NLMS problem was introduced in [1,2,3], corresponds to a special case of the *Generalized Maintenance Scheduling Problem (GMSP)* without broadcast costs, and admits a simple and practical 2-approximation algorithm [4]. The golden ratio sequence algorithm (*GRSA*) was developed in [11,10] and subsequently proven to be 9/8-approximation [4] for NLMS. A polynomial-time approximation scheme *(PTAS)* [12] partitions the data set by popularity and exhaustively searches all schedules for the popular data. Several variants of NLMS have also been investigated [16,13,14,5]. The NLMS problem can be solved in polynomial time in the special case when the server multicasts only two data items [6]. Certain NLMS variants are NP-hard [4,13,16], but the hardness of the original NLMS is an open problem.

Layered Multicast Scheduling (LMS). Previous work on LMS extensively focuses on the arrangement of a file with erasure-correcting codes across multiple layers in a network with packet losses. Two data arrangements, *SO* and *CO*, are proposed and use reception efficiency as the performance metric [17]. The multirate scheme *PO* further improves over SO for some performance measures [9]. Another approach breaks a file into equal size groups, encodes individual groups, and carefully interleaves the intra- and inter-group data [7].

Previous work on LMS ignores data popularity, which was a central factor in the original NLMS problem. If data popularity is uniform, the optimal schedule is flat and can be incrementally extended to higher layers if we shift it by a number of transmission slots proportional to the period of an optimal schedule. Otherwise, there is no clear way to obtain a provable performance guarantee by shifting a schedule from the lower to the upper layers.

Theoretical Significance. The original NLMS problem is at heart similar to a natural problem of combinatorial flavor: find an infinite sequence of the integers $\{1, 2, \ldots, n\}$ that is "regular", in the sense that, given a set of real numbers $\{\tau_1, \tau_2, \ldots, \tau_n\}$, the integer i appears in the sequence (approximately) every τ_i positions. The LMS problem is a stronger version in which we wish to find a highly regular sequence that contains highly regular subsequences. Specifically, suppose that we start with a regular sequence and we extract a subsequence by taking one position out of 2 $(4, 8, \ldots)$. Then, we wish to ensure that the subsequences are also regular in the sense that the integer i appears (approximately) every τ_i positions. An LMS solution can be regarded as being "more regular" than an NLMS solution in that regularity also holds if we "zoom in" on the subsequences. In terms of layered multicast applications, subsequences correspond to layers, and the problem becomes to find a good overall schedule whose sub-schedules are also good.

An LMS algorithm is fundamentally different depending on whether its approximation ratio is greater than 2 or not. Specifically, a $(2 + \epsilon)$-approximation algorithm can be easily obtained by running the previously known PTAS [12] for NLMS on each layer independently of the other layers. However, this $(2+\epsilon)$-approximation algorithm is intuitively unsatisfactory, since it completely disregards the fact that layers are cumulative so that the way how the data are

scheduled across layers has an influence on the performance. By contrast, a 2-approximation algorithm requires cross-layer interaction. Although a 2-approximation algorithm would shave off only an arbitrarily small factor from the $(2+\epsilon)$-approximation algorithm, it would highlight a qualitatively important construction in that it makes the layers interdependent.

A schedule could in principle have satisfactory performance for some layers but not for others. The multiple per-layer objectives can be reconciled by taking the weighted sum of per-layer costs using the user subscription profiles (i.e., the distribution of users subscribing to a layer) as the weighting coefficients. However, subscription profiles are not always available in practice and, in these cases, we adopt the L_∞ norm by taking the maximum of per-layer approximation ratios. Both objectives can arise in practice. The L_1 objective corresponds to the case where the user subscription profiles are known, and the L_∞ objective may be the only possible objective if the user subscription profiles are absent. For the L_1 objective, we obtain a 2-approximation algorithm by randomization and derandomization via conditional expectations. Conditional expectations become weaker or inapplicable under progressively more stringent objectives, so that we turn to combinatorial constructions to exploit cross-layer dependence. Combinatorial constructions differ from conditional expectations primarily in that they use more structural information. For example, our 1.6875-approximation (ignoring an additive constant) algorithm captures the "regularity" characteristics of the golden ratio sequence schedule. For the strictest L_∞ objective, this algorithm establishes a combinatorial construction for the LMS problem by compressing and stretching the golden ratio sequence schedule.

Our Results. Compared to the previous work on multicast scheduling, this paper addresses the data popularity and the interaction among layers simultaneously. To the best of our knowledge, this paper is the first to address LMS with data popularity taken into account.

We provide a simple 2-approximation randomized algorithm for both objectives of the LMS problem. After derandomization via conditional expectations, we obtain a deterministic algorithm for the L_1 objective and prove its performance guarantee of 2 by the potential method.

The L_∞ objective is substantially more complicated. We present a 1.6875-approximation (ignoring an additive constant) algorithm in the two-layer case. This algorithm is simple, but its performance analysis is complicated and can not use known techniques. In this algorithm, higher layer benefits from lower layer either by sharing the transmissions of hot pages on the lower layer or by borrowing bandwidth from the lower layer depending on the existence of hot pages.

Organization. In Section 2, we formalize the LMS problem and introduce our notations. Section 3 gives a lower bound to the cost of an optimal schedule, which is the basis for the performance analysis in this paper. In Section 4, we present a simple 2-approximation randomized algorithm for the LMS problem. Section 5 introduces a 2-approximation deterministic algorithm for the L_1 objective.

For the L_∞ objective, we present a 1.6875-approximation (ignoring an additive constant) algorithm for the two-layer LMS problem and analyze its performance in Section 6. Finally, Section 7 concludes this paper.

2 Preliminaries

A multicast server maintains a database that contains N equal size *pages* numbered $1,\ldots,N$. Each page is associated with an access probability p_i. The server continuously and repeatedly transmits these pages on L layers numbered $0,\ldots,L-1$ ($L\geq 2$), where each layer corresponds to a multicast rate. The time axis is divided into unit length *time-slots*, which is the duration for the server to transmit one page on layer 0. During each time-slot, the number of pages transmitted on layer l is defined as:

$$r_l = \begin{cases} 1, & l=0; \\ 2^{l-1}, & 1\leq l\leq L-1. \end{cases}$$

Let $R_l = \sum_{j=0}^{l} r_j = 2^l$ denote the cumulative rate for layer l. Without loss of generality, we suppose $R_{L-1} = 2^{L-1} \leq N$. A *schedule* is a sequence $S = \{A_{l,t} : 0\leq l\leq L-1, t\geq 1\}$, where $A_{l,t}$ denotes the set of pages multicast on layer l during time slot t. We then have $|A_{0,t}| = r_0 = 1$ and $|A_{l,t}| = r_l = 2^{l-1}$, $(1\leq l\leq L-1)$.

We assume that a client receives every page the server multicasts. A client is said to *subscribe* to layer l if he receives $A_{l,t}$ during time-slot t. When a client subscribes to layer l, he also subscribes to layers $0,1,\ldots,l-1$, since layers are cumulative. The *subscription level* of a client is the highest layer that he subscribes to. Let q_l be the fraction of clients whose subscription levels are l.

Consider the scenario where a client accesses a page. Suppose, at the beginning of time-slot t, a request for page i is raised by a client with subscription level l. This request will be satisfied at the end of time-slot t' ($t\leq t'$) during which page i is multicast on at least one of layers $0,1,\ldots,l$, i.e., $i\in\bigcup_{h=0}^{l} A_{h,t'}$ and $i\notin\bigcup_{h=0}^{l} A_{h,j}$ for all $t\leq j<t'$. Then, the client has to wait $w_{l,t}(i) = t'-t+1$ time-slots before page i can be accessed. In the long run, the average waiting time for the client to receive a page is $Cost(S,l) = \limsup_{n\to\infty} \frac{1}{n}\sum_{t=1}^{n}\sum_{i=1}^{N} p_i w_{l,t}(i)$.

An ideal schedule should satisfy the following requirements:

Reliable: Each page must be transmitted on layer 0. A consequence of reliability is that clients can receive all pages independently of their subscription levels.

Efficient: If the user subscription profiles (q_l's) are present, we look for a schedule S to minimize the average per-client access cost (the L_1 objective), i.e., to minimize

$$Cost(S) = \sum_{l=0}^{L-1} q_l Cost(S,l) = \sum_{l=0}^{L-1} q_l \limsup_{n\to\infty}\frac{1}{n}\sum_{t=1}^{n}\sum_{i=1}^{N} p_i w_{l,t}(i)\ .$$

Otherwise, let OPT_l denote the optimal value of $Cost(S, l)$. Our goal is to minimize the maximum of per-layer approximation ratios (the L_∞ objective), i.e., to minimize

$$\max_{0 \leq l \leq L-1} \{Cost(S, l)/OPT_l\} .$$

3 Lower Bound

The set of feasible schedules for LMS is a proper subset of the set of feasible schedules for the GMSP problem [4] due to the reliability requirement. Consequently, a lower bound for GMSP is also a lower bound for LMS. By referring to GMSP, the optimal value of the following programs (1)-(3) is a lower bound to $Cost(S, l)$:

$$\min \frac{1}{2} \sum_{i=1}^{N} p_i \left(\tau_{l,i} + 1\right) \tag{1}$$

$$\text{s.t.} \quad \sum_{i=1}^{N} \frac{1}{\tau_{l,i}} \leq R_l \tag{2}$$

$$\tau_{l,i} \geq 1 \qquad\qquad 1 \leq i \leq N \tag{3}$$

The optimal solution to (1)-(2), i.e., the non-linear programs above without the constraint $\tau_{i,l} \geq 1$, is given by $\tau_{l,i} = \sum_{j=1}^{N} \sqrt{p_j} / \left(R_l \sqrt{p_i}\right)$ [4].

Let $\tau^*_{l,i}$ be an optimal solution to (1)-(3) and computed by an algorithm in [4]. Suppose that LB_L is a lower bound to $Cost(S)$. By applying the GMSP lower bound to each layer, we have $LB_L = 1/2 \sum_{l=0}^{L-1} q_l \sum_{i=1}^{N} p_i \left(\tau^*_{0,i}/R_l + 1\right)$.

4 A Simple Randomized Algorithm

In spite of its simplicity, the randomized algorithm RA1 is our starting point to explore the cross-layer dependence. More importantly, it lays the foundation for the deterministic algorithm DA1, which is obtained by derandomizing RA1. The algorithm is as follows:

Randomized algorithm 1 (RA1)
At time slot t ($t = 1, 2, \ldots$):
for $l = 0 \ldots L-1$ do
for $j = 1 \ldots r_l$ do
schedule page i on layer l with probability $1/\tau^*_{0,i}$;

In RA1, page i is receivable to layer l with the probability $1 - \left(1 - 1/\tau^*_{0,i}\right)^{R_l}$, and this probability increases as l increases. Therefore, lower layers benefit higher layers by increasing the probability that a request is satisfied on higher layers.

Theorem 1. *There exists a 2-approximation randomized algorithm for both objectives of the LMS problem.*

Proof (Sketch). In RA1, the expected waiting time for page i on layer l is $1/(1-(1-1/\tau^*_{0,i})^{R_l})$, which is less than $\tau^*_{0,i}/R_l+1$ by induction. Then, by comparing the expected waiting time $E[Cost(S,l)]$ of a schedule S by RA1 with the lower bound $\left(1/2\left(\tau^*_{0,i}/R_l+1\right)\right)$, we conclude that RA1 is 2-approximation for each layer. □

5 A Deterministic Algorithm for the L_1 Objective

At time-slot t ($t \geq 0$), let $s^t_{l,i}$ denote the number of time-slots elapsed from the most recent transmission of page i on any of the layers $0,\ldots,l$. We assume that all pages are transmitted on layer 0 at time-slot 0, i.e., $s^0_{l,i}=0$. The state vector is defined as $\overrightarrow{s^t}=\left(\overrightarrow{s^t_0},\ldots,\overrightarrow{s^t_{L-1}}\right)$, where $\overrightarrow{s^t_l}=\left(s^t_{l,1},\ldots,s^t_{l,N}\right)$. Then, the cost of a schedule S incurred at time-slot t can be rewritten as $\sum_{l=0}^{L-1} q_l \sum_{i=1}^{N} p_i\left(s^t_{l,i}+1\right)$.

Deterministic algorithm 1 (DA1)
At time slot t:

$U=\emptyset$;
for $l=0$ *to* $L-1$
  for $i=1$ *to* r_l
    On layer l, *schedule page* j ($j\notin U$) *that maximizes*

$$\sum_{k=l}^{L-1} q_k p_j\left(s^{t-1}_{k,j}+1\right)\frac{\left(1-\frac{1}{\tau^*_{0,j}}\right)^{R_k-R_{l-1}-i}}{1-\left(1-\frac{1}{\tau^*_{0,j}}\right)^{R_k}};$$

    $U=U\cup\{j\}$;

We now give some intuitions for algorithm DA 1. First, when scheduling a layer, we consider its influence on higher layers by taking higher layers' state into account. Second, it is natural not to reschedule a page which is already scheduled on that layer or lower layers at the same time-slot. Therefore, for each time-slot, we keep track of the set (U) of pages that have been scheduled so far.

Theorem 2. *There exists a 2-approximation deterministic algorithm for the L_1 objective of the LMS problem.*

Proof (Sketch). Although a standard technique (conditional expectations) is used to obtain DA1 from derandomizing RA1, it is not suitable for the performance proof of DA1, since the cost is defined in the long-run average form and the choice at an individual time-slot has no remarkable influence on the average cost. Hence, we apply the potential method as in [4,12]. The potential function of schedule S at time-slot t is defined by

$$\Phi(t,S)=\sum_{l=0}^{L-1} q_l \sum_{i=1}^{N}\frac{p_i s^t_{l,i}}{1-\left(1-\frac{1}{\tau^*_{0,i}}\right)^{2^l}}.$$

Note that $\Phi(t,S)\geq 0$ ($t\geq 1$) and $\Phi(0,S)=0$, since $s^0_{l,i}=0$ ($0\leq l\leq L-1, 1\leq i\leq N$). □

6 An Approximation Algorithm for the L_∞ Objective in the Two Layer Case

For the strictest L_∞ objective, several complications arise. First, conditional expectations become weaker or inapplicable. Second, although the golden ratio sequence algorithm (GRSA) is 9/8-approximation for the NLMS problem [4], its performance for the LMS problem is unknown, since the "regularity" property (i.e., the three-gap property as described later) is not guaranteed to hold for the subschedules on each layer. However, we can establish a combinatorial construction by stretching and compressing the golden ratio sequence schedule, where the stretching and compressing ratio is determined by carefully balancing the provable performance guarantee on each layer such that the L_∞ objective is minimized. The algorithm is simple and elegant, but the performance analysis is substantially more complicated.

Deterministic algorithm 2 (DA2)
– If there are hot pages (the classification is specified later), we ignore them on layer 1, and use GRSA to schedule the remaining non-hot pages on layer 1. The schedule on layer 0 is generated by GRSA for all (hot and non-hot) pages.
– Otherwise, 1 out of 3 time-slots on layer 0 is reserved for layer 1. We apply GRSA to generate an original schedule S' for NLMS, and fill S' in the time-slots available to layer 0 and 1 respectively (see Fig. 1).

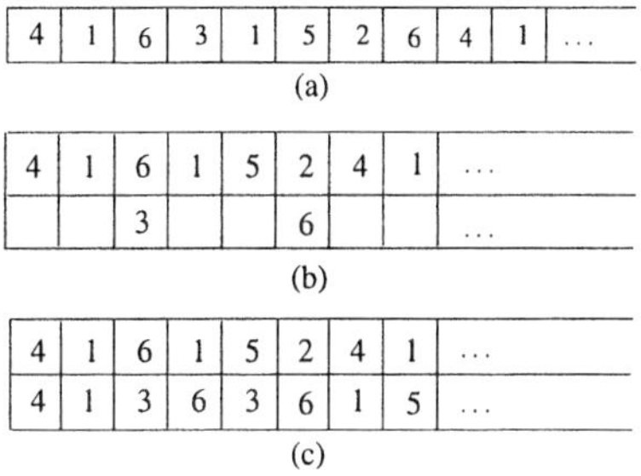

Fig. 1. An illustration of the algorithm DA2 when there are no hot pages. (a) An original schedule S' generated by GRSA for NLMS; (b) A partial schedule after filling S' into layer 1 and the reserved time-slots on layer 0; (c) A complete schedule after filling S' into the non-reserved time-slots on layer 0.

The main idea behind the algorithm DA2 is that hot pages are transmitted frequently on layer 0, and layer 1 may ignore them without considerable sacrifice to the performance since layers are cumulative so that layer 1 can receive pages transmitted on layer 0. However, if there are no such hot pages, layer 1 will suffer from neglecting pages. In this case, layer 1 may still benefit from layer 0 by reserving a fraction of transmission capacity on layer 0 for layer 1, and this fraction can be determined by carefully balancing the approximation ratios of both layers so that the maximum ratio is minimized.

In DA2, the pages are categorized into hot pages and non-hot pages. The classification of a hot page is done by examining if there is an inter-transmission gap of 1 for this page in the original GRSA schedule. There are two reasons for this classification. First, the pages with inter-transmission gaps of 1 are indeed pages with high access probability and there are at most two such pages due to the three-gap property of GRSA. Second, it is not suitable for these pages to use reserved time-slots because two adjacent transmissions of these pages may fall into the same time-slot after filling the original GRSA schedule on layer 1, and consequently may invalidate the benefit of reserved time-slots.

We now briefly recall GRSA. Let $f_i = 1/\tau^*_{0,i}$ be the optimal transmission frequency of page i for layer 0. GRSA generates a schedule of length F_k ($k \geq 8$), where each page i ($1 \leq i \leq N$) appears N_i ($N_i = \lfloor f_i F_k \rfloor$ or $\lceil f_i F_k \rceil$) times. Let $N_i = F_{k_i} + S_i$ ($0 \leq S_i < F_{k_i - 1}$), i.e., F_{k_i} is the largest Fibonacci number which is no more than N_i. GRSA generates a schedule where there are at most three types of inter-transmission gaps for each page i: i) S_i gaps of length F_{k-k_i}; ii) $F_{k_i-2} + S_i$ gaps of length F_{k-k_i+1}; iii) $F_{k_i-1} - S_i$ gaps of length F_{k-k_i+2}.

Due to the three-gap property above and our page classification, the hot pages are those pages with $N_i = F_{k_i} + S_i$, where $k - 2 \leq k_i \leq k - 1$. We have the following lemmas in the presence of hot pages.

Lemma 1. *If there is a hot page i with $k_i = k-1$, then $Cost(S, 1) \leq 1.39 OPT_1$.*

Proof. Consider an optimal solution to programs (1)-(3) for this case. Without the constraint (3), page i has $\tau^*_{0,i} < F_k/(F_{k-1}+S_i-1) < 2$, and $\tau^*_{1,i} = \tau^*_{0,i}/2 < 1$, which is set to 1 by an argument in [1]. Consequently, the optimal cost incurred by page i is F_k in a schedule of length F_k. Then, for the remaining pages, we obtain $\tau^*_{1,j}$ ($j \neq i$) by computing an optimal solution to programs (1)-(3) with R_1 replaced by 1.

In the analysis, we only count the transmissions of hot pages i on layer 0 as the contribution made to layer 1 by layer 0. Due to GRSA on layer 0, page i has two types of gaps: $F_{k-3} + 2S_i$ gaps of length 1 and $F_{k-2} - S_i$ gaps of length 2. Thus, the total cost incurred by page i for layer 1 is $F_{k-3} + 2S_i + 3(F_{k-2} - S_i) = F_k + F_{k-2} - S_i$. Therefore, the approximation ratio for page i is $(F_k + F_{k-2} - S_i)/F_k$, which is no more than 1.39. For the remaining pages, the approximation ratio is 9/8 by GRSA on layer 1. Hence, the approximation ratio for layer 1 is no more than the maximum ratio of pages, i.e., $\max\{1.39, 9/8\} = 1.39$. □

Lemma 2. *If there is a hot page i with $k_i = k-2$, then $Cost(S, 1) \leq 1.628 OPT_1$.*

Proof (Sketch). On layer 0, pages i have 3 types of gaps: S_i gaps of length 1, $F_{k-4} + S_i$ gaps of length 2, and $F_{k-3} - S_i$ gaps of length 3. When there is a page i with $\tau^*_{1,i} \leq 1$, we have a constraint for S_i such that $S_i \geq F_k/2 - F_{k-2} - 1$. By the application of this constraint and similar calculation as in the proof of Lemma 1, we can bound the approximation ratio for page i from above by 1.619. Then, the approximation ratio of layer 1 is at most $\max\{1.619, 9/8\} = 1.619$.

If pages i satisfies $\tau^*_{1,i} > 1$, there must be a constraint $S_i < F_k/2 - F_{k-2} - 1$. Then, the approximation ratio r for page i is $r = 4(3F_{k-1} - 2S_i)(F_{k-2} + S_i)$

$/ (F_k (F_k + 2F_{k-2} + 2S_i))$. By looking at r as a function of S_i and solving the equation $r'(S_i) = 0$, we have

$$r_{max} = 8 + \frac{2F_{k-1}}{F_k} - 4\sqrt{3 + \frac{F_{k-1}}{F_k}} < 1.628 .$$

For the remaining pages j $(j \neq i)$, let $\tau_{1,j}$ be the average inter-transmission gap of page j. Then, we have $\tau_{1,j} \leq 9\sum_{m=1,m\neq i}^{N} \sqrt{p_m}/(8\sqrt{p_j})$ due to GRSA. We can prove that $\tau_{1,j}/\tau^*_{1,j} < 1.393$ using the constraint $\tau^*_{0,i} = \sum_{m=1}^{N} \sqrt{p_m}/\sqrt{p_i} \leq F_k/(F_{k-2} + S_i - 1)$. Then, the approximation ratio for layer 1 is no more than $\max\{1.628, 1.393\} = 1.628$. □

Lemma 3. *If there are hot pages, the algorithm DA2 generates a schedule S such that* $\max_{0\leq l\leq 1}\{Cost(S,l)/OPT_l\} \leq 1.628$.

Proof. This lemma immediately follows from Lemma 1, Lemma 2, and the fact that $Cost(S,0)/OPT_0 \leq 9/8$ due to GRSA on layer 0. □

When no hot pages exist (i.e., $k_i \leq k - 3$ for all $1 \leq i \leq N$), the following lemma gives an upper bound to $Cost(S,l)$.

Lemma 4. *If there are no hot pages, the algorithm DA2 generates a schedule S such that $Cost(S,l) \leq 1.6875OPT_l + 53/64$ for $0 \leq l \leq L-1$.*

Proof (Sketch). In the analysis, we only consider the transmissions in the reserved time-slots on layer 0 as the benefit that layer 1 can gain from layer 0. Obviously, for layer 1, any inter-transmission gap of length v in the original GRSA schedule will be reduced in length when $v \geq 4$, and may remain the same if $v < 4$ after schedule "compression" for layer 1. To get a precise approximation analysis, we individually discuss the performances for the pages i with gap length $v < 4$ (i.e., $k_i = k-3$ or $k_i = k-4$), and $v \geq 4$ (i.e., $k_i \geq k-5$). Meanwhile, an original gap of length v, after being "stretched" on layer 0, becomes $3/2\ v$ if v is even, and $3/2\ v \pm 1/2$ if v is odd. This lemma can be proved by carefully calculating the cost on each layer using the numeric properties of Fibonacci numbers and applying appropriate constraints implied by DA2 in various cases.

Theorem 3. *For the L_∞ objective in the two layer case ($L = 2$), DA2 generates a schedule S where $Cost(S,l) \leq 1.6875OPT_l + 53/64$ for $0 \leq l \leq L-1$.*

Proof. The proof follows from Lemma 3 and 4. □

7 Concluding Remarks

We have studied the LMS problem with or without the knowledge of the user subscription profiles, i.e., the L_1 and L_∞ objectives respectively. A deterministic 2-approximation algorithm is obtained for the L_1 objective. However, the L_∞ objective is substantially more complicated than the L_1 objective. For the L_∞ objective, the algorithm DA2 has a good performance guarantee (1.6875 ignoring an additive constant) only for the two-layer case. Therefore, a natural direction for further investigation would be to achieve an efficient algorithm to the LMS problem with any number of layers for the L_∞ objective.

Acknowledgement

The authors acknowledge Kirk Pruhs from University of Pittsburgh for discussion on this problem, and anonymous reviewers for helpful comments.

References

1. M.H. Ammar, J.W. Wong: The Design of Teletext Broadcast Cycles. Performance Evaluations. **5**:4 (1985) 235–242
2. M.H. Ammar, J.W. Wong: On the Optimality of Cyclic Transmission in Teletext Systems. IEEE Transactions on Communications.**35**:1 (1987) 68–73
3. S. Anily, C. A. Glass, R. Hassin: The scheduling of maintenance service. Discrete Applied Mathematics. **80** (1998) 27–42
4. A. Bar-Noy, R. Bhatia, J. Naor, B. Schieber: Minimizing Service and Operation Costs of Periodic Scheduling. In Mathematics of Operations Research. **27**:3 (2002) 518–544
5. A. Bar-Noy, B. Patt-Shamir, I. Ziper: Broadcast Disks with Polynomial Cost Functions. Proc. of IEEE INFOCOM'00. 575–584
6. A. Bar-Noy, Y. Shilo: Optimal Broadcasting of Two Files over an Asymmetric Channel. Journal of Parallel and Distributed Computing. **60**:4 (2000) 474–493
7. Y. Birk, D. Crupnicoff: A Multicast Transmission Schedule for Scalable Multi-Rate Distribution of Bulk Data using Non-Scalable Erasure-Correcting Codes. Proc. of IEEE INFOCOM'03. 1033–1043
8. J. Byers, M. Luby, M. Mitzenmacher: Fine-Grained Layered Multicast. Proc. of IEEE INFOCOM'01. 1143–1151
9. M. J. Donahoo, M. H. Ammar, E. W. Zegura: Multiple-Channel Multicast Scheduling for Scalable Bulk-data Transport. Proc. of IEEE INFOCOM'99. 847–855.
10. M. Hofri, Z. Rosberg: Packet Delay under the Golden Ratio Weighted TDM Policy in a multiple-access channel. IEEE Transactions on Information Theory. **33**:3 (1987) 341–349
11. A. Itai, Z. Rosberg: A golden ratio control policy for a multiple-access channel. IEEE Transactions on Automatic Control. **29**:8 (1984) 712–718
12. C. Kenyon, N. Schabanel, N. Young: Polynomial-Time Approximation Scheme for Data Broadcast. Proc. of the 32nd Annual ACM Symp. on Theory of Computing (STOC 2000). 659–666
13. C. Kenyon, N. Schabanel: The Data Broadcast Problem with non-uniform transmission times. Proc. of the 10th Annual ACM-SIAM Symp. on Discrete Algorithms (SODA 1999). 547–556
14. V. Liberatore: Multicast scheduling for list requests. Proc. of IEEE INFOCOM'02. 1129–1137
15. S. McCanne, V. Jacobson, M. Vetterli: Receiver-driven Layered Multicast. ACM Special Interest Group on Data Communication (SIGCOMM 1996). 117–130
16. N. Schabanel: The Data Broadcast Problem with Preemption. Proc of the 17th International Symp. on Theoretical Computer Science (STACS 2000). 181–192
17. L. Vicisano: Notes on a cumulative layered organisation of data packets across multiple streams with different rates. Unpublished notes.

Minimum Weight Triangulation by Cutting Out Triangles

Magdalene Grantson[1], Christian Borgelt[2], and Christos Levcopoulos[1]

[1] Department of Computer Science,
Lund University, Box 118, 221 Lund, Sweden
{magdalene, christos}@cs.lth.se

[2] Department of Knowledge Processing and Language Engineering,
University of Magdeburg, Universitätsplatz 2, 39106 Magdeburg, Germany
borgelt@iws.cs.uni-magdeburg.de

Abstract. We describe a fixed parameter algorithm for computing the minimum weight triangulation (MWT) of a simple polygon with $(n-k)$ vertices on the perimeter and k hole vertices in the interior, that is, for a total of n vertices. Our algorithm is based on cutting out empty triangles (that is, triangles not containing any holes) from the polygon and processing the parts or the rest of the polygon recursively. We show that with our algorithm a minimum weight triangulation can be found in time at most $O(n^3 k!\, k)$, and thus in $O(n^3)$ if k is constant. We also note that $k!$ can actually be replaced by b^k for some constant b. We implemented our algorithm in Java and report experiments backing our analysis.

1 Introduction

A *triangulation* of a set S of n points in the plane is a maximal set of non-intersecting edges connecting the points in S. A *minimum weight triangulation* (MWT) of S is a triangulation of minimum total edge length. (Note that a MWT need not be unique.) Although it is unknown whether the problem of computing a MWT of a set S of points is NP-complete or solvable in polynomial time [3], it is surely a non-trivial problem, for which no efficient (i.e. polynomial time) algorithm is known. The MWT problem has been studied extensively in computational geometry and has applications in computer graphics [11], image processing [10], database systems [8], and data compression [9].

In this paper we consider the slightly more general problem of finding a MWT of a simple polygon with $(n-k)$ vertices on the perimeter and k hole vertices in the interior, that is, for a total of n vertices. In this case a triangulation is a maximal set of non-intersecting edges (in addition to the perimeter edges), all of which lie inside the polygon. Note that the problem of finding the MWT of a set S of points can be reduced to this problem by finding the convex hull of S, which is then treated as a (convex) polygon, while all vertices not on the convex hull are treated as holes. Here, however, we do *not* require the polygon to be convex and thus solve a more general problem. Note also that for $k = 0$ (no holes) a MWT can be found by dynamic programming in time $O(n^3)$ [4, 7].

X. Deng and D. Du (Eds.): ISAAC 2005, LNCS 3827, pp. 984–994, 2005.

Recent attempts to give exact algorithms for computing a MWT in the general case exploit the idea of parameterization. The basis of such approaches is the notion of a so-called *fixed parameter algorithm*. Generally, such an algorithm has a time complexity of $O(n^c \cdot f(k))$, where n is the input size, k is a (constrained) parameter, c is a constant independent of k, and f is an arbitrary function [2]. The idea is that an algorithm with such a time complexity can be tractable if the parameter k is constrained. For example, for constant k the problem becomes efficiently tractable, because then the time complexity is polynomial in n.

W.r.t. a MWT of a simple polygon with holes the total number n of vertices is the size of the input and we may choose the number k of hole vertices as the constrained parameter. An algorithm based on such an approach was presented in [6] and analyzed to run in $O(n^5 \log(n)\, 6^k)$ time. In [5] we presented an improved algorithm, which we analyzed to run in $O(n^4 4^k k)$ time. We implemented our algorithm in Java and performed experiments backing our analysis.

In this paper we also present a fixed parameter algorithm for the MWT problem, which, however, is based on a different idea. The algorithm developed in [5] repeatedly splits a pointgon with lexi-monotone paths and processes the parts recursively. In contrast to this, our new algorithm repeatedly cuts empty triangles (that is, triangles not containing any holes) from the polygon and processes the parts or the remaining polygon recursively. We analyze that our algorithm runs in at most $O(n^3 k!\, k)$ time, and thus in $O(n^3)$ time if k is constant. We also note that $k!$ can actually be replaced by b^k for some constant b. Again we implemented our algorithm in Java and report experiments backing our analysis.

2 Preliminaries and Basic Idea

As already pointed out, we consider as input a simple polygon with $(n-k)$ vertices on the perimeter and k hole vertices, thus a total of n input vertices. Following [1], we call such a polygon with holes a *pointgon* for short. We denote the set of perimeter vertices by $V_p = \{v_0, v_1, \ldots, v_{n-k-1}\}$, assuming that they are numbered in counterclockwise order starting at an arbitrary vertex. The set of hole vertices we denote by $V_h = \{v_{n-k}, v_{n-k+1}, \ldots, v_{n-1}\}$. The set of all vertices is denoted by $V = V_p \cup V_h$, the pointgon formed by them is denoted by G.

The core idea of our algorithm is based on the following simple observation:

Observation 1. *With the definition* $s(i) = (i+1) \bmod (n-k)$, *let* $e_i = (v_i, v_{s(i)})$, $1 \le i < n-k$, *be an arbitrary edge on the perimeter of a pointgon* G. *Then in every triangulation* T *of* G *there exists a vertex* $v \in V \backslash \{v_i, v_{s(i)})\}$, *adjacent to* v_i *and* $v_{s(i)}$ *that together with* v_i *and* $v_{s(i)}$ *forms an empty triangle, that is, a triangle without any hole vertices in its interior.*

As a consequence of this observation, we can try to find the MWT of a given pointgon G—which is not a triangle without hole vertices itself—by checking all possible empty triangles that can be built over an arbitrarily chosen perimeter edge $(v_i, v_{s(i)})$. Each such empty triangle splits the pointgon into at most two sub-pointgons (see Figure 1), which are then processed recursively.

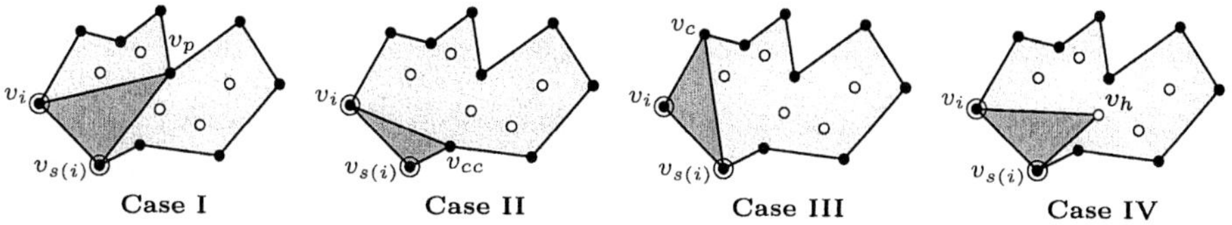

Fig. 1. The four possible ways of forming a triangle from a perimeter edge $(v_i, v_{s(i)})$ (encircled points). A black point indicates a vertex belonging to V_p, a white point a vertex belonging to V_h.

We consider four cases I, II, III, IV in our recursion, as shown in Figure 1. Note that cases II and III can be included as special cases into case I. In the following description, v_c is the vertex adjacent to v_i in clockwise direction, and v_{cc} is the vertex adjacent to $v_{s(i)}$ in counterclockwise direction.

I In this case we consider all empty triangles formed by v_i, $v_{s(i)}$, and a perimeter vertex $v_p \in V_p \backslash \{v_i, v_{s(i)}, v_c, v_{cc}\}$. Each such triangle $(v_i, v_{s(i)}, v_p)$ splits the pointgon G into two sub-pointgons, one to the left and one to the right of the triangle. We denote the set of all such triangles by $\Theta_p(G, v_i)$, where the second argument v_i indicates the base edge $e_i = (v_i, v_{s(i)})$ of the triangle. For each $\theta \in \Theta_p(G, v_i)$, we denote by $L(G, \theta)$ and $R(G, \theta)$ the sub-pointgons to the left and to the right of θ, respectively.

II In this case we consider the triangle $(v_i, v_{s(i)}, v_{cc})$, provided that it is empty. Cutting such a triangle from G yields the sub-pointgon $L(G, (v_i, v_{s(i)}, v_{cc}))$ to the left of the triangle. (There is no sub-pointgon to the right.)

III In this case we consider the triangle $(v_i, v_{s(i)}, v_c)$, provided that it is empty. Cutting such a triangle from G yields the sub-pointgon $R(G, (v_i, v_{s(i)}, v_c))$ to the right of the triangle. (There is no sub-pointgon to the left.)

IV In this case we consider all empty triangles formed by v_i, $v_{s(i)}$, and any hole vertex $v_h \in V_h$. Cutting any such triangle from the pointgon G leaves a sub-pointgon. We denote the set of all such triangles by $\Theta_h(G, v_i)$, where the second argument v_i indicates the base edge $e_i = (v_i, v_{s(i)})$ of the triangle. For each $\theta \in \Theta_h(G, v_i)$, we denote by $L(G, \theta)$ the remaining sub-pointgon.

Given this, the weight of a MWT of G can be computed recursively as

$$\begin{aligned} \mathrm{MWT}(G) = \min \Big\{ & \min_{\theta \in \Theta(G, v_i)} \{\mathrm{MWT}(L(G,\theta)) + \mathrm{MWT}(R(G,\theta)) + |(v_i, v_{s(i)}|\}, \\ & \mathrm{MWT}(L(G, (v_i, v_{s(i)}, v_{cc}))) + |(v_i, v_{s(i)})| + |(v_{s(i)}, v_{cc})|, \\ & \mathrm{MWT}(R(G, (v_i, v_{s(i)}, v_c))) + |(v_i, v_{s(i)})| + |(v_i, v_c)|, \\ & \min_{\theta_h \in \theta(G, v_h)} \{\mathrm{MWT}(L(G, \theta_h)) + |(v_i, v_{s(i)})|\} \Big\}, \end{aligned}$$

where i is an arbitrary integer number in $\{1, \ldots, n-k-1\}$. The four terms in the outer minimum refer to the cases I, II, III, and IV, respectively.

Although the above recursive formula only computes the weight of a MWT, it is easy to see how it can be extended to yield the edges of a MWT. For this, each recursive call also has to return the set of edges that is added in order to achieve a triangulation. The union of these sets of edges for the term that yields the minimum weight is a MWT for the original pointgon G.

Technically, one selects an arbitrary initial perimeter edge $(v_i, v_{s(i)})$ of the pointgon G to be triangulated. Then all triangles listed in the above four cases I, II, III, and IV are generated. Each of these is cut out of the pointgon and the parts or the rest is processed recursively. The minimum weight triangulation is obtained as the minimum over all resulting triangulations.

3 Dynamic Programming

The basic idea, as it was outlined in the preceding section, solves the MWT problem of an input pointgon G in a recursive manner. However, it is immediately clear that it should not be implemented in this way, because several branches of the tree recursion lead to the same sub-pointgon. Hence a direct implementation would lead to considerable redundant computations. A better approach consists in using dynamic programming, which ensures that each possible subproblem is solved at most once, thus rendering the computation much more efficient.

To apply dynamic programming, we have to identify the classes of subproblems that we encounter in the recursion, and we have to find a representation for them. The basic idea is to exploit the fact that we can choose freely, which perimeter edge of a sub-pointgon we want to use as the base of the triangles. By always choosing an edge that was a side of the triangle chosen in the preceding recursion step (preferring the left side if both are in the new subproblem), we can achieve that the sides of the triangles chosen in consecutive recursion steps connect into a path through the holes (of the input pointgon). As a consequence, every subproblem we encounter can be described by a path π through zero or more hole vertices and a possible coherent perimeter piece δ_p of G (see Figure 2). We show later that we actually encounter only subproblems of this type.

We represent a subproblem by an index word over an alphabet with n characters, which uniquely identifies each subproblem. This index word has the general form (v_i, π) and describes a counterclockwise walk round the perimeter of the subproblem. The first element is the so-called *anchor* v_i of the subproblem, which may be a perimeter vertex or a hole vertex of the input pointgon. π describes a sequence of hole vertices of the input pointgon, i.e., all elements of π are in V_h, with the possible exception of the last element, which may be a perimeter vertex $s_\pi \in V_p$ of the input pointgon. Note that in this case a possible coherent perimeter piece between the anchor v_i and the end vertex s_π (if it exists) is not part of the subproblem representation, but is left implicit. Note that in the case where the path π bounding a subproblem loops back to the anchor, this end vertex is not contained in π to avoid duplicate entries. As a consequence we can have at most $O(n^2 k!)$ index words, where the n^2 comes from the possible choices of the anchor and the end vertex, and the $k!$ from the possible sequences of holes.

The general idea of using such an index word is the following: in order to avoid redundant computations, we have to be able to efficiently store and retrieve the solutions of already solved subproblems. For the problem at hand it is most convenient to use a trie structure, which is accessed through the index word representing a subproblem. That is, for our implementation, we do not use the standard, table-based form of dynamic programming, but a version of implementing the recursion outlined above, which is sometimes called "memorized". In each recursive call, we first access the trie structure (using the index word of a sub-pointgon) in order to find out whether the solution of the current subproblem is already known. If it is, we simply retrieve and return the solution. Otherwise we actually carry out the split computations and in the end store the found solution in the trie. Although this approach is slightly less efficient than a true table-based version (since there are superfluous accesses to the trie, namely the unsuccessful ones), its additional costs do not worsen the asymptotic time complexity. The following pseudocode describes our algorithm:

```
function MWT (word key) : real
begin
    if    key is in trie
    then return trie.getweight(key); fi;
    if    polygon(key) is a triangle
    and   polygon(key) contains no holes
    then wgt = perimeter_length(polygon(key));
          trie.add(key, wgt, ⊥);
          return wgt;
    fi;
    min = +∞; best = ⊥;
    for all triangles θ ∈ Θ_p(polygon(key), key.v_i)              (* case I *)
                      ∪ {(key.v_i, key.v_s(i), v_c(key.v_i))}     (* case II *)
                      ∪ {(key.v_i, key.v_s(i), v_cc(key.v_i))}    (* case III *)
                      ∪ Θ_h(polygon(key), key.v_i) do             (* case IV *)
        wgt = MWT(L(key, θ)) + MWT(R(key, θ)) + |v_i, v_s(i)|;
        if    (wgt < min)
        then min = wgt; best = θ; fi;
    done;
    trie.add(key, min, best);
    return min;
end;
```

In this pseudocode the symbol $\perp$ is used to indicate that there is no triangle θ that can be cut out. (This is, of course, only the case if the sub-pointgon is itself a triangle without holes.) $key.v_i$ and $key.v_{s(i)}$ are the anchor and the first vertex on the path representing a subproblem. v_c and v_{cc} are to be seen as functions yielding the clockwise or the counterclockwise neighbor on the perimeter of the subproblem. Note that if we are in one of the cases II, III, or IV, one of $\mathrm{L}(key, \theta)$ or $\mathrm{R}(key, \theta)$ is empty. It should be clear that in this case the corresponding term is dropped from the sum computing the triangulation weight.

Fig. 2. The two types of pointgons we encounter. A black point indicates a vertex belonging to V_p, a white point a vertex belonging to V_h, a grey point a vertex either in V_p or in V_h. The encircled and labeled points are the anchors. Thick lines indicate pieces of the perimeter of the input pointgon.

To collect the edges of the solution, the following function is used:

function collect (word *key*) : set of edges
begin
  $\pi = \mathit{trie}.\text{getpath}(\mathit{key})$;
  if $\pi = \perp$ **then return** $\emptyset$; **fi**;
  return $\text{collect}(\text{L}(\mathit{key}, \theta)) \cup \text{collect}(\text{R}(\mathit{key}, \theta)) \cup \text{edges}(\theta)$;
end;

This function is called with the index word describing the input pointgon, that is, an index word consisting only of the initial anchor vertex v_i.

3.1 Types of Pointgons

Apart from the input pointgon, which is of neither of these types, we encounter two types of sub-pointgons (see Figure 2 for sketches):

A Sub-pointgons of this type are bounded by a coherent perimeter piece of the original pointgon and a path π through zero or more hole vertices. Its anchor is the vertex at the clockwise end of the path (perimeter vertex).

B Sub-pointgons of this type are bounded by a path π through hole vertices and zero or one perimeter vertex $v_p \in V_p$ of the input pointgon. Its anchor is the lexicographically smallest vertex (hole or perimeter vertex).

We show now that we only encounter these two types of pointgons in the recursion. For this, recall that in each step of the recursion, one of the four cases I, II, III, and IV is used to cut a triangle from a pointgon (see Section 2).

Type A Pointgons. (Note that the input pointgon, although strictly speaking it is not of this type, behaves exactly in the same way.) The different splits of type A pointgons are sketched in Figure 3. Note that the triangle is always formed over the edge at the clockwise end of the path through hole vertices. Therefore, if we have a triangle of case I, we obtain either two sub-pointgons of type A or one sub-pointgon of type A and B each. If the triangle is of case II, then depending on the number of vertices in the coherent perimeter section the result is either a sub-pointgon of type A or of type B. Type B results if there are only two perimeter vertices in the section, type A otherwise. If the triangle is of case III or case IV, the result is necessarily of type A again. Note from the sketches in Figure 3 how the anchors of the sub-pointgons are selected.

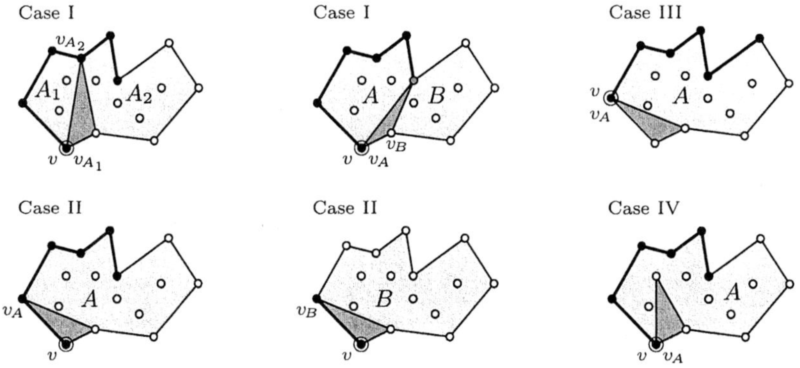

Fig. 3. Behavior of Type A pointgons in the recursion (possible splits). v denotes the original anchor and v_*, $* \in \{A, B\}$, denotes the anchor of a sub-pointgon of type $*$ due to the split.

Type B Pointgons. Since type B pointgons do not have a consecutive perimeter piece of the input pointgon (they may contain at most one perimeter vertex), it is immediately clear that regardless of the case we consider, the resulting sub-pointgon(s) must be of type B again (cf. Figure 1, which shows the four cases). It is only important to notice that in a sub-pointgon to the right of the triangle, the anchor has to be set to the lexicographically smallest vertex on the perimeter of the sub-pointgon. (Note that in a sub-pointgon to the left of the triangle the old anchor must still be the lexicographically smallest vertex.)

4 Analysis

In order to estimate the time complexity of our algorithm, we multiply the (worst case) number of subproblems by the (worst case) time it takes to process one subproblem. This time is computed as the (worst case) number of possible triangles that can be built over a given perimeter edge multiplied with the (worst case) time for processing one such triangle.

As discussed in Section 3, there are at most $O(n^2k!)$ possible subproblems if we proceed in the described manner. The factor $k!$ is actually an overestimate as the path through the holes must not intersect itself. Drawing on bounds for planar configurations of a set of points [12], it can be shown that the actual number of valid paths is only $O(b^k)$ for some constant b. However, the best known asymptotic bounds on the constant b are fairly large, so that for practical purposes, for which k has to be very small, $O(k!)$ is a better estimate.

Given a subproblem and a perimeter edge of this subproblem, there are at most $n-2$ possible triangles that can be built over the given edge, as there are at most $n-2$ possible choices for the tip of the triangle.

For each triangle, we have to check whether its two additional sides (the base of the triangle, of course, need not be checked) intersects the perimeter of the sub-pointgon. This check can be made very efficient by a preprocessing step in

which we determine for each edge that could be part of a triangle whether it intersects the perimeter of the input pointgon or not. The resulting table has a size of at most n^2, which is negligible compared to the number of subproblems, and its computation takes at most $O(n^3)$ time, because each of the edges has to be checked against the $n - k$ perimeter edges. With this table we can check in $O(1)$ whether a given triangle intersects a (possibly existing) perimeter piece.

In addition, we have to check for an intersection of the two added triangle sides with the path through hole vertices, which contains at most $k + 1$ edges. Since we have to check for each of the edges on the path whether it intersects with one of the two triangle sides, this check can be carried out in $O(k)$.

Finally we have to check whether the triangle is empty (i.e., contains no hole vertices). This also takes at most $O(k)$ time, because we have to check the location of at most k hole vertices w.r.t. the triangle sides.

Once a triangle is found to be valid, the sub-pointgons have to be constructed by collecting their at most $k+2$ defining vertices, and their solutions have to be looked up using the describing index words. Both operations obviously take $O(k)$ time. Finally the length of the base of the triangle and maybe also the length of one of the added sides have to be computed, which takes constant time. As a consequence processing one triangle takes in all $O(k)$ time.

As a consequence the overall time complexity is at most

$$O(\underbrace{n^2 k!}_{\text{subproblems}} \cdot \underbrace{n}_{\text{triangles}} \cdot \underbrace{k}_{\text{time/triangle}}) = O(n^3 k!\, k).$$

5 Implementation

As already pointed out above, we implemented our algorithm in Java. Example results for different numbers of holes and perimeter vertices are shown in Table 1. The test system was an Intel Pentium 4C@2.6GHz with 1GB of main memory running S.u.S.E. Linux 9.3 and Sun Java 1.5.0_01. All execution times are averages of 20 runs, carried out on randomly generated convex pointgons. We used convex pointgons, because they seem to represent the worst case. Non-convex pointgons, due to intersections of paths through holes with the perimeter, are usually processed faster (there are fewer possible sub-pointgons). The left table shows the results for the algorithm as it was described in the preceding sections, while the right table shows the results for the algorithm we developed in [5]. As these tables show, our new algorithm beats the previous one by an order of magnitude, making it the method of choice for few hole vertices.

To check our theoretical result about the time complexities, we computed the ratios of the measured execution times to the theoretical values (see last columns of both tables; note that the two algorithms have different theoretical time complexities). As can be seen, these ratios are decreasing for increasing values of n and k, indicating that the theoretical time complexity is actually a worst case, while average results in practice are considerably better. As pointed out above, it can even be shown that the dependence on k is actually only $O(b^k)$ for some constant b, and thus asymptotically better than $O(k!) = O(k^k)$.

Table 1. Results obtained with our Java implementations of the described MWT algorithm (left) and the path-based algorithm of [5] (right, version without hole distribution). All results are averages over 20 runs, with randomly generated convex pointgons.

$n-k$	k	time in seconds	time/$n^3 k!\, k$
3	1	0.008± 0.000	$1.250 \cdot 10^{-4}$
6	1	0.009± 0.000	$2.624 \cdot 10^{-5}$
9	2	0.014± 0.001	$2.630 \cdot 10^{-6}$
12	3	0.039± 0.005	$4.280 \cdot 10^{-7}$
15	4	0.060± 0.004	$3.417 \cdot 10^{-8}$
18	5	0.092± 0.017	$2.420 \cdot 10^{-9}$
21	6	0.272± 0.102	$2.962 \cdot 10^{-10}$
24	7	0.885± 0.285	$3.607 \cdot 10^{-11}$
27	8	2.849± 0.818	$3.961 \cdot 10^{-12}$
30	9	12.791± 5.480	$5.566 \cdot 10^{-13}$

$n-k$	k	time in seconds	time/$n^4 4^k k$
3	1	0.010± 0.000	$9.766 \cdot 10^{-6}$
6	1	0.011± 0.000	$1.145 \cdot 10^{-6}$
9	2	0.049± 0.004	$1.046 \cdot 10^{-7}$
12	3	0.076± 0.006	$7.819 \cdot 10^{-9}$
15	4	0.151± 0.029	$1.132 \cdot 10^{-9}$
18	5	0.535± 0.144	$3.734 \cdot 10^{-10}$
21	6	2.115± 0.869	$1.619 \cdot 10^{-10}$
24	7	9.142± 3.982	$8.631 \cdot 10^{-11}$
27	8	43.548±19.198	$5.535 \cdot 10^{-11}$
30	9	176.588±71.473	$3.235 \cdot 10^{-11}$

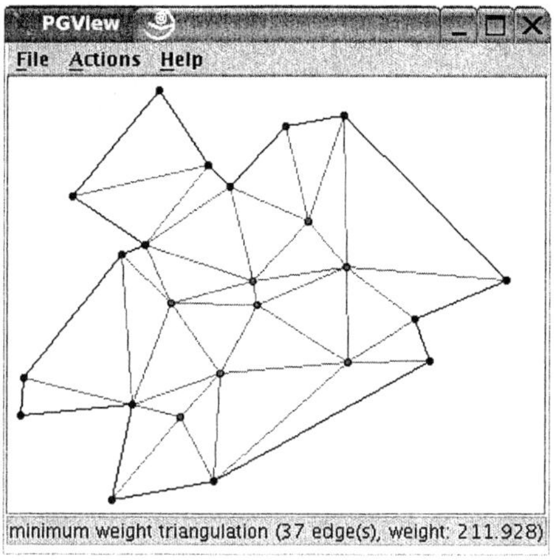

Fig. 4. A screen shot of the graphical user interface to our implementation of the described algorithm. The program allows for loading pointgons from a text file, generating random pointgons, finding their minimum weight triangulation, and modifying a triangulation as well as recomputing its weight in order to check whether it is actually minimal. When a MWT is computed, search statistics are printed about the number of triangles, subproblems, and separating paths considered. This screen shot shows the MWT of a randomly generated (star-like) pointgon with 16 vertices on the perimeter and 8 holes, which has 37 edges (excluding the perimeter edges). It was computed in about 0.3 seconds (on an Intel Pentium 4C@2.6GHz system with 1GB of main memory running S.u.S.E. Linux 9.3 and Sun Java 1.5.0_01).

To give an impression of the graphical user interface (GUI) of the program, Figure 4 shows a screen shot of the main window. With this user interface it is possible to load pointgons from text files, to generate random pointgons, to find their minimum weight triangulation, and to modify a triangulation as well as to recompute its weight in order to check whether it is actually minimal. Apart from

the GUI version, the program can be invoked on the command line, a feature we exploited to script the test runs reported above. The Java source code of our implementation as well as an executable Java archive (jar) can be downloaded free of charge at `http://fuzzy.cs.uni-magdeburg.de/~borgelt/pointgon.html`.

6 Conclusions

We described a fixed parameter algorithm for computing the minimum weight triangulation of a simple polygon with hole vertices, which is based on recursively cutting out empty triangles from the input pointgon. As we showed in our analysis, the time complexity of our algorithm is $O(n^3 k!\, k)$. (Note that we use $k!$ instead of b^k, for some constant b, because in known upper bounds on the number of crossing-free paths of a set of points the constant b is so large that b^k is worse than $k!$ for the small values of k that are practically relevant.) W.r.t. the total number of vertices it is therefore much better than the algorithm we developed in [5] (time complexity $O(n^4 4^k k)$). An important advantage of our algorithm is that for a constant number of holes (and in particular for no holes, i.e. for $k = 0$) it achieves the best known $O(n^3)$ time bound for the minimum weight triangulation of a simple polygon. In addition, we presented a Java implementation of our algorithm, and reported experiments that were carried out with this implementation. These experiments indicate that the actual time complexity may be considerably better than the result of our theoretical analysis (presumably due to an overestimate of the number of subproblems).

References

1. O. Aichholzer, G. Rote, B. Speckmann, and I. Streinu. The Zigzag Path of a Pseudo-Triangulation. *Proc. 8th Workshop on Algorithms and Data Structures (WADS 2003), LNCS 2748*, 377–388. Springer-Verlag, Berlin, Germany 2003
2. R. Downey and M. Fellows. *Parameterized Complexity.* Springer-Verlag, New York, NY, USA 1999
3. M.R. Garey and D.S. Johnson. *Computers and Intractability: A Guide to Theory of NP-Completeness.* Freeman, New York, NY, USA 1979
4. P.D. Gilbert. New Results in Planar Triangulations. *Report R-850.* University of Illinois, Coordinated Science Lab, 1979
5. M. Grantson, C. Borgelt and C. Levcopoulos. A Fixed Parameter Algorithm for Minimum Weight Triangulation: Analysis and Experiments. Technical Report LU-CS-TR:2005-234, ISSN 1650-1276 Report 154. Lund University, Sweden 2005
6. M. Hoffmann and Y. Okamoto. The Minimum Triangulation Problem with Few Inner Points. *Proc. 1st Int. Workshop on Parameterized and Exact Computation (IWPEC 2004), LNCS 3162*, 200–212. Springer-Verlag, Berlin, Germany 2004
7. G.T. Klincsek. Minimal Triangulations of Polygonal Domains. *Annals of Discrete Mathematics* 9:121–123. ACM Press, New York, NY, USA 1980
8. E. Lodi, F. Luccio, C. Mugnai, and L. Pagli. On Two-Dimensional Data Organization, Part I. *Fundaments Informaticae* 2:211–226. Polish Mathematical Society, Warsaw, Poland 1979

9. A. Lubiw. The Boolean Basis Problem and How to Cover Some Polygons by Rectangles. *SIAM Journal on Discrete Mathematics* 3:98–115. Society of Industrial and Applied Mathematics, Philadelphia, PA, USA 1990
10. D. Moitra. Finding a Minimum Cover for Binary Images: An Optimal Parallel Algorithm. *Algorithmica* 6:624–657. Springer-Verlag, Heidelberg, Germany 1991
11. D. Plaisted and J. Hong. A Heuristic Triangulation Algorithm. *Journal of Algorithms* 8:405–437 Academic Press, San Diego, CA, USA 1987
12. M. Sharir and E. Welzl. On the Number of Crossing-Free Matchings (Cycles and Partitions), *Proc. 17th Annual ACM-SIAM Symposium on Discrete Algorithms (SODA 2006)*, to appear.

Multi-directional Width-Bounded Geometric Separator and Protein Folding*

Bin Fu[1,2], Sorinel A Oprisan[3], and Lizhe Xu[1,2]

[1] Dept. of Computer Science, University of New Orleans, LA 70148, USA
[2] Research Institute for Children, 200 Henry Clay Avenue, LA 70118
[3] Dept. of Psychology, University of New Orleans, LA 70148
fu@cs.uno.edu, soprisan@uno.edu, lxu@chnola-research.org

Abstract. We introduce the concept of multi-directional width-bounded geometric separator and get improved separator for the grid graph, which improves exact algorithm for the protein folding problem in the HP-model. For a grid graph G with n grid points P, there exists a separator $A \subseteq P$ such that A has less than or equal to $1.02074\sqrt{n}$ points, and $G - A$ has two disconnected subgraphs with less than or equal to $\frac{2}{3}n$ nodes on each of them. We also derive $0.7555\sqrt{n}$ lower bound for such a separator on grid graph. The previous upper bound record for the grid graph $\frac{2}{3}$-separator is $1.129\sqrt{n}$ [6].

1 Introduction

Lipton and Tarjan [11] showed that every n vertices planar graph has at most $\sqrt{8n}$ vertices whose removal separates the graph into two disconnected parts of size at most $\frac{2}{3}n$. Their $\frac{2}{3}$-separator was improved by a series of papers [4, 8, 1, 5] with best record $1.97\sqrt{n}$ by Djidjev and Venkatesan [5]. Spielman and Teng [14] found a $\frac{3}{4}$-separator with size $1.82\sqrt{n}$ for planar graphs. Some other forms of the geometric separators were studied by Miller, Teng, Thurston, and Vavasis [12] and by Smith and Wormald [13]. If each of n input points is covered by at most k regular geometric object such as circles, rectangles, etc, then there exist $O(\sqrt{k \cdot n})$ size separators [12, 13]. In particular, Smith and Wormald obtained the separator of size $4\sqrt{n}$ for the case $k = 1$. The lower bounds $1.555\sqrt{n}$ and $1.581\sqrt{n}$ for the $\frac{2}{3}$-separator for the planar graph were proven by Djidev [5], and by Smith and Wormald [13], respectively.

Each edge in a grid graph connects two grid points of distance 1 in the set of vertices. Thus a grid graph is a special planar graph. Fu and Wang [6] developed a method for deriving sharper upper bound separator for grid graphs by controlling the distance to the separator line. Their separator is determined by a straight line on the plane and the set of grid points with distance less than or equal to $\frac{1}{2}$ to the line. They proved that for an n vertices grid graph on the plane, there is a separator that has less than or equal to $1.129\sqrt{n}$ grid points and each of

* This research is supported by Louisiana Board of Regents fund under contract number LEQSF(2004-07)-RD-A-35.

X. Deng and D. Du (Eds.): ISAAC 2005, LNCS 3827, pp. 995–1006, 2005.

two disconnected subgraphs has at most $\frac{2}{3}n$ grid points. Using this separator and their approximation to the separator line, they obtained the first $n^{O(n^{1-\frac{1}{d}})}$-time exact algorithm for the d-dimensional protein folding problem of the HP-model. The method of Fu and Wang [6] was further developed and generalized by Fu [7]. The notion of width-bounded geometric separator was introduced by Fu [7]. For a positive constant a and a set of points Q on the plane, an a-wide separator is the region between two parallel lines of distance a that partitions Q into Q_1 (on the left side of the separator's region), S (inside the separator's region), and Q_2 (on the right side of the separator's region). The width-bounded geometric separators were applied to many other problems, which include the disk covering problem on the plane, the maximum independent set problem, the vertex covering problem, and dominating set problem on disk graphs. Fu [7] derived $2^{O(\sqrt{n})}$-time exact algorithms for all of them, whose previous algorithms need $n^{O(\sqrt{n})}$ time.

This paper introduces the concept of a multi-directional width-bounded separator. For a set of points P on the plane and two vectors v_1 and v_2, the (a, b)-wide separator (along the directions v_1 and v_2) is the region of points that have distance less than or equal to a to L along v_1 or distance less than or equal to b to L along v_2, where L is a straight line (separator line) on the plane. The separator size is measured by the number of points from P in the region and the line L partitions the set P into two balanced subsets. In this paper we use this new method to improve the separator for the grid graph. The multi-directional width approach is different from that used in [6, 7], which only controls the regular distance to the middle line in the separator area. Pursuing smaller and more balanced separator is an interesting problem in combinatorics and also gives more efficient algorithms for the divide and conquer applications. For a grid graph G with n grid points P, there exists a separator subset $A \subseteq P$ such that A has up to $1.02074\sqrt{n}$ points, and $G - A$ has two disconnected subgraphs with up to $\frac{2}{3}n$ nodes on each of them. This improves the previous $1.129\sqrt{n}$ size separator for the grid graph [6]. We also prove a $0.7555\sqrt{n}$ lower bound for the size of the separators for grid graphs. Our lower bound is based on a result that the shortest curve partitioning an unit circle into two areas with ratio $1 : 2$ is a circle arc. Its length is less than that of the straight line partitioning the circle with the same ratio.

Protein structure prediction with computational technology is one of the most significant problems in bioinformatics. A simplified representation of proteins is a lattice conformation, which is a self-avoiding sequence in Z^3. An important representative of lattice models is the HP-model, which was introduced by Lau and Dill [9, 10]. In this model, the 20 amino acids are reduced to a two letter alphabet by H and P, where H represents hydrophobic amino acids, and P, polar or hydrophilic amino acids. Two monomers form a contact in some specific conformation if they are not consecutive, but occupy neighboring positions in the conformation (i.e., the distance vector between their positions in the conformation is a unit vector). A conformation with minimal energy is a conformation with the maximal number of contacts between non-consecutive H-monomers (see

Figure 3). The folding problem in the HP-model is to find the conformation for any HP-sequence with minimal energy. This problem was proved to be NP-hard in both 2D and 3D [2, 3]. We will apply our new separator to the protein folding problem in the 2D HP model to get an $O(n^{5.563\sqrt{n}})$ time exact algorithm, improving the previous $O(n^{6.145\sqrt{n}})$-time algorithm [6].

2 Separators Upper Bound for Grid Graphs

We first give a series of notations. For a set A, $|A|$ denotes the number of elements in A. For two points p_1, p_2 in the d-dimensional space (R^d), $\text{dist}(p_1, p_2)$ is the Euclidean distance. For a set $A \subseteq R^d$, $\text{dist}(p_1, A) = \min_{q \in A} \text{dist}(p_1, q)$. The integer set is represented by $Z = \{\cdots, -2, -1, 0, 1, 2, \cdots\}$. For integers x_1 and x_2, (x_1, x_2) is a *grid point.* A *grid square* is an 1×1 square that has four grid points as its four corner points. For a set V of grid points on the plane, let E_V be the set of edges (v_i, v_j) (straight line segments) such that $v_i, v_j \in V$ and $\text{dist}(v_i, v_j) = 1$. Define $G = (V, E_V)$ as the *grid graph.* For $0 < \alpha < 1$, an α-separator for a grid graph $G = (V, E_V)$ is a subset $A \subseteq V$ such that $G - A$ has two disconnected areas $G_1 = (V_1, E_{V_1})$ and $G_2 = (V_2, E_{V_2})$ with $|V_1|, |V_2| \leq \alpha|V|$. Define $C(o, r) = \{(x, y)|\text{dist}((x, y), o) \leq r\}$, which is the disc area with center at point o and radius r. For $r > 0$, define $D(r)$ to be the union region of 4 discs $C((0, -r), r) \cup C((0, r), r) \cup C((-r, 0), r) \cup C((r, 0), r)$ (see the left of Figure 1). For a region R on the plane, define $G(R)$ to be the set of all grid points in the region R. For a 2D vector v, a *line* L in R^2 through a fixed point $p_0 \in R^2$ along the direction v corresponds to the equation $p = p_0 + tv$ that characterizes all the points p on L, where the parameter $t \in (-\infty, +\infty)$. For a point p_0 and a line L, the distance of p_0 to L along direction v is $\text{dist}(p_0, q)$, where q is the intersection between $p = p_0 + tv$ and L. Let $v_1, v_2, \cdots, v_k$ be k fixed vectors. A point p has distance $\leq (a_1, \cdots, a_k)$ to L along directions $v_1, v_2, \cdots, v_k$ if p has distance $\leq a_i$ along direction v_i for some $i = 1, \cdots, k$.

Definition 1. *Let P be a set of points in R^2, $v_1, \cdots, v_m$ be 2D vectors, and $w_1, \cdots, w_m$ be positive real numbers. A $(w_1, \cdots, w_m)$-wide separator for the set P along the directions $v_1, \cdots, v_m$ is a region R determined by a line L. The region R consists all points with distance $\leq (w_1, \cdots, w_m)$ along $v_1, \cdots, v_m$. The separator size is measured by the number of points of P in the region R. Its balance number is the least number α such that each side of L has at most $\alpha|P|$ points from P.*

In the rest of this paper, we use two vectors $v_1 = (1, 0)$ an $v_2 = (0, 1)$ to represent the horizontal and vertical directions, respectively. If a point p has distance $\leq (a, a)$ from a line L, it means that the point p has distance $\leq a$ from L along either direction $(1, 0)$ or $(0, 1)$ in the rest of this paper.

Lemma 1. *([15]) For an n-element set P in a d-dimensional space, there is a point q with the property that any half-space that does not contain q covers at most $\frac{d}{d+1}n$ elements of P. (Such a point q is called a centerpoint of P).*

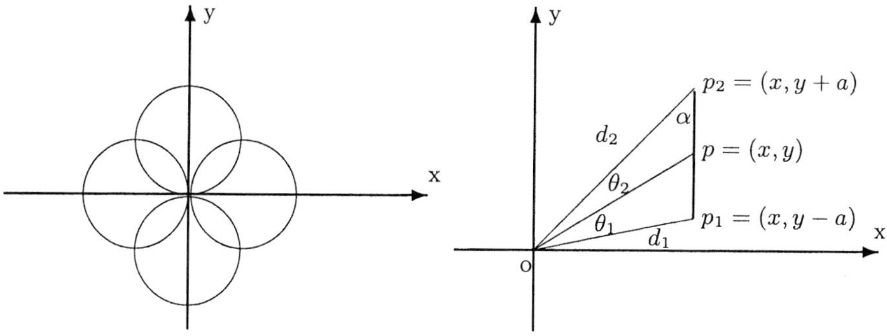

Fig. 1. Left: Area of grid points with maximal expectation. Right: Probability analysis.

Lemma 2. *Let P be a set of grid points on the plane and $(0,0) \notin P$. The sum $\sum_{p=(x,y)\in P} \max(\frac{|x|}{x^2+y^2}, \frac{|y|}{x^2+y^2})$ is maximal when $P \subseteq G(D(R))$, where R is the least radius with $|G(D(R))| \geq |P|$.*

Proof. Let L be the line segment connecting $o = (0,0)$ and $p = (x,y)$. If $p' = (x', y')$ is another point between o and p on the line L, we have $\frac{\max(|x|,|y|)}{\text{dist}(o,p)} = \frac{\max(|x'|,|y'|)}{\text{dist}(o,p')}$. Since $\text{dist}(o,p) > \text{dist}(o,p')$, we have $\frac{\max(|x|,|y|)}{\text{dist}(o,p)^2} < \frac{\max(|x'|,|y'|)}{\text{dist}(o,p')^2}$. For the constant c, let $\frac{|x|}{x^2+y^2} = c$ or $\frac{|y|}{x^2+y^2} = c$. We have $x^2 + y^2 - \frac{1}{c}|x| = 0$ or $x^2 + y^2 - \frac{1}{c}|y| = 0$. The two equations characterize the four circles of $D(\frac{1}{2c})$. All points on the external boundary $D(r)$ have the same value $\frac{\max(|x|,|y|)}{\text{dist}(o,p)^2}$. □

Let a be a constant > 0, p and o be two points on the plane, and P be a set of points on the plane. We define the function $f_{p,o,a}(L) = 1$ if p has $\leq (a,a)$ distance to the line L and L is through o; and 0 otherwise. Define $F_{P,o,a}(L) = \sum_{p\in P} f_{p,o,a}(L)$, which is the number of points of P with $\leq (a,a)$ distance to L for the line L through o. The expectation $\text{E}(F_{P,o,a})$ is the expected number of points in P with distance $\leq (a,a)$ to the random line L through o.

Lemma 3. *Let $a > 0$ be a constant and $\delta > 0$ be a small constant. Let P be a set of n grid points on the plane and o be a point on the plane. Then $\text{E}(F_{P,o,a}) \leq \frac{(4\pi+8)(1+\delta)a\sqrt{n}}{\pi\sqrt{4+2\pi}}$.*

Proof. Without loss generality, we assume that $o = (0,0)$ (Notice that $F_{P,o,a}$ is invariant under translation). Let $\epsilon > 0$ be a small constant that will be fixed later. Let us consider a grid point $p = (x,y) \in P$ on the plane and let $p_1 = (x, y-a)$ and $p_2 = (x, y+a)$. The angle between the two lines op_1 and op_2 will be estimated (Figure 1). Let $d = \text{dist}(o,p)$, $d_1 = \text{dist}(o,p_1)$ and $d_2 = \text{dist}(o,p_2)$. Define the angles $\theta_1 = \angle pop_1, \theta_2 = \angle pop_2$ and $\alpha = \angle op_2p$.

From $\frac{a}{\sin\theta_2} = \frac{d}{\sin\alpha}$, we have $\sin\theta_2 = \frac{a}{d} \cdot \sin\alpha = \frac{a}{d} \cdot \frac{|x|}{d_2} = \frac{a|x|}{dd_2}$. Similarly, $\sin\theta_1 = \frac{a|x|}{dd_1}$. If $d > a$, then $\frac{a|x|}{d(d+a)} \leq \sin\theta_1, \sin\theta_2 \leq \frac{a|x|}{d(d-a)}$.

Let $\beta_1 = \angle poq_1$ ($\beta_2 = \angle poq_2$) be the angle between the line segments op and oq_1 (oq_2 respectively), where $q_1 = (x - a, y)$ and $q_2 = (x + a, y)$. If $d > a$, then we also have $\frac{a|y|}{d(d+a)} \leq \sin\beta_1, \sin\beta_2 \leq \frac{a|y|}{d(d-a)}$. There is a constant d_0 such that if $d > d_0$, then we have the following inequalities: (i) $\frac{a|y|}{d^2}(1-\epsilon) \leq \beta_1, \beta_2 \leq \frac{a|y|}{d^2}(1+\epsilon)$, (ii) $\frac{a|x|}{d^2}(1-\epsilon) \leq \theta_1, \theta_2 \leq \frac{a|x|}{d^2}(1+\epsilon)$, and (iii) $\frac{(1-\epsilon)a\max(|x'|,|y'|)}{d'^2} < \frac{a\max(|x|,|y|)}{d^2} < \frac{(1+\epsilon)a\max(|x'|,|y'|)}{d'^2}$ for any (x', y') with $\mathrm{dist}((x,y),(x',y')) \leq \sqrt{2}$, where $d' = \mathrm{dist}((x',y'),o)$.

Let $Pr(o,p,a)$ be the probability that the point p has distance $\leq (a,a)$ to a random line L through o. If $d \leq d_0$, then $Pr(o,p,a) \leq 1$. Otherwise, $Pr(o,p,a) \leq \frac{\max(2\max(\beta_1,\beta_2), 2\max(\theta_1,\theta_2))}{\pi} \leq \frac{2}{\pi}\max\left(\frac{a|y|}{d^2}, \frac{a|x|}{d^2}\right)(1+\epsilon)$. The number of grid points with distance $\leq d_0$ to o is $\leq \pi(d_0+\sqrt{2})^2$. Then $\mathrm{E}(F_{P,o,a}) = \sum_{p\in P} E(f_{o,p,a}) = \sum_{p\in P} Pr(o,p,a) \leq \sum_{p\in P \text{ and } \mathrm{dist}(p,o)>d_0} Pr(o,p,a) + \sum_{p\in P \text{ and } \mathrm{dist}(p,o)\leq d_0} Pr(o,p,a) \leq \frac{2(1+\epsilon)}{\pi}\sum_{p\in P \text{ and } \mathrm{dist}(p,o)>d_0} \max\left(\frac{|x|}{d^2}, \frac{|y|}{d^2}\right) + \pi(d_0+\sqrt{2})^2$.

We only consider the case to make $\sum_{p\in P \text{ and } d>d_0} \max(\frac{|x|}{d^2}, \frac{|y|}{d^2})$ maximal. By Lemma 2, it is maximal when the points of P are in the area $D(R)$ with the smallest R.

For a grid point $p = (i,j)$, define $\mathrm{grid}_1(p) = \{(x,y) | i - \frac{1}{2} < x < i + \frac{1}{2} \text{ and } j - \frac{1}{2} < y < j + \frac{1}{2}\}$, and $\mathrm{grid}_2(p) = \{(x,y) | i - \frac{1}{2} \leq x \leq i + \frac{1}{2} \text{ and } j - \frac{1}{2} \leq y \leq j + \frac{1}{2}\}$. If the grid point $p \notin D(R)$, then $\mathrm{grid}_1(p) \cap D(R - \frac{\sqrt{2}}{2}) = \emptyset$. The area size of $D(R)$ is $2\pi R^2 + 4R^2$. Assume R is the minimal radius such that $D(R)$ contains at least n grid points. The region $D(R-\epsilon)$ contains $< n$ grid points for every $\epsilon > 0$. This implies $D(R - \epsilon - \frac{\sqrt{2}}{2}) \subseteq \cup_{\text{grid point } p\in D(R-\epsilon)} \mathrm{grid}_2(p)$. Therefore, $2\pi(R - \frac{\sqrt{2}}{2} - \epsilon)^2 + 4(R - \frac{\sqrt{2}}{2} - \epsilon)^2 \leq n$. Hence, $R \leq \frac{\sqrt{n}}{\sqrt{4+2\pi}} + \frac{\sqrt{2}}{2} + \epsilon < \frac{\sqrt{n}}{\sqrt{4+2\pi}} + \sqrt{2}$ (the constant ϵ will be $\leq \frac{\sqrt{2}}{2}$).

Let $A_1 = \{p = (x,y) \in D(R) | \text{the angle between } op \text{ and } x\text{-axis is in } [0, \frac{\pi}{4}]\}$, which is the $\frac{1}{8}$ area of $D(R)$. The probability that a point $p(= (x,y))$ has distance $\leq (a,a)$ to the random line L is $\leq \frac{2(1+\epsilon)ax}{\pi d^2}$ for p in A_1 with $\mathrm{dist}(p,o) > d_0$. The expectation of the number of points (with distance $\leq (a,a)$ to L and distance $> d_0$ to o) of P in the area A_1 is $\sum_{p\in A_1\cap P \text{ and } \mathrm{dist}(p,o)>d_0} Pr(o,p,a) \leq \sum_{p\in A_1\cap P \text{ and } \mathrm{dist}(p,o)>d_0} \frac{2(1+\epsilon)ax}{\pi d^2} \leq \int\int_{A_1} \frac{2(1+\epsilon)^2 ax}{\pi d^2} d_x d_y = \frac{2(1+\epsilon)^2 a}{\pi}\int_0^{\frac{\pi}{4}}\int_0^{2R\cos\theta} \frac{r\cos\theta}{r^2} \cdot r d_r d_\theta = \frac{2(1+\epsilon)^2 a}{\pi}\int_0^{\frac{\pi}{4}}\int_0^{2R\cos\theta} \cos\theta d_r d_\theta = \frac{2(1+\epsilon)^2 aR}{\pi}\int_0^{\frac{\pi}{4}} 2(\cos\theta)^2 d_\theta = \frac{2(1+\epsilon)^2 aR}{\pi}\cdot(\frac{\pi}{4}+\frac{1}{2}) = \frac{(1+\epsilon)^2 aR}{\pi}\cdot(\frac{\pi}{2}+1)$.

Since $R \leq \frac{\sqrt{n}}{\sqrt{4+2\pi}} + \sqrt{2}$, the total expectation is

$\mathrm{E}(F_{P,o,a}) \leq 8\sum_{p\in A_1\cap P \text{ and } \mathrm{dist}(p,o)>d_0} Pr(o,p,a) + \pi(d_0+\sqrt{2})^2 \leq \frac{8(1+\epsilon)^2 aR}{\pi} \cdot (\frac{\pi}{2}+1) + \pi(d_0+\sqrt{2})^2 \leq \frac{(4\pi+8)(1+3\epsilon)a\sqrt{n}}{\pi\sqrt{4+2\pi}} \leq \frac{(4\pi+8)(1+\delta)a\sqrt{n}}{\pi\sqrt{4+2\pi}}$ for all large n. We assign to the constant ϵ the value $\min(\frac{\delta}{3}, \frac{\sqrt{2}}{2})$. □

Theorem 1. *Let $a > 0$ be a constant and P be a set of n grid points on the plane. Let $\delta > 0$ be a small constant. There is a line L such that the number of points in P with $\leq (a, a)$ distance to L is $\leq \frac{(4\pi+8)(1+\delta)a\sqrt{n}}{\pi\sqrt{4+2\pi}}$, and each half plane has $\leq \frac{2n}{3}$ points from P for all large n.*

Proof. Let o be the center point of set P (by Lemma 1). The theorem follows from Lemma 3. □

The following corollary shows that for each grid graph of n nodes, its $\frac{2}{3}$-separator size is bounded by $1.02074\sqrt{n}$. For two grid points of distance 1, if they stay on different sides of separator line L, one of them has $\leq (\frac{1}{2}, \frac{1}{2})$ distance to L.

Corollary 1. *Let P be a set of n grid points on the plane. There is a line L such that the number of points in P with $\leq (1/2, 1/2)$ distance to L is $\leq 1.02074\sqrt{n}$, and each half plane has $\leq \frac{2n}{3}$ points from P.*

Proof. By Theorem 1 with $a = \frac{1}{2}$, we have, $\frac{8(1+\epsilon)}{\pi}\frac{1}{2} \cdot (\frac{\pi}{2} + 1) \cdot \frac{1}{\sqrt{4+2\pi}} < 1.02074$ when ϵ is small enough. □

3 Separator Lower Bound for Grid Graphs

In this section we prove the existence of a lower bound of $0.7555\sqrt{n}$ for the grid graph separator. We calculate the length of the shortest curve partitioning the unit circle into two areas with ratio $1 : 2$ (Theorem 2). A simple closed curve in the plane does not cross itself. Jordan's theorem states that every simple closed curve divides the plane into two compartments, one inside the curve and one outside of it, and that it is impossible to pass continuously from one to the other without crossing the curve.

Theorem 2. *The shortest curve that partitions an unit circle into two regions with ratio $1 : 2$ has length > 1.8937.*

Proof. (Sketch) Using the standard method of variational calculus, the shortest curve partitioning the unit circle with the fixed area ratio between two regions is a circle arc. Using the numerical method, we can calculate the length of the circle arc. □

Definition 2. *A graph is connected if there is a path between every two nodes in the graph. For a connected grid graph $G = (V, E_V)$, a contour of G is a circular path $C = v_1 v_2 \cdots v_k v_1$ such that 1) $(v_i, v_{i+1}) \in E_V (i = 1, 2, \cdots, k-1)$ and $(v_k, v_1) \in E_V$; 2) all points of V are in the one side of C; and 3) for any $i \leq j$, $v_1 \cdots v_{i-1} v_{j+1} \cdots v_k v_1$ does not satisfy both 1) and 2). A point $v \in V$ is a boundary point if $d(v, u) = 1$ for some grid point $u \notin V$. A contour C separates w from all grid points V if every path from w to a node in V intersects C.*

Example: Let V be the set of all dotted grid points in Figure 2.

$C = v_1v_2v_3v_4v_5v_6v_7v_8v_9v_{10}v_{11}v_{12}v_{13}v_{14}v_1$ is a contour for V. The condition 3) prevents $C' = v_1v_2v_{15}v_2v_3v_4v_5v_6v_7v_8v_9v_{10}v_{11}v_{12}v_{13}v_{14}v_1$ from being a contour.

Lemma 4. *Let $G = (V, E_V)$ be a connected grid graph. If the grid point $v \in V$ and grid point $w \notin V$ have the distance* $\text{dist}(v, w) = 1$*, then there is a contour C such that C contains v and separates w from all grid points of V.*

Proof. Imagine that a region starting from the grid point w grows until it touches all of the reachable edges of G (but never crosses any of them). Since G is a connected grid graph, the boundary forms a contour that consists of edges of G. As $\text{dist}(w, v) = 1$, the vertex v should appear in the contour. □

Lemma 5. *Let $G = (V, E_V)$ be a grid graph and C be a contour of G. Let $U = \{u | u$ is a grid point not in V with $\text{dist}(u, v) = 1$ for some $v \in V$ and C separates u from $V\}$. Then there is a list of grid points $u_1, u_2, \cdots, u_{m+1}$ in U such that $u_{m+1} = u_1$, $dist(u_i, u_{i+1}) \le \sqrt{2}$ for $i = 1, 2, \cdots, m$ and all points of P are on one side of the circle path $u_1u_2 \cdots u_{m+1}$ (the edge connecting every two consecutive points u_1, u_2 is straight line).*

Proof. Walking along the contour $C = v_1 \cdots v_kv_1$, we assume that only the left side has the points from V. A point v_i on C is called *special point* if $v_{i-1} = v_{i+1}$. The point v_9 is a special point at the contour $v_1v_2 \cdots v_{14}v_1$ in Figure 2. For each edge (v_i, v_{i+1}) in C, the grid square, which is on the right side of (v_i, v_{i+1}) and contains (v_i, v_{i+1}) as one of the four boundary edges, has at least one point not in V. Let $S_1, S_2, \cdots, S_k$ be those grid squares for $(v_1, v_2), (v_2, v_3), \cdots, (v_k, v_1)$, respectively. For each special point v_i on C, it has two special grid squares S_i' and S_i'' that share the edge (v_i, u) for some $u \in U$ with $\text{dist}(u, v_i) = 1$ and $\text{dist}(u, v_{i-1}) = 2$ (for example, S_9' and S_9'' on Figure 2). Insert S_i' and S_i'' between S_i and S_{i+1}. We get a new list of grid squares $H_1, H_2, \cdots, H_m$. We claim that for every two consecutive H_i and H_{i+1}, there are grid points $u_i \in H_i \cap U$ and $u_{i+1} \in H_{i+1} \cap U$ with $\text{dist}(u_i, u_{i+1}) \le \sqrt{2}$. The lemma is verified by checking the following cases:

Case 1. $H_i = S_j$ and $H_{i+1} = S_{j+1}$ for some $j < k$.

Subcase 1.1. S_j and S_{j+1} share one edge $v_{j+1}u$. An example of this subcase is the grid squares S_1 and S_2 on Figure 2. It is easy to see that $u \in U$ since u is on the right side when walking along the cycle path C.

Subcase 1.2. $S_j = S_{j+1}$. An example of this subcase is the grid squares S_5 and S_6 on Figure 2. This is a trivial case.

Subcase 1.3. S_j and S_{j+1} only share the point v_{j+1}. An example of this subcase is the grid squares S_{11} and S_{12} on Figure 2. We have grid points $u_1 \in U$ and $u_2 \in U$ such that $\text{dist}(u_1, v_{j+1}) = 1$, $\text{dist}(u_2, v_{j+1}) = 1$. Furthermore, $\text{dist}(u_1, u_2) = \sqrt{2}$.

Case 2. $H_i = S_j''$ and $H_{i+1} = S_j$ for some $j < m$. An example of this subcase is the grid squares S_9'' and S_9 on Figure 2. The two squares share the edge v_ju for some $u \in U$.

Case 3. $H_i = S'_j$ and $H_{i+1} = S''_j$. An example of this subcase is the grid squares S'_9 and S''_9 on Figure 2. The two squares share the edge $u_j u$ for some $u \in U$.

Case 4. $H_i = S_{j-1}$ and $H_i = S'_j$. An example of this subcase is the grid squares S_8 and S'_9 on Figure 2. The two squares share $v_j u$ for some $u \in U$. □

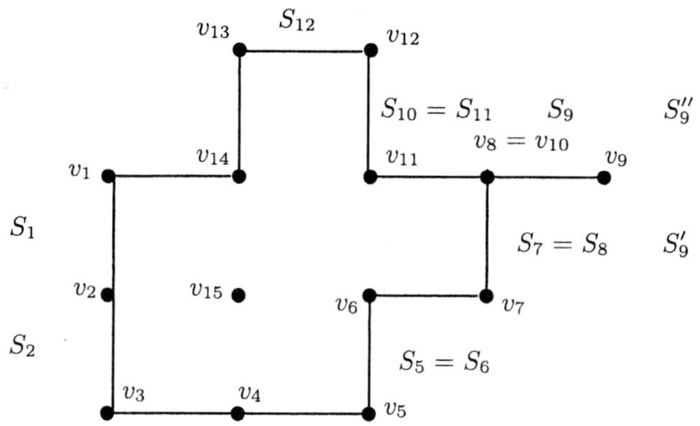

Fig. 2. Contour $C = v_1 v_2 \cdots v_{14} v_1$. The node v_9 is a special point. When walking along $v_1 \cdots v_{14} v_1$, we see that each S_i is the grid square on the right of $v_i v_{i+1}$.

Definition 3. *For a region R on the plane, define $A(R)$ to be the area size of R. An unit circle has radius 1. For a region R in the unit circle, $L(R)$ is the length of the boundary of R inside the internal area of the unit circle. A region R inside a unit circle is type 1 region if part of its boundary is from the unit circle boundary. Otherwise, it is called type 2 region, which does not share any boundary with the unit circle.*

Lemma 6. *Assume $s > 0$ is a constant and p_1, p_2 are two points on the plane. We have 1) the area with the shortest boundary and area size s on the plane is a circle with radius $\sqrt{\frac{s}{\pi}}$; and 2) the shortest curve that is through both p_1 and p_2, and forms an area of size s with the line segment $p_1 p_2$ is a circle arc.*

The proof of Lemma 6 can be found in regular variational calculus textbooks (e.g. [16]). Let R be a type 1 region of area size s. Let C be the part of R boundary that is an unit circle arc with p_1 and p_2 as two end points. Let C' be the rest of the boundary of R. Let R' be the region with the boundary C and line segment $p_1 p_2$. Assume the length of C' is minimal. If $A(R) = A(R')$, then C' is the same as the line segment $p_1 p_2$. If $A(R) < A(R')$, then C' is a circle arc inside R' (between C and $p_1 p_2$). If $A(R) > A(R')$, then C' is also a circle arc outside R'. Those facts above follow from Lemma 6.

Lemma 7. *Let $s \leq \pi$ be a constant. Let $R_1, R_2, \cdots, R_k$ be k regions inside an unit circle (they may have overlaps), $\sum_{i=1}^{k} A(R_i) = s$ and $\sum_{i=1}^{k} L(R_i)$ is minimal. Then $k = 1$ and R_1 is a type 1 region.*

Proof. We consider the regions $R_1, \cdots, R_k$ that satisfy $\sum_{i=1}^k A(R_i) = s$ and $\sum_{i=1}^k L(R_i)$ is minimal for $k \geq 1$. Each $R_i (i = 1, \cdots, k)$ is either type 1 or type 2 region. The part of boundary of R_i that is also the boundary of the unit circle is called *old boundary.* Otherwise it is called type *new boundary.*

A type 2 region has to be a circle (by Lemma 6). For a type 1 region, its new boundary inside the unit circle is also a circle arc (otherwise, its length is not minimal by part 2 of Lemma 6). If we have both type 1 region R_1 and type 2 region R_2. Move R_1 to R_1^* and R_2 to R_2^* on the plane so that R_1^* and R_2^* have some intersection (not a circle) at their new boundaries. Let R_2' be the circle with the same area size as $R_1^* \cap R_2^*$. The boundary length of R_2' is less than that of $R_1^* \cap R_2^*$. So, $L(R_1) + L(R_2)$ reduces to $L(R_1^* \cup R_2^*) + L(R_2')$ if R_1 and R_2 are replaced by $R_1^* \cup R_2^*$ and R_2' (Notice that $A(R_1) + A(R_2) = A(R_1^* \cup R_2^*) + A(R_1^* \cap R_2^*) = A(R_1^* \cup R_2^*) + A(R_2')$). This contradicts that $\sum_{i=1}^k L(R_i)$ is minimal. Therefore, there is no type 2 region. We only have type 1 regions left. Assume that R_1 and R_2 are two type 1 regions. Let R_1 and R_2 have the unit circle arcs p_1p_2 and p_2p_3 respectively. They can merge into another type 1 region R with the unit circle arc $p_1p_2p_3$ and the same area size $A(R) = A(R_1) + A(R_2)$. Furthermore, $L(R) < L(R_1) + L(R_2)$. A contradiction again. Therefore, $k = 1$ and R_1 is a type 1 region. □

Definition 4. *Let q be a positive real number. Partition the plane into $q \times q$ squares by the horizontal lines $y = iq$ and vertical lines $x = jq$ $(i, j \in Z)$. Each point (iq, jq) is a (q, q)-grid point, where $i, j \in Z$.*

Lemma 8. *Let V be the set of all (q, q) grid points in the unit circle C. Let $G = (V, E_V)$ be the grid graph on V, where $E_V = \{(v_i, v_j) | \text{dist}(v_i, v_j) = 1 \text{ and } v_i, v_j \in V\}$. Let t be a constant ≥ 1. Assume that l is a curve that partitions a unit circle C into two regions U_1 and U_2 with $\frac{A(U_1)}{A(U_2)} = \frac{1}{t}$. If the minimal length of l is c_0, then every $\frac{t}{t+1}$-separator for the grid graph G has a size $\geq \frac{(c_0 - o(1))(\sqrt{n} - \sqrt{2\pi})}{\sqrt{2}\sqrt{\pi}}$.*

Proof. Assume that the unit circle C area has n (q, q)-grid points. We have $\pi(1 + q\sqrt{2})^2 \geq n \cdot q^2$. It implies $q \leq \frac{1}{\frac{\sqrt{n}}{\sqrt{\pi}} - \sqrt{2}}$. Assume $S \subseteq V$ is the smallest separator for $G = (V, E_V)$ such that $G - S$ has two disconnected subgraphs $G_1 = (V_1, E_{V_1})$ and $G_2 = (V_2, E_{V_2})$, which satisfy $|V_1|, |V_2| \leq \frac{tn}{t+1}$. By Corollary 1, $|S| \leq 1.021\sqrt{n}$. Let G_1 have connected components $F_1, \cdots, F_m$. By Lemma 5, each F_i is surrounded by a circular path H_i with grid points not from G_1. Actually, the grid points of H_i inside C are from the separator A. Let $P_1, \cdots, P_k$ be the parts of $H_1, \cdots, H_m$ inside the C. They consist of vertices in A and the distance between every two consecutive vertices in each P_i is $\leq \sqrt{2}q$ (by Lemma 5 and scaling (q, q) grid points to $(1, 1)$ grid points).

The number of (q, q)-grid points with distance $\leq 2q$ to the unit circle boundary is $O(\sqrt{n})$. For a (q, q)-grid point $p = (iq, jq)$ and $h > 0$, define $\text{grid}_h(p) = \{(x, y) | iq - \frac{h}{2} \leq x \leq iq + \frac{h}{2} \text{ and } jq - \frac{h}{2} \leq y \leq jq + \frac{h}{2}\}$. Let V_H be the set of all (q, q)-grid points in $H_1, \cdots, H_m$ and V_P be the set of all (q, q)-grid points in

$P_1, \cdots, P_k$. Let $S_1 = \cup_{p \in V_1} \text{grid}_q(p)$. Since $|V_1| + |V_2| + |S| = n$, $|V_1|, |V_2| \leq \frac{tn}{t+1}$, and $|S| \leq 1.021\sqrt{n}$, we have $\frac{tn}{t+1}q^2 \geq A(S_1) \geq (\frac{n}{t+1} - 1.021\sqrt{n})q^2$.

Assume that $P_1, \cdots, P_k$ together with the circle boundary partition the unit circle into two parts X_1 and X_2, where X_1 contains all grid points of V_1 and X_2 contains all grid points of V_2. It is easy to see that $S_1 - \cup_{p \in V_H} \text{grid}_{2q}(p) \subseteq X_1 \subseteq S_1 \cup (\cup_{p \in V_H} \text{grid}_{2q}(p))$. Therefore, $A(S_1) - O(q^2\sqrt{n}) \leq A(X_1) \leq A(S_1) + O(q^2\sqrt{n})$. Thus, $A(S_1) - O(\frac{1}{\sqrt{n}}) \leq A(X_1) \leq A(S_1) + O(\frac{1}{\sqrt{n}})$.

For the variable $x \leq \frac{\pi}{2}$, define the function $g(x)$ to be the length of the shortest curve that partitions the unit circle into regions Q_1 and Q_2 with $A(Q_1) = x$. Then $g(x)$ is an increasing continuous function. For a small real number $\delta > 0$, let D be the disk of area size δ. Therefore, D has radius $\sqrt{\frac{\delta}{\pi}}$. Put D into the region Q_2 and let D be tangent to the boundary of Q_1. The length of the boundary of $Q_1 \cup D$ inside the unit circle is $g(x) + 2\pi\sqrt{\frac{\delta}{\pi}} = g(x) + 2\sqrt{\delta\pi}$. Thus, $g(x+\delta) \leq g(x) + 2\sqrt{\delta\pi}$.

Assume that total length of $P_1, \cdots, P_k$ is minimal, then $k = 1$ by Lemma 7. The length of P_1 is at least $g(\frac{1}{3}) - o(1) = c_0 - o(1)$ by the analysis in the last paragraph. Since every two consecutive grid points in P_1 has distance $\leq \sqrt{2}q$, there are at least $\frac{c_0 - o(1)}{q\sqrt{2}} \geq \frac{(c_0 - o(1))(\frac{\sqrt{n}}{\sqrt{\pi}} - \sqrt{2})}{\sqrt{2}} = \frac{(c_0 - o(1))(\sqrt{n} - \sqrt{2\pi})}{\sqrt{2}\sqrt{\pi}}$ grid points of A along P_1. □

Theorem 3. *There exists a grid graph $G = (V, E_V)$ such that for any $A \subseteq V$ if $G - A$ has two disconnected graphs G_1 and G_2, and $G_i (i = 1, 2)$ has $\leq \frac{2|V|}{3}$ nodes, then $|A| \geq 0.7555\sqrt{n}$ when n is large.*

Proof. By Theorem 2, the length of the shortest curve partitioning the unit circle into $1:2$ ratio is ≥ 1.8937. By Lemma 8 with $c_0 = 1.8937$ and $k = 1$, we have $|A| \geq 0.7555\sqrt{n}$. □

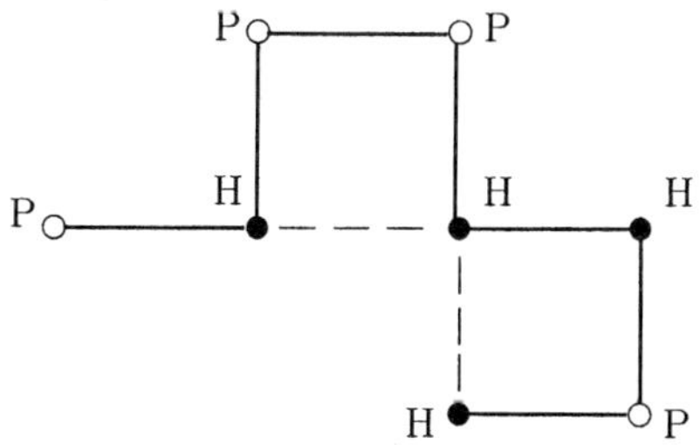

Fig. 3. The sequence PHPPHHPH is put on the 2 dimensional grid. There are 2 H-H contacts marked by the dotted lines.

4 Application to Protein Folding in the HP-Model

We have shown that there is a size $O(\sqrt{n})$ separator line to partition the folding problem of n letters into 2 problems in a balanced way. The 2 smaller problems

are recursively solved and their solutions are merged to derive the solution to the original problem. As the separator has only $O(\sqrt{n})$ letters, there are at most $n^{O(\sqrt{n})}$ cases to partition the problem. The major revision from the algorithm in [6] is the approximation of the optimal separator line.

Theorem 4. *There is a $O(n^{5.563\sqrt{n}})$ time algorithm for the 2D protein folding problem in the HP-model.*

Proof. (Sketch) The algorithm is similar to that in [6]. We still use approximation to the separator line. Lemma 3 shows that we can avoid the separator line that is almost parallel or vertical to the x-axis. Let $\delta > 0$ be a small constant. Let $a = \frac{1}{2}, c = \frac{2}{3} + \delta$ and $d = k_0(a + \delta)$, where $k_0 = \frac{(4\pi+8)(1+\delta)}{\pi\sqrt{4+2\pi}}$ such that $k_0 a\sqrt{n}$ is the upper bound for the number of grid points with $\leq (a, a)$ distance to the separator line (by Theorem 1). With the reduced separator, its computational time is $(\frac{n}{\delta})^{O(\log n)} 2^{O(\sqrt{n})} n^{d(\frac{1}{1-\sqrt{c}})\sqrt{n}} = O(n^{5.563\sqrt{n}})$. □

Acknowledgement. We are very grateful to Seth Pincus and Sansan Lin for their valuable comments on the earlier version of this paper. We are also grateful to the reviewers from ISAAC 2005 for their helpful comments.

References

1. N. Alon, P.Seymour, and R.Thomas, Planar Separator, SIAM J. Discr. Math. 7,2(1990) 184-193.
2. B. Berger and T. Leighton, Protein folding in the hydrophobic-hydrophilic (HP) model is NP-complete, Journal of Computational Biology, 5(1998), 27-40.
3. P. Crescenzi and D. Goldman and C. Papadimitriou and A. Piccolboni and M. Yannakakis,On the complexity of protein folding, Journal of computational biology, 5(1998), 423-465.
4. H.N. Djidjev, On the problem of partitioning planar graphs. SIAM Journal on Discrete Mathematics, 3(2) June, 1982, pp. 229-240.
5. H. N. Djidjev and S. M. Venkatesan, Reduced constants for simple cycle graph separation, Acta informatica, 34(1997), pp. 231-234.
6. B. Fu and W. Wang, A $2^{O(n^{1-1/d}\log n)}$-time algorithm for d-dimensional protein folding in the HP-model, Proceedings of 31st International Colloquium on Automata, Languages and Programming, 2004, pp.630-644.
7. B. Fu, Theory and application of width bounded geometric separator, Electronic Colloquium on Computational Complexity 2005, TR05-13.
8. H.Gazit, An improved algorithm for separating a planar graph, manuscript, USC, 1986.
9. K. F. Lau and K. A. Dill, A lattice statistical mechanics model of the conformational and sequence spaces of proteins, Macromolecules, 22(1989), 3986-3997.
10. K. F. Lau and K. A. Dill, Theory for protein mutability and biogenesis, Proc. Natl. Acad. Sci, 87(1990), 638-642.
11. R. J. Lipton and R. Tarjan, A separator theorem for planar graph, SIAM J. Appl. Math. 36(1979) 177-189.

12. G. L. Miller, S.-H. Teng, W. P. Thurston, S. A. Vavasis: Separators for sphere-packings and nearest neighbor graphs. J. ACM 44(1): 1-29 (1997)
13. W. D. Smith and N. C. Wormald, Application of geometric separator theorems, FOCS 1998, 232-243.
14. D. A. Spielman and S. H. Teng, Disk packings and planar separators, 12th Annual ACM Symposium on Computational Geometry, 1996, pp.349-358.
15. J. Pach and P.K. Agarwal, Combinatorial Geometry, Wiley-Interscience Publication, 1995.
16. R. Weinstock, Calculus of variations, McGraw-Hill, 1952.

Shortest Paths and Voronoi Diagrams with Transportation Networks Under General Distances
[Extended Abstract]

Sang Won Bae and Kyung-Yong Chwa

Division of Computer Science, Department of EECS,
Korea Advanced Institute of Science and Technology, Daejeon, Korea
{swbae, kychwa}@jupiter.kaist.ac.kr

Abstract. Transportation networks model facilities for fast movement on the plane. A transportation network, together with its underlying distance, induces a new distance. Previously, only the Euclidean and the L_1 distances have been considered as such underlying distances. However, this paper first considers distances induced by general distances and transportation networks, and present a unifying approach to compute Voronoi diagrams under such a general setting. With this approach, we show that an algorithm for convex distances can be easily obtained.

1 Introduction

Transportation networks model facilities for fast movement on the plane. They consist of roads and nodes; *roads* are assumed to be segments along which one can move at a certain fixed speed and *nodes* are endpoints of roads. We assume that there are no crossings among roads but roads can share their endpoints as nodes. We thus define a transportation network as a plane graph whose vertices are nodes and whose edges are roads with speeds assigned. Also, we assume that one can access or leave a road through any point on the road. This, as Aichholzer et al. [3] pointed out, makes the problem distinguishable from and more difficult than those in other similar settings such as the airlift distance.

In the presence of a transportation network, the distance between two points is defined to be the shortest elapsed time among all possible paths joining the two points using the roads of the network. We call such an induced distance a *transportation distance.* (In other literature [2, 3], it is called a *time distance* or a *city metric.*) More precisely, a transportation distance is induced on the plane by a transportation network and its underlying distance that measures the distance between two points without roads.

Since early considerations for roads [7, 17, 20], fundamental geometric problems, such as shortest paths and Voronoi diagrams, under transportation distances have been receiving much attention recently [1, 2, 3, 4, 8, 18]. However, underlying distances considered in the literature were only the Euclidean distance [2, 4] and the L_1 distance [1, 3, 8]. Since transportation distances have

X. Deng and D. Du (Eds.): ISAAC 2005, LNCS 3827, pp. 1007–1018, 2005.

quite different properties depending on their underlying distances, there has not been a common approach to extend such problems to more general underlying distances. This paper thus considers geometric problems, in particular, shortest paths and Voronoi diagrams under transportation distances induced from general distances.

The Results. This paper presents, to the best of our knowledge, the first result that studies general underlying distances and gives algorithms for computing Voronoi diagrams with transportation networks, in Section 3. More precisely, we classify *transportable* distance functions where transportation networks and transportation distances are well-defined. Transportable distances include asymmetric convex distances, nice metrics, and even transportation distances. In this general setting, a unifying approach to compute Voronoi diagrams is presented.

As a special case of transportable distances, we take the convex distances into account in Section 4. Based on the approach in Section 3 together with geometric and algorithmic observations on convex distances, we first obtain an efficient and practical algorithm that computes the Voronoi diagram with a transportation network under a convex distance.

For the L_1 metric, previous work considered only isothetic networks and a single or a constant number of speeds for roads [3, 8]. Our results first deal with more general transportation networks, which have no restriction except for straightness; the roads can have arbitrarily fixed speeds and directions. For the Euclidean metric, we obtain the same time and space bounds as those of the previously best results [4].

Note that a resulting diagram of our algorithm is in fact a refined diagram of the real Voronoi diagram so that it consists of shortest-paths information in each cell and it can also serve as a shortest path map structure.

2 Preliminaries

2.1 Transportation Networks Under General Distances

Here, we let $d : \mathbb{R}^2 \times \mathbb{R}^2 \to \mathbb{R}$ be a total distance function. For the Euclidean and the L_1 distances, a transportation network can be sufficiently represented as a planar straight-line graph. If, however, we consider more general distances, the meaning of "straight" should be reconsidered. Note that a straight segment is a shortest path or a geodesic on the Euclidean plane or on the L_1 plane. Geodesics, in general, naturally generalize straight segments, and a road can be defined to be a segment along a geodesic. Thus, in order to build a transportation network under d, d needs to admit a geodesic between any two points on the plane.

In this step, we define a *transportable distance* that satisfies several axioms. It is easy to observe that distances that admit geodesics and are possibly asymmetric are transportable. We will show that transportable distances admit a geodesic between any two points on the plane. We call a distance d over $\mathbb{R}^2$ *transportable* if the following properties hold:

1. d is non-negative and $d(p,q) = d(q,p) = 0$ iff $p = q$ and d satisfies the triangle inequality.
2. The backward topology induced by d on X induces the Euclidean topology.
3. The backward d-balls are bounded with respect to the Euclidean metric.
4. For any two points p and r, there exists a point $q \notin \{p,r\}$ such that $d(p,q) + d(q,r) = d(p,r)$.

Distances with Condition 1 are called *quasi-metrics* and they induce two associated topologies by two families of open balls, $B_d^+(x,\epsilon) = \{y|d(x,y) < \epsilon\}$ and $B_d^-(x,\epsilon) = \{y|d(y,x) < \epsilon\}$, called *forward* and *backward*, respectively [9, 16]. Here, we consider only the backward case since we shall take only the *inward* Voronoi diagrams into account; Voronoi sites are supposed to be static and fixed. In fact, Conditions 2-4 of the definition of transportable distances mimic those of *nice metrics* in the sense of Klein and Wood [13]. By these conditions, $(\mathbb{R}^2, d)$ is known to be backward-complete [16]. Also, we can find a geodesic, whose length is the same as the distance, between two points in $\mathbb{R}^2$.

Lemma 1. *Let d be a transportable distance. Then, for any two points p and r, there exists a path π from p to r such that for each point q on π the equality $d(p,r) = d(p,q) + d(q,r)$ holds.*

Such paths are called *d-straight* and they generalize straight line segments with respect to the Euclidean metric, indeed. This lemma can be shown by Menger's *Verbindbarkeitssatz* that implies the existence of d-straight paths in a complete metric space [19].

Now, we are able to define a transportation network under a transportable distance d. Since d can be asymmetric, roads in a transportation network may have an orientation. Thus, throughout this paper, a transportation network under d is defined to be a directed plane graph $G = (V,E)$ such that any edge e in E is a segment of a d-straight line and has its own weight $v(e) > 1$, called *speed.* We note that an edge has an orientation so that it can be regarded as a one-way road. And we call edges in E *roads* and vertices in V *nodes*, and roads and nodes may denote d-straight paths and points by themselves referenced. Two incident nodes of a road e is identified by $p_1(e)$ and $p_2(e)$, where e has the orientation toward $p_2(e)$ from $p_1(e)$. Note that any crossings among roads can be removed by introducing additional nodes. An anomaly occurs when we think of a two-way road. Thus, we allow coincidence only for two roads having the same incident nodes.

Now, we consider a distance d_G induced by a transportable distance d and a transportation network $G = (V,E)$ under d, which can be defined as follows:

$$d_G(p,q) = \min_{P=(p_1,\cdots,p_\ell)\in\mathcal{P}(p,q)} \sum_{i=1}^{\ell-1} \frac{1}{v_i} d(p_i, p_{i+1}),$$

where $\mathcal{P}(p,q)$ is the set of all piecewise d-straight paths from p to q and $v_i = v(e)$ if there exists an oriented road $e \in E$ such that the path from p_i to p_{i+1} passes along e, otherwise, $v_i = 1$. We call d_G the *transportation distance* induced by d

and G. Note that transportable distances include nice metrics, convex distances, and even transportation distances.

2.2 Needles

Bae and Chwa [4, 4] defined a needle as a generalized Voronoi site, which is very useful for a transportation network. In fact, the concept of needles was first proposed by Aichholzer, Aurenhammer, and Palop [3] but needles were not thought of as Voronoi sites in their results.

Here, we consider a needle under a transportable distance d. A needle $\mathbf{p}$ under d can be represented by a 4-tuple $(p_1(\mathbf{p}), p_2(\mathbf{p}), t_1(\mathbf{p}), t_2(\mathbf{p}))$ with $t_2(\mathbf{p}) \geq t_1(\mathbf{p}) \geq 0$, where $p_1(\mathbf{p}), p_2(\mathbf{p})$ are two endpoints and $t_1(\mathbf{p}), t_2(\mathbf{p})$ are additive weights of the two endpoints, respectively. In addition, a needle under d is d-straight in a sense that it can be viewed as a set of weighted points on the d-straight path from $p_2(\mathbf{p})$ to $p_1(\mathbf{p})$. Other terms associated with a needle under d are determined in a similar way with the Euclidean case. Thus, let $s(\mathbf{p})$ be the set of points on the d-straight path from $p_2(\mathbf{p})$ to $p_1(\mathbf{p})$, and $v(\mathbf{p})$ be the *speed* of $\mathbf{p}$, defined by $d(p_2(\mathbf{p}), p_1(\mathbf{p}))/(t_2(\mathbf{p}) - t_1(\mathbf{p}))$.

The distance from any point x to a needle $\mathbf{p}$ is measured as $d(x, \mathbf{p}) = \min_{y \in s(\mathbf{p})}\{d(x, y) + w_{\mathbf{p}}(y)\}$, where $w_{\mathbf{p}}(y)$ is the weight assigned to y on $\mathbf{p}$, given as $w_{\mathbf{p}}(y) = t_1(\mathbf{p}) + d(y, p_1(\mathbf{p}))/v(\mathbf{p})$, for all $y \in s(\mathbf{p})$.

For the Euclidean case, the Voronoi diagram for pairwise non-piercing needles has been shown to be an abstract Voronoi diagram. Two needles are called *non-piercing* if, and only if, the bisector between them contains at most one connected component. For more details, we refer to [4, 4].

3 Voronoi Diagrams Under Transportation Distances

3.1 d_G-Straight Paths and Needles

As noted in the previous section, a transportable distance d and a transportation network G induce a new distance d_G and d_G-straight paths. The structure of any d_G-straight path can be represented by a string of {S, T}, where S denotes a d-straight path without using any road and T denotes that along a road.

Let us consider a single road e as a simpler case. Given a transportation network G with only one road e, a d_G-straight path is of the form STS or its substring except for SS. This is quite immediate; paths represented by longer strings than STS can be reduced since a road is d-straight and d satisfies the triangle inequality. Thus, any d_G-straight path P from p to q using a road e can be represented as $P = (p, p', q', q)$, where p' is the entering point to e and q' is the exiting point to q. We then call q' a *footpoint* of q on e. A point may have several or infinitely many footpoints on a road. Let $FP_e(q)$ be the set of footpoints of q on e for all d_G-straight paths from any point to q using e. Let us consider a total order $\prec_e$ on the points on a road e, where $x \prec_e y$ for $x, y \in e$ if the orientation of the d-straight path from x to y is equivalent to that of e. Then, the following property of footpoints can be shown.

Lemma 2. *Let $F \subseteq e$ be connected. Then, the following are equivalent.*

1. *F is a connected component of $FP_e(q)$.*
2. *F has the least point q_0 with respect to $\prec_e$ such that $q_0 \in FP_e(q)$, and $d(q_1, q) = d(q_1, q_2)/v(e) + d(q_2, q)$ for any $q_1, q_2 \in F$ with $q_1 \prec_e q_2$.*

For such a shortest path $P = (p, p', q', q)$, we can find a needle $\mathbf{q}$ such that $d(p, \mathbf{q}) = d_G(p, q)$. Such a needle $\mathbf{q}$ is said to be *produced on a road e from a point q for a footpoint q'*, and can be defined by setting parameters as follows; $p_1(\mathbf{q}) = q'$, $p_2(\mathbf{q}) = p_1(e)$, $t_1(\mathbf{q}) = d(q', q)$, and $t_2(\mathbf{q}) = d(p'q')/v(e) + d(q', q)$. We let $\sigma_e(q)$ be the set of needles produced on e from q for the least footpoint of every connected component in $FP_e(q)$. Also, both of $FP_e(\cdot)$ and $\sigma_e(\cdot)$ can be naturally extended for needles since shortest paths to a needle under d_G are also d_G-straight paths.

Lemma 3. *If a transportation network G under d contains only one road e, for a point x and a needle $\mathbf{p}$, $d_G(x, \mathbf{p}) = d(x, \sigma_e(\mathbf{p}) \cup \{\mathbf{p}\})$.*

Now, we consider multiple roads. Let $\sigma_G(\mathbf{p}) = \bigcup_{e \in E} \sigma_e(\mathbf{p})$ and $\sigma_G(A) = \bigcup_{\mathbf{p} \in A} \sigma_G(\mathbf{p})$ for a set A of needles. Since d_G-straight paths may pass through several roads, we apply $\sigma_G(\cdot)$ repeatedly. We thus let $\sigma_G^k(\mathbf{p}) = \sigma_G(\sigma_G^{k-1}(\mathbf{p}))$ and $\sigma_G^0(\mathbf{p}) = \{\mathbf{p}\}$. Also, we let $\mathcal{S}_\mathbf{p}^k$ denote $\bigcup_{i=0}^k \sigma_G^i(\mathbf{p})$ and $\mathcal{S}_\mathbf{p}$ denote $\mathcal{S}_\mathbf{p}^\infty$.

Theorem 4. *Given a transportable distance d and a transportation network G under d, for a point x and a needle $\mathbf{p}$,*

$$d_G(x, \mathbf{p}) = d(x, \mathcal{S}_\mathbf{p}).$$

Proof. We first define $d_G^k(p, q)$ be the length of a shortest path from p to q where the path passes through at most k roads in G. Surely, $d_G(p, q) = d_G^\infty(p, q)$.

We claim that $d_G^k(x, \mathbf{p}) = d(x, \mathcal{S}_\mathbf{p}^k)$, which directly implies the theorem. We prove this by induction. Lemma 3 gives us an inductive basis. We have $d(x, \mathcal{S}_\mathbf{p}^{\ell+1}) = \min\{d(x, \mathcal{S}_\mathbf{p}^\ell), d(x, \sigma_G^{\ell+1}(\mathbf{p}))\} = \min\{d(x, \mathcal{S}_\mathbf{p}^\ell), d(x, \sigma_G(\sigma_G^\ell(\mathbf{p})))\}$. By inductive hypothesis and Lemma 3, the equation is evaluated as $d(x, \mathcal{S}_\mathbf{p}^{\ell+1}) = \min\{d_G^\ell(x, \mathbf{p}), d_G^1(x, \sigma_G^\ell(\mathbf{p}))\}$.

As pointed out in the proof of Lemma 3, $d_G^1(x, \sigma_G^\ell(\mathbf{p}))$ implies a shortest path to a needle in $\sigma_G^\ell(\mathbf{p})$ using exactly one road in G, and further a shortest path to $\mathbf{p}$ using exactly $\ell + 1$ roads. Therefore, we conclude $d(x, \mathcal{S}_\mathbf{p}^{\ell+1}) = d_G^{\ell+1}(x, \mathbf{p})$, implying the theorem. □

Theorem 4 says a nice relation between needles and roads. Furthermore, it directly implies that the Voronoi diagram $\mathcal{V}_d(\mathcal{S})$ under d for $\mathcal{S}$ induces the Voronoi diagram $\mathcal{V}_{d_G}(S)$ under d_G for S, where $\mathcal{S}$ denotes $\bigcup_{p \in S} \mathcal{S}_p$. In other words, any Voronoi region in $\mathcal{V}_d(\mathcal{S})$ is completely contained in a Voronoi region in $\mathcal{V}_{d_G}(S)$, i.e., $\mathcal{V}_{d_G}(S)$ is a sub-diagram of $\mathcal{V}_d(\mathcal{S})$.

Corollary 5. *$\mathcal{V}_{d_G}(S)$ can be extracted from $\mathcal{V}_d(\mathcal{S})$ in time linear in the size of $\mathcal{V}_d(\mathcal{S})$.*

3.2 Computing Effective Needles

In this subsection, we present an algorithm for computing the Voronoi diagram $\mathcal{V}_{d_G}(S)$ for a given set S of sites under d_G. The algorithm consists of three phases; it first computes the set $\mathcal{S}$ of needles from G and S, secondly, the Voronoi diagram $\mathcal{V}_d(\mathcal{S})$ for $\mathcal{S}$ under d is constructed, and the Voronoi diagram $\mathcal{V}_{d_G}(S)$ for S under d_G is finally obtained from $\mathcal{V}_d(\mathcal{S})$.

The second phase, computing $\mathcal{V}_d(\mathcal{S})$, would be solved by several technics and general approaches to compute Voronoi diagrams, such as the abstract Voronoi diagram [10]. Also, the third phase can be done by Corollary 5. We therefore focus on the first phase, computing $\mathcal{S}$—in fact, its finite subset—from G and S.

Recall that $\mathcal{S}$ is defined as all the needles recursively produced from given sites and contains infinitely many useless needles, that is, needles that do not constitute the Voronoi diagram $\mathcal{V}_d(\mathcal{S})$. We call a needle $\mathbf{p} \in \mathcal{S}$ *effective with respect to* $\mathcal{S}$ if the Voronoi region of $\mathbf{p}$ in $\mathcal{V}_d(\mathcal{S})$ is not empty. Let $\mathcal{S}^* \subseteq \mathcal{S}$ be the largest set of effective needles with respect to $\mathcal{S}$, i.e., $\mathcal{V}_d(\mathcal{S}^*) = \mathcal{V}_d(\mathcal{S})$. Our algorithm computes $\mathcal{S}^*$ from G and S.

The algorithm works with handling events, which are defined by a certain situation at a time. Here, at each time t, we implicitly maintain the (backward) d_G-balls of the given sites, where the backward d_G-ball of a site p is defined as the set $B^-_{d_G}(p,t) = \{x | d_G(x,p) < t\}$, and d_G-balls expand as time t increases. Here, we have only one kind of events, called *birth events* which occur when a d_G-ball touches any footpoint on a road during their expansions; at that time a new needle will be produced in the algorithm. We can determine a birth event associated with a footpoint of a needle on a road. In order to handle events, we need two data structures: Let $\mathcal{Q}$ be an event queue implemented as a priority queue such that the priority of an event e is its occurring time and $\mathcal{Q}$ supports inserting, deleting, and extracting-minimum in logarithmic time with linear space. And, let $\mathcal{T}_1, \mathcal{T}_2, \cdots, \mathcal{T}_m$ be balanced binary search trees, each associated with e_i, where the road set E is given as $\{e_1, e_2, \cdots, e_m\}$. Each $\mathcal{T}_i$ stores needles on e_i in order and the precedence for a needle $\mathbf{p}$ follows from that of $p_1(\mathbf{p})$ with respect to $\prec_{e_i}$. $\mathcal{T}_i$ supports inserting and deleting of a needle in logarithmic time, and also a linear scan for needles currently in $\mathcal{T}_i$ in linear time and space.

Now, we are ready to describe the algorithm ComputeEffectiveNeedles. First, the algorithm computes $\sigma_G(S)$ and the associating birth events, and insert events into $\mathcal{Q}$. Then, while the event queue $\mathcal{Q}$ is not empty, repeat the following procedure: (1)Extract the next upcoming event b, say that b is a birth event on a road e_i associated with a needle $\mathbf{p}$. (2)Test the effectiveness of $\mathbf{p}$ and, (3)if the test has passed, compute birth events associated with $\sigma_G(\mathbf{p})$, and insert the events into $\mathcal{Q}$.

ComputeEffectiveNeedles returns exactly $\mathcal{S}^*$ by the effectiveness test in (2). This test can be done by checking if the associating footpoint of the current event has been already dominated by d_G-balls of other sites. Thus, if the test is passed, we decide that the new needle should be effective and insert it into $\mathcal{T}_i$. The following lemma shows that the effectiveness test is necessary and sufficient to compute $\mathcal{S}^*$.

Lemma 6. *For every birth event and its associating needle* $\mathbf{p}$, $\mathbf{p}$ *is effective with respect to* $\mathcal{S}$ *if and only if it passes the effectiveness test of Algorithm* COMPUTE-EFFECTIVENEEDLES.

The algorithm always ends, which can be shown by the finiteness of $\mathcal{S}^*$. The effectiveness test always fails if all the roads are covered with the d_G-balls after some large time T and there exists such T since distances between any two points in the space are always defined. Hence, $\mathcal{S}^*$ contains a finite number of needles. Moreover, the number of events handled while running the algorithm is also bounded by the following lemma.

Lemma 7. $\mathcal{S}^*$ *is finite and the number of handled events is* $O(s \cdot |\mathcal{S}^*|)$*, where* s *is the maximum cardinality of* $\sigma_G(\mathbf{p})$ *for any needle* $\mathbf{p}$.

We end this section with the following conclusion.

Theorem 8. *Given a transportable distance* d*, a transportation network* G *under* d*, and a set* S *of sites,* $\mathcal{S}^*$ *can be computed in* $O(\epsilon(\log \epsilon + T_{ef}))$ *time, where* ϵ *is the number of events handled while running the algorithm, the same as* $O(s \cdot |\mathcal{S}^*|)$*, and* T_{ef} *denotes time taken to test the effectiveness.*

In most natural cases, such as the Euclidean metric and convex distances, sufficiently $T_{ef} = O(\log \epsilon)$ so that the total running time becomes $O(\epsilon \log \epsilon)$.

4 Transportation Networks Under Convex Distances

In this section, we deal with convex distances as a special case of transportable distances. We thus investigate geometric and algorithmic properties of the induced distance by a convex distance and a transportation network, and construct algorithms that compute the Voronoi diagram for given sites under the induced distance.

In order to devise such an algorithm, we apply the abstract scheme described in the previous section; a bundle of properties have to be shown: how to compute needles produced from a needle, how to check the effectiveness of needles, how many needles and events to handle, how to compute the Voronoi diagram for needles, and some technical lemmas to reduce the complexity.

A *convex distance* is defined by a compact and convex body C containing the origin, or the *center*, and is measured as the factor that C centered at the source should be expanded or contracted for its boundary to touch the destination. Note that a convex distance is symmetric, i.e. being a metric, if and only if C is symmetric at its center.

We consider the convex C as a black box which supports some kinds of elementary operations. These are finding the Euclidean distance from the center to the boundary in a given direction, finding two lines which meet at a given point and are tangent to C, finding the footpoint for a needle and a road, and computing the bisecting curve between two sites under the convex distance based on C. Here, we assume that these operations consume reasonable time bounds.

Throughout this section, for a convex body C, we denote by $C+p$ a translation of C by a vector p and by λC an expansion or a contraction of C by a factor λ. And, we may denote by C' the reflected body of C at its center and by ∂C the boundary of C.

4.1 Roads and Needles Under a Convex Distance

For the Euclidean distance, to take a shortest path with a road e, one should enter or exit e with angle $\pi/2 \pm \alpha$, where $\sin\alpha = 1/v(e)$ [2]. For a convex distance d based on C, there also exist such entering or exiting angles for a road that they lead to a shortest path. We can obtain these angles by simple operations on C. First, we pick a point x on e not $p_2(e)$ and consider $(d(x, p_2(e))/v(e))C' + p_2(e)$, say D. We then compute two lines which meet at x and are tangent to D. Each of the two lines is above or below e. We let α_e^+ and α_e^- be two inward directions perpendicular to the two lines. And, let us consider two directions from $p_2(e)$ to two meeting points between the two lines and D. We then let β_e^+ and β_e^- be reflections of the two directions at the center of D. The symbols + and − mean "above" and "below" with respect to e, respectively. See Figure 1(a).

Lemma 9. *Given a road e under a convex distance d, any shortest path from p to q passing through e has the following properties:*

- *If p is above e, either the access direction is α_e^+ or the access point is $p_1(e)$.*
- *If p is below e, either the access direction is α_e^- or the access point is $p_1(e)$.*
- *If q is above e, either the leaving direction is β_e^+ or the exiting point is $p_2(e)$.*
- *If q is below e, either the leaving direction is β_e^- or the exiting point is $p_2(e)$.*

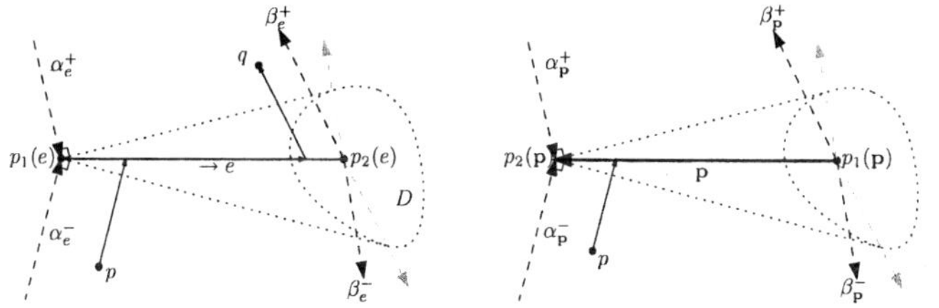

(a) A road e and a shortest path from p to q

(b) A needle $\mathbf{p}$ and a shortest path from p to $\mathbf{p}$

Fig. 1. Directions defined for a road and for a needle

By the observation of Lemma 9, we can easily find a shortest path with one road. For instance, Figure 1(a) shows a shortest path from p to q using road e.

We can define similar terms for a needle as we did for a road. The (backward) d-ball $B_d^-(\mathbf{p}, t)$ of $\mathbf{p}$ can be computed as follows: If $0 < t \le t_1(\mathbf{p})$, $B_d^-(\mathbf{p}, t)$ is empty. If $t_1(\mathbf{p}) < t \le t_2(\mathbf{p})$, $B_d^-(\mathbf{p}, t)$ is the convex hull of $(t - t_1(\mathbf{p}))C' + p_1(\mathbf{p})$ and the point x on $\mathbf{p}$ such that $d(x, p_1(\mathbf{p}))/v(\mathbf{t}) = w_{\mathbf{p}}(x) - t_1(\mathbf{p})$. And, if $t > t_2(\mathbf{p})$, $B_d^-(\mathbf{p}, t)$ is the convex hull of $(t - t_1(\mathbf{p}))C' + p_1(\mathbf{p})$ and $(t - t_2(\mathbf{p}))C' + p_2(\mathbf{p})$. Thus,

when $t > t_1(\mathbf{p})$, $\partial B_d^-(\mathbf{p}, t)$ contains two line segments tangent to two scaled C' and the slopes of these line segments do not change even if t changes. Hence, the meeting points between the line segments and the convex bodies scaled from C' make 4 rays, and they have two directions $-\beta_\mathbf{p}^+$ and $-\beta_\mathbf{p}^-$. Then, we can define $\alpha_\mathbf{p}^+$, $\alpha_\mathbf{p}^-$, $\beta_\mathbf{p}^+$, and $\beta_\mathbf{p}^-$, equivalently as for a road, see Figure 1(b). Note that if $\mathbf{p} \in \sigma_e(\mathbf{q})$ for any needle $\mathbf{q}$, then $\alpha_\mathbf{p}^+ = \alpha_e^+$, $\alpha_\mathbf{p}^- = \alpha_e^-$, $\beta_\mathbf{p}^+ = \beta_e^+$, and $\beta_\mathbf{p}^- = \beta_e^-$, by definition of $\sigma_e(\mathbf{q})$.

4.2 Computing $\mathcal{S}^*$

In this subsection, we discuss how to compute $\mathcal{S}^*$ for given sites S from the algorithm ComputeEffectiveNeedles described in the previous section.

The footpoints. Under a convex distance, any needle has at most one connected component of footpoints on a road. Actually, Lemma 9 tells us how to compute the least footpoint of a point p on a road e; it can be obtained as either the intersecting point of the road e and the ray with direction $-\beta_e^+$ or $-\beta_e^-$ from p, or just $p_2(e)$.

Lemma 10. *Let d be a convex distance based on a convex C and G be a transportation network under d. For a road e in G and a needle $\mathbf{p}$, the number of connected components of footpoints of $\mathbf{p}$ on e is at most one and the least footpoint is either the least footpoint of $p_1(\mathbf{p})$ or $p_2(\mathbf{p})$, or just $p_2(e)$.*

The effectiveness test. At every time a birth event occurs, the effectiveness test is done by testing if the point where the event occurs is dominated by other already produced effective needles at that time. Under a convex distance, d-balls of any needle are convex so that we can test the effectiveness in logarithmic time by maintaining the tree structures $\mathcal{T}_i$ and doing a couple of operations on them.

The number of needles and events. By convexity of convex distances, we can show a couple of lemmas that prove the number of needles and events we handle.

Lemma 11. *Let $\mathbf{p}$ be a needle produced on a road e from a needle $\mathbf{q}$. For another road $e' \in E$, if $\mathbf{p}$ does not dominate any node of e or e', no needles in $\sigma_{e'}(\mathbf{p})$ are effective with respect to S.*

Lemma 12. *$\mathcal{S}^*$ contains at most $O(m(n+m))$ needles, where n is the number of sites in S and m is the number of roads in G. Further, the number of handled events while the algorithm* ComputeEffectiveNeedles *running is $O(m^2(n+m))$.*

Remarks. Consequently, we have an algorithm to compute $\mathcal{S}^*$ in $O(m^2(n+m)(\log(n+m) + T_{op}(C)))$ time with $O(m^2(n+m))$ space by Theorem 8, where $T_{op}(C)$ is time taken during a simple operation on C. This complexity, however, can be reduced by small modifications on the algorithm. For the Euclidean case, Bae and Chwa [4] additionally maintains *node events*, which occur when $B_{d_G}^-(p, t)$

first touches a node, to reduce the number of events to $O(m(n+m))$. Here, we also can apply the approach by Lemma 11.

The authors also introduced primitive paths; a path is called *primitive* if the path contains no nodes in its interior and passes through at most one road. We can show the lemma of primitive paths in our setting, following from Lemma 11.

Lemma 13. *Given a transportation network G under a convex distance d, for two points p and q, there exists a d_G-straight path P from p to q such that P is a sequence of shortest primitive paths whose endpoints are p, q, or nodes in V.*

By Lemma 13 together with node events, we can improve our algorithm to be more efficient. Indeed, we can compute a shortest path for two given points by constructing an edge-weighted complete graph such that vertices are nodes and two given points, and edges are shortest primitive paths among vertices. Furthermore, we can use the graph to avoid useless computations during the algorithm. For more details, we refer to [4].

Lemma 14. *One can compute $\mathcal{S}^*$ in $O(m(n+m)(m+\log n+T_{op}(C)))$ time with $O(m(n+m))$ space.*

4.3 Voronoi Diagrams for Needles

In general, bisectors between two needles under a convex distance can be parted into two connected components. However, it will be shown that $\mathcal{S}^*$ can be replaced by such nice needles, so called *non-piercing*, that the Voronoi diagram for them is an abstract Voronoi diagram which can be computed in the optimal time and space.

Computing the Voronoi diagrams for non-piercing needles. The *abstract Voronoi diagram* is a unifying approach to define and compute general Voronoi diagrams, introduced by Klein [10]. In this model, we deal with not a distance but bisecting curves $J(p,q)$ defined in an abstract fashion between two sites p and q. A system $(S, \{J(p,q)|p,q \in S, p \neq q\})$ of bisecting curves for S is called *admissible* if the following conditions are fulfilled: (1) $J(p,q)$ is homeomorphic to a line or empty, (2) $R(p,q) \cap R(q,r) \subset R(p,r)$, (3) for any subset $S' \subseteq S$ and $p \in S'$, $R(p,S')$ is path-connected if it is nonempty, and (4) the intersection of any two bisectors consists of finitely many components, where $R(p,q) = \{x \in \mathbb{R}^2 | d(x,p) < d(x,q)\}$, $R(p,S) = \bigcap_{q \in S, p \neq q} R(p,q)$, and $J(p,q)$ is the bisector between p and q.

In fact, the first three conditions are enough to handle abstract Voronoi diagrams theoretically but the fourth one is necessary in a technical sense [11]. Though all convex distances satisfy the first three ones, there exist convex distances violating the fourth one. We guarantee the fourth condition by postulating that ∂C is semialgebraic [5]. We also note that two-dimensional bisectors can be avoided by a total order on given sites [13].

Lemma 15. *Let S be a set of pairwise non-piercing needles under a convex distance d based on C whose boundary is semialgebraic. Then, the system $(S, \{J(\mathbf{p},\mathbf{q})|\ \mathbf{p},\mathbf{q} \in S,\ \mathbf{p} \neq \mathbf{q}\})$ of bisecting curves for S is admissible.*

There are several optimal algorithms computing abstract Voronoi diagrams [10, 15, 12, 6]. These algorithms assume that the bisector between two sites can be computed in constant time, and construct the Voronoi diagram in $O(n \log n)$ time with $O(n)$ space when n sites are given. In our setting, we assume that a representation of the bisector between two needles has the complexity $S_b(C)$, and can be computed in $T_b(C)$ time.

Corollary 16. *Let S be a set of n pairwise non-piercing needles under a convex distance d based on a convex C. Then, the Voronoi diagram $\mathcal{V}_d(S)$ for S can be computed in $O(T_b(C) \cdot n \log n)$ time and $O(S_b(C) \cdot n)$ space.*

Making $\mathcal{S}^*$ pairwise non-piercing. We can make $\mathcal{S}^*$ pairwise non-piercing by the procedure introduced in the proof of Lemma 2 in [4]. The only difference arises when we consider non-piercing needles as input sites; the produced needles on the roads may pierce the original needles. This problem can be solved by cutting a pierced original needle into two non-pierced needles. Since these piercing cases can occur only between original needles and produced needles dominating a node, we can check all the cases in $O(T_{op}(C) \cdot mn)$ time and the asymptotic number of needles does not increase. We denote by $\mathcal{S}^*_{np}$ the resulting set of pairwise non-piercing needles for a given set S of pairwise non-piercing needles. Note that $\mathcal{V}_d(\mathcal{S}^*_{np})$ is a refined diagram of $\mathcal{V}_d(\mathcal{S}^*)$.

4.4 Putting It All Together

From the previous discussions, we finally conclude the following theorem.

Theorem 17. *Let d be a convex distance based on a convex C whose boundary is semialgebraic, G be a transportation network with m roads under d, and S be a set of n sites. Then, the Voronoi diagram $\mathcal{V}_{d_G}(S)$ under d_G can be computed in $O(m(n+m)(m+T_b(C)\log(n+m)+T_{op}(C)))$ time with $O(S_b(C)m(n+m))$ space, where $T_{op}(C)$, $T_b(C)$, and $S_b(C)$ are defined as before.*

If C is a k-gon, we can see that $T_{op}(C) = O(\log k)$, $T_b(C) = S_b(C) = O(k)$ [14].

Corollary 18. *Let d be a convex distance based on a convex k-gon C, G be a transportation network with m roads under d, and S be a set of n sites. Then, the Voronoi diagram $\mathcal{V}_{d_G}(S)$ under d_G can be computed in $O(m(n+m)(m + k\log(n+m)))$ time with $O(km(n+m))$ space.*

References

1. M. Abellanas, F. Hurtado, C. Icking, R. Klein, E. Langetepe, L. Ma, B. Palop, and V. Sacristán. Proximity problems for time metrics induced by the l_1 metric and isothetic networks. *IX Encuetros en Geometria Computacional*, 2001.
2. M. Abellanas, F. Hurtado, C. Icking, R. Klein, E. Langetepe, L. Ma, B. Palop, and V. Sacristán. Voronoi diagram for services neighboring a highway. *Information Processing Letters*, 86:283–288, 2003.

3. O. Aichholzer, F. Aurenhammer, and B. Palop. Quickest paths, straight skeletons, and the city voronoi diagram. In *Proceedings of the 8th SoCG*, pages 151–159, 2002.
4. S. W. Bae and K.-Y. Chwa. Voronoi diagrams with a transportation network on the euclidean plane. Technical report, Korea Advanced Institute of Science and Technology, 2005. A preliminary version appeared in proceedings of ISAAC 2004.
5. A. G. Corbalan, M. Mazon, and T. Recio. Geometry of bisectors for strictly convex distances. *International Journal of Computertational Gemoetry and Applications*, 6(1):45–58, 1996.
6. F. Dehne and R. Klein. "the big sweep": On the power of the wavefront approach to Voronoi diagrams. *Algorithmica*, 17:19–32, 1997.
7. L. Gewali, A. Meng, Joseph S. B. Mitchell, and S. Ntafos. Path planning in $0/1/\infty$ weighted regions with applications. *ORSA J. Comput.*, 2(3):253–272, 1990.
8. R. Görke and A. Wolff. Computing the city voronoi diagram faster. In *Proc. 21st Euro. Workshop on Comput. Geom.*, pages 155–158, 2005.
9. J. C. Kelly. Bitopological spaces. *Proc. London Math. Soc.*, 13(3):71–89, 1963.
10. R. Klein. *Concrete and Abstract Voronoi Diagrams*, volume 400 of *LNCS*. Springer-Verlag, Berlin, Germany, 1989.
11. R. Klein. Private communication, 2005.
12. R. Klein, K. Mehlhorn, and S. Meiser. Randomized incremental construction of abstract voronoi diagrams. *Computational Geometry: Theory and Applications*, 3:157–184, 1993.
13. R. Klein and D. Wood. Voronoi diagrams based on general metrics in the plane. In *Proc. 5th STACS*, volume 294 of *LNCS*, pages 281–291. Springer-Verlag, 1988.
14. L. Ma. *Bisectors and Voronoi Diagams for Convex Distance Functions*. PhD thesis, Fern Unversität Hagen, 2000.
15. K. Mehlhorn, S. Meiser, and C. O'Dunlaing. On the construction of abstract Voronoi diagrams. *Discrete Comput. Geom.*, 6:211–224, 1991.
16. A. C. G. Mennucci. On asymmetric distances. *Preprint*, 2004. available at `http://cvgmt.sns.it/cgi/get.cgi/papers/and04/`.
17. J. S. B. Mitchell. Shortest paths among obstacles, zero-cost regions, and "roads". Technical Report 764, School Oper. Res. Indust. Engrg., Cornell Univ., Ithaca, NY, 1987.
18. B. Palop. *Algorithmic problems on proximity and location under metric constraints.* PhD thesis, U. Politécnica de Catalunya, 2003.
19. W. Rinow. *Die innere Geometrie der Metrischen Räume*, volume 105 of *Grundlehren der Mathematischen Wissenschaften in Einzeldarstellungen.* Springer-Verlag, Berlin, 1961.
20. N. C. Rowe. Roads, rivers, and obstacles: optimal two-dimensional path planning around linear features for a mobile agent. *Internat. J. Robot. Res.*, 9:67–73, 1990.

Approximation Algorithms for Computing the Earth Mover's Distance Under Transformations

Oliver Klein[1,*] and Remco C. Veltkamp[2]

[1] Department of Computer Science, FU Berlin
oklein@inf.fu-berlin.de
[2] Department of Computer Science, Universiteit Utrecht
remco.veltkamp@cs.uu.nl

Abstract. The Earth Mover's Distance (EMD) on weighted point sets is a distance measure with many applications. Since there are no known exact algorithms to compute the minimum EMD under transformations, it is useful to estimate the minimum EMD under various classes of transformations. For weighted point sets in the plane, we will show a 2-approximation algorithm for translations, a 4-approximation algorithm for rigid motions and an 8-approximation algorithm for similarity transformations. The runtime for translations is $O(T^{EMD}(n,m))$, the runtime of the latter two algorithms is $O(nmT^{EMD}(n,m))$, where $T^{EMD}(n,m)$ is the time to compute the EMD between two fixed weighted point sets with n and m points, respectively. All these algorithms are based on a more general structure, namely on reference points. This leads to elegant generalizations to higher dimensions. We give a comprehensive discussion of reference points for weighted point sets with respect to the EMD.

1 Introduction

The Earth Mover's Distance on weighted point sets is a very useful distance measure for e.g. shape matching, colour-based image retrieval and music score matching, see [5], [6], [7] and [11] for more information. For these applications it is useful to have a quick estimation on the minimum distance between two weighted point sets which can be achieved under a considered class of transformations $\mathcal{T}$. Thus we want to find algorithms to compute an approximation where $EMD^{apx}(A,B) \leq \alpha \cdot \min\{EMD(A,\Phi(B)) : \Phi \in \mathcal{T}\}$. This problem was first regarded by Cohen ([5]). He constructed an iterative Flow-Transformation algorithm, which he proved to converge, but not necessarily to the global minimum. In this paper we will take a different approach and use reference points to get an approximation on the problem. These points have already been introduced in [1] and [2] to construct approximation algorithms for matching compact subsets of $\mathbb{R}^d$ under translations, rigid motions and similarity transformations with respect

* This research was supported by the Deutsche Forschungsgemeinschaft within the European graduate program 'Combinatorics, Geometry and Computation' (No. GRK 588/2).

X. Deng and D. Du (Eds.): ISAAC 2005, LNCS 3827, pp. 1019–1028, 2005.

to the Hausdorff-distance. Also approximation algorithms using reference points for matching with respect to the Area of Symmetric Difference have been given, see [3] and [12]. A general discussion of reference point methods for matching according to the Hausdorff-distance has been given in [1]. Here we will extend the definition of reference points to weighted point sets and get fast constant factor approximation algorithms for matching weighted point sets under translations, rigid motions and similarity transformations with respect to the EMD. Quite recently, Cabello et al. ([4]) have been working on similar problems. The advantage of our approach is that the results given can be applied to arbitrary dimension and distance measure on the ground set, even on more than the in this abstract mentioned L_p-distances. Therefore the results are widely applicable.

2 Basic Definitions

Definition 1 (Weighted Point Set). *([6]) Let $A = \{a_1, a_2, ..., a_n\}$ be a weighted point set such that $a_i = (p_i, \alpha_i)$ for $i = 1, ..., n$, where p_i is a point in $\mathbb{R}^d$ and $\alpha_i \in \mathbb{R}_0^+$ its corresponding weight. Let $W^A = \sum_{i=1}^n \alpha_i$ be the total weight of A. Let $\mathbb{W}^d$ be the set of all weighted point sets in $\mathbb{R}^d$ and $\mathbb{W}^{d,G}$ be the set of all weighted point sets in $\mathbb{R}^d$ with total weight $G \in \mathbb{R}^+$.*

In the following we will use a considered class of transformations on both weighted point sets and discrete subsets of $\mathbb{R}^d$. By a transformation on a weighted point set we mean to transform the coordinates of the weighted points and leave their weights unchanged.

We now introduce the center of mass, which plays an important role in our approximation algorithms. The computation time of this point is linear, so it does not affect the runtime of the presented algorithms.

Definition 2 (Center of Mass). *Let $A = \{(p_i, \alpha_i)_{i=1,...,n}\} \in \mathbb{W}^{d,G}$ be a weighted point set for some $G \in \mathbb{R}^+$. The center of mass of A is defined as $C(A) = \frac{1}{W^A} \sum_{i=1}^n \alpha_i p_i$.*

As we will see, the center of mass is an instance of a more general class of mappings, namely reference points. Later we will prove the correctness of abstract algorithms based on this class of mappings. By plugging in the center of mass we will get concrete and implementable algorithms.

Definition 3 (Reference Point). *([1]) Let $\mathcal{K}$ be a subset of $\mathbb{W}^d$ and $\delta : \mathcal{K} \times \mathcal{K} \to \mathbb{R}_0^+$ be a distance measure on $\mathcal{K}$. Let $||\cdot|| : \mathbb{R}^d \to \mathbb{R}_0^+$ be any norm on $\mathbb{R}^d$. A mapping $r : \mathcal{K} \to \mathbb{R}^d$ is called a δ-reference point for $\mathcal{K}$ with respect to a set of transformations $\mathcal{T}$ on $\mathcal{K}$, if the following two conditions hold:*

1. *Equivariance with respect to $\mathcal{T}$: For all $A \in \mathcal{K}$ and $\Phi \in \mathcal{T}$ we have*

$$r(\Phi(A)) = \Phi(r(A)).$$

2. *Lipschitz-continuity: There is a constant $c \geq 0$, such that for all $A, B \in \mathcal{K}$,*

$$||r(A) - r(B)|| \leq c \cdot \delta(A, B).$$

We call c the quality of the δ-reference point r.

In section 4.3 we will construct approximation algorithms for similarities. For this reason we will have to rescale one of the weighted point sets. Unfortunately, rescaling in a way that the diameters of the underlying point sets in $\mathbb{R}^d$ are equal, does not work. The key to a working algorithm is to rescale the set in a way that the normalized first moments with respect to their reference points coincide. Here we give the well known definition of the normalized first moment of a weighted point set with respect to an arbitrary point $p \in \mathbb{R}^d$.

Definition 4 (Normalized First Moment). *Let $A = \{(p_i, \alpha_i)_{i=1,\dots,n}\} \in \mathbb{W}^{d,G}$ be a weighted point set for some $G \in \mathbb{R}^+$ and let $p \in \mathbb{R}^d$ be an arbitrary point. We call $m_p(A) = \frac{1}{W^A}\sum_{i=1}^{n} \alpha_i ||p_i - p||$ the normalized first moment of A with respect to p.*

Note that the normalized first moment of a weighted point set with respect to an arbitrary point can be calculated efficiently in linear time.

Next we will introduce the EMD, a distance measure on weighted point sets.

Definition 5 (Earth Mover's Distance). *([5]) Let $A = \{(p_i, \alpha_i)_{i=1,\dots,n}\}$, $B = \{(q_j, \beta_j)_{j=1,\dots,m}\} \in \mathbb{W}^d$ be weighted point sets with total weights $W^A, W^B \in \mathbb{R}^+$. Let $D : \mathbb{R}^d \times \mathbb{R}^d \to \mathbb{R}_0^+$ be a distance measure on the ground set $\mathbb{R}^d$. The Earth Mover's Distance between A and B is defined as*

$$EMD(A, B) = \frac{\min_{F \in \mathcal{F}} \sum_{i=1}^{n} \sum_{j=1}^{m} f_{ij} D(p_i, q_j)}{\min\{W^A, W^B\}}$$

where $F = \{f_{ij}\}$ is a feasible flow, i.e.

1. $f_{ij} \geq 0, i = 1, ..., n, j = 1, ..., m$
2. $\sum_{j=1}^{m} f_{ij} \leq \alpha_i, i = 1, ..., n$
3. $\sum_{i=1}^{n} f_{ij} \leq \beta_j, j = 1, ..., m$
4. $\sum_{i=1}^{n} \sum_{j=1}^{m} f_{ij} = \min\{W^A, W^B\}$

For the rest of the paper the distance measure D used in the definition of the EMD will be the metric induced by the norm used in the definition of the EMD-reference point. When working with weighted point sets in $\mathbb{R}^d$ we will call $\mathbb{R}^d$ the ground set and $D : \mathbb{R}^d \times \mathbb{R}^d \to \mathbb{R}_0^+$ the ground distance. If D is the Euclidean Distance, we will also use EEMD as a notation for the Euclidean Earth Mover's Distance. If D is any L_p-distance for $1 \leq p \leq \infty$ we will write EMD_p to denote the Earth Mover's Distance based on this distance measure.

3 EMD-Reference Points

In this section we discuss the existence of EMD-reference points. We start with a negative result.

3.1 Non-existence of EMD-Reference Points for Non-equal Total Weights

Let r be any reference point with respect to translations. Regarding the weighted point sets $A := \{(p, 1)\}$, $B := \{(q, 1)\}$ and $C := A \cup B$, $p \neq q \in \mathbb{R}^d$, we can easilysee that $EMD(A, C) = EMD(B, C) = 0$ and therefore $r(A) = r(C) =$

$r(B)$. On the other hand, B is a translation of A by $q - p \neq 0$ and therefore, by equivariance, $r(B)$ has to be a translation of $r(A)$ by $q - p$, which leads to a contradiction. Thereby we have proven the following theorem:

Theorem 1. *There is no EMD-reference point for weighted point sets with unequal total weights with respect to all transformation sets that include translations.*

Unfortunately, Theorem 1 has a deep impact on the usability of the reference point approach for shape matching since it makes it impossible to use this approach for partial matching applications. For a more detailed discussion on partial matching using Mass Transportation Distances, see [6].

3.2 The Center of Mass as a Reference Point

In the next section we will present approximation algorithms for the EMD under transformations using EMD-reference points. Since this would be useless if there was no EMD-reference point, we will restrain the consideration to weighted point sets with equal total weight. In this case, the equivariance of the center of mass under affine transformations is well known and the proof of the Lipschitz-continuity appeared already in [5]. Therefore we can formulate the following theorem:

Theorem 2. *The center of mass is an EMD-reference point for weighted point sets with equal total weight with respect to affine transformations. Its quality is* 1*. This holds for any dimension d and any distance measure on the ground set.*

3.3 Lower Bound on the Quality of an EMD-Reference Point

Using the center of mass to construct implementable algorithms raises the question if there is a better reference point. We can prove that the center of mass is optimal in the sence that there is no reference point inducing a better quality. But we do not know so far if there is a reference point inducing approximation algorithms with a better approximation factor.

Theorem 3. *Let $r : \mathbb{W}^{d,G} \to \mathbb{R}^d$ be an EMD-reference point with respect to any transformation set including the set of translations for some $G \in \mathbb{R}^+$ and some dimension d, and let c be its quality. Then $c \geq 1$. This holds for any distance measure on the ground set.*

4 Approximation Algorithms Using EMD-Reference Points

The following three sections are organized as follows: In each section we consider a class of transformations, construct an approximation algorithm for matching under these transformations for general EMD-reference points and finally use the center of mass to get a concrete algorithm.

For the rest of the paper let $A = \{(p_i, \alpha_i)_{i=1,\dots,n}\}$, $B = \{(q_j, \beta_j)_{j=1,\dots,m}\} \in \mathbb{W}^{d,G}$ be two weighted point sets in dimension d with positive equal total weight $G \in \mathbb{R}^+$. Please be reminded that the following results do not hold for weighted point sets with unequal total weight. Further, let $r : \mathbb{W}^{d,G} \longrightarrow \mathbb{R}^d$ be an EMD-reference point for weighted point sets with respect to the considered class of transformations with quality c. Let $T^{ref}(n)$ be the time to compute the EMD-reference point of A, $T^{EMD}(n,m)$ and $T^{EEMD}(n,m)$ be the time to compute the EMD and $EEMD$ between A and B and $T^{rot}(n,m)$ be the time needed to find a rotation R around a fixed point minimizing $EMD(A, R(B))$.

An upper bound on $T^{EMD}(n,m)$ and $T^{EEMD}(n,m)$ is $O((nm)^2 \log(n+m))$ using a strongly polynomial minimum cost flow algorithm by Orlin ([9]). In practice, an algorithm using the simplex method to solve the linear program will be faster. Since we are developing approximation algorithms anyway, one can consider using an $(1+\varepsilon)$-approximation algorithm for the Earth Mover's Distance by Cabello et al. ([4]) with runtime $O(\frac{n^2}{\varepsilon^2}\log^2(\frac{n}{\varepsilon}))$.

4.1 Translations

The first algorithm will find an approximation for the EMD under translations:

> Algorithm *TranslationApx*:
> 1. Compute $r(A)$ and $r(B)$ and translate B by $r(A) - r(B)$. Let B' be the image of B.
> 2. Output B' together with the approximate distance $EMD(A, B')$.

Theorem 4. *Algorithm TranslationApx finds an approximately optimal matching for translations with approximation factor* $c+1$ *in time* $O(T^{ref}(\max\{n,m\}) + T^{EMD}(n,m))$. *This holds for arbitrary dimension d and distance measure on the ground set.*

Applying the center of mass leads to the following corollary. The approximation factor of 2 is tight, a proof for this can also be found in [8].

Corollary 1. *Algorithm TranslationApx using the center of mass as an EMD-reference point induces an approximation algorithm with approximation factor* 2. *Its runtime is* $O(T^{EMD}(n,m))$.

4.2 Rigid Motions

The following algorithm is a first approach to get an approximation on the EMD under rigid motions, i.e. combinations of translations and rotations:

> Algorithm *RigidMotionApx*:
> 1. Compute $r(A)$ and $r(B)$ and translate B by $r(A) - r(B)$. Let B' be the image of B.
> 2. Find an optimal matching of A and B' under rotations of B' around $r(A)$. Let B'' be the image of B' under this rotation.
> 3. Output B'' and the approximate distance $EMD(A, B'')$.

Theorem 5. *Algorithm RigidMotionApx finds an approximately optimal matching for rigid motions with approximation factor* $c+1$ *in time* $O(T^{ref}(\max\{n,m\}) + T^{EMD}(n,m) + T^{rot}(n,m))$. *This holds for arbitrary dimension* d *and distance measure on the ground set.*

Corollary 2. *Algorithm RigidMotionApx using the center of mass as EMD-reference point induces an approximation algorithm with approximation factor* 2 *in time* $O(T^{rot}(n,m) + T^{EMD}(n,m))$. *This holds for arbitrary dimension* d *and distance measure on the ground set .*

Since the position of the EMD-reference point as rotation center is fixed, several degrees of freedom have been eliminated and the problem to find the optimal rotation should be easier than the one finding the optimal rigid motion itself. Unfortunately, even for this problem no efficient algorithm is known so far. Therefore it would be nice to have at least an approximation algorithm for this problem. In the next lemma we will give an approximation for the Euclidean Distance as the ground distance. This result was already published in [4]. Next we will use this lemma to extend the result to all L_p-distances, $1 \leq p \leq \infty$. Unfortunately, the approximation factor will be worse than 2 for $p \neq 2$.

Lemma 1. *Let* $A, B \in \mathbb{W}^d$ *be two weighted point sets and* p^* *be any point. Let* $Rot(p^*)$ *be the set of all rotations around* p^*. *Then there is a rotation* $R' \in Rot(p^*)$ *such that* R' *aligns* p^* *and any two points of* A *and* B, *and*

$$EEMD(A, R'(B)) \leq 2 \cdot \min_{R \in Rot(p^*)} EEMD(A, R(B)).$$

As mentioned above, we will now use the last lemma to extend the result to all L_p-distances, $1 \leq p \leq \infty$. The proof is based on the fact that for $1 \leq p, q \leq \infty$ and any vector $v \in \mathbb{R}^d$ it holds that $||v||_p \leq \sqrt{d}||v||_q$.

Lemma 2. *Let* $A, B \in \mathbb{W}^d$ *be two weighted point sets and* p^* *be any point. Let* $Rot(p^*)$ *be the set of all rotations around* p^*. *Then there is a rotation* $R' \in Rot(p^*)$ *such that* R' *aligns* p^* *and any two points of* A *and* B, *and*

$$EMD_p(A, R'(B)) \leq 2\sqrt{d} \cdot \min_{R \in Rot(p^*)} EMD_p(A, R(B)).$$

An Applicable Algorithm in the Plane. Based on the last two lemmata we are able to construct an approximation algorithm for the problem of finding an optimal rotation of a weighted point set around their coinciding reference points. In this section we will discuss the case of weighted point sets in the plane.

> Algorithm *RotationApx*
> 1. Compute the minimum EMD over all possible alignments of the coinciding reference points and any two points of A and B.

Since there are $O(nm)$ possibilities to align the reference point and any two points of A and B, the runtime of this algorithm is $O(nmT^{EMD}(n,m))$. Using

this algorithm, we now get an easy to implement and fast approximation algorithm for rigid motions. Unfortunately, the fact that we now constructed an implementable algorithm must be paid by an increased approximation factor. Figure 1 shows an illustration of this algorithm.

Algorithm *RigidMotionApxUsingRotationApx*

1. Compute $r(A)$ and $r(B)$ and translate B by $r(A) - r(B)$. Let B' be the image of B.
2. Find a best matching of A and B' under rotations of B' around $r(A) = r(B')$ where $r(A)$ and any two points in A and B' are aligned. Let B'' be the image of B' under this rotation.
3. Output B'' and the approximate distance $EMD(A, B'')$.

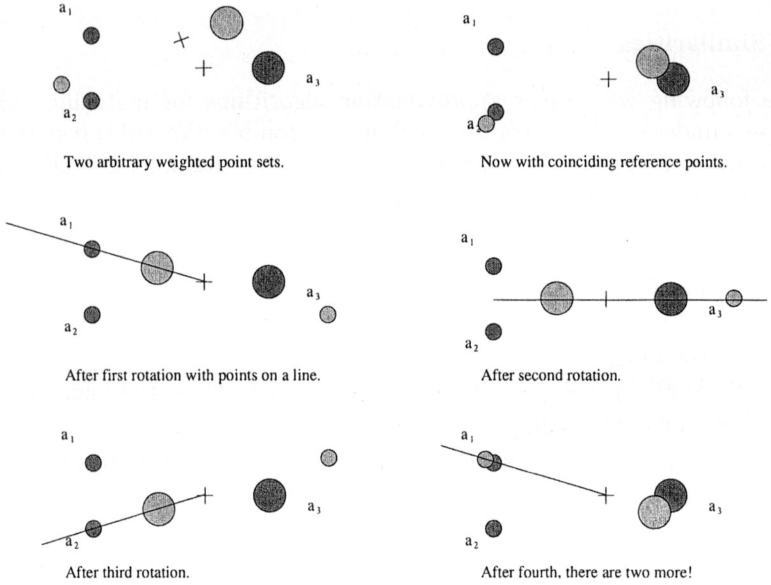

Fig. 1. Illustration of algorithm *RigidMotionApxUsingRotationApx*

Theorem 6. *Regarding EEMD in the plane, Algorithm RigidMotionApxUsing-RotationApx finds an approximately optimal matching for rigid motions with approximation factor* $2(c+1)$ *in time* $O(T^{ref}(\max\{n,m\}) + nmT^{EEMD}(n,m))$.

Using Lemma 2 we can extend the result to all L_p-distances:

Theorem 7. *Regarding* EMD_p *in the plane,* $1 \leq p \leq \infty$*, Algorithm Rigid-MotionApxUsingRotationApx finds an approximately optimal matching for rigid motions with approximation factor* $2\sqrt{2}(c+1)$ *in time* $O(T^{ref}(\max\{n,m\}) + nmT^{EMD_p}(n,m))$.

Application of the center of mass as an EMD-reference point leads to the following corollary:

Corollary 3. *Algorithm RigidMotionApxUsingRotationApx using the center of mass as EMD-reference point induces an approximation algorithm with approximation factor* 4 *in case of the Euclidean Distance in the plane and* $4\sqrt{2}$ *for any other* L_p *distance,* $1 \leq p \leq \infty$. *Its runtime is* $O(nmT^{EMD_p}(n,m))$.

In this section we have constructed approximation algorithms to minimize the EMD of weighted point sets under rigid motions in the plane. These algorithms have elegant generalizations to higher dimensions. In case of dimension $d \geq 3$, the approximation factor of the implementable algorithm using the center of mass as a reference point is $2^{d-1}\sqrt{d}(c+1)$ and the runtime is $O(n^d T^{EMD})$ if $m = O(n)$, see [8].

4.3 Similarities

In the following we present approximation algorithms for matching weighted point sets under similarity transformations, i.e. combinations of translations, rotations and scalings. More precisely, we want to compute $\min_S EMD(A, S(B))$, where the minimum is taken over all similarity transformations S. Note that in this case exchanging A and B makes a difference.

Algorithm *SimilarityApx*:
1. Compute $r(A)$ and $r(B)$ and translate B by $r(A) - r(B)$. Let B' be the image of B.
2. Scale B' by $\frac{m_{r(A)}(A)}{m_{r(B')}(B')}$ around $r(A)$ and let B'' be the image of B' under this scaling.
3. Find an optimal matching of A and B'' under rotations of B'' around $r(A)$. Let B''' be the image of B'' under this rotation.
4. Output B''' and the approximate distance $EMD(A, B''')$.

To show the correctness of this algorithm we use the following two lemmata:

Lemma 3. *Let* $A \in \mathbb{W}^{d,G}$ *for some* $G \in \mathbb{R}^+$ *and let* $m_p(A)$ *be its normalized first moment with respect to some point* $p \in \mathbb{R}^d$. *Let* τ_1, τ_2 *be scalings around the same center* p *and ratios* γ_1 *and* γ_2, *respectively. Then*

$$EMD(\tau_1(A), \tau_2(A)) \leq |(\gamma_1 - \gamma_2)| m_p(A).$$

This lemma gives a new lower bound for the EMD of weighted point sets:

Lemma 4. *Let* $A, B \in \mathbb{W}^{d,G}$ *for some* $G \in \mathbb{R}^+$. *Then*

$$|m_{r(A)}(A) - m_{r(B)}(B)| \leq (1+c)EMD(A, B).$$

Using Lemmata 3 and 4 we can prove the following:

Theorem 8. *Algorithm SimilarityApx finds an approximately optimal matching for similarities with approximation factor* $2(c+1)$ *in time* $O(T^{ref}(\max\{n,m\}) + T^{EMD}(n,m) + T^{rot}(n,m))$. *This holds for arbitrary dimension d and distance measure on the ground set.*

Corollary 4. *Algorithm SimilarityApx using the center of mass as EMD-reference point induces an approximation algorithm with approximation factor* 4. *Its runtime is* $O(T^{EMD}(n,m) + T^{rot}(n,m))$. *This holds for any dimension of the ground set and every distance measure defined on it.*

As for *RigidMotionApx*, *SimilarityApx* depends on finding the optimal rotation, which is impractical. Again, we make this algorithm practical and efficient by using *RotationApx* and again we have to pay by a worse approximation factor.

Algorithm *SimilarityApxUsingRotationApx*
1. Compute $r(A)$ and $r(B)$ and translate B by $r(A) - r(B)$. Let B' be the image of B.
2. Scale B' by $\frac{m_{r(A)}(A)}{m_{r(B')}(B')}$ around $r(A) = r(B')$ and let B'' be the image of B' under this scaling.
3. Find a best matching of A and B'' under rotations of B'' around $r(A) = r(B'')$ where $r(A)$ and any two points in A and B'' are aligned. Let B''' be the image of B'' under this rotation.
4. Output B''' and the approximate distance $EEMD(A, B''')$.

Theorem 9. *Regarding EEMD in the plane, Algorithm SimilarityApxUsingRotationApx finds an approximately optimal matching for similarities with approximation factor* $4(c+1)$ *in time* $O(T^{ref}(\max\{n,m\}) + nmT^{EEMD}(n,m))$.

Using Lemma 2 we can easily see:

Theorem 10. *Regarding* EMD_p *in the plane,* $1 \leq p \leq \infty$, *Algorithm SimilarityApxUsingRotationApx finds an approximately optimal matching for similarities with approximation factor* $4\sqrt{2}(c+1)$ *in time* $O(T^{ref}(\max\{n,m\}) + nmT^{EMD_p}(n,m))$.

Application of the center of mass as an EMD-reference point leads to the following corollary:

Corollary 5. *Algorithm SimilarityApxUsingRotationApx using the center of mass as an EMD-reference point induces an approximation algorithm with approximation factor* 8 *in case of the Euclidean Distance in the plane and* $8\sqrt{2}$ *for any other* L_p *distance,* $1 \leq p \leq \infty$. *Its runtime is* $O(nmT^{EMD_p}(n,m))$.

5 Conclusion

In this paper we introduced EMD-reference points for weighted point sets and constructed efficient approximation algorithms for matching under various classes

of transformations. In contrast to previous work, this approach allows elegant extension to higher dimensions and more general ground distances. Additionally, we presented the center of mass as an EMD-reference point for weighted point sets with equal total weight. This reference point, in fact, turns out to be an optimal reference point in the sence that there is none with a Lipschitz-constant smaller than 1. Unfortunately, the center of mass is no EMD-reference point if you consider the set of all weighted point sets, including those with different total weights. Even worse, there is no EMD-reference point for all weighted point sets. A variation of the EMD is the Proportional Transportation Distance (PTD). We can show that the center of mass is a PTD-reference point even for weighted point sets with different total weights and all theorems and corollaries mentioned in this paper carry over.

References

1. H. Alt, O. Aichholzer, G. Rote. Matching Shapes with a Reference Point. In *Proc. 10th Annual Symposium on Computational Geometry*, pages 85–92, 1994.
2. H. Alt, B. Behrends, J. Blömer. Approximate Matching of Polygonal Shapes. In *Proc. 7th Ann. Symp. on Comp. Geometry*, pages 186–193, 1991.
3. H. Alt, U. Fuchs, G. Rote, G. Weber. Matching Convex Shapes with Respect to the Symmetric Difference. In *Proc. 4th Ann. Eur. Symp. on Algorithms, Barcelona, LNCS Vol. 1136*, pages 320–333, Springer Verlag, 1996.
4. S. Cabello, P. Giannopoulos, C. Knauer, G. Rote. Matching Point Sets with respect to the Earth Mover's Distance. In *Proc. ESA*, 2005.
5. S. Cohen. Finding Color and Shape Patterns in Images. PhD thesis, Stanford University, Department of Compute Science, 1999.
6. P. Giannopoulos, R. Veltkamp. A pseudo-metric for weighted point sets. In *Proc. 7th European Conf. Comp. Vision*, LNCS 2352, pages 715–731, 2002.
7. K. Graumann, T. Darell. Fast contour matching using approximate Earth Mover's Distance. In *Proc. of the IEEE Conf. Comp. Vision and Pattern Recognition*, LNCS 2352, pages I: 220–227, 2004.
8. O. Klein, R. C. Veltkamp. Approximation Algorithms for the Earth Mover's Distance Under Transformations Using Reference Points. Technical Report UU-CS-2005-003, http://ftp.cs.uu.nl/pub/RUU/CS/techreps/CS-2005/2005-003.pdf, 2005.
9. J. B. Orlin. A Faster Strongly Polynomial Minimum Cost Flow Algorithm. In *Operations Research*, vol.41,no.2, pages 338–350, 1993.
10. Y. Rubner, C. Tomasi, L. J. Guibas. The Earth Mover's Distance as a Metric for Image Retrieval. In *Int. J. of Comp. Vision 40(2)*, pages 99–121, 2000.
11. R. Typke, P. Giannopoulos, R. C. Veltkamp, F. Wierking, R. Oostrum. Using transportation distances for measuring melodic similarity. In *Proc. of the 4th Int. Conf. Music Inf. Retrieval*, pages 107–114, 2003.
12. G. Weber. The Centroid is a Reference Point for the Symmetric Difference in d Dimensions. *Tech. Rep. UoA-SE-2004-1*, The University of Auckland, 2004.

Fast k-Means Algorithms with Constant Approximation

Mingjun Song and Sanguthevar Rajasekaran

Computer Science and Engineering,
University of Connecticut,
Storrs CT 06269, USA
{mjsong, rajasek}@engr.uconn.edu

Abstract. In this paper we study the k-means clustering problem. It is well-known that the general version of this problem is $\mathcal{NP}$-hard. Numerous approximation algorithms have been proposed for this problem. In this paper, we proposed three constant approximation algorithms for k-means clustering. The first algorithm runs in time $O((\frac{k}{\epsilon})^k nd)$, where k is the number of clusters, n is the size of input points, d is dimension of attributes. The second algorithm runs in time $O(k^3 n^2 \log n)$. This is the first algorithm for k-means clustering that runs in time polynomial in n, k and d. The run time of the third algorithm $\left(O(k^5 \log^3 kd)\right)$ is independent of n. Though an algorithm whose run time is independent of n is known for the k-median problem, ours is the first such algorithm for the k-means problem.

1 Introduction

Among the different clustering techniques known, k-means is very popular. In this problem, given a set $P \subset \Re^d$ of n data points and a number k, we are required to partition P into k subsets (i.e., clusters). Each such cluster has a center defined by the centroid (i.e.,mean) of the points in that cluster. The partitioning should minimize the following cost function:

$$\triangle^P(K) = \sum_{x \in P} \|x - K(x)\|^2,$$

Where $K(x)$ denotes the nearest centroid to x, and $\|x-y\|$ denotes the Euclidean distance between two points x and y.

One of the most popular heuristic algorithms for k-means is Lloyd's algorithm [1], which initially chooses k centers randomly. For each input point, the nearest center is identified. Points that choose the same center belong to a cluster. Now new centers are calculated for the clusters. Each input point identifies its nearest center; and so on. This process is repeated until no changes occur. The process of identifying the nearest center for each input point and recomputing centers is refered to as an *iteration*. The number of iterations taken by Lloyd's algorithm is unknown. This algorithm may converge to a local minimum with an arbitrarily bad distortion with respect to the optimal solution [2].

X. Deng and D. Du (Eds.): ISAAC 2005, LNCS 3827, pp. 1029–1038, 2005.

Researches have been conducted to find algorithms with bounded quality, either $(1+\epsilon)$-approximation or constant approximation. Matousek [3] has presented a $(1+\epsilon)$-approximation algorithm with a run time of $O(n \log^k n\epsilon^{-2k^2 d})$ for any fixed $\epsilon > 0$, k, and d using the approximate centroid set idea. The centroid set was constructed by recursively subdividing the 3-enlargement cube of the bounding box of the point set P. Then the algorithm generates all well-spread k-tuples and returns the k-tuple with the minimum cost.

Kanungo et al. [2] have given a $(9+\epsilon)$-approximation algorithm. This algorithm uses an ϵ-approximate centroid set generated from the algorithm of [3] as the candidate centers. The algorithm starts with k initial centers selected from the candidate centers, and iteratively removes p centers (for some appropriate value of p) and replaces them with another p centers from the candidate centers if the resulting cost decreases. The running time is $O(n \log n + n\epsilon^{-d} \log(1/\epsilon) + n^2 k^3 \log n)$.

The algorithm of Har-Peled Mazumdar [4] takes time $O(n + k^{k+2}\epsilon^{-(2d+1)k} \log^{k+1} n \log^k \frac{1}{\epsilon})$ to find a $(1+\epsilon)$-approximate solution to the k-means problem. If k and d are fixed, the run time is $O(n)$. The algorithm constructed a corset by sampling in an exponential grid. The authors achieved the linear time solution by combining many other known algorithms.

Kumar et al. [5] propose a simple $(1+\epsilon)$-approximation algorithm with a run time of $O(2^{(\frac{k}{\epsilon})^{O(1)}} dn)$. The idea of the algorithm is to approximate the centroid of the largest cluster by trying all subsets of constant size from the sample, and doing the same on the smaller cluster by pruning points from the larger cluster.

A problem closely related to k-means clustering is the k-median clustering problem. In this problem the objective is to minimize the sum of the distances to the nearest median. Also, the cluster centers should form a subset of the input points. Finding optimal solutions to k-means and k-medians problems are NP-hard. Jain et. al. [8] even showed that it is NP-hard to obtain an approximation within a factor of $1 + \frac{2}{e}$. Thus most of the research focusses on approximation algorithms. In this paper, we focus on constant approximations to the k-means problem. None of the previous ($O(1)$-approximation) algorithms for the k-means problem run in time polynomial on n, k and d at the same time. We present three algorithms in this paper. Run time of the first one is polynomial on n and d, of the second one is polynomial on n, k and d, of the third one is polynomial on k and d while being independent of n.

2 Algorithm1

This algorithm is inspired by the following facts: the centroid of one cluster can be approximated by the centroid of a random sample from this cluster. Also the centroid of the sample can be approximated by the closest point to the centroid of the samples. Inaba et. al. [9] showed the first approximation by the following lemma.

Lemma 1. *[9] Let P be the set of input points, T be a random sample with size of $|T|$ from P, μ_P be the centroid of P, μ_T be the centroid of T, then with probability at least $1-\delta$ $(\delta > 0)$,*

$$\sum_{x_i \in P} \|x_i - \mu_T\|^2 \leq (1 + \frac{1}{\delta|T|}) \sum_{x_i \in P} \|x_i - \mu_P\|^2 .$$

Let $\delta = \frac{1}{4}$. Then if we choose $|T|$ to be $\frac{4}{\epsilon}$, with a probability at least $\frac{3}{4}$, the cost computed using the centroid of the sample is $1+\epsilon$ approximation to the real cost.

We show the second approximation by the following lemma.

Lemma 2. *Let C_T be the closest point within the sample to the centroid of the sample, then with probability greater than $\frac{1}{12}$,*

$$\sum_{x_i \in P} \|x_i - C_T\|^2 \leq (5 + 2\epsilon) \sum_{x_i \in P} \|x_i - \mu_P\|^2 .$$

Proof. By the doubled triangle inequality,

$$\|x_i - C_T\|^2 \leq 2(\|x_i - \mu_T\|^2 + \|C_T - \mu_T\|^2).$$

With respect to the second term on the right side of the above inequality,

$$\sum_{x_i \in P} \|C_T - \mu_T\|^2 = |P| \, \|C_T - \mu_T\|^2 \leq \frac{|P|}{|T|} \sum_{x_i \in P} \|x_i - \mu_T\|^2 = |P| Var(T),$$

Where $Var(T)$ is the variance of the sample and is defined as $\frac{1}{|T|} \sum_{x_i \in P} \|x_i - \mu_T\|^2$.

Let $Var(P)$ denote the variance of P, then we have[7],

$$E(Var(T)) = \frac{|T|-1}{|T|} Var(P).$$

By Markov's inequality,

$$Pr[Var(T) \leq 1.5 Var(P)] \geq 1 - \frac{|T|-1}{1.5|T|} > \frac{1}{3}.$$

Thus, with a probability greater than $\frac{1}{3}$,

$$\sum_{x_i \in P} \|C_T - \mu_T\|^2 \leq 1.5 |P| Var(P) = 1.5 \sum_{x_i \in P} \|x_i - \mu_P\|^2 .$$

Let A represents this event, B be the event of satisfying the statement of Lemma 1 (with $\delta = \frac{1}{4}$), then $Pr(AB) = 1 - Pr(\bar{A} \bigcup \bar{B}) \geq 1 - (Pr(\bar{A}) + Pr(\bar{B})) = Pr(A) + Pr(B) - 1 > \frac{3}{4} + \frac{1}{3} - 1 = \frac{1}{12}$.

Therefore, with a probability greater than $\frac{1}{12}$,

$$\sum_{x_i \in P} \|x_i - C_T\|^2 \leq 2 \sum_{x_i \in P} \|x_i - \mu_T\|^2 + 2 \sum_{x_i \in P} \|C_T - \mu_T\|^2$$
$$\leq 2(1+\epsilon) \sum_{x_i \in P} \|x_i - \mu_P\|^2 + 3 \sum_{x_i \in P} \|x_i - \mu_P\|^2$$
$$= (5+2\epsilon) \sum_{x_i \in P} \|x_i - \mu_P\|^2 .$$

□

Next, we will figure out the sample size $|T|$ such that the sample would includes $\frac{4}{\epsilon}$ points for each cluster with high probability. Let n_s be the size of the smallest cluster, and assume $n_s = \alpha\frac{|P|}{k}$ (The similar assumption is found in [6]). By Chernoff Bounds, we have the following inequality with respect to the number of points (X_s) falling in the smallest cluster:

$$Pr[X_s \geq \beta|T|\frac{n_s}{|P|}] \geq 1 - exp(-\frac{(1-\beta)^2}{2}|T|\frac{n_s}{|P|}),$$

then

$$Pr[X_s \geq \beta|T|\frac{\alpha}{k}] \geq 1 - exp(-\frac{(1-\beta)^2}{2}|T|\frac{\alpha}{k}).$$

Let $\beta|T|\frac{\alpha}{k} = \frac{4}{\epsilon}$, and $\beta = \frac{1}{2}$, then $|T| = \frac{8}{\epsilon\alpha}k$. The probability is greater than $1 - exp(-\frac{1}{\epsilon})$.

Therefore, we get algorithm1:

1) Draw a random sample of size $\frac{8}{\epsilon\alpha}k$, where $\alpha = \frac{n_s k}{n}$, n_s is the size of the smallest cluster, $n = |P|$.

2) Using each k-subset of sample points as centers, calculate the cost of clustering with respect to all the original input points.

3) Retrieve the k-subset that results in the minimum cost.

Theorem 1. *The output of algorithm1 is a $(5+2\epsilon)$-approximation to the optimal clustering with a probability greater than $\frac{1}{12}$. Algorithm1 runs in time $O((\frac{k}{\epsilon})^k nd)$.*

Proof. The cost of clusters from algorithm1 is less than the cost of the following clustering: Each center of the cluster is the closest point within the sample to the centroid of the sample. By Lemma 2 and simple summation, $(5+2\epsilon)$-approximation holds. Obviously, The running time is $O((\frac{k}{\epsilon})^k nd)$. □

3 Algorithm2

In this section we present an algorithm with a running time that is polynomial on n, k and d. Kanungo et al. [2]'s local search algorithm is polynomial on n and k, but exponential on d because they used the candidate centroid sets constructed by the algorithm of [3]. In our algorithm, we employ the local search algorithm,

but we use all the input points as the candidate centers instead of just the candidate centroid sets. The algorithm is described as follows:

1) Initially select an arbitrary set of k centers (S) from the input points.

2) For some integer p, swap between any subset of p' ($p' \leq p$) centers from S and p' elements from the input points if the new centers decrease the cost.

3) Repeat step 2 until there is no cost change.

Theorem 2. *The local search algorithm using all input points as candidate centers yields an $O(1)$-approximation to the optimal k-means clustering problem.*

To prove this theorem, we prove some related lemmas.

Lemma 3. *Let C_P be the closest input point to the mean of the input points P. Then,*

$$\sum_{x_i \in P} \|x_i - C_P\|^2 \leq 2 \sum_{x_i \in P} \|x_i - \mu_P\|^2 .$$

Proof.

$$\begin{aligned}
\sum_{x_i \in P} \|x_i - C_P\|^2 &\leq \sum_{x_i \in P} ((x_i - \mu_P) + (\mu_P - C_P))^2 \\
&= \sum_{x_i \in P} \|x_i - \mu_P\|^2 + 2 \sum_{x_i \in P} ((x_i - \mu_P)(\mu_P - C_P)) \\
&\quad + \sum_{x_i \in P} \|C_P - \mu_P\|^2 \\
&\leq \sum_{x_i \in P} \|x_i - \mu_P\|^2 + 2(\mu_P - C_P) \sum_{x_i \in P} (x_i - \mu_P) \\
&\quad + \sum_{x_i \in P} \|x_i - \mu_P\|^2 \\
&= 2 \sum_{x_i \in P} \|x_i - \mu_P\|^2 . \square
\end{aligned}$$

Lemma 4. *The algorithm that enumerates all sets of k points from the input, uses them as centers, computes the clustering cost for each such set, and identifies the best set yields a 2-approximation to the optimal k-means clustering problem.*

Proof. The cost of the algorithm described in the lemma is less than the cost of the following algorithm: The center of each cluster is taken to be the closest point to the centroid of this cluster. By Lemma 3, this lemma follows. □

Next, we prove theorem 2.

Proof. We use the same construction of the set of swap pairs as [2]. The readers are referred to [2] for details. Here, we redescribe the representation of some symbols. S is a local optimal set of k centers resulting from the local search

algorithm, O is the optimal set of k centers from the input points. $\triangle(O)$ denotes the cost using the optimal centers O, $\triangle(S)$ denotes the cost using the heuristic centers S. For any optimal center $o \in O$, s_o represents the closest heuristic center in S to o, $N_O(o)$ represents the neighborhood of o. For any point $q \in P$, s_q denotes the closest heuristic center to q, o_q denotes the closest optimal center to q, s_{o_q} denotes the closest heuristic center to o_q. We use $d(x, y)$ to denote the Euclidean distance between two points x and y, i.e. $\|x - y\|$, and $\triangle(x, y)$ to denote $\|x - y\|^2$.

The following two lemmas adapted from [2] will be used.

Lemma 5. *[2]*
$$0 \le \triangle(O) - 3\triangle(S) + 2R,$$
where $R = \sum_{q \in P} \triangle(q, s_{o_q})$.

Lemma 6. *[2] Let* $\alpha > 0$ *and* $\alpha^2 = \frac{\sum_i s_i^2}{\sum_i o_i^2}$ *for two sequences of reals* $< o_i >$ *and* $< s_i >$*, then*
$$\sum_{i=1}^{n} o_i s_i \le \frac{1}{\alpha} \sum_{i=1}^{n} s_i^2.$$

First, consider the 1-swap case. By the triangle inequality and lemma 6, we have

$$\begin{aligned}
R &= \sum_{o \in O} \sum_{q \in N_O(o)} \triangle(q, s_o) \\
&= \sum_{o \in O} \sum_{q \in N_O(o)} (\triangle(q, o) + \triangle(o, s_o) + 2d(q, o)d(o, s_o)) \\
&\le \sum_{o \in O} \sum_{q \in N_O(o)} (\triangle(q, o) + \triangle(o, s_q) + 2d(q, o)d(o, s_q)) \\
&= \sum_{q \in P} (\triangle(q, o_q) + \triangle(o_q, s_q) + 2d(q, o_q)d(o_q, s_q)) \\
&\le \sum_{q \in P} (\triangle(q, o_q) + \sum_{q \in P} (d(o_q, q) + d(q, s_q))^2 + 2 \sum_{q \in P} d(q, o_q)(d(o_q, q) + d(q, s_q)) \\
&= 4 \sum_{q \in P} \triangle(q, o_q) + \sum_{q \in P} \triangle(q, s_q) + 4 \sum_{q \in P} d(q, o_q)d(q, s_q) \\
&\le 4\triangle(O) + \triangle(S) + \frac{4}{\alpha}\triangle(S) \\
&= 4\triangle(O) + (1 + \frac{4}{\alpha})\triangle(S).
\end{aligned}$$

By lemma 5, we have

$$0 \le \triangle(O) - 3\triangle(S) + 2(4\triangle(O) + (1 + \frac{4}{\alpha})\triangle(S)),$$

$$0 \le 9\triangle(O) - (1 - \frac{8}{\alpha})\triangle(S),$$

$$\frac{9}{1-\frac{8}{\alpha}} \geq \frac{\Delta(S)}{\Delta(O)} = \alpha^2,$$

$$(\alpha+1)(\alpha-9) \leq 0.$$

We get $\alpha \leq 9$. Therefore, $\Delta(S) \leq 81\Delta(O)$.

Second, for p-swap case, by the replacement of $2R$ with $(1+\frac{1}{p})$ in lemma 5, we have

$$0 \leq \Delta(O) - (2+\frac{1}{p})\Delta(S) + (1+\frac{1}{p})(4\Delta(O) + (1+\frac{4}{\alpha}\Delta(S)))$$
$$= (5+\frac{4}{p})\Delta(O) - (1-\frac{4}{\alpha}(1+\frac{1}{p}))\Delta(S),$$

$$\frac{5+\frac{4}{p}}{1-\frac{4}{\alpha}(1+\frac{1}{p}} \geq \frac{\Delta(S)}{\Delta(O)} = \alpha^2,$$

$$(\alpha+1)(\alpha-(5+\frac{4}{p})) \leq 0.$$

We get

$$\alpha \leq 5+\frac{4}{p}.$$

Therefore,

$$\Delta(S) \leq (5+\frac{4}{p})^2\Delta(O).$$

As p increases, $\frac{\Delta(S)}{\Delta(O)}$ approaches 25. Further, using lemma 4, the output of algorithm2 is a 50-approximation to the optimal k-means clustering.

The number of swaps the algorithm takes is proportional to $\log(\frac{\Delta(S_O)}{\Delta(O)})$, where S_O is the initial solution. Because $\Delta(S_O)$ is polynomial in n, the algorithm terminates after $O(k \log n)$ swaps. Each swap involves nk candidate sets of centers in the worst case. For each set, computing the cost of clusters requires $O(nk)$ time. Therefore, the running time of the algorithm is $O(k^3 n^2 \log nd)$.

4 Algorithm3

In this algorithm, we apply the sampling approach of [6] for k-median clustering. These samples resulting from this approach are processed by algorithm2 to yield a solution. The algorithm has a run time that is polynomial in k and d while being independent of n. A description of the algorithm follows.

1) Draw a random sample T of size $\frac{512k}{\alpha}\log(32k)$, where $\alpha = \frac{n_s k}{n}$, n_s is the size of the smallest cluster.

2) Do k-means clustering on the sample using algorithm2

3) Use the centers from step 2 as the centers for all the input points P.

Replacing n in the run time of algorithm2 with the sample size, we see that the run time of Algorithm3 is $O(k^5 \log^3 kd)$.

Theorem 3. *The output of Algorithm3 is an $O(1)$-approximation to the optimal k-means clustering with probability greater than $1/32$.*

We precede the proof of this theorem with the following lemma.

Lemma 7. *For any subset of k centers $K_T \subseteq T$, and any subset of k centers $K_P \subseteq P$,*

$$\Delta^T(K_T) \leq 4\Delta^T(K_P),$$

where $\Delta^T(K_T) = \sum_{x\in T} \|x - K_T(x)\|^2$, $\Delta^T(K_P) = \sum_{x\in T} \|x - K_P(x)\|^2$, $K_T(x)$ is the closest point in K_T to x, $K_P(x)$ is the closest point in K_P to x.

Proof. Let $q_T(K_P(x))$ denote the closest point in T to $K_P(x)$. For any $x \in T$,

$$\|x - K_T(x)\|^2 \leq \|x - q_T(K_P(x))\|^2 \leq 2(\|x - K_P(x)\|^2 + \|K_P(x) - q(K_P(x))\|^2).$$

By the definition of $q_T(K_P(x))$,

$$\|K_P(x) - q_T(K_P(x))\| \leq \|x - K_P(x)\|.$$

Thus we have,

$$\|x - K_T(x)\|^2 \leq 4\|x - K_P(x)\|^2.$$

Therefore,

$$\Delta^T(K_T) \leq 4\Delta^T(K_P).\square$$

Next, we prove the theorem. We adapt the proof from [6] given for the k-medians problem.

Proof. Let $K = K_1, ..., K_k$ denote the set of centers obtained by algorithm3, $K^* = K_1^*, ..., K_k^*$ denote the optimal centroid set, $K(K_i^*)$ denote the closest center in K to K_i^*, $N(K_i^*)$ denote the neighborhood of K_i^*, $n_i = |N(K_i^*)|$, $n_i^T = |N(K_i^*) \cap T|$. For any $x \in P$, let $K(x)$ denote the closest center in K to x, $K^*(x)$ denote the closest center in K^* to x. Let $Q_i = \sum_{x\in N(K_i^*)} \|x - K_i^*\|^2$, $R_i = \sum_{x\in N(K_i^*)} \|x - K(x)\|^2$, $Q_i^T = \sum_{x\in N(K_i^*)\cap T} \|x - K_i^*\|^2$, $R_i^T = \sum_{x\in N(K_i^*)\cap T} \|x - K(x)\|^2$. It follows that $\sum_{1\leq i\leq k} Q_i = \Delta^P(K^*)$, $\sum_{1\leq i\leq k} R_i = \Delta^P(K)$, $\sum_{1\leq i\leq k} Q_i^T = \Delta^T(K^*)$, and $\sum_{1\leq i\leq k} R_i^T = \Delta^T(K)$.

By the doubled triangle inequality,

$$\begin{aligned}
\|K_i^* - K(K_i^*)\|^2 &\leq min_{x\in N(K_i^*)\cap T} \|K_i^* - K(x)\|^2 \\
&\leq min_{x\in N(K_i^*)\cap T} 2(\|x - K_i^*\|^2 + \|x - K(x)\|^2 \\
&\leq \frac{2}{n_i^T} \sum_{x\in N(K_i^*)\cap T} (\|x - K_i^*\|^2 + \|x - K(x)\|^2 \\
&= \frac{2}{n_i^T}(Q_i^T + R_i^T).
\end{aligned}$$

For $x \in N(K_i^*)$,

$$\begin{aligned}\|x - K(x)\|^2 &\le \|x - K(K_i^*)\|^2 \\ &\le 2(\|x - K^*(x)\|^2 + \|K^*(x) - K(K_i^*)\|^2).\end{aligned}$$

Therefore,

$$\begin{aligned}\Delta^P(K) &\le 2\sum_{x\in P} \|x - K^*(x)\|^2 + 2\sum_{i=1}^{k} \sum_{x\in N(K_i^*)} \frac{2}{n_i^T}(Q_i^T + R_i^T) \\ &= 2\sum_{x\in P} \|x - K^*(x)\|^2 + 4\sum_{i=1}^{k} \frac{n_i}{n_i^T}(Q_i^T + R_i^T).\end{aligned}$$

By Chernoff Bounds,

$$Pr[n_i^T < \beta n_i(s/n)] < exp(-sn_i(1-\beta)^2/(2n)).$$

Assume $n_i \ge \frac{\alpha n}{k}$, then if $s = \frac{2k}{(1-\beta)^2\alpha}\log(32k)$, $Pr[n_i^T < \beta n_i(s/n)] < 1/32$. Therefore,

$$\begin{aligned}\Delta^P(K) &\le 2\Delta^P(K^*) + \frac{4n}{s\beta}\sum_{i=1}^{k}(Q_i^T + R_i^T) \\ &= 2\Delta^P(K^*) + \frac{4n}{s\beta}(\Delta^T(K) + \Delta^T(K^*)).\end{aligned}$$

By Lemma 4,

$$\Delta^T(K) \le c_2\Delta^T(K_T^*),$$

where $c_2 = 50$, and K_T^* is the optimal clustering for sample T.

By Lemma 7,

$$\Delta^T(K_T^*) \le 4\Delta^T(K^*).$$

Thus

$$\Delta^T(K) \le 4c_2\Delta^T(K^*),$$

and

$$\Delta^P(K) \le 2\Delta^P(K^*) + \frac{4n}{s\beta}(1 + 4c_2)\Delta^T(K^*).$$

It is known that

$$E(\Delta^T(K^*)) = \frac{s-1}{n}\Delta^P(K^*).$$

By Markov's inequality,

$$Pr[\Delta^T(K^*) \le \frac{16s}{15n}\Delta^P(K^*)] > \frac{1}{16}.$$

Let $\beta = \frac{15}{16}$ (the corresponding sample size is $\frac{512k}{\alpha}\log(32k)$). We have

$$Pr[\Delta^P(K) \le 2\Delta^P(K^*) + 4(1+4c_2)\Delta^P(K^*)] > 1 - \frac{1}{32} - \frac{15}{16} = \frac{1}{32}.$$

Therefore, with probability greater than 1/32,

$$\Delta^S(K^*) \le (6 + 16c_2)\Delta^P(K^*).$$

□

5 Conclusion

In this paper we have proposed three $O(1)$-approximation algorithms for the k-means clustering problem. Algorithm2 is the first algorithm for the k-means problem whose run time is polynomial in n, k, and d. There is a trade-off between the approximation ratio and the running time. Although the constant seems big, it is the bound in the worst case. In practice, especially when n is large, we could get better approximations. Also, the run time of Algorithm3 is independent of n. No prior algorithm for the k-means problem had this property.

References

1. Lloyd, S.P.: Least sqares quantization in PCM. IEEE Transactions on Information Theory, 28:129-137(1982).
2. Kanungo,T., Mount,D.M., Netanyahu, N.S., Piatko, C.D., Silverman, R., and Wu, A.Y.: A local search approximation algorithm for k-means clustering. In: Proceedings of the 18th Annual ACM Symposium on Computational Geometry, (2002)10-18.
3. Matousek, J.: On approximate geometric k-clustering. Discrete and Computational Geometry, 24:61-84, (2000).
4. Har-Peled, S., Mazumdar, S.: Coresets for k-means and k-median clustering and their applications, To appear in: Proceedings of the 36th Annual Symposium on Theory of Computing, (2004).
5. Kumar,A., Sabharwal, Y. and Sen, S.: A simple linear time $(1 + \epsilon)$-approximation algorithm for k-means clustering in any dimensions. To appear in: Proceedings of the 45th Annual IEEE Symposium on Foundations of Computer Science (FOCS '04), (2004).
6. Meyerson, A., O'callaghan, A., and Plotkin, S.: A k-median algorithm with running time independent of data size. Machine Learning, 56:61-87, (2004).
7. Milton J., and Arnold, J.: Introduction to Probability and Statistics, 3rd edition, McGraw Hill, (1994).
8. Jain, K., Mahdian, M. and Saberi, A.: A new greedy approach for facility location problems. In Proceedings of the 34th ACM Symposium on Theory of Computation, (2002)731-740.
9. Inaba, M., Katoh,N., and Imai, H.: Applications of weighted Voronoi diagrams and randomization to variance-based k-clustering. In: Proceedings of the Tenth Annual ACM Symposium on Computational Geometry, Stony Brook, NY, (1994)332-339.

On Efficient Weighted Rectangle Packing with Large Resources*

Aleksei V. Fishkin[1], Olga Gerber[2], and Klaus Jansen[2]

[1] University of Liverpool, Chadwick Building, Peach Street, Liverpool, L69 7ZF, UK
avf@csc.liv.ac.uk
[2] Institute of Computer Science, University of Kiel, 24118 Kiel, Germany
{oge, kj}@informatik.uni-kiel.de

Abstract. We address the problem of packing of a set of n weighted rectangles into a single rectangle so that the total weight of the packed rectangles is maximized. We consider the case of large resources, that is, the single rectangle is $\Omega(1/\varepsilon^3)$ times larger than any rectangle to be packed, for small $\varepsilon > 0$. We present an algorithm which finds a packing of a subset of rectangles with the total weight at least $(1-\varepsilon)$ times the optimum. The running time of the algorithm is polynomial in n and $1/\varepsilon$. As an application we present a $(2+\varepsilon)$-approximation algorithm for a special case of the advertisement placement problem.

Keywords: rectangle packing, approximation, resources.

1 Introduction

There has recently been increasing interest in solving 2D packing problems such as 2D strip packing [20, 23, 26], 2D bin packing [5, 6, 7, 24], and 2D rectangle packing [2, 3, 18]. These problems play an important role in a variety of applications in Computer Science and Operations Research, e.g. cutting stock, VLSI design, image processing, and multiprocessor scheduling, just to name a few.

In this paper we address the problem of packing a set of weighted rectangles into a single rectangle so that the total weight of the packed rectangles is maximized. More precisely, we are given a single rectangle R of width $a > 0$ and height $b > 0$, and a set L of n rectangles R_i $(i = 1, \dots, n)$ of widths $a_i \in (0, a]$ and heights $b_i \in (0, b]$. Each rectangle R_i has a positive weight $w_i > 0$. For any subset of rectangles $L' \subseteq L$, a *packing* of L' into R is a positioning of the rectangles from L' within the area $[0, a] \times [0, b]$ of R, so that all the rectangles of L' have disjoint interiors. Rectangles are not allowed to rotate. The goal is to find a subset of rectangles, $L' \subseteq L$, and a packing of L' into R, of maximum weight, $\sum_{R_i \in L'} w_i$.

This problem is known to be strongly NP-hard even for the case of packing squares with unit weights [22]. Hence it is very unlikely that any polynomial

* Supported by EU-Project CRESCCO IST-2001-33135, and by EU-Project AEOLUS 015964.

X. Deng and D. Du (Eds.): ISAAC 2005, LNCS 3827, pp. 1039–1050, 2005.

time algorithm for this problem exists. So, we look for efficient heuristics with good performance guarantees.

A polynomial time algorithm A is said to be a *ρ-approximation algorithm* for a maximization (minimization) problem Π if on every instance I of Π algorithm A outputs a feasible solution with a value $A(I) \geq \frac{1}{\rho} \cdot \mathrm{OPT}(I)$ (respectively $A(I) \leq \rho \cdot \mathrm{OPT}(I)$), where $\mathrm{OPT}(I)$ is the optimum. The value of $\rho \geq 1$ is called the *approximation ratio* or *performance ratio.* If ρ is achieved on instances I with $\mathrm{OPT}(I)$ tending to infinity, then A is said to be an *asymptotic ρ-approximation algorithm.* A *polynomial time approximation scheme* (PTAS) is a family of approximation algorithms $\{A_\varepsilon\}_{\varepsilon>0}$ such that A_ε is a $(1+\varepsilon)$-approximation algorithm and its running time is polynomial in the size of I. If the running time of each A_ε is polynomial in the size of I and $1/\varepsilon$, then $\{A_\varepsilon\}_{\varepsilon>0}$ is called a *fully polynomial time approximation scheme* (FPTAS).

The 1-dimensional version of our packing problem is the knapsack problem: given a knapsack capacity B and a set of items with profits and sizes, pack the items into a knapsack of size B so that the total profit of the packed items is maximized. The knapsack is weakly NP-hard [12], and it admits an FPTAS [19, 21]. In contrast, our 2D version is strongly NP-hard, and, hence, it admits no FPTAS unless $P = NP$.

One can find a relationship between our packing problem and the 2D bin packing: given a set L of rectangles of specified size (width, height), pack the rectangles into N square bins of unit area such that N is minimized. The problem is strongly NP-hard [22] and there is no better than a 2-approximation algorithm for it [10], unless $P = NP$. A long history of approximation results exists for this problem and its special cases [5, 6, 7, 24]. Recently, it has been shown that the general version with rectangles does not admit an asymptotic FPTAS unless $P = NP$, and there is one if all rectangles are squares [5]. In [9] a polynomial algorithm was presented which packs any set of rectangles into a sequence of square bins of side length $(1+\varepsilon)$, and the number of that bins is at most the minimum number of unit bins required to pack the rectangles.

Finally, one can also find a relationship between our problem and 2D strip packing [13]: given a set of rectangles pack the rectangles into a vertical strip $[0,1] \times [0,+\infty)$ so that the height of the packing is minimized. The problem is strongly NP-hard since it includes the classical bin packing problem as a special case. In fact many simple "shelves" heuristics come from that, e.g. Bottom-Left, First-Fit, First-Fit-Decreasing-Height [4, 8, 14, 25], with the best asymptotic performance ratio of 5/4 [2]. There are two algorithms having the same absolute performance ratio, 2 [23, 26], and there is an asymptotic FPTAS if all rectangles have side lengths at most 1 [20].

In contrast to all the above mentioned problems, there are just few results known for packing rectangles into a rectangle so as to maximize the total weight. For a long time the only known result has been an asymptotic (4/3)-approximation algorithm for packing squares with unit weights into a rectangle [3]. Only very recently this algorithm has been improved to a PTAS [17]. For packing rectangles, two approximability results have been presented in [18]:

a simple $(3+\varepsilon)$-approximation algorithm whose running time is polynomial in the number of rectangles n and $1/\varepsilon$ from one side, and a sophisticated $(2+\varepsilon)$-approximation algorithm whose running time is double exponential in $1/\varepsilon$.

In this paper we consider the so-called case of large resources, that is, the single rectangle R has a width $a > 0$ and height $b > 0$, whereas each rectangle R_i in L has a width $a_i \in (0, a]$ and height $b_i \in (0, \varepsilon^3 \cdot b]$, for some small enough $\varepsilon > 0$. We present an algorithm which finds a packing of a subset of L into R whose total weight is at least $(1-\varepsilon)$ times the optimum. The running time of the algorithm is polynomial in n and $1/\varepsilon$.

There are three main steps of our algorithm: LP approximation, Rounding, and Shifting. In the first step we relax our problem to *fractional packing*: any rectangle is allowed to be cut by horizontal lines into several fractional rectangles of the same width, and then some of that fractions are independently packed. In order to find a solution to the relaxation we formulate a linear program (LP) which consists of an exponential number of variables. Since we cannot solve it directly, we reformulate the LP as an instance of the *resource-sharing* problem and then make use of some recent approximation tools for it (see [15, 16] for details). By approximating a sequence of $O(n/\varepsilon^2)$ instances of the resource-sharing problem, that is performed in quite an elegant way, we find an approximate LP solution. In the second step of our algorithm, we round this LP solution to a "near-optimal" solution. Namely, by solving and rounding $O(1/\varepsilon^2)$ instances of the fractional knapsack problem we find a subset of rectangles which can be packed within the area $[0, a] \times [0, (1+O(\varepsilon))b + O(1/\varepsilon^2)]$ by using a strip packing algorithm from [20], and its total weight is $(1 - O(\varepsilon))$ times the optimum. In the third step of the algorithm, we apply a shifting technique. We cut the packing into small pieces of roughly equal heights $O(\varepsilon^2 b)$, and then remove some less weighted part. This is done so that the final packing fits within the area $[0, a] \times [0, b]$, and its total weight still remains within $(1 - O(\varepsilon))$ times the optimum. By an appropriate combination of these ideas and a careful analysis of the algorithm, we prove the following result.

Theorem 1. *Fro any $\varepsilon \in (0, 1/35]$ and any single rectangle R of width $a > 0$ and height $b > 0$, and a set L of rectangles R_i $(i = 1, \ldots, n)$ of widths $a_i \in (0, a]$ and heights $b_i \in (0, \varepsilon^3 \cdot b]$, there is an algorithm A_ε which outputs a packing of a subset of L within the area $[0, a] \times [0, b]$ of R whose total weight $A_\varepsilon(L) \geq (1 - 72\varepsilon)\mathrm{OPT}(L)$, where $\mathrm{OPT}(L)$ is the optimum. The running time of A_ε is polynomial in n and $1/\varepsilon$.*

Remark. The $O(1/\varepsilon^2)$ bound given in [20] comes within the shifting step of the algorithm, which only works if all $b_i = O(\varepsilon^3 b)$, and then makes some play in $(1 - O(\varepsilon))$ bound. In order to get an $(1-\varepsilon)$-approximation, we must scale all values in an appropriate way. This leads to an upper bound $1/\beta$ on the value of ε. So far, we can bound β by 2.6×10^3 if all $b_i \in (0, \varepsilon^3 \cdot b]$, and by 2.5×10^2 if all $b_i \in (0, \varepsilon^4 \cdot b]$. It is believed that all bounds can be further improved.

Interestingly, by considering a weakly restricted case we are able to achieve the best possible approximation result, in terms of trade-off between approximation ratio and running time. From theoretical point of view this makes a significant

step in understanding the approximation properties of the problem. In order to cope with the problem we design several new approximation techniques, some of those are nice combinations of various classical techniques used for knapsack problems, strip packing, and, surprisingly, for the resource-sharing problem.

More generally, it should be noted that – although phrased in terms of "packing" – our result really is about dynamic storage, i.e., given a set of tasks L and a resource pool R, we fix the resources R and attempt to maximize the amount of L serviced. Thus, we can reformulate our Theorem in terms of dynamic storage, e.g., if the resources of R are large, $\Omega(1/\varepsilon^3)$, then one can efficiently serve at least a fraction $(1-\varepsilon)$ of the maximum amount of tasks L.

Due to the difference in side lengths the number of the packed rectangles is large indeed. However, that case occurs widely in practice. As an example consider the advertisement placement problem for newspapers and the Internet [1, 11]: we are given k identical pages on which advertisements may be placed, where each page appears as a rectangle of size (a, b), and a set of n advertisements where each ith $(i = 1, \ldots, n)$ advertisement appears as a small rectangle of size (a_i, b_i) with an associated profit p_i; overlapping is not allowed; and, it is required to maximize the total profit of the advertisements placed on all k pages. We can find an approximate solution as follows. We first take all k pages together, as a rectangle of size $(a, k \cdot b)$, and run our packing algorithm. This outputs a packing within $[0, a] \times [0, k \cdot b]$ whose profit is at least $(1-\varepsilon)$OPT. Next, we draw $(k-1)$ vertical lines that cut this packing into k "pages" of equal size (a, b). Now one can obtain two feasible solutions for the original problem: one with those rectangles that lie inside the k "pages", and one with those rectangles that are cut by the $(k-1)$ lines (for each line the rectangles fit on a page of size (a, b)). We pick up the one with maximum profit, that is least $(1-\varepsilon)$OPT$/2$.

Theorem 2. *If the number of pages $k = \Omega(1/\varepsilon^3)$ for some small enough $\varepsilon > 0$, then there is a $(2+\varepsilon)$-approximation algorithm. The running time of the algorithm is polynomial in n and $1/\varepsilon$.*

Organization. The paper is organized as follows. Section 2 introduces notations and presents some preliminary results. Section 3 describes the steps of our algorithm. Due to space limitations, proofs and technical details are omitted.

2 Preliminaries

We will use the following notations. We write (p, q) to denote a rectangle whose width $p > 0$ and height $q > 0$. In the input, we are given a dedicated rectangle $R = (a, b)$, a set L of rectangles $R_i = (a_i, b_i)$ $(i = 1, \ldots, n)$ with positive weights $w_i > 0$, and an accuracy $\varepsilon \in (0, 1]$ such that all $a_i \in (0, a]$ and $b_i \in (0, \varepsilon^3 \cdot b]$. We write $w_{\max} = \max_{i=1}^n w_i$ to denote the maximum rectangle weight, and OPT to denote the optimum weight. For simplicity, we scale all widths by a and all heights by $\max_{R_i \in L} b_i$. Hence, throughout of the paper we assume w.l.o.g. that each R_i has side lengths $a_i, b_i \in (0, 1]$, whereas R has unit width $a = 1$ and height $b \geq 1/\varepsilon^3$. In addition, we also assume w.l.o.g. that $w_{max} \in [\varepsilon, 1]$,

OPT $\in [w_{\max}, n \cdot w_{\max}]$, and ε is selected such that $\varepsilon \in (0, 1/4]$ and $1/\varepsilon$ is integral.

2.1 The LP Formulation

Here we relax the problem to fractional packing: any rectangle R_i from L is allowed to be cut by horizontal lines into fractional several rectangles of the same width, and then some of them are independently packed within the area of R. We formulate the relaxation as an LP. We find an approximate LP solution in the first step of our algorithm.

First, we need some definitions. We start with *configurations.* Let L be a set of rectangles. A configuration is a sebset of rectangles $C \subseteq L$ whose total width is at most 1, i.e. they are able to occur at the same level. Without loss of generality, the configurations are assumed to be arbitrary ordered.

Let $\#C$ be the number of distinct configurations. (Notice that $\#C$ is $O(2^n)$.) Then, for each configuration C_j we define a variable $y_j \geq 0$, whose interpretation will be the height of C_j. For simplicity, we use y to denote the vector of all $y_j \geq 0$ $(j = 1, \ldots, \#C)$.

Let $(1, b)$ be a rectangle of height b. A *fractional* packing of L into $(1, b)$ is a packing within the area $[0, 1] \times [0, b]$ of any set of rectangles obtained from L by subdividing some of the rectangles by horizontal cuts: each rectangle (a_i, b_i) is replaced by a sequence $(a_i, x_{i_1} \cdot b_i), (a_i, x_{i_2} \cdot b_i), \ldots, (a_i, x_{i_k} \cdot b_i)$ of rectangles with the same width b_i such that the total sum $x_i = \sum_{j=1}^{k} x_{i_j}$ is a *fraction* in $[0, 1]$.

For simplicity, we write x to denote the vector of all x_i $(i = 1, \ldots, n)$, and $L(x)$ to denote the set of all $(a_i, x_i \cdot b_i)$ $(i = 1, \ldots, n)$. The weight of $L(x)$ is defined as $\sum_{i=1}^{n} x_i \cdot w_i$. We say that $L(x)$ is *integral* if all $x_i \in \{0, 1\}$ $(i = 1, \ldots, n)$, i.e. $L(x)$ is a subset of L that consists of all rectangles R_i with $x_i = 1$, and *fractional* otherwise.

Now we can define the values of y_j $(j = 1, \ldots, \#C)$ in the vector y as follows. We scan the area $[0, 1] \times [0, b]$ bottom-up with a horizontal sweep line $y = h$, $0 \leq h \leq b$. (Here y means the ordinate axis, or Y-line.) Every such line canonically associates to a configuration, that consists of all the rectangles of L whose fractions' interior is intersected by the sweep line. The value of y_j, $1 \leq j \leq \#C$, is equal to the measure of the h's such that the sweep line $y = h$ is associated to configuration C_j. Thus, the sum of y_j over all configurations C_j is at most b.

So, we are ready to combine the two ideas togeter. We look for a fractional packing of L within the area $[0, 1] \times [0, b]$ whose total weight is maximum. This relaxation can be formulated as the following linear program $LP(L, b)$:

$$
\begin{aligned}
\text{maximize } & \textstyle\sum_{i=1}^{n} x_i \cdot w_i \\
\text{subject to } & \textstyle\sum_{j: R_i \in C_j} y_j \geq x_i \cdot b_i, \text{ for all } i = 1, \ldots, n, \\
& \textstyle\sum_{j=1}^{\#C} y_j \leq b, \\
& y_j \geq 0, \quad \text{for all } j = 1, \ldots, \#C, \\
& x_i \in [0, 1], \quad \text{for all } i = 1, \ldots, n.
\end{aligned}
\tag{1}
$$

Each x_i defines a fraction of rectangle R_i. Each y_j defines the height value of configuration C_j. The objective value defines the total fractional weight. In the first line, the sum of y_j over all configurations C_j that include rectangle R_i is at least x_i times its height b_i. In the second line, the sum of y_j over all configurations C_j is bounded by b. In the last two lines, all y_j are non-negative and all x_i are fractions in $[0, 1]$. We can conclude the following result.

Lemma 1. *Let* $\overline{\mathrm{OPT}}$ *be the optimum of* $LP(L, b)$*. Then,* $\overline{\mathrm{OPT}}$ *is an upper bound on* OPT.

2.2 Separating Rectangles

Let $\varepsilon' = \varepsilon/(2 + \varepsilon)$. Let (p, q) be a rectangle of width $p > 0$ and height $q > 0$. We say that (p, q) *narrow* if its width p is at most ε', and *wide* otherwise. For any given set of rectangles Q, we will write Q_{wide} to denote the set of all wide rectangles in Q, and Q_{narrow} to denote the set of all narrow rectangles in Q, respectively. So, Q is partitioned into Q_{narrow} and Q_{wide}. This will be used later in the LP rounding step of our algorithm.

2.3 The KR-Algorithm

By adopting the result in [20] we can move from a fractional packing to a "non-fractional" packing.

Theorem 3. *Let* $h \geq b$ *and* $L(x)$ *be a (possibly fractional) solution for* $LP(L, h)$*. Then, there is an algorithm which, given an accuracy* $\varepsilon \in (0, 1]$*, finds a positioning of all rectangles* $(a_i, x_i \cdot b_i)$ $(i = 1, \ldots, n)$ *from* $L(x)$ *within the vertical strip* $[0, 1] \times [0, \infty)$ *of unit width such that all the rectangles of* $L(x)$ *have disjoint interiors and the height to which the strip is filled is bounded by* $h' \leq h(1 + 1/(m\varepsilon'))/(1 - \varepsilon') + 4m + 1$*, where* $m = \lceil (1/\varepsilon')^2 \rceil$ *and* $\varepsilon' = \varepsilon/(2 + \varepsilon)$*. The running time of the algorithm is polynomial in* n *and* $1/\varepsilon$*.*

Remark. For simplicity, such an algorithm is called the KR-algorithm, and its running time is denoted by $KR(n, \varepsilon)$.

2.4 An Outline of the Algorithm

In overall we can describe an outline of our algorithm as follows:
ALGORITHM A_ε:
Input: A set L of rectangles, rectangle $R = (1, b)$, and accuracy $\varepsilon > 0$.
Output: A packing of a subset of L in the area $[0, 1] \times [0, b]$ of R.
[**LP approximation**] Define $\bar{\varepsilon} = \varepsilon^2/n$. Perform a modified linear search over all $\bar{\varepsilon}$-approximate solutions for a sequence of n/ε^2 instances of the resource-sharing problem. This defines an LP solution $L(x)$ for $LP(L, (1 + 2\varepsilon)b)$ whose weight is at least $(1 - 3\varepsilon)$OPT.

[Rounding] Define $\varepsilon' = \varepsilon/(2+\varepsilon)$ and $m = \lceil 1/(\varepsilon')^2 \rceil$. Perform the partition $L_{wide} = L \setminus L_{narrow}$ to set aside the rectangles of width less than ε'. Round $L(x)$ to a subset of rectangles $L' \subseteq L$ which is given as the union of m non-intersecting groups $\cup_{k=1}^{m} L'_{wide,k}$ of wide rectangles and a subset L'_{narrow} of narrow rectangles. The weight of L' is at least $(1-3\varepsilon)$OPT. Given accuracy ε, using the KR-algorithm on the set L' of rectangles outputs a strip-packing of L' within the area $[0,1] \times [0,(1+O(\varepsilon))b + O(1/\varepsilon^2)]$.
[Shifting] Cut the strip-packing of L' into small pieces of roughly equal heights $O(\varepsilon^2 b)$, and then remove some less weighted part. This gives a packing within the area $[0,1] \times [0,b]$ of R. The weight of the rectangles in the packing is at least $(1-O(\varepsilon))$OPT.

Remark. In order to obtain a $(1-\varepsilon)$-approximation, we need to define bounds on b and ε, use A_ε together with Lemmas 4, 5, 6, and then scale ε in an appropriate way. If $b \geq 1/\varepsilon^4$ and $\varepsilon \in (0, 1/10]$, the algorithm outputs a packing whose weight is at least $(1-22\varepsilon)$OPT. If $b \geq 1/\varepsilon^3$ and $\varepsilon \in (0, 1/35]$, the algorithm outputs a packing whose weight is at least $(1-72\varepsilon)$OPT. Hence, we can obtain a required algorithm either for $b \geq 1/\varepsilon^4$ and $\varepsilon \in (0, 1/220]$, or for $b \geq 1/\varepsilon^3$ and $\varepsilon \in (0, 1/2520]$. In the first case $b \approx 2.4 \times 10^9$. In the second case $b \approx 1.6 \times 10^{10}$. Notice also that some steps of our algorithm can be performed in a more efficient way. For example, one can improve the search procedure in Step 1. Here we mainly concentrate our attention on the polynomial time efficiency of the algorithm.

3 The Packing Algorithm

As described in Section 2.4, our algorithm consists of the three main steps: LP approximation, Rounding, and Shifting. The first step is described in Section 3.1, and the next two steps are described in Sections 3.2, 3.3 respectively.

3.1 LP Approximation

Here we work with our LP formulation $LP(L,b)$. Since the number of configurations $\#C = O(2^n)$, i.e. the LP is large, we look for an LP approximation. We transform the LP to the resource-sharing problem. By performing a modified linear search over approximate solutions for the latter problem, we find a solution $L(x)$ for $LP(L,(1+2\varepsilon)b)$, whose weight is at least $(1-3\varepsilon)$OPT. In order to resolve the resource-sharing problem in a required time we use the results from [15, 16]. We formulate the block-problem and present a block solver for it, due to space limitations we omit proofs.

Resource-sharing problem. We can assume w.l.o.g. that the LP optimum $\overline{\text{OPT}}$ is lower bounded by the maximum weight $w_{\max}$ and upper bounded by $n \cdot w_{\max}$, i.e. $\overline{\text{OPT}} \in [w_{\max}, nw_{\max}]$. Then, for each value $w \in [w_{\max}, nw_{\max}]$ we introduce the following resource-sharing problem:

$$
\begin{array}{lll}
\text{maximize} & \lambda & \\
\text{subject to} & \sum_{i=1}^{n} x_i \cdot (w_i/w) \geq \lambda, & \\
& \sum_{j:R_i \in C_j} \lfloor y_j/b_i \rfloor - x_i + 1 \geq \lambda, & \text{for all } i = 1, \ldots, n, \\
& \sum_{j=1}^{\#C} y_j/b \leq 1, & \\
& y_j \geq 0, & \text{for all } j = 1, \ldots, \#C, \\
& x_i \in [0,1], & \text{for all } i = 1, \ldots, n.
\end{array} \tag{2}
$$

Lemma 2. *Let λ^* be the optimum. If $\lambda^* < 1$, then the value of w is larger than* $\overline{\text{OPT}}$.

Linear search. Assume that we can solve any instance of the above resource-sharing problem to the optimum. Then, we can perform a search at each value $w \in \{(1 + \varepsilon^2 \cdot \ell)w_{max} | \ell = 0, 1, \ldots, (n-1)/\varepsilon^2\}$, and simply take a solution (x, y) given by the maximum value of w whose optimum $\lambda^* \geq 1$. First, this solution (x, y) is feasible for $LP(L, b)$. Second, we know that $\overline{\text{OPT}} \geq w_{\max}$, and, due to the search procedure, $w + \varepsilon^2 w_{max} > \overline{\text{OPT}}$. Hence, the objective value at (x, y) is at least $w \geq \overline{\text{OPT}} - \varepsilon^2 w_{max} \geq (1 - \varepsilon^2)\overline{\text{OPT}}$. Thus, this solution (x, y) is "near" optimal for $LP(L, b)$. Unfortunately, we cannot resolve the above resource-sharing problem, and so we need to look for approximate solutions.

Approximating the general resource-sharing problem. Let M and N be two positive integers. Let B be a non-empty compact convex set in R^N. Let $f_m : B \to \mathrm{R}_+$ $(m = 0, \ldots, M)$ be non-negative linear functions over B. Let $\lambda(z) = \max_{m=1}^{M} f_m(z)$. Then, the *general resource-sharing problem* can be formulated as the following linear program: $\max\{\lambda(z) | z \in B\}$.

A *price vector* is a vector p of non-negative values $p_m \geq 0$ $(m = 0, \ldots, M)$ such that $\sum_{m=0}^{M} p_m = 1$. Let $\Lambda(p, z) = \sum_{m=0}^{M} p_m f_m(z)$. Then, for any fixed p, the *block problem* is defined as the following linear program: $\max\{\Lambda(p, z) | z \in B\}$.

Let λ^* be the optimum. For an accuracy $\bar{\varepsilon} \in (0, 1]$, a solution $z \in B$ is called an $\bar{\varepsilon}$-approximate solution if $\lambda(z) \geq (1 - \bar{\varepsilon})\lambda^*$. Let $\Lambda^*(p)$ be the optimum. For an accuracy $\bar{t} \in (0, 1]$, a solution $z(p) \in B$ is called a $(p, \bar{t})$-approximate solution if $\Lambda(p, z(p)) \geq (1 - \bar{t})\Lambda^*(p)$.

If N is polynomial in M, then we can solve the two LPs in time polynomial in M. If $N = O(2^M)$, i.e. N can be exponential in M, then the LPs are large. In order to cope with that case, we will use the following result.

Theorem 4 (Grigoriadis et al. [15], Jansen [16]). *For any given $\bar{\varepsilon} > 0$, there is a resource sharing algorithm $RSA(\bar{\varepsilon})$ which finds an $\bar{\varepsilon}$-approximate solution for the resource-sharing problem, provided that given any $\bar{t} = \Theta(\bar{\varepsilon})$, any price vector p there is a block solver algorithm $BSA(p, \bar{t})$ which finds a $(p, \bar{t})$-approximate solution for the block problem. The algorithm $RSA(\varepsilon)$ runs as a sequence of $O(M(\ln M + \bar{\varepsilon}^{-2} \ln \bar{\varepsilon}^{-1}))$ iterative steps, each of those requires a call to $BSA(p, \bar{t})$ and incurs an overhead of $O(M \ln\ln(M\bar{\varepsilon}^{-1}))$ elementary operations.*

Remark. The algorithm proposed in [16] uses price vectors p whose positive coordinates $p_m = \Omega([\bar{\varepsilon}/M]^q)$ $(m = 0, 1, \ldots, M)$, for some constant q. We use this important fact in the analysis of our algorithm.

Solving the block problem. In order to approximate (2) we turn to the block problem. Let $w \in [w_{\max}, nw_{\max}]$. Let x and y denote the vectors of all x_i's and y_j's. Let $B(x)$ be the set of all x such that $x_i \in [0,1]$ for all $i = 1,\dots,n$. Let $B(y)$ be the set of all y such that $\sum_{j=1}^{\#C} y_j/b \leq 1$ and $y_j \geq 0$, for all $j = 1,\dots,\#C$. Then, $B(x)$ and $B(y)$ are non-empty, compact and convex.

For any given price vector p of non-negative values $p_i \geq 0$ $(i = 0,\dots,n)$ such that $\sum_{i=0}^{n} p_i = 1$ we define the objective function of the block problem as $\Lambda(p,x,y) = p_0[\sum_{i=1}^{n} x_i \cdot (w_i/w)] + \sum_{i=1}^{n} p_i[\sum_{j:R_i \in C_j} y_j/b_i - x_i + 1]$. For simplicity, we combine the coefficients for each of the variables. For x_i and y_j we get $c_i = p_0(w_i/w) - p_i \sum_{j:R_i \in C_j} 1$ and $d_j = \sum_{R_i \in C_j} p_i/b_i$, respectively. Let $\Lambda(p,x) = \sum_{i=1}^{n} c_i \cdot x_i$ and $\Lambda(p,y) = \sum_{j=1}^{\#C} d_j \cdot y_j$. Then, x and y are independent.

So, we reformulate the block problem as two LPs: $\max\{\Lambda(p,x) | x \in B(x)\}$ and $\max\{\Lambda(p,y) | y \in B(y)\}$. Now it is quite an easy task to define optimal solutions. Let x^* and y^* be defined such that (i) $x_i^* = 0$ if c_i is non-positive, and $x_i^* = 1$ otherwise $(i = 1,\dots,n)$, (2) $y_k^* = b$, and $y_j^* = 0$ for all $C_j \neq C_k$ $(j = 1,\dots,\#C)$, where C_k is a configuration with $d_k = \max_{j=1}^{\#C} d_j$. Then, x^* and y^* define an optimal solution for the block problem.

Recall now that the number of configurations $\#C$ can be exponential. Hence, we cannot find an optimal solution as defined above in a straightforward way. Our idea is to look for an approximation. In overall, we prove the following result.

Lemma 3. *Let q be some positive constant. Then, for any accuracy $\bar{\varepsilon} > 0$ and any price vector p whose positive coordinates $p_i = \Omega([\bar{\varepsilon}/M]^q)$ $(i = 0,\dots,n)$, there is a block solver algorithm $BSA(p,\bar{\varepsilon})$ which finds a $(p,\bar{\varepsilon})$-approximate solution for the block problem in $O(n^2 \cdot [M/\bar{\varepsilon}]^q) + KS(n,\bar{\varepsilon})$ time, where $KS(n,\bar{\varepsilon})$ is time required to find an $(1-\bar{\varepsilon})$-approximate solution for an instance of the knapsack problem with n items (for example $O(n/\bar{\varepsilon}^3)$).*

Using $\bar{\varepsilon}$-approximate solutions. Now we can use $\bar{\varepsilon}$-approximate solutions of our resource-sharing problem. Let $w \in [w_{\max}, n \cdot w_{\max}]$. Let λ^* be the optimum of the resource-sharing problem for given w. Let (x,y) be an $\bar{\varepsilon}$-approximate solution that satisfies the constrains of (2) for some $\lambda \geq \lambda^*(1-\bar{\varepsilon})$. Assume that $\lambda < (1-\bar{\varepsilon})$. Then $\lambda^* < 1$, and by Lemma 2, we can conclude that $\overline{\mathrm{OPT}}$ is smaller than w. Now assume that $\lambda \geq (1-\bar{\varepsilon})$ and $\bar{\varepsilon} = \varepsilon^2/n$. For such values of λ and $\bar{\varepsilon}$, we can see the following three facts. First, consider all $x_i < \varepsilon/n$. Then, $\sum_{R_i \in L\,:\,x_i \leq \varepsilon/n} x_i \cdot w_i < (\varepsilon/n) \cdot [\sum_{i=1}^{n} w_i] \leq \varepsilon \cdot w_{\max}$. Second, for each $x_i \geq \varepsilon/n$, we have that $x_i - \bar{\varepsilon} = x_i - \varepsilon^2/n \geq x_i - \varepsilon x_i = (1-\varepsilon)x_i$. Third, for $\varepsilon \in (0,1/4]$ we have that $(1-\varepsilon)(1+2\varepsilon) = 1 + \varepsilon - 2\varepsilon^2 > 1$. Hence, $\sum_{j=1}^{\#C} y_j/(1-\varepsilon) \leq b/(1-\varepsilon) \leq b(1+2\varepsilon)$.

Using this $\bar{\varepsilon}$-approximate solution (x,y) we can create a new solution as follows. For each $x_i < \varepsilon/n$, we set the value of x_i to 0. Then, from $w \in [w_{\max}, n \cdot w_{\max}]$ the objective function value can be bounded as $\sum_{i=1}^{n} x_i \cdot w_i \geq \lambda \cdot w - \varepsilon \cdot w_{\max} \geq (1-\varepsilon^2/n)w - \varepsilon \cdot w_{\max} \geq (1-\varepsilon^2/n-\varepsilon)w \geq (1-2\varepsilon)w$. Next, notice the following. If $x_i = 0$, then using $y_j \geq 0$ we obviously get $\sum_{j:R_i \in C_j} y_j/b_i \geq 0 = (1-\varepsilon)x_i$. If $x_i \geq \varepsilon/n$, then using $\lambda \geq (1-\bar{\varepsilon})$ we can

bound $\sum_{j:R_i \in C_j} y_j/b_i \geq \lambda + x_i - 1 \geq x_i - \bar{\varepsilon} = x_i - \varepsilon^2/n \geq x_i - \varepsilon x_i = (1-\varepsilon)x_i$. By scaling the values of all y_j ($j = 1, \ldots, \#C$) by $1/(1-\varepsilon)$, the new values of all x_i and y_j satisfy $\sum_{i=1}^{n} x_i \cdot w_i \geq (1-2\varepsilon)w$, $\sum_{j:R_i \in C_j} y_j \geq b_i \cdot x_i$ for all $i = 1, \ldots, n$, and $\sum_{j=1}^{\#C} y_j \leq b(1+2\varepsilon)$, where all $y_j \geq 0$ and all $x_i \in [0,1]$.

Modified linear search. Now we can modify our linear search. We define $\bar{\varepsilon} = \varepsilon^2/n$. Then, we perform a search at each value $w \in \{(1+\varepsilon^2 \cdot \ell)w_{max} | \ell = 0, 1, \ldots, (n-1)/\varepsilon^2\}$. Each time we find an $\bar{\varepsilon}$-approximate solution, and then modify it as shown above. We take the modified $\bar{\varepsilon}$-approximate solution (x, y) given by the maximum value of w, that is feasible for $LP(L, (1+2\varepsilon)b)$. Furthermore, the objective function value at (x, y) is at least $(1-3\varepsilon)\overline{\text{OPT}}$. So, we conclude with the following result.

Lemma 4. *Let $\varepsilon \in (0, 1/4]$ and $\bar{\varepsilon} = \varepsilon^2/n$. Then, by performing a linear search over $\bar{\varepsilon}$-approximate solutions for a sequence of n/ε^2 instances of the resource-sharing problem, one can determine a feasible (possibly fractional) solution $L(x)$ for $LP(L, (1+2\varepsilon)b)$ whose objective function is at least $(1-3\varepsilon)\overline{\text{OPT}}$.*

3.2 Rounding

Here we round our LP solution $L(x)$. We separate the rectangles of L into narrow, L_{narrow}, and wide, L_{wide}. First, we round $L_{narrow}(x)$. We find a subset, $L'_{narrow} \subseteq L_{narrow}$, of narrow rectangles whose weight is at least the weight of $L_{narrow}(x)$, and total area is at most the area of $L_{narrow}(x)$ plus 1. This can be done by selecting the rectangles $(a_i, x_i \cdot b_i)$ appearing in $L_{narrow}(x)$ in order of non-increasing ratio $w_i/(a_i \cdot b_i)$.

Next, we round $L_{wide}(x)$. We define $H(x) = \sum_{R_i \in L_{wide}} x_i \cdot b_i$, that is the total height of $L_{wide}(x)$. Here, in the beginning we select the fractional rectangles $(a_i, x_i \cdot b_i)$ form $L_{wide}(x)$ in order of non-increasing widths a_i one by one, and working in a greedy way create $m = \lceil 1/(\varepsilon')^2 \rceil$ non-intersecting groups $L_{wide,1}(x), L_{wide,2}(x), \ldots, L_{wide,m}(x)$ of roughly equal heights $(\varepsilon'^2) \cdot H(x)$. (Ties are broken arbitrarily.) Then, for each kth group $L_{wide,k}(x)$ we find a subset, $L'_{wide,k} \subseteq L_{narrow}$, of narrow rectangles whose weight is at least the weight of $L_{wide,k}(x)$, and total height is at most the height of $L_{wide,k}(x)$ plus 1. This can be done by selecting the rectangles $(a_i, x_i \cdot b_i)$ appearing in $L_{wide,k}(x)$ in order of non-increasing ratio w_i/b_i.

The main idea behind the above rounding procedure is that to keep preserved a "nice area" for narrow rectangles, and "nice heights" for all $m = O(1/\varepsilon^2)$ groups of wide rectangles. In overall, we can show that the subset of rectangles $L' \subseteq L$ given as the union of $\cup_{k=1}^{m} L'_{wide,k}$ and L'_{narrow}, can be nicely packed within the area $[0,1] \times [0, (1+O(\varepsilon))b + O(1/\varepsilon^2)]$.

Lemma 5. *Let $\alpha \geq 1$ and $\beta \geq 4$. Let $b \geq (\alpha/\varepsilon^3)$ and $\varepsilon \in (0, 1/\beta]$. Then, a fractional solution, $L(x)$, can be rounded to a subset $L' \subseteq L$ of rectangles so that the weight of L' is at least $(1 - \delta_1 \cdot \varepsilon)$OPT, where $\delta_1 = 3$, and the KR-algorithm outputs a packing of L' within the area $[0,1] \times [(1+\delta_2 \cdot \varepsilon)b]$, where $\delta_2 = 4 + (33/\beta) + (82/\beta^2)$ if $\alpha = 1/\varepsilon$, and $\delta_2 = 32 + (45/\beta) + (42/\beta^2)$ if*

$\alpha = 1$. *The complete rounding and packing procedure requires at most* $O([1/\varepsilon^2] \cdot (n \log n) + KR(n, \varepsilon))$ *running time, where* $KR(n, \varepsilon)$ *is the running time of the KR-algorithm.*

Remark: One can also obtain slightly different bounds by taking $\alpha \in \{10, 20\}$.

3.3 Shifting

Assume that we are given a strip packing of L' within the area $[0, 1] \times [0, (1+\delta_2 \cdot \varepsilon)b]$, and the weight of L' is at least $(1 - \delta_1 \cdot \varepsilon)\mathrm{OPT}$, for some $\delta_1, \delta_2 = O(1)$. The idea of our shifting technique is to cut this strip packing into $k = \left\lfloor \frac{(1+\delta_2 \cdot \varepsilon)b+2}{(\delta_2 \cdot \varepsilon)b+2} \right\rfloor$ pieces of roughly equal height $(\delta_2 \cdot \varepsilon)b$. Putting some bound together we can show that there must exit one piece whose weight is at most $[2\mathrm{OPT}](1/k) \leq 2\varepsilon(\delta_2 + 1)\mathrm{OPT}$. So, throwing away this piece can decrease the weight of L' to $(1 - (\delta_1 + 2\delta_2 + 2)\varepsilon)\mathrm{OPT}$, but it reduces the height of the strip packing to b. This gives an approximate packing within the required area $[0, 1] \times [0, b]$. Omitting all technical details, we use Lemma 5 and obtain the following result.

Lemma 6. *Let* $\alpha \geq 1$ *and* $\beta \geq 4$. *Let* $b \geq \alpha/\varepsilon^3$ *and* $\varepsilon \in (0, 1/\beta]$. *Then, given a packing of* $L(\bar{x})$ *in the area* $[0, 1] \times [0, (1 + \delta_2 \cdot \varepsilon)b]$ *whose weight is at least* $(1-\delta_1 \cdot \varepsilon)\mathrm{OPT}$, *in time* $O(n+1/\varepsilon)$ *one can obtain a packing in the area* $[0, 1] \times [0, b]$ *whose weight is at least* $(1 - 22\varepsilon)\mathrm{OPT}$ *if* $\alpha = 1/\varepsilon$ *and* $\beta = 10$, *and at least* $(1 - 72\varepsilon)\mathrm{OPT}$ *if* $\alpha = 1$ *and* $\beta = 35$.

References

1. M. Adler, P. Gibbons, and Y. Matias. Scheduling space-sharing for internet advertising. Journal of Scheduling, 1998.
2. B.S. Baker, D.J. Brown, and H.P. Katseff. A 5/4 algorithm for two dimensional packing. *Journal of Algorithms*, 2:348–368, 1981.
3. B.S. Baker, A.R. Calderbank, E.G. Coffman, and J.C. Lagarias. Approximation algorithms for maximizing the number of squares packed into a rectangle. *SIAM Journal on Algebraic and Discrete Methods*, 4:383–397, 1983.
4. B.S. Baker, E.G. Coffman, and R.L. Rivest. Orthogonal packings in two dimensions. *SIAM Journal on Computing*, 9:846–855, 1980.
5. N. Bansal and M. Sviridenko. New approximability and inapproximability results for 2-dimensional bin packing. In *Proceedings 15th Annual ACM-SIAM Symposium on Discrete Algorithms (SODA)*, pages 189–196, 2004.
6. A. Caprara. Packing 2-dimensional bins in harmony. In *Proceedings 43rd Annual Symposium on Foundations of Computer Science (FOCS)*, pages 490–499, 2002.
7. F.R.K. Chung, M.R. Garey, and D.S. Johnson. On packing two-dimensional bins. *SIAM Journal on Algebraic and Discrete Methods*, 3:66–76, 1982.
8. E.G.Jr. Coffman, M.R. Garey, D.S. Johnson, and R.E. Tarjan. Performance bounds for level-oriented two-dimensional packing algorithms. *SIAM Journal on Computing*, 9:808–826, 1980.
9. J.R. Correa and C. Kenyon. Approximation schemes for multidimensional packing. In *Proceedings 15th Annual ACM-SIAM Symposium on Discrete Algorithms (SODA)*, pages 179–188, 2004.

10. C.E. Ferreira, F.K. Miyazawa, and Y. Wakabayashi. Packing squares into squares. *Perquisa Operacional*, 19:349–355, 1999.
11. A. Freund and J. Naor. Approximating the advertisement placement problem. In *Proceedings 9th Conference on Integer Programming and Combinatorial Optimization (IPCO)*, LNCS 2337, pages 415–424, 2002.
12. M. R. Garey and D. S. Johnson. *Computers and intractability: A guide to the theory of NP-completeness.* Freeman, San Francisco, CA, 1979.
13. P.C. Gilmore and R.E. Gomory. Multistage cutting stock problems of two and more dimensions. *Operations Research*, 13:94–120, 1965.
14. I. Golan. Performance bounds for orthogonal, oriented two-dimensional packing algorithms. *SIAM Journal on Computing*, 10:571–582, 1981.
15. M.D. Grigoriadis, L.G. Khachiyan, L. Porkolab, and J. Villavicencio. Approximate max-min resource sharing for structured concave optimization. *SIAM Journal on Optimization*, 41:1081–1091, 2001.
16. K. Jansen. Approximation algorithms for the general max-min resource sharing problem: faster and simpler. In *Proceedings 9th Scandinavian Workshop on Algorithm Theory (SWAT)*, 2004. To appear.
17. K. Jansen and G. Zhang. Maximizing the number of packed rectangles. In *Proceedings 9th Scandinavian Workshop on Algorithm Theory (SWAT)*, 2004.
18. K. Jansen and G. Zhang. On rectangle packing: maximizing benefits. In *Proceedings 15th Annual ACM-SIAM Symposium on Discrete Algorithms (SODA)*, pages 197–206, 2004.
19. Hans Kellerer, U. Pferschy, and D. Pisinger. *Knapsack problems.* Springer, 2004.
20. C. Kenyon and E. Rémila. Approximate strip-packing. In *Proceedings 37th Annual Symposium on Foundations of Computer Science (FOCS)*, pages 31–36, 1996.
21. E. Lawler. Fast approximation algorithms for knapsack problems. *Mathematics of Operations Research*, 4:339–356, 1979.
22. J.Y.-T. Leung, T.W. Tam, C.S. Wong, G.H. Young, and F.Y.L. Chin. Packing squares into a square. *Journal of Parallel and Distributed Computing*, 10:271–275, 1990.
23. I. Schiermeyer. Reverse fit: a 2-optimal algorithm for packing rectangles. In *Proceedings 2nd European Symposium on Algorithms (ESA)*, pages 290–299, 1994.
24. S. Seiden and R. van Stee. New bounds for multi-dimentional packing. *Algorithmica*, 36(3):261–293, 2003.
25. D.D. Sleator. A 2.5 times optimal algorithm for bin packing in two dimensions. *Informatin Processing Letters*, (10):37–40, 1980.
26. A. Steinberg. A strip-packing algorithm with absolute performance bound 2. *SIAM Journal on Computing*, 26(2):401–409, 1997.

On Routing in VLSI Design and Communication Networks*

Tamás Terlaky[1], Anthony Vannelli[2], and Hu Zhang[1]

[1] AdvOL, Department of Computing and Software,
McMaster University, Hamilton, Canada
{terlaky, zhanghu}@mcmaster.ca
[2] Department of Electrical and Computer Engineering,
University of Waterloo, Canada
a.vannelli@ece.uwaterloo.ca

Abstract. In this paper, we study the global routing problem in VLSI design and the multicast routing problem in communication networks. We first propose new and realistic models for both problems. Both problems are $\mathcal{NP}$-hard. We present the integer programming formulation of both problems and solve the linear programming (LP) relaxations approximately by the fast approximation algorithms for min-max resource-sharing problems in [10]. For the global routing problem, we investigate particular properties of lattice graphs and propose a combinatorial technique to overcome the hardness due to the bend-dependent vertex cost. Finally we develop asymptotic approximation algorithms for both problems depending on the best known approximation ratio for the minimum Steiner tree problem. They are the first known theoretical approximation bound results for these problems.

1 Introduction

Hardness of many combinatorial problems is represented by two characteristics in their mathematical programming formulation: First, the integrality of variables leads to $\mathcal{NP}$-hardness for many problems. Second, in general the limited resources represented by the constraints increase the complexity of the problems. In fact the occurrence of constraints changes the feasible set of the problem, i.e., the polytope of the mathematical program. Thus the characteristics of the problem differs much from the original one. Furthermore, in many problems arising in graph theory, when trees are required to be generated instead of paths, $\mathcal{NP}$-hardness appears again.

In this paper, we shall study routing problems in two different engineering applications: in VLSI design and in multicast communication networks. In both problems, there are only limited resources shared by multiple entities. Furthermore, the solutions of both problems consist of trees generated for these entities. Specifically, we shall study the global routing problem in VLSI design and

* Research supported by a MITACS grant for all the authors, the NSERC Discovery Grant #5-48923 for the first and third author, the NSERC Grant #15296 for the second author, and the Canada Research Chair Program for the first author.

X. Deng and D. Du (Eds.): ISAAC 2005, LNCS 3827, pp. 1051–1060, 2005.

the multicast routing problem in communication networks. There exist capacity constraints on all edges. The purpose of both problems is to minimize the total cost (e.g., the overall edge length of the trees or with some additional cost). As described above, both problems are hard. We propose the first asymptotic approximation algorithms for both problems. In addition, as a byproduct, we develop a combinatorial technique for lattice graphs with bend-dependent vertex cost, which can be applied to many optimization problems in VLSI design and transportation networks.

Global routing in VLSI design: In the global routing stage of VLSI design, the circuits are assumed to be in a one-layer frame. The channels on a chip form the edge set of a lattice graph, and their crosses consist of the vertex set. A group of pins to be connected is called a *net.* Wires are to be routed along the channels to connect the nets with respect to the channel capacity, and certain objective functions (e.g., overall wirelength or maximum edge congestion) are optimized.

The global routing problem is $\mathcal{NP}$-hard [14]. Therefore, heuristics tend to be used to find reasonable solutions in fast time. In sequential global routing, nets are ordered to their routing importance, then each net is routed separately. This method is called the *maze runner* method and was initially introduced by Lee [13]. Enhancements to the maze runner heuristics were shown in [9]. The quality of a sequential global router depends on the ordering of the nets.

Integer programming methodologies allow global routing problems to be solved in a "global" sense by routing all the nets simultaneously. This class of techniques is used to formulate the global routing problem as a 0-1 integer programming problem where the objective is to select one Steiner tree for each net such that the total wirelength (tree length) is minimized and channel capacity (edge capacity) constraints are not violated. Three traditional models of the global routing by integer programming are listed in [3]. In the first model [20], the goal is to minimize the overall wirelength with respect to the capacity constraints. The goal of the second model is to minimize the maximum tree length [15]. The third model is to minimize the maximum edge congestion [17]. Based on the above traditional models, two new models are proposed in [3,4]. In the first model, the goal is to minimize a linear combination of overall wirelength, overall number of *vias* (bends of the routed paths or trees) and the maximum edge congestion. The weights of these three terms depends on a pre-specified set of trees. A big improvement is that the impact of vias is counted, while vias increase the hardness of manufacturing and leads to more heat generation in general. In the second model, besides the objective in the first model, the maximum edge congestion is also to be minimized.

Multicast routing in communication networks: We notice that one model of the global routing problem in VLSI design [17] is to minimize the maximum edge congestion, which is essentially equivalent to the multicast congestion problem [2,11] or multicast packing problem [7] in communication networks. In fact there are many similarities between the routing problems in VLSI design and routing problems in communication networks.

A multicast communication network consists of a number of processors (usually routers) and other devices. The processors can receive, duplicate and deliver packets of data, and are linked by infrastructure backbone. The signals are transmitted among processors through fixed cables. Furthermore, for each cable there is a bandwidth (capacity) constraint. There exist a certain number of *requests* in a multicast communication network. Each request utilizes some of the processors. One processor in a request is the source processor to send the data packets, and others are destinations of the signal flows. Thus, it is required to find a routing tree for each request in the network structure such that the transmission can be realized.

From algorithmic point of view, a communication network can be represented by an undirected graph, where the vertices are the processor and the edges are the cables. A request is a subset of the vertex set. A solution of the multicast routing problem is a set of trees spanning the requests with respect to the edge capacity constraints. And the goal is to minimize the total cost. In the multicast congestion/multicast packing problem, there is no edge capacity nor cost of the routing. The goal is to minimize the maximum edge congestion [2,11]. Another related problem is the so-called group multicast routing problem [6,12]. Recently, the problem to maximize the number of successfully routed requests with respect to the edge capacity is studied in [18].

Our contribution: In this paper, we propose generalized models for the global routing problem in VLSI design and the multicast routing in communication networks. For the global routing problem, our model generalizes the previous models developed in [3,4,17,20]. Here we consider the three important factors in global routing: the total wirelength, the edge congestion, and the number of bends. The edge capacity can vary according to the local requirement instead of posing extra edge length as penalty for the high congested edges. The goal is to optimize a combination of the total wirelength and total number of bends. Similarly, for the multicast routing problem, our model also generalizes the previous models in [2,6,11,12]. It is a special case of the global routing problem, but in arbitrary graphs. In our approximation algorithms for the routing problems, we first apply binary search strategy to reformulate the linear relaxations to min-max resource-sharing problems or packing problems. Then we use the approximation algorithm in [10] as a subroutine to solve the packing problems. We show that the block problem of the multicast routing problem is the minimum Steiner tree problem in graphs. We also develop a combinatorial technique for lattice graphs, with which we also convert the block problem of the global routing problem to the minimum Steiner tree problem. Finally we obtain asymptotic approximation algorithms for the routing problems. To our best knowledge, no approximation results have been previously reported for these problems. In addition, with the combinatorial technique for lattice graphs, we also present polynomial time algorithms for some generalized optimization problems in urban transportation networks.

The remaining part of this paper is organized as follows: We give the formal models of the routing problems in Section 2 and reformulate them to the packing problems in Section 3. The approximation algorithms are presented in Section

4, and the approximation guarantee is shown in Section 5. Finally we address some more applications of the combinatorial technique in Section 6. Due to the limit of space we do not give proofs of our results in this version. We refer the readers to the full version of our paper [19] for details.

2 Mathematical Formulation

The global routing problem in VLSI design we study is defined as follows: Given an edge-weighted lattice graph and a set of nets (subsets of the vertex set), the edge set is associated with a nonnegative length function and a positive capacity function. Furthermore, there also exists a vertex cost function. The solution is a set of trees spanning the given nets. The total cost of a solution consists of two parts: the edge cost and the bend-dependent vertex cost. The edge cost is the sum of edge length of the generated trees, and the bend-dependent vertex cost is caused by the structure of the generated trees. For a tree in a solution in the given lattice graph, if the tree has a bend at vertex v, then its bend-dependent vertex cost on v is the given vertex cost. Otherwise, its bend-dependent vertex cost on v is 0. In fact a bend on a vertex corresponds a via in VLSI design, which leads to extra cost for manufacturing and results in risk of hot spots. The goal of the problem is to minimize the overall cost, which is a linear combination of the total edge cost and the total bend-dependent vertex cost, while the edge capacity constraints are fulfilled. In the multicast routing problem in communication networks, we are given an edge-weighted undirected graph and requests. There are a nonnegative cost function and a positive capacity function associated to the edge set. The goal is to find a set of trees spanning the requests with respect to the edge capacity constraints, such that the overall cost is minimized. Indeed the minimum Steiner tree problem in graphs, which is $\mathcal{APX}$-hard [5], is a special case of our global routing problem, where the vertex cost is 0 and all edge capacities are 1.

Given an planar edge-weighted lattice graph $G = (V, E)$ (with some rectangular holes) and nets $S_1, \ldots, S_K \subseteq V$, the edge set is associated with a length function $l : E \to \mathbb{R}^+ \cup \{0\}$ and a capacity function $c : E \to \mathbb{R}^+$. Without loss of generality, we assume that $|S_k|$ is bounded by some constant for all $k = 1, \ldots, K$. Furthermore, there also exists a vertex cost $w : V \to \mathbb{R}^+ \cup \{0\}$. A feasible solution is a set of K trees spanning $S_1, \ldots, S_K$ with respect to the edge capacity constraints. The goal of the global routing problem in VLSI design is to minimize the overall cost defined as a linear combination $\alpha l_{\text{total}} + \beta v_{\text{total}}$, where l_{total} is the sum of edge length of all K trees and v_{total} is the sum of numbers of bends of all K trees, while $\alpha, \beta \geq 0$ are artificial weights corresponding to the impact of the total wirelength and the total number of vias. For simplicity, we denote by c_i the capacity of edge $e_i \in E$ from now on.

We now develop the integer linear program formulation of our generalized model. Denote by $\mathcal{T}_k$ the set of all trees in G connecting the vertices in S_k. It is worth noting that $|\mathcal{T}_k|$ can be exponentially large. We also denote by $x_k(T)$ the indicator variable as follows: If $T \in \mathcal{T}_k$ is selected for the net S_k, then $x_k(T) = 1$;

Otherwise $x_k(T) = 0$. In addition, we define by $l(T)$ and $v(T)$ the length of tree T and the number of bends of tree T, respectively. Therefore the integer linear program formulation of the global routing problem in VLSI design is as follows:

$$\begin{array}{lll} \min \alpha \sum_{k=1}^{K} \sum_{T \in \mathcal{T}_k} l(T) x_k(T) + \beta \sum_{k=1}^{K} \sum_{T \in \mathcal{T}_k} v(T) x_k(T) & & \\ \text{s.t. } \sum_{T \in \mathcal{T}_k} x_k(T) = 1, & \forall k; & \\ \quad \sum_{k=1}^{K} \sum_{T \in \mathcal{T}_k \& e_i \in T} x_k(T) \leq c_i, & \forall e_i \in E; & (1) \\ \quad x_k(T) \in \{0, 1\}, & \forall T \& k. & \end{array}$$

Here the first set of constraints means that for any set $\mathcal{T}_k$ we choose exactly one tree for S_k, and the second set of constraints are capacity constraints.

In the multicast routing problem in communication networks, G can be an arbitrary graph. Formally, we are given an edge-weighted undirected graph $G = (V, E)$ and requests $S_1, \ldots, S_K \subseteq V$. There is a cost function $l : E \rightarrow \mathbb{R}^+ \cup \{0\}$ and a capacity function $c : E \rightarrow \mathbb{R}^+$ associated to the edge set. The goal is to find K trees spanning $S_1, \ldots, S_K$ with respect to the edge capacity constraints, such that the overall cost is minimized. The integer linear programming formulation of the multicast routing problem in communication networks is similar to (1) except that the weight $\beta = 0$ and $\alpha = 1$. Accordingly, we just need to study the integer linear program (1) for both problems.

3 Packing Formulation of the Generalized Model

We can show that the optimum value of the integer linear program (1) is equivalent to the optimum value of the following integer linear program:

$$\begin{array}{lll} \min g & & \\ \text{s.t. } \sum_{k=1}^{K} \sum_{T \in \mathcal{T}_k} (\alpha l(T) + \beta v(T)) x_k(T) \leq g, & & \\ \quad \sum_{k=1}^{K} \sum_{T \in \mathcal{T}_k \& e_i \in T} x_k(T) \leq c_i, & \forall e_i \in E; & (2) \\ \quad \sum_{T \in \mathcal{T}_k} x_k(T) = 1, & \forall k; & \\ \quad x_k(T) \in \{0, 1\}, & \forall T \& k. & \end{array}$$

Furthermore, we can show that any ρ-approximate solution of (2) is also a ρ-approximate solution of (1) for any $\rho \geq 1$. We shall study (2) instead of (1).

As usual, we shall study the LP-relaxation of (2), and then apply rounding techniques to obtain a feasible solution. It is worth noting that there may be exponentially many variables in (2) and its LP-relaxation. Thus many exact algorithms for LPs such as standard interior point methods can not be applied. The LP-relaxation of (2) may be solved by the volumetric-center [1] or the ellipsoid methods with separation oracle. However, those approaches will lead to a large running time. Therefore, we shall study approximation algorithms for the LP-relaxation of (1).

Now we consider the LP-relaxation of (2). For any given $\varepsilon \in (0, 1)$, we can show if we can solve the following linear program (see [19] for details)

$$\begin{array}{lll} \min \lambda & & \\ \text{s.t.} \sum_{k=1}^{K} \sum_{T \in \mathcal{T}_k} (\alpha l(T) + \beta v(T)) x_k(T)/g \leq \lambda, & & \\ \sum_{k=1}^{K} \sum_{T \in \mathcal{T}_k \& e_i \in T} x_k(T)/c_i \leq \lambda, & \forall e_i \in E; & (3) \\ \sum_{T \in \mathcal{T}_k} x_k(T) = 1, & \forall k; & \\ x_k(T) \in [0,1], & \forall T \& k, & \end{array}$$

then we can find a $(1+\varepsilon)$-approximate solution to the LP-relaxation of (2). The remaining task is to solve (3). Here we call the linear program (3) the packing formulation of the global routing problem in VLSI design.

4 Approximation Algorithms for the LP-Relaxations

We shall present approximation algorithms for the LP-relaxations of the multicast routing problem in communication networks and the global routing problem in VLSI design in this section. Here we call the linear relaxation of the multicast routing problem (respectively, the global routing problem) the *fractional* multicast routing problem (respectively, the *fractional* global routing problem).

4.1 Approximation Algorithms for Fractional Multicast Routing

In the fractional multicast routing problem in communication networks, we shall minimize the overall edge length of the trees spanning requests with respect to the capacity constraints (e.g., $\alpha = 1$ and $\beta = 0$). We notice that (3) is indeed the formulation of the *convex min-max resource-sharing problems* (*packing problems* in the linear case) in [8,10,21]. The convex min-max resource-sharing problems is defined as follows:

$$\min\{\lambda | f_m(x) \leq \lambda, m = 1, \ldots, M, x \in B\}, \tag{4}$$

where $B \subseteq \mathbb{R}^N$ is a nonempty convex compact set and $f_m : B \to \mathbb{R}_+$ are nonnegative continuous convex functions on B for $m \in \{1, \ldots, M\}$. Let $f(x) = (f_1(x), \ldots, f_M(x))^T$ and $\lambda(x) - \max_{m \in \{1,\ldots,M\}} f_m(x)$ for $x = (x_1, \ldots, x_N)^T \in B$. Denote by x^* the optimal solution and by $\lambda^* = \lambda(x^*)$ the optimal value of (4). In [8,21], algorithms are proposed to find $(1+\varepsilon)$-approximate solutions to (4) provided an oracle to deliver a $(1 + O(\varepsilon))$-approximate solution to the corresponding block problem (which depends on the specified $f(x)$). However, in the fractional multicast routing problem in communication networks (also the fractional global routing in VLSI design), we shall show that there is no PTAS for the block problems. Therefore there only exist r-approximation algorithms for the block problem with $r > 1$. In such a case the algorithms in [8,21] are not applicable. An approximation algorithm for convex min-max resource-sharing problems with only weak block solvers is proposed in [10].

The algorithm is based on the Lagrangian relaxation. The algorithm maintains a pair of a feasible solution x and a *price vector* (dual vector) $p \in P = \{p \in \mathbb{R}^M | \sum_{m=1}^{M} p_m = 1, p_m \geq 0\}$. The block problem $ABS(p, t, r)$ of (4) is to compute $\hat{x} = \hat{x}(p) \in B$ such that $p^T f(\hat{x}) \leq r(1+t) \min\{p^T f(y) | y \in B\}$ [10], where $t = O(\varepsilon)$ and $r \geq 1$ is the approximation ratio. Furthermore, the

algorithm is an iterative method, where in each iteration an approximate block solver is called as an oracle once. More details of packing problems or convex min-max resource-sharing problems can be found in [22].

Proposition 1. *[10] For any $\varepsilon \in (0,1)$, there exists an Algorithm $\mathcal{L}$ that finds an $r(1+\varepsilon)$-approximate solution to (4) in $O(M(\log M + \varepsilon^{-2}\log \varepsilon^{-1}))$ iterations and an $r(1+O(\varepsilon))$-approximate block solver is called once in each iteration.*

In the fractional multicast routing problem in communication networks, the packing constraints are: $f_i = \sum_{k=1}^{K} \sum_{T \in \mathcal{T}_k \& e_i \in T} x_k(T)/c_i$, for $i = 1, \ldots, m$, and $f_{m+1} = \sum_{k=1}^{K} \sum_{T \in \mathcal{T}_k} l(T) x_k(T)/g$. Then we can directly apply algorithm $\mathcal{L}$ here and the following theorem holds (see [19] for proof).

Theorem 1. *For any accuracy $\varepsilon \in (0,1)$, there exists an $r(1+\varepsilon)$-approximation algorithm for the fractional multicast routing problem in communication networks provided an r-approximate minimum Steiner tree solver.*

4.2 Approximation Algorithms for Fractional Global Routing

The fractional global routing problem is formulated as the linear program (3), where the weight of overall number of bends (vias) $\beta \neq 0$. We shall also apply the approximation algorithm $\mathcal{L}$ in [10] for this problems. Here the block problem is different from that for the multicast routing problem due to the additional cost caused by the tree bends. However, we still have the following result for the fractional global routing problem:

Theorem 2. *For any accuracy $\varepsilon \in (0,1)$, there exists an $r(1+\varepsilon)$-approximation algorithm for the fractional global routing problem in VLSI design (3) provided an r-approximate minimum Steiner tree solver.*

We only show the main idea of the proof in this version. For the original lattice graph $G = (V, E)$, we construct a two-layer graph H, where on each layer the vertex set is identical to V, and all horizontal edges are in the lower layer, while all vertical edges are in the upper layer. In addition, Each pair of vertices in the

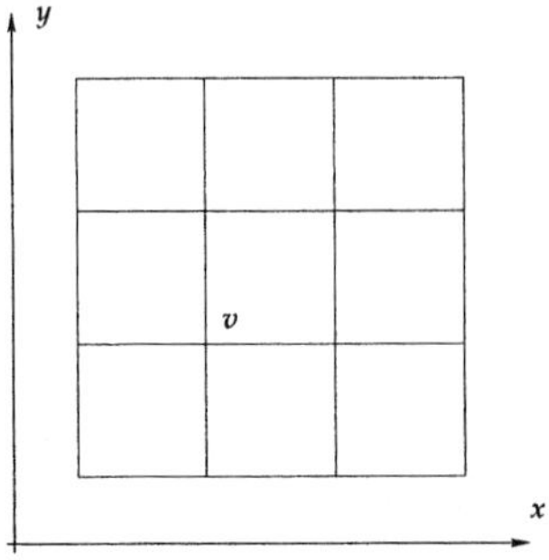

Fig. 1. Original lattice graph G

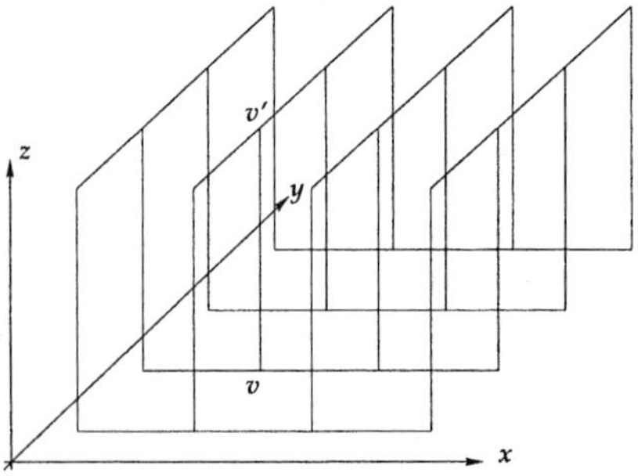

Fig. 2. Two-layer graph H

two layers corresponding to the same vertex in V is connected by a new edge (see Figure 1) and 2). In this way, a bend in G corresponds to crossing the new edge in H once. We assign the vertex cost in G to the new edges in H. Thus the block problem is shown the minimum Steiner tree problem in H, and the theorem follows (see [19] for details). This technique is called the *virtual layer method.*

5 Approximation Algorithms

We have presented $r(1+\varepsilon)$-approximation algorithms for the fractional global routing problem in VLSI design and the multicast routing problem in communication networks in Section 4. Then we use the randomized rounding in [16,17] to obtain feasible solutions. Let OPT denote the optimal objective value of the LP-relaxation (3). Directly from [16] we have the following theorem:

Theorem 3. *There are approximation algorithms for the global routing problem in VLSI design and the multicast routing problem in communication networks such that the objective value is bounded by*

$$\begin{cases} r(1+\varepsilon)OPT + (\exp(1)-1)(1+\varepsilon)\sqrt{rOPT\ln m}, & \text{if } rOPT > \ln m; \\ r(1+\varepsilon)OPT + \dfrac{\exp(1)(1+\varepsilon)\ln m}{1+\ln(\ln m/(rOPT))}, & \text{otherwise.} \end{cases}$$

If $rOPT > \ln m$, the objective value generated by our algorithm is at most $\exp(1)r(1+\varepsilon)OPT$. In the other case, the objective value generated by our algorithm is at most $r(1+\varepsilon)OPT + O(\ln m)$. Therefore, we obtain asymptotic approximation algorithms for the routing problems.

In (1) there are exponential number of variables. However, by applying Algorithm $\mathcal{L}$, we just need to generate K minimum Steiner trees for the K nets/requests in each iteration corresponding to the current price vector. So there are only a polynomial number of Steiner trees generated in Algorithm $\mathcal{L}$ in total.

Corollary 1. *The approximation algorithms for the global routing problem in VLSI design and the multicast routing problem in communication networks only generates at most $O(Km(\log m + \varepsilon^{-2}\log\varepsilon^{-1}))$ Steiner trees.*

This is similar to the column generation technique for linear programs, and the hardness due to exponential number of variables in (1) is overcome.

6 More Applications of the Virtual Layer Method

We consider an optimization problem as follows: In a given lattice graph $G = (V, E)$, there is a source-destination pair (s, t) and an amount of demand $d \leq \min_e c(e)$ between the pair, where $c(e)$ is the capacity of edge e. There is an edge cost function $l : E \rightarrow \mathbb{R}^+ \cup \{0\}$. Furthermore, for each vertex there is a vertex cost $w : V \rightarrow \mathbb{R}^+ \cup \{0\}$. A feasible solution is a routed path connecting the source and the destination. The overall cost of the path is the sum of edge costs

and bend-dependent vertex costs. The goal of the problem is to find a routing path between the source and the destination such that the overall cost of the path is minimized.

This problem is a variant of the classical shortest path problem in lattice graphs, which has a realistic background in urban transportation networks. In many cases the urban transportation networks are lattice graphs or can be represented as lattice graphs. When a vehicle reaches an intersection, it can turn left or right, or drive straightly through the intersection. The delays corresponding to these three choices are different. We assume that the delays to turn left and right are identical due to the traffic situation, pedestrians and traffic signal lights. The delay to pass straightly is much less than the delays of turns and can also be merged to the edge cost. To deliver a certain amount commodity from a source to a destination in such a transportation network, a routing path for the vehicle with minimum overall time is to be designed. To solve this problem, we can directly apply the virtual layer method and the following theorem holds:

Theorem 4. *The shortest path problem in lattice graphs, where the total cost is the sum of edge costs and the bend-dependent vertex costs, is polynomial time solvable.*

Corollary 2. *The splitable min-cost flow problem (or the splitable min-cost multicommodity flow problem) in lattice graphs, where the total cost is the sum of edge costs and the bend-dependent vertex costs, is polynomial time solvable.*

7 Conclusions and Future Research

In this paper we have presented approximation algorithms for the global routing problem in VLSI design and the multicast routing problem in communication networks. Our models generate many previous models for these routing problems. In fact, with binary search technique one can study a bicriteria version of the minimization problem with two objective functions: the overall cost and the maximum edge congestion. There are many interesting topics for further study. First, we intend to apply the virtual layer method to more optimization problems in lattice graphs. Second, we would like to propose more general models for the global routing problem in VLSI design and the multicast routing problems in communication networks such that the impact of more factors are included. Nice problems arise when one studies the 3-dimensional lattice graphs for the global routing problem. We shall also implement our algorithm to explore the power of our approximation algorithm in computational practice.

References

1. K. M. Anstreicher, Towards a practical volumetric cutting plane method for convex programming, *SIAM Journal on Optimization*, 9 (1999), 190-206.
2. A. Baltz and A. Srivastav, Fast approximation of minimum multicast congestion - implementation versus theory, *RAIRO Operations Research*, 38 (2004), 319-344.

3. L. Behjat, New modeling and optimization techniques for the global routing problem, *Ph.D. Thesis*, University of Waterloo, 2002.
4. L. Behjat, A. Vannelli, W. Rosehart, Linear programming models for the global routing problem, *to appear in Informs Journal on Computing*, 2004.
5. M. Bern and P. Plassmann, The Steiner problem with edge lengths 1 and 2, *Information Professing Letters*, 32 (1989), 171-176.
6. M. Cai, X. Deng and L. Wang, Minimum k arborescences with bandwidth constraints, *Algorithmica*, 38 (2004), 529-537.
7. S. Chen, O. Günlük and B. Yener, The multicast packing problem, *IEEE/ACM Transactions on Networking*, 8(3) (2000), 311-318.
8. M. D. Grigoriadis and L. G. Khachiyan, Coordination complexity of parallel price-directive decomposition, *Mathematics of Operations Research*, 2 (1996), 321-340.
9. F. Hadlock, Finding a maximum cut of a planar graph in polynomial time, *SIAM Journal on Computing*, 4(3) (1975), 221-225.
10. K. Jansen and H. Zhang, Approximation algorithms for general packing problems with modified logarithmic potential function, *Proceedings of TCS 2002*, 255-266.
11. K. Jansen and H. Zhang, An approximation algorithm for the multicast congestion problem via minimum Steiner trees, *Proceedings of ARACNE 2002*, 77-90.
12. X. Jia and L. Wang, A group multicast routing algorithm by using multiple minimum Steiner trees, *Computer Communications*, 20 (1997), 750-758.
13. C. Y. Lee, An algorithm for path connection and its application, *IRE Transactions on Electronic Computers*, 10 (1961), 346-365.
14. T. Lengauer, *Combinatorial algorithms for integrated circuit layout*, J. Wiley, New York, 1990.
15. T. Lengauer and M. Lungering, Provably good global routing of integrated circuits, *SIAM Journal on Optimization*, 11(1) (2000), 1-30.
16. P. Raghavan, Probabilistic construction of deterministic algorithms: approximating packing integer programs, *Journal of Computer and System Sciences*, 37 (1988), 130-143.
17. P. Raghavan and C. D. Thompson, Randomized rounding: a technique for provably good algorithms and algorithmic proofs, *Combinatorica*, 7(4) (1987), 365-374.
18. M. Saad, T. Terlaky, A. Vannelli, and H. Zhang, Packing trees in communication networks, *to appear in Proceedings of WINE 2005*, LNCS.
19. T. Terlaky, A. Vannelli and H. Zhang, On routing in VLSI design and communication networks, *Technical Report*, AdvOL #2005-14, Advanced Optimization Lab., McMaster University, `http://www.cas.mcmaster.ca/~oplab/research.htm`.
20. A. Vannelli, An adaptation of the interior point method for solving the global routing problem, *IEEE Transactions on Computer-Aided Design*, 10(2) (1991), 193-203.
21. J. Villavicencio and M. D. Grigoriadis, Approximate Lagrangian decomposition with a modified Karmarkar logarithmic potential, *Network Optimization, P. Pardalos, D. W. Hearn and W. W. Hager (Eds.), Lecture Notes in Economics and Mathematical Systems 450, Springer-Verlag, Berlin,* (1997), 471-485.
22. H. Zhang, Approximation algorithms for min-max resource sharing and malleable tasks scheduling, *Ph.D. Thesis*, University of Kiel, 2004.

The Capacitated Traveling Salesman Problem with Pickups and Deliveries on a Tree

Andrew Lim, Fan Wang, and Zhou Xu*

Department of Industrial Engineering and Engineering Management,
The Hong Kong University of Science and Technology,
Clear Water Bay, Kowloon, Hong Kong
{iealim, fanwang, xuzhou}@ust.hk

1 Introduction

The Capacitated Traveling Salesman Problem with Pickups and Deliveries (CTSP-PD) [1] can be defined on an undirected graph $T = (V, E)$, where V is a set of n vertices and E is a set of edges. A nonnegative weight $d(e)$ is associated with each edge $e \in E$ to indicate its length. Each vertex is either a pickup point, a delivery point, or a transient point. At each pickup point is a load of unit size that can be shipped to any delivery point which requests a load of unit size. Hence we can use $a(v) = 1, 0, -1$ to indicate v to be a pickup, a transient, or a delivery point, and $a(v)$ is referred to as the volume of v. The total volumes for pickups and for deliveries are usually assumed to be balanced, i.e., $\sum_{v \in V} a(v) = 0$, which implies that all loads in pickup points must be shipped to delivery points [1]. Among V, one particular vertex $r \in V$ is designated as a depot, at which a vehicle of limited capacity of $k \geq 1$ starts and ends. The problem aims to determine a minimum length feasible route that picks up and delivers all loads without violating the vehicle capacity.

The CTSP-PD on a general graph is $\mathcal{NP}$-hard in the strong sense because it coincides with the TSP. It was first studied by [6] together with a polynomial time algorithm which achieved an approximation factor of $(5\alpha + 2 - 5\alpha/k)$, where α is the approximation factor of the TSP algorithm. Later, [1] proposed the other two approximation algorithms for the CTSP-PD on a general graph, and they achieved approximation factors of $(4\alpha + 1 - 2\alpha/k)$ and $\alpha + \lceil \log_2 k \rceil/2 + (2\lceil k/2 \rceil - 1)/2^{\lceil \log_2 k \rceil}$ respectively. Other related routing problems include the *capacitated vehicle routing problem* [4], the *swapping problem* [3, 2], the *stacker crane problem* [9], and etc. See [12] for a detailed survey.

Routing problems on a tree-shaped network have also been studied in a great deal of literatures [7, 8, 13, 11], because of its wide application in transportation industry. For example, Labbe *et al* demonstrated a lot of real applications of the tree network, such as river networks, some railway networks, pit mine railways, and mining or logging areas in Northern Canada [14]. Basnet and *et al.* proposed heuristics for vehicle routing problem on tree-like networks, especially for those

* Corresponding Author.

X. Deng and D. Du (Eds.): ISAAC 2005, LNCS 3827, pp. 1061–1070, 2005.

rural systems where supply (or delivery) nodes are located on rural roads leading off from a few highways which form the "trunks" of a tree-like network [5].

We thus focus our interest in the CTSP-PD on a tree. We have proved its $\mathcal{NP}$-hardness in a strong sense, and proposed a 2-approximation algorithm with a time complexity of $O[n^2/\min(n,k)]$ for the CTSP-PD on a tree. Since the TSP on a tree can be solved optimally in a trivial way, a straightforward approach to solve the CTSP-PD on a tree is to apply the approximation algorithms devised for the general graph case in [6, 1]. Since $\alpha = 1$ in this case, their approximation factors are $(7 - 5/k)$ for [6], $(5 - 2/k)$ and $1 + \lceil \log_2 k \rceil /2 + (2\lceil k/2 \rceil - 1)/2^{\lceil \log_2 k \rceil}$ for [1] respectively. Comparing with them, our method has a much smaller approximation factor.

The approximation algorithm presented in this paper is based on a recurrent construction process, which is differently from those approaches for the CTSP-PD in a general graph in [6, 1]. The rest of the paper is organized as follows. Section 2 introduces basic notation and derives a lower bound of the length of the shortest route for the CTSP-PD on a tree. Afterwards a 2-approximation algorithm is presented and analyzed in Section 3. Due to the lack of space, some statements here are presented without details of proofs. (Refer to [15] for details.)

2 Preliminaries

Without loss of generality, we let the tree T rooted at the depot r. For each vertex $v \in V$, let $child(v)$ denote its children set, $p(v)$ denote its parent, $e(v)$ indicate the edge between $p(v)$ and v, $T(v)$ represent the sub-tree rooted by v, and $b(v)$ represent the number of vertices in $T(v)$. Let $A(v) = \sum_{i \in T(v)} a(v)$ indicate the total volume of products in sub-tree $T(v)$.

Let $L(H)$ denote the length of a route H of the CTSP-PD on the T. Let H^* denote the shortest feasible solution to the CTSP-PD on the T, and therefore, $L(H^*)$ is the length of the shortest feasible solution and often referred to as the optimal length. Furthermore, we define the length of a route set by summing up all the lengths of routes in the set. In other words, let $L(A)$ denote the length of a route set A and $L(A) = \sum_{H \in A} L(H)$.

The following theorem reveals that the CTSP-PD on a tree is $\mathcal{NP}$-hard in the strong sense, and therefore, no efficient algorithm exists unless $\mathcal{NP} = \mathcal{P}$.

Theorem 1. *Finding a shortest feasible solution to the CTSP-PD on a tree T is $\mathcal{NP}$-hard in the strong sense.*

Proof. It can be proved by a reduction from the **3-Partition** problem [10]. For the lack of space, we omit the details here. □

2.1 Assumptions About T

To simplify the presentation along the paper, we will always assume that the tree T is standard according to the following definition.

Definition 1. *The tree T is standard if T satisfies the following conditions.*

1. *Each leaf of T is either pickup point or delivery point.*
2. *Each non-leaf of T must be transit-point.*
3. *For each non-leaf $v \in V$, the $child(v)$ satisfies*
   (a) *either every $u \in child(v)$ has $A(u) \geq 0$;*
   (b) *or every $u \in child(v)$ has $A(u) < 0$;*
   (c) *or v has exactly two children, i.e., $child(v) = \{u_1, u_2\}$, where $A(u_1) \geq 0$ and $A(u_2) < 0$.*

It is easy to see the assumption is safe, since any tree T can be transformed to a standard tree $\widetilde{T}$ so that each feasible route to the CTSP in $\widetilde{T}$ corresponds to a route to the CTSP in T with the same length. Details are omitted for the lack of space.

2.2 Lower Bound of $L(H^*)$

Since the vehicle can carry at most k units of products once at a time and the route forms a close loop starting and ending at the root r, each edge $e(v)$ must be traversed at least $2n(v)$ times, where $n(v) = \max\{\lceil |A(v)|/k \rceil, 1\}$. Hence, the length of the shortest route is at least $LB(r)$, where

$$LB(v) = 2d(e(v))n(v), \text{ for } v \in V \text{ and } v \text{ is a leaf;}$$
$$LB(v) = \sum_{i \in T(v)} 2d(e(v))n(v), \text{ for } v \in V \text{ and } v \text{ is not a leaf.}$$

Lemma 1. *Each edge $e(v)$ is visited at least $2n(v)$ times, and $LB(r) \leq L(H^*)$.*

3 A 2-Approximation Algorithm

Briefly speaking, our approximation algorithm constructs a series of route sets for all vertices, recurrently from leaves to the root of the tree. For each $v \in V$, the route set is denoted by H_v and contains $h(v)$ routes, i.e., $H_v = \{H_{v,j} | 1 \leq j \leq h(v)\}$. In the rest of this section, we will explain the details about the construction of H_v.

3.1 w-Structured Route Set

In order to ensure its total length close to the $LB(v)$, the route set H_v that we are going to construct holds a w-structure, which will be defined later. To ease the presentation, let

$$\alpha(x) = \begin{cases} k, & \text{if } x \geq k \text{ and } x \bmod k = 0 \\ x \bmod k, & \text{if } x \leq k-1 \end{cases} \tag{1}$$

We use $s^-(v,j)$ and $s^+(v,j)$ respectively denote the units of products that are initially carried from $p(v)$ to v and the units that are finally carried from v to $p(v)$ by the route $H_{v,j} \in H_v$ for $1 \leq j \leq h(v)$.

Definition 2. *For $v \in V$ and $0 \le w \le k$, a route set $H_v = \{H_{v,j} | 1 \le j \le h(v)\}$ holds a w-structure if it satisfies the following five conditions.*

1. *Routes in H_v satisfy all the pickup and the delivery requests for vertices in $T(v)$ and visits each edge $e(u)$ for u in $T(v)$ at most $2[n(u)+1]$ times.*
2. *Each route $H_{v,j}$ for $1 \le j \le h(v)$ starts from $p(v)$ to v through $e(v)$, ends from v to $p(v)$ through $e(v)$, and never traverses $e(v)$ in between, and therefore, routes in H_v traverse $e(v)$ totally at most $2h(v)$ times.*
3. *The route $H_{v,1}$ starts with w units of products from $p(v)$ to v, i.e., $s^-_{v,1} = w$.*
4. *H_v can be divided into two disjoint subsets, i.e., $Y_v = \{H_{v,j} | 1 \le j \le y(v)\}$ and $X_v = \{H_{v,j} | y(v)+1 \le j \le h(v)\}$ where $0 \le y(v) \le h(v)$, satisfying:*
   (a) If $y(v) > 1$, then $h(v) = y(v)$ implying $X_v = \emptyset$; otherwise either $h(v) = y(v)$ implying $X_v = \emptyset$, or $h(v) = y(v)+1$ and $X_v = \{H_{v,h(v)}\}$ with $s^+_{v,h(v)} = s^-_{v,h(v)} = 0$;
   (b) If $A(v) > 0$, then $y(v) = \lceil (A(v)+w)/k \rceil$, $s^-(v,j) = 0$ for $2 \le j \le y(v)$, $s^+(v,j) = k$ for $1 \le j \le y(v)-1$, and $s^+(v,y(v)) = \alpha(A(v)+w)$.
   (c) If $A(v) < 0$, then $y(v) = \lceil -(A(v)+w)/k \rceil + 1$, $s^-(v,j) = k$ for $2 \le j \le y(v)$, $s^+(v,j) = 0$ for $1 \le j \le y(v)-1$, and $s^+(v,y(v)) = \alpha(A(v)+w)$;
   (d) If $A(v) = 0$, then $y(v) = 1$ and $s^+_{v,1} = w$;

Condition 2 implies that H_v totally traverses $e(v)$ at most $2h(v)$ times. Condition 4 claims that the w-structured route set H_v can be participated into two disjoint subsets Y_v and X_v, and Condition 4(a) says that X_v is either empty set, or may contain a single route $H_{v,h(v)}$, which carries nothing into and out of $T(v)$ if $Y(v)$ contains more than one route. The X_v may be redundant, but such a redundant provides us a flexibility to construct a heuristic solution with a certain quality guarantee in polynomial time. The Conditions 4(b) and 4(c) restrict the routes in Y_v to carry products out of (or into) the $T(v)$ as many as possible, if $A(v) > 0$ (or $A(v) < 0$ respectively). The Condition 4(d) forces the vehicle to carry equal units of loads when it moves into and out of the $T(v)$. Based on the above definition of the w-structured route set, the following proposition can be easily observed.

Proposition 1. *A w-structured route set $H_v = Y_v \cup X_v$ for $v \in V$ where $0 \le w \le k$, we have $1 \le y(v) \le h(v) \le y(v)+1$, $h(v) \le n(v)+1$, and thus routes in H_v visit $e(v)$ at most $2(n(v)+1)$ times in total.*

Notice that $A(r) = 0$ because of the balance between pickup and delivery points. Thus, a 0-structured route set H_r must be able to satisfy all the pickup and the delivery requests in T and to visit each edge $e(u)$ for $u \in V$ at most $2[n(u)+1]$ times, and thus has a total length at most two times of the optimal length.

3.2 Route Construction

To construct a 0-structured route set as a special case for the root r, we are now going to present a recurrent process to generate a w-structured route set H_v together with Y_v and X_v, for any vertex $v \in V$ and $0 \le w \le k$.

Algorithm 1 illustrates the recurrent framework, and thus, the construction of the 0-structured route set for the root r can be started by calling $Construct(r, 0)$. According to the assumption that T is standard, five cases, including Case 0, 1, 2, 3a, and 3b, as shown in Algorithm 1, need to be considered for any execution of $Construct(v, w)$ where $v \in V$ and w. The recurrence happens since sub-procedures, $ConstructCase1(v, w)$, $ConstructCase2(v, w)$, $ConstructCase3a(v, w)$, and $ConstructCase3b(v, w)$, are based on route sets for children of v, and these route sets are obtained by calling $ConstructCase()$ recurrently.

Algorithm 1 $Construct(v, w)$: to construct a w-structured route set $H_v = Y_v \cup X_v$ for subtree $T(v)$ where $v \in V$

---

if v is a leaf of T **then**
  Call $ConstructCaseLeaf(v, w)$ to get H_v together with Y_v and X_v;
else
  Consider the following three cases to get H_v together with Y_v and X_v;
  - Case 1: when $A(u) \geq 0$ for all $u \in child(v)$, call $ConstructCase1(v, w)$;
  - Case 2: when $A(u) < 0$ for all $u \in child(v)$, call $ConstructCase2(v, w)$;
  - Case 3a: when $A(u_1) \geq 0$ and $A(u_2) < 0$ and $A(v) \geq 0$ where $child(v) = \{u_1, u_2\}$, call $ConstructCase3a(v, w)$;
  - Case 3b: when $A(u_1) \geq 0$ and $A(u_2) < 0$ and $A(v) < 0$ where $child(v) = \{u_1, u_2\}$, call $ConstructCase3b(v, w)$;

end if
Return $H_v = Y_v \cup X_v$;

To fill the details for Algorithm 3.2 and prove the correctness of Algorithm 3.2, we will show the Claim 1 is true for all vertices $v \in V$ through a constructive proof by induction from leaves to the root.

Claim 1. *For a vertex $v \in V$, it can take only $O[b^2(v)/\min(b(v), k)]$ time to construct a w-structured route set H_v for each $0 \leq w \leq k$.*

Lemma 2. *Case 0: Claim 1 is true for all leaves $v \in V$.*

Proof. Consider a leaf $v \in V$. Since T is standard, either $A(v) = a(v) = 1$ or $A(v) = a(v) = -1$. For any w with $0 \leq w \leq k$, a w-structured route set $H_v = Y_v \cup X_v$ can be easily constructed by Algorithm 2. □

Lemma 3. *Case 1: For a non-leaf $v \in V$ with $A(u) \geq 0$ for all $u \in child(v)$, if Claim 1 is true for all $u \in child(v)$, it is still true for v.*

Proof. Consider a non-leaf $v \in V$ with $A(u) \geq 0$ for all $u \in child(v)$. Let $child(v) = \{u_1, u_2, ..., u_t\}$ denote the set of t children of v. We have that $A(v) = \sum_{i=1}^{t} A(u_i)$ and $b(v) = \sum_{i=1}^{t} b(u_i) + 1$.

Algorithm 2 ConstructCaseLeaf(v, w): to construct a w-structured route set $H_v = Y_v \cup X_v$ for a leaf v

$X_v \leftarrow \emptyset$;
if $a(v) = 1$ **then**
  if $w < k$ **then**
    $Y_v = \{H_{v,1}\}$, where $H_{v,1}$ starts from $p(v)$ with w unit products, picks up one unit from v, and carries $w + 1$ units to $p(v)$;
  else
    $Y_v = \{H_{v,1}, H_{v,2}\}$, where $H_{v,1}$ starts from $p(v)$ to v with k units and comes back to $p(v)$, while $H_{v,2}$ starts from $p(v)$ with nothing, but picks up one unit from v, and carries it to $p(v)$.
  end if
else
  if $w > 0$ **then**
    $Y_v = \{H_{v,1}\}$, where $H_{v,1}$ starts from $p(v)$ with w unit products, delivers one unit to v, and carries $w - 1$ units to $p(v)$;
  else
    $Y_v = \{H_{v,1}, H_{v,2}\}$, where $H_{v,1}$ starts from $p(v)$ to v with nothing and comes back to $p(v)$, while $H_{v,2}$ starts from $p(v)$ with k units, delivers one unit to v, and carries $k - 1$ units to $p(v)$.
  end if
end if
Return $H_v = Y_v \cup X_v$;

By the induction assumption, for each $u \in child(v)$ it can take only $O[b^2(u)/\min(b(u), k)]$ time to construct a s-structured route set $H_u = Y_u \cup X_u$ for each $0 \leq s \leq k$. We are now going to construct a w-structured route set $H_v = Y_v \cup X_v$ for any $0 \leq w \leq k$ in the following way as shown in Algorithm 3.

Since every $u_i \in child(v)$ has $A(u_i) \geq 0$, we know $A(v) \geq 0$. Our construction of a w-structured route set for v is based on a series of w_i-structured route sets $H_{u_i} = Y_{u_i} \cup X_{u_i}$ for $i = 1, ..., t$, where $s^-(u_1, 1) = w_1 = w$, and $s^-(u_i, 1) = w_i = s^+(u_{i-1}, y(u_{i-1}))$ for $2 \leq i \leq t$. In other words, the first route in H_{u_i} carries into $T(v)$ the same units of products as that are carried out of $T(v)$ by the last route in $H_{u_{i-1}}$. By $A(v) = \sum_{i=1}^{t} A(u_i)$ and $s^-(u_1, 1) = w$, we know that $s^+(u_t, y(u_t)) = \alpha(A(v)+w)$. Because of the induction assumption, each route set H_{u_i} can be obtained in $O(b^2(u_i)/min(b(u_i), k))$ time. Furthermore, each route $H_{u_i,j}$, for $1 \leq i \leq t$ and $1 \leq j \leq y(u_i)$, can extend its start point and end point from v to $p(v)$. Since $s^-(u_i, 1) = s^+(u_{i-1}, y(u_{i-1})) = w_i$ for $2 \leq i \leq t$, every $H_{u_i,1}$ can then be appended to $H_{u_{i-1}, y(u_{i-1})}$.

After the above appending process, we can decide $H_v = Y_v \cup X_v$ as follows. For Y_v, the first $y(u_1)$ routes in it equals routes $H_{u_1,j}$ for $1 \leq j \leq y(u_1)$, the next $[y(u_2) - 1]$ routes equals routes $H_{u_2,j}$ for $2 \leq j \leq y(u_2)$, ..., and the last $[y(u_t) - 1]$ routes of H_v equals routes $H_{u_t,j}$ for $2 \leq j \leq y(u_t)$. In total, there are $y(v) = [\sum_{i=1}^{t} y(u_i) - t + 1]$ routes in Y_v. For X_v, it can be decided based on Y_v and S, where $S = \cup_{i=1}^{t} X_{u_i}$, as shown in Algorithm 3.2. It can be seen that the $H_v = Y_v \cup X_v$ constructed satisfies all the constraints of Definition 2 and must

Algorithm 3 ConstructCase1(v, w): to construct a w-structured route set $H_v = Y_v \cup X_v$ for a non-leaf v with $A(u_i) \geq 0$ for all $u_i \in child(v) = \{u_1, u_2, ..., u_t\}$

1: For $i = 1, 2, ..., t$, call $Construct(v, w_i)$ to construct a w_i-structured route set H_{u_i}, where $w_1 = w$ and $w_i = s^+(u_{i-1}, y(u_{i-1}))$.
2: For $i = t, t-1, ..., 2$, append $H_{u_{i-1}, y(u_{i-1})}$ to $H_{u_i,1}$ so that vehicle starts from $p(v)$ to v, follows $H_{u_{i-1}, y(u_{i-1})}$ until it visits v again, then the vehicle changes to follow $H_{u_i,1}$ and comes back to $p(v)$;
3: Let $Y_v \leftarrow \{H_{u_1,j} : 1 \leq j \leq y(u_1)\} \cup \{H_{u_i,j} : 2 \leq i \leq t, 2 \leq j \leq y(u_i)\}$, and let $S \leftarrow \cup_{i=1}^{t} X_{u_i}$;
4: **if** $S = \emptyset$ **then**
5: Let $X_v \leftarrow \emptyset$;
6: **else**
7: Link routes in S one by one into a single route $H_{v,y(v)+1}$ and extend its start and end points to $p(v)$;
8: Let $X_v \leftarrow \{H_{v,y(v)+1}\}$ if $y(v) = 1$, and otherwise, let $X_v \leftarrow \emptyset$ and insert $H_{v,y(v)+1}$ before $H_{v,2}$.
9: **end if**
10: Return $H_v = Y_v \cup X_v$;

Algorithm 4 ConstructCase2(v, w): to construct a w-structured route set $H_v = Y_v \cup X_v$ for a non-leaf v with $A(u_i) < 0$ for all $u_i \in child(v) = \{u_1, u_2, ..., u_t\}$

1: For $i = 1, 2, ..., t$, call $Construct(v, w_i)$ to construct a w_i-structured route set H_{u_i}, where $w_1 = w$ and $w_i = s^+(u_{i-1}, y(u_{i-1}))$.
2: For $i = t, t-1, ..., 2$, append $H_{u_{i-1}, y(u_{i-1})}$ to $H_{u_i,1}$ so that vehicle starts from $p(v)$ to v, follows $H_{u_{i-1}, y(u_{i-1})}$ until it visits v again, then the vehicle changes to follow $H_{u_i,1}$ and comes back to $p(v)$;
3: Let $Y_v \leftarrow \{H_{u_1,j} : 1 \leq j \leq y(u_1)\} \cup \{H_{u_i,j} : 2 \leq i \leq t, 2 \leq j \leq y(u_i)\}$;
4: Let $S \leftarrow \cup_{i=1}^{t} X_{u_i}$;
5: **if** $S = \emptyset$ **then**
6: Let $X_v \leftarrow \emptyset$;
7: **else**
8: Link routes in S one by one into a single route $H_{v,y(v)+1}$ and extend its start and end points to $p(v)$;
9: Let $X_v \leftarrow \{H_{v,y(v)+1}\}$ if $y(v) = 1$, and otherwise, let $X_v \leftarrow \emptyset$ and append $H_{v,y(v)+1}$ after $H_{v,1}$.
10: **end if**
11: Return $H_v = Y_v \cup X_v$;

be a w-structured route set. Finally, the time complexity of the above can also be shown as $O[b^2(v)/\min(b(v), k)]$, because each route set H_{u_i} can be obtained in $O[b^2(v)/\min(b(u_i), k)]$ for $1 \leq i \leq t$ by the induction assumption. □

Lemma 4. *Case 2: For a non-leaf $v \in V$ with $A(u) < 0$ for all $u \in child(v)$, if Claim 1 is true for all $u \in child(v)$, it is still true for v.*

Proof. The construction for Case 2 is very similar to Case 1, as shown in Algorithm 4. Details are omitted for the lack of space.

Algorithm 5 ConstructCase3a(v, w): to construct a w-structured route set $H_v = Y_v \cup X_v$ for a non-leaf v with $A(u_1) \geq 0$ and $A(u_2) < 0$ and $A(v) \geq 0$ where $child(v) = \{u_1, u_2\}$

For $i = 1, 2$, call $Construct(u_i, w_i)$ to construct a w_i-structured route set $H_{u_i} = Y_{u_i} \cup X_{u_i}$, where $w_1 = w$ and $w_2 = s^+(u_1, y(u_1))$, and extend each route $H_{u_i,j}$ to start and end at $p(v)$ instead of v, for $1 \leq i \leq 2$ and $1 \leq j \leq h(u_i)$;
Let $S \leftarrow X_{u_1} \cup X_{u_2}$;
if $y(u_1) = 1$ **then**
    Construct $H_{v,1}$ that follows $H_{u_1,1}$ and $H_{u_2,1}$ consecutively, and let $Y_v \leftarrow \{H_{v,1}\}$;
else
    Let $q \leftarrow y(u_1) - y(u_2)$, $H_{v,j} \leftarrow H_{u_1,j}$ for $1 \leq j \leq q$, and $Y_v \leftarrow \{H_{v,j} : 1 \leq j \leq q\}$;
    Construct a route A that follows $H_{u_1,y(u_1)}$ and $H_{u_2,1}$;
    if $s^+(u_2, 1) > 0$ **then**
        Let $H_{v,q+1} = A$ and $Y_v \leftarrow Y_v \cup \{H_{v,q+1}\}$;
    else
        Let $S \leftarrow S \cup \{A\}$;
        if $y(u_2) \geq 2$ **then**
            Construct a route B that follows $H_{u_i,q+1}, H_{u_2,2}, H_{u_1,q+2}, \ldots, H_{u_2,y(u_2)}$;
            Let $S \leftarrow S \cup \{B\}$ if $s^-(u_1, q+1) = s^+(u_2, y(u_2)) = 0$, and otherwise, let $Y_{v,q+1} \leftarrow B$ and $Y_v \leftarrow Y_v \cup \{H_{v,q+1}\}$;
        end if
    end if
end if
Let $Y_v \leftarrow \{H_{v,1}\}$ and $S \leftarrow S - H_{v,1}$, if $Y_v = \emptyset$;
Decide X_v based on Y_v and S in the same way as that in $ConstructCase1(v, w)$;
Return $H_v = Y_v \cup X_v$;

Lemma 5. *Case 3a: For a non-leaf $v \in V$ with $A(u_1) \geq 0$ and $A(u_2) < 0$ and $A(v) \geq 0$ where $child(v) = \{u_1, u_2\}$, if Claim 1 is true for u_1 and u_2, it is still true for v.*

Proof. Similarly to Case 1, a w_i-structured route set $H_{u_i} = Y_{u_i} \cup X_{u_i}$ can be constructed in $O(b^2(u_i)/\min(b(u_i), k))$ time for $i = 1, 2$ sequentially, where $w_1 = w = s^-(u_1, 1)$ and $w_2 = s^+(u_1, y(u_1)) = s^-(u_2, 1)$. Since $A(u_1) + A(u_2) = A(v)$ and $s^+(u_1, y(u_1)) = s^-(u_2, 1) = \alpha(A(u_1) + w) \leq k$ and $A(u_2) < 0$, we know $s^-(u_1, 1) = w$, and $s^+(u_2, y(u_2)) = [A(v) + w] \bmod k \leq k - 1$. Furthermore, since $A(u_1) + A(u_2) \geq 0$ and $w_1 - s^+(u_2, y(u_2)) \geq -(k-1)$, we can derive that $y(u_1) \geq y(u_2)$, which implies that the number of routes in Y_{u_1} is at least as large as that in Y_{u_2}. Based on this, the route set $H_v = Y_v \cup X_v$ can be constructed by Algorithm 5. It first decides Y_v and S, where the route set S contains routes that starts and ends with zero unit products, so that routes in $Y_v \cup S$ can together serve all the requests in $T(v)$, and each route in Y_v carries as many products from $T(v)$ as the vehicle can, and therefore satisfy Condition 4(b), 4(c), and 4(d) of Definition 2. Then it decides X_v, based on Y_v and S, through the same approach as that for Case 1, so that all the conditions of Definition 2 are satisfied, and

Algorithm 6 ConstructCase3b(v, w): to construct a w-structured route set $H_v = Y_v \cup X_v$ for a non-leaf v with $A(u_1) \geq 0$ and $A(u_2) < 0$ and $A(v) < 0$ where $child(v) = \{u_1, u_2\}$

For $i = 1, 2$, call $Construct(u_i, w_i)$ to construct a w_i-structured route set $H_{u_i} = Y_{u_i} \cup X_{u_i}$, where $w_1 = w$ and $w_2 = s^+(u_1, y(u_1))$, and extend each route $H_{u_i,j}$ to start and end at $p(v)$ instead of v, for $1 \leq i \leq 2$ and $1 \leq j \leq h(u_i)$;
Let $S \leftarrow X_{u_1} \cup X_{u_2}$;
if $y(u_1) = 1$ **then**
  Construct a route $H_{v,1}$ that follows $H_{u_1,1}$ and $H_{u_2,1}$ consecutively, let $H_{v,j} \leftarrow H_{u_2,j}$ for $2 \leq j \leq y(u_2)$, and $Y_v \leftarrow \{H_{v,j} : 1 \leq j \leq y(u_2)\}$;
else
  Construct a route $H_{v,1}$ that follows $H_{u_1,j}$ and $H_{u_2,1+j}$ alternatively for $j = 1, 2, ..., y(u_1) - 1$, and let $Y_v \leftarrow \{H_{v,1}\}$;
  Construct a route B that follows $H_{u_1,y(u_1)}$ and $H_{u_2,1}$ consecutively, and let $S \leftarrow S \cup \{B\}$.
  Let $H_{v,j} \leftarrow H_{u_2,j}$ for $2 \leq j \leq y(u_2)$, and $Y_v \leftarrow Y_v \cup \{H_{v,j}, 2 \leq j \leq y(u_2)\}$
end if
Decide X_v based on Y_v and S in the same way as that in $ConstructCase2(v, w)$;
Return $H_v = Y_v \cup X_v$;

$H_v = Y_v \cup X_v$ must be a w-structured route set. the time consuming for the construction of $H_v = Y_v \cup X_v$ can also been seen to be $O[b^2(v)/\min(b(v), k)]$. □

Lemma 6. *Case 3b: For a non-leaf $v \in V$ with $A(u_1) \geq 0$ and $A(u_2) < 0$ and $A(v) < 0$ where $child(v) = \{u_1, u_2\}$, if Claim 1 is true for u_1 and u_2, it is still true for v.*

Proof. The construction for Case 3b is similar to Case 3, as shown in Algorithm 6. Details are omitted for the lack of space. □

3.3 Approximation Factor

Theorem 2. *The CTSP-PD on a tree can be approximatively solved in $O[n/\min(n, k)]$ time to achieve an approximation factor of 2.*

Proof. By Lemma 2–Lemma 6, we know Claim 1 is true for all vertices $v \in V$. Hence, we can construct a 0-structured route set $H_r = Y_r \cup X_r$ in $O[n/\min(n, k)]$ time. The Y_r must contain exactly one route $H_{r,1}$. If X_r is not empty but contains an extra route $H_{r,2}$, we can construct a feasible H by following $H_{r,2}$ and $H_{r,1}$ consecutively, otherwise let $H = H_{r,1}$. Hence, H becomes a feasible solution and $L(H) = L(H_r)$. Since routes in H_r traverses each edge $e(v)$ for $v \in V$ at most $2[n(v) + 1]$ time, we have $L(H) = L(H_r) \leq \sum_{v \in V} 2[n(v) + 1]d(e(v)) \leq LB(r) + 2L(T)$, where $L(T)$ is the total edge length of the tree T. By $LB(r) \geq 2L(T)$ and Lemma 1, we have $L(H) = L(H_r) \leq 2LB(r) \leq 2L(H^*)$, which completes the proof. Remark that the approximation factor of two can be shown to be tight to our algorithm. (Refer [15] for details) □

4 Conclusion

We presented a 2-approximation algorithm for the capacitated traveling salesman problem with pickup and delivery (CTSP-PD) on a tree. Although only a unit requests is considered for each pickup or delivery point, our methods can be extended to solve multiple unit case. We also remark here that the algorithm proposed in this paper can be further adapted to optimally solve some special cases of the CTSP-PD, including the case that k is infinity or unit. However, due to the lack of space, we have omitted these extensions here but presented them in [15]. Our future research will focus on how to design approximation algorithms for the CTSP-PD on a general graph by employing the results from cases on a line and a tree.

References

1. S. Anily and J. Bramel. Approximation algorithms for the capacitated traveling salesman problem with pickups and deliveries. *Naval Research Logistics*, 56:654–670, 1999.
2. S. Anily, M. Gendreau, and G. Laporte. The swapping problem on a line. *SIAM Jounral on Computing.*, 29(1):327–335, 1999.
3. S. Anily and R. Hassin. The swapping problem. *Networks*, 22:419–433, 1992.
4. T. Asano, N. Katoh, and K. Kawashima. A new approximation algorithm for the capacitated vehicle routing problems on a tree. *Journal of Combinatorial Optimization*, 5(2):213–231, 2000.
5. C. Basnet, L.R. Foulds, and J.M. Wilson. Heuristics for vehicle routing on tree-like networks. *Journal of the Operational Research Society*, 50:627–635, 1999.
6. P. Chalasani and R. Motwani. Approximating capacitated routing and delivery problems. *SIAM J. Comput.*, 28(6):2133–2149, 1999.
7. G. N. Frederickson. Notes on the complexity of a simple transportation problem. *SIAM J. Comput.*, 22:57–61, 1993.
8. G. N. Frederickson and D. Guan. Preemptive ensemble motion planning on a tree. *SIAM J. Comput.*, 21:1130–1152, 1992.
9. G. N. Frederickson, M. S. Hecht, and C. E. Kim. Approximation algorithms for some routing problems. *SIAM J. Comput.*, 7:178–193, 1978.
10. M. R. Garey and D. S. Johnson. *Computers and Intractability – A Guide to the Theory of NP-Completeness.* Freeman, San Francisco, 1979.
11. Shinya Hamaguchi and Naoki Katoh. A capacitated vehicle routing problem on a tree. *Proc. of ISAAC98, Lecture Notes in Computer Science*, pages 397–407, 1998.
12. H. Hernandez-Perez and J.J. Salazar-Gonzalez. Heuristics for the one-commodity pickup-and-delivery traveling salesman problem. *Transportation Science*, 38(2):245–255, 2004.
13. Y. Karuno, H. Nagamochi, and T. Ibaraki. Vehicle scheduling on a tree with release and handling times. *Proc. of ISAAC93, Lecture Notes in Computer Science*, pages 486–495, 1993.
14. M. Labbe, G. Laporte, and H. Mercure. Capacitated vehicle routing problems on trees. *Operations Research*, 39(2):616–622, 1991.
15. A. Lim, F. Wang, and Z. Xu. A 2-approximation algorithm for capacitated traveling salesman problem with pickup and delivery on a tree (complete version), Sept. 2005. Working Paper, Hong Kong University of Science and Technology.

Distance Labeling in Hyperbolic Graphs

Cyril Gavoille* and Olivier Ly

LaBRI – Bordeaux University

Abstract. A graph G is δ-hyperbolic if for any four vertices u, v, x, y of G the two larger of the three distance sums $d_G(u,v) + d_G(x,y), d_G(u,x) + d_G(v,y), d_G(u,y) + d_G(v,x)$ differ by at most δ, and the smallest $\delta \geqslant 0$ for which G is δ-hyperbolic is called the hyperbolicity of G.

In this paper, we construct a distance labeling scheme for bounded hyperbolicity graphs, that is a vertex labeling such that the distance between any two vertices of G can be estimated from their labels, without any other source of information. More precisely, our scheme assigns labels of $O(\log^2 n)$ bits for bounded hyperbolicity graphs with n vertices such that distances can be approximated within an additive error of $O(\log n)$. The label length is optimal for every additive error up to n^{ε}. We also show a lower bound of $\Omega(\log \log n)$ on the approximation factor, namely every s-multiplicative approximate distance labeling scheme on bounded hyperbolicity graphs with polylogarithmic labels requires $s = \Omega(\log \log n)$.

Keywords: Distance queries, distance labeling scheme, hyperbolic graphs.

1 Introduction

It is well-known that a metric space (V, d) embeds into a tree metric if and only if the 4-point condition holds, that is, for any 4 points u, v, x, y of V the two larger of the sums $d(u,v) + d(x,y), d(u,x) + d(v,y), d(u,y) + d(v,x)$ are equals [1]. More generally, if the two larger sums differ by at most δ, then the metric space is said to be *δ-hyperbolic*. Introduced by Gromov [21, 20], δ-hyperbolic spaces arise naturally in the area of geometric group theory. In a certain extend hyperbolicity measures the deviation from tree-likeness. And thus, it appears in a natural way as a generalization of the study of trees in metric graph theory [4, 5, 13], classification theory [8], phylogenetic analysis [29], and Gauber dynamics also known as Gibbs samplers [25].

A graph $G = (V, E)$ is *δ-hyperbolic* if (V, d_G) is a δ-hyperbolic metric space, where d_G is the shortest-path metric of G, associating to each pair of vertices the length of a shortest path connecting them. The *hyperbolicity* of G is the smallest $\delta \geqslant 0$ for which G is δ-hyperbolic. The graphs considered in this paper are unweighted, simple, and connected.

0-hyperbolic graphs are precisely the block graphs [5, 12, 24], i.e., graphs in which every 2-connected subgraph is a clique, and chordal graphs, i.e., the graphs containing no induced cycles of length larger than three, are 2-hyperbolic [7]. It is not difficult to see from the definition that graphs of diameter D are $(2\lfloor D/2 \rfloor)$-hyperbolic.

* Supported by the project "PairAPair" of the ACI Masses de Données.

X. Deng and D. Du (Eds.): ISAAC 2005, LNCS 3827, pp. 1071–1079, 2005.

1-hyperbolic graphs have been partially characterized in [26], and recently a full characterization has been given in terms of a convexity condition and forbidden isometric subgraphs [3].

This paper deals with the problem of the distance computation and distributed abilities of δ-hyperbolic graphs. Commonly, when we make a query concerning a set of nodes in a graph (adjacency, distance, connectivity, etc.), we need to make a global access to the structure. In our approach, the compromise is to store the maximum of information in a label associated with a vertex to have directly what we need with a local access. Motivation of localized data-structures in distributed computing is survey and widely discussed in [18].

We are especially interested in the distance labeling problem, introduced by Peleg in [30]. The problem consists in labeling the vertices of a graph in order to compute or estimate the distance between any two of its vertices x and y using only the information stored in the labels of x and y, without any other source of information. The main parameters taken into account when designing a solution is the maximum label length (in bits) assigned by the labeling. More formally, an (s,r)*-approximate distance labeling scheme* on a given graph family $\mathfrak{F}$ is a pair $\langle L, f\rangle$, L is called the *labeling* function and f the *distance decoder*, such that, for every $G \in \mathfrak{F}$ and for all $x, y \in V(G)$: $L(x,G) \in \{0,1\}^*$, and $d_G(x,y) \leqslant f(L(x,G), L(y,G)) \leqslant s \cdot d_G(x,y) + r$. If $s = 1$ and $r = 0$, then we shortly deal with a distance labeling scheme (or DLS). Also, an $(s,0)$-approximate DLS is called s-multiplicative, and a $(1,r)$-approximate DLS is called r-additive.

Related works for distance labeling. The main results on the field are that general graphs support an (exact) distance labeling scheme with labels of $O(n)$ bits [19], and that trees [2, 30], bounded tree-width graphs [19], distance-hereditary graphs [16], bounded clique-width graphs [10], some non-positively curved plane graphs [9], all support distance labeling schemes with $O(\log^2 n)$ bit labels. Since 0-hyperbolic graphs are block graphs, which are distance-hereditary, it follows that this class supports a $O(\log^2 n)$ bit label DLS.

The $O(n)$ bit upper bound is tight for general graphs (simply by counting the number of n-vertex graphs), and a lower bound of $\Omega(\log^2 n)$ bit on the label length is known for trees [19], implying that all the results mentioned above (including 0-hyperbolic graphs) are tight as well since all of them contains trees. Recently, [17, 6] showed an optimal bound of $O(\log n)$ bits for interval graphs, permutation graphs, and their generalizations (circular-arc graphs and cicurlar permutation graphs).

Other results concern approximated distance labeling schemes. For arbitrary graphs, the best scheme in date is due to Thorup and Zwick [34]. They propose a $(2k-1)$-multiplicative DLS, for each integral parameter $k \geqslant 1$, with labels of $O(n^{1/k}\log^2 n)$ bits. Moreover, $\Omega(n^{1/k})$ bit labels are required in the worst-case for every s-multiplicative DLS with $s < 2k+1$. In fact, this result relies to a 1963 girth conjecture of Erdös [14] proved for $k = 1, 2, 3$ and 5. However, for all the other values of k, the results of [27] implies that the $\Omega(n^{1/k})$ lower bound is true for $s < 4k/3 + 2$.

In [15], it is proved that trees (and bounded tree-width graphs as well) enjoy a $(1 + 1/\log n)$-multiplicative DLS with labels of $O(\log n \cdot \log\log n)$ bits, and this is tight in terms of label length and approximation. They also design some $O(1)$-additive

DLS with $O(\log^2 n)$ bit labels for several families of graphs including: the graphs with bounded longest induced cycle, and, more generally, the graphs of bounded tree-length, i.e., that admit a Robertson-Seymour tree-decomposition in bags of bounded diameter (see [11]). Interestingly, it is easy to show that every exact DLS for these families of graphs needs labels of $\Omega(n)$ bits in the worst-case (e.g., considering some chordal graphs, namely the split graphs [15]).

Recently, the graphs with doubling dimension α have been considered, i.e., the graphs for which, for every r, each ball of radius $2r$ can be covered by at most 2^α balls of radius r. It generalizes Euclidian metrics and bounded growth graphs, and includes many realistic networks. After several successive improvements [22, 32, 28], the best scheme in date, due to Slivkins [31], is a $(1+\varepsilon)$-multiplicative DLS with $O(\varepsilon^{-O(\alpha)} \log n \cdot \log\log n)$ bit labels. This is optimal for bounded α by combining the results of [28] and the lower bound of [15] for trees. Finally, in [33], it is shown that planar graphs enjoy a $(1+\varepsilon)$-multiplicative DLS with labels of $O(\varepsilon^{-1} \log^3 n)$ bits (see also [23]).

Our results. From the above list of results, it is clear that 0-hyperbolic graphs enjoy an (exact) DLS with $O(\log^2 n)$ bit labels, and that moreover every DLS for 2-hyperbolic graphs requires some labels of $\Omega(n)$ bits. Again, some chordal graphs, that are all 2-hyperbolic, require $\Omega(n)$ bit labels [15].

Our first contribution is a lower bound on an s-multiplicative DLS for bounded hyperbolicity graphs. We construct a family of bounded hyperbolic graphs for which, for every integer $k \geqslant 1$, every s-multiplicative DLS with $s < 2\log k + O(1)$ requires some labels of $\Omega(n/\log k)^{1/k}$ bits. In particular, for $k = \Theta(\log n/\log\log n)$, it implies that any s-multiplicative DLS using labels of any poly-logarithmic length requires $s = \Omega(\log\log n)$.

On the positive side, we construct for δ-hyperbolic graphs an $\delta\log n$-additive DLS with labels of $O(\log^2 n)$ bits. The label length is optimal since every r-additive DLS for trees, and thus for δ-hyperbolic graphs for every $\delta \geqslant 0$, requires $\Omega(\log^2(n/r))$ bit labels [19]. In the full version, we show that any poly-log label DLS for bounded hyperbolic graphs requires $r = \Omega(\log n)$, proving the optimality of the approximation of our scheme.

Due to the lack of space, proofs appear in the full version.

2 Pyramidal Construction

Our lower bound combines several ingredients. First we show how to construct from any graph G a graph P, called the *pyramid* of G, such that: 1) G is a subgraph of P; 2) P has bounded hyperbolicity (i.e., bounded by some constant independent of G); and 3) $d_P(x,y) \geqslant 2\log d_G(x,y) - O(1)$ for all x,y in G. In particular the girth of P is at least the log of the girth of G.

Let G be a graph. Let D denote the diameter of G. Let us consider $\lceil \log D \rceil + 1$ copies of G denoted by $G_0, G_1, \ldots, G_{\lceil \log D \rceil}$. Let us consider a new graph constructed as follows: we start from the disjoint union of the G_i's. We add some new edge as follows: First, for any vertex v of G, let us denote v^i the copy of v in G_i. For any $i = 0, \ldots, \lceil \log D \rceil - 1$, let us add an edge between v^i and v^{i+1}. Such an edge shall

be said to be vertical. Second, by induction, let us add a new edge between any two vertices of G_i, let say v_1^i and v_2^i, if their copies in G_{i-1}, v_1^{i-1} and v_2^{i-1}, are at distance 2. Such an edge but also any orginal edge of some G_i's shall be said to be transversal. The graph that we obtain is denoted by $P(G)$, it is called the *pyramid* graph of G, G_0 is called the *base* of $P(G)$.

Lemma 1. *There exists a constant K such that for any G, the hyperbolicity of $P(G)$ is at most K.*

Geodesics of $P(G)$. Here we consider the shape of geodesics of $P(G)$ in order to prove that $d_{P(G)}(x, y) \geqslant 2\log d_G(x, y) - O(1)$ for all x,y in G. The successive steps of this study are presented here along the following propositions. For any vertex v of $P(G)$, let us call the *height* of v the unique i such that $v \in G_i$, it shall be denoted by $h(v)$. For any two vertices v_1 and v_2 of the same height h, we denote $d_{G_h}(v_1, v_2)$ the distance between v_1 and v_2 in the subgraph of $P(G)$ generated by the vertices of G_h. We denote by D_{G_h} the maximum of d_{G_h}. If p is a geodesic, i.e., a shortest path, then $\ell(p)$ denotes its length.

Proposition 1. *Let v_1 and v_2 be two vertices of G. Then $d_{G_h}(v_1^h, v_2^h) = \lceil d_G(v_1, v_2)/2^h \rceil$. In particular $D_{G_h} \leqslant \lceil D/2^h \rceil$.*

Proposition 2. *Let p be a geodesic of $P(G)$ which only uses transversal edges. Then $\ell(p) \leqslant 5$.*

Let us consider a path $p = v_0 v_1 \ldots v_t$ of length t. Let us consider the sequence of respective heights : $h_0 h_1 \ldots h_t$. We say that p is *increasing* (resp. *decreasing*) if the sequence of heights is increasing (resp. decreasing).

Proposition 3. *Let $p = v_0 v_1 \ldots v_t$ be a geodesic of $P(G)$. Let us assume that $h(v_0) \geqslant h(v_t)$. Then there exists a vertex v_i of p such that $v_0 \ldots v_i$ is increasing and $v_i \ldots v_t$ is decreasing.*

We consider a special kind of geodesic that we call *straight* geodesic. These are those having the following shape: first, it starts by using a sequence of vertical edges; second, it carries on by a sequence of transversal edges; and finally it uses a sequence of vertical edges.

Proposition 4. *For any geodesic p, there exists a straight geodesic p' with same extremities. Moreover, p is totally included into a 5-neighbourhood of p' and conversely.*

Proposition 5. *Let x and y be two vertices of $P(G)$. Let p be such a straight geodesic between x and y. Let us assume that $h(x) \leqslant h(y)$. Let x' be the copy of x in $G_{h(y)}$. Let h be the minimal of the lengths of the vertical parts of p Then $\log(d_{G_{h(y)}}(x', y)) - 3 \leqslant h \leqslant \log(d_{G_{h(y)}}(x', y)) - 1$*

The following proposition compares distances of $P(G)$ with those of the G_i's.

Proposition 6. *Let x and y be two vertices of $P(G)$ with $h(x) \leqslant h(y)$. Let x' be the copy of x in $G_{h(y)}$. Then $h(y) - h(x) + 2\log(d_{G_{h(y)}}(x', y)) - 3 \leqslant d_{P(G)}(x, y) \leqslant h(y) - h(x) + 2\log(d_{G_{h(y)}}(x', y)) + 4$.*

In particular, for all $x, y \in G$, $2\log(d_G(x, y)) - 3 \leqslant d_{P(G)}(x, y) \leqslant 2\log(d_G(x, y)) + 4$.

Proposition 7. *If p and p' are two geodesics with same extremities, then p is totally included into a 11-neighbourhood of p' and conversely.*

Sketch of the Proof of Lemma 1. Let us be given with 3 vertices x, y and z of $P(G)$ (see Fig. 1). We consider 3 geodesics p_{xy}, p_{yz} and p_{xz} connecting respectively x and y, y and z, and x and z. By the criterion of Rips (cf. [20]), it suffices to show that there exists a constant K', independent of x, y and z, such that p_{xz} is included into the K'-neighbourhood of $p_{xy} \cup p_{yz}$. First, let us assume that p_{xy}, p_{yz} and p_{xz} are straight.

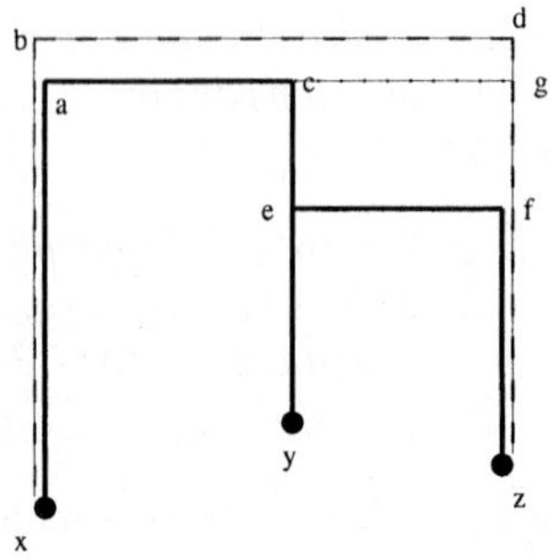

Fig. 1. Rips's Criterion

We claim that in this case p_{xz} is included into a 5-neighbourhood of $p_{xy} \cup p_{yz}$. Let us consider the notations indicated in Figure 1. Let us look at vertices of p_{xz} case by case:

- Vertices of p_{xz} which are located between x and a belong also to p_{xy}.
- Without loss of generality, let us suppose that p_{xy} is higher than p_{yz}. Vertices between a and b are at distance at most 3 from p_{xy}. Indeed, if a is higher than b, it is true seeing that the segment ab is totally included into p_{xy}. If b is higher than a, one can verify the previous claim by applying Proposition 5.
- By Proposition 2, vertices between b and d are at distance at most 5 from b, and therefore at most 8 from a.
- Vertices between d and g are within a distance at most 3 from d and therefore at most 11 from a.
- Vertices between g and f are at most at distance 5 from the segment ec, because of the length of ef which is at most 5.
- Finally, vertices between f and z belong to p_{yz}.

We conclude that p_{xz} is totally included into the 11-neighbourhood of $p_{xy} \cup p_{yz}$.

The general case where p_{xy}, p_{yz} and p_{xz} are not straight can be obtained from the above discussion by applying Proposition 4: we get that in general, p_{xz} is included into the 21-neighbourhood of $p_{xy} \cup p_{yz}$. □

3 Distance Labeling Lower Bound

We consider the conjecture of Erdös according to which for any pair of integers $k \geqslant 1$ and $n \geqslant 1$, the maximal number of edges of a graph of girth $2k + 2$ with n vertices is

$\Omega(n^{1+1/k})$ (see [14]). It is true for $k = 1, 2, 3, 5$; it is also true if we consider graphs of girth $4k/3 + 3$ (see [27]). In the following, for any k and n we shall consider a graph $G_{n,k}$ of girth $4k/3 + 3$ with n vertices and with maximal number of edges equal to $\Omega(n^{1+1/k})$.

We consider subgraphs defined by subsets of edges: given a graph G, a subset E of edges of G defines a subgraph H whose vertices are the vertices of G and whose edges are the elements of E.

Proposition 8. *Let us fix $k \geqslant 1$ and $n \geqslant 1$, and let us consider a subgraph H of $G_{n,k}$. Let us consider $P(H)$ the pyramid graph of H, and a pair (x, y) made of two vertices of the base of $P(H)$ which are connected by an edge in $G_{n,k}$. Then either $d_{P(H)}(x, y) = 1$ or $d_{P(H)}(x, y) \geqslant 2\log(4k/3 + 2) - 3$.*

Theorem 1. *For $n \geqslant 1$ and $k \geqslant 1$, there exists a family $\mathfrak{F}_{n,k}$ of graphs of bounded hyperbolicity with $O(n \log k)$ vertices for which every (s, r)-approximated distance labeling scheme such that $s + r < 2\log(4/3k + 2) - 3$ requires labels of $\Omega(n^{1/k})$ bits.*

In particular, for $k = \Theta(\log n/ \log\log n)$, every s-multiplicative DLS on n-vertex bounded hyperbolic graphs with poly-log label length requires $s = \Omega(\log\log n)$.

Proof. Let us consider the family $\mathfrak{F}_{n,k}$ of the pyramid graphs of the connected subgraphs of $G_{n,k}$. By maximality of the number of edges, it is not difficult to see that $G_{n,k}$ has diameter $O(k)$. We restrict ourself to connected subgraphs of diameter $O(k)$ by fixing some shortest path spanning tree in $G_{n,k}$. Observe that pyramid graphs that we obtain have $O(n \log k)$ vertices. By Lemma 1, $\mathfrak{F}_{n,k}$ is of bounded hyperbolicity. Let us be given with an (s, r)-approximated distance labeling scheme $\langle L, f\rangle$ for $\mathfrak{F}_{n,k}$.

For each $H \in \mathfrak{F}_{n,k}$, let us denote by S_H the word $L(1, H)\#L(2, H)\# \ldots \#L(n, H)$ obtained by concatenation of the labels of all the vertices of its base. We suppose that the vertex set of $G_{n,k}$ is $\{1, 2, \ldots, n\}$. Besides, we use a special symbol $\#$ as delimiter.

Let us assume that $\max_{H\in\mathfrak{F}_{n,k}, x\in V(H)}\{|L(x, H)|\} < c \cdot n^{1/k}$ for some constant $c > 0$. It follows that the number of words for $\mathfrak{F}_{n,k}$ is at most $2^{c\cdot n^{1+1/k}}$. Because $|\mathfrak{F}_{n,k}| = 2^{|E(G_{n,k})-(n-1)|} \geqslant 2^{c'\cdot n^{1+1/k}}$ for some suitable constant $c' > 0$. This implies, for $c < c'$ that there exists a pair H_1 and H_2 of distinct graphs of $\mathfrak{F}_{n,k}$ such that L does not distinguish H_1 and H_2, i.e., $S_{H_1} = S_{H_2}$. Let us choose a pair of vertices (x, y) of $G_{n,k}$ such that (x, y) is an edge of the base of H_1 but not of the base of H_2. If such a pair does not exist, we exchange H_1 and H_2. If we cannot find such a pair, this means that $H_1 = H_2$ which is a contradiction.

$S_{H_1} = S_{H_2}$ implies $L(x, H_1) = L(x, H_2)$ and $L(y, H_1) = L(y, H_2)$; and thus $f(L(x, H_1), L(y, H_1)) - f(L(x, H_2), L(y, H_2))$.

Besides, by definition of $\langle L, f\rangle$, we have $d_{H_1}(x, y) \leqslant f(L(x, H_1), L(y, H_1) \leqslant s \cdot d_{H_1}(x, y) + r$ and $d_{H_2}(x, y) \leqslant f(L(x, H_2), L(y, H_2)) \leqslant s \cdot d_{H_2}(x, y) + r$.

All together we get $d_{H_2}(x, y) \leqslant s \cdot d_{H_1}(x, y) + r$. But $d_{H_1}(x, y) = 1$ by assumption, and $d_{H_2}(x, y) \geqslant 2\log(4k/3 + 2) - 3$ by Proposition 8. Finally we get $s + r \geqslant 2\log(4k/3 + 2) - 3$.

By contraposition, we have thus proved that for any k and any n, $s + r < 2\log(4k/3 + 2) - 3$ implies that $\max_{H \in \mathfrak{F}_{n,k}, x \in V(H)}\{|L(x, H)|\} \geqslant c \cdot n^{1/k}$. □

4 Tree Approximation and Distance Labeling

This section is devoted to the proof of Theorem 2. It is based on the classical result about approximation of hyperbolic metric spaces by real trees (cf. e.g. [20–Thm. 12, p. 33]. We set up a combinatorial version of this result based on the same method of proof.

We use the characterization of hyperbolicity in terms of Gromov product. Let G be a connected finite graph. Let x, y and w be vertices of G. One defines the *Gromov product* of x and y regarding w to be $(x|y)_w = \frac{1}{2}(|x - w| + |y - w| - |x - y|)$ where $|u - v|$ denotes $d_G(u, v)$. Let G be a connected undirected finite graph. Then the hyperbolicity of G is equal to $2\max_{x,y,z,w \in G}\{\min\{(x|z)_w, (z|y)_w\} - (x|y)_w\}$ (see [20]).

Proposition 9. *Let X be a finite 0-hyperbolic metric space with integral distances; let D be the diameter of X. Then there exists a mapping $\sigma : X \longrightarrow T$ where T is a tree of at most $2(|X| - 1) \cdot D$ nodes such that for any pair (x, y) of elements of X, $d_T(x, y) = 2d_X(x, y)$.*

Let G be a connected undirected finite graph. Let us fix a vertex w_0 of G. In the following, $|x - w_0|$ shall be denoted by $|x|$ for any vertex x of G, it shall be called the length of x (regarding w_0). Following [20], let us define $(x|y)' = \max\{\min_{2 \leqslant j \leqslant \ell}\{(x_{j-1}|x_j)_{w_0}\}$ where $x_1, \ldots, x_\ell$ denotes any sequence of vertices. And from this, let $|x - y|' = |x| + |y| - 2(x|y)'$.

Lemma 2. *Let δ be the hyperbolicity of G. Then for any pair of vertices x and y of G, we have $|x - y| - \delta\log n \leqslant |x - y|' \leqslant |x - y|$*

Then we consider the equivalence relation defined by $x \equiv y$ if and only if $|x - y|' = 0$. And the metric space whose elements are $G/\equiv$ provided with the distance $d'([x]_\equiv, [y]_\equiv) = |x - y|'$. We have the following property:

Lemma 3. *$(G/\equiv, d')$ is a 0-hyperbolic metric space.*

Theorem 2. *The family of δ-hyperbolic graphs with n vertices have a $\delta\log n$-additive distance labeling scheme with $O(\log^2 n)$ bit labels.*

Proof. Let us be given with G a δ-hyperbolic graph with n vertices. We consider the mapping chain $G \xrightarrow{\pi} G/\equiv \xrightarrow{\sigma} T$ where $G \xrightarrow{\pi} G/\equiv$ is defined as above and $G/\equiv \xrightarrow{\sigma} T$ as in Proposition 9 (let us recall that $G/\equiv$ is 0-hyperbolic).

Since T is a tree, there exists an exact distance labeling scheme $\langle L_T, f_T\rangle$ using labels of length $0(\log^2|T|)$ (cf. [19]). By Proposition 9, $|T| \leqslant 2(n-1)^2$ because $|G/\equiv| \leqslant n$. So, labels used by $\langle L_T, f_T\rangle$ are of length $O(\log^2 n)$. Besides we have $|x - y| - \delta\log n \leqslant \frac{1}{2}d_T(\sigma \circ \pi(x), \sigma \circ \pi(y)) \leqslant |x - y|$.

Finally, let us define $L(x, G) = L_T(\sigma \circ \pi(x))$ and $f(\ell_1, \ell_2) = \frac{1}{2}f_T(\ell_1, \ell_2) + \delta\log n$. Then $\langle L, f\rangle$ satisfies the conditions of the Theorem. □

References

1. Richa Agarwala, Vineet Bafna, Martin Farach, Mike Paterson, and Mikkel Thorup. On the approximability of numerical taxonomy (fitting distances by tree metrics). In 7th *Symposium on Discrete Algorithms (SODA)*, pages 365–372. ACM-SIAM, January 1996.
2. Stephen Alstrup, Philip Bille, and Theis Rauhe. Labeling schemes for small distances in trees. In 14th *Symposium on Discrete Algorithms (SODA)*, pages 689–698. ACM-SIAM, January 2003.
3. Hans-Jürgen Bandelt and Victor D. Chepoi. 1-hyperbolic graphs. *SIAM Journal on Discrete Mathematics*, 16(2):323–334, 2003.
4. Hans-Jürgen Bandelt, A. Henkmann, and F. Nicolai. Powers of distance-hereditary graphs. *Discrete Mathematics*, 145:37–60, 1995.
5. Hans-Jürgen Bandelt and Henry Martyn Mulder. Distance-hereditary graphs. *Journal of Combinatorial Theory, Series B*, 41:182–208, 1986.
6. Fabrice Bazzaro and Cyril Gavoille. Localized and compact data-structure for comparability graphs. Research Report RR-1343-05, LaBRI, University of Bordeaux 1, 351, cours de la Libération, 33405 Talence Cedex, France, February 2005.
7. Gunnar Brinkmann, Jack H. Koolen, and Vincent Moulton. On the hyperbolicity of chordal graphs. *Annals of Combinatorics*, 5(1):61–65, 2001.
8. Peter Buneman. The recovery of trees from measures of dissimilarity. *Mathematics in Archaeological and Historical Sciences*, pages 387–395, 1971.
9. Victor D. Chepoi, Feodor F. Dragan, and Yann Vaxes. Distance and routing labeling schemes for non-positively curved plane graphs. *Journal of Algorithms*, 2005. To appear.
10. Bruno Courcelle and Rémi Vanicat. Query efficient implementation of graphs of bounded clique-width. *Discrete Applied Mathematics*, 131:129–150, 2003.
11. Yon Dourisboure and Cyril Gavoille. Tree-decomposition of graphs with small diameter bags. In J. Fila, editor, 2nd *European Conference on Combinatorics, Graph Theory and Applications (EUROCOMB)*, pages 100–104, September 2003.
12. Andreas W. M. Dress, Vincent Moulton, and Michael A. Steel. Trees, taxonomy, and strongly compatible multi-state characters. *Advances in Applied Mathematics*, 19:1–30, 1997.
13. Andreas W. M. Dress, Vincent Moulton, and Werner Terhalle. T-theory: an overview. *European Journal of Combinatorics*, 17:161–175, 1996.
14. Paul Erdös. Extremal problems in graph theory. In *Publ. House Cszechoslovak Acad. Sci., Prague*, pages 29–36, 1964.
15. Cyril Gavoille, Michal Katz, Nir A. Katz, Christophe Paul, and David Peleg. Approximate distance labeling schemes. In F. Meyer auf der Heide, editor, 9th *Annual European Symposium on Algorithms (ESA)*, volume 2161 of Lecture Notes in Computer Science, pages 476–488. Springer, August 2001.
16. Cyril Gavoille and Christophe Paul. Distance labeling scheme and split decomposition. *Discrete Mathematics*, 273(1-3):115–130, 2003.
17. Cyril Gavoille and Christophe Paul. Optimal distance labeling schemes for interval and circular-arc graphs. In G. Di Battista and U. Zwick, editors, 11th *Annual European Symposium on Algorithms (ESA)*, volume 2832 of Lecture Notes in Computer Science, pages 254–265. Springer, September 2003.
18. Cyril Gavoille and David Peleg. Compact and localized distributed data structures. *Journal of Distributed Computing*, 16:111–120, May 2003. PODC 20-Year Special Issue.
19. Cyril Gavoille, David Peleg, Stéphane Pérennès, and Ran Raz. Distance labeling in graphs. *Journal of Algorithms*, 53(1):85–112, 2004.
20. Etienne Ghys and Pierre de La Harpe. *Sur les Groupes Hyperboliques d'après Mikhael Gromov*. Birkhäuser, 1990.

21. Mikhael Gromov. Hyperbolic groups. *Essays in Group Theory*, Mathematical Sciences Research Institute Publications, 8:75–263, 1987.
22. Anupam Gupta, Robert Krauthgamer, and James R. Lee. Bounded geometries, fractals, and low-distortion embeddings. In 44th *Annual IEEE Symposium on Foundations of Computer Science (FOCS)*, pages 534–543. IEEE Computer Society Press, October 2003.
23. Anupam Gupta, Amit Kumar, and Rajeev Rastogi. Traveling with a pez dispenser (or, routing issues in mpls). *SIAM Journal on Computing*, 34(2):453–474, 2005.
24. Edward Howorka. On metric properties of certain clique graphs. *Journal of Combinatorial Theory, Series B*, 27:67–74, 1979.
25. Haim Kaplan and Tova Milo. Parent and ancestor queries using a compact index. In 20th *ACM Symposium on Principles of Database Systems (PODS)*. ACM-SIAM, May 2001.
26. Jack H. Koolen and Vincent Moulton. Hyperbolic bridged graphs. *European Journal of Combinatorics*, 23(6):683–699, 2002.
27. Felix Lazebnik, Vasiliy A. Ustimenko, and Andrew J. Woldar. A new series of dense graphs of high girth. *Bulletin of the American Mathematical Society (New Series)*, 32(1):73–79, January 1995.
28. Manor Mendel and Sariel Har-Peled. Fast construction of nets in low dimensional metrics, and their applications. In 21st *Annual ACM Symposium on Computational Geometry (SoCG)*, pages 150–158, 2005.
29. Vincent Moulton and Michael A. Steel. Retractions of finite distance functions onto tree metrics. *Discrete Applied Mathematics*, 91:215–233, 1999.
30. David Peleg. Proximity-preserving labeling schemes. *Journal of Graph Theory*, 33:167–176, 2000.
31. Aleksandrs Slivkins. Distance estimation and object location via rings of neighbors. In 24th *Annual ACM Symposium on Principles of Distributed Computing (PODC)*, pages 41–50. ACM Press, 2005.
32. Kunal Talwar. Bypassing the embedding: Algorithms for low dimensional metrics. In 36th *Annual ACM Symposium on Theory of Computing (STOC)*, pages 281–290, June 2004.
33. Mikkel Thorup. Compact oracles for reachability and approximate distances in planar digraphs. *Journal of the ACM*, 51(6):993–1024, November 2004.
34. Mikkel Thorup and Uri Zwick. Approximate distance oracles. *Journal of the ACM*, 52(1): 1–24, January 2005.

Multi-source Trees: Algorithms for Minimizing Eccentricity Cost Metrics

Paraskevi Fragopoulou[1], Stavros D. Nikolopoulos[2], and Leonidas Palios[2]

[1] Department of Applied Informatics and Multimedia,
Technological Educational Institute of Crete,
P.O.Box 1939, GR-71004 Heraklion-Crete, Greece
fragopou@epp.teiher.gr

[2] Department of Computer Science, University of Ioannina,
GR-45110 Ioannina, Greece
{stavros, palios}@cs.uoi.gr

Abstract. We consider generalizations of the k-source sum of vertex eccentricity problem (k-SVET) and the k-source sum of source eccentricity problem (k-SSET) [1], which we call SDET and SSET, respectively, and provide efficient algorithms for their solution. The SDET (SSET, resp.) problem is defined as follows: given a weighted graph G and sets S of source nodes and D of destination nodes, which are subsets of the vertex set of G, construct a tree-subgraph T of G which connects all sources and destinations and minimizes the SDET cost function $\sum_{d \in D} max_{s \in S}\, d_T(s,d)$ (the SSET cost function $\sum_{s \in S} \max_{d \in D} d_T(s,d)$, respectively). We describe an $O(nm \log n)$-time algorithm for the SDET problem and thus, by symmetry, to the SSET problem, where n and m are the numbers of vertices and edges in G. The algorithm introduces efficient ways to identify candidates for the sought tree and to narrow down their number to $O(m)$. Our algorithm readily implies $O(nm \log n)$-time algorithms for the k-SVET and k-SSET problems as well.

Keywords: Multi-source trees, eccentricity, weighted graphs, networks, communication, algorithms, complexity.

1 Introduction

The work in this paper is motivated by problems in collective communication on networks modeled by graphs. In the general case, a group of network nodes, defined as the sources, wish to consistently transmit information to another group of network nodes, the destinations. This type of collective communication is served by the establishment of a tree T connecting the sources to the destinations which minimizes certain criteria in order to guarantee efficient communication. The criteria to optimize are diverse and various cases have been considered in the literature [1, 3, 6, 7, 9]. The main reason is that different applications pose different cost requirements; note also that some of these cost requirements lead to intractable problems in general graphs. The problem of collective communication from a single source node is a well studied problem [4, 5, 8, 10]. Multiple sources have also been considered [3, 6], although to a lesser degree.

X. Deng and D. Du (Eds.): ISAAC 2005, LNCS 3827, pp. 1080–1089, 2005.

In its general form, the construction of optimum communication spanning trees was initiated in [4] where the problem was defined on the complete graph with a length and a requirement on its edges; the cost measure that had to be minimized was the sum of vertex distances weighted by the requirements between all vertices of the graph (network). By setting the requirement equal to 0 and parameterizing the number of source nodes, we obtain the k-SPST problem or *sum of distances from every source to every destination*, which was studied in [3] and was shown to be NP-complete. Some exact solutions were provided for the 2-source SPST problem on restricted classes of graphs, such as unicycles and cactuses [3]. Furthermore, approximation algorithms for both the 2-source SPST problem on general graphs [3] and the all source SPST problem [10] are available. In [9], heuristic algorithms were given for the minimum partial spanning trees taking into consideration the delay between a single source node and a group of destination nodes, while trying to bound the maximum difference in these delays. Another problem, the k-MEST problem, is defined in terms of the *maximum distance from a source to a destination*, known as the *maximum eccentricity*. This cost function was studied in [3] for some special types of graphs; in [6], the problem in its general form (arbitrary sets of source and destination nodes) was shown to be tractable and an efficient polynomial algorithm which for a graph on n vertices runs in $O(n^3)$ time was given. The same was independently established in [7] considering all graph vertices as destinations via an $O(n^3 + nm \log n)$-time algorithm, where m is the number of edges of the graph.

The k-SPST problem takes into consideration all source-destination distances but defines an intractable problem, whereas the k-MEST problem takes into consideration only the maximum source-destination distance and thus does not give any indication for the distances for the rest of the source-destination pairs. Two cost functions that fill the gap between these two extreme cases were introduced in [1]: the k-SVET cost function or *sum of vertex eccentricities spanning tree* is defined as the sum of distances from each destination to its most distant graph vertex in the constructed spanning tree, and the k-SSET cost function or *sum of source eccentricities spanning tree* is defined as the sum of distances from each source to its most distant vertex in the constructed tree.

In this paper, we consider the following generalizations of the k-SVET and the k-SSET problems: for a set of source nodes and a set of destination nodes, which are arbitrary subsets of the vertex set of a graph, we are interested in minimizing the sum of distances from each destination (source, respectively) to its most distant source (destination, respectively) node in the constructed tree. Under this generalization, the two problems, the generalized k-SVET and k-SSET, become symmetric (simply exchange S and D) and thus the results derived for one of them apply to the other in a straightforward manner. We call the generalized k-SVET problem *SDET* and the generalized k-SSET problem *SSET*. We derive an $O(nm \log n)$-time algorithm for the SDET problem and thus, by symmetry, to the SSET problem, where n and m are the numbers of vertices and edges of the given graph (network). The algorithm introduces efficient ways to identify candidates for the sought tree and to narrow down their

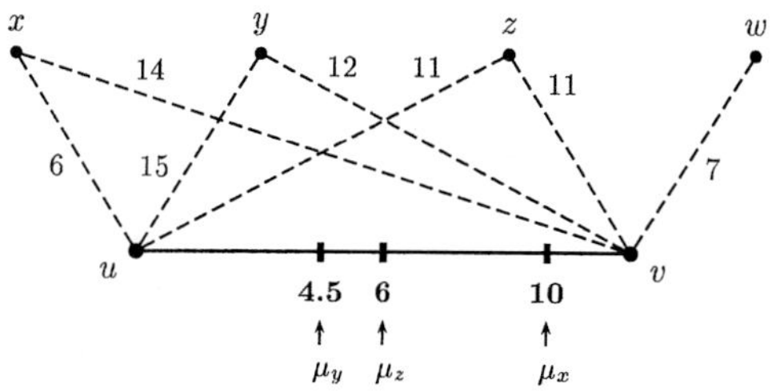

Fig. 1. The midpoints of the vertices x, y, z on an edge uv of length 12

number to $O(m)$. Our algorithm readily implies $O(nm \log n)$-time algorithms for the k-SVET and k-SSET problems as well.

2 Theoretical Framework

Let G be a simple weighted graph which models a network and has vertex set $V(G)$, edge set $E(G)$, and non-negative symmetric weights on the edges, and let $S, D \subseteq V(G)$ be the set of source nodes and the set of destination nodes, respectively; the sets S and D do not need to be disjoint. By $d(u,v)$ we denote the length (weight) of the shortest path from vertex u to vertex v in the graph G, and by $d_T(u,v)$ the length (weight) of the path from node u to node v in a spanning tree T of G.

Let uv be an edge of G. A *point* on uv is either a vertex-endpoint or a location on uv. For any two points α, β on uv, we denote by $\ell(\alpha,\beta)$ the length from α to β along uv. Thus, the length (weight) of the edge uv is denoted by $\ell(u,v)$. (Note that $d(u,v) \le \ell(u,v)$, although $\ell(u,v), d(u,v)$ are not necessarily equal.) With respect to the edge uv, we partition the vertex set $V(G)$ as follows:

$$V_u = \{u\} \cup \{\, w \mid \text{the shortest paths from } v \text{ to } w \text{ all go along the edge } vu \,\}$$
$$V_v = \{v\} \cup \{\, w \mid \text{the shortest paths from } u \text{ to } w \text{ all go along the edge } uv \,\}$$
$$V_{uv} = V(G) - (V_u \cup V_v).$$

(For example, in Figure 1, $V_u = \{u\}$, $V_v = \{v, w\}$, and $V_{uv} = \{x, y, z\}$.) For each vertex $x \in V_{uv}$, we define the midpoint μ_x of x with respect to uv as the point on uv such that $\ell(v, \mu_x) = 1/2 \cdot \big(d(u,x) - d(v,x) + \ell(u,v)\big)$; note that for each such vertex x, any shortest path from μ_x to x through u and any shortest path from μ_x to x through v are of equal length (see Figure 1).

Let T be a spanning tree of the graph G and let u, v be any two distinct vertices of G. The removal of the path ρ connecting u, v in T produces two subtrees of T, containing u and v, respectively. Then, we say that a vertex x that does not belong to the path ρ is *connected to* u (v, resp.) in T if x and u (v, resp.) belong to the same subtree. Then, the definitions of V_u, V_v, V_{uv}, and of the midpoint of a vertex imply the following observation:

Observation 1. *Let p be a point on an edge uv and T_p a shortest paths tree rooted at p. Then, for any vertex $x \in V(G)$, we have:*

(i) If $x \in V_u$ then vertex x is connected to u in T_p; if $x \in V_v$ then vertex x is connected to v in T_p.
(ii) If $x \in V_{uv}$ then: if the midpoint μ_x of x with respect to uv is located in the interval $[u, p)$ then vertex x is connected to v in T_p; if μ_x is located in the interval $(p, v]$ then x is connected to u in T_p; if μ_x coincides with p, then x may be connected either to u or to v in T_p.

Observation 1 implies the following corollary.

Corollary 1. *If a vertex x is connected to u in a shortest paths tree T_p rooted at a point p of an edge uv, then x is connected to u in any shortest paths tree rooted at any point in the interval $[u, p]$; if x is connected to v in T_p, then x is connected to v in any shortest paths tree rooted at any point in the interval $[p, v]$.*

Given a spanning tree T of the graph G and a vertex v, a *critical source* for vertex v in T is an element of the set S of source nodes at maximum distance from v in T. For the tree T, two sources at maximum intrasource distance in T (i.e., their distance in T is no less than the distance of any other pair of sources) form a *pair of critical sources.* It is important to note that a tree T may have more than one pair of critical sources; if this is the case, then we can show the following:

Lemma 1. *Let T be a spanning tree of a graph, a set S of source nodes, and suppose that T has more than one pair of critical sources. Then:*

(i) The midpoints of all the paths in T connecting pairs of critical sources coincide.
(ii) Let a, b and c, d be two pairs of critical sources. Then, a, c and b, d or a, d and b, c are also pairs of critical sources.

The following lemma, established in [1], gives properties of trees which are important both for the k-SVET and the SDET problems.

Lemma 2. [1] *Let s_1 and s_2 be two sources with maximum intrasource distance in a tree T. For any vertex $d \in V(T)$ and any source $s_i \in S - \{s_1, s_2\}$, either $d_T(d, s_i) \leq d_T(d, s_1)$ or $d_T(d, s_i) \leq d_T(d, s_2)$.*

In other words, for each vertex v in a tree T with a pair s_1, s_2 of critical sources, s_1 or s_2 is a critical source for v in T. Based on this and other results, Connamacher and Proskurowski [1] showed the following theorem (the same technique had been used to show that the maximum eccentricity problem is polynomial [6]):

Theorem 1. [1] *Given a weighted graph G, there exists a point χ such that any shortest paths tree rooted at χ is an optimal tree for the k-SVET problem.*

Theorem 1 shows that the tree minimizing the k-SVET cost function is a shortest paths tree rooted at a vertex or at a point on an edge of G. In a similar fashion, we can show the following:

Theorem 2. *Given a weighted graph G, a set S of source nodes, and a set D of destination nodes, there exists an optimal tree for the SDET problem which is a shortest paths tree rooted at a point on an edge of the graph G.*

3 The Algorithm

Our algorithm for the SDET problem relies on Theorem 2. It receives as input a weighted undirected graph G with non-negative symmetric weights and outputs a shortest paths tree that minimizes the SDET cost function. In high level, it works as follows:

Algorithm SDET

1. `for` each vertex v of the graph G `do`
   1.1 compute the distances $d(v, x)$ in G from v to every other vertex $x \in V(G)$;
   1.2 associate each pair v, x with either a neighbor u of v if all the shortest paths from v to x go along the edge uv, or with v otherwise;
2. $mincost \leftarrow +\infty$;
   `for` each edge uv of G `do`
   2.1 determine the description of a shortest paths tree T of G rooted at a point on the edge uv which minimizes the SDET cost function over all shortest paths trees rooted at points of uv and let $\text{cost}(T)$ be the value of the SDET cost function for T;
   2.2 `if` $\text{cost}(T) < mincost$
   `then` save the description of T as it gives the currently optimal tree T_{opt};
   $mincost \leftarrow \text{cost}(T)$;
3. Construct the tree T_{opt} from its description and clip it by repeatedly removing leaves that do not belong to $S \cup D$.

Clearly, the correctness of Step 2.1 implies the correctness of the entire algorithm. For the execution of Step 2.1 on an edge uv, we first compute a list L_{uv} of shortest paths trees rooted at points on uv, which includes a tree exhibiting the minimum of the SDET cost function over all shortest paths trees rooted on uv, and then, among the trees in L_{uv}, we select one with the minimum value of the SDET cost; these are discussed in the following subsections.

3.1 Finding Candidates (for the Optimal Tree) Rooted on an Edge

Let us consider an edge uv of the input graph G. In order to guarantee the correctness of our algorithm, we should consider all structurally different shortest paths trees of G rooted at points on uv. Observation 1 narrows down the possibilities and determines which vertices are connected to u and which to v in a tree: it implies that if we walk along the edge uv from u to v and compute the shortest paths tree rooted at the current point of uv, the tree is unique and remains the same for as long as we do not cross any midpoint; when we cross the midpoint μ_x of a vertex x, then x, which has been connected to u in the shortest paths trees considered so far, gets now connected to v. Since the vertices in V_{uv} are those contributing midpoints on the edge uv, we consider the partition of the set S of source nodes into the following three sets:

$$S_u = S \cap V_u \qquad S_v = S \cap V_v \qquad S_{uv} = S \cap V_{uv}$$

(for example, if $S = \{x, y, z, w\}$ in Figure 1, then $S_u = \emptyset$, $S_v = \{w\}$, and $S_{uv} = \{x, y, z\}$).

The following lemma, which we establish, helps us avoid additional unnecessary work as well.

Lemma 3. *Let G be a graph, and $S, D \subseteq V(G)$ be sets of source and destination nodes, respectively. For the computation of an optimal solution for the SDET problem for G, S, D, it suffices that we consider only shortest paths trees of G rooted at points r satisfying both following conditions:*

(i) the root r of the tree lies on the path connecting a pair of critical sources in the tree;
(ii) the root r and the (common) midpoint of all the paths connecting pairs of critical sources in the tree are located on (the closure of) an edge of G.

Lemma 3 has the following very important implications:

Corollary 2. *Let G, S, D be as described in Lemma 3. For the computation of an optimal solution for the SDET problem for G, S, D, it suffices that we consider only shortest paths trees T of G such that if σ_u (σ_v, resp.) is a source connected to u (v, resp.) in T at maximum distance from u (v, resp.), the midpoint of the path connecting σ_u and σ_v in T belongs to the edge uv. Then, in any such tree:*

(i) the sources σ_u, σ_v form a pair of critical sources in T;
(ii) the source σ_u (σ_v, resp.) is critical for every vertex x connected to v (u, resp.) in T.

Finally, if the midpoints of k source nodes with respect to the edge uv coincide at a point p, then there are 2^k structurally different shortest paths trees rooted at p (and in fact, even more if midpoints of other vertices also coincide with p). Yet, in such a case, we can show the following:

Lemma 4. *Let G be a graph, $S, D \subseteq V(G)$ be sets of source and destination nodes, respectively, and $A \subseteq S$ be a set of source nodes whose midpoints all fall at a point r on an edge uv of G. Then, for the computation of an optimal solution for the SDET problem for G, S, D, among all the shortest paths trees of G rooted at r, it suffices that we consider only those in which the sources in A are either all connected to u or all connected to v.*

The details of the processing of an edge uv of the input graph G are given in the Algorithm TREES_ROOTED_ON_EDGE presented below. For an edge uv of G, the algorithm produces a list L_{uv} of shortest paths trees rooted at points on uv, which is guaranteed to include a tree minimizing the SDET cost function over all such trees with their roots on uv; the trees are listed in L_{uv} in the order their roots are met along uv from u to v, and each such tree T is represented by its root r, the distance $\delta_u(r)$ of u to the critical sources in T connected to u, the distance $\delta_v(r)$ of v to the critical sources in T connected to v, and a number $k_S(r)$ such that the sources stored in a subarray $\Sigma[1..k_S(r)]$ of an array Σ of size $|S|$

are those connected to v in T whereas the remaining ones are those connected to u. We also note that in the case that the midpoints of several sources coincide, Algorithm TREES_ROOTED_ON_EDGE considers more possibilities than the two specified in Lemma 4; nevertheless, it certainly considers those two.

Algorithm TREES_ROOTED_ON_EDGE

1. Compute the sets S_u, S_v, and S_{uv}, and the midpoints of the source nodes in S_{uv} as well as their distances from vertex u on the edge uv;
2. Construct a sorted list $(s_1, s_2, \ldots, s_t)$ of the source nodes in S_{uv} in order of non-decreasing distance of their midpoints (on the edge uv) from u;
   Construct a sorted array Σ of the source nodes which stores the elements of S_v, followed by the sources $s_1, s_2, \ldots, s_t$ in that order, followed by the elements in S_u;
   $s_0 \leftarrow$ source node in S_v (if any) at maximum distance from v;
   $s_{t+1} \leftarrow$ source node in S_u (if any) at maximum distance from u;
3. Construct two arrays: $A_u = [d(u, s_1), d(u, s_2), \ldots, d(u, s_t), d(u, s_{t+1})]$ and $A_v = [d(v, s_0), d(v, s_1), d(v, s_2), \ldots, d(v, s_t)]$;
4. Compute the suffix-maxima $[a_1, a_2, \ldots, a_{t+1}]$ on the array A_u and the prefix-maxima $[b_0, b_1, \ldots, b_t]$ on the array A_v, i.e., $a_i = \max\{d(u, s_i), d(u, s_{i+1}), \ldots, d(u, s_{t+1})\}$ and $b_i = \max\{d(v, s_0), d(v, s_1), \ldots, d(v, s_i)\}$;
5. $L_{uv} \leftarrow$ an empty list;
   `for` each $i = 0, 1, 2, \ldots, t$ `do`
       {*consider the shortest paths tree rooted at* u *if* $i = 0$ *or at* μ_{s_i} *otherwise, in which the sources in* $S_v \cup \{s_1, s_2, \ldots, s_i\}$ *are connected to* v *and the sources in* $\{s_{i+1}, s_{i+2}, \ldots, s_{t+1}\} \cup S_u$ *are connected to* u}
       `if` ($i = 0$ `and` $S_v = \emptyset$) `or` ($i = t$ `and` $S_u = \emptyset$)
       `then` do nothing; {b_0 *or* a_{t+1} *are not well defined*}
       `else if` $|\, a_{i+1} - b_i \,| \leq \ell(u, v)$
           `then` {*the midpoint of the paths connecting pairs of critical sources belongs to* uv}
               `if` $i = 0$ `then` $r \leftarrow u$;
                   `else` $r \leftarrow$ midpoint μ_{s_i} of s_i on the edge uv;
               $\delta_u(r) \leftarrow a_{i+1}$;
               $\delta_v(r) \leftarrow b_i$;
               $k_S(r) \leftarrow |S_v| + i$;
               insert at the end of L_{uv} a record for a shortest paths tree represented by its root r, $\delta_u(r)$, $\delta_v(r)$, and $k_S(r)$;
6. Return the list L_{uv} of shortest paths trees and the array Σ;

The correctness of the algorithm follows from Observation 1, Theorem 2, Corollary 2, and Lemma 4. It is not difficult to see that the algorithm runs in $O(|S| \log |S|)$ time. Thus, we have:

Lemma 5. *Given a weighted graph* G*, a set of source nodes* $S \subseteq V(G)$*, and an edge* uv *of* G*, Algorithm* TREES_ROOTED_ON_EDGE *computes in* $O(|S| \log |S|)$ *time a collection of shortest paths trees rooted at points on* uv *among which there is one that minimizes the SDET cost function over all shortest paths trees rooted on* uv.

3.2 Selecting a Shortest Paths Tree of Minimum Cost Among Those Rooted at Points of an Edge

Let uv be an edge of the given graph G. In light of Lemma 5, finding a shortest paths tree of minimum SDET cost among those rooted at points on uv reduces to finding a tree of minimum SDET cost among those computed by Algorithm TREES_ROOTED_ON_EDGE (see Section 3.1). We process these trees in the order their roots are met on the edge uv from u to v; similarly, we process the destination nodes in the order their midpoints are met on the edge uv from u to v. If we consider the partition of the set D of destination nodes into

$$D_u = D \cap V_u, \qquad D_v = D \cap V_v, \qquad D_{uv} = D \cap V_{uv},$$

then, for any shortest paths tree T rooted at a point r of uv and any destination node d, Observation 1 specifies whether d is connected to u or to v in T. Additionally, the critical sources connected to u (v, resp.) are critical for each destination node connected to v (u, resp.); see Corollary 2. Finally, Lemma 6 (similar to Lemma 4) addresses the case of destination nodes whose midpoints coincide with the root of a shortest paths tree.

Lemma 6. *Let G be a graph, $D \subseteq V(G)$ the set of destination nodes, $B \subseteq D$ be a set of destination nodes whose midpoints all fall at a point r on an edge uv of G, and let σ_u, σ_v be sources connected to u and v, repsectively, forming a critical pair in a shortest paths tree rooted at r. Then, for the computation of an optimal solution for the SDET problem for G, S, D, among all the shortest paths trees of G rooted at r, it suffices that we consider only the one in which the destinations in B are either all connected to v if $d(u,\sigma_u) + \ell(u,r) < \ell(r,v) + d(v,\sigma_v)$, or all connected to u otherwise.*

The details of the computation are given below:

Algorithm EDGE_MIN_COST

1. Compute the sets D_u, D_v, and D_{uv}, and the midpoints of the destination nodes in D_{uv} as well as their distances from vertex u on the edge uv;
2. Construct a sorted array Δ of the destination nodes which stores the elements of D_v, followed by the elements of D_{uv} in order of non-decreasing distance of their midpoints (on the edge uv) from u, followed by the elements in D_u;
3. Execute Algorithm TREES_ROOTED_ON_EDGE on the edge uv: in addition to an ordered array Σ of the sources, the algorithm returns a list L_{uv} of shortest paths trees rooted at points on uv in the order their roots $r_1, r_2, \ldots, r_{t'}$ appear along uv from u to v; the tree rooted at r_i is also associated with the distances $\delta_u(r_i)$ and $\delta_v(r_i)$, and the number $k_S(r_i)$ (see Section 3.1);
4. $c_u \leftarrow \sum_{x \in D - D_v} d(u,x)$ and $c_v \leftarrow \sum_{x \in D_v} d(v,x)$;
   $j \leftarrow |D_v| + 1$ and $mincost \leftarrow +\infty$;
   `for` each candidate root location $r_i \in L_{uv}$, $1 \le i \le t'$, `do`
   4.1 `while` $j \le |D| - |D_u|$ and $\mu_{\Delta[j]}$ is to the left of r_i on the edge uv `do`
       subtract the value $d(u, \Delta[j])$ from c_u;
       add the value $d(v, \Delta[j])$ to c_v;
       $j \leftarrow j + 1$;

4.2 **if** $\delta_u(r_i) + \ell(u, r_i) < \ell(r_i, v) + \delta_v(r_i)$ {*apply Lemma 6*}
then while $j \leq |D| - |D_u|$ and $\mu_{\Delta[j]}$ coincides with r_i **do**
subtract the value $d(u, \Delta[j])$ from c_u;
add the value $d(v, \Delta[j])$ to c_v;
$j \leftarrow j + 1$;
$k_D(r_i) \leftarrow j - 1$; {*number of destination nodes connected to v*}
4.3 $cost \leftarrow c_v + k_D(r_i) \cdot (\ell(u, v) + \delta_u(r_i)) +$
$c_u + (|D| - k_D(r_i)) \cdot (\ell(u, v) + \delta_v(r_i))$;
if $cost < mincost$
then $\widehat{r}_{uv} \leftarrow r_i$;
$mincost \leftarrow cost$;

5. return a description of the computed shortest paths tree consisting of its root $\widehat{r}_{uv}$, its cost *mincost*, the ordered pair (u, v), and the source nodes stored in $\Sigma[1..k_S(\widehat{r}_{uv})]$ and the destination nodes stored in $\Delta[1..k_D(\widehat{r}_{uv})]$ which are to be connected to v whereas the remaining source and destination nodes are to be connected to u;

The correctness of the algorithm follows from the discussion preceding the description of the algorithm. Regarding the complexity of the algorithm, we have: Step 1 requires $O(|D|)$ time, Step 2 $O(|D| \log |D|)$ time, Step 3 $O(|S| \log |S|)$ time (Lemma 5), Step 4 $O(|S| + |D|)$ time (for each $d_i \in D$, the distances $d(u, d_i)$ and $d(v, d_i)$ are available in constant time thanks to Step 1 of Algorithm SDET and we spend $O(1)$ time for each candidate root r_i and each destination node), and Step 5 $O(|S| + |D|)$ time. In total, the algorithm runs in $O(|S| \log |S| + |D| \log |D|)$ time. Thus, we have the following result.

Lemma 7. *Given a weighted graph G, a set of source nodes $S \subseteq V(G)$, a set of destination nodes $D \subseteq V(G)$, and an edge uv of G, then Algorithm* EDGE_MIN_COST *runs in $O(|S| \log |S| + |D| \log |D|)$ time and produces the description of a shortest paths tree of G rooted at a point of uv which minimizes the SDET cost function over all shortest paths trees rooted at points on uv.*

3.3 Time Complexity of Algorithm SDET

We assume that the input graph G has n vertices and m edges and is given in adjacency list representation.

Step 1: The distances $d(v, x)$ from v to all other vertices x of the input graph G can be computed using the well known Dijkstra's algorithm in $O((n + m) \log n)$ time [2]. Since $|S| + |D| = O(n)$, we have that Step 1.1 is executed in $O(n(n + m) \log n)$ time. Finding whether the shortest paths from vertex v to any source or destination node x all go along the same edge incident on v or not can be easily carried out by executing a slightly modified version of Dijkstra's algorithm for v which maintains this information; the modified version runs in $O(n(n + m) \log n)$ time for all pairs of vertices of G as well. In total, Step 1 requires $O(n(n + m) \log n)$ time.

Step 2: Since Algorithm EDGE_MIN_COST runs in $O(|S| \log |S| + |D| \log |D|)$ time for each edge of G (Lemma 7), the step is executed in $O(nm \log n)$ time.

Step 3: The shortest paths tree T_{opt} can be constructed from its description in $O((n+m)\log n)$ time by using Dijkstra's algorithm to determine the desired shortest paths, while the clipping of unnecessary leaves takes $O(n)$ time.

Therefore, we obtain the following result.

Theorem 3. *The SDET problem on a weighted graph G on n vertices and m edges is solved in $O(nm\log n)$ time.*

4 Concluding Remarks

We described an algorithm for the SDET problem, which runs in $O(nm\log n)$ time and, to the best of our knowledge, is the first one for the problem in question. The algorithm also provides $O(nm\log n)$-time algorithms for the k-SVET and k-SSET problems for which it has only been proven that they are polynomial. The obvious open question is whether a faster algorithm can be obtained for the SDET problem; a potential improvement would arise if the set of edges contributing candidate roots for the sought shortest paths tree could be narrowed to only $o(m)$ edges.

References

1. H.S. Connamacher and A. Proskurowski, "The complexity of minimizing certain cost metrics for k-source spanning trees", *Discrete Applied Mathematics* **131** (2003) 113–127.
2. T.H. Cormen, C.E. Leiserson, R.L. Rivest, and C. Stein, *Introduction to Algorithms* (2nd edition), MIT Press, Inc., 2001.
3. A.M. Farley, P. Fragopoulou, D.W. Krumme, A. Proskurowski, and D. Richards, "Multi-source spanning tree problems", *Journal of Interconnection Networks* **1** (2000) 61–71.
4. T.C. Hu, "Optimum communication spanning trees", *SIAM Journal on Computing* **3** (1974) 188–195.
5. D.S. Johnson, J.K. Lenstra, and A.H.G. Rinnoy Kan, "The complexity of the network design problem", *Networks* **8** (1978) 279–285.
6. D.W. Krumme and P. Fragopoulou, "Minimum eccentricity multicast trees", *Discrete Mathematics and Theoretical Computer Science* **4** (2001) 157–172.
7. B. McMahan and A. Proskurowski, "Multi-source spanning trees: algorithms for minimizing source eccentricities", *Discrete Applied Mathematics* **137** (2004) 213–222.
8. R. Ravi, R. Sundaram, M.V. Marathe, D.J. Rosenkrantz, and S.S. Ravi, "Spanning trees - short or small", *SIAM Journal of Discrete Mathematics* **9** (1996) 178–200.
9. G.N. Rouskas and I. Baldine, "Multicast routing with end-to-end delay and delay variation constraints", *IEEE Journal on Selected Areas in Communications* **15** (1997) 346–356.
10. B.Y. Wu, G. Lancia, V. Bafna, K.-M. Chao, R. Ravi, and C.Y. Tang, "A polynomial time approximation scheme for minimum routing cost spanning trees", *Proc. 9th Annual ACM-SIAM Symposium on Discrete Algorithms – SODA'98* (1998) 21–32.

Edge-Pancyclicity of Twisted Cubes

Jianxi Fan[1], Xiaola Lin[2], Xiaohua Jia[1], and Rynson W.H. Lau[1]

[1] Department of Computer Science,
City University of Hong Kong, Kowloon,
Hong Kong, China
{fanort, csjia, rynson}@cityu.edu.hk
[2] College of Information Science and Technology,
Sun Yat-sen University,
Guangzhou, China
issxlin@zsu.edu.cn

Abstract. Twisted cubes are attractive alternatives to hypercubes. In this paper, we study a stronger pancyclicity of twisted cubes. We prove that the n-dimensional twisted cube is edge-pancyclic for $n \geq 3$. That is, for any $(x, y) \in E(TQ_n)(n \geq 3)$ and any integer l with $4 \leq l \leq 2^n$, a cycle C of length l can be embedded with dilation 1 into TQ_n such that (x, y) is in C. It is clear that an edge-pancyclic graph is also a node-pancyclic graph. Therefore, TQ_n is also a node-pancyclic graph for $n \geq 3$.

1 Introduction

Interconnection networks take a key role in parallel computing systems. An interconnection network can be represented by a graph $G = (V, E)$, where V represents the node set and E represents the edge set. In this paper, we use graphs and interconnection networks interchangeably.

Graph embedding is to embed a graph into another graph. This operation is required in interconnection networks. Graph embedding can be formally defined as: Given two graphs $G_1 = (V_1, E_1)$ and $G_2 = (V_2, E_2)$, an embedding from G_1 to G_2 is an injective mapping ψ: $V_1 \rightarrow V_2$. An important performance metric of embedding is dilation. The dilation of embedding ψ is defined as

$$\mathrm{dil}(G_1, G_2, \psi) = \max\{\mathrm{dist}(G_2, \psi(u), \psi(v)) | (u, v) \in E_1\},$$

where $\mathrm{dist}(G_2, \psi(u), \psi(v))$ denotes the distance between the two nodes $\psi(u)$ and $\psi(v)$ in G_2. The smaller the dilation of an embedding is, the shorter the communication delay that the graph G_2 simulates the graph G_1. We call ψ the optimal embedding from G_1 to G_2, if ψ has the smallest dilation in all the embeddings from G_1 to G_2. Clearly, the dilation of the optimal embedding is at least 1. In fact, under this circumstance, G_1 is a subgraph of G_2. Finding the optimal embedding of graphs is NP-hard.

Many graph embeddings take cycles, trees, meshes, paths, etc. as guest graphs [5], [9], [11], [13], [15], [18], [19], because these interconnection networks are widely used in parallel computing systems.

X. Deng and D. Du (Eds.): ISAAC 2005, LNCS 3827, pp. 1090–1099, 2005.

Twisted cubes [1], [12] are attractive alternatives to hypercubes. It possesses some desirable features for interconnection networks [12]. The diameters, wide diameters, and faulty diameters in twisted cubes are about half of those in comparable hypercubes [7]. A complete binary tree can be embedded into a twisted cube [2]. Recently, it was also proved that twisted cubes have the same diagnosability as hypercubes under the t/k-diagnosis strategy based on the well-known PMC diagnostic model [8]. In particular, it is shown, respectively, that twisted cubes are pancyclic graphs [7] and that the n-dimensional twisted cube is $(n-2)$-Hamiltonian $(n \geq 2)$ and $(n-3)$-Hamiltonian connected $(n \geq 3)$ [14].

In this paper, we will study a stronger result on the pancyclicity of twisted cubes. We will prove that the n-dimensional twisted cube is an edge-pancyclic graph when $n \geq 3$. Obviously, an edge-pancyclic graph is also a node-pancyclic graph. The issues on node-pancyclicity and edge-pancyclicity have been discussed in [3], [4], [6], [10], [16], [17].

2 Preliminaries

In this section, we give some definitions and notations used in the paper.

Let $G = (V, E)$ be a graph. A path P from node u to node v in G is denoted by P: $u = u^{(0)}, u^{(1)}, \ldots, u^{(k)} = v$. The number k of edges in path P is called the *length* of path P. Nodes u and v are called the two *end nodes* of path P. If $u = v$, then P is called a *cycle*.

If (x, y) is an edge in a cycle C, then the path between x and y in C is denoted by $C - (x, y)$, which is a path after deleting the edge (x, y) in C.

G is called a *pancyclic* graph, if G contains any cycle of length l with $3 \leq l \leq |V|$, i. e., any cycle of length l with $4 \leq l \leq |V|$, can be embedded into G with dilation 1. However, there is no cycle of length 3 in twisted cubes. For convenience of discussion in this paper, we call G a *pancyclic* graph, if G contains any cycle of length l with $4 \leq l \leq |V|$. Similarly, we define *node-pancyclic* graphs and *edge-pancyclic* graphs as follows:

G is called a node-pancyclic graph if, for every node u and any integer l with $4 \leq l \leq |V|$, there exists a cycle C of length l in G such that u is in C. G is called an *edge-pancyclic* graph if, for every edge (u, v) and any integer l with $4 \leq l \leq |V|$, there exists a cycle C' of length l in G such that (u, v) is in C'. Obviously, if G is an edge-pancyclic graph, then it is also a node-pancyclic graph.

Let G_1 and G_2 be two subgraphs of G. We use $G_1 \bigcup G_2$ to denote the subgraph induced by $V(G_1) \bigcup V(G_2)$ of G. The *cartesian product* of G_1 and G_2 is defined as the graph $G_1 \times G_2$, where $V(G_1 \times G_2) = V(G_1) \times V(G_2)$, and for any $x, y \in V(G_1 \times G_2)$, $x = (u, u')$, $y = (v, v')$, $(x, y) \in E(G_1 \times G_2)$ if and only if $u = v$ and $(u', v') \in E(G_2)$, or $u' = v'$ and $(u, v) \in E(G_1)$.

A binary string u of length n will be denoted by $u_{n-1}u_{n-2}\ldots u_0$. The complement of u_i will be denoted by $\overline{u_i} = 1 - u_i$. In [12], the n-dimensional twisted cube TQ_n was defined. It is an n-regular graph with 2^n nodes and $n2^{n-1}$ edges, where n is an odd integer. We label all the nodes of TQ_n by binary strings of

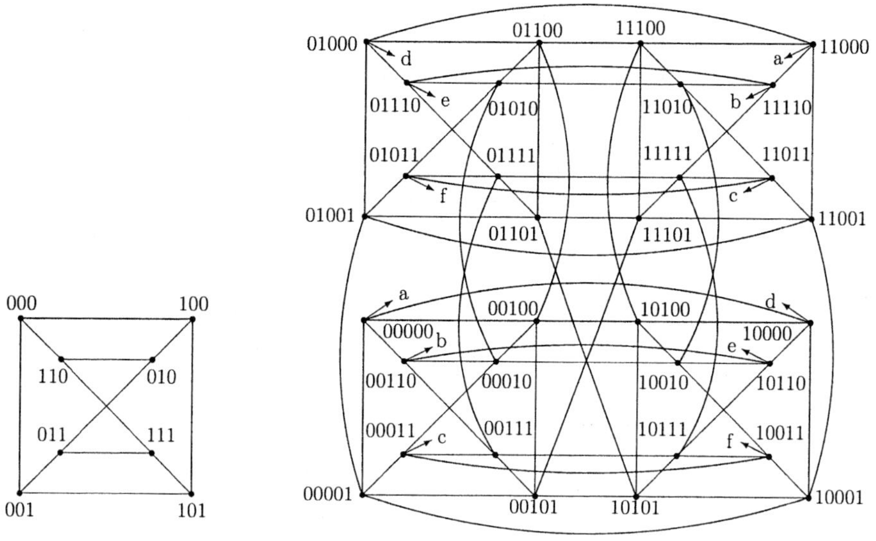

(a) (b)

Fig. 1. (a) The 3-dimensional twisted cube TQ_3. (b) The 5-dimensional twisted cube TQ_5, where the end nodes of a missing edge are marked with arrows labeled with the same letter.

length n. In this paper, we will not distinguish between the nodes of TQ_n and their labels. If $u = u_{n-1}u_{n-2}\ldots u_0 \in V(TQ_n)$, for $0 \leq i \leq n-1$, we define $f(u,i) = u_i \bigoplus u_{i-1} \bigoplus \ldots \bigoplus u_0$, where $\bigoplus$ is the exclusive operation. According to the definition of TQ_n in [12], we may give a recursive definition of TQ_n for any odd integer $n \geq 1$ as follows:

Definition 1. The 1-dimensional twisted cube TQ_1 is defined as the complete graph with two nodes labelled 0 and 1. For an odd integer $n \geq 3$, TQ_n consists of four subcubes TQ_{n-2}^{00}, TQ_{n-2}^{01}, TQ_{n-2}^{10}, TQ_{n-2}^{11}, where TQ_{n-2}^{ab} is isomorphic to TQ_{n-2} and $V(TQ_{n-2}^{ab}) = \{\, abx|\ x \in V(TQ_{n-2})\,\}$ and $E(TQ_{n-2}^{ab}) = \{\, (abx,\ aby)|\ (x,\ y) \in E(TQ_{n-2})\,\}$ for $a,\ b \in \{\,0,\ 1\,\}$; and $V(TQ_n) = \bigcup_{a,b\in\{0,1\}} V(TQ_{n-2}^{ab})$, $E(TQ_n) = \bigcup_{a,b\in\{0,1\}} E(TQ_{n-2}^{ab}) \bigcup E'$, where for the nodes $u = u_{n-1}u_{n-2}\ldots u_0$, $v = v_{n-1}v_{n-2}\ldots v_0 \in V(TQ_n)$, $(u,v) \in E'$ if u and v satisfy one of the following three conditions:

(1) $u = \overline{v_{n-1}}v_{n-2}v_{n-3}\ldots v_0$;

(2) $u = \overline{v_{n-1}}\ \overline{v_{n-2}}v_{n-3}\ldots v_0$ and $f(u,n-3) = 0$;

(3) $u = v_{n-1}\ \overline{v_{n-2}}v_{n-3}\ldots v_0$ and $f(u,n-3) = 1$.

According to Definition 1, Fig. 1 shows TQ_3 and TQ_5.

Notation 1. For $n \geq 3$ and $a,b \in \{0,1\}$, the four subcubes of TQ_n are denoted by TQ_{n-2}^{ab}, $TQ_{n-2}^{(1-a)b}$, $TQ_{n-2}^{a(1-b)}$, and $TQ_{n-2}^{(1-a)(1-b)}$, respectively.

3 Edge-Pancyclicity of Twisted Cubes

In this section, we will prove that TQ_n is an edge-pancyclic graph for $n \geq 3$. To prove this result, we introduce the following two results associated with crossed cubes. In the following, the parameter n always denotes an odd integer.

Crossed cubes are also variants of hypercubes [9], [15]. The n-dimensional crossed cube, denoted by CQ_n, is recursively defined as follows.

Definition 2. CQ_1 is the complete graph on two nodes whose addresses are 0 and 1. CQ_n consists of two subcubes CQ^0_{n-1} and CQ^1_{n-1}. The most significant bit of the addresses of the nodes of CQ^0_{n-1} and CQ^1_{n-1} are 0 and 1, respectively. The nodes $u = u_{n-1}u_{n-2}\ldots u_1u_0$ and $v = v_{n-1}v_{n-2}\ldots v_1v_0$, where $u_{n-1} = 0$ and $v_{n-1} = 1$, are joined by an edge in CQ_n if and only if

(1) $u_{n-2} = v_{n-2}$ if n is even, and

(2) $(u_{2i+1}u_{2i}, v_{2i+1}v_{2i}) \in \{(00,00),(10,10),(01,11),(11,01)\}$, for $0 \leq i < \lfloor \frac{n-1}{2} \rfloor$.

CQ_3 is shown in Fig 2. From Fig 1 and Fig 2, we can easily verify the following result.

Lemma 1. TQ_3 is isomorphic to CQ_3.

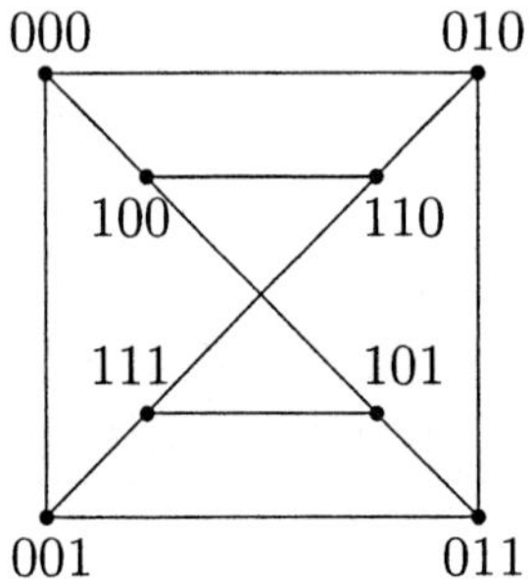

Fig. 2. The 3-dimensional crossed cube CQ_3

Lemma 2 [10]. If $n \geq 2$, for any $(x, y) \in E(CQ_n)$ and any integer l with $4 \leq l \leq 2^n$, there exists a cycle C of length l such that (x, y) is in C.

Theorem 1. TQ_n is an edge-pancyclic graph for $n \geq 3$.

Proof. We use induction on n. By Lemma 1, TQ_3 is isomorphic to CQ_3. Further, by Lemma 2, the theorem holds in TQ_3.

Supposing that the theorem holds for $n = \tau - 2$ ($\tau \geq 5$), we will prove the theorem still holds for $n = \tau$. For any an edge $(u, v) \in E(TQ_\tau)$ and any $a, b \in \{0, 1\}$, we separately deal with the following cases.

Case 1. $u, v \in V(TQ^{ab}_{\tau-2})$. For $4 \leq l \leq 2^\tau$, we have the following sub-cases.

Case 1.1. $4 \leq l \leq 2^{\tau-2}$. By the induction hypothesis, there is a cycle of length l that contains (u, v) in $TQ^{ab}_{\tau-2}$.

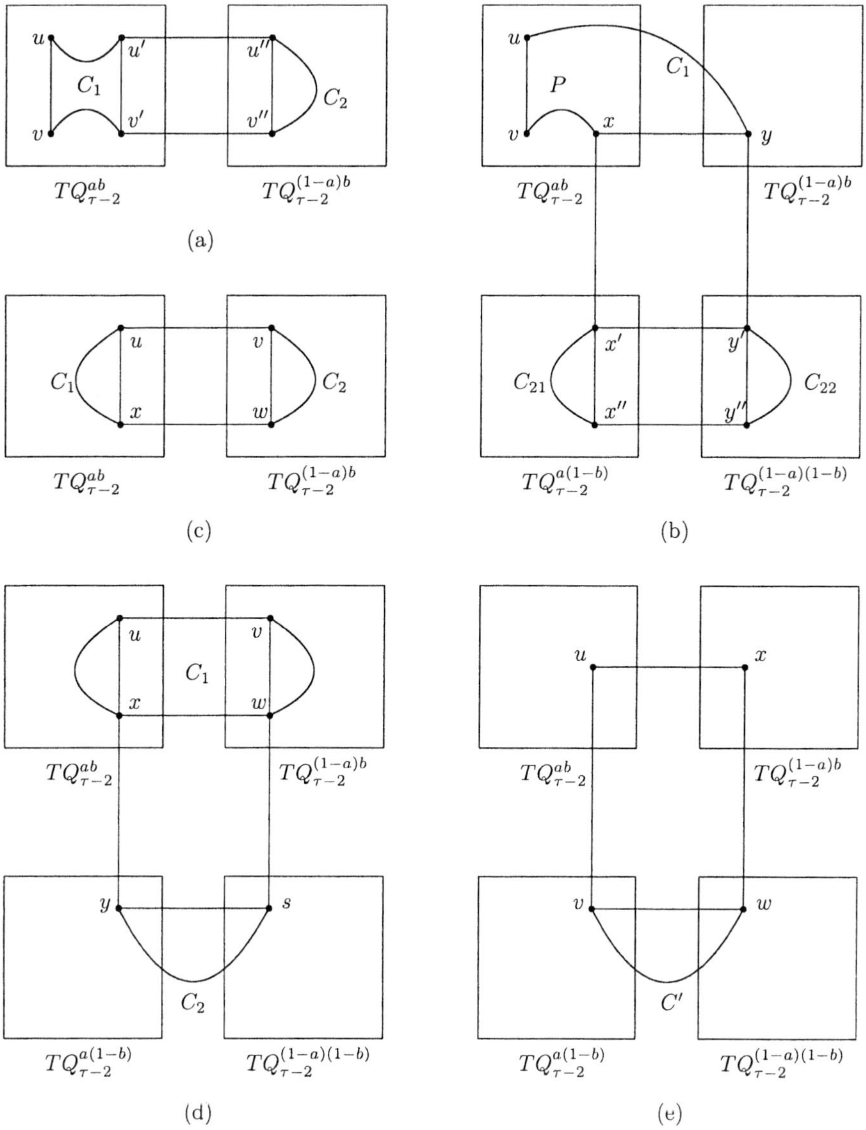

Fig. 3. The cycle of length l that contains (u, v) in TQ_τ, where a straight line represents an edge and a curve line represents a path between two nodes

Case 1.2. $2^{\tau-2} + 1 \leq l \leq 2^{\tau-1}$. Let $l_1 = \lfloor \frac{l}{2} \rfloor$, $l_2 = l - l_1$. Then $l_1 + l_2 = l$ and $4 \leq 2^{\tau-3} \leq l_1 \leq l_2 \leq 2^{\tau-2}$. By the induction hypothesis, there is a cycle C_1 of length l_1 that contains (u, v) in $TQ^{ab}_{\tau-2}$ (See Fig.3 (a)). Select an edge (u', v') in C_1. Further, by Definition 1, we can respectively select the neighbors u'' and v'', in $TQ^{(1-a)b}_{\tau-2}$, of the nodes u' and v'. By the induction hypothesis, there is a cycle C_2 of length l_2 that contains (u'', v'') in $TQ^{(1-a)b}_{\tau-2}$. Then

$$C_1 - (u', v'), C_2 - (v'', u''), u'$$

is a cycle of length $(l_1-1)+(l_2-1)+2=l$ that contains (u,v) in $TQ^{ab}_{\tau-2} \bigcup TQ^{(1-a)b}_{\tau-2}$ and thus in TQ_τ.

Case 1.3. $2^{\tau-1}+1 \leq l \leq 2^\tau$. Let $l_1 = \lfloor \frac{l}{2} \rfloor$, $l_2 = l - l_1$. Then $l_1 + l_2 = l$ and $2^{\tau-2} \leq l_1 \leq l_2 \leq 2^{\tau-1}$.

We can prove the following Claim:

Claim. If $(u,v) \in E(TQ^{ab}_{\tau-2})$, then there exists a cycle C of length l' with $2^{\tau-2} \leq l' \leq 2^{\tau-1}$ in $TQ^{ab}_{\tau-2} \bigcup TQ^{(1-a)b}_{\tau-2}$, such that C contains (u,v) and there exists an edge (x,y) in C with $x \in V(TQ^{ab}_{\tau-2})$ and $y \in V(TQ^{(1-a)b}_{\tau-2})$.

Now we prove the above claim. Let u' and v' be the neighbors, in $TQ^{(1-a)b}_{\tau-2}$, of u and v, respectively. Then, $(u',v') \in E(TQ^{(1-a)b}_{\tau-2})$. By the induction hypothesis, there is a cycle C' of length $2^{\tau-2}-2$ that contains (u',v') in $TQ^{(1-a)b}_{\tau-2}$. Let $x = u$ and $y = u'$. Then, $u, v, C' - (v', u'), u$ is a cycle C of length $2^{\tau-2}$ in $TQ^{ab}_{\tau-2} \bigcup TQ^{(1-a)b}_{\tau-2}$, such that C' contains (u,v) and (x,y) with $x \in V(TQ^{ab}_{\tau-2})$ and $y \in V(TQ^{(1-a)b}_{\tau-2})$. Thus, the claim holds for $l' = 2^{\tau-2}$.

For $2^{\tau-2}+1 \leq l' \leq 2^{\tau-1}$, by Case 1. 2, there is a cycle C'' of length l' that contains (u,v) in $TQ^{ab}_{\tau-2} \bigcup TQ^{(1-a)b}_{\tau-2}$. Considering that $l' \geq 2^{\tau-2}+1 \geq V(TQ^{ab}_{\tau-2})$, there exists an edge (x,y) in C'' with $x \in V(TQ^{ab}_{\tau-2})$ and $y \in V(TQ^{(1-a)b}_{\tau-2})$ (Otherwise, all the nodes of C'' are in $TQ^{ab}_{\tau-2}$ and the length of C'' is at most $2^{\tau-2}$).

In summary, the above claim holds.

Since $2^{\tau-2} \leq l_1 \leq 2^{\tau-1}$, according to the above claim, there always exists a cycle C_1 of length l_1 in $TQ^{ab}_{\tau-2} \bigcup TQ^{(1-a)b}_{\tau-2}$, such that C_1 contains (u,v) and there exists an edge (x,y) in C_1 with $x \in V(TQ^{ab}_{\tau-2})$ and $y \in V(TQ^{(1-a)b}_{\tau-2})$. Suppose that x is between v and y in C_1 (See Fig.3 (b)).

Let x' and y' be the neighbors, in $TQ^{a(1-b)}_{\tau-2} \bigcup TQ^{(1-a)(1-b)}_{\tau-2}$, of x and y, respectively. Then we have $x' \in V(TQ^{a(1-b)}_{\tau-2})$ and $y' \in V(TQ^{(1-a)(1-b)}_{\tau-2})$, or $x' \in V(TQ^{(1-a)(1-b)}_{\tau-2})$ and $y' \in V(TQ^{a(1-b)}_{\tau-2})$; and $(x', y') \in E(TQ^{a(1-b)}_{\tau-2} \bigcup TQ^{(1-a)(1-b)}_{\tau-2})$. Without loss of generality, we assume that $x' \in V(TQ^{a(1-b)}_{\tau-2})$, $y' \in V(TQ^{(1-a)(1-b)}_{\tau-2})$. Select a neighbor x'', in $TQ^{(1-a)b}_{\tau-2}$, of x'. Further, select the neighbor y'', in $TQ^{(1-a)(1-b)}_{\tau-2}$, of x''. By Definition 1, we have $(y', y'') \in E(TQ^{(1-a)(1-b)}_{\tau-2})$. Let $l_{21} = \lfloor \frac{l_2}{2} \rfloor$, $l_{22} = l_2 - l_1$. Then $l_{21} + l_{22} = l_2$ and $4 \leq 2^{\tau-3} \leq l_{21} \leq l_{22} \leq 2^{\tau-2}$. By the induction hypothesis, there is a cycle C_{21} of length l_{21} that contains (x', x'') in $TQ^{a(1-b)}_{\tau-2}$ and a cycle C_{22} of length l_{22} that contains (y', y'') in $TQ^{(1-a)(1-b)}_{\tau-2}$. Then,

$$C_1 - (x, y), C_{22} - (y', y''), C_{21} - (x'', x'), x$$

is a cycle of length $(l_1 - 1) + (l_{22} - 1) + (l_{11} - 1) + 3 = l$ that contains (u, v) in TQ_τ.

Case 2. $u \in V(TQ^{ab}_{\tau-2})$ and $v \in V(TQ^{(1-a)b}_{\tau-2})$. For $4 \le l \le 2^\tau$, we have the following sub-cases.

Case 2.1. $l = 4$. Select a neighbor x, in $TQ^{ab}_{\tau-2}$, of u. Let w be the neighbor, in $TQ^{(1-a)b}_{\tau-2}$, of x. By Definition 1, $(v, w) \in E(TQ^{(1-a)b}_{\tau-2})$. Thus,

$$u, v, w, x, u$$

is a cycle of length 4 that contains (u, v) in $TQ^{ab}_{\tau-2} \bigcup TQ^{(1-a)b}_{\tau-2}$ and thus in TQ_τ.

Case 2.2. $l = 5$. Let $u = u_{n-1}u_{n-2}\ldots u_0$. Then $v = \overline{u_{n-1}}u_{n-2}\ldots u_0$. Let $y = u_{n-1}u_{n-2}\ldots\overline{u_0}$. Then,$f(v, \tau-3) = \overline{f(y, \tau-3)}$. Further, let w and x be the neighbors, in $TQ^{a(1-b)}_{\tau-2} \bigcup TQ^{(1-a)(1-b)}_{\tau-2}$, of v and y, respectively. According to Definition 1, $(w, x) \in E(TQ_\tau)$. Thus,

$$u, v, w, x, y, u$$

be a cycle of length 5 that contains (u, v) in TQ_τ.

Case 2.3. $6 \le l \le 2^{\tau-1}$. Let u, v, w, x, u be a cycle, found in Case 2. 1, of length 4 in $TQ^{ab}_{\tau-2} \bigcup TQ^{(1-a)b}_{\tau-2}$.

Case 2.3.1. $6 \le l \le 2^{\tau-2} + 2$. By the induction hypothesis, there is a cycle C of length $l-2$ that contains (v, w) in $TQ^{(1-a)b}_{\tau-2}$. Then

$$u, C - (v, w), x, u$$

is a cycle of length $(l-3)+3 = l$ that contains (u, v) in $TQ^{ab}_{\tau-2} \bigcup TQ^{(1-a)b}_{\tau-2}$ and thus in TQ_τ.

Case 2.3.2. $2^{\tau-2} + 3 \le l \le 2^{\tau-1}$. Let $l_1 = \lfloor \frac{l}{2} \rfloor$, $l_2 = l - l_1$. Then $l_1 + l_2 = l$ and $5 \le 2^{\tau-3} + 1 \le l_1 \le l_2 \le 2^{\tau-2}$. By the induction hypothesis, there is a cycle C_1 of length l_1 that contains (u, x) in $TQ^{ab}_{\tau-2}$ and a cycle C_2 of length l_2 that contains (v, w) in $TQ^{(1-a)b}_{\tau-2}$ (See Fig. 3 (c)). Then

$$C_1 - (u, x), C_2 - (w, v), u$$

is a cycle of length $(l_1-1)+(l_2-1)+2-l$ that contains (u, v) in $TQ^{ab}_{\tau-2} \bigcup TQ^{(1-a)b}_{\tau-2}$ and thus in TQ_τ.

Case 2.4. $2^{\tau-1} + 1 \le l \le 2^\tau$. Let $l_1 = \lfloor \frac{l}{2} \rfloor$, $l_2 = l - l_1$. Then $l_1 + l_2 = l$ and $8 \le 2^{\tau-2} \le l_1 \le l_2 \le 2^{\tau-1}$. By Case 2. 3, there is a cycle C_1 of length l_1 that contains (u, v) in $TQ^{ab}_{\tau-2} \bigcup TQ^{(1-a)b}_{\tau-2}$. Clearly, there is an edge (x, w) in C_1 such that $x \in V(TQ^{ab}_{\tau-2}) - \{u\}$ and $w \in V(TQ^{(1-a)b}_{\tau-2}) - \{v\}$ (See Fig. 3 (d)). Let y and s be the neighbors, in $TQ^{a(1-b)}_{\tau-2} \bigcup TQ^{(1-a)(1-b)}_{\tau-2}$, of x and w, respectively. Then we have $s \in V(TQ^{a(1-b)}_{\tau-2})$ and $y \in V(TQ^{(1-a)(1-b)}_{\tau-2})$, or $s \in V(TQ^{(1-a)(1-b)}_{\tau-2})$

and $y \in V(TQ_{\tau-2}^{a(1-b)})$; and $(y,s) \in E(TQ_{\tau-2}^{a(1-b)} \bigcup TQ_{\tau-2}^{(1-a)(1-b)})$. Without loss of generality, we assume that $y \in V(TQ_{\tau-2}^{a(1-b)})$, $s \in V(TQ_{\tau-2}^{(1-a)(1-b)})$. Since $8 \le l_2 \le 2^{\tau-1}$, by Case 2. 3, there is a cycle C_2 of length l_2 that contains (y,s) in $TQ_{\tau-2}^{a(1-b)} \bigcup TQ_{\tau-2}^{(1-a)(1-b)}$. Then

$$C_1 - (x,w), C_2 - (s,y), x$$

is a cycle of length $(l_1-1)+(l_2-1)+2 = l$ that contains (u,v) in in TQ_τ.

Case 3. $u \in V(TQ_{\tau-2}^{ab} \bigcup TQ_{\tau-2}^{(1-a)b})$ and $v \in V(TQ_{\tau-2}^{a(1-b)} \bigcup TQ_{\tau-2}^{(1-a)(1-b)})$. Without loss of generality, we have the following sub-cases.

Case 3.1. $u \in V(TQ_{\tau-2}^{ab})$ and $v \in V(TQ_{\tau-2}^{a(1-b)})$. Let x be the neighbor, in $TQ_{\tau-2}^{(1-a)b}$, of u and let w be the neighbor, in $TQ_{\tau-2}^{(1-a)(1-b)}$, of v. By Definition 1, $(x,w) \in E(TQ_\tau)$. Thus,

$$u, v, w, x, u$$

is a cycle of length 4 that contains (u,v) in TQ_τ. For $5 \le l \le 2^\tau$, we have the following sub-cases.

Case 3.1.1. $l=5$. Let $x = x_{\tau-1}x_{\tau-2}\dots x_0$ and $y = x_{\tau-1}x_{\tau-2}\dots x_1\overline{x_0}$. Then, $u = \overline{x_{\tau-1}}x_{\tau-2}\dots x_0$ and $v = \overline{x_{\tau-1}x_{\tau-2}}x_{\tau-3}x_{\tau-4}\dots\overline{x_0}$. Further, let s be the neighbor, in $TQ_{\tau-2}^{a(1-b)} \bigcup TQ_{\tau-2}^{(1-a)(1-b)}$, of y. According to Definition 1, $(x,y), (v,s) \in E(TQ_\tau)$. Thus,

$$u, v, s, y, x, u$$

be a cycle of length 5 that contains (u,v) in TQ_τ.

Case 3.1.2. $6 \le l \le 2^{\tau-1}+2$. By Cases 2. 1, 2. 2, and 2. 3, there is a cycle C' of length $l-2$ that contains (v,w) in $TQ_{\tau-2}^{a(1-b)} \bigcup TQ_{\tau-2}^{(1-a)(1-b)}$ (See Fig. 4 (a)). Then

$$u, C' - (v,w), x, u$$

is a cycle of length $(l-3)+3 = l$ that contains (u,v) in TQ_τ.

Case 3.1.3. $2^{\tau-1}+3 \le l \le 2^\tau$. Let $l_1 = \lfloor \frac{l}{2} \rfloor$, $l_2 = l - l_1$. Then $l_1 + l_2 = l$ and $9 \le 2^{\tau-2}+1 \le l_1 \le l_2 \le 2^{\tau-1}$. By Case 2. 3, there is a cycle C_1' of length l_1 that contains (u,x) in $TQ_{\tau-2}^{ab} \bigcup TQ_{\tau-2}^{(1-a)b}$ and a cycle C_2' of length l_2 that contains (w,v) in $TQ_{\tau-2}^{a(1-b)} \bigcup TQ_{\tau-2}^{(1-a)(1-b)}$ (See Fig. 4 (b)). Then,

$$C_1' - (u,x), C_2' - (w,v), u$$

is a cycle of length $(l_1-1)+(l_2-1)+2 = l$ that contains (u,v) in TQ_τ.

Case 3.2. $u \in V(TQ_{\tau-2}^{ab})$ and $v \in V(TQ_{\tau-2}^{(1-a)(1-b)})$. Let x be the neighbor, in $TQ_{\tau-2}^{(1-a)b}$, of u and let w be the neighbor, in $TQ_{\tau-2}^{(1-a)(1-b)}$, of v. By Definition 1, $(x,w) \in E(TQ_\tau)$. Thus,

$$u, v, w, x, u$$

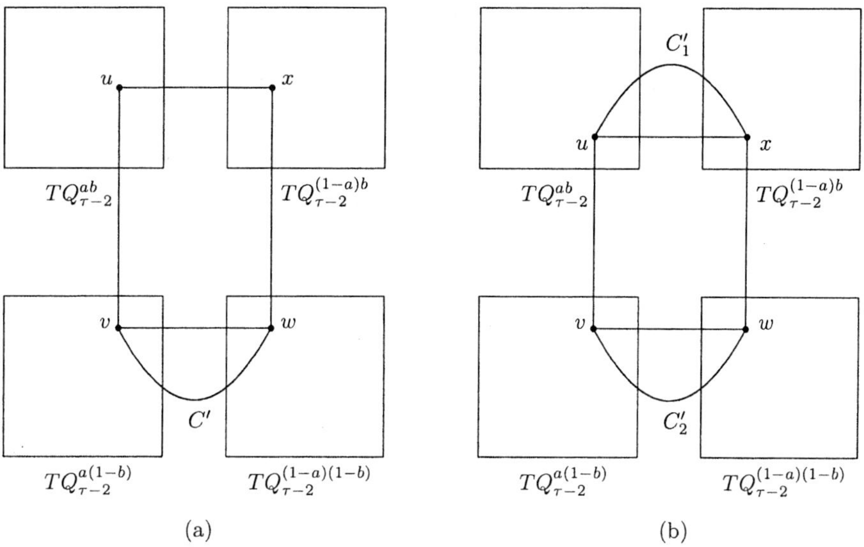

Fig. 4. The cycle of length l that contains (u, v) in TQ_τ, where a straight line represents an edge and a curve line represents a path between two nodes

is a cycle of length 4 that contains (u, v) in TQ_τ. For $5 \le l \le 2^\tau$, we have the following sub-cases.

Case 3.2.1. $l = 5$. Let $u = u_{\tau-1}u_{\tau-2}\ldots u_0$. Then $x = \overline{u_{\tau-1}}u_{\tau-2}u_{\tau-3}\ldots u_0$ and $v = \overline{u_{\tau-1}u_{\tau-2}}u_{\tau-3}\ u_{\tau-4}\ldots u_0$. Further, let $y = \overline{u_{\tau-1}}u_{\tau-2}u_{\tau-3}\ldots u_1\overline{u_0}$ and $s = \overline{u_{\tau-1}u_{\tau-2}}u_{\tau-3}u_{\tau-4}\ldots u_1\overline{u_0}$. According to Definition 1, $(v, s), (x, y), (y, s) \in E(TQ_\tau)$. Thus,

$$u, v, s, y, x, u$$

is a cycle of length 5 that contains (u, v) in TQ_τ.

Case 3.2.2. $6 \le l \le 2^{\tau-1} + 2$. Similar to Case 3. 1. 2, we can get a cycle of length l that contains (u, v) in TQ_τ.

Case 3.2.3. $2^{\tau-1} + 3 \le l \le 2^\tau$. Similar to Case 3. 1. 3, we can get a cycle of length l that contains (u, v) in TQ_τ.

So far, we have proven that the theorem also holds for $n = \tau$. Hence, the theorem holds. □

Acknowledgments

This research is partially supported by Hong Kong grants CERG under No. 9040816 and No. 9040909.

References

1. S. Abraham, K. Padmanabhan, The Twisted Cube Topology for Multiprocessors: A Study in Networks Asymmetry, J. Parallel and Distributes Computing, vol. 13, no. 1, pp. 104–110, 1991.
2. E. Abuelrub and S. Bettayeb, Embedding of Complete Binary Trees in Twisted Hypercubes, Proc. Int'l Conf. Computer Applications in Design, Simulation, and Analysis, pp. 1–4, 1993.
3. B. Alspach, D. Hare, Edge-pancyclic block-intersection graphs, Discrete Mathematics, vol. 97, pp. 17–24, 1991.
4. T. Araki, Edge-pancyclicity of recursive circulants, Information Processing Letters, vol. 88, pp. 287–292, 2003.
5. L. Auletta, A. A. Rescigno, V. Scarano, Embedding Graphs onto the Supercube, IEEE Trans. Computers, vol. 44, no. 4, pp. 593–597, 1995.
6. J. Bang-Jansen, Y. Guo, A note on vertex pancyclic oriented graphs, J. Graph Theory, vol. 31, pp. 313–318, 1999.
7. C. -P. Chang, J. -N. Wang, L. -H. Hsu, Topological Properties of Twisted Cubes, Inform. Sci., vol. 113, pp. 147–167, 1999.
8. J. Fan, X. Lin, The t/k-Diagnosability of the BC Graphs, IEEE Trans. Computers, vol. 54, no. 2, pp. 176–184, 2005.
9. J. Fan, X. Lin, X. Jia, Optimal Path Embedding in Crossed Cubes, IEEE Trans. Parallel and Distributed Systems, accepted.
10. J. Fan, X. Lin, X. Jia, Node-Pancyclicity and Edge-Pancyclicity of Crossed Cubes, Info. Process. Lett., vol. 93, pp. 133–138, 2005.
11. J. -S. Fu, Fault-Tolerant Cycle Embedding in the Hypercube, Parallel Computing, vol. 29, no. 6, pp. 821–832, 2003.
12. P. A. J. Hilbers, M. R. J. Koopman, J. L. A. Van de Snepscheut, The Twisted Cube, PARLE: Parallel Architectures and Languages Europe, Parallel Architectures, vol. 1, J. deBakker, A. Numan, and P. Trelearen, Berlin: Springer-Verlag, pp. 152–158, 1987.
13. H. -C. Hsu, T. -K. Li, J. J. M. Tan, L. -H. Hsu, Fault Hamiltonicity and Fault Hamiltonian Connectivity of the Arrangement Graphs, IEEE Trans. Computers, vol. 53, no. 1, pp. 39–53, 2004.
14. W. -T. Huang, J. J. M. Tan, C. -N. Hung, L. -H. Hsu, Fault-Tolerant Hamiltonicity of Twisted Cubes, J. Parallel and Distributes Computing, vol. 62, pp. 591–604, 2002.
15. P. Kulasinghe and S. Bettayeb, Embedding Binary Trees into Crossed Cubes, IEEE Trans. Computers, vol. 44, no.7, pp. 923–929, 1995.
16. K. -W. Lih, S. Zengmin, W. Weifan, Z. Kemin, Edge-pancyclicity of coupled graphs, Discrete Applied Mathematics, vol. 119, pp. 259–264, 2002.
17. B. Randerath, I. Schiermeyer, M. Tewes, L. Volkmann, Vertex pancyclic graphs, Discrete Applied Mathematics, vol. 120, pp. 219–237, 2002.
18. M. -C. Yang, T. -K. Li, J. J. M. Tan, L. -H. Hsu, Fault-Tolerant Cycle-Embedding of Crossed Cubes, Info. Process. Lett., vol. 88, no. 4, pp. 149–154, 2003.
19. P. -J. Yang, S. -B. Tien, C. S. Raghavendra, Embedding of Rings and Meshes onto Faulty Hypercubes Using Free Dimensions, IEEE Trans. Computers, vol. 43, no. 5, pp. 608–613, 1994.

Combinatorial Network Abstraction by Trees and Distances

Stefan Eckhardt[1], Sven Kosub[1], Moritz G. Maaß[1,⋆], Hanjo Täubig[1,⋆⋆], and Sebastian Wernicke[2,⋆⋆⋆]

[1] Fakultät für Informatik, Technische Universität München, Boltzmannstraße 3, D-85748 Garching, Germany
{eckhardt, kosub, maass, taeubig}@in.tum.de

[2] Institut für Informatik, Friedrich-Schiller-Universität Jena, Ernst-Abbe-Platz 2, D-07743 Jena, Germany
wernicke@minet.uni-jena.de

Abstract. This work draws attention to combinatorial network abstraction problems which are specified by a class $\mathcal{P}$ of pattern graphs and a real-valued similarity measure ϱ based on certain graph properties. For fixed $\mathcal{P}$ and ϱ, the optimization task on any graph G is to find a subgraph G' which belongs to $\mathcal{P}$ and minimizes $\varrho(G, G')$. We consider this problem for the natural case of trees and distance-based similarity measures. In particular, we systematically study spanning trees of graphs that minimize distances, approximate distances, and approximate closeness-centrality with respect to some standard vector and matrix norms. The complexity analysis shows that all considered variants of the problem are NP-complete, except for the case of distance-minimization with respect to the L_∞ norm. We further show that unless P = NP, there exist no polynomial-time constant-factor approximation algorithms for the distance-approximation problems if a subset of edges can be forced into the spanning tree.

1 Introduction

Motivation. Network analysis aims at algorithmically exposing certain meaningful structures and characteristics of a complex network that can be considered essential for its functionality (see, e.g., [3] for a recent survey). A (simple) subnetwork containing only the essential parts of a given network is what we refer to as a *network abstraction*.

In this work, we formalize the combinatorial network abstraction problem by specifying a class $\mathcal{P}$ of admissible pattern graphs and a real-valued similarity measure ϱ that rates the degree of correct approximation of a given graph G

⋆ Supported by DFG, grant Ma 870/5-1 (Leibnizpreis Ernst W. Mayr).

⋆⋆ Supported by DFG, grant Ma-870/6-1 (DFG-SPP 1126 Algorithmik großer und komplexer Netzwerke).

⋆⋆⋆ Supported by Deutsche Telekom Stiftung and Studienstiftung des deutschen Volkes.

X. Deng and D. Du (Eds.): ISAAC 2005, LNCS 3827, pp. 1100–1109, 2005.

by a subgraph $G' \subseteq G$ based on certain graph properties. For a fixed pattern class $\mathcal{P}$ and a fixed measure ϱ, the optimization task is to find for any input graph G a subgraph G' which belongs to $\mathcal{P}$ such that $\varrho(G, G')$ is minimal.

Here, we restrict ourselves to trees as the class of pattern graphs (although some results seem to easily carry over to related structures such as spanning subgraphs with a restricted number of edges) because they are the sparsest and simplest subgraphs that may connect all vertices of a network. Moreover, for several applications the use of spanning trees as an approximation of the network has some promising advantages:

1. *Understanding network dynamics.* A recent study [15] of communication kernels (which handle the majority of network traffic) shows that the organization of many complex networks is heavily influenced by their scale-free spanning trees.
2. *Guiding graph-layout for large networks.* We can use elegant tree-layout algorithms for drawing a tree that closely reflects the main characteristics of a given network.
3. *Compressing networks.* Even with most complex networks being sparse themselves, abstraction by trees reduces network sizes significantly.

In search of suitable graph properties for which a high amount of similarity between a network and its abstraction is desirable, we concentrate in this paper on distances as an inherent graph property. To quantify this degree of similarity, we use standard vector and matrix norms $\|\cdot\|_r$ (see Sect. 2 for a review and definitions) on the distance matrix D_G of an input graph G and the distance matrices of its spanning trees. To this end, we consider the following three optimization problems:

1. *Find a spanning tree that minimizes distances.* This corresponds to a similarity measure $\varrho_r(G,T) = \|D_T\|_r$. As an example, for the L_1 norm, the tree realizing the minimum is known as the minimum average distance tree (or, MAD-tree for short) [14, 8]. For the L_∞ matrix norm, the tree realizing the minimum is known as the minimum diameter spanning tree [6, 12].
2. *Find a spanning tree that approximates distances.* This corresponds to a similarity measure $\varrho_r(G,T) = \|D_T - D_G\|_r$. As an example, for the L_∞ matrix norm, we seek a tree that, for all vertex pairs, does not exceed a certain amount of additive increase in distance. Such trees are known as additive tree-spanners [17]. With the L_1 norm, we are again looking for a MAD-tree.
3. *Find a spanning tree that approximates centralities.* In this paper, we consider the popular notion of closeness centrality [2, 23] which, for any graph $G = (V, E)$ and vertex $v \in V$, is defined as $c_G(v) = (\sum_{t \in V} d_G(v,t))^{-1}$. The optimization problem is then based on the similarity measure $\varrho_r(G,T) = \|c_G - c_T\|_r$ for some vector norm $\|\cdot\|_r$.

Note that—except for the L_1 matrix norm—distance-minimizing spanning trees and optimal distance-approximating spanning trees typically cannot be used to

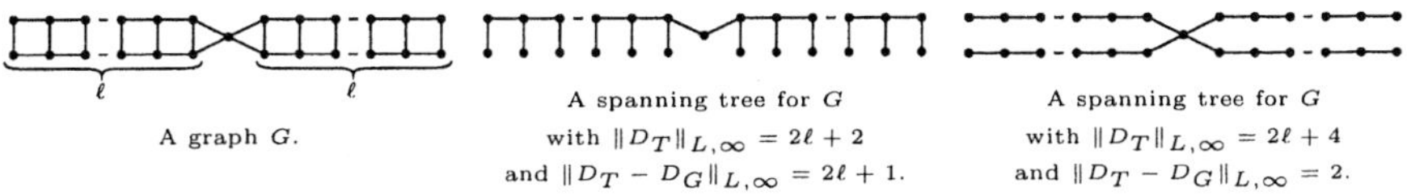

Fig. 1. Distance-minimization and distance-approximation do not provide good approximate solutions for each other with respect to the norm L_∞

provide good approximate solutions for each other. An example for this (with respect to L_∞) is given in Fig. 1.

Results. We study the impact of the norm on the computational complexity of the above-mentioned network abstraction problems. For computing distance-minimizing spanning trees, two results have already been known, namely that there exists a polynomial-time algorithm for computing a minimum diameter spanning tree [6, 12] and that it is NP-complete to decide on input (G, γ) whether there is a spanning tree T of G such that $\|D_T\|_{L,1} \leq \gamma$ [14]. For distance-approximating spanning trees, even for L_1 and L_∞, no such results have so far been established to the best of our knowledge.[1]

In Sect. 3.2, we prove that deciding whether there exists a spanning tree T such that $\|D_T\|_r \leq \gamma$ for any given instance (G, γ) is NP-complete for all matrix norms within our framework where complexity has been unknown so far. We also consider fixed-edge versions (as, e.g., in [5]) where problem instances additionally specify a set of edges E_0 that must be contained in the spanning tree. If we allow arbitrary edge sets for E_0, then even Minimum Diameter Spanning Tree becomes NP-complete.

In Sect. 3.3, we prove that deciding whether there is a spanning tree T of G such that $\|D_T - D_G\| \leq \gamma$ for any given instance (G, γ) is NP-complete for all matrix norms within our framework, i.e., essentially for all standard norms (with exception of the spectral norm, a case which is left open). This is somewhat surprising, since at least in the case of L_∞ one might have hoped for a polynomial-time algorithm based on the polynomial-time algorithms for computing minimum diameter spanning trees. We also prove that the fixed-edge versions of finding optimal distance-approximating spanning trees cannot be approximated in polynomial-time within constant factor unless P = NP.

Finally, in Sect. 3.4, we prove that with respect to closeness centrality, deciding for a given instance (G, γ) whether there is a spanning tree T such that $\|c_G - c_T\|_r \leq \gamma$ is NP-complete for the L_1 vector norm.

Related work. Besides the already mentioned minimum diameter spanning trees [6, 12] and MAD-trees [14, 8], several notions of distance-approximability by trees have been considered in the literature. One variant is obtained by considering the *stretch* $d_T(u, v)/d_G(u, v)$ over all distinct vertices $u, v \in V$. If the stretch is at most γ, then the tree is called γ-multiplicative tree spanner

[1] Note that in contrast to some claims in the literature the results in [18] do *not* provide a proof for the NP-completeness of deciding whether there is a spanning tree T with $\|D_T - D_G\|_{L,\infty} \leq \gamma$, neither does an easily conceivable adaption.

(see, e.g., [22]—recently, also combinations of additive and multiplicative tree-spanners have been studied [10]). Finding a minimum maximum-stretch tree is NP-hard even for unweighted planar graphs [11] and cannot be approximated by a factor better than $(1+\sqrt{5})/2$ unless P = NP [19]. The problem of finding a minimum *average*-stretch tree is also NP-hard [14].

Spanning *subgraphs* (not only trees) with certain bounds on distance increases have been intensively studied since the pioneering work in [1, 21, 7]. The most general formulation of a spanner problem is the following [18]: A spanning subgraph H of G is an $f(x)$-spanner for G if and only if $d_H(u,v) \leq f(d_G(u,v))$ for all $u, v \in V(G)$. The computational problem then is to find an $f(x)$-spanner with the minimum number of edges, a problem somewhat dual to ours since it fixes a bound on the distance increase and tries to minimize the size of the subgraphs, whereas we fix the size of the subgraph and try to minimize the bounds. In a series of papers, the hardness of the spanner problems has been exhibited (see, e.g., [20, 5, 4, 16]). The version closest to our problem is to ask for a given graph G and two given parameters m, t if there exists an additive t-spanner for G with no more than m edges. This problem is NP-complete [18]. In the case that $m = n - 1$ is fixed, it becomes the problem of finding the best possible distance-approximating spanning tree with respect to $\|\cdot\|_{L,\infty}$. However, the corresponding NP-completeness proof for the general case relies heavily on the number of edges in the instance and hence a translation to an NP-completeness proof for the tree case is not obvious.

2 Notation

We consider simple, undirected, and unweighted graphs G with vertex set V and edge set E. For two vertices $v, w \in V$, the distance between v and w (i.e., the minimum number of edges in a path between u to v) in G is denoted by $d_G(v, w)$. The corresponding distance matrix is denoted by D_G. Clearly, D_G is symmetric with all entries being non-negative. Moreover, for any spanning tree T of a graph G, we have $D_T[i,j] \geq D_G[i,j]$ for all $v_i, v_j \in V$. We use the following well-known norms to evaluate a matrix A in $\mathbb{R}^{n \times n}$:

- The L_p norms $\|A\|_{L,p} \stackrel{\text{def}}{=} \left(\sum_{i=1}^n \sum_{j=1}^n |a_{i,j}|^p\right)^{1/p}$ for $1 \leq p < \infty$.[2]
- The L_∞ norm $\|A\|_{L,\infty} \stackrel{\text{def}}{=} \max_{i,j \in \{1,\ldots,n\}} |a_{i,j}|$.
- The maximum-column-sum norm $\|A\|_1 \stackrel{\text{def}}{=} \max_{j \in \{1,\ldots,n\}} \sum_{i=1}^n |a_{i,j}|$.
- The maximum-row-sum norm $\|A\|_\infty \stackrel{\text{def}}{=} \max_{i \in \{1,\ldots,n\}} \sum_{j=1}^n |a_{i,j}|$.

Trivially, for symmetric matrices we have $\|A\|_1 = \|A\|_\infty$. Therefore, we only consider the maximum-column-sum norm in our results to avoid redundancy.

[2] In the last part of the paper, we use L_p norms for vectors as well: for any $1 \leq p < \infty$ and vector $x \in \mathbb{R}^n$, define $\|x\|_p \stackrel{\text{def}}{=} \left(\sum_{i=1}^n |x_i|^p\right)^{1/p}$.

3 Hardness Results

All our theorems establish hardness results that rely on similar constructions (which, however, depend on parameters that must be tuned in a non-trivial manner). We gather these essential constructions in the next section, followed by our results.

Note that, due to lack of space, we defer the proofs for our results to [9].

3.1 Gadgets

Graph representation of X3C instances. Given a family $\mathcal{C} = \{C_1, \ldots, C_s\}$ of 3-element subsets of a set $L = \{l_1, \ldots, l_{3m}\}$, the NP-complete problem EXACT-3-COVER (X3C) asks whether there exists a subfamily $\mathcal{S} \subseteq \mathcal{C}$ of pairwise disjoint sets such that $\bigcup_{A \in \mathcal{S}} = L$. A subfamily $\mathcal{S}$ satisfying this property is called an *admissible solution* to $(\mathcal{C}, L)$. Suppose we are given an X3C instance $(\mathcal{C}, L)$ and let a, b be arbitrary natural numbers. Following a construction from [14], we define the graph $G_{a,b}(\mathcal{C}, L)$ to consist of the vertex set

$$V \stackrel{\text{def}}{=} \mathcal{C} \cup L \cup \underbrace{\{r_1, \ldots, r_a\}}_{\stackrel{\text{def}}{=} R} \cup \underbrace{\{x\}}_{\stackrel{\text{def}}{=} X} \cup \underbrace{\{k_{1,1}, \ldots, k_{1,b}, \;\; \cdots \;\; , k_{3m,1}, \ldots, k_{3m,b}\}}_{\stackrel{\text{def}}{=} K}$$

and the edge set

$$\begin{aligned} E \stackrel{\text{def}}{=} & \{ \; \{r_\mu, x\} \mid \mu \in \{1, \ldots, a\} \; \} \cup \{ \; \{C_\mu, x\} \mid \mu \in \{1, \ldots, s\} \; \} \cup \\ & \cup \{ \; \{l_\mu, C_\nu\} \mid l_\mu \in C_\nu \; \} \cup \{ \; \{l_\mu, l_\nu\} \mid \mu, \nu \in \{1, \ldots, 3m\} \; \} \cup \\ & \cup \{ \; \{k_{\mu,\nu}, l_\mu\} \mid \mu \in \{1, \ldots, 3m\} \text{ and } \nu \in \{1, \ldots, b\} \; \}. \end{aligned}$$

This construction is illustrated in Fig. 2. Given an admissible solution $\mathcal{S}$ to an X3C instance $(\mathcal{C}, L)$, we can identify a corresponding spanning subgraph $T_{\mathcal{S}}$ called *solution tree* in $G_{a,b}(\mathcal{C}, L)$ through the edge set

$$\begin{aligned} E(T_{\mathcal{S}}) = & \{ \; \{r_\mu, x\} \mid \mu \in \{1, \ldots, a\} \; \} \cup \{ \; \{C_\mu, x\} \mid \mu \in \{1, \ldots, s\} \; \} \cup \\ & \cup \{ \; \{l_\mu, C_\nu\} \mid l_\mu \in C_\nu \text{ and } C_\nu \in \mathcal{S} \; \} \cup \\ & \cup \{ \; \{k_{\mu,\nu}, l_\mu\} \mid \mu \in \{1, \ldots, 3m\} \text{ and } \nu \in \{1, \ldots, b\} \; \}. \end{aligned}$$

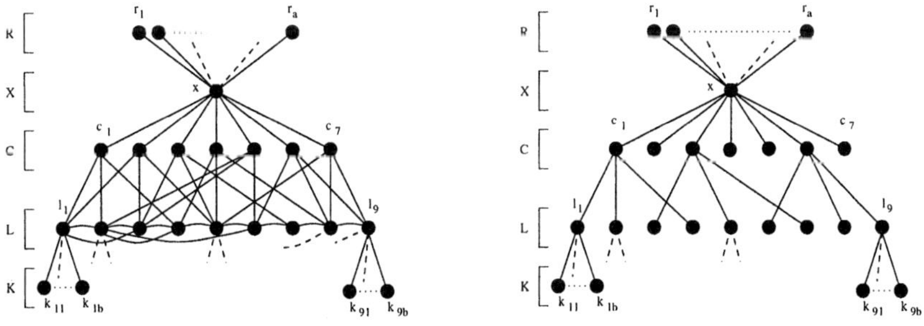

Fig. 2. Graph representation of an X3C instance and a corresponding solution tree

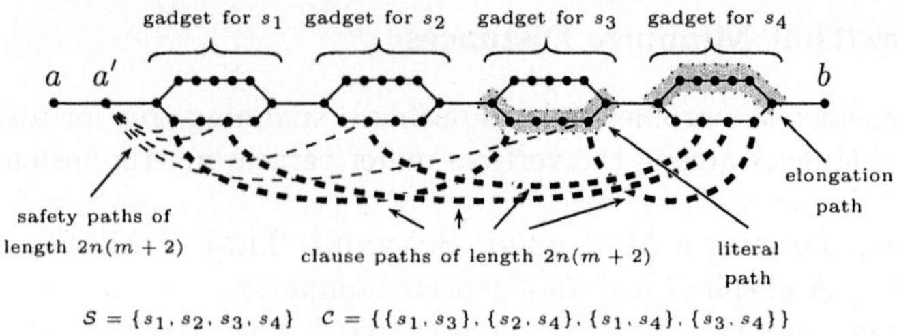

Fig. 3. Construction of a 2HS gadget $G(\mathcal{C}, \mathcal{S}, k)$. The dashed paths that are drawn bold consist solely of edges that must be contained in a spanning tree for the graph.

Lemma 1. *Let $(\mathcal{C}, L)$ be an X3C instance, $a, b \in \mathbb{N}$. Let T be any spanning tree of the graph $G_{a,b}(\mathcal{C}, L)$. There exists an admissible solution $\mathcal{S} \subseteq \mathcal{C}$ such that $T = T_{\mathcal{S}}$ if and only if the following conditions are satisfied:*

1. *For all $\mu \in \{1, \ldots, s\}$, the tree T contains the edge $\{C_\mu, x\}$.*
2. *For all $\mu \in \{1, \ldots, 3m\}$, there is a $\nu \in \{1, \ldots, s\}$ such that T contains the edge $\{l_\mu, C_\nu\}$.*
3. *For all $\mu \in \{1, \ldots, s\}$, vertex C_μ has either four neighbors in T or one.* □

Graph representation of 2HS instances. Given a family $\mathcal{C} = \{C_1, \ldots, C_m\}$ of 2-element subsets of a set $\mathcal{S} = \{s_1, \ldots, s_n\}$ and a natural number k, the NP-complete 2-HITTING SET (2HS) problem asks whether there exists a subset $\mathcal{S}' \subseteq \mathcal{S}$ such that $\|\mathcal{S}'\| \leq k$ and $\mathcal{S}' \cap C_\mu \neq \emptyset$ for all $\mu \in \{1, \ldots, m\}$.[3] A subset $\mathcal{S}' \subseteq \mathcal{S}$ having this property is called an *admissible solution* to a 2HS instance $(\mathcal{C}, \mathcal{S}, k)$. Suppose we are given an instance $(\mathcal{C}, \mathcal{S}, k)$ of 2HS where $\|\mathcal{C}\| = m$ and $\|\mathcal{S}\| = n$. We define the graph $G(\mathcal{C}, \mathcal{S}, k)$ to consist of

- two vertices a, a', and b,
- for each $s_\mu \in \mathcal{S}$, consisting of vertices $v_\mu, v'_\mu, u^\mu_1, \ldots, u^\mu_{m+1}$ and $v^\mu_1, \ldots, v^\mu_m$,
- for each clause $C_\mu = \{s_\nu, s_\kappa\} \in \mathcal{C}$, *clause paths* of length $2n(m+2)$ connecting v^ν_μ with v^κ_μ and *safety paths* of length $2n(m+2)$ connecting v^ν_μ with a'.

For each $s_\mu \in \mathcal{S}$ the *literal gadget* G_μ consists of two vertices v_μ and v'_μ called *connection vertices*. Both v_μ and v'_μ are connected via a path $(v_\mu, u^\mu_1, \ldots, u^\mu_{m+1} v'_\mu)$ of length $m+2$ called *elongation path* and a path $(v_\mu, v^\mu_1, \ldots, v^\mu_m v'_\mu)$ of length $m+1$ called the *literal path.* The construction is illustrated in Fig. 3.

Lemma 2. *Let $(\mathcal{C}, \mathcal{S}, k)$ be an instance of* 2HS. *Then we have $d_{G(\mathcal{C},\mathcal{S},k)}(a, b) = 2 + n(m+2)$. Moreover, there exists an admissible solution $\mathcal{S}' \subseteq \mathcal{S}$ to $(\mathcal{C}, \mathcal{S}, k)$ if and only if there exists a spanning tree T of $G(\mathcal{C}, \mathcal{S}, k)$ containing all edges in the clause paths such that $d_T(a, b) \leq d_{G(\mathcal{C},\mathcal{S},k)}(a, b) + k$.* □

[3] 2HS is better known as VERTEX COVER. For the sake of readability (i.e., to avoid an overuse of the terms "vertices" and "edges"), we use the 2HS formulation.

3.2 Trees That Minimize Distances

Here we consider the problem of computing a spanning tree for a graph which minimizes distances among the vertices under certain matrix norms $\|\cdot\|_r$.

> *Problem:* DISTANCE-MINIMIZING SPANNING TREE (DMST)
> *Input:* A graph G and an algebraic number γ.
> *Question:* Does G contain a spanning tree T with $\|D_T\|_r \leq \gamma$?

Using the graph representation $G_{a,b}(\mathcal{C}, L)$ for any X3C instance $(\mathcal{C}, L)$, the following theorem can be shown.

Theorem 3. DMST *with respect to the norms* $\|\cdot\|_1$ *and* $\|\cdot\|_{L,p}$ *is* NP*-complete for all* $p \in \mathbb{N}_+$*, even when restricted to planar graphs.* □

It is known that a minimum-diameter spanning tree in a graph—i.e., DMST with respect to $\|\cdot\|_{L,\infty}$)—can be found in polynomial time [6, 12]. However, the next theorem shows that the *fixed-edge* version of this problem is intractable. This version additionally contains an edge set $E_0 \subseteq E(G)$ and we seek a spanning tree T such that $\|D_T\|_r \leq \gamma$ and $E_0 \subseteq E(T)$. Using the graph representation $G(\mathcal{C}, \mathcal{S}, k)$ for any given instance $(\mathcal{C}, \mathcal{S}, k)$ of 2HS gives us the following theorem.

Theorem 4. *The fixed-edge version of* DMST *with respect to the norm* $\|\cdot\|_{L,\infty}$ *is* NP*-complete.* □

3.3 Trees That Approximate Distances

We now turn to the problem of finding spanning trees that approximate the distances in a graph under a given matrix norm $\|\cdot\|_r$. We also examine the fixed-edge version of this problem, which is specified in the same way as for DMST.

> *Problem:* DISTANCE-APPROXIMATING SPANNING TREE (DAST)
> *Input:* A graph G and an algebraic number γ
> *Question:* Does G contain a spanning tree T with $\|D_T - D_G\|_r \leq \gamma$?

NP*-completeness results.* Using the graph representation $G_{a,b}(\mathcal{C}, L)$ for any X3C instance $(\mathcal{C}, L)$, the following theorem can be shown.

Theorem 5. DAST *with respect to the norms* $\|\cdot\|_1$ *and* $\|\cdot\|_{L,p}$ *is* NP*-complete for all* $p \in \mathbb{N}_+$. □

For proving the NP-completeness for the L_∞ matrix norm, it is helpful to first establish the result for the fixed-edge version (by reduction from 2HS).

Lemma 6. *The fixed-edge version of* DAST *with respect to the norm* $\|\cdot\|_{L,\infty}$ *is* NP*-complete.* □

To get rid of the fixed edges, we replace them by cycles such that deleting a fixed edge will cause the distance between two cycle vertices to increase by more than the allowed threshold γ, which then gives us the hardness result for the norm $\|\cdot\|_{L,\infty}$.[4]

Lemma 7. *Let $G = (V, E)$ be any graph and let $\{v, w\}$ be an arbitrary non-bridge edge in G. For $k > 3$, let G' be the graph resulting from adding a path $(v, u_1, \ldots, u_k, w)$ to G where $u_\mu \notin V$ for all $\mu \in \{1, \ldots, k\}$. There exists a spanning tree T of G which includes the edge $\{v, w\}$ and satisfies $\|D_T - D_G\|_{L,\infty} \leq k$ if and only if there exists a spanning tree T' of G' with $\|D_{T'} - D_{G'}\|_{L,\infty} \leq k$.* □

Theorem 8. DAST *with respect to the norm $\|\cdot\|_{L,\infty}$ is* NP*-complete.* □

Inapproximability results. We now show that—independent of the norm—it is hard to approximate the fixed-edge version of DAST in polynomial time within a constant factor. Let G be a graph and $E' \subseteq E(G)$ be the set of fixed edges. We say that an algorithm $\mathcal{A}$ is a *polynomial-time constant-factor approximation algorithm* for the fixed-edge version of DAST with respect to $\|\cdot\|_r$ if, for some constant $\delta > 0$, it computes a spanning tree $T_\mathcal{A}$ of G with $E' \subseteq E(T)$ in polynomial time such that $\|D_{T_\mathcal{A}} - D_G\|_r \leq \delta \cdot \|D_{T_{\text{opt}}} - D_G\|_r$. Here, T_{opt} is the *optimal tree*, i.e., $\|D_{T_{\text{opt}}} - D_G\|_r = \min_T \|D_T - D_G\|_r$ for all spanning trees T of G.

2HS is not polynomial-time approximable within a constant factor better than 7/6 unless P = NP [13]. To make use of this in our context, note that the main idea behind the graph representation $G(\mathcal{C}, \mathcal{S}, k)$ for any given instance $(\mathcal{C}, \mathcal{S}, k)$ of 2HS is that choosing an element from $\mathcal{S}$ into the solution corresponds to opening a literal path in $G(\mathcal{C}, \mathcal{S}, k)$. This opening is "penalized" by the elongation path, increasing the distance between the vertices a and b. The key idea now is to increase this penalty in a super-linear way by recursively replacing elongation paths with graph representations of the given 2HS instance.

More formally, suppose we are given an instance $(\mathcal{C}, \mathcal{S}, k)$ of 2HS where $\|\mathcal{C}\| = m$ and $\|\mathcal{S}\| = n$. For $j \in \mathbb{N}_+$ we define the graph $G(\mathcal{C}, \mathcal{S}, k, j)$ recursively as follows: For $j = 1$, the graph $G(\mathcal{C}, \mathcal{S}, k, j)$ is just the graph $G(\mathcal{C}, \mathcal{S}, k)$. For $j > 1$, define $l_{j-1} \stackrel{\text{def}}{=} d_{G(\mathcal{C},\mathcal{S},k,j-1)}(a, b)$. Then $G(\mathcal{C}, \mathcal{S}, k, j)$ consists of

- three vertices a, a', and b,
- *literal paths* P_μ for each $s_\mu \in \mathcal{S}$, consisting of vertices $v_1^\mu, \ldots, v_{l_{j-1}-1}^\mu$,
- *elongation gadgets* G_μ for each $s_\mu \in \mathcal{S}$, consisting of a copy of $G(\mathcal{C}, \mathcal{S}, k, j-1)$, with the vertices a, a', and b in $G(\mathcal{C}, \mathcal{S}, k, j-1)$ relabeled as a_μ, a'_μ and b_μ,
- for each $C_\mu = \{s_\nu, s_\kappa\} \in \mathcal{C}$, *clause paths* and *safety paths* of length $2nl_{j-1}$ connecting v_μ^ν with v_μ^κ and v_μ^ν with a', respectively.

For each clause $C_\mu \in \mathcal{C}$ the vertices a_μ and b_μ are connected via a literal path $(a_\mu, v_1^\mu, \ldots, v_{l_{j-1}-1}^\mu, b_\mu)$. Furthermore, the edges $\{a, a'\}$, $\{a', a_1\}$, $\{b_m, b\}$, and the

[4] A similar technique with two cycles was used in [5–Lemma 3] to guarantee that any minimum t-spanner (i.e., a spanning subgraph with smallest number of edges such that $d_G(u, v) \leq t \cdot d_T(u, v)$ for all $u, v \in V$) contains a certain edge. However, this construction does not work in the context of additive distance growth and trees.

edges $\{b_i, a_{i+1}\}$ for all $1 \leq i \leq m-1$ are in $G(\mathcal{C}, \mathcal{S}, k, j)$. Note that the graph size is polynomial in the size of the instance $(\mathcal{C}, \mathcal{S}, k)$ and j. The following analogue to Lemma 2 can be established for our new graph representation of 2HS:

Lemma 9. *Let* $(\mathcal{C}, \mathcal{S}, k)$ *be a given instance of* 2HS *and* $j \in \mathbb{N}_+$. *Then, we have that* $d_{G(\mathcal{C},\mathcal{S},k,j)}(a,b) = 3\frac{n^{j+1}-1}{n-1} + n^j(m-1) - 1$. *Moreover, there exists an admissible solution* $\mathcal{S}' \subseteq \mathcal{S}$ *to* $(\mathcal{C}, \mathcal{S}, k)$ *if and only if there exists a spanning tree* T *of* $G(\mathcal{C}, \mathcal{S}, k, j)$ *containing all edges in the clause paths of all instances* $G(\mathcal{C}, \mathcal{S}, k, j')$, *for* $j' < j$, *of which* $G(\mathcal{C}, \mathcal{S}, k, j)$ *is composed, including the clause paths in* $G(\mathcal{C}, \mathcal{S}, k, j)$ *such that* $d_T(a,b) \leq d_{G(\mathcal{C},\mathcal{S},k,j)}(a,b) + k^j$. □

Lemma 10. *Unless* P = NP, *there is no polynomial-time algorithm* $\mathcal{A}$ *that, given a* 2HS *instance* $(\mathcal{C}, \mathcal{S}, k)$ *and parameter* $j \in \mathbb{N}_+$, *computes a spanning tree* $T_\mathcal{A}$ *of* $G(\mathcal{C}, \mathcal{S}, k, j)$ *such that* $d_{T_\mathcal{A}}(a,b) - d_{G(\mathcal{C},\mathcal{S},k,j)}(a,b) \leq \delta \cdot d_{T_{opt}}(a,b) - d_{G(\mathcal{C},\mathcal{S},k,j)}(a,b)$ *for any* $\delta > 0$. *Here,* T_{opt} *is a spanning tree of* $G(\mathcal{C}, \mathcal{S}, k, j)$ *such that* $d_{T_{opt}}(a,b) - d_{G(\mathcal{C},\mathcal{S},k,j)}(a,b) = \min_T d_T(a,b) - d_{G(\mathcal{C},\mathcal{S},k,j)}(a,b)$ *where the minimum is taken over all spanning trees that include all clause paths.* □

Theorem 11. *Unless* P = NP, *there is no polynomial-time constant-factor approximation algorithm for the fixed-edge version of* DAST *with respect to the norms* $\|\cdot\|_{L,\infty}$, $\|\cdot\|_{L,p}$, *and* $\|\cdot\|_1$. □

3.4 Trees That Approximate Centralities

Closeness centrality $c_G : V \to \mathbb{R}$ for a graph $G = (V, E)$ is defined for all $v \in V$ as $c_G(v) \stackrel{\text{def}}{=} (\sum_{t \in V} d_G(v,t))^{-1}$ [2, 23]. Here we consider the problem of computing a spanning tree such that its centrality function is as close as possible to the centrality function of the original graph with respect to some vector norm $\|\cdot\|_r$.

Problem: Closeness-Approximating Spanning Tree (CAST)
Input: A graph G and an algebraic number γ
Question: Does G contain a spanning tree T with $\|c_G - c_T\|_r \leq \gamma$?

By a reduction from our X3C gadget, we can show the following theorem.

Theorem 12. CAST *with respect to the norm* $\|\cdot\|_1$ *is* NP-*complete, even when restricted to planar graphs.* □

4 Conclusion

We have introduced the problem of combinatorial network abstraction and systematically studied it for the natural case of trees and distance-based similarity measures. This provides the first computational complexity study in this area, presented in a unifying framework.

As an interesting problem left open here, future research might consider the presented problems with respect to the spectral norm—in the light that NP-completeness appears with coarser norms and the value of the spectral norm is always smaller than that of the norms considered here, there might even be a chance for polynomial-time solvability.

References

1. B. Awerbuch. Complexity of network synchronization. *J. ACM*, 32(4):804–823, 1985.
2. M. A. Beauchamp. An improved index of centrality. *Behavioral Science*, 10:161–163, 1965.
3. U. Brandes and T. Erlebach, editors. *Network Analysis: Methodological Foundations*, volume 3418 of *LNCS*. Springer-Verlag, 2005.
4. U. Brandes and D. Handke. NP-completeness results for minimum planar spanners. *Discr. Math. & Theor. Comp. Sci.*, 3(1):1–10, 1998.
5. L. Cai. NP-completeness of minimum spanner problems. *Discr. Appl. Math.*, 48(2):187–194, 1994.
6. P. M. Camerini, G. Galbiati, and F. Maffioli. Complexity of spanning tree problems: Part I. *Europ. J. Oper. Res.*, 5(5):346–352, 1980.
7. L. P. Chew. There are planar graphs almost as good as the complete graph. *J. Comp. Sys. Sci.*, 39(2):205–219, 1989.
8. E. Dahlhaus, P. Dankelmann, W. Goddard, and H. C. Swart. MAD trees and distance-hereditary graphs. *Discr. Appl. Math.*, 131(1):151–167, 2003.
9. S. Eckhardt, S. Kosub, M. G. Maaß, H. Täubig, and S. Wernicke. Combinatorial network abstraction by trees and distances. Technical Report TUM-I0502, Technische Universität München, Institut für Informatik, 2005.
10. M. Elkin and D. Peleg. $(1+\epsilon, \beta)$-spanner constructions for general graphs. *SIAM J. Comp.*, 33(3):608–631, 2004.
11. S. P. Fekete and J. Kremer. Tree spanners in planar graphs. *Discr. Appl. Math.*, 108(1–2):85–103, 2001.
12. R. Hassin and A. Tamir. On the minimum diameter spanning tree problem. *Inform. Proc. Lett.*, 53(2):109–111, 1995.
13. J. Håstad. Some optimal inapproximability results. *J. ACM*, 48(4):798–859, 2001.
14. D. S. Johnson, J. K. Lenstra, and A. H. G. Rinnooy Kan. The complexity of the network design problem. *Networks*, 8:279–285, 1978.
15. D.-H. Kim, J. D. Noh, and H. Jeong. Scale-free trees: The skeletons of complex networks. *Phys. Rev. E*, 70(046126), 2004.
16. G. Kortsarz. On the hardness of approximating spanners. *Algorithmica*, 30(3):432–450, 2001.
17. D. Kratsch, H.-O. Le, H. Müller, E. Prisner, and D. Wagner. Additive tree spanners. *SIAM J. Discr. Math.*, 17(2):332–340, 2003.
18. A. L. Liestman and T. C. Shermer. Additive graph spanners. *Networks*, 23(4):343–363, 1993.
19. D. Peleg and E. Reshef. Low complexity variants of the arrow distributed directory. *J. Comp. Sys. Sci.*, 63(3):474–485, 2001.
20. D. Peleg and A. A. Schäffer. Graph spanners. *J. Graph Theory*, 13(1):99–116, 1989.
21. D. Peleg and J. D. Ullman. An optimal synchronizer for the hypercube. *SIAM J. Comp.*, 18(4):740–747, 1989.
22. E. Prisner. Distance approximating spanning trees. In *Proc. STACS'97*, volume 1200 of *LNCS*, pages 499–510. Springer-Verlag, 1997.
23. G. Sabidussi. The centrality index of a graph. *Psychometrica*, 31:581–603, 1966.

Drawing Phylogenetic Trees*
(Extended Abstract)

Christian Bachmaier, Ulrik Brandes, and Barbara Schlieper

Department of Computer & Information Science, University of Konstanz, Germany
{christian.bachmaier, ulrik.brandes, barbara.schlieper}@uni-konstanz.de

Abstract. We present linear-time algorithms for drawing phylogenetic trees in radial and circular representations. In radial drawings given edge lengths (representing evolutionary distances) are preserved, but labels (names of taxons represented in the leaves) need to be adjusted, whereas in circular drawings labels are perfectly spread out, but edge lengths adjusted. Our algorithms produce drawings that are unique solutions to reasonable criteria and assign to each subtree a wedge of its own. The linear running time is particularly interesting in the circular case, because our approach is a special case of Tutte's barycentric layout algorithm involving the solution of a system of linear equations.

1 Introduction

Phylogeny is the study of the evolutionary relationships within a group of organisms. A phylogenetic tree represents the evolutionary distances among the organisms represented by its leaves. Due to the increasing size of data sets, drawings are essential for exploration and analysis. In addition to the usual requirements for arbitrary tree structures, drawings of phylogenetic trees should also depict given edge lengths and leaf names. Standard approaches [3,12,18] do not take these criteria into account (see [2,7] for an overview of tree drawing algorithms). Popular software tools in computational biology such as TreeView [10], PAUP* [14], or PHYLIP [11] also provide drawings of phylogenetic trees, but the underlying algorithms are not documented.

There are essentially two forms of representation for phylogenetic trees. For an overview see, e. g., [1]. Both are variations of dendrograms, since many algorithms for the construction of phylogenetic trees are based on clustering (see, e. g., [15]). They differ in that leaf labels are either placed monotonically along one axis or around the tree structure. While the first class of representations is very similar to standard dendrograms and easy to layout, it is somewhat difficult to understand the nesting of subtrees from the resulting drawings. We focus on the algorithmically more challenging and graphically more appealing second class of representations.

In radial tree drawings edges extend radially monotonic away from the root, and we give a linear-time algorithm that preserves all edge lengths exactly. In

* Partially supported by DFG under grant Br 2158/1-2.

X. Deng and D. Du (Eds.): ISAAC 2005, LNCS 3827, pp. 1110–1121, 2005.

circular tree drawings leaves are placed equidistantly on the perimeter of a circle and the tree is confined to the inside of the circle. Note that it may not be possible to preserve edge lengths in this representation. We give an algorithm that heuristically minimizes length deviations and, even though based on solving a system of linear equations, runs in linear time as well. Both algorithms yield drawings that are unique in a well-defined sense up to scaling, rotation, and translation. Since each subtree is confined to a wedge rather than an interval of its own, their nesting structure is more apparent than in vertical or horizontal representations. While, typically, phylogenetic trees are cases extended binary trees, our algorithms apply to general trees.

This paper is organized as follows. Basic notation and some background is provided in Sect. 2. In Sects. 3 and 4 we present our algorithms for radial and circular representations, and conclude in Sect. 5.

2 Preliminaries

Throughout the paper let $T = (V, E, \delta)$ denote a phylogenetic tree with $n = |V|$ vertices, $m = |E|$ edges, and positive edge lengths $\delta\colon E \to \mathbb{R}^+$. The leaves of a phylogenetic tree represent the species, molecules, or DNA sequences (*taxons*) under study and its inner vertices represent virtual or hypothetical ancestors. The length of an edge represents the evolutionary divergence between its incident vertices, and the entire tree represents a tree metric fitted to a (potentially noisy and incomplete) dissimilarity matrix defined over all taxons. Since we want the length of an edge $e \in E$ to resemble $\delta(e)$ as closely as possible, only positive values are allowed. If a method for tree construction, e. g., [5, 6, 9, 13], assigns negative or zero length to an edge, we set it to a small positive constant, e. g., to a fraction of the smallest positive edge length in the tree.

Let $\deg\colon V \to \mathbb{N}$ denote the number of edges incident to a vertex and note that typical tree reconstruction methods yield rooted trees in which most inner vertices have two children. We use $\mathsf{root}(T)$ to refer to the root of a tree T. Each vertex $v \in V \setminus \{\mathsf{root}(T)\}$ has an unique parent $\mathsf{parent}(v)$, and we denote the set of children by $\mathsf{children}(v)$. For a vertex $v \in V$ let $T(v)$ be the induced subtree, i. e., the subtree of all descendants of v (including v itself). Clearly, $T(\mathsf{root}(T)) = T$. For a subtree $T(v)$ of T we use $\mathsf{leaves}(T(v))$ containing all vertices v in $T(v)$ which have $\deg(v) = 1$ in T to denote the set of its leaves. Tree edges are directed away from the root, i. e., for an edge (u, v) we have $u = \mathsf{parent}(v)$. We sometimes use $\{u, v\}$ to refer to the underlying undirected edge.

Finally, we assume that in ordered trees the outgoing edges are to be drawn in counterclockwise order, i. e., for an inner vertex other than the root the counterclockwise first edge after the incoming edge is that of the first child.

3 Radial Drawings

In this section, we describe a drawing algorithm that yields a planar radial drawing of a phylogenetic tree $T = (V, E, \delta)$. It is $\mathcal{NP}$-complete to decide whether a

graph can be drawn in the plane with prescribed edge lengths, even if the graph is planar and all edges have unit lengths [4], but for trees we can represent any assignment of positive edge lengths exactly.

3.1 Basic Algorithm

The main idea is to assign to each subtree $T(v)$ a wedge of angular width proportional to the number of leaves in $T(v)$. The wedge of an inner vertex is divided among its children, and tree edges are drawn along wedge angle bisectors, so that they can have any length without violating disjointness. See Fig. 1 for illustration. Algorithm 1 therefore traverses the input tree twice:

- in a postorder traversal, the number $l_v = |\mathsf{leaves}(T_v)|$ of leaves in each subtree $T(v)$ is determined, and
- in a subsequent preorder traversal, a child w of an inner vertex v is placed at distance $\delta(v, w)$ on the angular bisector of the wedge reserved for w.

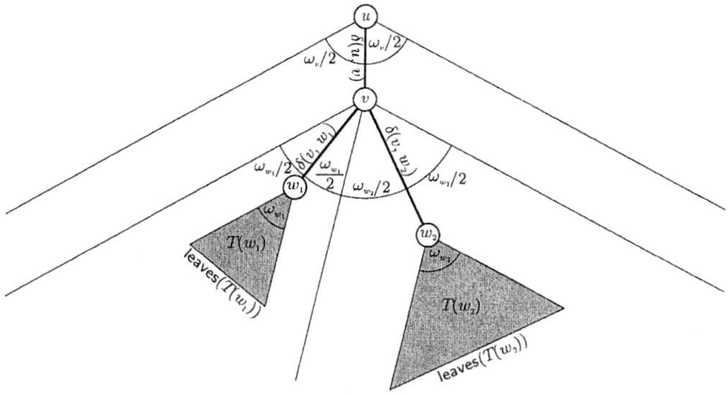

Fig. 1. Wedges of vertex v's neighbors

The following theorem shows that the layouts determined by Algorithm 1 are essentially the only ones that fulfill all natural requirements for radial drawings of trees with given edge lengths.

Theorem 1. *For an unrooted ordered phylogenetic tree $T = (V, E, \delta)$, there is a unique planar radial drawing up to rotation, translation and scaling, that satisfies the following properties:*

1. *Relative edge lengths are preserved exactly.*
2. *Disjoint subtrees are confined to disjoint wedges.*
3. *Subtrees are centered at the bisectors of their wedges.*
4. *The angular width of the wedge of a subtree is proportional to the number of leaves in that subtree.*

Moreover, it can be computed in linear time.

Algorithm 1: RADIAL-LAYOUT

Input: Rooted tree $T = (V, E, \delta)$
Data: Vertex arrays l (number of leaves in subtree), ω (wedge size), and τ (angle of right wedge border)
Output: Coordinates x_v for all $v \in V$

```
begin
    postorder_traversal (root(T))
    x_root(T) ← (0,0)
    ω_root(T) ← 2π
    τ_root(T) ← 0
    preorder_traversal (root(T))
end

procedure postorder_traversal(vertex v)
    if deg(v) = 1 then
        l_v ← 1
    else
        l_v ← 0
        foreach w ∈ children(v) do
            postorder_traversal(w)
            l_v ← l_v + l_w

procedure preorder_traversal(vertex v)
    if v ≠ root(T) then
        u ← parent(v)
        x_v ← x_u + δ(u,v) · (cos(τ_v + ω_v/2), sin(τ_v + ω_v/2))
    η ← τ_v
    foreach w ∈ children(v) do
        ω_w ← (l_w / l_root(T)) · 2π
        τ_w ← η
        η ← η + ω_w
        preorder_traversal(w)
```

Proof. Choose any vertex as the root. By Property 4, the angular width of the wedge of a subtree $T(v)$ is

$$\omega_v = \alpha \cdot \frac{|\mathsf{leaves}\,(T(v))\,|}{|\mathsf{leaves}(T)|} \ . \tag{1}$$

If the root is altered such that $u = \mathsf{parent}(v)$ becomes a child of v in the newly rooted tree T', then

$$\frac{|\mathsf{leaves}\,(T'(u))\,|}{|\mathsf{leaves}(T')|} = \frac{|\mathsf{leaves}(T)| - |\mathsf{leaves}\,(T(v))\,|}{|\mathsf{leaves}(T)|} \ , \tag{2}$$

so that the proportionality factor $\alpha = 2\pi$ and Properties 4, 3 and 2 imply that angles between incident edges are independent of the actual choice of the root. Since relative lengths of edges need to be preserved as well, layouts are unique up to rotation, translation and scaling. Planarity is implied. Clearly, Algorithm 1 determines the desired layouts in linear time. □

3.2 Extensions

Labels of leaves are placed on the angle bisector of the respective wedge. Since the angle of a leaf wedge is $\frac{2\pi}{|\mathsf{leaves}(T)|}$, labels placed close to their leaf may not fit into the wedge. When using a font of height h, non-overlapping labels are guaranteed if they are placed at distance at least

$$\frac{h}{2 \cdot \tan\left(\pi \,/\, |\mathsf{leaves}(T)|\right)} \tag{3}$$

from the parent of their associated leaf.

While the number of leaves is a good indicator of how much angular space is required by a subtree, other scaling schemes can be used to emphasize different aspects such as height, size, or importance of subtrees.

Since child orders are respected by the algorithm, we may sort children according to, say, the size of their subtrees in a preprocessing step. This serves to modify the general appearance of the final layout.

While the confinement of subtrees to wedges of their own nicely separates them, it also results in poor angular resolution and a fair amount of wasted drawing space. We may thus wish to relax this requirement and increase angles between incident edges where possible. This can be achieved during post-processing using a bottom-up traversal, in which the angle between outgoing edges of a vertex v are scaled to the maximum value possible within the wedge of v rooted at $\mathsf{parent}(v)$ (see Fig. 2). Note that labels need to be be taken into account in this angular spreading step, and that the resulting drawings depend on the choice of root. Our experiments suggest that placing the root at the center of the tree yields favorable results.

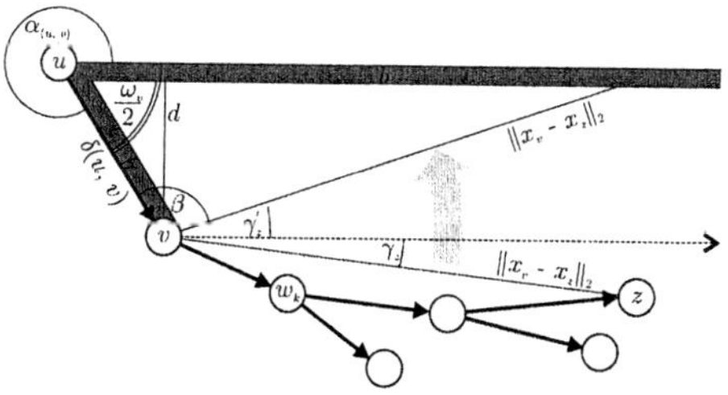

Fig. 2. Increasing angles to fill empty strips around subtrees

4 Circle Drawings

Since it can be difficult to place labels in radial tree drawings, we describe a second method to draw phylogenetic trees. Here, leaves are placed equidistantly along the perimeter of a circle. Again, each leaf thus obtains a wedge of angular width $\frac{2\pi}{|\mathsf{leaves}(T)|}$, but now the radius of the circle determines the maximum height of the font according to (3). It is easy to see that, with this constraint, it might not be possible to draw all edges $e \in E$ with length proportional to $\delta(e)$. Edge length preservation therefore turns into an optimization criterion.

4.1 Basic Algorithm

We use a variant of the weighted version of Tutte's barycentric layout algorithm [16, 17]. The general idea is to fix some vertices to the boundary of a convex polygon and place all other vertices in the weighted barycenter of their neighbors, i. e. v_i is positioned at $x_i = \sum_{v_j \in V}(a_{ij}x_j)$ where a_{ij} is the relative influence of v_j on v_i.

For circular drawings, the leaves of a tree T are fixed to a circle and we define weights a_{ij} by (4). These weights reflect the desired edge lengths, i. e., the shorter an edge $\{v_i, v_j\}$ should be, the more influence has x_j on the resulting coordinate x_i. In the original Tutte algorithm $a_{ij} = \frac{1}{\deg(v_i)}$. Note that the weights a_{ij} sum up to 1 for each v_i, so that v_i is placed inside of the convex hull of its neighbors.

Since we fix leaves to the perimeter of a circle, we can expect in general that an inner vertex of a tree is placed between its children on the one side and its parent on the other side. To counterbalance the accumulated radial influence of the children, their weight is scaled down.

$$s_{ij} = \begin{cases} \dfrac{1}{\delta(v_i, v_j) \cdot (\deg(v_i) - 1)} & \text{if } v_i = \mathsf{parent}(v_j), \\ \dfrac{1}{\delta(v_i, v_j)} & \text{if } v_j = \mathsf{parent}(v_i) \end{cases}$$
$$a_{ij} = \begin{cases} \dfrac{s_{ij}}{\sum_{\{v_i, v_{j'}\} \in E} s_{ij'}} & \text{if } \{v_i, v_j\} \in E, \\ 0 & \text{otherwise} \end{cases} \tag{4}$$

In the following, let $V = \{v_1, \ldots, v_n\}$ with $\mathsf{leaves}(T) = \{v_1, \ldots, v_k\}$ for $k = \lceil \frac{n}{2} \rceil$. After fixing the leaves equidistantly around the circle, we need to solve

$$\begin{pmatrix} a_{k+1,1} & \cdots & a_{k+1,n} \\ \vdots & & \vdots \\ a_{n,1} & \cdots & a_{n,n} \end{pmatrix} \cdot \begin{pmatrix} x_1 \\ \vdots \\ x_n \end{pmatrix} = \begin{pmatrix} x_{k+1} \\ \vdots \\ x_n \end{pmatrix} . \tag{5}$$

By the following lemma, this can be done in linear time by traversing the tree first in postorder to resolve the influence of leaves and then in preorder passing down positions of parents.

Lemma 1. *For $v_i \in V$ and $v_p = \mathit{parent}(v_i)$ define coefficients*

$$c_i = \begin{cases} 0 & \text{if } v_i \in \mathsf{leaves}(T) \cup \{\mathsf{root}(T)\}, \\ \dfrac{a_{pi}}{1 - \sum_{(v_i,v_j)\in E}(a_{ij}c_j)} & \text{otherwise} \end{cases} \tag{6a}$$

and offsets

$$d_i = \begin{cases} x_i & \text{if } v_i \in \mathsf{leaves}(T), \\ \dfrac{\sum_{(v_i,v_j)\in E}(a_{ij}d_j)}{1 - \sum_{(v_i,v_j)\in E}(a_{ij}c_j)} & \text{otherwise .} \end{cases} \tag{6b}$$

Then, (5) has a unique solution with

$$x_i = \begin{cases} d_i & \text{if } v_i = \mathsf{root}(T), \\ c_i x_p + d_i & \text{otherwise} \end{cases} \tag{7}$$

for all inner vertices $v_i \in V \setminus \mathsf{leaves}(T)$.

Proof. We use induction over the vertices. For the base case let v_i be an inner vertex having only leaves as children. Then in case $v_i \neq \mathrm{root}(T)$ let v_p be the parent of v_i and thus

$$\begin{aligned} x_i &= a_{pi}x_p + \sum_{(v_i,v_j)\in E} (a_{ij} \underbrace{x_j}_{=d_j}) = \frac{a_{pi}}{1-0}x_p + \frac{\sum_{(v_i,v_j)\in E}(a_{ij}d_j)}{1-0} = \\ &= \frac{a_{pi}}{1 - \sum_{(v_i,v_j)\in E}(a_{ij}c_j)}x_p + \frac{\sum_{(v_i,v_j)\in E}(a_{ij}d_j)}{1 - \sum_{(v_i,v_j)\in E}(a_{ij}c_j)} = c_i x_p + d_i \ . \end{aligned} \tag{8}$$

The case $v_i = \mathrm{root}(T)$ is a special case of (8) which uses only the second addend. For the inductive step let $v_i \neq \mathrm{root}(T)$ be an inner vertex and v_p the parent of v_i. Then

$$\begin{aligned} x_i &= a_{pi}x_p + \sum_{(v_i,v_j)\in E} (a_{ij}x_j) \overset{\text{i. h.}}{=} a_{pi}x_p + \sum_{(v_i,v_j)\in E} (a_{ij}(c_j x_i + d_j)) = \\ &= a_{pi}x_p + x_i \sum_{(v_i,v_j)\in E} (a_{ij}c_j) + \sum_{(v_i,v_j)\in E} (a_{ij}d_j) = \\ &= \frac{a_{pi}}{1 - \sum_{(v_i,v_j)\in E}(a_{ij}c_j)}x_p + \frac{\sum_{(v_i,v_j)\in E}(a_{ij}d_j)}{1 - \sum_{(v_i,v_j)\in E}(a_{ij}c_j)} = c_i x_p + d_i \ . \end{aligned} \tag{9}$$

The proof for $v_i = \mathrm{root}(T)$ is a special case of (9) which uses only the second addend. □

Algorithm 2: CIRCLE-LAYOUT

Input: Ordered rooted tree $T = (V, E, \delta)$
Data: Vertex arrays c (coefficient), d (offset), and edge array s (weighting)
Output: Coordinates x_v in/on the unit circle for each vertex $v \in V$

```
begin
    i ← 0
    k ← 0
    foreach v ∈ V do
        if deg(v) = 1 then k ← k + 1

    postorder_traversal (root(T))
    preorder_traversal (root(T))
end

procedure postorder_traversal(vertex v)
    foreach w ∈ children(v) do
        postorder_traversal(w)            // opt. ordered by h(w) + δ(v, w)
    if is_leaf(v) or (v = root(T) and deg(root(T)) = 1) then
        c_v ← 0
        d_v ← (cos(2πi/k), sin(2πi/k))    // fix vertex on circle
        i ← i + 1
    else
        S ← 0
        foreach adjacent edge e ← {v, w} do
            if v = root(T) or w = parent(v) then
                s_e ← 1/δ(e)
            else
                s_e ← 1/(δ(e)·(deg(v)−1))
            S ← S + s_e
        t ← t' ← 0
        foreach outgoing edge e ← (v, w) do
            t ← t + (s_e/S) · c_w
            t' ← t' + (s_e/S) · d_w
        if v ≠ root(T) then
            e ← (parent(v), v)
            c_v ← s_e/(S·(1−t))
        d_v ← t'/(1−t)

procedure preorder_traversal(vertex v)
    if v = root(T) then
        x_v ← d_v
    else
        u ← parent(v)
        x_v ← c_v · x_u + d_v

    foreach w ∈ children(v) do
        preorder_traversal(w)
```

Theorem 2. *For a rooted ordered phylogenetic tree $T = (V, E, \delta)$, there is a unique planar circle drawing up to rotation, translation, and scaling, that satisfies the following properties:*

1. *Leaves are placed equidistantly on the perimeter of a circle.*
2. *Disjoint subtrees are confined to disjoint wedges.*
3. *Inner vertices are placed in the weighted barycenter of their neighbors with weights defined by Eq. (4).*

Moreover, it can be computed in linear time.

The most general version of Tutte's algorithm for arbitrary graphs fixes at least one vertex of each component and simply places each vertex in the barycenter of its neighbors, which yields a unique solution. The running time corresponds to solving n symmetric equations, which can be done in $\mathcal{O}(n^3)$ time. For planar graphs $\mathcal{O}(n \log n)$ time can be achieved [8], but it is not known whether this is also a lower bound. We showed that for trees with all leaves in convex position, the running time is in $\mathcal{O}(n)$. Giving up planarity this result can be generalized to trees having arbitrary fixed vertices (at least one). Consider a fixed inner vertex v and let each other vertex w in the subtree $T(v)$ be free. Then $T(v)$ collapses to the position of v. On the other hand, if v has a fixed ancestor u, then all positions of vertices in $T(v)$ are computed and for the rest a restart on $T \backslash T(v) \cup \{v\}$ computes all remaining positions. It follows by induction that our algorithm computes in $\mathcal{O}(n)$ time the unique solution.

Note that the desirable property of disjoint subtree wedges together with the circular leaves constraint further restricts the class of edge lengths that can are represented exactly. Nevertheless, our experiments indicate that typical phylogenetic tree metrics are represented fairly accurately.

4.2 Extensions

Our weights depend on the choice of the root, since re-rooting the tree changes weights along the path between the previous and the new root. The rationale behind reducing the influence weights of children suggests that the tree should be rooted at its center (minimum eccentricity element). Using weights independent of the parent-child relation the layout can be made independent of the root just like the radial layouts discussed in the previous section.

It is easy to see that by fixing the order of leaves we are also fixing the child order of all inner vertices. If no particular order is given, we can permute the children of inner vertices to improve edge lengths preservation. Ordering the children of each vertex v according to ascending height $h(w)$ of the subtrees $T(w)$ plus $\delta(v, w)$ ensures that shallow and deep subtrees are never placed alternatingly. See Fig. 3. Though sorting the children leads to $\mathcal{O}(n \log n)$ preprocessing

time in general, most phylogenetic trees have bounded degree, so that sorting can be performed in linear time.

Since correct edge lengths cannot be guaranteed in circle drawings, we use the following coloring scheme to depict the error: Let $\sigma = \frac{\sum_{e=(u,v)\in E} \|x_v - x_u\|_2}{\sum_{e\in E} \delta(e)}$ be the mean resolution of the drawing, i. e., the scaling factor between drawn units and length units of δ. Then we obtain the (multiplicative) error of an edge $e = (u, v)$ by $f_e = \frac{\|x_u - x_v\|_2}{\sigma \cdot \delta(e)}$, which we encode into a color rgb: $\mathbb{R}^+ \to [0; 1]^3$ by

$$\mathrm{rgb}(f_e) = \begin{cases} (0,0,1) & \text{if } f_e \leq \frac{1}{2}, \\ (0,0,-\log_2(f_e)) & \text{if } \frac{1}{2} < f_e < 1, \\ (\log_2(f_e),0,0) & \text{if } 1 \leq f_e < 2, \\ (1,0,0) & \text{if } 2 \leq f_e \ . \end{cases} \tag{10}$$

so that blue and red signify edges that are too short and too long, respectively. 100% red means that the edge is at least twice as long as desired whereas 100% blue means that the edge should be at least twice as long. Weaker saturation reflects intermediate values. Black edges have the correct lengths.

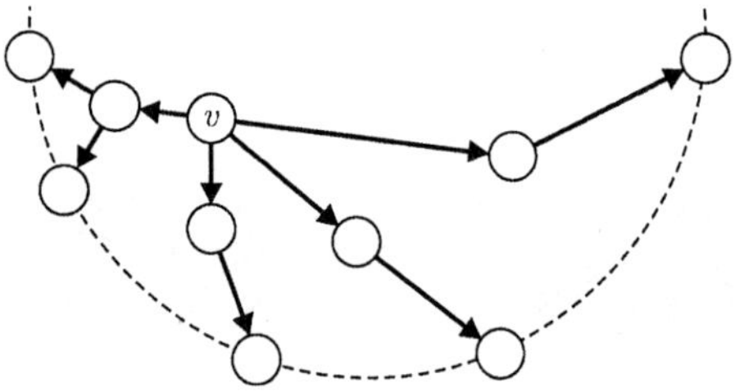

Fig. 3. Ordering children according to subtree hight supports postulated edge lengths

5 Discussion

We have presented two linear-time algorithms for drawing phylogenetic trees. Example drawings are shown in Figs. 4 and 5. Both are easy to implement and scale very well. While the algorithm for radial drawings preserves edge lengths exactly, the algorithm for circle drawings is constrained by having leaves fixed on the perimeter of a circle. Since each inner vertex is positioned in the weighted barycenter of its neighbors, it would be interesting to devise a weighting scheme that, in a sense to be defined, is provably optimal with respect to the pre-specified edge lengths. A related open question is the complexity status of deciding whether a circle drawing preserving the edge lengths exists.

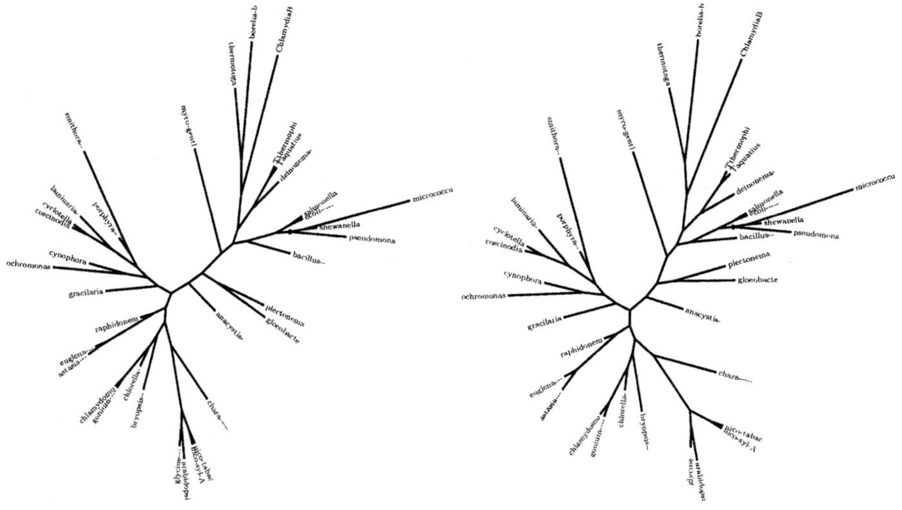

(a) Our basic algorithm (b) Angle-spread extension

Fig. 4. Radial drawing examples

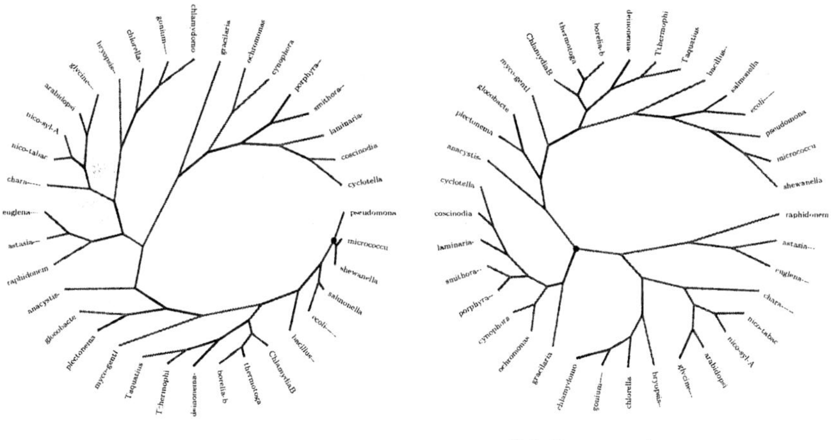

(a) Our basic algorithm (b) Re-rooted at center

Fig. 5. Circular drawing examples

Acknowledgment. We wish to thank Lars Volkhardt for implementing our algorithms in Java using yFiles version 2.3 [19], and Falk Schreiber from the Institute of Plant Genetics and Crop Plant Research in Gatersleben for providing real-world data.

References

1. S. F. Carrizo. Phylogenetic trees: An information visualization perspective. In Y.-P. Phoebe Chen, editor, *Asia-Pacific Bioinformatics Conference (APBC 2004)*, volume 29 of *CRPIT*, pages 315–320. Australian Compter Science, 2004.
2. G. Di Battista, P. Eades, R. Tamassia, and I. G. Tollis. *Graph Drawing: Algorithms for the Visualization of Graphs.* Prentice Hall, 1999.
3. P. Eades. Drawing free trees. *Bulletin of the Institute of Combinatorics and its Applications*, 5:10–36, 1992.
4. P. Eades and N. C. Wormald. Fixed edge-length graph drawing is $\mathcal{NP}$-hard. *Discrete Applied Mathematics*, 28:111–134, 1990.
5. J. Felsenstein. Maximum likelihood and minimum-steps methods for estimating evolutionary trees from data on discrete characters. *Systematic Zoology*, 22:240–249, 1973.
6. W. M. Fitch. Torward defining the course of evolution: Minimum change for a specified tree topology. *Systematic Zoology*, 20:406–416, 1971.
7. M. Kaufmann and D. Wagner. *Drawing Graphs*, volume 2025 of *LNCS*. Springer, 2001.
8. R. J. Lipton, D. J. Rose, and R. E. Tarjan. Generalized nested dissection. *SIAM Journal on Numerical Analysis*, 16:346–358, 1979.
9. C. D. Michener and R. R. Sokal. A quantitative approach to a problem in classification. *Evolution*, 11:130–162, 1957.
10. R. D. M. Page. TreeView. `http://taxonomy.zoology.gla.ac.uk/rod/treeview.html`. University of Glasgow.
11. PHYLIP.Phylogeny inference package. `http://evolution.genetics.washington.edu/ phylip.html`.
12. E. M. Reingold and J. S. Tilford. Tidies drawing of trees. *IEEE Transactions on Software Engineering*, 7(2):223–228, 1981.
13. N. Saitou and M. Nei. The neighbor-joining method: A new method for reconstructing phylogenetic trees. *Molecular Biology and Evolution*, 4(4):406–425, 1987.
14. D. L. Swofford. PAUP*. Phylogenetic analysis using parsimony (and other methods). `http://paup.csit.fsu.edu/`. Florida State University.
15. D. L. Swofford, G. J. Olsen, P. J. Waddel, and D. M. Hillis. Phylogenetic inference. In D. Hillis, C. Moritz, and B. Mable, editors, *Molecular Systematics*, pages 407–514. Sinauer Associates, 2nd edition, 1996.
16. W. T. Tutte. Convex representations of graphs. In *Proc. London Mathematical Society, Third Series*, volume 10, pages 304–320, 1960.
17. W. T. Tutte. How to draw a graph. In *Proc. London Mathematical Society, Third Series*, volume 13, pages 743–768, 1963.
18. J. Q. Walker. A node-positioning algorithm for general trees. *Software Practice & Experience*, 20(7):685–705, 1990.
19. R. Wiese, M. Eiglsperger, and M. Kaufmann. yFiles: Visualization and automatic layout of graphs. In P. Mutzel, M. Jünger, and S. Leipert, editors, *Proc. Graph Drawing 2001*, volume 2265 of *LNCS*, pages 453–454. Springer, 2002.

Localized and Compact Data-Structure for Comparability Graphs

(Extended Abstract)

Fabrice Bazzaro* and Cyril Gavoille*

LaBRI UMR CNRS 5800, Université Bordeaux 1, 33405 Talence Cedex, France
{bazzaro, gavoille}@labri.fr

Abstract. We show that every comparability graph of any two-dimensional poset over n elements (a.k.a. permutation graph) can be preprocessed in $O(n)$ time, if two linear extensions of the poset are given, to produce an $O(n)$ space data-structure supporting distance queries in constant time. The data-structure is localized and given as a distance labeling, that is each vertex receives a label of $O(\log n)$ bits so that distance queries between any two vertices are answered by inspecting on their labels only. As a byproduct, our data-structure supports all-pair shortest-path queries in $O(d)$ time for distance-d pairs, and so identifies in constant time the first edge along a shortest path between any source and destination. More fundamentally, we show that this optimal space and time data-structure cannot be extended for higher dimension posets (we prove that for comparability graphs of three-dimensional posets, every distance labeling scheme requires $\Omega(n^{1/3})$ bit labels).

1 Introduction

The dimension of a partially ordered set (or poset for short) $P = (V, <)$ is a fundamental invariant. It is the minimum number d of totally ordered sets $(V, <_1), \ldots, (V, <_d)$ whose intersection is P, i.e., $x < y$ if and only if $x <_i y$ for all $i \in \{1, \ldots, d\}$. Each total order $<_i$ is called linear extension of P, and a k-dimensional poset is a poset of dimension at most k.

The comparability graph of a poset $P = (V, <)$ is the graph $G = (V, E)$ such that $\{x, y\} \in E$ if and only if $x < y$ or $y < x$. The important point is that all posets with the same comparability graph have the same dimension [5]. So the comparability graph is definitively a fundamental tool for the study of posets.

Of special interest are the two-dimensional posets because they can be characterized in term of a single ordering [15]. It is NP-complete to recognize posets of dimension three [33] whereas linear time (linear in the size of the relation) algorithms exist for two-dimensional posets [26]. Actually, the comparability graphs of two-dimensional posets are exactly the permutation graphs, namely the intersection graphs of straight segments between two parallel lines [3]. Intersection

* The two authors are supported by the project "PairAPair" of the ACI Masses de Données.

X. Deng and D. Du (Eds.): ISAAC 2005, LNCS 3827, pp. 1122–1131, 2005.

graphs are graphs in which vertices are mapped to objects, with the vertices defined to be adjacent if and only if the corresponding objects have nonempty intersection. See [27] for a comprehensive introduction to the intersection graphs.

This paper deals with the problem of the distance computation and distributed abilities of comparability graphs. Commonly, when we make a query concerning a set of nodes in a graph (adjacency, distance, connectivity, etc.), we need to make a global access to the structure. In our approach, the compromise is to store the maximum of information in a label associated with a vertex to have directly what we need with a local access. Motivation of localized data-structures in distributed computing is surveyed and widely discussed in [19].

We are especially interested in the distance labeling problem, introduced in [28]. The problem consists in labeling the vertices of a graph to compute the distance between any two of its vertices x and y using only the information stored in the labels of x and y, without any other source of information. The main parameters taken into account when designing a solution are: (1) length (in bits) of the labels; (2) time complexity to decode the distance from the labels; and (3) time complexity to preprocess the graph and to compute all the labels.

Related works. Distance computation in graphs is one of the most fundamental graph algorithmic problem. Computing the distance matrix of a general graph is strongly related to Boolean matrix multiplication [13], and achieving this task as quickly as possible is a widely open problem.

However, the time complexity of this problem is known, and can be reduced significantly from the naive $O(n^3)$ upper bound, for many families of graphs: planar graphs [12, 25, 31], bounded tree-width graphs [6], interval graphs [2, 7, 18], etc.

Beyond the classical all-pair distance problem, whose goal is to preprocess (possibly linearly) a graph and to produce a data-structure supporting distance or shortest-path queries in the minimum time complexity, the distance labeling problem is a variant in which the queries must be answered *locally*, by looking at the information related to the concerned vertices only. Introduced in [28], it generalizes adjacency labeling [1, 23] whose goal is only to decide whether the distance is 1 or not between any two vertices. At this point, it is worth to mention that any distance labeling scheme on a family $\mathcal{F}$ of graphs with $\ell(n)$ bit labels converts trivially into a non-distributed data-structure for $\mathcal{F}$ of $O(\ell(n) \cdot n/\log n)$ space supporting distance queries within the same time complexity, being assumed that a cell of space can store $\Omega(\log n)$ bits of data.

The main results on the field are that general graphs support a distance labeling scheme with labels of $O(n)$ bits [20], and that trees [1, 28], bounded tree-width graphs [20], distance-hereditary graphs [17], bounded clique-width graphs [11], some non-positively curved plane graphs [8], interval and permutation graphs, all support distance labeling schemes with $O(\log^2 n)$ bit labels. There are also deep results concerning approximated distance labeling schemes we will not discuss here, e.g., see [16, 30, 31, 32].

The $O(n)$ bit upper bound is tight for general graphs, and a lower bound of $\Omega(\log^2 n)$ bit on the label length is known for trees [20], implying that all the

results mentioned above are tight as well, excepted for interval and permutation graphs that does not contain trees. Recently, [18] showed an optimal bound of $O(\log n)$ bits for interval graphs and circular-arc graphs.

Concerning permutation graphs, we should mention the related work of [29]. It is shown how to compute one fixed BFS tree and DFS tree in $O(n)$ and $O(n \log\log n)$ time respectively when the permutation of the graph is given. Although original, the technique is limited; it does not allow us to choose arbitrarily the root of the trees.

Our results. In this paper, we are only interested in comparability graphs, and in particular the permutation graphs, the comparability graphs of two-dimensional posets. This latter family is well-known and have a lot of valuable properties due to its characterizations as a partially ordered sets [15], as an intersection graph of segments between two parallel lines [21] or as an Asteroidal Triple-free graph [10].

We show that permutation graphs with n vertices enjoy a distance labeling with labels of length $O(\log n)$ bits, more precisely $9\lceil \log n \rceil + 6$ bits. The distance can be computed in constant time, and all the labels are computed in $O(n)$ time if a realizer (the two linear extensions of the poset) is given. (Such realizer can be obtained be in $O(n+m)$ time [26] otherwise, m the number of edges). This result has optimal complexities (label length, distance decoder time complexity, and preprocessing time complexity). It improves within a $\log n$ factor on the label length the previous result of Katz, Katz and Peleg [24] in the STACS '00.

We also show that $3\log n$ bits for the label length of *any* labeling distance scheme is inevitably in the worst-case. As intermediate combinatorial result to prove this latter Information-Theoretic lower bound, and of independent interest, we prove that there are $2^{\Omega(n \log n)}$ unlabeled permutation graphs with n vertices.

As remarked previously, this leads to an optimal $O(n)$ space data-structure supporting distance queries in constant time, and computable in $O(n)$ time. We also show how to adapt our data-structure such that a length-d shortest path can be extracted in $O(d)$ time for any distance-d pair of vertices. One can also use our localized data-structure to identify the first shortest-path edge, and we therefore present a new shortest-path compact routing scheme for permutation graphs, improving the result of [14]. All these results can be extended to circular permutation graphs, a natural generalization of permutation graphs.

Looking for a generalized scheme with $O(f(d) \log n)$ bit labels for all comparability graphs of d-dimensional posets, we have proved that unfortunately no such function $f(d)$ can exist. More precisely, for every distance labeling scheme, there are comparability graphs of posets of dimension three that requires $\Omega(n^{1/3})$ bit labels. This makes a difference between comparability graphs of two- and three-dimensional posets for distance computation. This could not be observed when only adjacency is required, since $O(\log n)$ bit labels suffices for comparability graphs of bounded dimension posets.

Outline of the paper. Let us sketch our distance labeling scheme. We use a 2D geometric representation of the permutation graph, each vertex being associated with a point of $\mathbb{N}^2$, and u is adjacent to v if and only if v is in the South-East or North-West quadrant around u (cf. Fig. 2). For our purpose, two sets of points (or vertices)

are of special interests: those with no neighbors in their North-West quadrant (call A), and those with no neighbors in their South-East quadrant (call B). Intuitively, one can always construct a shortest path between non-adjacent vertices by using only edges alternating between A and B. The main difficulty resides in deciding whether the first edge must be incident to a vertex of A or of B.

The first trick is to treat differently short and long distances. We entirely characterize distances $\leqslant 3$ by comparing the coordinates of six specific vertices spread around each vertex (three belongs to A and three to B). Then, for long distances, i.e., distances $\geqslant 4$, we show that they reduce to the distance computation between vertices of two intermediate graphs, G_A and G_B, defined on respectively the vertices of A and of B. The edges of G_A and G_B are entirely determined by an asymmetric relation between the points of A and of B, and we show that a distance labeling for these graphs with $O(\log n)$ bit labels is possible. The distance is then computed by evaluating the distance in the graphs G_A and G_B between the six points associated with the source and the six points associated with the destination.

Finally, after several optimizations, the resulting data-structure is extremely simple. It is composed of 9 integers of $\{0, \ldots, n-1\}$ plus 6 bits. The distance decoder simply consists of a constant number of additions and comparisons on these integers.

Background and preliminaries are presented in Section 2, and the implementation of the scheme is presented in Section 3. An extension of the data-structure for all-pair shortest-path queries, compact routing, and to circular permutation graphs is discussed in Section 4.

Due to space limitation, the proofs can be found in [4].

2 Preliminaries

We consider simple undirected graphs. Moreover, along this paper, we will assume that graphs are connected. We denote by $d_G(u, v)$ the distance between u and v in G. For a vertex u of graph G, we denote by $N(u)$ the set of neighbors of u, and $N[u] := N(u) \cup \{u\}$.

A *permutation graph* is an intersection graph of straight segments between two parallel lines [21]. On Fig. 1, Segment 1 intersects the segments 2,5,6,7, so in the permutation graph vertex 1 is adjacent to the vertices 2,5,6,7. More formally, a permutation graph is isomorphic to a graph $G = (V, E)$ with $V = \{1, \ldots, n\}$ such that there exists a permutation π of V, called *realizer* of G, satisfying: u is adjacent to v if and only if $u < v$ and $\pi^{-1}(u) > \pi^{-1}(v)$.

It is not difficult to see that the isomorphism and the permutation π can be combined to form two linear extensions of a 2-dimensional poset. Actually, permutation graphs are exactly the comparability graphs of 2-dimensional posets [3].

We can draw in the plane a permutation graph G with the realizer π, by associating with each vertex $u \in \{1, \ldots, n\}$ the point of coordinates $(u, \pi^{-1}(u))$. Within this graphic representation, the neighbors of u are the points located in the North-West and South-East quadrants around u (see Fig. 2 for an example).

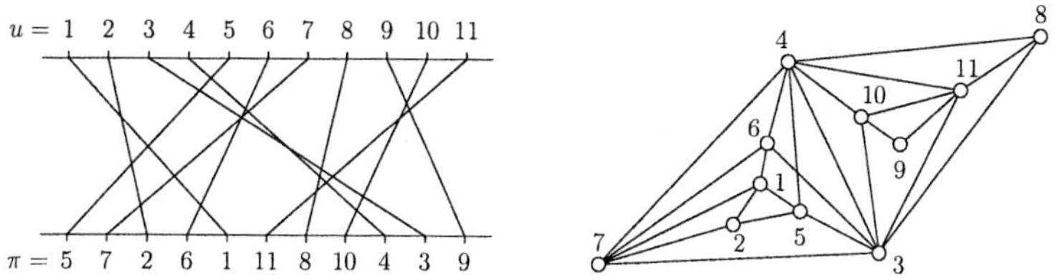

Fig. 1. A permutation graph G with the realizer $\pi = (5, 7, 2, 6, 1, 11, 8, 10, 4, 3, 9)$

Hereafter, we assume that $G = (V, E)$ is a given connected permutation graph, and π a realizer of G. Since $V = \{1, \ldots, n\}$, we use the total ordering on natural numbers as total ordering of the vertices of G. We partition the neighbors of u in the subsets $N^+(u) := \{v \in N(u) \mid v > u\}$ and $N^-(u) := \{v \in N(u) \mid v < u\}$ (see left side of Fig. 2).

We distinguish two particular subsets of vertices, A and B depicted in black and gray on Fig. 2, defined by $A := \{u \in V \mid N^-(u) = \varnothing\}$ and $B := \{u \in V \mid N^+(u) = \varnothing\}$. Note that A and B are nonempty stables of G. If G is connected and has at least one edge, then $A \cap B = \varnothing$. Moreover, we check that G is bipartite if and only if $V = A \cup B$.

Lemma 1. *For every $u \in V$, there exist $a^-(u), a^+(u) \in A$ such that $a^-(u) \leqslant a^+(u)$ and $N[u] \cap A = [a^-(u), a^+(u)] \cap A$. Similarly, there exist $b^-(u), b^+(u) \in B$ such that $b^-(u) \leqslant b^+(u)$ and $N[u] \cap B = [b^-(u), b^+(u)] \cap B$.*

According to Lemma 1, $N[u] \cap A$ and $N[u] \cap B$ are never empty and are consecutive in A and B respectively. Observe also that necessarily $a^-(u) = \min\{N[u] \cap A\}$ and $a^+(u) = \max\{N[u] \cap A\}$, and similarly for $b^-(u)$ and $b^+(u)$.

Hereafter, we denote by G_A the intersection graph of the family $\{[b^-(u), b^+(u)] \mid u \in A\}$. Similarly, we denote by G_B the intersection graph of the family $\{[a^-(u), a^+(u)] \mid u \in B\}$. By construction, G_A and G_B are interval graphs with vertex sets A and B respectively. As we will see later in Lemma 4, G_A and G_B are proper interval graphs.

The interesting connection between G and G_A, G_B is the following:

Lemma 2. *For all $u, v \in A$, $d_G(u, v) = 2d_{G_A}(u, v)$, and for all $u, v \in B$, $d_G(u, v) = 2d_{G_B}(u, v)$.*

E.g., in our example, $d_G(1, 9) = 4$ because $1, 9 \in A$ and $d_{G_A}(1, 9) = 2$. By Lemma 2, since G is connected, G_A and G_B are connected too. We are now ready to prove the main result of this section that is the heart of our distance decoder.

Theorem 1. *Let u, v be two vertices with $u < v$. Then,*

1. *if $\pi^{-1}(u) > \pi^{-1}(v)$, then $d_G(u, v) = 1$;*
2. *otherwise, if $a^-(v) \leqslant a^+(u)$ or $b^-(v) \leqslant b^+(u)$, then $d_G(u, v) = 2$;*
3. *otherwise, if $a^-(v) \leqslant a^+(b^+(u))$ or $b^-(v) \leqslant b^+(a^+(u))$, then $d_G(u, v) = 3$;*

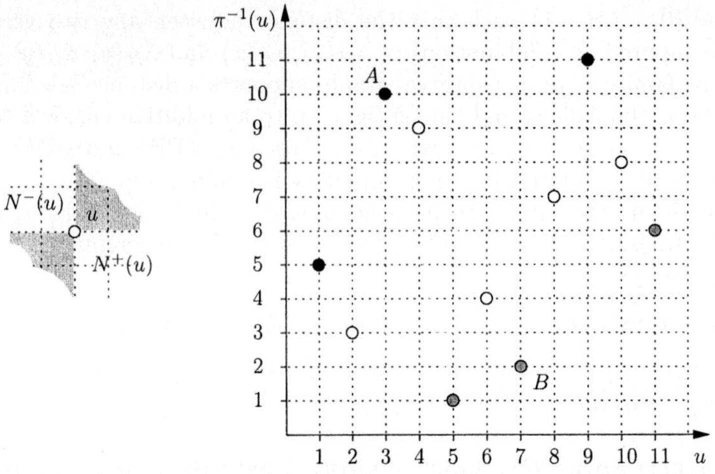

Fig. 2. Graphic representation of the permutation graph of Fig. 1

4. otherwise, $d_G(u,v)$ *is the minimum between the four distances:*

- $2 + 2d_{G_B}(b^+(u), b^-(v))$
- $3 + 2d_{G_A}(a^+(b^+(u)), a^-(v))$
- $2 + 2d_{G_A}(a^+(u), a^-(v))$
- $3+2d_{G_B}(b^+(a^+(u))), b^-(v))$

3 Implementation of the Scheme

Distance labeling for proper interval graphs. An interval graph is the intersection graph of a family of intervals of the real line. The *layout* of an interval graph is the set of intervals associated with each vertex of the graph. A layout is *proper* if no intervals are strictly contained in another one, i.e., there are no intervals $[a,b]$ and $[c,d]$ with $a < c$ and $d < b$ (however $a = c$, or $b = d$ is possible). The graphs having such layouts are called proper interval graphs.

Motivated by Lemma 2, we will use as a sub-routine the distance labeling scheme of [18] for proper interval graphs that we need to detail. To compute all the labels in $O(n)$ time, the scheme of [18] takes as input a proper layout in a special form: a normalized proper layout. A layout is *normalized* if: (1) the left boundaries of the intervals are distinct; (2) the intervals are sorted according to their left boundaries (so there is a way to list them in $O(n)$ time by increasing left boundary); (3) the boundaries are integers bounded by some polynomials of n (so that boundaries can be manipulated in constant time on RAM-word computers). Note that the layouts provided by the linear time recognizing algorithms of [9, 22] have such properties. Observe also that a layout satisfying properties (2) and (3) can be easily transformed in $O(n)$ time in a normalized layout by scanning all the intervals by increasing left boundaries.

Consider any connected proper interval graph H with n vertices with a normalized proper layout $\mathcal{L}_H = \{[\mathrm{L}_H(u), \mathrm{R}_H(u)] \mid u \in V(H)\}$. In [18], it has been proved that in $O(n)$ time it is possible to compute two mappings λ_H, σ_H :

$V(H) \to \{0,\ldots,n-1\}$ such that the distance between any two vertices x, y can be computed as combinations of $\lambda_H(x), \sigma_H(x)$ and $\lambda_H(y), \sigma_H(y)$. In other words, the family of proper interval graphs supports a distance labeling scheme with $2\lceil \log n \rceil$ bit labels, and this is tight up to an additive $\log\log n$ term [18]. The mappings λ_H and σ_H are respectively based on a BFS and a DFS initiated from the vertex x_0 of H whose left boundary is minimum in $\mathcal{L}_H$.

Let us define the binary relation $\mathrm{adj}_H(x,y) \in \{0,1\}$ by: $\mathrm{adj}_H(x,y) = 1$ if and only if $\lambda_H(x) < \lambda_H(y)$ and $\sigma_H(x) > \sigma_H(y)$. These mappings satisfy the following properties:

Lemma 3 ([18]). *For all distinct vertices x and y of H with $\lambda_H(x) \leqslant \lambda_H(y)$:*

1. $d_H(x,y) = \lambda_H(y) - \lambda_H(x) + 1 - \mathrm{adj}_H(x,y)$.
2. $\lambda_H(x) = d_H(x_0, x)$.
3. *σ_H is bijective.*
4. *If $\lambda_H(x) = \lambda_H(y)$, then $\mathrm{L}_H(x) < \mathrm{L}_H(y)$ if and only if $\sigma_H(x) < \sigma_H(y)$.*

Lemma 4. *The families $\mathcal{I}_A := \{[b^-(u), b^+(u)] \mid u \in A\}$ and $\mathcal{I}_B := \{[a^-(u), a^+(u)] \mid u \in B\}$ are proper layouts of the graphs G_A and G_B.*

Distance decoder. At this step, we have enough material to prove that permutation graphs enjoy an $O(\log n)$ bit distance labeling scheme. Indeed, Theorem 1 can be easily implemented with short labels. Each vertex u could store: (1) the integers u and $\pi^{-1}(u)$ in order to compute distances $\leqslant 1$; (2) the x-coordinates about the six vertices $a^-(u), a^+(u), b^-(u), b^+(u), a^+(b^+(u)), b^+(a^+(u))$ for distances 2 and 3; and (3) the distance labels (two integers per label) in the proper interval graphs G_A and G_B about the same six vertices for distances $\geqslant 4$. The resulting label is composed of a total of $2 + 6 + 6 \times 2 = 20$ integers in $O(n)$. Thus such a label is of length $20 \log n + O(1)$ bits. However, we will show how to significantly reduce this length.

In order to apply the result of [18] (Lemma 3) we need to transform the initial layouts $\mathcal{I}_A$ and $\mathcal{I}_B$ into the normalized layouts that we denote hereafter by $\mathcal{L}_A = \{[\mathrm{L}_A(u), \mathrm{R}_A(u)] \mid u \in A\}$ and $\mathcal{L}_B = \{[\mathrm{L}_B(u), \mathrm{R}_B(u)] \mid u \in B\}$ respectively. Clearly, all the boundaries of $\mathcal{I}_A$ and of $\mathcal{I}_B$ are in $O(n)$, so sorting them can be done in $O(n)$ time. Actually, we check that the left boundaries of $\mathcal{I}_A$ and $\mathcal{I}_B$ are already sorted if A and B are sorted. To produce normalized layouts, the left boundaries are ordered and distinguished with the following rule:

$$\forall u, v \in A, \quad \mathrm{L}_A(u) < \mathrm{L}_A(v) \text{ iff } b^-(u) < b^-(v), \text{ or } b^-(u) = b^-(v) \text{ and } u < v.$$

Similary, left boundaries of B are ordered such that $\mathrm{L}_B(u) < \mathrm{L}_B(v)$ if and only if $a^-(u) < a^-(v)$, or $a^-(u) = a^-(v)$ and $u < v$. This can be done in linear time by scanning the left boundaries of $\mathcal{I}_A$ (and of $\mathcal{I}_B$) by increasing order, and by shifting them to produce $\mathcal{L}_A$. We check that the resulting boundaries are in $O(n)$. We can prove the following properties for the layouts $\mathcal{L}_A$ and $\mathcal{L}_B$:

Lemma 5. *For all $u, v \in A$, $u < v$ if and only if $\mathrm{L}_A(u) < \mathrm{L}_A(v)$. Similarly, for all $u, v \in B$, $u < v$ if and only if $\mathrm{L}_B(u) < \mathrm{L}_B(v)$.*

Let λ_A, σ_A and λ_B, σ_B be the mappings obtained after application of Lemma 3 for the graphs G_A and G_B with normalized proper layouts $\mathcal{L}_A$ and $\mathcal{L}_B$. We are now ready to define the label of u, which is composed of the 14 following fields:

$$\text{label}(u) := \langle u, \pi^{-1}(u),\ \lambda_A(a), \sigma_A(a),\ \lambda_B(b), \sigma_B(b),\ \ldots \rangle$$

for all $a \in \{a^-(u), a^+(u), a^+(b^+(u))\}$ and $b \in \{b^-(u), b^+(u), b^+(a^+(u))\}$.

The information about the six vertices needed to compute the 4 distances involved at Step 4 of Theorem 1 is available in the labels. So, when implementing the distance decoder the only problem concerns the comparison tests between vertices, involved at Steps 2 and Steps 3 of Theorem 1. For these two steps, we need to make comparison tests between some vertices of A or some vertices of B. Typically, in Step 2 for instance, we need to test whether $a^-(v) \leqslant a^-(u)$ or not. The problem is solved thanks to the next lemma.

Lemma 6.
For all $a_u, a_v \in A$,
$a_u \leqslant a_v$ if and only if:

- $\sigma_A(a_u) = \sigma_A(a_v)$, *or*
- $\lambda_A(a_u) < \lambda_A(a_v)$, *or*
- $\lambda_A(a_u) = \lambda_A(a_v)$ *and* $\sigma_A(a_u) < \sigma_A(a_v)$.

For all $b_u, b_v \in B$,
$b_u \leqslant b_v$ if and only if:

- $\sigma_B(b_u) = \sigma_B(b_v)$, *or*
- $\lambda_B(b_u) < \lambda_B(b_v)$, *or*
- $\lambda_B(b_u) = \lambda_B(b_v)$ *and* $\sigma_B(b_u) < \sigma_B(b_v)$.

Therefore, according to Lemma 3, Lemma 6 and Theorem 1, the distance decoder takes a constant number of integer additions and comparisons.

Label size. Since each field of label(u) ranges in $\{0, \ldots, n-1\}$ (formally we need to decrease by 1 the two first fields), the label size is a priori $14 \lceil \log n \rceil$. However we will work a little and use correlations between some values of label(u) to reduced it to $9 \log n + O(1)$. As we will see all the six λ's values involved in label(u) are strongly related.

The precise description of the final implementation of label(u) is given in the proof of Lemma 7, which is based on the following fact: Let $u \in V$. For all $a \in N[u] \cap A$ and $b \in N[u] \cap B$, $|\lambda_A(a) - \lambda_B(b)| \leqslant 1$.

Lemma 7. *Labels are of* $9 \lceil \log n \rceil + 6$ *bits at most, and it takes constant time to decode the distance from the labels.*

In [4] we show that label(u) for all $u \in V$ can be computed in linear time once the realizer π is given. Combining this with Lemma 7, we obtain:

Theorem 2. *Permutation graphs with n vertices enjoy a distance labeling scheme using labels of* $9 \lceil \log n \rceil + 6$ *bits. The distance decoder has constant time complexity, and given a realizer of the graph, i.e., a permutation of $\{1, \ldots, n\}$, all the labels can be computed in $O(n)$ time.*

4 Extensions

The full explanations of the following results are presented in [4].

Theorem 3. *Given a realizer of a permutation graph of n vertices, one can construct in $O(n)$ time an $O(n)$ space data-structure supporting all-pair shortest path extraction in linear time (linear in the length of the shortest path returned).*

Theorem 4. *Permutation graphs have a shortest-path routing scheme with constant time protocol, and with $O(\log n)$ bit addresses, and, for every vertex u, the routing table of u is of size $O(\deg(u) \log n)$ bits. If the graph is bipartite then the size of the local routing tables can be reduced to $O(\log n)$ bit per vertex. Moreover, given a realizer of the graph, addresses and routing tables can be constructed in $O(n)$ time.*

Theorem 5. *Circular permutation graphs with n vertices enjoy a distance labeling scheme with $O(\log n)$ bit labels and constant time distance decoder.*

Theorem 6. *There are comparability graphs of three-dimensional posets for which every distance labeling scheme requires $\Omega(n^{1/3})$ bit labels.*

Theorem 7. *The number $P(n)$ of labeled n-vertex connected permutation graphs satisfies $\frac{1}{n} \log P(n) \geqslant 2 \log n - O(\log \log n)$. It follows that there are $2^{\Omega(n \log n)}$ unlabeled n-vertex permutation graphs.*

Theorem 8. *Any distance labeling scheme on the family of n-vertex permutation graphs requires a label of length at least $3 \log n - O(\log \log n)$ bits.*

References

1. S. Alstrup, P. Bille, and T. Rauhe, *Labeling schemes for small distances in trees*, in 14th Symp. on Disc. Algorithms (SODA), ACM-SIAM, 2003, pp. 689–698.
2. M. Atallah, D. Chen, and D. Lee, *An optimal algorithm for shortest paths on weighted interval and circular-arc graphs, with applications*, Algorithm., 14 (1995).
3. K. A. Baker, P. C. Fishburn, and F. S. Roberts, *Partial orders of dimension 2*, Networks, 2 (1971), pp. 11–28.
4. F. Bazzaro and C. Gavoille, *Localized and compact data-structure for comparability graphs*, RR-1343-05, LaBRI, University of Bordeaux 1, Feb. 2005.
5. A. Brandstädt, V. B. Le, and J. P. Spinrad, *Graph Classes – A survey*, SIAM Monographs on Discrete Mathematics and Applications, Philadelphia, 1999.
6. S. Chaudhuri and C. D. Zaroliagis, *Shortest paths in digraphs of small treewidth. Part I: Sequential algorithms*, Algorithmica, 27 (2000), pp. 212–226.
7. D. Chen, D. Lee, R. Sridhar, and C. Sekharan, *Solving the all-pair shortest path query problem on interval and circular-arc graphs*, Networks, 31 (1998).
8. V. D. Chepoi, F. F. Dragan, and Y. Vaxes, *Distance and routing labeling schemes for non-positively curved plane graphs*, J. of Algorithms, (2005).
9. D. G. Corneil, H. Kim, S. Natarajan, S. Olariu, and A. P. Sprague, *Simple linear time algorithm of unit interval graphs*, Inf. Proc. Letters, 55 (1995).

10. D. G. Corneil, S. Olariu, and L. Stewart, *Asteroidal triple-free graphs*, SIAM Journal on Discrete Mathematics, 10 (1997), pp. 399–430.
11. B. Courcelle and R. Vanicat, *Query efficient implementation of graphs of bounded clique-width*, Discrete Applied Mathematics, 131 (2003), pp. 129–150.
12. H. N. Djidjev, *Efficient algorithms for shortest path queries in planar digraphs*, in 22nd WG, vol. 1197 of LNCS, Springer, June 1996, pp. 151–165.
13. D. Dor, S. Halperin, and U. Zwick, *All-pairs almost shortest paths*, SIAM Journal on Computing, 29 (2000), pp. 1740–1759.
14. F. F. Dragan and I. Lomonosov, *New routing schemes for interval graphs, circular-arc graphs, and permutation graphs*, in 14th PDCS, Nov. 2002, pp. 78–83.
15. B. Dushnik and E. W. Miller, *Partially ordered sets*, American Journal of Mathematics, 63 (1941), pp. 600–610.
16. C. Gavoille, M. Katz, N. A. Katz, C. Paul, and D. Peleg, *Approximate distance labeling schemes*, in 9th ESA, vol. 2161 of LNCS, 2001, pp. 476–488.
17. C. Gavoille and C. Paul, *Distance labeling scheme and split decomposition*, Discrete Mathematics, 273 (2003), pp. 115–130.
18. ——, *Optimal distance labeling schemes for interval and circular-arc graphs*, in 11th ESA, vol. 2832 of LNCS, Springer, Sept. 2003, pp. 254–265.
19. C. Gavoille and D. Peleg, *Compact and localized distributed data structures*, J. of Distributed Computing, 16 (2003), pp. 111–120. PODC 20-Year Special Issue.
20. C. Gavoille, D. Peleg, S. Pérennès, and R. Raz, *Distance labeling in graphs*, Journal of Algorithms, 53 (2004), pp. 85–112.
21. M. C. Golumbic, *Algorithmic Graph Theory and Perfect Graphs*, Academic Press, Harcourt Brace Jovanovich, Academic Press ed., 1980.
22. C. M. Herrera de Figueiredo, J. a. Meidanis, and C. Picinin de Mello, *A linear-time algorithm for proper interval recognition*, Information Processing Letters, 56 (1995), pp. 179–184.
23. S. Kannan, M. Naor, and S. Rudich, *Implicit representation of graphs*, SIAM Journal on Discrete Mathematics, 5 (1992), pp. 596–603.
24. M. Katz, N. Katz, and D. Peleg, *Distance labeling schemes for well-separated graph classes*, in 17th STACS, vol. 1770 of LNCS, 2000, pp. 516–528.
25. P. N. Klein, S. Rao, M. R. Henzinger, and S. Subramanian, *Faster shortest-path algorithms for planar graphs*, J. of Comp. and Sys. Sci., 55 (1997), pp. 3–23.
26. R. M. McConnell and J. P. Spinrad, *Linear-time transitive orientation*, in 8th Symp. on Discrete Algorithms (SODA), ACM-SIAM, Jan. 1997, pp. 19–25.
27. T. A. McKee and F. R. McMorris, *Topics in Intersection Graph Theory*, SIAM Monographs on Discrete Mathematics and Applications, 1999.
28. D. Peleg, *Proximity-preserving labeling schemes*, Journal of Graph Theory, 33 (2000), pp. 167–176.
29. C. J. Rhee, Y. D. Liang, S. K. Dhall, and S. Lakshmivarahan, *Efficient algorithms for finding depth-first and breadth-first search trees in permutation graphs*, Information Processing Letters, 49 (1994), pp. 45–50.
30. K. Talwar, *Bypassing the embedding: Algorithms for low dimensional metrics*, in 36th STOC, ACM Press, June 2004, pp. 281–290.
31. M. Thorup, *Compact oracles for reachability and approximate distances in planar digraphs*, in 42nd FOCS, IEEE Computer Society Press, Oct. 2001, pp. 242–251.
32. M. Thorup and U. Zwick, *Approximate distance oracles*, in 33rd Annual ACM Symp. on Theory of Computing (STOC), ACM Press, July 2001, pp. 183–192.
33. M. Yannakakis, *The complexity of the parital order dimension problem*, SIAM Journal on Algebraic and Discrete Methods, 3 (1982), pp. 351–358.

Representation of Graphs by OBDDs

Robin Nunkesser[1] and Philipp Woelfel[2,⋆]

[1] FB Informatik, LS2, Universität Dortmund, 44221 Dortmund, Germany
robin.nunkesser@udo.edu
[2] University of Toronto, 10 King's College Road, Toronto M5S 3G4, Canada
pwoelfel@cs.toronto.edu

Abstract. In this paper, the space requirements for the OBDD representation of certain graph classes, specifically cographs, several types of graphs with few P_4s, unit interval graphs, interval graphs and bipartite graphs are investigated. Upper and lower bounds are proven for all these graph classes and it is shown that in most (but not all) cases a representation of the graphs by OBDDs is advantageous with respect to space requirements.

1 Introduction

Some modern applications require such huge graphs that the usual, explicit representation by adjacency lists or adjacency matrices is infeasible. E.g., a typical state transition graph arising in the process of verification and synthesis of sequential circuits may consist of 10^{27} vertices and 10^{36} edges. Such huge graphs appear also if the basic graph as e.g. the street network of a city is interlinked with other components as e.g. traffic amount and time slots.

In order to be able to store huge graphs, implicit representations of graphs can be used. In the standard implicit representation the vertices of a graph are labeled in such a way that adjacency of two vertices is uniquely determined by their labels (a prominent example is the representation of interval graphs, where the nodes are labeled by intervals and two nodes are adjacent if and only if the corresponding intervals intersect). However, such an implicit representation is not a generic graph representation because by each representation only a small number of graphs can be described (e.g. most graphs are not interval graphs). Hence, for different graph classes different implicit representations and different algorithms are needed.

Another approach is to use a generic graph representation (i.e. a representation which can represent *all* graphs) which requires less space for sufficiently structured graphs than for unstructured graphs. One idea is to store the characteristic function of the vertex and the edge set by a generic data structure for boolean functions. A data structure which is well suited for this task is the *Ordered Binary Decision Diagram* (OBDD), because for all important operations on boolean functions (e.g. the synthesis operation, substitution by constants or

⋆ Supported in part by DFG grants We 1066/10-1 and Wo 1232/1-1.

X. Deng and D. Du (Eds.): ISAAC 2005, LNCS 3827, pp. 1132–1142, 2005.

satisfiability test) efficient algorithms are known. Such OBDD operations allow to efficiently convert any graph represented by e.g. an adjacency list or an adjacency matrix to a corresponding OBDD representation. Moreover, it has turned out that several graph problems can be solved with a polylogarithmic number of OBDD operations, if the input graphs are represented by OBDDs. Recently, several OBDD algorithms for fundamental graph problems have been devised, as e.g. for network flow maximization [8, 15], topological sorting [19] or for finding shortest paths in networks [16].

Since an OBDD representation is a generic representation of graphs, most graphs cannot be represented in less space by OBDDs than by adjacency matrices or adjacency lists. But one hopes (and we show in this paper) that for sufficiently structured inputs as they may appear in practical applications, a good compression can be achieved. Experimental studies have already shown that in many practical relevant cases OBDD representations of graphs may be advantageous. However, it has not yet been theoretically investigated for which general graph classes this is the case. While it is obvious that very simply structured graphs as e.g. grid networks have a small OBDD representation, this is not clear for more complicated graph classes. In this paper, we start filling this gap by investigating several important graph classes with respect to their space requirements in an OBDD representation.

In the following, let B_n denote the class of boolean functions $f : \{0,1\}^n \to \{0,1\}$. OBDDs have been introduced by Bryant in 1986 [3] as a representation type for boolean functions.

Definition 1. Let $X_n = \{x_1, \ldots, x_n\}$ be a set of boolean variables. A *variable ordering* π on X_n is a bijection $\pi : \{1, \ldots, n\} \to X_n$, leading to the ordered list $\pi(1), \ldots, \pi(n)$ of the variables. A π-OBDD on X_n for a variable ordering π is a directed acyclic graph with one root, two sinks labeled with 0 and 1, respectively, and the following properties: Each inner node is labeled by a variable from X_n and has two outgoing edges, one of them labeled by 0, the other by 1. If an edge leads from a node labeled by x_i to a node labeled by x_j, then $\pi^{-1}(x_i) < \pi^{-1}(x_j)$.

A π-OBDD is said to *represent* a boolean function $f \in B_n$, if for any $a = (a_1, \ldots, a_n) \in \{0,1\}^n$, the path starting at the root and leading from any x_i node over the edge labeled by the value of a_i, ends at a sink with label $f(a)$. The *size* of a π-OBDD G is the number of its nodes and is denoted by $|G|$. The π-OBDD *size* of a boolean function f (π-OBDD(f)) is the size of the minimum π-OBDD computing f. The OBDD size of a boolean function f (OBDD(f)) is the size of the minimum π-OBDD computing f for some variable ordering π.

Let $G = (V, E)$ be a graph. We can use an OBDD for representing the graph G by letting it represent the characteristic function $\chi_E : E \to \{0,1\}$ of the edge set, where $\chi_E(v_1, v_2) = 1 \Leftrightarrow \{v_1, v_2\} \in E$. We denote the binary value represented by the n bit string $z_{n-1} \ldots z_0 \in \{0,1\}^n$ by $|z| := \sum_{i=0}^{n-1} z_i 2^i$. Conversely we denote by $[k]_n$, $n \geq \lceil \log(k+1) \rceil$, the n bit string representing the integer $k \geq 0$, i.e. the string $z_{n-1} \ldots z_0 \in \{0,1\}^n$ with $k = |z|$. In order to encode the vertices by boolean variables, we use the convention that $V = \{[0]_n, \ldots, [N-1]_n\}$, where

$N \geq 0$ and $n = \lceil \log N \rceil$. Thus, we can store V using n bits or by a π-OBDD for the characteristic function χ_V of V, and the minimal π-OBDD for χ_V has a size of $\mathcal{O}(n \log n) = \mathcal{O}(\log N \log \log N)$ for all variable orderings π.

Definition 2. The π-OBDD *size* of a graph $G = (V, E)$ (π-OBDD(G)) is the size of the minimum π-OBDD computing χ_E for some labeling of the vertices. Analogously, the OBDD *size* of G is the minimum of π-OBDD(G) for all variable orderings π. The *worst-case OBDD size* of a graph class $\mathcal{G}$ is the maximum of OBDD(G) over all $G \in \mathcal{G}$.

Breitbart, Hunt III and Rosenkrantz have shown in [2] that any function $f \in B_n$ can be represented by a π-OBDD of size $(2 + \mathcal{O}(1))\, 2^n/n$ (for any variable ordering π). Hence, the π-OBDD size of any graph with N vertices is bounded by $(4 + \mathcal{O}(1))\, N^2/\log N$, because $n < 2 + 2\log N$ for appropriately chosen vertex labels. In order to obtain a general bound which is better for not very dense graphs, we may use the fact that the π-OBDD size of any function $f \in B_n$ is bounded by roughly $n \cdot |f^{-1}(1)|$. This yields the following straight forward proposition.

Proposition 1. *Let $G = (V, E)$ be a graph with $N := |V|$ and $M := |E|$. The* OBDD *size of G is bounded above by*

$$\min\left\{(4 + \mathcal{O}(1))\, N^2/\log N,\ 2M \cdot (2\lceil \log N \rceil - \lfloor \log M \rfloor) + 1\right\} .$$

Note that if a graph G with N edges and M vertices is given by an adjacency matrix or adjacency list, then the OBDD for G satisfying the size bound of the above Proposition can be constructed in time $\mathcal{O}(M \log N \log^2 M)$. Hence, OBDDs are a truly generic graph representation.

Obviously, an OBDD B can be uniquely described by $\mathcal{O}\left(|B|\,(\log |B| + \log n)\right)$ bits. Consider a graph G with N vertices and M edges. The above proposition shows that an OBDD representation for dense graphs needs at most $\mathcal{O}(N^2)$ bits and thus is at most a constant factor larger than the representation by adjacency matrices. Since $\mathcal{O}\left(M \log N (\log M + \log\log N)\right)$ bits are sufficient to store G an OBDD representation is more space efficient than an adjacency matrix for not very dense graphs. An adjacency list representation of G needs $\mathcal{O}\left((N + M)\log N\right)$ bits. However, an adjacency test may take linear time in the worst case. Here, OBDDs are more efficient because adjacency can be tested in $\mathcal{O}(\log N)$ time (start at the root of the OBDD and traverse the graph according to the input until the sink is found). Hence, the space requirements of OBDDs are only little worse than that of adjacency lists and much better than that of adjacency matrices for sparse graphs, while the adjacency test is less efficient than that of a matrix representation but much better than that of a list representation in the worst case.

But Proposition 1 covers only the worst case. In the following we will show that several very natural and large graph classes have OBDD sizes which yield a much better space behaviour than that of explicit representations. We start in Sect. 2 with the investigation of several types of graph classes which do not

contain many P_4s, most prominently cographs. We show that the worst-case OBDD size of these graphs is between $\Omega(N/\log N)$ and $\mathcal{O}(N \log N)$. In Sect. 3 we investigate interval graphs. We show that the worst-case OBDD size of unit interval graphs is between $\Omega(N/\log N)$ and $\mathcal{O}(N/\sqrt{\log N})$, while that of general interval graphs is between $\Omega(N)$ and $\mathcal{O}(N^{3/2}/\log^{3/4} N)$. Finally, in Sect. 4 we try to find a natural graph class which is very hard to represent for OBDDs. We show that the representation of some bipartite graphs requires OBDDs of size $\Omega(N^2/\log N)$. Hence, the OBDD representation of bipartite graphs is not necessarily more space efficient than that of an adjacency matrix representation.

2 Graphs with Few Induced P_4s

Hereinafter P_4 denotes a chordless path with four vertices and three edges. Many graphs with few induced P_4s have common properties such as a unique (up to isomorphism) tree representation. Starting from the tree representation developed by Lerchs [4] of the well-known class of *cographs* (graphs with no induced P_4), Jamison and Olariu have developed and studied tree representations for various graph classes with few induced P_4s such as *P_4-reducible graphs* [9], *P_4-extendible graphs* [10] and *P_4-sparse graphs* [11]. *P_4-reducible graphs* contain no vertex, that belongs to more than one induced P_4, *P_4-extendible graphs* contain at most one additional vertex for each induced P_4 p that induces a different P_4 together with three vertices from p and in *P_4-sparse graphs* every set of five vertices induces one P_4 at most. In the following discussion we omit P_4-reducible graphs, because their class is the intersection of the classes of P_4-extendible and P_4-sparse graphs.

All these graph classes have in common that they can be constructed from single vertex graphs by graph operations which join several vertex disjoint graphs together. Consider for example two graphs $G_1 = (V_1, E_1)$ and $G_2 = (V_2, E_2)$ on two disjoint vertex sets. Then the *union* of G_1 and G_2 is $G_1 \cup G_2 := (V_1 \cup V_2, E_1 \cup E_2)$ and the *join* is $G_1 + G_2 := (V_1 \cup V_2, E_1 \cup E_2 \cup E')$, where E' contains all edges $\{v_1, v_2\}$ with $v_1 \in V_1$ and $v_2 \in V_2$. It is well-known that any *cograph* can be obtained from single vertex graphs by a sequence of $\cup$ and $+$ operations.

Now consider a set Ω of operations, each of them joining several vertex disjoint graphs together. Further assume that a graph G can be constructed from single vertex graphs by using only the joining operations in Ω. Then G has a natural representation as a rooted tree $T(G)$ which can be constructed recursively as follows: If G is a single vertex graph, consisting of the vertex v, then $T(G) = v$. If $G = \omega(G_1, \dots, G_k)$ for an operation $\omega \in \Omega$, then the root of $T(G)$ is a vertex labeled with ω whose children are the roots of the trees $T(G_1), \dots, T(G_k)$ (note that the order of the children may be important).

In order to obtain upper bounds for the OBDD-size of graph classes which have a tree representation, we devise an algorithm which has the following property: In each step of the algorithm a variable is queried (in the order determined by the variable ordering π). In the ith step the variable $\pi(i)$ is queried and after the query the algorithm stores a state value q_i which depends only on the previ-

ous stored state value and the result of the variable query. Each possible stored state value q_i of the algorithm corresponds to an OBDD node labeled with the variable $\pi(i+1)$ and thus the sum of the number of possible state values q_i over all $0 \le i \le n$ is the number of OBDD nodes (q_0 is the unique starting state corresponding to the root of the OBDD and the two possible final state values $q_{n+1} \in \{0,1\}$ corresponding to the sinks of the OBDD). It is obvious how to construct the π-OBDD corresponding to such an algorithm.

A tree representation of a graph is helpful if we want to devise an algorithm deciding adjacency which can then be turned into an OBDD for the graph. E.g. if G is a cograph, then two vertices v_1 and v_2 are adjacent if and only if the least common ancestor of v_1 and v_2 in $T(G)$, $\text{lca}(v_1, v_2)$, is labeled with $+$. Hence, for the algorithm it suffices to determine the lca of v_1 and v_2. For P_4-extendible graphs and P_4-sparse graphs adjacency is not so simple to determine. However, we develop a new tree representation for these graphs such that adjacency of two vertices can be determined by computing the lca of two vertices and some additional information.

Recall that $V = \{[0]_n, \ldots, [N-1]_n\}$. We label the vertices for our representation in such a way that $|v_1|$ is less than $|v_2|$ for two vertices v_1, v_2, if a preorder traversal of $T(G)$ traverses the leaf corresponding to v_1 first. Furthermore, for two vertices v_1, v_2 of a graph $G = (V, E)$ with a tree representation $T(G)$ let $\delta_d(v_1, v_2)$ be $||v_1| - |v_2||$, if $||v_1| - |v_2|| \le d$ and 0 otherwise. Let $c : V \to \mathbb{N}$ be the function with $c(v) = i$ if the vertex v is the ith child of its parent in $T(G)$.

Lemma 1. *Let $\mathcal{G}$ be the class of either cographs, P_4-sparse graphs or P_4-extendible graphs. Then there is a tree representation $T(G)$ for all graphs $G = (V, E) \in \mathcal{G}$ such that for any two vertices $v_1, v_2 \in V$ the characteristic function $\chi_E(v_1, v_2)$ is uniquely determined by*

(a) $lca(v_1, v_2)$, in the case of cographs,
(b) $lca(v_1, v_2)$, $\delta_1(v_1, v_2)$, $|v_1| \bmod 2$, $|v_2| \bmod 2$ and the information whether $|v_1| < |v_2|$, in the case of P_4-sparse graphs,
(c) $lca(v_1, v_2)$, $\delta_4(v_1, v_2)$, $c(v_1)$, $|v_1| \bmod 2$, $|v_2| \bmod 2$ and the information whether $|v_1| < |v_2|$, in the case of P_4-extendible graphs.

The proof of part (a) follows right away from the definition of cographs. The proofs of parts (b) and (c) are omitted due to space restrictions and can be found in the full version of the paper.

The following algorithm determines the lowest common ancestor (lca) of two nodes in a tree representation. The idea of the algorithm is to search the lca starting in the leaf corresponding to v_1 and ascending successively while reading the vertex coding of v_2.

Algorithm 1. The algorithm is defined for a fixed tree T with N leaves labeled with values from $\{0,1\}^n$, $n = \lceil \log N \rceil$, in such a way that if $v_0, \ldots, v_{N-1}$ are the leaves found in a preorder traversal then $|v_i| = i$. The inputs of the algorithm are $x, y \in \{0,1\}^n$ and the output is the lca of x and y if x and y are both leaves in the tree. If either x or y is not a leaf of T, then the output is $-\infty$. The algorithm queries all input variables once in the order $x_{n-1}, \ldots, x_0, y_{n-1}, \ldots, y_0$ and after

each query two values b and c are stored. The value of c is one of the relations "$<$", "$>$" or "$=$" and b is a node in T.

We describe two invariants which are true after each step of the algorithm unless the algorithm terminated. Consider a situation in which all variables up to y_i, $0 \le i \le n-1$, have been queried and the algorithm has not terminated. The first invariant is that c is the relation between $|x_{n-1} \dots x_i|$ and $|y_{n-1} \dots y_i|$ (e.g., if c="$<$", then $|x_{n-1} \dots x_i| < |y_{n-1} \dots y_i|$). Now assume c = "$<$" and let y_i^0 be the leaf $y_{n-1} \dots y_i 0 \dots 0$. The second invariant is that $b = \text{lca}(x, y_i^0)$ and $y_i^1 = y_{n-1} \dots y_i 1 \dots 1$ is not in the subtree rooted at b. For the case c = "$>$", the invariant is analogous, but the roles of y_i^0 and y_i^1 are exchanged. In the case c = "$=$", we have $b = x$.

Note that if these invariants are true, then by knowing c and b, the value of $y_{n-1} \dots y_i$ is uniquely determined. For c = "$=$" this is obvious. For c = "$<$", this follows because due to the enumeration of the leaves, in the right subtree of the tree rooted at b, there can only be one leaf $a_{n-1} \dots a_i 0 \dots 0$ such that $a_{n-1} \dots a_i 1 \dots 1$ is not in this subtree. The case c = "$>$" is analogous. Hence, it suffices to describe an algorithm for which these invariants remain true after each query of a y variable and which – under the assumption that the invariant remains true – outputs the correct result.

Step 1: Store "$=$" in c. Query all x variables and let b be the corresponding leaf of T. If there is no corresponding leaf: output $-\infty$. Clearly, the invariants remain true after this step unless the algorithm terminates.
Step 2: Query the next y variable, say y_i. Since the invariants were true before querying y_i, by knowing b and c we now know $y_{n-1} \dots y_i$. If c = "$=$" and $y_i = b_i$, we can proceed with querying the next variable because the invariants remain true. Hence, we continue with Step 2, again. If c = "$=$" and $y_i \ne b_i$, then we have found the most significant bit in which x and y differ. We can change the value of c to "$<$" or "$>$" such that it reflects the relation between $|x_{n-1} \dots x_i|$ and $|y_{n-1} \dots y_i|$. Hence, if we reach this point in any case $c \ne$ "$=$". Let b' be $\text{lca}(b, y_i^0)$ ($= \text{lca}(x, y_i^0)$) in the case c = "$<$" and $b' = \text{lca}(b, y_i^1)$ ($= \text{lca}(x, y_i^1)$) in the case c = "$>$". Since we know $y_{n-1} \dots y_i$ and b, b' is uniquely determined, if it exists. However, it may happen that such a b' does not exist. In this case y cannot be a leaf of the tree and thus we output $-\infty$. Assume c = "$<$" (the case c = "$>$" is analogous with the roles of y_i^1 and y_i^0 exchanged). If the leaf y_i^1 is not in the subtree rooted at b', then we replace b with b'. Clearly, the invariants are now true again and we proceed with the next y variable by going to Step 2. If on the other hand, y_i^1 is in the subtree rooted at b', then obviously all leaves $y_{n-1} \dots y_i a_{i-1} \dots a_0$ for $a_{i-1} \dots a_0 \in \{0,1\}^i$ are in this subtree. Hence, b' is the lca of x and y and we output b'.

Note that after querying the last y variable, the algorithm terminates in Step 2, because either an appropriate b' is not found (and the algorithm outputs $-\infty$) or the found b' is in fact the lca of x and y.

Theorem 1. *The* OBDD *size of a cograph with N vertices is at most $3N \lceil \log N \rceil$.*

Proof. Let $V = \{v_0, \ldots, v_{N-1}\}$ and let $G = (V, E)$ be a cograph. Let $T(G)$ be the tree representation of G as described before Lemma 1. We have shown above that Algorithm 1 computes for two vertices $u, v \in V$ (given by n-bit strings x, y) the corresponding $\text{lca}(u, v)$ in $T(G)$ by querying each x- and y-variable at most once in the order $x_{n-1}, \ldots, x_0, y_{n-1}, \ldots, y_0$. According to Lemma 1 the adjacency of u and v in G is uniquely determined by their lca, and thus the algorithm describes a π-OBDD for the variable ordering π with $(\pi(1), \ldots, \pi(2n)) = (x_{n-1}, \ldots, x_0, y_{n-1}, \ldots, y_0)$. In the following we bound the number of possible states the algorithm has to store after each variable query.

For Step 1 it suffices to store $x_{n-1} \ldots x_i$ once theses variables have been queried. If it turns out that $|x_{n-1} \ldots x_i 0 \ldots 0|$ is at least N, the algorithm outputs $-\infty$. Then the π-OBDD has at most as many nodes as a complete binary tree in the first n levels and the $(n+1)$th level has at most N nodes. Hence, there are at most $2^n - 1 < 2N - 1$ x nodes and at most N y_{n-1} nodes

Now consider Step 2. As long as b is a leaf, the value of c is "=" and once b is no leaf anymore, the value of c is either "<" or ">". Hence, by knowing c, we can conclude on whether b is a leaf or not. Since there are N leaves and at most $N-1$ inner nodes, $3N$ states suffice for storing c and b. Therefore, there are at most $3N$ y_i nodes for $0 \leq i < n-1$. To conclude, the total number of y nodes of the OBDD is bounded by $N + (n-1) \cdot 3N = 3N \cdot \lceil \log N \rceil - 2N$. □

In order to obtain OBDDs for P_4-sparse and P_4-extendible graphs, our Algorithm 1 has to be modified in such a way that it computes in addition to the lca of two leaves the other information which is needed in order to decide adjacency between the vertices v_1 and v_2, as described in Lemma 1. The necessary modifications of the algorithm can be found in the full version of the paper.

Theorem 2. *The* OBDD *size of P_4-sparse graphs and P_4-extendible graphs is $\mathcal{O}(N \log N)$.*

We contrast the above results by a lower bound for cographs, which also applies to its superclasses of P_4-reducible, P_4-extendible and P_4-sparse graphs.

Theorem 3. *The worst-case* OBDD *size of cographs is at least* $1.832 \cdot N / \log N - \mathcal{O}(1)$.

We prove this with counting arguments. Let in the following $N_{\mathcal{G}}(N)$ denote the number of graphs with N vertices in a graph class $\mathcal{G}$. Note that unless stated otherwise the graphs in the considered graph classes $\mathcal{G}$ are unlabeled.

Proposition 2. *Consider functions $s_N : \mathbb{N} \to \mathbb{R}$ for $N \in \mathbb{N}$ and let $\mathcal{G}$ be a graph class that allows the addition of isolated vertices. If $\lim_{N\to\infty}(2^{s_N \log s_N + s_N \log\log N + \mathcal{O}(s_N)} \cdot (N_{\mathcal{G}}(N))^{-1}) < 1$, then for large enough N there are graphs with N or more vertices in $\mathcal{G}$ such that the* OBDD *size of these graphs is more than s_N.*

Proof. Wegener has shown in [18] that OBDDs of size s can compute at most $sn^s (s+1)^{2s} / s! = 2^{s \log s + s \log n + \mathcal{O}(s)}$ different functions $f \in B_n$. It is easy to see that if the limit in the claim is less than 1, then for large enough N there are more graphs than OBDDs for functions in $n = 2 \cdot \lceil \log N \rceil$ variables. □

An asymptotic formula due to Finch [6] states that the number of graphs in the class of cographs $\mathcal{C}$ satisfies $N_{\mathcal{C}}(N) \sim \lambda\kappa^N N^{-\frac{3}{2}}$ for the constants $\lambda = 0.4127\ldots$ and $\kappa = 3.5608\ldots$. Since $\kappa^N = 2^{N\cdot 1.8322\ldots}$, Theorem 3 follows directly from Proposition 2.

3 Interval Graphs

An interval graph is defined by a set of closed intervals $I \subseteq \mathbb{R}^2$, each of them corresponding to a vertex in the graph. Two vertices are adjacent if and only if the corresponding two intervals intersect.

We first analyse *unit interval graphs*, i.e. interval graphs where the underlying intervals have unit length. Therefore, we can identify the intervals with just one endpoint. We assume w.l.o.g. that no two intervals have the same endpoints and label the vertices in such way that if the interval represented by a vertex v_1 starts further left than the interval represented by a vertex v_2, then $|v_1| < |v_2|$.

Theorem 4. *The* OBDD *size of unit interval graphs with* N *vertices is bounded above by* $\mathcal{O}(N/\sqrt{\log N})$.

In order to proof this result we use the characterization of minimal OBDDs due to Sieling and Wegener [17]: The minimal π-OBDD representing a function f on $x_1, \ldots, x_n$ has as many nodes labeled with the variable x_i, $1 \le i \le n$, as there are different subfunctions of f essentially depending on x_i when all variables x_j with $\pi^{-1}(x_j) < \pi^{-1}(x_i)$ are set to constants (a subfunction essentially depends on a variable x_i, if the substitution $x_i = 0$ leads to a different subfunction than the substitution $x_i = 1$).

Proof (of Theorem 4). Let $G = (V, E)$ be a unit interval graph labeled as described above. For $x \in \{0,1\}^n$ let the interval corresponding to the vertex x be denoted by $I(x) = [a, a+1]$, where $a \in \mathbb{R}$. Let π be the variable ordering where $(\pi(1), \ldots, \pi(2n)) = (x_{n-1}, y_{n-1}, \ldots, x_0, y_0)$. Further, let $f := \chi_E$ and $s_{k,\ell}$, $1 \le k < n$, $\ell \in \{k-1, k\}$, be the number of non-constant subfunctions $f_{|\alpha,\beta}$ of f, where α is an assignment to the variables $x_{n-1}, \ldots, x_{n-k}$ and β is an assignment to the variables $y_{n-1}, \ldots, y_{n-\ell}$. Then $s_{k,k}$ is an upper bound on the number of x_k nodes and $s_{k,k-1}$ is an upper bound on the number of y_k nodes as stated in Sect 1. For the sake of simplicity we assume $k = \ell$ using the simple observation that $s_{k,k+1}$ is at most $2s_{k,k}$ and denote $s_{k,k}$ by s_k.

Since there are 2^{2^m} boolean functions in m variables, we have

$$s_k \le 2^{2^{2n-2k}} . \tag{1}$$

If k is small, we need a better bound. We derive an upper bound for the number of non-constant subfunctions $f_{|\alpha,\beta}$, where α and β are assignments to the variables $x_n \ldots x_{n-k}$ and $y_n \ldots y_{n-k}$, respectively, and $|\alpha| \le |\beta|$. Then s_k is at most twice the result. Let $(\alpha_1, \beta_1), \ldots, (\alpha_p, \beta_p)$ be different pairs of assignments to the x and y variables such that $|\alpha_i| \le |\beta_i|$ and $f_{|\alpha_i,\beta_i} \notin \{0,1\}$ (i.e. these

subfunctions are not constant) for all $1 \leq i \leq p$. Furthermore, assume that $(|\alpha_1|, |\beta_1|), \ldots, (|\alpha_p|, |\beta_p|)$ are ordered lexicographically. We prove below that

$$\forall 1 \leq i \leq p: \quad |\beta_i| \leq |\beta_{i+1}| \; . \tag{2}$$

Using the fact that $(|\alpha_1|, |\beta_1|), \ldots, (|\alpha_p|, |\beta_p|)$ are ordered lexicographically, it is easy to see that the number of these pairs, p, is bounded by $|\alpha_p| + |\beta_p| + 1$. Hence, we obtain $p \leq 2^{k+1} - 1$ and thus $s_k \leq 2^{k+2} - 2$. Using the upper bound of (1) it follows that $s_k \leq \min\{2^{k+2} - 2, 2^{2^{2n-2k}}\}$. We plug this inequality into the upper bound $2 + \sum_{k=0}^{n-1}(s_k + 2s_k)$ on the OBDD size of G, and using simple algebra obtain an upper bound of $\mathcal{O}(N/\sqrt{\log N})$.

It remains to prove the claim (2). If $|\alpha_i| = |\alpha_{i+1}|$, this follows right away from the lexicographical ordering of the pairs $(|\alpha_j|, |\beta_j|)$. Hence, assume $|\alpha_i| < |\alpha_{i+1}|$. If (2) is not true, i.e. $|\beta_{i+1}| < |\beta_i|$, then we have $|\alpha_i| < |\alpha_{i+1}| \leq |\beta_{i+1}| < |\beta_i|$ (recall that we only count the pairs (α_j, β_j) where $|\alpha_j| \leq |\beta_j|$). Since $f_{|\alpha_i,\beta_i}$ is not the constant 0-function, there is an assignment c to the remaining x variables $x_{n-k-1}, \ldots, x_0$ and an assignment d to the remaining y variables $y_{n-k-1}, \ldots, y_0$ such that $f_{|\alpha_i,\beta_i}(c, d) = 1$. Hence, $\chi_E(\alpha_i c, \beta_i d) = 1$ and the intervals $I(\alpha_i c)$ and $I(\beta_i d)$ intersect. Now consider additional arbitrary assignments c' to the remaining x variables and d' to the remaining y variables. Obviously, then $|\alpha_i c| < |\alpha_{i+1} c'| < |\beta_i d|$ and $|\alpha_i c| < |\beta_{i+1} d'| < |\beta_i d|$. Hence, the intervals $I(\alpha_{i+1} c')$ and $I(\beta_{i+1} d')$ are neither right of $I(\beta_i d)$ nor left of $I(\alpha_i c)$. But since the latter intervals intersect, obviously $I(\alpha_{i+1} c')$ and $I(\beta_{i+1} d')$ intersect, too. Because this is true for all c' and d' we obtain that $f_{|\alpha_{i+1},\beta_{i+1}} = 1$, which contradicts the assumption that this subfunction is not constant. □

Using a similar idea yields an upper bound for general interval graphs stated in the following theorem (the proof can be found in the full version of the paper).

Theorem 5. *Interval graphs with N vertices have* OBDD*s of size* $\mathcal{O}(N^{3/2}/\log^{3/4} N)$.

Finch [5] provided an asymptotic formula for the size of the class of unit interval graphs $\mathcal{U}$, $N_{\mathcal{U}}(N) \sim \frac{1}{8e^{\kappa}\sqrt{\pi}} \frac{4^N}{N^{\frac{3}{2}}}$, and Gavoille and Paul [7] obtained an asymptotic formula for the size of the class of general interval graphs $\mathcal{I}$: $N_{\mathcal{I}}(N) \geq 2^{N \log N - \mathcal{O}(N)}$. Thus, the following lower bounds follow directly from Proposition 2.

Theorem 6. *For all $\varepsilon > 0$, the worst-case* OBDD *size of unit interval graphs with N vertices is at least $(2-\varepsilon)N/\log N - \mathcal{O}(1)$ and the worst-case* OBDD *size of interval graphs with N vertices is at least $(1-\varepsilon)N - \mathcal{O}(1)$.*

4 Bipartite Graphs

The goal of this section is to show for a specific graph class that a representation by OBDDs is not necessarily more space efficient than a representation by adjacency matrices.

Theorem 7. *For all $\varepsilon > 0$, the worst-case* OBDD *size of bipartite graphs with N vertices is at least $(\frac{1}{8} - \varepsilon)N^2/\log N - \mathcal{O}(1)$.*

Proof. We consider the class of labeled 2-*coloured graphs*, where different colourings of 2-*colourable (bipartite) graphs* lead to different graphs. The asymptotic relation between the size of the class of labeled 2-coloured graphs $\mathcal{C}_\ell$ and the class of labeled 2-colourable graphs $\mathcal{B}_\ell$ has been proven by Prömel and Steger in [13] to be as follows: $\lim_{N\to\infty} N_{\mathcal{C}_\ell}(N)/N_{\mathcal{B}_\ell}(N) = 2$. Therefore, a random bipartite graphs has almost surely only one 2-colouring (and the inverse 2-colouring). Asymptotics for the number of labeled 2-coloured graphs were given by Wright [20, 14] as $N_{\mathcal{C}_\ell}(N) \sim \kappa 2^{\frac{N^2}{4}} 2^N \sqrt{\left(\frac{2}{N \ln 2}\right)}$ where $\kappa = 1 \pm 0.0000013097\ldots$ is a constant. According to Prömel [12] the relation between the number of labeled and unlabeled bipartite graphs is bounded by $N!$. Combining all these results leads to the following relation for the size of the class of unlabeled bipartite graphs $\mathcal{B}$: $\lim_{N\to\infty} 2(N!)N_{\mathcal{B}}(N)/2^{\frac{N^2}{4}} 2^N \sqrt{\left(\frac{2}{N \ln 2}\right)} \leq 1$. The lower bound now follows directly from Proposition 2. □

The disadvantage of proving lower bounds with counting arguments is that they only show the existence of graphs which are hard to represent. However, such graphs might for large N never appear in applications because e.g. they are not computable in polynomial time. A statement showing how to construct such a graph or at least telling us that such a graph is computable in polynomial time has much more relevance. In order to achieve such results, we show how any boolean function can be represented by a bipartite graph. This way, we can conclude from known lower bounds for the OBDD size of boolean function on lower bounds for the OBDD size of the corresponding bipartite graphs.

Definition 3. Let $f \in B_n$, n even, be a boolean function. The bipartite graph $G_f = (V_1 \cup V_2, E)$ is given by the vertex sets $V_1 := \{v_1 \in \{0,1\}^{\frac{n}{2}} \mid |v_1| < 2^{\frac{n}{2}}\}$ and $V_2 := \{v_2 \in \{0,1\}^{\frac{n}{2}+1} \mid 2^{\frac{n}{2}} \leq |v_2| < 2^{\frac{n}{2}+1}\}$ and the edge set $E := \{\{v_1, v_2\} | v_1 \in V_1, v_2 \in V_2, f(v_1[|v_2| - 2^{\frac{n}{2}}]_{\frac{n}{2}}) = 1\}$.

Theorem 8. *For each function $f \in B_n$ there is a bipartite graph $G_f = (V, E)$ such that the* OBDD *size of χ_E is not smaller than the* OBDD *size of f.*

Proof. Let us assume that an OBDD B with smaller size exists for χ_E. Let $\{x_0, \ldots, x_{\frac{n}{2}}, y_0, \ldots, y_{\frac{n}{2}}\}$ be the set of variables of χ_E, then

$$f\left(x_1, \ldots, x_{\frac{n}{2}}, y_1, \ldots, y_{\frac{n}{2}}\right) = \chi_E\left(0, x_1, \ldots, x_{\frac{n}{2}}, 1, y_1, \ldots, y_{\frac{n}{2}}\right)$$

follows from Definition 3. We therefore can construct an OBDD for f from B by redirecting all edges leading to a node labeled with x_0 or y_0 to the appropriate 0-successor or 1-successor of this node, respectively. We represent f with this OBDD of smaller size, which is a contradiction. □

Andreev, Baskakov, Clementi and Rolim [1] presented a boolean function which is computable in polynomial time and has an OBDD size of $2^{n-\mathcal{O}(\log^2 n)}$. According to the knowledge of the authors this is the best known lower bound for the OBDD size of a function in P.

Corollary 1. *There is a bipartite graph G_f, $f \in B_{2k}$, with $N = 2^{k+1}$ vertices which is computable in polynomial time and for which the* OBDD *size of χ_E is at least $N^2/(\log N)^{\mathcal{O}(\log\log N)}$.*

Acknowledgment. The authors thank Ingo Wegener for helpful comments on the paper.

References

1. A. Andreev, J. Baskakov, A. Clementi, and J. Rolim. Small pseudo-random sets yield hard functions: New tight explicit lower bounds for branching programs. In *26th ICALP*, volume 1644 of *LNCS*, pp. 179–189. 1999.
2. Y. Breitbart, H. Hunt III, and D. Rosenkrantz. On the size of binary decision diagrams representing boolean functions. *Theor. Comp. Sci.*, 145:45–69, 1995.
3. R. E. Bryant. Graph-based algorithms for boolean function manipulation. *IEEE Trans. Comput.*, C-35:677–691, 1986.
4. D. G. Corneil, H. Lerchs, and L. Stewart Burlingham. Complement reducible graphs. *Discrete Appl. Math.*, 3:163–174, 1981.
5. S. R. Finch. *Mathematical Constants.* Cambridge University Press, 2003.
6. S. R. Finch. Series-parallel networks, 2003. Supplementary Material for [5].
7. C. Gavoille and C. Paul. Optimal distance labeling schemes for interval and circular-arc graphs. In G. Di Battista and U. Zwick, editors, *11th ESA*, volume 2832 of *LNCS*. Springer, 2003.
8. G. D. Hachtel and F. Somenzi. A symbolic algorithm for maximum flow in 0-1 networks. *Formal Methods in System Design*, 10:207–219, 1997.
9. B. Jamison and S. Olariu. p_4-reducible graphs - a class of uniquely tree representable graphs. *Stud. Appl. Math.*, 81:79–87, 1989.
10. B. Jamison and S. Olariu. On a unique tree representation for p_4-extendible graphs. *Discrete Appl. Math.*, 34:151–164, 1991.
11. B. Jamison and S. Olariu. A tree representation for p_4-sparse graphs. *Discrete Appl. Math.*, 35:115–129, 1992.
12. H. J. Prömel. Counting unlabeled structures. *J. Combin. Theory Ser. A*, 44:83–93, 1987.
13. H. J. Prömel and A. Steger. Random l-colorable graphs. *Random Structures Algorithms*, 6:21–37, 1995.
14. R. C. Read and E. M. Wright. Coloured graphs: A correction and extension. *Canad. J. Math.*, 22:594–596, 1970.
15. D. Sawitzki. Implicit flow maximization by iterative squaring. In P. Van Emde Boas, J. Pokorny, M. Bielikova, and J. Stuller, editors, *30th SOFSEM*, volume 2932 of *LNCS*, pp. 301–313. Springer, 2004.
16. D. Sawitzki. A symbolic approach to the all-pairs shortest-paths problem. In *30th WG*, volume 3353 of *LNCS*, pp. 154 – 116. 2004.
17. D. Sieling and I. Wegener. NC-algorithms for operations on binary decision diagrams. *Parallel Processing Letters*, 3:3–12, 1993.
18. I. Wegener. *Branching Programs and Binary Decision Diagrams.* SIAM, 2000.
19. P. Woelfel. Symbolic topological sorting with OBDDs. *Journal of Discrete Algorithms*, to appear.
20. E. M. Wright. Counting coloured graphs. *Canad. J. Math.*, 13:683–693, 1961.

Space-Efficient Construction of LZ-Index*

Diego Arroyuelo and Gonzalo Navarro

Dept. of Computer Science, University of Chile,
Blanco Encalada 2120, Santiago, Chile
{darroyue, gnavarro}@dcc.uchile.cl

Abstract. A *compressed full-text self-index* is a data structure that replaces a text and in addition gives indexed access to it, while taking space proportional to the compressed text size. The LZ-index, in particular, requires $4uH_k(1+o(1))$ bits of space, where u is the text length in characters and H_k is its k-th order empirical entropy. Although in practice the LZ-index needs 1.0-1.5 times the text size, its construction requires much more main memory (around 5 times the text size), which limits its applicability to large texts. In this paper we present a practical space-efficient algorithm to construct LZ-index, requiring $(4+\epsilon)uH_k+o(u)$ bits of space, for any constant $0<\epsilon<1$, and $O(\sigma u)$ time, being σ the alphabet size. Our experimental results show that our method is efficient in practice, needing an amount of memory close to that of the final index.

1 Introduction and Previous Work

A *full-text database* is a system providing fast access to a large mass of textual data. The simplest (yet realistic and rather common) scenario is as follows. The text collection is regarded as a unique sequence of characters $T_{1\ldots u}$ over an alphabet Σ of size σ, and the search pattern $P_{1\ldots m}$ as another (short) sequence over Σ. Then the text search problem consists of finding all the *occ* occurrences of P in T. To provide fast access, data structures called *indexes* are built on the text. Typical text databases contain natural language texts, DNA or protein sequences, MIDI pitch sequences, program code, etc.

Until a short time ago, the smallest indexes available in practice were the suffix arrays [21], requiring $u \log u$ bits (log means $\log_2$ in this paper). Since the text requires $u \log \sigma$ bits to be represented, this index is usually much larger than the text (typically 4 times the text size). To handle huge texts like the Human Genome (about 3×10^9 base pairs), one solution is to store the indexes on secondary memory. However, this has significant influence on the running time of an application, as access to secondary memory is considerably slower.

Several attempts to reduce the space of the suffix trees [2] or arrays [13,17,19,1] focused on reducing the size of the data structures but not the text, and did not relate text compressibility with the size of its index.

A parallel track started at about the same time [15,14,10,28,4,5,6,8,9,20,7], with the distinguishing feature of providing *compressed* indexes, whose sizes are proportional to the compressed text size. Moreover, in most cases, those indexes *replace* the text by

* Supported in part by CONICYT PhD Fellowship Program (first author) and Fondecyt Grant 1-050493 (second author).

X. Deng and D. Du (Eds.): ISAAC 2005, LNCS 3827, pp. 1143–1152, 2005.

being able to reproduce any text substring. This is called *self-indexing*. Taking space proportional to the compressed text, replacing it, and providing efficient indexed access to it, is an unprecedented breakthrough in text indexing and compression.

The LZ-index [24,25,26] is a full-text self-index on these lines, based on the Ziv-Lempel parsing of the text. If the text is parsed into n *phrases* by the LZ78 algorithm [29], then the LZ-index takes $4n \log n(1 + o(1))$ bits of space, which is 4 times the size of the compressed text and also 4 times the k-th order text entropy, i.e. $4uH_k + o((1 + H_k)u)$, for any $k = o(\log n / \log^2 \sigma)$ [16,6]. See the original article for details on its search algorithms, as we focus only in construction in this paper.

However, all these works do not consider the space-efficient construction of the indexes. For example, construction of *CS-array* [28] and *FM-index* [4] involves building first the suffix array of the text. Similarly, the LZ-index is constructed over a non-compressed intermediate representation. In both cases, one needs about 5 times the text size. For example, the Human Genome may fit in 1 GB of main memory using these indexes (and thus it can be operated entirely in RAM on a desktop computer), but 15 GB of main memory are needed to build them! Using secondary memory for the construction is usually rather inefficient.

The works of T.-W. Lam et al. [18] and W.-K.Hon et al. [12] deal with the space (and time) efficient construction of *CS-array*. The former work presents an algorithm that uses $(2H_0 + 1 + \epsilon)u$ bits of space to build the *CS-array*, where H_0 is the 0-th order empirical entropy of the text, and ϵ is any positive constant; the construction time is $O(\sigma u \log u)$, which is good enough if the alphabet is small (as in the case of DNA), but may be impractical in the case of proteins and Oriental languages, such as Chinese or Japanese. The second work [12] addresses this problem by requiring $(H_0 + 2 + \epsilon)u$ bits of space and $O(u \log u)$ time to build the *CS-array*. Also, they show how to build the *FM-index* from *CS-array* in $O(u)$ time.

Our work follows this line of research. We present a practical and efficient algorithm to construct the LZ-index using little space. Our idea is to replace the (non-compressed) intermediate representations of the tries that conform the index by space-efficient counterparts. Basically, we use the balanced parentheses representation of Munro and Raman [23] as an intermediate representation for the tries, but we modify such representation to allow fast incremental construction directly from the text. The resulting intermediate data structure consists of a tree whose nodes are small subsequences of balanced parentheses, which are easier and cheaper to update. The idea is inspired in the work of Clark and Munro [3], yet ours differs in numerous technical aspects and practical considerations (structuring inside the nodes, overflow management policies, etc.).

Our algorithm requires $(4+\epsilon)uH_k + o(u)$ bits to build the LZ-index, for any constant $0 < \epsilon < 1$. This is very close to the space the final LZ-index requires to operate. This is the *first* construction algorithm for a self index requiring space proportional to H_k instead of H_0. In practice our algorithm also requires about the same memory as the final index. That is, wherever the LZ-index can be used, we can build it. The indexing speed is approximately 5 sec/MB in a 2GHz machine, which is competitive with the (non-space-efficient) construction of *FM-index* and much faster than *CS-array* construction [26]. We argue that our method outperforms (in time) previous work [12] when indexing the Human Genome, using about the same indexing space.

2 The LZ-Index Data Structure

Assume that the text $T_{1...u}$ has been partitioned using the LZ78 [29] algorithm into $n+1$ blocks $T = B_0 \ldots B_n$, such that $B_0 = \varepsilon$; $\forall k \neq \ell, B_k \neq B_\ell$; and $\forall k \geqslant 1, \exists \ell < k$, $c \in \Sigma$, $B_k = B_\ell \cdot c$. To ensure that B_n is not a prefix of another B_i, we append to T a special character "\$" $\notin \Sigma$. The data structures that conform LZ-index are [26,25]:

1. *LZTrie*: is the trie formed by all the blocks $B_0 \ldots B_n$. Given the properties of LZ78 compression, this trie has exactly $n+1$ nodes, each one corresponding to a string.
2. *RevTrie*: is the trie formed by all the reverse strings $B_0^r \ldots B_n^r$. In this trie there could be internal nodes not representing any block. We call these nodes *empty*.
3. *Node*: is a mapping from block identifiers to their node in *LZTrie*.
4. *RNode*: is a mapping from block identifiers to their node in *RevTrie*.

Each of these 4 structures requires $n \log n(1 + o(1)) = uH_k(1 + o(1))$ bits of space.

For the construction of *LZTrie* we traverse the text and at the same time build a *normal trie* (using one pointer per parent-child relation) of the strings represented by Ziv-Lempel blocks. At step k (assume $B_k = B_i \cdot c$), we read the text that follows and step down the trie until we cannot continue. At this point we create a new trie leaf (child of the trie node of block i, by character c, and assigning the leaf block number k), go to the root again, and go on with step $k+1$ reading the rest of the text. The process completes when the last block finishes with the text terminator "\$".

Then we compact the normal trie, essentially using the parentheses representation of Munro and Raman [23]. We traverse the normal trie in preorder, writing an opening parenthesis when going down to a node, and a closing parenthesis when going up.

The *LZTrie* structure consists of the above sequence of parentheses plus a sequence *lets* of characters that label each trie edge and a sequence *ids* of block identifiers, both in preorder. We identify a trie node x with its opening parenthesis in the representation. The subtree of x contains those nodes (parentheses) enclosed between the opening parenthesis representing x and its matching closing parenthesis.

Once the *LZTrie* is built we free the space of the normal trie, and build *Node*. This is just an array with the n nodes of *LZTrie*, using $\lceil \log n \rceil$ bits for each. It is built as the inverse of permutation *ids*.

To construct *RevTrie* we traverse *LZTrie* in depth-first order, generating each string stored in *LZTrie* in constant time, and then inserting it into a *normal trie of reversed strings*. We then traverse the trie and represent it using a sequence of parentheses and block identifiers, *rids*. Empty unary nodes are removed only at this step. Finally, we build the normal reverse trie and build *RNode* as the inverse permutation of *rids*.

In the experiments of the original LZ-index [26,25], the largest extra space needed to build *LZTrie* is that of the normal trie, which is 1.7–2.0 times the text size. The indexing space for the normal reverse trie is, in some cases, 4 times the text size. This is, mainly, because of the empty unary nodes. This space dictates the maximum indexing space of the algorithm (note that the text itself can be processed by buffers and hence does not require significant space). The overall indexing space was 4.8–5.8 times the text size for English text, and 3.4–3.7 times the text size for DNA. As a comparison, the construction of a plain suffix array without any extra data structure requires 5 times the text size.

3 Space-Efficient Construction of LZTrie

The main memory requirement to build the LZ-index comes from the normal tries used to build *LZTrie* and *RevTrie*. We focus on building those tries in little memory, by replacing them with space-efficient data structures that support insertions. These can be seen as hybrids between normal tries and their final parentheses representations.

Let us start with *LZTrie*. In its final representation as a linear sequence of balanced parentheses [23], the insertion of a new node at any position of the sequence may force rebuilding the sequence from scratch. To avoid that cost, we work on a *hierarchical representation of balanced parentheses* (hrbp for short), such that we rebuild only a small part of the entire sequence to insert a new node.

In a hrbp we cut the trie in *pages*, that is, in subsets of trie nodes such that if a node x is stored in page q, then node y, the parent of x, is: (1) also stored in q (enclosing x), or (2) stored in a page p, the *parent page* of q, and hence y is ancestor of all nodes stored in q. We store in p information indicating that node y encloses all nodes in q. In a hrbp we arrange the pages in a tree, thus the entire trie is represented by a tree of pages.

We represent a page as a contiguous block of memory. Let $N_1 < \ldots < N_t$ be even integers. We say that a page has size N_i if it can store N_i parentheses ($N_i/2$ nodes), although its physical size is larger than N_i bits. Each page p of size N_i consists of: N_i bits to represent a subsequence of balanced parentheses; $N_i/2$ bits (the *flags*) indicating which opening parentheses (nodes) in a page have their subtree stored in a child page; $\lceil \log N_i/2 \rceil$ bits to tell the number of nodes stored in the page; $(N_i/2)\lceil \log u \rceil$ bits to store the block identifiers (*ids*) in the page (in preorder); $(N_i/2)\lceil \log \sigma \rceil$ bits to store the characters (*lets*) in the page (in preorder); and a variable number of pointers to child pages. The number of pointers varies from 0 to $N_i/2$, and it corresponds to the number of flags with value 1 in p. To maintain a constant physical page size, these pointers are stored in a separately allocated array, and we store a pointer to them in the page.

As in the parentheses representation of *LZTrie*, in the hrbp a node encloses its subtree, but not necessarily a node and its subtree are stored both in the same page. If the subtree of the j-th opening parenthesis of page p is stored in page q, then q is a child page of p and the j-th flag in p has the value 1. If the number of flags in 1 before the j-th flag (not including it) is h, then the h-th pointer of p points to q. An important property we enforce is that sibling nodes must be stored all in the same page.

Initially, the data structure consists of a sole empty page (the *root page*) of size N_1. The construction of *LZTrie* proceeds as explained in Section 2, but this time the nodes are inserted in a hrbp of *LZTrie*, instead of a normal trie. The insertion of a new node $B_k = B_i \cdot c$ in the hrbp implies to recompute the page in which the insertion is done. If the new leaf must become j-th opening parenthesis in the page (counting from left to right), then we insert "()" at the corresponding parentheses position and the j-th flag is set to 0. Also, c is inserted at position j within the characters, and k is inserted at the same position within the identifiers.

We do not store information to traverse the parentheses sequence in the pages of the hrbp. Instead, all the navigation inside each page is done sequentially, in a single $O(N_t)$ time pass: the first child of an opening parenthesis starts at the next position (unless that contains a closing parenthesis, in which case the node is a leaf in the page), and the next sibling starts right after the closing parenthesis matching the current position, which is

found sequentially. As we traverse the page, we maintain the current position j in *flags*, *ids* and *lets*, as well as the count h of 1-bits seen in *flags*.

A *page overflow* occurs when inserting a new node in a full page p. If the size of p is N_i, $1 \leqslant i < t$, we perform a `grow` operation on p, which allocates a new page p' of size N_{i+1}, copies the content of p to p', frees p, and replaces the pointer to p by a pointer to p' in the parent of p. If the size of p is N_t, instead, we select a subset of nodes to be copied to a new child page of p and then deleted from p.

We only allow the selection of the whole subtree of a node in the page (without selecting the node itself). This simple way ensures that sibling nodes are always stored in the same page. As the maximum number of siblings is σ, we must have $N_t \geqslant 2\sigma$ so that a page with children always has space for its top-level nodes at least. We choose the subtree of maximum size not exceeding $N_t/2$ nodes. It is easy to see that this guarantees that the size of the new leaf p' is at least $\lfloor N_t/(2\sigma) \rfloor - 1$ nodes.

Assume we have selected in this way the subtree of the j-th opening parenthesis in the page. The selected subtree is copied to a new page p', whose size is the minimum N_i suitable to hold the subtree. As p' is a new leaf page, all its flags are initialized to 0. Next we add in p a pointer to p', inserted at the current (j-th) position, and set to 1 the j-th bit in *flags*. Finally, we delete from p the selected subtree. After that, if the number of parentheses in p does not exceed N_i for some $i < t$, we perform a `shrink` operation, which is the opposite of `grow`.

Once we solved the overflow, the insertion of the new node may have to be done in p or in p', but we are sure that there is room for the new pair of parentheses in either page. The following lemma states the condition to achieve a minimum fill ratio α in the pages of the data structure, thus controlling the wasted space. The proof is obvious.

Lemma 1. *Let $0 < \alpha < 1$ be a real number. If each page has the smallest posible size N_i to hold its parentheses, and we define $N_i = N_{i-1}/\alpha$, $i = 2, \ldots, t$, and $2 \leqslant N_1 \leqslant 2/\alpha$, then all pages of the data structure have a fill ratio of at least α.*

As the trie has n nodes, we need $2n + n + n \log u + n \log \sigma + n \log u(2\sigma/N_t)$ bits of storage to represent the parentheses, flags, identifiers, characters and pointers to child pages, respectively. The last bound holds because leaves are created with at least $N_t/(2\sigma)$ nodes and thus there is at worst one pointer for each $N_t/(2\sigma)$ nodes (except the root). If, in addition, we define the N_is as in Lemma 1, in the worst case the construction algorithm needs $\frac{n}{\alpha}(3 + \log \sigma + (1 + 2\sigma/N_t) \log u)$ bits of storage. We can relate this space requeriment to H_k: as $n \log u = uH_k + O(kn \log \sigma)$ for any k [16], and since $n \leqslant u/\log_\sigma u$, the space is $\frac{1+2\sigma/N_t}{\alpha} uH_k + o(u)$ for any $k = o(\log n/\log^3 \sigma)$.

When constructing *LZTrie*, the navigational cost per character of the text is $O(N_t)$ in the worst case. Hence, the overall navigational cost is $O(N_t u)$. On the other hand, the cost of rebuilding pages after an insertion is $O(N_t)$, with or without page overflows. As there are n insertions, the total cost is $O(N_t n)$. However, the constant involved with page overflows is greater than that of simple insertions, thus in practice we expect that larger values of α yield a greater construction time (and a smaller space requirement). In general, choosing $N_t = 2\sigma/\gamma$ for any constant $0 < \gamma < 1$, we get $\frac{1+\gamma}{\alpha} uH_k + o(u)$ bits of space, which can be made $(1 + \epsilon)uH_k + o(u)$ for any constant $0 < \epsilon < 1$ by properly choosing γ and α. The construction time is $O(\frac{1}{\gamma}\sigma u) = O(\sigma u)$.

Once we construct the hrbp for *LZTrie*, we build the final version of *LZTrie* in $O(n)$ time. We perform a preorder traversal on the hrbp, writing an opening parenthesis each time we reach a node, then checking the corresponding flag, traversing the subtree of the node recursively in preorder (which, depending of the flag, may be stored in the same or in a child page), and then writing a closing parenthesis.

4 Space-Efficient Construction of *RevTrie*

For the space-efficient construction of *RevTrie*, we use a hrbp to represent not the original reverse trie but its *PATRICIA tree* [22], which compresses *empty* unary paths of the reverse trie. This yields an important saving of space. We do not store the skips in each node since they can be computed using the connection with *LZTrie*. We store, in the nodes of the reverse trie, pointers to nodes of *LZTrie*, instead of the corresponding block identifiers. Each pointer uses $\lceil \log 2n \rceil$ bits. This is done to avoid access to *Node* when constructing the reverse trie, so we can build *Node* after both tries have been built (thus reducing the indexing space). The empty non-unary nodes are marked by storing in them the same pointer to *LZTrie* stored in their first child.

To construct the reverse trie we traverse *LZTrie* in depth-first order, generating each string B_i stored in *LZTrie* in constant time, and then inserting its reverse B_i^r into the reverse trie. As we compress empty unary paths, the edges of the trie are labeled with strings instead of simple characters. The *PATRICIA* tree stores only the first character of the string that labels the edge. Given a node v in the reverse trie, the position of the character in B_i^r on which v discriminates is 1 plus the length of the string represented by v. If v is not an empty node, then it stores a pointer to *LZTrie* node n_v. The length of the string is the same as the depth of n_v in *LZTrie*, which can be computed in constant time [26]. On the other hand, if v is an empty node, we we must use instead a procedure similar to that used in [26] to compute the string that labels an edge of the trie.

The hrbp of the reverse trie requires at least $\frac{1}{\alpha}(2n' + n' + n' \log 2n + n' \log \sigma + (2\sigma/N_t)n' \log n')$ bits of storage to represent the parentheses, flags, pointers to *LZTrie*, characters and pointers to child pages, respectively, where $n' \geqslant n$ is the number of nodes in the reverse trie. As we compress unary paths, $n' \leqslant 2n$, and thus the space is upper bounded by $\frac{2(1+2\sigma/N_t)}{\alpha} uH_k + o(u)$. However, in practice we expect that n' will be much less than $2n$ (see Section 5 for empirical results).

For each string B_i^r to be inserted in the reverse trie, $1 \leqslant i \leqslant n$, the navigational cost is $O(N_t|B_i^r| + |B_i^r|^2)$ in the worst case (when we work $O(N_t)$ per character, and every traversed node is empty). The total construction cost is $\sum_{i=1}^{n} (N_t|B_i^r| + |B_i^r|^2)$. The sum $\sum_{i=1}^{n} |B_i^r|^2$ is usually $O(u \log_\sigma u)$, but in pathological cases it is $O(u^{3/2})$. To have a better theoretical bound, we can explicitly store the skips, using $2 \log \log u$ extra bits per node (inserting *empty* unary nodes when the skip is exceeded). In this way, one of each $\log^2 u$ empty unary nodes could be explicitly represented. In the worst case there are $O(u)$ empty unary nodes, of which $O(\frac{u}{\log u})$ are explicitly represented. This means $o(u)$ extra bits in the hrbp, and the construction cost is reduced to $\sum_{i=1}^{n} (N_t|B_i^r| + |B_i^r|)$. As $\sum_{i=1}^{n} |B_i^r| = u$, the total cost is $O(N_t u)$.

After we construct the hrbp for the reverse trie, we construct *RevTrie* directly from the hrbp in $O(n')$ time, replacing the pointers to *LZTrie* by the corresponding block

identifiers (*rids*), and then we free the space of the hrbp. If we use $n' \log n$ bits for the *rids* array, *RevTrie* requires $2uH_k + o(u)$ bits of storage, and the whole index requires $5uH_k(1+o(1))$ bits. Instead, we can represent the *rids* array with $n \log n$ bits (i.e., only the non-empty nodes), plus a bitmap of $2n(1 + o(1))$ bits supporting $rank$ queries in $O(1)$ time [27]. The j-th bit of the bitmap is 1 if the node represented by the j-th open parenthesis is not an empty node, otherwise the bit is 0. The *rids* index corresponding to the j-th opening parenthesis is $rank(j)$. Using this representation, *RevTrie* requires $uH_k + o(u)$ bits of storage. This was unclear in the original LZ-index paper [26,25].

We finally build *Node* mapping from *ids* array in time $O(n)$ and $n \log n = uH_k + o(u)$ bits of space, and *RNode* from *rids* in $O(n')$ time and $n \log n' = uH_k + o(u)$ bits.

Now we summarize the construction process, and show in parentheses the total space requeriment and the time in each step. Then, we conclude with a theorem.

1. We build the hrbp of *LZTrie* from the text ($(1 + \epsilon)uH_k + o(u)$ bits, $O(\sigma u)$ time).
2. We build *LZTrie* from its hrbp ($(1 + \epsilon)uH_k + uH_k + o(u)$ bits, $O(n)$ time).
3. We free the hrbp of *LZTrie* and build the hrbp of the reverse trie from *LZTrie* ($(2 + \epsilon)uH_k + uH_k + o(u)$ bits, $O(\sigma u)$ time).
4. We build *RevTrie* from its hrbp ($(2+\epsilon)uH_k + uH_k + uH_k + o(u)$ bits, $O(n)$ time).
5. We free the hrbp of *RevTrie* and build *Node* from *ids* ($uH_k + uH_k + uH_k + o(u)$ bits, $O(n)$ time).
6. We build *RNode* from *rids* ($uH_k + uH_k + uH_k + uH_k + o(u)$ bits, $O(n)$ time).

Theorem 1. *Our space-efficient algorithm to construct LZ-index requires* $(4+\epsilon)uH_k + o(u)$ *bits of space, reached at step 4 above, and* $O(\sigma u)$ *time. This holds for any constant* $0 < \epsilon < 1$ *and any* $k = o(\log n / \log^3 \sigma)$.

5 Experimental Results

For the experiments we use the file `ohsumed.88-91` from the *OHSUMED* collection [11], as a representative of other text collections we tested, such as DNA, music, and others. We use incremental subsets of the file, ranging from 10MB to 100MB. We run our experiments on an AMD Athlon processor at 2GHz, 1024MB of RAM and 512Kb of cache, running version 2.6.7-gentoo-r11 of Linux kernel. We compiled the code with `gcc 3.3.4` using optimization option `-O9`. Times were obtained using 10 repetitions.

In Fig. 1 we show the construction space for *LZTrie* and *RevTrie*. As expected, the construction space is smaller as we use a greater value of α. On average, the construction space of *LZTrie* ranges from approximately 0.5 ($\alpha = 0.95$) to 0.64 ($\alpha = 0.5$) times the text size, and from approximately 1.00 ($\alpha = 0.95$) to 1.27 ($\alpha = 0.5$) times the size of the final version of *LZTrie*. For construction of *RevTrie* the space varies from 0.52 ($\alpha = 0.95$) to 0.65 ($\alpha = 0.5$) times the text size, and from 1.47 ($\alpha = 0.95$) to 1.85 ($\alpha = 0.5$) times the size of the final *RevTrie*. The greater difference among *RevTrie* and its hrbp is due to the fact that the final version of the trie does not store the characters. As a comparison, the original construction algorithm [26] (labeled "Original" in the plots) needs on average 1.82 times the text size to hold the normal trie and 3.30 times to hold the normal reverse trie. The sizes as a fraction of the final tries are 3.62 and 9.87 times, respectively.

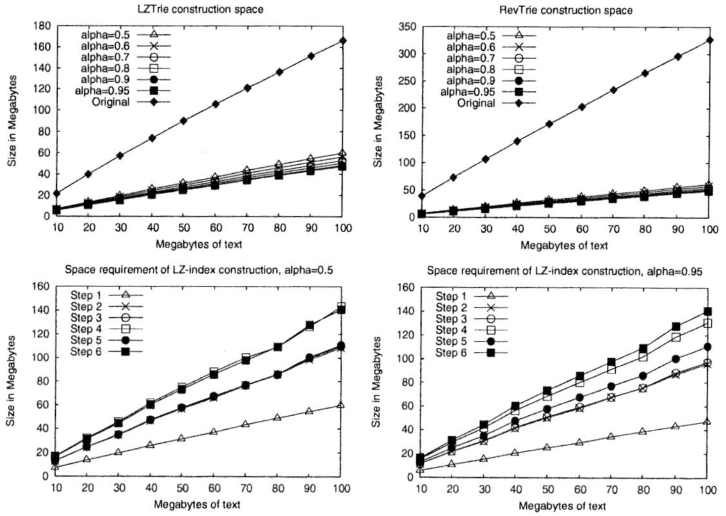

Fig. 1. Size of the hrbps of *LZTrie* and *RevTrie*, $N_1 = 2$, $N_t = 512$

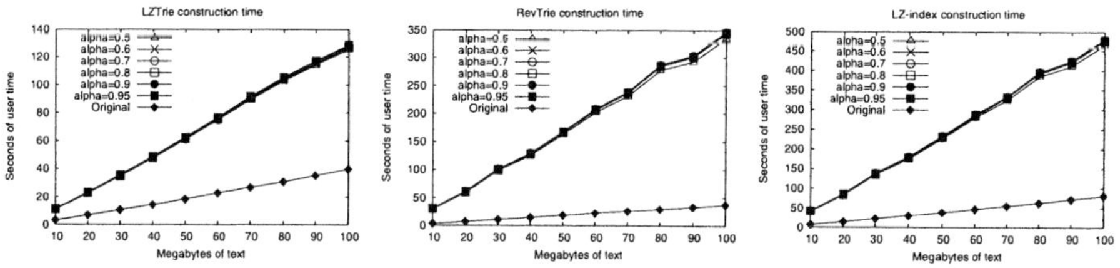

Fig. 2. Average user time to build *LZTrie*, *RevTrie* and the whole LZ-index, $N_1 = 2$, $N_t = 512$

In the same Fig. 1 (below) we show the space requirements in each step of the construction. The space to construct LZ-index varies from 1.46 ($\alpha = 0.95$) to 1.49 ($\alpha = 0.5$) times the text size, and from 1.00 ($\alpha = 0.95$) to 1.02 ($\alpha = 0.5$) times the size of the final index (labeled "Step 6" in the plots). This confirms that the indexing space is about the same to that needed by the final index. For $\alpha = 0.5$ the maximum is reached in step 4, as predicted in the analysis. However, for $\alpha = 0.95$ the maximum is reached in step 6, mainly because in the experiments the number of nodes of the reverse trie is (on average) $n' \approx 1.032n$, which is much less than the pesimistic theoretic bound $n' \leqslant 2n$ we used in the space requirement analysis.

In Fig. 2 we show the indexing time for the tries and the whole index. The average indexing rate for *LZTrie* varies from 0.805MB/sec ($\alpha = 0.95$) to 0.828MB/sec ($\alpha = 0.5$). For *RevTrie* it varies from 0.302MB/sec ($\alpha = 0.95$) to 0.309MB/sec ($\alpha = 0.5$). The whole indexing rate varies from 0.217MB/sec ($\alpha = 0.95$) to 0.223MB/sec ($\alpha = 0.5$). As we expected, the indexing rate is greater as α becomes smaller. The original construction has an average indexing rate of 2.745MB/sec for *LZTrie*, 2.752MB/sec for *RevTrie*, and 1.310MB/sec for the whole indexing process. Thus the succinct construction is 6 times slower in practice, as the upper bound $O(\sigma u)$ is too pessimistic.

We also tested our algorithm on DNA data [1], where the indexing rate is about 0.197MB/sec ($\alpha = 0.95$, $N_1 = 2$, $N_t = 192$), using on average 1.19 times the text size of main memory to index. Extrapolating these results we can argue that the human genome can be indexed in approximately 4.23 hours and using less than 4 GB of main memory. As a comparison, W.-K. Hon et al. [12] argued that they can index the human genome in 20 hours (although they do not describe the CPU of the machine used).

6 Conclusions and Future Work

In this paper we proposed a practical space-efficient algorithm to construct LZ-index. The basic idea is to construct the tries of LZ-index using space-efficient intermediate representations that allow fast incremental insertion of nodes. The algorithm requires at most $(4+\epsilon)uH_k + o(u)$ bits $(0 < \epsilon < 1)$ to construct LZ-index for the text $T_{1...u}$ in time $O(\sigma u)$, being σ the alphabet size. This is the first construction algorithm of a compressed full-text index whose space requirement is related to H_k (the k-th order empirical entropy of the text). In our experiments the construction required approximately 1.45 times the text size, or 1.02 times the final index size, which is much better than the original LZ-index construction algorithm (4–5 times the text size), and the indexing speed was approximately of 5sec/MB.

We believe that our intermediate hrbp representation could be made searchable, so that it could be taken as the final index. The result would be a LZ-index supporting efficient insertion of new text. Those pages could also be handled in secondary memory, so as to have an efficient disk-based LZ-index. Furthermore, the hrbp might have independent interest as a practical technique to represent dynamic general trees in little space, so we plan to work on making them fully dynamic. For the near future, we plan to compare our method against previous work [12], both in time and space.

References

1. M. Abouelhoda, E. Ohlebusch, and S. Kurtz. Optimal exact string matching based on suffix arrays. In *Proc. SPIRE'02*, LNCS 2476, pages 31–43, 2002.
2. A Apostolico. The myriad virtues of subword trees. In *Combinatorial Algorithms on Words*, NATO ISI Series, pages 85–96. Springer-Verlag, 1985.
3. D. Clark and J. I. Munro. Efficient suffix trees on secondary storage. In *Proc. SODA'96*, pages 383–391, 1996.
4. P. Ferragina and G. Manzini. Opportunistic data structures with applications. In *Proc FOCS'00*, pages 390–398, 2000.
5. P. Ferragina and G. Manzini. An experimental study of an opportunistic index. In *Proc. SODA'01*, pages 269–278, 2001.
6. P. Ferragina and G. Manzini. On compressing and indexing data. Technical Report TR-02-01, Dipartamento di Informatica, Univ. of Pisa, 2002.
7. P. Ferragina, G. Manzini, V. Mäkinen, and G. Navarro. An alphabet-friendly FM-index. In *Proc.SPIRE'04*, LNCS 3246, pages 150–160. Springer, 2004.
8. R. Grossi, A. Gupta, and J. S. Vitter. High-order entropy-compressed text indexes. In *Proc. SODA'03*, pages 841–850. SIAM, 2003.

[1] 52.71MB from *GenBank* (Homo Sapiens DNA, `http://www.ncbi.nlm.nih.gov`).

9. R. Grossi, A. Gupta, and J.S. Vitter. When indexing equals compression: experiments with compressing suffix arrays and applications. In *Proc. SODA'04*, pages 636–645. SIAM, 2004.
10. R. Grossi and J.S. Vitter. Compressed suffix arrays and suffix trees with applications to text indexing and string matching. In *Proc. STOC'00*, pages 397–406, 2000.
11. W. Hersh, C. Buckley, T. Leone, and D. Hickam. Ohsumed: An interactive retrieval evaluation and new large test collection for research. In *Proc. SIGIR'94*, pages 192–201, 1994.
12. W. K. Hon, T. W. Lam, K. Sadakane, and W. K. Sung. Constructing compressed suffix arrays with large alphabets. In *Proc. ISAAC'03*, LNCS 2906, pages 240–249, 2003.
13. J. Kärkkäinen. Suffix cactus: a cross between suffix tree and suffix array. In *Proc. CPM'95*, LNCS 937, pages 191–204, 1995.
14. J. Kärkkäinen. *Repetition-based text indexes*. PhD thesis, Dept. of Computer Science, University of Helsinki, Finland, 1999.
15. J. Kärkkäinen and E. Ukkonen. Lempel-Ziv parsing and sublinear-size index structures for string matching. In *Proc. WSP'96*, pages 141–155. Carleton University Press, 1996.
16. R. Kosaraju and G. Manzini. Compression of low entropy strings with Lempel-Ziv algorithms. *SIAM Journal on Computing*, 29(3):893–911, 1999.
17. S. Kurtz. Reducing the space requeriments of suffix trees. Technical Report 98-03, Technische Kakultät, Universität Bielefeld, Germany, 1998.
18. T. W. Lam, K. Sadakane, W. K. Sung, and S. M. Yiu. A space and time efficient algorithm for constructing compressed suffix arrays. In *Proc. COCOON 2002*, pages 401–410, 2002.
19. V. Mäkinen. Compact suffix array - a space-efficient full-text index. *Fundamenta Informaticae*, 56(1–2):191–210, 2003.
20. V. Mäkinen and G. Navarro. Succinct suffix arrays based on run-length encoding. In *Proc. CPM'05*, LNCS 3537, pages 45–56, 2005.
21. U. Manber and G. Myers. Suffix arrays: A new method for on–line string searches. *SIAM Journal on Computing*, 22(5):935–948, 1993.
22. D. R. Morrison. Patricia – practical algorithm to retrieve information coded in alphanumeric. *Journal of the ACM*, 15(4):514–534, 1968.
23. I. Munro and V. Raman. Succinct representation of balanced parentheses, static trees and planar graphs. In *Proc. FOCS'97*, pages 118–126, 1997.
24. G. Navarro. Indexing text using the Ziv-Lempel trie. In *Proc. SPIRE'04*, LNCS 2476, pages 325–336, 2002.
25. G. Navarro. Indexing text using the Ziv-Lempel trie. Technical Report TR/DCC-2002-2, Dept. of Computer Science, Univ. of Chile, 2002. `ftp://ftp.dcc.uchile.cl/pub/users/gnavarro/lzindex.ps.gz`.
26. G. Navarro. Indexing text using the Ziv-Lempel trie. *Journal of Discrete Algorithms (JDA)*, 2(1):87–114, 2004.
27. V. Raman and S. Rao. Static dictionaries supporting rank. In *Proc. ISAAC '99*, LNCS 1741, pages 18–26, 1999.
28. K. Sadakane. Compressed text databases with efficient query algorithms based on the compressed suffix array. In *Proc. ISAAC'00*, LNCS 1969, pages 410–421, 2000.
29. J. Ziv and A. Lempel. Compression of individual sequences via variable–rate coding. *IEEE Trans. Inform. Theory*, 24(5):530–536, 1978.

Longest Increasing Subsequences in Windows Based on Canonical Antichain Partition

Erdong Chen, Hao Yuan, and Linji Yang

Department of Computer Science and Engineering,
Shanghai Jiao Tong University,
200030 Shanghai, P.R. China
{edchen, hyuan, ljyang}@cs.sjtu.edu.cn

Abstract. We consider the LISW problem, which is to find the longest increasing subsequences (LIS) in a sliding window of fixed-size w over a sequence $\pi_1\pi_2 \dots \pi_n$. Formally, it is to find a LIS for every window in a set $S_{\text{FIX}} = \{\pi\langle i+1, i+w\rangle \mid 0 \le i \le n-w\} \cup \{\pi\langle 1, i\rangle, \pi\langle n-i, n\rangle \mid i < w\}$, where a window $\pi\langle l, r\rangle$ is a subsequence $\pi_l\pi_{l+1}\dots\pi_r$. By maintaining a *canonical antichain partition* in windows, we present an optimal *output-sensitive* algorithm to solve this problem in $O(\text{OUTPUT})$ time, where OUTPUT is the sum of the length of the $n+w-1$ longest increasing subsequences in those windows of S_{FIX}. In addition, we propose a more generalized problem called LISSET , which is to find the LIS for every window in a set S_{VAR} containing *variable-size* windows. By applying our algorithm, we provide an efficient solution for LISSET problem which is better than the straight forward generalization of classical LIS algorithms. An upper bound of our algorithm on LISSET is discussed.

1 Introduction

Given a sequence $\pi = \pi_1\pi_2\dots\pi_n$ of n elements, the longest increasing subsequences (LIS) problem is to find a longest subsequence $\sigma = \pi_{i_1}\pi_{i_2}\dots\pi_{i_T}$ such that $1 \le i_1 < i_2 < \dots < i_T \le n$ and $\pi_{i_1} < \pi_{i_2} < \dots < \pi_{i_T}$. The LISSET problem proposed in this paper, is to find the longest increasing subsequences in a set of variable-size windows, which is to solve the LIS problem in a subsequence $\pi\langle l, r\rangle = \pi_l\pi_{l+1}\dots\pi_r$ for different pairs of indices l and r.

In [1], Albert et al. defined the problem LISW , which is to find the longest increasing subsequences in sliding windows over a sequence of n elements. A w-size window is a subsequence $\pi\langle i+1, i+w\rangle$ for some $0 \le i \le n-w$. Additionally, all the truncated windows $\pi\langle 1, j\rangle$ for $j < w$ and $\pi\langle j, n\rangle$ for $j > n-w+1$ are also regarded as w-size windows(see Fig. 1 for example). Albert et al. proposed an algorithm to solve the LISW problem in $O(n \log\log n + \text{OUTPUT})$ time, where OUTPUT is the total size of the output. In this paper, we will give a faster algorithm for this problem, which runs in $O(\text{OUTPUT})$ time. Our algorithm solves LISW problem in optimal time, linear on the size of the output.

The remainder of this paper is organized as follows. In Section 2, some problems and techniques related to LISSET are reviewed. In Section 3, the LISSET

X. Deng and D. Du (Eds.): ISAAC 2005, LNCS 3827, pp. 1153–1162, 2005.

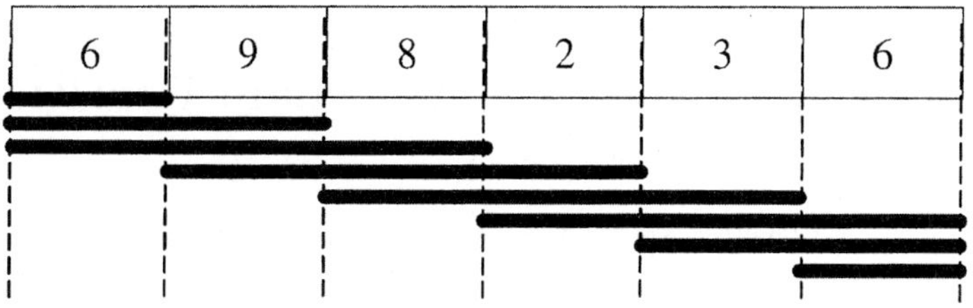

Fig. 1. Sliding windows of size 3 over sequence 6, 9, 8, 2, 3, 6 (LISW problem)

problem is defined, which takes LISW as a subcase. In Section 4, the Canonical Antichain Partition is discussed, which is the basic data structure of our algorithm. In Section 5, we give design and analysis of a sweep algorithm on LISSET problem and LISW problem, and summarize our contributions. Finally, conclusions and future work are discussed in Section 6.

2 Related Work

The longest increasing subsequences problem is a fundamental combinatorial problem which has been widely studied[2, 3, 4]. Knuth proposed an $O(n \log n)$ algorithm whose mechanism is to maintain the first row of Young tableau [5]. Fredman proved an $\Omega(n \log n)$ lower bound under the decision tree model [6]. However, an $O(n \log \log n)$ algorithm is possible by using van Emde Boas tree [7] on a permutation.

David et al.[8] proposed two algorithms for LIS problem in data streaming model[9]. One is to decide whether the LIS of a given stream of integers drawn from $\{1 \ldots M\}$ has length at least k using $O(k \log M)$ space and update time $O(\min\{\log k, \log\log k\})$, and the other is a *multipass* data streaming algorithm to return the actual LIS itself using space $O(k^{1+\varepsilon} \log M)$. If $n \gg w$(average window size), then our algorithm is a data streaming algorithm which makes only a single pass over the input sequence with $O(w)$ space.

Longest Increasing Subsequence has been widely used in bioinformatics[10]. On the topic of alignment of sequences, Zhang proposed a BLAST+LIS strategy to find the correct longest consecutive list of high scoring segment pairs(HSPs) in the BLAST output, if the BLAST output contains multiple HSPs for a pair of sequences; hence reduced the redundant HSPs in each hit and filtered out the redundant genomic hits[11].

3 Problem Definition

The longest increasing subsequence (LIS) in a window $W = \pi\langle l, r\rangle$ is a longest subsequence $\sigma = \pi_{i_1}\pi_{i_2}\ldots\pi_{i_t}$ such that $l \le i_1 < i_2 < \cdots < i_t \le r$ and $\pi_{i_1} < \pi_{i_2} < \cdots < \pi_{i_t}$. Given a window W, let $\omega(W) = |\sigma|$ denote the length of such LIS in window W.

Let $S = \{W_i \mid W_i = \pi\langle l_i, r_i\rangle\}$ be a set of m variable-size windows. The LISSET problem is to calculate $\omega(W_i)$ and find out a corresponding LIS in W_i for each $i(1 <= i <= m)$.

4 Canonical Antichain Partition

Given a sequence $\pi = \pi_1\pi_2 \dots \pi_n$, each element π_i may be represented by a point (i, π_i) in the plane. For example, the sequence $6, 9, 8, 2, 3, 5, 1, 4, 7$, is represented by $p_1p_2 \dots p_9$(See Fig. 2). Let P be the planar point set of n elements in the form of (i, π_i). For any point $p \in P$, let p_x and p_y denote its x- and y-coordinates. Following Felsner and Wernisch[12], for two points $p, q \in$ P, the *dominance order* is given by the relation $p \prec q$(say q dominates p) if $p_x < q_x$ and $p_y < q_y$. The *shadow*[12] of a point p is defined as the area of all points (u, v) *dominating* p, i.e. with $u > p_x$ and $v > p_y$. A chain $C \subseteq P$ is an ordered points list of the dominance order on P, i.e. $C = \langle p_1, p_2, \dots, p_t\rangle$ is a chain if and only if $p_j \prec p_{j+1}$ for $1 \leq j < t$. For instance, the chain $\langle p_4, p_5, p_6, p_9\rangle$ is a *longest chain* in Fig. 2.

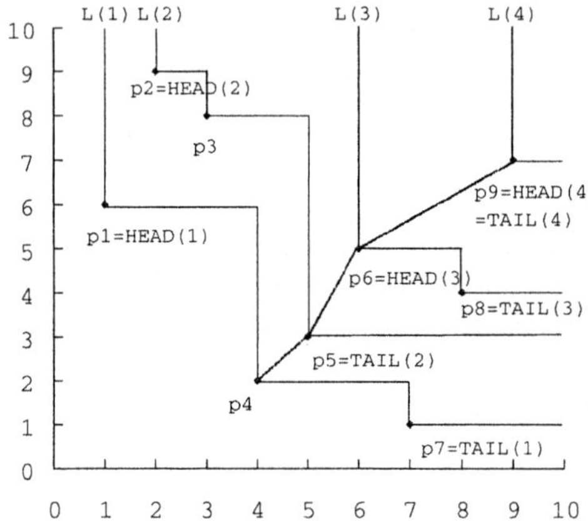

Fig. 2. The canonical antichain partition of $6, 9, 8, 2, 3, 5, 1, 4, 7$, and a longest chain: $\langle(4, 2), (5, 3), (6, 5), (9, 7)\rangle$

The *height* of a point $p \in P$, denoted $h(p)$, is the length of a longest chain with p as maximal point. Hence, given a point $p_i = (i, \pi_i)$, the height $h(p_i)$ is also the length of the longest increasing subsequence with π_i as the maximal element[12]. In addition, the points with the same height are not *comparable* by *dominance order*. Thus, all points with the same height in the same set yield a partition of P into *antichains*, the *canonical antichain partition*. All points with the same height h in P form an antichain L_h. Theorem 1 is a well-known result about antichains[12], which is a key issue in designing our algorithm.

Theorem 1. *All points with the same height h in P form an antichain L_h. Points in an* antichain L_h *are sorted in increasing order by x-coordinate and in non-increasing order by y-coordinate.*

Since all the points with height h are stored in *linked list* L_h, we define HEAD(h) and TAIL(h) to be the first and last point in L_h respectively. Obviously, HEAD(h) is the point with the smallest x-coordinate in antichain L_h, and TAIL(h) is the point with the largest x-coordinate in antichain L_h. Given a point $p \in L_h$, the point following p in L_h is denoted by NEXT(p). By definition, NEXT(TAIL(h)) = $\emptyset$ for any h. A point q is said to be a *preceding point* of p, if $h(q) = h(p) - 1$ and $q \prec p$. In order to return a chain with p as the maximal point, the rightmost preceding point of p, denoted PRED(p), is defined to be the one with the largest x-coordinate among all p's preceding points. By definition, if $h(p) = 1$, then we have PRED(p) = $\emptyset$.

5 Sweep Algorithm

In order to output the longest increasing subsequence in a window, we have to design a data structure to maintain structural information about all LIS's in a window. Our data structure needs to support the following operations: remove the first element of the sequence, insert an element to the end of the sequence, output a longest increasing subsequence with constraints. Formally, if $\mathscr{W} = \pi\langle l, r\rangle$ is a subsequence, our data structure needs to support:

1. remove π_l from $\mathscr{W}$;
2. insert π_{r+1} to $\mathscr{W}$;
3. output a longest increasing subsequence $\sigma = \pi_{i_1}\pi_{i_2}\dots\pi_{i_T}$ in $\mathscr{W}$ satisfying $i_T \le X_{\text{QRY}}$ for a fixed X_{QRY} between l and r.

Instead of dealing subsequences directly, our data structure maintains a *canonical antichain partition* in a window, for there is a mapping between a sequence and a point set in the plane. Let $P = [p_0, p_1, \dots, p_{r-l+1}]$ be the ordered points set that represents $\mathscr{W}$, i.e. $p_i = (l+i, \pi_{l+i})$, for $0 \le i \le r-l$, and $\omega(P) = \omega(\mathscr{W})$ is the number of antichains in the canonical antichain partition. We design three operations on the point set P corresponding to the three operations on the subsequence $\mathscr{W}$:

1. DELETE: remove from P a point p_{DEL} with the smallest x-coordinate
2. INSERT: insert into P a new point p_{INS} with a larger x-coordinate than that of any point in P
3. QUERY(X_{QRY}): output a longest chain σ satisfying that the largest x-coordin ate of points in σ is equal to or less than X_{QRY}

For any h, let L'_h be the new antichain with height h after an operation, and for any $p \in P$, the point PRED$'(p)$ is the new rightmost preceding point of p after an operation. In the following, we will provide the details and complexity analysis for each operation.

DELETE operation. The DELETE operation is to delete the point p_{DEL} with the smallest x-coordinate in P. After deleting p, the height of some points may decrease. For any $p \in P$, the new height $h'(p)$ is defined to be the height of p after deleting p_{DEL}. Let $D_i = \{p|h(p) = i \text{ and } h'(p) = h(p) - 1\}$, so we have $D_1 = \{p_{\text{DEL}}\}$ and $D_{\omega(\mathscr{W})+1} = \emptyset$. Since D_i is a set of consecutive points with lower x-coordinate in L_i, we have $D_i = \{\text{HEAD}(i), \text{NEXT}(\text{HEAD}(i)), \ldots, \text{PMAX}(i)\}$, (PMAX($i$) is defined to be the point with the largest x-coordinate in D_i). Since the height of each point may decrease at most by one, our method is simple: for each L_i, first delete D_i from L_i, then insert D_{i+1} to head part of L_i.

Algorithm 1 Algorithm for DELETE operation

initialize $\text{PMAX}(1) \leftarrow p_{\text{DEL}}, D_1 \leftarrow \{p_{\text{DEL}}\}$, and $D_i \leftarrow \emptyset$ for $i > 1$
for $i = 2$ to $\omega(P)$ **do**
  $\text{PMAX}(i) \leftarrow \max\{p_x | \neg(\text{NEXT}(\text{PMAX}(i-1)) \prec p)\}$
  if $\text{PMAX}(i) = \emptyset$ **then**
    EXIT LOOP
  end if
  $D_i = \{p | p \in L_i \text{ and } p_x \le \text{PMAX}(i)_x\}$
end for
$\text{PRED}'(p) = \emptyset$ for $p \in D_2$
for $i = 1$ to $\omega(P)$ **do**
  if $\text{PMAX}(i) = \emptyset$ **then**
    EXIT LOOP
  end if
  $L'_i \leftarrow D_{i+1} + (L_i \setminus D_i)$
end for

We will illustrate our method based on the example in Fig. 3 and 4. By definition, $\text{PMAX}(1) = p_1$ and $D_1 = \{p1\}$. The point p_2 is the point with the largest x-coordinate that does not *dominate* $\text{NEXT}(\text{PMAX}(1)) = p_3$. In Fig. 3, the points in the *shadow* area *dominate* D_1 but not any point in $L_1 \setminus D_1$, so $D_2 = \{p_2\}$. The point p_5 is the point with the largest x-coordinate that does not *dominate* $\text{NEXT}(\text{PMAX}(2)) = p_6$, so $\text{PMAX}(3) = p_5$ and $D_3 = \{p_4, p_5\}$. Next step is computing L'_i. $L'_1 = D_2 + (L_1 \setminus D_1) = \{p_2, p_3\}$, $L'_2 = \{p_4, p_5, p_6, p_7\}$, and $L'_3 = \{p_8\}$. In Fig. 4, the *dashed line* represents removing the D_i, and the *broad-brush line* represents the *concatenation* of D_{i+1} and $L_i \setminus D_i$.

Lemma 1. *For any $i > 1$, the height of a point in D_i decreases if and only if it does not dominate any point in $L_{i-1} \setminus D_{i-1}$.*

Lemma 2. *The x-coordinate of any point in D_{i+1} is smaller than that of any point in $L_i \setminus D_i$.*

Proof. Suppose $\exists p \in D_{i+1}$, $q \in L_i \setminus D_i$ for a particular i, and $p_x > q_x$. Because $h(p)$ decreases, the point p does not dominate q, that means $p_y \le q_y$. Before delete p_{DEL}, there exits $v \in L_i$ that p dominates v, i.e. $p_x > v_x$ and $p_y > v_y$.

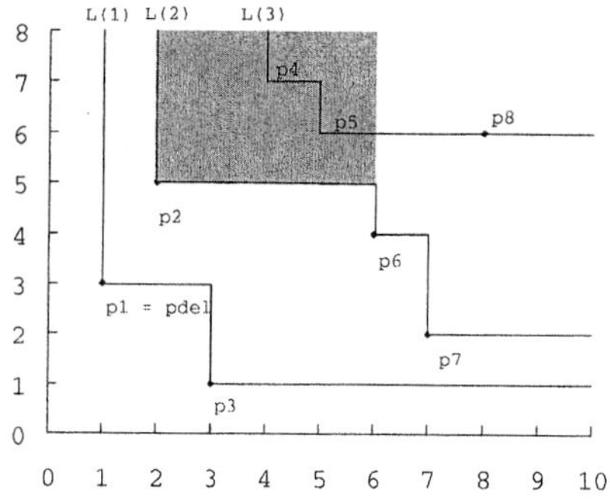

Fig. 3. Before DELETE operation

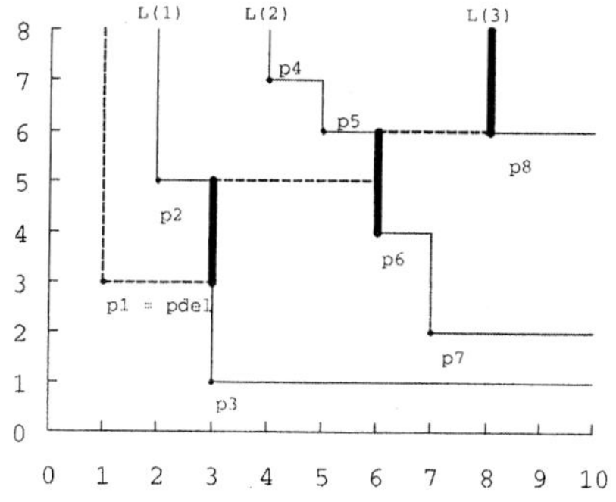

Fig. 4. After DELETE operation

Since v precedes q in L_i, we will have $v_x < q_x$, $v_y \geq q_y$ according to Theorem 1. Therefore, $p_y > v_y \geq q_y$ which contradicts with the inequity $p_y \leq q_y$. Thus, no such p, q exists. □

By lemma 1, to compute PMAX$(i)(i > 1)$, we scan L_i from the head until the first point which is *dominating* NEXT(PMAX$(i-1)$). By lemma 2, for a fixed i, we can move D_{i+1} to the head part of $L_i \setminus D_i$ in $O(1)$ time without destroying the relative orders of the points in D_{i+1} and $L_i \setminus D_i$. Let $\mathscr{D}$ be the union of D_i, i.e. $\mathscr{D} = \bigcup_{i=1}^{\omega(\mathscr{W})} D_i$. Thus, the time complexity of maintaining the data structure after deleting p_{DEL} is $O(|\mathscr{D}|)$.

Theorem 2. *The cost of one DELETE operation equals the total number of points whose height decreases, i.e.* $O(|\mathscr{D}|)$.

INSERT operation. The INSERT operation is to insert a new point p_{INS} with a larger x-coordinate than that of any point in P. The main problem of INSERT operation is to find an *antichain* which p_{INS} belongs to. There are two cases:

1. if $h(p_{\text{INS}}) = i(1 \leq i \leq \omega(\mathscr{W}))$, then insert p_{INS} to the end of L_i, i.e $L_i' = L_i \cup \{p_{\text{INS}}\}$.
2. if $h(p_{\text{INS}}) = \omega(\mathscr{W}) + 1$, then create a new *antichain* $L'_{\omega(\mathscr{W})+1} = \{p_{\text{INS}}\}$.

By Theorem 1, if $\langle a_1, a_2, \ldots, a_t, p_{\text{INS}} \rangle$ is a longest chains, there always be another longest chain by replacing a_t by TAIL$(h(a_t))$. As in Fig. 5, we can always replace the *solid line* $\langle p_1, p_2, p_5, p_{\text{INS}} \rangle$ by the *dashed line* $\langle p_1, p_2, p_8, p_{\text{INS}} \rangle$.

Therefore, to compute $h(p_{\text{INS}})$ is to check TAIL(k) for $k = 1, 2, \ldots, \omega(P)$, until we find the highest TAIL(k) that p_{INS} *dominates*. If $k = 0$, PRED$'(p_{\text{INS}}) = \emptyset$, otherwise PRED$'(p_{\text{INS}})$ = TAIL(k). If $k = \omega(\mathscr{W})$, we create a new *antichain* $L'_{\omega(\mathscr{W})+1}$ with only a point p_{INS}. Since finding the antichain which p_{INS} belongs to requires $h(p_{\text{INS}})+1$ steps, the time complexity of INSERT opreation is $O(h(p_{\text{INS}}))$.

Theorem 3. *The cost of INSERT operation equals* $O(h(p_{INS}))$.

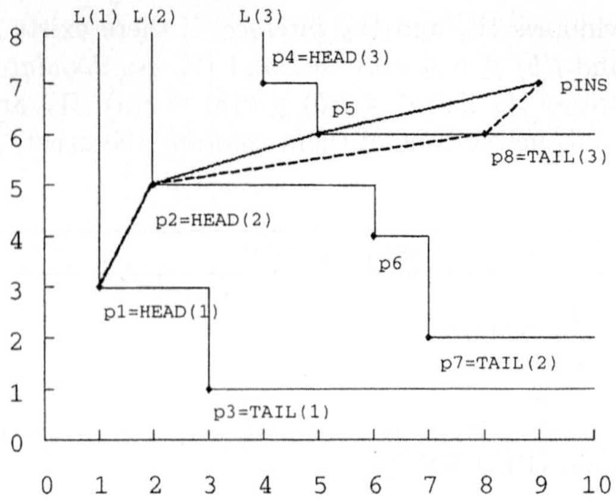

Fig. 5. INSERT operation

QUERY operation. Suppose $\sigma = \langle p_1, p_2, \ldots, p_t \rangle$ is a longest chain in P that σ satisfies the x-coordinate of p_t is equal to or less than X_{QRY}. Let P' be the set of points in P whose x-coordinate is equal to or less than X_{QRY}, i.e. $P' = \{p \mid p \in P \text{ and } p_x \leq X_{\text{QRY}}\}$. Let H be the length of a longest chain in P', which is the size of QUERY(X_{QRY})'s output.

By Theorem 1, since x-coordinate of HEAD(H) is the smallest among that of all points in L_H, if a longest chain in P' with any point $p \in L_H$ as the maximal point, there always be another longest chain in P' with HEAD(H) as the maximal point. Therefore, to compute H is to check HEAD(i) for $i = 1, 2, \ldots, \omega(P)$, until we find the highest HEAD(i) that the x-coordinate of HEAD(i) is equal to or less than X_{QRY}, i.e. $H = \max\{i \mid \text{HEAD}(i)_x \leq X_{\text{QRY}}\}$. We can find a longest chain C in P' satisfying that $C = \langle c_1, c_2, \ldots, c_H \rangle$, $c_H = \text{HEAD}(H)$ and $c_i = \text{PRED}(c_{i+1})$ for $i = 1, 2, \ldots, H-1$. By careful analysis, computing H requires $H+1$ steps of searching, and outputting the sequence requires H steps. In short, the total time complexity of QUERY(X_{QRY}) is $O(H)$.

Theorem 4. *The cost of outputting a longest chain σ satisfying that the largest x-coordinate of points in σ is equal to or less than X_{QRY}, is the length of σ i.e. $O(|\sigma|)$.*

5.1 Algorithm

Our algorithm is based on the data structure proposed above. Firstly, windows are sorted in increasing order, and then we slide the window $\mathscr{W}$ from left to right to output the LIS in each window. In order to make sure that points are inserted in to $\mathscr{W}$ only once, the windows are ordered by their left endpoints (if two windows share the same left endpoint, the longer window comes first), i.e. $i < j$ if and only if $l_i < l_j$ or ($l_i = l_j$ and $r_i > r_j$). Since the endpoints are drawn from $\{1, 2, \ldots, n\}$, the preprocessing sorting can be completed in $O(n+m)$ time.

Two distinct windows W_a and W_b *intersect* if there exists a number c that $l(a) \leq c \leq r(a)$ and $l(b) \leq c \leq r(b)$; W_a and W_b are *disjointed* if they do not *intersect*; W_a *contains* W_b if $l(a) \leq l(b) \leq r(b) \leq r(a)$; W_a and W_b *overlap*, if W_a *intersects* W_b but neither one of them *contains* the other.

Algorithm 2 Algorithm for LISSET Problem

```
initialize 𝒲 ← ∅
INSERT into 𝒲 the elements in π⟨l_1, r_1⟩ separately
QUERY(r_1) to output the LIS in 𝒲
for j = 1 to m − 1 do
  slide the window from W_j to W_{j+1}, let W_a = W_j, and W_b = W_{j+1}
  if (l_b > r_a) then {DISJOINT}
    reset 𝒲 ← ∅
    INSERT into 𝒲 the elements in π⟨l_b, r_b⟩ separately
  else if (l_b ≤ r_a and r_b > r_a) then {OVERLAP}
    DELETE from 𝒲 the elements in π⟨l_a, l_b − 1⟩ separately
    INSERT into 𝒲 the elements in π⟨r_a + 1, r_b⟩ separately
  else if (l_a ≤ l_b and r_b ≤ r_a) then {CONTAIN}
    if (l_b = l_a) then {SAME LEFT ENDPOINT}
      NO OPERATIONS
    else if l_b > l_a then {DIFFERENT LEFT ENDPOINTS}
      DELETE from 𝒲 the elements in π⟨l_a, l_b − 1⟩ separately
    end if
  end if
  QUERY(r_b) to output a LIS σ = π_{i_1} π_{i_2} ... π_{i_T} in 𝒲 satisfying i_T ≤ r_b
end for
```

Before analyzing the algorithm, depth_i is defined to be the *largest* height that π_i achieved in the m windows. In other words, among all increasing subsequences in m windows, depth_i is the length of the longest one with π_i as the maximal element.

Theorem 5 (LISSET Problem). *The algorithm described above computes the m longest increasing subsequences, one for each window, in total time $O(n + \text{OUTPUT} + \sum_{i=1}^{n} \text{depth}_i)$.*

Proof. Most of the time required comes from three operations : INSERT, DELETE, QUERY. By Theorem 4, the cost of QUERY operation equals the total length of the LIS's in all m windows, so $T_{QUERY} = O(\text{OUTPUT})$. By Theorem 3, the cost of inserting π_i is equal to the length of the longest LIS with π_i as the maximal element in the m windows, i.e. $T_{INS} = O(\sum_{i=1}^{n} \text{depth}_i)$. For DELETE operations, it is difficult to analyse the cost of each DELETE operation, but we can calculate the overall cost of DELETE operations. By Theorem 2, the cost equals to the sum of the number of points which decrease after each operation, and a point $p = (i, \pi_i)$ may decrease at most depth_i times. Therefore, $T_{DEL} = O(\sum_{i=1}^{n} \text{depth}_i)$. Thus, $T = T_{QUERY} + T_{INS} + T_{DEL} = O(n + \text{OUTPUT} + \sum_{i=1}^{n} \text{depth}_i)$. □

Theorem 6 (LISW Problem). *Our algorithm finds the longest increasing subsequence in a sliding window over a sequence of n elements in* $O(\text{OUTPUT})$ *time.*

Proof. By Theorem 5, the time complexity is $O(n+\text{OUTPUT}+\sum_{j=1}^{n}\text{depth}_j)$. By definition, for $1 \le i < w$, depth_i is equal to or less than the length of the LIS in the window $\pi\langle 1, i\rangle$. For $w \le i \le n$, depth_i is equal to or less than the length of the LIS in the window $\pi\langle i-w+1, i\rangle$. Therefore ,$\sum_{i=1}^{n}\text{depth}_i = O(\sum_{i=1}^{m}\omega(W_i)) = O(\text{OUTPUT})$. In short, our time complexity is $O(n+\text{OUTPUT}+\sum_{j=1}^{n}\text{depth}_j) = O(n+\text{OUTPUT}) = O(\text{OUTPUT})$. □

LISSET problem has a *straight forward* approach, which is to find all LIS's for each window separately. In the worst case that all m windows are disjointed, our algorithm does not give any asymptotic improvement over the straight forward method. However, similar to the analysis of Albert et al., our algorithm gives a better performance in average cases.

If average window size is $O(w)$, the optimal algorithm finds LIS in a single window in time $O(\min\{w\log w, w\log\log n\})$, so the time complex of finding LIS's in m windows is $O(\min\{mw\log w, mw\log\log n\})$. On the other hand, it was shown in[3] that average length of LIS of a permutation of length n is asymptotically $2\sqrt{n}$. Suppose the permutation π is randomly chosen from $n!$ different permutations, so the relative ordering of the elements in each window is also randomly chosen. Thus, the expected length of LIS in any window is asymptotically $2\sqrt{w}$. By Theorem 2, we can prove that $\text{depth}_i = O(\sqrt{w})$. Thus, our time complexity is $O(n+\text{OUTPUT}+\sum_{i=1}^{n}\text{depth}_i) = O(n+\text{OUTPUT}+n\sqrt{w}) = O((n+m)\sqrt{w})$. If $m = O(n)$, then our algorithm would certainly perform better than the *straight forward* approach.

6 Conclusions Remarks

We investigate the problem of finding the longest increasing subsequences in a set of variable-size sliding windows over a given sequence. By maintaining a canonical antichain partition, we propose an approach that solve the problems in time $O(n+\text{OUTPUT}+\sum_{i=1}^{n}\text{depth}_i)$ for m windows over a sequence of n elements. This algorithm is able to solve the problem significantly better than straight forward methods. In addition, we show that LISW problem is a subcase of LISSET problem. Our algorithm solve LISW in time $O(\text{OUTPUT})$, while the time complexity of the best solution for LISW previously achieved is $O(\text{OUTPUT}+n\log\log n)$.

Some problems related to LISSET problem are still open problems. Firstly, the time complexity of finding *global maximal* LIS among all LIS's in all windows is particularly interesting. Because our algorithm only finds the global maximal LIS after computing all LIS's in all m window individually, it is expected to design an algorithm to find global maximal LIS *directly* in the future. The second problem is to design a space-efficient *online* algorithm to output the LIS in a window with size w in *linear* time $O(w)$. Our algorithm requires $O(n^2)$ time and $O(n^2)$ space for preprocessing to solve this problem.

Acknowledgements

Thanks to Hong Zhu, Binhai Zhu and the reviewers for their helpful comments. Thanks to Prof. Yong Yu[1] for his support on this work.

References

1. Michael H. Albert, Alexander Golynski, Angèle M. Hamel, Alejandro López-Ortiz, S. Srinivasa Rao, and Mohammad Ali Safari. Longest increasing subsequences in sliding windows. *Theor. Comput. Sci.*, 321(2-3):405–414, 2004.
2. C. Schensted. Longest increasing and decreasing subsequences. *Canadian Journal of Mathematics*, 1961.
3. Sergei Bespamyatnikh and Michael Segal. Enumerating longest increasing subsequences and patience sorting. *Inf. Process. Lett.*, 76(1-2):7–11, 2000.
4. Alberto Apostolico, Mikhail J. Atallah, and Susanne E. Hambrusch. New clique and independent set algorithms for circle graphs (discrete applied mathematics 36 (1992) 1-24). *Discrete Applied Mathematics*, 41(2):179–180, 1993.
5. D. E. Knuth. *The Art of Computer Programming. Vol 3, Sorting and Searching. Second Edition.* Addison-Wesley, 1998.
6. M. L. Fredman. On computing the length of longest increasing subsequences. *Discrete Mathematics*, 11:29–35, 1975.
7. Peter van Emde Boas. Preserving order in a forest in less than logarithmic time and linear space. *Inf. Process. Lett.*, 6(3):80–82, 1977.
8. Erik Vee David Liben-Nowell and An Zhu. Finding longest increasing and common subsequences in streaming data. Technical Report MIT-LCS-931, Cambridge, MA 02139, November 2003.
9. P. Raghavan M. R. Henzinger and S. Rajagopalon. Computing on data streams. Technical Report 1998-011, Digital Equipment Corporation,Systems Research Center, May 1998.
10. T.F. Smith and M.S. Waterman. Comparison of biosequences. *Adv. in Appl. Math.*, 2:482–489, 1981.
11. Hongyu Zhang. Alignment of blast high-scoring segment pairs based on the longest increasing subsequence algorithm. *Bioinformatics*, 19(11):1391–1396, 2003.
12. Stefan Felsner and Lorenz Wernisch. Maximum k-chains in planar point sets: Combinatorial structure and algorithms. *SIAM J. Comput.*, 28(1):192–209, 1998.

[1] The advisor of the undergraduate program in Department of Computer Science and Engineering, Shanghai Jiao Tong University.

Pareto Optimality in House Allocation Problems*

David J. Abraham[1,**], Katarína Cechlárová[2],
David F. Manlove[3,***], and Kurt Mehlhorn[4]

[1] Computer Science Department, Carnegie-Mellon University, 5000 Forbes Ave, Pittsburgh PA 15213-3890, USA
dabraham@cs.cmu.edu

[2] Institute of Mathematics, P.J. Šafárik University in Košice, Faculty of Science, Jesenná 5, 040 01 Košice, Slovakia
cechlarova@science.upjs.sk

[3] Department of Computing Science, University of Glasgow, Glasgow G12 8QQ, UK
davidm@dcs.gla.ac.uk

[4] Max-Planck-Institut für Informatik, Stuhlsatzenhausweg 85, 66123 Saarbrücken, Germany
mehlhorn@mpi-sb.mpg.de

Abstract. We study Pareto optimal matchings in the context of house allocation problems. We present an $O(\sqrt{n}m)$ algorithm, based on Gale's Top Trading Cycles Method, for finding a maximum cardinality Pareto optimal matching, where n is the number of agents and m is the total length of the preference lists. By contrast, we show that the problem of finding a minimum cardinality Pareto optimal matching is NP-hard, though approximable within a factor of 2. We then show that there exist Pareto optimal matchings of all sizes between a minimum and maximum cardinality Pareto optimal matching. Finally, we introduce the concept of a signature, which allows us to give a characterization, checkable in linear time, of instances that admit a unique Pareto optimal matching.

1 Introduction

We study the problem of allocating a set H of heterogeneous indivisible goods among a set A of agents [14,8,3,4]. We assume that each agent $a \in A$ ranks in order of preference a subset of H (the *acceptable* goods for a) and that monetary compensations are not possible. In the literature the situation in which each agent initially owns one good is known as a *housing market* [14,12,11].

* Due to mistakes during data conversion, this paper was originally published in LNCS 3341 (ISAAC 2004) with several special characters missing in text and figures.

** Work done whilst at Department of Computing Science, University of Glasgow, and Max-Planck-Institut für Informatik.

*** Supported by grant GR/R84597/01 from the Engineering and Physical Sciences Research Council and RSE/Scottish Executive Personal Research Fellowship.

X. Deng and D. Du (Eds.): ISAAC 2005, LNCS 3827, pp. 1163–1175, 2005.

When there are no initial property rights, we obtain the *house allocation problem* [8,16,1]. A mixed model, in which a subset of agents initially owns a good has also been studied [2]. Yuan [15] describes a large-scale application of these problems in the allocation of families to government-subsidized housing in China.

Following convention we refer to the elements of H as *houses*, though the class of problems under consideration could equally be formulated in terms of allocating graduates to trainee positions, professors to offices, clients to servers, etc. For ease of exposition we begin by assuming that there are no initial property rights, though we later show how to take account of such a situation.

Given such a problem instance, the task is to construct a *matching*, i.e. a subset M of $A \times H$ such that $(a, h) \in M$ implies that a finds h acceptable, each agent is assigned to at most one house and vice versa. Furthermore one seeks a matching that is *optimal* in a precise sense, taking into account the agents' preferences. Various notions of optimality have been considered in the literature, but a criterion that has received much attention, particularly from economists, is *Pareto optimality.* A matching M is Pareto optimal if there is no other matching M' such that no agent is worse off in M' than in M, whilst some agent is better off in M' than in M. For example, a matching M is not Pareto optimal if two agents could improve by swapping the houses that they are assigned to in M.

There is a straightforward greedy algorithm, which we denote by Greedy-POM, for finding a Pareto optimal matching [1]: consider each agent a in turn, giving a his/her most-preferred vacant house (assuming such a house exists). This algorithm is also known as a *serial dictatorship* mechanism [1]. Roth and Sotomayor [13, Example 4.3] remark that a similar mechanism is used by the United States Naval Academy in order to match graduating students to their first posts as Naval Officers (in this context however, the algorithm considers each student in non-decreasing order of graduation results). However one may construct an example instance (see Section 2 for further details) in which Pareto optimal matchings may have different cardinalities and Greedy-POM could fail to produce a Pareto optimal matching of maximum size. Yet in many applications, one wishes to match as many agents as possible.

Stronger notions of optimality have been considered in the literature. For example a matching M is *rank-maximal* [10] if, in M, the maximum number of agents are matched to their first-choice house, and subject to this condition, the maximum number of agents are matched to their second-choice house, and so on. Irving et al. [10] describe two algorithms for finding a rank-maximal matching, with complexities $O(\min\{n + C, C\sqrt{n}\}m)$ and $O(Cnm)$, where $n = |A| + |H|$, m is the total length of the preference lists and C is the maximum k such that some agent is assigned to his/her kth-choice house in the constructed matching. Clearly a rank-maximal matching is Pareto optimal, however a rank-maximal matching need not be a maximum cardinality Pareto optimal matching (henceforth a maximum Pareto optimal matching). Alternatively, one may consider a *maximum cardinality maximum utility* matching M, in which we maximise $\sum_{(a,h)\in M} u_{a,h}$ over all maximum cardinality matchings, where $u_{a,h}$ indicates the utility of house h being allocated to agent a. If one defines $u_{a,h} = l - rank_{a,h}$,

where $rank_{a,h}$ is the rank of house h in agent a's preference list and l is the maximum length of an agent's list, then a maximum cardinality maximum utility matching is in turn a maximum Pareto optimal matching. Since all utilities are integral, a maximum cardinality maximum utility matching may be found in $O(\sqrt{n}m \log n)$ time [5]. However if one only requires to find a maximum cardinality matching that satisfies the weaker condition of being Pareto optimal, it is of interest to consider whether faster algorithms exist.

The next two sections of this paper work towards answering this question. In Section 2 we give a formal definition of the problem model, and present necessary and sufficient conditions for a matching to be Pareto optimal. In Section 3 we use these conditions as the basis for an $O(\sqrt{n}m)$ algorithm for finding a maximum Pareto optimal matching. This algorithm extends the Top Trading Cycles Method due to Gale [14], which has been the focus of much attention, particularly in the game theory and economics literature [14,12,11,15,2]. We then show that any improvement to the complexity of our algorithm would imply an improved algorithm for finding a maximum matching in a bipartite graph. We also demonstrate how to modify our algorithm in order to take account of initial property rights, guaranteeing that those who own a good initially will end up with a good that is either the same or better.

In the remainder of the paper, we prove several related results. In Section 4 we consider the problem of finding a minimum Pareto optimal matching, showing that this problem is NP-hard, though approximable within a factor of 2. In Section 5 we prove an interpolation result, showing that there exist Pareto optimal matchings of all sizes between a minimum and maximum Pareto optimal matching. Finally, in Section 6 we give a characterization, checkable in linear time, of instances that admit a unique Pareto optimal matching.

2 Preliminaries

We begin with a formal definition of the problem model under consideration. An instance I of the PARETO OPTIMAL MATCHING problem (POM) comprises a bipartite graph $G = (A, H, E)$, where $A = \{a_1, a_2, \ldots, a_r\}$ is the set of *agents* and $H = \{h_1, h_2, \ldots, h_s\}$ is the set of *houses*. For each $a_i \in A$, we denote by $A_i \subseteq H$ the vertices adjacent to a_i – these are referred to as the *acceptable* houses for a_i. Moreover a_i has a linear order over A_i. We let $n = r + s$ and $m = |E|$. Henceforth we assume that G contains no isolated vertices.

An *assignment* M is a subset of $A \times H$ such that $(a_i, h_j) \in M$ only if a_i finds h_j acceptable (i.e. $h_j \in A_i$). If $(a_i, h_j) \in M$, we say that a_i and h_j are *assigned* to one another. For each $q \in A \cup H$, let $M(q)$ denote the assignees of q in M. A *matching* is an assignment M such that $|M(q)| \leq 1$ for each $q \in A \cup H$. If $M(q) = \emptyset$, we say that q is *unmatched* in M, otherwise q is *matched* in M.

Let M be a matching in I. M is *maximal* if there is no (agent,house) pair (a_i, h_j) such that a_i and h_j are both unmatched in M and $h_j \in A_i$. Also M is *trade-in-free* if there is no (agent,house) pair (a_i, h_j) such that a_i is matched in M, h_j is unmatched in M, and a_i prefers h_j to $M(a_i)$. Finally M is *coalition-*

free if M admits no *coalition*, which is a sequence of matched agents $C = \langle a_0, a_1, \ldots, a_{k-1}\rangle$, for some $k \geq 2$, such that a_i prefers $M(a_{i+1})$ to $M(a_i)$ $(0 \leq i \leq k-1)$ (here, and in the remainder of this paper, all subscripts are taken modulo k when reasoning about coalitions). The matching

$$M' = (M \backslash \{(a_i, M(a_i)) : 0 \leq i \leq k-1\}) \cup \{(a_i, M(a_{i+1})) : 0 \leq i \leq k-1\}$$

is defined to be the matching obtained from M by *satisfying* C.

The preferences of an agent extend to matchings as follows. Given two matchings M and M', we say that an agent a_i *prefers* M' *to* M if either (i) a_i is matched in M' and unmatched in M, or (ii) a_i is matched in both M and M' and prefers $M'(a_i)$ to $M(a_i)$. Given this definition, we may define a relation $\prec$ on the set of all matchings as follows: $M' \prec M$ if and only if no agent prefers M to M', and some agent prefers M' to M. It is straightforward to then prove the following.

Proposition 1. *Given an instance I of POM, the relation $\prec$ forms a strict partial order over the set of matchings in I.*

A matching is defined to be *Pareto optimal* if and only if it is $\prec$-minimal. Intuitively a matching is Pareto optimal if no agent a_i can be better off without requiring another agent a_j to be worse off. The following proposition gives necessary and sufficient conditions for a matching to be Pareto optimal.

Proposition 2. *Let M be a matching in a given instance of POM. Then M is Pareto optimal if and only if M is maximal, trade-in-free and coalition-free.*

Proof. Let M be a Pareto optimal matching. If M is not maximal, then there exists an agent a_i and a house h_j, both unmatched in M, such that $h_j \in A_i$. Let $M' = M \cup \{(a_i, h_j)\}$. If M is not trade-in-free, then there exist an agent a_i and a house h_j, such that a_i is matched in M, h_j is unmatched in M, and a_i prefers h_j to $M(a_i)$. Let $M' = (M \backslash \{(a_i, M(a_i))\}) \cup \{(a_i, h_j)\}$. Finally if M admits some coalition C, let M' be the matching obtained by satisfying C. In all three cases, $M' \prec M$, a contradiction.

Conversely let M be a matching that is maximal, trade-in-free and coalition-free, and suppose for a contradiction that M is not Pareto optimal. Then there exists some matching M' such that $M' \prec M$. Let a_0 be any agent matched in M who prefers M' to M. Note that such an agent must exist, since M is maximal and at least one agent prefers M' to M.

It follows that $M'(a_0)$ is matched in M, say to a_1, for otherwise M is not trade-in-free. Therefore, $M'(a_1) \neq M(a_1)$, and so a_1 must also prefer M' to M. Using this same argument, $M'(a_1)$ is matched in M, say to a_2. We can continue in this manner finding a sequence of agents $\langle a_0, a_1, a_2, \ldots\rangle$, where a_i prefers $M(a_{i+1})$ to $M(a_i)$. Since the number of agents is finite, this sequence must cycle, thereby contradicting the assumption that M is coalition-free. □

Henceforth we will establish the Pareto optimality of a given matching by showing that the conditions of the above proposition are satisfied. For a given matching M, we can trivially check whether M satisfies the maximality and trade-in-free properties in $O(m)$ time. To check for the absence of coalitions, we construct

the *envy graph* of M. This is a directed graph, denoted by $G(M)$, consisting of one vertex for each agent, with an edge from a_i to a_j whenever a_j is matched in M and either (i) a_i is unmatched in M and finds $M(a_j)$ acceptable, or (ii) a_i is matched in M and prefers $M(a_j)$ to $M(a_i)$. It is clear that M is coalition-free if and only if $G(M)$ is acyclic. So we can perform this last check in $O(m)$ time by using a cycle-detection algorithm on $G(M)$. Putting these observations together, we have the following result.

Proposition 3. *Let M be a matching in a given instance of POM. Then we may check whether M is Pareto optimal in $O(m)$ time.*

It is easy to construct an instance of POM in which the Pareto optimal matchings are of different sizes. For example let $A = \{a_1, a_2\}$ and let $H = \{h_1, h_2\}$. Suppose that a_1 prefers h_1 to h_2, whilst a_2 finds only h_1 acceptable. Then both $M_1 = \{(a_1, h_1)\}$ and $M_2 = \{(a_1, h_2), (a_2, h_1)\}$ are Pareto optimal. Given this observation it is natural to consider the complexity of each of the problems of finding a maximum and minimum Pareto optimal matching. (Note that Greedy-POM produces M_1 given the agent ordering $\langle a_1, a_2 \rangle$, and produces M_2 given the agent ordering $\langle a_2, a_1 \rangle$.)

3 Maximum Pareto Optimal Matchings

In this section, we describe a three-phase algorithm for finding a maximum Pareto optimal matching, mirroring the three necessary and sufficient conditions in Proposition 2. We let I be an instance of POM, and we assume the notation and terminology introduced in Section 2. Phase 1 involves using the Hopcroft-Karp algorithm [7] to compute a maximum matching M in G. This phase, which guarantees that M is maximal, takes $O(\sqrt{n}m)$ time and dominates the runtime. The final two phases transform M into a trade-in-free and coalition-free matching respectively. We describe these phases in more detail below.

3.1 Phase 2 of the Algorithm

In this phase, we transform M into a trade-in-free matching by repeatedly identifying and promoting agents that prefer an unmatched house to their existing assignment. Each promotion breaks the existing assignment, thereby freeing a house, which itself may be a preferred assignment for a different agent. With the aid of suitable data structures, we can ensure that the next agent and house can be identified efficiently.

For each house h, we maintain a linked list L_h of pairs (a, r), where a is a matched agent who finds h acceptable, and r is the rank of h in a's preference list. Initially the pairs in L_h involve only those matched agents a who prefer h to $M(a)$, though subsequently the pairs in L_h may contain agents a who prefer $M(a)$ to h. The initialization of these lists can be carried out using one traversal of the agent preference lists, which we assume are represented as doubly linked lists or arrays, in $O(m)$ time.

For each matched agent a, we also use this traversal to initialize a variable, denoted by $curr_a$, which stores the rank of $M(a)$ in a's preference list. This variable is maintained during the execution of the algorithm. We also assume that, for each matched agent a we store $M(a)$. One final initialization remains: construct a stack S of all unmatched houses h where L_h is non-empty. We now enter the loop described in Figure 1.

```
while S ≠ ∅
    h := S.pop();
    (a, r) := L_h.removeHead();
    if r < curr_a
        // h is unmatched in M, a is matched in M and prefers h to M(a)
        h' := M(a);
        M := (M\{(a, h')}) ∪ {(a, h)};
        curr_a := r;
        h := h';
    if L_h ≠ ∅
        S.push(h);
```

Fig. 1. Phase 2 loop

During each loop iteration we pop an unmatched house h from S and remove the first pair (a, r) from the list L_h (which must be non-empty). If a prefers h to $M(a)$ (i.e. $r < curr_a$) then a is promoted from $h' = M(a)$ to h, also M and $curr_a$ are updated, and finally h', which is now unmatched, is pushed onto S if $L_{h'}$ is non-empty. Otherwise h is pushed back onto S if L_h is non-empty.

Each iteration of the loop removes a pair from a list L_h. Since agent preference lists are finite and no new pair is added to a list L_h during a loop iteration, the while loop must eventually terminate with S empty. At this point no matched agent a would trade $M(a)$ for an unmatched house, and so M is trade-in-free. Additionally, M remains a maximum matching, since any agent matched at the end of Phase 1 is also matched at the end of Phase 2. Finally, it is clear that this phase runs in $O(m)$ time given the data structures described above.

3.2 Phase 3 of the Algorithm

In this phase, we transform M into a coalition-free matching. Recall that coalitions in M correspond to cycles in the envy graph $G(M)$. So a natural algorithm involves repeatedly finding and satisfying coalitions in $G(M)$ until no more coalitions remain. This algorithm has a runtime of $O(m^2)$, since there are $O(m)$ coalitions, and cycle-detection takes $O(m)$ time.

A better starting point for an efficient algorithm is Gale's Top Trading Cycles Method [14]. This method is also based on repeatedly finding and satisfying coalitions, however the number of iterations is reduced by the following observation: no agent assigned to his/her first choice can be in a coalition. We remove such agents from consideration, and since the houses assigned to them are no

```
for each matched agent a such that p(a) ≠ M(a)
    P := {a};   // P is a stack of agents
    c(a) := 1;   // counters record the number of times an agent is in P
    while P ≠ ∅
        a' := P.pop();
        p(a') := most-preferred unlabelled house on preference list of a';
        if c(a') = 2
            C := coalition in P containing a';
            satisfy C;
            for each a'' ∈ C
                label M(a'');
                c(a'') := 0;
                P.pop();
        else if p(a') = M(a')
            label M(a');
            c(a') := 0;
        else
            P.push(a');
            a'' := M(p(a'));
            c(a'') := c(a'') + 1;
            P.push(a'');
```

Fig. 2. Phase 3 loop

longer exchangeable, they can be deleted from the preference lists of the remaining agents. This observation can now be recursively applied to the reduced preference lists. At some point, either no agents remain, in which case the matching is coalition-free, or no agent is assigned to his/her reduced first choice (i.e. the first choice on his/her reduced preference list).

In this last case, it turns out that there must be a coalition C in M, which can be found in $O(r)$ time by searching the envy graph restricted to reduced first-choice edges. After satisfying C, each agent in C is assigned to his/her reduced first choice. Therefore, no agent is in more than one coalition, giving $O(r)$ coalitions overall. The runtime of this preliminary implementation then is $\Omega(m + r^2)$. We now present a linear-time extension of Yuan's description of the Top Trading Cycles Method [15].

In our implementation, deletions of houses from agents' preference lists are not explicitly carried out. Instead, a house that is no longer exchangeable is labelled (all houses are initially unlabelled). For each agent a we maintain a pointer $p(a)$ to the first unlabelled house on a's preference list – this is equivalent to the first house on a's reduced preference list. Initially $p(a)$ points to the first house on a's preference list, and subsequently $p(a)$ traverses left to right. Also, in order to identify coalitions, we initialize a counter $c(a)$ to 0 for each agent a. Then, we enter the main body of the algorithm, as given in Figure 2.

This algorithm repeatedly searches for coalitions, building a path P of agents (represented by a stack) in the (simulated) envy graph restricted to reduced first-choice edges. At each iteration of the while loop, we pop an agent a' from the stack and move up $p(a')$ if necessary. If P cycles (i.e. we find that $c(a') = 2$),

there is a coalition C – the agents involved in C are removed from consideration and the houses assigned to these agents are labelled (in practice the agents in C can be identified and C can be satisfied during the stack popping operations). Alternatively, if P reaches a dead-end (a' is already assigned to his/her first choice), this agent is removed from consideration and his/her assigned house is labelled. Otherwise, we keep extending the path by following the reduced first-choice edges.

At the termination of this phase we note that M is coalition-free by the correctness of the Top Trading Cycles Method [14]. Also M remains a maximum trade-in-free matching, since each agent and house matched at the end of Phase 2 is also matched at the end of Phase 3. Finally, it is clear this phase runs in $O(m)$ time given the data structures described above. We summarize the preceding discussion in the following theorem.

Theorem 1. *A maximum Pareto optimal matching can be found in $O(\sqrt{n}m)$ time. Such a matching is also a maximum matching of agents to houses.*

We now show that any improvement to the complexity of the above algorithm would imply an improved algorithm for finding a maximum matching in a bipartite graph. Without loss of generality, let $G = (A, H, E)$ be an arbitrary bipartite graph with no isolated vertices. Construct an instance I of POM with bipartite graph G, where each agent a's preference list in I is an arbitrary permutation over a's neighbours in G. By Theorem 1, any maximum Pareto optimal matching in I is also a maximum matching in G. Since I may be constructed from G in $O(m)$ time, the complexity of finding a maximum matching in a bipartite graph is bounded above by the complexity of finding a maximum Pareto optimal matching.

3.3 Initial Property Rights

Suppose that a subset A' of the agents already own a house. We now describe an *individually rational* modification of our algorithm, which ensures that every agent in A' ends up with the same house or better.

We begin with a matching M that pre-assigns every agent $a \in A'$ to his/her existing house h. We then truncate the preference list of each such a by removing all houses less preferable than $M(a)$. Now, we enter Phase 1, where we use the Hopcroft-Karp algorithm to exhaustively augment M into some matching M'. Members of A' must still be matched in M', and since their preference lists were truncated, their new assignments must be at least as preferable as those in M. Note that M' may not be a maximum matching of A to H, however M' does have maximum cardinality among all matchings that respect the initial property rights. The remaining two phases do not move any agent from being matched to unmatched, and so the result follows immediately.

In the special case that all agents own a house initially (i.e. I is an instance of a housing market), it is clear that Phases 1 and 2 of the algorithm are not required. Moreover it is known that Phase 3 produces the unique matching that belongs to the core of the market [12], a stronger notion than Pareto optimality.

4 Minimum Pareto Optimal Matchings

In this section, we consider the problem of finding a minimum Pareto optimal matching. Let MIN-POM denote the problem deciding, given an instance I of POM and an integer K, whether I admits a Pareto optimal matching of size at most K. We firstly prove that MIN-POM is NP-complete via a reduction from MMM, which is the problem of deciding, given a graph G and an integer K, whether G admits a maximal matching of size at most K.

Theorem 2. *MIN-POM is NP-complete.*

Proof. By Proposition 3, MIN-POM belongs to NP. To show NP-hardness, we give a reduction from the NP-complete restriction of MMM to subdivision graphs [6] (given a graph G, the *subdivision graph* of G is obtained by subdividing each edge $e = \{u, w\}$ into two edges $\{u, v_e\}, \{v_e, w\}$, where v_e is a new vertex corresponding to e).

Let $G = (V, E)$ (a subdivision graph) and K (a positive integer) be given as an instance of MMM. Then V is a disjoint union of two sets U and W, where each edge $e \in E$ joins a vertex in U to a vertex in W. Assume that $U = \{u_1, u_2, \ldots, u_r\}$ and $W = \{w_1, w_2, \ldots, w_s\}$. Without loss of generality assume that each vertex $u_i \in U$ has degree 2, and moreover assume that p_i and q_i are two sequences such that $p_i < q_i$, $\{u_i, w_{p_i}\} \in E$ and $\{u_i, w_{q_i}\} \in E$.

We create an instance I of MIN-POM as follows. Let A be the set of agents and let H be the set of houses, where $A = A^1 \cup A^2$, $A^t = \{a_1^t, a_2^t, \ldots, a_r^t\}$ ($t = 1, 2$), $H = W \cup X$ and $X = \{x_1, x_2, \ldots, x_r\}$. For each i ($1 \le i \le r$), we create preference lists for agents a_i^1 and a_i^2 as follows:

$$a_i^1 : x_i \quad w_{p_i} \quad w_{q_i} \qquad\qquad a_i^2 : x_i \quad w_{q_i} \quad w_{p_i}$$

We claim that G has a maximal matching of size at most K if and only if I has a Pareto optimal matching of size at most $K + r$.

For, suppose that M is a maximal matching in G of size at most K. We construct a set M' as follows. For any $u_i \in U$ that is unmatched in M, add the pair (a_i^1, x_i) to M'. Now suppose that $(u_i, w_j) \in M$. If $j = p_i$, add the pairs (a_i^1, w_j) and (a_i^2, x_i) to M'. If $j = q_i$, add the pairs (a_i^1, x_i) and (a_i^2, w_j) to M'. Clearly M' is a matching in I, and $|M'| = |M| + r \le K + r$. It is straightforward to verify that, by the maximality of M in G, M' is Pareto optimal in I.

Conversely suppose that M' is a Pareto optimal matching in I of size at most $K + r$. For each i ($1 \le i \le r$), either $(a_i^1, x_i) \in M'$ or $(a_i^2, x_i) \in M'$, for otherwise M' is not trade-in-free. Hence we may construct a matching M in G as follows. For each i ($1 \le i \le r$), if $(a_i^t, w_j) \in M'$ for some t ($1 \le t \le 2$), add (u_i, w_j) to M. Then $|M| = |M'| - r \le K$. The maximality of M' clearly implies that M is maximal in G. □

For a given instance I of POM with bipartite graph G, we denote by $p^-(I)$ and $p^+(I)$ the sizes of a minimum and maximum Pareto optimal matching in I respectively. Similarly, we denote by $\beta_1^-(G)$ and $\beta_1(G)$ the sizes of a minimum maximal

and a maximum matching in G respectively. It is known that $\beta_1^-(G) \geq \beta_1(G)/2$ [9]. By Proposition 2, Pareto optimal matchings in I are maximal matchings in G. Hence, by Theorem 1, we have that $\beta_1^-(G) \leq p^-(I) \leq p^+(I) = \beta_1(G)$. It is therefore immediate that, for a given instance I of POM, the problem of finding a minimum Pareto optimal matching is approximable within a factor of 2.

5 Interpolation of Pareto Optimal Matchings

In this section, we prove that, for a given instance I of POM, there are Pareto optimal matchings of all sizes between $p^-(I)$ and $p^+(I)$.

Given a matching M, an augmenting path P for M is an alternating sequence of distinct agents and houses $\langle a_1, h_1, a_2, \ldots, a_k, h_k \rangle$, where a_1 and h_k are unmatched in M, $h_i \in A_i$, and $M(a_{i+1}) = h_i$ $(1 \leq i \leq k-1)$. We associate with each such augmenting path a vector $rank_P$, whose ith component contains the rank of a_i for h_i. Given two augmenting paths P and Q for M, we say that $P \lhd Q$ if (i) both P and Q begin from the same agent, and (ii) $rank_P$ is lexicographically less than $rank_Q$. Also for paths P and Q, we define three operations: $Prefix_P(v)$ is the substring of P from a_1 to $v \in P$, $Suffix_P(v)$ is the substring of P from $v \in P$ to h_k, and $P \cdot Q$ denotes the concatenation of P and Q.

Theorem 3. *For a given instance I of POM, there exist Pareto optimal matchings of size k, for each k such that $p^-(I) \leq k \leq p^+(I)$.*

Proof. Let M be any Pareto optimal matching such that $|M| < p^+(I)$, and let M' be the matching that results from augmenting M along some $\lhd$-minimal augmenting path P. We will show in turn that M' is maximal, trade-in-free and coalition-free; the result then follows by induction.

If M' is not maximal, then clearly we contradict the maximality of M. Now suppose that M' is not trade-in-free. Then there exists an agent a and house h, where a is matched in M', h is unmatched in M', and a prefers h to $M'(a)$. Since h is also unmatched in M, a must be in P, for otherwise $M(a) = M'(a)$, and M is not trade-in-free. But then $P' = Prefix_P(a) \cdot \langle h \rangle$ is an augmenting path for M, contradicting the $\lhd$-minimality of P.

Finally suppose for a contradiction that M' is not coalition-free. Then there exists a coalition $C = \langle a_0, a_1, \ldots, a_{k-1} \rangle$ with respect to M'. At least one agent in P must also be in C, for otherwise M is not coalition-free. Let a_i be the first such agent in P. We establish some properties of $M'(a_{i+1})$. Firstly, $M'(a_{i+1})$ must be matched in M, for otherwise M admits the augmenting path $Prefix_P(a_i) \cdot \langle M'(a_{i+1}) \rangle$, contradicting the $\lhd$-minimality of P. Also, $M'(a_{i+1})$ cannot appear before a_i in P, for otherwise a_i is not the first agent in P to be in C. Lastly, $M'(a_{i+1})$ cannot appear after a_i in P, for otherwise M admits the augmenting path $Prefix_P(a_i) \cdot Suffix_P(M'(a_{i+1}))$, contradicting the $\lhd$-minimality of P. So, it must be the case that $M'(a_{i+1})$ is matched in M and does not appear in P. Let a_{i+j} be the first agent in C after a_i, such that a_{i+j} is in P. Note that $a_{i+j} \neq a_{i+1}$ by the above properties of $M'(a_{i+1})$, but since C is a cycle, $a_{i+j} = a_i$ is possible. It follows that the subsequence $S = \langle M'(a_{i+1}), a_{i+1}, \ldots, M'(a_{i+j-1}), a_{i+j-1} \rangle$ of

C is disjoint from P, and so $P' = \mathit{Prefix}_P(a_i) \cdot S \cdot \mathit{Suffix}_P(M'(a_{i+j}))$ is a valid augmenting path of M. But then P' contradicts the $\lhd$-minimality of P, since a_i prefers $M'(a_{i+1})$ to $M'(a_i)$. □

Corollary 1. *Given an instance I of POM and a Pareto optimal matching M in I of size k, we can construct a Pareto optimal matching M' of size $k+1$, or determine that no such matching exists, in $O(m)$ time.*

Proof. Let G be the bipartite graph in I, with edges in M directed from H to A, and edges not in M directed from A to H. Also associate with each non-matching edge (a_i, h_j) the rank of a_i for h_j. We search for a $\lhd$-minimal augmenting path by performing an ordered depth first search of G starting from the set of unmatched agents, where for each agent a in the search, we explore outgoing edges from a in increasing order of rank. In general, ordered depth-first search is asymptotically slower than depth-first search. However, the $O(m)$ result holds, since each preference list is already given in increasing order of rank. □

We remark that the results of this section extend to the case where a subset of the agents have initial property rights.

6 Uniqueness of Pareto Optimal Matchings

In this section, we give a characterization of instances with no initial property rights that admit a unique Pareto optimal matching. This is based on the concept of a *signature* of a Pareto optimal matching.

If a matching M is Pareto optimal, the envy graph $G(M)$ contains no cycles, and therefore admits a topological ordering. We say that a reversed topological ordering of $G(M)$, denoted by $\sigma(M)$, is a *signature* of M. The next lemma will help us establish that the signature of a matching is unique for that matching. This lemma is similar to [1, Lemma 1], though the proof here, which uses the concept of a signature, is much simpler.

Lemma 1. *Given an instance I of POM, the algorithm Greedy-POM can generate any Pareto optimal matching in I.*

Proof. Let M be an arbitrary Pareto optimal matching in I. We claim that by processing the agents in order of $\sigma(M)$, the greedy algorithm returns M.

Suppose for a contradiction that Greedy-POM returns a matching $M' \neq M$. It follows that since M' is Pareto optimal, some agent must prefer M' to M. Let a be the first such agent in $\sigma(M)$.

Now, $M'(a)$ must be matched in M, say to a', for otherwise M is not maximal (if a is unmatched in M), or M is not trade-in-free (if a is matched in M). $G(M)$ must therefore contain an edge from a to a', meaning that a' precedes a in $\sigma(M)$. At the time a' is processed by Greedy-POM, $M'(a)$ is unmatched (since it is assigned later to a). So, a' must prefer $M'(a')$ to $M(a') = M'(a)$, contradicting the assumption that a was the first such agent in $\sigma(M)$. □

Corollary 2. *Given an instance I of POM, every agent permutation is a signature of exactly one Pareto optimal matching in I.*

We can now present a necessary and sufficient condition, checkable in linear time, for a POM instance to admit a unique Pareto optimal matching.

Theorem 4. *An instance I of POM admits a unique Pareto optimal matching M if and only if every agent is matched in M with his/her first choice.*

Proof. Let M be the unique Pareto optimal matching in I. Since every agent permutation is a signature of M, $G(M)$ contains no edges. Then every agent must be matched to his/her first choice.

Conversely, let M be a matching in I in which every agent is matched with his/her first choice. Then if M' is any matching in I such that $M' \neq M$, it follows that $M \prec M'$. Hence M is the unique Pareto optimal matching in I. □

7 Concluding Remarks

We conclude with an open problem. The basic POM definition given in Section 2 can be generalized by permitting agents to contain *ties* in their preference lists (i.e. to rank equally two or more houses). In this context the definition of the relation $\prec$ is the same as that given in Section 2, and hence the definition of Pareto optimality remains unchanged. A maximum Pareto optimal matching can be found in $O(\sqrt{n}m \log n)$ time using a similar reduction to the Assignment problem as described in Section 1 (in this case $rank_{a,h}$ is the number of houses that a prefers to h). However is the problem of finding a maximum Pareto optimal matching solvable in $O(\sqrt{n}m)$ time?

References

1. A. Abdulkadiroğlu and T. Sönmez. Random serial dictatorship and the core from random endowments in house allocation problems. *Econometrica*, 66(3):689–701, 1998.
2. A. Abdulkadiroğlu and T. Sönmez. House allocation with existing tenants. *Journal of Economic Theory*, 88:233–260, 1999.
3. X. Deng, C. Papadimitriou, and S. Safra. On the complexity of equilibria. *Journal of Computer and System Sciences*, 67(2):311–324, 2003.
4. S.P. Fekete, M. Skutella, and G.J. Woeginger. The complexity of economic equilibria for house allocation markets. *Inf. Proc. Lett.*, 88:219–223, 2003.
5. H.N. Gabow and R.E. Tarjan. Faster scaling algorithms for network problems. *SIAM Journal on Computing*, 18(5):1013–1036, 1989.
6. J.D. Horton and K. Kilakos. Minimum edge dominating sets. *SIAM Journal on Discrete Mathematics*, 6:375–387, 1993.
7. J.E. Hopcroft and R.M. Karp. A $n^{5/2}$ Algorithm for Maximum Matchings in Bipartite Graphs. *SIAM Journal on Computing*, 2:225–231, 1973.
8. A. Hylland and R. Zeckhauser. The efficient allocation of individuals to positions. *Journal of Political Economy*, 87(2):293–314, 1979.

9. B. Korte and D. Hausmann. An analysis of the greedy heuristic for independence systems. *Annals of Discrete Mathematics*, 2:65–74, 1978.
10. R.W. Irving, T. Kavitha, K. Mehlhorn, D. Michail, and K. Paluch. Rank-maximal matchings. *Proceedings of SODA '04*, pages 68–75. ACM-SIAM, 2004.
11. A.E. Roth. Incentive compatibility in a market with indivisible goods. *Economics Letters*, 9:127–132, 1982.
12. A.E. Roth and A. Postlewaite. Weak versus strong domination in a market with indivisible goods. *Journal of Mathematical Economics*, 4:131–137, 1977.
13. A.E. Roth and M.A.O. Sotomayor. *Two-sided matching: a study in game-theoretic modeling and analysis.* Cambridge University Press, 1990.
14. L. Shapley and H. Scarf. On cores and indivisibility. *Journal of Mathematical Economics*, 1:23–37, 1974.
15. Y. Yuan. Residence exchange wanted: a stable residence exchange problem. *European Journal of Operational Research*, 90:536–546, 1996.
16. L. Zhou. On a conjecture by Gale about one-sided matching problems. *Journal of Economic Theory*, 52(1):123–135, 1990.

Generalized Geometric Approaches for Leaf Sequencing Problems in Radiation Therapy*,**

Danny Z. Chen[1], Xiaobo S. Hu[1], Shuang Luan[2,***], Shahid A. Naqvi[3], Chao Wang[1], and Cedric X. Yu[3]

[1] Department of Computer Science and Engineering,
University of Notre Dame,
Notre Dame, IN 46556, USA
{chen, hu, cwang1}@cse.nd.edu

[2] Department of Computer Science,
University of New Mexico,
Albuquerque, NM 87131, USA
sluan@unm.edu

[3] Department of Radiation Oncology,
University of Maryland School of Medicine,
Baltimore, MD 21201-1595, USA
{snaqv001, cyu002}@umaryland.edu

Abstract. The 3-D *static leaf sequencing* (SLS) problem arises in radiation therapy for cancer treatments, aiming to deliver a prescribed radiation dose to a target tumor accurately and efficiently. The treatment time and machine delivery error are two crucial factors of a solution (i.e., a treatment plan) for the SLS problem. In this paper, we prove that the 3-D SLS problem is NP-hard, and present the first ever algorithm for the 3-D SLS problem that can determine a tradeoff between the treatment time and machine delivery error (also called the "tongue-and-groove" error in medical literature). Our new 3-D SLS algorithm with error control gives the users (e.g., physicians) the option of specifying a machine delivery error bound, and subject to the given error bound, the algorithm computes a treatment plan with the minimum treatment time. We formulate the SLS problem with error control as computing a k-weight shortest path in a directed graph and build the graph by computing g-matchings and minimum cost flows. Further, we extend our 3-D SLS algorithm to the popular radiotherapy machine models with different constraints. In our extensions, we model the SLS problems for some of the radiotherapy systems as computing a minimum g-path cover of a directed acyclic graph. We implemented our new 3-D SLS algorithm suite and conducted

* Due to mistakes during data conversion, this paper was originally published in LNCS 3341 (ISAAC 2004) with several special characters missing in text and figures.

** This research was supported in part by the National Science Foundation under Grant CCR-9988468.

*** Corresponding author. The research of this author was supported in part by a faculty start-up fund from the Department of Computer Science, University of New Mexico.

X. Deng and D. Du (Eds.): ISAAC 2005, LNCS 3827, pp. 1176–1186, 2005.

an extensive comparison study with commercial planning systems and well-known algorithms in medical literature. Some of our experimental results based on real medical data are presented.

1 Introduction

In this paper, we study the 3-D *static leaf sequencing* (SLS) problem in *intensity-modulated radiation therapy* (IMRT). IMRT is a modern cancer therapy technique and aims to deliver a high radiation dose that is as conformal to a tumor as possible. Performing IMRT is based on the ability to *accurately* and *efficiently* deliver prescribed discrete dose distributions of radiation, called *intensity maps (IMs)*. An IM is a dose prescription specified by a set of nonnegative integers on a 2-D grid (see Figure 1(a)). The value in each grid cell indicates the amount (in units) of radiation to be delivered to the body region corresponding to that IM cell.

One of the most advanced tools today for delivering intensity maps is the *multileaf collimator* (MLC) [11]. An MLC consists of up to 60 pairs of tungsten leaves (see Figure 1(b)). The leaves can move up and down to form a rectilinear region, called an *MLC-aperture*. To reduce radiation leakage, two pairs of backup metal diaphragms (along the x and y axes) are used to form a "bounding box" of each rectilinear MLC-aperture.

Currently, there are three popular MLC systems in clinical use. They are the Elekta, the Siemens, and the Varian MLC systems [11]. Depending on the specific MLC in use, there are some differences among the geometric shapes of the "deliverable" MLC-apertures. In this abstract, we will focus on the Elekta MLC system, and leave the details of Siemens and Varian MLCs to the full paper. (Section 2 describes the geometric properties of an Elekta MLC-aperture.)

There are several techniques for delivering IMRT [18, 19]. The most popular one is called the *static leaf sequencing* (SLS) or the "*step-and-shoot*" technique [9, 19, 20]. In the SLS approach, an IM is delivered as follows: Form an MLC-aperture, turn on the beam source to deliver radiation to the area of the IM exposed by the MLC-aperture, turn off the beam source, reposition MLC leaves to form another MLC-aperture, and repeat until the entire IM is done. In this setting, the boundary of each MLC-aperture does not intersect the interior of any IM cell. In delivering a beam shaped by an MLC-aperture, all the cells inside the region of the MLC-aperture receive the same integral amount of radiation dose (say, one unit), i.e., the numbers in all such cells are decreased by the same integer value. The IM is done when each cell has a value zero.

A *treatment plan* for a given IM consists of a set of MLC-apertures for delivering the IM. (Each MLC-aperture is associated with an integral value that represent the amount of radiation prescribed to it.) Two key criteria are used to evaluate the quality of an IMRT treatment plan:

(1) **Treatment time (the efficiency):** Minimizing the treatment time is important because it not only lowers the patients' costs but also enables more patients to be treated. Since the overhead associated with turning the beam

0	0	1	1	0
0	0	1	1	0
1	1	2	1	0
0	0	1	1	0
1	1	1	1	1
0	0	1	0	0
0	0	1	1	0

(*a*)

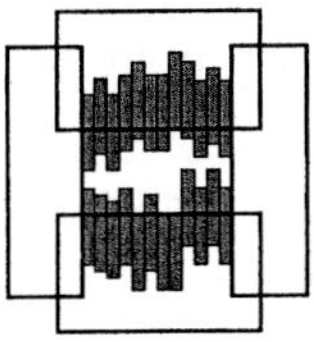

(*b*)

Fig. 1. (a) An IM. (b) Illustrating the Elekta MLC systems. The shaded rectangles represent the MLC leaves, and the unshaded rectangles represent the backup diaphragms.

source on/off and repositioning MLC leaves dominates the total treatment time [9, 20], we like to minimize the number of MLC-apertures used for delivering an IM.

(2) **Machine delivery error (the accuracy):** Ideally, an IM is partitioned into a set of MLC-apertures that delivers the IM perfectly. However, in reality, due to the special geometric shape of the MLC leaves [21], an MLC-aperture cannot be delivered as perfectly as its geometry. This gives rise to an error between the prescribed dose and actually delivered dose. We call this error the *machine delivery error*, which is also called the *"tongue-and-groove" error* in medical literature [17, 20, 21] (Section 2 discusses more on its nature). Minimizing the machine delivery error is important because the maximum machine delivery error can be up to 10% [10], which is well beyond the allowed 3–5% limit. Carefully choosing MLC-apertures to use can reduce the error.

The 3-D **static leaf sequencing (SLS)** problem is: Given an IM and an error bound $\mathcal{E}$, find a set S of MLC-apertures for delivering the IM such that the total machine delivery error incurred by S is $\leq \mathcal{E}$, and the size $|S|$ is minimized. Note that the key to the problem is to find a good tradeoff between the accuracy (error) and efficiency (treatment time).

The 3-D SLS problem has received a great deal of attention from medical physics community [4, 5, 6, 9, 17, 20], computational geometry [7, 8, 15], and operations research [1, 3, 13]. An influential paper is by Xia and Verhey [20] for solving the 3-D SLS problem **without** error control (i.e., the machine delivery error is ignored). Their algorithm has been implemented by many researchers and used as a widely accepted benchmark program for leaf sequencing software. However, none of the known approaches so far can solve the 3-D SLS problem **with** machine delivery error control (i.e., determining a good tradeoff between the error and time).

The main results of this paper are summarized as follows:

(1) We prove that the 3-D SLS problem is NP-hard under all three popular MLC models, and with or without machine delivery error control.
(2) We present the first ever modeling of the 3-D SLS problem with a tradeoff between the error and treatment time and an efficient algorithm for the prob-

lem on the Elekta model. In our solution, the 3-D SLS problem is formulated as a k-weight shortest path problem on a directed graph, in which each edge is defined by a minimum weight g-cardinality matching. Every such k-weight path specifies a set S of k MLC-apertures for delivering the given IM, and the cost of the path indicates the machine delivery error of the set S of MLC-apertures.
(3) We extend our 3-D SLS algorithm to other MLC models, such as Siemens and Varian. Our extensions are based on computing a minimum g-path cover of a directed acyclic graph.
(4) We also solve the 2-D case of the 3-D SLS problem in which the given IM consists of entries of only 1 or 0. Our solution is based on computing a minimum cost flow in a special graph.
(5) We implemented our algorithm suite and carried out experimental studies using medical data.

Due to space limit, we will focus on results (2) and (5) in this extended abstract, and leave the details of results (1), (3) and (4) to the full paper.

2 Preliminaries and Problem Statement

2.1 Constraints of Multileaf Collimators and Their Geometry

The mechanical constraints of an MLC preclude certain aperture shapes from being used [11]. One such constraint is called the *minimum leaf separation*, which requires the distance between the opposite leaves of any MLC leaf pair to be $\geq$ a given value δ (say, $\delta = 1cm$). Another constraint is called the *no-interleaf motion*, which forbids the tip of each MLC leaf to surpass those of its neighboring leaves on the opposite leaf bank. These constraints prevent opposite MLC leaves from colliding into each other and being damaged. The Elekta MLC (i.e., the default MLC system in this paper) is subject to both the leaf separation and the no-interleaf motion constraint, hence geometrically, each Elekta MLC-aperture is a rectilinear x-monotone *simple* polygon whose minimum vertical "width" is $\geq$ the minimum separation value δ (see Figure 1(b)).

2.2 Machine Delivery Errors

On most current MLCs, the sides of the leaves are designed to have a "tongue-and-groove" (TG) feature (see Figure 2(a)). This design reduces the radiation leakage through the gap between two neighboring MLC leaves [21]. But, it also causes an unwanted underdose and leakage situation when an MLC leaf is used for blocking radiation (see Figures 2(b) and 2(c)). Geometrically, the underdose and leakage error caused by the tongue-and-groove feature associating with an MLC-aperture is a set of 3-D axis-parallel boxes $w \cdot l_i \cdot h$, where w is the (fixed) width of the tongue or groove side of an MLC leaf, l_i is the length of the portion of the i-th leaf that is actually involved in blocking radiation, and $h = \alpha \cdot r$ is the amount of radiation leakage with α being the (fixed) leakage ratio and r being the amount of radiation delivered by that MLC-aperture. Figure 2(b) illustrates

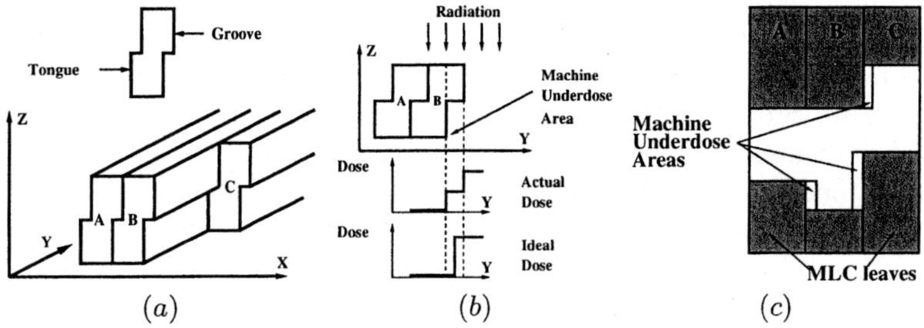

Fig. 2. (a) Illustrating the tongue-and-groove (TG) interlock feature of the MLC in 3-D, where leaf B is used for blocking radiation. (b) When leaf B is used for blocking radiation, there is an underdose and leakage in the tongue or groove area. (c) The underdose and leakage areas of an MLC-aperture region.

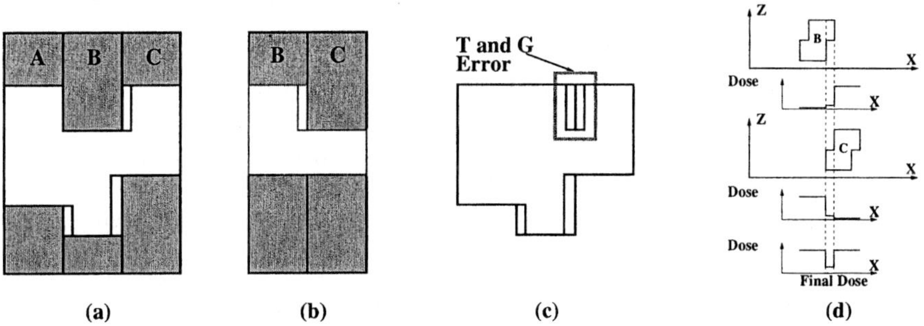

Fig. 3. Illustrating the tongue-and-groove error. (a) and (b): Two MLC-apertures (the shaded rectangles represent MLC leaves). (c) When delivering the two MLC-apertures in (a) and (b) (one by one), the groove side of leaf B and tongue side of leaf C are both used for blocking radiation, causing a tongue-and-groove error in the area between the leaves B and C. (d) Illustrating a dose "dip" in the final dose distribution where a tongue-and-groove error occurs.

the height of the underdose and leakage error, and Figure 2(c) illustrates the width and length of the underdose and leakage error.

We distinguish two types of errors caused by the tongue-and-groove feature of MLC leaves:

Tongue-or-groove error: The tongue-**or** groove error of an MLC-aperture is defined as the amount of underdose and leakage error occurred whenever the tongue side or groove side of an MLC leaf is used for blocking radiation. The tongue-or-groove error of an IMRT plan (i.e., a set of MLC-apertures) is the sum of the errors of all its MLC-apertures.

Tongue-and-groove error: Unlike the tongue-or-groove error which is defined on each individual MLC-aperture, the tongue-**and**-groove error is defined by the

relations between different MLC-apertures. The tongue-and-groove error occurs whenever the tongue side of an MLC leaf and the corresponding groove side of its neighboring leaf are both used for blocking radiation in any two different MLC-apertures of an IMRT plan (see Figure 3). Clearly, the tongue-and-groove error is a subset of the tongue-or-groove error. Note that minimizing the tongue-and-groove error is also important because it usually occurs in the middle of the delivered intensity maps [20], and is actually the most damaging part of the tongue-or-groove error.

Geometrically, for a set of MLC-apertures, the errors caused by using the tongue sides of the MLC leaves for blocking radiation are a set of 3-D axis-parallel boxes, denoted by V_T. Similarly, the errors by using the groove sides of the leaves for blocking radiation are another set of 3-D axis-parallel boxes, denoted by V_G. Then the tongue-or-groove error for the MLC-aperture set is the sum of the volume values of these two sets (i.e., $|V_T| + |V_G|$), and the tongue-and-groove error is equal to twice the volume value of the intersection between V_T and V_G (i.e., $2 \cdot |V_T \cap V_G|$). We can also view the given IM as a 3-D rectilinear terrain (mountain), denoted by V^*. Then the magnitude of the tongue-or-groove (resp., tongue-and-groove) error can be quantified by the percentage $\frac{|V_T|+|V_G|}{|V^*|}$ (resp., $\frac{|V_T \cap V_G|}{|V^*|}$).

If we view each MLC-aperture as a rectilinear polygonal region on the xy-plane, then the tongue-or-groove error occurs along every vertical boundary edge of this region, except its leftmost and rightmost vertical edges. The leftmost and rightmost vertical edges are excluded (i.e., no error on them) since they are defined by the backup diaphragms along the x-axis, not by the MLC leaves.

2.3 The Static Leaf Sequencing Problem

The 3-D **static leaf sequencing (SLS)** problem is: Given an IM and an error bound $\mathcal{E}$, find a set S of MLC-apertures for delivering the IM, such that the machine delivery error (i.e., either the tongue-or-groove error or the tongue-and-groove error) is $\leq \mathcal{E}$ and the size $|S|$ is minimized.

Interchangeably, we also call such MLC-apertures the *B-segments*[7] (for *block-segments*). Each B-segment is of a rectilinear x-monotone polygonal shape of a uniform height $h \geq 1$ (h is the number of dose units delivered by the MLC-aperture).

A key special case of the 3-D SLS problem is the **basic 3-D SLS** problem [4,5,6]. This case is similar to the general 3-D SLS problem, except that the height of each of its B-segments must be one. Note that in the general 3-D SLS problem, the uniform height of each B-segment can be any integer ≥ 1. Studying the basic case is important because the maximum heights of the majority of IMs used in the current clinical treatments are around 5, and an optimal solution for the basic case on such an IM is often very close to an optimal solution for the general case.

3 3-D SLS Algorithms with Error Control

This section presents our SLS algorithms with error control. Due to space limit, we will use our basic 3-D SLS algorithms with error control for the Elekta model to illustrate some of the key ideas of our approach, and leave the details of our complete SLS algorithm suite to the full paper.

3.1 Algorithm for Basic 3-D SLS Problem with Tongue-or-Groove Error Control

Let S be a solution for the basic 3-D SLS problem with tongue-or-groove error control ($S = \{S_i \mid i \in I\}$ is a set of B-segments of height 1 for the given IM, where I is an index set). Consider a B-segment $S_i \in S$. Observe that since each B-segment S_i has a unit height and an x-monotone rectilinear simple polygonal shape, S_i actually builds a continuous block of a unit height on every IM column C_j that intersects the projection of S_i on the IM grid [7]. We denote such a continuous block by an interval $B_{i,j}$ on the column C_j. (Interchangeably, we also call $B_{i,j}$ a block). Note that when delivering a B-segment S_i, each of its blocks $B_{i,j}$ is delivered by the pair of MLC leaves that is aligned with C_j.

Let $B(S_i) = \{B_{i,j} \mid B_{i,j} = S_i \cap C_j, j = e_i, e_i+1, \ldots, k_i\}$ be the set of blocks that form S_i, where S_i "runs" consecutively from the IM columns C_{e_i} to C_{k_i}. As discussed in Section 2.2, the tongue-or-groove error of S_i occurs along every vertical boundary edge of the polygonal region of S_i, except its leftmost and rightmost vertical edges. Let $|B_{i,j} \oplus B_{i,j+1}|$ denote the length of the symmetric difference between the two intervals of $B_{i,j}$ and $B_{i,j+1}$, $j = e_i, e_i+1, \ldots, k_i - 1$. Thus, the total tongue-or-groove error $TorG(S_i)$ of S_i is the sum of a set of 3-D error volumes, $TorG(S_i) = \sum_{j=e_i}^{k_i-1} w \cdot l_{i,j} \cdot h_i$, where w is the (fixed) width of the tongue or groove side of an MLC leaf, $l_{i,j} = |B_{i,j} \oplus B_{i,j+1}|$ is the length of the leaf portion that is actually used for blocking radiation between blocks $B_{i,j}$ and $B_{i,j+1}$, and $h_i = \alpha \cdot r_i$ is the amount of radiation leakage associated with S_i with r_i being the "height" of S_i. Since in the basic 3-D SLS problem, the B-segments are all of a height one (i.e., $r_i = 1, \forall i \in I$), the tongue-or-groove error of S is $TorG(S) = w \cdot \alpha \cdot \sum_{i \in I} \sum_{j=e_i}^{k_i-1} l_{i,j}$. Observe that $\sum_{j=e_i}^{k_i-1} l_{i,j}$ is the sum of the lengths of all non-extreme vertical edges of the B-segment S_i (e.g., see Figure 2(c)).

Thus, we have the following geometric version of the basic 3-D SLS problem with tongue-or-groove error control: Given an IM and an error bound $\mathcal{E}^*$, find a set $S = \{S_i \mid i \in I\}$ of B-segments of height one, such that the value $\mathcal{C}(S) = \sum_{i \in I} \sum_{j=e_i}^{k_i-1} l_{i,j} \leq \mathcal{E}^*$ and the size $|S|$ is minimized. Note that here, $\mathcal{E}^* = \mathcal{E}/(w \cdot \alpha)$ for the error bound $\mathcal{E}$ in the definition of the SLS problems in Section 2.3.

For each B-segment $S_i \in S$, now let $B(S_i) = \{B_{i,j} \mid j = 1, 2, \ldots, n\}$, such that n is the number of columns of the given IM and some intervals $B_{i,j}$ may be empty. Then we have $\mathcal{C}(S) = \sum_{i \in I} \sum_{j=e_i}^{k_i-1} |B_{i,j} \oplus B_{i,j+1}| = \sum_{j=1}^{n-1} \sum_{i \in I} |B_{i,j} \oplus B_{i,j+1}|$. Note that for each $j = 1, 2, \ldots, n-1$, the value $\sum_{i \in I} |B_{i,j} \oplus B_{i,j+1}|$ is actually the tongue-or-groove error for "stitching" the two block-sets $BS(C_j)$ and $BS(C_{j+1})$

for the IM columns C_j and C_{j+1} to form the B-segments as defined by S. Suppose g pairs of blocks are stitched together by S between $BS(C_j)$ and $BS(C_{j+1})$. To minimize the error of S, the error incurred for such a stitching configuration (i.e., $\sum_{i \in I} |B_{i,j} \oplus B_{i,j+1}|$) must be the smallest among all possible stitching configurations with exactly g stitched block pairs between $BS(C_j)$ and $BS(C_{j+1})$ defined by S.

Now we can relate the tongue-or-groove error to the number of B-segments, by associating an error to each stitching configuration between the block-sets for any two consecutive IM columns. Specifically, for any two block-sets $BS(C_j)$ and $BS(C_{j+1})$ for columns C_j and C_{j+1} and each $g = 1, 2, \ldots, |M_j|$, where M_j is a maximum size matching between $BS(C_j)$ and $BS(C_{j+1})$, we stitch together exactly g pairs of blocks with the minimum total error. Note that every stitching configuration of exactly g block pairs between $BS(C_j)$ and $BS(C_{j+1})$ with the minimum error corresponds to a matching of exactly g pairs of intervals with the minimum total weight between the two interval sets of $BS(C_j)$ and $BS(C_{j+1})$. We call such a bipartite matching of intervals (subject to the MLC constraints) an *optimal g-matching*. Hence, an optimal solution for the basic 3-D SLS problem with error control is specified by a list of block-sets (one for each IM column) and an optimal g_j-matching between two such block-sets $BS(C_j)$ and $BS(C_{j+1})$ for any consecutive columns C_j and C_{j+1} (for some value g_j).

To find the sought block-sets and g_j-matchings, we construct the following directed acyclic graph G^*: (1) Generate all distinct block-sets for each IM column, and let every vertex of G^* correspond to exactly one such block-set. (2) For any two vertices of G^* corresponding to two block-sets $BS(C_j)$ and $BS(C_{j+1})$ for two consecutive IM columns C_j and C_{j+1}, compute a minimum weight g-matching for each $g = 1, 2, \ldots, |M_j|$, where M_j is a maximum size matching between $BS(C_j)$ and $BS(C_{j+1})$. For each such g-matching, put a left-to-right edge between the two vertices in G^*, and assign the edge a weight of $(|BS(C_{j+1})| - g)$ and a cost equal to the minimum weight of the g-matching. (3) Add a source vertex s to G^* and connect s to all block-sets for the first IM column; add a sink t and connect all block-sets for the last IM column to t (all edges added here have weight zero and cost zero).

Lemma 1. *An s-to-t path in the graph G^* with a weight k and a cost $\mathcal{C}$ specifies a set of k B-segments of a unit height for the given IM whose total tongue-or-groove error is $\mathcal{C}$.*

Lemma 1 implies that for the basic 3-D SLS problem with a given error bound $\mathcal{E}^*$, we can obtain a minimum B-segment set for the given IM subject to this error bound, by finding a minimum weight s-to-t path in G^* with a total cost $\leq \mathcal{E}^*$. This is called the *constrained shortest path problem* [12].

To compute such a set of B-segments, several issues must be resolved: (1) How to generate all distinct block-sets for each IM mountain slice? (2) How to compute an optimal g-matching between two block-sets? (3) How to find a minimum weight s-to-t path in G^* with a total cost $\leq \mathcal{E}^*$?

To generate all distinct block-sets for an IM column, we use the algorithm in [14], whose running time is linear in terms of the number of output block-sets.

To find a desired k-weight shortest s-to-t path in G^*, we use Lawler's dynamic programming algorithm for constrained shortest paths [12]. The sought path can be easily obtained from the dynamic programming table once it becomes available.

Now we show how to compute the optimal g-matchings. Given two block-sets BS_r and BS_b for any two consecutive IM columns, we construct a bipartite graph $G = (R \cup B, E)$ as follows: Each block in BS_r (resp., BS_b) corresponds to a red (resp., blue) vertex in G, and every *stitchable block pair* corresponds to an edge between the corresponding red and blue vertices. Here, a red block and a blue block are *stitchable* if they satisfy the (machine model specific) MLC constraints. For each edge $e(u, v)$ in G, let their corresponding intervals be $I_u = [l_u, r_u]$ and $I_v = [l_v, r_v]$; then the cost of $e(u, v)$ is assigned as $|l_u - l_v| + |r_u - r_v|$, i.e., the length of the symmetric difference between I_u and I_v.

To compute an optimal g-matching in the bipartite graph G for each $g = 1, 2, \ldots, |M|$, where M is the maximum size matching of G, we transform G into a unit-capacity flow network and formulate the task as a minimum cost flow problem with a flow value g. The $|M|$ optimal g-matchings can be computed efficiently by the successive shortest path algorithm [2], i.e., at the end of the g-th stage of the algorithm ($1 \leq g \leq |M|$), a desired optimal g-matching is produced. Since the single source shortest paths in G at each stage can be computed in $O(m+n \log n)$ time, the total time for computing all the $|M|$ optimal g-matchings ($1 \leq g \leq |M|$) is $O(|M|(m + n \log n))$, where $m = |E|$ and $n = |R \cup B|$.

Theorem 1. *The basic 3-D SLS problem with tongue-or-groove error control is solvable in* $O(\sum_{j=1}^{n-1} \Pi_j \cdot \Pi_{j+1} \cdot \Gamma + \mathcal{K})$ *time, where* n *is the number of columns of the input IM,* Π_j *is the number of block-sets used for column* C_j*,* Γ *is the time for computing all optimal* g*-matchings between two block-sets, and* $\mathcal{K}$ *is the time for computing the minimum cost* k*-weight* s*-to-*t *paths in* G^**.*

3.2 3-D Basic SLS Algorithms with Tongue-and-Groove Error Control

As discussed in Section 2.2, the tongue-and-groove error for an IMRT plan is the intersection $|V_T \cap V_G|$, where V_T (resp., V_G) is the error set (i.e., a set of 3-D axis-parallel boxes) caused by the tongue sides (resp., groove sides) of the MLC leaves. At first sight, due to the nonlinearity, it might appear that the tongue-and-groove error is much harder to handle than the tongue-or-groove error. Interestingly, we are able to show that handling the tongue-and-groove error is in fact equivalent to handling the tongue-or-groove error. The key to our solution is the next lemma which implies that our SLS algorithms for tongue-or-groove error control are also applicable to the case of tongue-and-groove error control.

Lemma 2. *Let* $\mathcal{M}$ *be the input IM and* S *be a B-segment set that builds the IM* $\mathcal{M}$*. Then the difference between the values of the tongue-or-groove error and tongue-and-groove error of* S *is merely a value* $\mathcal{F}(\mathcal{M})$ *that depends only on the input IM* $\mathcal{M}$*.*

4 Implementation and Experiments

To further examine the clinical feasibility of our new 3-D SLS algorithms, we implemented them using C on the Linux systems and conducted extensive experimental studies using real medical data. We performed Monte Carlo simulations [16] on the treatment plans computed by our software, which proved the correctness of our algorithms/software. We also compared our new 3-D SLS algorithms with the current most popular commercial planning system CORVUS (version 5.0), which showed significant improvements. E.g., on a head and neck cancer case consisting of 7 IMs, to eliminate tongue-and-groove error, CORVUS would need 262 B-segments, in contrast to the 142 B-segments computed by our new SLS program.

References

1. R.K. Ahuja and H.W. Hamacher. Minimizing Beam-on Time in Radiation Therapy Treatment Planning Using Network Flows. *submitted to Networks.*
2. R.K. Ahuja, T.L. Magnanti, and J.B. Orlinr. *Network Flows: Theory, Algorithms, and Applications.* Prentice Hall, Inc., 1993.
3. N. Boland, H.W. Hamacher, and F. Lenzen. Minimizing Beam-on Time in Cancer Radiation Treatment Using Multileaf Collimators. Report, Department of Mathematics, University Kaiserslautern, 2002.
4. T.R. Bortfeld, A.L. Boyer, W. Schlegel, D.L. Kahler, and T.L. Waldron. Realization and Verification of Three-Dimensional Conformal Radiotherapy with Modulated Fields. *Int. J. Radiat. Oncol. Biol. Phys.*, 30:899–908, 1994.
5. T.R. Bortfeld, D.L. Kahler, T.J. Waldron, and A.L. Boyer. X-ray Field Compensation with Multileaf Collimators. *Int. J. Radiat. Oncol. Biol. Phys.*, 28:723–730, 1994.
6. A.L. Boyer. Use of MLC for Intensity Modulation. *Med. Phys.*, 21:1007, 1994.
7. D.Z. Chen, X.S. Hu, S. Luan, C. Wang, and X. Wu. Geometric Algorithms for Static Leaf Sequencing Problems in Radiation Therapy. In *Proc. of 19th ACM Symposium on Computational Geometry (SoCG'03)*, pages 88–97, 2003.
8. D.Z. Chen, X.S. Hu, S. Luan, X. Wu, and C.X. Yu. Optimal Terrain Construction Problems and Applications in Intensity-Modulated Radiation Therapy. In *Lecture Notes in Computer Science, Springer-Verlag, Proc. 10th Annual European Symposium on Algorithms (ESA'02)*, volume 2461, pages 270–283, 2002.
9. J. Dai and Y. Zhu. Minimizing the Number of Segments in a Delivery Sequence for Intensity-Modulated Radiation Therapy with Multileaf Collimator. *Med. Phys.*, 28(10):2113–2120, 2001.
10. J. Deng, T. Pawlicki, Y. Chen, J. Li, S.B. Jiang, and C.-M. Ma. The MLC Tongue-and-Groove Effect on IMRT Dose Distribution. *Physics in Medicine and Biology*, 46:1039–1060, 2001.
11. T.J. Jordan and P.C. Williams. The Design and Performance Characteristics of a Multileaf Collimator. *Phys. Med. Biol.*, 39:231–251, 1994.
12. E. Lawler. *Combinatorial Optimization: Networks and Matroids.* Holt, Rinehart and Winston, 1976.
13. F. Lenzen. An Integer Programming Approach to the Multileaf Collimator Problem. Master's thesis, University of Kaiserslautern, June 2000.

14. S. Luan, D.Z. Chen, L. Zhang, X. Wu, and C.X. Yu. An Optimal Algorithm for Computing Configuration Options of One-dimensional Intensity Modulated Beams. *Phys. Med. Biol.*, 48(15):2321–2338, 2003.
15. S. Luan, C. Wang, S.A. Naqvi, D.Z. Chen, X.S. Hu, C.L. Lee, and C.X. Yu. A New MLC Segmentation Algorithm/Software for Step and Shoot IMRT. *Med. Phys.*, 31(4):695–707, 2004.
16. S.A. Naqvi, M. Earl, and D. Shepard. Convolution/Superposition Using the Monte Carlo Method. *Phys. Med. Biol.*, 48(14):2101–2121, 2003.
17. R.A.C. Siochi. Minimizing Static Intensity Modulation Delivery Time Using an Intensity Solid Paradigm. *Int J. Radiat. Oncol. Biol. Phys.*, 43(3):671–680, 1999.
18. S. Webb. *The Physics of Three-Dimensional Radiation Therapy.* Bristol, Institute of Physics Publishing, 1993.
19. S. Webb. *The Physics of Conformal Radiotherapy — Advances in Technology.* Bristol, Institute of Physics Publishing, 1997.
20. P. Xia and L.J. Verhey. MLC Leaf Sequencing Algorithm for Intensity Modulated Beams with Multiple Static Segments. *Med. Phys.*, 25:1424–1434, 1998.
21. C.X. Yu. Design Considerations of the Sides of the Multileaf Collimator. *Phys. Med. Biol.*, 43(5):1335–1342, 1998.

Author Index

Lecture Notes in Computer Science

For information about Vols. 1–3739

please contact your bookseller or Springer

Vol. 3789: A. Gelbukh, Á. de Albornoz, H. Terashima-Marín (Eds.), MICAI 2005: Advances in Artificial Intelligence. XXVI, 1198 pages. 2005. (Subseries LNAI).

Vol. 3788: B. Roy (Ed.), Advances in Cryptology - ASIACRYPT 2005. XIV, 703 pages. 2005.

Vol. 3785: K.-K. Lau, R. Banach (Eds.), Formal Methods and Software Engineering. XIV, 496 pages. 2005.

Vol. 3784: J. Tao, T. Tan, R.W. Picard (Eds.), Affective Computing and Intelligent Interaction. XIX, 1008 pages. 2005.

Vol. 3783: S. Qing, W. Mao, J. Lopez, G. Wang (Eds.), Information and Communications Security. XIV, 492 pages. 2005.

Vol. 3781: S.Z. Li, Z. Sun, T. Tan, S. Pankanti, G. Chollet, D. Zhang (Eds.), Advances in Biometric Person Authentication. XI, 250 pages. 2005.

Vol. 3780: K. Yi (Ed.), Programming Languages and Systems. XI, 435 pages. 2005.

Vol. 3779: H. Jin, D. Reed, W. Jiang (Eds.), Network and Parallel Computing. XV, 513 pages. 2005.

Vol. 3778: C. Atkinson, C. Bunse, H.-G. Gross, C. Peper (Eds.), Component-Based Software Development for Embedded Systems. VIII, 345 pages. 2005.

Vol. 3777: O.B. Lupanov, O.M. Kasim-Zade, A.V. Chaskin, K. Steinhöfel (Eds.), Stochastic Algorithms: Foundations and Applications. VIII, 239 pages. 2005.

Vol. 3776: S.K. Pal, S. Bandyopadhyay, S. Biswas (Eds.), Pattern Recognition and Machine Intelligence. XXIV, 808 pages. 2005.

Vol. 3775: J. Schönwälder, J. Serrat (Eds.), Ambient Networks. XIII, 281 pages. 2005.

Vol. 3774: G. Bierman, C. Koch (Eds.), Database Programming Languages. X, 295 pages. 2005.

Vol. 3773: A. Sanfeliu, M.L. Cortés (Eds.), Progress in Pattern Recognition, Image Analysis and Applications. XX, 1094 pages. 2005.

Vol. 3772: M. Consens, G. Navarro (Eds.), String Processing and Information Retrieval. XIV, 406 pages. 2005.

Vol. 3771: J.M.T. Romijn, G.P. Smith, J. van de Pol (Eds.), Integrated Formal Methods. XI, 407 pages. 2005.

Vol. 3770: J. Akoka, S.W. Liddle, I.-Y. Song, M. Bertolotto, I. Comyn-Wattiau, W.-J. van den Heuvel, M. Kolp, J. Trujillo, C. Kop, H.C. Mayr (Eds.), Perspectives in Conceptual Modeling. XXII, 476 pages. 2005.

Vol. 3769: D.A. Bader, M. Parashar, V. Sridhar, V.K. Prasanna (Eds.), High Performance Computing – HiPC 2005. XXVIII, 550 pages. 2005.

Vol. 3768: Y.-S. Ho, H.J. Kim (Eds.), Advances in Multimedia Information Processing - PCM 2005, Part II. XXVIII, 1088 pages. 2005.

Vol. 3767: Y.-S. Ho, H.J. Kim (Eds.), Advances in Multimedia Information Processing - PCM 2005, Part I. XXVIII, 1022 pages. 2005.

Vol. 3766: N. Sebe, M.S. Lew, T.S. Huang (Eds.), Computer Vision in Human-Computer Interaction. X, 231 pages. 2005.

Vol. 3765: Y. Liu, T. Jiang, C. Zhang (Eds.), Computer Vision for Biomedical Image Applications. X, 563 pages. 2005.

Vol. 3764: S. Tixeuil, T. Herman (Eds.), Self-Stabilizing Systems. VIII, 229 pages. 2005.

Vol. 3762: R. Meersman, Z. Tari, P. Herrero (Eds.), On the Move to Meaningful Internet Systems 2005: OTM 2005 Workshops. XXXI, 1228 pages. 2005.

Vol. 3761: R. Meersman, Z. Tari (Eds.), On the Move to Meaningful Internet Systems 2005: CoopIS, DOA, and ODBASE, Part II. XXVII, 653 pages. 2005.

Vol. 3760: R. Meersman, Z. Tari (Eds.), On the Move to Meaningful Internet Systems 2005: CoopIS, DOA, and ODBASE, Part I. XXVII, 921 pages. 2005.

Vol. 3759: G. Chen, Y. Pan, M. Guo, J. Lu (Eds.), Parallel and Distributed Processing and Applications - ISPA 2005 Workshops. XIII, 669 pages. 2005.

Vol. 3758: Y. Pan, D.-x. Chen, M. Guo, J. Cao, J.J. Dongarra (Eds.), Parallel and Distributed Processing and Applications. XXIII, 1162 pages. 2005.

Vol. 3757: A. Rangarajan, B. Vemuri, A.L. Yuille (Eds.), Energy Minimization Methods in Computer Vision and Pattern Recognition. XII, 666 pages. 2005.

Vol. 3756: J. Cao, W. Nejdl, M. Xu (Eds.), Advanced Parallel Processing Technologies. XIV, 526 pages. 2005.

Vol. 3754: J. Dalmau Royo, G. Hasegawa (Eds.), Management of Multimedia Networks and Services. XII, 384 pages. 2005.

Vol. 3753: O.F. Olsen, L.M.J. Florack, A. Kuijper (Eds.), Deep Structure, Singularities, and Computer Vision. X, 259 pages. 2005.

Vol. 3752: N. Paragios, O. Faugeras, T. Chan, C. Schnörr (Eds.), Variational, Geometric, and Level Set Methods in Computer Vision. XI, 369 pages. 2005.

Vol. 3751: T. Magedanz, E.R. M. Madeira, P. Dini (Eds.), Operations and Management in IP-Based Networks. X, 213 pages. 2005.

Vol. 3750: J.S. Duncan, G. Gerig (Eds.), Medical Image Computing and Computer-Assisted Intervention – MICCAI 2005, Part II. XL, 1018 pages. 2005.

Vol. 3749: J.S. Duncan, G. Gerig (Eds.), Medical Image Computing and Computer-Assisted Intervention – MICCAI 2005, Part I. XXXIX, 942 pages. 2005.

Vol. 3748: A. Hartman, D. Kreische (Eds.), Model Driven Architecture – Foundations and Applications. IX, 349 pages. 2005.

Vol. 3747: C.A. Maziero, J.G. Silva, A.M.S. Andrade, F.M.d. Assis Silva (Eds.), Dependable Computing. XV, 267 pages. 2005.

Vol. 3746: P. Bozanis, E.N. Houstis (Eds.), Advances in Informatics. XIX, 879 pages. 2005.

Vol. 3745: J.L. Oliveira, V. Maojo, F. Martín-Sánchez, A.S. Pereira (Eds.), Biological and Medical Data Analysis. XII, 422 pages. 2005. (Subseries LNBI).

Vol. 3744: T. Magedanz, A. Karmouch, S. Pierre, I. Venieris (Eds.), Mobility Aware Technologies and Applications. XIV, 418 pages. 2005.

Vol. 3742: J. Akiyama, M. Kano, X. Tan (Eds.), Discrete and Computational Geometry. VIII, 213 pages. 2005.

Vol. 3740: T. Srikanthan, J. Xue, C.-H. Chang (Eds.), Advances in Computer Systems Architecture. XVII, 833 pages. 2005.

Knaur

meine Familie & ich

Das Backbuch

Knaur

meine Familie & ich

Das Backbuch

400 Rezepte • 400 Bilder

Aus dem Backstudio von „meine Familie & ich“

Inhalt

Liebe Leserinnen und Leser!

Was haben Kuchen und Musik gemeinsam? Zunächst dachte ich: nichts! Außer, dass man natürlich zur gleichen Zeit sowohl ein Stück Kuchen als auch ein Musikstück genießen kann. Zum Gaumenschmaus der Ohrenschmaus! Doch dann fiel mir auf, dass, genauso wie mich manche Melodie durch das Leben begleitet, auch immer wieder bestimmtes Backwerk eine Rolle spielt. So wie bei einigen Songs Erinnerungen an das erste Verliebtsein auftauchen, ein trauriger Moment sich ins Bewusstsein schleicht, ganze Lebensphasen lebendig werden, so ist beispielsweise der Duft von Butterkuchen für mich unweigerlich mit Hamburg verbunden, meiner Heimatstadt, in der ich schon lange nicht mehr lebe. Bei Rosinenkuchen werde ich gleich wieder zum Kind, Kuchen mit Schokoladenplätzchen versetzt mich augenblicklich an die sonntägliche Kaffeetafel bei meiner Großmutter.

Kuchen, Torten, Plätzchen und vieles mehr finden Sie in diesem Buch für jede erdenkliche Lebenssituation – und jedes Rezept ist eine Erinnerung wert! Von der großen Hochzeitstorte bis zum Küchlein zum Verschenken. Vom wunderbaren Alltagskuchen ohne großen Aufwand bis zu eindrucksvollen Liebhaberstücken. Von internationalen Spezialitäten bis zu lieb gewonnenen Bekannten. Mit uns können Sie sogar backen ohne zu backen. Auf „kaltem" Wege entstehen Köstlichkeiten wie eine Caipirinha-Torte oder eine Eistorte mit Pflaumen.

Wenn Sie sich für etwas ganz Besonderes begeistern können, ist vielleicht die Tiramisu-Torte das Richtige für Sie. Oder die Panna-cotta-Torte oder, oder, oder ... In diesem Buch gibt es über 400 Rezepte zu entdecken, zu probieren und zu genießen. Alle basieren auf dem Know-how der Zeitschrift „meine Familie & ich".

Birgitt Micha, Chefredakteurin „meine Familie & ich"

Als langjährige Chefredakteurin von Europas größtem Food-Magazin kann ich Ihnen versichern, dass sämtliches Gebäck – von der Torte bis zum Brot – in unserem eigenen Koch- und Backstudio getestet worden ist.

Damit Sie bei jedem Rezept genau wissen, woran Sie sind, verwenden wir Symbole. So können Sie sich ganz einfach orientieren, ob es sich um aufwändiges oder raffiniertes Backwerk handelt, ob es besonders preiswert oder schnell gebacken ist. Das bedeutet, denke ich, eine große Hilfe. Denn wir alle, die wir gerne backen, wissen ja: Kochen ist oft die Pflicht, Backen aber ist die Kür. Das kleine Extra, das das Leben im besten Sinne versüßt und Menschen, Ereignisse und schöne Stunden in Erinnerung bleiben lässt.

Viel Spaß wünscht Ihnen

Ihre

Warenkunde

Getreide, Mehlsorten und Backtreibmittel

Getreide

Botanisch betrachtet gehören alle Getreidearten zur Familie der Süßgräser. Die Früchte dieser Pflanzen nennt man Körner und sie spielen seit Anbeginn der Menschheit eine wichtige Rolle in unserer Ernährung. Aus allen alten Kulturen dieser Welt sind archäologische Funde bekannt, die die Verarbeitung von Getreide belegen. Getreide ist heute noch das wichtigste Erzeugnis der Weltagrarwirtschaft. Es wird überall zur Herstellung von Brot, Gebäck und Teigwaren verarbeitet.

Die Arten

Weizen zählt seit der Frühzeit zu den bedeutendsten Getreidearten. Durch Kultivierung verschiedener Wildweizen sind Sorten entstanden, deren unterschiedliche Eigenschaften vielfältig genutzt werden können. Von allen Getreidearten hat Weizen die besten Backeigenschaften. Sie beruhen auf dem ausgewogenen Verhältnis von Stärke, Eiweiß und Kleber, einem Eiweißstoff, der den Teig bindet. Je nach Ausmahlungsgrad schmeckt es mild bis nussig. Für die Herstellung von Nudeln wird **Hartweizen,** der sogenannte Durum-Weizen, verwendet. Er liefert ein kleberreiches Mehl, aus dem man einen stabilen gut formbaren Teig kneten kann.

Dinkel ist ein enger Verwandter des Weizens und war in unseren Breiten das beliebteste Getreide. Sein kleberreiches Mehl wird heute noch vor allem in Schwaben für Mehlspeisen und Gebäcke verwendet. Das Dinkelkorn fällt beim Dreschen nicht nackt aus der Ähre wie der Weizen, sondern ist von Spelzen umschlossen. Im Lebensmittelgeschäft ist er vorwiegend als Mehlltype 630 zu bekommen. Grünkern ist der unreif geerntete – und deshalb grüne – Dinkel, der ein herzhaft-nussiges Aroma hat.

Roggen ist nach dem Weizen das zweitwichtigste Getreide in Europa. Er hat zwar einen geringeren Eiweißgehalt, dafür aber wesentlich mehr Mineralien und Vitalstoffe als Weizen und er ist besonders würzig. Roggenmehl wird bei uns vor allem zum Brotbacken genutzt, jedoch muss es durch Zugabe von Säure backfähig gemacht werden. Deshalb wird dem Teig immer Sauerteig (siehe Seite 11) zugesetzt.

Buchweizen ist keine Getreidesorte, sondern ein Knöterichgewächs und ein enger Verwandter von Sauerampfer und Rhabarber. Er bringt rotbraune, dreikantige, 4-6 mm große Kerne hervor, die wie winzige Bucheckern aussehen. Das daraus gewonnene Mehl wird z. B. in Südtirol gerne zu Fladen und Pizzoccheri verwendet. In den USA ist es zur Herstellung von Pfannkuchen besonders beliebt. Auch die kleinen russischen Pfannkuchen, die Blinis, sind aus Buchweizenmehl gebacken.

Gerste ist für das Backen von untergeordneter Bedeutung. Gerstenmehl wird in einigen Mittelmeerländern zu Fladenbrot verarbeitet. Auch für Nudeln und ähnliche Erzeugnisse wird Gerstenmehl manchmal mit Hartweizen vermischt verarbeitet, denn es hat keinen Kleber. Weltweit spielt Braugerste als Grundzutat für Bier eine große Rolle. Gerste wird vorwiegend zu Malz, Grütze und Graupen verarbeitet.

Hafer ist zwar kein wichtiges Getreide für Gebäck, im Handel gibt es jedoch viele Vollkornbrote und -produkte, in denen Hafer enthalten ist. Weil er keinen Kleber hat, lässt er sich nur mit Weizenmehl vermischt backen. Hafer enthält jedoch viel Fett, die V tamine B1 und E und die wichtigen Mineralstoffe Eisen und Kalium. Wegen seiner cholesterinsenkenden Eigenschaften wird Hafermehl immer häufiger verwendet.

Mais ist das Grundnahrungsmittel für Millionen von Menschen, vor allem in seiner ursprünglichen Heimat Lateinamerika. Heute wird er in vielen Ländern, vor allem in Süd- und Südosteuropa, angebaut. Die vielen Maisvarianten werden je nach Stärkegehalt der Körner unterschieden. Aus Weich- oder Stärkemais, Zucker- und Wachsmais werden Maismehl, Maisstärke und Maisgrieß gewonnen. In Italien wird aus letzterem die berühmte Polenta hergestellt. Ähnliche Verwendung findet er auch in anderen Mittelmeerländern. Und in den USA gehört *cornbread* – eine Art Maisbrot – zu jedem Thanksgiving-Dinner.

Mehltypen und Backtreibmittel

Mehltypen und Mahlstufen

Vollkornmehl ist ernährungsphysiologisch gesehen die wertvollste Mehlsorte, weil sich vor allem in den Randschichten Vitamine, Mineral- und Ballaststoffe befinden. Zwar lässt sich jedes Getreide zu Mehl vermahlen, aber nicht jedes enthält alle Bestandteile des vollen Korns. Wie viele es enthält, sagt der Ausmahlungsgrad bzw. die Mehl-Type. Je höher die Zahl, desto mehr Nährstoffe sind im Mehl. Neben dem Ausmahlungsgrad gibt es die Feinheitsgrade, in denen das Korn gemahlen wird:

Schrot ist die gröbste Mahlstufe.

Grieß ist das körnig gemahlene Produkt aus dem Mehlkern. Er hat einen hohen Eiweißgehalt und gute Quellfähigkeit.

Dunst ist feinkörniger als Grieß, man verwendet ihn gerne für Hefegebäck.

Griffiges oder **doppelgriffiges Mehl** ist zu noch feineren Körnchen vermahlen.

Mehl ist die feinste Mahlstufe.

Spezialmehle werden durch bestimmte Verfahren oder mit bestimmten Zusätzen hergestellt:

Instantmehl ist Weizenmehl, das nicht mit Wasser verklumpt, weil sich die Mehlkörnchen schon vor dem Quellen gleichmäßig in der Flüssigkeit verteilt haben.

Fertigmehle sind Back-, Brot- und Kuchenmischungen, die das Backen im Haushalt erleichtern sollen und daher bereits Backzutaten wie Zucker, Gewürze, Aroma, Treibmittel, Ei- und Milchpulver, enthalten.

Nicht alle Mehle eignen sich zum Backen gleich gut. Das ist abhängig vom Stärke-, Kleber- und Eiweißan-

Dinkelmehl: eiweißreiches und besonders aromatisches Mehl.

Weizenmehl Typ 550: höher ausgemahlenes Mehl, an der dunkleren Farbe zu erkennen.

Weizenmehl Typ 405: gängigste Mehlart zur Herstellung von Gebäck.

Roggenmehl Type 1370: ein nahezu vollwertiges Mehl.

Hafermehl wird vor allem zur Herstellung von Säuglingsnahrung verwendet.

Maismehl: mit kleberhaltigem Mehl gemischt, eignet es sich auch für Gebäck.

teil des jeweiligen Getreides. Für das Backen sind die Eigenschaften vor allem eines dieser Eiweißstoffe wichtig, nämlich des Klebers, auch Gluten genannt. Der Kleber ist dafür verantwortlich, dass sich im Teig eine Art Gerüst ausbildet, an dem sich die durch Treibmittel erzeugten Gasbläschen festsetzen und so eine lockere Krume ergeben.

Die besten Backeigenschaften besitzt der Kleber des Weizenmehls; er braucht jedoch etwas Zeit, bevor er richtig arbeiten kann. Daher müssen Mürbteige vor der Verarbeitung ruhen und Hefeteige eine Weile kräftig durchgeknetet werden. Pur kann außer Weizen- auch Roggenmehl verbacken werden, dann allerdings mit Sauerteig als Treibmittel. Für alle übrigen Getreide empfiehlt sich eine Beimischung von Weizenmehl, wenn der Teig schön aufgehen soll.

Weizengrieß aus Hartweizen ist ideal für Teigwaren aller Art.

Weizen-Vollkornmehl ist gut für Vollkorn-Backwaren aller Art und Gebäck.

Treibmittel

Hefe und Sauerteig sind vor allem für das Brotbacken unverzichtbar. Beide Treibmittel sind seit dem Altertum bekannt. Hefe (siehe auch Seite 125) wird frisch oder getrocknet angeboten. Sauerteig entsteht durch in der Luft vorhandene Essig- und Milchsäurebakterien. Es gibt ihn flüssig und getrocknet, oder man setzt ihn selbst an.

Grundrezept Sauerteig: 100 g Roggenmehl Type 997 mit 100 ml Wasser (Temperatur 40-45 Grad) verrühren. Mit Folie und einem Tuch bedeckt 1-2 Tage an einem warmen Ort gehen lassen. Dann den Vorgang wiederholen: Den Teig mit 100 g Mehl und 100 ml Wasser vermischen und wieder zugedeckt 24 Stunden gehen lassen. Ein drittes Mal Mehl und Wasser zufügen, verrühren, doch diesmal die Mengen verdoppeln. Teig einen weiteren Tag gehen lassen. Jetzt ist er gebrauchsfertig.

Backhefe gibt es frisch zu 42 g-Würfeln gepresst und getrocknet in 10 g-Tütchen im Lebensmittelhandel zu kaufen.

Eier: Eiweiss & Eigelb

Eier

Eier sind von Natur aus ideal verpackt und haben einen Nährwert, der kaum zu schlagen ist: Eiweiß, Fett, Lecithin, Vitamine, Mineralstoffe und Spurenelemente. Das Ei ist für das Backen unverzichtbar. Ob zur Lockerung, Stabilisierung oder Bindung – kaum ein Rezept dieses Buches kommt ohne Eier aus. Was es außerdem zu einer so wichtigen Zutat macht, sind die leichte Trennbarkeit von Eigelb und Eiweiß sowie deren unterschiedliche Eigenschaften, die man sich beim Backen gezielt zunutze machen kann.

Eiweiß oder Eiklar verfügt über ein hervorragendes Schaumbildungsvermögen, es erhöht den Lockerungsgrad von Massen und Teigen und vermag während des Backens die entstehenden Luftbläschen gut festzuhalten. Und es enthält Klebstoffe, die Teigschichten verbinden, und Glanzstoffe, die zur Verschönerung des Gebäcks beitragen.

Eigelb bindet die Zutaten sehr gut, besonders bei fett- und zuckerreichen Massen oder Cremes. Zudem macht Eigelb Gebäck mürbe und sorgt für einen guten Geschmack des Gebäcks. Der schönen Farbe wegen bestreicht man Blätterteig häufig auch mit verquirltem Eigelb, bevor er in den Ofen kommt.
Vom Gesamtgewichts eines Eis machen das Eiklar etwa 57 % aus, das Eigelb 33 % und die Schale 10 %. Sein Kaloriengehalt liegt zwischen 75 und 90 kcal (315 bis 390 kJ).

Lagerung

Die Luftkammer, die sich am runden Ende des Eis befindet, ist sein schwacher Punkt. Hier können Keime und Gerüche am ehesten durch die poröse Schale dringen. Trotzdem sollte man Eier so aufbewahren, dass die Luftkammer nach oben zeigt. Liegt sie unten, dringt die Luft langsam nach oben und die innere Haut kann sich von der Schale lösen.

Frisch-Eier sollten immer im geschlossenen Eierfach des Kühlschranks und nicht in der Vorratskammer gelagert werden. Im Gefriergerät bleiben Eier ohne Schale als Eimasse oder getrennt in Eigelb und Eiweiß etwa 4 Monate frisch. Werden für Cremes oder zum Backen von Baisers sehr frische Eier benötigt, sollte man einen der beiden unten gezeigten Tests durchführen, denn von außen sieht man einem Ei leider nicht an, wie frisch es wirklich ist.

Frische Eier verursachen beim Schütteln kein Geräusch. Generell gilt: Je älter das Ei, um so mehr Flüssigkeit verdunstet durch die poröse Schale. Dadurch wird die Luftkammer größer und das Ei leichter. Nach 2 bis 3 Wochen steht es beim Schwimmtest nahezu auf der Spitze, später schwimmt es ganz oben (siehe unten). Beim Aufschlagtest (siehe unten rechts) wird der Dotter mit zunehmendem Alter flacher, das Eiklar verliert an Spannung und fließt auseinander.

Eier gehören in den Kühlschrank

Sie können als Masse oder getrennt in Eigelb und Eiklar eingefroren werden. Eigelb, mit etwas Öl bedeckt, kann etwa 2 Wochen im Kühlschrank lagern.

Sicherheit & Qualität

Wer sicher sein will, dass ein Ei höchsten Qualitäts-Anforderungen entspricht, achtet auf die Herkunftssicherung und das CMA-Gütezeichen, das nur nach strengen Kontrollen (Schlupf der Küken, Aufzucht der Legehennen und Eierablage) vergeben wird. Eier werden in vier Gewichtsklassen eingeteilt:

- **XL** ist ein besonders großes Ei, das mindestens 73 Gramm wiegt;
- **L** sind große Eier, die zwischen 63 und 72 Gramm wiegen;
- **M** sind mittelgroße Eier mit 53 bis 62 Gramm;
- **S** sind kleine Eier, mit weniger als 53 Gramm.

Frischetest: Ein frisches Ei sinkt im Wasserglas zu Boden. Das Ei in der Mitte ist ca. 3 Wochen, das linke ca. 5 Wochen alt.

So sieht ein frisches, 1-2 Tage altes Ei aus, wenn es aufgeschlagen wird.

Zucker

Man kann es sich kaum vorstellen: Über Jahrhunderte hinweg wurde Zucker wie Gewürze nur in Apotheken verkauft und war entsprechend teuer. Schließlich gewann man ihn aus Zuckerrohr und das wuchs nur in exotischen Ländern.

Ein Berliner Apotheker fand 1747 heraus, dass man aus der Runkelrübe einen Stoff gewinnen konnte, der mit dem Zucker aus Zuckerrohr völlig identisch ist. Damit änderte sich alles: Das Zuckermonopol der Kolonialmächte war gebrochen und Zucker wurde ein Grundnahrungsmittel, das aus unserer Ernährung nicht mehr wegzudenken ist.

Zucker wird in den verschiedensten Formen sowie in den Farben weiß und braun angeboten. Brauner Zucker, auch Rohzucker oder Farinzucker genannt, ist nicht vollständig von der bei der Herstellung entstehenden Melasse gereinigt. Im Vergleich zum völlig gereinigten, weißen Zucker hat er einen etwas kräftigeren Geschmack. Zum Süßen von gewürztem dunklem Gebäck wird er wegen seines Karamell-Aromas und seiner Bräunungsstoffe gerne verwendet.

Für Feingebäck nimmt man grundsätzlich weißen Zucker. Brauner und weißer Zucker kommen in unterschiedlicher Körnung in den Handel. Zum Backen eignet sich am besten eine feinkörnige Sorte, denn je kleiner die Kristalle sind, desto leichter lösen sie sich auf und verbinden sich mit dem Teig. Die größten Zuckerkristalle bildet der braune Krustenkandis, die kleinsten der fein gemahlene Puderzucker. Er ist immer weiß und wird daher auch gerne zum Bestäuben von Gebäck verwendet.

Doch ganz gleich, welchen Zucker man bevorzugt, alle Sorten müssen kühl, vor allem trocken aufbewahrt werden, denn Feuchtigkeit lässt Zucker verklumpen.

Zuckersorten

1 **Brauner Kandis** – zum Backen und für Desserts
2 **Raffinade** – Zucker von höchster Reinheit
3 **Grümmelkandis** – zum Süßen von Teigen und Obst
4 **Brauner Würfelzucker** – für Kaffee oder Espresso
5 **Weißzucker** – es ist die einfachste Zuckersorte
6 **Weißer Kandis** – für Tee
7 **Brauner Zucker** (Kandisfarin) – beliebter Backzucker
8 **Würfelzucker** – für alle heißen Getränke
9 **Einmachzucker** – zum Konservieren von Früchten
10 **Kluntjekandis** – in Ostfriesland klassisch für Tee
11 **Hagelzucker** – Dekorierzucker für Gebäck
12 **Krustenkandis** – entsteht in sog. Kandiswannen
13 **Gelierzucker** – süßt und geliert zugleich
14 **Zuckerhut** – für die berühmte Feuerzangenbowle
15 **Glückswürfelzucker** – wird feucht in Form gepresst
16 **Puderzucker** – für Eischnee und Glasuren

Ahornsirup

wird aus dem Saft des in Nordamerika beheimateten Zuckerahorns gewonnen. Der Baum wird angezapft und die Flüssigkeit eingekocht. Ahornsirup hat ein ganz besonderes Aroma.

Zuckerrüben-Sirup

erhält man durch das langsame Einkochen von Zuckerrübenschnitzeln. Dabei erhält er seine typische dunkle Farbe und seinen Karamellgeschmack. Man kann ihn wie braunen Zucker zum Backen verwenden.

Honig

wird aus Bienenwaben „geschleudert", ausgetropft oder gepresst. Seine Süßkraft ist höher als die des Zuckers. Man bekommt ihn in vielen Sorten, je nach der Pflanzenart, von der er gesammelt wurde. Grob wird zwischen Blüten- und Waldhonig unterschieden. Letzterer ist weniger süß, dunkler und aromatischer.

Süßstoffe

Wichtig für Diabetiker und für alle, die abnehmen wollen. Man kann auch mit Süßstoffen backen, allerdings müssen sie hitzebeständig sein. Vier Süßstoffe, die allesamt so gut wie keine Kalorien haben, befinden sich bei uns im Handel: **Saccharin** ist sehr stabil und daher zum Backen und Kochen geeignet. Es ist etwa 450-mal süßer als Zucker. **Cyclamat** ist nur etwa 35-mal süßer und wird daher häufig mit Saccharin kombiniert. Es ist lange lagerfähig und ausgesprochen hitzebeständig. Auch mit **Acesulfam** kann man backen, es ist ebenso wie **Aspartam** etwa 200-mal so süß wie Zucker. Aspartam ist allerdings nur bis 180 Grad erhitzbar. Deshalb sollte man es lieber für Cremes verwenden. Achten Sie beim Dosieren auf die Packungsaufschrift. Zuckeraustauschstoffe wie **Fruchtzucker, Isomalt, Sorbit, Xylit** u. ä. können Sie wie Zucker verwenden, sie haben aber ebenso viele oder fast ebenso viele Kalorien wie Zucker.

Zucker und Fruchtzucker: beide sind gut zum Backen geeignet

Butter und Fette

Das wichtigste Backfett ist **Butter.** Sie wird aus der Sahne der Milch hergestellt. Sie ist als Sauerrahmbutter und Süßrahmbutter erhältlich. Beide eignen sich gleich gut zum Backen. Auch gesalzene Butter findet man im Handel, die man besonders gut in Mürbeteig oder Hefeteig einsetzen kann, dann lässt man allerdings die im Rezept angegebene Salzmenge weg.

Butterschmalz ist reines Butterfett, das von allen Wasser- und Eiweißbestandteilen befreit wurde. Dadurch lässt es sich höher erhitzen als Butter und eignet sich vor allem zum Braten. Andere tierische Fette wie Schweineschmalz oder Rindertalg spielen ihres ausgeprägten Eigengeschmacks wegen beim Backen heute keine Rolle mehr; jedoch für manches traditionelle oder regionale Rezepte wird noch Schmalz verwendet.

Heute werden zum Backen häufig **Pflanzenfette** eingesetzt, weil sie kein Cholesterin enthalten. Pflanzenöle eignen sich vor allem für Hefeteige, in Mürbe- oder Rührteigen kann man ohne weiteres die Butter durch die gleiche Menge Pflanzenmargarine ersetzen, sollte sich jedoch darüber im Klaren sein, dass der Teig dann ein anderes Aroma bekommt.

Zum Einfetten von Formen oder Blechen eignen sich Butter und Pflanzenfette gleich gut. Darauf verzichten kann man nur bei Teigen, deren Fettanteil sehr hoch ist, beispielsweise bei vielen Mürbeteigen.

1 **Butter –** ist das traditionelle Fett zum Backen
2 **Kokosfett –** für Schokoladenglasuren
3 **Pflanzenfett –** ideal zum Frittieren

Milchprodukte

Ob Sahne, Schichtkäse und Quark oder Joghurt, Frischkäse und gereifter Käse – alle Milchprodukte lassen sich beim Backen verwenden. Und wenn nicht für Süßes, dann für die pikante Bäckerei wie Käsestangen aus Blätterteig, Quiche und Pizza. Die Palette der Milchprodukte im Supermarkt bietet eine riesige Auswahl und wird darüber hinaus von Spezialitäten aus Frankreich wie **Crème fraîche** und **Crème double,** oder aus Italien wie **Ricotta, Mascarpone** und **Mozzarella** bereichert.

Sahne findet steif geschlagen oft pur Verwendung und lockert zarte Cremes für Füllungen auf. **Joghurt und Quark** sind ebenfalls beliebt für Füllungen, oft in Kombination mit Früchten. **Schichtkäse,** mit dem Quark eng verwandt, wird bei seiner Herstellung schichtweise in Formen gefüllt. Er schmeckt säuerlich-aromatisch, ist geschmeidiger und trockener als Quark. Auch wenn er etwas teurer ist, bleibt er bei Käsekuchen die beste Wahl.

Was den Käse für Pikantes betrifft, so lässt sich hier zwischen frischem und gereiftem Käse unterscheiden. Frischkäse, etwa der leicht gesalzene, sahnige Doppelrahmfrischkäse, oder auch Mozzarella bringen relativ wenig Eigengeschmack mit und eignen sich daher gut zum Kombinieren mit Pikantem. Gereifte Hartkäse dagegen, vom Gouda über Emmentaler bis hin zum Parmesan, besitzen jede Menge Würze.

Milch & Co.

1 Sahne
2 Crème fraîche
3 Dickmilch
4 Kondensmilch
5 Joghurt
6 Sahnequark

Edle Backzutaten

Nüsse

Walnuss Sie gehört in Asien und Europa seit Jahrtausenden zu den Grundnahrungsmitteln. Sie hat einen Fettgehalt von etwa 62 Prozent und ist daher wie alle Nüsse, weil sie leicht ranzig wird, nicht lange lagerfähig.

Pecannuss Sie stammt aus Nordamerika und und ist eng verwandt mit der Walnuss. Sie schmeckt angenehm lieblich und findet bei uns bisher viel zu wenig Beachtung.

Paranuss Sie kommt aus Südamerika und hat eine besonders harte Schale. Man kann sie als Mandelersatz verwenden. Ihr Fettgehalt beträgt etwa 66 Prozent.

Kokosnuss Sie wächst in den Tropen und dient vor allem als Rohstofflieferant für Öl und Fett. Zum Backen werden Kokosraspel und für ausländische Spezialitäten auch Kokosmilch verwendet.

Macademia-Nuss Mit 76 Prozent hat die australische Nuss den höchsten Fettgehalt. Ihre Kerne sind besonders aromatisch und werden geröstet, gesalzen oder ungesalzen angeboten.

Esskastanie oder Marone gibt es bei uns frisch und geröstet in Dosen oder als Püree.

Cashewnuss Sie enthält etwa 45 Prozent Fett und viel Vitamin E. Sie schmeckt süßlich-aromatisch und ist bei uns geschält, geröstet, gesalzen und ungesalzen zu haben.

Haselnuss Ihre ursprüngliche Heimat ist zwar Kleinasien, heute ist sie aber in allen gemäßigten Klimazonen zu Hause. Ihr Fettanteil beträgt etwa 50 Prozent. Nüsse in der Schale halten länger frisch als geschälte Kerne.

Pistatzie Ihre grasgrüne Farbe zeigt ihren Frischegrad an. Alte Nüsse werden weiß. Wegen ihres hohen Fettgehaltes (über 60 Prozent) ist sie nicht lange lagerfähig.

Erdnuss Sie ist die bei uns bekannteste exotische Nuss. Als Backzutat wird sie wie andere Nüsse verwendet. Ihr Fettgehalt beträgt „nur" etwa 30 Prozent.

Pinienkern Die süßlich schmeckenden, schlanken Samenkerne einer mediterranen Kiefernart haben etwa 60 Prozent Fett. Man kann sie wie Mandeln verwenden.

Trockenfrüchte

Sultaninen sind getrocknete Weinbeeren mit heller Schale ohne Kerne.

Korinthen kommen aus Griechenland und stammen von einer kleinbeerigen dunklen Traube. Sie schmecken intensiver als Rosinen und Sultaninen.

Rosinen kommen aus dem gesamten Mittelmeerraum mit Kernen und Stiel in den Handel.

Getrocknete Aprikosen sind saftig, schmecken süß-säuerlich und haben ein kräftiges Fruchtaroma. 6 bis 8 Kilo frische Aprikosen ergeben 1 kg Trockenfrüchte.

Getrocknete Feigen werden nach dem Trocknen mit Dampf behandelt und haben im Gegensatz zur Naturalfeige keinen weißen Überzug aus Zucker, sie sind dunkel und glänzend.

Getrocknete Pflaumen schmecken zuckersüß, verlieren mit der Zeit Wasser und werden hart. Dann entsteht ein weißgrauer Belag aus Zuckerkristallen auf der Pflaumenhaut. Sie werden gerne roh gegessen oder zu Pflaumenmus verarbeitet.

Getrocknete Birnen werden ganz oder in Schnitzen gedörrt. In den Alpenländern finden sie vor allem für das berühmte Kletzenbrot Verwendung.

Getrocknete Äpfel kommen in Scheiben geschnitten in den Handel. Dafür werden vorwiegend spätreife säuerliche Apfel mit kleinem Kerngehäuse genommen.

Kandierte Früchte

Kandierte Kirschen werden vorwiegend für Dekoration von Gebäck verwendet oder sie machen als Cocktailkirsche in alkoholischen Drinks Karriere. Allerdings gehören sie obligatorisch in den englischen Teekuchen.

Kandierte Ananas Auch sie finden hauptsächlich als Dekoration für Torten und Kuchen Verwendung.

Kandierte Orangen Wie Kirschen und Ananas werden sie als dekoratives und besonders aromatisches Element für Kuchen und Torten eingesetzt.

Orangeat wird nicht aus herkömmlichen Orangen hergestellt, sondern aus der spanischen Bitterorange, die man auch „Pomeranze" nennt. Aus den kleinen dunkel-orangefarbenen Früchten wird auch die englische Orangenmarmelade gemacht.

Zitronat stammt von der Zedratzitrone. Die unreifen Früchte der Pomeranze und der Zedratzitrone werden in Salzlake gelegt, dann wieder entsalzt und schließlich in hochprozentiger Zuckerlösung kandiert.

1
2
3
4
5
6
7
8
9
10
11
12
13
14
15
16
17
18
19
20
21
22
23
24
25

Die Geräte

Damit Torten, Kuchen und Kleingebäck gelingen, ist es nicht nur wichtig, dass alle Rezepte genau nachvollzogen werden, auch das nötige Werkzeug ist oft von entscheidender Bedeutung. Wenn Sie beispielsweise eine Crememasse aufschlagen, gelingt das am besten in einem Schneekessel. Oder wenn Sie statt der angebenen Form eine größere oder kleiner verwenden, verändert sich auch die Backzeit. Links sehen Sie eine Übersicht abgebildet, die es Ihnen erleichtern soll, Ihre persönliche Auswahl zu treffen.

1 **Springformen**
2 **Tortenringe**
3 **Rührschüssel**
4 **Schneekessel mit Holzspatel**
5 **Schneebesen**
6 **Kuchengitter**
7 **Mehlsieb**
8 **Spritzbeutel mit Stern- und Lochtüllen**
9 **Alufolie**
10 **Backpapier**
11 **Teigroller**
12 **Schmale Winkelpalette**
13 **Breite Winkelpalette**
14 **Lange Palette**
15 **Kurze Palette**
16 **Gummispatel**
17 **Backpinsel**
18 **Teigrädchen mit glattem Rand**
19 **Teigrädchen mit gezacktem Rand**
20 **Zesteur/Zestenreißer**
21 **Feine Reibe für Zitrusschale**
22 **Tortenunterlage**
23 **Teigschaber**
24 **Ausstecher mit gewelltem Rand**
25 **Ausstecher mit glattem Rand**

Der Ofen

Erfolg oder Misserfolg beim Backen hängt sehr häufig auch von Ihrem Ofen ab. Ober- und Unterhitze, Heißluft oder ein Kombination von beidem – moderne Öfen bieten die Auswahl. Die Temperaturen in unseren Rezepten sind für die gängigen Modelle mit Ober- und Unterhitze angegeben. Für Gasöfen stehen sie in Klammern daneben.

Alle angegebenen Hitzegrade sind nur ungefähre Werte, denn auch mit unserer modernen Technik hat jeder Ofen seine eigenen Temperaturen. Deshalb: Lesen Sie die Gebrauchsanweisung Ihres Ofens genau durch.

Backt man Plätzchen, ist Heißluft sehr praktisch, weil man mehrere Bleche auf einmal backen kann. Mit einem Testbacken verschiedener Teige machen Sie sich am ehesten mit Ihrem Gerät vertraut. Es kann sehr nützlich sein, die Backergebnisse aufzuschreiben, um zukünftige Fehler zu vermeiden.

Beim Kauf eines neuen Gerätes achten Sie darauf, dass die Ofenklappe ein Fenster hat, denn beispielsweise Plätzchen, Kekse und dünne Teigböden backt man am besten nach Sicht. Handelt es sich um Kuchen, macht man die Stäbchenprobe: Man sticht mit einem langen Holzstäbchen hinein und zieht es wieder heraus. Bleibt kein weicher Teig mehr daran hängen, ist der Kuchen fertig.

Kleine Material-Info

Backgeräte, wie links abgebildet, gibt es aus verschiedenen Materialien. Selbstverständlich können Sie das Material verwenden, das Ihnen am sympathischsten ist. Grundsätzlich gilt aber: Geräte aus Metall sind besonders hygienisch, weil sie gut zu reinigen sind. Plastikbretter, -löffel, -spatel etc. bekommen häufig Kerben, in denen sich Bakterien ansiedeln können. Dasselbe gilt für Holz, das beim Reinigen auch noch gerne den Geschmack der jeweiligen Reinigungsmittel annimmt.

Backformen: Klassiker und Neuheiten

Herzformen Nicht nur am Valentinstag, sondern einfach auch mal so: Sag's mit einem Kuchen in Herzform.

Gugelhupfform Der Klassiker als Variante mit Antihaftbeschichtung. Er lässt sich besonders gut stürzen.

1 Weißblechform

Hier lässt sich der Boden samt Kuchen herausheben. Die Weißblechform ist vor allem für Gasherde geeignet – auf dunklen Blechen wird Kuchen schnell zu braun.

2 Für Biskuitböden

Durch den geraffelten Boden löst sich Biskuitkuchen besonders leicht aus der Form.

3 Schnittfest

Form mit schnittfester Antihaftbeschichtung, d. h., Sie können den Kuchen nach dem Backen gleich in der Form aufschneiden, ohne dass Kratzer entstehen.

4 Rechteck

Warum nicht mal eckige Kuchen backen? Ideal dafür: diese Form mit heraushebbarem Boden.

5 Achteckige Forn

Originell und praktisc zugleich: Springform mit acht Ecken. So lassen sich Torten gerecht in gleichmäßige Stücke teilen.

Gut zu wissen

- Für alle Kuchenformen – egal, ob antihaftbeschichtet oder nicht – gilt: Vor dem Backen immer dünn mit Butter oder Margarine einstreichen – das verleiht dem Boden schöne Bräune.
- Fertigen Kuchen vor dem Lösen aus der Form erst 10 Minuten abkühlen lassen. Und dann nie mit einem Messer herauslösen. Wenn nötig, lieber einen Kunststoffschaber benutzen.

6 Springform mit Superboden
Das Besondere an dieser Springform ist der überstehende Boden. Beim Backen dient er als Auslaufschutz und später als Servierplatte für den fertigen Kuchen.

7 Mal was anderes: ovale Form
Für Abwechslung auf der Kaffeetafel sorgen ovale Kuchen aus dieser ungewöhnlichen Form mit Rohrboden.

1 Mit Glasboden
Schön und appetitlich: Springform mit Glasboden. Darauf können Kuchen und Torten auch serviert werden.

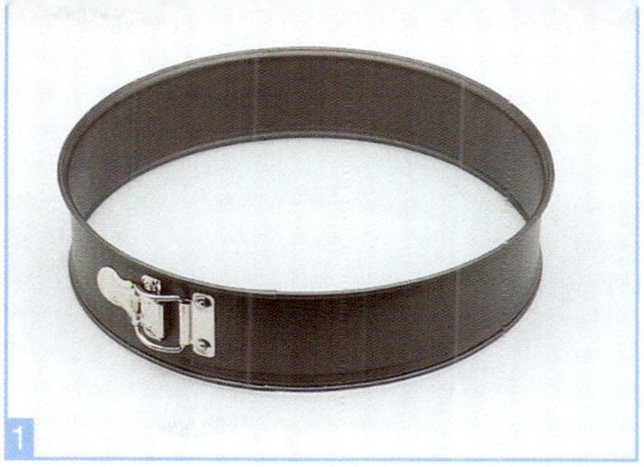
1

2 Für Muffins
Original-Bleche für Muffins: Den Teig in die Vertiefungen füllen und backen. Wer mag, kann vorher auch Papierförmchen hineinsetzen.

2

3 Praktisch mit Griffen
Springform mit Rohrbodeneinsatz. Das Besondere: die Edelstahlgriffe an den Seiten zum Hineinstellen und Herausheben aus dem Ofen.

3

4 Blech mit Einsatz
Links im Bild der Klassiker für Blechkuchen, daneben eine Variante mit Einsätzen für verschiedene Beläge.

4

5 Für Torten im Doppelpack
Wer beim Kuchenbacken hoch hinaus will: Diese Backringe mit breitem Boden sind auch für zweistöckige Exemplare (z. B. für unsere Hochzeitstorte auf Seite 258) gedacht.

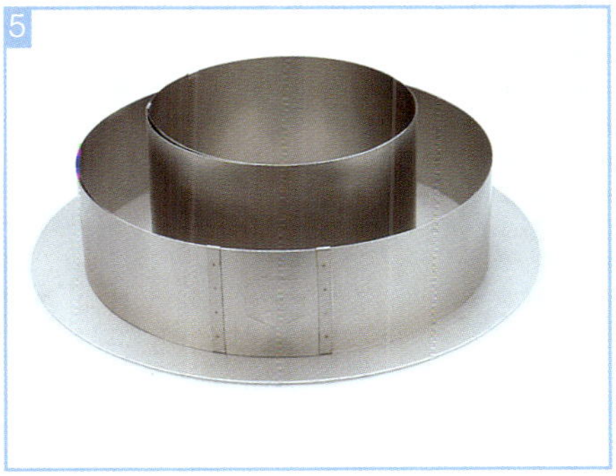
5

6 Form mit Haube
Etwas teurer, aber für leidenschaftliche Tortenbäcker bequemer Luxus: Springform auf schnittfester Servierplatte aus Edelstahl und – zum Kühlstellen – eine Tortenhaube mit Griff.

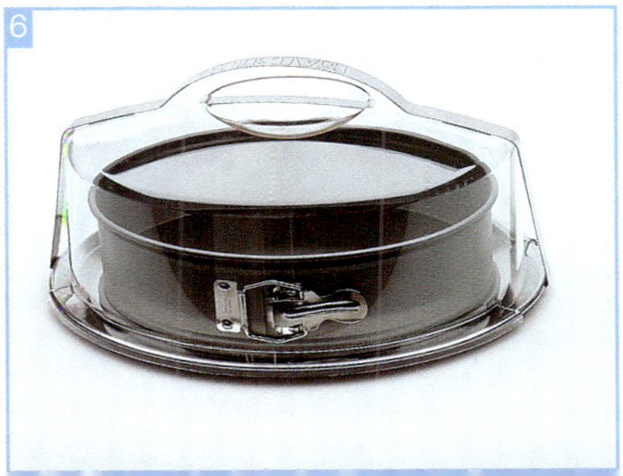
6

Mengen, Masse und Gewichte

Schnell auslöffeln

Lebensmittel	1 Teelöffel (gestrichen)	1 Esslöffel (gestrichen)	1 Esslöffel (gehäuft)
	Menge in g		
Wasser, Sahne, Milch	5	15	
Öl	4	10	
Quark, Crème fraîche	10	20	30
Butter, Margarine	6	15	25
Mehl, Stärkepulver	3	10	15
Kakao, Backpulver	3	5	8
Salz	4	10	20
Zucker	5	15	25
Grieß	5	10	20
Haferflocken, Semmelbrösel	3	6	10
Reis	4	10	15
Käse (gerieben)	3	5	8
Mandeln, Nüsse (gemahlen)	3	8	16
Konfitüre	17	30	50
Honig	7	15	20

Für Flüssiges – Liter & Co.

1 ml (Milliliter)	=	1 ccm*	=	1/1000 l
1 cl (Zentiliter)	=	10 ccm*	=	1/100 l
1 dl (Deziliter)	=	100 ccm*	=	1/10 l

*1 ccm = 1 Kubikzentimeter

1 l	=	1000 ml	=	100 cl	=	10 dl
3/4 l	=	750 ml	=	75 cl	=	7,5 dl
1/2 l	=	500 ml	=	50 cl	=	5 dl
3/8 l	=	375 ml	=	37,5 cl	=	3,75 dl
1/3 l	=	333 ml	=	33,3 cl	=	3,33 dl
1/4 l	=	250 ml	=	25 cl	=	2,5 dl
1/8 l	=	125 ml	=	12,5 cl	=	1,25 dl
1/10 l	=	100 ml	=	10 cl	=	1 d

1 Teelöffel	=	etwa 4 ml	=	0,4 cl
1 Esslöffel	=	etwa 10 ml	=	1 cl
1 Schnapsglas	=	20 ml	=	2 cl
1 Tasse	=	125–150 ml	=	12,5–15 cl

Die richtige Temperatur

Fahrenheit	*Celsius*	*Gas Stufe*
225°	107°	0-1
300°	149°	1
325°	163°	1-2
350°	177°	2
375°	191°	2-3
400°	204°	3
425°	218°	3-4
450°	232°	4
475°	246°	4-5
500°	260°	5

Für Festes – Pfund(s)sache

1 kg	=	1000 g	=	2 Pfund
1/2 kg	=	500 g	=	1 Pfund
1/4 kg	=	250 g	=	1/2 Pfund
1/8 kg	=	125 g	=	1/4 Pfund

Maße aus England und Amerika

1 lb (pound)	=	0,454 kg	=	454 g
1 oz (ounze)	=	0,028 kg	=	28 g
1 pt (pint)	=	0,568 l	=	568 ml
1 cup (amerik. Tasse)	=	0,24 l	=	240 ml
1 cup (engl. Tasse)	=	0,2 l	=	200 ml

Alle Ofentemperaturen in diesem Buch beziehen sich auf Elektroherde mit Ober- und Unterhitze. Bei Gasherden bitte die Gebrauchsanweisung beachten.

Glossar

abbrennen Wenn man Brandteig herstellt, wird Mehl in kochende Flüssigkeit gerührt, bis sich der Teig als Kloß vom Topfboden löst und dieser mit einer weißen Haut überzogen ist.

aprikotieren Gebäck, Kuchen oder Plätzchen mit Aprikosenkonfitüre, die vorher durch ein Sieb gestrichen wurde, überziehen. So ergibt sich ein glatte Oberfläche für Schokoladenüberzüge oder Zuckerglasuren.

aufschlagen Eischnee, Sahne, Saucen oder Cremes mit Schneebesen, Handrührgerät oder Mixstab bearbeiten, bis sie locker und luftig sind und ihr Volumen deutlich vergrößert haben.

Baiser Schaummasse aus Eiweiß und Zucker, die im Ofen bei geringer Hitze mehr getrocknet als gebacken wird. In Italien nennt man sie Meringue.

blanchieren Früchte oder Mandeln mit kochendem Wasser übergießen, um die Frucht- bzw. die Mandelhaut anschließend besser abziehen zu können.

blind backen Mürbeteigböden (ohne Belag) vorbacken. Damit er schön flach bleibt und sich nicht aufwirft, beschwert man ihn mit Hülsenfrüchten.

erstarren Festwerden von Cremes oder Gelees, aber auch von Schokoladenüberzug und Zuckerglasuren.

gehen lassen Hefeteig mit einem Küchentuch zudecken und an einen warmen Ort stellen, damit der Teig aufgehen kann und dabei sein Volumen vergrößert.

Gelatine Geliermittel tierischen Ursprungs, das Cremes, Fruchtpürees und Sahne „Stand" verleiht. Es ist geschmacklos und transparent.

glasieren Gebäck, Plätzchen oder Kuchen mit einem Überzug aus Puderzuckerglasur (weiß oder mit Lebensmittelfarbe eingefärbt) versehen.

karamelliesieren Nüsse oder Früchte mit geschmolzenem und leicht gebräuntem Zucker überziehen.

melieren Vorsichtiges Unterziehen von Zutaten mit unterschiedlicher Konsistenz. Am besten geht das mit einem Spatel. Die Masse darf dabei nicht an Volumen verlieren.

passieren Früchte oder Konfitüren durch ein Sieb streichen, um sie von Fruchthaut oder Kernen zu befreien.

reduzieren Flüssigkeit (meist mit Zucker) einkochen, um die Konsistenz dickflüssiger und den Geschmack intensiver zu machen.

Stäbchenprobe Garprobe, bei der ein Holzstäbchen in die dickste Stelle des Gebäcks gestochen und wieder herausgezogen wird. Haften keine feuchten Teigreste daran, kann der Kuchen aus dem Ofen genommen werden.

temperieren Schokolade schmelzen, abkühlen lassen und langsam wieder auf 32 Grad erwärmen.

tourieren Teige (z. B. Plunder- oder Blätterteig), in die Butterschichten eingearbeitet werden sollen, mehrfach zusammenlegen und wieder ausrollen.

tränken Luftiges Gebäck (Biskuit oder Rührkuchen) mit Likör, Frucht- oder Zuckersirup beträufeln.

Wasserbad In einen mit Wasser gefüllten Behälter wird eine Schüssel eingesetzt, in der unter Hitzezufuhr eine Masse mit dem Rührbesen aufgeschlagen wird, um ihr Volumen zu vergrößern oder um Schokolade zu schmelzen.

Zesten Feine Streifen von Zitrusschalen, die entweder mit dem Messer fein abgeschält oder mit dem sogenannten Zestenreißer abgezogen sind.

Diese Symbole finden Sie bei unseren Rezepten

Die kleinen Vignetten, die über den Rezepten stehen, sollen Ihnen bei der Orientierung helfen.

Die rote Kochmütze heißt: Das Rezept ist besonders „raffiniert", etwas für Gäste oder z. B. eine Torte für einen speziellen Anlass.

Die roten Mützen bedeuten: ein „preisgünstiges Gebäck", für das Sie pro Stück nicht mehr als 1 Euro ausgeben müssen.

Das um die Taille gebundene Maßband in Rot zeigt an: Dieser Kuchen ist „schlank" und hat pro Stück nie mehr als 250 Kalorien.

Die rote Uhr sagt Ihnen: Dieses Rezept geht „schnell". Sie haben den Kuchen in weniger als 2 Stunden auf dem Kaffeetisch.

Ist eines oder sind mehrere Symbole grau, trifft oder treffen diese Aussage(r) nicht zu.

★★★ Großer Aufwand
★★ Mittlerer Aufwand
★ Wenig Aufwand

Knusprig & zart.

Mürbeteig

Pies und Torten mit süßen Sommerfrüchten belegt sind einfach unwiderstehlich – und Mürbeteig ist die ideale Basis für diese Köstlichenkeiten

Erdbeer-Kokos-Tarte

Rezept siehe folgende Seite

Erdbeer-Kokos-Tarte

FÜR 1 PIEFORM VON 26 cm Ø
1 Eiweiß (Ei: Größe M)
30 g Zucker
80 g Butter, 150 g Mehl
FÜR DEN BELAG
3 Eigelbe (Eier: Größe M)
40 g Speisestärke
300 ml Milch, 1/2 Vanilleschote
200 ml ungesüßte Kokoscreme (Dose), 60 g Zucker
750 g Erdbeeren
1 EL Kokoschips
einige Zweige Zitronenmelisse
AUSSERDEM
Fett für die Form
1 Pck. heller Tortenguss

- Eiweiß mit 30 g Zucker, Fett, Mehl und 1-2 EL Wasser verkneten. Teig zur Kugel formen und in Klarsichtfolie gewickelt 30 Minuten kalt stellen.
- Elektro-Ofen auf 200 Grad heizen. Teig ausrollen. Eine gefettete Pieform damit auslegen. Teig mehrmals mit einer Gabel einstechen. Im Ofen (Gas: Stufe 3) 10-15 Minuten backen.
- Eigelbe, Stärke und 200 ml Milch verrühren. Vanilleschote aufschlitzen, Mark herausschaben und mit übriger Milch, Kokoscreme und Zucker aufkochen. Angerührte Stärke zugeben, nochmals aufkochen. Etwas abkühlen lassen, auf den Mürbeteig streichen.
- Erdbeeren waschen, putzen und halbieren. Auf der Creme verteilen.
- Tortenguss nach Packungsangabe zubereiten. Auf den Erdbeeren verteilen. Mit Kokoschips bestreuen. Mit Melisseblättchen verzieren.

Aufwand:	★
Zubereitungszeit:	1 1/4 Stunden
Backzeit:	15 Minuten
Pro Stück:	285 kcal/1200 kJ

Beeren-Stracciatella-Kuchen

FÜR 1 BACKBLECH (30 x 40 cm)
250 g Mehl
125 g weiche Butter
60 g Zucker, 1 Prise Salz
1 Ei (Größe M)
FÜR DIE CREME
2 Orangen (1 davon unbehandelt)
350 ml Milch
200 g Schlagsahne
2 Packungen Stracciatella-Mousse (Fertigprodukt; für 300 ml Flüssigkeit)
FÜR DEN BELAG
1 kg gemischte Beeren (z. B. Erdbeeren, Himbeeren, Heidelbeeren, Brombeeren)
50 g Schokoraspel
AUSSERDEM
Fett für das Blech, Mehl zum Arbeiten, Backpapier, Hülsenfrüchte zum Blindbacken

- Aus Mehl, Butter, Zucker, Salz und Ei rasch einen glatten Teig kneten. Teig in Folie gewickelt 30 Minuten im Kühlschrank ruhen lassen.
- Unbehandelte Orange heiß abwaschen, die Schale – am besten mit einem Zestenreißer – in feinen Streifen abziehen. Saft aus beiden Orangen pressen. Saft und die Hälfte der Schalen in einen Topf geben, sirupartig einkochen.
- Elektro-Ofen auf 200 Grad vorheizen. Eine Hälfte des Backblechs einfetten. Teig auf bemehlter Fläche in halber Backblechgröße ausrollen. Gefettete Blechhälfte mit der Teigplatte auslegen, dabei den Teig an den Rändern hochziehen. Teig mit Backpapier belegen und Hülsenfrüchte zum Blindbacken darauf verteilen. Im Ofen (Gas: Stufe 3) 15-20 Minuten backen, auskühlen lassen.
- Milch, Sahne und Mousse-Pulver in einen hohen Rührbecher geben. Die Masse nach Packungsangabe schaumig-fest aufschlagen. Orangensirup unterrühren. Mousse auf dem Teigboden verstreichen und etwa 30 Minuten kalt stellen.
- Früchte waschen, putzen, gut trockentupfen. Je nach Größe in mundgerechte Stücke teilen. Früchte dekorativ auf der Creme verteilen, mit Schokoraspeln und restlichen Orangenschalen bestreuen.

Aufwand:	★★
Zubereitungszeit:	1 1/4 Stunden
Backzeit:	20 Minuten
Pro Stück:	365 kcal/1530 kJ

Cappuccino-Pflaumenkuchen

FÜR 1 BACKBLECH (30 x 40 cm)
250 g Mehl, 1 Msp. Backpulver
75 g gemahlene Mandeln
3 Tropfen Mandelaroma
125 g Butter, 65 g Zucker
1 Prise Salz, 2 Eier (Größe M)

FÜR DEN BELAG
5 EL Cappuccino-Kaffeepulver (Instant)
750 g Pflaumen
150 g Schlagsahne
3 Eier (Größe M), 50 g Zucker
750 g Sahnequark

AUSSERDEM
Fett für das Blech
Mehl zum Arbeiten
Puderzucker zum Bestäuben

- Mehl, Backpulver, Mandeln, Mandelaroma, Butter, Zucker und Eier zu einem glatten Teig kneten, in Folie gewickelt 30 Minuten kalt stellen.
- Cappuccinopulver in 3-4 EL heißem Wasser auflösen, abkühlen lassen. Pflaumen waschen, halbieren, entsteinen. Sahne steif schlagen. Eier mit Zucker schaumig rühren. Mit Quark und Cappuccino verrühren. Sahne unterheben. Elektro-Ofen auf 200 Grad vorheizen.
- Backblech einfetten. Teig ausrollen auf bemehlter Fläche ausrollen, Backlech damit auslegen. Einen Rand hochziehen. Quark-Masse auf den Teig streichen. Pflaumen mit der Wölbung nach oben dicht an dicht darauf legen.
- Kuchen im Ofen (Gas: Stufe 3) etwa 30 Minuten backen. Kuchen weitere 10 Minuten im ausgeschalteten Ofen ruhen lassen. Herausnehmen, mit Puderzucker bestäuben.

Aufwand:	★
Zubereitungszeit:	1 Stunde
Backzeit:	40 Minuten
Pro Stück:	430 kcal/1810 kJ

Französische Apfeltarte

FÜR 1 PIEFORM VON 26 cm Ø
200 g Mehl
1 Prise Salz
50 g Zucker
1 Eigelb (Ei: Größe M)
100 g kalte Butter
FÜR DEN BELAG
2 Eier (Größe M)
50 g Puderzucker
200 g Crème fraîche
1 EL Zitronensaft
4 säuerliche Äpfel (ca. 800 g)
1-2 EL Aprikosenkonfitüre
1 EL Pinienkerne
AUSSERDEM
Fett für die Form
Mehl zum Arbeiten
Puderzucker zum Bestäuben

- Mehl, Salz, Zucker, Eigelb, Butter in Flöckchen und 1-2 EL eiskaltes Wasser rasch zu einem Mürbeteig verkneten. Form einfetten. Den Teig auf bemehlter Arbeitsfläche dünn ausrollen, in die Form geben, dabei einen 2-3 cm hohen Rand formen. Kühl stellen.
- Elektro-Ofen auf 200 Grad vorheizen. Eier mit Puderzucker mischen, schaumig rühren. Crème fraîche und Zitronensaft unterrühren.
- Äpfel waschen, schälen und vierteln, Kerngehäuse entfernen. Jedes Viertel längs in drei Spalten schneiden. Die Hälfte der Creme auf den Teigboden gießen. Apfelspalten dachziegelartig überlappend darauf verteilen. Restliche Creme darüber geben. Im Ofen (Gas: Stufe 3) etwa 15 Minuten backen.
- Aprikosenkonfitüre erwärmen, durch ein Sieb streichen. Auf die Äpfel pinseln. Pinienkerne darauf streuen, mit Puderzucker bestäuben. Kuchen in 10-15 Minuten fertig backen.

Aufwand:	★
Zubereitungszeit:	25 Minuten
Backzeit:	25 Minuten
Pro Stück:	250 kcal/970 kJ

Orangen-Tarte

FÜR 1 PIEFORM VON 26 cm Ø
250 g Mehl, 175 g Butter
2 EL Zucker, 1 Eigelb
1 Prise Salz
abgeriebene Schale von
1 unbehandelten Orange
3 EL Orangensaft
50 g geriebene dunkle Kuvertüre
FÜR DIE ORANGENCREME
6 Eigelbe (Eier: Größe M)
125 g Zucker
Saft von 2 Orangen
1 EL Speisestärke
250 g Butter
abgeriebene Schale von
1 unbehandelten Orange
AUSSERDEM
Butter und Mehl für die Form
25 g zerlassene Butter
Backpapier, Mehl zum Arbeiten
Trockenerbsen zum Backen
FÜR DIE 1. DEKO-VARIANTE
5 Eiweiße (Eier: Größe M)
1 Prise Salz
Saft und abgeriebene Schale
von 1 unbehandelten Orange
200 g Zucker
FÜR DIE 2. DEKO-VARIANTE
500 g Schlagsahne
Mark von 1 Vanilleschote
abgeriebene Schale von
1 unbehandelten Orange
1 Päckchen Schoko-Weihnachts-dekor, Puderzucker

- Mehl, Butter, Zucker, Eigelb und Salz verkneten. Orangenschale, -saft und Kuvertüre untermischen und alles zu einem glatten Teig verarbeiten. Den Teig zur Kugel formen, in Klarsichtfolie wickeln und 1 Stunde kühlen.
- Form einfetten, mit Mehl ausstäuben. Elektro-Ofen auf 200 Grad vorheizen. Teig auf bemehlter Fläche zu 4 mm dicker Platte (ca. 26 cm Ø) ausrollen. Form damit auslegen, dabei Rand (3 cm) bilden. Mit Butter bestreichen, mit Backpapier belegen, mit Erbsen auffüllen. Im Ofen (Gas: Stufe 3) etwa 20 Minuten backen.
- Erbsen und Papier entfernen. Kuchen weitere 15 Minuten backen. Herausnehmen.
- Eigelbe, Zucker und Saft aufschlagen. Stärke dazusieben. Butter in Stücken und Orangenschale zugeben. Masse über dem heißen Wasserbad unter Rühren erhitzen, bis eine dicke Creme entsteht.
- Creme abkühlen lassen, auf die Tarte gießen und etwa 15 Minuten kalt stellen.

Aufwand:	★★
Zubereitungszeit:	1 1/4 Stunden
Backzeit:	35 Minuten
Pro Stück:	680 kcal/2860 kJ

Deko-Variationen

- Die Eiweiße mit Salz, Orangensaft und -schale und Zucker steif schlagen. In einen Spritzbeutel mit Sterntülle (6 mm Ø) füllen, Tupfen auf die Oberfläche spritzen. Unter dem Grill überbacken. Nach Belieben mit Puderzucker bestäuben.
- Sahne mit Vanillemark und Orangenschale steif schlagen. In Spritzbeutel (Sterntülle: 11 mm Ø) füllen, spiralförmig aufspritzen.

Cheese-Cake mit Früchten

FÜR 1 SPRINGFORM VON 24 cm Ø
50 g Zucker, 100 g Butter, 1 Ei
1 Msp. Salz, 150 g Mehl
Mark von 1/4 Vanilleschote
abgeriebene Schale von
1/2 Zitrone
FÜR DIE KÄSEMASSE
1 1/4 kg Doppelrahmfrischkäse
3 EL Mehl, 200 g Zucker
1 Pck. Vanillezucker
je 2 TL abgeriebene Zitronen- und Orangenschale
5 Eier, 2 Eigelb (Eier: Größe M)
FÜR DEN BELAG
250 g Erdbeeren, 2 Kiwis
1/2 Mango, 1 Dose Mandarinen
(Abtropfgewicht: 175 g)
1/2 Päckchen weißer Tortenguss
AUSSERDEM
Fett für die Form, Mehl zum Arbeiten, Backpapier

- Zucker, Butter, Ei, Salz, Mehl, Vanillemark und Zitronenschale zu einem glatten Teig verkneten und zu einer Kugel formen. Die Teigkugel in Klarsichtfolie wickeln und für 30 Minuten in den Kühlschrank legen.
- Elektro-Ofen auf 175 Grad vorheizen. Frischkäse, Mehl, Zucker, Vanillezucker, Zitronen- und Orangenschale glatt rühren und nach und nach Eier und Eigelbe unterziehen.
- Springform fetten. Teig auf bemehlter Fläche etwas größer als die Form etwa 3 mm dick ausrollen. Einen 10 cm breiten Backpapierstreifen an den Springformrand legen. Die Form mit dem Teig auskleiden und den Teigrand hochziehen.
- Käsemasse in die Springform füllen. Kuchen im Ofen (Gas: Stufe 2) etwa 75 Minuten backen.
- Inzwischen die Erdbeeren waschen, halbieren. Kiwis und Mango schälen, in Spalten teilen. Mandarinen abtropfen lassen. Nach dem Backen die Torte gut abkühlen lassen, mit den Früchten belegen. Tortenguss nach Packungsangabe zubereiten und die Torte damit dünn überziehen.

Aufwand:	★
Zubereitungszeit:	45 Minuten
Backzeit:	1 1/4 Stunden
Pro Stück:	415 kcal/1750 kJ

Tipp

Käsekuchen können Sie auch mal mit anderen Milchprodukten backen, z. B. den Quark durch 200 g Schmand und 300 g Kefir oder die entsprechende Menge Ricotta ersetzen. So bekommt der Kuchen einen fein säuerlichen Geschmack.

Käse-Aprikosen-Kuchen

FÜR 1 SPRINGFORM VON 24 cm Ø
400 g Mehl
3/4 Päckchen Backpulver
200 g Zucker
2 Eier (Größe M)
125 g Butter
FÜR DEN BELAG
1 Dose Aprikosen
(Abtropfgewicht: 480 g)
500 g Quark (20 % F. i. Tr.)
1 Päckchen Vanille-
Puddingpulver
abgeriebene Schale von
1 Zitrone (unbehandelt)
50 g Zucker
3 Eier (Größe M)
nach Belieben Puderzucker
AUSSERDEM
Fett für die Form
Puderzucker zum Bestreuen

- Mehl und Backpulver mischen und mit dem Zucker, den Eiern und der Butter verkneten. Den Teig 30 Minuten kalt stellen.
- Elektro-Ofen auf 200 Grad vorheizen. Aprikosen gut abtropfen lassen. Den Quark, das Puddingpulver, die Zitronenschale, der Zucker und die Eier verrühren.
- Eine Springform einfetten. 2/3 des Mürbeteiges in die Form geben und andrücken. Die Quark-Masse auf dem Teig verteilen und die Aprikosen auf die Quark-Masse setzen. Den restlichen Teig krümelig kneten und die Streusel gleichmäßig auf den Aprikosen verteilen.
- Der Kuchen im Ofen auf der mittleren Schiene bei 200 Grad (Gas: Stufe 3) etwa 45 Minuten backen. Kuchen herausnehmen, auskühlen lassen und nach Belieben mit Puderzucker bestäubt servieren.

Aufwand:	★
Zubereitungszeit:	40 Minuten
Backzeit:	45 Minuten
Pro Stück:	305 kcal/1280 kJ

Orangen-Anis-Tarte

FÜR 1 PIEFORM VON 30 cm Ø
200 g Mehl
100 g Butter
40 g Zucker
1 Ei (Größe M)
FÜR DIE FÜLLUNG
4-5 Orangen
50 g Butter
50 g Puderzucker
2 Eier (Größe M)
1-2 TL Speisestärke
300 g Crème fraîche
1/2-1 TL Anispulver
AUSSERDEM
Fett für die Form
Mehl zum Arbeiten
Puderzucker zum Bestäuben

- Aus Mehl, Butter, Zucker und Ei einen glatten Teig kneten. Teig in Folie wickeln und 30 Minuten kühl stellen. Pieform einfetten.
- Elektro-Ofen auf 175 Grad vorheizen. Den Mürbeteig auf bemehlter Fläche dünn ausrollen und die Pieform damit auslegen. Kalt stellen.
- Die Orangen dick abschälen, in dünne Scheiben teilen oder filetieren. Butter, Puderzucker, Eier, Speisestärke und Crème fraîche verrühren und mit Anispulver abschmecken. Die Masse auf den Mürbeteig streichen. Die Orangenscheiben oder -filets auflegen und die Tarte im Ofen (Gas: Stufe 2) etwa 35 Minuten backen. Abkühlen lassen und mit Puderzucker bestäuben.

Aufwand:	★
Zubereitungszeit:	1 Stunde
Backzeit:	35 Minuten
Pro Stück:	330 kcal/1390 kJ

Tipp

Die Orangen-Anis-Tarte lässt sich prima auch als Dessert servieren. Am besten schmeckt der Kuchen lauwarm, mit einer Kugel Nuss-, Vanille- oder Schokoladeneis. Oder: Sie reichen geschlagene Sahne zum Kuchen, verfeinert mit dem Mark von 1 Vanilleschote, 2-3 EL geraspelter Schokolade oder 1-2 EL Kaffeelikör.

Quark-Rhabarber-Kuchen

FÜR 1 BACKBLECH (30 x 40 cm)
80 g Zucker
150 g Butter
300 g Mehl
1 Ei (Größe M)
FÜR DEN BELAG
1 kg Rhabarber
150 g Zucker
4 Eier (Größe M)
800 g Magerquark
800 g Milchreis (Fertigprodukt aus dem Kühlregal)
2 TL abgeriebene Zitronenschale
etwa 100 ml Zitronensaft
1 EL Speisestärke
1 Päckchen heller Tortenguss
AUSSERDEM
Fett für das Blech

- Für den Mürbeteig die Zutaten zu einem glatten Teig verkneten, in Folie wickeln, im Kühlschrank 30 Minuten ruhen lassen. Backblech einfetten.
- Elektro-Ofen auf 200 Grad vorheizen. Den Teig auf dem Backblech ausrollen, einen Rand hochziehen.
- Rhabarber putzen, waschen und entfädeln. In 2 cm große Stücke schneiden. Mit 50 g Zucker bestreuen und mit 100 ml Wasser im geschlossenen Topf etwa 3 Minuten weich dünsten. Den Rhabarber in ein Sieb geben, den Sud auffangen.
- Eier trennen. Eigelbe mit Quark, Milchreis, Zitronenschale und -saft, Zucker und Stärke verrühren. Eiweiße steif schlagen und unterheben.
- Masse auf den Mürbeteig streichen im Ofen (Gas: Stufe 3) etwa 30 Minuten backen. Ofen auf 175 Grad (Gas Stufe 2) herunterschalten, weitere 15 Minuten backen.
- Rhabarberstücke auf dem Quarkkuchen verteilen. Den aufgefangenen Sud mit Wasser auf 1/4 l auffüllen und mit dem Tortengusspulver nach Packungsangabe zubereiten. Über den Rhabarber gießen und abkühlen lassen.

Info

Wer statt eckiger Kuchenstücke lieber Tortenstücke sevieren möchte, kann den Blechkuchen auch in einer Springform (28 cm Ø!) backen, wobei die Backzeit dann allerdings etwa 20 Minuten länger ist.

Aufwand:	★★
Zubereitungszeit:	1 Stunde
Backzeit:	45 Minuten
Pro Stück:	210 kcal/880 kJ

Pflaumentarte mit Zimtguss

FÜR TARTEFORM VON 26 cm Ø
200 g Mehl, 100 g Butter
50 g Zucker, 1 Msp. Backpulver
1 Prise Salz, 1 Ei (Größe: M)
FÜR DEN BELAG
1/2 l Milch
50 g Weichweizengrieß
1 Msp. Nelkenpulver
750 g Pflaumen
3 Eiweiße (Eier: Größe M)
100 g Zucker, 1/2 TL Zimt
AUSSERDEM
Fett für die Form
Mehl zum Arbeiten
Puderzucker zum Bestäuben

● Aus den Teigzutaten rasch einen glatten Teig kneten. Zur Kugel formen und in Folie gewickelt mindestens 30 Minuten kühl stellen.

● In der Zwischenzeit für den Belag die Milch zum Kochen bringen. Grieß mit Nelkenpulver vermischen und in die kochende Milch einrieseln lassen. Herdplatte abschalten und den Grieß 5-10 Minuten auf der abgeschalteten Kochstelle ausquellen lassen.

● Elektro-Ofen auf 200 Grad vorheizen. Die Tarteform (ersatzweise Springform) fetten. Den Teig auf bemehlter Fläche dünn ausrollen. Die Form damit auslegen. Bei einer Springform den Rand nur halbhoch mit Teig bedecken. Den Teigboden mit einer Gabel mehrfach einstechen.

● Die Pflaumen waschen, vierteln und entsteinen. Viertel mit der Hautseite nach unten dekorativ auf dem Teig verteilen.

● Die Eiweiße zu steifem Schnee schlagen. Den Zucker einrieseln lassen und weiterschlagen, bis er

sich gelöst hat. Die Eiweißmasse unter die lauwarme Grieß-Creme rühren und mit Zimt abschmecken. Die Masse gleichmäßig auf den Pflaumen verteilen.

- Kuchen im Ofen (Gas: Stufe 3) in 30-35 Minuten auf der untersten Schiene goldbraun backen. Vor dem Servieren mit Puderzucker überstäuben.

Aufwand:	★ ★
Zubereitungszeit:	45 Minuten
Backzeit:	35 Minuten
Pro Stück:	220 kcal/920 kJ

Tipp

So gelingen Ihnen Grieß-Hauben: Mit Weichweizengrieß wird der Guss lockerer als mit Hartweizengrieß. Der gekochte Grießbrei muss noch lauwarm sein, wenn der Eischnee untergehoben wird. Ist er zu heiß, zerläuft der Eischnee, ist er zu kalt, lässt sich der Eischnee nicht mehr unterheben.

Quark-Himbeer-Schiffchen

FÜR 12 SCHIFFCHEN-FORMEN,
(10-12 cm Länge)
200 g Mehl
60 g Zucker
1 Pck. Vanillezucker
75 g gemahlene Mandeln
2 Eigelbe (Eier: Größe M)
1 EL Milch
120 g kalte Butter
FÜR DIE FÜLLUNG
100 g dunkle Kuvertüre
400 g frische Himbeeren
200 g Quark (Magerstufe)
2 EL Puderzucker
2 EL Himbeergeist
4 Blatt weiße Gelatine
200 g Schlagsahne
AUSSERDEM
Fett für die Formen
Mehl zum Arbeiten

- Aus Mehl, Zucker, Vanillezucker, Mandeln, Eigelben, Milch und Butterflöckchen schnell einen Mürbeteig kneten, in Frischhaltefolie wickeln, 30 Minuten kalt stellen.
- Den Elektro-Ofen auf 200 Grad vorheizen. Förmchen mit Fett ausstreichen. Mürbeteig auf leicht bemehlter Arbeitsfläche messerrückendick ausrollen. Die Förmchen damit auskleiden, überstehenden Teig abschneiden. Den Teig im Ofen (Gas: Stufe 3) in etwa 8 Minuten goldbraun backen.
- Kuvertüre im warmen Wasserbad schmelzen, Teigschiffchen damit ausstreichen, trocknen lassen. 200 g Himbeeren durch ein Sieb streichen. Das Fruchtmark mit Quark, Puderzucker und Himbeergeist verrühren.
- Gelatine in kaltem Wasser einweichen. Tropfnass bei milder Hitze auflösen, mit 1 EL Creme verrühren und unter die übrige Creme rühren. Im Kühlschrank angelieren lassen.
- Die Sahne steif schlagen und unter die Creme ziehen. Schiffchen mit ganzen Himbeeren auslegen. Die Quark-Himbeer-Sahne in einen Spritzbeutel mit Sterntülle füllen und spiralenartig in die Schiffchen spritzen.

Aufwand:	★★
Zubereitungszeit:	45 Minuten
Backzeit:	8 Minuten
Pro Stück:	250 kcal/1050 kJ

Käse-Früchte-Torte

FÜR 1 SPRINGFORM VON 24 cm Ø
50 g Zucker, 100 g Butter
1 Msp. Salz, 150 g Mehl
Mark von 1/4 Vanilleschote
abger. Schale von 1/2 Zitrone
FÜR DIE QUARK-MASSE
500 g Magerquark
6 Eier und 3 Eigelbe (Größe M)
abger. Schale und Saft von
2 unbehandelten Zitronen
600 g Schlagsahne
6 EL Mehl, 3 EL Speisestärke
300 g Zucker, 2 Msp. Salz
AUSSERDEM
Fett für die Form, Mehl zum
Arbeiten, Backpapier
250 g Erdbeeren, 2 Kiwis
1/2 Mango
1 Dose Mandarin-Orangen
(Abtropfgewicht: 175 g)
1/2 Päckchen weißer Tortenguss

- Zucker, Butter, Salz, Mehl, Vanillemark und Zitronenschale zu einem glatten Teig verarbeiten, in Folie wickeln und 30 Minuten kalt stellen.
- Elektro-Ofen auf 175 Grad heizen. Quark, Eier, Eigelbe, Zitronenschale, Zitronensaft und Sahne vermischen. Mehl, Speisestärke, Zucker und Salz unterrühren.
- Springform einfetten. Teig auf bemehlter Fläche etwas größer als die Form etwa 3 mm dick ausrollen. Einen 10 cm breiten Backpapierstreifen an den Formrand legen. Teig in die Form legen, Rand hochziehen.
- Quark-Masse in die Form füllen. Kuchen im Ofen (Gas: Stufe 2) etwa 75 Minuten backen. Erdbeeren waschen, halbieren. Kiwis und Mango schälen, in Spalten teilen.
- Torte abkühlen lassen, mit Früchten und Mandarin-Orangen belegen. Tortenguss nach Packungsangabe zubereiten, Torte damit dünn überziehen.

Aufwand:	★
Zubereitungszeit:	45 Minuten
Backzeit:	1 1/4 Stunden
Pro Stück:	415 kcal/1750 kJ

Himbeer-Kokos-Käsekuchen

FÜR 1 SPRINGFORM VON 24 cm Ø
250 g Mehl
1 Msp. Backpulver
125 g Butter
65 g Zucker
1 Prise Salz
1 Ei (Größe M)
FÜR DEN BELAG
4 Eier (Größe M)
750 g Magerquark
200 g Creme fraîche
2 Pck. Vanillezucker
1 Dose gesüßte Kokoscreme
3-4 EL Speisestärke
100 g Zucker
250 g Tiefkühl-Himbeeren
AUSSERDEM
Fett für die Form
Mehl zum Arbeiten
Puderzucker zum Bestäuben

- Mehl und Backpulver mischen, mit Butterflöckchen, Zucker, Salz und Ei verkneten. Den Teig 30 Minuten kalt stellen.
- Für den Belag die Eier trennen. Quark mit Crème fraîche, Vanillezucker, Eigelben, Kokoscreme und Speisestärke glatt rühren. Den Elektro-Ofen auf 175 Grad vorheizen.
- Die Springform einfetten. Den Teig auf einer bemehlten Fläche etwas größer als die Form etwa 3 mm dick ausrollen. Form mit dem Teig auskleiden, dabei einen Rand formen.
- Die Eiweiße steif schlagen. Den Zucker zugeben, nochmals 4-5 Minuten weiterschlagen. Zusammen mit den gefrorenen Himbeeren vorsichtig unter die Quark-Masse heben und auf dem Mürbeteigboden verstreichen. Kuchen im Ofen (Gas: Stufe 2) in 60-70 Minuten goldgelb backen. Nach Belieben mit Puderzucker bestäuben.

Aufwand:	★
Zubereitungszeit:	45 Minuten
Backzeit:	1 1/4 Stunden
Pro Stück:	298 kcal/1250 kJ

Pfirsich-Makronen-Kuchen

FÜR 1 BACKBLECH (30 x 40 cm)
350 g Mehl
250 g kalte Butter
100 g Zucker
1 Ei, 1 Eigelb (Größe M)
FÜR DEN BELAG
1 1/2 kg Pfirsiche
500 g Sauerkirschen
3 Eiweiße (Eier: Größe M)
1 Prise Salz
150 g Zucker
2 Eigelbe
1 Päckchen-Vanillezucker
abgeriebene Schale von
1 unbehandelten Zitrone
150 g geschälte, gemahlene Mandeln
AUSSERDEM
Fett für die Form
4 EL Orangenmarmelade

Ausrollen leicht gemacht

Mürbeteig macht oft Probleme beim Ausrollen: Er klebt oder reißt. Da hilft ein kleiner Trick: Den Teig zwischen 2 Lagen Klarsichtfolie legen und mit dem Wellholz darüber rollen.

- Aus Mehl, Butter, Zucker, Ei und Eigelb rasch einen glatten Teig kneten. 1 Stunde kalt stellen. Das Blech fetten.
- Elektro-Ofen auf 175 Grad heizen. Pfirsiche überbrühen, häuten, halbieren, entsteinen. Kirschen waschen, entsteinen.
- Mürbeteig auf dem Blech ausrollen. Mit einer Gabel mehrmals einstechen und im Ofen (Gas: Stufe 2) etwa 20 Minuten backen. Auskühlen lassen. Ofen auf 200 Grad schalten.
- Eiweiße mit Salz steif schlagen, 100 g Zucker einrühren. Eigelbe mit restlichem Zucker, Vanillezucker und Zitronenschale schaumig rühren, Mandeln und Eischnee unterheben.
- Makronenmasse auf die vorgebackene Teigplatte streichen. Obst darauf verteilen und im Ofen (Gas: Stufe 3) etwa 15 Minuten backen. Früchte mit Marmelade überglänzen.

Aufwand:	★★
Zubereitungszeit:	1 1/4 Stunden
Backzeit:	35 Minuten
Pro Stück:	310 kcal/1300 kJ

Früchte-Dreieck

FÜR 1 BACKBLECH (30 X 40 cm)
200 g Mehl
1 Msp. Backpulver
1 Prise Salz
60 g Zucker
1/4 TL abgeriebene Zitronenschale
2 Eigelbe (Eier: Größe M)
120 g kalte Butter
FÜR DEN BELAG
je 150 g Brombeeren und Himbeeren
200 g grünes Melonen-Fruchtfleisch (Galia- oder Ogen-Melone)
2 EL Aprikosenkonfitüre
1 Päckchen weißer Tortenguss
1/4 l halbtrockener Riesling
2 EL Zucker
nach Belieben 1 EL Zitronenmelisse-Streifen
AUSSERDEM
Mehl zum Arbeiten
Backpapier

- Mehl mit Backpulver, Salz, Zucker, Zitronenschale, Eigelben und Butter rasch zu einem glatten Teig verkneten. Zugedeckt 1 Stunde kalt stellen.
- Beeren abbrausen. Melone in ganz feine Scheiben schneiden.
- Elektro-Ofen auf 200 Grad vorheizen. Teig ausrollen zu einem gleichseitigen Dreieck (Länge = 32 cm) zuschneiden und auf ein mit Backpapier belegtes Blech legen. Teigreste zu 3 je 32 cm langen, gleich dicken Rollen formen und als Ränder an die Dreiecksseiten drücken.
- Teig-Dreieck im Ofen (Gas: Stufe 3) etwa 20 Minuten backen und auskühlen lassen.
- Dreieck mit Konfitüre bestreichen, mit Obst belegen. Guss mit Wein und Zucker nach Packungsangabe zubereiten, eventuell Melisse zufügen. Guss gleichmäßig auftragen. Kalt stellen.

Aufwand:	★★
Zubereitungszeit:	35 Minuten
Backzeit:	35 Minuten
Pro Stück:	250 kcal/1050 kJ

Apfel-Soufflé-Kuchen

FÜR TARTEFORM VON 26 cm Ø
200 g Mehl
2 Msp. Backpulver
45 g Zucker
1 Eiweiß (Ei: Größe M)
120 g Butter
FÜR DIE FÜLLUNG
1 kg Äpfel
3 EL Zitronensaft
5 EL Orangenlikör
6 Eier (Größe M)
75 g Zucker, 75 g Mehl
2 Msp. Backpulver
AUSSERDEM
Fett für die Form
3-4 EL Mandelblättchen
Puderzucker zum Bestäuben

- Mehl, Backpulver, Zucker, Eiweiß und Butter verkneten. Teig in Folie wickeln und 30 Minuten kalt stellen.
- Teig ausrollen, die gefettete Tarteform damit auslegen. Teig mehrmals einstechen und 30 Minuten kalt stellen.

Info

Mürbeteig lässt sich sehr gut tiefgekühlt lagern und ist für Überraschungsgäste verfügbar. Er hält sich etwa drei Monate. Bereiten Sie einfach die doppelte oder dreifache Menge vor und frieren Sie den Teig portionsweise ein.

- Elektro-Ofen auf 175 Grad heizen. Äpfel schälen, achteln, entkernen, mit Zitronensaft und 2 EL Likör beträufeln. Eier trennen. Eigelbe mit 2 EL warmem Wasser aufschlagen. Eiweiße steif schlagen, Zucker zugeben, 3-4 Minuten weiterschlagen. Eigelbe, Mehl, Backpulver und Eischnee vorsichtig vermischen. Übrigen Likör unterheben.
- Die Hälfte der Soufflé-Masse auf den Teig geben, Äpfel darauf verteilen, dann die übrige Masse. Mit Mandeln bestreut im Ofen (Gas: Stufe 2) etwa 50 Minuten backen. Mit Puderzucker bestäuben.

Aufwand:	★★
Zubereitungszeit:	70 Minuten
Backzeit:	50 Minuten
Pro Stück:	400 kcal/1680 kJ

Apfeltorte mit Weingelee

BITTE BEACHTEN: TORTE ÜBER NACHT KALT STELLEN

FÜR 1 SPRINGFORM VON 26 cm Ø

250 g Mehl, 125 g kalte Butter

70 g Zucker, Salz, 1 Ei (Größe M)

FÜR DEN BELAG

1 1/2 kg säuerliche Äpfel

1 haselnussgroßes Stück Ingwer

3/4 l halbtrockener Roséwein

1-2 Nelken, 2 Zitronenscheiben

75 g Zucker

2 Pck. Sahnepuddingpulver

FÜR DAS BAISER

3 Eiweiße (Eier: Größe M)

1 Prise Salz, 150 g Zucker

AUSSERDEM

Fett für die Form, Mehl

- Für den Mürbeteig Mehl mit Butterflöckchen, Zucker, 1 Prise Salz und Ei zu einem glatten Teig verkneten. 30 Minuten kalt stellen.
- Äpfel waschen, vierteln, putzen, schälen und klein würfeln. Ingwer schälen, in Scheiben schneiden. In einem Sud aus 650 ml Wein, Ingwer, Nelken, Zitronen und 50 g Zucker etwa 2 Minuten köcheln lassen. Gewürze entfernen.
- Elektro-Ofen auf 150 Grad vorheizen. Topf vom Herd nehmen. Restlichen Zucker und Puddingpulver mit restlichem Wein anrühren, unter die Äpfel mischen, einmal aufkochen, vom Herd ziehen.
- Teig auf bemehlter Fläche ausrollen, die gefettete Form damit auskleiden, dabei einen Teigrand von maximal 4,5 cm formen. Apfelmasse einfüllen. Im Ofen (Gas: Stufe 1) etwa 60 Minuten backen. Auskühlen lassen, über Nacht zugedeckt kalt stellen.
- Ofen auf 200 Grad (Gas: Stufe 3) vorheizen. Eiweiße mit Salz steif schlagen, Zucker einrieseln lassen, weiterschlagen, bis die Masse fest und glänzend ist. In Spritzbeutel mit Sterntülle füllen, auf den Kuchen spritzen. Unterm Grill auf mittlerer Schiene 3-4 Minuten leicht bräunen. Sofort servieren.

Aufwand:	★★
Zubereitungszeit:	50 Minuten
Backzeit:	1 Stunde
Pro Stück:	400 kcal/1680 kJ

Campari-Zitrus-Torte

FÜR 1 SPRINGFORM VON 24 cm Ø
150 g Mehl
40 g Zucker
1 Prise Salz
1 Eiweiß (Ei: Größe M)
100 g Butter
FÜR DEN BISKUITTEIG
50 g Butter
3 Eier (Größe M)
1 Prise Salz
150 g Zucker
100 g Mehl
FÜR DIE FÜLLUNG
10 rosa Grapefruits
3-4 Orangen, 300 g Joghurt
7 Blatt weiße Gelatine
125 g Puderzucker
300 g Mascarpone
8 EL Campari (Bitterlikör)
600 ml Pink-Grapefruit-Nektar
350 g Schlagsahne
2 EL Orangenmarmelade
1 Päckchen weißer Tortenguss
AUSSERDEM
Backpapier für die Form
Mehl zum Arbeiten
100 g Mandelblättchen

● Mürbeteig-Zutaten verkneten, 30 Minuten kalt stellen. Elektro-Ofen auf 200 Grad vorheizen. Form mit Backpapier auslegen. Teig auf Mehl ausrollen, in die Form legen. Im Ofen (Gas: Stufe 3) etwa 10 Minuten backen. Abkühlen lassen, aus der Form nehmen.
● Temperatur auf 175 Grad reduzieren. Butter zerlassen, Eier trennen. Eiweiße mit Salz und 80 g Zucker steif schlagen. Eigelbe mit 3 EL Wasser und übrigem Zucker aufschlagen. Mehl, Eischnee und Butter unterziehen. Teig in die Form füllen. Im Ofen (Gas: Stufe 2) etwa 30 Minuten backen. Abkühlen lassen und quer halbieren.
● Grapefruits und Orangen dick abschälen und filetieren. Gelatine einweichen. Puderzucker, Joghurt, Mascarpone und 5 EL Campari verrühren. Gelatine in 350 ml Nektar erwärmen, auflösen. Unter die Creme rühren, kalt stellen.
● Sahne steif schlagen, 5 EL abnehmen. Rest unter die Creme ziehen. Marmelade erwärmen, auf den Mürbeteig streichen. 1 Biskuitboden darauflegen, Springformring darumlegen. Die Hälfte der Grapefruits und Orangen darauf legen, Creme darauf verteilen. Mit zweitem Biskuitboden abdecken und 2 Stunden kalt stellen.
● Übrige Orangen auflegen. Guss mit restlichem Campari und Nektar nach Packungsangabe zubereiten. Auf die Torte gießen und kalt stellen. Den Tortenrand mit übriger Sahne und Mandeln verzieren.

Aufwand:	★★
Zubereitungszeit:	1 Stunde
Backzeit:	40 Minuten
Pro Stück:	550 kcal/2310 kJ

Aprikosen-Himbeer-Tarte

FÜR TARTEFORM VON 26 cm Ø
225 g Mehl, 70 g Zucker
125 g Butter
1 Eigelb (Ei: Größe M)
1 EL Schlagsahne
FÜR DEN BELAG
500 g Aprikosen
50 g Amarettini (ital. Mandelgebäck; ersatzweise Mürbeteigplätzchen)
3 Eigelb (Eier: Größe M)
25 g Zucker
125 g Schlagsahne
einige Himbeeren
AUSSERDEM
Fett für die Form
Mehl zum Arbeiten
Hülsenfrüchte (z. B. getrocknete Bohnen) zum Blindbacken

- Für den Teig Mehl mit Zucker und Butter krümelig kneten. Eigelb mit Sahne verrühren, dazugeben und alles zu einem glatten Teig kneten. Eventuell etwas Wasser zugeben. Den Teig in Folie wickeln und 30 Minuten kalt stellen.
- Teig auf einer leicht bemehlten Fläche ausrollen und in eine gefettete Tarte- oder Pieform legen. Den Rand hochziehen und gerade schneiden. Den Teig mit einer Gabel mehrmals einstechen und 30 Minuten kalt stellen.
- Elektro-Ofen auf 175 Grad vorheizen. Auf den Teig Backpapier legen, darauf die Hülsenfrüchte streuen und den Kuchen im Ofen (Gas: Stufe 2) etwa 20 Minuten backen. Hülsenfrüchte und Papier entfernen und den Kuchenboden in weiteren 5 Minuten goldbraun backen. Herausnehmen, in der Form abkühlen lassen. Ofen nicht ausschalten.
- Aprikosen waschen, halbieren, entsteinen. Fruchtfleisch in Spalten schneiden. Kekse zerbröseln, auf dem Kuchenboden verteilen. Die Aprikosen kreisförmig darauf anordnen. Eigelbe mit Zucker und Sahne verrühren und darüber gießen. Die Tarte im Ofen (Gas: Stufe 2) etwa 5 Minuten backen.
- Tarte auskühlen lassen. Die Himbeeren verlesen, waschen und auf dem Kuchen verteilen. Die Tarte servieren.

Aufwand:	★★
Zubereitungszeit:	1 1/2 Stunden
Backzeit:	30 Minuten
Pro Stück:	450 kcal/1890 kJ

Variationen

- Diese Tarte lässt sich ebenso gut mit der entsprechenden Menge anderer Früchte, wie Pfirsiche, Äpfel, Pflaumen, Orangenfilets oder Heidelbeeren, belegen
- Statt der Himbeeren nehmen Sie auch mal Rosinen, geröstete Walnüsse oder Mandeln.

Badischer Apfelkuchen

FÜR 1 SPRINGFORM VON 26 cm Ø
200 g Mehl
1 Prise Salz, 50 g Zucker
1 Eigelb (Ei: Größe M)
100 g kalte Butter
FÜR DEN BELAG
750 g mittelgroße säuerliche Äpfel, 50 g Butter
50 g Zucker, 1/4 TL Zimt
2 EL Apfelgelee, 50 g Rosinen
50 g Mandelblättchen
AUSSERDEM
Fett für die Form
Mehl zum Arbeiten

- Elektro-Ofen auf 200 Grad heizen. Springform fetten.
- Mehl mit Salz, Zucker, Eigelb, Butter in Flöckchen und 1-2 EL eiskaltem Wasser rasch zu einem Mürbeteig verkneten. Den Teig auf bemehlter Arbeitsfläche dünn ausrollen, in die Form geben, dabei einen 2-3 cm hohen Rand formen. 30 Minuten kühl stellen.
- Äpfel waschen, schälen, vierteln, Kerngehäuse entfernen. Äpfel in dünne Scheiben hobeln. Butter zerlassen. Zucker mit Zimt vermischen. Teigboden mit Apfelgelee bestreichen. Apfelscheiben in die Form schichten. Rosinen dazwischen streuen.
- Die Äpfel mit zerlassener Butter bestreichen. Mandelblättchen und Zimtzucker überstreuen.
- Apfelkuchen im unteren Drittel des Ofens (Gas: Stufe 3) 30-35 Minuten backen. In der Form abkühlen lassen.

Aufwand:	★
Zubereitungszeit:	40 Minuten
Backzeit:	35 Minuten
Pro Stück:	260 kcal/1090 kJ

Himbeer-Soufflé-Torte

FÜR 1 SPRINGFORM VON 24 cm Ø
125 g Mehl
1 Msp. Backpulver
65 g Butter, 2 EL Zucker
1 Prise Salz
1 Eigelb (Ei: Größe M)
FÜR DEN BELAG
10 Eier (Größe M)
1 unbehandelte Zitrone
130 g saure Sahne
550 g Quark (20 % Fett i. Tr.)
1 Pck. Vanille-Puddingpulver
40 g Speisestärke
3 Pck. Vanillezucker
1 EL Rum
80 g Butter, Salz
200 g Zucker
3-4 EL Himbeerkonfitüre
500 g Himbeeren (frisch oder tiefgekühlt)
AUSSERDEM
Fett für die Form
Mehl zum Arbeiten, Backpapier
1/2 Päckchen roter Tortenguss

- Mehl und Backpulver mischen, mit Butterflöckchen, Zucker, Salz und Eigelb verkneten. Den Teig 30 Minuten kalt stellen.
- Elektro-Ofen auf 200 Grad vorheizen.
- Springform fetten. Teig auf bemehlter Fläche ca. 3 mm dick ausrollen, in die Form legen, mit einer Gabel mehrmals einstechen. Im Ofen (Gas: Stufe 3) etwa 12 Minuten backen, herausnehmen. Die Hitze auf 175 Grad (Gas: Stufe 2) reduzieren.
- Eier trennen. Die Zitrone heiß waschen, Schale abreiben. Eigelbe mit saurer Sahne, Quark, Puddingpulver, Stärke, Vanillezucker, Rum und Zitronenschale glatt rühren. Butter schmelzen. Eiweiße mit Salz steif schlagen, Zucker einrieseln lassen, unter die Quarkmasse heben. Butter unterziehen.
- Aus Backpapier einen ca. 80 cm langen, 10 cm breiten Rand formen, in die Form legen, einfetten. Quark-

masse einfüllen. Kuchen im Ofen (Gas: Stufe 2) etwa 30 Minuten vorbacken, bei 150 Grad (Gas: Stufe 1) in etwa 60 Minuten fertig backen.

● Torte abkühlen lassen, mit Konfitüre bestreichen und mit Beeren belegen. Tortenguss nach Packungsangabe zubereiten, darüber träufeln. Teigrand eventuell mit Puderzucker bestäuben.

Aufwand:	★★
Zubereitungszeit:	50 Minuten
Backzeit:	1 3/4 Stunden
Pro Stück:	305 kcal/1280 kJ

Variation

● Mit frischen Mangospalten, aber auch Früchten aus der Dose, wie Ananas oder Mandarinen, können Sie diese köstliche Torte immer wieder verändern.

Frisches Obst für Kuchen und Torten

Mit den vielen wertvollen Inhaltsstoffen, die in Früchten enthalten sind, sollte man schonend umgehen

Mangos schälen

Die Mangoschale ist ungenießbar, deshalb wird sie abgeschält.

1 Das Messer oben ansetzen und die Frucht streifenweise dünn schälen. Dazu kann man auch einen Sparschäler verwenden.
2 Da der Kern fest mit der Frucht verwachsen ist, wird das Fruchtfleisch auf beiden Seiten nahe dem Kern abgeschnitten.
3 Das Fruchtfleisch kann dann in Scheiben oder Würfel geschnittenen werden.

Äpfel und Birnen vorbereiten

Äpfel und Birnen werden vor dem Backen meist geschält und das Kerngehäuse wird entfernt.

1 So wird der Apfel im Ganzen verwendet: Zuerst das Kerngehäuse ausstechen, dann spiralförmig schälen.
2 Oder man viertelt oder achtelt die Frucht, schneidet das Kerngehäuse heraus und schält die Segmente anschließend.
3 Birnen kann man auch ungeschält in dünne Scheiben schneiden, das Kerngehäuse ausstechen und mit den Birnenscheiben eine Torte belegen.

Pfirsiche blanchieren

Die Pfirschhaut schmeckt bitter und sollte deshalb entfert werden:

1 Das Wasser zum Kochen bringen und die Pfirsiche darin 1-2 Minuten blanchieren.
2 Die Früchte anschließend in Eiswasser abschrecken.
3 Jetzt lässt sich die Haut ganz leicht abziehen.
Mit Aprikosen und Nektarinen können Sie genauso verfahren.

Beeren mögen sanften Umgang

Beeren haben eine dünne Haut und sind besonders empfindlich. Sie werden daher nach dem Abzupfen nur abgebraust und vorsichtig trocken getupft. Man kann sie entweder im Teig mitbacken oder aber den gebackenen Kuchen damit garnieren. Will man sie zuckern, dann erst kurz vor dem Servieren Puderzucker darüber streuen. Zucker will schmelzen und entzieht den Früchten Wasser, d. h. sie werden matschig.

Würfel und Streifen

Rhabarber

1 Rhabarberschalen enthalten Oxalsäure, einen Kalziumräuber, deshalb wird Rhabarber immer geschält.
2 Die Stangen anschließend in 1-2 cm breite Streifchen schneiden.

Papaya

1 Mit einem Löffel die Kerne aus den halbierten Früchten lösen.
2 Papaya zuerst in Streifen, dann in Würfel schneiden.

Orangen filetieren

1 Die Schale rund herum mit einem scharfen Messer in Segmente schneiden. Die Schalensegmente samt der weißen, bitteren Haut herunterschneiden.
2 Die Orangenfilets anschließend mithilfe des Messers herauslösen und in ein Schälchen legen.
3 Damit die Filets feucht bleiben, den Saft aus der restlichen Orange drücken und darüberträufeln.

Rührkuchen

heiß geliebt –

Auch wenn man wenig Backerfahrungen hat – an Rührteig kann sich jeder wagen. Sein einziges Geheimnis: Er muss gründlich gerührt werden. Dann wird ein lockerer, wunderbarer Kuchen daraus

und schnell gemacht

Walnuss-Orangen-Kuchen

Rezept siehe folgende Seite

Walnuss-Orangen-Kuchen

FÜR NAPFKUCHENFORM VON 2 l
3 unbehandelte Orangen
150 g Walnusskerne
100 g Orangenkekse (zartes Orangen-Mürbeteiggebäck, z. B. von Bahlsen), 7 Eier (Größe M)
250 g Zucker
100 g weiche Butter
150 g Mehl
1/2 Päckchen Backpulver
100 g Orangenmarmelade
AUSSERDEM
Fett und Brösel für die Form
kandierte Orangenstücke

- Die Orangen heiß abwaschen, die Schale abreiben und den Saft auspressen. Saft bei milder Hitze auf etwa die Hälfte einköcheln lassen. Walnusskerne fein hacken, Kekse fein zerbröseln.
- Den Elektro-Ofen auf 175 Grad vorheizen. Die Form gut einfetten und mit Semmelbröseln ausstreuen.
- Die Eier trennen. Die Eigelbe mit 125 g Zucker schaumig aufschlagen. Die Butter unterrühren. Orangensaft und -schale, Mehl, Backpulver, Keksbrösel und Nüsse unterheben.
- Eiweiße zu steifem Schnee schlagen, restlichen Zucker einrieseln lassen und weiterschlagen, bis er sich gelöst hat. Den Eischnee unter den Teig ziehen.
- Teig in die vorbereitete Form geben, glatt streichen und im Ofen (Gas: Stufe 2) etwa 60 Minuten backen.
- Kuchen zum Abkühlen auf ein Gitter stürzen. Orangenmarmelade in einem kleinen Topf sirupartig einkochen lassen und den noch heißen Kuchen damit überziehen. Kandierte Orangen in feine Streifen schneiden und den Kuchen damit garnieren.

Aufwand:	★
Zubereitungszeit:	30 Minuten
Backzeit:	1 Stunde
Pro Stück:	605 kcal/2540 kJ

Deko-Tipp

Die Schalen von unbehandelten Zitrusfrüchten sind eine schöne Verzierung für Kuchen und Desserts: Frucht heiß abwaschen, trockenreiben und mit einem Zestenreißer Streifen abziehen.

Limetten-Mohn-Kuchen

FÜR 1 GUGELHUPF-FORM (2 l)
250 g weiche Butter
150 g Zucker, 1 Prise Salz
4 Eier (Größe M), 400 g Mehl
1 Päckchen Backpulver
100 g Schlagsahne
abgeriebene Schale und Saft von 2 Limetten
250 g Mohnback (Fertigprodukt)
2-3 EL Orangenlikör
AUSSERDEM
Fett für die Form
3 Limetten
150 g Puderzucker

- Elektro-Ofen auf 175 Grad vorheizen. Die Form einfetten. Butter, Zucker, Salz und Eier schaumig rühren. Mehl und Backpulver mischen und mit der Sahne unterheben. Teig in zwei Hälften teilen. Unter die eine Teighäfte Limettenschale und -saft rühren, unter die andere Teighälfte Mohnback und Likör heben.
- Erst den Mohn-Teig in die Form füllen und darauf den Limetten-Teig. Mit einer Gabel durch die Teige ziehen, so dass eine Marmorierung entsteht. Im Ofen auf der untersten Schiene (Gas: Stufe 2) etwa 50 Minuten backen.
- In der Zwischenzeit 2 Limetten heiß abwaschen und mit einem Zestenreißer einige Streifen abziehen.
- Den Kuchen etwas abkühlen lassen und aus der Form stürzen. Übrige Limette auspressen und den Saft mit dem Puderzucker zu einem Guss rühren. Guss auf dem Kuchen verteilen, mit Limettenstreifen verzieren.

Aufwand:	★
Zubereitungszeit:	25 Minuten
Backzeit:	50 Minuten
Pro Stück:	200 kcal/840 kJ

Gugelhupf mit Nougat

FÜR GUGELHUPF-FORM (1 1/2 l)
150 g Nuss-Nougat (im Backregal)
2 Vanilleschoten
5 Eier (Größe M)
200 g weiche Butter
200 g Zucker
150 g Mehl
200 g gemahlene Haselnüsse
4 EL Eierlikör, Salz

AUSSERDEM
Fett für die Form
150 g Puderzucker
2-3 EL Eierlikör

- Elektro-Ofen auf 175 Grad vorheizen. Die Form einfetten. Nougat würfeln, über dem warmen Wasserbad schmelzen. Vanilleschoten längs halbieren, Mark herauskratzen. Eier trennen.
- Butter mit 150 g Zucker cremig aufschlagen. Eigelbe nach und nach unterrühren. Mehl und Nüsse unter die Buttercreme heben. Masse halbieren. Unter die eine Hälfte Vanillemark und 4 EL Eierlikör, unter die andere Hälfte die Nougatmasse rühren. Eiweiße mit übrigem Zucker und 1 Prise Salz steif schlagen. Ebenfalls halbieren, jeweils unter die Teige heben.
- Nougatteig in die Form füllen, Vanilleteig darauf geben, mit Gabel marmorieren. Im Ofen (Gas: Stufe 2) 50-55 Minuten backen. Kuchen etwas abkühlen lassen und aus der Form nehmen.
- Puderzucker mit dem übrigen Eierlikör verrühren, den kalten Kuchen damit gleichmäßig beträufeln und den Guss fest werden lassen.

Aufwand:	★
Zubereitungszeit:	30 Minuten
Backzeit:	55 Minuten
Pro Stück:	340 kcal/1430 kJ

Schokoladenkuchen mit Beeren

FÜR 1 SPRINGFORM MIT KAMIN VON 26 cm Ø
300 g tiefgekühlte gemischte Beeren, 2 EL Puderzucker
150 g weiße Schokolade
200 g Butter, 5 Eier (Größe M)
150 g Zucker, 100 g geschälte gemahlene Mandeln, 250 g Mehl
1 Päckchen Backpulver
1-2 EL Zitronensaft
40 ml Crème de Cassis (Johannisbeerlikör; ersatzweise Johannisbeersaft)
AUSSERDEM
Fett für die Form
2-3 EL Johannisbeergelee

- Beeren auftauen und gut abtropfen lassen, 200 g mit Puderzucker pürieren. 100 g Schokolade fein raspeln. Die Butter in einem Topf schmelzen und lauwarm abkühlen lassen. Elektro-Ofen auf 175 Grad vorheizen. Form fetten.
- Eier mit Zucker in einer Schüssel schaumig schlagen. Butter unterrühren. Mandeln, Mehl und Backpulver mischen, unter die Eiercreme heben. Den Teig halbieren. Unter die eine Teighälfte übrige Beeren, Beerenpüree, Zitronensaft und Likör rühren, unter die andere Hälfte die geraspelte Schokolade geben.
- Beerenteig in die Springform füllen. Schokoteig darauf geben, beide Teige mit einer Gabel marmorieren. Kuchen im Ofen (Gas: Stufe 2) etwa 60 Minuten backen und aus der Form nehmen. Auf einem Kuchengitter abkühlen lassen.
- Johannisbeergelee in einem kleinen Topf erwärmen und auf den Kuchen streichen. Übrige Schokolade in Späne hobeln und über das Gelee streuen.

Aufwand:	★
Zubereitungszeit:	30 Minuten
Backzeit:	1 Stunde
Pro Stück:	270 kcal/1130 kJ

Schoko-Nuss-Kuchen

FÜR 1 KASTENFORM VON 30 cm
200 g Mokka-Schokolade
250 g Butter
200 g brauner Zucker
Salz, 5 Eier (Größe M)
300 g Mehl
200 g gemahlene Haselnüsse
1 Päckchen Backpulver
4 cl Whisky-Sahne-Likör
AUSSERDEM
Fett und gemahlene Nüsse
oder Brösel für die Form
150 g Zartbitter-Kuvertüre

- Elektro-Ofen auf 175 Grad vorheizen. Die Kastenform fetten und mit Nüssen oder Semmelbröseln ausstreuen.
- Die Mokka-Schokolade grob hacken. Butter, Zucker und 1 Prise Salz schaumig schlagen. Eier einzeln unterrühren. Mehl, Nüsse und Backpulver mischen, nach und nach unter die Butter-Zucker-Masse rühren. Schokoladenstücke und Likör unterheben.
- Den Teig in die Form geben, glatt streichen und im Ofen (Gas: Stufe 2) 65-70 Minuten backen.
- Kuchen aus dem Ofen nehmen, 5 Minuten in der Form ruhen lassen, auf ein Kuchengitter stürzen und auskühlen lassen.
- Kuvertüre grob zerkleinern und über dem Wasserbad schmelzen. Kuchen damit überziehen und am Rand herunter laufen lassen.

Aufwand:	★
Zubereitungszeit:	20 Minuten
Backzeit:	1 1/4 Stunden
Pro Stück:	255 kcal/1070 kJ

Info

Rührteig lässt sich in allen Formen backen: Wer z. B. keine kleinen Gugelhupfformen hat, gibt den Teig in eine große Form mt 2 1/2 l Inhalt (oder in eine Spring- oder Kastenform). Der Kuchen muss dann allerdings 10-15 Minuten länger backen. Machen Sie die Stäbchenprobe: ein Holzstäbchen in die Mitte des Kuchens stechen, kommt es trocken wieder heraus, ist der Kuchen fertig.

Mini-Mohn-Gugelhupf

FÜR 12 KLEINE GUGELHUPF-FORMEN (150 ml INHALT)
6 Eier (Größe M)
250 g Butter, 175 g Zucker
1 Päckchen Vanillezucker
1 Prise Salz, 500 g Mehl
1 Päckchen Backpulver
etwa 150 ml Milch
250 g Mohnback (Fertigprodukt)
1-2 TL abgeriebene Schale und Saft von 1/2 Limette
3 Msp. Anispulver
AUSSERDEM
Fett für die Form
Puderzucker zum Bestäuben

- Elektro-Ofen auf 175 Grad vorheizen. Gugelhupfförmchen einfetten.
- Eier, Butter, Zucker, Vanillezucker und Salz schaumig rühren. Mehl und Backpulver unterrühren. Nach und nach die Milch zugeben, bis der Teig schwer reißend vom Löffel fällt.
- Mohnback, Limettenschale, Limettensaft und Anis unterrühren. Teig in die Gugelhupfförmchen verteilen.
- Die Mini-Kuchen im Ofen (Gas: Stufe 2) etwa 35 Minuten backen.
- Die kleinen Kuchen stürzen und abkühlen lassen. Mit Puderzucker bestäuben.

Aufwand:	★
Zubereitungszeit:	20 Minuten
Backzeit:	35 Minuten
Pro Stück:	605 kcal/2540 kJ

Zitronenkuchen

FÜR 1 KASTENFORM VON 30 cm
200 g Butter
200 g Zucker
1 Prise Salz
60 g Zitronengelee
(ersatzweise Orangenmarmelade)
3 Eier und 2 Eigelbe
(Eier: Größe M)
250 g Mehl, 2 TL Backpulver
AUSSERDEM
Fett für die Form
1/8 l frisch gepresster
Zitronensaft (aus etwa
3 Zitronen)
100 g Puderzucker
Zitronen-Geleebonbons zum
Verzieren

- Elektro-Ofen auf 175 Grad vorheizen. Die Kastenform fetten. Butter, Zucker, Salz, Zitronengelee, Eier und Eigelbe dickschaumig aufschlagen. Mehl und Backpulver mischen und unter den Teig heben.
- Den Teig in die Form füllen und im Ofen (Gas: Stufe 2) etwa 50 Minuten backen.
- Kuchen in der Form etwas auskühlen lassen. Mit einem Holzstäbchen mehrmals einstechen. Zitronensaft und Puderzucker verrühren und den Kuchen damit beträufeln, bis die ganze Flüssigkeit aufgesogen ist. Den Kuchen mindestens 1 Stunde ruhen lassen. Dann vorsichtig aus der Form stürzen und eventuell mit Geleebonbons verzieren.

Aufwand:	★
Zubereitungszeit:	1 1/4 Stunden
Backzeit:	50 Minuten
Pro Stück:	180 kcal/670 kJ

Bratapfelkuchen

FÜR 1 SPRINGFORM VON 26 cm Ø
150 g Marzipan-Rohmasse
150 g Butter
70 g Zucker, 2 Eier (Größe M)
150 g Mehl
1/2 TL Backpulver, 2 EL Milch
FÜR DEN BELAG
10 kleine säuerliche Äpfel
1 EL Zitronensaft, 1 EL Butter
AUSSERDEM
Fett für die Form
2 EL Aprikosenkonfitüre
2 EL Calvados (Apfelschnaps)
2 EL Mandelblättchen

- Für den Rührteig Marzipan-Rohmasse und Butter in kleine Stücke teilen. Mit Zucker und Eiern schaumig rühren. Mehl mit Backpulver mischen, unter den Teig heben. Milch unterrühren.
- Den Elektro-Ofen auf 175 Grad vorheizen. Die Springform einfetten. Die Äpfel schälen und die Kerngehäuse großzügig entfernen. 1/2 l Wasser mit dem Zitronensaft in einem Topf zum Kochen bringen. Äpfel darin etwa 3 Minuten auf der abgeschalteten Herdplatte ziehen lassen.
- Teig in die Form füllen. 4 EL abnehmen. Äpfel eng aneinander in den Teig setzen, mit dem übrigen Teig füllen. Die Oberfläche mit zerlassener Butter bestreichen. Im Ofen (Gas: Stufe 2) in etwa 45 Minuten goldbraun backen.
- Die Konfitüre erwärmen, durch ein Sieb streichen und mit Calvados verrühren. 10 Minuten vor Ende der Backzeit die Äpfel damit bestreichen. Mandeln überstreuen und den Kuchen fertig backen.

Aufwand:	★★
Zubereitungszeit:	40 Minuten
Backzeit:	45 Minuten
Pro Stück:	390 kcal/1640 kJ

Schweizer Apfelkuchen

FÜR 1 SPRINGFORM VON 26 cm Ø
125 g Butter
2 Eier (Größe M)
75 g Zucker, 150 g Mehl
FÜR DIE FÜLLUNG
1 unbehandelte Zitrone
750 g säuerliche Äpfel
nach Belieben 3-4 EL Calvados (Apfelschnaps)
125 g Butter, 50 g Puderzucker
1 EL Vanille-Puddingpulver
3 Eier (Größe M)
AUSSERDEM
Fett für die Form
Puderzucker zum Bestäuben

- Elektro-Ofen auf 175 Grad heizen. Springform fetten.
- Butter schmelzen, leicht abkühlen lassen. Eier mit Zucker cremig rühren. Mehl rasch unter die Eiercreme heben. Flüssige Butter unterrühren. Teig in die Form streichen, dabei mit einem Löffel einen 4 cm hohen Rand formen. Den Teig in der Form kühl stellen.
- Für die Füllung die Zitrone waschen. Schale abreiben, Saft auspressen. Äpfel waschen, schälen, vierteln. Kerngehäuse entfernen. Apfelviertel längs in dünne Schnitze schneiden oder hobeln. Mit Zitronensaft und nach Belieben mit Calvados vermischen. Butter mit Puderzucker und Puddingpulver cremig rühren.
- Eier trennen. Eigelbe und Zitronenschale unter die Buttermischung rühren. Eiweiße steif schlagen, unterheben. Marinierte Apfelscheiben mit der Creme vermischen und in die Form auf den Rührteig füllen.
- Kuchen im Ofen (Gas: Stufe 2) etwa 55 Minuten backen. Den fertigen Kuchen abkühlen lassen, aus der Form lösen und mit Puderzucker bestäuben.

Aufwand:	★
Zubereitungszeit:	30 Minuten
Backzeit:	55 Minuten
Pro Stück:	325 kcal/1370 kJ

Apfel-Mohn-Kuchen

FÜR 1 SPRINGFORM VON 26 cm Ø
1 kg säuerliche Äpfel
1 EL Zitronensaft
FÜR DEN RÜHRTEIG
3 Eier (Größe M)
130 g Butter, 130 g Zucker
250 g Mehl, 1 TL Backpulver
etwa 5 EL Milch
100 g Mohnback (Fertigprodukt)
2 EL Apfelsaft
AUSSERDEM
Fett für die Form
2 EL Butter
2 EL Aprikosenkonfitüre
Puderzucker zum Bestäuben

● Äpfel putzen, waschen, schälen, Kerngehäuse entfernen. Äpfel vierteln, längs 3-5 mm tief einschneiden. Mit Zitronensaft beträufeln.
● Elektro-Ofen auf 175 Grad heizen. Die Springform einfetten.
● Eier trennen. Eiweiße steif schlagen. Fett mit Zucker und Eigelben schaumig rühren. Mehl mit Backpulver sieben und unter die Eigelbmasse heben. Eischnee vorsichtig unterziehen. So viel Milch zufügen, dass der Teig schwer reißend vom Löffel fällt.
● Mohnmasse mit Apfelsaft verrühren. Teig halbieren. Unter eine Hälfte die Mohnmasse rühren und in die Form füllen. Hellen Teig darauf geben, mit einer Gabel unterziehen.
● Butter schmelzen. Äpfel auf der Teigoberfäche verteilen, leicht eindrücken, mit Butter bestreichen. Den Kuchen im Ofen (Gas: Stufe 2) in 50-55 Minuten goldgelb backen.
● Den heißen Kuchen mit Konfitüre überglänzen. Rand vor dem Servieren mit Puderzucker bestäuben.

Aufwand:	★
Zubereitungszeit:	30 Minuten
Backzeit:	55 Minuten
Pro Stück:	310 kcal/1300 kJ

Variation

● Den Kuchen können Sie auch mit der entsprechenden Menge Birnen belegen, mit denen Sie genauso verfahren wie mit den Äpfeln. Aprikosen, Kirschen oder Pflaumen schmecken ebenso gut. Diese Früchte werden allerdings nicht eingeschnitten.

Kokos-Kirsch-Kuchen

FÜR FETTPFANNE (30 x 40 cm)
5 Eier (Größe M), 350 g Zucker
1 Prise Salz
1/2 TL Ingwerpulver
175 g Kokoscreme (Fertigprodukt; im Asienregal; ersatzweise 150 g Butter)
175 g Butter
400 g Mehl
1/2 Päckchen Backpulver
1/4 l Milch
1 1/2 kg Sauerkirschen
100 g Kokos-Chips (Reformhaus; ersatzweise Kokosraspel)
AUSSERDEM
Fett für das Blech

● Elektro-Ofen auf 175 Grad vorheizen. Eier mit 200 g Zucker, Salz, Ingwer, Kokoscreme und 75 g Butter verrühren. Mehl und Backpulver unterheben, nach und nach die Milch zugießen.

● Die Fettpfanne des Backofens einfetten, den Teig darin glatt verstreichen. Kirschen waschen, putzen, entsteinen und auf dem Teig verteilen. Mit den Kokos-Chips, der übrigen Butter in Flöckchen und dem restlichen Zucker bestreuen.

● Kuchen im Ofen (Gas: Stufe 2) etwa 40 Minuten backen. Abkühlen lassen und in gleich große Stücke schneiden.

Aufwand:	★
Zubereitungszeit:	25 Minuten
Backzeit:	40 Minuten
Pro Stück:	275 kcal/1160 kJ

Johannisbeertorte mit Mandelbaiser

FÜR 2 SPRINGFORMEN (24 cm Ø)
150 g weiche Butter
100 g Zucker, 2 EL Zitronensaft
4 Eigelbe (Eier: Größe M)
150 g Mehl, 1 TL Backpulver
FÜR DIE BAISER-MASSE
4 Eiweiße (Eier: Größe M)
180 g Zucker
50 g gehackte Mandeln
FÜR DIE FÜLLUNG
500 g Johannisbeeren
1 EL Puderzucker
500 g Schlagsahne
2 Päckchen Sahnesteif
AUSSERDEM
Fett für die Formen
nach Belieben Zitronenmelisse

● Elektro-Ofen auf 175 Grad heizen. Die Springformen einfetten.

● Butter mit Zucker, Zitronensaft und den Eigelben schaumig rühren. Mehl mit dem Backpulver mischen, darüber sieben und unter die Eigelb-Masse ziehen.

● Eiweiße mit Zucker steif schlagen. Je eine Teighälfte in eine Backform geben, glatt streichen. Baiser-Masse wolkenartig darauf verteilen. Mandeln darüber streuen.

● Beide Kuchen im Ofen (Gas: Stufe 2) etwa 25 Minuten backen. Herausnehmen, abkühlen lassen und aus der Form lösen.

● Johannisbeeren waschen, einige Rispen beiseite legen. Beeren abstreifen, mit Puderzucker mischen. Sahne mit Sahnesteif steif schlagen.

● Einen Kuchen auf eine Platte legen. Die Hälfte der Sahne und die Beeren darauf verteilen. Mit übriger Sahne bedecken. Den zweiten Boden in 12 Tortenstücke teilen, fächerartig darauf setzen. Johannisbeer-Rispen und Zitornenmelisse obenauf setzen.

Aufwand:	★
Zubereitungszeit:	25 Minuten
Backzeit:	25 Minuten
Pro Stück:	370 kcal/1560 kJ

Maikäfer-Schoko-Torte

FÜR 1 SPRINGFORM VON 26 cm Ø
6 Eier, 250 g weiche Butter
250 g Zucker, Salz, 300 g Mehl
50 g Speisestärke
1/2 TL Backpulver, 3 TL Kakao
FÜR DIE FÜLLUNG
1 Päckchen Mandelpudding
400 ml Milch, 40 g Zucker
2 Eigelbe, 1 EL Puderzucker
Bittermandelöl, 1 Dose Pfirsiche
(Abtropfgewicht: 470 g)
3 EL Aprikosenkonfitüre
4 cl Sherry (süß)
FÜR DIE DEKORATION
Kakao, 100 g Marzipan-Rohmasse
100 g Puderzucker
50 g Vollmilch-Kuvertüre
24 Mandelhälften, Puderzucker
zum Bestäuben, Schokospäne

1 Aus Marzipan 12 ovale Käferkörper formen, eine Seite in flüssige Kuvertüre tauchen und auf einem Abtropf-Gitter fest werden lassen.
2 Mit feinem Pinsel Rücken und Punkte markieren, seitlich Mandelhälften als Flügel ins Marzipan stecken.

- Elektro-Ofen auf 200 Grad vorheizen. Springformboden einfetten.
- Eier trennen. Butter mit Eigelben und Zucker schaumig rühren. Eiweiße mit 1 Prise Salz steif schlagen. Mehl mit Stärke, Backpulver und Kakao mischen, mit Eischnee unter die Eigelbcreme ziehen.
- 2 EL Teig mit einer Palette auf den Springformboden streichen. Im Ofen (Gas: Stufe 3) etwa 10 Minuten backen, vorsichtig ablösen. Aus dem übrigen Teig auf die gleiche Art weitere 5 Böden backen und auskühlen lassen.
- Für die Füllung Puddingpulver mit 4 EL Milch anrühren. Übrige Milch mit Zucker aufkochen. Angerührtes Pulver einrühren, aufkochen. Topf vom Herd nehmen, Eigelbe, Puderzucker und 2 Tropfen Mandelöl unterziehen. Masse kalt rühren.
- Pfirsiche in hauchdünne Spalten schneiden. Konfitüre erhitzen, durch ein Sieb streichen. Die Kuchenböden mit Sherry beträufeln. 3 Böden mit Konfitüre bestreichen und mit Früchten belegen. Die anderen dünn mit Creme bestreichen. Abwechselnd aufeinander schichten.
- Torte mit der übrigen Creme überziehen. Rand mit Kakao bestäuben. Auf der Oberfläche mit einer grob gezackten Teigkarte Rillen ziehen.
- Marzipan mit Puderzucker verkneten. Maikäfer formen, mit Kuvertüre und Mandeln verzieren und auf die Torte setzen. Torte mit Puderzucker bestäuben und mit Schoko-Spänen dekorieren.

Aufwand:	★★★
Zubereitungszeit:	40 Minuten
Backzeit:	50 Minuten
Pro Stück:	605 kcal/2540 kJ

Marzipan-Früchte-Kuchen

FÜR 1 SPRINGFORM VON 26 cm Ø
BITTE BEACHTEN: MUSS 4 STUNDEN KÜHLEN
50 g Butter
1/2 unbehandelte Zitrone
5 Eier (Größe M), 120 g Zucker
Salz, 100 g Mehl
40 g Speisestärke

FÜR DIE FÜLLUNG
1 1/4 kg Obst (z. B. Nektarinen Melonen, Weintrauben, Feigen)
200 ml Marsala (Dessertwein)
3 Blatt weiße Gelatine
2 Eier (Größe M)
3 Päckchen Vanillezucker
50 g Zucker, 300 ml Milch
200 g Schlagsahne

AUSSERDEM
Backpapier, 1 Marzipandecke (400 g; etwa 38 cm Ø)
1 Päckchen weißer Tortenguss
1/4 l Apfelsaft
2 EL Aprikosenkonfitüre
gehackte Pistazienkerne

- Boden der Springform mit Backpapier auslegen. Elektro-Ofen auf 200 Grad vorheizen.
- Butter zerlassen. Von der Zitrone 1/2 TL Schale abreiben, Saft auspressen. Eier, Zucker, Salz, Zitronenschale etwa 5 Minuten schlagen. Mehl, Stärke, Butter und Zitronensaft unterrühren. In die Form geben und im Ofen (Gas: Stufe 3) etwa 30 Minuten backen .
- Inzwischen das Obst waschen, putzen, in Stücke schneiden und 30 Minuten in Marsala marinieren. Im Sieb gut abtropfen lassen. Marsala auffangen.
- Gelatine einweichen. Eier trennen. Eigelbe mit Vanillezucker, 2 EL Zucker und heißer Milch über dem Wasserbad schaumig schlagen. Ausgedrückte Gelatine darin auflösen.
- Eiweiße und Sahne mit dem übrigen Zucker getrennt steif schlagen. Zuerst Sahne, dann Eischnee unter die abgekühlte Eigelb-Masse heben.
- Kuchenboden waagerecht halbieren, mit Marsala tränken. Tortenring um unteren Boden stellen, 1/3 der Früchte darauf verteilen, Creme einfüllen, glatt streichen. 1/3 Früchte daraufgeben, Kuchendeckel aufsetzen. Torte 4 Stunden kalt stellen.
- Tortenring lösen. Torte mit Marzipan überziehen. Tortenmitte mit restlichen Früchten dekorieren.
- Guss nach Packungsangabe mit Apfelsaft kochen, Früchte damit überziehen. Marzipan mit Konfitüre einpinseln. Tortenrand eventuell mit Pistazien bestreuen.

Aufwand:	★★★
Zubereitungszeit:	1 1/2 Stunden
Backzeit:	30 Minuten
Pro Stück:	490 kcal/2080 kJ

Marzipan-Sandkuchen

FÜR 1 KASTENFORM VON 30 cm
250 g Butter
160 g Zucker, 1 Prise Salz
abgeriebene Schale von 1 unbehandelten Orange
200 g Marzipan-Rohmasse
3-4 Tropfen Mandelaroma
5 Eier (Größe M)
200 g Mehl
80 g Speisestärke
100 g gemahlene geschälte Mandeln
1-2 TL Backpulver
AUSSERDEM
Fett für die Form
250 g Schlagsahne
abgeriebene Schale von 1 unbehandelten Orange
2 EL Zucker

● Kastenform einfetten. Elektro-Ofen auf 175 Grad heizen. Butter, Zucker, Salz und Orangenschale schaumig rühren. Marzipan grob hacken, unterrühren. Mandelaroma und Eier darunter schlagen. Mehl, Stärke, Mandeln und Backpulver schnell unter die Butter-Eier-Masse rühren. Teig in die Form füllen.

● Kuchen im Ofen (Gas: Stufe 2) etwa 60 Minuten backen. Aus der Form nehmen, abkühlen lassen.

● Sahne mit Zucker und Orangenschale schlagen und zum Kuchen servieren.

Aufwand:	★
Zubereitungszeit:	20 Minuten
Backzeit:	1 Stunde
Pro Stück:	310 kcal/1300 kJ

Variation

● Sie wollen den Kuchen in Gläsern servieren? So geht's: 200-ml-Gläser mit möglichst dicken Wänden innen einfetten. Teig einfüllen, im vorgeheizten Ofen bei 175 Grad (Gas: Stufe 2) etwa 35 Minuten backen.

Möhren-Muffins

FÜR 1 MUFFINFORM MIT
12 VERTIEFUNGEN
250 g Möhren, 200 g Mehl
Saft und Schale von
1 unbehandelten Zitrone
150 g Walnusskerne
2 gehäufte TL Backpulver
1/2 TL gemahlener Ingwer
1 Ei (Größe M)
150 g brauner Zucker
100 ml Öl, 200 g saure Sahne
AUSSERDEM
100 g Puderzucker
1 EL Zitronensaft
12 Marzipanmöhren (vom
Konditor)
Öl für die Muffinform

- Möhren waschen, schälen und fein reiben. Mit 1 EL Zitronensaft und Zitronenschale mischen. Walnusskerne fein hacken.
- Elektro-Ofen auf 175 Grad vorheizen. Die Vertiefungen der Muffinform einölen.
- Mehl, Backpulver und Ingwer mischen. Ei mit Zucker schaumig rühren. Öl, saure Sahne, Möhren und Nüsse zufügen. Mehlmischung unterrühren.
- Den Teig in die Vertiefungen füllen und im Ofen (Gas: Stufe 2) in 25 Minuten goldbraun backen.
- Muffins herausnehmen, etwa 5 Minuten ruhen lassen. Die Muffinform auf ein Kuchengitter legen, die Muffins stürzen und wieder wenden.
- Puderzucker mit Zitronensaft zu einer nicht zu dicken Glasur verrühren. Die Muffins mit Glasur bestreichen und mit den Möhrchen dekorieren.

Aufwand:	★
Zubereitungszeit:	20 Minuten
Backzeit:	25 Minuten
Pro Stück:	360 kcal/1510 kJ

Aprikosen-Muffins

FÜR 1 MUFFINFORM MIT
12 VERTIEFUNGEN
10 frische Aprikosen
2 EL Ahornsirup
50 g Pinienkerne
220 g Mehl
2 gehäufte TL Backpulver
1 Ei (Größe M)
180 g brauner Zucker
100 ml Öl
200 g Buttermilch
2 cl Aprikosenlikör
AUSSERDEM
Öl für die Muffinform

- Die Aprikosen blanchieren und häuten. Dann halbieren, die Steine entfernen. 6 Aprikosenhälften nochmals halbieren und fächerförmig einschneiden. Die restlichen Aprikosen würfeln und mit Ahornsirup vermischen. Pinienkerne in einer trockenen Pfanne anrösten.
- Den Elektro-Ofen auf 175 Grad vorheizen. Die Vertiefungen der Muffinform einölen.
- Das Mehl mit Backpulver mischen. Ei mit 150 g Zucker, Öl, Buttermilch und Likör glatt rühren. Mehlmischung schnell unterrühren. Aprikosenwürfel und Pinienkerne unterziehen.
- Den Teig in die Vertiefungen füllen und im Ofen (Gas: Stufe 2) etwa 15 Minuten vorbacken. Muffins mit Aprikosenfächern belegen, mit übrigem braunem Zucker bestreuen und in 5-10 Minuten fertig backen.
- Muffins aus dem Ofen, 5 Minuten ruhen lassen. Dann herauslösen und am besten lauwarm servieren.

Aufwand:	★
Zubereitungszeit:	20 Minuten
Backzeit:	25 Minuten
Pro Stück:	250 kcal/1060 kJ

Schoko-Muffins

FÜR 1 MUFFINFORM MIT
12 VERTIEFUNGEN
100 g Zartbitter-Schokolade
100 ml Milch
100 g Mehl
100 g Kokosraspel
2 gehäufte TL Backpulver
100 g weiche Butter
100 g Puderzucker
1 Ei (Größe M)
AUSSERDEM
Öl für die Muffin-Form
1 EL Puderzucker
60 g Vollmilch-Kuvertüre
1 TL Butter, 50 g Kokos-Chips

- Schokolade hacken, in der erwärmten Milch schmelzen und abkühlen lassen.
- Elektro-Ofen auf 175 Grad vorheizen. Vertiefungen der Muffinform einölen.
- Mehl mit Kokosraspeln und Backpulver mischen. Butter mit Puderzucker und Ei cremig schlagen. Schokoladen-Milch unterrühren. Mehlmischung unterheben.
- Teig in die Vertiefungen füllen und im Ofen (Gas: Stufe 2) 25 Minuten backen. Muffins herausnehmen und abkühlen lassen.
- Für den Guss 2 EL Wasser mit Puderzucker erhitzen, bis sich dieser gelöst hat. Etwas abkühlen lassen. Die Kuvertüre hacken und in der Zuckerlösung schmelzen. Glasur vom Herd nehmen und die Butter einrühren.
- Muffins aus den Vertiefungen lösen und auf ein Kuchengitter setzen. Den Guss auf die Muffin-Oberflächen geben, leicht antrocknen lassen. Muffins mit Kokos-Chips dekorieren.

Aufwand:	★
Zubereitungszeit:	30 Minuten
Backzeit:	25 Minuten
Pro Stück:	260 kcal/1090kJ

Er ist wohl der variationsreichste aller Teige, kinderleicht

Torte im Marzipankleid

Mit fertig ausgerollten Marzipandecken (im Handel) kann man Torten süß verhüllen: Aus Backpapier einen Kreis schneiden, der so groß ist wie die Form und auf den fertigen Kuchen legen. Darüber die Marzipandecke legen, die man wie Tortenstücke einschneidet. Die Segmente von der Mitte zum Rand hin einrollen und das Backpapier vorsichtig herausziehen. Dann die Marzipandecke am Rand rundherum andrücken.

Becher-Kuchen

Das ist das einfachste Rezept der Welt: Becherkuchen ist ein Rührkuchen vom Blech. Das Maß ist ein 250-g-Joghurt- oder Sahne-Becher. Für ein Blech von 30 x 40 cm verrühren Sie 1 Becher Schlagsahne, 4 Eier (Größe M), 1 Becher Zucker, 1 Päckchen Vanillezucker, 2 Becher Mehl, 1 Päckchen Backpulver und 1 Prise Salz zu einem Rührteig. Teig auf dem gefetteten Blech verteilen. 1 Becher Zucker, vermischt mit 4 EL Milch oder Sahne, daraufgeben und 400 g Mandelblättchen darüber streuen. Im vorgeheizten Ofen wird der Kuchen bei 175 Grad (Gas: Stufe 2) etwa 15 Minuten gebacken.

Puderzucker-Zaubereien

Natürlich können Sie jeden Kuchen ganz einfach mit Puderzucker überstreuen, aber Sie können sein Aussehen auch individuell, ja künstlerisch gestalten. Das ist ganz einfach: Schneiden Sie aus dünner Pappe Streifen, Ornamente, Figuren oder Ähnliches aus und legen Sie sie, wie oben die Zickzack-Streifen, auf den Kuchen. Anschließend sieben Sie Puderzucker darüber. Dann nehmen Sie die Streifen sehr vorsichtig weg und der Kuchen hat ein Zickzack-Muster. Übrigens, wer Tortendeckchen aus Papier zur Hand hat, kann auch diese als Schablone benutzen und so ein filigranes Muster auf den Kuchen stäuben.

u Rührkuchen

ken und lässt sich vielfältig verzieren und belegen

Puder-zucker Glasur

Man kann alle Kuchen damit überziehen und sie ist ganz leicht herzustellen: Für 250 g Puderzucker braucht man 3-5 EL Wasser, je nachdem, wie dickflüssig die Glasur sein soll. Das Wasser in eine Schale geben, den Puderzucker portionsweise daraufsieben und mit einem Schneebesen glatt rühren. Es dürfen keine Klümpchen entstehen. Statt Wasser können Sie auch Zitronensaft verwenden.

Grundrezept für Gugelhupf

FÜR 1 GUGELHUPFFORM
250 g weiche Butter
250 g Zucker
1 Päckchen Vanillezucker
6 Eier (Größe M), 500 g Mehl
1 Päckchen Backpulver
1/8 l Milch
AUSSERDEM
Fett und Semmelbrösel für die Form

- Elektro-Ofen auf 175 Grad vorheizen. Form einfetten und mit Bröseln ausstreuen. Fett, Zucker, Vanillezucker und Eier mit dem Schneebesen des Handrührgerätes schlagen, bis eine cremige Masse entsteht (s. Step 1). Der Zucker soll sich ganz auflösen.
- Mehl und Backpulver mischen. Mehl-Mischung und Milch abwechselnd in den Teig rühren. Schnell rühren, sonst fängt das Mehl an zu quellen und der Kuchen wird trocken. Der Teig hat die richtige Konsistenz, wenn er schwer reißend vom Löffel fällt (s. Step 2). Ist er zu flüssig oder zu fest, etwas mehr Mehl oder Milch zugeben.
- Teig in die Form füllen (s. Step 3), sofort in die Ofenmitte (Gas: Stufe 2) stellen und etwa 60 Minuten backen. Auskühlen lassen.
- Noch warm aus der Form stürzen (s. Step 4). Nach Belieben mit Puderzucker bestäuben.

Variation

- Für Marmorkuchen die Hälfte des Teiges mit 100 g geriebener Halbbitter-Schokolade und 2 EL Kakao mischen und beide Teige abwechselnd in die Form geben, damit eine Marmorierung entsteht. Weiter verfahren wie im Rezept angegeben.

leicht

Der Edelste unter den Teigen besteht vorwiegend aus Eiern. Das steif geschlagene Eiweiß wird untergehoben und macht die Masse fast schwerelos – ideal für Torten

Biskuit
ocker & luftig

Planter's Punch-Torte

Rezept siehe folgende Seite

Planter's Punch-Torte

BITTE BEACHTEN: TORTE MUSS 4 STUNDEN KÜHLEN

FÜR 1 BACKBLECH (30 x 40 cm)
5 Eier (Größe M), 130 g Zucker
100 g Mehl, 1 EL Grenadine-Sirup, 100 g Johannisbeergelee
FÜR DIE CREME
4 Eigelbe (Eier: Größe M)
40 g Puderzucker
10 Blatt weiße und
2 Blatt rote Gelatine
9 EL Zitronensaft
400 ml Orangensaft
10 EL Mango-Sirup
10 EL weißer Rum
300 g Schlagsahne
AUSSERDEM
Backpapier

● Elektro-Ofen auf 200 Grad heizen. Backblech mit Backpapier auslegen.
● Eier trennen, Eiweiße steif schlagen. Eigelbe mit 100 g Zucker schaumig rühren. Eischnee auf die Eigelb-Masse geben, das Mehl darüber sieben, locker mischen. Teig auf das Blech streichen, im Ofen (Gas: Stufe 3) etwa 15 Minuten backen.
● Zwei feuchte Küchentücher mit übrigem Zucker bestreuen. Biskuit halbieren, darauf stürzen, Backpapier abziehen. Beide Biskuits fest mit den Tüchern einrollen, mit Klarsichtfolie bedeckt abkühlen lassen.
● Biskuit wieder aufrollen, Grenadine-Sirup mit Johannisbeergelee glatt rühren, auf den Biskuit streichen. Biskuit ohne Tuch wieder einrollen, in Klarsichtfolie wickeln und kühl stellen.
● Eigelbe mit Puderzucker im heißen Wasserbad schaumig schlagen. Gelatine einweichen, ausdrücken und in der Eiercreme auflösen, Zitronen-, Orangensaft, Mango-Sirup und Rum einrühren. Im kalten Wasserbad abkühlen lassen. Creme kühl stellen, bis sie zu gelieren beginnt.
● Biskuit-Rouladen aus der Folie nehmen, in 1 cm dicke Scheiben schneiden. Eine Schüssel (etwa 21 cm Ø und 1 1/2 l Inhalt) mit Klarsichtfolie auskleiden. Am Boden beginnend eng mit Biskuitscheiben auslegen.
● Sahne steif schlagen und unter die Creme rühren. Die Masse in die Schüssel füllen und mindestens 4 Stunden kalt stellen. Die Torte vor dem Servieren stürzen und die Folie abziehen.

Aufwand:	★★
Zubereitungszeit:	45 Minuten
Backzeit:	15 Minuten
Pro Stück:	185 kcal/780 kJ

Variationen

● Wenn Sie diese Torte ohne Alkohol zubereiten möchten, ersetzen Sie den Rum durch 10 EL Kokosmilch (entfettete, homogenisierte Kokosmilch gibt es im Asialaden) oder durch 10 EL Ananassaft.
● Mit den Biskuitscheiben können Sie auch viele andere Torten verzieren z. B.: Füllen Sie die Roulade mit Schokocreme (Fertigprodukt) und bringen Sie die Scheiben am Rand der Tiramisu-Torte mit Erdbeeren (siehe Seite 78) an.

Tipp

Für alle, die wenig Zeit haben, empfehlen wir für Obst-, Creme-, Quarktorten und Co. Tortenböden aus Löffelbiskuit: 175 g weiche Butter mit 1 Prise Salz schaumig rühren. 250 g Löffelbikuits zerkrümeln und untermischen. In eine Springform geben und glatt andrücken. Form etwa 30 Minuten kalt stellen und dann belegen oder füllen.

Beerenkuchen mit Mandelbaiser

FÜR 1 BACKBLECH (30 x 40 cm)
6 Eier (Größe M), 150 g Zucker
Salz, 125 g Mehl
180 g geschälte gemahlene Mandeln
FÜR DEN BELAG
700 g Himbeeren
500 g Brombeeren
75 g Himbeergelee
FÜR DAS BAISER
5 Eiweiße (Eier: Größe M)
175 g Zucker
3-4 Tropfen Bittermandel-Aroma
AUSSERDEM
Backpapier
2-3 EL Mandelblättchen
Puderzucker zum Bestäuben

● Elektro-Ofen auf 200 Grad vorheizen. Eier trennen. Eigelbe mit 2 EL Wasser und 75 g Zucker schaumig schlagen. Eiweiße mit 1 Prise Salz steif schlagen. Übrigen Zucker einrieseln lassen. 4 Minuten weiterschlagen, dann Eischnee auf die Eigelb-Masse geben.
● Mehl und Mandeln mischen, auf den Eischnee geben und unterheben. Biskuit-Masse auf ein mit Backpapier belegtes Blech streichen und im Ofen (Gas: Stufe 3) in etwa 12 Minuten goldgelb backen. Backofen-Temperatur auf 250 Grad erhöhen.
● Beeren waschen und verlesen. Gelee erwärmen und auf die abgekühlte Biskuitplatte streichen. Beeren darauf verteilen.
● Eiweiße steif schlagen. Zucker esslöffelweise zugeben, Mandel-Aroma zufügen. Weiterschlagen, bis sich der Zucker gelöst hat. Eiweiß-Masse auf den Beeren verstreichen. Mandeln darüber streuen. Baiser im Ofen (Gas: Stufe 5) in 4 Minuten goldbraun übergrillen. Kuchen abkühlen lassen und mit Puderzucker überstäuben.

Aufwand:	★
Zubereitungszeit:	30 Minuten
Backzeit:	16 Minuten
Pro Stück:	170 kcal/ 710 kJ

1 **Erdbeerscheiben** vor dem Verzieren von Sahnetorten mit Küchenpapier abtupfen, damit sie nicht abfärben.

2 **Erdbeerkranz:** Sahne mit glatter Tülle als Tupfen aufspritzen und Erdbeerscheiben an die Tupfen stellen.

1

2

Erdbeer-Marzipan-Torte

BITTE BEACHTEN: TORTE MUSS 4 STUNDEN KÜHLEN

FÜR SPRINGFORM VON 26 cm Ø

3 Eier (Größe M)
1 Prise Salz
90 g Zucker
90 g Mehl

FÜR DIE FÜLLUNG

250 g Erdbeeren
70 g Marzipan-Rohmasse
70 g Puderzucker
30 g gemahlene Pistazien
1 Zitrone
8 Blatt weiße Gelatine
450 g Sahne-Joghurt
150 g Zucker
2 Päckchen Vanillezucker
4 cl weißer Kokoslikör
200 g Schlagsahne
3-4 EL Quittengelee

FÜR DEN BELAG

500 g kleine Erdbeeren
300 g Schlagsahne
1 Päckchen Sahnesteif
100 g Mandelblättchen

AUSSERDEM

Backpapier

● Elektro-Ofen auf 175 Grad heizen. Eier trennen. Eiweiße mit Salz und 1 EL Zucker steif schlagen. Eigelbe mit übrigem Zucker und 2 EL lauwarmem Wasser einige Minuten lang cremig aufschlagen. Eischnee und gesiebtes Mehl unterheben.

● Boden der Springform mit Backpapier auslegen. Biskuit einfüllen. Im Ofen (Gas: Stufe 2) etwa 25 Minuten backen. Auskühlen lassen.

● Erdbeeren waschen, putzen, vierteln. Marzipan mit Puderzucker und Pistazien verkneten. Zwischen Klarsichtfolien zu 2 dünnen Platten (Ø 26 cm) ausrollen. Zitrone auspressen.

● Gelatine einweichen. Joghurt mit Zucker, Vanillezucker, Zitronensaft und Likör glatt rühren. Gelatine bei milder Hitze auflösen. Erst 2 EL Creme, dann die restliche Creme in die Gelatine rühren. Sahne steif schlagen, unterziehen.

● Biskuit waagerecht halbieren. Gelee aufkochen. Boden und Deckel dünn damit bestreichen, mit je 1 Marzipanplatte belegen. Um den Boden einen Tortenring legen. Creme einfüllen, Deckel aufsetzen. 4 Stunden kalt stellen.

● Erdbeeren waschen, putzen, in dünne Scheiben teilen. Sahne mit Sahnesteif schlagen. Mandeln leicht rösten.

● Torte mit Sahne überziehen, Rand mit Mandeln bedecken. Mit Erdbeeren und Sahnetupfen dekorieren.

Aufwand:	★
Zubereitungszeit:	45 Minuten
Backzeit:	25 Minuten
Pro Stück:	405 kcal/1700 kJ

Panna-cotta-Torte

BITTE BEACHTEN: TORTE MUSS 3 STUNDEN KÜHLEN
FÜR SPRINGFORM VON 24 cm Ø
3 Eier (Größe M), 130 g Zucker
Schale und Saft von 1/2 Zitrone
40 g Mehl, 40 g Speisestärke
40 g gemahlene Mandeln
1/2 TL Backpulver
5 Tropfen Bittermandel-Aroma
FÜR DIE FÜLLUNG
6 Blatt weiße und 3 rote Gelatine
300 ml Milch, 400 g Schlagsahne
2 Päckchen Vanillezucker
100 g Zucker, 300 g Joghurt
500 g Erdbeeren
AUSSERDEM
Backpapier
200 g Schlagsahne, 100 g Mandelblättchen, 25 g gehackte Pistazien, 250 g Erdbeeren

● Elektro-Ofen auf 175 Grad heizen. Eier trennen. Eigelbe mit 3 EL Wasser, 65 g Zucker, Zitronenschale dickschaumig schlagen. Eiweiße mit übrigem Zucker sehr steif schlagen.
● Mehl, Stärke, Mandeln, Backpulver mischen. Mit Eigelb-Masse, Eischnee, Zitronensaft und Aroma vermengen. In die mit Backpapier ausgelegte Springform füllen und im Ofen (Gas: Stufe 2) etwa 25 Minuten backen. Abkühlen lassen, waagerecht halbieren.
● Gelatine einweichen. Milch mit 100 g Sahne, Vanillezucker und Zucker etwa 5 Minuten köcheln. Vom Herd nehmen. Gelatine ausdrücken und darin auflösen. Creme angelieren lassen. Übrige Sahne steif schlagen und mit dem Joghurt unterheben.
● Tortenring um einen Boden legen. Erdbeeren waschen, putzen, vierteln, unter die Creme heben. Erdbeercreme einfüllen, zweiten Boden auflegen und 3 Stunden kalt stellen.
● Tortenring entfernen, Sahne steif schlagen. Mandeln rösten. Torte mit Sahne, Mandeln und Pistazien dekorieren. Die Erdbeeren waschen, putzen, in Scheiben teilen und als Herz auf die Torte legen.

Aufwand:	★★
Zubereitungszeit:	45 Minuten
Backzeit:	25 Minuten
Pro Stück:	455 kcal/1905 kJ

Variation

● Sie können die Torte ebensogut mit in Scheiben geschnittenen frischen Kiwis oder frischen Himbeeren belegen. Auch Ananas macht sich gut, und sie darf ruhig aus der Dose kommen.

Tiramisu-Torte mit Erdbeeren

BITTE BEACHTEN: TORTE MUSS 4 STUNDEN KÜHLEN

FÜR SPRINGFORM VON 24 cm Ø

4 Eier (Größe M), 100 g Zucker
100 g Mehl, 80 g Speisestärke

FÜR DIE FÜLLUNG

150 g Kaffeebohnen
400 g Schlagsahne
4 Eigelbe (Eier: Größe M)
50 g Zucker, 500 g Mascarpone
250 g Sahnequark, 2 EL Zitronensaft, 2 Pck. Sahnesteif
800 g Erdbeeren
3 EL Amaretto, etwa 30 Löffelbiskuits, Kakaopulver

- Kaffeebohnen in die Sahne geben und über Nacht kalt stellen.
- Elektro-Ofen auf 200 Grad (Gas: Stufe 3) heizen.
- Eier trennen, Eigelbe mit 100 g Zucker schaumig rühren. Eiweiße steif schlagen, übrigen Zucker einrieseln lassen, auf die Eigelb-Masse geben. Mehl, Stärke zufügen, alles verrühren. In einer mit Backpapier ausgelegten Springform etwa 15 Minuten backen.
- Eigelbe mit Zucker schaumig schlagen. Mascarpone, Quark und Zitronensaft unterziehen. Kaffeebohnen aus der Sahne nehmen. 200 g Sahne mit Sahnesteif steif schlagen. Erdbeeren waschen, vierteln, 8 Viertel beiseite legen. Restliche mit der Sahne unter die Masse geben.
- Biskuit waagerecht halbieren. Um eine Hälfte einen Tortenring legen. Biskuit mit Amaretto tränken. Hälfte der Creme einfüllen, 2. Boden auflegen. Übrige Creme einfüllen.
- Restliche Sahne steif schlagen. Tortenring entfernen. Tortenrand dünn mit Sahne bestreichen. Löffelbiskuits auf Tortenhöhe abschneiden und an den Rand drücken. Torte mit Kakao bestäuben, mit Sahne und übrigen Erdbeeren verzieren.

Aufwand:	★★
Zubereitungszeit:	1 Stunde
Backzeit:	15 Minuten
Pro Stück:	590 kcal/ 2580 kJ

Topfentorte

BITTE BEACHTEN: TORTE MUSS 4 STUNDEN KÜHLEN

FÜR SPRINGFORM VON 26 cm Ø

8 Eier (Größe M), 190 g Zucker
3 EL Zitronensaft
1 Pck. Bourbon-Vanillezucker
225 g Mehl, 50 g Butter

FÜR DIE FÜLLUNG

8 Blatt weiße Gelatine
8 Eigelbe (Eier: Größe M)
3 EL Zitronensaft, 180 g Zucker
1 Pck. Bourbon-Vanillezucker
3 EL Rum, 400 g Schlagsahne
500 g Quark (20 % Fett i.Tr.)
4 EL Johannisbeerkonfitüre
300 g Früchte (z. B. abgetropfte Mandarinen und/oder Ananas aus der Dose)

AUSSERDEM

Backpapier, Puderzucker
gehackte Mandeln

- Elektro-Ofen auf 175 Grad heizen. Eier und Zucker etwa 5 Minuten schaumig schlagen. Zitronensaft und Vanillezucker unterrühren. Mehl und zerlassene Butter unterheben.
- Springform mit Backpapier auslegen. Biskuit-Masse in die Form füllen, im Ofen (Gas: Stufe 2) etwa 40 Minuten backen. Herausnehmen, auskühlen lassen.
- Gelatine in kaltem Wasser einweichen. Eigelbe mit Zitronensaft, Zucker, Vanillezucker und Rum verrühren, über dem heißen Wasserbad etwa 5 Minuten aufschlagen. Abgetropfte Gelatine unterrühren.
- Sahne steif schlagen. Quark und Sahne verrühren, unter die Eigelb-Masse ziehen. Biskuit waagerecht halbieren. Boden und Deckel dünn mit Konfitüre bestreichen. Tortenring um den Boden legen. Früchte unter die Creme heben, einfüllen. Biskuitdeckel aufsetzen. 4 Stunden kalt stellen.
- Vor dem Servieren Tortenring entfernen und die Torte mit Puderzucker bestäuben. Rand nach Belieben mit gerösteten, gemahlenen Mandeln bestreuen.

Aufwand:	★★
Zubereitungszeit:	30 Minuten
Backzeit:	40 Minuten
Pro Stück:	410 kcal/ 1720 kJ

Cappuccino-Torte

BITTE BEACHTEN: TORTE MUSS 2 STUNDEN KÜHLEN

FÜR SPRINGFORM VON 22 cm Ø
- *40 g gemahlene Mandeln*
- *2 Eier (Größe M), Salz*
- *60 g Zucker, 20 g Speisestärke*
- *40 g Mehl, 2 Msp. Backpulver*
- *2 TL Kakaopulver*

FÜR DIE FÜLLUNG
- *80 g Mokka- oder Zartbitter-Schokolade, 4 Blatt weiße Gelatine, 400 g Schlagsahne*
- *80 ml Espresso*
- *80 ml Kaffeelikör*
- *2 Pck. Bourbon-Vanillezucker*
- *1 EL Kakaopulver*

AUSSERDEM
- *Backpapier*
- *Puderzucker, Kakaopulver*

● Mandeln rösten, beiseite stellen. Elektro-Ofen auf 175 Grad heizen.

● Eier trennen. Eiweiße mit 1 Prise Salz steif schlagen, 40 g Zucker unterziehen. Eigelbe mit übrigem Zucker und 4 EL warmem Wasser aufschlagen. Stärke, Mehl, Backpulver, Kakao und Mandeln mit Eischnee und Eigelb verrühren.

● Teig in die mit Backpapier ausgelegte Springform füllen. Im Ofen (Gas: Stufe 2) etwa 20 Minuten backen. Abkühlen lassen, waagerecht in zwei Böden teilen.

● Schokolade hacken. Gelatine einweichen. Sahne steif schlagen. 50 ml Espresso und 2 EL Likör mischen, erwärmen. Ausgedrückte Gelatine darin auflösen, mit 1 EL Sahne verrühren. Unter die übrige Sahne ziehen. Menge halbieren.

● Unter die eine Hälfte Vanillezucker ziehen, unter die andere Schokolade und Kakao. Tortenring um einen Biskuitboden legen. Mit dem übrigen Kaffee und Likör tränken.

● Dunkle und helle Creme (2 EL beiseite stellen) in zwei Spritzbeutel mit großer Lochtülle füllen. Abwechselnd Ringe auf den Boden spritzen. Dabei außen mit der hellen Creme beginnen. Zweiten Boden darauf setzen, 2 Stunden kalt stellen.

● Den Ring entfernen. Rand mit den übrigen 2 EL Creme bestreichen. Die Torte mit Puderzucker bestäuben. Mehrere 2 cm breite Papierstreifen auf die Torte legen, mit Kakao bestäuben. Streifen entfernen.

Aufwand:	★
Zubereitungszeit:	1 Stunde
Backzeit:	20 Minuten
Pro Stück:	320 kcal/ 1340 kJ

Eierlikör-Kirsch-Torte

BITTE BEACHTEN: TORTE MUSS 4 STUNDEN KÜHLEN

FÜR SPRINGFORM VON 26 cm Ø
150 g Halbbitter-Kuvertüre
6 Eier (Größe M)
180 g Zucker
Mark von 1 Vanilleschote
3 TL Speisestärke
3 TL Kakaopulver
120 g Mehl
3/4 TL Backpulver

FÜR DIE FÜLLUNG
1 Glas Kirschen (Abtropfgewicht: 350 g)
3 TL Speisestärke
1 EL Zucker
1 Pck. Vanille-Puddingpulver
500 ml Milch, 1 Prise Salz
70 g Zucker, 250 g Mascarpone
125 ml Eierlikör

FÜR DIE DEKORATION
100 g gelbe Marzipan-Rohmasse
Mandelblättchen und -stifte (für Flügel und Schnäbel)
500 g Schlagsahne
je 2 Päckchen Sahnesteif und Vanillezucker
50 g geröstete Mandelblättchen
25 g gehackte Pistazien (ungesalzen), bunte Zuckereier

AUSSERDEM
Backpapier

● Elektro-Ofen auf 175 Grad vorheizen. Springform mit Backpapier auslegen. Kuvertüre hacken, bei kleiner Hitze schmelzen. Eier und Zucker schaumig rühren. Vanillemark und <abgekühlte flüssige Kuvertüre zugeben. Stärke, Kakao, Mehl, Backpulver einrühren. Teig in die Form füllen. Im Ofen (Gas: Stufe 2) etwa 25 Minuten backen. Stürzen, Papier abziehen. Biskuit erkalten lassen, waagerecht halbieren. 1 Boden mit Tortenring umstellen.

● Kirschen abtropfen, dabei Saft auffangen. Stärke mit etwas Saft anrühren. Übrigen Saft mit Zucker aufkochen. Stärke zugießen, aufkochen. Kirschen unterheben. Abgekühlt auf den Biskuitboden verteilen.

● Puddingpulver mit etwas Milch verquirlen. Übrige Milch mit Salz, Zucker, Mascarpone aufkochen. Puddingmilch einrühren, unter Rühren aufkochen. Vom Herd nehmen, Likör unterheben. Warm auf die abgekühlten Kirschen streichen. Übrigen Boden auflegen. 4 Stunden kalt stellen.

● Aus Marzipan je 14 große und kleine Kugeln formen, aneinander setzen. Mandeln als Schnäbel und Flügel ansetzen.

● Den Tortenring entfernen. Sahne, Sahnesteif und Vanillezucker steif schlagen. Die Torte damit rundum bestreichen. Tortenrand mit Mandeln bestreuen. Mit Marzipanküken, Pistazien, Zuckereiern verzieren.

Aufwand:	★★
Zubereitungszeit:	1 Stunde
Backzeit:	25 Minuten
Pro Stück:	585 kcal/ 2460 kJ

Variationen

● Die Frühlingstorte schmeckt auch mit einem hellen Biskuitteig sehr gut: Geben Sie statt der Kuvertüre 120 g geriebene Mandeln in den Teig und lassen Sie das Kakaopulver weg.

● Wer auf Alkohol verzichten möchte, nimmt statt des Eierlikörs 125 g mehr Mascarpone und gibt 1 TL geriebene Schale von1 unbehandelten Orange unter die Füllmasse.

Malakow-Torte

BITTE BACHTEN: TORTE MUSS 2 STUNDEN KÜHLEN
FÜR SPRINGFORM VON 26 cm Ø
7 Eier (Größe M)
190 g Zucker, 90 g Speisestärke
80 g Mehl, 2 EL Puderzucker
FÜR DIE CREME
6 Blatt weiße Gelatine
250 g Erdbeeren
2 Eier (Größe M)
50 ml Milch, 90 g Zucker, Salz
Mark von 1 Vanilleschote
2 EL Orangenlikör
Saft von 1 Orange
600 g Schlagsahne
ca. 20 Löffelbiskuits
AUSSERDEM
Backpapier
Mandelblättchen

- Elektro-Ofen auf 175 Grad vorheizen. Eier trennen. Eigelbe mit 50 g Zucker schaumig rühren. Eiweiße mit übrigem Zucker steif schlagen. Stärke und Eischnee unter die Eigelb-Masse rühren, dann das Mehl vorsichtig unterheben.
- 1/3 der Teig-Masse in die mit Backpapier ausgelegte Springform füllen. Im Ofen (Gas: Stufe 2) etwa 20 Minuten backen, abkühlen lassen.
- Übrigen Teig in Spritzbeutel mit Lochtülle (Größe 7) geben. Auf ein mit Backpapier belegtes Blech etwa 8 cm lange Zungen (in Form von Biskuitstäbchen) spritzen. Mit Puderzucker bestäuben. Im Ofen (Gas: Stufe 2) 8-10 Minuten backen.
- Gelatine einweichen. Erdbeeren waschen, putzen. 5 Beeren für die Deko beiseite legen, Rest je nach Größe halbieren oder vierteln. Eier trennen. Eigelbe mit Milch, der Hälfte des Zuckers, Salz und Vanillemark über heißem Wasserdampf dickflüssig rühren. Likör, Saft und ausgedrückte Gelatine einrühren, abkühlen lassen. Eiweiße mit übrigem Zucker steif schlagen, unter die Creme heben. Sahne schlagen, 100 g beiseite stellen, Rest mit Erdbeer-Stücken unterheben.
- Um den Boden einen Tortenring stellen. 1/3 der Creme auf dem Boden verstreichen, mit einer Lage Löffelbiskuits abdecken. Wieder 1/3 der Creme einfüllen und mit Löffelbiskuits abschließen. Restliche Creme darauf verteilen. 2 Stunden kalt stellen.
- Vor dem Servieren die Tortenoberfläche mit 16 Sahnetuffs, den restlichen Erdbeeren und Biskuitstäbchen verzieren. Rand mit Mandelblättchen bestreuen.

Aufwand:	★★
Zubereitungszeit:	45 Minuten
Backzeit:	30 Minuten
Pro Stück:	360 kcal/ 1510 kJ

Info

Im Jahr 1855 während des Krimkrieges erstürmte der französische General Pelissier die Bastion Malakow. Als Auszeichnung erhielt er den Titel „Herzog von Malakow“. Ein Pariser Patissier kreierte ihm zu Ehren diese Torte. – Servieren Sie sie immer gut durchgezogen und garnieren Sie sie erst zuletzt mit Sahne.

Schwarzwälder Kirschtorte

BITTE BEACHTEN: TORTE MUSS 4 STUNDEN KÜHLEN

FÜR SPRINGFORM VON 26 cm Ø
- *100 g Zartbitter-Schokolade*
- *75 g Butter*
- *6 Eier (Größe M)*
- *180 g Zucker, 100 g Mehl*
- *50 g Speisestärke*
- *3 EL Kakaopulver*
- *1 1/2 TL Backpulver*

FÜR DIE FÜLLUNG
- *2 Gläser Sauerkirschen (Abtropfgewicht: à 350 g)*
- *150 g Zucker*
- *1 Zimtstange*
- *2-3 TL Speisestärke*
- *2 Päckchen Sahnesteif*
- *750 g Schlagsahne*
- *6-8 cl Kirschwasser*
- *100 g Vollmilch-Kuvertüre oder 50 g Schoko-Raspel*

AUSSERDEM
- *Backpapier*

● Elektro-Ofen auf 175 Grad vorheizen. Den Boden der Springform mit Backpapier auslegen.

● Schokolade und Butter bei milder Hitze im heißen Wasserbad schmelzen. Eier trennen. Eigelbe mit Zucker cremig rühren, Schokoladen-Butter-Masse zufügen. Eiweiße steif schlagen, unter die Masse heben. Mehl mit Stärke, Kakao und Backpulver mischen, vorsichtig unter die Masse heben. Teig in die Form füllen, im Ofen (Gas: Stufe 2) etwa 35 Minuten backen. Aus der Form lösen, auskühlen lassen.

● Kirschen abtropfen lassen, Saft dabei auffangen. 1/4 l Kirschsaft mit 40 g Zucker und der Zimtstange aufkochen, mit angerührter Stärke binden. 16 schöne Kirschen beiseite legen, übrige Kirschen zum Saft geben, aufkochen. Zimtstange entfernen, Kirsch-Mischung abkühlen lassen.

● Biskuit zweimal waagerecht durchschneiden. Restlichen Zucker mit Sahnesteif mischen. Sahne damit steif schlagen. Einen Tortenboden dünn mit Sahne bestreichen, darauf die abgekühlten Kirschen verteilen, mit Sahne bestreichen. Zweiten Boden auflegen, mit etwas Kirschwasser tränken und mit einem Drittel der übrigen Sahne bestreichen. Letzten Boden auflegen.

● 5 EL Sahne in Spritzbeutel mit Sterntülle füllen. Torte rundum mit übriger Sahne bestreichen. 16 Rosetten aufspritzen, Kirschen darauf setzen. Die Mitte mit Schoko-Locken (siehe Seite 94) oder fertigen Schoko-Raspeln verzieren. Torte etwa 2 Stunden kühlen.

Aufwand:	★★
Zubereitungszeit:	40 Minuten
Backzeit:	35 Minuten
Pro Stück:	560 kcal/ 2350 kJ

Tipp

Wer für die etwas aufwändige Torte zu wenig Zeit hat, findet im Handel schon fertig gebackene Schoko-Tortenböden aus Biskuitteig. Sie werden von mehreren Herstellern angeboten und sind auch geschmacklich recht unterschiedlich. Probieren Sie einfach aus, welche Ihnen am besten schmecken.

Nektarinen-Kuchen mit Kokoscreme

FÜR 1 BACKBLECH (30 x 40 cm)
50 g Butter, 6 Eier (Größe M)
150 g Zucker
1 Päckchen Bourbon-Vanillezucker
Saft und abgeriebene Schale von 1 unbehandelten Zitrone
125 g Mehl
75 g Speisestärke
1 TL Backpulver
Fett für das Blech
FÜR DEN BELAG
2 Päckchen Puddingpulver mit Sahnegeschmack
150 g Zucker
3/4 l Milch
6 cl Kokoscreme-Likör
600 g Nektarinen
350 g Brombeeren
2 Pck. weißer Tortenguss
AUSSERDEM
Fett für das Blech

Info

Biskuit als Blechkuchen ist ungewöhnlich, aber lecker. Wichtig ist, dass Sie für den Belag „saftarme" Früchte nehmen, damit der Teig nicht aufweicht und sicherheitshalber Kuchenoblaten (im Handel) zwischen Biskuit und Früchte legen. Oder: Sie bestreichen den kalten Kuchen mit flüssiger Kuvertüre, lassen sie erstarren und geben dann die Früchte darauf. Auch gut: Die Früchte unter Sahnecremes mischen und auf dem Kuchen verteilen.

- Elektro-Ofen auf 175 Grad heizen. Butter schmelzen, abkühlen lassen. Eier trennen, Eiweiße mit 2 EL Zucker zu steifem Schnee schlagen. Eigelbe mit restlichem Zucker, Vanillezucker, Zitronensaft und -schale schaumig rühren. Butter untermischen. Mehl mit Speisestärke und Backpulver mischen und locker unterziehen. Den Eischnee daraufgeben, mit dem Schneebesen unterheben.
- Teig auf gefettetem Backblech verteilen und glatt streichen, im Ofen (Gas: Stufe 2) etwa 20 Minuten backen.
- Das Puddingpulver nach Packungsanweisung mit 70 g Zucker, aber nur 3/4 l Milch zubereiten. Mit etwas Zucker dünn bestreuen, auskühlen lassen und den Likör unterrühren.
- Nektarinen waschen, halbieren, entsteinen und die Früchte in dünne Spalten schneiden. Brombeeren kurz abbrausen und abtropfen lassen.
- Puddingmasse auf den abgekühlten Biskuit streichen, mit Nektarinenspalten und Brombeeren belegen. Tortenguss nach Packungsangabe mit Wasser und 80 g Zucker zubereiten, gleichmäßig über die Früchte ziehen.

Aufwand:	★
Zubereitungszeit:	40 Minuten
Backzeit:	20 Minuten
Pro Stück:	230 kcal/ 920 kJ

Variation

- Mischen Sie 1 TL Zimt unter den Teig und verzichten Sie auf die Zitronenschale. Belegen Sie den Kuchen mit abgetropftem Pflaumenkompott. Den aufgefangenen Saft verwenden Sie für den Tortenguss.

Schmetterlings-Torte

BITTE BEACHTEN: TORTE MUSS 3 STUNDEN KÜHLEN

FÜR SPRINGFORM VON 26 cm Ø

- *150 g Butter*
- *4 Eier (Größe M)*
- *150 g Zucker*
- *50 g Kokosraspel*
- *150 g Mehl*

FÜR DIE FÜLLUNG

- *250 g Erdbeeren, Zitronensaft*
- *1 EL Erdbeer-Konfitüre*
- *6 Blatt weiße Gelatine*
- *125 g Schlagsahne*
- *500 g Magerquark*
- *3 EL Puderzucker*
- *250 g Mascarpone*

FÜR DIE VERZIERUNG

- *100 g Marzipan-Rohmasse*
- *100 g Puderzucker*
- *gelbe Lebensmittelfarbe*
- *4 Erdbeeren*
- *500 g Schlagsahne*
- *dünne Limettenschalenstreifen*

AUSSERDEM

- *Backpapier*

● Elektro-Ofen auf 150 Grad heizen. Butter schmelzen. Eier trennen. Eiweiße steif schlagen, 2 EL Zucker und Kokosraspel unterheben. Eigelbe mit übrigem Zucker schaumig rühren. Mehl unterrühren. Erst die Eiweißmasse, dann die Butter unterziehen.

● Den Boden der Springform mit Backpapier auslegen, Masse einfüllen und im Ofen (Gas: Stufe 1) etwa 45 Minuten backen.

● Erdbeeren waschen, putzen und klein schneiden, mit Zitronensaft und Konfitüre mischen und fein pürieren.

● Gelatine einweichen. Sahne steif schlagen. Quark, Puderzucker und Mascarpone glatt rühren, mit etwas Zitronensaft abschmecken. Sahne unterziehen. Gelatine bei milder Hitze auflösen, unterziehen. Masse halbieren. Unter eine Hälfte das Erdbeermus mischen.

● Kuchen zweimal waagerecht durchschneiden. Um den unteren Boden einen Tortenring legen. Erdbeercreme einfüllen. Mittleren Boden auflegen. Weiße Creme einfüllen, mit dem oberen Boden belegen. Mindestens 3 Stunden kalt stellen.

● Für die Dekoration Marzipan mit Puderzucker und ein paar Tropfen Lebensmittelfarbe glatt kneten. Zwischen Klarsichtfolie 2 mm dünn ausrollen und Herzen ausstechen. Erdbeeren putzen, waschen, längs in 1/2 cm dicke Spalten schneiden.

● Sahne steif schlagen, auf die ganze Torte auftragen. Herzen, Erdbeeren und Limettenschalenstreifen zu Schmetterlingen geformt auf die Sahne setzen.

Aufwand:	★★
Zubereitungszeit:	1 Stunde
Backzeit:	45 Minuten
Pro Stück:	585 kcal/ 2460 kJ

Waldmeister-Sekt-Torte

BITTE BACHTEN: TORTE MUSS 4 STUNDEN KÜHLEN

FÜR SPRINGFORM VON 24 cm Ø

2 Eier (Größe M), 50 g Zucker
1/2 Päckchen Vanillezucker
etwas abgeriebene Zitronenschale, 1 EL Speisestärke
40 g Mehl, Salz

FÜR DIE FÜLLUNG

300 g TK-Himbeeren
750 ml trockener Sekt
2 cl Waldmeistersirup
80 g Speisestärke
180 g Zucker, 200 g Schlagsahne
1 Päckchen Sahnesteif
1-2 TL Sofort-Gelatine-Pulver

FÜR DIE VERZIERUNG

1/2 Beutel Götterspeise (Waldmeistergeschmack), Apfelsaft und Zucker nach Packungsangabe
200 g Schlagsahne

AUSSERDEM

Backpapier
Minzeblättchen

- Elektro-Ofen auf 175 Grad heizen. Eier trennen. Eiweiße mit 1 EL Zucker steif schlagen. Eigelbe mit übrigem Zucker, Vanillezucker und der Zitronenschale schaumig rühren. Mehl mit 1 Prise Salz und Speisestärke sieben, unter die Masse ziehen.
- Boden der Form mit Backpapier auslegen. Masse einfüllen. Im Ofen (Gas: Stufe 2) 20-30 Minuten backen.
- Himbeeren auftauen lassen, gut trockentupfen. Sekt mit Waldmeistersirup mischen und in einen Topf geben. Stärke und Zucker mischen, einrühren, aufkochen und erkalten lassen. Sahne mit Sahnesteif steif schlagen und unterheben.
- Einen Tortenring um den Biskuit legen. Himbeeren darauf verteilen, 12 Beeren beiseite legen. Die Sofort-Gelatine darüber streuen. Die Creme einfüllen und die Torte mindestens 4 Stunden kalt stellen.
- Götterspeise nach Packungangabe mit Apfelsaft und Zucker zubereiten. Dünn auf die Torte auftragen. 1 Stunde kalt stellen, bis die Gelatine fest geworden ist.
- Sahne steif schlagen. Torte mit kleinen Tupfen, Beeren und Minzeblättchen garnieren.

Aufwand:	★★
Zubereitungszeit:	1 Stunde
Backzeit:	30 Minuten
Pro Stück:	300 kcal/ 1260 kJ

Variationen

- Wer Waldmeistergeschmack nicht mag, nimmt Himbeersirup und rote Götterspeise.
- Das funktioniert auch mit Aprikosen, Mandarinen aus der Dose, Erdbeeren und Kirschen. Dabei kann man mit der Götterspeise (rot oder gelb) variieren.

Sachertorte

FÜR SPRINGFORM VON 22 cm Ø
120 g Zartbitter-Kuvertüre
100 g Butter
110 g Puderzucker
5 Eigelbe (Eier: Größe M)
Mark von 1/2 Vanilleschote
4 Eiweiße (Eier: Größe M)
50 g Mehl
120 g gemahlene Mandeln
FÜR DIE GLASUR
200 g Aprikosenkonfitüre
300 g Zucker
100 g Zartbitter-Kuvertüre
60 g Kakaopulver
AUSSERDEM
Butter und Brösel für die Form

- Elektro-Ofen auf 175 Grad vorheizen. Nur den Boden der Form fetten und mit Bröseln ausstreuen.
- Kuvertüre über warmem Wasserbad schmelzen, warm halten (etwa 40 Grad). Butter mit 40 g Puderzucker und Eigelben schaumig rühren. Vanillemark und lauwarme Kuvertüre unterrühren. Eiweiße mit restlichem Puderzucker steif schlagen und mit dem gesiebten Mehl und den Mandeln locker unterheben. Teig in die Form füllen, glatt streichen. Im Ofen (Gas: Stufe 2) 40–50 Minuten backen.
- Ausgekühlten Biskuit einmal durchschneiden. Einen Boden mit der Hälfte der Aprikosenkonfitüre bestreichen, zweiten Boden darauf setzen. Restliche Konfitüre mit 2–3 EL Wasser kurz kochen, durch ein Sieb streichen, gleichmäßig auf die Torte streichen.
- Für die Glasur Zucker, klein geschnittene Kuvertüre und Kakaopulver mit 1/8 l Wasser mischen und aufkochen. Vom Herd nehmen, in eine Schüssel gießen, mit einem Teigschaber ständig in Bewegung halten, bis die Glasur dickflüssig wird. Torte mit der Schokoladenglasur überziehen und trocknen lassen. Nach Belieben mit Schokospänen verzieren.

Aufwand:	★★
Zubereitungszeit:	1 Stunde
Backzeit:	50 Minuten
Pro Stück:	480 kcal/2020 kJ

Rehrücken

FÜR 1 REHRÜCKEN-FORM
4 Eier (Größe M), 100 g Zucker
1 Prise Salz, 75 g Speisestärke
1 TL Backpulver
75 g geriebene Mandeln
100 g geriebene Zartbitter-Kuvertüre, 1 TL Kakaopulver
100 g Butter oder Margarine
FÜR DIE GLASUR
80 g Schlagsahne
5 EL Butter, 1 Prise Salz
250 g Kakaopulver
300 g Zucker, Vanille-Essenz (in Feinkostläden; ersatzweise Vanillearoma), 125 g Mandelstifte
AUSSERDEM
Fett für die Form

- Elektro-Ofen auf 180 Grad vorheizen. Die Form einfetten.
- Eier schaumig rühren. Zucker mit Salz mischen und nach und nach einrieseln lassen. Masse so lange rühren, bis sie cremig und glänzend ist.
- Speisestärke mit Backpulver mischen, auf die Masse sieben und unterrühren. Mandeln, Kuvertüre, Kakaopulver und Fett zugeben und ebenfalls unterrühren.
- Teig in die Form füllen und im Ofen auf der unteren Schiene (Gas: Stufe 2) 40-50 Minuten backen.
- Rehrücken herausnehmen, in der Form etwa 5 Minuten abkühlen lassen, auf einen Rost setzen.
- Für die Glasur Sahne, Butter, Salz, Kakao und Zucker bei kleiner Hitze unter Rühren etwa 5 Minuten kochen. Vom Herd nehmen, mit Vanille-Essenz abschmecken, abkühlen lassen.
- Die Mandelstifte in Reihen mit etwa 3 cm Abstand in den Rehrücken stecken. Die Glasur gleichmäßig über den abgekühlten Kuchen gießen.

Aufwand:	★★
Zubereitungszeit:	40 Minuten
Backzeit:	30 Minuten
Pro Stück:	625 kcal/2625 kJ

Beerentorte mit weißer Schokocreme

BITTE BEACHTEN: TORTE MUSS 2 STUNDEN KÜHLEN
FÜR SPRINGFORM VON 26 cm Ø
50 g Zartbitter-Kuvertüre
3 Eier (Größe M), 125 g Zucker
50 g Mehl, 1/4 TL Backpulver
FÜR DIE MOUSSE
150 g weiße Kuvertüre
4 Blatt weiße Gelatine
2 Eier (Größe M), 200 g Sahnequark, 200 g Schlagsahne
FÜR DEN BELAG
600 g gemischte Beeren
oder 1 kg Aprikosen
1 Päckchen weißer Tortenguss
1/8 l Johannisbeernektar
AUSSERDEM
Fett für die Form

- Eine Springform (26 cm Ø) fetten. Den Elektro-Ofen auf 175 Grad vorheizen.
- Zartbitter-Kuvertüre in einer Schüssel im warmen Wasserbad schmelzen. Eier trennen, Eiweiße steif schlagen. Zucker nach und nach unterrühren. Eigelbe mit Kuvertüre vermischen, unter die Eiweiß-Masse rühren. Mehl und Backpulver unterheben. Masse in die Form füllen, im Ofen (Gas: Stufe 2) etwa 30 Minuten backen. Biskuit auskühlen lassen.
- Inzwischen für die Mousse weiße Kuvertüre in Stücke brechen, in einer Schüssel im warmen Wasserbad schmelzen. Gelatine nach Packungsangabe einweichen und auflösen. Eier in einer Schüssel im warmen Wasserbad schaumig rühren. Kuvertüre und Gelatine einrühren. Quark untermischen. Etwas abkühlen lassen.
- Sahne steif schlagen und vorsichtig unter die Schokocreme heben.
- Biskuitboden mit einem Tortenring umschließen. Schokocreme darauf geben und glatt streichen. 2 Stunden zum Festwerden in den Kühlschrank stellen.
- Beeren waschen und verlesen. Oder Aprikosen brühen, häuten und in Schnitze schneiden. Torte damit belegen. Guss mit 1/8 l Wasser und Nektar nach Packungsangabe zubereiten und dünn über die Früchte ziehen. Erstarren lassen. Tortenring entfernen und die Torte bis zum Servieren kühl stellen.

Aufwand:	★★
Zubereitungszeit:	1 Stunde
Backzeit:	30 Minuten
Pro Stück:	300 kcal/1260 kJ

Espresso-Torte

BITTE BEACHTEN: TORTE MUSS 2 STUNDEN KÜHLEN

FÜR SPRINGFORM VON 20 cm Ø

FÜR DIE KAFFEE-CREME
- *250 g Crème double*
- *60 g Espresso-oder Kaffeebohnen*
- *Mark von 1 Vanilleschote*
- *200 g Schlagsahne*
- *150 g weiße Kuvertüre*
- *1 Päckchen Sahnesteif*

FÜR DEN MÜRBETEIG
- *100 g Mehl, 30 g Zucker*
- *1 Eiweiß (Ei: Größe M)*
- *50 g Butter*

FÜR DEN BISKUIT
- *2 Eier (Größe M), Salz*
- *50 g Zucker*
- *20 g Speisestärke*
- *40 g Mehl*
- *2 Msp. Backpulver*

AUSSERDEM
- *Fett, Backpapier, 6 EL Orangenmarmelade, 7 cl Orangenlikör, 50 g Mandelblättchen, kandierte Orangenscheiben*

● Crème double mit Kaffeebohnen verrühren, über Nacht ziehen lassen.

● Vanillemark mit Sahne und Kuvertüre erhitzen, bis diese geschmolzen ist. Über Nacht kalt stellen.

● Für den Mürbeteig die Zutaten zu einem glatten Teig verkneten, in Folie wickeln, 30 Minuten kalt stellen.

● Elektro-Ofen auf 175 Grad vorheizen. Boden der gefetteten Form mit Mürbeteig auslegen. Im Ofen (Gas: Stufe 2) etwa 15 Minuten backen. Herausnehmen, abkühlen lassen. Ofen eingeschaltet lassen.

● Für den Biskuit die Eier trennen. Eiweiße mit 1 Prise Salz steif schlagen. Eigelbe mit Zucker und 2 EL warmem Wasser aufschlagen. Eischnee und Stärke-Mehl-Backpulver-Mischung unter die Eigelb-Masse heben.

● Form mit Backpapier auslegen, Teig hineinfüllen, im Ofen (Gas: Stufe 2) etwa 30 Minuten backen. Herausnehmen, Papier abziehen, waagerecht halbieren.

● Creme double durch ein Sieb streichen. Crème double mit Sahnesteif und Schoko-Sahne getrennt steif schlagen. Dann beides mischen.

● Die Hälfte der Marmelade erwärmen, auf den Mürbeteig streichen. 1 Biskuitboden darauf legen. Mit 4 cl Likör beträufeln. Tortenring herumlegen. Die Hälfte der Kaffee-Creme darauf streichen. Zweiten Boden darauf legen, mit übrigem Likör beträufeln, mit restlicher Creme bestreichen. 2 Stunden kalt stellen. Tortenring entfernen.

● Mandeln rösten, abkühlen. Übrige Marmelade erwärmen, auf Tortenrand estreichen, mit Mandeln bestreuen. Mit Orangenscheiben verzieren.

Aufwand:	★★
Zubereitungszeit:	1 Stunde
Backzeit:	45 Minuten
Pro Stück:	555 kcal/2330 kJ

Ananas-Mandel-Torte

BITTE BEACHTEN: TORTE MUSS 3 STUNDEN KÜHLEN
FÜR SPRINGFORM VON 26 cm Ø
75 g geschälte, gemahlene Mandeln, 4 Eier, 1 Prise Salz
80 g Zucker, 2 EL Speisestärke
1/2 TL abgeriebeneZitronenschale
3 EL Mehl, 1/4 TL Backpulver
FÜR DIE ANANASCREME
2 reife Ananas (1,5 kg), 100 g Zucker, 1 Pck. Vanillezucker
8 Blatt weiße Gelatine
500 g Quark (20 % Fett i.Tr.)
5 EL Rum (weiß oder braun, 40%ig)
500 g Schlagsahne
FÜR DIE DEKORATION
2 EL Mandelblättchen
1 Zweig Zitronenmelisse
AUSSERDEM
Fett und Mehl für die Form

- Elektro-Ofen auf 175 Grad vorheizen. Nur den Boden der Form fetten und mit Mehl bestäuben.
- Mandeln anrösten. Eier trennen. Eiweiße mit Salz steif schlagen. Eigelbe mit Zucker, Zitronenschale und 3 EL lauwarmem Wasser cremig rühren. Eischnee auf die Eigelb-Masse setzen. Mehl, Speisestärke, Backpulver mischen, darüber sieben. Mandeln zufügen und alles vorsichtig mischen.
- Teig in Form füllen, im Ofen (Gas: Stufe 2) etwa 30 Minuten backen. Auf Kuchengitter auskühlen lassen.
- Ananas schälen, vierteln, die holzigen Strünke herausschneiden. Viertel quer in 1 cm dicke Stücke schneiden. In einem Topf mit Zucker und Vanillezucker 30 Minuten ziehen lassen. Kurz aufkochen, in ein Sieb schütten, Saft auffangen. Den aufgefangenen Saft sirupartig einkochen lassen.
- Biskuit zweimal quer durchschneiden, so dass 3 Böden entstehen.
- Gelatine in kaltem Wasser 5 Minuten einweichen. 12 Ananasstücke für die Dekoration beiseite legen, Rest klein schneiden. Quark mit Ananassirup und Rum glatt rühren. Gelatine tropfnass bei milder Hitze auflösen, unter die Quarkmasse rühren, im Kühlschrank 15 Minuten angelieren lassen. Sahne steif schlagen, unterheben. Von der Creme 1/4 beiseite stellen, unter die übrige Creme die Ananasstückchen mengen.
- Auf 1 Boden die Hälfte der Creme verteilen. Mittleren Boden darauf legen, mit Creme bestreichen. Darauf oberen Boden setzen, leicht andrücken. Creme ohne Ananasstücke auf die Torte streichen. Mit übriger Ananas garnieren. 3 Stunden kühlen.
- Mandeln rösten, abkühlen lassen. Den Tortenrand damit dekorieren. Torte mit Zitronenmelisse verzieren.

Aufwand:	★★
Zubereitungszeit:	2 Stunden
Backzeit:	30 Minuten
Pro Stück:	300 kcal/1260 kJ

Apfel-Biskuit-Kuchen

BITTE BEACHTEN: KUCHEN MUSS 2 STUNDEN KÜHLEN
FÜR SPRINGFORM VON 26 cm Ø
3 Eier (Größe M), 90 g Zucker
90 g Mehl
FÜR DEN BELAG
750 g säuerliche Äpfel
1 Pck. Puddingpulver
450 ml Apfelsaft, 50 g Zucker
200 g Schlagsahne
100 g Magerquark
30 g Puderzucker
etwa 50 g Zartbitter-Kuvertüre
AUSSERDEM: Backpapier

- Elektro-Ofen auf 175 Grad vorheizen. Den Boden der Springform mit Backpapier auslegen.
- Eier mit Zucker etwa 5 Minuten cremig schlagen. Mehl unterheben. Masse in die Form geben. Im Ofen (Gas: Stufe 2) etwa 15 Minuten backen.
- Äpfel waschen, schälen, vierteln, putzen. Apfelviertel quer in dünne Scheiben schneiden.
- Puddingpulver mit 5 EL Apfelsaft verrühren. Restlichen Apfelsaft mit Zucker aufkochen lassen, angerührtes Puddingpulver einrühren. Äpfel unterheben, 2 Minuten ziehen lassen. Vom Herd nehmen.
- Springformrand vom Biskuit lösen, Backpapier abziehen. Biskuit mit einem Tortenring umschließen, Apfelmasse darauf verteilen. 2 Stunden kalt stellen.
- Sahne steif schlagen. Quark mit Puderzucker verrühren. Die Sahne unterziehen. Creme wolkenartig auf dem Kuchen verteilen oder als Gitter darauf spritzen. Von der Kuvertüre mit dem Sparschäler Späne abziehen, den Kuchen damit verzieren.

Aufwand:	★
Zubereitungszeit:	35 Minuten
Backzeit:	15 Minuten
Pro Stück:	240 kcal/1010 kJ

Deko-Variationen und Trick

Auch die feinste Torte schmeckt noch mal so gut, wenn sie hübs

1 Torte überziehen
Mittels einer Palette die Kuvertüre auf der Tortenoberfläche verstreichen.

2 Stücke markieren
Tortenstücke auf der fast erkalteten Kuvertüre ca. 1 mm tief einschneiden.

3 Torte anschneiden
Jetzt läßt sich jedes einzelne Stück schneiden, ohne die Glasur zu „sprengen".

Schokotorte verzieren

Für unser Tortenmuster brauchen Sie eine kleine Spritztüte aus Pergamentpapier (im Handel), in die Sie grün eingefärbte Puderzuckerglasur füllen. Spritzen Sie dünne Linien als Blütenstengel auf die Torte und setzen Sie an das jeweilige Ende kleine Zuckerblüten (Fertigprodukt).

Schokolade schmelzen

Starke Hitze macht Kuvertüre grau und stumpf, daher beim Schmelzen behutsam vorgehen: Metallschüssel in einen Topf mit warmem Wasser setzen, klein geschnittene Kuvertüre hineingeben und unter Rühren langsam zum Schmelzen bringen. Schüssel vom Topf nehmen, weiter rühren, bis die Kuvertüre abgekühlt und dickflüssig ist.

Zart und fein: Schoko-Locken

Aus einfachen Torten können Sie wahre Prachtstücke machen. So geht's: Geschmolzene Kuvertüre auf einem Marmorbrett ausgießen und gleichmäßig dünn verstreichen. Kurz bevor die Masse ganz erstarrt ist, mit einem Spatel vorsichtig papierdünne Röllchen abschaben und beliebig auf der Torte verteilen.

ür

koriert

Biskuit-Torten

ist – und das ist ganz einfach

Sahne-Zopfmuster
Spritzbeutel in Zick-zack-Bewegungen über die Torte führen.

Sahne-Sterne
Spritzbeutel punktuell nur kurz drücken, und hochziehen.

Sahne-Spiralen
Spritzbeutel ohne abzusetzen in fortlaufenden Kreis-Bewegungen führen.

Aber bitte mit Sahne

Sie ist die einfachste und beliebteste Füllung oder Garnierung für Torten. Damit sie nach dem Schlagen ihren Stand behält, müssen alle verwendeten Geräte wie Schüssel und Schneebesen gekühlt sein. Sahne wird in flüssigem Zustand gezuckert (30-40 g auf 150 g Sahne). Zuerst mit dem Handrührgerät auf mittlerer Stufe aufschlagen, dann mit dem Schneebesen fertigschlagen. Nur so kann man die gewünschte Konsistenz genau feststellen. Wer will, kann natürlich auch „Sahnesteif" (Fertigprodukt) verwenden.

Puder-zucker-Glasur

250 g Puderzucker sorgfältig durchsieben, mit 2-3 EL Wasser glatt rühren. Anschließend kann man sie mit Lebensmittelfarbe (Fertigprodukt) färben und wie hier Petit fours (kleine Biskuitschnitten) damit glasieren.

Schnelle Torten-Garnierung mit Fruchtcreme

1 **Spiral-Muster**
Die Torte mit der Sahne rund herum einstreichen und glätten.

2 Spritzbeutel mit kleiner Lochtülle mit der Fruchtcreme füllen und in der Mitte der Torte ansetzen.

3 Spritzbeutel ohne abzusetzen in einer gleichmäßigen Spiralbewegung führen.

4 Mithilfe eines Holzstäbchens von der Mitte aus in gleichen Abständen gerade Linien nach außen führen.

Würziger Gugelhupf

Rezept siehe folgende Seite

Hefeteig

aromatisch & vielseitig

Ob als Boden für Obstkuchen vom Blech, als Gugelhupf mit Nüssen und Rosinen oder als Plundergebäck mit unterschiedlichen Füllungen – Hefeteig macht immer eine gute Figur

Würziger Gugelhupf

FÜR GUGELHUPF-FORM (2 1/2 l)
100 g Rosinen
8 EL Orangenlikör
30 g Hefe, Zucker
100 ml Milch
500 g Mehl, 1 TL Backpulver
1/2 TL gem. Muskatblüte (Macis)
1/2 TL Bourbon-Vanille
abger. Schale von je 1 unbehandelten Orange und Zitrone
2 Eier und 3 Eigelbe (Größe M)
200 g weiche Butter
100 g Zucker, Salz
100 g Mandelblättchen
FÜR DEN LÄUTERZUCKER
100 g Zucker, 1 Sternanis
2 Gewürznelken
AUSSERDEM
Fett für die Form

- Rosinen in Orangenlikör einweichen. Hefe mit 1 Prise Zucker in lauwarmer Milch auflösen. Mehl in einer Schüssel mit Backpulver, Macis und Vanille mischen. Mulde ins Mehl drücken, Hefemilch hineingießen. Mit etwas Mehl vermischen, 30 Minuten gehen lassen. Rosinen abgießen, Likör auffangen.
- Elektro-Ofen auf 175 Grad vorheizen. Übrige Teigzutaten (von den Mandeln nur 50 g) und Rosinen mit dem Vorteig zu einem glatten Teig verkneten. Zugedeckt 1 Stunde an einem warmen Ort gehen lassen.
- Form einfetten, übrige Mandeln einstreuen. Teig erneut durchkneten, einfüllen, 30 Minuten gehen lassen.
- Gugelhupf im Ofen (Gas: Stufe 2) etwa 1 Stunde backen.
- Zutaten für den Läuterzucker mit 100 ml Wasser aufkochen. Beiseite stellen, 10 Minuten ziehen lassen. Durch ein Sieb gießen, Likör untermischen.
- Den Gugelhupf aus der Form stürzen, abkühlen, mit Läuterzucker bestreichen.

Aufwand:	★★
Zubereitungszeit:	1 1/4 Stunden
Backzeit:	1 Stunde
Pro Stück:	480 kcal/2020 kJ

Mandel-Butterkuchen

FÜR 1 BACKBLECH (30 x 40 cm)
1 Würfel Hefe (42 g)
225 ml Milch
200 g Zucker, Salz
270 g Butter
2 Päckchen Vanillezucker
abgeriebene Schale von
1 unbehandelten Zitrone
150 g gehobelte Mandeln
AUSSERDEM
Fett für die Form
Mehl zum Arbeiten

- Mehl in eine Schüssel geben, in die Mitte eine Vertiefung drücken und Hefe hineinbröckeln. Mit 4 EL lauwarmer Milch, 1 TL Zucker und etwas Mehl vom Rand einen Vorteig rühren. Diesen zugedeckt an einem warmen Ort 20 Minuten gehen lassen.
- 70 g Butter würfeln, mit übriger Milch, 50 g Zucker, 1 Päckchen Vanillezucker, 1 Prise Salz und der Zitronenschale unter den Teig kneten. So lange kneten, bis er sich vom Schüsselrand löst. Weitere 40 Minuten gehen lassen.
- Elektro-Ofen auf 200 Grad vorheizen. Das Backblech einfetten. Teig auf einer bemehlten Arbeitsfläche ausrollen, das Blech damit auslegen und mindestens 30 Minuten gehen lassen, bis er sein Volumen verdoppelt hat. In gleichmäßigen Abständen mit einem Kochlöffelstiel Vertiefungen in den Teig drücken.
- Restliche Butter würfeln und in die Vertiefungen drücken. Mandeln auf der Oberfläche verteilen. Übrigen Zucker und Vanillezucker mischen und darüberstreuen. Den Butterkuchen im Ofen (Gas: Stufe 3) etwa 25 Minuten backen.

Aufwand:	★★
Zubereitungszeit:	2 Stunden
Backzeit:	25 Minuten
Pro Stück:	220 kcal/920 kJ

Ananas-Gugelhupf

FÜR GUGELHUPF-FORM (2 1/2 l)
500 g Mehl
2 Päckchen Trockenhefe
1/4 l Milch
100 g Zucker
125 g kalte Butter, Salz
3 Eigelbe (Eier: Größe M)
Saft und abgeriebene Schale
von 1 unbehandelten Zitrone
100 g in Sirup eingelegter
Ingwer (Glas)
100 g kandierte Ananas
100g kandierte Papaya
AUSSERDEM
Fett für die Form
200 g Puderzucker
etwas Grenadinesirup
2 EL weißer Rum

- Mehl in eine Schüssel geben und in die Mitte eine Vertiefung drücken. Hefe hineingeben, mit 4 EL lauwarmer Milch, 1 TL Zucker und etwas Mehl vom Rand einen Vorteig rühren. Diesen zugedeckt an einem warmen Ort 20 Minuten gehen lassen.
- Butter würfeln und mit übriger Milch, Zucker, 1 Prise Salz, Eigelben, Zitronenschale und 3 EL Ingwersirup verrühren und unter den Teig kneten. Weitere 40 Minuten gehen lassen. Ingwer fein hacken, Ananas und Papaya klein würfeln und jeweils 1 EL davon beiseite legen.
- Elektro-Ofen auf 175 Grad vorheizen. Die Gugelhupfform einfetten. Die gewürfelten Früchte unter den Teig kneten und in die Form geben. Etwa 20 Minuten gehen lassen.
- Im Ofen (Gas: Stufe 2) etwa 1 Stunde backen. Auf ein Gitter stürzen und abkühlen lassen.
- Puderzucker mit Zitronensaft, Grenadinesirup und Rum glatt rühren. Gugelhupf damit überziehen. Ananas- und Papayawürfel obenauf streuen.

Aufwand:	★★
Zubereitungszeit:	1 3/4 Stunden
Backzeit:	1 Stunde
Pro Stück:	270 kcal/1130 kJ

Rosenkuchen

FÜR 1 SPRINGFORM VON 24 cm Ø

- *500 g Mehl*
- *1 Würfel Hefe (42 g)*
- *1/4 l Milch, 100 g Zucker*
- *100 g Butter, Salz*
- *abgeriebene Schale von*
- *1 unbehandelten Zitrone*
- *150 g getrocknete Aprikosen*
- *100 ml Aprikosenlikör*
- *4 Eiweiße*
- *100 g Marzipan-Rohmasse*
- *50 g Pistazien*

AUSSERDEM

- *Fett für die Form*
- *4 EL Aprikosenkonfitüre*
- *50 g gehackte Pistazien*

● Das Mehl in eine Schüssel geben und in die Mitte eine Vertiefung drücken. Die Hefe hineinbröckeln, mit 4 EL lauwarmer Milch, 1 TL Zucker und etwas Mehl vom Rand verrühren. Zugedeckt an einem warmen Ort 20 Minuten gehen lassen.

● Danach die in Stücke geschnittene Butter, übrige Milch, Zucker, 1 Prise Salz und Zitronenschale unterkneten. Weitere 40 Minuten gehen lassen.

● Aprikosen klein würfeln. Im Likör 30 Minuten marinieren. Marzipan raffeln. Eiweiße steif schlagen. Mit Aprikosen, Likör und Marzipan verrühren.

● Den Teig kneten und zu einem 1/2 cm dünnen Rechteck ausrollen. Mit der Masse bestreichen und mit Pistazien bestreuen. Den Teig von der Längsseite her aufrollen und in 14 Scheiben von etwa 4 cm Dicke schneiden. Diese Scheiben mit der Schnittfläche nach oben kreisförmig in die gefettete Springform setzen. 20 Minuten gehen lassen, zwischenzeitlich Elektro-Ofen auf 200 Grad heizen.

● Kuchen im Ofen (Gas: Stufe 3) etwa 40 Minuten backen. Konfitüre erhitzen, durch ein Sieb streichen und den Kuchen damit bepinseln. Den Rand mit Pistazien bestreuen.

Aufwand:	★★
Zubereitungszeit:	2 Stunden
Backzeit:	25 Minuten
Pro Stück:	430 kcal/1800 kJ

Variationen

● Eine italienische Variante des Rosenkuchens: Der Teig bleibt derselbe, allerdings ersetzen Sie die getrockneten Aprikosen durch Orangeat und Zitronat. Auf Aprikosenlikör wird ganz verzichtet und statt der Aprikosenkonfitüre nehmen Sie Orangenmarmelade.

● Oder Sie verwenden Rosinen statt Aprikosen und bestreuen den Teig in der Form mit einer Mischung aus 4 EL Zucker, 1/2 TL Zimt und 2 EL geriebenen Mandeln. Pistazien und Aprikosenkonfitüre weglassen.

Apfel-Käse-Kuchen

FÜR 1 BACKBLECH (30 x 40 cm)
350 g Mehl, 1 Prise Salz
1/2 Päckchen Trockenhefe
1 TL abgeriebene Zitronenschale, 40 g Zucker
175 ml lauwarme Milch
75 g flüssige Butter
1 EL Quark (20 % Fett i. Tr.)
FÜR DEN BELAG
1 kg Quark (20 % Fett i. Tr.)
2 Pck. Vanille-Puddingpulver
100 g Zucker, 5 Eier (Größe M)
Schale und Saft von
1/2 unbehandelten Zitrone
1 Packung Mohnback
1 1/2-2 kg kleine Äpfel
3 EL flüssige Butter
AUSSERDEM
Fett für das Blech
Mehl zum Arbeiten

Info

Hefeteig lässt sich sehr gut einfrieren: Nach zweimaligem Gehen den Teig nochmals durchkneten, portionieren und in Beuteln verpackt – maximal 2 Monte – tiefkühlen. Vor dem Weiterverwenden im Kühlschrank auftauen lassen, nochmals gehen lassen, dann wie gewohnt belegen und backen.

- Blech einfetten. Mehl, Hefe, Salz und Zitronenschale mischen. Zucker, Milch, Butter, und Quark hinzufügen, alles zu einem glatten Teig verarbeiten. Zugedeckt 30 Minuten ruhen lassen.
- Quark, Puddingpulver, Zucker, Eier und Zitronenschale verrühren. Masse halbieren. Unter eine Hälfte die fertige Mohnmischung rühren, mit Zitronensaft abschmecken.
- Elektro-Ofen auf 175 Grad vorheizen. Hefeteig auf bemehlter Arbeitsfläche ausrollen und auf das Blech legen. Quark- und Mohnmasse abwechselnd auf den Teig geben und ineinander ziehen, so dass eine Marmorierung entsteht.
- Äpfel schälen, halbieren, Kerngehäuse entfernen. Die runden Seiten gitterförmig einschneiden. Mit der Wölbung nach oben in die Quarkmasse setzen und mit flüssiger Butter bestreichen. Kuchen im Ofen (Gas: Stufe 2) 45-50 Minuten backen.

Aufwand:	★
Zubereitungszeit:	45 Minuten
Backzeit:	50 Minuten
Pro Stück:	400 kcal/1680 kJ

Aprikosen-Mohn-Kuchen

FÜR 1 BACKBLECH (30 x 40 cm)
500 g Mehl, 2 EL Zucker
1 Würfel Hefe (42 g)
150 ml Milch, 100 g Butter
2 Eier (Größe M)
2 Pck. Vanillezucker, Salz
FÜR DEN BELAG
2 Päckchen Mohnback
2 EL Marillenlikör
100 g Sultaninen
100 g Butter, 4 Eier (Größe M)
750 g Quark (20 % Fett i.Tr.)
250 ml Kefir, 250 g Zucker
2 Pck. Vanille-Puddingpulver
(für je 500 ml Flüssigkeit)
Saft und Schale von
1/2 unbehandelten Zitrone
2 Dosen Aprikosen-Hälften
(Abtropfgewicht: à 500 g)
AUSSERDEM
Fett für das Blech, Mehl zum
Arbeiten, Puderzucker

● Mehl in eine Schüssel geben, Mulde eindrücken, 1 TL Zucker und die zerbröckelte Hefe mit etwas lauwarmer Milch verrühren, in die Mulde gießen. Mit Mehl bestäuben, 15 Minuten an einem warmen Ort gehen lassen. Übrigen Zucker, Butter, Eier, Milch, Vanillezucker und 1 Prise Salz unterkneten. Teig erneut gehen lassen.

● Elektro-Ofen auf 175 Grad heizen. Mohnback mit Likör und Sultaninen mischen. Butter schmelzen, abkühlen lassen. Eier trennen. Quark mit Kefir, Zucker, Butter, Puddingpulver, Eigelben, Zitronensaft und -schale glatt rühren. Aprikosen abgießen, abtropfen lassen.

● Backblech fetten. Hefeteig ausrollen, Blech damit auskleiden. Mohncreme darauf verteilen. Aprikosen darauf setzen. Eiweiße steif schlagen, unter die Quarkcreme ziehen und auf die Früchte streichen. Im Ofen (Gas: Stufe 2) 1 Stunde backen. Mit Puderzucker bestäuben.

Aufwand:	★★
Zubereitungszeit:	45 Minuten
Backzeit:	1 Stunde
Pro Stück:	395 kcal/1660 kJ

Käse-Mohn-Kuchen

FÜR 1 BACKBLECH (30 x 40 cm)
400 g Mehl, 50 g Butter
50 g Zucker, 1/2 Würfel Hefe
200 ml lauwarme Milch, Salz
FÜR DIE MOHN-MASSE
1/2 l Milch, 1 EL Honig
350 g Mohn (Reformhaus, fein
mahlen lassen),100 g Zucker
1 Päckchen Vanillezucker
200 g Marzipan-Rohmasse
FÜR DIE QUARK-MASSE
4 Eigelbe und 7 Eiweiße
(Größe M), 100 g Zucker
2 Päckchen Vanillezucker
350 g Quark (20 % Fett i.Tr.)
140 g Speisestärke, 450 ml Milch
AUSSERDEM
Fett für das Blech
Mehl zum Arbeiten

● Mehl, Butter, Zucker, Salz in eine Schüssel geben. Hefe in Milch auflösen, mit den anderen Zutaten zu einem glatten Teig kneten. 30 Minuten an einem warmen Ort gehen lassen.

● Milch mit Honig zum Kochen bringen. Mohn, Zucker und Vanillezucker dazugeben, einmal aufkochen, von der Kochstelle ziehen und kurz ausquellen lassen. Marzipan hacken, unter die Masse heben und mit dem Handrührgerät glatt rühren.

● Elektro-Ofen auf 175 Grad vorheizen. Blech fetten. Teig in Blechgröße ausrollen, aufs Blech legen, Teigränder hochziehen. An den Teigrand einen 3 cm breiten Alustreifen setzen.

● Abgekühlte Mohn-Masse auf dem Teigboden verteilen. Eigelbe, Zucker und Vanillezucker schaumig rühren. Quark, Stärke und Milch unterrühren. Eiweiße steif schlagen, unterheben. Auf die Mohn-Masse streichen.

● Im Ofen (Gas: Stufe 2) etwa 45 Minuten backen. Kuchen bei leicht geöffneter Ofentüre auskühlen lassen.

Aufwand:	★★
Zubereitungszeit:	2 Stunden
Backzeit:	45 Minuten
Pro Stück:	430 kcal/1800 kJ

Stachelbeer-Streuselkuchen

FÜR 1 BACKBLECH (30 x 40 cm)
350 g Mehl, Salz
1 Würfel frische Hefe (42 g)
1/8 l lauwarme Milch
60 g Zucker, 150 g Butter

FÜR DEN BELAG
1 kg rote Stachelbeeren oder
2 kg Nektarinen
2 EL Puderzucker
1 unbehandelte Zitrone
3 Eier (Größe M)
150 g Zucker, 750 g Quark
1 Pck. Vanille-Puddingpulver

FÜR DIE STREUSEL
200 g kalte Butter
150 g Zucker
300 g Mehl, 1 Pck. Vanillezucker

AUSSERDEM
Fett für das Blech

- Mehl mit Salz in eine Schüssel geben, in die Mitte eine Mulde drücken. Hefe hineinbröckeln, mit 2 EL Milch und 1 TL Zucker verrühren. Etwas Mehl darüber stäuben, zugedeckt 15 Minuten gehen lassen. Mit übrigem Zucker, übriger Milch und Butter zu einem glatten Teig verkneten. Zugedeckt 30 Minuten bis zur doppelten Größe aufgehen lassen.
- Beeren putzen und waschen, mit Puderzucker vermischen. Zitrone heiß abwaschen, Schale abreiben, Saft auspressen. Eier mit Zucker, Zitronensaft und -schale, Quark und Puddingpulver verrühren.
- Für die Streusel Mehl mit Butterflöckchen, Zucker und Vanillezucker verkneten.
- Elektro-Ofen auf 200 Grad vorheizen. Blech fetten, Teig darauf ausrollen. Quark-Masse aufstreichen. Stachelbeeren darauf verteilen, Streusel darüber streuen.
- Kuchen im Ofen (Gas: Stufe 3) etwa 45 Minuten backen.

Aufwand:	★ ★
Zubereitungszeit:	45 Minuten
Backzeit:	45 Minuten
Pro Stück:	320 kcal/134 kJ

Zwetschgen-Birnen-Kuchen

FÜR 1 BACKBLECH (30 x 40 cm)
375 g Mehl, 50 g Zucker
1/2 Würfel frische Hefe (20 g)
1/8 l lauwarme Milch
1 Prise Salz, 1 Pck. Vanillezucker, 50 g Butter
FÜR DIE STREUSEL
200 g Butter, 250 g Mehl
150 g Zucker
FÜR DEN BELAG
750 g Zwetschgen, 100 g Zucker
750 g Birnen, Zitronensaft
4 Eigelbe (Eier: Größe M)
250 g Crème fraîche
3-4 EL Zucker,
150 g Kokosraspel
AUSSERDEM
Fett für das Blech

- Mehl in eine Schüssel geben, in die Mitte eine Mulde drücken. Hefe hineinbröckeln und mit 2 EL Milch, 1 TL Zucker verrühren. Etwas Mehl darüber stäuben, zugedeckt 15 Minuten ruhen lassen. Dann mit übrigen Teigzutaten verkneten. Zugedeckt 30 Minuten gehen lassen.
- Für die Streusel Butter schmelzen. Mit Mehl und Zucker krümelig kneten. Zugedeckt kalt stellen.
- Elektro-Ofen auf 175 Grad vorheizen. Früchte waschen, halbieren, entsteinen bzw. entkernen. Früchte in Spalten teilen, mit Zitronensaft beträufeln.
- Eigelbe mit Zucker schaumig aufschlagen. Crème fraîche und Kokosraspel unterrühren.
- Hefeteig ausrollen. Ein tiefes Backblech oder eine Fettpfanne einfetten, mit dem Teig auslegen. Mit Kokoscreme bestreichen. Obst schachbrettartig in 10 cm großen Quadraten auf die Creme legen. Die Zwischenräume mit Streuseln ausstreuen. Kuchen im Ofen (Gas: Stufe 2) ca. 40 Minuten backen.

Aufwand:	★★
Zubereitungszeit:	1 Stunde
Backzeit:	40 Minuten
Pro Stück:	340 kcal/1430 kJ

Aprikosen-Käsekuchen mit Butterstreuseln

FÜR 1 BACKBLECH (30 x 40 cm)
300 g Mehl
40 g frische Hefe
90 ml lauwarme Milch
60 g Zucker, Salz
1 Ei (Größe M), 1 EL Butter
1 Päckchen Vanillezucker
FÜR DEN BELAG
1 Dose Aprikosen
(Abtropfgewicht: 480 g)
6 Eier (Größe M), 200 g Butter
1 1/2 kg Magerquark
300 g Zucker
2 Päckchen Sahne-Pudding-pulver (für je 500 ml Flüssigkeit)
2 TL Backpulver
Saft und Schale von
1 unbehandelte Zitrone
100 g Sultaninen
FÜR DIE STREUSEL
250 g Mehl, 150 g Zucker
1 Päckchen Vanillezucker
1 Prise Salz
120 g Butter
AUSSERDEM
Fett und Mehl für das Blech

- Mehl in eine Schüssel geben. Eine Mulde eindrücken. Hefe hineinbröckeln und mit 2 EL Milch und 1 TL Zucker verrühren. Etwas Mehl überstäuben, 15 Minuten gehen lassen.
- Übrigen Zucker, 1 Prise Salz, Ei, Butterflöckchen und Vanillezucker unterkneten. Den Teig an einem warmen Ort gehen lassen, bis sich sein Volumen verdoppelt hat.
- Aprikosen abgießen. Eine Fettpfanne oder ein Blech mit hohem Rand einfetten mit Mehl bestäuben. Elektro-Ofen auf 175 Grad vorheizen.
- Eier trennen. Butter schmelzen. Quark mit Eigelben und 150 g Zucker verrühren. Abgekühlte Butter und das mit Backpulver gemischte Puddingpulver unterrühren. Mit Zitronensaft und -schale würzen. Eiweiße mit übrigen Zucker steif schlagen. Eischnee und Sultaninen unterheben.
- Hefeteig auf dem Backblech ausrollen, Rand hochziehen. Quark-Masse darauf verteilen. Aprikosen darauf setzen. Im Ofen (Gas: Stufe 2) etwa 30 Minuten vorbacken.
- Für die Streusel Mehl, Zucker, Vanillezucker und Salz vermischen. Butter schmelzen, abühlen lassen mit, einer Gabel untermischen, bis kleine Streusel entstehen. Noch 1 TL Mehl überstäuben und durchschütteln.
- Streusel zwischen die Aprikosen auf dem Käsekuchen verteilen. Im Ofen (Gas: Stufe 3) in etwa 25 Minuten fertig backen.

Aufwand:	★★
Zubereitungszeit:	1 1/2 Stunden
Backzeit:	55 Minuten
Pro Stück:	380 kcal/1590 kJ

Tipp

Auch gebackenen Hefekuchen kann man mühelos einfrieren: Übrig gebliebene Stücke einzeln in Plastiktüten verpacken und rasch tiefkühlen. Vor dem Servieren bei Zimmertemperatur etwa 2 Stunden auftauen lassen.

Aprikosen-Mandel-Kuchen

FÜR 1 BACKBLECH (30 x 40 cm)
500 g Mehl, 1 Würfel Hefe (42 g)
1/4 l Milch, 80 g Zucker
2 Eier, 100 g Butter
1 Prise Salz
FÜR DEN BELAG
1 kg reife, aromatische Aprikosen
50 g Mandelstifte, 2 EL Zucker
AUSSERDEM
Backpapier

- Mehl in eine Schüssel geben, eine Mulde eindrücken, die Hefe hineinbröckeln und mit etwas lauwarmer Milch, 1 EL Zucker und 3 EL Mehl verrühren. Mit Mehl bestäubt zugedeckt 20 Minuten gehen lassen.
- Mit Eiern, Butter, restlicher Milch, übrigem Zucker verkneten und zugedeckt 1 Stunde gehen lassen.
- Elektro-Ofen auf 175 Grad heizen. Backblech mit Backpapier auslegen. Teig durchkneten, auf dem Blech ausrollen. 20 Minuten ruhen lassen.
- Aprikosen waschen, halbieren und entsteinen, mit den Schnittflächen nach oben nebeneinander in den Teig drücken. Mit Mandeln und Zucker bestreuen. Im Ofen (Gas: Stufe 2) etwa 40 Minuten backen.

Aufwand:	★★
Zubereitungszeit:	1 1/2 Stunden
Backzeit:	40 Minuten
Pro Stück:	245 kcal/1030 kJ

Variation

- Mit Baiserhaube wird der Kuchen noch feiner: 3 Eiweiße mit 125 g Zucker steif schlagen, die Masse in einen Spritzbeutel füllen. Den Kuchen nach 25 Minuten Backzeit aus dem Ofen nehmen und kleine Häubchen oder ein beliebiges Muster aufspritzen. Den Kuchen wieder in den Ofen schieben und in weiteren 15 Minuten fertig backen.

Kirsch-Käsekuchen

FÜR 1 BACKBLECH (30 x 40 cm)
300 g Mehl, 1 Würfel Hefe (42 g)
100 ml lauwarme Milch
60 g Zucker, 1 Prise Salz
1 Ei (Größe M), 2 EL Butter
1 Päckchen Vanillezucker
FÜR DEN BELAG
2 Gläser Sauerkirschen (à 350 g)
6 Eier (Größe M), 250 g Zucker
500 g Mascarpone
1 1/4 kg Magerquark
2 Pck. Grieß-Puddingpulver (für je 500 ml Flüssigkeit)
2 EL Kakaopulver
AUSSERDEM
Fett und Mehl für das Blech
Puderzucker

● Mehl in eine Schüssel geben, eine Mulde ins Mehl drücken. Hefe hineinbröckeln, mit 2 EL Milch und 1 TL Zucker verrühren. Etwas Mehl überstäuben, 15 Minuten gehen lassen. Übrigen Zucker, restliche Milch, Salz, Ei, Butter und Vanillezucker unterkneten. Teig an warmem Ort gehen lassen, bis er doppelt so groß ist

● Kirschen abtropfen lassen. Eier trennen. Eigelbe mit 150 g Zucker und Mascarpone aufschlagen. Quark und Puddingpulver untermischen. Eiweiße steif schlagen, übrigen Zucker zugeben und etwa 4 Minuten weiterschlagen, unter die Käse-Masse heben.

● Elektro-Ofen auf 175 Grad vorheizen. Etwa 1/3 der Masse abnehmen, den Kakao untermischen. Kirschen unter die helle Masse heben. Ein tiefes Blech fetten, mit Mehl bestäuben. Hefeteig auf dem Backblech ausrollen, Ränder hochziehen. Die dunkle Käse-Masse auf dem Teig verteilen. Helle Käse-Masse darüber streichen. Mit einer Gabel marmorieren.

● Im Ofen (Gas: Stufe 2) etwa 50 Minuten backen. Mit Puderzucker bestäuben.

Aufwand:	★★
Zubereitungszeit:	1 1/4 Stunden
Backzeit:	50 Minuten
Pro Stück:	325 kcal/1370 kJ

Kirschkuchen mit Sauerrahmguss

FÜR 1 BACKBLECH (30 x 40 cm)
400 g Mehl
1 Würfel Hefe (42 g)
50 g Zucker
200 ml lauwarme Milch
1 Ei (Größe M), 50 g Butter
FÜR DEN BELAG
1 1/2 kg Sauerkirschen
150 g Marzipan-Rohmasse
2 EL Puderzucker
FÜR DEN GUSS
4 Eier (Größe M), 50 g Zucker
450 g saure Sahne
AUSSERDEM
Fett und Mehl fürs Blech
Puderzucker zum Bestäuben

- Mehl in eine Schüssel geben, eine Mulde ins Mehl drücken. Hefe hineinbröseln, mit 1 EL Zucker bestreuen. Mit etwas Milch verrühren und zugedeckt an einem warmen Ort 15 Minuten gehen lassen.
- Übrigen Zucker, Ei, übrige Milch und Butter zum Vorteig geben. Alles zu einem geschmeidigen Teig verkneten. Zugedeckt 30 Minuten gehen lassen
- Kirschen waschen und entsteinen. Marzipan-Rohmasse grob zerteilen, Puderzucker darüber sieben und das Marzipan fein würfeln. Backblech fetten mit Mehl bestäuben. Elektro-Ofen auf 200 Grad vorheizen. Teig auf dem Blech ausrollen, mit Marzipan bestreuen, die Kirschen darauf verteilen.
- Für den Guss die Eier trennen. Die Eigelbe mit Zucker so lange rühren, bis sich der Zucker gelöst hat. Die saure Sahne unterrühren. Die Eiweiße zu steifem Schnee schlagen und vorsichtig unterheben. Guss auf den Kirschen verteilen.
- Kuchen im Ofen (Gas: Stufe 3) etwa 45 Minuten backen. Nach Belieben mit Puderzucker bestäuben.

Aufwand:	★★
Zubereitungszeit:	1 Stunde
Backzeit:	45 Minuten
Pro Stück:	210 kcal/880 kJ

Zwetschgen-Nuss-Pie

FÜR 1 PIEFORM VON 30 cm Ø
250 g Mehl
20 g frische Hefe
1/8 l lauwarme Milch
30 g Zucker
1 TL abgeriebene Schale einer unbehandelten Zitrone
1 Prise Salz, 70 g weiche Butter
FÜR DEN BELAG
1 1/4 kg Zwetschgen
150 g gemahlene Haselnüsse
50 g Löffelbiskuits oder Zwieback
2 Eier (Größe M), 60 g Zucker
1/4 TL gemahlener Zimt
nach Belieben 2 cl Zwetschgen-Schnaps
50 g Haselnussblättchen
AUSSERDEM
Fett für das Blech
Mehl zum Arbeiten

- Mehl in eine Schüssel geben und in die Mitte eine Mulde drücken. Hefe hineinbröckeln, mit 2 EL Milch und 1 TL Zucker verrühren. Etwas Mehl darüber stäuben, zugedeckt 15 Minuten gehen lassen.
- Zitronenschale mit Salz, übrigem Zucker, übriger Milch und der Butter zum Vorteig geben. Alles zu einem Teig verkneten. Mit Mehl bestäuben und 30 Minuten zugedeckt gehen lassen.
- Zwetschgen waschen, längs einschneiden, Steine herauslösen. Die Hälften nochmals längs einschneiden. Gemahlene Haselnüsse in einer Pfanne ohne Fett leicht anrösten. Löffelbiskuits in einen Gefrierbeutel geben und mit einer Teigrolle fein zerbröseln.
- Elektro-Ofen auf 175 Grad heizen. Pie-Form einfetten. Hefeteig auf bemehlter Arbeitsfläche ausrollen. Die Form damit auskleiden.
- Die Eier trennen und die Eiweiße mit Zucker steif schlagen. Eigelbe mit Zimt und Biskuitbröseln verrühren, unter die Eischnee-Masse ziehen. Die Haselnüsse unterheben und nach Belieben mit Zwetschgen-Schnaps aromatisieren. Die Nuss-Masse in die Form füllen. Zwetschgen dachziegelartig dicht an dicht nebeneinander setzen. Nussblättchen darüber streuen.
- Pie im Ofen (Gas: Stufe 2) etwa 40 Minuten backen.

Aufwand:	★★
Zubereitungszeit:	1 1/4 Stunden
Backzeit:	40 Minuten
Pro Stück:	350 kcal/1470 kJ

Variationen

- Fruchtwechsel: Belegen Sie den Kuchen mit Äpfeln, Aprikosen oder Birnen.
- Sie können die Nussmasse ebenso gut auch auf die Pflaumen geben und bekommen so eine Knusperkruste.

Zwetschgen-Quark-Teilchen

Für 16 Stück
250 g Mehl, 20 g frische Hefe
40 g Zucker
1/8 l Milch
40 g Butter
1 Prise Salz
1 TL abgeriebene Schale einer unbehandelten Zitrone
1 Ei (Größe S)
FÜR DEN BELAG
600 g Zwetschgen
1 Ei
100 g Sahnequark
2 EL Crème fraîche
50 g Zucker
2 EL Aprikosenkonfitüre
AUSSERDEM
Fett für die Bleche

- Mehl in eine Schüssel sieben, eine Vertiefung eindrücken. Die Hefe hineinbröckeln, mit dem Zucker und lauwarmer Milch verrühren. Zugedeckt 14 Minuten gehen lassen.
- Butter zerlassen, mit übriger Milch, Salz, Zitronenschale und dem Ei vermengen. Zum Vorteig geben, alles verkneten. Zugedeckt gehen lassen, bis sich die Teigmenge verdoppelt hat. Nochmals kräftig durchkneten und weitere 15 Minuten gehen lassen.
- Inzwischen 2 Backbleche fetten. Zwetschgen waschen, entsteinen und in schmale Spalten schneiden. Das Ei trennen. Eiweiß steif schlagen. Eigelb mit Quark, Crème fraîche und Zucker verrühren, Eischnee unterziehen.
- Den Backofen auf 200 Grad vorheizen.
- Den Teig 1/2 cm dick ausrollen und Kreise von etwa 8 cm Ø ausstechen. Die Mitte flach drücken, die Ränder etwas hochbiegen. Quark-Masse in die Teilchen füllen, darauf die Zwetschgenspalten legen. Teilchen auf die Bleche setzen, nacheinander im Ofen (Gas: Stufe 3) etwa 15 Minuten backen.
- Die Konfitüre erhitzen und durch ein Sieb streichen, dann die noch heißen Zwetschgen damit bepinseln.

Aufwand:	★
Zubereitungszeit:	1 1/2 Stunden
Backzeit:	30 Minuten
Pro Stück:	245 kcal/1030 kJ

Variationen

- Auch hier lassen sich verschiedene Früchte wie Aprikosen, Pfirsiche oder Kirschen verwenden.
- Heben Sie unter die Quarkmasse gehackte Pistazien und süßen Sie die Mischung mit Vanillezucker.

Rhabarberkuchen mit Schmand

FÜR 1 BACKBLECH (30 x 40 cm)
200 g Mehl
20 g frische Hefe
50 ml lauwarme Milch
40 g Zucker
1 Prise Salz
1 Ei, 40 g weiche Butter
FÜR DEN BELAG
1 kg Rhabarber
2 TL Zimt
4 EL Zucker, 30 g Ingwer
400 g Schmand
AUSSERDEM
Fett für das Blech
Mehl zum Arbeiten

● Mehl in eine Schüssel sieben, in die Mitte eine Mulde drücken, Hefe hineinbröckeln. Mit Milch, 1 EL Zucker und wenig Mehl vom Rand einen Vorteig rühren. Zugedeckt an einem warmen Ort 15 Minuten gehen lasen.

● Restlichen Zucker, Salz, Ei und die Butter unter den Vorteig mischen. Alles zu einem glatten Teig verkneten. Zugedeckt nochmals 30 Minuten gehen lassen.

● Den Rhabarber putzen und in ca. 2 cm lange Stücke schneiden. 1 TL Zimt mit 2 EL Zucker mischen. Den Rhabarber damit bestreuen und etwa 20 Minuten ziehen lassen.

● Backofen auf 200 Grad vorheizen. Das Blech einfetten. Den Teig auf der bemehlten Arbeitsfläche ausrollen und auf das Blech legen.

● Rhabarber abtropfen lassen. Ingwer schälen und fein reiben oder ganz fein würfeln. Restlichen Zimt und Zucker mischen. Den Teig mit Rhabarber belegen, mit Zimtzucker und Ingwer bestreuen.

● Kuchen im Ofen (Gas: Stufe 3) etwa 20 Minuten backen. Schmand daraufgeben und in etwa 10 Minuten fertig backen.

Aufwand:	★★
Zubereitungszeit:	1 1/4 Stunden
Backzeit:	30 Minuten
Pro Stück:	325 kcal/1370kJ

Pflaumenkuchen mit Baisergitter

FÜR 1 BACKBLECH (30 x 40 cm)
400 g Mehl
40 g frische Hefe
100 ml lauwarme Milch
80 g Zucker
1 Prise Salz
2 Eier
80 g Butter
FÜR DEN BELAG
1 1/2-2 kg Pflaumen
(oder Zwetschgen)
125 g Zucker
3 Eiweiße
2 Msp. Ingwerpulver
AUSSERDEM
Fett für das Blech
Mehl zum Arbeiten

- Mehl in eine Schüssel sieben, eine Vertiefung eindrücken. Die Hefe hineinbröckeln. Mit Milch, 1 EL Zucker und etwas Mehl verrühren. Den Vorteig 15 Minuten gehen lassen.
- Übrigen Zucker, Salz, Eier und Butter dazugeben, alles gut vermischen und zu einem glatten Teig verkneten. Auf einer bemehlten Fläche durchkneten, zugedeckt gehen lassen, bis sich sein Volumen verdoppelt hat.
- Backofen auf 200 Grad vorheizen. Pflaumen waschen, halbieren oder vierteln, entsteinen.
- Hefeteig auf dem gefetteten Blech dünn ausrollen, mit etwa 40 g Zucker bestreuen und mit den Früchten, die Schnittflächen nach oben, belegen.
- Den Kuchen im Ofen (Gas: Stufe 3) etwa 20 Minuten backen. Eiweiße mit übrigem Zucker und Ingwerpulver steif schlagen, in einen Spritzbeutel mit dünner Tülle füllen.
- Den Kuchen aus dem Ofen nehmen, die Baisermasse gitterförmig aufspritzen, in 10-15 Minuten fertig backen.

Aufwand:	★★
Zubereitungszeit:	1 1/4 Stunden
Backzeit:	35 Minuten
Pro Stück:	245 kcal/1030 kJ

Apfel-Rosenkuchen mit Ingwer

FÜR SPRINGFORM VON 26 cm Ø
500 g Mehl
1 Würfel Hefe (42 g)
200 ml Milch, 1 Ei (Größe M)
100 g Zucker
100 g Kokoscreme (Fertigprodukt; ersatzweise 75 g Butter)
FÜR DIE FÜLLUNG
750 g säuerliche Äpfel
2 EL Zitronensaft
50 g Kokosraspel
50 g gehackte Pistazien
40 g in Sirup eingelegter Ingwer (Glas)
AUSSERDEM
Mehl zum Arbeiten
Fett für die Form
50 g Apfel- oder Quittengelee
Puderzucker zum Bestäuben

● Mehl in eine Schüssel geben, eine Mulde hineindrücken. Hefe hineinbröckeln, mit 50 ml lauwarmer Milch und 1 EL Zucker verrühren. Mehl darüber stäuben, 30 Minuten zugedeckt an einem warmen Ort ruhen lassen. Übrige Teigzutaten unterkneten und zugedeckt noch 30 Minuten ruhen lassen.
● Elektro-Ofen auf 175 Grad vorheizen. Äpfel schälen, grob reiben, sofort mit Zitronensaft, Kokosraspeln und Pistazien mischen. Ingwer würfeln und mit 2 EL Ingwersirup untermischen.
● Teig noch mal durchkneten und auf bemehlter Arbeitsfläche 1/2 cm dick zu einem Rechteck (50 x 40 cm) ausrollen. Füllung darauf verteilen. Teig von der Seite her einrollen, in 10 Stücke schneiden. Teigstücke mit der Schnittfläche nach oben in die gefettete Springform setzen. Im Ofen (Gas: Stufe 2) etwa 40 Minuten backen.
● Gelee erwärmen, warmen Kuchen damit bepinseln. Abkühlen lassen. Mit Puderzucker bestäuben.

Aufwand:	★★
Zubereitungszeit:	1 1/2 Stunden
Backzeit:	40 Minuten
Pro Stück:	350 kcal/1470 kJ

Aprikosen-Küchlein

Für 10 Stück
1 Packung tiefgekühlter Hefeteig (450 g)
240 g Aprikosenhälften (Dose)
250 g Quark (20% Fett i. Tr.)
1 Ei (Größe M)
1-2 EL Speisestärke
1-2 EL Amaretto (Mandellikör)
100 g Zucker, 50 g Butter
75 g Mehl
50 g gemahlene Mandeln
AUSSERDEM
Backpapier

● Hefeteig auftauen lassen. Elektro-Ofen auf 175 Grad vorheizen. Teigplatten auf ein mit Backpapier belegtes Blech legen.
● Aprikosen abtropfen lassen und in Achtel schneiden. Quark in einem Küchentuch gut ausdrücken und mit Ei, Speisestärke, Amaretto und 20 g Zucker verrühren. Übrigen Zucker, Fett, Mehl und Mandeln zu Streuseln verkneten.
● Auf die Hefeteig-Quadrate jeweils 2 EL Quarkmasse verteilen, dabei einen 1/2 cm breiten Rand frei lassen. Ecken leicht bis zur Mitte klappen. Aprikosen darauf setzen und die Streusel darüber streuen. Im Ofen (Gas: Stufe 2) etwa 25 Minuten backen. Abkühlen lassen.

Aufwand:	★
Zubereitungszeit:	40 Minuten
Backzeit:	25 Minuten
Pro Stück:	340 kcal/1430 kJ

Grundrezept Plunderteig

125 ml Milch
50 g Butter
1/2 Würfel Hefe (21 g)
2 Eier (Größe M)
250 g Mehl
50 g Zucker
1/2 TL abgeriebene Schale einer unbehandelten Zitrone
450 g Tiefkühl-Blätterteig (6 rechteckige Scheiben)
Mehl zum Arbeiten

● Für den Hefeteig Milch mit Butter erhitzen, bis das Fett geschmolzen ist. Auf Handwärme abkühlen lassen und die zerbröckelte Hefe darin auflösen.
● Eier trennen. Mehl mit Zucker und Zitronenschale mischen. Eigelbe und Hefemischung zum Mehl geben, alles zum glatten Teig kneten. An einem warmen Ort 20 Minuten gehen lassen. Blätterteigscheiben nebeneinander auftauen lassen. Hefeteig noch mal gut durchkneten.
● 3 Blätterteigplatten auf einer bemehlten Arbeitsfläche zu einem Rechteck zusammenlegen: Die Seiten, die aneinander stoßen, mit wenig Eiweiß bepinseln und anschließend 1 cm überlappend aufeinander legen. Zu einem Rechteck (45 x 25 cm) ausrollen und mit etwas Eiweiß bestreichen.
● Die Hälfte des Hefeteiges zu einem ebenso großen Rechteck ausrollen, auf den Blätterteig legen und etwas andrücken. Diese doppelte Teigplatte etwa 10 Minuten ruhen lassen, bevor sie – wie in den jeweiligen Rezepten beschrieben – in Stücke geteilt wird. Mit den restlichen Teigen ebenso verfahren.

Gittertasche mit exotischer Füllung

Für 10 Plunderteilchen
3 Dosen tropischer Fruchtcocktail (Abtropfgewicht: à 250 g)
2 Orangen (1 unbehandelt)
400 g Marzipan-Rohmasse
2 EL Puderzucker
1 Grundrezept Plunderteig
1 Ei (Größe M), 1 Päckchen weißer Tortenguss, 2 EL Milch
AUSSERDEM
Backpapier

● Elektro-Ofen auf 200 Grad vorheizen. Früchte abtropfen lassen, Saft auffangen. Von der unbehandelten Orange 1 TL Schale abreiben. Beide Orangen schälen, filetieren. Obst mischen.
● Marzipan hacken und mit Puderzucker verkneten. Masse halbieren und jeweils zwischen Klarsichtfolie auf 40 x 10 cm ausrollen.
● Jede Plunderteigplatte quer fünfteln, längs halbieren. Das Ei trennen. 10 Teigrechtecke mit Eiweiß bepinseln, übrige mit Gitterroller einschneiden.
● Marzipanmasse auf die glatten Teigstücke legen. 2-3 EL Obst darauf geben, dabei 2 cm Rand frei lassen, mit Eiweiß bepinseln.
● 1/4 l Saft mit Orangenschale und Tortenguss nach Packungsangabe kochen, über dem Obst verteilen. Teiggitter auflegen, etwas auseinander ziehen, am Rand festdrücken.
● Eigelb mit Milch verquirlen, auf Gebäck pinseln. Auf einem mit Backpapier ausgelegten Blech im Ofen (Gas: Stufe 3) 20-30 Minuten backen.

Aufwand:	★★
Zubereitungszeit:	1 1/4 Stunden
Backzeit:	30 Minuten
Pro Stück:	430 kcal/1800 kJ

Lochmuster: Gitterroller auf das Teigrechteck setzen, kräftig eindrücken, darüber rollen.

Hörnchen rollen: Füllung auf unteres Teigdrittel geben, Teig überklappen, zur Spitze einrollen.

Schoko-Kirsch-Hörnchen

Für 16 Stück
100 g weiße Kuvertüre
1 Glas Schattenmorellen (Abtropfgewicht: 175 g)
1-2 TL Speisestärke
1 Msp. Kardamom
1 Grundrezept Plunderteig
1 Eigelb (Ei: Größe M)
2 EL Milch
AUSSERDEM
Backpapier

- Elektro-Ofen auf 200 Grad heizen. 50 g Kuvertüre grob hacken. Kirschen in ein Sieb abgießen, Saft dabei auffangen.
- 100 ml Kirschsaft erhitzen. Stärke in wenig kaltem Wasser anrühren und in die heiße Flüssigkeit einrühren. Aufkochen lassen, dann mit Kardamom abschmecken. Vom Herd ziehen, Kirschen darunter mengen und abkühlen lassen.
- Jede Plunderteigplatte quer vierteln. Jedes Viertel diagonal halbieren, so dass je 2 Dreiecke entstehen. Je 1-2 EL Kirschen und Schokolade auf das untere Drittel der Teigdreiecke setzen. Teig darüber klappen und zur Spitze hin einrollen (s. Bild unten). Zu Hörnchen formen. Eigelb mit Milch verquirlen und die Hörnchen damit bepinseln.
- Die Plunderhörnchen auf ein mit Backpapier ausgelegtes Backblech legen und im Ofen (Gas: Stufe 3) etwa 20 Minuten backen.
- Die übrige Kuvertüre über dem heißen Wasserbad schmelzen. Plunderstücke damit nach Belieben verzieren.

Aufwand:	★★
Zubereitungszeit:	1 1/4 Stunden
Backzeit:	20 Minuten
Pro Stück:	260 kcal/1090 kJ

Apfel-Teilchen mit Kastanien

Für 10 Stück
4 Äpfel
200 g Kastanien (Dose)
1 Vanilleschote
100 g Zucker
20 g Butter
1 Grundrezept Plunderteig
(siehe Seite 116)
1 Eigelb (Ei: Größe M)
2 EL Milch
75 g Zitronenmarmelade
(ersatzweise Aprikosenkonfitüre)
AUSSERDEM
Backpapier

● Die Äpfel schälen, halbieren, entkernen und in 3 mm dünne Scheiben schneiden. Die Kastanien abtropfen lassen und fein würfeln. Vanilleschote aufschneiden, Mark herauskratzen.
● Zucker in einem Topf mit dickem Boden karamellisieren lassen. Wenn er braun ist, Kastanien und 2 EL Wasser zugeben, umrühren, kurz aufkochen.
● Inzwischen den Elektro-Ofen auf 200 Grad heizen. Butter zerlassen, Vanillemark zugeben.
● Aus den Plunderteigplatten Kreise von ca. 10 cm ø ausstechen (s. Bild rechts).
● Apfelscheiben kreisförmig auf den Teig legen. Mit Vanillebutter bepinseln. Karamellisierte Kastanien in die Mitte setzen. Teig-Ränder etwas hochklappen.
● Eigelb mit Milch verquirlen, die Teigränder damit bepinseln. Plunderteilchen auf einem mit Backpapier ausgelegten Blech im Ofen (Gas: Stufe 3) 20-30 Minuten backen.
● Marmelade erwärmen, auf die heißen Plunderteilchen streichen.

Aufwand:	★★
Zubereitungszeit:	1 1/4 Stunden
Backzeit:	30 Minuten
Pro Stück:	300 kcal/1260 kJ

Kreise ausstechen: Mit einer Tasse oder Ausstechform dicht an dicht den Teig ausstechen.

Taschen formen: Füllung auf den Teig geben, zwei Ecken zur Spitze nach oben klappen.

Cheese-cake-Schiffchen

Für 16 Stück
1 Ei (Größe M)
50 g Zucker
1 unbehandelte Zitrone
400 g Doppelrahm-Frischkäse
1 Grundrezept Plunderteig
(siehe Seite 116)
1 Eigelb (Ei: Größe M)
4 EL Milch
4 EL Johannisbeergelee
AUSSERDEM
Backpapier

● Elektro-Ofen auf 200 Grad vorheizen. Ei mit Zucker schaumig schlagen. Zitrone heiß abwaschen, trockenreiben. Schale abreiben, Saft auspressen. Frischkäse, Zitronenschale und -saft unter die Ei-Masse rühren.
● Jede Plunderteigpeigplatte quer in 2 Rechtecke teilen, dann die Rechtecke vierteln. Die Viertel noch mal diagonal teilen, so dass pro Teigplatte 16 kleine Dreiecke entstehen. Eigelb mit Milch verquirlen. Alle Teigdreiecke damit bepinseln.
● Je 1 EL Füllung auf die Teig-Dreiecke verteilen. Rechte und linke Ecke zur Spitze hochklappen und fest andrücken. Oberfläche ebenfalls mit verquirltem Eigelb einpinseln.
● Die Plundertäschchen auf ein mit Backpapier ausgelegtes Backblech setzen und im Ofen (Gas: Stufe 3) 20-30 Minuten backen.
● Das Gelee erhitzen und nach dem Backen auf die Ränder der noch heißen Füllung streichen.

Aufwand:	★★
Zubereitungszeit:	1 1/4 Stunden
Backzeit:	30 Minuten
Pro Stück:	325 kcal/1370 kJ

Gefüllte Buchteln

FÜR 1 OFENFESTE FORM
1 Päckchen Trockenhefe
1/4 l lauwarme Milch
70 g Zucker
500 g Mehl
1 Ei und 3 Eigelbe
(Eier: Größe M)
130 g weiche Butter, Salz
Mark von 1 Vanilleschote
abgeriebene Schale von
1 unbehandelten Zitrone
FÜR DIE FÜLLUNG
60 g Konfitüre
(z. B. Waldfrucht)
1 TL gem. Zimt
1 TL Kakaopulver
AUSSERDEM
Fett für die Form
3 EL Puderzucker
nach Belieben Vanillesauce

- Hefe, Milch, Zucker, Mehl, Ei, Eigelbe, 80 g Butter, 1 Prise Salz, Vanillemark und Zitronenschale zu einem Teig verrühren. An einem warmen Ort zugedeckt 30 Minuten gehen lassen, bis sich das Volumen verdoppelt hat.
- Noch mal gut durchkneten und mit einem Küchentuch bedeckt 15 Minuten ruhen lassen.
- Elektro-Ofen auf 175 Grad heizen. Konfitüre mit Zimt und Kakao abschmecken. Teig zu 16 Kugeln formen. Mulde eindrücken, je 1/4 TL Konfitüre hineinsetzen. Teig darüber zusammendrücken, zur Kugel formen.
- Eine ofenfeste Form einfetten, Buchteln nebeneinander hineinsetzen. Mit 2 EL Puderzucker bestäuben und 15 Minuten gehen lassen.
- Übrige Butter zerlassen, die Buchteln damit bestreichen. Im Ofen (Gas: Stufe 2) etwa 40 Minuten backen.
- Buchteln mit übrigem Puderzucker bestäuben. Nach Belieben Vanillesauce dazureichen.

Aufwand:	★★
Zubereitungszeit:	1 1/4 Stunde
Backzeit:	40 Minuten
Pro Stück:	225 kcal/940 kJ

Grundrezept Hefeteig

500 g Mehl
1/2 Würfel Hefe (21 g)
etwa 300 ml lauwarme Milch
100 g Zucker
50 g Butter
1 Prise Salz

- Mehl in die Schüssel geben, in die Mitte des Mehls eine Mulde drücken. Hefe mit 4 EL Milch und 1 TL Zucker verrühren und in die Vertiefung geben. Mit etwas Mehl vom Rand verrühren. Hefeansatz zugedeckt an einem warmen Ort 20 Minuten gehen lassen.
- Vorteig mit den übrigen Zutaten verkneten und weitere 30 Minuten ruhen lassen. Teig wie in den Rezepten beschrieben weiter verarbeiten.

Tipp

Bestreuen Sie den Zopf vor dem Backen mit Mandeln oder Hagelzucker.

Schweizer Hefezopf

FÜR 1 BACKBLECH (30 x 40 cm)
1 Grundrezept Hefeteig (s. links)
1 Eigelb (Ei: Größe M)
1-2 EL Milch
AUSSERDEM
Backpapier

- Elektro-Ofen auf 200 Grad vorheizen. Hefeteig halbieren, zu 2 Strängen (Länge: 50-60 cm) rollen. Diese wie eine Kordel drehen. In der Mitte umbiegen. Wieder ineinander drehen. Die Zopfenden andrücken.
- Zopf auf das mit Backpapier ausgelegtes Blech setzen. Eigelb mit Milch verquirlen und den Hefezopf damit bepinseln. Im Ofen (Gas: Stufe 3) etwa 35 Minuten backen.

Aufwand:	★
Vorbereitungszeit:	1 1/4 Stunden
Backzeit:	35 Minuten
Pro Stück:	180 kcal/760 kJ

Korianderzopf

FÜR 1 BACKBLECH (30 x 40 cm)
1 Grundrezept Hefeteig
(siehe Seite 121)
300 g entsteinte Backpflaumen
200 ml Apfelsaft
2 EL Milch
100 g Mandelstifte
je 30 g gewürfeltes Orangeat
und Zitronat
je 1 Msp. Zimtpulver, gemahlener Kardamom und Anis
Mehl zum Arbeiten
eventuell Fett für das Blech
1 Eigelb (Ei: Größe M)
1 EL zerstoßene Korianderkörner
AUSSERDEM
Mehl zum Arbeiten
Fett oder Backpapier für das Blech

● Den Hefeteig nach dem Grundrezept (siehe Seite 121) zubereiten.
● Die Pflaumen 30 Minuten im Apfelsaft einweichen, abtropfen lassen und würfeln. Mit Mandeln, Orangeat, Zitronat, Zimt, Kardamom und Anis vermischen. Elektro-Ofen auf 200 Grad vorheizen.
● Den Teig auf einer bemehlten Fläche zu einem Rechteck (etwa 25 x 30 cm) ausrollen und auf ein gefettetes oder mit Backpapier ausgelegtes Blech legen. Die Pflaumen-Masse in die Mitte des Teiges geben. Den Teig zu beiden Seiten der Füllung einschneiden und die Teigstreifen abwechselnd über der Füllung über Kreuz legen.
● Das Eigelb und die Milch verquirlen, den Zopf damit einpinseln und mit Koriander bestreuen. Zopf im Ofen (Gas: Stufe 3) 25-30 Minuten backen.

Aufwand:	★★
Zubereitungszeit:	1 1/2 Stunden
Backzeit:	30 Minuten
Pro Stück:	270 kcal/1130 kJ

Nuss-Mandel-Brot

FÜR 1 KASTENFORM (30 cm LÄNGE)
1 Grundrezept Hefeteig
(siehe Seite 121)
3 Eiweiße (Eier: Größe M)
3-4 EL Zucker
100 g gemahlene Mandeln
100 g gemahlene Walnusskerne
8 Tropfen Bittermandelöl
2 EL Mandelblättchen
1 EL Butter
AUSSERDEM
Mehl zum Arbeiten
Fett für die Form

- Den Hefeteig nach dem Grundrezept (siehe Seite 121) zubereiten.
- Die Eiweiße steif schlagen und dabei den Zucker einrieseln lassen. Die Mandeln und Walnüsse mischen und mit dem Bittermandelöl unter den Eischnee heben.
- Vom Teig etwas weniger als die Hälfte abnehmen, den übrigen Teil auf einer bemehlten Fläche zu einem Rechteck (25 x 35 cm) ausrollen. Die Nuss-Masse darauf verteilen und den Teig von der Längsseite her einrollen.
- Den Elektro-Ofen auf 200 Grad vorheizen. Die Form einfetten und mit 1 EL Mandelblättchen ausstreuen. Die Teigrolle in die Form legen.
- Aus dem übrigen Teig 3 Stränge (Länge: jeweils 40 cm) formen und zu einem Zopf flechten. Den Zopf auf das Brot legen. Butter in einem Topf zerlassen und dünn auf die Oberfläche des Brotes pinseln. Mit den restlichen Mandelblättchen bestreuen. Brot im Ofen (Gas: Stufe 3) 30 Minuten backen.
- Das Brot in der Form auskühlen lassen, damit es nicht reißt.

Aufwand:	★★
Zubereitungszeit:	1 1/2 Stunden
Backzeit:	30 Minuten
Pro Stück:	220 kcal/920 kJ

Rund um

Hefeteig lässt sich nicht nur gut formen, sonde

Mandeln abziehen

Die stumpfe, bittere Schale von Mandeln lässt sich ganz leicht entfernen: Mandeln mit heißem Wasser überbrühen und etwa 10 Minuten ziehen lassen. Wasser abgießen und die Mandeln kurz unter kaltem Wasser abschrecken. Anschließend lassen sie sich ganz leicht aus der Schale drücken.

Korianderzopf

Es gibt verschiedene Variationen, Zopfbrote herzustellen. Hier die Anleitung für den Korinaderzopf von Seite 122.

Die freien Teigstücke an beiden Seiten diagonal in 2 cm breite Streifen schneiden.

Abwechselnd linke und rechte Streifen über die Füllung legen und andrücken.

Wenn's um die Nüsse geht

Nüsse lieber ganz kaufen und selber hacken oder mahlen. Nicht nur, weil es billiger ist: Sie schmecken einfach besser, weil sie frischer sind. Außerdem sind Nüsse in der Schale bis zu 3 Monaten haltbar. Wenn gehackte oder gemahlene Nüsse vorher angeröstet werden, geben sie ein intensiveres Aroma ab. Goldbraune Mandelblättchen machen sich auch hübsch als Dekoration. Geröstet wird in einer beschichteten Pfanne ohne Fett.

Haselnü

Haselnüsse knacken, die harten Schalen entfernen und die Nüsse

den Hefeteig

ch füllen: Nüsse und Mandeln sind dafür die idealen Zutaten

So flechten Sie den Zopf

Hefeteig zubereiten, gehen lassen, nochmals gut durcharbeiten und anschließend mit bemehlten Händen in 3 gleichmäßige Stränge teilen. Nebeneinander legen. Abwechselnd den inneren Strang über den jeweils äußersten legen. Die Stränge an den Enden festdrücken und leicht nach unten einschlagen.

Für Eilige

Hefeteig kann man auch schon am Vortag kneten und über Nacht im Kühlschrank gehen lassen. Er geht zwar nicht ganz so auf wie bei Zimmertemperatur, aber für Blechkuchen reicht es. Am nächsten Tag einfach auf bemehlter Fläche ausrollen und wie gewünscht belegen.

Was ist Hefe?

Der typische, leicht säuerliche Geschmack und die Luftigkeit des Teiges werden durch die Hefebakterien verursacht, die sich im feuchten und warmen Teig vermehren. Mithilfe bestimmter Fermente im Mehl wird Stärke in Traubenzucker verwandelt und dieser in Alkohol und Kohlendioxyd. Dabei entstehen zahllose kleine gasgefüllte Bläschen, die das Volumen des Teiges ständig vergrößern. Frische Hefe riecht und schmeckt angenehm säuerlich. Sie hält im Kühlschrank mehrere Wochen. Trockenbackhefe behält mindestens ein Jahr lang ihre Triebkraft, ist also praktisch für den Vorrat.

rden kuchenfein

00 Grad 20 Minuten trocknen, in uch geben und reiben.

Die geschälten Nüsse auf einem Brett mit einem scharfen Messer hacken.

Blätterteig & Co
für beliebte Klassiker

Strudelteig, Brandteig und Blätterteig sind zwar nicht so vielseitig wie andere Teige, aber was man aus ihnen machen kann, bleibt in guter Erinnerung: zum Beispiel Wiener Apfelstrudel, Flockentorte und knusprige Windräder

Apfel-Blätterteig-Torte

Rezept siehe folgende Seite

Apfel-Blätterteig-Torte

FÜR 1 BACKBLECH (30 x 40 cm)
450 g TK-Blätterteig (6 Platten)
1 EL Puderzucker, 4 EL Zucker
100 g Marzipan-Rohmasse
600 g Schlagsahne, 1 kg Äpfel
1 roter Apfel, 200 ml Apfelsaft
50 ml Orangensaft
1 TL abgeriebene Orangenschale
1 Zimtstange, 3-4 TL Speisestärke
je 2 Pck. Sahnesteif und Vanillezucker
200 g Preiselbeerkonfitüre
etwas Melisse
AUSSERDEM
Mehl zum Arbeiten

● Teig auftauen lassen. Elektro-Ofen auf 200 Grad vorheizen. Je 2 Teigscheiben aufeinander legen, auf Mehl zum Quadrat (25x25 cm) ausrollen. 1 Platte mit Puderzucker bestreuen. Marzipan zerbröckeln, mit 50 g Sahne verrühren, auf die übrigen Platten streichen und alle auf einem kalt abgespülten Blech im Ofen (Gas: Stufe 3) je 15 Min. backen. Aus den heißen Platten mit einem Tortenring Kreise (24 cm Ø) ausstechen.

● Äpfel – bis auf den roten – schälen, vierteln, entkernen, quer in Schnitze teilen. Roten Apfel entkernen, in Spalten teilen. Zucker mit Säften, Schale und Zimt aufkochen. Spalten darin 5 Minuten dünsten, herausheben. Apfelschnitze im Sud etwa 5 Minuten köcheln, mit angerührter Stärke binden, abkühlen lassen.

● Übrige Sahne, -steif und Vanillezucker steif schlagen. Konfitüre mit übrigem Orangensaft 2 Minuten kochen.

● Marzipan-Blätterteig mit Konfitüre bestreichen, mit Kompott und zwei Drittel Sahne belegen, aufeinander setzen. Übrige Teigplatte auflegen, mit Sahne bestreichen, mit Apfelspalten und Melisse verzieren.

Aufwand:	★★
Zubereitungszeit:	1 1/4 Stunde
Backzeit:	45 Minuten
Pro Stück:	480 kcal/2020 kJ

Fruchtige Blätterteigküchlein

FÜR 1 BACKBLECH (30 x 40 cm)
5 Platten TK-Blätterteig
1 Eigelb
FÜR DIE FÜLLUNG
700 g gemischte Früchte (z. B. Banane, Honigmelone, Erdbeeren)
4 EL Zitronensaft
2 EL Honig
1 Ei (Größe L)
50 g Butter
75 g Puderzucker
100 g Quark (20 % Fett i. Tr.)
1/2 TL abgeriebene Zitronenschale, 1 Bund Minze
AUSSERDEM
Mehl zum Arbeiten

● Teigplatten auftauen lassen und zu 9 Rechtecken dritteln (10 x 7 cm). Aus jedem Rechteck einen Rahmen mit 1 cm breiten Rand schneiden. Rechtecke zu Rahmengröße ausrollen. Ränder mit Wasser bestreichen. Rahmen auf die Rechtecke legen, so dass die Ränder abschließen. Leicht andrücken.

● Elektro-Ofen auf 200 Grad vorheizen. Küchlein auf ein kalt abgespültes Blech legen. Mit Eigelb einpinseln.

● Rechtecke im Ofen (Gas: Stufe 3) 15-20 Minuten backen. Wölbung in der Mitte herunterdrücken.

● Banane schälen. Aus der Melone Kugeln ausstechen. Erdbeeren waschen, in Scheiben schneiden. 3 EL Zitronensaft und Honig verquirlen, darüber träufeln und ziehen lassen.

● Ei trennen. Eigelb, Butter, 50 g Puderzucker, Quark, 1 EL Zitronensaft und die Zitronenschale schaumig rühren. Eiweiß mit 25 g Puderzucker steif schlagen und unterheben.

● 15 Minzeblättchen beiseite legen. Restliche Blätter streifig schneiden, unter die Creme mischen. Creme auf die abgekühlten Küchlein verteilen, das Obst darauflegen. Mit Minze garnieren und Puderzucker bestäuben.

Aufwand:	★★
Zubereitungszeit:	50 Minuten
Backzeit:	40 Minuten
Pro Stück:	360 kcal/1510 kJ

Apfeltarte

FÜR 1 TARTEFORM VON 26 cm Ø
300 g TK-Blätterteig (5 Platten)
100 g kalte Butter
6-7 gleich große Äpfel
Zitronensaft
75 g Zucker
40 g Mandelblättchen
4 EL Aprikosenkonfitüre
AUSSERDEM
Mehl zum Arbeiten

- Den Blätterteig auftauen lassen. 2 EL Butter schmelzen und dünn auf 4 Teigplatten streichen. Aufeinander stapeln und die restliche Teigplatte obenauf legen. Die Platten auf Mehl zu einem dünnen Boden ausrollen. In die kalt ausgespülte Form legen und rundum einen kleinen Rand hochziehen.
- Elektro-Ofen auf 250 Grad vorheizen. Äpfel waschen und trockenreiben, schälen, vierteln, entkernen und längs in gleichmäßig dünne Scheiben schneiden. In Zitronensaft wenden.
- Die Apfelscheiben rosettenförmig überlappend auf dem Teigboden anordnen. Mit Butterflöckchen belegen, mit Zucker und Mandelblättchen bestreuen.
- Die Tarte im Ofen (Gas: Stufe 5) etwa 35 Minuten backen, bis die Kanten der Äpfel hellbraun sind.
- Die Tarte aus dem Ofen nehmen. Konfitüre erhitzen, durch ein Sieb streichen und die Tarte sofort damit gleichmäßig bestreichen.

Aufwand:	★★
Zubereitungszeit:	40 Minuten
Backzeit:	35 Minuten
Pro Stück:	295 kcal/1240 kJ

Variation

- Aus diesem Rezept kann auch eine Pflaumentarte werden. Dafür brauchen Sie 1 1/2 kg Pflaumen, die nach dem Waschen entsteint und rosettenförmig auf die Tarte gelegt werden.

Gestürzter Birnenkuchen

FÜR 1 TARTEFORM VON 24 cm Ø
250 g TK-Blätterteig
750 g fast reife Birnen
50 g Butter
30 g Zucker
2 EL Mandelblättchen
2 cl weißer Rum
AUSSERDEM
Mehl zum Arbeiten
Puderzucker zum Bestäuben

- Blätterteig auftauen lassen. Elektro-Ofen auf 200 Grad vorheizen.
- Birnen schälen, halbieren, entkernen und in Spalten schneiden.
- Butter in der Form auf dem Herd schmelzen, den Zucker darin goldbraun karamellisieren lassen. Die Mandeln einstreuen. Die Birnenspalten rosettenförmig einlegen, im Karamell leicht bräunen lassen. Mit Rum beträufeln und kurz dünsten, vom Herd nehmen.
- Die Teigplatten auf Mehl aufeinander legen, in der Größe der Form ausrollen, die Birnen damit bedecken. Teig fest andrücken und mit einer Gabel mehrmals einstechen. Tarte im Ofen (Gas: Stufe 3) etwa 30 Minuten backen. Sofort vorsichtig stürzen, mit Puderzucker bestäuben und warm servieren.

Aufwand:	★
Zubereitungszeit:	30 Minuten
Backzeit:	30 Minuten
Pro Stück:	165 kcal/690 kJ

Variation

- Das Original-Rezept, allerdings mit Äpfeln, haben die Franzosen erfunden: die berühmte Tarte Tatin. Sehr gut schmeckt die Tarte auch mit Ananas. Dafür können Sie ruhig Ananas aus der Dose verwenden und so verfahren wie im Rezept angegeben.

Apfel-Aprikosen-Kuchen

FÜR 1 BACKBLECH (30 x 40 cm)
8 Platten TK-Blätterteig
75 g Butter
700 g säuerliche Äpfel
4 EL Zitronensaft
500 g Aprikosen, 75 g Zucker
1 Päckchen Vanillezucker
1 Msp. Zimtpulver
6 EL Aprikosenkonfitüre
3 cl Orangenlikör
AUSSERDEM
Mehl zum Arbeiten

- Blätterteig auftauen lassen. Vier Platten auf einer bemehlten Fläche nebeneinander zu einem Rechteck legen, dünn mit etwas Butter bestreichen, die restlichen Platten obenauf legen und mit etwas Mehl bestäuben, zu Backblechgröße ausrollen. Auf das kalt abgespülte Blech legen, rundherum einen kleinen Rand formen. Teigboden mit einer Gabel mehrmals einstechen.
- Elektro-Ofen auf 250 Grad vorheizen.
- Äpfel schälen, vierteln, entkernen und in dünne Spalten schneiden. In Zitronensaft wenden. Aprikosen waschen, überbrühen, häuten, halbieren, entsteinen, in Spalten schneiden.
- Zucker, Vanillezucker und Zimt mischen, einige Esslöffel auf den Teig streuen. Äpfel und Aprikosen dachziegelartig auflegen. Mit übriger Zucker-Zimt-Mischung bestreuen, mit restlicher Butter in Flöckchen belegen. Den Kuchen im Ofen (Gas: Stufe 5) etwa 35 Minuten backen, bis die Oberfläche zu karamellisieren beginnt.
- Konfitüre mit Likör erwärmen, durch ein Sieb streichen, den heißen Kuchen damit bepinseln.

Aufwand:	★
Zubereitungszeit:	40 Minuten
Backzeit:	35 Minuten
Pro Stück:	125 kcal/530 kJ

Quarktorte mit Erdbeergrütze

BITTE BEACHTEN: TORTE MUSS 2 STUNDEN KÜHLEN
FÜR 1 BACKBLECH (30 x 40 cm)
4 rechteckige Platten TK-Blätterteig (300 g)
300 g Rhabarber
4 EL Puderzucker
75 g weiße Kuvertüre
1 EL Speisestärke
300 g Erdbeeren
8 Blatt weiße Gelatine
500 g Quark (20 % Fett i. Tr.)
100 g Zucker
1 Päckchen Vanillezucker
50 ml Orangensaft
400 g Schlagsahne
AUSSERDEM
Mehl zum Arbeiten
75 g Puderzucker
1 EL Grenadinesirup oder
1 Tropfen rote Speisefarbe
100 g Schlagsahne
Zitronenmelisse

- Blätterteigplatten nebeneinander auftauen lassen. Rhabarber putzen, schälen und in 1 cm große Stücke schneiden, Puderzucker überstäuben, etwa 1 Stunde ziehen lassen.
- Elektro-Ofen auf 225 Grad vorheizen. 2 Blätterteigplatten mit Wasser bestreichen, mit den übrigen Platten belegen, auf Mehl 28 x 28 cm groß ausrollen. Mit einem Springformrand (28 cm Ø) 2 Böden ausstechen und auf ein mit Wasser bestrichenes Backblech legen. Böden mehrmals einstechen und einzeln im Ofen (Gas: Stufe 4) etwa 15 Minuten backen. Auf Gittern auskühlen lassen. Kuvertüre schmelzen, 1 Boden damit bestreichen.
- Rhabarber etwa 2 Minuten kochen. Speisestärke in etwas Sud anrühren und einrühren. Nochmal aufkochen und abkühlen lassen.
- Erdbeeren überbrausen, putzen und in 2 cm große Würfel schneiden.
- Gelatine in Wasser einweichen. Quark, Zucker, Vanillezucker und Saft miteinander verrühren.
- Gelatine tropfnass in einem Töpfchen bei schwacher Hitze auflösen, 2 EL Quarkcreme unterrühren, alles unter die übrige Creme rühren. Kalt stellen.
- Sahne steif schlagen, unter die angelierte Quarkmasse heben. Erdbeeren unter den Rhabarber mischen.
- Boden auf eine Kuchenplatte legen, einen Tortenring darum legen. Obstmasse darauf geben, glatt streichen. Quarkcreme darüber streichen, 2 Stunden kalt stellen.
- Puderzucker mit Sirup oder Speisefarbe und 2 EL Wasser glatt rühren. Übrigen Boden in 16 gleich große Tortenstücke teilen und mit dem Guss bestreichen, trocknen lassen.
- Tortenring lösen. Die Sahne steif schlagen, in einen Spritzbeutel mit Sterntülle füllen und Tupfen auf die Tortenoberfläche spritzen. Blätterteigecken schräg darauf setzen. Mit übriger Sahne Tortenrand bestreichen und mit Melisseblättchen dekorieren.

Aufwand:	★★
Zubereitungszeit:	1 1/2 Stunden
Backzeit:	30 Minuten
Pro Stück:	300 kcal/1260 kJ

Variationen

- Wandeln Sie die Torte ab mit anderem Obst, z. B. Beeren, die Sie je nach Konsistenz ein paar Minuten wie den Rhabarber zubereiten.
- Fein schmeckt es auch, wenn Sie Obst wie Mango in Würfel schneiden und unter die Creme mischen, die dann direkt auf den Boden gegeben wird.
- Oder pürieren Sie 100 g Erdbeeren und etwas Zucker zu einem Fruchtmus, das Sie zusammen mit 200 g gewürfelten Erdbbeeeren unter den Quark mischen.

Bratapfelkuchen vom Blech

FÜR 1 BACKBLECH (30 x 40 cm)	
1 1/2 Packungen TK-Blätterteig (9 Platten)	
18 kleine oder 9 große Äpfel (z. B. Boskop, Jonathan, Berlepsch)	
Zitronensaft	
250 g Zucker	
je 1 Prise Zimt- und Nelkenpulver	
300 g Marzipan-Rohmasse	
3 Eigelbe (Eier: Größe M)	
je 40 g Orangeat und Zitronat	
2 EL geschälte Mandeln	
2 EL Rum	
2 EL Schlagsahne	
AUSSERDEM	
Mehl zum Arbeiten	

- Blätterteigplatten nebeneinander legen, auftauen lassen.
- Die Äpfel schälen und größere Äpfel quer halbieren. Das Kerngehäuse großzügig herausstechen, so dass breite Öffnungen entstehen. Die Äpfel mit Zitronensaft einpinseln, damit sie sich nicht verfärben.
- Die Apfelschalen und das Kerngehäuse mit 3/8 l Wasser aufkochen, etwa 10 Minuten köcheln lassen. Leicht durch ein Sieb pressen und dabei den Saft auffangen. 1/4 l Saft mit Zucker unter Rühren aufwallen lassen, zu einem dickflüssigen Gelee kochen. Die Gewürze unterrühren, das Gelee abkühlen lassen.
- Jede Blätterteigplatte halbieren und auf bemehlter Fläche dünn zu einem Kreis ausrollen. Auf jeden Teigkreis 1 Apfel (oder Apfelhälfte) setzen.
- Den Elektro-Ofen auf 220 Grad vorheizen. Die Marzipan-Rohmasse mit 2 Eigelben verkneten. Orangeat, Zitronat und Mandeln fein hacken, mit dem Rum unter das Marzipan mischen. Masse in die Äpfel füllen.
- Den Teig über den Äpfeln wie einen Beutel locker zusammenlegen und gut festdrücken. Überstehenden Teig abschneiden, übereinander legen (nicht kneten!) und ausrollen. Aus dem Teig schmale Streifen schneiden und kleine Kreise ausstechen.
- Übriges Eigelb mit Sahne glatt rühren. Die Oberfläche der Beutel damit bestreichen, die Mitte mit Teigstreifen und Kreisen verzieren. Nochmals mit Eigelb bestreichen.
- Apfelbeutel dicht nebeneinander auf das kalt abgespülte Blech setzen, dabei die Kontaktstellen dünn mit Wasser bestreichen. Im Ofen (Gas: Stufe 3-4) 30-40 Minuten backen.
- Die noch heißen Apfelpäckchen mit Apfelgelee bestreichen, leicht abkühlen lassen und nochmals bestreichen. Bratapfelkuchen am besten lauwarm mit leicht gesüßter Sahne servieren.

Aufwand:	★★
Zubereitungszeit:	1 1/4 Stunden
Backzeit:	40 Minuten
Pro Stück:	310 kcal/1300 kJ

Apfel-Pecannuss-Pie

FÜR 6 TORTELETT-FÖRMCHEN (12 cm Ø) ODER 1 PIEFORM (30 cm Ø)
200 g Mehl, 1/4 TL Salz
200 g kalte Butter
200 g Magerquark (10 % Fett i.Tr.)
FÜR DEN BELAG
1 kg Äpfel (Boskop oder Elstar)
300 ml klarer Apfelsaft
40 g Butter
3 EL brauner Zucker
2 TL Speisestärke
100 g Pecannusshälften
AUSSERDEM
Fett für die Förmchen
Mehl zum Arbeiten

- Mehl mit Salz und Butter in Flöckchen auf die Arbeitsfläche häufen. Quark gut ausdrücken, hinzufügen und rasch zu einem glatten Teig verkneten. In Klarsichtfolie wickeln, 20 Minuten kalt stellen. Teig zu einem Rechteck ausrollen, von beiden Seiten zur Mitte einschlagen, zusammenlegen 20 Minuten kalt stellen.
- Äpfel waschen, schälen, vierteln, die Kerngehäuse entfernen. Fruchtfleisch klein würfeln. Apfelsaft, Butter und Zucker in einem Topf aufkochen. Apfelstücke zufügen und bissfest köcheln. Stärke in etwas Saft anrühren, dazugeben und aufkochen.
- Elektro-Ofen auf 175 Grad vorheizen. Die Backförmchen einfetten. Teig auf bemehlter Arbeitsfläche dünn ausrollen. Förmchen mit Teig auslegen, der Teig sollte etwas über den Rand hinausragen.
- Apfelkompott in Förmchen geben. Pecannüsse klein hacken, überstreuen. Teigrand nach Belieben verzieren.
- Im Ofen (Gas: Stufe 2) 35-40 Minuten backen. 5 Minuten vor Ende der Backzeit mit Alufolie zudecken. Lauwarm servieren.

Aufwand:	★
Zubereitungszeit:	1 Stunde
Backzeit:	40 Minuten
Pro Stück	355 kcal/1490 kJ

Knusprige Windräder

Für 10 Stück
10 TK-Blätterteigplatten (450 g, quadratisch), 150 g Mehl
90 g Butter, 75 g Zucker
1 TL abgeriebene unbehandelte Orangenschale
1 Eigelb (Ei: Größe S)
2 EL Schlagsahne
10 Belegkirschen
250 g Puderzucker
AUSSERDEM
Backpapier für das Blech

- Die Teigplatten auftauen lassen. Den Elektro-Ofen auf 200 Grad vorheizen.
- Mehl, Butter, Zucker und Orangenschale krümelig kneten. Blätterteigquadrate von den Ecken aus diagonal einschneiden (siehe unten). Auf ein mit Backpapier belegtes Blech legen, Orangenstreusel darauf verteilen.
- Teigecken so falten (siehe unten), dass eine Windradform entsteht. Eigelb und Sahne verquirlen, Oberfläche der Windräder damit bestreichen. In die Mitte je 1 Kirsche drücken.
- Im Backofen (Gas: Stufe 3) etwa 20 Minuten backen. Auf einem Kuchengitter abkühlen lassen.
- Den Puderzucker mit 2 EL Wasser verrühren und dünne Linien über die Windräder ziehen, trocknen lassen.

Aufwand:	★
Zubereitungszeit:	35 Minuten
Backzeit:	20 Minuten
Pro Stück:	265 kcal/ 1110 kJ

1 Blätterteig von den Ecken diagonal zur Mitte hin einschneiden. Den Teig nach Rezept belegen. Jede zweite Ecke nach innen einlegen, gut andrücken.

2 Oberfläche nach Belieben mit Eigelb oder Sahne bestreichen und belegen, z. B. mit Kirschen.

Äpfel auf Blätterteig mit Sabayon

__Für 4 Stück__
250 g TK-Blätterteig
4 Äpfel
1 EL Zitronensaft
60 g Marzipan-Rohmasse
3 EL Calvados oder Weinbrand
30 g gehackte Walnusskerne
Zimtpulver, Zucker
100 g Schlagsahne
50 g heller Honig
3 Eigelbe (Eier: Größe M)
2 EL Aprikosenkonfitüre
AUSSERDEM
Backpapier für das Blech
Mehl zum Arbeiten
Puderzucker zum Bestäuben

- Blätterteig auftauen lassen. Elektro-Ofen auf 200 Grad vorheizen. Backblech mit Backpapier belegen. Äpfel schälen, vierteln, entkernen und in Spalten teilen. Mit Zitronensaft beträufeln.
- Marzipan mit 2 EL Calvados und Nüssen verkneten. Teigplatten aufeinander legen, auf Mehl 3 mm dünn ausrollen und 4 runde Kreise (14 cm Ø) ausstechen. Kreise auf das Blech legen, Marzipan darauf streichen. Äpfel fächerförmig darauf legen. Mit Zimt und Zucker bestreuen. Im Ofen (Gas: Stufe 3) etwa 15 Minuten backen.
- Sahne, Honig, Eigelbe, Zimt und übrigen Calvados in einer Schüssel über dem warmen Wasserbad dickschaumig aufschlagen. Schüssel in ein kaltes Wasserbad stellen. Weiterschlagen, bis die Sauce abgekühlt ist.
- Aprikosenkonfitüre erwärmen. Apfeltörtchen mit Konfitüre bestreichen. Nach Belieben mit Puderzucker bestreuen, mit Zimt-Sabayon umgießen und sofort servieren.

Aufwand:	★
Zubereitungszeit:	35 Minuten
Backzeit:	15 Minuten
Pro Stück:	595 kcal/2500 kJ

Apfel-Baiser-Kuchen

FÜR 1 BACKBLECH (30 x 40 cm)
250 g TK-Blätterteig
FÜR DEN BELAG
750 g säuerliche Äpfel
1 EL Zucker
125 ml Orangensaft
1 EL Zitronensaft
1 kleines Bund frische Minze
100 g Zartbitter-Kuvertüre
FÜR DAS BAISER
3 Eiweiße (Eier: Größe M)
150 g Zucker
1 Prise Salz
AUSSERDEM
Mehl zum Arbeiten

- Die Blätterteigplatten nebeneinander legen und auftauen lassen.
- Äpfel putzen, waschen, schälen, mit dem Apfelausstecher Kerngehäuse entfernen. Äpfel halbieren und längs in 1/2 cm dicke Spalten schneiden.
- Zucker mit Orangen- und Zitronensaft und gewaschenen Minzestängeln in einem breiten Topf aufkochen. Apfelspalten hineinlegen und zugedeckt etwa 10 Minuten auf der abgeschalteten Kochplatte ziehen lassen.
- Elektro-Ofen auf 225 Grad vorheizen. Blätterteigplatten aufeinander legen, auf Mehl etwa 1/2 cm dick ausrollen. Eine runde Platte (28 cm Ø) ausschneiden. Mit einer Gabel mehrmals einstechen und im Ofen (Gas: Stufe 4) in etwa 15 Minuten knusprig backen. Herausnehmen und Ofen auf Grill stellen.
- Kuvertüre im warmen Wasserbad schmelzen. Blätterteigboden dünn damit bestreichen. Fest werden lassen. Die Apfelspalten abtropfen lassen und dachziegelartig auf den Kuchen legen.
- Eiweiße mit Zucker und 1 Prise Salz steif schlagen. In eine Spritztüte mit mittelgroßer Sterntülle füllen und in engen Schlangenlinien auf den Kuchen spritzen. Im Ofen goldbraun übergrillen. Unbedingt frisch servieren.

Aufwand:	★
Zubereitungszeit:	50 Minuten
Backzeit:	20 Minuten
Pro Stück:	220 kcal/920 kJ

Kiwi-Kuppeltorte

BITTE BEACHTEN: TORTE MUSS 1 STUNDE KÜHLEN

Für 1 BACKBLECH (30 x 40 cm)
450 g TK-Blätterteig (9 Platten), 120 g Nougat
9 Blatt weiße Gelatine
3 Eigelbe (Eier: Größe M)
75 g Zucker
1 Päckchen Vanillezucker
150 ml Cidre (Apfelwein)
2 EL Orangenlikör
9 Kiwis, 4 EL Orangensaft
200 g Schlagsahne

FÜR DAS BAISER
3 Eiweiße (Eier: Größe M)
1 Prise Salz, 150 g Zucker

AUSSERDEM
Mehl zum Arbeiten
Backpapier für das Blech

- Blätterteig auftauen lassen. Elektro-Ofen auf 200 Grad vorheizen. Backblech mit Backpapier auslegen.
- Je 3 Teigplatten aufeinander legen und dünn auf Mehl ausrollen, 3 Kreise (20 cm Ø) ausschneiden. Auf das Blech legen, mit einer Gabel mehrmals einstechen. Die Böden im Ofen (Gas: Stufe 3) je etwa 10 Minuten backen.
- Gleichzeitig Gelatine einweichen. Eigelbe, Zucker und Vanillezucker im heißen Wasserbad aufschlagen.
- Cidre erhitzen, ausgedrückte Gelatine darin auflösen. Leicht abgekühlt unter die Eiercreme geben, Likör zufügen.
- 5 Kiwis schälen, pürieren, mit Orangensaft etwa 4 Min. köcheln, abkühlen lassen. Zur Eiercreme geben, kalt stellen.
- Die Sahne steif schlagen und unter die Creme heben. Eine kuppelförmige Schüssel (Eisbombengefäß, 16 cm Ø) mit Klarsichtfolie auskleiden. Die Creme einfüllen und 1 Stunde ins Gefrierfach stellen.
- Den Nougat im Wasserbad schmelzen. 2 Teigböden damit bestreichen und aufeinander setzen. Mit dem übrigen Boden abdecken.
- Elektro-Ofen auf 250 Grad (Gas: Stufe 5) vorheizen. Die feste Creme auf die Böden stürzen.
- Eiweiße mit Salz und Zucker schnittfest schlagen. Dekorativ auf die Creme spritzen. Im Ofen 1-2 Minuten leicht bräunlich überbacken.
- Übrige Kiwis schälen, in dünne Scheiben schneiden und halbieren. Den überstehenden Teigrand damit garnieren.

Aufwand:	★★
Zubereitungszeit:	1 1/2 Stunden
Backzeit:	30 Minuten
Pro Stück:	375 kcal/1580 kJ

Zwetschgen-strudel

FÜR 1 BACKBLECH (30 x 40 cm)
1 Grundrezept Strudelteig
(s. Seite 155)
FÜR DIE FÜLLUNG
600 g frische Zwetschgen
1 Dose Aprikosen (Abtropfgewicht: 480 g), 80 g Zucker
1 TL Vanillezucker
2 Msp. Zimtpulver
2 EL Zwetschgenwasser
4 EL zerlassene Butter
2 EL gemahlene Walnüsse
2 EL Semmelbrösel
FÜR DIE SCHOKOSAUCE
200 g weiße Schokolade
100 ml Milch, 100 g Schlagsahne, 1 EL Puderzucker
AUSSERDEM
Backpapier für das Blech
Mehl zum Arbeiten

- Zwetschgen waschen, entsteinen, sechsteln. Aprikosen abgießen, ca. 1 1/2 cm groß würfeln, abtropfen lassen. Früchte mit Zucker, Vanillezucker, Zimt, Zwetschgenwasser mischen.
- Elektro-Ofen auf 200 Grad vorheizen. Backblech mit Backpapier belegen. Strudelteig halbieren. Eine Hälfte auf einem bemehlten Tuch ausrollen und hauchdünn (ca. 30x40 cm) ausziehen.
- Teig mit 1 EL Butter einpinseln. Nüsse mit Bröseln mischen, 2 EL auf den Teig streuen. Die Hälfte der Früchte darauf geben, Teig einrollen. Mit der Nahtstelle nach unten auf das Blech geben. Zweite Teighälfte ebenso füllen.
- Strudel mit Butter einpinseln. Im Ofen (Gas: Stufe 3) 35-40 Min. backen, dabei mit Butter bestreichen.
- Schokolade hacken. Milch, Sahne und Puderzucker erwärmen. Schoko unter Rühren auflösen. Sauce leicht abgekühlt servieren.

Aufwand:	★★
Zubereitungszeit:	50 Minuten
Backzeit:	40 Minuten
Pro Stück:	265 kcal/1110 kJ

Möhren-Orangen-Strudel

FÜR 1 BACKBLECH (30 x 40 cm)
1 Grundrezept Strudelteig
(s. Seite 155)
FÜR DIE FÜLLUNG
100 g Mandelblättchen
400 g Möhren, 1 Dose Ananas
(Abtropfgewicht: 260 g)
5 Orangen, 1 Msp. Ingwerpulver
2 EL Orangenmarmelade
2 Eier (Größe M), 100 g Zucker
250 g Quark (40 % Fett i. Tr.)
3 EL zerlassene Butter
1 Päckchen Vanillezucker
AUSSERDEM
Backpapier für das Blech
Mehl zum Arbeiten
Puderzucker zum Bestäuben

- Mandeln rösten. Möhren schälen, raspeln, Saft ausdrücken. Ananas abgießen, in 1 cm breite Stücke schneiden. 1 Orange auspressen, übrige schälen, Fruchtfilets herauslösen. Marmelade mit Ingwer, Orangensaft sirupartig einköcheln, mit Möhren mischen.
- Eier trennen. Eigelbe mit 50 g Zucker und Quark cremig rühren. Eiweiße mit übrigem Zucker steif schlagen. Erst Möhren, Orangen, Mandeln, dann Eischnee locker unter die Quarkmasse heben.
- Elektro-Ofen auf 200 Grad heizen. Backblech mit Backpapier auslegen. Strudelteig halbieren. Eine Hälfte auf einem bemehlten Tuch ausrollen, hauchdünn (ca. 30 x 40 cm) ausziehen.
- Strudelteig mit 1 EL Butter einpinseln. Die Hälfte der Füllung darauf geben, Teig einrollen. Mit der Nahtstelle nach unten auf das Blech legen. Übrigen Teig genauso füllen. Strudel mit übriger Butter bestreichen. Mit Vanillezucker bestreuen. Im Ofen (Gas: Stufe 3) 35-40 Min. backen. Mit Puderzucker bestreuen.

Aufwand:	★★
Zubereitungszeit:	40 Minuten
Backzeit:	40 Minuten
Pro Stück:	240 kcal/1010 kJ

Wiener Apfelstrudel

FÜR 1 BACKBLECH (30 x 40 cm)
1 Grundrezept Strudelteig
(s. Seite 155)
FÜR DIE FÜLLUNG
1 kg säuerliche Äpfel
Saft von 1/2 Zitrone
125 g Rum-Rosinen
(Fertigprodukt)
50 g grob gehackte Haselnüsse
1/2 TL Zimtpulver
100 g Zucker
4 EL Semmelbrösel
etwa 5 EL zerlassene Butter
100 g saure Sahne
AUSSERDEM
Backpapier für das Blech
Mehl zum Arbeiten
3 EL Zucker, vermischt mit
1/4 TL Zimtpulver

- Äpfel putzen, schälen, vierteln, blättrig schneiden. Mit Zitronensaft, Rosinen, Nüssen, Zimt und Zucker mischen. Semmelbrösel in 1 EL Butter rösten. Elektro-Ofen auf 200 Grad vorheizen. Das Blech mit Backpapier auslegen.
- Den Teig halbieren. Jede Hälfte auf einem bemehlten Tuch ausrollen, hauchdünn (ca. 30 x 40 cm) ausziehen. Die Strudel mit je 1 EL Butter einpinseln.
- Jeden Strudel mit Apfelfüllung belegen, dann mit je 50 g saurer Sahne bestreichen, einrollen.
- Die Strudel mit der Nahtstelle nach unten auf das Blech legen. Mit etwas flüssiger Butter bestreichen und im Ofen (Gas: Stufe 3) 30-35 Minuten backen. Zwischendurch immer wieder mit Butter bestreichen. Nach dem Backen den heißen Apfelkuchen mit Zimtzucker bestreuen.

Aufwand:	★★
Zubereitungszeit:	50 Minuten
Backzeit:	35 Minuten
Pro Stück:	210 kcal/880 kJ

Kokos-Kirsch-Strudel

FÜR 1 BACKBLECH (30 x 40 cm)
1 Grundrezept Strudelteig
(s. Seite 155)
FÜR DIE FÜLLUNG
2 Gläser Sauerkirschen
(Abtropfgewicht: à 370 g)
2 EL Honig
Mark von 1 Vanilleschote
2 Eier und 1 Eigelb (Größe M)
50 g Puderzucker
250 g Mascarpone
etwa 20 g Vanille-Puddingpulver
3 EL zerlassene Butter
2 EL Kokosraspel
3 EL Milch
AUSSERDEM
Backpapier für das Blech
Mehl zum Arbeiten

● Kirschen in einem Sieb abtropfen lassen. Honig mit Vanillemark erhitzen. Noch warm mit den gut abgetropften Kirschen mischen. Die Eier trennen. 2 Eigelbe, 25 g Puderzucker, Mascarpone und Puddingpulver verrühren. Eiweiße mit übrigem Puderzucker steif schlagen und unterheben.

● Elektro-Ofen auf 200 Grad vorheizen. Backblech mit Backpapier belegen. Teig halbieren. Eine Hälfte auf einem bemehlten Tuch ausrollen, hauchdünn (ca. 30 x 40 cm) ausziehen. Strudelteig mit 1 EL Butter einpinseln. Die Hälfte der Creme und die Hälfte der Kirschen daraufgeben. Mit 1 EL Kokosraspeln bestreuen.

● Teig einrollen. Mit der Nahtstelle nach unten aufs Blech legen. Übrigen Teig genauso füllen. Mit restlicher Butter einstreichen.

● Strudel im Ofen (Gas: Stufe 3) 35-40 Minuten backen. Übriges Eigelb mit Milch verrühren, Strudel nach der Hälfte der Backzeit damit bestreichen.

Aufwand:	★★
Zubereitungszeit:	30 Minuten
Backzeit:	40 Minuten
Pro Stück:	210 kcal/880 kJ

Kirsch-Mandel-Strudel

FÜR 1 BACKBLECH (30 x 40 cm)
1 Grundrezept Strudelteig (s. Seite 155)
50 g zerlassene Butter
FÜR DIE FÜLLUNG
150 g Mandelmakronen
1 1/4 kg Kirschen, 100 g Zucker
2 Päckchen Vanillezucker
100 g blättrige oder gehackte Mandeln
AUSSERDEM
Backpapier für das Blech
Mehl zum Arbeiten
Puderzucker zum Bestäuben

- Für die Füllung Makronen in einer Plastiktüte mit dem Teigroller sehr fein zerbröseln. Kirschen waschen, abtrocknen, entstielen und entsteinen. Zucker mit Vanillezucker mischen.
- Elektro-Ofen auf 200 Grad vorheizen. Das Backblech mit Backpapier auslegen. Den Strudelteig halbieren, jede Hälfte auf einem bemehlten Tuch hauchdünn ausziehen und dünn mit zerlassener Butter bepinseln.
- Erst die Makronenbrösel, dann die Kirschen auf dem Teig verteilen. Mit Zuckermischung und Mandeln bestreuen.
- Den Strudel mithilfe des Tuchs zusammenrollen und mit der Nahtstelle nach unten auf das Blech legen. Mit Butter bepinseln. Die Strudel im Ofen (Gas: Stufe 3) 35-40 Minuten backen. Mit Puderzucker bestäuben.

Aufwand:	★★
Zubereitungszeit:	40 Minuten
Backzeit:	40 Minuten
Pro Stück:	250 kcal/1050 kJ

Variation

- Die Makronenmischung harmoniert auch sehr gut mit Äpfeln: Dafür 1 1/2 kg Äpfel schälen, Kerngehäuse entfernen und Äpfel kleinschneiden. Etwa 4 Minuten in 1/4 l Weißwein und 5 EL Zucker dünsten und auf der Makronenmischung verteilen. Wie im Rezept beschrieben fortfahren.

Millirahmstrudel

FÜR 1 BACKBLECH (30 x 40 cm)
1 Ei (Größe M)
1 Prise Salz
2 EL Öl
125 g Mehl (Type 405)
125 g doppelgriffiges Mehl
FÜR DIE FÜLLUNG
3 Eier (Größe M)
80 g Butter
50 g Puderzucker
40 g Mehl
1 Prise Salz
Mark von 1 Vanilleschote
abgeriebene Schale und Saft von 1/2 unbehandelten Zitrone
300 g Schichtkäse
200 g saure Sahne
100 g Rum-Rosinen
80 g Zucker
FÜR DIE SAUCE
300 ml Milch
2 Eigelbe (Eier: Größe M)
30 g Zucker
1 Päckchen Vanillezucker
1 TL Zitronenschale
AUSSERDEM
Mehl zum Arbeiten
50 g Butter zum Fetten der Form und zum Bestreichen

- Für den Strudelteig 100 ml lauwarmes Wasser mit Ei, Salz und Öl verrühren. Gesiebtes Mehl mit den Knethaken des Handrührgerätes etwa 3 Minuten unterkneten. Teig auf einer bemehlten Arbeitsfläche durch Auseinanderziehen und Zusammenfalten etwa 10 Minuten bearbeiten, bis er elastisch ist.
- Den Teig 30 Minuten ruhen lassen.
- Für die Füllung die Eier trennen. Butter, Puderzucker und Eigelbe cremig rühren. Mehl und Gewürze zufügen. Gut abgetropften Schichtkäse und saure Sahne unterschlagen. Rosinen einarbeiten. Die Eiweiße mit Zucker steif schlagen und unterziehen.
- Elektro-Ofen auf 200 Grad vorheizen. Form einfetten.
- Teig halbieren. Jede Hälfte zur Kugel formen, flach drücken und auf bemehlter Fläche so dünn wie möglich zu einer rechteckigen Platte ausrollen. Dann auf ein bemehltes Küchentuch legen, mit den Handrücken darunterfahren und so dünn ausziehen, dass die Finger durchscheinen.
- Teigflächen mit flüssiger Butter bestreichen. Füllung jeweils von der langen Seite aus auf die Hälfte der Teigfläche streichen, dabei links und rechts einen Rand frei lassen. Die Teige mithilfe des Küchentuches von der bestrichenen Längsseite her aufrollen und die Enden nach innen umschlagen. Die beiden Strudel mit den Nahtstellen nach unten in die Form legen.
- Zutaten für die Sauce verquirlen und ein Drittel über die Strudel gießen. Strudel im Ofen (Gas: Stufe 3) 40-45 Minuten backen. Nach und nach die übrige Sauce darüber gießen.

Aufwand:	★★
Zubereitungszeit:	1 1/4 Stunden
Backzeit:	45 Minuten
Pro Stück:	325 kcal/1370 kJ

Variation

- Sie können diesen Klassiker leicht verändern, indem Sie statt der Rosinen 100 g klein geschnittene Trockenpflaumen oder -aprikosen verwenden.

Birnen-Preiselbeer-Strudel

FÜR 1 BACKBLECH (30 x 40 cm)
1 Grundrezept Strudelteig (s. Seite 155)
30 g zerlassene Butter
FÜR DIE FÜLLUNG
75 g Amaretti (Mandelkekse)
400 g Schmand (saure Sahne mit 24 % Fett)
2 unbehandelte Zitronen
1 1/2 kg feste Birnen
200 g frische Preiselbeeren (oder 150 g aus dem Glas)
50 g Zucker
AUSSERDEM
Backpapier für das Blech
Mehl zum Arbeiten
Puderzucker zum Bestäuben

- Für die Füllung Kekse zerbröseln. Schmand mit Saft und abgeriebener Schale von 1 Zitrone glatt rühren.
- Birnen schälen, vierteln, entkernen, quer in dünne Scheiben schneiden. Mit dem Saft der übrigen Zitrone (Schale vorher abreiben und beiseite stellen) beträufeln. Preiselbeeren waschen, verlesen und mit 4 EL Wasser kurz aufkochen, abkühlen lassen.
- Den Elektro-Ofen auf 200 Grad vorheizen. Das Blech mit Backpapier auslegen. Den Strudelteig ausziehen (s. Seite 155).
- Teig mit Schmand bestreichen und mit Birnen belegen. Preiselbeeren darüber verteilen. Abgeriebene Zitronenschale mit Kekskrümeln und Zucker vermischen, über das Obst streuen.
- Strudel zusammenrollen und mit der Nahtseite nach unten auf das Blech legen, mit Butter bepinseln. Im Ofen (Gas: Stufe 3) 40-50 Minuten backen. Mit Puderzucker bestäuben.

Aufwand:	★★
Zubereitungszeit:	55 Minuten
Backzeit:	50 Minuten
Pro Stück:	230 kcal/970 kJ

Quitten-Pfirsich-Strudel

FÜR 1 BACKBLECH (30 x 40 cm)
1 Grundrezept Strudelteig (s. Seite 155)
30 g Butter
200 ml Milch
FÜR DIE FÜLLUNG
1 3/4 kg Quitten
80 g Zucker
400 g Pfirsiche
250 g Mascarpone
1/2 TL gemahlenes Zimtpulver
Saft und Schale von 1/2 unbehandelten Orange
50 g Zwieback
AUSSERDEM
Mehl zum Arbeiten
Backpapier für das Blech
2 EL Aprikosenkonfitüre

- Für die Füllung Quitten schälen, entkernen und feinblättrig schneiden. Quitten mit 50 g Zucker in 1/2 l Wasser in etwa 25 Minuten weich dünsten. Pfirsiche überbrühen, häuten, entsteinen und würfeln.
- Mascarpone mit Zimt, 3 EL Orangensaft und Orangenschale leicht aufschlagen, übrigen Zucker einrühren.
- Elektro-Ofen auf 200 Grad vorheizen. Strudelteig wie beschrieben (s. Seite 155) ausziehen.
- Teig mit Mascarpone bestreichen. Abgetropfte Quitten und Pfirsiche darauf verteilen. Zwieback zerbröseln und darüberstreuen.
- Strudel formen, mit der Naht nach unten in einen Bräter setzen. Mit zerlassener Butter bestreichen und einen Teil der Milch angießen. Im Ofen (Gas: Stufe 3) 40-45 Minuten backen, dabei nach und nach die übrige Milch nachgießen.
- Konfitüre erwärmen, durch ein Sieb streichen und den Strudel damit bepinseln.

Aufwand:	★★
Zubereitungszeit:	55 Minuten
Backzeit:	45 Minuten
Pro Stück:	240 kcal/1010 kJ

Zwetschgen-Kokos-Strudel

FÜR 1 BACKBLECH (30 x 40 cm)
1 Grundrezept Strudelteig (s. Seite 155)
50 g zerlassene Butter
FÜR DIE FÜLLUNG
1 1/4 kg Zwetschgen
125 g Kokosraspel
75 g Zucker
1 Päckchen Vanillezucker
AUSSERDEM
Backpapier für das Blech
Mehl zum Arbeiten

- Für die Füllung die Zwetschgen waschen, vierteln und entkernen. Kokosraspel in einer Pfanne ohne Fett anrösten.
- Backofen auf 200 Grad vorheizen. Backblech mit Backpapier auslegen. Den Strudelteig wie auf Seite 155 beschrieben dünn ausziehen.
- Teig mit gut 3/4 der Kokosraspel, dem Zucker und Vanillezucker bestreuen. Zwetschgen darauf verteilen.
- Strudel – wie beschrieben – formen, mit etwas Butter bepinseln. Im Ofen (Gas: Stufe 3) etwa 45 Minuten backen. Den heißen Strudel mit übriger Butter bestreichen und mit den restlichen Kokosraspeln bestreuen.

Aufwand:	★★
Zubereitungszeit:	55 Minuten
Backzeit:	45 Minuten
Pro Stück:	200 kcal/840 kJ

Tipp

Strudelteig herzustellen ist nicht ganz einfach. Wem es zu aufwändig ist, findet fertigen Strudelteig im Kühlregal der Supermärkte. Man sollte ihn aber möglichst rasch verwenden, da er schnell brüchig wird.

Erdbeersahne-Ringe

FÜR 1 BACKBLECH (30 x 40 cm)
50 g Butter, Salz
150 g Mehl, 30 g Speisestärke
4-6 Eier (Größe M)
FÜR DIE FÜLLUNG
500 g Schlagsahne
1 EL Orangenlikör
2 Päckchen Vanillezucker
500 g Erdbeeren
30 g Mandelblättchen
AUSSERDEM
Backpapier für das Blech
Puderzucker zum Bestäuben

- Elektro-Ofen auf 225 Grad vorheizen. Blech mit Backpaier auslegen.
- 1/4 l Wasser, Butter und 1 Prise Salz aufkochen, Topf vom Herd ziehen. Mehl und Stärke mischen, unter Rühren auf einmal in den Topf schütten.
- Topf auf den Herd stellen, Teig unter Rühren in 1-2 Minuten zu einem Kloß abbrennen. Am Topfboden muss eine dünne, weiße Schicht sichtbar sein.
- Teig in eine Schüssel geben. Eier einzeln unterrühren. Das nächste erst unterrühren, wenn das vorige vom Teig aufgenommen wurde. Nur so viele Eier verwenden, bis der Teig glänzt und in Spitzen vom Löffel hängt.
- Teig in einen Spritzbeutel mit großer Sterntülle füllen, 16 Kreise auf das Blech spritzen. Im Ofen (Gas: Stufe 4) etwa 25 Minuten backen.
- Mit einer Schere quer aufschneiden, auskühlen lassen.
- Sahne steif schlagen, mit Likör und Vanillezucker aromatisieren. Erdbeeren waschen, einige beiseite legen. Übrige putzen, klein schneiden, unter die Sahne heben.
- Ringe mit Sahne füllen. Geröstete Mandeln und Puderzucker aufstreuen, mit Erdbeeren dekorieren.

Aufwand:	★★
Zubereitungszeit:	40 Minuten
Backzeit:	25 Minuten
Pro Stück:	205 kcal/860 kJ

Flockentorte

FÜR 1 SPRINGFORM VON 26 cm Ø
1 Prise Salz
50 g Butter
100 g Mehl
3 Eier (Größe M)

FÜR DIE FÜLLUNG
1 Dose Ananas (Abtropfgewicht 260 g)
300 g Himbeeren
750 g Schlagsahne
50 g Zucker
1 Päckchen Vanillezucker
3 Päckchen Sahnesteif

AUSSERDEM
Fett und Mehl für die Form
Puderzucker zum Bestäuben

● 1/8 l Wasser und Salz aufkochen, das Mehl zufügen und alles mit einem Holzlöffel bei mittlerer Hitze so lange rühren, bis ein Teigkloß entsteht und sich am Topfboden eine weiße Haut zeigt. Teig in eine Schüssel geben und 1 Ei untermischen. Teig auf Handwärme abkühlen lassen und dann die restlichen Eier unter den Teig heben.
● Elektro-Ofen auf 225 Grad vorheizen. Den Springformboden einfetten und mit Mehl bestäuben. Ein Drittel des Teiges darauf verstreichen und im Ofen (Gas Stufe 4) etwa 20 Minuten backen. Den Boden auf einem Kuchengitter erkalten lassen. Aus dem restlichen Teig 2 weitere Böden ebenso backen und abkühlen lassen.
● Ananas abtropfen lassen, in kleine Stifte schneiden und trockentupfen. Himbeeren waschen und vorsichtig trockentupfen.
● Sahne mit Zucker, Vanillezucker und Sahnesteif sehr steif schlagen und die Früchte darunter heben. Die Hälfte davon auf einem der Böden verteilen. Den zweiten Boden auflegen und die restlichen Sahnemischung darauf geben. Zum Schluss den dritten darauflegen und alles mit Puderzucker bestäuben.

Aufwand:	★★
Zubereitungszeit:	1 Stunde
Backzeit:	1 Stunde
Pro Stück:	270 kcal/1130 kJ

Variation

● Praktisch ist es, den dritten Boden in 12 Tortenstücke zu schneiden und obenauf zu legen. Oder Sie brechen den Boden in kleine Stücke und streuen sie auf die Sahnemischung.

Windbeutel mit Kirschen

FÜR 1 BACKBLECH (30 x 40 cm)
30 g Butter, 1/2 TL Zucker
1 Prise Salz, 100 g Mehl
2 Eier (Größe M)
FÜR DIE FÜLLUNG
500 g Kirschen
250 g Quark (20% Fett i. Tr.)
1 Päckchen Sahnesteif
3 cl Kirschwasser
200 g Schlagsahne
2 EL Zucker
2 Päckchen Vanillinzucker
2 TL abgeriebene
Zitronenschale
AUSSERDEM
Backpapier für das Blech
Puderzucker zum Bestäuben

● 1/8 l Wasser, Butter, Zucker und Salz in einem Topf aufkochen lassen. Mehl zugeben und alles bei mittlerer Hitze so lange rühren, bis ein Teigkloß entsteht, der sich leicht vom Topfboden löst. Teig in eine Schüssel geben, etwas abkühlen lassen und nacheinander die Eier unterrühren.

● Elektro-Ofen auf 225 Grad vorheizen. Backblech mit Backpapier auslegen. Den Teig in einen Spritzbeutel mit großer Sterntülle füllen und 8-10 große Tupfen im Abstand von 3-4 cm auf das Backblech spritzen.

● Windbeutel im Ofen (Gas: Stufe 4) in 30 Minuten goldgelb backen. Herausnehmen und etwa 10 Minuten ruhen lassen, dann mit einer Küchenschere waagerecht aufschneiden und auskühlen lassen.

● Kirschen waschen, abtropfen lassen, entstielen und entsteinen. Die Hälfte der Kirschen vierteln. Quark in einem Tuch auspressen. Kirschwasser unter den Quark mischen. Schlagsahne mit Sahnesteif, Zucker und Vanillezucker steif schlagen. Sahne unter den Quark heben und Kirschen und Zitronenschale untermischen.

● Die Füllung auf die unteren Windbeutelhälften geben, die Deckel aufsetzen und die Windbeutel mit Puderzucker bestäuben.

Aufwand:	★
Zubereitungszeit:	1 Stunde
Backzeit	30 Minuten
Pro Stück:	330 kcal/970 kJ

Mandel-Karamell-Torte

FÜR 2 BACKBLECHE (30 x 40 cm)
75 g Butter, 180 g Mehl
6 Eier (Größe M)
1 EL Zucker
FÜR DIE CREME
3 Blatt weiße Gelatine
200 g Zucker
175 g Mandelmus
(Reformhaus oder Bioladen)
3-4 Tropfen Bittermandelaroma
700 g Schlagsahne
AUSSERDEM
Backpapier für die Bleche
4 EL Mandelblättchen

- Butter mit 375 ml Wasser aufkochen, Mehl unterrühren, bis sich ein weißlicher Belag bildet, abkühlen lassen. Eier und Zucker unterrühren.
- Elektro-Ofen auf 200 Grad vorheizen. Bleche mit Backpapier belegen. 3 Kreise (20 cm Ø) aufzeichnen.
- Teig in Spritzbeutel mit Lochtülle geben, 10-12 haselnussgroße Tupfen aufs Papier spritzen. Übrigen Teig auf die Kreise streichen, im Ofen (Gas: Stufe 3) etwa 15 Minuten backen.
- Gelatine einweichen. Zucker in einem breiten Topf bei mittlerer Hitze ohne Rühren goldbraun schmelzen, 75 ml Wasser zufügen. Rühren, bis der Karamell flüssig ist. Abkühlen lassen. Die Hälfte mit Mandelmus, Mandelaroma und 200 g Sahne verrühren. Gelatine darin auflösen. Restliche Sahne steif schlagen, unter die angelierte Masse heben.
- Brandteigböden mit Creme bestreichen und aufeinander setzen. Windbeutel kreisförmig oben auf die Creme setzen. Übrigen Karamell nochmal aufkochen, bis er dickflüssig wird, mit einem Löffel in Linien über die Torte ziehen. Geröstete Mandelblättchen überstreuen.

Aufwand:	★★
Zubereitungszeit:	1 Stunde
Backzeit:	15 Minuten
Pro Stück:	385 kcal/1620 kJ

Prosecco-Torte

BITTE BEACHTEN: TORTE MUSS 3 STUNDEN KÜHLEN

FÜR 2 BACKBLECHE (30 x 40 cm)
- *1 Prise Salz*
- *50 g Butter*
- *75 g Mehl*
- *2 Eier (Größe M)*
- *1-2 TL Zucker*

FÜR DIE FÜLLUNG
- *8 Blatt weiße Gelatine*
- *250 g helle, kernlose Trauben*
- *250 ml Prosecco*
- *200 ml Traubensaft*
- *1 EL Zitronensaft*
- *3 EL Zucker*
- *250 g Mascarpone (ital. Frischkäse)*

AUSSERDEM
- *Backpapier für die Bleche*
- *Kakaopulver zum Bestäuben*

● 50 ml Wasser mit Salz und Butter aufkochen. Mehl einrühren, unter Rühren erhitzen, bis sich ein Kloß bildet und der Topfboden einen weißen Belag bekommt. Teig in eine Schüssel umfüllen, etwas abkühlen lassen. Eier einzeln unterrühren.

● Elektro-Ofen auf 200 Grad vorheizen. Backbleche mit Backpapier auslegen, 3 Kreise (je etwa 18 cm Ø) aufzeichnen, Teig darauf verstreichen (pro Kreis etwa 75 g Teig). 1 Teigboden mit Zucker bestreuen. Im Ofen (Gas: Stufe 3) 15-20 Minuten backen. Noch warm die Kreise mit Hilfe eines Springformrandes (18 cm Ø) sauber ausstechen, auskühlen lassen.

● Die Gelatine kalt einweichen. Trauben waschen, abzupfen und halbieren. Prosecco, Trauben- und Zitronensaft mit Zucker und Mascarpone glatt rühren. Tropfnasse Gelatine bei milder Hitze auflösen, mit 2 EL Prosecco-Mischung verrühren, unter Rühren in die Creme gießen. Creme angelieren lassen, halbierte Trauben unterheben.

● Den mit Zucker bestreuten Boden in 8 Stücke teilen. Einen Boden mit einem Tortenring umstellen. Die Hälfte der Creme darauf verteilen. Zweiten Boden auflegen, gut andrücken und restliche Creme darauf verstreichen. Leicht mit Kakao überstäuben. Die Stücke des dritten Bodens schräg in die Creme setzen. Im Kühlschrank in etwa 3 Stunden fest werden lassen. Tortenring lösen.

Aufwand:	★★
Zubereitungszeit:	50 Minuten
Backzeit:	40 Minuten
Pro Stück:	500 kcal/2100 kJ

Variationen

● Statt der Weintrauben können Sie auch andere Früchte nehmen, wie Himbeeren, Erdbeeren oder klein geschnittene Pfirsiche.

● Wer keinen Prosecco zur Hand hat, kann ebensogut einen trockenen Weißwein, wie Chardonnay oder Riesling, nehmen.

Grundrezepte: Blätte

Es gibt sie zwar fertig zu kaufen, aber auch wenn es etwas Mü

Und so geht's

400 g Mehl
1 TL Salz
1 Eigelb
FÜR DIE BUTTERMISCHUNG
500 g kalte Butter
100 g Mehl
AUSSERDEM
Mehl für Arbeitfläche

- 400 g Mehl mit 1 TL Salz, 150 ml Wasser und dem Eigelb zu einem glatten Teig verkneten und 30 Minuten kühlen.
- 500 g Butter mit 100 g Mehl verkneten, zu einem flachen Ziegel formen und etwa 30 Minuten kühlen.
- Den Teig auf einer bemehlten Arbeitsfläche etwa 3 mm dünn ausrollen. Butterziegel in die Mitte des Teiges geben (s. Step 1) und die vier Teigecken straff über das Butterstück spannen. Das Teigstück dann etwa fingerdick zu einem Rechteck ausrollen .
- Vom Teigstück erst ein Drittel des Teiges über das mittlere Drittel klappen, dann das übrige Drittel darüber legen (s. Step 2). Die Teigschichten wiederum zu einem Rechteck ausrollen.
- Die Teigplatte dann vierteilig zusammenlegen (s. Step 3). Teig in Frischhaltefolie wickeln und etwa 30 Minuten im Kühlschrank ruhen lassen (s. Step 4).
- Den Teig dann wieder zu einem Rechteck ausrollen und dreiteilig (wie Step 2) zusammenlegen. Nachmals ausrollen und vierteilig zusammenklappen (wie Step 3). Den Teig über Nacht im Kühlschrank ruhen lassen, dann entsprechend verarbeiten.

Grundrezept Blätterteig

Tipps

- Blätterteig immer in zwei Richtungen ausrollen (quer und längs), sonst schrumpft er beim Backen.
- Blätterteig mit einem scharfen Messer oder Teigrad schneiden, sonst gehen die Ränder beim Backen nicht auf.
- Wenn Sie den Teig mit Eigelb bepinseln, darf kein Eigelb an die Schnittkanten gelangen, sonst gehen diese beim Backen ebenfalls nicht auf.

Gelatine für glatte Cremes

Blattgelatine in kaltem Wasser 5 Minuten einweichen, dann tropfnass in einen kleinen Topf geben. Bei milder Hitze auf dem Herd auflösen. Mit etwas Creme verrühren und zur restlichen Creme geben. Gelatine nie direkt in Cremes schütten: Sonst entstehen die gefürchteten Klümpchen.

So kom

teig und Strudelteig

acht: Selbst gemacht schmecken Teige immer noch am besten

Grundrezept Strudelteig

1

2

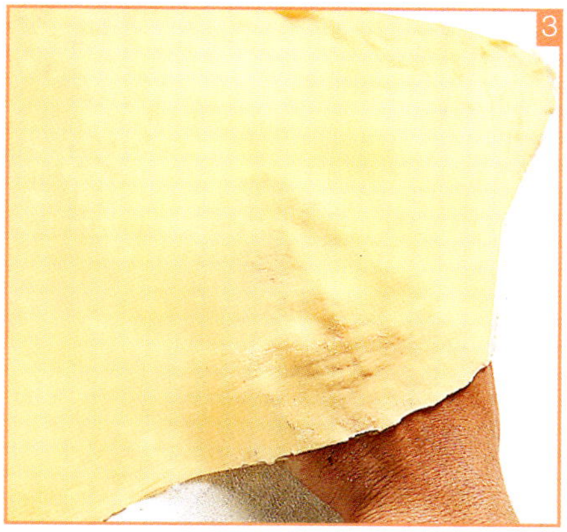
3

4

Und so geht's

250 g Mehl, 1 Prise Salz
2 EL Öl, 1 Ei (Größe M)
etwas zerlassene Butter

- Mehl mit Salz in eine Schüssel geben, in die Mitte eine Vertiefung drücken (s. Step 1).
- Öl, Ei und 80-100 ml lauwarmes Wasser zugeben. Alles kräftig zu einem glatten, elastischen Teig durchkneten. Teig zu einer Kugel formen, in Folie wickeln und bei Zimmertemperatur mindestens 30 Minuten ruhen lassen.
- Teigkugel mit zerlassener Butter bestreichen (s. Step 2) und auf einem bemehlten Tuch dünn ausrollen. Mit den Händen von innen nach außen ziehen (s. Step 3).
- Den ausgezogenen Teig nochmals mit Butter bestreichen (s. Step 4).

Tipps

- Alle Zutaten für einen Strudelteig sollten Raumtemperatur haben, also rechtzeitig aus dem Kühlschrank nehmen.
- Wichtig ist es, den Teig gut und kräftig zu kneten, er soll sehr geschneidig werden.
- Die Ruhezeit für den fertigen Teig sollte 30 Minuten bis 1 Stunde betragen.
- Bleiben nach dem Ausziehen noch dicke Teigränder, schneiden Sie diese am besten ab.

e Fülle in die Hülle

3

1 Teigdreiecke belegen. Von der breiten Seite her aufrollen.

2 Blätterteigquadrat diagonal belegen. Ränder mit Wasser oder Eiweiß bepinseln.

3 Teig über die Füllung klappen, Rand andrücken, eventuell mit einer Gabel ein Muster einkerben.

Brandteig: Gut zu wissen

- Zubereiteter Brandteig wird vom Stehen grau und zäh, also gleich nach der Zubereitung backen.
- Brandteig-Gebäck verliert an der Luft schnell an Knusprigkeit. Mit Sahne oder Creme gefüllte Windbeutel deshalb rasch servieren.
- Ohne Füllung lassen sich fertige Gebäckstücke gut einfrieren; bei Zimmertemperatur müssen sie 30 Minuten auftauen.

Kalte Torten
wenig Mühe & viel Erfolg

Bei diesen Torten bleibt der Ofen kalt, dafür werden köstliche Sahne- und Käse-Cremes gerührt und mit frischen Früchten belegt. Genau das Richtige für den Sommer

Joghurttorte mit Obstguss

Rezept siehe folgende Seite

Joghurttorte mit Obstguss

FÜR 1 SPRINGFORM VON 20 cm Ø
150 g Eiswaffeln
50 g Amaretti
75 g weiche Butter
FÜR DEN BELAG
4 Blatt weiße Gelatine
250 g Speisequark (20 % F. i. Tr.)
500 g Vanillejoghurt
2 EL Zitronensaft, 5 EL Zucker
1 Päckchen Vanillezucker
1 Mango (ca. 500 g)
250 g Himbeeren
1 1/2 EL Speisestärke
AUSSERDEM
Backpapier für die Form

- Formboden mit Backpapier auslegen. Waffeln und Kekse fein zerbröseln, mit Butter verkneten. In der Form verteilen, andrücken, kalt stellen.
- Gelatine einweichen. Quark mit Joghurt, 1 EL Zitronensaft, 3 EL Zucker und Vanillezucker glatt rühren. Gelatine bei kleiner Hitze auflösen. Mit 2 EL Quarkmasse mischen. Gelatine in übrige Masse rühren, auf dem Kuchenboden verteilen. 2 Stunden kühl stellen.
- Mango schälen, Fruchtfleisch vom Stein schneiden, mit übrigem Zitronensaft und 1/2 EL Zucker pürieren. Beeren durch ein feines Sieb streichen, mit restlichem Zucker mischen. Beide Pürees eventuell mit Wasser auf 1/4 l auffüllen.
- Von jedem Püree 2 EL abnehmen. Mangopüree mit 1/2 EL Stärke, Himbeerpüree mit 1 EL anrühren. Pürees getrennt aufkochen, angerührte Stärke einfließen lassen und nochmal aufkochen.
- Pürees in großen Klecksen auf den Kuchen geben, verstreichen. Mit einem Stäbchen marmorieren. 1 Stunde kalt stellen.

Aufwand:	★★
Zubereitungszeit:	50 Minuten
Kühlzeit:	3 Stunden
Pro Stück:	260 kcal/1090 kJ

Aprikosen-Schoko-Torte

FÜR 1 SPRINGFORM VON 26 cm Ø
100 g Halbbitter-Kuvertüre
100 g Vollmilch-Kuvertüre
2 EL Butter
100 g Cornflakes
25 g Mandelblättchen
FÜR DEN BELAG
5 Blatt weiße Gelatine
600 g Aprikosen
600 g Doppelrahm-Frischkäse
300 g Magermilchjoghurt
100 g Zucker
etwas abgeriebene, unbehandelte Zitronenschale
3 EL Zitronensaft
1 Päckchen heller Tortenguss
2-3 EL gelber Fruchtsirup
(z. B. Pfirsich-Maracuja)
AUSSERDEM
Backpapier für die Form
einige Melisseblättchen für die Dekoration

- Springform mit Backpapier auslegen. Beide Kuvertüren grob hacken. Mit der Butter in einen Topf geben und über dem heißen Wasserbad bei kleinster Hitze schmelzen lassen.
- Cornflakes und Mandeln locker unterheben, die Masse auf dem Springformboden gleichmäßig verteilen. Für 30 Minuten in den Kühlschrank stellen und fest werden lassen.
- Die Gelatine in kaltem Wasser einweichen. Einige Aprikosen für die Dekoration beiseite legen, die übrigen Aprikosen mit kochendem Wasser überbrühen, häuten, halbieren und entsteinen. Das Fruchtfleisch in kleine Stücke schneiden.
- Den Frischkäse mit Joghurt, Zucker und Zitronenschale verrühren. Die Gelatine ausdrücken und in einem Töpchen mit Zitronensaft bei kleinster Hitze auflösen.
- 3 EL Frischkäsecreme in die Gelatine rühren, dann alles unter die übrige Frischkäsecreme mischen. Aprikosenstücke unterheben.
- Die Frischkäse-Aprikosen-Masse auf den Schokoladenboden streichen und die Torte mindestens 4 Stunden kalt stellen.
- Die beiseite gelegten Aprikosen waschen, halbieren, entsteinen und in dünne Spalten schneiden. Auf der fest gewordenen Torte rosettenförmig verteilen.
- Den Tortenguss nach Packungsangabe mit Fruchtsirup (anstelle des angegebenen Zuckers) und Wasser zubereiten. Etwas abgekühlt gleichmäßig über die Torte geben und fest werden lassen. Zum Schluss nach Belieben mit Melisseblättchen verzieren.

Aufwand:	★
Zubereitungszeit:	1 Stunde
Kühlzeit:	4 Stunden
Pro Stück:	470 kcal/1970 kJ

Variationen

- Das Raffinierte an dieser Torte ist der Schokoladenboden, den Sie beliebig mit Nüssen, Puffreis oder Keksstückchen variieren können.
- Die Aprikosencreme schmeckt natürlich auch mit einem Boden aus Löffelbiskuits oder zerbröselten Butterkeksen.
- Fruchtwechsel: Anderes Steinobst wie Pfirsiche, Kirschen, Mirabellen oder Pflaumen, lassen sich für die Creme ebensogut verwenden.

Florentiner Erdbeerkuchen

FÜR 1 SPRINGFORM VON 24 cm Ø
100 g Halbbitter-Kuvertüre
150 g Schoko-Sandkekse
10 Belegkirschen
100 g gehackte Mandeln
80 g weiche Butter
FÜR DEN BELAG
1 Pck. Sahne-Puddingpulver
1 gestrichener EL Speisestärke
300 ml Milch, 80 g Zucker
200 g Schlagsahne
100 ml Orangensaft
200 g Doppelrahm-Frischkäse
750 g Erdbeeren
50 g Halbbitter-Kuvertüre
AUSSERDEM
Backpapier für die Form
50 g Mandelblättchen

- Springform mit Backpapier auslegen. Elektro-Ofen auf 175 Grad vorheizen. Kuvertüre im heißen Wasserbad schmelzen.
- Kekse fein zerdrücken. Kirschen würfeln. Mandeln, Kirschen, Kekskrümel und Butter in einer Schüssel mit der flüssigen Kuvertüre mischen. Masse in die Form drücken, kalt stellen.
- Puddingpulver und Stärke mit 6 EL Milch anrühren. Übrige Milch mit Zucker und Sahne aufkochen. Angerührtes Puddingpulver einrühren, aufkochen. Topf vom Herd nehmen, Masse abkühlen lassen.
- Orangensaft und Frischkäse verrühren und unter den Pudding heben. Erdbeeren waschen, 250 g beiseite legen. Restliche Beeren putzen, würfeln und unter die Creme heben.
- Die Creme auf den Boden geben und 4-5 Stunden kalt stellen. Kuvertüre im heißen Wasserbad schmelzen. Die übrige Erdbeeren in die Kuvertüre tauchen, auf den Kuchen legen. Mandeln rösten und Tortenrand damit verzieren.

Aufwand:	★
Zubereitungszeit:	45 Minuten
Kühlzeit:	5 Stunden
Pro Stück:	420 kcal/1760 kJ

Eistorte „Avanti“

FÜR 1 SPRINGFORM VON 24 cm Ø
1 l Himbeereiscreme
50 g Mandelstifte
100 g Zartbitterschokolade
150 g Cantuccini (italienische Mandelkekse, ersatzweise Vollkorn-Butterkekse)
200 g Schlagsahne
1 EL Zucker
1 Päckchen Sahnesteif
AUSSERDEM
250 g frische Himbeeren
2-4 EL Schokospäne zum Verzieren

- Das Eis antauen lassen und die Form mit einer Frischhaltefolie auslegen. Die Eiscreme in der Form glatt streichen und die Form für etwa 15 Minuten in das Tiefkühlfach stellen.
- Inzwischen die Mandelstifte in einer Pfanne ohne Fett hellbraun rösten. Die Schokolade grob hacken und über dem heißen Wasserbad schmelzen. Kekse im Blitzhacker grob zermahlen oder in einem Gefrierbeutel mit dem Nudelholz zerdrücken. Mit den Mandeln und der Schokolade vermischen.
- Die Mandel-Schokoladen-Mischung gleichmäßig auf das Eis streichen und für etwa 30 Minuten ins Tiefkühlfach stellen. Torte nicht länger im Eisfach lassen, da der Boden sonst zu fest wird.
- Inzwischen Sahne steif schlagen und dabei langsam Zucker und Sahnesteif einrieseln lassen. Die Himbeeren waschen und vorsichtig trockentupfen.
- Die Eistorte aus der Form stürzen, die Folie entfernen und die Torte rundum mit Sahne bestreichen. Mit Himbeeren und Schokospänen garnieren und gleich servieren.

Aufwand:	★
Zubereitungszeit:	30 Minuten
Kühlzeit:	30 Minuten
Pro Stück:	315 kcal/1320 kJ

Sahnige Apfeltorte

FÜR 1 SPRINGFORM VON 20 cm Ø
4 Äpfel (z. B. Elstar)
1 EL Zitronensaft, 100 g Zucker
Saft von 1 Orange
2-3 EL Apfellikör nach Belieben
4 Blatt weiße Gelatine
2-3 eingelegte Ingwerpflaumen (Glas), 3 Eier (Größe M)
250 g Mascarpone
100 g Löffelbiskuits
ZUM DEKORIEREN
Schokoherzen
AUSSERDEM
Fett zum Bestreichen
Backpapier für die Form
Puderzucker zum Bestäuben

- Elektro-Ofen auf 225 Grad vorheizen. Einen Apfel waschen, etwa 15 hauchdünne Scheiben abschneiden. Mit Zitronensaft beträufeln, mit 1 EL Zucker bestreuen und die Apfelscheiben auf ein gefettetes Stück Alufolie legen. Im Ofen (Gas: Stufe 4-5) etwa 10 Minuten rösten. Für die Dekoration beiseite legen.
- Alle Äpfel, auch den angefangenen, schälen, entkernen, in kleine Blättchen schneiden. Mit Orangensaft und Likör aufkochen, etwa 2 Minuten dünsten, in einem Sieb abtropfen lassen, Saft auffangen.
- Gelatine einweichen. Ingwer fein würfeln. Eier trennen, Eiweiße steif schlagen, 2 EL Zucker einrieseln lassen. Eigelbe mit 60 g Zucker aufschlagen, Mascarpone, Ingwerwürfel und 1-2 EL Ingwersirup einrühren. Gelatine bei kleiner Hitze auflösen, 3 EL Mascarponecreme einrühren, dann alles unter die Creme rühren. Erst Apfelstücke, dann Eischnee unterheben.
- Boden der Form mit Backpapier auslegen, mit Biskuits auslegen, mit dem aufgefangen Apfelsaftsud beträufeln. Mascarponecreme einfüllen, glatt streichen. 3 Stunden kalt stellen.
- Torte aus der Springform nehmen, mit Apfelchips und Schokoherzen garnieren, mit Puderzucker bestäuben.

Aufwand:	★
Zubereitungszeit:	40 Minuten
Kühlzeit:	3 Stunden
Pro Stück:	300 kcal/1260 kJ

Kirsch-Bananen-Torte

FÜR 1 SPRINGFORM VON 26 cm Ø
200 g Orangenkekse
100 g flüssige Butter
4 Blatt weiße Gelatine
300 g Schlagsahne
400 g Doppelrahm-Frischkäse
300 g Bananenjoghurt (Fertigprodukt)
3 EL Zitronensaft
80 g Zucker
750 g Süßkirschen
AUSSERDEM
Backpapier für die Form
Orangenkekse, Kirschen und Melisseblättchen zum Verzieren

- Form mit Backpapier auslegen. Die Kekse zerbröseln, mit Butter vermischen. In die Form drücken und kalt stellen.
- Gelatine in kaltem Wasser etwa 5 Minuten einweichen. 200 g Sahne steif schlagen.
- Frischkäse mit Joghurt, Zitronensaft und Zucker verrühren. Gelatine bei kleiner Hitze auflösen. Mit 3 EL Creme verrühren, dann in die übrige Creme rühren.
- Kirschen waschen, putzen, entsteinen und mit der geschlagenen Sahne unter die Creme heben. Die Masse auf den Keksboden geben und glatt streichen.
- Die Torte im Kühlschrank mindestens 4 Stunden, am besten über Nacht, fest werden lassen.
- Übrige Sahne steif schlagen. Torte mit Keksen, Kirschen, Melisseblättchen und Sahnetuffs verzieren.

Aufwand:	★
Zubereitungszeit:	30 Minuten
Kühlzeit:	4 Stunden
Pro Stück:	370 kcal/1550 kJ

Erdbeer-Mascarpone-Schnitten

FÜR 1 BACKBLECH (30 x 40 cm)
3 Päckchen Mini-Butterkekse mit Schokoglasur (à 125 g)
160 g Butter
FÜR DIE CREME
4 Blatt weiße Gelatine
5 eingelegte Ingwerpflaumen
350 g Mascarpone
350 g Vollmilchjoghurt
1 Päckchen Vanillezucker
70 g Zucker, 5 EL Zitronensaft
AUSSERDEM
Backpapier für das Blech
700 g Erdbeeren
1 Päckchen heller Tortenguss
50 g Halbbitter-Kuvertüre

- Blech zur Hälfte mit Backpapier auslegen. Kekse in Gefrierbeutel füllen, mit dem Nudelholz zerbröseln. Butter zerlassen. Beides vermischen und auf das Blech drücken. Ein Stück Alufolie zweimal falten, den freien Rand damit befestigen.
- Gelatine einweichen. Ingwer fein würfeln. Mascarpone mit Joghurt verrühren. Vanillezucker, Zucker, Zitronensaft, Ingwerwürfel und 6 TL Ingwersirup unterrühren. Gelatine bei milder Hitze auflösen, mit 2 EL Creme verrühren. Gelatine unter die Creme mischen, auf dem Keksboden verteilen und glatt streichen.
- Die Erdbeeren waschen, trockentupfen, putzen. In dünne Scheiben schneiden. Dachziegelartig auf der Creme verteilen. Tortenguss nach Packungsangabe anrühren, auf den Erdbeeren verteilen. Kuchen etwa 1 Stunde kalt stellen. Kuvertüre grob hacken, auf dem warmen Wasserbad schmelzen. In einen Gefrierbeutel füllen, eine Ecke abschneiden. Auf die Erdbeeren Linien spritzen, fest werden lassen.

Aufwand:	★
Zubereitungszeit:	40 Minuten
Kühlzeit:	1 Stunde
Pro Stück:	300 kcal/1260 kJ

Exotische Orangencremetorte

FÜR 1 SPRINGFORM VON 26 cm Ø
200 g Mürbeteig-Kekse
80 g weiche Butter
FÜR DIE CREME
4 Eigelbe (Eier: Größe M)
300 g Zucker
1/4 TL gemahlene Vanille
1 Prise Salz
3/8 l Orangensaft
3 EL Mehl, 3 EL Speisestärke
100 g Butter
FÜR DEN BELAG
1 Karambole (Sternfrucht)
1 Kiwi, 1 Orange, 2 Limetten
AUSSERDEM
Backpapier für die Form
Puderzucker zum Bestäuben

- Form mit Backpapier auslegen. Kekse zerbröseln, mit Butter vermischen. In die Form drücken und kalt stellen.
- Für die Creme Eigelbe mit 100 g Zucker und Vanille dickschaumig schlagen.
- Restlichen Zucker mit Salz und Orangensaft erhitzen. Das Mehl und die Stärke mit wenig Wasser glatt rühren und unter den Saft schlagen. Unter Rühren erhitzen, bis die Masse dick wird.
- Die Hälfte davon unter die Eigelb-Masse mischen, restliche Masse dazugeben. Alles unter Rühren leicht erhitzen, aber nicht kochen lassen. Vom Herd nehmen, die Butter in Flöckchen darunter schlagen. Creme abkühlen lassen, dabei immer wieder umrühren.
- Creme in die Springform auf den Tortenboden füllen, glatt streichen und die Torte zum Festwerden mindestens 4 Stunden kalt stellen.
- Für den Belag Karambole waschen, Kiwi schälen, beides in Scheiben schneiden. Die Orange und 1 Limette schälen, in Scheiben schneiden.
- Übrige Limette waschen und mit der Schale zu einer hauchdünnen Spirale aufschneiden. Früchte auf der Torte dekorativ anrichten und vor dem Servieren mit Puderzucker bestäuben.

Aufwand:	★
Zubereitungszeit:	1 Stunde
Kühlzeit:	4 Stunden
Pro Stück:	280 kcal/1180 kJ

Joghurt-Kirsch-Torte

FÜR 1 HERZ-BACKRING
100 g Butterkekse
75 g Bananen-Chips
125 g Butter
7 Blatt weiße Gelatine
1 Vanilleschote
600 g Sahnejoghurt
Saft und abgeriebene Schale von 1/2 unbehandelten Zitrone
100 g Zucker
1 Päckchen Vanillezucker
2 Gläser Schattenmorellen (Abtropfgewicht: à 175 g)
200 g Schlagsahne
AUSSERDEM
5 Blatt Gelatine
Melissezweige

● Kekse und Bananen-Chips im Blitzhacker fein zerbröseln. Die Butter zerlassen und unter die Brösel mischen. Den Backring auf eine Tortenplatte stellen. Bröselmischung hineingeben, festdrücken und kalt stellen.

● Gelatine einweichen. Vanilleschote längs aufschneiden, Mark herausschaben. Vanillemark, Joghurt, Zitronensaft, -schale, Zucker und Vanillezucker verrühren. Schattenmorellen abtropfen lassen, Saft auffangen und beiseite stellen. Früchte trockentupfen.

● Gelatine bei milder Hitze auflösen. 2-3 EL Joghurtcreme in die Gelatine rühren, dann die Gelatinemasse in den übrigen Joghurt rühren. Kalt stellen. Wenn die Masse zu gelieren beginnt, Kirschen unterheben.

● Sahne steif schlagen, unter die Creme rühren. Masse in die Form geben, mit einem Löffelrücken Wellenmuster eindrücken und die Torte mindestens 2 Stunden kalt stellen.

● 5 Blatt Gelatine im Abtropfsaft der Kirschen auflösen, aufkochen. 1-2 Stunden kalt stellen, bis die Masse dickflüssig wird. Mit einem Teelöffel große Tropfen auf einen geölten Teller gießen, fest werden lassen. Tropfen mit einer Palette vorsichtig abnehmen, auf die Torte legen. Mit Melissezweigen „Kirschstängel" anlegen.

Aufwand:	★★
Zubereitungszeit:	45 Minuten
Kühlzeit:	3 Stunden
Pro Stück:	300 kcal/1260 kJ

Caipirinha-Torte

FÜR 1 SPRINGFORM VON 18 cm Ø
125 g Mürbeteig-Kekse
50 g weiche Butter
FÜR DIE CREME
2 Limetten
6 Blatt weiße Gelatine
3 Eigelbe (Eier: Größe M)
150 g Puderzucker
250 g Sahnejoghurt
6 EL Cachaça (Zuckerrohrschnaps)
300 g Schlagsahne
AUSSERDEM
30 g Pistazienkerne
1 Limette
100 g Schlagsahne

● Die Kekse portionsweise in einen Gefrierbeutel füllen, auf eine Arbeitsfläche legen und mit dem Teigroller fein zermahlen. Die Brösel mit der Butter verkneten. Mischung auf den Boden der Springform geben, fest andrücken und kalt stellen.

● Die Limetten heiß abwaschen und trockenreiben. Die Schale von 1 Limette abreiben und diese und 1 weitere Limette auspressen. Gelatine in lauwarmem Wasser einweichen.

● Eigelbe mit Puderzucker im heißen Wasserbad cremig aufschlagen. Die Gelatine ausdrücken, in der Eiercreme geben und darin auflösen. Alles gründlich verrühren und die Creme anschließend im kalten Wasserbad abkühlen lassen.

● Abgeriebene Limettenschale und -saft mit Joghurt und Cachaça glatt rühren. Nach und nach unter die Eigelbcreme ziehen. Die Masse bis zum Angelieren etwa 30 Minuten kühlen.

● Die Sahne sehr steif schlagen und gleichäßig unter die Limettencreme ziehen. Die Springform mit der Keksgrundlage auf eine Arbeitsplatte stellenund die Creme einfüllen. Etwa 2 Stunden kalt stellen.

● Den Springformrand vorsichtig von der Torte entfernen. Die Pistazien sehr fein hacken und den Tortenrand rundherum damit verzieren, dabei leich andrücken.
● Die Limette heiß abwaschen und trockenreiben. Mit einem scharfen Messer 5 hauchdünne Scheiben abschneiden und halbieren. Die Schlagsahne sehr steif schlagen und in einen Spritzbeutel mit gezackter Tülle füllen.
● Auf die Tortenoberfläche am Rand 10 kleine Tupfen spritzen und mit je 1/2 Limettenscheibe dekorieren.

Aufwand:	★
Zubereitungszeit:	50 Minuten
Kühlzeit:	3 1/2 Stunden
Pro Stück:	460 kcal/1930 kJ

Variation

● Die Torte kann auch mit anderen Zitrusfrüchten zubereitet werden: Die Limette durch 4 kleine Orangen ersetzen und die Creme statt mit Cachaça mit 6 EL Orangenlikör, z. B. Cointreau, aromatisieren.

Eistorte mit Pflaumen

FÜR 1 SPRINGFORM VON 18 cm Ø
1 l Pfirsich-Joghurt-Eis
FÜR DEN TEIG
100 g Mehl, 30 g Zucker
1 Eigelb (Ei: Größe M)
60 g kalte Butter
FÜR DEN BELAG
250 g Pflaumen, 2 Minzezweige
1 EL Puderzucker
1 EL Zitronensaft
1 EL Weinbrand
AUSSERDEM
2 EL gehackte Pistazienkerne

- Eis 1 Stunde im Kühlschrank antauen lassen, bis es streichfähig ist. Elektro-Ofen auf 200 Grad (Gas: Stufe 3) vorheizen.
- Für den Mürbteig die Zutaten verkneten, 30 Minuten kühl stellen.
- Die Pflaumen waschen, entsteinen. Minze waschen, die Blättchen von 1 Zweig hacken. 1 Pflaume beiseite legen, restliche klein würfeln, mit Zucker, Zitronensaft, Weinbrand und Minzeblättchen marinieren, etwa 15 Minuten ziehen lassen.
- Teig in die Form geben und im Ofen etwa 10 Minuten backen. Gut auskühlen lassen.
- Pflaumen abtropfen lassen. Auf den Boden eine dünne Schicht Eis und dann die Pflaumen geben.
- Eis bis auf 3 EL mit den Pistazien vermengen und in die Form füllen. Torte und übriges Eis 3 Stunden in das Gefrierfach stellen.
- Die Torte 30 Minuten vor dem Servieren in den Kühlschrank stellen, aus der Form nehmen. Aus dem übrigen Eis mit einer Spritztülle kleine Tupfen auf die Torte spritzen. Die Pflaume in Spalten schneiden, zwischen die Tupfen setzen, mit Minze garnieren.

Aufwand:	★
Zubereitungszeit:	1 1/4 Stunden
Backzeit:	10 Minuten
Kühlzeit:	3 1/2 Stunden
Pro Stück:	295 kcal/1240 kJ

Schoko-Cheese-Cake

FÜR 1 SPRINGFORM VON 26 cm Ø
80 g Butter
1 TL Spekulatiusgewürz
200 g Butterkekse
FÜR DEN BELAG
12 Blatt Gelatine
350 g Zartbitter-Kuvertüre
4 Eier (Größe M)
140 g Zucker
700 g Doppelrahm-Frischkäse
2 Päckchen Vanillezucker
250 g Crème fraîche
1 EL Espressopulver (Instant)
125 g Schlagsahne
AUSSERDEM
Backpapier für die Form
Schokospäne zum Verzieren

- Form mit Backpapier auslegen. Butter mit dem Spekulatiusgewürz schmelzen. Die Kekse fein zerkleinern, mit der Buttermischung verkneten. Masse in der Form verteilen, einen etwa 2 cm hohen Rand formen und die Form kalt stellen.
- Gelatine in lauwarmem Wasser einweichen. Kuvertüre hacken, über dem heißen Wasserbad schmelzen. Eier und Zucker ebenfalls über einem heißen Wasserbad schaumig schlagen und die ausgedrückte Gelatine darin auflösen. Frischkäse, 1 Päckchen Vanillezucker und Crème fraîche unterrühren.
- Von der Masse etwa die Hälfte abnehmen, das Espressopulver und den übrigen Vanillezucker unterrühren.
- Die Kuvertüre gleichmäßig unter die andere Hälfte rühren. Die Sahne steif schlagen und darunter heben. Zuerst die Schokomasse auf den Teigboden geben und glatt streichen. Dann die Espressomasse vorsichtig darauf verteilen und ebenfalls glatt streichen.
- Den Kuchen in der Form in den Kühlschrank stellen und 4-5 Stunden kühlen.
- Danach den Kuchen aus der Form lösen und nach Belieben mit gehobelter Schokolade verzieren.

Aufwand:	★
Zubereitungszeit:	35 Minuten
Kühlzeit:	5 Stunden
Pro Stück:	660 kcal/2770 kJ

Variation

- Statt Espressopulver können Sie unter die zweite Cremehälfte auch 100 g pürierte Himbeeren ziehen. Vanillezucker durch 2 EL Zucker ersetzen.

Kefirtorte mit Sauerkirschen

FÜR 1 SPRINGFORM VON 26 cm Ø
150 g weiche Butter
Salz, 80-100 g Zucker
300 g Pumpernickel
3 EL Kirschwasser
FÜR DEN BELAG
350 g Sauerkirschen
2 EL Puderzucker
500 g Kefir, 150 g Zucker
200 g Schlagsahne
1 Päckchen Vanillezucker
1 unbehandelte Orange
10 Blatt weiße Gelatine
FÜR DIE GARNITUR
1/4 l Rotwein oder Kirschsaft
3 EL Puderzucker, 5 Blatt rote
Gelatine, 200 g Schlagsahne
2 EL Zucker

● Butter, Salz und Zucker cremig rühren. Pumpernickel fein zerbröckeln, mit Kirschwasser beträufeln und unter die Buttermischung rühren. In der Form glatt drücken. Kirschen entsteinen, mit Puderzucker marinieren.
● Kirschen im eigenen Saft etwa 5 Minuten dünsten. Kefir, Zucker, Sahne, Vanillezucker, Saft und abgeriebene Schale der Orange verrühren.
● Weiße Gelatine einweichen und bei milder Hitze auflösen. 2 EL Kefircreme einrühren, unter die restliche Creme mischen. Kalt stellen, bis die Masse zu gelieren beginnt.
● Hälfte der Creme auf den Boden streichen. Abgetropfte Kirschen darauf verteilen, restliche Creme darüberstreichen. 2 1/2 Stunden kalt stellen.
● Wein mit Puderzucker verrühren. Rote Gelatine einweichen, bei milder Hitze auflösen, mit Wein verrühren, als dünne Schicht über die Torte gießen. Kalt stellen.
● Sahne mit Zucker steif schlagen, in einen Spritzbeutel mit Sterntülle füllen und die Torte verzieren.

Aufwand:	★
Zubereitungszeit:	55 Minuten
Kühlzeit:	3 Stunden
Pro Stück:	320 kcal/1340 kJ

Himbeer-Melonen-Torte

FÜR 1 SPRINGFORM VON 24 cm Ø
150 g weiche Butter
1 Prise Salz
200 g Löffelbiskuits
FÜR DEN BELAG
18 Blatt weiße Gelatine
2 Honigmelonen (etwa 1 1/2 kg)
Saft von 1 Zitrone
80-100 g Zucker
500 g saure Sahne
150 g Crème fraîche
150 g Himbeeren
200 g Schlagsahne
1 EL Zucker
Melisseblättchen

- Butter mit Salz cremig rühren. Biskuits zerbröseln, unter die Butter heben. Als Boden in die Form drücken, kalt stellen.
- Gelatine einweichen. Melonen halbieren, Kerne entfernen, dabei den Saft auffangen. Aus dem Fruchtfleisch 8 Kugeln ausstechen, beiseite legen.
- Fruchtfleisch herausheben, mit dem aufgefangenen Saft und Zitronensaft pürieren. Durch ein Sieb streichen, mit Zucker, saurer Sahne und Crème fraîche verrühren.
- Tropfnasse Gelatine bei milder Hitze auflösen, mit 2 EL Creme verrühren, unter die übrige Creme schlagen. Kühl stellen.
- Die Himbeeren verlesen. Sahne mit Zucker steif schlagen. Die Hälfte mit 100 g Himbeeren mischen. Unter die gelierende Creme heben, auf dem Boden glatt streichen. Die Torte und die Sahne 2 Stunden kalt stellen.
- Sahne in einen Spritzbeutel mit großer Sterntülle füllen und Tupfen auf die Torte spritzen. Mit Melonenkugeln, Himbeeren und Melisse verzieren.

Aufwand:	★
Zubereitungszeit:	40 Minuten
Kühlzeit:	2 1/2 Stunden
Pro Stück:	395 kcal/1660 kJ

Johannisbeer-Eistorte

FÜR 1 SPRINGFORM VON 24 cm Ø
2 Eiweiße (Eier: Größe M)
120 g Puderzucker
10 g Speisestärke
70 g Kokosraspel
FÜR DAS EIS
2 Eigelbe (Eier: Größe M)
70 g Puderzucker
1 EL abgeriebene Zitronenschale, 2 cl Johannisbeerlikör
500 g rote und 100 g schwarze Johannisbeeren
250 g Schlagsahne
AUSSERDEM
Backpapier für die Form

- Elektro-Ofen auf 150 Grad vorheizen. Eiweiße steif schlagen. Puderzucker, Stärke und Kokosraspel unterziehen.
- Gut die Hälfte der Eiweiß-Masse in die Form füllen. Den Boden im Ofen (Gas: Stufe 1) 30-35 Minuten backen. Mit dem Papier herausnehmen. Dann die zweite Teighälfte backen. Auskühlen lassen, bis die Böden fest sind.
- Eigelbe und Zucker im heißen Wasserbad aufschlagen, erkalten lassen. Zitronenschale und Likör unterrühren.
- Beeren waschen, trockentupfen, von den Rispen zupfen. Die roten Beeren durch ein Sieb passieren. Die Sahne steif schlagen, mit dem Mus unter die Eicreme rühren. Die schwarzen Beeren unterheben.
- 1 Boden (ohne Backpapier) in die Form legen. Creme darauf geben, 3-4 Stunden in das Gefrierfach stellen. Zweiten Boden stürzen und 1 cm dicke Streifen abschneiden. Diese fein zerkrümeln und nach 2 Stunden auf die Eistorte streuen.
- Torte 30 Minuten vor dem Servieren in den Kühlschrank stellen.

Aufwand:	★★
Zubereitungszeit:	40 Minuten
Backzeit:	1 1/4 Stunden
Kühlzeit:	4 Stunden
Pro Stück:	205 kcal/860 kJ

Fruchtige Waffeltorte

500 g Johannisbeeren
250 g Himbeeren
1-2 Limetten
4 Blatt weiße Gelatine
200 g Frischkäse
300 g Schlagsahne
80 g Zucker
100 g Mehl
150 g Buttermilch
1 Päckchen Vanillezucker
3 Eier (Größe M)
1 Prise Salz
AUSSERDEM
Fett für das Waffeleisen
Puderzucker zum Bestäuben

- Beeren waschen und putzen. Einige Beeren für die Dekoration beiseite legen. Limetten heiß waschen, trockenreiben. Die Schale von 1/2 Limette abreiben. Limetten auspressen. Himbeeren pürieren und durch ein Sieb streichen. Himbeermark mit 3-4 EL Limettensaft verrühren.
- Gelatine einweichen und tropfnass bei kleiner Hitze auflösen. Mit 1 EL Himbeermark glatt rühren, dann unter das restliche Mark rühren. Mit Frischkäse vermischen.
- Sahne steif schlagen, 50 g Zucker langsam einrieseln lassen. Unter die Frischkäse-Himbeer-Masse heben. Johannisbeeren – bis auf einige für die Dekoration – unter die Masse heben.
- Mehl, Buttermilch, restlichen Zucker, Vanillezucker, Limettenschale und 1-2 EL Limettensaft gut mischen. Eier trennen. Eigelbe unter die Mehl-Masse rühren. Eiweiße mit dem Salz steif schlagen, dann ebenfalls unter die Mehl-Masse heben.
- Waffeleisen, falls nötig, leicht einfetten. Nacheinander aus je 2 EL Teig 8 goldbraune Waffeln backen, abkühlen lassen.
- Erste Waffel auf eine Platte legen, mit Himbeercreme bestreichen, mit der nächsten Waffel abdecken. So weiter verfahren, bis alles verbraucht ist. Mit einer Waffel abschließen. (Oder aus Waffeln und Himbeercreme zwei Stapel machen.) Torte 1 Stunde zum Festwerden kühl stellen. Mit restlichen Beeren und Puderzucker dekorieren.

Aufwand:	★
Zubereitungszeit:	1 Stunde
Kühlzeit:	1 Stunde
Pro Stück:	245 kcal/1030 kJ

Heidelbeer-Quarktorte

BITTE BEACHTEN: TORTE MUSS ÜBER NACHT KÜHLEN
FÜR 1 SPRINGFORM VON 26 cm Ø

- *150 g Mandelkekse (z. B. Amarettini)*
- *1 unbehandelte Orange*
- *125 g Butter*
- *750 ml heller Traubensaft*
- *80 g Speisestärke*
- *180 g Zucker*
- *400 g Schlagsahne*
- *1 Päckchen Sahnesteif*
- *450 g Heidelbeeren (frisch oderTK)*
- *1 Päckchen heller Tortenguss*

AUSSERDEM

- *Backpapier für die Form*
- *Puderzucker zum Bestäuben*

● Die Kekse zerbröseln. Orange heiß waschen, 1 TL Schale abreiben und die Frucht auspressen. Butter schmelzen. Mit Bröseln und Schale mischen. Die Springform mit Backpapier auslegen. Brösel einfüllen, festdrücken, 30 Minuten kühlen.

● Inzwischen Orangen- und 700 ml Traubensaft mischen. Etwas Flüssigkeit abnehmen, mit Stärke verrühren. Säfte mit Zucker und Stärke einmal aufkochen, erkalten lassen.

● Sahne mit Sahnesteif steif schlagen, unter die abgekühlte Saftmasse heben. Creme auf dem Boden verstreichen. Über Nacht kühl stellen.

● TK-Beeren auftauen, frische abbrausen, verlesen, trockentupfen. Guss mit übrigem Traubensaft und Wasser nach Packungsangabe zubereiten. Die Beeren auf die Torte verteilen, mit Guss überziehen. Abkühlen lassen. Mit Puderzucker bestäuben.

Aufwand:	★
Zubereitungszeit:	50 Minuten
Kühlzeit:	über Nacht
Pro Stück:	380 kcal/1600 kJ

Limetten-Torte

BITTE BEACHTEN: TORTE MUSS ÜBER NACHT KÜHLEN
FÜR 1 SPRINGFORM VON 26 cm Ø

- *150 g Butterkekse*
- *200 g weiße Kuvertüre*
- *6 Limetten*
- *9 Blatt weiße Gelatine*
- *300 g Joghurt*
- *400 ml Kokosmilch (ungesüßt)*
- *250 g Puderzucker*
- *400 g Schlagsahne*
- *4-5 EL Ingwerkonfitüre*

AUSSERDEM

- *Backpapier für die Form*
- *brauner Zucker zum Bestreuen*

● Kekse in Gefrierbeutel geben und mit dem Nudelholz fein zerbröseln. Kuvertüre grob hacken und über heißem Wasserbad schmelzen. Geschmolzene Kuvertüre mit den Keksbröseln vermischen. Springform mit Backpapier auslegen. Keksmischung hineinfüllen, festdrücken, kalt stellen.

● 3 Limetten heiß abwaschen und trockenreiben. 1 für die Deko beiseite legen, von den anderen die Schale abreiben. Die Limetten auspressen.

● Gelatine einweichen. Limettenschale, -saft, Joghurt, Kokosmilch und Zucker verrühren. Gelatine tropfnass in einem Topf bei milder Hitze schmelzen. Unter die Kokosmischung rühren. Masse bis zum Angelieren kalt stellen.

● Sahne steif schlagen, unter die Limettencreme heben. Teig aus der Form lösen, auf eine Kuchenplatte legen. Mit Ingwermarmelade bestreichen. Einen Tortenring um den Boden legen. Creme auf dem Boden verteilen. Kuchen über Nacht kühl stellen.

● Übrige Limette halbieren und in dünne Scheiben schneiden. Kurz vorm Servieren Zucker auf der Torte verstreuen. Limettenscheiben paarweise auf der Torte verteilen.

Aufwand:	★★
Zubereitungszeit:	45 Minuten
Kühlzeit:	über Nacht
Pro Stück:	520 kcal/2180 kJ

Tipps & Tricks für Füllungen

Gewusst wie! Und Füllungen und Optik von Torten gelingen mühelo

Blütenblätter zuckern

1 TL Gummi arabicum und 1 TL Zucker in 1/4 l Wasser auflösen. Blütenblätter mit einer Pinzette eintauchenund abtropfen lassen. Dann in Dekorzucker wenden, bis sie davon überzogen sind. Die Blätter vor dem Verwenden auf Küchenpapier einige Stunden trocknen lassen.

Karamell herstellen

1 10 EL Zucker mit 4 EL Wasser in einem kleinen Topf bei kleiner Hitze verrühren.

2 Bei starker Hitze zum Kochen bringen – dabei ständig rühren, bis das Wasser vollständig verdampft ist.

3 Danach färbt sich der Karamell langsam braun. Er muss immer weiter gerührt werden, damit er nicht anbrennt. Dann die Hitze zurück schalten.

4 Wenn er die typische goldbraune Farbe angenommen hat, kann man ihn zum Zuckerspinnen oder als Deckel für Cremetorten verwenden.

Schokoladenblätter

Gewaschene und trockengetupfte Blätter bereit legen. Blätter mit ihrer Oberseite flach in geschmolzene Kuvertüre tauchen und mit der Schokoladenseite nach oben trocknen lassen. Die Blätter von der festgewordenen Schokolade vorsichtig abziehen.

und Dekorationen

nn oft sind es nur die kleinen Kniffe, die zum Erfolg führen

Cremes zur „Rose" abziehen

Erst das Eigelb verrühren **1**, Zucker zufügen und die Mischung mit dem Schneebesen cremig rühren **2**. Erhitzte Milch oder Saft zugießen und unterrühren **3**. Die Masse unter Rühren erhitzen, aber nicht kochen! Wenn die Creme auf dem Kochlöffel beim Daraufblasen Kringel zeigt **4**, die an eine „Rose" erinnern, ist sie fertig.

Zitrusfrüchte

1 Das Aroma von Zitrusfrüchten ist in der Schale verborgen. Deshalb wird sie auf einer feinen Reibe abgerieben und unter Cremes oder Sahne gemischt.

2 Zur Dekoration schneidet man hauchdünne Zitronen-, Orangen- oder Limettenschale in sehr feine Streifchen, die man zuckern (siehe links) und auf Torten streuen kann.

Mit Sahne dekorieren

1 Die Sahne in eine Rührschüssel geben. Zucker und aufgelöste Gelatine zufügen.

2 Die Sahne anschließend mit dem Schneebesen des Handrührgerätes sehr steif schlagen.

3 Jetzt hat sie die richtige Konsistenz, um fest zu bleiben. Die Sahne in einen Spritzbeutel füllen und die Torte damit dekorieren.

Vanillemark herauslösen

Vanille hat ein wunderbares Aroma. Schoten mit einem spitzen Messer aufschlitzen, das Mark herausschaben und in Sahne oder Cremes rühren.

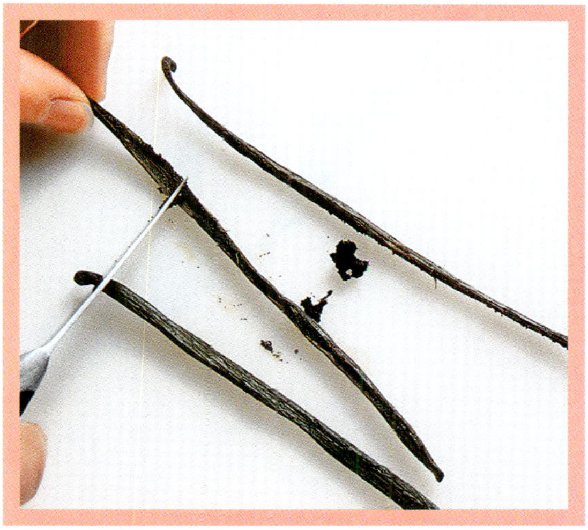

Kuchen
und Gebäck aus aller Welt

Überall auf der Welt wird Kuchen geschätzt und jedes Land hat seine eigenen Traditionen. Üppige Torten, Tartes und Gewürzkuchen – lassen Sie sich überraschen

Feigentorte
Rezept siehe folgende Seite

Feigentorte

Italien

FÜR 1 SPRINGFORM VON 24 cm Ø

200 g Mehl, 2 EL Zucker

150 g kalte Butter, 1 Prise Salz

FÜR DEN BELAG

3 Eigelbe (Eier: Größe M)

4 EL Zucker

250 g Mascarpone

1 EL Orangenlikör

14-16 reife, blaue Feigen

150 g Aprikosenkonfitüre

AUSSERDEM

Fett für die Form

● Aus Mehl, Zucker, Butter und Salz rasch einen glatten Teig kneten. In Folie gewickelt 1 Stunde kalt stellen.
● Elektro-Ofen auf 225 Grad vorheizen. Die Form einfetten. Mürbeteig zwischen Klarsichtfolie ausrollen, Form damit auslegen. Den Boden mit einer Gabel mehrmals einstechen. Teig im Ofen (Gas: Stufe 4) etwa 25 Minuten backen. Auskühlen lassen.
● Für den Belag Eigelbe mit Zucker dickschaumig schlagen, Mascarpone und Likör unterrühren. Creme auf dem Tortenboden verteilen, glatt streichen.
● Feigen waschen, abtrocknen, längs vierteln und dekorativ auf die Creme setzen. Konfitüre aufkochen, durch ein Sieb streichen und noch lauwarm hauchdünn auf die Feigen pinseln. Torte ganz frisch servieren.

Aufwand:	★
Zubereitungszeit:	40 Minuten
Backzeit:	25 Minuten
Pro Stück:	815 kcal/3410 kJ

Piña-Colada-Biskuit

Karibik

FÜR 1 BACKFORM (25 x 25 cm)

300 g Löffelbiskuits

1 Dose Ananas in Scheiben (Einwaage: 800 g)

100 ml Ananassaft

6 Eigelbe (Eier: Größe M)

100 g Zucker, 8 EL weißer Rum

10 EL Kokoslikör

500 g Mascarpone

50 g Kokos-Chips für die Dekoration

● Die Hälfte der Löffelbiskuits in die Form schichten. Ananas abgießen, Sud auffangen und mit dem Ananassaft mischen. Die Biskuits mit der Hälfte der Saftmischung beträufeln. Die Ananasscheiben vierteln und die Hälfte davon auf den Biskuits in der Form verteilen.
● Die Eigelbe mit Zucker schaumig rühren. Rum, Likör und Mascarpone zufügen und locker-cremig schlagen.
● Die Hälfte der Masse glatt auf die Ananasschicht streichen. Restliche Biskuits darauf schichten. Mit der übrigen Saftmischung beträufeln, restliche Ananasstücke darauf legen. Mit der übrigen Creme bestreichen. Den Kuchen zugedeckt 1 Stunde kalt stellen.
● Die Kokos-Chips in einer beschichteten Pfanne leicht anrösten und den Biskuit vor dem Servieren damit bestreuen.

Aufwand:	★
Zubereitungszeit:	35 Minuten
Kühlzeit:	1 Stunde
Pro Stück:	290 kcal/1210 kJ

Mein liebster Käsekuchen

Australien

FÜR 1 SPRINGFORM VON 26 cm Ø
250 g frische oder tiefgekühlte Himbeeren
100 g weiße Schokolade
2 Bananen, 2 EL Apfelsaft
1 TL Zitronensaft
150 g weiche Butter
200 g Zucker, 4 Eier (Größe M)
1 kg Magerquark, 1 Päckchen Vanille-Puddingpulver
AUSSERDEM:
Butter für die Form
Puderzucker zum Bestäuben
2 EL Honig

- Tiefgekühlte Himbeeren auftauen. Die Schokolade fein reiben, für die Verzierung ein paar Späne grob hobeln. Bananen schälen und mit Apfel- und Zitronensaft gründlich zerdrücken oder mixen.
- Elektro-Ofen auf 200 Grad vorheizen. Die Form fetten. Die Butter mit dem Zucker cremig rühren, nach und nach die Eier darunter schlagen. Geriebene Schokolade, Bananen, abgetropften Quark und Puddingpulver unterrühren.
- Quarkmasse in die Form füllen, Kuchen im Ofen (Gas: Stufe 3) etwa 40 Minuten backen, bis er aufgegangen und goldgelb ist.
- Inzwischen die frischen Himbeeren verlesen, waschen. Etwa 10 Beeren für die Dekoration beiseite legen. Übrige Beeren mit dem Honig pürieren und durch ein Sieb streichen.
- Den fertigen Kuchen in der Form auskühlen lassen, dann herauslösen. Käsekuchen mit Schokospänen, Himbeeren und Puderzucker verzieren und mit der Himbeersauce servieren.

Aufwand:	★
Zubereitungszeit:	50 Minuten
Backzeit:	40 Minuten
Pro Stück:	340 kcal/1440 kJ

Erdbeer-Kuppeltorte

Frankreich

BITTE BEACHTEN: CREME MUSS 3-4 STUNDEN KÜHLEN

FÜR 1 SPRINGFORM VON 22 cm Ø

30 g Butter
2 Eier (Größe M)
60 g Zucker, 50 g Mehl

FÜR CREME UND BELAG

400 g Erdbeeren
1/2 l Dickmilch
2 EL ungesalzene Pistazienkerne
50 g Puderzucker
2 EL Amaretto
2 EL Zitronensaft
2 Eiweiße (Eier: Größe M)
10 Blatt weiße Gelatine

AUSSERDEM

Backpapier und Fett für die Form
Vollmilchkuvertüre
2 EL Aprikosenkonfitüre
nach Belieben Puderzucker

● Elektro-Ofen auf 180 Grad vorheizen. Butter in einem Topf zerlassen. Eier trennen. Eigelbe mit der Hälfte des Zuckers cremig rühren. Eiweiße steif schlagen, restlichen Zucker einrieseln lassen. Mehl über die Eigelbcreme sieben und mit dem Eischnee vorsichtig unterheben. Zuletzt die flüssige Butter hinzufügen.

● Den Boden der Springform mit Backpapier auslegen, Ränder der Form einfetten. Teig einfüllen, im Ofen (Gas: Stufe 2) 15-20 Minuten backen. Form herausnehmen, Biskuit auf ein Kuchengitter stürzen, abkühlen lassen.

● Erdbeeren waschen und putzen. Die Hälfte in kleine Würfel, die andere Hälfte in dünne Scheiben schneiden. Erdbeeren kalt stellen. Pistazien grob hacken.

● Dickmilch mit der Hälfte des Puderzuckers, Amaretto und Zitronensaft verrühren. Eiweiße mit übrigem Puderzucker steif schlagen, unter die Dickmilch-Mischung heben.

● Eine Schüssel (20 cm Ø; Inhalt: 1 l) von innen anfeuchten und mit den Erdbeerscheiben auslegen.

● Gelatine in kaltem Wasser einweichen. In einer Tasse über dem heißen Wasserbad auflösen und unter die Creme rühren. Erdbeerwürfel und Pistazien untermischen.

● Creme in die mit Erdbeeren ausgelegte Schüssel füllen und 4 Stunden kühlen. Creme auf den Biskuitboden stürzen.

● Von der Kuvertüre mit einem Sparschäler dünne Schokospäne abhobeln. Den Teigrand mit der Konfitüre bestreichen. Die Torte am Rand und oben mit den Schokospänen verzieren und nach Belieben die oberen Schokospäne mit etwas Puderzucker bestäuben.

Aufwand:	★★
Zubereitungszeit:	1 Stunde
Backzeit:	20 Minuten
Pro Stück:	265 kcal/1120 kJ

Mürbeteig-Tortelettes

Frankreich

Für 8 Tortelettes

50 g Butter
100 g Zartbitter-Kuvertüre
100 g Schlagsahne
50 g Zucker
1 Ei und 1 Eigelb (Eier: Größe M)
Mark von 1/2 Vanilleschote
8 vorgebackene Tortelettes (Fertigprodukt)

AUSSERDEM

125 g Schlagsahne
rote Früchte der Saison (z. B. Erdbeeren, Johannisbeeren, Himbeeren)

- Elektro-Ofen auf 125 Grad vorheizen. Butter, Kuvertüre und Sahne in einer Schüssel über dem warmen Wasserbad erwärmen, bis die Butter und die Kuvertüre geschmolzen sind. In einer anderen Schüssel Zucker, Ei, Eigelb und Vanillemark mit dem Schneebesen cremig rühren. Die Schokoladenmischung darunter rühren. Die Masse in die vorgebackenen Tortelettes füllen.
- Tortelettes auf ein Backblech stellen und im Ofen (Gas: niedrigste Stufe) 15-20 Minuten backen. Herausnehmen und etwas abkühlen lassen.
- Sahne steif schlagen, nach Belieben auf die Törtchen spritzen oder getrennt dazu servieren. Die Früchte verlesen, waschen und putzen. Gut trockentupfen und auf den Törtchen anrichten.

Aufwand:	★
Zubereitungszeit:	20 Minuten
Backzeit:	20 Minuten
Pro Stück:	130 kcal/550 kJ

Gefüllter Hefestrudel

Italien

FÜR 1 BACKBLECH (30 x 40 cm)
300 g Mehl, 50 ml Milch
20 g Hefe (etwa 1/2 Würfel)
60 g Butter, 50 g Zucker
1 EL Vanillezucker
2 Eier (Größe M), 1 Prise Salz
FÜR DIE FÜLLUNG
150 g Rosinen, getrocknete Feigen, Aprikosen oder anderes Dörrobst, 100 ml Süßwein (Passito, Marsala) oder Wasser
100 g grob gemahlene Nüsse (Mandeln, Hasel- oder Walnüsse)
70 g Pinienkerne (ersatzweise Pistazien- oder Mandelkerne)
etwas abgeriebene Schale von je 1 unbehandelten Zitrone und Orange
1 TL Zimtpulver, 1 Prise Pfeffer
40 g Zartbitter-Schokolade oder 2 EL Kakaopulver
1 Ei (Größe M), 40 g Amaretti-Kekse (ca. 20 Stück; ersatzweise trockener Biskuit, Zwieback oder Semmelbrösel)
50 g Butter
AUSSERDEM
Mehl zum Arbeiten
Backpapier für das Blech
1 Eigelb (Ei: Größe M)
Puderzucker zum Bestäuben

● Mehl in eine Schüssel sieben, in der Mitte eine Mulde machen. Milch in einem Topf leicht erwärmen und die Hefe darin auflösen. In die Mulde gießen und mit etwas Mehl verrühren. Zugedeckt 30 Minuten an einem warmen Ort gehen lassen.

● Butter zerlassen, Zucker mit Vanillezucker mischen. Vorteig mit Eiern, Salz, Butter und Zucker zu einem glatten weichen Teig verkneten. Sobald der Teig nicht mehr klebt und sich vom Schüsselrand löst, zu einer Kugel formen. Mit etwas Mehl bestäuben, die Schüssel mit einem Küchentuch bedecken und an einem warmen, zugfreien Ort 1-2 Stunden gehen lassen, bis sich das Volumen verdoppelt hat.

● Inzwischen Rosinen oder Dörrobst in kleine Stücke schneiden, mindestens 30 Minuten in Süßwein oder Wasser einweichen.

● Kurz bevor der Teig fertig gegangen ist, die Füllung zubereiten. Die gemahlenen Nüsse und die grob gehackten Pinienkerne in eine Schüssel geben. Mit abgeriebener Zitronen- und Orangenschale, Gewürzen und Schokolade oder Kakao vermischen. Rosinen oder Dörrobst mit der Einweichflüssigkeit zugeben.

● Das Ei trennen. Eigelb in die Schüssel geben, Eiweiß zu Schnee schlagen. Amaretti-Kekse mit einem Teigroller oder im Mörser zerbröseln. Die Butter in einer Pfanne erhitzen und die Brösel darin hellbraun anrösten. Zu den übrigen Zutaten rühren und den Eischnee unterheben.

● Elektro-Ofen auf 175 Grad vorheizen. Blech mit Backpapier auslegen. Den Teig nochmal durchkneten und auf einem bemehlten Küchentuch quadratisch (ca. 40 x 40 cm) ausrollen (oder auf einer bemehlten Arbeitsfläche ausrollen und dann auf das Tuch legen). Die Füllung mit einem Spatel auf dem Teig verteilen, einen kleinen Rand lassen. Mithilfe des Tuchs den Teig samt Füllung zu einem Strudel einrollen und auf das Blech gleiten lassen.

● Strudel vorsichtig schneckenförmig eindrehen. Mit verquirltem Eigelb bepinseln und im Ofen (Gas: Stufe 2) etwa 45 Minuten backen. Wenn die Oberfläche zu braun wird, mit Alufolie bedecken.

● Zum Servieren den Strudel abkühlen lassen, bis er lauwarm ist, in Scheiben schneiden und mit Puderzucker bestäubt auftragen.

Aufwand:	★★
Zubereitungszeit:	55 Minuten
Backzeit:	45 Minuten
Pro Stück:	750 kcal/3150 kJ

Zimtkuchen

Spanien

BITTE BEACHTEN: KUCHEN MUSS 1 TAG ZIEHEN

FÜR 1 SPRINGFORM VON 24 cm Ø

4 Eier (Größe M)
200 g Zucker
200 ml Olivenöl
200 ml Milch
2 TL Zimtpulver
350 g Mehl
150 g ungeschälte gemahlene Mandeln
1 gehäufter 1 TL Backpulver

AUSSERDEM
Öl für die Form
Puderzucker zum Bestäuben

- Eier und Zucker mit dem Schneebesen des elektrischen Handrührgeräts schaumig rühren. Öl unter Rühren nach und nach zugeben, dann Milch und Zimt untermischen.
- Den Elektro-Ofen auf 180 Grad vorheizen. Mehl, Mandeln und Backpulver mischen und vorsichtig, aber gründlich unter die Ei-Zucker-Öl-Masse heben. Es soll ein glatter, geschmeidiger Teig entstehen.
- Springform mit Öl ausfetten und den Teig einfüllen. In den Ofen (Gas: Stufe 2-3) schieben. Den Kuchen 45-50 Minuten backen, dann noch einige Minuten im ausgeschalteten Ofen nachziehen lassen.
- Den Kuchen vor dem Servieren mit Puderzucker bestäuben, in Stücke schneiden und zu Kaffee oder Wein reichen.

Aufwand:	★
Zubereitungszeit:	20 Minuten
Backzeit:	50 Minuten
Pro Stück:	430 kcal/1800 kJ

Tipp

Wer sich scheut, dieses Rezept mit Olivenöl zu backen, nimmt stattdessen 220 g Butter. Noch besser schmeckt der Kuchen, wenn Sie ihn mit einer Schokoladenglasur überziehen oder gehackte Schokolade unter den Teig mischen.

Zitronen-Mandel-Tarte

Spanien

FÜR 1 TARTEFORM VON 28 cm Ø
150 g Mehl
1 Prise Salz
100 g eiskalte Butter
FÜR DEN BELAG
150 g Butter
3 Eier (Größe M)
100 g Zucker
2 unbehandelte Zitronen
100 g gemahlene Mandeln
AUSSERDEM
Mehl zum Arbeiten
Fett für die Form
Hülsenfrüchte zum Blindbacken
nach Belieben kandierte Zitronen und 50 g geschälte ganze Mandeln zum Garnieren

● Mehl in eine Schüssel sieben und mit dem Salz vermischen. Butter kurz im Mehl wenden und auf einer groben Reibe in das Mehl raspeln. Nach und nach etwas eiskaltes Wasser dazugeben und die Masse zusammendrücken, nicht kneten! Nur so viel Wasser zufügen, dass sich der Teig zu einer Kugel formen lässt. Teigkugel in Folie wickeln und mindestens 1 Stunde im Kühlschrank ruhen lassen.

● Elektro-Ofen auf 220 Grad vorheizen. Tarteform einfetten. Teig auf einer bemehlten Fläche in der Größe der Form dünn ausrollen. Teig in die Form legen. Rand etwas andrücken. Überstehende Ränder abschneiden.

● Hülsenfrüchte einfüllen und die Tarte im Ofen auf der unteren Schiene (Gas: Stufe 4) in 10-15 Minuten knusprig braun backen. Hülsenfrüchte entfernen.

● Inzwischen Butter in einem kleinen Topf schmelzen. Die Eier trennen. Die Eigelbe mit dem Zucker schaumig aufschlagen. Zitronen heiß abwaschen, trockenreiben. Auf einer feinen Raspel die Schale zu den Eiern reiben. Beide Zitronen auspressen und den Saft dazugießen.

● Die gemahlenen Mandeln und die flüssige Butter unterrühren. Die Eiweiße schaumig, aber nicht steif schlagen und unter die Masse ziehen. Die Zitronen-Mandel-Masse in den vorbereiteten Teigboden füllen. Die Tarte mit den kandierten Zitronen und nach Belieben mit den ganzen Mandeln belegen.

● Die Temperatur des Backofens auf 200 Grad (Gas Stufe 3) reduzieren und die Tarte etwa 30 Minuten backen. Falls sie zu dunkel werden sollte, mit Alufolie bedecken. Die Tarte aus der Form nehmen und vor dem Anschneiden abkühlen lassen.

Aufwand:	★
Zubereitungszeit:	40 Minuten
Backzeit:	45 Minuten
Pro Stück:	330 kcal/1380 kJ

Möhrenkuchen vom Blech

Schweiz

FÜR 1 TIEFES BACKBLECH VON 30 x 40 cm
500 g Möhren
300 g Datteln
250 g Mehl
2 TL Soda (Apotheke)
2 TL Zimtpulver
250 g Mandelblättchen
350 g brauner Zucker
4 Eier (Größe M)
1/4 l Olivenöl

FÜR DIE CREME
250 g Mascarpone
150 g Puderzucker
80 g weiche Butter

AUSSERDEM
Backpapier für das Blech
Zimtpulver zum Bestreuen

- Die Möhren putzen, schälen und raspeln. Datteln halbieren, entsteinen und hacken. Das Mehl mit Soda und Zimt vermischen. Möhren, Datteln, Mandeln, Mehlmischung und Zucker miteinander vermengen.
- Elektro-Ofen auf 175 Grad vorheizen. Die Eier schaumig schlagen, nach und nach das Öl dazugeben und weiterschlagen. Diese Creme mit der Möhrenmischung vermengen.
- Backblech mit Backpapier auslegen. Die Möhrenmasse darin glatt streichen und im Ofen (Gas: Stufe 2) 40-50 Minuten backen. 5 Minuten ausdampfen lassen, auf ein Kuchengitter stürzen und das Papier abziehen. Abkühlen lassen.
- Für die Creme alle Zutaten schaumig schlagen. Den Kuchen wenden und wellig mit der Creme bestreichen. Zimt dekorativ darüber sieben. Zum Servieren den Kuchen in Stücke teilen.

Aufwand:	★
Zubereitungszeit:	40 Minuten
Backzeit:	50 Minuten
Pro Stück:	370 kcal/1550 kJ

Tipp

Dekorieren Sie die Kuchenstücke ruhig mal mit ein paar jungen, geschälten Möhren (wie auf unserem Bild) – zum Anknabbern! Natürlich können Sie auch Marzipanmöhren verwenden.

Zitronen-Tarte

Frankreich

FÜR 1 SRINGFORM VON 26 cm Ø
200 g Mehl, 1 Prise Salz
80 g Zucker, 1 Ei (Größe M)
150 g Butter
FÜR DEN BELAG
3 Eier und 1 Eigelb (Eier: Größe M)
180 g Zucker
Saft und abgeriebene Schale von 3 unbehandelten Zitronen
125 g Crème fraîche
50 g flüssige Butter
AUSSERDEM
Fett und Backpapier für die Form, Mehl zum Arbeiten
Hülsenfrüchte zum Blindbacken
2 EL Zitronengelee
2 EL Zitronensaft
einige schöne Minzeblättchen
nach Belieben Puderzucker

● Die Teigzutaten mit den Knethaken des Handrührgeräts zu einem glatten Mürbeteig verarbeiten, in Folie gewickelt 1 Stunde kühlen. Elektro-Ofen auf 200 Grad vorheizen.

● Form einfetten. Den Teig auf Mehl kreisförmig dünn ausrollen und den Boden sowie 3 cm Rand der Form damit auskleiden. Backpapier auf den Teig legen und mit Hülsenfrüchten bedecken. Im Ofen (Gas: Stufe 3) etwa 15 Minuten backen. Papier und Hülsenfrüchte entfernen. Den Ofen auf 170 Grad (Gas: Stufe 2) schalten.

● Eier, Eigelb und Zucker schaumig rühren. Nach und nach Zitronensaft, die Hälfte der Zitronenschale, Crème fraîche und flüssige Butter zufügen. Creme auf dem Kuchen verteilen, mit Zitronenschale bestreuen. Im Ofen (Gas: Stufe 2) 45 Minuten backen.

● Zitronengelee mit Zitronensaft erhitzen, Oberfläche des heißen Kuchens damit bestreichen. Kuchen mit Minzeblättchen belegen und leicht glasieren. Mit Puderzucker bestäuben.

Aufwand:	★
Zubereitungszeit:	1 1/2 Stunden
Backzeit:	1 Stunde
Pro Stück:	355 kcal/1490 kJ

Zucchinikuchen

Italien

FÜR 1 SPRINGFORM VON 24 cm Ø

200 g kleine Zucchini
2 gestrichene TL Salz
125 g Sonnenblumenkerne
200 g Zucker, 4 Eier (Größe M)
1 TL Bourbon-Vanillezucker
Saft und abgeriebene Schale von 1/2 unbehandelten Zitrone
100 g Mehl
1 gehäufter TL Backpulver
je 1 Handvoll fein gehackte Petersilie und Basilikumstreifen

AUSSERDEM

Backpapier für die Form
Puderzucker zum Bestäuben

- Die Zucchini waschen, putzen und die Enden abschneiden. Die Zucchini grob in eine Schüssel raspeln, mit Salz bestreuen und Wasser ziehen lassen. Die Sonnenblumenkerne in einer Pfanne ohne Fett leicht anrösten, abkühlen lassen und in der Mandelmühle (oder im Mixer) mahlen.
- Zucker und Eier weißschaumig schlagen. Vanillezucker, Zitronensaft und -schale dazugeben. Das Mehl, das Backpulver, die gemahlenen Sonnenblumenkerne, die gut ausgedrückten Zucchiniraspel und die Kräuter unterrühren.
- Den Elektro-Ofen auf 180 Grad vorheizen. Die Springform mit Backpapier auslegen. Die Zucchini-Masse einfüllen. Kuchen im Ofen (Gas: Stufe 2) auf der untersten Schiene 45-50 Minuten backen. Kuchen herausnehmen und aus der Form lösen. Mit Puderzucker bestreuen und lauwarm oder kalt servieren.

Aufwand:	★
Zubereitungszeit:	30 Minuten
Kühlzeit:	50 Minuten
Pro Stück:	190 kcal/800 kJ

Variation

- Eine andere Geschmacksnote bekommt der Kuchen, wenn Sie statt Sonnenblumenkernen Kürbiskerne nehmen und je 1 Prise Nelkenpulver und gemahlenen Anis sowie 1 TL Zimtpulver in den Teig geben.

Zitronen-Kokos-Kuchen

England

FÜR 1 SPRINGFORM VON 24 cm Ø
250 g Mehl
80 g Zucker
180 g kalte Butter
1 Ei

FÜR DIE FÜLLUNG
7 EL Kokosraspel
350 g Clementinen
4 Eier
750 g Magerquark
abgeriebene Schale und Saft von 1 unbehandelten Zitrone
250 g Zucker
3 EL Speisestärke
2 EL Zitronenmarmelade

AUSSERDEM
Fett für die Form

- Mehl, Zucker, Butter in Flöckchen und das Ei zu einem glatten Teig verkneten. In Folie gewickelt etwa 30 Minuten kalt stellen. Den Backofen auf 175 Grad (Gas Stufe 2) vorheizen.
- Die Springform fetten. 2/3 des Teigs auf den Springformboden rollen und 10 Minuten vorbacken. Dann den Teigrest zur Rolle drehen, in die Form als Rand drücken.
- Für die Füllung 5 EL Kokosraspel in der Pfanne anrösten. Die Clementinen schälen, in Spalten teilen. Eier trennen. Eiweiße steif schlagen. Eigelbe, Quark, Zitronensaft, Zitronenschale, Zucker, Stärke und geröstete Kokosraspel verrühren. Eischnee unterheben.
- Auf dem Teigboden 3 EL Quarkmasse verteilen, die Clementinen auflegen, die übrige Quarkmasse darüber streichen. Im Ofen (Gas: Stufe 2) 60-70 Minuten backen.
- Den warmen Kuchen mit Marmelade bestreichen, mit Kokosraspeln bestreuen.

Aufwand:	★
Zubereitungszeit:	1 Stunde
Backzeit:	1 Stunde
Pro Stück:	290 kcal/1220 kJ

Orangenkuchen

Spanien

FÜR 1 SPRINGFORM VON 28 cm Ø

4 Eier (Größe M)
200 g Zucker
200 ml Sonnenblumen- oder Maiskeimöl
abgeriebene Schale von 1 unbehandelten Orange
1/4 l frisch gepresster Orangensaft
1 TL Zimtpulver
500 g Mehl (Type 405)
1 Päckchen Backpulver
1 Glas Orangenkonfitüre

AUSSERDEM

Öl und Mehl für die Form
Puderzucker zum Bestäuben

● Die Eier trennen. Eigelbe in einer Schüssel mit dem Zucker schaumig rühren. Nach und nach das Öl dazugießen und rühren, bis eine dickcremige Masse entstanden ist. Orangenschale und Orangensaft unterrühren.

● Den Backofen auf 180 Grad vorheizen. Zimt, Mehl und Backpulver vermischen. Nach und nach das Mehl über die Eimasse sieben und unterrühren. Die Eiweiße zu steifem Schnee schlagen und unter den Teig heben.

● Die Form einölen und mit etwas Mehl ausschwenken. Den Teig einfüllen, im Ofen (Gas: Stufe 2-3) etwa 50 Minuten backen und kurz in der Form abkühlen lassen. Den Kuchen aus der Form auf ein Backgitter stürzen und erkalten lassen.

● Die Orangenkonfitüre in einem kleinen Topf erwärmen. Den Kuchen zweimal quer durchschneiden und die beiden oberen Platten abheben. Den Boden mit der Hälfte der Orangenkonfitüre einstreichen, den zweiten Boden auflegen, ebenfalls mit Konfitüre einstreichen und mit der oberen Lage bedecken. Den Kuchen mit Puderzucker überstäuben.

Dazu Orangen- oder Mandeleis und etwas Sahne servieren.

Aufwand:	★
Zubereitungszeit:	30 Minuten
Backzeit:	50 Minuten
Pro Stück:	375 kcal/1580 kJ

Variation

● Aus einem Stück Backpapier einen Kreis in Kuchengröße ausschneiden, zweimal falten und ein dekoratives Muster einschneiden. Das Papier auseinanderfalten, als Schablone auf die Oberfläche des Kuchens legen und dick Puderzucker aufsieben. Das Papier vorsichtig abnehmen, dann erscheint das Muster auf dem Kuchen.

Mandelkuchen

Spanien

FÜR 1 SPRINGFORM VON 26 cm Ø

8 Eier (Größe M)
250 g Puderzucker
1 unbehandelte Orange
1 Msp. Zimtpulver
Mark von 1 Vanilleschote
250 g geschälte gemahlene Mandeln
50 g flüssige Butter
2 cl Orangenlikör

AUSSERDEM

Butter und Semmelbrösel für die Form
Puderzucker zum Bestäuben

● Elektro-Ofen auf 180 Grad vorheizen. Eier trennen. Die Eigelbe und den Puderzucker mit den Schneebesen des Handrührgeräts dickschaumig schlagen. Orange heiß waschen, abreiben. Saft auspressen. Orangenschale, Zimt und Vanillemark zu den Eiern geben. Abwechselnd Mandeln und flüssige Butter unterrühren. In einer zweiten Schüssel die Eiweiße steif schlagen, vorsichtig und gleichmäßig unter die Schaummasse ziehen.

● Die Springform einfetten und mit Bröseln ausstreuen. Die Mandel-Masse einfüllen, Oberfläche glatt streichen. Kuchen im Ofen (Gas: Stufe 2) 45-50 Minuten backen.

● Den Kuchen auf ein Kuchengitter stürzen und noch warm mit einer Mischung aus Orangensaft und -likör beträufeln. Dick mit Puderzucker bestäuben und möglichst am Backtag genießen.

Aufwand:	★
Zubereitungszeit:	30 Minuten
Backzeit:	50 Minuten
Pro Stück:	225 kcal/950 kJ

Info

Der Mandelkuchen kommt von der Insel Mallorca und heißt dort „Gato". In den Restaurants dort wird er auch als Dessert angeboten. Der Kuchen fällt beim Auskühlen wieder zusammen!

Haselnusskuchen

Italien

FÜR 1 SPRINGFORM VON 28 cm Ø

250 g Haselnusskerne (ersatzweise 250 g gemahlene Haselnüsse)
5 Eier (Größe M)
200 g Zucker
1 TL Backpulver
200 g flüssige Butter
150 g Mehl
80 g Semmelbrösel

AUSSERDEM

Fett für die Form
nach Belieben Puderzucker zum Bestäuben

- Elektro-Ofen auf 220 Grad vorheizen. Die Haselnüsse auf einem Backblech verteilen und so lange im heißen Ofen rösten, bis sich die braune Haut leicht abreiben lässt. Etwas abkühlen lassen, dann 2/3 davon fein mahlen, den Rest grob hacken. Die Temperatur auf 180 Grad (Gas: Stufe 2) reduzieren.
- Form fetten. Eier, Zucker und Backpulver mit dem elektrischen Handrührgerät cremig rühren. Nach und nach abwechselnd zerlassene Butter, Mehl, Semmelbrösel und die Nüsse unterrühren. Die Masse in die Form füllen und im Ofen etwa 1 Stunde backen. Auf einem Kuchengitter auskühlen lassen. Nach Belieben mit Puderzucker bestäuben und mit Schlagsahne servieren.

Aufwand:	★
Zubereitungszeit:	30 Minuten
Backzeit:	1 Stunde
Pro Stück:	365 kcal/1530 kJ

meine Familie & ich

Tipp

Das Rösten der Haselnüsse verstärkt das nussige Aroma. Achten Sie aber darauf, dass die Nüsse nicht zu braun werden, sonst schmecken sie bitter.

Linzer Törtchen

Österreich

FÜR 8 TORTELETTEFORMEN
250 g Butter, 1 Prise Salz
150 g Zucker
1 Päckchen Vanillezucker
2 Eigelbe (Eier: Größe M)
abgeriebene Schale von 1/2 unbehandelten Zitrone
je 1 Prise Zimt- und Nelkenpulver
250 g Mehl
150 g gemahlene Mandeln
1 Glas Johannisbeerkonfitüre
AUSSERDEM
Fett und Mehl für die Förmchen
Mehl zum Arbeiten
1 Eigelb, 3 EL Schlagsahne
8 Backoblaten von 6 cm Ø
Puderzucker zum Bestäuben

- Butter mit Salz, Zucker und Vanillezucker cremig schlagen. Eigelbe, Gewürze und Zitronenschale unterrühren. Mehl und Mandeln unterheben und den Teig verkneten. In Folie gewickelt 30 Minuten kalt stellen.
- Elektro-Ofen auf 200 Grad vorheizen. Förmchen einfetten und mit Mehl bestäuben. 3/4 des Teiges in 8 Stücke teilen, jedes Stück rund (Ø 12 cm) ausrollen und in die Förmchen legen. Boden der Tortelettes mehrfach einstechen. Eigelb mit Sahne verrühren und damit den Teigrand bestreichen.
- Die Oblaten auf die Tortelettes legen und großzügig mit Johannisbeerkonfitüre bestreichen. Restlichen Teig ausrollen, in 1 cm breite Streifen schneiden, gitterartig auf die Tortelettes legen und mit der restlichen Eimischung bestreichen.
- Tortelettes im Ofen (Gas: Stufe 3) etwa 5 Minuten backen, bei 160 Grad (Gas: Stufe 1-2) etwa 35 Minuten weiterbacken.
- Törtchen herausnehmen, sofort mit Puderzucker bestäuben und auf einem Kuchengitter auskühlen lassen.

Aufwand:	★★
Zubereitungszeit:	1 Stunde
Backzeit:	40 Minuten
Pro Stück:	605 kcal/2540 kJ

Chocolate Cake

USA

BITTE BEACHTEN: TORTENBODEN AM VORTAG BACKEN

FÜR 1 SPRINGFORM VON 22 cm Ø

100 g Zartbitter-Kuvertüre
6 Eier (Größe M), 120 g Butter
40 g Puderzucker, 1 Msp. Salz
2 EL Kakaopulver
180 g Zucker, 130 g Mehl

FÜR DIE CREME

230 g Butter, 125 g Kakaopulver
1 Prise Salz, 200 g Puderzucker
Mark von 1/2 Vanilleschote
200 g Schlagsahne

AUSSERDEM

Backpapier für die Form
Schokospäne zum Dekorieren

- Am Vortag für den Biskuit Kuvertüre hacken, über dem warmen Wasserbad schmelzen, abkühlen lassen. Eier trennen. Butter mit Puderzucker und Salz schaumig rühren. Nach und nach Eigelbe, lauwarme Kuvertüre und Kakaopulver darunter rühren.
- Elektro-Ofen auf 175 Grad vorheizen. Eiweiße steif schlagen, dabei den Zucker einrieseln lassen. 1/3 des Eischnees unter die Eigelbmasse rühren. Übrigen Eischnee und das gesiebte Mehl unterheben. Boden der Springform mit Backpapier belegen, den Teig einfüllen und glatt streichen.
- Biskuit im Ofen (Gas: Stufe 2) 45-50 Minuten backen. Am nächsten Tag Boden aus der Form lösen, dreimal quer durchschneiden, so dass 4 dünne Böden entstehen.
- Butter cremig rühren, nach und nach Kakaopulver, Salz, Puderzucker und Vanillemark zugeben. Die Sahne steif schlagen und unterrühren.
- Cremenge halbieren, auf 3 Böden streichen. Böden aufeinander setzen, mit dem 4. Boden abschließen. Torte mit der restlichen Creme ringsherum dick einstreichen. Nach Belieben mit Schokospänen garnien.

Aufwand:	★★
Zubereitungszeit:	1 Stunde
Backzeit:	50 Minuten
Pro Stück:	485 kcal/2040 kJ

Vanille-Schoko-Kuchen

Spanien

FÜR 1 SPRINGFORM VON 24 cm Ø
120 g dunkle Kuvertüre
120 g Marzipan-Rohmasse
150 g Zucker, 50 ml Milch
150 g zimmerwarme Butter
Mark von 2 Vanilleschoten
5 Eier (Größe M)
1 Prise Salz, 130 g Mehl
50 g gemahlene Mandeln

FÜR DIE GLASUR
200 g Aprikosenkonfitüre
350 g dunkle Kuvertüre
50 ml Milch, 50 g Schlagsahne
30 g Honig, 1 1/2 EL Zucker

AUSSERDEM
Backpapier für die Form
Speisestärke zum Bestreuen

- Backofen auf 175 Grad vorheizen. Den Springformboden mit Backpapier auslegen. Die Kuvertüre in kleine Stücke hacken.
- Marzipan-Rohmasse würfeln, mit 50 g Zucker und der Milch mit dem Handrührgerät glatt rühren. Butter hinzufügen, nicht zu schaumig schlagen.
- Eier trennen. Vanillemark unterrühren, nach und nach die Eigelbe. Eiweiße mit Salz und restlichem Zucker zu cremigem Schnee aufschlagen. Vorsichtig erst 1/3 unter die Marzipanmasse heben, dann den Rest.
- Mehl, Mandeln und Kuvertürestücke vermischen und unter den Teig ziehen. In die Form füllen und im Ofen (Gas: Stufe 2) 40-45 Minuten backen. Ein Tuch mit Speisestärke bestreuen, den Kuchen nach dem Backen aus der Form darauf stürzen und auskühlen lassen.
- Die Konfitüre aufkochen lassen und durch ein Sieb streichen. Den Kuchen damit bestreichen.
- Für die Glasur die Kuvertüre fein hacken. Alle übrigen Zutaten und 3 EL Wasser einmal aufkochen lassen, vom Herd nehmen, die Kuvertüre unter sanftem Rühren darin auflösen. Mit dem Mixstab einmal kurz durchmixen, so dass nicht zu viel Luft untergerührt wird. Die Masse auf 30 Grad abkühlen lassen und den Kuchen damit überziehen.
- Kuchen auf eine Platte heben und die Glasur fest werden lassen, eventuell mit Marzipan-Rosen dekorieren. Bis zum Servieren kühl (nicht im Kühlschrank) aufbewahren.

Aufwand:	★★
Zubereitungszeit:	1 Stunde
Backzeit:	45 Minuten
Pro Stück:	490 kcal/2060 kJ

Kürbis-Pie

USA	
FÜR 1 PIEFORM VON 26 cm Ø	
200 g Mehl	
100 g kalte Butter	
50 g Zucker	
1 Prise Salz	
1 Eiweiß (Ei: Größe M)	
FÜR DIE FÜLLUNG	
700 g Kürbis (500 g geputzt)	
100 g Schlagsahne	
100 ml Ahornsirup	
1 Pck. Vanille-Puddingpulver	
1/2 TL abgeriebene Zitronenschale	
4 Eier und 1 Eigelb (Eier: Größe M)	
AUSSERDEM	
Fett und Mehl für die Form	
Mehl zum Arbeiten	
Backpapier und Hülsenfrüchte zum Blindbacken	

- Alle Zutaten für den Boden rasch zum glatten Teig verkneten, zur Kugel formen. In Folie gewickelt 30 Minuten kühl stellen.
- Für die Füllung Kürbis schälen und würfeln. Kürbis mit etwas Sahne in einem Topf zugedeckt in etwa 30 Minuten weich dünsten. Den Kürbis auskühlen lassen.
- Ahornsirup, die restliche Sahne und das Puddingpulver zufügen und fein pürieren. Mit Zitronenschale würzen. Die ganzen Eier und das Eigelb unterrühren.
- Elektro-Ofen auf 175 Grad vorheizen. Die Pieform einfetten und mit Mehl bestäuben. Den Teig auf bemehlter Fläche dünn ausrollen und die Form damit auslegen, einen Rand hochziehen. Den Boden mit einer Gabel mehrmals einstechen. Backpapier auf den Teigboden legen und darauf die Hülsenfrüchte verteilen.
- Den Boden im Ofen (Gas: Stufe 2) etwa 15 Minuten vorbacken. Herausnehmen und die Temperatur auf 200 Grad (Gas: Stufe 3) erhöhen.
- Das Papier samt den Hülsenfrüchten entfernen. Die Kürbis-Füllung auf dem Boden verteilen und glatt streichen.
- Die Pie in weiteren 45 Minuten im Ofen fertig backen. Pie herausnehmen, auf einem Gitter auskühlen lassen.

Aufwand:	★
Zubereitungszeit:	45 Minuten
Backzeit:	1 Stunde
Pro Stück:	290 kcal/1220 kJ

Info

Gut geeignet für diesen Kuchen ist der aromatische Moschuskürbis, z. B. Muskat oder Butternut. Gartenkürbis schmeckt für dieses Rezept zu fad.

Servieren Sie zum Kuchen eiskalte Crème fraîche.

Orangen-Nougat-Kuchen

Spanien

FÜR 1 SPRINGFORM VON 26 cm Ø

- 5 Orangen (2 davon unbehandelt)
- 200 g Nougat am Stück
- 2 gestrichene TL Instant-Kaffee
- 4 Eier (Größe M)
- 250 g Butter
- 250 g Zucker
- 250 g Mehl
- 2 gestrichene TL Backpulver
- 3 EL Orangenmarmelade

AUSSERDEM

- Fett für die Form
- 1 Päckchen klarer Tortengus
- 20 g Zucker
- 4 cl Orangenlikör oder weißer Rum
- 2 EL Mandelblättchen

● Die unbehandelten Orangen heiß waschen und Schale abreiben. Den Saft von 1/2 Orange auspressen. Nougat zerbröckeln, im Wasserbad schmelzen. Kaffee in 1 EL heißem Wasser auflösen, ins Nougat rühren. Abkühlen lassen.

● Backofen auf 175 Grad vorheizen. Die Springform einfetten. 2 Eier trennen. Eiweiße steif schlagen. Butter mit 250 g Zucker cremig rühren, nach und nach die Eigelbe und die 2 ganzen Eier dazugeben. Zuletzt Mehl mit Backpulver mischen und unterrühren. Ist der Teig zu fest, Orangensaft dazugeben.

● Den Teig halbieren, in eine Hälfte flüssiges Nougat rühren, in die andere Hälfte die Orangenschale und den Saft von 1/2 Orange. Jeweils die Hälfte des Eischnees unterheben.

● Erst dunklen, dann hellen Teig in die Form füllen, mit einer Gabel marmorieren. Im Ofen (Gas: Stufe 2) 60-70 Minuten backen. Auf das Gitter stürzen, abkühlen lassen.

● Restliche Orangen so schälen, dass auch die weiße Haut entfernt wird. In dünne Scheiben schneiden und entkernen. Marmelade erwärmen, durch ein Sieb streichen, den Kuchen damit rundum einpinseln. Die Orangenscheiben überlappend auf den Kuchen legen.

● Den Tortenguss, wie auf der Packung beschrieben, mit 20 g Zucker und der angegebenen Wassermenge (minus 4 cl) zubereiten. Likör oder Rum einrühren. Guss über die Orangen auf dem Kuchen verteilen. Den Rand mit Mandelblättchen verzieren.

Aufwand:	★★
Zubereitungszeit:	20 Minuten
Backzeit:	1 Stunde 10 Minuten
Pro Stück:	425 kcal/1790 kJ

Früchtekuchen

England

BITTE BEACHTEN: KUCHEN MUSS 1 WOCHE DURCHZIEHEN

FÜR KASTENFORM 28 cm LÄNGE

je 100 g Rosinen und Korinthen, 150 g Früchte-Mix
4 cl Rum oder Weinbrand
250 g Butter (zimmerwarm)
200 g brauner Zucker
1 Prise Salz
abgeriebene Schale von 1 unbehandelte Zitrone
2 EL Zitronensaft
5 Eier (Größe M)
250 g Weizenmehl
100 g Weizenvollkornmehl
1 gehäufter TL Backpulver

ZUM VERZIEREN

125 g Puderzucker
3-4 EL Zitronensaft
50 g kandierte Kirschen
2 EL Mandelstifte

AUSSERDEM

Fett und Mehl für die Form

- Die Früchte in einer Schüssel mit dem Alkohol übergießen und gut durchmischen. Zugedeckt mindestens 2 Stunden durchziehen lassen.
- Inzwischen den Backofen auf 150 Grad vorheizen. Die Form einfetten und sorgfältig mit Mehl ausstäuben.
- Butter cremig rühren. Zucker, Salz, Zitronenschale und -saft hinzufügen und unterrühren. Die Eier einzeln dazugeben und alles zu einer cremigen Masse schlagen.
- Mehl und Backpulver mischen, auf die Masse häufen und mit einem Rührlöffel untermischen. Die eingeweichten Früchte unterrühren.
- Den Teig in die Form geben, glatt streichen. Im Ofen (Gas: Stufe 1) auf der unteren Schiene gut 2 Stunden backen. Falls die Oberfläche zu dunkel wird, mit Alufolie abdecken.
- Den Kuchen aus der Form lösen und auf einem Gitter auskühlen lassen.
- Für die Glasur den Puderzucker mit Zitronensaft zu einem dünnflüssigen Guss verrühren. Kuchen damit bestreichen und mit Kirschen und Mandeln verzieren. Gut trocknen lassen. Fest in Folie verpacken, 1 Woche durchziehen lassen. Kühl aufbewahren.

Aufwand:	★
Zubereitungszeit:	20 Minuten
Marinierzeit:	2 Stunden
Backzeit:	2 Stunden
Pro Stück:	130 kcal/550 kJ

Knödelhupf mit Kirschen

Österreich

BITTE BEACHTEN: KUCHEN MUSS ÜBER NACHT AUSKÜHLEN

FÜR 1 GUGELHUPFFORM

500 g Mehl, 20 g frische Hefe
1/4 l lauwarme Milch
60 g Zucker, 2 Eigelbe (Größe M)
80 g weiche Butter

FÜR FÜLLUNG UND GLASUR

400 g frische Kirschen (ersatzweise 1 Glas Sauerkirschen)

AUSSERDEM

Fett für die Form
200 g Puderzucker
6 EL Kirschwasser

- Mehl in eine große Schüssel geben, eine Mulde hineindrücken und die Hefe hineinbröckeln. Die Hefe in der Mulde mit etwas Milch und Mehl sowie 1 EL Zucker zum Vorteig verrühren und zugedeckt an einem warmen Ort zu doppelter Größe aufgehen lassen. Frische Kirschen waschen, trockentupfen und entsteinen. (Kirschen aus dem Glas abtropfen lassen.)
- Elektro-Ofen auf 180 Grad vorheizen. Vorteig mit übrigem Mehl, übriger Milch, restlichem Zucker, Eigelben und Butter zu einem glatten Teig verkneten. Daraus eine Rolle formen, diese in 30 Scheiben schneiden. Scheiben flach drücken, Kirschen darauf geben, Teig darumschlagen und zu Knödeln formen. Die Gugelhupfform gut einfetten und die Teigkugeln in zwei Lagen hineinschichten.
- Den Kuchen im Ofen (Gas: Stufe 2-3) auf der untersten Schiene 60-70 Minuten backen. Den Kuchen in der Form 5 Minuten ruhen lassen, auf ein Gitter stürzen und über Nacht auskühlen lassen.
- Puderzucker mit dem Kirschwasser verrühren und den Kuchen damit bestreichen.

Aufwand:	★★
Zubereitungszeit:	1 1/2 Stunden
Kühlzeit:	1 1/4 Stunden
Pro Stück:	340 kcal/1430 kJ

Toskanischer Weihnachtskuchen

Italien

BITTE BEACHTEN: KUCHEN MUSS 1 WOCHE DURCHZIEHEN

FÜR 1 SPRINGFORM VON 26 cm Ø

- *300 g Mandeln*
- *100 g getrocknete Feigen*
- *je 150 g gewürfeltes Orangeat und Zitronat*
- *je 1 TL Zimt- und Korianderpulver*
- *1 Prise Muskat*
- *1 unbehandelte Orange*
- *8 Stück Würfelzucker*
- *250 g Zucker, 200 g Mehl*

AUSSERDEM

- *Backpapier für die Form*
- *Puderzucker zum Bestäuben*

● Elektro-Ofen auf 200 Grad vorheizen. Die Mandeln kurz in kochendes Wasser geben, kalt abschrecken. Die Kerne aus den Häuten drücken und auf dem Blech im Ofen (Gas: Stufe 3) goldbraun rösten, grob hacken. Den Ofen auf 150 Grad (Gas: Stufe 1) schalten.

● Die Feigen würfeln. Mit den Mandeln, dem Orangeat und Zitronat sowie den Gewürzen mischen. Die Orange heiß waschen, abtrocknen, mit dem Würfelzucker die Schale abreiben. Die Orange auspressen.

● Den Zucker bei milder Hitze schmelzen, den Würfelzucker darin auflösen und so lange rühren, bis die Masse hellbraun ist. Die Mandel-Früchte-Mischung und den Orangensaft unterrühren, auf der Arbeitsfläche mit dem gesiebten Mehl verkneten.

● Springform mit Backpapier auskleiden. Teig hineindrücken und im Ofen (Gas: Stufe 1) etwa 30 Minuten backen. Auf einem Kuchengitter auskühlen lassen.

● Kuchen 1 Woche, in einer Dose oder in Alufolie luftdicht verpackt, durchziehen lassen. Vor dem Servieren mit Puderzucker bestäuben. Der Kuchen hält sich 1 1/2 bis 2 Monate.

Aufwand:	★
Zubereitungszeit:	1 Stunde
Backzeit:	30 Minuten
Pro Stück:	335 kcal/1410 kJ

Honigkuchen

Holland

BITTE BEACHTEN: KUCHEN MUSS 1 WOCHE ZIEHEN

FÜR KASTENFORM 30 cm LÄNGE

200 g zimmerwarme Butter
125 g brauner Zucker
3 Eier (Größe M), 1 Prise Salz
1 Fläschchen Rum-Aroma
250 g Honig
100 g gemahlene Haselnusskerne
400 g Roggenmehl
1 Päckchen Backpulver
1 TL Lebkuchengewürz
1 TL gemahlener Ingwer
2 EL Hagelzucker
2 EL Krokant

AUSSERDEM

Fett für die Form
Puderzucker zum Bestäuben

● Elektro-Ofen auf 175 Grad vorheizen. Die Form einfetten. Die Butter cremig rühren, dabei Zucker, Eier, Salz und Rum-Aroma zufügen. Honig löffelweise darunter ziehen.

● Haselnüsse in einer trockenen Pfanne anrösten. Abgekühlt mit Mehl, Backpulver, Lebkuchengewürz und Ingwer mischen, unter den Teig kneten.

● Den Teig in die Form füllen, mit Hagelzucker und Krokant bestreuen. Im Ofen (Gas: Stufe 2) auf der untersten Schiene 1 1/2 Stunden backen.

● Kuchen kurz in der Form ausdampfen lassen, auf ein Gitter stürzen und abkühlen lassen. In Folie luftdicht verpacken und mindestens 1 Woche ruhen lassen.

● Vor dem Servieren mit Puderzucker bestäuben. Der Kuchen hält sich gut 2 Wochen.

Aufwand:	★
Zubereitungszeit:	30 Minuten
Backzeit:	1 1/2 Stunden
Pro Stück:	335 kcal/1410 kJ

Lime Pie

USA

FÜR 1 PIEFORM VON 26 cm Ø

125 g kalte Butter
1 Ei, Salz
150 g Mehl

FÜR DIE CREME

6 Eigelbe (Eier: Größe M)
1 Dose gezuckerte Kondensmilch (400 g)
3 Tropfen Bittermandel-Aroma
6 Limetten

FÜR DAS BAISER

6 Eiweiße (Eier: Größe M)
150 g Zucker
2 Päckchen Vanillezucker

AUSSERDEM

Fett für die Form
Hülsenfrüchte zum Blindbacken

● Butterwürfel, Ei, Mehl und 1 Prise Salz zu einem glatten Teig verkneten. In Klarsichtfolie gewickelt etwa 2 Stunden kühlen.

● Elektro-Ofen auf 200 Grad vorheizen. Form einfetten. Teig zwischen Klarsichtfolie zu einem Kreis (28 cm Ø) ausrollen, die Form damit auslegen.

● Teigboden mit Hülsenfrüchten bedecken, im Ofen (Gas: Stufe 3) etwa 15 Minuten blindbacken. Herausnehmen, abkühlen lassen. Ofen eingeschaltet lassen.

● Die Eigelbe mit Kondensmilch und Mandel-Aroma cremig schlagen. Die Limetten auspressen und Saft unterrühren. Creme auf den Boden gießen und die Pie nochmals etwa 20 Minuten backen. Abkühlen lassen.

● Eiweiße steif schlagen, dabei nach und nach Zucker und Vanillezucker einrieseln lassen, bis der Eischnee schnittfest ist und glänzt. Baiser auf der Pie verteilen, mit dem Löffelrücken Spitzen ziehen, in etwa 5 Minuten goldbraun überbacken und sofort servieren.

Aufwand:	★
Zubereitungszeit:	20 Minuten
Backzeit:	40 Minuten
Pro Stück:	230 kcal/970 kJ

Joghurt-Grießkuchen

Syrien

FÜR 1 SPRINGFORM VON 26 cm Ø
125 g weiche Butter
150 g Puderzucker
abgeriebene Schale und Saft von 1 unbehandelten Orange
3 Eier (Größe M)
500 g Weichweizengrieß
1 TL Backpulver
150 g Vollmilchjoghurt
75 g gemahlene Mandeln
300 g Zucker
2 TL Zitronensaft
1/2 TL Orangenblütenwasser

AUSSERDEM
Fett für die Form
exotische Früchte (z. B. 2 Granatäpfel, 1 Baby-Ananas, 1 Papaya, 4 Feigen)
Orangensaft zum Beträufeln
Puderzucker zum Bestäuben
nach Belieben Minzeblättchen

- Elektro-Ofen auf 175 Grad heizen. Form einfetten. Butter mit Puderzucker, Orangenschale und Eiern schaumig aufschlagen. Grieß und Backpulver mischen, unterrühren. Joghurt und Mandeln vermischen, locker unterheben.
- Den Teig in die Form füllen und im Ofen (Gas: Stufe 2) etwa 40 Minuten backen. Herausnehmen und etwas abkülen lassen
- Inzwischen den Zucker mit 1/4 l Wasser und Zitronensaft aufkochen und abkühlen lassen. Dann das Blütenwasser zufügen.
- Den leicht abgekühlten Kuchen mit einem Holzstäbchen mehrmals einstechen und mit dem Sirup tränken. Zugedeckt beiseite stellen.
- Die Granatäpfel halbieren und die Kerne mit einem Löffel herauslösen. Die Ananas schälen, vierteln, den Strunk herausschneiden und das Fruchtfleisch in Stücke schneiden.
- Die Papaya schälen, halbieren und Kerne mit einem Löffel herausschaben. Das Fruchtfleisch in dünne Spalten schneiden. Die Feigen waschen und vierteln.
- Alle Früchte auf eine Schale legen und mit Orangensaft beträufeln. Zugedeckt 30 Minuten kalt stellen.
- Den Kuchen nach Belieben mit Puderzucker bestäuben, in Stücke schneiden und mit dem Obst und auf Tellern anrichten.Eventuell mit Minzeblättchen bestreuen.

Aufwand:	★
Zubereitungszeit:	40 Minuten
Backzeit:	40 Minuten
Pro Stück:	355 kcal/1480 kJ

Cheese Cake

USA

FÜR 1 SPRINGFORM VON 26 cm Ø
200 g Mehl, 50 g Zucker
1 Päckchen Vanillezucker
130 g kalte Butter
1 Eiweiß (Ei: Größe M)
FÜR DIE FÜLLUNG
600 g Doppelrahm-Frischkäse
1 TL abgeriebene Zitronenschale
150 g Zucker, 1 Päckchen Vanillezucker, 2 EL Mehl
3 Eier (Größe M)
3 EL Schlagsahne
AUSSERDEM
Fett und Backpapier für die Form, Hülsenfrüchte zum Blindbacken, 125 g Heidelbeeren (frisch oder TK)
250 g Erdbeeren, 1 Päckchen heller Tortenguss, 1 EL Zucker

- Aus Mehl, Zucker, Vanillezucker, Butter und Eiweiß einen glatten Teig kneten. In Folie gewickelt mindestens 1 Stunde kalt stellen.
- Elektro-Ofen auf 200 Grad vorheizen. Die Form einfetten. Teig ausrollen und die Form damit auskleiden. Rand 2-3 cm hochziehen. Teig mit Backpapier belegen, Hülsenfrüchte darauf streuen. Im Ofen (Gas: Stufe 3) etwa 10 Minuten vorbacken.
- Backpapier und Hülsenfrüchte entfernen. Temperatur auf 225 Grad (Gas: Stufe 4) erhöhen.
- Frischkäse mit Zitronenschale, Zucker, Vanillezucker, Mehl, Eiern und Sahne gut verrühren. Auf den gebackenen Teig streichen, im Ofen etwa 15 Minuten backen. Temperatur dann auf 100 Grad (Gas: niedrigste Stufe) reduzieren und Kuchen weitere 30-45 Minuten backen. Kalt stellen.
- Heidelbeeren verlesen. Oder Früchte auftauen. Erdbeeren waschen, putzen, vierteln oder halbieren, zuckern. Zugedeckt beiseite stellen.
- Erdbeeren und Heidelbeeren auf dem Kuchen verteilen. Aufgefangenen Fruchtsaft mit Saft oder Wasser auffüllen. Tortengusspulver mit Zucker mischen und nach Packungsangabe aufkochen. Über die Früchte gießen. Kalt stellen. Kuchen aus der Form lösen und servieren.

Aufwand:	★
Zubereitungszeit:	30 Minuten
Back- und Kühlzeit:	2 1/4 Stunden
Pro Stück:	380 kcal/1600 kJ

Peach Upside-Down Cake

USA

BITTE BEACHTEN: KUCHEN MUSS ÜBER NACHT ZIEHEN

FÜR 1 SPRINGFORM VON 26 cm Ø

6-8 Pfirsiche
100 g Pecannüsse oder Haselnüsse, 100 g Butter
120 g Zucker
1 Päckchen Vanillezucker
2 Eier (Größe M)
200 g Mehl, Salz
1/2 Päckchen Backpulver
1/4 l Buttermilch

AUSSERDEM
Backpapier für die Form
Puderzucker zum Bestäuben

- Pfirsiche erst in kochendes Wasser legen, dann in Eiswasser tauchen und die Haut abziehen. Früchte vierteln und entsteinen. Den Boden einer Springform mit Backpapier auslegen. Früchte darauf verteilen.
- Einige Nüsse zur Seite legen, restliche grob hacken. Elektro-Ofen auf 175 Grad vorheizen. Butter, Zucker, Vanillezucker und Eier schaumig rühren. Mehl mit 1 Prise Salz und Backpulver mischen und abwechselnd mit der Buttermilch unterrühren. Teig auf den Pfirsichen verteilen und mit gehackten Nüssen bestreuen.
- Kuchen im Ofen (Gas: Stufe 2) etwa 55 Minuten backen. Einige Minuten ruhen lassen, dann auf ein Kuchengitter stürzen, Backpapier abziehen. Zugedeckt über Nacht kühl stellen.
- Zum Servieren mit restlichen Nüssen garnieren und mit Puderzucker bestäuben.

Aufwand:	★
Zubereitungszeit:	25 Minuten
Backzeit:	55 Minuten
Pro Stück:	270 kcal/1130 kJ

Joghurtkuchen

Türkei

FÜR 1 SPRINGFORM VON 26 cm Ø

125 g Butter
500 g Joghurt
4 Eier (Größe M)
600 g Zucker
Mark von 1 Vanilleschote
1 Päckchen Backpulver
500 g Hartweizengrieß

AUSSERDEM

Butter und Brösel für die Form
Saft von 1 Zitrone
50 g Kokosraspel

- Butter schmelzen. Joghurt, Eier, 200 g Zucker und Vanillemark mischen. Butter zugeben. Alles mit dem Schneebesen gründlich verrühren, bis sich der Zucker vollständig aufgelöst hat. Elektro-Ofen auf 175 Grad vorheizen.
- Backpulver unter den Grieß mischen, über die Zutaten in der Schüssel sieben und gründlich untermischen. Die Form einfetten und mit Bröseln ausstreuen. Den Teig einfüllen und den Kuchen im Ofen (Gas: Stufe 2) etwa 45 Minuten backen.
- Bei starker Hitze den restlichen Zucker mit 3/4 l Wasser aufkochen und ohne Deckel etwa 5 Minuten kochen lassen. Zitronensaft durch ein Sieb dazugießen, unterrühren, den Sirup kalt werden lassen.
- Den Kuchen 1 Minute abkühlen lassen. In der Form auf eine größere Platte stellen, um auslaufende Zuckerlösung aufzufangen. Den Kuchen nach und nach mit der Zuckerlösung tränken. Den Kuchen abkühlen lassen und noch in der Form in Quadrate schneiden. Die Kuchenstücke auf eine Kuchenplatte legen, mit den Kokosraspeln bestreuen und servieren.

Aufwand:	★
Zubereitungszeit:	40 Minuten
Kühlzeit:	45 Minuten
Pro Stück:	420 kcal/1760 kJ

Brownies

USA

FÜR 1 BACKBLECH (40 x 40 cm)
200 g Zartbitter-Schokolade
160 g Pecannuss- oder Walnusskerne
150 g Mehl
50 g Speisestärke
320 g Butter, 210 g Zucker
Mark von 1 Vanilleschote
1 Prise Salz
5 Eier (Größe M)
200 g Zartbitter- oder Vollmilch-Kuvertüre
3 EL Kakaopulver

AUSSERDEM
Fett für das Blech

- Elektro-Ofen auf 150 Grad vorheizen. Das Blech einfetten. Die Schokolade in kleine Stücke brechen und über dem warmen Wasserbad schmelzen.
- Die Nüsse grob hacken. Mehl, Nüsse und Speisestärke vermischen. Butter, Zucker, Vanillemark, Salz und die lauwarme Schokolade cremig rühren. Nach und nach die Eier unterrühren.
- Die Mehl-Nuss-Mischung nach und nach unter die Schokoladenmasse heben. Den Teig auf das Blech füllen und im Ofen (Gas: Stufe 1) etwa 50 Minuten backen.
- Den Kuchen aus dem Ofen nehmen, etwas abkühlen lassen.
- Die Kuvertüre über dem warmen Wasserbad schmelzen und die Kuchenoberfläche damit bestreichen.
- Sobald die Glasur fest geworden ist, Papierstreifen kreuz und quer darauf legen, den Kuchen mit Kakaopulver bestäuben. Den Kuchen mit einem scharfen Messer, das immer wieder in heißes Wasser getaucht wird, in Stücke schneiden.

Aufwand:	★
Zubereitungszeit:	40 Minuten
Kühlzeit:	50 Minuten
Pro Stück:	480 kcal/2020 kJ

Venezianische Krapfen

Italien

Für etwa 15 Stück

60 g Rosinen, 4 EL Grappa
250 g Mehl
20 g Hefe (1/2 Würfel)
150 ml Milch
40 g feiner Zucker
1 Ei (Größe L)
1 Prise Zimtpulver, 1 Prise Salz
abgeriebene Schale von
1 unbehandelten Zitrone

AUSSERDEM

Backpapier, Öl zum Ausbacken
3 Päckchen Vanillezucker zum
Bestäuben

- Die Rosinen in Grappa etwa 1 Stunde einweichen.
- Das Mehl in eine Schüssel geben, eine Mulde ins Mehl drücken. Die Hefe hineinbröckeln, mit etwas erwärmter Milch verrühren und den Vorteig 30 Minuten an einem warmen Ort gehen lassen.
- Alle übrigen Zutaten sowie die Rosinen samt Grappa zufügen und mit den Knethaken des Handrührgeräts zu einem glatten Teig verkneten. Schüssel mit einem Tuch bedecken und den Teig 1-2 Stunden gehen lassen, bis sich sein Volumen verdoppelt hat. Teig noch mal durchkneten.
- Aus dem Teig mit 2 Esslöffeln kleine, gleichmäßige Bällchen formen, auf ein mit Backpapier ausgelegtes Blech setzen. Noch mal 10-15 Minuten gehen lassen.
- Inzwischen Öl auf 160 Grad erhitzen. Die Bällchen darin portionsweise in 4-5 Minuten ausbacken. Fertige Bällchen mit einer Schaumkelle herausnehmen, auf Küchenpapier abtropfen lassen. Mit Vanillezucker bestreuen. Sofort servieren.

Aufwand:	★
Zubereitungszeit:	1 Stunde
Ruhezeit:	3 Stunden
Pro Stück:	140 kcal/590 kJ

Teigkringel

Italien

Für etwa 16 Stück

250 g Weizenmehl
1/2 Päckchen Trockenhefe
50 g Zucker
100 ml Milch
100 g Butter
1 Ei (Größe M)
1 Prise Salz
AUSSERDEM
Mehl zum Arbeiten
Backpapier

● Mehl in eine Schüssel sieben, mit Hefe und Zucker vermischen. In die Mitte eine Mulde drücken. Milch in einem Topf erwärmen und die Hälfte der Butter darin schmelzen. In die Mulde gießen und mit Ei und Salz zu einem glatten, weichen Teig verkneten. Eventuell noch etwas Mehl einarbeiten. Sobald der Teig nicht mehr klebt, zu einer Kugel formen. Mit etwas Mehl bestäuben, die Schüssel mit einem Küchentuch bedecken und an einem warmen Ort 3-4 Stunden gehen lassen.

● Elektro-Ofen auf 160 Grad vorheizen. Das Backblech mit Backpapier auslegen. Den Teig durchkneten und kleine Portionen abstechen. Auf einer bemehlten Arbeitsfläche fingerdicke Rollen von etwa 20 cm Länge rollen. Die Teigrollen zu einem Kringel (Kreis) formen und die beiden Enden zusammendrücken. Auf das Blech legen. Mit einem Tuch bedecken und nochmal 30 Minuten gehen lassen.

● Teigkringel im Ofen (Gas: Stufe 1-2) etwa 20 Minuten backen. Abkühlen lassen. Restliche Butter schmelzen und die Kringel damit einpinseln. Erneut in den Ofen schieben und in etwa 5 Minuten goldbraun, aber nicht zu dunkel werden lassen. Auf dem Blech auskühlen lassen.

Aufwand:	★
Zubereitungszeit:	1 Stunde
Ruhezeit:	4 1/2 Stunden
Backzeit:	25 Minuten
Pro Stück:	120 kcal/500 kJ

Honigkugeln

Griechenland

FÜR ETWA 60 STÜCK

1 TL Trockenhefe

70 g Zucker, 450 ml Milch

70 g Butter, 1 Prise Salz

3 Eier (Größe M), 600 g Mehl

FÜR DEN SIRUP

375 g Honig, 375 g Zucker

AUSSERDEM

Pflanzenöl zum Frittieren

100 g gehackte Pistazienkerne

- Hefe mit 1 Prise Zucker in 75 ml warmem Wasser auflösen, an einen warmen Ort stellen. Milch aufkochen, vom Herd nehmen.
- Butter, übrigen Zucker und Salz in einer Schüssel mischen, mit heißer Milch übergießen. Umrühren, bis alles aufgelöst ist und abkühlen lassen.
- Eier cremig schlagen. Sobald die Milchmischung lauwarm ist, die Hefemischung, dann die Eier unterrühren und das Mehl einarbeiten, bis ein dicker Teig entsteht. Den Teig zugedeckt mindestens 1 Stunde bei Zimmertemperatur gehen lassen.
- Honig und Zucker mit 190 ml Wasser in einem Topf etwa 5 Minuten sprudelnd kochen. Umrühren, abschäumen und abkühlen lassen.
- Öl auf 170 Grad erhitzen. (Es ist heiß genug, wenn an einem ins Fett getauchten Holzlöffelstiel Bläschen aufsteigen.) Einen Teelöffel in kaltes Wasser tauchen. 1 gestrichenen TL Teig abstechen und mit einem zweiten angefeuchteten Löffel in das Öl geben. Teigkugeln im Öl von allen Seiten goldbraun backen.
- Die Kugeln mit der Schaumkelle herausnehmen, auf Küchenpapier abtropfen lassen. Noch warm in den Sirup tauchen, mit Pistazien bestreuen und servieren.

Aufwand:	★★
Zubereitungszeit:	2 Stunden
Pro Stück:	110 kcal/460 kJ

Nusspastetchen

Griechenland

BITTE BEACHTEN: PASTETEN MÜSSEN ÜBER NACHT ZIEHEN

FÜR 1 BACKBLECH (30 x 40 cm)

FÜR DEN SIRUP

250 g Honig, 500 g Zucker

abgeriebene Schale und Saft von 2 unbehandelten Orangen

1 Zimtstange

FÜR DIE PASTETEN

200 g Walnusskerne

160 g Mandelkerne

140 g grüne Pistazienkerne

500 g Fillo-Teig (Fertigprodukt)

150 g zerlassene Butter

2 Eier (Größe M)

AUSSERDEM

Backpapier für das Blech

- 400 ml Wasser, Honig, Zucker, Orangenschale, Orangensaft und die Zimtstange bei großer Hitze kochen, bis der Sirup eindickt. Umrühren und abkühlen lassen.
- Elektro-Ofen auf 200 Grad vorheizen. Walnüsse, Mandeln und Pistazien auf ein Backblech geben und im Ofen (Gas: Stufe 3) 3-4 Minuten rösten. Sobald die Nüsse goldbraun sind, aus dem Ofen nehmen. Fein hacken und in einer Schüssel mischen. Nussmischung in 12 Portionen aufteilen. Ofen auf 150 Grad (Gas: Stufe 1) stellen. Blech mit Backpapier auslegen.
- 1 Fillo-Teigblatt auf der Arbeitsfläche ausbreiten. Mit etwas zerlassener Butter, dann mit etwas verquirltem Ei bestreichen. Ein zweites Teigblatt darauf legen, mit Butter, dann mit Ei bestreichen.
- Eine der 12 Nussportionen auf das obere Teigblatt streuen, dabei an einer Schmalseite einen Streifen von 1 cm frei lassen. Den Teig wie eine Biskuitrolle einrollen, dabei nach jeder Umdrehung mit Ei und Butter bestreichen. Die Außenseite ebenfalls mit Ei bepinseln und die Rolle vorsichtig auf das Blech setzen.
- Mit den übrigen Teigblättern ebenso verfahren, bis alle Zutaten aufgebraucht sind.
- Die Rollen im Ofen (Gas: Stufe 1) etwa 5 Minuten backen, dann die Temperatur auf 130 Grad (Gas: niedrigste Stufe) reduzieren. Weiterbacken, bis die Pastetchen goldbraun sind. Rollen abkühlen lassen.
- Jede Rolle in 3 Teile schneiden. Die Teile mit dem Sirup übergießen und über Nacht darin tränken.

Aufwand:	★★
Zubereitungszeit:	1 Stunde
Backzeit:	30 Minuten
Pro Stück:	250 kcal/1050 kJ

Frauennabel

Türkei

FÜR ETWA 20 STÜCK
350 g Zucker
Saft von 1/2 Zitrone
50 g Butter
1 Prise Salz
150 g Mehl
2 Eier (Größe M)
AUSSERDEM
1/2 l Sonnenblumenöl zum Ausbacken
etwas Öl zum Arbeiten
50 g fein gehackte Pistazien

- Zucker mit 1/2 l Wasser aufkochen, bis sich der Zucker aufgelöst hat. Den Zitronensaft durch ein Sieb dazugießen, unterrühren und den Sirup erkalten lassen.
- 1/4 l Wasser, Butter und Salz in einem Topf aufkochen. Das Mehl auf einen Teller sieben und dann auf ein Mal zur Butter geben und unterrühren, so dass ein dicker, zäher Teig entsteht. Diesen Teig 3-4 Minuten rühren, bis er sich als Kloß vom Topfboden löst.
- Den Teig in eine Schüssel geben und leicht abkühlen lassen. Mit den Schneebesen des Handrührgeräts nacheinander die Eier darunter rühren.
- Das Öl in einem großen Topf stark erhitzen, bis am Stiel eines Kochlöffels Bläschen aufsteigen.
- Jeweils 1 knappen EL Teig abnehmen und mit eingeölten Händen daraus Bällchen formen. Die Bällchen etwas flach drücken und in die Mitte mit dem Finger einen „Nabel" eindrücken.
- Die Teigteile im Öl von beiden Seiten goldgelb frittieren, kurz im Sirup wenden und auf eine Platte legen. In die Mitte der Plätzchen die Pistazien streuen.

Aufwand:	★★
Zubereitungszeit:	45 Minuten
Pro Stück:	160 kcal/670 kJ

Vogelnester

Türkei

FÜR 1 BACKBLECH (30 x 40 cm)
500 g Fillo-Teigblätter
(1 Paket; aus dem griechischen Lebensmittelladen)
6 EL Butterschmalz
1 TL Speisestärke
1 Eiweiß (Ei: Größe M)
50 g Zucker
je 70 g gehackte Walnüsse und geschälte gehackte Mandeln
1 EL Rosenwasser (Gül suyu; aus türkischem Lebensmittelladen oder aus der Apotheke)
25 g gehackte Pistazienkerne
FÜR DEN SIRUP
200 g Zucker, 1 EL Zitronensaft
1 EL Orangenblütenwasser (aus Reformhaus oder Feinkostladen)
1 EL Rosenwasser (Gül suyu)
AUSSERDEM
Fett für das Blech

- Den Teig auf der Arbeitsfläche ausbreiten und mit einem trockenen Küchentuch bedecken. Ein zweites Tuch in warmes Wasser tauchen, gut auswringen und auf das trockene legen. 10 Minuten ruhen lassen. Butterschmalz schmelzen. Die Arbeitsfläche neben dem Teigstapel leicht mit Speisestärke einreiben.
- Das erste Fillo-Teigblatt vorsichtig auf die Seite ziehen und leicht mit Butterschmalz einpinseln. Das zweite Teigblatt vorsichtig auf das erste ziehen und einpinseln. Bis auf das letzte Fillo-Teigblatt so fortfahren.
- Mit einem scharfen, langen Messer den Stapel Teigblätter in Quadrate von 10 cm Kantenlänge schneiden. Alle Quadrate bis zur Verwendung mit einem trockenen Tuch bedecken. Das Blech einfetten. Elektro-Ofen auf 175 Grad vorheizen.
- Das Eiweiß zu Eischnee schlagen, dabei nach und nach den Zucker einrieseln lassen. Nüsse, Mandeln und Rosenwasser untermischen. In die Mitte jedes Quadratstapels 1 gehäuften TL Füllung geben. Die Teigecken darüber legen und leicht auf die Füllung drücken. Auf diese Weise alle Quadratstapel füllen.
- Stapel dicht nebeneinander auf das Blech setzen, im Ofen (Gas: Stufe 2) etwa 30 Minuten backen. Hitze auf 150 Grad (Gas: Stufe 1) reduzieren und die Nester weitere 15 Minuten backen.
- Inzwischen Zucker mit 375 ml Wasser bei starker Hitze etwa 10 Minuten kochen. Zitronensaft durch ein Sieb dazugießen, Orangenblüten- und Rosenwasser unterrühren. Den Zuckersirup kalt werden lassen und über das heiße Gebäck träufeln. Nach dem Abkühlen in die Mitte jedes Nestchens Pistazien streuen.

Aufwand:	★★
Zubereitungszeit:	40 Minuten
Kühlzeit:	45 Minuten
Pro Stück:	500 kcal/2100 kJ

Zitronenschnitten

Spanien

FÜR 1 BACKBLECH (30 x 40 cm)
200 g Butter
200 g Zucker
4 Eier (Größe M)
abgeriebene Schale von 1
unbehandelten Zitrone
200 g Mehl
1/2 Päckchen Backpulver
150 g Puderzucker und
Saft von 1 Zitrone
AUSSERDEM
Backpapier für das Blech
Puderzucker zum Bestäuben

- Elektro-Ofen auf 200 Grad vorheizen. Backblech mit Backpapier belegen. Die Butter mit dem Zucker und den Eiern schaumig schlagen. Zitronenschale, Mehl und Backpulver unterheben. Den Teig dünn auf das Blech streichen und im Ofen (Gas: Stufe 3) etwa 15 Minuten backen.
- 150 g Puderzucker mit dem Zitronensaft verrühren. Den Kuchen aus dem Ofen nehmen, das Backpapier abziehen und den Kuchen noch heiß mit der Zitronenglasur bepinseln.
- Den Kuchen mit Puderzucker bestäuben, in gleich große Stücke schneiden und servieren.

Aufwand:	★
Zubereitungszeit:	15 Minuten
Backzeit:	15 Minuten
Pro Stück:	100 kcal/420 kJ

Baklava

Türkei

FÜR 1 VIERECKIGE FORM
60 g Pistazienkerne
100 g Walnusskerne
100 g gemahlene Mandeln
100 g Butter
400 g Yufka-Teig
(10-12 Blätter; in türkischen
Lebensmittelläden; ersatzweise
300 g TK-Blätterteig)
300 g Zucker, Saft von
je 1/2 Orange und Zitrone
10 Aprikosen
Melissezweige zum Verzieren

- Pistazien mahlen, Walnüsse fein hacken. Mit Mandeln vermischen.
- Butter schmelzen. Form damit auspinseln. Yufka-Teig passend zur Form zuschneiden. 5-6 Blätter dünn mit Butter bepinseln, übereinander in die Form legen. Nüsse darauf verteilen. Restliche Teigblätter bepinseln, übereinander auf die Nüsse legen. Obere Teigschicht mit einem scharfen Messer in Rauten teilen. Mit einem feuchten Küchentuch bedecken, 30 Minuten ruhen lassen.
- Elektro-Ofen auf 200 Grad vorheizen. Kuchen im Ofen (Gas Stufe 3) auf der unteren Schiene 25-30 Minuten backen.
- In der Zwischenzeit Zucker mit Orangen- und Zitronensaft und 1/4 l Wasser aufkochen, bis sich der Zucker gelöst hat. 2-3 Minuten einköcheln lassen.
- Kuchen herausnehmen. Kurz ruhen lassen. Vom Sirup 4 EL abnehmen, Rest über den Kuchen träufeln. Beiseite stellen.
- Aprikosen waschen, vierteln und entsteinen. Im übrigen Sirup 1-2 Minuten dünsten. Zugedeckt beiseite stellen.
- Kuchen in Stücke teilen. Mit Früchten und Melisse anrichten.

Aufwand:	★ ★
Zubereitungszeit:	50 Minuten
Backzeit:	30 Minuten
Pro Stück:	880 kcal/ 3700 kJ

Kaffee, Tee, Kakao – die genus

Ohne sie würden Kuchen, Torten und Gebäc

Beim Röstvorgang werden die Kaffeebohnen unter Druck trocken erhitzt, wodurch das in den Bohnen enthaltene Wasser verdampft und sich das spezifische Kaffee-Aroma bildet. Der Geschmack hängt von der Bohnensorte bzw. der jeweiligen Mischung der Bohnensorten und dem Grad der Röstung ab. Für Espresso oder Cappuccino zum Beispiel röstet man die Bohnen stärker und mahlt sie besonders fein.

Übrigens: Für den so genannten Schonkaffee wird den Kaffeebohnen beim Rösten das Koffein weitgehend entzogen, so dass sich der Kaffee auch für koffeinempfindliche Konsumenten eignet.

Kaffee – nicht nur für Müßiggänger

Afrika, genauer gesagt die Hochebene Äthiopiens, gilt als die Ur-Heimat des Kaffees. Heute wird er in vielen tropischen Ländern rund um die Welt angebaut. Brasilien gilt inzwischen als der größte Kaffee-Produzent. Nach der Ernte werden die Bohnen enthülst und getrocknet. In Säcken verpackt werden sie verschifft und erst in den jeweiligen Importländern geröstet.

Schwarzer Tee – das Ri

Schwarzer Tee, der bei uns angeboten wird, wie „Assam", „Darjeeling" oder „Ceylon", stammt vorwiegend aus Indien und Sri Lanka, wo er in großem Stil angebaut wird. Auf den Teeplantagen wird dreimal pro Jahr geerntet und das grundsätzlich von Hand. Die erste Pflückung wird „First Flush" genannt, diese Ernte gilt als die qualitativ hochwertigste.
Nach dem Trocknen wird der Tee „gerollt", wobei die Blattzellen aufgebrochen werden, anschließend fermentiert und nochmals getrocknet. Schließlich werden die Blätter über Rüttelsiebe nach Blattgrößen sortiert.
In Teekisten verpackt, tritt er seine Reise per Schiff in alle Welt an, natürlich auch zu uns. Vor der Freigabe prüft man ihn u. a. noch auf

ollen Begleiter

ur halb so gut schmecken

e für alle Genießer

Rückstände von Pflanzenschutzmitteln. Anschließend wird er abgepackt oder vorher noch aromatisiert, wie beispielsweise der berühmte Klassiker „Earl Grey", dem man die zitrusduftenden Bergamotte-Aromen zufügt. Auch Zitronen-, Orangen- oder Schwarze-Johannisbeer-Tee sind sehr beliebte Sorten.

So wird schwarzer Tee gebrüht: Für sechs Tassen Tee nimmt man einen gehäuften Teelöffel Teeblätter. Das Wasser sprudelnd kochen und sofort über den Tee gießen. Nach 3-4 Minuten entfaltet er seine belebende Wirkung. Danach lösen sich die Gerbstoffe aus den Blättern, die beruhigend wirken, doch der Tee schmeckt bitter.

Heißer Kakao – für kleine und große Kinder

Man nennt sie das braune Gold der Tropen: Kakaobohnen, die wertvollen Kerne der Kakaofrucht. Gleich nach der Ernte werden die Früchte geöffnet und die Bohnen mit dem Fruchtfleisch, in das sie eingebettet sind, per Hand herausgelöst. In Kästen zwischen Bananenblättern geschichtet, fermentieren sie etwa 5-7 Tage in der Tropensonne und entwickeln dabei das typische Schokoladenaroma. Nun erst befreit man die Kakaobohnen vom Fruchtfleisch und trocknet sie etwa 1 Woche wiederum in der Sonne. Bei diesem Prozess verlieren die „Gold-Nuggets" bis zu 94 Prozent ihrer Feuchtigkeit, gewinnen in gleichem Maß aber an Aroma. In Jutesäcken gehen die duftenden Bohnen dann auf die Reise zu Kakao- und Schokoladenproduzenten in aller Welt. Hier werden die verschiedenen Sorten gemischt, geröstet und zu staubfeinem Kakaopulver vermahlen – die Grundlage für den köstlichen Kakao.

Pikante Bäckerei

Wer sagt denn, dass Kuchen immer süß sein muss? Wie köstlich gesalzener Hefe-, Mürbe- oder Blätterteig mit knackigem Gemüse, würzigem Käse und herzhaftem Schinken belegt oder gefüllt sein kann, erfahren Sie auf den folgenden Seiten

Südstaaten-Pie

Rezept siehe folgende Seite

Südstaaten-Pie

FÜR 1 PIEFORM VON Ø 26 cm
400 g Mehl
200 g kalte Butter
2 Eiweiße (Eier: Größe M), Salz
FÜR DIE FÜLLUNG
1 Zwiebel, 1 Knoblauchzehe
1 rote Peperoni
1 Bund Lauchzwiebeln
1/2 TL getrockneter Thymian
1 EL Butter
400 g stückige Tomaten (Fertigprodukt)
2 EL Tomatenmark
350 g ausgelöste Flusskrebsschwänze (Kühlregal)
Salz, Cayennepfeffer
60 g Schlagsahne
AUSSERDEM
Fett für die Form

- Mehl, Butter, Eiweiße und etwas Salz zu einem Teig verkneten, 30 Minuten kühl stellen.
- Zwiebel und Knoblauch abziehen, fein würfeln. Peperoni putzen, entkernen, abbrausen. Lauchzwiebeln putzen und waschen. Beides in Streifen schneiden.
- Zwiebel, Knoblauch und Thymian in Butter andünsten. Peperoni und Tomaten zufügen, etwa 10 Minuten dünsten. Das Tomatenmark einrühren. Elektro-Ofen auf 200 Grad vorheizen. Die Form einfetten.
- 1/3 Teig beiseite legen. Rest ausrollen, in die Form legen, einen Rand formen. Übriges Teigdrittel ausrollen, in 1 cm breite Streifen schneiden.
- Krebse und Lauchzwiebeln zur Sauce geben. Mit Salz und Cayennepfeffer würzen und auf den Teig in die Form geben.
- Teigstreifen gitterförmig auf die Füllung legen und dünn mit Sahne bestreichen. Pie im Ofen (Gas: Stufe 3) etwa 30 Minuten backen.

Aufwand:	★
Zubereitungszeit:	50 Minuten
Backzeit:	30 Minuten
Pro Stück:	905 kcal/3800 kJ

Würzige Käsewähe

FÜR 1 PIEFORM VON 26 cm Ø
10 g frische Hefe
1/8 l Milch
300 g Mehl
40 g Butter, 1/2 TL Salz
FÜR DEN BELAG
1 großes Bund Lauchzwiebeln
2 EL Butter, 3 Eier (Größe M)
150 g Joghurt
2 EL Mehl, je 100 g geriebener Emmentaler und Appenzeller
Salz, Pfeffer
Paprikapulver (edelsüß)
AUSSERDEM
Fett für die Form
Mehl zum Arbeiten

- Hefe in lauwarmer Milch auflösen. Mit Mehl, Butter, Salz zu einem glatten Teig verkneten. Teig an einem warmen Ort zugedeckt 30 Min. gehen lassen, bis sich das Volumen verdoppelt hat.
- Lauchzwiebeln putzen, in Ringe schneiden. 1 EL Butter erhitzen, Lauchzwiebeln darin glasig dünsten.
- Eier trennen. Eigelbe mit Joghurt, Mehl und dem Käse verquirlen. Die Masse kräftig mit Salz, Pfeffer und Paprika würzen. Die Eiweiße steif schlagen und unter die Eigelb-Masse heben.
- Elektro-Ofen auf 200 Grad vorheizen. Pieform einfetten. 1/4 des Hefeteigs abnehmen. Rest auf einer bemehlten Arbeitsfläche ausrollen, in die Springform legen und den Teig über den Rand hochziehen.
- Zwiebeln auf dem Teigboden verteilen, Käse-Masse darüber gießen. Teigrand nach innen klappen. Übrigen Teig zu einer Rolle (in der Größe des Springform-Durchmessers) ausrollen, auf den Teigrand legen, Rolle einkerben. Restliche Butter in Flöckchen auf der Wähe verteilen.
- Die Käsewähe im Ofen (Gas: Stufe 3) etwa 30 Minuten backen.

Aufwand:	★★
Zubereitungszeit:	50 Minuten
Backzeit:	30 Minuten
Pro Stück:	245 kcal/1030 kJ

Pikante Käsetorte

FÜR 1 SPRINGFORM VON 20 cm Ø
300 g Mehl, Salz
150 g kalte Butter in Würfeln
1 Ei, 1 Eigelb (Eier: Größe M)
FÜR DEN BELAG
300 g milder Ziegen-Frischkäse
je 40 g Butter und Mehl
200 ml Milch
200 g Schlagsahne
75 g junger Gouda, gerieben
3 Eier und 3 Eigelbe (Eier: Größe M)
2 Bund Schnittlauch, in Röllchen geschnitten
geriebene Muskatnuss
Pfeffer
100 g Salat (z.B. Rucola, Lollo rosso und Lollo bionda)
2 EL Sherry-Essig
4 EL Olivenöl
AUSSERDEM
Fett für die Form
Mehl zum Arbeiten

● Für den Boden Mehl, Salz, Butter, Ei und Eigelb rasch zu einem Teig verkneten. In Frischhaltefolie wickeln, 30 Minuten kalt stellen. Form einfetten.

● Für den Belag Ziegenkäse zerdrücken. Butter zerlassen, Mehl darin anschwitzen. Mit Milch und Sahne ablöschen, unter Rühren etwas köcheln lassen. Ziegenkäse und Gouda unterrühren. Vom Herd nehmen, Eier, Eigelbe und Schnittlauch unterrühren. Käsemasse mit Pfeffer und Muskat abschmecken.

Tipp

Dieser herzhafte Torte eignet sich besonders gut für Partys: Verdoppeln Sie die Mengen und backen Sie den Kuchen in der Fettpfanne. Die Backzeit bleibt gleich. Um zu testen, ob die Käsemasse fest ist, in die Mitte ein Messer einstechen, kommt es sauber wieder heraus, ist sie gar.

● Elektro-Ofen auf 175 Grad vorheizen. Den Teig auf bemehlter Fläche ausrollen, Formboden mit 2/3 des Teiges auslegen, mehrmals mit einer Gabel einstechen. Rest zu einem 5 cm breiten Streifen formen, eine Längsseite mit dem Teigrad nachfahren. Formrand mit dem Teigstreifen auskleiden, gezackte Seite nach oben. Im Backofen (Gas: Stufe 2) etwa 15 Minuten vorbacken. Käsemasse einfüllen, Torte bei 150 Grad (Gas: Stufe 1) in etwa 50 Minuten fertig backen.

● Rucola und den übrigen Salat waschen, trockentupfen, Salat zerpflücken. Mit einer Marinade aus Essig, Salz, Pfeffer und Öl anmachen. Fertige Torte kurz im ausgeschalteten Backofen ruhen lassen. Torte mit Salat servieren.

Aufwand:	★
Zubereitungszeit:	50 Minuten
Backzeit:	1 Stunde 5 Minuten
Pro Stück:	640 kcal/2690 kJ

Pizza mit Käse und Oliven

FÜR 1 BACKBLECH (30 x 40 cm)
300 g Weizenmehl (Type 550)
1/2 Päckchen Hefe (20 g)
Olivenöl, Salz
400 g stückige Tomaten (Dose)
6 Tomaten, Pfeffer
125 g Mozzarella
100 g geriebener Emmentaler
75 g geriebener Parmesan-Käse
je 1 TL getrockneter Thymian
und Oregano
85 g schwarze Oliven
AUSSERDEM
Olivenöl für das Blech
Mehl zum Arbeiten

- Mehl auf eine Arbeitsfläche häufen und in die Mitte eine Mulde drücken. Die Hefe hineinbröckeln, 50 ml lauwarmes Wasser dazugießen, mit der Hefe und etwas Mehl leicht verrühren und 15 Minuten ruhen lassen.
- 3 EL Olivenöl und 1 TL Salz zugeben und alles zu einem geschmeidigen, glatten Teig verkneten. Zugedeckt in einer Schüssel an einem warmen Ort etwa 1 Stunde gehen lassen.
- Backblech mit Öl bestreichen. Elektro-Ofen auf 225 Grad vorheizen. Den Teig auf Mehl zu Backblechgröße ausrollen.
- Die stückigen Tomaten salzen und pfeffern. Die frischen Tomaten waschen und in Spalten schneiden. Mozzarella würfeln und mit den übrigen Käsesorten mischen.
- 2 EL Olivenöl auf den Teig pinseln, dann die stückigen Tomaten auf dem Teig verteilen. Die Hälfte der Käse-Mischung darüber geben und mit Tomaten, Kräutern und Oliven belegen, mit restlichem Käse bestreuen.
- Pizza im Ofen (Gas: Stufe 3-4) auf der untersten Schiene etwa 25 Minuten backen.

Aufwand:	★★
Zubereitungszeit:	45 Minuten
Backzeit:	25 Minuten
Pro Stück:	380 kcal/1600 kJ

Variation

- Sie können statt des rechteckigen Blechs ebenso gut (wie im Foto) 2 runde Pizzableche von 30 cm Ø verwenden, dann den Teig bzw. die Zutaten für den Belag auf beide Pizzen aufteilen. Die Backzeit bleibt die gleiche.

Kartoffel-Pizza

BITTE BACHTEN: TEIG MUSS ÜBER NACHT RUHEN
FÜR 2 BACKBLECHE (30 x 40 cm)
350 g mehligkochende Kartoffeln
Salz, 450 g Mehl
1 Würfel Hefe (42 g)
150 ml Milch, Zucker, 1 Zwiebel
3 Stiele Thymian, 3 EL Olivenöl
750 g stückige Tomaten
(Fertigprodukt), Pfeffer
300 g Schafskäse
4 Bund Rucola
300 g Parma-Schinken
AUSSERDEM
Olivenöl fürs Blech

- Kartoffeln in Salzwasser garen, pellen, durch die Presse drücken, ausdampfen lassen.
- Eine Mulde ins Mehl drücken. Hefe in 4 EL warmer Milch mit 1 Prise Zucker auflösen, in die Mulde gießen, mit Mehl bestäuben. 30 Minuten gehen lassen. Übrige Milch, 1/2 TL Salz und Kartoffeln zugeben, alles zum glatten Teig verkneten. Zugedeckt über Nacht kühl stelllen.
- Zwiebel abziehen und fein hacken. Thymian waschen, Blättchen grob hacken. Zwiebeln in Öl andünsten, Tomaten und Thymian zugeben. Mit Salz, Pfeffer und Zucker würzen. Bei milder Hitze sämig einkochen lassen. Schafskäse zerbröckeln. Alles kühl stellen.
- Elektro-Ofen auf 200 Grad vorheizen. 2 Bleche ölen. Teig halbieren, jeweils auf Blechgröße ausrollen, Bleche damit auslegen. Sauce auf dem Teig verteilen und mit Käse bestreuen.
- Pizzen nacheinander im Ofen (Gas: Stufe 3) auf der untersten Schiene etwa 40 Minuten backen.
- Rucola waschen, trockentupfen. Pizza noch heiß mit Rucola und Schinken belegen, in kleine Stücke teilen und warm servieren.

Aufwand:	★★
Zubereitungszeit:	50 Minuten
Backzeit:	1 Stunde 20 Minuten
Pro Stück:	405 kcal/1700 kJ

Pizza mit Hack
Pizza Hawaii

FÜR 2 BACKBLECHE (30 x 40 cm)
600 g Weizenmehl (Type 550)
1 Päckchen Trockenhefe
4 EL Olivenöl, 2 TL Salz
FÜR DEN HACK-BELAG
1 Aubergine, Salz
1 rote Paprikaschote
3 Knoblauchzehen
1 Schalotte, 4 Stängel Minze
150 g Bergkäse, 2 EL Öl
250 g Hackfleisch (gemischt)
Pfeffer, Kreuzkümmel
Paprikapulver (edelsüß)
1-2 Msp. Sambal Oelek
1 TL Aprikosenkonfitüre
FÜR DEN HAWAII-BELAG
1/2 Ananas (ersatzweise
5 Ananasringe, ungezuckert)
3 Stangen Staudensellerie
1 EL Öl, Salz
Pfeffer, Currypulver
150 g geräucherte Putenbrust
75 g Gorgonzola, 100 g Bergkäse
400 g stückige Tomaten
(Fertigprodukt)
AUSSERDEM
Olivenöl für die Bleche

- Mehl, Hefe und 100 ml lauwarmes Wasser vermischen, Öl und Salz zugeben, zu einem geschmeidigen Teig kneten. Zugedeckt an einem warmen Ort etwa 1 Stunde gehen lassen.
- 2 Backbleche ölen. Elektro-Ofen auf 225 Grad vorheizen. Teig nochmals verkneten, halbieren und jeweils auf Backblechgröße ausrollen, Bleche damit auslegen.
- Für den Hack-Belag Aubergine putzen, waschen, in 1 cm große Würfel schneiden, salzen, 15 Minuten ziehen lassen.
- Paprika putzen, halbieren, entkernen, abbrausen, in schmale Spalten schneiden. Knoblauch abziehen, in dünne Scheiben schneiden. Schalotte abziehen, in Ringe teilen. Minze abbrausen, trockenschütteln, Blättchen abzupfen und in feine Streifen schneiden. Käse reiben.

● Auberginen mit Küchenkrepp trockentupfen. Schalotte und Knoblauch in 1 EL heißem Öl andünsten. Fleisch zugeben und bei starker Hitze krümelig braten. Mit Salz, Pfeffer, Kreuzkümmel und Paprika würzen, Sambal Oelek und Konfitüre unterrühren. Fleisch aus der Pfanne nehmen, Auberginenwürfel im übrigen Fett rundum anbraten. Minze unter das Fleisch mischen.

● Für den Hawaii-Belag Ananas schälen, längs achteln. Mittelstrunk entfernen, Achtel in dünne Scheiben teilen. Dosen-Ananas abtropfen lassen, in kleine Stücke schneiden. Sellerie putzen, waschen, in dünne Scheiben teilen, im heißen Öl kurz andünsten, mit Salz, Pfeffer, Curry würzen. Putenbrust halbieren oder vierteln. Gorgonzola fein würfeln, Bergkäse reiben.

● Stückige Tomaten salzen, pfeffern, auf dem Teig verteilen. Fleisch auf einer Teighälfte verteilen, mit Aubergine und Paprika belegen. Käse darüber streuen. Auf der zweiten Teighälfte Gorgonzola verteilen, dann mit Putenbrust, Sellerie und Ananas belegen und mit Bergkäse bestreuen.

● Die Pizzen nacheinander im Ofen (Gas: Stufe 4) auf der unteren Schiene etwa 25 Minuten backen. In Stücke teilen, servieren.

Aufwand:	★★
Zubereitungszeit:	55 Minuten
Backzeit:	50 Minuten
Pro Stück:	485 kcal/2040 kJ

Tipp

Alle, die gerne Pizza essen, aber wenig Zeit haben, stellen die doppelte Menge Teig her und frieren die Hälfte ein. Den Teig nach dem Aufgehen und dem zweiten Knetvorgang in eine Gefriertüte geben und in das Gefrierfach legen. Er braucht dann etwa 2 Stunden zum Auftauen.

Spinat-Pizza mit Schafskäse

FÜR 2 BACKBLECHE (30 x 40 cm)
200 g TK-Blattspinat
1 Knoblauchzehe
1 kleine Zwiebel
1 EL Butter, Salz, Pfeffer
1 Bund Minze oder
glatte Petersilie
150 g Schafskäse
250 g Pizzateig (Kühlregal)
2 EL Pinienkerne
(ersatzweise Mandelblättchen)
AUSSERDEM
Backpapier für die Bleche
Olivenöl zum Beträufeln

● Spinat auftauen lassen. Knoblauch und Zwiebel abziehen, hacken, in heißer Butter andünsten. Spinat ausdrücken, hacken, unterrühren. Kräuter waschen, Blättchen hacken, unterheben. Käse würfeln oder zerbröckeln. Spinat mit dem Käse mischen. Mit Salz und Pfeffer würzen.

● Elektro-Ofen auf 200 Grad (Gas: Stufe 3) vorheizen. Die Backbleche mit Backpapier auslegen. Den Pizzateig falls nötig ausrollen, 4 Kreise von 20 cm Ø ausschneiden, auf die Bleche legen. Mehrmals mit der Gabel einstechen, mit Öl beträufeln. Im Ofen etwa 8 Minuten backen. Spinat-Käse-Mischung und Pinienkerne darauf verteilen und weitere 2-3 Minuten backen. Pizzen halbieren und heiß servieren.

Aufwand:	★
Zubereitungszeit:	30 Minuten
Backzeit:	22 Minuten
Pro Stück:	175 kcal/ 740 kJ

Mini-Pizza mit Tomaten

FÜR 1 BACKBLECH (30 x 40 cm)
450 g TK-Blätterteig
4 kleine Tomaten
2 mittelgroße weiße Zwiebeln
2 EL Olivenöl
250 ml pürierteTomaten (Fertigprodukt)
Salz
1 TL getrockneter Thymian
150 g Mozzarella
AUSSERDEM
Backpapier für das Blech

- Blätterteigscheiben zum Auftauen nebeneinander legen.
- Tomaten waschen, putzen, abtrocknen und in Scheiben schneiden. Zwiebeln abziehen und in dünne Ringe schneiden.
- Die Zwiebeln in heißem Olivenöl etwa 2 Minuten andünsten und beiseite stellen.
- Elektro-Ofen auf 220 Grad vorheizen. Backblech mit Backpapier auslegen. Die Blätterteigplatten quer halbieren und auf das Backblech legen.
- Je 2 EL pürierte Tomaten auf den Platten verstreichen, dabei einen kleinen Rand frei lassen. Salzen und etwas Thymian darüber streuen, mit den Tomatenscheiben belegen.
- Mozzarella in dünne Scheiben schneiden, diese halbieren und auf die Tomaten legen. Mini-Pizzen im Ofen (Gas: Stufe 3) etwa 15 Minuten backen. Vor dem Servieren etwas abkühlen lassen.

Aufwand:	★
Zubereitungszeit:	30 Minuten
Backzeit:	15 Minuten
Pro Stück:	80 kcal/340 kJ

Info

Kaum ein anderes pikantes Gebäck ist so variabel wie Pizza. Man kann sie nicht nur auf verschiedenste Art belegen (neben Tomaten und diversen Käsesorten mit Pilzen, Salami, Schinken, Krabben, Zwiebeln, Zucchini etc.), man kann auch unterschiedliche Teige als Grundlage verwenden. Klassisch ist zwar der Hefeteig, aber auch Blätterteig oder Mürbeteig sind gute Altenativen. Und die findet man – wenn es mal eilig ist – auch fertig im Kühlregal des Supermarkts.

Kürbis-Tarte

FÜR 1 TARTEFORM VON 30 cm Ø
200 g Mehl, 100 g Butter
1 Ei (Größe M)
geriebene Muskatnuss, Salz
FÜR DEN BELAG
1 kg Kürbis (z. B. Muskat-Kürbis; ergibt etwa 800 g Fruchtfleisch)
200 ml Milch
2 Zweige Salbei
Salz, Pfeffer, 100 g Schlagsahne
3 Eier (Größe M)
150 g Gorgonzola
50 g geriebener Parmesan
50 g Butter
AUSSERDEM
Fett für die Form
getrocknete Hülsenfrüchte zum Blindbacken

● Für den Teig Mehl, Butter, Ei, Muskat und Salz verkneten. Den Teig in Folie wickeln und mindestens 30 Minuten kühl stellen.

● Elektro-Ofen auf 200 Grad vorheizen. Tarteform einfetten. Teig ausrollen, Form damit auslegen, Rand hochziehen. Boden erst mit Backpapier, dann mit Hülsenfrüchten belegen. Im Ofen (Gas: Stufe 3) etwa 10 Minuten backen. Hülsenfrüchte und Papier entfernen, Tarte 3-5 Minuten weiterbacken. Temperatur auf 175 Grad (Gas: Stufe 2) reduzieren.

● Für den Belag Kürbis in Spalten teilen, entkernen, schälen. 500 g Fruchtfleisch quer in Scheiben teilen. Übriges Fruchtfleisch 2 cm groß würfeln, in die Milch geben. Salbei, Salz und Pfeffer zugeben, etwa 20 Minuten köcheln. Sahne einrühren, Masse pürieren. Eier verquirlen, vorsichtig unterrühren, abschmecken.

● Die Tarte mit den Kürbisscheiben belegen. Gorgonzola zerbröseln, darauf streuen. Kürbispüree darüber gießen. Parmesan und Butter in Flöckchen darauf setzen. Im Ofen (Gas: Stufe 2) etwa 35 Minuten backen.

Aufwand:	★★
Zubereitungszeit:	35 Minuten
Backzeit:	35 Minuten
Pro Stück:	310 kcal/1300 kJ

Variation

● Besonders edel wird die Tarte, wenn Sie 100 g Lachsschinken in feinen Streifchen oder 100 g kleine Krabben auf die Kürbismasse geben. Statt Gorgonzola nehmen Sie dann 200 g Gouda oder Edamer Käse und verzichten auf den Parmesan.

Bunte Pizza

FÜR 2 PIZZABLECHE VON 30 cm Ø
250 g Mehl
1/2 Päckchen Hefe (20 g)
1 Prise Zucker, 2 EL Olivenöl
1/2 TL Salz
FÜR DEN BELAG
2 Kugeln Mozzarella (etwa 400 g)
500 g Flaschentomaten
500 g Zucchini, 1 große Zwiebel
je 1 rote, grüne und gelbe
Paprikaschote, 1-2 EL Kapern
2-3 grüne Pfefferschoten
1 Dose Maiskörner (212 ml)
20 schwarze Oliven
3-4 Zweige Basilikum
1 EL Sardellenpaste
3 EL Olivenöl, Salz, Pfeffer
2 EL getrockneter Oregano
AUSSERDEM
Olivenöl zum Einfetten
und Beträufeln
Mehl zum Bestäuben und
zum Arbeiten

● Für den Teig in das Mehl eine Mulde drücken. Die Hefe hineinbröckeln, mit Zucker, 2-3 EL lauwarmes Wasser und wenig Mehl verrühren und mit einem Tuch bedeckt gehen lassen.

● 100 ml lauwarmes Wasser mit Salz und Öl zufügen und den Teig schlagen, bis er elastisch ist und Blasen wirft. Mit etwas Mehl bestäubt 30 Minuten gehen lassen.

● Elektro-Ofen auf 225 Grad vorheizen. Die Bleche einfetten. Für den Belag Mozzarella abtropfen lassen. Tomaten und Zucchini waschen und in Scheiben schneiden. Tomaten im Sieb abtropfen lassen. Die geschälte Zwiebel in Ringe teilen. Paprikaschoten putzen und in Streifen schneiden. Mais, Oliven und Kapern abtropfen lassen, Mozzarella in Scheiben teilen. Basilikum abbrausen und größere Blätter in Streifen schneiden.

● Den Teig auf bemehlter Fläche durchkneten, halbieren, zu flachen Fladen ausrollen und auf die Bleche legen. Sardellenpaste mit Öl verrühren und auf den Teig streichen.

● Das Gemüse mit Mozzarella, den Oliven und Kapern gleichmäßig auf den Fladen verteilen und mit Salz, Pfeffer, Oregano und Basilikumstreifen würzen.

● Die Pizzen mit Olivenöl beträufeln und im Ofen (Gas: Stufe 4) jeweils 20-25 Minuten backen, dann nochmals etwa 5 Minuten bei Unterhitze backen. Mit Basilikumblättchen bestreuen, heiß servieren.

Aufwand:	★★
Zubereitungszeit:	1 1/4 Stunden
Backzeit:	50 Minuten
Pro Stück:	370 kcal/1550 kJ

Variationen

● Schmeckt auch gut: Statt der Kapern 100 g geputzte und in feine Scheiben geschnittenene Champignons auf die Pizza geben.

● Oder 125 g gekochten Schinken in Streifen schneiden und zusätzlich auf der Pizza verteilen.

● Oder 125 g Garnelen unter das Gemüse mischen und auch dann wieder auf die Kapern verzichten.

Würzige Käse-Tarte

FÜR 1 PIEFORM VON 26 cm Ø
1 Grundrezept Mürbeteig
(s. Seite 223)
FÜR DEN BELAG
2 rote Paprikaschoten
1 Zwiebel, 2 Knoblauchzehen
1 Bund Schnittlauch
2 EL Olivenöl
3 Eier (Größe M)
200 g Doppelrahmfrischkäse
200 g geriebener Gruyère
100 g Crème double
Salz, Pfeffer, Muskatnuss
Cayennepfeffer
AUSSERDEM
Fett für die Form
Mehl zum Arbeiten

- Mürbeteig zubereiten, kalt stellen.
- Elektro-Ofen auf 175 Grad vorheizen. Form einfetten. Teig etwas größer als die Form ausrollen. Form damit auslegen, dabei einen kleinen Rand bilden. Im Ofen (Gas: Stufe 2) etwa 15 Minuten vorbacken, Ofen nicht ausschalten.
- Paprikaschoten waschen, putzen und klein würfeln oder in Streifen schneiden. Zwiebel und Knoblauch abziehen. Zwiebel in Achtel schneiden, Knoblauch fein hacken. Schnittlauch waschen, trockenschütteln, in Röllchen schneiden.
- Paprika, Zwiebel und Knoblauch im erhitzten Olivenöl andünsten.
- Eier trennen, Eiweiße steif schlagen. Frischkäse, Gruyère, Crème double, Eigelbe und Gewürze verrühren. Paprika-Zwiebel-Mischung und den Schnittlauch untermischen. Eischnee unterheben.
- Käsemasse auf den Boden geben und etwa 30 Minuten backen, bis die Oberfläche leicht gebräunt ist.

Aufwand:	★
Zubereitungszeit:	1 Stunde
Backzeit:	30 Minuten
Pro Stück:	350 kcal/1470 kJ

Flammekuche Elsässer Art

FÜR 2 PIZZABLECHE VON 30 cm Ø
300 g Mehl
1/2 Würfel Hefe (20 g)
1 Msp. Zucker, Salz
2 TL weiches Schweine- oder Butterschmalz, 2 EL Butter
1 Gemüsezwiebel, 1 EL Öl
150 g magerer Räucherspeck
250 g Quark (20 % Fett i. Tr.)
100 g Schlagsahne
2 Eier (Größe M), weißer Pfeffer
geriebene Muskatnuss
AUSSERDEM
Mehl zum Arbeiten
Fett für das Blech

- Mehl in eine Schüssel geben, eine Mulde ins Mehl drücken. Hefe hineinbröseln und mit Zucker, etwas lauwarmem Wasser und wenig Mehl zum Vorteig verrühren. Den Teig zugedeckt 20 Minuten gehen lassen.
- Etwa 100 ml lauwarmes Wasser mit Salz und Schmalz in Flöckchen zum Teig geben und so lange kneten, bis sich der Teig vom Schüsselrand löst. Teig weitere 30 Minuten gehen lassen.
- Elektro-Ofen auf 250 Grad vorheizen. Das Blech einfetten. Zwiebel abziehen, in dünne Ringe teilen. Butter und Öl in einer Pfanne erhitzen, die Zwiebel darin glasig dünsten.
- Speck in sehr feine Streifen teilen. Quark mit Sahne, Eiern, Salz, Pfeffer und Muskat verrühren.
- Den Teig nochmal durchkneten. Auf einer leicht bemehlten Arbeitsfläche zu einer rechteckigen Platte in Backblechgröße oder zu 2 runden Platten ausrollen. Die Teigplatten auf ein Blech legen. Gleichmäßig mit Zwiebeln und Speck bestreuen. Den Kuchen im Ofen (Gas: Stufe 5) in 10-12 Minuten knusprig und goldgelb backen. Sofort servieren.

Aufwand:	★
Zubereitungszeit:	1 Stunde
Backzeit:	12 Minuten
Pro Stück:	260 kcal/1090 kJ

Tomaten-Tarte

FÜR 1 TARTEFORM VON 28 cm Ø
200 g Mehl
100 g kalte Butter
3 Eier (Größe M)
600 g Flaschentomaten
4 EL Olivenöl
Salz, Pfeffer
2 TL Honig
2 EL weißer Balsamico-Essig
200 g gereifter Ziegenfrischkäse
50 g Ricotta
1-2 TL Dijon-Senf
1 Bund Rucola
AUSSERDEM
Mehl zum Arbeiten
Fett für die Form

- Mehl, Butter in Flöckchen und 1 Ei verkneten. Teig 1 Stunde zugedeckt kühl stellen.
- Elektro-Ofen auf 175 Grad vorheizen. Tomaten waschen, putzen, in 1 cm dicke Scheiben teilen. Öl, Salz, Pfeffer, Honig und Essig verrühren. Tomaten darin marinieren.
- Käse entrinden, 100 g zerdrücken. Mit Ricotta, übrigen Eiern und Senf vermengen, würzen. Restlichen Käse 1 cm groß würfeln. Rucola putzen, waschen, trockenschütteln.
- Teig ausrollen. Die Tarteform mit übriger Butter einfetten. Teig hineinlegen und im Ofen (Gas Stufe2) etwa 15 Minuten vorbacken, abkühlen lassen. Ofen nicht ausschalten.
- Käse-Masse auf den abgekühlten Boden streichen. Mit Tomaten belegen und im Ofen etwa 20 Minuten backen. Nach 15 Minuten Käsewürfel überstreuen. Vorm Servieren mit Rucola belegen.

Aufwand:	★
Zubereitungszeit:	1 1/4 Stunden
Backzeit:	35 Minuten
Pro Stück:	270 kcal/ 1130 kJ

Zwiebel-Kürbis-Kuchen

FÜR 1 BACKBLECH (30 x 40 cm)
1 Würfel Hefe (42 g), 1/4 l Milch
Zucker, 1 kleine Messerspitze
Safranpulver
500 g Mehl, 1 Ei (Größe M)
Salz, 3-4 EL Olivenöl
FÜR DIE ZWIEBELMASSE
800 g Zwiebeln
500 g eingelegter Kürbis (Glas)
3 Eier (Größe M), 3 EL Butter
200 g Schlagsahne, Salz
250 g Mascarpone
1 Bund Schnittlauch, Pfeffer
AUSSERDEM
Mehl zum Arbeiten
Fett fürs Blech

- Hefe in lauwarmer Milch auflösen. 1 Prise Zucker, Safran und 3 EL Mehl einrühren. Übriges Mehl in eine Schüssel geben. Mulde ins Mehl drücken, Vorteig hineingeben. 20 Minuten gehen lassen.
- Zwiebeln abziehen, halbieren und quer in Ringe schneiden. Kürbis abtropfen lassen und in Spalten oder 1 cm große Würfel schneiden.
- Vorteig mit Ei, 1 TL Salz und Öl zu einem Teig verkneten, zugedeckt 30 Minuten gehen lassen.
- Butter in einem Topf erhitzen, Zwiebeln darin etwa 10 Minuten dünsten. Mit Salz und Pfeffer abschmecken. Eier, Sahne und Mascarpone verrühren, mit Salz und Pfeffer würzen. Elektro-Ofen auf 175 Grad vorheizen. Backblech einfetten.
- Teig ausrollen. Backblech damit auslegen. Zwiebeln mit Kürbis mischen, auf dem Teig verteilen. Eier-Sahne-Mischung darauf gießen. Im Ofen (Gas: Stufe 2) 45 Minuten backen.
- Schnittlauch abbrausen, trockenschütteln und in Röllchen schneiden. Zwiebelkuchen mit Schnittlauch bestreuen und servieren.

Aufwand:	★★
Zubereitungszeit:	45 Minuten
Backzeit:	45 Minuten
Pro Stück:	820 kcal/ 3440 kJ

Schinken-Käse-Tarte

FÜR 1 SPRINGFORM VON 26 cm Ø
300 g Mehl, 1 TL Salz
150 g kalte Butter
200 g Crème fraîche
FÜR DEN BELAG
1 Stange Lauch (etwa 200 g)
200 g gekochter Schinken
150 g geriebener Emmentaler
200 g Crème fraîche
1 Ei (Größe M), 1 TL Salz
Pfeffer, geriebene Muskatnuss
AUSSERDEM
Fett für die Form
Mehl zum Arbeiten

- Elektro-Ofen auf 200 Grad vorheizen. Springform einfetten. Mehl, Salz, Butter und Crème fraîche rasch zu einem glatten Mürbeteig verkneten. 3/4 davon zu einem Kreis ausrollen und die Form damit auslegen. Dabei den Teigrand bis fast ganz nach oben ziehen.
- Lauch putzen, waschen, in dünne Ringe teilen. Schinken in feine Streifen schneiden. Beides auf dem Mürbeteigboden verteilen. Käse darüber streuen.
- Crème fraîche, Ei, Salz, Pfeffer und Muskat verquirlen, in die Form gießen. Übrigen Mürbeteig dünn auf etwas Mehl ausrollen, in 1/2 cm breite Streifen teilen, als Gitter über den Belag legen.
- Kuchen im Ofen (Gas: Stufe 3) etwa 30 Minuten backen, bis er goldbraun ist.

Aufwand:	★
Zubereitungszeit:	45 Minuten
Backzeit:	30 Minuten
Pro Stück:	395 kcal/1660 kJ

Variationen

- Dieses Rezept bietet zahlreiche Variationsmöglichkeiten, z. B. können Sie den Schinken durch die gleiche Menge geräuchte Putenbrust oder Kassler ersetzen.
- Schmeckt auch gut: Den Lauch durch 2 mittelgroße, in Würfel geschnitttene und blanchierte Zwiebeln ersetzen und statt Schinken 150 g gewürfelten Räucherspeck zufügen
- Vegetarier nehmen statt des Schinkens 200 g Champignons oder Egerlinge, die nach dem Säubern in feine Scheiben geschnitten werden.

Sauerkraut-Wähe

FÜR 1 QUICHEFORM VON 26 cm Ø
250 g Mehl, 125 g Butter
60 ml Milch, Salz, Pfeffer
2 Eigelbe (Eier: Größe M)
FÜR DEN BELAG
1 Zwiebel, 1 Apfel, 75 g Speck
1 Dose Sauerkraut (Abtropfgew.: 520 g), 2 Wacholderbeeren, 1/2-1 TL Kümmel
1 Lorbeerblatt, Salz, Pfeffer
1/8 l trockener Weißwein (z. B. Gewürztraminer)
1/8 l Gemüsebrühe (Instant)
nach Belieben 20 ml Kirschwasser, 200 g Schlagsahne
4 Eier (Größe M)
100 g Emmentaler
AUSSERDEM
Fett für die Form
Mehl zum Arbeiten

- Zwiebel abziehen, fein würfeln. Apfel schälen, entkernen, in dünne Scheiben schneiden. Speck würfeln.
- Speck in einem Topf auslassen. Zwiebel und Apfel zugeben, etwa 5 Minuten mitdünsten. Sauerkraut, Wacholder, Kümmel, Lorbeer, Salz und Pfeffer zugeben. So viel Wein angießen, dass die Oberfläche bedeckt ist. Etwa 20 Minuten köcheln, eventuell Brühe nachgießen. Kurz vor Ende der Garzeit Kirschwasser zugeben.
- Alle Teigzutaten zu einem glatten Teig verkneten. 30 Minuten zugedeckt kühl stellen.
- Elektro-Ofen auf 175 Grad vorheizen. Form einfetten. Teig ausrollen, in die Form legen, einen Rand formen. Lorbeer und Wacholder aus dem Kraut nehmen, Kraut gut abtropfen lassen, auf dem Teig verteilen. Sahne und Eier verquirlen, darüber gießen. Käse würfeln, darauf verteilen. Wähe im Ofen (Gas: Stufe 2) etwa 35 Minuten backen.

Aufwand:	★
Zubereitungszeit:	50 Minuten
Backzeit:	35 Minuten
Pro Stück:	520 kcal/2180 kJ

Käse-Tarte mit Schalotten

FÜR 1 TARTEFORM VON 22 cm Ø
90 g kalte Butter, 175 g Mehl
1 Eigelb (Ei: Größe M), Salz
FÜR DIE FÜLLUNG
100 g Schalotten, 1 EL Olivenöl
150 g zweierlei würziger Schnittkäse, z. B. Greyère und Comté
150 g Ziegenfrischkäse
100 ml Milch
3 Eier (Größe M)
je 1 EL frischer gehackter Thymian, Salbei, Rosmarin
Pfeffer, geiebene Muskatnuss
AUSSERDEM
Mehl zum Arbeiten
Fett für die Form
Hülsenfrüchte zum Blindbacken

- Butter würfeln, mit Mehl, Eigelb, etwas Salz und 2 EL eiskaltem Wasser rasch zu einem glatten Teig verkneten. Zugedeckt 1 Stunde kalt stellen.
- Schalotten abziehen, fein würfeln. Öl in einer Pfanne erhitzen, Schalotten darin glasig dünsten, beiseite stellen.
- Elektro-Ofen auf 225 Grad vorheizen. Form einfetten. Schnittkäse eventuell entrinden, längs in dünne Scheiben schneiden.
- Frischkäse mit Milch glatt rühren. Eier und Kräuter unterziehen, mit Pfeffer, Muskat wüzen.
- Teig auf bemehlter Fläche ausrollen, in die Form legen, Rand formen. Boden mit Gabel einstechen. Mit Hülsenfrüchten belegen, im Ofen (Gas: Stufe 4) etwa 10 Minuten backen. Temperatur auf 200 Grad (Gas: Stufe 3) reduzieren. Hülsenfrüchte entfernen und den Boden weitere 20 Minuten backen.
- Käsestreifen und Schalotten auf dem Boden verteilen, Ziegenkäsecreme darüber geben. Im Ofen in 30-40 Minuten fertig backen.

Aufwand: ★
Zubereitungszeit: 1 1/4 Stunden
Backzeit: 1 Stunde 10 Minuten
Pro Stück: 430 kcal/1810 kJ

Gemüsestrudel mit Hackfleisch

FÜR 1 BACKBLECH (30 x 40 cm)
200 g Mehl, 2 EL Maiskeimöl
1 EL Weißweinessig
FÜR DEN BELAG
200 g Zucchini, 60 g Butter
200 g Möhren, 1/2 TL Zucker
6 EL Crème fraîche, Salz, Pfeffer
1 Zwiebel, 300 g gemischtes Hackfleisch, 3 EL Olivenöl
1 Ei (Größe M), 3 EL Paniermehl
je 1 EL frischer oder 1 TL getrockneter Thymian und Majoran
AUSSERDEM
Mehl zum Arbeiten
Backpapier für das Blech
1 Eigelb, je 1 TL Sesamkörner und Schwarzkümmel (aus dem türkischen Lebensmittelladen)
Öl zum Bestreichen

● Aus Mehl, 100 ml lauwarmem Wasser, Öl und Essig einen elastischen Strudelteig kneten. Zur Kugel formen, mit Öl einpinseln, in Folie wickeln und 1 Stunde ruhen lassen.

● Zucchini waschen, putzen und trockenreiben. Zucchini schräg in dünne Scheiben schneiden, in 20 g Butter andünsten, beiseite stellen.

● Möhren schälen und raspeln. 20 g Butter in der Pfanne erhitzen, den Zucker darin leicht karamellisieren lassen. Möhren unter Rühren etwa 5 Minuten dünsten, dabei 2 EL Wasser unterrühren und einkochen lassen. Abkühlen. 3 EL Crème fraîche unterrühren,salzen und pfeffern.

● Zwiebel schälen und hacken. Fleisch und Zwiebel in Olivenöl in der Pfanne anbraten und abkühlen lassen. Ei, Paniermehl und 3 EL Crème fraîche unterrühren. Mit Salz, Pfeffer und Kräutern abschmecken.

● Ein großes Küchentuch mit Mehl bestreuen. Den Teig darauf ausrollen, dann von innen nach außen auf eine Größe von 40x55 cm dünn ausziehen. Elektro-Ofen auf 200 Grad vorheizen. Backblech mit Backpapier auslegen.

● 20 g Butter schmelzen und auf den Teig damit pinseln. Rundum einen 5 cm breiten Rand frei lassen. Ein Drittel der Fläche mit Zucchinischeiben belegen, ein Drittel mit Möhren und das letzte Drittel mit Hackfleisch.

● Den Rand an der Fleischseite über das Fleisch klappen und von dort her den Strudel mithilfe des Tuches einrollen.

● Den Strudel mithilfe des Tuches auf das Blech (Naht nach unten) wälzen. Die Enden unter den Strudel schieben. Oberfläche mit Eigeb bepinseln, mit Sesam und Schwarzkümmel bestreuen. Im Ofen (Gas: Stufe 3) etwa 30 Minuten backen .

Aufwand:	★★
Zubereitungszeit:	1 3/4 Stunden
Backzeit:	30 Minuten
Pro Stück:	240 kcal/1760 kJ

Gefüllte Teigtaschen

FÜR 1 BACKBLECH (30 x 40 cm)
300 g Mehl, 1 TL Salz
75 g zimmerwarmes Schweine- oder Butterschmalz
6-8 EL Orangensaft
FÜR DIE FÜLLUNG
30 g Rosinen
3 EL Amontillado-Sherry
1 Knoblauchzehe
500 g frische Spinatblätter
3 EL Olivenöl, Salz, Muskatnuss
2 EL Sherryessig, 2 EL Pinienkerne
AUSSERDEM
Mehl zum Arbeiten
Backpapier für das Blech
1 Eigelb (Ei: Größe M)
2 EL Schlagsahne

- Mehl, Salz und Schmalz in einer Schüssel mit den Knethaken des Handrührgeräts zu einem geschmeidigen Teig verkneten, dabei so viel Orangensaft wie nötig hinzufügen. Kurz durchkneten und in Folie gewickelt 30 Minuten im Kühlschrank ruhen lassen.
- Elektro-Ofen auf 170 Grad vorheizen. Backblech mit Backpapier auslegen. Für die Füllung Rosinen mit Sherry begießen, etwa 15 Minuten einweichen. Knoblauch abziehen. Spinat waschen, tropfnass mit Knoblauch im heißen Öl bei mittlerer Hitze andünsten, bis die Blätter zusammengefallen sind. Mit Salz, Muskat würzen, mit Essig ablöschen. Rosinen mit Einweichflüssigkeit dazugeben. Kurz einkochen lassen. Pinienkerne in der Pfanne ohne Fett anrösten, zum Spinat geben. Abkühlen lassen.
- Teig in 4 Portionen teilen, dünn zu Fladen (16 cm Ø) ausrollen. Füllung auf je 1 Hälfte geben, andere Hälfte darüber klappen. Ränder fest zusammendrücken, auf das Blech geben. Eigelb und Sahne verquirlen, auf die Fladen streichen, im Ofen (Gas: Stufe 2) etwa 30 Minuten goldbraun backen.

Aufwand:	★★
Zubereitungszeit:	1 3/4 Stunden
Backzeit:	30 Minuten
Pro Stück:	600 kcal/2520 kJ

Pikantes Kuchen-Quartett

FÜR 1 BACKBLECH (30 x 40 cm)
1 doppeltes Grundrezept Mürbeteig (s. Seite 223)
FÜR DEN ANANASBELAG
je 2 EL Tomatenmark und Tomatenketchup
1 EL Olivenöl
1 kleine Dose Ananasringe (Abtropfgew.: 255 g)
10 Kirschtomaten
150 g gekochter Schinken in Scheiben
100 g junger Gouda
FÜR DEN THUNFISCHBELAG
je 1 gelbe und rote Tomate
1 kleine Zwiebel
1 EL Öl
1 Dose Thunfisch in Öl (Abtropfgew.: 150 g)
1 EL Zitronensaft
4 Zweige Thymian
1 TL Kapern
100 g geriebener Pizzakäse
FÜR DEN LACHSBELAG
200 g Crème fraîche
1 Ei (Größe M)
11/2 TL mittelscharfer Senf
200 g Lachsfilet
1 Bund Lauchzwiebeln
FÜR DEN GEMÜSEBELAG
100 g Ziegenfrischkäse
100 g Schlagsahne
2 Gläser in Öl eingelegtes Gemüse (Abtropfgew.: je 120 g)
AUSSERDEM
Salz, Pfeffer
Fett für das Blech

● Mürbeteig zubereiten, kalt stellen.
● Elektro-Ofen auf 175 Grad vorheizen. Backblech einfetten. Vom Teig 1/5 abnehmen. Restlichen Teig auf dem Blech ausrollen, dabei einen Rand hochziehen. Übrigen Teig zu 2 Strängen formen und damit die Teigfläche in 4 Rechtecke unterteilen. Teig im Ofen (Gas: Stufe 2) etwa 15 Minuten vorbacken.
● Für den Ananasbelag Tomatenmark, -ketchup, Olivenöl und Salz, Pfeffer und 2 EL Wasser verrühren. Abgetropfte Ananasringe halbieren. Kirschtomaten waschen, halbieren. Ein Teigviertel mit der Tomatenpaste bestreichen, Schinken vierteln, darauf geben. Ananas darauf legen. Tomatenhälften, Rundung nach oben, darauf setzen. Käse überstreuen.
● Für den Thunfischbelag Tomaten waschen, in Scheiben schneiden, auf ein Teigviertel legen, salzen und pfeffern.
● Zwiebel abziehen, in Ringe hobeln, in 1 EL Öl andünsten auf die Tomaten legen. Die Hälfte des Thunfischs abtropfen lassen, mit einer Gabel zerpflücken, auf dem Teig verteilen.
● Übrigen Thunfisch mit dem Thunfischöl pürieren. Mit Zitronensaft, Salz und Pfeffer würzen. Thymianblättchen und die Hälfte der Kapern hacken, unterrühren. Thunfischcreme aufstreichen. Kapern und Käse überstreuen.
● Für den Lachsbelag Crème fraîche mit Ei, Senf, Salz und Pfeffer verquirlen. Lachs kleinschneiden.
● Lauchzwiebeln waschen, putzen, in feine Ringe schneiden. Ein Teigviertel mit der Hälfte der Creme bestreichen, Lachs darauf verteilen. Restliche Crème fraîche darüber geben, die Lauchzwiebeln bis auf 2 EL überstreuen.
● Für den Gemüsebelag Ziegenkäse zerdrücken, mit Sahne glatt rühren. Die Hälfte der Creme auf ein Teigviertel streichen. Abgetropftes Gemüse darauf geben, den Rest der Creme darauf verteilen.
● Kuchen im Ofen (Gas: Stufe 2) etwa 30 Minuten backen. Zwiebelringe auf den Lachsbelag streuen.

Aufwand:	★
Zubereitungszeit:	1 1/4 Stunden
Backzeit:	30 Minuten
Pro Stück:	420 kcal/1760 kJ

Käse-Trauben-Tartes

FÜR 8 TARTEFÖRMCHEN VON 12 cm Ø
400 g Mehl
200 g Butter
2 Eiweiße und 4 Eier
(Eier: Größe M)
Salz
1 Stange Lauch (etwa 450 g)
250 g rosé Trauben
200 g Sahne-Gorgonzola
250 g Milch oder Schlagsahne
Pfeffer
geriebene Muskatnuss
AUSSERDEM
Mehl zum Arbeiten
Fett für die Tarte-Formen

- Aus Mehl, Butter, 2 Eiweißen und 1 Prise Salz einen glatten Teig kneten. In Folie gewickelt 30 Minuten kalt stellen.
- Lauch putzen, waschen, in dünne Ringe teilen. In kochendem Salzwasser etwa 1 Minute vorkochen. Abschrecken, abtropfen lassen.
- Trauben waschen, halbieren und entkernen. Gorgonzola in kleine Würfel schneiden. Übrige Eier und Milch oder Sahne verrühren. Mit Salz, Pfeffer und Muskat abschmecken.
- Elektro-Ofen auf 200 Grad vorheizen. Die Formen einfetten.
- Den Teig auf leicht bemehlter Fläche ausrollen, in Größe der Tarte-Formen ausschneiden. Formen damit auslegen, Rand hochziehen. Trauben, Käse und Lauch darin verteilen. Mit Eiermilch begießen. Im Ofen (Gas: Stufe 3) 20-30 Minuten backen.

Aufwand:	★
Zubereitungszeit:	40 Minuten
Backzeit:	30 Minuten
Pro Stück:	550 kcal/ 2310 kJ

Herbstlicher Kartoffelkuchen

FÜR 1 TARTEFORM (20 x 30 cm)
300 g Mehl
Salz, 1 Ei (Größe M)
150 g kalte Butter
FÜR DEN BELAG
500 g Kartoffeln (vorwiegend fest kochend)
500 g Rosenkohl, Salz
250 g Champignons
1 1/2 EL Butter, 250 g Crème fraîche, 3-4 EL Senf (mittelscharf), 2 Eier (Größe M)
1 EL gehackte Petersilie
Pfeffer, geriebene Muskatnuss
100 g geriebener Käse (z. B. Emmentaler)
AUSSERDEM
Mehl zum Arbeiten
Fett für die Form

- Mehl, 1/4 TL Salz, Ei und Butter in Würfeln bröselig kneten, zu einer Kugel formen, in Folie gewickelt 1 Stunde kalt stellen.
- Inzwischen Kartoffeln mit Schale in kochendem Wasser in etwa 25 Minuten garen. Kalt abschrecken, pellen, auskühlen lassen und in Scheiben teilen. Rosenkohl putzen, die größeren Röschen halbieren. In leicht gesalzenem Wasser etwa 5 Minuten köcheln lassen. Abschrecken und abtropfen lassen.
- Pilze putzen und in Scheiben schneiden. In 1 EL Butter kurz anbraten. Die Crème fraîche mit Senf, Eiern, Petersilie verrühren, mit Salz, Pfeffer, Muskat würzen. Elektro-Ofen auf 200 Grad heizen. Die Form einfetten.
- Teig auf bemehlter Arbeitsfläche 1/2 cm dick ausrollen. Form damit auskleiden. Mit Kartoffeln, Pilzen und Rosenkohl belegen. Eiercreme angießen, mit Käse bestreuen. Im Ofen (Gas: Stufe 3) 35-40 Minuten backen.

Aufwand:	★★
Zubereitungszeit:	1 Stunde
Backzeit	40 Minuten
Pro Stück:	700 kcal/ 2940 kJ

Brioche

BITTE BEACHTEN: TEIG MUSS 2 1/2 STUNDEN UND ÜBER NACHT RUHEN

Für 10 Briocheförmchen

- *10 g Hefe*
- *3 EL lauwarme Milch*
- *250 g Mehl*
- *2 Eier und 2 Eigelbe (Größe M)*
- *20 g Zucker, 1 Prise Salz*
- *120 g weiche Butter*

AUSSERDEM

- *Fett für die Formen*
- *1 Eigelb*

- Die Hefe mit der Milch vermischen. Mehl in eine Rührschüssel sieben. Eine Mulde in die Mitte drücken und die aufgelöste Hefe hineingeben. Mit einer Gabel etwas Mehl unter die Hefemilch mischen. Diesen Vorteig so lange gehen lassen, bis sich Blasen bilden.
- Zuerst Eier, Eigelbe, Zucker und Salz mit den Knethaken des elektrischen Rührgeräts unter das Mehl mischen. Dann nach und nach die Butter unterkneten. Die Masse so lange kräftig – zuerst mit den Knethaken, dann mit den Händen – durchkneten, bis ein elastischer Teig entsteht. Zur Kugel formen, leicht mit Mehl überstäuben, mit einem Tuch abdecken und 2 Stunden an einem warmen Ort gehen lassen.
- Den aufgegangenen Hefeteig kurz durchschlagen und über Nacht im Kühlschrank ruhen lassen.
- Die Förmchen einfetten. Den Teig einfüllen und – wieder zugedeckt – an einem warmen Ort gehen lassen, bis sich das Volumen fast verdoppelt hat. Den Backofen auf 200 Grad (Gas: Stufe 3-4) vorheizen.
- Den Teig mit verquirltem Eigelb bestreichen. Förmchen in den Ofen schieben. Die Hitze auf 180 Grad (Gas: Stufe 2-3) reduzieren. Brioches 25-30 Minuten backen.
- Die Brioches sind fertig, wenn sich ein hineingestochenes Holzstäbchen sauber wieder herausziehen lässt. Die Brioches als Beilage servieren. Oder den Deckel der Brioches abschneiden, sie etwas aushöhlen und füllen.

Aufwand:	★★
Zubereitungszeit:	45 Minuten
Backzeit:	30 Minuten
Pro Stück:	220 kcal/920 kJ

Tipp

Falls Sie eine große Brioche backen wollen: Die angegebene Teigmenge reicht aus für eine Form von 1,2 l Inhalt. Dann muss die Backzeit um 10-15 Minuten verlängert werden. Wenn Teig übrig bleibt: Er lässt sich sehr gut einfrieren.

Camembert-Muffins

Für 8 Muffinförmchen
5 Brötchen vom Vortag
300 ml Milch, Salz, Pfeffer
geriebene Muskatnuss
2 Eier (Größe M)
300 g reifer Camembert
1/2 Bund Petersilie
1 Zwiebel, 1 TL Butter
AUSSERDEM
Butter und 2-3 EL Semmelbrösel für die Förmchen

● Brötchen in kleine Stücke teilen. Milch erwärmen, mit Salz, Pfeffer und Muskatnuss abschmecken. Mit Eiern und Brötchen vermengen und etwa 20 Minuten ziehen lassen.
● Elektro-Ofen auf 200 Grad vorheizen. Käse (mit Rinde) etwa 1 cm groß würfeln. Petersilie waschen, trockenschütteln, Blättchen fein hacken. Petersilie und Käse unter die Brötchenmasse mischen.
● Zwiebel abziehen und in kleine Würfel schneiden. Butter in einer Pfanne schmelzen, Zwiebel dazugeben und glasig dünsten. Zur Brötchenmasse geben. Nochmal mit Salz, Pfeffer und Muskat abschmecken.
● Förmchen mit Butter einfetten und mit Semmelbröseln ausstreuen. Die Masse einfüllen. Die Förmchen in die Fettpfanne des Ofens stellen und so viel kochendes Wasser zugießen, dass die Förmchen zur Hälfte im Wasserbad stehen. Im Ofen (gas Stufe 3) in etwa 30 Minuten garen.

Aufwand:	★
Zubereitungszeit:	30 Minuten
Backzeit:	30 Minuten
Pro Stück:	240 kcal/1010 kJ

500g. Mehl
21g. Hefe
300ml Lauwarme Milch
100g. Zucker
50g. Butter
1 Prise Salz

Gemüsebrot

FÜR 1 KASTENFORM 24 cm LÄNGE
250 g Mehl
1/8 l lauwarme Milch
1/2 Päckchen Hefe (20 g)
1/2 TL Salz, 1 Prise Zucker
2 Eier, 1 Eigelb (Eier: Größe M)
200 g Gemüsewürfel (z. B. von Möhren, Stangensellerie, Paprika, Zucchini)
50 g TK-Erbsen
50 g Butter, 100 g Magerquark
100 g Fetakäse in Würfeln
3 EL gehackte Kräuter (z. B. Thymian, Petersilie, Rosmarin, Basilikum)
Pfeffer, geriebene Muskatnuss
AUSSERDEM
Fett für die Form

● Mehl auf eine Arbeitsfläche häufeln, ein Mulde eindrücken und die Hefe hineinbröckeln. Milch zufügen und mit der Hefe und etwas Mehl leicht verrühren. 15 Minuten ruhen lassen.

● Salz, Zucker, Eier und Eigelb zufügen und alles zu einem glatten Teig verkeneten. 1 Stunde gehen lassen.

● Form gut fetten. Gemüsewürfel und Erbsen in Butter andünsten und abkühlen lassen. Mit Quark, Käse und Kräutern unter den Teig rühren und würzen. In die Form füllen und erneut 20 Minuten gehen lassen.

● Elektro-Ofen auf 180 Grad vorheizen. Das Brot im Ofen (Gas: Stufe 2-3) etwa 45 Minuten backen.

Aufwand:	★★
Zubereitungszeit:	2 Stunden
Backzeit:	45 Minuten
Pro Stück:	325 kcal/1370 kJ

Strudelecken mit Spinat

FÜR 2 BACKBLECHE (30 x 40 cm)
30 g Korinthen
300 g Blattspinat
1 kleine Zwiebel
3 EL Olivenöl
300 g gehacktes Lamm- oder Rindfleisch
Salz, Pfeffer
1/2 TL Zimtpulver
2 EL gehackte Petersilie
1 EL gehackter Dill
1 Ei (Größe M)
50 g Schafskäse
25 g Pinienkerne
40 g Butter
50 ml Milch
2 EL Crème fraîche
3 Blätter Yufka-Teig (aus dem türkischen Lebensmittelladen)
AUSSERDEM
Backpapier für die Bleche
1 EI (Größe M)
2 EL Sesamsamen zum Bestreuen

● Korinthen in Wasser einweichen. Spinat putzen, waschen und tropfnass in der Pfanne bei starker Hitze dünsten. Spinat im Sieb ausdrücken und hacken.

● Zwiebel schälen, klein würfeln und im erhitzten Olivenöl glasig dünsten. Hackfleisch unterrühren und braten, bis der Saft verdampft ist.

● Spinat und die abgetropften Korinthen unterrühren. Mit Salz, Pfeffer und Zimt würzen. Kräuter und 1 Ei unterrühren. Käse zerbröckeln. Pinienkerne in einer trockenen Pfanne anrösten. Käse, Kerne und Crème fraîche unter die Füllung mischen.

● Butter schmelzen, Milch unterrühren. Den Teig auseinander falten und leicht mit Wasser besprühen. Elektro-Ofen auf 220 Grad vorheizen. Backbleche mit Backpapier belegen.

● Teigblätter der Länge nach in 3 gleich breite Streifen schneiden und mit Butter-Milch einpinseln. Die Streifen halbieren. Auf den unteren Teil jedes Streifens 1 EL Füllung geben. Die Streifen zickzackartig von unten nach oben falten, so dass 18 Dreiecke entstehen, die an allen Seiten geschlossen sind.

- Dreiecke mit verquirltem Ei bepinseln, auf die Bleche legen, mit Sesam bestreuen. Im Ofen (Gas: Stufe 4) in je 8-10 Minuten goldgelb backen.
- Dazu passt ein bunter Salat: Die Dreiecke am besten auf einem Salatbett, beträufelt mit einer Marinade aus Olivenöl und Balsamico-Essig, anrichten.

Aufwand:	★★
Zubereitungszeit:	1 Stunde
Backzeit:	20 Minuten
Pro Stück:	440 kcal/1850 kJ

Info

Den türkischen Blätterteig Yufka gibt es bei uns eingeschweißt in Folie zu kaufen. In jedem Paket befinden sich acht runde, etwa 38 cm große, zusammengerollte Scheiben. Sie sind nicht so elastisch wie frisch ausgerollt. Deshalb werden sie kurz vor der Verwendung mit etwas Wasser eingesprüht. Danach zügig weiterarbeiten, weil Yufka-Teig sich sonst wellt und leicht reißt. Für süßes Gebäck lieber den hauchdünnen Baklava-Teig (beim Griechen heißt er Fillo-Teig) nehmen. Ihn gibt es ebenfalls in Folie.

Olivenbrot mit Käsecreme

FÜR 1 BACKBLECH (30 x 40 cm)
325 g Mehl
1/2 Würfel Hefe (20 g)
1/2 TL Zucker
75 ml Olivenöl, 1 TL Salz
75 g schwarze, entsteinte Oliven
AUSSERDEM
Mehl zum Arbeiten
Öl für das Blech

● Mehl in eine Schüssel geben, in die Mitte eine Mulde drücken. Hefe mit 100 ml warmem Wasser und Zucker verrühren, in die Mulde gießen und mit Mehl bestäuben. 15 Minuten ruhen lassen.

● Öl und Salz zugeben, alles zu einem glatten Teig kneten. Zugedeckt an einem warmen Ort 40 Minuten gehen lassen.

● Elektro-Ofen auf 200 Grad vorheizen. Backblech einölen. 50 g Oliven hacken, in 1 TL Mehl wälzen, unter den Teig kneten. Den Teig zu einem länglichen Laib formen und die übrigen Oliven hineindrücken. Den Laib auf das Backblech setzen und Olivenbrot im Ofen (Gas: Stufe 3) etwa 30 Minuten backen.

Aufwand:	★
Zubereitungszeit:	1 1/4 Stunden
Backzeit:	30 Minuten
Pro Stück:	315 kcal/1320 kJ

Käsecreme

1 EL gehackte Basilikumblätter
200 g Doppelrahm-Frischkäse
3 EL Schlagsahne
Salz
1/2 TL getrocknete italienische Kräuter

● Basilikum waschen, trocken schütteln und fein hacken. Den Frischkäse mit der Sahne verrühren. Basilikum untermischen und mit Salz und Kräutern würzig abschmecken.

● Die Creme zum Brot servieren.

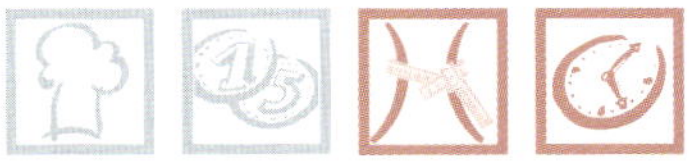

Maisbrot

FÜR 1 BACKBLECH (30 x 40 cm)
500 g Maismehl
500 g Weizenmehl (Type 550)
1 1/2 Würfel Hefe (60 g)
1 TL Zucker
125 g Frühstücksspeck (Bacon)
1 EL Salz, 1 Ei (Größe L)
500-600 g zimmerwarme Buttermilch
200 g Maiskörner (Dose)
AUSSERDEM
Backpapier für das Blech
Mehl zum Bestäuben

● Beide Mehlsorten in einer Schüssel mischen, in die Mitte eine Mulde drücken. Die Hefe mit 75 ml lauwarmem Wasser glatt rühren, in die Mulde gießen, mit Zucker bestreuen. Zugedeckt 10 Minuten gehen lassen. Inzwischen den Speck fein würfeln.

● Speck, Salz, Ei, Buttermilch und die abgetropften Maiskörner zugeben und alles zu einem Brotteig verkneten. Zugedeckt 30 Minuten gehen lassen.

● Backblech mit Backpapier auslegen. Teig nochmals durchkneten und zu 2 runden Broten formen. Diese über Kreuz einschneiden. Auf das Blech legen, mit Mehl bestäuben. Brote in den kalten Ofen schieben, bei 200 Grad (Gas: Stufe 3) etwa 45 Minuten backen.

Aufwand:	★
Zubereitungszeit:	50 Minuten
Backzeit:	45 Minuten
Pro Stück:	165 kcal/695 kJ

Variation

● Wer keinen Mais mag, kann 500 g Dinkelmehl oder Roggenmehl verwenden, auf die Maiskörner verzichten und die doppelte Menge Frühstücksspeck nehmen.

Pfälzer Weinbrot mit Zwiebeln

Im Bild rechts unten

FÜR 1 BACKBLECH (30 x 40 cm)
150 g geräucherter, durchwachsener Speck
1 große Zwiebel
20 g Butter
1 Pck. Brotbackmischung für ein Roggenmischbrot
1 Pck. Trockenhefe
120 ml zimmerwarmer, trockener Pfälzer Weißwein
1/4 TL Paprikapulver, edelsüß
1 EL Kümmelsamen
AUSSERDEM
Mehl zum Arbeiten
Kümmelsamen zum Bestreuen
Fett für das Backblech

- Speck würfeln und in einer Pfanne auslassen. Zwiebel abziehen, klein würfeln, in heißer Butter glasig dünsten.
- Backmischung mit der Trockenhefe in einer Schüssel vermengen. Wein und 240 ml lauwarmes Wasser zugießen.
- Alles mit den Knethaken des Handrührgerätes locker vermischen. Zwiebel- und Speckwürfel, Paprikapulver und Kümmel zufügen und zu einem glatten Teig verarbeiten.
- Den Teig abdecken und an einem warmen Ort etwa 30 Minuten gehen lassen.
- Teig auf einer bemehlten Fläche kräftig durchkneten und zu einem ovalen Laib formen. Zugedeckt noch 20 Minuten gehen lassen. Elektro-Ofen auf 200 Grad vorheizen. Das Backblech einfetten.
- Den Laib mehrmals quer etwa 2 cm tief einschneiden und mit Kümmel bestreuen. Auf das Blech setzen. Im Ofen (Gas: Stufe 3) etwa 50 Minuten backen.

Aufwand:	★
Zubereitungszeit:	1 1/4 Stunden
Backzeit:	50 Minuten
Pro Stück:	95 kcal/400 kJ

Rosinen-Haselnuss-Brot

FÜR KASTENFORM 2 l INHALT
650 g Dinkelmehl
1/2 Würfel Hefe (20 g)
1/2 TL Salz
1/2 TL gemahlener Koriander
75 g Haselnusskerne
75 g Rosinen
AUSSERDEM
Fett und Mehl für die Form

- Dinkelmehl in eine Schüssel geben. Die Hefe in einer Tasse lauwarmem Wasser auflösen. In das Mehl eine Mulde drücken und die Hefe hineingeben. Abgedeckt etwa 30 Minuten an einem warmem Ort gehen lassen.
- 100 ml lauwarmes Wasser zugießen und alles mit dem Knethaken des Handrührgerätes vermischen. Salz und Koriander zufügen und nochmals 30 Minuten gehen lassen.
- Elektro-Ofen auf 175 Grad vorheizen. Kastenform einfetten und mit Mehl bestäuben.
- Die Haselnusskerne grob hacken und mit den Rosinen unter den Teig mischen. Den Teig in die Kastenform einfüllen und gleichmäßig verteilen. (Oder: Einen Brotlaib formen, kreuzweise einschneiden und auf ein gefettes Backblech setzen.) Im Ofen (Gas: Stufe 2) etwa 1 Stunde backen. In der Form auskühlen lassen.

Aufwand:	★★
Zubereitungszeit:	1 1/4 Stunde
Backzeit:	1 Stunde
Pro Stück:	205 kcal/860 kJ

Knuspriges Nussbrot

Im Bild rechts Mitte

FÜR 1 KASTENFORM VON 1 1/2-2 l INHALT
500 g Weizenmehl
1/2 Pck. Trockenhefe
1 TL Honig
200 g ungesalzene Nussmischung (Haselnüsse, Walnüsse, Mandeln, Paranüsse, Cashewnüsse)
1/2 Pck. Natur-Sauerteig
1 1/2 TL Salz
AUSSERDEM
Mehl zum Arbeiten
Fett für die Form

- Mehl in eine Schüssel sieben, eine Mulde eindrücken und die Hefe hineinstreuen. Honig mit 200 ml lauwarmem Wasser verrühren, in die Mehlmulde gießen, abdecken und etwa 20 Minuten ruhen lassen. Die Nüsse klein hacken.
- Sauerteig durchkneten, klein schneiden und in die Schüssel geben. Salz in 200 ml Wasser verrühren und zufügen.
- Alles mit den Knethaken des Handrührgerätes zu einem Teig verarbeiten. Zuletzt die Nüsse einarbeiten. Schüssel abdecken und Teig an einem warmen Ort etwa 30 Minuten gehen lassen.
- Elektro-Ofen auf 200 Grad vorheizen. Teig auf einer bemehlten Arbeitsfläche kräftig durchkneten. Kastenform einfetten, mit Mehl bestäuben, Teig einfüllen, abdecken und noch 10 Minuten gehen lassen.
- Teigoberfläche mit Wasser bestreichen und den Teig im Ofen (Gas: Stufe 3) etwa 50 Minuten backen.

Aufwand:	★
Zubereitungszeit:	1 1/2 Stunden
Backzeit:	50 Minuten
Pro Stück:	300 kcal/1260 kJ

Gewürztes Kürbiskernbrot

Im Bild links

FÜR 1 BACKBLECH (30 x 40 cm)
100 g Kürbiskerne (grün, geschält)
1 Pck. Brotbackmischung für ein Vollkornbrot
1/2 TL gemahlener Kardamom
2 Msp. gemahlene Nelken
1/2 TL gemahlener Koriander
AUSSERDEM
Mehl zum Arbeiten
Fett für das Backblech
Kürbiskerne zum Bestreuen

● Kürbiskerne in einer trockenen Pfanne rösten.

● Die Backmischung mit der Trockenhefe, Kürbiskernen und Gewürzen in einer Schüssel vermischen. 365 ml lauwarmes Wasser darüber gießen. Mit den Knethaken des Handrührgerätes einen geschmeidigen Teig herstellen. Teig abgedeckt an einem warmen Ort 30 Minuten gehen lassen.

● Teig auf einer bemehlten Arbeitsfläche kurz durchkneten und zu einem runden Laib formen. Laib abgedeckt nochmals 30 Minuten gehen lassen. Elektro-Ofen auf 250 Grad vorheizen. Backblech einfetten.

● Laib mit Kürbiskernen bestreuen, diese leicht andrücken, auf das Blech setzen und im Ofen (Gas: Stufe 5) auf der untersten Schiene etwa 20 Minuten backen. Dann die Temperatur auf 175 Grad (Gas: Stufe 2) herunterschalten und das Brot noch etwa 55 Minuten backen.

Aufwand:	★
Zubereitungszeit:	1 1/4 Stunden
Backzeit:	1 1/4 Stunden
Pro Stück:	310 kcal/1300 kJ

Schweizer Rüblibrötchen

Im Bild rechts Mitte

2 große Möhren
1 Pck. Brotbackmischung für ein Bauernbrot mit Bio-Vollkorn-Sauerteig
1 Pck. Trockenhefe
3 EL Walnussöl, 1 EL Honig
50 g Sonnenblumenkerne
AUSSERDEM
Mehl zum Arbeiten
Fett für das Backblech
Sonnenblumenkerne zum Bestreuen

● Möhren waschen, schälen und fein reiben.

● Die Backmischung mit Trockenhefe in einer Schüssel vermischen. Etwa 200 ml lauwarmes Wasser, Walnussöl, Honig, Sonnenblumenkerne und Möhrenraspel zufügen. Alles mit den Knethaken des Handrührgerätes zu einem geschmeidigen Teig verarbeiten. Teig abgedeckt an einem warmen Ort 30 Minuten gehen lassen.

● Teig auf einer bemehlten Arbeitsfläche mehrmals durchkneten, zu einer Rolle formen und davon 16 gleich große Stücke abschneiden. Elektro-Ofen auf 200 Grad vorheizen.

● Jedes Teigstück durchkneten, zu einem runden Brötchen formen und die Oberfläche kreuzweise einschneiden. Mit Sonnenblumenkernen bestreuen und diese etwas andrücken. Mit einem Tuch abdecken und nochmals 30 Minuten gehen lassen.

● Brötchen auf ein gefettetes Backblech legen, mit lauwarmem Wasser bestreichen und im Ofen (Gas: Stufe 3) etwa 30 Minuten backen.

Aufwand:	★★
Zubereitungszeit:	1 1/2 Stunden
Backzeit:	30 Minuten
Pro Stück:	285 kcal/1200 kJ

Pizzahörnchen

Im Bild rechts unten

Für 6 Stück
Fertigteig-Croissants aus der Kühltheke (Knack & Back)
80 g Parmaschinken
50 g Mozzarella
30 g grüne entsteinte Oliven
2 EL Olivenöl
1/2 TL getrockneter Oregano
Salz, Pfeffer
AUSSERDEM
Mehl für die Arbeitsfläche
Backpapier für das Backblech
Olivenöl zum Bestreichen

● Die Teigdreiecke einzeln voneinander lösen und auf einer bemehlten Arbeitsfläche auslegen. Die Teigstücke leicht mit Olivenöl bepinseln.

● Den Parmaschinken und Mozzarella sehr fein würfeln. Die Oliven klein hacken.

● Parma- und Mozzarellawürfel mit Oliven, Olivenöl und Oregano locker vermischen. Nur leicht mit Salz und Pfeffer würzen. Elektro-Ofen auf 200 Grad vorheizen. Backblech mit Backpapier auslegen.

● Die vorbereitete Füllung auf den Teigstücken verteilen. Teigstücke zu Hörnchen aufrollen und die Teigränder leicht andrücken.

● Die Pizzahörnchen jeweils mit der Teignaht nach unten das Blech setzen und im Ofen (Gas: Stufe 3) in etwa 15 Minuten goldbraun backen.

Aufwand:	★
Zubereitungszeit:	30 Minuten
Backzeit:	15 Minuten
Pro Stück:	490 kcal/2060 kJ

Feiner Käse, edles Öl un

Auch beim pikanten Backen kommt es auf die richtigen Zutaten

Käse: Reife ist wichtig

Butterkäse oder **Fontina** sind Käsesorten, die man dann verwendet, wenn ein zartes, milderes Käsearoma gewünscht ist. Gut für Füllungen mit Kräutern oder Gemüse.

Parmesan oder **Grana Padano**: Je nach Reifezustand sind sie von unterschiedlicher Würze. Ältere Käse schmelzen beim Überbacken nicht richtig, sind dafür aber um so aromatischer. Deshalb immer mit Mozzarella mischen.

Alter und **mittelalter Gouda** sind ideal für würzige Tartes oder Pies. Junger Gouda sorgt für mildes Aroma und eignet sich gut zum Überbacken von Gemüse oder Pilzen.

Bergkäse, Emmentaler oder **Greyerzer** sind ein fester Bestandteil in der Schweizer Küche, sie sind besonders gut für Käsewähen und herzhafte Tartes. Alle Käsesorten lassen sich auch gut unter Teige mischen.

Olivenöl zum Backen

In den Mittelmeerländern ist es ganz selbstverständlich mit Olivenöl zu backen, häufig auch süßes Gebäck. Man verwendet es dort wie bei uns Butter. Da sich natives Olivenöl bis zu 210 Grad erhitzen lässt, ohne seine wertvollen Inhaltsstoffe einzubüßen, eignet es sich besonders für Gebäck, das bei hohen Temperaturen gebacken wird.

Für knusprige Pizzen & Co

So bleibt der Teigboden schön knusprig: Antihaft-beschichtete und gelochte Bleche sorgen dafür, dass die Feuchtigkeit gut entweichen kann. Es gibt sie mit flachem und hohem Rand im Fachhandel.

Mit dem Pizzaschneider bekommt jeder das richtige Stück. Er schneidet auch Wähen, Tartes, Flammekuchen, Fladen und Blätterteig.

aromatische Gewürze

Sie verleihen den Rezepten das ganz spezielle Aroma

Back-Gewürze aus aller Welt

Kurkuma
Im Geschmack ist die Wurzelknolle dem Ingwer ähnlich, doch etwas bitterer. Man findet Kurkuma meistens als gelbes Pulver im Handel, er heißt auch Gelbwurz. Er würzt orientalisches Gebäck und Brot.

Koriander
Die Samen des Blattkorianders sind als ganze Kerne oder als Pulver zu bekommen. Er ist eine besonders raffinierte Würze für Brotteige, und ist aber auch in der Weihnachtsbäckerei bekannt.

Anis
Das uralte Würzmittel ist vor allem Zutat für feines Gebäck, aber auch für Brot. Die Samen der Doldenpflanze sind ganz und gemahlen zu kaufen. Anis ist auch als Heilmittel für Magenleiden und Erkältungen bekannt.

Kardamom
Es gibt ihn als geschlossene Fruchtkapseln, als Samen und gemahlen. Am intensivsten schmecken die Samen, die in den Fruchtkapseln stecken. Er ist unerlässlich für die Weihnachtsbäckerei.

Sternanis
Der Gourmet denkt an Glühwein – der Fromme an den Stern von Bethlehem. Doch Sternanis kommt ursprünglich aus China. Er riecht und schmeckt fast wie Anis, ist feiner und würzt Plätzchen und Lebkuchen.

Sesam
Er ist eine der älteste Kulturpflanzen der Welt. Sesam ist sehr ölhaltig und schmeckt leicht süßlich, nussartig. Es gibt ihn als getrocknete Samen im Handel und er wird zum Bestreuen von Gebäck verwendet.

Safran
„Safran macht den Kuchen gelb", heißt es in einem Kinderreim. Das teuerste Gewürz der Welt gibt Gebäck ein feines Aroma. Bei uns wird er als Safranfäden oder gemahlen verkauft.

Kreuzkümmel
Er wird auch Kumin genannt und ist als Samen und gemahlen im Handel. Als scharf-aromatisches, fein-bitteres Gewürz aromatisiert er Brotspezialitäten und Käsegebäck.

Geburtstag und Hochzeit
Fasching, Ostern und Muttertag
oder nur ein Kaffeeklatsch –
es gibt genügend Anlässe, Ihrer
Backleidenschaft nachzugeben

Backen für besondere Gelegenheiten

Hochzeitstorte

Rezept siehe folgende Seite

Biskuit und Füllung für Hochzeits- & Herztorte

FÜR 3 SPRINGFORMEN VON 26, 18, 12 cm Ø
7 frische Eier (Größe M)
350 g Zucker
1 Päckchen Vanillezucker
350 g Mehl
2 TL Backpulver
4-6 cl Orangenlikör (z. B. Cointreau)
FÜR DIE FÜLLUNG
4 unbehandelte Limetten
5 EL Puderzucker
8 Blatt weiße Gelatine
700 g Schlagsahne
AUSSERDEM
Backpapier für die Formen

- Den Elektro-Ofen auf 175 Grad heizen. Eier trennen. Die Eigelbe mit 7 EL Wasser, Zucker und Vanillezucker schaumig schlagen. Eiweiße steif schlagen und auf die Eigelbmasse geben. Mehl mit Backpulver darüber sieben, vorsichtig unterheben. Böden der 3 Formen mit Backpapier auslegen, Biskuit jeweils bis zur Hälfte einfüllen.
- Teig in kleiner und mittlerer Form gleichzeitig im Ofen (Gas: Stufe 2) auf dem Rost 30-35 Minuten, den in der großen Form extra etwa 35 Minuten backen. Falls Sie die Etagen-Form verwenden: 40-45 Minuten backen .
- Füllung: Die Limetten waschen und die Schale abziehen. Saft auspressen, mit Schale, 4 EL Wasser, Puderzucker sirupartig einköcheln. Die Gelatine einweichen, im Sirup auflösen. Schale entfernen. Sahne steif schlagen und den abgekühlten Sirup unterziehen.
- Die Böden waagerecht halbieren, für die Herztorte aus den 2 kleinsten halben Böden mit einer Karton-Schablone je ein Herz schneiden. Alle Böden mit Likör tränken. Je 2 passende mit Limettensahne füllen, dabei eine Alu-Manschette um die Biskuits legen. Kalt stellen.

Dekoration für die Hochzeitstorte

400 g Marzipan-Rohmasse
250 g Puderzucker
2-3 EL Aprikosenkonfitüre
gelbe Speisefarbe
400 g weiße Kuvertüre
25 g Kokosfett
FÜR DIE GLASUR
1 Eiweiß (Ei: Größe M)
250 g Puderzucker

So entstehen Marzipanrosen

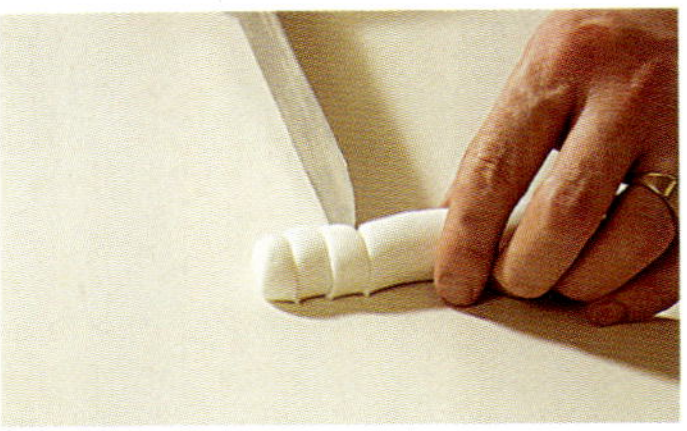

Die Marzipanrolle in 1/2 cm dicke Scheiben teilen. Pro Rose braucht man ca. 7 Stücke.

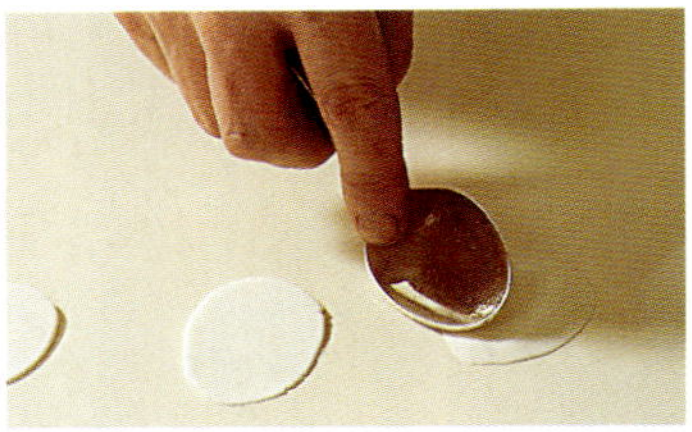

Scheiben mit dem Löffel zu dünnen Blättern flach drücken oder mit dem Wellholz ausrollen.

Für die Blütenmitte je 1 Blatt zur Tüte rollen, dann rundum 6 Blätter leicht versetzt andrücken.

- Marzipan-Rohmasse mit Puderzucker zu einer glatten Masse verkneten. 2/3 davon abnehmen, Rest für die Rosen in Folie einwickeln.
- Marzipan portionsweise zwischen Klarsichtfolie dünn ausrollen. Mit jedem Springformrand einen Kreis ausstechen. Marzipanreste zusammenkneten, nochmals dünn ausrollen und 3 Streifen ausschneiden: 70 cm, 50 cm, 30 cm lang und so breit, wie die gefüllten Biskuitkuchen hoch sind.
- Konfitüre erwärmen, durch ein Sieb streichen. Manschette von den Biskuits lösen, Kuchenränder mit der Konfitüre bestreichen und mit Marzipanstreifen umhüllen. Nun die Kuchenoberflächen mit Konfitüre bestreichen, Marzipankreise darauflegen, andrücken.
- Das Marzipan für die Rosen mit ein paar Tropfen gelber Speisefarbe zartgelb verkneten. Eine Rolle (1 cm Ø) formen und, wie in der Anleitung linksbeschrieben, Rosen formen.
- Kuvertüre mit Kokosfett im warmen Wasserbad schmelzen. Die 3 Biskuits mithilfe einer Palette mit Kuvertüre überziehen, antrocknen lassen, dann Kuchen pyramidenförmig aufeinandersetzen.
- Eiweiß steif schlagen, Puderzucker nach und nach unterrühren. In einen Gefrierbeutel füllen, eine winzige Ecke davon abschneiden und die Ränder der Torte mit kleinen Bögen und Punkten verzieren. Trocknen lassen. Zuletzt die Marzipanrosen auf der Torte verteilen, leicht andrücken.

Aufwand:	★★
Zubereitungszeit:	2 1/2 Stunden
Backzeit:	1 Stunde 20 Minuten
Pro Stück:	607 kcal/ 2540 kJ

Dekoration für die Herztorte

800 g Schlagsahne
2 Päckchen Sahnesteif
500 g reife Erdbeeren

● Die Sahne mit Sahnesteif steif schlagen und die 3 gefüllten Biskuits mithilfe einer Palette dünn mit Sahne überziehen. Biskuits und übrige Sahne kalt stellen.

● Die Erdbeeren waschen, putzen und längs in dünne Scheiben schneiden. Die Erdbeeren mit Küchenpapier trockentupfen.

● Die Biskuits aus der Manschette lösen und aufeinandersetzen. Übrige Sahne in einen Spritzbeutel mit Sterntülle (Größe 4) füllen und auf die einzelnen Terrassen Wellenlinien spritzen (siehe Bild links).

● Die Erdbeerscheiben auf der unteren und mittleren Stufe leicht überlappend rund um die Torte legen. Herz völlig mit den Erdbeerscheiben auslegen.

Für die Verzierung der Torte den Spritzbeutel in halbkreisförmigen Bewegungen führen.

Aufwand:	★★
Zubereitungszeit:	1 1/2 Stunden
Backzeit:	1 Stunde 20 Minuten
Pro Stück:	595 kcal/ 2500 kJ

Schweizer Hasenöhrl

Im Bild rechts

BITTE BEACHTEN: TEIG MUSS 2 STUNDEN KÜHLEN

Für 40 Stück

75 g Butter
150 g Zucker
1 Prise Salz
2 Eier (Größe M)
100 g Schlagsahne
2 EL Kirschwasser
500 g Mehl

AUSSERDEM

Mehl zum Arbeiten
Frittierfett
Zimtzucker zum Bestreuen

- Butter mit Zucker weiß-schaumig rühren. Salz, Eier, Sahne und Kirschwasser unterrühren. Mehl zugeben, alles zu einem glatten, geschmeidigen Teig verkneten. Mit Folie abgedeckt etwa 2 Stunden kühl stellen.
- Teig auf leicht bemehlter Arbeitsfläche messerrückendick ausrollen. Mit einem Teigrädchen oder einem Messer Rhomben (6 cm Seitenlänge) ausschneiden. Jedes Teigstück in der Mitte längs einschneiden und einen Zipfel der Rhombe durch die Öffnung ziehen (s. Seite 311).
- Fett auf 180 Grad erhitzen. Die Hasenöhrl darin portionsweise je 4-5 Minuten backen, dabei einmal wenden. Sie sollen hellbraun und innen durchgebacken sein.
- Die Hasenöhrl auf Küchenpapier abtropfen lassen. Mit Zimtzucker bestreuen. Heiß servieren.

Aufwand:	★★
Zubereitungszeit:	45 Minuten
Frittierzeit:	20 Minuten
Pro Stück:	525 kcal/2205 kJ

Berliner Pfannkuchen

Im Bild links oben

Für 20 Stück

500 g Mehl
1 Würfel Hefe, 50 g Zucker
1/4 l lauwarme Milch
2 EL Öl, 1 Ei, 1 Eigelb (Größe M)
1/2 TL Salz

AUSSERDEM

Mehl zum Arbeiten
Himbeerkonfitüre
1 Eiweiß zum Bestreichen
Frittierfett
Puderzucker zum Bestäuben

- Mehl in eine Schüssel geben, eine Mulde eindrücken. Hefe hineinbröckeln. 1 TL Zucker, Hälfte der Milch hineingeben und mit etwas Mehl zu einem Vorteig verrühren. 15 Minuten ruhen lassen.
- Restlichen Zucker, übrige Milch, Öl, Ei, Eigelb und Salz zum Vorteig geben und mit den Knethaken des Handrührgerätes verkneten. Dann den Teig mit einem Holzlöffel so lange schlagen, bis er sich vom Schüsselrand löst. Zugedeckt an einem warmen Ort 40 Minuten gehen lassen.
- Den Teig 1 cm dick ausrollen und halbieren. Auf einer Hälfte der Teigplatte 20 Kreise (6-7 cm Ø) markieren und jeweils in die Mitte einen Klecks Konfitüre geben. Markierungslinien mit Eiweiß bestreichen. Zweite Teigplatte darauflegen und Krapfen ausstechen (s. Seite 310). Ränder andrücken. Krapfen noch 15-20 Minuten gehen lassen.
- Frittierfett auf 180 Grad erhitzen. Krapfen portionsweise darin 3 Minuten backen, wenden und weitere 3 Minuten backen.
- Wenn sie hellbraun sind, herausnehmen und auf Küchenpapier abtropfen lassen. Abkühlen lassen und mit Puderzucker bestäuben.

Aufwand:	★★
Zubereitungszeit:	2 Stunden
Frittierzeit:	45 Minuten
Pro Stück:	185 kcal/780 kJ

Rheinische Muzen

Im Bild links Mitte

Für 60 Stück

75 g Butter
50 g Zucker
1 Päckchen Vanillezucker
1 Ei (Größe M)
1/2 TL abgeriebene, unbehandelte Zitronenschale
2 EL Rum
4 EL Weißwein
etwa 250 g Mehl
1 Prise Salz

AUSSERDEM

Mehl zum Arbeiten
Frittierfett
Puderzucker zum Bestäuben

- Butter in einem Topf zerlassen. Topf vom Herd nehmen. Zucker, Vanillezucker, Ei, Zitronenschale und Rum einrühren.
- Mehl in eine Schüssel sieben, Buttermischung, Wein und Salz zufügen und alles zu einem glatten Teig verkneten. Sollte der Teig kleben, noch etwas Mehl unterkneten.
- Teig auf einer bemehlten Arbeitsfläche zu einer dicken Platte ausrollen, zusammenschlagen, in Klarsichtfolie wickeln und 1 Stunde in den Kühlschrank legen.
- Teig messerrückendick ausrollen. Mit dem Teigrädchen Rhomben (etwa 4 cm Seitenlänge) ausrädeln.
- Das Fett auf 180 Grad erhitzen und die Muzen darin portionsweise 1-2 Minuten goldgelb ausbacken. Dabei einmal wenden.
- Die Muzen aus dem Fett heben, auf Küchenpapier abtropfen und erkalten lassen. Zum Schluss mit etwas Puderzucker bestäuben.

Aufwand:	★★
Zubereitungszeit:	1 1/4 Stunden
Frittierzeit:	15 Minuten
Pro Stück:	110 kcal/460 kJ

Schwäbische Fasnachtsküchle

Im Bild rechts unten

Für 50-60 Stück

500 g Mehl
1/2 Würfel Hefe (21 g)
50 g Zucker
knapp 1/4 l lauwarme Milch
1/2 TL Salz
2 Eier (größe M)
75 g weiche Butter

AUSSERDEM
Mehl zum Arbeiten
Frittierfett
Puderzucker zum Bestäuben

- Mehl in eine Schüssel schütten, in die Mitte eine Mulde drücken. Hefe hineinbröckeln, mit 1 TL Zucker, der Hälfte der Milch und etwas Mehl zu einem Vorteig verrühren. Vorteig zugedeckt an einem warmen Ort etwa 15 Minuten gehen lassen.
- Die restliche Milch, den übrigen Zucker, Salz, Eier und Butter in die Schüssel geben. Alles zuerst mit den Knethaken des Handrührgerätes, dann mit einem Holzlöffel zu einem glatten Teig verarbeiten: Er soll Blasen werfen und sich vom Schüsselrand lösen. Teig noch 30-40 Minuten gehen lassen.
- Den Teig auf leicht bemehlter Arbeitsfläche etwa fingerdick ausrollen. Mit einem Glas oder einer runden Ausstechform (3-4 cm Ø) Plätzchen ausstechen. Jeweils in der Mitte mit einer kleinen Form (1-2 cm Ø) ein Loch ausstechen.
- Frittierfett auf 175-180 Grad erhitzen. Kringel darin portionsweise ausbacken, bis sie hellbraun sind. Auf Küchenpapier abtropfen lassen. Mit Puderzucker bestäuben.

Aufwand:	★★
Zubereitungszeit:	1 1/2 Stunden
Frittierzeit:	50 Minuten
Pro Stück:	100 kcal/420 kJ

Crailsheimer Horaffen

Im Bild rechts Mitte

BITTE BEACHTEN: TEIG MUSS 2 STUNDEN RUHEN

Für 15 Stück

500 g Mehl, 1/2 Würfel Hefe (21 g)
60 g Zucker
200 ml lauwarme Milch
1 Päckchen Vanillezucker
1/2 TL Salz, 60 g weiche Butter
80 g kalte Butter in Flocken

ZUM VERZIEREN
150 g Puderzucker
2 EL Milch
2-3 EL gehackte Haselnüsse

So werden Horaffen geformt

Den gut gekühlten Teig ausrollen und die Teigplatte in gleichmäßige längliche Dreiecke schneiden, an den kurzen Seiten einschneiden.

Dann jeweils die Spitzen und Teigstränge kordelartig verdrehen. Die Schenkel der Dreiecke nach den Seiten ziehen und wie ein Widdergehörn ausformen.

AUSSERDEM
Mehl zum Arbeiten
Backpapier für das Blech

- Mehl in eine Schüssel schütten, in die Mitte eine Mulde drücken. Hefe hineinbröckeln, mit 1 TL Zucker, der Hälfte der Milch und etwas Mehl zu einem Vorteig verrühren. Vorteig zugedeckt an einem warmen Ort etwa 15 Minuten gehen lassen.
- Restliche Milch, übrigen Zucker, Vanillezucker, Salz und weiche Butter in die Schüssel geben. Alles mit den Knethaken des Handrührgerätes zu einem glatten Teig verarbeiten. Teig noch 30-40 Minuten gehen lassen, bis er sein Volumen verdoppelt hat.
- Dann den Teig mit einem Holzlöffel schlagen, bis er sich vom Schüsselrand löst. Teig auf bemehlter Arbeitsfläche zu rechteckiger Platte ausrollen. Butterflocken auf die Hälfte der Teigplatte streuen, die andere Hälfte darüberklappen. In Klarsichtfolie gewickelt etwa 20 Minuten kühl legen.
- Teig auf bemehlter Arbeitsfläche zur dreifachen Länge ausrollen, linke und rechte Seiten über die Mitte schlagen, Teig 20 Minuten kühlen. Vorgang wiederholen. Dann Teig zu einem Rechteck (15 x 30 cm) ausrollen. Elektro-Ofen auf 200 Grad vorheizen.
- Die Teigplatte in gleichmäßige Dreiecke (5 x 15 cm) schneiden (siehe links) und von den kurzen Seiten her einschneiden. Jeweils Spitze und Teigstränge kordelartig verdrehen und wie ein Gehörn formen (siehe links). Horaffen auf ein mit Backpapier belegtes Blech setzen und im Ofen auf der 2. Schiene von unten (Gas: Stufe 3) etwa 20 Minuten backen.
- Den Puderzucker mit der Milch zu einem Guss verrühren. Nüsse in einer beschichteten Pfanne anrösten. Die Horaffen auf ein Gitter setzen, noch heiß mit Zuckerguss bestreichen und mit Nüssen bestreuen. Die Horaffen schmecken am besten warm.

Aufwand:	★★★
Zubereitungszeit:	1 1/2 Stunden
Backzeit:	20 Minuten
Pro Stück:	265 kcal/1110 kJ

Spritzkuchen

1/4 l Milch
100 g Butter
1 Msp. Salz, 1 TL Zucker
230 g Mehl
7 Eier (Größe M)
AUSSERDEM
Fett für das Papier
Frittierfett
100 g Aprikosenkonfitüre
100 g Puderzucker
nach Belieben 2 EL Arrak
oder Rum

● Milch, Butter, Salz und Zucker unter Rühren aufkochen. Mehl auf einmal dazuschütten, dabei kräftig weiterrühren. Masse weiter erhitzen, dabei ständig rühren, so lange, bis sie sich als Kloß vom Topf löst und eine weiße Haut den Topfboden überzieht.
● Masse in eine Schüssel füllen, abkühlen lassen. 1 Ei unterrühren, bis es sich völlig mit der Masse verbunden hat. Die übrigen Eier nacheinander zugeben, dabei jedes einzelne gut unterrühren. Die Masse soll glatt sein und weich vom Löffel fallen.
● Pergamentpapier in 15 x 15 cm breite Stücke teilen, einfetten und das Fett im Kühlschrank erstarren lassen. Den Teig in einen Spritzbeutel mit großer Sterntülle füllen und Ringe aufs Papier spritzen.
● Zum Frittieren Fett in einem Topf oder einer Fritteuse erhitzen. Die Kuchen auf dem Papier ins heiße Fett gleiten lassen. Das Papier entfernen. Die Kuchen etwa 6 Minuten ausbacken, nach 3 Minuten wenden. Herausheben und auf Küchenpapier abtropfen lassen.
● Konfitüre mit 2 EL Wasser erwärmen. Warmen Kuchen damit bestreichen, trocknen lassen. Puderzucker mit Arrak oder Rum oder Wasser glatt rühren. Kuchen mit der Oberseite in die Glasur tauchen. Abtropfen lassen. Frisch servieren.

Aufwand:	★★
Zubereitungszeit:	40 Minuten
Frittierzeit:	20 Minuten
Pro Stück:	80 kcal/340 kJ

Reiskrapfen

Für 15 Stück
30 g kandierte Kirschen
50 g Walnusskerne, 60 g Rosinen
20 ml Madeira oder
trockener Sherry
1/2 l Milch
1/4 TL Salz, 40 g Zucker
200 g Arborio-Reis oder
Milchreis, 50 g Butter
2 Eier (Größe M)
abgeriebene Schale von je 1/2
unbehandelten Zitrone und Orange
10 g frische Hefe, 50 g Mehl
AUSSERDEM
Frittierfett
Puderzucker zum Bestäuben

- Kirschen und Nüsse hacken. Mit Rosinen in Madeira oder Sherry etwa 1 Stunde marinieren.
- Inzwischen die Milch aufkochen, Salz und 2 TL Zucker zugeben. Reis in die kochende Milch geben, unter Rühren 4-5 Minuten kochen lassen. Butter zugeben, umrühren. Bei kleiner Hitze köcheln lassen, bis die Flüssigkeit völlig aufgesogen ist und ein fast trockener Reisbrei entstanden ist. Abkühlen lassen, bis die Masse lauwarm ist.
- Die Eier trennen. Eiweiße steif schlagen, restlichen Zucker langsam einrieseln lassen.
- Zitrusschalen, Eigelbe und zerbröckelte Hefe unter den Reisbrei mischen. Mehl, eingeweichte Kirschen und Nüsse unterheben. Eischnee vorsichtig unterziehen.
- Zum Frittieren Fett in einem Topf oder einer Fritteuse erhitzen. Zwei Esslöffel in Wasser tauchen und damit von der Reismasse Nocken abstechen. Die Reiskrapfen im heißen Fett 3-5 Minuten knusprig ausbacken, sie sollen innen aber noch weich sein.
- Die Krapfen herausnehmen und auf Küchenpapier abtropfen lassen. Warm mit Puderzucker bestäuben.

Aufwand:	★★
Zubereitungszeit:	1 Stunde
Frittierzeit:	20 Minuten
Pro Stück:	150 kcal/630 kJ

Valentinstorte

BITTE BEACHTEN: TORTE MUSS 2 STUNDEN KÜHLEN
FÜR SPRINGFORM VON 18 cm Ø
2 Eier (Größe M), Salz
60 g Zucker, 20 g Speisestärke
40 g Mehl, 2 Msp. Backpulver
40 g gemahlene Mandeln
FÜR DIE FÜLLUNG
5 Blatt rote Gelatine
150 g TK-Himbeeren
250 g Magerquark, 30 g Zucker
1/2 unbehandelten Zitrone
150 g Schlagsahne
FÜR DIE VERZIERUNG
200 g Puderzucker
2 EL Zitronensaft
rote Speisefarbe
100 g Marzipan-Rohmasse
AUSSERDEM
Backpapier für die Form

So entstehen Herzblätter

1

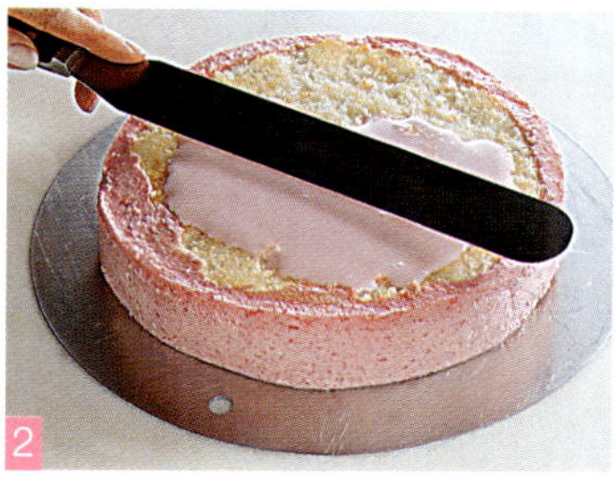

2

1 Marzipan auf Puderzucker 3 mm dünn ausrollen und verschiedene große Herzen ausstechen.

2 Zuckerguss auf die Torte streichen, das geht am besten mit einer Palette. Herzen auf den Guss setzen und mit Puderzucker bestäuben.

- Elektro-Ofen auf 175 Grad vorheizen.
- Eier trennen. Eiweiße mit 1 Prise Salz steif schlagen, 40 g Zucker langsam einrieseln lassen, dabei auf niedrigster Stufe weiterschlagen. Eigelbe mit übrigem Zucker und 4 EL lauwarmem Wasser cremig aufschlagen. Eischnee auf die Eigelb-Masse geben. Speisestärke, Mehl und Backpulver mischen, darüber sieben und mit den Mandeln unterziehen.
- Boden der Form mit Backpapier auslegen, Teig in die Form füllen und im Ofen (Gas: Stufe 2) etwa 25 Minuten backen. Herausnehmen, Papier abziehen, abkühlen lassen und waagerecht halbieren.
- Gelatine einweichen. Zitrone heiß waschen, abtrocken, Schale abreiebn und Saft auspressen. Himbeeren auftauen, pürieren und durch ein Sieb streichen. Mit Quark, Zucker, Zitronensaft und -schale verrühren. Die Gelatine bei milder Hitze auflösen, mit 2-3 EL Quarkmasse vermischen und unter die Quarkcreme rühren.
- Sahne steif schlagen und, wenn die Quarkcreme zu gelieren beginnt, unterheben. Tortenboden auf eine Platte setzen, leicht geöffneten Springformrand darumlegen. Die Creme so auf den Boden streichen, dass sie über den Tortenrand bis zum Boden läuft. Zweiten Boden darauf legen, etwas andrücken und mindestens 2 Stunden kalt stellen.
- Den Tortenring entfernen. 150 g Puderzucker mit wenig Wasser, Zitronensaft und Lebensmittelfarbe glatt rühren, auf die Torte streichen. Trocknen lassen. Marzipan mit übrigem Puderzucker und Speisefarbe verkneten. Etwa 3 mm dünn ausrollen und unterschiedlich große Herzen ausstechen. Die Torte mit den Marzipanherzen dekorieren.

Aufwand:	★★
Zubereitungszeit:	1 1/2 Stunden
Backzeit:	25 Minuten
Pro Stück:	360 kcal/1510 kJ

Giotto-Torte

FÜR SPRINGFORM VON 18 cm Ø
50 g Butter
40 g Zucker
1 Ei (Größe M)
60 g Mehl
80 g gemahlene Mandeln
1 TL Backpulver
FÜR DIE FÜLLUNG
1 Glas Kirschen
(Abtropfgewicht: 370 g)
etwa 60 Giotto-Kugeln
1 EL Zucker
1 1/2 EL Speisestärke
AUSSERDEM
40 g gehackte Haselnusskerne
250 g Schlagsahne
1 Päckchen Vanillezucker
1 Päckchen Sahnesteif
Backpapier und Fett für die Form

● Elektro-Ofen auf 200 Grad vorheizen. Butter und Zucker cremig rühren, Ei unterrühren. Mehl, Mandeln und Backpulver vermischen und unter die Butter-Masse rühren. Springformboden mit Backpapier auslegen, Rand fetten. Teig einfüllen, glatt streichen. Im Ofen (Gas: Stufe 3) etwa 20 Minuten backen. Aus der Form stürzen, Backpapier abziehen, Kuchen erkalten lassen.

● Die Kirschen in ein Sieb schütten, Saft auffangen. Giotto-Kugeln halbieren, 18 Hälften für die Deko beiseite legen. Zucker und Stärke mit etwas Kirschsaft glatt rühren. Übrigen Saft aufkochen, Stärke unter Rühren zugießen. Nochmals aufkochen. Boden dünn mit angedicktem Saft bestreichen. 200 g Kirschen mit restlichem Saft vermischen.

● Kuchen waagerecht teilen, unteren Boden mit Tortenring umstellen. Giotto-Hälften dicht aneinander, Schnittfläche nach unten, auf den Boden legen. Abgekühlte Kirschen darüber verteilen, mit übrigen Giottos (bis auf die für das Herz!) und Kuchendeckel belegen. 30 Minuten kühlen.

● Gehackte Nüsse in einer Pfanne ohne Fett rösten. Sahne mit Vanillezucker und Sahnesteif schlagen, Torte damit bestreichen. Rand mit Nüssen bestreuen, Oberfläche in Herzform mit Sahne-Tuffs und beiseite gelegten Giotto-Hälften verzieren.

Aufwand:	★★
Zubereitungszeit:	1 Stunde
Backzeit:	20 Minuten
Pro Stück:	500 kcal/2100 kJ

Schokoherz mit Vanillecreme

BITTE BEACHTEN: TORTE MUSS 2 STUNDEN KÜHLEN

FÜR 1 HERZBACKFORM

- 1 Packung Backmischung Schokokuchen sowie Fett, Eier und Flüssigkeit nach Packungsangabe
- 2 TL Instant-Kaffeepulver
- 1 Vanilleschote
- 350 ml Milch
- 1 Päckchen Vanille-Puddingpulver (für 1/2 l Flüssigkeit)
- 1 Päckchen Vanillezucker
- 250 g Mascarpone
- 40 g Zucker
- nach Belieben 2 EL Portwein oder Orangenlikör
- 3-4 EL Orangenmarmelade

AUSSERDEM

- Fett für die Form
- 50 g Mandelblättchen
- Puderzucker und Kakaopulver

● Nach Packungsangabe Elektro-Ofen vorheizen und Teig zubereiten. Die Form einfetten. Kaffeepulver in den Teig rühren. Den Teig in die Form füllen und im Ofen wie angegeben backen.

● Vanilleschote längs aufschneiden, das Mark herauschaben. 300 ml Milch mit Vanillemark aufkochen. Die übrige Milch mit Puddingpulver, Zucker und Vanillezucker glatt rühren. In die kochende Milch geben, einmal aufkochen. Mascarpone unterrühren und abkühlen lassen. Nach Belieben mit Wein oder Likör aromatisieren.

● Den Kuchen aus der Form nehmen, abkühlen lassen. Marmelade erwärmen. Kuchen zweimal waagerecht halbieren, so dass 3 Böden entstehen. Den unteren Boden mit 1-2 EL Marmelade bestreichen, zweiten Boden darauf legen. Vanillecreme darauf streichen, mit dem dritten Boden bedecken. Torte 2 Stunden kalt stellen.

● Mandeln in einer Pfanne ohne Fett rösten. Kuchenrand mit übriger Marmelade einstreichen, mit Mandeln bestreuen. Kuchen mit Puderzucker bestäuben. Mithilfe einer Deko-Schablone (Haushaltswarenladen) mit Kakaopulver Muster darauf stäuben.

Aufwand:	★★
Zubereitungszeit:	1 Stunde
Backzeit:	20 Minuten
Pro Stück:	500 kcal/2100 kJ

Variation

● Schokoladen-Puddingpulver statt Vanille-Puddingpulver nehmen, auf Vanilleschote und Vanillezucker verzichten, statt dessen 1/2 TL Zimt und 1 EL Zucker unterrühren.

Mambo-Torte

BITTE BEACHTEN: TORTE MUSS 3 STUNDEN KÜHLEN

FÜR SPRINGFORM VON 18 cm Ø

100 g Zartbitter-Kuvertüre
2 Eier (Größe M), 60 g Zucker
1 Päckchen Vanillezucker
1 TL Speisestärke, 40 g Mehl
1 TL Kakaopulver
1 Msp. Backpulver

FÜR DIE CREME

6 Blatt weiße Gelatine
2 Limetten, 3 EL Zucker
3 Päckchen Vanillezucker
450 g Sahnejoghurt
200 g Schlagsahne
3 EL Blue Curaçao

AUSSERDEM

1 Mango, 1 Papaya, 2 Kiwis
1 Pck. heller Tortenguss
1/4 l heller Traubensaft
Backpapier für die Form

Kuvertüre nach Noten

1 Flüssige Kuvertüre in Tütchen füllen. Spitze abschneiden, Noten und Violinschlüssel damit auf Pergamentpapier „aufmalen".
2 Schokolade trocknen lassen, mit einem Spatel abheben und auf den Tortenrand setzen.

- Kuvertüre schmelzen. Die Hälfte in eine kleine Pergamentpapier-Tüte füllen, Noten und Notenschlüssel auf Papier spritzen (s. unten). Erstarren lassen.
- Elektro-Ofen auf 175 Grad vorheizen. Eier, Zucker und Vanillezucker schaumig schlagen. Übrige Kuvertüre unterrühren. Stärke, Mehl, Kakao und Backpulver mischen, unterheben. Springform mit Backpapier auslegen. Teig einfüllen und glatt streichen. Im Ofen (Gas: Stufe 2) etwa 25 Minuten backen. Aus der Form stürzen, Backpapier abziehen und erkalten lassen.
- Gelatine einweichen. Limetten auspressen. Beide Zuckersorten mit Joghurt verrühren. Gelatine bei milder Hitze auflösen. Mit Limettensaft vermischt in den Joghurt rühren. Sahne steif schlagen. Unterziehen, wenn der Joghurt zu gelieren beginnt. 1/4 der Creme abnehmen und mit Curaçao verrühren.
- Biskuit waagerecht halbieren. Unteren Boden mit einem Tortenring umstellen. 2/3 der hellen Creme aufstreichen, Likör-Creme löffelweise darauf geben und mit dem Löffelstiel marmorieren. Oberen Boden auflegen, mit der übrigen Creme rundum bestreichen. 3 Stunden kalt stellen.
- Mango schälen, Fruchtfleisch in Scheiben vom Stein schneiden. Papaya schälen, halbieren, Kerne herauslösen, Fruchtfleisch in Scheiben schneiden. Kiwis schälen, in Scheiben schneiden. Aus dem Fruchtfleisch Herz-Formen ausstechen.
- Den Tortenguss mit Saft nach der Packungsangabe zubereiten. Früchte mit Guss bestreichen, mit der Guss-Seite nach unten auf die Torte geben. Torte mit Guss überziehen. Rand mit Schoko-Noten verzieren.

Aufwand:	★★★
Zubereitungszeit:	55 Minuten
Backzeit:	25 Minuten
Pro Stück:	360 kcal/1510 kJ

Osterlämmer

Für 8 oder 16 Stück	
500 g Mehl	
1/2 Würfel Hefe (21 g)	
etwa 300 ml lauwarme Milch	
100 g Zucker, 50 g Butter, Salz	
1 Dose Aprikosen	
(Abtropfgewicht: 480 g)	
1 EL Zitronensaft	
150 g Butter	
150 g feste Kokoscreme	
(Fertigprodukt)	
300 g Mehl, 150 g Zucker	
1 Eigelb (Ei: Größe M)	
2 EL Milch	
AUSSERDEM	
Backpapier für das Blech	
Puderzucker zum Bestäuben	

- Mehl in Schüssel geben, in die Mitte des Mehls eine Mulde drücken. Hefe mit 4 EL Milch und 1 TL Zucker verrühren und in die Vertiefung geben. Mit etwas Mehl vom Rand verrühren. Hefeansatz zugedeckt an einem warmen Ort 20 Minuten gehen lassen.
- Vorteig mit Milch, Zucker, Butter und 1 Prise Salz verkneten und 30 Minuten ruhen lassen. Teig in 8 oder 16 Teile teilen und 1/2 cm dick ausrollen. Nach einer Schablone Lämmer ausschneiden (z. B. von einem Buch abmalen). Auf ein mit Backpapier ausgelegtes Backblech legen. Elektro-Ofen auf 175 Grad vorheizen.
- Aprikosen abtropfen lassen, mit Zitronensaft beträufeln. Früchte in Spalten schneiden, auf den Lämmern verteilen (Kopf und Beine aussparen). Butter, Kokoscreme, Mehl und Zucker zu Streuseln kneten, auf den Aprikosen verteilen. Eigelb mit Milch verquirlen, Kopf und Beine der Lämmchen damit bepinseln. Lämmer im Ofen auf unterster Schiene (Gas: Stufe 2) etwa 15 Minuten backen. Auskühlen lassen, mit Puderzucker bestäuben.

Aufwand:	★★★
Zubereitungszeit:	2 Stunden
Backzeit:	15 Minuten
Pro Stück:	430 kcal/1810 kJ

Hefezopf mit Marzipan

FÜR 1 BACKBLECH (30 x 40 cm)
500 g Mehl, 1 Pck. Trockenhefe
75 g Zucker, 180 ml Milch
1 Ei (Größe M)
100 g weiche Butter
FÜR DIE FÜLLUNG
200 g Marzipan-Rohmasse
50 g Pistazienkerne
200 g getrocknete Aprikosen
1 Päckchen Vanillezucker
100 ml Orangensaft
100 g geschälte gemahlene Mandeln
AUSSERDEM
Backpapier für das Blech
1 Eigelb (Ei: Größe M)
2 EL Milch
2 EL Mandelblättchen
1 EL Puderzucker

● Mehl, Trockenhefe und Zucker mischen. Lauwarme Milch, Ei und Butter unterrühren, gut verkneten und zugedeckt an einem warmen Ort 20 Minuten gehen lassen.
● Marzipan zerkrümeln, Pistazien und Aprikosen fein hacken. Mit Vanillezucker, Orangensaft und Mandeln mischen. Hefeteig durchkneten und weitere 20 Minuten gehen lassen.
● Teig dritteln, zu je 50 cm langen Strängen formen, zu 10 cm breiten Rechtecken ausrollen, mit jeweils 1/3 Füllung bestreichen, dabei einen Rand frei lassen. Von der Längsseite her einrollen, gut festdrücken.
● Aus den gefüllten Strängen einen Zopf flechten, auf ein mit Backpapier belegtes Blech setzen. Zugedeckt noch mal 30 Minuten gehen lassen. Elektro-Ofen auf 175 Grad vorheizen.
● Eigelb mit Milch verquirlen, Zopf damit bestreichen. Im Ofen (Gas: Stufe 2) etwa 45 Minuten backen. Nach Belieben mit Mandelblättchen bestreuen, mit Puderzucker bestäuben.

Aufwand:	★★
Zubereitungszeit:	1 1/2 Stunden
Backzeit:	45 Minuten
Pro Stück:	210 kcal/880 kJ

Vanillekranz

FÜR 1 BACKBLECH (30 x 40 cm)
400 g Mehl, 1 Würfel Hefe (42 g)
175 ml Milch, 100 g Zucker
2 Vanilleschoten
1-2 Rosmarinzweige
2 Eier (Größe M), 1 Prise Salz
100 g weiche Butter
100 g gemahlene Mandeln
AUSSERDEM
1 Eigelb, 1-2 EL Milch
100 g Puderzucker
1 EL Zitronensaft
einige Rosmarinnadeln, etwas Zitronensaft und 1 EL Zucker

- Mehl in eine Schüssel geben. Hefe zerbröckeln, mit 50 ml lauwarmer Milch, 2 EL Zucker und etwas Mehl verrühren und in eine Mehlmulde geben. Zugedeckt 20 Minuten gehen lassen.
- Vanilleschoten aufschlitzen, Mark herauskratzen. Schoten, Mark und Rosmarinzweige in übriger Milch erhitzen, 5 Minuten köcheln lassen. Abkühlen lassen, Vanilleschoten und Rosmarin entfernen.
- Eier, Salz, Butter, Mandeln und restlichen Zucker zum Vorteig geben, gut verkneten. Dabei so viel Vanille-Milch zugeben, bis ein elastischer Teig entsteht. Zugedeckt 30 Minuten gehen lassen.
- Elektro-Ofen auf 175 Grad vorheizen. Teig in 2 Hälften teilen und zu etwa 50 cm langen Strängen ausrollen. Spiralförmig miteinander verdrehen, einen Kranz formen. Enden aneinander drücken. Auf ein mit Backpapier belegtes Blech setzen, 15 Minuten gehen lassen. Eigelb mit Milch verquirlen, auf den Kranz streichen. Im Ofen (Gas: Stufe 2) etwa 35 Minuten backen.
- Puderzucker mit Zitronensaft verrühren, auf den Kranz streichen. Rosmarinnadeln mit Zitronensaft bepinseln, in Zucker wälzen, aufstreuen.

Aufwand:	★★
Zubereitungszeit:	1 1/2 Stunden
Backzeit:	35 Minuten
Pro Stück:	240 kcal/1010 kJ

Gefüllte Ostereier

Für 24 Stück	
500 g Mehl	
1/2 Würfel Hefe (21 g)	
etwa 300 ml lauwarme Milch	
100 g Zucker, 50 g Butter	
1 Prise Salz, 100 g gemischte getrocknete Früchte	
200 g Marzipan-Rohmasse	
etwas Ingwerpulver	
Saft und abgeriebene Schale von 1 unbehandelten Zitrone	
AUSSERDEM	
Backpapier für das Blech	
200 g Puderzucker	
1 frisches Eiweiß (Ei: Größe M)	
Lebensmittelfarben	

- Mehl in Schüssel geben, eine Mulde eindrücken. Hefe mit 4 EL Milch, 1 TL Zucker und etwas Mehl verrühren und in die Mulde geben. Vorteig zugedeckt an einem warmen Ort etwa 20 Minuten gehen lassen.
- Vorteig mit Milch, Zucker, Butter und Salz verkneten und weitere 30 Minuten ruhen lassen. Teig in 2 Teile teilen und zu dünnen Strängen ausrollen, jeden in 12 Teile schneiden.
- Früchte und Marzipan fein hacken. Miteinander verkneten, mit Ingwer, Zitronensaft und -schale abschmecken. Marzipan zu 24 Kugeln formen. Elektro-Ofen auf 175 Grad vorheizen.
- Teigkugeln etwas flach drücken. Marzipan in die Mitte legen und mit Teig umhüllen. Eier formen, auf ein mit Backpapier ausgelegtes Backblech setzen und im Ofen (Gas: Stufe 2) 15-20 Minuten backen. Abkühlen.
- Puderzucker mit Eiweiß glatt rühren. Teilweise mit Lebensmittelfarben leicht einfärben. Jedes Ei glasieren (siehe links), trocknen lassen. Übrigen Guss in einen Gefrierbeutel geben, eine kleine Ecke abschneiden, so dass ein Spritzloch entsteht, Eier verzieren.

Aufwand:	★★
Zubereitungszeit:	45 Minuten
Backzeit:	15-20 Minuten
Pro Stück:	195 kcal/820 kJ

Eier verzieren

Die gebackenen Eier auf einen Kuchenrost legen und mit rosa Guss bestreichen. Trocknen lassen und anschließend mit weißem Guss Schlangenlinien oder Blüten aufspritzen.

Biskuit-Lämmchen

FÜR 1 LÄMMCHEN-FORM
160 g Zucker
1 Päckchen Vanillezucker
4 Eier (Größe M)
160 g Mehl
40 g Speisestärke
abgeriebene Zitronenschale

AUSSERDEM
Fett und Mehl für die Form
Puderzucker zum Bestäuben
1 TL geriebene Kuvertüre

● Elektro-Ofen auf 175 Grad vorheizen. Zucker und Vanillezucker mischen. Eier mit 4 EL heißem Wasser verquirlen, dabei Zucker-Vanillezucker-Mischung esslöffelweise zufügen. 2 Minuten schaumig schlagen.

● Mehl, Speisestärke und Zitronenschale mischen und nach und nach unter den Teig heben.

● Die Form gründlich einfetten und mit Mehl ausstäuben. Den Teig einfüllen, die Form schließen und den Kuchen im Ofen (Gas: Stufe 2) 30-40 Minuten backen.

● Biskuit in der Form abkühlen lassen, dann herausnehmen. Kuvertüre schmelzen und die Augen damit markieren. Das Lämmchen zum Schluss mit Puderzucker bestreuen.

Aufwand:	★★
Zubereitungszeit:	15 Minuten
Backzeit:	40 Minuten
Insgesamt:	110 kcal/460 kJ

Hasenbrötchen

Für 16 Stück
500 g Mehl
1/2 Würfel Hefe (21 g)
etwa 300 ml lauwarme Milch
100 g Zucker
50 g Butter
1 Prise Salz
AUSSERDEM
Backpapier für das Blech
1 Eigelb (Ei: Größe M)
3–4 EL Schlagsahne
Rosinen
Mandelstifte

- Mehl in die Schüssel geben, in die Mitte des Mehls eine Mulde drücken. Hefe mit 4 EL Milch, 1 TL Zucker und etwas Mehl verrühren und in die Vertiefung geben. Hefeansatz zugedeckt an einem warmen Ort etwa 20 Minuten gehen lassen.
- Vorteig mit den übrigen Zutaten verkneten und weitere 30 Minuten ruhen lassen. Teig in 16 Teile teilen. Jedes Teil zu einer Kugel formen und etwas flach drücken. Mit einem Messer bis zur Hälfte einschneiden und einen Hasenkopf mit Ohren formen. Auf ein mit Backpapier belegtes Blech setzen. Elektro-Ofen auf 175 Grad vorheizen.
- Eigelb mit Sahne verquirlen, auf die Brötchen streichen. Für die Gesichter je 3 Rosinen als Augen und Mund sowie 2 Mandelstifte als Zähne an den entsprechenden Stellen eindrücken. Im Ofen (Gas: Stufe 2) etwa 15 Minuten backen. Abkühlen lassen.

Aufwand:	★★
Zubereitungszeit:	2 Stunden
Backzeit:	15 Minuten
Pro Stück:	190 kcal/800 kJ

Variation

- Die Brötchen schmecken besonders gut, wenn Sie noch je 50 g fein gehacktes Zitronat und Orangeat, 100 g gehackte Mandeln und 50 g Rosinen unter den Teig mischen.

Osterkranz

BITTE BEACHTEN: TEIG MUSS 2 STUNDEN RUHEN
FÜR 1 BACKBLECH (30 x 40 cm)
100 g Butter
1 Döschen Safran
450 g Mehl
1 Päckchen Trockenhefe
60 g Zucker, etwa 100 ml Milch
2 Eier (Größe M)
AUSSERDEM
Mehl zum Bestäuben
Backpapier für das Blech
1 Eiweiß und 2 EL Milch
bunte Eier zur Dekoration

- Butter zerlassen, Safran einrühren, abkühlen lassen. Mehl und Hefe mischen. Zucker, Milch und Eier mit der Butter verrühren, unter das Mehl mischen. Teig kneten, bis er nicht mehr klebt. Falls nötig, etwas Milch oder Mehl zugeben.
- Teig zur Kugel formen, mit Mehl bestäuben. Entweder über Nacht kalt stellen oder an einem warmen Ort 2 Stunden gehen lassen, bis sich sein Volumen verdoppelt hat.
- Blech mit Backpapier auslegen. Teig in 3 Teile teilen, Rollen von etwa 2 cm Ø formen und zum lockeren Zopf flechten. Zopf auf dem Backblech zum Kranz legen. Weitere 20 Minuten gehen lassen.
- Elektro-Ofen auf 200 Grad vorheizen. Eiweiß mit 2 EL Milch verquirlen und den Kranz damit bestreichen. Im Ofen (Gas: Stufe 3) etwa 35 Minuten backen. Abgekühlt mit bunten Eiern anrichten.

Aufwand:	★★
Zubereitungszeit:	30 Minuten
Backzeit:	35 Minuten
Pro Stück:	250 kcal/1220 kJ

Variation

- Aus der 1 1/2-fachen Teig-Menge kleine Kränze formen. Kränze mit Eiweiß bestreichen, 20-25 Minuten backen, je 1 buntes Ei hineinlegen.

Blumen-Torte

BITTE BEACHTEN: TEIG MUSS
3 STUNDEN KÜHLEN

FÜR SPRINGFORM VON 18 cm Ø
- *1 Ei (Größe M)*
- *Salz*
- *30 g Zucker*
- *10 g Speisestärke*
- *20 g Mehl*
- *1 Msp. Backpulver*
- *20 g gemahlene Mandeln*

FÜR DIE FÜLLUNG
- *4 Blatt Gelatine*
- *1 Vanilleschote*
- *300 g Vollmilch-Joghurt*
- *Saft und abgeriebene Schale von 1 unbehandelten Orange*
- *40 g Zucker*

FÜR DIE VERZIERUNG
- *exotische Früchte, z. B. 1 Mango, 1 Kiwi, 6 Physalis (Kapstachelbeeren)*
- *1 Stange kandierte Angelika (Feinkostladen)*
- *1 Päckchen heller Tortenguss*
- *50 g gehackte Pistazienkerne*

AUSSERDEM
- *Backpapier für die Form*

● Elektro-Ofen auf 175 Grad vorheizen.

● Ei trennen. Eiweiß mit 1 Prise Salz steif schlagen, 20 g Zucker unterschlagen. Das Eigelb mit übrigem Zucker und 2 EL warmem Wasser aufschlagen. Stärke, Mehl, Backpulver mischen, mit Eischnee und Mandeln unter die Eigelb-Masse ziehen.

● Den Boden der Form mit Backpapier auslegen, Teig einfüllen, im Ofen (Gas: Stufe 3) etwa 25 Minuten backen. Herausnehmen und das Papier abziehen.

● Die Gelatine einweichen. Vanilleschote aufschlitzen und das Mark herausschaben. Mit Joghurt, Orangensaft, -schale und Zucker verrühren. Die Gelatine bei milder Hitze auflösen, mit 3 EL Joghurtmasse vermischen, in die übrige Masse einrühren. Den Tortenring um den Boden legen. Wenn die Masse zu gelieren beginnt, auf den Boden geben. 3 Stunden kühl stellen.

● Mango und Kiwi schälen, Physalis von den Hüllen befreien. Von der Mango mit dem Sparschäler Streifen abziehen. Kiwi in Scheiben schneiden. Mangostreifen um die Physalis zu Blüten legen. Mit Kiwi-„Blättern" und in Streifen geschnittener Angelika auf die Torte legen. Tortenguss bereiten, Früchte beträufeln. Den Tortenring entfernen. Pistazien an den Rand drücken.

Aufwand:	★★
Zubereitungszeit:	1 Stunde
Backzeit:	25 Minuten
Pro Stück:	170 kcal/710 kJ

So wird die Torte dekoriert

1

2

1 Mango schälen, mit Sparschäler Streifen abziehen, die Streifen zu Blüten rollen. Aus Kiwischeiben Blätter ausstechen (Formen gibt es in Haushaltswarenläden).

2 Pistazien auf eine breite Messerklinge legen. Torte leicht schräg halten und Pistazien an den Rand drücken.

Mandelherz mit Mango-Sahne

FÜR 1 HERZFORM (20 x 25 cm)	
3 Eier (größe M)	
75 g Zucker	
80 g Mehl	
1 Msp. Backpulver	
50 g gemahlene Mandeln	
FÜR FÜLLUNG UND ÜBERZUG	
4 Blatt weiße Gelatine	
1 kleine reife Mango	
175 g Sahnedickmilch	
1-2 EL Zucker	
225 g Schlagsahne	
1 EL Puderzucker	
1 EL Zitronensaft	
AUSSERDEM	
Fett und Brösel für die Form	

● Backofen auf 175 Grad vorheizen. Herzform fetten, mit Bröseln ausstreuen.

● Eier trennen. Die Eiweiße steif schlagen. Eigelbe mit Zucker dickschaumig rühren. Den Eischnee darauf geben. Mehl mit Backpulver auf die Masse sieben, leicht unterrühren. Teig in die Form füllen, im Ofen (Gas: Stufe 2) etwa 25 Minuten backen.

● 3 Blatt Gelatine einweichen. Mango schälen und das Fruchtfleisch in Spalten vom Stein schneiden. 9 dünne Scheiben vom Fruchtfleisch abschneiden, den Rest pürieren.

● Dickmilch, Zucker und das Mangopüree verrühren. Gelatine in einem Töpfchen bei milder Hitze auflösen, mit etwas Mango-Milch verrühren, zur übrigen Dickmilch geben. 10 Minuten kalt stellen. 75 g Sahne steif schlagen.

● Kuchenherz quer halbieren. Sahne unter die angelierte Creme heben und das Herz mit der Creme füllen.

● Restliche Gelatine auflösen. Übrige Sahne mit Puderzucker und Zitronensaft steif schlagen. Gelatine unter die Sahne rühren. Das Herz damit leicht wolkig überziehen, kalt stellen. Vor dem Servieren mit den Mangoscheiben fächerförmig belegen.

Aufwand:	★
Zubereitungszeit:	1 Stunde
Backzeit:	25 Minuten
Pro Stück:	220 kcal/920 kJ

Zitronen-Himbeer-Kuchen

BITTE BEACHTEN: KUCHEN MUSS 2 STUNDEN KÜHLEN

FÜR 1 HERZ-BACKFORM

1 Packung Backmischung Zitronenkuchen (Rührteig) sowie Fett, Eier und Flüssigkeit nach Packungsangabe
300 g Himbeeren
400 g Schlagsahne
2 Päckchen Sahnesteif
4-5 EL Zucker

AUSSERDEM

Fett für die Form
eventuell Schokosrapel zum Verzieren

● Nach Packungsangabe Elektro-Ofen vorheizen und Teig zubereiten. Teig in gefettete Form füllen. Im Ofen nach Angabe backen. Kuchen aus der Form nehmen, abkühlen lassen.

● Die Himbeeren verlesen, waschen. 100 g pürieren, durch ein Sieb streichen. Sahne mit Sahnesteif steif schlagen, Zucker unterrühren und 1/3 Sahne mit Himbeerpüree verrühren.

● Kuchen waagerecht halbieren. 1 Boden mit der puren Sahne bestreichen, mit den ganzen Himbeeren belegen. Mit dem zweiten Boden bedecken. 2 Stunden kalt stellen.

● Kuchen mit der Himbeersahne rundum einstreichen. Mit einer gezackten Teigkarte Muster auf der Torte streichen und Torte eventuell mit Schokoraspeln verzieren.

Aufwand:	★
Zubereitungszeit:	30 Minuten
Backzeit:	45 Minuten
Pro Stück:	365 kcal/1530 kJ

Muttertags-Herz

BITTE BEACHTEN: KUCHEN MUSS 3 STUNDEN KÜHLEN

FÜR 1 HERZFORM VON 23 cm Ø

30 g Butter
3 Eier (Größe M)
1 Prise Salz
150 g Zucker
50 g geschälte gemahlene Mandeln, 50 g Mehl

FÜR DIE FÜLLUNG

500 g Rhabarber, Zitronensaft
1 EL Himbeer-Gelee
200 ml roter Traubensaft
3 EL Zucker
1 Päckchen Sahne-Puddingpulver
1 TL Sofort-Gelatine-Pulver
200 g Schlagsahne
1 Päckchen Sahnesteif

FÜR DIE VERZIERUNG

2 EL Aprikosen-Konfitüre
Erdbeeren, Kiwis und Mango
1/2 Päckchen weißer Tortenguss
300 g Schlagsahne

AUSSERDEM

Backpapier für das Blech

- Elektro-Ofen auf 175 Grad vorheizen. Herztortenform ohne Boden auf ein mit Backpapier belegtes Blech stellen.
- Butter schmelzen. Eier trennen. Eiweiße mit Salz steif schlagen. Zucker und Eigelbe unter Rühren dazugeben, die Masse schaumig rühren. Mandeln und Mehl unterheben. Flüssige Butter unterziehen. In die Form füllen. Im Ofen (Gas: Stufe 2) etwa 30 Minuten backen. Abkühlen lassen.
- Rhabarber putzen, waschen und klein würfeln. In Zitronensaft, Himbeergelee und 150 ml Traubensaft in etwa 2 Minuten bissfest garen. Zucker und Puddingpulver mit übrigem Saft verrühren, unter den Rhabarber rühren, aufkochen und erkalten lassen. Gelatine unterrühren. Sahne mit Sahnesteif steif schlagen.
- Kuchenherz waagerecht halbieren. Herzform um den Kuchenboden legen. Zuerst Rhabarbergrütze einfüllen und glatt streichen. Sahne darauf geben, verstreichen. Kuchendeckel darauf legen. 2-3 Stunden kalt stellen.
- Die Konfitüre erhitzen, durch ein Sieb streichen und Kuchenoberfläche mit Konfitüre bestreichen. Die Früchte waschen und putzen. Erdbeeren und Kiwis längs in Spalten, Mango in dünne Scheiben schneiden. Mango kurz blanchieren, abschrecken, trockentupfen. Die Früchte in der Kuchenmitte herzförmig auflegen.
- Tortenguss nach Packungsangabe zubereiten. Mit einem Löffel dünn über die Früchte ziehen.
- Sahne steif schlagen, in einen Spritzbeutel mit Sterntülle füllen und dekorativ aufspritzen (siehe unten).

Aufwand:	★★
Zubereitungszeit:	1 1/2 Stunden
Backzeit:	30 Minuten
Pro Stück:	605 kcal/ 2540 kJ

So wird die Torte dekoriert

1 Etwa 4 cm vom Rand der herzförmigen Torte mit Erdbeerspalten herzförmig belegen, mit Kiwi und Mango ausfüllen.

2 Den freien Rand mit Sahne verzieren, dabei den Spritzbeutel spiralförmig bewegen.

Bären-Torte

BITTE BEACHTEN: TORTE MUSS 2 STUNDEN KÜHLEN

FÜR SPRINGFORM VON 18 cm Ø

2 Eier (Größe M), Salz
60 g Zucker
20 g Speisestärke
40 g Mehl
2 Msp. Backpulver
40 g gemahlene Mandeln

FÜR DIE FÜLLUNG

1 Pck. Sahne-Puddingpulver
400 ml Milch, 1-2 EL Zucker
1 Eigelb (Ei: Größe M)
200 g Nuß-Nougat
200 g Butter

FÜR DIE DEKORATION

300 g Marzipan-Rohmasse
300 g Puderzucker
50 g Vollmilch-Kuvertüre
rote Speisefarbe, Zuckerperlen

AUSSERDEM

Backpapier für die Form

● Elektro-Ofen auf 175 Grad vorheizen.

● Eier trennen. Eiweiße mit 1 Prise Salz steif schlagen, dabei 40 g Zucker zufügen. Die Eigelbe mit übrigem Zucker und 4 EL warmem Wasser aufschlagen. Stärke, Mehl und Backpulver mischen, mit Eischnee und Mandeln unterziehen.

● Den Boden der Form mit Backpapier auslegen, Teig einfüllen. Im Ofen (Gas: Stufe 3) etwa 25 Minuten backen. Herausnehmen, Papier abziehen, waagerecht halbieren.

● Puddingpulver nach Packungsangabe mit Milch, Zucker und Eigelb zubereiten. Gehacktes Nougat im heißen Pudding auflösen, kalt stellen. Butter aufschlagen. Nach und nach den Pudding in die Butter rühren.

● Um den Tortenboden einen Tortenring legen. Mit 2/3 der Creme bestreichen. Zweiten Boden darauf legen, übrige Creme darauf glatt streichen. 2 Stunden kalt stellen. Ring entfernen.

● Das Marzipan mit 200 g Puderzucker verkneten, ausrollen, Bären ausstechen. Kuvertüre hacken, im Wasserbad schmelzen. Bären mit Kuvertüre verzieren. Etwas Marzipan zu Schleifen formen. Übriges Marzipan zu einem großen Kreis ausrollen, über die Torte legen.

● Übrigen Puderzucker mit Speisefarbe und 2 EL Wasser verrühren. Schleifen bemalen. Die Bären und Schleifen auf die Torte legen. Mit Zuckerperlen bestreuen.

Aufwand:	★★★
Zubereitungszeit:	1 1/2 Stunden
Backzeit:	25 Minuten
Pro Stück:	870 kcal/3660 kJ

So wird die Torte dekoriert

1

2

1 Aus der Marzipanplatte mit Ausstechform (im Küchenfachhandel) Bären ausstechen.

2 Streifen für Schleifen formen. Mit Zuckerguss und Kuvertüre bestreichen und trocknen lassen. Torte dekorieren.

Marienkäfertorte

BITTE BEACHTEN: TORTE MUSS 4 STUNDEN KÜHLEN

FÜR 1 SPRINGFORM VON 26 cm Ø

3 Eier (Größe M), 1 Prise Salz
90 g Zucker, 90 g Mehl

FÜR DIE FÜLLUNG

30 g Pistazienkerne
250 g Erdbeeren
70 g Marzipan-Rohmasse
70 g Puderzucker, 1 Zitrone
8 Blatt weiße Gelatine
450 g Sahne-Joghurt
150 g Zucker
2 Päckchen Vanillezucker
4 cl weißer Kokoslikör
200 g Schlagsahne
3-4 EL Quittengelee

FÜR DEN BELAG

500 g kleine Erdbeeren
300 g Schlagsahne
1 Päckchen Sahnesteif
100 g Mandelblättchen
1/2 Päckchen weißer Tortenguss
2 Stücke Konfekt (z. B. Toffifee)
2 Gebäckstangen mit Schokoüberzug
1 Stück Vanilleschote

AUSSERDEM

Backpapier für die Form

- Elektro-Ofen auf 175 Grad vorheizen.
- Für den Biskuit Eier trennen. Eiweiße mit Salz und 1 EL Zucker steif schlagen. Eigelbe mit übrigem Zucker und 2 EL lauwarmem Wasser 3-4 Minuten cremig aufschlagen. Eischnee und gesiebtes Mehl unterheben.
- Boden der Form mit Backpapier auslegen. Biskuit einfüllen. Im Ofen (Gas: Stufe 2) etwa 25 Minuten backen. Auskühlen lassen.
- Pistazien fein mahlen. Erdbeeren waschen, putzen, vierteln und pürieren. Marzipan mit Puderzucker und Pistazien verkneten. Zwischen Klarsichtfolien zu 2 dünnen Platten (26 cm Ø) ausrollen. Zitrone auspressen.
- Gelatine einweichen. Joghurt mit Zucker, Vanillezucker, Erdbeerpüree, Zitronensaft und Likör glatt rühren. Gelatine bei milder Hitze auflösen. Erst 2 EL Creme, dann die restliche Creme in die Gelatine rühren. Sahne steif schlagen, unterziehen.
- Biskuit waagerecht halbieren. Gelee aufkochen. Boden und Deckel dünn damit bestreichen, mit je 1 Marzipanplatte belegen. Um den Boden einen Tortenring legen. Creme einfüllen, Deckel aufsetzen. 4 Stunden kalt stellen.
- Die Erdbeeren waschen, putzen, in dünne Scheiben teilen. Sahne mit Sahnesteif schlagen. Mandeln ohne Fett leicht rösten.
- Torte kuppelartig mit Sahne überziehen. Rand mit Mandeln bestreuen. Erdbeerscheiben dachziegelartig zu Flügeln legen (genug Platz für das Gesicht lassen). Tortenguss nach Angabe zubereiten, auf die Erdbeeren mpinseln. Gesicht mit Konfekt, Sticks und dem Stück Vanilleschote „modellieren".

Aufwand:	★★
Zubereitungszeit:	2 Stunden
Backzeit:	25 Minuten
Pro Stück:	605 kcal/2540 kJ

Info

Erdbeeren verderben schnell, deshalb kaufen Sie sie immer erst am Verwendungstag. Früchte, die für den Belag gedacht sind, müssen besonders sorgfältig behandelt werden: Erdbeeren putzen, waschen und sorgfältig trockentupfen. Dann mit einem scharfen Messer in dünne Scheiben schneiden und den Kuchen damit belegen. Den Tortenguss sofort darüber verteilen, denn er hält die Erdbeeren einige Stunden frisch.

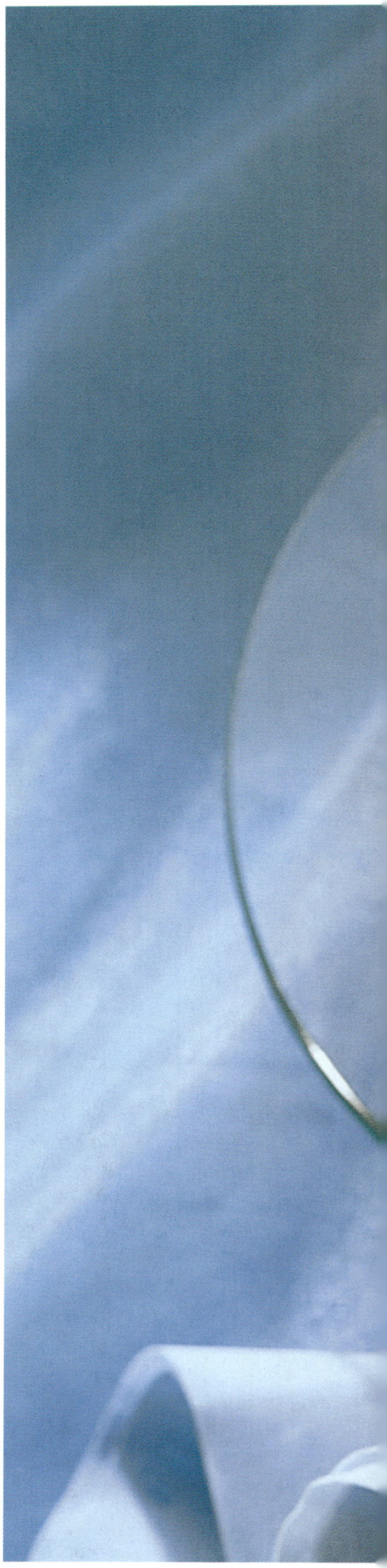

Himbeer-Kokos-Tarte

BITTE BEACHTEN: TARTE MUSS 1 STUNDE KÜHLEN

FÜR 1 TARTEFORM VON 28 cm Ø

300 g Tiefkühl-Blätterteig
150 ml Kokosmilch (Dose, aus dem Asienladen)
100 g Sahnejoghurt
40 g Puderzucker
1/2 Limette oder Zitrone
3 Blatt weiße Gelatine
2 EL Kokoslikör nach Belieben
150 g Schlagsahne
500 g Himbeeren oder
1 kg Kirschen

AUSSERDEM

Mehl zum Ausrollen
Fett für die Form

● Den Elektro-Ofen auf 200 Grad vorheizen. Dann Blätterteig auftauen, aufeinander legen. Auf bemehlter Arbeitsfläche etwas größer als die Backform ausrollen (etwa 30 cm Ø).

● Die Form einfetten, mit dem Teig auskleiden. Im Ofen (Gas: Stufe 3) 15-20 Minuten backen. Tarte kurz abkühlen lassen, vorsichtig aus der Form nehmen.

● Die Kokosmilch mit Joghurt und Puderzucker verrühren. Limette auspressen. Gelatine nach Packungsangabe einweichen, tropfnass bei milder Hitze auflösen. Mit 3 EL Kokoscreme verrühren, dann unter die übrige Creme mischen. Den Limettensaft und nach Belieben Likör dazugeben. Creme im Kühlschrank angelieren lassen.

● Sahne steif schlagen, Kokoscreme esslöffelweise unterrühren. Kokossahne auf den Tarteboden geben und glatt streichen. Himbeeren verlesen und rosettenartig auf die Creme setzen. Tarte mindestens 1 Stunde kühl stellen. Nach Belieben mit Limettenschalen-Streifen garnieren.

Aufwand:	★
Zubereitungszeit:	45 Minuten
Backzeit:	20 Minuten
Pro Stück:	190 kcal/800 kJ

Aprikosen-Sekt-Torte

BITTE BEACHTEN: TORTE MUSS 4 STUNDEN KÜHLEN

FÜR 1 SPRINGFORM VON 20 cm Ø
- *60 g Mehl, 1 EL Zucker*
- *1 Eigelb (Ei: Größe M)*
- *30 g Butter*

FÜR DEN BISKUIT
- *2 Eier (Größe M), 50 g Zucker*
- *1 Päckchen Vanillezucker*
- *20 g Speisestärke*
- *40 g Mehl, 2 Msp. Backpulver*

AUSSERDEM
- *Fett für die Form*
- *Backpapier für die Form*
- *250 g Aprikosen*
- *200 g Erdbeeren*
- *2 EL Erdbeerkonfitüre*
- *5 Blatt weiße Gelatine*
- *2 Eier (Größe M)*
- *60 g Zucker*
- *125 ml halbtrockener Sekt*
- *250 g Schlagsahne*

- Mehl mit Zucker, Eigelb, Butter und 2 EL Wasser verkneten. Kalt stellen.
- Elektro-Ofen auf 175 Grad heizen. Knetteig ausrollen, Boden der gefetteten Form damit auslegen. Bei 175 Grad (Gas: Stufe 2) ca. 10 Minuten backen.
- Eier trennen. Eiweiße steif schlagen, Zucker einrieseln lassen. Eigelbe mit Vanillezucker und 4 EL Wasser cremig schlagen. Stärke, Mehl, Backpulver und Eischnee unterziehen. Biskuitmasse in die mit Backpapier ausgelegte Form geben. Im Ofen (Gas: Stufe 2) etwa 25 Minuten backen.
- Aprikosen überbrühen, häuten, entsteinen. Erdbeeren waschen, putzen. Einige Früchte beiseite legen, übrige würfeln. Biskuit herausnehmen, waagerecht teilen. Konfitüre auf den Mürbeteigboden streichen, 1 Biskuit darauf legen. Springformring drum herum stellen.
- Gelatine einweichen. Eier trennen, Eigelbe mit Zucker aufschlagen, Sekt einrühren. Gelatine bei kleiner Hitze auflösen. 3 EL Sektcreme und Gelatine verrühren, dann übrige Creme unterheben. 1 Stunde kalt stellen.
- 200 g Sahne und Eiweiße getrennt steif schlagen. Sobald die Creme zu gelieren beginnt, beides unter Eiercreme ziehen und die Fruchtstücke unterheben. Auf den Biskuit streichen. Mit zweitem Boden bedecken. Etwa 3 Stunden kalt stellen.
- Die übrige Sahne ziemlich steif schlagen. Den Rand damit bestreichen und mit Tortenkamm verzieren. Obst in die Mitte legen, Sahneherzen aufspritzen.

Aufwand:	★★
Zubereitungszeit:	1 1/2 Stunden
Backzeit:	35 Minuten
Pro Stück:	705 kcal/2960 kJ

Schoko-Himbeer-Torte

BITTE BEACHTEN: TORTE MUSS 3 STUNDEN KÜHIEN

FÜR 1 SPRINGFORM VON 20 cm Ø
- 50 g Zartbitter-Kuvertüre
- 2 Eier (Größe M)
- 60 g Zucker
- 20 g Speisestärke
- 1 EL Kakaopulver
- 40 g Mehl
- 2 Msp. Backpulver

FÜR DIE FÜLLUNG
- 8 Blatt weiße Gelatine
- 80 g flüssiger Honig
- Saft und abgeriebene Schale von 1/2 unbehandelten Zitrone
- 4 EL Zucker
- 500 g Sahnejoghurt
- 400 g Himbeeren (frisch oder TK)
- 200 g Schlagsahne

AUSSERDEM
- Backpapier für die Form
- 3-4 EL Schokoraspel
- Puderzucker zum Verzieren

● Elektro-Ofen auf 175 Grad vorheizen. Kuvertüre grob hacken, in einer Schüssel über dem warmen Wasserbad schmelzen. Abkühlen lassen. Eier trennen, Eigelbe mit Zucker schaumig schlagen. Eiweiße steif schlagen.

● Die Schokomasse unter die Eigelbmasse rühren, dann Stärke, Kakao und Mehl, Backpulver und Eischnee vorsichtig unterziehen. Den Teig in eine mit Backpapier ausgelegte Springform füllen und im Ofen (Gas: Stufe 2) etwa 25 Minuten backen. In der Form auskühlen lassen, waagerecht halbieren.

● Um 1 Tortenboden den Springformring legen. Gelatine einweichen. Honig, Zitronensaft, -schale und Zucker erwärmen. Gelatine ausdrücken und bei milder Hitze in der Honigmischung auflösen.

● 3 EL Joghurt darunter rühren, dann diese Mischung unter den restlichen Joghurt rühren. Kalt stellen.

● Die Himbeeren verlesen, einige beiseite legen. Sobald die Masse zu gelieren beginnt, die Sahne steif schlagen und unterziehen. Himbeeren unterziehen. Creme auf den Boden streichen, mit dem zweiten Schokoboden bedecken. 3 Stunden kalt stellen.

● Tortenring entfernen. Die Torte dick mit Schokoraspeln bestreuen, mit Puderzucker bestäuben und mit den beiseite gelegten Himbeeren verziert servieren.

Aufwand:	★★
Zubereitungszeit:	1 Stunde
Backzeit:	25 Minuten
Pro Stück:	705 kcal/2960 kJ

Johannisbeer-Torte mit Erdnüssen

BITTE BEACHTEN: TORTE MUSS 4 STUNDEN KÜHLEN

FÜR 1 SPRINGFORM VON 20 cm Ø

50 g Erdnussbutter (ersatzweise Butter)
25 g Zucker
1 Ei (Größe M)
50 g Mehl
1/2 TL Backpulver
1-2 EL Milch

FÜR DIE FÜLLUNG

250 g Johannisbeeren (frisch oder tiefgekühlt)
200 ml Johannisbeer-Nektar
5 EL Zucker
2 EL Speisestärke
4 Blatt weiße Gelatine
1 Vanilleschote
Saft von 1 Orange (80 ml)
200 g Doppelrahm-Frischkäse
400 g Sahnejoghurt

AUSSERDEM

Backpapier für die Form
3-4 EL gehackte ungesalzene Erdnüsse
einige Johannisbeer-Rispen
feiner Zucker
Minzeblättchen zum Garnieren

● Elektro-Ofen auf 175 Grad vorheizen. Die Springform mit Backpapier auslegen. Erdnussbutter mit Zucker und Ei schaumig rühren. Mehl und Backpulver unterrühren. Nach und nach die Milch dazugeben, bis der Teig schwer reißend vom Löffel fällt.

● Den Teig in die Springform füllen und im Ofen (Gas: Stufe 2) etwa 20 Minuten backen. Aus dem Ofen nehmen und in der Form abkühlen lassen.

● Die Johannisbeeren waschen und verlesen. Johannisbeersaft mit 2 EL Zucker aufkochen. Die Stärke mit wenig Wasser anrühren und dazu geben, die Saftmischung noch mal aufkochen lassen. Johannisbeeren unterheben und abkühlen lassen. Wenn die Beeren-Saft-Mischung dick wird, die Masse auf dem Kuchen verteilen. Den Kuchen in der Form abkühlen lassen.

● Gelatine in kaltem Wasser einweichen. Vanilleschote längs aufschneiden. Das Mark herauskratzen, in den Orangensaft rühren. 3 EL Zucker zugeben und aufkochen. Vom Herd nehmen, Gelatine darin auflösen. Frischkäse und Joghurt verrühren. 2 EL Frischkäsecreme in die Gelatinemasse einrühren, dann alles unter die übrige Creme rühren.

● Creme auf auf den Beeren verteilen und glatt streichen. Torte für 3-4 Stunden kalt stellen.

● Springformrand entfernen, den Tortenrand mit gehackten Erdnüssen verzieren. Die Johannisbeer-Rispen waschen, in restlichem Zucker wälzen und mit Minzeblättchen dekorativ auf dem Kuchen verteilen.

Aufwand:	★★
Zubereitungszeit:	40 Minuten
Backzeit:	20 Minuten
Pro Stück:	305 kcal/1280 kJ

Variationen

● Nach dem gleichen Rezept kann man diese Torte auch mit der selben Menge Weintrauben (weiße oder blaue, ganz nach Belieben), Erdbeeren oder Orangen und dem entsprechenden Saft herstellen.

● Mit Äpfeln: Statt Sahnejoghurt nehmen Sie Sahne und statt der Erdnüsse 100 g Amaretti, die in einer Plastiktüte mithilfe einer Kuchenrolle fein zerkrümelt werden.

Käsekuchen mit Brombeeren

FÜR 1 SPRINGFORM VON 26 cm Ø
4 Eier (Größe M)
125 g Zucker
1 unbehandelte Zitrone
500 g Quark
75 g Speisestärke
400 g Brombeeren
AUSSERDEM
Fett und Zucker für die Form
1 TL Puderzucker zum Bestäuben
Melisseblättchen

- Backofen auf 175 Grad vorheizen. Boden und Rand der Springform fetten, Form mit Zucker ausstreuen.
- Eier trennen. Die Eigelbe und 75 g Zucker hellschaumig rühren. Zitronenschale abreiben, den Saft auspressen. Quark, Zitronenschale und 3 EL Zitronensaft unter die Eigelbe rühren.
- Eiweiße steif schlagen. Den übrigen Zucker zufügen und weiterschlagen, bis der Eischnee schnittfest ist. Eischnee und Speisestärke unter die Quarkmasse heben.
- Brombeeren verlesen. Die Hälfte der Quarkmasse in die Form füllen. 300 g Beeren darauf verteilen, die übrige Quarkmasse darauf geben und glatt streichen. Im Ofen (Gas: Stufe 2) auf der 2. Schiene von unten in 1 1/4 Stunden goldbraun backen. Nach etwa 40 Minuten mit Alufolie abdecken. Kuchen im ausgeschalteten Ofen abkühlen lassen.
- Kuchen aus der Form nehmen, die restlichen Beeren obenauf legen. Puderzucker darüber stäuben und mit den Melisseblättchen verzieren.

Aufwand:	★
Zubereitungszeit:	30 Minuten
Backzeit:	1 1/4 Stunden
Pro Stück:	150 kcal/630 kJ

Sahnige Pflaumentorte

FÜR 1 SPRINGFORM VON 26 cm Ø
4 Eier (Größe M)
130 g Zucker
1 Päckchen Vanillezucker
abgeriebene Schale von
2 unbehandelten Orangen
1 Prise Salz
80 g Mehl
40 g Speisestärke
1 TL Backpulver
FÜR DIE FÜLLUNG
1 kg Pflaumen
(oder Zwetschgen)
100 g Zucker
8 Blatt rote Gelatine
300 g Frischkäse
5 EL Orangenlikör
500 g Schlagsahne
1 unbehandelte Orange
AUSSERDEM
Backpapier für die Form

- Den Boden der Form mit Backpapier auslegen. Elektro-Ofen auf 175 Grad vorheizen.
- Eier trennen. Eigelbe mit 4 EL lauwarmem Wasser, Zucker und Vanillezucker schaumig rühren und Orangenschale zufügen.
- Eiweiße mit Salz steif schlagen und auf die Eigelbmasse geben. Mehl, Stärke und Backpulver darüber sieben und unterheben. Den Teig in die Form füllen, im Ofen (Gas: Stufe 2) 30-35 Minuten backen. Den Kuchen auf ein Kuchengitter stürzen und auskühlen lassen.
- Pflaumen waschen und 5 besonders schöne Früchte beiseite legen. Restliche Früchte waschen, entsteinen und vierteln. In einem Topf geben und mit 60 g Zucker 3-4 Minuten köcheln lassen.
- Gelatine einweichen und ausdrücken und in einem Topf bei geringer Hitze flüssig werden lassen. Dann unter das noch heiße Kompott rühren. Kalt stellen.
- Den Biskuitkuchen in der Mitte quer durchschneiden. Den unteren Teil auf eine Tortenplatte setzen, drum herum den Springformrand legen.
- Frischkäse mit Likör und übrigem Zucker gut verrühren. 250 g Sahne steif schlagen und unterheben. Erst die gelierte Pflaumenmasse auf den Tortenboden streichen, darauf dann die Käse-Creme. Torte für 30 Minuten kalt stellen.
- Springformrand entfernen, die obere Teigplatte auflegen. Restliche Sahne steif schlagen, die Torte rundum damit bestreichen. Übrige Pflaumen halbieren, in dünne Spalten schneiden als kleine Fächer auf die Torte legen. Von der Orange mit einem Zestenschneider feine Schalenspäne abziehen, über die Torte streuen.

Aufwand:	★★
Zubereitungszeit:	1 1/2 Stunden
Backzeit:	35 Minuten
Pro Stück:	400 kcal/1680 kJ

Kirsch-Mokka-Torte

BITTE BEACHTEN: TORTE MUSS 1 STUNDE KÜHLEN

FÜR 1 SPRINGFORM VON 26 cm Ø

1 TL Kaffeepulver (Instant)
80 g Mehl
1/2 TL Kakaopulver
1 EL Zucker, 1 Prise Salz
1 Eiweiß (Ei: Größe M)
40 g Butter

FÜR DEN BISKUIT

4 Eier (Größe M), 175 g Zucker
1 Päckchen Vanillezucker
50 g Mehl, 50 g Speisestärke
1 TL Backpulver

AUSSERDEM

750 g Sauerkirschen
Saft von 1 Limette oder 1/2 Zitrone
3 EL brauner Zucker
2 TL Speisestärke
4 EL Anisschnaps (z. B. Pernod)
50 g Mokka-Sahne-Schokolade
800 g Schlagsahne
70 g Zucker, 3 Pck. Sahnesteif

AUSSERDEM

Backpapier für die Form
Schokoröllchen, Kirschen Mokkabohnen und Minzeblättchen zum Verzieren

- Für den Mürbeteigboden Kaffeepulver, Mehl, Kakao, Zucker, Salz, Eiweiß und Butter zum glatten Teig verkneten. 30 Minuten kalt stellen.
- Elektro-Ofen auf 175 Grad vorheizen. Für den Biskuit die Eier mit 4 EL warmem Wasser schaumig schlagen. Zucker und Vanillezucker einrieseln lassen. Mehl, Stärke und Backpulver mischen, unterziehen. In eine mit Backpapier ausgelegte Springform füllen und im Ofen (Gas: Stufe 2) etwa 25 Minuten backen. Abkühlen lassen. 2-mal waagerecht teilen, so dass 3 Böden entstehen.
- Elektro-Ofen auf 200 Grad einstellen. Mürbeteig ausrollen, den Boden der mit Backpapier belegten Springform damit auskleiden. Im Ofen (Gas: Stufe 3) etwa 10 Minuten backen. Abkühlen lassen.
- Kirschen waschen, entsteinen und mit Zitrussaft aufkochen, zuckern. Stärke mit Schnaps oder 4 EL Wasser anrühren. Zugießen, umrühren und aufkochen. Abkühlen lassen. Die Schokolade hacken, bei milder Hitze schmelzen.
- Den Mürbeteigboden auf eine Platte legen, mit Schokolade bestreichen. 1 Biskuit darauf drücken und Tortenring drum herum legen. Die Sahne mit Zucker und Sahnesteif steif schlagen.
- Etwa 6 EL Sahne als Kleckse auf dem Biskuit verteilen. Die Hälfte der Kirschen in die Zwischenräume geben. Zweiten Biskuit darauf leicht andrücken. Die übrigen Kirschen und etwa 5 EL Sahne ebenso darauf verteilen. Mit der dritten Platte abdecken. Die Torte ringsum mit der restlichen Sahne bestreichen und mindestens 1 Stunde kalt stellen.
- In die Torten-Oberfläche mit einer gezackten Teigkarte dekorative Rillen ziehen und die Torte nach Belieben mit Schokoröllchen, Kirschen, Mokkabohnen und Minze verzieren.

Aufwand:	★★
Zubereitungszeit:	1 1/2 Stunden
Backzeit:	35 Minuten
Pro Stück:	390 kcal/1640 kJ

Tipp

Dunkler Biskuit

Eine Idee für Schoko-Fans: Verfeinern Sie den Biskuitteig zusätzlich mit 2 TL Kakaopulver. Einfach mit dem Mehl und der Stärke zugeben, dann den Teig wie beschrieben backen.

Trüffel-Sahne-Torte

BITTE BEACHTEN: TORTE MUSS 5 STUNDEN KÜHLEN

FÜR 1 SPRINGFORM VON 26 cm Ø
- 30 g Butter
- 4 Eier (Größe M)
- 110 g Zucker, 1 Prise Salz
- 70 g Mehl, 20 g Kakaopulver
- 20 g Speisestärke

ZUM TRÄNKEN
- 80 g Zucker, 4 cl Rum

FÜR DIE SCHOKO-SAHNE
- 6 Blatt weiße Gelatine
- 150 g Zartbitter-Kuvertüre
- 100 ml Milch
- herausgeschabtes Mark von 1 Vanilleschote
- 4 cl Orangenlikör
- 4 cl Rum, 1 kg Schlagsahne

AUSSERDEM
- Backpapier für die Form
- 100 g Schokoladen-Raspel
- 100 g Zartbitter-Kuvertüre
- Kakaopulver zum Bestäuben

● Den Elektro-Ofen auf 175 Grad vorheizen. Form mit Backpapier auslegen.

● Butter schmelzen. Eier, Zucker, Salz im warmen Wasserbad schaumig aufschlagen. Aus dem Wasserbad nehmen, weiterschlagen bis die Masse abgekühlt ist. Mehl, Kakao und Speisestärke mischen, über die Masse sieben, vorsichtig unterheben. Butter zugießen und unterziehen. Masse in die Form füllen. Im Ofen (Gas: Stufe 2) etwa 20 Minuten backen.

● Kuchen stürzen, Papier abziehen. Den Kuchen zweimal waagerecht in 3 gleich dicke Böden schneiden und um den unteren Boden einen Springformring legen.

● Zum Tränken 200 ml Wasser mit Zucker aufkochen, abkühlen lassen, Rum unterrühren. Den unteren Boden mit 1/3 der Flüssigkeit tränken.

● Für die Schoko-Sahne Gelatine einweichen. Kuvertüre hacken, im heißen Wasserbad schmelzen. Die Milch mit Vanillemark aufkochen. Vom Herd nehmen, ausgedrückte Gelatine darin auflösen. Kuvertüre, Likör und Rum unterrühren. Sahne steif schlagen, unter die abgekühlte Creme heben. 6 EL für den Rand abnehmen.

● Ein Drittel der Schoko-Sahne auf den getränkten Boden streichen. Mittleren Boden auflegen, ebenfalls tränken, mit dem zweiten Drittel der Sahne bestreichen. Dritten Boden auflegen, tränken, restliche Sahne darauf verteilen, 5 Stunden kalt stellen. Ring entfernen.

● Rand mit 2 EL Schoko-Sahne und Raspeln verzieren. Die übrige Schoko-Sahne in Spritzbeutel mit Sterntülle füllen, Rosetten aufspritzen. Für die Schoko-Röllchen Kuvertüre schmelzen, daraus Röllchen herstellen (s. S. 94). Torte damit dekorieren. Mit Kakao bestäuben.

Aufwand:	★
Zubereitungszeit:	1 1/2 Stunden
Backzeit:	20 Minuten
Pro Stück:	430 kcal/550 kJ

Schokoladen-Cappuccino-Torte

BITTE BACHTEN: TORTE MUSS 4 STUNDEN KÜHLEN

FÜR 1 SPRINGFORM VON 26 cm Ø

4 Eier (Größe M)
120 g Puderzucker
100 ml Pflanzenöl
90 g Zucker, Salz
160 g Mehl, 50 g Kakaopulver

FÜR DIE FÜLLUNG

50 g Johannisbeergelee
2 Blatt weiße Gelatine
250 g Zartbitter-Kuvertüre
1 TL Espresso-Pulver (Instant)
3 cl Weinbrand
herausgeschabtes Mark von 1 Vanilleschote, Salz
4 Eigelbe, 1 Ei (Eier: Größe M)
500 g Schlagsahne

FÜR DIE DEKO

300 g Zartbitter-Kuvertüre
50 g Schlagsahne

AUSSERDEM

Backpapier für die Form

- Elektro-Ofen auf 175 Grad vorheizen. Form mit Backpapier auslegen.
- Die Eier trennen. Eigelbe mit Puderzucker schaumig schlagen. 200 ml Wasser und das Öl leicht erwärmen und schnell unter die Masse arbeiten, nicht mehr schlagen.
- Die Eiweiße steif schlagen. Dabei Zucker und 1 Prise Salz einrieseln lassen. Eischnee mit dem Schneebesen unter die Eigelbmasse ziehen. Mehl und Kakaopulver darauf sieben und unterheben. Masse in die Form füllen. Im Ofen (Gas: Stufe 2) 35-40 Minuten backen. Kuchen auf ein Gitter stürzen, Papier abziehen.
- Kuchen waagerecht halbieren, Schnittflächen der Hälften mit Gelee bestreichen. Unteren Boden mit Tortenring umlegen.
- Gelatine einweichen. Kuvertüre hacken, im Wasserbad schmelzen. Weinbrand erwärmen, Espresso-Pulver darin auflösen, mit Vanillemark und 1 Prise Salz unter die Kuvertüre rühren. Eigelbe und Ei im Wasserbad schaumig schlagen. Sahne steif schlagen, unter die ageкühlte Eicreme heben.
- Für die helle Creme ein Drittel der Ei-Sahne-Masse abnehmen. Gelatine bei milder Hitze auflösen, erst mit 1 EL Creme verrühren, dann die Gelatine unter die übrige Creme ziehen.
- Für die dunkle Creme die abgekühlte, flüssige Kuvertüreunter die übrige Ei-Sahne-Masse ziehen. Schoko-Sahne auf den Boden streichen, 15 Minuten im Gefriergerät erstarren assen.
- Helle Creme auf die dunkle Creme streichen. Zweiten Boden darauf legen, ancrücken und 4 Stunden kalt stellen.
- Für den Guss 150 g Kuvertüre hacken in der Sahne schmelzen. Mit dem Pürierstab kurz durchmixen. Die Masse abkühlen lassen. Die Tortenoberfläche damit überziehen und fest werden lassen. Den Tortenring entfernen.
- Für die Schoko-Fächer die übrige Kuvertüre schmelzen und daraus Fächer herstellen (ähnlich wie die Schoko-Röllchen, s. Seite 94). Torte damit dekorieren.

Aufwand:	★★
Zubereitungszeit:	2 Stunden
Backzeit:	40 Minuten
Pro Stück:	455 kcal/1910 kJ

Apfelkuchen mit Pecannuss-Kruste

FÜR 1 SPRINGFORM VON 26 cm Ø
150 g Butter
1 Prise Salz
1 Päckchen Vanillezucker
150 g brauner Zucker
1/2 TL abgeriebene Orangenschale (unbehandelt)
4 Eier (Größe M)
1 TL Backpulver
250 g Mehl
750 g säuerliche Äpfel
FÜR DEN BELAG
100 g kalte Butter
75 g gehackte und
75 g halbe Pecannüsse (ersatzweise Haselnüsse)
3 EL Zucker
AUSSERDEM
Fett für die Form
Puderzucker zum Bestäuben

- Den Elektro-Ofen auf 175 Grad vorheizen. Die Form einfetten. Butter, Salz, Vanillezucker, Zucker und die Orangenschale schaumig rühren. Die Eier unterrühren. Dann das Backpulver und das Mehl mischen und unterheben.
- Die Äpfel schälen, halbieren und entkernen. 1 Apfel klein würfeln und unter den Teig rühren. Den Teig in die Form füllen. Die übrigen Äpfel an der gewölbten Seite mehrmals längs einritzen, mit der gewölbten Seite nach oben in den Teig setzen. Butter in Flöckchen und gehackte Nüsse auf den Äpfeln verteilen. Mit Zucker bestreuen. Halbe Pecannüsse an den Rand setzen (siehe Bild).
- Kuchen im Ofen (Gas: Stufe 2) etwa 50 Minuten backen. Eventuell mit Alufolie abdecken.
- Kuchen aus der Form nehmen, abkühlen lassen und nach Belieben mit Puderzucker bestäuben.

Aufwand:	★
Zubereitungszeit:	45 Minuten
Backzeit:	50 Minuten
Pro Stück:	420 kcal/1760 kJ

Apfel-Cidre-Kuchen

FÜR 1 SPRINGFORM VON 26 cm Ø
300 g Mehl
150 kalte Butter
240 g Zucker
1 Ei
1 Eigelb (Ei: Größe M)
Salz
1 kg säuerliche Äpfel
250 ml herber Cidre (Apfelwein)
1 EL Zitronensaft
1-2 EL Zucker
2 EL Speisestärke
2 EL Sultaninen
FÜR DAS MARZIPAN-GITTER
100 g Marzipan-Rohmasse
2 Tropfen Mandelaroma
1 Eigelb (Ei: Größe M)
AUSSERDEM
Fett für die Form

● Das Mehl mit den Butterflöckchen, Zucker, Ei, Eigelb und 1 Prise Salz rasch zu einem glatten Teig verkneten und mindestens 30 Minuten kalt stellen.

● Inzwischen die Äpfel waschen und schälen, das Kerngehäuse entfernen. Die Äpfel einen cm groß würfeln und in einem Sud aus Cidre, Zitronensaft und Zucker kurz andünsten. Den Sud abgießen. Die Speisestärke in 2 EL Sud anrühren und in dem übrigen Sud aufkochen lassen. Die Sultaninen und die Äpfel unterheben.

● Den Elektro-Ofen auf 200 Grad vorheizen. 3/4 des Teiges abnehmen, ausrollen und die Form damit auskleiden. Marzipan zerkleinern, mit Mandelaroma unter den übrigen Mürbeteig kneten, dünn ausrollen. Mit einem Messer oder Teigrädchen mit Wellenrand 10 etwa 1 1/2 cm breite Streifen ausschneiden.

● Die Äpfel auf den Teig geben. Mit Streifen gitterartig belegen. Mit Eigelb bepinseln.

● Kuchen im Ofen (Gas: Stufe 3) etwa 30 Minuten backen.

Aufwand:	★
Zubereitungszeit:	1 Stunde
Backzeit:	30 Minuten
Pro Stück:	395 kcal/1660 kJ

Apfelkuchen mit Möhren

FÜR 1 SPRINGFORM VON 26 cm Ø
200 g Möhren
600 g säuerliche Äpfel
3 EL Zitronensaft
4 Eier (Größe M)
100 g Butter
100 g Zucker
100 g Maismehl (ersatzweise Weizenmehl)
125 g geschälte gemahlene Mandeln
1/2 TL Backpulver
AUSSERDEM
Fett für die Form
50 g Orangenmarmelade oder Aprikosenkonfitüre
Puderzucker zum Bestäuben

● Die Möhren putzen, schälen und fein reiben. Die Äpfel schälen, 100 g fein reiben und mit den Möhren vermengen. Übrige Äpfel entkernen und in Spalten teilen, sofort mit Zitronensaft beträufeln.

● Den Elektro-Ofen auf 175 Grad vorheizen. Die Form einfetten. Die Eier trennen. Eigelbe mit Butter und 50 g Zucker schaumig rühren. Die Eiweiße steif schlagen, dabei den übrigen Zucker unterrühren. Mehl, Mandeln und das Backpulver unter die Butter-Zucker-Masse rühren. Danach die Apfel-Möhren-Mischung unterheben. Den Eischnee vorsichtig unterheben.

● Die Teig-Masse in die Form gießen, glatt streichen. Teig mit Apfelspalten rosettenförmig belegen und im Ofen (Gas: Stufe 2) etwa 45 Minuten backen.

● Marmelade erwärmen, auf den noch warmen Kuchen streichen, abkühlen lassen. Kuchen aus der Form nehmen und vor dem Servieren mit Puderzucker bestäuben.

Aufwand:	★
Zubereitungszeit:	45 Minuten
Backzeit:	45 Minuten
Pro Stück:	260 kcal/1090 kJ

Mandelstollen

FÜR 1 BACKBLECH (30 x 40 cm)
500 g Mehl
1 Würfel frische Hefe
1/8 l lauwarme Milch
125 g Zucker, 1 Prise Salz
abgeriebene Schale von
1/2 unbehandelten Zitrone
je 1 Prise gemahlene
Muskatblüte und Gewürznelke
200 g zimmerwarme Butter
je 100 g geschälte gemahlene
und gehackte Mandeln
3 geschälte und gemahlene
Bittermandeln
AUSSERDEM
Mehl zum Arbeiten
Backpapier für das Blech
50 g zerlassene Butter
2 EL Puderzucker zum
Bestäuben

● Mehl in eine Schüssel geben, in die Mitte die Hefe geben, mit Milch und etwas Zucker glatt rühren. Zugedeckt 30 Minuten gehen lassen.
● Alle übrigen Zutaten zufügen und so lange schlagen, bis der Teig Blasen wirft und sich vom Rand löst. Zugedeckt etwa 1 Stunde gehen lassen.
● Den Teig auf leicht bemehlter Arbeitsfläche durchkneten. Zu einem Rechteck (etwa 20 x 30 cm) ausrollen. Eine Teigseite so über die andere schlagen, dass die typische Stollenform entsteht. Auf ein mit Backpapier belegtes Blech setzen, 30 Minuten gehen lassen. Elektro-Ofen auf 175 Grad vorheizen.
● Stollen im Ofen (Gas: Stufe 2-3) etwa 1 Stunde backen. Dann mit Butter bestreichen und Puderzucker bestäuben. 1-2 Wochen durchziehen lassen.

Aufwand:	★★
Zubereitungszeit:	2 Stunden
Backzeit:	1 Stunde
Pro Stück:	215 kcal/900 kJ

Quarkstollen

FÜR 1 BACKBLECH (30 x 40 cm)
FÜR DIE FÜLLUNG
100 g Marzipan-Rohmasse
50 g Puderzucker
1/2 Päckchen Vanillezucker
1 TL Kirschwasser
50 g kandierte Kirschen
30 g Pistazienkerne
FÜR DEN STOLLEN
50 g Schweineschmalz
100 g Butter
500 g Mehl
1 Päckchen Backpulver
je 1 Msp. gemahlener Zimt
Kardamom und Piment
2 Eier (Größe M)
250 g Magerquark
1 Päckchen Vanillezucker
150 g Zucker
abgeriebene Schale von
1/2 unbehandelten Zitrone
4 Tropfen Bittermandelöl
3 EL Rum, 250 g Korinthen
150 g gewürfeltes Zitronat
100 g Mandelstifte
AUSSERDEM
Mehl zum Bestäuben
Backpapier für das Blech
etwa 50 g zerlassene Butter
zum Bepinseln
Zucker und Puderzucker
zum Bestreuen

● Marzipan mit Puder-, Vanillezucker und Alkohol verkneten. Kirschen fein hacken, mit den Pistazien untermengen. Alles zur 4 cm dicken, knapp 30 cm langen Rolle formen.
● Schmalz mit Butter schmelzen und auf Handwärme abkühlen lassen. Mehl mit Backpulver und Gewürzen mischen. Eier, Quark, Vanillezucker, Zucker, Zitronenschale, Mandelöl und Rum cremig aufschlagen und samt dem Fett in das Mehl kneten.
● Die Korinthen waschen, trockentupfen, mit Zitronat und Mandeln in einer trockenen Pfanne anwärmen. Mit Mehl bestäuben und unter den Teig mengen.

● Elektro-Ofen auf 175 Grad vorheizen. Teig zur etwa 30 cm langen Rolle formen, mit der Handkante der Länge nach in der Mitte eindrücken.

● Marzipanrolle hineinlegen und den Teig darüber schließen, so dass die Marzipanfüllung etwa in der Mitte des Stollens liegt.

● Den Stollen auf ein mit Backpapier belegtes Blech setzen. Mit einem mehrfach gefalteten Streifen Alufolie umlegen und die Enden feststecken. Den Stollen im Ofen (Gas: Stufe 2-3) etwa 1 1/4 Stunden auf der unteren Schiene backen.

● Mehrmals mit zerlassener Butter bestreichen. Ausgekühlt dick mit Zucker und Puderzucker bestreuen.

● Den Stollen zuerst in Alufolie, dann in einen Frischhaltebeutel verpacken. Am besten 2-3 Wochen an einem kühlen Ort, aber nicht im Kühlschrank, durchziehen lassen. Der Stollen ist 2-3 Monate haltbar.

Aufwand:	★★
Zubereitungszeit:	1 Stunde
Backzeit:	1 1/4 Stunden
Pro Stück:	280 kcal/1180 kJ

Lebkuchentorte

FÜR 1 STERNFORM VON 26 cm Ø
75 g Honig, 1 EL Butter
125 g Zucker
1 geh. TL Hirschhornsalz
1/2 TL Pottasche
1 Ei (Größe M), 170 g Mehl
30 g gemahlene Mandeln
1/2 EL Kakaopulver
1/2 TL Lebkuchengewürz
30 g Mandelblättchen
1 unbehandelte Orange
125 g Mascarpone
1 Blatt weiße Gelatine
100 g Schlagsahne, gem. Zimt
AUSSERDEM
Fett für die Form

- Elektro-Ofen auf 175 Grad vorheizen. Honig, Butter und 75 g Zucker in einem kleinen Topf unter Rühren erwärmen, bis sich der Zucker aufgelöst hat. Die Masse etwas auskühlen lassen. Hirschhornsalz und Pottasche mit 3 EL Wasser verrühren.
- Das Ei kräftig in die Honigmasse einarbeiten. Mehl, Mandeln, Kakao, Lebkuchengewürz und Pottasche-Hirschhornsalz-Mischung zufügen und alles zu einem glatten Teig verrühren.
- Form einfetten und mit Mandelblättchen ausstreuen. Teig hineinfüllen, und glatt streichen. Kuchen im Ofen (Gas: Stufe 2) 20-25 Minuten backen.
- Orange heiß abwaschen und trockenreiben. Mit einem scharfen Messer ein großes Stück Schale abschneiden und in ganz feine Streifen schneiden. Übrige Schale fein abreiben. Orange auspressen.
- Kuchen noch heiß auf ein Kuchengitter stürzen. Mit einem Holzstäbchen mehrmals einstechen, mit 3 EL Orangensaft beträufeln. Kuchen auskühlen lassen.
- Mascarpone, restlichen Zucker, abgeriebene Orangenschale glatt rühren. Gelatine in Wasser 10 Minuten einweichen, ausdrücken. Übrigen Orangensaft erwärmen, Gelatine darin auflösen. Erst 1 EL Orangen-Mascarpone-Creme in die Gelatine rühren, dann die Gelatine in die übrige Creme. Etwa 10 Minuten kalt stellen.
- Sahne steif schlagen und unter die gekühlte Creme heben. Auf der Sterntorte verteilen. Bis zum Servieren kalt stellen. Mit den Orangenschalenstreifen verzieren und mit etwas Zimt bestäuben.

Aufwand:	★★
Zubereitungszeit:	1 Stunde
Backzeit:	25 Minuten
Pro Stück:	215 kcal/900 kJ

Deko-Tipp

Eine unbehandelte Orange heiß abwaschen, trockenreiben und mit einem Sparschäler die Schale in möglichst großen Stücken abziehen. Fruchtfleisch anderweitig verwenden. Mit Förmchen aus der Schale Sterne ausstechen, Torte damit verzieren.

Mandel-Gewürzkuchen

BITTE BEACHTEN: BROTMISCHUNG MUSS 2 STUNDEN ZIEHEN

FÜR 1 SPRINGFORM VON 24 cm Ø
- *250 g Weißbrot vom Vortag*
- *3/4 l Milch*
- *ausgeschabtes Mark von 1 Vanilleschote*
- *100 g Amaretti (italienische Mandelkekse)*
- *3 Eier (Größe M)*
- *180 g Zucker*
- *Salz*
- *1/2 EL abgeriebene Zitronenschale (unbehandelt)*
- *50 g Zitronat*
- *50 g Orangeat*
- *100 g gemahlene Mandeln*
- *1 Msp. gerieben Muskatnuss*
- *1 TL Zimtpulver*
- *80 g Zartbitterschokolade*

AUSSERDEM
- *Fett für die Form*
- *1 EL Paniermehl*
- *30 g geschälte ganze Mandeln*
- *2 EL Mandelblättchen*

- Brot würfeln und in eine Schüssel geben. Milch mit Vanillemark einmal aufkochen lassen, über das Brot gießen und 2 Stunden ziehen lassen.
- Amaretti in einen Gefrierbeutel geben und mit einer Teigrolle so lange darüber rollen, bis die Kekse grob zerkrümelt sind.
- Elektro-Ofen auf 175 Grad vorheizen. Die Eier mit Zucker und 1 Prise Salz schaumig schlagen. Zitronenschale, Zitronat, Orangeat, gemahlene Mandeln, Muskat und Zimt unterrühren. Schokolade fein hacken oder raspeln und dazugeben.
- Eingeweichte Brotwürfel zu einer glatten Masse verquirlen und mit den Amaretti-Stücken unter die Eischaummasse rühren.
- Springform einfetten und den Boden mit Paniermehl ausstreuen. Teig einfüllen und glatt streichen. Ganze Mandeln auf den Teig am Rand entlang nebeneinander legen. Die Mandelblättchen in die Mitte streuen.
- Den Kuchen im Ofen (Gas: Stufe 3) etwa 1 1/2 Stunden backen. Nach etwa 1 Stunde Backzeit den Kuchen mit Backpapier bedecken. Den Kuchen herausnehmen, aus der Form lösen, abkühlen lassen und servieren.

Aufwand:	★
Zubereitungszeit:	45 Minuten
Backzeit	1 1/2 Stunden
Pro Stück:	330 kcal/1390 kJ

Variationen

- Dieses einfache Rezept kann man ebenso gut mit der entsprechenden Menge Walnüsse oder Haselnüsse zubereiten.
- Schmeckt besonders fein: Man rührt statt der gemahlenen Mandeln 100 g Marzipan unter den Teig und nimmt dann nur 150 g Zucker.

Im heißen Fett gebacken

Schmalzgebäck schmeckt zwar zur Faschingszeit besonders gut, a
Man sollte sie allerdings immer so frisch wie mög

Reiskrapfen

So werden die Reiskrapfen schön locker

1 Reismasse vorbereiten: Unter den Milchreis werden gehackte Früchte, Mehl und Eigelb gerührt. Wichtig ist dabei, dass der Milchreis nur lauwarm ist, sonst gerinnt das Ei.

2 Eischnee unterziehen: Das Eiweiß steif schlagen, bis es mittelfest ist. Dann wird es vorsichtig unter die Reismasse gehoben. Jetzt hat der Teig die richtige Konsistenz und man kann die Nocken formen.

Das fertige Backwerk mit der Siebkelle aus dem Fett fischen und auf Küchenpapier abtropfen lassen. Danach sofort mit Puderzucker bestäuben.

Kneten & formen

1 Kräftig kneten: Das ist das A und O beim Hefeteig. Denn dadurch gelangt viel Luft in den Teig und er wird schön locker.

2 Gleich große Krapfen: Die Kuchen werden gleichmäßig, wenn Sie den Teig rollen und mit einem Ausstecher oder Glas Kreise ausstechen.

Krapfen, Kücherl & Co

liner oder Spritzkuchen finden das ganze Jahr über ihre Fans.
ießen, denn dann schmecken sie am besten

Das richtige Frittierfett

Außen knusprig, innen zart – so schmeckt frittiertes Backwerk allen. Damit dies gelingt, brauchen Sie das richtige Fett, denn längst nicht jedes ist geeignet.

Stark erhitzbar muss es sein

Frittierfett muss geschmacksneutral und bis zu 200 Grad erhitzbar sein. Diese Voraussetzungen erfüllen z. B. die 100 Prozent reinen Pflanzenfette, die als spezielle Frittierfette gekennzeichnet sind.

Butter verbrennt

Nicht geeignet sind Butter und Margarine: Die Eiweißanteile verbrennen ab 100 Grad.

Öle eignen sich gut

Öle sind hocherhitzbar und zum Frittieren deshalb besonders gut geeignet, wie beispielsweise Sojaöl oder natives Olivenöl und alle als Frittieröle ausgewiesenen Öle.

Fette nicht mischen

Niemals verschiedene Fettsorten, z. B. Frittierfett mit Sojaöl, mischen: Wegen der unterschiedlichen Siedepunkte können sie spritzen.

Nach Gebrauch filtern

Frittierfett kann 2- bis 3-mal verwendet werden, es muss aber grundsätzlich nach jedem Gebrauch gefiltert werden, um alle Rückstände zu entfernen. Wer keine Fritteuse mit Fettfilter hat, gießt das noch warme Fett durch ein Sieb, das mit Küchenpapier ausgelegt ist, in ein Schraubglas.

Entsorgen, aber wie?

Nach 3- bis 4-maligem Gebrauch sollten Sie Frittierfett entsorgen. Gießen Sie es auf keinen Fall in den Ausguss! Gebrauchtes Öl in Milchkartons füllen, Plattenfett in Papier einwickeln und in den Hausmüll werfen.

chweizer Hasenöhrl

Teig kühlen, dünn ausrollen. Mit einem Teigrädchen Rhomben mit 6 cm eitenlänge ausschneiden. Jedes Teigstück in der Mitte längs einschneiden. Nun braucht es etwas Fingerspitzengefühl: Jeweils einen Zipfel der Teigomben vorsichtig durch den Schlitz in der Mitte ziehen. Sofort ausbacken.

Spritzgebäck

So wird's gemacht

1 Wasser und Gewürze aufkochen, das Mehl unterrühren. Den Teig so lange rühren, bis er dick wird und sich als Kloß vom Topfboden löst.

2 Perfekte Form: Den Teig in einen Beutel mit Sterntülle füllen und auf Pergamentpapier Kreise spritzen. Mithilfe des Papiers können Sie die Kuchen leicht in das heiße Fett gleiten lassen, ohne dass sie ihre Form verlieren.

3 Glasieren: Spritzkuchen mit Konfitüre bepinseln, dann mit der Oberseite in die Zuckerglasur tauchen und trocknen lassen.

Weihnachts-gebäck

Einmal im Jahr kommen auch kleine Bäcker groß raus: In der Adventszeit, wenn es ans Plätzchen-Backen geht

Plätzchen-Rezepte siehe folgende Seiten

Gewürzplätzchen

Für etwa 70 Stück

100 g weiche Butter
80 g brauner Zucker
1 Ei (Größe M)
80 g Zuckerrohrsirup (ungeschwefelt)
260 g Mehl
2 Msp. Natronpulver
2 Msp. Backpulver
1 1/2 TL Ingwerpulver
1/2 TL Zimtpulver
1/2 TL Nelkenpulver
1/2 TL Salz
1 Prise Pfeffer

FÜR DIE DEKORATION

1 Eiweiß (Ei: Größe M)
150 g Puderzucker
Lebensmittelfarben

AUSSERDEM

Mehl zum Arbeiten
Backpapier für die Bleche

● Butter und Zucker schaumig schlagen. Ei und Sirup unterrühren.

● Mehl, Natron, Backpulver und die Gewürze vermischen und unter die Creme kneten. Teig in Folie einwickeln, 1 Stunde ruhen lassen.

● Den Elektro-Ofen auf 175 Grad vorheizen. Backbleche mit Backpapier auslegen. Teig auf bemehlter Arbeitsfläche 3 mm dick ausrollen. Formen (z. B. Eisbären, Robben und Sterne) ausstechen. Auf einem mit Backpapier belegtem Blech (Gas: Stufe 2) 8-10 Minuten backen.

● Eiweiß mit Puderzucker aufschlagen, teilweise mit Lebensmittelfarbe färben. Kekse einfarbig bestreichen. Weißen Guss in einen Gefrierbeutel geben, eine winzige Ecke abschneiden, Konturen und Muster damit aufspritzen.

Aufwand:	★
Zubereitungszeit:	1 1/2 Stunden
Backzeit pro Blech:	8-10 Minuten
Pro Stück:	40 kcal/170 kJ

Minztaler

BITTE BEACHTEN: MUSS EINIGE STUNDEN ANFRIEREN

Für etwa 65 Stück

300 g Mehl
150 g Zucker
1 Päckchen Vanillezucker
2 Eigelbe (Ei: Größe M)
200 g kalte Butter
12 Tropfen Pfefferminzöl (aus der Apotheke)
grüne und gelbe Lebensmittelfarbe

AUSSERDEM

Mehl zum Arbeiten
Backpapier für die Bleche

● Mehl, Zucker, Vanillezucker mit Eigelben und Butterflöckchen rasch zu einem glatten Teig verarbeiten.

● Den Teig in 3 Portionen teilen. Die erste Portion mit Minzöl und einigen Tropfen grüner Lebensmittelfarbe verkneten, bis sich die Farbe gleichmäßig verteilt hat. Die zweite Portion mit ein paar Tropfen gelber Lebensmittelfarbe verkneten. Alle 3 Portionen in Folie gewickelt 1 Stunde kalt stellen.

● Grünen und gelben Teig jeweils zu 4 Rollen von je 30 cm Länge formen.

● Den hellen Teig halbieren. Jede Hälfte zu einem Rechteck von etwa 30 x 15 cm ausrollen. Jeweils 1 grüne und gelbe Rolle in die Mitte legen und die übrigen Rollen farbig versetzt darauf legen. Mit dem hellen Teig umwickeln und die Nähte zusammendrücken. Im Gefrierbeutel einige Stunden anfrieren lassen.

● Elektro-Ofen auf 200 Grad vorheizen. Rollen in 1/2 cm dicke Scheiben teilen, auf mit Backpapier belegte Bleche legen und im Ofen (Gas: Stufe 3) etwa 8 Minuten hell backen.

Aufwand:	★
Zubereitungszeit:	50 Minuten
Backzeit pro Blech:	8-10 Minuten
Pro Stück:	50 kcal/210 kJ

Pekan-Nuss-Würfel

BITTE BEACHTEN: TEIG MUSS ÜBER NACHT KÜHLEN

Für etwa 75 Stück

200 g weiche Butter
120 g Margarine
180 g Zucker
2 Eier (Größe M)
3 Päckchen Vanillezucker
400 g Mehl, 1 TL Backpulver
1 TL Salz

FÜR DEN BELAG

225 g Butter
120 g Honig, 60 g Zucker
200 g brauner Zucker
60 g Schlagsahne
450 g gehackte Pekan-Nüsse

AUSSERDEM

Mehl zum Arbeiten
Fett für das Blech

● Butter und Margarine schaumig schlagen. Erst den Zucker, dann Eier und Vanillezucker unterrühren. Mit Mehl, Backpulver und Salz verkneten. Teig über Nacht kühl stellen.

● Elektro-Ofen auf 200 Grad vorheizen. Ein tiefes Backblech fetten. Teig zu einer Platte (26 x 40 cm) ausrollen, auf das Blech legen. Mehrmals einstechen. Im Ofen (Gas: Stufe 3) 12 Minuten goldgelb backen. Ofen eingeschaltet lassen.

● Butter, Honig, Zucker und braunen Zucker bei mittlerer Hitze unter ständigem Rühren etwa 5 Minuten köcheln, bis der Zucker zu schmelzen beginnt und sich die Masse dunkel färbt. Von der Herdplatte nehmen, in die Sahne einrühren. Nüsse untermischen.

● Belag gleichmäßig auf dem gebackenen Boden verteilen und etwa 15 Minuten backen (Gas: Stufe 3). Abgekühlt in 3 cm große Würfel teilen.

Aufwand:	★
Zubereitungszeit:	40 Minuten
Backzeit:	30 Minuten
Pro Stück:	150 kcal/630 kJ

Aprikosen-Häufchen

Für etwa 80 Stück	
275 g getrocknete Aprikosen	
8 EL Apricot Brandy	
3 EL Öl	
60 g weiche Butter	
200 g Zucker	
2 Päckchen Vanillezucker	
2 Eier (Größe M)	
270 g Mehl	
1 TL Natronpulver	
1 Prise Salz	
1 TL Zimtpulver	
4 geh. EL Weizenkeime	
100 g Aprikosenkonfitüre	
AUSSERDEM	
Backpapier für die Bleche	

- Aprikosen 1/2 cm groß würfeln und in Brandy einweichen. Elektro-Ofen auf 175 Grad vorheizen. Backbleche mit Backpapier auslegen.
- Öl, Butter, Zucker, Vanillezucker und Eier schaumig rühren. Mehl, Natron, Salz, Zimt und Weizenkeime vermischen und unter die Butter-Zucker-Ei-Creme rühren. Konfitüre und 2/3 der Aprikosenwürfel unterrühren.
- Mit einem Teelöffel kleine Häufchen (Ø 3 cm) auf die Backbleche setzen, dabei Abstand halten. Die Plätzchen mit den übrigen Aprikosenwürfeln bestreuen und im Ofen (Gas: Stufe 2) in 12-15 Minuten goldbraun backen.
- Aprikosenkonfitüre mit 2 EL Wasser erwärmen und durch ein Sieb streichen. Die noch warmen Plätzchen damit bestreichen und abkühlen lassen.

Aufwand:	★
Zubereitungszeit:	40 Minuten
Backzeit:	15 Minuten
Pro Stück:	50 kcal/210 kJ

Winter-Küchlein

Für etwa 6 Stück

1 großer säuerlicher Apfel
1 EL Zitronensaft
2 Eier (Größe M), 150 g Butter
150 g brauner Zucker
300 g saure Sahne
250 g Mehl
1 Prise Salz, 1/2 TL Zimtpulver
1 Msp. abgeriebene Zitronenschale (unbehandelt)
1/2 Päckchen Backpulver

AUSSERDEM

Fett und Mehl für die Formen
1 Eiweiß (Ei: Größe M)
250 g Puderzucker
Lebensmittelfarben nach Belieben
eventuell bunte Zuckerperlen, Streusel oder Ähnliches als Dekoration

● Elektro-Ofen auf 175 Grad vorheizen. Formen fetten und mit Mehl ausstäuben.

● Apfel waschen, vierteln und das Kerngehäuse entfernen. Apfelviertel grob raspeln. Raspel mit Zitronensaft vermischen.

● Die Eier mit Butter und Zucker schaumig rühren, die saure Sahne unterziehen. Mehl mit Salz, Zimt, Zitronenschale und Backpulver vermischen. Die Apfelraspel etwas ausdrücken, dann mit der Mehl-Mischung unter den Teig rühren.

● Teig in die Formen verteilen. Im Ofen (Gas: Stufe 2) in etwa 25 Minuten goldbraun backen. Küchlein vorsichtig aus den Formen lösen, auf ein Gitter stürzen, vollständig auskühlen lassen.

● Zum Verzieren das Eiweiß zu steifem Schnee schlagen, nach und nach den Puderzucker unterrühren. Guss in Portionen teilen und mit Speisefarbe beliebig einfärben. Kuchen nach Geschmack mit buntem Guss verzieren, eventuell zusätzlich mit Zuckerperlen, bunten Streuseln oder Ähnlichem dekorieren.

Aufwand:	★
Zubereitungszeit:	20 Minuten
Backzeit:	25 Minuten
Pro Stück:	520 kcal/2180 kJ

Weihnachtsglocke

FÜR 1 GLOCKENFORM
120 g Früchte-Mix (Fertigprodukt aus Zitronat Orangeat und kandierten Kirschen)
je 50 g Rosinen und Korinthen
75 ml Kirschwasser
250 g Butter
200 g Zucker
1 Päckchen Vanillezucker
6 Eier (Größe M)
200 g gemahlene Mandeln
300 g Mehl
2 TL Backpulver
100 ml Milch
Saft und Schale von 1/2 unbehandelten Zitrone
ZUM VERZIEREN
25 g gehackte Pistazienkerne
1 Eiweiß (Größe M)
200 g Puderzucker
blaue Lebensmittelfarbe
nach Belieben Puder- und Kristallzucker
AUSSERDEM
Fett und Mehl für die Form

- Früchte-Mix hacken. Alles mit Rosinen, Korinthen und Kirschwasser vermischen und zugedeckt ziehen lassen.
- Glockenform fetten, mit Mehl ausstäuben. Elektro-Ofen auf 175 Grad vorheizen. Butter mit Zucker und Vanillezucker schaumig schlagen. Eier einzeln unterrühren. Mandeln, Mehl und Backpulver mischen, mit Milch, Zitronensaft und -schale unter die Fettmasse rühren. Die eingeweichten Früchte mit dem Kirschwasser unterheben.
- Den Teig in die vorbereitete Form geben, glatt streichen. Kuchen im Ofen (Gas: Stufe 2) 55-60 Minuten backen. Aus dem Ofen nehmen, kurz ruhen lassen, auf ein Kuchengitter stürzen.
- Zum Verzieren Pistazien fein hacken. Das Eiweiß steif schlagen und nach und nach den Puderzucker unterrühren. Einen Teil des Gusses mit der Lebensmittelfarbe blau färben. Schleife, Konturen und Klöppel mit dem blauen Guss überziehen, den restlichen Kuchen mit weißem Guss.
- Den Rand mit Pistazien bestreuen. Wenn der Guss ganz getrocknet ist, nach Belieben den Kuchen mit Puderzucker und feinem Kristallzucker überstäuben.

Aufwand:	★
Zubereitungszeit:	25 Minuten
Backzeit:	1 Stunde
Insgesamt:	7120 kcal/29 900 kJ

Pfefferkuchen-Spitzen & -Bären

BITTE BEACHTEN: MUSS ÜBER NACHT KÜHLEN

Für etwa 50 Stück

125 g Rübensirup oder Honig
60 g Zucker
50 g Butter oder Margarine
250 g Mehl
2 TL Ingwerpulver
2 TL Zimtpulver
1/2 TL Nelkenpulver
1 gestr. TL Anispulver
1 Prise Salz
1 Msp. Pottasche
1 Msp. Hirschhornsalz
50 g Orangeat

AUSSERDEM

Fett und Mehl für die Formen bzw. das Blech
je 200 g Zartbitter-Kuvertüre und weiße Kuvertüre

● Sirup mit Zucker und Fett erhitzen. Das gesiebte Mehl und Ingwer, Zimt, Nelke, Anis und Salz mischen. Die Pottasche und das Hirschhornsalz in der Sirup-Masse auflösen. Die Sirupmasse abkühlen, mit dem Mehlgemisch verarbeiten. Orangeat ganz fein hacken und unterheben. Über Nacht kalt stellen.

● Den Elektro-Ofen auf 175 Grad vorheizen. Die kleinen Formen (10 cm Ø) und das Blech mehlen und fetten. Für die Spitzen den Teig etwa 1 1/2 cm hoch in die Formen füllen und im Ofen (Gas: Stufe 2) 18-20 Minuten backen. Für die Bären Teig ausrollen und mit der Schablone Bären ausschneiden, aufs Blech legen und backen.

● Kuvertüren schmelzen. Jeden Pfefferkuchen noch warm in 8 Ecken teilen und in die geschmolzene weiße bzw. Zartbitter-Kuvertüre tauchen, dann verzieren. Bären wie auf dem Foto verzieren.

Aufwand:	★★
Zubereitungszeit:	1 1/4 Stunden
Backzeit:	18-20 Minuten
Pro Stück:	80 kcal/340 kJ

Nürnberger Lebkuchen

BITTE BEACHTEN: MUSS ÜBER NACHT RUHEN

Für etwa 20 Stück

200 g gehackte Mandeln
60 g Zitronat, 60 g Orangeat
4 Eier (Größe M), 250 g Zucker
1 Msp. weißer Pfeffer
1 TL gemahlener Kardamom
1 Msp. Hirschhornsalz
3 TL Zimtpulver
2 TL Nelkenpulver, 250 g Mehl
1 TL abgeriebene, unbehandelte Zitronenschale

AUSSERDEM

20 Oblaten (Ø 8 cm)
Backpapier für das Blech
Mandelhälften
20 halbe rote Belegkirschen
4 EL Puderzucker

● Elektro-Ofen auf 200 Grad vorheizen. Gehackte Mandeln auf einem Backblech verteilen und im Ofen (Gas: Stufe 3) 7 Minuten goldbraun rösten. Zitronat und Orangeat fein hacken.

● Eier trennen. Eigelbe und Zucker 5 Minuten cremig rühren. Eiweiße steif schlagen, zur Ei-Creme geben und 5 Minuten weiterschlagen. Gewürze mit Mehl mischen und portionsweise mit Zitronenschale, Mandeln, Zitronat und Orangeat unter die Masse heben. Zugedeckt über Nacht kühl stellen.

● Elektro-Ofen auf 175 Grad vorheizen. Lebkuchenteig mit einem in warmem Wasser vorgewärmten Messer auf Oblaten streichen oder mit der Lebkuchenmühle darauf pressen.

● Lebkuchen auf ein mit Backpapier belegtes Blech legen und mit Mandelhälften und Belegkirschen verzieren. Im Ofen (Gas: Stufe 2) etwa 15 Minuten backen.

● Puderzucker mit 2 EL Wasser verrühren. Die Lebkuchen noch heiß damit glasieren.

Aufwand:	★★
Zubereitungszeit:	45 Minuten
Backzeit:	15 Minuten
Pro Stück:	200 kcal/840 kJ

Zimtsterne

Für etwa 65 Stück

6 Eiweiße (Eier: Größe M)
Salz
500 g Puderzucker
600 g geschälte gemahlene Mandeln
1 geh. EL Zimtpulver
abgeriebene Schale von 1/2 unbehandelten Zitrone
1 EL Quittengelee (ersatzweise Apfelgelee)
AUSSERDEM
Zucker zum Ausrollen
Backpapier für die Bleche

- Die Eiweiße mit 1 Prise Salz steif schlagen, dabei den Puderzucker einrieseln lassen. Die Masse 10 Minuten auf höchster Stufe weiterschlagen. 5 EL Eischneemasse beiseite stellen.
- Mandeln mit Zimt und Zitronenschale mischen, mit dem Quittengelee unter den Eischnee ziehen. 30 Minuten kühl stellen.
- Masse portionsweise auf mit Zucker ausgestreuter Arbeitsfläche 1/2 cm dick ausrollen. Sterne ausstechen, Ausstecher dabei immer wieder in Zucker tauchen.
- Sterne auf ein mit Backpapier ausgelegtes Blech setzen. Übrigen Eischnee auf die Sterne streichen. Plätzchen 1 Stunde ruhen lassen, bis die Schnittflächen trocken sind.
- Elektro-Ofen auf 150 Grad vorheizen. Zimtsterne im Ofen (Gas: Stufe 1) etwa 17 Minuten backen.

Aufwand:	★★
Zubereitungszeit:	2 1/2 Stunden
Backzeit pro Blech:	17 Minuten
Pro Stück:	90 kcal/380 kJ

Variation

- Man kann die Zimtsterne statt mit Mandeln auch mit der selben Menge geriebenen Haselnüssen backen oder mit jeweils der Hälfte Mandeln und Haselnüsse.

Walnusstaler

Für etwa 45 Stück

200 g Walnusskerne
200 g Butter
150 g Zucker
Mark von 1 Vanilleschote
1 Ei (Größe M)
1 Prise Salz
100 g Mehl
100 g Speisestärke
AUSSERDEM
4 EL Preiselbeerkonfitüre
200 g Puderzucker
3-4 EL Zitronensaft
Backpapier für die Bleche

- 45 Walnusshälften beiseite legen, Rest fein mahlen. Mit Butter, Zucker, Vanillemark, Ei, Salz, Mehl und Stärke verkneten. In Folie gewickelt 30 Minuten kalt stellen.
- Elektro-Ofen auf 175 Grad vorheizen. Den Teig 1/2 cm dick ausrollen. Mit runden Austechern mit gewelltem Rand (3-4 cm Ø) Plätzchen ausstechen und auf mit Backpapier belegte Bleche setzen. Im Ofen (Gas: Stufe 2) etwa 12 Minuten backen. Gut auskühlen lassen.
- Die Konfitüre erwärmen und durch ein Sieb streichen. Die Hälfte der Plätzchen damit bestreichen und jeweils mit 1 unbestrichenen Plätzchen zusammensetzen. Puderzucker mit Zitronensaft verrühren, gitterartig über die Plätzchen ziehen. Jeweils 1 Nusshälfte darauf setzen.

Aufwand:	★
Zubereitungszeit:	1 1/4 Stunden
Backzeit pro Blech:	12 Minuten
Pro Stück:	100 kcal/420 kJ

Mandel-Honig-Kipferln

BITTE BEACHTEN: TEIG MUSS ÜBER NACHT KÜHLEN

Für etwa 80 Stück

400 g Mehl, 280 g Butter
80 g Zucker
1 Päckchen Vanillezucker
80 g flüssiger Honig
150 g geschälte gemahlene Mandeln
2 Tropfen Bittermandelaroma
1 Ei (Größe M)
AUSSERDEM
150 g brauner Zucker
Backpapier für die Bleche

- Mehl mit Butter, weißem Zucker, Vanillezucker, 50 g Honig, Mandeln, Aroma und Ei verkneten. Über Nacht kalt stellen.
- Elektro-Ofen auf 175 Grad vorheizen. Backbleche mit Backpapier auslegen. Den Teig in 8 Stücke teilen. Jedes Stück zu einem 1 1/2 cm dicken Strang ausrollen. Jeden Strang in 7 cm lange Stücke teilen. Stücke zu Hörnchen formen und im Abstand von 3 cm auf das Blech legen.
- Die Kipferln im Ofen (Gas: Stufe 2) etwa 15 Minuten backen. Die Plätzchen noch heiß mit übrigem Honig bestreichen, im braunen Zucker wälzen.

Aufwand:	★
Zubereitungszeit:	50 Minuten
Backzeit pro Blech:	15 Minuten
Pro Stück:	70 kcal/290 kJ

Tipp

Statt Bittermandelaromakönnen Sie auch 2-3 bittere Mandeln (man bekommt sie in der Apotheke) fein reiben und unter den Teig mischen. Sie haben ein besonders feines Aroma.

Neros

Für etwa 40 Stück
250 g Butter
150 g Puderzucker
5 Eigelbe (Eier: Größe M)
Mark von 1 Vanilleschote
abgeriebene Schale von
2 unhandelten Orangen
220 g Mehl
50 g Kakaopulver
FÜR DIE FÜLLUNG
100 g Marzipan-Rohmasse
1-2 EL Orangenlikör
1-2 EL Orangenmarmelade
AUSSERDEM
Backpapier für das Blech
50 g weiße Kuvertüre
40 Kaffeebohnen

● Elektro-Ofen auf 200 Grad vorheizen. Butter, Puderzucker und Eigelbe mit Vanillemark und der Hälfte der Orangenschale schaumig rühren. Mehl und Kakao mischen und unterrühren.

● Backblech mit Backpapier auslegen. Teig in einen Spritzbeutel mit Lochtülle (0,5 cm Ø) füllen. Kaffeebohnenförmige, 3-4 cm lange Stücke aufs Blech spritzen. Im Ofen (Gas: Stufe 3) etwa 10 Minuten backen.

● Für die Füllung Marzipan, restliche Orangenschale, Likör und Marmelade verkneten. Die Unterseite der Kekse damit bestreichen, Kekse zusammensetzen, nach Belieben mit geschmolzener weißer Kuvertüre und mit Kaffeebohnen verzieren.

Aufwand:	★★
Zubereitungszeit:	35 Minuten
Backzeit:	10 Minuten
Pro Stück:	110 kcal/460 kJ

Odenwälder Dreispitz

Für etwa 40 Stück
125 g gekochte Kartoffeln
(mehlig kochend)
60 g Butter, 60 g Zucker
1 Ei (Größe M), 250 g Mehl
1 TL Backpulver
1 Prise Salz
Mark von 1 Vanilleschote
1 TL abgeriebene
Zitronenschale (unbehandelt)
150 g Pflaumenmus
1 Eigelb (Ei: Größe M)
2 EL Milch
2 EL Hagelzucker
Backpapier für die Bleche

● Kalte Kartoffeln pellen, fein reiben. Mit Butter, Zucker, Ei, und Mehl, Backpulver, Salz, Vanillemark und Zitronenschale zu einem glatten Teig verkneten. Teig etwa 30 Minuten kalt stellen.

● Elektro-Ofen auf 175 Grad vorheizen. Backbleche mit Backpapier auslegen.

● Teig etwa 3 mm dick ausrollen, Kreise (6 cm Ø) ausstechen. Auf jeden Kreis etwas Pflaumenmus aufstreichen, dabei am Rand etwa 1 cm frei lassen. Kreise jeweils an drei Seiten hochklappen. Die Teigseiten zu einem Dreispitz fest zusammendrücken (s. Seite 335).

● Eigelb mit Milch verrühren. Plätzchen damit bepinseln, mit Hagelzucker bestreuen und im Ofen (Gas: Stufe 2) in etwa 15 Minuten goldgelb backen.

Aufwand:	★★
Zubereitungszeit:	1 1/2 Stunden
Backzeit po Blech:	15 Minuten
Pro Stück:	110 kcal/460 kJ

Michelstädter Wölkchen

Für etwa 30 Stück
2 Eiweiße (Eier: Größe M)
125 g Puderzucker
20 g Speisestärke
rote Lebensmittelfarbe
50 g fein gemahlene Mandeln
FÜR DIE FÜLLUNG
2 Eier (Größe M)
80 g Zucker
250 g Butter
100 g Himbeeren (frisch oder
tiefgekühlt)
1-2 EL Himbeergeist
Backpapier für die Bleche

● Elektro-Ofen auf 175 Grad vorheizen. Eiweiße mit Puderzucker und Stärke steif schlagen. Nach Belieben die Masse mit Lebensmittelfarbe rosa einfärben. Mandeln unterheben.

● Masse in einen Spritzbeutel mit Lochtülle (1cm Ø) füllen, Tupfen auf ein mit Backpapier belegtes Blech spritzen. Im Ofen 15-20 Minuten backen.

● Für die Füllung die Eier mit Zucker in einer Schüssel über dem heißen Wasserbad zu einer cremigen Masse aufschlagen, danach kalt schlagen.

● Butter mit Himbeeren und Himbeergeist schaumig schlagen und dann die Eimasse unterheben. Jeweils 2 abgekühlte Baiser-Tupfen mit etwas Buttercreme zusammensetzen.

Aufwand:	★★
Zubereitungszeit:	40 Minuten
Backzeit pro Blech:	20 Minuten
Pro Stück:	55 kcal/230 kJ

Engels-Plätzchen

Für etwa 60 Stück
100 g Pinienkerne
250 g Mehl
2 Päckchen Vanillezucker
50 g Zucker
150 g Butter
1 Eigelb (Ei: Größe M)
2 EL süßer Sherry (ersatzweise
Orangensaft)
AUSSERDEM
Mehl zum Arbeiten
Backpapier für das Blech
1 frisches Ei (Größe M)
1 EL Milch, 150 g Puderzucker
2-3 EL Zitronensaft oder
gelbe Lebensmittelfarbe
farbiger Dekor-Zucker

- Pinienkerne in einer beschichteten Pfanne ohne Fett goldbraun rösten, abkühlen lassen und fein mahlen.
- Pinienkerne mit Mehl, Vanillezucker, Zucker, Butter, Eigelb und Sherry zu einem glatten Teig verkneten. Zugedeckt etwa 1 Stunde kühl stellen.
- Elektro-Ofen auf 175 Grad vorheizen. Teig auf einer bemehlten Arbeitsfläche etwa 1/2 cm dick ausrollen. Mit Ausstechern „himmlische" Formen ausstechen (Engel, Trompeten, Sterne oder Sternschnuppen). Plätzchen auf mit Backpapier belegte Bleche legen.
- Ei trennen. Eigelb mit Milch verrühren und die Plätzchen damit bestreichen. Im Ofen (Gas: Stufe 2) in etwa 10 Minuten goldgelb backen. Abkühlen lassen.
- Puderzucker mit Zitronensaft oder steif geschlagenem Eiweiß glatt rühren. Guss eventuell einfärben und in einen Einweg-Spritzbeutel oder Gefrierbeutel füllen (eine kleine Ecke abschneiden). Konturen der Plätzchen mit Zuckerguss nachziehen. Plätzchen nach Belieben mit Dekor-Zucker dekorieren, trocknen lassen.

Aufwand:	★
Zubereitungszeit:	2 Stunden
Backzeit pro Blech:	10 Minuten
Pro Stück:	60 kcal/250 kJ

Wolken-Cookies

Für etwa 35 Stück
250 g Mehl
1 TL Backpulver
1/4 TL Salz
2 EL Butter
200 g Zucker
1 Ei (Größe M)
50 g feste Kokoscreme
(ungesüßt; aus der Dose,
ersatzweise Butter)
1-2 TL Zitronensaft
100 ml Buttermilch
50 g Kokosflocken
AUSSERDEM
Backpapier für die Bleche
200 g Puderzucker
Zitronensaft oder Kokoslikör
zum Glattrühren
Kokosflocken zum Bestreuen

- Elektro-Ofen auf 175 Grad vorheizen. Mehl mit Backpulver und Salz mischen. Butter, Zucker, Ei, Kokoscreme und Zitronensaft mit den Schneebesen des Handrührgeräts schaumig aufschlagen. Abwechselnd die Mehlmischung und Buttermilch zufügen, alles gut verrühren. Zum Schluss Kokosflocken unterheben.
- Von dem Teig mithilfe eines Teelöffels 3 kleine Kleckse auf einem mit Backpapier belegten Blech wolkenförmig aneinander setzen. Diesen Vorgang mit dem restlichen Teig wiederholen, dabei einen größeren Abstand zwischen den einzelnen Wolken einhalten.
- Die Plätzchen im Ofen (Gas: Stufe 2) etwa 12 Minuten backen. Auskühlen lassen.
- Puderzucker mit Zitronensaft oder Likör glatt rühren. Auf die Plätzchen streichen und mit Kokosflocken bestreuen.

Aufwand:	★
Zubereitungszeit:	45 Minuten
Backzeit pro Blech:	12 Minuten
Pro Stück:	90 kcal/380 kJ

1001-Nacht-Plätzchen

Für etwa 40 Stück
300 g Mehl
250 g Butter
150 g Zucker
100 g gemahlene Haselnüsse
3 EL Orangenblütenwasser (Apotheke)
1 Ei (Größe M)
je 1 TL Zimt- und Kardamompulver
1/2 TL gemahlene Nelken
50 g kandierte Früchte (Früchte-Mix)
3 EL Milch
0,2 g Safranfäden (2 Döschen)
100 g Honig, 75 g Sesamsamen
30 g gehackte Pistazienkerne
AUSSERDEM
Backpapier für das Blech
3 EL Aprikosenkonfitüre

- Mehl mit 200 g Butter, Zucker, Haselnüssen, Orangenwasser, Ei und Gewürzen verkneten. Teig 1 Stunde kühl stellen.
- Elektro-Ofen auf 175 Grad vorheizen. Den Teig 1/2 cm dick ausrollen, Halbmonde ausstechen. Auf ein mit Backpapier belegtes Blech legen. Früchte-Mix feiner hacken.
- Milch erwärmen, Safran darin auflösen. Übrige Butter, Honig, Sesam, Früchte-Mix und Pistazien zufügen, erwärmen, auf die Monde streichen. Im Ofen (Gas: Stufe 2) etwa 15 Minuten backen.
- Konfitüre erwärmen, durch ein Sieb streichen und auf die Plätzchen streichen. Eventuell mit Trockenfrüchten, zu Mini-Figuren ausgestochen, verzieren.

Aufwand:	★
Zubereitungszeit:	1 1/2 Stunden
Backzeit:	15 Minuten
Pro Stück:	130 kcal/550 kJ

Glamour-Sterne

Für etwa 55 Stück

BITTE BEACHTEN: TEIG MUSS 4 STUNDEN KÜHLEN

- *100 g Vollmilch- und 250 g Halbbitter-Kuvertüre*
- *100 g Schlagsahne*
- *2 EL Ahornsirup*
- *225 g Butter*
- *50 ml Whisky*
- *100 g Bananenchips (Reformhaus)*
- *100 g Zucker*
- *4-5 Tropfen Buttervanille-Aroma*
- *1 Ei (Größe M), 250 g Mehl*
- *1 Msp. Backpulver*

AUSSERDEM

- *Backpapier für die Bleche*
- *Blattgold nach Belieben*

● Kuvertüren hacken. Sahne und Sirup aufkochen, Topf vom Herd ziehen. Kuvertüren und 100 g Butter darin schmelzen, Whisky unterrühren. 4 Stunden kalt stellen.

● Inzwischen 50 g Bananenchips im Mixer oder mit der Teigrolle fein zerbröseln und mit übriger Butter, Zucker, Vanille-Aroma, Ei, Mehl und Backpulver verkneten. In Folie gewickelt 1 Stunde kalt stellen.

● Elektro-Ofen auf 200 Grad vorheizen. Teig dünn ausrollen. Sterne ausstechen, auf mit Backpapier belegte Bleche setzen. Im Ofen (Gas: Stufe 3) etwa 10 Minuten backen, abkühlen lassen.

● Kuvertüre-Masse mit dem Handrührgerät aufschlagen. In Spritzbeutel mit Sterntülle füllen, auf die Sterne spritzen. Mit Blattgold und übrigen Bananenchips verzieren.

Aufwand:	★★
Zubereitungszeit:	1 Stunde
Backzeit pro Blech:	10 Minuten
Pro Stück:	100 kcal/420 kJ

Walnuss-Plätzchen

150 g weiche Butter
125 g Zucker
2 Eier (Größe M)
200 g gemahlene Walnusskerne
100 g Mehl
1-2 EL Kakaopulver
1 Prise Salz
50 g Puderzucker
1 TL Zitronensaft
Beleg-Kirschen
AUSSERDEM
Backpapier für die Bleche

- Butter und Zucker schaumig schlagen, Eier unterrühren. Walnüsse, Mehl, Kakao und Salz unterheben. Teig in Folie gewickelt 45 Minuten kühl stellen.
- Elektro-Ofen auf 175 Grad vorheizen. Aus dem Teig kleine Kugeln formen. Mit etwas Abstand auf mit Backpapier ausgelegte Bleche setzen, sie zerlaufen etwas beim Backen.
- Im Ofe (Gas: Stufe 2) etwa 15 Minuten backen. Aus dem Ofen nehmen und einige Minuten auf dem Backblech ruhen lassen. Auf einem Kuchengitter erkalten lassen.
- Aus Puderzucker und Zitronensaft einen festen Guss rühren. Die Beleg-Kirschen halbieren. Mit etwas Zuckerguss auf jedes Plätzchen 1 Kirschenhälfte (mit der Schnittfläche nach unten) kleben.

Aufwand:	★
Zubereitungszeit:	1 1/4 Stunden
Backzeit pro Blech:	15 Minuten
Pro Stück	75 kcal/315 kJ

Glühwein-Schnitten

FÜR ETWA 80 STÜCK	
1/8 l Glühwein (Fertigprodukt)	
200 g Butter	
200 g Zucker	
1 EL Vanillezucker	
3 Eier (Größe M)	
300 g Mehl	
2 TL Backpulver	
150 g Raspel-Schokolade	
200 g Puderzucker	
AUSSERDEM	
Backpapier fürs Blech	

- Elektro-Ofen auf 200 Grad vorheizen. Blech mit Backpapier auslegen.
- 4 EL Glühwein für die Glasur abnehmen, aufheben. Aus Butter, Zucker, Vanillezucker, Eiern, Mehl, Backpulver und Glühwein einen glatten Teig rühren. Raspel-Schokolade unterheben.
- Den Teig gleichmäßig auf dem Blech verstreichen. Im Ofen (Gas: Stufe 3) etwa 25 Minuten backen.
- Puderzucker mit restlichem Glühwein verrühren. Kuchen damit glasieren. Ist die Glasur fest, die Kuchenplatte in kleine Dreiecke schneiden.

Aufwand:	★
Zubereitungszeit:	30 Minuten
Backzeit:	25 Minuten
Pro Stück:	65 kcal/270 kJ

Cappuccino-Plätzchen

Für etwa 60 Stück
2-3 TL Instant-Espressopulver
75 g Mokkaschokolade
200 g Butter
120 g brauner Zucker
2 Eier (Größe M)
200 g Mehl
50 g gemahlene Haselnüsse
2 Msp. abgeriebene
Zitronenschale (unbehandelt)
2 Msp. Kardamom
1 TL Backpulver
2 Eiweiße (Eier: Größe M)
125 g Zucker
AUSSERDEM
60 Pralinenförmchen aus Metall
Kakaopulver zum Bestäuben

- Das Kaffeepulver in 1-2 TL heißem Wasser auflösen. Die Schokolade grob hacken oder reiben. Elektro-Ofen auf 175 Grad vorheizen.
- Die Butter mit braunem Zucker, 2 Eiern und Kaffee verrühren. Mehl, Nüsse, Zitronenschale, Kardamom, Backpulver und Schokolade nach und nach unter den Teig rühren.
- Den Teig in einen Spritzbeutel mit Lochtülle füllen und in kleine Metall-Pralinenformen spritzen.
- Die Eiweiße steif schlagen, dabei den weißen Zucker einrieseln lassen. Jeweils 1 kleinen Klecks auf die Plätzchen setzen.
- Die Förmchen auf Backbleche setzen und im Ofen (Gas: Stufe 2) etwa 15 Minuten backen, eventuell zudecken. Abgekühlt mit Kakao bestäuben.

Aufwand:	★
Zubereitungszeit:	30 Minuten
Backzeit pro Blech:	15 Minuten
Pro Stück:	70 kcal/290 kJ

Vanille-Monde

Für etwa 50 Stück

2 Vanillerschoten
250 g Mehl
1 Prise Salz
210 g kalte Butter
70 g Zucker
100 g geschälte, fein gemahlene Mandeln

AUSSERDEM

Backpapier für die Bleche
Puderzucker zum Bestäuben

● Die Vanilleschoten in der Länge halbieren und das Mark herausschaben. Das Mehl mit Salz, der Butter in Flöckchen, Zucker und Mandeln rasch zu einem glatten Teig verkneten. Den Teig zugedeckt etwa 1 Stunde kühl stellen.

● Elektro-Ofen auf 175 Grad vorheizen. Den Teig zu einer Rolle formen (3-4 cm Ø), walnussgroße Stücke abschneiden und zu Monden biegen.

● Plätzchen auf mit Backpapier belegte Backbleche legen. Im Ofen (Gas: Stufe 2) 10-12 Minuten backen.

● Die Vanillle-Monde vorsichtig vom Blech nehmen und noch warm mit Puderzucker bestäuben. Auskühlen lassen.

Aufwand:	★
Zubereitungszeit:	2 Stunden
Backzeit pro Blech:	12 Minuten
Pro Stück:	65 kcal/270 kJ

Variationen

● Dieses Rezept lässt sich sehr leicht abwandeln. Wer z. B. Schokoladen-Monde backen möchte, lässt die Vanilleschoten weg und gibt statt dessen 100 g geriebene Zartbitterschokolade unter den Teig.

● Oder Haselnuss-Monde: Statt der Mandeln 100 g geriebene Haselnüsse unter den Teig heben.

● Oder feine Marzipan-Monde: Nur 50 g Mandeln und 100 g Marzipan unter den Teig mischen.

● Oder statt sie mit Puderzucker zu bestreuen, überziehen Sie die Vanille-Monde mit 100 g geschmolzener dunkler Kuvertüre.

Heidesand

Für etwa 50 Stück
250 g Butter
200 g feiner Zucker
1 Päckchen Vanillezucker
250 g Mehl, 125 g Speisestärke
1 gestrichener TL Backpulver
je 25 rote und grüne Belegkirschen
AUSSERDEM
Backpapier für die Bleche
1 EL Aprikosenkonfitüre
2 EL Kokosraspel

- Butter in einem Topf schmelzen und leicht bräunen, dann abkühlen lassen. Mit Zucker und Vanillezucker schaumig rühren. Mehl, Stärke und Backpulver mischen, unterkneten. Teig 1 Stunde kalt stellen.
- Aus dem Teig Rollen formen (2 cm Ø), Teigrollen in Folie gewickelt nochmal 1 Stunde kalt stellen.
- Elektro-Ofen auf 175 Grad vorheizen. Backblech mit Backpapier auslegen. Rollen in 1/2 cm dicke Scheiben schneiden, aufs Blech legen und im Ofen (Gas: Stufe 2) etwa 15 Minuten hellgelb backen.
- Kekse aus dem Ofen nehmen, je 1/2 Belegkirsche in die Mitte setzen. Auf einem Kuchengitter leicht abkühlen lassen. Die Ränder mit erwärmter Konfitüre bestreichen und in Kokosraspeln wälzen.

Aufwand:	★★
Zubereitungszeit:	2 1/2 Stunden
Backzeit pro Blech:	15 Minuten
Pro Stück:	80 kcal/340 kJ

meine Familie & ich

Deko-Tipp

Aus den Belegkirschen kleine Herzen und Sterne ausstechen und die Plätzchen damit verzieren. Das macht zwar etwas Arbeit, sieht aber sehr hübsch aus.

Weihnachts-Amerikaner

Für etwa 12 Stück
100 g Marzipan-Rohmasse
200 g saure Sahne, 50 g Zucker
1 Päckchen Vanillezucker
2 Eier (Größe M)
250 g Mehl, 1 TL Backpulver
1 TL Lebkuchengewürz
2 EL Milch
AUSSERDEM
Backpapier für das Blech
200 g Puderzucker
Zitronensaft oder
1 Eiweiß (Ei: Größe M)
rote Lebensmittelfarbe

- Elektro-Ofen auf 200 Grad vorheizen. Marzipan in Würfel schneiden, mit saurer Sahne, Zucker, Vanillezucker und Eiern glatt rühren. Mehl, Backpulver, Lebkuchengewürz und Milch zugeben und zu einem glatten Teig rühren.
- Jeweils 1-2 EL Teig als kleines flaches Häufchen auf ein mit Backpapier belegtes Backblech setzen. Dabei etwas Abstand halten, da die Kekse beim Backen auseinander laufen. Im Ofen (Gas: Stufe 3) etwa 10 Minuten backen. Danach auf einem Kuchengitter auskühlen lassen.
- Puderzucker mit Zitronensaft oder steif geschlagenem Eiweiß glatt rühren. Guss teilweise einfärben und getrennt in Einweg-Spritzbeutel (gibt es in Haushaltsläden) oder Gefrierbeutel füllen. Eine kleine Ecke abschneiden und die Amerikaner auf einer Seite mit Zuckerguss zu Weihnachtsmann-Gesichtern verzieren und gut trocknen lassen.

Aufwand:	★
Zubereitungszeit:	30 Minuten
Backzeit:	10 Minuten
Pro Stück:	230 kcal/970 kJ

Tipps, Tricks und Ideen fü

Jetzt hat Backen Hochkonjunktur. Selbst wer d
Damit es Ihnen und Ihren Kindern auch richtig Spaß macht, habe

Je frischer, um so würziger

Je frischer die Gewürze sind, um so intensiver duften und schmecken Gewürzkuchen. Deshalb immer auf das Haltbarkeitsdatum achten! Lieber nur kleine Mengen kaufen. Und: Frisch gemahlen in der Gewürzmühle, zerstampft im Mörser oder gerieben entfalten sie ihr Aroma am besten.

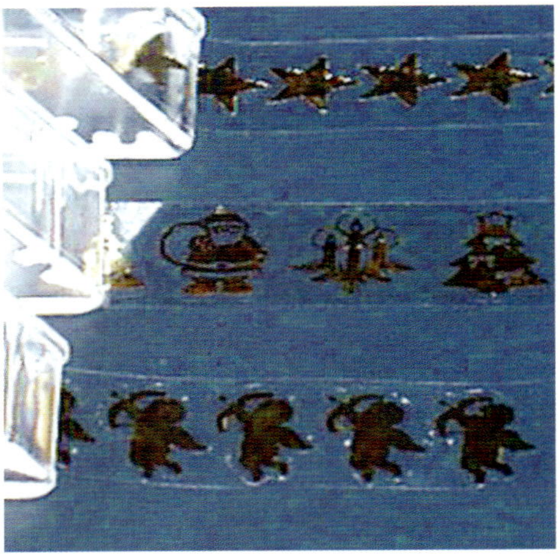

Goldene Plätzchen-Deko

Besonders edel werden Plätzchen, z. B. die Glamour-Sterne (Rezept Seite 327), Pralinen oder Kuchen mit Motiven aus essbarem, hauchdünnem Gold. Das geschmacksneutrale Gold gibt es als Pulver, Blattgold oder als kleine Figuren. Sie sind beim Hobbybäcker-Versand (Tel. 0 73 06 / 92 59 00) zu bestellen. Wesentlich preisgünstiger, aber nicht so filigran, sind goldene Zuckerperlen (im Backregal des Supermarkts).

In die Form gebracht

Glockenformen, Sternformen u. ä. gibt es in unterschiedlichen Größen, antihaftbeschichtet oder aus Weißblech (besonders geeignet für Gasöfen) in Haushaltsläden.

die Weihnachtsbäckerei

anze Jahr nicht backt, greift im Advent zum Handrührer.
ir hier ein paar hilfreiche Informationen zusammengetragen

Zimtsterne ausstechen

Der Mandel-Zucker-teig ist ziemlich klebrig, tauchen Sie die Ausstecher daher immer wieder in lauwarmes Wasser. Mit einem auseinander klappbaren Ausstecher geht es besonders gut (im Fachhandel).

Spritzige Ideen

Ganz neu und superpraktisch: Einmal-Spritzbeutel. Einfach Creme oder Sahne in den Klarsichtbeutel mit einer der drei Tüllen füllen, Plätzchen oder Torten dekorieren. Beutel hinterher wegwerfen, Tülle abwaschen und wieder verwenden.

Plätzchen glasieren

Egal, ob Sie Plätzchen, Teilchen oder ganze Kuchen glasieren wollen, am besten geht das auf einem Kuchengitter. So kann die überflüssige Glasur einfach abtropfen (in Haushaltsfachgeschäften erhältlich).

Dreispitze formen

In die Mitte der ausgestochenen Teigkreise einen kleinen Klecks Konfitüre geben. Den Teigrand mit zwei Fingern fast bis zur Mitte hochheben und eine Ecke einkneifen. Die flache Seite einfach hochklappen und die Teigränder festdrücken.

Plätzchen richtig aufbewahren

Bewahren Sie jede Plätzchensorte getrennt in Blechdosen auf und stellen Sie sie an einen möglichst kühlen Ort (nicht in den Kühlschrank). Bei knusprigem Gebäck wie Vanillekipferln und Mürbeteigplätzchen (ohne Verzierung) Dose gleich nach dem Abkühlen luftdicht verschließen, damit es keine Feuchtigkeit aus der Luft aufnimmt. Verzierte und glasierte Kekse erst trocknen lassen, bevor Sie sie in Dosen verpacken.

Dem besseren Geschmack zuliebe

Gemahlene Nüsse zu kaufen erspart zwar Arbeit, aber jedes Gebäck schmeckt einfach besser, wenn man frisch gemahlene verwendet. Am besten ganze Nüsse kaufen und sie je nach Bedarf knacken, mahlen und dann dem Rezept entsprechend abwiegen. Im Handel gibt es Nussmühlen mit verschiedenen Mahltrommeln von fein bis grob.

Hübsch verpackt

Pralinenförmchen kann man auch als Backförmchen verwenden, z. B. für die Cappuccino-Plätzchen. Es gibt sie aus Metall-Folie oder Papier (weiß und farbig) mit weihnachtlichen Motiven.

Feines Konfekt
selbst gemacht

Nichts geht über Pralinen und Schokoladentrüffel, die aus Ihrer eigenen Küche kommen. Denn in ihnen steckt Liebe, die man schmecken kann

Pralinenrezepte
siehe folgende Seiten

Schwarze Sesam-Trüffel

Für etwa 70 Stück	
650 g weiße Schokolade	
50 g Zucker	
150 g Schlagsahne	
30 g Sesamöl	
AUSSERDEM	
100 g schwarze Sesamsamen	
(Reformhaus),	
ersatzweise weißer Sesam	

- Die Schokolade grob hacken. Den Zucker in einem kleinen Topf goldgelb schmelzen lassen, ohne umzurühren. Sahne zufügen und so lange kochen, bis sich der Zucker aufgelöst hat.
- 450 g gehackte Schokolade und das Sesamöl einrühren. Form (z. B. eine Auflaufform, ca. 27 x 17 cm) mit Klarsichtfolie auslegen. Die handwarme Masse etwa 1,5 cm hoch einfüllen. Im Kühlschrank mindestens 2-3 Stunden (am besten über Nacht) kalt stellen.
- Die restliche Schokolade im warmen Wasserbad schmelzen.
- Erkaltete Masse mit Hilfe der Folie aus der Form heben, in Würfel, Kugeln oder Rechtecke teilen. Pralinen in geschmolzener Schokolade wenden und mit Sesam bestreuen. Auf einem Kuchengitter auskühlen lassen. Nach Belieben mit Spänen von weißer Schokolade verzieren.

Aufwand:	★
Zubereitungszeit:	35 Minuten
Kühl-/Trockenzeit:	3 Stunden
Pro Stück:	70 kcal/290 kJ

Champagner-Trüffel

Für etwa 30 Stück
300 g weiße Kuvertüre
50 g Butter, 50 g Puderzucker
2 EL Champagner
oder trockener Sekt
1 TL Zitronensaft
AUSSERDEM
Backpapier für das Blech
150 g weiße Kuvertüre oder
weiße Schokoraspel
(Fertigprodukt)

- 125 g Kuvertüre grob hacken, in einer Schüssel über dem warmen Wasserbad schmelzen. Abkühlen lassen. Butter und Puderzucker schaumig aufschlagen. Lauwarme flüssige Kuvertüre unter die Buttermischung heben. Champagner oder Sekt und Zitronensaft zugeben, rühren, bis eine spritzfähige Masse entsteht.
- Creme in einen Spritzbeutel mit Lochtülle (etwa 1 1/2 cm Ø) füllen, kirschgroße Tupfen auf ein mit Backpapier belegtes Blech spritzen. 2 Stunden kühl stellen.
- Trüffeln zu Kugeln formen und 30 Minuten tiefkühlen. Die übrigen 175 g Kuvertüre hacken, in einer Schüssel über dem warmen Wasserbad schmelzen. Kugeln einzeln in die Kuvertüre tauchen, auf ein Gitter legen. Zum Verzieren entweder 150 g weiße Kuvertüre auf einer Reibe in Späne hobeln und die Trüffeln darin wenden oder in weißen Schokoraspeln wälzen, trocknen lassen.

Aufwand:	★
Zubereitungszeit:	1 1/2 Stunden
Kühl-/Trockenzeit:	2 Stunden
Pro Stück:	100 kcal/420 kJ

Marzipan-Konfekt

Für etwa 70 Stück
150 g getrocknete Aprikosen
100 g getrocknete Ananas
100 g getrocknete Pflaumen
(ohne Stein)
300 g Marzipan-Rohmasse
30 ml Orangenlikör
150 g Puderzucker
2 EL Orangenmarmelade
300 g Zartbitter-Kuvertüre

- Aprikosen, Ananas und Pflaumen fein würfeln. Marzipan würfeln, mit Likör, Puderzucker, Marmelade und Früchten verkneten. Zwischen aufgeschnittenen Gefrierbeuteln 2 cm dick ausrollen. Dann etwa 1 Stunde tiefkühlen.
- Aus der Marzipanmasse Früchte ausstechen oder Masse in Würfel teilen. Konfekt etwa 1 Stunde bei Zimmertemperatur trocknen lassen.
- Kuvertüre grob hacken und in einer Schüssel über dem warmen Wasserbad schmelzen. Konfekt mit Kuvertüre überziehen, auf Folie oder Gitter setzen, trocknen lassen. Nach Belieben mit getrockneten Früchten, z. B. mit Zitronatstreifen, verzieren.

Aufwand:	★
Zubereitungszeit:	30 Minuten
Kühl-/Trockenzeit:	4 Stunden
Pro Stück:	65 kcal/270 kJ

Variation

- Wer lieber auf den Orangenlikör verzichten möchte, nimmt statt dessen die gleiche Menge Orangensaft-Konzentrat.

Walnuss-Pralinen

Für etwa 80 Stück
100 g Vollmilch-Schokolade
2 EL Honig
250 g gemahlene Walnusskerne
80 g geschälte, gemahlene
Mandeln
6-7 EL Birnengeist
300 g Vollmilch-Kuvertüre
AUSSERDEM
Puderzucker zum Bestäuben
etwa 80 Walnusskernhälften
(oder entsprechend der
Pralinenzahl ersatzweise
geschälte Mandelhälften)

- Schokolade fein hacken. Honig mit Schokolade und Nüssen unter Rühren erhitzen, bis Honig und Schokolade geschmolzen sind und sich die Masse gut vermischt hat. Topf vom Herd nehmen, Mandeln und Birnengeist unter die Schokoladenmasse mischen.
- Masse portionsweise auf ein großes Stück Klarsichtfolie geben und mit Hilfe der Folie zu 1-2 cm dicken Stangen formen. Die Marzipanrollen rundherum mit Puderzucker bestäuben, in Alufolie wickeln und im Kühlschrank fest werden lassen.
- Kuvertüre grob hacken und in einer Schüssel über dem warmen Wasserbad schmelzen.
- Die Schokostangen in 1 1/2 cm lange Stücke teilen, diese in die Kuvertüre tauchen und auf ein Gitter oder auf Folie legen. Kurz bevor die Kuvertüre fest wird, je 1 Walnusshälfte auf die Pralinen legen. Wer möchte, überzieht die Walnusskerne vorher mit Karamell.

Aufwand:	★★
Zubereitungszeit:	50 Minuten
Kühl-/Trockenzeit:	3 Stunden
Pro Stück:	70 kcal/295 kJ

Knusper-Pralinen

Für etwa 30 Stück
75 g gemischte Nusskerne und
getrocknete Früchte
(z. B. Haselnusskerne, Mandeln,
Aprikosen und Feigen
oder Fertig-Mix)
200 g weiße Kuvertüre
50 g Cornflakes

- Nüsse und Früchte hacken. Kuvertüre grob hacken und in einer Schüssel über dem warmen Wasserbad schmelzen.
- Die Früchte, Nüsse und Cornflakes vorsichtig unter die flüssige Kuvertüre heben. Mit einem Teelöffel kleine Häufchen abnehmen und auf Pergamentpapier setzen. Gut trocknen lassen. Sie halten etwa 2-3 Wochen.

Aufwand:	★
Zubereitungszeit:	30 Minuten
Kühl-/Trockenzeit:	2 Stunden
Pro Stück:	55 kcal/230 kJ

Variationen

- Die Cornflakes kann man durch Rice-Crispies oder andere gemischte Knusperflocken ersetzen.
- Wer es vollwertig mag, nimmt die gleiche Menge Haferflocken und röstet sie in einer unbeschichteten Pfanne ohne Fett.
- Und statt der weißen kann man auch helle oder dunkle Kuvertüre verwenden.

Cointreau-Trüffel

Für etwa 65 Stück
BITTE BEACHTEN: MUSS ÜBER NACHT KÜHLEN
125 g Schlagsahne
60 g Zucker, 1 Msp. Salz
Mark von 1 Vanilleschote
200 g Zartbitter-Kuvertüre
250 g weiße Kuvertüre
10 ml Cointreau
250 g zimmerwarme Butter
AUSSERDEM
Kakaopulver zum Wälzen
300 g Zartbitter-Kuvertüre zum Überziehen, Backpapier

- Sahne mit Zucker, Salz und Vanillemark aufkochen. Kuvertüren klein hacken und unter die Sahnemischung rühren. Abkühlen lassen. Cointreau unterheben. Die Butter schaumig rühren und flöckchenweise mit der Schokoladencreme verrühren (dabei darauf achten, dass Butter und Creme Zimmertemperatur haben).
- Schokocreme in einen Spritzbeutel mit großer Lochtülle geben und Kugeln auf Backpapier spritzen. Im Kühlschrank über Nacht erstarren lassen.
- Trüffeln aus dem Kühlschrank nehmen. Einige Zeit bei Raumtemperatur stehen lassen. Kuvertüre über dem warmen Wasserbad schmelzen, auf Körpertemperatur abkühlen lassen, die Fingerprobe (s. Info im Kasten Mitte) machen. Trüffeln in die Kuvertüre tauchen, gut abtropfen lassen, in Kakao wälzen. Überschüssigen Kakao abschütteln, Trüffeln trocknen lassen.

Aufwand:	★★
Zubereitungszeit:	40 Minuten
Kühl-/Trockenzeit:	40 Minuten
Pro Stück:	100 kcal/420 kJ

Honig-Trüffel

Für etwa 50 Stück
BITTE BEACHTEN: MUSS ÜBER NACHT KÜHLEN
75 g Schlagsahne, 150 g Honig
Mark von 1 Vanilleschote
250 g Zartbitter-Kuvertüre
175 g Vollmilch-Kuvertüre
250 g zimmerwarme Butter
AUSSERDEM
Backpapier
150 g Zartbitter-Kuvertüre

- Sahne mit Honig und Vanillemark aufkochen. Die Kuvertüren hacken, unter Rühren in der Sahnemischung schmelzen. Abkühlen lassen.
- Die Butter schaumig rühren. Nach und nach die Schokoladenmasse zugeben (dabei darauf achten, dass Butter und Schokomasse Zimmertemperatur haben) und zu einer glatten Masse rühren. Masse in einen Spitzbeutel mit großer Lochtülle füllen, auf Backpapier kleine Kugeln spritzen. Im Kühlschrank über Nacht erstarren lassen.
- Kuvertüre über dem warmen Wasserbad schmelzen, auf Körpertemperatur abkühlen lassen und die Fingerprobe (s. unten) machen. Trüffeln darin eintauchen. Abkühlen lassen. Nach Belieben mit geschmolzener Vollmilch-Kuvertüre feine Linien auf den Trüffeln ziehen.

Aufwand:	★★
Zubereitungszeit:	40 Minuten
Kühl-/Trockenzeit:	1 Stunde
Pro Stück:	100 kcal/460 kJ

Info

Fingerprobe: Kuvertüre wird bei etwa 37 Grad verarbeitet. Dippen Sie zum Testen einen Finger in die geschmolzene Schokolade. Die Masse darf sich nicht zu kalt und nicht zu heiß anfühlen – eben körperwarm.

Muskat-Trüffel

Für etwa 50 Stück
BITTE BEACHTEN: MUSS ÜBER NACHT KÜHLEN
50 g Zucker
200 g Schlagsahne, 1 Msp. Salz
50 g Läuterzucker
1/2 TL geriebene Muskatnuss
200 g Zartbitter-Kuvertüre
200 g Vollmilch-Kuvertüre
100 g weiche Butter
AUSSERDEM
Backpapier
200 g Zartbitter-Kuvertüre
Puderzucker zum Wälzen

- Zucker mit 50 ml Wassser unter Rühren aufkochen. Ist der Zucker gelöst, 1 Minute kochen lassen. Schaum abnehmen und den Läuterzucker abkühlen lassen.
- Sahne mit Läuterzucker aufkochen. Muskat und Salz zugeben. Kuvertüren klein hacken und unter die Sahnemischung rühren, abkühlen lassen. Sobald die Creme fest zu werden beginnt, erneut leicht cremig rühren und nach und nach die Butter in Flöckchen unterrühren.
- Creme in einen Spritzbeutel mit großer Lochtülle füllen und Kugeln auf Backpapier spritzen. Eventuell mithilfe einer Gabel etwas rollen, damit die Oberfläche angeraut wird. Über Nacht im Kühlschrank erstarren lassen.
- Zartbitter-Kuvertüre über dem warmen Wasserbad schmelzen, auf Körpertemperatur abkühlen lassen und die Fingerprobe (s. Info im Kasten Mitte) machen. Trüffeln mit Kuvertüre überziehen, in Puderzucker wälzen und auf Backpapier trocknen lassen.

Aufwand:	★
Zubereitungszeit:	20 Minuten
Kühl-/Trockenzeit:	2 Stunden
Pro Stück:	95 kcal/400 kJ

Ingwer-Pecan-nuss-Marzipan

Für etwa 40 Stück	
90 g kandierter Ingwer	
170 g Marzipan-Rohmasse	
90 g geriebene Pecannüsse	
30 ml Curaçao	
AUSSERDEM	
Backpapier	
200 g Zartbitter-Kuvertüre	
50 g helle Kuvertüre	

- Den Ingwer sehr fein hacken und mit Marzipan, Nüssen und Curaçao verkneten.
- Die Masse auf Backpapier geben und mit einer Kuchenrolle rechteckig etwa 1 cm dick ausrollen. Die Kanten mit einem Messer begradigen. Oder geben Sie die Masse in einen 20 x 20 cm großen Holzrahmen und rollen sie darin aus.
- Kuvertüre über dem warmen Wasserbad schmelzen, auf Körpertemperatur abkühlen lassen. Fingerprobe (s. Seite 343) machen. Die Oberfläche der Marzipanmasse mit Kuvertüre bestreichen, trocknen lassen.
- Die Masse in 2 x 2 cm große Quadrate teilen. Die Quadrate in die restliche Kuvertüre tauchen und auf Backpapier setzen. Wenn sie getrocknet sind, jedes Quadrat nach Belieben mit geschmolzener Vollmilch-Kuvertüre verzieren.

Aufwand:	★
Zubereitungszeit:	30 Minuten
Kühl-/Trockenzeit:	2 Stunden
Pro Stück:	70 kcal/295 kJ

Venus Nipples

Für etwa 85 Stück	
BITTE BEACHTEN: MUSS ÜBER NACHT KÜHLEN	
600 g Zartbitter-Kuvertüre	
100 g Schlagsahne	
3 EL weiche Butter	
100 g weiße Kuvertüre	

- 300 g Zartbitter-Kuvertüre klein hacken, mit der Sahne in einer Schüssel über dem warmen Wasserbad schmelzen lassen. Butter zufügen, alles zu einer glatten Creme verrühren, abkühlen lassen.
- Creme mit den Schneebesen des Handrührgeräts aufschlagen. In einen Spritzbeutel mit runder Lochtülle (Nr. 12) füllen, gleich große Häufchen auf Pergamentpapier spritzen. Im Kühlschrank über Nacht fest werden lassen.
- Übrige Zartbitter-Kuvertüre über dem warmen Wasserbad schmelzen, auf Körpertemperatur abkühlen lassen und die Fingerprobe (s. Seite 343) machen. Die Cremehäufchen in die Kuvertüre tauchen und auf Pergamentpapier trocknen lassen.
- Die weiße Kuvertüre ebenfalls schmelzen und erwärmen und nur jeweils die Spitzen der fest gewordenen Pralinen hineintauchen. Trocknen lassen.

Aufwand:	★★
Zubereitungszeit:	40 Minuten
Kühl-/Trockenzeit:	2 Stunden
Pro Stück:	50 kcal/210 kJ

Info

Das Rezept „Venus Nipples" stammt aus dem Film „Chocolat", in dem die junge Vianne eine kleine Stadt irgendwo in der französischen Provinz mit ihren köstlichen Pralinen verzaubert.

Williams-Trüffel

Für etwa 60 Stück
200 g Schlagsahne
1 Vanilleschote
300 g Vollmilch-Kuvertüre
300 g dunkle Kuvertüre
125 g zimmerwarme Butter
4-5 EL Williams-Christ-Birnengeist
AUSSERDEM
Backpapier
Puderzucker oder Kakaopulver zum Bestäuben
600 g Vollmilch-Kuvertüre

- Sahne mit der aufgeschlitzten Vanilleschote aufkochen, vom Herd nehmen, kurz ziehen lassen. Die Schote entfernen.
- Die Kuvertüre hacken, zusammen mit der Butter in Flöckchen zur Sahne geben und unter Rühren schmelzen lassen. Birnengeist tropfenweise unterrühren. Die Masse zugedeckt kühl stellen, bis sie sich formen lässt.
- Die Trüffelmasse mit dem Schneebesen aufschlagen. In einen Spritzbeutel mit Lochtülle füllen, Kugeln auf ein mit Backpapier ausgelegtes Blech spritzen. Kurz antrocknen lassen.
- Die Kugeln dünn mit Puderzucker oder Kakao übersieben. 400 g Vollmilch-Kuvertüre schmelzen, die Kugeln damit überziehen. Die restliche Vollmilch-Kuvertüre zu feinen Röllchen formen, den noch weichen Überzug damit bestreuen.

Aufwand:	★★
Zubereitungszeit:	40 Minuten
Kühl-/Trockenzeit:	2 Stunden
Pro Stück:	130 kcal/550 kJ

Variation

- Aus den Williams-Trüffeln werden Himbeergeist-, Kirschgeist- oder Aprikosengeist-Trüffel, wenn Sie den Schnaps entsprechend austauschen.

Mailänder Nuss-Herzen

Für etwa 80 Stück
350 g Zucker, 50 g Traubenzucker
80 g fein gemahlene Haselnüsse
80 g fein gemahlene Mandeln
4-5 EL Kirschwasser
150 g Puderzucker
rote Lebensmittelfarbe
etwa 80 Walnuss- oder Pistazienhälften
150 g Zartbitter-Kuvertüre
AUSSERDEM
Öl zum Arbeiten
Puderzucker zum Ausrollen

- Zucker, Traubenzucker und 1/8 l kaltes Wasser in einem Topf unter Rühren langsam erhitzen. Sobald der Zuckersirup kocht, nicht mehr rühren.
- Alle 3 Minuten nach dem Aufkochen zur Probe einige Tropfen Sirup in eine Schüssel mit eiskaltem Wasser fallen lassen. Der Sirup ist fertig, wenn die Tropfen, die man zwischen den Fingern zusammenpresst, ihre Form behalten. Sirup zum Abkühlen in kaltes Wasser stellen. Sofort Nüsse und Mandeln gründlich darunter mischen.
- Ein großes, gekühltes Arbeitsbrett (Marmor oder Kunststoff) einölen. Die Masse dünn aufstreichen, erstarren lassen. Mit dem Teigroller zu Bröseln zerdrücken. Mit 2-3 EL Kirschwasser zu einer festen Paste vermischen.
- Auf einer mit gesiebtem Puderzucker bestäubten Arbeitsfläche etwa 1 cm dick ausrollen, kleine Herzen ausstechen. Trocknen lassen.
- Puderzucker mit 1-2 EL Kirschwasser glatt rühren, mit 1-2 Tropfen Lebensmittelfarbe rosa färben.
- Herzen eintauchen, auf einem Gitter abtropfen lassen. Mit je 1 Nusshälfte verzieren. Fest werden lassen.
- Zartbitter-Kuvertüre schmelzen. Die Herzen mit der Unterseite hineintauchen und trocknen lassen.

Aufwand:	★★
Zubereitungszeit:	1 Stunde
Kühl-/Trockenzeit:	2 Stunden
Pro Stück:	65 kcal/270 kJ

Florentiner Honig-Pralinen

Für etwa 45 Stück
100 g Schlagsahne
75 g Honig
200 g extrabittere Kuvertüre
50 g Orangeat
3 EL Orangenlikör
40 g Nougat
150 g weiße Kuvertüre
150 g dunkle Kuvertüre
AUSSERDEM
etwa 45 Pralinen-Kapseln
etwas Blattgold zum Verzieren (aus der Confiserie) oder kleine Metallic-Perlen

- Sahne und Honig langsam erhitzen, vom Herd nehmen. Extrabittere Kuvertüre hacken, dazugeben und schmelzen lassen. Die Masse glatt rühren, zugedeckt abkühlen lassen.
- Orangeat sehr fein hacken. Orangenlikör, Nougat und das Orangeat unter die Schokolade mischen.
- Helle und dunkle Kuvertüre grob hacken, gemeinsam im Wasserbad schmelzen. Pralinen-Kapseln damit ausgießen. Kapseln sofort wenden, damit die überflüssige Kuvertüre gut ablaufen kann.
- Sobald die Kuvertüre erstarrt ist, die Kapseln mit der Pralinenmasse füllen, etwas antrocknen lassen. Oberfläche mit der restlichen Kuvertüre überziehen. Pralinen mit winzigen Blattgold-Stückchen oder mit je 1 Perle verzieren.

Aufwand:	★★
Zubereitungszeit:	30 Minuten
Kühl-/Trockenzeit:	1 Stunde
Pro Stück:	80 kcal/340 kJ

Schokoladen-Kirschen

Für etwa 26 Stück
26 Kirschen mit Stielen
200 g Zartbitter-Kuvertüre
AUSSERDEM
Backpapier
Kakaopulver nach Belieben

- Kirschen waschen und trockentupfen. Kuvertüre hacken, über dem warmen Wasserbad schmelzen, auf Körpertemperatur abkühlen lassen (Fingerprobe, s. Seite 343).
- Die Kirschen in die Kuvertüre tauchen und auf Backpapier trocknen lassen. Eventuell noch mit Kakaopulver bestreuen

Aufwand:	★
Zubereitungszeit:	15 Minuten
Kühl-/Trockenzeit:	1 Stunde
Pro Stück:	45 kcal/190 kJ

Orangen-Schoko-Pralinen

Für etwa 60 Stück
1 große unbehandelte Orange
150 g Zucker
3 Eigelbe (Eier: Größe M)
100 g Butter, 100 g Schlagsahne
400 g Zartbitter-Kuvertüre
300 g Vollmilch-Kuvertüre
AUSSERDEM
Backpapier
100 g Zartbitter-Kuvertüre

- Orange heiß waschen, abtrocknen. Schale fein abreiben, Orange halbieren, auspressen.
- Zucker und Eigelbe im Wasserbad gut verrühren. Die Butter in Flöckchen hinzufügen. Sahne, Orangenschale und -saft dazugeben. Unter Rühren erhitzen. Vom Herd nehmen.
- Zartbitter-Kuvertüre hacken, in der Eigelbmasse schmelzen, glatt rühren. Masse 1 cm hoch auf ein mit Backpapier ausgelegtes Blech gießen. Kalt stellen, erstarren lassen.
- Mit einem runden Ausstecher von 2-3 cm Ø Kreise ausstechen. Die Vollmilch-Kuvertüre schmelzen, Kreise eintauchen, mit Kuvertüre überziehen und auf einem Gitter erstarren lassen.
- Für die Dekoration die übrige Kuvertüre schmelzen, in einen Mini-Spritzbeutel füllen und die Pralinen mit feinen Dekor-Linien verzieren.

Aufwand:	★★
Zubereitungszeit:	1 Stunde
Kühl-/Trockenzeit:	2 Stunden
Pro Stück:	100 kcal/420 kJ

Tipp

Lassen Sie Ihrer Phantasie freien Lauf! Auch andere Früchte schmecken lecker, wenn man sie in Schokolade taucht. Z. B. Birnen, Äpfel, Bananen, Mandarinen oder Trockenfrüchte wie Aprikosen.

Spanisches Mandel-Konfekt

Für etwa 30 Stück
30 geschälte Mandeln
1 Eiweiß (Ei: Göße M)
50 g Puderzucker
100 g Vollmilch-Kuvertüre
250 g Marzipan-Rohmasse
25 g zimmerwarme Butter
25 g Nougat
AUSSERDEM
Backpapier
400 g Vollmilch-Kuvertüre

- Die Mandeln in einer großen Pfanne ohne Fett unter Rühren leicht rösten. Herausnehmen und abkühlen lassen. Eiweiß verquirlen, Mandeln hineintauchen, kurz antrocknen lassen. In Puderzucker wälzen.
- Vollmilch-Kuvertüre im Wasserbad schmelzen und wieder leicht abkühlen lassen.
- Marzipan-Rohmasse würfeln, mit der Butter und dem Nougat zur Kuvertüre geben. Alles zu einer Paste vermischen. Abkühlen lassen und mit einem Spachtel durcharbeiten.
- Konfektmasse auf ein mit Backpapier ausgelegtes Blech 1,5 cm dick aufstreichen, erstarren lassen. Mit einem ovalen oder runden Ausstecher Formen ausstechen.
- Vollmilch-Kuvertüre schmelzen, das Konfekt damit überziehen. Die Oberfläche mit je 1 gezuckerten Mandel belegen. Erstarren lassen. Um jede Mandel einen Rand aus der übrigen, leicht abgekühlten Kuvertüre spritzen. Fest werden lassen.

Aufwand:	★★
Kühl-/Trockenzeit:	1 1/2 Stunden
Zubereitungszeit:	3 Stunden
Pro Stück:	150 kcal/630 kJ

Marzipankartoffeln

Für etwa 50 Stück

100 g Walnusskerne
400 g Marzipan-Rohmasse
60 ml Cointreau
100 g Kakaopulver zum Wälzen

- Die Walnüsse fein hacken und das Marzipan grob würfeln.
- Marzipan, Nüsse und Cointreau miteinander verkneten. Die Marzipanmasse zu Kugeln mit 2 cm Ø formen.
- Das Kakaopulver in eine Schüssel geben und die Marzipankugeln darin wälzen.

Aufwand:	★
Zubereitungszeit:	20 Minuten
Pro Stück:	60 kcal/250 kJ

Schokoladen-Pflaumen

Für etwa 25 Stück

250 g entsteinte Trockenpflaumen
200 ml Weinbrand (Armagnac)
200 g Zartbitter-Kuvertüre

- Pflaumen mit einer Nadel mehrmals einstechen und 1 Stunde in Weinbrand einlegen.
- Die Pflaumen in einem Sieb gut abtropfen lassen und trockentupfen. Die Kuvertüre hacken, über dem warmen Wasserbad schmelzen und auf Körpertemperatur abkühlen lassen. Die Pflaumen in die Kuvertüre tauchen, auf Pergamentpapier trocknen lassen.

Aufwand:	★
Zubereitungszeit:	30 Minuten
Marinierzeit:	1 Stunde
Pro Stück:	85 kcal/360 kJ

Gefüllte Walnüsse im Schokomantel

Für etwa 30 Stück

50 g Rosinen, 50 ml weißer Rum
200 g Zartbitter-Kuvertüre
100 g Marzipan-Rohmasse
250 g Walnusskerne

● Rosinen kurz in Rum einlegen. Abgießen, Flüssigkeit auffangen. Kuvertüre hacken, im Wasserbad erwärmen und abkühlen lassen. Rosinen trockentupfen, fein hacken und mit Marzipan vermischen. Eventuell von der Rosinen-Rum-Abtropfflüssigkeit zufügen.
● Jeweils 1 kleines Stück Marzipanmischung zwischen 2 Walnusshälften geben. Die Walnusshälften leicht zusammendrücken und in die Kuvertüre tauchen. Auf Pergamentpapier setzen und trocknen lassen.

Aufwand:	★
Zubereitungszeit:	1 Stunde
Kühl-/Trockenzeit:	1 Stunde
Pro Stück:	115 kcal/480 kJ

Schokoladen-Datteln

Für etwa 40 Stück

200 g Marzipan-Rohmasse
50 g gehackte Pistazienkerne
20 ml Grand Marnier
250 g entsteinte Trockendatteln
200 g weiße Kuvertüre

● Marzipan hacken, mit Pistazien und Grand Marnier verkneten. Masse 1/2 cm dick ausrollen und in kleine Streifen teilen. Die Marzipanstreifen in die Kernhöhlung der Datteln stecken.
● Kuvertüre hacken, über dem Wasserbad schmelzen, auf Körpertemperatur abkühlen lassen. Datteln hineintauchen, auf Pergamentpapier trocknen lassen. Nach Belieben mit dunkler Kuvertüre verzieren.

Aufwand:	★★
Zubereitungszeit:	30 Minuten
Kühl-/Trockenzeit:	1 Stunde
Pro Stück:	80 kcal/340 kJ

Pfefferminz-Sterne

BITTE BEACHTEN: MUSS 2 STUNDEN TROCKNEN

Für etwa 40 Stück
450 g Puderzucker
1 Eiweiß (Ei: Größe M)
4 TL Zitronensaft
1/2 TL Pfefferminzöl (erhältlich in Apotheken)
150 g weiße Kuvertüre
50 g Zartbitter-Kuvertüre
2 TL Butterschmalz
AUSSERDEM
etwas Puderzucker zum Ausrollen

- Puderzucker sieben und mit Eiweiß, Zitronensaft und Pfefferminzöl verkneten.
- Die Pfefferminzmasse zwischen zwei aufgeschnittenen Gefrierbeuteln 5-6 mm dick ausrollen. Eventuell mit etwas Puderzucker bestäuben, falls die Masse zu stark klebt.
- Die Sterne (4-5 cm Ø) ausstechen. Entweder 2 Stunden an der Luft oder (wenn vorhanden) 1 Stunde im Umluft-Ofen bei etwa 50 Grad trocknen.
- Die Kuvertüren hacken, getrennt schmelzen. Schmalz in die weiße Kuvertüre rühren. Sterne damit überziehen. Dunkle Kuvertüre in einen Gefrierbeutel füllen, eine Ecke abschneiden. Die Sterne mit der Kuvertüre verzieren.

Aufwand:	★★
Zubereitungszeit:	45 Minuten
Kühl-/Trockenzeit:	2 Stunden
Pro Stück:	75 kcal/315 kJ

Variation

- Für Experimentierfreudige: Tauschen Sie das Pfefferminzöl gegen Rosenöl oder Veilchenessenz (Apotheke) aus. Statt Zitronensaft nehmen Sie 4 EL Wasser. Rosen- und Veilchen-Pralinen sind die neuesten Kreationen in der Konfiserie.

Marzipan-Pasteten

Für etwa 30 Stück
550 g Marzipan-Rohmasse
225 g Puderzucker
2 EL Orangenlikör
1 TL abgeriebene Schale von 1 unbehandelten Orange
100 g gehackte Pistazienkerne
2 EL Cocktail-Kirschen (Glas)
1 Eiweiß (Ei: Größe S)
AUSSERDEM
Backpapier

- 400 g Marzipan mit 150 g Puderzucker, Orangenlikör, Orangenschale und Pistazien vermischen. Abgetropfte Kirschen vorsichtig unterkneten, Marzipanmasse halbieren und 2 Rollen von etwa 17 cm Länge formen. Rollen längs durchschneiden.
- Übriges Marzipan mit restlichen Puderzucker verkneten. Zwischen aufgeschnittenen Gefrierbeuteln dünn ausrollen. Pasteten darin einwickeln, Rand gerade abschneiden.
- Das abgeschnittene Marzipan erneut ausrollen und etwa 30 Sterne (2-3 cm Ø) ausstechen. Auf ein mit Backpapier belegtes Blech legen. Unter dem Backofengrill 1 Minute bräunen. Rückseite mit Eiweiß bepinseln, auf die Pasteten kleben.

Aufwand:	★★
Zubereitungszeit:	1 Stunde
Pro Stück:	140 kcal/590 kJ

Schokolade-Minze-Pralinen

Für etwa 50 Stück	
1 Hand voll frische Minzeblätter	
200 g Zucker	
500 g Schokolade mit 60 % Kakaoanteil	
200 g Schlagsahne	
35 g weiche Butter	
AUSSERDEM	
Backpapier	
etwa 50 Papiermanschetten	

● Minze waschen, in feine Streifen schneiden. 1/8 l Wasser zum Kochen bringen und über die Minze gießen. Etwa 10 Minuten ziehen lassen und durchseihen. Minzetee erneut aufkochen und auf 1/4 einkochen lassen.

● Zucker in einer Pfanne bei milder Hitze schmelzen (nicht karamellisieren!). 1-2 EL Minzetee zufügen und die Zuckermasse auf ein mit Backpapier belegtes Blech gießen. Trocknen lassen.

● Zuckermasse vom Papier trennen, sehr fein hacken und 1 Stunde in den Kühlschrank stellen.

● Schokolade hacken und über dem warmen Wasserbad schmelzen. Nach und nach Sahne und Butter unterrühren. Die Masse auf Körpertemperatur abkühlen lassen, Fingerprobe machen (s. Info Seite 343).

● Zuckermasse unter die Schokoladenmasse rühren. Masse noch weiter abkühlen lassen, bis sie spritzfähig ist und sie dann in einen Spritzbeutel mit großer Sterntülle füllen.

● Masse in die Papiermanschetten spritzen und nach Belieben mit kandierten Veilchenblättern dekorieren.

Aufwand:	★★
Zubereitungszeit:	45 Minuten
Kühl-/Trockenzeit:	2 Stunden
Pro Stück:	90 kcal/380 kJ

Dreierlei Trüffel

BITTE BEACHTEN: MÜSSEN ÜBER NACHT KÜHLEN	
Für etwa 60 Stück	
je 100 g Vollmilch- und Zartbitter-Kuvertüre	
100 g Schlagsahne	
2 EL Zucker	
100 g Butter	
herausgeschabtes Mark von 2 Vanilleschoten	
1 EL Mandellikör	
1-2 EL Haselnuss-Krokant	
AUSSERDEM	
je 50 g gehackte Mandeln, Puderzucker und Kakaopulver	

● Kuvertüren hacken. Mit Sahne, Zucker und Butter bei milder Hitze schmelzen. Masse dritteln. 1 Teil mit Vanillemark, 1 Teil mit Mandellikör und 1 Teil mit Krokant verrühren. 2 Stunden kühl stellen.

● Trüffelmasse jeweils getrennt aufschlagen. Entweder mit den Händen kleine Kugeln formen oder die Masse in einen Spritzbeutel mit Lochtülle (Nr. 6) füllen und auf Pergamentpapier kleine Tupfen setzen. Abgedeckt über Nacht kalt stellen.

● Mandeln in einer Pfanne ohne Fett rösten, fein hacken. Mandel-Trüffeln in Mandeln, Vanille-Trüffeln in Puderzucker und Krokant-Trüffeln in Kakao wälzen. Kühl aufbewahren.

Aufwand:	★
Zubereitungszeit:	40 Minuten
Kühl-/Trockenzeit:	2 Stunden
Pro Stück:	50 kcal/210 kJ

Info

Alle selbst hergestellten Trüffel sind höchstens eine Woche haltbar. Bewahren Sie sie immer dunkel, trocken und vor allem kühl auf, allerdings nicht im Kühlschrank, denn dort „beschlagen" sie.

Tipps, Tricks und Infos zu

Konfekt kann man zwar fertig kaufen, doch Selbstgemachtes ist bei Familie u
Die Herstellung erfordert allerdings ein bisschen Zeit, Geduld und etw

Toffee-Pralinen

In einem Topf 500 g Schlagsahne, 300 g Krümelkandis, 50 g Butter und 200 g geschälte, gehackte Erdnüsse (ungesalzen) aufkochen und bei mittlerer Hitze unter ständigem Rühren in etwa 40 Minuten einkochen. Konsistenz-Test: Etwas Masse auf einen Teller geben. Wird sie gleich fest, auf ein Backblech gießen und abkühlen lassen. Zerläuft sie, weiterkochen. Wird sie zu fest, noch etwas Sahne zufügen. Wenn die Toffee-Masse abgekühlt ist, in 2 x 2 cm große Würfel schneiden.

So gelingen Trüffeln perfekt

1 Schlagsahne aufkochen und die Vanilleschote kurz darin ziehen lassen, dann herausnehmen.
2 Gehackte Kuvertüre und Butter zur Sahne geben. Unter ständigem Rühren schmelzen lassen.

3 Trüffelmasse kühl stellen. Anschließend mit einem Schneebesen aufschlagen und in einen Spritzbeutel füllen. Kugelförmig auf Pergament spritzen.
4 Mit Kakao oder Puderzucker bestäuben oder in Schokospänen wälzen.

Kuvertüre

Die Sorten

Ganz dunkle, extrabittere Kuvertüre hat über 60 % Kakaoanteile. Sie gibt es vor allem in Confiserien und Patisserien.

Dunkle Kuvertüre hat einen Kakaoanteil von 60 %.

Helle Vollmilch-Kuvertüre besteht nur aus 36 % Kakaomasse.

Weißer Kuvertüre fehlen die festen Kakaobestandteile, die für Farbe und den typischen Schokoladengeschmack zuständig sind. Sie enthält nur etwa 30 % Kakaobutter.

Kuvertüre richtig temperieren

Die wichtigste Regel beim Umgang mit Kuvertüre: Sie darf nur langsam und nicht zu heiß geschmolzen werden.

- Kuvertüre klein schneiden. Die halbe Menge in einer Metallschüssel ins warme (etwa 36-38 Grad), keinesfalls ins heiße Wasserbad setzen und unter ständigem Rühren schmelzen. Sobald alles flüssig ist, die Schüssel aus dem Wasserbad nehmen.
- Über 40 Grad erwärmt, verliert Kuvertüre ihre Bindung. Kakaomasse, -butter und Zucker trennen sich dann voneinander ab. Statt zart zu schmelzen, klumpt die Masse, lässt sich nicht verstreichen.
- Auch wichtig: Kuvertüre nie mit Wasser oder Wasserdampf in Berührung bringen, sie gerinnt sonst.
- Dann erst die restliche Kuvertüre einrühren und warten, bis auch sie geschmolzen ist. Anschließend kalt stellen, bis die Kuvertüre dickflüssig geworden ist.
- Zum Überziehen von Pralinen, Plätzchen, Kuchen und Torten erneut sanft im Wasserbad erwärmen, aber nur auf 30-32 Grad, das entspricht in etwa unserer Handwärme.
- Mit einer einfachen Methode können Sie feststellen, ob Kuvertüre den idealen Punkt erreicht hat: Ein breites Messer hineintauchen und wieder herausziehen. Beginnt der Schokobelag nach kurzer Zeit fest und glatt zu werden, ist die Kuvertüre fertig. Zeigt sie Streifen, noch einmal bei der erforderlichen Temperatur von 30-32 Grad gleichmäßig durchrühren.

Pralinen und Trüffeln

unden besonders beliebt – ganz einfach, weil es am besten schmeckt.
schick. Damit alles gelingt, zeigen wir Ihnen hier, wie es geht

Das Aroma bringt es

Schokolade ist ein idealer Partner für die unterschiedlichsten Aromen. Neben Gewürzen wie Zimt und Muskat sind Hochprozentiges, Vanille, Kaffee, Zitronen- oder Orangenschale sehr beliebt. Mutige sollten auch schwarzen Pfeffer probieren.

Schokoraspel herstellen

1 Späne
Schokolade oder Kuvertüre mit Butterbrot- oder Küchenpapier festhalten. Mit Sparschäler Späne abziehen.

2 Raspel, Pulver
Ideal ist hier eine Vierkantreibe. Je nach gewünschter Form (Pulver oder Raspel) wird die Schokolade oder Kuvertüre in Papier geklemmt und über die entsprechende Seite gerieben.

Trüffel in Schokoraspeln wälzen

1 Ob Sie feine oder große Raspel mögen, ist Geschmacksache. Kühlen Sie die Kuverüre jedenfalls vor dem Raspeln etwa 30 Minuten im Kühlschrank.
2 Geben Sie die Schokoraspel auf einen Teller und rollen Sie die ebenfalls gekühlten Trüffelkugeln darin, indem Sie den Teller hin und her bewegen.

Gesamtregister

Backpraxis

Raffinierte Rezepte

Preisgünstige Rezepte

Kalorienarme Rezepte

Schnelle Rezepte

Bildnachweis

Barbara Bonisolli 125 M., 154 u. l.
Michael Brauner 146, 308.
Walter Cimbal 251, 253.
Achim Deimling-Ostrinsky 4, 28, 49 o. l., 50, 56, 65, 66, 67, 70 o., 152, 156, 173, 180, 198, 206, 207, 216, 217, 224, 225,226, 227, 231, 233, 240, 249, 279.
Werner Deisenroth 153, 268, 270, 271.
StockFood/Eising 4, 5, 30, 32, 33, 34, 35, 36, 38, 39, 40, 41, 42, 44, 46, 47, 48 (8) l., M. o., r., 49 (7) u., r., 53, 58, 60, 61, 62, 64, 70 (4) M., u., 71 o. l., 75, 76 (2) u. l., u. r., 77, 78, 79, 81, 86, 87, 92, 93, 94 (6) M. l., o. r., 95 (6) o. r., M. l., u., 101, 102, 103, 105, 106, 107, 108, 109, 111, 114, 121, 122, 123, 124 (6) o. l., o. r., u. r., 125 (4) o., u., 126, 128, 138, 154 (4) o., 155 (4) o., 160, 161, 162, 163, 164, 166, 174, 175, 176, 177, 181, 182, 187, 190, 199, 201, 204, 220, 222, 228, 236, 237, 243, 248, 256, 259, 269, 272, 274, 276 (2), 278, 282, 284, 285, 288, 296, 297, 298, 302, 304, 305, 306, 309, 310 (2) o. r., 312, 315, 316, 317, 318, 320, 223, 324, 326, 327, 330, 331, 332, 333, 335 (3) M., 338, 339, 341, 356 (4) u., 357 u. M.
C. P. Fischer 71 (4) o. r., M.
Maria Harder 4 (2), 5 (2), 26, 71 u. r., 80, 88, 89, 91, 94 u., 95 (4) o. l., M. r., 96, 124 u. l., 134, 136 (2) u., 137, 144, 145, 148, 150, 155 (3) u. l., r., 156 u. r., 165, 178, 183, 184, 185, 186, 188, 189, 192, 193, 194, 195, 196, 197, 200, 202, 203, 208, 209, 210, 211, 214, 215, 219 o., 230, 232, 234, 238, 239, 242,244, 246, 247, 255, 254 (3) u. l., 266, 267, 280, 281, 286, 287, 300, 301, 307, 328, 336, 342, 344, 345, 346, 348, 349, 350, 351, 352, 353, 354, 355, 357 (3) o. l., o. r.,u. l.
Ulrich Kerth 129, 357 (2) u. r.
Adalbert Linz 275, 283, 292, 293, 294.
Chris Meier 48 u. M., 52.
Adam Pentos 22, 23.
Helmut Peters 43, 82, 83, 98, 99, 100, 260, 262, 263, 311 (2) u.
Ch. Rosenfeld 218 u. l.
Schinharl 63, 90, 104, 110, 135, 290.
Amos Schliack 205.
Studio R. Schmitz 218 o.
O. Schwarzwald 254 o. r.
Elke Seeger 4, 29, 37, 69, 72, 115, 116, 117, 118 unten, 119, 132, 140, 141, 142, 143, 167, 245.
M. Spachmann 95 o.
Teubner Foodfoto 8, 9, 10, 11, 12, 13, 15, 16. 17, 18, 19, 20, 258, 264, 265, 310 (2) u., l., 311 (3) r.
Martina Urban 154 u. r.
Misha Vetter 45, 159, 229.
J. P. Westermann 139, 223.
B. Winkelmann 57, 169.
Michael Wissing 31, 212, 213, 254 o. l.
B. Wohlgemuth 76 unten, 84, 113, 130, 131, 168, 170, 171, 172, 206, 207.
Zimmermann 54, 55.
Werkfotos: CMA, Easy, Kaiser Backform, Wirths PR, Zyliss Backform.
Titelfotos: Walter Cimbal, Stockfood/Eising, Maria Harder (2).
Alle übrigen Fotos aus Hubert Burda Media-Information Services.

Impressum

Bibliografische Information –
Die Deutsche Bibliothek

Die Deutsche Bibliothek verzeichnet diese Publikation in der Deutschen Nationalbibliografie; detaillierte bibliografische Daten sind im Internet über **http://dnb.ddb.de** abrufbar.

Idee, Konzeption und Redaktion:
Gabriele Redden

Projektleitung:
Kathrin Gritschneder, Bettina Huber

Grafische Gestaltung:
Hans-Jürgen Riegel

Bildredaktion, Koordination, Layout, Satz und Herstellung:
Heidi Paal

Schlussredaktion:
Adelheid Schmidt-Thomé

Fotos:
siehe Bildnachweis

Gesamtverantwortung:
Birgitt Micha, Chefredakteurin
„meine Familie & ich"

Litho:
Premedia GmbH, Wels/Österreich

Druck und Bindung:
MOHN Media • Mohndruck GmbH, Gütersloh

Printed in Germany
Gedruckt auf umweltfreundlich chlorfrei gebleichtem Papier.
ISBN 3-426-66855-6
Bitte besuchen Sie uns im Internet:
www.droemer-knaur.de
Weitere kulinarische Titel finden Sie im Internet unter **www.genussvoll-leben.de**